Reagents for Organic Synthesis

Fieser and Fieser's

Reagents for Organic Synthesis

COLLECTIVE INDEX FOR VOLUMES 1–12

Janice G. Smith
Mount Holyoke College

Mary Fieser
Harvard University

A WILEY-INTERSCIENCE PUBLICATION
John Wiley & Sons, Inc.
NEW YORK / CHICHESTER / BRISBANE / TORONTO / SINGAPORE

Library of Congress Catalog Card Number: 88-659800
ISBN 0-471-84091-2
ISSN 0271-616X

Printed in the United States of America

10 9 8 7 6 5 4 3 2 1

PREFACE

This index was prepared in response to suggestions from various chemists for an easier access to the information in the published volumes of Reagents. It covers the material in Volumes 1–12. We have made an attempt to systematize the nomenclature, particularly of the organometallic compounds. Several new sections have been added. Surprisingly, this includes a comprehensive list of reagents, both those singled out as reagents previously, as well as those only cited in the text. For example, the carbenes are now indexed separately as individual reagents. The original "Index According to Type" has now been subdivided and expanded into three separate indices. There is a "Type of Compound" Index in which we have grouped together similar kinds of reagents, for example, Diels-Alder Dienes, Dienophiles, Ylides, and most especially, many of the organometallic compounds whose complex names often make them difficult to locate. Secondly, there is an Index by reaction type. Whereas in the original indices for volumes 1–12 all oxidizing agents were grouped together, they are now divided on the basis of the substrate oxidized as well as the product obtained. Finally, we have included an expanded Synthesis Index, based on both the product and starting material.

We are deeply indebted to Dr. Daniel C. Smith who singlehandedly indexed the over 50,000 author citations in this index. We realize that a cumulative listing of references of chemists with large research groups will be of only limited value. However, reference to a co-worker will provide easy access to specific papers. We have followed Chemical Abstracts' lead in deleting foreign accents, and in indexing the German ä, ö, and ü, as ae, oe, and ue respectively. In both the Type of Reaction and Synthesis Indices, the page numbers refer either to the page on which the citation for a "boldface" reagent begins, or to a page on which a reagent in the text is actually cited.

We want to extend thanks to Jerry Lotto, a member of E. J. Corey's computer group. His expertise was invaluable in both the purchase of our computer hardware and in the execution of the project. We also wish to thank Mr. Bill Farrington and the Department of Electronic Services at Mount Holyoke who were instrumental in maintaining the computer equipment needed to complete this index.

JANICE G. SMITH
MARY FIESER

South Hadley, MA
Cambridge, MA
May, 1990

CONTENTS

Reagents for Organic Synthesis

TYPE OF REACTION INDEX

ACETALIZATION (*see* SYNTHESIS INDEX)
ACETOACETIC ESTER CONDENSATION
Potassium hydride, **1,** 935
Tricaprylylmethylammonium chloride, **6,** 404
ACETOACETYLATION
Acetoacetyl fluoride, **1,** 4
Diketene–Iodotrimethylsilane, **10,** 151
Mercury(II) acetate, **2,** 264
ACETONYLATION
OF RH, ArH
π-Allylnickel halides, **8,** 324
O,O-*t*-Butyl O-isopropenyl peroxycarbonate, **11,** 100
3-Chloro-2-trimethylsilyloxy-1-propene, **11,** 129
2,3-Dichloro-1-propene, **6,** 332
Isopropenyl acetate, **6,** 356
Manganese(III) acetate, **6,** 355
Mercury(II) trifluoroacetate, **9,** 294
2-Methoxyallyl bromide, **8,** 322
Tetrakis(triphenylphosphine)-palladium(0), **12,** 468
OF RX
Acetoacetic acid, **12,** 1
π-Allylnickel halides, **5,** 437
Lithium diethylamide, **6,** 332
S-(2-Methoxyallyl)-N,N-dimethyldithiocarbamate, **6,** 364
2-Oxo-3-(triphenylphosphoranylidene)-propyllithium, **5,** 375
Phenylsulfinylacetone, **5,** 524
ACETOXYLATION (*see also* ACYLOXYLATION, ALLYLIC REACTIONS)
OF RH, ArH
Acetic anhydride–Pyridine, **6,** 4
Bis(2,2-dipyridyl)silver(II) peroxydisulfate, **6,** 51
Borane–Tetrahydrofuran, **11,** 156
Cerium(IV) ammonium nitrate, **1,** 120; **4,** 71
m-Chloroperbenzoic acid, **6,** 110; **8,** 97
N,N-Dimethylhydrazine, **7,** 126
Iodine, **1,** 495
Lead tetraacetate, **1,** 537; **2,** 234; **3,** 168; **5,** 365; **6,** 313; **7,** 185; **12,** 270
Manganese(III) acetate, **4,** 318; **12,** 292
Mercury(II) acetate, **1,** 644
Palladium(II) acetate, **2,** 303; **5,** 496; **7,** 274
Palladium(II) chloride, **2,** 303; **10,** 300
Thallium(III) acetate, **2,** 406
Trimethyloxonium tetrafluoroborate, **3,** 314
OF C=C–H
Palladium(II) acetate, **1,** 778; **5,** 496
Palladium(II) chloride, **2,** 303
ACETOXYMERCURATION (*see* ADDITION REACTIONS)
ACETYLATION (*see also* FRIEDEL–CRAFTS ACYLATION, ENOL ACYLATION)
OF C–H
Acetaldoxime, **1,** 3
Acetyl hexafluoroantimonate, **1,** 692
Acetyl methanesulfonate, **5,** 5
Acetyl tetrafluoroborate, **11,** 6
Aluminum chloride, **3,** 7
Boron trifluoride, **1,** 68
Boron trifluoride–Acetic acid, **1,** 69
Dimethylacetamide, **1,** 270
Hydrogen fluoride, **1,** 455
Ion-exchange resins, **1,** 511
Ketene, **4,** 275
Lithium 2,2,6,6-tetramethylpiperidide, **4,** 310
Magnesium, **1,** 627
N-Methoxy-N-methylacetamide, **12,** 314
Methyllithium, **5,** 448
Trifluoroacetic anhydride, **1,** 1221
Zirconium(IV) chloride, **1,** 1295
OF N–H (*see* SYNTHESIS INDEX—AMIDES)
OF O–H (*see also* SYNTHESIS INDEX—ESTERS)
Hydrobromic acid, **3,** 154
Lithium dimethylcuprate, **5,** 234

ACETYLATION (*Continued*)
Perchloric acid, **1,** 796; **2,** 309
Potassium acetate, **1,** 906
Sodium hydride, **2,** 382
Sulfoacetic acid, **1,** 1117
Sulfosalicylic acid, **1,** 1118
p-Toluenesulfonic acid, **1,** 1172; **4,** 508
OF S–H
1,3,4,6-Tetraacetylglycouril, **6,** 563
Thiolacetic acid, **1,** 1154
ACYLATION (*see also* ACETYLATION, ALLYLIC REACTIONS, BENZOYLATION, ENOL ACYLATION, FORMYLATION, FRIEDEL–CRAFTS ACYLATION)
OF ACTIVE C–H
Acetic anhydride–Sodium acetate, **6,** 5
Bis(trimethylsilyl) lithiomalonate, **9,** 55
Boron trifluoride, **3,** 32
Boron trifluoride etherate, **2,** 35
t-Butyl lithioacetate, **9,** 82
t-Butyl α-lithioisobutyrate, **6,** 84
Butyllithium–Tetramethylethylenediamine, **10,** 73
t-Butyl trimethylsilylacetate, **7,** 51
N,N′-Carbonyldiimidazole, **9,** 96
Chlorotrimethylsilane, **2,** 435; **4,** 537
Copper(I) iodide, **7,** 81
Dibenzyl malonate, **1,** 198
Di-*t*-butyl malonate, **1,** 210
Dichlorobis(trifluoromethanesulfonato)titanium(IV), **12,** 173
(N,N-Diethylamino)methyloxosulfonium methylide, **5,** 210
Diethyl oxalate, **7,** 106
Diethyl phosphorocyanidate, **9,** 163
(Dimethoxyphosphinyl)methyllithium, **6,** 339
N,N-Dimethylhydrazine, **9,** 184
Dimethyl sulfone, **1,** 296; **2,** 157
Dimethyl sulfoxide, **1,** 296
Dimethylsulfoxonium methylide, **1,** 315
Ethyl ethylthiomethyl sulfoxide, **5,** 299
Ethyl malonate, **10,** 182
Ethyl phenyl sulfoxide, **6,** 257
Ethyl trichloroacetate, **4,** 233
Grignard reagents, **9,** 229
Isobutene, **1,** 522
Lithium dibutylcuprate, **6,** 163
Lithium diethylamide, **6,** 331
Lithium diisopropylamide, **5,** 400
Lithium hexamethyldisilazide, **7,** 197
Magnesium, **1,** 627
Magnesium ethyl malonate, **8,** 311
Meldrum's acid, **10,** 252
Mesityllithium, **8,** 317; **10,** 234
N-Methanesulfinyl-*p*-toluidine, **2,** 269
4-Methoxy-3-buten-2-one, **10,** 258
Methylthiomethyl *p*-tolyl sulfone, **12,** 327
1-Morpholino-1-cyclohexene, **1,** 707
β-Nitropropionyl chloride, **8,** 363
Phase-transfer catalysts, **9,** 356
Polyphosphoric acid, **5,** 540
Potassium *t*-butoxide, **5,** 544
Pyridine, **1,** 958
Sodium amide, **1,** 1034
Sodium hydride, **1,** 1075; **2,** 382
Sodium methylsulfinylmethylide, **1,** 310; **2,** 166
Thallium(I) ethoxide, **2,** 407
Tosylmethyl isocyanide, **4,** 514; **8,** 493
Triethylamine, **1,** 1198; **5,** 689
Trimethylsilyl 2-lithio-2-ethoxycarbonylacetate, **5,** 722; **9,** 493
Triphenylbis(2,2,2-trifluoroethoxy)phosphorane, **10,** 43
Triphenylmethylpotassium(or -sodium), **1,** 1258, 1259
Zinc chloride, **11,** 602
OF ArH (*see also* FRIEDEL–CRAFTS ACYLATION)
Boron trichloride, **9,** 62
Boron trifluoride etherate, **6,** 65
t-Butyl hydroperoxide, **3,** 37
1-Chloro-N,N,2-trimethylpropenylamine, **4,** 94
Chromium carbonyl, **9,** 117
Iron carbonyl, **8,** 265
Lithium 2,2,6,6-tetramethylpiperidide, **6,** 345
Nafion-H, **9,** 320
Polyphosphoric acid, **5,** 540
OF C=C–H, C≡C–H
Alkylaluminum halides, **11,** 7
Alumina, **11,** 22
Aluminum chloride, **3,** 7
Boron trifluoride etherate, **12,** 66
Di-μ-carbonylhexacarbonyldicobalt, **8,** 148; **9,** 144
Dichlorobis(triphenylphosphine)palladium(II), **8,** 151

Dichloromethyl methyl ether, **8,** 485; **9,** 154
Dimethylformamide–Phosphoryl chloride, **5,** 251
1- or 2-Lithio-3,3,6,6-tetramethoxy-1,4-cyclohexadiene, **7,** 191; **8,** 280
Lithium diisopropylamide, **10,** 241
Nitromethane, **4,** 357
Polyphosphoric acid, **12,** 399
Simmons–Smith reagent, **12,** 437
Tetrakis(triphenylphosphine)-palladium(0), **12,** 468
Tibutylvinyltin, **9,** 42
Trifluoroacetic anhydride, **1,** 1221
Vilsmeier reagent, **9,** 514

INTRAMOLECULAR REACTIONS
Aluminum bromide, **2,** 19
Aluminum chloride, **5,** 10; **10,** 9; **12,** 26
Boric acid, **1,** 63
Boron trifluoride etherate, **2,** 35; **3,** 33
sec-Butyllithium, **10,** 75
γ-Butyrolactone, **1,** 101
Carbonyl sulfide, **11,** 112
Copper(II) bromide, **6,** 138
Copper(II) trifluoromethanesulfonate, **10,** 110
Diethyl ethoxymethylenemalonate, **5,** 216
Dimethylaluminum benzenethiolate, **11,** 194
Dimethylsulfoxonium methylide, **1,** 315
Fluorosulfuric acid, **1,** 396
Grignard reagents, **9,** 229
Hydrogen fluoride, **1,** 455; **9,** 240
Lithium *o*-lithiobenzoate, **10,** 246
Methanesulfonic acid, **11,** 321
Molecular sieves, **11,** 350
Phosphorus(V) oxide, **1,** 871
Phosphoryl chloride, **2,** 330
Polyphosphate ester, **2,** 333; **11,** 430
Polyphosphoric acid, **1,** 894; **3,** 231; **4,** 395; **5,** 540; **7,** 294; **12,** 399
Potassium *t*-butoxide, **5,** 544
Pyridinium chloride, **6,** 497; **11,** 449
Sodium hexamethyldisilazide, **6,** 529
Sodium methylsulfinylmethylide, **1,** 310
Sodium tetrachloroaluminate, **1,** 1027; **4,** 438
Succinic anhydride, **4,** 468
Tetrabutylammonium fluoride, **12,** 458
Thionyl chloride, **1,** 1158
Tin(IV) chloride, **1,** 1111; **10,** 370; **12,** 486
Titanium(IV) chloride, **9,** 468
Trifluoroacetic anhydride, **1,** 1221
Trifluoromethanesulfonic acid, **12,** 531
2-Trimethylsilylethanol, **8,** 510

OF N–H (*see* SYNTHESIS INDEX—AMIDES)

OF RM TO FORM C=O COMPOUNDS
Aluminum chloride, **7,** 7; **12,** 26
Arene(tricarbonyl)chromium complexes, **12,** 34
1,3-Benzodithiolylium perchlorate, **8,** 34
Benzylchlorobis(triphenylphosphine)-palladium(II), **8,** 35; **9,** 41; **12,** 44
[1,2-Bis(diphenylphosphine)ethane]-(dichloro)nickel(II), **12,** 171
Bis(trimethylsilyl)acetylene, **5,** 44
2-(2-Bromoethyl)-1,3-dioxane, **7,** 37
(2-Bromovinyl)trimethylsilane, **11,** 82
Butyllithium, **6,** 85; **8,** 65
Carbonylchlorobis(triphenylphosphine)-rhodium(I), **6,** 105
Carbonylphenylbis(triphenylphosphine)-rhodium(I), **5,** 45
Cesium fluoride, **11,** 115
μ-Chlorobis(cyclopentadienyl)-(dimethylaluminum)-μ-methylene-titanium, **12,** 110
Chlorobis(cyclopentadienyl)-hydridozirconium(IV), **8,** 84
[Chloro(*p*-methoxyphenyl)methylene]-diphenylammonium chloride, **11,** 220
1-Chloro-N,N,2-trimethyl-propenylamine, **7,** 66; **12,** 123
Chlorotrimethylsilane, **10,** 96; **12,** 126
Copper, **12,** 140
Copper(I) bromide–Dimethyl sulfide, **8,** 117
Copper(I) trimethylsilylacetylide, **6,** 148
Dicarbonylcyclopentadienylcobalt, **7,** 84
Dimethylaluminum methylselenolate, **12,** 197
N,N-Dimethylformamide, **8,** 189; **11,** 198
N,O-Dimethylhydroxylamine, **11,** 201
Dimethylphenylsilyllithium, **10,** 162
Dimethylzinc, **3,** 128
Disodium tetracarbonylferrate, **4,** 461; **5,** 624
Grignard reagents, **1,** 415; **5,** 321; **8,** 235; **9,** 229; **10,** 189; **11,** 245

ACYLATION (*Continued*)
Imidazole, **1,** 492
Iron(III) acetylacetonate, **12,** 557
Iron carbonyl, **8,** 265
Lithium, **5,** 376
Lithium alkyl(*t*-butoxy)cuprates, **5,** 395
Lithium *t*-butyl(phenylthio)cuprates, **5,** 414; **7,** 211
Lithium dibutylcuprate, **4,** 127
Lithium dimethylcuprate, **3,** 106; **4,** 177; **5,** 234
Lithium iodide, **5,** 410
Lithium *o*-lithiobenzoate, **10,** 246
Lithium methyl(vinyl)cuprate, **6,** 342
Manganese(II) chloride, **9,** 288
Manganese(II) iodide, **8,** 312; **9,** 289; **10,** 290
Mercury(II) acetate, **7,** 222
4-Methoxy-3-buten-1-ynylcopper, **9,** 297
3-Methoxy-3-methylbutynylcopper, **9,** 300
Methylcopper, **4,** 334
Methyllithium, **1,** 686; **2,** 274
Monochloroborane diethyl etherate, **5,** 465
Nickel carbonyl, **4,** 353
Nickel chloride–Zinc, **10,** 277
Organocopper reagents, **9,** 328; **11,** 365; **12,** 345
Organolithium reagents, **9,** 5; **12,** 350
Organotitanium reagents, **12,** 110
Organozinc reagents, **3,** 128
Phenylthiocopper, **6,** 465
1-Phenylthio-1-trimethylsilylethylene, **11,** 583
Silver tetrafluoroborate, **12,** 434
Sodium diisopropylamide, **1,** 1064
Tetrakis(triphenylphosphine)-palladium(0), **6,** 571; **11,** 503; **12,** 468
N,N,N′,N′-Tetramethylsuccinamide, **4,** 490
Titanium(IV) chloride, **8,** 483
Trifluorovinyllithium, **6,** 622
3-Trimethylsilyl-1-cyclopentene, **8,** 509
2-Trimethylsilylmethyl-1,3-butadiene, **9,** 493; **12,** 539
Trimethylsilylmethyllithium, **6,** 635; **9,** 495; **11,** 581
Trimethylsilylmethylmagnesium chloride, **5,** 724
Trimethylsilyl trifluoromethane-sulfonate, **12,** 543
Triphenylphosphine dihalides, **12,** 554
Vinyllithium, **5,** 748
Vinyltrimethylsilane, **9,** 498; **10,** 444
OF O–H (*see* ENOL ACYLATION, SYNTHESIS INDEX—ESTERS)

ACYLOIN CONDENSATION
3-Benzyl-5-(2-hydroxyethyl)-4-methyl-1,3-thiazolium chloride, **10,** 27
Chlorotrimethylsilane, **2,** 435; **3,** 310; **4,** 537; **5,** 709; **6,** 626; **8,** 107
Chlorotrimethylsilane–Sodium, **2,** 435; **3,** 311
Sodium naphthalenide, **7,** 340
Sodium–Potassium alloy, **1,** 1102

ACYLOXYLATION (*see also* ACETOXYLATION, ALLYLIC REACTIONS)
t-Butyl acetoacetate, **1,** 83
t-Butyl perbenzoate, **4,** 66; **9,** 90
Cobalt(III) trifluoroacetate, **5,** 146
Dibenzoyl peroxide, **1,** 196; **6,** 160
Iodine–Silver carboxylates, **11,** 268
Lead tetraacetate, **5,** 365
Lead tetrabenzoate, **7,** 189
Lead tetrakis(trifluoroacetate), **2,** 238; **4,** 282
Thallium(III) acetate, **4,** 492
Thallium(III) trifluoroacetate, **3,** 286

ACYLOXYSELENYLATION, ACYLOXYSULFENYLATION (*see* ADDITION REACTIONS)

ADDITION REACTIONS TO C = C, C ≡ C (*see also* CONJUGATE ADDITION, CYCLOADDITION, EPOXIDATION, HALOCARBAMATION, HALOLACTAMIZATION, HALOLACTONIZATION, HYDROMETALLATION, REDUCTION REACTIONS, SYNTHESIS INDEX—THREE-MEMBERED RINGS)
ACETOXYMERCURATION (AcO, HgX)
Mercury(II) acetate, **7,** 222; **12,** 298
ACYLOXYSELENYLATION, ALKOXYSELENYLATION (*see* OXYSELENYLATION)
ACYLOXYSULFENYLATION, ALKOXYSULFENYLATION (*see* OXYSULFENYLATION)

AMIDOMERCURATION (RCONH, HgX)
Mercury(II) acetate, **3,** 194; **12,** 298
Mercury(II) nitrate, **3,** 197; **11,** 317
AMINATION TO FORM AZIRIDINES
N-Aminophthalimide, **5,** 14
Benzenesulfonyl azide, **3,** 17; **4,** 24
N-Bromosuccinimide–Sodium azide, **5,** 66
Cyanamide, **3,** 71
N,N-Dichlorourethane, **2,** 121
Diethyl dibromophosphoramidate, **7,** 103
Diethyl N-phenylamidophosphate, **7,** 106
2,4-Dinitrobenzenesulfenamide, **9,** 267
Ethyl azidoformate, **2,** 191
Iodine azide, **1,** 500; **3,** 160; **5,** 350
Iodine isocyanate, **1,** 501; **2,** 223; **3,** 161
Lead tetraacetate, **3,** 168; **5,** 365
O-Mesitylenesulfonylhydroxylamine, **10,** 256
Methyl azidoformate, **5,** 441
p-Nitrobenzenesulfonoxyurethane, **1,** 736
Nitrosyl chloride, **1,** 748
Phenyl azide, **5,** 513
Picryl azide, **1,** 885
Potassium hydroxide, **4,** 409
Sodium azide–Triphenylphosphine, **9,** 418
Tetrabutylammonium azide, **10,** 377
Tin(II) chloride, **1,** 1113
Triphenylphosphine–Carbon tetrachloride, **8,** 516
AMINOMERCURATION (RNH, HgX)
Mercury(II) acetate, **9,** 291; **11,** 315
AMINOMETHYLATION (H, CH_2NR_2)
Rhodium(III) oxide, **4,** 420
AZASELENYLATION (NR_2, SeR)
Benzeneselenenyl halides, **10,** 16; **12,** 39
Chloramine-T, **9,** 101
N,N-Dimethylbenzeneselenenamide, **7,** 119
N-Phenylselenophthalimide, **9,** 366; **12,** 390
AZASULFENYLATION (NR_2, SR)
Benzenesulfenyl chloride, **12,** 42
Chloramine-T, **9,** 101
Dimethyl(methylthio)sulfonium tetrafluoroborate, **11,** 204
AZIDOMERCURATION (N_3, HgX)
Mercury(II) azide, **3,** 196; **4,** 323
Mercury(II) trifluoroacetate, **9,** 294
CARBOMETALLATION (C, M)
Bis(cyclopentadienyl)diiodozirconium, **12,** 53
Carbon dioxide, **5,** 93
Copper(I) bromide–Dimethyl sulfide, **8,** 117
Dichlorobis(cyclopentadienyl)titanium–Alkylaluminum reagents, **9,** 471, 488
Dichlorobis(cyclopentadienyl)-zirconium–*t*-Butylmagnesium chloride, **12,** 171
Grignard reagents, **6,** 269; **8,** 235
Iodine, **9,** 248
Nickel(II) acetylacetonate, **8,** 42
Nickel(II) acetylacetonate–Trimethylaluminum, **9,** 52
Organocopper reagents, **10,** 282
Tetrakis(triphenylphosphine)-palladium(0), **11,** 503
N,N,N′,N′-Tetramethylethylenediamine, **3,** 284; **6,** 576
Trimethylaluminum-Dichlorobis-(cyclopentadienyl)zirconium, **11,** 567
Trimethylsilyllithium, **11,** 575
Vinylcopper reagents, **6,** 662
CYANOSELENYLATION (CN, SeR)
Phenyl selenocyanate, **11,** 416
CYANOSULFENYLATION (CN, SR)
Dimethyl(methylthio)sulfonium tetrafluoroborate, **11,** 204
DIAMINATION (NR_2, NR_2)
Bis(cyclopentadienyl)bis(μ-nitrosyl-N:N)dicobalt, **10,** 116
Bis(*p*-toluenesulfonyl)seleniumdiimide, **7,** 316
Cyanamide, **12,** 147
Dioxobis(*t*-butylimido)osmium(VIII), **8,** 206
Mercury(II) oxide–Tetrafluoroboric acid, **9,** 293
HALOACYLATION (X, COR)
Aluminum chloride, **4,** 10
Methylene chloride, **4,** 337
Tin(IV) chloride, **6,** 553
HALOAMINATION, HALONITRATION, ETC. [X, NR_2 (NO_2, etc.)]
Bromine azide, **3,** 34
Bromonium dicollidine perchlorate, **12,** 78

ADDITION REACTIONS TO C=C, C≡C (*Continued*)
N-Bromosuccinimide–Sodium azide, **5,** 66
Chromium(II) chloride, **3,** 60
Cyanamide, **3,** 71; **12,** 147
N,N-Dichlorourethane, **2,** 121; **4,** 139
Iodine, **8,** 256; **11,** 261
Iodine azide, **1,** 500; **2,** 222; **4,** 262; **5,** 350; **6,** 297; **8,** 260
Iodine isocyanate, **1,** 501; **2,** 223; **3,** 161
Iodine nitrate, **2,** 224
Levulinic acid, **2,** 239
Nitrogen dioxide–Iodine, **3,** 130; **8,** 205
Nitrosyl chloride, **1,** 748; **2,** 298
Nitryl chloride, **1,** 756
Nitryl iodide, **1,** 757; **2,** 300
Pyridinium poly(hydrogen fluoride), **5,** 538
Tin(II) chloride, **1,** 1113
Trichloroacetonitrile, **12,** 526
HALOGENATION (X, X)
Chlorination
Antimony(V) chloride, **4,** 21; **7,** 12
Benzeneselenenyl halides, **12,** 39
Chlorodimethylsulfoxonium chloride, **4,** 200
Copper(II) chloride, **6,** 139
Hydrogen peroxide + co-reagent, **8,** 248; **11,** 256
Molybdenum(V) chloride, **6,** 412
Phenyliodine(III) dichloride, **1,** 505; **2,** 225; **3,** 164; **6,** 298
Tetrabutylammonium iodotetrachloride, **1,** 1132
Tetrabutylammonium octamolybdate, **9,** 447
Trichloramine, **3,** 295; **4,** 520
Bromination
Bromine + co-reagent, **1,** 1025; **9,** 66; **11,** 156
N-Bromoacetamide, **1,** 74
Copper(II) bromide, **1,** 161
Hydrogen peroxide–Hydrobromic acid, **8,** 248
Pyridine perbromide, **5,** 568
Pyridinium bromide perbromide, **1,** 967; **5,** 568
Silver(I) nitrate, **2,** 366
2,4,4,6-Tetrabromo-2,5-cyclohexadienone, **7,** 351
Tetramethylammonium tribromide, **1,** 1144
Addition of X, Y
Bromine chloride, **2,** 38
Bromine fluoride, **2,** 365; **5,** 351; **8,** 53
N-Bromoacetamide–Hydrogen halides, **1,** 74, 75; **2,** 39; **4,** 48
Chloroiodomethane, **10,** 89
Copper(II) chloride, **6,** 139
Hydrogen fluoride, **5,** 336
Iodine fluoride, **2,** 365; **5,** 351
Iodine monochloride, **8,** 261
Iodine–Pyridinium chlorochromate, **12,** 256
Methyliodine(III) difluoride, **7,** 242
Pyridinium poly(hydrogen fluoride)–N-Bromosuccinimide, **5,** 539; **11,** 454
Tellurium(IV) chloride, **9,** 443
HALOHYDRINATION [X, OH(OR)]
Acetyl hypohalites, **1,** 12; **10,** 1
N-Bromoacetamide, **1,** 74; **4,** 47; **5,** 61
N-Bromosuccinimide, **1,** 78; **2,** 40; **3,** 34; **6,** 74; **7,** 182
t-Butyl hydroperoxide–Titanium(IV) chloride, **12,** 94
t-Butyl hypobromite, **1,** 90
Chloramine-T, **10,** 85
N-Chlorosuccinimide, **5,** 127
Chromium(II) sulfate–Dimethylformamide, **5,** 144
Chromyl chloride, **1,** 151; **5,** 144; **8,** 112
Dimethylformamide–Chlorine, **5,** 249
Dimethyl sulfoxide–N-Bromosuccinimide, **2,** 159; **9,** 71
Hexamethylphosphorous triamide, **10,** 199
Hydrogen peroxide–1,3-Dibromo-5,5-dimethylhydantoin, **6,** 287
Hypobromous acid, **1,** 800; **4,** 49
Hypochlorous acid, **10,** 208
Iodine, **3,** 159; **8,** 256; **9,** 248; **11,** 261; **12,** 253
Iodine + co-reagent, **4,** 261; **8,** 260
N-Iodosuccinimide–Carboxylic acids, **7,** 182
Lithium–Ammonia, **8,** 282
Mercury(II) chloride, **1,** 652
Mercury(II) trifluoroacetate, **11,** 320
Oxygen difluoride, **1,** 772
Peracetic acid, **2,** 220

Silver imidazolate, **10,** 349
Silver(I) trifluoroacetate, **1,** 1018
Sodium hypobromite, **1,** 1083
Sodium methoxide, **5,** 617
2,4,4,6-Tetrabromo-2,5-cyclohexadienone, **12,** 457
Thallium(I) acetate–Iodine, **5,** 654; **9,** 458
Titanium(IV) chloride, **6,** 590
Tributyltin hydride, **4,** 518
Trifluoroacetyl chloride, **10,** 419
Trifluoromethyl hypofluorite, **3,** 146; **7,** 156
Urea, **1,** 1262

HALOSELENYLATION (X, SeR)
Benzeneselenenyl halides, **5,** 518; **8,** 25; **9,** 25; **10,** 16
Selenium(I) chloride, **7,** 318

HALOSULFENYLATION (X, SR)
Benzenesulfenyl chloride, **8,** 32; **9,** 35; **10,** 24; **11,** 39
2,4-Dinitrobenzenesulfenyl chloride, **6,** 231
Phthalimide-N-sulfenyl chloride, **9,** 375
Sulfur dichloride, **1,** 1121; **2,** 391; **4,** 469

HALOSULFONYLATION (X, SO_2R)
Bromomethanesulfonyl bromide, **12,** 75
Copper(II) chloride, **4,** 105; **5,** 158

HALOTELLURATION (X, TeX_3)
Tellurium(IV) chloride, **12,** 457

HYDRATION OF C=C (*see also* HYDROMETALLATION–OXIDATION)
Chlorobis(cyclopentadienyl)-hydridozirconium(IV), **6,** 175
m-Chloroperbenzoic acid, **8,** 97
Diethoxymethylsilane, **12,** 182
Hydrogen peroxide, **12,** 242
Lithium aluminum hydride, **9,** 274
Thallium(III) acetate, **9,** 459
Tin(IV) chloride–Sodium borohydride, **9,** 438

HYDRATION OF C≡C
Aluminum, **2,** 19
9-Borabicyclo[3.3.1]nonane, **9,** 57
Catecholborane, **4,** 69
Chlorosulfonyl isocyanate, **3,** 51
Dichloroborane diethyl etherate, **5,** 191
1,3-Dichloro-2-butene, **2,** 111
Diethoxymethylsilane, **12,** 182
Di-2-mesitylborane, **12,** 195
Ion-exchange resins, **1,** 511
Lithium acetylide, **6,** 324
Mercury(II) acetate, **1,** 644; **12,** 303
Mercury(II) oxide, **1,** 655; **6,** 360
Mercury(II) sulfate, **1,** 658
Mercury(II) *p*-toluenesulfonamide, **1,** 660
Monochloroborane diethyl etherate, **5,** 465
Nafion-H, **9,** 320
Phenylmercuric hydroxide, **11,** 415
Phosphorus(V) oxide, **1,** 871
Thallium(III) acetate, **7,** 360
Trimethylamine N-oxide, **10,** 423

HYDROACYLATION (H, COR)
Azobisisobutyronitrile, **1,** 45
Chlorobis(cyclopentadienyl)-hydridozirconium(IV), **6,** 175; **8,** 84
Chlorotris(triphenylphosphine)-rhodium(I), **4,** 559; **8,** 109; **9,** 113; **10,** 98; **12,** 130
Dibenzoyl peroxide, **1,** 196
Disodium tetracarbonylferrate, **6,** 550
Hydridodinitrogentris(triphenyl-phosphine)cobalt(I), **5,** 331
Monoisopinocampheylborane, **11,** 350
Potassium hydroxide, **4,** 409
Potassium tetracarbonylhydridoferrate, **6,** 483
Thexylborane, **2,** 148; **4,** 175
Tributyltin hydride, **4,** 518

HYDROACYLOXYLATION (H, OOCR)
Bis(aceto- or benzonitrile)-dichloropalladium(II), **7,** 21; **12,** 51
Lithium aluminum hydride, **9,** 274
Mercury(II) oxide, **1,** 655; **12,** 305
Methoxyacetylene, **1,** 668
Sulfuric acid, **9,** 441

HYDROALKOXYLATION (H, OR)
Butadiyne, **1,** 185
N,N-Diethylaniline, **5,** 212
Dimethyl sulfoxide, **5,** 263
Mercury(II) trifluoroacetate, **11,** 320
N-Methyl-N-phenylaminoethynyl-lithium, **10,** 267
Methyl vinyl ketone, **1,** 697
Potassium hydroxide, **5,** 557
Silver(I) nitrate, **12,** 433

HYDROAMINATION (H, NR_3)
Acetic acid, **8,** 1

ADDITION REACTIONS TO C=C, C≡C (*Continued*)
Butadiyne, **1,** 185
Cadmium acetate–Zinc acetate, **1,** 103
Chloramine, **1,** 122
Chloramine-T–Diphenyl disulfide, **9,** 101
Dimethylacetamide, **4,** 165
Dimethylformamide diethyl acetal, **7,** 125
Dimethyl(methylthio)sulfonium tetrafluoroborate, **11,** 204
Ethyllithium, **5,** 306
Hydroxylamine-O-sulfonic acid, **1,** 481
Mercury(II) acetate, **9,** 291; **11,** 315
Mercury(II) nitrate, **11,** 317
O-Mesitylenesulfonylhydroxylamine, **5,** 430
Nickel(II) bromide, **4,** 351
Norbornene, **1,** 757
Sodium naphthalenide, **4,** 349; **5,** 468

HYDROCARBOXYLATION AND RELATED REACTIONS [H, COOH (COOR, etc.)]
Carbon monoxide, **7,** 53; **9,** 95
Chlorobis(cyclopentadienyl)-hydridozirconium(IV), **6,** 175
Di-*t*-butyl peroxide, **1,** 211
Di-μ-carbonylhexacarbonyldicobalt, **1,** 224
Dichlorobis(cyclopentadienyl)titanium, **9,** 146
Dichlorobis(triphenylphosphine)-palladium(II), **3,** 81
N,N-Dimethylformamide, **1,** 278
Dimethylformamide dialkyl acetals, **7,** 125
Ethyl bromoacetate, **2,** 192
Ethyl chloroformate, **7,** 147
Formamide, **1,** 402
Formic acid, **4,** 239
Hydrogen hexachloroplatinate(IV)–Tin(IV) chloride, **4,** 87
Nickel carbonyl, **1,** 720; **2,** 290; **5,** 472
Norbornene, **1,** 757
Organopentafluorosilicates, **11,** 373
Palladium(II) chloride + co reagent, **5,** 503; **6,** 450; **7,** 278; **8,** 385
Trifluoromethanesulfonic acid, **9,** 485
Tris(methylthio)methyllithium, **7,** 412

HYDROCYANATION (H, CN) (*see also* CONJUGATE ADDITION)
Chlorotrimethylsilane, **4,** 537
Di-μ-carbonylhexacarbonyldicobalt, **1,** 224
Diisobutylaluminum hydride, **2,** 140
Disodium tetracyanonickelate(II), **9,** 207
Hydrogen cyanide, **9,** 239
Tetrakis(triphenyl phosphite)nickel(0), **11,** 503

HYDROFORMYLATION (H, CHO)
9-Borabicyclo[3.3.1]nonane, **3,** 24
Carbon monoxide, **2,** 253
Carbonylchlorobis(triphenylphosphine)-rhodium(I), **5,** 46
Carbonylhydridotris(triphenyl-phosphine)rhodium(I), **5,** 331; **7,** 53; **9,** 259
Chlorobis(cyclopentadienyl)-hydridozirconium(IV), **6,** 175; **7,** 101
Chlorotris(triphenylphosphine)-rhodium(I), **1,** 1252
Cobalt tetracarbonyl hydride, **2,** 80
Di-μ-carbonylhexacarbonyldicobalt, **1,** 224; **5,** 204
Lithium trimethoxyaluminum hydride, **2,** 252
Rhodium(II) carboxylates, **12,** 423
Rhodium(III) oxide, **4,** 420; **8,** 437

HYDROHALOGENATION (H, X)
Hydrofluorination
Hydrogen chloride, **2,** 215
Hydrogen fluoride, **5,** 336
Phenyliodine(III) difluoride, **6,** 300
Pyridinium poly(hydrogen fluoride), **5,** 538; **9,** 399
Hydrochlorination
Chlorobis(cyclopentadienyl)-hydridozirconium(IV), **6,** 175
N-Chlorosuccinimide, **8,** 103
Copper(II) chloride, **6,** 259
Diisobutylaluminum hydride, **11,** 185
Ferric chloride, **6,** 259
Hydrogen chloride, **5,** 335
Lithium aluminum hydride, **9,** 274
Tin(IV) chloride, **12,** 486
Trifluoroacetic acid, **5,** 695
Trimethylaluminum, **5,** 707
Hydrobromination
Bromine, **3,** 160; **6,** 259
N-Bromosuccinimide, **8,** 54
Catecholborane, **5,** 100

Chlorobis(cyclopentadienyl)-hydridozirconium(IV), **6,** 175
N-Chlorosuccinimide, **8,** 103
Dibenzoyl peroxide, **1,** 196
1,3-Dibromo-2-pentene, **7,** 90
Diisobutylaluminum hydride, **2,** 140; **11,** 185
B-Halo-9-borabicyclo[3.3.1]nonane, **12,** 236
Hydrogen bromide, **1,** 196, 453; **6,** 284
Lithium aluminum hydride, **9,** 274
Sodium bromide–Chloramine-T, **11,** 489

Hydroiodination
Bromine, **6,** 70
Chlorobis(cyclopentadienyl)-hydridozirconium(IV), **6,** 175; **8,** 84
N-Chlorosuccinimide, **8,** 103
Dichlorobis(cyclopentadienyl)-titanium, **10,** 130; **11,** 163
Diisobutylaluminum hydride, **11,** 185
B-Halo-9-borabicyclo[3.3.1]nonane, **12,** 236
Iodine, **3,** 159; **5,** 346; **6,** 259, 293; **7,** 179
Iodine monochloride, **10,** 212; **11,** 268
3-Lithio-1-trimethylsilyl-1-propyne, **2,** 239; **3,** 173
Lithium aluminum hydride, **4,** 295; **9,** 274
Mercury(II) chloride, **6,** 259
Phosphorus(V) oxide–Phosphoric acid, **1,** 872
Sodium iodide–Chloramine-T, **11,** 488
Trifluoroacetic acid, **5,** 695

HYDROSULFENYLATION AND RELATED REACTIONS [H, SR (SO_2R, etc.)]
π-Allylpalladium complexes, **9,** 132
2,4-Dinitrobenzenesulfenyl chloride, **1,** 319
Lithium naphthalenide, **8,** 305
Palladium(II) chloride, **8,** 384; **11,** 393
Silver carbonate, **4,** 425
Thiolacetic acid, **1,** 1154
Thiophenol, **12,** 485

HYDROXYLATION (OH, OH)
anti-
Ammonium persulfate, **12,** 33
Dimethyl sulfoxide, **2,** 157
Dioxygentetrakis(triphenylarsine)-rhodium(I) hexafluorophosphate, **8,** 206
Hydrogen peroxide, **1,** 457
Hydrogen peroxide + co-reagent, **1,** 477; **5,** 340
Iodine–Potassium iodate, **7,** 180
Perbenzoic acid, **1,** 791
Perchloric acid, **1,** 796
Periodic acid, **1,** 815
Persuccinic acid, **1,** 820
Potassium peroxomonosulfate, **1,** 952
Silver iododibenzoate, **1,** 1007; **9,** 411
o-Sulfoperbenzoic acid, **4,** 469
Thallium(I) acetate–Iodine, **10,** 394
Thallium(III) sulfate, **4,** 498
Trifluoroperacetic acid, **1,** 821

syn-
N-Bromoacetamide–Silver acetate–Acetic acid, **7,** 36
t-Butyl hydroperoxide, **1,** 88
Cyanoacetic acid, **11,** 145
Hydrogen peroxide, **1,** 457
Iodine + co-reagent, **5,** 349; **6,** 296; **12,** 256
Methyltriphenylphosphonium permangate, **10,** 446
Osmium tetroxide, **1,** 759; **10,** 290; **12,** 358
Osmium tetroxide + co-reagent, **1,** 88, 475, 764; **2,** 301, 363; **4,** 361; **7,** 255, 256; **9,** 334; **10,** 291, 292
Perbenzoic acid, **1,** 791
Potassium manganate, **1,** 938
Potassium osmiate, **4,** 412
Potassium permanganate, **1,** 942; **4,** 30; **9,** 388; **12,** 413
Silver acetate, **1,** 1002; **2,** 362
Silver chlorate, **1,** 1005
Sodium nitrite–Acetic acid, **6,** 548
Thallium(I) acetate–Iodine, **5,** 654; **7,** 360; **9,** 458; **10,** 394
Trifluoroacetic acid trianhydride, with iodous acid, **4,** 263

NITROMERCURATION (NO_2, HgX)
Mercury(II) nitrite, **9,** 292

NITROSELENYLATION (NO_2, SeR)
Benzeneselenenyl bromide–Silver nitrate, **11,** 33
Silver nitrite–Mercury(II) chloride, **11,** 467

OXYAMINATION [HO, NHTs (NHR)]

ADDITION REACTIONS TO C=C, C≡C (*Continued*)
Iodonium di-*sym*-collidine perchlorate, **11,** 269
Mercury(II) oxide–Tetrafluoroboric acid, **10,** 254
Osmium tetroxide + co-reagent, **7,** 256; **8,** 365
Trioxo(*t*-butylimido)osmium(VIII), **6,** 641; **10,** 445
OXYMERCURATION [HO(RO), HgX]
Dimethyl sulfoxide, **5,** 263
Mercury(II) acetate, **1,** 644; **2,** 264; **3,** 194; **4,** 319; **5,** 424; **6,** 358; **10,** 252; **12,** 298
Mercury(II) chloride, **9,** 291
Mercury(II) nitrate, **7,** 223
Mercury(II) pivalate, **11,** 319
Mercury(II) trifluoroacetate, **5,** 429
OXYSELENYLATION [HO(RO,RCOO), SeR]
Acetoxymethyl methyl selenide, **8,** 5
Benzeneselenenic acid, **8,** 24; **9,** 24
Benzeneselenenyl halides, **8,** 25; **9,** 25; **10,** 16; **12,** 39
Benzeneselenenyl hexafluorophosphate, **10,** 21
Benzeneselenenyl trifluoroacetate, **5,** 522
Benzeneseleninic acid, **8,** 28
Copper(II) chloride, **8,** 119; **10,** 106
Dimethyl selenoxide, **6,** 224
Diphenyl diselenide–Copper(II) acetate, **9,** 199
Methaneselenenic acid, **8,** 319
Phenyl selenocyanate–Copper(II) chloride, **9,** 34
N-Phenylselenophthalimide, **9,** 366; **10,** 312; **11,** 417
OXYSULFENYLATION [HO(RO,RCOO), SR]
Benzenesulfenyl chloride, **8,** 32
Dimethyl(methylthio)sulfonium tetrafluoroborate, **11,** 204
Lead tetrakis(trifluoroacetate)–Diphenyl disulfide, **9,** 269
Methyl bis(methylthio)sulfonium hexachloroantimonate, **11,** 335
PEROXYMERCURATION (HOO, HgX)
t-Butyldimethylsilyl hydroperoxide–Mercury(II) trifluoroacetate, **12,** 85
Hydrogen peroxide–Mercury(II) trifluoroacetate, **8,** 248
Mercury(II) nitrate–Hydrogen peroxide, **7,** 224
Mercury(II) trifluoroacetate, **8,** 316
Tributyltin hydride, **11,** 545
SELENOSULFONYLATION (SeR, SO_2R)
Phenylselenenyl benzenesulfonate, **11,** 407; **12,** 390
Phenylselenenyl *p*-toluenesulfonate, **10,** 315; **12,** 394
SULFONAMIDOMERCURATION (RSO_2NH, HgX)
Mercury(II) nitrate, **11,** 317
SULFURIZATION
Iodine thiocyanate, **4,** 263
Methyl bis(methylthio)sulfonium hexachloroantimonate, **6,** 375
Phthalimide-N-sulfenyl chloride, **9,** 375
Sulfur monochloride, **3,** 275
ADDITION REACTIONS TO C=O—HETEROATOMS
N NUCLEOPHILES (*see also* SYNTHESIS INDEX—ENAMINES, HYDRAZONES, IMINES, OXIMES)
Azidotrimethylsilane, **9,** 21
O,N-Bis(trifluoroacetyl)hydroxylamine, **1,** 60
N-Chlorosuccinimide–Dimethyl sulfide, **9,** 111
Cyanic acid, **1,** 170
Dicyclohexyl-18-crown-6, **5,** 207
Difluoramine, **2,** 134
Hydrazoic acid, **1,** 446
Hydroxylamine, **1,** 478
Hydroxylamine-O-sulfonic acid, **1,** 481
Lead(II) oxide, **1,** 536
O-Mesitylenesulfonylhydroxylamine, **5,** 430
Methyl carbazate, **4,** 333
Silver(I) nitrate, **5,** 582
Sodium hexamethyldisilazide, **1,** 1046
O NUCLEOPHILES (*see also* SYNTHESIS INDEX—ACETALS)
Acetic anhydride–Ferric chloride, **12,** 1
Tetrabutylammonium fluoride, **12,** 458
X NUCLEOPHILES
Boron trichloride, **1,** 67
Carbonyl difluoride, **1,** 116
Dibromomethyllithium, **7,** 90

Dichloromethyl methyl ether, **1,** 220
(Diethylamino)sulfur trifluoride, **6,** 183; **8,** 166; **10,** 142
1,1-Diiodoethane, **6,** 254
Iodotrimethylsilane, **9,** 251
Lithium chloride, **1,** 609
Molybdenum(VI) fluoride, **5,** 465; **6,** 412
Oxalyl chloride, **1,** 767; **5,** 481; **7,** 257
Phenylsulfur trifluoride, **1,** 849
Phosphorus(III) chloride, **1,** 875
Phosphorus(V) chloride, **4,** 388
Selenium(IV) fluoride, **5,** 576
Sulfur tetrafluoride, **1,** 1123; **2,** 392
2,2,2-Trichloro-1,3,2-benzodioxaphosphole, **1,** 120; **2,** 63
Triphenylphosphine–Carbon tetrachloride, **4,** 551; **6,** 644
Triphenylphosphine dihalides, **1,** 1247; **6,** 646, 647
Vilsmeier reagent, **9,** 465
Zinc chloride, **9,** 522
S NUCLEOPHILES (*see also* SYNTHESIS INDEX—THIOACETALS)
Chlorotrimethylsilane, **7,** 66
Hexamethyldisilazane, **7,** 167
Iodotrimethylsilane, **10,** 216
Methylthiotrimethylsilane, **8,** 352
Morpholine, **1,** 705
Phosphorus(V) sulfide, **6,** 470
Potassium cyanide, **7,** 299
Sodium hydrogen sulfite, **1,** 1047
Se NUCLEOPHILES
Benzeneselenol, **6,** 28
Butyllithium, **6,** 85
Methaneselenol, **6,** 361
Phenyl trimethylsilyl selenide, **9,** 373
Thionyl chloride–Triethylamine, **7,** 367
Tris(phenylseleno)borane, **8,** 522; **9,** 511
Zinc chloride, **8,** 536
Si NUCLEOPHILES
Dimethylphenylsilyllithium, **11,** 209
Organoaluminum reagents, **12,** 339
Tetrabutylammonium fluoride, **12,** 458
Trimethylsilyllithium, **9,** 493
ADDITION REACTIONS TO C=O—UNFUNCTIONALIZED CARBON REAGENTS (*see also* ASYMMETRIC REACTIONS, BARBIER REACTION, ENE REACTIONS)
Al REAGENTS
Aluminum, **2,** 19
Diisobutylaluminum hydride, **4,** 158
Formaldehyde–Dimethylaluminum chloride, **10,** 186
Tetrakis(triphenylphosphine)-palladium(0), **10,** 384
Trialkylaluminums, **6,** 622; **11,** 539
Triisobutylaluminum–Bis(N-methylsalicylaldimine)nickel, **11,** 562
B REAGENTS
B-1-Alkenyl-9-borabicyclo[3.3.1]-nonanes, **8,** 6
B-Allyl-9-borabicyclo[3.3.1]nonane, **12,** 15
Allyl phenyl selenide–Trialkylborane, **12,** 20
9-Borabicyclo[3.3.1]nonane, **9,** 57
2-Chloromethyl-4,4,5,5-tetramethyl-1,3,2-dioxaborolane, **12,** 397
B-Crotyl-9-borabicyclo[3.3.1]nonane, **12,** 81
2-Crotyl-4,4,5,5-tetramethyl-1,3,2-dioxaborolane, **11,** 86
Fluorodimethoxyborane, **11,** 240
Lithium 3-chloropropargylide, **9,** 279
Triethylborane, **10,** 416
Ce REAGENTS
Cerium amalgam, **11,** 114
Cerium(III) iodide, **11,** 114
Organocerium reagents, **12,** 345
Cr REAGENTS
Chromium(II) chloride, **11,** 132; **12,** 136
Chromium(III) chloride–Lithium aluminum hydride, **8,** 110; **9,** 119; **10,** 101
Cu REAGENTS
Dilithium trimethylcuprate, **6,** 386; **7,** 115
Dilithium tris(1-pentynyl)cuprate, **6,** 203
Grignard reagents, **10,** 189
Lithium dibutylcuprate, **3,** 79; **5,** 187
Lithium dimethylcuprate, **6,** 209
Li REAGENTS
Boron trifluoride etherate, **12,** 66
n- or *t*-Butyllithium, **4,** 60; **5,** 80; **7,** 47
Chromium(II)–Amine complexes, **3,** 57
1-Cyclobutenylmethyllithium, **6,** 151
gem-Dichloroallyllithium, **5,** 188
1,3-Dilithiopropyne, **6,** 202
Dilithium trimethylcuprate, **6,** 386
β-Furyllithium, **10,** 188

ADDITION REACTIONS TO C=O—UNFUNCTIONALIZED CARBON REAGENTS *(Continued)*
3-Lithio-1-triisopropylsilyl-1-propyne, **11,** 566
1-Lithio-2-vinylcyclopropane, **7,** 192
Lithium, **1,** 570; **4,** 286
Lithium acetylide, **1,** 573, 574; **6,** 324; **7,** 195; **8,** 285; **12,** 272
Lithium dibutylcuprate, **5,** 187
Lithium diisopropylamide, **6,** 334
Lithium naphthalenide, **2,** 288; **4,** 348
Methyllithium, **1,** 686; **9,** 311
Selenium(IV) oxide, **4,** 422
Sodium methylsulfinylmethylide, **2,** 166
N,N,N',N'-Tetramethylethylenediamine, **4,** 485; **5,** 652
2,4,6-Triisopropylbenzenesulfonylhydrazide, **9,** 486
Vinyllithium, **1,** 1273
Mg REAGENTS
Aluminum chloride, **12,** 26
Cyanotrimethylsilane, **7,** 397
1,3-Dichloroacetone, **2,** 109
4,4-Dichloro-3-buten-1-ol, **5,** 192
Dilithium trimethylcuprate, **6,** 386
Dimethylformamide dialkyl acetals, **6,** 222
Grignard reagents, **1,** 415; **10,** 189
3-Lithio-1-trimethylsilyl-1-propyne, **2,** 239
Lithium dibutylcuprate, **5,** 187
3-Methoxyisoprene, **4,** 330
Methylenemagnesium bromide, **2,** 273
Pentane-1,5-di(magnesium bromide), **9,** 355
2-Phenylthiocyclobutanone, **11,** 422
Titanium(IV) chloride, **1,** 1169
Na REAGENTS
Benzylsodium, **6,** 40
Methyl vinyl ketone, **1,** 697
Sodium acetylide, **1,** 1027
Sodium amide, **1,** 1034
Sodium methylsulfinylmethylide, **2,** 166
Si REAGENTS
Allyltrimethylsilane, **10,** 6
Aluminum chloride, **7,** 7
2-Bromo-3-trimethylsilyl-1-propene, **11,** 80
Cesium fluoride, **11,** 115
Crotyltrimethylsilane, **12,** 146
Iodotrimethylsilane, **12,** 259
(E)-(2,4-Pentadienyl)trimethylsilane, **10,** 303
Tetrabutylammonium fluoride, **9,** 444; **10,** 378
Titanium(IV) chloride, **7,** 370; **12,** 494
Trimethylsilylallene, **10,** 428
3-Trimethylsilyl-1-cyclopentene, **8,** 509
2-Trimethylsilylmethylenecyclobutane, **9,** 494
Sm REAGENTS
Samarium(II) iodide, **8,** 439; **11,** 464
Sn REAGENTS
Allyltin difluoroiodide, **10,** 6
Allyltributyltin, **12,** 21
Allyltrimethyltin, **9,** 8
Aluminum chloride, **12,** 26
Bis(diethylaluminum) sulfate, **9,** 46
1,3-Dimethyl-2-imidazolidinone, **11,** 202
Tin, **11,** 519
Tin–Aluminum, **12,** 486
Tin(II) halides, **10,** 374; **11,** 521, 524
Tributylcrotyltin, **10,** 411; **11,** 143, 542; **12,** 146, 513
Ti REAGENTS
Dichlorobis(cyclopentadienyl)titanium, **10,** 130; **12,** 168
Organotitanium reagents, **10,** 138, 270, 422; **11,** 374; **12,** 353
Tetrabutylammonium fluoride, **12,** 458
Zn REAGENTS
Aluminum amalgam, **11,** 24
Organozinc reagents, **12,** 356
Propargyl bromide, **10,** 332
Zinc, **1,** 1285, 1286; **5,** 753; **11,** 598
Zr REAGENTS
Dichlorobis(cyclopentadienyl)zirconium(II), **11,** 51
Organozirconium reagents, **12,** 358
INTRAMOLECULAR REACTIONS
Alkylaluminum halides, **10,** 177; **11,** 7
Allyltrimethylsilane, **10,** 6
2-Bromo-3-trimethylsilyl-1-propene, **11,** 80
Chloromethyllithium, **12,** 77
Chlorotrimethylsilane–Zinc, **12,** 568
Chromium carbonyl, **7,** 71
Chromium(II) chloride, **12,** 136
Dicarbonylbis(triphenylphosphine)nickel, **10,** 125
N-Ethyl-2-pyrrolidone, **9,** 220

2-Hydroxymethyl-3-trimethylsilyl-1-propene, **12,** 252
2-Iodomethyl-3-trimethylsilyl-1-propene, **11,** 579
Lithium dibutylcuprate, **3,** 79
Lithium dimethylcuprate, **7,** 120
Nickel carbonyl, **11,** 356
Potassium–Ammonia, **9,** 377
Samarium(II) iodide, **12,** 429
Sodium naphthalenide, **11,** 490
Tin–Aluminum, **12,** 486
Tin(IV) chloride, **9,** 436
Zinc-copper couple, **7,** 428
ALDOL REACTIONS (*see also* ASYMMETRIC REACTIONS)
GENERAL CONSIDERATIONS
Aluminum *t*-butoxide, **1,** 23
Bis(*p*-methoxyphenyl) telluroxide, **11,** 55
Boric acid, **2,** 32
Calcium carbide, **3,** 40
Glycine, **1,** 412
Ion-exchange resins, **1,** 511
Nafion-H–Palladium on carbon, **10,** 275
Potassium fluoride, **12,** 406
2,4,6-Trimethylphenoxymagnesium bromide, **9,** 489
Triphenylmethyllithium, **1,** 1256
CLAISEN–SCHMIDT REACTIONS
Bis(*p*-methoxyphenyl) telluroxide, **11,** 55
Boric acid, **2,** 32
Boron trifluoride etherate, **11,** 72
Bromomagnesium N-methylanilide, **5,** 440
Dialkylboryl trifluoromethanesulfonates, **7,** 91
Lithium iodide, **10,** 245
Potassium carbonate, **9,** 382
INTRAMOLECULAR REACTIONS
3-Benzyl-5-(2-hydroxyethyl)-4-methyl-1,3-thiazolium chloride, **6,** 38; **10,** 27
Bis(*p*-methoxyphenyl) telluroxide, **11,** 55
Bromoacetylmethylenetriphenylphosphorane, **9,** 66
2-(2-Bromoethyl)-1,3-dioxane, **11,** 78
2-(2-Bromoethyl)-1,3-dioxolane, **9,** 70
2-(2-Bromoethyl)-2-methyl-1,3-dioxolane, **1,** 75
Bromomagnesium diisopropylamide, **12,** 74
t-Butoxybis(dimethylamino)methane, **5,** 71
Copper(I) acetylacetonate, **4,** 100
1-Cyclopropyl-1-trimethylsilyloxyethylene, **8,** 140
Dibenzylammonium trifluoroacetate, **9,** 136
1,3-Dichloro-2-butene, **1,** 214; **2,** 111; **4,** 472
Diethyl acetylmethylmalonate, **7,** 102
Diethyl lithiomorpholinomethylphosphonate, **7,** 105
Diisobutylaluminum phenoxide, **9,** 172; **10,** 150
Ethyl phenyl sulfoxide, **6,** 257
Ion-exchange resins, **10,** 220
Isopropenyl acetate, **6,** 356
Ketene dimethyl thioacetal monoxide, **8,** 268
2-Lithiobenzothiazole, **8,** 274
Lithium–Ammonia, **3,** 179
Lithium diisopropylamide, **8,** 292; **10,** 241
Lithium dimethylcuprate, **6,** 209
Lithium 2,2,6,6-tetramethylpiperidide, **9,** 285
2-Methoxyallyl bromide, **8,** 322
5-Methoxycarbonyl-1-penten-3-one, **1,** 1271
Methyl 4-methylthiocrotonate, **6,** 389
Methyl 3-oxo-4-pentenoate, **1,** 1272
Methyl vinyl ketone, **1,** 697; **7,** 247
2-Methyl-6-vinylpyridine, **6,** 409
Morpholine–Camphoric acid, **9,** 317; **11,** 352
2-Nitropropene, **7,** 253
Phenylsulfinylacetone, **5,** 524
Piperidine, **1,** 886; **7,** 293
Potassium *t*-butoxide, **1,** 911; **9,** 382
Potassium carbonate, **9,** 382
Potassium fluoride, **8,** 410
Potassium hydride, **8,** 412
Potassium hydroxide, **11,** 439
(S)-(–)-Proline, **7,** 307; **12,** 414
Pyrrolidine, **1,** 972; **2,** 354; **7,** 309
Sodium bicarbonate, **5,** 595
Tetrakis(triphenylphosphine)palladium(0), **10,** 384
2,4,4,6-Tetramethyl-5,6-dihydro-1,3-(4H)-oxazine, **4,** 481
Titanium(IV) chloride, **9,** 468; **12,** 494
Triethylamine, **1,** 1198
Triphenyltin hydride, **12,** 555

ALDOL REACTIONS (*Continued*)
Zirconium(IV) propoxide, **11,** 605
STEREOSELECTIVE ALDOLS, DIRECTED ALDOLS, ETC.
with Al enolates
Alkylaluminum halides, **8,** 164
Diisobutylaluminum phenoxide, **9,** 172
Dimethylaluminum benzenethiolate, **10,** 163
Tributyl(diethylaluminum)tin, **12,** 339
Tributyltinlithium–Diethylaluminum chloride, **12,** 523
with B reagents
Benzeneboronic acid, **11,** 408
9-Borabicyclononyl trifluoromethanesulfonate, **8,** 504; **9,** 131
2-Chloro-1,3,2-dioxaborolane, **12,** 224
Dialkylboryl trifluoromethanesulfonates, **7,** 91; **8,** 504; **9,** 140; **10,** 118
Dichloro(diethylamino)borane, **9,** 152
Dichlorophenylborane, **12,** 178
Dicyclopentylboryl trifluoromethanesulfonate, **9,** 131
with Imines
(S)-1-Amino-2-methoxymethyl-1-pyrrolidine, **8,** 16
Boron trichloride, **9,** 62
N,N-Dimethylhydrazine, **7,** 126
Hexamethylphosphoric triamide, **11,** 253
Isobornylamine, **9,** 62
Lithio propylidene-*t*-butylimine, **4,** 283
Lithium diisopropylamide, **2,** 249; **3,** 184
Organotitanium reagents, **11,** 374
(S)-(−)-Proline, **12,** 414
with Li enolates
t-Butyllithium, **11,** 103
Dichlorobis(cyclopentadienyl)-zirconium(II), **10,** 131
Formaldehyde, **6,** 264; **12,** 232
Lithium 1,1-bis(trimethylsilyl)-3-methylbutoxide, **7,** 198
Lithium diisopropylamide, **6,** 334; **8,** 292; **9,** 280
Lithium hexamethyldisilazide, **7,** 197
Lithium iodide, **10,** 245
4-Methoxy-3-penten-2-one, **7,** 231
2-Methyl-2-trimethylsilyloxypentan-3-one, **11,** 348
Phenylselenoacetaldehyde, **10,** 310
Phenylsulfinylacetone, **5,** 524
Trimethylsilylmethylmagnesium chloride, **12,** 540
with Mg enolates
Bromomagnesium diisopropylamide, **9,** 173; **12,** 74
Bromomagnesium N-methylanilide, **5,** 440
with Silyl enol ethers
Boron trifluoride etherate, **8,** 51
Cesium fluoride, **10,** 81
Iodotrimethylsilane, **12,** 259
Lithium diisopropylamide, **8,** 292
Mandelic acid, **12,** 291
1-Methoxy-2-methyl-3-trimethylsilyloxy-1,3-pentadiene, **11,** 326
Organotitanium reagents, **12,** 353
Tetrabutylammonium fluoride, **8,** 467; **12,** 458
Tin(IV) chloride, **12,** 486
Titanium(IV) chloride, **6,** 590; **12,** 494
Tributyltin fluoride, **12,** 515
Trimethylsilyl trifluoromethanesulfonate, **10,** 438
Triphenylcarbenium perchlorate, **12,** 549
Tris(diethylamino)sulfonium difluorotrimethylsilicate, **10,** 452
with Sn enolates
3-Acylthiazolidine-2-thiones, **12,** 4
2,3-Butanediol, **11,** 84
(−)-Norephedrine, **12,** 338
Tin, **11,** 519
Tin(II) fluoride, **11,** 524
Tin(II) trifluoromethanesulfonate, **11,** 525; **12,** 490
Tributyltin fluoride, **12,** 515
Tributyltinlithium–Diethylaluminum chloride, **12,** 523
Triphenyltin chloride, **10,** 451
with Ti enolates
Organotitanium reagents, **11,** 374; **12,** 54
Titanium(IV) chloride, **12,** 494
with Zn enolates
Formaldehyde, **6,** 264
Lithium dimethylcuprate, **6,** 209

Zinc, **7,** 426
Zinc chloride, **5,** 763; **6,** 676

ALDOL-TYPE REACTIONS (*see also* ASYMMETRIC REACTIONS)
HOMOALDOLS (*see* SYNTHESIS INDEX—HOMOALLYLIC ALCOHOLS)
TO FORM β-HYDROXY ACIDS, ESTERS, ETC. OR α,β-UNSATURATED ACIDS, ESTERS, ETC.
Acetic anhydride–Triethylamine, **6,** 6
Aceto(carbonyl)cyclopentadienyl-(triphenylphosphine)iron, **12,** 1
(R)-2-Acetoxy-1,1,2-triphenylethanol, **12,** 3
3-Acylthiazolidine-2-thiones, **12,** 4
Alkylaluminum halides, **8,** 164
Benzyltrimethylammonium hydroxide, **1,** 1252
Bis[methoxy(thiocarbonyl)] disulfide, **10,** 39
Bromomagnesium diethylamide, **5,** 209
t-Butyl 2-chloro-2-lithiotrimethyl-silylacetate, **8,** 60
t-Butyldimethylchlorosilane, **11,** 88
t-Butyl lithioacetate, **5,** 371; **6,** 84
t-Butyl lithiobis(trimethylsilyl)acetate, **7,** 42
t-Butyl lithio(trimethylsilyl)acetate, **5,** 371
2-Chloro-1,3,2-dioxaborolane, **12,** 224
Chloromethyldiphenylsilane, **12,** 321
Chloromethyl methyl ether, **7,** 61
Dialkylboryl trifluoromethanesulfonates, **9,** 131; **10,** 118
1,8-Diazabicyclo[5.4.0]undecene-7, **5,** 177
Dichlorobis(cyclopentadienyl)-zirconium(II), **10,** 131
Diethyl ethylidenemalonate, **9,** 161
3,4-Dihydro-2H-pyran, **10,** 147
N,N-Dimethylhydrazine, **7,** 126
2,6-Dimethylphenyl propionate, **10,** 162
O-Ethyl-S-ethoxycarbonylmethyl dithiocarbonate, **7,** 148
Ethyl formate, **12,** 226
Ethyl lithioacetate, **3,** 172; **8,** 225
Ethyl lithio(trimethylsilyl)acetate, **5,** 373
Ethyl 2-phenylsulfinylacetate, **6,** 256
Ethyl trimethylsilylacetate, **7,** 150; **12,** 228
Formaldehyde, **1,** 397; **5,** 312; **7,** 158
Hexamethylphosphoric triamide, **8,** 240
Isopropenyl acetate, **6,** 308
Ketene derivatives, **10,** 261, **12,** 268
Lithium amide, **1,** 600
Lithium 1,1-bis(trimethylsilyl)-3-methylbutoxide, **7,** 198
Lithium α-ethoxycarbonylvinyl-(1-hexynyl)cuprate, **6,** 329
Lithium dicyclohexylamide, **5,** 397
Lithium diethylamide, **6,** 331
Lithium diisopropylamide, **4,** 298; **6,** 334; **7,** 204; **8,** 292; **12,** 277
Lithium N-isopropylcyclohexylamide, **4,** 306
Lithium naphthalenide, **3,** 208
Magnesium methoxide, **11,** 309
α-Mercapto-γ-butyrolactone, **8,** 314
p-Methoxybenzyl itaconate, **6,** 365
Methyl O-methyllactate, **11,** 341
2-Methyl-2-trimethylsilyl-oxypentan-3-one, **11,** 348
Organoaluminum reagents, **12,** 339
2-Oxazolidones, chiral, **11,** 379
Potassium carbonate, **5,** 552; **9,** 382
Sodium ethoxide, **1,** 1065
Tetrabutylammonium fluoride, **12,** 458
Tin(II) trifluoromethanesulfonate, **12,** 490
Titanium(IV) alkoxides, **6,** 11
p-Tolylsulfinylacetic acid, **10,** 405; **11,** 538
Tributylcrotyltin, **10,** 411; **11,** 143
Trimethylsilylacetic acid, **6,** 631
Trimethylsilylacetyl chloride, **8,** 509
(Trimethylsilyl)acetyltrimethylsilane, **11,** 572
Trimethylsilylmethyllithium, **6,** 635
Tris(methylthio)methyllithium, **7,** 412
Tris(trimethylsilyloxy)ethylene, **9,** 512
Zinc chloride, **11,** 602
Zinc–Graphite, **12,** 570

INTRAMOLECULAR ALDOLS
Bis(1,5-cyclooctadiene)nickel(0), **9,** 45
t-Butylmagnesium chloride, **4,** 63
Carbon monoxide, **12,** 105
Crown ethers, **6,** 133
Cryptates, **12,** 404
Diethylaluminum chloride, **8,** 164
Dimethyl sulfone, **2,** 157

ALDOL-TYPE REACTIONS (*Continued*)
Tetrabutylammonium fluoride, **11,** 499
NITROALDOLS
Ion-exchange resins, **1,** 511
Lithium aluminum hydride, **12,** 272
Nitromethane, **1,** 739; **9,** 323; **12,** 337
β-Nitropropionyl chloride, **8,** 363
Sodium ethoxide, **1,** 1065
Tetrabutylammonium fluoride, **9,** 444; **10,** 378; **11,** 499; **12,** 458
Titanium(III) chloride, **5,** 669
ALKOXYCARBONYLATION
OF RH, ArH, C=C–H
Ammonium persulfate–Silver nitrate, **9,** 18
Arylthallium bis(trifluoroacetates), **10,** 300
Bromine–Antimony(V) chloride–Sulfur dioxide, **5,** 57
Carbon disulfide, **5,** 95
Carbonyl sulfide, **11,** 112
Chloromethyl methyl ether, **6,** 109
Diethyl carbonate, **1,** 247; **2,** 129; **5,** 213
Diethyl oxalate, **1,** 250
Dimethyl carbonate, **2,** 149; **5,** 234; **12,** 201
Ethyl chloroformate, **1,** 364; **5,** 294
Ethyl diazoacetate, **6,** 252
Ethyl diethoxyphosphinylformate, **1,** 370
Ferrous sulfate, **4,** 237
Hexafluoroantimonic acid, **2,** 216
Lithium–Ammonia, **8,** 282
Lithium diisopropylamide, **6,** 334
Magnesium methyl carbonate, **1,** 631; **6,** 354
Mercury(II) oxide–Boron trifluoride, **11,** 318
Methyl cyanoformate, **12,** 321
Oxalyl chloride, **3,** 216
Palladium(II) chloride, **10,** 300
Palladium(II) chloride–Copper(II) chloride, **10,** 302
Potassium amide, **3,** 232
Trichloroacetyl chloride, **4,** 521
Trimethylsilylmethylpotassium, **12,** 541
OF RX, RM
Arene(tricarbonyl)chromium complexes, **10,** 13
Boron trifluoride etherate, **11,** 72
Butyllithium, **9,** 83
Carbon monoxide, **7,** 53
Dihalobis(triphenylphosphine)-palladium(II), **6,** 59, 60
Disodium tetracarbonylferrate, **4,** 461
Ethyl chloroformate, **1,** 364; **7,** 147
Grignard reagents, **1,** 415
Iron carbonyl, **8,** 265
Mercury(II) acetate, **12,** 298
Methyl chloroformate, **9,** 306
Methyl methylthiomethyl sulfone, **11,** 242
Nickel carbonyl, **1,** 720; **3,** 210
Organopentafluorosilicates, **11,** 373
Palladium(II) chloride, **11,** 393; **12,** 371
Sodium tetracarbonylcobaltate, **1,** 1058
OF OTHER SUBSTRATES
Alkyl chloroformates, **5,** 117; **12,** 15
Benzyltrimethylammonium fluoride, **10,** 29
Bromine–Antimony(V) chloride–Sulfur dioxide, **5,** 57
t-Butyl azidoformate, **2,** 44; **6,** 77
1-Chlorocarbonylbenzotriazole, **8,** 87
Copper(I) alkoxides, **9,** 122
N,N-Dimethylformamide, **5,** 247
Dimethyl sulfate, **4,** 188
Ethyl chloroformate, **1,** 364
Ethyl diazoacetate, **1,** 367
Hexamethylphosphorous triamide–Dialkyl azodicarboxylates, **6,** 280
Lithium 2,2,6,6-tetramethylpiperidide, **8,** 307
Methyl chloroformate, **6,** 376; **11,** 336
Nickel carbonyl, **1,** 720
Rhodium catalysts, **12,** 426
Triphenylphosphine–Diethyl azodicarboxylate, **11,** 589
ALKOXYLATION (*see also* METHOXYLATION)
t-Butyl perbenzoate, **1,** 98
Chlorotris(triphenylphosphine)-rhodium(I), **4,** 559
Silver(I) oxide, **5,** 583
ALKOXYSELENYLATION, ALKOXYSULFENYLATION (*see* ADDITION REACTIONS)
ALKYLATION REACTIONS (*see also* ARYLATION, ASYMMETRIC REACTIONS, COUPLING REACTIONS, METHYLATION, PROPARGYLATION)
OF RCOX

Arene(tricarbonyl)chromium complexes, **6,** 125
2-(3-Bromopropyl)-2-methyl-1,3-dioxolane, **6,** 74
t-Butyl lithioacetate, **10,** 67
Butyllithium, **2,** 51
Butyllithium–Nickel(II) bromide, **8,** 66
Chromium carbonyl, **6,** 125
Copper(I) iodide, **8,** 121
Diisopropylethylamine, **10,** 151
Dimethylformamide dimethyl acetal, **8,** 191
Ethoxycarbonylmethylcopper, **4,** 224
Ethyl 4,4-dimethoxy-2-phenylthiobutyrate, **8,** 222
Ethyl trichloroacetate, **4,** 233
Lead tetraacetate, **8,** 269
Lithium diethylamide, **6,** 331; **7,** 201; **11,** 295
Lithium diisopropylamide, **5,** 400, 406; **7,** 204; **8,** 292; **10,** 241
Lithium dimethylcuprate, **8,** 301
Lithium N-isopropylcyclohexylamide, **4,** 306
4-Methoxycarbonylthiolane-3-one, **11,** 324
Methyl α-benzalaminocrotonate, **9,** 305
Methyl 3-(N,N-dimethylamino)-propionate, **9,** 306
Methyl phenylsulfinylacetate, **6,** 393
Phase-transfer catalysts, **9,** 356; **11,** 403
Potassium *t*-butoxide, **1,** 911
Potassium 2,6-di-*t*-butylphenoxide, **3,** 234
Potassium–Graphite, **8,** 405
Reformatsky reagent, **12,** 422
Silver(I) oxide, **3,** 252
Triphenylmethylpotassium, **1,** 1258
Zinc chloride, **8,** 536
OF RCHO, R_2CO
via Enolates
Arene(tricarbonyl)chromium complexes, **6,** 125
Benzeneselenenyl halides, **6,** 459; **8,** 25
Benzoin, **6,** 34
Benzyltrialkylammonium halides, **4,** 27; **6,** 44
Bis(dibenzylideneacetone)-palladium(0), **11,** 53
Bis(methylthio)(trimethylsilyl)-methyllithium, **6,** 53
9-Borabicyclo[3.3.1]nonane, **3,** 24
1-Butanethiol, **2,** 53; **4,** 64
2-(2-Buten-2-yl)-1,3-dithiane, **11,** 447
Butyllithium–Tetramethylethylene-diamine, **10,** 73
Chromium carbonyl, **6,** 125
Copper(I) cyanide, **9,** 124
Cryptates, **10,** 112
Dichlorobis(tri-*o*-tolylphosphine)-palladium(II), **12,** 173
1,3-Dichloro-2-butene, **2,** 111
4,4-Dichloro-3-buten-1-ol, **5,** 192
Dicyclohexyl-18-crown-6, **6,** 174
Dimethylphenylsilyllithium, **12,** 210
Dimethyl sulfoxide, **1,** 296
Diphenyl disulfide, **6,** 235
Ethyl ethylthiomethyl sulfoxide, **5,** 299
Ethyl formate, **2,** 197
Hexamethylphosphoric triamide, **3,** 149
Lithium diallylcuprate, **7,** 86
Lithium diisopropylamide, **4,** 298
Lithium 2-dimethylaminoethoxide, **10,** 244
Lithium dimethylcuprate, **5,** 234
Lithium methyl(vinyl)cuprate, **6,** 342
Methyl α-(methyldiphenylsilyl)-acetate, **12,** 325
2-Oxo-3-(triphenylphosphoranyl-idene)propyllithium, **5,** 375
Phase-transfer catalysts, **9,** 356
2-Phenylthiocyclobutanone, **11,** 422
Potassium *t*-amyloxide, **9,** 379
Potassium *t*-butoxide, **1,** 911; **2,** 336
Potassium hydride, **5,** 557; **8,** 412; **9,** 386; **12,** 407
Simmons–Smith reagent, **3,** 255
Sodium, **8,** 445
Sodium amide, **1,** 1034
Tetrabutylammonium iodide, **5,** 646
Tetrakis(triphenylphosphine)-palladium(0), **10,** 384; **11,** 503; **12,** 468
Tetramethylurea, **1,** 1146
Trichloroethylene, **11,** 552
Triethanolamine borate, **8,** 500
Triethylaluminum, **3,** 299
Triethylborane, **9,** 482
Trimethyloxonium 2,4,6-trinitrobenzenesulfonate, **2,** 438
Trimethylsilylmethyllithium, **9,** 495

ALKYLATION REACTIONS (*Continued*)
Triphenylcarbenium tetrafluoroborate, **8,** 524
Triphenylmethylsodium, **1,** 1259
Tris(diethylamino)sulfonium difluorotrimethylsilicate, **10,** 452
Zinc, **2,** 459
via Silyl enol ethers
1-*t*-Butylthio-1-trimethylsilyloxyethylene, **9,** 92
Chlorotrimethylsilane, **2,** 435
Iodotrimethylsilane, **12,** 259
Methyl α-chloro-α-phenylthioacetate, **12,** 320
Simmons–Smith reagent, **9,** 415
Titanium(IV) chloride, **8,** 483; **10,** 401
Zinc halides, **9,** 520; **11,** 600; **12,** 574
Zinc–Copper–Isopropyl iodide, **10,** 460
via Other intermediates
Allyl chloroformate, **12,** 15
sec-Butyllithium, **12,** 97
Dichloroketene, **12,** 176
Ethyl diazoacetate, **5,** 295
Grignard reagents, **10,** 189
Lithium alkyl(*t*-butoxy)cuprates, **5,** 395
Lithium *t*-butyl(phenylthio)cuprate, **5,** 414
Lithium dimethylcuprate, **4,** 177; **5,** 234; **6,** 209
Potassium 3-aminopropylamide, **8,** 406
Potassium carbonate, **10,** 323
Potassium 2,6-di-*t*-butylphenoxide, **3,** 234
Potassium tetracarbonylhydridoferrate, **6,** 483
OF C=C–H, C≡C–H (*see* COUPLING REACTIONS)
OF AMINES AND OTHER N COMPOUNDS (*see also* SYNTHESIS INDEX—AMINES)
N-Alkylhydroxylamines, **9,** 42
N-Benzoyl-N′-trifluoromethanesulfonylhydrazide, **6,** 36
Benzyldodecyldimethylammonium bromide, **7,** 16
Benzyltriethylammonium chloride, **6,** 41
N-Benzyl trifluoromethanesulfonamide, **6,** 43
Bisbenzenesulfenimide, **3,** 20
t-Butyl methyl iminodicarboxylate potassium salt, **8,** 72
Chloroacetyl chloride, **3,** 46
Crown ethers, **10,** 110
Di-*t*-butyl iminodicarboxylate, **1,** 210
Diethoxycarbenium tetrafluoroborate, **4,** 144
Dimethylformamide dibenzyl acetal, **6,** 221
Dimethyl sulfoxide, **5,** 263
Disodium tetrachloropalladate(II), **10,** 175
Ethyl chloroformate, **4,** 228
Ethylene oxide, **5,** 297
Formamide, **1,** 402
Hexamethylphosphoric triamide, **4,** 244
Hydrazine, **5,** 327
Lithium diisopropylamide, **5,** 400
Lithium methylsulfinylmethylide, **6,** 342
Organocopper reagents, **11,** 365
1,2,2,6,6-Pentamethylpiperidine, **4,** 370
Phase-transfer catalysts, **8,** 387; **10,** 305; **11,** 403
Polyphosphate ester, **6,** 474
Potassium *t*-butoxide, **1,** 911
Potassium fluoride, **10,** 325
Potassium hydride, **9,** 387
Potassium hydroxide, **4,** 409; **9,** 388
Sodium amide, **1,** 1034
Sodium borohydride, **6,** 530
Sodium hydride, **1,** 1075; **4,** 452; **5,** 610
Sodium hydroxide–Potassium carbonate, **9,** 429
Tetrabutylammonium salts, **6,** 565; **7,** 353; **8,** 467
Tetramethylene sulfone, **1,** 1144
Thallium(I) ethoxide, **2,** 407
Tributylhexadecylphosphonium bromide, **7,** 166
Tricaprylylmethylammonium chloride, **7,** 246
Triethyloxonium tetrafluoroborate, **3,** 303; **4,** 527; **5,** 691
Trifluoromethanesulfonic anhydride, **5,** 702
Trimethylaluminum, **8,** 506
Trimethyl phosphate, **5,** 716
Trimethylsilylmethyl trifluoromethanesulfonate, **10,** 434
OF ArH (*see* COUPLING REACTIONS,

FRIEDEL–CRAFTS ACYLATION)
OF RCOOH
Bis(chlorodimethylsilyl)ethane, **10,** 140
Bis(ethylthio)acetic acid, **9,** 228
Butyllithium, **7,** 45
Lithium diisopropylamide, **3,** 184; **8,** 292; **11,** 296
Methylthioacetic acid, **6,** 395
2-Phenylselenopropionic acid, **8,** 397
Phenylthioacetic acid, **6,** 463
Sodium amide, **1,** 1034
Trimethylsilylacetic acid, **6,** 631
OF DICARBONYLS AND RELATED COMPOUNDS
Barium hydroxide, **4,** 23
Benzyltriethylammonium chloride, **6,** 41
Bis(acetonitrile)dichloropalladium(II), **8,** 39; **11,** 46
Bis(dibenzylideneacetone)palladium(0), **10,** 34
Boron trifluoride, **1,** 68
t-Butyl cyanoacetate, **1,** 87
Butyllithium, **6,** 85
4-Chloromethyl-3,5-dimethylisoxazole, **2,** 150
Cobalt(II) chloride, **10,** 101
Cryptates, **5,** 156
Cyanotrimethylsilane, **12,** 148
1,8-Diazabicyclo[5.4.0]undecene-7, **8,** 141
Dibenzoyl peroxide, **4,** 122
(Z)-1,2-Dichloro-4-phenylthio-2-butene, **7,** 98
S,S′-Diethyl dithiomalonate, **9,** 160
Diethyl oxalate, **2,** 132
Dilithium tetrachloropalladate(II), **8,** 176
6,6-Dimethyl-5,7-dioxaspiro[2.5]-octane-4,8-dione, **6,** 216
N,N-Dimethylformamide, **1,** 278
Dimethyl sulfoxide, **1,** 296; **2,** 157; **3,** 119
Disodium tetrachloropalladate(II), **8,** 217
Ethyl 3-ethyl-5-methyl-4-isoxazolecarboxylate, **4,** 232
Ethyl formate, **2,** 197
Ethyl malonate, **6,** 255
Ethyl 2-methylsulfinylacetate, **6,** 255
Guanidines, **11,** 249
Hexamethylphosphoric triamide, **6,** 273; **10,** 196
Ion-exchange resins, **8,** 263
Iron carbonyl, **12,** 266
Isobutene, **1,** 522
Lithium diisopropylamide, **4,** 298
Magnesium methyl carbonate, **1,** 631
Meldrum's acid, **1,** 526
Methyl iodide, **1,** 682
5-Methylisoxazole, **9,** 309
Molybdenum carbonyl, **12,** 330
Nickel(II) acetylacetonate, **6,** 417
Palladium(II) chloride, **4,** 369; **5,** 500, 503; **6,** 447; **7,** 277
Pentaethoxyphosphorane, **2,** 305
Phase-transfer catalysts, **9,** 356
Phenylsulfinylacetone, **5,** 524
(Phenylsulfonyl)nitromethane, **11,** 419
Potassium carbonate, **5,** 552
Potassium 2,6-di-*t*-butylphenoxide, **3,** 234
Potassium tetracarbonylhydridoferrate, **6,** 483
Pyrrolidine, **1,** 972
Raney nickel, **9,** 405
Sodium ethoxide, **1,** 1065
Sodium methylsulfinylmethylide, **2,** 166; **7,** 338
Tetrakis(triphenylphosphine)-palladium(0), **8,** 472; **9,** 451; **11,** 503
Thallium(I) ethoxide, **2,** 407
Thallium(I) hydroxide, **6,** 578
Triethylamine, **1,** 1198
Triphenylmethylpotassium, **9,** 502
Triphenylphosphine–Diethyl azodicarboxylate, **4,** 553
Tris(acetonitrile)tricarbonyltungsten, **12,** 556
OF ENAMINES
1,1-Dichloro-3-bromopropene, **5,** 191
1,3-Dichloro-2-butene, **1,** 214
2,3-Dichloro-1-propene, **8,** 158
Diethyl lithio-N-benzylideneamino-methylphosphonate, **9,** 161
Diisopropylethylamine, **7,** 148
Hexamethylphosphoric triamide, **4,** 244
2-Methyl-6-vinylpyridine, **6,** 409
Morpholine, **1,** 705
Palladium(II) acetate, **4,** 365
β-Propiolactone, **1,** 957
Pyrrolidine, **1,** 972
Tetrakis(triphenylphosphine)-palladium(0), **11,** 503

ALKYLATION REACTIONS *(Continued)*
GEMINAL ALKYLATION
1-Butanethiol, **4,** 64
t-Butoxybis(dimethylamino)methane, **5,** 71
2-Chloroethyl(dimethyl)sulfonium iodide, **12,** 113
1,8-Diazabicyclo[5.4.0]undecene-7, **7,** 87
1,4-Dichloro-2-butene, **10,** 134
Diethyl N-benzylideneaminomethylphosphonate, **11,** 178
Diphenyl disulfide, **6,** 235; **7,** 137
Diphenylsulfonium cyclopropylide, **4,** 211; **6,** 242
1-Lithiocyclopropyl phenyl sulfide, **6,** 319
Lithium diisopropylamide, **5,** 400, 406
Lithium dimethylcuprate, **4,** 177
Nickel(II) acetylacetonate, **6,** 417
Organotitanium reagents, **10,** 138
Phase-transfer catalysts, **9,** 356
Phenylthioacetic acid, **6,** 463
Potassium amide, **5,** 543
Trimethylaluminum, **6,** 622
β-Trimethylsilylethylidenetriphenylphosphorane, **9,** 492
OF IMINES AND OTHER C=N
t-Butyldimethylchlorosilane, **11,** 88
Butyllithium, **11,** 101
N,N-Dimethylhydrazine, **7,** 126
Lithium diethylamide, **7,** 201
2-Methyl-2-thiazoline, **6,** 403
INTRAMOLECULAR ALKYLATION
Aluminum chloride, **12,** 26
Arene(tricarbonyl)chromium complexes, **6,** 125
Benzyltriethylammonium chloride, **6,** 41
t-Butyllithium, **10,** 76
t-Butyl perbenzoate, **7,** 49
4-Chloro-1-butenyl-2-lithium, **12,** 113
Chromium carbonyl, **6,** 125
1,8-Diazabicyclo[5.4.0]undecene-7, **7,** 87
S,S′-Diethyl dithiomalonate, **9,** 160
Diethylthallium *t*-butoxide, **7,** 108
Diisopropylethylamine, **7,** 148
1,3-Dithiane, **2,** 182
Ethylmagnesium bromide, **9,** 218
Gases, **4,** 241
Guanidines, **12,** 477
2-Hydroxymethyl-3-trimethylsilyl-1-propene, **11,** 258
Iodine, **3,** 159; **4,** 258
4-Iodobutyltrimethyltin, **10,** 219
Ion-exchange resins, **1,** 511; **4,** 267
Lithium amide, **3,** 178; **5,** 392
Lithium chloride, **10,** 240
Lithium diethylamide, **2,** 247
Lithium diisopropylamide, **5,** 400; **6,** 334; **8,** 292
Lithium dimethylcuprate, **5,** 234
Methyl 2-bromocrotonate, **7,** 235
Methyllithium, **9,** 311
Methyl methylthiomethyl sulfone(or -sulfoxide), **6,** 390; **11,** 242
Methyl 2-trimethylsilylacrylate, **11,** 346
Palladium(II) acetate, **4,** 365; **10,** 297
Phenylthioacetyl chloride, **9,** 370
Phosphorus(V) oxide, **10,** 319
Potassium amide, **5,** 543
Potassium *t*-butoxide, **4,** 399
Potassium fluoride, **2,** 346
Potassium hexamethyldisilazide, **4,** 407; **9,** 385
Potassium trifluoromethanesulfinate, **5,** 564
Sodium amide, **5,** 591
Sodium ethoxide, **1,** 1065
Sodium hexamethyldisilazide, **6,** 529; **12,** 446
Sodium hydride, **1,** 1075; **4,** 452; **5,** 610
Sodium methylsulfinylmethylide, **2,** 166; **4,** 195
Tetrakis(triphenylphosphine)-palladium(0), **8,** 472; **9,** 451; **11,** 503; **12,** 468
2,4,4,6-Tetramethyl-5,6-dihydro-1,3-(4H)-oxazine, **3,** 280; **4,** 481
Tin(IV) chloride, **11,** 522
Tosylmethyl isocyanide, **10,** 409; **11,** 539
Trifluoromethanesulfonyl chloride, **9,** 485
2,4,6-Triisopropylbenzenesulfonylhydrazide, **12,** 533
2,4,4-Trimethyl-2-oxazoline, **6,** 629
Triphenylphosphine, **3,** 317; **4,** 548
Triphenylphosphine–Diethyl azodicarboxylate, **7,** 404
OF RCN
Arene(tricarbonyl)chromium complexes, **9,** 21; **12,** 34
Benzyltriethylammonium chloride, **3,** 19; **5,** 26; **7,** 18

Cyanomethylcopper, **4,** 111
Cyanotrimethylsilane, **5,** 720; **6,** 632; **9,** 127
Dibenzo-18-crown-6, **6,** 159
N,N-Diethylaminoacetonitrile, **9,** 159
Diisobutylaluminum hydride, **11,** 185
N,N-Dimethyldithiocarbamoylacetonitrile, **7,** 123
2-(2,6-Dimethylpiperidino)acetonitrile, **11,** 212
Lithium diethylamide, **1,** 610
Lithium diisopropylamide, **4,** 298
Lithium N-isopropylcyclohexylamide, **5,** 411
Magnesium naphthalenide, **1,** 711
Methyl phenylsulfinylacetate, **6,** 393
Phase-transfer catalysts, **8,** 387
Potassium amide, **1,** 907
Potassium 2,6-di-*t*-butylphenoxide, **3,** 234
Potassium–Graphite, **8,** 405
Potassium hexamethyldisilazide, **4,** 407
Potassium tetracarbonylhydridoferrate, **6,** 483
Sodium hexamethyldisilazide, **6,** 529
Sodium naphthalenide, **1,** 711

OF O–H (*see also* ENOL ETHERIFICATION, SYNTHESIS INDEX—ESTERS, ETHERS)
Adogen 464, **6,** 10
Chloromethyl methyl ether, **6,** 109
Ion-exchange resins, **1,** 511
Isopropyl bromide, **9,** 69
Lithium diisopropylamide, **5,** 400
Methylene chloride, **7,** 239
Methylene chloride–Cesium fluoride, **7,** 239
Silver tetrafluoroborate, **1,** 1015
Thallium(I) ethoxide, **10,** 395
Triethyloxonium tetrafluoroborate, **1,** 1210; **3,** 303; **5,** 691
Triphenylphosphine–Diethyl azodicarboxylate, **7,** 404

OF α,β- OR β,γ-UNSATURATED SUBSTRATES
Bis(dibenzylideneacetone)palladium(0), **9,** 46
2-(2-Bromoethyl)-2-methyl-1,3-dioxolane, **1,** 75
Butyllithium, **9,** 83, 87; **10,** 73
4-Chloromethyl-3,5-dimethylisoxazole, **1,** 276
Copper(I) halides, **7,** 81; **10,** 104, 107
N,N-Dimethylhydrazine, **8,** 192
Dimethyl sulfoxide, **9,** 189
Lithium diisopropylamide, **4,** 298; **5,** 400, 406; **10,** 241
Lithium dimethylcuprate, **6,** 209
Lithium α-ethoxycarbonylvinyl-(1-hexynyl)cuprate, **6,** 329
Lithium hexamethyldisilazide, **5,** 393
Lithium N-isopropylcyclohexylamide, **4,** 306; **5,** 411
Lithium 2,2,6,6-tetramethylpiperidide, **6,** 345
Methyl iodide, **1,** 682
Methyl 4-methylthiocrotonate, **6,** 389
Potassium *t*-amyloxide, **1,** 939
Potassium *t*-butoxide, **1,** 911
Potassium 2,6-di-*t*-butylphenoxide, **3,** 234
Potassium hydride, **7,** 301; **8,** 412
Simmons–Smith reagent, **6,** 521
Sodium *t*-amyloxide, **1,** 1096
Titanium(IV) chloride, **11,** 529
Trichloroethylene, **9,** 479; **10,** 414
Trimethylene bis(thiotosylate), **1,** 956
Zinc bromide, **9,** 520

ALKYLSULFONYLATION (*see* SULFONYLATION)

ALKYLTHIOALKYLATION
Benzyl bromomethyl sulfide, **5,** 25
Chlorodimethylsulfonium chloride, **5,** 114
Chloromethyl phenyl sulfide, **9,** 107
N-Chlorosuccinimide + co-reagent, **5,** 129; **6,** 119
Dimethyl(phenylthio)methylamine, **10,** 164
Dimethyl sulfide, **4,** 190; **5,** 260
Dimethyl sulfoxide, **1,** 296
Dimethyl sulfoxide + co-reagent, **2,** 163; **12,** 213
S,S-Dimethyl-N-*p*-toluenesulfonylsulfilimine, **9,** 193
2,4-Dinitrobenzenesulfenyl chloride, **1,** 319
Iodomethyl methyl sulfide, **5,** 353
Iodotrimethylsilane, **12,** 259
Ketene *t*-butyldimethylsilyl methyl acetal, **11,** 279
N-Methylthiomethylpiperidine, **12,** 326

ALKYLTHIOALKYLATION (*Continued*)
Thioanisole, **5,** 660
Tin(IV) chloride, **10,** 370
Zinc bromide, **9,** 520
ALLYLIC REACTIONS
ACETOXYLATION
Benzeneselenenyl bromide, **5,** 518
Bromine–Silver acetate, **12,** 72
t-Butyl perbenzoate, **2,** 54
Cobalt(III) acetate, **9,** 119
Lead tetraacetate, **2,** 238; **3,** 168; **9,** 265
Manganese(III) acetate, **4,** 318
Mercury(II) acetate, **1,** 644; **4,** 319
Palladium(II) acetate, **1,** 778
Palladium(II) chloride–Silver(I) acetate, **11,** 396
Palladium(II) trifluoroacetate, **12,** 373
ACYLATION
Acetyl hexachloroantimonate, **8,** 5
Silver iododibenzoate, **1,** 1007
Simmons–Smith reagent, **12,** 437
ACYLOXYLATION
t-Butyl perbenzoate, **1,** 98; **2,** 54
ALKOXYLATION
Mercury(II) acetate, **6,** 358
Mercury(II) oxide–Tetrafluoroboric acid, **10,** 254
ALKYLATION (*see* COUPLING REACTIONS)
AMINATION
Bis(methoxycarbonyl)sulfur diimide, **12,** 57
Bis(N-*p*-toluenesulfonyl)sulfodiimide, **7,** 317; **8,** 43
4-Methyl-1,2,4-triazoline-3,5-dione, **2,** 282
Nitrosocarbonylmethane, **9,** 326
N-Phenyl-1,2,4-triazoline-3,5-dione, **5,** 528; **9,** 372
Selenium(IV) chloride–*p*-Toluenesulfonamide, **7,** 316
CARBOXYLATION
Diethyl oxomalonate, **10,** 143
COUPLING (*see* COUPLING REACTIONS)
HALOGENATION
Chlorination
t-Butyl hypochlorite, **1,** 90
Chlorine oxide, **12,** 110
N-Chloro-N-cyclohexylbenzenesulfonamide, **2,** 67
N-Chlorosuccinimide, **9,** 111, 367
Hypochlorous acid, **10,** 208; **11,** 260
Trichloroisocyanuric acid, **3,** 297; **6,** 605
Bromination
N-Bromocaprolactam, **1,** 76
N-Bromosuccinimide, **1,** 78; **2,** 40; **5,** 65; **8,** 54
Dibenzoyl peroxide, **1,** 196
1,3-Dibromo-5,5-dimethylhydantoin, **1,** 208
1,2-Dibromotetrachloroethane, **1,** 209
Iodination
N-Bromosuccinimide, **2,** 40
OXIDATION TO ALLYLIC ALCOHOLS
Ammonium persulfate, **5,** 15
Benzeneselenenic acid, **8,** 24
t-Butyl hydroperoxide–Selenium(IV) oxide, **8,** 64; **10,** 65; **11,** 96
t-Butyl perbenzoate, **7,** 49
m-Chloroperbenzoic acid, **6,** 46
Chlorotrimethylsilane, **10,** 96
Chromium(VI) oxide–3,5-Dimethylpyrazole complex, **8,** 110
Fluorodimethoxyborane, **6,** 261; **10,** 186
Hydrogen peroxide–Selenium(IV) oxide, **2,** 362; **8,** 440
Lead tetraacetate–N-Bromosuccinimide, **2,** 238
Mercury(II) trifluoroacetate, **6,** 360
Oxygen, **5,** 482; **10,** 293
Oxygen, singlet, **4,** 362; **6,** 431; **7,** 261; **8,** 367; **9,** 338; **10,** 294; **11,** 385
Perbenzoic acid, **1,** 791
Potassium perchromate, **4,** 412
Selenium(IV) oxide, **1,** 992; **2,** 360; **3,** 245; **7,** 319; **8,** 439; **10,** 345
Sodium methoxide, **5,** 617
Triphenyl phosphite ozonide, **3,** 323
OXIDATION TO α,β-UNSATURATED C=O's
N-Bromosuccinimide, **3,** 34; **4,** 49; **7,** 37
t-Butyl chromate, **1,** 86; **4,** 54; **7,** 42
t-Butyl hydroperoxide, **12,** 88
t-Butyl hydroperoxide–Chromium carbonyl, **12,** 89
Chromium(VI) oxide, **3,** 54; **7,** 70; **9,** 110; **10,** 99
2,3-Dichloro-5,6-dicyano-1,4-benzoquinone, **3,** 83

Ethyl trimethylsilylacetate, **11,** 234
Hexakis(acetato)trihydrato-μ_3-oxotrisrhodium acetate, **11,** 252
Mercury(II) acetate, **4,** 319
Nitrosyl fluoride, **3,** 214
Oxygen, **5,** 482
Oxygen, singlet, **12,** 363
Palladium(II) chloride, **11,** 393
Pyridinium chlorochromate, **9,** 397; **12,** 417
Selenium(IV) oxide, **4,** 422; **5,** 575; **6,** 509
Sodium chromate, **1,** 1059
Sodium peroxide, **11,** 492

AMIDATION (RH, RX, RM → RNHCOR′)
Lead tetraacetate–Trifluoroacetic acid, **6,** 317
Nitronium tetrafluoroborate, **9,** 324
Nitrosocarbonylmethane, **9,** 326
Nitrosonium hexafluorophosphate, **9,** 326
Silver carbonate, **12,** 432

AMIDOMERCURATION (*see* ADDITION REACTIONS)

AMINATION (*see also* ALLYLIC REACTIONS)
OF RH
Chloramine, **5,** 103
Hydroxylamine-O-sulfonic acid, **4,** 256
Trichloramine, **1,** 1193; **2,** 424; **3,** 295
OF ArH, C=C–H
Benzeneseleninic anhydride, **11,** 37
Chloramine, **5,** 103
Dimethyl(methylthio)sulfonium tetrafluoroborate, **11,** 204
Hydrazoic acid, **1,** 446
Hydroxylamine, **1,** 478
Hydroxylamine-O-sulfonic acid, **1,** 481; **5,** 343
Silver carbonate, **12,** 432
Thallium(III) trifluoroacetate, **3,** 286
Trichloramine, **1,** 1193; **2,** 424
OF RM, ArM
Ammonia, **1,** 1290
Azidomethyl phenyl sulfide, **10,** 14; **12,** 37
Benzeneseleninic anhydride, **8,** 29; **11,** 37
N,O-Dimethylhydroxylamine, **12,** 205
O-(Diphenylphosphinyl)hydroxylamine, **11,** 221
Diphenyl phosphoroazidate, **12,** 217
Hydroxylamine, **1,** 478; **12,** 251
Hydroxylamine-O-sulfonic acid, **1,** 481
Methoxyamine, **11,** 322
Silver carbonate, **12,** 432
Thallium(III) trifluoroacetate, **3,** 286
Trichloramine, **2,** 424
Trimethylsilylmethyl azide, **12,** 538
OF OTHER SUBSTRATES
Chloramine, **2,** 65
Hydroxylamine-O-sulfonic acid, **1,** 481; **5,** 343
O-Mesitoylhydroxylamine, **1,** 660
O-Mesitylenesulfonylhydroxylamine, **5,** 430
Triphenylphosphine–Diethyl azodicarboxylate, **7,** 404

AMINOALKYLATION (*see also* MANNICH REACTION)
Butadiyne, **1,** 185
Chloroacetaldehyde, **5,** 106
Copper(II) acetate, **2,** 84
Formaldehyde, **9,** 225; **10,** 186
Methoxymethylbis(trimethylsilyl)amine, **12,** 62
Tri-μ-carbonylhexacarbonyldiiron, **5,** 221

AMINOMERCURATION (*see* ADDITION REACTIONS)

ARBUZOV REACTION
Triisopropyl phosphite, **1,** 1229

ARNDT–EISTERT REACTION
t-Butyllithium, **11,** 103
Copper(I) iodide, **1,** 169
Copper(II) sulfate, **6,** 141
Diazomethane, **1,** 191; **2,** 102; **3,** 74
Trimethylsilyldiazomethane, **10,** 431

AROMATIC SUBSTITUTION (*see* SUBSTITUTION AT AROMATIC CARBON)

AROMATIZATION
OF SIX-MEMBERED CARBOCYCLES
Acetic anhydride–Phosphoric acid, **6,** 3
Anthraquinone, **4,** 20
Butyllithium–Tetramethylethylenediamine, **5,** 80
Chloranil, **1,** 125; **3,** 46
o-Chloranil, **1,** 128
1,8-Diazabicyclo[5.4.0]undecene-7, **7,** 87
2,3-Dichloro-5,6-dicyano-1,4-benzoquinone, **1,** 215; **2,** 112; **8,** 153
Diethyl azodicarboxylate, **4,** 148

AROMATIZATION (*Continued*)
Dimethyl sulfoxide–Acetic anhydride, **2,** 163
Diphenylpicrylhydrazyl, **1,** 347
Formic acid, **4,** 239
Hydrogen peroxide–Salt and oxide catalysts, **1,** 472
Iodine, **1,** 495
N-Lithioethylenediamine, **1,** 567
Lithium aluminum hydride, **5,** 382
Manganese(IV) oxide, **2,** 257; **3,** 191; **11,** 311
Nitric acid, **1,** 733
Palladium catalysts, **1,** 778; **4,** 367; **8,** 382
Palladium(II) trifluoroacetate, **10,** 302
Phenyliodine(III) bis(trifluoroacetate), **6,** 301
Phosphorus(V) sulfide, **3,** 226; **4,** 389
Potassium *t*-butoxide, **1,** 911; **6,** 477; **7,** 298
Potassium fencholate, **8,** 228
Potassium hydride, **4,** 409
Potassium permanganate, **9,** 389
Pyrene, **4,** 414
Pyridinium chloride, **8,** 424
Rhodium catalysts, **1,** 979
Rhodium(III) chloride, **8,** 435
Selenium, **1,** 990
Silver perchlorate, **5,** 585
Sodium bis(2-methoxyethoxy)aluminum hydride, **6,** 528
Sulfur, **1,** 1118
Tetracyanoethylene, **1,** 1133
N,N,N′,N′-Tetramethylethylenediamine, **4,** 485
Trifluoroacetic acid, **4,** 530
Triphenylcarbenium salts, **1,** 1256; **6,** 657

OF CYCLOHEXANONES, -ENONES TO PHENOLS
Benzeneselenenyl halides, **11,** 34
o-Chloranil, **9,** 102
Copper(II) bromide, **5,** 158
Diphenyl disulfide, **8,** 210
o-Ethoxycarbonylbenzyl phenyl sulfoxide, **8,** 75
Hydrobromic acid, **6,** 282
Lead tetraacetate–Copper(II) acetate, **5,** 365
N-Lithioethylenediamine, **1,** 567
Lithium–Biphenyl, **1,** 612
Lithium carbonate–Lithium bromide, **2,** 245
Methyl phenylsulfinylacetate, **8,** 347
Methyl vinyl ketone, **6,** 407
Palladium catalysts, **1,** 778; **7,** 274
Palladium(II) chloride, **5,** 500
Potassium *t*-butoxide, **2,** 336
Pyridinium bromide perbromide, **6,** 499
Rhodium(III) chloride, **8,** 435; **11,** 460
Sodium hydride, **5,** 610
Sulfuric acid, **6,** 558
Tetrabutylammonium hydroxide, **11,** 500
Tetrakis(triphenylphosphine)-palladium(0), **6,** 571

OF HETEROCYCLES, OTHER CARBOCYCLES
Amidines, bicyclic, **4,** 16
Barium manganate, **10,** 16
Benzeneseleninic anhydride, **11,** 37
o-Chloranil, **7,** 355
1,5-Diazabicyclo[4.3.0]nonene-5, **4,** 116
2,3-Dichloro-5,6-dicyano-1,4-benzoquinone, **4,** 130; **9,** 148; **11,** 166
Ferric nitrate/K10 Bentonite, **12,** 231
Manganese(IV) oxide, **2,** 257; **4,** 317; **6,** 357
Nickel peroxide, **8,** 357; **9,** 322
Palladium(II) chloride, **7,** 277
Phenanthrenequinone, **1,** 827
Potassium nitrosodisulfonate, **5,** 562; **10,** 329
Sulfur, **1,** 1118; **3,** 273
Sulfuryl chloride, **10,** 375
Vinyltriphenylphosphonium bromide, **6,** 666

ARYLATION (*see also* COUPLING REACTIONS)
OF ALLYLIC ALCOHOLS TO RCHO, R_2CO
Palladium(II) acetate, **7,** 274; **8,** 378; **12,** 367
Phenylpalladium acetate, **4,** 379

α TO C=O, C≡N, ETC.
π-Allylnickel halides, **8,** 324
Arene(tricarbonyl)chromium complexes, **6,** 27, 103; **8,** 21
Aryllead triacetates, **11,** 31
Benzenesulfenyl chloride, **12,** 42
9-Borabicyclo[3.3.1]nonane, **3,** 24
Butyllithium–Nickel(II) bromide, **8,** 66
α-Chloro-N-cyclohexylpropanal-

donitrone, **6,** 106
Chromium carbonyl, **9,** 117; **10,** 100; **11,** 131
Copper(I) bromide, **5,** 163; **6,** 143; **11,** 140
Dimethyl sulfide, **5,** 260
Diphenyliodonium chloride, **1,** 341
Ethyl 3-bromo-2-hydroxyimino-propanoate, **9,** 212
Ethyl diazoacetate, **5,** 295
Lithium diphenylcuprate, **6,** 234
Phenylcopper, **7,** 282
Tributyltin fluoride, **11,** 544
Triphenylbismuth carbonate, **10,** 446
Tris(dimethylamino)sulfonium difluorotrimethylsilicate, **12,** 558
OF O–H
Diphenyliodonium bromide, **1,** 340
Hexamethylphosphorous triamide, **6,** 279
Tetraphenylbismuth trifluoroacetate, **10,** 393; **11,** 515
Triphenylbismuth diacetate, **12,** 548
ASYMMETRIC REACTIONS
ACYLATION
2-Oxazolidones, chiral, **12,** 359
(S)-(+)-*p*-Tolyl *p*-tolylthiomethyl sulfoxide, **12,** 510
Zinc borohydride, **12,** 572
ADDITION TO C=C
t-Butyl hydroperoxide + co-reagent, **12,** 90
Dimethyl sulfoxide, **5,** 263
Hydrogen cyanide, **9,** 239
2,3-O-Isopropylidene-2,3-dihydroxy-1,4-bis(diphenylphosphine)butane, **9,** 259
ADDITION TO C=O, C=N
B-Allyldiisopinocampheylborane, **12,** 17
Allyltrimethylsilane, **11,** 16; **12,** 23
(S)-2-(Anilinomethyl)pyrrolidine, **9,** 19
(S,S)-(+)-1,4-Bis(dimethylamino)-2,3-dimethoxybutane, **9,** 47
2-Crotyl-4,4,5,5-tetramethyl-1,3,2-dioxaborolane, **11,** 86
Diethyl tartrate, **11,** 181
Diisobutylaluminum hydride, **12,** 191
(3S,6S)-(+)-2,5-Dimethoxy-3,6-dimethyl-3,6-dihydropyrazine, **10,** 151
(–)-N,N-Dimethylephedrinium bromide, **6,** 219
N,S-Dimethyl-S-phenylsulfoximine, **11,** 210
Ephedrine, **11,** 230
Hexahydro-4,4,7-trimethyl-4H-1,3-benzothiin, **12,** 237
(2S,2′S)-2-Hydroxymethyl-1-[(1-methylpyrrolidin-2-yl)methyl]-pyrrolidine, **8,** 253; **9,** 246; **10,** 207
Lithium tetrabutylaluminate–(–)-N-Methylephedrine, **7,** 240
(–)-10-Mercaptoisoborneol, **9,** 290
(γ-Methoxypropyl)-α-phenylethylamine, **12,** 318
α-Methylbenzylamine, **8,** 393
N-Methylephedrinium chloride, **8,** 340
(S)-(–)-Methyltri(2-methyl-1-butoxy)-titanium, **10,** 422
Organocopper reagents, **12,** 345
Organotitanium reagents, **11,** 374; **12,** 353
(2R,4R)-Pentanediol, **12,** 375
(+)-3-Phenylbornane-2,3-diol, **9,** 363
8-Phenylmenthol, **11,** 412; **12,** 389
Quinine, **8,** 430
p-Tolylsulfinylacetic acid, **11,** 538
p-(Tolylsulfinyl)methyllithium, **4,** 513; **5,** 682
(S)-(+)-*p*-Tolyl p-tolylthiomethyl sulfoxide, **9,** 474; **10,** 408
Tributylcrotyltin, **12,** 513
4,6,6- or 4,4,6-Trimethyl-1,3-oxathiane, **8,** 508; **12,** 534
ALDOL AND RELATED REACTIONS
(S)-1-Amino-2-methoxymethyl-1-pyrrolidine, **8,** 16; **9,** 17
Boron trichloride, **9,** 62
Bromomagnesium diethylamide, **5,** 209
2,3-Butanediol, **11,** 84
(R)-(+)-*t*-Butyl (*p*-tolylsulfinyl)acetate, **11,** 106
Cobaltacyclopentan-2-ones, **11,** 136
Dialkylboryl trifluoromethanesulfonates, **10,** 118
N,N-Dimethyl-α-*p*-tolyl-sulfinylacetamide, **12,** 509
(4S,5S)-2-Ethyl-4-(methoxymethyl)-5-phenyloxazoline, **11,** 232
Lithium tri-*sec*-butylborohydride, **7,** 307
Mandelic acid, **12,** 291
Menthyl *p*-toluenesulfinate, **12,** 295
2-Methyl-6-vinylpyridine, **6,** 409

ASYMMETRIC REACTIONS (*Continued*)
(−)-Norephedrine, **12,** 338
2-Oxazolidones, chiral, **11,** 379
(2R,4R)-Pentanediol, **12,** 375
(S)-(−)-Proline, **8,** 421; **9,** 393; **12,** 414
Tin(II) trifluoromethanesulfonate, **11,** 525; **12,** 490
p-Tolylsulfinylacetic acid, **10,** 405; **11,** 538
ALKYLATION
(S)-1-Alkoxymethyl-2-aminopropane, **9,** 3
(S)-1-Amino-2-methoxymethyl-1-pyrrolidine, **7,** 10; **8,** 16; **9,** 17; **12,** 30
(S)-(−)-2-Amino-1-methoxy-3-phenylpropane, **10,** 11
(1S,2S)-2-Amino-3-methoxy-1-phenyl-1-propanol, **7,** 229
(1S,2S)-2-Amino-1-phenyl-1,3-propanediol, **12,** 32, 383
(S)-2-Aminopropyl benzyl ether, **11,** 30
(Benzylmethoxymethyl)methylamine, **7,** 17; **9,** 42
(−)-N-Benzyl-N-methylephedrinium bromide, **6,** 39
trans-2,5-Bis(methoxymethyl)-pyrrolidine, **12,** 59
7-Chloro-5-phenyl-1-[(S)-α-phenylethyl]-1,3-dihydro-2H-1,4-benzodiazepine-2-one, **11,** 124
Cobaltacyclopentan-2-ones, **11,** 136
(3S,6S)-(+)-2,5-Dimethoxy-3,6-dimethyl-3,6-dihydropyrazine, **10,** 151; **11,** 192
(S)-1-(Dimethoxymethyl-2-methoxymethyl)pyrrolidine, **10,** 152
trans-2,4-Dimethoxymethyl-5-phenyloxazoline, **6,** 204
(+)-*trans*-2,5-Dimethylpyrrolidine, **8,** 197
Ephedrine, **9,** 209
Hexamethylphosphoric triamide, **12,** 239
(1S,2S,5S)-2-Hydroxypinan-3-one, **7,** 177
Leucine *t*-butyl esters, **8,** 272; **10,** 229
Lithium diisopropylamide, **10,** 241
erythro-2-Methoxy-1,2-diphenylethylamine, **12,** 313
(4S,5S)-4-Methoxymethyl-2-methyl-5-phenyl-2-oxazoline, **6,** 386; **7,** 229
(S)-(+)-2-Methoxymethylpyrrolidine, **10,** 259
α-Methylbenzylamine, **8,** 393; **9,** 363
[(2S)-(2α,3aα,4α,7α,7aα)]-2,3,3a,4,5,6,7,7a-Octahydro-7,8,8-trimethyl-4,7-methanobenzofuran-2-ol, **12,** 339
2-Oxazolidones, chiral, **11,** 379
Palladium(II) chloride, **5,** 500
Phase-transfer catalysts, **12,** 379
(S)-Prolinol, **10,** 332
(S)-Valine *t*-butyl ester, **12,** 563
L-Valinol, **12,** 563
CONJUGATE ADDITION
(S)-1-Amino-2-methoxymethyl-1-pyrrolidine, **12,** 30
(1S,2S)-2-Amino-3-methoxy-1-phenyl-1-propanol, **7,** 229; **9,** 312; **12,** 310, 381
(2S,4S)-(Anilinomethyl)-1-ethyl-4-hydroxypyrrolidine, **10,** 12
(S,S)-(+)-1,4-Bis(dimethylamino)-2,3-dimethoxybutane, **9,** 47
(4S)-*t*-Butylthio-(2S)-methoxymethyl-N-pivaloylpyrrolidine, **12,** 315
Crown ethers, **11,** 143
Diisobutylaluminum hydride, **12,** 191
Ephedrine, **11,** 230
(1R)-(−)-*cis*-3-Hydroxyisobornyl neopentyl ether, **12,** 249
(S)-2-Hydroxymethyl-1-methyl-pyrrolidine, **10,** 266
Leucine *t*-butyl esters, **7,** 189; **10,** 229
Menthyl *p*-toluenesulfinate, **11,** 312; **12,** 295
(4S,5S)-4-Methoxymethyl-2-methyl-5-phenyl-2-oxazoline, **6,** 386; **7,** 229; **9,** 312
(S)-(+)-2-Methoxymethylpyrrolidine, **11,** 326
8-Phenylmenthol, **11,** 412
Quinine, **6,** 501; **8,** 430; **9,** 403
Tetramethyltartaric acid diamide, **12,** 480
p-Tolylsulfinylacetic acid, **10,** 405
(S)-(+)-*p*-Tolyl *p*-tolylthiomethyl sulfoxide, **10,** 408
COUPLING
Allyltrimethylsilane, **11,** 16
(1S,2S)-2-Amino-3-methoxy-1-phenyl-1-propanol, **12,** 310, 381
trans-2,3-Bis(diphenylphosphine)-bicyclo[2.2.1]hept-5-ene, **10,** 36
Ferrocenylphosphines, **9,** 200; **11,** 237

Hydrogen peroxide, **12,** 242
2,3-O-Isopropylidene-2,3-dihydroxy-1,4-bis(diphenylphosphine)butane, **5,** 360
CYCLOADDITION
Bis(1,5-cyclooctadiene)nickel(0), **4,** 33
Cinchona alkaloids, **11,** 134
Dichloroketene, **12,** 176
Titanium(IV) chloride, **10,** 401
p-Tolyl vinyl sulfoxide, **11,** 538
CYCLOPROPANATION
(−)-N,N-Dimethylephedrinium bromide, **6,** 219
N,S-Dimethyl-S-phenylsulfoximine, **11,** 210
Dimethylsulfoxonium methylide, **9,** 186
Ephedrine, **11,** 230
Menthol, **12,** 294
DEPROTONATION
Lithium L-α-,α′-dimethyldibenzylamide, **10,** 245
(S)-(+)-2-(1-Pyrrolidinyl)-methylpyrrolidine, **12,** 421
DIELS–ALDER REACTION
Camphor-10-sulfonic acid, **12,** 103
Dichlorodiisopropoxytitanium(IV), **11,** 164
Ethyl (S)-lactate, **12,** 226
(R)-Ethyl *p*-tolylsulfinylmethylene-propionate, **12,** 228
(S)-(+)-α-Hydroxy-β,β-dimethyl-propyl vinyl ketone, **12,** 248
(1R)-(−)-*cis*-3-Hydroxyisobornyl neopentyl ether, **12,** 249
Menthoxyaluminum dichloride, **9,** 289
trans-1-Methoxy-3-trimethylsilyloxy-1,3-butadiene, **12,** 318
2-Oxazolidones, chiral, **12,** 359
8-Phenylmenthol + esters, **10,** 402; **11,** 412
(S)-(−)-Proline, **8,** 421
Titanium(IV) chloride, **10,** 401
Tris(6,6,7,7,8,8,8-heptafluoro-2,2-dimethyl-3,5-octanedionato)-europium, **12,** 559
ENE REACTION
Alkylaluminum halides, **10,** 177
8-Phenylmenthol, **11,** 412; **12,** 389
EPOXIDATION
(−)-Benzylquininium chloride, **10,** 27
t-Butyl hydroperoxide, **10,** 64
t-Butyl hydroperoxide + co-reagent, **8,** 393; **11,** 92; **12,** 90, 91
Dibenzyltartaric acid diamide, **12,** 91
(S)-(2-Hydroxy-N,N-dimethylpropan-amide-O,O′)-oxodiperoxymolyb-denum(VI), **10,** 206
α-Methylbenzylamine, **3,** 199; **6,** 457
N-Phenylcampholylhydroxamic acid, **8,** 392
Quinine, **7,** 311; **8,** 430; **10,** 338
p-Tolylsulfinylacetic acid, **10,** 405
HALOLACTONIZATION
(S)-(−)-Proline, **8,** 421; **9,** 393; **10,** 331; **11,** 446
HYDROBORATION
Borane–Tetrahydrofuran, **1,** 199; **4,** 124
Diisopinocampheylborane, **1,** 262; **4,** 161; **5,** 283; **6,** 202; **11,** 188, 225; **12,** 193
Dilongifolylborane, **11,** 191
Iodine, **7,** 179
Monoisopinocampheylborane, **8,** 267; **9,** 317; **10,** 224; **11,** 350; **12,** 333
HYDROGENATION, CATALYTIC
N-Acryloyl-4-(diphenylphosphine)-2-[(diphenylphosphine)methyl]-pyrrolidine, **11,** 6
trans-2,3-Bis(diphenylphosphine)bicyclo-[2.2.1]hept-5-ene, **9,** 48
2,2′-Bis(diphenylphosphine)-1,1′-binaphthyl, **10,** 36
Bis[2-(diphenylphosphine)ethyl]amine hydrochloride, **11,** 54
(2S,4S)-N-(*t*-Butoxycarbonyl)-4-(diphenylphosphino)-2-[(diphenyl-phosphino)methyl]pyrrolidine, **8,** 57; **11,** 87
Chlorotris(neomenthyldiphenyl-phosphine)rhodium(I), **6,** 416
Ferrocenylphosphines, **9,** 200
2,3,-O-Isopropylidene-2,3-dihydroxy-1,4-bis(diphenylphosphine)butane, **4,** 273; **6,** 309; **9,** 259; **11,** 277
α-Methylbenzylamine, **11,** 411
Methyl 2,3-bis-O-diphenylphosphine-4,6-O-benzylidene-α-D-glucopyranoside, **8,** 333
Neomenthyldiphenylphosphine, **6,** 416
Palladium catalysts, **9,** 353
(S)-(−)-Proline, **6,** 492
Tetrakis(triphenylphosphine)-palladium(0), **9,** 451

ASYMMETRIC REACTIONS (*Continued*)
HYDROXYLATION
t-Butyl perbenzoate, **1,** 98
[(−)-Camphor-10-ylsulfonyl]-3-aryloxaziridines, **11,** 108
Osmium tetroxide–Dihydroquinine acetate, **10,** 291
OXIDATION
t-Butyl hydroperoxide + co-reagent, **12,** 90
t-Butyl perbenzoate, **1,** 98
PROTONATION
(S)-2-Aminopropyl benzyl ether, **11,** 30
(2R,3R)-Dipivaloyltartaric acid, **9,** 204
REDUCTION OF C=N
Borane–Tetrahydrofuran, **6,** 161
2,3-O-Isopropylidene-2,3-dihydroxy-1,4-bis(diphenylphosphine)butane, **5,** 360
(S)-(−)-Proline, **11,** 446
Sodium tris[(S)-N-benzyloxycarbonyl-prolyoxy]hydroborate, **12,** 454
REDUCTION OF C=O
(S)-4-Anilino-3-methylamino-1-butanol, **12,** 33
1-Benzyl-1,4-dihydronicotinamide, **6,** 36
(R)-1-(S)-1′,2-Bis(diphenylphosphine)-ferrocenylethanol, **11,** 239
Borane–(S)-(−)-2-Amino-3-methyl-1,1-diphenyl-1-butanol, **12,** 31
(2S,4S)-N-(*t*-Butoxycarbonyl)-4-(diphenylphosphino)-2-[(diphenylphosphino)methyl]-pyrrolidine, **8,** 57; **11,** 87
Darvon alcohol, **6,** 206; **10,** 154
Diisobutylaluminum hydride, **12,** 191
Diisopinocampheylborane, **1,** 262; **8,** 174
(−)-N-Dodecyl-N-methylephedrinium bromide, **11,** 337
NB-Enantrane, **11,** 229
NB-Enantride, **11,** 229
Hexahydro-4,4,7-trimethyl-4H-1,3-benzothiin, **12,** 237
Lithium aluminum hydride + co-reagent, **2,** 243; **5,** 231, 387; **8,** 184, 423; **9,** 169, 308; **10,** 148, 238, 265, 458; **11,** 153, 289, 292; **12,** 60, 190, 191, 322
Lithium B-isopinocampheyl-9-borabicyclo[3.3.1]nonyl hydride, **8,** 303
Monoisopinocampheylborane, **11,** 350
(2R,4R)-Pentanediol, **12,** 375
8-Phenylmenthol, **12,** 389
B-3-Pinanyl-9-borabicyclo[3.3.1]nonane, **8,** 403; **10,** 320; **11,** 429; **12,** 397
(S)-(−)-Proline, **8,** 421
Raney nickel, **9,** 405
Sodium borohydride + co-reagent, **7,** 144, 239; **8,** 431; **9,** 422; **12,** 441
RING-FORMING REACTIONS
Alkylaluminum halides, **10,** 177
2,3-Butanediol, **11,** 84
Lithium tri-*sec*-butylborohydride, **7,** 307
Menthyl *p*-toluenesulfinate, **12,** 295
2-Methyl-6-vinylpyridine, **6,** 409
Palladium(II) acetate, **9,** 344
(S)-(−)-Proline, **12,** 414
Pyridinium (*d*)-camphor-10-sulfonate, **11,** 449
Rhodium(II) carboxylates, **12,** 423
Sulfuric acid, **4,** 470
Tin(IV) chloride, **11,** 522
WITTIG REACTION
(R)-(+)-*o*-Anisylcyclohexyl-methylphosphine, **11,** 31
AZASELENYLATION, AZASULFENYLATION, AZIDOMERCURATION (*see* ADDITION REACTIONS)
AZO COUPLING
Arenediazonium halides, **4,** 24; **10,** 306
Copper, **5,** 146
Diphenylzinc, **1,** 349
Phase-transfer catalysts, **10,** 305
Silica, **11,** 466
p-Toluenesulfonyl azide, **1,** 1178

BAEYER–VILLIGER REACTION
GENERAL REACTIONS
Bis(trimethylsilyl) monoperoxysulfate, **9,** 56
Bis(trimethylsilyl) peroxide, **11,** 67
Caro's acid, **1,** 118
m-Chloroperbenzoic acid, **1,** 135; **2,** 68; **5,** 120; **11,** 124
3,5-Dinitroperbenzoic acid, **9,** 196
Hydrogen peroxide–Boron trifluoride, **1,** 463
2-Hydroperoxyhexafluoro-2-propanol, **3,** 149
p-Methoxycarbonylperbenzoic acid, **4,** 68
Permaleic acid, **12,** 379

Permonophosphoric acid, **10,** 304
Perphthalic acid, **3,** 220
Potassium persulfate, **5,** 563
o-Sulfoperbenzoic acid, **4,** 469
Trifluoroperacetic acid, **1,** 821; **10,** 421
Triphenylsilyl hydroperoxide, **9,** 509
SPECIFIC EXAMPLES OF LACTONE SYNTHESIS
Benzeneperoxyseleninic acid, **8,** 22
Bis(trimethylsilyl) peroxide, **11,** 67
Caro's acid, **1,** 118
Cerium(IV) ammonium sulfate, **7,** 56
Cerium(III) sulfate, **9,** 99
m-Chloroperbenzoic acid, **2,** 68; **4,** 85; **5,** 120; **8,** 97
1,5-Diazabicyclo[4.3.0]nonene-5, **6,** 157
Grignard reagents, **10,** 189
Hydrogen peroxide, **4,** 254
Hydrogen peroxide–Acetic acid, **3,** 155; **6,** 286
2-Hydroperoxyhexafluoro-2-propanol, **3,** 149
Lead tetraacetate, **10,** 228; **12,** 270
Peracetic acid, **1,** 787; **3,** 155; **4,** 372; **6,** 286
Permaleic acid, **1,** 819; **12,** 379
Potassium dichromate, **6,** 124
Potassium peroxomonosulfate, **1,** 952
Potassium persulfate, **2,** 348; **10,** 331
Ruthenium tetroxide, **10,** 343
Silver carbonate–Celite, **3,** 247; **6,** 511
Sodium bromite, **12,** 445
Sodium dichromate, **7,** 68
Trifluoroperacetic acid, **1,** 821
Zinc permanganate, **12,** 576

BAKER–VENKATARAMAN REARRANGEMENT
Potassium *t*-butoxide–*t*-Butyl alcohol, **12,** 402

BAMFORD–STEVENS REACTION (*see* ELIMINATION REACTIONS)

BARBIER REACTION
Magnesium, **11,** 307
Samarium(II) iodide, **12,** 429

BECKMANN FRAGMENTATION
Methanesulfonyl chloride, **5,** 435
Methyl nitrite, **1,** 691
S-Methyl *p*-toluenethiosulfonate, **6,** 400; **7,** 245
Tetrakis(triphenylphosphine)-palladium(0), **6,** 571
Triethyl orthoformate, **6,** 610
Trimethylsilyl trifluoromethane-sulfonate, **12,** 543

BECKMANN REARRANGEMENT
p-Acetamidobenzenesulfonyl chloride, **1,** 3
Alkylaluminum halides, **12,** 5
Boron trifluoride, **1,** 68
Dichloromethylenedimethylammonium chloride, **6,** 170
N,N-Dimethylformamide, **1,** 278
Formic acid, **1,** 404
Hexamethylphosphoric triamide, **5,** 323
Hydrazoic acid, **1,** 446
Hydrochloric acid, **5,** 333
Hypochlorous acid, **1,** 487
Iodine pentafluoride, **1,** 503
O-Mesitylenesulfonylhydroxylamine, **5,** 430
Phosphorus(V) oxide–Methanesulfonic acid, **5,** 535
Phosphoryl chloride, **1,** 876
Polyphosphate ester, **3,** 229
Polyphosphoric acid, **1,** 894; **10,** 321
Potassium *t*-amyloxide, **1,** 939
Pyridinium chloride, **2,** 352
Silica, **11,** 466
Sodium azide, **1,** 901
p-Toluenesulfonyl chloride, **1,** 1179; **9,** 472
Trialkylaluminums, **11,** 539
Trifluoroacetic anhydride, **1,** 1221
Trimethylsilyl polyphosphate, **10,** 437; **11,** 427
Triphenylphosphine, **2,** 443
Triphenylphosphine–Carbon tetrachloride, **5,** 727

BENZILIC ACID REARRANGEMENT
Manganese(IV) oxide, **3,** 191
Potassium *t*-butoxide, **1,** 911
Sodium bromate, **1,** 1055

BENZOANNELATION
1-Acetoxybutadiene, **1,** 7
2-Acetoxy-1-methoxy-3-trimethyl-silyloxy-1,3-butadiene, **11,** 2
Aluminum chloride, **11,** 25
1-Aminobenzotriazole, **3,** 10
1,2-Bis(methylthio)-1,3-butadiene, **10,** 40
1,3-Bis(trimethylsilyloxy)-1,3-butadiene, **10,** 44
2,5-Bis(trimethylsilyloxy)furans, **10,** 45

BENZOANNELATION (*Continued*)
t-Butyl azidoformate, **2,** 44
(*o*-Carboxyphenyl)phenyliodonium hydroxide inner salt, **1,** 341; **4,** 207
p-Chlorobenzoyl nitrite, **4,** 79
1-Chloro-1-dimethylamino-2-methyl-1,3-butadiene, **9,** 104
Chromium carbene complexes, **11,** 397; **12,** 132
(E,E)-1,4-Diacetoxybutadiene, **1,** 183
2,5-Di-*p*-anisyl-3,4-diphenylcyclopentadienone, **1,** 188
Diels–Alder solvents, **1,** 236
1(E),3-Dimethoxybutadiene, **11,** 192
4,6-Dimethoxy-2-pyrone, **9,** 175
5,5-Dimethoxy-1,2,3,4-tetrachlorocyclopentadiene, **1,** 270
1,2-Dimethoxy-1-trimethylsilyloxy-1,3-pentadiene, **12,** 197
Dimethyl acetylenedicarboxylate, **1,** 272; **3,** 103; **4,** 168; **7,** 117
4-Dimethylaminopyridine, **10,** 155
4-Dimethylamino-1,1,2-trimethoxybutadiene, **11,** 197
Dimethyl methoxymethylenemalonate, **12,** 206
o-Ethoxycarbonylbenzyl phenyl sulfoxide, **8,** 75
Ethyl diphenylphosphinite, **3,** 139
Ethyl β-phenylsulfonylpropiolate, **11,** 233
Isoamyl nitrite, **5,** 358
Ketene dialkyl acetals, **9,** 262; **11,** 279
Lithium 2,2,6,6-tetramethylpiperidide, **6,** 345
1-Methoxy-1,3-bis(trimethylsilyloxy)-1,3-butadiene, **10,** 257; **11,** 66
3-Methoxymethylene-2,4-bis(trimethylsilyloxy)-1,4-pentadiene, **12,** 315
6-Methoxy-4-methyl-2-pyranone, **8,** 325
1-Methoxy-3-methyl-1-trimethylsilyloxy-1,3-butadiene, **9,** 302
trans-1-Methoxy-3-trimethylsilyloxy-1,3-butadiene, **6,** 370; **8,** 328
Methyl vinyl ketone, **6,** 407; **10,** 272
Nickel carbonyl, **4,** 353
3-Phenylsulfonyl-1(3H)-isobenzofuranone, **9,** 368
3-Phenylthio-3-buten-2-one, **11,** 421
Potassium *t*-butoxide, **1,** 911
Potassium persulfate, **1,** 952
Sodium methylsulfinylmethylide, **2,** 166
3-Sulfolene, **2,** 389
Tetraphenylcyclopentadienone, **1,** 1149; **4,** 490
Titanium(IV) chloride, **11,** 529
N-*p*-Tolylvinylmethylketenimine, **5,** 683
1-Trimethylsilyloxy-1,3-butadiene, **8,** 512
2-Trimethylsilyloxy-1,3-cyclohexadiene, **8,** 512

BENZOIN CONDENSATION
3-Benzyl-5-(2-hydroxyethyl)-4-methyl-1,3-thiazolium chloride, **6,** 38; **7,** 16
3-Benzyl-4-methylthiazolium chloride, **7,** 18
Potassium cyanide, **7,** 299

BENZOYLATION (*see also* SYNTHESIS INDEX—AMIDES, ESTERS)
Aluminum chloride, **1,** 24
t-Butyl perbenzoate, **1,** 98
Dibenzoyl peroxide, **4,** 122
Lithium 2,2,6,6-tetramethylpiperidide, **6,** 345
Pyridine, **1,** 958
Sodium hexamethyldisilazide, **3,** 261
Zirconium(IV) chloride, **1,** 1295

BIRCH REDUCTION AND RELATED REACTIONS
TO FORM CYCLOHEXADIENES
Birch reduction, **1,** 54; **5,** 30; **8,** 38
Calcium hexamine, **1,** 104
Hexamethylphosphoric triamide, **4,** 244
Lithium–Methylamine, **1,** 574; **5,** 378
Sodium amalgam, **3,** 259
Sodium–Ammonia, **3,** 259; **10,** 355
Sodium borohydride, **6,** 530
Ytterbium–Ammonia, **9,** 517
TO FORM CYCLOHEXENONES
Birch reduction, **1,** 54; **2,** 27
Butyllithium, **9,** 87
Lithium–Alkylamines, **5,** 377
Lithium–Ammonia, **1,** 54; **2,** 245; **7,** 21; **8,** 282
2-Methyl-6-vinylpyridine, **6,** 409
Oxalic acid, **1,** 764
Potassium *t*-butoxide, **1,** 911
Ytterbium–Ammonia, **9,** 517
TO FORM TETRAHYDROARENES
Calcium–Amines, **9,** 94; **12,** 103
Lithium–Alkylamines, **1,** 614, 618; **2,** 241, 250; **3,** 175; **4,** 287
Lithium–Ammonia, **2,** 28; **8,** 282

Sodium–Ammonia, **2,** 374
BISCHLER–NAPIERALSKI REACTION
Diethyl N-benzylideneamino-methylphosphonate, **12,** 185
Phosphorus(V) oxide, **1,** 871
Polyphosphate ester, **1,** 892; **2,** 333; **3,** 229; **4,** 394; **5,** 539
Polyphosphoric acid, **5,** 540
BOHN–SCHMIDT REACTION
Boric acid, **1,** 63
BOUVEAULT ALDEHYDE SYNTHESIS
N,N-Dimethylformamide, **11,** 198
BOUVEAULT–BLANC REDUCTION
Phenol, **1,** 828
Sodium–Ammonia–Ethanol, **1,** 1041
BROMINATION (*see also* ADDITION REACTIONS)
OF RCHO, R_2CO AND RELATED SUBSTRATES
Boron trifluoride etherate, **5,** 52
Bromine, **1,** 32, 927; **5,** 452, 712; **6,** 70; **7,** 33; **11,** 75
2-Bromo-2-cyano-N,N-dimethylacetamide, **6,** 73
N-Bromosuccinimide, **7,** 37
2-Carboxyethyltriphenyl-phosphonium perbromide, **6,** 97
N-Chlorosuccinimide, **6,** 115
Copper(II) bromide, **1,** 161; **5,** 158
2,4-Diamino-1,3-thiazole hydrotribromide, **10,** 119
5,5-Dibromo-2,2-dimethyl-1,3-dioxane-4,6-dione, **8,** 143
Dibromomethyllithium, **5,** 403
Dimethyl bromomalonate, **12,** 200
Dimethyl sulfoxide, **2,** 157
Dioxane–Bromine, **1,** 333; **5,** 58
Epichlorohydrin, **1,** 355
Ethanolamine, **4,** 222
Iodine monobromide, **2,** 224; **6,** 297
Lead tetraacetate–Metal halides, **11,** 283
Magnesium bromide etherate–Hydrogen peroxide, **7,** 220
Phenyltrimethylammonium perbromide, **1,** 855; **5,** 531
Potassium acetate, **1,** 906
Pyridinium bromide perbromide, **1,** 967
Pyrrolidone-2 hydrotribromide, **3,** 240
Sulfur, **1,** 1118
2,4,4,6-Tetrabromo-2,5-cyclohexadienone, **5,** 643
p-Toluenesulfonic acid, **1,** 1172
ω-Tribromoacetophenone, **1,** 1188
2-Trimethylsilyloxy-1,3-butadiene, **7,** 401
OF RH
Aluminum bromide, **9,** 11
Boron tribromide, **1,** 66; **2,** 33
Bromine, **5,** 55; **6,** 70
Bromine chloride, **6,** 102
N-Bromopolymaleimide, **4,** 49
N-Bromosuccinimide + co-reagent, **4,** 90; **5,** 66
Bromotrichloromethane, **1,** 80; **7,** 40
Chromium carbonyl, **11,** 131
Dibenzoyl peroxide, **1,** 196
1,3-Dibromo-5,5-dimethylhydantoin, **1,** 208; **12,** 158
Dioxane–Bromine, **3,** 130
Mercury(II) oxide–Bromine, **4,** 324
Phenyltrimethylammonium perbromide, **11,** 426
Pyridine perbromide, **1,** 966
Pyridinium bromide perbromide, **1,** 967
Silver hexafluoroantimonate, **5,** 577
Tetramethylammonium tribromide, **1,** 1144
ω-Tribromoacetophenone, **1,** 1188
Trichloromethanesulfonyl bromide, **1,** 1195
OF ALLYLIC C–H (*see* ALLYLIC REACTIONS)
OF ArH, C=C–H
Benzoyl nitrate, **5,** 25
Bromine, **1,** 499; **7,** 33
Bromine + co-reagent, **1,** 32, 960; **2,** 429; **4,** 508; **5,** 672; **6,** 17
Bromine chloride, **1,** 73
Bromine on Graphite, **5,** 58
N-Bromoacetamide, **2,** 39
N-Bromosuccinimide, **1,** 78; **2,** 40; **7,** 33
N-Bromosuccinimide + co-reagent, **6,** 510; **9,** 72
t-Butylamine, **2,** 43
Butyllithium–Potassium *t*-butoxide, **10,** 72
Copper(II) bromide, **1,** 161; **2,** 84; **5,** 159
2,4-Diamino-1,3-thiazole hydrotribromide, **10,** 119
Dibromoisocyanuric acid, **2,** 109; **10,** 122
Dimethyl sulfoxide–Bromine, **6,** 229
Dioxane–Bromine, **1,** 333; **3,** 130
Hexafluoroantimonic acid, **10,** 195

BROMINATION (*Continued*)
Hydrobromic acid–Dimethyl sulfoxide, **5,** 333
Iodine monobromide, **1,** 501
Iron, **1,** 519
Mercury(II) acetate, **1,** 644
Palladium(II) acetate, **5,** 496
Potassium bromate, **11,** 431
Pyridinium bromide perbromide, **1,** 967
Silver sulfate, **1,** 1015
Sulfur monochloride, **1,** 1122
2,4,4,6-Tetrabromo-2,5-cyclohexadienone, **4,** 476; **7,** 351
Tetramethylammonium tribromide, **1,** 1144
Thallium(III) acetate–Bromine, **3,** 286; **4,** 492
Triethylamine, **2,** 427
Trifluoroacetic acid, **1,** 1219
Trifluoroacetyl hypobromite, **1,** 1227
Trimethyl phosphate, **7,** 393
OF RCOOH, RCOX
Barium hydroxide, **4,** 23
N-Bromosuccinimide, **3,** 34; **5,** 65; **6,** 74; **7,** 37
Chlorine–Chlorosulfuric acid, **10,** 86
1,2-Dibromoethane, **4,** 126
Diethyl dibromomalonate, **8,** 169
Lithium N-isopropylcyclohexylamide, **4,** 306
Magnesium bromide etherate–Hydrogen peroxide, **7,** 220
Phosphorus(III) bromide, **1,** 873
Phosphorus, red, **1,** 861
Silver(I) oxide, **5,** 583
OF $ARN_2^+X^-$, $RN_2^+X^-$
Copper(I) bromide, **1,** 165
Hexamethylphosphoric triamide, **5,** 323
Pyridinium poly(hydrogen fluoride), **6,** 473; **12,** 419
Sodium bromide, **6,** 285
Tin(II) bromide, **1,** 1112
BROMOLACTAMIZATION, BROMOLACTONIZATION (*see* HALO-)
BUCHNER REACTION
Ethyl diazoacetate, **1,** 367
Rhodium(II) carboxylates, **10,** 340; **12,** 423

CANNIZZARO REACTION
Formaldehyde, **1,** 397
CARBAMOYLATION
OF RH
Benzyltrimethylammonium fluoride, **10,** 29
Dichlorobis(triphenylphosphine)-palladium(II), **8,** 151
Dichloromethylenedimethylammonium chloride, **4,** 135
N-Phenyl-1,2,4-triazoline-3,5-dione, **4,** 381
Tetramethylene sulfone, **2,** 402
OF RX
Dihalobis(triphenylphosphine)-palladium(II), **6,** 60
Disodium tetracarbonylferrate, **4,** 461
Lithium bis(N,N-diethylcarbamoyl)-cuprate, **9,** 279
Lithium tricarbonyl(dimethyl-carbamoyl)nickelate, **4,** 302
Nickel carbonyl, **5,** 93
Organocopper reagents, **11,** 365
CARBOALKOXYLATION (*see* ALKOXYCARBONYLATION)
CARBONYL TRANSPOSITION
1,2-
Chlorotrimethylsilane, **10,** 96
Chromium(II) acetate, **3,** 59
Diphenyl disulfide, **6,** 235
Lithium aluminum hydride–Aluminum chloride, **8,** 289
Phenyl benzenethiosulfonate, **8,** 391; **9,** 362
p-Toluenesulfonylhydrazide, **6,** 598; **8,** 489; **9,** 472; **11,** 537
1,3-
Benzeneselenenyl halides, **10,** 16
Benzenesulfenyl chloride, **6,** 30
Tetrafluoroboric acid, **12,** 465
CARBONYLATION (*see also* FORMYLATION)
OF ALKYNYL, ALKENYL ALCOHOLS
2-Bromo-3-trimethylsilyl-1-propene, **11,** 80
Dicarbonylbis(triphenylphosphine)-nickel, **10,** 125
Dimethylaluminum acetylide, **8,** 385
Disodium tetracarbonylferrate, **9,** 205; **10,** 174
Lithium acetylide, **6,** 451

Nickel carbonyl, **6,** 451
Palladium(II) chloride–Thiourea, **6,** 450; **7,** 278; **11,** 397
α-Trimethylsilylvinylmagnesium bromide, **8,** 515
OF BORANES
Carbon monoxide, **2,** 60; **4,** 69; **8,** 76
Chlorothexylborane, **10,** 95
Dichloromethyl methyl ether, **5,** 200; **7,** 98
Lithium triethylmethoxide, **4,** 314
Monochloroborane diethyl etherate, **7,** 249
Monoisopinocampheylborane, **11,** 350
Potassium triisopropoxyborohydride, **9,** 393
Sodium cyanide, **4,** 446; **7,** 333
Thexylborane, **2,** 148; **4,** 175; **10,** 397
Tris(phenylthio)methyllithium, **11,** 305
OF DIENES
Bis(cyclopentadienyl)diiodozirconium, **12,** 53
Bromine, **5,** 55
Carbon monoxide, **3,** 41; **8,** 76
Di-μ-carbonylhexacarbonyldicobalt, **12,** 163
Disodium tetracarbonylferrate, **6,** 550
Monochloroborane diethyl etherate, **7,** 249
Nickel carbonyl, **2,** 290
Palladium(II) chloride, **2,** 303
Sodium cyanide, **6,** 535
Tetracarbonyldi-μ-chlorodirhodium, **7,** 59
Thexylborane, **2,** 148
OF N COMPOUNDS
N,N′-Carbonyldiimidazole, **2,** 61
Selenium, **6,** 507
Tellurium, **11,** 498
Tetracarbonyldi-μ-chlorodirhodium, **12,** 112
OF RM TO FORM R_2CO
Dicarbonylbis(triphenylphosphine)-nickel, **10,** 125
Di-μ-carbonylhexacarbonyldicobalt, **3,** 89
Disodium tetracarbonylferrate, **4,** 461; **10,** 174
Lithium aluminum hydride, **9,** 274
Nickel carbonyl, **4,** 353; **12,** 335
CARBOXYLATION (*see also* ORGANOMETALLIC REAGENTS AND REACTIONS)
OF RH, ArH
Aluminum chloride, **1,** 24
Bromine–Antimony(V) chloride–Sulfur dioxide, **5,** 57
Carbon dioxide, **3,** 40; **5,** 93; **6,** 94
Catechol dichloromethylene ether, **1,** 119; **4,** 70
Copper(I) carbonyl ion, **5,** 149
Cyclodextrins, **12,** 151
Diphenylcarbamoyl chloride, **1,** 337; **2,** 177
Formic acid, **1,** 404
Hexafluoroantimonic acid, **11,** 252
Iron carbonyl, **8,** 265
Lithium 2,6-di-*t*-butyl-4-methylphenoxide, **5,** 412
Lithium diisopropylamide, **7,** 204
Magnesium methyl carbonate, **1,** 631; **3,** 190; **5,** 420; **6,** 354; **11,** 310
Oxalyl chloride, **1,** 28, 767
Pentafluorophenylcopper, **5,** 504
Sodium amide, **1,** 1034
Sodium phenoxide, **1,** 1102
Triphenylmethylpotassium, **1,** 1258
OF $ArN_2^+X^-$, $RN_2^+X^-$
Arenediazonium tetrahaloborates, **1,** 43
Nickel carbonyl, **1,** 720
Palladium(II) acetate, **10,** 297
OF RM, RX
Arene(tricarbonyl)chromium complexes, **10,** 13
n- or *t*-Butyllithium, **1,** 95; **6,** 109; **7,** 47
Carbon dioxide, **3,** 40; **5,** 93; **6,** 94
Copper(I) phenylacetylide–Tributylphosphine, **6,** 130
1,2-Dibromoethane, **1,** 374
gem-Dichloroallyllithium, **8,** 150
Dichlorobis(cyclopentadienyl)titanium, **9,** 146
Diisobutylaluminum hydride, **2,** 140; **7,** 147
Disodium tetracarbonylferrate, **4,** 461
1,3-Dithiane, **2,** 182
Ethyl chloroformate, **7,** 147
Grignard reagents, **1,** 415
Lithium–Ammonia, **3,** 179
Lithium diisobutylmethylaluminum hydride, **2,** 248
Lithium naphthalenide, **8,** 305

CARBOXYLATION (*Continued*)
Lithium nitride, **1,** 618
Magnesium, **7,** 218; **10,** 251; **11,** 307
Magnesium methyl carbonate, **3,** 190
Methylenemagnesium bromide, **2,** 273
Nickel(II) acetylacetonate–Trimethylaluminum, **8,** 42; **9,** 52
Nickel carbonyl, **1,** 720; **5,** 472
Phenylpotassium, **1,** 848
Sodium amalgam, **1,** 1033
Sodium amide, **1,** 1034
Sodium naphthalenide, **1,** 711
1,1,3,3-Tetramethylbutyl isocyanide, **3,** 279; **4,** 480
N,N,N′,N′-Tetramethylethylenediamine, **3,** 284; **4,** 485; **5,** 652
2,4,6-Triisopropylbenzenesulfonylhydrazide, **9,** 486
Triphenylmethyllithium, **1,** 1256
Triphenylphosphine–Carbon tetrabromide, **4,** 550
Vinylcopper reagents, **6,** 662
Vinyllithium, **1,** 1273
CATION–OLEFIN REACTION (*see* POLYENE CYCLIZATION)
CHLORINATION (*see also* ADDITION REACTIONS)
OF RCHO, R_2CO AND RELATED COMPOUNDS
N-Chlorosuccinimide, **6,** 115; **7,** 39; **8,** 103
Chlorotrimethylsilane, **5,** 709
Copper(II) halides, **1,** 161; **2,** 84; **5,** 158; **10,** 106
N,N-Dimethylformamide, **9,** 182
Ethanolamine, **4,** 222
Hexachloroacetone, **8,** 239
2,3,4,5,6,6-Hexachloro-2,4-cyclohexadien-1-one, **11,** 251
Lead tetraacetate–Metal halides, **11,** 283
Lithium chloride–Hydrogen peroxide, **7,** 220
Selenium oxychloride, **1,** 1000
Sulfuryl chloride, **1,** 1128
Tetrabutylammonium iodotetrachloride, **1,** 1132
Trifluoromethanesulfonyl chloride, **9,** 485
OF RH
Antimony(V) chloride, **5,** 18; **6,** 22
t-Butyl hypochlorite, **1,** 90; **2,** 50
Chlorine oxide, **11,** 119
N-Chloro-N-cyclohexylbenzenesulfonamide, **2,** 67
Chromium carbonyl, **11,** 131
N,N-Dichlorourethane, **8,** 161; **9,** 156
Molybdenum(V) chloride, **6,** 412
Nitrosyl chloride, **3,** 214
Phenyliodine(III) dichloride, **1,** 505; **4,** 264; **5,** 352; **6,** 298; **9,** 249; **10,** 215
Phosphorus(V) chloride, **1,** 866
Potassium hydroxide–Carbon tetrachloride, **5,** 96
Silver hexafluoroantimonate, **5,** 577
Sulfuryl chloride, **1,** 1128; **3,** 276; **4,** 474; **5,** 641
N,2,4,6-Tetrachloroacetanilide, **1,** 1132
Trichloroisocyanuric acid, **3,** 297; **6,** 605
Trichloromethanesulfonyl chloride, **1,** 1195
AT ALLYLIC C–H (*see* ALLYLIC REACTIONS)
OF ArH, C=C–H
Aluminum chloride, **7,** 252
Antimony(V) chloride–Graphite, **6,** 22
Benzoyl nitrate, **5,** 25
t-Butyl hypochlorite, **1,** 90
Chloramine, **5,** 103
Chlorine, **6,** 101
Chlorine oxide, **11,** 119
1-Chlorobenzotriazole, **4,** 78
N-Chloropolymaleimide, **4,** 87
N-Chlorosuccinimide, **1,** 139
N-Chlorosuccinimide + co-reagent, **6,** 510; **9,** 111
Chlorosulfonyl isocyanate, **5,** 132
N-Chlorotriethylammonium chloride, **12,** 122
Copper(II) chloride, **1,** 163; **2,** 84; **5,** 158; **8,** 120
N,N-Dichlorourethane, **2,** 121
2,3,4,5,6,6-Hexachloro-2,4-cyclohexadien-1-one, **11,** 251
Hydrogen peroxide, **1,** 457; **7,** 174
Palladium(II) acetate, **5,** 496
Sodium hypochlorite, **6,** 543
Sulfuryl chloride, **1,** 1128; **2,** 394; **6,** 510; **7,** 349
Thionyl chloride, **5,** 663
Titanium(IV) chloride, **4,** 507
Trichloroisocyanuric acid, **3,** 297
Triethylamine, **2,** 427

Trimethyl phosphate, **7,** 393
OF RCOOH, RCOX
N-Bromosuccinimide, **6,** 74
Chlorine–Chlorosulfuric acid, **8,** 83; **10,** 86
N-Chlorosuccinimide, **6,** 115
7,7,8,8-Tetracyanoquinodimethane, **12,** 464
Thionyl chloride, **5,** 663
OF $ArN_2^+X^-$, $RN_2^+X^-$
Copper(I) chloride, **1,** 166
Pyridinium poly(hydrogen fluoride), **12,** 419
Sodium chloride, **6,** 285
Tin(II) bromide, **1,** 1112
OF N COMPOUNDS
t-Butyl hypochlorite, **1,** 90; **4,** 58
N-Chlorosuccinimide, **1,** 139
Hypochlorous acid, **1,** 487
Potassium superoxide, **11,** 442
Sodium hypochlorite, **6,** 543
Sulfuric acid, **5,** 633
CHLORO- (*see* HALO-)
CHUGAEV REACTION (*see* ELIMINATION REACTIONS)
CINE REARRANGEMENT
Triethylamine, **9,** 481
CLAISEN CONDENSATION
Dichlorobis(trifluoromethanesulfonato)-titanium(IV), **12,** 173
Ethyl acetate, **1,** 1075
Lithium tri-*sec*-butylborohydride, **6,** 348
Magnesium naphthalenide, **1,** 711
Molecular sieves, **5,** 465
Polyphosphoric acid, **4,** 395
Potassium hydride, **1,** 935; **6,** 482
Sodium–Dimethyl sulfoxide, **2,** 158
Sodium alkoxide, **1,** 1065; **5,** 617
Sodium hydride, **1,** 1075
Triphenylmethylpotassium, **1,** 1258
CLAISEN REARRANGEMENT (*see* SIGMATROPIC REARRANGEMENTS)
CLAISEN–SCHMIDT CONDENSATION (*see* ALDOL REACTION)
CLEAVAGE REACTIONS (*see also* HYDROLYSIS, OXIDATIVE CLEAVAGE)
AZIRIDINES
Aziridine, **1,** 378
Bis(ethylthio)acetic acid, **9,** 228
Carbon dioxide, **7,** 52
Ethyl azidoformate, **2,** 191
Ethyl chloroformate, **12,** 223
Phenylcarbonimidic dichloride, **6,** 458
Pyridinium poly(hydrogen fluoride), **11,** 453
CYCLOBUTANES
Borane–Tetrahydrofuran, **1,** 199
Cyanogen bromide, **1,** 174
Diphenyl disulfide, **7,** 137
Diphenylsulfonium cyclopropylide, **4,** 211; **6,** 242; **8,** 212
Lead tetraacetate, **6,** 313
1-Lithiocyclopropyl phenyl sulfide, **6,** 319; **9,** 271
Sodium dichromate, **7,** 68
Zinc bromide, **2,** 463
CYCLOPROPANES
Acetic anhydride–Boron trifluoride etherate, **6,** 3
Acetyl methanesulfonate, **9,** 2
Benzeneselenenyl halides, **9,** 25
Bis(benzonitrile)dichloroplatinum(II), **11,** 49
1,1-Bis(phenylsulfonyl)cyclopropane, **12,** 61
Borane–Tetrahydrofuran, **1,** 199
Bromine, **6,** 70
Butyllithium, **6,** 85
Cerium(IV) ammonium nitrate, **2,** 63
Copper(I) acetylacetonate, **4,** 100
Copper(0)–Isonitrile complexes, **9,** 122
Di-μ-carbonylhexacarbonyldicobalt, **7,** 99
Diethyl(2-chloro-1,1,2-trifluoroethyl)-amine, **4,** 149; **5,** 214
Dimethyl diazomalonate, **8,** 187
6,6-Dimethyl-5,7-dioxaspiro[2.5]-octane-4,8-dione, **6,** 216; **7,** 122
6,6-Dimethyl-2-vinyl-5,7-dioxaspiro[2.5]octane-4,8-dione, **6,** 230
Diphenylsulfonium cyclopropylide, **8,** 212
Diphosphorus tetraiodide, **6,** 243; **12,** 218
Fluorosulfuric acid, **5,** 310
Formic acid, **5,** 316; **7,** 160
Hydrobromic acid, **4,** 249
Hydrogen chloride, **7,** 172
Iodotrimethylsilane, **9,** 251; **10,** 216

CLEAVAGE REACTIONS (*Continued*)
Lead tetraacetate, **1,** 537; **8,** 269
1-Lithiocyclopropyl phenyl sulfide, **9,** 271; **11,** 284
1-Lithio-2-methoxycyclopropane, **6,** 366
Lithium–Ammonia, **4,** 288
Lithium benzeneselenoate, **8,** 306
Lithium dimethylcuprate, **6,** 209; **7,** 120
Lithium divinylcuprate, **4,** 219
Magnesium iodide, **6,** 353
Mercury(II) acetate, **4,** 319; **5,** 424; **7,** 222; **12,** 298
Palladium catalysts, **7,** 275
Palladium(0)–Phosphines, **9,** 344
Potassium thiophenoxide, **8,** 420
Pyridinium chloride, **8,** 424; **10,** 333
Pyridinium chlorochromate, **9,** 397
Silver acetate, **2,** 362; **6,** 511
Silver perchlorate, **10,** 354
Silver(I) trifluoroacetate, **11,** 471
Simmons–Smith reagent, **3,** 255; **5,** 588; **12,** 437
Sodium benzeneselenoate, **10,** 356
Sodium chlorodifluoroacetate, **5,** 603
Sodium cyanide, **8,** 453
N,N,N′,N′-Tetramethylethylenediamine, **4,** 485
Thallium(III) acetate, **1,** 1150; **5,** 655
Titanium(IV) chloride, **12,** 494
Tributyltin hydride, **3,** 294; **7,** 379
Triethylammonium fluoride, **10,** 415
Zinc bromide, **2,** 463
Zinc–Zinc chloride, **6,** 675
EPISULFIDES
Ethylene sulfide, **1,** 378
Silver(I) nitrate, **5,** 582
Sodium borohydride, sulfurated, **5,** 601
EPOXIDES
with N nucleophiles
Alumina, **8,** 9; **9,** 8
Benzylamine, **1,** 51
Boron trifluoride etherate, **6,** 65
Calcium ethoxide, **7,** 52
Cyanotrimethylsilane, **11,** 147; **12,** 148
Dialkylaluminum amides, **10,** 117
Dimethyl sulfoxide, **3,** 119; **7,** 133
Iodonium di-*sym*-collidine perchlorate, **10,** 212
Sodium azide, **1,** 1041; **9,** 418
Titanium(IV) alkoxides, **12,** 504
Triphenylphosphine–Carbon tetrachloride, **8,** 516
with O nucleophiles
Alumina, **6,** 16; **8,** 9; **9,** 8; **10,** 8
Bis(triphenylphosphine)nickel(0), **4,** 41
(+)-Camphanic acid, **8,** 74
Carbon dioxide, **9,** 94
Copper(I) chloride, **12,** 141
Copper(I) cyanoacetate, **7,** 74
Copper(II) sulfate, **8,** 125; **9,** 125
Di-*t*-butylperoxyoxalate, **7,** 93
Dimethyl sulfoxide, **2,** 157; **4,** 192; **5,** 263
Dimethyl sulfoxide–Trifluoroacetic acid, **6,** 226
Ferric chloride, **8,** 229
Formic acid, **5,** 316
Hydrogen peroxide, **4,** 253; **7,** 174
Iodine, **11,** 261
Lithium hydroxide, **7,** 208
Molybdenum(V) chloride–Triphenylphosphine, **10,** 274
Nafion-H, **10,** 275
Perchloric acid, **1,** 796
Periodic acid, **1,** 815
Potassium hydroxide, **12,** 411
Sodium borohydride, **8,** 449
Tetramethylammonium hydroxide, **11,** 514
Thallium(III) nitrate, **10,** 395
Titanium(IV) alkoxides, **12,** 504
Trimethyl phosphonoacetate, **7,** 394
Trimethylsilyl trifluoromethanesulfonate, **9,** 497; **11,** 584
with X nucleophiles
Acetic anhydride–Magnesium bromide, **6,** 352
Bis(benzonitrile)dichloropalladium(II), **9,** 44
Bromine, **8,** 52
Bromodimethylsulfonium bromide, **9,** 66
Bromotrimethylsilane, **10,** 59, 216
2-Chloro-3-ethylbenzoxazolium tetrafluoroborate, **8,** 90
Chlorotrimethylsilane, **11,** 125
Copper(II) sulfate–Pyridine, **9,** 125
Dilithium tetrabromonickelate, **12,** 195
Dimethylformamide–Phosphoryl chloride, **1,** 284
Hydriodic acid, **8,** 246

Hydrobromic acid, **1,** 450
Hydrogen iodide, **7,** 173
Iodotrimethylsilane, **10,** 216
Magnesium bromide etherate, **7,** 218
Pyridinium chloride, **5,** 566; **7,** 308; **10,** 333
Sodium iodide–Acetic acid, **1,** 881; **5,** 595
Sulfuryl chloride, **2,** 394
Trifluoroacetyl chloride, **10,** 419
Triphenylphosphine dihalides, **6,** 646; **12,** 554
Vilsmeier reagent, **1,** 286
with S nucleophiles
Alumina, **6,** 16; **8,** 9
t-Butyl hydroperoxide + co-reagent, **11,** 92
Lithium diethylamide, **5,** 398
Organoaluminum reagents, **10,** 281
Polyphosphoric acid, **1,** 894
Sodium borohydride, **4,** 443
Titanium(IV) alkoxides, **12,** 504
Tributylphosphine–Diphenyl disulfide, **9,** 200
Triphenylphosphine–Thiocyanogen, **9,** 507
Zinc iodide, **10,** 462
with other Heteroatoms
Alumina, **8,** 9
Dimethylaluminum methylselenolate, **8,** 182
Diphenyl diselenide, **5,** 272
(Diphenylphosphine)lithium, **5,** 408
Diphenylsulfonium cyclopropylide, **8,** 212
o-Nitrophenyl selenocyanate, **10,** 278
Phenyl trimethylsilyl selenide, **9,** 496
Sodium benzeneselenoate, **9,** 432
Trimethylsilylpotassium, **7,** 402
Triphenyltin hydride, **10,** 451
Tris(phenylseleno)borane, **10,** 454
with Organocopper reagents
Copper(I) iodide, **5,** 167; **9,** 124
Dilithium dialkyl(cyano)cuprates, **12,** 349
Grignard reagents, **6,** 269
Lithium dimethylcuprate, **3,** 106; **4,** 177; **5,** 234; **6,** 209
Lithium dipropylcuprate, **6,** 245
Lithium divinylcuprate, **5,** 288
Lithium methyl(vinyl)cuprate, **6,** 342
Organocopper reagents, **9,** 328; **10,** 282; **11,** 365; **12,** 345
Phenylcopper, **7,** 282
α-Trimethylsilylvinylmagnesium bromide, **8,** 515
with Organolithium reagents,
1-(Alkylthio)vinyllithium, **5,** 6
1,3-Bis(methylthio)allyllithium, **4,** 38
Boron trifluoride etherate, **12,** 66
t-Butyl dilithioacetoacetate, **8,** 60
n- or *sec*-Butyllithium, **6,** 85; **7,** 45; **8,** 65; **12,** 97
Copper(I) iodide, **9,** 124
Copper(I) trifluoromethanesulfonate, **8,** 125
gem-Dichloroallyllithium, **8,** 150
2-(2,2-Dimethoxyethyl)-1,3-dithiane, **4,** 164
N,N-Dimethylhydrazine, **7,** 126
1,3-Dithiane, **2,** 182; **3,** 135
Hexamethylphosphoric triamide, **10,** 196
Isoprene epoxide, **6,** 307
1-Lithiocyclopropyl phenyl sulfide, **11,** 284
α-Lithiomethylselenocyclobutane, **10,** 232
3-Lithio-1-triisopropylsilyl-1-propyne, **11,** 566
Lithium, **1,** 570
Lithium acetylide, **8,** 285
Lithium diethylamide, **2,** 247; **6,** 332; **7,** 201
Lithium diisopropylamide, **6,** 334; **7,** 207; **8,** 292
Lithium α-lithioacetate, **7,** 113; **8,** 175
Lithium naphthalenide, **6,** 415
S-(2-Methoxyallyl)-N,N-dimethyldithiocarbamate, **6,** 364
(4S,5S)-4-Methoxymethyl-2-methyl-5-phenyl-2-oxazoline, **6,** 386
Methylenetriphenylphosphorane, **11,** 338
Organocopper reagents, **10,** 282
Organolithium reagents, **10,** 3
Palladium(II) chloride–Thiourea, **6,** 450
Phenyllithium, **8,** 396
2-Phenylselenopropionic acid, **8,** 397
Phenylsulfinylacetone, **5,** 524
2-(2-Phenylsulfonylethyl)-1,3-

CLEAVAGE REACTIONS (*Continued*)
dioxolane, **6,** 461
Phenylthioacetic acid, **6,** 463
2-Phenylthiopropionic acid, **6,** 466
Propiolic acid, **6,** 495
2,4,4,6-Tetramethyl-5,6-dihydro-1,3-(4H)-oxazine, **3,** 280
2,4,4-Trimethyl-2-oxazoline, **5,** 714
Trimethylsilylacetic acid, **6,** 631
Trimethylsilylacetonitrile, **10,** 427
Trimethylstannylmethyllithium, **12,** 547
with Organomagnesium reagents
Butyllithium–Magnesium bromide, **10,** 71
Dimethylmagnesium, **1,** 292
Dimethyl N-(*p*-toluenesulfonyl)-sulfoximine, **12,** 216
Ethylene oxide, **1,** 377
Grignard reagents, **1,** 415
Phenylcopper, **7,** 282
with other Carbon reagents
Alkylaluminum halides, **4,** 144
1,3-Butadiene monoxide, **4,** 53
Butyllithium–Tetramethylethylenediamine, **10,** 73
Chloro(methoxy)methylalane, **5,** 438
Cyanotrimethylsilane, **5,** 720; **11,** 147
Di-μ-carbonylhexacarbonyldicobalt, **1,** 224; **8,** 148; **12,** 163
Diethylcarbo-*t*-butoxymethylaluminum, **7,** 103
Diethyl cyanomethylphosphonate, **1,** 250
Diethyl methylmalonate, **5,** 216
Diisobutylaluminum hydride, **6,** 198; **7,** 111; **8,** 173
Dimethyl malonate, **8,** 193
Dodecamethylcyclohexasilane, **12,** 219
Ethylene oxide, **5,** 297
Organoaluminum reagents, **1,** 244; **11,** 364; **12,** 339
Oxygen, **5,** 482
Sodium ethoxide, **4,** 451
Tetrakis(triphenylphosphine)-palladium(0), **11,** 503
Trialkylaluminums, **1,** 1197; **11,** 539; **12,** 512
Triethyl phosphonoacetate, **1,** 1217
Trifluoromethanesulfonic anhydride, **4,** 533
Trimethylaluminum–Dichlorobis-(cyclopentadienyl)zirconium, **10,** 423
Trimethylsilylmethylpotassium, **6,** 637
Tris(tribenzylideneacetylacetone)-tripalladium(chloroform), **12,** 561

ETHERS (*see also* DEALKYLATION, DEMETHYLATION, REDUCTION OF BENZYL ETHERS)
Acetic anhydride + co-reagent, **2,** 36; **6,** 352
Acetyl methanesulfonate, **5,** 5
Acetyl *p*-toluenesulfonate, **2,** 14
Aluminum chloride–Sodium iodide, **12,** 29
Aluminum iodide, **12,** 30
Boron trifluoride etherate, **2,** 35
Bromodimethylborane, **12,** 199
Bromotrimethylsilane, **10,** 59
Butyllithium, **2,** 51; **8,** 65
3-Chloro-4,5-dihydrofuryl-2-copper, **11,** 119
Copper(I) bromide, **7,** 79
Copper–Chromium oxide, **1,** 157
Di-μ-carbonylhexacarbonyldicobalt, **9,** 144
3,4-Dihydro-2H-pyran, **7,** 109
1,1-Diphenylhexyllithium, **5,** 277
Hydrobromic acid, **1,** 450
Hydrogen iodide, **7,** 173
Iodotrimethylsilane, **9,** 251
Lithium–Ammonia, **9,** 273
Lithium–Ethylamine, **10,** 236
Lithium naphthalenide, **4,** 348
Lithium trialkoxyaluminum hydrides, **4,** 312
Lithium triethylborohydride–Aluminum *t*-butoxide, **9,** 287
Molybdenum carbonyl, **4,** 346
Phase-transfer catalysts, **9,** 356
Phosphorus(V) oxide–Phosphoric acid, **1,** 872
Phosphoryl chloride, **1,** 876
Polyphosphoric acid, **1,** 894
Pyridinium chloride, **6,** 497
Sodium, **4,** 437; **7,** 324
Sodium iodide–Trimethylacetyl chloride, **11,** 489
Tetrachlorosilane–Sodium iodide, **11,** 466
Tetrahydrofuran, **1,** 1140

Thallium(III) nitrate, **10,** 395
Trichloro(methyl)silane, **11,** 553
Triphenylphosphine dibromide, **1,** 1247; **3,** 320; **5,** 729
Zinc chloride, **1,** 1289
OXETANES
Acetic anhydride–Pyridinium chloride, **4,** 4
Boron trifluoride etherate, **12,** 66
Bromotrimethylsilane, **10,** 59
Copper(I) bromide, **7,** 79
Dialkylaluminum amides, **8,** 164
Dimethyl N-(*p*-toluenesulfonyl)-sulfoximine, **12,** 216
Lithium aluminum hydride–Aluminum chloride, **4,** 293
Lithium–Ethylenediamine, **4,** 291
Lithium triethylborohydride–Aluminum *t*-butoxide, **9,** 287
Trimethylene oxide, **1,** 1232
CLEMMENSEN REDUCTION (*see* DEOXYGENATION)
COLLINS OXIDATION (*see* REAGENT INDEX)
CONJUGATE ADDITION (*see also* ASYMMETRIC REACTIONS)
WITH HETEROATOMS
Alumina, **10,** 8; **11,** 22
Aluminum chloride, **11,** 25
N-Benzylhydroxylamine, **9,** 42
Dimethyl acetylenedicarboxylate, **2,** 145
Dimethylaluminum benzenethiolate, **10,** 163
Dimethylaluminum methylselenolate, **8,** 182
N,N-Dimethylbenzeneselenenamide, **7,** 119
Dimethylphenylsilyllithium, **8,** 196; **12,** 210
Hydrogen bromide, **10,** 200
Hydroxylamine, **1,** 478
Iodotrimethylsilane, **9,** 251; **11,** 271
Ion-exchange resins, **8,** 263
Lithium diisopropylamide, **5,** 406
Methoxyamine, **1,** 670
Methyl 3-oxo-4-pentenoate, **6,** 662
Methylthiotrimethylsilane, **6,** 399
Methyl vinyl ketone, **6,** 407
Organoaluminum reagents, **12,** 339
Organocopper reagents, **10,** 282
Oxalic acid, **1,** 764
Phenyl trimethylsilyl selenide, **9,** 373
Phosphorus(V) sulfide, **9,** 374
Piperidine, **1,** 886
Potassium hydride, **12,** 407
Sodium acetate, **1,** 1024
Sodium benzeneselenoate, **9,** 432; **10,** 368
Sodium thiophenoxide, **6,** 552
Tetrabutylammonium fluoride, **7,** 353
Tetrachlorosilane, **12,** 463
Tetraethylammonium iodide, **12,** 465
Thiolacetic acid, **1,** 1154
Tributyltinlithium, **8,** 495; **12,** 523
Triethyl orthoformate, **6,** 610
Trimethylsilyllithium, **7,** 400
Vinyltriphenylphosphonium bromide, **1,** 1274
HYDROCYANATION
Acetone cyanohydrin, **1,** 5; **8,** 409
t-Butyl isocyanide, **11,** 99
Crown ethers, **6,** 133
Cyanotrimethylsilane–Triethylaluminum, **10,** 113
Diethylaluminum cyanide–Chlorotrimethylsilane, **8,** 130
Hydrogen cyanide, **1,** 1197, 1198; **5,** 689
N-Methyl-2-pyrrolidone, **2,** 281
Organoaluminum cyanides, **1,** 244; **2,** 127; **4,** 146; **6,** 180; **9,** 158, 328
Potassium cyanide–Acetone cyanohydrin, **8,** 409
Triethylaluminum, **2,** 427; **4,** 526; **5,** 688
INTRAMOLECULAR REACTIONS
Alkylaluminum halides, **12,** 5
Allytrimethylsilane, **12,** 23
Butyllithium, **12,** 96
Cryptates, **12,** 404
Dimethyl 1,3-acetonedicarboxylate, **9,** 176
Disodium tetracarbonylferrate, **8,** 216
Ethyl vinyl ether, **11,** 235
4-Iodobutyltrimethyltin, **10,** 219
Iodotrimethylsilane, **11,** 271
Mercury(II) acetate, **11,** 315
Phenyl vinyl sulfone, **10,** 315
Potassium carbonate, **8,** 408
Sodium hydride, **12,** 447
Tetrabutylammonium fluoride, **10,** 378
Titanium(IV) chloride, **12,** 494
MICHAEL REACTIONS
Allylidenetriphenylphosphorane, **6,** 14
Aniline, **6,** 21

CONJUGATE ADDITION (*Continued*)
Benzyltrimethylammonium hydroxide, **1,** 1252
Bis(methylthio)(trimethylsilyl)-methyllithium, **6,** 53
t-Butyl cyanoacetate, **1,** 87
t-Butyllithium, **8,** 70
1-*t*-Butylthio-1-trimethylsilyloxy-ethylene, **9,** 92
Cesium fluoride, **10,** 81; **11,** 117; **12,** 108
Chromium carbonyl, **7,** 71
Crown ethers, **8,** 128
1-Cyclopropyl-1-trimethylsilyloxy-ethylene, **8,** 140
1,4-Diazabicyclo[2.2.2]octane, **11,** 153
1,5-Diazabicyclo[5.4.0]undecene-5, **11,** 155
Diethyl acetylmethylmalonate, **7,** 102
Diethyl lithiomorpholinomethyl-phosphonate, **7,** 105
Dimethyl 1,3-acetonedicarboxylate, **9,** 176
2-(2,6-Dimethylpiperidino)acetonitrile, **11,** 212
Dimethyl sulfoxide, **1,** 296
Dimethylsulfoxonium methylide, **4,** 197
Dimethyl(tetrahydro-2-oxo-3-furanyl)-sulfonium tetrafluoroborate, **5,** 262
o-Ethoxycarbonylbenzyl phenyl sulfoxide, **8,** 75
2-Ethoxycarbonyl-1,3-dithiolane, **5,** 304
Ethyl acrylate, **6,** 251
Ethyl diethoxyacetate, **6,** 253
Ethyl ethylthiomethyl sulfoxide, **5,** 299
Ethyl 2-methylsulfinylacetate, **6,** 255
Ethyl phenyl sulfoxide, **6,** 257
1-Ethylsulfinyl-3-pentanone, **9,** 221
Grignard reagents, **11,** 245
Hexamethylphosphoric triamide, **8,** 240; **9,** 235; **10,** 196; **12,** 239
Ion-exchange resins, **1,** 511
Ketene alkyl trialkylsilyl acetals or ketals, **10,** 261; **11,** 279
Ketene dimethyl thioacetal monoxide, **6,** 311; **8,** 268; **9,** 261
3-Lithiobenzothiazole, **8,** 274
Lithium 1,3-butadien-1-olate, **8,** 291
Lithium N-isopropylcyclohexylamide, **6,** 411
Lithium naphthalenide, **10,** 247
Lithium 2,2,6,6-tetramethylpiperidide, **9,** 285
Magnesium ethyl malonate, **8,** 311
Magnesium methoxide, **3,** 189
2-(2-Methoxy)allylidene-1,3-dithiane, **7,** 226
5-Methoxycarbonyl-1-penten-3-one, **1,** 1271
Methyl bis(ethylthio)acetate, **5,** 444
Methyl 2-bromocrotonate, **7,** 235
3-Methyl-1,3-butadienyl phenyl sulfoxide, **8,** 333
Methyl lithiodithioacetate, **11,** 340
Methyl α-(methyldiphenylsilyl)acetate, **12,** 325
Methyl 2-methylthioacrylate, **5,** 455
Methyl 4-methylthiocrotonate, **6,** 389
Methyl methylthiomethyl sulfoxide, **8,** 344
Methyl 3-oxo-4-pentenoate, **1,** 1272
Methyl phenylsulfinylacetate, **8,** 347; **11,** 343
Methyl vinyl ketone, **1,** 697; **5,** 464; **7,** 247; **12,** 329
Nickel(II) acetylacetonate, **10,** 42
1-Nitro-1-(phenylthio)propene, **10,** 279
2-Nitro-3-pivaloyloxypropene, **11,** 360
2-Nitropropene, **7,** 253
(2-Oxo-2-phenylethylide)-dimethylselenonium ylide, **5,** 257
Ozone, **8,** 374
Phase-transfer catalysts, **9,** 356
3-Phenylsulfonyl-1(3H)-isobenzofuranone, **9,** 368
Phenylthiotrimethylsilylmethane, **11,** 423
Phenyl 2-(trimethylsilyl)ethynyl sulfone, **11,** 426
Phenyl vinyl sulfone(or sulfoxide), **6,** 468; **10,** 315
Phthalide, **8,** 402
Piperidine, **1,** 886; **7,** 293
Potassium fluoride, **1,** 933; **6,** 481; **12,** 406
Potassium permanganate, **11,** 440
β-Propiolactone, **1,** 957
Sodium cyanide, **4,** 446; **5,** 606; **6,** 535
Sodium dicarbonyl(cyclopentadienyl)-ferrate, **6,** 538
Sodium ethoxide, **1,** 1065
(+)-(R)-*trans*-β-Styryl *p*-tolyl sulfoxide, **4,** 466
Tetrabutylammonium fluoride, **9,** 446; **11,** 499

2,4,4,6-Tetramethyl-5,6-dihydro-1,3-(4H)-oxazine, **4,** 481
1,1,3,3-Tetramethylguanidine, **1,** 1145; **4,** 489; **12,** 477
Titanium(IV) chloride, **6,** 590; **7,** 370; **10,** 401; **12,** 494
Tributylphosphine, **5,** 685
Tributyltin hydride, **11,** 545
Trichloroisocyanuric acid, **6,** 605
Triethylamine, **1,** 1198
Triphenylphosphine, **12,** 550
Tris(dimethylamino)sulfonium difluorotrimethylsilicate, **12,** 558
Tris(methyl- or phenylthio)-methyllithium, **6,** 650; **7,** 412; **9,** 511
β-Vinylbutenolide, **6,** 661; **9,** 515; **11,** 594
Zirconium(IV) propoxide, **11,** 605
WITH ORGANOALUMINUM REAGENTS
Alkylaluminum halides, **4,** 144
Diisobutylaluminum hydride, **5,** 224
Nickel(II) acetylacetonate–Diisobutylaluminum hydride, **8,** 42
μ-Oxybis(chlorotriphenylbismuth), **11,** 381
Trialkylaluminums, **5,** 734; **6,** 622
WITH ORGANOBORON REAGENTS
Acrolein, **2,** 15; **3,** 5
Acrylonitrile, **7,** 4
B-1-Alkynyl-9-borabicyclo[3.3.1]-nonanes, **8,** 6
9-Borabicyclo[3.3.1]nonane, **7,** 29
Bromine, **5,** 55
3-Butyn-2-one, **3,** 38
Dimethyl(methylene)ammonium halides, **4,** 186
Diphenylborane, **9,** 198
Methyl vinyl ketone, **2,** 283
Titanium(IV) chloride, **10,** 401
WITH ORGANOCOPPER REAGENTS
Benzeneselenenyl halides, **6,** 459
trans-1,2-Bis(tributylstannyl)ethylene, **5,** 43
Boron trifluoride etherate, **10,** 52
2-(2-Bromoethyl)-1,3-dioxane, **11,** 78
2-(2-Bromoethyl)-1,3-dioxolane, **9,** 70
3-Bromopropyl 1-ethoxyethyl ether, **4,** 226
2-Bromo-3-trimethylsilyl-1-propene, **11,** 80
1-Butanethiol, **4,** 64
4-Chloro-1-butenyl-2-lithium, **12,** 113
Copper(II) acetate, **1,** 159
Copper(I) bromide, **1,** 165; **4,** 108; **8,** 116, 117
Copper(I) chloride, **1,** 166; **2,** 91
Copper(I) iodide, **2,** 92; **3,** 278; **4,** 516
Copper(I) phenoxides, **5,** 148
Dichlorobis(cyclopentadienyl)titanium, **6,** 48
N,N-Dimethylformamide, **8,** 189
N,N-Dimethylhydrazine, **7,** 126; **9,** 184
1-Ethoxy-1-trimethylsilyloxy-cyclopropane, **12,** 221
(Z)-2-Ethoxyvinyllithium, **8,** 221
Formaldehyde, **6,** 264
Grignard reagents, **8,** 235
Hydrogen peroxide, **12,** 242
Ketene dimethyl thioacetal monoxide, **9,** 261
Lithium alkyl(*t*-butoxy)cuprates, **5,** 395
Lithium alkyl(phenylthio)cuprates, **5,** 414; **7,** 212
Lithium–Ammonia, **10,** 234
Lithium bis(2-vinylcyclopropyl)cuprate, **7,** 199
Lithium diallylcuprate, **5,** 175; **7,** 86
Lithium dibutylcuprate, **6,** 163
Lithium (1,1-diethoxy-2-propenyl)-(3,3-dimethyl-1-butynyl)cuprate, **8,** 300
Lithium di(α-methoxyvinyl)cuprate, **6,** 204
Lithium dimethylcuprate, **2,** 151; **3,** 106; **4,** 177; **5,** 234; **6,** 209; **8,** 301
Lithium di[(E)-1-propenyl]cuprate, **8,** 302
Lithium dipropylcuprate, **5,** 283
Lithium divinylcuprate, **4,** 219, 221
Lithium methyl(vinyl)cuprate, **6,** 342
Lithium α-ethoxycarbonylvinyl (1-hexynyl)cuprate, **6,** 329
Lithium 2,2,6,6-tetramethylpiperidide, **6,** 345
3-Methoxy-3-methylbutynylcopper, **8,** 324
Methylcopper, **2,** 282, 441; **6,** 377
Organocopper reagents, **9,** 328; **10,** 282; **11,** 365; **12,** 345
Organolithium reagents, **12,** 350
Phenylthiocopper, **6,** 465
3-Tetrahydropyranyloxy-1-

CONJUGATE ADDITION (*Continued*)
tributylstannyl-1-propene, **6,** 602
Titanium(IV) chloride, **12,** 494
2,4,6-Triisopropylbenzene-sulfonylhydrazide, **10,** 422
3-Trimethylsilyl-3-buten-2-one, **5,** 461
α-Trimethylsilyl-2-methyl-benzothiazole, **8,** 511
1-Trimethylsilylpropynylcopper, **6,** 638
Vinylcopper, **5,** 747
WITH ORGANOLITHIUM REAGENTS
2-Lithiobenzothiazole, **8,** 274
3-Lithio-1-triisopropylsilyl-1-propyne, **11,** 566
2-Lithio-2-trimethylsilyl-1,3-dithiane, **4,** 284
Lithium acetylides, **11,** 288
Methyllithium, **9,** 310; **12,** 324
Methyl 2-trimethylsilylacrylate, **11,** 346
Organolithium reagents, **9,** 5
WITH ORGANOMAGNESIUM REAGENTS
Grignard reagents, **1,** 415; **9,** 229; **10,** 189
Hexamethylphosphoric triamide, **5,** 323
WITH ORGANOSILANES
Alkylaluminum halides, **12,** 5
Allyltrimethylsilane, **12,** 23
Cesium fluoride, **11,** 115
2-Hydroxymethyl-3-trimethylsilyl-1-propene, **11,** 258
Organocopper reagents, **12,** 345
Titanium(IV) chloride, **7,** 370; **11,** 529; **12,** 494
Trimethylsilylallene, **10,** 428
3-Trimethylsilyl-1-cyclopentene, **8,** 509
WITH ORGANOTIN REAGENTS
Allyltrimethyltin, **11,** 20
Bis(diethylaluminum) sulfate, **9,** 46
WITH YLIDES
Allylidenetriphenylphosphorane, **5,** 7
2-Butenylidenetriphenylphosphorane, **5,** 69
Isopropylidenetriphenylphosphorane, **5,** 361
CONJUGATE REDUCTION (*see* REDUCTION REACTIONS)
COPE ELIMINATION (*see* ELIMINATION REACTIONS)
COPE REARRANGEMENT (*see* SIGMATROPIC REARRANGEMENTS)
COREY–KIM OXIDATION
N-Chlorosuccinimide–Dimethyl sulfide, **4,** 87
Dimethyl sulfoxide–Oxalyl chloride, **8,** 200
Formaldehyde, **7,** 158
Phenylselenoacetaldehyde, **9,** 365
Thioanisole, **6,** 582
COREY–WINTER REACTION (*see* ELIMINATION REACTIONS)
COUPLING REACTIONS (*see also* ASYMMETRIC REACTIONS, OXIDATIVE AND REDUCTIVE COUPLING)
HOMOCOUPLING, ALKYL GROUPS
Chromium(II) chloride, **11,** 132
Chromium(III) chloride–Lithium aluminum hydride, **8,** 110
Chromium(II) sulfate, **1,** 150
Copper, **12,** 140
Copper(II) acetylacetonate, **2,** 81
Copper(I) iodide–Tributylphosphine, **2,** 400
Copper(0)–Isonitrile complexes, **9,** 122
Dicyclopentadienyltitanium methylene–Zinc iodide, **12,** 506
Difluoramine, **1,** 253
Dilithium tetrachlorocuprate, **4,** 163
N,N-Dimethylformamide, **1,** 278
Ferric chloride, **9,** 222
Hydrogen peroxide–Salt and oxide catalysts, **1,** 472
Lead, **2,** 233
Lithium aluminum hydride, **9,** 274
Lithium dibutylcuprate, **3,** 79
Magnesium, **4,** 315
Samarium(II) iodide, **10,** 344
Silver(I) nitrate, **2,** 366; **4,** 429
Sodium naphthalenide, **1,** 711
Tetraphenylethylene, **2,** 404
Thallium(I) bromide, **2,** 405
Titanium(III) chloride, **2,** 415
Titanium(IV) chloride, **1,** 1169
Titanium(III) chloride–Lithium aluminum hydride, **6,** 588; **7,** 369; **10,** 401
Tungsten carbonyl, **8,** 525
Vanadium(II) chloride tetrapyridine complex, **5,** 744

HOMOCOUPLING, ALKYNYL GROUPS
t-Butylamine, **1,** 84
Copper(II) acetate, **1,** 159; **4,** 105; **6,** 138
Copper(I) chloride, **1,** 166; **4,** 109
Copper(I) chloride + co-reagent, **1,** 165; **5,** 165
Copper(I) cyanide, **6,** 146
Hydroxylamine, **1,** 478
Thionyl chloride, **5,** 663
HOMOCOUPLING, ALLYL GROUPS
Allylmagnesium chloride, **1,** 420
Chlorotris(triphenylphosphine)cobalt, **12,** 128
Chromium(II) chloride, **11,** 132
Chromium(III) chloride–Lithium aluminum hydride, **8,** 110
Copper(0)–Isonitrile complexes, **9,** 122
Crotyl chloride, **2,** 290
Iron, **3,** 167
Lithium bis(alkylamino)cuprates, **6,** 327
Magnesium, **4,** 315
Nickel carbonyl, **1,** 720; **2,** 290; **3,** 210
Palladium(II) chloride, **10,** 300; **12,** 371
Samarium(II) iodide, **10,** 344
Titanium(III) chloride, **2,** 415
Titanium(III) chloride–Lithium aluminum hydride, **6,** 588; **7,** 369; **10,** 401
Tris(triphenylphosphine)nickel(0), **6,** 653
HOMOCOUPLING, ARYL GROUPS
Antimony(III) chloride, **8,** 19
Bis(1,5-cyclooctadiene)nickel(0), **4,** 33; **11,** 51
trans-Bromobis(triethylphosphine)-*o*-tolylnickel(II), **10,** 58
Cerium(IV) trifluoroacetate, **5,** 102
Chromium(VI) oxide, **2,** 72
Copper, **1,** 157; **5,** 146; **7,** 73; **9,** 122
Copper(II)–Amine complexes, **8,** 114
Copper bronze, **1,** 155; **4,** 100
Copper(II) chloride, **2,** 84
Copper(I) iodide–Triethyl phosphite, **8,** 122
Copper(I) trifluoromethanesulfonate, **6,** 130
1,4-Dichloro-2-butene, **10,** 134
Diethyltrimethylsilylamine, **6,** 184
N,N-Dimethylformamide, **1,** 278
Ferric chloride, **10,** 185
Magnesium, **7,** 218
Nickel, **12,** 335
Nickel(II) bromide–Zinc, **9,** 322
Nickel catalysts, **1,** 723
Palladium(II) acetate, **6,** 442; **10,** 297
Periodic acid, **2,** 313
Potassium–Graphite, **7,** 296
Pyridine–Aluminum chloride, **2,** 351
Tellurium(IV) chloride, **5,** 643
Tetracarbonyldi-μ-chlorodirhodium, **8,** 469
Tetramethylurea, **1,** 1146
Thallium(I) bromide, **2,** 405; **3,** 286; **5,** 655; **7,** 361
Thallium(III) trifluoroacetate, **8,** 478; **10,** 397
Thallium(III) trifluoroacetate–Palladium(II) acetate, **11,** 516; **12,** 484
Tris(*p*-bromophenyl)ammoniumyl hexachloroantimonate, **5,** 735
Tris(triphenylphosphine)nickel(0), **6,** 653; **8,** 523
Vanadium(IV) chloride, **3,** 332
Vanadyl trifluoride, **8,** 528
HOMOCOUPLING, PROPARGYL GROUPS
Copper(I) chloride, **6,** 145
HOMOCOUPLING, VINYL GROUPS
Bis(1,5-cyclooctadiene)nickel(0), **4,** 33
Chlorobis(cyclopentadienyl)-hydridozirconium(IV), **8,** 84
Copper, **4,** 102
Copper(I) bromide–Dimethyl sulfide, **10,** 104
Copper(I) chloride, **2,** 91
Copper(I) trifluoromethanesulfonate, **6,** 130
Diisobutylaluminum hydride, **4,** 158
Manganese(II) chloride, **7,** 222
Methylcopper, **6,** 377
(R)-4-Methylcyclohexylidene-methylcopper, **7,** 238
Monochloroborane diethyl etherate, **5,** 465
Nickel(II) chloride, **12,** 336
Organopentafluorosilicates, **11,** 373
Palladium(II) acetate, **2,** 303
Palladium(II) chloride, **8,** 384
Tetracarbonyldi-μ-chlorodirhodium, **8,** 469
Tetrakis(triphenylphosphine)-palladium(0), **6,** 571

COUPLING REACTIONS (*Continued*)
Thionyl chloride, **5,** 663
Tris(triphenylphosphine)nickel(0), **6,** 653
Vinylcopper reagents, **6,** 662
CROSS-COUPLING, TWO ALKYL GROUPS
Butyllithium, **6,** 85
N-Chlorosuccinimide–Triethylamine, **6,** 120
Copper(I) methyltrialkylborates, **7,** 75
Dichlorodimethyltitanium, **10,** 138
Diethyl sulfate, **1,** 253
Dilithium tetrachlorocuprate(II), **4,** 163; **5,** 226; **6,** 203; **7,** 114
1,1-Diphenylhexyllithium, **5,** 277
Grignard reagents, **1,** 415
2,3-O-Isopropylidene-2,3-dihydroxy-1,4-bis(diphenylphosphine)butane, **5,** 360
Lithium alkyl(*t*-butoxy)cuprates, **5,** 395
Lithium alkyl(phenylthio)cuprates, **5,** 414
Lithium dibutylcuprate, **7,** 92; **8,** 291
Lithium diisopropylamide, **3,** 184
Lithium dimethylcuprate, **2,** 151; **3,** 106; **4,** 177; **5,** 234; **8,** 301
Lithium trimethylferrate and related reagents, **3,** 112, 312
3-Methoxy-3-methylbutynylcopper, **8,** 324
Methyl fluoride–Antimony(V) fluoride, **3,** 201
Methyllithium, **5,** 448
Organocopper reagents, **11,** 365; **12,** 345
Sodium bis(2-methoxyethoxy)aluminum hydride, **4,** 441; **6,** 528
Sodium methoxide, **1,** 1091
N,N,N′,N′-Tetramethylethylenediamine, **5,** 652
Trimethylaluminum, **5,** 707; **6,** 622
CROSS-COUPLING, TWO ALKYNYL GROUPS
Copper(I) chloride–Ammonium chloride, **1,** 165
Dicyclohexylmethylthioborane, **7,** 101
3,3,-Diethoxy-1-propyne, **2,** 126
Iodine, **7,** 179
Lithium aluminum hydride, **5,** 382
Palladium catalysts, **7,** 275
CROSS-COUPLING, TWO ALLYL GROUPS
Allyl phenyl ether, **11,** 14
Allyltrimethylsilane, **11,** 16
Benzylchlorobis(triphenylphosphine)-palladium(II), **10,** 26
1-Bromo-3-methyl-2-butene, **5,** 84
B-Butyl-9-borabicyclo[3.3.1]nonane, **9,** 76
Copper, **9,** 122
Copper(I) bromide–Dimethyl sulfide, **10,** 104
Copper(I) iodide, **8,** 121
1-Cyclobutenylmethyllithium, **6,** 151
Hexamethylphosphoric triamide, **3,** 149
Lithium–Alkylamines, **5,** 377
Nickel(II) acetylacetonate–Trimethylaluminum, **9,** 52
Palladium(II) chloride, **12,** 371
Tetrakis(triphenylphosphine)-palladium(0), **10,** 384
N,N,N′,N′-Tetramethylethylenediamine, **4,** 485
p-Toluenesulfonyl chloride–Lithium chloride, **3,** 293
Zinc chloride, **12,** 574
CROSS-COUPLING, TWO ARYL GROUPS
(1S,2S)-2-Amino-3-methoxy-1-phenyl-1-propanol, **12,** 310, 381
Bis(1,5-cyclooctadiene)nickel(0), **11,** 51
Bis-3,4-dichlorobenzoyl peroxide, **3,** 21
[1,3-Bis(diphenylphosphine)propane]-(dichloro)nickel(II), **11,** 167
t-Butyl thionitrate, **9,** 91
Cobalt(II) chloride, **1,** 155
Copper, **8,** 113
Copper(I) alkoxides, **9,** 122
Copper(I) iodide–Triethyl phosphite, **8,** 122
Crown ethers, **8,** 128
Dichlorobis(triphenylphosphine)-nickel(II), **6,** 59; **9,** 147
1,3-Dichloro-2-butene, **10,** 134
Diethyltrimethylsilylamine, **6,** 184
Dinitrogen trioxide–Boron trifluoride, **1,** 329
Diphenyl selenoxide, **9,** 201
Grignard reagents–Nickel(II) reagents, **7,** 250
Isoamyl nitrite, **1,** 40
Methanesulfonyl chloride, **6,** 362
2-(*o*-Methoxyphenyl)-4,4-dimethyl-2-oxazoline, **6,** 368

Nickel(II) acetylacetonate, **11,** 58
Pentafluorophenylcopper, **5,** 504
Phase-transfer catalysts, **9,** 356; **12,** 379
Phenylsodium, **1,** 848
Potassium *t*-butoxide, **5,** 544
Tetrakis(triphenylphosphine)nickel(0), **9,** 450
Tetrakis(triphenylphosphine)-palladium(0), **8,** 472; **11,** 503
Tetraphenylbismuth trifluoroacetate, **11,** 515
Thallium(III) trifluoroacetate, **3,** 286; **12,** 481
p-Toluenesulfonic acid, **1,** 1172
Vanadyl trifluoride, **6,** 660; **7,** 418; **9,** 513

CROSS-COUPLING, TWO VINYL GROUPS
B-1-Alkenyl-9-borabicyclo[3.3.1]-nonanes, **12,** 5
Benzylchlorobis(triphenylphosphine)-palladium(II), **12,** 44
Bis(acetonitrile)dichloropalladium(II), **11,** 46
Chlorobis(cyclopentadienyl)-hydridozirconium(IV), **8,** 84
Copper(I) iodide–Tributylphosphine, **2,** 400
Diacetatobis(triphenylphosphine)-palladium(II), **6,** 156
Dichlorobis(triphenylphosphine)-palladium(II)–Diisobutylaluminum hydride, **7,** 95
1,2-Dichloroethylene, **12,** 175
Dihalobis(triphenylphosphine)-palladium(II), **6,** 60
Grignard reagents, **10,** 189
Nickel(II) acetylacetonate–Trimethylaluminum, **9,** 52
Organocopper reagents, **10,** 282
Palladium(II) acetate, **12,** 367
Palladium catalysts, **10,** 297; **11,** 392
Palladium(II) chloride, **8,** 384
Tetrakis(triphenylphosphine)-palladium(0), **8,** 472; **9,** 451; **12,** 468
Thexylborane, **5,** 232
Trimethylaluminum, **8,** 506
Zinc chloride, **12,** 574

CROSS-COUPLING, ALKYL + ALKYNYL GROUPS
Acetylene, **1,** 11
2-Chloro-1,1-diethoxyethane, **5,** 107
Diiodo(methyl)bis(triphenylphosphine)-rhodium, **10,** 149
Hexamethylphosphoric triamide, **2,** 208; **4,** 244
Iodine, **5,** 346; **6,** 293; **8,** 256
Lithium acetylide–Ethylenediamine, **1,** 574; **5,** 382
Lithium amide, **9,** 278
Methanesulfinyl chloride, **5,** 434
Organocopper reagents, **9,** 328
Organolithium reagents, **9,** 5
Sodium amide, **1,** 1034; **5,** 591
Sodium methylsulfinylmethylide, **2,** 166
Trialkynylalanes, **6,** 600

CROSS-COUPLING, ALKYL + ALLYL GROUPS (*see also* SYNTHESIS INDEX–ALKENES)
Alkylcopper + co-reagent, **8,** 334; **9,** 333; **10,** 282, 286
Allyl phenyl ether, **11,** 14
Allyltributyltin, **11,** 15
Allyltrimethylsilane, **11,** 16
[1,2-Bis(diphenylphosphine)ethane]-(dichloro)nickel(II), **10,** 36
[1,1′-Bis(diphenylphosphine)ferrocene]-(dichloro)palladium(II), **12,** 171
Boron trifluoride etherate, **12,** 66
Butyllithium–Tetramethyl-ethylenediamine, **5,** 80
α-Chloroallyllithium, **10,** 87
1-Chloro-2-methyl-N,N-tetramethyl-enepropenylamine, **12,** 124
Copper(I) iodide, **8,** 121
Di-μ-bromobis(3-methyl-2-butenyl)-dinickel, **8,** 181
2,3-Dibromopropene, **1,** 420
gem-Dichloroallyllithium, **8,** 150
Dichlorobis(triphenylphosphine)-nickel(II), **9,** 147
Dilithium tetrachlorocuprate(II), **5,** 226; **8,** 176; **11,** 190
1,1-Diphenylhexyllithium, **5,** 277
Ferrocenylphosphines, **10,** 37
Grignard reagents, **2,** 110; **6,** 269; **10,** 189
Iron carbonyl, **12,** 266
(1-Lithiovinyl)trimethylsilane, **7,** 193; **9,** 498
Lithium aluminum hydride, **9,** 274
Lithium dibutylcuprate, **8,** 291
Lithium dimethylcuprate, **2,** 151; **3,** 106; **7,** 120

COUPLING REACTIONS (*Continued*)
Lithium trimethylferrate, **3,** 312
N,N-Methylphenylaminotriphenylphosphonium iodide, **8,** 346
Molybdenum carbonyl, **12,** 330
Nickel carbonyl, **2,** 290; **6,** 417
Organocopper reagents, **9,** 328; **10,** 282; **11,** 365; **12,** 345
Palladium(II) chloride, **5,** 500
Silver(I) nitrate, **1,** 1008
Silver tetrafluoroborate, **12,** 434
Titanium(IV) chloride, **6,** 590
Tributyl(methylphenylamino)phosphonium iodide, **8,** 345; **10,** 268; **11,** 342
Triisobutylaluminum, **7,** 391
β-Trimethylsilylethylidenetriphenylphosphorane, **9,** 492

CROSS-COUPLING, ALKYL + ARYL GROUPS (*see also* FRIEDEL–CRAFTS ALKYLATION)
Ammonium persulfate, **5,** 15
Ammonium persulfate–Silver nitrate, **5,** 16
Arene(tricarbonyl)chromium complexes, **6,** 27
[1,1′-Bis(diphenylphosphine)ferrocene](dichloro)palladium(II), **12,** 171
[1,3-Bis(diphenylphosphine)propane](dichloro)nickel(II), **11,** 167
t-Butyldimethylsilyl trifluoromethanesulfonate, **11,** 90
Butyllithium, **1,** 95; **9,** 83
Butyllithium–Potassium *t*-butoxide, **10,** 72
t-Butyl trifluoroacetate, **5,** 87
Cobalt(II) chloride, **10,** 101
Copper, **12,** 140
Copper(I) iodide, **8,** 121
Dibromobis(triethylphosphine)nickel(II), **6,** 57
Di-*t*-butyl peroxide, **1,** 211
Dichlorobis(triphenylphosphine)nickel(II), **11,** 165
Dilithium tetrachlorocuprate, **6,** 203
Dimethylbromonium hexafluoroantimonate, **5,** 231
Dimethylsulfonium methylide, **2,** 169
Dimethylsulfoxonium methylide, **1,** 315
Diphenyliodonium bromide, **1,** 340
Grignard reagents, **1,** 415; **6,** 50
Lead tetraacetate–Trifluoroacetic acid, **6,** 317
Lithium aluminum hydride–Pyridine, **4,** 294
Lithium dimethylcuprate, **2,** 151; **3,** 106
Lithium trimethylferrate and related reagents, **3,** 112, 312
Methoxycarbenium hexafluoroantimonate, **4,** 328
2-(*o*-Methoxyphenyl)-4,4-dimethyl-2-oxazoline, **6,** 368
Methyl fluoride–Antimony(V) fluoride, **3,** 201
Methyl iodide, **1,** 682
Methyltris(triphenylphosphine)rhodium(I), **5,** 463
Nafion-H, **9,** 320
Nickel(II) acetylacetonate, **11,** 58
Nickel carbonyl, **4,** 353
Nitrosonium hexafluorophosphate, **1,** 747
Organocopper reagents, **12,** 345
Organolithium reagents, **9,** 5
Palladium(II) acetate–Sodium chloride, **6,** 443
Pentafluorophenylcopper, **5,** 504
Peracetic acid, **11,** 402
Sodium hydroxide, **5,** 616
Sodium methylsulfinylmethylide, **1,** 310; **2,** 166; **5,** 621
Tetrahydrofuran, **6,** 570
Thallium(III) trifluoroacetate, **8,** 478
Trifluoromethanesulfonic-alkanesulfonic anhydrides, **6,** 618

CROSS-COUPLING, ALKYL + PROPARGYL GROUPS
n- or *t*-Butyllithium, **5,** 80; **6,** 85; **9,** 89
1-Chloro-N,N,2-trimethylpropenylamine, **12,** 123
Copper(I) bromide, **6,** 143; **7,** 79
3,3-Diethoxy-1-methylthiopropyne, **5,** 207
1,3-Dilithiopropyne, **6,** 202
Ferric chloride, **7,** 153
Grignard reagents, **6,** 147; **7,** 163
3-Lithio-1-trimethylsilyl-1-propyne, **2,** 239
Lithium aluminum hydride, **9,** 274
Lithium 3-chloropropargylide, **5,** 397
Lithium dimethylcuprate, **2,** 151; **3,** 106; **7,** 120

3-Methoxy-1-methylthio-1-propyne, **6,** 397
Organocopper reagents, **12,** 345
Phenyl trimethylsilylpropargyl ether, **9,** 373
Propargyltriphenyltin, **12,** 415
Tributyl(methylphenylamino)-phosphonium iodide, **10,** 268
CROSS-COUPLING, ALKYL + VINYL GROUPS (*see also* SYNTHESIS INDEX—ALKENES)
Ammonium persulfate–Silver nitrate, **8,** 18
trans-2,3-Bis(diphenylphosphine)bicyclo-[2.2.1]hept-5-ene, **10,** 36
[1,1′-Bis(diphenylphosphine)ferrocene]-(dichloro)palladium(II), **9,** 147; **12,** 171
[1,3-Bis(diphenylphosphine)propane]-(dichloro)nickel(II), **10,** 191; **11,** 167
trans-1,2-Bis(tributylstannyl)ethylene, **5,** 43
n- or *t*-Butyllithium, **6,** 85; **7,** 47
μ-Chlorobis(cyclopentadienyl)-(dimethylaluminum)-μ-methylenetitanium, **8,** 83
Copper(I) bromide–Dimethyl sulfide, **10,** 104
Copper(I) iodide, **12,** 141, 143
Dichlorobis(cyclopentadienyl)titanium, **11,** 163
Dichlorobis(cyclopentadienyl)titanium–Trialkylaluminum, **9,** 146
Diisobutylaluminum hydride, **7,** 111
Dilithium tetrachlorocuprate, **12,** 195
Dimethylformamide diethyl acetal, **9,** 182
1,1-Diphenylhexyllithium, **5,** 277
(S)-α-(R)-2-Diphenylphosphine-ferrocenyl ethyldimethylamine, **9,** 200; **11,** 237
Ferric chloride, **5,** 307; **6,** 259
Grignard reagents + co-reagents, **4,** 236; **6,** 50; **7,** 82, 411; **10,** 189
Hydrogen peroxide, **12,** 242
3-Lithio-1-trimethylsilyl-1-propyne, **2,** 239
(1-Lithiovinyl)trimethylsilane, **5,** 374
Lithium dibutylcuprate, **5,** 187; **7,** 92
Lithium dimethylcuprate, **2,** 151; **3,** 106; **4,** 177, 295; **6,** 209
Lithium di[(E)-1-propenyl]cuprate, **7,** 141
Lithium phenylthio(alkyl)cuprates, **6,** 343; **7,** 211
Lithium trimethylferrate, **3,** 312
Lithium trimethylmanganate, **3,** 112
3-Methoxy-3-methylbutynylcopper, **8,** 324
Methylcopper, **7,** 236
Methyltris(triphenylphosphine)-rhodium(I), **5,** 463
Nickel(II) chloride–Triphenylphosphine, **11,** 358
Organoaluminum reagents, **12,** 339
Organocopper reagents, **9,** 328; **10,** 282; **11,** 365; **12,** 345
Organolithium reagents, **12,** 350
Organotitanium reagents, **8,** 83
Palladium(II) acetate, **5,** 496; **8,** 378
Pentane-1,5-di(magnesium bromide), **9,** 355
Pentylsodium, **7,** 400
Phenylselenenyl *p*-toluenesulfonate, **12,** 394
Potassium diisopropylamide–Lithium *t*-butoxide, **9,** 383
(E)-1-Propenyllithium, **6,** 308, 495
Sodium amide, **5,** 591
Sodium methylsulfinylmethylide, **1,** 310
3-Tetrahydropyranyloxy-1-tributylstannyl-1-propene, **6,** 602
Tetrakis(triphenylphosphine)nickel(0), **12,** 467
Tetrakis(triphenylphosphine)-palladium(0), **6,** 571; **8,** 472; **9,** 451; **10,** 384; **12,** 468
p-Toluenesulfonylhydrazide, **9,** 472
2,4,6-Triisopropylbenzene-sulfonylhydrazide, **9,** 486
Trimethylaluminum–Dichlorobis-(cyclopentadienyl)zirconium, **11,** 567
Trimethyltinlithium, **12,** 547
Triphenylcarbenium tetrafluoroborate, **8,** 524
Tris(dimethylamino)methane, **7,** 411
Vinylcopper reagents, **6,** 662
Vinyllithium, **2,** 456; **6,** 665
CROSS-COUPLING, ALLYL + ALKYNYL GROUPS
Copper(I) chloride, **1,** 166
4-Methoxy-3-buten-1-ynylcopper, **9,** 297

COUPLING REACTIONS *(Continued)*
Palladium(II) chloride, **7,** 277
CROSS-COUPLING, ALLYL + ARYL GROUPS
Boron trifluoride etherate, **12,** 66
1-Bromo-3-methyl-2-butene, **5,** 64
1-Chloro-N,N,2-trimethylpropenylamine, **12,** 123
Copper, **12,** 140
Dichlorobis(triphenylphosphine)nickel(II), **9,** 147; **11,** 165
Ferrocenylphosphines, **10,** 37
Lead tetraacetate, **9,** 265
N,N-Methylphenylaminotriphenylphosphonium iodide, **8,** 346
Nickel carbonyl, **4,** 353
Oxalic acid, **1,** 764
Palladium(II) acetate, **5,** 497; **10,** 297; **12,** 367
Potassium bisulfate, **1,** 909
Silver(I) oxide, **6,** 515
Thallium(III) trifluoroacetate, **11,** 515
Tributyl(methylphenylamino)phosphonium iodide, **8,** 345
Zinc chloride, **10,** 461; **12,** 574
CROSS-COUPLING, ALLYL + VINYL GROUPS
Allyltrimethylsilane, **10,** 6
Benzylchlorobis(triphenylphosphine)palladium(II), **12,** 44
Bis(dibenzylideneacetone)palladium(0), **12,** 56
Copper(I) bromide–Dimethyl sulfide, **8,** 117
Diisobutylaluminum hydride, **6,** 198
Dilithium tetrachloropalladate (II), **9,** 174
Lithium (1,1-diethoxy-2-propenyl)(3,3-dimethyl-1-butynyl)cuprate, **8,** 300
Lithium dimethylcuprate, **4,** 177
Lithium α-ethoxycarbonylvinyl-(1-hexynyl)cuprate, **6,** 329
Methylcopper, **7,** 236
Nickel(II) acetylacetonate–Trimethylaluminum, **9,** 52
Nickel carbonyl, **4,** 353; **7,** 250
Palladium(II) acetate, **8,** 378
Palladium(II) chloride, **11,** 393
Tetrakis(triphenylphosphine)palladium(0), **10,** 384; **11,** 503
Tributyltin trifluoromethanesulfonate, **12,** 524
Trimethyltinlithium, **9,** 499
CROSS-COUPLING, ALKYNYL + ARYL GROUPS
Cobalt(II) chloride, **1,** 155
Diacetatobis(triphenylphosphine)palladium(II), **6,** 156; **10,** 117
Dicarbonylcyclopentadienylcobalt, **12,** 160
Dihalobis(triphenylphosphine)palladium(II), **6,** 60
Iodoethynyl(trimethyl)silane, **4,** 265
Lithium 2,2,6,6-tetramethylpiperidide, **4,** 310
4-Methoxy-3-buten-1-ynylcopper, **9,** 297
Organoaluminum reagents, **12,** 339
Tetrakis(triphenylphosphine)palladium(0), **6,** 571; **8,** 472
Zinc chloride, **12,** 574
CROSS-COUPLING, ALKYNYL + PROPARGYL GROUPS
Copper(I) chloride, **3,** 67
Di-μ-carbonylhexacarbonyldicobalt, **12,** 163
3,3-Diethoxy-1-propyne, **2,** 126
Lithium methoxyaluminum hydride, **6,** 341
Tetrahydrofuran, **1,** 1140
CROSS-COUPLING, ALKYNYL + VINYL GROUPS
B-1-Alkynyl-9-borabicyclo[3.3.1]nonanes, **8,** 6
1-Bromo-3-tetrahydropyranyloxy-1-propyne, **6,** 576
Diacetatobis(triphenylphosphine)palladium(II), **6,** 156
1,2-Dichloroethylene, **12,** 175
Dihalobis(triphenylphosphine)palladium(II), **6,** 60
B-Halo-9-borabicyclo[3.3.1]nonane, **12,** 236
Tetrakis(triphenylphosphine)palladium(0), **6,** 571; **8,** 472; **12,** 468
Tetrakis(triphenylphosphine)palladium(0)–Copper(I) iodide, **12,** 475
Trimethylaluminum, **8,** 506
CROSS-COUPLING, ARYL + VINYL GROUPS
Benzeneselenenyl halides, **5,** 518
1,8-Bis(dimethylamino)naphthalene, **6,** 50

[1,3-Bis(diphenylphosphine)propane]-(dichloro)nickel(II), **11,** 167
Chloranil, **10,** 85
Chlorobis(cyclopentadienyl)-hydridozirconium(IV), **8,** 84
Diacetatobis(triphenylphosphine)-palladium(II), **6,** 156
2,6-Di-*t*-butyl-4-methylpyridine, **8,** 145; **9,** 141
Dichlorobis(triphenylphosphine)-nickel(II), **9,** 147
1,2-Dichloroethylene, **12,** 175
Dihalobis(triphenylphosphine)-palladium(II), **6,** 60
Dilithium tetrachloropalladate(II), **5,** 413
Palladium(II) acetate, **5,** 496, 497; **8,** 378; **9,** 350; **11,** 389
Palladium catalysts, **10,** 297
Tetrakis(triphenylphosphine)nickel(0), **7,** 357
Tetrakis(triphenylphosphine)-palladium(0), **9,** 451
Trimethylaluminum, **8,** 506
Vinyllithium, **2,** 456

CRIEGEE REACTION
Ozone, **12,** 365

CURTIUS REARRANGEMENT
Azidotrimethylsilane, **7,** 394; **10,** 14
2-Chloroacrylyl chloride, **4,** 77
Diethyl oxomalonate, **8,** 170
Diphenyl phosphoroazidate, **4,** 210; **5,** 280; **11,** 222
Hydrazoic acid, **1,** 446; **5,** 329
Hypochlorous acid, **1,** 487
Sodium azide, **1,** 901, 1041, 1225; **2,** 376; **4,** 440
Sodium nitrite, **1,** 1097
Tetrabutylammonium azide, **6,** 563
Tributyltin azide, **7,** 377
Trifluoroacetic acid, **12,** 529

CYANATION
Arylthallium bis(trifluoroacetates), **12,** 483
Chlorosulfonyl isocyanate, **2,** 70; **3,** 51; **5,** 132
Copper(I) cyanide, **5,** 166
Cyanogen chloride, **1,** 176
Cyanotrimethylsilane, **12,** 148
Diethyl phosphorocyanidate, **10,** 145
Hydroxylamine-O-sulfonic acid, **9,** 245
Palladium(II) acetate, **5,** 496
Phenyl cyanate, **10,** 307
Potassium cyanide, **7,** 299
Thallium(III) trifluoroacetate, **3,** 286
p-Toluenesulfonyl cyanide, **11,** 536
Trichloroacetyl isocyanate, **7,** 382
Triphenylphosphine–Thiocyanogen, **8,** 518

CYANOSELENYLATION, CYANOSULFENYLATION (*see* ADDITION REACTIONS)

CYCLOADDITION REACTIONS (*see also* ASYMMETRIC REACTIONS)
[1 + 2]
N,N-Diethylaminopropyne, **2,** 133
[1 + 3]
N,N-Diethylaminopropyne, **2,** 133
[1 + 4]
Alkylaluminum halides, **11,** 7
t-Butyl isocyanide, **11,** 99
Chlorosulfonyl isocyanate, **4,** 90
Hydroxylamine-O-sulfonic acid, **2,** 217
Sulfur dioxide, **5,** 633; **6,** 558; **9,** 440
[2 + 2] (*see* SYNTHESIS INDEX—FOUR-MEMBERED RINGS, β-LACTAMS, ETC.)
[2 + 2 + 2]
Aluminum chloride, **2,** 21
Bis(benzonitrile)dichloropalladium(II), **1,** 56; **7,** 95
o-Bromoiodobenzene, **1,** 77
Chlorotris(triphenylphosphine)-rhodium(I), **5,** 736; **11,** 130
π-Cyclopentadienyl(triphenyl-phosphine)cobaltatetraphenyl-cyclopentadiene, **5,** 173
Dicarbonylbis(triphenylphosphine)-nickel, **1,** 61
Dicarbonylcyclopentadienylcobalt, **5,** 172; **6,** 153; **7,** 84, 94; **8,** 146; **9,** 142; **10,** 126; **11,** 161; **12,** 160
Di-μ-carbonyldecacarbonyltri-*triangulo*-iron, **7,** 144
Di-μ-carbonylhexacarbonyldicobalt, **1,** 224
Dicyclopentadienylcobalt, **6,** 128; **12,** 180
Iron carbonyl, **1,** 519
Nickel cyanide, **4,** 356
Tetrakis(trichlorophosphine)nickel(0), **1,** 1142
Tetrakis(triphenylphosphine)-palladium(0), **5,** 649

CYCLOADDITION REACTIONS
(*Continued*)
N,N'-Thiocarbonyldiimidazole, **6,** 583
Titanium(IV) chloride, **1,** 1169
Tri-μ-carbonylhexacarbonyldiiron, **12,** 525
[3 + 2]
N-Alkylhydroxylamines, **7,** 241; **12,** 13, 322
Aluminum azide, **5,** 9
Arenesulfonyl azides, **3,** 17
Azidotrimethylsilane, **1,** 1236
Benzenesulfonylnitrile oxide, **9,** 39; **11,** 40
Bis(acrylonitrile)nickel(0), **3,** 20
3-Bromo-3-methyl-2-trimethylsilyloxy-1-butene, **9,** 68
2-Bromo-3-trimethylsilyl-1-propene, **11,** 80
t-Butyl azidoformate, **6,** 77
t-Butyldimethylsilyl ethylnitronate, **12,** 84
Carbon disulfide, **6,** 95
Copper(II) acetylacetonate, **2,** 81
Cyanogen chloride N-oxide, **11,** 146
Cyclopropenone 1,3-propanediyl ketal, **12,** 152
Diazoacetaldehyde, **4,** 120
2-Diazo-1,1-dimethoxyethane, **10,** 120
Diazomethane, **4,** 120; **5,** 179; **9,** 133
2,4-Dibromo-3-pentanone, **4,** 158
gem-Dichloroallyllithium, **8,** 150
Dimethyl acetylenedicarboxylate, **5,** 227
Diphenyldiazomethane, **1,** 338; **4,** 204
Diphenylketene, **4,** 210
Diphenyl phosphoroazidate, **7,** 138
Ethoxycarbonylformonitrile oxide, **7,** 145
Ethyl azidoformate, **1,** 363; **4,** 225
Ethyl diazoacetate, **1,** 367
Ethyl β-(1-pyrrolidinyl)acrylate, **8,** 226
Hydrogen peroxide–Sodium tungstate, **12,** 246
Iron carbonyl, **8,** 533
Isocyanomethyllithium, **10,** 231; **11,** 285
Lithium aluminum hydride, **11,** 289
Lithium diisopropylamide, **10,** 241
Lithium divinylcuprate, **12,** 345
Lithium 2,2,6,6-tetramethylpiperidide, **9,** 285
Nickel catalysts, **11,** 457
Nitrobenzene, **7,** 251
Nitrogen dixide–Iodine, **8,** 205
Organocopper reagents, **12,** 345
Phenyl azide, **1,** 829; **4,** 375; **5,** 513
Phenyldiazomethane, **1,** 834; **5,** 515
Phenyl isocyanate, **10,** 309; **12,** 386
Potassium diazomethanedisulfonate, **1,** 928
1-Pyrroline-1-oxide, **9,** 401
Sodium azide, **5,** 593
Tetrakis(triphenylphosphine)-palladium(0), **9,** 451
p-Toluenesulfonic acid, **12,** 507
Tri-μ-carbonylhexacarbonyldiiron, **4,** 157; **5,** 221; **8,** 498; **9,** 477
Triethyl phosphonoacetate, **6,** 612
Trimethylamine N-oxide, **12,** 533
Trimethylsilylallene, **10,** 428
Trimethylsilyldiazomethane, **4,** 543
Trimethylsilylmethyl azide, **12,** 538
Trimethylsilylmethyl trifluoromethanesulfonate, **10,** 434; **12,** 542
Vinyldiazomethane, **6,** 664
Vinyltrimethylsilane, **12,** 566
Zinc chloride, **9,** 522; **10,** 461; **11,** 602
[3 + 4]
3-Bromo-3-methyl-2-trimethylsilyloxy-1-butene, **9,** 68
Dimethyl(methylthio)sulfonium tetrafluoroborate, **11,** 204
2-Methoxyallyl bromide, **4,** 327; **5,** 437; **6,** 364
Potassium fluoride, **6,** 481
Silver perchlorate, **11,** 469
Sodium iodide–Copper, **6,** 544
Tri-μ-carbonylhexacarbonyldiiron, **4,** 157; **5,** 221; **6,** 195; **8,** 498; **9,** 477
Zinc-copper couple, **8,** 533
Zinc–Silver couple, **9,** 519
[4 + 2] (*see* TYPE OF COMPOUND INDEX—DIELS–ALDER DIENES, DIENOPHILES)
Dicyanoacetylene, **1,** 230
[6 + 2]
Dimethyl acetylenedicarboxylate, **7,** 117
Titanium(IV) chloride–Diethylaluminum chloride, **12,** 502
CYCLODEHYDRATION (*see also* BISCHLER–NAPIERALSKI REACTION)
Aluminum chloride, **1,** 24
Benzeneselenenyl halides, **10,** 16

Benzenesulfonyl chloride, **1,** 46
Boron trifluoride etherate, **7,** 31
3-Bromopropyl 1-ethoxyethyl ether, **4,** 226
2,3-Dichloro-5,6-dicyano-1,4-benzoquinone, **5,** 193
Dicyclohexylcarbodiimide, **3,** 91
Diisopropylethylamine, **4,** 230
Dimethylformamide dimethyl acetal, **3,** 115
Dimethyl sulfoxide, **1,** 296
Diphenyldi(1,1,1,3,3,3-hexafluoro-2-phenyl-2-propoxy)sulfurane, **5,** 270
Ethyl formate, **1,** 380
Fluorosulfuric acid, **1,** 396
Formic acid, **4,** 239
Hydrobromic acid, **2,** 214
Hydrogen fluoride, **1,** 455; **4,** 252
Hydroxylamine sulfate, **1,** 481
Ion-exchange resins, **4,** 266; **5,** 355
Magnesium methyl carbonate, **6,** 354
Magnesium sulfate, **5,** 421
N-Methyl-N,N′-di-*t*-butylcarbodiimium tetrafluoroborate, **6,** 378
Nafion-H, **11,** 354
Oxalic acid, **5,** 481
Palladium(II) acetate, **12,** 367
Perchloric acid, **1,** 796
Phosphorus(V) oxide, **10,** 319
Phosphorus(V) oxide–Phosphoric acid, **5,** 535
Phosphoryl chloride, **5,** 535
Polyphosphate ester, **1,** 892
Polyphosphoric acid, **1,** 894; **2,** 334; **5,** 540; **10,** 321
Potassium bisulfate, **1,** 909
Sulfur trioxide–Dimethylformamide, **1,** 1125
Tetramethylene sulfone, **2,** 402
Thionyl chloride, **4,** 503
p-Toluenesulfonic acid, **1,** 1172
Trifluoroacetic acid, **1,** 1219
Triphenylphosphine + co-reagent, **9,** 504; **12,** 551
Triphenyl phosphite ozonide, **8,** 519
Vilsmeier reagent, **1,** 286
CYCLODEHYDROGENATION
Ammonium persulfate, **6,** 20
Bis(benzonitrile)dichloropalladium(II), **4,** 129
Bromine–Silver(I) salts, **1,** 73; **5,** 60
t-Butyl hypoiodite, **1,** 94
Cerium(IV) ammonium nitrate, **3,** 44
Chloranil, **4,** 75
Chromyl chloride, **2,** 79
Cobalt(III) acetate, **6,** 127
Dichlorobis(triphenylphosphine)-palladium(II), **5,** 190
2,3-Dichloro-5,6-dicyano-1,4-benzoquinone, **2,** 112; **3,** 83; **9,** 148
Hydrobromic acid, **6,** 282
Iodine, **8,** 256; **9,** 248
Lead tetraacetate, **1,** 537; **2,** 234; **12,** 270
Mercury(II) acetate, **4,** 319
Mercury(II) oxide–Iodine, **1,** 658
Nickel peroxide, **1,** 731
Palladium(II) acetate, **9,** 344
Palladium catalysts, **1,** 778
Platinum catalysts, **8,** 405
Polyphosphoric acid, **2,** 334
Potassium dichromate, **9,** 383
Potassium–Graphite, **12,** 400
Pyridinium bromide perbromide, **6,** 499
Selenium(IV) oxide, **4,** 422; **5,** 575
Silver(I) oxide, **3,** 252
Sodium tetrachloroaluminate, **1,** 1027; **6,** 524; **8,** 446
Trifluoromethanesulfonyl chloride, **9,** 485
Triphenylcarbenium tetrafluoroborate, **12,** 550
CYCLOPROPANATION (*see* ASYMMETRIC REACTIONS, SYNTHESIS INDEX—THREE-MEMBERED RINGS)
CYCLOREVERSION REACTIONS (*see also* RETRO-DIELS–ALDER REACTION)
Bis(acrylonitrile)nickel(0), **6,** 45
1,2-Bis(trimethylsilyloxy)cyclobutene, **10,** 45
Copper(I) trifluoromethanesulfonate, **10,** 108
1-Cyclobutenylmethyllithium, **6,** 151
1,2-Dicyanocyclobutene, **6,** 172
α-Lithiomethylselenocyclobutane, **10,** 232
Methyl cyclobutenecarboxylate, **8,** 335
Trichloro(phenyl)tungsten–Aluminum chloride, **7,** 288
2-Trimethylsilylmethylenecyclobutane, **9,** 494

CYCLOREVERSION REACTIONS (*Continued*)
Vinyl acetate, **1,** 1271

DAKIN REACTION
Hydrogen peroxide, **1,** 466; **2,** 216; **5,** 337
DAKIN-WEST REACTION
4-Dimethylaminopyridine, **3,** 118
Hydrogen peroxide, **5,** 337
DARZENS REACTION
Benzyltriethylammonium chloride, **4,** 27; **6,** 41
1-Butadienyldimethylsulfonium tetrafluoroborate, **9,** 74
t-Butyl chloroacetate, **1,** 86; **3,** 36
Chloroacetonitrile, **4,** 30; **6,** 102, 159
1-Chloro-3-diazoacetone, **5,** 114
Dibromoacetonitrile, **5,** 186
Ethyl chloroacetate, **1,** 911, 1039, 1077
Ethyl fluoroacetate, **5,** 304
Hexamethylphosphoric triamide, **7,** 168
Lithium dicyclohexylamide, **5,** 397
Lithium hexamethyldisilazide, **4,** 296
Methyl bromoacetate, **5,** 398
N-Methylephedrinium chloride, **8,** 340
Phase-transfer catalysts, **9,** 356
Potassium *t*-butoxide, **5,** 544
Potassium hydroxide, **12,** 411
Sodium amide, **1,** 1034
Sodium hydride, **1,** 1075; **6,** 542; **12,** 447
DEACETALIZATION (*see* HYDROLYSIS)
DEACETOXYLATION (*see* REDUCTION REACTIONS)
DEACETYLATION
Barium hydroxide, **4,** 23
Boron trifluoride–Ethanethiol, **10,** 51
Dichlorotris(triphenylphosphine)-ruthenium(II), **5,** 740
Magnesium methoxide, **7,** 220
Potassium tetracarbonylhydridoferrate, **6,** 483; **8,** 419
Sodium methoxide, **5,** 617
Trimethylene bis(thiotosylate), **6,** 628
DEACYLATION (*see also* SYNTHESIS INDEX—CARBOXYLIC ACIDS)
Formaldehyde, **9,** 224
Meldrum's acid, **10,** 252
Sodium amide, **8,** 446
Triethyloxonium tetrafluoroborate, **1,** 1210
DEACYLOXYLATION (*see* REDUCTION REACTIONS)
DEALKOXYCARBONYLATION
Alumina, **7,** 5
Birch reduction, **4,** 31
Boric acid, **5,** 48; **9,** 59
t-Butyl acetoacetate, **1,** 83
Chlorosulfuric acid, **2,** 70
Chlorotrimethylsilane, **4,** 537
Crown ethers, **8,** 128
1,5-Diazabicyclo[4.3.0]nonene-5, **4,** 116
1,4-Diazabicyclo[2.2.2]octane, **5,** 176; **7,** 86
1,8-Diazabicyclo[5.4.0]undecene-7, **4,** 119
Dibenzyl malonate, **1,** 198
Di-*t*-butyl malonate, **1,** 210
Dimethyl sulfoxide, **5,** 263
Hexamethylphosphoric triamide, **5,** 323
Hydrochloric acid, **6,** 283
Iodotrimethylsilane, **8,** 261; **9,** 251
Ion-exchange resins, **1,** 511
Isobutene, **1,** 522
Lithium halides, **1,** 617; **7,** 200; **9,** 283; **10,** 240; **11,** 300; **12,** 282
Magnesium, **1,** 627
Magnesium chloride, **10,** 251
Palladium(II) acetate–1,2-Bis(diphenylphosphine)ethane, **11,** 391
Palladium(II) chloride, **6,** 447
Piperidine, **6,** 472
Propionic acid, **12,** 415
Quinoline–Acetic acid, **6,** 501
Sodium chloride, **4,** 445; **9,** 422
Sodium cyanide, **2,** 381; **4,** 447; **6,** 536
DEALKOXYLATION (*see also* REDUCTION REACTIONS—ACETALS, C=C–OR)
Allyllithium, **1,** 18
Benzyllithium, **1,** 52
Dichlorobis(cyclopentadienyl)titanium, **6,** 179
Iodotrimethylsilane, **9,** 251
Lithium–Alkylamines, **1,** 574
Lithium aluminum hydride, **5,** 382
Lithium aluminum hydride–Aluminum chloride, **4,** 293
Lithium triethylborohydride, **11,** 304
Sodium–Potassium alloy, **1,** 1102
Titanium(III) chloride, **10,** 400
Titanium(IV) chloride–Lithium aluminum hydride, **7,** 372

Zinc, **4,** 574
DEALKYLATION (*see also* DEMETHYLATION)
ArR, R-R′ → ArH, RH
Nickel–Alumina, **9,** 321
Trifluoroacetic acid, **5,** 695
ArOR → ArOH
Aluminum bromide, **1,** 22
Aluminum iodide, **12,** 30
Borane–Tetrahydrofuran, **1,** 199
Boron trihalides, **1,** 66; **4,** 42; **9,** 61
Boron trifluoride–Ethanethiol, **9,** 63
5-Chloro-1-phenyltetrazole, **7,** 64
(Diphenylphosphine)lithium, **1,** 345
Hexamethylphosphoric triamide, **7,** 168
Iodotrimethylsilane, **8,** 261
Lead tetraacetate, **9,** 265
Lithium–Ethylenediamine, **2,** 250
Lithium iodide, **4,** 304
Phase-transfer catalysts, **9,** 356
Phosphorus(V) chloride, **1,** 866
Selenium(IV) oxide, **3,** 245
Sodium bis(2-methoxyethoxy)aluminum hydride, **7,** 327
Sodium–Potassium alloy, **1,** 1102
Triethyloxonium tetrafluoroborate, **7,** 386
ROR′ → ROH
Acetic anhydride–Ferric chloride, **6,** 260
Borane–Tetrahydrofuran, **1,** 199
Boron tribromide, **1,** 66
Boron trifluoride–Ethanethiol, **9,** 63
sec-Butyllithium, **9,** 87
Diethyl azodicarboxylate, **5,** 212
Hydrogen bromide, **1,** 453
Iodotrimethylsilane, **8,** 261
Lithium bromide–Boron trifluoride etherate, **1,** 604
Phenylthiotrimethylsilane, **10,** 426
Selenium(IV) oxide, **3,** 245
Sodium borohydride, **2,** 377
Sodium thiophenoxide, **3,** 268
Trichloro(methyl)silane, **11,** 553
Trimethylammonium formate, **1,** 1231
Uranium(VI) fluoride, **7,** 417
N COMPOUNDS
Benzeneselenol, **9,** 35
t-Butyldimethylsilyl trifluoromethanesulfonate, **12,** 86
α-Chloroethyl chloroformate, **12,** 112
1,4-Diazabicyclo[2.2.2]octane, **4,** 119
Diethyl azodicarboxylate, **5,** 212
Manganese(III) acetate, **6,** 355
2-Nitroprop-2-yl hydroperoxide, **3,** 213
Phenyl chloroformate, **2,** 318
Pyridinium chloride, **1,** 964; **3,** 239
Triphenylphosphine, **6,** 643
Vinyl chloroformate, **8,** 530
DEAMINATION (*see also* DEDIAZOTIZATION, ELIMINATION REACTIONS, REDUCTION REACTIONS)
Amyl nitrite, **5,** 18
Arenediazonium salts, **1,** 42, 43
Difluoramine, **1,** 253
Hexamethylphosphoric triamide, **5,** 323
Hydroquinone, **5,** 341
Hydroxylamine-O-sulfonic acid, **1,** 481
Hypophosphorous acid, **1,** 489; **8,** 255
Nickel–Alumina, **9,** 321
Nitrogen dioxide, **8,** 361
Sodium borohydride, **6,** 530, 531
Sodium cyanoborohydride, **9,** 424
Sodium nitrite, **11,** 491
Tetramethylurea, **1,** 1146
2,4,6-Triphenylpyrylium perchlorate, **7,** 408
Triphenyltin hydride, **1,** 1250
DEBENZYLATION (*see also* REDUCTION REACTIONS—BENZYL ETHERS)
Aluminum chloride, **4,** 10
Lithium 1-propanethiolate, **6,** 496
Oxygen, **12,** 362
Phosphoric acid, **7,** 255
Sodium thiophenoxide, **3,** 268
Sulfuryl chloride, **1,** 1128
2,2,2-Trichloroethyl chloroformate, **5,** 686
DEBROMINATION (*see* ELIMINATION REACTIONS, REDUCTION REACTIONS)
DECARBOALKOXYLATION (*see* DEALKOXYCARBONYLATION)
DECARBONYLATION (*see also* ELIMINATION REACTIONS)
OF RCOCl → RCl
Carbonylchlorobis(triphenylphosphine)-rhodium(I), **2,** 67
Chlorotris(triphenylphosphine)-rhodium(I), **1,** 140; **2,** 448
Palladium catalysts, **11,** 392
3-Pyridinethiol-1-oxide, **12,** 417

DECARBONYLATION (*Continued*)
OF RCHO → RH
Bis[1,3-bis(diphenylphosphine)propane]-chlororhodium(I), **12,** 111
Chlorotris(methyldiphenylphosphine)-rhodium(I), **7,** 67
Chlorotris(triphenylphosphine)-rhodium(I), **1,** 140, 1252; **2,** 448; **3,** 325; **4,** 559; **8,** 109
Lithium dimethylcuprate, **6,** 209
Nickel–Alumina, **9,** 321
Palladium catalysts, **1,** 778
OF RCOOH → ROH
m-Chloroperbenzoic acid, **1,** 135
DECARBOXYLATION
Alumina, **7,** 5
Aniline, **6,** 21
Benzoic anhydride, **1,** 49
Bis[1,3-bis(diphenylphosphine)propane]-chlororhodium(I), **12,** 111
N-Bromosuccinimide, **1,** 78
t-Butyl hydroperoxide, **1,** 88; **2,** 49; **6,** 81
Copper, **1,** 157; **2,** 82; **3,** 63; **5,** 146
Copper carbonate, basic, **1,** 163
Copper chromite, **1,** 156
Copper salts, **1,** 158
Crown ethers, **8,** 128
Cryptates, **5,** 156
Dibenzo-18-crown-6, **6,** 159
Dicyclohexylcarbodiimide, **3,** 91
Dimethylacetamide, **1,** 270
N,N-Dimethylaniline, **1,** 274
2,4-Dimethylpyridine, **6,** 224
Dimethyl sulfoxide, **2,** 157
Ferrous perchlorate, **6,** 260
Iodosylbenzene, **10,** 213
Lithium nitride, **1,** 618
Lithium phenylethynolate, **6,** 343
Mercury(II) oxide, **7,** 224
Morpholine, **11,** 352
Nitric acid, **1,** 733; **4,** 356
Osmium tetroxide–*t*-Butyl hydroperoxide, **1,** 88
3,3,6,9,9-Pentamethyl-2,10-diazabicyclo-[4.4.0]decene, **9,** 354
Phosphoryl chloride, **5,** 535; **7,** 292; **9,** 374
Potassium fluoride, **1,** 933
Potassium permanganate, **1,** 942
Potassium persulfate, **3,** 238
2-Pyridinethiol-1-oxide, **12,** 417
Quinoline, **1,** 975
Sodium chloride, **6,** 534
Tributyltin hydride, **10,** 411
DECARBOXYLATIVE BROMINATION (*see* HUNSDIECKER REACTION)
DECARBOXYLATIVE DEHYDRATION (*see* ELIMINATION REACTIONS)
DECARBOXYLATIVE DIMERIZATION (*see* KOLBE REACTION)
DECHLOROCARBONYLATION
m-Chloroperbenzoic acid, **5,** 120
Zinc chloride, **3,** 338
DECONJUGATION (*see* ISOMERIZATION REACTIONS)
DECYANATION
α-Chloro-N-cyclohexylpropanal-donitrone, **4,** 80
Hexamethylphosphoric triamide, **4,** 244
Iron(III) acetylacetonate, **4,** 268
Nickel–Alumina, **9,** 321
Potassium, **4,** 245; **5,** 543; **10,** 322
Sodium–Ammonia, **7,** 324
Sodium hydroxide, **5,** 616
Sodium naphthalenide, **4,** 349
Zinc, **4,** 574
DEDIAZOTIZATION
Amyl nitrite, **5,** 18
Arenediazonium salts, **1,** 42, 43
Chlorotris(triphenylphosphine)-rhodium(I), **4,** 559
Hexamethylphosphoric triamide, **5,** 323
Hypophosphorous acid, **1,** 489; **8,** 255
Sodium borohydride, **1,** 1049
Tetramethylurea, **1,** 1146
Thiophenol, **9,** 465
Tributyltin hydride, **3,** 294
DEFLUORINATION (*see* ELIMINATION REACTIONS, REDUCTION REACTIONS)
DEFORMYLATION (*see* DECARBONYLATION)
DEHALOCARBONYLATION (*see* DECHLOROCARBONYLATION)
DEHALOGENATION (*see* ELIMINATION REACTIONS, REDUCTION REACTIONS)
DEHYDRATION, DEHYDRATIVE DECARBOXYLATION, DEHYDROAMINATION, DEHYDROCYANATION (*see* ELIMINATION REACTIONS)

DEHYDROGENATION (*see also* AROMATIZATION, CYCLODEHYDROGENATION)
OF ALKANES → ALKENES
Benzeneseleninic anhydride, **11,** 37
1,4-Benzoquinone, **1,** 49
Bis(benzonitrile)dichloropalladium(II), **11,** 48
t-Butyl hypochlorite, **6,** 82
Butyllithium–Tetramethylethylenediamine, **5,** 80; **8,** 67
Chloranil, **1,** 125; **2,** 66
Copper(II) acetate, **12,** 140
Diborane–Lithium borohydride, **2,** 108
2,3-Dichloro-5,6-dicyano-1,4-benzoquinone, **2,** 112; **3,** 83; **4,** 130; **6,** 168; **7,** 96; **8,** 153
N,N-Dichlorourethane, **8,** 161
Diphenylpicrylhydrazyl, **1,** 347
Mercury(II) acetate, **1,** 644
Phenyliodine(III) dichloride, **4,** 264; **5,** 352; **6,** 298; **10,** 215
Potassium superoxide, **11,** 442
Selenium(IV) oxide, **1,** 992
Sulfur, **6,** 556
Thionyl chloride, **1,** 1158
Triphenylcarbenium tetrafluoroborate, **2,** 454; **6,** 657
OF KETONES, ENONES, ETC. → ENONES, DIENONES, ETC.
Acetic anhydride, **5,** 3
Allyl chloroformate, **12,** 15
Benzeneselenenyl halides, **11,** 34
Benzeneseleninic anhydride, **8,** 29; **11,** 37
Bis(acetonitrile)dichloropalladium(II), **12,** 50
Chloranil, **1,** 125
2,3-Dichloro-5,6-dicyano-1,4-benzoquinone, **1,** 215; **2,** 112; **3,** 83; **4,** 130; **5,** 193; **8,** 153; **10,** 135
Dichloroketene, **12,** 176
Dimethyl sulfoxide–Iodine, **9,** 190
Iodine–Potassium *t*-butoxide, **5,** 349
Iodylbenzene, **11,** 275
Lithium tri-*sec*-butylborohydride, **10,** 248
Methyl vinyl ketone, **7,** 247
Palladium(II) acetate, **11,** 391; **12,** 367
Palladium(II) chloride, **5,** 500; **8,** 384
Potassium nitrosodisulfonate, **3,** 238
Pyridine, **2,** 349
Pyridine N-oxide, **9,** 396
Pyridinium bromide perbromide, **1,** 967
Selenium(IV) oxide, **1,** 992
Sulfur, **3,** 273; **5,** 632
Thallium(III) acetate, **7,** 360
p-Toluenesulfinyl chloride, **12,** 507
DEHYDROHALOGENATION, DEHYDROMETALLATION DEHYDROSULFONYLATION (*see* ELIMINATION REACTIONS)
DEHYDROXYLATION (*see* DEOXYGENATION)
DEIODINATION (*see* ELIMINATION REACTIONS, REDUCTION REACTIONS)
DEKETALIZATION (*see* HYDROLYSIS)
DELEPINE REACTION
Hexamethylenetetramine, **1,** 427
DEMETHYLATION (*see also* DEALKYLATION)
$ROCH_3 \rightarrow ROH$
Acetic anhydride–Boron trifluoride etherate, **1,** 72
Aluminum chloride–Sodium iodide, **12,** 29
Boron tribromide, **11,** 71
Boron trichloride, **1,** 67
Boron trifluoride–Ethanethiol, **7,** 33; **9,** 63
Boron trifluoride etherate, **11,** 72
Bromodimethylborane, **12,** 199
Chlorotrimethylsilane–Sodium iodide, **10,** 97
Chromium(VI) oxide, **2,** 72
Phenylthiotrimethylsilane, **10,** 426
Pyridinium chloride, **4,** 415
Quinuclidine, **4,** 417
2,4,4,6-Tetrabromo-2,5-cyclohexadienone, **11,** 498
Trichloro(methyl)silane, **11,** 553
$ArOCH_3 \rightarrow ArOH$
Aluminum bromide, **1,** 22
Aluminum chloride, **1,** 24; **3,** 7; **4,** 10
Boron tribromide, **1,** 66; **2,** 33; **3,** 30; **10,** 50
Boron trichloride, **2,** 34; **4,** 42; **12,** 65
Boron trifluoride etherate, **7,** 33; **12,** 66
Boron triiodide, **9,** 65
Bromodimethylborane, **12,** 199
Chlorotrimethylsilane–Acetic anhydride, **12,** 126

DEMETHYLATION (*Continued*)
Copper(II) chloride, **6,** 139
Diisobutylaluminum hydride, **6,** 198
(Diphenylphosphine)lithium, **1,** 345; **5,** 408; **8,** 302
Hydriodic acid, **1,** 449
Hydriodic acid–Red phosphorus, **1,** 864
Hydrobromic acid, **1,** 450; **4,** 249
Hydrochloric acid, **4,** 250
Iodotrimethylsilane, **9,** 251
Lithium iodide, **11,** 300
Lithium methanethiolate, **8,** 303
Lithium 2-methylpropane-2-thiolate, **7,** 210
Magnesium iodide etherate, **12,** 290
Methylmagnesium iodide, **1,** 689; **2,** 278; **3,** 204
Pyridinium chloride, **1,** 964; **2,** 352; **6,** 497
Sodium–Ammonia, **2,** 27
Sodium cyanide, **9,** 423
Sodium ethanethiolate, **3,** 115; **4,** 465
Sodium iodide, **1,** 1087; **2,** 384
Sodium methaneselenoate, **12,** 450
Sodium N-methylanilide, **10,** 367
Sodium nitrite, **7,** 169
Sodium 2-propanethiolate, **11,** 473
Sodium *p*-thiocresolate, **7,** 342
Thioanisole–Trifluoromethanesulfonic acid, **9,** 464
N COMPOUNDS
Alkyl chloroformates, **5,** 117
N,N-Dimethylformamide, **4,** 184; **7,** 124
Lithium alkanethiolates, **5,** 412, 415
Lithium tri-*sec*-butylborohydride, **10,** 248
Lithium triethylborohydride, **6,** 348
Phenylcarbonimidic dichloride, **6,** 458
Potassium ferricyanide, **1,** 929
Silver nitrite, **6,** 515
Sodium thiophenoxide, **1,** 1106
2,2,2-Trichloroethyl chloroformate, **5,** 686; **7,** 383
DENITRATION (*see also* REDUCTION REACTIONS)
1-Benzyl-1,4-dihydronicotinamide, **12,** 47
Lithium aluminum hydride, **12,** 272
Sodium borohydride, **5,** 597
Tributyltin hydride, **10,** 411; **11,** 545; **12,** 516
DENITROSATION
Lithium diisopropylamide, **7,** 204
Molybdenum carbonyl, **3,** 206
Raney nickel, **7,** 312
Tin(II) chloride, **1,** 1113
Titanium(IV) chloride–Sodium borohydride, **10,** 404
DEOXIMATION (*see* HYDROLYSIS)
DEOXYGENATION
OF ROH → RH
Bis(dimethylamino) phosphorochloridate, **4,** 480
Borane–Pyridine, **9,** 59
Chlorotrimethylsilane–Sodium iodide, **10,** 97; **11,** 127
Dicyclohexylcarbodiimide, **5,** 206
Diiododimethylsilane, **9,** 170
N,N-Dimethylsulfamoyl chloride, **9,** 187
N,N-Dimethylthiocarbamoyl chloride, **3,** 127
Formic acid, **1,** 404
Hydrazine, **1,** 434
Hydriodic acid–Red phosphorus, **1,** 865
Iodine, **1,** 495
Iodotrimethylsilane, **12,** 259
Iron carbonyl, **10,** 221
Lithium aluminum hydride, **4,** 473; **5,** 326
Lithium aluminum hydride–Aluminum chloride, **1,** 595
Lithium triethylborohydride, **9,** 297
Nickel catalysts, **1,** 718; **9,** 321; **11,** 355
Palladium catalysts, **1,** 778; **2,** 303
Phenyl chlorothionocarbonate, **10,** 306
Potassium, **9,** 377; **10,** 322
Sodium borohydride, **12,** 441
Sodium borohydride + co-reagent, **8,** 451; **11,** 479
Sodium–Hexamethylphosphoric triamide, **9,** 416
Sulfur trioxide–Pyridine, **3,** 275
Titanium(IV) chloride–Lithium aluminum hydride, **7,** 372
Tributyltin hydride, **10,** 411; **11,** 545; **12,** 516
Triethylsilane–Boron trifluoride, **7,** 387; **8,** 501
Trifluoroacetic acid–Alkylsilanes, **6,** 616
Zinc, **4,** 574
Zinc amalgam, **8,** 534
OF C=O → CH_2 (*see also* DESULFURIZATION, DESELENYLATION)
Clemmensen reduction

Zinc, **2,** 459; **3,** 334; **5,** 753
Zinc amalgam, **1,** 1287; **2,** 462
Wolff–Kishner reduction
Dimethyl sulfoxide, **1,** 296
Hydrazine, **1,** 434; **2,** 211; **4,** 248; **5,** 327; **9,** 236
Lithium–Ammonia, **7,** 195
via Arenesulfonylhydrazones
Bis(benzoyloxy)borane, **10,** 32
Bis(triphenylphosphine)copper(I) borohydride, **10,** 47
Sodium borohydride, **6,** 530
Sodium cyanoborohydride, **6,** 537; **7,** 334
p-Toluenesulfonylhydrazide, **1,** 1051; **2,** 377; **4,** 443; **5,** 607; **6,** 98
other Routes
Aluminum isopropoxide, **1,** 35
Bis(dimethylamino) phosphorochloridate, **4,** 480
Borane–Tetrahydrofuran, **2,** 106
Chloromethyl methyl ether, **4,** 83
2,6-Di-*t*-butyl-4-methylpyridine, **11,** 160
Di-μ-carbonylhexacarbonyldicobalt, **1,** 224
Dicyclopentadienyltitanium, **5,** 672
Diphenylsilane, **1,** 348
Hydriodic acid–Red phosphorus, **7,** 171
Lithium aluminum hydride + co-reagent, **1,** 595; **2,** 243; **3,** 176; **12,** 275
Lithium–Ammonia, **4,** 288
Nickel on alumina, **11,** 355
Palladium catalysts, **1,** 778; **2,** 303; **7,** 275
Phosphorus(V) sulfide, **7,** 290
Platinum catalysts, **12,** 338
Polymethylhydrosiloxane, **4,** 393
Potassium *t*-butoxide, **1,** 911
Pyrene, **4,** 414
Raney nickel, **10,** 339
1:1 Raney nickel alloy, **1,** 718
Sodium bis(2-methoxyethoxy)-aluminum hydride, **3,** 260; **4,** 441
Sodium borohydride, **3,** 262
Sodium borohydride + co-reagent, **9,** 421; **11,** 479
Tin(II) chloride, **1,** 1113
Triethylsilane–Boron trifluoride, **8,** 501
Trifluoroacetic acid–Alkylsilanes, **5,** 695; **8,** 502
Zinc, **1,** 1276
OF C=O → ALKENES
Chlorotrimethylsilane–Zinc, **5,** 714; **6,** 628
Diborane, **6,** 161
Sodium borohydride–Palladium(II) chloride, **11,** 479
OF DIOLS (AND RELATED COMPOUNDS) → C=C, C≡C
Diols
Bis(1,5-cyclooctadiene)nickel(0), **5,** 34
Butyllithium, **2,** 51
Chlorotrimethylsilane–Sodium iodide, **11,** 127
Dichloromethylenedimethyl-ammonium chloride, **5,** 195
Diethylaluminum amides, **6,** 181
Dimethylformamide dimethyl acetal, **4,** 184; **8,** 191
1,3-Dimethyl-2-phenyl-1,3,2-diazaphospholidine, **11,** 208
N,N-Dimethylphosphoramidic dichloride, **6,** 215
Diphosphorus tetraiodide, **1,** 349
Dipotassium hexachloro-tungstate(IV), **4,** 407
Formic acid, **1,** 404
Iron carbonyl, **4,** 268
Methyl iodide, **6,** 384
N,N′-Thiocarbonyldiimidazole, **5,** 661
Titanium(0), **7,** 368; **9,** 466; **11,** 526
Titanium(III) chloride–Lithium aluminum hydride, **5,** 391
Tributyltin hydride, **8,** 497
Triethyl orthoformate, **7,** 385; **11,** 555
Trimethyl phosphite, **1,** 1233; **2,** 439; **3,** 315; **4,** 541
Triphenylphosphine + co-reagent, **9,** 507; **10,** 450
Disulfonates
Methanesulfonyl chloride, **1,** 662
Sodium iodide, **1,** 1087; **4,** 457
Sodium naphthalenide, **4,** 349; **8,** 464
Tributyltinlithium, **12,** 523
OF EPOXIDES → C=C
Stereospecific methods
Chlorotrimethylsilane–Sodium iodide, **11,** 127
Dimethylphenylsilyllithium, **7,** 133

DEOXYGENATION (*Continued*)
(Diphenylphosphine)lithium, **4,** 303; **5,** 408
Lithium, **10,** 234
Lithium diethylamide, **5,** 398
3-Methyl-2-selenoxo-1,3-benzothiazole, **7,** 245
Phosphorus(III) iodide, **10,** 318
Phosphoryl chloride, **1,** 876
Potassium selenocyanate, **6,** 487
Sodium dicarbonyl(cyclopentadienyl)-ferrate, **5,** 610; **6,** 538
Sodium iodide, **1,** 1116
Tributylphosphine, **1,** 1191
Trifluoroacetic anhydride–Sodium iodide, **8,** 504
Trimethylsilylpotassium, **7,** 402; **8,** 513
Triphenylphosphine selenide, **4,** 556
Triphenylphosphonium iodide–Triphenylphosphine diiodide, **10,** 450
Tris(phenylseleno)borane, **10,** 454
Tungsten(VI) chloride, **4,** 569; **6,** 658
Zinc, **1,** 1276
Zinc-copper couple, **5,** 758
Nonstereospecific methods
Chromium(II)–Amine complexes, **3,** 57
Chromium(II) chloride, **1,** 149
Dimethyl diazomalonate, **12,** 203
Diphosphorus tetraiodide, **10,** 174
Ferric chloride–Butyllithium, **6,** 260
Iron carbonyl, **8,** 265
3-Methyl-2-selenoxo-1,3-benzothiazole, **10,** 270
3-Methyl-2-thioxobenzothiazole, **7,** 245
Methyltriphenoxyphosphonium iodide, **8,** 354
Potassium iodide–Zinc–Phosphorus(V) oxide, **10,** 329
Samarium(II) iodide, **10,** 344
Sodium O,O-diethyl phosphorotelluroate, **8,** 455; **10,** 362
Titanium(III) chloride–Lithium aluminum hydride, **6,** 588
Triethyl phosphite, **1,** 1212
Triphenylphosphine, **1,** 1238
Triphenylphosphine–Iodine, **12,** 554
OF EPOXY C=O's
Di-μ-carbonylhexacarbonyldicobalt, **5,** 204
Dimethyl diazomalonate, **12,** 203
Diphosphorus tetraiodide, **10,** 174
Iron carbonyl, **8,** 265
Lithium dimethylcuprate, **5,** 234
Triphenylphosphine, **1,** 1238
Zinc, **1,** 1276
OF α-HYDROXY C=O's → C=O's
Iodine, **1,** 495
Iodotrimethylsilane, **9,** 251; **12,** 259
Phosphorus, red, **1,** 861; **6,** 470
Sodium iodide, **1,** 1087
Thionyl chloride, **1,** 1158
Tin, **1,** 1168
Zinc, **1,** 1276
OF N, P COMPOUNDS (*see* REDUCTION REACTIONS)
OF ArOH → ArH
5-Chloro-1-phenyltetrazole, **2,** 319; **4,** 377; **8,** 102; **10,** 93
Cyanogen bromide, **5,** 169
Cyanuric chloride, **6,** 149
Dicyclohexylcarbodiimide, **5,** 206
Diethyl phosphite, **1,** 251; **2,** 132
Diethyl phosphorochloridate, **1,** 248
2,4-Dinitrofluorobenzene, **1,** 321
Hydriodic acid, **9,** 238
Nonafluorobutanesulfonyl fluoride, **12,** 338
Phenyl isocyanate, **4,** 378
Raney nickel, **6,** 502
Sodium–Ammonia, **1,** 1041
Titanium(0), **9,** 466
OF S COMPOUNDS (*see* REDUCTION REACTIONS)
OF UNSATURATED C=O's → C=C
Catecholborane, **6,** 98; **8,** 79; **10,** 79
Diethyl phosphorochloridate, **3,** 98
Hydrazine, **1,** 434
Lithium aluminum hydride–Aluminum chloride, **1,** 595; **5,** 389
Sodium cyanoborohydride, **6,** 537; **7,** 334
Zinc, **1,** 1276; **7,** 426
DEPROTECTION (*see* HYDROLYSIS, PROTECTION)
DESELENYLATION (*see also* REDUCTION REACTIONS)
OF RSeR → RH
Lithium–Alkylamines, **7,** 194
Lithium triethylborohydride, **11,** 304
Nickel catalysts, **12,** 335
Triphenyltin hydride, **8,** 521; **10,** 451

Tris(phenylseleno)borane, **9,** 511
OF $R_2C(SeR)_2 \rightarrow R_2CH_2$
Lithium–Ethylamine, **7,** 194
Nickel catalysts, **12,** 335
Triphenyltin hydride, **8,** 521; **10,** 451
Tris(phenylseleno)borane, **9,** 511
DESILYLATION (*see also* PROTECTION, REDUCTION REACTIONS)
Boron trifluoride etherate, **12,** 66
Butadiynyl(trimethyl)silane, **5,** 68
Cesium fluoride, **9,** 100
Chlorotrimethylsilane, **5,** 709
Dicarbonylcyclopentadienylcobalt, **6,** 153
Hydrogen chloride, **9,** 239
Hydrogen fluoride, **9,** 240
Methoxy(trimethylsilyl)methyllithium, **9,** 284
Methyllithium–Lithium bromide, **9,** 311
Potassium *t*-butoxide, **11,** 432
Potassium fluoride, **10,** 325
Sodium methoxide, **5,** 617
Tetraalkylammonium fluoride, **7,** 356; **10,** 378
Tributyltin fluoride, **12,** 515
1-Trimethylsilyl-1,3-butadiene, **7,** 395
3-Trimethylsilyl-3-buten-2-one, **5,** 461
β-Trimethylsilylethylidenetriphenylphosphorane, **9,** 492
2-Trimethylsilylmethyl-1,3-butadiene, **10,** 432
Trimethylsilylmethyllithium, **9,** 495
Trimethylsilylmethylmagnesium chloride, **5,** 724
Trimethylsilylmethyl trifluoromethanesulfonate, **10,** 434
DESULFONYLATION
Aluminum amalgam, **5,** 9
Chlorotris(triphenylphosphine)-rhodium(I), **4,** 559
Diethyl phosphite, **2,** 132
Dimethyl sulfone, **2,** 157
Hexamethylphosphoric triamide, **6,** 273
Hydrogen bromide–Acetic acid, **7,** 171
Lithium dimethylcuprate, **4,** 177
Lithium–Ethylamine, **5,** 377; **11,** 287
Lithium triethylborohydride, **11,** 304
Nickel catalysts, **1,** 723
Organolithium reagents, **10,** 3
Potassium *t*-butoxide, **1,** 911
Potassium–Crown ether, **11,** 431
Potassium–Graphite, **10,** 326
Sodium amalgam, **1,** 1030; **7,** 326; **11,** 473
Sodium benzylate, **1,** 1045
Sodium bis(2-methoxyethoxy)aluminum hydride, **4,** 441
Sodium–Ethanol, **11,** 472
Sodium naphthalenide, **1,** 711
Sulfuric acid, **5,** 633
Tetrakis(triphenylphosphine)-palladium(0), **11,** 503
Tributyltin hydride, **9,** 476
Zinc, **2,** 459
DESULFURIZATION (*see also* ELIMINATION REACTIONS, REDUCTION REACTIONS)
OF CYCLIC SULFIDES $\rightarrow$ C=C
Lithium–Ethylamine; Raney nickel, **5,** 381
Raney nickel, **5,** 263
OF $R_2C(SR)_2 \rightarrow R_2CH_2$
Copper(II) chloride–Zinc chloride–Lithium aluminum hydride, **5,** 161
1,2-Ethanedithiol, **1,** 356
Hydrazine, **1,** 434
Lithium–Ethylamine, **1,** 574
Nickel catalysts, **1,** 720, 723; **5,** 471
Titanium(IV) chloride–Lithium aluminum hydride, **8,** 486
Tributyltin hydride, **10,** 411
OF RSOR $\rightarrow$ RH
Nickel catalysts, **1,** 723
Organolithium reagents, **10,** 3
OF $R_2C{=}S \rightarrow R_2CH_2$
Iron carbonyl, **8,** 265
Potassium tetracarbonylhydridoferrate, **6,** 483; **8,** 266
OF RSH, RSR, RSSR $\rightarrow$ RH
Aluminum amalgam, **1,** 20
Bis(3-dimethylaminopropyl)phenylphosphine, **5,** 36
Chlorotrimethylsilane–Zinc, **8,** 532
Lithium–Alkylamines, **1,** 574
Lithium aluminum hydride, **5,** 382
Lithium naphthalenide, **8,** 305
Lithium triethylborohydride, **11,** 304
Molybdenum carbonyl, **10,** 273
Nickel catalysts, **1,** 723; **2,** 289
Potassium thiocyanate, **1,** 954
Sodium triethylborohydride–Ferrous chloride, **10,** 369

DESULFURIZATION (*Continued*)
Triethyl phosphite, **1,** 1212; **6,** 612
DETHIOACETALIZATION (*see* HYDROLYSIS)
DIALKYLATION (*see* ALKYLATION REACTIONS)
DIAMINATION (*see* ADDITION REACTIONS)
DIAZO TRANSFER REACTIONS
Arenesulfonyl azides, **2,** 62; **10,** 422
Benzyltriethylammonium chloride, **7,** 18
Dimethyl diazomalonate, **5,** 244
Sodium azide, **1,** 1041
p-Toluenesulfonyl azide, **1,** 1178; **2,** 415; **3,** 291; **4,** 510; **5,** 675; **6,** 597; **7,** 19; **11,** 535
Tricaprylylmethylammonium chloride, **5,** 460
DIAZONIATION
Dinitrogen trioxide–Boron trifluoride, **1,** 329
DIAZONIUM COUPLING (*see* AZO COUPLING)
DIAZOTIZATION
p-Aminoacetophenone, **4,** 18
Amyl nitrite, **5,** 18
Arenediazonium hexafluorophosphates, **1,** 42
Butyl nitrite, **1,** 97
t-Butyl nitrite, **1,** 97
Copper(I) chloride, **1,** 166
Hypophosphorous acid, **1,** 489
Isoamyl nitrite, **1,** 520; **5,** 358
Nitrosylsulfuric acid, **1,** 755
Sodium nitrite, **1,** 1097
Tetrafluoroboric acid, **1,** 394
Tetrahydrofuran, **1,** 1140
Trichloroacetic acid, **1,** 1194
Urea, **1,** 1262
DIECKMANN REACTION
Dichlorobis(trifluoromethanesulfonato)-titanium(IV), **12,** 173
Dimethyl sulfoxide, **2,** 157
Ethyl acetate, **1,** 1075
Hydrogen fluoride, **6,** 285
Magnesium methoxide, **3,** 189
Organocopper reagents, **12,** 345
Potassium, **12,** 400
Potassium *t*-butoxide, **1,** 911; **2,** 336
Potassium ethoxide, **1,** 928
Potassium hydride, **6,** 482
Sodium hydride, **1,** 1075; **9,** 427
Sodium methylsulfinylmethylide, **1,** 310
Triphenylmethylpotassium, **9,** 502
DIELS–ALDER REACTION, INTRAMOLECULAR (*see also* TYPE OF COMPOUND INDEX—DIENES, DIENOPHILES; SYNTHESIS INDEX—ARENES, SIX-MEMBERED RINGS)
Alkylaluminum halides, **10,** 177; **12,** 5
Aluminum chloride, **11,** 25
Benzyl *trans*-1,3-butadiene 1-carbamate, **11,** 43
O-*t*-Butyldimethylsilylhydroxylamine, **12,** 85
Cesium carbonate, **11,** 114
Cesium fluoride, **11,** 115; **12,** 108
Chlorotrimethylsilane, **11,** 125
Cyclodextrins, **11,** 150
Dichlorobis(cyclopentadienyl)titanium, **12,** 168
1,3-Dihydrobenzo[c]thiophene 2,2-dioxide, **9,** 168
Dimethyl 1,2,4,5-tetrazine-3,6-dicarboxylate, **12,** 214
Florisil, **9,** 223
Methyl 4-diethoxyphosphinylcrotonate, **8,** 336
Molybdenum carbonyl, **12,** 330
2-Oxazolidones, chiral, **12,** 359
Potassium hydride, **11,** 435
Tetraethylammonium periodate, **9,** 448
3-Triethylsilyloxypentadienyllithium, **11,** 556
Trifluoromethanesulfonic acid, **12,** 531
1-Trimethylsilyl(pentadienyl)lithium, **10,** 436
DIENONE–PHENOL REARRANGEMENT
Cyanotrimethylsilane, **5,** 720
Trifluoroacetic acid, **5,** 695; **11,** 557
Trifluoroacetic anhydride, **1,** 1221
DIHYDROXYLATION (*see* ADDITION REACTIONS)
DIMERIZATION (*see* COUPLING REACTIONS)
DISSOLVING-METAL REDUCTION (*see* BIRCH REDUCTION, REDUCTION OF SPECIFIC FUNCTIONAL GROUPS)
DOEBNER–MILLER REACTION
Dimethyl 2-oxoglutaconate, **11,** 206

DUFF REACTION
Hexamethylenetetramine, **1,** 427; **4,** 243

ELBS OXIDATION
Potassium persulfate, **1,** 952; **8,** 417
ELECTROCYCLIC REARRANGEMENTS
Chlorotrimethylsilane, **3,** 310
Dimethyl acetylenedicarboxylate, **4,** 168
Silver tetrafluoroborate, **3,** 250
ELIMINATION REACTIONS TO FORM C=C, C≡C (*see also* AROMATIZATION, DEHYDROGENATION, DEOXYGENATION, EXTRUSION REACTIONS)
BAMFORD–STEVENS REACTION
Sodium hydroxide, **8,** 460
p-Toluenesulfonylhydrazide, **1,** 1185; **2,** 417
2,4,6-Triisopropylbenzenesulfonylhydrazide, **11,** 563
CHUGAEV AND RELATED REACTIONS
N,N-Dimethylthiocarbamoyl chloride, **3,** 127
Sodium methylsulfinylmethylide, **3,** 123; **4,** 195
p-Tolylthionochlorocarbonate, **4,** 342; **5,** 457; **9,** 474
CLEAVAGE OF BISHYDRAZONES TO C≡C
Copper(I) chloride, **7,** 80
Copper(I) chloride–Oxygen, **5,** 165
Hydrazine, **1,** 434
Mercury(II) oxide, **1,** 655
Oxygen, **6,** 426
Silver(I) oxide, **3,** 252
Silver(II) oxide, **4,** 431
Silver(I) trifluoroacetate, **1,** 1018
CLEAVAGE OF ETHERS (–H, OR)
Alumina, **6,** 16
Iodotrimethylsilane, **11,** 271
Molecular sieves, **6,** 411
Palladium(II) acetate, **8,** 378; **9,** 349
Pentacarbonyl(trimethylsilyl)manganese, **11,** 401
Potassium *t*-butoxide, **5,** 544
Sodium amide, **1,** 1034; **2,** 373; **5,** 591
Sodium hydride–Dimethyl sulfoxide, **10,** 364
COPE ELIMINATION
Dimethyl(methylene)ammonium salts, **8,** 194
Dimethyl sulfoxide, **1,** 296
Hydrogen peroxide, **1,** 471
COREY–WINTER REACTION
Bis(1,5-cyclooctadiene)nickel(0), **5,** 34
Dichloromethylenedimethylammonium chloride, **5,** 195
1,3-Dimethyl-2-phenyl-1,3,2-diazaphospholidine, **11,** 208
Iron carbonyl, **4,** 268
Methyl iodide, **6,** 384
N,N′-Thiocarbonyldiimidazole, **5,** 661; **6,** 583
Trimethyl phosphite, **1,** 1233; **2,** 439; **3,** 315; **4,** 541; **7,** 393
DEAMINATION OF AZIRIDINES
Difluoramine, **1,** 253
3-Nitro-N-nitrosocarbazole, **1,** 741
Nitrosyl chloride, **1,** 748
DECARBONYLATION OF RCOCl (–H,COCl)
Chlorotris(triphenylphosphine)rhodium(I), **2,** 448
DECARBOXYLATIVE DEHYDRATION (–OH,COOH)
Benzenesulfonyl chloride, **5,** 22
Copper, **1,** 157
Dimethylformamide dialkyl acetals, **6,** 221, 222; **11,** 198; **12,** 204
Sodium methylsulfinylmethylide, **2,** 166
Tetrakis(triphenylphosphine)palladium(0), **10,** 384; **11,** 503
Triphenylphosphine–Diethyl azodicarboxylate, **8,** 517; **9,** 504; **12,** 552
DEHALOGENATION (–X,X)
Aluminum, **4,** 9
Benzyltrimethylammonium mesitoate, **1,** 54
1,8-Bis(dimethylamino)naphthalene, **6,** 50
Bis(trimethylsilyl)mercury, **9,** 55
Chromium(II)–Amine complexes, **3,** 57
Chromium(II) chloride, **1,** 149; **12,** 136
Chromium(II) sulfate, **1,** 150
(Diphenylphosphine)lithium, **6,** 340
Diphenyltelluride, **1,** 348
Disodium phenanthrene, **1,** 350; **2,** 182
Disodium tetracarbonylferrate, **6,** 550
Hexamethylphosphoric triamide, **6,** 273
Iron–Graphite, **11,** 276

ELIMINATION REACTIONS TO FORM C=C, C≡C (*Continued*)
Lithium amalgam, **3,** 177; **4,** 287
Magnesium–Iodine–Ether, **1,** 630
Potassium *t*-butoxide, **8,** 407
Potassium hydroxide, **1,** 938
Sodium + solvent, **5,** 589; **6,** 523; **7,** 325
Sodium dithionite, **7,** 336
Sodium hydrogen telluride, **8,** 459
Sodium iodide, **1,** 1087; **6,** 543; **7,** 338
Sodium methaneselenoate, **10,** 368
Sodium methylsulfinylmethylide, **1,** 310
Sodium naphthalenide, **2,** 289
Sodium phenanthrenide, **4,** 375
Sodium sulfide, **11,** 492; **12,** 453
Sodium thiosulfate, **4,** 466
Tetracarbonyl(cyclopentadienyl)-vanadium, **8,** 468
Thiourea, **1,** 1164
Titanium(III) chloride–Lithium aluminum hydride, **7,** 369
Trialkyl phosphites, **1,** 1212, 1229, 1233; **4,** 529
Tributyltin hydride, **5,** 685
Trichloroethylene, **10,** 414
Trimethyltinsodium, **9,** 500
Triphenylphosphine, **1,** 1238; **2,** 443
Vanadium(II) chloride tetrapyridine complex, **5,** 744
Zinc, **1,** 1276, 1286; **2,** 459; **3,** 334; **7,** 426; **11,** 598
Zinc-copper couple, **12,** 569
Zinc–Dimethylformamide, **3,** 338
Zinc–Silver couple, **5,** 760; **12,** 571
Zinc–Titanium(IV) chloride, **11,** 599

DEHYDRATION (–H,OH)
Alumina, **1,** 19, 958; **2,** 17
Bis(benzonitrile)dichloropalladium(II), **11,** 48
Boric acid, **1,** 63
Boron trifluoride etherate, **1,** 70
N-Bromoacetamide–Pyridine–Sulfur dioxide, **1,** 75
2-Chloro-3-ethylbenzoxazolium tetrafluoroborate, **9,** 105
Chloromethyl methyl ether, **1,** 132
Copper(II) sulfate, **2,** 89; **10,** 107
Dicyclohexylcarbodiimide, **2,** 126; **4,** 141; **11,** 173
Diethoxyaluminum chloride, **6,** 180
(Diethylamino)sulfur trifluoride, **8,** 166
Diketene, **1,** 264
Dimethylformamide dialkyl acetals, **7,** 125; **9,** 183
Dimethylformamide–Sodium acetate, **1,** 285
1,3-Dimethyl-2-imidazolidinone, **11,** 202
N,N-Dimethylphosphoramidic dichloride, **9,** 187
Dimethyl sulfoxide, **1,** 296; **4,** 192; **10,** 166
N,N-Dimethylthiocarbamoyl chloride, **3,** 127
Diphenyldi(1,1,1,3,3,3-hexafluoro-2-phenyl-2-propoxy)sulfurane, **4,** 205
N,N′-Disuccinimidyl carbonate, **11,** 226
Ethyl(carboxysulfamoyl)triethyl-ammonium hydroxide inner salt, **4,** 227; **5,** 442
Ferric chloride, **8,** 229
Florisil, **1,** 393
Formaldehyde, **5,** 312; **6,** 264
Girard reagent T, **1,** 410
Hexamethylphosphoric triamide, **4,** 244
Hydrobromic acid, **1,** 450
Iodine, **1,** 495; **2,** 220; **4,** 258
Lithium diisopropylamide, **4,** 298
Lithium piperidide, **6,** 344
Magnesium methyl carbonate, **5,** 420
Methanesulfonyl chloride–Sulfur dioxide, **1,** 664
Methyl chlorosulfinate, **1,** 675
Methyl N-sulfonylurethane triethylamine complex, **4,** 343
Methyltriphenoxyphosphonium iodide, **4,** 557; **6,** 649
Methyl vinyl ketone, **1,** 697
Naphthalene-β-sulfonic acid, **1,** 712
Oxalic acid, **1,** 764
Phenylene orthosulfite, **8,** 393
Phenyl isocyanate, **1,** 842
Phosphorus(V) oxide, **1,** 871; **7,** 291
Phosphoryl chloride, **1,** 876
Phthalic anhydride, **1,** 882
Potassium bisulfate, **1,** 909
Silver perchlorate, **9,** 413
Sulfuryl chloride–Pyridine, **5,** 642
N,N′-Sulfuryldiimidazole, **11,** 496
Tetrafluoropyrrolidine, **9,** 449
Thionyl chloride, **1,** 1158
Thorium oxide, **1,** 1167; **5,** 669
Titanium(IV) chloride, **11,** 529

p-Toluenesulfonic acid, **1,** 1172
p-Toluenesulfonyl chloride, **1,** 1179
p-Tolylthionochlorocarbonate, **5,** 457
Triphenylphosphine bis(trifluoromethanesulfonate), **6,** 648
Triphenylphosphine + co-reagent, **7,** 404; **9,** 504
Vilsmeier reagent, **7,** 292
DEHYDROAMINATION + RELATED REACTIONS [−H,NR_2(NX_2,ETC.)]
Copper(I) chloride, **6,** 145
N,N-Dimethylaniline, **1,** 274
N,N-Dimethylbenzeneselenenamide, **7,** 119
Dimethyl(methylene)ammonium iodide, **8,** 194
Formaldehyde, **5,** 312
Isoamyl nitrite, **5,** 358
Methyl vinyl ketone, **1,** 697
Oxalic acid, **1,** 764
Phenyllithium, **1,** 845
Potassium amide, **1,** 907
5,6,8,9-Tetrahydro-7-phenyldibenzo-[c,h]xanthylium tetrafluoroborate, **11,** 502
Trifluoromethanesulfonic anhydride, **5,** 702
DEHYDROCYANATION (−H,CN)
Ascarite, **4,** 21
Dimethyl sulfoxide, **1,** 296
Potassium *t*-butoxide, **9,** 380; **10,** 323
DEHYDROHALOGENATION (−H,X)
Alkylaluminum halides, **12,** 5
Alumina, **2,** 18; **8,** 9; **9,** 8
Aluminum chloride, **6,** 17
Benzyltrimethylammonium hydroxide, **5,** 29; **7,** 20
Bis(dimethylamino) phosphorochloridite, **6,** 575
t-Butylamine, **1,** 84
Calcium carbonate, **1,** 103; **5,** 89
m-Chloroperbenzoic acid, **8,** 97
Collidine, **1,** 155
Crown ethers, **5,** 152, 207; **6,** 133; **10,** 110
1,5-Diazabicyclo[4.3.0]nonene-5, **1,** 189; **2,** 98; **4,** 16; **5,** 176; **6,** 157
1,4-Diazabicyclo[2.2.2]octane, **2,** 99
1,8-Diazabicyclo[5.4.0]undecene-7, **2,** 101; **7,** 87
1,2-Dibromoethane, **4,** 126
1,3-Dichloro-2-butene, **2,** 111
Dichloro(diethoxyphosphinyl)-methyllithium, **6,** 188
Dichloromethyllithium, **9,** 154
Dicyclohexylethylamine, **1,** 370
Diethylamine, **4,** 148
Diisopropylethylamine, **1,** 371
Dimethylaminotrimethyltin, **2,** 147
N,N-Dimethylaniline, **1,** 274
Dimethyl bromomalonate, **12,** 200
N,N-Dimethylformamide, **1,** 278
N,N-Dimethylhydrazine, **8,** 192
1,3-Dimethyl-2-imidazolidinone, **11,** 202
Dimethyl sulfoxide, **1,** 296; **2,** 157; **4,** 192
2,4-Dinitrophenylhydrazine, **1,** 330
(Diphenylphosphine)lithium, **6,** 340
Ethyl carbazate, **1,** 360
Ethylene glycol, **1,** 375
Ethylene oxide, **5,** 297
Hexamethylphosphoric triamide, **2,** 208; **4,** 244; **8,** 240
Ion-exchange resins, **1,** 511
Lithium carbonate, **1,** 606; **3,** 183
Lithium carbonate–Lithium halide, **1,** 606; **2,** 245; **4,** 298; **5,** 395; **7,** 200
Lithium chloride, **1,** 609; **2,** 246; **12,** 277
Lithium dicyclohexylamide, **2,** 246
Magnesium oxide, **2,** 256
Methyllithium, **5,** 448
Pentaisopropylguanidine, **11,** 249
Phase-transfer catalysts, **9,** 356; **11,** 403
N-Phenylbenzamidine, **5,** 513
N-Phenylmorpholine, **1,** 846
Phosphoryl chloride, **2,** 330
Potassium amide, **1,** 907
Potassium *t*-butoxide, **1,** 911; **2,** 336; **3,** 233; **4,** 358, 401; **5,** 544; **6,** 477; **7,** 296, 298; **8,** 407; **9,** 126, 380
Potassium fluoride, **2,** 346; **5,** 153, 555; **12,** 406
Potassium hydroxide, **6,** 486
Potassium superoxide, **6,** 488; **11,** 442
Potassium triethylmethoxide, **2,** 349
Pyridine, **1,** 958; **11,** 448
Quinoline, **1,** 975; **9,** 403
Silica, **9,** 410
Silver fluoride, **2,** 364; **10,** 348
Silver(I) nitrate, **1,** 1008
Silver(I) oxide, **1,** 1011
Sodium acetate, **1,** 1024
Sodium amide, **1,** 1034; **2,** 373; **5,** 591; **6,** 525

ELIMINATION REACTIONS TO FORM C=C, C≡C (*Continued*)
Sodium amide–Sodium *t*-butoxide, **4,** 439; **5,** 593; **9,** 417
Sodium 2-butylcyclohexoxide, **1,** 1056
Sodium dicarbonyl(cyclopentadienyl)-ferrate, **9,** 426
Sodium hydride, **1,** 1075
Sodium iodide, **1,** 1087
Sodium isopropoxide, **2,** 385
Sodium methoxide, **1,** 1091; **6,** 545
Sodium methylsulfinylmethylide, **3,** 123
Tetrabutylammonium salts, **4,** 477; **7,** 354
Tetraethylammonium chloride, **1,** 1137
Tetramethylammonium dimethyl phosphate, **6,** 573
Tributylamine, **1,** 1189
Triethylamine, **1,** 1198; **2,** 427; **4,** 527
Trimethyl phosphite, **1,** 1233
Triphenylmethylpotassium, **12,** 550
DEHYDROMETALLATION (–H,M)
trans-1,2-Bis(tributylstannyl)ethylene, **5,** 43
Nickel(II) chloride–Ethylene, **10,** 276
Tributyltinlithium, **11,** 551
Triphenylmethylpotassium, **8,** 524
DEHYDROSULFONYLATION ($-H,OSO_2R$)
Alumina, **7,** 5; **8,** 9
Amidines, bicyclic, **4,** 16
Dimethyl sulfoxide, **1,** 296
Formaldehyde, **5,** 312; **6,** 264
Guanidines, **11,** 105, 249
o-Nitrobenzenesulfonyl chloride, **6,** 419
Potassium acetate, **6,** 475
Potassium *t*-butoxide, **1,** 911; **2,** 336; **3,** 233; **4,** 399; **5,** 544
Potassium iodide–Zinc-copper couple, **4,** 411
Potassium trifluoromethanesulfinate, **5,** 564
Sodium dicarbonyl(cyclopentadienyl)-ferrate, **9,** 426
N,N′-Sulfuryldiimidazole, **11,** 496
Tetrabutylammonium formate, **4,** 478
Trifluoromethanesulfonic anhydride, **5,** 702; **6,** 618
DEOXYGENATION OF DIOLS, EPOXIDES (*see* DEOXYGENATION)
DESULFURIZATION OF EPISULFIDES AND RELATED COMPOUNDS
Butyllithium, **4,** 60
Chromium(II)–Amine complexes, **3,** 57
Diazomethane, **2,** 102
4,4-Dimethyl-2-methylthio-2-oxazoline, **6,** 584
Diphosphorus tetraiodide, **12,** 218
Methyl iodide, **2,** 274
3-Methyl-2-selenoxo-1,3-benzothiazole, **7,** 245
Potassium thiocyanate, **1,** 954
Triethyl phosphite, **1,** 1212
Triphenylphosphine, **1,** 1238
ELIMINATION OF ESTERS AND RELATED COMPOUNDS [–H,OCOR(OCSR)]
1,8-Diazabicyclo[5.4.0]undecene-7, **9,** 132
Dibenzoyl peroxide, **4,** 122
Hexamethylphosphoric triamide, **7,** 168
Molybdenum carbonyl, **12,** 330
Palladium(II) acetate, **8,** 378; **9,** 349
Pyridine, **4,** 414
Sodium methylsulfinylmethylide, **4,** 195
Tetrakis(triphenylphosphine)-palladium(0), **10,** 384
Thiobenzoyl chloride, **6,** 582
Zinc oxide, **1,** 1294
ELIMINATION OF H, SR AND RELATED
Benzyl phenyl sulfoxide, **6,** 394
m-Chloroperbenzoic acid, **10,** 92
Chloro(phenylsulfinyl)methyllithium, **9,** 271
Copper(I) trifluoromethanesulfonate, **6,** 130; **8,** 125
Dialkoxycarbenium tetrafluoroborates, **4,** 114
Dimethyl disulfide, **5,** 246; **6,** 217
Dimethyl sulfoxide, **1,** 296
Diphenyl disulfide, **5,** 276; **6,** 235
Diphosphorus tetraiodide, **11,** 224
2,2′-Dipyridyl disulfide, **8,** 214
Ethyl 2-phenylsulfinylacetate, **6,** 256
Ethyl α-trifluoromethyl-sulfonyloxyacetate, **7,** 149
Iodomethyl methyl sulfide, **5,** 353
Lithium diisopropylamide, **9,** 280
Methyl phenylsulfinylacetate, **6,** 393
Phenylsulfinylacetone, **5,** 524

Potassium *t*-butoxide, **1,** 911; **8,** 407; **12,** 401
Potassium trifluoromethanesulfinate, **5,** 564
Sodium methylsulfinylmethylide, **7,** 338
Tetrakis(triphenylphosphine)-palladium(0), **9,** 451
Thiophenol, **10,** 399

ELIMINATION OF H, SeR AND RELATED
Benzeneselenenic acid, **8,** 24
Benzeneselenenyl halides, **5,** 518; **6,** 459; **7,** 286; **8,** 25; **10,** 16
Benzeneseleninic anhydride, **8,** 29
Benzeneselenol, **6,** 28
t-Butyl hydroperoxide, **8,** 64
4,4′-Dichlorodiphenyl diselenide, **6,** 421
Dimethyl disulfide, **5,** 246
Diphenyl diselenide, **5,** 272; **6,** 235; **7,** 136
Diphosphorus tetraiodide, **11,** 224
Lithium diisopropylamide, **6,** 334; **9,** 280
Methyl fluorosulfonate, **9,** 307
o-Nitrophenyl selenocyanate, **6,** 420
Ozone, **10,** 295
Potassium 3-aminopropylamide, **8,** 406
2-Pyridineselenenyl bromide, **11,** 455
Selenium, **11,** 465
Sodium benzeneselenoate, **8,** 447
Sodium pyridylselenate, **10,** 368

ELIMINATION OF H, TeR
Chloramine-T, **10,** 85

ELIMINATION OF OH(OR),X
Chromium(II)–Amine complexes, **3,** 57
Chromium(II) chloride, **1,** 149
Dibromomethyllithium, **11,** 158
Hexamethylphosphorous triamide, **10,** 199
Methanesulfonyl chloride, **4,** 326
Phosphoryl chloride, **1,** 881; **4,** 390; **5,** 535
Sodium iodide, **1,** 1116
Tin(II) chloride, **1,** 1113
Titanium(III) chloride–Lithium aluminum hydride, **6,** 588
Trimethyltinsodium, **9,** 500
Zinc, **1,** 1276
Zinc–Silver couple, **5,** 760

ELIMINATION OF OH, SiR_3 AND RELATED
Bis(cyclopentadienyl)(η^3-1-trimethylsilylallyl)titanium, **11,** 174
α-Bromovinyltriphenylsilane, **5,** 68
t-Butyldimethylchlorosilane, **11,** 88
t-Butyl lithioacetate, **6,** 84
Chloromethyldiphenylsilane, **12,** 321
Diisobutylaluminum hydride, **6,** 198
Lithium dipropylcuprate, **6,** 245
Methoxy(trimethylsilyl)methyllithium, **10,** 246; **11,** 331
Methyllithium, **7,** 242
Organotitanium reagents, **11,** 174
Potassium fluoride, **5,** 555
[(Trimethylsilyl)allyl]lithium, **11,** 572
Trimethylsilylmethyllithium, **6,** 635; **11,** 581
Trimethylsilylmethylmagnesium chloride, **5,** 724

ELIMINATION OF OH, SR AND RELATED
Diethyl phosphorochloridate, **8,** 171
Dimethyl(methylthio)sulfonium tetrafluoroborate, **11,** 204
Diphenyl disulfide, **6,** 235
Diphosphorus tetraiodide, **9,** 203
Lithium diethylamide, **5,** 398
N-Methanesulfinyl-*p*-toluidine, **2,** 269
Methoxymethyl phenyl sulfide, **12,** 316
Phenylthioacetic acid, **6,** 463
Sodium amalgam, **9,** 416; **11,** 473
Sodium methylsulfinylmethylide, **4,** 195
Sulfuryl chloride, **6,** 561
Titanium(0), **11,** 526
Tributyltin hydride, **8,** 497

ELIMINATION OF OH, SeR AND RELATED
Chlorotrimethylsilane–Sodium iodide, **10,** 97
Phenylselenoacetaldehyde, **10,** 310
Phosphorus(III) iodide, **10,** 318
Phosphoryl chloride, **8,** 401
Sodium benzeneselenoate, **6,** 548
Thionyl chloride–Triethylamine, **7,** 367
Zinc chloride, **8,** 536

ELIMINATION OF NO_2,NO_2
Calcium amalgam, **8,** 74
Sodium sulfide, **4,** 460
Tin(II) chloride, **6,** 554
Tributyltin hydride, **11,** 545

ELIMINATION OF X, SiR_3 AND RELATED
Alumina, **8,** 9
Bromine, **11,** 75

ELIMINATION REACTIONS TO FORM C=C, C≡C (*Continued*)
α-Bromovinyltrimethylsilane, **8,** 56
Cesium fluoride, **6,** 100; **10,** 81; **11,** 115
Chloro[(trimethylsilyl)methyl]ketene, **12,** 127
Iodine monochloride, **8,** 261
(1-Lithiovinyl)triphenylsilane, **5,** 375
Peracetic acid–Sodium acetate, **7,** 279
1-Phenylsulfonyl-2-trimethylsilylethane, **11,** 41
Potassium fluoride, **5,** 555; **6,** 481
Tetrabutylammonium fluoride, **9,** 444; **10,** 378
Vinyltrimethylsilane, **10,** 444
HOFMANN ELIMINATION
Hydrogen peroxide, **1,** 471
Methyl 3-(N,N-dimethylamino)-propionate, **9,** 306
Methylenetriphenylphosphorane, **8,** 339
Methyl iodide, **5,** 447
Phenyllithium, **1,** 845
Potassium amide, **1,** 907
Potassium hydroxide, **5,** 557
Potassium 2-methylcyclohexoxide, **5,** 560
Silver(I) oxide, **3,** 252; **5,** 583
OXIDATIVE DECARBOXYLATION (*see* OXIDATIVE . . .)
RAMBERG–BACKLUND REACTION
Bromomethanesulfonyl bromide, **12,** 75
Copper(II) chloride, **5,** 158
Lithium aluminum hydride, **5,** 382
Nickel(II) acetylacetonate, **11,** 58
Organocopper reagents, **11,** 365
Potassium hydroxide–Carbon tetrachloride, **5,** 96, 560
Sodium *t*-amyloxide, **6,** 546
Sulfuryl chloride, **5,** 641
SHAPIRO REACTION
Catecholborane, **9,** 443
Lithium diisopropylamide, **7,** 204
Lithium hydride, **4,** 304
p-Toluenesulfonyl chloride, **5,** 676
p-Toluenesulfonylhydrazide, **1,** 1185; **2,** 417; **4,** 511; **5,** 678; **6,** 598; **7,** 375; **8,** 460, 489

EMDE DEGRADATION
Sodium amalgam, **1,** 1030

ENE REACTION (*see also* ASYMMETRIC REACTIONS)
Acetyl hexachloroantimonate, **8,** 5
Alkylaluminum halides, **10,** 177; **11,** 7; **12,** 5
Aluminum chloride, **5,** 10; **7,** 7; **8,** 13; **9,** 11, 13
Butyllithium–Tetramethylethylenediamine, **8,** 67
Butyl N-(*p*-toluenesulfonyl)-iminoacetate, **7,** 50
Chloral, **8,** 82
p-Chlorobenzoyl nitrite, **4,** 79
Diethyl azodicarboxylate, **11,** 177
Diethyl oxomalonate, **10,** 143
Formaldehyde, **11,** 8
Grignard reagents, **11,** 245
Methyl cyanodithioformate, **7,** 237
Methyl cyclobutenecarboxylate, **10,** 263
Methyl fluoride–Antimony(V) fluoride, **6,** 381
Nitrosocarbonylmethane, **11,** 362
Oxygen, singlet, **6,** 431; **7,** 261; **9,** 338; **11,** 385
Phenyliodine(III) dichloride, **6,** 298
N-Phenyl-1,2,4-triazoline-3,5-dione, **4,** 381; **5,** 528; **6,** 467; **7,** 287; **9,** 372
Selenium(IV) oxide, **5,** 575
Sodium α-(N,N-dimethylamino)-naphthalenide, **10,** 362
N-Sulfinylbenzenesulfonamide, **9,** 439
Tetracyanoethylene, **5,** 647
Tin(IV) chloride, **9,** 436; **11,** 522; **12,** 486
Triphenyl phosphite ozonide, **3,** 323
Vinyltrichlorosilane, **5,** 749

ENOL ACYLATION
4-Dimethylaminopyridine, **9,** 178
Dimethylformamide diethyl acetal, **9,** 182
Hydrobromic acid, **3,** 154
Isopropenyl acetate, **1,** 524
Ketene, **1,** 528
Lithium dimethylcuprate, **4,** 177; **5,** 234
Perchloric acid, **1,** 796; **2,** 309
Potassium acetate, **1,** 906
Potassium hydride, **9,** 386
Sodium hexamethyldisilazide, **3,** 261
Sodium hydride, **2,** 382
Sulfoacetic acid, **1,** 1117
Sulfosalicylic acid, **1,** 1118
Thallium(I) ethoxide, **2,** 407
p-Toluenesulfonic acid, **1,** 1172; **4,** 508
Trifluoroacetyl trifluoromethanesulfonate, **9,** 485

ENOL ETHERIFICATION
Chloromethyl methyl ether, **4,** 83
2,2-Dimethoxypropane, **1,** 268
Ion-exchange resins, **5,** 355
Lithium diisopropylamide, **12,** 277
Methyl fluorosulfonate, **4,** 339; **6,** 381
Potassium fluoride, **10,** 325
Thallium(I) ethoxide, **4,** 501
p-Toluenesulfonic acid, **1,** 1172
Trialkyl orthoformate, **1,** 1204; **5,** 714
Triethyloxonium tetrafluoroborate, **2,** 430; **6,** 611
ENOL LACTONIZATION
Acetic anhydride, **8,** 1
Mercury(II) acetate, **8,** 315
Mercury(II) oxide, **8,** 316; **12,** 305
Mercury(II) trifluoroacetate, **11,** 320
Phosphorus(V) oxide–Methanesulfonic acid, **7,** 291
Potassium hydride, **12,** 407
Silver(I) nitrate, **12,** 433
Thallium(III) trifluoroacetate, **12,** 481
EPOXIDATION (*see also* ASYMMETRIC REACTIONS)
OF C=C
Benzeneperoxyseleninic acid, **8,** 22
N-Benzoylperoxycarbamic acid, **6,** 35
O-Benzylmonoperoxycarbonic acid, **6,** 40
(−)-Benzylquininium chloride, **7,** 311; **10,** 27
Bis(acetonitrile)chloronitropalladium(II), **11,** 45
3-Bromo-4,5-dihydro-5-hydroperoxy-4,4-dimethyl-3,5-diphenyl-3H-pyrazole, **11,** 76
2,3-Butanedione, **7,** 21
t-Butyl hydroperoxide, **2,** 49; **9,** 78
t-Butyl hydroperoxide + co-reagent, **2,** 287; **3,** 331; **4,** 346; **7,** 63; **8,** 62; **11,** 95
[(−)-Camphor-10-ylsulfonyl]-3-aryloxaziridines, **11,** 108
m-Chloroperbenzoic acid, **1,** 135; **3,** 49; **4,** 85; **5,** 120; **6,** 110; **7,** 62; **8,** 97; **9,** 108; **11,** 124
3,5-Dinitroperbenzoic acid, **9,** 196; **11,** 217
O-Ethylperoxycarbonic acid, **9,** 220
Hydrogen peroxide, **1,** 457
Hydrogen peroxide + co-reagent, **1,** 475; **3,** 156; **5,** 307; **6,** 35; **8,** 248; **9,** 242; **10,** 204; **11,** 256
2-Hydroperoxyhexafluoro-2-propanol, **9,** 244
(S)-(2-Hydroxy-N,N-dimethylpropanamide-O,O′)-oxodiperoxymolybdenum(VI), **10,** 206
Iodine–Silver oxide, **7,** 180
Iodosylbenzene, **10,** 213
p-Methoxycarbonylperbenzoic acid, **4,** 68
2-Nitrobenzeneseleninic acid, **8,** 359
p-Nitroperbenzoic acid, **1,** 743
Oxodiperoxymolybdenum(pyridine)-(hexamethylphosphoric triamide), **4,** 203
μ_3-Oxohexakis(μ-trimethylacetato)-trimethanoltriiron(III) chloride, **11,** 382
Oxomethoxymolybdenum(V) 5,10,15,20-tetraphenylprophyrin, **11,** 383
Oxygen, **5,** 482
Oxygen, singlet, **5,** 486; **6,** 431
Ozone, **5,** 491
Peracetic acid, **1,** 785, 787; **2,** 307; **3,** 219; **4,** 372
Perbenzoic acid, **1,** 791; **6,** 453; **7,** 279
Perlauric acid, **3,** 220
Permaleic acid, **1,** 819
Permonophosphoric acid, **10,** 304
Peroxyacetimidic acid, **1,** 470; **8,** 387
Peroxyacetyl nitrate, **5,** 510
Peroxybenzimidic acid, **4,** 375; **5,** 121, 511
Peroxytrichloroacetimidic acid, **12,** 379
Perphthalic acid, **1,** 819
Potassium *o*-nitrobenzeneperoxysulfonate, **12,** 412
Potassium peroxomonosulfate, **11,** 442; **12,** 413
Potassium superoxide, **12,** 413
Sodium hypochlorite, **8,** 461
Sodium perborate, **2,** 387
Succinoyl peroxide, **1,** 1117
o-Sulfoperbenzoic acid, **4,** 469
Tetralin hydroperoxide, **3,** 278
Trifluoroperacetic acid, **1,** 821
Triphenylsilyl hydroperoxide, **9,** 509
OF C=C SUBSTITUTED BY C=O, CN, TMS, ETC.
O-Benzylmonoperoxycarbonic acid, **6,** 40

EPOXIDATION (*Continued*)
(−)-Benzylquininium chloride, **7,** 311; **10,** 27
1-Bromo-1-trimethylsilyl-1(Z),4-pentadiene, **11,** 80
t-Butyl hydroperoxide, **1,** 88; **2,** 49; **7,** 43; **9,** 78
t-Butyl hydroperoxide + co-reagent, **5,** 338; **9,** 80, 81; **11,** 97; **12,** 90
m-Chloroperbenzoic acid, **1,** 135; **2,** 68; **6,** 110
Hydrogen peroxide, **1,** 51, 466; **3,** 155; **5,** 337; **9,** 241; **12,** 246
(E)-1-Iodo-3-trimethylsilyl-2-butene, **5,** 355
(1-Lithiovinyl)trimethylsilane, **5,** 374
Lithium aluminum hydride–Aluminum chloride, **8,** 289
α-Methoxyvinyllithium, **7,** 233
Osmium tetroxide–*t*-Butyl hydroperoxide, **1,** 88
Oxygen, **5,** 482
Peracetic acid, **1,** 787; **7,** 279
Peroxytrichloroacetimidic acid, **12,** 379
Potassium hydroxide, **6,** 486
Quinine, **10,** 338
Silver(I) oxide, **6,** 515
Sodium hypochlorite, **1,** 1084; **7,** 337; **8,** 430, 461
Sodium perborate, **2,** 387
[(Trimethylsilyl)allyl]lithium, **8,** 273
OF ALLYLIC, HOMOALLYLIC ALCOHOLS
t-Butyldimethylsilyl hydroperoxide–Mercury(II) trifluoroacetate, **12,** 85
t-Butyl hydroperoxide, **10,** 64
t-Butyl hydroperoxide + co-reagent, **5,** 75; **7,** 44, 63; **8,** 393; **9,** 81; **11,** 92, 97, 99; **12,** 90, 91
m-Chloroperbenzoic acid, **1,** 135; **4,** 85; **7,** 44, 62; **9,** 81, 108
Dibenzyltartaric acid diamide, **12,** 91
Hydrogen peroxide–Iron(III) acetylacetonate, **6,** 304
2-Hydroperoxyhexafluoro-2-propanol, **9,** 244
Lead tetrakis(trifluoroacetate), **6,** 318
Lithium alkyl(cyano)cuprates, **9,** 329
Oxodiperoxymolybdenum(pyridine)-(hexamethylphosphoric triamide), **9,** 197
Oxoperoxobis(N-phenylbenzohydroxamato)molybdenum(VI), **10,** 292
Perbenzoic acid, **1,** 791; **5,** 76
Tetraethylammonium fluoride, **7,** 356
Triphenylsilyl hydroperoxide, **9,** 509

ESCHENMOSER FRAGMENTATION
N-Bromosuccinimide, **10,** 57
2,4-Dinitrobenzenesulfonylhydrazide, **6,** 232
Hydrazine, **5,** 327; **7,** 170
Hydroxylamine-O-sulfonic acid, **2,** 217; **5,** 343
Mesitylenesulfonylhydrazide, **10,** 255
p-Toluenesulfonylhydrazide, **2,** 417

ESCHWEILER–CLARKE REACTION
Formaldehyde, **4,** 238

ESTERIFICATION (*see* SYNTHESIS INDEX—ESTERS)

EXTRUSION REACTIONS
−CO
Alumina, **10,** 8
Benzylsulfonyldiazomethane, **11,** 43
$-CO_2$
1,8-Diazabicyclo[5.4.0]undecene-7, **7,** 87
Lithium phenylethynolate, **6,** 343; **7,** 210
$-SO_2$
N-Chlorosuccinimide, **5,** 127; **6,** 115
1,3-Dihydrobenzo[*c*]thiophene 2,2-dioxide, **9,** 168; **10,** 146
1,3,3a,4,7,7a-Hexahydro-4,7-methanobenzo[*c*]thiophene 2,2-dioxide, **12,** 236
Lithium aluminum hydride, **8,** 286
Sodium acetate, **5,** 591
3-Sulfolene, **3,** 272; **12,** 455
Sulfur dioxide, **5,** 633; **9,** 440
Vinyltriphenylphosphonium bromide, **6,** 666
TWO-FOLD EXTRUSION
Selenium, **6,** 507
Triethyl phosphite, **5,** 693
Triphenylphosphine, **3,** 317; **4,** 548; **5,** 725; **6,** 643

FAVORSKII REARRANGEMENT
1,2-Dimethoxyethane, **1,** 267
Dimethylsulfoxonium methylide, **3,** 125
Potassium hydroxide–Carbon tetrachloride, **5,** 96
Sodium methoxide, **1,** 1091; **4,** 457
Triethylamine, **1,** 1198

FINKELSTEIN REACTION
Dicyclohexyl-18-crown-6, **6,** 174
Dimethyl sulfoxide, **7,** 133
FISHER INDOLE SYNTHESIS
Ethylene glycol, **1,** 375; **5,** 296
Phenylhydrazine, **1,** 838; **6,** 457; **7,** 284; **8,** 424
Phosphorus(III) chloride, **10,** 318
Polyphosphate ester, **1,** 892; **3,** 229
Polyphosphoric acid, **1,** 894; **10,** 321
Zinc chloride, **1,** 1289
FLUORINATION (*see also* ADDITION REACTIONS)
OF RCHO, R_2CO AND RELATED COMPOUNDS
Acetyl hypofluorite, **12,** 3
N-Alkyl-N-fluoro-*p*-toluenesulfonamides, **12,** 231
N-Bromosuccinimide, **7,** 37
Cesium fluoroxysulfate, **11,** 118
Fluorine–Sodium trifluoroacetate, **9,** 223
Perchloryl fluoride, **1,** 802
Sodium hexamethyldisilazide, **3,** 261
Trifluoromethyl hypofluorite, **2,** 200; **4,** 237; **9,** 224; **10,** 420
OF RH
Carbonyl difluoride, **12,** 106
Cesium fluoroxysulfate, **10,** 84; **11,** 118
Sulfur tetrafluoride, **1,** 1123
Trifluoromethyl hypofluorite, **7,** 156
AT ALLYLIC C–H (*see* ALLYLIC REACTIONS)
OF ArH, C=C–H
Acetyl hypofluorite, **11,** 5
Bis(fluoroxy)difluoromethane, **4,** 36
Cesium fluoroxysulfate, **10,** 84; **11,** 118
Crown ethers, **7,** 76
Fluorine, **5,** 309
Potassium fluoride, **8,** 480
Trifluoromethyl hypofluorite, **9,** 224
Xenon(II) fluoride, **6,** 669
Xenon(VI) fluoride, **7,** 425
OF RCOOH, RCOX
N-Alkyl-N-fluoro-*p*-toluenesulfonamides, **12,** 231
Perchloric acid, **7,** 279
Perchloryl fluoride, **1,** 802; **2,** 310
OF $ArN_2^+X^-$, $RN_2^+X^-$
Arenediazonium salts, **1,** 42, 43
Dinitrogen trioxide–Boron trifluoride, **1,** 329
Pyridinium poly(hydrogen fluoride), **6,** 285, 473; **9,** 399; **11,** 453
Tetrafluoroboric acid, **1,** 394
Tetrahydrofuran, **1,** 1140
OF RM, ArM
N-Alkyl-N-fluoro-*p*-toluenesulfonamides, **12,** 231
Perchloryl fluoride, **1,** 802
FLUOROLACTAMIZATION, FLUOROLACTONIZATION (*see* HALO-)
FORMYLATION (*see also* CARBONYLATION)
AT ACTIVE C–H
t-Butoxybis(dimethylamino)methane, **5,** 71
2-Chloro-1,3-dithiane, **8,** 88
Diethoxycarbenium tetrafluoroborate, **11,** 175
1,2-Dimethoxyethane, **1,** 267
1,3-Dithienium tetrafluoroborate, **11,** 227
Ethyl formate, **1,** 380, 1077; **2,** 197; **4,** 233; **12,** 226
N-Formylimidazole, **8,** 233
Lithium diisopropylamide, **3,** 184
Methoxymethylenetriphenylphosphorane, **5,** 438
Methyl formate, **5,** 446
Sodium alkoxide, **1,** 1065, 1091
Sodium hydride, **1,** 1075
Titanium(IV) chloride, **6,** 590
Trimethyl orthoformate, **11,** 568
Vilsmeier reagent, **1,** 284; **7,** 422
Zinc chloride, **11,** 602
OF C=C–H
Dimethylformamide–Phosphoryl chloride, **5,** 251
Vilsmeier reagent, **8,** 529; **9,** 514
OF ArH
Vilsmeier reaction
Dichloromethylenedimethylammonium chloride, **5,** 195
Dimethylthioformamide–Acetic anhydride, **3,** 128
N-Methylformanilide, **1,** 680
N-Methylformanilide–Phosphoryl chloride, **1,** 681, 878
Phosphoryl chloride, **1,** 876
Triphenylphosphine dibromide, **3,** 320
Vilsmeier reagent, **1,** 284, 286

FORMYLATION (*Continued*)
Related reactions
Acetic–formic anhydride, **2,** 10
Acetone cyanohydrin, **11,** 1
Aluminum chloride, **1,** 24
Bis(1,3-diphenyl-2-imidazolidinylidene), **7,** 22
n- or *sec*-Butyllithium, **6,** 85; **12,** 97
Chloroform–Sodium hydroxide, **12,** 151
N-Chlorosuccinimide + co-reagent, **6,** 119, 120
Copper(I) chloride, **1,** 166
Dichloromethyl methyl ether, **1,** 220; **2,** 120; **5,** 671; **9,** 154
Dimethyl(methylthio)sulfonium tetrafluoroborate, **12,** 207
N,N-Dimethyl-4-nitrosoaniline, **1,** 746
N,N′-Diphenylformamidine, **1,** 339
1,3-Dithiane, **6,** 248
Formaldehyde, **8,** 242
Formaldoxime, **1,** 402
Formyl fluoride, **1,** 407, 1018
Hexamethylenetetramine, **1,** 427, 898; **4,** 243
Hydrogen cyanide, **1,** 454
Iron carbonyl, **8,** 265
N-Methyl-N-phenylcarbamoyl chloride, **1,** 694
Oxalyl chloride, **7,** 257; **11,** 379
Polyphosphoric acid, **1,** 894
Thioanisole, **5,** 660
Titanium(IV) chloride, **2,** 414
1,3,5-Triazine, **2,** 423
Triethyl orthoformate, **1,** 1204; **2,** 430
Zinc cyanide, **1,** 1293
OF N–H (*see* SYNTHESIS INDEX—AMIDES)
OF RM, RX
Acetic—formic anhydride, **2,** 10
2-Alkoxy-1,3-benzodithiolanes, **8,** 236
1,3-Benzodithiolylium perchlorate, **8,** 34
gem-Dichloroallyllithium, **8,** 150
Dichloromethyl methyl ether, **9,** 154
Diethyl phenyl orthoformate, **3,** 97
Dihalobis(triphenylphosphine)-palladium(II), **6,** 60
p-Dimethylaminobenzaldehyde, **2,** 146
N,N-Dimethylformamide, **1,** 278; **5,** 247; **7,** 124; **8,** 189; **11,** 198; **12,** 98
4,4-Dimethyl-1,3-oxathiolane-3,3-dioxide, **9,** 186
Dipotassium(or Disodium) tetracarbonylferrate, **3,** 267; **8,** 214; **10,** 174
Ethoxymethyleneaniline, **1,** 362
2-(N-Formyl-N-methyl)aminopyridine, **8,** 341
N-Formylpiperidine, **11,** 244
Grignard reagents, **1,** 415; **10,** 189
Lithium, **5,** 376
Methyl methylthiomethyl sulfoxide, **5,** 456; **8,** 344
3-Methylthio-1,4-diphenyl-5-triazolium iodide, **6,** 396
Methylthiomethyl-N,N′-dimethyldithiocarbamate, **6,** 398
Methylthiomethyl *p*-tolyl sulfone, **12,** 327
1-Phenylthio-1-trimethylsilylethylene, **12,** 394
Tetrabutylammonium iodide, **7,** 355
Tetrakis(triphenylphosphine)-palladium(0), **12,** 468
1,1,3,3-Tetramethylbutyl isocyanide, **3,** 279; **4,** 480
Thionyl chloride, **4,** 503
Titanium(IV) chloride, **8,** 483
Triethyl orthoformate, **1,** 1204
N,4,4-Trimethyl-2-oxazolinium iodide, **4,** 540; **6,** 630
s-Trithiane, **4,** 564
OF O–H (*see* SYNTHESIS INDEX—ESTERS)
FRAGMENTATION REACTIONS (*see also* BECKMANN, ESCHENMOSER FRAGMENTATIONS)
Aluminum isopropoxide, **8,** 15
Birch reduction, **8,** 38
Borane–Tetrahydrofuran, **3,** 76
N-Bromosuccinimide, **10,** 57
Cesium fluoride, **11,** 115
α-Chloro-N-cyclohexylpropanaldonitrone, **4,** 80
m-Chloroperbenzoic acid, **11,** 122
Copper(II) acetate–Ferrous(II) sulfate, **10,** 103
Copper(I) trifluoromethanesulfonate, **8,** 125
Lead(IV) acetate azides, **4,** 276
Lead tetraacetate, **9,** 265
Lithium aluminum hydride, **9,** 274

Methyllithium, **8,** 342
3,3,6,9,9-Pentamethyl-2,10-diazabicyclo-[4.4.0]decene, **9,** 354
Phenyliodine(III) diacetate–Azidotrimethylsilane, **6,** 297
Potassium *t*-butoxide, **7,** 296
1,3-Propanedithiol, **4,** 413
Sodium methylsulfinylmethylide, **7,** 338

FRIEDEL–CRAFTS ACYLATION
2-Acetoxypyridine, **1,** 9
Aluminum halides, **1,** 24, 35; **2,** 19; **5,** 10; **10,** 9; **11,** 25
Boric acid, **1,** 63
Boron trifluoride etherate, **11,** 72
γ-Butyrolactone, **1,** 101
Carbonyl sulfide, **11,** 112
Chloroacetylium hexafluoroantimonate, **6,** 103
Chloroform, **1,** 130
Copper(II) trifluoromethanesulfonate, **10,** 110
Diethyl ethoxymethylenemalonate, **5,** 216
Fluorosulfuric acid, **1,** 396
Grignard reagents, **9,** 229
Hydrogen fluoride, **1,** 455; **6,** 285; **9,** 240
Lithium *o*-lithiobenzoate, **10,** 246
Methanesulfonic acid, **11,** 321
Methylene chloride, **1,** 676
Nitroethane, **4,** 357
Oxalyl chloride, **1,** 767
Phosphorus(V) oxide, **1,** 871
Phosphoryl chloride, **2,** 330
Polyphosphate ester, **2,** 333
Polyphosphoric acid, **1,** 894; **3,** 231; **4,** 395; **7,** 294
Pyridinium chloride, **6,** 497
Sodium tetrachloraluminate, **1,** 1027; **4,** 438
Succinic anhydride, **4,** 468
Tetrachloroethane–Nitrobenzene, **4,** 479
Thionyl chloride, **1,** 1158
Tin(IV) chloride, **1,** 1111
Trichloroacetyl chloride, **4,** 521
Trifluoroacetic anhydride, **1,** 1221; **9,** 484
Trifluoromethanesulfonic acid, **5,** 701; **12,** 531
Trifluoromethanesulfonic-carboxylic anhydrides, **4,** 533; **10,** 420
2-Trimethylsilylethanol, **8,** 510
Trimethylsilyl trifluoromethanesulfonate, **11,** 584
Zinc chloride, **1,** 1289
Zirconium(IV) chloride, **1,** 1295

FRIEDEL–CRAFTS ALKYLATION
Aluminum chloride, **1,** 24; **4,** 10; **5,** 10; **11,** 25; **12,** 26
Boron trifluoride, **1,** 68
γ-Butyrolactone, **1,** 101
Ethylmagnesium bromide, **9,** 218
Hydrobromic acid, **6,** 282
Iodine, **2,** 220
Iodotrimethylsilane, **10,** 216
Ion-exchange resins, **1,** 511
Iron, **2,** 229
Methyl chloride, **1,** 674
Methylene chloride, **4,** 337
Methyl fluoride–Antimony(V) fluoride, **3,** 201
2-Methylfuran, **1,** 682
Nafion-H, **9,** 320
Nickel(II) acetylacetonate, **9,** 51
Oxalic acid, **1,** 764
Phosphorus(V) oxide, **10,** 319
Polyphosphoric acid, **1,** 894; **5,** 540
Potassium bisulfate, **1,** 909
Silver perchlorate, **11,** 469
Titanium(IV) chloride, **1,** 1169

FRIEDLANDER QUINOLINE SYNTHESIS
Cobalt(II) phthalocyanine, **10,** 102

FRIES REARRANGEMENT
Aluminum chloride, **1,** 24; **5,** 10; **8,** 13
Boric acid, **1,** 63
Boron trifluoride etherate, **7,** 31
Cyclodextrins, **6,** 151
Hydrogen fluoride, **5,** 336
Ion-exchange resins, **1,** 511
Sodium tetrachloroaluminate, **1,** 1027
Titanium(IV) chloride, **1,** 1169
Trifluoromethanesulfonic acid, **5,** 701

GABRIEL SYNTHESIS
N-Benzyl trifluoromethanesulfonamide, **5,** 29; **6,** 43
t-Butyl methyl iminodicarboxylate potassium salt, **8,** 72
Copper(I) bromide, **3,** 67
Di-*t*-butyl iminodicarboxylate, **1,** 210
Tributylhexadecylphosphonium bromide, **7,** 166

GATTERMAN SYNTHESIS
Acetone cyanohydrin, **11,** 1

GATTERMAN SYNTHESIS (*Continued*)
Aluminum chloride, **1,** 24
Copper(I) chloride, **1,** 166
Hydrogen cyanide, **1,** 454
1,3,5-Triazine, **2,** 423
GLASER AND RELATED COUPLINGS
t-Butylamine, **1,** 84
Copper(II) acetate, **1,** 159; **4,** 105
Copper(I) chloride, **1,** 166; **4,** 109
Copper(I) chloride–Ammonium chloride, **1,** 165
GOMBERG–BACKMANN REACTION
Crown ethers, **8,** 128
Dinitrogen trioxide–Boron trifluoride, **1,** 329
Phase-transfer catalysts, **9,** 356; **12,** 379

HALLER–BAUER REACTION
Sodium amide, **6,** 525; **8,** 446
HALOALKYLATION
Aluminum halides, **1,** 22; **6,** 17
Benzyltriethylammonium chloride, **4,** 27
1,4-Bis(chloromethoxy)butane, **7,** 22
1-Chloro-4-bromomethoxybutane, **6,** 104
Chloromethyl methyl ether, **1,** 132; **4,** 83
Copper, **5,** 146
Dimethoxymethane, **1,** 671
Formaldehyde, **1,** 397, 399; **4,** 238
Methoxyacetyl chloride, **12,** 310
Molybdenum(VI) fluoride, **5,** 465
Sodium trichloroacetate, **1,** 1107
Zinc chloride, **2,** 464
HALOAMINATION (*see* ADDITION REACTIONS)
HALOCARBAMATION
Iodine, **11,** 261; **12,** 253
HALOCARBONYLATION
Oxalyl halides, **1,** 767; **3,** 216; **11,** 379
HALODECARBOXYLATION (*see also* HUNSDIECKER REACTION)
t-Butyl hypoiodite, **1,** 94
Iodine–Lead tetraacetate, **1,** 552
Lead tetraacetate, **1,** 537; **4,** 278
Lead tetraacetate + co-reagent, **1,** 557; **5,** 370; **6,** 317
Mercury(II) oxide, **1,** 655; **4,** 323; **5,** 428; **9,** 293
Mercury(II) oxide + halogen, **1,** 657; **5,** 428
Potassium chloride, **6,** 480
2-Pyridinethiol-1-oxide, **12,** 417
Thallium(I) carbonate, **11,** 515
Thallium(I) ethoxide, **2,** 407
HALOFORM REACTION
Potassium hydroxide–Carbon tetrachloride, **5,** 96
Sodium hypochlorite, **1,** 1084
HALOGENATION (*see* FLUORINATION, ETC.)
HALOLACTAMIZATION
Bromine, **11,** 75
HALOLACTONIZATION (*see also* ASYMMETRIC REACTIONS)
Bromine, **9,** 65; **12,** 70
Chloramine-T, **11,** 118
1,8-Diazabicyclo[5.4.0]undecene-7, **7,** 87
Iodine, **8,** 256; **9,** 248; **11,** 261; **12,** 253
Lithium α-ethoxycarbonylvinyl (1-hexynyl)cuprate, **6,** 329
Molybdenum carbonyl, **12,** 330
Thallium(I) acetate-Iodine, **10,** 394
HALOSELENYLATION (*see* ADDITION REACTIONS)
HALOSULFENYLATION (*see also* ADDITION REACTIONS)
Thionyl chloride, **5,** 663
HALOSULFONATION
Sulfuryl chloride, **4,** 474
HALOSULFONYLATION, HALOTELLURATION (*see* ADDITION REACTIONS)
HELL–VOLHARD–ZELINSKY REACTION
Phosphorus(III) bromide, **1,** 873
Phosphorus, red, **1,** 861
HINSBERG REACTION
Benzenesulfonyl chloride, **1,** 46
HOESCH REACTION
Zinc chloride, **1,** 1289
HOFMANN CARBYLAMINE REACTION
Benzyltriethylammonium chloride, **4,** 27; **5,** 26; **7,** 18
HOFMANN DEGRADATION
Iodosylbenzene, **12,** 258
Lead tetraacetate, **1,** 537; **2,** 234; **4,** 278; **6,** 313
Methyl hypobromite, **5,** 446
Phenyliodine(III) bis(trifluoroacetate), **9,** 54
Sodium hypobromite, **1,** 1083
HOFMANN ELIMINATION (*see* ELIMINATION REACTIONS)

HOFMANN–LOFFLER–FREYTAG REACTION
Sulfuric acid, **5,** 633
Triethylamine, **7,** 385
HOMOALDOL REACTION (*see* ALDOL-TYPE REACTIONS)
HOOKER REACTION
Hydrogen peroxide, **1,** 466
HUNSDIECKER REACTION
Bromotrichloromethane, **1,** 80
Mercury(II) oxide, **1,** 655; **4,** 323; **5,** 428; **7,** 224; **9,** 293
Thallium(I) carbonate, **11,** 515
Thallium(I) ethoxide, **2,** 407
HYDRIDE ABSTRACTION
Antimony(V) halide, **1,** 42; **5,** 19
t-Butyl chloride, **2,** 46
o-Chloranil, **1,** 128
2,3-Dichloro-5,6-dicyano-1,4-benzoquinone, **1,** 215
Phosphorus(V) chloride, **1,** 866
Sodium methylsulfinylmethylide, **4,** 195
Tetrafluoroboric acid, **1,** 394
Triphenylcarbenium salts, **1,** 1256; **3,** 330
Tropylium salts, **4,** 568
HYDROACYLATION (*see* ADDITION REACTIONS)
HYDROALUMINATION (*see* HYDROMETALLATION REACTIONS)
HYDROAMINATION (*see* ADDITION REACTIONS)
HYDROBORATION (*see* HYDROMETALLATION REACTIONS)
HYDROBROMINATION, HYDROCARBOXYLATION, HYDROCHLORINATION (*see* ADDITION REACTIONS)
HYDROCYANATION (*see also* ADDITION REACTIONS, CONJUGATE ADDITION REACTIONS)
Cyanotrimethylsilane, **6,** 632
Hydrogen cyanide, **4,** 252
HYDROFLUORINATION, HYDROFORMYLATION (*see* ADDITION REACTIONS)
HYDROGENATION (*see* REDUCTION REACTIONS, TYPE OF COMPOUND INDEX—HYDROGENATION CATALYSTS)
HYDROHALOGENATION, HYDROIODINATION (*see* ADDITION REACTIONS)
HYDROLYSIS (INCLUDING OXIDATIVE AND REDUCTIVE HYDROLYSIS)
OF ACETALS, KETALS
Acetic acid, **2,** 6
Acetone, **3,** 4
Aluminum chloride–Sodium iodide, **12,** 29
Benzoyl trifluoromethanesulfonate, **12,** 44
Boric acid, **1,** 63
Boron trihalides, **1,** 67; **5,** 49
3-Bromo-1,2-dihydroxypropane, **4,** 154
Bromodimethylborane, **12,** 199
Bromotrimethylsilane, **12,** 79
Ethylene glycol, **1,** 375
Formic acid, **2,** 202; **5,** 316
Hydrogen iodide, **7,** 173
Iodotrimethylsilane, **8,** 261
Magnesium sulfate, **1,** 634
Nafion-H, **12,** 335
Perchloric acid, **1,** 796
Periodic acid, **1,** 815
Tetrachlorosilane–Sodium iodide, **11,** 466
Titanium(IV) chloride, **12,** 494
Trichloroacetic acid, **4,** 520
Trichloro(methyl)silane–Sodium iodide, **12,** 527
Trifluoroacetic acid, **2,** 433
Triphenylcarbenium tetrafluoroborate, **4,** 565
OF RCOX → RCOOH (*see* SYNTHESIS INDEX—CARBOXYLIC ACIDS)
OF C=N → C=O
Arenesulfonylhydrazones
Benzeneseleninic anhydride, **8,** 29; **10,** 22
Bromine, **9,** 65
N-Bromosuccinimide, **6,** 74
N,N-Dimethylhydrazine, **7,** 416
Hydrogen peroxide–Potassium carbonate, **9,** 243
Molybdenum(V) trichloride oxide, **7,** 248
Sodium hypochlorite, **6,** 543
Sodium nitrite, **9,** 432
Sodium peroxide, **7,** 341

HYDROLYSIS (*Continued*)
Thallium(III) acetate, **10,** 393
Titanium(III) chloride, **6,** 587
p-Toluenesulfonylhydrazide, **7,** 31, 416
Hydrazones
Acetone, **6,** 9
Benzeneseleninic anhydride, **10,** 22
Boron trifluoride etherate, **11,** 72
m-Chloroperbenzoic acid, **12,** 118
Chromium(II) chloride, **1,** 149
Cobalt(III) fluoride, **8,** 113
N,N-Dimethylhydrazine, **3,** 117; **7,** 126, 416
2,4-Dinitrophenylhydrazine, **2,** 176
Ferric nitrate/K10 Bentonite, **12,** 231
Levulinic acid, **1,** 564
Molybdenum(V) trichloride oxide, **7,** 248
Nitrosonium tetrafluoroborate, **7,** 253
α-Oxoglutaric acid, **1,** 531
Oxygen, singlet, **8,** 367
2,4-Pentanedione, **1,** 10
Peracetic acid, **1,** 785
Picryl azide, **1,** 885
Sodium nitrite, **1,** 1097; **9,** 432
Thallium(III) nitrate, **4,** 492
Tin(II) chloride, **1,** 1113
Titanium(III) chloride, **6,** 587
p-Toluenesulfonylhydrazide, **7,** 416
Uranium(VI) fluoride, **7,** 417
Vanadium(II) chloride, **11,** 593
Imines
α-Chloro-N-cyclohexylpropanaldonitrone, **4,** 80
Lithium diethylamide, **6,** 332
Oximes
Acetone, **6,** 9
Aluminum isopropoxide, **5,** 14
Benzeneseleninic anhydride, **8,** 29; **10,** 22
Benzyl chloroformate, **3,** 59
Bispyridinesilver permanganate, **12,** 62
Bromine, **9,** 65
Cerium(IV) ammonium nitrate, **2,** 63; **3,** 44
Chromium(II) acetate, **3,** 59; **5,** 143
Cobalt(III) fluoride, **8,** 113
Collins reagent, **6,** 124
Dioxygen bis(triphenylphosphine)-palladium, **5,** 510
Hydrogen peroxide, **10,** 201
Iron carbonyl, **2,** 229
Lead tetraacetate, **2,** 234
Levulinic acid, **1,** 564
Molybdenum(V) trichloride oxide, **7,** 248
Nitrosonium tetrafluoroborate, **7,** 253
Nitrosyl chloride, **8,** 364
Perchloryl fluoride, **1,** 802
Potassium persulfate, **1,** 952
Pyridinium chlorochromate, **8,** 425; **10,** 335
Pyridinium dichromate, **11,** 453
Raney nickel, **12,** 422
Sodium dichromate, **6,** 123
Sodium dithionite, **10,** 363
Sodium hydrogen sulfite, **2,** 377
Sodium nitrite, **1,** 1097
Thallium(III) nitrate, **4,** 492
Titanium(III) chloride, **8,** 482
Tributylphosphine–Diphenyl disulfide, **12,** 514
Tri-μ-carbonylhexacarbonyldiiron, **12,** 525
Triphenylbismuth carbonate, **9,** 501
Vanadium(II) chloride, **10,** 457
Semicarbazones
Benzeneseleninic anhydride, **10,** 22
Cerium(IV) ammonium nitrate, **3,** 44
Pyruvic acid, **1,** 974
Sodium nitrite, **1,** 1097
Thallium(III) nitrate, **4,** 492
Titanium(III) chloride, **8,** 482
OF CYANOHYDRINS AND RELATED COMPOUNDS
Benzyltriethylammonium chloride, **5,** 26
2-Chloroacrylonitrile, **4,** 76; **5,** 107
Copper(II) sulfate, **8,** 125
Copper(II) tetrafluoroborate, **3,** 66; **6,** 142
Cyanotrimethylsilane, **6,** 632; **9,** 127
N,N-Diethylaminoacetonitrile, **9,** 159
N,N-Dimethyldithiocarbamoylacetonitrile, **7,** 123
2-(2,6-Dimethylpiperidino)acetonitrile, **11,** 212
Silver carbonate–Celite, **4,** 425
Sodium hexamethyldisilazide, **6,** 529
Tosylmethyl isocyanide, **11,** 539
OF GEMINAL DIHALIDES
Chlorotrifluoroethylene, **4,** 94
Dichloromethyllithium, **5,** 199

Morpholine, **1,** 705
Silver acetate, **1,** 1002
Silver(I) trifluoroacetate, **1,** 1018; **7,** 323
OF HEMITHIOACETALS
Chloramine-T, **4,** 445
Chlorotrimethylsilane, **7,** 66
Mercury(II) oxide–Tetrafluoroboric acid, **10,** 254
Nickel boride, **5,** 471
Phenyl(phenylthio)trimethylsilylmethane, **10,** 314
OF ORTHOESTERS, ORTHOTHIOESTERS
Sulfur dioxide, **7,** 346
Tris(methylthio)methyllithium, **7,** 412
OF PROTECTED ALCOHOLS (*see* PROTECTION)
OF SELENOACETALS
Benzeneseleninic anhydride, **9,** 32
Chloramine-T, **4,** 445
OF C=C–X → C–C=O
Enamines
Ion-exchange resins, **1,** 511
Enol ethers, enol esters, etc.
Benzeneselenenyl trifluoroacetate, **5,** 522
Chlorotrimethylsilane–Sodium iodide, **10,** 97
Dimethyl sulfoxide, **10,** 166
2-Ethoxyallyl vinyl sulfide, **5,** 292
α-Lithio-α-methoxyallene, **9,** 272
Lithium di(α-methoxyvinyl)cuprate, **6,** 204
α-Methoxyvinyllithium, **6,** 372
Oxalic acid, **1,** 764
Potassium thiolacetate, **10,** 325
Tributyltin fluoride, **12,** 515
3-Triethylsilyloxypentadienyllithium, **11,** 556
Vinyl halides
1,3-Dibromo-2-pentene, **7,** 90
1,3-Dichloro-2-butene, **1,** 214; **2,** 111
Ethyl trichloroacetate, **4,** 233
Lithium diethylamide, **6,** 332
Mercury(II) acetate, **8,** 315
Mercury(II) perchlorate, **5,** 428
Mercury(II) trifluoroacetate, **9,** 294
Potassium *t*-butoxide, **5,** 544
Sodium amalgam, **7,** 326
Sulfuric acid, **4,** 470
Titanium(IV) chloride, **5,** 80
Vinylsilanes
Hydrogen peroxide, **12,** 242
(E)-1-Iodo-3-trimethylsilyl-2-butene, **5,** 355
(1-Lithiovinyl)trimethylsilane, **5,** 374
Lithium aluminum hydride–Aluminum chloride, **8,** 289
Vinyl sulfides
1-(Alkylthio)vinyllithium, **5,** 6
S-Allyl N,N-dimethyldithiocarbamate, **6,** 11
Benzenesulfenyl chloride, **6,** 30
Diethyl (1-methylthio)ethylphosphonate, **4,** 151
Diethyl methylthiomethylphosphonate, **3,** 97
1-Lithiocyclopropyl phenyl sulfide, **7,** 190
Mercury(II) chloride, **6,** 359
Methyl 4-methylthiocrotonate, **6,** 389
3-Methylthio-2-propenyl *p*-tolyl sulfone, **12,** 327
2-(2-Phenylsulfonylethyl)-1,3-dioxolane, **6,** 461
Titanium(IV) chloride, **5,** 80
Trichloroisocyanuric acid, **6,** 605
OF THIOACETALS
Benzeneseleninic anhydride, **8,** 29
Bis(dimethylaluminum) 1,3-propanedithiolate, **6,** 49
Bromodimethylsulfonium bromide, **9,** 66
N-Bromosuccinimide, **6,** 74
Cerium(IV) ammonium nitrate, **4,** 71; **10,** 79
Chloramine-T, **4,** 75, 445
1-Chlorobenzotriazole, **4,** 78
Copper(II) chloride–Copper(II) oxide, **5,** 160
7,8-Dimethyl-1,5-dihydro-2,4-benzodithiepin, **6,** 216
Dimethyl sulfoxide–Iodine, **5,** 347
Diphenyl disulfide, **6,** 235
1,3-Dithiane, **2,** 182; **3,** 135; **4,** 216; **5,** 287
1,2-Ethanedithiol, **2,** 189
Ethyl ethylthiomethyl sulfoxide, **5,** 299
Ferric nitrate/K10 Bentonite, **12,** 231
Iodine–Pyridine N-oxide, **6,** 296
Iodylbenzene, **11,** 275
Isoamyl nitrite, **9,** 258
Ketene dimethyl thioacetal monoxide, **6,** 311; **8,** 268

HYDROLYSIS (*Continued*)
Lead(IV) oxide–Boron trifluoride etherate, **11,** 283
2-Lithio-4,5-dihydro-5-methyl-[4H]-1,3,5-dithiazine, **8,** 305
Mercury(II) chloride, **1,** 652
Mercury(II) oxide–Tetrafluoroboric acid, **10,** 254
O-Mesitylenesulfonylhydroxylamine, **5,** 430
Methyl bis(methylthio)sulfonium hexachloroantimonate, **11,** 335
Methyl fluorosulfonate, **4,** 339
Methyl iodide, **4,** 341; **9,** 308
Methyl methylthiomethyl sulfoxide, **4,** 341; **5,** 456; **6,** 390
Nitronium tetrafluoroborate, **9,** 324
Oxygen, **8,** 366
Periodic acid, **10,** 304
Pyridinium bromide perbromide, **10,** 333
Silver(I) oxide, **4,** 430
Sulfuric acid, **4,** 470
Sulfuryl chloride, **7,** 349
Thallium(III) nitrate, **7,** 362; **9,** 460
Trialkyloxonium tetrafluoroborate, **4,** 527; **5,** 691; **9,** 482; **10,** 426
OF 1,1,1-TRIHALIDES
Brij 35, **6,** 70
Carbon tetrabromide–Tin(II) fluoride, **11,** 111
Hexamethyldisiloxane, **8,** 240
HYDROMETALLATION OF C = C, C ≡ C
HYDROALUMINATION
Chlorobis(cyclopentadienyl)-hydridozirconium(IV), **8,** 84
N-Chlorosuccinimide, **8,** 103
Dichlorobis(cyclopentadienyl)titanium, **10,** 130
Dichlorobis(cyclopentadienyl)titanium–Bis(dialkylamino)alanes, **9,** 46
Dichlorobis(cyclopentadienyl)-zirconium(II), **10,** 131
Dichlorobis(triphenylphosphine)-palladium(II), **7,** 95
Diisobutylaluminum hydride, **2,** 140; **4,** 158; **5,** 224; **6,** 198; **7,** 111; **8,** 173
Ethyl chloroformate, **7,** 147
3-Lithio-1-trimethylsilyl-1-propyne, **3,** 173
Lithium aluminum hydride, **8,** 286; **9,** 274
Lithium diisobutylmethylaluminum hydride, **2,** 248
Titanium(IV) chloride–Lithium aluminum hydride, **9,** 276
HYDROBORATION (*see also* ASYMMETRIC REACTIONS, CARBONYLATION)
General methods
9-Borabicyclo[3.3.1]nonane, **8,** 47
Borane + co-reagent, **1,** 1229; **4,** 191; **10,** 50; **11,** 70
Catecholborane, **4,** 25; **6,** 33; **7,** 54
Diborane, **1,** 199; **5,** 184
Dibromoborane–Dimethyl sulfide, **8,** 143
Dichloroborane diethyl etherate, **7,** 96
Dicyclohexylborane, **3,** 90
Monoisopinocampheylborane, **9,** 317
Sodium borohydride–Aluminum chloride, **1,** 1053
Thexylborane, **6,** 207
followed by Amination
Dichlorophenylborane, **4,** 377
Hydroxylamine-O-sulfonic acid, **1,** 481
O-Mesitylenesulfonylhydroxylamine, **5,** 430
Norbornene, **1,** 757
followed by Halogenation
Bromine, **6,** 70
Catecholborane, **5,** 100
Ferric chloride, **6,** 259
Iodine, **3,** 159; **5,** 346; **6,** 293; **7,** 179
Iodine monochloride, **10,** 212
Sodium iodide–Chloramine-T, **11,** 488
followed by Oxidation
Bis(3,6-dimethyl)borepane, **4,** 35
9-Borabicyclo[3.3.1]nonane, **2,** 31; **5,** 46; **6,** 62; **8,** 47; **9,** 57
Borane + co-reagent, **1,** 199, 963; **4,** 124; **5,** 47; **9,** 136; **11,** 69
Catecholborane, **4,** 69; **9,** 97
Chromium(VI) oxide, **3,** 54
Dichlorobis(cyclopentadienyl)-titanium–Lithium borohydride, **9,** 146
Dichloroborane diethyl etherate, **5,** 191
Dicyclohexylborane, **6,** 62; **8,** 162; **11,** 172
Diisopinocampheylborane, **6,** 202

Di-2-mesitylborane, **12,** 195
Disiamylborane, **1,** 57; **5,** 39; **6,** 62; **11,** 226
Iodine monochloride, **11,** 268
Lithium acetylide, **6,** 324
Lithium 2,2,6,6-tetramethylpiperidide, **5,** 417
Lithium triethylborohydride, **8,** 309
Monochloroborane diethyl etherate, **4,** 346; **5,** 465
Monochloroborane–Dimethyl sulfide, **8,** 354
Sodium borohydride–Acetic acid, **5,** 601
Sodium cyanide, **4,** 446
Sodium dichromate, **2,** 70
Thexylborane, **1,** 276; **10,** 397; **11,** 516; **12,** 484
Triisopinocampheylborane, **1,** 1228
Trimethylamine N-oxide, **3,** 309; **10,** 423
HYDROMAGNESIATION
Dichlorobis(cyclopentadienyl)titanium, **9,** 146; **10,** 130; **11,** 163; **12,** 168
Pentane-1,5-di(magnesium bromide), **9,** 355
Titanium(IV) chloride, **1,** 1169
HYDROSILYLATION
N-Bromosuccinimide, **8,** 54
Carbon monoxide, **9,** 95
m-Chloroperbenzoic acid, **8,** 97
Chlorotris(triphenylphosphine)-rhodium(I), **4,** 559
Dichlorosilane, **4,** 139
Diethoxymethylsilane, **12,** 182
Dimethylphenylsilyllithium, **10,** 162
Hydrogen hexachloroplatinate(IV)–Diethoxymethylsilane, **12,** 243
Hydrogen hexachloroplatinate(IV)–Trichlorosilane, **8,** 98
Hydrogen peroxide, **12,** 242
2,3-O-Isopropylidene-2,3-dihydroxy-1,4-bis(diphenylphosphine)butane, **5,** 360
Organomagnesium reagents, **12,** 352
Organopentafluorosilicates, **11,** 373
Palladium(II) acetate, **8,** 378
HYDROSTANNATION
Organoaluminum reagents, **12,** 339
Tin(IV) chloride–Sodium borohydride, **9,** 438
Tributyltin hydride, **6,** 604
Tributyltin trifluoromethanesulfonate, **12,** 524
HYDROZIRCONATION
Chlorobis(cyclopentadienyl)-hydridozirconium(IV), **6,** 175; **7,** 101; **8,** 84
Dichlorobis(cyclopentadienyl)-zirconium–*t*-Butylmagnesium chloride, **12,** 171
HYDROPEROXYLATION (*see also* ALLYLIC REACTIONS)
Benzyltrimethylammonium hydroxide, **1,** 1252
Hydrogen peroxide–Iron salts, **5,** 340
Oxygen, **5,** 482; **6,** 426; **7,** 258; **8,** 366
Potassium *t*-butoxide, **1,** 911; **2,** 336
HYDROSILYLATION, HYDROSTANNATION (*see* HYDROMETALLATION)
HYDROSULFENYLATION (*see* ADDITION REACTIONS)
HYDROXYALKYLATION
Benzyl chloromethyl ether, **1,** 52
Ethyl formate, **4,** 233
Hydroxylamine-O-sulfonic acid, **3,** 156
Tributyltinmethanol, **7,** 378
HYDROXYLATION (*see also* ADDITION REACTIONS, ASYMMETRIC REACTIONS)
OF RH
Benzyl(triethyl)ammonium permanganate, **9,** 43
t-Butyl hydroperoxide–Selenium(IV) oxide, **9,** 79
Chlorotris(triphenylphosphine)-rhodium(I), **2,** 448
Chromium(VI) oxide, **3,** 54
Chromyl acetate, **2,** 78
Di-*t*-butyl diperoxycarbonate, **6,** 166
2,3-Dichloro-5,6-dicyano-1,4-benzoquinone, **6,** 168
Dimethyl sulfoxide, **1,** 296
Ferrous perchlorate, **9,** 258
Ferrous sulfate–Oxygen, **5,** 308
Hydrogen peroxide, **6,** 286
Hydrogen peroxide–Iron salts, **5,** 340
Lead(IV) acetate azides, **5,** 363
Lead tetrakis(trifluoroacetate), **2,** 238
Nitrobenzene, **8,** 358
p-Nitroperbenzoic acid, **9,** 324
Oxygen, **6,** 426; **11,** 384

HYDROXYLATION (*Continued*)
Ozone–Silica gel, **6,** 440; **7,** 271; **9,** 343; **10,** 296
Palladium(II) chloride, **2,** 303
Peracetic acid, **1,** 787; **5,** 505
Potassium permanganate, **1,** 942
Silver hexafluoroantimonate, **11,** 467
Silver(II) oxide, **4,** 431
Silver(II) picolinate, **4,** 432
Sulfur trioxide, **4,** 473
Trifluoroacetic acid trianhydride, with iodous acid, **7,** 181
Trifluoroperacetic acid, **7,** 281
OF ArH
Borane–Tetrahydrofuran, **11,** 156
Boric acid, **1,** 63
t-Butylperoxy isopropyl carbonate, **2,** 55
m-Chloroperbenzoic acid, **12,** 118
Cobalt(III) trifluoroacetate, **5,** 146
Copper(II)–Amine complexes, **8,** 115
Diisopropyl peroxydicarbonate, **1,** 263
Hydrogen peroxide, **10,** 201
Hydrogen peroxide + co-reagent, **1,** 472; **4,** 255; **9,** 244; **11,** 256
Hypofluorous acid, **8,** 254
Lead tetraacetate, **6,** 313
Lead tetrakis(trifluoroacetate), **2,** 238; **4,** 282
Nitrobenzene, **8,** 358
m-Nitrobenzenesulfonyl peroxide, **2,** 296
Osmium tetroxide–Sodium chlorate, **6,** 424
Oxalyl chloride, **5,** 481
Oxygen, **5,** 482
Perchloryl fluoride, **1,** 802
Permonophosphoric acid, **10,** 304
Potassium nitrosodisulfonate, **1,** 940
Potassium persulfate, **1,** 952; **8,** 417
Potassium superoxide, **11,** 442
Tetralin hydroperoxide, **3,** 278
Thallium(III) trifluoroacetate, **3,** 286; **6,** 579
Trifluoroacetyl nitrate, **12,** 530
Trifluoroperacetic acid, **1,** 821
AT ACTIVE C–H
2-Arylsulfonyl oxaziridines, **12,** 392
Benzeneseleninic anhydride, **9,** 32; **10,** 22
t-Butyl hydroperoxide, **12,** 88
[(−)-Camphor-10-ylsulfonyl]-3-aryloxaziridines, **11,** 108
m-Chloroperbenzoic acid, **6,** 110; **8,** 97; **9,** 108; **11,** 122; **12,** 118
Chlorotrimethylsilane, **3,** 310
Copper(II) chloride, **6,** 139
Dibenzoyl peroxide, **12,** 157
2,3-Dichloro-5,6-dicyano-1,4-benzoquinone, **6,** 168
Dioxygentetrakis(triphenylarsine)-rhodium(I) hexafluorophosphate, **8,** 206
Iodosylbenzene, **10,** 213; **11,** 270
o-Iodosylbenzoic acid, **12,** 259
Osmium tetroxide–N-Methylmorpholine N-oxide, **10,** 291
Oxodiperoxymolybdenum(pyridine)-(hexamethylphosphoric triamide), **5,** 269; **7,** 136; **9,** 197; **10,** 169; **11,** 382
Oxygen, **6,** 426; **8,** 366; **9,** 335; **10,** 293; **11,** 384
Oxygen, singlet, **11,** 385
Ozone, **7,** 269
Peracetic acid, **1,** 785
Perchloryl fluoride, **10,** 304
Platinum oxide, **2,** 332
Potassium *t*-butoxide, **1,** 911
Potassium fluoride–Oxygen, **11,** 434
Potassium peroxomonosulfate, **11,** 442
Sodium methoxide, **10,** 366
Thallium(III) nitrate, **4,** 492
p-Toluenesulfonic acid, **1,** 1172
Triethyl phosphite, **2,** 432
HYDROXYMETHYLATION (*see* HYDROXYALKYLATION)
HYDROZIRCONATION (*see* HYDROMETALLATION)

INVERSION REACTIONS
OF ROH
Benzeneboronic acid, **11,** 408
Cesium propionate, **11,** 118; **12,** 109
3-Ethyl-2-fluorobenzothiazolium tetrafluoroborate, **8,** 223
Potassium hydroxide, **5,** 557
Potassium nitrite, **10,** 329
Potassium superoxide, **6,** 488; **11,** 442
Raney nickel, **9,** 405
Sulfene, **11,** 494
Tetraalkylammonium salts, **1,** 1137; **2,** 397; **3,** 277; **4,** 478
Triphenylphosphine–Diethyl azodicarboxylate, **5,** 727
Zinc *p*-toluenesulfonate, **11,** 605

OF RNH_2
Ammonium tetrafluoroborate, **5,** 17
IODINATION (*see also* ADDITION REACTIONS, ASYMMETRIC REACTIONS)
OF RCHO, R_2CO AND RELATED COMPOUNDS
1,3-Diiodo-5,5-dimethylhydantoin, **1,** 258
Iodine, **1,** 495; **4,** 258
Iodine + co-reagent, **9,** 249; **10,** 211; **12,** 256
N-Iodosuccinimide, **1,** 510
Lead tetraacetate–Metal halides, **11,** 283
Magnesium bromide–Hydrogen peroxide, **7,** 220
Potassium hydride, **9,** 386
Thallium(I) acetate–Iodine, **8,** 260
OF RH
t-Butyl hypoiodite, **2,** 50
AT ALLYLIC C–H (*see* ALLYLIC REACTIONS)
OF ArH, C=C–H
Bis(cyclopentadienyl)methyltitanium, **8,** 40
Butyllithium, **10,** 68
Butyllithium–Potassium *t*-butoxide, **10,** 72
1-Chloro-2-iodoethane, **9,** 106
1,3-Diiodo-5,5-dimethylhydantoin, **1,** 258
Ethylenediamine, **1,** 372
Iodine, **1,** 495; **4,** 105; **10,** 210
Iodine + co-reagent, **1,** 502, 504, 1019; **2,** 220; **3,** 163; **4,** 259; **11,** 267; **12,** 256
Iodine monochloride, **1,** 502
Mercury(II) oxide–Tetrafluoroboric acid, **12,** 306
Organotitanium reagents, **8,** 40
Thallium(I) acetate–Iodine, **7,** 359
Thallium(III) trifluoroacetate, **3,** 286; **4,** 498; **6,** 579; **7,** 365
Trifluoroacetyl hypoiodite, **1,** 1227; **10,** 419
OF RCOOH, RCOX
N-Bromosuccinimide, **6,** 74
Chlorine–Chlorosulfuric acid, **10,** 86
N-Chlorosuccinimide, **6,** 115
Iodine–Copper(II) acetate, **12,** 256
Lithium N-isopropylcyclohexylamide, **4,** 306
OF $ArN_2^+X^-$, $RN_2^+X^-$
Potassium iodide, **11,** 440
Sodium iodide, **6,** 285
OF N COMPOUNDS
t-Butyl hypoiodite, **1,** 94
IODOLACTAMIZATION, IODOLACTONIZATION (*see* HALO-)
ISOMERIZATION
OF ALKENES (*cis–trans*)
Copper(I) chloride, **4,** 109
Di-μ-carbonylhexacarbonyldicobalt, **5,** 204
Diisobutylaluminum hydride, **11,** 185
Diphenyl diselenide, **9,** 199
(Diphenylphosphine)lithium, **4,** 303; **5,** 408
Gallium oxide, **4,** 241
Hydrogen bromide–Acetic acid, **5,** 335
Iodine, **1,** 495
Iodine isocyanate, **3,** 161
Lithium diethylamide, **5,** 398
Mercury(II) acetate, **2,** 264
Palladium catalysts, **9,** 351
Palladium(II) chloride, **9,** 352
Polyphosphoric acid, **7,** 294
Potassium–Graphite, **7,** 296
Potassium selenocyanate, **6,** 487
Selenium, **1,** 990
Siloxene, **6,** 510
Sodium iodide, **7,** 338
Tellurium(IV) chloride, **10,** 377
Tetrakis(triphenylphosphine)-palladium(0), **12,** 468
Thioglycolic acid, **1,** 1153
Thiophenol, **4,** 505; **6,** 585; **7,** 367
p-Toluenesulfinic acid, **7,** 374
Trifluoroacetyl chloride, **10,** 419
Trimethylsilylpotassium, **8,** 513
Trimethyltinsodium, **9,** 500
Triphenylphosphine dihalides, **7,** 407
OF ALKENES (1- → 2-ALKENES AND RELATED REACTIONS)
Borane–Tetrahydrofuran, **11,** 69
Boron trifluoride etherate, **10,** 52
Dimethyl sulfoxide, **1,** 296
Ethylenebis(triphenylphosphine)-platinum(0), **9,** 216
Hydrochloric acid, **6,** 283
N-Lithioethylenediamine, **1,** 567; **2,** 239
Methyl fluorosulfonate, **5,** 445
Oxygen, singlet, **8,** 367

ISOMERIZATION (*Continued*)
Palladium(II) chloride, **6,** 447
Potassium amide, **5,** 543
Potassium 3-aminopropylamide, **12,** 400
Potassium *t*-butoxide, **3,** 233
Rhodium(III) chloride, **3,** 242; **8,** 435
Sodium, **1,** 1022
Sulfur dioxide, **8,** 464
Tri-μ-carbonylhexacarbonyldiiron, **6,** 195
OF ALKYNES, ALLENES
Boron trifluoride–Hydrogen fluoride, **4,** 252
Butyllithium–Potassium *t*-butoxide, **11,** 103
Copper(I) chloride, **2,** 91; **8,** 119
1,3-Dichloro-2-butene, **2,** 111
Ethylenediamine, **4,** 231
Oxygen, **5,** 482
Potassium 3-aminopropylamide, **6,** 476; **7,** 296; **8,** 406; **11,** 431
Potassium *t*-butoxide, **1,** 911; **2,** 336; **4,** 399; **11,** 432
Potassium–Graphite, **7,** 296
Pyridinium *p*-toluenesulfonate, **12,** 420
Silver perchlorate, **2,** 369
Silver(I) trifluoroacetate, **10,** 355
Sodium methylsulfinylmethylide, **3,** 123
Tetrafluoroboric acid, **4,** 252
OF ALLYLIC ALCOHOLS → RCHO, R_2CO
Butyllithium, **5,** 80
Diisobutylaluminum hydride, **12,** 191
Iridium catalysts, **8,** 135
Ruthenium(III) chloride, **10,** 343
Tetracarbonyldi-μ-chlorodirhodium, **10,** 382
p-Toluenesulfonic acid, **1,** 1172
Trimethylsilylmethylmagnesium chloride, **12,** 540
Tris(aquo)hexa-μ-acetato-μ_3-oxotriruthenium(III,III,III) acetate, **6,** 425, 650
OF DIENES
Chlorotris(triphenylphosphine)-rhodium(I), **2,** 448
Chromium carbonyl, **9,** 117
Dichloromaleic anhydride, **4,** 135; **5,** 195
Diethyl azodicarboxylate, **11,** 177
Dimethyl sulfoxide, **1,** 296
N-Lithioethylenediamine, **1,** 567
Lithium–Alkylamines, **1,** 574
Lithium diethylamide, **1,** 610
Phenylpotassium, **1,** 848
Potassium *t*-butoxide, **1,** 911
Rhodium(III) chloride, **8,** 435
Sulfur dioxide, **1,** 1122
OF EPOXIDES → ALLYLIC ALCOHOLS
Aluminum isopropoxide, **8,** 15; **9,** 14; **11,** 29
Bromomagnesium isopropylcyclo-hexylamide, **10,** 115
t-Butyldimethyliodosilane, **10,** 63
Diethylaluminum 2,2,6,6-tetramethylpiperidide, **6,** 181; **8,** 164
Diisobutylaluminum hydride, **1,** 260
Hydrobromic acid, **4,** 249
Iodotrimethylsilane, **10,** 216
Iron carbonyl, **6,** 304
Lithium diethylamide, **2,** 247; **4,** 298; **7,** 201
Lithium diisopropylamide, **6,** 334; **11,** 296
Lithium L-α-,α′-dimethyldibenzyl-amide, **10,** 245
Lithium phosphate, **4,** 309; **6,** 334
Perbenzoic acid, **1,** 791
Phenyl trimethylsilyl selenide, **9,** 496
(S)-(+)-2-(1-Pyrrolidinyl)-methylpyrrolidine, **12,** 421
Sodium benzeneselenoate, **5,** 272
Sodium hydride, **4,** 455
Tetrakis(triphenylphosphine)-palladium(0), **9,** 451
Titanium(IV) isopropoxide, **10,** 404; **12,** 504
Trifluoroacetic acid, **2,** 433
Trimethylsilyl trifluoromethane-sulfonate, **9,** 497
Tris(phenylseleno)borane, **10,** 454
OF EPOXIDES → C=O's
Aluminum isopropoxide, **11,** 29
Boron trifluoride, **1,** 68
Boron trifluoride etherate, **1,** 70; **2,** 35; **5,** 52
t-Butyl hydroperoxide–Molybdenum carbonyl, **11,** 95
Chloroacetonitrile, **6,** 102
Diazo(trimethylsilyl)methyllithium, **7,** 216
Di-μ-carbonylhexacarbonyldicobalt, **1,** 224

Dimethyl sulfoxide–Sulfur trioxide, **2,** 165
Lithium bromide, **4,** 297; **5,** 395
Lithium diethylamide, **1,** 610
Magnesium bromide etherate, **1,** 629
Magnesium iodide, **6,** 353
Molybdenum carbonyl, **7,** 247
o-Nitrophenyl selenocyanate, **10,** 278
Perchloric acid, **5,** 506
Sodium hydride, **4,** 455
Sulfuric acid, **6,** 558; **7,** 347
Tetrakis(triphenylphosphine)-palladium(0), **10,** 384
Tin(IV) chloride, **5,** 627
Tributylphosphine oxide, **1,** 1192
OF α,β- → β,γ-UNSATURATED C=O's
Lithium diisopropylamide, **5,** 406; **12,** 277
Lithium hexamethyldisilazide, **5,** 393
Lithium N-isopropylcyclohexylamide, **4,** 306
Potassium *t*-butoxide, **1,** 911; **5,** 544
Potassium hexamethyldisilazide, **12,** 407
Pyridine, **12,** 416
OF β,γ- → α,β-UNSATURATED C=O's
Alumina, **11,** 22
Lithium–Ammonia, **1,** 601
Lithium diisopropylamide, **11,** 296
2-Methoxypropene, **2,** 230
Molybdenum carbonyl, **7,** 247
Oxalic acid, **1,** 764
2,2,2-Trifluoroethylamine, **6,** 617

JAPP–KLINGEMANN REACTION
Arenediazonium tetrahaloborates, **8,** 22
JONES OXIDATION (*see* REAGENT INDEX)

KETALIZATION (*see* SYNTHESIS INDEX)
KNOEVENAGEL CONDENSATION
β-Alanine, **1,** 16; **5,** 6; **6,** 10
Alumina, **11,** 22
Ammonium acetate, **1,** 38
Boron trioxide, **3,** 33
t-Butyl acetoacetate, **1,** 83
t-Butyl cyanoacetate, **1,** 87
Cesium fluoride, **1,** 121
1,8-Diazabicyclo[5.4.0]undecene-7, **6,** 158
N,N-Dimethylformamide, **5,** 247
Dimethyl sulfoxide, **1,** 296
Ethyl 2-phenylsulfinylacetate, **10,** 183
Glycine, **1,** 412
Ion-exchange resins, **1,** 511
Meldrum's acid, **1,** 526; **8,** 313
Molecular sieves, **11,** 350
Palladium(II) chloride, **6,** 447
Phase-transfer catalysts, **9,** 356
α-(Phenylsulfinyl)acetonitrile, **12,** 391
Piperidine, **1,** 886; **7,** 293
Potassium carbonate, **9,** 382
Potassium fluoride, **1,** 933
Potassium tetracarbonylhydridoferrate, **6,** 483
Pyridine, **1,** 958
Rubidium fluoride, **1,** 983
Sodium acetate, **1,** 1024
Titanium(IV) chloride, **3,** 291; **4,** 507; **6,** 590; **8,** 483
Triethanolamine, **1,** 1196
KOCH–HAAF CARBOXYLATION
Formic acid, **1,** 404; **5,** 316
Trifluoromethanesulfonic acid, **9,** 485
KOENIGS–KNORR SYNTHESIS
Bromine, **12,** 70
Cadmium carbonate, **4,** 67
Mercury(II) oxide, **1,** 655
Mercury(II) oxide–Mercury(II) bromide, **6,** 360
Nitromethane, **1,** 739
Silver carbonate, **6,** 511
Silver carbonate–Celite, **6,** 511
Tetraethylammonium bromide, **6,** 568
2-Trimethylsilylethanol, **11,** 574
KOLBE REACTION
N,N-Dimethylformamide, **1,** 278
Sodium methoxide, **1,** 1091
KRONKE REACTION
N,N-Dimethyl-4-nitrosoaniline, **1,** 746
Pyridine, **1,** 958

LACTAMIZATION
Aceto(carbonyl)cyclopentadienyl-(triphenylphosphine)iron, **12,** 1
Ammonium acetate, **7,** 11
Catecholborane, **9,** 97
B-Chlorocatecholborane, **9,** 98
Dibutyltin oxide, **10,** 123; **12,** 160
Dicyclohexylcarbodiimide, **1,** 231
Dowtherm A, **1,** 353
Grignard reagents, **10,** 189

LACTAMIZATION (*Continued*)
2-Halopyridinium salts, **12,** 116
Hexamethyldisilazane, **9,** 234
2-Mesitylenesulfonyl chloride, **8,** 318
Methoxymethylbis(trimethylsilyl)amine, **12,** 62
o-Nitrophenyl thiocyanate, **9,** 325
Phase-transfer catalysts, **10,** 305
4-Pyrrolidinopyridine, **10,** 336
Raney nickel, **1,** 723
p-Tolylsulfinylacetic acid, **10,** 405
Triphenylphosphine–2,2′–Dipyridyl disulfide, **10,** 449
Tris(dimethylamino)borane, **11,** 590

LACTONIZATION (*see also* ASYMMETRIC REACTIONS, ENOL LACTONIZATION)
Alkylaluminum halides, **12,** 5
Benzenesulfonyl chloride, **5,** 22
Bis(1,5-cyclooctadiene)nickel(0), **9,** 45
N,N-Bis(2-oxo-3-oxazolidinyl)-phosphorodiamidic chloride, **11,** 57
Borane–Dimethyl sulfide, **10,** 49
Boron trifluoride etherate, **7,** 31; **10,** 52
t-Butyl dilithioacetoacetate, **8,** 60
Butyllithium–Tetramethyl-ethylenediamine, **10,** 73
(R)-(+)-*t*-Butyl (*p*-tolylsulfinyl)acetate, **11,** 106
N,N′-Carbonyldiimidazole, **5,** 97
Cesium carbonate, **9,** 99; **11,** 114
Chloromethyl methyl sulfide, **12,** 117
Copper(I) iodide, **11,** 141
Copper(I) trifluoromethanesulfonate, **10,** 108
Cyanotrimethylsilane–Triethylaluminum, **10,** 113
Cyanuric chloride, **10,** 114
1-Cyclohexyl-3-(2-morpholinoethyl)-carbodiimide metho-*p*-toluene-sulfonate, **11,** 151
Dialkylboryl trifluoromethanesulfonates, **12,** 155
4,4′-Di(2-amino-6-methylpyrimidyl) disulfide, **9,** 413
Dibutyltin oxide, **10,** 123; **12,** 160
2,3-Dichloro-5,6-dicyano-1,4-benzoquinone, **1,** 215
2,3-Dichloro-1-propene, **8,** 158
Dicyclohexylcarbodiimide, **1,** 231
Diethylaluminum ethoxyacetylide, **7,** 103
Diethylcarbo-*t*-butoxymethylaluminum, **7,** 103
Diethyl succinate, **8,** 296
2,2′-Di(1-isopropyl-4-*t*-butylimidazolyl) disulfide, **7,** 141
4-Dimethylamino-3-butyn-2-one, **12,** 198
Diphenyl phosphorochloridate, **11,** 223
2,2′-Dipyridyl disulfide, **5,** 285; **6,** 246
Ethyl lithioacetate, **6,** 255; **8,** 225
Formaldehyde, **7,** 158
2-Halopyridinium salts, **8,** 95; **9,** 234; **12,** 116
Iodotrimethylsilane, **12,** 259
Ion-exchange resins, **6,** 302
Iron, **5,** 357
Lithium aluminum hydride–(−)-N-Methylephedrine-3,5-Dimethyl-phenol, **10,** 265
Lithium (1,1-diethoxy-2-propenyl)(3,3-dimethyl-1-butynyl)cuprate, **8,** 300
Lithium (3,3-diethoxy-1-propen-2-yl)-phenylthiocuprate, **7,** 212
Lithium α-lithioacetate, **8,** 175
Lithium β-lithiopropionate, **8,** 303
Lithium naphthalenide, **6,** 415
2-Mesitylenesulfonyl chloride, **6,** 625
(4R)-Methoxycarbonyl-1,3-thiazolidine-2-thione, **11,** 323
(4S,5S)-4-Methoxymethyl-2-methyl-5-phenyl-2-oxazoline, **6,** 386; **7,** 229; **9,** 312
(γ-Methoxypropyl)-α-phenylethylamine, **12,** 318
Oxygen, singlet, **10,** 294; **12,** 363
Ozone, **7,** 269
Perchloric acid, **7,** 279
1-Phenyl-2-tetrazoline-5-thione, **11,** 420
2-Phenylthiopropionic acid, **6,** 466
Phosgene, **2,** 328
B-3-Pinanyl-9-borabicyclo[3.3.1]nonane, **11,** 429
Potassium carbonate, **6,** 479
Potassium hydride, **12,** 407
Potassium tri-*sec*-butylborohydride, **6,** 490
Pyridinium chlorochromate, **12,** 417
Silica, **12,** 431
Silver(II) oxide, **9,** 99
Silver perchlorate, **6,** 519; **9,** 413
Silver tetrafluoroborate, **6,** 519
Silver(I) trifluoroacetate, **8,** 444

Tetrabutylammonium fluoride, **12,** 458
N,N,N′,N′-Tetramethylchloroformamidinium chloride, **12,** 477
Thallium(I) 2-methyl-2-propanethiolate, **6,** 581
Thionyl chloride, **5,** 663
p-Toluenesulfonic acid, **1,** 1172
p-Tolylsulfinylacetic acid, **10,** 405
Tributylcrotyltin, **11,** 143; **12,** 513
2,4,6-Trichlorobenzoyl chloride, **9,** 478
Triethyl orthoacetate, **6,** 607
Trimethyl orthoformate, **3,** 313
2,4,4-Trimethyl-2-oxazoline, **5,** 714
Triphenylcarbenium tetrafluoroborate, **9,** 512
Triphenylphosphine + co-reagent, **5,** 285; **6,** 246; **7,** 404; **8,** 215, 517; **9,** 167; **12,** 552
Zinc-copper couple, **1,** 1292; **7,** 428

LEMIEUX–JOHNSON, LEMIEUX–RUDLOFF OXIDATIONS (*see* OXIDATIVE CLEAVAGE)

LEUCKART–WALLACH REACTION
N,N-Dimethylformamide, **2,** 153

LITHIATION (*see* METALLATION)

LOSSEN REARRANGEMENT
Triphenylphosphine–Diethyl azodicarboxylate, **5,** 727

MALONIC ESTER SYNTHESIS
Dibromodifluoromethane, **12,** 158
Sodium ethoxide, **1,** 1065
Tetrahydrofuran, **1,** 1140

MANNICH REACTION (*see also* AMINOALKYLATION)
Bis(dimethylamino)methane, **3,** 21; **7,** 359
Chloroacetaldehyde, **5,** 106
Dichloromethylenedimethylammonium chloride, **5,** 195
Dimethyl(methylene)ammonium salts, **3,** 114; **4,** 186; **7,** 130; **8,** 194; **10,** 160
Dimethyl oxomalonate, **11,** 207
Formaldehyde, **1,** 397; **5,** 312; **8,** 231; **9,** 224
Glyoxylic acid, **5,** 320
Iodotrimethylsilane, **11,** 271
Magnesium methyl carbonate, **3,** 190
Methoxymethylbis(trimethylsilyl)amine, **12,** 62
Methylenetriphenylphosphorane, **8,** 339
Methyl vinyl ketone, **1,** 697
Oxalic acid, **1,** 764
Titanium(IV) chloride, **10,** 401; **12,** 494
Tri-μ-carbonylhexacarbonyldiiron, **5,** 221
Trifluoroacetic anhydride, **3,** 308

MARSCHALK REACTION
Benzeneboronic acid, **9,** 23
Benzyltrimethylammonium hydroxide, **12,** 48
Boron trichloride, **9,** 62
Glyoxylic acid, **9,** 228
Piperidinium acetate, **9,** 376
Sodium dithionite, **8,** 456; **10,** 363

MCFADYEN–STEVENS REDUCTION
p-Toluenesulfonylhydrazide, **5,** 678
2,4,6-Triisopropylbenzenesulfonylhydrazide, **10,** 422

MEERWEIN ARYLATION
p-Aminoacetophenone, **4,** 18
Copper(I) bromide, **2,** 90

MEERWEIN–PONNDORF–VERLEY REDUCTION
Aluminum *t*-butoxide, **2,** 21
Aluminum isopropoxide, **1,** 35; **4,** 15

MERCURATION (*see* METALLATION)

MESYLATION (*see* SULFONYLATION, SYNTHESIS INDEX—SULFONIC ACID ESTERS)

METALLATION OF C–H
LITHIATION
1-(Alkylthio)vinyllithium, **5,** 6
2-Amino-2-methyl-1-propanol, **6,** 20
Arene(tricarbonyl)chromium complexes, **10,** 13
Butyllithium, **1,** 95; **2,** 51; **4,** 60; **5,** 80; **7,** 45; **8,** 67; **9,** 87; **10,** 68, 72
sec-Butyllithium, **5,** 78; **10,** 75; **12,** 97
Chloromethyl methyl ether, **6,** 109
Cryptates, **5,** 156
1,4-Diazabicyclo[2.2.2]octane, **1,** 1203; **2,** 99
Ethyllithium, **5,** 306
Hexamethylphosphoric triamide, **8,** 240
Lithium dicyclohexylamide, **5,** 397
Lithium diisopropylamide, **4,** 298; **5,** 400; **6,** 334
Lithium naphthalenide, **3,** 208
Lithium 2,2,6,6-tetramethylpiperidide, **5,** 417; **6,** 345
Organolithium reagents, **11,** 13

METALLATION OF C–H (*Continued*)
Phenylthiomethyllithium, **2,** 324
Potassium *t*-butoxide, **2,** 336
Sodium methylsulfinylmethylide, **2,** 166
Tetrahydrofuran, **5,** 649
N,N,N′,N′-Tetramethylethylenediamine, **2,** 403; **4,** 485
MERCURATION
Borane–Tetrahydrofuran, **11,** 156
Mercury(II) acetate, **1,** 644
Tri-μ-carbonylhexacarbonyldiiron, **6,** 195
Trifluoroacetic acid, **1,** 1219
BY POTASSIUM
Butyllithium–Potassium *t*-butoxide, **5,** 552
Crown ethers, **8,** 128
Cryptates, **5,** 156
Fluorodimethoxyborane, **6,** 261
Phenylpotassium, **1,** 848
Potassium, **1,** 905
Potassium amide, **5,** 543
Potassium diisopropylamide, **10,** 324
Potassium–Graphite, **8,** 405
Potassium hydride, **5,** 557; **6,** 482; **7,** 301
Trimethylsilylmethylpotassium, **6,** 637; **7,** 400
BY SODIUM
Benzyltriethylammonium chloride, **6,** 41
Copper, **1,** 157
Cryptates, **5,** 156
Sodium hexamethyldisilazide, **3,** 261
Sodium methylsulfinylmethylide, **2,** 166
Sodium naphthalenide, **1,** 711
N,N,N′,N′-Tetramethylethylenediamine, **5,** 652; **7,** 358
Trimethylsilylmethylpotassium, **7,** 400
THALLATION
Palladium(II) chloride, **10,** 300
Thallium(I) ethoxide, **2,** 407
Thallium(I) hydroxide, **6,** 578
Thallium(III) trifluoroacetate, **3,** 286; **4,** 498; **5,** 658; **6,** 579; **7,** 365; **8,** 478; **9,** 462; **12,** 481
METATHESIS
Tungsten carbonyl, **6,** 658
Tungsten(VI) chloride, **4,** 569; **5,** 742
Tungsten(VI) chloride–Tetramethyltin, **7,** 415; **10,** 455
METHOXYLATION (*see also* ALKOXYLATION)
t-Butyl hypochlorite, **4,** 58
2,3-Dichloro-5,6-dicyano-1,4-benzoquinone, **6,** 168
N,N-Dimethylhydrazine, **7,** 126
Methanol–Sodium tetraborate, **11,** 322
Methyl methanethiosulfonate, **5,** 454
Phenyliodine(III) diacetate, **11,** 411
Phenyl isocyanide, **10,** 309
Silver(I) oxide, **5,** 583
Tris(tetrabutylammonium)-hexacyanoferrate(III), **6,** 656
METHYLATION (*see also* ALKYLATION)
OF N–H (*see also* SYNTHESIS INDEX—AMINES)
Chloromethyl methyl sulfide, **6,** 109
Dimethylchloronium chloropentafluoroantimonate, **8,** 186
Dimethyl sulfate, **1,** 293
Dimethylsulfonium methylide, **4,** 196
Dimethylsulfoxonium methylide, **1,** 315; **3,** 125
Methyl fluorosulfonate, **3,** 202; **5,** 445; **6,** 381; **7,** 240
Methyl iodide, **7,** 242
Methyl iodide–Potassium bicarbonate, **7,** 242
Methyl trifluoromethanesulfonate, **12,** 329
Silver perchlorate, **2,** 369
Sodium hydride, **2,** 382
Tetrabutylammonium fluoride, **9,** 444
Triethyl orthoformate, **5,** 690
Trimethyloxonium tetrafluoroborate, **4,** 541
Trimethylselenonium hydroxide, **9,** 491
Trimethylsulfoxonium hydroxide, **8,** 509
Trimethylsulfoxonium iodide, **1,** 1236
OF O–H (*see also* SYNTHESIS INDEX—ESTERS, ETHERS)
Benzenesulfonyl chloride, **1,** 46
2,6-Di-*t*-butyl-4-methylpyridine, **9,** 141
Dimethyl sulfate + co-reagent, **1,** 295
Dimethyl sulfoxide, **1,** 296
O-Methyldibenzofuranium tetrafluoroborate, **4,** 336
Methyl fluoride–Antimony(V) fluoride, **3,** 201
Methyl fluorosulfonate, **3,** 202; **4,** 339; **6,** 381; **8,** 340
Methyl trifluoromethanesulfonate, **6,** 406

Phase-transfer catalysts, **10,** 305
Potassium methylsulfinylmethylide, **10,** 329
Sodium methylsulfinylmethylide, **2,** 166
p-Toluenesulfonyl chloride, **1,** 1179
Trimethyloxonium tetrafluoroborate, **1,** 1232
Trimethyloxonium 2,4,6-trinitrobenzenesulfonate, **1,** 1232
OF S–H
Dialkoxycarbenium tetrafluoroborates, **4,** 114
Diazomethane, **9,** 135
Dimethylformamide dimethyl acetal, **4,** 184
Dimethylsulfoxonium methylide, **5,** 254
Methyl fluoride–Antimony(V) fluoride, **3,** 201
Methyl fluorosulfonate, **6,** 381
Pentamethoxyphosphorane, **9,** 353
Trimethyloxonium tetrafluoroborate, **6,** 630
Trimethylselenonium hydroxide, **9,** 491
Trimethylsulfonium iodide, **1,** 1236
METHYLENATION (*see also* SYNTHESIS INDEX—THREE-MEMBERED RINGS)
AT ACTIVE C–H
Benzyl bromomethyl sulfide, **5,** 25
Bis(dimethylamino)methane, **7,** 359
Bis(dimethylamino)methoxymethane, **10,** 34
Bromine, **11,** 75
Bromomethanesulfonyl bromide, **12,** 75
Bromomethyl methyl ether, **10,** 56
Carbon dioxide, **5,** 93
Chloromethyl phenyl sulfide, **9,** 107, 521
1,5-Diazabicyclo[4.3.0]nonene-5, **6,** 157
Dibenzoyl peroxide, **4,** 122
1,2-Dibromoethane, **5,** 187
Dimethoxymethane, **11,** 193
Dimethyl(methylene)ammonium trifluoroacetate, **7,** 132
Dimethyl sulfoxide, **1,** 296
Diphenyl diselenide, **5,** 272; **7,** 136
Diphenyl disulfide, **5,** 276
Ethyl formate, **12,** 226
Ethyl α-trifluoromethylsulfonyloxyacetate, **7,** 149
Formaldehyde, **5,** 312; **6,** 264; **7,** 158; **8,** 231; **9,** 224; **12,** 232
Iodomethyl methyl sulfide, **5,** 353
Lead tetraacetate, **5,** 365
Lithium diisopropylamide, **4,** 298
Magnesium methyl carbonate, **3,** 190; **5,** 420; **11,** 310
N-Methylanilinium trifluoroacetate, **8,** 331
Methyl formate, **5,** 446
N-Methylthiomethylpiperidine, **12,** 326
Potassium carbonate, **9,** 382
Simmons–Smith reagent, **9,** 415
Sodium benzeneselenoate, **10,** 368
Sodium cyanoborohydride, **5,** 607; **6,** 537
Sodium dicarbonyl(cyclopentadienyl)-ferrate, **11,** 483
Trimethylsilyl dichloroacetate, **9,** 491
Tris(methylthio)methyllithium, **7,** 412
Zinc bromide, **9,** 520
OF $C=O \rightarrow C=CH_2$
by Peterson reaction
Trimethylsilylmethyllithium, **6,** 637; **10,** 433; **11,** 581
Trimethylsilylmethylmagnesium chloride, **5,** 724; **6,** 636
Trimethylsilylmethylsodium, **5,** 617
by Wittig and related reactions
Benzyl chloride, **2,** 158
Butyllithium, **1,** 95
Dimethyl methylphosphonate, **11,** 203
Dimethyl methylphosphonothioate, **2,** 155
Ethylene oxide, **5,** 297
Formaldehyde, **7,** 158
α-Lithiomethylenetriphenylphosphorane, **11,** 338
Methylenetriphenylphosphorane, **6,** 380; **8,** 244, 339; **9,** 307; **10,** 264
Potassium–Dimethyl sulfoxide, **2,** 158
Sodium hexamethyldisilazide, **1,** 1046
Sodium methylsulfinylmethylide, **1,** 310
N,N,P-Trimethyl-P-phenylphosphinothioic amide, **11,** 569; **12,** 534
Trimethylsilylmethylenetriphenylphosphorane, **5,** 723
Trimethylsilylmethyl trifluoromethanesulfonate, **10,** 434
Triphenylphosphine, **1,** 1238
by other Methods
Alkyldimesitylboranes, **12,** 12
μ-Chlorobis(cyclopentadienyl)-

METHYLENATION (*Continued*)
(dimethylaluminum)-μ-methyl-enetitanium, **8,** 83; **10,** 87; **12,** 110
Diazomethane, **5,** 569
Dibromomethane–Zinc–Titanium(IV) chloride, **11,** 337; **12,** 322
Dicyclopentadienyltitanium methylene–Zinc iodide complex, **12,** 506
Diiodomethane–Zinc–Trimethylaluminum, **8,** 339
4,4-Dimethyl-2-methylthio-2-oxazoline, **6,** 584
N,S-Dimethyl-S-phenylsulfoximine, **6,** 395
N-Methanesulfinyl-*p*-toluidine, **2,** 269
B-Methyl-9-borabicyclo[3.3.1]nonane, **4,** 310
Methylenemagnesium bromide, **2,** 273; **3,** 189
N-Methylphenylsulfonimidoylmethyllithium, **5,** 458; **11,** 343
Organomolybdenum reagents, **12,** 352
Phenylselenomethyllithium, **6,** 549
Phenylthiomethyllithium, **4,** 379; **5,** 527; **6,** 596
Simmons–Smith reagent, **3,** 255
Sodium methylsulfinylmethylide, **4,** 195
Titanium(0), **11,** 526
Titanium(IV) chloride–Lithium aluminum hydride, **6,** 596
Trimethyl(or -phenyl)stannylmethyllithium, **9,** 509; **12,** 547
OF RCOOR → ENOL ETHERS
μ-Chlorobis(cyclopentadienyl)-(dimethylaluminum)-μ-methyl-enetitanium, **10,** 87; **11,** 52
METHYLTHIOMETHYLATION (*see* ALKYLTHIOALKYLATION)
MICHAEL REACTION (*see* CONJUGATE ADDITION)
MITSUNOBU REACTION (*see* REAGENT INDEX)
MUKAIYAMA ALDOL REACTION (*see* ALDOL REACTIONS)

NAZAROV AND RELATED REACTIONS
Aluminum chloride, **12,** 26
Benzylchlorobis(triphenylphosphine)-palladium(II), **12,** 44
Boron trifluoride etherate, **12,** 66
(2-Bromovinyl)trimethylsilane, **11,** 82
Chlorotrimethylsilane, **10,** 96
gem-Dichloroallyllithium, **8,** 150
Iodotrimethylsilane, **11,** 271
Manganese(IV) oxide, **1,** 637
1-Phenylthio-1-trimethylsilylethylene, **11,** 583
Phosphoric acid–Formic acid, **1,** 860; **5,** 534; **10,** 317
Propargyl alcohol, **9,** 394
Sodium tetrachloroaluminate, **1,** 1027
Tetrakis(triphenylphosphine)-palladium(0), **12,** 468
1-Trimethylsilyl-2-trimethylstannyl-ethylene, **12,** 469
Vinyltrimethylsilane, **9,** 498; **10,** 444; **12,** 26
NEBER REARRANGEMENT
O-Mesitylenesulfonylhydroxylamine, **5,** 430
NEF REACTION (*see also* OXIDATION OF NITRO GROUPS)
Mercury(II) nitrite, **9,** 292
Nitroethylene, **5,** 476
Potassium permanganate, **11,** 440
Sodium methoxide, **6,** 545
NENCKI REACTION
Zinc chloride, **1,** 1289
NITRATION
OF RH
Acetone cyanohydrin nitrate, **1,** 5
Isoamyl nitrate, **1,** 40; **2,** 25
Lithium diisopropylamide, **3,** 184
Methyl nitrate, **1,** 691
Nitryl chloride, **1,** 756
Polyphosphoric acid, **1,** 894
Potassium amide, **2,** 336
Potassium *t*-butoxide, **1,** 911
Sodium hydride, **1,** 1075
Trifluoroacetyl nitrate, **12,** 530
OF C=C–H
Benzeneselenenyl bromide–Silver nitrate, **11,** 33
Mercury(II) nitrite, **9,** 292
Nitrogen dioxide, **1,** 324; **2,** 175; **3,** 130
Nitrogen dioxide–Iodine, **3,** 130
Nitrosyl chloride, **1,** 748; **2,** 298
Nitryl iodide, **1,** 757
Silver nitrite–Mercury(II) chloride, **11,** 467
Tetranitromethane, **10,** 392

OF ArH
Benzoyl nitrate, **5,** 25
Copper(II) bromide–Nitric oxide, **2,** 87
Copper(II) nitrate–Acetic anhydride, **1,** 163
Lanthanum nitrate, **11,** 283
Methyl nitrate, **5,** 51
Nafion-H, **9,** 320
Nitric acid, **3,** 212; **6,** 284
Nitrogen dioxide–Boron trifluoride, **1,** 329
Nitrogen oxides, **1,** 322; **4,** 202
Nitronium tetrafluoroborate, **1,** 742, 1144; **2,** 296; **5,** 476
Nitronium trifluoromethanesulfonate, **5,** 477; **6,** 420
Nitryl chloride, **1,** 1018
Palladium(II) acetate, **5,** 496
Sodium nitrate, **7,** 393
Tetranitromethane, **1,** 1147
Thallium(III) trifluoroacetate, **6,** 579
Trifluoroacetic acid, **1,** 1219
Trifluoroacetyl nitrate, **12,** 530
OF N COMPOUNDS
Nitrogen oxides, **1,** 322
Trifluoroacetic anhydride, **1,** 1221
NITROALDOL REACTION (*see* ALDOL-TYPE REACTIONS)
NITROMERCURATION (*see* ADDITION REACTIONS)
NITROSATION
OF RH, RNH_2
2-Hydroperoxy-2-nitropropane, **4,** 256
Nitrogen oxides, **1,** 324; **2,** 175; **7,** 252
3-Nitro-N-nitrosocarbazole, **1,** 741
Nitrosonium tetrafluoroborate, **1,** 747
Nitrosyl chloride, **1,** 748; **7,** 254
Nitrosylsulfuric acid, **2,** 299
Sodium dithionite, **1,** 1081
Sodium nitrite, **1,** 1097
Thallium(III) trifluoroacetate, **5,** 658
OF $RCH_2Z \rightarrow RZC{=}NOH$
Butyl nitrite, **1,** 97
Isoamyl nitrite, **1,** 40
Methyl nitrite, **1,** 691
Nitrosyl chloride, **1,** 748; **6,** 422
Sodium nitrite, **1,** 1097
NITROSELENYLATION (*see* ADDITION REACTIONS)

OPPENAUER OXIDATION
Aluminum *t*-butoxide, **1,** 23
Aluminum isopropoxide, **1,** 35
1,4-Benzoquinone, **1,** 49
2-(Diethoxyphosphinyl)propionitrile, **7,** 106
Potassium *t*-butoxide, **1,** 911; **9,** 382
Rochelle salt, **1,** 983
ORGANOMETALLIC REAGENTS AND THEIR REACTIONS (*see also* ASYMMETRIC REACTIONS, CARBONYLATION, REFORMATSKY REACTION)
WITH ACETALS, THIOACETALS
Allyltributyltin, **12,** 21
Allyltrimethylsilane, **11,** 16
1,4-Bis(trimethylsilyl)-2-butyne, **11,** 64
Boron trifluoride etherate, **10,** 52
Grignard reagents, **11,** 245
Iodotrimethylsilane, **10,** 216
3-Methyl-3-trimethylsilyl-1-butene, **8,** 181
Organocopper reagents, **12,** 345
Titanium(IV) chloride, **7,** 370; **11,** 529
2-Trimethylsilylmethyl-1,3-butadiene, **12,** 539
Trimethylsilyl trifluoromethane-sulfonate, **10,** 438
WITH RCOOH, RCOX
RCOCl
Aluminum chloride, **7,** 7; **12,** 26
Benzylchlorobis(triphenylphosphine)-palladium(II), **8,** 35; **9,** 41; **12,** 44
Bis(trimethylsilyl)acetylene, **5,** 44
2-(2-Bromoethyl)-1,3-dioxane, **7,** 37
(2-Bromovinyl)trimethylsilane, **11,** 82
Butyllithium, **6,** 85
Carbonylchlorobis(triphenyl-phosphine)rhodium(I), **6,** 105
Carbonylphenylbis(triphenyl-phosphine)rhodium(I), **5,** 45
μ-Chlorobis(cyclopentadienyl)-(dimethylaluminum)-μ-methyl-enetitanium, **12,** 110
Chlorobis(cyclopentadienyl)-hydridozirconium(IV), **8,** 84
Chlorotrimethylsilane, **10,** 96
Copper, **12,** 140
Copper(I) bromide–Dimethyl sulfide, **8,** 117
Copper(I) trimethylsilylacetylide, **6,** 148

ORGANOMETALLIC REAGENTS AND THEIR REACTIONS *(Continued)*
Dicarbonylcyclopentadienylcobalt, **7,** 84
Dichloromethyllithium, **6,** 170
Dimethylphenylsilyllithium, **10,** 162
Grignard reagents, **1,** 415; **8,** 235; **9,** 229; **10,** 189
Iodo(methyl)calcium, **5,** 442
Iron(III) acetylacetonate, **12,** 557
Lithium *t*-butyl(phenylthio)cuprate, **5,** 414; **7,** 211
Lithium dibutylcuprate, **4,** 127
Lithium dimethylcuprate, **4,** 177
Lithium iodide, **5,** 410
Lithium methyl(vinyl)cuprate, **6,** 342
Manganese(II) iodide, **8,** 312; **9,** 289; **10,** 290
Mercury(II) acetate, **7,** 222
4-Methoxy-3-buten-1-ynylcopper, **9,** 297
Methylcopper, **4,** 334
Organocopper reagents, **9,** 328; **11,** 365
Organolithium reagents, **12,** 350
Organotitanium reagents, **12,** 110
Organozinc reagents, **3,** 128
Phenylthiocopper, **6,** 465
1-Phenylthio-1-trimethylsilylethylene, **11,** 583
Silver tetrafluoroborate, **12,** 434
Tetrakis(triphenylphosphine)-palladium(0), **6,** 571; **11,** 503; **12,** 468
Trifluorovinyllithium, **6,** 622
3-Trimethylsilyl-1-cyclopentene, **8,** 509
2-Trimethylsilylmethyl-1,3-butadiene, **9,** 493; **12,** 539
Trimethylsilylmethyllithium, **11,** 581
Trimethylsilyl trifluoromethane-sulfonate, **12,** 543
Vinyltrimethylsilane, **9,** 498; **10,** 444

$RCONR_2$
Bis(chlorodimethylsilyl)ethane, **12,** 179
Butyllithium, **6,** 85
1,4-Dichloro-1,4-dimethoxybutane, **12,** 175
N,N-Dimethylformamide, **1,** 278; **5,** 247; **7,** 124; **8,** 189
N,O-Dimethylhydroxylamine, **11,** 201
2-(N-Formyl-N-methyl)amino-pyridine, **8,** 341; **10,** 265
N-Formylpiperidine, **11,** 244
Grignard reagents, **1,** 415
Imidazole, **1,** 492
Lithium, **5,** 376
N,N,N′,N′-Tetramethylsuccinamide, **4,** 490

RCOOH
Aluminum chloride, **12,** 26
[1,2-Bis(diphenylphosphine)ethane]-(dichloro)nickel(II), **12,** 171
[Chloro(*p*-methoxyphenyl)methylene]-diphenylammonium chloride, **11,** 220
1-Chloro-N,N,2-trimethylpropenyl-amine, **12,** 123
Chlorotrimethylsilane, **12,** 126
Grignard reagents, **10,** 189
Methoxymethylenetriphenyl-phosphorane, **6,** 368
Methyllithium, **1,** 686; **5,** 448; **6,** 384
Trimethylaluminum, **5,** 707; **6,** 622
Trimethylsilylmethyllithium, **6,** 635
Vinyllithium, **5,** 748

RCOOR′, Lactones
Allyltrimethylsilane, **11,** 16
Cesium fluoride, **11,** 115
Chlorotrimethylsilane, **12,** 126
Copper(I) iodide, **11,** 141
gem-Dichloroallyllithium, **8,** 150
Ethylene glycol, **9,** 217
Ethyl formate, **1,** 380
Grignard reagents, **1,** 415; **7,** 163; **10,** 189; **11,** 245
Iodo(methyl)calcium, **5,** 442
2-Lithio-1,3-dithianes, **11,** 285
Lithium, **4,** 286
Lithium borohydride, **12,** 276
Lithium 2,2,6,6-tetramethylpiperidide, **6,** 345
Methyllithium, **5,** 448
Nickel chloride–Zinc, **10,** 277
Organocopper reagents, **10,** 282; **12,** 345
Sodium diisopropylamide, **1,** 1064
Sodium methylsulfinylmethylide, **2,** 166
Trimethylsilylmethyllithium, **9,** 495

other RCOX
Acetic-formic anhydride, **2,** 10

Bis(1,5-cyclooctadiene)nickel(0), **10,** 33
Dimethylaluminum methylselenolate, **12,** 197
Grignard reagents, **5,** 321; **9,** 229; **10,** 189; **11,** 245
Lithium dimethylcuprate, **5,** 234
Manganese(II) chloride, **9,** 288
3-Methoxy-3-methylbutynylcopper, **9,** 300
Organocopper reagents, **11,** 365
Organolithium reagents, **9,** 5
Triethylamine, **1,** 1198
Triphenylphosphine dihalides, **12,** 554
WITH RCHO, R_2CO (*see* ADDITION REACTIONS)
WITH RX, C=C (*see* COUPLING REACTIONS)
WITH CARBONATES, HALOFORMATES, ETC.
Ethyl chloroformate, **1,** 364; **7,** 147
Grignard reagents, **1,** 415
Iodo(methyl)calcium, **5,** 442
Methyl chloroformate, **9,** 306
WITH CO_2
Arene(tricarbonyl)chromium complexes, **10,** 13
n- or *t*-Butyllithium, **1,** 95; **6,** 109; **7,** 47
Carbon dioxide, **5,** 93; **6,** 94
Chloromethyl methyl ether, **6,** 109
Copper(I) phenylacetylide–Tributylphosphine, **6,** 130
1,2-Dibromoethane, **1,** 374
gem-Dichloroallyllithium, **8,** 150
Dichlorobis(cyclopentadienyl)titanium, **9,** 146
Diisobutylaluminum hydride, **2,** 140; **7,** 147
1,3-Dithiane, **2,** 182
Ethyl chloroformate, **7,** 147
Grignard reagents, **1,** 415
Lithium diisobutylmethylaluminum hydride, **2,** 248
Lithium naphthalenide, **8,** 305
Lithium nitride, **1,** 618
Magnesium, **7,** 218; **10,** 251; **11,** 307
Methylenemagnesium bromide, **2,** 273
Nickel(II) acetylacetonate–Trimethylaluminum, **8,** 42; **9,** 52
Phenylpotassium, **1,** 848
Sodium amalgam, **1,** 1033
Sodium amide, **1,** 1034
Sodium naphthalenide, **1,** 711
N,N,N′,N′-Tetramethylethylenediamine, **3,** 284; **5,** 652
2,4,6-Triisopropylbenzenesulfonylhydrazide, **9,** 486
Triphenylmethyllithium, **1,** 1256
Triphenylphosphine–Carbon tetrabromide, **4,** 550
Vinylcopper reagents, **6,** 662
Vinyllithium, **1,** 1273
WITH CS_2
Carbon disulfide, **5,** 94; **9,** 94
Grignard reagents, **11,** 245
Lithium diisopropylamide, **8,** 292
Organocopper reagents, **9,** 328
WITH EPOXIDES (*see* CLEAVAGE REACTIONS)
WITH C=N
B-Allyl-9-borabicyclo[3.3.1]nonane, **12,** 15
N-Benzenesulfonylformimidic acid ethyl ester, **3,** 18
Dimethyl(methylene)ammonium salts, **7,** 130; **8,** 194
Ethoxymethyleneaniline, **1,** 362
Grignard reagents, **1,** 415
6-Methoxy-7-hydroxy-3,4-dihydroisoquinolinium methiodide, **4,** 329
2,4,4,6-Tetramethyl-5,6-dihydro-1,3-(4H)-oxazine, **3,** 280; **4,** 481
p-Toluenesulfonyl chloride, **5,** 676
p-Toluenesulfonylhydrazide, **9,** 472
p-(Tolylsulfinyl)methyllithium, **5,** 682
N,4,4-Trimethyl-2-oxazolinium iodide, **4,** 540
Zinc, **5,** 753
Zinc-copper couple, **10,** 459
INTRAMOLECULAR REACTIONS
Butyllithium, **8,** 65
2-Hydroxymethyl-3-trimethylsilyl-1-propene, **11,** 258
Iodosylbenzene, **11,** 270
Lithium diisopropylamide, **8,** 292
Lithium dimethylcuprate, **7,** 120
Magnesium, **6,** 351; **11,** 307
Titanium(IV) chloride, **11,** 529
Zinc-copper couple, **7,** 428
WITH RCN
t-Butyllithium, **10,** 76
1,3-Dithiane, **2,** 182

ORGANOMETALLIC REAGENTS AND THEIR REACTIONS (*Continued*)
Grignard reagents, **1,** 415; **12,** 235
Magnesium, **6,** 351
5-Methylisoxazole, **9,** 309
Methyllithium, **8,** 342
Methyl methylthiomethyl sulfoxide, **8,** 344
Organolithium reagents, **10,** 3
p-(Tolylsulfinyl)methyllithium, **5,** 682
Trialkylaluminum, **1,** 1197; **6,** 622
4-Trimethylsilyloxyvaleronitrile, **8,** 513
Triphenylmethyl isocyanide, **6,** 642
Vinyltriphenylphosphonium bromide, **6,** 666
Zinc, **12,** 567
Zinc–Silver couple, **10,** 460
WITH O_2
Chlorobis(cyclopentadienyl)-hydridozirconium(IV), **6,** 175
Grignard reagents, **1,** 415
Lithium, **3,** 174
Sodium borohydride, **5,** 597
OXIDATION REACTIONS (*see also* ADDITION REACTIONS, ALLYLIC OXIDATION, AROMATIZATION, BAEYER–VILLIGER REACTION, DEHYDROGENATION, EPOXIDATION, OXIDATIVE ..., OZONOLYSIS)
ACETALS → RCOOR′
Ozone, **4,** 363; **6,** 436
Peracetic acid, **1,** 787
ALCOHOLS → RCHO, R_2CO
using Cr reagents
Bis(tetrabutylammonium) dichromate, **10,** 42
sec-Butyl chloroformate, **1,** 86
Chromium(VI) oxide, **1,** 144; **5,** 140; **10,** 306
Chromium(VI) oxide + co-reagent, **1,** 278; **4,** 97; **5,** 142; **7,** 69, 71; **9,** 115, 116
Chromium(VI) oxide–Pyridine, **1,** 145; **2,** 74; **4,** 96, 215; **9,** 121; **10,** 99
Chromyl chloride + co-reagent, **6,** 126; **8,** 112
Cornforth reagent, **1,** 146
Dimethyl sulfide–Trifluoroacetic anhydride, **6,** 225
Fieser reagent, **1,** 145
Jones reagent, **1,** 142; **6,** 123
Manganese(IV) oxide, **1,** 637
Molecular sieves, **10,** 273
Picolinic acid, **5,** 538
Potassium chromate, **9,** 357
Potassium dichromate, **8,** 410; **10,** 324; **12,** 405
Pyridinium chlorochromate, **6,** 498; **7,** 308; **8,** 425; **9,** 397; **10,** 334, 335; **12,** 417
Pyridinium dichromate, **3,** 239; **9,** 399; **12,** 419
Pyridinium fluorochromate, **11,** 453; **12,** 419
Sodium dichromate, **1,** 1059; **3,** 265; **4,** 95; **6,** 225; **8,** 110; **9,** 114
Tetrabutylammonium chromate, **9,** 443
Trimethyl borate, **9,** 398
using Mn reagents
Barium manganate, **8,** 21
Copper(II) permanganate, **9,** 389; **11,** 142
Manganese(IV) oxide, **1,** 637
Potassium permanganate, **1,** 942; **8,** 416; **9,** 389
Sodium permanganate, **10,** 368
using DMSO-based and related reagents
Chlorodimethylsulfonium chloride, **4,** 191
Chlorodimethylsulfoxonium chloride, **4,** 200
N-Chlorosuccinimide–Dialkyl sulfide, **4,** 87, 192; **5,** 129; **8,** 200; **12,** 195
N,N-Diethylaminopropyne, **3,** 98
Dimethyl sulfide–Trifluoroacetic anhydride, **6,** 225
Dimethyl sulfoxide–Acetic anhydride, **2,** 163; **3,** 121
Dimethyl sulfoxide–Chlorosulfonyl isocyanate, **10,** 167
Dimethyl sulfoxide–Cyanuric chloride, **5,** 266
Dimethyl sulfoxide–1-Cyclohexyl-3-(2-morpholinoethyl)carbodiimide metho-*p*-toluenesulfonate, **6,** 227
Dimethyl sulfoxide–Dicyclohexyl-carbodiimide, **1,** 304; **2,** 162; **3,** 121; **6,** 228; **8,** 199
Dimethyl sulfoxide–Diphenylketene-*p*-tolylimine, **3,** 122

Dimethyl sulfoxide–Methanesulfonic anhydride, **5,** 266
Dimethyl sulfoxide–4-Nitrobenzene-sulfonyl chloride, **4,** 193
Dimethyl sulfoxide–Oxalyl chloride, **8,** 200; **9,** 192; **10,** 167; **11,** 215
Dimethyl sulfoxide–Phosgene, **1,** 856; **6,** 226
Dimethyl sulfoxide–Sulfur trioxide, **2,** 165; **4,** 200; **7,** 159; **8,** 200; **11,** 216
Dimethyl sulfoxide–Sulfur trioxide–Pyridine, **2,** 394
Dimethyl sulfoxide–Sulfuryl chloride, **8,** 200
Dimethyl sulfoxide–Thionyl chloride, **8,** 200
Dimethyl sulfoxide–*p*-Toluenesulfonyl chloride, **5,** 266
Dimethyl sulfoxide–Trifluoroacetic anhydride, **7,** 136; **8,** 200; **10,** 166
Polymeric carbodiimide reagent, **5,** 539
Thioanisole, **6,** 582; **7,** 159

of $ArCH_2OH$
Alumina, **6,** 16
Benzeneseleninic anhydride, **9,** 32
Bis(benzyltriethylammonium) dichromate, **11,** 50
Bispyridinesilver permanganate, **11,** 61
Bis(tetrabutylammonium) dichromate, **10,** 42
Cerium(IV) ammonium nitrate, **1,** 120
Chloramine, **4,** 74
o-Chloranil, **1,** 128
Chromium(VI) oxide–Diethyl ether, **9,** 115
Copper(I) chloride, **8,** 118
1,4-Diazabicyclo[2.2.2]octane–Bromine, **8,** 141
1,3-Dibromo-5,5-dimethylhydantoin, **2,** 108
2,3-Dichloro-5,6-dicyano-1,4-benzoquinone, **1,** 215; **10,** 135
4-(Dimethylamino)pyridinium chlorochromate, **11,** 196
3,5-Dinitrobenzoyl *t*-butyl nitroxyl, **8,** 204
Ferric nitrate/K10 Bentonite, **11,** 237
Iodylbenzene, **11,** 275
Jones reagent, **6,** 123
Manganese(IV) oxide, **1,** 636; **2,** 257
Nickel peroxide, **1,** 731
Nitric acid, **5,** 474
Nitrosonium tetrafluoroborate, **7,** 253
Periodates, **4,** 373
Phenyliodine(III) bis(trifluoroacetate), **6,** 301
N-Phenyl-1,2,4-triazoline-3,5-dione, **7,** 287
Potassium dichromate, **10,** 324
Potassium hypochlorite, **1,** 938
Silver carbonate–Celite, **7,** 319
Silver(II) oxide, **2,** 369
Sodium hypochlorite, **7,** 337
Sodium persulfate, **1,** 1102
Tetrabutylammonium salts, **8,** 468; **12,** 458
2,2,6,6-Tetramethylpiperidinyl-1-oxy, **12,** 479

other Reagents
Alumina, **7,** 5
1,1′-(Azodicarbonyl)dipiperidine, **8,** 19; **9,** 22
Benzeneseleninic anhydride, **9,** 32
Bis(tributyltin) oxide, **7,** 26; **8,** 43
Bis(trimethylsilyl)peroxide, **12,** 63
Bromine + co-reagent, **5,** 60; **7,** 35
N-Bromoacetamide, **1,** 74
N-Bromosuccinimide, **1,** 78; **9,** 70
t-Butyl hydroperoxide + co-reagent, **9,** 79; **10,** 65; **12,** 89, 94
t-Butyl hypohalite, **1,** 90, 94
Calcium hypochlorite, **11,** 107
Cerium(IV) ammonium nitrate, **2,** 63; **11,** 114
Chlorine–Pyridine, **5,** 106
1-Chlorobenzotriazole, **2,** 67; **3,** 46
m-Chloroperbenzoic acid–2,2,6,6-Tetramethylpiperidine, **6,** 110
N-Chlorosuccinimide, **6,** 115
Copper(I) chloride, **8,** 118
Copper–Chromium oxide, **1,** 157
Copper(II) oxide, **4,** 108
Copper(II) permanganate, **11,** 142
1,4-Diazabicyclo[2.2.2]octane–Bromine, **8,** 141
Dibenzoyl peroxide–Nickel(II) bromide, **9,** 136
1,3-Dibromo-5,5-dimethylhydantoin, **1,** 208
Dibutyltin oxide, **9,** 141
2,3-Dichloro-5,6-dicyano-1,4-benzoquinone, **5,** 193

OXIDATION REACTIONS (*Continued*)
Dichlorotris(triphenylphosphine)-ruthenium(II), **4,** 564; **7,** 244; **10,** 141
2-(Diethoxyphosphinyl)propionitrile, **7,** 106
Diethyl azodicarboxylate, **1,** 245
3,5-Dinitrobenzoyl *t*-butyl nitroxyl, **8,** 204
Dodecacarbonyltri-*triangulo*-ruthenium, **7,** 244
Ethyl nitroacetate–Diethyl azodicarboxylate–Triphenyl-phosphine, **10,** 182
Ferric nitrate/K10 Bentonite, **11,** 237
Hydridotetrakis(triphenylphosphine)-rhodium(I), **11,** 255
Hydrogen peroxide–Ammonium heptamolybdate, **12,** 245
Iodine–Silver salts, **1,** 504
N-Iodosuccinimide–Tetrabutyl-ammonium iodide, **10,** 216
Iodosylbenzene–Ruthenium catalysts, **11,** 270
N-Lithioethylenediamine, **1,** 567
Nickel catalysts, **1,** 723; **10,** 339
Nickel peroxide, **2,** 294
Nitrogen dioxide, **1,** 324
Nitrosonium tetrafluoroborate, **6,** 226
Oxoperoxobis(N-phenylbenzo-hydroxamato)molybdenum(VI), **10,** 292
μ-Oxybis(chlorotriphenylbismuth), **9,** 335
Oxygen, singlet, **5,** 486
Ozone, **7,** 269
Palladium(II) acetate–Triphenylphosphine, **10,** 298
Palladium black, **4,** 365
Perbenzoic acid–2,2,6,6-Tetramethylpiperidine, **6,** 111
Periodinane, **12,** 378
Phenyliodine(III) bis(trifluoroacetate), **6,** 301
Phenyliodine(III) diacetate, **11,** 271
Phenyliodine(III) dichloride, **5,** 352
N-Phenyl-1,2,4-triazoline-3,5-dione, **7,** 287
Potassium *t*-butoxide, **1,** 911; **9,** 382
Potassium ferrate(VI), **4,** 405
Potassium persulfate–Silver nitrate, **2,** 348
Potassium ruthenate, **9,** 391
Pyruvyl chloride, **7,** 310
Ruthenium(III) chloride, **7,** 244; **8,** 437
Ruthenium(IV) oxide–Sodium periodate, **2,** 358
Ruthenium tetroxide, **1,** 986; **2,** 357; **3,** 243; **4,** 420
Silver carbonate–Celite, **2,** 363; **4,** 425; **5,** 577; **6,** 511; **7,** 319
Silver(II) picolinate, **3,** 16; **5,** 20
Sodium hypochlorite, **7,** 337; **10,** 365; **11,** 487
Sodium ruthenate, **5,** 622
2,2,6,6-Tetramethylpiperidinyl-1-oxy, **12,** 479
Trialkylaluminums, **11,** 539
Triethyltin methoxide, **6,** 613
Triphenylcarbenium tetrafluoroborate, **8,** 524
Urushibara catalysts, **4,** 571

ALCOHOLS → RCOOH
m-Chloroperbenzoic acid–2,2,6,6-Tetramethylpiperidine, **6,** 110
Chromium(VI) oxide–Diethyl ether, **9,** 115
Copper(II) permanganate, **11,** 142
Heyn's catalyst, **1,** 432
Iodosylbenzene–Ruthenium catalysts, **11,** 270
Nickel peroxide, **1,** 731; **2,** 294
Nitric acid, **1,** 733
Nitrogen dioxide, **1,** 324
Oxygen, singlet, **5,** 486
Platinum catalysts, **1,** 432
Potassium ruthenate, **9,** 391
Pyridinium dichromate, **9,** 399
Silver(II) oxide, **2,** 369
Silver(II) picolinate, **3,** 16
Sodium ruthenate, **5,** 622; **9,** 432
Tetrabutylammonium salts, **5,** 644; **8,** 468

ALCOHOLS → RCOOR′
t-Butyl hydroperoxide–Benzyltrimethylammonium tetrabromooxomolybdate, **12,** 89
t-Butyl hypochlorite, **1,** 90
m-Chloroperbenzoic acid–2,2,6,6-Tetramethylpiperidine, **8,** 99
Collins reagent, **12,** 139
Dibenzoyl peroxide–Nickel(II) bromide, **9,** 136

Dihydridotetrakis(triphenylphosphine)-ruthenium(II), **11,** 182
p-Nitroperbenzoic acid, **9,** 324
Sodium bromite, **12,** 445
Sodium dichromate, **1,** 1059

ALDEHYDES, KETONES → RCOX, RCN
Aluminum chloride–Ethanol, **8,** 15
N-Amino-4,6-diphenylpyridone, **7,** 10
Chloramine, **5,** 103
Dicyclohexylcarbodiimide, **5,** 206
S,S-Diphenylsulfilimine, **10,** 174
Hydroxylamine, **2,** 217; **7,** 176; **9,** 245
Hydroxylamine-O-sulfonic acid, **6,** 290; **9,** 245
N-Imino-N,N-dimethyl-2-hydroxy-propanaminium ylide, **8,** 256
Manganese(IV) oxide, **4,** 317
N-Methyl-2-pyrrolidone, **11,** 346
Nickel peroxide, **1,** 731
Nitroethane–Pyridinium chloride, **11,** 359
1-Nitropropane, **1,** 745
Selenium(IV) oxide, **9,** 409
Uranium(VI) fluoride, **7,** 417

ALDEHYDES → RCOOH
t-Butyl hydroperoxide–Benzyltrimethylammonium tetrabromooxomolybdate, **12,** 89
Calcium hypochlorite, **11,** 107
Caro's acid, **3,** 43
Cetyltrimethylammonium bromide, **6,** 70
Copper(II) permanganate, **11,** 142
1,3-Dithiane, **4,** 216
Hydrogen peroxide + co-reagent, **1,** 477; **12,** 245
2-Hydroperoxyhexafluoro-2-propanol, **9,** 244; **10,** 206
Hypohalite solution, **1,** 488
Nickel peroxide, **8,** 357
Nitric acid, **1,** 733
Peroxyacetyl nitrate, **5,** 510
Potassium hydroxide, **1,** 935
Potassium permanganate, **1,** 942; **4,** 143; **6,** 70
Potassium ruthenate, **9,** 391
Pyridinium dichromate, **9,** 399; **10,** 335
Silver(I) nitrate, **1,** 1008
Silver(I) oxide, **1,** 1011; **11,** 468
Silver(II) oxide, **2,** 369
Silver(II) picolinate, **3,** 16
Sodium chlorite, **5,** 603; **9,** 423
Sodium permanganate, **10,** 368
Tetrabutylammonium permanganate, **8,** 468
Tollens reagent, **1,** 1171

ALDEHYDES → RCOOR′, ROCOH
3-Benzylthiazolium bromide, **10,** 27
N-Bromosuccinimide, **7,** 37; **8,** 54; **9,** 70
t-Butyl hydroperoxide–Benzyltrimethyl-ammonium tetrabromooxomolybdate, **12,** 89
Caro's acid, **3,** 43
m-Chloroperbenzoic acid, **5,** 120; **11,** 124
Crown ethers, **8,** 128
2-Lithio-2-methylthio-1,3-dithiane, **4,** 217
Manganese(IV) oxide, **2,** 257; **3,** 191
Ozone, **8,** 374
Peracetic acid, **1,** 787
Sodium hypochlorite, **11,** 487

1-ALKENES → RCH_2CHO
N-Chlorosuccinimide, **7,** 368
Chromyl chloride, **2,** 79; **3,** 62
Disiamylborane, **10,** 40
Lead tetraacetate, **5,** 365
Palladium(II) chloride–Copper(I) chloride, **12,** 372
Thallium(III) nitrate, **4,** 492

1-ALKENES → $RCOCH_3$
Allyltrimethylsilane, **10,** 6
Bis(acetonitrile)chloronitro-palladium(II), **10,** 30; **11,** 45
[Bis(salicylidene-γ-iminopropyl)]-methylaminecobalt(II), **11,** 138
Hydrogen peroxide–Palladium acetate, **10,** 203
Jones reagent, **6,** 123
Mercury(II) acetate, **4,** 319
Oxotris(*t*-butylimido)osmium(VIII), **8,** 206
Palladium(or Platinum) *t*-butyl peroxide trifluoroacetate, **10,** 299; **11,** 430
Palladium(II) chloride, **10,** 300
Palladium(II) chloride + co-reagent, **7,** 278; **11,** 396; **12,** 380
Thallium(III) trifluoroacetate, **5,** 658

ALKENES → 1,2-DICARBONYLS
Potassium permanganate, **4,** 412; **5,** 563; **6,** 487; **9,** 388

ALKENES → α-HYDROXY C=O's (AND RELATED COMPOUNDS)

OXIDATION REACTIONS (*Continued*)
Chromyl chloride, **4,** 98
Hydrogen peroxide–N-Methylmorpholine oxide, **1,** 690; **3,** 204
Osmium tetroxide + co-reagent, **1,** 475; **2,** 363
Oxygen difluoride, **1,** 772
Phenyliodine(III) diacetate, **1,** 508
Potassium permanganate, **9,** 388
Zinc permanganate, **12,** 576

ALKENES → R_2CO
Allyltrimethylsilane, **10,** 6
Benzyl chloroformate, **3,** 59
Bis(acetonitrile)chloronitropalladium(II), **10,** 30; **11,** 45
[Bis(salicylidene-γ-iminopropyl)]-methylaminecobalt(II), **11,** 138
Borane–Tetrahydrofuran, **1,** 199
Bromine–Silver nitrate, **5,** 59
Chromium(II) acetate, **3,** 59
Chromyl chloride, **2,** 79; **3,** 62
Dimethyl(methylthio)sulfonium tetrafluoroborate, **11,** 204
Disodium tetrachloropalladate(II)–*t*-Butyl hydroperoxide, **10,** 175
Hydrogen peroxide–Palladium acetate, **10,** 203
Jones reagent, **6,** 123
Lead tetraacetate, **5,** 365
Mercury(II) acetate, **4,** 319
Nitrosyl chloride, **2,** 298
Oxotris(*t*-butylimido)osmium(VIII), **8,** 206
Palladium(or Platinum) *t*-butyl peroxide trifluoroacetate, **10,** 299; **11,** 430
Palladium(II) chloride, **10,** 300
Palladium(II) chloride + co-reagent, **11,** 396; **12,** 380
Picryl azide, **1,** 885
Pyridinium chlorochromate, **9,** 397
Thallium(III) nitrate, **4,** 492; **5,** 656
Trifluoroperacetic acid, **1,** 821; **2,** 316

ALKYLARENES (ArR → ArCOOH)
Benzyl(triethyl)ammonium permanganate, **10,** 28
Chromium(VI) oxide, **1,** 144
Cobalt(II) acetate, **1,** 154; **4,** 99
Hexamethylphosphoric triamide, **1,** 430
Lead(IV) oxide, **1,** 533
Nitric acid, **1,** 733
Palladium(II) chloride, **6,** 447
Potassium *t*-butoxide, **1,** 911
Potassium ferricyanide, **1,** 929
Potassium hypochlorite, **1,** 938
Potassium permanganate, **1,** 942; **4,** 143
Potassium peroxomonosulfate, **1,** 952
Selenium, **4,** 422
Selenium(IV) oxide, **10,** 345
Sodium dichromate, **1,** 1059; **7,** 68
Sodium hypochlorite, **1,** 1084

ALKYLARENES (ArR → RCOOH)
Ozone, **6,** 436
Potassium *t*-butoxide, **8,** 128
Potassium permanganate, **1,** 942
Ruthenium(IV) oxide–Sodium hypochlorite, **6,** 506
Ruthenium tetroxide, **2,** 357; **6,** 504; **11,** 462

ALKYLARENES → ArCHO
Benzeneseleninic anhydride, **9,** 32
Cerium(IV) ammonium nitrate, **1,** 120; **2,** 63
Chromium(VI) oxide–Acetic anhydride, **1,** 146
Chromyl chloride, **1,** 151
2,3-Dichloro-5,6-dicyano-1,4-benzoquinone, **3,** 83; **10,** 135; **11,** 166
N,N-Dimethyl-4-nitrosoaniline, **1,** 746
Oxygen, **6,** 477
Potassium *t*-butoxide, **6,** 477
Potassium persulfate, **11,** 441
Selenium(IV) oxide, **2,** 360
Silver(II) oxide, **2,** 369
Sodium persulfate, **1,** 1102
Thallium(III) nitrate, **12,** 481

ALKYL HALIDES → RCHO, R_2CO
Bis(tetrabutylammonium) dichromate, **9,** 53
N-Chlorosuccinimide, **7,** 65
N,N-Dimethyl-4-nitrosoaniline, **1,** 746
Dimethyl sulfoxide, **1,** 296; **2,** 157; **3,** 119, 161
Hexamethylenetetramine, **1,** 427; **4,** 243
Mercury(I) nitrate, **6,** 360
Potassium chromate, **7,** 298
Pyridine, **1,** 958
Silver chromate, **1,** 1005
Silver tetrafluoroborate, **1,** 1015; **5,** 264
Silver *p*-toluenesulfonate, **1,** 1018
Sodium benzeneselenoate, **7,** 341
Sodium 4,6-diphenyl-1-oxido-2-pyridone, **8,** 456

Sodium ethoxide, **1,** 1065
Sodium 2-nitropropanenitronate, **1,** 1101
Trimethylamine N-oxide, **1,** 1230; **2,** 434
ALKYNES → α-DICARBONYLS
Dimethyl sulfoxide–N-Bromosuccinimide, **4,** 51
Iodosylbenzene–Ruthenium catalysts, **11,** 270
Osmium tetroxide–Potassium chlorate, **9,** 334
Oxygen, **9,** 335
Potassium permanganate, **1,** 942; **9,** 388
Ruthenium tetroxide, **4,** 420
Tetracyanoethylene, **5,** 647
Thallium(III) nitrate, **4,** 492
Zinc permanganate, **12,** 576
ALKYNES → α-HYDROXY C=O's
Thallium(III) nitrate, **4,** 492
ALKYNES → OTHER COMPOUNDS
Chromium(VI) oxide, **4,** 96
Collins reagent, **9,** 121
Iodosylbenzene–Ruthenium catalysts, **11,** 270
Oxygen difluoride, **1,** 772
ALLYLIC ALCOHOLS → α,β-UNSATURATED C=O's
Barium manganate, **9,** 23
2,2′-Bipyridinium chlorochromate, **11,** 44
Bis(tetraalkylammonium) dichromate, **10,** 42; **11,** 50
Bis(tributyltin) oxide, **7,** 26
Bis(trimethylsilyl) peroxide, **12,** 63
t-Butyl hydroperoxide–Diaryl diselenides, **9,** 79; **10,** 65
Chloranil, **1,** 125
o-Chloranil, **1,** 128
N-Chlorosuccinimide, **6,** 115
N-Chlorosuccinimide–Dialkyl sulfide, **8,** 200; **12,** 195
Chromium(VI) oxide, **1,** 144; **4,** 96; **7,** 70
Chromium(VI) oxide + co-reagent, **7,** 71; **8,** 110
Collins reagent, **3,** 55; **4,** 215
Collins reagent–Celite, **5,** 285
Copper(I) chloride, **8,** 118
Copper(II) permanganate, **11,** 142
2,3-Dichloro-5,6-dicyano-1,4-benzoquinone, **1,** 215; **2,** 112; **9,** 148; **12,** 174
Dichlorotris(triphenylphosphine)-ruthenium(II), **11,** 171
4-(Dimethylamino)pyridinium chlorochromate, **11,** 196
Dimethyl sulfoxide, **1,** 296
Dimethyl sulfoxide + co-reagent, **2,** 165; **8,** 200; **10,** 167; **11,** 215
N,N-Dimethylthiocarbamoyl chloride, **8,** 200
3,5-Dinitrobenzoyl *t*-butyl nitroxyl, **8,** 204
Dodecacarbonyltri-*triangulo*-ruthenium, **7,** 244
Hexakis(acetato)trihydrato-μ_3-oxotrisrhodium acetate, **11,** 252
Hexamethylphosphoric triamide, **3,** 149
Jones reagent, **6,** 123
Manganese(IV) oxide, **1,** 637; **2,** 257; **5,** 422; **6,** 357
Nickel peroxide, **1,** 731; **2,** 294
μ-Oxybis(chlorotriphenylbismuth), **9,** 335; **11,** 381
Palladium(II) acetate–Triphenylphosphine, **10,** 298
Peracetic acid, **4,** 372
Phase-transfer catalysts, **9,** 356
Potassium dichromate, **10,** 324
Potassium ferrate(VI), **4,** 405
Potassium permanganate–Copper sulfate, **11,** 441
Pyridinium chlorochromate, **6,** 498; **7,** 308; **8,** 425; **10,** 334; **12,** 419
Pyridinium dichromate, **9,** 399
Pyruvyl chloride, **7,** 310
Ruthenium(III) chloride, **7,** 244
Ruthenium(IV) oxide, **12,** 428
Selenium(IV) oxide, **4,** 422
Silver carbonate–Celite, **2,** 363; **3,** 247; **6,** 511
Silver(II) oxide, **2,** 369
Sodium dichromate, **2,** 70
Sodium permanganate, **10,** 368
Sulfur trioxide–Pyridine, **2,** 393
Tetrabutylammonium chlorochromate, **12,** 458
2,2,6,6-Tetramethylpiperidinyl-1-oxy, **12,** 479
Triphenylbismuth carbonate, **9,** 501
Tris(aquo)hexa-μ-acetato-μ_3-oxotriruthenium(III,III,III) acetate, **6,** 425, 650
ALLYLIC ALCOHOLS OR

OXIDATION REACTIONS (*Continued*)
RCH=CHCHO → RCH=CHCOOH
Hydrogen peroxide–Selenium(IV) oxide, **1,** 477
Lithium bis(3,3-diethoxy-1-propen-2-yl)-cuprate, **6,** 331
Silver(II) oxide, **2,** 369
Sodium chlorite, **9,** 423; **11,** 481
Sodium ruthenate, **5,** 622
ALLYLIC HALIDES → α,β-UNSATURATED C=O's
Potassium chromate, **7,** 298
Silver tetrafluoroborate–Dimethyl sulfoxide, **5,** 264
ALLYLIC SILANES → ALLYLIC ALCOHOLS
Benzeneselenenyl halides, **11,** 34
(Chloromethyl)dimethylsilane, **12,** 27
Dimethylphenylsilyllithium, **12,** 210
Hydrogen peroxide, **12,** 242
1-Trimethylsilyl-1,3-butadiene, **7,** 395
ALLYLIC SILANES → α,β-UNSATURATED C=O's
Iodosylbenzene, **12,** 258
AMINES → RCHO, R_2CO
Aluminum chloride, **4,** 10
Benzeneseleninic anhydride, **7,** 139
Benzo-1,3-thiazole-2-carboxaldehyde, **5,** 24
5-Bromo-3-methylthio-1,4-diphenyl-5-triazolium bromide, **8,** 53
t-Butyl hypochlorite, **1,** 90
t-Butyl nitrite, **7,** 48
m-Chloroperbenzoic acid, **6,** 110
Copper halide nitrosyls, **7,** 73
3,5-Di-*t*-butyl-1,2-benzoquinone, **3,** 78
Dimethyl sulfoxide, **7,** 133
4-Formyl-1-methylpyridinium benzenesulfonate, **11,** 244
Hexamethylenetetramine, **1,** 427; **2,** 208
Hexamethylphosphoric triamide, **5,** 323
Hydrogen tetrachloroaurate(III), **7,** 277
Hypochlorous acid, **1,** 487
Manganese(IV) oxide, **1,** 636
p-Nitrobenzenesulfonyl peroxide, **7,** 251
Palladium(II) chloride, **7,** 277
Potassium ferrate(VI), **4,** 405
Potassium permanganate, **2,** 348
Potassium superoxide, **8,** 417
4-Pyridinecarboxaldehyde, **11,** 448
Ruthenium(III) chloride, **8,** 437
Silver(II) oxide, **2,** 369
Sodium hypochlorite, **7,** 337
Sodium nitrite, **2,** 386
Sodium persulfate, **3,** 267
Sodium ruthenate, **5,** 622
N-Sulfinylaniline, **6,** 556
2,2,6,6-Tetramethylpiperidinyl-1-oxy, **12,** 479
Trifluoromethanesulfonic anhydride, **11,** 560
AMINES → $RCONR_2$
Acetic anhydride–Manganese(III) acetate, **6,** 355
Benzyl(triethyl)ammonium permanganate, **10,** 28
Bromine, **6,** 70
Chromium(VI) oxide–Pyridine, **2,** 74
2,6-Di-*t*-butyl-*p*-benzoquinone, **9,** 139
Manganese(IV) oxide, **1,** 637; **2,** 257
Mercury(II) acetate–Ethylenediaminetetraacetic acid, **5,** 427
Ruthenium tetroxide, **6,** 504
Sarett reagent, **2,** 74
AMINES → –N=N–
Benzeneseleninic acid, **10,** 22
Chloramine, **4,** 74
Hydrogen peroxide, **1,** 457
Iodine pentafluoride, **1,** 503
Manganese(IV) oxide, **1,** 637
Nickel peroxide, **1,** 731
Phenyliodine(III) diacetate, **1,** 508; **3,** 166
Potassium superoxide, **9,** 391
Silver carbonate–Celite, **6,** 511
Silver(II) oxide, **4,** 431
Sodium hypochlorite, **4,** 456
Sodium perborate, **1,** 1102
AMINES → C=N
t-Butyl hypochlorite, **1,** 90
Diphenyldi(1,1,1,3,3,3-hexafluoro-2-phenyl-2-propoxy)sulfurane, **6,** 239
Diphenylselenium bis(trifluoroacetate), **11,** 223
Manganese(IV) oxide, **1,** 637
p-Nitrobenzenesulfonyl peroxide, **7,** 251
Potassium superoxide, **8,** 417
Sodium persulfate, **3,** 267
AMINES → RCN
Copper(I) chloride, **8,** 118
Diphenyldi(1,1,1,3,3,3-hexafluoro-2-phenyl-2-propoxy)sulfurane, **6,** 239

Hydridotris(triisopropylphosphine)-rhodium(I), **9,** 238
Iodine pentafluoride, **1,** 503
Lead tetraacetate, **1,** 537
Nickel peroxide, **1,** 731
Ruthenium(III) chloride, **8,** 437
Silver(II) oxide, **2,** 369
Silver(II) picolinate, **5,** 20
Sodium hypochlorite, **7,** 337
2,2,6,6-Tetramethylpiperidinyl-1-oxy, **12,** 479
AMINES → RNO_2
m-Chloroperbenzoic acid, **1,** 135
Ozone–Silica gel, **8,** 375
Potassium permanganate, **1,** 942
AMINES → RNO
Caro's acid, **1,** 118; **6,** 97
m-Chloroperbenzoic acid, **11,** 122
Hydrogen peroxide, **1,** 457
Oxygen difluoride, **1,** 772
Potassium peroxomonosulfate, **1,** 952
ANILINES, ArNO → $ArNO_2$
t-Butyl hydroperoxide + co-reagent, **3,** 331; **5,** 77
Hydrogen peroxide, **1,** 457
2-Hydroperoxyhexafluoro-2-propanol, **3,** 149
Nitric acid, **1,** 733
Peracetic acid, **1,** 787
Permaleic acid, **1,** 819
Sodium perborate, **12,** 452
Trifluoroperacetic acid, **1,** 821; **2,** 316; **3,** 221
ARENES → *o*- or *p*-QUINONES
Cerium(IV) ammonium salts, **5,** 101; **8,** 80; **12,** 107
Chromium(VI) oxide, **1,** 144
Hexamethylphosphoric triamide, **3,** 149
Hydrogen peroxide, **1,** 457
Lithium, **3,** 150
Manganese(III) sulfate, **9,** 289
Periodic acid, **2,** 313
Pyridinium fluorochromate, **11,** 453
Sodium chlorate, **1,** 1056
Thallium(III) trifluoroacetate, **5,** 658
AZO → AZOXY COMPOUNDS
Hydrogen peroxide, **1,** 457
CARBOXYLIC ACIDS → RCO_3H
Hydrogen peroxide, **1,** 457
Methanesulfonic acid, **1,** 666
CATECHOLS (AND RELATED COMPOUNDS) → *o*-QUINONES
Bis(*p*-methoxyphenyl) telluroxide, **9,** 50
Cerium(IV) ammonium nitrate, **4,** 71
Cerium(IV) sulfate, **2,** 65
Chloranil, **4,** 76
o-Chloranil, **1,** 128; **5,** 104
N-Chlorosuccinimide–Triethylamine, **6,** 120
2,3-Dichloro-5,6-dicyano-1,4-benzoquinone, **2,** 112
Diphenyl selenoxide, **5,** 280; **9,** 201
Ferric chloride, **1,** 390
Lead(IV) oxide, **1,** 533; **2,** 233
Nickel peroxide, **1,** 731
Periodates, **1,** 809
Phenyliodine(III) diacetate, **4,** 266
Potassium ferricyanide, **4,** 406
Silver carbonate–Celite, **5,** 577
Silver(I) oxide, **1,** 1011
Succinimidodimethylsulfonium tetrafluoroborate, **6,** 555
CYCLIC ETHERS → LACTONES
Bromine, **3,** 34
N-Bromosuccinimide, **7,** 37
t-Butyl chromate, **3,** 37
Copper(I) trifluoromethanesulfonate, **11,** 142
Ruthenium tetroxide, **1,** 986; **3,** 243; **10,** 343
Zinc permanganate, **12,** 576
1,2-DIOLS (AND RELATED COMPOUNDS) → α-SUBSTITUTED C=O's
Alumina, **8,** 9
Bis(tributyltin) oxide, **7,** 26
N-Bromosuccinimide, **1,** 78
t-Butyl hydroperoxide–Diaryl diselenides, **10,** 65
Chlorine–Pyridine, **5,** 106
N-Chlorosuccinimide–Dimethyl sulfide, **5,** 129
Chromium(VI) oxide–Graphite, **5,** 732
Dibutyltin oxide, **5,** 188; **9,** 141
Dimethyl sulfoxide + co-reagent, **2,** 163; **9,** 192; **10,** 168
Silver carbonate–Celite, **3,** 247; **4,** 425; **5,** 577; **6,** 511
Sodium dichromate–Manganese(II) nitrate, **2,** 72
Triphenylcarbenium tetrafluoroborate, **4,** 565; **8,** 524

OXIDATION REACTIONS (*Continued*)
1,4- OR 1,5-DIOLS → LACTONES
Barium manganate, **12,** 38
Bromine–Nickel(II) alkanoates, **11,** 358; **12,** 72
N-Bromosuccinimide, **9,** 70
Butyllithium–Potassium *t*-butoxide, **8,** 67
Copper chromite, **1,** 156
Dibenzoyl peroxide–Nickel(II) bromide, **10,** 121
Dichlorotris(triphenylphosphine)-ruthenium(II), **10,** 141
Dihydridotetrakis(triphenylphosphine)-ruthenium(II), **11,** 182
Formic acid, **11,** 243
Heyn's catalyst, **5,** 326; **7,** 320
Platinum catalysts, **5,** 326
Pyridinium chlorochromate, **11,** 450
Silver carbonate–Celite, **3,** 247; **6,** 511; **7,** 319; **8,** 441; **12,** 433
Sodium bromite, **12,** 445
Tetra-μ_3-carbonyldodecacarbonyl-hexarhodium, **5,** 326
Thexylborane, **11,** 516
ENOL ETHERS (AND RELATED COMPOUNDS) → α-HYDROXY C=O's
t-Butyl hydroperoxide, **12,** 88
Chloramine-T, **9,** 101
m-Chloroperbenzoic acid, **6,** 110; **8,** 97; **10,** 92
Chlorotrimethylsilane, **3,** 310
Chromyl chloride, **11,** 134
1-(Methoxymethoxy)styrene, **12,** 99
N-Methylanilinium trifluoroacetate, **8,** 331
Ozone, **8,** 374
Peracetic acid, **1,** 785
2-Trimethylsilyloxy-1,3-butadiene, **7,** 401
ENOL ETHERS (AND RELATED COMPOUNDS) → α,β-UNSATURATED C=O's
Allyl chloroformate, **12,** 15
Boron trifluoride, **11,** 71
Chlorodimethyl(2,4,6-tri-*t*-butylphenoxy)silane, **11,** 217
Chloromethylcarbene, **10,** 90
2,3-Dichloro-5,6-dicyano-1,4-benzoquinone, **9,** 148
Palladium(II) acetate, **8,** 378; **9,** 344; **12,** 367
Potassium hydride, **9,** 386
Pyridinium chlorochromate, **11,** 450
Triphenylcarbenium tetrafluoroborate, **8,** 524
ETHERS, $RCH_2OSiR'_3$, $RCH_2OSnR'_3$ → RCHO, R_2CO
Bromine, **3,** 34
N-Bromosuccinimide, **7,** 37; **8,** 54
Butyl azide, **1,** 84
Cerium(IV) ammonium nitrate, **10,** 79
2,3-Dichloro-5,6-dicyano-1,4-benzoquinone, **12,** 174
Nitronium tetrafluoroborate, **8,** 361
Nitrosonium tetrafluoroborate, **7,** 253
Trichloroisocyanuric acid, **2,** 426
Triethyltin methoxide, **6,** 613
Triphenylcarbenium salts, **4,** 548, 565; **7,** 414
Uranium(VI) fluoride, **7,** 417
ETHERS, $RCH_2OSiR'_3$ → RCOOR′ (*see also* CYCLIC ETHERS → LACTONES)
Benzyl(triethyl)ammonium permanganate, **9,** 43
N-Bromosuccinimide, **8,** 54
Chromium(VI) oxide, **2,** 72; **3,** 54
Pyridinium chlorochromate, **8,** 425
Ruthenium(IV) oxide, **12,** 428
Ruthenium tetroxide, **11,** 462
Trichloroisocyanuric acid, **2,** 426
HOMOALLYLIC ALCOHOLS → α,β-UNSATURATED C=O's
Aluminum *t*-butoxide, **1,** 23
Aluminum isopropoxide, **1,** 35
Raney nickel, **1,** 723
Rochelle salt, **1,** 983
HOMOALLYLIC ALCOHOLS → β,γ-UNSATURATED C=O's
Chromium(VI) oxide–Diethyl ether, **9,** 115
Diisobutylaluminum hydride, **7,** 111
Dimethyl sulfoxide, **2,** 157
N-Iodosuccinimide–Tetrabutylammonium iodide, **10,** 216
Jones reagent, **1,** 142
Palladium(II) acetate–Triphenylphosphine, **10,** 298
Pyruvyl chloride, **7,** 310
HYDRAZO COMPOUNDS → –N=N–
Benzeneseleninic anhydride, **8,** 29; **10,** 22
N-Bromosuccinimide, **6,** 74

1-Chlorobenzotriazole, **2,** 67; **3,** 46
Copper(II) chloride, **2,** 84; **3,** 66
Iodine pentafluoride, **1,** 503
Manganese(IV) oxide, **1,** 637; **8,** 312
Oxygen-Palladium, **5,** 491
Periodic acid, **3,** 220
Phenyliodine(III) bis(trifluoroacetate), **6,** 301
Potassium superoxide, **8,** 417
Sodium periodate, **11,** 492
Triphenylbismuth carbonate, **9,** 501

HYDRAZONES → DIAZO COMPOUNDS
N-Chlorosuccinimide–Triethylamine, **6,** 120
Copper(I) chloride, **5,** 164
Manganese(IV) oxide, **1,** 637
Mercury(II) oxide, **1,** 655
Mercury(II) trifluoroacetate, **1,** 659
Nickel peroxide, **2,** 294
Silver carbonate–Celite, **6,** 511
Silver(I) oxide, **2,** 368
Triphenylbismuth carbonate, **9,** 501

HYDROQUINONES → *p*-QUINONES
Ammonium metavanadate, **1,** 39
Benzeneseleninic anhydride, **10,** 22
Bis(*p*-methoxyphenyl) telluroxide, **9,** 50
Boron trifluoride etherate, **11,** 72
Cerium(IV) ammonium nitrate, **4,** 71; **7,** 55; **9,** 99; **10,** 79
o-Chloranil, **5,** 104
Chlorine, **6,** 101
N-Chlorosuccinimide–Triethylamine, **6,** 120
Cyclodextrins, **8,** 133
2,3-Dichloro-5,6-dicyano-1,4-benzoquinone, **1,** 215; **7,** 96
3,5-Dinitrobenzoyl *t*-butyl nitroxyl, **8,** 204
4,4'-Dinitrodiphenylnitroxide, **2,** 174
Ferric chloride, **1,** 390; **11,** 237
Hydrogen peroxide, **1,** 466
Hydrogen tetrachloroaurate(III), **1,** 414
Lead(IV) oxide, **1,** 533
Lead tetraacetate, **1,** 537
Manganese(IV) oxide, **11,** 311
Mercury(II) oxide, **8,** 316
Nitric acid, **1,** 733
Nitrogen dioxide, **1,** 324, 737
Phenyliodine(III) bis(trifluoroacetate), **6,** 301
Phosphatolead(IV) acids, **3,** 225
Pyridinium chlorochromate, **11,** 450
Silver carbonate, **2,** 363; **4,** 425
Silver(II) dipicolinate, **10,** 348
Silver(I) oxide, **1,** 1011
Silver(II) oxide, **4,** 431; **7,** 322; **9,** 412; **10,** 352; **11,** 469
Silver(II) picolinate, **3,** 16
Sodium nitrite, **1,** 1097
Sodium periodate, **11,** 492
Succinimidodimethylsulfonium tetrafluoroborate, **6,** 555
Thallium(III) nitrate, **7,** 362
Thallium(III) trifluoroacetate, **3,** 286
Vanadium(V) oxide, **1,** 1057

α-HYDROXY C=O's (AND RELATED COMPOUNDS) → 1,2-DICARBONYLS
Bismuth(III) oxide, **1,** 59
Bromine, **4,** 46, 537
N-Bromosuccinimide, **2,** 40
t-Butyl hydroperoxide, **12,** 88
N-Chlorosuccinimide, **8,** 103
Chlorotrimethylsilane, **4,** 537
Chromium(VI) oxide–Diethyl ether, **9,** 115
Copper(II) acetate, **1,** 159
Copper carbonate, basic, **1,** 163
Copper(II) sulfate, **1,** 164; **6,** 141
Dimethyl sulfide–Trifluoroacetic anhydride, **6,** 225
Dimethyl sulfoxide, **1,** 296; **6,** 225
Dimethyl sulfoxide + co-reagent, **2,** 163; **6,** 227
Epichlorohydrin, **4,** 222
Ferric chloride, **1,** 390
Jones reagent, **4,** 95
Lead tetraacetate, **2,** 234
Manganese(IV) oxide, **1,** 637
Mercury(II) acetate, **5,** 424
Nickel(II) acetate, **5,** 470
Nitric acid, **1,** 733
Oxygen, **7,** 258
Phenyl benzenethiosulfonate, **8,** 391
Potassium ferricyanide, **8,** 410
Selenium(IV) oxide, **1,** 992; **3,** 245
Sodium acetate, **1,** 1024
Sodium bromate, **1,** 1055
Sulfuryl chloride, **1,** 1128
Thallium(III) nitrate, **9,** 460
Thionyl chloride, **1,** 1158

OXIDATION REACTIONS (*Continued*)
Triphenylphosphine dibromide, **4,** 555
Ytterbium(III) nitrate, **6,** 671
HYDROXYLAMINES → RNO
Benzeneseleninic anhydride, **8,** 29; **10,** 22
Bis(*p*-methoxyphenyl) telluroxide, **9,** 50
1-Chlorobenzotriazole, **3,** 46
Manganese(IV) oxide, **1,** 637
Silver carbonate–Celite, **4,** 425
Tetraethylammonium periodate, **2,** 397; **9,** 448
KETONES (AND RELATED COMPOUNDS) → 1,2-DICARBONYLS
t-Butoxybis(dimethylamino)methane, **7,** 41
t-Butyl hydroperoxide, **1,** 88
m-Chloroperbenzoic acid, **10,** 92
N,N-Dimethyl-4-nitrosoaniline, **1,** 746
Dimethyl sulfoxide, **6,** 225
Dimethyl sulfoxide–Iodine, **8,** 200; **9,** 190
Iodine–Silver salts, **1,** 504
Methyl 2-nitrophenyl disulfide, **9,** 314
S-Methyl *p*-toluenethiosulfonate, **6,** 400
Nitrosyl chloride, **1,** 748
Osmium tetroxide–*t*-Butyl hydroperoxide, **1,** 88
Oxygen, **1,** 921
Oxygen, singlet, **6,** 431; **9,** 338; **12,** 363
Ozone, **11,** 387
Phenyl benzenethiosulfonate, **8,** 391
Phenyliodine(III) diacetate, **12,** 384
Potassium permanganate, **1,** 942
Pyridine, **1,** 958
Selenium(IV) oxide, **1,** 992; **7,** 319
Sodium methoxide, **8,** 463
Sodium nitrite, **1,** 1097
Thionyl chloride, **8,** 481
LACTOLS (AND RELATED COMPOUNDS) → LACTONES
Bromocarbamide, **1,** 76
m-Chloroperbenzoic acid, **8,** 97
Cobaloxime(I), **11,** 135
B-Crotyl-9-borabicyclo[3.3.1]nonane, **12,** 81
Crotyl carbamates, **12,** 82
Lead tetraacetate, **1,** 537
1-Methoxy-1,3-butadiene, **10,** 258
Ozone, **6,** 436
Pyridinium dichromate, **11,** 453
Sodium tris(3,5-di-*t*-butylphenoxy)-borohydride, **10,** 369
Tributyltin hydride, **11,** 545; **12,** 516
METHYLENE GROUPS → C=O's (*see also* ArR → ArCHO, ALLYLIC OXIDATION)
Ammonium persulfate–Silver nitrate, **3,** 15
Benzeneseleninic anhydride, **9,** 32
Benzyl(trialkyl)ammonium salts, **1,** 1252; **9,** 43; **10,** 28
Cerium(IV) ammonium nitrate, **2,** 63; **7,** 55
Chlorotris(triphenylphosphine)-rhodium(I), **2,** 448
Chromium(VI) oxide, **5,** 140; **7,** 70; **9,** 115
Chromyl acetate, **2,** 78
Cobalt(II) acetate–Hydrogen bromide, **1,** 154
2,3-Dichloro-5,6-dicyano-1,4-benzoquinone, **3,** 83; **8,** 153; **10,** 135
Hexafluoroantimonic acid, **8,** 239
Hexamethylphosphoric triamide, **3,** 149
Manganese(III) acetylacetonate, **3,** 194
Nitrogen dioxide, **1,** 737
Oxygen, **1,** 1253; **7,** 258; **8,** 366
Ozone–Silica gel, **8,** 375; **9,** 343
Potassium permanganate, **1,** 942
Potassium persulfate, **11,** 441
Pyridinium chlorochromate, **12,** 417
Selenium(IV) oxide, **2,** 360; **6,** 509
Sodium dichromate, **1,** 1059
Sodium hypochlorite, **1,** 1084
Thallium(III) perchlorate, **8,** 478
Thionyl chloride, **8,** 481
NITRO COMPOUNDS → C=O's
t-Butyl hydroperoxide, **8,** 62
N″-(*t*-Butyl)-N,N,N′,N′-tetramethylguanidinium *m*-iodylbenzoate, **12,** 102
Cerium(IV) ammonium nitrate, **10,** 79
Hydrogen peroxide, **10,** 201
Mercury(II) nitrite, **9,** 292
Nitroethylene, **5,** 476
Nitrogen dioxide–Iodine, **8,** 205
Oxodiperoxymolybdenum(pyridine)-(hexamethylphosphoric triamide), **11,** 218
Oxygen, singlet, **8,** 367
Ozone, **5,** 491; **8,** 374
Potassium permanganate, **1,** 942; **10,** 330; **11,** 440

Propyl nitrite–Sodium nitrite, **5,** 565
Sodium hydroxide, **8,** 461
Sodium methoxide, **6,** 545
1,1,3,3-Tetramethylbutyl isocyanide, **5,** 650
Titanium(III) chloride, **4,** 506; **5,** 669
Vanadium(II) chloride, **7,** 418

NITROSO COMPOUNDS → RNO_2
t-Amyl hydroperoxide, **4,** 20
t-Butyl hydroperoxide, **5,** 75
Hypochlorous acid, **10,** 208
Nitric acid, **1,** 733; **3,** 212
Trifluoroperacetic acid, **1,** 821

PHENOLS (AND RELATED COMPOUNDS) → 6-SUBSTITUTED-1,4-CYCLOHEXADIEN-3-ONES
Acetic anhydride–Nitric acid, **5,** 475
Antimony(V) chloride, **6,** 22
Benzeneseleninic anhydride, **6,** 240
Chromium(VI) oxide, **5,** 140
2,3-Dichloro-5,6-dicyano-1,4-benzoquinone, **9,** 148
Hydrogen peroxide–Cerium(IV) oxide, **6,** 99
Lead tetraacetate, **6,** 313; **7,** 185
Manganese(IV) oxide, **1,** 637
Oxygen, **6,** 426
Oxygen, singlet, **11,** 385
Perchloryl fluoride, **1,** 802
Salcomine, **3,** 245
Thallium(III) nitrate, **7,** 362
Thallium(III) perchlorate, **5,** 657; **8,** 478
Trifluoromethyl hypofluorite, **2,** 200
Tris(tetrabutylammonium)-hexacyanoferrate(III), **6,** 656

PHENOLS → *o,p*-QUINONES
Benzeneseleninic anhydride, **7,** 139; **9,** 32; **10,** 22
Bis(tricaprylylmethyl)ammonium nitrosodisulfonate, **10,** 42
Copper(II)–Amine complexes, **8,** 115
2,3-Dichloro-5,6-dicyano-1,4-benzoquinone, **6,** 168; **7,** 96; **10,** 135
3,5-Dinitrobenzoyl *t*-butyl nitroxyl, **8,** 204
4,4′-Dinitrodiphenylnitroxide, **2,** 174
Ferric chloride, **7,** 153
Hydrogen peroxide, **1,** 457
Hydrogen peroxide + co-reagent, **6,** 99; **7,** 174
Iodylbenzene, **11,** 275
Lead(IV) oxide, **3,** 168
Mercury(II) oxide, **8,** 316
Oxygen, singlet, **7,** 261
Peracetic acid, **8,** 386
Potassium nitrosodisulfonate, **1,** 940; **2,** 347; **4,** 411; **10,** 329
Salcomine, **2,** 360; **3,** 245; **6,** 507; **12,** 429
Sodium dichromate, **12,** 131
Thallium(III) nitrate, **7,** 362; **9,** 460; **10,** 395
Thallium(III) perchlorate, **8,** 478
Thallium(III) trifluoroacetate, **3,** 286; **4,** 498; **9,** 462

P COMPOUNDS
Bis(trimethylsilyl) peroxide, **11,** 67
Dimethyl selenoxide, **8,** 197
Nitrogen oxides, **1,** 324
Oxygen, singlet, **6,** 431

PROPARGYL ALCOHOLS → ACETYLENIC C=O's
Lead(IV) oxide, **2,** 233
Manganese(IV) oxide, **1,** 637; **4,** 317
Nickel peroxide, **2,** 294
Pyridinium chlorochromate, **6,** 498

Se COMPOUNDS
Benzeneselenenyl halides, **5,** 518; **6,** 459
t-Butyl hydroperoxide, **8,** 64
t-Butyl hypochlorite, **10,** 66
m-Chloroperbenzoic acid, **11,** 122; **12,** 118
Oxygen, singlet, **7,** 261
2-Pyridineselenenyl bromide, **11,** 455

SULFIDES → RSOR
Acetyl nitrate, **7,** 3
Benzoyl nitrate, **7,** 3
Bis(tributyltin) oxide, **8,** 43
Bromine, **9,** 65
3-Bromo-4,5-dihydro-5-hydroperoxy-4,4-dimethyl-3,5-diphenyl-3H-pyrazole, **11,** 76
N-Bromosuccinimide, **3,** 34
t-Butyl hydroperoxide + co-reagent, **12,** 90
t-Butyl hypochlorite, **1,** 90
[(−)-Camphor-10-ylsulfonyl]-3-aryloxaziridines, **11,** 108
Cerium(IV) ammonium nitrate, **4,** 71
1-Chlorobenzotriazole, **3,** 46
1,4-Diazabicyclo[2.2.2]octane–Bromine, **2,** 99

OXIDATION REACTIONS (*Continued*)
Dimethyl sulfoxide, **1,** 296
Diphenyl sulfide, **6,** 241
Hydrogen peroxide, **1,** 471
Hydrogen peroxide + co-reagent, **8,** 248; **9,** 244
Hydrogen tetrachloroaurate(III), **5,** 647
2-Hydroperoxyhexafluoro-2-propanol, **10,** 206
Iodylbenzene, **11,** 275
Manganese(III) acetylacetonate, **3,** 194
Manganese(IV) oxide, **1,** 637
Nitrogen tetroxide, **1,** 324
Oxygen, **9,** 335
Perbenzoic acid, **3,** 219; **7,** 279
Periodates, **1,** 809; **2,** 311
Perlauric acid, **2,** 315
Phenyliodine(III) diacetate, **3,** 166
Phenyliodine(III) dichloride, **3,** 164
Potassium peroxomonosulfate, **10,** 328
Ruthenium tetroxide, **1,** 986
Sodium perborate, **12,** 452
Sodium permanganate, **10,** 368
Sulfuryl chloride, **5,** 641; **7,** 350
Tetrabutylammonium periodate, **10,** 381
Titanium(III) chloride–Hydrogen peroxide, **10,** 401
Trifluoroperacetic acid, **11,** 562
SULFIDES, RSOR → RSO_2R
Benzyl(triethyl)ammonium permanganate, **11,** 44
Bis(trimethylsilyl) peroxide, **11,** 67
N-Bromosuccinimide, **3,** 34
Hydrogen peroxide, **1,** 471, 475; **10,** 205
2-Hydroperoxyhexafluoro-2-propanol, **10,** 206
Iodosylbenzene, **1,** 507
Manganese(III) acetylacetonate, **3,** 194
Oxygen, **9,** 335
Potassium permanganate, **8,** 416
Potassium peroxomonosulfate, **10,** 328
Ruthenium catalysts, **7,** 314
Ruthenium tetroxide, **2,** 357
Sodium perborate, **12,** 452
Sodium periodate, **11,** 492
Trifluoroperacetic acid, **11,** 562
THIOLS → RSSR
Alumina, **9,** 8
Bis(*p*-methoxyphenyl) telluroxide, **9,** 50
o-Carboxyphenyl *o*-carboxybenzenethiosulfonate, **5,** 100
Cyanogen bromide, **5,** 169
Diethyl azodicarboxylate, **2,** 128
Dimethyl dithiobis(thioformate), **5,** 247
Dimethyl sulfoxide, **1,** 296; **5,** 263; **12,** 212
2,4-Dinitrobenzenesulfenyl chloride, **6,** 231
Ferric chloride, **1,** 390
Ferric nitrate/K10 Bentonite, **12,** 231
Hydrogen peroxide, **1,** 471
Lead(II) acetate, **4,** 276
Lead tetraacetate, **1,** 537
Manganese(IV) oxide, **1,** 637; **8,** 312
Nickel peroxide, **10,** 277
Phenyliodine(III) bis(trifluoroacetate), **6,** 301
Potassium superoxide, **9,** 391
1,3-Propanedithiol, **9,** 394
2-Pyridinesulfenyl chloride, **9,** 397
Sodium trithiocarbonate, **9,** 435
2,4,4,6-Tetrabromo-2,5-cyclohexadienone, **6,** 563
Tetramethylene sulfoxide, **1,** 1145
Triphenylbismuth carbonate, **9,** 501
THIOLS → OTHER S COMPOUNDS
Barium permanganate, **1,** 46
m-Chloroperbenzoic acid, **5,** 120
Hexamethylphosphoric triamide, **1,** 430
Hydrofluoric acid, **6,** 284
N-Hydroxymethylphthalimide, **8,** 253
Potassium peroxomonosulfate, **1,** 952
OXIDATIVE CLEAVAGE
OF ALCOHOLS
Cerium(IV) ammonium nitrate, **3,** 44; **5,** 101
2,3-Dichloro-5,6-dicyano-1,4-benzoquinone, **10,** 135
Fieser reagent, **1,** 145
Hypohalite solution, **1,** 488
Lead tetrakis(trifluoroacetate), **6,** 318
Mercury(II) oxide–Iodine, **5,** 347; **11,** 267
Nitric acid, **1,** 733
Potassium *t*-butoxide, **1,** 911
Pyridinium dichromate, **10,** 335
Silver carbonate–Celite, **5,** 577
Sodium dichromate, **3,** 54; **5,** 138; **7,** 68
OF C=C → RCHO, R_2CO
Chromyl chloride, **4,** 98
Chromyl trichloroacetate, **1,** 152
Dimethyl sulfide, **2,** 156; **5,** 260

Dioxygentetrakis(triphenylarsine)-rhodium(I) hexafluorophosphate, **8,** 206
Ethoxycarbonylformonitrile oxide, **7,** 145
Ferric chloride, **12,** 230
Glyoxylic acid, **5,** 320
Lindlar catalyst, **3,** 171
Osmium tetroxide, **2,** 301
Oxygen, **5,** 482; **9,** 335
Oxygen, singlet, **6,** 431
Ozone, **1,** 773; **4,** 363; **5,** 491; **8,** 374; **9,** 341; **10,** 295; **11,** 387
Phosphatolead(IV) acids, **3,** 225
Potassium permanganate, **5,** 562; **9,** 388
Ruthenium tetroxide, **1,** 986; **2,** 357
Sodium periodate + co-reagent, **1,** 810, 812; **2,** 312
Thallium(III) nitrate, **5,** 656
Thiourea, **11,** 519
Trimethyl phosphite, **1,** 1233
Triphenylphosphine, **1,** 1238
OF C=C → RCOOH, R_2CO
t-Butyl hydroperoxide–Molybdenyl acetylacetonate, **11,** 91; **12,** 89
2,3-Dichloro-1-propene, **8,** 158
Jones reagent, **3,** 54
Ozone, **6,** 436; **7,** 269; **9,** 341
Potassium permanganate, **1,** 942; **4,** 30, 143, 412; **6,** 175
Potassium superoxide, **9,** 391
Ruthenium tetroxide, **11,** 462
Ruthenium tetroxide–Sodium periodate, **1,** 813
Sodium periodate–Potassium permanganate, **1,** 810; **2,** 312; **6,** 455
Sodium ruthenate, **5,** 622
Tetrabutylammonium salts, **5,** 644; **8,** 468
OF ArR (*see* OXIDATION REACTIONS)
OF C≡C
Oxygen, **9,** 335
Ruthenium(IV) oxide–Sodium hypochlorite, **4,** 421
Thallium(III) nitrate, **4,** 492
OF ArH → DICARBONYLS
N-Benzoylperoxycarbamic acid, **6,** 35
Dimethyl sulfide, **2,** 156
Glyoxal, **1,** 413
Hydrogen peroxide, **1,** 457
Oxygen, **6,** 426
Ozone, **1,** 773; **6,** 436; **9,** 341
Peracetic acid, **1,** 785
Potassium permanganate, **1,** 942
Ruthenium(IV) oxide–Sodium hypochlorite, **6,** 506
Ruthenium tetroxide, **2,** 357
BICYCLIC ALKENES → MACROLIDES
m-Chloroperbenzoic acid, **1,** 135; **2,** 68
Lithium–Alkylamines, **2,** 241
Molybdenum carbonyl, **3,** 206
Oxygen, singlet, **8,** 367; **9,** 9
Ozone, **6,** 436; **12,** 365
OF CYCLOALKANEDIONES → DIACIDS
Dicyclopentadienylcobalt–Oxygen adduct, **5,** 145
Hydrogen peroxide, **1,** 457, 466
Perbenzoic acid, **1,** 791
Periodates, **1,** 809
Phenyl benzenethiosulfonate, **8,** 391
Potassium superoxide, **7,** 304
Sodium hypochlorite, **1,** 1084; **9,** 430
OF CYCLOALKENES → DIACIDS (*see also* OXIDATIVE CLEAVAGE OF C=C)
Brij 30, **1,** 892
t-Butyl hydroperoxide–Molybdenyl acetylacetonate, **11,** 91
Ozone, **1,** 773; **9,** 341
Ruthenium(IV) oxide–Sodium periodate, **5,** 508
Sodium dichromate, **3,** 54
Sodium periodate–Potassium permanganate, **1,** 810
OF α-DICARBONYLS (*see also* OXIDATIVE CLEAVAGE—CYCLOALKANEDIONES)
Calcium hypochlorite, **11,** 107
Copper(I) chloride, **11,** 140
Dicyclopentadienylcobalt–Oxygen adduct, **5,** 145
Dimethyl sulfoxide, **7,** 133
Hydrogen peroxide, **1,** 466
Iodosylbenzene, **10,** 213
Nickel peroxide, **1,** 731
Oxygen, **9,** 335
Oxygen, singlet, **7,** 261; **11,** 385
Perbenzoic acid, **1,** 791
Perchloric acid, **1,** 796

OXIDATIVE CLEAVAGE (*Continued*)
Potassium superoxide, **7,** 304
Trimethyl phosphite, **1,** 1233
OF β-DICARBONYLS (*see also* OXIDATIVE CLEAVAGE—CYCLOALKANEDIONES)
Hydrogen peroxide, **1,** 457
Periodates, **1,** 809
OF DIOLS (AND RELATED COMPOUNDS) → RCHO, R_2CO
Benzoin, **6,** 34
N″-(*t*-Butyl)-N,N,N′,N′-tetramethylguanidinium *m*-iodylbenzoate, **12,** 102
Calcium hypochlorite, **11,** 107
Cerium(IV) ammonium nitrate, **4,** 71
Chlorotrimethylsilane, **6,** 626
Chromium(VI) oxide, **2,** 72; **4,** 96
Cobalt(II) acetate, **2,** 80
N-Iodosuccinimide, **10,** 216
Iodylbenzene, **11,** 275
Lead tetraacetate, **1,** 537
Lead tetrakis(trifluoroacetate)–Diphenyl disulfide, **9,** 269
Manganese(IV) oxide, **5,** 422
Oxygen, **9,** 335
Periodates, **2,** 311
Periodic acid, **1,** 815; **2,** 313; **4,** 374
Phenyliodine(III) bis(trifluoroacetate), **6,** 301
Phenyliodine(III) diacetate, **1,** 508
Phosphatolead(IV) acids, **3,** 225
Potassium persulfate–Silver nitrate, **1,** 954
Pyridinium chlorochromate, **11,** 450
Silver carbonate–Celite, **5,** 577; **6,** 511
Silver iododibenzoate, **3,** 251
Sodium bismuthate, **1,** 1045
Sodium periodate, **11,** 492
Thallium(III) nitrate, **4,** 492
Triphenylbismuth carbonate, **9,** 501; **11,** 587
Zinc, **2,** 459
OF DIOLS (AND RELATED COMPOUNDS) → RCOOH, R_2CO
Cobalt(II) acetate, **2,** 80
Kiliani reagent, **1,** 144
Lead tetraacetate, **6,** 313
Periodic acid–Chromium(VI) oxide, **2,** 315
Ruthenium tetroxide, **11,** 462
Xenic acid, **2,** 458
OF EPOXIDES
2-Butanone, **1,** 678
Dimethyl sulfoxide, **1,** 296; **2,** 157
Hydrogen peroxide, **7,** 174
Periodic acid, **1,** 815; **5,** 508
OF α-HYDROXY C=O's (AND RELATED COMPOUNDS) → RCOOH, R_2CO
Calcium hypochlorite, **11,** 107
Cerium(IV) ammonium nitrate, **4,** 71
Hydrogen peroxide, **8,** 247
Lead tetraacetate, **7,** 185
Nickel peroxide, **1,** 731
Potassium superoxide, **7,** 304
Sodium bismuthate, **1,** 1045
Tetrabutylammonium periodate, **10,** 381
OF α-HYDROXY C=O's (AND RELATED COMPOUNDS) → RCHO, R_2CO
Benzeneseleninic anhydride, **8,** 29
N-Bromosuccinimide, **1,** 78
Calcium hypochlorite, **11,** 107
Cerium(IV) ammonium nitrate, **4,** 71
2-Chloro-3-ethylbenzoxazolium tetrafluoroborate, **8,** 90; **9,** 105
2,4-Dinitrofluorobenzene, **1,** 321
Dipotassium tetrachloroplatinate(II), **4,** 215
Hydrogen peroxide, **1,** 472; **8,** 247
N-Iodosuccinimide, **12,** 258
Lead tetraacetate, **1,** 537; **5,** 365; **8,** 269
Methylthioacetic acid, **6,** 395
Periodic acid, **1,** 815; **2,** 313
Phosphonitrilic chloride trimer, **6,** 469
Phosphoryl chloride, **1,** 876
Potassium *t*-butoxide, **2,** 336
Pyridine N-oxide, **2,** 353
Sodium bismuthate, **1,** 1045
Sodium hypochlorite, **1,** 1084
N-Sulfinylaniline, **6,** 556
Tetrabutylammonium periodate, **10,** 381
Tributyl(iodoacetoxy)tin, **12,** 258
OF KETONES
Butyl azide, **1,** 84
Caro's acid, **1,** 118
Chloramine, **1,** 122
Cobalt(II) acetate, **4,** 99
4,4-Dichloro-3-buten-1-ol, **5,** 192
Diphenyl phosphoroazidate, **10,** 173
Ferric chloride, **12,** 230
Hexamethylphosphoric triamide, **1,** 430

Hydrogen peroxide–Selenium(IV) oxide, **1,** 477
Hypohalite solution, **1,** 488
Iodine–Pyridine, **1,** 503
Nitric acid, **1,** 733
Potassium *t*-butoxide, **1,** 911; **2,** 336; **9,** 380
Potassium hydroxide–Carbon tetrachloride, **5,** 96
Potassium permanganate, **1,** 942
Potassium superoxide, **9,** 391; **11,** 442
Silver perchlorate, **2,** 369
Sodium dichromate, **1,** 1059
Sodium hypohalite, **1,** 1083, 1084
Tetra-μ_3-carbonyldodecacarbonylhexarhodium, **6,** 504
Zinc permanganate, **12,** 576
OF *o*-PHENYLENEDIAMINES
Copper(I) chloride, **5,** 164; **7,** 80; **8,** 118
Lead tetraacetate, **1,** 537
Nickel peroxide, **1,** 731
OXIDATIVE COUPLING (*see also* COUPLING REACTIONS)
OF CARBANIONS, RM
Copper, **12,** 140
Copper(I) iodide–Tributylphosphine, **2,** 400
Copper(II) tetrafluoroborate, **12,** 144
Hydrogen peroxide, **8,** 247
Lithium aluminum hydride, **9,** 274
Lithium dibutylcuprate, **3,** 79
(R)-4-Methylcyclohexylidenemethylcopper, **7,** 238
Molybdenum carbonyl, **7,** 247
Nickel(II) acetylacetonate, **11,** 58
Silver(I) nitrate, **2,** 366; **4,** 429
Triphenyl phosphite ozonide, **7,** 408
OF ArH (*see* COUPLING REACTIONS)
OF ArOH
Bis(N-propylsalicylideneaminate)cobalt(II), **11,** 61
Chloramine, **2,** 65
Copper(II)–Amine complexes, **8,** 114
Di-μ-chlorodimethoxybis(pyridine)dicopper, **10,** 136
Diphenyl selenoxide, **9,** 201
Ferric chloride, **1,** 390; **2,** 199; **3,** 145; **4,** 236; **10,** 185
Lead(IV) oxide, **1,** 533
Manganese(III) acetylacetonate, **2,** 264
Manganese(IV) oxide, **2,** 257; **3,** 191
Palladium(II) chloride, **4,** 369
Potassium ferricyanide, **1,** 929; **2,** 345; **4,** 406; **6,** 480; **7,** 300
Salcomine, **7,** 316
Silver carbonate–Celite, **3,** 247; **4,** 425
Silver(I) oxide, **6,** 515
Thallium(III) trifluoroacetate, **4,** 498; **6,** 579; **9,** 462
Vanadyl trichloride, **3,** 331; **5,** 744; **8,** 527; **11,** 593
Vanadyl trifluoride, **5,** 745; **7,** 418; **9,** 513
OF SILYL ENOL ETHERS, ENOLATES, AND RELATED COMPOUNDS
Copper(I) chloride, **6,** 145
Copper(II) chloride, **6,** 139; **9,** 123
Copper(II) trifluoromethanesulfonate, **8,** 126; **10,** 110
2,3-Dichloro-5,6-dicyano-1,4-benzoquinone, **1,** 215
Iodine, **1,** 495
Lithium diisopropylamide, **12,** 277
Manganese(III) acetate, **6,** 355
Manganese(IV) oxide, **4,** 317
Nickel peroxide, **5,** 474
2,4-Pentanedione, **1,** 10
Silver(I) oxide, **6,** 515; **7,** 321
Titanium(IV) chloride, **8,** 483; **11,** 529
Zinc-copper couple, **7,** 428
OXIDATIVE DEALKYLATION
Acetic anhydride–Boron trifluoride etherate, **1,** 72
t-Butyl chromate, **5,** 73
Cerium(IV) ammonium nitrate, **9,** 99; **10,** 79
Chromium(VI) oxide, **2,** 72; **5,** 140
Copper(II) chloride, **2,** 84
Ferric chloride, **7,** 153
Lead tetraacetate, **1,** 537
Manganese(IV) oxide, **1,** 637
Silver(II) dipicolinate, **10,** 348
Silver(II) oxide, **4,** 431; **7,** 322; **9,** 412; **10,** 352
OXIDATIVE DECARBONYLATION
Copper(II)–Amine complexes, **3,** 65; **5,** 157
Oxygen, **11,** 384
Oxygen, singlet, **8,** 367
Periodic acid, **1,** 815
OXIDATIVE DECARBOXYLATION
BIS-DECARBOXYLATION OF

OXIDATIVE DECARBOXYLATION (*Continued*)
DIACIDS
t-Butyl hydroperoxide, **3,** 37
Dicarbonylbis(triphenylphosphine)-nickel, **7,** 94; **8,** 147
Lead(IV) oxide, **1,** 533
Lead tetraacetate, **1,** 537; **2,** 234; **3,** 168; **4,** 278; **8,** 269
Potassium ferricyanide, **1,** 929
OF CARBOXYLIC ACIDS → RCHO, R_2CO
m-Chloroperbenzoic acid, **8,** 97
N-Chlorosuccinimide, **6,** 115; **8,** 103
Copper carbonate, basic, **4,** 101
3-Methylthio-1,4-diphenyl-5-triazolium iodide, **8,** 351
Oxygen, **6,** 426; **7,** 258
Oxygen, singlet, **8,** 367
Pyridine N-oxide, **1,** 966
Tetrabutylammonium periodate, **10,** 381
OF CARBOXYLIC ACIDS → C=C (–H,COOH)
Copper(II) acetate, **2,** 84
Lead(IV) oxide, **1,** 533; **2,** 233
Lead tetraacetate, **1,** 537; **2,** 234; **3,** 168; **5,** 365; **6,** 313; **7,** 185
Palladium(II) acetate–1,2-Bis(diphenylphosphine)ethane, **11,** 391
Potassium persulfate, **3,** 238
p-Toluenesulfonyl chloride, **1,** 1179
OF α-HYDROXY ACIDS (AND RELATED COMPOUNDS)
N-Bromosuccinimide, **1,** 78
2-Chloro-3-ethylbenzoxazolium tetrafluoroborate, **8,** 90; **9,** 105
Dipotassium tetrachloroplatinate(II), **4,** 215
Hydrogen peroxide–Iron salts, **1,** 472
N-Iodosuccinimide, **12,** 258
Lead tetraacetate, **1,** 537; **5,** 365
Methylthioacetic acid, **6,** 395
Periodic acid, **2,** 313
Pyridine N-oxide, **2,** 353
Sodium bismuthate, **1,** 1045
Sodium hypochlorite, **1,** 1084
Tetrabutylammonium periodate, **10,** 381
OXIDATIVE DECYANATION
Benzenesulfenyl chloride, **5,** 523
t-Butyl chromate, **5,** 73
t-Butyldimethylchlorosilane, **5,** 74
Oxygen, **6,** 426; **9,** 335; **10,** 293
Polyphosphoric acid, **1,** 894
Potassium *t*-butoxide, **10,** 323
OXIDATIVE DESULFONYLATION
Bis(trimethylsilyl) peroxide, **12,** 63
Potassium *t*-butoxide, **12,** 401
OXIDATIVE ESTERIFICATION
N-Bromosuccinimide, **7,** 37
OXIDATIVE HYDROLYSIS
(*see* HYDROLYSIS)
OXIDATIVE LACTONIZATION
Dimethyl sulfoxide–Methanesulfonic anhydride, **9,** 191
Lead tetraacetate, **9,** 265; **10,** 228
OXO REACTION, OXYAMINATION
(*see* ADDITION REACTIONS)
OXY-COPE REARRANGEMENT
(*see* SIGMATROPIC REARRANGEMENTS)
OXYMERCURATION, OXYSELENYLATION, OXYSULFENYLATION (*see* ADDITION REACTIONS)
OZONOLYSIS
2,3-Dichloro-1-propene, **8,** 158
Hexamethylphosphorous triamide, **2,** 210
Ozone, **1,** 773, 1234; **2,** 156; **4,** 363; **5,** 320, 491; **6,** 436; **7,** 269; **8,** 374; **9,** 341; **11,** 387
Potassium hypochlorite, **1,** 938
Sodium borohydride, **1,** 1049
Tetracyanoethylene, **5,** 647
Thiourea, **11,** 519
Trimethyl phosphite, **1,** 1233
Triphenylphosphine, **1,** 1238

PERKIN REACTION
Acetic anhydride, **5,** 4; **9,** 1
Lead(II) oxide, **1,** 536
Potassium acetate, **1,** 906
Sodium acetate, **1,** 1024
Triethylamine, **1,** 1198
PERKOW REACTION
Triethyl phosphite, **3,** 304
PEROXYMERCURATION (*see* ADDITION REACTIONS)
PETERSON OLEFINATION (*see also* METHYLENATION)
Bis(methylthio)(trimethylsilyl)-methyllithium, **6,** 53

1,3-Bis(triisopropylsilyl)propyne, **11,** 63
t-Butyl 2-chloro-2-lithiotrimethylsilylacetate, **8,** 60
t-Butyl lithiobis(trimethylsilyl)acetate, **7,** 42
t-Butyl lithio(trimethylsilyl)acetate, **5,** 371
N-Chlorosuccinimide, **6,** 115
N,N-Dimethylhydrazine, **7,** 126
Ethyl lithio(trimethylsilyl)acetate, **5,** 373
Ethyl trimethylsilylacetate, **12,** 228
2-Lithio-2-trimethylsilyl-1,3-dithiane, **5,** 374; **6,** 320
Lithium 1-(dimethylamino)-naphthalenide, **12,** 279
Lithium naphthalenide, **11,** 302
Methoxy(phenylthio)trimethylsilylmethane, **12,** 317
Methoxy(phenylthio)trimethylsilylmethyllithium, **12,** 317
N-Methyl-N-trimethylsilylmethyl-N′-*t*-butylformamidine, **11,** 347
Organotitanium reagents, **11,** 374
Phenylsulfonyl(trimethylsilyl)methane, **12,** 393
Phenylthiotrimethylsilylmethane, **11,** 423
Sodium methoxide, **5,** 617
Titanium(IV) alkoxides, **12,** 504
Trimethylsilylacetic acid, **6,** 631
Trimethylsilylacetonitrile, **11,** 571
[(Trimethylsilyl)allyl]lithium, **8,** 273
Trimethylsilylbenzyllithium, **6,** 637
Trimethylsilylmethylmagnesium chloride, **6,** 636
Tris(trimethylsilyl)ketenimine, **11,** 591

PFITZNER–MOFFATT REACTION
Dimethyl sulfoxide, **1,** 304; **2,** 162; **3,** 119; **6,** 225; **8,** 198
Formaldehyde, **7,** 158
Polymeric carbodiimide reagent, **5,** 539

PHOSPHORYLATION
Acetoinenediolcyclopyrophosphate, **6,** 6
Benzyltriethylammonium chloride, **6,** 41
Bis(*p*-nitrophenethyl) phosphorochloridate, **11,** 56
Bis(2,2,2-trichloro-1,1-dimethylethyl) phosphorochloridate, **11,** 62
Bis(2,2,2-trichloroethyl) phosphorochloridate, **2,** 187
α-Bromocyanoacetamide, **2,** 40
Cesium fluoride, **7,** 57; **8,** 81
(2-Chloromethyl-4-nitrophenyl) dichlorophosphate, **3,** 48
2-Chloro-2-oxobenzo-1,3,2-dioxaphosphole, **1,** 837; **2,** 321; **3,** 223
2-Chlorophenyl dichlorophosphite, **6,** 114
2-Cyanoethyl phosphate, **1,** 172
Cyclohexyl isocyanide, **6,** 152
Dibenzyl phosphorochloridate, **1,** 198
Dibromomalonamide, **1,** 208
Di(2-*t*-butylphenyl) phosphorochloridate, **6,** 167
Di-*t*-butyl phosphorobromidate, **8,** 146
2-(N,N-Dimethylamino)-4-nitrophenyl phosphate, **6,** 207
1-(4,5-Dimethyl-1,3,2-dioxaphosphol-2-yl)-1H-imidazole, **6,** 218
Dimorpholinophosphinyl chloride, **1,** 861
Di-*p*-nitrobenzyl phosphorochloridate, **1,** 321
Dioxane diphosphate, **1,** 334
Diphenyl phosphorochloridate, **1,** 345
Ethoxyacetylene, **1,** 357
Lithium dibutylcuprate, **7,** 92
Lithium diisopropylamide, **10,** 241
Lithium dimethylcuprate, **4,** 177
N-Methylimidazole, **5,** 447
p-Nitrophenyl dichlorophosphate, **1,** 744
4-Nitrophenyl phenyl phosphorochloridate, **8,** 362
Organocopper reagents, **10,** 282
Phenyl N-phenylphosphoramidochloridate, **7,** 286
Phenyl 8-quinolyl phosphate, **5,** 517
Phosphoric acid, **1,** 860
Phosphorus(V) oxide–Phosphoric acid, **1,** 872
Phosphoryl chloride, **1,** 876; **3,** 228, 229
Polyphosphoric acid, **1,** 894
Pyrophosphoryl chloride, **1,** 971; **2,** 353
Silver dibenzyl phosphate, **1,** 1006
Sodium methyl phosphorazidate, **2,** 386
Tetrabutylammonium bromide, **7,** 353
Tetrakis(triphenylphosphine)-palladium(0), **10,** 384
Tetra-*p*-nitrophenyl pyrophosphate, **1,** 1148
Trichloroacetonitrile, **1,** 1194
2,2,2-Trimethoxy-4,5-dimethyl-1,3-dioxaphospholene, **4,** 535

PHOSPHORYLATION (*Continued*)
Triphenylphosphine–Diethyl azodicarboxylate, **4,** 553
PICTET–SPENGLER ISOQUINOLINE SYNTHESIS
Diisobutylaluminum hydride, **11,** 185
Dimethyl(methylene)ammonium salts, **10,** 160
Methyl iodide, **9,** 308
PINACOL COUPLING (*see* REDUCTIVE COUPLING)
PINACOL REARRANGEMENT
Boron trifluoride dibutyl etherate, **4,** 43
Calcium carbonate, **4,** 67
Ferric chloride, **8,** 229
Iodine, **1,** 495; **2,** 220
Nafion-H, **9,** 320
Potassium *t*-butoxide, **2,** 336
Potassium hydroxide, **4,** 409
Sulfuric acid, **5,** 633
Trialkylaluminums, **12,** 512
POLONOVSKI REACTION
Acetic anhydride, **2,** 7; **5,** 3
Trichloroacetic anhydride, **7,** 380
Trifluoroacetic anhydride, **3,** 308; **7,** 389; **9,** 484
POLYENE CYCLIZATION
Acetic acid, **7,** 1
Alumina, **9,** 8
Benzeneselenenyl halides, **10,** 16; **11,** 34
Boron trifluoride etherate, **4,** 44; **9,** 64; **11,** 72
Bromine–Silver salts, **7,** 36
N-Bromosuccinimide, **10,** 57
Camphor-10-sulfonic acid, **11,** 107
Chlorotris(triphenylphosphine)-rhodium(I), **4,** 559
Darvon alcohol, **8,** 184
1,1-Dichloro-3-bromopropene, **5,** 191
4,4-Dichloro-3-buten-1-ol, **5,** 192
2,3-Dichloro-1-propene, **2,** 120
Dimethyl(methylthio)sulfonium tetrafluoroborate, **11,** 204
3-Ethyl-2-fluorobenzothiazolium tetrafluoroborate, **8,** 223
Formaldehyde, **11,** 240
Formic acid, **5,** 316; **7,** 160; **8,** 232
Hexafluoroantimonic acid, **7,** 166
Ion-exchange resins, **2,** 227; **6,** 302
2-Lithio-2-trimethylsilyl-1,3-dithiane, **6,** 320
Mercury(II) acetate, **3,** 194; **5,** 424
Mercury(II) chloride, **9,** 291
Mercury(II) trifluoroacetate, **4,** 325; **9,** 294
Mercury(II) trifluoromethanesulfonate–N,N-Dimethylaniline, **12,** 307
Perchloric acid, **1,** 796
Phosphoric acid–Boron trifluoride, **7,** 289
Phosphorus(V) oxide–Methanesulfonic acid, **11,** 428
Pyridinium chlorochromate, **7,** 308; **8,** 425
Sulfuric acid, **4,** 470; **5,** 633
2,4,4,6-Tetrabromo-2,5-cyclohexadienone, **10,** 377
Tetrakis(2-methylpropyl)-μ-oxodialuminum, **8,** 471
Tin(IV) chloride, **5,** 627; **7,** 342; **9,** 436
Titanium(IV) alkoxides, **10,** 404
Titanium(IV) chloride–N-Methylaniline **9,** 470
Trifluoroacetic acid, **3,** 305; **4,** 530; **5,** 695; **6,** 613; **7,** 388; **8,** 503
Trifluoromethanesulfonic anhydride, **6,** 618
Trimethylsilylmethyl trifluoromethane-sulfonate, **10,** 434
Zinc bromide, **8,** 535
POMERANZ–FRITSCH REACTION
Boron trifluoride–Trifluoroacetic anhydride, **4,** 45
PONZIO REACTION
Nitrogen dioxide, **1,** 324
PREVOST REACTION
Iodine–Silver acetate, **12,** 256
Silver iododibenzoate, **1,** 1007; **9,** 411
PRINS REACTION
Alkylaluminum halides, **11,** 7; **12,** 5
Formaldehyde, **1,** 397; **7,** 158
Hexakis(acetato)trihydrato-μ_3-oxotrisrhodium acetate, **11,** 252
PROPARGYLATION
Di-μ-carbonylhexacarbonyldicobalt, **8,** 148; **10,** 129
PROTECTION OF
ALCOHOLS
Benzyl ethers
Benzyl bromide, **5,** 25
Benzyl trichloroacetimidate, **11,** 44
2,3-Dichloro-5,6-dicyano-1,4-

benzoquinone, **11,** 166
Palladium catalysts, **10,** 299
Esters
Acetic-formic anhydride, **1,** 4
Crotonic anhydride, **8,** 127
2-Dibromomethylbenzoyl chloride, **9,** 138
Levulinic acid, **10,** 230
Levulinic anhydride, **6,** 318
Sodium triphenylmethoxyacetate, **5,** 627
Tigloyl chloride, **1,** 1167
Trifluoroacetic anhydride, **1,** 1221
Vinyl chloroformate, **8,** 530
Zinc–Ammonium chloride, **5,** 758
Zinc-copper couple, **11,** 599
MEM ethers
Bromodimethylborane, **12,** 199
2-Chloro-1,3,2-dithioborolane, **12,** 112
Chlorotrimethylsilane–Sodium iodide, **12,** 127
Mercury(II) chloride, **12,** 304
β-Methoxyethoxymethyl chloride, **7,** 227; **9,** 299
Pyridinium *p*-toluenesulfonate, **12,** 420
MOM ethers
Bromodimethylborane, **12,** 199
Bromotrimethylsilane, **12,** 79
Chloromethyl methyl ether, **5,** 120
Dimethoxymethane, **6,** 374; **7,** 115
Formaldehyde, **1,** 397
Mercury(II) chloride, **12,** 304
Methoxymethyl methanesulfonate, **3,** 198
2-Methyl-2-thiazoline, **6,** 403
Nafion-H, **11,** 354
Pyridinium *p*-toluenesulfonate, **12,** 420
MTM ethers
Chloromethyl methyl sulfide, **9,** 107
Dimethyl sulfoxide–Acetic anhydride, **7,** 135; **9,** 190
Iodomethyl methyl sulfide, **6,** 301
Silyl ethers
Acetic anhydride–Ferric chloride, **6,** 260
Allyltrimethylsilane, **10,** 6
Bis(trimethylsilyl)acetamide, **1,** 61; **3,** 23
N,O-Bis(trimethylsilyl) sulfamate, **7,** 28
Bis(trimethylsilyl)urea, **11,** 68
Boron trifluoride etherate, **7,** 31; **9,** 64
Bromo-*t*-butylmethoxyphenylsilane, **12,** 100
Bromotrimethylsilane, **12,** 79
t-Butyldimethylchlorosilane, **4,** 57, 176; **5,** 74; **6,** 78; **8,** 59; **9,** 77; **10,** 62
N-(*t*-Butyldimethylsilyl)-N-methyltrifluoroacetamide, **11,** 336
t-Butyldimethylsilyl perchlorate, **9,** 78
t-Butyldiphenylchlorosilane, **6,** 81; **12,** 87
N-Chlorosuccinimide, **10,** 94
Chlorotrimethylsilane, **1,** 1232; **9,** 112; **10,** 96; **11,** 125
Diethyltrimethylsilylamine, **3,** 317; **4,** 544
4-Dimethylaminopyridine, **9,** 178
Ethyl trimethylsilylacetate, **7,** 150; **11,** 234
Guanidines, **12,** 477
Hexamethyldisilazane, **1,** 427; **2,** 207
Hexamethyldisiloxane, **9,** 55
Hydrofluoric acid, **9,** 238
Ketene alkyl trialkylsilyl acetals or ketals, **9,** 310; **11,** 279
Potassium superoxide, **9,** 391
Tetrabutylammonium fluoride, **4,** 477
Triisopropylsilyl trifluoromethanesulfonate, **11,** 567
N-Trimethylsilylacetamide, **1,** 1235
Trimethylsilyl trifluoromethanesulfonate, **11,** 584
THP ethers
Alkylaluminum halides, **12,** 5
Bis(trimethylsilyl) sulfate, **11,** 67
Boron trifluoride etherate, **4,** 44
Bromotrimethylsilane, **12,** 79
3,4-Dihydro-2H-pyran, **1,** 256
Ion-exchange resins, **9,** 256
Magnesium sulfate, **1,** 634
Pyridinium *p*-toluenesulfonate, **8,** 427
p-Toluenesulfonic acid, **5,** 673
Trichloro(methyl)silane, **11,** 553
other Ethers
Alkylaluminum halides, **11,** 7
2-Benzyloxy-1-propene, **12,** 48
Boron trifluoride–Methanol, **9,** 64
Bromobis(isopropylthio)borane, **12,** 57
Bromotrimethylsilane, **12,** 79
t-Butyl chloromethyl ether, **9,** 77

PROTECTION OF (*Continued*)
9-Chloro-9-phenylxanthene, **9,** 110
2-Chlorotetrahydrofuran, **9,** 112
Chlorotris(triphenylphosphine)-rhodium(I), **5,** 736
(1,5-Cyclooctadiene)bis(methyl-diphenylphosphine)iridium(I) hexafluorophosphate, **8,** 135; **10,** 115
2,3-Dichloro-5,6-dicyano-1,4-benzoquinone, **12,** 174
Diethoxymethane, **11,** 241
5,6-Dihydro-4-methoxy-2H-pyran, **2,** 271; **3,** 197
4-Dimethylaminopyridine, **9,** 178; **10,** 155
Dimethyl sulfoxide, **1,** 296
Guaiacylmethyl chloride, **10,** 194
9-Hydroxy-9-phenylanthrone, **4,** 378
Iridium catalysts, **8,** 135
Isobutene, **1,** 522
Mercury(II) chloride, **12,** 304
2-Methoxypropene, **5,** 360
N-Methyl-N-phenylaminoethynyl-lithium, **10,** 267
α-Naphthyldiphenylmethyl chloride, **6,** 415
Palladium catalysts, **7,** 275; **10,** 299
Selenium(IV) oxide, **3,** 245
Sodium bis(2-methoxyethoxy)-aluminum hydride, **7,** 327
Sulfuryl chloride, **7,** 356
2-Tetrahydrothienyl diphenylacetate, **9,** 450
Tri(2-chloroethyl) orthoformate, **3,** 297
Trichloro(methyl)silane, **11,** 553
β-(Trimethylsilyl)ethoxymethyl chloride, **10,** 431
Trimethyl(triphenylmethoxy)silane, **10,** 444
Triphenylmethyl chloride, **1,** 1254; **2,** 453; **6,** 656
Triphenylmethylpyridinium tetrafluoroborate, **5,** 741
Tris(*p*-bromophenyl)ammoniumyl hexachloroantimonate, **9,** 510
4,4′,4″-Tris(4,5-dichlorophthalamido)-trityl bromide, **12,** 557
Zinc bromide, **10,** 461; **11,** 600
other Methods
Boric acid, **3,** 29
Disiamylborane, **3,** 22
Methanesulfonyl chloride, **6,** 362
Nickel carbonyl, **5,** 472
2-Phenylsulfonylethyl chloroformate, **11,** 419
O-(Phenylthiomethyl)hydroxylamine, **5,** 527
2,2,2-Trichloroethyl chloroformate, **2,** 426; **3,** 296
2-(Trimethylsilyl)ethyl chloroformate, **10,** 431

ALDEHYDES, KETONES (*see also* SYNTHESIS INDEX—ACETALS, THIOACETALS)
Cyanotrimethylsilane, **4,** 542
Hydroxylamine, **7,** 176
Lithium morpholide, **12,** 284
Methoxyamine, **1,** 670
Organotitanium reagents, **12,** 353
Piperidine, **1,** 886
Sodium hydrogen sulfite, **11,** 487
Triphenylmethyllithium, **4,** 567

ALKENES
Cyclopentadiene, **8,** 137
Diazomethane, **7,** 88
Dicarbonylcyclopentadienyl(2-methyl-1-propene)iron tetrafluoroborate, **6,** 154
Ozone–Silica gel, **7,** 271
Pyridinium bromide perbromide, **9,** 399
Trimethyltinsodium, **9,** 500

AMINES
Alkyl chloroformates, **10,** 431
9-Anthrylmethyl *p*-nitrophenyl carbonate, **8,** 18
2-Benzoyloxymethylbenzoyl chloride, **7,** 14
Bis(chlorodimethylsilyl)ethane, **10,** 140
1,4-Bis(N,N-dimethylamino)-1,1,4,4-tetramethyldisilethylene, **12,** 476
Borane–Tetrahydrofuran, **8,** 141
Chlorotris(triphenylphosphine)-rhodium(I), **10,** 98
Cobalt(II) phthalocyanine, **9,** 119
Crown ethers, **7,** 76
Di-*t*-butyl dicarbonate, **12,** 159
2,3-Dichloro-5,6-dicyano-1,4-benzoquinone, **11,** 166
3,4-Dihydro-2H-pyran, **1,** 256
Dilithium tetrachloropalladate(II), **8,** 470

4,4'-Dimethoxybenzhydrylamine, **9,** 174
4-Dimethylaminopyridine, **10,** 155
Diphenyldi(1,1,1,3,3,3-hexafluoro-2-phenyl-2-propoxy)sulfurane, **6,** 239
Diphenylethylene carbonate, **4,** 25; **5,** 23
Iodotrimethylsilane, **11,** 271
p-Methoxybenzenesulfonyl chloride, **9,** 297
4-Methoxy-2,3,6-trimethylbenzenesulfonyl chloride, **11,** 330
Nickel carbonyl, **5,** 472
o-Nitrobenzenesulfenyl chloride, **8,** 360
p-Phenylbenzoyl chloride, **7,** 282
2-Picolyl chloride N-oxide, **5,** 538
2-(2-Pyridyl)ethyl-*p*-nitrophenyl carbonate, **12,** 420
Rhodium(III) chloride, **8,** 435
Sodium borohydride, **12,** 441
Tetraethylammonium fluoride, **8,** 470
2,2,2-Trichloroethyl chloroformate, **7,** 383
Trifluoroacetic anhydride, **1,** 1221
Trifluoromethanesulfonic anhydride, **5,** 702
2-(Trimethylsilyl)ethyl chloroformate, **10,** 431
Vinyl chloroformate, **8,** 530
AMINO ACIDS (*see* SYNTHESIS INDEX—AMINO ACIDS)
CARBOXYL GROUPS
Allyltrimethylsilane, **10,** 6
2-Amino-2-methyl-1-propanol, **3,** 14; **6,** 20
Bis(trimethylsilyl)urea, **11,** 68
5-Bromo-7-nitroindoline, **11,** 79
p-Bromophenacyl bromide, **3,** 34
t-Butyl acetate, **1,** 82
p-Chlorobenzyl alcohol, **7,** 59
Chlorodimethylsulfonium chloride, **9,** 105
Chloromethyl methyl sulfide, **4,** 84; **8,** 94; **9,** 107; **12,** 117
2-Chlorotetrahydrofuran, **9,** 112
Chlorotrimethylsilane, **5,** 709; **10,** 96; **11,** 125
Cobalt(II) phthalocyanine, **9,** 119
Crown ethers, **5,** 152
5,6-Dihydrophenanthridine, **11,** 184
3,4-Dihydro-2H-pyran, **1,** 256
N,N'-Diisopropylhydrazine, **4,** 162
Dimethoxymethane, **7,** 115
Dimethyl sulfoxide–*t*-Butyl bromide, **9,** 190
Hexamethyldisiloxane, **10,** 196
2-Hydroxymethylanthraquinone, **8,** 252
Isobutene, **1,** 522
Ketene alkyl trialkylsilyl acetals or ketals, **9,** 310
Lithium dimethylcuprate, **8,** 301
Lithium methylsulfinylmethylide–Tributylborane, **11,** 302
Mercury(II) acetate, **8,** 315
p-Methoxyphenacyl bromide, **5,** 439
3-Methyl-3-hydroxymethyloxetane, **12,** 323
p-Nitrobenzyl bromide, **8,** 360
Organolithium reagents, **9,** 5
Phenacyl bromide, **12,** 381
2-Pyridineethanol, **12,** 416
Sodium–Ammonia, **9,** 415
Sodium ethanedithiolate, **6,** 540
Sodium hydrogen selenide, **8,** 459
Sodium methanethiolate, **5,** 626
Sodium trithiocarbonate, **6,** 553
Sulfuryl chloride–Tetrahydrofuran, **7,** 356
Tributyltinlithium, **8,** 495
2,2,2-Trichloroethanol, **3,** 295
4-Trimethylsilyl-2-buten-1-ol, **12,** 538
2-Trimethylsilylethanol, **8,** 510
N-Trimethylsilyl-2-oxazolidinone, **11,** 582
3-Trimethylsilyl-2-oxazolidone, **11,** 383
Trimethylsilyl trifluoromethanesulfonate, **11,** 584
Trimethyltinlithium, **9,** 499
DICARBONYL COMPOUNDS
Dicyanodimethylsilane, **8,** 161
Isopropyl bromide, **9,** 69
1,3-DIENES
Benzylideneacetone(tricarbonyl)iron, **6,** 39
Di-μ-carbonyldecacarbonyltri-*triangulo*-iron, **5,** 706
Iron carbonyl, **10,** 221
N-Phenyl-1,2,4-triazoline-3,5-dione, **3,** 223; **9,** 372; **12,** 395
Tri-μ-carbonylhexacarbonyldiiron, **7,** 110
DIOLS
1,2-
Benzeneboronic acid, **3,** 221

PROTECTION OF (*Continued*)
Bis(1-methoxy-2-methyl-1-propenyloxy)dimethylsilane, **12,** 58
Boronic acid resins, **9,** 59
t-Butyldimethylchlorosilane, **9,** 77
N,N′-Carbonyldiimidazole, **6,** 97
Copper(II) sulfate, **8,** 125
Dibromomethane, **11,** 157
Di-*t*-butyldichlorosilane, **11,** 159
Dichlorodimethylsilane, **3,** 114
Diisopropylbis(trifluoromethanesulfonato)silane, **11,** 189
2,2-Dimethoxypropane, **5,** 226
N,N-Dimethylacetamide dimethyl acetal, **4,** 166
Dimethyl sulfoxide, **3,** 119
Dimethyl sulfoxide–N-Bromosuccinimide, **4,** 199
2-Methoxypropene, **11,** 329; **12,** 317
Perchloric acid, **3,** 220
Phase-transfer catalysts, **8,** 387
2-Phenyl-1,3-dithiolane, **6,** 26
Tetramethyl orthocarbonate, **3,** 285
2-Trimethylsilyloxy-1-propene, **5,** 718
1,3-
Acetone, **3,** 4
Benzaldehyde, **6,** 26
Benzeneboronic acid, **5,** 513; **7,** 284
Bis(1-methoxy-2-methyl-1-propenyloxy)dimethylsilane, **12,** 58
Boronic acid resins, **7,** 284
Diacetoxydimethylsilane, **3,** 113
Dibromomethane, **11,** 157
Di-*t*-butyldichlorosilane, **11,** 159
2,2-Diethoxypropane, **5,** 208
Diisopropylbis(trifluoromethanesulfonato)silane, **11,** 189
2-Methoxypropene, **11,** 329
2-Phenyl-1,3-dithiolane, **6,** 26
LACTONES
Bis(dimethylaluminum) 1,2-ethanedithiolate, **5,** 35
m-Chloroperbenzoic acid, **8,** 97
α-METHYLENE LACTONES
Bromine, **11,** 75
Dimethylamine, **7,** 119
Diphenyl diselenide, **5,** 272
Sodium thiophenoxide, **6,** 552
PHENOLS
Benzyl *p*-toluenesulfonate, **11,** 44
p-Bromophenacyl bromide, **3,** 34
Chloromethyl methyl ether, **1,** 132; **7,** 61
Chloromethyl methyl sulfide, **8,** 94
2-Chlorotetrahydrofuran, **9,** 112
Crown ethers, **5,** 152
3,4-Dihydro-2H-pyran, **1,** 256
Fluorene-9-carboxylic acid, **1,** 394
Hexamethyldisilazane, **6,** 273
Ion-exchange resins, **9,** 256
Isopropyl bromide, **9,** 69
Ketene alkyl trialkylsilyl acetals or ketals, **9,** 310
Phase-transfer catalysts, **8,** 387
Sodium methanethiolate, **5,** 626
Sulfuryl chloride, **7,** 356
Trimethylacetyl chloride, **8,** 404
Triphenylmethyl chloride, **1,** 1254
Vinyl chloroformate, **8,** 530
Zinc, **10,** 459
PHOSPHATES
t-Butylamine, **10,** 62
9-Fluorenylmethanol, **5,** 308
2-Phenylthioethanol, **5,** 516
Tetrabutylammonium fluoride, **7,** 353
2-Trimethylsilyl-2-propen-1-ol, **9,** 497
2-(*p*-Triphenylmethylphenyl)-thioethanol, **7,** 414
QUINONES
Cyanotrimethylsilane, **5,** 720; **7,** 397
Hexamethyldisilane, **11,** 253
Silver(II) oxide, **4,** 431
THIOLS, THIOPHENOLS
o-Carboxyphenyl *o*-carboxybenzenethiosulfonate, **5,** 100
Diethyl methylenemalonate, **3,** 96
3,4-Dihydro-2H-pyran, **1,** 256
Methoxymethyl isocyanate, **5,** 439
o-Nitrobenzenesulfenyl chloride, **10,** 277
2-Picolyl chloride 1-oxide hydrochloride, **6,** 472
Sodium methanethiolate, **5,** 626
Sulfuryl chloride, **7,** 356
Trifluoroacetic anhydride, **12,** 530
Triphenylmethyl chloride, **1,** 1254
PSCHORR RING CLOSURE
6-Methoxy-7-hydroxy-3,4-dihydroisoquinolinium methiodide, **4,** 329
Phase-transfer catalysts, **12,** 379
Sodium iodide, **3,** 267; **12,** 449
Zinc carbonate, **4,** 579
PUMMERER REARRANGEMENT
Acetic anhydride, **5,** 3; **11,** 1

Acetic anhydride + co-reagent, **6,** 3, 5; **7,** 2; **8,** 2
Acetyl chloride, **4,** 5
Cerium(IV) ammonium nitrate, **4,** 71
Chlorotrimethylsilane, **6,** 626
2,2-Dimethyl-6-(*p*-chlorophenylthiomethylene)cyclohexanone, **6,** 208
Iodine, **2,** 220
Methoxy(phenylthio)methyllithium, **6,** 369
Methyl methylthiomethyl sulfone, **11,** 242
Phenylthiotrimethylsilylmethane, **10,** 313
Sodium methylsulfinylmethylide, **1,** 310
Tin(II) trifluoromethanesulfonate, **12,** 490
Trifluoroacetic anhydride, **12,** 530

RAMBERG–BACKLUND REACTION (*see* ELIMINATION REACTIONS)
REDUCTION REACTIONS (*see also* ACYLOIN REACTION, ASYMMETRIC REACTIONS, DEOXYGENATION, REDUCTIVE...)

ACETALS, KETALS → ROR
Diborane, **5,** 184
Diisobutylaluminum hydride, **1,** 260
Lithium aluminum hydride, **1,** 595; **5,** 382, 389; **9,** 277
Rhodium catalysts, **1,** 979
Sodium bis(2-methoxyethoxy)aluminum hydride, **9,** 418
Sodium cyanoborohydride, **8,** 454; **11,** 481
Triethylsilane, **1,** 1218
Trimethylsilyl trifluoromethanesulfonate, **9,** 497; **11,** 584

ACETATES, OTHER ROCOR′ → RH
Calcium–Ammonia, **1,** 106
Chromium(II) chloride, **1,** 149; **2,** 76
Lithium–Ethylamine, **8,** 284
Palladium(II) acetate, **12,** 367
Palladium catalysts, **2,** 303
Potassium, **9,** 377
Sodium, **9,** 416
Tetrakis(triphenylphosphine)palladium(0), **10,** 384
Trichlorosilane, **6,** 606
Zinc, **1,** 1276; **4,** 574

ACETYLENIC C=O's → PROPARGYL ALCOHOLS
Darvon alcohol, **10,** 154
2,2′-Dihydroxy-1,1′-binaphthyl, **12,** 190
NB-Enantrane, **11,** 229
Lithium aluminum hydride, **1,** 581
Lithium aluminum hydride + co-reagent, **8,** 184; **9,** 308; **10,** 148
Phosphorus(V) sulfide, **10,** 320
B-3-Pinanyl-9-borabicyclo[3.3.1]nonane, **11,** 429; **12,** 397
Potassium tri-*sec*-butylborohydride, **7,** 307

ACYL HALIDES → ROH
9-Borabicyclo[3.3.1]nonane, **7,** 29
Borane–Alkylamines, **1,** 16
Diisobutylaluminum hydride, **12,** 191
Lithium aluminum hydride, **1,** 581; **12,** 272
Lithium aluminum hydride–Aluminum chloride, **1,** 595
Lithium borohydride, **1,** 603
Sodium bis(2-methoxyethoxy)aluminum hydride, **3,** 260
Sodium borohydride, **1,** 1049
Sodium borohydride + co-reagent, **1,** 1053; **9,** 421; **12,** 444
Sodium cyanoborohydride, **4,** 448
Sodium trimethoxyborohydride, **1,** 1108
Tetrabutylammonium octahydrotriborate, **11,** 501

ACYL HALIDES → RCHO
Aziridine, **1,** 378
μ-Bis(cyanotrihydroborato)tetrakis-(triphenylphosphine)dicopper(I), **10,** 33
Bis(triphenylphosphine)copper(I) borohydride, **8,** 47; **9,** 57; **10,** 47
Cadmium borohydride, **12,** 574
N,O-Dimethylhydroxylamine, **11,** 201
Disodium tetracarbonylferrate, **4,** 461
Hydrido(tributylphosphine)copper(I), **3,** 154
Imidazole, **1,** 492
Lithium aluminum hydride, **1,** 581
Lithium tri-*t*-butoxyaluminum hydride, **1,** 620; **2,** 251
Methanesulfonic acid, **4,** 326
S-Methyl-1,4-diphenylisothiosemicarbazide, **5,** 445
3-Methyl-1-phenyl-2-phospholene, **6,** 392
Palladium catalysts, **1,** 975; **4,** 368; **9,** 351

REDUCTION REACTIONS (*Continued*)
Palladium on barium sulfate, **7,** 275
Pyridine, **11,** 448
Rosenmund catalyst, **1,** 975; **4,** 367
Sodium borohydride, **10,** 357
Sodium borohydride–Cadmium chloride–DMF, **8,** 451; **10,** 359
Tetracarbonyl(cyclopentadienyl)-vanadium, **8,** 468
Tetramethylammonium tetracarbonylhydridoferrate, **8,** 476
Tetramethylthiourea, **1,** 1145
1,3-Thiazolidine-2-thione, **11,** 518
Tributyltin hydride, **1,** 1192; **10,** 411; **11,** 545
Triethylsilane, **3,** 304
Zinc borohydride, **12,** 574
ALCOHOLS → RH (*see* DEOXYGENATION)
ALDEHYDES, KETONES → ROH
Al-based reagents
Alumina, **8,** 9
Aluminum amalgam, **1,** 20
Aluminum *t*-butoxide, **2,** 21
Aluminum hydride, **5,** 13
Aluminum isopropoxide, **1,** 35; **4,** 15
N,N-(Diethylethanamine)-trihydridoaluminum, **6,** 607
Diisobutylaluminum hydride, **6,** 198; **12,** 191
Isobornyloxyaluminum dichloride, **1,** 522
Lithium aluminum hydride, **1,** 581; **12,** 272
Lithium aluminum hydride + co-reagent, **1,** 595, 599; **11,** 293
Lithium *n*- or *t*-butyldiisobutyl-aluminum hydride, **10,** 239; **11,** 296; **12,** 276
Lithium trialkoxyaluminum hydrides, **1,** 620, 625; **2,** 251; **3,** 188; **4,** 312; **5,** 392
Sodium bis(2-methoxyethoxy)-aluminum hydride, **3,** 260; **5,** 596; **9,** 418
Tetra-μ-hydridotetrahydro-aluminummagnesium, **4,** 316
Tetramethylammonium borohydride, **1,** 1143
Trialkylaluminums, **1,** 1188; **9,** 486; **11,** 539
B-based reagents
μ-Bis(cyanotrihydroborato)tetrakis-(triphenylphosphine)dicopper(I), **10,** 33
Bis(triphenylphosphine)copper(I) borohydride, **10,** 47
9-Borabicyclo[3.3.1]nonane, **7,** 29
9-Borabicyclo[3.3.1]nonane ate complexes, **7,** 30
Borane–Alkylamines, **1,** 16, 963, 1229; **9,** 59; **10,** 12; **11,** 87, 188, 219
Borane–Tetrahydrofuran, **1,** 199; **2,** 106; **3,** 76
Butyllithium–Pyridine, **2,** 351
Calcium borohydride, **2,** 57
Catecholborane, **8,** 79
Chlorobis(cyclopentadienyl)tetra-hydroboratozirconium(IV), **9,** 103
Dicyclohexylborane, **5,** 39
Diisopinocampheylborane, **5,** 39
Disiamylborane, **1,** 57; **3,** 22; **5,** 39
Lithium borohydride, **1,** 603
Lithium butylborohydride, **11,** 294
Lithium cyanoborohydride, **3,** 183
Lithium 9,9-dibutyl-9-borabicyclo-[3.3.1]nonanate, **7,** 30
Lithium dimesitylborohydride bis-(dimethoxymethane), **5,** 408
Lithium perhydro-9b-boraphenalyl-hydride, **3,** 187; **4,** 309
Lithium tri-*sec*-butylborohydride, **4,** 312; **12,** 286
Lithium triethylborohydride, **8,** 309; **11,** 288; **12,** 289
Lithium trisiamylborohydride, **7,** 216
Potassium borohydride–(–)-N-Dodecylmethylephedrinium bromide, **6,** 249
Potassium 9-(2,3-dimethyl-2-butoxy)-9-boratabicyclo[3.3.1]nonane, **12,** 406
Potassium tri-*sec*-butylborohydride, **5,** 557; **6,** 490; **11,** 445
Potassium triisopropoxyborohydride, **5,** 565
Sodium acetanilidoborohydride, **7,** 325
Sodium borohydride, **1,** 1049; **3,** 262; **5,** 597; **10,** 357; **12,** 441
Sodium borohydride + co-reagent, **1,** 1053; **6,** 175, 249; **10,** 198, 359; **11,** 479

Sodium borohydride, sulfurated, **3,** 264; **4,** 444
Sodium cyanoborohydride, **4,** 448; **5,** 607
Sodium triacetoxyborohydride, **12,** 453
Sodium trimethoxyborohydride, **1,** 1108
Tetrabutylammonium borohydride, **7,** 352; **10,** 378
Tetrabutylammonium cyanoborohydride, **5,** 645
Tetrabutylammonium octahydrotriborate, **11,** 501
Tetramethylammonium borohydride, **1,** 1143
Thexylborane, **4,** 175
Thexylborane–N,N-Diethylaniline, **9,** 464
Tributylborane, **10,** 410
Zinc borohydride, **3,** 337; **10,** 460; **11,** 599; **12,** 572, 574
Zinc bromide, **11,** 600

Catalytic hydrogenation
Chlorotris(triphenylphosphine)-rhodium(I), **3,** 325
Iridium catalysts, **2,** 434
Nickel(II) acetate–Sodium hydride–*t*-Amyloxide, **10,** 365
Nickel catalysts, **1,** 718; **7,** 312; **11,** 356
Palladium catalysts, **3,** 218
Platinum catalysts, **1,** 1220
Raney nickel, **5,** 570; **7,** 312
Rhenium heptaselenide, **1,** 979
Rhodium catalysts, **1,** 982; **6,** 503; **8,** 433
Ruthenium catalysts, **1,** 983
Ruthenium–Silica, **10,** 342
Tin(II) chloride, **1,** 1113
Trihydridobis(triphenylphosphine)-iridium(III), **2,** 434
Urushibara catalysts, **4,** 571; **5,** 743; **6,** 659; **7,** 417

Metals + solvents
Lithium + solvent, **4,** 246; **5,** 543; **7,** 195
Potassium–Graphite, **4,** 397
Potassium–Hexamethylphosphoric triamide, **4,** 245; **5,** 543
Sodium + solvent, **4,** 246; **5,** 543; **11,** 472
Sodium amalgam, **1,** 1030

using other Methods
Butyllithium–Pyridine, **2,** 351
Butylphenyltin dihydride, **6,** 92
Chlorotris(triphenylphosphine)-rhodium(I), **3,** 325; **6,** 652
Copper(I) bromide–Lithium trimethoxyaluminum hydride, **8,** 120
Copper–Chromium oxide, **1,** 157
Copper hydride ate complexes, **5,** 330; **6,** 492
Dichlorobis(cyclopentadienyl)-titanium, **10,** 130
Dichlorotris(triphenylphosphine)-ruthenium(II), **6,** 654; **11,** 171
Dicyclopentadienyltitanium, **5,** 672
Dihydridotetrakis(triphenyl-phosphine)ruthenium(II), **7,** 109
2,6-Diisopropylphenoxymagnesium hydride, **8,** 175
Dimethylphenylsilane, **12,** 209
Diphenyltin dihydride, **1,** 349; **11,** 224
Ferric chloride–Sodium hydride, **7,** 155
Formamidinesulfinic acid, **4,** 506; **5,** 668; **6,** 586
Hydrogen hexachloroiridate(IV), **1,** 131; **2,** 67; **3,** 47; **4,** 83; **5,** 119
Hydrogen telluride, **10,** 205
Iridium(IV) chloride, **2,** 228; **3,** 166
Lithium butyl(hydrido)cuprate, **5,** 330
Lithium dibutylcuprate, **5,** 187
Lithium pyrrolidide, **5,** 416
Nickel(II) acetate–Sodium hydride–*t*-Amyloxide, **10,** 365
Polymethylhydrosiloxane, **4,** 393
Potassium hydroxide, **3,** 238
Samarium(II) iodide, **8,** 439; **10,** 344
Sodium dithionite, **8,** 456; **9,** 426
Tributyltin hydride, **1,** 1192; **3,** 294; **9,** 477; **12,** 525
Triethoxysilane, **11,** 554
Trifluoroacetic acid–Alkylsilanes, **6,** 616
Ytterbium(II) iodide, **8,** 439

Reduction to axial alcohols
Aluminum *t*-butoxide, **2,** 21
Butyllithium–Pyridine, **2,** 351
Chlorotris(triphenylphosphine)-rhodium(I), **3,** 325
2,6-Diisopropylphenoxymagnesium hydride, **8,** 175

REDUCTION REACTIONS (*Continued*)
Hydrogen hexachloroiridate(IV), **1,** 131; **2,** 67; **3,** 47; **4,** 83; **5,** 119
Iridium(IV) chloride, **2,** 228; **3,** 166
Isobornyloxyaluminum dichloride, **1,** 522
Lithium *t*-butyldiisobutylaluminum hydride, **10,** 239; **11,** 296
Lithium perhydro-9b-boraphenalylhydride, **3,** 187
Lithium trialkoxyaluminum hydrides, **1,** 620; **5,** 392
Lithium tri-*sec*-butylborohydride, **4,** 312
Lithium trisiamylborohydride, **7,** 216
Nickel catalysts, **5,** 570
Potassium 9-(2,3-dimethyl-2-butoxy)-9-boratabicyclo[3.3.1]nonane, **12,** 406
Potassium tri-*sec*-butylborohydride, **5,** 557
Potassium triisopropoxyborohydride, **5,** 565
Rhodium catalysts, **8,** 433
Triisobutylaluminum, **1,** 1188
Urushibara catalysts, **5,** 743; **6,** 659
Reduction to equatorial alcohols
μ-Bis(cyanotrihydroborato)tetrakis-(triphenylphosphine)dicopper(I), **10,** 33
Chlorotris(triphenylphosphine)-rhodium(I), **3,** 325
N,N-(Diethylethanamine)-trihydridoaluminum, **6,** 607
Lithium aluminum hydride, **1,** 581
Lithium aluminum hydride–Aluminum chloride, **1,** 595
Lithium butylborohydride, **11,** 294
Lithium perhydro-9b-boraphenalylhydride, **3,** 187
Sodium–Ammonia, **11,** 472
Sodium borohydride, **3,** 262
Sodium borohydride–Palladium(II) chloride, **11,** 479
Sodium borohydride, sulfurated, **4,** 444
Sodium cyanoborohydride, **5,** 607
Selective reduction of RCHO
Alumina, **6,** 16; **8,** 9
Bis(triphenylphosphine)copper(I) borohydride, **10,** 47
9-Borabicyclo[3.3.1]nonane–Pyridine, **8,** 49
Catecholborane, **8,** 79
Dichlorotris(triphenylphosphine)-ruthenium(II), **8,** 159
Formaldehyde, **1,** 397
8-Hydroxyquinolinedihydroboronite, **12,** 252
Lithium cyanoborohydride, **1,** 618; **3,** 183
Lithium 9,9-dibutyl-9-borabicyclo-[3.3.1]nonanate, **7,** 30
Lithium tris(3-ethyl-3-pentyloxy)-aluminum hydride, **11,** 305
B-(3-Methyl-2-butyl)-9-borabicyclo-[3.3.1]nonane, **8,** 334
Samarium(II) iodide, **8,** 439; **10,** 344
Sodium borohydride–Alkanethiols, **8,** 449
Sodium borohydride, sulfurated, **4,** 444
Sodium triacetoxyborohydride, **6,** 553
Sodium tris(3,5-dimethylphenoxy)-borohydride, **10,** 369
Tetrabutylammonium borohydride, **10,** 378; **11,** 499
Tetrabutylammonium cyanoborohydride, **5,** 645
Tin(II) chloride, **1,** 1113
Tributyltin hydride, **9,** 477; **12,** 525
Trihydridobis(triphenylphosphine)-iridium(III), **2,** 434
Ytterbium(II) iodide, **8,** 439
Selective reduction of R_2CO
Borane–Tetrahydrofuran, **11,** 487
Erbium(III) chloride, **11,** 232
Lithium dibutylcuprate, **5,** 187
Lithium pyrrolidide, **5,** 416
Lithium tri-*t*-butoxyaluminum hydride, **10,** 248
Sodium borohydride–Cerium(III) chloride, **9,** 422
Reduction of ArCHO
Borane–Tetrahydrofuran, **2,** 106
Chromium(II) chloride, **1,** 149
Diimide, **3,** 99
Reduction of Ar_2CO
Lithium diisopropylamide, **8,** 292
Sodium–Ammonia, **4,** 290
Sodium bis(2-methoxyethoxy)-aluminum hydride, **4,** 441

Sodium hydride, **1,** 1075; **4,** 452
Zinc, **1,** 1276
ALDEHYDES, KETONES → C=C OR $-CH_2-$ (*see* DEOXYGENATION)
ALKENES → RH
Catalytic hydrogenation
Chloro(hexamethylbenzene)-hydridotriphenylphosphine-rhodium, **8,** 153
Chlorohydridotris(triphenyl-phosphine)ruthenium(II), **7,** 60
Chlorotris(triphenylphosphine)-rhodium(I), **1,** 140, 1252; **2,** 448; **4,** 559; **6,** 652; **9,** 113
(1,5-Cyclooctadiene)bis(trialkyl- or arylphosphine)iridium(I) hexafluorophosphate, **8,** 135; **10,** 116; **12,** 151
Dicarbonyldichlorobis(triphenyl-phosphine)ruthenium(II), **4,** 421
Di-μ-chlorodichlorobis(pentamethyl-cyclopentadienyl)dirhodium, **8,** 152
Hydrogen hexachloroplatinate(IV) + co-reagent, **1,** 890; **3,** 51
Iridium catalysts, **2,** 228; **8,** 135; **9,** 257
Lindlar catalyst, **4,** 283
Nickel(II) acetate–Sodium hydride–*t*-Amyloxide, **10,** 365
Nickel boride, **1,** 720; **3,** 208; **4,** 351; **5,** 471
Nickel catalysts, **1,** 720; **2,** 293; **7,** 312; **10,** 339
Palladium catalysts, **6,** 443, 445; **9,** 351
Palladium(II) chloride–Diphenylbenzylphosphine, **6,** 450
Perchloric acid, **1,** 796
Polymethylhydrosiloxane, **4,** 393
Potassium–Graphite, **7,** 296
Raney nickel, **1,** 723; **2,** 293; **5,** 570; **7,** 312; **10,** 339
Rhodium catalysts, **1,** 979; **12,** 426
Ruthenium catalysts, **1,** 983
Tetrahydrofuran, **1,** 1140
Tributylborane, **1,** 1190
Urushibara catalysts, **4,** 571; **6,** 659
Diimide reduction
Diimide, **1,** 42, 257; **2,** 139; **4,** 535; **6,** 195; **8,** 172
Potassium azodicarboxylate, **1,** 909
Hydrometallation–protonation
Borane–Tetrahydrofuran, **1,** 199
Borane–Trimethylamine, **1,** 1229
Chlorobis(cyclopentadienyl)-hydridozirconium(IV), **6,** 175
Dichlorobis(cyclopentadienyl)-titanium, **9,** 46; **10,** 130
Dichlorobis(cyclopentadienyl)-zirconium–*t*-Butylmagnesium chloride, **12,** 171
Lithium aluminum hydride, **9,** 274
Lithium aluminum hydride–Dichlorobis(cyclopentadienyl)-zirconium, **10,** 237
Lithium triethylborohydride, **8,** 309
Sodium bis(2-methoxyethoxy)-aluminum hydride, **6,** 528
Thexylborane, **1,** 276
Tributylborane, **1,** 1190
Metals + acid
Birch reduction, **4,** 31
Lithium–Ethylenediamine, **1,** 614; **9,** 274
Magnesium–Methanol, **10,** 251
Sodium + solvent, **2,** 27; **3,** 151
Zinc–copper couple, **12,** 569
other Methods
Chlorotris(triphenylphosphine)-rhodium(I), **6,** 652
Chromium(II) sulfate, **2,** 77
Cobalt(II) phthalocyanine, **11,** 138
Dicyclopentadienyltitanium, **5,** 672
Lithium aluminum hydride, **2,** 242
Lithium aluminum hydride–Cobalt(II) chloride, **8,** 291
Lithium tri-*sec*-butylborohydride, **12,** 286
Palladium(II) chloride, **12,** 371
Sodium hydrazide, **1,** 1074
Sodium hydrogen telluride, **8,** 459
Titanium(IV) chloride–Lithium aluminum hydride, **7,** 372
Trifluoroacetic acid–Alkylsilanes, **6,** 616
ALKYL HALIDES → RH
Aluminum chloride, **7,** 7
Aluminum hydride, **6,** 19
Ammonium acetate, **1,** 38
1-Benzyl-1,4-dihydronicotinamide, **7,** 15
9-Borabicyclo[3.3.1]nonane, **6,** 62
9-Borabicyclo[3.3.1]nonane ate complexes, **11,** 295
Borane + co-reagent, **3,** 76, 299

REDUCTION REACTIONS (*Continued*)
B-Butyl-9-borabicyclo[3.3.1]nonane, **11,** 88
Butylphenyltin dihydride, **6,** 92
Chromium(II) acetate, **1,** 147; **2,** 75; **5,** 143
Chromium(II)–Amine complexes, **3,** 57
Chromium(II) chloride, **1,** 149; **3,** 60; **5,** 144
Chromium(II) perchlorate–Butanethiol, **10,** 101
Chromium(II) sulfate, **1,** 150
Cobalt(II) phthalocyanine, **11,** 138
Copper hydride ate complexes, **3,** 154; **5,** 168, 330; **6,** 492; **8,** 120
Dichlorobis(cyclopentadienyl)titanium, **6,** 179
Dicyclopentadienyltitanium, **5,** 672; **6,** 596
Diethyl phosphite, **12,** 187
2,6-Diisopropylphenoxymagnesium hydride, **8,** 175
Diphenyltin dihydride, **3,** 134
Grignard reagents, **1,** 415
Hexaethylphosphorous triamide, **3,** 148
Hydrazine, **5,** 327
Iodoform, **3,** 165
Lithium + solvent, **1,** 604; **4,** 291; **6,** 322
Lithium aluminum hydride, **1,** 581; **4,** 291; **5,** 382; **10,** 236; **11,** 289; **12,** 272
Lithium aluminum hydride–Cobalt(II) chloride, **8,** 291
Lithium butyldiisobutylaluminum hydride, **12,** 276
Lithium 4,4′-di-*t*-butylbiphenylide, **7,** 200
Lithium 9,9-dibutyl-9-borabicyclo-[3.3.1]nonanate, **11,** 295
Lithium triethylborohydride, **4,** 313; **10,** 249
Magnesium, **1,** 627
Mercury(II) oxide, **7,** 224
Methyllithium, **5,** 448
Methylmagnesium bromide, **1,** 689
Nickel catalysts, **9,** 321; **11,** 355
Potassium–Crown ether, **11,** 431
Potassium tetracarbonylhydridoferrate, **6,** 483
Potassium thiocyanate, **1,** 954
Raney nickel, **1,** 723
Samarium(II) iodide, **10,** 344
Sodium–Alcohol, **2,** 378; **9,** 416
Sodium–Ammonia, **5,** 31
Sodium bis(2-methoxyethoxy)aluminum hydride, **3,** 260
Sodium borohydride, **1,** 1049; **3,** 120, 262; **4,** 443; **6,** 530; **11,** 477; **12,** 444
Sodium cyanoborohydride, **4,** 448; **12,** 446
Sodium 9-cyano-9-hydrido-9-borabicyclo[3.3.1]nonane, **11,** 483
Sodium methylsulfinylmethylide, **1,** 310
Tetrabutylammonium cyanoborohydride, **5,** 645
Tetracarbonyl(cyclopentadienyl)-vanadium, **8,** 468
Tin(II) chloride, **1,** 1113
Titanium(III) chloride–Tetrahydrofuran–Magnesium, **6,** 590
Tributyltin hydride, **1,** 1192; **2,** 424; **3,** 294; **4,** 518; **5,** 685; **6,** 604; **7,** 379; **8,** 497; **11,** 545
Trichlorosilane, **3,** 298
Triethylamine, **2,** 427
Triethyl phosphite, **11,** 555
Trifluoroacetic acid–Alkylsilanes, **6,** 616
Triphenyltin hydride, **1,** 1250; **3,** 324
Urushibara catalysts, **5,** 743
Zinc, **1,** 1276
Zinc borohydride, **12,** 572
Zinc-copper couple, **8,** 533

ALKYL SULFONATES → RH
Copper hydride ate complexes, **5,** 168, 330
Lithium aluminum hydride, **1,** 581; **8,** 286
Lithium triethylborohydride, **7,** 215; **10,** 249; **12,** 289
Nickel–Alumina, **9,** 321
Nickel catalysts, **1,** 723
Palladium catalysts, **5,** 499
Potassium thiocyanate, **1,** 954
Samarium(II) iodide, **10,** 344
Sodium borohydride, **1,** 1049; **3,** 262; **11,** 477
Sodium borohydride + co-reagent, **8,** 449; **12,** 443, 444
Sodium cyanoborohydride, **4,** 448
Sodium 9-cyano-9-hydrido-9-borabicyclo[3.3.1]nonane, **11,** 483
Sodium iodide, **1,** 1087
Sodium iodide–Zinc, **7,** 338

ALKYNES → $-CH_2CH_2-$
Chlorotris(triphenylphosphine)-rhodium(I), **1,** 140, 1252; **2,** 448; **4,** 559
Cobalt boride–Borane–*t*-Butylamine, **11,** 138
Diimide, **4,** 154
Hexamethylphosphoric triamide, **3,** 149
Nickel boride, **3,** 208
Palladium(II) acetate–Sodium hydride–*t*-Amyloxide, **12,** 370
Ruthenium catalysts, **1,** 983
Titanium(IV) chloride–Lithium aluminum hydride, **7,** 372
Urushibara catalysts, **4,** 571
ALKYNES → $-CH{=}CH_2$
Chromium(II)–Amine complexes, **9,** 117
Dichloroborane diethyl etherate, **5,** 191
Disiamylborane, **1,** 57
Lithium aluminum hydride–Nickel(II) chloride, **8,** 291
Monochloroborane diethyl etherate, **5,** 465
Zinc-copper couple, **10,** 459
ALKYNES → *cis* C=C
Bis(cyclopentadienyl)trihydridoniobium, **6,** 47
9-Borabicyclo[3.3.1]nonane, **9,** 57
Borane–Tetrahydrofuran, **1,** 199
Catecholborane, **4,** 69; **8,** 79
Chromium(II)–Amine complexes, **9,** 117
Copper hydride ate complexes, **6,** 492; **7,** 80
Dichlorobis(cyclopentadienyl)titanium, **10,** 130; **11,** 163
Dichloroborane diethyl etherate, **5,** 191
Dichlorotris(triphenylphosphine)-ruthenium(II), **2,** 121
Dicyclohexylborane, **4,** 141
Diisobutylaluminum hydride, **1,** 260
Disiamylborane, **3,** 22; **4,** 37
Ethylmagnesium bromide–Copper(I) iodide, **7,** 149
Gold, **5,** 321
Hydrogen hexachloroplatinate(IV)–Triethylsilane, **3,** 51
Iodine, **6,** 293
Lindlar catalyst, **6,** 319; **9,** 270
Lithium, **4,** 286
Lithium aluminum hydride–Nickel(II) chloride, **8,** 291
Magnesium hydride–Copper(I) iodide, **8,** 311
Monochloroborane diethyl etherate, **5,** 465
Nickel(II) acetate–Sodium hydride–*t*-Amyloxide, **10,** 365
Nickel boride, **1,** 720; **5,** 471
Nickel–Graphite, **11,** 356
Palladium(II) acetate, **8,** 378
Palladium(II) acetate–Sodium hydride–*t*-Amyloxide, **12,** 370
Palladium catalysts, **2,** 356; **7,** 275; **9,** 351; **10,** 297; **11,** 392
Palladium(II) chloride, **9,** 352
Palladium on barium sulfate, **1,** 778
Palladium, poisoned catalyst, **1,** 566; **2,** 356
Raney nickel, **1,** 723
Sodium borohydride–Palladium(II) chloride, **12,** 399
Titanium(IV) chloride–Lithium aluminum hydride, **7,** 372
Triethylammonium formate–Palladium, **9,** 481
Tris(dimethylphenylphosphine)-(norbornadiene)rhodium(I) hexafluorophosphate, **7,** 411
Zinc, **6,** 672
Zinc–1,2-Dibromoethane, **12,** 570
ALKYNES → *trans* C=C
Borane–Tetrahydrofuran, **7,** 321
Chromium(II) sulfate, **1,** 150
Copper hydride ate complexes, **7,** 80
Dichlorobis(cyclopentadienyl)titanium, **11,** 163
Lithium aluminum hydride, **2,** 242; **8,** 286; **9,** 274
Lithium–Ammonia, **1,** 601; **8,** 282
Lithium bronze, **11,** 293
Lithium diisobutylmethylaluminum hydride, **8,** 292
Lithium–Ethylamine, **1,** 574
Palladium catalysts, **4,** 366
Sodium–Ammonia, **5,** 589
Ytterbium–Ammonia, **9,** 517
ALLENES → C=C
Chlorotris(triphenylphosphine)-rhodium(I), **4,** 559
Diimide, **3,** 99
Diisobutylaluminum hydride, **10,** 149
Disiamylborane, **5,** 39
Sodium–Ammonia, **2,** 374; **4,** 438; **5,** 589; **6,** 523

REDUCTION REACTIONS (*Continued*)
AMIDES → ROH
9-Borabicyclo[3.3.1]nonane, **7,** 29
Lithium triethylborohydride, **8,** 309
Sodium (dimethylamino)borohydride, **12,** 446
Sodium–Hexamethylphosphoric triamide, **5,** 324
AMIDES → RCHO
Aziridine, **1,** 378
Bis(N-methylpiperazinyl)aluminum hydride, **6,** 52
N,N′-Carbonyldiimidazole, **1,** 114, 591
1,4-Dichloro-1,4-dimethoxybutane, **12,** 175
N,O-Dimethylhydroxylamine, **11,** 201
Disiamylborane, **1,** 57; **3,** 22
Lithium aluminum hydride, **1,** 581
Lithium butyldiisobutylaluminum hydride, **12,** 276
Lithium diethoxyaluminum hydride, **1,** 610
Lithium–Methylamine, **1,** 574
Lithium triethoxyaluminum hydride, **1,** 625
N-Methyl-N-phenylcarbamoyl chloride, **1,** 694
Phosphorus(V) chloride, **1,** 866
Sodium aluminum hydride, **3,** 259
Tin(II) chloride, **1,** 1116
Trifluoromethanesulfonic anhydride, **5,** 702
Zinc, **7,** 426
AMIDES → RNH_2
Aluminum hydride, **2,** 23
Borane–Dimethyl sulfide, **6,** 64; **11,** 69
Borane–Tetrahydrofuran, **1,** 199; **2,** 106; **4,** 124
N-Ethylmorpholine, **1,** 383
Iodosylbenzene, **12,** 258
Lead tetraacetate, **6,** 313
Lithium aluminum hydride, **1,** 581; **4,** 291; **5,** 382
Lithium 9-boratabicyclo[3.3.1]nonane, **12,** 275
Lithium borohydride, **12,** 276
Monochloroalane, **12,** 333
Sodium acetoxyborohydride, **7,** 325
Sodium borohydride, **3,** 262; **4,** 443; **7,** 329
Sodium borohydride + co-reagent, **1,** 1053; **3,** 264; **7,** 330; **11,** 479
Sodium (dimethylamino)borohydride, **12,** 446
Sodium trifluoroacetoxyborohydride, **7,** 326
Tetrabutylammonium borohydride, **10,** 378
Thexylborane–N,N-Diethylaniline, **9,** 464
Tin(IV) chloride–Sodium borohydride, **8,** 452; **9,** 438
Triethyloxonium tetrafluoroborate, **2,** 430; **5,** 691
AMINE OXIDES → RNH_2
Boron trifluoride etherate, **6,** 65
Chlorotrimethylsilane–Sodium iodide–Zinc, **11,** 128
Chromium(II) chloride, **7,** 73
9-Diazofluorene, **1,** 190
Diphosphorus tetraiodide, **10,** 174
Hexachlorodisilane, **10,** 195
Iron carbonyl, **4,** 268
Molybdenum(V) chloride–Zinc, **10,** 274
Phenyl(trichloromethyl)mercury, **1,** 851
Phosphorus(III) chloride, **1,** 875
Sodium azide, **5,** 593
Sodium borohydride, sulfurated, **7,** 331
Sulfur dioxide, **2,** 392
Titanium(III) chloride, **6,** 587
Titanium(IV) chloride–Sodium borohydride, **10,** 404
Triphenylphosphine, **1,** 1238
ANHYDRIDES → ROH
Diborane, **5,** 184
Diisobutylaluminum hydride, **12,** 191
Lithium aluminum hydride, **1,** 581
Sodium bis(2-methoxyethoxy)aluminum hydride, **3,** 260
Sodium borohydride–Aluminum chloride, **1,** 1053
Sodium trimethoxyborohydride, **1,** 1108
Tin(IV) chloride–Sodium borohydride, **9,** 438
ANHYDRIDES → RCHO
Disodium tetracarbonylferrate, **5,** 624; **6,** 550
Pyridine N-oxide, **5,** 567
ANHYDRIDES → LACTONES
Diborane, **5,** 184
Dichlorotris(triphenylphosphine)-ruthenium(II), **6,** 654; **7,** 99

2,3-O-Isopropylidene-2,3-dihydroxy-1,4-bis(diphenylphosphine)butane, **11,** 277
Sodium borohydride, **3,** 262; **5,** 597

ARENES → DIHYDRO-, TETRAHYDROARENES (*see also* BIRCH REDUCTION)
1-Benzyl-1,2-dihydroisonicotinamide, **12,** 46
Borane–Amines, **5,** 708; **8,** 50
Dichlorotris(triphenylphosphine)-ruthenium(II), **12,** 179
Lithium trialkoxyaluminum hydrides, **12,** 286
Methyl chloroformate, **6,** 376
Palladium catalysts, **8,** 382
Potassium borohydride, **8,** 407
Sodium amalgam, **2,** 373
Sodium borohydride, **3,** 262
Sodium cyanoborohydride, **9,** 424
Sodium dithionite, **1,** 1081
Sodium hydride, **5,** 610
Tetrabutylammonium borohydride, **10,** 378
Tin, **1,** 1168
Triethylsilane–Boron trifluoride, **9,** 483
Trifluoroacetic acid, **1,** 1219; **6,** 613
Trimethylsilane, **1,** 1235

ARENES → HEXAHYDROARENES
trihapto-Allyltris(trimethylphosphite)-cobalt(I), **6,** 15
Calcium hexamine, **1,** 104
Chloro(hexamethylbenzene)hydrido-triphenylphosphinerhodium, **8,** 153
Di-μ-chlorobis(1,5-hexadiene)-dirhodium, **12,** 172
Lithium–Ethylenediamine, **1,** 614
Nickel catalysts, **1,** 723, 731
Palladium catalysts, **8,** 382
Palladium(II) chloride, **9,** 352
Palladium hydroxide, **8,** 385
Rhodium catalysts, **1,** 979, 982; **4,** 418; **6,** 503
Rhodium oxide–Platinum oxide, **9,** 408
Ruthenium catalysts, **1,** 983; **5,** 574
Sodium borohydride–Rhodium(III) chloride, **11,** 480
Sodium sulfite, **2,** 387

ARYL HALIDES → ArH
Aluminum chloride–Ethanethiol, **12,** 29
Butyllithium, **6,** 85
Butylphenyltin dihydride, **6,** 92
Chromium(II)–Amine complexes, **3,** 57; **4,** 97
Copper, **12,** 140
Copper–Ascorbic acid, **1,** 155
Copper–Benzoic acid, **1,** 158
Copper(II) bromide, **1,** 161
Copper hydride ate complexes, **3,** 154; **5,** 168; **6,** 492
Copper(I) trifluoromethanesulfonate, **6,** 130
Cyclodextrins, **9,** 129
Dichlorobis(cyclopentadienyl)titanium, **6,** 179
Dicyclopentadienyltitanium, **6,** 596
Dimethyl sulfoxide, **3,** 119
Hydrido(tributylphosphine)copper(I), **3,** 154
Hydriodic acid, **1,** 449
Hypophosphorous acid, **1,** 489
Lithium aluminum hydride, **3,** 176; **11,** 289; **12,** 272
Lithium aluminum hydride–Cobalt(II) chloride, **8,** 291
Lithium–*t*-Butanol–Tetrahydrofuran, **1,** 604
Magnesium, **1,** 627
Magnesium oxide, **1,** 633
Manganese(II) chloride, **7,** 222
Nickel catalysts, **1,** 718; **6,** 502
Palladium catalysts, **11,** 392
Phosphorus, red, **1,** 861
Potassium *t*-butoxide, **2,** 336
Raney nickel, **1,** 723; **6,** 502
Sodium acetate, **1,** 1024
Sodium bis(2-methoxyethoxy)aluminum hydride, **3,** 260
Sodium borohydride, **5,** 597, 602; **11,** 479
Sodium formate, **8,** 458
Sodium hydride, **5,** 610
Sodium naphthalenide, **9,** 431
Sodium telluride, **2,** 388
Tetracarbonyl(cyclopentadienyl)-vanadium, **8,** 468
Tin, **2,** 414
Titanium(III) chloride–Tetrahydrofuran–Magnesium, **6,** 590
p-Toluenesulfonylhydrazide, **1,** 1185
Tributyltin hydride, **3,** 294
Trimethyltinsodium, **9,** 500
Triphenylphosphine, **1,** 1238
Triphenyltin hydride, **1,** 1250

REDUCTION REACTIONS (*Continued*)
Zinc, **1,** 1276; **5,** 753
AZIDES → RNH_2, $ArNH_2$
Bis(triphenylphosphine)copper(I) borohydride, **10,** 47
1,4-Dimercapto-2,3-butanediol, **9,** 394
Lindlar catalyst, **6,** 319
1,3-Propanedithiol, **9,** 394
Raney nickel, **8,** 433
Sodium borohydride, **1,** 1049; **11,** 477
Sodium hydride, **5,** 610
Titanium(III) chloride, **7,** 418; **8,** 482
Triphenylphosphine, **12,** 550
AZO COMPOUNDS → RNH_2
Palladium catalysts, **8,** 382
Sodium dithionite, **1,** 1081
AZO COMPOUNDS → –NHNH–
Dicyclopentadienyltitanium, **6,** 596
Diimide, **1,** 257; **5,** 220
Hydrazine, **1,** 434
Hydrogen hexachloroplatinate(IV)–Triethylsilane, **3,** 51
Hydrogen telluride, **10,** 205
Palladium catalysts, **6,** 445
Phenylhydrazine, **1,** 838
Potassium azodicarboxylate, **1,** 909
Sodium borohydride–Palladium on charcoal, **1,** 1054
2,4,6-Triisopropylbenzenesulfonylhydrazide, **7,** 392
AZOXY COMPOUNDS → –N=N–
Hexamethylphosphorous triamide–Iodine, **9,** 236
Trialkyl phosphites, **1,** 1212; **5,** 717
Triphenylphosphine, **1,** 1238
BENZYL ETHERS → ROH
Boron trifluoride–Dimethyl sulfide, **10,** 51
sec-Butyllithium, **9,** 87
Chromium(VI) oxide, **3,** 54
(Diphenylphosphine)lithium, **1,** 345
Ferric chloride, **11,** 237
Lithium aluminum hydride–Boron trifluoride etherate, **1,** 599
Palladium catalysts, **10,** 299
Perchloric acid, **1,** 796
Phenylthiotrimethylsilane, **10,** 426
Sodium–Ammonia, **9,** 415
Sodium bis(2-methoxyethoxy)aluminum hydride, **7,** 327
Sodium–Potassium alloy, **1,** 1102
Titanium(IV) chloride, **11,** 529
Trifluoroacetic acid, **1,** 1219
Uranium(VI) fluoride, **7,** 417
CARBOXYLIC ACIDS → ROH
9-Borabicyclo[3.3.1]nonane, **7,** 29
Borane + co-reagent, **6,** 64; **11,** 219
Borane–Tetrahydrofuran, **1,** 199; **3,** 76; **5,** 48; **11,** 69
Diborane, **5,** 184
Diisobutylaluminum hydride, **1,** 260
N-Ethyl-5-phenylisoxazolium-3′-sulfonate, **5,** 306
Lithium aluminum hydride, **1,** 581; **12,** 272
Magnesium methyl carbonate, **5,** 420
Peracetic acid, **1,** 787
Rhenium catalysts, **1,** 978
Sodium bis(2-methoxyethoxy)aluminum hydride, **3,** 260; **5,** 596
Sodium borohydride + co-reagent, **1,** 1053; **11,** 479; **12,** 565
Tetra-μ-hydridotetrahydroaluminum-magnesium, **4,** 316
Thexylborane–N,N-Diethylaniline, **9,** 464
Tin(IV) chloride–Sodium borohydride, **9,** 438
Titanium(IV) chloride–Sodium borohydride, **10,** 404
CARBOXYLIC ACIDS → RCHO
Bis(N-methylpiperazinyl)aluminum hydride, **6,** 52; **12,** 60
Borane–Dimethyl sulfide, **9,** 398
N,N′-Carbonyldiimidazole, **1,** 114
Chlorothexylborane–Dimethyl sulfide, **12,** 485
Dichlorobis(cyclopentadienyl)titanium, **11,** 163
Disodium tetracarbonylferrate, **6,** 550
Ethyl formate, **3,** 185
Lithium aluminum hydride, **1,** 581
Lithium diisopropylamide, **3,** 184
Lithium–Methylamine, **3,** 175; **4,** 288
Pyridinium chlorochromate, **9,** 397
Raney nickel, **2,** 293
Tetrabutylammonium borohydride, **6,** 564
1,3-Thiazolidine-2-thione, **11,** 518
p-Toluenesulfonylhydrazide, **5,** 678
Trifluoromethanesulfonic anhydride, **5,** 702

2,4,6-Triisopropylbenzene-
sulfonylhydrazide, **10,** 422
Vilsmeier reagent–Lithium
tri-*t*-butoxyaluminum hydride, **12,** 201
CARBOXYLIC ACIDS, RCOX → RCH_3
Borane–Tetrahydrofuran, **6,** 161
Dicyclopentadienyltitanium, **5,** 672
Iodotrimethylsilane, **9,** 251
Lead tetraacetate, **5,** 365
Lithium aluminum hydride–Aluminum
chloride, **5,** 389
Nickel on alumina, **11,** 355
Sodium bis(2-methoxyethoxy)aluminum
hydride, **3,** 260; **9,** 418
Trichlorosilane, **3,** 298; **4,** 525
Trichlorosilane–*t*-Amines, **4,** 525; **5,** 688;
8, 499
DIAZONIUM SALTS → ArH, RH (*see*
DEDIAZOTIZATION)
DIENES → C=C
Aluminum amalgam, **1,** 20
Arene(tricarbonyl)chromium complexes,
12, 34
9-Borabicyclo[3.3.1]nonane, **8,** 47
Chlorotris(triphenylphosphine)-
rhodium(I), **3,** 325; **5,** 736
Chromium carbonyl, **5,** 142
Dicarbonyldichlorobis(triphenyl-
phosphine)ruthenium(II), **5,** 193
Dichlorotris(triphenylphosphine)-
ruthenium(II), **8,** 159
Diimide, **3,** 99; **4,** 154; **6,** 195; **7,** 121
Gold, **5,** 321
Lithium aluminum hydride, **2,** 242
Lithium-Amines, **1,** 618; **11,** 287
Magnesium oxide, **7,** 221
Nickel boride, **3,** 208
Palladium black, **6,** 443
Palladium(II) chloride, **9,** 352
Ruthenium(III) chloride, **4,** 421
Sodium–Ammonia, **2,** 374
Triethylamine, **5,** 689
Triethylammonium formate–Palladium,
9, 481
Urushibara catalysts, **6,** 659
Zinc, **1,** 1276
DISULFIDES → RSR
Hexaethylphosphorous triamide, **2,** 207;
4, 242
Hexamethylphosphorous triamide, **6,**
279; **9,** 235
Sodium cyanide, **5,** 606
Triethyl phosphite, **1,** 1212
Triphenylphosphine, **3,** 317
EPOXIDES → ROH
with Metal hydrides
9-Borabicyclo[3.3.1]nonane, **7,** 29
9-Borabicyclo[3.3.1]nonane ate
complexes, **7,** 201
Borane–Tetrahydrofuran, **1,** 199; **9,**
136
t-Butyl hydroperoxide–Dialkyl
tartrate–Titanium(IV)
isopropoxide, **11,** 92
Copper hydride ate complexes, **5,** 168
Diborane + co-reagent, **2,** 107, 108
Diisobutylaluminum hydride, **4,** 158
Lithium aluminum hydride, **1,** 581; **7,**
196
Lithium aluminum hydride–
Aluminum chloride, **1,** 595; **8,** 289
Lithium borohydride, **1,** 603; **12,** 276
Lithium 9,9-dibutyl-9-borabicyclo-
[3.3.1]nonanate, **7,** 201
Lithium triethylborohydride–
Aluminum *t*-butoxide, **9,** 287
Lithium trimethoxyaluminum
hydride, **1,** 625
Sodium bis(2-methoxyethoxy)-
aluminum hydride, **10,** 357; **11,** 476
Sodium borohydride, **11,** 477
Sodium borohydride–Aluminum
chloride, **1,** 1053
Sodium cyanoborohydride, **9,** 424
with other Reagents
t-Butyl hydroperoxide, **7,** 43
Chromium(II) acetate, **1,** 147; **4,** 97
Dimethylsulfoxonium methylide, **10,**
168
Lithium–Alkylamines, **1,** 574; **3,** 186;
4, 291
Lithium–Ammonia, **4,** 288; **11,** 286
Lithium dimethylcuprate, **5,** 234
Raney nickel, **10,** 339
Sodium hydrogen telluride, **12,** 449
Sodium iodide, **6,** 544
Triphenyltin hydride, **10,** 451
Zinc–Acetic acid, **12,** 568
ESTERS → ROH
9-Borabicyclo[3.3.1]nonane, **7,** 29
Borane–Dimethyl sulfide, **6,** 64; **10,** 49;
11, 69; **12,** 64

REDUCTION REACTIONS (*Continued*)
Calcium borohydride, **2,** 57
Copper–Chromium oxide, **1,** 157
Dichlorobis(cyclopentadienyl)titanium, **10,** 130
Diisobutylaluminum hydride, **1,** 260; **12,** 191
Lithium aluminum hydride, **1,** 581; **12,** 272
Lithium–Ammonia, **1,** 601; **9,** 273
Lithium 9-boratabicyclo[3.3.1]nonane, **12,** 275
Lithium borohydride, **1,** 603; **11,** 293; **12,** 276
Lithium butyldiisobutylaluminum hydride, **12,** 276
Lithium triethylborohydride, **6,** 348; **9,** 286
Methylene chloride, **1,** 676
Phenol, **1,** 828
Platinum catalysts, **1,** 890
Sodium acetanilidoborohydride, **7,** 325
Sodium–Ammonia–Ethanol, **1,** 1041
Sodium bis(2-methoxyethoxy)aluminum hydride, **3,** 260; **5,** 596
Sodium borohydride, **1,** 1049; **11,** 477; **12,** 441
Sodium borohydride + co-reagent, **1,** 1053; **6,** 532; **11,** 479
Sodium (dimethylamino)borohydride, **12,** 446
Sodium trimethoxyborohydride, **3,** 268
Thexylborane, **4,** 175
Triethoxysilane, **11,** 554

ESTERS → RCHO, R_2CO
Bis(N-methylpiperazinyl)aluminum hydride, **6,** 52
Diisobutylaluminum hydride, **1,** 260; **2,** 140; **6,** 198
Lithium naphthalenide, **8,** 305
Lithium tri-*t*-butoxyaluminum hydride, **1,** 620
Sodium aluminum hydride, **1,** 1030
Sodium bis(2-methoxyethoxy)aluminum hydride, **3,** 260; **7,** 329
Tributyltin hydride, **10,** 411

ESTERS → RH (*see* REDUCTION OF ACETATES, OTHER ESTERS)

ESTERS → ROR
2,4-Bis(4-methoxyphenyl)-1,3-dithia-2,4-diphosphetane-2,4-disulfide, **10,** 39
Lithium aluminum hydride–Boron trifluoride etherate, **1,** 599
Sodium borohydride–Boron trifluoride, **1,** 1053
Trichlorosilane, **4,** 525; **6,** 606

HYDROPEROXIDES, ROOR, RCO_3R', ETC.
Dimethyl sulfoxide, **5,** 263
Hexamethylphosphorous triamide, **2,** 210
Palladium black, **4,** 365
Sodium iodide, **1,** 1087
Tetramethoxyethylene, **2,** 401
Thiourea, **6,** 586
Trialkyl phosphites, **1,** 1212, 1233; **2,** 432
Triphenylphosphine, **1,** 1238; **2,** 443; **5,** 725
Zinc, **1,** 1276

α-HYDROXY(ALKOXY) C=O's → 1,2-DIOLS
Borane–Dimethyl sulfide, **12,** 64
Calcium–Ammonia, **1,** 106
Copper chromite, **1,** 156
Dimethylphenylsilane, **12,** 209
Dimethyl sulfoxide–Trifluoroacetic anhydride, **10,** 168
Hexahydro-4,4,7-trimethyl-4H-1,3-benzothiin, **12,** 237
Lithium aluminum hydride, **6,** 206; **11,** 289
Lithium cyanoborohydride, **3,** 183
Potassium formate, **5,** 556
Sodium bis(2-methoxyethoxy)aluminum hydride, **12,** 440
Sodium borohydride, **10,** 357
Tetrabutylammonium fluoride, **12,** 458
Triisobutylaluminum, **4,** 535
Zinc borohydride, **12,** 572

IMINES → RNH_2
(S)-N-Benzyloxycarbonylproline, **11,** 447
Borane + co-reagent, **1,** 273, 1229; **6,** 161
Chlorotris(triphenylphosphine)-rhodium(I), **5,** 736
Cobalt(II) phthalocyanine, **11,** 138
Di-μ-carbonyldecacarbonyltri-*triangulo*-iron, **4,** 534
(S)-1-(Dimethoxymethyl-2-methoxymethyl)pyrrolidine, **10,** 152
Hydrogen hexachloroplatinate(IV)–Triethylsilane, **3,** 51

2,3,-O-Isopropylidene-2,3-dihydroxy-1,4-bis(diphenylphosphine)butane, **5,** 360
Lithium pyrrolidide, **5,** 416
Lithium tri-*sec*-butylborohydride, **12,** 286
Potassium–Graphite, **9,** 378
Sodium 9-cyano-9-hydrido-9-borabicyclo[3.3.1]nonane, **11,** 483
Sodium dithionite, **10,** 363
Trifluoroacetic acid–Alkylsilanes, **6,** 616

LACTONES → CYCLIC ETHERS
Diborane, **5,** 184
Lithium aluminum hydride–Boron trifluoride etherate, **1,** 599
Perchloric acid, **1,** 796
Sodium borohydride–Boron trifluoride, **1,** 1053
Triethylsilane–Boron trifluoride, **10,** 418

LACTONES → LACTOLS
Borane–Tetrahydrofuran, **1,** 199
Diisobutylaluminum hydride, **1,** 260; **2,** 140; **6,** 198; **9,** 171
Disiamylborane, **1,** 57; **2,** 29; **3,** 22
Lithium triethoxyaluminum hydride, **2,** 252
Sodium amalgam, **1,** 1030
Sodium bis(2-methoxyethoxy)aluminum hydride, **6,** 528; **7,** 327
Triethylsilane–Boron trifluoride, **10,** 418

NITRILES → RCHO
Diisobutylaluminum hydride, **1,** 260; **4,** 158; **6,** 198; **11,** 185
Lithium aluminum hydride, **1,** 581
Lithium triethoxyaluminum hydride, **1,** 625
Raney nickel, **1,** 723; **2,** 293; **9,** 405
1:1 Raney nickel alloy, **4,** 350
Semicarbazide, **2,** 98
Sodium bis(2-methoxyethoxy)aluminum hydride, **8,** 448; **9,** 418
Sodium hypophosphite, **1,** 1087
Sodium triethoxyaluminum hydride, **1,** 1108
Tin(II) chloride, **1,** 1116
Triethylsilane, **5,** 694

NITRILES → RNH_2
Aluminum hydride, **2,** 23
Borane + co-reagent, **1,** 199; **11,** 69
Cobalt boride–Borane–*t*-Butylamine, **11,** 138
Cobalt(II) phthalocyanine, **11,** 138
Diborane, **5,** 184
Hydridotris(triisopropylphosphine)-rhodium(I), **9,** 238
Lithium aluminum hydride–Aluminum chloride, **1,** 595
Nickel catalysts, **1,** 723; **4,** 351
1:1 Raney nickel alloy, **1,** 718
Rhodium catalysts, **1,** 979
Sodium bis(2-methoxyethoxy)aluminum hydride, **3,** 260; **5,** 596
Sodium borohydride + co-reagent, **1,** 1053; **3,** 264
Sodium trifluoroacetoxyborohydride, **7,** 342
Tetrabutylammonium borohydride, **10,** 378
Tin(IV) chloride–Sodium borohydride, **9,** 438

NITRILE OXIDES → RCN
Trimethyl phosphite, **1,** 1233
Triphenylphosphine, **1,** 1238

NITRO COMPOUNDS → RNH_2
Borane–Tetrahydrofuran, **12,** 65
Cobalt(II) phthalocyanine, **10,** 102; **11,** 138
Hydrogen hexachloroplatinate(IV)–Triethylsilane, **3,** 51
Iron–Acetic acid, **8,** 264
Molybdenum(V) chloride–Zinc, **10,** 274
Palladium–Graphite, **10,** 297
Raney nickel, **1,** 723; **7,** 312; **12,** 422
Sodium borohydride, **4,** 443
Sodium borohydride, sulfurated, **4,** 444
Sulfides, **5,** 632
Tin(IV) chloride + co-reagent, **9,** 438; **12,** 503
Tributyltin hydride, **11,** 545

NITRO COMPOUNDS → $ArNH_2$
Aluminum, **4,** 9
Ammonium sulfide, **3,** 270
Borane–Dimethylamine, **1,** 273
Cobalt(II) phthalocyanine, **10,** 102; **11,** 138
Di-μ-carbonyldecacarbonyltri-*triangulo*-iron, **4,** 534; **8,** 130
Dichlorotris(triphenylphosphine)-ruthenium(II), **6,** 654; **12,** 179
Ferrous sulfate, **1,** 393
Hydrazine, **1,** 434; **6,** 280; **7,** 170; **9,** 236; **11,** 255

REDUCTION REACTIONS (*Continued*)
Hydrogen hexachloroplatinate(IV)–Triethylsilane, **3,** 51
Hydrogen telluride, **10,** 205
Hydroxylamine, **12,** 251
Iron, **1,** 519
Iron–Acetic acid, **8,** 264
Lithium aluminum hydride, **5,** 382
Lithium–Ethylamine, **1,** 574
Molybdenum(V) chloride–Zinc, **10,** 274
Palladium(II) acetylacetonate, **6,** 45
Palladium catalysts, **6,** 445; **10,** 297; **11,** 392
Phase-transfer catalysts, **8,** 387
Phenylhydrazine, **1,** 838
Platinum oxide, **1,** 890
Platinum sulfide-on-carbon, **1,** 892
Polymethylhydrosiloxane, **4,** 393
Rhodium(III) chloride, **7,** 313
Sodium borohydride, **7,** 329
Sodium borohydride + co-reagent, **1,** 1054; **3,** 264; **5,** 602; **9,** 421
Sodium borohydride, sulfurated, **7,** 331
Sodium disulfide, **1,** 1064
Sodium dithionite, **1,** 1081
Sodium hydrogen sulfide, **3,** 266
Sodium sulfide, **1,** 1104
Sodium sulfide–Sulfur, **3,** 270
Tin, **1,** 1168
Tin(II) bromide, **1,** 1112
Tin(II) chloride, **1,** 1113; **3,** 271
Tin(II) chloride–Sodium borohydride, **11,** 480
Titanium(III) chloride, **5,** 669; **10,** 400
Titanium(IV) chloride–Magnesium amalgam, **12,** 503
Urushibara catalysts, **4,** 571
Zinc, **1,** 1276

NITRO COMPOUNDS → RNHOH
Aluminum amalgam, **1,** 20
Iridium catalysts, **2,** 228
Nitromethane, **1,** 903
Phenylhydrazine, **1,** 838
Zinc, **1,** 1276

NITRO COMPOUNDS → OTHER N COMPOUNDS
Hexamethylditin, **6,** 273
Lead, **2,** 233
Phosphine, **1,** 859
Ruthenium catalysts, **1,** 983
Sodium borohydride, **4,** 443
Sodium borohydride–Cobalt(II) chloride, **3,** 264
Thallium, **3,** 285
Tributylphosphine–Diphenyl disulfide, **12,** 514
Zinc, **1,** 1276

NITROSO COMPOUNDS → RNH_2
Iron carbonyl, **4,** 268
Tin(II) chloride, **1,** 1113
Zinc, **1,** 1276

OXIMES → RNH_2
Aluminum hydride, **2,** 23
Borane–Tetrahydrofuran, **3,** 76; **4,** 124
Diisobutylaluminum hydride, **12,** 191
Lithium aluminum hydride, **12,** 272
Lithium aluminum hydride–Aluminum chloride, **1,** 595
Lithium–Ethylamine, **4,** 288
1:1 Raney nickel alloy, **1,** 718; **2,** 289
Rhodium catalysts, **1,** 979; **6,** 503
Sodium amalgam, **1,** 1030
Sodium bis(2-methoxyethoxy)aluminum hydride, **3,** 260; **5,** 596
Sodium borohydride, sulfurated, **3,** 264
Tributylphosphine–Diphenyl disulfide, **12,** 514
Zinc, **1,** 1276

OXIMES → RNHOH
Borane–Tetrahydrofuran, **1,** 199; **3,** 76
Hydroxylamine, **11,** 257
Organoaluminum reagents, **12,** 339
Sodium cyanoborohydride, **4,** 448; **5,** 607; **6,** 537; **11,** 481

PHENOLS → ArH (*see* DEOXYGENATION)

P COMPOUNDS
Hexachlorodisilane, **3,** 148
Phenylsilane, **5,** 523
Trichlorosilane, **5,** 687

PROPARGYL ALCOHOLS (AND RELATED COMPOUNDS) → ALLYLIC ALCOHOLS
Butyllithium–Diisobutylaluminum hydride, **12,** 96
Chlorotris(triphenylphosphine)-rhodium(I), **6,** 652
Chromium(II) sulfate, **1,** 150
Dichlorobis(cyclopentadienyl)titanium, **11,** 163
Diisobutylaluminum hydride, **3,** 174; **9,** 171

Lithium aluminum hydride, **1,** 581; **4,** 295
Lithium methoxyaluminum hydride, **6,** 341
Sodium–Ammonia, **5,** 589
Sodium bis(2-methoxyethoxy)aluminum hydride, **12,** 440
Zinc–Potassium cyanide, **8,** 534
QUINONES → HYDROQUINONES
Chromium(II) chloride, **3,** 60
Polymethylhydrosiloxane, **4,** 393
Tetramethoxyethylene, **2,** 401
Titanium(III) chloride, **7,** 418
Trimethyl phosphite, **1,** 1233
QUINONES → ArOH
Borane–Tetrahydrofuran, **3,** 76
Chromium(II) chloride, **3,** 60
N,N-Diethylhydroxylamine, **6,** 187
Lithium tri-*t*-butoxyaluminum hydride, **7,** 329
Polymethylhydrosiloxane, **4,** 393
Sodium borohydride, **7,** 329
Sodium dithionite, **1,** 1081
Trimethyl phosphite, **12,** 536
SUBSTITUTED C=C → C=C
Dealkoxylation
Borane–Tetrahydrofuran, **6,** 161
2,6-Di-*t*-butyl-4-methylpyridine, **8,** 145
Diethyl phosphorochloridate, **3,** 98
Diisobutylaluminum hydride, **1,** 260
Titanium(0), **8,** 481
Triethyl phosphite, **3,** 304
Deamination, Denitration
Aluminum hydride, **3,** 9
Borane–Tetrahydrofuran, **1,** 199
Diethyl phosphite, **12,** 187
Lithium aluminum hydride, **11,** 289
Lithium aluminum hydride–Aluminum chloride, **1,** 595
Sodium sulfide–Thiophenol, **9,** 434
Dehalogenation
Chromium(II)–Amine complexes, **3,** 57
Dichlorobis(cyclopentadienyl)-titanium, **6,** 179
Difluoromethylenetriphenyl-phosphorane, **9,** 166
Iron carbonyl, **4,** 268
Lithium aluminum hydride, **4,** 291
Lithium–*t*-Butanol–Tetrahydrofuran, **4,** 291
Magnesium, **11,** 307
Manganese(II) chloride, **7,** 222
Potassium (or Sodium) tetracarbonyl-hydridoferrate, **6,** 483; **9,** 435
Sodium–*t*-Butanol–Tetrahydrofuran, **2,** 378
Tin(II) chloride, **1,** 1116
Tributyltin hydride, **2,** 424
Triethyl phosphite, **11,** 555
Zinc–Silver couple, **5,** 760; **9,** 519
Deselenylation
Diphosphorus tetraiodide, **11,** 224
Desilylation
Hydriodic acid, **6,** 281
(1-Lithiovinyl)trimethylsilane, **7,** 193; **9,** 498
Potassium fluoride, **6,** 481
Tetraethylammonium fluoride, **7,** 356
p-Toluenesulfinic acid, **8,** 488
Desulfurization
Allyl phenyl sulfone, **8,** 405
Aluminum amalgam, **5,** 9
Dichlorobis(triphenylphosphine)-nickel(II), **11,** 165
Nickel(II) chloride–Triphenylphosphine, **11,** 358
Reduction of C=C–OPO(OR)$_2$
Diethyl phosphorochloridate, **3,** 98
Lithium dimethylcuprate, **4,** 177
Titanium(0), **8,** 481
Triethyl phosphite, **3,** 304
α-SUBSTITUTED C=O's → C=O's
Aluminum chloride–Ethanethiol, **12,** 29
Cerium(III) sulfate, **9,** 99
Chlorotrimethylsilane + co-reagent, **7,** 429; **10,** 97
Chromium(II) acetate, **1,** 147
Chromium(II) chloride, **1,** 149
Chromium(III) chloride–Lithium aluminum hydride, **8,** 110
Di-μ-carbonylhexacarbonyldicobalt, **8,** 148
Dicyclopentadienyltitanium, **6,** 596
Diisobutylaluminum hydride, **11,** 185
N,N-Dimethylaniline, **2,** 147
Hydriodic acid, **1,** 449; **2,** 213
Hydrobromic acid, **2,** 214
Iron carbonyl, **4,** 268
Lithium dimethylcuprate, **5,** 234
Lithium iodide–Boron trifluoride, **4,** 305
Molybdenum carbonyl, **7,** 247; **9,** 317

REDUCTION REACTIONS (*Continued*)
Phosphorus(III) iodide, **10,** 318
Potassium tetracarbonylhydridoferrate, **6,** 483
Pyridine, **6,** 497
Pyridine–Sodium hydrosulfite, **5,** 566
Sodium O,O-diethyl phosphorotelluroate, **11,** 485
Sodium α-(N,N-dimethylamino)-naphthalenide, **10,** 362
Sodium dithionite, **11,** 485
Sodium iodide, **10,** 365
Sulfur trioxide–Pyridine, **9,** 396
Titanium(III) chloride, **5,** 669
Tributylphosphine, **1,** 1191
Tributyltin hydride, **9,** 476; **10,** 411
Trichloro(methyl)silane–Sodium iodide, **12,** 527
Triethylamine, **2,** 427
Triphenylphosphine, **1,** 1238
Vanadium(II) chloride, **7,** 418
Zinc, **1,** 1276; **2,** 459; **5,** 753; **6,** 672
Zinc–Acetic acid, **5,** 757
α-SUBSTITUTED C=O's → 1,2-DISUBSTITUTED COMPOUNDS
Diisobutylaluminum hydride, **6,** 198
Isobutylaluminum dichloride, **6,** 307
Lithium diisopropylamide, **8,** 292
Lithium tri-*t*-butoxyaluminum hydride, **1,** 620
Sodium borohydride, **1,** 1049
SULFIDES, RSSR → RSH
Formamidinesulfinic acid, **6,** 586
Lithium–Ethylamine, **1,** 574
o-Nitrobenzenesulfenyl chloride, **10,** 277
Sodium hydrogen selenide, **6,** 542
Triphenylphosphine, **5,** 725
SULFONES → RSOR
Arenediazonium tetrahaloborates, **9,** 102
SULFONIC ACIDS, RSO_2X → RSSR
Hydriodic acid, **1,** 449
Iodotrimethylsilane, **11,** 271
Molybdenum carbonyl, **3,** 206
Potassium iodide–Boron triiodide, **10,** 329
Trichlorosilane, **3,** 298
SULFONIC ACIDS, RSO_2X → RSH
Benzenesulfonyl chloride, **1,** 46
Phosphorus(V) sulfide, **11,** 428
Triphenylphosphine–Iodine, **10,** 220
Zinc, **1,** 1276
Zinc amalgam, **1,** 1287
SULFOXIDES, RSO_2R → RSR
Acetyl chloride, **4,** 5; **5,** 4
Benzeneselenol, **9,** 35
Bromine, **5,** 55
Bromodimethylborane, **12,** 199
Bromotrimethylsilane, **8,** 262
2-Chloro-1,3,2-benzodioxaphosphole, **7,** 58
Chlorotrimethylsilane–Zinc, **10,** 98
Cyanuric fluoride, **10,** 115
Dichloroborane, **4,** 129
O,O-Diethyl hydrogenphosphoroselenoate, **8,** 169
Diisobutylaluminum hydride, **5,** 224
Dimethyl sulfide–Trifluoroacetic anhydride, **9,** 188
Diphosphorus tetraiodide, **9,** 203
Ethylmagnesium bromide–Copper(I) iodide, **9,** 219
Formamidinesulfinic acid, **8,** 232
Hexamethylcyclotrisilathiane, **8,** 240
Hexamethyldisilathiane, **8,** 240
Hexamethylphosphorous triamide–Iodine, **9,** 236
Iodine–Pyridine–Sulfur dioxide, **6,** 296
Iodotrimethylsilane, **8,** 261; **9,** 251
Iron carbonyl, **3,** 167
Methanesulfinyl chloride, **7,** 224
Molybdenum(V) trichloride oxide, **7,** 248
Oxalyl chloride–Sodium iodide, **9,** 335
Palladium catalysts, **6,** 445
2-Phenoxy-1,3,2-benzodioxaphosphole, **5,** 512
Phenyl trimethylsilyl selenide, **9,** 496
Phosphorus(III) iodide, **10,** 318
Phosphorus(V) sulfide, **7,** 290; **8,** 401
Samarium(II) iodide, **10,** 344
Sodium bis(2-methoxyethoxy)aluminum hydride, **6,** 528
Sodium borohydride–Cobalt(II) chloride, **11,** 479
Sodium cyanoborohydride, **5,** 607
Thionyl chloride, **6,** 585
Thiophosphoryl bromide, **9,** 466
Tin(II) chloride, **5,** 631
Titanium(III) chloride, **5,** 669
Titanium(IV) chloride + co-reagent, **7,** 372; **8,** 487; **10,** 404
Trichloro(methyl)silane–Sodium iodide, **12,** 527

Trifluoroacetic anhydride–Sodium iodide, **8,** 504
Triphenylphosphine, **1,** 1238
Triphenylphosphine–Iodine, **8,** 518
Tripotassium tri-μ-chlorohexachloroditungstate, **8,** 522
Tris(phenylseleno)borane, **9,** 511
THIOL ESTERS → RCHO
1,2-Dianilinoethane, **2,** 98
Raney nickel, **2,** 293
α,β-UNSATURATED C=O's → ROH
Di-μ-carbonylhexacarbonyldicobalt, **1,** 224
Lithium aluminum hydride, **1,** 581
Lithium borohydride, **4,** 296
Osmium-on-carbon, **3,** 216
Potassium–Graphite, **4,** 397
Raney nickel, **6,** 502
Rhodium catalysts, **1,** 982
Sodium borohydride–Pyridine, **10,** 360
Sodium cyanoborohydride, **5,** 607; **6,** 537
Trifluoroacetic acid–Alkylsilanes, **5,** 695
Urushibara catalysts, **7,** 417
α,β-UNSATURATED C=O's → ALLYLIC ALCOHOLS
Alumina, **6,** 16
Aluminum hydride, **1,** 34; **3,** 9
Aluminum isopropoxide, **1,** 35; **6,** 19
Bis(triphenylphosphine)copper(I) borohydride, **10,** 47
9-Borabicyclo[3.3.1]nonane, **6,** 62; **7,** 29; **8,** 47
Borane–Dimethyl sulfide, **8,** 49
Calcium borohydride, **5,** 89
Chlorotris(triphenylphosphine)-rhodium(I)–Hydrosilanes, **11,** 130
Corey's reducing agent, **4,** 103
Dichlorotris(triphenylphosphine)-ruthenium(II), **6,** 654
N,N-(Diethylethanamine)-trihydridoaluminum, **6,** 607
Diisobutylaluminum 2,6-di-*t*-butyl-4-methylphenoxide, **9,** 171
Diisobutylaluminum hydride, **3,** 101; **6,** 198; **12,** 191
2,6-Diisopropylphenoxymagnesium hydride, **8,** 175
Dimethylformamide dialkyl acetals, **6,** 222
Dimethylphenylsilane, **12,** 209
Erbium(III) chloride, **11,** 232
Ethane-1,2-diaminoborane, **1,** 355
Lithium aluminum hydride, **1,** 581; **4,** 291; **9,** 274; **11,** 289
Lithium aluminum hydride + co-reagent, **5,** 389; **9,** 169; **10,** 148, 238; **12,** 33, 190
Lithium butylborohydride, **11,** 294
Lithium *n*- or *t*-butyldiisobutylaluminum hydride, **10,** 239; **12,** 276
Lithium tri-*t*-butoxyaluminum hydride, **1,** 620, 983; **4,** 312
B-(3-Methyl-2-butyl)-9-borabicyclo-[3.3.1]nonane, **8,** 334
Nitromethane, **4,** 357
B-3-Pinanyl-9-borabicyclo[3.3.1]nonane, **11,** 429
Platinum catalysts, **1,** 890
Polymethylhydrosiloxane, **4,** 393
Potassium tri-*sec*-butylborohydride, **6,** 490; **7,** 307
Raney nickel, **7,** 312
Sodium bis(2-methoxyethoxy)aluminum hydride, **12,** 440
Sodium borohydride + co-reagent, **8,** 451; **9,** 422; **12,** 444
Sodium cyanoborohydride, **5,** 607; **6,** 538
Sodium hydride–Sodium *t*-amyloxide–Zinc chloride, **11,** 486
Triisobutylaluminum–Bis(N-methylsalicylaldimine)nickel, **11,** 562
Zinc borohydride, **3,** 337; **5,** 761
α,β-UNSATURATED C=O's, C≡N → C=O's, C≡N
Aluminum chloride, **8,** 13
Birch reduction, **3,** 19
Chlorotris(triphenylphosphine)-rhodium(I), **2,** 448; **3,** 325; **4,** 562; **6,** 652; **11,** 130
Chromium(II)–Amine complexes, **5,** 142
Chromium(II) chloride, **3,** 60
Chromium(II) sulfate, **2,** 77
Cobalt boride–Borane–*t*-Butylamine, **11,** 138
Copper hydride ate complexes, **5,** 330; **6,** 144; **8,** 120; **10,** 237
Cryptates, **9,** 127
(1,5-Cyclooctadiene)(pyridine)-(tricyclohexylphosphine)iridium(I) hexafluorophosphate, **12,** 151
Di-μ-carbonylhexacarbonyldicobalt, **1,** 224; **6,** 172

REDUCTION REACTIONS (*Continued*)
Dichlorotris(triphenylphosphine)-ruthenium(II), **4,** 564
Diimide, **5,** 220
Diisobutylaluminum hydride, **11,** 185
Ethoxydiethylsilane, **5,** 293
Fluorosulfuric acid, **5,** 310
Hexafluoroantimonic acid, **6,** 272
Hexamethylphosphoric triamide, **2,** 208; **3,** 149
Hydrazine, **1,** 434
Hydridotetrakis(triphenylphosphine)-rhodium(I), **11,** 255
Hydrogen sulfide, **12,** 247
Hydrogen telluride, **10,** 205
Iron carbonyl, **4,** 268
Lithium aluminum hydride, **11,** 289
Lithium aluminum hydride–Copper(I) iodide, **6,** 326; **7,** 196; **10,** 237
Lithium–Ammonia, **1,** 601; **5,** 379
Lithium bronze, **11,** 293
Lithium tri-*sec*-butylborohydride, **6,** 348
Lithium trimethoxyaluminum hydride, **1,** 625
Magnesium, **6,** 351
Magnesium–Methanol, **10,** 251
Nickel catalysts, **1,** 718; **8,** 433; **11,** 356
Palladium catalysts, **5,** 499
2-Phenylbenzothiazoline, **12,** 384
Platinum catalysts, **1,** 890
Polymethylhydrosiloxane, **4,** 393
Potassium di-μ-carbonyloctacarbonyl-μ-hydridochromate, **7,** 299
Potassium–Graphite, **9,** 378
Potassium tetracarbonylhydridoferrate, **8,** 419
Potassium tri-*sec*-butylborohydride, **6,** 490; **7,** 307
Rhodium catalysts, **11,** 460
Ruthenium catalysts, **1,** 983
Samarium(II) iodide, **8,** 439; **10,** 344
Sodium amalgam, **1,** 1030
Sodium bis(2-methoxyethoxy)aluminum hydride, **5,** 596
Sodium borohydride, **2,** 377; **12,** 441
Sodium borohydride, sulfurated, **7,** 331
Sodium cyanoborohydride, **7,** 334
Sodium di-μ-carbonylhexacarbonyl-μ-hydridoferrate, **7,** 340
Sodium dithionite, **11,** 485
Sodium hydrogen telluride, **8,** 459; **10,** 365
Sodium iodide–Hydrochloric acid, **10,** 366
Tetrabutylammonium borohydride, **10,** 378
Tin, **1,** 1168
Titanium(III) chloride, **5,** 669
Tributyltin hydride, **11,** 545
Tributyltin trifluoromethanesulfonate, **12,** 524
Triethylammonium formate, **3,** 300; **9,** 481
Trifluoroacetic acid–Alkylsilanes, **5,** 695
Triisobutylaluminum–Bis(N-methylsalicylaldimine)nickel, **11,** 562
Trimesitylborane, **3,** 308
Triphenyltin hydride, **2,** 448; **3,** 324; **4,** 559; **5,** 734; **6,** 649
Urushibara catalysts, **6,** 659
Ytterbium, **9,** 517
Zinc, **1,** 1276; **5,** 753
Zinc–copper couple, **12,** 569
Zinc–Zinc chloride, **7,** 430

REDUCTIVE ACYLATION
Acetic anhydride, **1,** 3; **6,** 4
Di-μ-carbonylhexacarbonyldicobalt, **11,** 162
Formic acid–Formamide, **1,** 407
Lead tetraacetate, **3,** 168
Sodium acetate, **1,** 1024
Tetramethylammonium bromide, **1,** 1143
Triethylamine, **1,** 1198
Triphenyltin hydride, **2,** 448
Zinc–Acid anhydride–Catalyst, **1,** 1143; **4,** 577

REDUCTIVE ALKYLATION
$R_2C=O \rightarrow R_2CHR'$
Lithium–Ammonia, **5,** 379
Methyllithium, **5,** 448
Potassium tetracarbonylhydridoferrate, **6,** 483
Sodium bis(2-methoxyethoxy)aluminum hydride, **4,** 441; **6,** 528
p-Toluenesulfonyl chloride, **5,** 676
p-Toluenesulfonylhydrazide, **8,** 489
$R_2C=O \rightarrow R_2CHOR'$
Palladium hydroxide, **2,** 303
Triethylsilane, **4,** 530
ArH
Birch reduction, **7,** 21; **8,** 38
Lead tetraacetate, **7,** 185

Lithium–Ammonia, **10,** 234
α,β-UNSATURATED C=O's
t-Butyllithium, **8,** 70
Hexamethylphosphoric triamide, **3,** 149
Lithium–Ammonia, **1,** 601; **3,** 179; **8,** 282; **10,** 234
Lithium N-isopropylcyclohexylamide, **4,** 306
Lithium tri-*sec*-butylborohydride, **6,** 348; **8,** 308
Nickel carbonyl, **7,** 250
Sodium–Ammonia, **5,** 589
Titanium(III) chloride, **11,** 529

REDUCTIVE AMINATION
Benzylamine, **1,** 51
Borane–Amines, **1,** 273; **12,** 65
Borane–Tetrahydrofuran, **1,** 199
N,N-Dimethylformamide, **2,** 153
5,5-Dimethylhydantoin, **7,** 126
Ephedrine, **11,** 230
Formaldehyde, **1,** 397; **4,** 238; **6,** 305; **12,** 232
Hexamethylphosphoric triamide, **4,** 244
Iron carbonyl, **5,** 357; **6,** 304
Isovaline, **4,** 274
Lithium aluminum hydride, **6,** 325
Lithium cyanoborohydride, **3,** 183
α-Methylbenzylamine, **11,** 411
Morpholine, **1,** 705
Nickel catalysts, **1,** 723
Organotitanium reagents, **11,** 374
Potassium tetracarbonylhydridoferrate, **6,** 483
Sodium borohydride, **6,** 530; **11,** 477
Sodium cyanoborohydride, **4,** 448; **5,** 607; **6,** 537; **10,** 360; **12,** 445
Sodium hydrogen telluride, **12,** 449
Tetrabutylammonium cyanoborohydride, **11,** 499
Titanium(IV) chloride, **5,** 671

REDUCTIVE COUPLING (*see also* ACYLOIN REACTION, COUPLING REACTIONS)
C=O's → C=C
Titanium(0), **7,** 368; **8,** 481; **9,** 466; **11,** 526
Titanium(III) chloride, **6,** 587
Titanium(III) chloride + co-reagent, **5,** 391, 671; **6,** 588; **7,** 369; **10,** 401; **11,** 529; **12,** 493
Titanium(IV) chloride + co-reagent, **7,** 372; **8,** 487
Tributyltinlithium, **10,** 413
Tungsten(VI) chloride, **4,** 569
C=O's → PINACOLS
Aluminum amalgam, **1,** 20; **3,** 7
Chlorotrimethylsilane–Lithium, **6,** 323
Copper chromite, **1,** 156
Dichlorodimethylsilane, **4,** 183
Lithium–Ammonia, **3,** 179
Magnesium, **1,** 627; **6,** 351
Magnesium–Mercuric chloride, **2,** 255
Mercury(II) chloride, **1,** 652
Samarium(II) iodide, **12,** 429
Tetrabutylammonium fluoride, **12,** 458
Titanium(0), **9,** 466
Titanium(III) chloride, **6,** 587; **12,** 492
Titanium(IV) chloride–Magnesium amalgam, **7,** 373
Trimethyl phosphite, **1,** 1233
Vanadium(II) perchlorate, **9,** 513
$R_2CX_2 \rightarrow R_2C=CR_2$
Copper(I) halides, **2,** 90; **3,** 67
Di-μ-carbonylhexacarbonyldicobalt, **4,** 139
Iron carbonyl, **1,** 519
Lithium dimethylcuprate, **4,** 177
Magnesium, **5,** 419
Molybdenum carbonyl, **7,** 247
Titanium(III) chloride–Lithium aluminum hydride, **7,** 369
Tungsten carbonyl, **8,** 525

REDUCTIVE CYANATION
Diethyl phosphorocyanidate, **12,** 187
Hydrogen cyanide, **4,** 512
Methyl carbazate, **4,** 333; **7,** 235
Potassium cyanide, **8,** 409
p-Toluenesulfonylhydrazide, **4,** 511
Tosylmethyl isocyanide, **5,** 684; **6,** 600; **10,** 409

REDUCTIVE CYCLIZATION (*see also* WURTZ REACTION)
3-Benzyl-4-methylthiazolium chloride, **7,** 18
Birch reduction, **5,** 30
Bis(1,5-cyclooctadiene)nickel(0), **4,** 33
1,8-Bis(dimethylamino)naphthalene, **6,** 50
n- or *t*-Butyllithium, **4,** 60; **11,** 103
Chlorotrimethylsilane–Magnesium–Hexamethylphosphoric triamide, **5,** 713

REDUCTIVE CYCLIZATION (*Continued*)
Chromium(II) sulfate, **5,** 144
Cobaloxime(I), **11,** 135
Cyclopentadienone ketals, **2,** 143
Ethylenediaminetetraacetic acid tetrasodium salt, **1,** 373
Lithium aluminum hydride, **4,** 291
Lithium amalgam, **3,** 177
Magnesium, **4,** 315
Nickel carbonyl, **1,** 720; **2,** 290; **3,** 210; **4,** 353; **5,** 472; **6,** 417
Potassium–Ammonia, **9,** 377
Silver(I) nitrate, **7,** 321
Silver(I) trifluoromethanesulfonate, **7,** 324
Sodium, **4,** 437
Sodium naphthalenide, **4,** 349; **11,** 490
Sodium telluride, **6,** 550
Tetrakis(triphenyl phosphite)nickel(0), **11,** 503
Titanium(0), **11,** 526
Titanium(III) chloride + co-reagent, **8,** 483; **9,** 468; **12,** 493
Triphenylphosphine, **5,** 725
Zinc–copper couple, **1,** 1292; **7,** 428
REDUCTIVE HYDROLYSIS (*see* HYDROLYSIS)
REDUCTIVE SILYLATION (*see also* HYDROMETALLATION)
Chlorotrimethylsilane, **5,** 709
Chlorotrimethylsilane–Sodium, **8,** 446
Chlorotris(triphenylphosphine)-rhodium(I), **6,** 652; **11,** 130
Dimethylphenylsilane, **12,** 209
Hexamethyldisilane, **11,** 253
Phenylthiotriethylsilane, **6,** 613
Sodium hydride–Sodium *t*-amyloxide–Zinc chloride, **11,** 486
Trichlorosilane, **3,** 298
Triethylsilane, **5,** 694
REFORMATSKY REACTION
Bis(chloromethyl) ether, **5,** 34
Chlorotrimethylsilane, **12,** 126
Diethylaluminum chloride, **8,** 164
3,4-Dihydro-2H-pyran, **10,** 147
Ethyl trichloroacetate, **4,** 233
Indium, **6,** 293
Iodine, **1,** 495
Magnesium, **1,** 627
Reformatsky reagent, **3,** 334; **5,** 753; **8,** 533
Samarium(II) iodide, **10,** 344
Trimethyl borate, **3,** 310
Zinc, **1,** 1285; **3,** 334; **5,** 753; **6,** 672; **11,** 598; **12,** 567
Zinc-copper couple, **7,** 428; **8,** 533
Zinc–Graphite, **12,** 570
REPPE REACTION
Nickel carbonyl, **1,** 720; **5,** 472
RESOLUTION REACTIONS
ROH
3β-Acetoxy-Δ^5-etienic acid, **1,** 9
Acetyl bromide, **6,** 9
Brucine, **11,** 83
t-Butyl hydroperoxide–Dialkyl tartrate–Titanium(IV) isopropoxide, **11,** 92
d- and *l*-10-Camphorsulfonyl chloride, **1,** 109
Dehydroabietylamine, **10,** 117
(S)-1-α-Naphthylethylamine, **8,** 355; **9,** 320
(R)-1-(1-Naphthyl)ethyl isocyanate, **6,** 416; **8,** 356
[(2S)-(2α,3aα,4α,7α,7aα)]-2,3,3a,4,5,6,7,7a-Octahydro-7,8,8-trimethyl-4,7-methanobenzofuran-2-ol, **12,** 339
RCHO, R_2CO
(S)-(–)-1-Amino-2-(silyloxymethyl)-pyrrolidines, **12,** 32
2,3-Butanediol, **1,** 81
Camphor-10-sulfonic acid, **2,** 58
4-(4-Carboxyphenyl)semicarbazide, **1,** 118
Ephedrine, **5,** 289
N-Methylphenylsulfonimidoyl-methyllithium, **11,** 343
(S)-(–)-5-(α-Phenylethyl)semioxamazide, **9,** 364
N,N,P-Trimethyl-P-phenyl-phosphinothioic amide, **11,** 569
C=C
1,3-Dibenzyl-2-methyl-1,3,2-diazaphospholidine, **2,** 105
α-Methylbenzylamine, **2,** 271
C=C=C
α-Methylbenzylamine, **3,** 199; **8,** 393
RNH_2
1,1′-Binaphthyl-2,2′-diyl hydrogen phosphate, **12,** 49
(–)-α-Bromocamphor-π-sulfonic acid, **2,** 40

t-Butyl hydroperoxide–Dialkyl tartrate–Titanium(IV) isopropoxide, **12,** 90
Camphor-10-sulfonic acid, **1,** 108
d- and *l*-10-Camphorsulfonyl chloride, **1,** 109
(−)-2,3;4,6-Di-O-isopropylidene-2-keto-L-gulonic acid hydrate, **7,** 113
[(2S)-(2α,3aα,4α,7α,7aα)]-2,3,3a,4,5,6,7,7a-Octahydro-7,8,8-trimethyl-4,7-methanobenzofuran-2-ol, **12,** 339
Resolving agents, **1,** 977
(+)- and (−)-α-(2,4,5,7-Tetranitro-9-fluorenylideneaminooxy)-propionic acid, **6,** 577
AMINO ACIDS
Chiral cyclic polyethers, **5,** 103
Copper(II) perchlorate, **7,** 79
2-Naphthol-6-sulfonic acid, **5,** 470
(S)-1-Nitroso-2-methylindoline-2-carboxylic acid, **3,** 214
L-Tyrosine hydrazide, **2,** 454; **3,** 330
ArH
Resolving agents, **1,** 977
(+)- and (−)-α-(2,4,5,7-Tetranitro-9-fluorenylideneaminooxy)propionic acid, **1,** 1147; **2,** 404
RCOOH
(+)-3-Aminomethylpinane, **8,** 17
Amphetamine, **5,** 18
Cyano-*t*-butyldimethylsilane, **6,** 80
Dehydroabietylamine, **1,** 183; **3,** 73
α-Methylbenzylamine, **1,** 838; **5,** 441
EPOXIDES
(R)-1-(1-Naphthyl)ethyl isocyanate, **9,** 321
RETRO-DIELS–ALDER REACTION
9-(Benzyloxy)methoxyanthracene, **12,** 47
Cyclopentadiene, **8,** 137
1,3,3a,4,7,7a-Hexahydro-4,7-methanobenzo[*c*]thiophene 2,2-dioxide, **12,** 236
4-Phenyloxazole, **12,** 389
Tetrabutylammonium fluoride, **10,** 378
RING CONTRACTION (*see also* FAVORSKII REACTION)
Arenesulfonyl azides, **5,** 19
Barium hydroxide, **12,** 38
Boron trifluoride dibutyl etherate, **4,** 43
Calcium carbonate, **2,** 57
Chloramine, **3,** 45
Copper(II) chloride, **4,** 105
Dimethylsulfoxonium methylide, **5,** 254
Diphenyl phosphoroazidate, **7,** 138; **10,** 173
Diphenyl-4-pyridylmethylpotassium, **9,** 384
Hydrogen peroxide–Selenium(IV) oxide, **1,** 477
Lead tetraacetate, **9,** 265; **10,** 228
Lithium bromide, **5,** 395
Methanesulfonyl chloride, **1,** 662
Oxygen, **7,** 258
Periodates, **5,** 507
Potassium *t*-butoxide, **1,** 911
Silver acetate, **8,** 440
Silver perchlorate, **6,** 518
Sodium chloride, **6,** 534
Sodium hydride, **4,** 452
Sodium hydroxide, **5,** 616
Sulfuric acid, **4,** 470
Thallium(III) nitrate, **4,** 492
Tin(IV) chloride, **3,** 269
Zinc–Silver couple, **9,** 519
RING EXPANSION (*see also* BUCHNER REACTION, SYNTHESIS INDEX—CYCLOBUTANONES)
π-Allylpalladium complexes, **7,** 5
Allyl trifluoromethanesulfonate, **8,** 8
Aluminum chloride, **12,** 26
Arenesulfonyl azides, **5,** 475
Benzylsulfonyldiazomethane, **11,** 43
Benzyltriethylammonium chloride, **4,** 27; **7,** 18
Bis(tributyltin) peroxide, **8,** 44
1-Bromo-1-ethoxycyclopropane, **12,** 73
t-Butyl hypochlorite, **5,** 77
n- or *t*-Butyllithium, **5,** 79; **9,** 83
Calcium carbonate, **1,** 103
Chloramine, **1,** 122
Chloromethylcarbene, **10,** 90
Chlorotrimethylsilane, **3,** 310
Chloro[(trimethylsilyl)methyl]ketene, **12,** 127
Chromium carbonyl, **8,** 110
Chromium(III) chloride–Lithium aluminum hydride, **8,** 110
Copper(I) halides, **1,** 165; **3,** 67
Copper(0)–Isonitrile complexes, **9,** 122
Copper(I) trifluoromethanesulfonate, **6,** 130
Cyanogen azide, **3,** 71; **5,** 169

RING EXPANSION (*Continued*)
Cyclopropyl phenyl sulfide, **8,** 138
Diazoalkanes, **1,** 191; **3,** 73; **9,** 133
1,1-Dibromoalkyllithium reagents, **6,** 162; **7,** 90; **9,** 138; **11,** 158
2,3-Dichloro-5,6-dicyano-1,4-benzoquinone, **9,** 148
Dichloroketene, **1,** 221; **10,** 139; **12,** 176
Dichloromethyllithium, **4,** 138
Diethyl (1,3-dithian-2-yl)phosphonate, **11,** 179
Diethylzinc–Haloform, **4,** 153; **6,** 194
Dimethyl acetylenedicarboxylate, **5,** 227
Dimethylsulfoxonium methylide, **5,** 254; **8,** 194; **10,** 168
Dimethyl N-(*p*-toluenesulfonyl)-sulfoximine, **9,** 193
Diphenyl-4-pyridylmethylpotassium, **9,** 384
Diphenylsulfonium cyclopropylide, **8,** 212
Diphosphorus tetraiodide, **6,** 243
Ethyl azidoformate, **7,** 146
Ethyl diazoacetate, **1,** 367; **2,** 464; **3,** 138; **6,** 252; **8,** 222; **12,** 223
Ferric chloride, **7,** 153; **10,** 185
Hydrogen bromide, **4,** 251
Hydrogen peroxide, **12,** 242
2-Hydroxymethyl-3-trimethylsilyl-1-propene, **11,** 258
Iodotrimethylsilane, **10,** 216
Isocyanomethyllithium, **4,** 272
1-Lithiocyclopropyl phenyl sulfide, **5,** 372; **9,** 271
Lithium aluminum hydride–Aluminum chloride, **4,** 293
Lithium diisopropylamide, **5,** 400
Lithium iodide, **11,** 300
Lithium naphthalenide, **4,** 348
Lithium piperidide, **6,** 344
Magnesium bromide etherate, **9,** 288
Methylene chloride–Butyllithium, **7,** 239
Methyllithium, **1,** 686; **5,** 448; **6,** 384
Nitromethane, **1,** 739
Norbornene, **1,** 757
Palladium catalysts, **7,** 275
Perchloric acid, **5,** 506; **6,** 453
Phenyldiazomethane, **1,** 366
Phenyl(trihalomethyl)mercury, **6,** 468
Potassium *t*-butoxide, **1,** 911; **5,** 544; **7,** 296
Potassium hydride, **12,** 407
Pyridine, **2,** 349; **6,** 497
Pyridinium chlorochromate, **9,** 397
Quinoline, **1,** 975
Silver acetate, **2,** 362
Silver perchlorate, **4,** 432; **6,** 518
Silver sulfate, **3,** 254
Silver tetrafluoroborate, **2,** 365; **4,** 428
Silver *p*-toluenesulfonate, **7,** 323
Simmons–Smith reagent, **6,** 521
Sodium–Ammonia, **4,** 438; **6,** 523
Sodium chlorodifluoroacetate, **5,** 603
Sodium ethoxide, **1,** 1065
Sodium hydrogen sulfite, **1,** 1047
Sodium methylsulfinylmethylide, **1,** 310
Sodium nitrite, **1,** 1097
Tetrabutylammonium fluoride, **12,** 458
Tetracarbonyldi-μ-chlorodirhodium, **5,** 572; **6,** 108
Tetrafluoroboric acid, **1,** 394; **10,** 382
Thallium(III) nitrate, **8,** 476
Thallium(III) perchlorate, **5,** 657
Tributyltin hydride, **12,** 516
Triethyloxonium tetrafluoroborate, **8,** 500
Trifluoroacetic acid, **5,** 695; **9,** 483
Tris(methylthio)methyllithium, **10,** 453
Zinc, **6,** 672
Zinc–Acetic acid, **5,** 757
Zinc–Copper–Isopropyl iodide, **10,** 460
Zinc halides, **2,** 464; **5,** 762, 765

RITTER REACTION
Arene(tricarbonyl)chromium complexes, **10,** 13
Hydrogen peroxide, **4,** 254
Iodine azide, **5,** 350
Sulfuric acid, **5,** 633

ROBINSON ANNELATION
2-(2-Bromoethyl)-2-methyl-1,3-dioxolane, **1,** 75
t-Butyl γ-iodotiglate, **6,** 82
4-Chloromethyl-3,5-dimethylisoxazole, **1,** 276
1-Chloro-3-pentanone, **3,** 49; **6,** 110
5-Chloro-1-penten-3-one, **3,** 47
1,3-Dibromo-2-pentene, **7,** 90
1,3-Dichloro-2-butene, **1,** 214
Ethyl phenyl sulfoxide, **6,** 257
Ethyl vinyl ketone, **1,** 388
2-Fluoro-1-buten-3-one, **6,** 263
Formaldehyde, **6,** 264

(E)-1-Iodo-3-trimethylsilyl-2-butene, **5,** 355
1-Methoxy-3-buten-2-one, **3,** 198
5-Methoxycarbonyl-1-penten-3-one, **1,** 1271
Methyl 5-methoxy-3-oxopentanoate, **9,** 314
Methyl 3-oxo-4-pentenoate, **1,** 1272; **5,** 746
Methyl vinyl ketone, **1,** 697; **6,** 407
2-Methyl-6-vinylpyridine, **6,** 409
1,7-Octadien-3-one, **9,** 327; **10,** 280
trans-3-Penten-2-one, **2,** 306; **3,** 218; **4,** 371
Piperidine, **1,** 886
Pyrrolidine, **1,** 972
Sodium dicarbonyl(cyclopentadienyl)-ferrate, **6,** 538
Sodium methylsulfinylmethylide, **2,** 166
3-Trimethylsilyl-3-buten-2-one, **5,** 461; **6,** 406

ROSENMUND REDUCTION
Palladium catalysts, **4,** 367, 368; **7,** 275
Quinoline, **1,** 975
Tetramethylthiourea, **1,** 1145

SAKURAI REACTION (*see* CONJUGATE ADDITION)
SANDMEYER REACTION
Copper(I) halides, **1,** 165, 166
Copper(II) chloride–Nitric oxide, **2,** 87
Phase-transfer catalysts, **9,** 356
Sodium nitrite, **1,** 1097
Tin(II) bromide, **1,** 1112
SARETT OXIDATION (*see* REAGENT INDEX)
SCHIEMANN REACTION
Arenediazonium salts, **1,** 42, 43
Tetrafluoroboric acid, **1,** 394
Tetrahydrofuran, **1,** 1140
SCHMIDT REACTION
Azidotrimethylsilane, **7,** 394; **10,** 14
Chloramine, **1,** 122
Diphenyl phosphoroazidate, **11,** 222
Hydrazoic acid, **1,** 446; **2,** 211; **5,** 329
Sodium azide, **1,** 901, 1041, 1225; **2,** 376; **4,** 440
Trifluoroacetic acid, **12,** 529
SCHOLL REACTION
Antimony(III) chloride, **8,** 19
SCHOTTEN–BAUMANN REACTION
Ethyl chlorothiolformate, **6,** 252
Formyl fluoride, **1,** 407
Sodium borohydride, **6,** 530
SELENYLATION
Benzeneselenenyl halides, **5,** 518; **6,** 459; **7,** 286; **8,** 25; **9,** 25; **10,** 16; **11,** 33
Benzeneselenenyl trifluoroacetate, **6,** 459
Benzeneseleninic acid, **8,** 28
N,N-Diethylbenzeneselenenamide, **11,** 178
Diphenyl diselenide, **5,** 272; **7,** 136; **11,** 219
Lithium diisopropylamide, **6,** 334
Phenylseleninyl chloride, **6,** 459
Phenyl selenocyanate, **11,** 416
1-Phenylthiovinyllithium, **8,** 281
2-Pyridineselenenyl bromide, **11,** 455
Selenium, **10,** 345; **11,** 465
p-Toluenesulfonylhydrazide, **11,** 537
SEMMLER–WOLFF REACTION
Tetramethylene sulfone, **5,** 651
SERINI REACTION
Zinc, **1,** 1276; **2,** 459; **3,** 334
SHAPIRO REACTION (*see* ELIMINATION REACTIONS)
SHARPLESS EPOXIDATION
t-Butyl hydroperoxide, **5,** 75; **7,** 43; **8,** 62; **10,** 64
t-Butyl hydroperoxide + co-reagent, **5,** 75; **9,** 81; **11,** 92, 97; **12,** 90
Oxodiperoxymolybdenum(pyridine)-(hexamethylphosphoric triamide), **9,** 197
p-Tolylsulfinylacetic acid, **10,** 405
SIGMATROPIC REARRANGEMENTS
[1,3]
Butyllithium, **9,** 83
2-(Chloroethoxy)carbene, **10,** 89
Diphenylsulfonium cyclopropylide, **4,** 211
1,3-Dithienium tetrafluoroborate, **4,** 218
1-Hydroxycyclopropanecarboxaldehyde, **12,** 247
Lead carbonate, **9,** 265
1-Lithiocyclopropyl phenyl sulfide, **7,** 190
(1-Lithiovinyl)trimethylsilane, **11,** 286
Lithium cyclopropyl(phenylthio)cuprate, **7,** 211
Organocopper reagents, **11,** 365
N-Phenyl-1,2,4-triazoline-3,5-dione, **4,** 381

SIGMATROPIC REARRANGEMENTS (*Continued*)
Potassium hydride–Hexamethylphosphoric triamide, **7,** 302
3-Tetrahydropyranyloxy-1-tributylstannyl-1-propene, **6,** 602
[2,3] (*see also* WITTIG REARRANGEMENT)
3-Acetoxy-2-trimethylsilylmethyl-1-propene, **11,** 578
S-Allyl N,N-dimethyldithiocarbamate, **6,** 11
Allyl trifluoromethanesulfonate, **8,** 8
Benzenesulfenyl chloride, **6,** 30; **9,** 35
Bis(methoxycarbonyl)sulfur diimide, **12,** 57
Cesium fluoride, **9,** 100
(3-Chloro-3-methyl-1-butynyl)lithium, **8,** 93
m-Chloroperbenzoic acid, **12,** 118
(Z)-1,2-Dichloro-4-phenylthio-2-butene, **7,** 98
Diethylzinc–Diiodomethane, **11,** 182
Dimethyl diazomalonate, **5,** 244
Dimethylformamide dimethyl acetal, **8,** 191
2,4-Dinitrobenzenesulfenyl chloride, **9,** 194
1,3-Dithiane, **4,** 216
1,3-Dithienium tetrafluoroborate, **4,** 218
Dodecamethylcyclohexasilane, **12,** 219
Lithium diisopropylamide, **5,** 400; **6,** 334
Lithium thiophenoxide, **5,** 418
(Z)-2-Methoxy-1-(phenylthio)-1,3-butadiene, **11,** 328
Methyl (allylthio)acetate, **10,** 261
Methyl bromide–Lithium bromide, **10,** 262
Methyl cyanodithioformate, **7,** 237
Methyl methanethiosulfonate, **7,** 243
p-Nitrophenyl selenocyanate, **9,** 325
Phase-transfer catalysts, **8,** 387
Phenylselenenyl benzenesulfonate, **11,** 407
Potassium carbonate, **10,** 323
Potassium thiophenoxide, **8,** 420
Silver tetrafluoroborate, **5,** 587
Sodium benzeneselenoate, **8,** 447
Sulfur dioxide, **8,** 464
p-Toluenesulfonyl-S-methylcarbazate, **5,** 681
Tributyl(iodomethyl)tin, **9,** 250
Trichloroacetonitrile, **12,** 526
Triethyloxonium tetrafluoroborate, **2,** 430
Trimethyl phosphite, **11,** 570
Trimethylsilylmethanethiol, **11,** 576
Trimethylsilylmethyl trifluoromethanesulfonate, **10,** 434
[3,3]
Claisen rearrangement
Acetic anhydride–Sodium acetate, **7,** 2
Alkylaluminum halides, **11,** 7
Benzeneselenenyl halides, **9,** 25
Benzylamine, **10,** 26
Boron trichloride, **5,** 50
Boron trifluoride etherate, **1,** 70
N-Bromosuccinimide, **4,** 49
(E)-(Carboxyvinyl)trimethylammonium betaine, **12,** 106
μ-Chlorobis(cyclopentadienyl)-(dimethylaluminum)-μ-methylenetitanium, **11,** 52
Chlorotrimethylsilane, **8,** 107
Claisen's alkali, **6,** 127
Copper(I) trifluoromethanesulfonate, **10,** 108
Dichlorotris(triphenylphosphine)-ruthenium(II), **8,** 159
3,3-Diethoxy-1-propene, **9,** 158
Diethylaluminum benzenethiolate, **12,** 343
N,N-Dimethylacetamide dimethyl acetal, **8,** 179
Ethyl vinyl ether, **1,** 386; **3,** 300; **4,** 234; **11,** 235
Iodine, **8,** 256
2-(E)-Lithiovinyltrimethylsilane, **10,** 443
Lithium methylsulfinylmethylide, **12,** 283
Mercury(II) acetate, **1,** 644
(S)-N-Methanesulfonylphenylalanyl chloride, **8,** 320
2- or 3-Methoxypropene, **2,** 230; **4,** 330
Organoaluminum reagents, **12,** 339
Organotitanium reagents, **11,** 52
Phenylselenoacetaldehyde, **10,** 310
Phenylthioacetylene, **9,** 370
Potassium hydride, **11,** 435
Sodium dithionite, **11,** 485

Sodium methylsulfinylmethylide, **12,** 451
1,1,1,3-Tetramethoxypropane, **10,** 424
Titanium(IV) chloride, **7,** 370; **11,** 529
p-Toluenesulfonic acid, **5,** 673
Trialkylaluminums, **11,** 539
Triethyl orthoacetate, **3,** 300; **4,** 527; **6,** 607; **10,** 417
Trifluoroacetic acid, **4,** 530; **11,** 557
1,1,1-Trimethoxy-3-phenylselenopropane, **9,** 489
Trimethyl orthoformate, **8,** 180
Zinc carbonate, **8,** 535
Zinc chloride, **10,** 461
amide-Claisen rearrangement
Chlorotris(triphenylphosphine)rhodium(I), **2,** 448
N,N-Dimethylacetamide dialkyl acetals, **1,** 271; **4,** 166; **5,** 226; **8,** 179
Hydrogen fluoride, **9,** 240
aza-Claisen rearrangement
Titanium(IV) chloride, **9,** 468
Zinc chloride, **5,** 763
ester enolate Claisen rearrangement
t-Butyldimethylchlorosilane, **6,** 78
Chlorotrimethylsilane, **6,** 626
Hexamethylphosphoric triamide, **6,** 273; **8,** 240
Lithium diisopropylamide, **11,** 296
Lithium hexamethyldisilazide, **12,** 280
Lithium N-isopropylcyclohexylamide, **4,** 306; **7,** 209
Triethyl orthoacetate, **6,** 607
Zinc, **4,** 574
thio-Claisen rearrangement
Aluminum chloride, **11,** 25
Diethyl allylthiomethylphosphonate, **3,** 94
2-Ethoxyallyl vinyl sulfide, **5,** 292
Potassium *t*-butoxide, **4,** 399
Sodium N,N-dimethyldithiocarbamate, **6,** 539
Tetraethylthiuram disulfide, **6,** 569
Triethylamine, **7,** 385
1-Vinylthioallyllithium, **5,** 749
Cope rearrangement
Antimony(V) fluoride, **6,** 23
Bis(aceto- or benzonitrile)dichloropalladium(II), **10,** 31; **11,** 46; **12,** 50
Chlorotrimethylsilane, **6,** 626
1-Lithio-2-vinylcyclopropane, **7,** 192; **9,** 273
Lithium bis(2-vinylcyclopropyl)cuprate, **7,** 199
Lithium cyclopropyl(phenylthio)cuprate, **7,** 211
Methyl 2-acetylacrylate, **11,** 334
2-Methylcyclopentenone-3-dimethylsulfoxonium methylide, **6,** 378
Methyl 4-methylthiocrotonate, **6,** 389
Methyl vinyl ketone, **5,** 464
Organocopper reagents, **9,** 328
Phenylselenoacetaldehyde, **10,** 310
Potassium hydride, **10,** 327
Sulfur dioxide, **4,** 469
p-Toluenesulfonylhydrazide, **5,** 678
Trifluoroacetic acid, **11,** 557
oxy-Cope rearrangement
Bis(benzonitrile)dichloropalladium(II), **12,** 51
Butyllithium, **11,** 101
Cyanotrimethylsilane, **7,** 397
Dimethylphenylsilyllithium, **12,** 210
Grignard reagents, **10,** 189
Mercury(II) trifluoroacetate, **11,** 320; **12,** 306
N-Methyl-2-pyrrolidone, **9,** 316
Potassium hydride, **7,** 302; **8,** 412; **10,** 327; **11,** 438; **12,** 407
Silver(I) trifluoromethanesulfonate, **12,** 435
other [3,3] Rearrangements
S-Allyl N,N-dimethyldithiocarbamate, **6,** 11
Bis(benzonitrile)dichloropalladium(II), **10,** 31
N-*t*-Butylhydroxylamine, **12,** 95
1,5-Diazabicyclo[4.3.0]nonene-5, **4,** 116
Dichloroketene, **9,** 152; **12,** 176
N,N-Dimethylthiocarbamoyl chloride, **8,** 200
α-Methoxyvinyllithium, **7,** 233
Pentadienyl-1-pyrrolidinecarbodithioate, **12,** 374
Tributyltin hydride, **10,** 411
Trichloroacetonitrile, **6,** 604; **7,** 381
Vinyl acetate, **12,** 565
Zinc oxide, **1,** 1294
SILYLATION (*see also* PROTECTION, SYNTHESIS INDEX—SILYL ENOL

SILYLATION (*Continued*)
ETHERS)
OF C–H
1,3-Bis(phenylseleno)propene, **11,** 59
t-Butyldimethylchlorosilane, **11,** 88
Butyllithiums, **5,** 80; **8,** 70; **10,** 76; **12,** 97
Chloromethyldiphenylsilane, **10,** 91; **12,** 321
Chlorotrimethylsilane, **6,** 626
Chlorotrimethylsilane–Sodium, **12,** 126
Chromium carbonyl, **9,** 117
N,N-Dimethylhydrazine, **7,** 126
1,3-Dithiane, **2,** 182
Ethyl trimethylsilylacetate, **11,** 234
Lithium 2,2,6,6-tetramethylpiperidide, **5,** 417
1-Phenylthiovinyllithium, **8,** 281
N,N,N′,N′-Tetramethylethylenediamine, **4,** 485
3-Triethylsilyloxypentadienyllithium, **11,** 556
2,4,6-Triisopropylbenzenesulfonylhydrazide, **9,** 486
Trimethylsilyl trifluoromethanesulfonate, **8,** 514
OF HETEROATOMS
Bis(trimethylsilyl)acetamide, **1,** 61
N,O-Bis(trimethylsilyl) sulfamate, **7,** 28
t-Butyldimethylchlorosilane, **5,** 74
t-Butyldimethylsilyl trifluoromethanesulfonate, **10,** 63
Chlorodimethyl(2,4,6-tri-*t*-butylphenoxy)silane, **11,** 217
Chlorotrimethylsilane, **2,** 435; **4,** 537; **5,** 709
Chlorotrimethylsilane–Sodium, **8,** 446
Chlorotris(triphenylphosphine)-rhodium(I), **5,** 736
Dimethyl sulfoxide, **2,** 157
Hexamethyldisilazane, **5,** 323
Hexamethyldisiloxane, **10,** 44, 196
Ketene alkyl trialkylsilyl acetals or ketals, **9,** 310; **11,** 279
Triethylsilyl perchlorate, **12,** 527
Trimethylsilyl trifluoromethanesulfonate, **11,** 584
SIMMONS–SMITH REACTION (*see* SYNTHESIS INDEX—THREE-MEMBERED RINGS)
SMILES REARRANGEMENT
Hexamethylphosphoric triamide, **8,** 240
SOMM–MULLER METHOD
Phosphorus(V) chloride, **1,** 866
Tin(II) chloride, **1,** 1116
SOMMELET REACTION
Hexamethylenetetramine, **1,** 427; **2,** 208; **4,** 243
SOMMELET AND RELATED REARRANGEMENTS
Butyllithium–Tetramethylethylenediamine, **5,** 80
SPIROANNELATION (*see* SYNTHESIS INDEX—SPIRO COMPOUNDS)
STEPHEN REDUCTION
Nickel catalysts, **4,** 350
Tin(II) chloride, **1,** 1116
STEVENS REARRANGEMENT
Butyllithium–Tetramethylethylenediamine, **5,** 80
Dialkoxycarbenium tetrafluoroborates, **4,** 114
Sodium methylsulfinylmethylide, **6,** 546
Sodium sulfide, **5,** 623
STOBBE CONDENSATION
1,8-Diazabicyclo[5.4.0]undecene-7, **5,** 177
Di-*t*-butyl succinate, **1,** 212
Lithium diisopropylamide, **8,** 292
5-Methoxycarbonyl-1-penten-3-one, **1,** 1271
Potassium *t*-butoxide, **1,** 911
Potassium hydride, **1,** 935
Sodium hydride, **1,** 1075
STORK ENAMINE REACTION (*see* ALKYLATION REACTIONS)
SUBSTITUTION REACTIONS AT AROMATIC CARBON (*see also* FRIEDEL–CRAFTS REACTION, NITRATION, REDUCTION REACTIONS, ULLMANN ETHER SYNTHESIS)
OF $ArN_2^+X^-$ (*see also* HALOGENATION)
Acetaldoxime, **1,** 3
Arenediazonium tetrahaloborates, **1,** 43
Copper(I) chloride, **1,** 166
Copper(I) oxide–Copper(II) nitrate, **8,** 124
Formaldoxime, **1,** 402
Mercury(II) chloride, **1,** 652
Potassium selenocyanate, **8,** 417
Sodium disulfide, **1,** 1064

Sodium nitrite, **1,** 1097
Sodium sulfide, **3,** 269
Sulfur dioxide–Copper(I) chloride, **11,** 495
Tetrafluoroboric acid, **1,** 394

ArX → ArCN
Copper(I) cyanide, **5,** 166
N,N-Dimethylformamide, **1,** 278
Ferric chloride, **1,** 390
N-Methyl-2-pyrrolidone, **1,** 696
Pyridine, **1,** 958
Sodium cyanide, **9,** 423

ArX → $ArNH_2$, $ArNR_2$
Aluminum chloride, **6,** 17
Azidomethyl phenyl sulfide, **12,** 37
Copper(I) chloride, **1,** 166
Copper(I) hexamethyldisilazide, **5,** 45; **6,** 57
N,N-Dimethylformamide, **4,** 184
N-Nitrosodimethylamine, **5,** 479
Phenylsodium, **1,** 848
Sodium amide, **4,** 439
Tetrakis(triphenylphosphine)-palladium(0), **12,** 468
Tris(trimethylsilyl)hydrazidocopper, **6,** 651
Urea, **1,** 1262

ArX → ArOH, ArOR
Barium oxide, **9,** 23
Borane–Tetrahydrofuran, **4,** 124
t-Butyl hydroperoxide, **1,** 88
2-*t*-Butylperoxy-1,3,2-dioxaborolane, **12,** 101
Copper(I) alkoxides, **5,** 148
Copper(I) iodide, **6,** 147
Copper(II) sulfate, **1,** 164
Copper(I) trifluoromethanesulfonate, **6,** 130
Crown ethers, **5,** 152
Dimethyl sulfoxide, **4,** 192
2,4-Dinitrofluorobenzene, **2,** 175
Dioxygentetrakis(triphenylarsine)-rhodium(I) hexafluorophosphate, **8,** 206
Osmium tetroxide–*t*-Butyl hydroperoxide, **1,** 88
Potassium superoxide, **7,** 304; **8,** 417
Silver(I) oxide, **5,** 583
Triaryl phosphates, **11,** 542
Trimethyl borate, **2,** 435

ArOH → $ArNH_2$
4-Chloro-2-phenylquinazoline, **4,** 86
Diethyl phosphorochloridate, **4,** 152; **5,** 217
Phenyl phosphorodiamidate, **5,** 517
Potassium amide, **4,** 398
Zinc chloride, **1,** 1289

ArOH → ArX
Phosphorus(V) bromide, **1,** 865
Triphenylphosphine halides, **1,** 1247; **2,** 446; **3,** 320; **12,** 554

SULFENYLATION
S-Allyl N,N-dimethyldithiocarbamate, **6,** 11
Aluminum thiophenoxide, **9,** 15
Benzenesulfenyl chloride, **6,** 30; **9,** 35, 429; **12,** 42
t-Butyllithium, **7,** 47
Dimethyl disulfide, **5,** 246; **6,** 217; **8,** 189
N,N-Dimethylhydrazine, **7,** 126
Diphenyl disulfide, **5,** 276; **6,** 235; **7,** 137; **9,** 429; **10,** 171; **11,** 220
2,2′-Dipyridyl disulfide, **8,** 214
Lead tetraacetate–Trifluoroacetic acid, **6,** 317
Lithium diisopropylamide, **6,** 334; **9,** 280
Methyl methanethiosulfonate, **5,** 454; **7,** 243
Methyl 2-nitrophenyl disulfide, **9,** 314
N-Methylthiophthalimide, **11,** 493
S-Methyl *p*-toluenethiosulfonate, **5,** 460; **6,** 400
Phenyl benzenethiosulfonate, **8,** 391; **9,** 362, 429
1-Phenylthiovinyllithium, **8,** 281
Potassium hydride, **9,** 386
Potassium hydroxide, **8,** 415
Sodium ethanethiolate, **11,** 492
Sodium hydride–Butyllithium, **9,** 429
Tetrabutylammonium fluoride, **9,** 446
Thiophenol, **9,** 447
p-Toluenesulfonylhydrazide, **9,** 472; **11,** 537
Trifluoromethylthiocopper, **6,** 621
Trimethylene bis(thiotosylate), **6,** 628

SULFONAMIDOMERCURATION (*see* ADDITION REACTIONS)

SULFONATION
Bis(trimethylsilyl) sulfate, **10,** 176
Butyllithium, **7,** 45
Methyl fluorosulfonate, **4,** 339
Phenyl chlorosulfate, **11,** 410

SULFONATION (*Continued*)
Sulfur trioxide + co-reagent, **1,** 1126, 1127, 1128; **2,** 393; **4,** 473
Thionyl chloride, **1,** 1158
SULFONYLATION (*see also* SYNTHESIS INDEX—SULFONIC ACID ESTERS, SULFONAMIDES)
Arenesulfonyl halide–Antimony(V) halide complexes, **5,** 20
Benzenesulfonic anhydride, **1,** 46
Benzenesulfonyl fluoride, **11,** 40
Chlorosulfuric acid, **1,** 139
2,6-Di-*t*-butyl-4-methylpyridine, **10,** 123
Methanesulfonic anhydride, **1,** 667
N-Methyl-N-tolylsulfonylpyrrolidinium perchlorate, **4,** 345
Nonafluorobutanesulfonyl fluoride, **5,** 479
Phenylselenenyl benzenesulfonate, **11,** 407; **12,** 390
Phenylselenenyl *p*-toluenesulfonate, **10,** 315
N-Phenyltrifluoromethanesulfonimide, **12,** 395
Polyphosphoric acid, **2,** 334
Silver(I) trifluoromethanesulfonate, **6,** 520
Sodium borohydride, **6,** 530
Thallium(I) ethoxide, **6,** 577
Thallium(III) trifluoroacetate, **9,** 462
p-Toluenesulfinylimidazolide, **11,** 535
p-Toluenesulfonyl chloride, **1,** 1179, 1184
p-Toluenesulfonyl perchlorate, **1,** 1187
Trifluoromethanesulfonic acid, **4,** 532
Trifluoromethanesulfonic-alkanesulfonic anhydrides, **6,** 618
Trifluoromethanesulfonic anhydride, **5,** 702; **6,** 618
Trifluoromethanesulfonylimidazole, **6,** 620
SULFURIZATION
Sulfur, **1,** 1118
Sulfur monochloride, **1,** 1122
SWERN OXIDATION
Dimethyl sulfoxide, **10,** 166
Dimethyl sulfoxide–Oxalyl chloride, **8,** 200; **9,** 192; **10,** 167; **11,** 215
Dimethyl sulfoxide–Trifluoroacetic anhydride, **5,** 266; **7,** 136; **10,** 168
THALLATION (*see* METALLATION)
THIELE REACTION
Acetic anhydride, **2,** 7
Acetic–phosphoric anhydride, **2,** 12
Boron trifluoride etherate, **1,** 70
2-Hydroxy-1,4-naphthoquinone, **1,** 484
Perchloric acid, **1,** 796
THIOALKYLATION (*see* ALKYLTHIOALKYLATION)
THIOCYANATION
Palladium(II) acetate, **5,** 496
Thiocyanogen, **1,** 1152
Thiocyanogen chloride, **1,** 1153
THIONATION
2,4-Bis(4-methoxyphenyl)-1,3-dithia-2,4-diphosphetane-2,4-disulfide, **8,** 327; **9,** 49; **12,** 59
Bis(tricyclohexyltin) sulfide–Boron trichloride, **11,** 63
Boron trisulfide, **5,** 55
Carbon disulfide, **5,** 94
Dioxane, **1,** 333
Hydrogen sulfide, **12,** 247
Lithium triethylborohydride, **11,** 304
3-Methylbenzothiazole-2-thione, **6,** 374
Phosphorus(V) sulfide, **1,** 870; **3,** 226; **5,** 534, 653; **9,** 374
Potassium thiocyanate, **1,** 954
Sodium hydrogen sulfide, **9,** 429; **11,** 487
Thiourea, **1,** 1164; **11,** 519
Triethyloxonium tetrafluoroborate, **10,** 417
Triphenylphosphine sulfide, **5,** 732
THORPE REACTION
Dimethyl sulfoxide, **1,** 296
Lithium diethylamide, **2,** 247
Potassium *t*-butoxide, **1,** 911
Ruthenium tetroxide, **6,** 545
Sodium hexamethyldisilazide, **1,** 1046
Sodium N-methylanilide, **1,** 1095; **4,** 459; **6,** 545
TIFFENEAU RING EXPANSION
Isocyanomethyllithium, **4,** 272
Sodium ethoxide, **1,** 1065
Sodium nitrite, **1,** 1097
TISHCHENKO REACTION
Boric acid, **5,** 48
TOSYLATION (*see* SULFONYLATION, SYNTHESIS INDEX—SULFONIC ACID ESTERS)

TRANSESTERIFICATION
Aluminum chloride, **5,** 10
Copper(I) acetate–*t*-Butyl isocyanide, **5,** 163
Molecular sieves, **2,** 286
Potassium cyanide, **5,** 553
Pyridinium *m*-nitrobenzenesulfonate, **12,** 419
Titanium(IV) alkoxides, **11,** 528
p-Toluenesulfonic acid, **1,** 1172
Triphenylphosphine–Diethyl azodicarboxylate, **6,** 645
TRANSETHERIFICATION
Silver(I) oxide, **4,** 430
TRANSLACTONIZATION
p-Toluenesulfonic acid, **8,** 488

UGI REACTION
t-Butyl isocyanide, **10,** 67
ULLMANN ETHER SYNTHESIS
Copper, **1,** 157
Copper(I) chloride, **3,** 67
Copper(I) oxide, **1,** 169
Copper(I) phenylacetylide, **12,** 143
Dimethyl sulfoxide, **1,** 296
Pentafluorophenylcopper, **6,** 451
ULLMANN REACTION
Bis(1,5-cyclooctadiene)nickel(0), **11,** 51
trans-Bromobis(triethylphosphine)-*o*-tolylnickel(II), **10,** 58
Copper, **1,** 157; **5,** 146; **7,** 73; **8,** 113; **9,** 122
Copper bronze, **1,** 155; **4,** 100
Copper(I) *t*-butoxide, **9,** 122
Copper(I) iodide, **8,** 122
Copper(I) trifluoromethanesulfonate, **6,** 130
Diethyltrimethylsilylamine, **6,** 184
N,N-Dimethylformamide, **1,** 278
Methanesulfonyl chloride, **6,** 362
Nickel(II) bromide–Zinc, **9,** 322
Pentafluorophenylcopper, **5,** 504
Tetramethylurea, **1,** 1146
Tris(triphenylphosphine)nickel(0), **6,** 653; **8,** 523

VILSMEIER REACTION (*see* FORMYLATION)
VON BRAUN DEGRADATION
Phosphorus(III) bromide, **1,** 873
Piperidine, **1,** 886
Thionyl chloride, **1,** 1158
VON BRAUN REACTION
Cyanogen bromide, **1,** 174; **5,** 169
2,2,2-Trichloroethyl chloroformate, **5,** 686
VON PECHMANN CONDENSATION
Hydrogen fluoride, **2,** 215
Ion-exchange resins, **1,** 511

WACKER OXIDATION
Allyltrimethylsilane, **10,** 6
Palladium(II) chloride, **10,** 300
Palladium(II) chloride–Copper(I) chloride, **11,** 396
Phase-transfer catalysts, **12,** 379
WAGNER–MEERWEIN REARRANGEMENT
Arenesulfonyl azides, **5,** 19
3,5-Dinitroperbenzoic acid, **11,** 217
p-Toluenesulfonic acid, **4,** 508; **9,** 471
Trifluoromethanesulfonic anhydride, **6,** 618
WALLACH REARRANGEMENT
Antimony(V) chloride, **7,** 12
WHARTON REACTION
Hydrazine, **1,** 434; **8,** 245
Tributyltin hydride, **11,** 545
WHITE DEAMINATION
Nitrogen dioxide, **1,** 324; **2,** 175
WICHTERLE REACTION
1,3-Dichloro-2-butene, **1,** 214; **2,** 111; **8,** 152; **10,** 134
Mercury(II) trifluoroacetate, **9,** 294
Sulfuric acid, **4,** 470
WILLGERODT AND SIMILAR REACTIONS
Ammonium polysulfide, **6,** 21
d- and *l*-10-Camphorsulfonyl chloride, **11,** 109
N,N-Dimethylformamide, **1,** 278
Iodine–Silver nitrate, **11,** 268; **12,** 257
Lead tetraacetate–Boron trifluoride etherate, **10,** 228
Morpholine, **1,** 705
Potassium hydroxide–Carbon tetrachloride, **5,** 96
Sulfur, **1,** 1118; **3,** 273
Thallium(III) nitrate, **4,** 492; **5,** 656; **7,** 362, 364; **12,** 481
WILLIAMSON ETHER SYNTHESIS (*see also* SYNTHESIS INDEX—ETHERS)

WILLIAMSON ETHER SYNTHESIS (*Continued*)
Crown ethers, **5,** 152; **6,** 133
Cryptates, **9,** 127
Dimethyl sulfoxide, **3,** 119
Thallium(I) ethoxide, **10,** 395

WITTIG REACTION (*see also* ASYMMETRIC REACTIONS, METHYLENATION, SYNTHESIS INDEX—ALKENES, TYPE OF COMPOUND INDEX)
GENERAL METHODS
Benzoic acid, **1,** 49
Crown ethers, **6,** 133
1,5-Diazabicyclo[4.3.0]nonene-5, **2,** 98
N,N-Dimethylformamide, **1,** 278
1,3-Dimethyl-2-oxohexahydropyrimidine, **11,** 207
p-Diphenylphosphinobenzoic acid, **2,** 179
Ethoxycarbonylcyclopropyltriphenylphosphonium tetrafluoroborate, **10,** 78
Ethylene oxide, **2,** 196; **5,** 297
Hexamethylphosphoric triamide, **6,** 273; **8,** 240
Hexamethylphosphorous triamide, **2,** 210
Methylenetriphenylphosphorane, **10,** 264
Potassium *t*-amyloxide, **9,** 379
Potassium *t*-butoxide, **11,** 432
Potassium carbonate, **11,** 433
Sodium *t*-amyloxide, **2,** 386
Sodium hexamethyldisilazide, **1,** 1046; **7,** 329
Sodium methoxide, **1,** 1091
Sodium methylsulfinylmethylide, **1,** 310
Tetrahydrofuran, **1,** 1140
Triphenylphosphine, **1,** 1238
Wittig reaction, **4,** 573; **5,** 752
INTRAMOLECULAR REACTIONS
Allylidenetriphenylphosphorane, **5,** 7
trans-1-Butadienyltriphenylphosphonium bromide, **6,** 76
Butane-1,4-bis(triphenylphosphonium) dibromide, **10,** 60
2-Butenylidenetriphenylphosphorane, **5,** 69
Cyclopropyltriphenylphosphonium bromide, **2,** 95
2-Ethoxyallylidenetriphenylphosphorane, **8,** 219
Ethoxycarbonylcyclopropyltriphenylphosphonium tetrafluoroborate, **5,** 90; **6,** 93; **8,** 76
Hexamethylphosphoric triamide, **6,** 273
Ketenylidenetriphenylphosphorane, **10,** 450; **12,** 268
N-Phenylketeniminyl(triphenyl)phosphorane, **12,** 387
1-Phenylthiocyclopropyltriphenylphosphonium tetrafluoroborate, **6,** 465
Potassium *t*-butoxide, **1,** 911
Tetrakis(triphenylphosphine)palladium(0), **10,** 384
Triphenyl(1-phenylthiovinyl)phosphonium iodide, **9,** 502; **11,** 425
Triphenylphosphine, **3,** 317; **4,** 548; **11,** 588
Triphenylpropargylphosphonium bromide, **6,** 494
Vinyltriphenylphosphonium bromide, **1,** 1274; **2,** 456; **5,** 750; **6,** 666

WITTIG REARRANGEMENT
Butyllithium, **6,** 85; **11,** 101; **12,** 96
Potassium amide, **1,** 907
Tributyl(iodomethyl)tin, **9,** 475

WITTIG–HORNER REACTION
GENERAL METHODS
Crown ethers, **10,** 110
Diethyl phenylsulfinylmethylphosphonate, **6,** 191
Lithium chloride–Diisopropylethylamine, **12,** 277
Methyl bis(trifluoroethoxy)phosphinylacetate, **12,** 320
N,N,N′,N′-Tetramethylethylenediamine, **12,** 477
INTRAMOLECULAR REACTIONS
1,3-Bis(dimethoxyphosphinyl)-2-propanone, **10,** 35
(Dimethoxyphosphinyl)methyllithium, **6,** 339
Dimethyl 3-bromo-2-ethoxypropenylphosphonate, **9,** 180
N-Ethyl(diethoxyphosphinyl)methylketenimine, **9,** 216
Lithium isopropoxide, **9,** 283
Potassium carbonate, **12,** 403
Trifluoroacetyl chloride, **11,** 560

WOLFF REARRANGEMENT

Chloramine, **3,** 45
Copper(II) acetylacetonate, **2,** 81
Copper(I) iodide, **1,** 169
Diazomethane, **1,** 191; **2,** 102; **3,** 74
Silver benzoate, **1,** 1004
Trimethylsilyldiazomethane, **4,** 543; **10,** 431
WOLFF–KISHNER REACTION (*see* DEOXYGENATION)
WOODWARD METHOD
Silver acetate, **1,** 1002
WURTZ REACTION (*see also* COUPLING REACTIONS)
Lithium amalgam, **3,** 177
Sodium, **4,** 437
Sodium naphthalenide, **1,** 711; **4,** 349

SYNTHESIS INDEX

ACETALS AND KETALS
(*see also* ACETONIDES)
GENERAL METHODS, CATALYSTS
Aluminum chloride, **6,** 17
Cerium(III) chloride, **9,** 99
Chlorotris(triphenylphosphine)-rhodium(I), **3,** 325
2,2-Dimethoxypropane, **1,** 268
Dimethylformamide–Dimethyl sulfate, **9,** 182
Erbium(III) chloride, **9,** 210
Iodotrimethylsilane, **12,** 259
Ion-exchange resins, **4,** 266
Molecular sieves, **4,** 345
Pyridinium chloride, **5,** 566
Selenium(IV) oxide, **1,** 992
Tin(IV) chloride–Dicyclohexyl-carbodiimide, **5,** 631
p-Toluenesulfonic acid, **5,** 673
Triethyl orthoformate, **1,** 1204
Trimethylsilyl trifluoromethane-sulfonate, **10,** 438
DIETHYL ACETALS(KETALS)
Triethyl orthoformate, **5,** 356
DIMETHYL ACETALS(KETALS)
m-Chloroperbenzoic acid, **12,** 118
Dimethylformamide–Dimethyl sulfate, **5,** 250
Nafion-H, **10,** 275
Oxalic acid, **5,** 481
Phenyliodine(III) diacetate, **11,** 411
Trimethyl orthoformate, **8,** 507
1,3-DIOXANES
Benzaldehyde, **6,** 26
Dimethyl sulfoxide, **2,** 157
2-Phenyl-1,3-dithiolane, **6,** 26
1,3-DIOXOLANES
from RCHO, RCOR′
2,3-Butanediol, **1,** 81
2-Chloroethanol, **5,** 117
2,4,6-Collidinium *p*-toluenesulfonate, **12,** 139
Copper(II) sulfate, **8,** 125
Cyclohexanone, **1,** 178
2-Dimethylamino-1,3-dioxolane, **1,** 283
Dimethylformamide–Dimethyl sulfate, **9,** 182
Ethylene glycol, **1,** 375; **9,** 217
Ethylene oxide, **3,** 140
2-Ethyl-2-methyl-1,3-dioxolane, **1,** 82
Fumaric acid, **5,** 319
Ion-exchange resins, **9,** 256
2-Methoxy-1,3-dioxolane, **9,** 298
2-Methyl-2-(2-methyl-1-propenyl)-1,3-dioxolane, **1,** 662
Oxalic acid, **1,** 764; **5,** 481
Pyridinium chloride, **1,** 964
Pyridinium *p*-toluenesulfonate, **9,** 400
1,4,6,9-Tetraoxaspiro[4.4]nonane, **6,** 186
p-Toluenesulfonic acid, **1,** 1172
from other starting materials
Ethylene glycol, **5,** 718; **6,** 27, 381
Mercury(II) acetate, **4,** 319
2-Phenyl-1,3-dithiolane, **6,** 26
FORMALS
Adogen 464, **6,** 10
Dibromomethane, **8,** 387; **11,** 157
Dimethyl sulfoxide + co-reagent, **4,** 199; **9,** 190
Methylene chloride, **7,** 239
p-Toluenesulfonic acid, **11,** 535
MONOACETALS(KETALS) FROM DICARBONYLS
2,3-Butanediol, **12,** 80
m-Chloroperbenzoic acid, **10,** 92
Diethoxycarbenium tetrafluoroborate, **11,** 175
N,N-Dimethylhydrazine, **7,** 126
Methyl iodide, **1,** 682
Palladium(II) chloride–Copper(I) chloride, **12,** 372
Sodium ethoxide, **1,** 1065
Thallium(III) nitrate, **7,** 362
Trialkyl orthoformate, **5,** 690; **11,** 568
SPIRO KETALS
Acrolein, **10,** 2
N,N-Dimethylhydrazine, **12,** 205
Mercury(II) chloride, **12,** 304
N-Phenylselenophthalimide, **11,** 417

ACETALS AND KETALS (*Continued*)
Potassium hydride, **12,** 407
α,β-UNSATURATED ACETALS (KETALS)
Adipic acid, **1,** 15
Ammonium nitrate, **1,** 39
Benzeneselenenyl bromide, **5,** 518
Bromine, **5,** 55
2,4,6-Collidinium *p*-toluenesulfonate, **12,** 139
1,5-Diazabicyclo[4.3.0]nonene-5, **4,** 116
Dimethylformamide–Dimethyl sulfate, **9,** 182
Dimethyl sulfoxide, **1,** 296
Fumaric acid, **5,** 319
2-Methyl-2,4-pentanediol, **1,** 693
Oxalic acid, **1,** 764
Potassium *t*-butoxide, **1,** 911
Potassium fluoride, **2,** 346
Tetracarbonyldi-μ-chlorodirhodium, **5,** 572
Thionyl chloride, **1,** 879
Triethyl orthoformate, **1,** 1204
OTHER ACETALS(KETALS)
3-Bromopropane-1,2-diol, **4,** 154
Mercury(II) acetate, **12,** 298
2-Methylene-1,3-propanediol, **6,** 379
2-Methyl-2,4-pentanediol, **1,** 693
Molecular sieves, **10,** 273
1,3-Propanediol, **5,** 718
Sulfosalicylic acid, **1,** 1118
2,2,2-Trichloroethanol, **4,** 521
Zinc iodide, **1,** 1294
ACETONIDES
Acetone, **3,** 4
Bis(*p*-nitrophenyl) hydrogen phosphate, **1,** 331
Copper(II) sulfate, **2,** 89; **8,** 125
2,2-Dialkoxypropane, **1,** 244, 268; **3,** 120; **5,** 208, 226
Ferric chloride, **8,** 228
Ion-exchange resins, **1,** 511
2-Methoxypropene, **11,** 329; **12,** 317
Perchloric acid, **1,** 796; **3,** 220
p-Toluenesulfonic acid, **1,** 1172
2-Trimethylsilyloxy-1-propene, **5,** 718
ACETYLENES (*see* ALKYNES)
ACETYLENIC CARBONYL COMPOUNDS
α,β-ACETYLENIC ACIDS
Acetylene, **1,** 712
Carbon dioxide, **1,** 1037
Lithium acetylides, **4,** 551
Lithium 2,6-di-*t*-butyl-4-methylphenoxide, **5,** 412
Pyridinium chlorochromate, **6,** 498
Sodium hydride, **1,** 1075
α,β-ACETYLENIC ALDEHYDES (KETONES)
from RCOX and alkynes
Bis(trimethylsilyl)acetylene, **5,** 44
Boron trifluoride etherate, **12,** 66
Copper(I) trimethylsilylacetylide, **6,** 148
Dichlorobis(triphenylphosphine)-palladium(II), **8,** 151
N,N-Dimethylformamide, **1,** 278
N,O-Dimethylhydroxylamine, **11,** 201
Lithium iodide, **5,** 410
Manganese(II) iodide, **9,** 289
Silver tetrafluoroborate, **12,** 434
by Oxidation at propargyl carbons
o-Chloranil, **1,** 128
Chlorotris(triphenylphosphine)-rhodium(I), **5,** 736
Chromium(VI) oxide–Pyridine, **4,** 96; **9,** 121
Lead(IV) oxide, **2,** 233
Manganese(IV) oxide, **1,** 637; **4,** 317
Nickel peroxide, **2,** 294
Other methods
Benzyltrimethylammonium hydroxide, **5,** 29; **7,** 20
[1,1′-Bis(diphenylphosphine)-ferrocene](dichloro)palladium(II), **12,** 171
Ferrocenylphosphines, **10,** 37
Formic acid, **5,** 316
Iodine, **8,** 256
Triethyl orthoformate, **1,** 1204
γ,δ-ACETYLENIC ALDEHYDES (KETONES)
β-1-Alkynyl-9-borabicyclo[3.3.1]-nonanes, **8,** 6
Copper(I) iodide, **8,** 121
Di-μ-carbonylhexacarbonyldicobalt, **10,** 129
Diethylaluminum chloride, **4,** 144
Diethylaluminum trimethylsilylacetylide, **8,** 42
Triethylborane, **9,** 482
ACETYLENIC ALDEHYDES

(KETONES) FROM α,β-EPOXY CARBONYLS
2,4-Dinitrobenzenesulfonylhydrazide, **6,** 232
Hydrazine, **5,** 327; **7,** 170
Hydroxylamine-O-sulfonic acid, **2,** 217; **5,** 343
Mesitylenesulfonylhydrazide, **10,** 255
p-Toluenesulfonylhydrazide, **2,** 417
α,β-ACETYLENIC AMIDES
Dimethylcarbamoyl chloride, **8,** 151
α,β-ACETYLENIC ESTERS
Benzyltrimethylammonium hydroxide, **5,** 29
Ethyl propiolate, **8,** 259
Methyl chloroformate, **9,** 306
Palladium(II) chloride–Copper(II) chloride, **10,** 302
Thallium(III) nitrate, **4,** 492; **7,** 362
Triethyl phosphonoiodoacetate, **1,** 1218
OTHER ACETYLENIC CARBONYLS
m-Chloroperbenzoic acid, **11,** 122
Di-μ-carbonylhexacarbonyldicobalt, **8,** 148
Ethyl diazoacetate, **4,** 228
Methyllithium, **8,** 342
Potassium 3-aminopropylamide, **8,** 406
Propargyl bromide, **6,** 493
Trichloroethylene, **9,** 479; **11,** 552
ACYL AZIDES
Dimethylformamide–Thionyl chloride, **12,** 204
Hydrazoic acid, **5,** 329
Nitrosyl chloride, **1,** 748
Phenyl dichlorophosphate, **12,** 384
Sodium azide, **1,** 1041
Sodium nitrite, **1,** 1097
Tetrabutylammonium azide, **6,** 563
ACYL HALIDES
ACYL FLUORIDES
Acyl fluorides, **1,** 14
Carbonic difluoride, **1,** 116
Cyanuric fluoride, **5,** 171
(Diethylamino)sulfur trifluoride, **6,** 183
2-Halopyridinium salts, **9,** 234
Potassium fluoride, **2,** 346; **5,** 153
Pyridinium poly(hydrogen fluoride), **9,** 399
Selenium(IV) fluoride, **5,** 576
Sodium fluoride, **2,** 382
Uranium(VI) fluoride, **7,** 417
ACYL CHLORIDES
from RCOOH
Amidines, bicyclic, **4,** 16
Benzoyl chloride, **1,** 50
N,N′-Carbonyldiimidazole, **1,** 114
Dichloromethyl methyl ether, **1,** 220
N,N-Diethyl-1,2,2-trichlorovinylamine, **1,** 253
Dimethylformamide–Thionyl chloride, **1,** 286
Hexamethylphosphoric triamide–Thionyl chloride, **3,** 153
Oxalyl chloride, **1,** 767; **8,** 365
N-Phenyltrimethylacetimidoyl chloride, **1,** 854
Phosgene, **5,** 532
Phosphorus(III) chloride, **1,** 875
Phosphorus(V) chloride, **1,** 866
Phthaloyl chloride, **1,** 882
Thionyl chloride, **1,** 1158
2,2,2-Trichloro-1,3,2-benzodioxaphosphole, **1,** 120
(Trichloromethyl)carbonimidic dichloride, **4,** 523
Triphenylphosphine–Carbon tetrachloride, **1,** 1247; **6,** 644; **9,** 503
Triphenylphosphine dichloride, **6,** 646
from Other sources
Chlorosulfuric acid, **1,** 139
Diisopropyl peroxydicarbonate, **1,** 263
Dimethylformamide–Oxalyl chloride, **9,** 335
Hexamethyldisiloxane, **8,** 240
Oxalyl chloride, **8,** 366; **11,** 379
Phthaloyl chloride, **1,** 882
Triphenylphosphine dihalide, **5,** 730; **6,** 645
Zinc chloride, **1,** 1289
ACYL BROMIDES
N-Bromosuccinimide, **9,** 70
Bromotrimethylsilane, **10,** 59
Thionyl bromide, **5,** 663
2,2,2-Tribromobenzo-1,3,2-dioxaphosphole, **2,** 63
Triphenylphosphine dibromide, **1,** 1247; **5,** 729
ACYL IODIDES
Iodotrimethylsilane, **10,** 59
Sodium iodide, **12,** 450
ACYL NITRILES
t-Butyl hydroperoxide, **12,** 88

ACYL NITRILES (*Continued*)
Copper(I) cyanide, **5,** 166
Cyanotrimethylsilane, **10,** 1
Iodophenylbis(triphenylphosphine)-palladium, **11,** 269
Phenyl selenocyanate, **11,** 416
Potassium cyanide, **11,** 433
Tetrabutylammonium bromide, **5,** 644
Thallium(I) cyanide, **8,** 476
Tributyltin cyanide, **10,** 411
ACYLOINS (*see* HYDROXY ALDEHYDES AND KETONES)
ACYL SILANES (*see* SILANES)
ALCOHOLS (*see also* ALLENIC ALCOHOLS, ALLYLIC ALCOHOLS, CHIRAL COMPOUNDS, DIOLS, HOMOALLYLIC ALCOHOLS, HYDROXY..., PROPARGYL ALCOHOLS)
BY ADDITION OF RM TO C=O
General methods
Cerium(III) iodide, **11,** 114
Cesium fluoride, **11,** 115
Grignard reagents, **1,** 415, 1170
Lithium, **4,** 286
Organocerium reagents, **12,** 345
Organolithium reagents, **10,** 3
Organozirconium reagents, **12,** 358
Pentane-1,5-di(magnesium bromide), **9,** 355
Samarium(II) iodide, **8,** 439; **10,** 344; **12,** 429
Tributyltinlithium, **8,** 495
Addition of CH_3M
Dilithium trimethylcuprate, **6,** 386; **7,** 115
Iodo(methyl)calcium, **5,** 442
Lithium dimethylcuprate, **6,** 209; **10,** 193
N-Methylphenylsulfonimidoyl-methyllithium, **5,** 458
Organotitanium reagents, **10,** 138, 270, 422
Addition of ArM
Benzeneboronic acid, **9,** 23
Boron trichloride, **9,** 62
Chromium carbonyl, **9,** 117
Chromium(II) chloride, **12,** 136
Dichlorophenylborane, **12,** 178
Lithium *o*-lithiophenoxide, **12,** 283
Methyl phenyl selenide, **6,** 86
Phenylytterbium iodide, **11,** 249
Phenylzinc bromide, **5,** 753
Selective additions ($RCHO \rightarrow R_2CHOH$)
2-(N-Formyl-N-methyl)-aminopyridine, **10,** 265
Selective additions ($R_2CO \rightarrow R_3COH$)
1,4-Dichloro-1,4-dimethoxybutane, **12,** 175
Diethyl carbonate, **1,** 247
Other RM additions
Benzylsodium, **6,** 40
Butyllithiums, **1,** 96; **4,** 60; **5,** 80
Chromium carbonyl, **11,** 131
Dichloromethyllithium, **1,** 223
gem-Difluoroallyllithium, **10,** 146
Dilithium tributylcuprate, **7,** 115
Trichloromethyllithium, **1,** 223
Zinc, **11,** 598
BY CLEAVAGE OF CYCLIC ETHERS
Copper(I) bromide, **7,** 79
1,1-Diphenylhexyllithium, **5,** 277
Grignard reagents–Copper(I) halides, **9,** 124
Lithium aluminum hydride–Aluminum chloride, **4,** 293
Lithium–Ethylenediamine, **4,** 291
Lithium naphthalenide, **4,** 348
Lithium triethylborohydride–Aluminum *t*-butoxide, **9,** 287
Phenyllithium–Boron trifluoride etherate, **12,** 68
Sodium, **4,** 437
Trimethylene oxide, **1,** 1232
BY CLEAVAGE OF $ROCH_3$, ROR′ (*see* TYPE OF REACTION INDEX, DEALKYLATION, DEMETHYLATION)
BY CLEAVAGE OF EPOXIDES WITH RM
Dimethylmagnesium, **1,** 292
Grignard reagents, **1,** 415; **9,** 124
2-Lithio-1,3-dithianes, **3,** 135; **12,** 573
3-Lithio-1-triisopropylsilyl-1-propyne, **11,** 566
Organoaluminum reagents, **12,** 339
Organocopper reagents, **3,** 106; **5,** 288; **11,** 365
Organolithium reagents, **1,** 571; **12,** 68
p-(Tolylsulfinyl)methyllithium, **4,** 513
Trialkylaluminums, **11,** 539

Trimethylsilylmethylpotassium, **6,** 637
BY HYDROLYSIS OF RX
m-Chloroperbenzoic acid, **8,** 97
Mercury(II) oxide–Tetrafluoroboric acid, **12,** 306
Potassium superoxide, **6,** 488
Silver sulfate, **4,** 435
BY HYDROLYSIS OF $ROSO_3R'$
Hexamethylphosphoric triamide, **6,** 273
Potassium *t*-butoxide, **1,** 911
Potassium nitrite, **10,** 329
Potassium superoxide, **6,** 488
Sodium amalgam, **1,** 1030
Sodium benzylate, **1,** 1045
Sodium naphthalenide, **1,** 711; **5,** 468
Tetrabutylammonium hydroxide, **5,** 645
BY HYDROMETALLATION-OXIDATION OF ALKENES
Hydroboration-oxidation
Bis(3,6-dimethyl)borepane, **4,** 35
9-Borabicyclo[3.3.1]nonane, **2,** 31; **5,** 46; **6,** 62; **8,** 47; **9,** 57
Borane + co-reagents, **1,** 199, 963; **5,** 47; **9,** 136; **11,** 69
Dichlorobis(cyclopentadienyl)-titanium-Lithium borohydride, **9,** 146
Dicyclohexylborane, **6,** 62
Diisopinocampheylborane, **6,** 202
Disiamylborane, **1,** 57; **6,** 62; **11,** 226
Lithium triethylborohydride, **8,** 309
Monochloroborane diethyl etherate, **4,** 346
Thexylborane, **11,** 516
Triisopinocampheylborane, **1,** 1228
Other routes
Diethoxymethylsilane, **12,** 182
Hydrogen hexachloroplatinate(IV) + co-reagents, **8,** 98; **12,** 243
Sodium borohydride–Acetic acid, **5,** 601
Tin(IV) chloride–Sodium borohydride, **9,** 438
Titanium(IV) chloride + co-reagents, **9,** 276; **10,** 404
BY HYDROXYLATION OF C–H BONDS (*see* TYPE OF REACTION INDEX)
FROM ORGANOBORANES
2-Alkyl-1,3-benzodithiolanes, **9,** 4
Bis(3,6-dimethyl)borepane, **4,** 35
Bis(phenylthio)methane, **7,** 25
Borane–Tetrahydrofuran, **1,** 199; **4,** 124
N-Bromosuccinimide, **4,** 49
Carbon monoxide, **2,** 60; **4,** 69; **9,** 393
Dichloromethyl methyl ether, **5,** 200
Hydrochloric acid, **6,** 283
Lithium triethylmethoxide, **4,** 314
α-Methoxyvinyllithium, **7,** 233
Oxygen, **4,** 362
Sodium cyanide, **4,** 446; **7,** 333
Tris(phenylthio)methyllithium, **11,** 305
Vinyllithium, **6,** 325
BY OXIDATION OF ALKYLSILANES (*see* TYPE OF REACTION INDEX)
BY OXYMERCURATION-DEMERCURATION INDEX
Mercury(II) acetate, **2,** 264; **3,** 194; **4,** 319; **5,** 424; **6,** 358; **10,** 252
BY REACTION OF RM WITH O_2
Chlorobis(cyclopentadienyl)-hydridozirconium(IV), **6,** 175
Sodium borohydride, **5,** 597
BY REDUCTION OF RCHO, R_2CO, RCOOH, RCOX, EPOXIDES, α,β-UNSATURATED CARBONYLS (*see* TYPE OF REACTION INDEX)
BY REDUCTION OF ROOH
Triethyl phosphite, **1,** 1212
OTHER METHODS
Allyl alcohol, **5,** 80
Aluminum hydride, **5,** 13
9-Borabicyclo[3.3.1]nonane, **9,** 57
sec-Butyllithium, **9,** 88
m-Chloroperbenzoic acid, **1,** 135
Chlorotrimethylsilane, **7,** 66
Diborane, **5,** 184
Di-μ-carbonylhexacarbonyldicobalt, **1,** 224
S,S′-Diethyl dithiomalonate, **9,** 160
N-Ethyl 2 pyrrolidone, **9,** 220
Ferrous perchlorate, **6,** 260
Hydroxylamine-O-sulfonic acid, **3,** 156
Lithium borohydride, **12,** 276
Nitrogen dioxide, **1,** 324
Sodium borohydride, **8,** 449
Sodium dithionite, **10,** 363
Tributyltinmethanol, **7,** 378
Trimethylamine N-oxide, **9,** 488
ALDEHYDES (*see also* α,β-ACETYLENIC CARBONYLS, ALLENIC CARBONYLS, CHIRAL

ALDEHYDES (*Continued*)
COMPOUNDS, CYANO CARBONYLS, DICARBONYLS, α,β-EPOXY CARBONYLS, UNSATURATED CARBONYLS)
FROM RX + CARBANIONS
Allyldimesitylborane, **12,** 12
Allylmercaptan, **5,** 84
3-Chloro-1-phenylthio-1-propene, **6,** 605
3,3-Diethoxy-1-propene, **6,** 270
N,N-Diethylaminoacetonitrile, **9,** 159
Diethyl phenyl orthoformate, **3,** 97
7,8-Dimethyl-1,5-dihydro-2,4-benzodithiepin, **6,** 216
4,4-Dimethyl-1,3-oxathiolane-3,3-dioxide, **9,** 186
2-(2,6-Dimethylpiperidino)acetonitrile, **11,** 212
2,4-Dimethylthiazole, **4,** 202
1,3-Dithiane, **5,** 287; **6,** 248
(Z)-2-Ethoxyvinyllithium, **8,** 221
Ethyl ethylthiomethyl sulfoxide, **5,** 299
2-Lithio-4,5-dihydro-5-methyl-[4H]-1,3,5-dithiazine, **8,** 305
Methoxymethyl phenyl sulfide, **12,** 316
Methyl methylthiomethyl sulfoxide, **4,** 341; **5,** 456
2-Methyl-2-thiazoline, **6,** 403
Methylthioacetic acid, **6,** 395
3-Methylthio-1,4-diphenyl-5-triazolium iodide, **6,** 396
Methylthiomethyl-N,N'-dimethyldithiocarbamate, **6,** 398
Phenylselenotrimethylsilylmethyllithium, **7,** 402
Phenylthiotrimethylsilylmethane, **10,** 313
2,4,4,6-Tetramethyl-5,6-dihydro-1,3-(4*H*)-oxazine, **3,** 280; **4,** 481; **5,** 651
3-Triethylsilyloxy-1-propene, **6,** 607
s-Trithiane, **3,** 329; **4,** 564
BY ARYLATION OF ALLYL ALCOHOLS (*see* TYPE OF REACTION INDEX)
BY CARBONYLATION OF RM
Dihalobis(triphenylphosphine)-palladium(II), **6,** 60
Dipotassium(or Disodium) tetracarbonylferrate, **3,** 267; **8,** 214; **10,** 174
Tetrakis(triphenylphosphine)-palladium(0), **12,** 468
BY FORMYLATION OF ArH (*see* TYPE OF REACTION INDEX)
BY FORMYLATION OF RM
Acetic-formic anhydride, **2,** 10
2-Alkoxy-1,3-benzodithiolanes, **8,** 236
1,3-Benzodithiolylium perchlorate, **8,** 34
p-Dimethylaminobenzaldehyde, **2,** 146
N,N-Dimethylformamide, **1,** 278; **11,** 198; **12,** 98
Ethoxymethyleneaniline, **1,** 362
2-(N-Formyl-N-methyl)aminopyridine, **8,** 341
N-Formylpiperidine, **11,** 244
Grignard reagents, **10,** 189
Lithium, **5,** 376
Methyl methylthiomethyl sulfoxide, **8,** 344
Methylthiomethyl *p*-tolyl sulfone, **12,** 327
1-Phenylthio-1-trimethylsilylethylene, **12,** 394
1,1,3,3-Tetramethylbutyl isocyanide, **3,** 279; **4,** 480
Triethyl orthoformate, **1,** 1204
N,4,4-Trimethyl-2-oxazolinium iodide, **4,** 540; **6,** 630
FROM HALOHYDRINS AND SIMILAR COMPOUNDS
Methanesulfonyl chloride, **1,** 662
Phosphoric acid, **4,** 387
Silver carbonate–Celite, **5,** 577
Sodium hydroxide, **5,** 616
BY HYDROFORMYLATION OF C=C
9-Borabicyclo[3.3.1]nonane, **3,** 24
Carbon monoxide, **2,** 253
Carbonylchlorobis(triphenylphosphine)-rhodium(I), **5,** 46
Carbonylhydridotris(triphenyl-phosphine)rhodium(I), **5,** 331; **7,** 53; **9,** 259
Chlorobis(cyclopentadienyl)-hydridozirconium(IV), **6,** 175
Di-μ-carbonylhexacarbonyldicobalt, **1,** 224; **5,** 204
Rhodium(II) carboxylates, **12,** 423
Rhodium(III) oxide, **4,** 420; **8,** 437
Tetracarbonylhydridocobalt, **2,** 80
BY HYDROLYSIS OF C=C–X, CX_2 (*see* TYPE OF REACTION INDEX)
BY HYDROMETALLATION-OXIDATION OF C≡C

Borane–Tetrahydrofuran, **1,** 199
Catecholborane, **4,** 69
Dichloroborane diethyl etherate, **5,** 191
Di-2-mesitylborane, **12,** 195
Disiamylborane, **1,** 57
Lithium acetylide, **6,** 324
Monochloroborane diethyl etherate, **5,** 465
Thiolacetic acid, **1,** 1154
FROM IMINES AND RELATED C=N (*see* TYPE OF REACTION INDEX)
BY ISOMERIZATION OF ALLYLIC ALCOHOLS
Diisobutylaluminum hydride, **12,** 191
Iridium catalysts, **8,** 135
FROM NITRO COMPOUNDS
N″-(*t*-Butyl)-N,N,N′,N′-tetramethylguanidinium *m*-iodylbenzoate, **12,** 102
Ozone, **5,** 491
Potassium permanganate, **1,** 942; **10,** 330; **11,** 440
Sodium hydroxide, **8,** 461
Sodium methoxide, **6,** 545
1,1,3,3-Tetramethylbutyl isocyanide, **5,** 650
Titanium(III) chloride, **4,** 506; **5,** 669
BY OXIDATION OF ROH, RX, RNH_2, 1-ALKENES, ArR, ROR′, and R_3SiOR' (*see* TYPE OF REACTION INDEX)
BY OXIDATIVE CLEAVAGE OF C=C, 1,2-DIOLS AND RELATED, EPOXIDES, OTHER SUBSTRATES (*see* TYPE OF REACTION INDEX)
BY REARRANGEMENT OF EPOXIDES
Boron trifluoride etherate, **1,** 70; **5,** 52
Dimethyl sulfoxide–Sulfur trioxide, **2,** 165
Lithium bromide, **4,** 297; **5,** 395
Lithium diethylamide, **1,** 610
Molybdenum carbonyl, **7,** 247
BY REDUCTION OF RCOOH, RCOCl, RCN (*see* TYPE OF REACTION)
BY VILSMEIER FORMYLATION (*see* TYPE OF REACTION INDEX)
BY WITTIG AND RELATED REAGENTS
[Bis(1,3,2-dioxaborola-2-nyl)methyl]-lithium, **6,** 328; **10,** 239
Bis(methylthio)methyllithium, **7,** 24
Chloro(trimethylsilyl)methyllithium, **8,** 277; **12,** 118
Diazo(trimethylsilyl)methyllithium, **7,** 216
Dibutyl tetrahydropyranyloxymethylphosphonate, **9,** 163
Diethyl N-benzylideneaminomethylphosphonate, **9,** 161; **11,** 178
Diethyl methoxyethoxymethylphosphonate, **9,** 162
Diethyl [(2-tetrahydropyranyloxy)-methyl]phosphonate, **11,** 181; **12,** 188
Dimesityl(trimethylsilylmethyl)borane, **12,** 12
Diphenylmethoxymethylphosphine, **10,** 172
Methoxymethyl(diphenyl)phosphine oxide, **9,** 301
Methoxymethylenetriphenylphosphorane, **1,** 671; **2,** 271; **6,** 368
Methoxy(trimethylsilyl)methyllithium, **9,** 284; **10,** 246; **11,** 331
Methyl methylthiomethyl sulfoxide, **6,** 446
N-Methyl-N-trimethylsilylmethyl-N′-*t*-butylformamidine, **11,** 347
N-Morpholinomethyldiphenylphosphine oxide, **9,** 318
Phenoxymethylenetriphenylphosphorane, **1,** 1240
Trimethylsilyllithium, **9,** 493
OTHER METHODS
Acetic anhydride, **5,** 3
Benzoic anhydride, **1,** 49
Bis(trimethylsilyl) peroxide, **12,** 63
Carbon monoxide, **9,** 393
α-Chloro-N-cyclohexylpropanaldonitrone, **5,** 110; **6,** 106
Diazoacetaldehyde, **3,** 73
2-Lithio-2-trimethylsilyl-1,3-dithiane, **4,** 284
Potassium fluoride, **10,** 325
Raney nickel, **12,** 422
α-SUBSTITUTED ALDEHYDES (*see also* AMINO CARBONYLS, HALO CARBONYLS, HYDROXY CARBONYLS)
α-Acetoxy
m-Chloroperbenzoic acid, **6,** 110
Lead tetraacetate, **1,** 537; **7,** 185
Sodium borohydride, **12,** 441
α-Acyloxy
N-*t*-Butylhydroxylamine, **12,** 95

ALDEHYDES (*Continued*)
Lead tetrabenzoate, **7,** 189
α-Alkoxy
(−)-10-Mercaptoisoborneol, **9,** 290
(S)-(+)-*p*-Tolyl *p*-tolylthiomethyl sulfoxide, **9,** 474; **10,** 408; **12,** 510
(−)-4,6,6-Trimethyl-1,3-oxathiane, **8,** 508
α-Alkyl(or Phenyl)seleno
Benzeneselenenic anhydride, **9,** 25
Benzeneselenenyl halides, **5,** 518; **9,** 25
Benzeneselenol, **11,** 537
N,N-Diethylbenzeneselenenamide, **11,** 178
Diphenyl diselenide–Selenium(IV) oxide, **11,** 219
α-Alkyl(or Phenyl)thio
Copper(I) trifluoromethanesulfonate, **6,** 130
Diphenyl disulfide, **9,** 386
Methoxy(phenylthio)methyllithium, **10,** 260
Thiophenol, **11,** 537
α-Silyl
t-Butyldimethylchlorosilane, **11,** 88
α-Trialkylsilyloxy
m-Chloroperbenzoic acid, **6,** 110
ALICYCLIC HYDROCARBONS—GENERAL METHODS
FROM ACYCLIC STARTING MATERIALS
1,3-Dithiane, **2,** 182
N-Phenylketeniminyl(triphenyl)-phosphorane, **12,** 387
Sodium hydride, **4,** 453
Titanium(0), **11,** 526
Titanium(III) chloride + co-reagent, **8,** 483; **12,** 493
Tosylmethyl isocyanide, **11,** 539
BY RING EXPANSION
Arenesulfonyl azides, **5,** 475
Diazomethane, **1,** 191
Dibromomethyllithium, **6,** 162
Ethyl diazoacetate, **8,** 222
Thallium(III) nitrate, **8,** 476
Thallium(III) perchlorate, **5,** 657
ALICYCLIC HYDROCARBONS—THREE-MEMBERED RINGS
(*see also* CHIRAL COMPOUNDS)
ALKYLIDENE CYCLOPROPANES
1-Chloro-2-methyl-1-trimethylsilyl-1-propene, **8,** 410
Cyclopropyltriphenylphosphonium bromide, **2,** 95
Isopropylidenecarbene, **3,** 118; **6,** 619; **7,** 358
1-(N-Nitrosoacetylaminomethyl)-cyclohexanol, **4,** 30
Phosphorus(III) iodide, **10,** 318
Trifluoromethanesulfonic anhydride, **5,** 702
CYCLOPROPANES
by Cyclization
Boron trifluoride etherate, **2,** 35
3-Butenyltributyltin, **4,** 53
t-Butyllithium, **11,** 103
Chromium(II) sulfate, **5,** 144
Grignard reagents, **9,** 229
Ion-exchange resins, **4,** 267
Lithium diethylamide, **2,** 247
Lithium diisopropylamide, **7,** 207; **8,** 292
Potassium fluoride, **2,** 346
Silver(I) nitrate, **4,** 429
Silver(I) oxide, **1,** 1011
Sodium–Ammonia, **5,** 31
Trimethylstannylmethyllithium, **12,** 547
Zinc–copper couple, **1,** 1292
Zinc–Silver couple, **4,** 436
by Cyclopropanation of C=C
Chloromethyl methyl ether, **1,** 672
Copper(I) iodide, **3,** 69
Diazomethane, **2,** 90; **3,** 68; **6,** 442, 577; **11,** 391
2,2-Dibromopropane, **11,** 102
Dicarbonyl(cyclopentadienyl)-[(dialkylsulfonium)methyl]iron tetrafluoroborate, **9,** 143; **10,** 128
Diethyl(iodomethyl)aluminum, **1,** 245
Diethylzinc–Diiodomethane(or –ethane), **1,** 253; **2,** 134; **3,** 141; **4,** 153; **8,** 172
Diiodomethane, **7,** 74; **9,** 170
Diphenylsulfonium methylide, **5,** 281
Ethyl(iodomethyl)zinc, **3,** 257
Iodo(iodomethyl)mercury, **1,** 506
Simmons–Smith reagent, **1,** 1019; **2,** 371; **3,** 255; **4,** 159; **5,** 588, 758; **8,** 445; **11,** 472
p-Toluenesulfonylhydrazide, **8,** 460
Zinc, **11,** 598

Other methods
9-Borabicyclo[3.3.1]nonane, **3,** 24
Borane–Tetrahydrofuran, **1,** 199
Chromium(II) chloride, **1,** 149; **2,** 76
Copper, **2,** 82
Diethyl cyanomethylphosphonate, **1,** 250
Difluoramine, **1,** 253
Hydrazine, **2,** 211
1-Lithiocyclopropyl phenyl sulfide, **10,** 231; **11,** 284
Lithium aluminum hydride, **1,** 581
Phenyllithium, **9,** 364
p-Toluenesulfonylhydrazide, **5,** 678
CYCLOPROPANOLS AND DERIVATIVES
from Acyclic precursors
9-Borabicyclo[3.3.1]nonane, **3,** 24
1,3-Dichloroacetone, **2,** 109
Methoxyallene, **10,** 257
Zinc, **7,** 426
by Cyclopropanation
2-Chloroethyl chloromethyl ether, **7,** 214
Dichloromethyl 2-chloroethyl ether, **2,** 118
Diethylzinc–Diiodomethane, **8,** 172
(Dimethoxymethyl)trimethoxysilane, **3,** 103
Simmons–Smith reagent, **5,** 588; **6,** 521
from Other routes
1-Bromo-1-ethoxycyclopropane, **12,** 73
Lead tetraacetate, **1,** 537
Methyllithium, **1,** 686
Oxygen, **3,** 174
Simmons–Smith reagent, **1,** 1019
Sodium borohydride, **6,** 530
CYCLOPROPANONES
Diazomethane, **2,** 102
Ketene, **1,** 528
Peracetic acid, **2,** 307
Phthalyl alcohol, **1,** 883
Potassium fluoride, **6,** 481
Sodium hexamethyldisilazide, **6,** 529
CYCLOPROPENES
by Cyclization
Crotyl chloride, **5,** 392
Lithium amide, **3,** 178
Sodium amide, **1,** 1034; **5,** 544
by Cyclopropanation of C≡C
Benzal chloride, **5,** 453
1,2-Dimethoxyethane, **1,** 267
Ethyl diazoacetate, **3,** 138
Lithium 2,2,6,6-tetramethylpiperidide, **4,** 310; **7,** 213
Methyl diazoacetate, **8,** 434
Rhodium(II) carboxylates, **11,** 458
Simmons–Smith reagent, **1,** 1019; **3,** 255; **4,** 436
from Halocyclopropanes
Cesium fluoride, **6,** 100
Potassium *t*-butoxide, **5,** 544; **7,** 296
Sodium trichloroacetate, **6,** 553
Other methods
Lithium aluminum hydride, **1,** 581
Sodium amide, **5,** 591
Titanium(III) chloride–Lithium aluminum hydride, **9,** 468
p-Toluenesulfonylhydrazide, **1,** 1185
CYCLOPROPENONES AND KETALS
(Bromodichloromethyl)phenylmercury, **2,** 326
Potassium amide, **4,** 398
Potassium *t*-butoxide, **4,** 399
Tributyltin hydride, **2,** 424; **4,** 518
Trichloromethyllithium, **4,** 524
Triethylamine, **1,** 1198
Triphenylphosphine, **5,** 725
CYCLOPROPYL ALDEHYDES FROM C=C + :CHCHO (*see also* CYCLOPROPYL CARBONYL COMPOUNDS)
Diazoacetaldehyde, **2,** 101; **4,** 5
2-Diazo-1,1-dimethoxyethane, **10,** 120
CYCLOPROPYL ALLENES
3-Chloro-3-methyl-1-butyne, **7,** 77; **8,** 93
Dimethylvinylidenecarbene, **5,** 27
Potassium *t*-butoxide, **1,** 911
CYCLOPROPYL CARBONYL COMPOUNDS—GENERAL METHODS
by Cyclization
1,2-Dibromoethane, **6,** 42
Dicyclohexylcarbodiimide, **3,** 91; **4,** 141
Guanidines, **12,** 477
Iodine, **3,** 159; **4,** 258
Potassium trifluoromethanesulfinate, **5,** 564
Sodium alkoxide, **1,** 1091; **4,** 451

ALICYCLIC HYDROCARBONS—THREE-MEMBERED RINGS (*Continued*)
Sodium hydride, **4,** 452; **5,** 610
by Cyclopropanation of $R_2C=CRCOR(H)$
Acetylmethylene(triphenyl)arsorane, **11,** 5
t-Amyl diazoacetate, **8,** 40
Antimony(V) fluoride, **9,** 20
Benzyltriethylammonium chloride, **7,** 18
Bis(1,5-cyclooctadiene)nickel(0), **5,** 34
Cerium(IV) ammonium nitrate, **10,** 79
Cyclopropenone 1,3-propanediyl ketal, **12,** 152
α-Diazoacetophenone, **10,** 274
Diazomethane, **4,** 120
2-Diazopropane, **2,** 105; **3,** 74
Di-*t*-butyl bromomalonate, **8,** 213
(N,N-Diethylamino)methyloxosulfonium methylide, **5,** 210
N,N-Dimethylaminocyclopropylphenyloxosulfonium tetrafluoroborate, **4,** 173
(Dimethylamino)phenyloxosulfonium methylide, **3,** 105
Dimethylsulfonium ethoxycarbonylmethylide, **10,** 164
Dimethylsulfoxonium methylide, **1,** 315; **2,** 171; **3,** 125; **4,** 197; **5,** 254
Dimethylvinylidenecarbene, **5,** 460
Diphenyldiazomethane, **1,** 338
Diphenylsulfonium isopropylide, **2,** 180
Ethyl (dimethylsulfuranylidine)-acetate, **2,** 196
Isopropylidenetriphenylphosphorane, **5,** 361; **7,** 183
1-Lithiocyclopropyl methyl selenide, **9,** 85
3-Methyl-2-butenylidenetriphenylphosphorane, **5,** 69
Methylphenyl-N-*p*-toluenesulfonylsulfoximine, **3,** 204
(2-Oxo-2-phenylethylide)-dimethylselenonium ylide, **5,** 257
Palladium(II) acetate, **6,** 442
Simmons–Smith reagent, **3,** 255; **6,** 521
Tetrabutylammonium iodide, **5,** 646
Triethyl phosphoenol pyruvate, **1,** 1216
Trimethylsilylmethylene dimethylsulfurane, **9,** 495
from α,β-Epoxy carbonyl compounds
Diethyl methylmalonate, **5,** 216
Triethyl phosphonoacetate, **1,** 1217
by Ring expansion
Boron trifluoride dibutyl etherate, **4,** 43
Sulfuric acid, **4,** 470
Other methods
2-Chloroethyl(dimethyl)sulfonium iodide, **12,** 113
Copper(II) sulfate, **5,** 162
Cyclopropyltrimethylsilane, **7,** 84
1-Cyclopropyl-1-trimethylsilyloxyethylene, **8,** 140
1,4-Dibromo-2-pentene, **9,** 358
Diethylzinc–Bromoform, **12,** 74
Diphenylsulfonium cyclopropylide, **7,** 140
Ferric chloride, **5,** 307
Iodine, **3,** 159
1-Lithio-2-vinylcyclopropane, **7,** 192
Methoxycarbonyltriphenylphosphorane, **1,** 112
Nickel carbonyl, **12,** 335
Nitromethane, **9,** 323
Potassium *t*-butoxide, **4,** 399
Silver acetate, **8,** 440
Simmons–Smith reagent, **2,** 371
Sodium chloride, **6,** 534
Titanium(IV) chloride, **11,** 529
Triethyl phosphoenol pyruvate, **1,** 1216

CYCLOPROPYL ESTERS FROM C=C + :CHCOOR (*see also* CYCLOPROPYL CARBONYL COMPOUNDS)
Benzyltriethylammonium chloride, **7,** 18
Bis(N-*t*-butylsalicylaldiminato)-copper(II), **12,** 52
t-Butyl diazoacetate, **7,** 313
Copper(I) *t*-butoxide, **6,** 144
Copper(0)–Isonitrile complexes, **6,** 128
Copper(I) oxide–*t*-Butyl isocyanide, **4,** 101; **5,** 150
Dibromomethoxycarbonylmethyl-(phenyl)mercury, **3,** 222
Di-μ-chlorobis(allyl)dipalladium, **2,** 109

2,5-Dichlorothiophenium bismethoxycarbonylmethylide, **8,** 159
Dimethyl diazomalonate, **5,** 244, 717; **8,** 187
Dimethyl malonate, **6,** 140
Dimethylsulfonium ethoxycarbonylmethylide, **10,** 164
Ethyl cyanoacetate, **6,** 140
Ethyl diazoacetate, **5,** 151, 496; **8,** 40; **9,** 406; **10,** 274, 382; **11,** 461
Ethyl (dimethylsulfuranylidine)acetate, **2,** 196
Methyl diazopropionate, **4,** 335
Rhodium(II) acetate, **10,** 340
Sodium hydride, **1,** 1075

CYCLOPROPYL KETALS
Chlorotrimethylsilane–Sodium, **11,** 127
Diazomethane, **1,** 165
1,3-Dithiane, **2,** 182
Ketene diphenyl thioacetal, **9,** 261
Magnesium, **4,** 315
Phthalyl alcohol, **1,** 883
Sodium methoxide, **5,** 617
Thiophenol, **9,** 312

CYCLOPROPYL KETONES (*see also* CYCLOPROPYL CARBONYL COMPOUNDS)
Acetylmethylene(triphenyl)arsorane, **11,** 5
t-Butyl diazomethyl ketone, **5,** 162
Copper, **2,** 82; **3,** 63; **4,** 102
Copper(I) chloride, **5,** 164
Copper(II) oxide, **5,** 161
Copper(I) oxide–*t*-Butyl isocyanide, **4,** 101
α-Diazoacetophenone, **6,** 577; **10,** 274
2,2,2-Trimethoxy-4,5-dimethyl-1,3-dioxaphospholene, **5,** 267

HALOCYCLOPROPANES
General methods
Crown ethers, **6,** 133
Haloform, **4,** 231; **5,** 656
Phenyl(trihalomethyl)mercury, **5,** 530
Addition of :CHF
Fluorodiiodomethane, **6,** 261
Addition of :CHCl
Methylene chloride, **1,** 95
Addition of :CHBr
Dibromomethane, **4,** 442
Diethylzinc–Bromoform, **6,** 194
Addition of :CHI
Chlorodiiodomethane, **7,** 74
Diethylzinc–Iodoform, **6,** 194
Iodoform, **3,** 165; **4,** 266
Addition of :CF_2
Bromodifluoromethyltriphenylphosphonium bromide, **5,** 62
Bromotrifluoromethane, **2,** 42
Phenyl(trifluoromethyl)mercury, **3,** 224; **4,** 384
Sodium chlorodifluoroacetate, **2,** 379; **3,** 265; **5,** 603
Thallium chlorodifluoroacetate, **2,** 411
Trimethyl(trifluoromethyl)tin, **1,** 1236
Addition of :CCl_2
Bis(trichloromethyl)mercury, **2,** 30
(Bromodichloromethyl)-phenylmercury, **1,** 851
Bromotrichloromethane, **1,** 80
Carbon tetrachloride, **6,** 596
Chloral, **3,** 45
Chloroform, **2,** 18; **4,** 27, 114; **6,** 38, 42, 159, 220, 349, 601; **7,** 19; **9,** 360
Ethyl trichloroacetate, **1,** 386; **3,** 143
Phenyl(trichloromethyl)mercury, **1,** 851; **2,** 326
Sodium trichloroacetate, **1,** 1107; **2,** 388; **6,** 553; **7,** 341
Tetramethylammonium chloride, **11,** 403
Thallium(I) trichloroacetate, **2,** 411
ω-Trichloroacetophenone, **3,** 45
Trichloromethyllithium, **1,** 1196; **2,** 119
Trifluoro(trichloromethyl)silane, **6,** 621
Addition of :CBr_2
Bis(tribromomethyl)mercury, **2,** 30
Bromoform–Potassium *t*-butoxide, **2,** 18; **3,** 174, 234; **5,** 26; **6,** 42, 220
Dibromocarbene, **6,** 601
Phenyl(tribromomethyl)mercury, **1,** 851
Addition of :CI_2
Iodoform, **5,** 27
Addition of :CClBr
Bromochlorocarbene, **8,** 129
Dibromochloromethyl(phenyl)-mercury, **1,** 851; **3,** 222
Addition of :CFCl
Dichlorofluoromethane, **6,** 175
Methyl dichlorofluoroacetate, **2,** 273

ALICYCLIC HYDROCARBONS—THREE-MEMBERED RINGS (*Continued*)
sym-Tetrachlorodifluoroacetone, **1,** 254
Addition of :CFI
Dibromofluoromethane, **4,** 29
Iodoform, **4,** 29
Addition of :CClI
Chlorodiiodomethane, **5,** 27
Addition of :CXY
Bis(dichlorotrimethylsilylmethyl)-mercury, **2,** 30
(1-Bromo-1-chloro-2,2,2-trifluoro-ethyl)phenylmercury, **4,** 376
(1-Bromo-1,2,2,2-tetrafluoroethyl)-phenylmercury, **5,** 515
Dibromomethoxycarbonylmethyl (phenyl)mercury, **3,** 222
α, α-Dichlorobenzyllithium, **1,** 1140
1,1-Dichloroethane, **10,** 90; **11,** 121
Dichloromethyl methyl sulfide, **5,** 194
Other methods
Antimony(III) fluoride, **6,** 553
Lithium aluminum hydride, **5,** 382
Silver fluoride, **5,** 581
VINYL CYCLOPROPANES
by Cyclopropanation
gem-Dichloroallyllithium, **9,** 145
1,3-Dichloro-1-propene, **6,** 129
Dimethylsulfonium methylide, **1,** 314
Lithium 2,2,6,6-tetramethylpiperidide, **9,** 83
3-Methyl-2-butenylidenetriphenyl-phosphorane, **5,** 69
Vinyldiazomethane, **6,** 664
using Cyclopropyl metal reagents
1-Lithio-2-vinylcyclopropane, **7,** 192; **9,** 273
Lithium bis(2-vinylcyclopropyl)-cuprate, **7,** 199
Lithium phenylthio(2-vinylcyclo-propyl)cuprate, **9,** 329
Other methods
3-Chloro-2-ethyl-1-propene, **5,** 544
1,4-Dibromo-2-pentene, **9,** 358
Dimethylsulfoxonium methylide, **8,** 194
Isopropylidenetriphenylphosphorane, **7,** 183
Lithium cyclopropyl(phenylthio)-cuprate, **7,** 211
Lithium diethylamide, **7,** 201
3-Tetrahydropyranyloxy-1-tributylstannyl-1-propene, **6,** 602
OTHER CYCLOPROPANES
Benzyl chloride, **4,** 310
Bromomalononitrile, **2,** 428
3-Chloro-2-ethyl-1-propene, **5,** 544
Chloromethyltrimethylsilane, **8,** 307
Chloromethyltrimethyltin, **8,** 308
Dicyanocarbene, **1,** 1202
Dicyanodiazomethane, **2,** 125
Dimethyl diazomethylphosphonate, **3,** 113
Dimethyl(tetrahydro-2-oxo-3-furanyl)-sulfonium tetrafluoroborate, **5,** 262
Phenylcarbene, **3,** 336
Phenylsulfinyldiazomethane, **6,** 457
Tris(phenylthio)methyllithium, **6,** 650

ALICYCLIC HYDROCARBONS—FOUR MEMBERED RINGS (*see also* CHIRAL COMPOUNDS)
CYCLOBUTANES
by Cyclization
9-Borabicyclo[3.3.1]nonane, **3,** 24
Butyllithiums, **11,** 103; **12,** 96
Chromium carbonyl, **6,** 125
Diethyl malonate, **1,** 1069
Hexamethylphosphoric triamide, **6,** 273
Lithium diisopropylamide, **6,** 334
Silver(I) trifluoromethanesulfonate, **7,** 324
Sodium methylsulfinylmethylide, **4,** 195
Tetrakis(triphenylphosphine)-palladium(0), **8,** 472
by [2 + 2]Cycloaddition
Aluminum chloride, **10,** 9; **12,** 26
1-Chloro-N,N,2-trimethylpropenyl-amine, **5,** 136
Copper(I) trifluoromethanesulfonate, **5,** 151; **11,** 142; **12,** 144
Di-μ-chlorobis(1,5-cyclooctadiene)-diiridium, **5,** 113
1,1-Dichloro-2,2-difluoroethylene, **1,** 220
Ethylaluminum dichloride, **6,** 251
Silver(I) nitrate, **7,** 321
Tetracyanoethylene, **1,** 1133; **5,** 647
Tetramethoxyethylene, **2,** 401; **5,** 649

Tribenzylcyclopentadienyltitanium, **4,** 112
Tricarbonyl(1,3-cyclobutadiene)iron, **6,** 149
Zinc chloride, **10,** 461; **12,** 574
by Ring contraction
Chloramine, **3,** 45
Potassium *t*-butoxide, **1,** 911
1,2-CYCLOBUTANEDIONES
Bromine, **4,** 46
Chlorotrimethylsilane, **3,** 310
Dichloroketene, **11,** 168
Dichlorovinylene carbonate, **2,** 122
1,3-Dithian-2-ylidenemethanone, **5,** 99
Methyl methylthiomethyl sulfoxide, **6,** 390
1,3-CYCLOBUTANEDIONES
Ethoxyacetylene, **1,** 357; **5,** 290
Triethylamine, **2,** 427
CYCLOBUTANONES AND RELATED COMPOUNDS
by Cyclization
1,3-Dithianes, **2,** 182; **4,** 241; **8,** 217; **10,** 380
Sodium hexamethyldisilazide, **6,** 529
Tosylmethyl isocyanide, **10,** 409
Trifluoroacetic acid, **3,** 305
Trifluoromethanesulfonic anhydride, **4,** 533; **6,** 618
by [2+2]Cycloaddition
t-Butylcyanoketene, **4,** 55
Chlorocyanoketene, **12,** 111
Chloro(methoxycarbonyl)ketene, **11,** 110
Chlorotrichloroethylketene, **9,** 112
1-Chloro-N,N,2-trimethylpropenylamine, **4,** 94; **5,** 136
Dichloroketene, **1,** 221; **2,** 118; **3,** 87; **4,** 134; **8,** 156; **9,** 152; **11,** 168
N,N-Diethylaminopropyne, **5,** 217; **7,** 107; **9,** 164
Dimethyl(or Diphenyl)ketene, **1,** 290; **3,** 131; **6,** 223
Ethoxyketene, **3,** 137
Isopropylidene diazomalonate, **9,** 258
Trifluoromethanesulfonic anhydride, **11,** 560
Trimethylsilyl tribromoacetate, **4,** 549
from Cyclopropyl epoxides
Chlorotrimethylsilane, **5,** 709
N,N-Dimethylaminocyclopropylphenyloxosulfonium tetrafluoroborate, **4,** 173
Diphenylsulfonium cyclopropylide, **3,** 133; **4,** 211; **5,** 281, 416; **8,** 212
Lithium iodide, **4,** 304
by Rearrangement of cyclopropanes
1-Bromo-1-cyclopropyllithium, **5,** 78
1-Bromo-1-ethoxycyclopropane, **12,** 73
Chlorotrimethylsilane, **5,** 709
Diphenylsulfonium cyclopropylide, **12,** 217
Hydrogen bromide, **4,** 251
1-Hydroxycyclopropanecarboxaldehyde, **12,** 247
1-Lithiocyclopropyl phenyl sulfide, **5,** 372; **6,** 319; **8,** 139
1-Lithio-1-methoxycyclopropane, **10,** 383
Simmons–Smith reagent, **6,** 521
Other routes
Boron trifluoride etherate, **6,** 65
Chlorotrimethylsilane–Sodium, **2,** 435; **5,** 711
Cobaltacyclopentan-2-ones, **11,** 136
2-Diazopropane, **3,** 74
Dimethylsulfoxonium methylide, **3,** 125
2-Phenylthiocyclobutanone, **11,** 422
p-Toluenesulfonylhydrazide, **4,** 511
CYCLOBUTENES
Aluminum chloride, **7,** 7; **11,** 25
Ethylaluminum dichloride, **7,** 146; **9,** 11
α-Lithiomethylselenocyclobutane, **10,** 232
Lithium aluminum hydride, **5,** 382; **8,** 286
Lithium aluminum hydride–Aluminum chloride, **4,** 293
Methyllithium, **9,** 311
Sodium dicarbonyl(cyclopentadienyl)ferrate, **11,** 483
Sodium hydride, **4,** 453
Tetrakis(triphenylphosphite)nickel(0), **11,** 503
Titanium(III) chloride + co-reagent, **7,** 369; **8,** 483
2-Trimethylsilylmethylenecyclobutane, **9,** 494
CYCLOBUTENEDIONES
Chlorotrifluoroethylene, **4,** 94

ALICYCLIC HYDROCARBONS—FOUR MEMBERED RINGS (*Continued*)
Dichloroketene, **12,** 176
CYCLOBUTENONES
Alumina, **6,** 16
t-Butylcyanoketene, **4,** 55
1-Chloro-N,N,2-trimethylpropenylamine, **6,** 122
Dichloroketene, **9,** 152
Ketene diethyl acetal, **10,** 226
Lithium aluminum hydride, **5,** 382
Methyl fluorosulfonate, **6,** 381
Trifluoromethanesulfonic anhydride, **11,** 560
ALICYCLIC HYDROCARBONS—FIVE-MEMBERED RINGS
(*see also* CHIRAL COMPOUNDS)
CYCLOPENTADIENONES
Dicarbonylcyclopentadienylcobalt, **10,** 126
Nickel carbonyl, **4,** 353
CYCLOPENTANES
by Cation–olefin reaction
Alumina, **9,** 8
4,4-Dichloro-3-buten-1-ol, **5,** 192
Trifluoroacetic acid, **4,** 530; **5,** 695; **8,** 503
Trimethylsilylmethyl trifluoromethanesulfonate, **10,** 434
by Intramolecular displacement
Butyllithium, **12,** 96
Lithium diethylamide, **2,** 247
Potassium hexamethyldisilazide, **4,** 407
Tetrakis(triphenylphosphine)-palladium(0), **8,** 472
2,4,4,6-Tetramethyl-5,6-dihydro-1,3-(4*H*)-oxazine, **3,** 280
2,4,4-Trimethyl-2-oxazoline, **6,** 629
Other routes
Acetyl methanesulfonate, **5,** 5
N-Alkylhydroxylamines, **12,** 13
Butyllithiums, **8,** 67; **11,** 103
Calcium carbonate, **2,** 57
2-Chloroacrylonitrile, **9,** 75
1,3-Diiodopropane, **6,** 129
Lithium dibutylcuprate, **6,** 163
Magnesium, **11,** 246
N-Phenylselenophthalimide, **11,** 417
Silver(I) trifluoromethanesulfonate, **7,** 324
Thallium(III) nitrate, **7,** 364
1,2-CYCLOPENTANEDIONES
Magnesium methoxide, **2,** 255
Sulfuric acid, **6,** 558
1,3-CYCLOPENTANEDIONES
Boron trifluoride etherate, **8,** 51
Copper(II) trifluoromethanesulfonate, **8,** 126
Sodium hexamethyldisilazide, **6,** 529
Zinc–Copper–Isopropyl iodide, **10,** 460
CYCLOPENTANOLS
by Intramolecular addition to carbonyls
t-Butylmagnesium chloride, **4,** 63
Chlorotrimethylsilane–Zinc, **12,** 568
Chromium carbonyl, **7,** 71
N-Ethyl-2-pyrrolidone, **9,** 220
Lithium dibutylcuprate, **3,** 79
Magnesium, **11,** 307
Potassium–Ammonia, **9,** 377
Sodium naphthalenide, **11,** 490
Other routes
Alkylaluminum halides, **11,** 7
2-(2-Bromoethyl)-1,3-dioxane, **11,** 78
1,5-Diazabicyclo[4.3.0]nonene-5, **2,** 98
Dichlorodimethylsilane, **4,** 183
Mercury(II) acetate, **2,** 264; **11,** 315
Mercury(II) chloride, **9,** 291
Tin(IV) chloride, **9,** 436
CYCLOPENTANONES
using Acyl anion equivalents
1,3-Dithianes, **2,** 182; **10,** 380
Methyl methylthiomethyl sulfoxide(or sulfone), **6,** 390; **11,** 242
by [3 + 2]Cycloaddition
3-Bromo-3-methyl-2-trimethylsilyloxy-1-butene, **9,** 68
Iron carbonyl, **8,** 533
Tri-μ-carbonylhexacarbonyldiiron, **5,** 221; **8,** 498; **9,** 477
from Diazoketones + C=C
Boron trifluoride etherate, **4,** 44; **10,** 52
Copper, **3,** 63
Copper(II) oxide, **5,** 161
Copper(II) sulfate, **5,** 162
Simmons–Smith reagent, **6,** 521
Trifluoroacetic acid, **11,** 557
by Dieckmann (and related) reactions
Dichlorobis(trifluoromethanesulfonato)titanium(IV), **12,** 173
Polyphosphoric acid, **4,** 395
Potassium *t*-butoxide, **1,** 911

Potassium fluoride, **1,** 933
Potassium (or Sodium) hydride, **1,** 1075; **6,** 482; **9,** 427
Sodium, **2,** 158
Sodium methylsulfinylmethylide, **1,** 310
from Dienes by carbonylation
Bromine, **5,** 55
Di-μ-carbonylhexacarbonyldicobalt, **12,** 163
Disodium tetracarbonylferrate, **6,** 550
Nickel carbonyl, **2,** 290
Thexylborane, **2,** 148
by Rearrangement of vinyl cyclopropanols
Diphenylsulfonium cyclopropylide, **4,** 211
1-Hydroxycyclopropane-carboxaldehyde, **12,** 247
1-Lithiocyclopropyl phenyl sulfide, **7,** 190
by Ring expansion
t-Butyl hypochlorite, **5,** 77
Diazomethane, **9,** 133
Dichloroketene, **10,** 139; **12,** 176
Lithium iodide, **11,** 300
Tetracarbonyldi-μ-chlorodirhodium, **6,** 108
Thallium(III) perchlorate, **5,** 657
from γ,δ-Unsaturated carbonyls
Alkylaluminum halides, **11,** 7
Chlorotris(triphenylphosphine)-rhodium(I), **4,** 559; **10,** 98; **12,** 130
Tin(IV) chloride, **9,** 436
Other routes
Alkylaluminum halides, **12,** 5
Bis[1,2-bis(diphenylphosphine)ethane]-palladium(0), **10,** 32
Boron trifluoride etherate, **12,** 66
5,5-Dimethoxy-1,2,3,4-tetrachloro-cyclopentadiene, **8,** 178
Dimethyl 1,3-acetonedicarboxylate, **11,** 193
Dimethyl sulfone, **2,** 157
Dimethylsulfoxonium methylide, **1,** 315
Disodium tetracarbonylferrate, **8,** 216; **10,** 174
(3-Iodopropyl)trimethylsilane, **12,** 26
Magnesium, **6,** 351
Organocopper reagents, **12,** 345
Palladium(II) acetate, **9,** 344; **10,** 297
1-Phenylthiocyclopropyltriphenyl-phosphonium tetrafluoroborate, **6,** 465
Polyphosphoric acid, **5,** 540
Propargyl alcohol, **9,** 394
Rhodium(II) carboxylates, **11,** 458
Sodium dichromate, **5,** 138
Sodium hydride, **1,** 1077; **12,** 447
Triphenyl(1-phenylthiovinyl)-phosphonium iodide, **9,** 502
CYCLOPENTENES
by Aldol reactions
2-(2-Bromoethyl)-1,3-dioxolane, **9,** 70
Dibenzylammonium trifluoroacetate, **9,** 136
Diethyl acetylmethylmalonate, **7,** 102
Methyl 4-methylthiocrotonate, **6,** 389
Morpholine–Camphoric acid, **9,** 317
Piperidine, **1,** 886
by [3+2]Cycloaddition
Lithium divinylcuprate, **12,** 345
Trimethylsilylallene, **10,** 428
Zinc chloride, **10,** 461; **11,** 602
by Displacements
1,3-Dichloro-2-butene, **10,** 134
1,4-Dichloro-2-butene, **7,** 88
Sodium hydride, **4,** 453
from Vinyl cyclopropanes
Lead carbonate, **9,** 265
(1-Lithiovinyl)trimethylsilane, **11,** 286
Lithium cyclopropyl(phenylthio)-cuprate, **7,** 211
Organocopper reagents, **11,** 365
by Wittig reaction
Ethoxycarbonylcyclopropyl-triphenylphosphonium tetrafluoroborate, **5,** 90; **6,** 93
Vinyltriphenylphosphonium bromide, **2,** 456
Other routes
Boron trifluoride etherate, **5,** 52; **10,** 52
Hydrazine, **8,** 245
Lithium bromide, **5,** 395
Methyllithium, **9,** 311
Palladium(II) acetate, **9,** 344
Potassium *t*-butoxide, **5,** 544
Tetracarbonyldi-μ-chlorodirhodium, **5,** 572
Tungsten(VI) chloride–Ethylaluminum dichloride, **5,** 742

ALICYCLIC HYDROCARBONS—FIVE-MEMBERED RINGS
(*Continued*)
Zinc iodide, **5,** 765
3-CYCLOPENTENOLS
2-(Chloroethoxy)carbene, **9,** 83; **10,** 89
Palladium(II) chloride–Copper(II) chloride, **9,** 353
2-CYCLOPENTENONES
by Acylation
2-(E)-Lithiovinyltrimethylsilane, **10,** 443
Nitromethane, **4,** 357
Polyphosphoric acid, **12,** 399
Tin(IV) chloride, **12,** 486
by Aldol reactions
3-Benzyl-5-(2-hydroxyethyl)-4-methyl-1,3-thiazolium chloride, **6,** 38
t-Butoxybis(dimethylamino)methane, **5,** 71
Copper(I) acetylacetonate, **4,** 100
1,3-Dichloro-2-butene, **4,** 472
Isopropenyl acetate, **6,** 356
2-Nitropropene, **7,** 253
Potassium *t*-butoxide, **1,** 911
Potassium fluoride, **8,** 410
Pyrrolidine, **7,** 309
Rexyn® ion-exchange resin, **10,** 220
2,4,4,6-Tetramethyl-5,6-dihydro-1,3-(4*H*)-oxazine, **4,** 481
by [3 + 2]Cycloaddition
Cyclopropenone 1,3-propanediyl ketal, **12,** 152
2,4-Dibromo-3-pentanone, **4,** 158
gem-Dichloroallyllithium, **8,** 150
Lithium 2,2,6,6-tetramethylpiperidide, **9,** 285
Tri-μ-carbonylhexacarbonyldiiron, **8,** 498
by Nazarov and related reactions
Benzylchlorobis(triphenylphosphine)-palladium(II), **12,** 44
(2-Bromovinyl)trimethylsilane, **11,** 82
Iodotrimethylsilane, **11,** 271
1-Phenylthio-1-trimethylsilylethylene, **11,** 583
Phosphoric acid–Formic acid, **1,** 860; **5,** 534; **10,** 317
Propargyl alcohol, **9,** 394
1-Trimethylsilyl-2-trimethylstannyl-ethylene, **12,** 469
Vinyltrimethylsilane, **9,** 498; **10,** 444; **12,** 26
from β,γ-Unsaturated diazoketones
Boron trifluoride etherate, **6,** 65; **10,** 52
by Wittig and related reactions
Bromoacetylmethylenetriphenyl-phosphorane, **9,** 66
(Dimethoxyphosphinyl)methyllithium, **6,** 339
Dimethyl 3-bromo-2-ethoxypropenyl-phosphonate, **9,** 180
Potassium carbonate, **12,** 403
Triphenyl(1-phenylthiovinyl)-phosphonium iodide, **11,** 425
Vinyltriphenylphosphonium bromide, **6,** 666
Other routes
Aluminum chloride, **11,** 25
Bis(alkylnitrile)dichloropalladium(II), **6,** 45; **12,** 50
Bis(cyclopentadienyl)diiodozirconium, **12,** 53
Copper(I) trifluoromethanesulfonate, **10,** 108
Di-μ-carbonylhexacarbonyldicobalt, **11,** 162; **12,** 163
Dilithium dialkyl(cyano)cuprates, **12,** 349
5,5-Dimethoxy-1,2,3,4-tetrachloro-cyclopentadiene, **10,** 153
2-(Dimethylamino)-3-pentenonitrile, **10,** 155
Dimethyl methylphosphonate, **3,** 117
Disodium tetracarbonylferrate, **9,** 205
Ethyl β-(1-pyrrolidinyl)acrylate, **8,** 226
Hydrobromic acid, **5,** 332
Hydrochloric acid, **6,** 283
2-Hydroxymethyl-3-trimethylsilyl-1-propene, **12,** 252
Mercury(II) acetate, **7,** 222
Methanesulfonic acid, **10,** 256
p-Nitroperbenzoic acid, **11,** 360
Organolithium reagents, **8,** 396, **11,** 13
Polyphosphoric acid, **6,** 474
Potassium hydroxide, **11,** 439
Silver tetrafluoroborate, **6,** 519
Trimethyl orthoformate, **10,** 425
Zinc–Zinc chloride, **7,** 430
3-CYCLOPENTENONES
1,3-Dithienium tetrafluoroborate, **4,** 218
Methyl methylthiomethyl sulfoxide, **6,** 390

Tetrakis(triphenylphosphine)-
palladium(0), **11,** 503
4-HYDROXY-2-CYCLOPENTENONES
Alumina, **8,** 9
Chloral, **6,** 100
Ethyl ethylthiomethyl sulfoxide, **5,** 299
Ketene dimethyl thioacetal monoxide, **8,** 268
β-Nitropropionyl chloride, **8,** 363
Pyridinium chlorochromate, **10,** 334
Sodium hypochlorite, **9,** 519
Thiophenol, **7,** 367
ALICYCLIC HYDROCARBONS—SIX-MEMBERED RINGS (*see also* CHIRAL COMPOUNDS)
CYCLOHEXADIENES
by Birch reduction
Birch reduction, **1,** 54; **5,** 30; **8,** 38
Calcium hexamine, **1,** 104
Lithium–Methylamine, **1,** 574; **5,** 378
Sodium–Ammonia, **3,** 259; **10,** 355
Ytterbium–Ammonia, **9,** 517
by Wittig reaction
Allylidenetriphenylphosphorane, **5,** 7, 70
trans-1-Butadienyltriphenyl-phosphonium bromide, **6,** 76
Butane-1,4-bis(triphenyl-phosphonium) dibromide, **10,** 60
2-Butenylidenetriphenylphosphorane, **5,** 69
Other routes
Bis(1,5-cyclooctadiene)nickel(0), **4,** 33
Diazadieneiron(0) complexes, **12,** 156
Ethynyl *p*-tolyl sulfone, **10,** 183
Methyl *trans*-2,4-pentadienoate, **2,** 279
1,4-CYCLOHEXADIEN-3-ONES
from Diazoketones + phenols
Copper(I) chloride, **8,** 118; **9,** 123
Trifluoroacetic acid, **6,** 613
by Diels–Alder reaction
4-Methoxy-1-phenylseleno-2-trimethylsilyloxy-1,3-butadiene, **8,** 398
trans-1-Methoxy-3-trimethylsilyloxy-1,3-butadiene, **8,** 328; **9,** 303
by Oxidation of phenols and related substrates (*see* TYPE OF REACTION INDEX)
by Oxidative phenol coupling
Copper(II)–Amine complexes, **8,** 114
Ferric chloride, **4,** 236
Potassium ferricyanide, **4,** 406; **6,** 480
Thallium(III) trifluoroacetate, **4,** 498; **9,** 462
Vanadyl trihalide, **3,** 331; **7,** 418; **8,** 527
2,4-CYCLOHEXADIEN-1-ONES
Chromium carbene complexes, **12,** 132
Hydrogen peroxide, **4,** 253
Lead tetraacetate, **7,** 185
Trifluoroacetic acid, **6,** 613
Trifluoroperacetic acid, **1,** 821
CYCLOHEXANES
by Cation–olefin routes
Perchloric acid, **1,** 796
Sulfuric acid, **5,** 633
Tin(IV) chloride, **5,** 627; **7,** 342; **9,** 436
Trifluoroacetic acid, **3,** 305; **4,** 530; **5,** 695; **6,** 613; **7,** 388; **8,** 503
Trimethylsilylmethyl trifluoromethanesulfonate, **10,** 434
by Cyclization
1,8-Diazabicyclo[5.4.0]undecene-7, **7,** 87
S,S′-Diethyl dithiomalonate, **9,** 160
Potassium hexamethyldisilazide, **4,** 407
from 1,5-Dienes
Benzeneselenenyl iodide, **11,** 36
Borane–Tetrahydrofuran, **7,** 321
N-Bromosuccinimide, **7,** 37
Mercury(II) trifluoroacetate, **9,** 294
by Hydrogenation of arenes
trihapto-Allyltris(trimethyl phosphite)-cobalt(I), **6,** 15
Calcium hexamine, **1,** 104
Chloro(hexamethylbenzene)hydrido-triphenylphosphinerhodium, **8,** 153
Di-μ-chlorobis(1,5-hexadiene)-dirhodium, **12,** 172
Palladium(II) chloride, **9,** 352
Palladium hydroxide, **8,** 385
Rhodium catalysts, **1,** 979; **4,** 418
Rhodium oxide–Platinum oxide, **9,** 408
Ruthenium catalysts, **1,** 983
Sodium borohydride–Rhodium(III) chloride, **11,** 480
Other methods
Acetyl *p*-toluenesulfonate, **6,** 10
Dibenzoyl peroxide, **4,** 122; **5,** 182

ALICYCLIC HYDROCARBONS—SIX-MEMBERED RINGS (*Continued*)
Dimethylphenylsilyl(methyl)-magnesium, **12,** 352
Grignard reagents, **11,** 245
Iodine, **8,** 256
Lithium diethylamide, **2,** 247
Nickel carbonyl, **3,** 210
Silver(I) trifluoromethanesulfonate, **7,** 324
1,3-CYCLOHEXANEDIONES
Dimethylaluminum benzenethiolate, **11,** 194
1,3-Dithiane, **2,** 182
Raney nickel, **1,** 723
Sodium ethoxide, **1,** 1065
1,4-CYCLOHEXANEDIONES
Copper(II) chloride, **9,** 123
p-Toluenesulfonic acid, **1,** 1172
CYCLOHEXANOLS
by Addition to carbonyls
Chlorotris(triphenylphosphine)-rhodium(I), **4,** 559
N-Ethyl-2-pyrrolidone, **9,** 220
Lithium dibutylcuprate, **3,** 79
Potassium–Ammonia, **9,** 377
Sodium naphthalenide, **11,** 490
Zinc bromide, **8,** 535
from 1,5-Dienes
Benzeneselenenyl halides, **9,** 25; **10,** 16
Mercury(II) acetate, **2,** 264; **5,** 424
from Epoxy–olefin cyclizations
Alumina, **9,** 8
Tin(IV) chloride, **5,** 627
Titanium(IV) isopropoxide, **10,** 404
Other routes
Alkylaluminum halides, **10,** 177; **11,** 7
Tributyltin hydride, **11,** 545
CYCLOHEXANONES
by Carbonylation
Carbon monoxide, **8,** 76
Disodium tetracarbonylferrate, **9,** 205; **10,** 174
Monochloroborane diethyl etherate, **7,** 249
Sodium cyanide, **6,** 535
from α-Diazoketones
Bis(N-propylsalicylideneaminate)-copper(II), **5,** 41
Copper, **3,** 63
Trifluoroacetic acid, **11,** 557
by Dieckmann (and related) condensations
1,3-Acetonedicarboxylic acid, **11,** 2
Bromomagnesium diisopropylamide, **12,** 74
4-Phenylseleno-1-penten-3-one, **8,** 397
Polyphosphoric acid, **4,** 395
Titanium(III) chloride–Lithium aluminum hydride, **12,** 493
Triphenylmethylpotassium, **9,** 502
by Displacements
1,3-Dithianes, **2,** 182; **10,** 380
Lithium diisopropylamide, **8,** 292
Magnesium, **6,** 351
Methyl methylthiomethyl sulfoxide, **6,** 390
Sodium methylsulfinylmethylide, **2,** 166
from Unsaturated carbonyls
Benzeneselenenyl chloride–Aluminum chloride, **11,** 33
Pyridinium chlorochromate, **7,** 308
Tetrakis(triphenylphosphine)-palladium(0), **11,** 503
Tin(IV) chloride, **11,** 522
Tributyltin hydride, **4,** 518
Other methods
t-Butyl hypochlorite, **5,** 77
2,3-Dichloro-1-propene, **2,** 120
Formic acid, **5,** 316
Polyphosphoric acid, **1,** 894
Pyridinium chlorochromate, **7,** 308
Sodium dichromate, **5,** 138
Thallium(III) perchlorate, **5,** 657
2,4,6-Triisopropylbenzene-sulfonylhydrazide, **11,** 563
2-Trimethylsilyloxy-1,3-butadiene, **7,** 401
CYCLOHEXENES (*see also* TYPE OF COMPOUND INDEX, DIELS–ALDER DIENES AND DIENOPHILES)
by Cation–olefin reactions
3-Ethyl-2-fluorobenzothiazolium tetrafluoroborate, **8,** 223
Sulfuric acid, **4,** 470
2,4,4,6-Tetrabromo-2,5-cyclohexadienone, **10,** 377
Tetrakis(2-methylpropyl)-μ-oxodialuminum, **8,** 471
Titanium(IV) chloride–

N-Methylaniline, **9,** 470
Trifluoroacetic acid, **3,** 305
by Reduction of benzenes
Calcium–Amines, **12,** 103
Lithium–Ammonia, Amines, **1,** 574; **2,** 28, 250; **4,** 287
Sodium–Ammonia, **2,** 374
Other routes
1-Butadienyldimethylsulfonium tetrafluoroborate, **9,** 74
Copper(I) iodide, **3,** 69
Dimethyl(methylthio)sulfonium tetrafluoroborate, **11,** 204
Hydrazine, **8,** 245
Lithium 1,3-butadiene-1-olate, **8,** 291
Sodium bicarbonate, **5,** 595
Sodium hydride–Dimethylformamide, **4,** 453
Titanium(III) chloride–Lithium aluminum hydride, **7,** 369
Vinyltriphenylphosphonium bromide, **2,** 456
2-CYCLOHEXENONES
Birch reduction
Birch reduction, **1,** 54; **2,** 27
Butyllithium, **9,** 87
Lithium–Ammonia, **1,** 54; **2,** 245; **7,** 21
Ytterbium–Ammonia, **9,** 517
from 1,5-Dicarbonyls
2-(2-Bromoethyl)-2-methyl-1,3-dioxolane, **1,** 75
4-Chloromethyl-3,5-dimethylisoxazole, **1,** 276
1-Chloro-3-pentanone, **3,** 49
5-Chloro-1-penten-3-one, **3,** 47
1,3-Dichloro-2-butene, **1,** 214; **2,** 111; **7,** 203
Diethyl lithiomorpholinomethylphosphonate, **7,** 105
2,2-Dimethyl-3(2*H*)-furanone, **10,** 159
Ethyl vinyl ketone, **1,** 388
2-Fluoro-1-buten-3-one, **6,** 263
Lithium–Ammonia, **3,** 179
Lithium diisopropylamide, **8,** 292; **10,** 241
Methyl 5-methoxy-3-oxopentanoate, **9,** 314
Methyl vinyl ketone, **1,** 697; **7,** 247; **12,** 329
2-Methyl-6-vinylpyridine, **6,** 409
1,7-Octadien-3-one, **9,** 327
trans-3-Penten-2-one, **2,** 306; **3,** 218; **4,** 371
Piperidine, **1,** 886; **7,** 293
(S)-(−)-Proline, **12,** 414
Pyrrolidine, **1,** 972; **2,** 354; **7,** 309
by Diels–Alder reactions
1,3-Bis(trimethylsilyloxy)-1,3-butadiene, **7,** 27
1-Chloro-1-dimethylamino-2-methyl-1,3-butadiene, **9,** 104
3-Hydroxy-2-pyrone, **6,** 291
Iron, **9,** 257
2-Lithio-2-trimethylsilyl-1,3-dithiane, **4,** 284
2-Methoxy-3-phenylthio-1,3-butadiene, **7,** 232
trans-1-Methoxy-3-trimethylsilyloxy-1,3-butadiene, **6,** 370; **8,** 328; **9,** 303; **11,** 475
(Z)-3-Methyl-1-phenylthio-2-trimethylsilyloxy-1,3-butadiene, **11,** 344
3-Nitro-2-cyclohexenone, **10,** 277
3-Trimethylsilyloxy-1,3-pentadiene, **10,** 271
Trimethylsilylvinylketene, **10,** 441
Other routes
Arene(tricarbonyl)chromium complexes, **9,** 21
Dimethyl methylphosphonate, **3,** 117
2-Ethoxyallylidenetriphenylphosphorane, **8,** 219
Ferric chloride, **7,** 153
Hexafluoroantimonic acid, **7,** 166; **8,** 239
Iron carbonyl, **12,** 266
Lead tetraacetate, **6,** 313
Methanesulfonic acid, **10,** 256
Methyl 4-methylthiocrotonate, **6,** 389
Piperidine, **1,** 886
Trimethyl orthoformate, **10,** 425
3-CYCLOHEXENONES
by Birch reduction
Lithium–Ammonia, **8,** 282
Oxalic acid, **1,** 764
by Diels–Alder reactions
2-Acetoxyacrylonitrile, **1,** 7
2-Chloroacrylonitrile, **4,** 76; **5,** 107
2-Chloroacrylyl chloride, **4,** 77
Copper(II) tetrafluoroborate, **3,** 66

ALICYCLIC HYDROCARBONS—SIX-MEMBERED RINGS (*Continued*)
Methyl methoxypropiolate, **11,** 340
Oxodiperoxymolybdenum(pyridine)-(hexamethylphosphoric triamide), **10,** 169
ALICYCLIC HYDROCARBONS—SEVEN-MEMBERED RINGS
(*see also* MEDIUM-LARGE RINGS)
BY CLEAVAGE OF BICYCLO[4.1.0]-HEPTANES
Allylidenetriphenylphosphorane, **5,** 7
Ferric chloride, **7,** 153; **10,** 185
Quinoline, **1,** 975
Silver sulfate, **3,** 254
CYCLIZATION
Dodecamethylcyclohexasilane, **12,** 219
Ethyl cyanoacetate, **7,** 406
Lithium diisopropylamide, **8,** 292
Methylaluminum bis(trifluoroacetate), **8,** 471
Palladium(II) acetate, **10,** 297
Sodium hexamethyldisilazide, **1,** 1046
Sodium hydride, **4,** 453
Titanium(IV) chloride–N-Methylaniline, **9,** 470
BY [3+4]CYCLOADDITION
3-Bromo-3-methyl-2-trimethylsilyloxy-1-butene, **9,** 68
Dimethyl(methylthio)sulfonium tetrafluoroborate, **11,** 204
2-Methoxyallyl bromide, **4,** 327
Silver perchlorate, **11,** 469
Tri-μ-carbonylhexacarbonyldiiron, **4,** 157; **9,** 477
Zinc-copper couple, **8,** 533
Zinc–Silver couple, **9,** 519
BY REARRANGEMENT OF DIVINYLCYCLOPROPANES
1-Lithio-2-vinylcyclopropane, **7,** 192; **9,** 273
Lithium phenylthio(2-vinylcyclopropyl)-cuprate, **7,** 212; **9,** 329
BY RING EXPANSION
Cyanogen azide, **3,** 71; **5,** 169
Diazoethane, **3,** 73
Diazomethane, **1,** 191
Diethylzinc–Haloform, **4,** 153; **6,** 194
Dimethyl acetylenedicarboxylate, **4,** 168
Ethyl diazoacetate, **1,** 367; **2,** 464; **8,** 222
Isocyanomethyllithium, **4,** 272
Nitromethane, **1,** 739
Phenyldiazomethane, **1,** 366
OTHER ROUTES
N-Bromosuccinimide, **4,** 49
Copper(II) acetylacetonate, **3,** 62
Sodium, **5,** 589
Zinc–Copper–Isopropyl iodide, **10,** 460
ALICYCLIC HYDROCARBONS—EIGHT-MEMBERED RINGS
(*see also* MEDIUM–LARGE RINGS)
Allene, **1,** 18
Bis(1,5-cyclooctadiene)nickel(0), **12,** 52
4,4′-Bis(dimethylamino)benzophenone, **2,** 285
Dicarbonylbis(triphenylphosphine)-nickel, **1,** 61
Dimethyl acetylenedicarboxylate, **4,** 168
Lithium naphthalenide, **4,** 348
Potassium hydride, **12,** 407
Silver perchlorate, **4,** 432
Sodium N-methylanilide, **1,** 1095
Tetrakis(trichlorophosphine)nickel(0), **1,** 1142
Zinc chloride, **9,** 272
ALICYCLIC HYDROCARBONS—MEDIUM–LARGE RINGS
BY ACYLOIN REACTIONS
Chlorotrimethylsilane–Sodium, **2,** 435; **3,** 311
Sodium–Potassium alloy, **1,** 1102
BY COUPLING OF ALLYLIC HALIDES
Nickel carbonyl, **1,** 720; **2,** 290; **5,** 472; **6,** 417
BY CYCLIZATION
Bis(*o*-methylaminophenyl) disulfide, **9,** 305
Butyllithium, **6,** 85; **7,** 46
Diisobutylaluminum phenoxide, **9,** 172; **10,** 150
Potassium *t*-butoxide, **1,** 911
Sodium hexamethyldisilazide, **12,** 446
Sodium N-methylanilide, **4,** 459
Tetrakis(triphenylphosphine)-palladium(0), **8,** 472; **11,** 503
Tin(IV) chloride, **6,** 553
Tosylmethyl isocyanide, **11,** 539
BY ESCHENMOSER FRAGMENTATION
N-Bromosuccinimide, **10,** 57
Mesitylenesulfonylhydrazide, **10,** 255

p-Toluenesulfonylhydrazide, **2,** 417
BY ONE-CARBON RING EXPANSION
Benzylsulfonyldiazomethane, **11,** 43
Chloromethylcarbene, **10,** 90
Dibromomethyllithium, **6,** 162
Methyllithium, **6,** 384
Perchloric acid, **5,** 506
Silver perchlorate, **6,** 518
Silver *p*-toluenesulfonate, **7,** 323
Sodium–Ammonia, **4,** 438; **6,** 523
BY OXY-COPE REACTION
Potassium hydride, **8,** 412; **10,** 327
BY REDUCTIVE COUPLING OF C=O
Titanium(0), **11,** 526
Titanium(III) chloride–Lithium aluminum hydride, **12,** 493
OTHER METHODS
Borane–Tetrahydrofuran, **3,** 76
Copper(II) acetylacetonate, **9,** 51
2-Hydroxymethyl-3-trimethylsilyl-1-propene, **11,** 258
Sodium methylsulfinylmethylide, **1,** 310

ALICYCLIC HYDROCARBONS—CYCLOALKYNES
BY DIENE COUPLING
Copper(II) acetate, **4,** 105; **6,** 138
Silver(I) nitrate, **1,** 1008
BY ELIMINATION
1,3-Diphenylisobenzofuran, **4,** 209
Phenyl azide, **1,** 829
Potassium *t*-butoxide, **2,** 336; **7,** 298
Selenium(IV) oxide, **9,** 409
Silver(I) nitrate, **2,** 366
Silver(I) oxide, **3,** 252
Sodium amide–Sodium *t*-butoxide, **5,** 593
BY ESCHENMOSER FRAGMENTATION
N-Bromosuccinimide, **10,** 57
Mesitylenesulfonylhydrazide, **10,** 255
p-Toluenesulfonylhydrazide, **2,** 417
OTHER ROUTES
Diethyl (diazomethyl)phosphonate, **12,** 186
Potassium *t*-butoxide, **1,** 911
Sodium hydride, **5,** 610

ALICYCLIC HYDROCARBONS—BICYCLO[2.2.1]HEPTANES
2-Acetoxyacrylonitrile, **1,** 7
Aluminum chloride, **2,** 21
Bis(acrylonitrile)nickel(0), **11,** 48
Boron trifluoride, **5,** 51
Butyllithium, **9,** 83
2-Chloroacrylyl chloride, **4,** 77
5,5-Dimethoxy-1,2,3,4-tetrachlorocyclopentadiene, **2,** 143
Ethynyl *p*-tolyl sulfone, **10,** 183
Methyl 2-acetylacrylate, **11,** 334
Methyl methoxypropiolate, **11,** 340
Methyl vinyl ketone, **1,** 697
Nitroethylene, **5,** 476
Tri-μ-carbonylhexacarbonyldiiron, **9,** 477

ALICYCLIC HYDROCARBONS—BICYCLIC [5,5] RING SYSTEMS
FROM ACYCLIC STARTING MATERIALS
2-Bromo-3-trimethylsilyl-1-propene, **11,** 80
Di-μ-carbonylhexacarbonyldicobalt, **12,** 163
Dimethyl 1,3-acetonedicarboxylate, **11,** 193
BY CYCLOPENTANNELATION
(R)-(+)-*o*-Anisylcyclohexylmethylphosphine, **11,** 31
Polyphosphoric acid, **12,** 399
FROM EIGHT-MEMBERED RINGS
Benzeneselenenyl iodide, **11,** 36
9-Borabicyclo[3.3.1]nonane, **6,** 62
Chloroform, **1,** 131
Lithium diethylamide, **2,** 247
Lithium diisopropylamide, **6,** 334
Lithium phosphate, **4,** 309
Potassium hydride, **2,** 346
Thallium(III) trifluoroacetate, **8,** 478

ALICYCLIC HYDROCARBONS—BICYCLIC [5,6] RING SYSTEMS
FROM ACYCLIC STARTING MATERIALS
Aluminum chloride, **11,** 25
Boron trifluoride etherate, **10,** 52
2-Bromo-3-trimethylsilyl-1-propene, **11,** 80
Carbon monoxide, **3,** 41
2-Oxazolidones, chiral, **12,** 359
3-Triethylsilyloxypentadienyllithium, **11,** 556
Trifluoroacetic acid, **4,** 530
Trifluoromethanesulfonic acid, **12,** 531
Zirconium(IV) propoxide, **11,** 605
BY CYCLOHEXANNELATION

ALICYCLIC HYDROCARBONS—
BICYCLIC [5,6] RING SYSTEMS
(*Continued*)
Aldol reaction
Ethyl phenyl sulfoxide, **6,** 257
2-Methyl-6-vinylpyridine, **6,** 409
(S)-Proline, **7,** 307; **12,** 414
Other routes
Acrolein, **11,** 11
1,2-Bis(trimethylsilyloxy)cyclobutene, **10,** 45
2-Bromomethyl-3-(trimethylsilylmethyl)-1,3-butadiene, **12,** 77
Dimethylphenylsilyllithium, **12,** 210
4-Iodobutyltrimethyltin, **10,** 219
Phenylselenenyl benzenesulfonate, **12,** 390
BY CYCLOPENTANNELATION
Aldol reaction
Bromoacetylmethylenetriphenylphosphorane, **9,** 66
2-(2-Bromoethyl)-1,3-dioxolane, **9,** 70
Isopropenyl acetate, **6,** 356
2-Lithiobenzothiazole, **8,** 274
2-Methoxyallyl bromide, **8,** 322
Morpholine–Camphoric acid, **9,** 317; **11,** 352
2-Nitropropene, **7,** 253
Phenylsulfinylacetone, **5,** 524
Potassium fluoride, **8,** 410
Potassium hydride, **8,** 412
Tetrakis(triphenylphosphine)palladium(0), **10,** 384
Titanium(IV) chloride, **12,** 494
Triphenyltin hydride, **12,** 555
Carbonyl addition
Chlorotrimethylsilane–Zinc, **12,** 568
Lithium dibutylcuprate, **3,** 79
Tin–Aluminum, **12,** 486
Nazarov reaction
(2-Bromovinyl)trimethylsilane, **11,** 82
gem-Dichloroallyllithium, **8,** 150
Phosphoric acid–Formic acid, **1,** 860
Propargyl alcohol, **9,** 394
Vinyltrimethylsilane, **9,** 498; **10,** 444
from Vinylcyclopropanes
Allyltrimethylsilane, **12,** 23
Lead carbonate, **9,** 265
(1-Lithiovinyl)trimethylsilane, **11,** 286
Organocopper reagents, **11,** 365
Simmons–Smith reagent, **6,** 521
Wittig reaction
(Dimethoxyphosphinyl)methyllithium, **6,** 339
Dimethyl 3-bromo-2-ethoxypropenylphosphonate, **9,** 180
Ethoxycarbonylcyclopropyltriphenylphosphonium tetrafluoroborate, **5,** 90
Other routes
Alkylaluminum halides, **10,** 177; **12,** 5
Boron trifluoride etherate, **6,** 65
Bromine, **5,** 55
4-Chloro-1-butenyl-2-lithium, **12,** 113
Chlorotris(triphenylphosphine)rhodium(I), **10,** 98
2-(Dimethylamino)-3-pentenonitrile, **10,** 155
Hydrazine, **8,** 245
Mercury(II) acetate, **11,** 315
Methanesulfonic acid, **10,** 256
Organocopper reagents, **9,** 328
Palladium(II) acetate, **9,** 344
Palladium(II) chloride–Copper(II) chloride, **9,** 353
1-Phenylthiocyclopropyltriphenylphosphonium tetrafluoroborate, **6,** 465
Potassium carbonate, **8,** 408
Potassium hexamethyldisilazide, **4,** 407
Tetrakis(triphenylphosphine)palladium(0), **8,** 472; **9,** 451
Thexylborane, **2,** 148
Titanium(IV) chloride–Magnesium amalgam, **7,** 373
Tributyltin hydride, **11,** 545; **12,** 516
2,4,6-Triisopropylbenzenesulfonylhydrazide, **12,** 533
Trimethylsilylallene, **10,** 428
FROM NINE-MEMBERED RINGS
Acetic anhydride–Zinc chloride, **4,** 4
Benzeneselenenyl chloride, **9,** 25
Organomagnesium reagents, **12,** 352
ALICYCLIC HYDROCARBONS—
BICYCLIC [5,7] RING SYSTEMS
BY CYCLOHEPTANNELATION
2-(2-Buten-2-yl)-1,3-dithiane, **11,** 447
Lead carbonate, **9,** 265
1-Lithio-2-vinylcyclopropane, **7,** 192
2-Methylcyclopentenone-3-dimethylsulfoxonium methylide, **6,** 378

BY CYCLOPENTANNELATION
Boron trifluoride etherate, **6,** 65; **11,** 72
Diphenylsulfonium cyclopropylide, **4,** 211
Titanium(IV) chloride, **9,** 468
Triethyloxonium tetrafluoroborate, **8,** 500
FROM TEN-MEMBERED RINGS
Acetic acid, **7,** 1
Boron trifluoride etherate, **9,** 64
Phosphoric acid–Formic acid, **10,** 317
OTHER ROUTES
Calcium carbonate, **1,** 103
Dibromomethyllithium, **11,** 158
Ethyl diazoacetate, **3,** 138
Thallium(III) nitrate, **9,** 460

ALICYCLIC HYDROCARBONS—BICYCLIC [6,6] RING SYSTEMS
FROM ACYCLIC STARTING MATERIALS
Alkylaluminum halides, **12,** 5
D-(−)- and L-(+)-2,3-Butanediol, **11,** 84
Carbon monoxide, **3,** 41
Ion-exchange resins, **2,** 227
Mercury(II) trifluoroacetate, **9,** 294
Mercury(II) trifluoromethanesulfonate–N,N-Dimethylaniline, **12,** 307
Potassium hydride, **11,** 435
Tin(IV) chloride, **5,** 627; **7,** 342
BY CYCLOHEXANNELATION
Benzeneselenenyl hexafluorophosphate, **10,** 21
Benzenesulfenyl chloride, **11,** 39
1,2-Bis(trimethylsilyloxy)cyclobutene, **10,** 45
1-Butadienyldimethylsulfonium tetrafluoroborate, **9,** 74
4-Chloro-1-butenyl-2-lithium, **12,** 113
1,1-Dichloro-3-bromopropene, **5,** 191
1,4 Dichloro-2-butanone, **9,** 75
Dimethyl 1,3-acetonedicarboxylate, **9,** 176
Disodium tetracarbonylferrate, **10,** 174
Formic acid, **5,** 316
4-Iodobutyltrimethyltin, **10,** 219
Lithium dimethylcuprate, **5,** 234; **6,** 209
Mercury(II) acetate, **5,** 424; **9,** 461
Methyl cyclobutenecarboxylate, **10,** 263
Methyl vinyl ketone, **5,** 464; **11,** 10
Potassium hexamethyldisilazide, **4,** 407
Potassium hydride, **7,** 302
Samarium(II) iodide, **12,** 429
Thexylborane, **2,** 148
Titanium(IV) chloride, **12,** 494
Triphenyltin hydride, **12,** 555
Vinyltriphenylphosphonium bromide, **5,** 750
FROM NAPHTHALENES
trihapto-Allyltris(trimethyl phosphite)-cobalt(I), **6,** 15
Calcium–Amines, **9,** 94
Chlorotris(triphenylphosphine)-rhodium(I), **3,** 325
Lithium–Alkylamines, **1,** 574, 614; **2,** 241
FROM TEN-MEMBERED RINGS
Boron trifluoride etherate, **4,** 44
N-Bromosuccinimide, **10,** 57
Formic acid, **5,** 316
Iodine, **8,** 256
Ion-exchange resins, **6,** 302
SOME SPECIFIC [6,6] SYSTEMS
$\Delta^{1(9)}$-2-Octalones
2-(2-Bromomagnesioethyl)-2-methyl-1,3-dioxolane, **7,** 148
t-Butyl γ-iodotiglate, **6,** 82
1-Chloro-3-pentanone, **3,** 49; **6,** 110
5-Chloro-1-penten-3-one, **3,** 47
1,3-Dibromo-2-pentene, **7,** 90
1,3-Dichloro-2-butene, **2,** 111; **8,** 152
4-Diethylamino-2-butanone, **1,** 698
Ethyl acetoacetate, **6,** 264
Ethyl 3-ethyl-5-methyl-4-isoxazolecarboxylate, **4,** 232
(E)-1-Iodo-3-trimethylsilyl-2-butene, **5,** 355
1-Methoxy-3-buten-2-one, **3,** 198
Methyl 5-methoxy-3-oxopentanoate, **9,** 314
Methyl 3-oxo-4-pentenoate, **1,** 1272; **5,** 746
1,7-Octadien-3-one, **10,** 280
trans-3-Penten-2-one, **2,** 306; **3,** 218; **4,** 371
Piperidine, **1,** 886
Pyrrolidine, **1,** 972; **2,** 354
Sodium dicarbonyl(cyclopentadienyl)-ferrate, **6,** 538
Triethylamine, **1,** 1198
3-Trimethylsilyl-3-buten-2-one, **5,** 461; **6,** 406
Δ^{1}-3-Octalones

ALICYCLIC HYDROCARBONS—BICYCLIC [6,6] RING SYSTEMS (*Continued*)
N,N-Dimethylhydrazine, **9,** 184
2-Ethoxyallylidenetriphenylphosphorane, **8,** 219
2-Lithiobenzothiazole, **8,** 274
trans-1-Methoxy-3-trimethylsilyloxy-1,3-butadiene, **6,** 370; **8,** 328
Δ^2-1-Octalones
N,N-Dimethylhydrazine, **9,** 184
Pyridinium chlorochromate, **8,** 425
$\Delta^{9(10)}$-1-Octalones
Methanesulfonic acid, **10,** 256
ALKANES (*see also* ALICYCLIC HYDROCARBONS, CHIRAL COMPOUNDS)
BY *gem*-ALKYLATION OF C=O
Dichlorodimethyltitanium, **10,** 138
Trimethylaluminum, **5,** 707; **6,** 622
BY CATALYTIC HYDROGENATION (*see* TYPE OF REACTION INDEX-REDUCTION)
BY COUPLING REACTIONS
Cross-coupling of R,R′
Dichlorodimethyltitanium, **10,** 138
Diethyl sulfate, **1,** 253
Dilithium tetrachlorocuprate(II), **4,** 163; **5,** 226
Grignard reagents, **1,** 415
Lithium alkyl(*t*-butoxy)cuprate, **5,** 395
Lithium alkyl(phenylthio)cuprates, **5,** 414
Lithium dibutylcuprate, **7,** 92; **8,** 291
Lithium dimethylcuprate, **2,** 151; **3,** 106; **5,** 234
Lithium trimethylferrate (and related), **3,** 112, 312
3-Methoxy-3-methylbutynylcopper, **8,** 324
Organocopper reagents, **11,** 365
Trimethylaluminum, **5,** 707; **6,** 622
Homocoupling of ROH
Titanium(III) chloride, **2,** 415
Titanium(IV) chloride, **1,** 1169
Titanium(III) chloride–Lithium aluminum hydride, **6,** 588
Tungsten carbonyl, **8,** 525
Homocoupling of RX
Chromium(II) chloride, **11,** 132
Chromium(III) chloride–Lithium aluminum hydride, **8,** 110
Magnesium, **4,** 315
Sodium naphthalenide, **1,** 711
Tetraphenylethylene, **2,** 404
Titanium(III) chloride–Lithium aluminum hydride, **7,** 369
Vanadium(II) chloride tetrapyridine complex, **5,** 744
Oxidative coupling of RM
Copper, **12,** 140
Copper(I) iodide, **2,** 400
Lithium aluminum hydride, **9,** 274
Lithium dibutylcuprate, **3,** 79
Silver(I) nitrate, **2,** 366; **4,** 429
Other couplings
Copper(II) acetylacetonate, **2,** 81
Difluoramine, **1,** 253
N,N-Dimethylformamide, **1,** 278
Sodium bis(2-methoxyethoxy)-aluminum hydride, **4,** 441
BY DEAMINATION (*see* TYPE OF REACTION INDEX)
BY DECARBONYLATION OF RCHO
Bis[1,3-bis(diphenylphosphine)propane]-chlororhodium(I), **12,** 111
Chlorotris(triphenylphosphine)-rhodium(I), **1,** 140, 1252; **2,** 448
Nickel–Alumina, **9,** 321
BY DECARBOXYLATION OF RCOOH AND DERIVATIVES
t-Butyl hydroperoxide, **1,** 88
2-Pyridinethiol-1-oxide, **12,** 417
Tributyltin hydride, **10,** 411
BY DECYANATION
Iron(III) acetylacetonate, **4,** 268
Nickel–Alumina, **9,** 321
Potassium, **4,** 245; **10,** 322
BY DEHALOGENATION (*see* TYPE OF REACTION INDEX-REDUCTION)
BY DENITRATION (*see* TYPE OF REACTION INDEX)
BY DEOXYGENATION OF ROH, C=O (*see* TYPE OF REACTION INDEX)
BY DESELENYLATION (*see* TYPE OF REACTION INDEX)
BY DESULFURATION (*see* TYPE OF REACTION INDEX)
BY HYDROMETALLATION-PROTONATION (*see* TYPE OF REACTION INDEX)

BY REDUCTION OF C=C, C≡C, RX, ROSO$_2$R′, ROCOR′ AND RELATED, RCOOH, RCOX (*see* TYPE OF REACTION INDEX)
BY REDUCTIVE ALKYLATION OF C=O (*see* TYPE OF REACTION INDEX)
OTHER ROUTES
Chlorotris(triphenylphosphine)-rhodium(I), **3,** 325
Dicyclopentadienyltitanium, **5,** 672
Methyl fluoride–Antimony(V) fluoride, **3,** 201
Nickel–Alumina, **9,** 321
ALKENES (*see also* CHIRAL COMPOUNDS, DIENES, α-METHYLENE CARBONYLS, METHYLENE CYCLOALKANES, UNSATURATED..., TYPE OF REACTION INDEX FOR ELIMINATION REACTIONS)
FROM ACYL DERIVATIVES + WITTIG REAGENTS
Methylenetriphenylphosphorane, **6,** 380; **9,** 307
BY ADDITIONS TO C≡C, C≡C–X
1-Alkenes
Grignard reagents, **6,** 269
Organocopper reagents, **10,** 282
Trimethylaluminum, **8,** 506
(E)-Alkenes
Chlorobis(cyclopentadienyl)-hydridozirconium(IV), **8,** 84
Copper(I) iodide, **12,** 143
Cyanogen bromide, **4,** 110
Dibromoborane–Dimethyl sulfide, **11,** 262
Diisobutylaluminum hydride, **7,** 111
Dilithium acetylide, **6,** 148
Palladium(II) acetate, **8,** 378
Tetrakis(triphenylphosphine)nickel(0), **7,** 357
Thexylborane, **4,** 175; **5,** 232
(Z)-Alkenes
Bis(*trans*-2-methylcyclohexyl)borane, **4,** 37
Dibromoborane–Dimethyl sulfide, **11,** 262
Iodine, **6,** 293; **11,** 261
Lithium acetylide, **6,** 324
Organocopper reagents, **10,** 282
Palladium(II) acetate, **8,** 378
Tributyltin chloride, **7,** 378
Trisubstituted alkenes
Cyanogen bromide, **4,** 110
Dichlorobis(cyclopentadienyl)-titanium, **9,** 488; **10,** 130
Iodine, **6,** 293; **11,** 261
Organocopper reagents, **9,** 328
Tetrakis(triphenylphosphine)-palladium(0), **9,** 451
Thexylborane, **6,** 207
Tributyltin chloride, **7,** 378
Vinylcopper reagents, **6,** 662
BY ALKYLATION (ARYLATION) OF C=C
Bis(cyclopentadienyl)isoprenezirconium, **11,** 174
μ-Chlorobis(cyclopentadienyl)-(dimethylaluminum)-μ-methylenetitanium, **8,** 83
Diacetatobis(triphenylphosphine)-palladium(II), **6,** 156
2,6-Di-*t*-butyl-4-methylpyridine, **9,** 141
Dichlorobis(cyclopentadienyl)titanium–Trialkylaluminum, **9,** 146
Palladium(II) acetate, **5,** 496
Palladium(II) acetate–Triphenylphosphine, **5,** 497
Rhodium(III) chloride, **6,** 504
FROM C=O AND WITTIG (OR RELATED) REAGENTS
Peterson olefination
Lithium 1-(dimethylamino)-naphthalenide, **12,** 279
Lithium naphthalenide, **11,** 302
Sodium methoxide, **5,** 617
Trimethylsilylmethyllithium, **6,** 637; **10,** 433; **11,** 581
Trimethylsilylmethylmagnesium chloride, **5,** 724; **6,** 636
Trimethylsilylmethylsodium, **5,** 617
Wittig (Wittig–Horner) reaction—General methods
Ethylene oxide, **5,** 297
Lithium chloride, **12,** 277
Pentamethylphosphonic diamide, **2,** 280
Phosphonamides, chiral, **12,** 396
Potassium *t*-butoxide, **11,** 432
Potassium carbonate, **11,** 433
Sodium hexamethyldisilazide, **1,** 1046

ALKENES (*Continued*)
Sodium methylsulfinylmethylide, **1,** 310
Triphenylphosphine, **1,** 1238
Wittig reaction, **4,** 573; **5,** 752
Wittig (Wittig–Horner) methylenation (*see* TYPE OF REACTION INDEX)
Wittig (Wittig–Horner) reaction, C=O → E-Alkenes
Alkyldiphenylphosphine oxides, **10,** 2
Crown ethers, **6,** 133
Diethyl benzylphosphonate, **1,** 1212; **2,** 432
Lithium hexamethyldisilazide, **11,** 300
Triphenylphosphine, **1,** 1238
Wittig (Wittig–Horner) reaction, C=O → Z-Alkenes
Alkyldiphenylphosphine oxides, **10,** 2
Benzyltriphenylphosphonium chloride, **1,** 279
Crown ethers, **6,** 133; **11,** 143
Hexamethylphosphoric triamide, **6,** 273
Sodium hexamethyldisilazide, **7,** 329
N,N,N′,N′-Tetramethylethylenediamine, **12,** 477
Zirconium carbene complexes, **12,** 577
Wittig (Wittig–Horner) reactions—Miscellaneous
Benzyltriphenylphosphonium chloride, **5,** 30
Benzyltris(dimethylamino)-phosphonium bromide, **2,** 210
Bis(trifluoromethyl)methylene-triphenylphosphorane, **9,** 54
Crown ethers, **10,** 110
Ethylidenetriphenylphosphorane, **3,** 141
N,N,P-Trimethyl-P-phenyl-phosphinothioic amide, **11,** 569
2-Trimethylstannylethylidene-triphenylphosphorane, **6,** 640
Vinyltriphenylphosphonium bromide, **5,** 750
Methylenation reactions (other methods) (*see* TYPE OF REACTION INDEX)
Other methods for $C=O \rightarrow C=CR_2$
Alkyldimesitylboranes, **12,** 12
N-Methylphenylsulfonimidoyl-1-ethyllithium, **5,** 459
Tris(2,2-dimethylpropyl)(2,2-dimethylpropylidene)niobium (or tantalum), **7,** 403
BY CLAISEN REARRANGEMENT (*see* γ,δ-UNSATURATED C=O's
BY CLEAVAGE OF CYCLIC ETHERS
Butyllithium, **2,** 51
Sodium, **4,** 437, 438; **7,** 324
BY CLEAVAGE OF CYCLOPROPYL CARBINYL HALIDES
Cyclopropyl methyl ketone, **1,** 676
Hydrobromic acid, **4,** 249
N,N,N′,N′-Tetramethylethylenediamine, **4,** 485
Zinc bromide, **2,** 463
BY COUPLING OF ALLYL + ALKYL
Allylic acetates + RM
Alkylcopper reagents–Aluminum chloride, **10,** 286
Boron trifluoride etherate, **12,** 66
Dilithium tetrachlorocuprate(II), **5,** 226
Lithium dimethylcuprate, **2,** 151; **3,** 106; **7,** 120
Lithium trimethylferrate, **3,** 312
Triisobutylaluminum, **7,** 391
Allylic alcohols + RM
Alkylcopper reagents–Boron trifluoride, **9,** 333; **10,** 282
[1,2-Bis(diphenylphosphine)ethane]-(dichloro)nickel(II), **10,** 36
1-Chloro-2-methyl-N,N-tetramethyl-enepropenylamine, **12,** 124
Grignard reagents–Nickel(II) reagents, **2,** 110
N,N-Methylphenylaminotriphenyl-phosphonium iodide, **8,** 346
Tributyl(methylphenylamino)-phosphonium iodide, **8,** 345
Allylic halides + RM
Alkylcopper reagents–Boron trifluoride, **8,** 334; **10,** 282
α-Chloroallyllithium, **10,** 87
2,3-Dibromopropene, **1,** 420
Dilithium tetrachlorocuprate(II), **11,** 190
Lithium aluminum hydride, **9,** 274
Allylic metals + RX
Butyllithium, **5,** 80
Copper(I) iodide, **8,** 121
Di-μ-bromobis(3-methyl-2-butenyl)-

dinickel, **8,** 181
Palladium(II) chloride, **5,** 500
Other allylic substrates + RM
Allyl phenyl ether, **11,** 14
[1,1′-Bis(diphenylphosphine)-ferrocene](dichloro)palladium(II), **12,** 171
Dichlorobis(triphenylphosphine)-nickel(II), **9,** 147
Ferrocenylphosphines, **10,** 37
Grignard reagents, **6,** 269
Organocopper reagents, **9,** 328; **10,** 282
BY COUPLING OF VINYL + ALKYL
Enol phosphates + RM
Lithium dibutylcuprate, **7,** 92
Tetrakis(triphenylphosphine)-palladium(0), **10,** 384
Vinyl halides + RM
trans-2,3-Bis(diphenylphosphine)-bicyclo[2.2.1]hept-5-ene, **10,** 36
[1,1′-Bis(diphenylphosphine)-ferrocene](dichloro)palladium(II), **9,** 147; **12,** 171
[1,3-Bis(diphenylphosphine)propane]-(dichloro)nickel(II), **11,** 167
Butyllithium, **6,** 85
(S)-α-(R)-2-Diphenylphosphino-ferrocenyl ethyldimethylamine, **9,** 200; **11,** 237
Ferric chloride, **6,** 259
Grignard reagents–Iron(III) reagents, **4,** 236; **6,** 50; **7,** 411
Lithium dimethylcuprate, **2,** 151; **3,** 106; **4,** 295; **6,** 209
Lithium trimethylferrate, **3,** 312
Lithium trimethylmanganate, **3,** 112
Pentane-1,5-di(magnesium bromide), **9,** 355
Tetrakis(triphenylphosphine)nickel(0), **12,** 467
Tetrakis(triphenylphosphine)-palladium(0), **6,** 571; **8,** 472; **9,** 451
Vinyl metal reagents + RX
n- or *t*-Butyllithium, **7,** 47
Copper(I) iodide, **12,** 141
Grignard reagents–Copper(I) halides, **7,** 82
Lithium di[(E)-1-propenyl]cuprate, **7,** 141
Methylcopper, **7,** 236
Organocopper reagents, **11,** 365; **12,** 345
Pentylsodium, **7,** 400
(E)-1-Propenyllithium, **6,** 308, 495
p-Toluenesulfonylhydrazide, **9,** 472
2,4,6-Triisopropylbenzene-sulfonylhydrazide, **9,** 486
Vinyllithium, **2,** 456; **6,** 665
Vinyl sulfides + RM
[1,3-Bis(diphenylphosphine)propane]-(dichloro)nickel(II), **10,** 191
Dichlorobis(triphenylphosphine)-nickel(II), **9,** 147
Nickel(II) chloride–Triphenylphosphine, **11,** 358
Other vinyl substrates + RM
Grignard reagents–Nickel(II) reagents, **10,** 189
Organocopper reagents, **10,** 282
BY DEHYDROGENATION (*see* TYPE OF REACTION INDEX)
BY DESULFURATION OF CYCLIC SULFIDES
Raney nickel, **5,** 263, 381
BY DOUBLE EXTRUSION (*see* TYPE OF REACTION INDEX)
BY ELIMINATION REACTIONS TO FORM SPECIFIC C=C (*see also* TYPE OF REACTION INDEX for a complete list of elimination reactions)
1-Alkenes
Chloramine-T, **10,** 85
Chloromethyldiphenylsilane, **12,** 321
Copper, **1,** 157
Dimethyl(methylene)ammonium iodide, **8,** 194
Dimethyl sulfoxide, **2,** 157
Diphenyl disulfide, **6,** 235
Ethyl α-trifluoromethyl-sulfonyloxyacetate, **7,** 149
Ethyl trimethylsilylacetate, **11,** 247
Hexamethylphosphoric triamide, **4,** 244
Methylselenomethyllithium, **9,** 307
o-Nitrophenyl selenocyanate, **6,** 420
Phenylthioacetic acid, **6,** 463
2-Pyridineselenenyl bromide, **11,** 455
Sodium dicarbonyl(cyclopentadienyl)-ferrate, **9,** 426
Sodium hydride, **1,** 1075
Sodium pyridylselenate, **10,** 368

ALKENES (*Continued*)
Tetrabutylammonium fluoride, **9,** 444
Thorium oxide, **1,** 1167; **5,** 669
(E)- or (Z)-Alkenes
Benzyl phenyl sulfoxide, **6,** 394
Bis(1,5-cyclooctadiene)nickel(0), **5,** 34
t-Butyldimethylchlorosilane, **11,** 88
Butyllithium, **2,** 51
N-Chlorosuccinimide, **5,** 127
Chlorotrimethylsilane–Sodium iodide, **10,** 97
Chromium(II) sulfate, **1,** 150
Copper(I) trifluoromethanesulfonate, **6,** 130
1,3-Dibenzyl-2-methyl-1,3,2-diazaphospholidine, **2,** 105
Diisobutylaluminum hydride, **6,** 198
Dimethylformamide dimethyl acetal, **8,** 191
N,N-Dimethylphosphoramidic dichloride, **6,** 215
Diphenyl diselenide, **6,** 235
Diphosphorus tetraiodide, **9,** 203
Dipotassium hexachlorotungstate(IV), **4,** 407
Ethyl(carboxysulfamoyl)triethylammonium hydroxide inner salt, **5,** 442
Lithium diethylamide, **5,** 398
Lithium dipropylcuprate, **6,** 245
N-Methanesulfinyl-*p*-toluidine, **2,** 269
Phosphoryl chloride, **1,** 881; **2,** 330
Sodium amalgam, **11,** 473
Sodium iodide, **1,** 1116; **6,** 543; **7,** 338
Sodium sulfide, **12,** 453
p-Toluenesulfonylhydrazide, **8,** 489
Tributyltin hydride, **8,** 497
Trimethyl phosphite, **1,** 1233; **2,** 439
Triphenylphosphine–Diethyl azodicarboxylate, **9,** 504; **12,** 552
Zinc, **1,** 1276
BY PROTONATION OF ALLYL METAL REAGENTS
Boron trifluoride etherate, **12,** 66
Hydrogen chloride, **9,** 239
Tributyltin hydride, **7,** 379; **9,** 476
β-Trimethylsilylethylidene-triphenylphosphorane, **9,** 492
2-Trimethylsilylmethyl-1,3-butadiene, **10,** 432
2-Trimethylstannylethylidene-triphenylphosphorane, **6,** 640
BY REDUCTION OF C≡C, C=C–X, DIENES (*see* TYPE OF REACTION INDEX)
BY REDUCTION OF ALLYLIC ETHERS AND RELATED
Allyl phenyl ether, **11,** 14
Lithium triethylborohydride, **11,** 304
Tributyltin hydride, **7,** 379; **9,** 476; **10,** 411
BY REDUCTIVE COUPLING (*see* TYPE OF REACTION INDEX)
BY 2,3-SIGMATROPIC REARRANGEMENT (*see also* TYPE OF REACTION INDEX)
S-Allyl N,N-dimethyldithiocarbamate, **6,** 11
Benzenesulfenyl chloride, **9,** 35
Dimethyl diazomalonate, **5,** 244
Lithium diisopropylamide, **5,** 400
FROM THREE-MEMBERED HETEROCYCLES (*see* TYPE OF REACTION INDEX)
OTHER ROUTES
Bismuth(III) oxide–Tin(IV) oxide, **6,** 55
(Bromodichloromethyl)phenylmercury, **1,** 851
Cerium(IV) ammonium nitrate, **4,** 71
Chlorotris(triphenylphosphine)-rhodium(I), **3,** 325
Dibromodifluoromethane, **1,** 207
Iodo(methyl)calcium, **5,** 442
Lithium naphthalenide, **11,** 302
Methyllithium, **7,** 242
Nickel(II) chloride–Ethylene, **10,** 276
Palladium(II) acetate, **12,** 367
Propargyl chloride, **5,** 565
p-Toluenesulfonylhydrazide, **6,** 537
Tributyltin hydride, **10,** 411
Tributyltinlithium, **10,** 413
β-[(Trimethylsilyl)ethyl]lithium, **11,** 574
Trimethylsilylmethyllithium, **6,** 635
Triphenylcarbenium salts, **1,** 1256; **8,** 524; **10,** 455
Triphenyl phosphite ozonide, **7,** 408
ALKYL HALIDES (*see also* ALLYLIC COMPOUNDS, CHIRAL COMPOUNDS, DIHALIDES,

HALO-, VINYL HALIDES)
GENERAL METHODS
2-Dimethylamino-N,N'-diphenyl-1,3,2-diazaphospholidine, **11,** 196
Sodium borohydride, **1,** 1049; **2,** 377
Triphenylphosphine–N-Halosuccinimides, **5,** 728
Triphenyl phosphite, **1,** 1249
ALKYL FLUORIDES
from ROH
N,N-Diethylaminopropyne, **2,** 133
(Diethylamino)sulfur trifluoride, **6,** 183; **8,** 166
Diethyl(2-chloro-1,1,2-trifluoroethyl)-amine, **1,** 249; **3,** 95
Phenyltetrafluorophosphorane, **5,** 526
Pyridinium poly(hydrogen fluoride), **6,** 473; **7,** 294
Selenium(IV) fluoride, **5,** 576
Sodium fluoride, **6,** 473
Sulfur tetrafluoride, **6,** 560
Triphenylphosphine difluoride, **2,** 135
from RH (*see* TYPE OF REACTION INDEX)
from RX'
Amberlyst ion-exchange resin, **7,** 182
Antimony(V) fluoride–Graphite, **5,** 19
Ethylene glycol, **1,** 375
Potassium fluoride, **1,** 933; **5,** 153, 322; **8,** 390
Silver fluoride, **6,** 514
Tributylmethylphosphonium fluoride, **9,** 223
from $ROSO_2R'$
Amberlyst ion-exchange resin, **7,** 182
Potassium fluoride, **5,** 153, 322; **6,** 405
Tributylmethylphosphonium fluoride, **9,** 223
Other methods
Potassium fluoride, **8,** 410
Pyridinium poly(hydrogen fluoride), **5,** 538; **9,** 399
2,4,6-Triphenylpyrylium fluoride, **9,** 508
ALKYL CHLORIDES
from ROH
Arsenic trichloride, **3,** 16
Benzyltriethylammonium chloride, **4,** 27
Bis(benzonitrile)dichloropalladium(II), **9,** 44
t-Butyl chloride, **1,** 86
t-Butyl hydroperoxide, **6,** 81
2-Chloro-3-ethylbenzoxazolium tetrafluoroborate, **8,** 90
N-Chlorosuccinimide–Dimethyl sulfide, **4,** 87; **5,** 129
Cyanuric chloride, **3,** 72; **4,** 522
N,N-Diethyl-1,2,2-trichlorovinylamine, **1,** 253; **2,** 134
N,N-Dimethylformamide, **12,** 203
N,N-Dimethylphosphoramidic dichloride, **9,** 187
Hexamethylphosphoric triamide–Thionyl chloride, **3,** 153
Hydrochloric acid, **5,** 322
Hydrogen chloride, **2,** 215; **6,** 278
Lithium chloride, **3,** 95; **8,** 223
Methanesulfonyl chloride, **2,** 268; **5,** 435
Oxalyl chloride, **1,** 767
Phosphorus(V) chloride, **5,** 534; **7,** 290
Phosphorus(V) sulfide, **6,** 470
Phosphoryl chloride, **2,** 335
Silver hexafluoroantimonate, **5,** 577
Sodium chloride, **6,** 473
Thionyl chloride, **1,** 1158; **4,** 245; **5,** 663; **6,** 676
Trichloroacetonitrile, **1,** 1194
Triphenylphosphine–Carbon tetrachloride, **1,** 1247; **4,** 551; **6,** 644; **9,** 503; **12,** 554
Triphenylphosphine dichloride, **1,** 1247; **6,** 646
Vilsmeier reagent, **6,** 220; **7,** 292, 422; **12,** 564
Zinc chloride, **1,** 1289
from RH (*see* TYPE OF REACTION INDEX)
from C=C
Copper(II) chloride, **6,** 259
Ferric chloride, **6,** 259
Hydrogen chloride, **5,** 335
Trimethylaluminum, **5,** 707
from RX'
Aluminum bromide, **4,** 10
Antimony(V) chloride–Graphite, **5,** 19; **6,** 22
Molybdenum(V) chloride, **6,** 412
Silver chlorodifluoroacetate, **3,** 249
Tetrabutylammonium chloride, **6,** 565

ALKYL HALIDES (*Continued*)
Tributyltin chloride, **6,** 604
from $ROSO_2R'$
Antimony(V) chloride–Graphite, **6,** 22
Lithium chloride, **7,** 169
Potassium chloride, **6,** 405
Pyridinium chloride, **1,** 964
Titanium(IV) chloride, **6,** 590
Vinylsulfonyl chloride–Trimethylamine, **9,** 211
from RCOOH
Lead tetraacetate–Metal halides, **1,** 557; **4,** 278; **5,** 370
Mercury(II) oxide, **1,** 655
2-Pyridinethiol 1-oxide, **12,** 417
from ROR′
Molybdenum carbonyl, **4,** 346
Benzylic chlorides
Aluminum chloride, **6,** 17
1,4-Bis(chloromethoxy)butane, **7,** 22
t-Butyl hypochlorite, **1,** 90
Chlorine oxide, **11,** 119
1-Chloro-4-chloromethoxybutane, **6,** 104
N-Chloro-N-cyclohexylbenzenesulfonamide, **2,** 67
Chloromethyl methyl ether, **1,** 132; **4,** 83
Dimethoxymethane, **1,** 671
Formaldehyde–Hydrochloric acid, **1,** 399
Trichloroisocyanuric acid, **3,** 297
Zinc chloride, **2,** 464
ALKYL BROMIDES
from ROH
Ammonium bromide, **6,** 473
Arsenic tribromide, **3,** 16
Benzoyl bromide, **5,** 249
Bromodimethylsulfonium bromide, **4,** 174
Bromotrimethylsilane, **9,** 73
t-Butyl hydroperoxide, **6,** 81
N,N′-Carbonyldiimidazole, **12,** 106
2-Chloro-3-ethylbenzoxazolium tetrafluoroborate, **10,** 204
Chlorotrimethylsilane–Lithium bromide, **10,** 96
2-Dimethylamino-N,N′-diphenyl-1,3,2-diazaphospholidine, **11,** 196
Dimethylbromomethyleneammonium bromide, **6,** 220; **7,** 422
Hydrobromic acid, **1,** 450
Hydrogen bromide, **1,** 453
Lithium bromide, **3,** 95; **5,** 326; **8,** 223
Phosphorus(III) bromide, **1,** 862, 873; **2,** 335; **7,** 292
Phosphorus(V) bromide, **1,** 865
Potassium bromide, **6,** 405
Thionyl bromide, **1,** 1157; **4,** 245
Tributylfluorophosphonium bromide, **11,** 543
Triphenylphosphine dihalides, **1,** 1247; **5,** 732; **12,** 554
Triphenyl phosphite dibromide, **1,** 1249
from RH (*see* TYPE OF REACTION INDEX)
from C=C
Bromine, **3,** 160; **6,** 259
N-Bromosuccinimide, **8,** 54
Chlorobis(cyclopentadienyl)-hydridozirconium(IV), **6,** 175
Hydrogen bromide, **1,** 196
Sodium bromide–Chloramine-T, **11,** 489
from RX′
Aluminum bromide, **4,** 10
N-Methyl-2-pyrrolidone, **11,** 346
Sodium bromide, **12,** 445
from $ROSO_2R'$
Lithium bromide, **4,** 297
Magnesium bromide, **2,** 254; **6,** 595
from RCOOH
Mercury(II) oxide, **9,** 293
Mercury(II) oxide–Bromine, **1,** 657; **4,** 323; **5,** 428
2-Pyridinethiol-1-oxide, **12,** 417
Thallium(I) carbonate, **11,** 515
Thallium(I) ethoxide, **2,** 407
from ROR′
Bromodimethylborane, **12,** 199
Bromotrimethylsilane, **10,** 59
Hydrobromic acid, **9,** 359
Triphenylphosphine dibromide, **3,** 320; **5,** 729
Other routes
Aluminum bromide, **1,** 22
Bromine, **7,** 33; **10,** 56
(Diphenylarsinyl)methyllithium, **8,** 280
Grignard reagents–Dilithium tetrachlorocuprate(II), **6,** 203

Trifluoromethanesulfonic anhydride, **5,** 702
Triphenylphosphine dibromide, **7,** 109; **9,** 112
Benzylic bromides
N-Bromopolymaleimide, **4,** 49
N-Bromosuccinimide–Dimethyl sulfide, **4,** 90
Bromotrichloromethane, **7,** 40
1-Chloro-4-bromomethoxybutane, **6,** 104
1,3-Dibromo-5,5-dimethylhydantoin, **1,** 208; **12,** 158
ω-Tribromoacetophenone, **1,** 1188
ALKYL IODIDES
from ROH
Aluminum iodide, **1,** 35
N,N′-Carbonyldiimidazole, **12,** 106
2-Chloro-1,3,2-benzodioxaphosphole, **2,** 321; **5,** 516
Cyanuric chloride, **4,** 522
2-Dimethylamino-N,N′-diphenyl-1,3,2-diazaphospholidine, **11,** 196
Diphosphorus tetraiodide, **9,** 203; **12,** 218
Iodotrimethylsilane, **8,** 261; **9,** 251
Mercury(II) oxide–Iodine, **1,** 657
N-Methyl-N,N′-dicyclohexylcarbodiimidium iodide, **4,** 336; **7,** 238
Methyltriphenoxyphosphonium iodide, **1,** 1249; **4,** 557
Phosphorus(III) iodide, **1,** 497
Phosphorus(V) oxide–Phosphoric acid, **1,** 872
Phosphorus(red)–Iodine, **1,** 862
Potassium iodide, **6,** 279, 405, 473
Sodium iodide, **8,** 223
Trichloro(methyl)silane–Sodium iodide, **12,** 527
Trimethylsilyl polyphosphate, **12,** 543
Triphenylphosphine + co-reagent, **7,** 407; **9,** 506, 507
Triphenyl phosphite, **4,** 556
from C=C
Dichlorobis(cyclopentadienyl)-titanium, **10,** 130
Iodine, **3,** 159; **6,** 259, 293; **7,** 179
Iodine monochloride, **10,** 212; **11,** 268
Mercury(II) chloride, **6,** 259
Sodium iodide–Chloramine-T, **11,** 488
from RX′
Aluminum bromide, **4,** 10
Potassium iodide, **6,** 174; **9,** 357
Sodium iodide, **1,** 1087
Tetrabutylammonium iodide, **6,** 565
Triphenylphosphine–Iodine, **10,** 450
from $ROSO_2R'$
Magnesium iodide, **5,** 420; **6,** 595
Sodium iodide, **1,** 1088; **4,** 456
Tetrabutylammonium iodide, **9,** 447
Vinylsulfonyl chloride–Trimethylamine, **9,** 211
from RCOOH
t-Butyl hypoiodite, **1,** 94
Iodine–Lead tetraacetate, **1,** 553
2-Pyridinethiol 1-oxide, **12,** 417
from ROR′
Aluminum iodide, **12,** 30
Phosphorus(V) oxide–Phosphoric acid, **1,** 872
Potassium iodide, **1,** 904
Other routes
t-Butyl hypoiodite, **2,** 50
Iodotrimethylsilane, **12,** 259
2-Methylthio-2-thiazoline, **5,** 459
Potassium iodide, **6,** 531
Sodium iodide, **2,** 324
Triphenylphosphine–Iodine, **10,** 450
2,4,6-Triphenylpyrylium iodide, **8,** 519
Tungsten(VI) chloride–Tetramethyltin, **10,** 455
ALKYNES (*see also* α,β-ACETYLENIC CARBONYLS, DIYNES, ENYNES, HALOALKYNES, HOMOPROPARGYL ALCOHOLS, PROPARGYL ALCOHOLS)
GENERAL METHODS
Acetic anhydride–Pyridine, **9,** 1
Benzyltrimethylammonium hydroxide, **5,** 29; **7,** 20
Dichloro(diethoxyphosphinyl)-methyllithium, **6,** 188
(Diphenylphosphine)lithium, **6,** 340
Phase-transfer catalysts, **11,** 403
Potassium fluoride, **5,** 555
Potassium hydroxide, **6,** 486
Sodium amide, **1,** 1034; **2,** 373
Tetrabutylammonium hydrogen sulfate, **7,** 354
1-ALKYNES
using Acetylide anions
Acetylene, **1,** 11

ALKYNES (*Continued*)
Lithium acetylide, **1,** 574; **5,** 345, 382
Sodium acetylide, **1,** 1037
from RCHO
Bromomethylenetriphenylphosphorane, **10,** 56
Dibromomethyllithium, **11,** 158
Dimethyl diazomethylphosphonate, **9,** 181
Trimethylsilyldiazomethane, **4,** 543
Triphenylphosphine–Carbon tetrabromide, **4,** 550
from Allenes
Bromoallene, **8,** 288
Ethyllithium, **5,** 306
Lithium aluminum hydride, **9,** 274
Methoxyallene, **6,** 146; **7,** 163
by Elimination
trans-1,2-Bis(tributylstannyl)ethylene, **5,** 43
Lithium diisopropylamide, **10,** 241; **11,** 296
Potassium 3-aminopropylamide, **8,** 406
Potassium *t*-butoxide, **2,** 336; **4,** 401; **5,** 544; **6,** 477
Sodium amide, **1,** 1034
Sodium methylsulfinylmethylide, **3,** 123
Trichloroethylene, **10,** 414
Trifluoromethanesulfonic anhydride, **5,** 702
using Propargyl anions
1,3-Dilithiopropyne, **6,** 202
3-Lithio-1-trimethylsilyl-1-propyne, **2,** 239; **3,** 173
Other routes
Butyllithium, **6,** 85
Di-μ-carbonylhexacarbonyldicobalt, **8,** 148
Dichloromethyllithium, **9,** 154
Organocopper reagents, **9,** 328
Potassium 3-aminopropylamide, **6,** 476
p-Toluenesulfonylhydrazide, **3,** 293
Tributyltinlithium, **11,** 551
ARYL ALKYNES
1-Alkynes
Chloromethylenetriphenylphosphorane, **8,** 96
Diacetatobis(triphenylphosphine)palladium(II), **10,** 117
Dimethyl diazomethylphosphonate, **9,** 181
Iodoethynyl(trimethyl)silane, **4,** 265
3-Methyl-5(4*H*)-isoxazolone, **9,** 309
Tetrakis(triphenylphosphine)palladium(0), **6,** 571
Aryl alkyl alkynes
Butylamine, **4,** 53
2-Chloro-3-ethylbenzoxazolium tetrafluoroborate, **9,** 105
Diacetatobis(triphenylphosphine)palladium(II), **6,** 156
Dimethyl diazomethylphosphonate, **9,** 181
Lithium 2,2,6,6-tetramethylpiperidide, **4,** 310
Mercury(I) trifluoroacetate, **2,** 268
Tetrakis(triphenylphosphine)palladium(0), **8,** 472
Tris(phenylethynyl)aluminum, **6,** 601
DIARYL ALKYNES
from Diaryl diketones
Copper(I) chloride, **7,** 80
Mercury(II) oxide, **1,** 655
Oxygen, **6,** 426
Silver(I) trifluoroacetate, **1,** 1018
from Diaryl ketones
Crown ethers, **10,** 110
Dimethyl diazomethylphosphonate, **8,** 188; **9,** 181
Trimethylsilyldiazomethane, **4,** 543; **8,** 188
Other routes
Acetylene, **6,** 61
Alumina, **10,** 8
Isoamyl nitrite, **1,** 40
Sodium methoxide–Dimethyl sulfoxide, **6,** 545
Tetrakis(triphenylphosphine)palladium(0), **6,** 571
Tetramethylphosphorodiamidous chloride, **6,** 575
Triethyl phosphite, **1,** 1212
Trifluoromethanesulfonic anhydride, **5,** 702
Triphenylphosphine, **1,** 1238
Triphenylphosphine–Potassium *t*-butoxide, **5,** 729
INTERNAL ALKYNES
from α-Dicarbonyls

Copper(I) chloride–Oxygen, **5,** 165
Grignard reagents–Cobalt(II) chloride, **1,** 155
Hydrazine, **1,** 434
Triethyl phosphite, **1,** 1212
Trimethyl phosphite, **7,** 393
by Elimination reactions
Diethyl phosphorochloridate, **8,** 171
Potassium *t*-butoxide, **5,** 557; **8,** 407
Sodium amalgam, **9,** 416; **12,** 439
Trifluoromethanesulfonic anhydride, **6,** 618
using Alkynyl metals
Dimethyl(methylthio)sulfonium tetrafluoroborate, **12,** 207
Lithium acetylides, **2,** 166, 208; **5,** 346; **8,** 285; **12,** 70
Lithium amide, **9,** 278
Methanesulfinyl chloride, **5,** 434
Sodium acetylides, **2,** 166
Trialkynylalanes, **6,** 600
Other routes
Chloromethylenetriphenylphosphorane, **5,** 119
Ethylidenetriphenylphosphorane, **3,** 141
Lithium chloroacetylide, **6,** 294
Organolithium reagents, **9,** 5
Potassium ferricyanide, **7,** 300
Sodium methylsulfinylmethylide, **3,** 123
Triphenylphosphine–Carbon tetrabromide, **4,** 550
ALKYNYL HALIDES
(*see* HALOALKYNES)
ALLENES, CUMULENES AND ALLENE DERIVATIVES (*see also* ALLENE OXIDES, ALLENIC ALCOHOLS, ALLENIC CARBONYLS, CHIRAL COMPOUNDS)
ALLENES (UNSTUBSTITUTED)
from 1,1-Dihalocyclopropanes
Butyllithium, **1,** 95; **2,** 51
Chromium(III) chloride–Lithium aluminum hydride, **8,** 110
Copper(0)–Isonitrile complexes, **9,** 122
Methyllithium, **1,** 686
Sodium methylsulfinylmethylide, **1,** 310
by Elimination
α-Bromovinyltrimethylsilane, **8,** 56
α-Bromovinyltriphenylsilane, **5,** 68, 375
1,3-Diphenylisobenzofuran, **2,** 178
Lithium diisopropylamide, **9,** 280
Potassium *t*-butoxide, **6,** 479
Potassium fluoride, **5,** 555
Sodium 2-methoxyethoxide, **5,** 620
Trifluoromethanesulfonic anhydride, **6,** 618
from Propargyl substrates + RM
1-Chloro-2-methyl-N,N-tetramethylenepropenylamine, **12,** 124
Grignard reagents, **6,** 143; **7,** 155, 163
Lithium acetylide, **8,** 285
Lithium 3-chloropropargylide, **5,** 397
Lithium dimethylcuprate, **2,** 151; **3,** 106; **7,** 120
(R)-1-(1-Naphthyl)ethyl isocyanate, **8,** 356
Organocopper reagents, **9,** 328
Tributyl(methylphenylamino)phosphonium iodide, **10,** 268
from Propargyl substrates by reduction
Chromium(III) chloride–Lithium aluminum hydride, **10,** 101
Disiamylborane, **3,** 22
Zinc, **4,** 574
Zinc–copper couple, **2,** 465
Other routes
Bis(cyclopentadienyl)titanacyclobutanes, **12,** 54
Carbon dioxide, **6,** 94
Catecholborane, **9,** 97
(3-Chloro-3-methyl-1-butynyl)lithium, **8,** 93
Dichlorobis(cyclopentadienyl)titanium–Trimethylaluminum, **9,** 488
Formaldehyde–Diisopropylamine–Copper(I) bromide, **9,** 225
Lithium aluminum hydride, **8,** 286
Lithium dimethylcuprate, **5,** 234
Lithium ethoxide, **1,** 612
Methyllithium, **2,** 274; **3,** 202
Organolithium reagents, **11,** 13
Potassium *t*-butoxide, **1,** 911
Propargyltriphenyltin, **12,** 415
Triethyl orthoacetate, **10,** 417
Trimethylaluminum, **8,** 506
Trimethylsilylmethylenetriphenylphosphorane, **5,** 723
1-Trimethylsilylpropynylcopper, **6,** 638
Triphenylphosphine, **1,** 1238
ALLENYL AMIDES
Potassium *t*-butoxide, **11,** 432

ALLENES, CUMULENES AND ALLENE DERIVATIVES (*Continued*)
ALLENYL ESTERS
Copper(I) chloride, **10,** 355
Silver perchlorate, **2,** 369
Silver tetrafluoroborate, **5,** 587
ALLENYL ETHERS
Potassium *t*-butoxide, **2,** 336
ALLENYL HALIDES
Copper(I) chloride, **8,** 119
Phosphorus(III) bromide, **2,** 330
Triphenyl phosphite dibromide, **1,** 1249
ALLENYL SULFUR COMPOUNDS
Benzenesulfenyl chloride, **9,** 35
Organolithium reagents, **10,** 3
CUMULENES
N,N-Dimethylformamide, **1,** 278
Diphosphorus tetraiodide, **1,** 349; **6,** 243
Methyllithium, **1,** 682
Organocopper reagents, **11,** 13
Potassium *t*-butoxide, **4,** 399
Sodium amide, **2,** 373
Thexylborane, **5,** 232
Zinc, **1,** 1276; **2,** 459; **7,** 426
ALLENE OXIDES
m-Chloroperbenzoic acid, **3,** 49
Peracetic acid, **2,** 307
Peracetic acid–Sodium acetate, **7,** 279
ALLENIC ALCOHOLS
α-ALLENIC ALCOHOLS
by Addition to carbonyls
Chromium(III) chloride–Lithium aluminum hydride, **9,** 119; **10,** 101
α-Lithio-α-methoxyallene, **9,** 272
Lithium 3-chloropropargylide, **9,** 279
Organotitanium reagents, **11,** 374
Tetrabutylammonium fluoride, **10,** 378
Tin(II) chloride, **11,** 521
by Cleavage of 1,3-diene monoxides
3,4-Epoxy-1-butyne, **5,** 484
Grignard reagents–Copper(I) halides, **5,** 167; **6,** 269
Lithium dimethylcuprate, **5,** 234
Other routes
Butyllithium, **5,** 80
3,4-Dihydro-2*H*-pyran, **3,** 99; **5,** 220
Formaldehyde, **9,** 225; **10,** 186
Formaldehyde–Dimethylaluminum chloride, **10,** 186
Lithium aluminum hydride, **5,** 382; **11,** 289
β-ALLENIC ALCOHOLS
3-Chloro-4,5-dihydrofuryl-2-copper, **11,** 119
Grignard reagents–Copper(I) halides, **6,** 147
Lithium aluminum hydride, **5,** 382
ALLENIC CARBONYL COMPOUNDS
α-ALLENIC CARBONYLS
Diethyl(2-chloro-1,1,2-trifluoroethyl)-amine, **4,** 149; **5,** 214
Diethyl(or Dimethyl)formamide diethyl acetal, **7,** 125; **8,** 169
N,N-Dimethylformamide, **8,** 189
Ethoxycarbonylmethylenetriphenyl-phosphorane, **10,** 78
Lithium N-isopropylcyclohexylamide, **4,** 306
Methyllithium, **7,** 242
Nickel carbonyl, **1,** 720
Nickel peroxide, **9,** 322
Thallium(III) nitrate, **4,** 492
Triethyl phosphonoacetate, **1,** 1217
Trimethylsilylketene, **6,** 635
β-ALLENIC CARBONYLS
Copper(I) iodide, **8,** 121
2-Methoxypropene, **2,** 230
Organocopper reagents, **10,** 282
Zinc, **4,** 574
ALLENYLSILANES (*see* SILANES)
ALLYLIC COMPOUNDS
ALLYLIC ACETATES
by Allylic acetoxylation (*see* TYPE OF REACTION INDEX)
by Displacements
Acetic anhydride + co-reagent, **5,** 470; **6,** 260
Lead(II) acetate, **2,** 233
Tetraalkylammonium acetate, **1,** 1136, 1142
Trioctylpropylammonium chloride, **6,** 640
Other routes
Acetic anhydride–Acetic acid, **6,** 1; **8,** 2
Lead tetraacetate, **1,** 537
4-Pyrrolidinopyridine, **4,** 416
Silver acetate, **2,** 362; **6,** 90
p-Toluenesulfonic acid, **5,** 673
ALLYLIC ALCOHOLS (*see also* CHIRAL COMPOUNDS)
by Addition to C=O
B-1-Alkenyl-9-borabicyclo[3.3.1]-

nonanes, **8,** 6
Allyl alcohol, O,2-dilithio derivative, **6,** 11
Benzeneselenol, **6,** 28, 85
α-Bromovinyltrimethylsilane, **8,** 56
t-Butyl hydroperoxide, **8,** 64
Butyllithium, **7,** 47
Chromium(II) chloride, **12,** 136
Dibromomethyllithium, **11,** 158
Diisobutylaluminum hydride, **4,** 158
(1-Lithiovinyl)trimethyl(or triphenyl)-silane, **5,** 368, 374, 375; **7,** 193; **9,** 498
Lithium diisopropylamide, **6,** 334; **9,** 280
Lithium naphthalenide, **9,** 284
3-Methoxyisoprene, **4,** 330
Nickel(II) acetylacetonate–Trimethylaluminum, **8,** 42; **9,** 52
Organocerium reagents, **12,** 345
1-Phenylsulfonyl-2-trimethyl-silylethane, **11,** 41
Phenyl vinyl selenide, **9,** 374
Potassium diisopropylamide–Lithium *t*-butoxide, **9,** 383
Potassium fluoride, **6,** 482
Tetrakis(triphenylphosphine)-palladium(0), **10,** 384
2,4,6-Triisopropylbenzene-sulfonylhydrazide, **9,** 486
Vinyllithium, **1,** 1273; **7,** 47
by Addition to α,β-unsaturated carbonyls
Boron trifluoride etherate, **12,** 66
Copper(I) chloride, **1,** 166
Dilithium tris(1-pentynyl)cuprate, **6,** 203
Grignard reagents, **1,** 415
Iodo(methyl)calcium, **5,** 442
Lithium, **1,** 570; **4,** 286
Organocopper reagents, **11,** 365
Organozirconium reagents, **12,** 358
by Allylic oxidation (*see* TYPE OF REACTION INDEX)
by Cleavage of 1,3-diene monoxides
1,3-Butadiene monoxide, **4,** 53; **5,** 483
Copper(I) bromide–Dimethyl sulfide, **10,** 104
Copper(I) chloride, **12,** 141
Diethylaluminum benzenethiolate, **10,** 281
Lithium alkyl(cyano)cuprates, **9,** 329; **10,** 287
Lithium dimethylcuprate, **3,** 106
Organolithium reagents, **10,** 3
Tetrakis(triphenylphosphine)-palladium(0), **11,** 503
Tris(tribenzylideneacetylacetone)-tripalladium(chloroform), **12,** 561
from Endoperoxides
Mercury(II) acetate, **1,** 644
Oxygen, singlet, **7,** 261
Palladium black, **4,** 365
Thiourea, **6,** 586
Zinc, **1,** 1276
by Isomerization of epoxides (*see* TYPE OF REACTION INDEX)
by Oxidation of allylsilanes (*see* TYPE OF REACTION INDEX)
by Reduction of 1,3-diene monoxides
Borane–Tetrahydrofuran, **9,** 136
Diisobutylaluminum hydride, **4,** 158
Lithium aluminum hydride, **7,** 196
Raney nickel, **10,** 339
by Reduction of propargyl alcohols, α,β-unsaturated carbonyls (*see* TYPE OF REACTION INDEX)
by Reductive cyclization of ynones
Potassium–Ammonia, **9,** 377
Sodium naphthalenide, **11,** 490
by 2,3-Sigmatropic rearrangement
Benzenesulfenyl chloride, **6,** 30; **9,** 35
Lithium diisopropylamide, **5,** 400; **6,** 334
Phenylseleneny! benzenesulfonate, **11,** 407
Potassium thiophenoxide, **8,** 420
Sodium benzeneselenoate, **8,** 447
by Wharton reaction
Hydrazine, **1,** 434; **8,** 245
Tributyltin hydride, **11,** 545
by Wittig reaction
Butyllithium, **11,** 101; **12,** 96
Potassium hexamethyldisilazide, **10,** 326
Other routes
Benzeneseleninic acid, **8,** 24
Butyllithium, **5,** 80
m-Chloroperbenzoic acid, **12,** 118
Copper(I) iodide, **12,** 141
Diethylzinc–Diiodomethane, **8,** 172
Ethylidenetriphenylphosphorane, **3,** 141

ALLYLIC COMPOUNDS (*Continued*)
Formic acid, **7,** 160
Hydrobromic acid, **4,** 249
Lithium dibutylcuprate, **3,** 79
Lithium 2,2,6,6-tetramethylpiperidide, **5,** 417
3-Methoxy-3-methylbutynylcopper, **8,** 324
Methylenetriphenylphosphorane, **11,** 338
Potassium superoxide, **6,** 488
Silver acetate, **6,** 511
Silver *p*-toluenesulfonate, **7,** 323
3-Tetrahydropyranyloxy-1-tributylstannyl-1-propene, **6,** 602
Trimethylstannylmethyllithium, **12,** 547
Trioctylpropylammonium chloride, **6,** 640
Triphenylphosphine, **4,** 548
Vinylcopper reagents, **6,** 662

ALLYLIC AMINES
by Allylic amination (*see* TYPE OF REACTION INDEX)
by Diels–Alder reaction
Benzyl *trans*-1,3-butadiene-1-carbamate, **8,** 34; **11,** 43
Ethyl *trans*-1,3-butadiene-1-carbamate, **7,** 147; **8,** 220
2,2,2-Trichloroethyl *trans*-1,3-butadiene-1-carbamate, **9,** 479
by Displacements
Chloramine-T, **9,** 367
N-Chlorosuccinimide, **12,** 121
4,4′-Dimethoxybenzhydrylamine, **9,** 174
Hexamethylenetetramine, **1,** 427
Tetrakis(triphenylphosphine)-palladium(0), **7,** 357; **9,** 451; **11,** 503; **12,** 468
Trichloroacetonitrile, **6,** 604; **7,** 381
by Wittig (and related) reactions
N-Alkyl-β-aminoethyltriphenylphosphonium bromides, **12,** 11
2-Dimethylaminoethylidenetriphenylphosphorane, **8,** 184
Vinyldiphenylphosphine oxide, **11,** 595
Vinyltriphenylphosphonium bromide, **11,** 596
Other routes
Diisobutylaluminum hydride, **9,** 171
Dimethyl(methylthio)sulfonium tetrafluoroborate, **11,** 204
Ethyl chloroformate, **12,** 223
Morpholine, **4,** 351
Palladium(II) chloride, **5,** 500
Sodium naphthalenide, **4,** 349; **5,** 468

ALLYLIC BENZOATES
Benzoic acid, **7,** 405
t-Butyl perbenzoate, **1,** 98; **4,** 66; **7,** 49

ALLYLIC ETHERS
by Allylic oxidation
Mercury(II) acetate, **6,** 358
Mercury(II) oxide–Tetrafluoroboric acid, **10,** 254
by Displacements
2,3-Bis(bromomethyl)-1,3-butadiene, **5,** 32
Dichlorobis(triphenylphosphine)-platinum(II)–Tin(II) chloride, **9,** 148
Ruthenium(III) chloride, **11,** 462
Other routes
Benzeneselenenyl bromide, **5,** 518
Chloromethyl ethyl ether, **7,** 112
Disiamylborane, **5,** 39
Grignard reagents, **6,** 594; **11,** 245
Mercury(II) acetate–Sodium trimethoxyborohydride, **10,** 253
Mercury(II) nitrate, **7,** 223
Silver perchlorate, **4,** 432; **10,** 354
Sodium borohydride, **7,** 329
Thallium(III) nitrate, **6,** 578
Vinyltriphenylphosphonium bromide, **2,** 456

ALLYLIC HALIDES
General methods
Triphenyl phosphite, **1,** 1249
Allylic fluorides
Diethyl(2-chloro-1,1,2-trifluoroethyl)-amine, **6,** 186
Allylic chlorides
2-Acetoxy-2-methylbutyryl chloride, **8,** 3
Boron trichloride, **6,** 65
α-Bromovinyltrimethylsilane, **7,** 193; **8,** 56
t-Butyl hypochlorite, **1,** 90
Chlorine oxide, **12,** 110
N-Chloro-N-cyclohexylbenzenesulfonamide, **2,** 67
N-Chlorosuccinimide + co-reagent, **4,** 87; **5,** 129; **6,** 608; **9,** 111, 367
1-Chloro-N,N,2-trimethyl-

propenylamine, **12,** 123
Disiamylborane, **3,** 22
Hypochlorous acid, **10,** 208
Lithium chloride, **6,** 233
Methanesulfonyl chloride, **4,** 326; **6,** 598; **7,** 225
Palladium(II) acetate, **12,** 367
p-Toluenesulfonyl chloride, **5,** 676
p-Toluenesulfonyl chloride–Lithium chloride, **3,** 152, 293; **5,** 677; **6,** 598
Trichloroisocyanuric acid, **3,** 297; **6,** 605
Triphenylphosphine–Carbon tetrachloride, **4,** 551; **6,** 644; **8,** 516
Triphenylphosphine–Hexachloroacetone, **8,** 516; **10,** 449
Allylic bromides
Benzeneselenenyl bromide, **5,** 518; **8,** 25; **10,** 16
N-Bromocaprolactam, **1,** 76
N-Bromosuccinimide, **1,** 78; **4,** 90; **8,** 54
Dibenzoyl peroxide, **1,** 196
1,3-Dibromo-5,5-dimethylhydantoin, **1,** 208
1,2-Dibromotetrachloroethane, **1,** 209
Methyl bromide–Lithium bromide, **10,** 262
N-Methyl-2-pyrrolidone, **11,** 346
Phosphorus(III) bromide, **1,** 873
Allylic iodides
Iodotrimethylsilane, **11,** 271
Methyl iodide, **5,** 447
2-Methylthio-2-thiazoline, **5,** 459
ALLYLIC HYDROPEROXIDES
Hydrogen peroxide, **8,** 247; **10,** 203
Oxygen, **10,** 293
Oxygen, singlet, **5,** 486; **6,** 431; **7,** 261; **8,** 367; **11,** 385
Potassium perchromate, **4,** 412
Sodium methoxide, **5,** 617
Triphenyl phosphite ozonide, **3,** 323
ALLYLIC PHOSPHORUS COMPOUNDS
(Diphenylphosphine)lithium, **6,** 340
Iron carbonyl, **5,** 96
ALLYLIC SELENIDES
Allyl phenyl selenide, **11,** 17
Benzeneselenenyl halides, **11,** 34
p-Nitrophenyl selenocyanate, **9,** 325
Phenylselenenyl benzenesulfonate, **11,** 407
N-Phenylselenophthalimide, **9,** 366
Sodium benzeneselenoate, **8,** 447
Tin(II) chloride, **11,** 521
Trimethylsilylmethyllithium, **11,** 581
ALLYLIC SILANES (*see also* CHIRAL COMPOUNDS)
by Coupling reactions
Allyltrimethylsilane, **11,** 16
Diethyl[dimethyl(phenyl)silyl]-aluminum, **12,** 342
Dimethylphenylsilyllithium, **12,** 210
Hydrogen peroxide, **12,** 242
Trimethylsilyllithium, **11,** 575; **12,** 538
Trimethylsilylmagnesium chloride, **11,** 58
Tris(trimethylsilyl)aluminum, **12,** 343
by Diels–Alder reactions
1-Acetoxy-4-trimethylsilyl-1,3-butadiene, **11,** 3
2,3-Bis(trimethylsilylmethyl)-1,3-butadiene, **11,** 64
1-Trimethylsilyl-1,3-butadiene, **7,** 395
2-Trimethylsilylmethyl-1,3-butadiene, **10,** 432
Other routes
(Chloromethyl)dimethylsilane, **12,** 27
Chlorotrimethylsilane, **5,** 709
Dodecamethylcyclohexasilane, **12,** 219
Hydrogen fluoride, **9,** 240
Methyl 2-trimethylsilylacrylate, **5,** 180
1-Phenylsulfonyl-2-trimethylsilylethane, **11,** 41
Tetrabutylammonium fluoride, **12,** 458
Tetrakis(triphenylphosphine)-palladium(0), **11,** 503
N,N,N′N′-Tetramethyl-ethylenediamine, **4,** 485
Tributyl(methylphenylamino)-phosphonium iodide, **11,** 342
[(Trimethylsilyl)allyl]lithium, **11,** 12
β-Trimethylsilylethylidene-triphenylphosphorane, **9,** 492
2-Trimethylsilylmethyl-enecyclobutane, **9,** 494
ALLYLIC SULFUR COMPOUNDS
Sulfides
Benzenesulfenyl chloride, **8,** 32
Diethylaluminum benzenethiolate, **10,** 281
Methyl cyanodithioformate, **7,** 237

ALLYLIC COMPOUNDS (*Continued*)
Potassium thiophenoxide, **8,** 420
Thiophenol, **10,** 399
p-Toluenesulfonic acid, **7,** 374
Tris(phenylthio)borane, **7,** 409
Sulfones
Benzenesulfinic acid, **9,** 132
Hexamethylphosphoric triamide, **10,** 196
Lithium–Ethylamine, **11,** 287
Sodium benzenesulfinate, **6,** 526
Tetrakis(triphenylphosphine)-palladium(0), **11,** 503
Sulfoxides
Benzenesulfenyl chloride, **6,** 30; **9,** 35
p-Toluenesulfinyl chloride, **11,** 8
Other S compounds
S-Allyl N,N-dimethyldithiocarbamate, **6,** 11
1,3-Dithienium tetrafluoroborate, **11,** 227
Methyl fluoride–Antimony(V) fluoride, **6,** 381
N-Sulfinylbenzenesulfonamide, **9,** 439
p-Toluenesulfonyl-S-methylcarbazate, **5,** 681
OTHER ALLYLIC COMPOUNDS
Lead(IV) acetate azides, **4,** 276; **5,** 363
Nitrosonium hexafluorophosphate, **9,** 326
Sodium hydride, **12,** 447
2-Trimethylstannylethylidene-triphenylphosphorane, **6,** 640
Tris(tetrabutylammonium)hydrogen pyrophosphate, **10,** 455
AMIDES
FROM RCOOH(X) + AMINES
RCOOH
Acetic anhydride, **1,** 3
Bis(*o*-nitrophenyl) phenylphosphonate, **10,** 41
N,N-Bis(2-oxo-3-oxazolidinyl)-phosphorodiamidic chloride, **10,** 41
Borane–Trimethylamine, **12,** 65
Boric acid, **11,** 70
Boron trifluoride etherate, **6,** 65
Catecholborane, **9,** 97
6-Chloro-1-*p*-chlorobenzene-sulfonyloxybenzotriazole, **6,** 106
2-Chloro-3-methyl-1,3-benzo-thiazolium trifluoromethane-sulfonate, **7,** 61
2-Chloro-2-oxobenzo-1,3,2-dioxaphosphole, **10,** 91
Dicyclohexylcarbodiimide, **1,** 231; **6,** 174
Diethylamidosulfurous acid, methyl ester, **1,** 1123
N,N-Diethylaminopropyne, **2,** 133
Diethyl phosphorobromidate, **8,** 170
Diethyl phosphorocyanidate, **5,** 217; **6,** 192; **7,** 107
5,6-Dihydrophenanthridine, **11,** 184
4-Dimethylaminopyridine, **9,** 178
N,N-Dimethylformamide, **2,** 153
Diphenyl 2-keto-3-oxazolinylphosphonate, **11,** 220
N-Ethoxycarbonyl-2-ethoxy-1,2-dihydroquinoline, **4,** 223
Ethyl chloroformate, **1,** 364
2-Ethyl-7-hydroxybenzisoxazolium tetrafluoroborate, **6,** 253
2-Fluoro-1,3,5-trinitrobenzene, **8,** 230
2-Halopyridinium salts, **7,** 110; **9,** 234
Ion-exchange resins, **1,** 511
Isopropyl isocyanate, **11,** 277
4(R)-Methoxycarbonyl-1,3-thiazolidine-2-thione, **11,** 323
N-Methyl-N-phenylbenzohydrazonyl bromide, **10,** 269
4-(4′-Methyl-1′-piperazinyl)-3-butyn-2-one, **8,** 348
o-Nitrophenyl thiocyanate, **9,** 325
Phenyl N-phenylphosphoramido-chloridate, **11,** 416
N-Phenylselenophthalimide, **10,** 312
Phosphonitrilic chloride trimer, **2,** 206
Sulfur trioxide–Dimethylformamide, **1,** 1125
Tetrachlorosilane, **3,** 277; **4,** 424
1,3-Thiazolidine-2-thione, **11,** 518
Thionyl chloride, **5,** 663
Triethoxydiiodophosphorane, **9,** 480
Triphenylbis(2,2,2-trifluoroethoxy)-phosphorane, **10,** 43
Triphenylphosphine bis(trifluoro-methanesulfonate), **6,** 648
Triphenylphosphine–Carbon tetrachloride, **9,** 503
Triphenyl phosphite, **4,** 556
Urea, **1,** 1262
Other RCOX

Ammonium acetate, **1,** 38
N,N-Bis(bromomagnesio)anilide, **4,** 33
Boron tribromide, **6,** 64
Chloro(methyl)aluminum amides, **11,** 121
Chlorotris(triphenylphosphine)-rhodium(I), **5,** 736
Dimethylaluminum amides, **8,** 182
Disodium tetracarbonylferrate, **7,** 341
Formamide, **2,** 201
Hexamethylphosphorous triamide, **4,** 247
Molecular sieves, **6,** 411
Morpholine, **6,** 429
2-Pyridone, **3,** 157
Sodium hydride, **4,** 456
Titanium(III) chloride, **10,** 400
N-(Trimethylsilyl)imidazole, **7,** 399

BY BECKMANN REARRANGEMENT
p-Acetamidobenzenesulfonyl chloride, **1,** 3
Boron trifluoride, **1,** 68
Dichloromethylenedimethylammonium chloride, **6,** 170
N,N-Dimethylformamide, **1,** 278
Formic acid, **1,** 404
Hexamethylphosphoric triamide, **5,** 323
Hydrazoic acid, **1,** 446
Iodine pentafluoride, **1,** 503
O-Mesitylenesulfonylhydroxylamine, **5,** 430
Polyphosphate ester, **3,** 229
Pyridinium chloride, **2,** 352
Silica, **11,** 466
p-Toluenesulfonyl chloride, **1,** 1179
Trifluoroacetic anhydride, **1,** 1221
Trimethylsilyl polyphosphate, **10,** 437; **11,** 427
Triphenylphosphine–Carbon tetrachloride, **5,** 727

BY HYDRATION OF RCN
Arene(tricarbonyl)chromium complexes, **10,** 13
[1,2-Bis(diphenylphosphine)ethane]-cyclohexyneplatinum(0), **5,** 171
Boron trifluoride, **1,** 68
Boron trifluoride–Acetic acid, **1,** 69
Chlorodiphenylcarbenium hexachloroantimonate, **5,** 115; **6,** 108
Copper(II) sulfate–Sodium borohydride, **11,** 142
Formic acid, **3,** 147; **5,** 316
Hydrogen peroxide, **1,** 466
Palladium(II) chloride, **6,** 447
Phase-transfer catalysts, **10,** 305
Potassium hydroxide, **7,** 303
Sodium methylsulfinylmethylide, **1,** 310
Sodium superoxide, **9,** 434
Vilsmeier reagent, **7,** 422

FROM ISOCYANATES
Methyllithium, **5,** 448
Triethylaluminum, **1,** 1197
Trimethylsilyl isocyanate, **6,** 634

BY OXIDATION OF AMINES (*see* TYPE OF REACTION INDEX)

FROM THIOAMIDES
Bis(*p*-methoxyphenyl) telluroxide, **9,** 50
m-Chloroperbenzoic acid, **12,** 118
Potassium superoxide, **11,** 442
Sodium nitrite, **11,** 491

OTHER ROUTES
Acetic anhydride, **2,** 7; **6,** 3
Aluminum chloride–Ethanol, **8,** 15
Benzeneselenenyl halides, **10,** 16
N-Bromosuccinimide, **9,** 70
Chloramine, **5,** 103
Chloroacetyl isocyanate, **6,** 634
Chlorobis(cyclopentadienyl)-hydridozirconium(IV), **11,** 119
Di-μ-carbonylhexacarbonyldicobalt, **1,** 224; **11,** 162
1,4-Dichloro-1,4-dimethoxybutane, **12,** 175
Dihalobis(triphenylphosphine)-palladium(II), **6,** 60
Dimethylketene, **1,** 290
Disodium tetracarbonylferrate, **4,** 461
Formamide, **1,** 212, 402
Lead tetraacetate–Trifluoroacetic acid, **6,** 317
Lithium bis(N,N-dialkylcarbamoyl)-cuprate, **9,** 279; **11,** 373
Lithium tricarbonyl(dimethyl-carbamoyl)nickelate, **4,** 302
Mercury(II) acetate, **3,** 194
Mercury(II) nitrate, **11,** 317
Methoxy(phenylthio)trimethylsilyl-methyllithium, **12,** 317
Nickel(II) acetate, **1,** 718
Nickel carbonyl, **3,** 210
Nickel peroxide, **1,** 731
Nitronium tetrafluoroborate, **4,** 358

AMIDES (*Continued*)
Phase-transfer catalysts, **11,** 403
Phosphoryl chloride, **1,** 876
Polyphosphoric acid, **1,** 894; **2,** 334
Sodium amide, **6,** 525
Sulfur, **1,** 1118
p-Toluenesulfonyl chloride, **6,** 598
METHODS SPECIFIC FOR
Acetamides
Acetic anhydride–Phosphoric acid, **6,** 3
Acetonitrile, **9,** 324
N-Acetoxyphthalimide, **1,** 9
2-Acetoxypyridine, **1,** 9
N-Acetylimidazde, **1,** 13
Bismuth(III) acetate, **4,** 40
Borane–Trimethylamine, **1,** 1229
Iodine azide, **10,** 211
Ketene, **1,** 528
Lead tetraacetate, **1,** 537
Lithium tricarbonyl(dimethylcarbamoyl)nickelate, **4,** 302
Manganese(III) acetate, **6,** 355
Peroxyacetyl nitrate, **5,** 510
Phase-transfer catalysts, **9,** 356
Silver perchlorate, **5,** 585
1,3,4,6-Tetraacetylglycouril, **6,** 563
Benzamides
Benzoic anhydride, **5,** 23
S-Benzoic O,O-diethyl phosphorodithioic anhydride, **6,** 34
2-Benzoylthio-1-methylpyridinium chloride, **9,** 40
sec-Butyllithium, **12,** 97
2,6-Di-*t*-butyl-*p*-benzoquinone, **9,** 139
Diethyl benzoylphosphonate, **11,** 178
Formamides
Acetic–formic anhydride, **2,** 10
Chromium(VI) oxide–Pyridine, **2,** 74
Ethyl formate, **1,** 380
Formic acid, **1,** 404
Formic acid–Formamide, **1,** 407
N-Formylimidazole, **1,** 407
Hydrogen selenide–Triethylamine, **5,** 341
Manganese(IV) oxide, **1,** 637; **2,** 257
p-Nitrophenyl formate, **1,** 744
Triethyl orthoformate, **1,** 1204
Triethylsilane, **5,** 694
Trimethylacetic formic anhydride, **11,** 567

AMIDINES
Aluminum chloride, **1,** 24
t-Butyl azidoformate, **6,** 77
3-(Dimethylamino)-2-azaprop-2-en-1-ylidenedimethylammonium chloride, **11,** 194
N,N-Dimethylthioformamide, **3,** 128
Diphenyl-N-*p*-tolylketenimine, **5,** 282
Ferric chloride, **5,** 307
Hexamethylphosphoric triamide, **4,** 244
Iron carbonyl, **6,** 304
Phosphorus(V) chloride, **1,** 866
Pyridine, **1,** 958
Silver chloride, **10,** 347
Triethyl orthoformate, **1,** 1204

AMINES (*see also* AMINO..., ANILINES, CHIRAL COMPOUNDS, ENAMINES)
GENERAL METHODS
by Alkylation of NH_3, amines
Aluminum *t*-butoxide–Raney nickel, **8,** 13
Dichlorotris(triphenylphosphine)-ruthenium(II), **10,** 141
Dicyclohexylethylamine, **1,** 370
Dihydridotetrakis(triphenylphosphine)ruthenium(II), **11,** 182
Diisopropylethylamine, **1,** 371
Lithium amide, **1,** 600
Lithium naphthalenide, **2,** 288
N,N-Methylphenylaminotriphenylphosphonium iodide, **6,** 392
Sodium hydride, **1,** 1075
Triethyloxonium tetrafluoroborate, **1,** 1210
Triethyl phosphate, **1,** 1212
by Reduction of RCONRR′, C=N (*see* TYPE OF REACTION INDEX)
by Reductive amination (*see* TYPE OF REACTION INDEX)
Other routes
Aluminum hydride, **2,** 23
Diethylaluminum iodide, **12,** 5
Grignard reagents, **1,** 415; **11,** 245
Hexachlorodisilane, **10,** 195
Lithium aluminum hydride, **5,** 382
Palladium black, **5,** 498; **6,** 443
Potassium tetracarbonylhydridoferrate, **6,** 483; **8,** 266
Raney nickel, **7,** 312
Sodium naphthalenide, **4,** 349

Tin(IV) chloride, **11,** 522
Zinc, **4,** 574

PRIMARY AMINES
by Alkylation of N compounds with RX
N-Benzylhydroxylamine, **9,** 42
N-Benzyl trifluoromethanesulfonamide, **5,** 29; **6,** 43
Bisbenzenesulfenimide, **3,** 20
Di-*t*-butyl iminodicarboxylate, **1,** 210
Hydrazine, **1,** 434
Lithium bisbenzenesulfenimide, **3,** 182
Potassium phthalimide, **7,** 166
Sodium hexamethyldisilazide, **12,** 441
Trifluoromethanesulfonic anhydride, **5,** 702
Triphenylphosphine–Diethyl azodicarboxylate, **4,** 553
Urea, **1,** 1262
by Amination of RH
Chloramine, **5,** 103
Trichloramine, **1,** 1193; **2,** 424; **3,** 295
by Amination of RM
Azidomethyl phenyl sulfide, **12,** 37
O-(Diphenylphosphinyl)-hydroxylamine, **11,** 221
Methoxyamine, **11,** 322
by Hydroamination of alkenes
Chloramine, **1,** 122
Chloramine-T–Diphenyl diselenide, **9,** 101
Dimethyl(methylthio)sulfonium tetrafluoroborate, **11,** 204
Ethyllithium, **5,** 306
Hydroxylamine-O-sulfonic acid, **1,** 481
Mercury(II) nitrate, **11,** 317
O-Mesitylenesulfonylhydroxylamine, **5,** 430
by Hofmann degradation
Iodosylbenzene, **12,** 258
Lead tetraacetate, **6,** 313
Phenyliodine(III) bis(trifluoroacetate), **9,** 54
by Reduction of $RCONH_2$, RN_3, C = N, RCN, RNO_2, C = NOH (*see* TYPE OF REACTION INDEX)
by Reductive amination (*see* TYPE OF REACTION INDEX)
by Schmidt (or Curtius) reaction
Azidotrimethylsilane, **10,** 14
Diphenyl phosphoroazidate, **11,** 222
Hydrazoic acid, **1,** 446
Sodium azide, **1,** 1041; **2,** 376; **4,** 440
Trifluoroacetic acid, **12,** 529
Other routes
N-Benzenesulfonylformimidic acid ethyl ester, **3,** 18
Borane–Tetrahydrofuran, **12,** 65
N-(Diphenylmethylene)methylamine, **8,** 210
Ethylene glycol, **1,** 375
Lithium aluminum hydride, **8,** 286
Methoxymethylbis(trimethylsilyl)-amine, **12,** 62

SECONDARY AMINES
ω-Acetophenonesulfonyl chloride, **3,** 221
t-Butyldimethylsilyl trifluoromethanesulfonate, **12,** 86
Dichlorophenylborane, **4,** 377
Diethoxycarbenium tetrafluoroborate, **4,** 144
Diisobutylaluminum hydride, **12,** 191
N,O-Dimethylhydroxylamine, **12,** 205
Dimethyl sulfate, **1,** 293
Lithium–Methylamine, **4,** 288
Methoxyacetonitrile, **5,** 436
N-Methyl-N-trimethylsilylmethyl-N′-*t*-butylformamidine, **11,** 347
Molybdenum carbonyl, **3,** 206
Organolithium reagents, **12,** 14
Phase-transfer catalysts, **8,** 387
Potassium–Graphite, **9,** 378
Potassium hydride, **9,** 387
Titanium(IV) chloride, **5,** 671
Titanium(IV) chloride–Sodium borohydride, **10,** 404
Trialkylaluminums, **11,** 539
Triethyloxonium tetrafluoroborate, **3,** 303; **5,** 691
Triphenylbis(2,2,2-trifluoroethoxy)-phosphorane, **10,** 43

TERTIARY AMINES
by Alkylation reactions
Dicyclohexylethylamine, **2,** 195
Formaldehyde, **8,** 231
Iodobenzene, **1,** 505
Piperidine, **1,** 886
by Reductive amination (*see* TYPE OF REACTION INDEX)
Other routes
Dimethyl(methylene)ammonium salts, **7,** 130

AMINES (*Continued*)
Hexamethylphosphoric triamide, **4,** 244
Iron carbonyl, **4,** 268
Lithium aluminum hydride, **6,** 325
Mercury(II) acetate, **4,** 319
Organotitanium reagents, **11,** 374
Rhodium(III) oxide, **4,** 420
Sodium methoxide, **11,** 489
Sodium nitrite, **11,** 491
Vilsmeier reagent, **1,** 284
AMINE OXIDES
t-Amyl hydroperoxide, **4,** 20
N-Benzoylperoxycarbamic acid, **6,** 35
3-Bromo-4,5-dihydro-5-hydroperoxy-4,4-dimethyl-3,5-diphenyl-3H-pyrazole, **11,** 76
t-Butyl hydroperoxide–Vanadyl acetylacetonate, **2,** 456; **3,** 331
Caro's acid, **8,** 79
m-Chloroperbenzoic acid, **3,** 49
Hydrogen peroxide, **1,** 457
2-Hydroperoxyhexafluoro-2-propanol, **10,** 206
Peracetic acid, **1,** 785; **4,** 372
Perdichloromaleic acid, **5,** 511
o-Sulfoperbenzoic acid, **4,** 469
Trifluoroperacetic acid, **1,** 821
AMINO ACIDS, AMINO ESTERS, AND DERIVATIVES (*see also* CHIRAL COMPOUNDS, PEPTIDES)
α-AMINO ACIDS (*see also* CHIRAL COMPOUNDS)
Barium hydroxide, **12,** 38
t-Butyldimethylchlorosilane, **11,** 88
t-Butyl isocyanide, **10,** 67
Diethyl carbonate, **5,** 213
Diphenyl phosphoroazidate, **5,** 280
Ethyl 3-bromo-2-hydroxyimino-propanoate, **9,** 212
Hippuric acid, **1,** 432
Hydrobromic acid, **1,** 450
Lithium cyanoborohydride, **3,** 183
Magnesium methyl carbonate, **1,** 631
Mercury(II) acetate, **1,** 644
Methyl methylthiomethyl sulfoxide, **5,** 456
Methyl nitrite, **1,** 691
Palladium catalysts, **4,** 368
Phase-transfer catalysts, **8,** 387; **11,** 403
Phenyl isocyanide, **10,** 309
Potassium tetracarbonylhydridoferrate, **6,** 483
Raney nickel, **5,** 570
Ruthenium tetroxide, **6,** 504
L-(−)-Serine, **12,** 430
Trichloroacetonitrile, **12,** 526
β-AMINO ACIDS
Hydroxylamine, **1,** 478
Iodine, **11,** 261
Methoxymethylbis(trimethylsilyl)amine, **12,** 62
Palladium hydroxide, **9,** 353
Ruthenium tetroxide, **6,** 504
BOC-AMINO ACIDS
t-Amyl chloroformate, **3,** 15
Boron tribromide, **5,** 49
1-*t*-Butoxycarbonyl-4-dimethyl-aminopyridinium salts, **4,** 173; **8,** 57
t-Butoxycarbonylimidazole, **1,** 98
1-(*t*-Butoxycarbonyloxy)benzotriazole, **12,** 83
2-*t*-Butoxycarbonyloxyimino-2-phenylacetonitrile, **6,** 91; **10,** 61
N-(*t*-Butoxycarbonyloxy)succinimide, **2,** 42
1-*t*-Butoxycarbonyl-1,2,4-triazole, **4,** 65
t-Butyl azidoformate, **1,** 84; **2,** 44; **3,** 36; **4,** 54; **6,** 77
t-Butylcarbonic diethylphosphoric anhydride, **2,** 46
t-Butyl chloroformate, **2,** 48
t-Butyl cyanoformate, **1,** 87
t-Butyl fluoroformate, **2,** 48, 54; **4,** 65
t-Butyl methyl iminodicarboxylate potassium salt, **8,** 72
t-Butyl *p*-nitrophenyl carbonate, **1,** 97
t-Butyl pentachlorophenyl carbonate, **2,** 54
t-Butyl phenyl carbonate, **5,** 86
t-Butyl 2-pyridyl carbonate, **12,** 102
t-Butyl 8-quinolyl carbonate, **4,** 66; **5,** 87
t-Butyl 2,4,5-trichlorophenyl carbonate, **2,** 55
Di-*t*-butyl dicarbonate, **4,** 128; **7,** 91; **10,** 122; **12,** 159
4,6-Diphenylthieno[3,4-*d*][1,3]-dioxol-2-one 5,5-dioxide, **9,** 202
Formic acid, **2,** 202
Thioanisole–Trifluoroacetic acid, **10,** 398
Toluenesulfonic acid, **6,** 597

Triethylsilane, **2,** 433
Trifluoroacetic acid, **5,** 695
Trifluoromethanesulfonic acid, **5,** 701
Trimethylsilyl perchlorate, **6,** 639
OTHER N-PROTECTED AMINO ACIDS
Acetic-formic anhydride, **1,** 4; **2,** 10
1-Adamantyl chloroformate, **1,** 15
t-Amyl chloroformate, **1,** 40; **2,** 24, 307; **3,** 15
Benzhydryl azidoformate, **1,** 48
5-Benzisoxazolemethylene chloroformate, **6,** 32
Benzyl chloroformate, **1,** 109; **2,** 59
Benzylthiocarbonyl chloride, **2,** 26
[2-(*p*-Biphenyl)isopropyl]phenyl carbonate, **2,** 179
Boron tribromide, **5,** 49
p-Bromobenzyl chloroformate, **1,** 76
t-Butyl 2,4,5-trichlorophenyl carbonate, **2,** 55
Chloroacetic anhydride, **1,** 129
Chloroacetyl chloride, **1,** 130
2-[(Chloroformyl)oxy]ethyltriphenylphosphonium chloride, **8,** 92
5-Chloro-2-hydroxybenzophenone, **5,** 118
5-Chlorosalicylaldehyde, **1,** 139
Cyano-*t*-butoxychloroformate, **5,** 168
Cyclopentyl chloroformate, **1,** 182
Dicyclohexylcarbodiimide, **2,** 126
p-Dihydroxyborylbenzyloxycarbonyl chloride, **6,** 194
Diketene, **1,** 264
Dimedone, **1,** 266
3,5-Dimethoxybenzyl *p*-nitrophenyl carbonate, **2,** 142
4-Dimethylaminopyridine, **3,** 118
Diphenyl-4-pyridylmethanol, **7,** 139
Ethoxyacetylene, **2,** 190
Ethoxycarbonylphthalimide, **1,** 111; **5,** 91; **9,** 212; **12,** 221
S-Ethyl trifluorothioacetate, **3,** 143
9-Fluorenylmethyl chloroformate, **3,** 145; **4,** 237; **8,** 230
Furfuryl chloroformate, **1,** 408
Hydrazine acetate, **1,** 445
Isobornyl chloroformate, **5,** 359
p-Methoxybenzyloxycarbonyl azide, **1,** 668
o-Methoxycarbonylbenzoyl chloride, **5,** 91
Methyl trifluoroacetate, **7,** 246
o-Nitrobenzenesulfenyl chloride, **1,** 745; **4,** 359
p-Nitrobenzyl chloroformate, **1,** 737
p-Nitrophenyl acetate, **2,** 297
o-Nitrophenylsulfenyl chloride, **4,** 359
Palladium black, **6,** 443
2,4-Pentanedione, **1,** 10
Phenylhydrazine, **1,** 838
Phenyl [2-(*p*-phenylazophenyl)-isopropyl] carbonate, **7,** 285
Phenyl trifluoroacetate, **1,** 850
Phthalic anhydride, **1,** 882, 1200
p-Toluenesulfonyl chloride, **1,** 1179
2-*p*-Tolylsulfonylethyl chloroformate, **1,** 1185
2,2,2-Trichloro-*t*-butoxycarbonylchloride, **8,** 499
1,1,1-Trichloro-3,3,3-trifluoroacetone, **6,** 606
Triphenylmethanesulfenyl chloride, **1,** 1261
Triphenylmethyl chloride, **1,** 1254
S-PROTECTED AMINO ACIDS
Acetamidomethanol, **2,** 4
Benzyl chloromethyl sulfide, **2,** 27
Diethyl methylenemalonate, **3,** 96
Diphenyl-4-pyridylmethanol, **7,** 139
p-Methoxybenzyl chloride, **1,** 668
Methoxymethyl isocyanate, **5,** 439
2-Pyridinesulfenyl dioxide, **9,** 397
Triphenylmethyl chloride, **5,** 741
AMINO ACID ESTERS USED FOR PROTECTION
General methods
Cesium carbonate, **8,** 81
O-(Diphenylphosphinyl)-hydroxylamine, **11,** 221
Ethyl acetoacetate, **12,** 222
o-Nitrobenzenesulfenyl chloride, **11,** 360
Phase-transfer catalysts, **9,** 356
Sulfuryl chloride, **1,** 1128
Vilsmeier reagent, **8,** 186
Benzyl esters
Benzyldimethylanilinium hydroxide, **6,** 37
p-Chlorobenzyl alcohol, **7,** 59
p-Methoxyphenacyl bromide, **5,** 439
p-Nitrobenzyl bromide, **1,** 736
Pentamethylbenzyl chloride, **2,** 305

AMINO ACIDS, AMINO ESTERS, AND DERIVATIVES (*Continued*)
Phenacyl bromide, **2,** 316
Phosgene, **1,** 856
4-Picolyl chloride hydrochloride, **4,** 391
t-Butyl esters
t-Butyl acetate, **1,** 82
Isobutene, **1,** 522
Methyl esters
2,2-Dimethoxypropane, **1,** 268
Dimethyl sulfite, **1,** 296
Ion-exchange resins, **1,** 511
Thionyl chloride, **1,** 1158; **2,** 412
Other esters
2-Bromoethanol, **11,** 77
N-Chloromethylphthalimide, **1,** 135
Hydrogen chloride, **5,** 335
2,2,2-Trichloroethanol, **4,** 521
Triethyloxonium tetrafluoroborate, **3,** 303
AMINO ACID ESTERS USED FOR PEPTIDE COUPLING
Bis(4-nitrophenyl) carbonate, **1,** 330
t-Butyl 2,4,5-trichlorophenyl carbonate, **2,** 55
Chloroacetonitrile, **1,** 129
N,N-Dimethylformamide, **1,** 278
Di-*p*-nitrophenyl sulfite, **2,** 176
Diphenyldiazomethane, **1,** 338; **2,** 178; **6,** 234
4-(Methylthio)phenol, **2,** 281
o-Nitrophenol, **5,** 477
p-Nitrophenol, **1,** 743; **2,** 297
p-Nitrophenyl trifluoroacetate, **1,** 745
Pentachlorophenol, **1,** 782; **2,** 305
Phosphonitrilic chloride trimer, **6,** 469
2,4,5-Trichlorophenol, **1,** 1196
N-Trifluoroacetylimidazole, **1,** 1227
Vinyl acetate, **1,** 1271
α-SUBSTITUTED α-AMINO ACIDS (*see also* CHIRAL COMPOUNDS)
α-Alkyl
Bis(chlorodimethylsilyl)ethane, **10,** 140
Diisopropylethylamine, **10,** 151
Dimethylformamide dimethyl acetal, **8,** 191
Lithium diethylamide, **11,** 295
Phase-transfer catalysts, **11,** 403
α-Halo
(Diethylamino)sulfur trifluoride, **11,** 177
DEHYDRO AMINO ACIDS
Butyl N-(*p*-toluenesulfonyl)-iminoacetate, **7,** 50
Cyanotrimethylsilane, **12,** 148
(Diethylamino)sulfur trifluoride, **12,** 183
N,N′-Disuccinimidyl carbonate, **11,** 226
Hydrogen chloride, **8,** 246
Oxygen, singlet, **12,** 363
Triphenylphosphine–Diethyl azodicarboxylate, **9,** 504
AMINO ALCOHOLS (*see also* CHIRAL COMPOUNDS)
1,2-AMINO ALCOHOLS
by Additions to C=O
sec-Butyllithium–Potassium *t*-butoxide, **12,** 99
N-(Diphenylmethylene)methylamine, **8,** 210
Lithio-N,N-dimethylthiopivalamide, **7,** 190
Lithium aluminum hydride, **9,** 444
Lithium 2,2,6,6-tetramethylpiperidide, **7,** 213
Sodium borohydride, **9,** 420
Tetrabutylammonium fluoride, **10,** 378
2,4,6-Tri-*t*-butylphenoxy-N,N-dimethylcarbamate, **8,** 494
from Allylic amines or alcohols
Bromonium dicollidine perchlorate, **12,** 78
Iodine, **12,** 253
Iodonium di-*sym*-collidine perchlorate, **10,** 212; **11,** 269
Thallium(III) acetate, **9,** 459
Trichloroacetonitrile, **12,** 526
from Epoxide cleavage
Alumina, **9,** 8
Benzylamine, **1,** 51
Cyanotrimethylsilane, **11,** 147; **12,** 148
Dialkylaluminum amides, **10,** 117
Dimethylsulfoxonium methylide, **5,** 254
from Oxyamination
Mercury(II) oxide–Tetrafluoroboric acid, **10,** 254
Osmium tetroxide + co-reagent, **7,** 256; **8,** 365
Trioxo(*t*-butylimido)osmium(VIII), **6,** 641; **10,** 445
from Reduction of α-amino carbonyls

Aluminum chloride, **10,** 9
(R)-1-(S)-1′,2-Bis(diphenylphosphine)-ferrocenylethanol, **11,** 237
Borane–Dimethyl sulfide, **12,** 64
Dimethylphenylsilane, **12,** 209
Dimethyl sulfoxide–Sulfur trioxide, **11,** 216
Isobutylaluminum dichloride, **6,** 307
Phenol, **1,** 828
Sodium bis(2-methoxyethoxy)-aluminum hydride, **5,** 596
from Reduction of cyanohydrins
Chlorotrimethylsilane, **4,** 537
Cyanotrimethylsilane, **4,** 542
Diborane, **5,** 184
(2R,4R)-Pentanediol, **12,** 375
Other routes
3-Acetyl-4-oxazoline-2-one, **9,** 2
Diphenyl-N-*p*-tolylketenimine, **5,** 282
Iodine isocyanate, **3,** 161
Lead tetraacetate, **4,** 278
Organolithium reagents, **10,** 3
Rhodium catalysts, **6,** 503
Sodium cyanoborohydride, **9,** 424
1,3-AMINO ALCOHOLS
Alkylaluminum halides, **12,** 5
Borane–Tetrahydrofuran, **11,** 156
Lithium aluminum hydride, **10,** 236; **11,** 289; **12,** 272
Lithium–Methylamine, **6,** 322
Phenyl isocyanate, **10,** 309
p-Tolyl vinyl sulfoxide, **11,** 538
AMINO CARBONYL COMPOUNDS (*see also* AMINO ACIDS)
α-AMINO CARBONYL COMPOUNDS
α-Amino aldehydes
t-Butyl hypochlorite, **1,** 90
Chloramine-T, **9,** 101
Diisobutylaluminum hydride, **6,** 198
Dimethyl sulfoxide, **11,** 214
Dimethyl sulfoxide–Sulfur trioxide, **11,** 216
α-Amino ketones
Aluminum chloride, **10,** 9
Chromium(II) chloride, **3,** 60
Hexamethylenetetramine, **1,** 427
Lithium diisopropylamide, **11,** 296
O-Mesitylenesulfonylhydroxylamine, **5,** 430
N-Morpholinomethyldiphenyl-phosphine oxide, **11,** 352
Silver carbonate–Celite, **4,** 425
Sodium methoxide, **1,** 1091
2,4,4,6-Tetramethyl-5,6-dihydro-1,3-(4*H*)-oxazine, **10,** 391
Zinc–Acetic acid, **12,** 568
β-AMINO CARBONYL COMPOUNDS
by Conjugate addition
Alumina, **10,** 8
Aziridine, **11,** 23
Dimethylamine, **7,** 119
Oxalic acid, **1,** 764
by Mannich reaction (*see* TYPE OF REACTION INDEX)
Other routes
Iodine–Potassium iodide, **4,** 260
Sodium bis(2-methoxyethoxy)-aluminum hydride, **5,** 596
AMINO NITRILES (*see also* CHIRAL COMPOUNDS)
α-AMINO NITRILES
Cyanotrimethylsilane, **6,** 632; **12,** 148
Diethyl phosphorocyanidate, **9,** 163
Formaldehyde, **1,** 397
α-Methylbenzylamine, **8,** 393; **12,** 319
Sodium hydrogen sulfite, **1,** 1047
Tributylphosphine–Diphenyl disulfide, **12,** 514
β-AMINO NITRILES
Acrylonitrile, **1,** 14
Boron trichloride, **9,** 62
3-Bromopropionitrile, **1,** 77
Copper(II) acetate, **1,** 159
N-CYANO COMPOUNDS
Cyanogen bromide, **5,** 169
ANHYDRIDES
FROM RCOOH OR RCOX
2-Benzoylthio-1-methylpyridinium chloride, **9,** 40
N,N-Bis(2-oxo-3-oxazolidinyl)-phosphorodiamidic chloride, **11,** 57
Carbon disulfide, **1,** 114
Cyanogen bromide, **5,** 169
trans-1,2-Dibenzoylethylene, **1,** 195
Dicyclohexylcarbodiimide, **1,** 231
Dicyclopentadienylcobalt–Oxygen adduct, **5,** 145
N,N-Diethylaminopropyne, **2,** 133
Dimethylketene, **1,** 290
Ethoxy(trimethylsilyl)acetylene, **12,** 221
Ethyl chloroformate, **1,** 364
Ketene, **1,** 528

ANHYDRIDES (*Continued*)
Methoxyacetylene, **1,** 668; **2,** 270
Phenyl cyanate, **2,** 320
Phenyl N-phenylphosphoramidochloridate, **10,** 310
Phosgene, **1,** 856; **3,** 225
Pyridine, **1,** 958
Sodium bicarbonate, **3,** 260
Sodium ethoxide, **1,** 1065
Sulfene, **12,** 454
Thallium(I) ethoxide, **2,** 407
Thionyl bromide, **5,** 663
Thionyl chloride, **2,** 412
Triethylamine, **2,** 427
Trifluoroacetic anhydride, **6,** 616
N-Trifluoroacetylimidazole, **1,** 1227
METHODS SPECIFIC FOR MIXED ANHYDRIDES
2-Benzoylthio-1-methylpyridinium chloride, **9,** 40
Thallium(I) ethoxide, **2,** 407
Thallium(III) trifluoroacetate, **12,** 481
Trifluoroacetic anhydride, **1,** 1221; **9,** 484
OTHER ROUTES
Arylthallium bis(trifluoroacetates), **10,** 300
Nitric acid, **5,** 474
Perbenzoic acid, **1,** 791
Thallium(III) trifluoroacetate, **12,** 481
Triethyl phosphite, **1,** 1212
Vanadyl acetylacetonate, **11,** 47
ANILINES
FROM RCOX BY DEGRADATION
Azidotrimethylsilane, **7,** 394
Hydroxylamine, **1,** 903
Lead tetraacetate, **2,** 234
Sodium azide, **1,** 901, 1225
Sodium hypobromite, **1,** 1083
BY AMINATION OF ArM
Ammonia, **1,** 1290
Azidomethyl phenyl sulfide, **10,** 14
Benzeneseleninic anhydride, **8,** 29; **11,** 37
Diphenyl phosphoroazidate, **12,** 217
Hydroxylamine, **1,** 478; **12,** 251
Hydroxylamine-O-sulfonic acid, **1,** 481
Silver carbonate, **12,** 432
Trichloramine, **2,** 424
Trimethylsilylmethyl azide, **12,** 538
FROM ArX
Aluminum chloride, **6,** 17
Azidomethyl phenyl sulfide, **12,** 37
Carbonylphenylbis(triphenylphosphine)rhodium(I), **5,** 45
Copper(I) chloride, **1,** 166
Copper(I) hexamethyldisilazide, **6,** 57
N,N-Dimethylformamide, **4,** 184
Phenylsodium, **1,** 848
Sodium amide, **4,** 439
Urea, **1,** 1262
FROM ArOH
4-Chloro-2-phenylquinazoline, **4,** 86
Diethyl phosphorochloridate, **4,** 152; **5,** 217
Potassium amide, **4,** 398
BY REDUCTION OF $ArNO_2$, ArN_3 (*see* TYPE OF REACTION INDEX)
OTHER ROUTES
Acetic anhydride–Phosphoric acid, **6,** 3
Bismuth(III) chloride, **7,** 24
t-Butyl azidoformate, **2,** 44
Copper(I) iodide, **8,** 121
5,5-Dimethylhydantoin, **7,** 126
Formaldehyde, **2,** 200; **12,** 232
Hexamethylphosphoric triamide, **8,** 240
Oxalyl chloride, **5,** 481
Piperidine, **1,** 1039
Rhodium(III) chloride, **8,** 435
Sodium dithionite, **1,** 1081
Tetramethylene sulfone, **5,** 651
N-*p*-Tolylvinylmethylketenimine, **5,** 683
Triethyl orthoformate, **5,** 690
ARENES (*see also* ANILINES, ARYL HALIDES, BIARYLS, PHENOLS, TYPE OF REACTION INDEX: DEDIAZOTIZATION, DEHYDROXYLATION, REDUCTION OF ArX)
ALKYLBENZENES
by Coupling of Ar,R
Arene(tricarbonyl)chromium complexes, **6,** 27
[1,1′-Bis(diphenylphosphine)ferrocene](dichloro)palladium(II), **12,** 171
[1,3-Bis(diphenylphosphine)propane](dichloro)nickel(II), **11,** 167
Butyllithium, **1,** 95
Dibromobis(triethylphosphine)nickel(II), **6,** 57
Grignard reagents, **1,** 415; **6,** 50
Lithium dimethylcuprate, **2,** 151; **3,** 106

Lithium trimethylcobaltate and related reagents, **3,** 112
Methyltris(triphenylphosphine)-rhodium(I), **5,** 463
Nickel(II) acetylacetonate, **11,** 58
Nickel carbonyl, **4,** 353
Pentafluorophenylcopper, **5,** 504
Tetrahydrofuran, **6,** 570
by Friedel–Crafts alkylation (*see* TYPE OF REACTION INDEX)
Other routes
t-Butyl trifluoroacetate, **5,** 87
Di-*t*-butyl peroxide, **1,** 211
Dimethylbromonium hexafluoroantimonate, **5,** 231
Dimethyl sulfide, **4,** 190
Dimethylsulfonium methylide, **2,** 169
Nitrosonium hexafluorophosphate, **1,** 747
Potassium *t*-butoxide, **6,** 228
Sodium methylsulfinylmethylide, **2,** 166
ANTHRACENES
Aluminum cyclohexoxide, **4,** 15
Diborane, **1,** 199; **2,** 117
Lithium 2,2,6,6-tetramethylpiperidide, **6,** 345
Phenyliodine(III) bis(trifluoroacetate), **6,** 301
Polyphosphoric acid, **6,** 474
Potassium *t*-butoxide, **1,** 911
Selenium, **1,** 990
Sodium borohydride, **5,** 597
Zinc, **1,** 1276; **8,** 532
AZULENES
Copper(I) chloride, **5,** 164
Dimethyl acetylenedicarboxylate, **1,** 272; **4,** 168
Ethyl diazoacetate, **1,** 367
Sulfur, **3,** 273
BENZENES AND GENERAL METHODS FOR ARENES
from Alkynes
Bis(benzonitrile)dichloro-palladium(II), **1,** 56; **7,** 95
Bis(trimethylsilyl)acetylene, **6,** 153; **7,** 94; **9,** 142
Chlorotris(triphenylphosphine)-rhodium(I), **11,** 130
Dicarbonylbis(triphenylphosphine)-nickel, **1,** 61
Di-μ-carbonyldecacarbonyltri-*triangulo*-iron, **7,** 144
Di-μ-carbonylhexacarbonyldicobalt, **1,** 224
Di-μ-chlorotetra(ethylene)dirhodium, **5,** 203
Nickel cyanide, **4,** 356
Titanium(IV) chloride–Triethylaluminum, **1,** 1170
by Dehydrogenation (*see* TYPE OF REACTION INDEX)
by Diels–Alder routes
1-Acetoxybutadiene, **1,** 7
2-Acetoxy-1-methoxy-3-trimethyl-silyloxy-1,3-butadiene, **11,** 2
1,2-Bis(methylthio)-1,3-butadiene, **10,** 40
1-Chloro-1-dimethylamino-2-methyl-1,3-butadiene, **9,** 104
(E,E)-1,4-Diacetoxybutadiene, **1,** 183
1(E),3-Dimethoxybutadiene, **11,** 192
4,6-Dimethoxy-2-pyrone, **9,** 175
5,5-Dimethoxy-1,2,3,4-tetrachloro-cyclopentadiene, **1,** 270
Dimethyl acetylenedicarboxylate, **1,** 272; **3,** 103; **7,** 117
4-Dimethylamino-1,1,2-trimethoxybutadiene, **11,** 197
Diphenylacetylene, **1,** 335
Ethyl formate, **1,** 380
Ethyl β-phenylsulfonylpropiolate, **11,** 233
Ketene dimethyl acetal, **11,** 279
Phenyl vinyl sulfoxide, **8,** 399
Tetraphenylcyclopentadienone, **1,** 236, 1149; **4,** 490
Titanium(IV) chloride–Lithium aluminum hydride, **11,** 534
Triphenylphosphine, **1,** 1238
Other routes
Chlorotris(triphenylphosphine)-rhodium(I), **5,** 736
Copper, **7,** 73
1,8-Diazabicyclo[5.4.0]undecene-7, **7,** 87
Dichloromethyllithium, **6,** 170
Ketene diethyl acetal, **5,** 362
Lead tetraacetate, **7,** 185
Potassium *t*-butoxide, **1,** 911; **4,** 399
Pyridinium chloride, **8,** 424
Silver perchlorate, **6,** 518

ARENES (*Continued*)
Tetracarbonyldi-μ-chlorodirhodium, **8,** 469
Triethyl phosphite, **2,** 432
BENZOCYCLOBUTENES
Bis(trimethylsilyl)acetylene, **6,** 153
Butyllithium–Magnesium bromide, **10,** 71
t-Butyllithium, **10,** 76
Dicarbonylcyclopentadienylcobalt, **5,** 172; **8,** 146; **9,** 142; **10,** 126
Hydrogen peroxide, **4,** 254
Lead tetraacetate, **7,** 185
Sodium amide, **2,** 373
Sodium telluride, **6,** 550
Tri-μ-carbonylhexacarbonyldiiron, **2,** 139
BENZOCYCLOPROPENES
Aluminum hydride, **6,** 19
Potassium *t*-butoxide, **4,** 399; **6,** 477; **7,** 298
BIPHENYLENES
Benzenediazonium 2-carboxylate, **2,** 26
1,5-Diazabicyclo[5.4.0]undecene-5, **12,** 156
Dicarbonylcyclopentadienylcobalt, **11,** 161; **12,** 160
Phosphorus(V) sulfide, **3,** 226
CYCLOPHANES
[1,3-Bis(diphenylphosphine)propane]-(dichloro)nickel(II), **6,** 50
Butyllithium, **6,** 85
1,3 Dithiane, **7,** 182
Phenyl(trichloromethyl)mercury, **2,** 326
Potassium *t*-butoxide, **2,** 336
Tetraphenylethylene, **2,** 404
Triethyl phosphite, **6,** 612
INDANS
Bis(benzonitrile)dichloropalladium(II), **6,** 45
Dicarbonylcyclopentadienylcobalt, **6,** 153; **8,** 146
Ethylmagnesium bromide, **9,** 218
Hexamethylphosphoric triamide, **6,** 273
Hydrobromic acid, **6,** 282
Ion-exchange resins, **4,** 266; **6,** 302
Phosphorus(V) oxide, **10,** 319
Polyphosphoric acid, **1,** 894
Sulfuric acid, **5,** 633
Tetracarbonyldi-μ-chlorodirhodium, **7,** 59
1-INDANONES
Aluminum chloride, **5,** 10; **11,** 25
Butyllithium, **8,** 65
Hydrogen fluoride, **1,** 455
Methanesulfonic acid, **11,** 321
Phosphoryl chloride, **2,** 330
Polyphosphoric acid, **3,** 231
Sodium amide–Sodium *t*-butoxide, **7,** 327
Trifluoromethanesulfonic acid, **12,** 531
NAPHTHALENES
by Diels–Alder routes
(*o*-Carboxyphenyl)phenyliodonium hydroxide inner salt, **1,** 341; **4,** 207
2,5-Di-*p*-anisyl-3,4-diphenylcyclopentadienone, **1,** 188
Ethyl diphenylphosphinite, **3,** 139
3-Hydroxy-2-pyrone, **6,** 291
Isoamyl nitrite, **5,** 358
Lithium naphthalenide, **6,** 415
trans-1-Methoxy-3-trimethylsilyloxy-1,3-butadiene, **6,** 370
Potassium persulfate, **1,** 952
Sodium borohydride, **11,** 477
Sodium methylsulfinylmethylide, **2,** 166
Zinc, **5,** 753
Other routes
Bis(trimethylsilyl)acetylene, **7,** 84
Boron trifluoride, **3,** 32
Chlorotris(triphenylphosphine)-rhodium(I), **5,** 736
2,3-Dichloro-5,6-dicyano-1,4-benzoquinone, **2,** 112; **9,** 148
Dimethyl acetylenedicarboxylate, **4,** 168
Dimethylformamide dimethyl acetal, **10,** 158
Magnesium, **1,** 627
4-Methoxy-6-methyl-2-pyranone, **9,** 302
Molecular sieves, **11,** 350
Phenyl(tribromomethyl)mercury, **6,** 468
Phosphorus(V) oxide, **10,** 319
Polyphosphoric acid, **5,** 540
Sodium bis(2-methoxyethoxy)-aluminum hydride, **6,** 528
Sulfur, **1,** 1118
Triphenylphosphine dibromide, **5,** 729
PHENANTHRENES

Camphor-10-sulfonic acid, **11,** 107
Chromyl chloride, **2,** 79
Dicarbonylcyclopentadienylcobalt, **12,** 160
Hydrobromic acid, **2,** 214
Iodine, **1,** 495
Ketene dimethyl acetal, **11,** 279
Magnesium, **11,** 307
Phosphorus(V) oxide, **8,** 400; **10,** 319; **11,** 428
Sodium iodide, **3,** 267; **12,** 449
Tetrakis(triphenylphosphine)nickel(0), **9,** 450
Tin(IV) chloride, **3,** 269
Triphenylcarbenium trifluoroacetate, **6,** 657
Triphenylphosphine, **3,** 317
Tris(*p*-bromophenyl)ammonium hexachloroantimonate, **5,** 735
Vanadyl trifluoride, **8,** 528

POLYCYCLIC AROMATICS
Aluminum cyclohexoxide, **1,** 34; **6,** 20
5-Carbomethoxy-1-penten-3-one, **1,** 1271
Chloranil, **4,** 75
Hydriodic acid, **9,** 238; **10,** 199; **11,** 255
Iodine, **9,** 248
2-(2-Lithio-4-methoxyphenyl)-4,4-dimethyl-2-oxazoline, **8,** 279
Maleic anhydride, **5,** 422
Phosphorus(V) oxide–Phosphoric acid, **5,** 535
Platinum catalysts, **8,** 405
Polyphosphoric acid, **1,** 894
Pyrene, **4,** 414
Pyridinium bromide perbromide, **6,** 499
Sodium tetrachloroaluminate, **1,** 1027
3-Sulfolene, **3,** 272
Tin(II) chloride, **1,** 1113
Tin(IV) chloride, **5,** 627

TETRALINS
Alkylaluminum halides, **12,** 5
Aluminum chloride, **12,** 26
2-Butenylidenetriphenylphosphorane, **5,** 69
Cesium fluoride, **10,** 81
Chromium(II) chloride, **12,** 136
Dicarbonylcyclopentadienylcobalt, **6,** 153; **8,** 146
Di-μ-chlorotetra(ethylene)dirhodium, **5,** 203
Dimethyl methoxymethylenemalonate, **12,** 206
Ethyl formate, **1,** 380
N-Lithioethylenediamine, **1,** 567
Manganese(IV) oxide, **2,** 257; **5,** 69
Methyl 2-trimethylsilylacrylate, **11,** 346
Phosphorus(V) oxide, **10,** 319
Sodium ethoxide, **1,** 1065
Tetrabutylammonium fluoride, **10,** 378
Trifluoroacetic acid, **4,** 530; **8,** 503
Zinc–Silver couple, **12,** 571

1-TETRALONES
Aluminum chloride, **3,** 7; **4,** 10
γ-Butyrolactone, **1,** 27, 101
Copper(II) trifluoromethanesulfonate, **10,** 110
Dimethylsulfonium ethoxycarbonylmethylide, **10,** 164
Hexafluoroantimonic acid, **8,** 239
Hydrogen fluoride, **1,** 455
Methanesulfonic acid, **11,** 321
Polyphosphate ester, **2,** 333
Polyphosphoric acid, **1,** 894; **3,** 231; **4,** 395
Sodium amide–Sodium *t*-butoxide, **7,** 327
Sodium tetrachloroaluminate, **9,** 435
Succinic anhydride, **4,** 468
Thionyl chloride, **1,** 1158
Tin(IV) chloride, **1,** 1111
Trifluoromethanesulfonic acid, **12,** 531

2-TETRALONES
Chromium carbene complexes, **12,** 132
Disodium tetracarbonylferrate, **6,** 550
Methylene chloride, **4,** 337
Rhodium(II) carboxylates, **12,** 423
Sodium amide, **5,** 591

TRIPHENYLENES
Boron trifluoride, **3,** 32
o-Bromoiodobenzene, **1,** 77
Palladium catalysts, **1,** 778

ARENE OXIDES
N-Bromosuccinimide, **9,** 70
m-Chloroperbenzoic acid, **8,** 97
Chlorotrimethylsilane, **5,** 709
(E,E)-1,4-Diacetoxybutadiene, **9,** 130
1,8-Diazabicyclo[5.4.0]undecene-7, **7,** 87
Dimethylformamide–Chlorine, **5,** 249
Hexamethylphosphorous triamide, **4,** 247
Hydrogen peroxide–Diisopropyl-

ARENE OXIDES (*Continued*)
carbodiimide, **8,** 248
(−)-Menthoxyacetyl chloride, **6,** 357
Oxygen, singlet, **7,** 261
Potassium peroxomonosulfate, **12,** 413
Potassium superoxide, **12,** 413
Sodium hypochlorite, **8,** 461
Sodium methoxide, **4,** 457
ARYL HALIDES
GENERAL METHODS
trans-Bromo-*o*-tolylbis(triethylphosphine)nickel(II), **10,** 58
Carbonylchlorobis(triphenylphosphine)rhodium(I), **2,** 67
ARYL FLUORIDES
by Fluorination of ArH (*see* TYPE OF REACTION INDEX)
by Fluorination of $ArN_2^+X^-$
Arenediazonium salts, **1,** 42, 43
Pyridinium poly(hydrogen fluoride), **6,** 285
Tetrafluoroboric acid, **1,** 394
Other routes
N-Alkyl-N-fluoro-*p*-toluenesulfonamides, **12,** 231
2,3-Dichloro-5,6-dicyano-1,4-benzoquinone, **9,** 148
Potassium fluoride, **1,** 296, 933; **5,** 556
Pyridinium poly(hydrogen fluoride), **9,** 399
ARYL CHLORIDES
by Chlorination of ArH (*see* TYPE OF REACTION INDEX)
by Chlorination of $ArN_2^+X^-$
Copper(I) chloride, **1,** 166
Sodium chloride, **6,** 285
from ArOH
Phenylphosphonic dichloride, **1,** 846
Triphenylphosphine dichloride, **1,** 1247
Other routes
Carbon tetrachloride, **3,** 43
Chlorotrimethylsilane, **8,** 107
Chlorotris(triphenylphosphine)rhodium(I), **1,** 140; **4,** 559
Copper(II) chloride–Nitric oxide, **2,** 87
Hydrogen bromide, **4,** 251
Palladium catalysts, **11,** 392
Sodium chlorate, **1,** 1056
ARYL BROMIDES
by Bromination of ArH (*see* TYPE OF REACTION INDEX)
by Bromination of $ArN_2^+X^-$
Copper(I) bromide, **1,** 165
Hexamethylphosphoric triamide, **5,** 323
Sodium bromide, **6,** 285
Other routes
Bromoform, **9,** 91
Chlorotris(triphenylphosphine)rhodium(I), **4,** 559
Dioxane–Bromine, **1,** 333
Hydrogen bromide, **4,** 251
Phenyl(tribromomethyl)mercury, **6,** 468
Phosphorus(V) bromide, **1,** 865
Triphenylphosphine dihalides, **1,** 1247; **2,** 446; **3,** 320; **12,** 554
ARYL IODIDES
by Iodination of ArH (*see* TYPE OF REACTION INDEX)
by Iodination of $ArN_2^+X^-$
Potassium iodide, **11,** 440
Sodium iodide, **6,** 285
Other routes
Arylthallium bis(trifluoroacetates), **4,** 498
Iodine, **9,** 91
Sodium iodide + co-reagent, **1,** 1087; **11,** 488
Thallium(III) trifluoroacetate, **3,** 286; **6,** 579
AZIDES
FROM ROH
Diphenyl phosphoroazidate, **8,** 211
Hydrazoic acid, **12,** 242
Silver azide, **9,** 296
Sodium azide, **5,** 326; **6,** 279
Triphenylphosphine + co-reagent, **6,** 644; **7,** 404
FROM C=C
Bromine azide, **2,** 37; **3,** 34
Iodine azide, **2,** 222; **4,** 262; **6,** 297
Mercury(II) azide, **3,** 196; **4,** 323
Sodium azide, **8,** 258
Tris(dimethylamino)mesityloxyphosphonium azide, **9,** 296
FROM RX OR $ROSO_2R'$
Azidotrimethylsilane, **11,** 32
2,3-Bis(bromomethyl)-1,3-butadiene, **5,** 32

Sodium azide, **2,** 209; **6,** 676; **7,** 246
Tetramethylguanidinium azide, **2,** 403
FROM RNH_2
Sodium azide, **1,** 1041
p-Toluenesulfonyl azide, **2,** 415; **3,** 291; **7,** 19
OTHER ROUTES
Azidotrimethylsilane, **5,** 207
Ferric azide, **7,** 153
Lead(IV) acetate azides, **4,** 276
Palladium(II) acetate, **5,** 496
Phenyliodine(III) diacetate–Azidotrimethylsilane, **5,** 354
Sodium nitrite, **1,** 1097
AZIRIDINES
FROM C=C (*see* TYPE OF REACTION INDEX)
FROM AMINO ALCOHOLS
Sodium hydride, **6,** 541
Triphenylphosphine + co-reagent, **5,** 727; **9,** 504
Triphenylphosphine dibromide, **3,** 320
FROM IMINES
(Bromodichloromethyl)phenylmercury, **5,** 514
N-Chlorosuccinimide, **6,** 115
Dichloromethyllithium, **1,** 223
(Dimethylamino)pheryloxosulfonium methylide, **3,** 105
Methyl diazoacetate, **7,** 76
Methylphenyl-N-*p*-toluenesulfonylsulfoximine, **3,** 204
Potassium cyanide, **11,** 433
Trichloromethyllithium, **1,** 223
OTHER ROUTES
Chloroacetyl chloride, **3,** 46
Diethyl N-phenylamidophosphate, **7,** 106
(E)-(Dimethylamino)phenyl-(2-phenylethenyl)sulfoxonium tetrafluoroborate, **4,** 174
Hydrazine, **1,** 434; **5,** 327
Lithium aluminum hydride, **1,** 581; **2,** 242
Sodium amide, **1,** 1034
Sodium azide, **9,** 418
Triethyl phosphonoacetate, **6,** 612
Triphenylphosphine–Carbon tetrachloride, **8,** 516
AZO COMPOUNDS
FROM AZOXY COMPOUNDS (*see* TYPE OF REACTION INDEX)
FROM DIAZONIUM COUPLING (*see* TYPE OF REACTION INDEX)
BY OXIDATION OF ANILINES, HYDRAZO COMPOUNDS (*see* TYPE OF REACTION INDEX)
OTHER ROUTES
t-Butyl hypochlorite, **4,** 58; **5,** 77
Hydroxylamine-O-sulfonic acid, **1,** 481
Lithium methanethiolate, **8,** 303
Nitrosobenzene, **5,** 478
Sodium bis(2-methoxyethoxy)aluminum hydride, **2,** 382
Sodium borohydride, **4,** 443
Sodium hydride–*t*-Butyl hypochlorite, **4,** 455
p-Toluenesulfonyl azide, **1,** 1178
Zinc, **1,** 1276
AZOXY COMPOUNDS
t-Butyl hydroperoxide–Titanium(IV) isopropoxide, **5,** 76
Copper(I) chloride, **6,** 145
Hexamethylditin, **6,** 273
Hydrogen peroxide, **1,** 457
Lead, **2,** 233
Nitrosobenzene, **5,** 478
Phosphine, **1,** 859; **2,** 329
Sodium bis(2-methoxyethoxy)aluminum hydride, **2,** 382
Sodium borohydride–Cobalt(II) chloride, **3,** 264
Thallium, **3,** 285
Triethyloxonium tetrafluoroborate, **4,** 527; **5,** 691
Triethyl phosphite, **1,** 1212

BENZOINS (*see* HYDROXY ALDEHYDES AND KETONES)
BIARYLS (*see also* CHIRAL COMPOUNDS)
BY COUPLING OF ArX
Bis(1,5-cyclooctadiene)nickel(0), **4,** 33; **11,** 51
trans-Bromo-*o*-tolylbis(triethylphosphine)nickel(II), **10,** 58
Copper, **1,** 157; **5,** 146; **7,** 73; **9,** 122
Copper bronze, **1,** 155; **4,** 100
Copper(II) chloride, **2,** 84
Copper(I) trifluoromethanesulfonate, **6,** 130
Diethyltrimethylsilylamine, **6,** 184

BIARYLS (*Continued*)
N,N-Dimethylformamide, **1,** 278
Nickel, **12,** 335
Nickel(II) bromide–Zinc, **9,** 322
Palladium(II) acetate, **6,** 442
Tetramethylurea, **1,** 1146
Thallium(I) bromide, **2,** 405; **7,** 361
Tris(triphenylphosphine)nickel(0), **6,** 653; **8,** 523
BY CROSS-COUPLING OF ArM + Ar'H
(1S,2S)-2-Amino-3-methoxy-1-phenyl-1-propanol, **12,** 310, 381
[1,3-Bis(diphenylphosphine)propane]-(dichloro)nickel(II), **11,** 167
Cobalt(II) chloride, **1,** 155
Copper(I) iodide, **8,** 122
Dichlorobis(triphenylphosphine)-nickel(II), **6,** 59; **9,** 147
1,3-Dichloro-2-butene, **10,** 134
Grignard reagents–Nickel(II) reagents, **7,** 250
2-(*o*-Methoxyphenyl)-4,4-dimethyl-2-oxazoline, **6,** 368
Pentafluorophenylcopper, **5,** 504
Tetrakis(triphenylphosphine)-palladium(0), **8,** 472
BY NONPHENOLIC OXIDATIVE COUPLING
Cerium(IV) trifluoroacetate, **5,** 102
Ferric chloride, **1,** 390; **10,** 185
Periodic acid, **2,** 313
Raney nickel, **1,** 723
Tellurium(IV) chloride, **5,** 643
Thallium(III) trifluoroacetate, **8,** 478; **10,** 397; **11,** 516; **12,** 481, 484
Vanadyl trifluoride, **5,** 745; **7,** 418; **8,** 528; **9,** 513
BY OXIDATIVE PHENOLIC COUPLING
Chloramine, **2,** 65
Copper(II)–Amine complexes, **8,** 114
Diphenyl selenoxide, **9,** 201
Ferric chloride, **10,** 185
Manganese(III) acetylacetonate, **2,** 264
Potassium ferricyanide, **7,** 300
Vanadium(IV) chloride, **3,** 332
OTHER ROUTES FOR SYMMETRICAL Ar–Ar
Arenediazonium tetrahaloborates, **12,** 380
Silver(I) nitrate, **4,** 429
Tetracarbonyldi-μ-chlorodirhodium, **8,** 469
Tetrakis(pyridine)copper(I) perchlorate, **4,** 479
Thallium(I) bromide, **3,** 286; **5,** 655
OTHER ROUTES FOR UNSYMMETRICAL Ar–Ar'
Arenediazonium halides, **1,** 1177
Bis-3,4-dichlorobenzoyl peroxide, **3,** 21
t-Butyl thionitrate, **9,** 91
Copper, **8,** 113
Copper(I) *t*-butoxide, **9,** 122
Dimethylformamide dimethyl acetal, **10,** 158
Isoamyl nitrite, **1,** 40
Nitrosylsulfuric acid, **2,** 299
Potassium acetate, **8,** 130
Tetrakis(triphenylphosphine)-palladium(0), **11,** 503
Tetraphenylbismuth trifluoroacetate, **11,** 515
Thallium(III) trifluoroacetate, **3,** 286
Titanium(IV) chloride, **11,** 529
Trifluoromethanesulfonic acid, **6,** 617
BUTENOLIDES (*see* UNSATURATED LACTONES)

CARBAMATES
FROM ROH
Benzenesulfonyl isocyanate, **1,** 48
3-Bromopropionyl isocyanate, **1,** 78
1-Chlorocarbonylbenzotriazole, **8,** 87
Chlorosulfonyl isocyanate, **1,** 117
Diphenyl phosphoroazidate, **4,** 210
Ethyl- or Methyl(carboxysulfamoyl)-triethylammonium hydroxide inner salt, **5,** 442; **8,** 334
Isopropyl isocyanate, **11,** 277
Methyl N-sulfonylurethane triethylamine complex, **4,** 343
α-Naphthyl isocyanate, **1,** 718
Phenyl isocyanate, **2,** 322; **8,** 396
Sodium azide, **1,** 1041
Sodium cyanate, **1,** 1059
Tin(II) octoate, **7,** 346
Tributyltin azide, **7,** 377
Trichloroacetyl isocyanate, **2,** 425
Trifluoroacetic acid, **1,** 1219
FROM RNH_2
Benzyl chloroformate, **1,** 109

t-Butyl isocyanide, **9,** 123
1-Chlorocarbonylbenzotriazole, **8,** 87
Cholesteryl chloroformate, **1,** 141
Copper(II) methoxide, **4,** 107
Ethyl chloroformate, **1,** 364; **12,** 223
Phenyl chloroformate, **2,** 318
Rhodium catalysts, **12,** 426
OTHER ROUTES
Bis(methoxycarbonyl)sulfodiimide, **12,** 57
Bis(*p*-methoxyphenyl) telluroxide, **9,** 50
Bromonium dicollidine perchlorate, **12,** 78
Chromium(II) chloride, **3,** 60
Dichloromethylenedimethylammonium chloride, **5,** 195
N,N-Dichlorourethane, **2,** 121
Diphenyl phosphoroazidate, **4,** 210
Iodine, **11,** 261; **12,** 253
Iodine isocyanate, **2,** 223; **3,** 161
Lead tetraacetate, **1,** 537; **6,** 313
Methyl hypobromite, **5,** 446
Phase-transfer catalysts, **10,** 305
Silver tetrafluoroborate, **1,** 1015
Thallium(III) nitrate, **4,** 492
CARBODIIMIDES
Butyllithium, **7,** 45
Diethyl azodicarboxylate, **4,** 148
Diethyl phosphite, **1,** 251
Diphenylcarbodiimide, **1,** 337
1-Ethyl-3-methyl-3-phospholene 1-oxide, **1,** 679
2-Halopyridinium salts, **8,** 95; **9,** 234
Iron carbonyl, **3,** 167
3-Methyl-1-phenyl-3-phospholene 1-oxide, **1,** 695
Palladium(II) chloride, **6,** 447
Phosphorus(V) oxide–*t*-Amine, **2,** 329
p-Toluenesulfonyl chloride, **1,** 1179
Triphenylphosphine–Carbon tetrachloride, **4,** 551
Triphenylphosphine dibromide, **3,** 320; **4,** 555; **10,** 60
CARBOHYDRATES AND DERIVATIVES
ALKYLATED CARBOHYDRATES
Benzyl bromide, **5,** 25
Bis(tributyltin) oxide, **9,** 53
Diazomethane, **2,** 102
2,6-Di-*t*-butyl-4-methylpyridine, **9,** 141
Dimethyl sulfate, **1,** 293, 295
Dimethyl sulfoxide, **2,** 157
Methyl trifluoromethanesulfonate, **6,** 406
Sodium hydride, **1,** 1075
Sodium methylsulfinylmethylide, **2,** 166
Triphenylmethyl chloride, **1,** 1254
AMINO SUGARS
Iodonium di-*sym*-collidine perchlorate, **10,** 212
Nitromethane, **1,** 739
Nitrosyl chloride, **1,** 748
Trifluoromethanesulfonic anhydride, **12,** 532
DISACCHARIDES, OLIGOSACCHARIDES
(Diethylamino)sulfur trifluoride, **12,** 183
Iodonium di-*sym*-collidine perchlorate, **10,** 212
Mercury(II) cyanide, **1,** 655
Polyphosphate ester, **1,** 892
Tetraethylammonium bromide, **6,** 568
Tin(IV) chloride, **7,** 342
C-GLYCOSIDES
Allyltributyltin, **12,** 21
Allyltrimethylsilane, **11,** 16; **12,** 544
Benzeneselenenyl halides, **12,** 39
(Diethylamino)sulfur trifluoride, **12,** 183
2-Lithio-1,3-dithianes, **11,** 285
Lithium naphthalenide, **12,** 284
Tetrakis(triphenylphosphine)-palladium(0), **11,** 503
Tributyltin hydride, **12,** 516
Trimethylsilyl trifluoromethane-sulfonate, **12,** 543
Triphenylphosphine, **12,** 550
O-GLYCOSIDES
Benzeneselenenyl halides, **10,** 16
Boron trifluoride etherate, **4,** 44
Bromotrimethylsilane, **10,** 59
Bromotrimethylsilane–Cobalt(II) bromide, **10,** 59
Ion-exchange resins, **1,** 511
Mercury(II) cyanide, **1,** 655
Mercury(II) nitrate, **10,** 254
Mercury(II) oxide–Mercury(II) bromide, **6,** 360
Nitromethane, **1,** 739
Phenylthiotrimethylsilane, **12,** 535
Silver carbonate, **6,** 511
Silver carbonate–Celite, **6,** 511
Silver imidazolate, **10,** 349; **11,** 467
Tetraethylammonium bromide, **6,** 568

CARBOHYDRATES AND DERIVATIVES (*Continued*)
Tetrafluorosilane, **12,** 466
Tin(IV) chloride, **7,** 342
Tin(II) chloride–Silver perchlorate, **10,** 374
2-Trimethylsilylethanol, **11,** 574
Trimethylsilyl trifluoromethanesulfonate, **11,** 584; **12,** 543
Triphenylcarbenium perchlorate, **12,** 549
GLYCOSYL HALIDES
Aluminum chloride, **2,** 21
Bromodimethylborane, **12,** 199
(Diethylamino)sulfur trifluoride, **12,** 183
Hydrogen bromide, **1,** 453
Mercury(II) oxide, **1,** 655
Pyridinium poly(hydrogen fluoride), **12,** 419
Silver tetrafluoroborate, **3,** 250
Triphenylphosphine–N-Halosuccinimides, **4,** 555
OTHER CARBOHYDRATES
Benzeneboronic acid, **1,** 833
Benzyl chloroformate, **1,** 109
Benzylthiocarbonyl chloride, **2,** 26
Bis(tributyltin) oxide, **8,** 43
Bromine–Silver acetate, **1,** 73
t-Butyl hydroperoxide–Dialkyl tartrate–Titanium(IV) isopropoxide, **11,** 92
Cadmium carbonate, **4,** 67
1,1-Dimethoxyethane, **2,** 4
Dimethyl sulfoxide–Sulfur trioxide, **4,** 200
N,N-Dimethylthiocarbamoyl chloride, **3,** 127
Diphenyl phosphorochloridate, **2,** 180
1,3-Dithiane, **4,** 216
Formaldehyde, **8,** 231
Hydrogen bromide–Acetic acid, **1,** 452
Levulinic acid, **10,** 230
Mercury(II) acetate, **1,** 644; **12,** 298
trans-1-Methoxy-3-trimethylsilyloxy-1,3-butadiene, **11,** 332
1-Methyl-1-phenylhydrazine, **1,** 694
Methyl tributylstannyl sulfide, **8,** 352
p-Nitrobenzoyl chloride, **1,** 736
Phenyl isocyanate, **2,** 322
Ruthenium tetroxide, **1,** 986; **3,** 243
Silver carbonate, **1,** 1005; **4,** 425
Silver fluoride, **2,** 364
Sodium amalgam, **1,** 1030
Sodium azide, **5,** 593
Sodium tetraborate, **4,** 461
Tetrabutylammonium iodide, **9,** 447
Tetrakis(triphenylphosphine)-palladium(0), **11,** 503
Trifluoroacetic anhydride, **5,** 701
Triphenylphosphine + co-reagent, **9,** 506, 507
Tris(dimethylamino)mesityloxyphosphonium azide, **9,** 296

CARBONATES
Benzeneseleninic anhydride, **8,** 29
Bis(triphenylphosphine)nickel(0), **4,** 41
Carbon dioxide, **9,** 94
N,N′-Carbonyldiimidazole, **6,** 97
Cerium(IV) ammonium nitrate, **4,** 71
Chlorosulfonyl isocyanate, **5,** 132
Copper(I) cyanoacetate, **7,** 74
Dichloromethylenedimethylammonium chloride, **4,** 135
Diisopropyl peroxydicarbonate, **1,** 263
Dimethyl sulfate, **4,** 188
Ethyl chloroformate, **1,** 364
Hexamethylphosphorous triamide–Dialkyl azodicarboxylates, **6,** 280
Molybdenum(V) chloride, **10,** 274
Performic acid, **4,** 373
Phosgene, **1,** 856; **2,** 328
Selenium, **5,** 575
Triphenylphosphine–Diethyl azodicarboxylate, **11,** 589

CARBOXYLIC ACIDS (*see also* AMINO ACIDS, CHIRAL COMPOUNDS, DICARBONYLS, HALO CARBOXYLIC ACIDS, HYDROXY CARBOXYLIC ACIDS, UNSATURATED CARBOXYLIC ACIDS)
BY CARBOXYLTION
of RH, ArH
Bromine–Antimony(V) chloride–Sulfur dioxide, **5,** 57
2,2-Dichloro-1,3-benzodioxole, **1,** 119; **4,** 70
Diphenylcarbamoyl chloride, **1,** 337; **2,** 177
Oxalyl chloride, **1,** 28, 767
of $ArN_2^+X^-$ (*see* TYPE OF REACTION INDEX)
of ArM
Arene(tricarbonyl)chromium

complexes, **10,** 13
Butyllithiums, **1,** 95; **6,** 109
Carbon dioxide, **6,** 94
Grignard reagents, **1,** 415
Lithium naphthalenide, **8,** 305
Lithium nitride, **1,** 618
Magnesium, **7,** 218; **10,** 251
Magnesium methyl carbonate, **3,** 190
Nickel carbonyl, **1,** 720; **5,** 472
Phenylpotassium, **1,** 848
Sodium amalgam, **1,** 1033
N,N,N′,N′-Tetramethylethylenediamine, **5,** 652
of Other substrates
Copper(I) carbonyl ion, **5,** 149
Cyclodextrins, **12,** 151
Disodium tetracarbonylferrate, **4,** 461
Formic acid, **1,** 404
Hexafluoroantimonic acid, **2,** 216
Iron carbonyl, **8,** 265
Nickel carbonyl, **1,** 720; **5,** 472
Tetrabutylammonium iodide, **7,** 355
BY HOMOLOGATION OF RCHO, R_2CO; $R_2CO \rightarrow R_2CHCOOH$
Carbon dioxide, **6,** 94
Chloroacetonitrile, **6,** 102
Cyanotrimethylsilane, **11,** 147
Diethyl *t*-butoxy(cyano)methylphosphonate, **8,** 167
1,3-Dithiocyclohexane-2-thione, **2,** 441
2-Lithio-2-trimethylsilyl-1,3-dithiane, **4,** 284
Tetraethyl dimethylaminomethylenediphosphonate, **3,** 277; **11,** 502
Tosylmethyl isocyanide, **4,** 272; **5,** 684
BY HYDROCARBOXYLATION OF C=C
Chlorobis(cyclopentadienyl)hydridozirconium(IV), **6,** 175
Dichlorobis(triphenylphosphine)palladium(II), **3,** 81
Formic acid, **4,** 239
Nickel carbonyl, **5,** 472
Palladium(II) chloride–Triphenylphosphine, **5,** 503
Trifluoromethanesulfonic acid, **9,** 485
Tris(methylthio)methyllithium, **7,** 412
BY HYDROLYSIS OF RCOX
Amides
α-Chloro-N-cyclohexylpropanaldonitrone, **4,** 80
Claisen's alkali, **1,** 153
Diphenyldi(1,1,1,3,3,3-hexafluoro-2-phenyl-2-propoxy)sulfurane, **6,** 239
Ephedrine, **9,** 209
Ion-exchange resins, **1,** 511; **11,** 276
Lithium hydroxide, **12,** 282
Nitrogen dioxide, **1,** 324
Nitrosonium tetrafluoroborate, **1,** 747
Nitrosyl chloride, **1,** 748
Potassium hydroxide, **8,** 415; **10,** 328
Sodium peroxide, **6,** 548
Esters (General methods)
Aluminum chloride–Ethanethiol, **9,** 13; **10,** 11
Boron tribromide, **5,** 49
Cinchonine, **6,** 501
Claisen's alkali, **1,** 153
Dimethylaluminum methylselenolate, **8,** 182
Dimethyl sulfoxide, **1,** 296
Iodotrimethylsilane, **8,** 261
Ion-exchange resins, **1,** 511
Lithium iodide, **7,** 209
Lithium methanethiolate, **8,** 303
Methanesulfonic acid, **1,** 666
Potassium hydroxide, **8,** 415; **10,** 328
Potassium superoxide, **7,** 304
Sodium benzeneselenoate, **8,** 447; **11,** 475
Sodium thiophenoxide, **1,** 1106
Tetrabutylammonium hydrogen sulfate, **9,** 356
Trichloro(methyl)silane–Sodium iodide, **12,** 527
Trifluoroacetic acid, **5,** 695
Hydrazides
Cerium(IV) ammonium nitrate, **4,** 71
Copper(II) acetate, **6,** 429; **10,** 103
Copper(I) chloride, **7,** 80
Ion-exchange resins, **11,** 276
Lead tetraacetate, **3,** 168
Manganese(IV) oxide, **1,** 637
Methyl esters
Boron trichloride, **4,** 42
1,8-Diazabicyclo[5.4.0]undecene-7, **4,** 119
Dimethylaluminum methylselenolate, **8,** 182
N,N-Dimethylhydrazine, **6,** 223
Lithium–Ammonia, **1,** 601

CARBOXYLIC ACIDS (*Continued*)
Lithium iodide, **1,** 615; **7,** 209
Lithium 1-propanethiolate, **3,** 188; **5,** 415
Potassium *t*-butoxide, **1,** 911
Potassium cryptate[2.2.2], **5,** 156
Pyridinium chloride, **11,** 449
Sodium cyanide, **5,** 606
Other specific esters
Lithium dimethylcuprate, **8,** 301
Mercury(II) acetate, **8,** 315
Nitrosonium tetrafluoroborate, **8,** 364
Sodium ethanedithiolate, **6,** 540
Sodium trithiocarbonate, **6,** 553
Trimethylsilyl trifluoromethanesulfonate, **10,** 438
BY KOCH–HAAF REACTION (*see* TYPE OF REACTION INDEX)
BY OXIDATION OF ROH, RCHO, ArR (*see* TYPE OF REACTION INDEX)
BY OXIDATIVE CLEAVAGE (*see* TYPE OF REACTION INDEX)
BY WILLGERODT REACTION (*see* TYPE OF REACTION INDEX)
OTHER ROUTES
Arenesulfonyl azides, **5,** 19
t-Butyl lithioacetate, **10,** 67
Cerium(IV) acetate, **4,** 71
Chloramine, **1,** 122
Cobalt(II) acetate, **4,** 99
Copper(I) carbonyl ion, **5,** 149
Diazomethane, **1,** 191
4,4-Dichloro-3-buten-1-ol, **5,** 192
1,1-Dichloroethylene, **1,** 220; **2,** 118
Dicyclohexylborane, **8,** 162
Dilithium tetrachlorocuprate, **7,** 114
Diphenyl phosphoroazidate, **10,** 173
Hexafluoroantimonic acid, **11,** 252
Hydrogen peroxide–Selenium dioxide, **1,** 477
2-(*o*-Methoxyphenyl)-4,4-dimethyl-2-oxazoline, **6,** 368
Oxodiperoxymolybdenum(pyridine)-(hexamethylphosphoric triamide), **11,** 218
Phenoxyacetic acid, **9,** 361
(Phenylsulfonyl)nitromethane, **11,** 419
Potassium *t*-butoxide, **10,** 323
Propyl nitrite–Sodium nitrite, **5,** 565
Sodium–Ethanol, **11,** 472
2,4,4,6-Tetramethyl-5,6-dihydro-1,3-(4*H*)-oxazine, **3,** 280
Thallium(III) nitrate, **8,** 476
2,4,4-Trimethyl-2-oxazoline, **3,** 313; **5,** 714
α-SUBSTITUTED CARBOXYLIC ACIDS (*see also* AMINO ACIDS, HALO CARBOXYLIC ACIDS, HYDROXY CARBOXYLIC ACIDS)
α-Acyloxy
Carbon tetrabromide–Tin(II) fluoride, **11,** 111
Pyridine N-oxide, **5,** 567
Thallium(III) acetate, **4,** 492
α-Aryl
d- and *l*-10-Camphorsulfonyl chloride, **11,** 109
Diphenyl phosphoroazidate, **8,** 211
Methyl methylthiomethyl sulfoxide, **9,** 314
Silver tetrafluoroborate, **11,** 471
Sodium benzeneselenoate, **12,** 452
α-Methoxy
Chloroform, **8,** 92
trans-2,4-Dimethoxymethyl-5-phenyloxazoline, **6,** 204
(−)-10-Mercaptoisoborneol, **9,** 290
α-Silyl
Trimethylsilylacetic acid, **6,** 631
CHIRAL COMPOUNDS
ALCOHOLS
by Additions to C=O
(S,S)-(+)-1,4-Bis(dimethylamino)-2,3-dimethoxybutane, **9,** 47
N,S-Dimethyl-S-phenylsulfoximine, **11,** 210
(2S,2′S)-2-Hydroxymethyl-1-[(1-methylpyrrolidin-2-yl)methyl]-pyrrolidine, **8,** 253; **9,** 246
Lithium tetrabutylaluminate–(−)-N-Methylephedrine, **7,** 240
(S)-(−)-Methyltri(2-methyl-1-butoxy)-titanium, **10,** 422
Organotitanium reagents, **11,** 374
p-Tolylsulfinylacetic acid, **11,** 538
p-(Tolylsulfinyl)methyllithium, **4,** 513
by Hydroboration
Diisopinocampheylborane, **5,** 283; **6,** 202; **11,** 188, 225; **12,** 193
Dilongifolylborane, **11,** 191
Monoisopinocampheylborane, **8,** 267; **9,** 317; **10,** 224

by Reduction of C=O
(R)-1-(S)-1′,2-Bis(diphenylphosphine)-ferrocenylethanol, **11,** 239
Borane–(S)-(–)-2-Amino-3-methyl-1,1-diphenyl-1-butanol, **12,** 31
Diisobutylaluminum hydride, **12,** 191
Diisopinocampheylborane, **1,** 262; **8,** 174
(–)-N-Dodecyl-N-methylephedrinium bromide, **11,** 337
NB-Enantride, **11,** 229
Lithium aluminum hydride + chiral co-reagent, **2,** 243; **5,** 231, 387; **8,** 184, 423; **9,** 169, 308; **10,** 458; **11,** 153, 289, 292; **12,** 60, 190, 191, 322
Lithium B-isopinocampheyl-9-borabicyclo[3.3.1]nonyl hydride, **8,** 303
Monoisopinocampheylborane, **11,** 350
B-3-Pinanyl-9-borabicyclo[3.3.1]-nonane, **8,** 403; **10,** 320; **11,** 429; **12,** 397
Sodium borohydride + chiral co-reagent, **7,** 144, 239; **8,** 431; **9,** 422
Other routes
1,4-[Bis(diphenylphosphine)butane]-(norbornadiene)rhodium(I) tetrafluoroborate, **12,** 426
2,3-Butanediol, **12,** 80
Dibromoalane, **12,** 377
Hexahydro-4,4,7-trimethyl-4*H*-1,3-benzothiin, **12,** 237
(S)-(+)-α-Hydroxy-β,β-dimethyl-propyl vinyl ketone, **12,** 248
Lithium dimethylcuprate–Boron trifluoride, **12,** 348
Organotitanium reagents, **12,** 353
2-Oxazolidones, chiral, **11,** 379
ALDEHYDES
Chiral at the α carbon
(S)-1-Amino-2-methoxymethyl-1-pyrrolidine, **8,** 16
(1S,2S)-2-Amino-3-methoxy-1-phenyl-1-propanol, **12,** 310
Dimethyl sulfoxide–Sulfur trioxide, **11,** 216
Leucine *t*-butyl esters, **10,** 229
(–)-10-Mercaptoisoborneol, **9,** 290
α-Methylbenzylamine, **9,** 363
(S)-(–)-Proline, **9,** 393
Sodium borohydride, **12,** 441
(S)-(+)-*p*-Tolyl *p*-tolylthiomethyl sulfoxide, **9,** 474; **10,** 408; **12,** 510
Trialkylaluminums, **12,** 512
(–)-4,6,6-Trimethyl-1,3-oxathiane, **8,** 508
Chiral at the β carbon
(1S,2S)-2-Amino-3-methoxy-1-phenyl-1-propanol, **12,** 310
Diisobutylaluminum hydride, **12,** 191
2,3-O-Isopropylidene-2,3-dihydroxy-1,4-bis(diphenylphosphine)butane, **9,** 259
Leucine *t*-butyl esters, **7,** 189; **10,** 229
(S)-(+)-2-Methoxymethylpyrrolidine, **10,** 259
ALICYCLIC HYDROCARBONS
Three-membered
Bis[(–)-camphorquinone-α-dioximato]cobalt(II) hydrate, **8,** 39
Bromoform (or Chloroform), **6,** 220
N,S-Dimethyl-S-phenylsulfoximine, **11,** 210
Dimethylsulfoxonium methylide, **9,** 186
Ephedrine, **11,** 230
Iron carbonyl, **10,** 221
d- and *l*-Menthol, **12,** 294
Four-membered
Bis(1,5-cyclooctadiene)nickel(0), **4,** 33
Cobaltacyclopentan-2-ones, **11,** 136
Five-membered
(R)-(+)-*o*-Anisylcyclohexylmethyl-phosphine, **11,** 31
(S)- or (R)-Menthyl *p*-toluenesulfinate, **11,** 312; **12,** 295
Rhodium(II) carboxylates, **12,** 423
Six-membered
D-(–)- and L-(+)-2,3-Butanediol, **11,** 84
Camphor-10-sulfonic acid, **12,** 103
Copper(II) acetate–Ferrous(II) sulfate, **10,** 103
Dichlorodiisopropoxytitanium(IV), **11,** 164
Ethyl (S)-lactate, **12,** 226
(R)-Ethyl *p*-tolylsulfinylmethylene-propionate, **12,** 228
(S)-Histidine, **9,** 393
(S)-(+)-α-Hydroxy-β,β-dimethyl-propyl vinyl ketone, **12,** 248

CHIRAL COMPOUNDS (*Continued*)
(1R)-(−)-*cis*-3-Hydroxyisobornyl neopentyl ether, **12,** 249
Menthoxyaluminum dichloride, **9,** 289
(S)- or (R)-Menthyl *p*-toluenesulfinate, **12,** 295
2-Oxazolidones, chiral, **12,** 359
8-Phenylmenthol and esters, **10,** 402; **11,** 412
(S)-(−)-Proline, **6,** 411; **7,** 307; **8,** 421; **9,** 393; **12,** 414
Tetrakis(triphenylphosphine)-palladium(0), **11,** 503
Tetramethyltartaric acid diamide, **12,** 480
Tin(IV) chloride, **11,** 522

ALKANES
Dichloro[2,3-O-isopropylidene-2,3-dihydroxy-1,4-bis(diphenylphosphine)-butane]nickel(II), **5,** 360
Neomenthyldiphenylphosphine, **6,** 416

ALKENES (HAVING A CHIRAL SP^3C)
Bis(acetonitrile)dichloropalladium(II), **11,** 46
trans-2,3-Bis(diphenylphosphine)-bicyclo[2.2.1]hept-5-ene, **10,** 36
Ferrocenylphosphines, **9,** 200
Phosphonamides, chiral, **12,** 396

ALKYL HALIDES
Iodine, **7,** 179

ALLENES
1-Chloro-2-methyl-N,N-tetramethylenepropenylamine, **12,** 124
Chromium(III) chloride–Lithium aluminum hydride, **10,** 101
Diisopinocampheylborane, **4,** 161
Methyllithium, **7,** 242
(R)-1-(1-Naphthyl)ethyl isocyanate, **8,** 356
Triethyl orthoacetate, **10,** 417

ALLYLIC ALCOHOLS
(S)-4-Anilino-3-methylamino-1-butanol, **12,** 33
Butyllithium, **12,** 96
2,2′-Dihydroxy-1,1′-binaphthyl, **9,** 169; **10,** 148; **12,** 190
Diisobutylaluminum hydride, **12,** 191
(+)-α-Ethyl camphorate, **1,** 100
Lithium aluminum hydride–(−)-N-Methylephedrine, **10,** 238
Lithium L-α,α′-dimethyldibenzylamide, **10,** 245
B-3-Pinanyl-9-borabicyclo[3.3.1]nonane, **11,** 429
(S)-(+)-2-(1-Pyrrolidinyl)-methylpyrrolidine, **12,** 421

ALLYLIC SILANES
α-(Trimethylsilyl)benzylmagnesium bromide, **11,** 19

AMINES
(S)-1-Amino-2-methoxymethyl-1-pyrrolidine, **12,** 30
(1S,2S)-2-Amino-1-phenyl-1,3-propanediol, **12,** 32, 383
(S)-N-Benzyloxycarbonylproline, **11,** 447
Borane–Tetrahydrofuran, **4,** 124
O-(Diphenylphosphinyl)-hydroxylamine, **11,** 221
Ephedrine, **11,** 230
Iron, **1,** 519
2,3-O-Isopropylidene-2,3-dihydroxy-1,4-bis(diphenylphosphine)butane, **5,** 360
α-Methylbenzylamine, **6,** 162; **11,** 411
2-Methylpyrrolidine, **4,** 471
8-Phenylmenthol, **11,** 412
Pyridinium (*d*)-camphor-10-sulfonate, **11,** 449
Tin(IV) chloride, **11,** 522
p-(Tolylsulfinyl)methyllithium, **5,** 682

AMINO ACIDS
2-Acetoxy-1-methoxy-3-trimethylsilyloxy-1,3-butadiene, **11,** 2
(2R,4R)- and (2S,4S)-N-Acryloyl-4-(diphenylphosphine)-2-[(diphenyl-phosphine)methyl]pyrrolidine, **11,** 6
1-Amino-(S)-2-[(R)-1-hydroxyethyl]-indoline, **3,** 11
(S)-1-Amino-2-hydroxymethylindoline, **3,** 12
(S)-15-Aminomethyl-14-hydroxy-5,5-dimethyl-2,8-dithia[9](2,5)-pyridinophane, **11,** 29
2,3-Bis(diphenylphosphine)bicyclo-[2.2.1]hept-5-ene, **9,** 49
2,2′-Bis(diphenylphosphine)-1,1′-binaphthyl, **10,** 36
(2S,4S)-N-(t-Butoxycarbonyl)-4-(diphenylphosphine)-2-[(diphenyl-phosphine)methyl]pyrrolidine, **8,** 57
7-Chloro-5-phenyl-1-[(S)-α-phenylethyl]-1,3-dihydro-2*H*-1,4-benzodiazepine-2-one, **11,** 124

(3S,6S)-(+)-2,5-Dimethoxy-3,6-dimethyl-3,6-dihydropyrazine, **10,** 151; **11,** 192
(S)-1-(Dimethoxymethyl-2-methoxymethyl)pyrrolidine, **10,** 152
(S)-α-(R)-Diphenylphosphineferrocenylethyldimethylamine, **9,** 200
(2R,3R)-Dipivaloyltartaric acid, **9,** 204
Ethylaluminum dichloride, **10,** 177
Hydrogen cyanide, **4,** 252
(1S,2S,5S)-2-Hydroxypinan-3-one, **7,** 177
2,3-O-Isopropylidene-2,3-dihydroxy-1,4-bis(diphenylphosphine)butane, **4,** 273; **9,** 259
α-Methylbenzylamine, **8,** 393
Methyl 2,3-bis-O-diphenylphosphine-4,6-O-benzylidene-α-D-glucopyranoside, **8,** 333
Palladium hydroxide, **9,** 353
Potassium tri-*sec*-butylborohydride–Copper(II) iodide, **6,** 492
1,1,3,3-Tetramethylbutyl isocyanide, **5,** 650
AMINO ALCOHOLS
Aluminum chloride, **10,** 9
(R)-1-(S)-1′,2-Bis(diphenylphosphine)-ferrocenylethanol, **11,** 237
p-Tolyl vinyl sulfoxide, **11,** 538
AMINO NITRILES
α-Methylbenzylamine, **8,** 393; **12,** 319
BIARYLS
(1S,2S)-2-Amino-3-methoxy-1-phenyl-1-propanol, **12,** 381
CARBOXYLIC ACIDS AND ACYL DERIVATIVES
Chiral at the α carbon
[1,2-Bis(diphenylphosphine)propanc]-(norbornadiene)rhodium perchlorate, **9,** 452
trans-2,5-Bis(methoxymethyl)-pyrrolidine, **12,** 59
(2S,4S)-N-(*t*-Butoxycarbonyl)-4-(diphenylphosphino)-2-[(diphenylphosphino)methyl]-pyrrolidine, **11,** 87
d- and *l*-10-Camphorsulfonyl chloride, **11,** 109
trans-2,4-Dimethoxymethyl-5-phenyloxazoline, **6,** 204
Ephedrine, **9,** 209
2,3-O-Isopropylidene-2,3-dihydroxy-1,4-bis(diphenylphosphine)butane, **6,** 309
(4S,5S)-4-Methoxymethyl-2-methyl-5-phenyl-2-oxazoline, **6,** 386
2-Oxazolidones, chiral, **11,** 379
(S)-Prolinol, **10,** 332
Chiral at the β carbon
(2S,4S)-N-(*t*-Butoxycarbonyl)-4-(diphenylphosphino)-2-[(diphenylphosphino)methyl]-pyrrolidine, **11,** 87
Chlorotris(neomenthyldiphenylphosphine)rhodium(I), **6,** 416
Ephedrine, **11**, 230
(1R)-(−)-*cis*-3-Hydroxyisobornyl neopentyl ether, **12,** 249
Lithium dibutylcuprate, **10,** 191
Mercury(II) acetate, **12,** 298
(4S,5S)-4-Methoxymethyl-2-methyl-5-phenyl-2-oxazoline, **7,** 229; **9,** 312
8-Phenylmenthol, **11,** 412
CYANOHYDRINS
(2R,4R)-Pentanediol, **12,** 375
DICARBONYLS WITH CHIRAL α OR β CARBONS
(S)-1-Amino-2-methoxymethyl-1-pyrrolidine, **12,** 30
Lithium diisopropylamide, **10,** 241
2-Oxazolidones, chiral, **12,** 359
Quinine, **9,** 403
(+)-(R)-*trans*-β-Styryl *p*-tolyl sulfoxide, **4,** 466
(S)-(+)-*p*-Tolyl *p*-tolylthiomethyl sulfoxide, **10,** 408
L-Valinol, **12,** 563
DIENES
(R)-4-Methylcyclohexylidene-methylcopper, **7,** 238
DIOLS
1,2-Diols
t-Butyl hydroperoxide–Dialkyl tartrate, **12,** 91
(+)-Camphanic acid, **8,** 74
(S)-N-Formyl-2-methoxymethyl-pyrrolidine, **11,** 243
Hexahydro-4,4,7-trimethyl-4*H*-1,3-benzothiin, **12,** 237
Lithium aluminum hydride–Darvon alcohol, **6,** 206
Osmium tetroxide–Dihydroquinine

CHIRAL COMPOUNDS (*Continued*)
acetate, **10,** 291
1,3-Diols
t-Butyl hydroperoxide–Dialkyl tartrate, **11,** 92
Organocopper reagents, **12,** 345
(2R,4R)-Pentanediol, **12,** 375
Raney nickel, **9,** 405
Titanium(IV) chloride, **12,** 494
Trichloromethyltitanium, **12,** 355
ENAMINES WITH CHIRAL CARBONS
2,2′-Bis(diphenylphosphine)-1,1′-binaphthyl, **11,** 53; **12,** 56
EPOXIDES
(−)-Benzylquininium chloride, **8,** 430
Chloromethyl *p*-tolyl sulfone, **8,** 340
2,2′-Dihydroxy-1,1′-binaphthyl, **12,** 190
Dimethylsulfonium methylide, **6,** 219
Hydrogen bromide–Acetic acid, **12,** 242
(S)-(2-Hydroxy-N,N-dimethylpropanamide-O,O′)-oxodiperoxymolybdenum(VI), **10,** 206
(R)-1-(1-Naphthyl)ethyl isocyanate, **9,** 321
Tellurium(IV) chloride, **12,** 457
EPOXY ALCOHOLS
t-Butyl hydroperoxide, **10,** 64
t-Butyl hydroperoxide + co-reagent, **8,** 393; **12,** 90
Dibenzyltartaric acid diamide, **12,** 91
EPOXY CARBONYL COMPOUNDS
(−)-Benzylquininium chloride, **7,** 311; **10,** 27
(S)-(−)-Proline, **11,** 446
Quinine, **10,** 338
HALOHYDRINS AND DERIVATIVES
Hydrogen bromide–Acetic acid, **12,** 242
Lithium aluminum hydride–(R)- and (S)-2,2′-Dihydroxy-1,1′-binaphthyl, **12,** 190
B-3-Pinanyl-9-borabicyclo[3.3.1]nonane, **12,** 397
Tellurium(IV) chloride, **12,** 457
HETEROCYCLES
(1S,2S)-2-Amino-1-phenyl-1,3-propanediol, **12,** 32, 383
(S)-N-Benzyloxycarbonylproline, **11,** 447
Dimethyl sulfoxide, **5,** 263
trans-1-Methoxy-3-trimethylsilyloxy-1,3-butadiene, **11,** 332
2-Methylpyrrolidine, **4,** 471
Palladium(II) acetate, **9,** 344
8-Phenylmenthol, **11,** 412
Pyridinium (*d*)-camphor-10-sulfonate, **11,** 449
HOMOALLYLIC ALCOHOLS
B-Allyldiisocaranylborane, **12,** 18
B-Allyl(or Methallyl)diisopinocampheylborane, **12,** 17
2-Crotyl-4,4,5,5-tetramethyl-1,3,2-dioxaborolane, **11,** 86
Diisopinocampheylborane, **4,** 161
Dilithium dialkyl(cyano)cuprates, **12,** 349
Hydrogen peroxide, **12,** 242
B-(3-Methyl-2-butenyl)-diisopinocampheylborane, **12,** 17
(2R,4R)-Pentanediol, **12,** 375
(+)-3-Phenylbornane-2,3-diol, **9,** 363
Titanium(IV) chloride, **12,** 494
Trialkylaluminums, **12,** 512
α-(Trimethylsilyl)benzylmagnesium bromide, **11,** 19
HOMOPROPARGYL ALCOHOLS
Diethyl tartrate, **11,** 181
HYDROXY ALDEHYDES AND KETONES
α-Hydroxy
(S)-2-(Anilinomethyl)pyrrolidine, **9,** 19
(S)-N-Formyl-2-methoxymethylpyrrolidine, **11,** 243
Hexahydro-4,4,7-trimethyl-4H-1,3-benzothiin, **12,** 237
(−)-10-Mercaptoisoborneol, **9,** 290
(S)-(+)-*p*-Tolyl *p*-tolylthiomethyl sulfoxide, **9,** 474; **10,** 408
(−)-4,6,6-Trimethyl-1,3-oxathiane, **8,** 508
β-Hydroxy
3-Acetylthiazolidine-2-thione, **12,** 490
1-Amino-2-methoxymethyl-1-pyrrolidine, **8,** 16; **9,** 17
2,3-Butanediol, **11,** 84
Isobornylamine, **9,** 62
Mandelic acid, **12,** 291
(−)-Norephedrine, **12,** 338
(S)-(−)-Proline, **6,** 411; **7,** 307
Tin(IV) chloride, **12,** 486
Tin(II) trifluoromethanesulfonate, **11,** 525
Titanium(IV) chloride, **12,** 494
HYDROXY CARBOXYLIC ACIDS

AND ACYL DERIVATIVES
α-Hydroxy
Allyltrimethylsilane, **12,** 23
[1,2-Bis(diphenylphosphine)propane]-(norbornadiene)rhodium perchlorate, **9,** 452
(2S,4S)-N-(*t*-Butoxycarbonyl)-4-(diphenylphosphino)-2-[(diphenylphosphino)methyl]-pyrrolidine, **8,** 57; **11,** 87
Dicyclohexylborane, **11,** 172
Hexahydro-4,4,7-trimethyl-4*H*-1,3-benzothiin, **12,** 237
8-Phenylmenthol, **11,** 412; **12,** 389
B-3-Pinanyl-9-borabicyclo[3.3.1]-nonane, **11,** 429; **12,** 397
(S)-(−)-Proline, **8,** 421
Titanium(IV) chloride, **12,** 494
Tributylcrotyltin, **12,** 513
β-Hydroxy
(R)-2-Acetoxy-1,1,2-triphenylethanol, **12,** 3
3-Acetylthiazolidine-2-thione, **12,** 490
Bromomagnesium diethylamide, **5,** 209
Dichlorobis(cyclopentadienyl)-zirconium(II), **10,** 131
N,N-Dimethyl-α-p-tolylsulfinyl-acetamide, **12,** 509
(4S,5S)-2-Ethyl-4-(methoxymethyl)-5-phenyloxazoline, **11,** 232
Menthyl *p*-toluenesulfinate, **12,** 295
trans-1-Methoxy-3-trimethylsilyloxy-1,3-butadiene, **11,** 332
2-Oxazolidones, chiral, **11,** 379; **12,** 359
Raney nickel, **9,** 405
Rhodium catalysts, **12,** 426
p-Tolylsulfinylacetic acid, **10,** 405; **11,** 538
Zinc borohydride, **12,** 572
γ-Hydroxy
Diisopinocampheylborane, **5,** 283
KETONES
Chiral at the α carbon
(S)-1-Alkoxymethyl-2-aminopropane, polymeric, **9,** 3
(S)-1-Amino-2-methoxymethyl-1-pyrrolidine, **7,** 10; **8,** 16; **9,** 17
(S)-(−)-2-Amino-1-methoxy-3-phenylpropane, **10,** 11
(S)-2-Aminopropyl-benzyl ether, **11,** 30
(Benzylmethoxymethyl)methylamine, **7,** 17; **9,** 42
(+)-*trans*-2,5-Dimethylpyrrolidine, **8,** 197
Ephedrine, **9,** 209
Leucine *t*-butyl esters, **8,** 272
erythro-2-Methoxy-1,2-diphenyl-ethylamine, **12,** 313
(S)-(+)-2-Methoxymethylpyrrolidine, **11,** 326
Monoisopinocampheylborane, **11,** 350
Trialkylaluminums, **12,** 512
Valine *t*-butyl ester, **12,** 563
Chiral at the β carbon
(2S,4S)-(Anilinomethyl)-1-ethyl-4-hydroxypyrrolidine, **10,** 12
(S,S)-(+)-1,4-Bis(dimethylamino)-2,3-dimethoxybutane, **9,** 47
(4S)-*t*-Butylthio-(2S)-methoxymethyl-N-pivaloylpyrrolidine, **12,** 315
N,N′-Dimethyltartaramide, **12,** 480
(S)-2-Hydroxymethyl-1-methylpyrrolidine, **10,** 266
(S)- or (R)-Menthyl *p*-toluenesulfinate, **11,** 312; **12,** 295
Methylmagnesium iodide, **10,** 191
Quinine, **9,** 403
Thiophenol, **8,** 431
(S)-(+)-*p*-Tolyl *p*-tolylthiomethyl sulfoxide, **10,** 408
LACTAMS
Dimethylketene methyl trimethylsilyl ketal, **10,** 401
Quinine, **11,** 456
LACTONES
(R)-(+)-*t*-Butyl (*p*-tolylsulfinyl)acetate, **11,** 106
Cinchona alkaloids, **11,** 134
Dimethylsulfoxonium methylide, **9,** 186
2,3-O-Isopropylidene-2,3-dihydroxy-1,4-bis(diphenylphosphine)butane, **11,** 277
Lithium aluminum hydride–(−)-N-Methylephedrine–3,5-Dimethylphenol, **10,** 265
(S)- or (R)-Menthyl *p*-toluenesulfinate, **12,** 295
4(R)-Methoxycarbonyl-1,3-thiazolidine-2-thione, **11,** 323

CHIRAL COMPOUNDS (*Continued*)
(4S,5S)-4-Methoxymethyl-2-methyl-5-phenyl-2-oxazoline, **6,** 386; **7,** 229; **9,** 312
(γ-Methoxypropyl)-α-phenylethylamine, **12,** 318
trans-1-Methoxy-3-trimethylsilyloxy-1,3-butadiene, **11,** 332
2-Oxazolidones, chiral, **11,** 379
B-3-Pinanyl-9-borabicyclo[3.3.1]nonane, **10,** 320; **11,** 429
p-Tolylsulfinylacetic acid, **10,** 405
NITRILES
Chlorosulfonyl isocyanate, **11,** 125
Hydrogen cyanide, **9,** 239
PEPTIDES
(–)-3-Hydroxy-5-methylhydantoin, **8,** 252
Thioglycolic acid, **7,** 366
PHOPHORUS COMPOUNDS
Boron trisulfide, **5,** 55
Camphor-10-sulfonic acid, **4,** 68
PROPARGYL ALCOHOLS
Darvon alcohol, **8,** 184; **10,** 154
NB-Enantrane, **11,** 229
Lithium aluminum hydride + co-reagent, **9,** 308; **10,** 148
(2R,4R)-Pentanediol, **12,** 375
B-3-Pinanyl-9-borabicyclo[3.3.1]nonane, **10,** 320; **11,** 429; **12,** 397
Trimethylsilylacetylene, **9,** 246
SULFUR COMPOUNDS
(2S,4S)-(Anilinomethyl)-1-ethyl-4-hydroxypyrrolidine, **10,** 12
t-Butyl hydroperoxide–Dialkyl tartrate, **12,** 90
(S)- or (R)-Menthyl *p*-toluenesulfinate, **11,** 312; **12,** 295
UNSATURATED CHIRAL CARBONYL COMPOUNDS
Lithium hexamethyldisilazide, **12,** 280
Palladium(II) chloride, **5,** 500
8-Phenylmenthol, **11,** 412
UNSATURATED LACTONES
1,3-Dimethoxy-1-trimethylsilyloxy-1,3-butadiene, **12,** 196
(2S,2′S)-2-Hydroxymethyl-1-[(1-methylpyrrolidin-2-yl)methyl]-pyrrolidine, **10,** 207
Lithium aluminum hydride–(–)-N-Methylephedrine–3,5-Dimethylphenol, **10,** 265
(R)-(+)-3-(*p*-Tolylsulfinyl)propionic acid, **12,** 509
CHLOROFORMATES (*see* HALOFORMIC ESTERS)
CYANO AMINES (*see* AMINO NITRILES)
CYANO CARBONYL COMPOUNDS (*see also* ACYL NITRILES)
α-CYANO CARBONYLS
O,N-Bis(trifluoroacetyl)hydroxylamine, **1,** 40
t-Butylcyanoketene, **4,** 55
Chlorosulfonyl isocyanate, **5,** 132
Diethyl carbonate, **1,** 247
5-Methylisoxazole, **9,** 309
Phase-transfer catalysts, **9,** 356
Sodium ethoxide, **1,** 1065
p-Toluenesulfonyl cyanide, **11,** 536
β-CYANO CARBONYLS
Acetone cyanohydrin, **1,** 5
3-Benzyl-5-(2-hydroxyethyl)-4-methyl-1,3-thiazolium chloride, **6,** 289; **7,** 16
t-Butyl isocyanide, **11,** 99
O,O-*t*-Butyl O-isopropenyl peroxycarbonate, **11,** 100
Cyanotrimethylsilane–Triethylaluminum, **10,** 113
Diethylaluminum cyanide–Chlorotrimethylsilane, **8,** 130
Hydrogen cyanide, **1,** 1197, 1198; **5,** 689
N-Methyl-2-pyrrolidone, **2,** 281
Organoaluminum cyanides, **1,** 244; **2,** 127; **4,** 146; **6,** 180; **9,** 158, 328
Potassium cyanide, **5,** 553; **8,** 409
Sodium cyanide, **4,** 446; **6,** 535; **8,** 453
Triethylaluminum, **2,** 427
ω-CYANO CARBONYLS
Formamide, **2,** 201
Mercury(II) acetate, **12,** 298
Nitrosyl chloride, **3,** 214
Phosphonitrilic chloride trimer, **6,** 469
OTHER CYANO CARBONYLS
Acrylonitrile, **1,** 973
Hydrogen cyanide–Triethylaluminum, **4,** 526
Methanesulfonyl chloride, **7,** 225
Phosphonitrilic chloride trimer, **6,** 469
Thallium(I) cyanide, **8,** 476
CYANOHYDRINS AND DERIVATIVES (*see also* CHIRAL COMPOUNDS)
CYANOHYDRINS

Acetonecyanohydrin, **1,** 5
Acetyl cyanide, **10,** 1
Formaldehyde, **1,** 397
Oxodiperoxymolybdenum(pyridine)-(hemamethylphosphoramic triamide), **7,** 136
(2R,4R)-Pentanediol, **12,** 375
Sodium cyanide, **5,** 606
Sodium hydrogen sulfite, **1,** 1047
CYANOHYDRIN ESTERS
Lithium iodide, **9,** 283
Phosphorus(III) chloride, **8,** 400
Potassium cyanide, **12,** 405
CYANOHYDRIN ETHERS
Arene(tricarbonyl)chromium complexes, **12,** 34
Benzyltriethylammonium chloride, **5,** 26
t-Butyl isocyanide, **12,** 95
Cyanotrimethylsilane, **11,** 147
Diethylaluminum cyanide, **12,** 182
(2R,4R)-Pentanediol, **12,** 375
Sodium hexamethyldisilazide, **6,** 529
CYANOHYDRIN SILYL ETHERS
Chlorotrimethylsilane, **4,** 537
Cyanotrimethylsilane, **4,** 542; **5,** 720; **6,** 80, 632; **9,** 127
Dicyclohexyl-18-crown-6, **5,** 207
Potassium cyanide, **8,** 409; **12,** 148
CYCLOALKANES (*see* ALICYCLIC HYDROCARBONS)

DIACETATES
1,1-DIACETATES
Acetic anhydride–Ferric chloride, **12,** 1
Lead tetraacetate, **4,** 278
Mercury(II) acetate, **4,** 319
1,2-DIACETATES
Ammonium persulfate, **12,** 33
Phenyliodine(III) diacetate, **1,** 508
Selenium(IV) oxide, **3,** 245
Silver acetate, **1,** 1002
Thallium(I) acetate, **5,** 654
Thallium(III) acetate, **1,** 1150
1,3-DIACETATES
Formaldehyde, **7,** 158
Lead tetraacetate, **1,** 537
Thallium(III) acetate, **1,** 1150
1,4-DIACETATES
Palladium(II) acetate, **12,** 367
DIALDEHYDES (*see* DICARBONYLS)
1,2-DIAMINES
Aziridine, **1,** 378
Bis(cyclopentadienyl)bis(μ-nitrosyl-N:N)dicobalt, **10,** 116
Bis(*p*-toluenesulfonyl)selenodiimide, **7,** 316
Cyanamide, **12,** 147
Dioxobis(*t*-butylimido)osmium(VIII), **8,** 206
Iodine isocyanate, **2,** 223
Mercury(II) oxide–Tetrafluoroboric acid, **9,** 293
DIAZO COMPOUNDS AND DIAZONIUM SALTS
DIAZO COMPOUNDS—GENERAL METHODS
Chloramine, **1,** 122
Hydrazine, **5,** 327
Lead tetraacetate, **1,** 537
Manganese(IV) oxide, **1,** 637
Mercury(II) oxide, **1,** 655
Phenyldiazomethane, **10,** 308
Silver(I) oxide, **1,** 1011
Sodium nitrite, **1,** 1097
p-Toluenesulfonyl azide, **1,** 1178; **2,** 415; **11,** 535
p-Toluenesulfonylhydrazide, **1,** 1185; **8,** 460
2,4,6-Triisopropylbenzene-sulfonylhydrazide, **11,** 563
Triphenylbismuth carbonate, **9,** 501
α-DIAZO ALDEHYDES AND KETONES
p-Carboxybenzenesulfonyl azide, **2,** 62
Chloramine, **1,** 122; **3,** 45
Copper(I) chloride–Oxygen, **5,** 165
α-Diazoacetophenone, **3,** 253
Diazomethane, **3,** 91; **4,** 120; **7,** 88
Lead tetraacetate, **8,** 269
Manganese(IV) oxide, **1,** 637
Mercury(II) trifluoroacetate, **1,** 659
Nitrosylsulfuric acid, **2,** 299
p-Toluenesulfonyl azide, **1,** 1178; **3,** 291; **4,** 510; **5,** 460, 675; **6,** 597
p-Toluenesulfonylhydrazide, **1,** 1185; **2,** 417
Triethylamine, **1,** 1198
2,4,6-Triisopropylbenzenesulfonyl azide, **10,** 422
α-DIAZO ESTERS
Bis(N-*t*-butylsalicylaldiminato)-copper(II), **12,** 52

DIAZO COMPOUNDS AND DIAZONIUM SALTS (*Continued*)
p-Carboxybenzenesulfonyl azide, **2,** 62
Diazoacetyl chloride, **9,** 133
Ethyl diazoacetate, **3,** 253
Isoamyl nitrite, **4,** 270; **6,** 307
Mercury bis(ethyl diazoacetate), **4,** 325
Silver(I) oxide, **2,** 368
Sodium nitrite, **1,** 1097
p-Toluenesulfonyl azide, **2,** 415; **3,** 291
DIAZONIUM SALTS
Dinitrogen trioxide-Boron trifluoride, **1,** 329
Isoamyl nitrite, **1,** 520
Nitrogen dioxide, **1,** 324
Nitrosylsulfuric acid, **1,** 755
p-Toluenesulfonyl azide, **1,** 1178; **2,** 415
Trichloroacetic acid, **1,** 1194
DICARBONYL COMPOUNDS (*see also* CHIRAL COMPOUNDS)
DIACIDS (*see also* CARBOXYLIC ACIDS, DIESTERS)
by Oxidative cleavage of arenes, cycloalkanediones, cycloalkanones, cycloalkenes, ketones (*see* TYPE OF REACTION INDEX)
Other routes
Copper(I) bromide, **5,** 163; **6,** 143
Formaldehyde, **1,** 397
Hypohalite solution, **1,** 488
Lithium diisopropylamide, **12,** 277
Methylenebis(magnesium bromide), **2,** 273
Potassium superoxide, **7,** 304
Sodium methoxide, **5,** 617
Vanadium(V) oxide, **1,** 733, 1057
1,3-DIALDEHYDES
2-Chloro-1,3-dithiane, **8,** 88
Methoxymethylenetriphenylphosphorane, **5,** 438
Thallium(III) nitrate, **7,** 362
1,4-DIALDEHYDES
Dimethyl sulfide, **2,** 156
Dimethyl sulfoxide–Oxalyl chloride, **11,** 215
Triphenylphosphine, **2,** 443
1,5-DIALDEHYDES
Raney nickel, **1,** 723
α,ω-DIALDEHYDES
Lead tetrakis(trifluoroacetate), **6,** 318; **9,** 269
Manganese(IV) oxide, **5,** 422
Ozone, **5,** 261
Periodic acid, **5,** 508
Potassium permanganate, **9,** 388
Sodium bismuthate, **1,** 1045
Triphenylbismuth carbonate, **9,** 501
DIAMIDES
Dichloromethylenedimethylammonium chloride, **4,** 135
DIESTERS (*see also* DIACIDS, ESTERS)
Diethyl carbonate, **1,** 247; **5,** 213
Diethyl oxalate, **1,** 929
Dimethyl carbonate, **12,** 201
Hexamethylphosphoric triamide, **12,** 239
Lead tetraacetate, **8,** 269
Lithium diisopropylamide, **6,** 334
Methyl 2-methylthioacrylate, **5,** 455
Ozone, **6,** 436; **9,** 341
Sodium ethoxide, **1,** 1065
(+)-(R)-*trans*-β-Styryl *p*-tolyl sulfoxide, **4,** 466
Titanium(IV) chloride, **8,** 483
1,2-DIKETONES
by Oxidation of C=C, C≡C, α-hydroxy ketones and related ketones (*see* TYPE OF REACTION INDEX)
by Rearrangement of epoxy carbonyl compounds
Acetic acid, **8,** 1
Boron trifluoride etherate, **1,** 70
Magnesium iodide, **6,** 353
Sulfuric acid, **7,** 347
Other routes
Acetic anhydride–Acetyl chloride, **7,** 1
Acyllithium reagents, **12,** 4
m-Chloroperbenzoic acid, **6,** 110
Dichlorotris(triphenylphosphine)-ruthenium(II), **4,** 564
Dichlorovinylene carbonate, **2,** 122
3,5-Dinitrobenzoyl *t*-butyl nitroxyl, **8,** 204
1,3-Dithianes, **2,** 182; **4,** 216; **10,** 231
Grignard reagents, **7,** 163
Iodosylbenzene, **10,** 213
Iron carbonyl, **8,** 265
Lead, **2,** 233
Lithium naphthalenide, **8,** 305
Lithium alkyl(phenylthio)cuprates, **7,** 211
α-Methoxyvinyllithium, **6,** 372
Nickel carbonyl, **1,** 720

Oxalyl chloride, **1,** 28
Oxygen, singlet, **6,** 431
Ozone, **8,** 374; **10,** 295
Samarium(II) iodide, **11,** 464
Silver(I) trifluoroacetate, **7,** 323
Sodium acetate, **5,** 591
Sodium dichromate–Cerium(III) acetate, **1,** 1062
1,1,3,3-Tetramethylbutyl isocyanide, **5,** 650
Thallium(III) nitrate, **4,** 492; **8,** 476
Titanium(III) chloride, **6,** 587
Tosylmethyl isocyanide, **8,** 493
Trimethylenebis(thiotosylate), **6,** 628

1,3-DIKETONES
by Acylation of C=O
Acetic anhydride, **1,** 69
Acetyl chloride, **6,** 164
Acetyl tetrafluoroborate, **11,** 6
Boron trifluoride, **3,** 32
Boron trifluoride–Acetic acid, **1,** 69
Butyllithium, **10,** 73
N,N-Dimethylhydrazine, **9,** 184
Dimethyl sulfoxide, **1,** 296
Grignard reagents, **9,** 229
Hydrogen fluoride, **6,** 285
Lithium hexamethyldisilazide, **7,** 197
Magnesium methoxide, **3,** 189
Mesityllithium, **8,** 317; **10,** 234
1-Morpholino-1-cyclohexene, **1,** 707
Potassium *t*-butoxide–*t*-Butyl alcohol, **12,** 402
Sodium amide, **1,** 1034
Sodium ethoxide, **1,** 1065
Sodium hydride, **1,** 1076; **2,** 382
Triphenylbis(2,2,2-trifluoroethoxy)-phosphorane, **10,** 43
Zinc chloride, **11,** 602
Other routes
Bis(3-dimethylaminopropyl)-phenylphosphine, **5,** 36
1-Diazolithioacetone, **11,** 155
Dimethylaluminum benzenethiolate, **11,** 194
Dimethyl sulfoxide–Oxalyl chloride, **11,** 215
Disodium tetrachloropalladate(II)–*t*-Butyl hydroperoxide, **10,** 175
Ferric chloride, **7,** 153
Hydroxylamine, **11,** 257
Sodium chromate, **5,** 605
Sodium methoxide, **1,** 1091
Tetrakis(triphenylphosphine)-palladium(0), **10,** 384
Trifluoroacetic anhydride, **1,** 1221
Triphenylphosphine–Thiocyanogen, **9,** 507
Zinc–Copper–Isopropyl iodide, **10,** 460

1,4-DIKETONES
by Acetonylation α to C=O
3-Chloro-2-trimethylsilyloxy-1-propene, **11,** 129
2,3-Dichloro-1-propene, **6,** 332
Isopropenyl acetate, **6,** 356
Mercury(II) trifluoroacetate, **9,** 294
2-Methoxyallyl bromide, **8,** 322
Tetrakis(triphenylphosphine)-palladium(0), **12,** 468
by Conjugate addition of acyl anions
3-Benzyl-5-(2-hydroxyethyl)-4-methyl-1,3-thiazolium chloride, **6,** 289; **7,** 16; **10,** 27
Bis(methylthio)(trimethylsilyl)-methyllithium, **6,** 53
Disodium tetracarbonylferrate, **8,** 216
Ethyl ethylthiomethyl sulfoxide, **5,** 299
Lithium di(α-methoxyvinyl)cuprate, **6,** 204
Methyl bis(ethylthio)acetate, **5,** 444
2-Methylfuran, **1,** 682
Methyl vinyl ketone, **6,** 38
Nickel carbonyl, **3,** 210
Ozone, **8,** 374
Potassium permanganate, **11,** 440
Sodium cyanide, **4,** 446
Titanium(III) chloride, **4,** 506
by Coupling of enolates and related compounds
Copper(I) chloride, **6,** 145
Copper(II) chloride, **6,** 139; **9,** 123
Copper(II) trifluoromethanesulfonate, **8,** 126; **10,** 110
Nickel peroxide, **5,** 474
Silver(I) oxide, **6,** 515
Zinc-copper couple, **7,** 428
by Reduction of enediones
Chromium(II) chloride, **3,** 60
Sodium iodide–Hydrochloric acid, **10,** 366
Tin, **1,** 1168
Titanium(III) chloride, **5,** 669

DICARBONYL COMPOUNDS (*Continued*)
Titanium(IV) chloride, **12,** 494
Zinc, **1,** 1276
Other routes
Chlorotrimethylsilane, **6,** 626
Copper(I) acetylacetonate, **4,** 100
1,3-Dichloro-2-butene, **4,** 472
(Dimethoxyphosphinyl)methyllithium, **6,** 339
Jones reagent, **7,** 68
S-(2-Methoxyallyl)-N,N-dimethyldithiocarbamate, **6,** 364
2-Nitropropene, **7,** 253
Oxygen, singlet, **9,** 338
Ozone, **7,** 269; **9,** 341
Palladium(II) chloride–Copper(I) chloride, **11,** 396
Phenylsulfinylacetone, **5,** 524
Phthaloyl peroxide, **6,** 471
1,3-Propanedithiol, **6,** 493
Silver carbonate–Celite, **3,** 247
N,N,N′,N′-Tetramethylsuccinamide, **4,** 490
Tin(II) trifluoromethanesulfonate, **12,** 490
4-Trimethylsilyloxyvaleronitrile, **8,** 513
Zinc-copper couple, **5,** 758
1,5-DIKETONES
by Alkylation of enolates and related compounds
1,3-Dichloro-2-butene, **7,** 203; **8,** 152
N,N-Dimethylhydrazine, **7,** 126
(E)-1-Iodo-3-trimethylsilyl-2-butene, **5,** 355
Mercury(II) trifluoroacetate, **9,** 294
Tetrakis(triphenylphosphine)-palladium(0), **11,** 503; **12,** 468
Triethylamine, **1,** 1198
by Conjugate addition of acetonyl anions
Acetoacetic acid, **6,** 21
Allyltrimethylsilane, **10,** 6
Cesium fluoride, **10,** 81; **12,** 108
1,4-Diazobicyclo[2.2.2]octane, **11,** 153
1-Ethylsulfinyl-3-pentanone, **9,** 221
Lithium diisopropylamide, **10,** 241
Methyl lithiodithioacetate, **11,** 340
Methyl vinyl ketone, **1,** 697
Titanium(IV) chloride, **6,** 590
Triphenylphosphine, **12,** 550
Other routes
2-(2-Bromomagnesioethyl)-2-methyl-1,3-dioxolane, **7,** 148
Chromium(VI) oxide, **2,** 72
Cobalt *meso*-tetraphenylporphine, **12,** 138
Hexafluoroantimonic acid, **8,** 239
6-Lithio-3,4-dihydro-(2*H*)-pyran, **8,** 70
Palladium(II) chloride–Copper(I) chloride, **7,** 278
N,N,N′,N′-Tetramethylglutaramide, **4,** 490
Titanium(IV) chloride, **10,** 401
Zinc–Zinc chloride, **6,** 675
1,6-DIKETONES
Chlorotrimethylsilane–Magnesium, **5,** 713
Chromium(VI) oxide, **2,** 72
Copper(II) tetrafluoroborate, **12,** 144
1,3-Dichloro-2-butene, **1,** 214
Manganese(IV) oxide, **5,** 422
Palladium(II) chloride–Copper(I) chloride, **7,** 278
Ruthenium tetroxide, **1,** 986
Vinylmagnesium chloride, **4,** 572
Zinc, **2,** 459
Zinc–Zinc chloride, **6,** 675
α-KETO ACIDS
Bis(ethylthio)acetic acid, **10,** 38
2,2-Difluoro-1-tolylsulfonyloxyvinyl-lithium, **9,** 166
1,3-Dithiane, **2,** 182
Potassium bisulfate, **1,** 909
Potassium permanganate, **1,** 942; **5,** 562
4-Pyridinecarboxaldehyde, **11,** 448
1,1,3,3-Tetramethylbutyl isocyanide, **3,** 279; **5,** 650
β-KETO ACIDS
Bis(trimethylsilyl) lithiomalonate, **9,** 55
t-Butyl α-lithioisobutyrate, **6,** 84
Carbon dioxide, **1,** 1102; **3,** 40; **6,** 94
Chlorotrimethylsilane, **4,** 537
Dimethylsulfoxonium methylide, **1,** 315
Lithium 2,6-di-*t*-butyl-4-methylphenoxide, **5,** 412
Lithium diisopropylamide, **7,** 204
Magnesium methyl carbonate, **3,** 190; **6,** 354
Sodium amide, **1,** 1034
Triethylamine, **1,** 1198
Triphenylmethylpotassium, **1,** 1258
γ-KETO ACIDS

O,O-*t*-Butyl O-isopropenyl peroxycarbonate, **11**, 100
Grignard reagents, **11**, 245
Ketene, **1**, 528
Oxygen, singlet, **11**, 385
Ruthenium tetroxide, **2**, 357
Tetrachloroethane–Nitrobenzene, **4**, 479

δ-KETO ACIDS
Periodates, **5**, 507
Potassium superoxide, **11**, 442
β-Propiolactone, **1**, 957
Ruthenium tetroxide, **2**, 357
Sodium dichromate, **5**, 138
Sodium periodate–Potassium permanganate, **2**, 312

KETO ACID HALIDES
Dichloromethyl methyl ether, **6**, 171; **9**, 154
Diketene, **1**, 264
Oxalyl bromide, **1**, 767

α-KETO ALDEHYDES
Diethyl azodicarboxylate, **1**, 245
Dimethyl sulfoxide, **1**, 296
Ethyl ethylthiomethyl sulfoxide, **5**, 299
Iodine, **2**, 220
Lithium diethylamide, **7**, 201
S-Methyl *p*-toluenethiosulfonate, **6**, 400
Nafion-H, **12**, 335
Nitrosyl chloride, **1**, 748
Selenium(IV) oxide, **1**, 992
Sodium acetate, **1**, 1024
Sodium methylsulfinylmethylide, **2**, 166
Sodium nitrite, **1**, 1097
Thallium(III) nitrate, **9**, 460
Triphenylphosphine, **1**, 1238

β-KETO ALDEHYDES
by Formylation α to C=O
2-Chloro-1,3-dithiane, **8**, 88
Diethoxycarbenium tetrafluoroborate, **11**, 175
1,2-Dimethoxyethane, **1**, 267
1,3-Dithienium tetrafluoroborate, **11**, 227
Ethyl formate, **1**, 380, 1077; **4**, 233
Sodium ethoxide, **1**, 1065
Titanium(IV) chloride, **6**, 590
Trimethyl orthoformate, **11**, 568
Other routes
1,3-Benzodithiolylium tetrafluoroborate, **10**, 24
Boron trifluoride etherate, **1**, 70
4-Methoxy-3-buten-1-ynylcopper, **9**, 297
3-Methoxy-1-methylthio-1-propyne, **6**, 397
Thallium(III) acetate, **2**, 406
Thallium(III) nitrate, **8**, 476

γ-KETO ALDEHYDES
by Conjugate addition
Ethyl ethylthiomethyl sulfoxide, **5**, 299
Ketene dimethyl thioacetal monoxide, **6**, 311; **8**, 268; **9**, 261
2-Lithio-1,3-dithianes, **9**, 235
Methyl vinyl ether, **8**, 276
(S)-(+)-*p*-Tolyl *p*-tolylthiomethyl sulfoxide, **10**, 408
Tris(phenylthio)methyllithium, **9**, 511
Other routes
2-(2-Bromoethyl)-1,3-dioxane, **7**, 37
2-Ethoxyallyl vinyl sulfide, **5**, 292
Mercury(II) acetate, **8**, 315
2,4,4,6-Tetramethyl-5,6-dihydro-1,3-(4*H*)-oxazine, **4**, 481
3-Trimethylsilyl-3-trimethylsilyloxy-1-propene, **8**, 485

δ-KETO ALDEHYDES
Acetone N,N-dimethylhydrazone, **8**, 276
Allyltrimethylsilane, **10**, 6
3,3-Diethoxy-1-propene, **9**, 158
2,2-Dimethyl-3(2*H*)-furanone, **10**, 159
1,3-Dithienium tetrafluoroborate, **11**, 227
(Z)-2-Ethoxyvinyllithium, **8**, 221
2,2,6-Trimethyl-1,3-dioxolenone, **10**, 424
α-Trimethylsilyl-2-methylbenzothiazole, **8**, 511

ε-KETO ALDEHYDES
Ozone, **1**, 773
Potassium hydride, **8**, 412; **10**, 327
Ruthenium tetroxide, **1**, 986

α-KETO AMIDES
Dichlorobis(methyldiphenylphosphine)-palladium(II), **11**, 165
Iodosylbenzene–Ruthenium catalysts, **11**, 270
Lithium bis(N,N-diethylcarbamoyl)-cuprate, **9**, 279
Methyl methylthiomethyl sulfoxide, **8**, 344
Oxygen, **7**, 258

β-KETO AMIDES
Acetoacetyl fluoride, **1**, 4

DICARBONYL COMPOUNDS (*Continued*)
Di-μ-carbonylhexacarbonyldicobalt, **11,** 162
Diketene, **1,** 264
Mercury(II) acetate, **2,** 264
2-Oxazolidones, chiral, **12,** 359
Polyphosphoric acid, **1,** 894
α-KETO ESTERS
Arene(tricarbonyl)chromium complexes, **12,** 34
Boron trifluoride, **1,** 68
t-Butyl hydroperoxide, **12,** 88
Cyanotrimethylsilane, **12,** 148
1,2-Diethoxy-1,2-bis(trimethylsilyloxy)-ethylene, **12,** 181
Diphenyl disulfide, **6,** 235
2-Ethoxycarbonyl-1,3-dithiane, **4,** 230
Ethyl 2-methylsulfinylacetate, **6,** 255
Grignard reagents, **11,** 245
Iodosylbenzene–Ruthenium catalysts, **11,** 270
Lead tetraacetate, **2,** 234
Lithium bromide, **2,** 245
Magnesium iodide, **6,** 353
Methyl methylthiomethyl sulfoxide, **8,** 344
S-Methyl *p*-toluenethiosulfonate, **5,** 456
Osmium tetroxide, **9,** 334
Oxygen, **7,** 258; **9,** 338
Ozone, **5,** 491
Potassium permanganate, **1,** 942
Selenium(IV) oxide, **1,** 992
N-Sulfinylaniline, **6,** 556
Trimethylphosphonoglycolate, **12,** 537
Triphenylphosphine, **1,** 1238
β-KETO ESTERS
from RCOX
Benzoyl chloride, **5,** 403
t-Butyl trimethylsilylacetate, **7,** 50
Ethoxycarbonylmethylenetriphenyl-phosphorane, **11,** 24
Isobutene, **1,** 522
Ketene alkyl trialkylsilyl acetals, **5,** 689
Lithium 2,2,6,6-tetramethylpiperidine, **4,** 310
Magnesium, **1,** 627
Meldrum's acid, **8,** 313; **10,** 252
Tetrakis(triphenylphosphine)-palladium(0), **11,** 503
Trimethylsilyl 2-lithio-2-ethoxy-carbonylacetate, **5,** 722; **9,** 493
Triphenylmethylsodium, **1,** 1259
Triphenylphosphine, **1,** 1238
by Alkoxycarbonylation
Carbon disulfide, **5,** 95
Diethyl (or Dimethyl) carbonate, **1,** 247; **2,** 129, 149; **5,** 234
Diethyl oxalate, **1,** 250
Ethyl diazoacetate, **6,** 252
Ethyl diethoxyphosphinylformate, **1,** 370
Magnesium methyl carbonate, **1,** 631; **6,** 354
Mercury(II) oxide–Boron trifluoride, **11,** 318
Methyl cyanoformate, **12,** 321
by Claisen (Dieckmann) reaction
t-Butyl lithioacetate, **9,** 82
Dichlorobis(trifluoromethane-sulfonato)titanium(IV), **12,** 173
Ethyl acetate, **1,** 1075
Magnesium naphthalenide, **1,** 711
Molecular sieves, **5,** 465
Potassium, **12,** 400
Potassium hydride, **6,** 482
Sodium, **2,** 158
Sodium ethoxide, **1,** 1065
Triphenylmethylpotassium, **1,** 1258; **9,** 502
Other routes
Acetoacetyl fluoride, **1,** 4
1,5-Diazabicyclo[4.3.0]nonene-5, **7,** 86
3,3-Diethoxy-1-methylthiopropyne, **5,** 207
Diethyl(2-chloro-1,1,2-trifluoroethyl)-amine, **4,** 149; **5,** 214
Diketene, **1,** 264; **6,** 202
Dimethyl sulfoxide–Oxalyl chloride, **11,** 215
Disodium tetrachloropalladate(II)–*t*-Butyl hydroperoxide, **10,** 175
Ethyl acetoacetate, **9,** 438
Ethyl diazoacetate, **1,** 367; **3,** 138
Ethyl 2-phenylsulfinylacetate, **6,** 256
Lithium–Ammonia, **3,** 179; **8,** 282
Potassium tetracarbonyl-hydridoferrate, **6,** 483
Rhodium(II) acetate, **9,** 406
Triethyloxonium tetrafluoroborate, **8,** 500
Zinc, **1,** 1285; **12,** 567

γ-KETO ESTERS (OR LACTONES)
by Conjugate additions
3-Benzyl-5-(2-hydroxyethyl)-4-methyl-1,3-thiazolium chloride, **7,** 16
Disodium tetracarbonylferrate, **8,** 216
Ethyl acrylate, **6,** 251
Ethyl ethylthiomethyl sulfoxide, **5,** 299
Lithium di(α-methoxyvinyl)cuprate, **6,** 204
Sodium cyanide, **4,** 446
Titanium(IV) chloride, **12,** 494
Tris(phenylthio)methyllithium, **6,** 650
Other routes
Copper(II) chloride, **6,** 139
Cyclopropenone 1,3-propanediyl ketal, **12,** 152
1-Ethoxy-1-trimethylsilyloxycyclopropane, **12,** 221
Ethyl diazoacetate, **4,** 228
2,2′-Ethylenebis(1,3-dithiane), **8,** 223
Hexamethylphosphoric triamide, **3,** 149
Lithium diisopropylamide, **11,** 296
Lithium N-isopropylcyclohexylamide, **7,** 209
2-Methoxyallyl bromide, **8,** 322
Methyl α-chloro-α-phenylthioacetate, **12,** 320
Palladium(II) chloride–Copper(I) chloride, **10,** 301; **11,** 396
Titanium(III) chloride, **5,** 669
Triethylammonium fluoride, **10,** 415
δ-KETO ESTERS (OR LACTONES)
(S)-1-Amino-2-methoxymethyl-1-pyrrolidine, **12,** 30
1-*t*-Butylthio-1-trimethylsilyloxyethylene, **9,** 92
Cesium fluoride, **10,** 81; **12,** 108
N,N-Dimethylhydrazine, **9,** 184
Ketene alkyl trialkylsilyl acetals, **7,** 371; **10,** 261; **11,** 279
Magnesium ethyl malonate, **8,** 311
Methyl α-(methyldiphenylsilyl)acetate, **12,** 325
Methyl 2-methylthioacrylate, **5,** 455
Nickel(II) acetylacetonate, **10,** 42
Quinine, **9,** 403
2,4,4,6-Tetramethyl-5,6-dihydro-1,3-(4*H*)-oxazine, **4,** 481
ε-KETO ESTERS (OR LACTONES)
6,6-Dimethyl-5,7-dioxaspiro[2.5]-octane-4,8-dione, **6,** 216
1-Ethoxy-1-trimethylsilyloxycyclopropane, **12,** 221
Ferric chloride, **12,** 230
Mercury(II) acetate, **12,** 298
KETO THIOLESTERS
N,N′-Carbonyldiimidazole, **9,** 96
Carbonyl sulfide, **11,** 112
S,S′-Dimethyl dithiocarbonate, **9,** 181
Sodium hydride, **9,** 427
1,2,3-TRICARBONYLS
t-Butyl hypochlorite, **4,** 58
2,3-Dichloro-5,6-dicyano-1,4-benzoquinone, **6,** 168
Dimethyl sulfoxide, **6,** 225; **8,** 200; **9,** 190
Oxygen, singlet, **12,** 363
Ozone, **11,** 387
Phenyliodine(III) diacetate, **12,** 384
Selenium(IV) oxide, **6,** 227
1,3,5-TRICARBONYLS
Butyllithium, **10,** 73
Magnesium methoxide, **7,** 220
Sodium hydride, **1,** 1075
DICARBOXYLIC ACIDS
(*see* DICARBONYLS)
DIENALS
2,2′-Bipyridinium chlorochromate, **11,** 44
trans-4-(*t*-Butylthio)-3-buten-2-one, **9,** 90
Dichloromethyl methyl ether, **9,** 154
4-Ethoxy-1,3-butadienyllithium, **8,** 278
1-Methoxy-1-buten-3-yne, **1,** 669
Molecular sieves, **6,** 411
Phenylthioacetylene, **9,** 370
Piperidine, **1,** 886
Potassium bisulfate, **1,** 909
Sodium borohydride, **1,** 1049
DIENES (*see also* ALLENES, CHIRAL COMPOUNDS)
1,3-DIENES
by Addition to C=O
Allyl phenyl selenide, **11,** 17
Lead(II) oxide, **1,** 536
Lithium diisopropylamide, **6,** 334; **9,** 280
Molybdenum carbonyl, **12,** 330
Organotitanium reagents, **11,** 174, 374
Tetrabutylammonium fluoride, **10,** 378
[(Trimethylsilyl)allyl]lithium, **8,** 273; **9,** 155; **11,** 12, 572
by Coupling (*see* TYPE OF

DIENES (*Continued*)
REACTION INDEX)
from Cyclobutenes
1-Cyclobutenylmethyllithium, **6,** 151
α-Lithiomethylselenocyclobutane, **10,** 232
2-Trimethylsilylmethylene-cyclobutane, **9,** 494
Vinyl acetate, **1,** 1271
by Elimination
Alumina, **1,** 19; **2,** 17
Bromomethanesulfonyl bromide, **12,** 75
Chlorotrimethylsilane–Sodium iodide, **11,** 127
Copper(I) cyanide, **5,** 166
1,5-Diazabicyclo[4.3.0]nonene-5, **1,** 189
Diethylaluminum 2,2,6,6-tetramethylpiperidide, **6,** 181
N,N-Dimethylaniline, **1,** 274
Dimethylformamide dialkyl acetals, **6,** 222; **11,** 198; **12,** 204
Dimethyl sulfoxide, **4,** 192
2,4-Dinitrobenzenesulfenyl chloride, **9,** 194
Ferric chloride, **8,** 229
Hydrobromic acid, **1,** 450
Iodine, **1,** 495; **4,** 258
Lithium amalgam, **4,** 287
Lithium carbonate–Lithium halide, **7,** 200; **8,** 244
Lithium chloride, **12,** 277
Methyltriphenoxyphosphonium iodide, **6,** 649
Palladium(II) acetate, **8,** 378; **9,** 349
Phthalic anhydride, **1,** 882
Potassium *t*-butoxide, **3,** 233; **12,** 401
Silver(I) oxide, **1,** 1011
Sodium amalgam, **11,** 473
Sodium isopropoxide, **2,** 385
Tetrabutylammonium fluoride, **9,** 444
Tetrakis(triphenylphosphine)-palladium(0), **10,** 384
Thiobenzoyl chloride, **6,** 582
Titanium(0), **11,** 526
Tributyl(iodomethyl)tin, **11,** 543
Tributyltin hydride, **11,** 545
Triethylammonium dimethylphosphate, **6,** 574
Trimethyl phosphite, **1,** 1233
Triphenylphosphine–Carbon tetrachloride, **7,** 404
Zinc-copper couple, **12,** 569
Zinc–Silver couple, **5,** 760
from 3-Sulfolenes
3,4-Dimethyl-3-sulfolene, **8,** 287
Grignard reagents, **8,** 235
1,3,3a,4,7,7a-Hexahydro-4,7-methanobenzo[*c*]thiophene 2,2-dioxide, **12,** 236
3-Sulfolene, **2,** 389; **12,** 455
Sulfur dioxide, **5,** 633
Vinyltriphenylphosphonium bromide, **5,** 750; **6,** 666
from Unsaturated arenesulfonyl-hydrazones
Lithium hydride, **4,** 304
p-Toluenesulfonylhydrazide, **2,** 417; **7,** 375; **10,** 400
by Wittig and related reactions
Allyldiphenylphosphine, **12,** 19
Diethyl benzylphosphonate, **1,** 1212
N,N′-Dimethyl-2-allyl-1,3,2-diazaphospholidine 2-oxide, **3,** 104
(Diphenylphosphine)lithium, **6,** 340
Formylmethylenetriphenyl-phosphorane, **8,** 234
Hexamethylphosphoric triamide, **12,** 239
Lithium ethoxide, **1,** 612
Sodium methylsulfinylmethylide, **6,** 546
Tetrabutylammonium iodide, **7,** 355
Other routes
Bis(cyclopentadienyl)-isoprenezirconium, **11,** 174
1,4-Bis(trimethylsilyl)-2-butyne, **11,** 64
N-Chlorosuccinimide, **5,** 127
Dichlorobis(cyclopentadienyl)-titanium, **12,** 168
Dicyclohexylborane, **4,** 141
Diethyl peroxide, **4,** 152
Dilithium tetrachlorocuprate(II), **12,** 195
Disiamylborane, **8,** 41
Iron carbonyl, **10,** 221
N-Lithioethylenediamine, **1,** 567
Lithium aluminum hydride, **1,** 581; **5,** 382; **6,** 325
Lithium dimethylcuprate, **2,** 151
Lithium divinylcuprate, **4,** 219

Phenyl-β-naphthylamine, **1,** 846
Potassium *t*-butoxide, **1,** 911; **2,** 336
Pyridinium chloride, **3,** 239
Pyridinium *p*-toluenesulfonate, **12,** 420
Silver perchlorate, **4,** 432
Sodium dicarbonyl(cyclopentadienyl)-ferrate, **11,** 483
Sodium methylsulfinylmethylide, **3,** 123
Thexylborane, **5,** 232
Titanium(0), **8,** 481
Trimethyltinlithium, **12,** 547

1,4-DIENES
Benzylchlorobis(triphenylphosphine)-palladium(II), **12,** 44
Benzyl 3-tributylstannylacrylate, **12,** 56
Bis(benzonitrile)dichloropalladium, **11,** 48
Bis(cyclopentadienyl)diiodozirconium, **12,** 53
Bis(cyclopentadienyl)isoprenezirconium, **11,** 174
Cesium fluoride, **11,** 115
Copper(I) bromide–Dimethyl sulfide, **8,** 117; **10,** 104
Dichlorobis(allyl)dipalladium, **5,** 189
Dichlorobis(cyclopentadienyl)titanium, **12,** 168
Diethyl azodicarboxylate, **11,** 177
Diisobutylaluminum hydride, **6,** 198
Dilithium tetrachloropalladate(II), **9,** 174
Diphosphorus tetraiodide, **6,** 243
Ethylaluminum dichloride, **9,** 11
Grignard reagents, **8,** 235; **9,** 229
Hydrogen hexachloroplatinate(IV)–Trichlorosilane, **8,** 378
Lead tetraacetate, **8,** 269
Lithium alkyl(phenylthio)cuprates, **7,** 212
Lithium-Ammonia, **8,** 282
Methylcopper, **7,** 236
Nickel(II) acetylacetonate–Trimethylaluminum, **9,** 52
Palladium(II) chloride, **11,** 393
Selenium(IV) oxide, **1,** 532, 993
Tetrakis(triphenylphosphine)-palladium(0), **10,** 384; **11,** 503
Tributyltin trifluoromethanesulfonate, **12,** 524
Zinc, **1,** 1276

1,5-DIENES
by Cope rearrangement
Bis(acetonitrile or benzonitrile)-dichloropalladium(II), **10,** 31; **11,** 46
Methyl 4-methylthiocrotonate, **6,** 389
by Coupling reactions (*see* TYPE OF REACTION INDEX)
by Dimerization of 1,3-dienes
Bis(1,5-cyclooctadiene)nickel(0), **12,** 52
Dicarbonylbis(triphenylphosphine)-nickel, **1,** 61
Lithium naphthalenide, **2,** 288
Other routes
Aluminum isopropoxide, **8,** 15
Butyllithium, **11,** 101
μ-Chlorobis(cyclopentadienyl)-(dimethylaluminum)-μ-methylenetitanium, **11,** 52
(2,6-Di-*t*-butyl-4-methylphenoxy)-methylaluminum trifluoromethane-sulfonate, **11,** 159
Dichlorobis(cyclopentadienyl)-titanium, **6,** 48
Hexamethylphosphoric triamide, **10,** 196
3-Lithio-1-trimethylsilyl-1-propyne, **2,** 239
Lithium aluminum hydride, **9,** 278
Lithium 2,2,6,6-tetramethylpiperidide, **6,** 345
Methyl cyclobutenecarboxylate, **8,** 335
Palladium(II) chloride, **6,** 447; **10,** 300; **12,** 371
Triethyloxonium tetrafluoroborate, **2,** 430
Tungsten(VI) chloride–Tetramethyltin, **7,** 415
Vinylcopper reagents, **6,** 662
Vinylmagnesium bromide, **10,** 190

TRIENES, TETRAENES
Allylidenetriphenylphosphorane, **6,** 14
1,5-Diazabicyclo[4.3.0]nonene-5, **4,** 16, 116
Hydrogen peroxide, **8,** 247
1-Lithiocyclopropyl phenyl sulfide, **9,** 271
4-Methoxy-3-penten-2-one, **7,** 231
Methyl 4-methylthiocrotonate, **6,** 389
N-Methyl-2-pyrrolidone, **11,** 346
Methyltriphenoxyphosphonium iodide, **6,** 649

DIENES (*Continued*)
Nickel(II) acetylacetonate, **11,** 58
Organocopper reagents, **11,** 365
Pentadienyl-1-pyrrolidinecarbodithioate, **12,** 374
N-Phenyl-1,2,4-triazoline-3,5-dione, **6,** 467
Potassium amide, **5,** 543
Potassium *t*-butoxide, **2,** 336
Potassium hydroxide, **5,** 96
Sodium amalgam, **11,** 473
Sodium benzenesulfinate, **6,** 526
Titanium(IV) chloride–Lithium aluminum hydride, **7,** 372
Zinc, **6,** 672

DIENOIC ACIDS (*see also* DIENOIC ESTERS)
Chlorotrimethylsilane, **6,** 626
Copper(I) iodide, **7,** 81; **8,** 121
Diethyl ethylidenemalonate, **9,** 161
Disodium tetrachloropalladate(II), **5,** 626
Ethyl (E)-3-methylthio-2-methyl-2-butenoate, **9,** 219

DIENOIC ESTERS (*see also* DIENOIC ACIDS)
2,4-DIENOIC ESTERS
Alumina, **11,** 22
Bis(1,5-cyclooctadiene)nickel(0), **4,** 33
Chlorobis(cyclopentadienyl)-hydridozirconium(IV), **8,** 84
Copper, **4,** 102
Copper(I) iodide, **8,** 121
Diacetatobis(triphenylphosphine)-palladium(II), **6,** 156
Di-μ-chlorobis(allyl)dipalladium, **6,** 394
Diethyl ethylidenemalonate, **9,** 161
Dihalobis(triphenylphosphine)-palladium(II), **6,** 60
Ethoxyacetylene, **1,** 357
Ethyl 2-phenylsulfinylacetate, **10,** 183
Lithium chloride, **12,** 277
Lithium divinylcuprate, **4,** 219
Methyl (allylthio)acetate, **10,** 261
Methyl 4-diethoxyphosphinylcrotonate, **8,** 336
Naphthalene-β-sulfonic acid, **1,** 712
Nickel carbonyl, **3,** 210
Palladium(II) acetate, **12,** 367
Palladium–Graphite, **10,** 297
Tetrakis(triphenylphosphine)-palladium(0), **11,** 503
Vinylcopper, **5,** 747

2,5-DIENOIC ESTERS
Allylcopper, **4,** 220
Benzyl 3-tributylstannylacrylate, **12,** 56
Ethylaluminum dichloride, **9,** 11

3,5-DIENOIC ESTERS
Lithium diisopropylamide, **12,** 277

OTHER ESTERS
(2-Carboxy-2-propenyl)triphenyl-phosphonium bromide, **9,** 97
1,4-Diazabicyclo[2.2.2]octane, **12,** 155
Lithium α-ethoxycarbonylvinyl-(1-hexynyl)cuprate, **6,** 329
Sodium benzeneselenoate, **9,** 432
Tetrakis(triphenylphosphine)-palladium(0), **9,** 451
1,1,1,3-Tetramethoxypropane, **10,** 424
1,1,1-Trimethoxy-3-phenylselenopropane, **9,** 489
Trimethylsilylmethylpotassium, **12,** 541

DIENONES
1,4-DIEN-3-ONES (*see also* CYCLOHEXDIENONES)
by Dehydrogenation of C=O, enones
Benzeneseleninic anhydride, **11,** 37
Chloranil, **1,** 125
2,3-Dichloro-5,6-dicyano-1,4-benzoquinone, **1,** 215; **2,** 112; **4,** 130
Methyl vinyl ketone, **7,** 247
Selenium(IV) oxide, **1,** 992

by Elimination
Calcium carbonate, **1,** 103
Di-μ-carbonylhexacarbonyldicobalt, **11,** 162
Lead tetraacetate, **5,** 365
Lithium carbonate–Lithium halides, **1,** 606; **5,** 395; **6,** 17
N-Methylthiomethylpiperidine, **12,** 326
Organolithium reagents, **11,** 13
Pyridine, **2,** 349
Pyridinium bromide perbromide, **1,** 967
Triphenylphosphine, **2,** 443

Other routes
Acetyl chloride–2-Trimethylsilyl-ethanol, **11,** 4
Benzylchlorobis(triphenylphosphine)-palladium(II), **12,** 44
(2-Bromovinyl)trimethylsilane, **11,** 82

Diallylzinc, **12,** 54
Dimethyl sulfoxide, **1,** 296
Iodylbenzene, **11,** 275
Manganese(IV) oxide, **1,** 637
Methoxyallene, **12,** 310
2-Methoxypropene, **2,** 230
3-Methyl-1,3-butadienyl phenyl sulfoxide, **8,** 333
Nickel carbonyl, **4,** 353
Oxygen, singlet, **5,** 486; **7,** 261
Perchloric acid, **11,** 402
1-Phenylthiovinyllithium, **9,** 371
Potassium *t*-butoxide, **1,** 911
Pyridinium chlorochromate, **9,** 397
Trimethyl(1-propenyl)tin, **12,** 470
Vinyltrimethylsilane, **12,** 26

2,4-DIEN-1-ONES
by Dehydrogenation of C = O, enones
Chloranil, **1,** 125
2,3-Dichloro-5,6-dicyano-1,4-benzoquinone, **1,** 215; **2,** 112; **4,** 130
by Elimination
Benzenesulfenyl chloride, **9,** 35
Copper(I) bromide–Lithium carbonate, **6,** 18
Lithium carbonate–Lithium bromide, **1,** 606
Tetrakis(triphenylphosphine)-palladium(0), **12,** 468
Other routes
B-1-Alkenyl-9-borabicyclo[3.3.1]-nonanes, **12,** 5
1,4-Benzoquinone, **1,** 49
Bis(acetonitrile)dichloropalladium(II), **11,** 46
Butyllithium, **10,** 68
Di-μ-carbonylhexacarbonyldicobalt, **8,** 148; **9,** 144
Dimethyl(methylene)ammonium halides, **10,** 160
Fluorosulfuric acid, **5,** 310
N-Methyl-2-pyrrolidone, **9,** 316
Pyridinium chlorochromate, **9,** 397
Selenium(IV) oxide, **5,** 575
2,4,6-Trimethylphenoxymagnesium bromide, **9,** 489
Zinc carbonate, **8,** 535

DIESTERS (*see* DICARBONYLS)

DI- AND POLYHALIDES

gem-DIFLUORIDES
Carbonic difluoride, **1,** 116
(Diethylamino)sulfur trifluoride, **6,** 183; **8,** 166
Lead(IV) fluoride, **1,** 563
Molybdenum(VI) fluoride, **5,** 465; **6,** 412
Phenyliodine(III) difluoride, **6,** 300
Phenylsulfur trifluoride, **1,** 849
Pyridinium poly(hydrogen fluoride), **5,** 538
Selenium(IV) fluoride, **5,** 576
Sulfur tetrafluoride, **1,** 1123; **2,** 392

gem-DICHLORIDES
(Bromodichloromethyl)phenylmercury, **5,** 514
t-Butyl nitrite, **7,** 48
Chlorodimethylsulfoxonium chloride, **4,** 200
Copper halide nitrosyls, **7,** 73
Dichloromethyl methyl ether, **1,** 220
Dimethylformamide–Thionyl chloride, **9,** 465
Molybdenum(V) chloride, **6,** 412
Phosphorus(V) chloride, **1,** 866
Sodium trichloroacetate, **1,** 1107
2,2,2-Trichloro-1,3,2-benzodioxaphosphole, **1,** 120

vic-DICHLORIDES
anti-
Hydrogen peroxide–Vilsmeier reagent, **11,** 256
Phenyliodine(III) dichloride, **2,** 225
Tetrabutylammonium iodotetrachloride, **1,** 1132
Trichloramine, **3,** 295; **4,** 520
syn-
Antimony(V) chloride, **4,** 21
Benzeneselenenyl halides, **12,** 39
2-Chloro-3-ethylbenzoxazolium tetrafluoroborate, **8,** 90
Phenyliodine(III) dichloride, **1,** 505
Sulfuryl chloride, **2,** 394
Tetrabutylammonium octamolybdate, **9,** 447
Triphenylphosphine–Carbon tetrachloride, **4,** 551
Triphenylphosphine dichloride, **6,** 646; **7,** 407
syn- and *anti*-
Copper(I) chloride, **3,** 67
Dimethylformamide–Thionyl chloride, **1,** 286
Hydrogen peroxide–Hydrochloric

DI- AND POLYHALIDES (*Continued*)
acid, **8,** 248
Phenyliodine(III) dichloride, **3,** 164
gem-DIBROMIDES
t-Butyl nitrite, **7,** 48
Copper halide nitrosyls, **7,** 73
Lithium diisopropylamide, **5,** 400
vic-DIBROMIDES
anti-
Bromine + co-reagent, **1,** 1025; **9,** 66; **11,** 156
N-Bromoacetamide, **1,** 74
Copper(II) bromide, **1,** 161
Pyridine perbromide, **5,** 568
Pyridinium bromide perbromide, **1,** 967
2,4,4,6-Tetrabromo-2,5-cyclohexadienone, **7,** 351
syn-
Triphenylphosphine dibromide, **7,** 407
syn- and *anti*-
Hydrogen peroxide–Hydrobromic acid, **8,** 248
Tetramethylammonium tribromide, **1,** 1144
Triphenylphosphine dibromide, **6,** 646
gem-DIIODIDES
Iodine, **4,** 258
Iodotrimethylsilane, **9,** 251
Sodium iodide, **5,** 323
OTHER DIODIDES
Phosphorus(V) oxide–Phosphoric acid, **1,** 872
Tetrahydrofuran, **1,** 1140
OTHER DIHALIDES
F, Cl
Hydrogen fluoride, **5,** 336
F, Br
Bromine fluoride, **2,** 365; **8,** 53
N-Bromoacetamide–Hydrogen fluoride, **1,** 75; **2,** 39
Pyridinium poly(hydrogen fluoride)–N-Bromosuccinimide, **5,** 539
F, I
Bromine fluoride, **5,** 351
Iodine fluoride, **2,** 365; **5,** 351
Methyliodine(III) difluoride, **7,** 242
Pyridinium poly(hydrogen fluoride)–N-Bromosuccinimide, **5,** 539
Cl, Br
Antimony(V) chloride, **5,** 18
Bromine chloride, **2,** 38
N-Bromoacetamide–Hydrogen chloride, **1,** 74
Cl, I
Chloroiodomethane, **10,** 89
Sodium iodide, **5,** 323
Triphenylphosphine–Diethyl azodicarboxylate, **7,** 404
POLYHALIDES
Polyfluorides
Phenylsulfur trifluoride, **1,** 849
Sulfur tetrafluoride, **1,** 1123; **2,** 392; **5,** 640
Trifluoroiodomethane, **5,** 148
Polychlorides
Dichlorotris(triphenylphosphine)-ruthenium(II), **5,** 740; **6,** 654
Trichloromethyllithium (or sodium), **1,** 1108; **2,** 119
DIKETONES (*see* DICARBONYLS)
DIOLS AND POLYOLS (*see also* CHIRAL COMPOUNDS)
1,2-DIOLS
from Epoxides
Alumina, **6,** 16; **8,** 9; **9,** 8; **10,** 8
(+)-Camphanic acid, **8,** 74
Dimethylsulfoxide, **1,** 157
Dimethyl sulfoxide–Trifluoroacetic acid, **6,** 226
Ferric chloride, **8,** 229
Nafion-H, **10,** 275
Perchloric acid, **1,** 796
Periodic acid, **1,** 815
Sodium borohydride, **8,** 449
from Epoxy alcohols
Sodium bis(2-methoxyethoxy)-aluminum hydride, **11,** 476
Titanium(IV) isopropoxide, **12,** 504
by *anti*- or *syn*-Hydroxylation of C=C (*see* TYPE OF REACTION INDEX)
by Pinacol coupling (*see* TYPE OF REACTION INDEX)
by Reduction of α-hydroxy(or alkoxy) carbonyls (*see* TYPE OF REACTION INDEX)
Other routes
Benzeneboronic acid, **11,** 408
Borane–Tetrahydrofuran, **1,** 199; **4,** 124
sec-Butyllithium, **8,** 69; **9,** 88
Cobalt(III) acetate, **9,** 119

Copper(II) sulfate–Pyridine, **9,** 125
Cyanoacetic acid, **11,** 145
Diethyl azodicarboxylate, **1,** 245
Dimethylphenylsilane, **12,** 209
Dimethyl sulfoxide, **4,** 192
(S)-N-Formyl-2-methoxymethyl-pyrrolidine, **11,** 243
Grignard reagents, **10,** 189
Hydrogen peroxide–Selenium(IV) oxide, **3,** 156
Iodine, **10,** 210
Lithium 1-(dimethylamino)-naphthalenide, **10,** 244
Mercury(II) acetate, **5,** 424
Mercury(II) trifluoroacetate, **5,** 429
Methoxy(trimethylsilyl)methyllithium, **9,** 284
Methyllithium, **1,** 686
Organoaluminum reagents, **11,** 364
Osmium tetroxide, **9,** 334
Phenyliodine(III) diacetate, **1,** 508
Potassium borohydride, **6,** 476
Sodium borohydride, **9,** 420
Tin, **11,** 519
Titanium(III) chloride, **4,** 506
Tributyltinmethanol, **7,** 378
Trifluoroacetic acid trianhydride, with iodous acid, **7,** 181

1,3-DIOLS
Borane, **4,** 124; **8,** 49; **9,** 136
t-Butyl hydroperoxide–Dialkyl tartrate, **11,** 92
Di-μ-carbonylhexacarbonyldicobalt, **12,** 163
Formaldehyde, **7,** 158
Hexakis(acetato)trihydrato-μ-oxotrisrhodium acetate, **11,** 252
Iodine, **10,** 210
Lithium–Ammonia, **5,** 379; **11,** 286
Lithium naphthalenide, **11,** 302
1-(Methoxymethoxy)styrene, **12,** 99
Organocopper reagents, **12,** 345
Sodium bis(2-methoxyethoxy)aluminum hydride, **11,** 476
Sodium borohydride, **12,** 441
Thexylborane, **12,** 484
Titanium(IV) chloride, **12,** 494
Tributylborane, **10,** 410
Tributyltin hydride, **12,** 516
Trichloromethyltitanium, **12,** 355
2,4,6-Trimethylphenoxymagnesium bromide, **9,** 489

1,4-DIOLS
3-Bromopropyl 1-ethoxyethyl ether, **4,** 226
Grignard reagents, **1,** 415
Hydrogen peroxide–Iron salts, **1,** 472
(4S,5S)-4-Methoxymethyl-2-methyl-5-phenyl-2-oxazoline, **6,** 386
Oxygen, singlet, **7,** 261
Thexylborane, **1,** 276; **10,** 397
Thiourea, **6,** 586

POLYOLS
9-(Benzyloxy)anthracene, **12,** 47
Formaldehyde, **1,** 397
Iodine, **11,** 261

DIOXETANES
Hydrogen peroxide, **6,** 287
Mercury(II) trifluoroacetate, **8,** 316
Oxygen, singlet, **5,** 486; **6,** 241
Tetramethylammonium hydroxide, **11,** 514
Triphenyl phosphite ozonide, **3,** 323

DISELENIDES
Lithium diselenide, **8,** 310
Sodium hydrogen selenide, **8,** 459
Sodium methaneselenoate, **12,** 450
Triphenylphosphine–Selenocyanogen, **9,** 506

DISULFIDES
BY OXIDATION OF THIOLS (*see* TYPE OF REACTION INDEX)
BY REDUCTION OF RSO_2X, RSO_2OH (*see* TYPE OF REACTION INDEX)
endo-DISULFIDES
Manganese(IV) oxide, **8,** 312
Sodium trithiocarbonate, **9,** 435
OTHER METHODS FOR RSSR
Bromodimethylsulfonium bromide, **9,** 66
Cyanogen bromide, **5,** 169
Sodium disulfide, **1,** 1064
Sodium sulfide–Sulfur, **3,** 272
METHODS FOR RSSR′
Cyanogen bromide, **5,** 169
2,4-Dinitrobenzenesulfenyl chloride, **6,** 231
Nitrogen dioxide, **8,** 204

DITHIOLS
gem-
Morpholine, **1,** 705
vic-
Sodium borohydride, sulfurated, **5,** 601

DITHIOLS (*Continued*)
Thiolacetic acid, **1,** 1154
Thiourea, **1,** 1164
DIYNES
1,3-DIYNES
by Coupling of C≡C (*see* TYPE OF REACTION INDEX)
Other routes
Alumina, **2,** 18
Butadiyne, **1,** 185
Chlorotris(triphenylphosphine)-rhodium(I), **3,** 325; **5,** 736; **6,** 652
Copper(I) chloride, **2,** 91
Copper(I) cyanide, **6,** 146
Dicyclohexylmethylthioborane, **7,** 101
3,3-Diethoxy-1-propyne, **2,** 126
Iodine, **6,** 293; **7,** 179
Tetrakis(triphenylphosphine)-palladium(0), **12,** 468
Thionyl chloride, **5,** 663
1,4-DIYNES
Copper(I) chloride, **2,** 91; **3,** 67
Di-μ-carbonylhexacarbonyldicobalt, **12,** 163
3,3-Diethoxy-1-propyne, **2,** 126

ENAMINES AND ENAMINONES
ENAMINES (*see also* CHIRAL COMPOUNDS)
from Allylic amines
2,2′-Bis(diphenylphosphine)-1,1′-binaphthyl, **11,** 53; **12,** 56
Butyllithium, **10,** 68
from C=O's + amines—General methods
Calcium chloride, **1,** 104
Diethyl lithiomorpholinomethyl-phosphonate, **7,** 105
Molecular sieves, **1,** 703; **4,** 345; **8,** 486
Morpholine, **1,** 705, 972, 1175
1-Morpholino-1-cyclohexene, **1,** 707
Piperidine, **1,** 886, 972
Pyrrolidine, **1,** 972; **2,** 354; **3,** 240; **7,** 309; **8,** 428
Tetrakis(dimethylamino)titanium, **2,** 399
Tin(IV) chloride, **6,** 553
Titanium(IV) chloride, **2,** 414; **12,** 494
from C=O + other reagents
Chlorotrimethylsilane, **5,** 709
N,N-Dimethylformamide, **5,** 247
Hexamethylphosphoric triamide, **4,** 244
N-Morpholinomethyldiphenyl-phosphine oxide, **9,** 318; **12,** 334
Organotitanium reagents, **11,** 374
Potassium *t*-butoxide, **9,** 380; **10,** 323
Other routes
Cadmium acetate–Zinc acetate, **1,** 103
N,N-Diethylaminopropyne, **5,** 217; **7,** 107
N,N-Dimethylacetamide diethyl acetal, **1,** 271
Mercury(II) acetate, **1,** 644
Morpholine, **11,** 352
ENAMINONES
Acetic anhydride, **5,** 3
Alkylaluminum halides, **12,** 5
N-Chlorosuccinimide–Dimethyl sulfide, **9,** 111
3-(Dimethylamino)-2-azaprop-2-en-1-ylidene dimethylammonium chloride, **11,** 194
Dimethylformamide diethyl acetal, **9,** 182
ENDOPEROXIDES (*see* PEROXIDES)
ENEDIONES
1,4-ENEDIONES
Bis(methylthio)methyllithium, **7,** 24
Copper(II) acetylacetonate, **3,** 62; **9,** 51
Copper(II) nitrate–Pyridine, **1,** 164
2,2-Diethoxyvinylidenetriphenyl-phosphorane, **3,** 93
2-Ethoxycarbonyl-1,3-dithiolane, **5,** 304
Methoxyallene, **8,** 320
Oxygen, singlet, **12,** 363
Phthaloyl peroxide, **6,** 471
Pyridinium chloride, **5,** 566
Pyridinium chlorochromate, **10,** 334; **11,** 450
Pyruvaldehyde 2-phenylhydrazone, **6,** 500
Ruthenium tetroxide, **2,** 357
Selenium(IV) oxide, **1,** 992
Sodium dichromate, **1,** 1059
Sodium peroxide, **11,** 492
Tin(II) trifluoromethanesulfonate, **12,** 490
1,5-ENEDIONES
1,3-Dithienium tetrafluoroborate, **11,** 227
α-Ketoketene dithioacetals, **11,** 282

Zinc bromide, **11,** 600
2-METHYLENE(OR ALKYLIDENE)-1,3-DICARBONYLS
Benzeneselenenyl halides, **5,** 518; **7,** 286
Boron oxide, **3,** 33
2,3-Dichloro-5,6-dicyano-1,4-benzoquinone, **1,** 215; **2,** 112
N-Methylthiomethylpiperidine, **12,** 326
Selenium, **11,** 465
Trimethylsilyl trifluoromethanesulfonate, **10,** 438
ENOL CARBAMATES
Benzyltrimethylammonium fluoride, **10,** 29
Crotyl carbamates, **12,** 82
ENOL CARBONATES
Benzyltrimethylammonium fluoride, **10,** 29
Ethyl(or Methyl) chloroformate, **8,** 307; **11,** 336
ENOL ESTERS
ENOL ACETATES
by Acetoxylation of C=C (*see* TYPE OF REACTION INDEX)
from C=O
Acetic anhydride, **4,** 509
Acetyl chloride, **2,** 383, 408; **9,** 387
4-Dimethylaminopyridine, **9,** 178
Dimethylformamide diethyl acetal, **9,** 182
Hydrobromic acid, **3,** 154
Isopropenyl acetate, **1,** 524; **3,** 103
Perchloric acid, **1,** 796; **2,** 309
Potassium acetate, **1,** 906
Sulfoacetic acid, **1,** 1117
Sulfosalicylic acid, **1,** 1118
p-Toluenesulfonic acid, **1,** 1172
Other routes
Acetyl methanesulfonate, **9,** 2
Copper(I) acetate, **6,** 142
Lithium dimethylcuprate, **4,** 177
Silver(I) trifluoroacetate, **10,** 355
Zinc–Acid anhydride–Catalyst, **4,** 577
OTHER ENOL ESTERS
Acetylene, **1,** 11
Benzoic anhydride, **3,** 262
Chloro(trimethylsilyl)methyllithium, **8,** 277
Dimethyl oxomalonate, **11,** 207
Divinylmercury, **1,** 352
Mercury(II) acetate, **1,** 644
Potassium persulfate, **5,** 563
Trifluoroacetyl trifluoromethanesulfonate, **9,** 485
ENOL ETHERS
FROM CARBONYLS, $RCOCH_2R' \rightarrow R(OR)C=CHR'$
Boron trichloride, **1,** 67
Chloromethyl methyl ether, **4,** 83
2,2-Dimethoxypropane, **1,** 268
Lithium diisopropylamide, **12,** 277
Methyl fluorosulfonate, **4,** 339; **6,** 381
Pentaethoxyphosphorane, **2,** 305
Phase-transfer catalysts, **9,** 356
Thallium(I) ethoxide, **4,** 501
p-Toluenesulfonic acid, **1,** 1172
Trialkyl orthoformate, **1,** 1204; **5,** 356, 714
Triethyloxonium tetrafluoroborate, **2,** 430; **6,** 611
Trimethylsilyldiazomethane, **12,** 538
FROM CARBONYLS, $R_2C=O \rightarrow R_2C=CHOR$
Carbon disulfide, **12,** 316
Diethyl methoxyethoxymethylphosphonate, **9,** 162
Dimethyl diazomethylphosphonate, **10,** 157
α-Methoxyethylidenetriphenylphosphorane, **1,** 669
Methoxymethyl(diphenyl)phosphine oxide, **7,** 229; **9,** 301
Methoxymethylenetriphenylphosphorane, **1,** 671; **2,** 271
Methoxy(trimethylsilyl)methyllithium, **9,** 284
BY ISOMERIZATION OF ALLYLIC ETHERS
Bis(benzonitrile)dichloropalladium(II), **5,** 31
Chlorotris(triphenylphosphine)rhodium(I), **5,** 736
Dimethyl sulfoxide, **1,** 296; **2,** 157
Iridium catalysts, **8,** 135
Palladium catalysts, **10,** 299
Potassium *t*-butoxide, **5,** 544
BY METHYLENATION OF ESTERS
μ-Chlorobis(cyclopentadienyl)-(dimethylaluminum)-μ-methylenetitanium, **10,** 87; **11,** 52
OTHER ROUTES
Benzenesulfonyl chloride, **5,** 22

ENOL ETHERS (*Continued*)
Bis(benzonitrile)dichloroplatinum(II), **11,** 49
Diisopropylethylamine, **1,** 371
Dimethylformamide dineopentyl acetal, **12,** 204
Divinylmercury, **1,** 352
(Z)-2-Ethoxyvinyllithium, **8,** 221
1,2-Ethylenebis(triphenylphosphonium) dibromide, **6,** 665
Grignard reagents, **6,** 269; **7,** 226; **10,** 190
Iodotrimethylsilane, **11,** 271
Magnesium–Iodine–Ether, **1,** 630
Mercury(II) acetate, **1,** 644
Methoxyallene, **7,** 163
Methoxy(phenylthio)methyllithium, **9,** 473
Pentacarbonyl(methoxyphenylmethylene)chromium(0) or -tungsten(0), **4,** 378; **11,** 397
Pentacarbonyl(trimethylsilyl)manganese, **11,** 401
Phase-transfer catalysts, **9,** 356
Phosphoryl chloride, **2,** 330
Potassium *t*-butoxide, **12,** 401
ENOL PHOSPHATES (*see* PHOSPHORUS COMPOUNDS)
ENYNES
1,3-ENYNES
by Coupling of vinyl + allyl substrates (*see* TYPE OF REACTION INDEX)
by Elimination
Chromium(II)–Amine complexes, **3,** 57
1,5-Diazabicyclo[4.3.0]nonene-5, **1,** 189; **4,** 16
Di-μ-carbonylhexacarbonyldicobalt, **7,** 99; **8,** 148
Diphosphorus tetraiodide, **1,** 349
Potassium bisulfate, **1,** 909
Potassium *t*-butoxide, **9,** 126; **12,** 401
Sodium amide, **2,** 373
Zinc oxide, **1,** 1294
by Reduction of diynes
Lithium diisobutylmethylaluminum hydride, **8,** 292
Palladium catalysts, **7,** 275
Zinc–1,2-Dibromoethane, **12,** 570
Other routes
Benzyltrimethylammonium hydroxide, **5,** 29
1,3-Bis(triisopropylsilyl)propyne, **11,** 63
t-Butyl hypochlorite, **6,** 82
Chlorotris(triphenylphosphine)rhodium(I), **8,** 109
Copper(I) oxide, **1,** 169
Diethyl phosphorochloridate, **8,** 171
Dihydridotetrakis(tributylphosphine)ruthenium(II), **10,** 148
Dimethyl(methylthio)sulfonium tetrafluoroborate, **12,** 207
Ethylene glycol, **1,** 375
Methyllithium, **5,** 448
Pyridazine 1-oxide, **9,** 395
Sodium amalgam, **9,** 416
Sodium methoxide, **5,** 617
Tributyl(methylphenylamino)phosphonium iodide, **10,** 268
Triethyl orthoacetate, **6,** 607
Triphenylpropargylphosphonium bromide, **2,** 447
1,4-ENYNES
Copper(I) chloride, **1,** 166
Dilithium tri(1-pentynyl)cuprate, **6,** 203
Lead tetraacetate, **8,** 269
Propargyl bromide, **6,** 493
1,5-ENYNES
Di-μ-carbonylhexacarbonyldicobalt, **10,** 129
3-Lithio-1-trimethylsilyl-1-propyne, **2,** 239
Lithium 3-chloropropargylide, **11,** 294
Lithium 2,2,6,6-tetramethylpiperidide, **6,** 345
Tetrakis(triphenylphosphine)palladium(0), **10,** 384
Trimethylaluminum–Dichlorobis(cyclopentadienyl)zirconium, **11,** 567
1-Trimethylsilylpropynylcopper, **6,** 638
Zinc chloride, **11,** 602
EPISULFIDES
Diazomethane, **5,** 179
4,4-Dimethyl-2-methylthio-2-oxazoline, **6,** 584
cis-3a,4,5,6,7,7a-Hexahydrobenzothiazolidine-2-thione, **11,** 304
Iodine thiocyanate, **4,** 263
3-Methylbenzothiazole-2-thione, **6,** 374
Phthalimide-N-sulfenyl chloride, **9,** 375
Potassium thiocyanate, **6,** 487
Sulfur monochloride, **3,** 275

Thiourea, **1,** 1164
Triphenylphosphine, **7,** 403
Triphenylphosphine sulfide, **5,** 732
EPOXIDES (*see also* CHIRAL COMPOUNDS, EPOXY..., GLYCIDIC ACIDS, ETC., HALO EPOXIDES)
BY ADDITION TO C=O
Adding CH_2
Bromomethyllithium, **6,** 74; **12,** 77
m-Chloroperbenzoic acid, **7,** 62
Diazomethane, **3,** 6
Dibromomethane–Lithium, **3,** 201
(N,N-Diethylamino)methyloxosulfonium methylide, **5,** 210
(Dimethylamino)phenyloxosulfonium methylide, **3,** 105
Dimethylsulfonium methylide, **1,** 314; **2,** 169; **3,** 124; **5,** 646; **6,** 219; **8,** 198; **9,** 188; **11,** 213
Dimethylsulfoxonium methylide, **1,** 315; **2,** 171; **3,** 125; **4,** 197; **5,** 646; **10,** 168
Dimethyl N-(*p*-toluenesulfonyl)-sulfoximine, **3,** 205; **9,** 193; **12,** 216
Lauryldimethylsulfonium chloride, **5,** 363
Methyl(phenyl)selenoniomethanide, **11,** 343
Methylphenyl-N-*p*-toluenesulfonylsulfoximine, **3,** 204
Phenylthiomethyllithium, **5,** 527
Adding CHR, CR_2
Benzeneselenol, **6,** 28, 362
Diphenyl selenide, **6,** 240
Diphenylsulfonium cyclopropylide, **4,** 211; **5,** 281; **8,** 212
Diphenylsulfonium isopropylide, **2,** 180
Hexamethylphosphoric triamide, **12,** 239
Methaneselenol, **6,** 361
Triphenylarsonium ethylide, **10,** 445
Zinc chloride, **8,** 536
Other routes
Hexamethylphosphorous triamide, **1,** 431
Nickel carbonyl, **2,** 290
BY CYCLIZATION
from 1,2-Diols
Dimethylformamide dimethyl acetal, **3,** 115
Diphenyldi(1,1,1,3,3,3-hexafluoro-2-phenyl-2-propoxy)sulfurane, **4,** 205; **5,** 270; **8,** 208
Hydrogen bromide–Acetic acid, **12,** 242
p-Toluenesulfonyl chloride, **8,** 489
Triphenylphosphine + co-reagent, **9,** 504; **12,** 551
from Halohydrins and related
Acetyl hypobromite, **1,** 12
N-Bromoacetamide, **5,** 61
N-Bromosuccinimide, **4,** 49; **5,** 65; **9,** 70
t-Butyl hydroperoxide–Titanium(IV) chloride, **12,** 94
Chlorotrimethylsilane, **5,** 709
2,2′-Dihydroxy-1,1′-binaphthyl, **12,** 190
Iodine, **3,** 159; **8,** 256
Silica, **10,** 346
Silver carbonate–Celite, **5,** 577
Silver(I) oxide, **1,** 1011
Tellurium(IV) chloride, **12,** 457
from β-Hydroxy sulfides
Lithium tri-*sec*-butylborohydride, **12,** 286
(R)-1-(1-Naphthyl)ethyl isocyanate, **9,** 321
Organotitanium reagents, **11,** 374
Phenylthioacetic acid, **6,** 463
Potassium *t*-butoxide, **9,** 380
Other routes
Methanesulfonyl chloride, **1,** 662
Potassium *t*-butoxide, **2,** 336; **4,** 399
BY EPOXIDATION (*see* TYPE OF REACTION INDEX)
BY OXYGENATION OF 1,3-DIENES
Cobalt *meso*-tetraphenylporphine, **12,** 138
Dichlorotris(triphenylphosphine)-ruthenium(II), **11,** 171
Oxygen, singlet, **4,** 362; **6,** 431; **8,** 367
OTHER ROUTES
Ozone, **8,** 375
Periodates, **4,** 373
Tetraethylammonium fluoride, **7,** 356
Trifluoromethanesulfonic anhydride, **10,** 419
Triphenylphosphine, **5,** 725
EPOXY ALCOHOLS (*see also* CHIRAL COMPOUNDS)
β,γ-EPOXY ALCOHOLS

EPOXY ALCOHOLS (*Continued*)
by Epoxidation of allylic alcohols (*see* TYPE OF REACTION INDEX)
by Reduction of α,β-epoxy carbonyls
Sodium borohydride–Cerium(III) chloride, **10,** 359
Sodium cyanoborohydride, **5,** 607
Triisobutylaluminum, **9,** 486
Zinc borohydride, **11,** 599
Other routes
Iodine, **11,** 261
Potassium permanganate, **9,** 388
Tetracarbonyl(cyclopentadienyl)-vanadium, **5,** 174
Vanadyl acetylacetonate–Azobisisobutyronitrile, **10,** 457
γ,δ-EPOXY ALCOHOLS
t-Butyl hydroperoxide–Dialkyl tartrate, **12,** 90
Lead tetrakis(trifluoroacetate), **6,** 318
α,β-EPOXY ALDEHYDES AND KETONES (*see also* CHIRAL COMPOUNDS)
BY EPOXIDATION OF α,β-UNSATURATED CARBONYLS (*see* TYPE OF REACTION INDEX)
BY OXIDATION OF EPOXY ALCOHOLS
m-Chloroperbenzoic acid–2,2,6,6-Tetramethylpiperidine, **8,** 99
Jones reagent, **2,** 70
OTHER ROUTES
1-Chloro-3-diazoacetone, **5,** 114
1,4-Dichloro-2-butanone, **9,** 75
Hydrogen peroxide, **10,** 201
(S)-(−)-Proline, **10,** 331; **11,** 446
Silver(I) oxide, **6,** 515
Sodium hydroxide, **8,** 430
Tin(II) fluoride, **11,** 524
Tin(II) trifluoromethanesulfonate, **11,** 525
α,β-EPOXY SILANES
Chloromethyltrimethylsilane, **12,** 118
m-Chloroperbenzoic acid, **8,** 290
Chloro(trimethylsilyl)methyl(or ethyl)-lithium, **8,** 277
(E)-1-Iodo-3-trimethylsilyl-2-butene, **5,** 355
(1-Lithiovinyl)trimethylsilane, **5,** 374
Peracetic acid–Sodium acetate, **7,** 279
α,β-EPOXY SULFONES
Chloromethyl *p*-tolyl sulfone, **6,** 41; **8,** 340
Hydrogen peroxide, **3,** 155; **6,** 486
Potassium *t*-butoxide, **5,** 544
Potassium hypochlorite, **6,** 486
ESTERS—GENERAL METHODS
(*see also* α,β-ACETYLENIC ESTERS, DICARBONYLS, ESTERS—SPECIFIC KINDS OF ESTERS, GLYCIDIC ESTERS, UNSATURATED CARBONYLS)
BY ACYLATION OF ROH
with ROCl
Alumina, **1,** 20
Bis(methylcyclopentadienyl)tin(II), **12,** 201
Dibutyltin oxide, **10,** 123
3-Dimethylaminopropylamine, **1,** 274
Silver cyanide, **7,** 320
Thallium(I) 2-methyl-2-propanethiolate, **6,** 581
with RCOOH
Aluminum chloride, **5,** 10
3-Benzylthiazolium bromide, **10,** 27
N,N-Bis(2-oxo-3-oxazolidinyl)-phosphorodiamidic chloride, **10,** 41
Boron trifluoride etherate, **3,** 33; **4,** 44
N,N′-Carbonyldiimidazole, **1,** 114
Cesium fluoride, **10,** 81
6-Chloro-1-*p*-chlorobenzene-sulfonyloxybenzotriazole, **6,** 106
Chlorotrimethylsilane, **11,** 125
Dicyclohexylcarbodiimide + co-reagent, **1,** 231; **8,** 163; **9,** 156
Diethyl phosphorocyanidate, **7,** 107
4-Dimethylaminopyridine, **10,** 155
4-(N,N-Dimethylamino)pyridinium chlorosulfite chloride, **12,** 199
N,N-Dimethylphosphoramidic dichloride, **9,** 187
2-Fluoro-1,3,5-trinitrobenzene, **8,** 230
Graphite bisulfate, **6,** 269
2-Halopyridinium salts, **6,** 375; **7,** 110; **8,** 95; **9,** 234
1-Methylsulfonyloxybenzotriazole, **6,** 106
Molecular sieves, **1,** 703; **2,** 286
3-Nitro-2-pyridinesulfenyl chloride, **9,** 325
Silver perchlorate, **6,** 519
Silver(I) trifluoroacetate, **8,** 444

N,N,N',N'-Tetramethylchloroformamidinium chloride, **12,** 477
Thionyl chloride, **5,** 663
p-Toluenesulfonic acid, **1,** 1172
p-Toluenesulfonyl chloride, **1,** 1179
Triphenylphosphine bis(trifluoromethanesulfonate), **6,** 648
Triphenylphosphine + co-reagent, **5,** 727; **6,** 246; **9,** 167; **11,** 589
Vilsmeier reagent, **8,** 186; **9,** 514
with other RCOX
Dimethylaluminum methylselenolate, **8,** 182
4-Dimethylaminopyridine, **3,** 118
Guanidines, **9,** 179
Ion-exchange resins, **11,** 276
Nickel carbonyl, **1,** 720
3-Nitro-2-pyridinesulfenyl chloride, **9,** 325
Oxygen, **6,** 426
Silver(I) trifluoroacetate, **8,** 444
p-Toluenesulfonic acid, **1,** 1172
FROM ADDITION OF ORGANOBORANES TO C=C
9-Borabicyclo[3.3.1]nonane, **3,** 24
Ethyl diazoacetate, **2,** 193; **4,** 228; **5,** 295
Ethyl dibromoacetate, **2,** 195
BY ALCOHOLYSIS OF RCN
Sodium borohydride, **5,** 597
p-Toluenesulfonic acid, **1,** 1172
Vilsmeier reagent, **7,** 422
BY ALKOXYCARBONYLATION
of RH, ArH
Arylthallium bis(trifluoroacetates), **10,** 300
Bromine–Antimony(V) chloride–Sulfur dioxide, **5,** 57
Ferrous sulfate, **4,** 237
Hexafluoroantimonic acid, **2,** 216
Oxalyl chloride, **3,** 216
Trichloroacetyl chloride, **4,** 521
of RX with loss of X
Dichlorobis(triphenylphosphine)-palladium(II), **6,** 59, 60
Disodium tetracarbonylferrate, **4,** 461
Ethyl chloroformate, **1,** 364
Grignard reagents, **1,** 415
Iron carbonyl, **8,** 265
Methyl methylthiomethyl sulfone, **11,** 242
Nickel carbonyl, **3,** 210
Sodium tetracarbonylcobaltate, **1,** 1058
BY ALKYLATION OF $RCOO^-$, RCOOH
Alkyl chloroformates, **12,** 12
Benzyltrimethylammonium chloride, **1,** 53
t-Butyl methyl ether, **4,** 333
Calcium sulfate, **4,** 266
Cesium propionate, **11,** 118
Chlorotrimethylsilane, **7,** 66
Copper(I) acetate–*t*-Butyl isocyanide, **5,** 163
Cryptates, **6,** 137
1,8-Diazabicyclo[5.4.0]undecene-7, **9,** 132
Dicyclohexylethylamine, **1,** 370
N,N-Dimethylformamide, **2,** 153
Dimethylformamide dialkyl acetals, **1,** 281, 283
Guanidines, **11,** 105, 249
Hexamethylphosphoric triamide, **6,** 273
Ion-exchange resins, **6,** 302
Phase-transfer catalysts, **11,** 403
Sodium hydroxide, **4,** 247
Triethylamine, **1,** 1198
Triphenylphosphine–Diethyl azodicarboxylate, **7,** 404
2,4,6-Triphenylpyrylium tetrafluoroborate, **8,** 520
Tris(2-hydroxypropyl)amine, **3,** 325
BY ARNDT–EISTERT REACTION OR WOLFF REARRANGEMENT
t-Butyllithium, **11,** 103
Copper(I) iodide, **1,** 169
Diazomethane, **2,** 102; **3,** 74
Trimethylsilyldiazomethane, **10,** 431
BY BAEYER-VILLIGER REACTION (*see* TYPE OF REACTION INDEX)
BY HYDROCARBOXYLATION OF C=C
Chlorobis(cyclopentadienyl)-hydridozirconium(IV), **6,** 175
Di-μ-carbonylhexacarbonyldicobalt, **1,** 224
Dichlorobis(triphenylphosphine)-palladium(II), **3,** 81
Ethyl bromoacetate, **2,** 192
Hydrogen hexachloroplatinate(IV)–Tin(II) chloride, **4,** 87
BY OXIDATION OF ACETALS,

ESTERS—GENERAL METHODS (*Continued*)
ALCOHOLS, ALDEHYDES, ETHERS, SILYL ETHERS (*see* TYPE OF REACTION INDEX)
BY OXIDATIVE CLEAVAGE OF C=C (*see* TYPE OF REACTION INDEX)
BY THIELE REACTION, TISHCHENCKO REACTION, TRANSESTERIFICATION, WILLGERODT–KINDLER REACTION (*see* TYPE OF REACTION INDEX)
OTHER ROUTES
Benzeneseleninic anhydride, **8,** 29
N-Bromosuccinimide, **7,** 37
Cobalt(III) acetate, **6,** 127
2-Diethylamino-4-phenylsulfonyl-2-butenenitrile, **12,** 182
Diphenyl phosphoroazidate, **7,** 138
Ethyl malonate, **6,** 255
Lead tetraacetate, **10,** 228
2-Lithio-2-methylthio-1,3-dithiane, **7,** 191
Methyl methylthiomethyl sulfoxide, **4,** 341; **9,** 314
Nitrogen dioxide, **1,** 324; **2,** 175
Pyridinium chlorochromate, **8,** 425
Silver tetrafluoroborate, **11,** 471
Sodium methoxide, **1,** 1091
Sulfur dioxide, **7,** 346
Tetramethoxyethylene, **2,** 401
Thallium(III) nitrate, **4,** 492
Triethyl orthoformate, **6,** 610
2,4,4-Trimethyl-2-oxazoline, **3,** 313; **5,** 714

ESTERS—SPECIFIC TYPES OF ESTERS (*see also* ESTERS—GENERAL METHODS)
ACETATES
by Acetoxylation of RH, ArH
Bis(2,2-dipyridyl)silver(II) peroxydisulfate, **6,** 51
Cerium(IV) ammonium nitrate, **1,** 120; **4,** 71
Lead tetraacetate, **1,** 537
Manganese(III) acetate, **4,** 318
Palladium(II) acetate, **2,** 303; **5,** 496; **7,** 274
by Acetylation of ROH
Acetic anhydride, **1,** 3, 797, 958, 1024, 1228, 1289; **2,** 353; **11,** 1
Acetic-phosphoric anhydride, **2,** 12
2-Acetoxypyridine, **1,** 9
Acetyl bromide, **6,** 9
Acetyl chloride, **9,** 357
Acetyl hexafluoroantimonate, **1,** 692
3-Acetyl-1,5,5-trimethylhydantoin, **3,** 4
Alumina, **11,** 22
Bismuth(III) acetate, **4,** 40
Calcium hydride, **2,** 58
Cyanogen bromide, **1,** 174
4-Dimethylaminopyridine, **9,** 178
Ketene, **1,** 528; **2,** 232
Molecular sieves, **9,** 316
Perchloric acid, **1,** 796
Phenyl acetate, **1,** 829
4-Pyrrolidinopyridine, **4,** 416
Silica, **11,** 466
1,3,4,6-Tetraacetylglycouril, **6,** 563
Tetraethylammonium hydroxide, **1,** 1138
by Alkylation of AcO^-, AcOH
Potassium acetate, **5,** 154; **6,** 137, 278, 640
Sodium acetate, **8,** 390
Tetraalkylammonium acetate, **1,** 1136, 1142; **3,** 277
by Reductive acylation
Triethylamine, **1,** 1198
Triphenyltin hydride, **2,** 448
Zinc–Acid anhydride–Catalyst, **1,** 1143; **4,** 577
Other routes
Acetic anhydride, **1,** 72
Acetyl *p*-toluenesulfonate, **2,** 14
Arenediazonium tetrahaloborates, **1,** 43
Cerium(IV) acetate, **4,** 71
Chlorotrimethylsilane–Acetic anhydride, **12,** 126
Lead tetraacetate, **3,** 168
Lithium bromide–Boron trifluoride etherate, **1,** 604
Silver(I) oxide, **5,** 583
Thallium(III) nitrate, **10,** 395
ARYL ESTERS
2-Benzoylthio-1-methylpyridinium chloride, **9,** 40
Boric acid, **4,** 41
Dibenzoyl peroxide, **6,** 160
N,N-Dimethylformamide, **5,** 247

Phosphoryl chloride, **1,** 876
Polyphosphate ester, **9,** 376
Polyphosphoric acid, **1,** 894
Silica, **11,** 466
1,3,4,6-Tetraacetylglycouril, **6,** 563
Thallium(I) ethoxide, **2,** 407
BENZOATES
Benzoic anhydride, **5,** 23
Benzoyl chloride, **1,** 959
Benzoyl cyanide, **6,** 35
2-Benzoylthio-1-methylpyridinium chloride, **9,** 40
Benzoyl trifluoromethanesulfonate, **12,** 44
t-Butyl perbenzoate, **9,** 90
Dibenzoyl peroxide, **6,** 160
Diethyl benzoylphosphonate, **11,** 178
Fieser reagent, **3,** 56
Heptafluoro-1-methylethyl phenyl ketone, **7,** 165
Organocopper reagents, **11,** 365
Polyphosphate ester, **9,** 376
Sodium benzoate, **1,** 1044
BENZYL ESTERS
Benzyldimethylanilinium hydroxide, **6,** 37
Dimethylformamide dineopentyl acetal, **1,** 283
Sodium acetate, **1,** 1024
Sulfuryl chloride, **1,** 1128
t-BUTYL ESTERS
t-Butyl acetate, **1,** 82
Collins reagent, **12,** 139
1,5-Diazabicyclo[5.4.0]undecene-5, **11,** 155
N,N-Dimethylaniline, **1,** 274
Dimethylketene, **1,** 290
Isobutene, **1,** 522
Lithium *t*-butoxide, **3,** 183; **4,** 297
Magnesium, **1,** 627
Potassium *t*-butoxide, **2,** 336
Silver benzoate, **1,** 1004
ETHYL ESTERS
Dicyclohexylethylamine, **1,** 370
2,2-Diethoxyvinylidenetriphenylphosphorane, **11,** 175
Pentaethoxyphosphorane, **2,** 305
Triethyl orthoformate, **1,** 1204
Triethyloxonium tetrafluoroborate, **1,** 1210; **4,** 527; **8,** 500; **9,** 482
FORMATES
Acetic-formic anhydride, **1,** 4; **2,** 10
Fieser reagent, **2,** 72
Formamide, **1,** 402
Formic acid, **1,** 404
Formyl fluoride, **1,** 407
N-Formylimidazole, **1,** 407
Hexamethylphosphorous triamide, **6,** 279
p-Toluenesulfonyl chloride, **1,** 1184
Triphenylphosphine dibromide, **5,** 732
METHYL ESTERS
Alumina, **10,** 8
Boron trifluoride–Methanol, **1,** 73
Copper(II) acetate, **10,** 103
Diazomethane, **1,** 191; **2,** 102; **3,** 74; **5,** 179; **6,** 159
1,4-Dichloro-1,4-dimethoxybutane, **12,** 175
Dicyclohexylethylamine, **1,** 370
Dicyclopentadienylcobalt–Oxygen adduct, **5,** 145
2-Diethylamino-4-phenylsulfonyl-2-butenenitrile, **12,** 182
Dimethylacetamide, **2,** 144
Dimethylformamide–Dimethyl sulfate, **5,** 250
Dimethyl sulfate, **2,** 153; **3,** 325; **4,** 188, 189
Dimethylsulfoxonium methylide, **1,** 315
Methanol, **1,** 678; **6,** 645
Methyl iodide, **1,** 682; **3,** 202
1-Methyl-3-*p*-tolyltriazene, **1,** 696
Methyl trichloroacetate, **12,** 329
Molecular sieves, **1,** 703
Pentamethoxyphosphorane, **9,** 353
Sodium methoxide, **1,** 1091
Triethyloxonium tetrafluoroborate, **9,** 482
2,2,2-Trimethoxy-5-methyl-Δ^4-oxaphospholene, **5,** 707
Trimethylanilinium hydroxide, **5,** 708
Trimethyloxonium tetrafluoroborate, **5,** 693
Trimethyl phosphate, **5,** 716
Trimethyl phosphite, **4,** 541
Trimethylselenonium hydroxide, **9,** 491
METHYLTHIOMETHYL ESTERS
Chlorodimethylsulfonium chloride, **9,** 105
Chloromethyl methyl sulfide, **4,** 84; **8,** 94
Dimethyl sulfoxide–*t*-Butyl bromide, **9,** 190

ESTERS—SPECIFIC TYPES OF ESTERS (*Continued*)
Lithium methylsulfinylmethylide–Tributylborane, **11,** 302
TRIFLUOROACETATES
Cobalt(III) trifluoroacetate, **5,** 146
Lead tetrakis(trifluoroacetate), **4,** 282; **6,** 318
Trifluoroacetic acid, **7,** 109
Trifluoroacetic anhydride, **1,** 1221
Trifluoroacetyl trifluoromethanesulfonate, **9,** 485
TRITYL ESTERS
Triphenylmethyl bromide, **1,** 1254
ESTERS—α-SUBSTITUTED (*see also* α-HALO ESTERS, α-HYDROXY ESTERS)
α-ACYLOXY
Lead tetraacetate, **3,** 168; **12,** 270
α-ALKOXY
Ethyl diazoacetate, **5,** 571
Phenyliodine(III) diacetate, **11,** 411
Thallium(III) nitrate, **7,** 362
α-ALKYL(OR PHENYL)SELENO
Benzeneselenenyl bromide, **5,** 518
Selenium, **10,** 345
Sodium benzeneselenoate, **5,** 519
α-ALKYL(OR PHENYL)THIO
Sodium ethanethiolate, **11,** 492
Thallium(I) ethanethiolate, **9,** 460
α-SILYL
Chloromethyldiphenylsilane, **10,** 91
Trimethylsilyl trifluoromethanesulfonate, **8,** 514
ETHERS (EXCLUDING PROTECTING GROUPS)—(*see also* ALLYLIC ETHERS, ENOL ETHERS, HOMOALLYLIC ETHERS)
GENERAL METHODS
by Displacements with RO^-, ROH
m-Chloroperbenzoic acid, **10,** 92
Cryptates, **9,** 127
Dimethyl sulfoxide, **3,** 119
Mercury(II) perchlorate, **5,** 428
Nickel(II) acetylacetonate, **8,** 42
Potassium fluoride, **11,** 434
Sodium methylsulfinylmethylide, **4,** 195
Tetrabutylammonium hydrogen sulfate, **6,** 565
Thallium(I) ethoxide, **6,** 577; **10,** 395
Triethyloxonium tetrafluoroborate, **1,** 1210
by Reduction of acetals and ketals, esters (*see* TYPE OF REACTION INDEX)
by Reductive alkylation of C=O (*see* TYPE OF REACTION INDEX)
Other routes
Aluminum chloride, **4,** 10
Borane–Pyridine, **9,** 59
Mercury(II) acetate, **3,** 194
Organocopper reagents, **12,** 345
Perchloric acid, **2,** 309
Tributyltin hydride, **4,** 518
ALKYL ARYL ETHERS
Aryl methyl ethers
Diazomethane, **5,** 179; **6,** 159; **9,** 133
Dimethyl sulfate, **1,** 293, 295; **2,** 153
Ion-exchange resins, **8,** 263
Methyl iodide, **1,** 682
Methyl trichloroacetate, **12,** 329
Pentamethoxyphosphorane, **9,** 353
Phenyltrimethylammonium ethoxide, **1,** 854
Potassium hydroxide, **9,** 388
Potassium methoxide, **5,** 154
Sodium hydride, **4,** 452; **5,** 614
Sodium methoxide, **1,** 1091
Trimethylselenonium hydroxide, **9,** 491
Trimethylsilyldiazomethane, **12,** 538
Trimethylsulfoxonium iodide, **1,** 1236
Alkyl aryl ethers
Barium oxide, **9,** 23
Benzyltriethylammonium chloride, **5,** 26
t-Butyl perbenzoate, **1,** 98
Copper(I) alkoxides, **5,** 148
Copper(I) oxide, **1,** 169
Dicyclohexylcarbodiimide, **1,** 231
N,N-Dimethylformamide, **2,** 153; **5,** 247
Grignard reagents, **1,** 415
Hexamethylphosphoric triamide, **6,** 273
Hexamethylphosphorous triamide, **6,** 279
Ion-exchange resins, **1,** 511
Pentaethoxyphosphorane, **2,** 305
Potassium *t*-butoxide, **1,** 298
Silver(I) oxide, **4,** 430; **5,** 583
Tetraethylammonium fluoride, **9,** 448

Triphenylbismuth diacetate, **12,** 548
Triphenylphosphine–Diethyl azodicarboxylate, **6,** 645
BENZYL ETHERS
Arene(tricarbonyl)chromium complexes, **12,** 34
Benzyl bromide, **5,** 25
Benzyl *p*-toluenesulfonate, **11,** 44
Benzyl trichloroacetimidate, **11,** 44
Benzyl trifluoromethanesulfonate, **6,** 44
Bis(tributyltin) oxide, **9,** 53
Ion-exchange resins, **1,** 511
Mercury(II) perchlorate, **5,** 428
Sodium hydride, **1,** 1075
Tetrabutylammonium iodide, **11,** 501
t-BUTYL ETHERS
Antimony(V) fluoride, **6,** 23
t-Butyl perbenzoate, **1,** 98
Grignard reagents, **1,** 45
Isobutene, **1,** 522
Potassium *t*-butoxide, **1,** 298; **5,** 544; **8,** 128
Silver carbonate, **8,** 441
DIARYL ETHERS
Copper, **1,** 157
Copper(I) chloride, **3,** 67
Copper(I) oxide, **1,** 169
Copper(I) phenylacetylide, **12,** 143
2,4-Dinitrofluorobenzene, **1,** 321
Diphenyliodonium bromide, **1,** 340
Pentafluorophenylcopper, **6,** 451
Tetraphenylbismuth trifluoroacetate, **10,** 393
Triaryl phosphates, **11,** 542
METHYL ETHERS
by Displacements with CH_3O^-
Sodium methoxide, **1,** 298
Thallium(III) nitrate, **7,** 362
Trifluoromethanesulfonic anhydride, **5,** 702
by Methylation of ROH
Diazomethane, **1,** 191; **2,** 102; **6,** 554; **9,** 135
Dimethyl phosphite, **4,** 188
Dimethyl sulfate, **1,** 293; **5,** 647
Iodine, **2,** 220
Mercury(II) perchlorate, **5,** 428
Methyl iodide, **2,** 274
Phase-transfer catalysts, **10,** 305
Potassium hydroxide, **9,** 388
Potassium methylsulfinylmethylide, **10,** 329
Ruthenium(III) chloride, **11,** 462
Sodium hydride, **4,** 452; **5,** 614; **6,** 541
Tetrafluoroboric acid, **1,** 394
2,2,2-Trimethoxy-5-methyl-Δ^4-oxaphospholene, **5,** 707
Trimethylanilinium hydroxide, **5,** 708
Trimethylsulfoxonium iodide, **1,** 1236
Other routes
m-Chloroperbenzoic acid, **12,** 118
Dimethyl sulfoxide–Acetic anhydride, **9,** 190
Sodium cyanoborohydride, **8,** 454
TRITYL ETHERS
4-Dimethylaminopyridine, **10,** 155
Trimethyl(triphenylmethoxy)silane, **10,** 444
Triphenylmethylpyridinium tetrafluoroborate, **5,** 741

FULVENES AND DERIVATIVES
Chloranil, **2,** 66
Copper, **7,** 73
Copper(I) bromide, **2,** 90
Crown ethers, **10,** 110
Dichlorocarbene, **4,** 130
Dimethylformamide diethyl acetal, **1,** 281
Dimethylformamide–Dimethyl sulfate, **1,** 282; **2,** 154
Hexabutylditin, **7,** 165
Ion-exchange resins, **1,** 511
Lithium bromide, **7,** 200
Molybdenum carbonyl, **7,** 247
Phase-transfer catalysts, **8,** 387
Tetracyanoethylene, **5,** 647
Triethyl phosphite, **1,** 1212; **4,** 529; **7,** 387
Triphenylcarbenium tetrafluoroborate, **2,** 454
Vilsmeier reagent, **7,** 422
FURANS
Benzeneselenenyl bromide, **6,** 459
Boron trifluoride etherate, **6,** 65
t-Butyllithium, **9,** 89
2-Carboxy-1-methoxycarbonyl-ethylidenetriphenylphosphorane, **8,** 77
m-Chloroperbenzoic acid, **8,** 97
1-Chloro-N,N,2-trimethyl-propenylamine, **12,** 123
Chlorotris(triphenylphosphine)-rhodium(I), **5,** 736
Copper bronze, **12,** 140

FURANS (*Continued*)
Copper(I) trifluoromethanesulfonate, **6,** 130
Cyclopropenone 1,3-propanediyl ketal, **12,** 152
Dichlorobis(cyclopentadienyl)titanium, **12,** 168
Diethylaluminum benzenethiolate, **10,** 281
Diisobutylaluminum hydride, **1,** 260; **2,** 140; **6,** 198
Dimethyl diazomalonate, **8,** 187
Dimethylformamide dimethyl acetal, **10,** 158
Dimethylsulfonium methylide, **4,** 196; **11,** 213
Dimethyl sulfoxide, **1,** 296
β-Ethoxyvinyltriphenylphosphonium iodide, **5,** 294
Ethyl diazoacetate, **2,** 193
Ethyl 4,4-dimethoxy-2-phenylthiobutyrate, **8,** 222
(α-Formylethylidene)triphenylphosphorane, **12,** 234
Ion-exchange resins, **5,** 355
Isocyanomethyllithium, **10,** 231; **11,** 285
Lithium diisopropylamide, **7,** 204
Methoxyallene, **7,** 225
1-Methoxy-1-trimethylsilylallene, **11,** 577
Nickel carbonyl, **2,** 290
Nickel peroxide, **8,** 357
1-Nitro-1-(phenylthio)propene, **10,** 279
2-Nitropropene, **6,** 481
Oxygen, singlet, **7,** 261
4-Phenyloxazole, **12,** 389
α-(Phenylsulfinyl)acetonitrile, **11,** 418
Pyridinium chlorochromate, **11,** 450
Rhodium(II) acetate, **5,** 571
Sulfur, **3,** 273
Tetrakis(triphenylphosphine)palladium(0), **10,** 384
Titanium(IV) chloride, **6,** 590; **9,** 468
Tributyltin hydride, **12,** 516
2,4,6-Trimethylpyrylium sulfoacetate, **11,** 571
Zinc, **2,** 459

GLYCIDIC ACIDS, ESTERS, NITRILES
BY DARZENS REACTION (*see* TYPE OF REACTION INDEX)
BY EPOXIDATION OF α,β-UNSATURATED SUBSTRATES (*see* TYPE OF REACTION INDEX)
OTHER ROUTES
2-Chloroacrylonitrile, **9,** 75
Chloromethyl methyl ether, **7,** 61
Dimethylsulfonium ethoxycarbonylmethylide, **10,** 164
Ethyl (dimethylsulfuranylidine)acetate, **2,** 196
Iodine, **8,** 256
Potassium carbonate, **9,** 382
Sodium cyanide, **8,** 430

HALIDES (*see* ALKYL HALIDES, ALLYLIC HALIDES, ARYL HALIDES, DIHALIDES, HALO...)
HALOACETALS
Bromine, **5,** 55; **6,** 70
t-Butyl hypobromite, **1,** 90
Dibromo(or Dichloro)carbene, **8,** 388
Dioxane–Bromine, **3,** 130
Phenyltrimethylammonium perbromide, **1,** 855; **11,** 426
Pyridine perbromide, **1,** 966
Pyridinium bromide perbromide, **1,** 967
Tetrachlorosilane, **12,** 463
α-HALO ACID HALIDES
Bromine–Phosphorus(III) bromide, **1,** 874
N-Bromosuccinimide, **3,** 34; **6,** 74
N-Chlorosuccinimide, **6,** 75, 115
Iodine, **6,** 75, 117
Thionyl chloride, **5,** 663
α-HALO ALDEHYDES AND KETONES
GENERAL METHODS
Lead tetraacetate–Metal halides, **11,** 283
α-FLUORO
from C=C
Nitrosyl fluoride, **1,** 755; **2,** 299; **3,** 214
Oxygen difluoride, **1,** 772
from Enol ethers, silyl enol ethers, etc. (*see* TYPE OF REACTION INDEX)
Other routes
Acetyl hypofluorite, **12,** 3
2-Fluoro-1-buten-3-one, **6,** 263
Lithium tri-*t*-butoxyaluminum hydride, **2,** 251
Perchloryl fluoride, **1,** 802; **7,** 39
Pyridinium poly(hydrogen fluoride), **6,** 473
Silver tetrafluoroborate, **9,** 414
α-CHLORO

by Chlorination α to C=O (*see* TYPE OF REACTION INDEX)
Other routes
Barium hydroxide, **4,** 23
t-Butyl hypochlorite, **1,** 90
Chloroacetylium hexafluoroantimonate, **6,** 103
Chlorodimethylsulfonium chloride, **9,** 67
Chloromethyl phenyl sulfoxide, **8,** 94
Chromyl chloride, **4,** 98
Dichloromethyllithium, **4,** 138; **5,** 199; **6,** 344
Ferric chloride, **12,** 230
Levulinic acid, **2,** 239
Sodium dichromate, **1,** 1059
Thionyl chloride, **1,** 1158
α-BROMO
by Bromination α to C=O (*see* TYPE OF REACTION INDEX)
Other routes
Bromine, **8,** 52
Bromoacetylium hexafluoroantimonate, **6,** 103
2-Bromoacrolein, **2,** 15
Bromodimethylsulfonium bromide, **9,** 66
Bromomethyl phenyl sulfoxide, **9,** 68
Dimethyl sulfoxide–N-Bromosuccinimide, **9,** 73
Magnesium bromide etherate, **7,** 218
Silver chromate, **1,** 1005
α-IODO
1,3-Diiodo-5,5-dimethylhydantoin, **1,** 258
Iodine, **1,** 495; **9,** 249, 386; **10,** 211; **12,** 256
N-Iodosuccinimide, **1,** 510
Iodotrimethylsilane, **10,** 216
Lithium iodide–Hydrogen peroxide, **7,** 220
Silver chromate–Iodine, **8,** 442
Thallium(I) acetate–Iodine, **8,** 260
β-HALO ALDEHYDES AND KETONES
β-CHLORO
Acetylene, **1,** 11
Aluminum chloride, **4,** 10
Ferric chloride, **7,** 153
Tetraethylammonium iodide, **12,** 465
β-BROMO
Bromine, **6,** 70
Dioxane–Bromine, **5,** 58
Tetraethylammonium iodide, **12,** 465
β-IODO
Iodotrimethylsilane, **9,** 251; **10,** 216
Tetraethylammonium iodide, **12,** 465
HALOALKYNES
CHLORO-
N-Chlorosuccinimide, **5,** 127; **9,** 111
Dichloro(diethoxyphosphinyl)-methyllithium, **6,** 188
Potassium superoxide, **11,** 442
Tetrabutylammonium hydrogen sulfate, **7,** 354
Trichloroethylene, **11,** 552
BROMO-
Benzyltrimethylammonium hydroxide, **7,** 20
N-Bromosuccinimide, **12,** 79
Bromotrifluoromethane, **2,** 42
Sodium hypobromite, **1,** 1083
IODO-
N-Bromosuccinimide, **12,** 79
Heptafluoro-1-iodopropane, **2,** 226
Iodine–Morpholine complex, **1,** 502
Methyltriphenoxyphosphonium iodide, **1,** 1249
HALOAMIDES AND -AMINES
N-HALOAMIDES
t-Butyl halites, **1,** 90, 94
Tetrafluorohydrazine, **1,** 1139
N-HALOAMINES
t-Butyl hypochlorite, **1,** 90
Carbonyl difluoride, **12,** 106
N-Chlorosuccinimide, **1,** 139
Difluoramine, **2,** 134
Hypochlorous acid, **1,** 487
Tetrafluorohydrazine, **1,** 1139
Trifluoromethyl hypofluorite, **5,** 312; **6,** 263
HALO CARBONYLS—POLYHALO CARBONYLS
α-DI- AND TRIFLUORO-
Diethyl(2-chloro-1,1,2-trifluoroethyl)-amine, **6,** 66
Oxygen fluoride, **1,** 772
Perchloryl fluoride, **1,** 802
Trifluoromethyl hypofluorite, **7,** 156; **9,** 224
α-DI- AND TRICHLORO-
N-Chlorosuccinimide, **6,** 115; **8,** 103, 463
N,N-Dimethylformamide, **9,** 182

HALO CARBONYLS—POLYHALO CARBONYLS (*Continued*)
Ethyl dichloroacetate, **6,** 331
Sodium trichloroacetate, **2,** 388
Trimethylsilyl dichloroacetate, **9,** 491
α-DI- AND TRIBROMO-
Bromine–Sulfur, **1,** 1120
N-Chlorosuccinimide, **6,** 115
α-HALO CARBOXYLIC ACIDS
α-FLUORO
N-Bromoacetamide–Hydrogen fluoride, **2,** 39
(Diethylamino)sulfur trifluoride, **10,** 142
Pyridinium poly(hydrogen fluoride), **6,** 473; **11,** 453
Trifluoromethyl hypofluorite, **10,** 420
α-CHLORO
Bromomethyllithium, **12,** 77
Chlorine–Chlorosulfuric acid, **8,** 83; **10,** 86
(−)-2-Chloromethyl-4-methoxymethyl-5-phenyloxazoline, **6,** 204
Ethyl dichloroacetate, **3,** 26
Pyridinium poly(hydrogen fluoride), **12,** 419
7,7,8,8-Tetracyanoquinodimethane, **12,** 464
α-BROMO
p-Aminoacetophenone, **4,** 18
Barium hydroxide, **4,** 23
Bromine–Phosphorus(III) chloride, **1,** 875
Bromomethyllithium, **12,** 77
N-Bromosuccinimide, **3,** 34; **5,** 65; **7,** 37
Chlorosulfuric acid, **10,** 86
Diethyl dibromomalonate, **8,** 169
Ethyl dibromoacetate, **3,** 26
Phosphorus, red, **1,** 861
Pyridinium poly(hydrogen fluoride), **12,** 419
α-IODO
Chlorosulfuric acid, **10,** 86
Iodine–Copper(II) acetate, **12,** 256
HALOEPOXIDES
t-Butyl hydroperoxide, **7,** 43
Dichloromethyllithium, **3,** 89; **5,** 199; **6,** 170
Lithium diisopropylamide, **5,** 400
α-HALO ESTERS AND LACTONES
α-FLUORO
N-Alkyl-N-fluoro-*p*-toluenesulfonamides, **12,** 231
Perchloryl fluoride, **1,** 802; **2,** 310; **7,** 280
Trifluoromethyl hypofluorite, **10,** 420
α-CHLORO
N-Chlorosuccinimide, **6,** 115
Ethylene oxide, **3,** 140
Lithium diethylamide, **6,** 331
Thionyl chloride, **1,** 1158
Trimethylsilyl dichloroacetate, **9,** 491
Triphenylphosphine–Carbon tetrachloride, **2,** 445
α-BROMO
Barium hydroxide, **4,** 23
Bromine, **1,** 1159; **4,** 306
Dibromoacetonitrile, **5,** 186
Diethyl dibromomalonate, **8,** 169
Dimethyl sulfoxide–N-Bromosuccinimide, **9,** 73
Ethylene oxide, **3,** 140
Magnesium bromide etherate–Hydrogen peroxide, **7,** 220
Triphenylphosphine–Carbon tetrabromide, **2,** 445
α-IODO
Iodine, **4,** 306
HALO ETHERS
t-Butyl hypohalites, **1,** 90; **2,** 50
Formaldehyde, **4,** 238
Sodium fluoride, **2,** 382
Sulfur tetrafluoride, **5,** 640
Sulfuryl chloride, **1,** 1128
2,4,4,6-Tetrabromo-2,5-cyclohexadienone, **7,** 351
Tetrahydrofuran, **1,** 1140
Trichloroisocyanuric acid, **3,** 297
HALOFORMIC ESTERS
Carbonic difluoride, **1,** 116
Phosgene, **1,** 856
Pyridinium poly(hydrogen fluoride), **6,** 473
HALOHYDRINS AND DERIVATIVES (*see also* CHIRAL COMPOUNDS)
1,2-FLUOROHYDRINS
Acetyl hypofluorite, **10,** 1
Ethyl fluoroacetate, **5,** 304
Oxygen fluoride, **1,** 772
Perchloryl fluoride, **1,** 802
Trifluoromethyl hypofluorite, **3,** 146; **7,** 156
1,2-CHLOROHYDRINS
from C=C

t-Butyl hydroperoxide–Titanium(IV) chloride, **12,** 94
Chloramine-T, **10,** 85
N-Chlorosuccinimide, **5,** 127
Chromyl chloride, **1,** 151; **5,** 144; **8,** 112
Dimethylformamide–Chlorine, **5,** 249
Hypochlorous acid, **10,** 208
Titanium(IV) chloride, **6,** 590
Trifluoroacetyl chloride, **10,** 419
Urea, **1,** 1262
from Epoxides
Bis(benzonitrile)dichloropalladium(II), **9,** 44
Chlorotrimethylsilane, **11,** 125
Dimethylformamide–Phosphoryl chloride, **1,** 284
Pyridinium chloride, **5,** 566; **10,** 333
Triphenylphosphine dihalides, **12,** 554
Other routes
2-Acetoxyisobutyryl chloride, **8,** 3
Bromomethyllithium, **12,** 77
Chlorotrimethylsilane, **5,** 709
B-3-Pinanyl-9-borabicyclo[3.3.1]-nonane, **12,** 397
Tetrahydrofuran, **1,** 1140
Triphenylmethyl chloride, **4,** 565
Triphenylphosphine–Carbon tetrachloride, **12,** 551
OTHER CHLOROHYDRINS
2-Acetylsalicylic acid chloride, **5,** 6
2,2-Dimethyl-1,3-propanediol, **5,** 259
Molybdenum carbonyl, **4,** 346
Triphenylphosphine–Carbon tetrachloride, **12,** 551
Zinc chloride, **1,** 1289
1,2-BROMOHYDRINS
from C=C
Acetyl hypobromite, **1,** 12
N-Bromoacetamide, **1,** 74; **4,** 47
N-Bromosuccinimide, **1,** 78; **3,** 34; **6,** 74; **7,** 182
t-Butyl hypobromite, **1,** 90
Dimethyl sulfoxide–N-Bromosuccinimide, **2,** 159; **9,** 71
Hypobromous acid, **1,** 800; **4,** 49
Sodium hypobromite, **1,** 1083
from Epoxides
Bromotrimethylsilane, **10,** 59, 216
Dilithium tetrabromonickelate, **12,** 195
Hydrobromic acid, **1,** 450
Triphenylphosphine dihalides, **12,** 554
Other routes
Hydrogen bromide–Acetic acid, **5,** 335
Lithium tri-*t*-butoxyaluminum hydride, **1,** 620
B-3-Pinanyl-9-borabicyclo[3.3.1]-nonane, **12,** 397
Sodium borohydride, **1,** 1049
OTHER BROMOHYDRINS
Acetic anhydride–Magnesium bromide, **6,** 352
Bromotrimethylsilane, **10,** 59
1,2-IODOHYDRINS
from C=C
Iodine, **3,** 159; **4,** 261; **8,** 256, 260
N-Iodosuccinimide–Carboxylic acids, **7,** 182
Peracetic acid, **2,** 220
Silver(I) trifluoroacetate, **1,** 1018
Thallium(I) acetate–Iodine, **5,** 654; **9,** 458
from Epoxides
Hydriodic acid, **8,** 246
Hydrogen iodide, **7,** 173
Iodotrimethylsilane, **10,** 216
Sodium iodide–Acetic acid, **1,** 881; **5,** 595
Triphenylphosphine dihalides, **12,** 554
OTHER IODOHYDRINS
Aluminum chloride–Sodium iodide, **12,** 29
Borane–Tetrahydrofuran, **1,** 199
Iodotrimethylsilane, **9,** 251
Sodium iodide–Trimethylacetyl chloride, **11,** 489
HALO SULFUR COMPOUNDS
HALO SULFIDES
Benzenesulfenyl chloride, **8,** 32
Diethyl phenylthiomethylphosphonate, **6,** 191
Sulfuryl chloride, **1,** 1128
Trifluoromethylthiocopper, **6,** 621; **7,** 391
Trimethylsilyl trifluoromethanesulfonate, **12,** 543
HALO SULFONES
Bromomethyl *p*-tolyl sulfone, **6,** 41
5-Bromo-2,2,5-trimethyl-1,3-dioxane-4,6-dione, **5,** 66
Copper(II) chloride, **4,** 105; **5,** 158

HALO SULFUR COMPOUNDS (*Continued*)
Dichloromethyl phenyl sulfone, **6,** 41
Potassium hydroxide, **5,** 96
Sulfuryl chloride, **4,** 474
HALO SULFOXIDES
Diazomethane, **4,** 120
Nitrosyl chloride, **3,** 214
Potassium iodide, **7,** 134
Sulfuryl chloride, **3,** 276; **4,** 474
HETEROCYCLES—THREE-MEMBERED RINGS (*see also* AZIRIDINES, EPISULFIDES, EPOXIDES)
1 N (*see also* AZIRIDINES)
1,4-Diazabicyclo[2.2.2]octane, **2,** 99; **6,** 157
Iodine azide, **1,** 500; **2,** 222; **8,** 260
Lead tetraacetate, **5,** 365
1 N, 1 O
t-Amyl hydroperoxide, **4,** 20
N-Benzoylperoxycarbamic acid, **6,** 35
m-Chloroperbenzoic acid, **2,** 68; **5,** 120; **6,** 110
α-Methylbenzylamine, **6,** 457
Peracetic acid, **1,** 787
2 N
Diaziridines
Chloramine, **2,** 65
Hydroxylamine-O-sulfonic acid, **1,** 481; **2,** 217
Potassium *t*-butoxide, **1,** 911; **3,** 233
Diazirines
Chloramine, **1,** 122
Dichloramine, **1,** 213
Difluoramine, **1,** 253; **2,** 134
Hydroxylamine-O-sulfonic acid, **1,** 481; **2,** 217
Sodium hypochlorite, **1,** 1084
1 S (*see also* EPISULFIDES)
Diazomethane, **2,** 102
Methylbis(methylthio)sulfonium hexachloroantimonate, **6,** 375
Perbenzoic acid, **3,** 219
Phenyldiazomethane, **5,** 515
Triphenylphosphine, **5,** 725
HETEROCYCLES—FOUR-MEMBERED RINGS (*see also* DIOXETANES, β-LACTAMS, β-LACTONES)
1 N—AZETIDINES
Benzenesulfonyl azide, **3,** 17
Dimethylsulfoxonium methylide, **10,** 168
Potassium *t*-butoxide, **1,** 911
Triphenylphosphine + co-reagent, **9,** 503; **12,** 552
Triphenylphosphine dibromide, **5,** 729; **6,** 645
1 N, 1 O
Acetyl chloride, **7,** 3
2 N
Boron trichloride, **3,** 31
1 O—OXETANES
Bis(tributyltin) oxide, **6,** 56
2,2-Dimethyl-1,3-propanediol, **5,** 259
Dimethylsulfoxonium methylide, **12,** 213
Dimethyl N-(*p*-toluenesulfonyl)-sulfoximine, **9,** 193; **12,** 216
Diphenyldi(1,1,1,3,3,3-hexafluoro-2-phenyl-2-propoxy)sulfurane, **5,** 270
Lead tetraacetate, **2,** 234
Lithium carbonate, **5,** 396
Lithium hydroxide, **7,** 208
Sodium methylsulfinylmethylide, **7,** 338
Triphenylphosphine–Diethyl azodicarboxylate, **9,** 504
Zinc chloride, **8,** 536
2 O (*see also* DIOXETANES)
Triphenyl phosphite ozonide, **8,** 519
1 S
p-Nitrophenyl α-toluenesulfonate, **5,** 477
Potassium 2-methylcyclohexoxide, **5,** 560
Sulfur dichloride, **4,** 469
Thionyl chloride, **5,** 663
2 S
2,4-Bis(4-methoxyphenyl)-1,3-dithia-2,4-diphosphetane-2,4-disulfide, **11,** 54
HETEROCYCLES—FIVE-MEMBERED RINGS (*see also* FURANS, LACTAMS, LACTONES, PEROXIDES)
1 N—CARBAZOLES
Triethyl phosphite, **1,** 1212
1 N—INDOLES
by Fisher indole synthesis (*see* TYPE OF REACTION INDEX)
Other routes
Benzeneseleninic anhydride, **11,** 37
Bis(acetonitrile)dichloropalladium(II), **7,** 21
Boron trifluoride–Trifluoroacetic anhydride, **4,** 45
t-Butyl isocyanide, **2,** 50

Butyllithium, **10,** 68
Chlorodimethylsulfonium chloride, **5,** 114
Copper(I) iodide, **12,** 141
Di-*t*-butyl nitroxide, **11,** 160
Di-μ-carbonylhexacarbonyldicobalt, **7,** 99
Diethyl oxalate, **1,** 250
Dimethylformamide dimethyl acetal, **11,** 198
Dimethyl sulfide, **4,** 190
Dimethylsulfonium methylide, **3,** 124
Ferric chloride, **3,** 145
Ion-exchange resins, **1,** 511
Lithium diisopropylamide, **8,** 292
Manganese(IV) oxide, **1,** 637
Meldrum's acid, **11,** 311
4-Methoxy-3-buten-1-ynylcopper, **9,** 297
Organoaluminum reagents, **12,** 339
Oxodiperoxymolybdenum(pyridine)-(hexamethylphosphoric triamide), **11,** 218
Phenacylidenedimethylsulfurane, **6,** 456
Potassium *t*-butoxide, **9,** 380
Potassium ferricyanide, **1,** 929
Potassium nitrosodisulfonate, **2,** 347
Sodium amide, **1,** 1034; **6,** 526
Sodium borohydride–Acetic acid, **5,** 599
Sodium cyanoborohydride, **9,** 424
Sulfuric acid, **5,** 633
Thallium(III) trifluoroacetate, **12,** 481
Titanium(III) chloride, **9,** 467
Triethyl phosphite, **1,** 1212
Trimethylsilyl polyphosphate, **11,** 427
Vinyl acetate, **12,** 565

1 N—INDOLINES
Bis(trifluoroacetoxy)borane, **8,** 45
Borane + co-reagent, **5,** 708; **8,** 50
Dichlorotris(triphenylphosphine)-ruthenium(II), **12,** 179
N,O-Dimethylhydroxylamine, **12,** 205
Dimethyl 1,2,4,5-tetrazine-3,6-dicarboxylate, **12,** 214
Palladium(II) chloride, **10,** 300
N-Phenylselenophthalimide, **12,** 390
Zinc, **2,** 459

1 N—OXINDOLES
Dimethyl sulfide, **5,** 260
Dimethyl sulfoxide, **7,** 133
Phenylhydrazine, **1,** 838
Potassium amide, **1,** 907
Potassium hydride, **8,** 412
Silver carbonate, **12,** 432

1 N—PYRROLES
from 1,4-Dicarbonyls
N-Aminophthalimide, **1,** 38
Ammonia, **5,** 15
Diaminomaleonitrile, **5,** 175
2,5-Dimethoxytetrahydrofuran, **2,** 144
Tributylphosphine–Diphenyl disulfide, **12,** 514
Other routes
2-Butene-1,4-diol, **6,** 444
1,8-Diazabicyclo[5.4.0]undecene-7, **6,** 158
Dimethyl acetylenedicarboxylate, **2,** 145; **5,** 227
Hydroxylamine-O-sulfonic acid, **4,** 256
Lithium, **5,** 376
Lithium diisopropylamide, **7,** 204
Methyl N-tolylsulfonylmethylthio-benzimidate, **7,** 245
Methyl vinyl ketone, **1,** 697
Nitrosylsulfuric acid, **1,** 755
2,4-Pentanedione, **1,** 10
Potassium hydroxide, **11,** 439
Tetrakis(triphenylphosphine)-palladium(0), **10,** 384
Tosylmethyl isocyanide, **7,** 377
Trifluoroacetic anhydride, **6,** 616

1 N—PYRROLIDINES (*see also* AMIDES → AMINES)
from Unsaturated N compounds
Benzeneselenenyl bromide, **10,** 16
Benzenesulfenyl chloride, **12,** 42
Formaldehyde, **7,** 158
Mercury(II) acetate, **12,** 298
N-Phenylselenophthalimide, **9,** 366
Tetrakis(triphenylphosphine)-palladium(0), **7,** 357
Other routes
2,3-Bis(bromomethyl)-1,3-butadiene, **5,** 32
Bis(chlorodimethylsilyl)ethane, **12,** 179
Copper(I) trifluoromethanesulfonate, **12,** 144
Dichlorotris(triphenylphosphine)-

HETEROCYCLES—FIVE-MEMBERED RINGS (*Continued*)
ruthenium(II), **10,** 141
Dimethylaluminum chloride, **10,** 177
Iodotrimethylsilane, **9,** 251
8-Phenylmenthol, **11,** 412
Sodium cyanoborohydride, **10,** 360
Trimethylamine N-oxide, **12,** 533
1 N—PYRROLINES
1-Pyrrolines
Diisobutylaluminum hydride, **12,** 191
Lithium trichloropalladate(II), **12,** 288
Titanium(III) chloride, **12,** 492
2-Pyrrolines
Ethoxycarbonylcyclopropyltriphenylphosphonium tetrafluoroborate, **10,** 78
Methyl vinyl ketone, **2,** 283
3-Pyrrolines
Hydroxylamine-O-sulfonic acid, **2,** 217
Mercury(II) oxide–Tetrafluoroboric acid, **11,** 318
Phosphonium iodide, **1,** 859
Trimethylsilylmethyl trifluoromethanesulfonate, **10,** 434
1 N—OTHERS
N-Chlorosuccinimide, **8,** 103
Di-μ-carbonylhexacarbonyldicobalt, **1,** 224
Dichloromethylenedimethylammonium chloride, **5,** 195
Dimethyl acetylenedicarboxylate, **4,** 168
N-Ethylmorpholine, **1,** 383
Methylamine, **5,** 598
Nitrosocarbonylmethane, **11,** 362
Tetra-μ_3-carbonyldodecacarbonylhexarhodium, **5,** 326
1 N, 1 O—AZLACTONES
Ethyl chloroformate, **12,** 223
Hippuric acid, **1,** 432
Polyphosphoric acid, **7,** 294
Sodium acetate, **1,** 1024
Sulfur trioxide–Dimethylformamide, **1,** 1125
1 N, 1 O—BENZISOXAZOLES
Hydroxylamine-O-sulfonic acid, **1,** 481
1 N, 1 O—BENZOXAZOLES
Dichloromethylenedimethylammonium chloride, **5,** 195
Nickel peroxide, **1,** 731
Phosphoryl chloride, **11,** 429
Polyphosphate ester, **2,** 333; **3,** 229
Triphenylpropargylphosphonium bromide, **6,** 494
1 N, 1 O—ISOXAZOLES
Dichlorobis(triphenylphosphine)palladium(II), **5,** 190
2,3-Dichloro-5,6-dicyano-1,4-benzoquinone, **9,** 148
Dichloroformoxime, **1,** 221
N,N-Dimethylformamide, **8,** 189
Ethylene glycol, **1,** 375
Hydroxylamine, **9,** 438; **11,** 257
Hydroxylamine sulfate, **1,** 481
Iodine–Potassium iodide, **4,** 260
Sodium nitrite, **4,** 459
p-Toluenesulfonic acid, **12,** 507
1 N, 1 O—ISOXAZOLINES
N-Alkylhydroxylamines, **7,** 241; **12,** 13, 322
Hydrogen peroxide–Sodium tungstate, **12,** 246
1-Pyrroline 1-oxide, **9,** 401
p-Tolyl vinyl sulfoxide, **11,** 538
1 N, 1 O—Δ^2-ISOXAZOLINES
Benzenesulfonylnitrile oxide, **9,** 39
Cyanogen chloride N-oxide, **11,** 146
Ethoxycarbonylformonitrile oxide, **7,** 145
Phenyl isocyanate, **8,** 205; **10,** 309; **12,** 386
Raney nickel, **11,** 457
Triphenylacetonitrile oxide, **7,** 146
1 N, 1 O—OXAZOLES
Benzonitrile, **5,** 742
(Bromodichloromethyl)phenylmercury, **5,** 514
Copper(II) acetylacetonate, **2,** 81
1,8-Diazabicyclo[5.4.0]undecene-7, **5,** 177
Diphenyl phosphoroazidate, **11,** 222
Hydrogen fluoride, **4,** 252
Isocyanomethyllithium, **10,** 231
Methyl N-tolylsulfonylmethylthiobenzimidate, **7,** 245
Nickel peroxide, **9,** 322
Tosylmethyl isocyanide, **4,** 514
1 N, 1 O—2-OXAZOLIDINONES
Calcium ethoxide, **7,** 52
Carbon dioxide, **7,** 52
N,N′-Carbonyldiimidazole, **2,** 61

Iodine isocyanate, **3,** 161
Lead tetraacetate, **4,** 278
Selenium, **6,** 507
Tin(II) octoate, **7,** 346
1 N, 1 O—OXAZOLINES
Acetonitrile, **6,** 69
2-Amino-2-methyl-1-propanol, **3,** 14, 314
Benzonitrile, **6,** 69
m-Chloroperbenzoic acid, **5,** 120
Copper(I) oxide, **10,** 107
Cyanotrimethylsilane, **11,** 147
Diethyl isocyanomethylphosphonate, **4,** 271; **6,** 187
Iodine, **11,** 261
Isocyanotolylthiomethyllithium, **4,** 516
Lithium 2,2,6,6-tetramethylpiperidide, **7,** 213
Sulfuric acid, **5,** 633
Trichloroacetonitrile, **12,** 526
Triphenylphosphine–Carbon tetrachloride, **11,** 588
1 N, 1 O—OTHERS
2-Aminoethanol, **6,** 225
Aziridine, **1,** 378
Chlorosulfonyl isocyanate, **5,** 132
1-Cyclohexyl-3-(2-morpholinoethyl)-carbodiimide metho-*p*-toluene-sulfonate, **11,** 151
Meldrum's acid, **11,** 311
Oxalyl chloride, **2,** 301
Phosgene, **5,** 532
Triethyl phosphite, **1,** 1212
Zinc, **5,** 753
1 N, 1 S—BENZOTHIAZOLES
Dichloromethylenedimethyl-ammonium chloride, **5,** 195
Ethyl ethoxyiminoacetate, **11,** 109
Polyphosphate ester, **3,** 229
Silver cyanide, **5,** 581
Sodium thiocyanate, **1,** 1105
Triethyl orthoformate, **1,** 1204
Trimethylsilyl polyphosphate, **11,** 427
1 N, 1 S—THIAZOLES
Ethoxycarbonyl isothiocyanate, **6,** 250
Nickel peroxide, **8,** 357; **9,** 322
Phosphorus(V) sulfide, **1,** 870
Thiourea, **1,** 1164
Tosylmethyl isocyanide, **8,** 493
1 N, 1 S—THIAZOLINES
Aziridine, **1,** 378
Sodium ethoxide, **3,** 265
Triphenylphosphine, **11,** 588
1 N, 1 S—OTHERS
Selenium, **6,** 507
Thionyl chloride, **3,** 290; **5,** 663; **10,** 399
2 N—BENZIMIDAZOLES
Cyanamide, **5,** 168
Cyanogen bromide, **2,** 93
Dimethylformamide dimethyl acetal, **5,** 254
Formic acid, **1,** 404
o-Phenylenediamine, **1,** 834
Polyphosphate ester, **1,** 892
Polyphosphoric acid, **1,** 894
Silver cyanide, **5,** 581
Trifluoroperacetic acid, **1,** 821
Trimethylsilyl polyphosphate, **11,** 427
2 N—IMIDAZOLES
Barium manganate, **10,** 16
Diaminomaleonitrile, **5,** 175
Formamide, **1,** 402
Formamidine acetate, **3,** 147
Imidazole, **1,** 492
Mercury(II) sulfate, **1,** 658
Methyl N-tolylsulfonylmethylthio-benzimidate, **7,** 245
Nickel peroxide, **8,** 357
Nitrosonium tetrafluoroborate, **4,** 360
Potassium cyanide, **10,** 324
Rhodium(III) oxide, **4,** 420
Silver chloride, **10,** 347
p-Tolylmethyl isocyanide, **7,** 377
Tosylmethyl isocyanide, **4,** 514; **8,** 493
Triethyloxonium tetrafluoroborate, **5,** 691
2 N—PYRAZOLES
Ammonium persulfate–Silver nitrate, **5,** 16
o-Chloranil, **7,** 355
Diazoacetyl azide, **4,** 120
2-Diazo-1,1-dimethoxyethane, **10,** 120
Dichloromethylenedimethylammonium chloride, **4,** 135; **5,** 195
Dimethyl acetylenedicarboxylate, **4,** 168; **5,** 227
Ethylene glycol, **1,** 375; **5,** 296
Hydrazine, **1,** 434; **4,** 248
Nickel peroxide, **8,** 357
2,4-Pentanedione, **1,** 10
Phenyl(trichloromethyl)mercury, **2,** 326
Sodium hydroxide, **8,** 460

HETEROCYCLES—FIVE-MEMBERED RINGS (*Continued*)
Thionyl chloride, **9,** 465
Vinyltriphenylphosphonium bromide, **5,** 750
2 N—PYRAZOLIDINES
dl-Dianilino-1,2-diphenylethane, **1,** 187
Manganese(IV) oxide, **1,** 637
2 N—PYRAZOLINES
1-Pyrazolines
Diazomethane, **4,** 120; **5,** 179; **9,** 133
Phenyldiazomethane, **1,** 834
2-Pyrazolines
Diphenyldiazomethane, **1,** 338
Hydrazine, **2,** 211
Manganese(IV) oxide, **2,** 257
Phenylhydrazine, **2,** 322
Potassium diazomethanedisulfonate, **1,** 928
p-Toluenesulfonylhydrazide, **2,** 417
Trimethylsilyldiazomethane, **4,** 543
2 N—OTHERS
Benzamidine, **6,** 27
N,N′-Carbonyldiimidazole, **2,** 61
Cyanotrimethylsilane, **6,** 632
Diiminosuccinonitrile, **4,** 155
Diketene, **1,** 264
Dimethylformamide diethyl acetal, **5,** 253
Diphenyldiazomethane, **4,** 204
Ethylene diisocyanate, **2,** 100
Ethyl formate, **1,** 380
Methylhydrazine, **4,** 340
Phenylhydrazine, **1,** 838
Polyphosphoric acid, **2,** 334
Sodium methoxide, **1,** 1091
Sodium nitrite, **1,** 1097
Tetrabutylammonium fluoride, **5,** 645
Triethyl phosphite, **1,** 1212
Triphenylphosphine, **11,** 588
Urea, **1,** 1262
2 N, 1 O
Benzonitrile oxide, **2,** 383
2,4-Bis(4-methoxyphenyl)-1,3-dithia-2,4-diphosphetane-2,4-disulfide, **11,** 54
(Bromodichloromethyl)phenylmercury, **5,** 514
Dichloromethylenedimethylammonium chloride, **5,** 195
Dimethylformamide–Thionyl chloride, **1,** 286
Nitrobenzene, **7,** 251
Nitrogen dioxide, **2,** 175
Phenyliodine(III) diacetate, **1,** 508
Sodium azide, **1,** 1041; **2,** 376
Sodium hypochlorite, **1,** 1084; **7,** 337
Sulfur trioxide–Dimethylformamide, **1,** 1125
Thionyl chloride, **4,** 503
(Trichloromethyl)carbonimidic dichloride, **4,** 523
Triphenylphosphine, **1,** 1238
2 N, 1 S
Diiminosuccinonitrile, **4,** 155
Dipotassium cyanodithioimidocarbonate, **4,** 215
Ethoxycarbonyl isothiocyanate, **6,** 250
3 N—BENZOTRIAZOLES
o-Phenylenediamine, **1,** 834
Sodium nitrite, **1,** 1097
3 N—1,2,3-TRIAZOLES
Azidotrimethylsilane, **1,** 1236
Chloramine, **4,** 74
Ethyl azidoformate, **3,** 138; **4,** 225
Phenyl azide, **1,** 829; **2,** 45; **4,** 375; **5,** 513
Sodium azide, **4,** 440; **5,** 593
p-Toluenesulfonyl azide, **2,** 415
Trimethylsilyldiazomethane, **11,** 573
Trimethylsilylmethyl azide, **12,** 538
3 N—1,2,4-TRIAZOLES
Dichloromethylenedimethylammonium chloride, **4,** 135; **5,** 195
Ethyl ethoxyiminoacetate, **11,** 109
Ethyl formate, **1,** 380
Formic acid, **1,** 404
Thiosemicarbazide, **1,** 1164
(Trichloromethyl)carbonimidic dichloride, **4,** 523
Triethyloxonium tetrafluoroborate, **5,** 691
3 N—OTHERS
Chloramine, **1,** 122
Dichloromethylenedimethylammonium chloride, **8,** 156
Dicyclohexylcarbodiimide, **3,** 91
4 N
Aluminum azide, **5,** 9
Hydrazoic acid, **1,** 446
Iodine azide, **5,** 350
Tetramethylguanidinium azide, **2,** 403
Trimethylsilyldiazomethane, **11,** 573
1 O—BENZOFURANS

Acetic anhydride–Acetic acid, **2,** 5
Bis(benzonitrile)dichloropalladium(II), **4,** 129
t-Butyl hypochlorite–Dialkyl sulfides, **6,** 118
α-Chloro-N-cyclohexylpropanal-donitrone, **5,** 110
N-Chlorosuccinimide–Dimethyl sulfide, **6,** 118
2,3-Dichloro-5,6-dicyano-1,4-benzoquinone, **11,** 166
Mercury(II) acetate, **12,** 298
3-Methyl-3-buten-1-ynylcopper, **8,** 123
Oxodiperoxymolybdenum(pyridine)-(hexamethylphosphoric triamide), **11,** 218
Phosphoryl chloride, **5,** 535
Silver tetrafluoroborate, **5,** 587
1 O—2,3-DIHYDROBENZOFURANS
sec-Butyllithium, **12,** 97
Butyllithium–Magnesium bromide, **10,** 71
Di-*t*-butyl nitroxide, **11,** 160
N,N-Diethylaniline, **5,** 212
Dilithium tetrachloropalladate(II), **7,** 114
Dimethyl sulfoxide, **5,** 263
Dimethylsulfoxonium methylide, **2,** 171
Methyl bis(methylthio)sulfonium hexachloroantimonate, **11,** 335
Palladium(II) acetate, **9,** 344
Titanium(IV) chloride, **11,** 529
1 O—DIHYDROFURANS
2,3-Dihydrofurans
t-Butyl perbenzoate, **1,** 98
(E)-(Dimethylamino)phenyl-(2-phenylethenyl)sulfoxonium tetrafluoroborate, **4,** 174
Ethoxycarbonylcyclopropyltriphenyl-phosphonium tetrafluoroborate, **6,** 93
Grignard reagents, **11,** 245
Manganese(III) acetate, **6,** 355
Perchloric acid, **5,** 506
Zinc, **5,** 753
Zinc chloride, **9,** 522
2,5-Dihydrofurans
Lithium–Ammonia, **6,** 322
Silver perchlorate, **11,** 469
Titanium(IV) chloride, **6,** 590
Vinyltriphenylphosphonium bromide, **2,** 456
1 O—TETRAHYDROFURANS
by Cyclization of unsaturated alcohols
Benzeneselenenyl chloride, **8,** 25
N-Bromosuccinimide, **4,** 49
t-Butyl hydroperoxide–Vanadyl acetylacetonate, **8,** 62
Collins reagent, **11,** 139
Ethyl 4-diphenylphosphinyl-3-oxobutanoate, **10,** 181
Formic acid, **4,** 239
Iodine, **8,** 256; **11,** 261; **12,** 253
Iodine–Potassium iodide, **9,** 249
Mercury(II) acetate, **3,** 194
Mercury(II) trifluoroacetate, **11,** 320
Palladium(II) chloride, **12,** 371
2,4,4,6-Tetrabromo-2,5-cyclohexadienone, **12,** 457
by Cyclodehydration of 1,4-diols
Benzeneselenenyl halides, **10,** 16
Benzenesulfonyl chloride, **1,** 46
3-Bromopropyl 1-ethoxyethyl ether, **4,** 226
Dimethyl sulfoxide, **1,** 296
Diphenyldi(1,1,1,3,3,3-hexafluoro-2-phenyl-2-propoxy)sulfurane, **5,** 270
Nafion-H, **11,** 354
Palladium(II) acetate, **12,** 367
p-Toluenesulfonic acid, **1,** 1172
Triphenylphosphine + co-reagent, **9,** 504; **12,** 551
by Cyclodehydrogenation of ROH
Bromine–Silver(I) salts, **1,** 73; **5,** 60
t-Butyl hypoiodite, **1,** 94
Cerium(IV) ammonium nitrate, **3,** 44
Lead tetraacetate, **1,** 537; **12,** 270
Mercury(II) oxide–Iodine, **1,** 658
Silver(I) oxide, **3,** 252
Trifluoromethanesulfonyl chloride, **9,** 485
Other routes
Benzeneselenenic acid, **9,** 24
2,3-Bis(bromomethyl)-1,3-butadiene, **5,** 32
Boron trifluoride etherate, **10,** 52
4-Chloro-1-butenyl-2-lithium, **12,** 113
N,N-Dimethylformamide, **12,** 203
Fluorosulfuric acid, **6,** 262
Iodosylbenzene, **11,** 270
Lithium aluminum hydride–Boron trifluoride etherate, **1,** 599
Lithium 1-(dimethylamino)-

HETEROCYCLES—FIVE-MEMBERED RINGS (*Continued*)
naphthalenide, **12,** 279
Mercury(II) oxide–Iodine, **11,** 267
3-Methyl-3-trimethylsilyl-1-butene, **8,** 181
Oxalic acid, **12,** 408
Phenyl selenocyanate–Copper(II) chloride, **9,** 34
N-Phenylselenophthalimide, **9,** 366
Potassium permanganate, **9,** 388
Silver(I) trifluoroacetate, **11,** 471
Sodium borohydride–Boron trifluoride, **1,** 1053
Triethylsilane–Boron trifluoride, **10,** 418
Triisobutylaluminum, **1,** 1188
Zinc chloride, **9,** 522
1 O—OTHERS
Bis(trimethylsilyl)acetylene, **5,** 44
Butyllithium, **8,** 67; **9,** 83
Chlorotris(triphenylphosphine)-rhodium(I), **11,** 130
Copper(II)–Amine complexes, **8,** 114
Copper(I) bromide, **11,** 140
Copper(I) iodide, **5,** 717
1,5-Diazabicyclo[4.3.0]nonene-5, **4,** 116
1,4-Diazabicyclo[2.2.2]octane, **2,** 99
2-(2,2-Dimethoxyethyl)-1,3-dithiane, **4,** 164
Dimethylsulfoxonium methylide, **5,** 254
Ferric chloride, **4,** 236
Hydrobromic acid, **6,** 282
Hydrogen chloride–Titanium(IV) chloride, **10,** 201
Hydrogen peroxide, **5,** 337
α-Lithio-α-methoxyallene, **9,** 272
Lithium morpholide, **12,** 284
Mercury(II) acetate, **6,** 358; **12,** 303
3-Methoxy-1-methylthio-1-propyne, **6,** 397
Osmium tetroxide, **4,** 361
Oxalic acid, **5,** 481
Oxalyl chloride, **6,** 424
Peroxybenzimidic acid, **7,** 281
Phenyliodine(III) diacetate, **12,** 384
Piperazine, **4,** 392
Potassium *t*-butoxide, **5,** 544
Potassium ferricyanide, **1,** 929
Potassium superoxide, **6,** 488; **7,** 304
Thallium(III) acetate, **10,** 393
Tri-μ-carbonylhexacarbonyldiiron, **5,** 221
1 O, 1 S
Adogen 464, **7,** 4
(N,N-Diethylamino)methyloxo-sulfonium methylide, **5,** 210
Diphenylketene, **4,** 210
2 O (*see also* PEROXIDES for ENDOPEROXIDES)
2-Acetoxyisobutyryl chloride, **8,** 3
Potassium superoxide, **6,** 488
1 S—BENZOTHIOPHENES
Boron trifluoride–Trifluoroacetic anhydride, **4,** 45
Polyphosphoric acid, **5,** 540
Tetraphenylcyclopentadienone, **4,** 490
Thionyl chloride, **5,** 663
1 S—THIOLANES
Hexaethylphosphorous triamide, **2,** 207; **4,** 242
Hexamethylphosphoric triamide, **2,** 208
Palladium catalysts, **8,** 382
Potassium thiolacetate, **10,** 325
Sulfur dichloride, **2,** 391
Trifluoroacetic acid–Alkylsilanes, **6,** 616
1 S—THIOPHENES
Hydrogen sulfide, **1,** 962
Phosphorus heptasulfide, **1,** 864
Phosphorus(V) sulfide, **6,** 470
Tetracyanoethylene, **1,** 1133
1 S—OTHERS
Alumina, **4,** 8
2,3-Bis(bromomethyl)-1,3-butadiene, **5,** 32
N-Chlorosuccinimide, **5,** 127
Sodium hydroxide, **7,** 336
Sulfur dioxide, **6,** 558; **9,** 440
Sulfuryl chloride, **10,** 375
Vinyltriphenylphosphonium bromide, **5,** 750; **6,** 666
2 S
Dimethyl sulfoxide, **5,** 263
Phosgene, **5,** 532
HETEROCYCLES—SIX-MEMBERED RINGS (*see also* LACTAMS, LACTONES)
1 N—DIHYDROPYRIDINES
Copper hydride ate complexes, **12,** 286
Formaldehyde, **1,** 397
Methyl chloroformate, **6,** 376
Organocopper reagents, **11,** 365

Sodium dithionite, **1,** 1081
Trimethylsilane, **1,** 1235
1 N—ISOQUINOLINES AND RELATED
Isoquinolines
[1,3-Bis(diphenylphosphine)propane]-(dichloro)nickel(II), **11,** 167
Boron trifluoride, **4,** 45; **6,** 65
Hydrochloric acid, **5,** 333
Ozone, **11,** 387
Potassium nitrosodisulfonate, **5,** 562
Thallium(III) trifluoroacetate, **10,** 397
Trimethylsilyl polyphosphate, **11,** 427
Dihydroisoquinolines
Benzeneseleninic anhydride, **11,** 37
t-Butyllithium, **10,** 76
Copper(I) iodide, **12,** 141
Diphenylselenium bis-(trifluoroacetate), **11,** 223
Isocyanomethyllithium, **10,** 231
Polyphosphate ester, **1,** 892; **2,** 333; **3,** 229; **4,** 394; **5,** 539
Potassium superoxide, **11,** 442
Sodium hydride, **5,** 610
Tetrahydroisoquinolines
Dihydridotetrakis(triphenyl-phosphine)ruthenium(II), **11,** 182
6-Methoxy-7-hydroxy-3,4-dihydro-isoquinolinium methiodide, **4,** 329
Polyphosphoric acid, **5,** 540
Sodium tris[(S)-N-benzyloxycarbonyl-prolyoxy]hydroborate, **12,** 454
1 N—PIPERIDINES
by Cyclization of unsaturated amines
Benzeneselenenyl halides, **10,** 16
Benzenesulfenyl chloride, **12,** 42
Mercury(II) acetate, **11,** 315; **12,** 298
Organoaluminum reagents, **12,** 339
Other routes
Dichlorotris(triphenylphosphine)-ruthenium(II), **10,** 141
Palladium nitrate, **6,** 451
Raney nickel, **7,** 312
Rhodium catalysts, **1,** 979, 982
Ruthenium catalysts, **5,** 574
Sodium borohydride–Rhodium(III) chloride, **11,** 480
Sodium cyanoborohydride, **12,** 445
Sulfuric acid, **5,** 633
Trialkylaluminum, **11,** 539
1 N—PYRIDINES
by Deoxygenation of N-oxides
Chlorotrimethylsilane–Sodium iodide–Zinc, **11,** 128
9-Diazofluorene, **1,** 190
Diphosphorus tetraiodide, **10,** 174
Phenyl(trichloromethyl)mercury, **1,** 851
Phosphorus(III) chloride, **1,** 875
Sulfur dioxide, **2,** 392
Titanium(IV) chloride–Sodium borohydride, **10,** 404
Triphenylphosphine, **1,** 1238
by Diels–Alder reaction
N-Butoxycarbonylmethylene-*p*-toluenesulfonamide, **1,** 98
1-Chloro-1-dimethylamino-2-methyl-1,3-butadiene, **9,** 104
3-Methylsulfonyl-2,5-dihydrofuran, **10,** 270
1,2,4-Triazine, **10,** 409
Trichloro-1,2,4-triazine, **9,** 480
Triethyl 1,2,4-triazine-3,5,6-tricarboxylate, **12,** 528
Other routes
Alkanoylketene dithioacetals, **11,** 286
O-Allylhydroxylamine, **10,** 5
t-Butyldimethylsilyl trifluoromethanesulfonate, **11,** 90
4-Chloromethyl-3,5-dimethyl-isoxazole, **2,** 150; **4,** 177
Cyanotrimethylsilane, **12,** 148
Cyclopentadienyl(triphenyl-phosphine)cobaltatetraphenyl-cyclopentadiene, **5,** 173
Dicarbonylcyclopentadienylcobalt, **12,** 160
Dicyclopentadienylcobalt, **6,** 128; **12,** 180
Ferric nitrate/K10 Bentonite, **12,** 231
Formaldehyde, **1,** 397
Hexamethyldisilazane, **12,** 239
Hydrazoic acid, **2,** 211
Lithium aluminum hydride–Pyridine, **4,** 294
Nitric acid, **1,** 733
2,4,6-Trimethylpyrylium sulfoacetate, **11,** 571
Triphenylcarbenium perchlorate, **1,** 1256
1 N—PYRIDONES
Azidotrimethylsilane, **5,** 719

HETEROCYCLES—SIX-MEMBERED RINGS (*Continued*)
Bis[1,2-bis(diphenylphosphine)ethane]-palladium(0), **10,** 32
1,3-Bis(*t*-butyldimethylsilyloxy)-2-aza-1,3-diene, **11,** 50
Dicarbonylcyclopentadienylcobalt, **12,** 160
Dimethyl sulfoxide, **5,** 263
Ethyl formate, **1,** 380
Mercury(II) acetate–Ethylenediamine-tetraacetic acid, **5,** 427
Molecular sieves, **10,** 273
Piperidine, **1,** 886
Potassium *t*-butoxide, **6,** 477
Potassium ferricyanide, **1,** 929
Sodium methoxide, **1,** 1091
1 N—QUINOLINES AND RELATED
Quinolines
Benzyltriethylammonium chloride, **7,** 18
Boron trifluoride etherate, **6,** 65
Cobalt(II) phthalocyanine, **10,** 102
Cyanotrimethylsilane, **12,** 148
Dichlorobis(norbornadiene)-dirhodium(I), **9,** 108
Dichlorotris(triphenylphosphine)-ruthenium(II), **11,** 171
N,N-Diethylaminopropyne, **5,** 217
Diisopropyl peroxydicarbonate, **12,** 194
Dimethyl acetylenedicarboxylate, **2,** 145
Ferrous sulfate, **1,** 393
Hexamethylphosphoric triamide, **6,** 273
Methyl vinyl ketone, **1,** 697
Polyphosphoric acid, **1,** 894; **4,** 395
Sodium acetate, **1,** 1024
N-*p*-Tolylvinylmethylketenimine, **5,** 683
Triphenylpropargylphosphonium bromide, **6,** 494
Vilsmeier reagent, **9,** 514; **12,** 564
Dihydroquinolines
1-Benzyl-1,2-dihydroisonicotinamide, **12,** 46
Diisobutylaluminum hydride, **1,** 260
Lithium–Ammonia, **5,** 379
Tetrahydroquinolines
Borane–Pyridine, **8,** 50
Cesium fluoride, **11,** 115
Dichlorotris(triphenylphosphine)-ruthenium(II), **12,** 179
Diisobutylaluminum hydride, **1,** 260
Trifluoroacetic acid, **6,** 613
1 N—OTHERS
Acetic acid, **5,** 3
Aluminum chloride, **1,** 24
Birch reduction, **5,** 30
1,3-Bis(*t*-butyldimethylsilyloxy)-2-aza-1,3-diene, **11,** 50
Butyllithium, **10,** 68
Carbon monoxide, **9,** 95
1-Chloro-N,N,2-trimethyl-propenylamine, **5,** 136
Chlorotrimethylsilane, **11,** 125
Diethyl ethoxymethylenemalonate, **5,** 216
Diisobutylaluminum hydride, **12,** 191
N,N-Dimethylhydrazine, **11,** 200
Dimethyl sulfoxide, **5,** 263
trans-1-Methoxy-3-trimethylsilyloxy-1,3-butadiene, **11,** 332
Methylene bis(urethane), **1,** 677
Methyl isothiocyanate, **5,** 173
5-Methylisoxazole, **9,** 309
Methyl 3-oxo-4-pentenoate, **1,** 1272
1-*o*-Nitrophenyl-1,3-butadiene, **1,** 743
Organocopper reagents, **11,** 365
Oxygen, **8,** 366
Palladium(II) acetate, **8,** 378
Phenyldiazomethane, **2,** 320
Phosphoric acid–Boron trifluoride complex, **7,** 289
Phosphoryl chloride, **1,** 876
Polyphosphoric acid, **1,** 894; **2,** 334
Potassium amide, **4,** 398
Sodium cyanoborohydride, **9,** 424
Sodium ethoxide, **1,** 1065
Sodium tetrachloroaluminate, **2,** 372
Thallium(III) trifluoroacetate, **12,** 481
Tin(IV) chloride, **7,** 342
N-*p*-Toluenesulfonyltrifluoro-methylimine, **5,** 681
Trimethylsilyl trifluoromethane-sulfonate, **12,** 543
Vilsmeier reagent, **7,** 422
1 N, 1 O
Acetyl chloride, **7,** 3
Azidotrimethylsilane, **5,** 719; **6,** 632
O-*t*-Butyldimethylsilylhydroxylamine, **12,** 85
t-Butyl isocyanide, **5,** 581

α-Chloro-N-cyclohexylpropanal-donitrone, **4,** 80; **5,** 110
Chlorosulfonyl isocyanate, **5,** 132; **6,** 122
N-Cyclohexylhydroxylamine, **8,** 135
Dimethyl acetylenedicarboxylate, **7,** 117
N-(2,3-Epoxypropylidene)-cyclohexylamine N-oxide, **9,** 209
Ethyl 3-bromo-2-hydroxyimino-propanoate, **9,** 212
Iodine, **11,** 261
Methyl α-phenylglycinate, **8,** 395
Potassium ferricyanide, **1,** 929
Sodium bicarbonate, **11,** 476
Trichloroacetyl isocyanate, **7,** 382
(Trichloromethyl)carbonimidic dichloride, **4,** 523
Triphenylphosphine–Thiocyanogen, **8,** 518

1 N, 2 O
Diphenylketene, **5,** 278

1 N, 1 S
Boron trifluoride etherate, **5,** 52
Dimethyl acetylenedicarboxylate, **6,** 206
Polyphosphate ester, **4,** 394
Triethyl phosphite, **1,** 1212

2 N—PYRAZINES
Diaminomaleonitrile, **5,** 175
Diiminosuccinonitrile, **4,** 155
Glyoxal, **1,** 413
Silver perchlorate, **7,** 322
Sodium nitrite, **4,** 459
Titanium(IV) chloride, **5,** 671

2 N—PYRIDAZINES
1,5-Diazabicyclo[4.3.0]nonene-5, **5,** 176
Di-*t*-butyl azodiformate, **1,** 209
Diethyl azodicarboxylate, **1,** 245
Hydrazine, **9,** 236
Lithium diisopropylamide, **7,** 204

2 N—PYRIMIDINES
Azidotrimethylsilane, **5,** 719
Benzyltriethylammonium chloride, **7,** 18
N,N-Bis(phenylthio)benzene-sulfenamide, **5,** 732
Carbonylisocyanatidic chloride, **5,** 109
Chlorosulfonyl isocyanate, **5,** 132
Dichloromethylenedimethylammonium chloride, **4,** 135; **8,** 156
Dicyclohexylcarbodiimide, **1,** 231
Diketene, **5,** 225
Dimethylformamide diethyl acetal, **10,** 157
Ethoxycarbonyl isothiocyanate, **4,** 223
Ethylene glycol, **1,** 375
Ethyl ethoxymethylenecyanoacetate, **1,** 380
Formamide, **1,** 402
Formamidine acetate, **1,** 403
Hydrogen chloride, **5,** 335
Malonaldehyde bis(diethyl acetal), **2,** 257
Palladium black, **5,** 498
Phosgene, **5,** 532
Polyphosphoric acid, **7,** 294
Raney nickel, **1,** 723
Sodium alkoxides, **1,** 1065, 1091
Thiourea, **1,** 1164
(Trichloromethyl)carbonimidic dichloride, **4,** 523
Tris(formamino)methane, **1,** 1251
Urea, **1,** 1262

2 N—QUINAZOLINES
Formamidinoyl isothiocyanates, **5,** 315
Phosphoryl chloride, **4,** 390
Tetrafluoroboric acid, **2,** 397
(Trichloromethyl)carbonimidic dichloride, **4,** 523

2 N—QUINOXALINES
Ethyl ethoxyiminoacetate, **11,** 109
Glyoxal, **1,** 413
Hydrogen peroxide–Iron salts, **1,** 472
o-Phenylenediamine, **1,** 834
Vilsmeier reagent, **9,** 514

2 N—OTHERS
Diphenyl-N-*p*-tolylketenimine, **5,** 282
Hydrazine, **1,** 434
Nitrobenzene, **2,** 295
Silver carbonate–Celite, **6,** 511

2 N, 1 S
2-Chloroethanesulfonyl chloride, **5,** 116
O-Mesitylenesulfonylhydroxylamine, **5,** 430
Potassium persulfate, **10,** 331

3 N
N-Cyanoguanidine, **1,** 229
Dichloromethylenedimethylammonium chloride, **8,** 156
Diketene, **5,** 225
Dimethylformamide diethyl acetal, **1,** 281
Dimethyl 1,2,4,5-tetrazine-3,6-dicarboxylate, **12,** 214
Formamidinoyl isothiocyanates, **5,** 315
Hydrazine, **5,** 327

HETEROCYCLES—SIX-MEMBERED RINGS (*Continued*)
Iron carbonyl, **1,** 519
Sodium azide, **5,** 593
Sodium methoxide, **1,** 1091
Tri-μ-carbonylhexacarbonyldiiron, **12,** 525
(Trichloromethyl)carbonimidic dichloride, **4,** 523
Tris(formamino)methane, **1,** 1251
1 O—CHROMENES
Copper(II) sulfate, **2,** 89
2,3-Dichloro-5,6-dicyano-1,4-benzoquinone, **2,** 112; **3,** 83
Grignard reagents, **8,** 235
Menthyl *p*-toluenesulfinate, **12,** 295
Mercury(II) oxide, **5,** 428
Potassium borohydride, **8,** 407
Potassium dichromate, **9,** 383
Silver tetrafluoroborate, **5,** 587
Sodium borohydride–Palladium(II) chloride, **11,** 479
Sodium hydride, **1,** 1075
Triphenylcarbenium tetrafluoroborate, **12,** 550
Vinyltriphenylphosphonium bromide, **1,** 1274; **2,** 456
1 O—COUMARINS
Acetic anhydride, **9,** 1
Diketene, **1,** 264
N,N-Dimethylformamide, **5,** 247
Ethyl acetoacetate, **1,** 29, 514
Hydrogen fluoride, **2,** 215
Piperidine, **1,** 886
Selenium, **11,** 465
Trifluoroacetic acid, **1,** 1219
Zinc carbonate, **4,** 579
Zinc chloride–Phosphoryl chloride, **1,** 1292
1 O—FLAVONES
Aluminum chloride, **4,** 10
2,3-Dichloro-5,6-dicyano-1,4-benzoquinone, **8,** 153
Dimethyl sulfoxide–Iodine, **9,** 190
Mercury(II) acetate, **12,** 298
Oxygen, singlet, **7,** 261
Selenium(IV) oxide, **1,** 422; **5,** 575
Thallium(III) nitrate, **9,** 460
1 O—ISOCOUMARINS
Butyllithium, **6,** 85
Di-μ-bromobis(2-methoxy-2-propenyl)-dinickel, **8,** 324
Dimethylsulfoxonium methylide, **12,** 213
4-Methoxy-6-methyl-2-pyranone, **9,** 302
Thallium(III) trifluoroacetate, **12,** 481
Trifluoroacetic acid, **10,** 418
Vilsmeier reagent, **7,** 422
1 O—ISOFLAVONES
N-Formylimidazole, **8,** 233
Silver hexafluoroantimonate, **7,** 320
1,3,5-Triazine, **10,** 410
1 O—PYRONES
2-Pyrones
Bromine, **12,** 70
sec-Butyllithium, **12,** 97
Chloroketene dimethyl acetal, **5,** 119
Dimethyl methoxymethyl-enemalonate, **12,** 206
Iron carbene complexes, **12,** 265
Ketene dimethyl acetal, **11,** 279
Palladium catalysts, **1,** 778
Sulfur, **3,** 273; **5,** 632; **6,** 556
Sulfuric acid, **5,** 633
Tetracarbonyl(ethoxyphenyl-methylidene)iron(0), **12,** 265
Thallium(I) hydroxide, **6,** 578
2,4,6-Trichlorophenylcyanoacetate, **7,** 384
Vinylketene, **11,** 596
4-Pyrones
Dibenzyl ketone, **1,** 198
4-Methoxy-3-buten-2-one, **10,** 258
Methyl fluorosulfonate, **6,** 381
Polyphosphoric acid, **1,** 894
Sodium ethoxide, **1,** 1065
1 O—PYRYLIUM SALTS
Alkanoylketene dithioacetals, **11,** 282
t-Butyl chloride, **2,** 46
2,3-Dichloro-5,6-dicyano-1,4-benzoquinone, **5,** 193
Lithium perchlorate, **5,** 413
Perchloric acid, **5,** 506
Sulfoacetic acid, **1,** 1117; **10,** 374
Tin(IV) chloride, **5,** 627
Triphenylcarbenium hexachloroantimonate, **3,** 330
1 O—TETRAHYDROPYRANS
by Cyclization of unsaturated ROH
Benzeneselenenyl chloride, **8,** 25
Mercury(II) acetate, **3,** 194
Mercury(II) trifluoroacetate, **9,** 294
Palladium(II) chloride, **12,** 371

Silver(I) nitrate, **12,** 433
2,4,4,6-Tetrabromo-2,5-cyclohexadienone, **7,** 351
by Cyclodehydration of 1,5-diols
Diphenyldi(1,1,1,3,3,3-hexafluoro-2-phenyl-2-propoxy)sulfurane, **5,** 270
Formic acid, **4,** 239
Ion-exchange resins, **5,** 355
Triphenylphosphine + co-reagent, **9,** 504; **12,** 551
by Reduction of γ-lactones
Lithium aluminum hydride–Boron trifluoride etherate, **1,** 599
Perchloric acid, **1,** 796
Triethylsilane–Boron trifluoride, **10,** 418
Other routes
Benzeneselenenic acid, **9,** 24
Benzeneselenenyl bromide, **9,** 25
Diborane, **5,** 184
Hexamethylphosphoric triamide, **6,** 273
Iodosylbenzene, **11,** 270
Lead tetraacetate, **6,** 313
Lithium 1-(dimethylamino)-naphthalenide, **12,** 279
Mercury(II) oxide–Iodine, **11,** 267; **12,** 305
N-Phenylselenophthalimide, **11,** 417
Silver(I) trifluoroacetate, **11,** 471
Titanium(IV) chloride, **11,** 529
Trifluoromethanesulfonyl chloride, **9,** 485

1 O—XANTHONES
Copper(II) bromide, **6,** 138
4-Dimethylaminopyridine, **9,** 178
Oxygen, **8,** 366
Phosphoryl chloride–Zinc chloride, **1,** 880
Polyphosphoric acid, **4,** 395
Potassium ferricyanide, **5,** 551
Potassium fluoride, **12,** 406
Pyridinium chloride, **4,** 415
Sodium tetrachloroaluminate, **9,** 435
Zinc chloride–Phosphoryl chloride, **1,** 1292

1 O—OTHERS
Alkylaluminum halides, **11,** 7
Allyltrimethyltin, **11,** 20
Aluminum chloride, **9,** 14; **11,** 25
Bornyloxyaluminum dichloride, **11,** 70
Bromine–Silver tetrafluoroborate, **7,** 36
1-Bromo-3-methyl-2-butene, **5,** 64
Butyllithium, **6,** 85; **10,** 71
Carbonyl cyanide, **2,** 60
Chlorotris(triphenylphosphine)cobalt, **12,** 128
Diethyl oxomalonate, **7,** 396
5,6-Dihydro-4-methoxy-2*H*-pyran, **5,** 438
Diisobutylaluminum hydride, **5,** 224; **11,** 185
2-(2,2-Dimethoxyethyl)-1,3-dithiane, **4,** 164
4-Dimethylaminopyridine, **10,** 155
Dimethylformamide dimethyl acetal, **4,** 184
Dimethylformamide–Phosphoryl chloride, **5,** 251
Diphenyl disulfide, **10,** 171
N-Ethyl(diethoxyphosphinyl)-methylketenimine, **9,** 216
Lithium trichloropalladate(II), **12,** 288
Mercury(II) trifluoroacetate, **4,** 325; **11,** 320
1-Methoxy-1,3-bis(trimethylsilyloxy)-1,3-butadiene, **11,** 66
1-Methoxy-1,3-butadiene, **10,** 258
1-Methoxy-2-methyl-3-trimethylsilyloxy-1,3-pentadiene, **11,** 326; **12,** 312
trans-1-Methoxy-3-trimethylsilyloxy-1,3-butadiene, **11,** 332
Methyl diformylacetate, **8,** 338
Methyl methylthiomethyl sulfoxide, **6,** 390
Nickel(II) acetylacetonate, **9,** 51
Peroxybenzimidic acid, **7,** 281
2-(2-Phenylsulfonylethyl)-1,3-dioxolane, **6,** 461
Ruthenium tetroxide, **1,** 986
Sodium borohydride, **5,** 597
Sodium dihydrogen phosphate, **4,** 451
Sulfuric acid, **4,** 470
Thallium(III) sulfate, **5,** 658
Titanium(IV) chloride, **12,** 494
Trifluoroacetic acid, **2,** 433
Trifluoroacetic anhydride, **9,** 484
Trimethylsilyl trifluoromethanesulfonate, **11,** 584
Triphenylcarbenium perchlorate, **1,** 1256
Tris(6,6,7,7,8,8,8-heptafluoro-2,2-dimethyl-3,5-octanedionato)-

HETEROCYCLES—SIX-MEMBERED RINGS (*Continued*)
europium(or -ytterbium), **12,** 559
1 O, 1 S
Acetyl chloride, **4,** 5
Crown ethers, **7,** 76
Sulfur, **1,** 1118
Sulfur dioxide, **7,** 346
2 O—(*see also* PEROXIDES for ENDOPEROXIDES)
Copper(II) sulfate, **8,** 125
Diphenyldi(1,1,1,3,3,3-hexafluoro-2-phenyl-2-propoxy)sulfurane, **5,** 270
Formaldehyde, **1,** 397
Lead(IV) oxide, **1,** 533
N-Methyl-N,N′-di-*t*-butylcarbodiimium tetrafluoroborate, **6,** 378
Oxygen, **5,** 482
Phosphoric acid, **10,** 317
Silver(I) oxide, **6,** 515; **10,** 350
Silver(II) oxide, **9,** 412
3 O
Hydrogen peroxide, **4,** 253
Tungsten(VI) oxide–Chlorosulfonic acid, **11,** 591
1 S
Carbon disulfide, **5,** 173
Polyphosphoric acid, **5,** 540
Sulfur dichloride, **1,** 1121; **2,** 391
2*H*-Thiopyran, **8,** 242
Triethylamine, **7,** 385
2 S
Sodium ethoxide, **1,** 1065
3 S
Formaldehyde, **1,** 397
HETEROCYCLES—SEVEN-MEMBERED RINGS
1 N
Aluminum chloride, **3,** 7
Chloramine, **1,** 122
Dichlorotris(triphenylphosphine)-ruthenium(II), **10,** 141
Diethylaluminum iodide, **12,** 5
Dimethyl acetylenedicarboxylate, **4,** 168
Ethyl azidoformate, **1,** 363; **7,** 146
Hexaethylphosphorous triamide, **5,** 323
Hydrazoic acid, **2,** 211
Iodine isocyanate, **2,** 223; **3,** 161
Mercury(II) oxide–Tetrafluoroboric acid, **11,** 318
Polyphosphate ester, **1,** 892
Triphenylphosphine, **2,** 443
1 O
m-Chloroperbenzoic acid, **1,** 135
1,5-Diazabicyclo[4.3.0]nonene-5, **1,** 189
Dimethylsulfoxonium methylide, **7,** 133
Ethoxycarbonylcyclopropyltriphenylphosphonium tetrafluoroborate, **5,** 90
Lithium aluminum hydride–Boron trifluoride etherate, **1,** 599
N-Methyl-N,N′-di-*t*-butylcarbodiimium tetrafluoroborate, **6,** 378
Potassium bisulfate, **1,** 909
Potassium *t*-butoxide, **3,** 233
Tri-μ-carbonylhexacarbonyldiiron, **5,** 221
Trichlorosilane, **4,** 525
Triphenylphosphine–Diethyl azodicarboxylate, **9,** 504
1 S
Di-μ-chlorobis(allyl)dipalladium, **7,** 5
Sulfur dioxide, **2,** 392; **4,** 469; **7,** 346
OTHER HETEROCYCLES
PYRROLIZIDINES
Benzenesulfenyl chloride, **12,** 42
Ethoxycarbonylcyclopropyltriphenylphosphonium tetrafluoroborate, **10,** 78
Iodotrimethylsilane, **9,** 251
Potassium nitrosodisulfonate, **10,** 329
Pyridinium (*d*)-camphor-10-sulfonate, **11,** 449
Rhodium catalysts, **1,** 979, 982
Tributyltin hydride, **11,** 545; **12,** 516
Trimethylsilylmethyl trifluoromethanesulfonate, **10,** 434
PURINES
Acetic acid, **8,** 1
Benzyltrimethylammonium iodide, **1,** 53
Diethoxymethyl acetate, **1,** 243
Diethyl azodicarboxylate, **2,** 128; **6,** 185
N,N-Dimethylhydrazine, **7,** 126
Formamidine acetate, **1,** 403
N-Nitrosodimethylamine, **5,** 479
Polyphosphoric acid, **1,** 894
Triethyl orthoformate, **5,** 690
Vilsmeier reagent, **1,** 284
HOMOALLYLIC ALCOHOLS (*see also* CHIRAL COMPOUNDS)
BY ADDITION OF ALLYL TO C=O
B-Allyl-9-borabicyclo[3.3.1]nonane, **12,** 15

Allyl bromide, **1,** 1287
B-Allyldiisopinocampheylborane + derivatives, **12,** 17
Allyllithium, **1,** 1250; **5,** 82
Allylmagnesium bromide, **1,** 415
Allyl phenyl selenide–Trialkylborane, **12,** 20
Allyltin difluoroiodide, **10,** 6, 374
Allyltrimethylsilane, **7,** 8, 370; **9,** 445
Allyltrimethyltin, **9,** 8, 47
9-Borabicyclo[3.3.1]nonane, **9,** 57
Cerium amalgam, **11,** 114
2-Chloromethyl-4,4,5,5-tetramethyl-1,3,2-dioxaborolane, **12,** 397
Chromium(II) chloride, **11,** 132; **12,** 136
Chromium(III) chloride–Lithium aluminum hydride, **8,** 110
B-Crotyl-9-borabicyclo[3.3.1]nonane, **12,** 81
Crotylmagnesium chloride, **12,** 28
2-Crotyl-4,4,5,5-tetramethyl-1,3,2-dioxaborolane, **11,** 86
Crotyltri(isopropoxy)titanium, **12,** 354
Crotyltrimethylsilane, **12,** 146
Di-μ-bromobis(2-methallyl)dinickel, **6,** 361
gem-Dichloroallyllithium, **5,** 188
Dichlorobis(cyclopentadienyl)titanium, **10,** 130; **12,** 168
Dichlorobis(cyclopentadienyl)-zirconium(II), **11,** 51
gem-Difluoroallyllithium, **12,** 188
Fluorodimethoxyborane, **11,** 240
Hypochlorous acid, **11,** 260
Manganese(II) chloride–Lithium aluminum hydride, **11,** 310
B-Methallyldiisopinocampheylborane, **12,** 18
Methallyllithium, **5,** 82
Organotitanium reagents, **11,** 374
(E)-(2,4-Pentadienyl)trimethylsilane, **10,** 303
(2R,4R)-Pentanediol, **12,** 375
(+)-3-Phenylbornane-2,3-diol, **9,** 363
Samarium(II) iodide, **11,** 464
Tetrakis(triphenylphosphine)-palladium(0), **11,** 503
N,N,N′,N′-Tetramethylethylenediamine, **4,** 485
Tin, **11,** 519
Tin–Aluminum, **12,** 486
Tin(II) fluoride, **11,** 202, 524
Titanium(IV) chloride, **12,** 494
Tributylcrotyltin, **10,** 411, 451; **11,** 143, 542; **12,** 146, 513
Triethylaluminum, **10,** 415
Triethylborane, **10,** 416
[(Trimethylsilyl)allyl]lithium, **9,** 155; **11,** 12
α-(Trimethylsilyl)benzylmagnesium bromide, **11,** 19
α-Trimethylsilylcrotyl-9-borabicyclo-[3.3.1]nonane, **10,** 430
3-Trimethylsilyl-1-cyclopentene, **8,** 509
Zinc iodide, **12,** 21
by Ene reactions
Alkylaluminum halides, **10,** 177; **12,** 5
Chloral, **8,** 82
Formaldehyde, **11,** 8
by Epoxide cleavage
Diisobutylaluminum hydride, **6,** 198; **7,** 111; **8,** 173
Dilithium dialkyl(cyano)cuprates, **12,** 349
Lithium divinylcuprate, **5,** 288
Methylenetriphenylphosphorane, **11,** 338
Trimethylaluminum–Dichlorobis-(cyclopentadienyl)zirconium, **10,** 423
α-Trimethylsilylvinylmagnesium bromide, **8,** 515
by Wittig and related reactions
2,2-Dihydro-2,2,2-triphenyl-1,2-oxaphospholane, **2,** 442
3-Hydroxy-3-methylbutylidene-triphenylphosphorane, lithium salt, **8,** 251
2-Methyl-2-thiazoline, **6,** 403
Other routes
Acetic anhydride, **6,** 3
Diethylaluminum N-methylanilide, **8,** 165
Diisopinocampheylborane, **4,** 161
Disiamylborane, **11,** 226
Grignard reagents, **6,** 147
Hydrogen peroxide, **12,** 242
Lithium aluminum hydride, **5,** 382; **9,** 274
Lithium bronze, **11,** 293
Lithium dimethylcuprate, **4,** 177
Lithium tri-*sec*-butylborohydride, **12,** 286

HOMOALLYLIC ALCOHOLS (*Continued*)
Peracetic acid, **2,** 307
Perchloric acid, **5,** 506
Phenylselenoacetaldehyde, **10,** 310
p-Toluenesulfonylhydrazide, **8,** 489
Trialkylaluminums, **12,** 512
Tributyl(iodomethyl)tin, **9,** 250, 475
Urushibara catalysts, **6,** 659
HOMOALLYLIC AMINES
B-Allyl-9-borabicyclo[3.3.1]nonane, **12,** 15
Palladium(II) acetate, **8,** 378
N-Sulfinyltoluenesulfonamide, **12,** 455
HOMOALLYLIC ETHERS
Allyltrimethylsilane, **7,** 370; **10,** 439
Iodotrimethylsilane, **10,** 216; **12,** 259
3-Methyl-3-trimethylsilyl-1-butene, **8,** 181
Peracetic acid, **2,** 307
Silver tetrafluoroborate, **12,** 434
Titanium(IV) chloride, **11,** 529
HOMOPROPARGYLIC ALCOHOLS
(*see also* CHIRAL COMPOUNDS)
Chloro(methoxy)methylalane, **5,** 438
Chromium(III) chloride–Lithium aluminum hydride, **10,** 101
Diethylaluminum ethoxyacetylide, **7,** 103
Diethyl tartrate, **11,** 181
Hexamethylphosphoric triamide, **10,** 196
Lithium acetylide, **8,** 285
Lithium amide, **9,** 278
Lithium 3-chloropropargylide, **9,** 279
Organoaluminum reagents, **4,** 144; **12,** 339
Organotitanium reagents, **11,** 374
Organozinc reagents, **11,** 25; **12,** 356
Organozirconium reagents, **11,** 378
Propargyl bromide, **1,** 1285; **2,** 19
Samarium(II) iodide, **11,** 464
Trimethylsilylallene, **10,** 428
HYDRAZIDES
N,N′-Diisopropylhydrazine, **4,** 162
N,N-Dimethylhydrazine, **5,** 254
Hydrazine, **6,** 280
Phosphonitrilic chloride trimer, **2,** 206
HYDRAZINES
1-Acetyl-1-methylhydrazine, **4,** 7
N-Aminophthalimide, **1,** 38
t-Butyl carbazate, **11,** 88
Hydrazine, **1,** 434
Hydroxylamine-O-sulfonic acid, **1,** 481
Lithium aluminum hydride, **5,** 382
O-Mesitoylhydroxylamine, **1,** 660
Sodium hydrazide, **1,** 1074
Sodium hypochlorite, **1,** 1084
Tetrabutylammonium chloride, **7,** 353
Tris(trimethylsilyl)hydrazidocopper, **6,** 651
Zinc–Acetic acid, **1,** 1276
HYDRAZONES
FROM C=O
1-Acetyl-1-methylhydrazine, **4,** 7
Azobenzene-4-carboxylic acid hydrazide, **1,** 44
Diethyl phosphite, **1,** 251
N,N-Dimethylhydrazine, **1,** 289; **2,** 154; **3,** 117
2,4-Dinitrophenylhydrazine, **1,** 330
1,1-Diphenylhydrazine, **1,** 340
Ethyl carbazate, **1,** 360
Hydrazine, **1,** 434; **3,** 153
5-Iodo-2,4-dinitrophenylhydrazine, **1,** 506
Phenylhydrazine, **1,** 838
Polyphosphoric acid, **4,** 395
Triethylamine, **1,** 1198
OTHER ROUTES
Arenediazonium salts, **2,** 62; **8,** 22
N-Benzoyl-N′-trifluoromethanesulfonylhydrazine, **6,** 36
p-Carboxybenzenediazonium chloride, **2,** 62
Lead tetraacetate, **1,** 537
ARENESULFONYLHYDRAZONES
N,N-Methyltoluenesulfonylhydrazide, **9,** 411
p-Toluenesulfonylhydrazide, **1,** 1185; **12,** 508
HYDROPEROXIDES (*see also* ALLYLIC COMPOUNDS, α-HYDROPEROXY CARBONYLS)
Alkyl hydroperoxides, **1,** 17
Hydrogen peroxide + co-reagent, **1,** 471; **4,** 253; **6,** 99; **7,** 174, 175; **8,** 250
Oxygen, **1,** 1253; **5,** 482; **7,** 261; **8,** 367
HYDROXAMIC ACIDS
Bis(N,N-dimethylformamido)-oxodiperoxymolybdenum(VI), **9,** 48
Diborane, **7,** 89
Hydroxylamine, **1,** 478
HYDROXY ALDEHYDES AND KETONES (*see also* CHIRAL COMPOUNDS)

α-HYDROXY
by Acyloin reaction (*see* TYPE OF REACTION INDEX)
by Addition of acyl anions to C=O
Acyllithium reagents, **12,** 4
1-(Alkylthio)vinyllithium, **5,** 6
3-Benzyl-5-(2-hydroxyethyl)-4-methyl-1,3-thiazolium chloride, **6,** 289
t-Butylhydrazine, **12,** 87
Cyanotrimethylsilane, **9,** 127
Dichloromethyllithium, **5,** 199
Diethyl 1-phenyl-1-trimethylsilyloxymethylphosphonate, **10,** 145
1,3-Dithiane, **7,** 142; **8,** 217
Ethyl ethylthiomethyl sulfoxide, **5,** 299
(S)-N-Formyl-2-methoxymethylpyrrolidine, **11,** 243
Grignard reagents, **12,** 235
Hexahydro-4,4,7-trimethyl-4*H*-1,3-benzothiin, **12,** 237
2-Lithio-4,5-dihyro-5-methyl-[4*H*]-1,3,5-dithiazine, **11,** 302
α-Methoxyvinyllithium, **6,** 372; **7,** 233; **8,** 331
Methyl methylthiomethyl sulfoxide, **4,** 341
Potassium cyanide, **7,** 299
Potassium fluoride, **10,** 325
Sodium bis(2-methoxyethoxy)aluminum hydride, **8,** 448; **9,** 418
Tosylmethyl isocyanide, **5,** 684
2,4,6-Triisopropylbenzenesulfonylhydrazide, **11,** 563
by Benzoin condensation (*see* TYPE OF REACTION INDEX)
from 1,2-Dicarbonyls
1-Benzyl-1,4-dihydronicotinamide, **6,** 36
Chlorotris(triphenylphosphine)rhodium(I), **6,** 652
Hexahydro-4,4,7-trimethyl-4*H*-1,3-benzothiin, **12,** 237
Hydroxylamine, **7,** 176
Magnesium iodide, **5,** 420
Titanium(III) chloride, **7,** 418
Trimethyl-1,3-oxathianes, **8,** 508; **12,** 534
Zinc, **4,** 574
by Hydroxylation α to C=O (*see* TYPE OF REACTION INDEX)
by Oxidation of alkenes, 1,2-diols, enol ethers and related compounds (*see* TYPE OF REACTION INDEX)
from Propargyl alcohols
Benzenesulfenyl chloride, **9,** 35
Ion-exchange resins, **1,** 511
Mercury(II) oxide, **1,** 655; **6,** 360
Mercury(II) sulfate, **1,** 658
Mercury(II) *p*-toluenesulfonamide, **1,** 660
Phenylmercuric hydroxide, **11,** 415
Other routes
Acetic acid, **8,** 1
Acetic anhydride, **5,** 3
t-Butyl acetoacetate, **1,** 83
Dibenzoyl peroxide, **1,** 196
Diethylmethylsilane, **9,** 145
Dimethyl sulfoxide, **1,** 296; **2,** 94, 157
Formic acid, **5,** 698
Hydrogen peroxide, **1,** 466
Methyltriphenylphosphonium permanganate, **10,** 446
Osmium tetroxide, **1,** 759
Palladium(II) acetate, **9,** 344
Periodic acid, **1,** 815
Potassium formate, **5,** 556
Potassium hydroxide, **12,** 411
Potassium permanganate, **1,** 942
Samarium(II) iodide, **12,** 429
Sodium dichromate, **1,** 678
Thallium(III) nitrate, **4,** 492
Tosylmethyl isocyanide, **12,** 511
2,2,2-Trimethoxy-4,5-dimethyl-1,3-dioxaphospholene, **2,** 97
Tris(trimethylsilyloxy)ethylene, **8,** 523
Zinc, **1,** 1276
β-HYDROXY
by Aldol reaction (*see* TYPE OF REACTION INDEX)
from Epoxy C=O
Chromium(II) acetate, **1,** 147; **4,** 97
Lithium alkyl(cyano)cuprates, **9,** 329
Lithium dimethylcuprate, **5,** 234; **6,** 209
Lithium tri-*t*-butoxyaluminum hydride, **2,** 251
Nickel–Graphite, **11,** 356
Phenylcopper, **7,** 282
Sodium hydrogen telluride, **12,** 449
Sodium iodide–Acetic acid–Sodium acetate, **6,** 544
by Oxidation of 1,3-diols

HYDROXY ALDEHYDES AND KETONES (*Continued*)
Bis(tributyltin) oxide, **7,** 26
Bromine, **7,** 35
Silver carbonate–Celite, **3,** 247
by Reduction of dicarbonyls
Ethyl formate, **4,** 233
Sodium bis(2-methoxyethoxy)-aluminum hydride, **9,** 418
Other routes
1,3-Bis(trimethylsilyloxy)-1,3-cyclohexadiene, **8,** 46
Di-μ-bromobis(2-methoxy-2-propenyl)dinickel, **5,** 437
Diethylmethylsilane, **9,** 145
Dimethylphenylsilyllithium, **12,** 210
1,3-Dithiane, **3,** 135
S-(2-Methoxyallyl)-N,N-dimethyldithiocarbamate, **6,** 364
2-Methyl-2-thiazoline, **4,** 344; **6,** 403
Oxygen, singlet, **10,** 294
Raney nickel, **11,** 457
Sodium dichromate, **3,** 54; **5,** 138
Tetrakis(triphenylphosphine)-palladium(0), **11,** 503
Triethylborane, **10,** 416
γ-HYDROXY
t-Butyl dilithioacetoacetate, **8,** 60
Butyllithium, **10,** 73
N,N-Dimethylhydrazine, **7,** 126
Disiamylborane, **1,** 57, 201; **3,** 22
Hydrogen peroxide, **12,** 242
Jones reagent, **7,** 68
5-Lithio-2,3-dihydrofuran, **8,** 70
Lithium diethylamide, **6,** 332
Lithium lithioacetate, **8,** 175
Silver carbonate–Celite, **3,** 247
Triphenyltin hydride, **10,** 451
δ-HYDROXY
2-(2-Bromomagnesioethyl)-2-methyl-1,3-dioxolane, **7,** 148
3,4-Dihydro-2*H*-pyran, **1,** 256
2-(2-Phenylsulfonylethyl)-1,3-dioxolane, **6,** 461
HYDROXY AMIDES (*see also* CHIRAL COMPOUNDS)
α-HYDROXY
Acyllithium reagents, **5,** 233
Allyltrimethylsilane, **12,** 23
Cyanotrimethylsilane, **10,** 112
N,N-Diisopropylformamide, **8,** 174
Dimethyl sulfoxide, **1,** 296
Hexamethylphosphoric triamide, **8,** 240
Lithium diisopropylamide, **7,** 204
N-Methyl-C-(trichlorotitanio)-formimidoyl chloride, **12,** 354
Oxygen, **6,** 426; **10,** 293
Titanium(IV) chloride, **12,** 494
Titanium(IV) isopropoxide, **12,** 504
β-HYDROXY
Dichlorobis(cyclopentadienyl)-zirconium(II), **10,** 131
N,N-Dimethyl-α-*p*-tolylsulfinyl-acetamide, **12,** 509
Menthyl *p*-toluenesulfinate, **12,** 295
2-Oxazolidones, chiral, **12,** 359
Rhodium catalysts, **12,** 426
Zinc borohydride, **12,** 572
HYDROXYAMINES (*see* AMINO ALCOHOLS)
HYDROXY CARBOXYLIC ACIDS (*see also* CHIRAL COMPOUNDS)
α-HYDROXY (*see also* α-HYDROXY ESTERS)
1-Benzyl-1,4-dihydronicotinamide, **6,** 36
Benzyltriethylammonium chloride, **6,** 41
Borane–*t*-Butylamine, **11,** 87
m-Chloroperbenzoic acid, **6,** 110
Cyano-*t*-butyldimethylsilane, **6,** 80
Dicyclohexylborane, **11,** 172
Dimethyl sulfoxide, **1,** 296
Hexahydro-4,4,7-trimethyl-4*H*-1,3-benzothiin, **12,** 237
Periodic acid, **1,** 815
(S)-(−)-Proline, **8,** 421; **10,** 331
1,1,3,3-Tetramethylbutyl isocyanide, **5,** 650
Tin(IV) chloride, **1,** 1111
Titanium(IV) isopropoxide, **12,** 504
Trialkylaluminums, **11,** 539
Triethyl phosphite, **8,** 501
Vanadium(II) perchlorate, **9,** 513
β-HYDROXY (*see also* β-HYDROXY ESTERS)
by Aldol reaction (*see* TYPE OF REACTION INDEX)
Other routes
Acetic anhydride–Sodium acetate, **6,** 5
Dicyclohexylborane, **11,** 172
Sodium bromate, **1,** 1055
1,1,3,3-Tetramethylbutyl isocyanide, **5,** 650

ω-HYDROXY
Dimethyl sulfoxide, **2,** 157
Hexahydro-4,4,7-trimethyl-4*H*-1,3-benzothiin, **12,** 237
Hydrogen peroxide, **9,** 241
Ozone, **7,** 269
Permaleic acid, **12,** 379
Sodium benzeneselenoate, **11,** 475
HYDROXY ESTERS AND LACTONES (*see also* CHIRAL COMPOUNDS)
α-HYDROXY
by Hydroxylation α to C = O (*see* TYPE OF REACTION INDEX)
by Reduction of α-keto esters
Aluminum amalgam, **1,** 20
(2S,4S)-N-(*t*-Butoxycarbonyl)-4-(diphenylphosphino)-2-[(diphenylphosphino)methyl]pyrrolidine, **8,** 57; **11,** 87
Lithium aluminum hydride, **11,** 293
8-Phenylmenthol, **11,** 412; **12,** 389
B-3-Pinanyl-9-borabicyclo[3.3.1]-nonane, **11,** 429; **12,** 397
Potassium *t*-butoxide, **1,** 911
Triethoxysilane, **11,** 554
Other routes
m-Chloroperbenzoic acid, **10,** 92
B-Crotyl-9-borabicyclo[3.3.1]nonane, **12,** 81
2-Lithio-2-methylthio-1,3-dithiane, **7,** 191
Lithium diisopropylamide, **8,** 292
Tin(IV) chloride, **11,** 522
Titanium(IV) chloride, **12,** 494
Tributylcrotyltin, **12,** 513
(−)-4,6,6-Trimethyl-1,3-oxathiane, **8,** 508
Tris(phenylthio)methyllithium, **11,** 591
β-HYDROXY
by Aldol-type reaction (*see* TYPE OF REACTION INDEX)
by Reformatsky reaction
Chlorotrimethylsilane, **12,** 126
Diethylaluminum chloride, **8,** 164
Indium, **6,** 293
Magnesium, **1,** 627
Reformatsky reagent, **3,** 334; **5,** 753; **8,** 533
Trimethyl borate, **3,** 310
Zinc, **1,** 1285; **6,** 672; **11,** 598
Other routes
Aluminum amalgam, **1,** 20
Di-μ-carbonylhexacarbonyldicobalt, **1,** 224
Dimethylphenylsilyllithium, **12,** 210
(4S,5S)-2-Ethyl-4-(methoxymethyl)-5-phenyloxazoline, **11,** 232
Ketene, **6,** 11
Lithium aluminum hydride–Silica, **11,** 293
Lithium dimethylcuprate, **3,** 106
Menthyl *p*-toluenesulfinate, **12,** 295
Periodates, **1,** 809
Raney nickel, **9,** 405
Titanium(IV) chloride, **12,** 494
Tributyltin hydride, **12,** 516
2,4,4-Trimethyl-2-oxazoline, **3,** 313; **5,** 714
Zinc borohydride, **10,** 460
Zinc chloride, **8,** 536
γ-HYDROXY
(*t*-Butoxycarbonylmethyl)-diethylaluminum, **7,** 107
Diethylaluminum ethoxyacetylide, **7,** 103
Diisopinocampheylborane, **5,** 283
1-Ethoxy-1-trimethylsilyloxy-cyclopropane, **8,** 219
Lithium lithioacetate, **7,** 113
Ozone, **6,** 436
Palladium *t*-butyl peroxide trifluoroacetate, **12,** 88
Titanium(IV) chloride, **12,** 494
HYDROXYLAMINES AND DERIVATIVES
BY REDUCTION OF RNO_2, $R_2C = NOH$ (*see* TYPE OF REACTION INDEX)
OTHER ROUTES
m-Chloroperbenzoic acid, **5,** 120
Dibenzoyl peroxide, **12,** 157
Hydrogen peroxide–Sodium tungstate, **12,** 246
ALKOXYAMINES
N-Hydroxyphthalimide, **7,** 177
Sodium cyanoborohydride, **11,** 481
Sodium hydroxide, **5,** 616
Triphenylphosphine–Diethyl azodicarboxylate, **7,** 404
HYDROXY NITRILES (*see also* CYANOHYDRINS)
β-HYDROXY
Benzenesulfonylnitrile oxide, **9,** 39
Cyanotrimethylsilane, **5,** 720; **11,** 147

HYDROXY NITRILES (*Continued*)
Dialkylboryl trifluoromethanesulfonates, **11,** 159
Diethylaluminum cyanide, **1,** 244
Formaldehyde, **1,** 1252
Hydrogen cyanide–Triethylaluminum, **1,** 1198
Lithium diethylamide, **7,** 201
Phase-transfer catalysts, **8,** 387
Tetrabutylammonium fluoride, **12,** 458
Zinc, **8,** 532
γ-HYDROXY
Ethyl 3-bromo-2-hydroxyiminopropanoate, **9,** 212
Lithium diethylamide, **7,** 201
α-(Phenylsulfinyl)acetonitrile, **11,** 418
β-HYDROXY SELENIDES
Alumina, **8,** 9
Benzeneselenenic acid, **8,** 24, 320
Benzeneselenenyl halides, **9,** 25; **10,** 16
Benzeneselenenyl trifluoroacetate, **5,** 522
Benzeneseleninic acid, **8,** 28
Dimethylaluminum methylselenolate, **8,** 182
Diphenyl diselenide–Copper(II) acetate, **9,** 199
Hydrogen peroxide–Acetoxymethyl methyl selenide, **8,** 5
Methaneselenenic acid, **8,** 319
Phenyl selenocyanate, **8,** 119
N-Phenylselenophthalimide, **9,** 366
Phenyl trimethylsilyl selenide, **9,** 496
Phenyl vinyl selenide, **9,** 374
Sodium benzeneselenoate, **9,** 432
Tris(phenylseleno)borane, **10,** 454
HYDROXY SULFUR COMPOUNDS
BISULFITE ADDUCTS
Sodium hydrogen sulfite, **1,** 1047
HYDROXY SULFIDES
Acetic anhydride–Trifluoroacetic anhydride, **7,** 2
Alumina, **8,** 9
Benzenesulfenyl chloride, **9,** 35
Dimethyl(methylthio)sulfonium tetrafluoroborate, **11,** 204
Ethanethiol, **6,** 16
Lead tetrakis(trifluoroacetate)–Diphenyl disulfide, **9,** 269
Lithium tri-*sec*-butylborohydride, **12,** 286
Phenylthioacetic acid, **6,** 463
Zinc iodide, **10,** 462
HYDROXY SULFONES
t-Butylmagnesium chloride, **4,** 63
HYDROXY SULFOXIDES
Lithium aluminum hydride, **11,** 289
(R)-(+)-Methyl *p*-tolyl sulfoxide, **4,** 513
HYDROXY THIOLS
Sodium borohydride, sulfurated, **4,** 444; **5,** 399

IMIDES
Acetic anhydride, **6,** 4; **8,** 228
Acetyl hexafluoroantimonate, **1,** 692
Bis(methylcyclopentadienyl)tin(II), **12,** 201
Diketene–Iodotrimethylsilane, **10,** 151
Diphenylketene, **5,** 278
Diphenyl-N-*p*-tolylketenimine, **1,** 345
Hexamethylphosphorous triamide, **9,** 235
1-Hydroxybenzotriazole, **5,** 342
N-Hydroxymethylphthalimide, **1,** 484
Manganese(III) acetylacetonate, **3,** 194
Molybdenum carbonyl, **12,** 330
Ruthenium tetroxide, **6,** 504; **7,** 315
Thallium(III) trifluoroacetate, **12,** 481
Triphenylphosphine–Diethyl azodicarboxylate, **7,** 404
IMINES
FROM C=O
p-Aminobenzoic acid, **2,** 24
Boron trifluoride etherate, **1,** 70
Butylamine, **2,** 286
Dibutyltin dichloride, **11,** 161
N,N-Dimethyl-*p*-phenylenediamine, **1,** 293
2,4-Dinitrobenzaldehyde, **1,** 318
α-Methylbenzylamine, **11,** 411
Molecular sieves, **3,** 206; **4,** 345
Titanium(IV) chloride, **2,** 414; **3,** 291
BY OXIDATION OF AMINES
(*see* TYPE OF REACTION INDEX)
OTHER ROUTES
Cadmium acetate–Zinc acetate, **1,** 103
Chromium(II) chloride, **1,** 149
Dicyclohexylcarbodiimide, **11,** 173
Diethyl N-benzylideneaminomethylphosphonate, **12,** 185
N-(Diphenylmethylene)methylamine, **8,** 210
Iron carbonyl, **6,** 304

Picryl azide, **1,** 885
Raney cobalt catalyst, **1,** 977
Titanium(III) chloride–
Diisobutylaluminum hydride, **9,** 467
p-Toluenesulfonyl chloride, **1,** 1179
Triphenylphosphine, **11,** 588

IMINO ETHERS, THIOETHERS
1,4-Diazabicyclo[2.2.2]octane, **2,** 99
Diazidotin dichloride, **10,** 120
Diazomethane, **9,** 135
Dimethyl sulfate, **1,** 294
Methyl fluorosulfonate, **6,** 381; **8,** 340
Sodium borohydride, **5,** 597
Triethyl orthoformate, **1,** 1204
Triethyloxonium tetrafluoroborate, **1,** 1210; **2,** 430

ISOCYANATES
FROM RNH_2
N,N'-Carbonyldiimidazole, **1,** 114
Oxalyl chloride, **4,** 361
Palladium(II) chloride, **1,** 782
Phosgene, **1,** 856
FROM RCOOH AND RCOX
Azidotrimethylsilane, **5,** 719; **10,** 14
2-Halopyridinium salts, **9,** 234
Lead tetraacetate, **1,** 537; **4,** 278; **6,** 313
Sodium azide, **1,** 1041, 1225
Tetrabutylammonium azide, **6,** 563
Tributyltin azide, **7,** 377
2,4,6-Triphenylpyrylium salts, **7,** 408; **8,** 520
OTHER ROUTES
Carbon monoxide, **3,** 41
2-Chloro-3-ethylbenzoxazolium tetrafluoroborate, **8,** 90
Chlorotrimethylsilane, **2,** 435
Cyanic acid, **1,** 170
Cyanogen chloride, **4,** 110
Dimethyl sulfoxide, **1,** 296; **3,** 119
Iodine isocyanate, **3,** 161
Oxalyl chloride, **1,** 767
Palladium(II) chloride, **5,** 500
Phosgene, **5,** 532
Sulfur dioxide, **8,** 464
Tetrahydro-2*H*-1,3-oxazine-2-one, **6,** 570
Toluene diisocyanate, **1,** 1171

ISOCYANIDES
BY DEHYDRATION OF FORMAMIDES
1,1'-Carbonylbis(3-imidazolium) bismethanesulfonate, **11,** 112
2-Chloro-3-ethylbenzoxazolium tetrafluoroborate, **8,** 90; **9,** 105
Diphosgene, **8,** 214
Ethyl formate, **1,** 380
Formic acid, **1,** 404
Phenyl isocyanide, **1,** 843
Phosgene, **1,** 856
Phosphoryl chloride, **1,** 876, 924
Thionyl chloride, **4,** 503
p-Toluenesulfonyl chloride, **1,** 1179
Triphenylphosphine, **4,** 548, 553
Triphenylphosphine dibromide, **3,** 320
Vilsmeier reagent, **4,** 186
BY HOFMANN CARBYLAMINE REACTION (*see* TYPE OF REACTION INDEX)
OTHER ROUTES
(Bromodichloromethyl)phenylmercury, **1,** 851
Cyanotrimethylsilane, **11,** 147; **12,** 148
Diethyl isocyanomethylphosphonate, **4,** 271
Methylketene diethyl acetal, **1,** 685
3-Methyl-2-phenyl-1,3,2-oxazaphospholine, **2,** 323
Silver cyanide, **1,** 1006; **8,** 442
Sodium hypobromite, **7,** 336
1,1,3,3-Tetramethylbutyl isocyanide, **9,** 458
Trichlorosilane–Triethylamine, **11,** 553
Triethyl phosphite, **1,** 1212

ISOTHIOCYANATES
Carbon disulfide, **5,** 94; **6,** 95
Dicyclohexylcarbodiimide, **3,** 91
2-Halopyridinium salts, **9,** 234
Phenyl isocyanide, **1,** 843
Sulfur, **5,** 632
Thiophosgene, **5,** 667
2,2,2-Trichloro-1,3,2-benzodioxaphosphole, **2,** 63

KETALS (*see* ACETALS AND KETALS)

KETENES
Copper(II) acetylacetonate, **2,** 81
Mercury(II) oxide, **1,** 655
Phosphorus, red, **1,** 861
Triethylamine, **1,** 1198; **4,** 527
Trifluoroacetic anhydride, **1,** 1221
Triphenylphosphine, **2,** 443
Zinc, **1,** 1286; **6,** 672

KETENE DERIVATIVES

KETENE DERIVATIVES (*Continued*)
O,O-ACETALS AND KETALS
(Diethoxymethyl)diphenylphosphine oxide, **12,** 181
3,3-Diethoxy-1-methylthiopropyne, **5,** 207
2,2-Diethoxyvinylidenetriphenylphosphorane, **5,** 209
Ethylene glycol, **1,** 375
Potassium *t*-butoxide, **1,** 911; **2,** 336
Tetracyanoethylene, **1,** 1133
O,S-ACETALS AND KETALS
Methoxy(phenylthio)trimethylsilylmethane, **12,** 317
S,S-ACETALS AND KETALS
from RCHO, R_2CO
Bis(methylthio)(trimethylsilyl)methyllithium, **6,** 53
Chloramine-T, **7,** 58
Diethyl (1,3-dithian-2-yl)phosphonate, **8,** 89
Dimethyl bis(methylthio)methylphosphonate, **7,** 125
1,3-Dithian-2-ylidenetriphenylphosphorane, **8,** 89
1,3-Dithiolan-2-yltriphenylphosphonium tetrafluoroborate, **10,** 176
Ethyl ethylthiomethyl sulfoxide, **5,** 299
2-Lithio-2-trimethylsilyl-1,3-dithiane, **4,** 284; **5,** 374; **6,** 320
Lithium iodide, **12,** 282
from RCOOR′
Aluminum thiophenoxide, **9,** 15
Bis(dimethylaluminum) 1,2-ethanedithiolate, **5,** 35
Bis(dimethylaluminum) 1,3-propanedithiolate, **6,** 49
Other routes
Carbon disulfide–Methyl iodide, **5,** 95; **9,** 94
Dimethylsulfonium methylide, **5,** 254
Diphosphorus tetraiodide, **11,** 224
Grignard reagents, **11,** 245
Lead tetraacetate, **9,** 265
Lithium diisopropylamide, **8,** 292
Lithium dimethylcuprate, **5,** 234
Methyl iodide–1,3-Propanedithiol, **9,** 308
1-Phenylthiovinyllithium, **8,** 281
Potassium *t*-butoxide, **4,** 399
KETENIMINES
Diphenyl-N-*p*-tolylketenimine, **5,** 282
Phosphorus(V) oxide–*t*-Amine, **2,** 329
Sodium iodide, **1,** 1087
Triphenylphosphine dibromide, **10,** 60
SELENOACETALS AND -KETALS
Diphosphorus tetraiodide, **11,** 224
SILYLACETALS AND -KETALS
Birch reduction, **4,** 31
Chlorotrimethylsilane, **2,** 435; **4,** 537
Methoxy(phenylthio)trimethylsilylmethyllithium, **12,** 317
Triethylsilyl perchlorate, **12,** 527
Trimethylsilyl trifluoromethanesulfonate, **7,** 390
KETIMINES (*see* IMINES)
KETOACIDS, -ALDEHYDES, ETC. (*see* DICARBONYLS)
KETONES—GENERAL METHODS (*see also* α,β-ACETYLENIC CARBONYLS, ALLENIC CARBONYLS, CHIRAL COMPOUNDS, DICARBONYLS, UNSATURATED ALDEHYDES AND KETONES)
FROM ACYL ANIONS + RX
2-Alkyl-1,3-benzodithiolanes, **9,** 4
1-(Alkylthio)vinyllithium, **5,** 6
Bis(phenylthio)methane, **6,** 267; **7,** 25; **10,** 42
t-Butylhydrazine, **12,** 87
N,N-Diethylaminoacetonitrile, **9,** 159
Diethyl 1-trimethylsilyloxyethylphosphonate, **9,** 165
N,N-Dimethyldithiocarbamoylacetonitrile, **7,** 123
Disodium tetracarbonylferrate, **4,** 461
1,3-Dithiane, **2,** 182
Ethyl ethylthiomethyl sulfoxide, **5,** 299
Lithium di(α-methoxyvinyl)cuprate, **6,** 204
α-Methoxyvinyllithium, **9,** 304
Methyl methylthiomethyl sulfone (or sulfoxide), **5,** 456; **11,** 242
N-Methyl-N-trimethylsilylmethyl-N′-*t*-butylformamidine, **11,** 347
Phenyl(phenylthio)trimethylsilylmethane, **10,** 314
Phenylthioacetic acid, **6,** 463
1-Phenylthio-1-trimethylsilylethylene, **12,** 394
Potassium fluoride, **10,** 325

Tosylmethyl isocyanide, **8,** 493
2,4,6-Triisopropylbenzenesulfonylhydrazide, **11,** 563
BY CARBONYLATION REACTIONS
Carbon monoxide, **2,** 60; **8,** 76
Chlorothexylborane, **10,** 95
Dicarbonylbis(triphenylphosphine)nickel, **10,** 125
Di-μ-carbonylhexacarbonyldicobalt, **3,** 89
Dichloromethyl methyl ether, **5,** 200
Iron carbonyl, **8,** 265
Lithium aluminum hydride, **9,** 274
Nickel carbonyl, **4,** 353; **12,** 335
Sodium cyanide, **4,** 446; **6,** 535; **7,** 333
Thexylborane, **2,** 148; **4,** 175
Tris(phenylthio)methyllithium, **11,** 305
FROM RCOOH(X)
RCOCl
N-Benzoyl imidazole, **1,** 424
Benzylchlorobis(triphenylphosphine)palladium(II), **8,** 35; **9,** 41; **12,** 44
t-Butyl α-lithioisobutyrate, **6,** 84
Butyllithium, **6,** 85
Cadmium chloride, **1,** 422
Carbonylchlorobis(triphenylphosphine)rhodium(I), **6,** 105
Carbonylphenylbis(triphenylphosphine)rhodium(I), **5,** 45
Chlorobis(cyclopentadienyl)hydridozirconium(IV), **6,** 175; **8,** 84
Copper, **12,** 140
Disodium tetracarbonylferrate, **4,** 461
Grignard reagents, **8,** 235; **9,** 229
Iron(III) acetylacetonate, **12,** 557
Isobutene, **1,** 522
Lithium *t*-butyl(phenylthio)cuprate, **5,** 414; **7,** 211
Lithium dibutylcuprate, **4,** 127
Manganese(II) iodide, **7,** 222; **8,** 312; **10,** 290
Nickel, **12,** 335
Organocopper reagents, **11,** 365
Organomanganese(II) iodides, **7,** 222; **8,** 312; **10,** 290
Organotitanium reagents, **12,** 54, 110
Phenylthiocopper, **6,** 465
Tetrakis(triphenylphosphine)palladium(0), **12,** 468
Tri-μ-carbonylhexacarbonyldiiron, **8,** 498
$RCONH_2$
N,O-Dimethylhydroxylamine, **11,** 201
Ephedrine, **9,** 209
Lithium, **5,** 376
RCOOH
[1,2-Bis(diphenylphosphine)ethane](dichloro)nickel(II), **12,** 171
[Chloro(*p*-methoxyphenyl)methylene]diphenylammonium chloride, **11,** 220
1-Chloro-N,N,2-trimethylpropenylamine, **12,** 123
Chlorotrimethylsilane, **12,** 126
Methyllithium, **1,** 686; **2,** 274
Triphenylphosphine dihalides, **12,** 554
RCOOR′
Grignard reagents, **10,** 189
Manganese(II) chloride, **9,** 288
Nickel chloride–Zinc, **10,** 277
Organocopper reagents, **12,** 345
Sodium diisopropylamide, **1,** 1064
other RCOX
Dimethylaluminum methylselenolate, **12,** 197
Grignard reagents, **5,** 321
Organocopper reagents, **5,** 234; **11,** 365
Organolithium reagents, **9,** 5
Tetrakis(triphenylphosphine)palladium(0), **6,** 571; **11,** 514
FROM CYANOHYDRINS
2-Acetoxyacrylonitrile, **1,** 7
Arene(tricarbonyl)chromium complexes, **6,** 103
Benzyltriethylammonium chloride, **5,** 26
Copper(II) sulfate, **8,** 125
Cyanotrimethylsilane, **5,** 720; **6,** 632; **9,** 127
Lithium diisopropylamide, **4,** 298
Sodium sulfide, **4,** 77
FROM DEGRADATION OF AMINO ACIDS
Lead tetraacetate, **8,** 269
Sodium hypochlorite, **1,** 1084
N-Sulfinylaniline, **6,** 556
BY FRIEDEL–CRAFTS ACYLATION AND RELATED REACTIONS
Acetic acid, **1,** 1223
Acetic anhydride, **1,** 514
Aluminum bromide (or chloride), **1,** 24; **2,** 19

KETONES—GENERAL METHODS (*Continued*)
1-Chloro-N,N,2-trimethylpropenylamine, **4,** 94
Copper(II) trifluoromethanesulfonate, **10,** 110
Hexafluoroantimonic acid, **2,** 216
Hydrogen fluoride, **9,** 240
Nafion-H, **9,** 320
Nitroethane, **4,** 357
Polyphosphoric acid, **7,** 294
Sodium tetrachloroaluminate, **1,** 1027
Tin(IV) chloride, **1,** 1111
Trifluoromethanesulfonic acid, **5,** 701
Trifluoromethanesulfonic-carboxylic anhydrides, **4,** 533; **10,** 420
BY FRIES REARRANGEMENT (*see* TYPE OF REACTION INDEX)
FROM HALOHYDRINS AND RELATED SUBSTRATES
Benzeneselenol, **6,** 28
Lithium aluminum hydride–Aluminum chloride, **8,** 289
Palladium(II) acetate, **11,** 389
Perchloric acid, **2,** 309
Pyridinium chloride, **1,** 964
Silver carbonate–Celite, **5,** 577
Silver(I) nitrate, **2,** 366
Zinc, **2,** 459; **3,** 334
BY HYDRATION OF C≡C
Aluminum, **2,** 19
9-Borabicyclo[3.3.1]nonane, **9,** 57
Catecholborane, **4,** 69
Chlorosulfonyl isocyanate, **3,** 51
Dichloroborane diethyl etherate, **5,** 191
Diethoxymethylsilane, **12,** 182
Di-2-mesitylborane, **12,** 195
Ion-exchange resins, **1,** 511
Mercury(II) acetate, **1,** 644
Monochloroborane diethyl etherate, **5,** 465
Nafion-H, **9,** 320
Phenylmercuric hydroxide, **11,** 415
Thallium(III) acetate, **7,** 360
BY HYDROACYLATION OF C=C
Azobisisobutyronitrile, **1,** 45
Chlorotris(triphenylphosphine)-rhodium(I), **9,** 113
Disodium tetracarbonylferrate, **6,** 550
Hydridodinitrogentris(triphenylphosphine)cobalt(I), **5,** 331
Monoisopinocampheylborane, **11,** 350
BY HYDROLYSIS OF KETALS AND RELATED (*see* TYPE OF REACTION INDEX)
BY HYDROLYSIS OF C=C–X (*see* TYPE OF REACTION INDEX)
FROM IMINES AND OTHER C=N (*see* TYPE OF REACTION INDEX)
BY ISOMERIZATION OF ALLYLIC AND HOMOALLYLIC ALCOHOLS
Butyllithium, **5,** 80
Chlorohydridotris(triphenylphosphine)-ruthenium(II), **6,** 425
Iridium catalysts, **8,** 135
Raney nickel, **10,** 339
Ruthenium(III) chloride, **10,** 343
Tetracarbonyldi-μ-chlorodirhodium, **10,** 382
Tris(aquo)hexa-μ-acetato-μ-oxotri-ruthenium(III,III,III) acetate, **6,** 425, 650
FROM RNC
Monochloroborane diethyl etherate, **5,** 465
1,1,3,3-Tetramethylbutyl isocyanide, **5,** 650
Triphenylmethyl isocyanide, **5,** 650; **6,** 642
FROM RCN
Acetonitrile, **1,** 1291
Boron trichloride, **9,** 62
Methoxyacetonitrile, **1,** 422
Methyllithium, **8,** 342
Triethyl(or Trimethyl)aluminum, **1,** 1197; **6,** 622
FROM RNO_2
t-Butyl hydroperoxide, **8,** 62
N″-(*t*-Butyl)-N,N,N′,N′-tetramethylguanidinium *m*-iodylbenzoate, **12,** 102
Hydrogen peroxide, **10,** 201
Oxodiperoxymolybdenum(pyridine)-(hexamethylphosphoric triamide), **11,** 218
Oxygen, singlet, **8,** 367
Ozone, **5,** 491; **8,** 374
Propyl nitrite–Sodium nitrite, **5,** 565
Sodium hydroxide, **8,** 461
Titanium(III) chloride, **4,** 506; **5,** 669
BY OXIDATION OF ROH, C=C, R_2CHX, R_2NH, $-CH_2-$, ROR, $ROSiR_3'$

(*see* TYPE OF REACTION INDEX)
BY OXIDATIVE CLEAVAGE OF C=C, 1,2-DIOLS AND RELATED GROUPS (*see* TYPE OF REACTION INDEX)
BY OXIDATIVE DECARBONYLATION OF RCHO
Copper(II)–Amine complexes, **3,** 65; **5,** 157
BY OXIDATIVE DECARBOXYLATION OF α-HYDROXY CARBOXYLIC ACIDS, DICARBOXYLIC ACIDS
2-Chloro-3-ethylbenzoxazolium tetrafluoroborate, **8,** 90; **9,** 105
N-Chlorosuccinimide, **8,** 103
Copper carbonate, basic, **4,** 101
Dipotassium tetrachloroplatinate(II), **4,** 215
N-Iodosuccinimide, **12,** 258
Lead tetraacetate, **1,** 537; **2,** 234; **5,** 365
Tetrabutylammonium periodate, **10,** 381
BY OXIDATIVE DECYANATION
Benzenesulfenyl chloride, **5,** 523
t-Butyl chromate, **5,** 73
Oxygen, **6,** 426; **10,** 293
Polyphosphoric acid, **1,** 894
BY OXIDATIVE DESULFONYLATION
Bis(trimethylsilyl) peroxide, **12,** 63
BY PINACOL REARRANGEMENT
Calcium carbonate, **4,** 67
Ferric chloride, **8,** 229
Iodine, **1,** 495
Nafion-H, **9,** 320
Trialkylaluminums, **12,** 512
BY REARRANGEMENT OF EPOXIDES, EPOXY SILANES, AND RELATED SUBSTRATES
Aluminum isopropoxide, **11,** 29
Boron trifluoride etherate, **2,** 35; **5,** 52
1-Chloro-1-(trimethylsilyl)ethyllithium, **8,** 277; **11,** 128
Di-μ-carbonylhexacarbonyldicobalt, **1,** 224
Lithium bromide, **4,** 297
Lithium diethylamide, **1,** 610
Magnesium bromide etherate, **1,** 629
o-Nitrophenyl selenocyanate, **10,** 278
Periodic acid, **1,** 815
Sodium hydride, **4,** 455
Tin(IV) chloride, **5,** 627
Tributylphosphine oxide, **1,** 1192
FROM THIONES
Benzeneseleninic anhydride, **8,** 29
Bis(*p*-methoxyphenyl) telluroxide, **9,** 50
t-Butyldimethylsilyl ethylnitronate, **12,** 84
Dimethyl selenoxide, **8,** 197
Dimethyl sulfoxide, **4,** 192
BY WITTIG REACTION
Bis(phenylthio)methane, **7,** 25
Diethyl lithio-N-benzylideneamino-methylphosphonate, **9,** 161
Diethyl methylthiomethylphosphonate, **3,** 97
Diphenyl-1-(phenylthiovinyl)phosphine oxide, **11,** 424
OTHER METHODS
Arene(tricarbonyl)chromium complexes, **6,** 27
α-Azidostyrene, **6,** 24
Benzenesulfonyl chloride, **5,** 22
1,3-Benzodithiolylium perchlorate, **8,** 34
Bismuth(III) chloride, **7,** 24
1,1-Dibromoalkyllithium reagents, **9,** 138
Diethylketene, **9,** 85
Ferric chloride, **1,** 390
Iron, **2,** 229
1-Lithiocyclopropyl phenyl sulfide, **11,** 284
Lithium aluminum hydride, **12,** 272
Lithium 2-dimethylaminoethoxide, **10,** 244
Nitrosyl chloride, **2,** 298
Palladium(II) acetate, **8,** 378
Raney nickel, **12,** 422
2,4,4,6-Tetramethyl-5,6-dihydro-1,3-(4*H*)-oxazine, **3,** 280; **4,** 481
Tributyltinlithium, **8,** 495
Trifluoroacetic acid, **5,** 695
Zinc, **1,** 1276
KETONES—SPECIFIC TYPES (*see also* ALICYCLIC HYDROCARBONS for CYCLOPROPYL KETONES)
DIARYL KETONES
Arene(tricarbonyl)chromium complexes, **6,** 103
Benzoyl chloride, **1,** 1295
Bis(1,5-cyclooctadiene)nickel(0), **10,** 33
Lithium, **3,** 150
Manganese(III) acetylacetonate, **3,** 194
Nafion-H, **9,** 320
Oxygen, **5,** 482; **8,** 366

KETONES—SPECIFIC TYPES (*Continued*)
Polyphosphoric acid, **1,** 894
Potassium cryptate[2.2.2], **5,** 156
Potassium superoxide, **8,** 417
Tetrabutylammonium permanganate, **8,** 468
METHYL KETONES
by Acetonylation of RX (RX → RCH_2COCH_3)
Acetoacetic acid, **12,** 1
Di-μ-bromobis(2-methoxyallyl)-dinickel, **5,** 437
Lithium diethylamide, **6,** 332
S-(2-Methoxyallyl)-N,N-dimethyl-dithiocarbamate, **6,** 364
2-Oxo-3-(triphenylphosphoranylidene)-propyllithium, **5,** 375
Phenylsulfinylacetone, **5,** 524
from Acyl anions + RX
Diethyl 1-trimethylsilyloxyethyl-phosphonate, **9,** 165
7,8-Dimethyl-1,5-dihydro-2,4-benzodithiepin, **6,** 216
(1-Lithiovinyl)trimethylsilane, **5,** 374
α-Methoxyvinyllithium, **6,** 372; **9,** 304
Phenylthioacetic acid, **6,** 463
from RCOOH
Acetyl chloride, **8,** 86
Chlorotrimethylsilane, **12,** 126
Dibenzyl malonate, **1,** 198
Di-*t*-butyl malonate, **1,** 210
Dimethylzinc, **3,** 128
Hydriodic acid, **1,** 449; **2,** 213
Lithium dimethylcuprate, **3,** 106; **4,** 177
Magnesium, **1,** 627
Meldrum's acid, **10,** 252
N-Methanesulfinyl-*p*-toluidine, **2,** 269
Methylcopper, **4,** 334
Methyllithium, **2,** 274
Organotitanium reagents, **12,** 54, 110
Phenylsulfonylmethylmagnesium bromide, **2,** 166
Sodium methylsulfinylmethylide, **1,** 310
Trimethylsilylmethyllithium, **9,** 495
Trimethylsilylmethylmagnesium chloride, **5,** 724
from Epoxy silanes
1-Chloro-1-(trimethylsilyl)-ethyllithium, **11,** 128
(1-Lithiovinyl)trimethylsilane, **5,** 374
from RCN
Methyllithium, **8,** 342
Trimethylaluminum, **6,** 622
by Oxidation of vinyl groups
Allyltrimethylsilane, **10,** 6
Bis(acetonitrile)chloronitro-palladium(II), **10,** 30; **11,** 45
[Bis(salicylidene-γ-iminopropyl)]-methylaminecobalt(II), **11,** 138
Hydrogen peroxide–Palladium acetate, **10,** 203
Jones reagent, **6,** 123
Oxotris(*t*-butylimido)osmium(VIII), **8,** 206
Palladium *t*-butyl peroxide trifluoroacetate, **10,** 299
Palladium(II) chloride, **10,** 300; **11,** 396; **12,** 380
Platinum *t*-butyl peroxide trifluoroacetate, **11,** 430
Other routes
Acetaldoxime, **1,** 3
Acetic acid, **1,** 1291
Acetone, **6,** 518
2-Acetoxypyridine, **1,** 9
Acetyl chloride–Aluminum chloride, **1,** 131, 1295
Acetyl hexafluoroantimonate, **1,** 692
Acetyl methanesulfonate, **5,** 5
Azobisisobutyronitrile, **1,** 45
Cyclodextrins, **6,** 151
Dicarbonylbis(triphenylphosphine)-nickel, **10,** 125
2,3-Dichloro-5,6-dicyano-1,4-benzoquinone, **10,** 135
Diethyl (1-methylthio)-ethylphosphonate, **4,** 151
Dimethylacetamide, **1,** 270
Diphenyl disulfide, **6,** 235
Ethyl 2-phenylsulfinylacetate, **6,** 256
Lead tetraacetate, **1,** 537
Lithium acetylide, **6,** 324
α-Methoxyethylidenetriphenyl-phosphorane, **1,** 669
Phenylmercuric hydroxide, **11,** 415
Potassium permanganate, **1,** 942
Potassium tetracarbonylhydrido-ferrate, **6,** 483; **8,** 419
Thallium(III) acetate, **7,** 360
KETONES—α-SUBSTITUTED (*see also* AMINO CARBONYLS, DIAZO

COMPOUNDS, HALO ALDEHYDES AND KETONES, HYDROXY ALDEHYDES AND KETONES, NITRO COMPOUNDS)
α-ACETOXY
m-Chloroperbenzoic acid, **8,** 97
N,N-Dimethylhydrazine, **7,** 126
Lead tetraacetate, **1,** 541; **2,** 234; **5,** 365; **6,** 4; **7,** 185
Manganese(III) acetate, **7,** 221; **12,** 292
Mercury(II) acetate, **1,** 644
Tetrabutylammonium fluoride, **9,** 446
Tetramethylammonium acetate, **1,** 1142
Thallium(III) acetate, **2,** 406
p-Toluenesulfonic acid, **1,** 1172
Triethylamine, **1,** 1198
Trimethyloxonium tetrafluoroborate, **3,** 314
α-ACYLOXY
Dibenzoyl peroxide, **1,** 196
Iodine–Silver carboxylates, **11,** 268
Lead tetrabenzoate, **7,** 186, 189
Lithium 2,2,6,6-tetramethylpiperidide, **6,** 345
Triethylamine, **1,** 1198
α-ALKOXY
Boron trifluoride etherate, **2,** 35
t-Butoxymethyllithium, **12,** 350
2,3-Dichloro-5,6-dicyano-1,4-benzoquinone, **6,** 168
N,N-Dimethylhydrazine, **7,** 126
Disiamylborane, **5,** 39
Pyridinium chlorochromate, **10,** 334
Thallium(III) nitrate, **4,** 492
α-ALKYL(OR ARYL)SELENO
by Selenylation of enolates
Benzeneselenenyl bromide, **5,** 518; **6,** 459; **8,** 25
Benzeneselenenyl trifluoroacetate, **6,** 459
Diphenyl diselenide–Selenium dioxide, **11,** 219
Phenylseleninyl chloride, **6,** 459
Selenium, **10,** 345
Other routes
Benzeneselenenic anhydride, **9,** 25
Benzeneselenenyl halides, **7,** 286; **9,** 25; **10,** 16
Benzeneselenenyl trifluoroacetate, **5,** 522
t-Butyl hydroperoxide–Diaryl diselenides, **10,** 65
Diphenyl diselenide–Bromine–Hexabutyldistannoxane, **8,** 209
Phenylselenoacetaldehyde, **9,** 365
p-Toluenesulfonylhydrazide, **11,** 537
α-ALKYL(OR ARYL)SULFINYL
Ethyl phenyl sulfoxide, **6,** 257
Methyl *p*-toluenesulfinate, **11,** 312
Methyl *p*-toluenesulfinate, **6,** 400
Sodium methylsulfinylmethylide, **6,** 6
α-ALKYL(OR ARYL)THIO
by Sulfenylation α to C=O
Benzenesulfenyl chloride, **9,** 429
Diphenyl disulfide, **6,** 235; **9,** 429; **10,** 171
N-Methylthiophthalimide, **11,** 493
S-Methyl *p*-toluenethiosulfonate, **6,** 400
Phenyl benzenethiosulfonate, **8,** 391; **9,** 429
Thiophenol, **9,** 447
Other routes
Bis(phenylthio)methane, **6,** 131; **7,** 25
Dimethyl disulfide, **8,** 189
Dimethyl(methylthio)sulfonium tetrafluoroborate, **11,** 204
2,2′-Dipyridyl disulfide, **8,** 214
2-Methoxy-3-phenylthio-1,3-butadiene, **7,** 232
(Z)-3-Methyl-1-phenylthio-2-trimethylsilyloxy-1,3-butadiene, **11,** 344
S-Methyl *p*-toluenethiosulfonate, **5,** 460
Phase-transfer catalysts, **11,** 403
Phenylthiotrimethylsilylmethane, **11,** 423
Sodium benzeneselenoate, **12,** 439
p-Toluenesulfonylhydrazide, **11,** 537
α-AZIDO
Lead(IV) acetate azides, **4,** 276; **5,** 363
Lithium azide, **6,** 327
Phenyliodine(III) diacetate–Azidotrimethylsilane, **5,** 354; **6,** 297
α-HYDROPEROXY
Benzyltrimethylammonium hydroxide, **1,** 1252
Oxygen, **1,** 921; **2,** 341; **7,** 258
α-SILYL
Butyllithiums, **10,** 76; **12,** 97
Grignard reagents, **9,** 229

KETONES—α-SUBSTITUTED (*Continued*)
Hydriodic acid, **8,** 246
Lithium 1-(dimethylamino)-naphthalenide, **10,** 244
Magnesium bromide etherate, **7,** 218
Trimethylsilylmethyllithium, **6,** 635; **9,** 495; **11,** 581
Trimethylsilylmethylmagnesium chloride, **12,** 540

LACTAMS (*see also* CHIRAL COMPOUNDS)
GENERAL METHODS
Bromine, **6,** 70
Iodine azide, **10,** 211
p-Toluenesulfonyl chloride, **6,** 598
α-LACTAMS
Catecholborane, **9,** 97
Potassium *t*-butoxide, **1,** 911
Sodium hydride, **1,** 1075
Trifluoromethanesulfonic anhydride, **11,** 560
β-LACTAMS
by [2+2]Cycloaddition and related reactions
Arenesulfonyl halide–Antimony(V) halide complexes, **5,** 20
Azidoacetyl chloride, **5,** 21
Chlorocyanoketene, **8,** 88
Chlorosulfonyl isocyanate, **1,** 117; **3,** 51; **4,** 90; **5,** 132; **6,** 122; **8,** 105; **11,** 125; **12,** 122
1-Chloro-N,N,2-trimethylpropenylamine, **5,** 136
Chlorotrimethylsilane, **5,** 709
Cyanuric chloride, **10,** 114
Dimethylketene, **9,** 185
Diphenyl-N-*p*-tolylketenimine, **5,** 282
Ketene alkyl trialkylsilyl acetals, **10,** 401
Ketene bis(trimethylsilyl)acetals, **12,** 268
Methyl(phenylthio)ketene, **8,** 348
Sulfur dioxide, **6,** 558
Tetramethoxyethylene, **2,** 401
Titanium(IV) chloride, **8,** 483
p-Tolylsulfinylacetic acid, **12,** 508
Trimethylsilylbromoketene, **7,** 395
Triphenylphosphine + co-reagent, **9,** 504; **10,** 447; **11,** 589; **12,** 552
Vilsmeier reagent, **12,** 204
by Cyclization
Benzenesulfenyl chloride, **11,** 39; **12,** 42
2-Chloro-1-methylpyridinium iodide, **12,** 116
Dicyclohexylcarbodiimide, **1,** 231
Diethylthallium *t*-butoxide, **7,** 108
Grignard reagents, **10,** 189
Methoxymethylbis(trimethylsilyl)-amine, **12,** 62
Phase-transfer catalysts, **9,** 356; **10,** 305; **11,** 403
Sodium hydride, **7,** 335; **9,** 427
Tetrabutylammonium iodide, **6,** 566
p-Toluenesulfonyl chloride, **11,** 536
Triphenylphosphine–2,2′-Dipyridyl disulfide, **10,** 449
by Oxidation of azetidines
m-Chloroperbenzoic acid, **8,** 97
Oxygen, **7,** 258
Oxygen, singlet, **8,** 367
Other routes
Aceto(carbonyl)cyclopentadienyl-(triphenylphosphine)iron, **12,** 1
Bromine, **11,** 75
Chromium carbene complexes, **11,** 397
Ion-exchange resins, **1,** 511; **10,** 220
Lithium phenylethynolate, **10,** 247
Manganese(IV) oxide, **5,** 422
Palladium(II) acetate, **9,** 344
Periodates, **5,** 507
Phenyl dichlorophosphate–Dimethylformamide, **11,** 410
Raney nickel, **8,** 433
Rhodium(II) acetate, **9,** 406
Silver(I) nitrate, **4,** 429
Sodium dicarbonyl(cyclopentadienyl)-ferrate, **8,** 454; **10,** 362
Tetracarbonyldi-μ-chlorodirhodium, **12,** 112
p-Toluenesulfonyl chloride, **5,** 676
Zinc, **12,** 567
γ-LACTAMS
from Amino acids
Catecholborane, **9,** 97
Dibutyltin oxide, **12,** 160
Hexamethyldisilazane, **9,** 234
o-Nitrophenyl thiocyanate, **9,** 325
Raney nickel, **1,** 723
Other methods
Benzeneselenenyl halides, **12,** 39

Calcium hydride, **1,** 105
Chlorosulfonyl isocyanate, **4,** 90
Dichloroketene, **11,** 168
6,6-Dimethyl-5,7-dioxaspiro[2.5]-octane-4,8-dione, **6,** 216
Ion-exchange resins, **1,** 511
Ruthenium tetroxide, **6,** 504
Sodium hydride, **8,** 458
Tetraethylammonium periodate, **9,** 448

δ-LACTAMS
from Amino acids
Ammonium acetate, **7,** 11
Dibutyltin oxide, **12,** 160
Dowtherm A, **1,** 353
Hexamethyldisilazane, **9,** 234
o-Nitrophenyl thiocyanate, **9,** 325
Other routes
Benzeneselenenyl halides, **12,** 39
Hydrazoic acid, **1,** 446
Mercury(II) acetate, **5,** 427
O-Mesitylenesulfonylhydroxylamine, **5,** 430
Ruthenium tetroxide, **6,** 504
Triethyl orthoformate, **6,** 610

ε-LACTAMS
from Amino acids
Dibutyltin oxide, **12,** 160
Hexamethyldisilazane, **9,** 234
o-Nitrophenyl thiocyanate, **9,** 325
by Beckmann rearrangement
Hydrochloric acid, **5,** 333
Phosphorus(V) oxide–Methanesulfonic acid, **5,** 535
Sodium azide, **1,** 901
p-Toluenesulfonyl chloride, **9,** 472
Triphenylphosphine, **2,** 443
Other routes
Hydrazoic acid, **1,** 446; **5,** 329
Hydroxylamine-O-sulfonic acid, **10,** 207
Methyl iodide, **9,** 308
N-Phenyl-1,2,4-triazoline-3,5-dione, **4,** 381
Sodium hydride, **8,** 458
p-Toluenesulfonyl chloride, **4,** 510

MACROLACTAMS
from Amino acids
Catecholborane, **9,** 97
Dibutyltin oxide, **10,** 123
2-Mesitylenesulfonyl chloride, **8,** 318
Tris(dimethylamino)borane, **11,** 590
Other routes
Ferrous sulfate, **11,** 276
Nitrosylsulfuric acid, **1,** 755
Oxygen, singlet, **9,** 338
Potassium 3-aminopropylamide, **9,** 378
Sodium borohydride, **11,** 477

LACTONES (*see also* CHIRAL COMPOUNDS, MACROLIDES, UNSATURATED LACTONES)
GENERAL METHODS—BAEYER–VILLIGER REACTION
Benzeneselenenyl chloride, **8,** 25
Bis(trimethylsilyl) peroxide, **11,** 67
Caro's acid, **1,** 118
Cerium(III) sulfate, **9,** 99
m-Chloroperbenzoic acid, **4,** 85
Grignard reagents, **10,** 189
Hydrogen peroxide–Acetic acid, **6,** 286
2-Hydroperoxyhexafluoro-2-propanol, **3,** 149
Lead tetraacetate, **10,** 228; **12,** 270
Peracetic acid, **1,** 787; **6,** 286
Potassium peroxomonosulfate, **1,** 952
Ruthenium tetroxide, **10,** 343
Silver carbonate–Celite, **3,** 247; **6,** 511
Sodium bromite, **12,** 445
Trifluoroperacetic acid, **1,** 821
Zinc permanganate, **12,** 576

α-LACTONES
Iodosylbenzene, **10,** 213

β-LACTONES
Benzenesulfonyl chloride, **5,** 22
t-Butyllithium, **11,** 103
Cinchona alkaloids, **11,** 134
Dichlorobis(triphenylphosphine)-palladium(II), **10,** 133
Dichloroketene, **3,** 87; **4,** 134
Iron carbonyl, **8,** 265
Ketene, **1,** 528
Lithium phenylethynolate, **6,** 343; **7,** 210
Thallium(I) acetate–Iodine, **10,** 394
Trimethyl orthoformate, **3,** 313

γ-LACTONES
by Addition of $-CH_2COO-$ to C=C
Cerium(IV) acetate, **4,** 71
α-Chloro-N-cyclohexylpropanal-donitrone, **4,** 80
Dichloroketene, **11,** 168; **12,** 176
Ethyl 3-bromo-2-hydroxyimino-

LACTONES (*Continued*)
propanoate, **9,** 212
Iodotrimethylsilane, **11,** 271
Manganese(III) acetate, **2,** 263; **6,** 355; **12,** 292
Tributyl(iodoacetoxy)tin, **12,** 258
by Baeyer–Villiger reaction
Bis(trimethylsilyl) peroxide, **11,** 67
m-Chloroperbenzoic acid, **5,** 120
Hydrogen peroxide, **3,** 155; **4,** 254
Peracetic acid, **3,** 155
Potassium dichromate, **6,** 124
Sodium dichromate, **7,** 68
from γ-Hydroxy acids and related compounds
Borane–Dimethyl sulfide, **10,** 49
(*t*-Butoxycarbonylmethyl)-diethylaluminum, **7,** 103
t-Butyl dilithioacetoacetate, **8,** 60
2-Chloro-1-methylpyridinium iodide, **12,** 116
Dicyclohexylcarbodiimide, **1,** 231
Diethyl succinate, **8,** 296
Formaldehyde, **7,** 158
Lithium aluminum hydride, **10,** 265
Lithium lithioacetate, **8,** 175
Lithium β-lithiopropionate, **8,** 303
Lithium naphthalenide, **6,** 415
(4S,5S)-4-Methoxymethyl-2-methyl-5-phenyl-2-oxazoline, **6,** 386
Potassium tri-*sec*-butylborohydride, **6,** 490
p-Toluenesulfonic acid, **1,** 1172
Triethyl orthoacetate, **6,** 607
2,4,4-Trimethyl-2-oxazoline, **5,** 714
Triphenylcarbenium tetrafluoroborate, **9,** 512
Zinc–copper couple, **1,** 1292
from Hydroxy nitriles
Diethylaluminum cyanide, **9,** 158
Lithium aluminum hydride, **11,** 289
Lithium diethylamide, **7,** 201
Organoaluminum cyanides, **9,** 158
Trimethylsilylacetonitrile, **10,** 427
from Oxidation of 1,4-diols, lactols and related compounds, tetrahydrofurans (*see* TYPE OF REACTION INDEX)
by Reduction of anhydrides (*see* TYPE OF REACTION INDEX)
from Unsaturated carboxylic acids
Benzeneselenenyl chloride, **8,** 25
Benzenesulfenyl chloride, **8,** 32
Bis(benzonitrile)dichloro-palladium(II), **12,** 51
Bromine, **9,** 65; **12,** 70
Chloramine-T, **11,** 118
Chloro(methoxycarbonyl)ketene, **11,** 110
Copper(II) acetate, **5,** 156
Dimethyl sulfoxide–Methanesulfonic anhydride, **9,** 191
Iodine, **8,** 256; **11,** 261; **12,** 253
Ion-exchange resins, **6,** 302
Lead tetraacetate, **2,** 234; **9,** 265; **10,** 228
Lithium–Ammonia, **10,** 234
Mercury(II) trifluoroacetate, **9,** 294
Potassium hydride, **12,** 407
Silver(I) nitrate, **12,** 433
Sulfuric acid, **9,** 441
Thallium(I) acetate–Iodine, **10,** 394
Other routes
Allyltrimethylsilane, **8,** 273; **10,** 5, 8, 430
Ammonium persulfate, **6,** 20
Benzeneselenenyl halides, **12,** 39
Bis(N-*t*-butylsalicylaldiminato)-copper(II), **12,** 52
Bis(3-dimethylaminopropyl)-phenylphosphine, **10,** 34
Bis(trimethylsilyl) sulfate, **11,** 67
Boron trifluoride etherate, **5,** 52
t-Butyl hypoiodite, **1,** 94
t-Butyl isocyanide, **11,** 99
Carbon tetrabromide–Tin(II) fluoride, **11,** 111
Cobalt(III) acetate, **6,** 127
Copper(I) bromide, **11,** 140
1,5-Diazabicyclo[4.3.0]nonene-5, **4,** 116
Dichlorobis(cyclopentadienyl)-titanium, **9,** 146
Dichlorobis(triphenylphosphine)-palladium(II), **10,** 133
2,3-Dichloro-1-propene, **8,** 158
Dihydridotetrakis(triphenyl-phosphine)ruthenium(II), **11,** 182
Dimethyl diazomalonate, **5,** 244
1-Ethoxy-1-trimethylsilyloxy-cyclopropane, **8,** 219
Ferric chloride, **9,** 222
Lead tetraacetate, **7,** 185

Magnesium bromide etherate, **9,** 288
Mercury(II) acetate, **12,** 298
Methyl isocyanide, **11,** 11
Molybdenum carbonyl, **12,** 330
Nickel carbonyl, **12,** 335
Oxygen, singlet, **4,** 362
Palladium(0)–Phosphines, **9,** 344
Perchloric acid, **2,** 309
Peroxybenzimidic acid, **7,** 281
Phase-transfer catalysts, **10,** 305
Potassium carbonate, **5,** 552
Potassium hydride, **8,** 412
Selenium(IV) oxide, **1,** 992
Silica, **12,** 431
Silver carbonate–Celite, **5,** 577; **7,** 319
Silver(I) trifluoroacetate, **11,** 471
Silver(I) trifluoromethanesulfonate, **12,** 435
Sulfuric acid, **5,** 633
Tetracarbonyldi-μ-chlorodirhodium, **7,** 59
Thallium(III) trifluoroacetate, **9,** 462; **12,** 481
p-Toluenesulfonyl-S-methylcarbazate, **5,** 681
Tributyl(1-methoxymethoxy-2-butenyl)tin, **11,** 325
Tri-μ-carbonylhexacarbonyldiiron, **6,** 195
Trifluoromethanesulfonyl chloride, **9,** 485
Trimethyl phosphonoacetate, **7,** 394
Zinc, **11,** 598
Zinc bromide, **11,** 600

δ-LACTONES
by Baeyer–Villiger reaction
Benzeneperoxyseleninic acid, **8,** 22
Cerium(IV) ammonium sulfate, **7,** 56
Hydrogen peroxide, **4,** 254
Peracetic acid, **1,** 787
Potassium persulfate, **2,** 348
Trifluoroperacetic acid, **1,** 821
from δ-Hydroxy acids
(R)-(+)-*t*-Butyl (*p*-tolylsulfinyl)-acetate, **11,** 106
2,3-Dichloro-5,6-dicyano-1,4-benzoquinone, **1,** 215
2,3-Dichloro-1-propene, **8,** 158
Ethyl lithioacetate, **6,** 255; **8,** 225
Iodotrimethylsilane, **12,** 259
4(R)-Methoxycarbonyl-1,3-thiazolidine-2-thione, **11,** 323
(4S,5S)-4-Methoxymethyl-2-methyl-5-phenyl-2-oxazoline, **7,** 229; **9,** 312
(γ-Methoxypropyl)-α-phenylethylamine, **12,** 318
Oxygen, singlet, **12,** 363
Ozone, **7,** 269
B-3-Pinanyl-9-borabicyclo[3.3.1]-nonane, **11,** 429
p-Tolylsulfinylacetic acid, **10,** 405
Tributylcrotyltin, **11,** 143; **12,** 513
from Hydroxy nitriles
Mercury(II) acetate, **11,** 315; **12,** 298
by Oxidation of 1,5-diols, lactols and related compounds (*see* TYPE OF REACTION INDEX)
by Reduction of anhydrides (*see* TYPE OF REACTION INDEX)
from Unsaturated acids
Benzeneselenenyl chloride, **8,** 25
Bis(acetonitrile or benzonitrile)-dichloropalladium(II), **7,** 21; **12,** 51
Bromine, **12,** 70
Iodine, **9,** 248
S-Methyl *p*-toluenethiosulfonate, **6,** 400
Pyridinium chlorochromate, **12,** 417
Other routes
Arylthallium bis(trifluoroacetates), **10,** 300
Bis(3-dimethylaminopropyl)phenyl-phosphine, **10,** 34
Caro's acid, **1,** 118
Dichlorobis(cyclopentadienyl)-titanium, **9,** 146; **12,** 168
Diethyl (1,3-dithian-2-yl)phosphonate, **11,** 179
Diisobutylaluminum hydride, **9,** 171
1,3-Dimethoxy-1-trimethylsilyloxy-1,3-butadiene, **12,** 196
Ethyl vinyl ether, **11,** 235
Iron carbonyl, **8,** 265
Jones reagent, **7,** 68
Lithium trichloropalladate(II), **12,** 288
Pyridinium chlorochromate, **8,** 425; **11,** 450
Rhodium(II) carboxylates, **12,** 423
Thallium(III) trifluoroacetate, **12,** 481
Trifluoroacetic acid, **11,** 557
Trimethyl phosphite, **11,** 570
Tris(tribenzylideneacetylacetone)-

LACTONES (*Continued*)
tripalladium(chloroform), **12,** 561
Zinc bromide, **11,** 600
ε-LACTONES
Aluminum chloride, **11,** 25
Benzeneperoxyseleninic acid, **8,** 22
Diethyl (1,3-dithian-2-yl)phosphonate, **11,** 179
Ozone, **5,** 491
Phosphotungstic acid, **5,** 537
Trifluoroperacetic acid, **10,** 421

MACROLIDES
FROM ACYCLIC ESTERS
Bis(acetonitrile)dichloropalladium(II), **11,** 46
Copper(II) acetate, **1,** 159
Diethylaluminum chloride, **8,** 164
Ketenylidenetriphenylphosphorane, **12,** 268
Nickel carbonyl, **4,** 353
Phenylthioacetyl chloride, **9,** 370
Potassium hexamethyldisilazide, **9,** 385
Tetrabutylammonium fluoride, **11,** 499
Tetrakis(triphenylphosphine)-palladium(0), **8,** 472; **9,** 451; **12,** 468
Thallium(III) trifluoroacetate, **11,** 515
BY BAEYER–VILLIGER REACTION
Peracetic acid, **1,** 787; **4,** 372
Permaleic acid, **1,** 819; **12,** 379
FROM CARBOXYLATES WITH ω-LEAVING GROUPS
Cerium(III) sulfate, **9,** 99
Cesium carbonate, **11,** 114
Tetrabutylammonium fluoride, **12,** 458
FROM ω-HYDROXY ACIDS AND RELATED COMPOUNDS
Alkylaluminum halides, **12,** 5
N,N-Bis(2-oxo-3-oxazolidinyl)-phosphorodiamidic chloride, **11,** 57
Boron trifluoride etherate, **7,** 31
N,N′-Carbonyldiimidazole, **5,** 97
Chloromethyl methyl sulfide, **12,** 117
Copper(I) iodide, **11,** 141
Copper(I) trifluoromethanesulfonate, **10,** 108
Cyanuric chloride, **12,** 114
1-Cyclohexyl-3-(2-morpholinoethyl)-carbodiimide metho-*p*-toluene-sulfonate, **11,** 151
Dialkylboryl trifluoromethanesulfonates, **12,** 155
4,4′-Di(2-amino-6-methylpyrimidyl) disulfide, **9,** 413
Dibutyltin oxide, **10,** 123; **12,** 160
2,2′-Di(1-isopropyl-4-*t*-butylimidazolyl) disulfide, **7,** 141
4-Dimethylamino-3-butyn-2-one, **12,** 198
Diphenyl phosphorochloridate, **11,** 223
2-Halopyridinium salts, **8,** 95; **9,** 234
2-Mesitylenesulfonyl chloride, **6,** 625
Oxygen, singlet, **10,** 294
1-Phenyl-2-tetrazoline-5-thione, **11,** 420
Phosgene, **2,** 328
Silver perchlorate, **6,** 519
Silver(I) trifluoroacetate, **8,** 444
N,N,N′,N′-Tetramethylchloro-formamidinium chloride, **12,** 477
Thallium(I) 2-methyl-2-propanethiolate, **6,** 581
Thionyl chloride, **5,** 663
2,4,6-Trichlorobenzoyl chloride, **9,** 478
Triphenylphosphine + co-reagent, **5,** 285; **6,** 246; **7,** 404; **8,** 215, 517; **9,** 167; **12,** 552
BY OXIDATIVE CLEAVAGE OF BICYCLIC ALKENES (*see* TYPE OF REACTION INDEX)
OTHER ROUTES
Benzeneselenenyl bromide, **9,** 25
Bis(3-dimethylaminopropyl)phenyl-phosphine, **10,** 34
Copper(II) acetate–Ferrous(II) sulfate, **10,** 103
Dichloroketene, **9,** 152; **12,** 176
N-Methyl-N-phenylaminoethynyl-lithium, **10,** 267
3,3,6,9,9-Pentamethyl-2,10-diazabicyclo-[4.4.0]decene, **9,** 354
N-Phenylselenophthalimide, **9,** 366
Pyridinium dichromate, **11,** 453
Sodium hydride, **6,** 541
1,1,6,6-Tetrabutyl-1,6-distanna-2,5,7,10-tetraoxycyclodecane, **12,** 462
p-Toluenesulfonic acid, **8,** 488
Tributyltin hydride, **12,** 516
MERCAPTANS (*see* THIOLS)
MESYLATES (*see* SULFONIC ACIDS, ESTERS)
α-METHYLENE ALDEHYDES AND KETONES
GENERAL METHODS

Acetyl chloride–2-Trimethylsilylethanol, **11,** 4
Benzyl bromomethyl sulfide, **5,** 25
Bis(dimethylamino)methane, **7,** 359
Bromine, **11,** 75
Bromomethanesulfonyl bromide, **12,** 75
Chloromethyl phenyl sulfide, **9,** 107
Dicyclohexylcarbodiimide, **4,** 141
Dimethylformamide dineopentyl acetal, **9,** 183
Dimethyl sulfoxide, **1,** 296
Formaldehyde, **5,** 314; **6,** 264; **8,** 231; **12,** 232
Lithium benzeneselenoate, **10,** 238
Lithium (3,3-diethoxy-1-propen-2-yl)-phenylthiocuprate, **7,** 212
Lithium methanethiolate, **8,** 303
Lithium alkyl(phenylthio)cuprates, **7,** 212
Methoxyallene, **8,** 320
Methylenetriphenylphosphorane, **8,** 339
N-Methylthiomethylpiperidine, **12,** 326
Organolithium reagents, **11,** 13
Simmons–Smith reagent, **9,** 415
s-Trioxane, **8,** 331
α-METHYLENE CYCLOALKANONES
α-Methylenecyclobutanones
Chloro[(trimethylsilyl)methyl]ketene, **12,** 127
Dimethyl(or Diphenyl)ketene, **5,** 278; **6,** 223
Methyl(phenylthio)ketene, **8,** 348; **9,** 315
α-Methylenecyclopentanones
Di-μ-carbonylhexacarbonyldicobalt, **11,** 162
Methoxyallene, **12,** 310
Raney nickel, **11,** 457
α-Methylenecyclohexanones
2-Methylthiomaleic anhydride, **8,** 351
METHYLENECYCLOALKANES
GENERAL METHODS
by Elimination
2-Bromoacrolein, **5,** 62
Hydrogen peroxide, **1,** 471
o-Nitrophenyl selenocyanate, **6,** 420
Phenylselenomethyllithium, **6,** 549
Phenylthioacetic acid, **6,** 463
Silver(I) oxide, **5,** 583
Sodium dicarbonyl(cyclopentadienyl)-ferrate, **9,** 426
Tetrabutylammonium fluoride, **9,** 444
Zinc, **1,** 1276
by Methylenation of C=O
Dibromomethane–Zinc–Titanium(IV) chloride, **11,** 337
Dichloroketene, **4,** 134
Methylenetriphenylphosphorane, **8,** 339
Organotitanium reagents, **8,** 83
Titanium(0), **11,** 526
Triphenylstannylmethyllithium, **9,** 509
Other routes
Bis(dimethylamino)methane, **3,** 21
9-Borabicyclo[3.3.1]nonane, **10,** 48
Bromomethanesulfonyl bromide, **12,** 75
Catecholborane, **10,** 79
Dimethoxymethane, **11,** 193
Lithium dibutylcuprate, **3,** 79
Nickel carbonyl, **6,** 417
Potassium–Ammonia, **9,** 377
2,4,6-Triisopropylbenzenesulfonyl-hydrazide, **12,** 533
METHODS FOR SPECIFIC RING SIZES
Methylenecyclopropanes
Potassium *t*-butoxide, **3,** 233; **5,** 544
Sodium amide, **5,** 591
Sodium hexamethyldisilazide, **6,** 529
Methylenecyclobutanes
Allene, **1,** 18; **6,** 251
Methylenecyclopentanes
3-Acetoxy-2-trimethylsilylmethyl-1-propene, **9,** 454; **11,** 578
Allyltrimethylsilane, **10,** 6; **12,** 23
Bis(acrylonitrile)nickel(0), **3,** 20
2-Bromomethyl-3-(trimethylsilyl-methyl)-1,3-butadiene, **12,** 77
2-Bromo-3-trimethylsilyl-1-propene, **11,** 80
4-Chloro-1-butenyl-2-lithium, **12,** 113
2-Chloromethyl-3-trimethylsilyl-1-propene, **11,** 259
Dibromobis(tributylphosphine)-nickel(II), **5,** 42
Diisobutylaluminum hydride, **7,** 111
2-Hydroxymethyl-3-trimethylsilyl-1-propene, **11,** 258; **12,** 252
2-Iodomethyl-3-trimethylsilyl-1-propene, **11,** 579
Sodium naphthalenide, **11,** 490

METHYLENELACTONES (*see* UNSATURATED LACTONES)

NITRILES (*see also* AMINO NITRILES, CHIRAL COMPOUNDS, CYANO CARBONYLS, CYANOHYDRINS)
ALKYL NITRILES
from RCHO (RCHO → RCN)
N-Amino-4,6-diphenylpyridone, **7,** 10
O,N-Bis(trifluoroacetyl)-hydroxylamine, **1,** 60
Dicyclohexylcarbodiimide, **5,** 206
N,N-Dimethylhydrazine, **1,** 289
S,S-Diphenylsulfilimine, **10,** 174
Hydrazoic acid, **1,** 446
Hydroxylamine, **1,** 478; **2,** 217; **7,** 176; **9,** 245
Hydroxylamine-O-sulfonic acid, **6,** 290; **9,** 245
N-Imino-N,N-dimethyl-2-hydroxypropanaminium ylide, **8,** 256
Manganese(IV) oxide, **4,** 317
N-Methyl-2-pyrrolidone, **11,** 346
Selenium(IV) oxide, **9,** 409
by Dehydration of $RCONH_2$
Chloroform, **5,** 27
Chlorosulfonyl isocyanate, **11,** 125
Chlorotris(triphenylphosphine)-rhodium(I), **3,** 325; **5,** 736
Dicyclohexylcarbodiimide, **1,** 231; **2,** 126
Dimethylformamide + co-reagent, **1,** 286; **3,** 228
Hexamethylphosphoric triamide, **4,** 244
Methanesulfonyl chloride, **11,** 322
Phosgene, **1,** 856
Phosphonitrilic chloride trimer, **4,** 386
Phosphorus(V) chloride, **1,** 866
Phosphorus(V) oxide, **1,** 871; **2,** 329
Phosphoryl chloride, **1,** 876
Polyphosphate ester, **3,** 229
Sodium borohydride, **3,** 262
Sodium tetrachloroaluminate, **1,** 1027
Thionyl chloride, **1,** 1158
p-Toluenesulfonyl chloride, **1,** 1179
2,2,2-Trichloro-1,3,2-benzodioxaphosphole, **1,** 120
(Trichloromethyl)carbonimidic dichloride, **4,** 523
Triethoxydiiodophosphorane, **9,** 480
Triethyloxonium tetrafluoroborate, **2,** 430
Trifluoroacetic anhydride–Pyridine, **8,** 504
Triphenylphosphine bis(trifluoromethanesulfonate), **6,** 648
Triphenylphosphine–Carbon tetrachloride, **3,** 320
by Dehydration of RNHOH are related substrates
N,N′-Carbonyldiimidazole, **5,** 97
Chlorodimethylsulfoxonium chloride, **6,** 229
Chloroform, **5,** 27
1-Chlorosulfinyl-4-dimethylaminopyridinium chloride, **12,** 122
Cyanuric chloride, **4,** 522
Dichloromethylenedimethylammonium chloride, **6,** 170
O-2,4-Dinitrophenylhydroxylamine, **6,** 233
Diphenyl phosphorochloridate, **3,** 133
N-Ethylacetonitrilium tetrafluoroborate, **6,** 250
Hexamethylphosphoric triamide, **5,** 323
Hydroxylamine, **1,** 478; **9,** 245
Hydroxylamine-O-sulfonic acid, **6,** 290; **9,** 245
Methyl isocyanate, **4,** 341
Methylketene diethyl acetal, **1,** 685
N-Methyl-2-pyrrolidone, **11,** 346
Phenyl chlorosulfite, **6,** 456
Phenyl isocyanate, **1,** 842
Phosphorus(III) iodide, **10,** 318
Selenium(IV) oxide, **9,** 409
Titanium(IV) chloride, **4,** 507
Triethoxydiiodophosphorane, **9,** 480
Triethyl orthoformate, **6,** 610
Trifluoroacetic anhydride–Pyridine, **8,** 504; **9,** 484
Trifluoromethanesulfonic anhydride, **7,** 390
Trifluoroperacetic acid, **1,** 821
Trimethylamine–Sulfur dioxide, **9,** 188
Triphenylphosphine–Carbon tetrachloride, **5,** 727
Vilsmeier reagent, **10,** 457; **12,** 204
by Displacements of RX, $ROSO_2R'$, etc.

Benzyltrimethylammonium cyanide, **1,** 53
18-Crown-6, **6,** 135
Cyanotrimethylsilane, **11,** 147
2-Halopyridinium salts, **9,** 234
N-Methyl-2-pyrrolidone, **1,** 696
Potassium cyanide, **6,** 406; **7,** 299; **10,** 324
Sodium cyanide, **1,** 17, 297, 516, 1088; **2,** 445; **6,** 38; **7,** 109, 170; **11,** 481
Tetraethylammonium cyanide, **6,** 569
Triphenylphosphine–Diethyl azodicarboxylate, **7,** 404
by Hydrocyanation of C=C
Di-μ-carbonylhexacarbonyldicobalt, **1,** 224
Hydrogen cyanide, **9,** 239
from RNO_2
Diphosphorus tetraiodide, **9,** 203
Iodotrimethylsilane, **12,** 259
Phosphorus(III) chloride, **8,** 400
Phosphorus(III) iodide, **10,** 318
Sodium borohydride, sulfurated, **4,** 444
Trimethylamine–Sulfur dioxide, **9,** 488
by Oxidation of RNH_2 (*see* TYPE OF REACTION INDEX)
by Reductive cyanation of C=O and related compounds (*see* TYPE OF REACTION INDEX)
Other routes
Acetonitrile, **2,** 209
Acrylonitrile, **7,** 4
Bis(tributyltin) oxide, **11,** 62
Bromine, **8,** 52
t-Butoxybis(dimethylamino)methane, **9,** 246
Chloroacetonitrile, **3,** 235
Chloroform, **5,** 27
Chlorosulfonyl isocyanate, **3,** 51
Cyanogen chloride, **1,** 176
Cyanotrimethylsilane, **12,** 148
Dimethylaluminum amides, **9,** 177
Dimethyl(methylthio)sulfonium tetrafluoroborate, **11,** 204
Disodium tetracyanonickelate(II), **9,** 207
Ethoxycarbonylformonitrile oxide, **7,** 145
Ethyl diazoacetate, **2,** 193
Hydroxylamine, **6,** 400
Lithium diethylamide, **7,** 201
Methanesulfonyl chloride, **5,** 435; **11,** 322
Phenyl isocyanate, **7,** 284
(Phenylsulfonyl)nitromethane, **11,** 419
Phosphonitrilic chloride trimer, **5,** 322; **6,** 469
Succinonitrile, **4,** 241
Trimethyl phosphite, **1,** 1233
Triphenylmethyl isocyanide, **5,** 650; **6,** 642
Triphenylphosphine, **1,** 1238
ARYL NITRILES
from ArCHO
Nitroethane–Pyridinium chloride, **11,** 359
1-Nitropropane, **1,** 745
from ArX
Copper(I) cyanide, **1,** 278, 391, 696, 960
Sodium cyanide, **9,** 423
Sodium dicyanocuprate, **3,** 265
by Cyanation of ArH
Arylthallium bis(trifluoroacetates), **12,** 483
Chlorosulfonyl isocyanate, **2,** 70
Copper(I) cyanide, **5,** 166
Palladium(II) acetate, **5,** 496
Thallium(III) trifluoroacetate, **3,** 286
Triphenylphosphine–Thiocyanogen, **8,** 518
Other routes
p-Chlorophenyl chlorothionoformate, **3,** 50
Cyanotrimethylsilane, **11,** 147; **12,** 148
Dicyanoacetylene, **4,** 140
N,N-Diethylaminopropyne, **8,** 165
Diethyl phosphorocyanidate, **10,** 145
Hydridotris(triisopropylphosphine)-rhodium(I), **9,** 238
Hydrogen peroxide, **2,** 217
Iron carbonyl, **5,** 357
Lead(IV) acetate azides, **4,** 276
Lead tetraacetate, **5,** 365
Nickel peroxide, **1,** 731
Silver(II) oxide, **2,** 369
Tetrakis(triphenylphosphine)-palladium(0), **6,** 571
Trichloroacetonitrile, **5,** 686
Triethyl orthoformate, **6,** 610

NITRILES (*Continued*)
Triphenylphosphine dichloride, **6,** 646
NITRILE OXIDES
N-Bromosuccinimide, **2,** 40
Phenyl isocyanate, **10,** 309; **12,** 386
Sodium hypobromite, **2,** 383
p-Toluenesulfonic acid, **12,** 507
NITRO COMPOUNDS
NITRO ALDOLS
Dicyclohexylcarbodiimide, **11,** 173
Nitromethane, **1,** 515, 739; **9,** 323; **12,** 337, 462
Sodium ethoxide, **1,** 1065
Tetrabutylammonium fluoride, **9,** 444; **10,** 378; **11,** 499
Titanium(IV) chloride, **4,** 507
NITROALKANES
by Addition of NO_2, X to C=C
Benzeneselenenyl bromide–Silver nitrate, **11,** 33
Nitrogen dioxide–Iodine, **2,** 225; **3,** 130; **8,** 205
Nitronium tetrafluoroborate, **4,** 358
Nitrosyl chloride, **2,** 298
Nitryl chloride, **1,** 756
Nitryl iodide, **1,** 757; **2,** 300
Pyridinium poly(hydrogen fluoride), **5,** 538
Silver nitrite–Mercury(II) chloride, **11,** 467
by Displacement
Hexafluoroantimonic acid, **6,** 272
Potassium nitrite, **6,** 135
Silver nitrite, **1,** 1011
by Oxidation of RNH_2, RNO, etc. (*see* TYPE OF REACTION INDEX)
Other routes
Hypochlorous acid, **10,** 208
Lithium diisopropylamide, **3,** 184
Lithium tri-*sec*-butylborohydride, **12,** 286
Methyl nitrate, **1,** 691
Nitric acid, **4,** 356
Nitrogen dioxide, **1,** 324
Ozone, **12,** 365
Sodium borohydride, **1,** 1049
Trifluoroperacetic acid, **1,** 821
NITROALKENES
from C=C
Benzeneselenenyl bromide–Silver nitrate, **11,** 33
Mercury(II) nitrite, **9,** 292
Nitrogen dioxide, **1,** 324; **2,** 175, 225; **3,** 130; **8,** 205
Nitrosyl chloride, **1,** 748; **2,** 298
Nitryl chloride, **1,** 756
Nitryl iodide, **1,** 757; **2,** 300
Pyridinium poly(hydrogen fluoride), **5,** 538
Silver nitrite–Mercury(II) chloride, **11,** 467
Other routes
Benzeneselenenyl halides, **10,** 16
Dicyclohexylcarbodiimide, **11,** 173
Nitromethane, **1,** 739; **12,** 337
Tetranitromethane, **10,** 392
Trifluoroperacetic acid, **8,** 505
NITROARENES
from $ArN_2^+X^-$
Arenediazonium tetrahaloborates, **1,** 43
Sodium nitrite, **1,** 1097
Tetrafluoroboric acid, **1,** 394
by Nitration of ArH (*see* TYPE OF REACTION INDEX)
by Oxidation of $ArNH_2$, ArNO (*see* TYPE OF REACTION INDEX)
NITRO CARBONYL COMPOUNDS
α-Nitro
Acetone cyanohydrin nitrate, **1,** 5
Amyl nitrite, **1,** 923
Isoamyl nitrate, **1,** 40; **2,** 25
Lithium aluminum hydride, **12,** 272
Nitryl chloride, **1,** 756
Polyphosphoric acid, **1,** 894
Potassium amide, **2,** 336
Potassium dichromate, **12,** 405
Pyridinium chlorochromate, **12,** 417
Sodium hydride, **1,** 1075
Sodium nitrite, **1,** 298
Trifluoroacetyl nitrate, **12,** 530
γ-Nitro
Benzyltrimethylammonium hydroxide, **1,** 1252
(S)-(+)-2-Methoxymethylpyrrolidine, **11,** 326
Nitromethane, **8,** 129, 430; **9,** 447
2-Nitropropane, **5,** 685
Potassium fluoride, **6,** 481
1,1,3,3-Tetramethylguanidine, **1,** 1145; **4,** 489
NITRONES

N,N-Diethylaminopropyne, **9,** 164
N,N-Dimethyl-4-nitrosoaniline, **1,** 746
Hydrogen peroxide–Sodium tungstate, **12,** 246
Methyl fluorosulfonate, **7,** 240
N-Methylhydroxylamine, **12,** 322
N-Methylhydroxylamine-O-sulfonic acid, **8,** 341
Pyridine, **1,** 958
Titanium(III) chloride, **12,** 492

NITROSO COMPOUNDS
BY NITROSATION
Sodium dithionite, **1,** 1081
Thallium(III) trifluoroacetate, **5,** 658
BY OXIDATION OF RNH_2, RNHOH (*see* TYPE OF REACTION INDEX)
N-NITROSO COMPOUNDS
2-Hydroperoxy-2-nitropropane, **4,** 256
Nitrogen dioxide, **1,** 324; **8,** 204
Nitrogen trioxide, **7,** 252
3-Nitro-N-nitrosocarbazole, **1,** 741
2-Nitroprop-2-yl hydroperoxide, **3,** 213
Nitrosonium tetrafluoroborate, **1,** 747
Nitrosyl chloride, **1,** 748; **7,** 254
Nitrosylsulfuric acid, **2,** 299
Silver nitrite, **6,** 515
Sodium nitrite, **1,** 1097
OTHER METHODS
O-*t*-Butyldimethylsilylhydroxylamine, **12,** 85
m-Chloroperbenzoic acid, **5,** 120
Lead tetraacetate, **1,** 537
Sodium bicarbonate, **11,** 476

NUCLEOSIDES, NUCLEOTIDES
NUCLEOSIDES
Bis(*p*-nitrophenyl) hydrogen phosphate, **2,** 29
Mercury(II) cyanide, **3,** 197
Phase-transfer catalysts, **11,** 403
Polyphosphate ester, **1,** 892
Tin(IV) chloride, **11,** 522
Trimethylsilyl perchlorate, **6,** 639
Trimethylsilyl trifluoromethanesulfonate, **10,** 438
NUCLEOTIDES
Bis(*p*-nitrophenethyl) phosphorochloridate, **11,** 56
Bis(*p*-nitrophenyl) hydrogen phosphate, **1,** 331
Bis(2,2,2-trichloro-1,1-dimethylethyl) phosphorochloridate, **11,** 62
Bis(2,2,2-trichloroethyl) phosphorochloridate, **2,** 187
2-Cyanoethyl phosphate, **1,** 172
Cyclohexyl isocyanide, **6,** 152
Dibenzyl phosphorochloridate, **1,** 198
Di(2-*t*-butylphenyl) phosphorochloridate, **6,** 167
Dicyclohexylcarbodiimide, **2,** 126
2-(N,N-Dimethylamino)-4-nitrophenyl phosphate, **6,** 207
N-Methylimidazole, **5,** 447
p-Nitrophenyl dichlorophosphate, **1,** 744
Phosphoryl chloride, **3,** 228, 229
Polyphosphoric acid, **1,** 894
Pyrophosphoryl chloride, **1,** 971
Tetra-*p*-nitrophenyl pyrophosphate, **1,** 1148
Triphenylphosphine–Diethyl azodicarboxylate, **4,** 553
OH PROTECTING GROUPS IN NUCLEOTIDE SYNTHESIS
Adamantane-1-carboxylic acid chloride, **2,** 15
Benzeneboronic acid, **3,** 221
3-Benzoylpropionic acid, **3,** 19
Benzoyltetrazole, **7,** 15
Bis(*p*-nitrophenyl) hydrogen phosphate, **1,** 331
Copper(II) sulfate, **2,** 89
Di-*p*-anisylphenylmethyl chloride, **1,** 188
2,2-Dimethoxypropane, **1,** 268
2,4-Dinitrobenzenesulfenyl chloride, **1,** 319
Dioxane–Bromine, **2,** 177
Ethyl vinyl ether, **1,** 386
Isobutyl chloroformate, **2,** 47
α-Naphthyldiphenylmethyl chloride, **6,** 415
p-Nitrophenyl chloroformate, **2,** 297
Sodium triphenylmethoxyacetate, **5,** 627
Tetrabutylammonium fluoride, **8,** 467
Tetraethylammonium acetate, **1,** 1136
Tetramethyl orthocarbonate, **3,** 285
2,2,2-Tribromoethyl chloroformate, **2,** 423
Tri(2-chloroethyl) orthoformate, **3,** 297
Triphenylmethyl chloride, **1,** 1254

ORTHO ESTERS
2,6-Dimethylpyridine, **1,** 626
Formamide, **3,** 147

ORTHO ESTERS (*Continued*)
3-Methyl-3-hydroxymethyloxetane, **12,** 323
Triethyl orthoformate, **1,** 1204; **4,** 527
OXIMES
FROM C=O
Hydroxylamine, **1,** 478; **5,** 342
O-(Phenylthiomethyl)hydroxylamine, **5,** 527
Potassium *t*-amyloxide, **1,** 939
OTHER ROUTES
Chromium(II) chloride, **2,** 76
Hydrogen peroxide–Sodium tungstate, **1,** 475
Hydroxylamine, **10,** 206
Iodotrimethylsilane, **12,** 259
Isoamyl nitrite, **1,** 40
Methyl nitrite, **1,** 691
Nitrosyl chloride, **1,** 748; **6,** 422
Sodium nitrite, **1,** 1097; **4,** 459
Vilsmeier reagent, **12,** 564

PEPTIDES—COUPLING AGENTS, REACTION CONDITIONS
(*see also* AMINO ACIDS AND DERIVATIVES, CHIRAL COMPOUNDS)
Azidotris(dimethylamino)phosphonium hexafluorophosphate, **6,** 25
Benzotriazolyloxytris(dimethylamino)-phosphonium hexafluorophosphate, **6,** 34; **7,** 14
O-Benzotriazolyl-N,N,N′,N′-tetra-methyluronium hexafluorophosphate, **12,** 44
Benzylcarbazate, **1,** 110
Bis(2,4-dinitrophenyl) carbonate, **1,** 56; **2,** 29
2,4-Bis(4-methoxyphenyl)-1,3-dithia-2,4-diphosphetane-2,4-disulfide, **11,** 54
Bis(*o*-nitrophenyl) phenylphosphonate, **9,** 50; **10,** 41
Bis-*o*-phenylene pyrophosphite, **1,** 60; **2,** 30
5-Bromo-7-nitroindoline, **11,** 79
sec-Butyl chloroformate, **2,** 47
t-Butyl isocyanide, **9,** 82
Butyl nitrites, **1,** 97
N,N′-Carbonyldiimidazole, **1,** 114
1,1-Carbonyldi-1,2,4-triazole, **1,** 116
3-Chloro-1,2-benzisothiazole 1,1-dioxide, **1,** 990
2-Chloro-1,3,2-dioxaphospholane, **1,** 372
Copoly(ethylene-N-hydroxymaleimide), **2,** 80
Copper(II) chloride, **5,** 158
Cyanogen bromide, **6,** 148
1-Cyclohexyl-3-(2-morpholinoethyl)-carbodiimide, **1,** 181
1,1-Dichloroethyl ethyl ether, **1,** 219
Dicyclohexylcarbodiimide, **1,** 231; **6,** 174; **10,** 142
N,N-Diethylaminopropyne, **2,** 133
Diethylcyanamide, **1,** 249
Diethyl phosphite, **1,** 251
Diethyl phosphorobromidate, **8,** 170
Diethyl phosphorochloridite, **1,** 248
Diethyl phosphorocyanidate, **5,** 217; **6,** 192; **7,** 107
2,4-Dimethoxybenzylamine, **4,** 164
1-(3-Dimethylaminopropyl)-3-ethylcarbodiimide, **1,** 274; **12,** 199
Dimethylformamide–Thionyl chloride, **1,** 286
Diphenylketene, **1,** 343
Diphenyl 2-keto-3-oxazolinyl-phosphonate, **11,** 220
Diphenylphosphinyl chloride, **7,** 138
Diphenyl phosphite, **4,** 210
Diphenyl phosphoroazidate, **4,** 210; **6,** 193
Diphenyl phosphorochloridate, **1,** 345
Diphenyl N-succinimidylphosphate, **10,** 173
4,6-Diphenylthieno[3,4-*d*][1,3]dioxol-2-one 5,5-dioxide, **7,** 140
Diphenyl-N-*p*-tolylketenimine, **1,** 345
N,N′-Disuccinimidyl carbonate, **9,** 208
Di(N-succinimidyl)oxalate, **12,** 219
Ethoxyacetylene, **1,** 357
N-Ethoxycarbonyl-2-ethoxy-1,2-dihydroquinoline, **2,** 191; **3,** 137; **4,** 223
2-Ethoxypyridine 1-oxide, **1,** 362
N-Ethylbenzisoxazolium tetrafluoroborate, **1,** 364
Ethyl chloroformate, **2,** 193
1-Ethyl-3-(3′-dimethylaminopropyl)-carbodiimide hydrochloride, **1,** 371
Ethyl ethylene pyrophosphite, **1,** 250
2-Ethyl-7-hydroxybenzisoxazolium tetrafluoroborate, **6,** 253
N-Ethyl-5-phenylisoxazolium-3′-

sulfonate, **1,** 384; **2,** 197; **8,** 226
Hexamethyldisilazane, **1,** 427
Hexamethylphosphoric triamide, **3,** 149
Hydrogen fluoride, **2,** 215
1-Hydroxybenzotriazole, **3,** 156; **5,** 342; **6,** 288
3-Hydroxy-4-keto-3,4-dihydrobenzo-1,2,3-triazine, **5,** 342
(−)-3-Hydroxy-5-methylhydantoin, **8,** 252
N-Hydroxy-5-norbornene-2,3-dicarboximide, **6,** 290
N-Hydroxyphthalimide, **1,** 485
N-Hydroxypiperidine, **1,** 486
3-Hydroxypyridine, **1,** 486
N-Hydroxysuccinimide, **1,** 487; **9,** 246
N-Hydroxysuccinimide trifluoroacetate, **2,** 219
Ion-exchange resins, **1,** 511; **2,** 227; **6,** 302
Isobutyl chloroformate, **1,** 86; **6,** 307
N-Methyl-N,N′-di-*t*-butylcarbodiimium tetrafluoroborate, **6,** 378
N-Methylmorpholine, **2,** 278
N-Methyl-N-phenylbenzohydrazonyl bromide, **10,** 269
4-(4′-Methyl-1′-piperazinyl)-3-butyn-2-one, **8,** 348
2-Morpholinoethyl isocyanide, **11,** 352
5-Nitro-[3*H*]-1,2-benzoxathiole S,S-dioxide, **11,** 359
(5-Nitropyridyl)diphenyl phosphinate, **11,** 361
Nitrosyl chloride, **1,** 748
Oxalyl chloride, **2,** 301
2,5-Oxazolidinedione, **5,** 99
μ-Oxybis[tris(dimethylamino)-phosphonium] bis(tetrafluoroborate), **6,** 425
Phenylphosphorodi(1-imidazolidate), **1,** 847
N-Phenyltrimethylacetimidoyl chloride, **1,** 854
Phosphorus(III) chloride, **1,** 875
Polyhexamethylenecarbodiimide, **2,** 333
Polyphosphate ester, **1,** 892
Propylphosphonic anhydride, **10,** 333
Pyrazole, **1,** 958
2-Pyridone, **3,** 157
4-Pyrrolidinopyridine, **10,** 336
Sulfur trioxide–Dimethylformamide, **1,** 1125
Tetrachlorosilane, **4,** 424
Tetraethyl pyrophosphite, **1,** 1138; **2,** 397
1,1,3,3-Tetramethylguanidine, **12,** 477
1,3-Thiazolidine-2-thione, **12,** 485
Thioanisole–Trifluoromethanesulfonic acid, **10,** 398
Thioglycolic acid, **1,** 1153; **7,** 366
N,N′-Thionyldiimidazole, **1,** 1163
Tin(II) chloride, **2,** 389
1,2,4-Triazole, **1,** 1188; **2,** 423
Triethoxydiiodophosphorane, **9,** 480
Triethylsilane, **1,** 1218
Trifluoroacetic acid, **1,** 1219
Trimethylacetyl chloride, **1,** 1229
Tri-*p*-nitrophenyl phosphorotrithioite, **1,** 1237
Triphenylphosphine + co-reagent, **2,** 443; **3,** 322; **4,** 551
Triphenyl phosphite, **3,** 322
Vilsmeier reagent, **9,** 514
Zinc chloride, **8,** 536

PEROXIDES (*see also* DIOXETANES, PEROXY ACIDS)
GENERAL METHODS
from 1,3-Dienes
Hydrogen peroxide–Cerium(IV) oxide, **6,** 99
Lithium diisopropylamide, **7,** 204
Oxygen, **5,** 482
Oxygen, singlet, **4,** 362; **5,** 486, 576; **6,** 431; **7,** 261; **8,** 173, 367; **9,** 338; **11,** 385
Triphenylcarbenium tetrafluoroborate, **4,** 565
Triphenyl phosphite ozonide, **3,** 323
Other routes
Di-*t*-butylperoxyoxalate, **7,** 93
Hydrogen peroxide, **1,** 457; **5,** 337; **7,** 175; **10,** 201
Hydrogen peroxide + co-reagent, **7,** 224; **8,** 248
Potassium superoxide, **6,** 488
Silver carbonate–Celite, **3,** 247
Tributyltin hydride, **11,** 545
DIACYL PEROXIDES
Bis(tributyltin) peroxide, **8,** 44
Diacyl peroxides, **1,** 186
Dicyclohexylcarbodiimide, **1,** 231
Hydrogen peroxide–Silver trifluoroacetate, **8,** 249; **10,** 203
Ketene, **1,** 528

PEROXIDES (*Continued*)
Perbenzoic acid, **8,** 386
Potassium superoxide, **7,** 304; **8,** 417
Silver acetate, **8,** 45
PEROXY ACIDS AND ESTERS
t-Butyl hydroperoxide, **1,** 88
N,N′-Carbonyldiimidazole, **1,** 114
Hydrogen peroxide, **1,** 457; **7,** 174
Ketene, **1,** 528
Methanesulfonic acid, **1,** 666
Silver diethyl phosphate, **4,** 428
Sodium methoxide, **1,** 1091
PHENOLS
BY AROMATIZATION OF CYCLOHEXANONES, -ENONES (*see* TYPE OF REACTION INDEX)
FROM ArX
Borane–Tetrahydrofuran, **4,** 124
t-Butyl hydroperoxide, **1,** 88
2-*t*-Butylperoxy-1,3,2-dioxaborolane, **12,** 101
Copper(II) sulfate, **1,** 164
Dimethyl sulfoxide, **4,** 192
Peroxytetrakis(triphenylarsine)-rhodium(I) hexafluorophosphate, **8,** 206
Potassium superoxide, **7,** 304; **8,** 417
Sodium methoxide, **6,** 147
Trimethyl borate, **1,** 673; **2,** 435
BY CLAISEN REARRANGEMENT
Acetic anhydride–Sodium acetate, **7,** 2
Benzylamine, **10,** 26
Boron trichloride, **5,** 50
Sodium dithionite, **11,** 485
Titanium(IV) chloride, **7,** 370
Trifluoroacetic acid, **4,** 530
Zinc chloride, **10,** 461
BY DAKIN REACTION (*see* TYPE OF REACTION INDEX)
BY DEALKYLATION OF ArOR (*see* TYPE OF REACTION INDEX)
BY DIELS–ALDER REACTION
1,4-Dihydronaphthalene-1,4-*endo*-oxide, **1,** 255
1,1-Dimethoxy-1-trimethylsilyloxy-1,3-butadiene, **8,** 178
Dimethyl methoxymethylenemalonate, **12,** 206
Homophthalic anhydride, **11,** 254
3-Methoxymethylene-2,4-bis(trimethyl-silyloxy)-1,4-pentadiene, **12,** 315
trans-1-Methoxy-3-trimethylsilyloxy-1,3-butadiene, **6,** 370; **8,** 328
1,1,2,3-Tetrakis(trimethylsilyloxy)-1,3-butadiene, **10,** 384
1-Trimethylsilyloxy-1,3-butadiene, **8,** 512
2-Trimethylsilyloxy-1,3-cyclohexadiene, **8,** 512
Trimethylsilylvinylketene, **10,** 441
BY DIENONE–PHENOL REARRANGEMENT (*see* TYPE OF REACTION INDEX)
BY FRIES REARRANGEMENT (*see* TYPE OF REACTION INDEX)
BY HYDROXYLATION OF ArH (*see* TYPE OF REACTION INDEX)
BY REDUCTION OF QUINONES (*see* TYPE OF REACTION INDEX)
BY RING-FORMING REACTIONS
Aluminum chloride, **11,** 25
Chromium carbene complexes, **12,** 132
Dibenzyl ketone, **4,** 123
Iron carbonyl, **1,** 519
Lithium diisopropylamide, **9,** 280
Malonyl dichloride, **12,** 291
1-Methoxy-1,3-bis(trimethylsilyloxy)-1,3-butadiene, **9,** 54; **10,** 46, 257; **11,** 66
Methyl phenylsulfinylacetate, **8,** 347
Methyl vinyl ketone, **6,** 407; **10,** 272
3-Phenylsulfonyl-1(3*H*)-isobenzofuranone, **9,** 368
3-Phenylthio-3-buten-2-one, **11,** 421
Phthalide, **8,** 402
Potassium persulfate, **1,** 952; **8,** 417
Potassium superoxide, **11,** 442
Tin(II) chloride, **1,** 1113
Titanium(IV) chloride, **6,** 590
OTHER ROUTES
Antimony(V) chloride, **7,** 12
m-Chloroperbenzoic acid, **5,** 120
Chlorotrimethylsilane, **8,** 107
Copper(I) oxide–Copper(II) nitrate, **8,** 124
Cyanotrimethylsilane, **7,** 397
Dimethyl phosphite, **1,** 293
N-Lithioethylenediamine, **1,** 567
Lithium biphenylide, **1,** 612
Nickel carbonyl, **4,** 353
Nitrosyl chloride, **1,** 748
Oxalyl chloride, **5,** 481
Peracetic acid, **1,** 787
Rhodium(III) chloride, **11,** 460

Tetracarbonyldi-μ-chlorodirhodium, **8,** 469
Triethylamine, **9,** 481
Trifluoroacetic anhydride, **1,** 1221
Trimethyl phosphite, **12,** 536
Zinc, **1,** 1276
PHOSPHORUS COMPOUNDS
(*see also* CHIRAL COMPOUNDS)
ENOL PHOSPHATES
Ethoxyacetylene, **1,** 357
Triethyl phosphite, **3,** 304
PHOSPHATE ESTERS
Monophosphates
(2-Chloromethyl-4-nitrophenyl) dichlorophosphate, **3,** 48
2-Chloro-2-oxobenzo-1,3,2-dioxaphosphole, **1,** 837; **2,** 321; **3,** 223
Di-*t*-butyl phosphorobromidate, **8,** 146
2-(N,N-Dimethylamino)-4-nitrophenyl phosphate, **6,** 207
Dimorpholinophosphinyl chloride, **1,** 861
Di-*p*-nitrobenzyl phosphorochloridate, **1,** 321
Diphenyl phosphorochloridate, **1,** 345
Metaphosphoric acid, **1,** 665
N-Methylimidazole, **5,** 447
Phosphoric acid, **1,** 860
Pyrophosphoryl chloride, **2,** 353
Silver diphenylphosphate, **1,** 1006
Tetrabutylammonium di-*t*-butylphosphate, **9,** 443
Tetra-*p*-nitrophenyl pyrophosphate, **1,** 1148
Trichloroacetonitrile, **1,** 1194
2,2,2-Trichloroethanol, **6,** 605
Diphosphates
Acetoinenediolcyclopyrophosphate, **6,** 6, 218
Cyclohexyl isocyanide, **6,** 152
1-(4,5-Dimethyl-1,3,2-dioxaphosphol-2-yl)-1*H*-imidazole, **6,** 218
p-Nitrophenyl(1,2-dimethylethylenedioxy)phosphate, **6,** 8
Phenyl dichlorophosphate, **1,** 847
Phenyl N-phenylphosphoramidochloridate, **7,** 286
Phenyl 8-quinolyl phosphate, **5,** 517
2,4,6-Triisopropylbenzenesulfonyl chloride, **3,** 308
2,2,2-Trimethoxy-4,5-dimethyl-1,3-dioxaphospholene, **4,** 535
Triphosphates
Cesium bicarbonate, **8,** 81
Cesium fluoride, **7,** 57
m-Chloroperbenzoic acid, **11,** 122
2-Cyanoethyl phosphate, **1,** 172
N,N-Dimethylformamide, **1,** 278
Dimethyl selenoxide, **8,** 197
Di-*p*-tolylcarbodiimide, **1,** 351
Nitrogen dioxide, **1,** 324
Oxygen, singlet, **6,** 431
Phosphoryl chloride, **1,** 876
Silver(II) oxide, **2,** 369
Tetrabutylammonium bromide, **7,** 353
Trimethyl phosphite, **1,** 1233
Triphenyl phosphite ozonide, **3,** 323
PHOSPHINES
Camphor-10-sulfonic acid, **4,** 68
Hexachlorodisilane, **3,** 148
Phenylsilane, **5,** 523
Trichlorosilane, **5,** 687
2,4,6-Triisopropylbenzenesulfonylhydrazide, **11,** 563
PHOSPHINE OXIDES
Bis(trimethylsilyl) peroxide, **11,** 67
Dimethyl selenoxide, **8,** 197
Dimethyl sulfoxide, **5,** 263; **6,** 225, 295
(Diphenylphosphine)lithium, **6,** 340
Nitrogen dioxide, **1,** 324
2,4,6-Triisopropylbenzenesulfonylhydrazide, **11,** 563
PHOSPHONATES
Copper(I) iodide, **7,** 81
Diethyl phosphite, **6,** 192
(Dimethoxyphosphinyl)methyllithium, **6,** 339
Dimethyl phosphite, **5,** 258
Diphenyl triphenylphosphoranylidenemethylphosphonate, **2,** 181
Methyl iodide, **1,** 682
Methylphosphonic dichloride, **1,** 666
Sodium borohydride, **6,** 530
Tetrakis(triphenylphosphine)palladium(0), **10,** 384
Triethyl phosphite, **1,** 1212
Triisopropyl phosphite, **1,** 1229
Tris(trimethylsilyl) phosphite, **7,** 412
PYROPHOSPHATES
2-Chloro-4,5-dimethyl-2-oxo-1,3,2-dioxaphosphole, **6,** 7

PHOSPHORUS COMPOUNDS (*Continued*)
Dibromomalonamide, **1,** 208
N,N-Dimethylformamide, **1,** 278
Pyrophosphoryl chloride, **1,** 971
POLYNUCLEOTIDES
Benzenesulfonyltetrazole, **7,** 13
3-Benzoylpropionic acid, **2,** 26
t-Butyldimethylchlorosilane, **8,** 59
Chloro-N,N-dimethylamino-methoxyphosphine, **10,** 88
m-Chloroperbenzoic acid, **11,** 122
2-Chlorophenyl dichlorophosphite, **6,** 114
Dicyclohexylcarbodiimide, **1,** 231
Dimethylformamide–Thionyl chloride, **1,** 286
9-Fluorenylmethanol, **5,** 308
2-Mesitylenesulfonyl chloride, **1,** 661
Mesitylenesulfonylimidazole, **5,** 434
1-(Mesitylenesulfonyl)-1,2,4-triazole, **5,** 434; **6,** 361
p-Nitrobenzenesulfonyl 4-nitroimidazole, **11,** 359
4-Nitrophenyl phenyl phosphorochloridate, **8,** 362
Picryl chloride, **1,** 885
Polyphosphate ester, **1,** 892
2,2,2-Trichloroethanol, **6,** 605
2,4,6-Triisopropylbenzenesulfonyl chloride, **1,** 1228; **6,** 622
PROPARGYL ALCOHOLS (*see also* CHIRAL COMPOUNDS)
BY ADDITIONS TO C=O
Acetylene, **1,** 11
Bis(trimethylsilyl)acetylene, **7,** 8; **9,** 311
Butadiyne, **1,** 185
3-Butyn-2-ol, **4,** 423
Chlorotris(triphenylphosphine)-rhodium(I), **5,** 736
1,3-Dilithiopropyne, **6,** 202
Dilithium tris(1-pentynyl)cuprate, **6,** 203
Ethoxyacetylene–Magnesium bromide, **7,** 145
Ethynylmagnesium bromide, **1,** 389, 415
(2S,2′S)-2-Hydroxymethyl-1-[(1-methylpyrrolidin-2-yl)methyl]-pyrrolidine, **10,** 207
Lithium acetylides, **1,** 573; **2,** 166, 288; **6,** 324; **7,** 195; **12,** 272
Lithium–Ethylamine, **11,** 287
Methyl vinyl ketone, **1,** 697
Organocerium reagents, **12,** 345
Phase-transfer catalysts, **11,** 403
1-Phenyl-2-trimethylsilylacetylene, **7,** 354
Sodium acetylides, **1,** 1027; **2,** 166
3-Tetrahydropyranyloxy-1-propynyllithium, **7,** 320
Trimethylsilylacetylene, **9,** 246
BY REDUCTION OF α,β-ACETYLENIC C=O's (*see* TYPE OF REACTION INDEX)
OTHER ROUTES
t-Butyl hydroperoxide–Selenium(IV) oxide, **9,** 79
Copper, **5,** 146
Dichloro(diethoxyphosphinyl)-methyllithium, **6,** 188
(2R,4R)-Pentanediol, **12,** 375
Sodium amide, **1,** 1034

QUINODIMETHANES
o-QUINODIMETHANES
Cesium fluoride, **10,** 81
Chromium(II) chloride, **12,** 136
1,3-Dihydrobenzo[c]thiophene 2,2-dioxide, **10,** 146
Lithium 2,2,6,6-tetramethylpiperidide, **12,** 285
Oxygen, **5,** 482
Potassium *t*-butoxide, **1,** 911
Tetrabutylammonium fluoride, **10,** 378
Tri-μ-carbonylhexacarbonyldiiron, **2,** 139
Zinc, **3,** 338; **11,** 598
Zinc-Silver couple, **12,** 571
p-QUINODIMETHANES
1,3-Dihydrobenzo[c]thiophene 2,2-dioxide, **9,** 168
Iodine, **1,** 495
Tetrabutylammonium fluoride, **10,** 378
QUINONES AND DERIVATIVES
GENERAL METHODS—*o*-QUINONES
by Oxidation of ArH, catechols, phenols and related compounds (*see* TYPE OF REACTION INDEX)
Other routes
Cyanotrimethylsilane, **5,** 720
2,3-Dichloro-5,6-dicyano-1,4-benzoquinone, **11,** 166
Dimethyl sulfoxide + co-reagent, **2,**

163; **4,** 201; **6,** 229
Potassium–Graphite, **12,** 400
Selenium(IV) oxide, **10,** 345
Silver(I) oxide, **8,** 442
GENERAL METHODS—
p-QUINONES
Annelation
2,5-Bis(trimethylsilyloxy)furans, **10,** 45
Boric acid, **1,** 63
Boron trifluoride etherate, **11,** 72
Chloranil, **10,** 85
Chlorotris(triphenylphosphine)cobalt, **12,** 128
Chromium carbene complexes, **12,** 132
Glyoxal, **1,** 413
2-Lithio-3,3,6,6-tetramethoxy-1,4-cyclohexadiene, **8,** 280
Magnesium methoxide, **2,** 255
Phosphorus(V) oxide, **1,** 871
Phthalide, **5,** 537
Sodium tetrachloroaluminate, **1,** 1027; **4,** 438
by Oxidation of ArH, hydroquinones phenols and related compounds (*see* TYPE OF REACTION INDEX)
Other routes
2,5-Diethoxy-1,4-benzoquinone, **8,** 163
Oxygen, **9,** 335
Thallium(III) trifluoroacetate, **5,** 658
ANTHRAQUINONES
by Diels–Alder routes
1,3-Bis(trimethylsilyloxy)-1,3-butadiene, **10,** 44
5,5-Dimethoxy-1,2,3,4-tetrachlorocyclopentadiene, **1,** 270
1,1-Dimethoxy-1-trimethylsilyloxy-1,3-butadiene, **8,** 178
1,2-Dimethoxy-1-trimethylsilyloxy-1,3-pentadiene, **12,** 197
N,N-Dimethylhydrazine, **11,** 200
6-Methoxy-4-methyl-2-pyranone, **8,** 325
1-Methoxy-3-methyl-1-trimethylsilyloxy-1,3-butadiene, **9,** 302
Naphthacene-9,10,11,12-diquinone, **1,** 709
Other routes
Boron trifluoride etherate, **7,** 31
N-Bromosuccinimide, **11,** 79
Butyllithiums, **8,** 67; **10,** 75
Chlorotris(triphenylphosphine)-rhodium(I), **5,** 736
Di-*t*-butylsilyl bis(trifluoromethanesulfonate), **11,** 189
Dicarbonylcyclopentadienylcobalt, **8,** 146
Dichlorobis(triphenylphosphine)-palladium(II), **5,** 190
Grignard reagents, **9,** 229
Ketene diethyl acetal, **5,** 362; **9,** 262
Lithium *o*-lithiobenzoate, **10,** 246
Lithium phthalide, **9,** 285
Organolithium reagents, **9,** 5
Oxygen, **8,** 366
3-Phenylsulfonyl-1(3*H*)-isobenzofuranone, **10,** 313
Potassium hydroxide, **6,** 486
1,4-BENZOQUINONES
Iron carbonyl, **5,** 357
1-Lithio-3,3,6,6-tetramethoxy-1,4-cyclohexadiene, **7,** 191
1,4-NAPHTHOQUINONES
Chromium carbene complexes, **11,** 397
Iron carbonyl, **10,** 221
Lead tetraacetate, **3,** 168
1-Methoxy-3-methyl-1-trimethylsilyloxy-1,3-butadiene, **9,** 302
Potassium *t*-butoxide, **2,** 336
Potassium superoxide, **12,** 413
Sulfur sesquioxide, **1,** 1123
Vanadium(V) oxide, **1,** 1057
1,5-NAPHTHOQUINONES
2,3-Dichloro-5,6-dicyano-1,4-benzoquinone, **6,** 168; **7,** 96
2,6-NAPHTHOQUINONES
t-Butyl chromate, **5,** 73
o-Chloranil, **1,** 128
Lead(IV) oxide, **1,** 533
QUINOLS
Antimony(V) chloride, **6,** 22
Bis(3-salicylideneaminopropyl)-aminecobalt(II), **3,** 245
Chromium(VI) oxide, **5,** 140
Hydrogen peroxide–Cerium(IV) oxide, **6,** 99
Lead tetraacetate, **6,** 313; **7,** 185
Magnesium, **7,** 398
Manganese(IV) oxide, **1,** 637
Organolithium reagents, **10,** 3; **11,** 13
Oxygen, **6,** 426
Oxygen, singlet, **6,** 431

QUINONES AND DERIVATIVES (*Continued*)
Thallium(III) perchlorate, **5,** 657; **8,** 478
Thallium(III) trifluoroacetate, **4,** 498
QUINONE KETALS
2,3-Dichloro-5,6-dicyano-1,4-benzoquinone, **9,** 148
Thallium(III) nitrate, **7,** 362; **9,** 460; **10,** 395

SELENIDES (*see also* KETENE DERIVATIVES)
GENERAL METHODS
Benzeneselenenyl chloride, **9,** 25
Bis(tricyclohexyltin) sulfide–Boron trichloride, **11,** 63
m-Chloroperbenzoic acid, **11,** 122
Dimethylaluminum methylselenolate, **8,** 182
Dimethyl selenoxide, **6,** 224
Diphosphorus tetraiodide, **9,** 203
Lithium selenide, **8,** 310
Phenyl selenocyanate, **9,** 34; **11,** 416
Phenyl trimethylsilyl selenide, **9,** 373, 496
Phenyl vinyl selenide, **9,** 282
Selenium(IV) oxide, **6,** 509
Sodium benzeneselenoate, **8,** 447; **10,** 356
PHENYL SELENIDES
Benzeneselenenyl bromide–Silver nitrate, **11,** 33
Lithium benzeneselenoate, **10,** 238
Nitrogen dioxide, **7,** 252
o-Nitrophenyl selenocyanate, **6,** 420
Phenyl selenocyanate, **7,** 253
N-Phenylselenophthalimide, **10,** 312
Phenyl trimethylsilyl selenide, **9,** 373, 496
Sodium benzeneselenoate, **9,** 432
Sodium pyridylselenate, **10,** 368
SELENO ACETALS AND KETALS
Potassium diisopropylamide–Lithium *t*-butoxide, **9,** 383
Tris(phenylseleno)borane, **8,** 522; **9,** 511
Zinc chloride, **8,** 536
SELENO CYANATES
Ferric thiocyanate, **7,** 155
Potassium selenocyanate, **8,** 417
Triphenylphosphine–Selenocyanogen, **9,** 506
SELENO ESTERS
Benzeneseleninic acid, **9,** 32
Bis(tricyclohexyltin) sulfide–Boron trichloride, **11,** 63
N,N′-Carbonyldiimidazole, **8,** 77
Dimethylaluminum methylselenolate, **8,** 182
4-Dimethylamino-3-butyn-2-one, **9,** 177
Lithium benzeneselenoate, **8,** 183
Phenyl selenocyanate, **8,** 396
VINYL SELENIDES
Benzeneselenenyl bromide, **8,** 25; **11,** 33
Diphosphorus tetraiodide, **11,** 224
Hydrazine, **12,** 241
Phenyl vinyl selenide, **8,** 294
Potassium diisopropylamide–Lithium t-butoxide, **9,** 383

SEMICARBAZONES
4-(4-Carboxyphenyl)semicarbazide, **1,** 118
Chlorotrimethylsilane, **3,** 310
Polyphosphoric acid, **4,** 395
Semicarbazide, **1,** 1000; **10,** 346

SILANES (*see also* ALLYLIC COMPOUNDS, SILYL ENOL ETHERS, VINYLSILANES)
ACYLSILANES
1,3-Bis(phenylseleno)propene, **11,** 59
t-Butyllithium, **12,** 100
Carbon monoxide, **11,** 111; **12,** 105
4,4-Dimethyl-1,3-oxathiolane 3,3-dioxide, **9,** 186
2-Lithio-2-trimethylsilyl-1,3-dithiane, **2,** 184
Palladium catalysts, **9,** 351
Trimethylamine N-oxide, **10,** 423
ALKYLSILANES
Dichlorosilane, **4,** 139
Dimethylphenylsilyllithium, **8,** 196; **12,** 210
Trimethylsilyllithium, **7,** 400
Tris(trimethylsilyl)aluminum, **12,** 343
Vinyltrichlorosilane, **5,** 749
ALKYNYLSILANES
Chlorotrimethylsilane, **2,** 435
Copper(I) trimethylsilylacetylide, **6,** 148
Ethyl trimethylsilylacetate, **7,** 150; **11,** 234
Organoaluminum reagents, **11,** 25
Phenyl trimethylsilylpropargyl ether, **9,** 373
Quinoline, **9,** 403
ALLENYLSILANES
Dimethylphenylsilyllithium, **12,** 210

Organoaluminum reagents, **11,** 25
Organocopper reagents, **12,** 345
Trimethylsilylketene, **6,** 635
Trimethylsilylmethylmagnesium chloride, **10,** 433
1-Trimethylsilylpropynylcopper, **6,** 638
ARYLSILANES
Bis(trimethylsilyl)acetylene, **6,** 153; **7,** 84, 94; **8,** 146; **9,** 142
sec-Butyllithium, **12,** 97
Dicarbonylcyclopentadienylcobalt, **10,** 126
Diphenyl disulfide, **11,** 220
1-Methoxy-1-trimethylsilylallene, **11,** 577
3-Methoxy-1-trimethylsilyl-1-propyne, **6,** 154
Trimethylsilyllithium, **8,** 514
Trimethylsilylpotassium, **8,** 513
Trimethylsilylsodium, **8,** 514
CYCLOPROPYLSILANES
Bis(dichlorotrimethylsilylmethyl)-mercury, **2,** 30
(Chloromethyl)dimethylsilane, **12,** 27
Chloromethyltrimethylsilane, **8,** 307
Trimethylsilylcarbene, **4,** 544
Trimethylsilylmethylene dimethylsulfurane, **9,** 495
PROPARGYLSILANES
Trimethylsilylmethyl trifluoromethanesulfonate, **10,** 434

SILYL ENOL ETHERS
FROM RCHO, R_2CO
Bis(trimethylsilyl)formamide, **6,** 58
Bromo-*t*-butylmethoxyphenylsilane, **12,** 100
t-Butyldimethylchlorosilane, **7,** 210; **12,** 83
t-Butyldimethylsilyl trifluoromethanesulfonate, **12,** 86
Chlorodimethyl(2,4,6-tri-*t*-butylphenoxy)silane, **11,** 217
Chlorotrimethylsilane, **2,** 435; **7,** 429; **9,** 387; **10,** 97; **12,** 127
Dodecamethylcyclohexasilane, **8,** 218
Ethyl trimethylsilylacetate, **7,** 150; **8,** 226
Hexamethylphosphoric triamide, **6,** 273
Iodotrimethylsilane, **9,** 251
Iron, **12,** 263
Ketene alkyl trialkylsilyl acetals, **11,** 340
Lithium *t*-butyl-*t*-octylamide, **12,** 285
Lithium diisopropylamide, **8,** 292
Lithium morpholide, **12,** 284
Phenylthiotriethylsilane, **6,** 613
Potassium hydride, **5,** 557
N-(Trimethylsilyl)imidazole, **7,** 399
Trimethylsilyl nonaflate, **9,** 496
3-Trimethylsilyl-2-oxazolidone, **11,** 383
Trimethylsilyl trifluoromethanesulfonate, **7,** 390; **10,** 438
OTHER ROUTES
Butyllithium, **9,** 83
Chlorotrimethylsilane, **6,** 675
Chlorotris(triphenylphosphine)-rhodium(I)–Hydrosilanes, **4,** 562
Di-μ-carbonylhexacarbonyldicobalt, **12,** 163
Diethylaluminum cyanide–Chlorotrimethylsilane, **8,** 130
Diethylmethylsilane, **9,** 145
Dihydridotetrakis(triphenylphosphine)-ruthenium(II), **9,** 167
Magnesium bromide etherate, **7,** 218
Methylthiotrimethylsilane, **6,** 399
Organocopper reagents, **10,** 282
Trimethylsilyl trifluoromethanesulfonate, **12,** 543

SPIRO COMPOUNDS
SPIRO[2.n]ALKANES
Alumina, **2,** 18
Bis(1,5-cyclooctadiene)nickel(0), **5,** 34
Bromoform, **5,** 26
Cyclobutylidene, **12,** 150
1,4-Dibromo-2-pentene, **9,** 358
Dichloromethyllithium, **6,** 170
Diethylzinc–Bromoform, **12,** 74
Dimethyl diazomalonate, **8,** 187
Dimethylsulfoxonium methylide, **3,** 125
Ethylenediaminetetraacetic acid tetrasodium salt, **1,** 373
Hydrazine, **5,** 327
Methyllithium, **3,** 202
Simmons–Smith reagent, **2,** 371
SPIRO[3.n]ALKANES
m-Chloroperbenzoic acid, **5,** 120
Cyclobutylidene, **12,** 150
Cyclopropyl phenyl sulfide, **8,** 138
N,N-Dimethylaminocyclopropyl-phenyloxosulfonium tetrafluoroborate, **4,** 173
Dimethylsulfoxonium methylide, **3,** 125
Diphenylsulfonium cyclopropylide, **3,** 133; **4,** 211; **5,** 281, 416

SPIRO COMPOUNDS (*Continued*)
Ethoxyacetylene, **5,** 290
1-Lithiocyclopropyl phenyl sulfide, **5,** 372; **6,** 319
Simmons–Smith reagent, **6,** 521
Tetracyanoethylene, **1,** 1133; **5,** 647
p-Toluenesulfonylhydrazide, **4,** 511
Triethylamine, **2,** 427
SPIRO[4.n]ALKANES
Arene(tricarbonyl)chromium complexes, **9,** 21
Bis(1,5-cyclooctadiene)nickel(0), **5,** 34
Boron trifluoride etherate, **8,** 51
t-Butoxybis(dimethylamino)methane, **5,** 71
Chlorotris(triphenylphosphine)-rhodium(I), **10,** 98
Chromium carbonyl, **10,** 100
1,4-Dibromobutane, **5,** 404
Dichloromethyllithium, **6,** 170
Diethyl allylthiomethylphosphonate, **3,** 94
Diisobutylaluminum hydride, **7,** 111
N,N-Dimethylaminocyclopropyl-phenyloxosulfonium tetrafluoroborate, **4,** 173
Dimethyl sulfoxide, **9,** 189
Diphenylsulfonium cyclopropylide, **6,** 242; **7,** 140
Disodium tetracarbonylferrate, **10,** 174
Ethoxycarbonylcyclopropyl-triphenylphosphonium tetrafluoroborate, **6,** 93; **8,** 76
Ethylaluminum dichloride, **12,** 5
Hydrazine, **5,** 327
Hydrogen chloride, **7,** 172
1-Hydroxycyclopropane-carboxaldehyde, **12,** 247
2-Hydroxymethyl-3-trimethylsilyl-1-propene, **12,** 252
4-Iodobutyltrimethyltin, **10,** 219
2-Lithiobenzothiazole, **8,** 274
2-(E)-Lithiovinyltrimethylsilane, **10,** 443
Lithium chloride–Hexamethyl-phosphoric triamide, **10,** 240
Lithium iodide, **11,** 300
Mercury(II) acetate, **8,** 315; **11,** 315
Nitromethane, **4,** 357
Palladium(II) acetate, **11,** 389
N-Phenylselenophthalimide, **10,** 312
Polyphosphoric acid, **4,** 395
Tin(IV) chloride, **12,** 486
Titanium(IV) chloride, **12,** 494
SPIRO[5.n]ALKANES
2-Acetoxy-1-methoxy-3-trimethyl-silyloxy-1,3-butadiene, **11,** 2
Allylidenecyclopropane, **12,** 19
Allylidenetriphenylphosphorane, **5,** 7
Alumina, **2,** 18
Arene(tricarbonyl)chromium complexes, **9,** 21
Bis(N-propylsalicylideneaminate)-copper(II), **5,** 41
Chromium carbonyl, **10,** 100
Copper(I) chloride, **8,** 118; **9,** 123
Diethyl allylthiomethylphosphonate, **3,** 94
Diethyl lithiomorpholinomethyl-phosphonate, **7,** 105
Dimethyl(methylthio)sulfonium tetrafluoroborate, **11,** 204
Dimethyl sulfoxide, **9,** 189
Ferric chloride, **4,** 236
Hexafluoroantimonic acid, **8,** 239
Iron carbonyl, **12,** 266
Isoprene, **5,** 630
Isopropylidene isopropylidenemalonate, **6,** 309
Lithium dimethylcuprate, **6,** 209; **7,** 120
4-Methoxy-5-methyl-*o*-benzoquinone, **6,** 363
trans-1-Methoxy-3-trimethylsilyloxy-1,3-butadiene, **9,** 303
N-Methyl-N-phenylaminoacetylene, **10,** 266
Methyl vinyl ketone, **7,** 247; **12,** 329
Potassium ferricyanide, **4,** 406
Pyrrolidine, **7,** 309
Samarium(II) iodide, **11,** 464
Tetrakis(2-methylpropyl)-μ-oxodialuminum, **8,** 471
Thallium(III) trifluoroacetate, **4,** 498
Tin(IV) chloride, **11,** 522
Trifluoroacetic acid, **6,** 613
α-Trimethylsilyl-2-methylbenzothiazole, **8,** 511
Vanadyl trichloride, **3,** 331; **8,** 527
SPIRO[6.n]ALKANES
Organocopper reagents, **9,** 328
SULFIDES (*see also* ALDEHYDES OR KETONES—α-SUBSTITUTED, VINYL SULFIDES)

BY CONJUGATE ADDITION OF RSH
Hydrogen sulfide, **1,** 1026
Ion-exchange resins, **8,** 263
Tetrabutylammonium fluoride, **7,** 353
Thiophenol, **8,** 431
DIARYL SULFIDES
Copper(I) oxide, **1,** 169
1,3-Dimethyl-2-imidazolidinone, **11,** 202
Diphenyl disulfide, **8,** 210
Lithium thiophenoxide, **4,** 311
Sulfur monochloride, **1,** 1122
Tetraphenylbismuth trifluoroacetate, **10,** 393
p-Toluenesulfenyl chloride, **6,** 510
BY DISPLACEMENTS USING S NUCLEOPHILES
Butyllithium, **5,** 80
Cesium carbonate, **12,** 108
Dimethylformamide dimethyl acetal, **4,** 184
N,N-Methylphenylaminotriphenylphosphonium iodide, **6,** 392
N-Phenylthiosuccinimide, **8,** 399
Sodium borohydride, sulfurated, **6,** 534
Sodium methanethiolate, **8,** 19
Sodium sulfide, **1,** 1104; **5,** 623; **6,** 271
Sodium thiophenoxide, **6,** 272
Tricaprylylmethylammonium chloride, **6,** 404
Trimethylselenonium hydroxide, **9,** 491
Triphenylbis(2,2,2-trifluoroethoxy)-phosphorane, **10,** 43
Triphenylphosphine–Diethyl azodicarboxylate, **7,** 404
Zinc iodide, **12,** 576
BY REDUCTION OF RSSR′, RSOR′, RSO_2R' (*see* TYPE OF REACTION INDEX)
OTHER ROUTES
Borane–Pyridine, **11,** 69
1-Butanethiol, **4,** 458
Chloromethyl phenyl sulfide, **6,** 510
N-Chlorosuccinimide–Dimethyl sulfide, **5,** 129
Dimethyl disulfide, **9,** 91
Dimethyl sulfoxide + co-reagent, **1,** 306; **12,** 213
S,S-Dimethyl-N-*p*-toluenesulfonylsulfilimine, **9,** 193
2,4-Dinitrobenzenesulfenyl chloride, **1,** 319
Ethanethiol, **2,** 209
Formamidinesulfinic acid, **6,** 586
Hexamethyldisilazane, **7,** 167
Lead tetraacetate–Trifluoroacetic acid, **6,** 317
Sodium ethanethiolate, **11,** 492
Sulfur, **1,** 1118
Tributylphosphine–Dialkyl(or aryl) disulfides, **6,** 602; **9,** 200
Trifluoromethylthiocopper, **6,** 621
SULFILIMINES
m-Chloroperbenzoic acid, **11,** 122
N-Chlorosuccinimide–Dialkyl sulfides, **6,** 118
Dimethyl sulfoxide + co-reagent, **5,** 266; **6,** 230; **9,** 192; **10,** 166
Diphenyldi(1,1,1,3,3,3-hexafluoro-2-phenyl-2-propoxy)sulfurane, **6,** 239
O-Mesitylenesulfonylhydroxylamine, **5,** 430
Sulfuric acid, **5,** 633
SULFINES
m-Chloroperbenzoic acid, **5,** 120
Ozone, **5,** 491
Sulfur dioxide, **8,** 464
SULFINIC ACIDS AND ESTERS
m-Chloroperbenzoic acid, **5,** 120
N-Hydroxymethylphthalimide, **8,** 253
Lead tetraacetate, **1,** 537
Lithium–Ethylamine, **1,** 574
Trifluoroacetic acid, **7,** 388
Zinc, **1,** 1276
SULFONAMIDES
Benzenesulfonyl chloride, **1,** 46
N-Benzenesulfonylformimidic acid ethyl ester, **3,** 18
Mercury(II) nitrate, **11,** 317
4-Methoxy-2,3,6-trimethylbenzenesulfonyl chloride, **11,** 330
N-Methyl-N-tolylsulfonylpyrrolidinium perchlorate, **4,** 345
Phase-transfer catalysts, **11,** 403
p-Phenylazobenzenesulfonyl chloride, **1,** 832
N-Phenyltrifluoromethanesulfonimide, **5,** 530
Titanium(III) chloride, **8,** 482
p-Toluenesulfonyl chloride, **1,** 1179, 1184; **5,** 676
Trifluoromethanesulfonic anhydride, **5,** 702

SULFONES
BY FRIEDEL–CRAFTS SULFONYLATION
Arenesulfonyl halide–Antimony(V) halide complexes, **5,** 20
Benzenesulfonic anhydride, **1,** 46
Methanesulfonic anhydride, **1,** 667
Silver(I) trifluoromethanesulfonate, **6,** 520
p-Toluenesulfonyl perchlorate, **1,** 1187
Trifluoromethanesulfonic–alkanesulfonic anhydrides, **6,** 618
BY OXIDATION OF SULFIDES, SULFOXIDES (*see* TYPE OF REACTION INDEX)
OTHER ROUTES
Benzenesulfonyl chloride, **1,** 46
Butyllithium, **7,** 45
Dimethyl sulfate, **1,** 295
Dimethyl sulfone, **1,** 296; **2,** 157
Hexamethylphosphorous triamide, **9,** 235
Palladium(II) chloride, **8,** 384; **11,** 393
Polyphosphoric acid, **2,** 334
Potassium trifluoromethanesulfinate, **5,** 564
Sulfur dioxide, **11,** 495
Sulfuric acid–Trifluoroacetic anhydride, **10,** 375
Tetrabutylammonium *p*-toluenesulfinate, **6,** 567
Thallium(III) trifluoroacetate, **9,** 462
p-Toluenesulfonyl chloride, **1,** 1179
SULFONIC ACIDS
Barium permanganate, **1,** 46
Bis(trimethylsilyl) sulfate, **10,** 176
N,N-Dimethylthiocarbamoyl chloride, **3,** 127
Hexamethylphosphoric triamide, **1,** 430
Hydrogen peroxide, **1,** 457
Potassium peroxomonosulfate, **1,** 952
Sodium hydrogen sulfite, **1,** 1047
Sulfur trioxide, **1,** 1125, 1126; **2,** 393; **4,** 473
Thiolacetic acid, **1,** 1154
SULFONIC ACID ESTERS
BENZENESULFONATES
Benzenesulfonic anhydride, **1,** 46
Benzenesulfonyl chloride, **1,** 46
Phase-transfer catalysts, **9,** 356
MESITYLENESULFONATES
2-Mesitylenesulfonyl chloride, **5,** 430
METHANESULFONATES
Diethylmethyl(methylsulfonyl)-ammonium fluorosulfonate, **6,** 190
Methanesulfonic acid, **4,** 326
Methanesulfonic anhydride, **1,** 667
Methanesulfonyl chloride, **1,** 17, 662; **4,** 326; **5,** 435; **6,** 362
Phase-transfer catalysts, **9,** 356
METHYL ALKANESULFONATES
Diazomethane, **1,** 191
Methyl fluorosulfonate, **4,** 339
NONAFLUOROBUTANE-SULFONATES
Nonafluorobutanesulfonyl fluoride, **5,** 479
p-TOLUENESULFONATES
m-Chloroperbenzoic acid, **3,** 49
Silver *p*-toluenesulfonate, **1,** 1018
Thallium(I) ethoxide, **2,** 407
p-Toluenesulfinylimidazolide, **11,** 535
p-Toluenesulfonyl chloride, **1,** 958, 1079, 1179; **3,** 292; **5,** 676
Zinc *p*-toluenesulfonate, **11,** 605
TRIFLUOROMETHANE-SULFONATES
N-Phenyltrifluoromethanesulfonimide, **5,** 530
Trifluoromethanesulfonic anhydride, **5,** 702; **6,** 618
Trifluoromethanesulfonylimidazole, **5,** 705
VINYL TRIFLUOROMETHANE-SULFONATES
2,6-Di-*t*-butyl-4-methylpyridine, **10,** 123
N-Phenyltrifluoromethanesulfonimide, **12,** 395
Trifluoromethanesulfonic acid, **4,** 532
Trifluoromethanesulfonic anhydride, **5,** 702; **6,** 618
Trifluoromethanesulfonylimidazole, **6,** 620
SULFONYL HALIDES
Chlorine, **7,** 58
Dichloromethyl methyl ether, **1,** 220
Hydrofluoric acid, **6,** 284
Phosphorus(V) chloride, **1,** 866, 870; **6,** 497
Phosphoryl chloride, **11,** 429
Sulfur dioxide–Copper(I) chloride, **11,** 495

Sulfuryl chloride, **5,** 641
Thionyl chloride, **1,** 1158; **4,** 503; **6,** 497

SULFOXIDES (*see also* CHIRAL COMPOUNDS)
BY OXIDATION OF SULFIDES (*see* TYPE OF REACTION INDEX)
BY REDUCTION OF SULFONES (*see* TYPE OF REACTION INDEX)
OTHER ROUTES
1-Lithiocyclopropyl phenyl sulfide, **9,** 271
Sodium methylsulfinylmethylide, **1,** 310

SULFOXIMINES
Dimethyl sulfoxide, **7,** 135; **10,** 166
O-Mesitylenesulfonylhydroxylamine, **5,** 430
Ruthenium(IV) oxide–Sodium periodate, **8,** 438

THIOACETALS AND KETALS
GENERAL METHODS
Aluminum chloride, **11,** 25
Glyoxylic acid, **7,** 162
Sulfur dioxide, **11,** 495
Titanium(IV) chloride, **12,** 494
1,3-DITHIANES
Diisobutylaluminum hydride, **11,** 185
1,3-Dithienium tetrafluoroborate, **11,** 227
1,3-Propanedithiol, **1,** 956; **4,** 413; **7,** 368
Phase-transfer catalysts, **11,** 403
Triethylsilane–Trifluoroacetic acid, **9,** 309
Trimethylene thiotosylate, **1,** 956; **4,** 539; **6,** 242, 628
1,3-DITHIOLANES
Bis(diisobutylaluminum) 1,2-ethanedithiolate, **12,** 342
Boron trifluoride etherate, **1,** 70
L(+)-Butane-2,3-dithiol, **1,** 82
Diethylene tetrathioorthocarbonate, **6,** 187
1,2-Ethanedithiol, **1,** 356, 1174; **2,** 433; **5,** 290
5,5′-Ethylene *p*-toluenethiosulfonate, **4,** 540; **6,** 628
Nafion-H, **10,** 275
Perchloric acid, **1,** 796
Phase-transfer catalysts, **11,** 403
Zinc trifluoromethanesulfonate, **12,** 577
1,3-OXATHIOLANES
Chlorotrimethylsilane, **6,** 626
2-Mercaptoethanol, **1,** 356, 643; **5,** 718; **6,** 381
Organolithium reagents, **9,** 5
OTHER THIOKETALS
Aluminum chloride–Ethanethiol, **11,** 28
Boron trifluoride–Ethanethiol, **10,** 51
Chlorotrimethylsilane, **7,** 66
N,N-Dimethylhydrazine, **7,** 126
Methylthiotrimethylsilane, **6,** 399; **8,** 352
Molybdenum carbonyl, **7,** 165
Phenylthiotrimethylsilane, **6,** 399
Tributylphosphine–Diphenyl disulfide, **9,** 200
Tricaprylylmethylammonium chloride, **6,** 404
Trimethylsilyl trifluoromethanesulfonate, **11,** 584
Tris(methylthio)methyllithium, **10,** 453

THIOALDEHYDES AND KETONES
2,4-Bis(4-methoxyphenyl)-1,3-dithia-2,4-diphosphetane-2,4-disulfide, **8,** 327; **9,** 49
Bis(tricyclohexyltin) sulfide–Boron trichloride, **11,** 63
Carbon disulfide, **5,** 94
Phosphorus(V) sulfide, **3,** 226; **5,** 653; **10,** 320
Sodium hydrogen sulfide, **5,** 251
Sulfur monochloride, **9,** 442; **11,** 495

THIOAMIDES
Bis(benzonitrile)dichloropalladium(II), **10,** 31
2,4-Bis(4-methoxyphenyl)-1,3-dithia-2,4-diphosphetane-2,4-disulfide, **9,** 49
Bis(trimethylsilyl)thioketene, **7,** 28
N,N-Dimethylformamide, **1,** 278
Dioxane, **1,** 333
Diphenylphosphinodithioic acid, **10,** 171
Ethoxycarbonyl isothiocyanate, **7,** 146
Phosphorus(V) sulfide, **9,** 374; **11,** 428
Sodium hydrogen sulfide, **11,** 487
Sulfur, **1,** 1118; **3,** 273
Thioacetamide, **4,** 502
Triphenylphosphine–Thiocyanogen, **9,** 507

THIOCYANATES
Cyanogen bromide, **1,** 174
Cyanotrimethylsilane, **9,** 127
Diethyl phosphorocyanidate, **11,** 181
Ferric thiocyanate, **6,** 259; **7,** 155

THIOCYANATES (*Continued*)
Mercury(II) thiocyanate, **9,** 293
Palladium(II) acetate, **5,** 496
Potassium thiocyanate, **1,** 954; **8,** 390
Sodium thiocyanate, **1,** 1105
Thiocyanogen, **1,** 1152
Thiocyanogen chloride, **1,** 1153; **5,** 661
Triphenylphosphine–Thiocyanogen, **8,** 518
2,4,6-Triphenylpyrylium thiocyanate, **8,** 521
Vinylsulfonyl chloride–Trimethylamine, **9,** 211

THIOLS
BY DISPLACEMENTS WITH S NUCLEOPHILES
1-Acetyl-2-thiourea, **4,** 7
Potassium ethylxanthate, **5,** 554
Potassium hydrogen sulfide, **1,** 935
Potassium thiolacetate, **1,** 955
Sodium amide, **1,** 1034
Sodium borohydride, sulfurated, **6,** 534
Sodium trithiocarbonate, **2,** 389; **11,** 493
Thiophenol, **6,** 279
Thiourea, **1,** 1164; **2,** 157
BY REDUCTION OF SULFIDES, DISULFIDES (*see* TYPE OF REACTION INDEX)
OTHER ROUTES
Butyllithium, **10,** 68
Ethylene sulfide, **1,** 378
Phosphorus(V) sulfide, **3,** 226; **11,** 428
Sodium sulfide–Sulfur, **9,** 434
Thiolacetic acid, **1,** 1154
Thiourea, **11,** 519

THIOL ESTERS
FROM RCOOH
2,4-Bis(methylthio)-2,4-dithioxo-cyclodiphosphathiane, **11,** 56
N,N′-Carbonyldiimidazole, **8,** 77
2-Chloro-3-methyl-1,3-benzothiazolium trifluoromethanesulfonate, **7,** 61
Dicyclohexylcarbodiimide, **7,** 100
Diethyl phosphorochloridate, **6,** 192
Diethyl phosphorocyanidate, **6,** 192
4-Dimethylamino-3-butyn-2-one, **9,** 177
4-Dimethylaminopyridine, **10,** 155
Diphenyl 2-keto-3-oxazolinyl-phosphonate, **11,** 220
Diphenyl phosphoroazidate, **11,** 222
Ethyl chlorothiolformate, **6,** 252
2-Halopyridinium salts, **9,** 234
Lithium thiophenoxide, **8,** 183
Polyphosphate ester, **11,** 430
Sodium borohydride, **11,** 477
2-Thiopyridyl chloroformate, **9,** 466
Triphenylphosphine–2,2′-Bis(3-cyano-4,6-dimethylpyridyl) disulfide, **9,** 167
Tris(ethylthio)borane, **8,** 522
OTHER ROUTES
Aluminum thiophenoxide, **9,** 15
Bis(methylthio)(trimethylsilyl)-methyllithium, **6,** 53
Carbon disulfide, **5,** 94
Carbonyl sulfide, **11,** 112
Cerium(IV) ammonium nitrate, **6,** 197
Chlorodiphenylphosphine, **11,** 120
Disodium tetracarbonylferrate, **9,** 205
Methoxy(phenylthio)trimethylsilyl-methyllithium, **12,** 317
1,3,4,6-Tetraacetylglycouril, **6,** 563
Tetrafluoroboric acid, **12,** 465
Thiolacetic acid, **1,** 1154; **5,** 326
2,4,6-Trichlorobenzoyl chloride, **11,** 552
Triethyl phosphite, **1,** 1212
Triphenylphosphine–Diethyl azodicarboxylate, **11,** 589
Tris(methylthio)methyllithium, **7,** 280

THIONOESTERS AND LACTONES
2,4-Bis(4-methoxyphenyl)-1,3-dithia-2,4-diphosphetane-2,4-disulfide, **8,** 327; **9,** 49
2,4-Bis(methylthio)-2,4-dithioxocyclodi-phosphathiane, **11,** 56
Bis(tricyclohexyltin) sulfide–Boron trichloride, **11,** 63
Dicyclohexylcarbodiimide–4-Dimethyl-aminopyridine, **9,** 156
Dimethylaluminum benzenethiolate, **5,** 36
Hydrogen sulfide, **12,** 247
Sodium hydrogen sulfide, **9,** 429
Thiobenzoyl chloride, **6,** 582
Triethyloxonium tetrafluoroborate, **10,** 417

THIOPHENOLS
Amberlyst ion-exchange resin, **5,** 355
Benzenesulfonyl chloride, **1,** 46
N,N-Dimethylthiocarbamoyl chloride, **2,** 173; **4,** 202
Hydrogen sulfide, **4,** 256
Sodium disulfide, **1,** 1064

Sodium, **11,** 473
Tetraisopropylthiuram disulfide, **12,** 466
Trifluoroacetic anhydride, **12,** 530
Triphenylphosphine, **5,** 725; **10,** 220
Zinc, **1,** 1276
Zinc amalgam, **1,** 1287

THIOUREAS
Benzoylisothiocyanate, **1,** 50
Dichloromethylenedimethylammonium chloride, **4,** 135
Phosphorus(V) sulfide, **5,** 534
Sodium thiocyanate, **5,** 626
Triphenylphosphine–Thiocyanogen, **8,** 518

TOSYLATES, TRIFLATES
(*see* SULFONIC ACID ESTERS)

TROPOLONES
Benzocyclopropene, **6,** 33
Copper(II) acetylacetonate, **3,** 62
Dichloroketene, **1,** 221
Dimethylsulfoxonium methylide, **8,** 194
Ethyl diazoacetate, **1,** 367
Pyridine, **2,** 349
Tri-μ-carbonylhexacarbonyldiiron, **5,** 221; **6,** 195
Zinc–Acetic acid, **5,** 757

TROPONES
Alumina, **4,** 8
Dimethyl sulfoxide, **9,** 189
Lithium carbonate, **3,** 183
Lithium dibutylcuprate, **5,** 187
Selenium(IV) oxide, **1,** 992
Silver tetrafluoroborate, **2,** 365
Sodium chlorodifluoroacetate, **5,** 603
Tributyltin hydride, **12,** 516
Tri-μ-carbonylhexacarbonyldiiron, **1,** 259; **4,** 157; **5,** 221; **6,** 195; **8,** 498
Zinc–Silver couple, **9,** 519

UNSATURATED ACID HALIDES
Boron trichloride, **3,** 31
Oxalyl bromide (or chloride), **1,** 767
Thionyl chloride, **3,** 290
Triphenylphosphine dichloride, **5,** 730

UNSATURATED ALDEHYDES AND KETONES (*see also* ACETYLENIC CARBONYLS, ALICYCLIC HYDROCARBONS—2-CYCLOPENTENONES, 2-CYCLOHEXENONES, ALLENIC CARBONYLS, CHIRAL COMPOUNDS, DIENALS, DIENONES, ENEDIONES, α-METHYLENE ALDEHYDES AND KETONES)

α,β-UNSATURATED ALDEHYDES AND KETONES
by Acylation of C=C or vinyl nucleophiles
Alumina, **11,** 22
Aluminum chloride, **3,** 7; **7,** 7
Dichloromethyl methyl ether, **8,** 485; **9,** 154
Dihalobis(triphenylphosphine)-palladium(II), **6,** 60
Dimethylformamide, **5,** 252; **7,** 124
3-Methoxy-3-methylbutynylcopper, **9,** 300
Simmons–Smith reagent, **12,** 437
Tetrakis(triphenylphosphine)-palladium(0), **12,** 468
Tributylvinyltin, **9,** 42
Trifluoroacetic anhydride, **1,** 1221
2,4,6-Triisopropylbenzenesulfonyl-hydrazide, **9,** 486
Vilsmeier reagent, **9,** 514
Vinyllithium, **5,** 748
by Aldol reaction, dehydration
Aluminum *t*-butoxide, **1,** 23
Bis(*p*-methoxyphenyl) telluroxide, **11,** 55
Boric acid, **2,** 32
Calcium carbide, **3,** 40
Cesium fluoride, **10,** 81
Diethyl oxalate, **7,** 106
Formaldehyde, **9,** 224
Lithium diisopropylamide, **2,** 249; **3,** 184
Lithium iodide, **10,** 245
Nickel carbonyl, **10,** 276
Potassium carbonate, **9,** 382
Sodium bicarbonate, **5,** 595
2,4,6-Trimethylphenoxymagnesium bromide, **9,** 489
Trimethylsilylmethylmagnesium chloride, **12,** 540
Triphenylmethyllithium, **1,** 1256
by Alkylation of β-acyl vinyl anion equivalents
S-Allyl N,N-dimethyldithiocarbamate, **6,** 11
1,3-Bis(methylthio)allyllithium, **4,** 38

UNSATURATED ALDEHYDES AND KETONES (*Continued*)
t-Butoxyallene, **8,** 322
Butyllithium, **10,** 68
Lithium diisopropylamide, **6,** 334
Methoxyallene, **8,** 320
3-Methoxy-1-phenylthio-1-propene, **8,** 327
3-Methylthio-2-propenyl *p*-tolyl sulfone, **12,** 327
2-(2-Phenylsulfonylethyl)-1,3-dioxolane, **6,** 461
Sodium N,N-dimethyldithiocarbamate, **6,** 539
from C=O ($R_2CO \rightarrow R_2C=CH(R)CHO$)
Acetaldehyde N-*t*-butylimine, **9,** 1
2,2-Diethoxyethylidenetriphenylphosphorane, **9,** 157
Diethyl β-(cyclohexylimino)ethylphosphonate, **2,** 131
N,N-Dimethylhydrazine, **7,** 126
N,N-Dimethylthiocarbonyl chloride, **8,** 200
1,3-Dioxolan-2-ylmethyltriphenylphosphonium bromide, **5,** 269
Ethoxyacetylene, **1,** 357
1-Ethoxy-1-propene, **1,** 385
(Z)-2-Ethoxyvinyllithium, **8,** 221
Ethyl vinyl ether, **1,** 386; **10,** 184
Formylmethylenetriphenylphosphorane, **6,** 267; **8,** 234
Lithio propylidene-*t*-butylimine, **4,** 283
Lithium ethoxyacetylide, **1,** 613
Oxotris(triphenylsilanolato)vanadium, **6,** 655
Pyridinium bromide perbromide, **1,** 967
2,4,4,6-Tetramethyl-5,6-dihydro-1,3-(4*H*)-oxazine, **3,** 280
(Z)-(Trimethylsilyloxy)vinyllithium, **11,** 582
Vinyltrimethylsilane, **12,** 566
from C=O ($R_2CO \rightarrow R_2C=CHCOR$)
Acetylmethylene(triphenyl)arsorane, **11,** 5
1,3-Dithiane, **3,** 135
Propargyl bromide, **10,** 332
Titanium(IV) chloride, **12,** 494
Triethyl phosphonoacetate, **12,** 527
from C=O [$R'COCH_2R \rightarrow RCH=CR'COH(R)$]
Chloromethyl phenyl sulfoxide, **8,** 94
1,3-Dithiane, **2,** 182
Formic acid, **2,** 202; **5,** 316
Ion-exchange resins, **1,** 511
2-Lithiobenzothiazole, **8,** 274
Mercury(II) sulfate, **1,** 658
Nafion-H, **11,** 354
Oxygen, **7,** 258; **8,** 367
Phosphorus(V) oxide, **1,** 871
Sodium methoxide, **8,** 461
Thiophenol, **12,** 485
Titanium(III) chloride, **5,** 669
Tosylmethyl isocyanide, **12,** 511
by Dehydration of aldols
t-Butyl acetoacetate, **1,** 83
t-Butyllithium, **11,** 103
Dicyclohexylcarbodiimide, **2,** 126
Diisobutylaluminum phenoxide, **9,** 172
Florisil, **1,** 393
Girard reagent T, **1,** 410
Iodine, **1,** 495
Pyridine, **4,** 414
Silver perchlorate, **9,** 413
Tetrafluoropyrrolidine, **9,** 449
p-Toluenesulfonic acid, **1,** 1172
by Dehydrogenation of C=O (*see* TYPE OF REACTION INDEX)
by Dehydrohalogenation
Benzyltrimethylammonium mesitoate, **1,** 54
Calcium carbonate, **1,** 103; **5,** 89
Copper(I) bromide–Lithium carbonate, **6,** 18
1,5-Diazabicyclo[4.3.0]nonene-5, **4,** 116
1,8-Diazabicyclo[5.4.0]undecene-7, **4,** 119
Dichloromethyllithium, **5,** 199
N,N-Dimethylhydrazine, **8,** 192
Dimethyl sulfoxide, **1,** 296; **4,** 192
2,4-Dinitrophenylhydrazine, **1,** 330
Hexamethylphosphoric triamide, **2,** 208
Lithium–Ammonia, **1,** 54
Lithium bromide, **3,** 75
Lithium carbonate, **1,** 606; **4,** 298; **5,** 452; **6,** 17
Lithium chloride, **1,** 609

Magnesium oxide, **2,** 256
Nitrosyl fluoride, **2,** 299
N-Phenylbenzamidine, **5,** 513
Potassium fluoride, **5,** 153
Silver(I) nitrate, **1,** 1008
Sodium carbonate, **7,** 332
Tetraethylammonium chloride, **1,** 1137
Tetrakis(triphenylphosphine)-palladium(0), **6,** 571
Tetramethylammonium dimethyl phosphate, **6,** 573
Tributylamine, **1,** 1189
by Deoxygenation of keto epoxides (*see* TYPE OF REACTION INDEX)
from β-Dicarbonyls
1-Butanethiol, **2,** 53
2-Cyclohexenone, **1,** 180
Lithium aluminum hydride, **1,** 581
Sodium borohydride, **6,** 530
N-(Trimethylsilyl)imidazole, **11,** 575
by Elimination of Se compounds
Benzeneselenenyl halides, **5,** 518; **6,** 459; **7,** 286; **8,** 25; **10,** 16
Benzeneseleninic anhydride, **8,** 29
Diphenyl diselenide, **5,** 272
2-Pyridineselenenyl bromide, **11,** 455
by Elimination of S compounds
m-Chloroperbenzoic acid, **10,** 92
Dimethyl disulfide, **5,** 246
Diphenyl disulfide, **6,** 235
2,2′-Dipyridyl disulfide, **8,** 214
Phenylsulfinylacetone, **5,** 524
Tetrakis(triphenylphosphine)-palladium(0), **9,** 451
by Oxidation of allylic OH, X, CH_2 (*see* TYPE OF REACTION INDEX)
by Oxidation of allyl Pd complexes
Chromium(VI) oxide–Dimethylformamide, **11,** 131
Palladium(II) chloride, **11,** 393
by Oxidation of allyltrimethylsilanes, homoallylic alcohols, silyl enol ethers (*see* TYPE OF REACTION INDEX)
by Oxidative decarboxylation of keto acids, keto esters
Lead(IV) oxide, **1,** 533; **2,** 233
Lead tetraacetate, **5,** 365
Palladium(II) acetate–1,2-Bis-(diphenylphosphine)ethane, **11,** 391
Other routes
1-(Alkyl or Phenylthio)vinyllithium, **5,** 6; **9,** 371
Benzeneselenenyl chloride, **9,** 25
Benzenesulfenyl chloride, **6,** 30
1,3-Benzodithiolylium tetrafluoroborate, **10,** 24
Benzylchlorobis(triphenylphosphine)-palladium(II), **9,** 41
Bis(dimethylaluminum) 1,3-propanedithiolate, **6,** 49
Bis(trimethylsilyl)acetylene, **5,** 44
1-Bromo-2-ethoxycyclopropyllithium, **9,** 67
Butyllithium, **4,** 60; **5,** 80
trans-4-(*t*-Butylthio)-3-buten-2-one, **9,** 90
3-Butyn-2-one, **3,** 38
Carbonylphenylbis(triphenylphosphine)rhodium(I), **5,** 45
Chlorobis(cyclopentadienyl)-hydridozirconium(IV), **9,** 104
α-Chloro-N-cyclohexylpropanal-donitrone, **5,** 110
(3-Chloro-3-methyl-1-butynyl)lithium, **8,** 93
Chloro(trimethylsilyl)methyllithium, **8,** 277
Chlorotris(triphenylphosphine)-rhodium(I), **1,** 1252
Copper(I) bromide–Dimethyl sulfide, **8,** 117
Cyanotrimethylsilane, **11,** 147
1,5-Diazabicyclo[5.4.0]undecene-5, **11,** 155
Di-μ-bromobis(2-methoxyallyl)-dinickel, **5,** 437
2,3-Dichloro-5,6-dicyano-1,4-benzoquinone, **12,** 174
(Z)-1,2-Dichloro-4-phenylthio-2-butene, **7,** 98
3,3-Diethoxy-1-propyne, **7,** 80
Dimethylformamide diethyl acetal, **9,** 182
2-(1,3-Dioxan-2-ylethylidene)-triphenylphosphorane, **9,** 196
Disodium tetracarbonylferrate, **9,** 205
Disodium tetrachloropalladate(II), **9,** 207
Ethylidenetriphenylphosphorane, **8,** 339
Grignard reagents, **10,** 189

UNSATURATED ALDEHYDES AND KETONES (*Continued*)
Hydridodinitrogentris(triphenylphosphine)cobalt(I), **5,** 331
Isopropylidenetriphenylphosphorane, **8,** 339
1-Lithiocyclopropyl phenyl sulfide, **11,** 284
2-Lithio-2-trimethylsilyl-1,3-dithiane, **6,** 320
Lithium bromide, **5,** 395
Lithium diisopropylamide, **7,** 204; **10,** 241
Lithium dimethylcuprate, **5,** 234
Magnesium bromide etherate, **1,** 629
Manganese(II) chloride, **9,** 288
3-Methoxyallylidenetriphenylphosphorane, **8,** 323
Methyllithium, **6,** 384
Organolithium reagents, **12,** 350
Oxalic acid, **1,** 764
Palladium(II) acetate, **10,** 297
Perchloric acid, **1,** 796
Phenylcopper, **7,** 282
N-Phenylketeniminyl(triphenyl)phosphorane, **12,** 387
Potassium fluoride, **2,** 346
Potassium hydride, **9,** 386
Selenium(IV) oxide, **9,** 409
Silver(I) oxide, **1,** 1011
Silver perchlorate, **10,** 354
Tetraethylthiuram disulfide, **6,** 569
Thexylborane, **5,** 232
Titanium(IV) isopropoxide, **12,** 504
3-Triethylsilyloxypentadienyllithium, **11,** 556
Trimethyl phosphite, **1,** 1233
3-Trimethylsilyl-1-cyclopentene, **8,** 509
Zinc halides, **9,** 520; **12,** 574

β,γ-UNSATURATED ALDEHYDES AND KETONES
by Acylation
Acetyl chloride, **11,** 11
Acetyl hexachloroantimonate, **8,** 5
Bis(1,5-cyclooctadiene)nickel(0), **10,** 33
Lithium di(1-ethoxyvinyl)cuprate, **6,** 205
Simmons–Smith reagent, **12,** 437
Zinc–Silver couple, **10,** 460
by Deconjugation of α,β-unsaturated carbonyls
Potassium *t*-butoxide, **1,** 911; **5,** 544
Pyridine, **12,** 416
by Oxidation of homoallylic alcohols (*see* TYPE OF REACTION INDEX)
by 2,3-Sigmatropic rearrangement
Lithium diisopropylamide, **5,** 400
Phase-transfer catalysts, **8,** 387
Trimethylsilylmethanethiol, **11,** 576
Other routes
Aluminum chloride, **7,** 7
t-Butyldimethylchlorosilane, **11,** 88
α-Chloro-N-cyclohexylpropanaldonitrone, **5,** 110
Chlorotrimethylsilane, **8,** 107
Chromium(II) acetate, **5,** 143
Dichlorobis(tri-*o*-tolylphosphine)palladium(II), **12,** 173
Diethyl phosphite, **12,** 187
Diethyl [(2-tetrahydropyranyloxy)methyl]phosphonate, **12,** 188
2-(1,3-Dioxan-2-ylethylidene)triphenylphosphorane, **9,** 196
1,3-Dithianes, **4,** 216; **8,** 346
1-Lithio-2-methoxycyclopropane, **6,** 366
Lithium–Ammonia, **1,** 601
Lithium dimethylcuprate, **8,** 301
Magnesium bromide etherate, **1,** 629
Manganese(II) chloride, **9,** 288
Mercury(II) acetate, **5,** 424
Methyllithium, **6,** 384
Methyl α-phenylglycinate, **8,** 395; **10,** 308
Oxotris(triphenylsilanolato)vanadium, **7,** 413
Phenylselenoacetaldehyde, **10,** 310
Phenylthiomethyllithium, **4,** 379
Phenyl vinyl sulfoxide, **6,** 468
Pyridinium chlorochromate, **9,** 397
Sodium–Ammonia, **5,** 589
Tetrakis(triphenylphosphine)palladium(0), **9,** 451
Titanium(IV) chloride, **6,** 590
Trialkylaluminums, **12,** 512

γ,δ-UNSATURATED ALDEHYDES AND KETONES
by Alkylation of C=O with allylic electrophiles
Bis(dibenzylideneacetone)palladium(0), **11,** 53

Diethyl lithio-N-benzylideneamino-methylphosphonate, **9,** 161
Lithium–Ammonia, **8,** 282
Potassium carbonate, **10,** 323
Potassium hydride, **8,** 412
Sodium amide, **1,** 1034
Tetrakis(triphenylphosphine)-palladium(0), **10,** 384; **11,** 503
Triethylborane, **9,** 482
by Claisen and related rearrangements
(E)-(Carboxyvinyl)trimethyl-ammonium betaine, **12,** 106
Dichlorotris(triphenylphosphine)-ruthenium(II), **8,** 159
Diethyl allylthiomethylphosphonate, **3,** 94
Diethylaluminum benzenethiolate, **12,** 343
Ethyl vinyl ether, **4,** 234
Lithium(or Sodium) methylsulfinyl-methylide, **12,** 283, 451
Mercury(II) acetate, **1,** 644
3-Methoxyisoprene, **4,** 330
2-Methoxypropene, **2,** 230
Organotitanium reagents, **11,** 52
Potassium *t*-butoxide, **4,** 399
Potassium hydride, **11,** 435
Titanium(IV) chloride, **9,** 468
Trialkylaluminums, **11,** 539
Triethylorthoacetate, **3,** 300
1-Vinylthioallyllithium, **5,** 749
by Conjugate addition of vinyl anions
9-Borabicyclo[3.3.1]nonane, **7,** 29
Chlorobis(cyclopentadienyl)-hydridozirconium(IV), **8,** 84
Copper(I) bromide–Dimethyl sulfide, **8,** 117
Diisobutylaluminum hydride, **5,** 224
Lithium dimethylcuprate, **5,** 234
Lithium di[(E)-1-propenyl]cuprate, **8,** 302
Lithium divinylcuprate, **4,** 219, 221
Lithium methyl(vinyl)cuprate, **6,** 342
Palladium(II) acetate, **4,** 365
3-Tetrahydropyranyloxy-1-tributylstannyl-1-propene, **6,** 602
2,4,6-Triisopropylbenzene-sulfonylhydrazide, **10,** 422
Vinylmagnesium bromide–Methylcopper, **8,** 237
Other routes
Allyl chloroformate, **12,** 15
1-Bromo-1-trimethylsilyl-1(Z),4-pentadiene, **11,** 80
Chlorobis(cyclopentadienyl)-hydridozirconium(IV), **7,** 101
(Z)-2-Ethoxyvinyllithium, **8,** 221
Hydrogen hexachloroplatinate(IV)–Ethoxydiethylsilane, **5,** 293
Lithium dimethylcuprate, **6,** 209
Methylenecyclopropane, **9,** 46
Nickel(II) bromide, **4,** 351
Palladium(II) chloride, **5,** 500
Palladium(II) chloride–P-Phenyl-1-phospha-3-methyl-3-cyclopentene, **5,** 503
Phenacyl bromide, **5,** 758
Potassium *t*-butoxide, **7,** 296
Vinyllithium, **8,** 275
Vinylmagnesium chloride, **4,** 572
δ,ε-UNSATURATED ALDEHYDES AND KETONES
Allyltrimethylsilane, **7,** 370
Allyltrimethyltin, **9,** 47
Aluminum chloride, **9,** 11
Bis(benzonitrile)dichloropalladium(II), **12,** 51
Cesium fluoride, **11,** 115
m-Chloroperbenzoic acid, **11,** 122
Lithium diallylcuprate, **7,** 86
Mercury(II) trifluoroacetate, **11,** 320; **12,** 306
Potassium hydride, **7,** 302; **8,** 412; **10,** 327
Sodium methylsulfinylmethylide, **7,** 338
Titanium(IV) chloride, **11,** 529
UNSATURATED AMIDES
Acetic anhydride, **6,** 1
Benzeneseleninic anhydride, **8,** 29
Butyllithium, **10,** 73
Carbon dioxide, **5,** 93
Chlorotris(triphenylphosphine)rhodium, **2,** 448
N,N-Diethylaminopropyne, **8,** 165
Dimethylacetamide, **6,** 84
N,N-Dimethylacetamide dialkyl acetals, **1,** 271; **4,** 166; **5,** 226; **8,** 180
Dimethylformamide dialkyl acetals, **5,** 253; **8,** 191
Lithium diallylcuprate, **5,** 175
Lithium tricarbonyl(dimethyl-carbamoyl)nickelate, **4,** 302

UNSATURATED AMIDES (*Continued*)
N-Phenyl-1,2,4-triazoline-3,5-dione, **4,** 381
Pyridine N-oxide, **9,** 396
Sodium cyanide, **4,** 317; **5,** 606
Tin(IV) chloride, **10,** 370
Trimethylsilylmethyllithium, **6,** 635
UNSATURATED CARBONYLS SUBSTITUTED BY HETEROATOMS
α-SUBSTITUTED-α,β-UNSATURATED CARBONYLS
α-Acyloxy, α-Alkoxy
Acetic anhydride–Triethylamine, **5,** 4
1,2-Dimethoxyvinyllithium, **7,** 116
N,N-Dimethylaniline, **1,** 274
Rhodium(II) carboxylates, **11,** 458
Sodium methoxide, **10,** 366
α-Alkyl(or Phenyl)seleno
Benzeneselenenic anhydride, **9,** 25
Benzeneselenenyl halides, **9,** 25; **10,** 16
N,N-Dimethylbenzeneselenenamide, **7,** 119
α-Alkyl(or Phenyl)thio, sulfinyl, etc.
Acetic anhydride, **6,** 3
Benzenesulfenyl chloride, **8,** 32; **12,** 42
Chlorotrimethylsilane–Sodium iodide, **10,** 97
N,N-Dimethylhydrazine, **7,** 126
Lead tetraacetate, **9,** 265
Lithium naphthalenide, **9,** 284
(S)- or (R)-Menthyl *p*-toluenesulfinate, **11,** 312
Polyphosphoric acid, **1,** 894
β-Vinylbutenolide, **11,** 594
α-Halo
Benzeneselenenyl halides, **10,** 16; **11,** 34
Bromine–Acetic acid, **1,** 1025
(Bromodichloromethyl)phenylmercury, **7,** 282
Bromodimethylsulfonium bromide, **11,** 198
t-Butyl 2-chloro-2-lithiotrimethylsilylacetate, **8,** 60
Chlorotrimethylsilane, **7,** 66
Copper(II) chloride, **2,** 84
Diethyl carboxychloromethylphosphonate, **8,** 168
Diethyl 2,2-dichloro-1-ethoxyvinyl phosphate, **9,** 160
Iodine azide, **4,** 262
Lithium dimethylcuprate, **3,** 106
Perchloryl fluoride, **1,** 802
Pyridinium dichromate, **10,** 335
Sodium chlorodifluoroacetate, **5,** 603
Sulfuryl chloride, **1,** 1128
Sulfuryl chlorofluoride, **6,** 562
Trifluorovinyllithium, **6,** 622; **8,** 505
α-N-Substituted
1-Chloro-N,N,2-trimethylpropenylamine, **7,** 66
Iodine azide, **4,** 262
Silver carbonate, **4,** 425
α-Silyl
t-Butyl lithiobis(trimethylsilyl)acetate, **7,** 42
Methylaluminum dichloride, **12,** 5
α-SUBSTITUTED-β,γ-UNSATURATED CARBONYLS
Benzenesulfenyl chloride, **12,** 42
Hypochlorous acid, **11,** 260
Perchloryl fluoride, **1,** 802
β-SUBSTITUTED-β,γ-UNSATURATED CARBONYLS
β-Acyloxy, Alkoxy
Chloromethyl methyl ether, **4,** 83
Dimethylformamide–Phosphoryl chloride, **1,** 284
β-Amino
Bis(3-dimethylaminopropyl)-phenylphosphine, **5,** 36
Molybdenum carbonyl, **11,** 350
Trifluoroacetic anhydride, **1,** 1221
β-Halo
Aluminum chloride, **1,** 24
Dichloromethylenedimethylammonium chloride, **4,** 135
Oxalyl chloride, **5,** 481; **7,** 257
Phosphoryl chloride, **1,** 609
Tin(IV) chloride, **12,** 486
Trifluoroacetic acid, **5,** 695
Triphenylphosphine + co-reagent, **6,** 644, 647
β-Thio
1-Butanethiol, **2,** 53; **4,** 64
3-Methoxy-1-methylthio-1-propyne, **6,** 397
γ-SUBSTITUTED-α,β-UNSATURATED CARBONYLS
γ-Halo
t-Butyl hypochlorite, **1,** 90
Copper(II) bromide, **1,** 161

Copper(II) chloride, **5,** 158
Dilithium tetrachlorocuprate(II), **7,** 79
Perchloryl fluoride, **1,** 802; **2,** 310
Trifluoromethyl hypofluorite, **4,** 237
γ-Hydroxy (*see also* 4-Hydroxy-2-cyclopentenones)
1,3-Bis(methylthio)allyllithium, **4,** 38; **6,** 53
Borane–Tetrahydrofuran, **11,** 69
t-Butyl chromate, **2,** 48
2-Diethylamino-4-phenylthio-2-butenenitrile, **11,** 176
1,3-Dithiane, **2,** 182
Magnesium methoxide, **11,** 309
Methyl methanethiosulfonate, **7,** 243
Methyl methylthiomethyl sulfoxide, **8,** 344
Perchloryl fluoride, **10,** 304
Potassium peroxomonosulfate, **11,** 442
Pyridinium chlorochromate, **10,** 334
Selenium(IV) oxide, **2,** 360

UNSATURATED CARBOXYLIC ACIDS (*see also* DIENOIC ACIDS, UNSATURATED ESTERS)
α,β-UNSATURATED CARBOXYLIC ACIDS
by Carboxylation of vinyl metal reagents
Butyllithium, **7,** 47
Carbon dioxide, **5,** 93
Diisobutylaluminum hydride, **2,** 140; **7,** 147
Lithium diisobutylmethylaluminum hydride, **2,** 248
Nickel(II) acetylacetonate–Trimethylaluminum, **8,** 42; **9,** 52
Nickel carbonyl, **1,** 720
N,N,N′,N′-Tetramethylethylenediamine, **3,** 284
Vinylcopper reagents, **6,** 662
Vinyllithium, **1,** 1273
from C=O ($R_2CO \rightarrow R_2C=CHCOOH$)
t-Butyllithium, **11,** 103
α-Dibenzylphosphonoacetic acid, **5,** 183
Diethyl carboxymethylphosphonate, **8,** 168
Diethyl trimethylsilyloxycarbonylmethylphosphonate, **8,** 171
N,N-Dimethylformamide, **5,** 247
Dimethyl phosphite–Chloroacetic acid, **11,** 211
Ethoxyacetylene, **1,** 357
Malonic acid, **1,** 887, 959
Meldrum's acid, **1,** 526
Potassium(or Sodium) acetate, **1,** 906, 1024
Triethylamine, **1,** 1198
2,4,4-Trimethyl-2-oxazoline, **5,** 714
Trimethylsilylacetic acid, **6,** 631
(Trimethylsilyl)acetyltrimethylsilane, **11,** 572
by Oxidation of allylic alcohols or α,β-unsaturated aldehydes (*see* TYPE OF REACTION INDEX)
Other routes
Benzyl bromomethyl sulfide, **5,** 25
Copper(I) iodide, **10,** 107
2,3-Dichloro-5,6-dicyano-1,4-benzoquinone, **4,** 130
Dicyclohexylborane, **8,** 162
2-Diethylamino-4-phenylthio-2-butenenitrile, **11,** 176
Diphenyltelluride, **1,** 348
Formic acid, **1,** 404
Lithium diisopropylamide, **4,** 298
Potassium *t*-butoxide, **1,** 911
Sodium hydrogen telluride, **8,** 459
Trichloroacetic acid, **2,** 425
β,γ-UNSATURATED CARBOXYLIC ACIDS
Benzeneselenenyl chloride, **9,** 25
Copper(I) iodide, **10,** 107
Dicyclohexylborane, **8,** 162
Diethyl oxomalonate, **10,** 143
Diethyl succinate, **1,** 911, 1076
Hydrogen fluoride, **9,** 240
Lithium methanethiolate, **8,** 303
Magnesium, **11,** 307
Sodium cyanide, **7,** 333
Sodium dithionite, **11,** 485
Sodium naphthalenide, **9,** 431
N,N,N′,N′-Tetramethylethylenediamine, **4,** 485
Triethanolamine, **1,** 1196
γ,δ-UNSATURATED CARBOXYLIC ACIDS
Benzeneselenenyl chloride, **9,** 25
t-Butyldimethylchlorosilane, **6,** 78
Copper(II) acetate, **5,** 156
Ethyl diazoacetate, **5,** 151
Hexamethylphosphoric triamide, **6,** 273; **8,** 240

UNSATURATED CARBOXYLIC ACIDS (*Continued*)
Lithium diisopropylamide, **11,** 296
Lithium hexamethyldisilazide, **12,** 280
Lithium N-isopropylcyclohexylamide, **4,** 306
Methoxycarbonylmethylene-triphenylphosphorane, **1,** 112
Tetrakis(triphenylphosphine)-palladium(0), **9,** 451
Zinc, **4,** 574
ω-UNSATURATED CARBOXYLIC ACIDS
Lead tetraacetate, **3,** 168
1,3-Propanedithiol, **4,** 413
UNSATURATED ESTERS (*see also* DIENOIC ESTERS, UNSATURATED CARBOXYLIC ACIDS)
α,β-UNSATURATED ESTERS
by Addition to α,β-acetylenic esters
Aluminum chloride, **8,** 13
Bromine, **5,** 55
Carbonylhydridotris(triphenyl-phosphine)rhodium(I), **4,** 249
Copper(I) iodide, **3,** 278
Lithium dialkylcuprates, **3,** 106; **5,** 283; **6,** 163
Organocopper reagents, **9,** 328
from C=O via Wittig, Wittig-Horner reagents
Benzoic acid, **1,** 49
Ethoxycarbonylcyclopropyltriphenyl-phosphonium tetrafluoroborate, **5,** 90; **6,** 93
Ethoxycarbonylmethylenetriphenyl-phosphorane, **1,** 111, 1241
Ethylene oxide, **5,** 297
1-Methoxycarbonylethyl(diphenyl)-phosphine oxide, **12,** 105
Methoxycarbonylmethylenetriphenyl-phosphorane, **1,** 112; **2,** 98
Methyl bis(trifluoroethoxy)-phosphinylacetate, **12,** 320
Methyl 2-bis(2,2,2-trifluoroethoxy)-phosphinylpropionate, **12,** 320
Tetraethyl phosphonosuccinate, **1,** 1138
Tetrahexylammonium bromide, **10,** 383
Triethyl(or Trimethyl) phosphonoacetate, **1,** 1217; **2,** 442; **6,** 566; **7,** 394; **12,** 527
Triethyl α-phosphonoproprionate, **5,** 693
Triphenylphosphine, **1,** 1238; **4,** 548
from C=O by other routes
Alumina, **11,** 22
t-Butyl lithio(trimethylsilyl)acetate, **5,** 371
Chloromethyldiphenylsilane, **12,** 321
1,8-Diazabicyclo[5.4.0]undecene-7, **5,** 177; **6,** 158
Dibutyltelluronium carboethoxymethylide, **12,** 159
Diethyl succinate, **1,** 911, 1076
Ethoxyacetylene, **1,** 357
Ethyl cyanoacetate, **1,** 121
O-Ethyl-S-ethoxycarbonylmethyl dithiocarbonate, **7,** 148
Ethyl trimethylsilylacetate, **5,** 373; **11,** 234; **12,** 228
Ketene alkyl trialkylsilyl acetals, **10,** 261
Malononitrile, **1,** 16
Meldrum's acid, **1,** 526
Piperidine, **1,** 886
Raney nickel, **1,** 723
Sodium ethoxide, **1,** 1065
Sulfuryl chloride, **6,** 561
Titanium(IV) chloride, **3,** 291; **4,** 507
2,4,4-Trimethyl-2-oxazoline, **5,** 714
by Coupling reactions
Diacetatobis(triphenylphosphine)-palladium(II), **6,** 156
Methyltris(triphenylphosphine)-rhodium(I), **5,** 463
Palladium(II) acetate, **8,** 378; **9,** 350
Palladium–Graphite, **10,** 297
Tetrakis(triphenylphosphine)-palladium(0), **10,** 384
by Deoxygenation of α,β-epoxy esters (*see* TYPE OF REACTION INDEX)
by Elimination
Benzeneselenenyl bromide, **5,** 518
Chloromethyl phenyl sulfide, **9,** 107
m-Chloroperbenzoic acid, **8,** 97
Diethoxyaluminum chloride, **6,** 180
Dimethyl bromomalonate, **12,** 200
Dimethyl disulfide, **5,** 246; **6,** 217
Ethyl 2-methylsulfinylacetate, **6,** 255
Ethyl α-trifluoromethylsulfonyl-oxyacetate, **7,** 149

Methyl 3-dimethylaminopropionate, **9,** 306
Methyl phenylsulfinylacetate, **6,** 393
Oxalic acid, **1,** 764
Ozone, **10,** 295
Phosphorus(III) bromide, **1,** 873
Phosphorus(V) oxide, **7,** 291
Phosphoryl chloride, **8,** 401
Quinoline, **1,** 975
Sulfuryl chloride–Pyridine, **5,** 642
Thionyl chloride, **1,** 1158
Trimethylsilylmethyl trifluoromethanesulfonate, **9,** 100
by Hydrocarboxylation of C≡C
Carbon monoxide, **7,** 53; **9,** 95
Ethyl chloroformate, **7,** 147
Organopentafluorosilicates, **11,** 373
from β-Keto esters
Lithium dimethylcuprate, **5,** 234
Organocopper reagents, **9,** 328; **10,** 282
Other routes
Bis(dimethylaluminum) 1,3-propanedithiolate, **6,** 49
Chloromethylcarbene, **11,** 121
Copper(I) bromide–Sodium bis-(2-methoxyethoxy)aluminum hydride, **6,** 144
2-Diethylamino-4-phenylsulfonyl-2-butenenitrile, **12,** 182
2-Diethylamino-4-phenylthio-2-butenenitrile, **11,** 176
Dihalobis(triphenylphosphine)-palladium(II), **6,** 60
Formaldehyde, **9,** 224
Grignard reagents, **11,** 245
Lindlar catalyst, **9,** 270
Manganese(IV) oxide, **2,** 257
4-Methoxycarbonylthiolane-3-one, **11,** 324
Organocopper reagents, **12,** 345
Oxygen, singlet, **8,** 367
Peracetic acid, **1,** 787
Rhodium(II) carboxylates, **11,** 458
Sodium methoxide, **4,** 457
Triphenylcarbenium perchlorate, **1,** 1256
Zinc bromide, **9,** 520
β,γ-UNSATURATED ESTERS
Bis(acetonitrile)dichloropalladium(II), **8,** 39
t-Butyldimethylchlorosilane, **11,** 88
t-Butyl lithioacetate, **6,** 84
Butyllithium–Nickel(II) bromide, **8,** 66
Ethyl 4-bromocrotonate, **3,** 236
Lithium diisopropylamide, **5,** 406; **11,** 296
Lithium dimethylcuprate, **8,** 301
Lithium N-isopropylcyclohexylamide, **4,** 306
Methyl methylthiomethyl sulfone, **11,** 242
Organocopper reagents, **12,** 345
Palladium(II) acetate–Triphenylphosphine, **11,** 392
Palladium(II) chloride, **11,** 393
Phenyl 2-(trimethylsilyl)ethynyl sulfone, **11,** 426
Potassium hexamethyldisilazide, **12,** 407
Sodium dithionite, **11,** 485
Tetraethyl phosphonosuccinate, **1,** 1138
p-Toluenesulfonylhydrazide, **8,** 489
Trimethylsilylmethyllithium, **6,** 635
Vinylcopper, **5,** 747
γ,δ-UNSATURATED ESTERS
by Alkylation of enolates with allylic electrophiles
Bis(dibenzylideneacetone)-palladium(0), **10,** 34
[1,2-Bis(diphenylphosphine)ethane]-dichloropalladium(II), **6,** 447
Ethoxycarbonylmethylcopper, **4,** 224
Palladium(II) chloride, **4,** 369; **5,** 500
Reformatsky reagent, **12,** 422
Tetrakis(triphenylphosphine)-palladium(0), **8,** 472; **11,** 503
by Claisen rearrangement
N,N-Dimethylacetamide dimethyl acetal, **4,** 166
Lithium N-isopropylcyclohexylamide, **7,** 209
Triethyl orthoacetate, **3,** 300; **4,** 527; **6,** 607
1,1,1-Trimethoxy-3-phenyl-selenopropane, **9,** 489
Trimethyl orthoformate, **8,** 180
Other routes
Copper(II) sulfate, **6,** 141
8-Phenylmenthol, **11,** 412
δ,ε-UNSATURATED ESTERS
Alkylaluminum halides, **11,** 7
Allyltrimethylsilane, **12,** 23

UNSATURATED ESTERS (*Continued*)
Aluminum chloride, **5,** 10; **7,** 7; **9,** 13
Lithium diallylcuprate, **5,** 175
UNSATURATED LACTONES
(*see also* CHIRAL COMPOUNDS)
α-ALKYLIDENE LACTONES
Bis[methoxy(thiocarbonyl)] disulfide, **10,** 39
Ethyl formate, **12,** 226
α-Mercapto-γ-butyrolactone, **8,** 314
Mercury(II) trifluoroacetate, **11,** 320
Phosphoryl chloride, **8,** 401
Potassium carbonate, **9,** 382
Tris(dimethylamino)methane, **7,** 411
α-METHYLENE LACTONES
by Carbonylation
Dimethylaluminum acetylide, **8,** 385
Lithium acetylide, **6,** 451
Nickel carbonyl, **6,** 451
Palladium(II) chloride–Thiourea, **7,** 278; **11,** 397
α-Trimethylsilylvinylmagnesium bromide, **8,** 515
from Cyclopropanes
6,6-Dimethyl-5,7-dioxaspiro[2.5]-octane-4,8-dione, **7,** 122
Iodotrimethylsilane, **9,** 251; **12,** 259
Trifluoroacetic acid, **9,** 483
Zinc bromide, **5,** 762
by Lactonization
Boron trifluoride etherate, **10,** 52
Butyllithium, **10,** 73
Lithium (1,1-diethoxy-2-propenyl)-(3,3-dimethyl-1-butynyl)cuprate, **8,** 300
Lithium (3,3-diethoxy-1-propen-2-yl)-phenylthiocuprate, **7,** 212
2-Phenylthiopropionic acid, **6,** 466
by Methylenation α to C=O's
Bis(dimethylamino)methoxymethane, **10,** 34
Bromine, **11,** 75
Bromomethyl methyl ether, **10,** 56
Carbon dioxide, **5,** 93
Chloromethyl phenyl sulfide, **9,** 107, 521
1,5-Diazabicyclo[4.3.0]nonene-5, **6,** 157
Dibenzoyl peroxide, **4,** 122
1,2-Dibromoethane, **4,** 126; **5,** 187
Dimethyl(methylene)ammonium trifluoroacetate, **7,** 132
Diphenyl diselenide, **5,** 272; **7,** 136
Diphenyl disulfide, **5,** 276
Ethyl formate, **12,** 226
Ethyl α-trifluoromethylsulfonyloxyacetate, **7,** 149
Formaldehyde, **5,** 312; **6,** 264; **9,** 224
Iodomethyl methyl sulfide, **5,** 353
Lead tetraacetate, **5,** 365
Lithium diisopropylamide, **4,** 298
Magnesium methyl carbonate, **3,** 190; **5,** 420; **11,** 310
Methyl formate, **5,** 446
Potassium carbonate, **9,** 382
Sodium benzeneselenoate, **10,** 368
Sodium cyanoborohydride, **5,** 607; **6,** 537
Sodium dicarbonyl(cyclopentadienyl)-ferrate, **11,** 483
Trimethylsilyl dichloroacetate, **9,** 491
Tris(methylthio)methyllithium, **7,** 412
Other routes
Bis(1,5-cyclooctadiene)nickel(0), **9,** 45
2-Bromo-3-trimethylsilyl-1-propene, **11,** 80
α-Chloro-N-cyclohexylpropanaldonitrone, **5,** 110
Di-μ-bromobis[(2-ethoxycarbonyl)-2-propenyl]dinickel, **6,** 93
Dicarbonylbis(triphenylphosphine)-nickel, **10,** 125
Dimethyl diazomalonate, **8,** 187
Dimethyl malonate, **8,** 193
N-(2,3-Epoxypropylidene)-cyclohexylamine N-oxide, **9,** 209
Formaldehyde, **7,** 158
Lithium diisopropylamide, **7,** 204
Lithium α-ethoxycarbonylvinyl-(1-hexynyl)cuprate, **6,** 329
Methallyl alcohol, **8,** 67
p-Methoxybenzyl itaconate, **6,** 365
Methoxy(phenylthio)methyllithium, **6,** 369
Nickel carbonyl, **11,** 356
2-Phenylselenopropionic acid, **8,** 397
Titanium(IV) chloride, **8,** 483
Tributyltin hydride, **12,** 516
2,4,6-Triisopropylbenzenesulfonylhydrazide, **9,** 486
1,1,1-Trimethoxy-3-phenylselenopropane, **9,** 489

Zinc, **3,** 334
Zinc–copper couple, **7,** 428
Zinc–Graphite, **12,** 570
γ-METHYLENE LACTONES
Dimethyl diazomalonate, **8,** 187
Iodine, **8,** 256
Mercury(II) acetate, **8,** 315
Mercury(II) oxide, **8,** 316
Palladium(0)–Phosphines, **9,** 344
Silver(I) nitrate, **12,** 433
α,β-UNSATURATED-γ-LACTONES (Δ^2-BUTENOLIDES)
by Additions to C=O
2-Carboxy-1-ethoxycarbonylethylidenetriphenylphosphorane, **8,** 77
Cyclopropenone 1,3-propanediyl ketal, **12,** 152
Ethyl lithiopropiolate, **9,** 217
Ethyl β-(1-pyrrolidinyl)acrylate, **8,** 226
Glyoxylic acid, **5,** 320
Iron, **5,** 357
Lithium ethoxyacetylide, **1,** 613
by Elimination
Acetic anhydride, **11,** 1
Benzeneselenenyl chloride, **6,** 459; **9,** 25
Chloroketene diethyl acetal, **9,** 522
1,2-Dibromoethane, **4,** 126
Dimethyl carbonate, **5,** 234
Diphenyl disulfide, **5,** 276
Ethyl α-trifluoromethylsulfonyloxyacetate, **7,** 149
Phenylthioacetic acid, **6,** 463
Pyridine, **11,** 448
(R)-(+)-3-(*p*-Tolylsulfinyl)propionic acid, **12,** 509
Triethylamine, **1,** 1198
Triethyl orthoformate, **11,** 555
Trimethylsilyl dichloroacetate, **9,** 491
Zinc bromide, **11,** 600
from Furans
2,5-Bis(trimethylsilyloxy)furans, **10,** 45
2-*t*-Butoxyfuran, **8,** 59
β-Furyllithium, **10,** 188
2-Methoxy-4-furyllithium, **11,** 325
1-Methoxy-1-trimethylsilylallene, **11,** 577
Organolithium reagents, **10,** 290
Peracetic acid, **11,** 402; **12,** 378
Phenylthiomethyllithium, **11,** 422
Pyridinium chlorochromate, **9,** 397
Titanium(IV) chloride, **11,** 529
Other routes
Acetic anhydride, **6,** 6
Bis[1,2-bis(diphenylphosphine)-ethane]palladium(0), **9,** 344
N-Bromosuccinimide, **7,** 37
t-Butyl diazomethyl ketone, **5,** 162
Carbon monoxide, **7,** 53
1-Chloro-N,N,2-trimethylpropenylamine, **12,** 123
Di-μ-carbonylhexacarbonyldicobalt, **8,** 148; **9,** 144
Dichlorobis(triphenylphosphine)-palladium(II), **10,** 133
3,3-Diethoxy-1-methylthiopropyne, **5,** 207
N,N-Diethylaminopropyne, **8,** 165
Dihalobis(triphenylphosphine)-palladium(II), **10,** 133
Dimethylformamide dimethyl acetal, **11,** 198
Dimethylsulfonium methylide, **11,** 213
Dimethyl sulfoxide–Acetic anhydride, **2,** 163; **4,** 199
Dimethylsulfoxonium methylide, **1,** 315
Grignard reagents, **11,** 245
α-Hydroxyhippuric acid, **6,** 289
Ketene, **1,** 528
Ketenylidenetriphenylphosphorane, **10,** 450; **11,** 281
Lithium aluminum hydride, **10,** 265
(S)- or (R)-Menthyl *p*-toluenesulfinate, **12,** 295
S-Methyl *p*-toluenethiosulfonate, **6,** 400
Phosphorus(V) oxide–Methanesulfonic acid, **7,** 291
Potassium phenylacetate, **6,** 137
Pyridinium dichromate, **10,** 335
Selenium(IV) oxide, **1,** 992; **2,** 360
Silver carbonate–Celite, **12,** 433
Sodium methylsulfinylmethylide, **7,** 338
Sulfuric acid, **5,** 633; **9,** 441
Sulfur trioxide–Dimethylformamide, **1,** 1125
Tributyltin hydride, **12,** 516
Triethyl orthoacetate, **6,** 607
Trifluoroacetyl chloride, **11,** 560
Trimethyl phosphonoacetate, **7,** 394

UNSATURATED LACTONES (*Continued*)
Tris(dimethylamino)methane, **7,** 411
β-Vinylbutenolide, **6,** 661; **9,** 515; **11,** 594
α,β-UNSATURATED-δ-LACTONES
Benzeneseleninic anhydride, **11,** 37
Boron trifluoride, **11,** 71
2-Butynoic acid, **8,** 308
2,3-Dichloro-5,6-dicyano-1,4-benzoquinone, **1,** 215
Diethyl oxomalonate, **6,** 188
1,3-Dimethoxy-1-trimethylsilyloxy-1,3-butadiene, **12,** 196
2,2′-Dipyridyl disulfide, **8,** 214
1-Methoxy-1,3-butadiene, **10,** 258
Methyl phenylsulfinylacetate, **11,** 343
Propiolic acid, **6,** 495
Pyridinium chlorochromate, **11,** 450; **12,** 417
Sodium dicarbonyl(cyclopentadienyl)-ferrate, **11,** 483
Zinc bromide, **11,** 600
β,γ-UNSATURATED-γ-LACTONES (Δ^3-BUTENOLIDES)
Acetic anhydride, **8,** 1
Bis(benzonitrile)dichloropalladium(II), **12,** 51
Peracetic acid, **12,** 378
β,γ-UNSATURATED-δ-LACTONES
Diethyl oxomalonate, **8,** 170
Ethyl lithioacetate, **8,** 225
γ,δ-UNSATURATED-δ-LACTONES
N,N-Dimethylhydrazine, **9,** 184
Mercury(II) oxide, **12,** 305
Silver(I) nitrate, **12,** 433
OTHER UNSATURATED LACTONES
Benzeneselenenyl bromide, **9,** 25
Ketenylidenetriphenylphosphorane, **12,** 268
Lithium isopropoxide, **9,** 283
UNSATURATED NITRILES
α,β-ACETYLENIC NITRILES
Copper(I) cyanide, **6,** 146
Lithium aluminum hydride, **9,** 274
Phenyl cyanate, **10,** 307
α,β-UNSATURATED NITRILES
by Adding CN to C=O
Cyanotrimethylsilane, **10,** 112
Diethylaluminum cyanide, **2,** 127; **4,** 146
Diethyl phosphorocyanidate, **12,** 187
o-Nitrophenyl selenocyanate, **8,** 362
by Adding anions α to CN to C=O
Acetonitrile, **5,** 621; **7,** 303
Alumina, **11,** 22
Benzyl nitrile, **1,** 1066
t-Butyl cyanoacetate, **1,** 87
Chlorocyanoketene, **9,** 103
Cyanoacetic acid, **1,** 39; **1,** 959
Cyanomethyldiphenylphosphine oxide, **8,** 133
Cyanomethylidenebis(triphenyl-phosphonium) dibromide, **4,** 112
2-(Diethoxyphosphinyl)propionitrile, **7,** 106
Diethyl cyanomethylphosphonate, **1,** 250; **2,** 130; **4,** 150; **6,** 566
Ethyl cyanoacetate, **1,** 121
O-Ethyl S-cyanomethyl dithiocarbonate, **9,** 215
Glycine, **1,** 412
Ion-exchange resins, **1,** 511
Malononitrile, **1,** 16; **5,** 6; **6,** 11
Methylthioacetonitrile, **8,** 349
Phosphoryl chloride, **1,** 876
Potassium hydroxide, **9,** 387
Sodium amide, **1,** 1034
Titanium(IV) chloride–Titanium(IV) isopropoxide, **12,** 503
Trimethylsilylacetonitrile, **10,** 427; **11,** 571
Triphenylphosphine, **4,** 548
Tris(trimethylsilyl)ketenimine, **11,** 591
by Oxidative cleavage of *o*-phenylenediamines
Copper(I) chloride, **5,** 165; **7,** 80; **8,** 118
Nickel peroxide, **1,** 731
from α,β-Unsaturated C=O
Chlorosulfonyl isocyanate, **2,** 70; **3,** 51
N,N-Dimethylhydrazine, **1,** 289
Lead tetraacetate, **1,** 537
from Vinyl halides
Dibromobis(triphenylphosphine)-nickel(II)–Zinc, **11,** 157
Nickel carbonyl, **3,** 210
Potassium cyanide, **8,** 473
Sodium dicyanocuprate, **3,** 265
Other routes
Ascarite, **4,** 21
2-(6-Bromopyridyl)lithium, **5,** 63
1-Chlorobenzotriazole, **2,** 67

Diisobutylaluminum hydride, **2,** 140
Dimethyl(methylthio)sulfonium tetrafluoroborate, **11,** 204
Diphenyl diselenide, **5,** 272
Hydrogen cyanide, **9,** 293
Phenyl selenocyanate, **11,** 416
α-(Phenylsulfinyl)acetonitrile, **6,** 395; **12,** 391
Phosphorus(III) chloride, **8,** 400
Potassium cyanide, **11,** 433
Tetrakis(triphenylphosphine)-palladium(0), **10,** 384; **11,** 503
Tetrakis(triphenyl phosphite)nickel(0), **11,** 503
Trichloroacetyl isocyanate, **7,** 382
β,γ-UNSATURATED NITRILES
Acetonitrile, **6,** 84
Benzyltriethylammonium chloride, **7,** 18
Copper(I) cyanide, **5,** 166
Iodoacetonitrile, **6,** 494
Methylthioacetonitrile, **8,** 349
Sodium cyanide, **7,** 333
Trimethylsilylmethyllithium, **6,** 635
γ,δ-UNSATURATED NITRILES
Cyanomethylcopper, **4,** 111
Palladium(II) chloride, **5,** 500
UREAS
Dichloromethylenedimethylammonium chloride, **4,** 135
Dicyclohexylcarbodiimide, **8,** 162
Dimethyl selenoxide, **8,** 197
Dimethyl sulfoxide–Iodine, **6,** 295
Iron carbonyl, **7,** 183
Isopropyl isocyanate, **11,** 277
Phenylcarbonimidic dichloride, **6,** 458
Phenyl isocyanate, **1,** 842
Potassium superoxide, **11,** 442
Selenium, **4,** 422
Silicon tetraisocyanate, **1,** 1002
Sodium azide, **1,** 1041
Sodium methoxide, **1,** 1091
Sodium naphthalenide, **5,** 468
Tellurium, **11,** 498
Tributyltin azide, **7,** 377
Urea, **1,** 1262
URETHANES (*see* CARBAMATES)

VINYL HALIDES
VINYL FLUORIDES
Alumina, **8,** 9; **9,** 8
Bromine fluoride, **8,** 53
Cesium fluoroxysulfate, **11,** 118
Difluoromethylenetriphenyl-phosphorane, **9,** 166
Fluoromethylenetriphenylphosphorane, **3,** 146; **6,** 262
Pyridinium poly(hydrogen fluoride)–N-Bromosuccinimide, **11,** 454
Silver acetate, **6,** 511
VINYL CHLORIDES
from RCHO, $RCOCH_2R'$ ($RCOCH_2R' \rightarrow R'CH=CRCl$ or $R'CH_2CR=CHCl$)
Chlorocyanoketene, **9,** 103
Chloromethylenetriphenyl-phosphorane, **1,** 1239; **7,** 62; **8,** 96
N-Chlorosuccinimide, **3,** 142
Oxalyl chloride, **1,** 767
Phosphorus(III) chloride, **1,** 875
Phosphorus(V) chloride, **4,** 388
2,2,2-Trichloro-1,3,2-benzodioxa-phosphole, **2,** 63
Triphenylphosphine–Carbon tetrachloride, **4,** 551
Triphenylphosphine dichloride, **6,** 646
from Alkynes
N-Chlorosuccinimide, **6,** 178; **8,** 235; **9,** 276
1,2-Dichloroethylene, **12,** 175
Diisobutylaluminum hydride, **11,** 185
Lithium aluminum hydride, **9,** 274
Trifluoroacetic acid, **5,** 695
by Elimination
Chloro(phenylsulfinyl)methyllithium, **9,** 271
N-Chlorosuccinimide, **8,** 103
Potassium *t*-butoxide, **9,** 380
Sodium amide–Sodium *t*-butoxide, **9,** 417
Other routes
Alumina, **8,** 9
Chlorine, **5,** 556
Grignard reagents, **10,** 189
Iodine monochloride, **8,** 261
Lithium 2-methoxyethoxide, **3,** 186
Potassium hydroxide, **5,** 560
Tetrabutylammonium fluoride, **10,** 378
Tetrakis(triphenylphosphine)nickel(0), **12,** 467
VINYL BROMIDES
from RCHO, R_2CO

VINYL HALIDES (*Continued*)
Bromomethylenetriphenylphosphorane, **10,** 56
Dibromomethyllithium, **11,** 158
from Alkynes
N-Bromosuccinimide, **6,** 178; **8,** 54, 235
Catecholborane, **5,** 100
Diisobutylaluminum hydride, **2,** 140; **11,** 185
B-Halo-9-borabicyclo[3.3.1]nonane, **12,** 236
Hydrogen bromide, **6,** 284
Mercury(II) bromide, **6,** 663
by Elimination
Benzeneselenenyl bromide, **5,** 518; **8,** 25
Potassium *t*-butoxide, **8,** 407
Sodium amide–Sodium *t*-butoxide, **4,** 439; **5,** 593; **9,** 417
Other routes
Bromine, **5,** 556
N-Bromosuccinimide, **2,** 40
Diphosphorus tetraiodide, **11,** 224
Ethylene oxide, **5,** 297
Lithium 2-methoxyethoxide, **3,** 186
Silver perchlorate, **4,** 432
Silver *p*-toluenesulfonate, **7,** 323
VINYL IODIDES
from C≡C
Chlorobis(cyclopentadienyl)hydridozirconium(IV), **8,** 84
Diallylzinc, **12,** 54
Dichlorobis(cyclopentadienyl)titanium, **10,** 130; **11,** 163; **12,** 168
Diisobutylaluminum hydride, **3,** 174; **11,** 185
B-Halo-9-borabicyclo[3.3.1]nonane, **12,** 236
Iodine, **5,** 346; **6,** 178, 662; **8,** 235; **9,** 248
Iodine monochloride, **10,** 212
Lithium aluminum hydride–Sodium methoxide, **4,** 295
Mercury(II) trifluoroacetate, **11,** 320
Nickel(II) acetylacetonate–Trimethylaluminum, **8,** 42; **9,** 52
Trifluoroacetic acid, **5,** 695
Trimethylaluminum–Dichlorobis(cyclopentadienyl)zirconium, **11,** 567
Other routes
Guanidines, **12,** 477
Hydrazine, **12,** 241
Iodine, **9,** 498
Iodine monochloride, **8,** 261
N-Iodosuccinimide, **2,** 40
Lithium 2-methoxyethoxide, **3,** 186
Potassium fluoride, **5,** 555
DIFLUOROALKENES
Chlorodifluoromethane, **7,** 60
Diethyl difluorotrimethylsilylmethylphosphonate, **11,** 180
Difluoromethylenetriphenylphosphorane, **1,** 1058, 1227; **9,** 166
Sodium chlorodifluoroacetate, **2,** 281, 379
Triphenylphosphine–Dibromodifluoromethane, **10,** 448
DICHLOROALKENES
Antimony(V) chloride, **7,** 12
Copper(II) chloride, **6,** 139
Dichloro(diethoxyphosphinyl)methyllithium, **6,** 188
Dichloromethylenetriphenylphosphorane, **1,** 222, 1239; **7,** 97
Lithium piperidide, **6,** 344
Molybdenum(V) chloride, **6,** 412
Phenyliodine(III) dichloride, **6,** 298
Potassium superoxide, **11,** 442
DIBROMOALKENES
Copper(II) bromide, **1,** 161
Diethyl dibromomethylphosphonate, **7,** 104
VINYL HALIDES WITH TWO DIFFERENT HALOGENS
N-Bromoacetamide–Hydrogen fluoride, **4,** 48
Chlorofluoromethylenetriphenylphosphorane, **2,** 443
Methyliodine(III) difluoride, **7,** 242
VINYL SILANES
FROM ARENESULFONYLHYDRAZONES
Chlorotrimethylsilane, **7,** 13; **8,** 492; **10,** 96
p-Toluenesulfonylhydrazide, **8,** 290
2,4,6-Triisopropylbenzenesulfonylhydrazide, **9,** 486
FROM SILYL-SUBSTITUTED ALKYNES
Bis(benzonitrile)dichloropalladium(II), **11,** 48

N-Chlorosuccinimide, **8,** 103
Copper(I) iodide, **7,** 81; **12,** 143
Dichlorobis(cyclopentadienyl)titanium, **10,** 130; **11,** 163; **12,** 168
Dichlorobis(cyclopentadienyl)titanium–Dialkylaluminum chloride, **9,** 471
Nickel(II) acetylacetonate–Trimethylaluminum, **8,** 42; **9,** 52
OTHER ROUTES
Chlorotrimethylsilane–Sodium, **10,** 356; **12,** 126
Diethyl[dimethyl(phenyl)silyl]aluminum, **12,** 342
Dimethylphenylsilyllithium, **10,** 162
(1-Lithiovinyl)trimethylsilane, **5,** 374; **7,** 193; **9,** 498; **11,** 286
Lithium diisopropylamide–Chlorotrimethylsilane, **12,** 279
Organocopper reagents, **12,** 345
Organomagnesium reagents, **12,** 352
Tetrakis(triphenylphosphine)-palladium(0), **11,** 503
Triethylsilane, **3,** 187
2-Triethylsilyl-1,3-butadiene, **8,** 502
Trimethylsilylallene, **10,** 428
α-Trimethylsilylcrotyl-9-borabicyclo-[3.3.1]nonane, **10,** 430
1-Trimethylsilyl(pentadienyl)lithium, **10,** 436
Tris(trimethylsilyl)methyllithium, **5,** 618
VINYL SULFUR COMPOUNDS
VINYL SULFIDES
Benzenesulfenyl chloride, **8,** 32
Butyllithiums, **5,** 79, 80
2-Chloroethyl thiocyanate, **9,** 106
Copper(I) bromide–Lithium trimethoxyaluminum hydride, **7,** 80
Copper(I) trifluoromethanesulfonate, **6,** 130; **7,** 75; **8,** 125
Diethyl allylthiomethylphosphonate, **3,** 94
Diethyl (1-methylthio)ethylphosphonate, **4,** 151
Diethyl methylthiomethylphosphonate, **3,** 97
Diethyl phenylthiomethylphosphonate, **6,** 191
Dimethylsulfonium methylide, **5,** 254
Diphenyl disulfide, **7,** 47
Divinylmercury, **1,** 352
Ethylmagnesium bromide–Copper(I) iodide, **9,** 219
Iodotrimethylsilane, **10,** 216; **12,** 259
Lithium aluminum hydride, **8,** 286
Lithium diisopropylamide–Chlorotrimethylsilane, **12,** 279
Lithium naphthalenide, **9,** 284
Organolithium reagents, **9,** 5
Phenyl phenylthiomethyl sulfoxide, **6,** 394
Phenylthiotrimethylsilylmethane, **11,** 423
Sodium methylsulfinylmethylide, **9,** 431
Tetrakis(triphenylphosphine)-palladium(0), **9,** 451
Trichloroisocyanuric acid, **6,** 605
Tris(phenylseleno)borane, **9,** 511
Tris(phenylthio)borane, **7,** 409
Vinyltrimethylsilane, **10,** 444
VINYL SULFONES
Allyl phenyl sulfone, **8,** 405
Benzenesulfenyl chloride, **8,** 32
Benzylsulfonyldiazomethane, **11,** 43
Benzyltriethylammonium chloride, **6,** 41
Butyllithium, **4,** 60
m-Chloroperbenzoic acid, **8,** 97
Copper(II) acetate, **12,** 140
Hexamethylphosphoric triamide, **10,** 196
Methyllithium, **9,** 310
Phenylselenenyl benzenesulfonate, **11,** 407
Phenylselenenyl *p*-toluenesulfonate, **10,** 315; **12,** 394
1-Phenylsulfonyl-2-trimethylsilylethane, **11,** 41
Phenylsulfonyl(trimethylsilyl)methane, **12,** 393
VINYL SULFOXIDES
Bis(phenylsulfinyl)methane, **6,** 394
Diethyl methylsulfinylmethyl-phosphonate, **6,** 193
Grignard reagents, **8,** 235
(S)- or (R)-Menthyl *p*-toluenesulfinate, **12,** 295
Methylcopper, **6,** 377
Organocopper reagents, **9,** 328

XANTHATES
p-Bromophenacyl bromide, **1,** 77
Carbon disulfide, **4,** 195; **9,** 94
Crown ethers, **11,** 143

TYPE OF COMPOUND INDEX

CARBENES AND METAL–CARBENE COMPLEXES

1,3-Benzodithiole-2-carbene
Benzoylcarbene
Bis(methoxycarbonyl)carbene
Bromocarbene
Bromochlorocarbene
Bromofluorocarbene
Bromo(methoxycarbonyl)carbene
Bromophenylcarbene
Chlorocarbene
2-(Chloroethoxy)carbene
Chloro(ethoxycarbonyl)carbene
Chlorofluorocarbene
Chloromethylcarbene
Chloro(methylthio)carbene
Chlorophenylcarbene
Chlorotrifluoromethylcarbene
Chlorotrimethylsilylcarbene
Chromium carbene complexes
Cyclobutylidene
Dibromocarbene
Dichlorocarbene
Dicyanocarbene
Difluorocarbene
Dihalocarbene
Dimethoxycarbene
Dimethylcarbene
Dimethylvinylidenecarbene
Diphenylcarbene
Ethoxycarbene
Fluoroiodocarbene
Fluorophenylcarbene
Fluoro(trifluoromethyl)carbene
Formylcarbene
Iron-carbene complexes
Isopropylidenecarbene
Methoxycarbene
Pentacarbonyl(methoxyphenylmethylene)chromium(0) or -tungsten(0)
Phenoxycarbene
Phenylcarbene
Phenylsulfinylcarbene
Tetracarbonyl(ethoxyphenylmethylidene)iron(0)
Trimethylsilylcarbene
Trimethylstannylcarbene
Tris(2,2-dimethylpropyl)(2,2-dimethylpropylidene)niobium or -tantalum
Zirconium-carbene complexes

CHIRAL COMPOUNDS

CHIRAL ACIDS

(2R,3R)-Dipivaloyltartaric acid

CHIRAL AUXILIARIES, LIGANDS, AND CATALYSTS

(R)-2-Acetoxy-1,1,2-triphenylethanol
(2R,4R)- and (2S,4S)-N-Acryloyl-4-(diphenylphosphine)-2-[(diphenylphosphine)methyl]pyrrolidine
(S)-1-Alkoxymethyl-2-aminopropane, polymeric
1-Amino-(S)-2-[(R)-1-hydroxyethyl]-indoline
(S)-1-Amino-2-hydroxymethylindoline
(R)- or (S)-1-Amino-2-methoxymethyl-1-pyrrolidine
(S)-(−)-2-Amino-1-methoxy-3-phenylpropane
(1S,2S)-2-Amino-3-methoxy-1-phenyl-1-propanol
(S)-(−)-2-Amino-3-methyl-1,1-diphenyl-1-butanol
(S)-15-Aminomethyl-14-hydroxy-5,5-dimethyl-2,8-dithia-[9](2,5)-pyridinophane
(1S,2S)-2-Amino-1-phenyl-1,3-propanediol
(S)-2-Aminopropyl benzyl ether
(S)-4-Anilino-3-methylamino-1-butanol
(2S,4S)-(Anilinomethyl)-1-ethyl-4-hydroxypyrrolidine
(S)-2-(Anilinomethyl)pyrrolidine
(R)-(+)-*o*-Anisylcyclohexylmethylphosphine
N-Benzylcinchoninium chloride
3-O-Benzyl-1,2-O-cyclohexylidene-α-D-glucofuranose
(Benzylmethoxymethyl)methylamine
(−)-N-Benzyl-N-methylephedrinium bromide

CHIRAL COMPOUNDS (*Continued*)
(S)-N-Benzyloxycarbonylproline
N-Benzylquinidinium chloride
(–)-Benzylquininium chloride or fluoride
Bis[(–)-camphorquinone-α-dioximato]-cobalt(II) hydrate
Bis(1,5-cyclooctadiene)nickel(0)–(–)-Benzylmethylphenylphosphine
trans-2,3-Bis(diphenylphosphine)bicyclo-[2.2.1]hept-5-ene
(R)-(+)- and (S)-(–)-2, 2′-Bis(diphenyl-phosphine)-1,1′-binaphthyl
[2,2′-Bis(diphenylphosphine)-1,1′-binaphthyl](cyclooctadiene)-rhodium(I) perchlorate
[2,2′-Bis(diphenylphosphine)-1,1′-binaphthyl](norbornadiene)-rhodium(I) perchlorate
[1,4-Bis(diphenylphosphine)butane]-(1,5-cyclooctadiene)iridium(I) tetrafluoroborate
1,4-[Bis(diphenylphosphine)butane]-(norbornadiene)rhodium(I) tetrafluoroborate
trans-2,5-Bis(methoxymethoxymethyl)-pyrrolidine
trans-2,5-Bis(methoxymethyl)pyrrolidine
(S,S)- or (R,R)-N,N′-Bis(α-methyl-benzyl)sulfamide
Bis(norbornadiene)rhodium(I) perchlorate–(R)-1-(S)-1′,2-Bis-(diphenylphosphine)ferrocenylethanol
(–)-Borneol
Bornyloxyaluminum dichloride
D-(–)- and L-(+)-2,3-Butanediol
L(+)-Butane-2,3-dithiol
[N-(*t*-Butoxycarbonyl)-4-diphenyl-phosphine]-2-[(diphenylphosphine)-methyl]pyrrolidine(cyclooctadiene)-rhodium chloride or perchlorate
(2S,4S)-N-(*t*-Butoxycarbonyl)-4-(diphenylphosphine)-2-[(diphenyl-phosphine)methyl]pyrrolidine
(4S)-*t*-Butylthio-(2S)-methoxymethyl-N-pivaloylpyrrolidine
(R)-(+)-*t*-Butyl (*p*-tolylsulfinyl)acetate
(R)-(+)-*t*-Butyl 2-(*p*-tolylsulfinyl)-propionate
Camphor-10-sulfonic acid
Carbonylhydridotris(triphenyl-phosphine)rhodium–2,3,O-Isopropylidene-2,3-dihydroxy-1,4-bis(diphenylphosphine)butane
(–)-2-Chloromethyl-4-methoxymethyl-5-phenyloxazoline
Chlorotris(neomenthyldiphenyl-phosphine)rhodium(I)
Cinchona alkaloids
Cinchonidine
Cinchonine
(1,5-Cyclooctadiene)bis(methyldiphenyl-phosphine)iridium (I)hexafluoro-phosphate
Darvon alcohol
Dibenzyltaramide
Dichlorobis(1,5-cyclooctadiene)-dirhodium + chiral co-reagents
Di-μ-chlorobis(1,5-hexadiene)dirhodium + chiral co-reagents
Dichlorobis(norbornadiene)dirhodium + chiral co-reagents
(–)-*trans*-Dichloro(ethylene)-α-methylbenzylamineplatinum(II)
Dichloro[2,3-O-isopropylidene-2,3-dihydroxy-1,4-bis(diphenyl-phosphine)butane]nickel(II)
Di-μ-chlorotetra(ethylene)dirhodium–2,3-O-Isopropylidene-2,3-dihydroxy-1,4-bis(diphenylphosphine)butane
Di-μ-chlorotetrakis(cyclooctene)-dirhodium–2,3-O-Isopropylidene-2,3-dihydroxy-1,4-bis(diphenyl-phosphine)butane
Diethyl tartrate
2,2′-Dihydroxy-1,1′-binaphthyl
(S)-(–)-10,10′-Dihydroxy-9,9′-biphenanthryl
Diisopropyl tartrate
Dimenthyl fumarate
(3S,6S)-(+)-2,5-Dimethoxy-3,6-dimethyl-3,6-dihydropyrazine
(S)-1-(Dimethoxymethyl-2-methoxymethyl)pyrrolidine
trans-2,4-Dimethoxymethyl-5-phenyloxazoline
R-(–)-(Dimethoxyphosphinyl)methyl *p*-tolyl sulfoxide
3-Dimethylamino-2-methyl-1-propanol
(S,R)-N,N-Dimethyl-1-[1′,2-bis-(diphenylphosphine)ferrocenyl]-ethylamine
(–)-N,N-Dimethylephedrinium bromide

N,N-Dimethyl-α-methylbenzylamine
N,S-Dimethyl-S-phenylsulfoximine
1,2-Dimethylpyrrolidine
(+)-*trans*-2,5-Dimethylpyrrolidine
N,N-Dimethyl-α-*p*-tolylsulfinyl-acetamide
(SαR)- or (RαS)-2-Diphenylphosphine-ferrocenyl ethyldimethylamine
(−)-N-Dodecyl-N-methylephedrinium bromide
Ephedrine
(R,R)-1,2-Ethanediylbis[(*o*-methoxy-phenyl)phenylphosphine]
Ethyl (S)-lactate
Ethyl mandelate
(4S,5S)-2-Ethyl-4-(methoxymethyl)-5-phenyloxazoline
(−)-N,N-Ethylmethylephedrinium bromide
(R)-Ethyl *p*-tolylsulfinylmethyl-enepropionate
Ferrocenylphosphines
(S)-N-Formyl-2-methoxymethyl-pyrrolidine
L-N-Hexadecyl-N-methylephedrinium bromide
Hexahydro-4,4,7-trimethyl-4*H*-1,3-benzothiin
(S)-Histidine
(S)-(+)-α-Hydroxy-β,β-dimethylpropyl vinyl ketone
(1R)-(−)-*cis*-3-Hydroxyisobornyl neopentyl ether
(−)-3-Hydroxy-5-methylhydantoin
(S)-2-Hydroxymethyl-1-methyl-pyrrolidine
(2S,2′S)-2-Hydroxymethyl-1-[(1-methylpyrrolidin-2-yl)methyl]-pyrrolidine
(1S,2S,5S)-2-Hydroxypinan-3-one
Isopropyl glyoxylate
2,3-O-Isopropylidene-2,3-dihydroxy-1,4-bis(diphenylphosphine)butane
[2,3-O-Isopropylidene-2,3-dihydroxy-1,4-bis(diphenylphosphine)butane]-palladium
2,3-O-Isopropylidene-2,3-dihydroxy-1,4-bis(di-*m*-tolylphosphine)butane
4-Isopropyl-2-oxazolidinone
2-Isopropylpyrrolidine
Leucine *t*-butyl esters
(S)-(+)- and (R)-(−)-Mandelic acid
d- and *l*-Menthol
(−)-Menthoxyacetyl chloride
Menthoxyaluminum dichloride
(S)-(−)-Menthyl *p*-methoxy-benzenesulfinate
(S)- or (R)-Menthyl *p*-toluenesulfinate
(−)-10-Mercaptoisoborneol
4-(R)-Methoxycarbonyl-1,3-thiazolidine-2-thione
erythro-2-Methoxy-1,2-diphenylethylamine
(4S,5S)-4-Methoxymethyl-2-methyl-5-phenyl-2-oxazoline
(S)-(+)-2-Methoxymethylpyrrolidine
α-Methylbenzylamine
Methyl 2,3-bis-O-diphenylphosphine-4,6-O-benzylidene-α-D-glucopyranoside
(−)-N-Methylephedrine
N-Methylephedrinium chloride
4-Methyl-5-phenyl-2-oxazolidinone
(R)-(−)-Methylphenylpropylphosphine
N-Methylphenylsulfonimidoyl-methyllithium
(S)-N-Methylprolinol
2-Methylpyrrolidine
(S)-1-[1-Methyl-2-pyrrolidinyl]-methylpiperidine
(R)-(+)-Methyl *p*-tolyl sulfoxide
Neomenthyldiphenylphosphine
Nickel(II) chloride–α-2-Diphenyl-phosphineferrocenyl ethyldimethyl-amine
(S)-1-Nitroso-2-methylindoline-2-carboxylic acid
(−)-Norephedrine
[(2S)-(2α,3aα,4α,7α,7aα)]-2,3,3a,4,5,6,7,7a-Octahydro-7,8,8-trimethyl-4,7-methano-benzofuran-2-ol
2-Oxazolidones, chiral
Palladium(II) acetate–(S,R)-N,N-Dimethyl-1-[1′,2-bis(diphenyl-phosphine)ferrocenyl]ethylamine
Palladium(II) chloride + chiral co-reagents
(2R,4R)-Pentanediol
(S)-Phenylalanine
(+)-3-Phenylbornane-2,3-diol
S-(+)-2-Phenylbutyryl chloride

CHIRAL COMPOUNDS (*Continued*)
N-Phenylcampholylhydroxamic acid
8-Phenylmenthol and esters
(+)-1-Phenyl-2-propylamine
(R)-(+)-Phenyl (*p*-tolylsulfinyl)acetate
Pinanediol
(S)-(−)-Proline
(S)-Prolinol
Pyridinium (*d*)-camphor-10-sulfonate
Quinine
(−)-Sparteine
(+)-(R)-*trans*-β-Styryl *p*-tolyl sulfoxide
Tartaric acid
Tartrate esters
(R,R)- and (S,S)-Tetramethyltartaric acid diamide
p-Tolylsulfinylacetic acid
(R)-(4-Tolylsulfinyl)-ethyl-N-methoxyacetimidate
p-(Tolylsulfinyl)methyllithium
(R)-(+)-3-(*p*-Tolylsulfinyl)propionic acid
(S)-(+)-*p*-Tolyl *p*-tolylthiomethyl sulfoxide
p-Tolyl vinyl sulfoxide
N-*p*-Trifluoromethylbenzylcinchoninium bromide
(−)-4,6,6-Trimethyl-1,3-oxathiane
4,4,6-Trimethyl-1,3-oxathiane
Tris[3-(heptafluoropropylhydroxy-methylene)-*d*-camphorato]-europium(III)
(S)-Valine *t*-butyl ester
L-Valinol
(S)-2-(2,6-Xylidinomethyl)pyrrolidine

CHIRAL BASES
L-α,α′-Dimethyldibenzylamine
Lithium L-α,α′-dimethyldibenzylamide
Lithium (γ-methoxypropyl)-α-phenylethylamide
Lithium 2-(1-pyrrolidinyl)methyl-pyrrolidide
(γ-Methoxypropyl)-α-phenylethylamine
(S)-(+)-2-(1-Pyrrolidinyl)methyl-pyrrolidine

CHIRAL BORANES
B-Allyldiisocaranylborane
B-Allyldiisopinocampheylborane
Borane–(S)-(−)-2-Amino-3-methyl-1,1-diphenyl-1-butanol
Diisopinocampheylborane
Dilongifolylborane
NB-Enantrane
NB-Enantride
Lithium B-isopinocampheyl-9-borabicyclo[3.3.1]nonyl hydride
B-Methallyldiisopinocampheylborane
B-(3-Methyl-2-butenyl)-diisopinocampheylborane
Monoisopinocampheylborane
B-3-Pinanyl-9-borabicyclo[3.3.1]nonane
Sodium tris[(S)-N-benzyloxycarbonyl-prolyoxy]hydroborate
Triisopinocampheylborane

CHIRAL METAL HYDRIDES
Lithium aluminum hydride–(S)-4-Anilino-3-methylamino-1-butanol
Lithium aluminum hydride–(S)-2-(Anilinomethyl)pyrrolidine
Lithium aluminum hydride–3-O-Benzyl-1,2-O-cyclohexylidene-α-D-glucofuranose complex
Lithium aluminum hydride–(S,S)- or (R,R)-N,N′-Bis(methylbenzyl)-sulfamide
Lithium aluminum hydride–Darvon alcohol
Lithium aluminum hydride–(R)- and (S)-2,2′-Dihydroxy-1,1′-binaphthyl
Lithium aluminum hydride–(S)-(−)-10,10′-Dihydroxy-9,9′-biphenanthryl
Lithium aluminum hydride–(−)-N-Methylephedrine
Lithium aluminum hydride–(−)-N-Methylephedrine–3,5-Dimethylphenol
Lithium aluminum hydride–(−)-*cis*-Pinanediol
Lithium aluminum hydride–(S)-2-(2,6-Xylidinomethyl)pyrrolidine
Lithium tetrabutylaluminate–(−)-N-Methylephedrine
Sodium borohydride–(−)-Benzylquininium chloride
Sodium borohydride–1,2;5,6-Di-O-isopropylidene-α-D-glucofuranose
Sodium borohydride–(−)-N-Dodecyl-N-methylephedrinium bromide
Sodium borohydride–L-N-Hexadecyl-N-methylephedrinium bromide
Sodium tris[(S)-N-benzyloxycarbonyl-prolyoxy]hydroborate

CHIRAL RESOLVING AGENTS

(S)-(−)-1-Amino-2-(*t*-butyldialkylsilyloxymethyl)pyrrolidine
(+)-3-Aminomethylpinane
Amphetamine
(S)-(+)- and (R)-(−)-1,1′-Binaphthyl-2,2′-diyl hydrogen phosphate
(−)-α-Bromocamphor-10-sulfonic acid
Brucine
(+)-Camphanic acid
Camphor-10-sulfonic acid
d- and *l*-10-Camphorsulfonyl chloride
(−)-*trans*-Dichloro(ethylene)[α-methylbenzylamine]platinum(II)
(−)-2,3;4,6-Di-O-isopropylidene-2-keto-L-gulonic acid hydrate
Di-*p*-toluoyl-D-tartrate
(S)-N-Methanesulfonylphenylalanyl chloride
(R)- and (S)-α-Methoxy-α-trifluoromethylphenylacetic acid
α-Methylbenzylamine
(S)-1-α-Naphthylethylamine
(R)-1-(1-Naphthyl)ethyl isocyanate
[(2S)-(2α,3aα,4α,7α,7aα)]-2,3,3a,4,5,6,7,7a-Octahydro-7,8,8-trimethyl-4,7-methanobenzofuran-2-ol
(S)-N-Phthaloylphenylalanyl chloride
L-Pyrrolidonecarboxylic acid
(+)- and (−)-α-(2,4,5,7-Tetranitro-9-fluorenylideneaminooxy)propionic acid

OTHER CHIRAL REAGENTS AND STARTING MATERIALS
(R)-(+)-*o*-Anisylcyclohexylmethylphosphine
N-Benzylcinchoninium chloride
(−)-N-Benzyl-N-methylephedrinium bromide
N-Benzylquinidinium chloride
(−)-Benzylquininium chloride or fluoride
(S,S)-(+)-1,4-Bis(dimethylamino)-2,3-dimethoxybutane
t-Butyl hydroperoxide–Dialkyl tartrate–Dichlorodiisopropoxytitanium(IV)
t-Butyl hydroperoxide–Dialkyl tartrate–Titanium(IV) *t*-butoxide
t-Butyl hydroperoxide–Dialkyl tartrate–Titanium(IV) isopropoxide
t-Butyl hydroperoxide–Vanadyl acetylacetonate–N-Phenylcampholylhydroxamic acid
Butyllithium–(S,S)-(+)-1,4-Bis-(dimethylamino)-2,3-dimethoxybutane
[(−)-Camphor-10-ylsulfonyl]-3-aryloxaziridines
7-Chloro-5-phenyl-1-[(S)-α-phenylethyl]-1,3-dihydro-2H-1,4-benzodiazepine-2-one
Cyclic polyethers, chiral
R-(−)-(Dimethoxyphosphinyl)methyl *p*-tolyl sulfoxide
(−)-N,N-Dimethylephedrinium bromide
Dimethyl tartrate
(−)-N-Dodecyl-N-methylephedrinium bromide
(−)-N,N-Ethylmethylephedrinium bromide
Grignard reagents–(S,S)-(+)-1,4-Bis-(dimethylamino)-2,3-dimethoxybutane
L-N-Hexadecyl-N-methylephedrinium bromide
2,3-O-Isopropylidene-2,3-dihydroxy-1,4-bis(diphenylphosphine)butane
(S)-(+)-2-(*p*-Methoxyphenylsulfinyl)-2-cyclopentenone
(R)-4-Methylcyclohexylidenemethylcopper
(S)-(−)-Methyltri(2-methyl-1-butoxy)-titanium
Osmium tetroxide–Dihydroquinidine (or Dihydroquinine) acetate
Phosphonamides, chiral
L-(−)-Serine
Tartrate esters
(S,S)-(−)-1,2,3,4-Tetramethoxybutane
(S)-(+)-2-(*p*-Tolylsulfinyl)-2-buten-4-olide
(S)-2-(*p*-Tolylsulfiny)-2-cyclopentenone
N-*p*-Trifluoromethylbenzylcinchoninium bromide

DIELS–ALDER CATALYSTS

Alkylaluminum halides
Alumina
Aluminum chloride
Bornyloxyaluminum dichloride
Boron triacetate
Boron trifluoride
Boron trifluoride etherate
Copper(II) acetate–Copper(II) tetrafluoroborate

DIELS–ALDER CATALYSTS (*Continued*)
Copper(II) tetrafluoroborate
β-Cyclodextrin
Diazadieneiron(0) complexes
Dichlorodiisopropoxytitanium(IV)
Dichloromaleic anhydride
Diethylaluminum chloride
Dimethylaluminum chloride
Ethylaluminum dichloride
Florisil
Iron
Menthoxyaluminum dichloride
Nafion-H
Silica
Tin(IV) chloride
Titanium(IV) chloride
Tris(*p*-bromophenyl)ammoniumyl hexachloroantimonate
Tris(6,6,7,7,8,8,8-heptafluoro-2,2-dimethyl-3,5-octanedionato)-europium(III) or -ytterbium(III)
Tris[3-(heptafluoropropylhydroxymethylene)-*d*-camphorato]europium(III)
Zinc chloride
Zinc iodide

DIELS–ALDER DIENES
1-Acetoxybutadiene
2-Acetoxy-3-*p*-methoxyphenylthio-1,3-butadiene
2-Acetoxy-1-methoxy-3-trimethylsilyloxy-1,3-butadiene
1-Acetoxy-4-trimethylsilyl-1,3-butadiene
3-Acetoxy-1-trimethylsilyl-1,3-butadiene
N-Acetylpyrrole
Allylidenecyclopropane
Benzene
Benzyl *trans*-1,3-butadiene 1-carbamate
5-Benzyloxymethyl-1,3-cyclopentadiene
2,3-Bis(azidomethyl)-1,3-butadiene
2,3-Bis(bromomethyl)-1,3-butadiene
1,3-Bis(*t*-butyldialkylsilyloxy)-2-aza-1,3-dienes
2,3-Bis(methoxymethyl)-1,3-butadiene
1,2-Bis(methylthio)-1,3-butadiene
2,5-Bis[(Z)-(2-nitrophenylsulfenyl)-methylene]-3,6-dimethylene-7-oxabicyclo[2.2.1]heptane
2,3-Bis(trimethylsilylmethyl)-1,3-butadiene
1,3-Bis(trimethylsilyloxy)-1,3-butadiene
2,3-Bis(trimethylsilyloxy)-1,3-butadiene
1,3-Bis(trimethylsilyloxy)-1,3-cyclohexadiene
2,3-Bis(trimethylsilyloxy)-1,3-cyclohexadiene
2,5-Bis(trimethylsilyloxy)furans
2,3-Bis(trimethylstannyl)-1,3-butadiene
5-Bromocyclopentadiene
2-Bromomethyl-3-(trimethylsilylmethyl)-1,3-butadiene
1,3-Butadiene
2-*t*-Butoxycarbonyl-1,3-butadiene
1-Chloro-1-dimethylamino-2-methyl-1,3-butadiene
Cyclopentadiene
Cyclopentadienone ketals
(E,E)-1,4-Diacetoxybutadiene
2,5-Di-*p*-anisyl-3,4-diphenyl-cyclopentadienone
2,3-Dicyano-1,3-butadiene
3,4-Diethoxycarbonyl-2,5-dimethylfuran
5,5-Diethoxycyclopentadiene
1(E),3-Dimethoxybutadiene
5,5-Dimethoxycyclopentadiene
4,6-Dimethoxy-2-pyrone
5,5-Dimethoxy-1,2,3,4-tetrachloro-cyclopentadiene
1,3-Dimethoxy-1-trimethylsilyloxy-1,3-butadiene
1,1-Dimethoxy-1-trimethylsilyloxy-1,3-butadiene
1,2-Dimethoxy-1-trimethylsilyloxy-1,3-pentadiene
4-Dimethylamino-1,1,2-trimethoxy-butadiene
5,5-Dimethyl-1,3-bis(trimethylsilyloxy)-1,3-cyclohexadiene
2,3-Dimethylbutadiene
2,5-Dimethylfuran
Dimethyl 1,2,4,5-tetrazine-3,6-dicarboxylate
1,5-Dioxaspiro[4.5]deca-7,9-diene
2,3-Diphenylbutadiene
(E,E)-1,4-Diphenylbutadiene
1,3-Diphenylisobenzofuran
1,3-Diphenylnaphtho[2.3-c]furan
1-Ethoxy-2-ethyl-3-trimethylsilyloxy-1,3-butadiene
1-Ethoxy-4-tributylstannyl-1,3-butadiene
Ethyl *trans*-1,3-butadiene-1-carbamate
Furan
1,3,4,6-Heptatetraene

Hexabromocyclopentadiene
Hexachlorocyclopentadiene
Homophthalic anhydride
Indanocyclone
Isobenzofuran
Isoprene
2-(2-Methoxy)allylidene-1,3-dithiane
4-Methoxy-1,2-bis(trimethylsilyloxy)-1,3-butadiene
1-Methoxy-1,3-bis(trimethylsilyloxy)-1,3-butadiene
1-Methoxy-1,3-butadiene
6-Methoxy-(E,E)-3,5-hexadienoic acid
1-Methoxy-3-methyl-1,3-butadiene
5-Methoxymethylcyclopentadiene
3-Methoxymethylene-2,4-bis(trimethylsilyloxy)-1,4-pentadiene
2-Methoxy-3-methyl-1-phenylthio-1,3-butadiene
6-Methoxy-4-methyl-2-pyranone
1-Methoxy-2-methyl-3-trimethylsilyloxy-1,3-butadiene
1-Methoxy-3-methyl-1-trimethylsilyloxy-1,3-butadiene
1-Methoxy-2-methyl-3-trimethylsilyloxy-1,3-pentadiene
4-Methoxy-1-phenylseleno-2-trimethylsilyloxy-1,3-butadiene
(Z)-2-Methoxy-1-(phenylthio)-1,3-butadiene
2-Methoxy-3-phenylthio-1,3-butadiene
trans-1-Methoxy-3-trimethylsilyloxy-1,3-butadiene
5-Methyl-1,3-bis(trimethylsilyl)-1,3-cyclohexadiene
5-Methyl-1,3-bis(trimethylsilyloxy)-1,3-cyclohexadiene
3-Methyl-1,2-butadiene
Methyl *trans*-2,4-pentadienoate
(Z)-3-Methyl-1-phenylthio-2-trimethylsilyloxy-1,3-butadiene
Methyl α-pyrone-6-carboxylate
Methyl N-sulfonylurethane
1-*o*-Nitrophenyl-1,3-butadiene
1,3-Pentadiene
Perfluorotetramethylcyclopentadienone
4-Phenyloxazole
1-Phenylthio-1,3-butadiene
N-Phenyl-1,2,4-triazoline-3,5-dione
α-Pyrone
Tetrachlorocyclopentadienone
6,7,8,9-Tetrachloro-1,4-dioxaspiro[4.4]-nona-6,8-diene
Tetrachlorofuran
1,1,2,3-Tetrakis(trimethylsilyloxy)-1,3-butadiene
2,3,5,6-Tetramethylidene-7-oxanorbornane
Tetraphenylcyclopentadiene
Tetraphenylcyclopentadienone
1,2,4-Triazine
Tricarbonyl(1,3-cyclobutadiene)iron
2,2,2-Trichloroethyl *trans*-1,3-butadiene-1-carbamate
Trichloro-1,2,4-triazine
2-Triethylsilyl-1,3-butadiene
1-Trimethylsilyl-1,3-butadiene
2-Trimethylsilylmethyl-1,3-butadiene
1-Trimethylsilyloxy-1,3-butadiene
2-Trimethylsilyloxy-1,3-butadiene
2-Trimethylsilyloxy-1,3-cyclohexadiene
3-Trimethylsilyloxy-1,3-pentadiene
Trimethylsilylvinylketene
2-Trimethylstannylmethyl-1,3-butadiene

DIELS–ALDER DIENOPHILES

2-Acetoxyacrylonitrile
5-Acetoxymethyl-4-methoxy-*o*-benzoquinone
Acetylenedicarboxylic acid
3-Acetyl-4-oxazoline-2-one
Acrolein
Benzocyclopropene
1,4-Benzoquinone
(Z)-1,2-Bis(phenylsulfonyl)ethylene
1,2-Bis(β-tolylsulfonylethoxycarbonyl)-diazene
Bis(2,2,2-trichloroethyl) azodicarboxylate
2-Bromoacrolein
1-Bromo-2-chlorocyclopropene
Carbonyl cyanide
2-Chloroacrylonitrile
2-Chloroacrylyl chloride
Chloroketene dimethyl acetal
Chloromaleic anhydride
Chloromethylmaleic anhydride
1-Chloro-N,N,2-trimethyl-propenylamine
Chromium carbene complexes
Cyclopropene
trans-1,2-Dibenzoylethylene
Di-*t*-butyl azodiformate

DIELS–ALDER DIENOPHILES (*Continued*)
Dichlorovinylene carbonate
Dicyanoacetylene
1,2-Dicyanocyclobutene
1,3-Diethoxycarbonylallene
Diethylacetylene dicarboxylate
Diethyl azodicarboxylate
Diethyl oxomalonate
1,4-Dihydronaphthalene-1,4-endooxide
Dimethyl acetylenedicarboxylate
2,6-Dimethylbenzoquinone
Dimethyl maleate
2,3-Dimethylmaleic anhydride
Diphenylacetylene
Ethyl acrylate
Ethylene
Ethyl β-phenylsulfonylpropiolate
(R)-Ethyl *p*-tolylsulfinylmethylenepropionate
Ethynyl *p*-tolyl sulfone
Hexafluoro-2-butyne
(S)-(+)-α-Hydroxy-β,β-dimethylpropyl vinyl ketone
2-Hydroxy-5-oxo-5, 6-dihydro-2H-pyran
3-Hydroxy-2-pyrone
Isopropylidene isopropylidenemalonate
Itaconic anhydride
Maleic anhydride
4-Methoxy-5-methyl-*o*-benzoquinone
Methyl 2-acetylacrylate
Methyl acrylate
Methylene bis(urethane)
4-Methyloxazole
3-Methylsulfonyl-2,5-dihydrofuran
2-Methylthiomaleic anhydride
4-Methyl-1,2,4-triazoline-3,5-dione
Methyl vinyl ketone
Naphthacene-9,10,11,12-diquinone
3-Nitro-2-cyclohexenone
3-Nitro-2-cyclopentenone
Nitroethylene
Nitrosobenzene
Oxygen, singlet
p-Phenylazomaleinanil
N-Phenylmaleimide
8-Phenylmenthyl acrylate
8-Phenylmenthyl crotonate
(E)-1-Phenylsulfonyl-2-trimethylsilylethylene
Phenyl vinyl sulfone
Phenyl vinyl sulfoxide
1,4-Phthalazinedione
3-Sulfolene
Sulfur dioxide
Tetracyanoethylene
1,3,4-Thiadiazoline-2,5-dione
2-Thioxo-1,3-dioxol-4-ene
1,2,4-Triazoline-3,5-dione
Vinylene carbonate
Vinyltriphenylphosphonium bromide

HYDROGENATION CATALYSTS
trihapto-Allyltris(trimethyl phosphite)cobalt(I)
Arene(tricarbonyl)chromium complexes
[2,2′-Bis(diphenylphosphine)-1,1′-binaphthyl](cyclooctadiene)rhodium(I) perchlorate
[2,2′-Bis(diphenylphosphine)-1,1′-binaphthyl](norbornadiene)rhodium(I) perchlorate
[1,4-Bis(diphenylphosphine)butane](1,5-cyclooctadiene)iridium(I) tetrafluoroborate
1,4-[Bis(diphenylphosphine)butane](norbornadiene)rhodium(I) tetrafluoroborate
[1,2-Bis(diphenylphosphine)propane](norbornadiene)rhodium perchlorate
Carbonylchlorobis(triphenylphosphine)iridium(I)
Carbonylchlorobis(triphenylphosphine)rhodium(I)
Carbonyldihydridotris(triphenylphosphine)ruthenium
Carbonylhydridotris(triphenylphosphine)rhodium(I)
Chloro(hexamethylbenzene)hydridotriphenylphosphinerhodium
Chlorohydridotris(triphenylphosphine)ruthenium(II)
Chlorotris(neomenthyldiphenylphosphine)rhodium(I)
Chlorotris(triphenylphosphine)rhodium(I)
Chromium carbonyl
Copper chromite
Copper–Chromium oxide
(1,5-Cyclooctadiene)bis(methyldiphenylphosphine)iridium(I) hexafluorophosphate

(1,5-Cyclooctadiene)(pyridine)-(tricyclohexylphosphine)iridium(I) hexafluorophosphate
(1,5-Cyclooctadiene)(pyridine)-(triphenylphosphine)iridium(I) hexafluorophosphate
Di-μ-carbonyldecacarbonyltri-*triangulo*-iron
Dichlorobis(1,5-cyclooctadiene)-dirhodium + co-reagents
Di-μ-chlorobis(1,5-hexadiene)dirhodium + co-reagents
Dichlorobis(norbornadiene)dirhodium + co-reagents
Di-μ-chlorodichlorobis(pentamethyl-cyclopentadienyl)diiridium or -dirhodium
Dichloro(dimethylformamido)bis-(pyridine)rhodium borohydride
Di-μ-chlorotetra(ethylene)dirhodium–2,3-O-Isopropylidene-2,3-dihydroxy-1,4-bis(diphenylphosphine)butane
Di-μ-chlorotetrakis(cyclooctene)-dirhodium–2,3-O-Isopropylidene-2,3-dihydroxy-1,4-bis(diphenyl-phosphine)butane
Dichlorotris(triphenylphosphine)-ruthenium(II)
(S)-α-(R)-2-Diphenylphosphine-ferrocenyl ethyldimethylamine
Hydridodinitrogentris(triphenyl-phosphine)cobalt(I)
Hydridotris(triisopropylphosphine)-rhodium(I)
Hydrogen hexachloroplatinate(IV)–Triethylsilane
Iridium catalysts
Iridium(IV) chloride
Iron carbonyl
[2,3-O-Isopropylidene-2,3-dihydroxy-1,4-bis(diphenylphosphine)butane]-palladium
Lindlar catalyst
Magnesium oxide
Nickel(II) acetate–Sodium hydride–*t*-Amyloxide
Nickel–Alumina
Nickel boride
Nickel catalysts
Nickel–Graphite
Nickel on alumina
Osmium-on-carbon
Palladium(II) acetate–Sodium hydride–*t*-Amyloxide
Palladium catalysts
Palladium(II) chloride
Palladium–Graphite
Palladium hydroxide
Palladium hydroxide on barium sulfate
Palladium on barium sulfate
Palladium on calcium carbonate
Palladium on carbon
Palladium on charcoal
Palladium(0)–Phosphines
Palladium, poisoned
Palladium–Poly(ethylenimine)
Platinum catalysts
Platinum on carbon
Platinum oxide
Platinum–Silica
Platinum sulfide-on-carbon
Platinum–Tin(II) chloride
Raney cobalt catalyst
Raney nickel
Raney/nickel alloy
Rhenium catalysts
Rhodium catalysts
Rhodium on carbon
Rhodium oxide–Platinum oxide
Rosenmund catalyst
Ruthenium catalysts
Ruthenium(III) chloride
Ruthenium hydrogenation catalyst
Ruthenium on alumina
Ruthenium on carbon
Ruthenium–Silica
Trihydridobis(triphenylphosphine)-iridium(III)
Tris(dimethylphenylphosphine)-(norbornadiene)rhodium(I) hexafluorophosphate
Urushibara catalysts

METAL CARBONYLS

Aceto(carbonyl)cyclopentadienyl-(triphenylphosphine)iron
Arene(tricarbonyl)chromium complexes
Benzenechromium tricarbonyl
Benzylideneacetone(tricarbonyl)iron
(Butadiene)tricarbonyliron
Carbonylchlorobis(triphenylphosphine)-iridium(I)

METAL CARBONYLS (*Continued*)
Carbonylchlorobis(triphenylphosphine)-rhodium(I)
Carbonyldihydridotris(triphenyl-phosphine)ruthenium
Carbonylhydridotris(triphenyl-phosphine)rhodium(I)
Carbonylphenylbis(triphenyl-phosphine)rhodium(I)
π-(Chlorobenzene)chromium tricarbonyl
Chromium carbonyl
Copper(I) carbonyl ion
Dicarbonylbis(triphenylphosphine)nickel
Dicarbonylbis(triphenyl phosphite)nickel
Dicarbonylcyclopentadienylcobalt
Dicarbonyl(cyclopentadienyl)-[(dimethylsulfonium)methyl]iron tetrafluoroborate
Dicarbonyl(cyclopentadienyl)-[(methylphenylsulfonium)ethyl]iron trifluoromethanesulfonate
Dicarbonylcyclopentadienyl(2-methyl-1-propene)iron tetrafluoroborate
Di-μ-carbonyldecacarbonyltri-*triangulo*-iron
Di-μ-carbonyldicarbonylbis-(cyclopentadienyl)diiron
Dicarbonyldichlorobis(triphenyl-phosphine)ruthenium(II)
Di-μ-carbonylhexacarbonyldicobalt
Dipotassium tetracarbonylferrate
Disodium tetracarbonylferrate
Disodium tetracarbonylferrate-sesquidioxanate
Dodecacarbonyltri-*triangulo*-ruthenium
Hexacarbonyltetra-μ-chlorodiruthenium
Hexacarbonyl[μ-(N,N′,N″-tricyclohexyl-guanidinato(2-)]-N,N′;N,N′-diiron
Hydridotris(triphenylphosphine)-carbonylrhodium(I)
Iron carbonyl
Lithium tricarbonyl(dimethyl-carbamoyl)nickelate
Molybdenum carbonyl
Nickel carbonyl
Nickel carbonyl–Butyllithium
Pentacarbonyl(methoxyphenyl-methylene)chromium(0) or -tungsten(0)
Pentacarbonyl(trimethylsilyl)manganese
Pentacarbonyl(triphenylphosphine)-tungsten
Potassium di-μ-carbonyloctacarbonyl-μ-hydridochromate
Potassium tetracarbonylhydridoferrate
Sodium decacarbonyl-μ-hydrido-dichromate(1-)
Sodium dicarbonyl(cyclopentadienyl)-ferrate
Sodium di-μ-carbonylhexacarbonyl-μ-hydridoferrate
Sodium tetracarbonylcobaltate
Sodium tetracarbonylhydridoferrate
Sodium tricarbonyl(cyclopentadienyl)-vanadate(2-)
Tetracarbonyl(cyclopentadienyl)-vanadium
Tetracarbonyldi-μ-chlorodirhodium
Tetracarbonyldi-μ-chlorodirhodium–Lithium chloride
Tetracarbonyldodecacarbonyl-hexarhodium
Tetracarbonyl(ethoxyphenyl-methylidene)iron(0)
Tetracarbonylhydridocobalt
Tetramethylammonium tetracarbonylhydridoferrate
Tricarbonyl(1,3-cyclobutadiene)iron
Tri-μ-carbonylhexacarbonyldiiron
Tricarbonyl(*p*-methoxybenzylidene-acetone)iron
Tri-μ-carbonylnonacarbonyl-tetrarhodium
Tris(acetonitrile)tricarbonyltungsten
Tungsten carbonyl
Zinc tetracarbonylcobaltate

METAL-CONTAINING COMPOUNDS
ALUMINUM COMPOUNDS
Aklylaluminum halides
Allyldiisobutylaluminum
Alumina
Aluminum
Aluminum amalgam
Aluminum amalgam–Trifluoroacetic acid–*t*-Butanol
Aluminum azide
Aluminum bromide
Aluminum bromide–Ethanethiol
Aluminum *t*-butoxide
Aluminum *t*-butoxide–Raney nickel
Aluminum chloride
Aluminum chloride + co-reagents

Aluminum cyclohexoxide
Aluminum ethoxide
Aluminum hydride
Aluminum iodide
Aluminum isopropoxide
Aluminum telluride
Aluminum thiophenoxide
Bis(diethylaluminum) sulfate
Bis(diisobutylaluminum) 1,2-ethanedithiolate
Bis(diisopropylamino)aluminum hydride
Bis(dimethylaluminum) 1,2-ethanedithiolate
Bis(dimethylaluminum) 1,3-propanedithiolate
Bis(N-methylpiperazinyl)aluminum hydride
Bornyloxyaluminum dichloride
Butyllithium–Diisobutylaluminum hydride
(Carbo-*t*-butoxymethyl)diethylaluminum
μ-Chlorobis(cyclopentadienyl)(dimethylaluminum)-μ-methylenetitanium
Chloro(methoxy)methylalane
Chloro(methyl)aluminum amides
Chromium(III) chloride–Lithium aluminum hydride
Copper(I) bromide + aluminum co-reagents
Copper(II) chloride–Zinc chloride–Lithium aluminum hydride
Copper(I) iodide–Lithium trimethoxyaluminum hydride
Crotyldimethylaluminum
Cyanotrimethylsilane–Triethylaluminum
Dialkylaluminum amides
Dibromoalane
(2,6-Di-*t*-butyl-4-methylphenoxy)-methylaluminum trifluoromethanesulfonate
Dichloroalane
Dichlorobis(cyclopentadienyl)titanium + aluminum co-reagents
Diethoxyaluminum chloride
Diethylaluminum amides
Diethylaluminum anilide
Diethylaluminum benzenethiolate
Diethylaluminum chloride
Diethylaluminum cyanide
Diethylaluminum ethoxyacetylide
Diethylaluminum iodide
Diethylaluminum N-methylanilide
Diethylaluminum phenylacetylide
Diethylaluminum propylamide
Diethylaluminum 2,2,6,6-tetramethylpiperidide
Diethylaluminum trimethylsilylacetylide
Diethyl[dimethyl(phenyl)silyl]aluminum
N,N-(Diethylethanamine)trihydridoaluminum
Diethyl(iodomethyl)aluminum
Diiodomethane–Zinc–Trimethylaluminum
Diisobutylaluminum chloride
Diisobutylaluminum 2,6-di-*t*-butyl-4-methylphenoxide
Diisobutylaluminum hydride
Diisobutylaluminum phenoxide
Dimethylaluminum acetylide
Dimethylaluminum amides
Dimethylaluminum benzenethiolate
Dimethylaluminum benzylamide
Dimethylaluminum chloride
Dimethylaluminum cyanide
Dimethylaluminum iodide
Dimethylaluminum 2-methyl-2-propanethiolate
Dimethylaluminum methylselenolate
Dimethylaluminum phenylmethanethiolate
Ethylaluminum dichloride
Hydrogen cyanide + aluminum co-reagents
Isobornyloxyaluminum dichloride
Isobutylaluminum dichloride
Lithium aluminum hydride
Lithium aluminum hydride + co-reagents
Lithium *n*- or *t*-butyldiisobutylaluminum hydride
Lithium diethoxyaluminum hydride
Lithium diisobutylmethylaluminum hydride
Lithium methoxyaluminum hydride
Lithium tetrabutylaluminate–(–)-N-Methylephedrine
Lithium trialkoxyaluminum hydrides
Lithium tri-*t*-butoxyaluminum hydride
Lithium tri-*t*-butoxyaluminum hydride + co-reagents

METAL-CONTAINING COMPOUNDS (*Continued*)
Lithium triethoxyaluminum hydride
Lithium triethylborohydride–Aluminum *t*-butoxide
Lithium trimethoxyaluminum hydride
Lithium tris(3-ethyl-3-pentyloxy)-aluminum hydride
Manganese(II) chloride–Lithium aluminum hydride
Menthoxyaluminum dichloride
Methylaluminum bis(trifluoroacetate)
Methylaluminum dichloride
Methylmagnesium bromide–Nickel(II) acetylacetonate–Trimethylaluminum
Monochloroalane
Monoiodoalane
Nickel(II) acetylacetonate + aluminum co-reagents
Organoaluminum cyanides
Organoaluminum reagents
Sodium aluminum hydride
Sodium bis(2-methoxyethoxy)aluminum hydride
Sodium bis(2-methoxyethoxy)aluminum hydride + co-reagents
Sodium diethylaluminum hydride
Sodium tetrachloroaluminate
Sodium triethoxyaluminum hydride
Tetra-μ-hydridotetrahydroaluminum-magnesium
Tetrakis(2-methylpropyl)-μ-oxodi-aluminum
Tetramethylcyclobutadiene–Aluminum chloride
Tin–Aluminum
Titanium(IV) chloride + aluminum co-reagents
Titanium(III) chloride + aluminum co-reagents
Trialkylaluminums
Trialkynylalanes
Tri-*t*-butylaluminum
Tributyl (diethylaluminum)plumbate
Tributyl(diethylaluminum)tin
Tributyltinlithium–Diethylaluminum chloride
Trichlorocyclopentadienyltitanium–Lithium aluminum hydride
Triethylaluminum
Triisobutylaluminum
Triisobutylaluminum–Bis(N-methyl-salicylaldimine)nickel
Trimethylaluminum
Trimethylaluminum + co-reagents
Triphenylaluminum
Tripropylaluminum
Tris(1-hexynyl)aluminum
Tris(phenylethynyl)aluminum
Tris(trimethylsilyl)aluminum
Tris(trimethylsilylethynyl)aluminum
Tungsten(VI) chloride + aluminum co-reagents
Vanadium(III) chloride–Lithium aluminum hydride
Vilsmeier reagent–Lithium tri-*t*-butoxyaluminum hydride
Zirconium(IV) chloride–Lithium aluminum hydride

BORON COMPOUNDS
B-1-Alkenyl-9-borabicyclo[3.3.1]-nonanes
Alkyldimesitylboranes
B-1-Alkynyl-9-borabicyclo[3.3.1]-nonanes
Allenylboronic acid
B-Allyl-9-borabicyclo[3.3.1]nonane
B-Allyldiisocaranylborane
B-Allyldiisopinocampheylborane
Allyldimesitylborane
Allyl phenyl selenide–Trialkylborane
Alpine Borane
Benzeneboronic acid
Bis(benzoyloxy)borane
μ-Bis(cyanotrihydroborato)tetrakis-(triphenylphosphine)dicopper(I)
Bis(3,6-dimethyl)borepane
[Bis(1,3,2-dioxaborola-2-nyl)methyl]-lithium
Bis(*trans*-2-methylcyclohexyl)borane
Bis(trifluoroacetoxy)borane
Bis(triphenylphosphine)copper(I) borohydride
9-Borabicyclo[3.3.1]nonanates
9-Borabicyclo[3.3.1]nonane
9-Borabicyclo[3.3.1]nonane–Pyridine
9-Borabicyclononyl trifluoromethanesulfonate
Borane + co-reagents
Boric acid
Boronic acid resins
Boron oxide

Boron triacetate
Boron tribromide
Boron tribromide + co-reagents
Boron trichloride
Boron trichloride–Dimethyl sulfide
Boron trifluoride
Boron trifluoride + co-reagents
Boron triiodide
Boron tris(trifluoroacetate)
Boron trisulfide
Bromobis(isopropylthio)borane
B-Bromo-9-borabicyclo[3.3.1]nonane
Bromodiethylborane
Bromodimethylborane
α-Bromoethyldiethylborane
B-Butyl-9-borabicyclo[3.3.1]nonane
2-*t*-Butylperoxy-1,3,2-dioxaborolane
Cadmium borohydride–Dimethylformamide
Calcium borohydride
p-Carboxyphenyldihydroxyborane
Catecholborane
B-Chloro-9-borabicyclo[3.3.1]nonane
B-Chlorocatecholborane
2-Chloro-1,3,2-dioxaborolane
2-Chloro-1,3,2-dithioborolane
2-Chloromethyl-4,4,5,5-tetramethyl-1,3,2-dioxaborolane
Chlorothexylborane
Chlorothexylborane–Dimethyl sulfide
Cobalt boride–Borane–*t*-Butylamine
Copper(I) methyltrialkylborates
Copper(II) sulfate–Sodium borohydride
Corey's reducing agent
B-Crotyl-9-borabicyclo[3.3.1]nonane
Crotyldimethoxyborane
2-Crotyl-4,4,5,5-tetramethyl-1,3,2-dioxaborolane
Cyclohexyl metaborate trimer
Cyclopentylthexylboryl trifluoromethanesulfonate
B-Cyclopropyl-9-borabicyclo[3.3.1]-nonane
Dialkylboryl trifluoromethanesulfonates
Diborane
Diborane + co-reagents
Dibromobis(tributylphosphine)-nickel(II)–Sodium borohydride
Dibromoborane–Dimethyl sulfide
Dibutylboryl trifluoromethanesulfonate
Di-*t*-butyl(*t*-butylthio)borane
Dibutylchloroborane
Dichlorobis(cyclopentadienyl)titanium–Lithium borobydride
Dichloroborane
Dichloroborane diethyl etherate
Dichloroborane–Dimethyl sulfide
Dichlorocyclohexylborane
Dichloro(diethylamino)borane
Dichloro(dimethylformamido)bis-(pyridine)rhodium borohydride
Dichloromethaneboronic acid
2-Dichloromethyl-4,5-dimethyl-1,3,2-dioxaborolane
Dichlorophenylborane
Dicyclohexylborane
Dicyclohexylmethylthioborane
Dicyclopentylboryl trifluoromethane-sulfonate
Diethylboryl trifluoromethanesulfonate
p-Dihydroxyborylbenzyloxycarbonyl chloride
Diiodoborane
Diisopinocampheylborane
Dilongifolylborane
Di-2-mesitylborane
Dimesityl(methyl)borane
B-3,3-Dimethyl-1-butynyl-9-borabicyclo[3.3.1]nonane
Diphenylborane
Dipropylboryl trifluoromethane-sulfonate
Disiamylborane
NB-Enantrane
NB-Enantride
Ethane-1,2-diaminoborane
B-Fluorodimesitylborane
Fluorodimethoxyborane
B-Halo-9-borabicyclo[3.3.1]nonane
B-Hexyl-9-borabicyclo[3.3.1]nonane
B-1-Hexynyl-9-borabicyclo[3.3.1]nonane
Hydrogen hexachloroplatinate(IV)–Sodium borohydride
10-Hydroxy-10,9-boroxarophenanthrene
8-Hydroxyquinolinedihydroboronite
B-Iodo-9-borabicyclo[3.3.1]nonane
Lithium 9-boratabicyclo[3.3.1]nonane
Lithium borohydride
Lithium butylborohydride
Lithium cyanoborohydride
Lithium 9, 9-dibutyl-9-borabicyclo-[3.3.1]nonanate

METAL-CONTAINING COMPOUNDS (*Continued*)

Lithium dimesitylborohydride bis(dimethoxymethane)
Lithium B-isopinocampheyl-9-borabicyclo[3.3.1]nonyl hydride
Lithium methylsulfinylmethylide–Tributylborane
Lithium perhydro-9b-boraphenalylhydride
Lithium tri-*t*-butoxyaluminum hydride–Triethylborane
Lithium tri-*sec*-butylborohydride
Lithium triethylborohydride
Lithium triethylborohydride–Aluminum *t*-butoxide
Lithium trisiamylborohydride
Lithium tris(*trans*-2-methylcyclopentyl)-borohydride
Mercury(II) acetate–Sodium trimethoxyborohydride
B-Methallyldiisopinocampheylborane
Methanol–Sodium tetraborate
3-Methoxycatecholborane
B-Methyl-9-borabicyclo[3.3.1]nonane
B-(3-Methyl-2-butenyl)diisopino-campheylborane
B-(3-Methyl-2-butyl)-9-borabicyclo-[3.3.1]nonane
Monochloroborane diethyl etherate
Monochloroborane–Dimethyl sulfide
Monoisopinocampheylborane
Nickel(II) acetylacetonate–Diisopropoxyphenylphosphine–Sodium borohydride
4-Nitrocatecholborane
B-3-Pinanyl-9-borabicyclo[3.3.1]nonane
Potassium borohydride
Potassium borohydride–(−)-N-Dodecylmethylephedrinium bromide
Potassium 9-(2,3-dimethyl-2-butoxy)-9-boratabicyclo[3.3.1]nonane
Potassium tri-*sec*-butylborohydride
Potassium tri-*sec*-butylborohydride–Copper(II) iodide
Potassium triisopropoxyborohydride
Sodium acetanilidoborohydride
Sodium acetoxyborohydride
Sodium acyloxyborohydrides
Sodium borohydride
Sodium borohydride + co-reagents
Sodium (*t*-butylamino)borohydride
Sodium cyanoborohydride
Sodium cyanoborohydride + co-reagents
Sodium 9-cyano-9-hydrido-9-borabicyclo[3.3.1]nonane
Sodium (dimethylamino)borohydride
Sodium perborate
Sodium tetraborate
Sodium triacetoxyborohydride
Sodium triethylborohydride + co-reagents
Sodium trifluoroacetoxyborohydride
Sodium triisopropoxyborohydride
Sodium–Trimesitylborane
Sodium trimethoxyborohydride
Sodium tris[(S)-N-benzyloxycarbonyl-prolyoxy]hydroborate
Sodium tris(3,5-di-*t*-butylphenoxy)-borohydride
Sodium tris(3,5-dimethylphenoxy)-borohydride
Tetrabutylammonium borohydride
Tetrabutylammonium cyanoborohydride
Tetrabutylammonium octahydrotriborate
Tetraethylammonium borohydride
Tetrafluoroboric acid
Tetralin hydroperoxide–Cyclohexyl metaborate
Tetramethylammonium borohydride
Thexylborane
Thexylborane–N,N-Diethylaniline
Tin(II) chloride–Sodium borohydride
Tin(IV) chloride–Sodium borohydride
Titanium(IV) chloride–Sodium borohydride
Tributylborane
Tributylboroxine
Tricaprylylmethylammonium borohydride
Tricyclohexylborane
Tricyclopentylborane
Triethanolamine borate
Tetraethylammonium borohydride
Triethylborane
Trihexylborane
Triisobutylborane
Triisopinocampheylborane
Trimesitylborane

Trimethyl borate
Trimethylboroxine
α-Trimethylsilylcrotyl-9-borabicyclo-[3.3.1]nonane
Trimethylsilylmethyldimesitylborane
Tri-2-norbornylborane
Triphenylborane
Tris(dimethoxyboryl)methane
Tris(dimethylamino)borane
Tris(1,3,2-dioxaborolan-2-yl)methane
Tris(ethylthio)borane
Tris(methylseleno)borane
Tris(phenylseleno)borane
Tris(phenylthio)borane
Zinc borohydride
Zinc borohydride–Dimethylformamide

CHROMIUM COMPOUNDS
Arene(tricarbonyl)chromium complexes
Benzenechromium tricarbonyl
2,2′-Bipyridinium chlorochromate
Bis(benzyltriethylammonium) dichromate
Bis(butylammonium) dichromate
t-Butyl chromate
π-(Chlorobenzene)chromium tricarbonyl
Chromium(II) acetate
Chromium(II) acetate–Ethylenediamine
Chromium(II)–Amine complexes
Chromium carbene complexes
Chromium carbonyl
Chromium(II) chloride
Chromium(III) chloride
Chromium(III) chloride–Lithium aluminum hydride
Chromium(VI) oxide
Chromium(VI) oxide + co-reagents
Chromium(II) perchlorate
Chromium(II) perchlorate + co-reagents
Chromium(II) sulfate
Chromyl acetate
Chromyl chloride
Chromyl chloride + co-reagents
Chromyl trichloroacetate
Collins reagent
Collins reagent–Celite
Copper–Chromium oxide
Cornforth reagent
4-(Dimethylamino)pyridinium chlorochromate
Fieser reagent
Jones reagent
Palladium(II) chloride–Potassium dichromate
Pentacarbonyl(methoxyphenyl-methylene)chromium(0)
Periodic acid–Chromium(VI) oxide
Potassium chromate
Potassium chromate–Dicyclohexyl-18-crown-6
Potassium di-μ-carbonyloctacarbonyl-μ-hydridochromate
Potassium dichromate
Potassium dichromate + co-reagents
Potassium perchromate
Pyridinium chlorochromate
Pyridinium chlorochromate + co-reagents
Pyridinium dichromate
Pyridinium dichromate–Acetic anhydride
Pyridinium fluorochromate
Pyridinium tetrafluoroboratochromate
Ratcliff's reagent
Sarett reagent
Silver chromate
Silver chromate–Iodine
Sodium chromate
Sodium decacarbonyl-μ-hydridodichromate(1-)
Sodium dichromate
Sodium dichromate + co-reagents
Tetrabutylammonium chlorochromate
Tetrabutylammonium chromate

COBALT COMPOUNDS
trihapto-Allyltris(trimethyl phosphite)-cobalt(I)
Bis[(−)-camphorquinone-α-dioximato]-cobalt(II) hydrate
Bis(cyclopentadienyl)bis(μ-nitrosyl-N:N)dicobalt
Bis(N-propylsalicylideneaminate)-cobalt(II)
Bis(3-salicylideneaminopropyl)-aminecobalt(II)
[Bis(salicylidene-γ-iminopropyl)]-methylaminecobalt(II)
Chloro(pyridine)cobaloxime(III)
Chlorotris(triphenylphosphine)cobalt
Cobaloxime(I)
Cobalt(II) acetate
Cobalt(III) acetate

METAL-CONTAINING COMPOUNDS (*Continued*)
Cobalt(II) acetate–Hydrogen bromide
Cobaltacyclopentan-2-ones
Cobalt boride–Borane–*t*-Butylamine
Cobalt(II) chloride
Cobalt(III) fluoride
Cobalt(II) phthalocyanine
Cobalt *meso*-tetraphenylporphine
Cobalt(III) trifluoroacetate
Cyclopentadienyl(diiodo)triphenylphosphinecobalt
Cyclopentadienyl(triphenylphosphine)cobaltatetraphenylcyclopentadiene
Dibromobis(triphenylphosphine)cobalt(II)
Dicarbonylcyclopentadienylcobalt
Di-μ-carbonylhexacarbonyldicobalt
Dicyclopentadienylcobalt
Dicyclopentadienylcobalt–Oxygen
Diiodotris(triphenylphosphine)cobalt
Grignard reagents–Cobalt(II) chloride
Hydridodinitrogentris(triphenylphosphine)cobalt(I)
4-Hydroxysalcomine
Lithium trimethylcobaltate
Phenylmagnesium bromide–Cobalt(II) chloride
Raney cobalt catalyst
Salcomine
Sodium tetracarbonylcobaltate
Tetracarbonylhydridocobalt
Zinc tetracarbonylcobaltate
COPPER COMPOUNDS
Alkylcopper reagents
Alkylcopper reagents + co-reagents
Allylcopper
Bis(N-*t*-butylsalicylaldiminato)copper(II)
μ-Bis(cyanotrihydroborato)tetrakis(triphenylphosphine)dicopper(I)
Bis(N-propylsalicylideneaminate)copper(II)
Bis(triphenylphosphine)copper(I) borohydride
Bromomagnesium alkyl(methyl)cuprates
Butylcopper
Butylcopper + co-reagents
3-Chloro-4,5-dihydrofuryl-2-copper
Chloromagnesium dimethylcuprate
Copper
Copper(I) acetate
Copper(II) acetate
Copper(I) acetate–*t*-Butyl isocyanide
Copper(II) acetate + co-reagents
Copper(I) acetylacetonate
Copper(II) acetylacetonate
Copper(I) alkoxides
Copper(II)–Amine complexes
Copper–Ascorbic acid
Copper–Benzoic acid
Copper(I) bromide
Copper(II) bromide
Copper(I) bromide + co-reagents
Copper(II) bromide–Nitric oxide
Copper bronze
Copper(I) *t*-butoxide
Copper *t*-butylacetylide
Copper(0)–*t*-Butyl isocyanide
Copper carbonate, basic
Copper(I) carbonyl ion
Copper(I) chloride
Copper(II) chloride
Copper(I) chloride + co-reagents
Copper(II) chloride + co-reagents
Copper chromite
Copper chromite–Quinoline
Copper–Chromium oxide
Copper(I) cyanide
Copper(II) cyanide
Copper(I) cyanoacetate
Copper(II) cyanoacetate
Copper halide nitrosyls
Copper(II) hexafluoroacetylacetonate
Copper(I) hexamethyldisilazide
Copper hydride ate complexes
Copper(II) hydroxide
Copper(I) iodide
Copper(I) iodide + co-reagents
Copper(0)–Isonitrile complexes
Copper(II) methoxide
Copper(I) methyltrialkylborates
Copper(II) nitrate + co-reagents
Copper(I) oxide
Copper(II) oxide
Copper(I) oxide–*t*-Butyl isocyanide
Copper(I) oxide–Copper(II) nitrate
Copper(II) perchlorate
Copper(II) permanganate
Copper(I) phenoxides
Copper(I) phenylacetylide
Copper(I) phenylacetylide–

Tributylphosphine
Copper(II) sulfate
Copper(II) sulfate + co-reagents
Copper(II) tetrafluoroborate
Copper(I) trifluoromethanesulfonate
Copper(II) trifluoromethanesulfonate
Copper(I) trimethylsilylacetylide
Cyanomethylcopper
Di-μ-chlorodimethoxybis(pyridine)-dicopper
(Dicyclohexylamido)copper(I)
3,3-Diethoxy-1-propenyl-2-copper
Dilithium dialkyl(cyano)cuprates
Dilithium pentamethyltricuprate
Dilithium tetrachlorocuprate(II)
Dilithium tributylcuprate
Dilithium trimethylcuprate
Dilithium tris(1-pentynyl)cuprate
Diphenylphosphidocopper(I)
1-Hexynylcopper
Hydrido(tributylphosphine)copper(I)
Lithium alkyl(*t*-butoxy)cuprates
Lithium alkyl(cyano)cuprates
Lithium alkyl(diethylamino)cuprates
Lithium alkyl(phenoxy)cuprates
Lithium alkyl(phenylthio)cuprates
Lithium anilidocyanocuprates
Lithium bis(alkylamino)cuprates
Lithium bis(3,3-diethoxy-1-propen-2-yl)-cuprate
Lithium bis(N,N-diethylcarbamoyl)-cuprate
Lithium bis(dimethylphenylsilyl)cuprate
Lithium bis(N-methyl-N-phenyl-carbamoyl)cuprate
Lithium bis(3-trimethylsilyl-1-propen-2-yl)cuprate
Lithium bis(2-vinylcyclopropyl)cuprate
Lithium bromo(methyl)cuprate–Diisobutylamine complex
Lithium *t*-butoxy(*t*-butyl)cuprate
Lithium butyl(cyano)cuprate
Lithium (*t*-butylethynyl)methylcuprate
Lithium *t*-butylethynyl(vinyl)cuprate
Lithium butyl(hydrido)cuprate
Lithium *n*- or *t*-butyl(3-methoxy-3-methylbutynyl)cuprate
Lithium *t*-butyl(phenylthio)cuprate
Lithium *t*-butylthio(alkyl)cuprates
Lithium cyano(dimethylphenylsilyl)-cuprate
Lithium cyano(isohexyl)cuprate
Lithium cyano(methyl)cuprate
Lithium cyano(phenyl)cuprate
Lithium cyano(trimethylsilyl)cuprate
Lithium cyclopropyl(phenylthio)cuprate
Lithium diallylcuprate
Lithium dibutylcuprate
Lithium di-*t*-butylcuprate
Lithium (1,1-diethoxy-2-propenyl)-(3,3-dimethyl-1-butynyl)cuprate
Lithium (3,3-diethoxy-1-propen-2-yl)-phenylthiocuprate
Lithium di(1-ethoxyvinyl)cuprate
Lithium di(2-ethoxyvinyl)cuprate
Lithium diethylcuprate
Lithium diethylcuprate–Tributyl-phosphine
Lithium di(α-methoxyvinyl)cuprate
Lithium dimethylcuprate
Lithium dimethylcuprate–Boron trifluoride
Lithium diphenylcuprate
Lithium di(2-propenyl)cuprate
Lithium di[(E)-1-propenyl]cuprate
Lithium dipropylcuprate
Lithium di(*p*-tolyl)cuprate
Lithium divinylcuprate
Lithium divinylcuprate–Tributyl-phoshine
Lithium α-ethoxycarbonylvinyl-(1-hexynyl)cuprate
Lithium (1-hexynyl)(2-tributyl-stannylvinyl)cuprate
Lithium α-methoxycarbonylvinyl-(1-hexynyl)cuprate
Lithium methyl(phenylseleno)cuprate
Lithium methyl(phenylthio)cuprate
Lithium methyl(vinyl)cuprate
Lithium (1-pentynyl)(3-tetrahydro-pyranyloxy-1-propenyl)cuprate
Lithium (1-pentynyl)(2-tributyl-stannylvinyl)cuprate
Lithium phenylthio(trimethylstannyl)-cuprate
Lithium phenylthio(2-vinylcyclopropyl)-cuprate
Magnesium hydride–Copper(I) *t*-butoxide
Magnesium hydride–Copper(I) iodide
4-Methoxy-3-buten-1-ynylcopper
Methoxycarbonylcopper

METAL-CONTAINING COMPOUNDS (*Continued*)
3-Methoxy-3-methylbutynylcopper
3-Methyl-3-buten-1-ynylcopper
Methylcopper
Methylcopper + co-reagents
(R)-4-Methylcyclohexylidenemethylcopper
Organocopper reagents
Organolithium reagents–Copper(I) halides
Palladium(II) chloride + co-reagents
Pentafluorophenylcopper
1-Pentynylcopper
1-Pentynylcopper–Hexamethylphosphoric triamide
Phenylcopper
Phenylcopper–Boron trifluoride
Phenylselenocopper
Phenylthiocopper
Sodium dicyanocuprate
Tetrakis(acetonitrile)copper(I) perchlorate or tetrafluoroborate
Tetrakis(pyridine)copper(I) perchlorate
Tributyltincopper
Trifluoromethylcopper
Trifluoromethylthiocopper
Trimethylsilylcopper
Trimethylsilylmethylcopper
1-Trimethylsilylpropynylcopper
Trimethylstannylcopper–Dimethyl sulfide
Tris(2-picoline)copper(I) perchlorate
Tris(trimethylsilyl)hydrazidocopper
Vinylcopper
IRIDIUM COMPOUNDS
[1,4-Bis(diphenylphosphine)butane]-(1,5-cyclooctadiene)iridium(I) tetrafluoroborate
Carbonylchlorobis(triphenylphosphine)-iridium(I)
(1,5-Cyclooctadiene)bis(methyldiphenylphosphine)iridium(I) hexafluorophosphate
(1,5-Cyclooctadiene)(pyridine)-(tricyclohexylphosphine)iridium(I) hexafluorophosphate
(1,5-Cyclooctadiene)(pyridine)-(triphenylphosphine)iridium(I) hexafluorophosphate
Di-μ-chlorobis(1,5-cyclooctadiene)-diiridium
Di-μ-chlorodichlorobis(pentamethylcyclopentadienyl)diiridium
Di-μ-chlorotetrakis(cyclooctene)-diiridium
Disodium hexachloroiridate(IV)
Hydrogen hexachloroiridate(IV)
Iridium(III) chloride
Iridium(IV) chloride
Iridium–Silica
Trihydridobis(triphenylphosphine)-iridium(III)
IRON COMPOUNDS
Aceto(carbonyl)cyclopentadienyl-(triphenylphosphine)iron
Benzylideneacetone(tricarbonyl)iron
1,1′-Bis(diphenylphosphine)ferrocene
[1,1′-Bis(diphenylphosphine)ferrocene]-dichloronickel(II)
[1,1′-Bis(diphenylphosphine)ferrocene]-(dichloro)palladium(II)
(R)-1-(S)-1′,2-Bis(diphenylphosphine)-ferrocenylethanol
Bis(norbornadiene)rhodium(I) perchlorate–(R)-1-(S)-1′,2-Bis-(diphenylphosphine)ferrocenyl-ethanol
(Butadiene)tricarbonyliron
Butylmagnesium bromide–Iron(III) reagents
Diazadieneiron(0) complexes
Dicarbonyl(cyclopentadienyl)-[(dimethylsulfonium)methyl]iron tetrafluoroborate
Dicarbonyl(cyclopentadienyl)-[(methylphenylsulfonium)ethyl]iron trifluoromethanesulfonate
Dicarbonylcyclopentadienyl(2-methyl-1-propene)iron tetrafluoroborate
Di-μ-carbonyldecacarbonyltri-*triangulo*-iron
Di-μ-carbonyldicarbonylbis-(cyclopentadienyl)diiron
Di-μ-chlorobis(1,5-hexadiene)-dirhodium–(SαR)-2-Diphenyl-phosphineferrocenyl ethyldimethylamine
(S,R)-N,N-Dimethyl-1-[1′,2-bis-(diphenylphosphine)ferrocenyl]-ethylamine
(SαR)- or (RαS)-2-Diphenylphosphine-

ferrocenyl ethyldimethylamine
Dipotassiumtetracarbonylferrate
Disodium tetracarbonylferrate
Disodium tetracarbonylferrate-sesquidioxanate
Ferric acetylacetonate
Ferric azide
Ferric chloride
Ferric chloride + co-reagents
Ferric ethoxide
Ferric nitrate
Ferric nitrate/K10 Bentonite
Ferric selenocyanate
Ferric sulfate
Ferric thiocyanate
Ferrocenylphosphines
Ferrous chloride
Ferrous chloride–Oxygen
Ferrous perchlorate
Ferrous sulfate
Ferrous sulfate–Oxygen
Grignard reagents–Iron(III) reagents
Hexacarbonyl[μ-(N,N′,N″-tricyclohexylguanidinato(2-)]N,N′;N,N′-diiron
Iron
Iron–Acetic acid
Iron(III) acetylacetonate
Iron-carbene complexes
Iron carbonyl
Iron–Graphite
Isopropylmagnesium chloride–Iron(III) reagents
Lithium trimethylferrate
Methylmagnesium bromide–Iron(III) reagents
Nickel(II) chloride–α-2-Diphenylphosphineferrocenyl ethyldimethylamine
μ_3-Oxohexakis(μ-trimethylacetato)-trimethanoltriiron(III) chloride
Palladium(II) acetate–(S,R)-N,N-Dimethyl-1-[1′,2-bis(diphenylphosphine)ferrocenyl]ethylamine
Palladium(II) chloride–α-2-Diphenylphosphineferrocenyl ethyldimethylamine
Potassium ferrate (VI)
Potassium ferricyanide
Potassium tetracarbonylhydridoferrate
Sodium dicarbonyl(cyclopentadienyl)-ferrate
Sodium di-μ-carbonylhexacarbonyl-μ-hydridoferrate
Sodium tetracarbonylhydridoferrate
Tetracarbonyl(ethoxyphenylmethylidene)iron(0)
Tetramethylammonium tetracarbonylhydridoferrate
Tricarbonyl(1,3-cyclobutadiene)iron
Tri-μ-carbonylhexacarbonyldiiron
Tricarbonyl(*p*-methoxybenzylideneacetone)iron
Tris(dibenzoylmethide)iron(III)
Tris(tetrabutylammonium)-hexacyanoferrate(III)
LEAD COMPOUNDS
Aryllead triacetates
(3,3-Dichloroallyl)triphenyllead
Lead
Lead(II) acetate
Lead(IV) acetate azides
Lead carbonate
Lead(IV) fluoride
Lead(II) hydroxide
Lead(II) oxide
Lead(IV) oxide
Lead(IV) oxide–Boron trifluoride etherate
Lead tetraacetate
Lead tetraacetate + co-reagents
Lead tetrabenzoate
Lead tetrakis(trifluoroacetate)
Lead tetrakis(trifluoroacetate)–Diphenyl disulfide
Lead(II) thiocyanate
Phosphatolead(IV) acids
Tetravinyllead
Tributyl(diethylaluminum) plumbate
LITHIUM COMPOUNDS (excluding ionic, inorganic salts)
Acyllithium reagents
Alkoxymethyllithium reagents
Alkyllithium reagents
1-(Alkylthio)vinyllithium
Allyl alcohol, O,2-dilithio derivative
Allyllithium
Benzyl lithioacetate
Benzyllithium
[Bis(1,3,2-dioxaborola-2-nyl)methyl]-lithium
1,3-Bis(methylthio)allyllithium

METAL-CONTAINING COMPOUNDS

(*Continued*)
Bis(methylthio)methyllithium
Bis(methylthio)(trimethylsilyl)-methyllithium
Bis(methylthio)(trimethylstannyl)-methyllithium
Bis(phenylthio)methyllithium
Bis(trimethylsilyl) lithiomalonate
1-Bromo-1-cyclopropyllithium
1-Bromo-2-ethoxycyclopropyllithium
1-Bromo-1-lithiocyclopropanes
Bromomethyllithium
2-(6-Bromopyridyl)lithium
t-Butoxymethyllithium
t-Butyl 2-chloro-2-lithiotrimethyl-silylacetate
t-Butyl dilithioacetoacetate
t-Butyldiphenylsilyllithium
t-Butyl lithioacetate
t-Butyl lithiobis(trimethylsilyl)acetate
t-Butyl α-lithioisobutyrate
t-Butyl lithio(trimethylsilyl)acetate
Butyllithium
sec-Butyllithium
t-Butyllithium
Butyllithium + co-reagents
sec-Butyllithium + co-reagents
t-Butyllithium–Tetramethyl-ethylenediamine
S-*t*-Butylthio lithioacetate
Carbon tetrabromide–Methyllithium
α-Chloroallyllithium
4-Chloro-1-butenyl-2-lithium
1-Chloro-2-butenyllithium
(3-Chloro-3-methyl-1-butynyl)lithium
Chloromethyllithium
3-Chloro-2-methyl-1-propenyl-3-lithium
5-Chloro-1-pentenyl-2-lithium
Chloro(phenylsulfinyl)methyllithium
Chlorotrimethylsilane–Lithium
1-Chloro-1-(trimethylsilyl)ethyllithium
Chloro(trimethylsilyl)methyllithium
Crotyllithium
1-Cyclobutenylmethyllithium
1-Diazolithioacetone
Diazo(trimethylsilyl)methyllithium
1,1-Dibromoalkyllithiums
Dibromomethane–Lithium
Dibromomethyllithium
1,3-Dibutoxy-1-propenyllithium
gem-Dichloroallyllithium
α, α-Dichlorobenzyllithium
Dichloromethyllithium
(Diethoxyphosphinyl)difluoro-methyllithium
3,3-Diethoxy-1-propenyl-2-lithium
3,3-Diethoxy-1-propynyllithium
Diethyl dichlorolithiomethyl-phosphonate
Diethyl lithio-N-benzylideneamino-methylphosphonate
Diethyl lithiomorpholinomethyl-phosphonate
Diethyl lithiosuccinate
gem-Difluoroallyllithium
2,2-Difluoro-1-tolylsulfonyloxy-vinyllithium
4,5-Dihydro-5-methyl-1,3,5-dithiazin-2-yllithium
Diiodomethyllithium
Diisopropylcarbamoyllithium
Dilithioacetate
1,3-Dilithiopropyne
Dilithium acetylide
2,2′-Dilithiumbiphenyl
2-(2′,2′-Dimethoxyethyl)-1,3-dithianyl-2-lithium
(Dimethoxyphosphinyl)methyllithium
1,2-Dimethoxyvinyllithium
(Dimethylcarbamoyl)lithium
Dimethylphenylsilyllithium
Dimethylsulfamoylmethyllithium
Dimethylthiocarbamoyllithium
(Diphenylarsinyl)methyllithium
1,1-Diphenylhexyllithium
Diphenylmethyllithium
(Diphenylphosphine)lithium
1-Diphenylphosphino-1-methoxymethyllithium
4-Ethoxy-1,3-butadienyllithium
4-Ethoxy-3-buten-1-ynyllithium
1-Ethoxycyclopropyllithium
(Z)-2-Ethoxyvinyllithium
α-Ethoxyvinyllithium
Ethyl diazolithioacetate
Ethyl lithioacetate
Ethyl lithiopropiolate
Ethyl lithio(trimethylsilyl)acetate
Ethyllithium
Ethyllithium–Tetramethyl-ethylenediamine

Ferric chloride–Butyllithium
β-Furyllithium
Isocyanomethyllithium
Isocyanotolylthiomethyllithium
Lithioacetonitrile
2-Lithiobenzothiazole
1-Lithiocyclopropyl methyl selenide
1-Lithiocyclopropyl phenyl sulfide
5-Lithio-2,3-dihydrofuran
6-Lithio-3,4-dihydro-(2*H*)-pyran
Lithio-N,N-dimethylacetamide
Lithio-N,N-dimethylthiopivalamide
2-Lithio-1,3-dithianes
α-Lithio-α-methoxyallene
1-Lithio-1-methoxycyclopropane
1-Lithio-2-methoxycyclopropane
2-(2-Lithio-4-methoxyphenyl)-4,4-dimethyl-2-oxazoline
1-Lithio-3-methoxy-3-trimethylsilylallene
2-Lithio-2-methyl-1,3-dithiane
α-Lithiomethylenetriphenylphosphorane
2-Lithio-2-methylpropionitrile
α-Lithiomethylselenocyclobutane
2-Lithio-2-methylthio-1,3-dithiane
1-Lithio-1-methyl-2-vinylcyclopropane
2-Lithiopropionitrile
2-Lithio-2-propyl-1,3-benzodithiolane
Lithio propylidene-*t*-butylimine
1-Lithio-3,3,6,6-tetramethoxy-1,4-cyclohexadiene
2-Lithio-3,3,6,6-tetramethoxy-1,4-cyclohexadiene
6-Lithio-2*H*-thiopyran
3-Lithio-1-triisopropylsilyl-1-propyne
Lithiotrimethylsilylacetonitrile
2-Lithio-2-trimethylsilyl-1,3-dithiane
3-Lithio-1-trimethylsilyl-1-propyne
1-Lithio-3-trimethylsilyl-1-propyne
2 Lithio 1,3,5-trithianc
1-Lithio-2-vinylcyclopropane
2-Lithiovinyltributyltin
2-(E)-Lithiovinyltrimethylsilane
(1-Lithiovinyl)trimethylsilane
(1-Lithiovinyl)triphenylsilane
Lithium
Lithium acetylide
Lithium acetylide–Ethylenediamine
Lithium acetylides
Lithium amalgam
Lithium 1,3-butadien-1-olate
Lithium chloroacetylide
Lithium 3-chloropropargylide
Lithium cryptates
Lithium ethoxyacetylide
Lithium *o*-lithiobenzoate
Lithium *o*-lithiophenoxide
Lithium α-lithiophenoxyacetate
Lithium β-lithiopropionate
Lithium methylsulfinylmethylide
Lithium methylsulfinylmethylide–Tributylborane
Lithium tributylmanganate
Lithium tri-*t*-butylzincate
Lithium triethylmanganate
Lithium trimethylcobaltate
Lithium trimethylferrate
Lithium trimethylmanganate
Lithium trimethylsilylacetylide
Lithium trimethylzincate
Mesityllithium
Methallyllithium
2-Methoxy-4-furyllithium
Methoxymethyllithium
Methoxy(phenylthio)methyllithium
Methoxy(phenylthio)trimethylsilylmethyllithium
2-Methoxy-2-propyllithium
Methoxy(trimethylsilyl)methyllithium
α-Methoxyvinyllithium
Methylene chloride–Butyllithium
Methylene chloride–Methyllithium
Methyl lithiodithioacetate
Methyl α-lithio-α-methyldiphenylsilylacetate
Methyl 1-lithio-1-methylselenoacetate
Methyl 1-lithio-1-phenylselenoacetate
Methyllithium
Methyllithium + co-reagents
N-Methyl-N-phenylaminoethynyllithium
N-Methylphenylsulfonimidoyl-1-ethyllithium
N-Methylphenylsulfonimidoylmethyllithium
Methylselenomethyllithium
Methylthiomethyllithium
1-(Methylthio)-3-triethylsilyloxypentadienyllithium
Nickel carbonyl–Butyllithium
Organolithium reagents
Organolithium reagents + co-reagents

METAL-CONTAINING COMPOUNDS (*Continued*)
2-Oxo-3-(triphenylphosphoranylidene)-propyllithium
Pentadienyllithium
Phenyllithium
Phenyllithium + co-reagents
Phenyl(phenylthio)methyllithium
Phenyl(phenylthio)trimethylsilyl-methyllithium
Phenylselenomethyllithium
Phenylselenotrimethylsilylmethyllithium
α-Phenylsulfonylethyllithium
Phenylsulfonyl(trimethylsilyl)-methyllithium
Phenylthiobenzyllithium
Phenylthiomethyllithium
Phenylthio(trimethylsilyl)methyllithium
Phenylthio(triphenylstannyl)-methyllithium
1-Phenylthiovinyllithium
2-Picolyllithium
4-Picolyllithium
Potassium hydride–*sec*-Butyllithium–Tetramethylethylenediamine
(E)-1-Propenyllithium
2-Propenyllithium
Sodium hydride–Butyllithium
2-(1,3-Thiazolinyl)thiomethyllithium
Titanium(III) chloride–Lithium
Titanium(III) chloride–Methyllithium
p-Tolyllithium
p-(Tolylsulfinyl)methyllithium
Tribromomethyllithium
2,4,6-Tri-*t*-butylphenyllithium
Tributyltinlithium
Tributyltinlithium–Diethylaluminum chloride
Trichloromethyllithium
3-Triethylsilyloxypentadienyllithium
Trifluorovinyllithium
[(Trimethylsilyl)allyl]lithium
Trimethylsilylbenzyllithium
α-[(Trimethylsilyl)ethyl]lithium
β-[(Trimethylsilyl)ethyl]lithium
1-Trimethylsilylhexyllithium
Trimethylsilyl 2-lithio-2-carboethoxyacetate
Trimethylsilyllithium
Trimethylsilylmethyllithium
(Z)-(Trimethylsilyloxy)vinyllithium
1-Trimethylsilyl(pentadienyl)lithium
Trimethylstannylmethyllithium
Trimethyltinlithium
Triphenylmethyllithium
Triphenylmethyllithium–Tetramethylethylenediamine
Triphenylstannylmethyllithium
Tris(methylthio)methyllithium
Tris(phenylthio)methyllithium
Tris(trimethylsilyl)methyllithium
Tungsten(VI) chloride–Butyllithium
Vinyllithium
1-Vinylthioallyllithium

MAGNESIUM COMPOUNDS
Acetylenedimagnesium bromide
(Allyldimethylsilyl)methylmagnesium chloride
Allylmagnesium bromide or chloride
Allylmagnesium bromide–Copper(I) iodide
Benzocyclobutenylmethylmagnesium bromide
Benzylmagnesium bromide or chloride
N,N-Bis(bromomagnesio)anilide
2-(2-Bromomagnesioethyl)-2-methyl-1,3-dioxolane
Bromomagnesium alkyl(methyl)cuprates
Bromomagnesium diethylamide
Bromomagnesium diisopropylamide
Bromomagnesium hexamethyldisilazide
Bromomagnesium isopropylcyclohexylamide
Bromomagnesium N-methylanilide
1,3-Butadienyl-2-magnesium chloride
Butane 1,4-(dimagnesium bromide)
Butylmagnesium bromide or chloride
t-Butylmagnesium bromide or chloride
Butylmagnesium bromide or chloride + co-reagents
4-Chloro-1-butenyl-2-magnesium bromide
p-Chlorophenylmagnesium bromide
Chlorotrimethylsilane + magnesium co-reagents
Crotylmagnesium bromide or chloride
Crotylmagnesium chloride–Dichlorobis-(cyclopentadienyl)titanium
Cyclopentadienylmagnesium hydride
Cyclopropylmagnesium bromide
Cyclopropylmethylmagnesium bromide
Dibutylmagnesium

Dichlorobis(cyclopentadienyl)titanium–Magnesium
Dichlorobis(cyclopentadienyl)zirconium–*t*-Butylmagnesium chloride
Dichlorodimethylsilane–Magnesium amalgam
3,3-Diethoxypropynylmagnesium bromide
Diethylmagnesium
(Diisopropoxymethylsilyl)methylmagnesium chloride
2,6-Diisopropylphenoxymagnesium hydride
4-Dimethylaminophenylmagnesium bromide
Dimethylmagnesium
Dimethylphenylsilyl(methyl)magnesium
Ethoxyethynylmagnesium bromide
Ethyl ethoxymagnesiomalonate
Ethylmagnesium bromide or chloride
Ethylmagnesium bromide + co-reagents
Ethynylmagnesium bromide
Grignard reagents
Grignard reagents + co-reagents
Isobutylmagnesium bromide
Isopropylmagnesium bromide or chloride
Isopropylmagnesium chloride–Iron(III) reagents
Magnesium
Magnesium acetate
Magnesium amalgam
Magnesium anthracenide
Magnesium bromide
Magnesium bromide etherate
Magnesium bromide etherate–Hydrogen peroxide
Magnesium carbonate
Magnesium chloride
Magnesium chloride–Potassium
Magnesium chloride–Sodium
Magnesium chloride–Sodium naphthalenide
Magnesium ethoxide
Magnesium ethyl malonate
Magnesium hydride
Magnesium hydride + co-reagents
Magnesium iodide
Magnesium iodide etherate
Magnesium–Mercuric chloride
Magnesium methoxide
Magnesium methyl carbonate
Magnesium naphthalenide
Magnesium oxide
Magnesium permanganate
Magnesium sulfate
Magnesium trifluoromethanesulfonate
2-Mesitylmagnesium bromide
Methallylmagnesium chloride–Dichlorobis(cyclopentadienyl)-titanium
Methallylmagnesium halides
p-Methoxybenzylmagnesium bromide
4-Methoxy-3-buten-1-ynylmagnesium bromide
Methoxymethylmagnesium chloride
3-Methyl-2-butenylmagnesium chloride
3-Methyl-2-butenylmagnesium halides–Copper(I) iodide
Methylenebis(magnesium bromide or iodide)
Methylmagnesium bromide(chloride, iodide)
Methylmagnesium bromide or iodide + co-reagents
1-Norbornylmagnesium chloride
Organomagnesium reagents
Pentane-1,5-di(magnesium bromide)
Phenylethynylmagnesium bromide
Phenylmagnesium bromide
Phenylmagnesium bromide + co-reagents
Phenylmagnesium fluoride
Phenylsulfonylmethylmagnesium bromide
2-Propenylmagnesium bromide
2-Propenylmagnesium bromide–Copper(I) iodide
Propynylmagnesium bromide
trans-Styrylmagnesium bromide
Tetra-μ-hydridotetrahydroaluminummagnesium
2-Thiophenylmagnesium bromide
Titanium(III) chloride + magnesium co-reagents
Titanium(IV) chloride + magnesium co-reagents
p-Tolylmagnesium bromide or chloride
Tributyltinmagnesium bromide
2,4,6-Trimethylphenoxymagnesium bromide
α-(Trimethylsilyl)benzylmagnesium

METAL-CONTAINING COMPOUNDS (*Continued*)
bromide
Trimethylsilylmagnesium chloride
Trimethylsilylmethylmagnesium chloride
2-(3-Trimethylsilyl-1-propenyl)-magnesium bromide
2-(3-Trimethylsilyl-1-propenyl)-magnesium bromide–Copper(I) iodide
α-Trimethylsilylvinylmagnesium bromide
β-Trimethylsilylvinylmagnesium bromide
α-Trimethylsilylvinylmagnesium bromide–Copper(I) iodide
Vinylmagnesium bromide or chloride
Vinylmagnesium bromide or chloride + copper co-reagents

MANGANESE COMPOUNDS
Barium manganate
Barium permanganate
Benzyl(triethyl)ammonium permanganate
Bispyridinesilver permanganate
Copper(II) permanganate
Ethylmanganese chloride
Grignard reagents–Manganese(II) chloride
Lithium tributylmanganate
Lithium triethylmanganate
Lithium trimethylmanganate
Magnesium permanganate
Manganese(III) acetate
Manganese(III) acetylacetonate
Manganese(II) chloride
Manganese(II) chloride–Lithium aluminum hydride
Manganese(II) iodide
Manganese(IV) oxide
Manganese(III) sulfate
3-Methyl-2-butenylmanganese chloride
Methyltriphenylphosphonium permangate
Organomanganese(II) chlorides or iodides
Palladium(II) chloride–Manganese(IV) oxide
Pentacarbonyl(trimethylsilyl)manganese
Potassium manganate
Potassium periodate–Potassium permanganate
Potassium permanganate
Potassium permanganate + co-reagents
Sodium periodate–Potassium permanganate
Sodium permanganate monohydrate
Tetrabutylammonium permanganate
Zinc permanganate
Zinc permanganate–Silica

MERCURY COMPOUNDS
Bis(dichlorotrimethylsilylmethyl)-mercury
Bis(2-propenyl)mercury
Bis(tribromomethyl)mercury
Bis(trichloromethyl)mercury
Bis(trifluoromethylthio)mercury
Bis(trimethylsilyl)mercury
Bis(trimethylsilylmethyl)mercury
(1-Bromo-1-chloro-2,2,2-trifluoroethyl)-phenylmercury
(Bromodichloromethyl)phenylmercury
(1-Bromo-1,2,2,2-tetrafluoroethyl)-phenylmercury
t-Butyldimethylsilyl hydroperoxide–Mercury(II) trifluoroacetate
Dibromochloromethyl(phenyl)mercury
Dibromo(methoxycarbonyl)methyl-(phenyl)mercury
Dichloro(methoxycarbonyl)methyl-(phenyl)mercury
Dihalo(methoxycarbonyl)methyl-(phenyl)mercury
Diphenylmercury
Divinylmercury
Iodo(iodomethyl)mercury
Mercury(II) acetate
Mercury(II) acetate + co-reagents
Mercury(II) azide
Mercury bis(ethyl diazoacetate)
Mercury(II) bis(*p*-toluenesulfonamide)
Mercury(II) bromide
Mercury(II) chloride
Mercury(II) chloride–Cadmium carbonate
Mercury(II) cyanide
Mercury(II) iodide
Mercury(II) methanesulfonate
Mercury(I) nitrate
Mercury(II) nitrate
Mercury(II) nitrate–Hydrogen peroxide
Mercury(II) nitrite
Mercury(II) oxide

Mercury(II) oxide + co-reagents
Mercury(II) perchlorate
Mercury(II) pivalate
Mercury(II) propionate
Mercury(II) sulfate
Mercury(II) tetrafluoroborate
Mercury(II) thiocyanate
Mercury(I) trifluoroacetate
Mercury(II) trifluoroacetate
Mercury(II) trifluoromethanesulfonate–N,N-Dimethylaniline
Phenylmercuric acetate
Phenylmercuric bromide
Phenylmercuric chloride
Phenylmercuric hydroxide
Phenylmercuric perchlorate
Phenyl(tribromomethyl)mercury
Phenyl(trichloromethyl)mercury
Phenyl(trifluoromethyl)mercury
Phenyl(trihalomethyl)mercury

MOLYBDENUM COMPOUNDS
Bis(N,N-dimethylformamido)-oxodiperoxymolybdenum(VI)
Cesium 1,1,1,2,2,2-hexachloro-μ-(hydrido)di-μ-(chloro)-dimolybdenum(III)
Chloromethyleneoxomolybdenum
Dichlorodimethylmolybdenum
(S)-(2-Hydroxy-N,N-dimethyl-propanamide-O,O')-oxodiperoxy-molybdenum(VI)
Molybdenum carbonyl
Molybdenum(V) chloride
Molybdenum(V) chloride + co-reagents
Molybdenum(VI) fluoride
Molybdenum(V) trichloride oxide
Molybdenum(V) trichloride oxide–Zinc
Organomolybdenum reagents
Oxodiperoxymolybdenum(pyridine)-(hexamethylphosphoric triamide)
Oxomethoxymolybdenum(V) 5,10,15,20-tetraphenylporphyrin
Oxoperoxobis(N-phenylbenzo-hydroxamato)molybdenum(VI)
Tetrabutylammonium octamolybdate
Tetrachlorooxomolybdenum
Trichloromethylenemolybdenum

NICKEL COMPOUNDS
π-Allylnickel halides
Bis(acrylonitrile)nickel(0)
Bis(1,5-cyclooctadiene)nickel(0)
Bis(1,5-cyclooctadiene)nickel(0) + co-reagents
[*trans*-2,3-Bis(diphenylphosphine)-bicyclo[2.2.1]hept-5-ene]-dichloronickel(II)
[1,2-Bis(diphenylphosphine)ethane]-(dichloro)nickel(II)
[1,1'-Bis(diphenylphosphine)ferrocene]-dichloronickel(II)
[1,3-Bis(diphenylphosphine)propane]-(dichloro)nickel(II)
Bis(triphenylphosphine)nickel(0)
trans-Bromo-*o*-tolylbis(triethyl-phosphine)nickel(II)
Butylmagnesium bromide–Nickel(II) reagents
Di-μ-bromobis(allyl)dinickel
Di-μ-bromobis[(2-ethoxycarbonyl)-2-propenyl]dinickel
Di-μ-bromobis(2-methallyl)dinickel
Di-μ-bromobis(2-methoxyallyl)dinickel
Di-μ-bromobis(3-methyl-2-butenyl)-dinickel
Dibromobis(tributylphosphine)nickel(II)
Dibromobis(tributylphosphine)-nickel(II)–Sodium borohydride
Dibromobis(triethylphosphine)nickel(II)
Dibromobis(triphenylphosphine)-nickel(II)
Dibromobis(triphenylphosphine)-nickel(II)–Zinc
Dicarbonylbis(triphenylphosphine)nickel
Dicarbonylbis(triphenyl phosphite)-nickel
Dichlorobis(tributylphosphine)nickel(II)
Dichlorobis(triphenylphosphine)-nickel(II)
Dichloro[2,3-O-isopropylidene-2,3-dihydroxy-1,4-bis(diphenyl-phosphine)butane]nickel(II)
Dicyanobis(triphenylphosphine)nickel(0)
Grignard reagents–Nickel(0) reagents
Grignard reagents–Nickel(II) reagents
Lithium tetrabromonickelate
Lithium tricarbonyl(dimethyl-carbamoyl)nickelate
Methylmagnesium bromide–Nickel(II) acetylacetonate–Trimethylaluminum
Methylmagnesium bromide–Nickel(II) reagents
Nickel

METAL-CONTAINING COMPOUNDS (*Continued*)
Nickel(II) acetate
Nickel(II) acetate–Sodium hydride–*t*-Amyloxide
Nickel(II) acetylacetonate
Nickel(II) acetylacetonate + co-reagents
Nickel–Alumina
Nickel boride
Nickel(II) bromide
Nickel(II) bromide + co-reagents
Nickel carbonyl
Nickel carbonyl–Butyllithium
Nickel catalysts
Nickel(II) chloride
Nickel(II) chloride + co-reagents
Nickel cyanide
Nickel–Graphite
Nickel on alumina
Nickel peroxide
Nickel(II) sulfate
Nickel tetraphenylporphine
Phenylmagnesium bromide–Nickel(II) reagents
Potassium hexacyanodinickelate(4-)
Raney nickel
Raney nickel alloy
Sodium tetracyanonickelate(II)
Tetrakis(trichlorophosphine)nickel(0)
Tetrakis(triphenylphosphine)nickel(0)
Tetrakis(triphenyl phosphite)nickel(0)
Triisobutylaluminum–Bis(N-methylsalicylaldimine)nickel
Tris(triphenylphosphine)nickel(0)
OSMIUM COMPOUNDS
Dioxobis(*t*-butylimido)osmium(VIII)
Dipotassium tetramethylosmiate
Osmium tetroxide
Osmium tetroxide + co-reagents
Oxotris(*t*-butylimido)osmium(VIII)
Potassium hexachloroosmiate(IV)–Zinc
Potassium osmiate
Potassium triacetylosmate
Sodium periodate–Osmium tetroxide
Trioxo(*t*-butylimido)osmium(VIII)
PALLADIUM COMPOUNDS
π-Allylpalladium complexes
Benzylchlorobis(triphenylphosphine)palladium(II)
Bis[μ-(acetato)]bis(2-propenyl)dipalladium
Bis(acetonitrile)chloronitropalladium(II)
Bis(acetonitrile)dichloropalladium(II)
Bis(benzonitrile)dibromopalladium(II)
Bis(benzonitrile)dichloropalladium(II)
Bis[1,2-bis(diphenylphosphine)ethane]palladium(0)
Bis(dibenzylideneacetone)palladium(0)
[1,2-Bis(diphenylphosphine)ethane]bis(triphenylphosphine)palladium(0)
[1,2-Bis(diphenylphosphine)ethane]dichloropalladium(II)
[1,1′-Bis(diphenylphosphine)ferrocene](dichloro)palladium(II)
Diacetatobis(triphenylphosphine)palladium(II)
Diacetatobis(tri-*o*-tolylphosphine)palladium(II)
Diammonium tetrachloropalladate(II)
Dibromobis(triphenylphosphine)palladium(II)
Di-μ-chlorobis(allyl)dipalladium
Dichlorobis(crotyl)dipalladium
Dichlorobis(cyclohexene)palladium(II)
Dichlorobis(methyldiphenylphosphine)palladium(II)
Dichlorobis(triphenylphosphine)palladium(II)
Dichlorobis(triphenylphosphine)palladium(II) + co-reagents
Dichlorobis(tri-*o*-tolylphosphine)palladium(II)
Dichloro(1,5-cyclooctadiene)palladium(II)
Dihalobis(triphenylphosphine)palladium(II)
Dilithium tetrachloropalladate(II)
Disodium tetrachloropalladate(II)
Grignard reagents–Palladium(0) reagents
Grignard reagents–Palladium(II) reagents
Iodophenylbis(triphenylphosphine)palladium
[2,3-O-Isopropylidene-2,3-dihydroxy-1,4-bis(diphenylphosphine)butane]palladium
Lithium trichloropalladate(II)
(Maleic anhydride)bis(triphenylphosphine)palladium(0)
Organolithium reagents–Palladium(0) reagents

Palladium(II) acetate
Palladium(II) acetate + co-reagents
Palladium(II) acetylacetonate
Palladium(II) acetylacetonate–Triphenylphosphine
Palladium black
Palladium *t*-butyl peroxide trifluoroacetate
Palladium catalysts
Palladium(II) chloride
Palladium(II) chloride + co-reagents
Palladium–Graphite
Palladium hydroxide
Palladium hydroxide on barium sulfate
Palladium nitrate
Palladium on barium sulfate
Palladium on calcium carbonate
Palladium on carbon
Palladium on charcoal
Palladium(0)-Phosphines
Palladium, poisoned
Palladium–Poly(ethylenimine)
Palladium(II) trifluoroacetate
Peroxybis(triphenylphosphine)palladium
Phenylpalladium acetate
Rosenmund catalyst
Tetrakis(acetonitrile)palladium tetrafluoroborate
Tetrakis(triisopropyl phosphite)-palladium(0)
Tetrakis(triphenylphosphine)-palladium(0)
Tetrakis(triphenylphosphine)-palladium(0) + co-reagents
Triethylsilane–Palladium(II) chloride
Triethylsilane–Palladium on carbon
Tris(dibenzylideneacetone)dipalladium-(chloroform)
Tris(tribenzylideneacetylacetone)-tripalladium(chloroform)
Zinc chloride–Benzylchlorobis-(triphenylphosphine)palladium(II)

PLATINUM COMPOUNDS
Bis(benzonitrile)dichloroplatinum(II)
[1,2-Bis(diphenylphosphine)ethane]-cyclohexyneplatinum(0)
Chlorohydridobis(triethylphosphine)-platinum
Diammonium hexachloroplatinate(IV)
Dichlorobis(triphenylphosphine)-platinum(II)–Tin(II) chloride
(–)-*trans*-Dichloro(ethylene)-[α-methylbenzylamine]platinum(II)
Dipotassium tetrachloroplatinate(II)
Disodium tetrachloroplatinate(II)
Ethylenebis(triphenylphosphine)-platinum(0)
Hydrogen hexachloroplatinate(IV) + co-reagents
Platinum
Platinum–Alumina catalyst
Platinum black
Platinum *t*-butyl peroxide trifluoroacetate
Platinum catalysts
Platinum on carbon
Platinum oxide
Platinum–Silica
Platinum sulfide on carbon
Platinum–Tin(II) chloride
Rhodium oxide–Platinum oxide

RHODIUM COMPOUNDS
Bis[1,3-bis(diphenylphosphine)propane]-chlororhodium(I)
[2,2′-Bis(diphenylphosphine)-1,1′-binaphthyl](cyclooctadiene)-rhodium(I) perchlorate
[2,2′-Bis(diphenylphosphine)-1,1′-binaphthyl](norbornadiene)-rhodium(I) perchlorate
1,4-[Bis(diphenylphosphine)butane]-(norbornadiene)rhodium(I) tetrafluoroborate
[1,2-Bis(diphenylphosphine)propane]-(norbornadiene)rhodium perchlorate
Bis(norbornadiene)rhodium(I) perchlorate–(R)-1-(S)-1′,2-Bis-(diphenylphosphine)ferrocenylethanol
[N-(*t*-Butoxycarbonyl)-4-diphenyl-phosphine]-2-[(diphenylphosphine)-methyl]pyrrolidine(cyclooctadiene)-rhodium chloride or perchlorate
Carbonylchlorobis(triphenylphosphine)-rhodium(I)
Carbonylhydridotris(triphenyl-phosphine)rhodium(I)
Carbonylhydridotris(triphenyl-phosphine)rhodium–2,3,O-Isopropylidene-2,3-dihydroxy-1,4-bis(diphenylphosphine)butane
Carbonylphenylbis(triphenyl-phosphine)rhodium(I)

METAL-CONTAINING COMPOUNDS
(*Continued*)
Chloro(hexamethylbenzene)hydridotriphenylphosphinerhodium
Chlorotris(methyldiphenylphosphine)rhodium(I)
Chlorotris(neomenthyldiphenylphosphine)rhodium(I)
Chlorotris(triphenylphosphine)rhodium(I)
Chlorotris(triphenylphosphine)rhodium(I)–Hydrosilanes
Dichlorobis(1,5-cyclooctadiene)dirhodium + co-reagents
Di-μ-chlorobis(1,5-hexadiene)dirhodium
Di-μ-chlorobis(1,5-hexadiene)dirhodium + co-reagents
Dichlorobis(norbornadiene)dirhodium(I)
Dichlorobis(norbornadiene)dirhodium(I) + co-reagents
Di-μ-chlorodichlorobis(pentamethylcyclopentadienyl)dirhodium
Dichloro(dimethylformamido)bis(pyridine)rhodium borohydride
Di-μ-chlorotetra(ethylene)dirhodium
Di-μ-chlorotetra(ethylene)dirhodium–2,3-O-Isopropylidene-2,3-dihydroxy-1,4-bis(diphenylphosphine)butane
Di-μ-chlorotetrakis(cyclooctene)dirhodium–2,3-O-Isopropylidene-2,3-dihydroxy-1,4-bis(diphenylphosphine)butane
Diiodo(methyl)bis(triphenylphosphine)rhodium
Hexakis(acetato)trihydrato-μ_3-oxotrisrhodium acetate
Hydridotetrakis(triphenylphosphine)rhodium(I)
Hydridotris(triisopropylphosphine)rhodium(I)
Hydridotris(triphenylphosphine)carbonylrhodium(I)
Methyltris(triphenylphosphine)rhodium(I)
Peroxytetrakis(triphenylarsine)rhodium(I) hexafluorophosphate or perchlorate
Rhodium(II) acetate
Rhodium(II) butanoate
Rhodium(II) carboxylates
Rhodium(III) chloride
Rhodium(III) chloride + co-reagents
Rhodium(II) methoxyacetate
Rhodium(III) oxide
Rhodium oxide–Platinum oxide
Rhodium(II) pivaloate
Rhodium(III) porphyrins
Rhodium(II) trifluoroacetate
Tetracarbonyldi-μ-chlorodirhodium
Tetracarbonyldi-μ-chlorodirhodium–Lithium chloride
Tetra-μ_3-carbonyldodecacarbonylhexarhodium
Tri-μ-carbonylnonacarbonyltetrarhodium
Tris(dimethylphenylphosphine)(norbornadiene)rhodium(I) hexafluorophosphate

RUTHENIUM COMPOUNDS
Acetatohydridotris(triphenylphosphine)ruthenium
Carbonyldihydridotris(triphenylphosphine)ruthenium
Chlorohydridotris(triphenylphosphine)ruthenium(II)
Dibromotris(triphenylphosphine)ruthenium(II)
Dicarbonyldichlorobis(triphenylphosphine)ruthenium(II)
Dichlorotetrakis(triphenylphosphine)ruthenium(II)
Dichlorotris(triphenylphosphine)ruthenium(II)
Dihydridotetrakis(tributylphosphine)ruthenium(II)
Dihydridotetrakis(triphenylphosphine)ruthenium(II)
Dihydridotris(triphenylphosphine)ruthenium
Dodecacarbonyltri-*triangulo*-ruthenium
Hexacarbonyltetra-μ-chlorodiruthenium
Potassium ruthenate
Ruthenium(III) acetylacetonate
Ruthenium(III) acetylacetonate–Alumina
Ruthenium catalysts
Ruthenium(III) chloride
Ruthenium(III) chloride + co-reagents
Ruthenium hydrogenation catalyst
Ruthenium on alumina
Ruthenium on carbon

Ruthenium(IV) oxide
Ruthenium(IV) oxide + co-reagents
Ruthenium–Silica
Ruthenium tetroxide
Ruthenium tetroxide–Sodium periodate
Sodium ruthenate
Tris(aquo)hexa-μ-acetato-μ_3-oxotriruthenium(III,III,III) acetate
SILICON COMPOUNDS
Acetone cyanohydrin trimethylsilyl ether
2-Acetoxy-1-methoxy-3-trimethylsilyloxy-1,3-butadiene
1-Acetoxy-4-trimethylsilyl-1,3-butadiene
3-Acetoxy-1-trimethylsilyl-1,3-butadiene
3-Acetoxy-2-trimethylsilylmethyl-1-propene
3-Acetoxy-1-trimethylsilyl-1-propene
3-Acetoxy-3-trimethylsilyl-1-propene
Acetyl chloride–2-Trimethylsilylethanol
Allyl-*t*-butyldimethylsilane
(Allyldimethylsilyl)methylmagnesium chloride
Allyltrimethylsilane
Allyltrimethylsilylzinc chloride
(S)-(−)-1-Amino-2-(*t*-butyldialkylsilyloxymethyl)pyrrolidines
Azidotrimethylsilane
Benzyltrimethylsilane
1,3-Bis(*t*-butyldimethylsilyloxy)-2-aza-1,3-diene
Bis(chlorodimethylsilyl)ethane
Bis(cyclopentadienyl)(1-trimethylsilylallyl)titanium
Bis(dichlorotrimethylsilylmethyl)mercury
1,4-Bis(N,N-dimethylamino)-1,1,4,4-tetramethyldisilethylene
Bis(1-methoxy-2-methyl-1-propenyloxy)dimethylsilane
Bis(methylthio)(trimethylsilyl)methane
Bis(methylthio)(trimethylsilyl)methyllithium
1,3-Bis(phenylseleno)-3-trimethylsilyl-1-propene
1,3-Bis(triisopropylsilyl)propyne
Bis(trimethylsilyl)acetamide
Bis(trimethylsilyl)acetylene
1,4-Bis(trimethylsilyl)butadiyne
1,4-Bis(trimethylsilyl)-2-butene
1,4-Bis(trimethylsilyl)-2-butyne
N,O-Bis(trimethylsilyl)carbamate
1,1-Bis(trimethylsilyl)ethylene
Bis(trimethylsilyl)formamide
O,N-Bis(trimethylsilyl)hydroxylamine
Bis(trimethylsilyl) lithiomalonate
Bis(trimethylsilyl) malonate
Bis(trimethylsilyl)mercury
Bis(trimethylsilyl)methane
2,3-Bis(trimethylsilylmethyl)-1,3-butadiene
Bis(trimethylsilylmethyl)mercury
Bis(trimethylsilyl) monoperoxysulfate
1,3-Bis(trimethylsilyloxy)-1,3-butadiene
2,3-Bis(trimethylsilyloxy)-1,3-butadiene
1,2-Bis(trimethylsilyloxy)cyclobutene
1,3-Bis(trimethylsilyloxy)-1,3-cyclohexadiene
2,3-Bis(trimethylsilyloxy)-1,3-cyclohexadiene
2,5-Bis(trimethylsilyloxy)furans
Bis(trimethylsilyl) peroxide
1,3-Bis(trimethylsilyl)propyne
N,O-Bis(trimethylsilyl) sulfamate
Bis(trimethylsilyl) sulfate
Bis(trimethylsilyl)thioketene
Bis(trimethylsilyl) trimethylsilyloxymethylphosphonite
Bis(trimethylsilyl)urea
Bromo-*t*-butylmethoxyphenylsilane
2-Bromoethyltrimethylsilane
Bromomagnesium hexamethyldisilazide
(Bromomethyl)chlorodimethylsilane
2-Bromomethyl-3-(trimethylsilylmethyl)-1,3-butadiene
3-Bromo-3-methyl-2-trimethylsilyloxy-1-butene
Bromotrimethylsilane
Bromotrimethylsilane-Cobalt(II) bromide
(Z)-2-Bromo-1-(trimethylsilyloxy)-ethylene
1-Bromo-3-trimethylsilyloxy-1-propyne
1-Bromo-1-trimethylsilyl-1(Z),4-pentadiene
3-Bromo-1-trimethylsilyl-1-propene
2-Bromo-3-trimethylsilyl-1-propene
1-Bromovinyltrimethylsilane
(2-Bromovinyl)trimethylsilane
(1-Bromovinyl)triphenylsilane
Butadiynyl(trimethyl)silane

METAL-CONTAINING COMPOUNDS (*Continued*)
t-Butoxytrimethylsilane
t-Butyl bis(trimethylsilyl)acetate
t-Butyl 2-chloro-2-lithiotrimethylsilylacetate
t-Butyl 2-chloro-2-trimethylsilylacetate
t-Butyldimethylchlorosilane
t-Butyldimethyliodosilane
t-Butyldimethylsilyl ethylnitronate
t-Butyldimethylsilyl hydroperoxide–Mercury(II) trifluoroacetate
O-*t*-Butyldimethylsilylhydroxylamine
N-(*t*-Butyldimethylsilyl)-N-methylacetamide
N-(*t*-Butyldimethylsilyl)-N-methyltrifluoroacetamide
4-*t*-Butyldimethylsilyloxy-3-penten-2-one
t-Butyldimethylsilyl perchlorate
t-Butyldimethylsilyl phenyl selenide
t-Butyldimethylsilyl trifluoromethanesulfonate
3-(*t*-Butyldimethylsilyl)-1-(trimethylsilyl)propyne
t-Butyldiphenylchlorosilane
t-Butyldiphenylsilyllithium
t-Butyl lithiobis(trimethylsilyl)acetate
t-Butyl lithio(trimethylsilyl)acetate
1-*t*-Butylthio-1-trimethylsilyloxyethylene
t-Butyl trimethylsilylacetate
Cesium fluoride–Tetraalkoxysilanes
Chlorodimethylphenylsilane
Chlorodimethyl(2,4,6-tri-*t*-butylphenoxy)silane
α-Chloroethyltrimethylsilane
(Chloromethyl)dimethylsilane
Chloro(methyl)diphenylsilane
Chloromethyltrimethylsilane
3-Chloro-3-methyl-2-trimethylsilyloxy-1-butene
1-Chloro-2-methyl-1-trimethylsilyl-1-propene
2-Chloromethyl-3-trimethylsilyl-1-propene
Chlorotriisopropylsilane
Chlorotrimethylsilane
Chlorotrimethylsilane + co-reagents
(E)-1-Chloro-3-trimethylsilyl-2-butene
Chlorotrimethylsilylcarbene
1-Chloro-1-(trimethylsilyl)ethyllithium
Chloro[(trimethylsilyl)methyl]ketene
Chloro(trimethylsilyl)methyllithium
1-Chloro-1-trimethylsilyl-2-methylpropene
3-Chloro-2-trimethylsilyloxy-1-propene
Chlorotris(triphenylphosphine)-rhodium(I)–Hydrosilanes
(2-Chlorovinyl)trimethylsilane
Cinnamyltrimethylsilane
Copper(I) hexamethyldisilazide
Copper(I) trimethylsilylacetylide
Crotyltrimethylsilane
Cyano-*t*-butyldimethylsilane
Cyanotrimethylsilane
Cyanotrimethylsilane–Triethylaluminum
Cyclopropyltrimethylsilane
1-Cyclopropyl-1-trimethylsilyloxyethylene
Diacetoxydimethylsilane
Diazo(trimethylsilyl)methyllithium
Di-*t*-butyldichlorosilane
Di-*t*-butylmethylsilyl perchlorate
Di-*t*-butylsilyl bis(trifluoromethanesulfonate)
Dichlorodimethylsilane
Dichlorodimethylsilane–Magnesium amalgam
1,3-Dichloro-3-methyl-2-trimethylsilyloxy-1-butene
Dichlorosilane
Dicyanodimethylsilane
1,2-Diethoxy-1,2-bis(trimethylsilyloxy)-ethylene
Diethoxymethylsilane
Diethylaluminum trimethylsilylacetylide
Diethyl difluorotrimethylsilylmethylphosphonate
Diethyl[dimethyl(phenyl)silyl]aluminum
Diethylmethylsilane
Diethyl 1-phenyl-1-trimethylsilyloxymethylphosphonate
Diethyltrimethylsilylamine
Diethyl trimethylsilyloxycarbonylmethylphosphonate
Diethyl 1-trimethylsilyloxyethylphosphonate
Diethyl trimethylsilyl phosphite
Diiododimethylsilane

(Diisopropoxymethylsilyl)methylmagnesium chloride
Diisopropylbis(trifluoromethanesulfonato)silane
Diketene–Iodotrimethylsilane
(Dimethoxymethyl)trimethoxysilane
1,3-Dimethoxy-1-trimethylsilyloxy-1,3-butadiene
1,1-Dimethoxy-1-trimethylsilyloxy-1,3-butadiene
1,2-Dimethoxy-1-trimethylsilyloxy-1,3-pentadiene
5,5-Dimethyl-1,3-bis(trimethylsilyloxy)-1,3-cyclohexadiene
Dimethylketene methyl trimethylsilyl ketal
Dimethylphenylsilane
Dimethylphenylsilyllithium
Dimethylphenylsilyl(methyl)magnesium
Dimethylsilylene
Dimethyl sulfoxide–Chlorotrimethylsilane
N,N-Dimethyl-α-trimethylsilylacetamide
Diphenylsilane
Dodecamethylcyclohexasilane
Ethoxydiethylsilane
1-(1-Ethoxyethoxyallyl)-trimethylsilane
1-Ethoxy-2-ethyl-3-trimethylsilyloxy-1,3-butadiene
Ethoxy(trimethylsilyl)acetylene
1-Ethoxy-1-trimethylsilyloxycyclopropane
Ethyl bis(2-trimethylsilylallyl) phosphate
Ethyl lithio(trimethylsilyl)acetate
Ethyl 2-(methyldiphenylsilyl)propionate
Ethyl potassio(trimethylsilyl)acetate
Ethylthiotrimethylsilane
Ethyl trimethylsilylacetate
Ethyl N-trimethylsilylcarbamate
Ethyl trimethylsilyl malonate
Ethyl 2-(trimethylsilylmethyl)acrylate
Fluorotrimethylsilane
Heptamethyldisilazane
Hexachlorodisilane
Hexamethylcyclotrisilathiane
Hexamethyldisilane
Hexamethyldisilane–Pyridinium bromide perbromide
Hexamethyldisilathiane
Hexamethyldisilazane
Hexamethyldisiloxane
Hydrogen hexachloroplatinate(IV) + silicon co-reagents
2-Hydroxymethyl-3-trimethylsilyl-1-propene
Iodoethynyl(trimethyl)silane
Iodomethyltrimethylsilane
2-Iodomethyl-3-trimethylsilyl-1-propene
(3-Iodopropyl)trimethylsilane
Iodotrimethylsilane
(E)-1-Iodo-3-trimethylsilyl-2-butene
3-Iodo-1-trimethylsilylpropyne
1-Isopropoxy-1-trimethylsilyloxycyclopropane
Ketene alkyl trialkylsilyl acetals or ketals
Ketene bis(trimethylsilyl) acetals or ketals
Ketene *t*-butyldimethylsilyl methyl acetal
1-Lithio-3-methoxy-3-trimethylsilylallene
3-Lithio-1-triisopropylsilyl-1-propyne
Lithiotrimethylsilylacetonitrile
2-Lithio-2-trimethylsilyl-1,3-dithiane
3-Lithio-1-trimethylsilyl-1-propyne
1-Lithio-3-trimethylsilyl-1-propyne
2-(E)-Lithiovinyltrimethylsilane
(1-Lithiovinyl)trimethylsilane
(1-Lithiovinyl)triphenylsilane
Lithium bis(dimethylphenylsilyl)cuprate
Lithium 1,1-bis(trimethylsilyl)ethoxide
Lithium 1,1-bis(trimethylsilyl)-3-methylbutoxide
Lithium bis(3-trimethylsilyl-1-propen-2-yl)cuprate
Lithium cyano(dimethylphenylsilyl)-cuprate
Lithium cyano(trimethylsilyl)cuprate
Lithium diisopropylamide–Chlorotrimethylsilane
Lithium hexamethyldisilazide
Lithium trimethylsilanolate
Lithium trimethylsilylacetylide
Methallyltrimethylsilane
4-Methoxy-1,2-bis(trimethylsilyloxy)-1,3-butadiene
1-Methoxy-1,3-bis(trimethylsilyloxy)-1,3-butadiene
Methoxymethylbis(trimethylsilyl)amine
3-Methoxymethylene-2,4-bis(trimethylsilyloxy)-1,4-pentadiene
(Methoxymethyl)trimethylsilane

METAL-CONTAINING COMPOUNDS (*Continued*)

1-Methoxy-2-methyl-3-trimethylsilyloxy-1,3-butadiene
1-Methoxy-3-methyl-1-trimethylsilyloxy-1,3-butadiene
1-Methoxy-2-methyl-3-trimethylsilyloxy-1,3-pentadiene
4-Methoxy-1-phenylseleno-2-trimethylsilyloxy-1,3-butadiene
Methoxy(phenylthio)trimethylsilylmethane
Methoxy(phenylthio)trimethylsilylmethyllithium
Methoxytrimethylsilane
1-Methoxy-2-trimethylsilylacetylene
1-Methoxy-1-trimethylsilylallene
1-Methoxy-1-trimethylsilylethylene
Methoxy(trimethylsilyl)methyllithium
trans-1-Methoxy-3-trimethylsilyloxy-1,3-butadiene
1-Methoxy-1-trimethylsilyloxycyclopropane
3-Methoxy-1-trimethylsilyl-1-propyne
5-Methyl-1,3-bis(trimethylsilyl)-1,3-cyclohexadiene
O-Methyl-C,O-bis(trimethylsilyl)ketene acetal
5-Methyl-1,3-bis(trimethylsilyloxy)-1,3-cyclohexadiene
3-Methyl-2-butenyltrimethylsilane
Methyl 3-*t*-butyldimethylsilyloxycrotonate
Methylketene bis(trimethylsilyl) acetal
Methylketene methyl trimethylsilyl acetal
Methyl α-lithio-α-methyldiphenylsilylacetate
Methyl α-(methyldiphenylsilyl)acetate
Methylneopentylphenylsilane–Boron trifluoride
(Z)-3-Methyl-1-phenylthio-2-trimethylsilyloxy-1,3-butadiene
1-Methylthio-3-triethylsilyloxy-1,4-pentadiene
1-(Methylthio)-3-triethylsilyloxypentadienyllithium
Methylthiotrimethylsilane
3-(Methylthio)-3-(trimethylsilyl)-1-propene
Methyl trimethylsilylacetate
Methyl 2-trimethylsilylacrylate
3-Methyl-1-trimethylsilyl-1,2-butadiene
3-Methyl-3-trimethylsilyl-1-butene
3-Methyl-1-trimethylsilyl-3-buten-2-one
N-Methyl-N-trimethylsilylmethyl-N′-*t*-butylformamidine
2-Methyl-1-(trimethylsilylmethyl)-pyridinium trifluoromethanesulfonate
2-Methyl-2-trimethylsilyloxypentan-3-one
Organopentafluorosilicates
Oxotris(triphenylsilanolato)vanadium
Pentacarbonyl(trimethylsilyl)manganese
(2,4-Pentadienyl)trimethylsilane
Phenyliodine(III) diacetate–Azidotrimethylsilane
Phenyl(phenylthio)trimethylsilylmethane
Phenyl(phenylthio)trimethylsilylmethyllithium
Phenylselenotrimethylsilylmethyllithium
Phenylsilane
(2-Phenylsulfonylethyl)trimethylsilane
1-Phenylsulfonyl-2-trimethylsilylethane
(E)-1-Phenylsulfonyl-2-trimethylsilylethylene
Phenylsulfonyl(trimethylsilyl)methane
Phenylsulfonyl(trimethylsilyl)methyllithium
Phenylthiotriethylsilane
Phenylthiotrimethylsilane
1-Phenylthio-1-trimethylsilylcyclopropane
1-Phenylthio-1-trimethylsilylethylene
1-Phenylthio-2-trimethylsilylethylene
Phenylthiotrimethylsilylmethane
Phenylthio(trimethylsilyl)methyllithium
2-Phenylthio-2-trimethylsilylpropane
1-Phenyl-2-trimethylsilylacetylene
Phenyl 2-(trimethylsilyl)ethynyl sulfone
Phenyl trimethylsilylpropargyl ether
Phenyl trimethylsilyl selenide
Polymethylhydrosiloxane
Potassium cyanide–Chlorotrimethylsilane
Potassium hexamethyldisilazide
Potassium trimethylsilanolate
Silica
Silicon tetraisocyanate
Siloxene

Sodium cyanide–Chlorotrimethylsilane–Sodium iodide
Sodium hexamethyldisilazide
Sodium trimethylsilanolate
Tetrabutylammonium trimethylsilylate
Tetrachlorosilane
Tetrachlorosilane–Sodium iodide
Tetraethoxysilane
Tetrafluorosilane
3-Tetrahydropyranyloxy-1-trimethylsilyl-1-propyne
1,1,2,3-Tetrakis(trimethylsilyloxy)-1,3-butadiene
Tetramethoxysilane
Tribenzylchlorosilane
Tri-*t*-butylsilyl perchlorate
(E)-1-Tributylstannyl-2-trimethylsilylethylene
Tributyl(trimethylsilylmethyl)tin
Trichloro(iodo)silane
Trichloro(methyl)silane
Trichloro(methyl)silane–Sodium iodide
Trichlorosilane
Trichlorosilane–*t*-Amines
Triethoxysilane
Triethyl(iodo)silane
Triethylsilane
Triethylsilane + co-reagents
2-Triethylsilyl-1,3-butadiene
3-Triethylsilyl-3-buten-2-one
3-Triethylsilyloxy-1-butene
1-Triethylsilyloxy-2-butene
3-Triethylsilyloxy-1,4-pentadiene
3-Triethylsilyloxypentadienyllithium
3-Triethylsilyloxy-1-propene
Triethylsilyl perchlorate
Trifluoroacetic acid–Alkylsilanes
1-Trifluoromethylsulfonyl-2-methyl-1-trimethylsilyl-1-propene
Trifluoro(trichloromethyl)silane
1-Triisopropylsilyl-1-propyne
Triisopropylsilyl trifluoromethanesulfonate
Trimethylphenylsilane–Iodine
Trimethylsilane
N-Trimethylsilylacetamide
N-Trimethylsilylacetanilide
Trimethylsilylacetic acid
Trimethylsilylacetone
Trimethylsilylacetonitrile
Trimethylsilylacetyl chloride
Trimethylsilylacetylene
(Trimethylsilyl)acetyltrimethylsilane
Trimethylsilylallene
[(Trimethylsilyl)allyl]lithium
2-Trimethylsilyl-1,3-benzothiazole
Trimethylsilylbenzyllithium
α-(Trimethylsilyl)benzylmagnesium bromide
Trimethylsilylbromoketene
3-Trimethylsilyl-1,2-butadiene
1-Trimethylsilyl-1,3-butadiene
3-Trimethylsilyl-1-butene
1-Trimethylsilyl-1-buten-3-ol
(E)-3-Trimethylsilyl-2-buten-1-ol
4-Trimethylsilyl-2-buten-1-ol
3-Trimethylsilyl-3-buten-2-one
Trimethylsilylcarbene
Trimethylsilylcopper
α-Trimethylsilylcrotyl-9-borabicyclo-[3.3.1]nonane
3-Trimethylsilyl-1-cyclopentene
Trimethylsilyldiazomethane
Trimethylsilyl dichloroacetate
Trimethylsilyl 2,2-dichloropropionate
2-Trimethylsilyl-1,3-dithiane
2-Trimethylsilylethanol
β-(Trimethylsilyl)ethoxymethyl chloride
2-(Trimethylsilyl)ethyl azidoformate
2-(Trimethylsilyl)ethyl chloroformate
β-Trimethylsilylethylidenetriphenylphosphorane
α-[(Trimethylsilyl)ethyl]lithium
β-[(Trimethylsilyl)ethyl]lithium
2-Trimethylsilylfuran
1-Trimethylsilylhexyllithium
N-(Trimethylsilyl)imidazole
Trimethylsilyl isocyanate
Trimethylsilylketene
Trimethylsilyl 2-lithio-2-ethoxycarbonylacetate
Trimethylsilyllithium
Trimethylsilylmagnesium chloride
Trimethylsilylmethanethiol
Trimethylsilylmethyl azide
α-Trimethylsilyl-2-methylbenzothiazole
2-Trimethylsilylmethyl-1,3-butadiene
3-Trimethylsilylmethyl-3-butenoic acid
Trimethylsilylmethylcopper
Trimethylsilylmethyldimesitylborane
2-Trimethylsilylmethylene-

METAL-CONTAINING COMPOUNDS (*Continued*)
cyclobutane
Trimethylsilylmethylene dimethylsulfurane
Trimethylsilylmethylene-trimethylphosphorane
Trimethylsilylmethylene-triphenylphosphorane
Trimethylsilylmethyllithium
Trimethylsilylmethylmagnesium chloride
Trimethylsilylmethylpotassium
Trimethylsilylmethylsodium
Trimethylsilylmethyl trifluoro-methanesulfonate
2-Trimethylsilylmethyl-3-trimethyl-silyl-1-propene
Trimethylsilyl nonaflate
N-Trimethylsilyl-2-oxazolidinone
3-Trimethylsilyl-2-oxazolidone
2-Trimethylsilyloxyallyl chloride
1-Trimethylsilyloxy-1,3-butadiene
2-Trimethylsilyloxy-1,3-butadiene
2-Trimethylsilyloxy-1,3-cyclohexadiene
1-(Trimethylsilyloxy)cyclohexene
3-Trimethylsilyloxy-1,3-pentadiene
3-Trimethylsilyloxy-1,4-pentadiene
2-Trimethylsilyloxy-1-propene
4-Trimethylsilyloxyvaleronitrile
1-Trimethylsilyloxy-1-vinylcyclopropane
(Z)-(Trimethylsilyloxy)vinyllithium
1-Trimethylsilyl(pentadienyl)lithium
Trimethylsilyl perchlorate
Trimethylsilyl polyphosphate
Trimethylsilylpotassium
2-Trimethylsilyl-2-propen-1-ol
3-Trimethylsilyl-2-propen-1-ol
2-(3-Trimethylsilyl-1-propenyl)-magnesium bromide
2-(3-Trimethylsilyl-1-propenyl)-magnesium bromide–Copper(I) iodide
1-Trimethylsilyl-1-propyne
3-Trimethylsilyl-1-propyne
3-Trimethylsilyl-2-propyn-1-ol
1-Trimethylsilylpropynylcopper
N-Trimethylsilylpyrrole
Trimethylsilylsodium
Trimethylsilyl tribromoacetate
Trimethylsilyl trichloroacetate
Trimethylsilyl trifluoroacetamide
Trimethylsilyl trifluoromethanesulfonate
3-Trimethylsilyl-3-trimethylsilyloxy-1-propene
1-Trimethylsilyl-2-trimethyl-stannylethylene
Trimethylsilylvinylketene
α-Trimethylsilylvinylmagnesium bromide
β-Trimethylsilylvinylmagnesium bromide
α-Trimethylsilylvinylmagnesium bromide–Copper(I) iodide
Trimethyl(triphenylmethoxy)silane
Triphenylsilane
Triphenylsilyl hydroperoxide
Triphenyl(trimethylsilylmethyl)-phosphonium trifluoromethane-sulfonate
Tris(diethylamino)sulfonium difluorotrimethylsilicate
Tris(dimethylamino)sulfonium difluorotrimethylsilicate
Tris(trimethylsilyl)aluminum
Tris(trimethylsilylethynyl)aluminum
Tris(trimethylsilyl)hydrazidocopper
Tris(trimethylsilyl)hydrazidolithium
Tris(trimethylsilyl)ketenimine
Tris(trimethylsilyl)methane
Tris(trimethylsilyl)methyllithium
Tris(trimethylsilyloxy)ethylene
Tris(trimethylsilyl) phosphite
Vinyltrichlorosilane
Vinyltrimethylsilane
Vinyltriphenylsilane
SILVER COMPOUNDS
Bis(2,2-dipyridyl)silver(II) peroxydisulfate
Bispyridinesilver permanganate
Palladium(II) chloride–Silver(I) acetate
Silver
Silver acetate
Silver azide
Silver benzoate
Silver carbonate
Silver carbonate–Celite
Silver chlorate
Silver chloride
Silver chlorodifluoroacetate
Silver chromate
Silver chromate–Iodine
Silver crotonate

Silver cyanide
Silver dibenzyl phosphate
Silver diethyl phosphate
Silver diphenylphosphate
Silver(II) dipicolinate
Silver fluoride
Silver(II) fluoride
Silver fluoride–Pyridine
Silver heptafluorobutanoate
Silver hexafluoroantimonate
Silver imidazolate
Silver iododibenzoate
Silver(I) nitrate
Silver nitrite
Silver nitrite–Mercury(II) chloride
Silver(I) oxide
Silver(II) oxide
Silver perchlorate
Silver(II) picolinate
Silver sulfate
Silver tetrafluoroborate
Silver tetrafluoroborate–Dimethyl sulfoxide
Silver *p*-toluenesulfonate
Silver(I) trifluoroacetate
Silver(II) trifluoroacetate
Silver(I) trifluoromethanesulfonate
Tollens reagent

THALLIUM COMPOUNDS
Arylthallium bis(trifluoroacetates)
Cyclopentadienylthallium(I)
Diethylthallium *t*-butoxide
Grignard reagents–Thallium(I) bromide
Thallium
Thallium(I) acetate
Thallium(III) acetate
Thallium(III) acetate–Bromine
Thallium(I) acetate–Iodine
Thallium(I) benzoate–Iodine
Thallium(I) bromide
Thallium(I) carbonate
Thallium chlorodifluoroacetate
Thallium(I) cyanide
Thallium cyclopentadienide
Thallium(I) ethoxide
Thallium(I) hydroxide
Thallium(I) 2-methyl-2-propanethiolate
Thallium(III) nitrate
Thallium(III) nitrate + co-reagents
Thallium(III) oxide
Thallium(III) perchlorate
Thallium 1-propanethiolate
Thallium 2-propanethiolate
Thallium(III) sulfate
Thallium(I) trichloroacetate
Thallium(III) trifluoroacetate
Thallium(III) trifluoroacetate + co-reagents

TIN COMPOUNDS
Allyltin difluoroiodide
Allyltributyltin
Allyltrimethyltin
Allyltriphenyltin
Benzyl 3-tributylstannylacrylate
Bis(methylcyclopentadienyl)tin(II)
Bis(methylthio)(trimethylstannyl)-methane
Bis(methylthio)(trimethylstannyl)-methyllithium
trans-1,2-Bis(tributylstannyl)ethylene
Bis(tributyltin) oxide
Bis(tributyltin) peroxide
Bis(tricyclohexyltin) selenide–Boron trichloride
Bis(tricyclohexyltin) sulfide–Boron trichloride
2,3-Bis(trimethylstannyl)-1,3-butadiene
3-Butenyltributyltin
Butylphenyltin dihydride, polymeric
Chloromethyltrimethyltin
4-Chloro-2-trimethylstannyl-1-butene
Cinnamyltriphenyltin
Diallyltin dibromide
Diazidotin dichloride
Dibutyldivinyltin
Dibutyltin dichloride
Dibutyltin dilaurate
Dibutyltin oxide
Dimethylaminotrimethyltin
Diphenyl diselenide–Bromine–Hexabutyldistannoxane
Diphenyltin dihydride
1-Ethoxy-4-tributylstannyl-1,3-butadiene
1-Ethoxy-2-tributylstannylethylene
Ethyl tributylstannyl sulfide
Geranyltrimethyltin
Hexabutylditin
Hexamethylditin
Hydrogen hexachloroplatinate(IV)–Tin(IV) chloride
1-Hydroxymethyl-2-tributylstannyl-

METAL-CONTAINING COMPOUNDS (*Continued*)
cyclopropane
4-Iodobutyltrimethyltin
Iodomethyltriphenyltin
2-Lithiovinyltributyltin
Lithium (1-hexynyl)(2-tributylstannylvinyl)cuprate
Lithium (1-pentynyl)(2-tributylstannylvinyl)cuprate
Lithium phenylthio(trimethylstannyl)cuprate
3-Methyl-2-butenyltrimethyltin
Methyl tributylstannyl sulfide
Palladium(II) chloride–Tin(II) chloride–Triphenylphosphine
Phenylthio(triphenylstannyl)methane
Phenylthio(triphenylstannyl)methyllithium
Propargyltriphenyltin
Tetraallyltin
1,1,6,6-Tetrabutyl-1,6-distanna-2,5,7,10-tetraoxycyclodecane
Tetrabutyldiacetoxytin oxide, dimer
3-Tetrahydropyranyloxy-1-tributylstannyl-1-propene
Tetramethyltin
Tetraphenyltin
Tetravinyltin
Tin
Tin–Aluminum
Tin amalgam
Tin(II) bromide
Tin(II) chloride
Tin(IV) chloride
Tin(IV) chloride + co-reagents
Tin(II) chloride–Silver perchlorate
Tin(II) fluoride
Tin(IV) fluoride
Tin(II) octoate
Tin(II) oleate
Tin(II) trifluoromethanesulfonate
Tribenzyltin chloride
Tributylcinnamyltin
Tributylcrotyltin
Tributyl(diethylaluminum)tin
Tributyl(ethynyl)tin
Tributyl(iodoacetoxy)tin
Tributyl(iodomethyl)tin
Tributylmethallyltin
Tributyl(1-methoxymethoxy-2-butenyl)tin
(E)-Tributyl(3-penten-2-yl)tin
Tributylstannyl 2-iodopropionate
3-Tributylstannyl-2-propen-1-ol
(E)-1-Tributylstannyl-2-trimethylsilylethylene
Tributyltin azide
Tributyltin chloride
Tributyltin chloride–Boron trifluoride
Tributyltincopper
Tributyltin cyanide
Tributyltin fluoride
Tributyltin hydride
Tributyltinlithium
Tributyltinlithium–Diethylaluminum chloride
Tributyltinmagnesium bromide
Tributyltinmethanol
Tributyltin methoxide
Tributyltin trifluoromethanesulfonate
Tributyl(trimethylsilylmethyl)tin
Tributylvinyltin
Triethyltin methoxide
Trimethyl(1-propenyl)tin
1-Trimethylsilyl-2-trimethylstannylethylene
Trimethylstannylcarbene
Trimethylstannylcopper–Dimethyl sulfide
2-Trimethylstannylethylidenetriphenylphosphorane
2-Trimethylstannylmethyl-1,3-butadiene
Trimethylstannylmethyllithium
Trimethyltin chloride
Trimethyltin hydride
Trimethyltinlithium
Trimethyltinsodium
Trimethyl(trifluoromethyl)tin
Trimethylvinyltin
Triphenylstannylmethyllithium
Triphenyltin chloride
Triphenyltin hydride
Tungsten(VI) chloride–Tetramethyltin

TITANIUM COMPOUNDS
Alkyltris(dialkylamino)titanium derivatives
Bis(cyclopentadienyl)(diiodozinc)-μ-methylenetitanium
Bis(cyclopentadienyl)methyltitanium
Bis(cyclopentadienyl)titanacyclobutanes
Bis(cyclopentadienyl)(1-trimethyl-

silylallyl)titanium
(E)-Bromocrotylbis(cyclopentadienyl)-titanium
Bromotriisopropoxytitanium
μ-Chlorobis(cyclopentadienyl)(dimethylaluminum)-μ-methylenetitanium
Chlorotriisopropoxytitanium
Chlorotriphenoxytitanium
Crotylbis(cyclopentadienyl)halotitanium reagents
Crotylmagnesium chloride–Dichlorobis-(cyclopentadienyl)titanium
Crotyltri(isopropoxy)titanium
Dichlorobis(cyclopentadienyl)titanium
Dichlorobis(cyclopentadienyl)titanium + co-reagents
Dichlorobis(trifluoromethanesulfonato)-titanium(IV)
Dichlorodiisopropoxytitanium(IV)
Dichlorodimethyltitanium
Dicyclopentadienyltitanium
Diisopropoxydimethyltitanium
Ethylmagnesium bromide–Dichlorobis-(cyclopentadienyl)titanium
Grignard reagents + titanium co-reagents
Methallylmagnesium chloride–Dichlorobis(cyclopentadienyl)-titanium
N-Methyl-C-(trichlorotitanio)-formimidoyl chloride
(S)-(–)-Methyltri(2-methyl-1-butoxy)-titanium
Organolithium reagents–Titanium(IV) chloride
Organotitanium reagents
Tetrakis(diethylamino)titanium
Tetrakis(dimethylamino)titanium
Titanium(0)
Titanium(III) acetate
Titanium(IV) alkoxides
Titanium(IV) butoxide
Titanium(III) chloride
Titanium(IV) chloride
Titanium(IV) chloride + co-reagents
Titanium(III) chloride + co-reagents
Titanium(IV) ethoxide
Titanium(IV) fluoride
Titanium(IV) isopropoxide
Titanium(IV) methoxide
Titanium(IV) propoxide
Tribenzylcyclopentadienyltitanium
Trichlorocyclopentadienyltitanium–Lithium aluminum hydride
Trichloromethyltitanium
Triisopropoxymethyltitanium
Trimethylaluminum–Titanium(IV) chloride
Tris(diethylamino)methyltitanium
Zinc–Titanium(IV) chloride
TUNGSTEN COMPOUNDS
Dipotassium hexachlorotungstate(IV)
Pentacarbonyl(methoxyphenyl-methylene)tungsten(0)
Pentacarbonyl(triphenylphosphine)-tungsten
Phosphotungstic acid
Sodium tungstate
Trichloro(phenyl)tungsten–Aluminum chloride
Tripotassium tri-μ-chlorohexa-chloroditungstate
Tris(acetonitrile)tricarbonyltungsten
Tungsten carbonyl
Tungsten carbonyl–Aluminum chloride
Tungsten(VI) chloride
Tungsten(VI) chloride + co-reagents
Tungsten(VI) fluoride
Tungsten(VI) oxide–Chlorosulfonic acid
VANADIUM COMPOUNDS
Ammonium metavanadate
Bis(trimethylsilyl) peroxide–Vanadyl acetylacetonate
Dicyclopentadienylvanadium
Oxotris(triphenylsilanolato)vanadium
Sodium tricarbonyl(cyclopentadienyl)-vanadate(2-)
Tetracarbonyl(cyclopentadienyl)-vanadium
Vanadium(II) chloride
Vanadium(III) chloride
Vanadium(IV) chloride
Vanadium(III) chloride–Lithium aluminum hydride
Vanadium(IV) chloride–N-Methylaniline
Vanadium(II) chloride tetrapyridine complex
Vanadium(V) oxide
Vanadium(II) perchlorate
Vanadyl acetylacetonate
Vanadyl acetylacetonate–

METAL-CONTAINING COMPOUNDS (*Continued*)
Azobisisobutyronitrile
Vanadyl trichloride
Vanadyl trifluoride
YTTERBIUM COMPOUNDS
Phenylytterbium iodide
Tris(6,6,7,7,8,8,8-heptafluoro-2,2-dimethyl-3,5-octanedionato)-ytterbium
Ytterbium
Ytterbium(II) chloride
Ytterbium(II) iodide
Ytterbium(III) nitrate
ZINC COMPOUNDS
Allyltrimethylsilylzinc chloride
Allylzinc bromide
Bis(cyclopentadienyl)(diiodozinc)-μ-methylenetitanium
Bromoform–Diethylzinc
Chloroiodomethane–Zinc-copper couple
Chlorotrimethylsilane + zinc co-reagents
Diallylzinc
Dibromobis(triphenylphosphine)-nickel(II)–Zinc
Dibromomethane–Zinc–Titanium(IV) chloride
Dibutylzinc
Dichlorobis(triphenylphosphine)-palladium(II)–Zinc
Diethylzinc
Diethylzinc + co-reagents
Diiodomethane–Zinc–Trimethylaluminum
Dimethylzinc
Diphenylzinc
Dipotassium hexachloroosmiate(IV)–Zinc
α-Ethoxyvinylzinc chloride
Ethyl iodide–Zinc-copper couple
Ethyl(iodomethyl)zinc
Ethylzinc iodide
Lithium bromide–Zinc
Lithium tri-*t*-butylzincate
Lithium trimethylzincate
Molybdenum(V) chloride–Zinc
Molybdenum(V) trichloride oxide–Zinc
Nickel bromide–Zinc
Nickel chloride–Zinc
Organozinc reagents
Phenylzinc bromide
Phenylzinc chloride
Phosphorus(V) chloride–Zinc chloride
Potassium iodide + zinc co-reagents
Reformatsky reagent
Simmons–Smith reagent
Sodium iodide–Zinc
Tetrakis(triphenylphosphine)-palladium(0)–Zinc
Titanium(IV) chloride–Zinc
Titanium(III) chloride–Zinc-copper couple
Vinylzinc chloride
Zinc
Zinc + co-reagents
Zinc amalgam
Zinc borohydride
Zinc borohydride–Dimethylformamide
Zinc bromide
Zinc carbonate
Zinc chloride
Zinc chloride + co-reagents
Zinc-copper couple
Zinc cyanide
Zinc dithionite
Zinc fluoride
Zinc iodide
Zinc nitrate
Zinc oxide
Zinc perchlorate–(S)-15-Amino-14-hydroxy-5,5-dimethyl-2,8-dithia[9]-(2,5)pyridinophane
Zinc permanganate
Zinc permanganate–Silica
Zinc-silver couple
Zinc tetracarbonylcobaltate
Zinc *p*-toluenesulfonate
Zinc trifluoromethanesulfonate
ZIRCONIUM COMPOUNDS
Bis(cyclopentadienyl)diiodozirconium
Bis(cyclopentadienyl)isoprenezirconium
Chlorobis(cyclopentadienyl)-hydridozirconium(IV)
Chlorobis(cyclopentadienyl)-tetrahydroboratozirconium(IV)
Dibromobis(cyclopentadienyl)zirconium
Dichlorobis(cyclopentadienyl)-zirconium(II)
Dichlorobis(cyclopentadienyl)-zirconium–*t*-Butylmagnesium chloride
Organozirconium reagents

Tributoxychlorozirconium
Trimethylaluminum–Dichlorobis-(cyclopentadienyl)zirconium
Zirconium(IV) butoxide
Zirconium carbene complexes
Zirconium(III) chloride
Zirconium(IV) chloride
Zirconium(IV) chloride–Lithium aluminum hydride
Zirconium(IV) propoxide

OTHER COMPOUNDS
Allyltrimethylgermanium
Bis(cyclopentadienyl)trihydridoniobium
Gold
Hydrogen tetrachloroaurate(III)
Tris(2,2-dimethylpropyl)(2,2-dimethylpropylidene)niobium
Tris(2,2-dimethylpropyl)(2,2-dimethylpropylidene)tantalum
Tris(6,6,7,7,8,8,8-heptafluoro-2,2-dimethyl-3,5-octanedionato)-europium(III)
Tris[3-(heptafluoropropylhydroxy-methylene)-*d*-camphorato]-europium(III)

WITTIG REAGENTS AND PHOSPHONIUM SALTS
Acetone-1,3-bis(triphenylphosphonium) dichloride
Acetylmethylenetriphenylphosphorane
N-Alkyl-β-aminoethyltriphenyl-phosphonium bromides
Allylidenetriphenylphosphorane
Allyltriphenylphosphonium bromide
N-Aminotriphenylphosphinimine
Benzoylmethylenetriphenylphosphorane
Benzylidenetriphenylphosphorane
Benzyltriphenylphosphonium chloride
Benzyltris(dimethylamino)phosphonium bromide
Bis(trifluoromethyl)methylenetriphenyl-phosphorane
Bromoacetylmethylenetriphenyl-phosphorane
Bromomethylenetriphenylphosphorane
3-Bromopropyltriphenylphosphonium bromide
trans-1-Butadienyltriphenyl-phosphonium bromide
Butane-1,4-bis(triphenylphosphonium) dibromide
2-Butenylidenetriphenylphosphorane
2-*t*-Butoxycarbonyl-1-methoxy-carbonylethylidenetriphenyl-phosphorane
4-Carboxybutyltriphenylphosphonium bromide
2-Carboxyethyltriphenylphosphonium perbromide
2-Carboxy-1-methoxycarbonyl-ethylidenetriphenylphosphorane
(2-Carboxy-2-propenyl)triphenyl-phosphonium bromide
Chlorofluoromethylenetriphenyl-phosphorane
2-[(Chloroformyl)oxy]ethyltriphenyl-phosphonium chloride
Chloromethylenetriphenylphosphorane
Chloromethyltriphenylphosphonium halides
Cinnamyltriphenylphosphonium chloride
Cyanomethylenetriphenylphosphorane
Cyanomethylidenebis(triphenyl-phosphonium) dibromide
Cyclopentylidenetriphenyl-phosphorane
Cyclopropylidenetriphenyl-phosphorane
Cyclopropyltriphenylphosphonium bromide
Dibromomethylenetriphenyl-phosphorane
Dichloromethylenetriphenyl-phosphorane
Dichloromethylenetris(dimethylamino)-phosphorane
2,2-Diethoxyethylidenetriphenyl-phosphorane
2,2-Diethoxyvinylidenetriphenyl-phosphorane
Difluoromethylenetriphenylphosphorane
2-Dimethylaminoethylidenetriphenyl-phosphorane
(2-Dimethylaminoethyl)triphenyl-phosphonium bromide
2-(1,3-Dioxan-2-ylethylidene)-triphenylphosphorane
[2-(1,3-Dioxan-2-yl)ethyl]triphenyl-phosphonium bromide
1,3-Dioxolan-2-ylmethylenetriphenyl-

WITTIG REAGENTS AND PHOSPHONIUM SALTS (*Continued*)
phosphorane
1,3-Dioxolan-2-ylmethyltriphenylphosphonium bromide
Diphenyl triphenylphosphoranylidenemethylphosphonate
1,3-Dithian-2-ylidenetriphenylphosphorane
1,3-Dithiolan-2-yltriphenylphosphonium tetrafluoroborate
2-Ethoxyallylidenetriphenylphosphorane
Ethoxycarbonylcyclopropyltriphenylphosphonium tetrafluoroborate
Ethoxycarbonylmethylenetriphenylphosphorane
Ethoxycarbonylmethyltriphenylphosphonium bromide
(2-Ethoxy-1-propenyl)triphenylphosphonium iodide
β-Ethoxyvinyltriphenylphosphonium iodide
1,2-Ethylenebis(triphenylphosphonium) dibromide
Ethylidenetriphenylphosphorane
Ethyltriphenylphosphonium salts
Fluoromethylenetriphenylphosphorane
(α-Formylethylidene)triphenylphosphorane
Formylmethylenetriphenylphosphorane
3-Hydroxy-3-methylbutylidenetriphenylphosphorane, lithium salt
Isopropylidenetriphenylphosphorane
Ketenylidenetriphenylphosphorane
α-Lithiomethylenetriphenylphosphorane
3-Methoxyallylidenetriphenylphosphorane
Methoxycarbonylmethylenetriphenylphosphorane
α-Methoxyethylidenetriphenylphosphorane
Methoxymethylenetriphenylphosphorane
Methoxymethyltriphenylphosphonium iodide
3-Methyl-2-butenylidenetriphenylphosphorane
Methylenetriphenylphosphorane
1-Methylthiovinyl(triphenyl)phosphonium chloride
Methyltriphenylphosphonium bromide
Methyltriphenylphosphonium permangate
2-Oxo-3-(triphenylphosphoranylidene)propyllithium
Phenacyltriphenylphosphonium bromide
Phenoxymethylenetriphenylphosphorane
N-Phenylketeniminyl(triphenyl)phosphorane
(3-Phenyl-2-propenyl)triphenylphosphonium chloride
1-Phenylthiocyclopropyltriphenylphosphonium tetrafluoroborate
2-Propenyltriphenylphosphonium bromide
β-Trimethylsilylethylidenetriphenylphosphorane
Trimethylsilylmethylenetrimethylphosphorane
Trimethylsilylmethylenetriphenylphosphorane
2-Trimethylstannylethylidenetriphenylphosphorane
Triphenyl(1-phenylthiovinyl)phosphonium iodide
Triphenylphosphoranylideneoxalacetonitrile
Triphenylpropargylphosphonium bromide
Triphenyl(trimethylsilylmethyl)phosphonium trifluoromethanesulfonate
Vinyltriphenylphosphonium bromide
Wittig reaction

WITTIG–HORNER REAGENTS
Alkyldiphenylphosphine oxides
Allyldiphenylphosphine
Allyldiphenylphosphine oxide
P-Allyl-N,N,N′,N′-tetramethylphosphonic diamide
Bis(diethoxyphosphinyl)methane
1,3-Bis(dimethoxyphosphinyl)-2-propanone
Bis(2,2,2-trifluoroethyl) methoxycarbonylmethylphosphonate
Cyanomethyldiphenylphosphine oxide
Dibenzyl carboxymethylphosphonate
Dibutyl tetrahydropyranyloxymethylphosphonate
(Diethoxymethyl)diphenylphosphine oxide
Diethoxyphosphinylacetic acid,

anhydride with trifluoroacetic acid
(Diethoxyphosphinyl)difluoromethyllithium
2-(Diethoxyphosphinyl)propionic acid
2-(Diethoxyphosphinyl)propionic acid, anhydride with trichloroacetic acid
2-(Diethoxyphosphinyl)propionitrile
Diethyl allylthiomethylphosphonate
Diethyl benzoylphosphonate
Diethyl N-benzylideneaminomethylphosphonate
Diethyl benzylphosphonate
Diethyl *t*-butoxy(cyano)methylphosphonate
Diethyl carboxychloromethylphosphonate
Diethyl carboxymethylphosphonate
Diethyl chloromethylphosphonate
Diethyl cyanomethylphosphonate
Diethyl β-(cyclohexylimino)-ethylphosphonate
Diethyl (diazomethyl)phosphonate
Diethyl dibromomethylphosphonate
Diethyl dichlorolithiomethylphosphonate
Diethyl dichloromethylphosphonate
Diethyl 2,2-diethoxyethylphosphonate
Diethyl difluoromethylphosphonate
Diethyl difluorotrimethylsilylmethylphosphonate
Diethyl (1,3-dithian-2-yl)phosphonate
Diethyl ethylthiomethylphosphonate
Diethyl formylmethylphosphonate
Diethyl isocyanomethylphosphonate
Diethyl lithio-N-benzylideneaminomethylphosphonate
Diethyl lithiomorpholinomethylphosphonate
Diethyl methoxyethoxymethylphosphonate
Diethyl methoxymethylphosphonate
Diethyl methylsulfinylmethylphosphonate
Diethyl methylsulfonylmethylphosphonate
Diethyl (1-methylthio)ethylphosphonate
Diethyl methylthiomethylphosphonate
Diethyl morpholinomethylphosphonate
Diethyl 2-oxobutylphosphonate
Diethyl (2-oxopropyl)phosphonate
Diethyl phenylsulfinylmethylphosphonate
Diethyl phenylthiomethylphosphonate
Diethyl 1-phenyl-1-trimethylsilyloxymethylphosphonate
Diethyl [(2-tetrahydropyranyloxy)-methyl]phosphonate
Diethyl trichloromethylphosphonate
Diethyl trimethylsilyloxycarbonylmethylphosphonate
Diethyl 1-trimethylsilyloxyethylphosphonate
Diisopropyl methylphosphonate
(Dimethoxyphosphinyl)methyllithium
R-(−)-(Dimethoxyphosphinyl)methyl *p*-tolyl sulfoxide
N,N′-Dimethyl-2-allyl-1,3,2-diazaphospholidine 2-oxide
Dimethyl bis(methylthio)methylphosphonate
Dimethyl 3-bromo-2-ethoxypropenylphosphonate
Dimethyl diazomethylphosphonate
Dimethyl methylphosphonate
Diphenylmethoxymethylphosphine
Diphenyl 2-(N-piperidinyl)ethylphosphine oxide
Ethyl 4-diethoxyphosphinylcrotonate
Ethyl 4-diethoxyphosphinyl-3-oxobutanoate
Ethyl 4-diphenylphosphinyl-3-oxobutanoate
P-Ethyl-N,N,N′,N′-tetramethylphosphonic diamide
P-Isopropyl-N,N,N′,N′-tetramethylphosphonic diamide
1-Methoxycarbonylethyl(diphenyl)-phosphine oxide
1-Methoxyethyldiphenylphosphine oxide
Methoxymethyl(diphenyl)phosphine oxide
Methyl bis(trifluoroethoxy)-phosphinylacetate
Methyl 2-bis(2,2,2-trifluoroethoxy)-phosphinylpropionate
Methyl 4-diethoxyphosphinylcrotonate
N-Morpholinomethyldiphenylphosphine oxide
Pentamethylphosphonic diamide
Phosphonamides, chiral
Tetraethyl dimethylaminomethylenediphosphonate

WITTIG–HORNER REAGENTS (*Continued*)
Tetraethyl phosphonosuccinate
Triethyl phosphonoacetate
Triethyl phosphonoiodoacetate
Triethyl α-phosphonopropionate
Trimethyl phosphonoacetate
Trimethylphosphonoglycolate

YLIDES (EXCLUDING WITTIG REAGENTS)
Acetylmethylene(triphenyl)arsorane
Dibutyltelluronium carboethoxymethylide
2,5-Dichlorothiophenium bismethoxycarbonylmethylide
(N,N-Diethylamino)methyloxosulfonium methylide
(N,N-Dimethylamino)methyloxosulfonium methylide
(Dimethylamino)phenyloxosulfonium methylide
Dimethyl(phenyl)selenonium methyl sulfate
Dimethylselenonium methylide
Dimethylsulfonium ethoxycarbonylmethylide
Dimethylsulfonium methylide
Dimethylsulfoxonium methylide
Dimethyl(tetrahydro-2-oxo-3-furanyl)sulfonium tetrafluoroborate
Diphenylsulfonium cyclopropylide
Diphenylsulfonium isopropylide
Diphenylsulfonium methylide
Ethyl (dimethylsulfuranylidine)acetate
N-Imino-N,N-dimethyl-2-hydroxypropanaminium ylide
Isopropylidenediphenylsulfurane
Lauryldimethylsulfonium chloride
2-Methylcyclopentenone-3-dimethylsulfoxonium methylide
Methyl(phenyl)selenoniomethanide
(2-Oxo-2-phenylethylide)dimethylselenonium ylide
Phenacylidenedimethylsulfurane
Trimethylsilylmethylene dimethylsulfurane
Triphenylarsonium cinnamylide
Triphenylarsonium ethylide

AUTHOR INDEX

Aalbersberg, W. G. L., 6, 154
Aanerud, T. U., **7,** 271
Abai, T., **10,** 21
Abatjoglou, A. G., **10,** 291
Abbas, S. A., **6,** 35
Abbott, D. E., **12,** 23, 146
Abbott, D. J., **11,** 539
Abbott, F. P., **1,** 518
Abbott, S. D., **4,** 295
Abbott, T. W., **1,** 12; **11,** 2
Abdallah, H., **10,** 121; **11,** 231
Abdel-Rahman, M. O., **1,** 44, 135, 611, 731; **9,** 520
Abderhalden, E., **1,** 489, 652; **8,** 272
Abdulla, R. F., **5,** 253; **9,** 183; **10,** 159
Abdullah, A. H., **12,** 286
Abdul-Mali, N. F., **11,** 109
Abe, H., **10,** 290
Abe, K., **12,** 446
Abe, O., **5,** 170
Abe, R., **10,** 142; **12,** 546
Abe, T., **9,** 87, 233
Abe, Y., **1,** 1000; **11,** 220
Abecassis, J., **12,** 237, 456
Abegg, V. P., **5,** 424
Abel, E. W., **8,** 240; **11,** 22
Abell, R. D., **1,** 1095
Aben, R. W. M., **8,** 537; **9,** 523; **11,** 71
Abend, P. G., **1,** 166
Abendroth, H. J., **1,** 484
Abenhaim, D., **7,** 240
Abeysekera, B., **9,** 181
Abiko, A., **9,** 198; **12,** 506
Abiko, S., **5,** 347; **6,** 149
Abjean, J. P., **12,** 36
Abley, P., **4,** 40; **5,** 658
Aboderin, A. A., **1,** 339
Abola, E., **5,** 394
Abott, M. T., **7,** 268
Abou-Donia, M. B., **6,** 456
Abou-Gharbia, M., **8,** 342
Abraham, N. A., **1,** 56, 158; **2,** 29
Abraham, W., **5,** 316
Abramovitch, A., **6,** 91; **8,** 41
Abramovitch, B., **1,** 1292
Abramovitch, R. A., **2,** 352
Abramow, W. S., **1,** 500
Abramson, N. L., **6,** 19
Abskharoun, G. A., **5,** 28
Abushanab, E., **10,** 450
Acar, M., **8,** 272
Acharya, S. P., **2,** 349; **6,** 468
Achaya, K. T., **2,** 368
Acher, A., **6,** 574
Acher, F., **10,** 91; **11,** 359
Acheson, R. M., **4,** 172; **5,** 230; **12,** 46
Achiwa, K., **3,** 65, 71, 79, 210, 276, 313; **5,** 293; **8,** 59; **11,** 87
Achmatowicz, B., **1,** 741
Achmatowicz, O., Jr., **6,** 291; **7,** 50
Achmatowicz, S., **6,** 583
Acker, D. S., **1,** 16, 500, 1136
Acker, M., **12,** 168
Acker, R. D., **11,** 373
Ackere, K. J., **9,** 435
Ackerman, J., **1,** 986
Ackerman, J. H., **2,** 208, 252
Ackerman, M. E., **4,** 252
Ackermann, K., **3,** 125
Ackman, R. G., **1,** 892
Ackrell, J., **3,** 87
Acott, B., **1,** 563, **2,** 238
Acton, E. M., **1,** 1045
Acton, N., **4,** 339, 419; **5,** 530
Adachi, I., **10,** 231
Adachi, K., **12,** 405
Adachi, M., **8,** 519; **9,** 63, 506, 507
Adam, G., **5,** 424
Adam, M. A., **12,** 513
Adam, R., **6,** 500
Adam, W., **1,** 409; **4,** 254; **5,** 23, 590; **7,** 47; **8,** 173, 249, 299, 317, 367, 373, 519; **9,** 56, 341
Adamcik, J. A., **1,** 1203
Adamczyk, M., **11,** 458; **12,** 412
Adamek, J. P., **5,** 677; **6,** 534
Adames, G., **5,** 464
Adams, B. R., **12,** 252
Adams, D. J. C., **4,** 282
Adams, D. R., **12,** 565
Adams, E. W., **1,** 629, 654
Adams, J., **9,** 236; **10,** 165
Adams, J. B., Jr., **1,** 49, 339
Adams, J. H., **9,** 377
Adams, J. L., **10,** 370
Adams, J. T., **1,** 69, 160, 251, 1040; **4,** 397; **7,** 106; **12,** 403
Adams, P. E., **8,** 356
Adams, R., **1,** 6, 23, 47, 48, 274, 343, 370, 402, 444, 455, 489, 536, 629, 654, 747, 771, 827, 870, 892, 982, 1056, 1073, 1087, 1284, 1294; **6,** 465; **7,** 110
Adams, R. C., **5,** 25
Adams, R. D., **12,** 557
Adams, R. E., **5,** 438
Adams, R. F., **2,** 182
Adams, R. M., **4,** 191
Adams, T., **3,** 244; **4,** 421
Adams, W., **5,** 491
Adams, W. R., **2,** 59
Adamson, D. W., **3,** 74
Adamson, J. R., **6,** 453
Adcock, W., **6,** 454
Addas, A. R., **7,** 102
Addison, C. C., **1,** 329
Adelakun, E., **12,** 481
Adelstein, S. J., **10,** 211
Adenis, J. C., **5,** 312
Adesogan, E. K., **9,** 414
Adhikary, P., **4,** 357; **5,** 614
Adickes, H. W., **3,** 284; **4,** 430, 484

Adinolfi, M., **5,** 349; **6,** 296; **7,** 181, 182
Adkins, H., **1,** 24, 157, 228, 465, 654, 710, 730, 828, 846, 963, 977; **4,** 267
Adkins, J. D., **3,** 175; **4,** 288
Adkins, R. L., **12,** 258
Adkins, S., **12,** 379
Adkins, T. J., **6,** 91
Adler, A. D., **3,** 81
Adler, E., **1,** 219, 643; **4,** 374; **10,** 136
Adler, R. W., **3,** 202
Adlercreutz, P., **10,** 265
Adlerova, E., **2,** 279
Adlington, M. G., **7,** 388; **8,** 501
Adlington, R. M., **9,** 488; **10,** 248; **11,** 536, 566; **12,** 87, 416
Adolph, H. G., **4,** 60
Adolphen, G., **3,** 74
Adriaens, P., **9,** 212
Advani, B. G., **2,** 178; **3,** 131; **6,** 556
Aerssens, M. H. P. J., **12,** 570
Aeschbacher, R., **1,** 660
Affrossman, S., **1,** 1145
Afonso, A., **1,** 603, 1273; **3,** 182; **8,** 365
Afza, N., **12,** 532
Afzali, A., **6,** 452; **12,** 144
Agami, C., **1,** 309; **2,** 143, 173; **3,** 285; **12,** 415
Agarwal, K. L., **7,** 414
Agarwal, S. K., **9,** 99
Agasimundin, Y. S., **5,** 339
Agata, I., **1,** 465
Agawa, T., **6,** 158; **8,** 348; **9,** 216, 316; **10,** 298, 391; **11,** 514, 555; **12,** 187, 336
Ager, D. J., **8,** 197; **10,** 42, 314; **11,** 299, 303, 424; **12,** 212, 229, 394
Aggarwala, V. P., **4,** 438
Agnello, E. J., **1,** 127, 1122
Agnihotri, R. K., **1,** 581; **4,** 288
Agosta, W. C., **2,** 33; **4,** 108, 512; **7,** 278, 383; **11,** 570
Agranat, I., **6,** 475, 670; **7,** 425
Aguiar, A. M., **5,** 410; **6,** 341, 416
Aguilar, D. A., **11,** 415
Ahern, M. F., **9,** 127
Ahern, S. P., **7,** 316
Ahfaenger, B., **6,** 201
Ahlbrecht, H., **9,** 381; **10,** 260, 323; **12,** 100
Ahlgren, G., **5,** 230
Ahluwalia, V. K., **10,** 317
Ahmad, A., **1,** 481; **7,** 285
Ahmad, I., **9,** 8
Ahmad, M. S., **12,** 294
Ahmad, N., **7,** 60
Ahmad, S. Z., **12,** 294
Ahmad, Y., **2,** 300
Ahmed, F. U., **6,** 20
Ahmed, M., **5,** 631
Ahmed, M. G., **3,** 202
Ahmed, Q. A., **2,** 311
Ahmed, S. S., **5,** 436; **7,** 24, 331
Ahn, K. H., **11,** 294, 296; **12,** 193, 276, 446
Ahond, A., **3,** 115, 308; **4,** 184
Ahouande, F. S., **8,** 486
Ahuja, H. S., **1,** 124
Aiba, K., **9,** 326
Aida, T., **4,** 175; **5,** 57, 639; **8,** 200; **9,** 191, 291; **10,** 219
Aigami, K., **6,** 526; **7,** 349
Aigner, H., **9,** 82; **11,** 352
Aikawa, Y., **7,** 110
Aiken, R. R., **2,** 305
Aimetti, J. A., **9,** 477
Aimi, N., **5,** 601
Ainscough, M. A., **7,** 345
Ainsworth, C., **1,** 382, 407, 1073, 1080, 1188; **4,** 33, 539; **6,** 114; **7,** 400; **8,** 299; **9,** 310; **12,** 181, 268
Aithie, G. C. M., **6,** 308
Aizawa, T., **6,** 96; **7,** 47
Aizpurua, J. M., **11,** 384, 410; **12,** 79, 204
Ajo, M. M., **5,** 700
Akabori, S., **1,** 668; **6,** 138
Akaboshi, S., **8,** 198; **9,** 188
Akada, M., **9,** 23
Akagi, H., **10,** 259
Akahori, Y., **1,** 269
Akai, S., **12,** 221
Akam, T. M., **10,** 459
Akano, M., **4,** 395
Akao, M., **2,** 147
Akasaka, K., **6,** 628; **9,** 138, 174
Akasaka, T., **8,** 200; **9,** 191
Akashi, C., **2,** 136
Akashi, K., **6,** 127; **7,** 244, 255
Akerkar, A. S., **1,** 364; **2,** 192
Akermark, B., **6,** 589; **7,** 22, 109; **9,** 14
Akhrem, A. A., **5,** 6
Akhtar, M., **1,** 95, 658; **5,** 61
Akhtar, M. N., **6,** 358
Akhunova, V. R., **9,** 343
Akiba, K., **11,** 91, 373; **12,** 546
Akimoto, H., **9,** 421
Akimoto, I., **10,** 306
Akimoto, K., **9,** 382
Akinde, B., **9,** 311
Akita, E., **11,** 322
Akita, H., **9,** 77
Akita, T., **9,** 297, 464; **10,** 398
Akita, Y., **7,** 73; **11,** 542
Akiyama, E., **8,** 377
Akiyama, F., **9,** 364
Akiyama, M., **12,** 527
Akiyama, S., **5,** 86, 408; **6,** 138; **9,** 91
Akiyama, T., **11,** 599; **12,** 491
Akssira, M., **11,** 249
Akuja, V. K., **5,** 472
Akutagawa, S., **6,** 49; **11,** 54; **12,** 57
Alais, J., **11,** 501
Alauddin, M., **5,** 436
Al-Azrak, A., **2,** 119
Albaugh, P., **9,** 103
Albaugh-Robertson, P., **11,** 533
Al-Bayati, R., **11,** 75
Albers, R. J., **2,** 59, 297, 427
Albert, A., **1,** 881, 1187; **3,** 229
Albert, A. H., **5,** 391
Albert, R., **5,** 115
Alberts, A. H., **5,** 603
Albertson, N. F., **1,** 182, 235,

1126; **4,** 521; **6,** 283
Albizati, K. F., **8,** 222; **11,** 488
Albonico, S. M., **4,** 105, 376, 502; **7,** 362
Albrecht, H. P., **1,** 741
Albrecht, R., **1,** 98; **5,** 682; **7,** 50
Albright, J. A., **2,** 217; **4,** 341
Albright, J. D., **1,** 310, 1185; **2,** 165; **5,** 266
Albright, J. O., **10,** 115, 199, 419
Alburn, H. E., **1,** 1203
Alcock, N. W., **6,** 301
Alden, L., **7,** 211
Alder, K., **1,** 445, 636, 832; **3,** 339; **8,** 47; **9,** 269
Alder, R. W., **1,** 1139, 1145; **2,** 403; **3,** 22; **4,** 172; **5,** 446
Alderidge, M., **11,** 34
Alderson, T., **2,** 357
Aldrich, P. E., **1,** 117
Aldridge, F. T., **1,** 121
Alemany, A. A., **2,** 45
Alexakis, A., **9,** 249, 289, 470; **10,** 289; **12,** 350
Alexander, E. C., **8,** 446
Alexander, H. E., **5,** 107
Alexander, K., **1,** 881, 1141
Alexander, M. L., **4,** 341
Alexander, R., **5,** 454
Alexandre, C., **3,** 93; **4,** 141
Alexanian, V., **9,** 157, 180
Alfieri, C. C., **1,** 1127; **3,** 221
Alford, D. O., **11,** 56
Alford, J. A., **5,** 495, 649
Alhaique, F., **6,** 541
Al-Hamdany, R., **8,** 443
Alheim, R. J., **1,** 904
Ali, A., **4,** 54; **7,** 247
Ali, A. E., **10,** 203
Ali, B., **8,** 443
Ali, S., **7,** 174
Ali, S. A., **9,** 382, 402; **10,** 119
Ali, S. M., **6,** 157
Ali, Y., **3,** 153
Aliminosa, L. M., **1,** 652
Alired, E. L., **5,** 590
Alkaysi, H. N., **6,** 646
Alkonyi, I., **1,** 652
Allain, L., **10,** 373
Allan, A. R., **6,** 385
Allan, J. L. H., **5,** 86
Allan, L. T., **1,** 407
Allcock, H. R., **4,** 387
Allegra, G., **5,** 115
Allen, C. F., **1,** 874, 926, 976, 1073
Allen, C. F. H., **1,** 33, 196, 276, 336, 373, 379, 383, 402, 430, 452, 500, 832, 876, 890, 963, 1040, 1041, 1043, 1049, 1058, 1064, 1065, 1084, 1095, 1100, 1106, 1150, 1163, 1167, 1177, 1209, 1219, 1270, 1292; **3,** 272; **9,** 159
Allen, D. S., Jr., **1,** 814, 986
Allen, G. F., **7,** 22; **10,** 302
Allen, G. R., Jr., **1,** 1000; **4,** 94; **5,** 553
Allen, H. R., **1,** 1131
Allen, J. C., **1,** 264
Allen, J. F., **12,** 89
Allen, J. M., **7,** 295
Allen, L. E., **3,** 329; **4,** 562
Allen, M. S., **9,** 115
Allen, P., **3,** 314
Allen, S., **1,** 68
Allen, T. G., **10,** 450
Allen, W. S., **1,** 150, 764, 881, 1054
Allenby, O. C. W., **1,** 195
Alleston, D. L., **1,** 1193; **3,** 222; **7,** 193
Allevi, P., **12,** 479
Alley, W. D., **6,** 576
Allinger, N. L., **1,** 567, 643, 782, 926; **2,** 356
Allphin, N. L., **1,** 1064
Allred, A. L., **6,** 399
Allred, E. L., **2,** 202
Almirante, L., **1,** 1226
Almog, J., **4,** 63
Alner, D. J., **2,** 412
Alneri, E., **8,** 367
Alo, B. I., **9,** 414
Alonso, C., **6,** 162
Alonso, J. H., **5,** 551
Alonso, M. E., **7,** 170; **8,** 188; **12,** 141
Alonso-Cires, L., **9,** 293; **10,** 255; **12,** 306
Alper, H., **2,** 230; **3,** 167, 207; **4,** 45, 270, 346, 534; **5,** 358, 675; **6,** 198, 485; **7,** 100, 166, 248, 323; **8,** 130, 150, 266, 267, 391; **9,** 119, 127, 145, 207, 317, 374, 435; **10,** 112, 223, 274, 369, 382; **11,** 119, 163; **12,** 112, 173, 333, 373, 381, 475
Alper, J. B., **7,** 122
Alt, G. H., **1,** 94; **2,** 425
Alt, H., **7,** 374
Altenbach, H. J., **9,** 66
Altevogt, R., **11,** 134
Altland, H. W., **3,** 289; **4,** 492; **6,** 581
Altman, J., **6,** 290
Altman, L. J., **1,** 224; **4,** 202
Altnau, G., **12,** 345
Alvarez, E., **11,** 445
Alvarez, F. S., **1,** 207, 329; **2,** 145, 201; **4,** 490
Alvarez, R. M., **12,** 338
Alvernhe, G., **10,** 459
Alvernhe, G. M., **11,** 454
Alward, S. J., **6,** 181
Alzerreca, A., **8,** 519
Amada, T., **6,** 500
Amann, A., **6,** 630
Amaratunga, S., **11,** 163
Amaro, A., **6,** 171
Amato, J. S., **9,** 107; **11,** 189
Ambasht, S., **10,** 436
Ambelang, T., **4,** 255
Ambles, A., **7,** 294
Amel, R. T., **6,** 456
Amende, J., **5,** 114
Amendolla, C., **1,** 642
Ameniya, T., **8,** 155
Ames, A., **8,** 183; **10,** 107, 110, 232
Ames, S. R., **1,** 814
Amiard, G., **1,** 465, 1256
Amib, B., **11,** 79
Amice, P., **3,** 258; **6,** 523; **7,** 35, 67, 402; **10,** 91
Amick, D. R., **5,** 132; **6,** 119,

120
Amiel, Y., **1,** 159; **4,** 107
Amin, E. S., **1,** 736
Amin, J. H., **2,** 40
Amin, N. V., **11,** 420
Ammons, A. S., **6,** 19
Amonoo-Neizer, E. H., **3,** 173
Amorosa, M., **1,** 562
Amos, R. A., **8,** 122, 316; **11,** 321
Amouroux, R., **9,** 498; **10,** 5; **11,** 600
Amundsen, L. H., **7,** 148
Amupitan, J., **9,** 87; **11,** 108
An, T. D., **9,** 381
Anagnostopoulos, C. E., **1,** 328
Anand, N., **6,** 284, 326
Anand, S. P., **6,** 670
Ananthanarayan, T. P., **11,** 465
Anantharamaiah, G. M., **8,** 383; **9,** 227
Anastasia, M., **7,** 70, 428; **9,** 391; **11,** 177, 288; **12,** 479
Anastassiou, A. G., **2,** 93; **5,** 127
Anbar, M., **1,** 155
Anciaux, A. J., **7,** 313; **8,** 435; **10,** 342
Anders, E., **4,** 384
Andersen, K. K., **4,** 514
Andersen, N. H., **2,** 99, 126; **3,** 338; **4,** 531; **5,** 285, 325, 481, 631; **6,** 19, 322; **10,** 381
Andersen, R. J., **6,** 316
Andersen, V. K., **1,** 169
Anderson, A. G., **2,** 36; **8,** 146; **9,** 141
Anderson, A. G., Jr., **1,** 164, 320, 730, 741, 982, 1073, 1100, 1148, 1152, 1226, 1249; **5,** 731; **7,** 292
Anderson, A. S., **1,** 685
Anderson, A. W., **1,** 755
Anderson, B. A., **2,** 355
Anderson, B. C., **1,** 1210
Anderson, B. T., **1,** 1271
Anderson, C. B., **1,** 1151
Anderson, C. D., **1,** 927; **6,** 600
Anderson, C. E., **1,** 1118
Anderson, D. G., **5,** 68, 376; **6,** 637; **7,** 194
Anderson, D. J., **3,** 171; **5,** 370
Anderson, D. K., **11,** 385
Anderson, D. M. W., **2,** 274
Anderson, D. R., **8,** 46
Anderson, E. L., **1,** 257; **7,** 338
Anderson, G. C., **6,** 32
Anderson, G. J., **1,** 310, 1018
Anderson, G. W., **1,** 98, 110, 116, 248, 250, 253, 487, 524, 1139; **2,** 47, 279
Anderson, H. A., **2,** 344
Anderson, H. H., **6,** 634; **9,** 170
Anderson, H. M., **5,** 601
Anderson, H. V., **1,** 787, 1118
Anderson, H. W., **4,** 162
Anderson, J. E., **2,** 322; **6,** 81
Anderson, J. M., **3,** 239
Anderson, L., **9,** 54; **10,** 427
Anderson, L. R., **9,** 245
Anderson, O. P., **11,** 416; **12,** 412
Anderson, P. C., **10,** 198
Anderson, P. H., **4,** 115; **5,** 624
Anderson, R., **1,** 320
Anderson, R. C., **8,** 428
Anderson, R. G., **1,** 982, 1226
Anderson, R. J., **3,** 112; **5,** 244; **6,** 165
Anderson, R. S., **1,** 40, 207
Anderson, R. T., **4,** 448
Anderson, R. W., **6,** 416
Anderson, W. K., **5,** 127, 663
Andersson, F., **12,** 419
Andersson, J. E., **5,** 421
Andisik, D., **12,** 242
Ando, K., **3,** 36; **12,** 563
Ando, M., **8,** 16; **12,** 144
Ando, R., **11,** 124
Ando, S., **12,** 25
Ando, T., **2,** 273, 424; **10,** 326; **11,** 434; **12,** 407
Ando, W., **6,** 436; **8,** 188, 218; **9,** 170; **11,** 489
Andoye, P., **11,** 346
Andrade, J. G., **8,** 476, 480; **9,** 463; **10,** 397; **11,** 355
Andrascheck, H. H., **9,** 56
Andre, D., **9,** 226; **10,** 186
Andreades, S., **1,** 16, 531, 755
Andrejevic, V., **1,** 563; **7,** 196
Andrews, D. A., **7,** 414
Andrews, G. C., **5,** 406; **6,** 32, 607, 615; **10,** 12; **11,** 126
Andrews, L. J., **3,** 59
Andrews, M. A., **10,** 30; **11,** 45
Andrews, S. A., **11,** 195
Andrews, S. B., **9,** 251, 509
Andrews, S. D., **3,** 76
Andriamialisoa, R. Z., **9,** 459
Andries, T., **11,** 360
Andrieux, J., **5,** 507; **8,** 407, 437
Andrus, A., **9,** 207; **10,** 175
Andrus, D. W., **1,** 257, 452
Andrus, W. A., **8,** 263, 311
Andruszkiewicz, R., **9,** 506
Aneja, R., **3,** 308; **4,** 552
Anet, E. F. L. J., **1,** 20
Anet, F. A. L., **6,** 15
Angell, E. C., **7,** 377
Angelo, B., **1,** 712; **3,** 208; **10,** 363
Angerbauer, R., **9,** 130
Angerer, J., **6,** 359
Angibeaud, P., **2,** 210
Angla, B., **1,** 1167
Anglesio, D., **5,** 595
Angres, I., **5,** 468
Angus, A. B., **1,** 1153
Angus, R. O., Jr., **12,** 570
Angyal, S. J., **1,** 180, 244, 257, 430, 509, 518, 562, 1102; **2,** 364, 377; **3,** 57, 122; **7,** 135; **9,** 190
Anh, N. T., **3,** 304
Anhoury, M. L., **5,** 186
Anliker, R., **1,** 1284
Annen, K., **9,** 324; **11,** 193; **12,** 79
Anner, G., **1,** 562, 1203; **2,** 77, 250, 422
Annis, G. D., **8,** 267; **11,** 236;

12, 127
Annunziata, R., **12,** 298, 509
Ansari, H. R., **5,** 639
Ansari, I. A., **12,** 294
Ansbacher, S., **1,** 169, 500
Anschuetz, R., **1,** 828
Ansell, J. M., **7,** 301
Ansell, L. L., **7,** 171
Ansell, M. F., **1,** 732, 735
Anselme, J. P., **1,** 834; **2,** 178, 321, 383, 417; **3,** 292; **4,** 54; **7,** 20
Anslow, W. K., **1,** 413
Anspon, H. D., **1,** 1163
Antal, E., **5,** 499
Anteunis, A., **6,** 579
Anteunis, M., **1,** 392; **2,** 94; **5,** 675; **7,** 160
Anthes, E., **1,** 657, 767
Anthony, J. M., **10,** 79
Anthony, K. V., **5,** 601
Antoine, J. P., **11,** 407
Anton, D. R., **12,** 550
Antonakis, K., **9,** 183; **10,** 273
Antonioletti, R., **10,** 336
Antonucci, R., **1,** 1178
Antropiusova, H., **12,** 503
Antus, S., **7,** 365; **11,** 188
Anwar, S., **12,** 568
Anwer, M. K., **9,** 352
Aoai, T., **9,** 35; **10,** 21, 106
Aoki, K., **1,** 500; **2,** 222
Aoki, S., **9,** 477; **11,** 551
Aoki, T., **1,** 1198; **5,** 370; **9,** 24
Aono, M., **11,** 155; **12,** 63
Aoshima, K., **5,** 447
Aoyama, H., **9,** 349
Aoyama, S., **8,** 223
Aoyama, T., **8,** 47; **10,** 431; **11,** 44, 472, 574; **12,** 217, 538
Aoyama, Y., **4,** 31
Apelgot, S., **5,** 60
Apparu, M., **7,** 204
Appel, H., **2,** 4
Appel, R., **1,** 484; **2,** 219; **4,** 550, 552; **5,** 727; **7,** 97, 106, 404; **8** 516; **11,** 170
Appelbaum, A., **1,** 101
Appelman, E. H., **8,** 255; **10,** 85
Applebaum, M., **1,** 139
Applegate, H. E., **7,** 244
Applequist, D. E., **1,** 374, 1041, 1125; **3,** 76, 178, 310; **5,** 593
Applequist, J., **1,** 783
Applewhite, T. H., **1,** 367
ApSimon, J., **6,** 288
Apt, K. E., **2,** 92
Aquila, H., **3,** 9; **4,** 248
Arai, M., **7,** 199; **9,** 87
Arai, T., **8,** 115
Arai, Y., **9,** 247
Arakawa, M., **6,** 95
Araki, K., **12,** 402
Araki, M., **5,** 321
Araki, N., **8,** 421
Araki, Y., **10,** 289; **11,** 103, 130; **12,** 540
Aranda, G., **6,** 501; **7,** 392
Arapakos, P. G., **1,** 574
Arase, A., **2,** 285; **3,** 5; **5,** 607; **6,** 259, 510; **7,** 156; **9,** 250
Aratani, M., **4,** 86; **6,** 611; **11,** 562
Araujo, H. C., **6,** 124, 441; **8,** 374
Arbiser, J., **12,** 131
Arbusow, B. A., **1,** 500; **3,** 304
Arce, J., **5,** 590
Archer, S., **1,** 708, 956; **2,** 17; **3,** 199, 308; **5,** 542; **6,** 283
Archer, W. L., **1,** 285, 904
Arco, M. J., **8,** 9
Arcoleo, J. P., **7,** 220
Arcus, C. L., **1,** 970, 1145
Ardao, M. I., **1,** 72; **5,** 756
Ardecky, R. J., **11,** 599; **12,** 159
Ardis, A. E., **2,** 390
Ardoin, N., **5,** 713
Arends, R., **8,** 431
Arens, J. F., **1,** 140, 220, 360, 668, 689; **2,** 344, 374, 438, 462; **3,** 97; **6,** 651; **7,** 226; **8,** 322; **12,** 310
Arentzen, R., **6,** 367; **8,** 128
Argabright, P. A., **1,** 309, 318, 927
Arhart, R. J., **4,** 206, 207; **5,** 272
Arhart, R. W., **5,** 96
Arias, L. A., **12,** 379
Arickx, M., **5,** 186
Arigoni, D., **1,** 562; **5,** 576
Ari-izumi, A., **8,** 201
Arimoto, M., **10,** 8; **11,** 270, 516; **12,** 259
Arison, B., **1,** 293, 1081
Arison, B. H., **10,** 11
Aristoff, P. A., **12,** 404
Arkell, A., **1,** 660
Arlt, H. G., Jr., **1,** 890
Armand, Y., **2,** 372; **3,** 258
Armitage, J. B., **1,** 186
Armour, A. G., **1,** 349
Armstrong, E. C., **1,** 1121
Armstrong, R. K., **2,** 110
Armstrong, R. W., **11,** 416; **12,** 437
Armstrong, V. W., **6,** 97; **8,** 79
Arnaboldi, M., **9,** 207
Arnarp, J., **6,** 406; **9,** 141
Arnason, B., **6,** 381
Arnaud, P., **2,** 372; **3,** 258
Arndt, F., **1,** 195; **3,** 74; **5,** 114, 182
Arndt, H. C., **5,** 262; **7,** 327; **8,** 140
Arnett, E. M., **1,** 1145
Arnett, J. F., **8,** 1
Arney, B. E., Jr., **12,** 73
Arnim, K. v., **1,** 974
Arnoff, M. S., **4,** 290
Arnold, C., **1,** 17, 309
Arnold, C., Jr., **5,** 10
Arnold, D. R., **3,** 234
Arnold, H. R., **1,** 156
Arnold, H. W., **1,** 654
Arnold, R., **4,** 411
Arnold, R. T., **1,** 12, 465, 567, 680, 927; **8,** 464; **11,** 2
Arnold, Z., **1,** 289; **2,** 102; **3,** 54, 88, 155
Arnoldi, A., **11,** 440
Arnone, A., **9,** 413
Arnost, M. J., **11,** 209
Arnould, D., **11,** 346
Arntzen, C. E., **1,** 573
Arora, K. K., **10,** 317
Arora, S., **3,** 178, 310; **4,** 405;

5, 392, 544, 551, 593; **6,** 530
Arora, S. K., **3,** 23; **12,** 259
Arpe, H. J., **5,** 622
Arreguin, M., **1,** 207
Arreguy, B., **7,** 9
Arrhenius, P., **10,** 40
Arrieta, A., **11,** 410; **12,** 122, 199, 204, 384
Arrieta, J. M., **11,** 111
Arrington, J. P., **2,** 309
Arsenijevic, L., **10,** 147
Arsenijevic, V., **10,** 147
Arters, A. A., **1,** 1292
Arth, G. E., **1,** 147, 360, 703, 802, 1000; **3,** 37; **5,** 640
Arthur, C, D., **10,** 85
Arthur, P., **1,** 705; **2,** 287
Arthur, P., Jr., **1,** 228
Arthur, S. D., **10,** 403
Artus, J. J., **5,** 486
Arudi, R. L., **8,** 419
Arunachalam, T., **2,** 246, 344
Arundale, E., **7,** 160
Arvanaghi, M., **9,** 67; **10,** 98, 167, 202, 251, 457; **11,** 244, 593
Arya, P. S., **12,** 569
Arz, A. A., **6,** 17
Arzoumanian, H., **1,** 1045; **4,** 38; **5,** 347, 429; **11,** 266
Asadorian, A. A., **1,** 23
Asami, M., **8,** 424; **9,** 20, 106; **10,** 458; **12,** 202, 421
Asao, K., **12,** 70
Asao, T., **8,** 530
Asaoka, M., **6,** 279
Asato, A. E., **2,** 117
Aschan, O., **8,** 75
Asensio, G., **9,** 293; **10,** 255; **11,** 319; **12,** 233, 306
Asgill, J. O., **8,** 238
Ash, A. B., **1,** 1131; **2,** 330; **3,** 161
Ash, B. D., **1,** 790
Ashby, E. C., **1,** 424, 1191, 1203, 1230; **2,** 205, 243; **3,** 160; **4,** 260, 316; **5,** 14, 419; **6,** 327; **7,** 197; **8,** 120, 175, 291, 311; **9,** 46; **10,** 131; **11,** 152, 373
Ashby, J., **5,** 631
Ashcraft, A. C., **1,** 1022
Ashe, A. J., III, **4,** 119; **5,** 507
Ashley, K. R., **5,** 510
Ashley, W. C., **1,** 841, 1292
Ashton, A. A., **1,** 1270
Ashworth, P. J., **1,** 1040
Ashworth, R. W., **8,** 470
Asinger, F., **2,** 292; **4,** 497
Askam, V., **6,** 285
Askani, R., **1,** 247
Askenasy, P., **1,** 510
Askin, D., **11,** 334; **12,** 313
Aslam, M., **9,** 212; **12,** 77
Aso, K., **4,** 520; **5,** 686
Aspburg, K., **1,** 1002
Assaf, Y., **6,** 645
Asscher, M., **5,** 160; **9,** 311
Asselineau, C., **1,** 838
Assercq, J. M., **12,** 217
Assony, S. J., **2,** 21
Assour, J., **3,** 81
Astill, B. D., **11,** 79
Astle, M. J., **1,** 518
Aston, J. G., **9,** 182
Astrab, D. P., **12,** 310
Asveld, E. W. H., **5,** 160
Atassi, M. Z., **5,** 687
Atherton, E., **8,** 230
Atherton, F. R., **5,** 323; **6,** 43; **7,** 106, 353
Atkins, G. M., Jr., **4,** 228
Atkins, K. E., **2,** 389
Atkins, T. J., **5,** 155, 574
Atkinson, J. G., **2,** 173; **12,** 200
Atkinson, R. E., **2,** 295
Atkinson, R. S., **9,** 269
Atlani, P., **6,** 441
Atsumi, K., **10,** 71; **11,** 322
Attah-Poku, S. K., **9,** 181, 406
Attanasi, O., **6,** 162; **7,** 175
Atta-ur-Rahman, G. M., **2,** 382; **7,** 331, 381, 428; **10,** 360
Attenbach, H. J., **5,** 491
Attenburrow, J., **1,** 595, 642, 951, 1295
Attridge, C. J., **5,** 514
Atwal, K. S., **11,** 382
Atwater, N. W., **1,** 394, 1059
Atwell, W. H., **3,** 103
Au, A., **6,** 567
Aube, J., **10,** 324
Aubke, F., **11,** 526
Aubrey, N. E., **1,** 465
Auchus, R. J., **10,** 89
Audeh, C. A., **9,** 361
Audette, R. J., **4,** 406
Audia, J. E., **12,** 11
Audrieth, L. F., **1,** 484
Aue, D. H., **2,** 379; **5,** 127, 136, 441
Auerbach, R. A., **6,** 676; **10,** 276
Augelmann, G., **11,** 362
Augustine, R. L., **1,** 333; **5,** 329
Auksi, H., **7,** 188
Ault, A., **2,** 397; **3,** 6; **9,** 364
Aumann, D., **4,** 184
Aumann, R., **6,** 306
Aumiller, J. C., **4,** 324, 437, 475
Aumueller, W., **1,** 488
Aune, J. P., **9,** 207
Aurnhammer, R., **1,** 1096
Austin, N. A., **1,** 1000
Austin, P. R., **1,** 563
Austin, W. B., **10,** 117
Autenrieth, W., **1,** 956
Autrey, R. L., **5,** 436; **6,** 403
Avaro, M., **3,** 117
Avasthi, K., **7,** 321; **12,** 145
Avendano, C., **11,** 110
Averbeck, H., **6,** 306
Averina, N. V., **4,** 472
Avery, M. A., **7,** 173; **11,** 281, 319
Avila, N. V., **1,** 219
Avram, M., **2,** 112; **4,** 57; **6,** 279; **7,** 94
Avramoff, M., **1,** 737, 1144, 1253, 1254
Avruch, L., **5,** 351
Awad, S. B., **11,** 108, 109
Awai, T., **10,** 21
Awang, D. V. C., **2,** 39; **3,** 241; **4,** 48
Axelrod, A. E., **1,** 46
Axelrod, E. H., **4,** 507

Axelrod, J., **1,** 476; **5,** 308
Axelrod, L. R., **1,** 802; **4,** 33; **5,** 508
Axen, U., **8,** 259; **9,** 249
Axiotis, G., **12,** 268
Axiotis, G. P., **10,** 5
Axon, B. W., **11,** 428
Aya, T., **5,** 589; **6,** 73
Ayad, K. N., **1,** 1132
Ayad, M., **5,** 631
Ayada, A., **1,** 53
Ayaguchi, Y., **12,** 399
Ayalon-Chass, D., **8,** 274
Ayer, D. E., **1,** 249
Ayer, R. P., **1,** 1011; **2,** 368
Ayer, W. A., **7,** 268
Ayers, D. C., **1,** 904; **2,** 252
Ayers, P. W., **1,** 57
Aylward, J. B., **3,** 171; **4,** 282
Ayral-Kaloustian, S., **11,** 570
Ayres, D. C., **6,** 506
Ayrey, G., **6,** 224, 241; **8,** 29
Ayyangar, N. R., **1,** 207, 263, 1228; **3,** 91; **11,** 255
Ayyar, K. S., **4,** 14; **6,** 19
Azizian, J., **3,** 297
Azoo, J. A., **2,** 233
Azran, J., **10,** 79
Azuma, H., **6,** 653
Azuma, K., **12,** 96
Azuma, S., **11,** 445
Azuma, Y., **10,** 151
Azzaro, M., **5,** 109

Baarda, D. G., **1,** 339
Baardman, F., **5,** 330; **6,** 252; **7,** 146
Baarschers, W. H., **7,** 47
Baba, S., **6,** 601; **7,** 96, 113, 357; **10,** 423, **12,** 474
Baba, Y., **5,** 224; **6,** 201; **8,** 499
Babad, E., **4,** 442
Babad, H., **1,** 926; **3,** 76
Babadjamian, A., **8,** 486
Babcock, G. S., **1,** 47
Babcock, J. C., **1,** 127, 394
Babcock, S. H., **1,** 433
Babers, F. H., **1,** 691
Babiak, K. A., **5,** 132, 329; **8,** 13; **11,** 23
Babiarz, J. E., **6,** 102
Babler, J. H., **5,** 675; **6,** 1; **7,** 208, 309; **10,** 359; **11,** 448; **12,** 445
Babler, J. J., **6,** 338
Babsch, H., **6,** 584
Babston, R. E., **12,** 528
Babu, J. R., **12,** 30
Bac, N. V., **1,** 731; **11,** 117
Baccolini, G., **4,** 387; **5,** 532; **6,** 534; **10,** 318
Bach, G., **2,** 21; **10,** 188
Bach, H., **2,** 412
Bach, M. D., **7,** 21
Bach, N. J., **4,** 318
Bach, R. D., **1,** 1275; **2,** 154, 368; **3,** 200, 333; **4,** 110, 322, 323; **8,** 387; **9,** 220, 324; **12,** 379
Bach, R. O., **2,** 241
Bach, S. R., **3,** 191
Bacha, J. D., **2,** 84; **3,** 171
Bachand, C., **8,** 143
Bacharach, G., **1,** 164
Bachelor, F. W., **1,** 1157; **3,** 31; **4,** 99; **5,** 145
Bachhawat, J. M., **4,** 469
Bachhuber, H., **4,** 155
Bachi, M. D., **2,** 76; **11,** 550
Bachman, G. B., **1,** 329, 484, 757, 905, 1049, 1209; **3,** 96; **4,** 357; **9,** 293, 324
Bachman, P. L., **4,** 526
Bachmann, J. P., **6,** 215; **8,** 143
Bachmann, W. E., **1,** 33, 93, 195, 251, 430, 500, 630, 1254, 1256; **4,** 544, 567; **5,** 333, 420; **11,** 105
Bachner, J., **11,** 533
Back, T. G., **6,** 509; **8,** 32; **9,** 32, 34, 361; **10,** 22, 315; **11,** 408, 493; **12,** 335, 394
Backeberg, O. G., **1,** 730, 1087; **2,** 294
Backer, H. J., **1,** 195, 1184; **2,** 390; **8,** 351, 392
Backes, J. V., **1,** 208
Backlund, S. J., **8,** 162; **9,** 280
Baclawski, L. M., **3,** 232
Bacon, C. C., **4,** 79
Bacon, R. G. R., **1,** 170, 1102, 1153; **3,** 67, 267; **5,** 20
Bacquet, C., **4,** 139; **5,** 406; **6,** 345; **10,** 254
Bacso, I., **5,** 389
Baczynski, L., **8,** 419
Baddeley, G. V., **2,** 344
Baddiley, J., **1,** 881, 1128
Badding, V. G., **1,** 598; **2,** 159; **5,** 186
Bader, A. R., **1,** 33, 102, 266, 471, 476, 904
Bader, F. E., **1,** 235, 764
Bader, H., **1,** 1157; **7,** 135
Badger, G. M., **1,** 731, 977; **2,** 305
Badhwar, I. S., **1,** 1292
Badshah, A., **5,** 571
Badstuebner, W., **1,** 890
Baeckvall, J. E., **7,** 22; **8,** 112; **10,** 377; **12,** 370, 457, 475
Baeckwell, J. E., **11,** 318
Baer, D. R., **4,** 272
Baer, E., **1,** 39, 335, 346, 847
Baer, H., **1,** 685
Baer, H. H., **1,** 741; **8,** 451
Baer, K., **1,** 854
Baer, T. A., **6,** 224, 586; **7,** 114
Baer, U., **4,** 516
Baese, H. J., **5,** 170
Baessler, T., **5,** 551
Baettig, K., **12,** 326
Baeyer, A., **1,** 17, 119
Baeza, J., **5,** 23
Bafus, D. A., **1,** 613
Baganz, H., **1,** 94; **2,** 119; **9,** 299
Bagby, M. O., **2,** 139
Baggaley, K. H., **2,** 238, 313
Bagheri, V., **10,** 122; **11,** 359; **12,** 72, 474
Bagli, J. F., **1,** 219, 562, 642
Bagnell, L., **6,** 624
Bagot, J., **5,** 107
Bahl, C. P., **5,** 309; **6,** 622
Bahr, M. L., **7,** 246
Bahurel, Y., **9,** 361
Baig, M. I., **2,** 300
Baigrie, B. D., **4,** 80, 172
Bailes, R. H., **3,** 245

Bailey, A. S., **1,** 289, 885
Bailey, D. J., **5,** 15
Bailey, D. M., **2,** 203; **3,** 264; **4,** 521; **5,** 542; **9,** 83
Bailey, E. J., **1,** 927; **3,** 96
Bailey, F., **10,** 359
Bailey, G. A., **1,** 764
Bailey, G. M., **1,** 1112
Bailey, J. R., **4,** 162
Bailey, P. S., **1,** 777, 1090; **2,** 156; **5,** 495
Bailey, R. B., **7,** 136; **10,** 167
Bailey, T. R., **9,** 341, 499; **10,** 96; **12,** 384
Bailey, W. F., **11,** 105, 124
Bailey, W. J., **1,** 80, 598; **8,** 396
Baillargeon, D. J., **8,** 415
Baillargeon, M., **9,** 89
Baillargeon, V. P., **12,** 475
Baillie, A. C., **2,** 344
Baily, T. R., **8,** 290
Baine, N. H., **11,** 550
Baine, O., **6,** 40
Baines, D., **5,** 542
Baiocchi, L., **8,** 425
Bair, K. W., **7,** 390; **8,** 69; **9,** 8
Baird, M. C., **6,** 665; **9,** 426; **10,** 455
Baird, M. S., **4,** 251, 435; **6,** 385; **9,** 361
Baird, R., **2,** 19
Baird, W. C., Jr., **2,** 139, 267, 304, 368; **4,** 107
Baizer, M., **1,** 5
Baizer, M. M., **4,** 85; **5,** 685
Bajgrowicz, J. A., **12,** 350
Bakeeva, R. S., **9,** 343
Baker, A. D., **5,** 482, 667
Baker, A. W., **1,** 652
Baker, B. R., **1,** 198, 310, 678, 754, 1045; **4,** 161; **10,** 187
Baker, C. S. L., **1,** 169, 1250
Baker, D. A., **6,** 569
Baker, E. B., **1,** 693; **8,** 6
Baker, E. M., **8,** 466
Baker, F. W., **1,** 81, 658
Baker, G. L., **11,** 7
Baker, G. P., **1,** 506, 510
Baker, J. D., Jr., **6,** 98; **7,** 55; **8,** 80
Baker, R., **3,** 277; **4,** 352; **10,** 112; **11,** 145; **12,** 289
Baker, R. H., **1,** 83, 754, 1064, 1292; **5,** 62
Baker, T. N., III, **1,** 566, 754; **4,** 346
Baker, W., **1,** 397, 430, 471, 782, 954; **10,** 36
Bakker, B. H., **11,** 198
Bakker, C. G., **9,** 523
Bakker, G. R., **1,** 904
Bakuzis, M. L., **7,** 368
Bakuzis, P., **7,** 368
Bal, B., **11,** 349
Bal, B. S., **9,** 55, 190; **11,** 275, 481; **12,** 1
Bal, K., **6,** 74
Balaban, A. T., **2,** 47; **4,** 266; **7,** 409; **8,** 521; **10,** 375; **11,** 571
Balabane, M., **12,** 475
Balachandran, K. S., **9,** 323
Balaczewski, P., **12,** 327
Balakrishnan, P., **11,** 466
Balani, S. K., **11,** 330
Balanson, R. D., **5,** 5; **6,** 319; **8,** 305
Balasubrahmanyam, S. N., **1,** 478
Balasubramanian, D., **1,** 783; **9,** 361
Balasubramanian, K. K., **5,** 428; **7,** 385
Balasubramanian, M., **7,** 320
Balavoine, G., **9,** 46; **10,** 32; **12,** 525
Balcells, J., **7,** 144
Balchi, M., **9,** 341
Balchunis, R. J., **8,** 189, 416; **9,** 381
Balci, M., **12,** 139
Balcom, D., **1,** 444
Balcom, D. M., **1,** 730
Bald, E., **7,** 372
Baldassarre, A., **7,** 13
Baldauf, H. J., **1,** 1195; **11,** 44
Balderman, D., **8,** 433
Baldinger, H., **11,** 222
Baldwin, D., **6,** 283
Baldwin, D. E., **1,** 37; **4,** 393
Baldwin, F. P., **2,** 251
Baldwin, J., **8,** 141
Baldwin, J. E., **2,** 84, 344, 431; **4,** 60, 577; **5,** 491, 620; **6,** 205, 373, 567; **7,** 234, 335, 404; **8,** 69; **9,** 8, 391; **11,** 439, 554, 566; **12,** 87, 416
Baldwin, J. J., **10,** 11
Baldwin, S., **1,** 860, 905, 926; **5,** 353
Baldwin, S. W., **6,** 606; **7,** 41, 384; **8,** 222; **10,** 160, 424
Baldwin, W. A., **6,** 517
Balenovic, K., **5,** 281; **6,** 486; **8,** 417
Bales, S. E., **5,** 419; **7,** 218; **10,** 251
Balfe, M. P., **12,** 160
Balgobin, N., **10,** 432; **11,** 419
Ball, D. H., **1,** 309
Ball, K. L., **11,** 241
Ball, S., **1,** 642, 951
Ballard, D. A., **1,** 1056
Ballard, D. H., **5,** 624
Ballatore, A., **9,** 122
Ballester, M., **1,** 1131; **12,** 546
Ballin, S. G., **1,** 215
Ballina, R., **1,** 783
Balling, W. J., **1,** 828
Ballini, R., **9,** 484; **12,** 275, 405, 418
Ballou, C. E., **1,** 346
Balls, D. M., **4,** 41; **5,** 681
Balme, G., **11,** 360; **12,** 502
Balmonte, F. G., **12,** 103
Balogh, M., **12,** 231
Balogh, V., **3,** 249
Balogh-Hergovich, E., **9,** 393; **11,** 141, 445
Balogh-Nair, V., **9,** 207
Balon, A. D. J., **1,** 1168
Balquist, J. M., **7,** 347
Balsamini, C., **6,** 357
Balsiger, R. W., **5,** 39
Balss, E., **12,** 168
Baltazzi, E., **1,** 433, 1126
Baltzly, R., **1,** 444, 657, 1116; **5,** 447
Bamberger, E., **1,** 119, 393, 562
Bamberger, F., **5,** 758

Bamford, W. R., **1,** 1187; **7,** 331
Ban, Y., **1,** 894; **3,** 231; **4,** 345, 529; **5,** 693; **6,** 312, 628; **7,** 225, 385; **9,** 349, 414; **10,** 378; **11,** 213, 427; **12,** 429
Bandlish, B. K., **3,** 284
Bandyopadhyay, B., **5,** 217
Banerjee, D., **5,** 217
Banerjee, D. K., **1,** 144
Banerjee, S., **9,** 151
Banerjee, S. K., **1,** 165; **2,** 294
Banfi, L., **11,** 315; **12,** 444, 509, 510
Bang, L., **5,** 631
Bangert, R., **1,** 494; **2,** 333
Banhidai, B., **5,** 234, 267
Bank, S., **1,** 712; **2,** 289; **4,** 350; **10,** 363
Banko, K., **6,** 43
Banks, C. M., **7,** 119; **10,** 58
Banks, C. V., **1,** 999; **2,** 94
Banks, D. F., **1,** 506
Banks, G. R., **2,** 190
Banks, R. B., **7,** 238
Bannard, R. A. B., **1,** 135, 518, 928
Banner, B. L., **8,** 15
Bannister, B., **1,** 926
Banno, K., **6,** 596
Banoub, B., **7,** 345
Banquer, C., **6,** 254
Bansal, R. C., **3,** 9
Bansal, S. R., **5,** 160
Banthorpe, D. V., **6,** 263
Bantjes, A., **1,** 1178
Banville, J., **5,** 362; **8,** 179; **9,** 264, 364; **11,** 373
Bao, L. Q., **5,** 66; **6,** 76, 118; **7,** 40
Bapat, C. P., **12,** 175
Bapat, D. S., **1,** 206
Bapat, J. B., **7,** 10, 409; **8,** 521
Barager, H. J., III, **4,** 122; **6,** 457
Barahona-Naranjo, S., **11,** 432
Barakat, M. Z., **1,** 80
Baraldi, P. G., **6,** 566; **9,** 151; **11,** 324
Baran, J. S., **1,** 60, 764; **2,** 142; **6,** 348
Baranova, V. A., **5,** 695
Baranton, F., **2,** 89
Barbachyn, M. R., **11,** 211
Barbadoro, S., **8,** 12
Barbara, C., **10,** 267, 268
Barbaro, G., **6,** 306
Barbee, T. G., Jr., **6,** 546
Barber, G. N., **7,** 215; **8,** 190; **10,** 89
Barber, J. J., **9,** 147
Barbier, M., **11,** 115; **12,** 523
Barbieri, G., **3,** 165
Barborak, J. C., **2,** 95; **6,** 198
Barcelos, F., **7,** 310
Barclay, L. R. C., **5,** 389
Barco, A., **4,** 490; **5,** 553; **6,** 497, 560, 566; **9,** 151; **11,** 324
Barden, T. C., **11,** 439
Bardos, T. J., **1,** 520
Bardwell, C. M., **7,** 271
Bare, T. M., **2,** 278
Barelski, P. M., **4,** 231
Baret, P., **2,** 42
Bargar, T., **9,** 119
Bargellini, G., **4,** 424
Barger, G., **1,** 1168
Bargiotti, A., **8,** 192
Barili, P. L., **5,** 568
Barker, G. R., **1,** 685
Barker, W. E., **1,** 876
Barket, T. P., **2,** 279
Barkhurst, R. C., **2,** 263
Barkley, L. B., **1,** 731, 802, 957, 1004
Barkovich, A. J., **6,** 154
Barlin, G. B., **1,** 430, 509; **5,** 585
Barlow, D. O., **2,** 36
Barlow, J. J., **9,** 54; **11,** 24
Barlow, M. G., **9,** 480
Barltrop, J. A., **1,** 110, 177
Barluenga, J., **9,** 233, 293, 465; **10,** 255; **11,** 303, 318, 319; **12,** 233, 306
Barna, P. M., **2,** 23, 344
Barnard, D., **6,** 224, 241
Barneis, Z., **6,** 645
Barner, B. A., **11,** 152; **12,** 312
Barner, R., **5,** 51
Barnes, C. S., **1,** 1000, 1284
Barnes, D. K., **1,** 39
Barnes, G. H., **5,** 294
Barnes, J. D., **1,** 172
Barnes, J. H., **1,** 500
Barnes, R. A., **1,** 96, 904
Barnes, R. K., **3,** 279
Barnes, R. P., **1,** 964
Barnett, E. B., **4,** 506
Barnett, W. E., **4,** 378; **7,** 88
Barnette, W. E., **8,** 28, 259; **9,** 24, 169, 237, 367; **10,** 21, 147; **12,** 232
Barnick, J. W. F. K., **9,** 55
Barnier, J. P., **4,** 44, 472; **5,** 713; **6,** 523; **7,** 402; **12,** 248
Barnikow, G., **5,** 316
Barnstorff, H. D., **1,** 361
Barnum, C., **10,** 345, 373; **11,** 14, 36, 466
Barnvos, D., **9,** 265
Baron, F. A., **1,** 431
Baron, W., **1,** 357
Barone, G., **5,** 349; **6,** 296; **7,** 181, 182
Barot, B. C., **11,** 535
Barr, D. E., **2,** 45
Barr, P. A., **4,** 247
Barreau, M., **6,** 141
Barreiro, E., **6,** 215; **7,** 122
Barrelle, M., **2,** 234; **7,** 204
Barrera, J., **1,** 808
Barreto, H. S., **2,** 366
Barrett, A. G. M., **8,** 285, 498; **9,** 199, 378, 488; **10,** 248, 322; **11,** 536, 566; **12,** 41
Barrett, A. W., **1,** 652
Barrett, C. B., **1,** 1011
Barrett, H. C., **2,** 17
Barrett, J. H., **1,** 139; **3,** 50, 163, 234, 259
Barrett, S. O., **1,** 1184
Barrish, J. C., **12,** 484
Barry, B. J., **4,** 253
Barry, C. N., **12,** 552
Barry, J., **11,** 407
Barsky, L., **9,** 89

Barstow, L. E., **9,** 503
Barta, M. A., **8,** 110, 252, 448
Bartel, J., **5,** 207
Bartels, A. P., **4,** 290; **5,** 381
Barth, J., **10,** 153
Barthel, J. W., **1,** 1284
Barthels, R., **12,** 421
Bartholomew, J. T., **6,** 102
Barti, G., **2,** 159
Bartke, M., **9,** 506
Bartlett, M. F., **1,** 563
Bartlett, P. A., **3,** 188; **5,** 55, 631; **7,** 339; **8,** 64, 260; **9,** 249, 417; **10,** 370; **11,** 267, 600; **12,** 256, 377, 411, 458
Bartlett, P. D., **1,** 7, 89, 109, 211, 220, 293, 348, 409, 573, 933; **2,** 118, 455; **3,** 324; **4,** 397; **5,** 406, 419, 551, 603, 727; **7,** 21; **10,** 347
Bartlett, R. S., **1,** 207; **3,** 77
Bartlett, W. R., **3,** 302
Bartley, J. P., **3,** 66; **5,** 157, 395
Bartling, G. J., **4,** 63; **5,** 406
Bartmann, W., **1,** 1184
Bartnik, R., **11,** 602
Bartok, W., **1,** 927; **2,** 344
Bartoletti, I., **6,** 61
Bartoli, G., **5,** 556; **8,** 122
Bartolini, G., **11,** 117
Barton, A. E., **10,** 419
Barton, D., **5,** 615
Barton, D. H. R., **1,** 95, 149, 150, 152, 310, 320, 394, 444, 506, 562, 652, 658, 764, 802, 841, 859, 926, 927, 933, 957, 1000, 1090, 1203, 1284; **2,** 77, 200, 222, 238, 263; **3,** 147, 171, 224, 262, 320, 332; **4,** 37, 43, 162, 262, 383, 511, 540, 550, 567, 568; **5,** 116, 349, 381, 437, 472, 486, 734, 758; **6,** 5, 100, 109, 124, 187, 229, 241, 264, 509, 581, 583, 587, 596, 598, 644, 657; **7,** 111, 139, 158, 380, 421; **8,** 32, 131, 285, 361, 432, 437, 460, 498; **9,** 34, 50, 102, 119, 199, 335, 361, 378, 501; **10,** 24, 307, 322, 393, 413, 446; **11,** 39, 49, 106, 250, 276, 515, 551, 587; **12,** 102, 241, 338, 395, 417, 479, 515, 523
Barton, D. L., **6,** 354
Barton, F. E., Jr., **2,** 182, 289
Barton, M. A., **1,** 828
Barton, T. J., **4,** 18; **5,** 624; **9,** 78
Bartroli, J., **11,** 216, 381, 518
Bartsch, R., **11,** 543
Bartsch, R. A., **5,** 207; **9,** 361
Barua, H. B., **1,** 1171
Barua, N. C., **11,** 128; **12,** 550
Barua, P., **12,** 550
Barua, R. K., **1,** 1171
Baruah, J. N., **12,** 126
Barus, N. C., **12,** 126
Basak, A., **10,** 429; **11,** 471; **12,** 350, 416
Basavaiah, D., **11,** 266, 267
Bascoul, J., **1,** 1284
Basedow, O. H., **1,** 1216
Basha, A., **7,** 24, 331, 381, 428; **8,** 182, 507; **9,** 177
Basha, F. Z., **12,** 179
Bashall, A. P., **6,** 10
Bashe, R. W., **1,** 606; **2,** 117; **3,** 112; **4,** 183; **5,** 175
Bashiardes, G., **12,** 241, 479
Bashir, M., **10,** 360
Basile, L. J., **10,** 85
Bass, J. D., **1,** 856
Bass, K. C., **1,** 1142
Bass, R. J., **7,** 424
Basselier, J. J., **10,** 461
Bassignani, L., **9,** 335
Bassilios, H. F., **1,** 131
Bassindale, A. R., **7,** 220
Bassova, G. I., **5,** 695
Bast, H., **1,** 652
Basta, S. J., **4,** 254
Bastani, B., **9,** 48; **11,** 326; **12,** 85
Bastard, J., **5,** 580
Bastian, B. N., **2,** 124
Bastini, B., **11,** 500
Baston, D. W., **5,** 105
Bastus, J. B., **1,** 413
Basu, N. K., **1,** 149; **2,** 93; **6,** 124
Basu, S. K., **10,** 396
Basus, V. J., **6,** 15
Batchco, A. D., **10,** 180
Batchelor, R. J., **11,** 526
Batcho, A. D., **11,** 12, 200
Batelaan, J. G., **7,** 221
Bates, A. J., **6,** 425
Bates, E. B., **1,** 595
Bates, G. S., **5,** 168, 331; **6,** 192; **7,** 61, 400; **8,** 444; **9,** 229; **10,** 38, 333
Bates, H. A., **9,** 375
Bates, R. B., **2,** 139; **9,** 87
Bateson, J. H., **5,** 415; **6,** 186
Bath, S. S., **2,** 156
Batig, K., **10,** 181
Bator, B., **1,** 83
Bats, J. P., **11,** 515
Batt, D. G., **8,** 503
Battaglia, J., **3,** 9
Batten, P. L., **3,** 171
Batten, R. J., **10,** 94
Battenberg, E., **1,** 1211
Battersby, A. R., **1,** 1169; **2,** 263; **9,** 349; **12,** 492
Battig, K., **11,** 249, 415
Battin, D. E., **2,** 389
Battioni, J. P., **5,** 507
Battioni, P., **6,** 165
Battiste, M. A., **5,** 206; **8,** 138; **12,** 117, 229
Battisti, A., **2,** 289
Baucher, R., **1,** 737
Baucom, K. B., **4,** 398
Baudat, R., **9,** 366; **10,** 312
Baude, F. J., **1,** 251
Baudin, J. B., **12,** 140
Baudouy, R., **5,** 389; **6,** 342; **10,** 101
Baudry, D., **8,** 137
Bauer, B., **12,** 227
Bauer, C. D., **1,** 333
Bauer, D. P., **6,** 229; **7,** 394
Bauer, H., **5,** 701; **6,** 422; **9,** 325

Bauer, L., **1,** 1184; **6,** 291
Bauer, M., **1,** 195; **9,** 127
Bauer, P., **11,** 255
Bauer, V. J., **1,** 235, 1178
Bauer, W., **4,** 46; **5,** 49
Baughman, G. L., **1,** 67
Bauld, N. L., **1,** 723; **10,** 452
Baum, A. A., **1,** 697; **2,** 282
Baum, J. S., **7,** 124
Baum, J. W., **6,** 224, 586
Baum, K., **2,** 135; **5,** 705
Bauman, B. A., **8,** 308
Bauman, L., **1,** 407
Baumann, A., **5,** 277
Baumann, M. E., **9,** 18
Baumann, R., **11,** 24
Baumann, W. J., **2,** 254, 313
Baumgarten, E., **1,** 524, 1258, 1261
Baumgarten, H. E., **1,** 94, 563, 629, 927, 1095; **6,** 317
Baumgarten, P., **1,** 1128
Baumgarten, R. J., **5,** 677; **6,** 534
Baumstark, A. L., **5,** 727; **7,** 370; **9,** 468; **11,** 77
Bausher, L. P., **7,** 138
Bavin, P. M. G., **1,** 444, 680
Bavley, A., **1,** 642
Baxter, A. G. W., **9,** 361
Baxter, G. J., **6,** 96
Baxter, I., **1,** 407
Baxter, J. F., **1,** 377
Baxter, S. L., **10,** 39
Bayer, C., **11,** 36
Bayer, E., **10,** 175
Bayer, H. O., **5,** 230
Bayer, M., **10,** 130
Bayer, O., **5,** 653
Bayer, R. P., **1,** 465
Bayer, W., **1,** 682, 1272
Bayet, P., **6,** 29, 90, 240; **7,** 184
Bayger, E., **5,** 701
Bayles, R., **8,** 245
Bayless, A. V., **7,** 407
Bayless, J. H., **2,** 422; **5,** 359
Bayley, H., **9,** 394
Baylor, M. R., **6,** 9
Bayon, A. M., **12,** 233
Baysdon, S. L., **10,** 223; **12,** 130, 178
Bazant, V., **4,** 283
Baze, M. E., **12,** 187
Bazikian, C. M., **5,** 266
Bazilevskii, M. V., **1,** 277, 675
Beach, F. E., **1,** 832
Beach, R. G., **3,** 160
Beach, R. L., **2,** 300; **3,** 215
Beachall, H. C., **8,** 162
Beachem, M. T., **1,** 904
Beachman, L. M., III, **4,** 45
Beadle, J. R., **12,** 381
Beak, P., **6,** 348, 383; **8,** 69, 341; **9,** 89, 302; **10,** 75; **11,** 323; **12,** 99, 205
Beal, D. A., **2,** 427; **3,** 297; **6,** 105; **7,** 22
Beal, G. D., **1,** 503
Beal, J. L., **4,** 395
Beal, P. F., III, **1,** 1005; **2,** 36, 104
Beal, P., III, **1,** 1203; **4,** 122
Beale, C. H., **2,** 222
Beale, W. J., **4,** 253
Beam, C. F., **8,** 190
Beames, A. N., **1,** 219
Beames, D. J., **3,** 65; **5,** 36, 44, 288; **6,** 615; **9,** 16; **11,** 373
Bean, F. R., **1,** 834
Bean, H. J., **1,** 452, 963
Beard, C., **1,** 1132; **2,** 201, 381; **3,** 265
Beard, C. D., **3,** 187; **5,** 705
Beard, J., **4,** 429
Bearder, J. R., **7,** 189
Beardsley, G. P., **4,** 282
Beasley, G. H., **5,** 474; **6,** 419
Beatty, H. R., **4,** 87
Beau, J. M., **10,** 21
Beaucage, S. L., **7,** 58, 354; **8,** 59, 82, 468; **9,** 77, 446; **10,** 88
Beaucaire, V. D., **5,** 6
Beaucourt, J. P., **5,** 723
Beaudoin, S., **12,** 396
Beaulieu, P. L., **11,** 373, 485; **12,** 41, 556
Beaumont, S. M., **1,** 486
Beaupere, D., **11,** 255
Beautement, K., **12,** 214
Beavers, L. E., **2,** 215
Beavers, W. A., **9,** 87
Becerra, R., **1,** 75
Bechara, E. J. H., **7,** 370
Bechtler, G., **1,** 295, 685
Beck, A. K., **2,** 187; **3,** 330; **4,** 216, 241, 285, 564; **6,** 29, 90; **8,** 318; **10,** 234, 381; **11,** 500; **12,** 85, 418
Beck, B. M., **2,** 412
Beck, B. R., **5,** 590
Beck, G., **4,** 256; **5,** 595
Beck, K. M., **2,** 93
Beck, P. E., **1,** 247, 309
Beck, W., **1,** 1058
Becke, F., **3,** 147; **5,** 319
Becke, J., **1,** 465
Becke-Goehring, M., **1,** 971
Becker, A., **1,** 814
Becker, D., **9,** 450; **12,** 401
Becker, D. A., **11,** 318
Becker, E. I., **1,** 33, 106, 172, 188, 424, 772, 841, 846, 1073, 1095, 1193, 1251; **2,** 201; **3,** 177
Becker, G., **12,** 416
Becker, H., **1,** 682; **3,** 320
Becker, H. D., **1,** 219, 296, 310, 318, 643, 732; **2,** 263; **4,** 374; **10,** 136
Becker, H. G. O., **1,** 1123
Becker, J., **2,** 422
Becker, K. B., **4,** 105; **5,** 336, 370; **9,** 135, 217; **10,** 61
Becker, K. H., **11,** 137
Becker, L., **7,** 193
Becker, M., **6,** 74
Becker, P. N., **10,** 116
Beckett, A. H., **6,** 538
Beckham, L. J., **1,** 710, 754
Beckhaus, H., **7,** 207
Beckler, J. D., **1,** 375, 424
Beckley, R. S., **4,** 435
Beckman, J. C., **8,** 486
Beckmann, T., **1,** 736
Beckwith, A. L. J., **1,** 95, 101, 212, 562, 563; **2,** 238; **8,** 377; **9,** 343; **11,** 161; **12,** 275
Becu, C., **5,** 675
Bedenbaugh, A. O., **3,** 175; **4,**

288
Bedenbaugh, J. H., **3,** 175; **4,** 288
Bederka, J. P. M., **1,** 1108
Bedford, G. R., **2,** 213
Bednarski, M., **12,** 313, 561
Bedoukian, P. Z., **1,** 494, 907, 1178
Beears, W. L., **1,** 601
Beebe, T. R., **10,** 216; **12,** 258
Beeby, P. J., **8,** 365
Beech, W. F., **1,** 3, 402
Beecham, A. F., **5,** 693
Beechan, C. M., **3,** 59; **9,** 442
Beel, J. A., **1,** 619; **8,** 276
Beereboom, J. J., **1,** 93
Beerman, C., **2,** 415
Beetham, D., **2,** 192
Beetz, T., **5,** 319
Beevor, P. S., **5,** 633; **6,** 547
Begg, G., **2,** 323
Begland, R. W., **4,** 157; **5,** 176, 220
Begue, J. P., **11,** 467
Beguin, F., **7,** 296
Behagel, A., **8,** 397
Behal, A., **2,** 12
Behan, J. M., **6,** 488
Behare, E. S., **3,** 191; **8,** 194
Behforouz, M., **8,** 399; **9,** 408, 468
Behn, N. S., **4,** 375; **5,** 127
Behnke, M., **12,** 25
Behr, L. D., **1,** 1015
Behrend, R., **1,** 494
Behrens, C. H., **11,** 94
Behringer, H., **1,** 1157
Beier, R., **9,** 257
Beijer, B., **4,** 555
Beiner, J. M., **5,** 182; **9,** 95
Beisler, J., **5,** 410
Beissner, C., **5,** 740
Bekersky, I., **1,** 787
Bekowies, P. J., **6,** 487
Belanger, A., **2,** 36; **3,** 33; **5,** 119
Belanger, D., **10,** 194
Belcher, A. I., **12,** 258
Belcher, R., **1,** 528, 951
Belen'kii, L. I., **1,** 334
Belew, J. S., **2,** 263, 295; **5,** 552
Belgaonkar, V. H., **7,** 424
Belin, B., **9,** 234
Belkind, B. A., **8,** 393
Bell, A., **1,** 276, 336, 383, 452, 1043, 1219, 1270, 1292
Bell, E. V., **1,** 319, 747
Bell, E. W., **1,** 93
Bell, F. A., **5,** 735; **9,** 510; **10,** 452
Bell, H. C., **6,** 581; **11,** 32
Bell, H. M., **1,** 1055; **3,** 77, 123, 264
Bell, I., **1,** 643
Bell, K. H., **3,** 264
Bell, L. T., **12,** 275
Bell, R. A., **3,** 269; **11,** 446
Bell, R. H., **3,** 128
Bell, R. P., **1,** 562
Bellaart, A. C., **1,** 859; **2,** 329
Bellamy, E., **1,** 966
Belleau, B., **2,** 191; **12,** 59
Beller, N. R., **10,** 227
Beller, R. N., **11,** 249
Bellesia, F., **7,** 1
Belletire, J. L., **4,** 285, 413; **9,** 458; **11,** 150; **12,** 263, 279, 282
Belli, A., **6,** 560; **11,** 471
Bellin, S. A., **2,** 48
Bellinger, O., **11,** 170
Bellino, A., **4,** 133
Bello, J., **1,** 80
Bellora, E., **5,** 299
Bellucci, G., **3,** 73; **5,** 568
Belluco, U., **6,** 450
Bellus, D., **6,** 173, 558; **7,** 9, 100; **9,** 112, 154, 481
Bellville, D. J., **10,** 452
Below, P., **10,** 243
Belsky, I., **8,** 130
Belzecki, C., **6,** 457
Bemis, A., **2,** 89
Benaim, J., **4,** 302; **7,** 130; **8,** 193; **9,** 207
Benckhuysen, C., **5,** 340
Ben-David, Y., **8,** 451
Bender, D. R., **5,** 508
Bender, M. L., **1,** 153, 494, 1073, 1101; **2,** 323; **9,** 139
Bender, P. E., **5,** 626
Bender, V., **1,** 518
Benechie, M., **9,** 480
Benecke, I., **11,** 278
Benedetti, E., **5,** 115
Benedikt, R., **4,** 477
Ben-Efraim, D. A., **1,** 926; **11,** 79
Benes, M. J., **2,** 172, 241
Benetti, S., **5,** 553; **6,** 497, 560, 566; **9,** 151; **11,** 324
Benewolenskaja, S. W., **7,** 139
Bengelsdorf, I. S., **1,** 1216
Bengen, M. F., **1,** 1269
Bengtsson, K. C., **10,** 391
Benigni, J., **2,** 107
Ben-Ishai, D., **1,** 110; **5,** 106; **6,** 290
Benjamin, B. M., **8,** 19
Benkeser, R. A., **1,** 348, 581, 609, 611, 754; **2,** 242; **3,** 175, 299; **4,** 139, 288, 526; **5,** 330, 379, 523, 688; **8,** 499; **9,** 94, 256; **11,** 554; **12,** 103
Benko, D. A., **6,** 486
Benn, M. H., **5,** 330
Benn, W. R., **1,** 444; **2,** 370
Benner, S. A., **10,** 172
Bennett, A., **1,** 140
Bennett, C. R., **2,** 250
Bennett, D. A., **7,** 410
Bennett, G., **6,** 610
Bennett, G. B., **5,** 227; **7,** 211
Bennett, G. E., **1,** 84
Bennett, G. M., **1,** 319, 329, 747
Bennett, J. M., **11,** 605
Bennett, M. A., **2,** 453; **5,** 172; **8,** 153
Bennett, M. J., **5,** 408
Bennett, R. P., **1,** 484; **3,** 43; **5,** 503
Bennett, S. G., **10,** 135
Bennett, S. M., **10,** 85
Bennett, S. W., **9,** 56
Bennett, W., **11,** 158
Bennington, F., **2,** 305
Bennua, B., **10,** 441
Benny, D., **12,** 400
Benoiton, L., **1,** 851; **2,** 297, 383; **3,** 36

Benoiton, N. L., **4,** 455; **5,** 615; **7,** 242
Bensel, N., **7,** 394
Ben-Shoshan, R., **5,** 358
Bensimon, Y., **5,** 677
Bensoam, J., **5,** 465; **6,** 412; **9,** 224
Benson, B. E., **3,** 171
Benson, R., **1,** 1027
Benson, R. E., **1,** 18, 295, 785
Benson, W. R., **1,** 333, 511
Benther, H., **10,** 316
Bentler, H., **8,** 281
Bentley, H. R., **3,** 54
Bentley, K. W., **1,** 176; **5,** 596
Bentley, M. D., **5,** 455
Bentley, P. H., **1,** 745, 1196
Bentley, R., **1,** 427
Benton, F. L., **1,** 67, 1055
Bentz, F., **1,** 1236
Bentz, H., **5,** 480
Benz, E., **1,** 1185
Benz, W., **1,** 116
Ben-Zvi, Z., **3,** 191
Beppu, K., **11,** 587
Bera, B. C., **1,** 122
Beranek, J., **1,** 4
Berberian, D. A., **2,** 17
Berchtold, G. A., **2,** 279; **5,** 491; **6,** 436; **8,** 56, 470; **10,** 135; **11,** 208
Berdahl, J. M., **2,** 234
Berenbaum, M. B., **1,** 445
Berenjian, N., **9,** 337
Berens, G., **6,** 552
Berenschot, D. R., **10,** 214
Beretta, E., **5,** 554
Berezin, G. H., **1,** 1022
Berezin, I. V., **11,** 516
Berg, C., **5,** 668
Berg, C. J., **1,** 506
Berg, H., **8,** 211
Berg, U., **7,** 124
Bergbreiter, D. E., **8,** 238, 299, 406
Berge, D. D., **9,** 410
Berge, J. M., **11,** 550
Bergelson, L. D., **1,** 281, 1231, 1246
Berger, A., **1,** 110, 654, 1015; **2,** 59
Berger, B., **11,** 37
Berger, C., **2,** 429; **3,** 76
Berger, C. R. A., **1,** 598
Berger, D. E., **10,** 180; **11,** 12
Berger, E., **8,** 537
Berger, G., **6,** 407
Berger, J., **1,** 77
Berger, J. G., **5,** 708
Berger, M., **2,** 156, 444
Berger, M. H., **7,** 62, 412; **8,** 392; **9,** 217, 363
Berger, M. R., **5,** 60
Berger, R., **5,** 19
Berger, S., **1,** 249, 802; **2,** 381; **3,** 265
Bergeron, R., **3,** 221; **5,** 29, 530, 705; **6,** 44
Berges, D. A., **3,** 199, 258
Bergin, W. A., **3,** 175; **4,** 288
Berglund, C., **1,** 101, 198
Bergman, J., **5,** 643; **9,** 50; **12,** 457
Bergman, R. G., **4,** 119, 130; **5,** 173; **8,** 469; **10,** 116; **11,** 137
Bergmann, E. D., **1,** 277, 504, 518, 1005; **2,** 130, 252, 453; **4,** 335; **5,** 680, 685; **6,** 216, 670
Bergmann, F., **1,** 772
Bergmann, K., **11,** 355
Bergmann, M., **1,** 110, 531, 820, 859; **2,** 389
Bergmann, W., **1,** 1284
Bergmark, W. R., **3,** 234
Bergstroem, F. W., **11,** 207
Bergstroem, S., **4,** 310
Bergthaller, P., **2,** 154; **4,** 189
Beringer, F. M., **1,** 188, 341, 342, 508, 509, 951
Berix, J. W. M., **5,** 268
Berkowitz, L. M., **1,** 986, 989
Berkowitz, W. F., **9,** 73
Berkoz, B., **1,** 104, 219, 444; **3,** 265
Berler, Z., **6,** 290
Berlin, J., **1,** 1090
Berlin, K. D., **1,** 1254; **6,** 475
Berlin, T., **1,** 66
Berlin, Y. A., **5,** 434; **7,** 414
Berliner, E., **6,** 475
Berlinguet, L., **1,** 412
Berlini, F., **2,** 274
Berlo, R. C., **1,** 444, 445
Berman, D. A., **5,** 720
Berman, E., **8,** 359; **9,** 76, 79, 156, 304
Bernad, P., **9,** 233
Bernard, D., **7,** 222; **8,** 29, 313
Bernardi, A., **12,** 226, 298
Bernardi, L., **6,** 110, 515
Bernardi, R., **4,** 237
Bernardinelli, G., **12,** 104
Bernardini, A., **12,** 350
Bernassau, J. M., **6,** 455
Bernatek, E., **5,** 320
Bernath, T., **5,** 146
Bernauer, K., **1,** 1284; **10,** 441
Berndt, H., **2,** 345; **5,** 700
Bernett, D. A., **6,** 606
Bernhard, H., **1,** 86, 1203
Bernhard, S. A., **1,** 527; **8,** 314
Bernhard, W., **12,** 212
Bernhardt, J. C., **7,** 9; **8,** 470; **9,** 174
Bernhart, C., **5,** 609; **8,** 166
Bernheim, M., **11,** 222
Berninger, C. J., **2,** 96
Bernreuther, E., **1,** 636
Bernstein, J., **11,** 341
Bernstein, M. D., **4,** 451
Bernstein, S., **1,** 16, 74, 80, 120, 150, 281, 634, 671, 764, 767, 881, 1000, 1054, 1178; **2,** 117, 281, 310; **4,** 67
Bernstein, S. C., **1,** 1286
Bernstein, Z., **6,** 254
Beroza, M., **1,** 814
Berrange, B., **6,** 577
Berrie, H., **8,** 265
Berris, B. C., **11,** 162; **12,** 163
Berry, C. M., **1,** 293
Berry, R. S., **1,** 364; **2,** 26
Berryhill, S. R., **10,** 362
Berscheid, J. R., Jr., **4,** 450
Berse, C., **1,** 737; **4,** 425
Bershas, J. P., **6,** 15
Berson, J. A., **1,** 343; **3,** 101
Bertele, E., **3,** 243
Bertelli, D. J., **1,** 396, 1258
Bertelo, C. A., **6,** 179; **7,** 102

Berther, C., **1,** 476
Berthet, D., **7,** 122
Bertho, A., **5,** 21
Berthold, R. V., **1,** 1167
Berti, G., **1,** 676; **3,** 73
Bertile, E., **1,** 742
Bertin, D., **1,** 215, 608
Bertin, J., **5,** 19; **6,** 23, 269
Bertini, F., **3,** 189, 201; **5,** 16; **6,** 74
Bertini, I., **11,** 138
Bertorello, H. E., **1,** 259
Bertrand, M., **5,** 389; **6,** 223; **7,** 281; **8,** 302; **9,** 227, 323; **12,** 51
Bertsch, R. J., **4,** 444
Bertz, R. T., **1,** 872
Bertz, S. H., **11,** 194, 373, 566
Berube, N., **7,** 332
Berwick, M. A., **1,** 633
Besendorf, H., **1,** 518
Beson, G., **2,** 173
Besozzi, D., **11,** 486
Besse, J. J., **10,** 198
Besserer, K., **5,** 507
Bessiere, Y., **10,** 392
Bessiere-Chretien, Y., **6,** 498
Bessodes, M., **9,** 183; **10,** 450
Besson, A., **1,** 971
Besters, J. S. M., **6,** 514
Bestian, H., **1,** 292, 380; **4,** 22, 93
Bestmann, H. J., **1,** 114, 1101, 1246, 1247, 1249; **2,** 98, 187; **3,** 94, 138, 320, 322; **4,** 550, 555; **5,** 209, 283; **6,** 95, 279, 359, 381, 509; **7,** 102, 329, 408; **8,** 234; **9,** 133, 158; **10,** 451; **11,** 147, 176, 281; **12,** 44, 269, 388, 439, 441
Betancor, C., **11,** 445
Beten'kii, L. I., **1,** 1103
Bethall, D., **2,** 166
Bethea, T. W., III, **2,** 406
Betkouski, M., **4,** 110
Betschart, C., **12,** 502
Betterton, K., **4,** 60
Bettoni, G., **7,** 315
Betts, B. E., **1,** 5
Betts, M. J., **10,** 248; **11,** 536
Betula, I., **3,** 218
Betz, W., **1,** 859
Betzing, H., **1,** 636
Beucker, H., **1,** 509
Beugelmans, R., **2,** 79; **3,** 152; **7,** 299; **8,** 432; **9,** 381; **12,** 533
Beumel, O. F., Jr., **1,** 349, 570, 574, 1193, 1251; **5,** 382, 649
Beverley, G. M., **5,** 455
Beverung, W. N., **1,** 982; **3,** 284; **4,** 418
Bevington, J. C., **1,** 14
Bevis, M. J., **4,** 46
Bey, P., **4,** 562
Beyer, E., **2,** 63, 120
Beyer, H., **1,** 16, 273
Beyer, R. D., **5,** 652
Beyerman, H. C., **1,** 116, 494, 524, 1188, 1220; **2,** 228; **3,** 157; **4,** 191; **5,** 92; **7,** 349; **8,** 103
Beyl, V., **5,** 480; **6,** 273
Beyler, R. E., **1,** 87, 147, 402, 703, 802, 926
Beynon, P. J., **1,** 989; **2,** 359
Bezman, I. I., **2,** 207
Bezman, S. A., **5,** 113
Bhagwat, M. M., **4,** 562
Bhagwatheeswaran, H., **7,** 11
Bhakuni, D. S., **6,** 581
Bhalerao, U. T., **3,** 71, 325; **4,** 424, 505
Bhandari, K. S., **6,** 515
Bhanu, S., **6,** 203
Bharadwaj, G. K., **4,** 15
Bharadwaj, M. M., **11,** 494
Bharucha, K. R., **1,** 927; **2,** 344
Bhat, H. B., **1,** 1226; **2,** 175, 359
Bhati, A., **2,** 334; **3,** 232
Bhatia, S. B., **2,** 98
Bhatnagar, I., **2,** 263
Bhatnagar, P. K., **6,** 255
Bhatnagar, S. P., **12,** 259
Bhatt, M. V., **1,** 207; **5,** 651; **6,** 283; **7,** 2; **8,** 81; **9,** 289; **11,** 441, 466; **12,** 30
Bhatt, R., **11,** 602
Bhatt, R. S., **10,** 249
Bhattacharjee, M. N., **11,** 453
Bhattacharju, D., **11,** 150
Bhattacharya, A., **5,** 614; **11,** 415; **12,** 389
Bhattacharya, R., **5,** 141
Bhattacharya, S. N., **10,** 183; **11,** 427
Bhattacharyya, B. K., **1,** 141, 796
Bhattacharyya, S., **8,** 216
Bhattacharyya, S. C., **1,** 570; **2,** 239
Bhooshan, B., **5,** 513
Bhushan, V., **11,** 426
Bianchi, G., **9,** 151
Biancini, B., **11,** 385
Bianco, E. J., **1,** 1184
Bible, R. H., Jr., **1,** 394
Bibler, J. P., **12,** 3
Bichan, D. J., **7,** 188
Bickart, P., **4,** 526
Bickel, A. F., **1,** 927, 1254
Bickel, H., **1,** 235, 764; **6,** 186
Bickelhaupt, F., **8,** 340; **9,** 55
Bickoff, E. M., **1,** 296
Bidd, I., **12,** 379
Biehl, E. R., **4,** 439; **9,** 489
Biekert, E., **1,** 933
Biellmann, J. F., **5,** 86; **9,** 77, 127
Bielski, R., **8,** 498
Biemann, K., **2,** 213, 422
Bien, S., **5,** 718
Biener, H., **4,** 93
Biere, H., **9,** 246
Biergans, H., **3,** 79
Bierling, V. B., **5,** 503
Bierman, T. F., **4,** 357
Biernat, J. F., **1,** 83, 1184
Bierwirth, E., **1,** 1218
Biffin, M. E. C., **1,** 195
Biftu, T., **8,** 529
Bigby, D. B., **1,** 263
Bigelow, H. E., **1,** 1284
Bigelow, L. A., **1,** 66
Bigelow, M. H., **1,** 186
Bigelow, S. S., **11,** 87
Biger, S., **8,** 383
Bigge, C. F., **10,** 62
Biggi, G., **4,** 417

Biggs, B. S., **1,** 1269
Biggs, J., **6,** 57
Bigham, E. C., **5,** 657; **8,** 481
Bigi, F., **10,** 462
Bigler, A. J., **2,** 98
Bigler, P., **9,** 523
Bigley, B. J., **1,** 269
Bigley, D. B., **1,** 59, 177, 207; **3,** 23
Bijsterveld, E. J. M., **7,** 349; **8,** 103
Bil, M. S., **3,** 233
Bilk, L., **2,** 43
Biller, S. A., **6,** 642; **8,** 365; **12,** 523
Billeter, O. C., **1,** 48
Billica, H. R., **1,** 730, 977; **4,** 267
Billigmeier, J. E., **5,** 507
Billimoria, J. D., **1,** 334, 518
Billington, D. C., **11,** 163; **12,** 167
Billman, J. H., **1,** 39, 273, 320, 345, 1230
Billman, W., **8,** 234
Billmers, J. M., **12,** 393
Bills, E. J., **1,** 430
Billups, W. E., **2,** 344; **4,** 405; **5,** 551, 587; **6,** 479; **12,** 73
Biloski, A. J., **9,** 245; **10,** 206; **11,** 76; **12,** 158
Bilow, N., **10,** 117
Biname, R., **5,** 99
Binder, V., **8,** 535
Bindra, J. S., **5,** 639
Bindra, R., **6,** 165
Binger, P., **3,** 178, 310; **4,** 405; **5,** 35, 392, 544, 551, 593, 649; **6,** 530
Bingham, E. M., **8,** 167; **10,** 142, 421
Bingham, K. D., **3,** 220; **4,** 204
Bingham, R., **2,** 129
Binkley, R. W., **7,** 310; **9,** 447
Binks, R., **1,** 135, 1169
bin Manas, A. R., **12,** 208
Binns, M. R., **10,** 198
Binovi, L. J., **1,** 80, 890
Binovic, K., **12,** 466
Binsch, G., **2,** 81; **4,** 383
Biollaz, M., **4,** 577
Biran, C., **5,** 713; **7,** 9
Birch, A. J., **1,** 56, 161, 166, 309, 598, 904, 926; **2,** 29, 331, 366, 453; **3,** 238, 254; **4,** 33, 135; **5,** 195, 381; **6,** 323, 475; **7,** 68, 188; **9,** 274
Birchall, J. M., **1,** 1058; **6,** 621
Birckenbach, L., **1,** 501
Bird, C. W., **1,** 260, 723; **5,** 55
Bird, J. G., **6,** 283
Bird, J. W., **2,** 65; **3,** 45
Bird, M. L., **9,** 430
Bird, P. H., **5,** 113
Bird, T. G. C., **9,** 169
Bird, W., **9,** 485
Birkeland, S. P., **1,** 195, 764
Birkhaeuser, A., **3,** 249
Birkhimer, C. A., **1,** 360
Birkofer, L., **1,** 427, 581, 1218, 1232, 1236; **2,** 412; **3,** 316; **4,** 66, 539; **5,** 713; **6,** 390; **7,** 28, 53, 400
Birladeanu, L., **1,** 563; **3,** 37
Birnbaum, S. M., **1,** 129
Birnberg, G. H., **4,** 520
Birstein, O., **1,** 1158
Birum, G. H., **1,** 1131; **2,** 447; **10,** 145
Bisacchi, G. S., **9,** 259
Bisaha, J., **9,** 119; **12,** 362
Bischoff, C., **1,** 472
Bishop, C. T., **1,** 1081; **2,** 315
Bishop, J. J., **10,** 38
Bishop, K. C., III, **5,** 574
Bishop, P. M., **11,** 309
Bishop, R., **5,** 639
Bishop, R. D., **10,** 107
Bishop, W. S., **1,** 872, 1163, 1269
Biskup, M., **2,** 344; **3,** 112, 278
Biss, J. W., **1,** 262
Bissell, E. R., **1,** 564, 927
Bissember, B. B., **10,** 335
Bissing, D. E., **1,** 223, 1192
Bissinger, W. E., **6,** 456
Bistrzycki, A., **3,** 320
Biswas, K. M., **2,** 107
Bither, T. A., **4,** 543; **6,** 81
Bithos, Z. J., **1,** 982
Bittler, K., **3,** 83
Bittman, R., **7,** 413
Bittner, S., **5,** 728; **6,** 645; **7,** 406
Bizzarro, F. T., **9,** 352
Bjeldanes, L. F., **5,** 508
Bjerrum, J., **1,** 629
Bjoerk, A., **10,** 136
Bjorklund, C., **4,** 268
Blaas, T. A. H., **2,** 390
Black, A. P., **1,** 691
Black, C. J., **6,** 76, 118
Black, D. K., **1,** 68, 1249; **2,** 330, 446; **5,** 460
Black, D. S. C., **6,** 114; **7,** 241; **11,** 277
Black, H. K., **1,** 155, 165
Black, T. H., **12,** 157
Blackburn, E. V., **11,** 551
Blackburn, G. M., **3,** 320; **9,** 256
Blackburn, T. F., **6,** 179
Blacklock, T., **8,** 261
Blacklock, T. J., **10,** 246, 354; **11,** 331, 391
Blackman, N. A., **6,** 114
Blackstone, R. C., **2,** 344
Blackwell, J., **1,** 598
Blackwood, J. L., **4,** 132
Blackwood, R. K., **1,** 4, 309; **5,** 566
Blade, R. J., **7,** 10, 409; **8,** 521; **9,** 135
Bladon, C. M., **11,** 562
Bladon, P., **1,** 144, 276, 329, 1090
Blagbrough, I. S., **11,** 461
Blair, H. S., **1,** 782; **8,** 383
Blair, I. A., **5,** 734; **11,** 559
Blair, L. K., **8,** 141
Blais, C., **10,** 274
Blaise, E. E., **12,** 568
Blake, A. B., **11,** 383
Blake, J., **1,** 890
Blake, M. R., **6,** 504
Blakeley, R. L., **5,** 617
Blakeney, A. J., **4,** 405; **10,** 347
Blaker, J. W., **1,** 69
Blakesley, C. N., **6,** 546

Blanc, J., **9,** 115
Blanc, P. Y., **1,** 8
Blanchard, A. A., **1,** 228
Blanchard, E. J., **1,** 618; **2,** 251
Blanchard, E. P., **1,** 104, 1022; **3,** 141
Blanchard, H. S., **1,** 154; **10,** 138
Blanchard, J. M., **10,** 265
Blanchard, M., **5,** 669
Blanchette, M. A., **12,** 277
Blanco, L., **7,** 35, 67; **10,** 91; **11,** 121
Blaney, F., **6,** 121
Blank, B., **1,** 989
Blank, R. H., **1,** 764, 881
Blankenhorn, G., **6,** 547
Blankenship, C., **12,** 462
Blankenship, R. M., **7,** 429; **9,** 520
Blankley, C. J., **2,** 84; **4,** 290; **12,** 52
Blanton, C. D., Jr., **5,** 691
Blardinelli, A. J., **1,** 697
Blarer, S. J., **11,** 326
Blaschek, W., **12,** 278
Blaschette, A., **9,** 56; **11,** 67
Blaschke, H., **1,** 364
Blasioli, C., **7,** 327; **8,** 316
Blaszczak, L. C., **5,** 370, 671; **7,** 93
Blatcher, P., **7,** 26; **9,** 5
Blatchford, J. K., **1,** 1034
Blatt, A. H., **1,** 491; **5,** 13; **6,** 240
Blatt, K., **4,** 82
Blatter, H. M., **1,** 746
Blazejewski, J. C., **5,** 219; **10,** 393
Blazevic, N., **9,** 234
Blazevich, J. N., **4,** 41
Blegen, J. R., **1,** 500
Bleicher, W., **11,** 355
Blicke, F. F., **1,** 347; **3,** 115, 308; **5,** 107, 320
Blickenstaff, R. T., **1,** 966
Blizzard, R. H., **1,** 595
Bloch, E., **1,** 707
Bloch, H. S., **1,** 629
Bloch, K., **1,** 1167
Bloch, R., **8,** 144; **12,** 237, 456
Block, E., **1,** 1122; **4,** 559; **7,** 380; **12,** 77
Block, H. D., **8,** 523; **9,** 511; **10,** 50
Block, P., Jr., **1,** 401, 1147; **2,** 404
Block, R., **8,** 459
Blodinger, J., **1,** 248, 253, 1139
Bloecker, H., **10,** 461
Blohm, H. W., **1,** 172
Blok, A. P., **5,** 55
Blom, J. H., **1,** 247
Blomberg, C., **10,** 5
Blomquist, A. T., **1,** 56, 156, 160, 231, 1101, 1136, 1157, 1187; **2,** 373; **6,** 634
Blomstrom, D. C., **7,** 237
Bloodworth, A. J., **7,** 223, 224; **8,** 173, 249; **11,** 551
Bloom, B. M., **1,** 808, 829
Bloom, S. H., **12,** 533
Bloomer, J. L., **5,** 551
Bloomfield, J. J., **1,** 309, 1080; **2,** 438; **4,** 33, 490, 539; **8,** 52, 109, 265; **12,** 181
Blosick, G. J., **4,** 541
Bloss, D. E., **5,** 46
Blossey, E. C., **2,** 462; **4,** 322; **5,** 13, 491; **6,** 19
Blotny, G., **1,** 1184
Blount, J. F., **4,** 250; **5,** 63; **7,** 255, 421; **9,** 31, 184; **10,** 223; **11,** 236
Blout, E. R., **2,** 80; **6,** 137
Bloy, V., **9,** 308; **10,** 265
Blucher, W. G., **5,** 477; **6,** 420
Bluemmel, F., **1,** 1187; **2,** 417
Blues, E. T., **1,** 629
Bluestein, B. A., **3,** 146; **5,** 309
Bluestein, B. R., **1,** 329
Bluhm, A. L., **1,** 1055
Bluhm, H. J., **3,** 194
Blum, D. M., **6,** 285, 407
Blum, J., **1,** 81, 140, 158, 1117; **2,** 453; **3,** 329; **4,** 254, 562, 564; **5,** 740, 741; **6,** 108, 542, 655; **7,** 106; **8,** 383, 516; **9,** 418; **10,** 79
Blum, L., **8,** 130, 255; **9,** 361
Blum, M. S., **10,** 361
Blum, S., **1,** 432; **4,** 248
Blum, Y., **11,** 184
Blumbergs, J. H., **1,** 139
Blumbergs, P., **5,** 200
Blume, E., **4,** 516; **7,** 215; **10,** 107
Blume, G., **4,** 31
Blume, R. C., **3,** 313
Blumenkopf, T. A., **9,** 255; **12,** 502
Bluthe, N., **11,** 321; **12,** 51, 307, 437
Boal, J. R., **8,** 465
Boar, R. B., **5,** 472; **6,** 5, 9, 587; **7,** 421; **8,** 285; **9,** 378
Boatman, R. J., **8,** 66
Boatman, S., **2,** 53, 197
Boatner, C. H., **5,** 333
Bobbitt, J. M., **1,** 518, 814; **7,** 148
Bobek, M., **8,** 167
Boberg, F., **5,** 195
Bobko, E., **1,** 952
Bobst, A., **1,** 1221
Bocchi, V., **7,** 246; **9,** 361
Boch, M., **6,** 141
Boche, G., **11,** 222, 455
Bock, M. G., **6,** 110, 302; **7,** 130; **8,** 319, 325
Bock, P. A., **11,** 527
Bockemueller, W., **1,** 564
Boctor, B., **4,** 411
Bodalski, R., **10,** 181
Bodanszky, M., **1,** 235, 360, 743; **5,** 477
Boddy, K., **11,** 486
Bode, K. D., **10,** 331
Boden, E. P., **11,** 127; **12,** 23, 146
Boden, H., **1,** 281, 696, 1064
Boden, R. M., **6,** 137, 436
Bodenbenner, K., **5,** 714
Bodenberger, A., **4,** 134
Bodesheim, F., **1,** 383, 881; **2,** 51; **10,** 67
Bodine, J. J., **11,** 487
Bodo, B., **4,** 416; **9,** 462
Bodrikov, I. V., **6,** 232
Bodurow, C., **12,** 372

Boechmann, G., **8,** 198
Boeckman, R. K., Jr., **4,** 409; **5,** 266, 463, 725, 732; **6,** 205, 285, 407; **7,** 320; **8,** 66, 72, 284, 300, 442; **9,** 11, 377; **10,** 403, 443; **12,** 303, 531
Boeder, C. W., **8,** 357
Boeeseken, J., **1,** 66, 764, 796
Boehm, I., **11,** 299
Boehm, M. C., **9,** 341
Boehm, P., **10,** 359; **12,** 574
Boehm, W., **1,** 465, 1195
Boehme, E., **3,** 118
Boehme, H., **5,** 552; **8,** 281; **10,** 316
Boehme, P., **4,** 272
Boehme, W. R., **1,** 1131
Boehmer, V., **4,** 84
Boehnke, E., **7,** 299
Boekelheide, V., **1,** 392, 444, 518, 685, 730, 1284; **2,** 75, 187, 405; **3,** 19; **4,** 115; **5,** 624; **6,** 90; **9,** 222; **11,** 79
Boelens, H., **5,** 253; **7,** 124; **8,** 530
Boelhouwer, C., **7,** 415; **10,** 456
Boell, W., **1,** 671, 1246; **10,** 270
Boell, W. A., **1,** 190; **4,** 579
Boennemann, H., **12,** 290
Boer, H., **1,** 105
Boerma, G. J. M., **4,** 516
Boerth, D. W., **6,** 554
Boerwinkle, F., **2,** 38
Boes, O., **1,** 472
Boese, A. B., Jr., **1,** 531
Boettger, S. D., **10,** 122
Boettner, E. F., **5,** 560
Boffi, C., **2,** 163
Bogard, T. D., **4,** 14
Bogavac, M., **10,** 147
Bogdanovic, B., **4,** 35; **12,** 290
Bogdanowicz, M. J., **3,** 134; **4,** 214, 255; **5,** 281, 373; **6,** 155, 243; **8,** 140, 214
Bogentoft, C., **5,** 389
Boger, D. L., **7,** 386; **8,** 276, 427, 511; **9,** 184; **10,** 272, 409; **11,** 281; **12,** 154, 207, 216, 474, 529
Bogert, M. T., **3,** 272
Bogert, V. V., **1,** 808
Boggio, R. J., **5,** 677; **6,** 534
Boggs, R. A., **5,** 243
Bognar, R., **8,** 132
Bohlen, D. H., **1,** 444
Bohlmann, F., **1,** 186, 350, 669, 1284; **8,** 365; **9,** 90; **10,** 451; **11,** 45, 281
Bohm, H., **5,** 205
Bohn, R., **1,** 66
Boigegrain, R., **6,** 280
Boigegrain, R. A., **9,** 297
Boikess, R. S., **1,** 1022
Boireau, G., **7,** 240
Boiselle, A. P., **8,** 94
Boisselle, A. P., **1,** 452
Boissonnas, R. A., **1,** 85, 86, 110, 112, 281, 445, 841, 1184, 1196, 1203, 1220, 1256
Bolan, E. N., **2,** 287
Boldrini, G. P., **6,** 306; **7,** 300; **12,** 571
Boldt, K. G., **7,** 236
Boldt, P., **2,** 117, 429; **3,** 86; **4,** 317; **6,** 170; **7,** 97, 175
Bolduc, P. R., **6,** 436
Bolhofer, W. A., **1,** 281
Bolker, H. J., **5,** 186
Boll, W. A., **5,** 454
Bollinger, F. W., **11,** 536; **12,** 220
Bollinger, M., **1,** 986
Bolon, D. A., **5,** 585
Bolourtchian, M., **5,** 713
Bolsman, T. A. B. M., **7,** 268
Bolt, W., **8,** 431
Bolton, G. L., **12,** 323
Bolton, M., **6,** 583
Bolzoni, L., **9,** 14
Bombala, M. U., **9,** 375
Bomse, D. S., **6,** 589
Bonadies, F., **12,** 418
Bonanomi, M., **8,** 425
Boraventura, M. M., **1,** 581
Bond, A. C., Jr., **1,** 594, 1251
Bond, F. T., **2,** 328, 422; **6,** 600; **7,** 48; **9,** 488
Bond, G. C., **5,** 321
Bondar, W. A., **5,** 668
Bondavalli, F., **1,** 500
Bondinell, W. E., **4,** 432
Boness, M., **4,** 119
Bonet, J. J., **5,** 486
Bonetti, G., **1,** 666
Bonetti, M., **9,** 355
Bonfiglio, J. N., **10,** 398
Bongers, S. L., **10,** 31
Bongini, A., **9,** 257; **11,** 267
Boniardi, O., **10,** 294
Boniece, J. M., **8,** 391
Boniface, P. J., **11,** 486
Bonini, B. F., **5,** 516
Bonini, C., **12,** 418
Bonitz, G. H., **10,** 161
Bonjouklian, R., **5,** 751; **6,** 188; **8,** 170; **11,** 275
Bonncr, R. M., **1,** 927
Bonner, T. G., **1,** 68
Bonner, W. A., **1,** 235, 1157
Bonnet, G., **10,** 260
Bonnet, M., **10,** 342
Bonnet, M. C., **11,** 383
Bonnet, Y., **1,** 1253
Bonnett, R., **2,** 287; **8,** 255; **12,** 492
Bonnier, J. M., **5,** 669
Bonse, G., **4,** 9
Bonso, O., **1,** 177
Bontekoe, J. S., **1,** 524, 1220
Bonthrone, W., **1,** 1258
Bonzougou, Y. **8,** 238
Boocock, J. R. B., **1,** 1132
Boon, W. R., **5,** 693
Boord, C. E., **4,** 135
Boorman, E. J., **1,** 1197
Booth, A. N., **1,** 296
Booth, B. L., **5,** 326; **9,** 485
Booth, H. S., **9,** 466
Booth, J., **1,** 562
Boothe, J. H., **1,** 39
Boots, S. G., **1,** 1193
Boozer, C. E., **1,** 348
Boozpeikar, B., **10,** 255
Bopp, R. J., **5,** 700
Borch, R., **1,** 76
Borch, R. F., **2,** 431; **3,** 184, 303; **4,** 115, 144, 239, 296, 451; **5,** 297, 694; **6,**

542
Borchardt, R. T., **12,** 284
Borcherdt, G. T., **2,** 292
Borchert, E. E., **1,** 162, 163
Borden, G. W., **1,** 66
Borden, M. R., **1,** 564
Borden, W. T., **5,** 198; **7,** 376; **10,** 336
Bordignon, E., **5,** 647
Bordner, J., **3,** 57; **5,** 689
Bordwell, F. G., **1,** 14, 83, 1127, 1131, 1157, 1216, 1292; **5,** 524, 601; **10,** 316
Borel, A. W., **7,** 237
Borel, E., **1,** 119, 147, 320, 735
Borer, R., **7,** 414
Borggrefe, G., **4,** 230
Borgman, R. J., **6,** 9
Borgogno, G., **6,** 586
Borgulya, J., **5,** 51; **10,** 441
Borgwardt, S., **1,** 281
Borisov, A. E., **5,** 44
Bork, S., **1,** 155
Borkenhagen, L., **1,** 581
Born, H., **1,** 180
Borner, P., **5,** 714
Bornstein, J., **1,** 244, 273, 452, 564
Borodulina-Shvetz, V. I., **3,** 222
Borovas, D., **1,** 745, 1256, 1261
Borowitz, I. J., **1,** 139, 268, 1247; **2,** 68, 242; **3,** 207; **6,** 267
Borraccini, A., **3,** 73
Borredon, E., **11,** 214
Borrmann, D., **3,** 88
Borromeo, P. S., **7,** 132; **8,** 194
Borrows, E. T., **1,** 500
Bortolussi, M., **8,** 459
Bortsch, R. A., **9,** 418
Borum, O. H., **1,** 963
Bory, S., **2,** 278; **8,** 245
Bos, H., **6,** 143
Bos, H. J. T., **1,** 360; **9,** 111; **10,** 96; **11,** 432
Bos, M. G. J., **3,** 57
Bos, W., **10,** 68
Boschelli, D., **11,** 5, 109
Boschi, T., **6,** 450
Boschman, F. E. H., **5,** 371
Boschung, A. F., **6,** 436; **7,** 268, 269, 428
Bose, A. K., **1,** 236, 250, 882, 1203, 1217; **2,** 131, 426; **5,** 127, 713, 728, 729; **6,** 645; **8,** 211; **10,** 115, 448; **12,** 554, 568
Bose, J. L., **1,** 882, 1292
Boshart, G. L., **1,** 274, 371, 1184
Bosin, T. R., **5,** 602
Bosisio, G., **6,** 515
Boskin, M. J., **1,** 114, 1192
Bosman, H. H., **7,** 349; **8,** 103
Bosnich, B., **9,** 458
Bosnjak, J., **10,** 229; **12,** 272
Boss, R., **7,** 277
Bosshard, H., **9,** 18
Bosshard, H. H., **1,** 289; **9,** 335
Bosshardt, H., **10,** 186
Bost, M., **6,** 141
Bost, R. W., **1,** 402
Boston, M. C., **12,** 568
Boswell, G. A., Jr., **1,** 755; **2,** 299, 393; **5,** 640; **9,** 11
Bothner-By, A. A., **1,** 56
Bott, K., **1,** 221; **2,** 118
Bott, R. W., **1,** 892
Botta, M., **8,** 13
Bottaccio, G., **1,** 1102; **8,** 367
Bottaro, J. C., **10,** 135; **11,** 554, 566; **12,** 87
Botteghi, C., **4,** 420; **5,** 331; **8,** 437; **9,** 260; **11,** 125
Botteron, D. G., **1,** 388
Bottini, A. T., **1,** 430, 1040; **5,** 560
Bottin-Strzalko, T., **12,** 105
Bottorff, E. M., **1,** 1073
Bouas-Laurent, H., **4,** 247; **5,** 167
Bouchard, R., **2,** 21; **3,** 275
Boucher, E. A., **3,** 148
Boucherot, D., **5,** 557
Bouchoule, C., **5,** 669
Boudakian, M. M., **1,** 935
Boudjouk, P., **6,** 58; **11,** 292, 599; **12,** 86
Bouffard, F. A., **9,** 408; **10,** 342
Bougault, J., **1,** 730
Bouket, J. L., **1,** 374
Boulch, N. L., **2,** 329
Bould, L., **4,** 43
Boulerice, M., **3,** 171
Boulette, B., **3,** 47
Boullier, P. A., **6,** 263
Boulos, A. L., **6,** 503
Boulton, A. J., **1,** 1044; **4,** 505; **7,** 10, 409; **8,** 521; **9,** 309
Bouma, R. J., **6,** 567
Bourdais, J., **7,** 240
Bourelle-Wargnier, F., **6,** 124; **7,** 69
Bourgain, M., **5,** 94, 411; **6,** 270, 331, 664
Bourgeois, P., **10,** 176
Bourgeois, R. C., **1,** 430
Bourguel, M., **1,** 1040
Bourne, E. J., **1,** 68, 1226
Bousquet, E. W., **1,** 481, 1229
Boussac, G., **11,** 216
Boussu, M., **8,** 238
Boutevin, B., **4,** 69
Boutigue, M. H., **5,** 609
Bouvier, P., **8,** 407
Bowden, E., **1,** 767
Bowden, K., **1,** 144, 165; **3,** 39
Bowen, P., **12,** 378
Bower, F. A., **1,** 94
Bower, J. D., **4,** 8
Bowers, A., **1,** 75, 104, 127, 144, 150, 219, 402, 562, 564, 624, 1000, 1090; **2,** 362
Bowers, C. W., **6,** 137
Bowes, C. M., **5,** 486
Bowlus, S. B., **4,** 424, 535; **5,** 285; **7,** 172
Bowman, C. M., **1,** 834
Bowman, H., **5,** 626
Bowman, P. S., **3,** 22
Bowman, R. E., **1,** 198, 257
Bowman, S. A., **4,** 394
Bowyer, P. M., **4,** 79
Bowyer, W. J., **1,** 598, 802
Boxberger, M., **12,** 151

Boxler, D., **5,** 246, 526; **6,** 422
Boya, M., **6,** 417
Boyd, D. R., **6,** 358; **11,** 330
Boyd, J. D., **12,** 139
Boyd, S. D., **4,** 461
Boyd, T., **1,** 864
Boyd, V., **1,** 34
Boyer, J., **10,** 84; **11,** 554; **12,** 109
Boyer, J. H., **1,** 84, 1044, 1065, 1247; **3,** 272; **5,** 632; **9,** 364
Boyland, E., **1,** 562, 954
Boyle, L. W., **4,** 5
Boyle, P. H., **4,** 55
Bozell, J. J., **11,** 401
Bozimo, H. T., **7,** 365
Bozzato, G., **6,** 215
Brace, N. O., **2,** 227
Bracho, R. D., **5,** 437
Brackman, W., **1,** 164; **2,** 87; **3,** 66
Bradbury, J. H., **1,** 445
Bradbury, S., **4,** 282
Bradley, A., **1,** 465
Bradley, D. C., **2,** 400; **11,** 185
Bradley, J. N., **3,** 74
Bradley, M. G., **7,** 60
Bradlow, H. L., **1,** 257
Bradshaw, C. P. C., **1,** 93
Bradshaw, J. S., **4,** 459; **6,** 137; **10,** 39, 112
Bradshaw, R. W., **2,** 404; **3,** 67; **7,** 166
Bradshaw, T. K., **6,** 624
Bradsher, C. K., **1,** 1145; **2,** 215, 336; **4,** 191; **5,** 617; **10,** 72, 246
Brady, S. F., **2,** 87, 331, 464; **4,** 23
Brady, W. T., **1,** 1287; **2,** 462; **4,** 135; **7,** 395
Braem, A., **5,** 538, 700
Braendstroem, A., **1,** 17; **6,** 41, 564, 565, 567; **10,** 378
Braeunling, H., **1,** 116, 494, 595
Braga de Oliveira, A., **3,** 87
Braitmaier, E., **10,** 194
Braitsch, D. M., **6,** 654
Bram, G., **4,** 66; **5,** 86; **8,** 311; **11,** 407
Bramblett, J. D., **4,** 152
Branca, M., **9,** 260
Branca, Q., **7,** 298
Branca, S. J., **6,** 70, 577; **9,** 87; **10,** 55
Branchaud, B., **6,** 110; **7,** 310
Branchli, J., **9,** 180
Brand, E., **1,** 859, 1167
Brand, F. C., **1,** 1167
Brand, K., **3,** 272
Brand, W. W., **8,** 101, 245
Brandenberg, W., **1,** 66
Brandes, D., **11,** 67
Brandi, A., **12,** 218
Brandsma, L., **2,** 344, 374, 462; **5,** 95, 208, 293, 306, 749; **6,** 144, 398; **7,** 80, 226; **8,** 245, 322, 407, 408; **9,** 95, 106, 334; **10,** 73, 96; **11,** 103, 432, 473; **12,** 310, 570
Brandt, A., **9,** 335
Brandt, D. R., **1,** 222
Brandt, E. V., **8,** 391
Brandt, S., **9,** 143
Brannen, C. G., **1,** 619
Brannock, K. C., **4,** 172; **5,** 647
Brant, J. H., **1,** 389
Brase, D., **5,** 60; **6,** 430
Brasen, S., **4,** 374
Brasen, W. R., **1,** 685, 1033, 1041
Brassard, P., **5,** 119, 362; **8,** 179; **9,** 264, 303; **10,** 44, 384; **12,** 196, 197, 407
Bratke, K., **1,** 119
Brattesani, D. N., **5,** 276
Brauchli, H., **4,** 191
Braude, E. A., **1,** 129, 147, 219, 281, 348, 642, 713, 744, 861
Brauer, G., **1,** 1122
Brault, J., **7,** 410
Brauman, J. I., **5,** 631; **7,** 340
Braun, C. E., **1,** 33, 870, 1292
Braun, D., **7,** 52
Braun, G., **1,** 158, 197, 340, 795, 864, 1005, 1058, 1095
Braun, H., **4,** 68; **6,** 630; **9,** 75
Braun, L. M., **4,** 191
Braun, M., **4,** 285; **5,** 78, 700; **9,** 233; **12,** 3
Braun, R. A., **1,** 336; **4,** 191
Braun, W., **1,** 1058
Braunschweig, E., **3,** 156
Braunstein, D. M., **3,** 77; **9,** 175
Braunworth, J. B., **1,** 904
Brausch, J. F., **12,** 389
Bravo, P., **1,** 221; **2,** 173; **3,** 125; **4,** 197; **7,** 133; **12,** 510
Braxmeier, H., **12,** 58
Braye, E. H., **2,** 140
Brechbuehler, H., **1,** 282, 283
Brecht, E. A., **1,** 1118
Breck, D. W., **1,** 705
Breddy, L. J., **1,** 782
Bredenberg, J. B., **1,** 603, 1273
Breder, C. V., **2,** 44, 429
Bredereck, H., **1,** 282, 283, 295, 403, 404, 494, 842, 1252; **4,** 157, 167, 173, 529; **5,** 73, 250, 254, 653; **6,** 58, 243; **12,** 204
Bredt, J., **8,** 75
Bregan, W. H., **1,** 1178
Bregant, N., **5,** 281
Bregeault, J. M., **11,** 430
Brehme, R., **6,** 43
Breitholle, E. G., **7,** 304; **10,** 136
Breitmaier, E., **10,** 194, 410
Breitner, E., **1,** 983, 986
Bremer, N. J., **4,** 114
Bremholt, T., **4,** 374
Bremmer, M. L., **11,** 115
Brendle, T., **5,** 653
Brener, L., **3,** 336; **4,** 74; **8,** 49
Brennan, M. E., **3,** 144; **6,** 577
Brennan, M. R., **1,** 561
Brennan, T. M., **5,** 415
Brenneisen, K., **1,** 329
Brenner, G. S., **6,** 412
Brenner, L., **7,** 56
Brenner, M., **1,** 98, 859, 978, 1163; **2,** 444
Brenner, S., **4,** 489

Brennerman, M. C., **1,** 905
Brenninger, W., **1,** 444, 707, 708, 1178
Bresciani-Pahor, N., **11,** 124
Breslauer, H., **1,** 652
Breslow, D. S., **1,** 155, 524, 1058, 1286
Breslow, R., **1,** 224, 595, 1203; **2,** 424, 427; **3,** 338; **4,** 265, 520; **5,** 19, 353; **6,** 289, 300; **7,** 324; **8,** 64; **9,** 129; **10,** 216; **11,** 151; **12,** 314
Bressel, B., **9,** 56
Bressel, U., **2,** 53; **4,** 510; **5,** 650
Brett, D., **2,** 445
Bretting, C., **1,** 169
Brettle, R., **10,** 251
Breuer, A., **1,** 970; **4,** 133
Breuer, E., **1,** 125, 484; **6,** 612
Breuer, S. W., **4,** 126, 430; **6,** 581
Brewer, H. W., **1,** 573
Brewster, A. G., **7,** 139; **8,** 32; **9,** 34; **10,** 24
Brewster, J., **9,** 71
Brewster, J. H., **1,** 1184
Brewster, R. Q., **1,** 158, 321, 500, 505, 845, 1152; **2,** 351; **5,** 148, 161; **9,** 486
Brey, W. S., Jr., **1,** 613
Brice, C., **1,** 562
Brich, Z., **8,** 448; **9,** 420
Brickner, J. S., **10,** 126
Brickner, S. J., **11,** 358; **12,** 475
Bridgeman, J. E., **2,** 243
Bridger, R. F., **3,** 69
Bridges, A. J., **5,** 410; **6,** 395; **7,** 233
Bridgwater, A., **1,** 46
Bridson, J. N., **4,** 122, 172, 187; **5,** 296; **10,** 85
Brieger, G., **3,** 243; **7,** 277
Brienne, M. J., **8,** 75
Briese, R., **5,** 620
Brigandat, Y., **11,** 383
Briggs, L. H., **1,** 448, 1008; **3,** 66; **5,** 157, 395
Bright, R. D., **1,** 909
Brigl, P., **1,** 346, 847
Brignole, A. B., **10,** 274
Brill, J. F., **12,** 422
Brimacombe, J. S., **2,** 274; **8,** 331
Brindle, J. R., **3,** 264; **4,** 444; **6,** 534; **7,** 332
Brine, G. A., **7,** 236
Briner, P. H., **11,** 557
Bringmann, G., **10,** 413
Brink, M., **6,** 357, 612
Brinker, U. H., **12,** 151
Brinkmann, R., **12,** 290
Brinkmeyer, R. S., **4,** 541; **6,** 630; **8,** 186; **9,** 334
Brintzinger, H., **1,** 755
Brion, F., **11,** 605
Brisson, C., **10,** 44
Britelli, D. R., **11,** 212
Britt, R. W., **7,** 99
Britten-Kelly, M. R., **6,** 509; **9,** 102
Britton, R. W., **2,** 131, 462; **3,** 65; **4,** 103, 183; **6,** 530
Britton, W. E., **5,** 470
Brixius, D. W., **5,** 102
Brizuela, C. L., **7,** 174
Brizzolara, A., **1,** 707, 974
Broadbent, H. S., **1,** 978, 979, 982; **6,** 493
Broadbent, W., **2,** 56
Broaddus, C. D., **4,** 489
Broadhurst, M. D., **5,** 474; **6,** 419
Broadhurst, M. J., **4,** 94; **11,** 410
Brocard, J., **11,** 132; **12,** 36
Brochamann-Hanssen, E., **8,** 310
Brock, D. J. H., **2,** 296
Brocker, U., **7,** 268
Brockhaus, R., **2,** 65
Brockington, R., **6,** 381; **7,** 240
Brockman, J. A., Jr., **1,** 1157
Brockmann, H., **1,** 20, 321; **4,** 407
Brockmann, W., **8,** 115
Brocksom, T. J., **3,** 302; **6,** 339
Brod, A. O., **5,** 693; **8,** 501; **9,** 483
Brodie, B. B., **1,** 476; **5,** 308
Brodowski, W., **6,** 263
Brodsky, L., **5,** 486
Broeckx, W., **4,** 291
Broekhof, N. L. J. M., **7,** 357; **8,** 90; **9,** 319; **11,** 180, 353; **12,** 334
Broekhuis, A. A., **11,** 71
Broekhuysen, M. P., **11,** 423
Broenneke, A., **12,** 83
Broger, E. A., **3,** 212
Brois, S. J., **1,** 755, 1116
Broline, B. M., **5,** 325
Bromley, D., **2,** 411; **3,** 286; **4,** 492
Brook, A. G., **1,** 129, 219, 329, 348, 738; **2,** 187; **5,** 68, 376; **6,** 637; **7,** 194, 220, 271; **9,** 166; **10,** 71
Brook, M. A., **12,** 175
Brook, P. R., **5,** 280, 617
Brooke, G. S., **6,** 234
Brooke, P. K., **4,** 529
Brookhart, M., **8,** 36
Brookhart, T., **10,** 180
Brooks, C. J. W., **1,** 1046; **2,** 317
Brooks, D. W., **9,** 96, 132
Brooks, L. A., **1,** 156, 251, 494, 1163; **4,** 437
Brooks, R. E., **1,** 444
Brooks, S. G., **1,** 954
Brooks, T. W., **1,** 1246
Broom, N. J. P., **8,** 402
Broome, J., **1,** 35, 598
Broos, R., **7,** 160
Brossi, A., **4,** 43, 250, 330; **9,** 256
Brossmer, R., **2,** 263, 349; **6,** 511
Broster, F. A., **4,** 126, 430
Brothers, D., **11,** 299
Brotherton, C. E., **12,** 154, 207
Brotherton, T. K., **1,** 1041, 1044
Broughton, H. B., **12,** 41
Brouillette, W. J., **10,** 113
Broussard, J., **6,** 479
Brousse, E., **2,** 163

Browder, L. E., **1,** 604
Brown, A. E., **1,** 186
Brown, A. G., **8,** 37
Brown, A. P., **2,** 462
Brown, B. K., **1,** 444, 489, 1087
Brown, B. R., **1,** 35, 269, 598; **5,** 391
Brown, C., **11,** 308
Brown, C. A., **1,** 720, 892; **2,** 215, 333; **3,** 210; **4,** 409; **5,** 472, 557, 565, 615; **6,** 476, 483, 492; **7,** 296; **8,** 407; **9,** 58, 235; **11,** 431; **12,** 401
Brown, D., **1,** 236
Brown, D. J., **4,** 431; **5,** 681, 740
Brown, D. M., **1,** 53, 199, 235, 346, 847, 881; **10,** 357
Brown, E., **8,** 114
Brown, E. V., **1,** 465; **6,** 21
Brown, F. C., **2,** 215; **4,** 191; **6,** 515
Brown, F. R., Jr., **10,** 35
Brown, G. B., **1,** 139
Brown, G. E., **2,** 179
Brown, G. L., **2,** 65
Brown, H. C., **1,** 35, 50, 59, 68, 72, 125, 130, 144, 197, 206, 207, 212, 263, 276, 380, 484, 595, 603, 610, 624, 625, 720, 771, 772, 892, 926, 1054, 1055, 1109, 1131, 1163, 1184, 1191, 1221, 1228, 1231; **2,** 15, 24, 31, 60, 108, 149, 193, 195, 215, 253, 267, 285, 310, 333, 344, 349, 368; **3,** 5, 10, 23, 28, 29, 39, 43, 77, 91, 160, 176, 186, 187, 196, 197, 237, 264, 295, 325; **4,** 10, 25, 36, 38, 53, 69, 70, 96, 126, 130, 162, 176, 230, 311, 312, 313, 314, 322, 347, 362, 377, 447; **5,** 41, 47, 48, 57, 101, 186, 191, 203, 233, 296, 347, 351, 467, 486; **6,** 34, 64, 73, 172, 208, 259, 284, 295, 325, 342, 350, 573; **7,** 30, 53, 96, 98, 181, 216, 217, 249, 321; **8,** 6, 7, 49, 50, 76, 143, 174, 267, 303, 310, 334, 354, 500; **9,** 58, 199, 278, 287, 317, 393, 398, 399; **10,** 40, 49, 50, 77, 95, 106, 224, 250, 398; **11,** 68, 69, 154, 188, 191, 194, 266, 267, 292, 293, 351, 430; **12,** 5, 18, 70, 194, 195, 236, 276, 334, 399, 406, 414, 485
Brown, I., **2,** 192
Brown, J., **4,** 223
Brown, J. B., **1,** 392, 759, 1040
Brown, J. D., **12,** 284
Brown, J. F., Jr., **3,** 157
Brown, J. H., **1,** 865
Brown, J. H., Jr., **1,** 269, 982
Brown, J. J., **1,** 16, 120, 634, 767
Brown, J. M., **8,** 39; **12,** 428
Brown, J. R., **9,** 439
Brown, J. W., **1,** 139, 581, 617
Brown, K. S., Jr., **2,** 387
Brown, L., **12,** 44
Brown, L. L., **8,** 19
Brown, L. R., **5,** 31, 447
Brown, M., **1,** 1040; **2,** 453
Brown, M. S., **1,** 1055
Brown, N. C., **1,** 954
Brown, P., **5,** 391
Brown, P. M., **1,** 859
Brown, R., **1,** 13, 1157
Brown, R. A., **1,** 1211; **9,** 89
Brown, R. F. C., **6,** 96
Brown, R. J., **11,** 95
Brown, R. K., **1,** 158, 445, 970; **5,** 186
Brown, R. L., **1,** 450
Brown, R. T., **1,** 652, 1284; **3,** 196; **12,** 415
Brown, R. W., **8,** 525
Brown, S. P., **11,** 28, 275
Brown, T. L., **1,** 613
Brown, T. M., **6,** 414
Brown, W., **3,** 86; **4,** 134
Brown, W. G., **1,** 582, 594, 595, 884, 1054; **3,** 146; **5,** 309
Brownbridge, P., **7,** 375; **9,** 55; **10,** 46, 47, 258
Browne, A. R., **4,** 405
Browne, E. N. C., **8,** 303; **10,** 373
Browne, L. J., **7,** 193, 199
Browne, L. M., **7,** 268
Browne, P. A., **3,** 48
Brownfain, D. S., **9,** 192
Brownfield, R. B., **1,** 173
Brownlee, P. J. E., **2,** 27
Brownlee, R. G., **2,** 187
Brownridge, J. R., **5,** 563
Broxton, T. J., **5,** 266
Brubaker, C. H., Jr., **4,** 107
Bruce, D. B., **1,** 1029
Bruce, J. M., **1,** 34; **11,** 311
Bruce, M. R., **10,** 251
Bruce, N. F., **1,** 808
Bruce, W. F., **1,** 22, 859, 892, 1064
Bruck, P., **1,** 606; **2,** 397
Bruckner, V., **1,** 34
Bruder, W. A., **11,** 407; **12,** 329
Brueckner, H., **1,** 353
Bruegel, W., **5,** 717
Bruenig, H., **2,** 397
Bruentrup, C., **9,** 288
Bruentrup, G., **8,** 517; **9,** 506; **11,** 200; **12,** 278
Bruggemann-Rotgans, I. E., **11,** 131
Brugger, M., **1,** 1018
Bruggink, A., **5,** 164; **6,** 144
Bruice, T. C., **1,** 494
Brun, P., **2,** 300; **4,** 234; **12,** 53
Brunelet, T., **10,** 306
Brunelle, D. J., **4,** 63, 183; **5,** 243, 244; **6,** 343, 465; **7,** 142; **8,** 87, 489; **11,** 258
Brunet, J. J., **10,** 365; **11,** 486; **12,** 371
Bruni, R. J., **1,** 207
Brunie, J. C., **1,** 905
Brunings, K. J., **1,** 1122
Brunken, J., **1,** 652
Brunner, H., **9,** 49; **10,** 36,

116; **12,** 3
Bruno, J. J., **1,** 262
Brunow, G., **6,** 656; **10,** 359
Bruns, L., **1,** 832
Bruson, H. A., **1,** 14, 952, 1254
Brussee, J., **12,** 191
Brutcher, F. V., Jr., **1,** 1004; **11,** 268
Brutschy, F. J., **1,** 37, 927
Bruza, K. J., **6,** 205; **8,** 72; **9,** 11
Bryan, C. A., **5,** 296
Bryan, R. F., **5,** 746
Bryant, D. R., **10,** 291
Bryant, R. J., **6,** 629
Bryce-Smith, D., **1,** 114, 465, 629, 1103; **2,** 47, 254
Brylikowska-Piotrowicz, J., **8,** 391
Brynestad, J., **8,** 19
Brynjolffssen, J., **6,** 467
Bryson, T. A., **4,** 531; **5,** 700; **6,** 354; **7,** 90, 302; **8,** 72, 76; **10,** 161; **11,** 53
Bubb, W. A., **7,** 188
Bubl, E. C., **1,** 814
Bublitz, D. E., **1,** 78, 182
Buc, S. R., **1,** 477; **10,** 205
Buch, V., **6,** 33
Buchan, G. M., **6,** 170
Buchanan, D. A., **5,** 321
Buchanan, G. L., **2,** 50, 422; **5,** 57, 257
Buchanan, J. B., **1,** 22
Buchanan, J. G., **1,** 1128
Buchanan, R. A., **12,** 178
Buchanan, R. L., **2,** 40
Buchardt, O., **11,** 147
Buchbauer, G., **11,** 533
Buchecker, C., **5,** 230
Bucher, J. E., **1,** 952
Buchholz, M., **11,** 78
Buchman, O., **10,** 79
Buchner, B., **1,** 33, 876
Buchner, E., **1,** 370; **4,** 325; **6,** 190
Buchner, W., **9,** 224
Buchschacher, P., **8,** 424
Buchta, E., **1,** 594, 682, 1272; **4,** 297, 364, 367
Buchwald, S. L., **7,** 125; **12,** 55, 111
Buck, A., **1,** 402; **11,** 255
Buck, H. M., **5,** 707; **7,** 323
Buck, J. S., **1,** 166, 471, 481, 870, 1027, 1113, 1116; **7,** 299; **10,** 306
Buck, K. T., **5,** 761
Buck, P., **8,** 359
Buck, R. R., **1,** 62
Buckingham, D. A., **1,** 1204
Buckland, P. R., **12,** 156
Buckle, D. R., **11,** 167
Buckler, S. A., **1,** 17, 89, 859
Buckles, R. E., **1,** 74, 75, 1004, 1132, 1163
Buckley, D. J., **10,** 21
Buckley, G. D., **7,** 253
Buckley, T. F., **11,** 244
Buckman, J. D., **1,** 718
Buckow, G., **7,** 60
Buckpitt, A. R., **5,** 602
Buckschewski, H., **1,** 1101, 1247
Buckwalter, B., **6,** 607
Buckwalter, B. L., **8,** 188; **12,** 141
Bucourt, R., **1,** 215, 518; **8,** 429
Buczel, C., **2,** 126
Budde, W. M., **1,** 1065; **3,** 272
Buddrus, J., **2,** 197; **3,** 140; **4,** 232, 264; **5,** 299; **6,** 263; **7,** 181
Budewitz, E. P., **1,** 785, 1284
Budziarek, R., **1,** 1284
Budzikiewicz, H., **1,** 444; **4,** 230
Buechi, G., **1,** 94, 266, 562, 652, 1163; **2,** 17, 154, 213, 400, 422; **3,** 71, 247; **4,** 65, 261, 283, 361, 427, 472, 579; **5,** 9, 97, 254, 293; **6,** 77, 286, 514; **7,** 122, 210; **8,** 125, 272, 338, 374, 375, 488; **9,** 1, 70, 151, 191, 361, 438, 458; **10,** 36, 58; **12,** 107
Buechi, H., **1,** 282, 283
Buechner, E., **1,** 562
Buechner, O., **2,** 219
Buechner, W., **1,** 484
Bueckel, C., **8,** 163
Buehler, C. A., **1,** 672, 827; **3,** 49; **5,** 552; **12,** 403
Buerger, H., **5,** 45; **11,** 378
Buerstinghaus, R., **6,** 55, 118
Buethe, I., **5,** 207
Buhler, J. D., **4,** 63
Buhr, G., **2,** 333
Buhro, W. E., **9,** 21; **10,** 382
Buijle, R., **2,** 134
Bui Khac, T., **8,** 194
Buisson, J. P., **6,** 498
Buist, G. J., **1,** 818
Bujnicki, B., **7,** 389
Bukala, J., **9,** 320
Bukownik, R. B., **11,** 479
Bukowski, P., **6,** 291
Bull, B. A., **1,** 1087
Bull, J. R., **5,** 244; **6,** 600
Bullee, R. J., **11,** 13
Bullock, M. W., **1,** 619, 1157
Bulman-Page, P. G., **12,** 502
Bulmer, G., **10,** 39
Bu'Lock, J. B., **2,** 256
Bu'Lock, J. D., **9,** 302
Bulten, E. J., **10,** 220
Bumgardner, C. L., **1,** 254, 742, 1140; **2,** 135; **11,** 284; **12,** 158
Bumpus, F. M., **1,** 235
Bunce, N. J., **4,** 324, 325
Buncel, E., **6,** 158; **7,** 302; **8,** 130, 524
Bundle, D. R., **6,** 569; **11,** 44
Bunds, J., **1,** 658
Bundy, G. L., **3,** 77; **4,** 545; **9,** 495
Bunes, L., **8,** 531
Bunge, W., **1,** 1103
Bunnell, C. A., **2,** 309; **8,** 492; **9,** 385
Bunnett, J. F., **1,** 329, 909, 1041, 1044, 1090; **2,** 344, 374; **4,** 153, 398; **5,** 86, 217; **6,** 192; **8,** 245
Buntin, S. A., **12,** 370
Bunton, C. A., **1,** 818; **2,** 363; **6,** 249, 598
Bunton, E. V., **7,** 239
Bunyan, P. J., **1,** 1216

Buono, G., **12,** 53
Burbage, J. J., **9,** 207
Burch, J. E., **1,** 270
Burchard, H., **2,** 139
Burckhalter, J. H., **3,** 9; **4,** 337
Burckhardt, G., **1,** 894
Burckhardt, U., **1,** 46, 360
Burczyk, B., **11,** 495
Burden, R. S., **8,** 337
Burdett, K. A., **6,** 675; **7,** 429; **9,** 520
Burdge, D. N., **1,** 581
Burditt, N. A., **2,** 403
Burdon, J., **4,** 533
Burdon, M. G., **1,** 310
Burfield, D. R., **9,** 130; **11,** 350
Burford, C., **8,** 278; **12,** 118
Burford, S. S., **10,** 437
Burg, A. B., **1,** 206, 432, 1230; **4,** 191
Burgard, A., **5,** 433
Burger, A., **1,** 452, 1049; **2,** 132; **3,** 74
Burger, H., **11,** 68
Burger, K., **5,** 537
Burger, T., **4,** 76, 251
Burger, U., **5,** 389; **7,** 74, 239; **11,** 472
Burgers, P. M. J., **8,** 362
Burgess, C., **2,** 42, 227
Burgess, E. M., **4,** 228, 333, 344; **5,** 93, 444; **8,** 334
Burgess, J. K., **5,** 389
Burgess, J. R., **2,** 182, 289; **5,** 495
Burgess, W. M., **3,** 212; **7,** 171
Burgher, R. D., **1,** 892
Burgmaier, G. J., **3,** 129, 143; **5,** 205
Burgoyne, E. E., **1,** 157
Burgstahler, A. W., **1,** 44, 209, 573, 581, 611, 624, 652, 731, 982, 983; **2,** 6, 58, 239, 263; **3,** 175, 302; **4,** 167; **5,** 92; **7,** 277, 386; **9,** 335, 520
Burham, R. L., **1,** 1196
Buriks, R. S., **1,** 1065; **3,** 272
Burk, C., **11,** 106
Burk, R. M., **12,** 14, 193
Burke, M. C., **8,** 299; **10,** 77
Burke, P. L., **5,** 47, 186
Burke, S. D., **6,** 460, 632; **7,** 160; **9,** 81; **10,** 156, 443
Burkett, A. R., **4,** 302
Burkhardt, G. N., **1,** 1128
Burkhardt, T. J., **4,** 379
Burkhart, J. P., **10,** 248
Burks, G., **12,** 61
Burks, S. R., **12,** 303
Burlant, W. J., **1,** 500
Burleson, F. R., **5,** 511
Burlitch, J. M., **1,** 854, 1237; **2,** 328; **5,** 514
Burman, D. L., **11,** 346
Burn, A. J., **1,** 1216
Burn, D., **1,** 219, 285, 767; **5,** 253
Burness, D. M., **1,** 158, 636, 976
Burnett, D. A., **12,** 523
Burnett, R. D., **5,** 370
Burnett, R. E., **1,** 454; **6,** 416
Burnham, J. W., **4,** 357, 391, 397
Burns, P. A., **4,** 363; **5,** 491; **7,** 268; **9,** 341
Burns, T. P., **11,** 308, 599
Burns, W., **6,** 473
Burpitt, R. D., **4,** 172
Burr, M., **2,** 70; **3,** 338
Burrell, J. W. K., **2,** 263
Burrous, M. L., **1,** 581, 611; **5,** 379
Burrows, G. J., **8,** 280
Burstein, S. H., **1,** 219, 333, 705, 926
Burtner, R. R., **1,** 56, 1145
Burton, A., **9,** 34
Burton, D. I., **10,** 448
Burton, D. J., **1,** 1058, 1227; **2,** 444; **3,** 308; **4,** 83; **5,** 63, 731; **6,** 261, 262, 646; **7,** 60; **9,** 54, 166
Burton, H., **1,** 716, 802
Burton, R., **6,** 596
Burton, R. B., **1,** 411
Burzynska, H., **7,** 406
Busch, M., **5,** 445
Busch, T., **1,** 198, 1141
Buschhoff, M., **10,** 121
Buse, C. T., **8,** 112, 299; **9,** 283; **10,** 133; **11,** 349
Bush, J. B., Jr., **2,** 264
Bush, W. M., **1,** 96
Bushey, D. F., **4,** 510
Bushweller, C. H., **2,** 303
Buss, A. D., **10,** 2; **12,** 477
Buss, D. H., **1,** 310
Buss, V., **10,** 237; **11,** 292
Busse, A., **1,** 321
Busser, U., **4,** 126
Buswell, R. L., **6,** 600
Butenandt, A., **1,** 736, 881, 933, 1284
Buter, J., **4,** 267; **5,** 160; **12,** 108
Butler, D., **5,** 62
Butler, D. E., **1,** 93
Butler, G., **10,** 85
Butler, G. B., **1,** 1246; **4,** 398; **12,** 154
Butler, K., **1,** 445, 1210
Butler, M. E., **1,** 772
Butler, P. E., **3,** 178
Butler, R. N., **2,** 238; **7,** 188; **10,** 394
Buton, C. A., **1,** 470
Butskus, R. F., **1,** 78
Butt, B. A., **5,** 285
Butterfield, R. O., **12,** 36
Butterworth, R., **3,** 78; **6,** 261
Butterworth, R. F., **9,** 121
Butula, I., **8,** 87
Butzlaff, B., **4,** 119
Buu-Hoi, N. P., **1,** 730, 731; **3,** 240
Buurman, E., **7,** 349; **8,** 103
Buyle, R., **1,** 130, 258
Buynak, J. D., **12,** 122
Buza, M., **2,** 267
Buzas, A., **1,** 235
Buzby, G. C., Jr., **3,** 204
Buzzolini, M. G., **3,** 251, 252
Bycroft, B. W., **6,** 493
Byers, J. D., **8,** 249; **10,** 203
Byers, J. R., **1,** 500, 864, 1163, 1167
Bynum, E., **2,** 94
Byram, S. K., **12,** 41
Byrd, D., **4,** 491
Byrd, J. E., **5,** 658

Byrd, L. R., **4,** 57
Byrne, B., **4,** 291, 294; **7,** 383
Byrne, M. P., **6,** 137
Bystroem, S. E., **12,** 370
Bywood, R., **6,** 453

Cabiddu, S., **7,** 4
Cable, J., **2,** 257
Cabre, J., **11,** 58
Cacchi, C., **10,** 306
Cacchi, S., **4,** 387, 512; **5,** 495; **6,** 470, 534, 607; **9,** 151, 352, 411, 443
Caciagli, V., **9,** 335; **11,** 385
Caddy, B., **5,** 436
Cade, J. A., **1,** 1141
Cadenbach, G., **1,** 15
Cadiot, P., **1,** 1022; **5,** 129; **9,** 365; **10,** 411; **11,** 143; **12,** 416
Cadogan, J. I. G., **1,** 41, 264, 1216; **2,** 433; **3,** 140; **4,** 80, 172; **5,** 18, 148; **6,** 419; **9,** 56
Cadogan, M., **3,** 9; **4,** 14, 337
Cady, G. H., **2,** 200, 458; **4,** 37, 325; **12,** 173
Caesar, G. V., **1,** 324
Caesar, P. D., **1,** 1289
Caglioti, L., **1,** 1055; **2,** 207, 378; **4,** 304, 444, 512; **5,** 495; **6,** 162; **7,** 175; **8,** 250; **9,** 432
Cagniant, D., **4,** 134
Cagniant, P., **4,** 134
Cahiez, G., **5,** 94; **6,** 270, 664; **7,** 83, 222; **8,** 313; **9,** 249, 289; **10,** 289, 290
Cahill, W. M., **1,** 531
Cahnmann, H. J., **6,** 436
Cain, A. M., **8,** 448
Cain, B. F., **1,** 1008; **7,** 15
Cain, E. N., **2,** 295; **3,** 38; **6,** 76
Cain, M. E., **1,** 705
Cain, P., **6,** 411, 606
Cain, P. A., **8,** 203, 401
Caine, D., **2,** 112; **6,** 492; **7,** 31; **8,** 284, 303
Cainelli, G., **1,** 562; **2,** 274; **3,** 189, 201; **4,** 134; **5,** 315; **6,** 74, 304, 486; **7,** 69, 182
Cairncross, A., **4,** 157; **5,** 220, 505; **10,** 304
Cairns, J., **10,** 304
Cairns, T. L., **1,** 295, 426; **5,** 589
Cais, M., **2,** 417; **12,** 36
Cajipe, G., **4,** 60
Calabrese, J. C., **12,** 171
Calama, M. A., **2,** 321
Calamai, E. C., **6,** 486
Calamai, E. G., **6,** 545
Calas, R., **5,** 713; **6,** 323; **7,** 9, 84, 199, 372; **9,** 372, 440; **10,** 176; **12,** 220
Calder, A., **12,** 95
Calder, I. C., **2,** 88
Calderon, J., **2,** 321
Calderon, N., **4,** 570; **7,** 289
Caldwell, C. G., **11,** 159
Caldwell, R. A., **4,** 129
Calf, G. E., **3,** 134; **4,** 215
Callahan, F. M., **1,** 487, 524; **2,** 47, 279
Callant, P., **12,** 204
Callen, G. R., **12,** 397
Callen, J. E., **1,** 379, 1209
Callery, P. S., **7,** 56
Callot, H. J., **11,** 462
Calo, V., **4,** 477; **5,** 24, 644; **6,** 73, 375; **7,** 35, 245; **8,** 53
Calvery, H. O., **1,** 349
Calvin, E. W., **12,** 85
Calvin, M., **1,** 894; **3,** 143, 245
Calzada, J. G., **4,** 552; **5,** 296; **6,** 645
Calzada, M. C., **1,** 402
Calzadilla, C. H., **2,** 117
Cama, L. D., **5,** 585
Camacho, C., **11,** 219
Cambie, R. C., **1,** 1008; **2,** 250; **5,** 655; **7,** 360; **8,** 260; **9,** 249, 458; **10,** 395; **11,** 305, 486, 515
Camble, R., **2,** 27; **4,** 391
Cambon, A., **5,** 527; **9,** 306
Camerino, B., **1,** 72, 927
Cameron, A. F., **8,** 79
Cameron, A. F. B., **1,** 595, 642, 951
Cameron, A. G., **11,** 426
Cameron, D. D., **1,** 777, 964
Cameron, D. W., **9,** 264
Cameron, M. D., **1,** 84
Cameron, P. J., **2,** 79
Cameron, T. S., **4,** 172; **11,** 494
Cameron-Wood, M., **1,** 1216
Cammarata, A., **1,** 101
Camp, R. L., **2,** 309; **5,** 424
Campagna, F., **8,** 504; **9,** 484
Campaigne, E., **1,** 80, 198, 285, 430, 1015, 1065, 1131; **3,** 272; **8,** 486
Campbell, B. K., **1,** 12, 329, 1040
Campbell, B. N., Jr., **1,** 35
Campbell, C. B., **5,** 429; **8,** 317
Campbell, C. D., **1,** 484, 563; **3,** 11
Campbell, D. E., **8,** 299
Campbell, H. M., **5,** 655
Campbell, I. D., **1,** 161
Campbell, I. M., **2,** 117
Campbell, J. A., **1,** 127, 394
Campbell, J. B., Jr., **10,** 106
Campbell, J. R., **2,** 226, 395; **3,** 9, 244; **4,** 283, 337
Campbell, J. R. B., **1,** 66, 1137
Campbell, K. N., **1,** 12, 329, 703, 1040
Campbell, M. M., **6,** 115; **8,** 83; **12,** 257
Campbell, R. G., **6,** 353
Campbell, R. H., **6,** 187
Campbell, R. W., **4,** 505
Campbell, S. F., **2,** 87, 464; **4,** 23
Campbell, T. W., **1,** 235, 337, 338, 613, 679, 695, 933, 1246; **8,** 208
Campbell, W. P., **1,** 1203; **4,** 579
Campos, M. M., **1,** 956
Campos, P. J., **12,** 306
Camps, F., **4,** 573; **8,** 440; **9,** 52, 427; **11,** 124, 486
Camps, P., **11,** 555
Canceill, J., **10,** 461
Cancienne, A. E., Jr., **2,** 34

Candy, C. F., **4,** 502
Cane, D. E., **3,** 105; **4,** 233, 473; **12,** 425
Cann, K., **6,** 198
Cannell, L. G., **4,** 113
Cannic, G., **5,** 246
Cannon, G. W., **1,** 676
Cannon, J. B., **5,** 332
Canonica, L., **6,** 518
Canonne, P., **2,** 215; **3,** 33; **5,** 542; **10,** 194; **11,** 249
Cantacuzene, D., **5,** 219
Cantoni, G. L., **1,** 1002
Cantrall, E. W., **1,** 615, 1261
Cantrell, G. L., **9,** 224
Cantrill, J. E., **1,** 409
Cao, Y. **11,** 387
Capiris, T., **5,** 345
Capitaine, J., **2,** 21
Capka, M., **3,** 261
Caplar, V., **6,** 474
Caple, R., **11,** 163
Caporusso, A. M., **11,** 563
Capozzi, G., **6,** 375; **11,** 336
Cappiello, V., **4,** 84
Capuano, L., **2,** 321; **6,** 600
Caputo, J. A., **1,** 333; **2,** 359; **3,** 244
Caputo, R., **6,** 586; **7,** 336; **11,** 128; **12,** 555
Carabateas, C. D., **2,** 252
Carayon-Gentil, A., **1,** 281; **6,** 574
Carbon, J. H., **1,** 1081
Carbonaro, A., **9,** 257
Carboni, R. A., **1,** 1125, 1136
Cardani, S., **12,** 226
Cardellach, J., **11,** 555
Cardenas, C. G., **1,** 309, 318, 926
Cardiff, E. A., **5,** 511
Cardillo, G., **2,** 117; **3,** 87; **4,** 134; **5,** 86; **6,** 43; **7,** 69, 71, 298; **8,** 442; **9,** 257, 383; **10,** 211; **11,** 267; **12,** 87
Cardin, D. J., **2,** 147
Cardwell, H. M. E., **1,** 767; **3,** 41
Cardy, D. E., **2,** 87; **5,** 160
Carey, F. A., **2,** 441; **4,** 285; **6,** 35; **9,** 309
Carg, H. G., **2,** 175
Carganico, G., **9,** 119
Cargill, R. L., **2,** 159; **4,** 4, 510
Carhart, H. W., **1,** 595, 1163
Cariati, F., **9,** 217
Carini, D. J., **10,** 429; **12,** 350
Carless, H. A. J., **10,** 300
Carleton, P. W., **1,** 129
Carletti, V., **8,** 367
Carli, B., **12,** 365
Carlin, K. J., **8,** 299
Carlin, R. B., **1,** 767
Carlisle, C., **4,** 521
Carlock, J. T., **9,** 506
Carlon, F. E., **2,** 232, 326, 332, 344, 433
Carlsen, P. H. J., **11,** 463
Carlsmith, A., **1,** 465, 787
Carlsohn, H., **2,** 39
Carlson, B. A., **4,** 314; **5,** 203; **6,** 172, 350; **7,** 98
Carlson, G. R., **5,** 535
Carlson, H. D., **1,** 531
Carlson, J. A., **2,** 154, 400; **4,** 65; **8,** 338
Carlson, R., **8,** 486; **12,** 502
Carlson, R. D., **1,** 1163
Carlson, R. G., **1,** 150; **2,** 344; **4,** 375; **5,** 127; **6,** 546; **7,** 178
Carlson, R. M., **1,** 8, 185; **2,** 99, 126; **3,** 73, 163, 338; **4,** 165, 522; **5,** 208; **6,** 366, 398, 496; **7,** 207; **8,** 67
Carlson, R. W., **8,** 466
Carlson, S. C., **11,** 551
Carlsson, R., **5,** 643
Carmack, M., **1,** 1027, 1121; **4,** 497
Carman, F., **5,** 245, 718
Carman, R. M., **7,** 290
Carmichael, J. F., **1,** 53
Carmody, D. R. **1,** 432
Carmody, M. F., **6,** 279
Carnahan, J. C., Jr., **4,** 350
Carney, R. L., **4,** 183, 220; **5,** 415; **7,** 114
Carney, R. W. J., **1,** 89
Caro, B., **12,** 36
Caro, H., **1,** 119
Caron, G., **5,** 23
Caron, M., **12,** 94, 506
Caronna, G., **1,** 448
Caronna, T., **3,** 38; **4,** 237; **5,** 16, 17
Carotti, A., **8,** 504, **9,** 484
Carpenter, B. G., **2,** 252
Carpenter, F. H., **1,** 737
Carpenter, P. D., **5,** 639
Carpenter, W., **6,** 300
Carpenter, W. R., **1,** 178; **4,** 510
Carpino, B. A., **1,** 85, 661; **5,** 433
Carpino, L. A., **1,** 85, 88, 210, 661; **2,** 45; **3,** 146, 194; **4,** 237; **5,** 433, 727; **7,** 301; **8,** 471; **9,** 446
Carpio, H., **1,** 104, 127, 219, 802; **4,** 150; **5,** 216
Carpita, A., **8,** 289; **9,** 277
Carr, D. B., **8,** 43, 87; **9,** 104
Carr, M., **4,** 45
Carr, M. D., **2,** 363; **4,** 253
Carr, R. V. C., **10,** 295, 316; **11,** 427
Carre, P., **1,** 676
Carrera, P., **12,** 510
Carretto, J., **5,** 194
Carrick, W. L., **3,** 333
Carrie, R., **6,** 472; **10,** 223; **11,** 231, 588; **12,** 551
Carroll, F. I., **6,** 577; **7,** 236; **9,** 156
Carroll, G. L., **4,** 543; **5,** 722
Carroll, M. F., **1,** 266
Carroll, P. M., **1,** 9, 332, 377
Carroll, S., **1,** 658
Carrupt, P. A., **10,** 392
Carruthers, W., **1,** 677
Carstens, E., **6,** 547
Carter, F. L., **3,** 139
Carter, H. E., **1,** 110, 652, 859; **2,** 208
Carter, J. P., **11,** 319
Carter, K. N., **1,** 377
Carter, L. G., **9,** 89
Carter, M. J., **7,** 397
Carter, P. H., **1,** 1169

Carty, D. T., **3,** 178
Caruso, A. J., **9,** 434
Caruso, T. C., **6,** 137
Caruthers, M. H., **2,** 26; **10,** 88, 461
Carvalho, C. F., **11,** 308; **12,** 65
Carver, D. R., **11,** 597
Casadevall, E., **4,** 298
Casagrande, F., **11,** 471
Casanova, J., Jr., **1,** 81, 562, 989; **2,** 376
Casanova, T. G., **3,** 299; **8,** 499
Casares, A. M., **10,** 154; **12,** 25
Casati, R., **12,** 458
Casensky, B., **2,** 382; **3,** 261
Caserio, M. C., **1,** 195; **4,** 57, 162; **11,** 206
Casey, C., **1,** 1220
Casey, C. P., **2,** 400; **3,** 81, 154; **4,** 183, 379; **5,** 243, 454; **6,** 495
Casey, M., **4,** 517
Casida, J. E., **11,** 445
Casini, G., **1,** 271; **8,** 504
Casiraghi, G., **8,** 245; **9,** 14; **10,** 462
Casnati, G., **7,** 246; **8,** 245; **9,** 14, 190, 361, 489; **10,** 462
Cason, J., **1,** 151, 309, 562, 611, 772, 926, 976, 1073, 1163, 1203; **2,** 117; **4,** 324; **5,** 186
Casper, E. W. R., **6,** 267
Caspi, E., **1,** 219, 478, 730, 1041; **2,** 117, 359, 453; **5,** 444; **6,** 202
Cass, M. W., **11,** 170
Cass, Q. B., **11,** 309
Cassar, L., **2,** 181; **3,** 251; **5,** 473, 474; **6,** 573; **7,** 355
Casselli, P., **4,** 304
Casser, L., **6,** 451
Casserly, E. W., **12,** 73
Cassidy, E. S., **11,** 330
Cassidy, H. G., **1,** 134
Castagnino, E., **9,** 408
Castagnoli, N., Jr., **7,** 56
Castaing, M., **8,** 445
Castaldi, G., **11,** 471
Castaner, J., **1,** 1131
Castedo, L., **5,** 540
Castell, J. V., **9,** 397
Castelli, P. P., **1,** 1210
Castellino, S., **12,** 197
Castells, J., **1,** 989; **2,** 19; **4,** 573
Castellucci, N. T., **1,** 689, 1022; **2,** 278, 376
Castle, J. E., **1,** 426
Castle, R. N., **1,** 1101
Castrillon, J. P. A., **1,** 1247; **3,** 166
Castro, B., **4,** 234; **5,** 259, 326; **6,** 25, 35, 280; **7,** 14; **9,** 297
Castro, C. E., **1,** 151, 162; **2,** 78; **3,** 59, 62; **4,** 97, 102; **5,** 44; **7,** 237; **9,** 298; **12,** 144
Casutt, M., **12,** 36
Catalane, D. B., **9,** 387
Catch, J. R., **1,** 928
Catel, J., **1,** 343
Catellani, M., **11,** 514
Cathcart, J. A., **1,** 33
Catlin, J., **4,** 354
Catlin, W. E., **1,** 253
Caton, M. P. L., **6,** 284; **8,** 1
Catsoulacos, P., **2,** 107
Catt, J. D., **5,** 700
Cattalini, L., **5,** 647
Cattelain, E., **1,** 730
Caubere, P., **2,** 210; **4,** 440, 455; **5,** 593; **6,** 70, 526; **7,** 327; **10,** 365; **11,** 486; **12,** 371
Cauble, R. L., **4,** 37
Caudle, E. C., **1,** 500, 864
Caunt, D., **1,** 41
Causa, A. G., **5,** 127
Cava, M. P., **1,** 93, 124, 125, 343, 465, 677, 828, 846, 1019, 1187; **2,** 66; **3,** 133, 177, 228, 231, 273; **4,** 9, 389; **5,** 551, 761; **6,** 152; **7,** 200, 260, 336, 387; **8,** 359, 391; **9,** 40, 169, 430; **10,** 147; **11,** 55, 56, 599; **12,** 159, 466
Cavagna, G., **1,** 531
Cavalla, D., **11,** 596
Cavallo, P., **4,** 164
Cavazza, M., **5,** 188; **6,** 27
Cave, A., **2,** 9, 75, 210; **3,** 115, 308; **4,** 184
Cavill, G. W. K., **1,** 561
Cawkill, E., **10,** 324
Cawley, J. J., **5,** 140
Cayen, C., **2,** 129
Cazeau, P., **10,** 98; **11,** 150
Cazes, B., **5,** 406; **8,** 391; **9,** 95; **10,** 101
Ceccherelli, P., **9,** 408
Cecere, M., **5,** 16
Cech, D., **5,** 309
Cedar, F. J., **5,** 408
Ceder, O., **5,** 421
Cekovic, Z., **3,** 254; **4,** 520; **5,** 61; **10,** 229; **12,** 272
Cella, J. A., **1,** 56, 1157; **5,** 359; **6,** 114; **8,** 102
Centini, F., **7,** 79; **11,** 385
Ceprini, M. Q., **1,** 783
Ceraso, J. M., **5,** 156
Cerefice, S., **3,** 329
Cernigliaro, G., **8,** 357
Cerny, M., **3,** 261; **4,** 442; **6,** 529
Cervantes, A., **4,** 415; **5,** 605
Cetenko, W., **1,** 872
Cha, D. Y., **7,** 257
Cha, J. K., **11,** 20; **12,** 359
Cha, J. S., **12,** 406, 485
Chaabouni, R., **12,** 223
Chabala, J. C., **4,** 247; **5,** 614
Chabardes, P., **11,** 346
Chabaud, B., **9,** 80; **10,** 81
Chabrier, P., **1,** 503, 730; **6,** 570, 574
Chackalamannil, S., **11,** 317
Chaco, M. C., **4,** 334
Chadenson, M., **5,** 620
Chadha, M. S., **5,** 411
Chadha, N. K., **4,** 162; **5,** 283; **11,** 226
Chadja, J. S., **3,** 308
Chadwick, D. H., **1,** 371
Chaikin, S. W., **1,** 1054
Chakhmakhcheva, O. G., **5,** 434

Chakrabarti, J. K., **4,** 523
Chakrabarti, P. M., **2,** 404
Chakrabarty, S., **7,** 387
Chakraborti, P. C., **5,** 161
Chakraborty, T. K., **12,** 433
Chakraborty, U. R., **11,** 47
Chakravarti, D., **2,** 294
Chakravarti, K. K., **2,** 239
Chakravarti, R. N., **2,** 294
Chalchat, J. C., **8,** 196
Chalk, A. J., **7,** 274
Chalkley, G. R., **2,** 403
Challenger, F., **8,** 397; **9,** 430
Challis, B. C., **7,** 252
Chalmers, W., **1,** 654
Chaltopadhyaya, J. B., **9,** 139
Chaly, T., **11,** 66
Chamberlain, P., **4,** 53
Chamberlain, T. R., **3,** 200
Chamberlin, A. R., **7,** 48; **9,** 488; **10,** 407; **11,** 267; **12,** 533
Chamberlin, E. M., **1,** 652, 676
Chamberlin, J. W., **2,** 143
Chambers, R., **8,** 88
Chambers, R. D., **1,** 827; **3,** 149; **9,** 245
Chambers, R. W., **1,** 335, 1148
Chambers, W. J., **1,** 601, 1040
Champagne, J., **2,** 310
Champion, J., **4,** 44, 305
Chan, A. S. K., **6,** 198
Chan, C. C., **11,** 50
Chan, C. M., **7,** 359; **8,** 299
Chan, D. M. T., **7,** 235; **9,** 458; **10,** 225; **11,** 82, 260
Chan, H. F., **6,** 385
Chan, H. W. S., **10,** 294
Chan, K. K., **2,** 301; **8,** 192
Chan, K. S., **12,** 136
Chan, T. H., **3,** 277, 299; **4,** 425; **5,** 68, 376, 725, 732; **6,** 100, 323, 482, 637; **7,** 13, 67, 194, 279, 356; **8,** 57, 60, 274, 486; **9,** 55, 497, 498; **10,** 46, 47, 219, 258; **11,** 66, 533; **12,** 175, 220, 279
Chan, W. H., **9,** 187; **11,** 373
Chan, W. K., **8,** 444
Chandani, B., **9,** 361
Chandhary, S. K., **10,** 156
Chandler, A., **10,** 348
Chandler, J. H., **9,** 99, 443
Chandrasekaran, S., **7,** 374; **9,** 99, 136, 193, 318, 398, 432, 467; **11,** 335, 426; **12,** 433, 454, 503
Chandrasekhar, B. P., **6,** 588
Chandy, M. J., **6,** 621
Chaney, A., **2,** 175
Chang, C. C., **7,** 428
Chang, C. S., **6,** 142
Chang, C. W., **8,** 69
Chang, C. Y., **9,** 224
Chang, D. C. K., **3,** 214
Chang, E., **5,** 725; **6,** 637
Chang, F. C., **1,** 563, 927, 966, 1187; **3,** 188; **5,** 436; **9,** 406; **11,** 88, 445
Chang, H., **12,** 83, 479
Chang, H. H., **4,** 194; **5,** 78
Chang, H. W., **3,** 338
Chang, L. W., **9,** 483
Chang, P. L., **5,** 561
Chang, S. M., **6,** 224, 586
Chang, T. C. T., **11,** 485
Chang, V. S., **8,** 391; **9,** 361, 391
Chang, V. S. K., **5,** 329
Chang, W. H., **1,** 129; **2,** 397
Chang, Y. C., **10,** 295
Chang, Y. H., **8,** 299
Chanley, J. D., **2,** 203
Channing, D. M., **1,** 76
Chanon, F., **6,** 228
Chanon, M., **6,** 228
Chanot, J. J., **6,** 70
Chao, K. H., **12,** 333
Chao, L. C., **5,** 419; **6,** 293
Chao, O., **1,** 125
Chao, S. T., **8,** 284
Chao, T. S., **1,** 832, 963
Chao, Y. L., **11,** 454, 593
Chapdelaine, M. J., **7,** 7; **8,** 12, 13; **9,** 241, 470; **11,** 315
Chapin, R. M., **1,** 214
Chapleur, Y., **5,** 326; **9,** 297; **11,** 339, 373
Chaplin, C. A., **12,** 160
Chapman, D., **12,** 25
Chapman, D. D., **6,** 409
Chapman, D. W., **8,** 403
Chapman, J. H., **1,** 107, 595, 642, 951, 955
Chapman, K. T., **10,** 365; **11,** 488; **12,** 362
Chapman, N. B., **1,** 80, 562, 1292; **3,** 131, 171
Chapman, O. L., **1,** 66, 695, 841; **2,** 59, 332; **4,** 370; **7,** 89, 428
Chapman, P. H., **6,** 234
Chapman, T. M., **2,** 80; **5,** 282
Chapovskaya, N. K., **4,** 441
Chappell, R. L., **1,** 633; **5,** 747
Chappelle, E. W., **1,** 80
Chapuis, C., **11,** 165, 414; **12,** 104, 251
Chapuis, J., **3,** 125
Charalambous, J., **7,** 118
Charbier, P., **1,** 281
Chard, S. J., **1,** 427, 828
Charles, G., **4,** 239; **8,** 232; **10,** 459; **12,** 570
Charles, H. C., **8,** 102
Charles, R. G., **1,** 11, 500; **3,** 123
Charleson, D. A., **11,** 124
Charlson, A. J., **1,** 562
Charlton, F. B., **4,** 241
Charney, W., **1,** 1000
Charpentier, R., **10,** 203, 299
Charpiot, B., **10,** 393; **11,** 515
Charrier, C., **9,** 207
Charrier, G., **1,** 833, 1123
Charubala, R., **4,** 407
Chary, P. K., **9,** 480
Chasar, D. W., **7,** 59
Chassin, C., **5,** 760
Chastrette, M., **11,** 600
Chatani, N., **12,** 167
Chatt, J., **1,** 652; **2,** 434; **3,** 83; **8,** 148; **12,** 111
Chattaway, F. D., **1,** 1144
Chatterjee, A., **5,** 217
Chatterjee, A. K., **11,** 185
Chatterjee, B. G., **1,** 518
Chatterjee, S., **1,** 230; **11,** 514; **12,** 411, 474

Chattopadhyaya, J. B., **5,** 347; **6,** 296; **9,** 110; **10,** 432; **11,** 419
Chatziiosifidis, I., **9,** 522; **11,** 150, 524; **12,** 150
Chau, L. V., **6,** 175, 511
Chaudhari, P. N., **9,** 307
Chaudhary, S. K., **9,** 180
Chaudhary, S. S., **1,** 1194; **2,** 425; **4,** 14
Chaudhuri, M. K., **11,** 453
Chauhan, V. S., **9,** 256
Chauncy, B., **3,** 267
Chauvet, F., **11,** 45
Chauvette, R. R., **8,** 360
Chavdarian, C. G., **6,** 205, 441; **7,** 64, 279
Chavez, L. G., Jr., **10,** 398
Chavis, C., **4,** 238; **6,** 583
Chawla, H. M., **7,** 268
Chawla, H. P. S., **12,** 272
Chayabunjonglerd, S., **9,** 438
Chaykovsky, M., **1,** 87, 106, 318, 612, 1246; **2,** 131, 173; **3,** 126, 142; **5,** 9, 591; **6,** 342; **7,** 141
Chebaane, K., **6,** 503
Cheer, C. J., **5,** 78
Cheikh, R. B., **12,** 223
Chelhot, N. C., **5,** 600
Chelucci, G., **11,** 125
Chemerda, J. M., **1,** 150, 293, 652, 676, 764, 1006, 1131; **2,** 377
Chen, A., **5,** 640
Chen, A. F. T., **4,** 383
Chen, B., **8,** 445; **12,** 267
Chen, C. H., **4,** 389, 565; **6,** 476
Chen, C. M., **7,** 70
Chen, E. Y., **4,** 459
Chen, F., **4,** 33, 539; **8,** 148; **9,** 310; **12,** 181
Chen, F. C. M., **7,** 242
Chen, H. Y., **5,** 127
Chen, J. P., **3,** 50
Chen, K. K. N., **3,** 9
Chen, K. L., **4,** 480
Chen, L. C., **9,** 111
Chen, M. M., **1,** 754
Chen, P., **7,** 220
Chen, R. H. K., **4,** 220; **5,** 244, 412, 748; **6,** 95
Chen, S., **10,** 335
Chen, S. C., **9,** 99
Chen, T. K., **5,** 617
Chen, T. S., **5,** 542
Chen, W. Y., **6,** 193
Chen, Y. S., **12,** 90
Chen, Y. Y., **11,** 562
Chenard, B. L., **12,** 414
Chenevert, R., **6,** 441; **11,** 145
Cheney, L. C., **1,** 183; **5,** 700
Cheng, C. C., **2,** 361
Cheng, C. W. F., **11,** 45
Cheng, K. F., **2,** 453
Cheng, L., **7,** 170; **8,** 74
Cheng, Y. M., **5,** 514
Cheng, Y. S., **10,** 335
Cheon, S. H., **12,** 531
Cherbas, P., **11,** 128
Cherbuliez, E., **1,** 905, 1269
Cheriyan, U. O., **4,** 99; **5,** 145
Cherkofsky, S. C., **4,** 21
Cherluck, R. M., **5,** 454
Chermprapai, A., **9,** 508
Chern, C. I., **6,** 490; **7,** 307; **8,** 112, 419
Chernick, C. L., **6,** 670
Chernyshev, V. O., **4,** 562
Chesick, J. P., **3,** 69
Cheung, H. C., **7,** 109, 414; **10,** 43
Cheung, J. J., **4,** 339
Cheung, L. D., **5,** 78
Cheung, N. H., **9,** 256
Cheung, Y. F., **9,** 71
Cheutin, A., **2,** 353; **5,** 60
Chew, L., **2,** 228
Chhabra, B. R., **11,** 96
Chianelli, D., **11,** 166, 473
Chiang, C. S., **7,** 365; **8,** 477, 508
Chiang, R. W., **2,** 117
Chiang, Y. H., **2,** 163; **7,** 135
Chiasson, B. A., **8,** 56
Chiba, K., **9,** 349
Chiba, M., **11,** 159
Chiba, T., **11,** 190
Chibata, I., **5,** 470
Chibber, S. S., **7,** 268
Chickashita, H., **12,** 384
Chiddix, M. E., **3,** 241
Chidgey, R., **6,** 544; **8,** 534
Chien, P. L., **1,** 44, 611
Chien, S. W., **2,** 192
Chihara, T., **10,** 9; **11,** 24, 466, 487
Chikami, E., **5,** 148
Childers, W. E., Jr., **11,** 481
Childs, J. D., **9,** 488
Childs, M. E., **7,** 79; **12,** 321
Childs, R. F., **5,** 311; **11,** 355
Chiles, H. M., **1,** 6
Chimiak, A., **1,** 83; **4,** 360; **10,** 277
Chinard, F. P., **1,** 510
Chinn, L. J., **1,** 926; **5,** 552
Chino, K., **11,** 550
Chino, N., **7,** 32
Chinoporos, E., **1,** 904
Chiou, B. L., **6,** 399; **7,** 379
Chip, G. K., **4,** 508; **5,** 659
Chiras, S. J., **1,** 619
Chirinko, J. M., **4,** 302
Chisholm, M., **1,** 1141
Chishti, N. H., **6,** 97
Chitrakorn, S., **12,** 419
Chittattu, G., **8,** 28, 522; **9,** 32, 325, 511; **10,** 452
Chitwood, J. L., **2,** 144
Chiu, S. K., **10,** 436
Chiusoli, G. P., **1,** 1102; **2,** 181; **5,** 474; **6,** 451; **11,** 514
Chivers, G. E., **4,** 372
Chladek, S., **1,** 388; **2,** 198
Chmielewski, M., **10,** 258
Cho, H., **4,** 489; **5,** 86; **6,** 316; **7,** 331; **10,** 407; **11,** 91, 567
Cho, I. H., **12,** 195
Chodkiewicz, W., **1,** 169, 481
Chodosh, D. F., **12,** 557
Chodowska-Palicka, J., **7,** 109
Chodroff, S., **1,** 500
Choi, J. K., **12,** 523
Choi, V. M. F., **12,** 378
Choi, Y. M., **9,** 501; **11,** 69, 293
Chokski, S. K., **4,** 520
Chollet, A., **11,** 559

Chong, A. O., **7,** 256; **8,** 206
Chong, B. P., **3,** 103
Chong, J. A., **4,** 542
Chong, J. M., **12,** 116, 506
Chono, M., **12,** 428
Chopard, P. A., **8,** 79; **11,** 70
Choppin, A. R., **1,** 85, 86
Chorev, M., **6,** 288; **7,** 79
Chorn, T. A., **10,** 81
Chortyk, O. T., **1,** 904, 1184
Chou, C. S., **12,** 480
Chou, K. J., **8,** 188
Chou, S. K., **5,** 765; **7,** 155
Chou, T. S., **4,** 357, 397; **12,** 111
Choudhry, S., **9,** 73
Choudhury, S. R. R., **5,** 614
Chow, D., **8,** 203
Chow, F., **6,** 549; **8,** 24, 25; **9,** 283, 365, 430
Chow, M. F., **8,** 519
Chow, T., **8,** 316
Chow, W. Y., **4,** 405; **5,** 551, 587; **6,** 479
Chow, Y. L., **1,** 320, 394; **9,** 421; **11,** 198
Chowdhary, M. S., **8,** 405
Chowdhury, P. K., **12,** 305
Choy, T., **12,** 258
Choy, W., **10,** 110, 119; **12,** 94, 249, 277
Chretien, F., **9,** 297
Chretien-Bessiere, Y., **1,** 207
Chrisope, D. R., **11,** 77
Christ, W. J., **12,** 359
Christen, K., **9,** 523
Christensen, A. T., **12,** 41
Christensen, B. G., **1,** 293; **4,** 60; **5,** 21, 585; **9,** 162, 408; **10,** 100, 342, 349
Christensen, J. E., **1,** 1055
Christensen, J. J., **6,** 137
Christensen, L. W., **5,** 478
Christensen, N. H., **1,** 1125
Christiaens, L., **12,** 451
Christian, J., **1,** 977
Christian, P. A., **6,** 670
Christian, S. D., **9,** 488
Christie, B., **9,** 304
Christie, B. J., **1,** 731
Christie, C. F., Jr., **4,** 203
Christie, J. J., **6,** 146
Christie, M. A., **6,** 567; **7,** 335
Christie, S. M. H., **1,** 199
Christl, M., **7,** 207
Christmann, K. F., **1,** 927
Christmann, O., **4,** 9
Christol, H., **6,** 665; **10,** 81
Christova, N., **5,** 129
Christy, K. J., **6,** 79
Christy, M. R., **12,** 412
Chruma, J. L., **4,** 85
Chu, C. C., **4,** 282
Chu, C. Y., **6,** 225
Chu, G. N., **12,** 350
Chu, J. Y. C., **5,** 639; **8,** 459
Chu, K. Y., **7,** 268
Chu, P. S., **8,** 284; **9,** 151, 438
Chu, S. H., **6,** 288
Chu, S. H. L., **10,** 427
Chu, V., **5,** 49
Chuang, C. P., **12,** 523
Chubb, F., **4,** 163
Chuche, J., **7,** 3
Chucholowski, A., **9,** 288, 506; **12,** 185, 278
Chudd, C. C., **1,** 785, 1284
Chuit, C., **2,** 110; **5,** 94; **6,** 270, 664; **7,** 83; **8,** 505; **10,** 289, 290, 326; **11,** 117
Chujo, Y., **6,** 130, 436; **7,** 74
Chum, P. W., **7,** 113, 373
Chung, B. C., **1,** 1249; **12,** 412
Chung, B. Y., **11,** 373
Chung, D., **1,** 385
Chung, S. K., **6,** 436; **11,** 479, 486
Chung, Y. W., **11,** 296
Chung-Tungshun, **1,** 444
Chupp, J. P., **2,** 415
Church, R. F., **1,** 144, 166
Church, R. F. R., **2,** 219
Chvalovsky, V., **3,** 261; **4,** 283
Chwang, T. L., **5,** 86; **10,** 125
Chys, J., **8,** 15
Ciabattoni, J., **2,** 279; **4,** 405; **5,** 127; **6,** 114, 479
Ciampolini, M., **5,** 42
Ciattini, P. G., **9,** 334
Cicala, G., **11,** 442
Cichowicz, M. B., **12,** 428
Ciecíuch, R. F. W., **5,** 581
Ciganek, E., **1,** 472, 563, 595, 892; **2,** 23, 125, 368; **4,** 547; **5,** 245
Cima, F. D., **4,** 417
Cimarusti, C. M., **2,** 238; **9,** 341
Ciminale, F., **4,** 477
Cinnamon, J. M., **1,** 247
Cinnamon, M., **2,** 163
Cinquini, M., **3,** 165; **5,** 554; **6,** 191; **7,** 79; **8,** 395; **12,** 298, 509
Ciochetto, L. J., **8,** 299
Ciotti, C. J., Jr., **1,** 1184
Cipullo, M. J., **11,** 282
Cisneros, A., **11,** 452
Cistone, F., **6,** 534; **12,** 288
Citron, J. D., **3,** 304
Citterio, A., **11,** 440, 529; **12,** 502
Ciuffreda, P., **11,** 288
Ciufolini, M. A., **11,** 4
Clader, J. W., **10,** 155
Claesson, A., **5,** 86, 389; **6,** 148, 326
Claeys, M., **5,** 389
Clagett, D. C., **1,** 1022
Claisen, L., **1,** 154, 383, 393, 842, 1073, 1101, 1209, 1210; **5,** 64
Clapp, M. A., **1,** 1090
Clapp, R. C., **1,** 563
Clar, E., **1,** 127; **5,** 422; **7,** 295
Clardy, J., **4,** 520; **7,** 268; **9,** 119, 194, 480; **11,** 315, 327
Clardy, J. C., **4,** 370
Claremon, D. A., **9,** 32, 367; **10,** 313
Claret, P. A., **2,** 412
Claringbold, P. J., **2,** 182
Clark, A. C., **5,** 86
Clark, B. C., Jr., **4,** 199; **5,** 631
Clark, C. T., **1,** 476; **5,** 308
Clark, D. A., **9,** 385, 466; **10,** 419
Clark, D. R., **4,** 465
Clark, D. S., **9,** 325
Clark, F. R. S., **6,** 443
Clark, G., **6,** 28; **8,** 22; **9,** 119;

12, 11
Clark, G. M., **4,** 161
Clark, H. C., **1,** 1237
Clark, J., **1,** 285; **2,** 290
Clark, J. C., **1,** 44, 723
Clark, J. H., **7,** 239; **8,** 264, 411; **9,** 447, 448; **11,** 6, 441
Clark, L. C., Jr., **2,** 305
Clark, M., **3,** 149; **9,** 245
Clark, M. C., **11,** 60
Clark, N. G., **10,** 324
Clark, P. D., **7,** 287
Clark, P. S., **10,** 170
Clark, P. W., **8,** 302
Clark, R. D., **1,** 48, 295, 742, 754, 1184; **3,** 163; **5,** 482, 508, 761; **6,** 215, 339; **7,** 258, 271, 279
Clark, S. D., **4,** 162
Clark, S. L., **2,** 274
Clark, T. H., **1,** 637
Clark, V. M., **1,** 199; **12,** 492
Clarke, C. T., **8,** 73
Clarke, F. H., Jr., **1,** 926
Clarke, H. T., **1,** 47, 165, 393, 452, 481, 767, 876, 951, 963, 976, 1015, 1064, 1073, 1169, 1237; **4,** 239, 309
Clarke, J. A., **6,** 73
Clarke, J. E., **6,** 307
Clarke, K., **1,** 1292
Clarke, M. F., **1,** 952
Clarke, R. L., **2,** 189, 330; **3,** 308
Clarke, S., **5,** 91
Clarke, S. J., **10,** 48
Clarke, T. G., **2,** 369; **3,** 16
Clasen, H., **1,** 1030
Classon, B., **12,** 71
Claudy, P., **5,** 14
Claus, C. J., **1,** 595
Claus, P., **5,** 132
Claus, P. K., **6,** 118
Claus, R. E., **11,** 388
Clausen, K., **8,** 327; **9,** 50; **11,** 55
Clauson-Kaas, N., **1,** 636; **2,** 144
Clauss, K., **2,** 415; **3,** 53; **4,** 93; **5,** 499; **6,** 122
Clavel, J. M., **4,** 416
Clawson, L., **12,** 111
Clayton, D. W., **1,** 1126
Clayton, F. J., **10,** 443
Clayton, J. C., **1,** 372
Clayton, R. B., **1,** 19; **5,** 351
Cleland, G. H., **2,** 91; **4,** 18
Cleland, J. H., **5,** 622
Clemans, G. B., **6,** 500
Clemens, D. H., **1,** 1073
Clemens, L. M., **1,** 1147
Clement, B. A., **7,** 97
Clement, H., **1,** 424
Clement, K. S., **12,** 70
Clement, R. A., **1,** 561, 850; **2,** 332
Clement, W. H., **3,** 123
Clemmensen, E., **1,** 1289
Clemo, G. R., **1,** 976
Cleophax, J., **5,** 595, 642
Clerici, A., **6,** 21; **12,** 492
Clerin, D., **7,** 285
Cleve, G., **6,** 443
Cleveland, E. A., **1,** 156
Clifford, A. A., **1,** 159
Clifford, A. F., **6,** 642
Clift, S. M., **12,** 578
Cline, L. J., **5,** 510
Clinet, J. C., **8,** 94, 190, 322; **11,** 14
Clinton, R. O., **1,** 72, 382, 678
Clive, D. L. J., **3,** 51; **4,** 43, 556; **5,** 522, 523; **8,** 28, 170, 439, 455, 522, 523; **9,** 32, 35, 325, 511; **10,** 21, 98, 312, 362, 452; **11,** 373, 485; **12,** 41, 556
Clizbe, L. A., **6,** 605; **7,** 147; **9,** 76
Cloke, J. B., **1,** 1131
Close, V., **2,** 228
Close, W. J., **1,** 172
Closs, G. L., **1,** 81, 96, 130, 604, 755, 1041, 1116, 1187; **5,** 454; **6,** 620
Closs, L. E., **1,** 81, 96, 1187; **5,** 454
Closson, R. D., **1,** 889
Closson, W. D., **1,** 712; **2,** 289, 368; **4,** 350; **5,** 614
Cloudsdale, I. S., **11,** 182
Clough, J. M., **12,** 214
Clough, R. L., **7,** 250
Coan, S. B., **1,** 188, 1073
Coates, G. E., **1,** 397; **6,** 362
Coates, I. H., **5,** 758
Coates, R. M., **3,** 50, 219; **4,** 65, 84, 380; **5,** 244, 528; **6,** 40, 238, 400; **9,** 158, 172, 426, 438; **12,** 95
Coates, W. M., **3,** 239
Coats, E. A., **8,** 440
Cobb, R. L., **11,** 150
Coblens, K. E., **11,** 551
Coburn, C. E., **11,** 385
Coburn, E. R., **1,** 276, 424, 452, 890
Coburn, R. A., **5,** 513
Cochand, C., **1,** 796
Cochy, R. E., **4,** 251
Cocker, J. D., **1,** 561, 1054
Cocker, W., **1,** 1121; **4,** 55
Cockerill, G. S., **12,** 502
Coderre, R. A., **1,** 956
Codington, J. F., **1,** 1045
Coe, D. G., **1,** 1249
Coffee, E. C. J., **6,** 284
Coffen, D. L., **2,** 434
Coffey, C. E., **1,** 520
Coffey, R. S., **2,** 434
Coffey, S., **1,** 34
Coffman, D. D., **1,** 12, 117, 1040, 1124, 1125; **2,** 382; **7,** 391
Coffman, K. J., **11,** 491
Coggins, J. R., **2,** 383; **5,** 615
Coghlan, M. J., **7,** 309
Cogins, K. A. Z., **11,** 253
Cogolli, P., **6,** 631; **11,** 203
Cohen, A., **2,** 130, 252; **4,** 411
Cohen, C. F., **2,** 368
Cohen, D., **1,** 360; **2,** 190
Cohen, F. L., **1,** 892
Cohen, G. M., **3,** 112
Cohen, G. S., **5,** 144
Cohen, H., **1,** 1210; **5,** 620
Cohen, H. M., **3,** 175
Cohen, J. S., **7,** 294
Cohen, M. L., **7,** 287; **9,** 35; **10,** 170
Cohen, N., **8,** 15, 192; **9,** 420;

10, 155
Cohen, S., **4,** 466
Cohen, S. G., **1,** 833; **4,** 466; **5,** 430
Cohen, T., **1,** 49, 310, 966; **2,** 94, 353; **3,** 65; **4,** 103, 480; **5,** 163, 282, 567, 587; **6,** 133, 359, 606; **7,** 76, 410; **8,** 125, 126, 306; **9,** 16, 262, 285, 312, 512; **10,** 98, 110, 245, 256, 383; **11,** 182, 220, 329, 340, 422; **12,** 61, 280
Cohen, Z., **6,** 441; **7,** 273; **8,** 161; **9,** 156; **10,** 296
Cohn, W. E., **1,** 982
Cohnen, E., **1,** 814
Coke, J. L., **8,** 344
Coker, W. P., **1,** 982
Colapret, J. A., **11,** 179
Colb, J., **11,** 124
Colburn, C. B., **1,** 254, 1140
Colby, T. H., **1,** 89
Cole, C. A., **6,** 17; **10,** 346
Cole, L. L., **6,** 279
Cole, R. A., **5,** 633
Cole, S. M., **8,** 189
Cole, T. E., **8,** 476
Cole, W., **1,** 150, 251, 500
Cole, W. J., **2,** 87; **5,** 160
Coleman, B. D., **1,** 1145
Coleman, D. R., **11,** 393
Coleman, G. H., **1,** 34, 47, 124, 169, 177, 274, 379, 382, 402, 654, 747, 754, 841, 1005, 1163, 1209; **8,** 132; **10,** 209
Coleman, J. E., **1,** 952, 1269; **9,** 391
Coleman, R. A., **2,** 253; **3,** 28; **4,** 312; **9,** 58, 287
Coleman, R. S., **12,** 216
Coles, J. A., **1,** 642, 861
Colin, G., **2,** 353; **5,** 58
Colinese, D. L., **1,** 260
Coll, J., **8,** 440; **9,** 52, 427; **11,** 486
Coll, J. C., **5,** 349
Collard, J., **4,** 138
Collet, A., **5,** 560; **8,** 75
Collett, A. R., **1,** 379
Collette, J., **4,** 163
Collier, J. R., **5,** 658
Collier, W. L., **5,** 551
Collin, G. J., **9,** 61
Collin, J., **9,** 207
Collington, E. W., **3,** 183; **4,** 327, 541, 552; **5,** 49, 651, 700; **6,** 630
Collins, D., **1,** 85
Collins, D. J., **1,** 72, 500; **7,** 277
Collins, J. C., **2,** 75; **4,** 216; **5,** 542
Collins, J. C., Jr., **1,** 1184
Collins, J. F., **6,** 10
Collins, P. F., **1,** 573
Collins, P. M., **1,** 989; **2,** 316, 359
Collins, R. F., **1,** 351, 518
Collins, S., **9,** 32; **10,** 315; **11,** 408; **12,** 394
Collinton, E. W., **2,** 246
Collman, J. P., **1,** 1204; **4,** 465; **5,** 625; **6,** 552; **7,** 340; **9,** 207
Collonges, F., **7,** 149
Collum, D. B., **9,** 99, 251, 476; **11,** 536
Colombo, L., **9,** 475; **10,** 409; **11,** 315; **12,** 226, 298, 510
Colomer, E., **2,** 267; **6,** 179
Colon, C., **11,** 195
Colon, I., **3,** 160; **4,** 260; **6,** 244
Colonge, J., **1,** 905
Colonius, H., **1,** 573
Colonna, S., **3,** 165; **5,** 554; **6,** 586; **7,** 144; **8,** 264, 340, 395, 431; **11,** 539
Colson, A., **1,** 15
Colter, A. K., **1,** 1015, 1147
Colvin, E., **5,** 355, 375; **10,** 381
Colvin, E. W., **4,** 544; **5,** 98; **8,** 188; **9,** 172, 181, 446; **11,** 500
Colyer, R. A., **2,** 309
Combret, J. C., **7,** 170
Comeau, A., **4,** 37
Comer, W. T., **5,** 621
Comi, R., **5,** 713
Comin, J., **5,** 88
Comins, D. L., **8,** 341; **9,** 22; **10,** 266; **12,** 276, 284, 286
Comisarow, M. B., **1,** 397; **2,** 200
Commercon, A., **6,** 143, 270, 577; **8,** 176
Comninellis, C., **6,** 284
Compere, E. L., Jr., **8,** 92
Conacher, H. B. S., **2,** 36, 248
Conant, J. B., **1,** 150, 500, 1064; **3,** 160
Conant, R., **12,** 363
Conard, J., **4,** 256
Conbere, J. P., **1,** 1006
Concannon, P. W., **5,** 127, 198; **6,** 114
Concilio, C., **10,** 142
Condom, R., **11,** 454
Condon, F. E., **3,** 266; **4,** 7; **10,** 320
Conduche, A., **5,** 343
Cone, L. H., **9,** 110
Confalone, D. L., **10,** 309
Confalone, P. N., **6,** 254; **8,** 383; **10,** 309, 400; **12,** 387
Conia, J., **8,** 8
Conia, J. M., **1,** 374, 1096; **2,** 372, 386; **3,** 258; **4,** 44, 47, 251, 305, 437, 472; **5,** 589, 593, 713; **6,** 523; **7,** 35, 67, 164, 402; **8,** 238, 340, 374, 459; **10,** 91, 93, 265, 460; **11,** 121
Conigliaro, P. J., **2,** 180; **3,** 143
Conley, R. A., **5,** 379; **7,** 365; **8,** 477
Conley, R. T., **1,** 904
Conlin, R. T., **6,** 577
Conlon, L. E., **1,** 652; **3,** 302
Conn, R. S. E., **8,** 42; **9,** 13, 53
Connell, R., **11,** 536
Conner, R. M., **1,** 329, 1090
Connolly, R. E., **1,** 1193, 1251
Connor, D. S., **2,** 26; **3,** 178; **4,** 239
Connor, D. T., **5,** 756
Connor, R., **1,** 1167; **4,** 69

Connor, R. E., **12,** 167
Connor, W. F., **8,** 198
Conover, L. H., **3,** 261
Conover, W. W., **5,** 495
Conrad, F., **1,** 757
Conrad, J., **3,** 214
Conrad, M., **1,** 677, 842
Conrad, P. C., **8,** 396
Conrad, R. A., **4,** 405
Conrow, K., **1,** 178, 396, 870, 1261
Conrow, R. B., **4,** 67
Conrow, R. E., **10,** 134
Considine, W. J., **3,** 294; **5,** 188; **12,** 463
Consiglio, G., **5,** 331; **10,** 37
Constable, E. W., **1,** 402
Constantin, J. M., **1,** 662
Constantine, D. A., **1,** 767
Constantinides, D., **6,** 390
Contento, M., **5,** 86; **9,** 378
Conti, F., **5,** 190
Contillo, L. G., Jr., **11,** 126
Contreras, R., **11,** 219
Controt, P., **5,** 187
Convery, J. R., **2,** 239
Conway, W. P., **6,** 395
Cook, A. F., **2,** 163, 423
Cook, A. G., **5,** 713; **7,** 310
Cook, A. H., **4,** 352
Cook, C. D., **1,** 870, 1292; **9,** 217
Cook, C. E., **1,** 318; **2,** 173, 216
Cook, E. L., **9,** 237
Cook, F., **2,** 422
Cook, F. L., **5,** 155, 266; **6,** 137
Cook, H. G., **1,** 248, 322; **4,** 481
Cook, J., **4,** 172
Cook, J. M., **10,** 296; **11,** 194
Cook, J. W., **1,** 872; **2,** 305
Cook, M. A., **5,** 375
Cook, N. C., **1,** 1235
Cook, P. F. E., **1,** 66, 1226
Cook, P. L., **1,** 720
Cook, S. L., **9,** 180
Cook, W. J., **9,** 383
Cooke, D. W., **3,** 329
Cooke, F., **8,** 278; **9,** 495, 499; **10,** 444; **11,** 83, 129, 584; **12,** 118
Cooke, G. B., **1,** 3
Cooke, H. G., Jr., **4,** 77
Cooke, M. J., **8,** 456
Cooke, M. P., Jr., **3,** 268; **5,** 285, 375, 410; **6,** 349, 552; **8,** 217; **11,** 25, 346; **12,** 96
Cookson, C. M., **6,** 628
Cookson, P. G., **7,** 175; **11,** 67
Cookson, R. C., **1,** 44, 231, 247, 260, 723, 850; **2,** 125, 282, 325; **3,** 224; **4,** 14, 383; **5,** 13; **6,** 19; **7,** 18, 288; **8,** 535; **9,** 371, 438; **10,** 314, 355; **11,** 299
Cooley, G., **2,** 42, 227
Coombs, M. M., **4,** 501
Coombs, R. V., **1,** 219; **3,** 182; **5,** 389
Coombs, W., **12,** 530
Coon, C. L., **5,** 477; **6,** 420
Cooper, B. E., **7,** 28; **9,** 113
Cooper, D., **9,** 481
Cooper, D. J., **2,** 383
Cooper, F. C., **1,** 33
Cooper, G. D., **1,** 1171; **2,** 415; **10,** 314; **12,** 394
Cooper, G. F., **4,** 227
Cooper, G. K., **7,** 368
Cooper, J. A., **1,** 1163
Cooper, J. E., **3,** 128
Cooper, R. D. G., **4,** 60
Cooper, S. R., **1,** 1292
Cooper, T. A., **5,** 744
Cope, A. C., **1,** 39, 266, 309, 335, 412, 445, 472, 562, 595, 611, 658, 685, 864, 892, 904, 905, 978, 982, 1049, 1096, 1177, 1284; **2,** 154, 248, 273, 368, 397; **3,** 185, 200, 237; **4,** 31, 110; **10,** 9
Copeland, P. G., **1,** 782
Copenhaver, J. E., **1,** 1292
Copenhaver, J. W., **1,** 186
Coppens, G., **1,** 1137
Coppinger, G. M., **1,** 281, 409, 933
Coppola, B. P., **11,** 82
Coppolino, A. P., **5,** 169
Copson, A. J., **4,** 336
Corbella, A., **3,** 158
Corbett, D. F., **8,** 37, 427
Corbett, G. E., **2,** 296
Corbett, J. F., **2,** 382
Corbett, M. D., **5,** 249
Corbett, R. E., **1,** 214
Corbin, N. C., **9,** 380
Corbin, T. F., **1,** 1293
Corbin, V. L., **6,** 165
Corby, N. S., **1,** 53
Corcoran, G. B., III, **4,** 541
Corcoran, R. J., **5,** 353; **6,** 300; **10,** 216
Cordes, E. H., **6,** 70
Cordes, R. E., **11,** 494
Cordova, R., **11,** 12; **12,** 71
Corell, M., **1,** 682
Corett, R., **1,** 518
Corey, E. J., **1,** 20, 42, 81, 82, 258, 318, 465, 527, 562, 608, 612, 615, 703, 723, 989, 999, 1022, 1122, 1184, 1234, 1246, 1261; **2,** 40, 99, 101, 106, 117, 126, 131, 150, 153, 155, 173, 181, 187, 241, 263, 270, 281, 292, 293, 322, 324, 369, 378, 441; **3,** 12, 13, 54, 60, 65, 67, 71, 79, 81, 88, 95, 97, 105, 112, 113, 118, 126, 136, 142, 155, 174, 210, 212, 214, 276, 313, 330, 338; **4,** 39, 53, 63, 67, 77, 90, 105, 112, 151, 154, 165, 177, 183, 192, 196, 200, 218, 220, 233, 295, 310, 327, 355, 376, 380, 473, 478, 479, 497, 551; **5,** 5, 9, 18, 36, 44, 55, 132, 142, 188, 244, 286, 288, 293, 339, 381, 394, 412, 415, 441, 445, 448, 454, 473, 476, 527, 528, 530, 552, 631, 639, 660, 706, 739, 748, 749, 761; **6,** 11, 50, 56, 76, 81, 95, 110, 133, 215, 233, 247, 248, 267, 292, 302, 319, 322, 339, 352,

367, 380, 383, 419, 490, 493, 499, 603, 604; **7,** 26, 91, 130, 142, 160, 229, 309, 339, 374; **8,** 87, 180, 216, 259, 276, 316, 319, 325, 337, 427, 489, 511; **9,** 16, 66, 68, 71, 136, 184, 193, 196, 220, 242, 243, 247, 251, 269, 293, 296, 301, 318, 399, 408, 420, 430, 432, 466, 467, 468, 493; **10,** 115, 117, 142, 144, 169, 172, 199, 209, 228, 278, 336, 393, 407, 419; **11,** 58, 64, 91, 146, 189, 207, 209, 335, 339, 373, 414, 423, 469, 566, 567; **12,** 52, 57, 85, 97, 139, 182, 247, 285, 294, 324, 350, 351, 366, 525, 569, 577
Corey, E. R., **7,** 380
Corigliano, M. A., **2,** 228
Cork, D. G., **11,** 441
Corley, R. C., **1,** 318; **2,** 173, 216
Corn, J. E., **5,** 337
Cornelia, R. H., **2,** 328
Cornelis, A., **5,** 649; **11,** 237; **12,** 231
Cornelius, J. E., **4,** 351
Cornell, D., **1,** 364
Cornforth, D. A., **3,** 336
Cornforth, J. W., **1,** 147, 703, 777, 814, 882, 937, 951, 1049, 1116, 1284; **3,** 15, 41, 160; **9,** 123; **11,** 110
Cornforth, R. H., **1,** 147, 814, 882, 1116, 1284; **11,** 110
Cornia, M., **8,** 245
Cornwell, D. G., **2,** 252
Corral, R. A., **1,** 74, 259; **2,** 108; **3,** 4
Corre, E., **5,** 268
Correia, J. S., **1,** 309
Corrie, J. F. T., **9,** 449
Corrigan, J. R., **3,** 239
Corriol, C., **12,** 420
Corriu, R., **6,** 179
Corriu, R. J. P., **4,** 562; **6,** 179; **10,** 84; **11,** 117, 554; **12,** 109
Corsano, S., **9,** 408
Corson, B. B., **1,** 537, 629, 841, 870, 1049, 1270; **4,** 317
Corson, F. P., **5,** 560
Cortes, D. A., **12,** 480
Cortes, G., **1,** 1158
Cortese, F., **1,** 407, 452
Cortese, N. A., **7,** 5; **8,** 382; **9,** 351, 482; **11,** 170
Cortez, C., **6,** 230, 326; **9,** 102
Cortez, H. V., **1,** 927
Cory, R. M., **7,** 235; **10,** 198, 225, 316
Cosgrove, R. E., **3,** 309
Cosmatos, A., **1,** 346
Cossement, E., **5,** 99
Cossy, J., **11,** 514; **12,** 61
Costa, A., **11,** 466
Costa, L., **7,** 164
Costisella, B., **3,** 278
Costopanagiotis, A., **1,** 445, 1220
Cota, D. J., **10,** 14
Cotter, R. J., **1,** 335, 367, 445, 658, 1096
Cotterrell, G., **6,** 165
Cottier, L., **9,** 361
Cottle, D. L., **1,** 1295
Cotton, F. A., **9,** 148; **10,** 274
Cotton, W. D., **5,** 188, 220, 756
Cottrell, D. A., **12,** 121
Coudert, G., **4,** 440; **5,** 593; **7,** 327; **10,** 194, 368
Couffignal, R., **11,** 465
Coughlin, D. J., **9,** 239; **10,** 110
Coulson, D. R., **1,** 670; **5,** 503; **6,** 573
Coulter, A. W., **1,** 125, 1284
Coulter, J. M., **3,** 10
Coulton, S., **8,** 79
Counsell, R. E., **1,** 986
Counts, K. M., **6,** 360
Courduvelis, C., **2,** 101
Court, A. S., **4,** 285; **9,** 309
Courtin, A., **2,** 50, 422
Courtneidge, J. L., **11,** 551
Courtney, T., **4,** 14, 126
Cousineau, T. J., **9,** 180
Cousseau, J., **9,** 293
Coustard, J. M., **6,** 273
Coutrot, P., **6,** 189, 191, 354; **7,** 104; **8,** 168
Coutsogeorgopoulos, C., **1,** 519
Coutts, I. G. C., **11,** 450
Couturier, D., **11,** 132; **12,** 36
Couvillion, J. L., **2,** 247, 344
Covert, L. W., **1,** 730
Covey, R. A., **1,** 1190
Cowan, D., **1,** 375, 424
Cowan, J. C., **4,** 77
Cowell, A., **10,** 133
Cowell, D. B., **5,** 646
Cowell, G. W., **3,** 74
Cowie, J. S., **3,** 99; **5,** 220
Cowles, C., **4,** 375
Cowles, E. J., **1,** 1148
Cowper, R. M., **1,** 34
Cox, A., **1,** 320, 394
Cox, D., **5,** 19
Cox, D. P., **12,** 462
Cox, E. F., **8,** 396
Cox, G. J., **1,** 1074
Cox, G. R., **6,** 165
Cox, J. L., **9,** 387
Cox, J. M., **11,** 536
Cox, J. R., Jr., **1,** 328
Cox, M. E., **2,** 27
Cox, R. F. B., **1,** 5, 859
Cox, R. H., **10,** 156
Cox, S. D., **10,** 116
Coxon, D. T., **10,** 294
Coxon, J. M., **2,** 173, 267, 362; **8,** 440; **12,** 425
Coyle, C., **5,** 66; **7,** 40
Coyle, J. J., **1,** 130
Coyle, S., **7,** 139
Cozzi, F., **8,** 395; **12,** 298, 509
Crabbe, P., **1,** 127; **2,** 153; **3,** 106; **4,** 150, 228, 390, 415, 577; **5,** 216, 244, 537, 605, 762; **6,** 215, 536; **7,** 7, 86, 122, 243; **8,** 452; **9,** 226, 436; **10,** 186
Crabtree, R. H., **8,** 137; **10,** 116; **12,** 152, 550
Cragg, J. E., **10,** 347
Cragg, R. H., **4,** 276

Cragoe, E. J., Jr., **1,** 1131; **3,** 229
Craig, A. P., **5,** 276
Craig, C. A., **1,** 61
Craig, D., **1,** 34, 169, 182, 382
Craig, J. C., **1,** 573, 1169; **3,** 50; **8,** 432
Craig, L. C., **6,** 2
Craig, T. A., **11,** 3
Craig, W. G., **10,** 172
Crain, D. L., **4,** 529
Cram, D. J., **1,** 195, 309, 320, 567, 782, 926, 927, 1004, 1027, 1145; **2,** 200, 318, 356, 393; **4,** 174, 281, 535; **5,** 103, 155, 329; **6,** 137; **8,** 509; **9,** 361; **11,** 145; **12,** 312, 383
Cram, J. M., **6,** 137
Cramer, F., **1,** 281, 848, 854, 885, 1195; **4,** 516; **6,** 152; **11,** 44
Cramer, R., **5,** 204
Cramer, R. D., **1,** 228, 892
Cramp, M. C., **10,** 355
Crandall, D. C., **5,** 491
Crandall, J., **1,** 144, 755
Crandall, J. K., **1,** 611; **2,** 203, 248, 309; **3,** 50; **4,** 298; **5,** 86, 495; **6,** 78, 165; **7,** 149; **8,** 440; **9,** 117; **11,** 309
Crane, K. F., **1,** 331
Cranford, G., **2,** 207
Crank, G., **7,** 386; **9,** 393
Crass, G., **6,** 578; **9,** 48; **10,** 395; **11,** 326; **12,** 193, 480
Crastes de Paulet, A., **1,** 1284
Cravador, A., **9,** 34; **10,** 454
Craven, B. M., **12,** 11
Craw, P. A., **11,** 486
Crawford, H. T., **6,** 91
Crawford, M., **1,** 782; **2,** 303; **8,** 383
Crawford, R., **4,** 163
Crawford, R. J., **4,** 489; **12,** 465
Crawford, T. C., **10,** 12; **11,** 126
Crawford, T. H., **2,** 61
Crawley, L., **5,** 394
Crawley, L. C., **4,** 298; **8,** 440
Crawshaw, A., **1,** 19
Creary, X., **6,** 192; **8,** 299; **10,** 308
Creaser, E. H., **5,** 596
Creasey, S. E., **5,** 430
Creasy, W. S., **4,** 405
Cree, G. M., **2,** 274; **4,** 201
Creese, M. W., **6,** 646
Creger, P. L., **3,** 185; **7,** 114
Cregge, R. J., **5,** 407, 445, 456; **9,** 423
Creighton, A. M., **1,** 219
Cremer, S. E., **1,** 164, 1081
Cremlyn, R. J. W., **1,** 1141; **4,** 390
Cresp, T. M., **4,** 508; **5,** 269, 672; **6,** 138; **7,** 119; **8,** 39; **9,** 69
Cress, J. P., **4,** 230
Cressman, H. W. J., **1,** 176, 500
Cretcher, L. H., **1,** 1090
Crews, P., **4,** 429
Cricchio, R., **2,** 117; **3,** 87
Crich, D., **12,** 417
Criegee, R., **1,** 185, 465, 509, 561, 562, 563, 764, 956, 1136; **3,** 178; **6,** 441, 548; **11,** 32; **12,** 245, 366
Crimmin, W. R. C., **1,** 518
Crimmins, M. T., **12,** 350
Crimmins, T. F., **7,** 359
Criner, G. X., **1,** 561
Cripps, H. N., **1,** 18
Crison, C., **6,** 498
Crisp, G. T., **12,** 474
Crissman, H. R., **4,** 191
Cristau, H. J., **4,** 68; **6,** 665; **10,** 81
Cristea, I., **6,** 133; **10,** 110
Cristensen, J. E., **2,** 434
Cristol, S. J., **1,** 152, 658; **2,** 226; **3,** 18
Criswell, T. R., **5,** 600
Crivello, J. V., **12,** 531
Crochet, R. A., **5,** 103, 691
Crociani, B., **6,** 450; **12,** 502
Crocker, R. E., **1,** 1095
Croft, R. C., **4,** 97
Crofts, P. C., **1,** 60; **2,** 322, 354; **5,** 512; **7,** 59
Crombie, L., **1,** 195, 642; **3,** 191; **5,** 257; **8,** 238, 337; **9,** 161, 396
Crombie, W. M. L., **9,** 161
Cromwell, N. H., **1,** 1011, 1027, 1137; **2,** 368; **3,** 330
Cronheim, G., **7,** 118
Cronin, C. J., **1,** 121, 935
Crooks, P. A., **5,** 257
Crooy, P., **5,** 186
Crosby, D. G., **1,** 791, 1167
Crosby, G. A., **6,** 92, 582; **9,** 207
Cross, A. D., **1,** 127, 219, 249, 267, 802; **2,** 34, 362, 372, 381; **3,** 265
Cross, B., **1,** 101
Cross, B. E., **1,** 1121; **2,** 301; **5,** 415; **6,** 186
Cross, G. W., **1,** 1058
Cross, L. C., **1,** 1157
Crossland, R. K., **4,** 327
Crossley, F., **1,** 692
Crossley, J., **1,** 642
Crossley, M. J., **9,** 264, 391
Crossley, N. S., **1,** 581
Croteau, R., **9,** 391
Crouch, R. K., **6,** 267
Croudace, M. C., **12,** 167
Crouse, D., **2,** 187
Crouse, D. J., **9,** 462; **10,** 396
Crouse, D. N., **6,** 81, 84, 636
Crouse, G. D., **12,** 390
Crouse, O. M., **4,** 573
Crout, D. H. G., **8,** 365
Crovetti, A. J., **1,** 465, 870
Crow, W. D., **1,** 41
Crowe, A. J., **7,** 27
Crowe, D. F., **2,** 344, 423; **3,** 262; **5,** 394, 552
Crowley, B. R., **1,** 212
Crowley, G. P., **1,** 383
Crowley, P. J., **1,** 85, 210
Crowshaw, K., **2,** 344
Crowther, G. P., **3,** 183; **4,** 297
Croxall, W. J., **2,** 231
Crozet, M. P., **9,** 122
Crozier, R. F., **7,** 241
Cruickshank, P. A., **1,** 274,

371, 956, 966, 1184; **2,** 196
Crumbie, R. L., **12,** 233
Crumbine, D. S., **6,** 676
Crumbliss, A. L., **1,** 1141
Crumrine, A. L., **7,** 83
Crumrine, D. S., **5,** 764
Crunden, E. W., **4,** 481
Cruz, A., **5,** 605; **7,** 7
Cruz de Maldonado, V., **12,** 322
Cuadriello, D., **1,** 249, 802; **2,** 381; **3,** 265
Cubbon, R. C. P., **2,** 344
Cue, B. W., Jr., **5,** 261; **8,** 105
Cuellar, L., **1,** 219
Cuellar Ibanez, L., **1,** 562, 624
Cueto, O., **8,** 367
Cuffe, J., **8,** 159; **10,** 342
Cuiffreda, P., **12,** 479
Culbertson, T. P., **1,** 772
Culhane, P. J., **1,** 652
Cullen, E., **1,** 730, 1041
Cullen, W. R., **8,** 333
Cullinane, N. M., **1,** 427, 828, 1116, 1171
Cullis, C. F., **1,** 951
Culvenor, C. C. J., **10,** 330
Cumas, P., **4,** 97
Cumbo, C. C., **1,** 139; **2,** 106
Cummins, C. H., **12,** 95
Cundasawmy, N. E., **5,** 334
Cunha, H., **8,** 407
Cunico, R. F., **5,** 144, 182, 376, 556; **7,** 358; **8,** 411; **10,** 443; **11,** 347, 582
Cunningham, J., **1,** 309
Cunningham, K. G., **1,** 885
Cunningham, R., **2,** 263
Cunningham, W. C., **4,** 389
Cuny, G., **4,** 120
Cuomo, J., **8,** 308; **10,** 29
Cupas, C., **2,** 17; **3,** 6, 7
Cupas, C. A., **1,** 397; **3,** 234; **8,** 361
Curci, R., **6,** 114, 487; **11,** 442
Cure, J., **3,** 336
Curini, M., **11,** 156
Curkovic, L., **8,** 87
Curphey, T. J., **2,** 353, 438; **3,** 113; **4,** 53; **5,** 716; **10,** 405
Curragh, E. F., **2,** 263
Curran, D. P., **8,** 481; **9,** 480; **10,** 328, 391; **11,** 31, 260, 458, 579; **12,** 422
Current, S., **9,** 32
Current, S. P., **8,** 534
Currie, J. K., **8,** 150; **9,** 145, 374
Currie, R. B., **11,** 249
Curry, D. C., **3,** 101
Curry, M. J., **5,** 119
Curtin, D. Y., **1,** 39, 41, 97, 349, 465, 677; **2,** 93; **6,** 517
Curtin, J. H., **1,** 518
Curtis, A. B., **6,** 490
Curtis, E. A., **5,** 454
Curtis, G. G., **1,** 125
Curtis, N. J., **9,** 325; **10,** 452
Curtis, O. E., Jr., **1,** 1095
Curtis, R. F., **2,** 295
Curtis, W. D., **6,** 137
Curtius, T., **1,** 1043, 1179; **2,** 195, 454
Cusack, N. J., **4,** 535; **5,** 707; **6,** 605; **7,** 392; **10,** 255
Cushing, I. B., **1,** 140
Cushing, M. L., **1,** 324
Cushmac, G. E., **5,** 140
Cushman, M., **5,** 254
Cussans, N. J., **8,** 32
Cutcliffe, A. B., **4,** 114
Cuthbertson, E., **6,** 550
Cutler, A., **8,** 455
Cutler, F. A., Jr., **1,** 150, 1006
Cutrufello, P. R., **4,** 503
Cutter, H. B., **1,** 150
Cutting, J. D., **5,** 77
Cuvigny, T., **1,** 431; **3,** 153; **4,** 247; **5,** 325, 543; **6,** 279, 323, 333, 352; **7,** 204; **8,** 299; **9,** 209, 217
Cvetkovic, M., **10,** 229
Cwirla, W. M., **8,** 69
Cymerman-Craig, J., **1,** 952, 1121
Cynkowski, T., **4,** 507; **9,** 123
Cyr, C. R., **5,** 552
Czaja, R. F., **6,** 19
Czarniecki, M. F., **9,** 129
Czarny, M., **9,** 119
Czarny, M. R., **7,** 139
Czebotar, V. J., **1,** 520
Czernecki, S., **6,** 233, 280; **9,** 296
Cziesla, M. J., **5,** 649
Czuba, L. J., **3,** 312
Czyewski, J., **7,** 20
Czyzewski, A., **1,** 1101

Dabard, R., **6,** 126; **8,** 110
Dabbagh, G., **11,** 373, 566
Dabby, R. E., **4,** 183
Dabdoub, A. M., **8,** 409
Dabescat, F., **12,** 86
Dabral, V., **6,** 284
Da Costa, R., **9,** 396
D'Adamo, A. F., Jr., **1,** 754
Daddona, P. E., **7,** 390
Daeniker, H. V., **1,** 929
Daggett, J. U., **12,** 14
Dagli, D. J., **6,** 252, 542
Dahill, R. T., Jr., **1,** 250, 1217
Dahl, H., **3,** 45
Dahl, O., **11,** 147
Dahle, N. A., **1,** 827
Dahlgren, G., **4,** 393
Dahm, K. H., **2,** 442
Dahman, F. J. M., **9,** 523
Dahmen, K., **4,** 264
Dahmen, W., **12,** 205
Dahn, H., **6,** 229
Daigle, D., **6,** 341
Daignault, R. A., **1,** 598; **2,** 243
Dailey, O. D., Jr., **10,** 168
Dainobu, Y., **11,** 552
Dains, F. B., **1,** 51, 340, 500, 505, 845
Dakin, H. D., **1,** 413, 470, 963, 977; **2,** 217; **3,** 119
Dalavoy, V. S., **6,** 475
Dale, J., **2,** 182, 293
Dale, J. A., **4,** 265; **5,** 353; **11,** 330; **12,** 282
Daley, J. W., **11,** 292
Daley, R. F., **5,** 735
Dali, B. B., **8,** 305
Dallacker, F., **1,** 882
Dallas, M. S. J., **1,** 1011
Dall'Asta, D., **9,** 257

Dalling, J., **9,** 441
Dalton, D. R., **2,** 42, 159; **3,** 36; **5,** 98, 343; **7,** 395
Dalton, J. C., **6,** 385
Dalton, J. R., **9,** 423
Daly, J., **7,** 298
Daly, J. W., **2,** 107
Dalzell, H. C., **9,** 274
D'Ambra, T. E., **11,** 218
Damiano, J., **5,** 109
Damiano, J. C., **7,** 243; **10,** 5
D'Amico, J. J., **6,** 187
Damin, B., **10,** 85; **11,** 118
Damle, S. B., **1,** 573
Damodaran, K. M., **7,** 395
Damodaran, N. P., **1,** 1011
Damon, R. E., **6,** 253; **7,** 280, 412, 421
Damon, T. P., **5,** 669
D'Amore, M. B., **4,** 130
Dampawan, P., **12,** 531
Damps, K., **6,** 9
Damrauer, R., **1,** 854; **2,** 328
Dams, R., **11,** 529
Damschroder, R. E., **1,** 837, 1101
Dance, J., **1,** 231; **2,** 125
Dane, E., **1,** 11
Danehy, J. P., **1,** 379
Danen, W. C., **6,** 419; **8,** 419
Daney, M., **5,** 167
Danforth, R. H., **3,** 289; **5,** 659
Dang, H. P., **8,** 475; **9,** 125, 458
Dang, T. P., **4,** 273; **5,** 361; **9,** 260
D'Angeli, F., **1,** 266, 310
d'Angelo, J., **6,** 267, 530; **7,** 50, 108; **8,** 222; **9,** 165
Dangieri, T. J., **3,** 123
Danheiser, R. L., **5,** 406, 412, 761; **7,** 24, 374; **9,** 87, 136, 193, 318, 432, 467; **10,** 89, 429, 442; **11,** 335, 471; **12,** 350
Danieli, B., **5,** 370; **7,** 56
Danieli, N., **1,** 614, 999, 1000; **2,** 361
Daniels, C. K., **6,** 146
Daniels, J., **6,** 64
Daniels, K., **11,** 99; **12,** 344
Daniels, P. J. L., **1,** 152; **4,** 363
Daniels, R., **1,** 185, 389, 477, 643
Daniels, R. G., **11,** 25
Daniels, W. E., **3,** 241
Daniewski, A. R., **6,** 10
Daniewski, W. M., **8,** 306; **9,** 285, 312; **12,** 61
Daniher, F. A., **2,** 392; **3,** 15
Daniil, D., **8,** 272
Danish, A. A., **2,** 206
Danishefsky, S., **1,** 643; **2,** 151, 263; **3,** 47; **5,** 315, 394; **6,** 43, 217, 230, 354, 364, 368, 372, 411, 606; **7,** 103, 114, 123, 132, 227, 233; **8,** 176, 179, 330, 397, 399; **9,** 75, 76, 79, 304; **10,** 161, 272, 403; **11,** 3, 20, 99, 317, 327, 334, 337; **12,** 303, 313, 391, 561
Danks, L. J., **3,** 147; **5,** 478
Danks, L. S., **7,** 158
Dann, A. E., **9,** 257
Dann, O., **1,** 942
Dannenberg, H., **1,** 127; **4,** 134
Dannley, R., **9,** 509
Dannley, R. L., **2,** 296; **7,** 252
d'Ans, J., **1,** 531
Dansette, P. M., **5,** 713; **6,** 477; **7,** 274
Dansted, E., **2,** 173, 362; **8,** 440
Dao, G. M., **11,** 165; **12,** 251
Dao, H. L., **6,** 229
Dao, L. H., **7,** 288; **8,** 503
Daotien, N., **11,** 568
Dappen, G. M., **1,** 689; **2,** 109, 119
Dappen, M. S., **12,** 86
Darby, N., **9,** 115
Darby, W. J., **1,** 163, 402
Dardis, R. E., **10,** 161
Dardoize, F., **7,** 116
Darling, S. D., **1,** 603; **2,** 435, 445; **3,** 309, 310; **4,** 406; **11,** 566
Darlington, W. H., **11,** 47
Darnall, K. R., **5,** 511
Darnbrough, G., **8,** 1
Darragh, K. V., **3,** 225
Darst, K. P., **10,** 327, 398, 445
Darzens, G., **1,** 1041, 1148, 1163; **3,** 9
Das, J., **9,** 296, 493; **10,** 55, 169, 373; **11,** 146, 469
Das, M. K., **11,** 209
D'Ascoli, R., **1,** 808
Das Gupta, A. K., **2,** 60, 303, 433; **3,** 40
Dasgupta, F., **8,** 228
Dasgupta, H. S., **11,** 453
Dasgupta, R., **3,** 65
Dasgupta, S. K., **3,** 65
DasGupta, T. K., **4,** 82
Dasler, W., **1,** 333
Dastur, K. P., **4,** 135; **5,** 195, 440, 591
Dathe, W., **3,** 123
Datta, A. P., **1,** 116
Datta, M. C., **6,** 45
Daub, G. H., **1,** 213, 881, 906, 926, 1080; **4,** 252; **8,** 299
Daub, J., **4,** 270
Daub, J. P., **12,** 282
Dauben, H. J., Jr., **1,** 80, 82, 178, 396, 730, 741, 1073, 1100, 1258; **4,** 568
Dauben, W. G., **1,** 257, 598, 754, 1022; **2,** 144, 236, 353, 422; **3,** 57, 251, 252; **4,** 235, 290, 304, 437; **5,** 9, 70, 391, 474, 574; **6,** 94, 310, 419, 620; **7,** 94, 161, 376; **8,** 76, 336, 412, 427; **10,** 180, 187; **11,** 194, 559
Daubert, R. G., **5,** 687
Daudon, M., **7,** 177
Daudt, W. H., **1,** 563
Dauernheim, L. W., **5,** 144
Daugherty, B. W., **12,** 329
Daughhetee, P. H., Jr., **1,** 310
Daulton, A. L., **1,** 757
Daum, H., **9,** 48; **11,** 326
Daum, S. J., **2,** 189; **3,** 308
Daum, U., **11,** 492
Daunis, J., **4,** 172

Dauphin, G., **5,** 589; **9,** 515
D'Auria, M., **8,** 427; **9,** 399; **10,** 335, 336, 366; **11,** 452; **12,** 256
Dauzonne, D., **11,** 360
Dave, K. G., **1,** 273
Dave, L. D., **12,** 202
Dave, V., **5,** 282; **6,** 284, 458; **7,** 284; **12,** 224
Daves, G. D., Jr., **1,** 1107
Davey, W., **1,** 5
Daviaud, G., **5,** 175
David, C. C., **2,** 165
David, L., **5,** 589
David, S., **5,** 188; **7,** 392; **9,** 141; **11,** 501; **12,** 548
Davidovich, Y. A., **6,** 634
Davidsen, S. K., **11,** 179; **12,** 186
Davidson, A. H., **6,** 341
Davidson, D., **1,** 401, 527; **8,** 314
Davidson, E. B., **1,** 1189
Davidson, J. G., **10,** 274
Davidson, J. M., **8,** 47
Davidson, L. H., **1,** 34
Davidson, N., **11,** 194
Davidson, R. S., **1,** 595
Davies, A. G., **7,** 40, 175; **11,** 67
Davies, A. P., **3,** 308; **4,** 552
Davies, C. W., **1,** 518
Davies, D. I., **5,** 454
Davies, J. E., **1,** 814; **11,** 492
Davies, J. S., **2,** 386
Davies, N., **12,** 553
Davies, S. G., **7,** 333; **9,** 119; **12,** 36
Davies, T. M., **1,** 561
Davies, V. H., **2,** 386
Davies, W. C., **1,** 1246
Davis, A. K., **5,** 646
Davis, A. P., **10,** 183
Davis, A. W., **1,** 393, 767, 976
Davis, B. A., **5,** 280
Davis, B. C., **5,** 370
Davis, B. R., **1,** 1008; **8,** 117; **11,** 428
Davis, C. C., **2,** 270; **4,** 564
Davis, C. S., **1,** 1131, 1210; **2,** 394
Davis, D. P., **8,** 148
Davis, E. A., **1,** 1126
Davis, F. A., **5,** 561, 582; **11,** 108, 109; **12,** 393
Davis, J. A., **1,** 658
Davis, J. B., **9,** 257
Davis, J. T., **12,** 70, 249, 277
Davis, K. E., **1,** 162
Davis, M., **1,** 1178; **3,** 101, 290; **5,** 667
Davis, N. R., **2,** 187
Davis, P., **12,** 381
Davis, P. D., **12,** 46
Davis, R., **2,** 229; **9,** 261; **11,** 132, 481
Davis, R. B., **1,** 134
Davis, R. C., **8,** 419
Davis, R. E., **1,** 1216, 1247; **8,** 454
Davis, R. G., **8,** 94
Davis, R. H., **1,** 818, 1011, 1157
Davis, R. V., **1,** 140
Davis, V. C., **7,** 241
Davison, A., **10,** 38
Davoust, S. G. **11,** 12
Davy, H., **11,** 56
Dawe, R. D., **12,** 195
Dawes, C. C., **6,** 187
Dawkins, C. W. C., **1,** 427, 828
Dawson, A. D., **6,** 230
Dawson, C. R., **1,** 1008, 1103
Dawson, D. J., **2,** 453; **4,** 167
Dawson, J. W., **9,** 63, 152; **11,** 590; **12,** 179
Dawson, M. I., **3,** 57, 174; **5,** 689
Dawson, N., **4,** 189
Dawson, N. D., **2,** 132
Dawson, R. L., **1,** 1022
Dawson, W., **1,** 885
Dax, K., **9,** 64
Dax, S. L., **12,** 108
Day, A. C., **1,** 177, 444, 933; **2,** 105; **3,** 76, 153
Day, M. J., **4,** 511; **6,** 124, 598
Day, R. J., **1,** 78
Dayan, J. E., **1,** 412
Dayer, F., **6,** 229
De, A., **5,** 542
De, B., **12,** 324
Deacon, B. D., **1,** 837
Deady, L. W., **4,** 184; **5,** 266
Deakin, S., **1,** 7
De Amici, M., **9,** 151
Dean, B. D., **11,** 219
Dean, E. W., **1,** 618
Dean, F. H., **1,** 808; **2,** 40
Dean, F. M., **1,** 989; **2,** 35; **7,** 115, 155
Dean, J., **8,** 320
Dean, J. M., **2,** 42, 384
Dean, P. D. G., **1,** 617
Dean, R. E., **1,** 782
Dean, R. T., **7,** 293
Deana, A. A., **1,** 343, 846; **9,** 169
De Angelis, F., **7,** 20; **8,** 13
Dear, R. E. A., **1,** 574, 603, 992; **2,** 53, 346; **4,** 48
Deardorff, D. R., **10,** 359
De Ath, G. C., **1,** 448
Deatherage, F. E., **1,** 392, 759, 1040
Debal, A., **6,** 352
De Belder, A. N., **1,** 1055; **2,** 378
DeBenneville, P. L., **1,** 707
Deber, C. M., **6,** 137
De Bernardini, S., **11,** 602
DeBernardis, J. F., **12,** 179
Debesse, J. J., **8,** 238
DeBoer, C. D., **2,** 286; **5,** 464
de Boer, J. J. J., **8,** 2
de Boer, T. J., **1,** 195, 1184; **4,** 108, 428; **8,** 187
Debon, A., **6,** 300
De Borger, L., **1,** 392; **2,** 94
Debost, J. L., **12,** 318
DeBruin, K. E., **3,** 148
DeBruyn, D. J., **4,** 530, 548
De Bucyk, L., **6,** 118; **8,** 15, 105, 463; **9,** 182; **11,** 434
Dec, J., **4,** 77
DeCamp, M. R., **11,** 145
DeCamp, V., **1,** 502
de Castiglione, R., **6,** 110
Dechatre, J. P., **6,** 514
De Chirico, G., **10,** 289
DeChristopher, P. J., **5,** 677; **6,** 534

De Cicco, G. J., **7,** 81
Deck, H. R., **1,** 625
Decker, C. A., **1,** 235
Declercq, G., **6,** 596
DeClercq, P. J., **9,** 23, 116, 223; **11,** 111
Decodts, G., **11,** 117, 407
Decorte, E., **11,** 124
Decorzant, R., **6,** 215, 675; **7,** 122; **9,** 361; **11,** 373
Dedericks, E., **12,** 205
Deebel, G. F., **1,** 672; **3,** 49
Deem, M. F., **3,** 101
Deets, G. L., **2,** 353
Defarrari, J. O., **2,** 104
Deffense, E., **11,** 407
Deflorin, A. M., **1,** 933
Defrenne, M. A., **4,** 138
Degani, I., **7,** 164; **8,** 34; **10,** 255
Degener, E., **4,** 256, 524
Degenhardt, C. R., **8,** 493; **11,** 502
Degering, E. F., **1,** 864
Deghenghi, R., **1,** 481, 526, 614, 772
Degl'Innocenti, A., **10,** 326
de Graaf, C., **5,** 168; **6,** 270
DeGrazia, C. G., **5,** 542
de Groot, A., **2,** 238; **3,** 124; **7,** 310; **8,** 225; **10,** 260; **11,** 423; **12,** 317
DeGuili, G., **4,** 490
Dehasse-Delombaert, C. G., **12,** 125
Dehesch, T., **4,** 533; **6,** 620
de Heus, J. G., **1,** 147
Dehm, D., **6,** 137
Dehmel, E., **4,** 357; **5,** 614
Dehmlow, E. V., **1,** 360; **4,** 31; **5,** 28, 292; **7,** 239, 342; **8,** 130; **9,** 127, 361, 393; **10,** 112; **11,** 337, 407, 432
Dehn, R. L., **2,** 423
Dehn, W. M., **1,** 1056
Dehnel, R. B., **9,** 82
Dehnicke, K., **11,** 174
Deichler, C., **1,** 66
Deiko, S. A., **11,** 516; **12,** 484
de Jeso, B., **11,** 161
DeJong, A. J., **2,** 344
de Jonge, C. R. H. I., **3,** 168; **6,** 507
Dejonghe, J. P., **7,** 66
de Kaifer, C. F., **12,** 322
de Keukeleire, D., **10,** 45
De Kimpe, N., **6,** 118; **8,** 15, 105, 463; **9,** 182; **11,** 434
Dekker, C. A., **1,** 802
Dekker, J., **6,** 675
Dekoker, A., **12,** 125
de Koning, A. J., **6,** 503
de Koning, H., **8,** 2
de Koning, J. H., **10,** 116
deKrassny, A. F., **3,** 275
DeLaitsch, D. M., **1,** 257
de la Mare, P. B. D., **1,** 1107; **2,** 395
DeLaMater, M. R., **8,** 208
de Lasalle, P., **12,** 278
de Lattre, J., **6,** 369
Delaumeny, M., **6,** 270
Delay, F., **4,** 293; **5,** 31
Delbaere, P., **1,** 1131
Delbecq, F., **7,** 164, 223; **9,** 119; **10,** 101
Del Buttero, P., **9,** 119
del Carmen Torre, M., **4,** 468
Del Cima, F., **6,** 27
de Leer, E. W. B., **2,** 228
De Leeuw, J. W., **5,** 319
Delepine, M., **1,** 430; **9,** 299
Deleris, G., **7,** 9, 372; **9,** 372, 440
del Fierro, J., **8,** 373
Delfin, I., **1,** 249
Delia, T. J., **1,** 139
de Liefde Meijer, H. J., **2,** 21
Della, E. W., **2,** 393; **6,** 624; **7,** 224
Dellacoletta, B. A., **11,** 393
Della Pergola, R., **12,** 449
Dellaria, J. F., Jr., **9,** 21
Della Vecchia, L., **5,** 527
Dellinger, M. H., **2,** 241
Delmas, M., **11,** 214, 433
Delmonte, D. W., **1,** 599
DeLoach, J. A., **12,** 350
De Lombaert, S., **11,** 176; **12,** 183
Delorme, D., **12,** 79, 396
Delphin, W. H., **5,** 742
Delpierre, G. R., **1,** 339
Delty, M. R., **9,** 497
DeLuca, H. F., **7,** 273
De Lucchi, O., **6,** 375; **11,** 61, 495
DeLue, N. R., **6,** 295; **7,** 181
DeLullo, G. C., **10,** 42
Deluzarche, A., **5,** 61
Demaecker, J., **1,** 1116
Demailly, G., **6,** 162; **11,** 292, 339, 524
de Man, J. H. M., **6,** 172; **7,** 310
de Marquez, L. A., **5,** 669
DeMaster, R. D., **4,** 468
de Mayo, P., **5,** 176; **9,** 337, 477
de Meijere, A., **5,** 681; **6,** 568; **7,** 163; **8,** 377
DeMember, J. R., **3,** 202; **5,** 232
Demerseman, P., **2,** 353; **4,** 416; **5,** 567; **11,** 360
Demerson, C., **2,** 301
Demetriou, B., **6,** 147
Deming, P. H., **1,** 471
Demmin, T. R., **6,** 611; **10,** 138
Demole, E., **4,** 53, 361, 452; **11,** 519
Demonceau, A., **10,** 342; **11,** 460
de Mos, J. C., **7,** 188
DeMott, D. N., **1,** 209
de Moura Campos, M., **1,** 349
Demoute, J. P., **5,** 266
Demuth, E., **1,** 345, 393
Demuth, M., **9,** 2, 495; **12,** 546
Demuth, M. R., **7,** 64, 200
Demuth, T. P., **10,** 136, 354
Demuynck, C., **11,** 113
Denekas, M. O., **1,** 890
DeNet, R. W., **1,** 281
De Neys, R., **5,** 186
Deng, C., **11,** 193
Den Hertog, H. J., **1,** 827
Den Hollander, C. W., **7,** 113
de Nijs, H., **5,** 389
Denis, J. N., **4,** 47, 437, 472;

5, 713; 8, 299; 9, 32, 204, 283; 10, 174, 219, 319, 368; 11, 225; 12, 219
Denis, P., 12, 53
Denivelle, L., 6, 514; 11, 251
Denkewalter, R. G., 2, 5
Denmark, S. E., 11, 83, 437; 12, 86, 283, 441, 451
Denney, D. B., 1, 101, 114, 139, 927, 1192; 2, 305; 9, 353
Denney, D. Z., 1, 101
Denney, R. C., 1, 904; 2, 252
Denniff, P., 7, 197
Denning, G. S., Jr., 1, 118, 743
Dennis, N., 6, 618
Denniston, A. D., 6, 322
Dennler, E. B., 2, 336
Denny, R. W., 5, 491
Deno, N. C., 1, 397, 1107; 3, 34; 4, 405; 7, 282
DeNoble, J. P., 8, 192
Denoon, C. E., Jr., 1, 11, 69, 160
Denot, E., 1, 75, 562, 564, 624, 1000
de Noten, L. J., 9, 111; 11, 432
Dent, W. T., 2, 110
Denton, D. A., 1, 904
Denyer, C. V., 4, 556
Denzel, T., 6, 95
Deorha, D. S., 1, 271
DePamphilis, M. L., 1, 518
De Pasquale, R. J., 4, 41; 5, 640; 10, 223
de Paulet, A. C., 2, 132
Depaye, N., 12, 231
de Perez, C., 5, 219
Depezay, J. C., 5, 132, 389; 6, 331, 530, 610
De Poortere, M., 5, 138, 513
Depres, J. P., 10, 140; 12, 178
DePuy, C. H., 1, 7, 566, 689, 906, 1179, 1187; 2, 109, 119; 5, 276; 6, 73
Derbyshire, D. H., 1, 1015
de Reinach Hirtzbach, F., 7, 220
Derevitskaya, V. A., 4, 333
Derguini-Boumechal, F., 7, 83; 8, 122; 9, 125
Dermer, O. C., 8, 130
Dernell, W., 10, 266
Derner, O. C., 3, 18
de Roch, I. S., 4, 204; 5, 537; 6, 430
Derome, A. E., 11, 554
De Rosa, M., 6, 543
de Rostolan, J., 3, 115
Dersch, F., 1, 1261
Dershowitz, S., 1, 1235, 1247; 6, 381; 8, 219
Dert, V., 9, 112
Dertouzos, H., 2, 442
de Ruggieri, P., 1, 5, 141
Dervan, P. B., 7, 403; 8, 197, 514
des Abbayes, H., 8, 130, 150, 391; 9, 145
Desai, H. K., 9, 426
Desai, M. C., 11, 188, 351
Desai, N. B., 1, 223, 1216, 1234; 2, 98, 447; 4, 552; 7, 97
Desai, R. C., 12, 109, 158
Desai, R. D., 1, 832
Desai, S. R., 8, 219
Desauvage, S., 11, 225
Deschamps-Vallet, C., 9, 462
De Schryver, F. C., 5, 513
Descoins, C., 4, 164; 7, 100; 8, 150, 176
Descotes, G., 9, 361
De Selms, R. C., 3, 116; 4, 57; 5, 758
Deshayes, H., 3, 153; 5, 667; 7, 170; 9, 378, 416
Deshmukh, M. N., 12, 94
DeShong, P. R., 9, 191; 12, 566
Deshpande, R. P., 12, 1
de Silva, S. O., 9, 8
Deslongchamps, P., 2, 36; 3, 33; 4, 364; 5, 234; 6, 441; 7, 102; 9, 429
Desmaele, D., 12, 183
De Smet, A., 6, 579
Desmond, R., 12, 25
deSolms, S. J., 7, 359
de Souza, J. P., 6, 3, 374
Despo, A. D., 10, 436
Des Roches, D., 7, 248; 8, 130
Dess, D. B., 12, 379
Dessau, R. M., 2, 264; 4, 71; 6, 127, 356, 357
Dessy, R. D., 2, 268
Dessy, R. E., 1, 353, 777, 938; 2, 53
de Stevens, G., 1, 746
DeTar, D. F., 1, 1184, 1203; 2, 45, 59, 297, 412, 427; 12, 450
Determann, H., 1, 97, 755
Detraz, P., 7, 50
Detre, G., 2, 423; 4, 270
Detty, M. R., 8, 245; 9, 373; 10, 63, 67
Detzer, N., 4, 9
Deuchert, K., 5, 722; 9, 129
Deuel, H., 1, 119, 147, 320, 735
Deulofeu, V., 1, 450, 864, 1033
Deutsch, C. J., 5, 563
Deutsch, E. A., 11, 12
Dev, S., 1, 1011; 3, 54; 4, 510; 5, 356, 486; 6, 475; 10, 346; 11, 519; 12, 272, 502
Dev, V., 1, 430
Devant, R., 12, 3
Devaprabhakara, D., 1, 309, 926; 2, 278, 376; 3, 101; 4, 438, 562; 6, 523; 7, 298, 321; 9, 99, 398
Deveaux, J., 4, 84
Dever, J. L., 4, 63
de Vera, N., 10, 332
Devgan, O. N., 3, 309
Deviny, E. I., 4, 290
DeVita, C., 1, 364
de Vlieger, J. J., 10, 321
Devlin, B. R. J., 9, 361
De Voe, S. V., 6, 494
de Voghel, G. J., 5, 198
de Vos, D., 5, 371; 6, 458
Devos, M. J., 7, 184
De Vos, M. J., 12, 295
de Vries, B., 1, 1011
de Vries, G., 2, 80
de Vries, J. G., 8, 458
DeVrieze, J. D., 11, 355

Devys, M., **12,** 523
de Waal, W., **2,** 131; **3,** 65; **4,** 103, 183
De Waard, E. R., **5,** 319, 560
DeWald, H. A., **1,** 1220
Dewar, J., **2,** 4
Dewar, J. H., **1,** 12, 266
Dewar, M. J. S., **1,** 68, 444; **2,** 32, 264; **5,** 337
Dewar, R. B. K., **1,** 68
Dewey, R. H., **5,** 663
Dewey, R. S., **1,** 258, 909; **4,** 535
Dewhirst, K. C., **5,** 740; **7,** 67
Dewhurst, B. B., **4,** 390
Dewhurst, P. B., **5,** 61
Dewick, P. M., **11,** 44
De Winter, M., **1,** 1000
DeWit, A. D., **9,** 165
de Wit, G., **10,** 321
de Wit, J., **3,** 337; **5,** 731
DeWitt, E. J., **1,** 207, 1190
de Wolf, W. H., **8,** 340
Dexheimer, E. M., **5,** 376, 556
De Xing, Y., **11,** 534
Dey, A. S., **6,** 477
Deyo, D., **12,** 389
Deyrup, A. J., **1,** 495
Deyrup, J. A., **1,** 224; **4,** 110; **5,** 551, 667; **9,** 3
Dezube, M., **11,** 267
Dhaenens, M., **10,** 265
Dhanoa, D. S., **11,** 373
Dhar, D. J., **8,** 106
Dhar, D. N., **2,** 397; **5,** 563
Dhawan, B., **11,** 493
Dhawan, K. L., **10,** 72
Dhingra, O. P., **7,** 421; **9,** 202, 514
Diago-Meseguer, J., **10,** 41
Diakur, J., **6,** 192; **7,** 400
Diamanti, J., **2,** 129
Diamond, P. M., **8,** 229
Diamond, S. E., **8,** 438
Diaper, D. G. M., **1,** 1055; **2,** 65, 79; **3,** 45, 208; **4,** 577
Dias, J. R., **4,** 83, 150
Diaz, G. E., **11,** 551
DiBase, S. A., **9,** 387
DiBattista, P., **12,** 449
Di Bello, N., **8,** 425
DiBiase, S. A., **7,** 304; **8,** 130; **10,** 323
DiCesare, P., **10,** 306
Dick, C. R., **4,** 82
Dick, K. F., **6,** 124
Dickason, W. C., **3,** 43, 187
Dickel, G. B., **1,** 465, 759
Dickerhoof, D. W., **1,** 613
Dickerson, R. E., **3,** 57
Dickey, J. B., **1,** 109, 424, 500, 864, 1123, 1167, 1169, 1269
Dickinson, C. L., **1,** 377, 1136, 1269
Dickinson, R. A., **6,** 70; **10,** 373
Dickinson, R. G., **4,** 19
Dickinson, W. B., **3,** 199
Dickopp, H., **1,** 1236
Dickore, K., **10,** 331
Dickson, I. E., **2,** 30
Dickson, R. S., **6,** 454
Diebold, J. L., **4,** 112
Dieck, H. A., **6,** 61, 157
Diederick, F., **8,** 155
Diehl, H., **1,** 874, 999; **2,** 94, 360
Diehl, H. W., **1,** 476, 955; **2,** 163
Diehl, J., **1,** 348
Diehl, J. F., **1,** 235
Diehl, L., **10,** 201
Diehr, H. J., **1,** 118, 1192
Diekman, J., **2,** 208
Diels, O., **1,** 247, 361, 489, 992, 1219, 1270, 1284
Diercks, R., **12,** 157
Dieter, J. W., **12,** 465
Dieter, R. K., **10,** 55, 56; **12,** 350, 465
Dieterich, D., **4,** 422
Dietl, H. K., **5,** 160, 647
Dietrich, B., **5,** 156
Dietrich, C. O., **7,** 318
Dietrich, M. W., **4,** 25
Dietrich, P., **1,** 361; **10,** 152
Dietrich, R., **1,** 1187
Dietrichs, H. H., **3,** 87
Dietsche, M., **11,** 421
Dietsche, T. J., **4,** 235; **5,** 502, 503; **6,** 395; **8,** 217
Dietsche, W., **12,** 182
Dietsche, W. H., **1,** 1132
Dietz, A. G., Jr., **8,** 125
Dietz, G., **6,** 547
Dietz, K. H., **2,** 348
Dietz, R., **2,** 32; **11,** 401
Dietz, S. E., **6,** 534
Di Fabio, R., **12,** 418
Differding, E., **12,** 125
DiFuria, F., **6,** 114, 487; **12,** 94
Digenis, G. A., **5,** 23, 249
DiGiacomo, P. M., **4,** 520; **5,** 686
DiGiorgio, J. B., **1,** 642; **4,** 363
DiGiorgio, P. A., **9,** 74
Dijkink, J., **8,** 232
Dijkstra, R., **8,** 351
Dike, M. S., **10,** 443
Dilaris, I., **1,** 321
Dilbeck, G. A., **6,** 475
Dilgen, Sr. St. F., **2,** 150
Dillon, J., **9,** 154
Dillon, T. E., **1,** 67
Dilthey, W., **1,** 243, 834, 1150
Dimant, E., **1,** 772
Dime, D. S., **9,** 499; **10,** 96; **11,** 388
Dimmel, D. R., **5,** 86; **7,** 182
Dimmig, D. A., **1,** 890
Dimroth, K., **2,** 345
Dimroth, O., **1,** 66, 561, 564, 832
Dimroth, P., **4,** 414; **11,** 32
Din, Z., **4,** 31
D'Incan, E., **10,** 378
Dincer, D., **1,** 269
Dinculescu, A., **10,** 375; **11,** 571
Dines, M. B., **4,** 358, 360
Dineshkumar, **6,** 252
Ding, J. Y., **6,** 288
Dingwall, J. G., **3,** 128
Dinizo, S. E., **6,** 114; **8,** 168, 229
Dinkes, L., **4,** 45; **5,** 675
Dinsmore, R., **1,** 874
Dinulescu, I. G., **2,** 112; **6,** 279
Di Nunno, L., **5,** 595

Dinwiddie, J. G., Jr., **1,** 703
Dinwoodie, A. H., **4,** 480
DiPasquale, F., **9,** 482
Dirksen, H. W., **1,** 1146
Dishart, K. T., **1,** 1227
Disher, J. M., **4,** 54
Dishong, D. M., **10,** 323
Disnar, J. R., **6,** 331
Di Stefano, M., **9,** 497
Distler, W., **7,** 408
Dithmar, K., **1,** 652
Ditrich, K., **12,** 101
Dittman, W. R., Jr., **8,** 501
Dittmann, W. D., **1,** 478
Dittmer, D. C., **5,** 561; **6,** 102; **10,** 205
Divaker, K. J., **5,** 370
Divanfard, H. R., **8,** 284; **10,** 67
Diveley, W. R., **1,** 270
Dix, D. T., **1,** 431
Dixit, S. M., **2,** 236
Dixit, V. M., **6,** 326; **10,** 455
Dixon, A. J., **10,** 94
Dixon, J., **3,** 59
Dixon, J. A., **1,** 1087
Dixon, J. R., **2,** 313
Dixon, J. S., **1,** 176
Djarmati, Z., **3,** 171; **6,** 201
Djerassi, C., **1,** 3, 9, 24, 37, 60, 80, 82, 93, 104, 107, 144, 150, 444, 500, 511, 526, 601, 608, 643, 707, 731, 802, 814, 967, 970, 989, 1000, 1090, 1284; **2,** 208, 378, 453; **3,** 329; **4,** 31, 265; **5,** 370, 381; **7,** 335; **8,** 374, 375
Djuric, S., **10,** 84, 141
Doak, G. O., **2,** 25
Doane, W. M., **10,** 39
Dobashi, S., **8,** 373
Dobinson, F., **2,** 156
Doboszewski, B., **12,** 420
Dobriner, K., **1,** 526
Dobson, T. A., **2,** 263
Docken, A. M., **12,** 200
Dockner, T., **1,** 11
Dockx, J., **5,** 28
Dodd, G. H., **6,** 79
Doddema, L. L., **11,** 423
Dodge, R. A., **1,** 1049
Dods, R. F., **3,** 16, 116
Dodson, R. M., **1,** 127, 162, 444, 680, 976, 1157; **2,** 351; **4,** 366
Dodsworth, D. J., **9,** 285
Doebel, K. J., **1,** 659
Doedens, R. J., **9,** 479
Doellscher, G., **4,** 35
Doering, W. von E., **1,** 96, 273, 329, 536, 688, 790, 832, 927, 1179; **2,** 84, 159, 166, 394; **3,** 69; **4,** 31, 336; **6,** 137; **7,** 369; **12,** 113
Doerk, E., **1,** 736
Doerr, H., **9,** 48; **11,** 326
Doerr, I. L., **1,** 1045
Doetz, K. H., **11,** 401; **12,** 136
Doetzer, R., **7,** 358
Doganges, P. T., **1,** 989
Doggart, J. R., **1,** 1102
Doherty, D. G., **1,** 982
Doherty, J. B., **11,** 524
Doherty, R. M., **6,** 649
Dohmaru, T., **4,** 526
Doi, E., **12,** 485
Doi, Y., **4,** 224
Doifode, K. B., **1,** 162
Dojo, H., **12,** 236, 502
Dolak, L. A., **3,** 307
Dolak, T. M., **8,** 72
Dolata, D. P., **5,** 677
Dolbier, W. R., Jr., **2,** 389; **5,** 551
Dolby, L. J., **2,** 462; **7,** 368
Doldouras, G. A., **8,** 251
Dolence, E. K., **12,** 412
Doleschall, G., **5,** 445; **6,** 397; **8,** 54, 351
Dolfini, J. E., **7,** 244
Dolinski, R. J., **1,** 935
Doll, L., **1,** 859
Doll, R. J., **10,** 216
Dollat, J. M., **7,** 122
Dolle, R. E., **12,** 185, 476
Dolling, U. H., **12,** 381
Dollinger, H., **12,** 100
Dolphin, D., **7,** 56
Dolphin, J. M., **9,** 473
Domaschke, L., **2,** 119; **9,** 299
Dominguez, D., **12,** 159
Dominguez, J. N., **12,** 565
Dominguez, X. A., **1,** 730, 1284
DoMinh, T., **3,** 138; **6,** 190; **8,** 222
Dominianni, S. J., **2,** 332
Domsch, D., **11,** 587
Donahoe, H. B., **1,** 1269
Donahue, K., **11,** 150
Donald, D. S., **4,** 157; **5,** 220
Donaldson, M. M., **2,** 17, 21; **3,** 6
Donaldson, R. E., **8,** 284
Donarume, L., **5,** 727
Donatelli, B. A., **10,** 9
Dondoni, A., **6,** 306
Donetti, A., **5,** 299; **10,** 294
Dongala, E. B., **5,** 210
Donges, R., **5,** 448
Donleavy, J. J., **1,** 1270
Donley, W. J., **5,** 419; **11,** 309
Donndelinger, P., **6,** 146
Donnelly, S. J., **5,** 695
D'Onofrio, F., **12,** 256
Donogues, J., **12,** 220
Donovan, D. J., **7,** 240
Donovan, S. F., **8,** 79; **11,** 281
Dooley, J. F., **2,** 328
Doolittle, R. E., **1,** 518; **10,** 198; **11,** 294; **12,** 97
Doorakian, G. A., **1,** 970
Doorenbos, N. L., **1,** 236, 904
Doornbos, T., **1,** 668; **4,** 455, 459
Dopper, J. H., **6,** 525, 575
D'Orchymont, H., **9,** 127
Dorfman, L., **1,** 80
Dorfman, R., **1,** 407
Dorfman, R. I., **2,** 75
Dormoy, J. R., **6,** 25, 35; **7,** 14
Dorn, C. P., **8,** 167
Dornfeld, C. A., **1,** 1209
Dorofeenko, G. N., **1,** 802
Dorow, R. L., **10,** 382
Dorsay, J. E., **2,** 69
Dorsch, D., **12,** 241
Dorsch, H. L., **2,** 368
Dorsch, M., **12,** 400
Dorsey, J. E., **3,** 50

Dorsky, J., **9,** 182
Dorta, H. L., **10,** 459
Doskotch, R. W., **4,** 395
Dossena, A., **7,** 246; **9,** 190, 361
Douchkine, N. D., **5,** 580
Dougherty, C. M., **4,** 311; **5,** 374, 398
Dougherty, G., **1,** 158
Doughty, D. H., **12,** 111
Doughty, J. B., **1,** 379
Douglas, D. E., **1,** 177
Douglas, G. H., **2,** 29; **6,** 258
Douglas, L. J., **2,** 256
Douglass, I. B., **1,** 51; **2,** 270; **4,** 122; **5,** 435, 455
Douglass, J. P., **1,** 322
Doukas, H. M., **1,** 598
Doumaux, A. R., Jr., **3,** 194
Dourtoglou, B., **7,** 14
Dourtoglou, V., **12,** 44
Douteau, M. H., **6,** 273
Douty, C. F., **1,** 1145
Dovell, F. S., **1,** 892
Dow, A. W., **4,** 544; **10,** 431
Dow, R. L., **10,** 122; **12,** 72
Dowalo, F., **2,** 270
Dowbenko, R., **1,** 131
Dowd, D., **1,** 595
Dowd, J. E., **2,** 25
Dowd, P., **1,** 348; **5,** 205; **11,** 192, 460
Dowd, S., **11,** 103
Dowd, S. R., **1,** 854
Dowlatshahi, H. A., **10,** 413
Downie, I. M., **1,** 1247; **2,** 354, 445; **3,** 149; **6,** 280
Downing, M., **8,** 466
Dox, A. W., **1,** 738
Doyle, M. P., **2,** 455; **4,** 530, 548; **5,** 695; **6,** 616, 620; **7,** 9, 49, 73; **9,** 21, 136; **10,** 122, 274, 382; **11,** 50, 359; **12,** 72, 224
Doyle, P., **5,** 639
Doyle, T. W., **1,** 132; **2,** 229; **3,** 166
Draber, W., **8,** 163
Drabowicz, J., **7,** 373, 389; **8,** 135, 232, 488, 504; **9,** 66, 244; **10,** 194, 335; **11,** 275
Dradi, E., **9,** 489
Draghici, C., **4,** 57
Drago, R. S., **1,** 124; **11,** 138
Drahowzal, F., **1,** 1011
Drake, A., **1,** 75
Drake, G. L., Jr., **1,** 379
Drake, H. A., **1,** 75
Drake, N. L., **1,** 3, 1011
Drake, S. S., **2,** 12
Drake, W. V., **1,** 502
Drecksler, K., **12,** 263
Drees, F., **1,** 11
Drefahl, G., **1,** 318, 329, 501, 618, 671; **3,** 184
Dreger, E. E., **1,** 165, 176, 251, 378, 1209
Dreger, L. H., **1,** 235, 1178
Dreiding, A. S., **1,** 93; **2,** 348; **3,** 336; **4,** 369; **5,** 57, 257, 357, 532, 651; **12,** 156
Drekter, I. J., **1,** 1128
Drenth, W., **11,** 567
Dressaire, G., **11,** 117
Dressler, R. L., **1,** 563
Dreux, J., **6,** 557
Dreux, M., **5,** 512; **6,** 189
Drew, H. D. K., **1,** 38; **5,** 15
Drew, M. G. B., **10,** 348
Drew, R. T., **4,** 84
Drewes, H. R., **5,** 661
Drexler, S. A., **11,** 315
Dreyer, G. B., **11,** 373
Driesen, H. E., **1,** 318
Driggs, R. J., **9,** 174
Driguez, H., **3,** 61; **6,** 569
Drischel, W., **1,** 224; **5,** 120
Driscoll, J. S., **2,** 447
Droege, H., **1,** 573
Droste, W., **2,** 123
Drouin, J., **8,** 238
Drozda, S. E., **12,** 538
Druckrey, E., **2,** 202
Drueger, D. S., **8,** 272
Druelinger, M., **7,** 17; **10,** 12
Drummond, G. I., **1,** 236
Dryden, H. L., Jr., **1,** 56, 412, 612
Du, P. C., **11,** 270
Duane, W. M., **5,** 247
Dubac, J., **5,** 749; **9,** 372
Dube, J., **8,** 429
Dube, S., **9,** 429
Dubeck, M., **2,** 310
Dubey, P. K., **11,** 493
DuBois, A. S., **7,** 424
DuBois, G. E., **7,** 389
Dubois, J. E., **4,** 183, 247; **5,** 244; **6,** 165, 215; **8,** 238, 299; **12,** 268
DuBois, J. E., **6,** 658
Dubois, M., **8,** 299
Dubois, R., **3,** 172
Dubois, R. A., **6,** 566
DuBois, R. H., **2,** 79; **3,** 62; **4,** 99
Dubois, R. H., **3,** 166
Duboudin, F., **10,** 98; **11,** 150; **12,** 86
Dubs, P., **5,** 39; **8,** 215
Dubuis, R., **1,** 567; **6,** 319
Duc, D. K. M., **9,** 474
Ducep, J. B., **6,** 345; **9,** 77
Duchaussoy, P., **10,** 306
Duck, E. W., **1,** 1154
Duckworth, A. C., **1,** 827, 1226; **6,** 617
Duclos, R. I., Jr., **12,** 450
Ducom, J., **2,** 210
Dudley, F. B., **2,** 458
Dudley, K. H., **2,** 42, 117
Dudman, C. C., **10,** 423; **11,** 566
Dudragne, F., **5,** 312
Dueber, T. E., **4,** 533; **6,** 620
Duerckheimer, W., **1,** 129, 1056; **3,** 101
Duerr, F. H., **1,** 116
Duerr, H., **4,** 122
Duerrfeld, R., **4,** 86
Dufay, P., **1,** 215
du Feu, E. C., **1,** 703
Duff, J. C., **1,** 430
Duff, J. M., **2,** 187; **5,** 68, 376; **6,** 637; **7,** 194
Duff, R. B., **1,** 1128
Duff, S. R., **3,** 41
Duffaut, N., **5,** 713; **6,** 323; **7,** 199; **10,** 176
Dufour, C., **1,** 74
Dufresne, Y., **12,** 79
Dugat, D., **9,** 291
Duggan, A. J., **6,** 323

Duggan, M. E., **12,** 143
Duhamel, L., **5,** 58, 377; **8,** 72, 193; **9,** 205; **11,** 583
Duhamel, P., **5,** 58, 377
Duhl-Emswiler, B. A., **11,** 300
Duinker, P. M., **1,** 105
Duisberg, C., **1,** 518; **2,** 216
Duke, A. J., **5,** 617
Dulcere, J. P., **7,** 281; **10,** 457; **11,** 126, 256
Dulenko, V. I., **1,** 802
Dulh, D. L., **11,** 330
Dull, D. L., **5,** 210; **12,** 282
Dull, M. F., **1,** 166
Dulow, R., **1,** 207
du Manoir, J. R., **6,** 190
Dumas, P., **4,** 397
Dumon, L., **9,** 212
Dumont, C., **2,** 372
Dumont, W., **6,** 29, 90, 240, 362; **7,** 35; **8,** 64, 199, 537; **9,** 34, 204, 235; **10,** 312
Dunach, E., **12,** 94
Dunathan, H. C., **5,** 631
Dunbar, B. I., **6,** 137
Duncan, D. M., **6,** 446
Duncan, G. R., **6,** 542
Duncan, J. H., **2,** 422; **4,** 512
Duncan, W. G., **1,** 1058; **2,** 187, 281; **4,** 57, 148; **6,** 78; **7,** 43
Duncan, W. P., **3,** 297; **6,** 304
Duncanson, F. D., **8,** 79
Dunford, J. A., **8,** 238
Dung, J. S., **12,** 437
Dunham, M., **5,** 700; **6,** 381
Dunitz, J. D., **11,** 500; **12,** 85
Dunkelblum, E., **4,** 126, 489; **5,** 741; **6,** 43
Dunkelmann, Z., **2,** 263
Dunkerton, L. V., **6,** 611; **11,** 47, 514
Dunlap, D. E., **6,** 586
Dunlop, N. K., **12,** 294
Dunn, A. D., **11,** 322
Dunn, G. E., **1,** 619
Dunn, G. L., **2,** 271; **3,** 198
Dunn, J. T., **1,** 710
Dunnavant, W. R., **1,** 909, 1040
Dunnenbergen, M., **1,** 411
Dunnigan, D. A., **1,** 281; **5,** 127
Dunny, S., **1,** 444
Dunogues, J., **5,** 713; **6,** 323; **7,** 9, 84, 199, 372; **8,** 512; **9,** 372, 440; **10,** 176
Duong, T., **8,** 377; **9,** 343
Dupeyre, R. M., **5,** 639
Dupont, J. A., **3,** 29
Duprat, P., **5,** 589
DuPree, L. E., Jr., **4,** 161
DuPreez, N., **5,** 700
DuPreez, N. P., **9,** 48; **11,** 326
Dupres, J. P., **9,** 135
Dupuis, J., **12,** 523
Dupuy, A. E., **2,** 34
Dupuy, C., **12,** 357, 400
Duraisamy, M., **12,** 121
Durandetta, J. L., **4,** 345; **6,** 404
Duranti, E., **4,** 460; **6,** 357
Durbut, P., **11,** 383
Dureja, P., **11,** 445
Durette, P. L., **5,** 702
Durham, L. J., **1,** 1163, 1203; **2,** 364
Durst, H. D., **3,** 184; **4,** 451; **5,** 155, 609; **6,** 121, 137, 406; **7,** 79
Durst, T., **2,** 270; **3,** 53, 276; **5,** 129; **6,** 118; **7,** 220, 248; **8,** 95; **9,** 291; **10,** 72
Dussault, P., **11,** 267
Dusza, J. P., **2,** 310
Dutcher, J. D., **1,** 22, 110
Dutcher, J. S., **9,** 314
Duthaler, R. O., **12,** 81, 537
Dutra, G. A., **5,** 243
Dutta, P. C., **1,** 940, 1096; **4,** 369
Dutta, P. K., **6,** 297
Dutta, V. P., **3,** 36
Duttmann, H., **9,** 97
Dutton, G. R., **1,** 881
Duvall, H. M., **1,** 782
du Vigneaud, V., **1,** 235, 407, 634, 731, 737, 743, 1139; **5,** 477; **6,** 445; **9,** 415
Dux, F. J., **5,** 415
Duynstee, E. F. J., **5,** 268
Dvolaitzky, M., **1,** 377, 856; **3,** 264; **5,** 651
Dvonch, W., **1,** 1203
Dvorken, L. V., **2,** 57
Dvornik, D., **1,** 814
Dworkin, A. S., **8,** 19
Dye, J. L., **5,** 156
Dyer, E., **6,** 570
Dyer, J. R., **1,** 310
Dyer, M. C. D., **8,** 190
Dykman, E., **6,** 471
Dynak, J., **7,** 103
Dyong, I., **9,** 102
Dyrkacz, G. R., **5,** 6
Dyrnesli, R., **4,** 247
Dyson, G. M., **1,** 718
Dyson, N. H., **2,** 381; **3,** 265
Dziemian, R. L., **1,** 904
Dzierzogowski, S., **7,** 374

Eaborn, C., **1,** 892; **3,** 51; **4,** 265; **5,** 375, 446; **6,** 640; **9,** 56
Eades, E. D. M., **1,** 309
Eadon, G., **4,** 108
Earborn, C., **2,** 438
Eardley, S., **1,** 671
Earl, R. A., **9,** 306, 435; **12,** 163
Earnshaw, C., **7,** 229; **9,** 301
Earp, S. E., **8,** 391
Easler, E. M., **10,** 110
Easley, W. K., **3,** 272
Easterby, D. G., **1,** 1187
Eastes, J. W., **3,** 212
Eastham, J. F., **1,** 37, 477, 598, 983, 1204; **2,** 53
Eastlick, D. T., **6,** 453
Eastman, R. H., **1,** 952; **11,** 324
Eastmond, R., **5,** 69
Eastwood, F. W., **4,** 185; **6,** 96; **7,** 386
Eaton, J. T., **6,** 534
Eaton, P., **1,** 34
Eaton, P. E., **1,** 309, 970; **3,** 251; **4,** 227; **5,** 495, 535
Eaton, R. E., **2,** 95
Eaves, D. E., **1,** 14
Ebata, T., **10,** 417

Eber, J., **5,** 621
Eberhard, I., **4,** 357; **5,** 614
Eberle, G., **7,** 366
Eberle, H., **1,** 1096
Eberle, M. K., **10,** 194
Eberlein, T. H., **9,** 387; **12,** 85
Ebersole, R. C., **5,** 436, 444
Eberson, L., **7,** 274
Eberstein, K., **6,** 545
Ebert, G. W., **12,** 140
Ebert, L., **7,** 425
Ebke, K., **1,** 125
Ebnoether, C., **1,** 666
Eby, C. J., **1,** 70
Echigo, Y., **8,** 92; **9,** 106
Eck, C., **5,** 542
Eck, C. R., **5,** 756
Eck, H., **4,** 275
Eck, J. C., **1,** 864, 963, 1015
Eck, R. R. v., **1,** 445
Eckert, H., **8,** 499; **9,** 121; **10,** 103; **11,** 63, 139
Eckes, L., **8,** 53
Eckhardt, H. J., **1,** 795
Eckhart, E., **2,** 23
Eckrich, T. M., **8,** 391; **12,** 351, 525
Eckstein, F., **2,** 188
Eckstein, Z., **3,** 284
Eda, S., **9,** 91
Edens, C. O., **1,** 373, 1209
Eder, U., **6,** 411; **10,** 5
Edgar, J. A., **10,** 330
Edgar, K. J., **10,** 246
Edinger, A., **3,** 164
Edington, C., **12,** 378
Edman, J. R., **2,** 270; **12,** 221
Edman, P., **1,** 845; **2,** 323
Edmison, M. T., **3,** 18; **8,** 130
Edward, J. T., **1,** 802, 1121; **2,** 230; **4,** 270; **5,** 508; **9,** 119
Edwards, B. E., **1,** 802
Edwards, B. F., R., **1,** 1116, 1171
Edwards, D., **1,** 643
Edwards, D. A., **6,** 143
Edwards, H. N., **9,** 380
Edwards, J., **1,** 808
Edwards, J. A., **1,** 402; **2,** 195, 344, 362, 381; **3,** 87, 118, 265
Edwards, J. D., **1,** 68
Edwards, J. D., Jr., **1,** 892, 904
Edwards, J. H., **8,** 69
Edwards, J. M., **11,** 194
Edwards, J. O., **1,** 355, 470; **11,** 442
Edwards, L. D., **5,** 4
Edwards, M., **7,** 365; **9,** 462; **11,** 450
Edwards, O. E., **1,** 814, 933; **2,** 192; **6,** 327
Edwards, P. D., **12,** 384
Edwards, P. N., **1,** 1284
Edwards, R. G., **5,** 436
Edwards, W. B., III, **6,** 339
Edwards, W. R., Jr., **2,** 12
Eekhof, J. H., **7,** 392
Effenberger, F., **1,** 282, 283, 404, 1252; **4,** 167, 173, 534; **5,** 250, 653, 702, 705; **6,** 420, 521, 618; **12,** 44, 204, 291
Efimov, V. A., **5,** 434
Efner, H. F., **5,** 419; **11,** 309
Egan, R., **11,** 255
Egan, R. S., **5,** 127, 514
Egawa, Y., **1,** 16, 289
Ege, G., **1,** 608; **4,** 186
Ege, S. N., **1,** 1141
Egge, H., **1,** 295, 685
Eggelte, H. J., **8,** 173, 249
Eggelte, T. A., **8,** 2
Eggensperger, H., **1,** 1015, 1056
Egger, B., **5,** 293; **9,** 361
Eggerichs, T. L., **5,** 198
Eggette, H. J., **9,** 341
Eggler, J., **5,** 315
Egli, C., **6,** 409
Egli, R. A., **5,** 602
Eglinton, G., **1,** 159, 161, 169, 360, 668, 926, 1040; **2,** 19
Egnell, C., **1,** 235
Eguchi, S., **3,** 260; **4,** 93; **5,** 28; **7,** 79
Eguchi, Y., **5,** 89
Egyed, J., **5,** 567
Ehler, D. F., **4,** 526; **5,** 688; **8,** 499
Ehlinger, E., **8,** 274, 278; **10,** 8, 430; **11,** 577
Ehmann, W. J., **3,** 153; **7,** 314
Ehrenfreund, J., **5,** 354; **6,** 297
Ehrenstein, M., **1,** 394, 411
Ehrhart, G., **1,** 488, 849
Ehrhart, O., **1,** 764
Ehrhart, W. A., **1,** 404; **2,** 202
Ehrig, V., **7,** 47; **9,** 48; **11,** 326
Ehrlich, K., **3,** 101
Eibach, F., **8,** 228
Eibeck, R. E., **9,** 429; **11,** 487
Eichel, W. F., **8,** 15
Eichen, R., **4,** 119
Eichenauer, H., **7,** 11; **8,** 16, 374; **9,** 17
Eicher, T., **1,** 1203, 1261
Eichler, S., **1,** 49, 966
Eichmeier, L. S., **4,** 74
Eickhoff, D. J., **12,** 365
Eickman, N., **7,** 268
Eickoff, D. J., **11,** 99
Eigenberger, E., **1,** 333
Eigendorf, G. K., **7,** 71
Eigenmann, E. W., **1,** 927
Eijsinga, H., **7,** 164, 226
Eilar, K. R., **1,** 152
Eilbracht, P., **12,** 168
Eilerman, R. G., **4,** 302; **10,** 240
Eilers, E., **7,** 215
Eilers, K. L., **1,** 689; **2,** 109, 119
Eilingsfeld, H., **1,** 289; **2,** 134; **7,** 424
Eimutis, E. C., **2,** 23
Einck, J. J., **9,** 277
Einhorn, A., **1,** 120; **2,** 5; **10,** 201
Einstein, F. W. B., **12,** 333
Eis, M. J., **12,** 70
Eisch, J. J., **1,** 19, 262, 612; **5,** 744; **7,** 47; **8,** 41, 290; **9,** 146, 233, 311, 471; **10,** 149; **12,** 507
Eisele, W., **3,** 320
Eisenbraun, E. J., **1,** 9, 144, 476, 531, 536, 705; **2,** 104, 150, 234, 287; **3,** 74; **4,** 267, 357, 378, 391, 397; **5,** 141; **6,** 9, 20, 304; **7,** 171; **9,** 256; **11,** 321

Eisenhart, E. K., **12,** 11
Eisenmann, J. L., **1,** 228
Eisenstadt, E., **9,** 71
Eisert, M. A., **1,** 507
Eisner, T., **2,** 208; **4,** 424
Eissenstat, M. A., **7,** 236; **8,** 72; **9,** 273, 398
Eistert, B., **2,** 63, 233; **3,** 220; **4,** 230; **9,** 259
Eitel, D. E., **1,** 697; **2,** 282
Eiter, K., **1,** 114, 190; **2,** 19, 211, 448; **4,** 18, 119
Eizember, R. F., **6,** 19
Ek, A., **7,** 268
Ekouya, A., **6,** 323
Ekwurite, N., **11,** 493
Elad, D., **1,** 102, 403, 764
Elam, E. U., **1,** 986
ElAmin, B., **9,** 227
El Antably, S., **5,** 249
Elberling, J. A., **6,** 174
Elbert, W., **1,** 673
El Bouz, M., **9,** 235
Elbs, K., **1,** 954
Elderfield, R. C., **1,** 230, 659, 1158; **4,** 75
Elena Cabezas, M., **1,** 562, 624
Eletti-Bianchi, G., **7,** 79; **11,** 385
El-Fekky, T. A., **9,** 485
Elgavi, A., **8,** 157
Elhafez, F. A. A., **2,** 318; **8,** 509
El Hallaoui, A., **12,** 350
El Heweihi, Z., **1,** 339
Eliaers, J., **5,** 186
Elian, M., **2,** 112
Eliel, E. L., **1,** 132, 402, 522, 595, 598, 599, 682, 866, 978, 1163; **2,** 68, 92, 229, 243, 352; **3,** 166; **4,** 231; **5,** 92, 186, 565; **6,** 615; **8,** 509; **9,** 139, 290; **11,** 71; **12,** 239, 534
Eling, T. E., **10,** 203
Eliot, C. G., **1,** 336
Elix, J. A., **1,** 1284; **3,** 339; **5,** 127, 672
El-Kholy, I. E. S., **7,** 10; **8,** 456
El-Khrisy, E. E. A. M., **11,** 55
Elkik, E., **2,** 346; **5,** 304
Elkobaisi, F. M., **1,** 1081
Elks, J., **1,** 107, 150, 500, 927, 1284
Ellegast, K., **1,** 1188
Ellestad, G. A., **5,** 675
Ellestad, R. B., **1,** 573
Ellingboe, E. K., **1,** 1157
Ellingssen, P. O., **10,** 326
Ellington, P. S., **1,** 1004, 1008
Elliot, D. F., **1,** 928
Elliot, H. W., **1,** 135
Elliott, A. J., **4,** 209
Elliott, D. F., **4,** 391
Elliott, J. D., **8,** 73; **11,** 106, 250; **12,** 377, 378
Elliott, R., **12,** 377
Elliott, R. C., **11,** 570
Ellis, A., **11,** 559
Ellis, A. I., **8,** 360
Ellis, B., **1,** 147, 219, 285
Ellis, G. P., **2,** 24
Ellis, J. E., **9,** 314
Ellis, J. W., **1,** 561; **2,** 237
Ellis, M. K., **12,** 242
Ellis, R. C., **1,** 676
Ellis, S. H., **1,** 431
Ellis, S. R., **1,** 448
Ellison, D. L., **6,** 676
Ellison, R. A., **3,** 81; **4,** 218; **6,** 255
Ellison, R. H., **6,** 454
Ellman, G. L., **1,** 351
Ellwood, M., **10,** 306
Ellzey, S. E., Jr., **1,** 1247
El-Meligy, M. S. A., **3,** 226
Elmes, P. S., **9,** 240
Elmore, D. T., **1,** 199, 345
El-Morey, S. S., **11,** 466
El-Namaky, H. M., **5,** 731
Eloy, F., **1,** 258
Elphimoff-Felkin, I., **3,** 337; **4,** 577; **8,** 374, 535
El-Sadr, M. M., **1,** 80
Else, M. J., **5,** 326
Elsinger, F., **1,** 617, 618, 1023
Elsom, L. F., **2,** 406; **3,** 286; **5,** 655; **7,** 361
ElSoukkary, O., **8,** 102
Elston, C. T., **1,** 905
Elsvier, G. J., **11,** 13
El-Telbany, F., **12,** 446
Elvidge, J. A., **1,** 195
El-Wahab, M. F. A., **1,** 80
Elwood, J. K., **6,** 409
El-Zaru, R. A., **8,** 410
Eman, A., **7,** 108
Emde, H., **1,** 1033; **8,** 515; **11,** 587
Emeleus, H. J., **1,** 1236; **2,** 54
Emerson, D. W., **4,** 75, 446; **5,** 276
Emerson, G. F., **3,** 101, 243; **7,** 111
Emerson, M., **6,** 573
Emerson, M. T., **1,** 1193
Emerson, O. H., **1,** 445, 1269
Emerson, T. R., **2,** 287
Emerson, W. S., **1,** 156, 248, 362, 412, 697, 730, 963; **5,** 499
Emery, A., **3,** 329
Emery, S. E., **9,** 421
Emge, D. E., **7,** 271
Emke, A., **6,** 467
Emmert, B., **4,** 100
Emmick, T. L., **1,** 904; **2,** 335
Emmons, W. D., **1,** 6, 250, 253, 323, 329, 790, 791, 819, 827, 1018, 1081, 1095, 1138, 1216, 1217, 1218, 1226; **2,** 316; **3,** 130, 144, 221; **8,** 505
Emoto, S., **11,** 30
Emptoz, G., **1,** 268
Emsley, J., **11,** 6
Enda, J., **9,** 216; **10,** 298; **11,** 514; **12,** 11
Enders, D., **5,** 267, 717; **6,** 339; **7,** 11, 130, 312; **8,** 16, 374; **9,** 17, 184; **10,** 260; **11,** 243, 299; **12,** 31, 33, 205
Enders, L., **2,** 385
Endicott, M., **4,** 87
Endler, A. S., **1,** 106, 841
Endo, J., **5,** 601
Endo, M., **7,** 372; **9,** 47
Endo, T., **6,** 232
Endres, G. F., **10,** 138
Eng, S. L., **12,** 416

Engberts, J., **5,** 472
Engel, C. R., **1,** 377, 526, 771, 974; **2,** 21; **3,** 275; **10,** 188
Engel, L. L., **1,** 625, 1000, 1006; **8,** 442
Engel, M. R., **4,** 370
Engel, N., **8,** 396
Engel, P., **9,** 217
Engel, R., **5,** 482, 667; **9,** 256
Engelbertz, P., **1,** 176
Engelhard, H., **11,** 147
Engelhardt, V. A., **1,** 228, 377, 1125, 1136, 1269
Enggist, P., **4,** 53, 452; **11,** 519
England, D. C., **1,** 228; **11,** 403
Engle, R. R., **1,** 144, 989
Englehardt, E. L., **9,** 264
Engler, D. A., **7,** 150; **8,** 9, 208
Engler, E. M., **7,** 176
Englert, G., **12,** 194
English, J., Jr., **1,** 134, 412, 654
Englund, B., **1,** 1073
Engman, A. M., **5,** 381
Engman, L., **10,** 377; **11,** 56; **12,** 457
Enholm, E. J., **12,** 23, 146
Ennakoua, C. M., **11,** 454
Ennen, B., **12,** 353
Ennis, M. D., **11,** 381; **12,** 362
Enokiya, M., **11,** 514
Ensley, H., **11,** 466
Ensley, H. E., **5,** 339; **7,** 309; **10,** 295; **11,** 414; **12,** 389
Enslin, P. R., **1,** 60
Entwistle, D. W., **7,** 354; **8,** 82
Entwistle, I. D., **6,** 446; **7,** 341; **10,** 93, 359; **12,** 574
Enz, W., **1,** 828
Enzell, C., **2,** 366
Ephritikhine, M., **8,** 137
Epple, G., **4,** 534; **5,** 702; **12,** 44
Eppner, J. B., **5,** 495
Epstein, J. W., **2,** 76; **7,** 21
Epstein, M., **1,** 859
Epstein, R., **9,** 89
Epstein, W. W., **3,** 123
Epsztajn, J., **7,** 10, 409; **8,** 521
Epsztein, R., **8,** 489
Epton, R., **1,** 38
Erbland, M. L., **6,** 244
Ercoli, A., **1,** 5, 141, 1210
Ercoli, R., **1,** 228
Erden, I., **8,** 173, 373
Erdik, E., **12,** 143
Erdtman, H., **1,** 72, 392, 966
Ergun, S., **1,** 228
Erhardt, F., **1,** 933
Erhardt, J. M., **10,** 289; **11,** 373
Erhardt, U., **4,** 270
Erickson, A. S., **10,** 331; **11,** 441
Erickson, B. W., **2,** 263; **3,** 113; **4,** 39, 53, 218; **5,** 381, 445, 527; **6,** 53; **8,** 230, 337; **9,** 251
Erickson, D., **7,** 292
Erickson, E. R., **1,** 244
Erickson, F. B., **1,** 198
Erickson, G. W., **10,** 12
Erickson, J., **9,** 337
Erickson, K., **2,** 208
Erickson, K. C., **3,** 33
Erickson, K. L., **1,** 926, 1217; **4,** 52, 520; **5,** 551
Erickson, R. E., **1,** 562, 777, 1090; **3,** 59
Erickson, R. L., **1,** 652
Erig, J., **1,** 1073
Eriksson, C., **11,** 416
Erker, G., **6,** 9
Erlanger, B. F., **1,** 859
Erman, W. F., **1,** 730, 966; **2,** 210; **3,** 30, 142; **4,** 45, 489; **8,** 352
Ermolenko, M. S., **12,** 198
Erne, K., **1,** 518
Erner, W. E., **1,** 1172
Ernest, I., **1,** 764; **5,** 511
Erntholt, D., **8,** 455
Errede, L. A., **1,** 926
Ertz, R., **1,** 445
Eruenlue, R. K., **4,** 76
Esayan, G. T., **5,** 478
Eschenfelder, V., **6,** 511
Eschenmoser, A., **1,** 147, 272, 282, 283, 574, 617, 927, 1018, 1212; **2,** 219, 422; **4,** 82, 187; **5,** 15, 39, 113, 329; **6,** 91, 222; **7,** 171; **8,** 135; **9,** 241, 349, 355
Eschenmoser, W., **8,** 209
Escher, S., **7,** 389; **8,** 186
Escher, S. D., **11,** 373
Escherich, C., **1,** 1146
Eschinasi, E. H., **9,** 15
Eschweiler, W., **4,** 239
Esclamadon, C., **1,** 1218
Escobar, C., **10,** 354
Escudero, J., **11,** 562
Eskenazi, C., **9,** 46
Esmail, R., **6,** 250
Esmay, D. L., **1,** 97, 573
Esmond, R., **11,** 453
Espejo de Ochoa, O., **4,** 430
Espinosa, J., **7,** 345
Ess, R. J., **1,** 1216
Essenfeld, A. P., **12,** 277
Essiz, M., **7,** 327
Estep, R. E., **5,** 552
Estes, J. M., **11,** 514
Estham, J. F., **5,** 649
Estreicher, H., **9,** 293; **10,** 209, 278, 393
Estrin, N. F., **2,** 45, 297
Eswarakrishnan, V., **12,** 77
Etard, M. A., **1,** 152
Etemad-Moghadam, G., **12,** 105
Etheredge, S. J., **4,** 146; **7,** 114; **11,** 99
Etheridge, S. J., **8,** 330; **10,** 272
Ethier, J. C., **4,** 431
Etienne, A., **5,** 117
Etournaud, G., **5,** 614
Etter, J. B., **12,** 430
Ettlinger, M., **9,** 158
Ettlinger, M. G., **1,** 471
Ettorre, R., **1,** 1054, 1284
Etzemueller, J., **2,** 429
Etzold, G., **5,** 309
Eugster, C. H., **5,** 538, 700; **8,** 209; **9,** 180; **10,** 418
Eustace, E. J., **5,** 14
Eustache, J. M., **11,** 209
Evanega, G. R., **2,** 326
Evans, B. E., **8,** 451; **9,** 421
Evans, B. J., **11,** 462

Evans, B. R., **7,** 250
Evans, D., **1,** 731; **5,** 331; **6,** 105
Evans, D. A., **4,** 77, 543; **5,** 207, 296, 406, 681, 722; **6,** 32, 399, 607; **7,** 181, 303, 399; **8,** 196, 203, 352, 401, 415; **9,** 11, 88, 140; **10,** 71, 88, 133, 328, 332; **11,** 215, 216, 381, 518; **12,** 362, 428
Evans, D. D., **1,** 274
Evans, D. E., **1,** 274
Evans, D. F., **12,** 202
Evans, D. H., **4,** 189
Evans, D. L., **8,** 358; **9,** 323
Evans, E. A., **1,** 281, 713
Evans, G., **6,** 39
Evans, G. E., **9,** 302
Evans, G. W., **1,** 101
Evans, I. A., **12,** 283
Evans, J. M., **2,** 313
Evans, M. B., **3,** 320
Evans, M. E., **2,** 269; **5,** 226
Evans, R. F., **1,** 827, 1163
Evans, R. M., **1,** 149, 595, 642, 802, 951, 954, 1090; **2,** 263
Evans, S., **5,** 582
Evans, S. A., Jr., **12,** 552
Evans, S. M., **6,** 455
Evans, T. E., **5,** 322
Evans, T. W., **1,** 1056
Evans, W. J., **8,** 69
Evans, W. L., **1,** 452, 454, 654, 952, 1015, 1256; **10,** 273
Evenhuis, B., **3,** 124
Evcrett, T. S., **12,** 158
Everhardus, R. H., **8,** 408
Evers, J. T., **11,** 143
Evers, M., **12,** 451
Evin, G., **6,** 35; **7,** 14
Evnin, A. B., **2,** 326
Eweiss, N. F., **8,** 520
Ewing, G. D., **9,** 222
Exon, C., **12,** 167
Exon, C. M., **12,** 94
Eylander, C., **7,** 80
Eyman, D. P., **12,** 60
Eyssen, H., **9,** 212
Eyton, W. B., **3,** 87
Ezhaya, A., **10,** 294

Faber, E. M., **1,** 1105; **3,** 272
Fabian, J., **1,** 707
Fabian, J. M., **1,** 144
Factor, A., **5,** 427
Faehrmann, U., **11,** 455
Faesel, J., **2,** 193
Fager, J. H., **1,** 966; **2,** 353
Fagerlund, U. H. M., **1,** 1246
Fagerson, I. S., **2,** 348
Fahey, D. R., **4,** 421; **5,** 193
Fahey, J., **5,** 345
Fahr, E., **1,** 247; **2,** 129; **3,** 32
Fahrbach, G., **4,** 476
Fahrenholtz, S. R., **1,** 101; **3,** 123
Fahrenholz, F., **4,** 54; **7,** 247
Fahrni, H. P., **8,** 349
Fahrni, P., **5,** 51
Faigenbaum, H. M., **2,** 360
Fainberg, A. H., **1,** 705
Fairhurst, J., **7,** 86
Fairlie, J. C., **5,** 52
Faisst, W., **2,** 67, 454
Fajkos, J., **1,** 444, 526, 624, 625
Fakhr, I. M. I., **1,** 45
Fakstorp, J., **1,** 636; **2,** 144; **9,** 439
Falbe, J., **5,** 205
Falck, J. R., **6,** 133; **7,** 76; **8,** 87, 126; **9,** 242, 243, 301; **10,** 115, 199; **11,** 329; **12,** 551
Falecki, J., **2,** 176
Faler, G. R., **5,** 491, 727
Fales, H. M., **10,** 361
Faller, J. W., **12,** 333, 557
Falling, S. N., **6,** 254
Fallis, A. G., **5,** 614; **7,** 304
Falmagne, J. B., **9,** 396; **11,** 562
Falou, S., **8,** 222
Falshaw, C. P., **1,** 933
Fancher, O., **2,** 12
Fanelli, J., **7,** 335
Fang, J. M., **10,** 34; **11,** 448
Fankhauser, J. E., **12,** 121
Fanta, G. F., **1,** 374; **5,** 593
Fanta, P. E., **1,** 154, 276, 1131; **2,** 146; **4,** 103; **5,** 148; **6,** 133
Fanta, W. I., **1,** 161, 703; **3,** 30
Fantazier, R. M., **5,** 268
Fanton, E., **11,** 329
Farachi, C., **9,** 421; **10,** 382
Farah, B. S., **1,** 255; **2,** 270; **4,** 206
Faraone, A., **2,** 29
Farasiu, D., **5,** 658
Farazmand, I., **4,** 533
Farber, M., **1,** 536
Farcasiu, D., **3,** 330; **4,** 199; **5,** 680
Farcasiu, M., **2,** 112
Farges, G., **11,** 346
Farhat, K., **8,** 405
Farina, F., **10,** 224, 354
Farina, J. S., **6,** 331
Farina, R., **9,** 119
Farina, V., **10,** 452; **11,** 373
Farkas, E., **1,** 970, 982, 986, 1131; **2,** 351
Farkas, E. R., **5,** 182
Farkas, L., **7,** 365
Farkas, L. V., **9,** 387
Farley, W. C., **2,** 66, 70
Farlow, M. M., **12,** 106
Farlow, M. W., **1,** 33, 117, 859
Farmer, M. L., **1,** 39, 677; **2,** 344
Farnham, W. B., **6,** 352, 568; **11,** 119
Farnsworth, D. W., **1,** 904
Farnum, D. G., **2,** 70; **3,** 338
Faro, H. P., **1,** 95, 562
Farona, M. F., **4,** 114
Farooqui, T. A., **7,** 24
Farrar, K. R., **1,** 19, 642, 846
Farrar, M. W., **1,** 802, 957, 1084, 1203; **4,** 85
Farrell, N., **5,** 446
Farrington, J. A., **1,** 1126, 1237
Farthing, A. C., **1,** 110
Fasman, G. D., **1,** 53, 1015
Fateen, A. E. K., **1,** 444
Fatiadi, A. J., **1,** 414, 1116; **2,**

13, 165, 263, 315; **3,** 220; **4,** 375; **5,** 510; **6,** 357
Fatma, W., **9,** 191
Faubl, H., **2,** 173, 307
Faulkner, D. J., **2,** 87, 464; **3,** 302; **4,** 23, 331; **6,** 316; **7,** 36; **8,** 160
Faulstich, H., **2,** 193
Faust, T., **1,** 66
Faust, W., **8,** 272
Faut, O. D., **9,** 148
Fauvarque, J. F., **3,** 152; **10,** 133, 194
Favara, D., **11,** 301
Favier, R., **4,** 498
Favorskaya, I. A., **3,** 302
Favre, A., **10,** 125
Favre, B., **7,** 365
Favre, H., **1,** 180; **2,** 310
Fawcett, F. S., **1,** 117, 1124
Fawcett, J. S., **1,** 147, 744
Fawzi, M. M., **2,** 84
Fazakerley, H., **3,** 96
Fazal, N., **11,** 67
Feather, P., **1,** 285; **2,** 42, 227
Fecher, R., **1,** 1045
Fedder, J. E., **1,** 772; **5,** 513; **7,** 383
Feder, H. M., **6,** 172
Fedorova, T. M., **5,** 157
Fedoruk, N. A., **1,** 195, 764
Fedorynski, M., **4,** 31; **5,** 28, 530; **6,** 601; **8,** 130, 391; **9,** 361
Feely, W., **1,** 730
Feger, H., **11,** 587
Fegley, M. F., **1,** 859
Feher, F., **6,** 288
Fehlhaber, H. W., **1,** 989
Fehn, J., **5,** 537
Fehnel, E. A., **1,** 1027
Fehr, C., **8,** 534; **10,** 58
Fehr, T., **4,** 577; **6,** 560
Feier, H., **1,** 45
Feigl, F., **1,** 274, 1044
Feiler, L. A., **3,** 132
Feinberg, R. S., **6,** 304
Feiring, A. E., **4,** 405; **6,** 479
Feitler, D., **5,** 675
Feldkamp, R. F., **1,** 865
Feldman, A. M., **1,** 211; **3,** 81
Feldman, D., **10,** 270
Feldman, L. I., **1,** 150, 764, 881
Feldstein, G., **8,** 357
Felfoeldi, K., **3,** 317
Felix, A. M., **5,** 49; **10,** 300
Felix, D., **1,** 272; **2,** 422; **4,** 82; **5,** 15, 113, 329; **7,** 171
Felix, S., **6,** 645
Felker, D., **10,** 113, 369
Felkin, H., **2,** 110; **6,** 270; **8,** 137; **10,** 37, 116; **12,** 152
Fell, B., **2,** 292
Fell, E., **8,** 471
Fellebun-Khan, T., **12,** 152
Fellman, J. H., **4,** 387
Fellman, P., **8,** 299
Fellows, C. A., **10,** 302; **12,** 483
Felman, S. W., **8,** 197; **10,** 245
Felsen-Reingold, D., **5,** 667
Felzmann, J. H., **2,** 63, 233
Fenical, W., **7,** 352
Fenlon, W. J., **7,** 42
Fennessey, P. V., **1,** 309
Fenselau, A. H., **1,** 767; **2,** 163
Fenton, D. M., **5,** 504
Fenton, H. S. H., **1,** 476
Fenton, S. W., **4,** 31
Fentrill, G. I., **9,** 264
Fenyes, J. G. E., **2,** 206
Ferber, E., **1,** 353
Ferchinger, D., **1,** 966
Ferebee, R., **1,** 703
Ferguson, A. N., **5,** 446
Ferguson, D. C., **6,** 553
Ferguson, L. N., **1,** 152; **4,** 243; **5,** 253, 680
Fergusson, S. B., **12,** 373
Feringa, B., **7,** 301; **8,** 115
Ferland, J. M., **1,** 802
Ferles, M., **6,** 2
Fernandez, A. H., **12,** 338
Fernandez, E., **9,** 274
Fernandez, S., **11,** 452
Fernandez, V. P., **3,** 10
Fernandez-Lizarbe, J. R., **10,** 41
Fernekess, H., **1,** 787
Fernelius, W. C., **9,** 207
Fernholtz, E., **1,** 1178
Fernholz, H., **1,** 465, 796, 820
Feron, A., **7,** 76
Ferraboschi, P., **10,** 42, 324; **11,** 157; **12,** 444
Ferracutti, N., **4,** 234
Ferrari, J. L., **1,** 952; **2,** 348
Ferraro, J. J., **1,** 601
Ferraz, H. M. C., **8,** 398
Ferreira, D., **7,** 62; **9,** 102
Ferreira, G. A. L., **6,** 124, 441
Ferreira, L., **9,** 255
Ferreira, T. W., **9,** 148; **11,** 358
Ferrel, R. E., **1,** 860, 873
Ferreri, C., **12,** 555
Ferretti, A., **6,** 465
Ferrier, R. J., **1,** 834, 1191, 1226; **2,** 365; **12,** 546
Ferrino, S., **10,** 366
Ferris, A. F., **1,** 692, 730, 827, 977, 1018, 1226; **6,** 423; **9,** 361
Ferry, C. W., **1,** 1116
Ferry, N., **1,** 1270
Fessenden, J. S., **5,** 374; **7,** 152
Fessenden, R. J., **5,** 374; **7,** 152
Fessler, D. C., **3,** 275
Fessler, W. A., **1,** 754
Fetizon, M., **2,** 243, 278, 310, 363; **3,** 249, 304; **4,** 341, 427; **5,** 60, 580, 693; **6,** 430, 455, 501, 514; **7,** 320; **9,** 474
Fetzer, V., **1,** 859
Feuer, B. I., **2,** 433; **5,** 61
Feuer, H., **1,** 40, 207, 329, 870, 927; **2,** 25, 336; **3,** 77; **9,** 175
Feuerstein, K., **1,** 134
Feulgen, R., **1,** 892
Feurer, M., **1,** 130, 360, 595
Feutrill, G. I., **3,** 115; **4,** 465
Fey, P., **12,** 31
Feyen, P., **7,** 42
Fiandanese, V., **10,** 289; **12,** 171, 557
Fiato, R. A., **5,** 206, 454
Fiaud, J. C., **5,** 28; **6,** 40; **10,**

34; **11,** 53
Fibiger, R., **12,** 242
Fichter, F., **3,** 239
Ficini, F., **9,** 274
Ficini, J., **2,** 84; **5,** 219, 389; **7,** 50, 108; **8,** 222; **9,** 165, 257; **10,** 267, 268; **12,** 183
Fick, H. H., **8,** 299
Fick, J., **2,** 390
Ficken, G. E., **1,** 859
Fidler, D. A., **2,** 274
Fiecchi, A., **7,** 428; **11,** 157, 177, 288; **12,** 444, 479
Fiedler, C., **1,** 672
Fiegenbaum, P., **12,** 353
Field, B. O., **1,** 329
Field, G. F., **7,** 255; **9,** 352
Field, K. W., **3,** 295; **4,** 520
Field, L., **1,** 46, 48, 295, 563, 667, 718, 872, 963, 1178, 1184; **3,** 153; **5,** 100; **6,** 374
Field, L. D., **10,** 329; **11,** 275; **12,** 263
Field, R. E., **1,** 890
Fields, D. B., **1,** 564
Fields, D. L., **1,** 383
Fields, E. K., **1,** 155, 777, 892, 1108; **7,** 105
Fields, E. S., **5,** 727
Fields, K. W., **10,** 77; **11,** 105
Fields, R., **5,** 326
Fields, T. L., **1,** 39
Fierz, G., **6,** 544
Fierz, H. E., **1,** 364
Fieser, L. F., **1,** 33, 66, 72, 80, 93, 106, 141, 147, 154, 164, 165, 187, 195, 243, 255, 328, 329, 333, 335, 339, 341, 357, 360, 367, 370, 392, 393, 394, 411, 444, 448, 456, 465, 470, 485, 489, 521, 526, 561, 562, 563, 573, 594, 606, 643, 682, 695, 708, 710, 713, 714, 716, 718, 731, 767, 785, 795, 796, 802, 819, 837, 872, 881, 929, 933, 937, 952, 962, 970, 971, 992, 999, 1000, 1004, 1011, 1015, 1023, 1027, 1029, 1049, 1054, 1056, 1064, 1073, 1082, 1096, 1101, 1108, 1116, 1118, 1121, 1123, 1127, 1128, 1131, 1143, 1150, 1152, 1163, 1203, 1246, 1269, 1284, 1286, 1289; **2,** 6, 10, 17, 117, 126, 381, 394, 464; **3,** 274; **4,** 21, 209, 416, 491, 579; **5,** 4, 64, 486, 532, 639, 756; **6,** 5, 525; **9,** 199
Fieser, M., **1,** 165, 360, 392, 444, 470, 695, 795, 937, 952, 992, 1004, 1029, 1049, 1056, 1082, 1116, 1118, 1284; **5,** 4, 486, 639; **6,** 5, 525; **9,** 199
Figuly, G. D., **10,** 71
Fijiwa, T., **11,** 59
Fikentscher, L., **1,** 81, 207; **2,** 42
Filby, J. E., **6,** 288
Filby, W. G., **5,** 127
Filini, L., **5,** 17
Filipek, S., **10,** 258
Filippini, L., **12,** 502
Filippo, J. S., Jr., **3,** 154; **5,** 175
Filira, F., **1,** 266, 310
Filler, R., **1,** 80, 630, 966, 1126; **2,** 119; **4,** 27; **6,** 6, 228, 670; **7,** 295
Filliatre, C., **11,** 101
Fillion, H., **9,** 226; **10,** 186
Filosa, M. P., **7,** 193; **9,** 273, 398
Finan, P. A., **1,** 39, 50
Finarelli, J. D., **3,** 81
Finch, A., **12,** 112
Finch, M. A. W., **9,** 239
Finch, N., **4,** 293; **5,** 527; **6,** 228, 229
Fincke, R. G., **10,** 175
Finckenor, L. E., **1,** 13, 75, 269
Findeisen, K., **4,** 524; **11,** 174
Findlay, J. A., **2,** 217
Findlay, J. W. A., **2,** 117; **3,** 86, 87; **4,** 134, 330; **6,** 170
Findlehor, R. S., **12,** 295
Fineberg, H., **1,** 23
Finelli, A. F., **1,** 248, 1073
Finer-Moore, J., **10,** 9
Finger, C., **2,** 197
Finger, G. C., **1,** 296, 322, 935
Finholt, A. E., **1,** 594, 1030, 1251
Fink, D. M., **12,** 350
Fink, M. L., **5,** 477
Finkbeiner, H., **1,** 61; **2,** 264
Finkbeiner, H. L., **1,** 633, 1171; **2,** 256, 415; **3,** 245
Finke, R. G., **7,** 340
Finkel, A. J., **7,** 173
Finkel, J. M., **2,** 37
Finkelhor, R. S., **5,** 244, 361, 418, 516
Finkelstein, H., **7,** 135
Finkelstein, M., **1,** 281, 536
Finkenbine, J. R., **5,** 732
Finkheiner, H., **10,** 138
Finklea, H. O., **12,** 568
Finn, J., **12,** 393
Finn, M. G., **12,** 94
Finnan, J. L., **6,** 115, 415
Finnegan, R. A., **1,** 89; **4,** 526
Finnegan, W. G., **1,** 426; **2,** 423; **5,** 10; **7,** 337
Finnerty, J. L., **1,** 322
Finney, J. A., **10,** 113
Finucane, B. W., **3,** 36
Fiorentino, M., **11,** 442
Fiorenza, M., **10,** 326; **11,** 117
Firestone, R. A., **4,** 60
Firouzabadi, H., **8,** 21; **11,** 62; **12,** 62, 144
Firth, W. C., Jr., **1,** 658
Fischer, A., **5,** 475; **10,** 5
Fischer, B., **12,** 546
Fischer, E., **1,** 340, 654, 841, 1033, 1169, 1292; **5,** 336
Fischer, E. O., **4,** 379
Fischer, G., **4,** 155
Fischer, G. J., **1,** 531
Fischer, H., **1,** 11, 22, 281, 367, 567, 1100; **8,** 280
Fischer, H. O. L., **1,** 39
Fischer, H. P., **1,** 367; **5,** 436; **6,** 403
Fischer, J., **10,** 203, 299
Fischer, J. L., **1,** 477

Fischer, K., **1,** 528; **2,** 104
Fischer, M., **2,** 378
Fischer, N., **2,** 104
Fischer, P., **11,** 103
Fischer, R., **11,** 587
Fischer, R. F., **1,** 694
Fischer, R. H., **2,** 119
Fischer, S., **4,** 248
Fischer, W., **2,** 417; **3,** 292
Fischer, W. F., **5,** 175
Fischer, W. F., Jr., **2,** 441; **3,** 112, 265, 278; **4,** 183; **6,** 203, 215
Fischli, A., **6,** 527; **7,** 289, 298; **10,** 103
Fish, M. S., **1,** 890
Fishbein, L., **1,** 519
Fishbein, P. L., **12,** 112
Fishel, D. L., **1,** 290, 1064; **2,** 155; **3,** 117; **8,** 366
Fisher, A., **3,** 329; **5,** 740
Fisher, B. S., **1,** 1136
Fisher, D., **2,** 130
Fisher, E. O., **11,** 401
Fisher, F., **1,** 1041; **3,** 178, 310
Fisher, J. F., **1,** 150, 1006; **2,** 245
Fisher, M. H., **2,** 173
Fisher, R. D., **4,** 14; **11,** 401
Fisher, R. P., **4,** 36, 38, 110; **6,** 195, 295; **7,** 181
Fishman, D., **10,** 184
Fishman, J., **1,** 731, 1041; **5,** 75
Fisk, M. T., **1,** 402, 682, 1163
Fisk, T. E., **10,** 289
Fitch, P. M., **10,** 107
Fitjer, L., **5,** 593; **12,** 187
Fitt, J. J., **5,** 527; **6,** 228, 229; **8,** 192; **10,** 75
Fittig, R., **1,** 870
Fitton, M., **9,** 274
Fitton, P., **6,** 573; **8,** 36; **9,** 42; **11,** 269
Fitzgibbon, W. E., **1,** 802
FitzJohn, S., **11,** 311
Fitzner, J. N., **12,** 121
Fitzpatrick, J. D., **6,** 150
Fitzpatrick, J. M., **3,** 284; **7,** 142
Fitzpatrick, J. T., **1,** 377
Fitzsimmons, B. W., **2,** 207
Fizet, C., **6,** 290; **9,** 246
Flahant, J., **9,** 279
Flaig, W., **1,** 8; **3,** 79
Flammang, M., **5,** 4, 601
Flann, C. J., **12,** 303
Flash, P. J., **3,** 45
Flaskamp, E., **5,** 389
Flatau, G. N., **5,** 527
Flatow, L., **1,** 954
Flautt, T. J., **1,** 730
Flavin, M., **1,** 351
Flechtner, T., **5,** 353
Flecker, P., **11,** 325
Fleet, G. W. J., **5,** 55, 142; **8,** 47; **9,** 57, 116; **10,** 48
Fleet, S. J., **9,** 116
Fleetwood, P., **1,** 619
Fleig, H., **5,** 319
Fleischmann, F. K., **5,** 649
Flematti, S. M., **2,** 104
Fleming, B., **5,** 186
Fleming, I., **5,** 334; **6,** 348; **7,** 375, 397; **8,** 197; **9,** 38, 108, 151, 363, 492, 522; **10,** 51, 162; **11,** 76; **12,** 212, 320
Fleming, M. P., **5,** 98, 392; **6,** 589; **7,** 369; **9,** 1, 293, 467
Fleming, N. B., **9,** 156
Fleming, S. A., **9,** 449
Fleming, W. C., **2,** 344
Fles, D., **1,** 982
Fletcher, A. S., **8,** 175
Fletcher, E. A., **8,** 500
Fletcher, G. A., **2,** 19
Fletcher, H. G., **2,** 163
Fletcher, H. G., Jr., **1,** 476; **5,** 226; **6,** 583
Fletcher, H., III, **7,** 116
Fletcher, J. H., **9,** 193
Fletcher, R. S., **1,** 1184
Fletcher, S. R., **9,** 361
Fletcher, T. L., **1,** 72, 444, 500, 1141, 1212; **2,** 159; **5,** 333; **6,** 17, 578; **10,** 346
Fleury, J. P., **7,** 285
Fliege, W., **5,** 148
Fligge, M., **1,** 756; **3,** 140
Flippin, L. A., **10,** 118
Fliri, H., **8,** 272
Fliszar, S., **1,** 49
Flitsch, W., **4,** 550; **10,** 78
Flitter, D., **1,** 1167, 1269
Flodin, P., **1,** 1001, 1002
Flood, D. T., **1,** 44, 396
Flood, S. H., **1,** 741, 1018, 1145
Flood, T., **10,** 436
Flood, T. C., **4,** 407, 544, 570; **8,** 499; **10,** 431
Floor, J., **6,** 600
Florent, J. C., **9,** 416
Flores, R. H., **1,** 125
Flores, R. J., **1,** 1203
Florey, A., **10,** 392
Florez, J., **11,** 303
Florian, W., **1,** 272, 282, 283; **4,** 167
Florio, S., **5,** 595
Flory, K., **1,** 224, 1196; **5,** 120
Flowers, H. M., **10,** 231
Floyd, D., **8,** 319, 325
Floyd, D. E., **1,** 1073
Floyd, D. M., **6,** 331; **9,** 333, 341; **11,** 373
Floyd, J. C., **5,** 748
Floyd, M. B., **3,** 139, 258, 338; **4,** 294; **10,** 295
Floyd, W. C., **8,** 22
Fludzinski, P., **9,** 480; **10,** 195, 243, 415; **11,** 70, 553
Flueckiger, E., **5,** 39
Flynn, D. L., **12,** 282
Flynn, F. F., **1,** 859
Flynn, G. A., **8,** 159
Foa, M., **5,** 473; **7,** 355
Fochi, R., **7,** 164; **8,** 34; **10,** 255
Fodor, G., **2,** 57
Foehlisch, B., **4,** 185, 568
Foell, T. J., **1,** 249
Foelsch, G., **1,** 235
Foersch, M., **5,** 475
Foettinger, W., **1,** 153
Foffani, A., **5,** 114
Fogel, E. R., **7,** 76; **8,** 126; **11,** 329
Fogel, J., **2,** 456
Fogelsong, W. D., **2,** 84
Foglia, T. A., **2,** 122; **4,** 247;

8, 161
Foldi, V. S., **1,** 235, 679
Foley, H. G., **5,** 98, 343
Foley, K. M., **3,** 299; **4,** 526
Foley, L. H., **12,** 150
Foley, P. J., Jr., **3,** 133
Folkers, K., **1,** 448, 730, 731, 1081, 1157; **2,** 213; **5,** 607
Follmann, H., **4,** 405; **7,** 299
Folsom, T. K., **3,** 33
Folting, K., **5,** 152
Foltz, C. M., **1,** 505
Fomum, Z. T., **12,** 570
Fones, W. S., **1,** 1081; **9,** 161
Fonken, A. E., **1,** 75
Fonken, G. J., **1,** 754; **2,** 353; **5,** 454
Fonken, G. S., **1,** 86, 93, 211, 213, 524
Font, J., **3,** 63; **4,** 573; **11,** 555
Fontaine, F., **1,** 320
Fontaine, G., **2,** 89
Fontaine, T. D., **1,** 598
Fontana, A., **6,** 419; **7,** 135; **8,** 51
Fookes, C. J. R., **12,** 492
Foos, J. S., **4,** 358, 457
Foote, C. S., **4,** 363; **5,** 491; **6,** 241, 436; **7,** 268, 307; **9,** 337, 341; **12,** 139
Foote, J. L., **1,** 484
Foote, R. S., **5,** 152
Forbes, C. P., **6,** 229
Forbes, E. J., **4,** 46
Forbes, G. S., **6,** 634
Forbes, W. F., **1,** 861
Forbus, T. R., Jr., **9,** 485
Ford, F. E., **12,** 547
Ford, M., **1,** 935
Ford, M. C., **1,** 1131
Ford, M. E., **5,** 429, 430, 475, 657; **6,** 360, 585; **7,** 231, 365; **8,** 477; **11,** 233; **12,** 383
Ford, S. H., **4,** 146
Ford, T. A., **1,** 228
Ford, T. M., **10,** 49
Fordham, W. D., **1,** 257; **6,** 263
Fordice, M. W., **2,** 273
Ford-Moore, A. H., **1,** 508, 685, 876, 1090, 1216, 1229; **5,** 517
Fordyce, C. R., **2,** 323
Foreman, G. M., **6,** 195
Forest, D. L., **7,** 101
Forestiere, A., **11,** 118
Forgione, P. S., **1,** 342; **2,** 268
Forkey, D. M., **4,** 230
Formaneck, H., **1,** 247
Formanek, K., **9,** 207
Formica, G., **7,** 312
Formo, M. W., **1,** 489, 508, 509, 511, 1087
Fornasier, R., **5,** 554; **6,** 586; **7,** 144; **8,** 340, 431
Fornefeld, E. J., **1,** 970; **2,** 351
Forrest, H. S., **1,** 881
Forrest, J., **1,** 50
Forrester, A. R., **2,** 174; **12,** 95
Forsek, J., **10,** 340
Forsman, J. P., **2,** 322
Forster, M. O., **1,** 124, 364
Fort, A. W., **1,** 162
Fort, D., **5,** 542
Fort, G., **2,** 4
Fort, R., **11,** 251
Fort, R. C., Jr., **2,** 21
Forte, P. A., **6,** 456
Fortenbaugh, R. B., **1,** 169
Fortes, C. C., **7,** 368
Fortey, E. C., **1,** 654
Fortin, R., **12,** 101, 576
Fortini, L., **10,** 120
Fortunak, J. M., **9,** 458; **10,** 390; **11,** 514
Fortunato, J. M., **6,** 348; **7,** 307
Fosbinder, R. T., **3,** 95
Fosker, A. P., **1,** 385, 736
Fossel, E. T., **2,** 84
Fosselius, G. A., **2,** 124
Foster, A. B., **1,** 68, 122, 1090
Foster, A. M., **4,** 512
Foster, C. H., **5,** 491; **6,** 436
Foster, D. G., **1,** 992; **9,** 35
Foster, D. J., **1,** 348, 353; **5,** 523
Foster, G. L., **1,** 654
Foster, H. M., **1,** 731, 1095, 1167
Foster, P. W., **1,** 1011; **2,** 368
Foster, R. T., **1,** 777
Foster, W. E., **1,** 1230
Fotadar, U., **5,** 214
Fothergill, G. A., **1,** 39, 50
Fouace, A. M., **1,** 503
Fouad, H., **6,** 422
Foucaud, A., **5,** 268; **8,** 416; **11,** 24
Foulon, J. P., **3,** 153; **6,** 622; **10,** 289, 290, 326
Fountain, K. R., **1,** 222; **9,** 480
Fouquet, G., **5,** 226
Fouquey, C., **12,** 49
Four, P., **10,** 413; **11,** 551; **12,** 525
Fournari, P., **8,** 59
Fourneron, J. D., **5,** 427
Fournier, P., **6,** 165
Fourrey, J. L., **12,** 241, 479
Fowler, D. L., **4,** 478
Fowler, F. W., **2,** 101, 223; **3,** 161; **6,** 376
Fowler, J. S., **3,** 289; **4,** 501
Fowler, K. W., **8,** 122
Fowler, R. B., **1,** 182
Fowles, C. W. A., **10,** 348
Fox, C. J., **3,** 116
Fox, D. P., **6,** 231; **8,** 412
Fox, F. P., **5,** 705
Fox, F. T., **1,** 673
Fox, G. J., **4,** 477; **7,** 352
Fox, H., **8,** 230
Fox, J. J., **1,** 1045; **2,** 412; **9,** 399
Fox, M. A., **11,** 203
Fox, W. B., **9,** 245
Foy, P., **2,** 310
Fracheboud, M. G., **3,** 127
Fraenkel, G., **1,** 431
Fraenkel-Conrat, H., **1,** 860, 873
Frahm, A. W., **11,** 411
Frahm, E. D. G., **1,** 139
Frahm, H., **1,** 333
Frahne, H. H., **11,** 359
Frainnet, E., **1,** 1218; **12,** 86
Frajerman, C., **2,** 110
Frampton, V. L., **1,** 892; **3,** 139

France, H., **1,** 754, 859
Francesch, C., **5,** 304
Franceschi, G., **9,** 491
Franchimont, M., **1,** 1118
Franchini, C., **7,** 315
Francillette, J., **9,** 274
Francis, T., **7,** 346
Francisco, C. G., **7,** 40
Franck, B., **1,** 67; **3,** 214; **4,** 256; **5,** 555
Franck, R. W., **3,** 53; **5,** 713; **6,** 415; **10,** 447
Franck-Neumann, M., **2,** 105, 429; **3,** 76; **5,** 182, 230; **6,** 407; **12,** 245
Franco, R., **1,** 730
Francotte, E., **11,** 476
Frank, B., **2,** 199, 263; **3,** 194; **12,** 257
Frank, F. J., **2,** 75
Frank, G. A., **1,** 49
Frank, R., **8,** 482
Frank, R. L., **1,** 110, 383, 500, 859, 1283
Frank, W. C., **8,** 382
Franke, A., **2,** 188
Franke, C., **3,** 254
Franke, G., **1,** 476
Franke, G. T., **8,** 340
Franke, K., **1,** 1015
Franke, W., **1,** 186, 285, 289
Franke, W. K., **4,** 120
Frankel, E. N., **12,** 36
Frankel, M., **2,** 43, 333
Frankel, M. B., **1,** 1226
Frankenfeld, J. W., **1,** 1184; **3,** 336
Franko, J. B., **10,** 361
Frankus, E., **1,** 581
Franz, J. A., **4,** 206, 207; **5,** 272; **6,** 239
Franz, J. E., **4,** 25
Franzen, H., **12,** 159
Franzen, V., **1,** 81, 207, 318, 1015, 1203, 1231; **2,** 42, 434; **5,** 282
Franzus, B., **1,** 247, 866; **2,** 139, 368
Frasca, A. R., **2,** 336; **5,** 266
Fraser, A. R., **5,** 113
Fraser, B. A., **10,** 329
Fraser, L. R., **11,** 383
Fraser, M., **1,** 129, 219
Fraser, M. M., **1,** 659
Fraser, R. R., **7,** 207; **9,** 364
Fraser, W. A., **1,** 911
Fraser-Reid, B., **1,** 73; **4,** 411; **5,** 389; **7,** 196; **8,** 331, 428; **9,** 114; **10,** 213; **11,** 269, 453; **12,** 527
Frasnelli, H., **3,** 223; **4,** 325; **5,** 296
Fratantoni, J. C., **1,** 332, 377
Frater, G., **5,** 212; **6,** 628; **10,** 243
Frauenrath, H., **10,** 56
Fray, G. I., **1,** 34
Frazee, W. J., **5,** 455; **8,** 214; **9,** 290; **12,** 239
Frazer, M. G., **7,** 393
Frazer, M. J., **1,** 1158, 1163; **7,** 118
Frazer, V. S., **7,** 393
Frazier, H. W., **1,** 1027; **6,** 229, 252
Frazier, J. O., **12,** 568
Frazier, K. A., **9,** 298; **10,** 219, 418
Frazza, M. S., **11,** 539
Freasier, B. F., **5,** 249
Frecchi, A., **9,** 391
Freche, A., **3,** 264; **4,** 444
Frechet, J. M. J., **6,** 26, 657; **7,** 33, 284; **9,** 60
Fredenhagen, K., **1,** 15
Fredga, A., **9,** 430
Freeborn, E., **1,** 1270
Freedman, H. H., **1,** 970; **5,** 652; **6,** 566; **7,** 338
Freedman, L., **1,** 386
Freedman, L. D., **2,** 25
Freedman, M. B., **7,** 391
Freedman, R., **2,** 187
Freeguard, G. F., **1,** 206, 207, 1055; **2,** 378
Freeman, D., **6,** 574
Freeman, F., **2,** 79; **3,** 62; **4,** 99
Freeman, J. P., **1,** 6, 254, 329, 561, 707, 734, 742, 808, 827, 1081, 1140; **2,** 135; **6,** 128, 646
Freeman, P. K., **4,** 41; **5,** 681; **7,** 201; **10,** 240; **12,** 290
Freeman, P. W., **5,** 257
Freeman, R. C., **1,** 253, 1192
Freeman, W. A., **12,** 386
Freeman, W. R., **1,** 1131
Freenor, F, J., **1,** 1249; **5,** 731; **7,** 292
Freer, A. A., **8,** 79
Freerks, R. L., **11,** 43
Freerksen, R. W., **8,** 168, 229
Frehel, D., **6,** 441
Frei, B., **10,** 321
Freiberg, L. A., **1,** 643
Freidinger, R. M., **5,** 97
Freidlander, B. T., **9,** 129
Freifelder, M., **1,** 982, 983, 986
Freire, R., **7,** 40; **11,** 445
Freitag, D., **4,** 258
Frejd, T., **8,** 29
Fremery, M., **3,** 339
Fremery, M. I., **1,** 777, 892
Fremont, S. L., **12,** 282
French, J. C., **1,** 345, 1090; **5,** 283
French, L. G., **11,** 188
Frenette, R., **10,** 463; **12,** 576
Frensdorff, H. K., **4,** 143
Freon, P., **1,** 235
Freppel, C., **4,** 498
Frering, R., **5,** 542
Fresco, J. R., **1,** 257
Freudenberg, K., **1,** 1187; **2,** 417
Freure, B. T., **1,** 636
Frey, A. J., **1,** 235, 764, 942
Frey, D. A., **1,** 1011
Frey, D. W., **2,** 344
Frey, H., **1,** 1000
Frey, H. O., **4,** 186
Frey, W., **1,** 531
Freyberger, G., **12,** 576
Freyer, W., **6,** 638
Freyermuth, H. B., **1,** 477; **10,** 205
Friary, R. J., **2,** 195; **3,** 182; **5,** 511
Frick, U., **11,** 587
Fricke, F., **1,** 195
Fricke, H., **1,** 166, 178; **3,** 69

Frickel, F., **5,** 262
Frickey, D. G., **2,** 182
Fridkin, M., **8,** 57; **11,** 347
Fried, J., **1,** 56, 74, 139, 150, 581, 617, 782, 802; **2,** 29; **4,** 146, 545; **5,** 326, 438; **12,** 167, 335
Fried, J. H., **1,** 671, 1000; **2,** 34, 195, 201, 362, 372, 381; **3,** 112, 118, 265, 278; **4,** 577; **5,** 120, 216, 360
Friedel, R. A., **1,** 228, 615
Friedemann, O., **1,** 561
Friediger, A., **2,** 62
Friedlander, B. T., **5,** 127
Friedli, H., **1,** 1005
Friedman, B. S., **1,** 1157
Friedman, C., **1,** 155
Friedman, L., **1,** 46, 131, 212, 281, 309, 373, 392, 445, 531, 696, 710, 1064, 1073, 1163, 1187, 1194; **2,** 26, 422; **5,** 359; **6,** 203
Friedman, N., **6,** 166
Friedman, O. M., **1,** 198, 563
Friedman, S., **1,** 228, 570, 1000; **5,** 205
Friedrich, E., **8,** 16, 374
Friedrich, E. C., **3,** 277; **6,** 254, 604
Friedrich, K., **4,** 389
Friedrich, L. E., **5,** 454; **10,** 332
Friege, H., **9,** 102
Friend, A. W., **1,** 595
Friend, N. A. C., **1,** 999
Fries, K., **2,** 414
Frimer, A., **6,** 490; **7,** 307
Frimer, A. A., **8,** 101, 208, 248; **9,** 393
Fricke, G. M., **11,** 388
Fringuelli, F., **5,** 624; **9,** 13
Fringuelli, R., **9,** 408; **11,** 156
Frisch, M. A., **1,** 235, 1178
Frisell, C., **1,** 101, 266
Frisque-Hesbain, A. M., **11,** 111
Fristad, W. E., **7,** 268; **8,** 290; **9,** 341, 499; **10,** 96; **12,** 33, 212
Fritsch, N., **11,** 327
Fritsch, W., **1,** 139, 488; **4,** 296; **5,** 119, 633
Fritschel, S. J., **11,** 7
Fritz, C. G., **1,** 1148
Fritz, E., **10,** 65
Fritz, H., **9,** 246; **11,** 362
Fritz, H. G., **6,** 317
Fritz, H. J., **7,** 414
Fritz, J. S., **1,** 802; **2,** 13
Fritze, P., **2,** 179; **4,** 354, 356
Fritzsche, H., **2,** 179; **6,** 509
Fritzsche, U., **5,** 165
Frobese, A. S., **8,** 303
Froehlich, H., **5,** 148
Froehlich, K., **6,** 306
Froeling, A., **6,** 651
Froemsdorf, D. H., **1,** 1187
Frohlich, H. O., **11,** 373
Froitzheim-Kuehlhorn, H., **1,** 832
Frolow, F., **12,** 463
Fromme, I., **1,** 45
Frommeld, H. D., **1,** 1235; **11,** 103
Frostick, F. C., Jr., **1,** 70, 790, 1118
Fruchey, O. S., **8,** 516; **10,** 450
Fruit, R. E., Jr., **1,** 397
Frulla, F. F., **1,** 619
Fruton, J. S., **1,** 339, 379
Fry, A. J., **1,** 1018; **3,** 251; **5,** 470; **9,** 291, 414
Fry, E. M., **1,** 1096, 1203; **4,** 579
Fry, J. L., **5,** 694; **7,** 388; **8,** 501; **12,** 462
Fryble, C. B., **10,** 50
Frye, L. L., **11,** 315; **12,** 298
Frye, S. V., **12,** 239
Fryzuk, M. D., **9,** 458
Fu, P. P., **6,** 658; **8,** 155, 383; **9,** 102, 238
Fu, S. C. J., **1,** 904
Fu, T. H., **7,** 277
Fu, Y. L., **8,** 167
Fuchs, G., **8,** 315
Fuchs, H. G., **1,** 818
Fuchs, O., **1,** 603; **2,** 57; **5,** 89
Fuchs, P. L., **3,** 136; **4,** 204, 220, 304, 551; **5,** 91, 445; **6,** 77, 234; **7,** 32, 283; **8,** 33, 284, 396, 492; **9,** 385; **10,** 168; **11,** 442; **12,** 402
Fuchs, R., **2,** 359; **3,** 244; **6,** 279
Fuegen-Koester, B., **11,** 401
Fueldner, H. W., **6,** 375
Fueno, T., **9,** 290
Fuentes, L. M., **9,** 58; **10,** 91; **11,** 249; **12,** 32, 384
Fuerst, A., **1,** 444, 445, 595, 664; **2,** 211, 329, 362; **8,** 424; **11,** 287
Fugger, J., **1,** 169
Fugiel, A. A., **9,** 408
Fugnitto, R., **1,** 1258; **5,** 414
Fuhr, H., **4,** 412; **5,** 511, 681
Fuhr, K. H., **4,** 290; **9,** 183; **10,** 159
Fuhrer, W., **10,** 71
Fuji, K., **6,** 374; **9,** 63, 258; **10,** 11, 51, 52; **11,** 29; **12,** 29, 30
Fuji, T., **9,** 297
Fujii, M., **12,** 233
Fujii, N., **5,** 702; **9,** 297; **10,** 398
Fujii, S., **5,** 433; **6,** 233; **7,** 155, 322; **10,** 185
Fujii, T., **1,** 329; **5,** 427, 691; **10,** 237; **11,** 279; **12,** 64
Fujikura, Y., **6,** 359, 526; **7,** 349
Fujimori, K., **8,** 345; **10,** 220
Fujimoto, J., **3,** 228
Fujimoto, K., **9,** 187; **10,** 450
Fujimoto, M., **1,** 1198; **11,** 29; **12,** 29
Fujimoto, T. T., **5,** 406; **6,** 32
Fujimoto, Y., **7,** 225, 338, 373; **11,** 283
Fujinami, T., **6,** 96; **7,** 47
Fujino, A., **5,** 308
Fujino, M., **1,** 40; **2,** 24, 54, 305, 307; **5,** 360; **6,** 291; **11,** 330
Fujioka, A., **11,** 168
Fujioka, M., **11,** 70
Fujisawa, T., **6,** 260; **7,** 155, 424; **9,** 431; **10,** 289; **11,** 141, 221, 235, 249, 293,

514; **12,** 25, 33, 125, 201, 282, 477, 555, 565
Fujishita, T., **6,** 662; **9,** 516
Fujita, E., **1,** 1168; **7,** 33, 365; **9,** 13, 63, 221, 258, 315, 462; **10,** 8, 11, 51, 52; **11,** 1, 29, 270, 324, 516, 518, 519, 544; **12,** 29, 30, 259, 485, 491, 523
Fujita, F., **6,** 374
Fujita, H., **12,** 432
Fujita, K., **7,** 361; **8,** 135; **11,** 240, **12,** 397
Fujita, M., **9,** 111; **12,** 209, 444
Fujita, N., **5,** 602
Fujita, S., **6,** 187; **11,** 429, 599
Fujita, T., **1,** 53; **2,** 289; **3,** 208; **4,** 240, 349, 350, 572; **5,** 470, 503; **6,** 415; **9,** 431
Fujita, Y., **9,** 316; **11,** 103
Fujiwara, H., **5,** 433; **11,** 427
Fujiwara, J., **12,** 11, 193, 344, 378
Fujiwara, K., **1,** 250, 425, 1217; **10,** 25
Fujiwara, Y., **2,** 389; **7,** 277; **8,** 526; **11,** 249
Fujiyama, R., **12,** 150
Fujiyoshi, K., **3,** 208; **4,** 349
Fujui, H., **8,** 172
Fukada, N., **4,** 395
Fukagawa, T., **11,** 249
Fukami, H., **2,** 263
Fukamiya, N., **7,** 307
Fukata, F., **12,** 59
Fukatsu, S., **9,** 317; **10,** 273
Fukaya, C., **12,** 350
Fuks, R., **2,** 134; **4,** 111; **5,** 308
Fukuda, H., **8,** 231
Fukuda, N., **11,** 496
Fukuda, T., **5,** 360; **6,** 291; **11,** 330
Fukui, K., **6,** 436; **9,** 138
Fukui, M., **7,** 225; **12,** 573
Fukui, T., **7,** 286
Fukumi, H., **8,** 115
Fukumoto, K., **4,** 407, 428; **5,** 157; **6,** 283; **7,** 329; **10,** 245; **11,** 40; **12,** 44, 477
Fukumoto, T., **4,** 526; **11,** 127
Fukunaga, K., **6,** 555; **11,** 255, 492
Fukunaga, T., **3,** 183; **12,** 559
Fukuoka, M., **2,** 108
Fukuoka, R., **12,** 291
Fukuoka, S., **4,** 302; **12,** 428
Fukure, R. T., **7,** 377
Fukushima, D., **8,** 204; **9,** 92
Fukushima, D. K., **1,** 1283; **5,** 75
Fukushima, H., **12,** 191
Fukushima, M., **11,** 239, 240
Fukutani, Y., **12,** 193, 344, 480
Fukuyama, M., **3,** 219
Fukuyama, T., **4,** 86; **5,** 183; **6,** 611; **7,** 135; **8,** 64; **9,** 110, 138, 174
Fukuyama, Y., **7,** 406
Fukuzaki, K., **12,** 490
Fukuzawa, A., **6,** 582
Fukuzawa, S., **11,** 397; **12,** 121, 452
Fukuzumi, K., **5,** 332; **6,** 653; **7,** 109, 314
Full, R., **10,** 454
Fuller, A. E., **12,** 258
Fuller, C. J., **8,** 47
Fuller, G., **1,** 1145
Fuller, G. B., **6,** 496
Fuller, S. E., **9,** 361; **12,** 123
Fullerton, D. S., **3,** 57; **4,** 304, 437; **7,** 70
Fullerton, T. J., **4,** 370; **5,** 503; **7,** 316; **8,** 217
Fulmer, R. W., **1,** 652
Fulmor, W., **1,** 40
Fulton, B. S., **12,** 403
Fulton, R. P., **10,** 390
Fumimoto, Y., **9,** 269
Funabashi, Y., **12,** 325
Funabiki, T., **9,** 207
Funahashi, T., **1,** 409
Funakoshi, K., **12,** 131
Funakoshi, S., **9,** 297; **10,** 398
Funakura, M., **4,** 158
Funamizu, M., **6,** 170
Funasaka, W., **2,** 273, 424
Fung, A. P., **9,** 244, 320, 396; **10,** 81, 115, 207, 275; **11,** 256, 355; **12,** 263
Fung, D. P. C., **1,** 1261
Fung, M. K., **6,** 469
Fung, N. Y. M., **9,** 477
Fung, S., **7,** 268; **10,** 421
Funk, H. R. L., **7,** 218
Funk, K. F., **1,** 1123
Funk, K. W., **5,** 477
Funk, M. O., **6,** 167; **7,** 94
Funk, R. L., **6,** 154; **7,** 84, 95; **8,** 147; **9,** 143; **12,** 14, 323
Funke, B., **5,** 250
Funke, P., **7,** 244
Funke, W., **4,** 22
Fuoss, R. M., **1,** 1145
Fuqua, S. A., **1,** 1058; **2,** 281
Furakawa, S., **10,** 323
Furlenmeier, A., **2,** 211, 362
Furrer, M., **1,** 747, 963
Furstoss, R., **3,** 252
Furui, S., **6,** 224
Furuichi, K., **10,** 279
Furukawa, H., **4,** 175; **12,** 167
Furukawa, J., **1,** 253; **2,** 134; **3,** 141; **6,** 451; **9,** 290
Furukawa, M., **5,** 226; **7,** 101; **9,** 353; **10,** 221
Furukawa, N., **5,** 57, 104, 639; **6,** 369; **8,** 200; **9,** 191; **10,** 174
Furukawa, S., **5,** 42; **6,** 392
Furukawa, Y., **12,** 256
Furusaki, A., **8,** 441
Furusato, M., **5,** 9
Furuta, K., **11,** 378, 571; **12,** 356
Furuya, Y., **1,** 500
Fuse, Y., **11,** 342
Fuson, R. C., **1,** 156, 402, 563, 904, 1087, 1294; **11,** 281; **12,** 207
Fuzesi, L., **7,** 237

Gaasbeek, C. J., **3,** 66
Gabbard, R. B., **1,** 808
Gabhe, S. Y., **8,** 208
Gabobardes, M. R., **11,** 219
Gabor, V., **1,** 603; **2,** 27, 57
Gabriel, S., **1,** 710; **8,** 73
Gabrijelcic, D., **9,** 472

Gadecki, F. A., **1,** 1258; **4,** 568
Gadelle, C., **5,** 537
Gadke, W., **1,** 992
Gadola, M., **5,** 576
Gadsby, B., **2,** 355; **3,** 204
Gadwood, R. C., **12,** 73, 411
Gaede, B., **9,** 491
Gaertner, V. R., **3,** 122
Gaffney, B. L., **6,** 84, 636
Gaffney, J. S., **5,** 511
Gaffron, H., **1,** 1128
Gagen, J. E., **7,** 252
Gagneux, A., **1,** 93, 276
Gagnier, R. P., **11,** 105
Gagolski, J., **2,** 23
Gaibel, Z. L. F., **1,** 68
Gaidis, J. M., **5,** 652
Gaiffe, A., **1,** 730
Gais, H. J., **5,** 198; **8,** 77, 184; **9,** 178; **12,** 199
Gait, M. J., **7,** 414
Gaj, B. J., **1,** 1256
Gajda, T., **7,** 20; **8,** 146; **11,** 407
Gajewski, R. P., **10,** 159
Gajewski, W., **10,** 359
Gal, G., **5,** 62
Galamb, V., **12,** 475
Galambos, G., **10,** 240
Galantay, E., **5,** 389
Galante, J. J., **5,** 677; **6,** 534
Galat, A., **1,** 66, 518
Galbraith, A. R., **1,** 159; **2,** 19
Galbraith, M. N., **3,** 24
Gale, D. M., **1,** 563; **4,** 21
Galeazzi, E., **5,** 605
Galeeva, R. I., **9,** 343
Gall, J. H., **9,** 441
Gall, M., **3,** 312; **4,** 302; **10,** 276
Gall, W. G., **11,** 79
Gallagher, G., **6,** 453
Gallagher, G., Jr., **5,** 694
Gallagher, T., **11,** 465; **12,** 434, 483
Gallagher, T. F., **1,** 74, 526, 1178, 1283, 1284
Gallais, F., **1,** 618
Galle, J. E., **4,** 323; **7,** 47; **9,** 233, 311
Galledou, B. S., **9,** 227; **12,** 490
Galli, C., **6,** 479; **9,** 484; **10,** 420
Galli, G., **7,** 70; **11,** 177
Galli, R., **4,** 237
Galliani, C., **9,** 491
Galliani, G., **11,** 445
Gallina, C., **9,** 334
Gallo, R., **7,** 124
Gallois, M., **8,** 150
Gallois, P., **10,** 365
Galloy, J., **11,** 108
Galls, J. E., **9,** 146
Galpin, I. J., **6,** 425
Galt, R. H. B., **5,** 639
Galton, S. A., **1,** 342
Galynker, I., **11,** 605
Gambacorta, A., **7,** 20
Gambale, R. J., **10,** 295
Gamble, E. L., **1,** 67
Gambos, G., **8,** 260
Gammill, R. B., **7,** 90, 302; **8,** 179; **11,** 187, 327; **12,** 275
Gamoh, K., **11,** 570
Gan, I., **5,** 709
Gancarz, R. A., **11,** 408
Gancher, E., **2,** 156
Ganellin, C. R., **1,** 978; **2,** 273
Ganem, B., **2,** 263, 369; **5,** 266, 406, 463, 732; **6,** 114, 260, 348, 407, 455, 492, 639; **7,** 45, 85, 88, 89, 216, 307; **8,** 56, 373, 503, 505; **9,** 99, 245, 300; **10,** 99, 122, 206, 259; **11,** 76, 138, 208, 460, 483, 551; **12,** 70, 158, 203, 444
Gange, D., **9,** 272
Ganguly, A. K., **2,** 200; **3,** 147; **7,** 158
Gani, V., **5,** 266
Ganis, P., **4,** 480
Gannon, J. J., **6,** 128
Gannon, W. F., **1,** 180, 195, 594, 1178
Ganschow, S., **6,** 36
Gante, J., **2,** 29
Gantseva, B. V., **11,** 341
Gao, Y., **12,** 94
Gaona, R. T., **5,** 88
Gaoni, Y., **1,** 159; **2,** 88, 278; **4,** 452; **5,** 33; **6,** 675; **7,** 429; **8,** 238, 289
Gapinski, D. M., **11,** 254
Gapinski, R. E., **9,** 16, 262
Garad, M. V., **12,** 13
Garapon, J., **10,** 85; **11,** 118
Garbarino, J. A., **6,** 109
Garbe, J. E., **8,** 310
Garbisch, E. W., Jr., **1,** 14; **3,** 131; **4,** 441, 459
Garbrecht, W. L., **1,** 1126; **7,** 337
Garcia, A., **5,** 762
Garcia, B. A., **6,** 224, 586; **11,** 346; **12,** 445
Garcia, E. E., **1,** 54, 1123
Garcia, G. A., **6,** 101, 339; **7,** 199
Garcia, H., **1,** 808
Garcia, J. L., **11,** 132
Garcia, T., **12,** 199
Garcia-Munoz, G., **1,** 982
Garcia-Sharp, F. J., **2,** 156
Garcia-Slanga, B. J., **12,** 381
Gard, G., **2,** 458
Gardano, A., **7,** 355
Garden, J. F., **1,** 685
Gardent, J., **1,** 407
Gardi, R., **1,** 1210
Gardiner, D., **2,** 359
Gardini, G. P., **3,** 38
Gardner, J. H., **1,** 951, 1293
Gardner, J. N., **2,** 232, 344, 355, 433; **5,** 225
Gardner, J. O., **4,** 409
Gardner, P. D., **1,** 309, 318, 444, 689, 904, 926, 927, 1197, 1253; **2,** 376
Gardner, T. S., **1,** 289
Garegg, P. J., **9,** 121, 508; **10,** 329, 350; **11,** 467, 483; **12,** 71, 555
Garfunkel, E., **9,** 189
Garg, C. P., **1,** 144, 207, 625; **4,** 96
Garg, S. P., **4,** 438
Gariano, A. L., **8,** 501
Gariano, P., **4,** 529
Gariano, P., Jr., **5,** 693; **8,** 501; **9,** 483

Gariboldi, P., **3,** 158
Garigipati, R. S., **12,** 455
Garin, D. L., **5,** 411
Garito, A. F., **7,** 165, 387
Garland, R., **12,** 564
Garland, R. B., **5,** 144; **9,** 269
Garmaise, D. L., **1,** 1178; **3,** 51; **6,** 585
Garnatt, J. L., **4,** 215
Garneau, F. X., **6,** 190
Garner, B. J., **1,** 310, 859
Garner, P., **11,** 460; **12,** 314
Garner, R., **2,** 334; **4,** 391
Garnett, J. L., **2,** 182; **3,** 134; **5,** 287; **6,** 504
Garnier, B., **5,** 713
Garnier, F., **6,** 658
Garrard, T. F., **4,** 53
Garratt, D. G., **10,** 22
Garratt, P. J., **5,** 471; **6,** 471
Garratt, S., **1,** 236
Garrett, E. R., **1,** 787, 1118
Garrett, P. E., **7,** 64, 200
Garrett, P. J., **2,** 88
Garrison, J., **1,** 320
Garrison, P. J., **8,** 323; **11,** 179
Garrison, W. E., Jr., **1,** 1095
Garry, R., **8,** 196
Garson, L. R., **3,** 128
Garst, J. F., **2,** 182, 289
Garst, M. E., **4,** 197; **5,** 294; **9,** 75; **10,** 40, 398; **12,** 234
Garver, L. C., **11,** 200, 214
Garvey, D. S., **10,** 119
Garwood, R. F., **1,** 170, 1132; **2,** 263
Gascoigne, J. A., **1,** 413
Gaset, A., **11,** 214, 346, 433
Gash, V. W., **1,** 652
Gasic, G. P., **8,** 259
Gasic, M., **3,** 171
Gaspar, P. P., **3,** 69; **6,** 577
Gasparrini, F., **6,** 162; **7,** 175; **8,** 250; **9,** 432
Gasquez, E. C., **5,** 427
Gass, J. D., **1,** 633
Gasser, P., **5,** 283
Gasser, R. J., **3,** 123
Gassman, P. G., **1,** 125, 247, 606, 814, 927; **2,** 87, 144, 173, 344, 379; **3,** 66, 69, 123; **4,** 191; **5,** 115, 132, 261, 262, 389, 574, 661; **6,** 91, 119, 120, 248; **7,** 289; **8,** 105, 189, 416; **9,** 128, 364, 381, 416; **11,** 150; **12,** 150, 532
Gastambide, B., **5,** 600
Gasteiger, J., **4,** 405; **6,** 24
Gaston, D. W., **4,** 223
Gaston, L. K., **1,** 658
Gates, J. W., Jr., **1,** 196, 1167; **2,** 10, 320; **4,** 377
Gates, M., **1,** 322, 392, 444, 927, 1203; **7,** 349
Gates, M. D., Jr., **4,** 579
Gati, A., **1,** 1171
Gatica, J., **1,** 150, 1090
Gattermann, L., **1,** 383, 455, 1283, 1294
Gatto, V. J., **12,** 147
Gattuso, M. J., **7,** 389
Gatzke, A. L., **2,** 79
Gaudemar, M., **1,** 1287; **3,** 336; **7,** 116; **12,** 423
Gaudemer, A., **5,** 413, 642
Gaudemer, F., **5,** 413
Gaudiano, G., **1,** 221; **2,** 173; **3,** 125; **4,** 197
Gaudino, J., **12,** 537
Gaudry, M., **7,** 35; **8,** 194
Gaudry, R., **1,** 219, 402, 412, 562, 614, 633, 772
Gaughan, E. J., **1,** 162, 222; **5,** 44; **9,** 298
Gaul, J. M., **3,** 299; **4,** 526
Gaule, A., **2,** 105
Gault, H., **1,** 74
Gault, R., **4,** 418
Gaur, S. P., **7,** 11
Gausing, W., **8,** 69
Gautechi, F., **2,** 422
Gauthier, J., **2,** 36
Gauthier, J. Y., **12,** 530
Gauthier, R., **4,** 425
Gautier, A., **1,** 844
Gaver, R. C., **2,** 208
Gavrilenko, V. V., **1,** 1030; **3,** 259
Gawish, A., **11,** 194
Gawley, R. E., **10,** 324; **11,** 75; **12,** 64
Gawne, G., **3,** 153
Gawronski, J. K., **12,** 126
Gay, R. L., **2,** 53
Gaylord, N. G., **1,** 582, 594, 595, 1171
Gazawa, K., **8,** 115
Gazdar, M., **1,** 472
Gazis, E., **1,** 745, 1256, 1261
Gebelein, C. G., **2,** 224; **3,** 163
Gebert, W., **1,** 966
Gebeyehu, G., **11,** 24
Geckeler, K., **10,** 175
Gedge, D. R., **9,** 431
Gee, M., **1,** 685
Gee, R. D., **2,** 254
Gee, V., **9,** 265
Geering, E. J., **1,** 342
Geevers, J., **9,** 177
Gehlhaus, J., **4,** 510; **5,** 650
Geib, J. R., **5,** 107
Geiger, K., **1,** 386
Geiger, R., **3,** 103, 156; **5,** 342; **6,** 288; **7,** 247
Geiger, U. P., **2,** 7
Geisel, M., **5,** 370, 580
Geiss, F., **9,** 259
Geiss, K., **5,** 86
Geiss, K. H., **4,** 285
Geisselmann, G., **9,** 167
Geissler, G., **1,** 1141
Geistlich, P., **1,** 652
Geke, J., **6,** 420
Gelas, G., **11,** 329
Gelas, J., **12,** 318
Gelas-Mialhe, Y., **10,** 174
Gelbard, G., **3,** 336; **8,** 264; **10,** 306
Gelblum, E., **1,** 211; **3,** 81
Gelboin, H. V., **11,** 501
Gelin, S., **8,** 311
Geller, H. H., **1,** 257
Gelleri, A., **5,** 601
Gellert, E., **3,** 267
Gellert, H. G., **1,** 1189, 1198
Gelotte, B., **1,** 1002
Geluk, H. W., **3,** 228
Gemal, A. L., **6,** 3; **9,** 99, 211, 422; **10,** 366; **11,** 198, 232, 373

Gemenden, C. W., **4,** 293
Gemmer, R. V., **8,** 534
Genco, N. A., **5,** 358
Gendler, P. L., **4,** 194
Gendreau, Y., **10,** 194
Geneste, J. M., **1,** 90
Geneste, P., **10,** 342
Genet, J. P., **7,** 358; **8,** 39; **9,** 257; **10,** 267, 268; **12,** 475
Gennari, C., **9,** 475; **10,** 409; **11,** 315, **12,** 226, 298, 320, 510
Gensler, W. J., **1,** 262, 1141; **2,** 311; **3,** 139, 258, 338; **4,** 46, 294; **5,** 532
Gent, P. A., **5,** 740
Geoffrey, G. L., **7,** 60
Geoghegan, P. J., Jr., **2,** 267; **3,** 196; **4,** 322
Georg, G. I., **12,** 154
George, A. V. E., **11,** 473
George, D. A., **5,** 285
George, J., **12,** 503
George, M. V., **2,** 263; **4,** 173; **9,** 323; **11,** 209
George, T. J., **5,** 749
Georges, L. W., **1,** 1101
Georges, M., **11,** 269
Georghiou, P. E., **5,** 331, 667
Georgiadis, G. M., **10,** 71
Georgian, V., **1,** 444
Georgoulis, C., **6,** 233, 280; **8,** 260; **9,** 296; **10,** 343
Geraghty, M. B., **6,** 530; **8,** 492
Gerdes, H. M., **7,** 79; **9,** 186
Gerdes, J. M., **8,** 94; **9,** 107
Gergel, W. C., **1,** 518
Gerhart, F., **3,** 223; **4,** 325; **7,** 215
Gerhart, H. L., **1,** 18, 1131, 1163
Geribaldi, S., **5,** 109
Gerin, G., **11,** 600
Gerkin, R. M., **4,** 297; **6,** 103; **7,** 220
Gerlach, H., **4,** 343, 397; **5,** 457, 542; **6,** 107, 247, 520, 625; **7,** 142; **8,** 511; **9,** 93
Germain, A., **6,** 103
Germain, G., **11,** 111
Germann, F. E. E., **1,** 350
Germeraad, P., **5,** 639
Gero, S. D., **1,** 68, 180, 257; **5,** 287, 595, 642; **11,** 106, 250
Gerold, C., **1,** 74, 1000, 1107; **9,** 380
Gerrard, W., **1,** 68, 673, 1158, 1163; **4,** 46; **8,** 60
Gershanov, F. B., **4,** 20
Gershon, H., **1,** 808; **5,** 667
Gersmann, H. R., **1,** 268, 927, 1254
Gerstl, R., **2,** 135
Gerstmans, A., **12,** 231
Gerszberg, S., **5,** 88
Gerzon, K., **1,** 15, 802; **2,** 17
Geschwinder, P. M., **12,** 450
Gesing, E. R. F., **10,** 127; **12,** 526
Gesson, J. P., **7,** 167; **10,** 136, 354
Geursen, H. S., **2,** 360
Geyer, U., **2,** 213
Ghaderi, E., **8,** 21
Ghadiri, M. R., **12,** 29
Ghali, N. I., **11,** 88
Gharia, M. M., **8,** 228
Gharpure, S. B., **5,** 86
Ghatak, U. R., **1,** 599; **3,** 65; **5,** 161; **6,** 70; **7,** 387; **10,** 426
Ghatge, B. B., **1,** 570
Ghattas, A. B. A. G., **11,** 491
Ghazala, M., **7,** 381; **10,** 360
Gheorghiu, M. D., **4,** 57
Ghera, E., **1,** 802, 927, 1254; **2,** 462; **3,** 337; **5,** 760; **6,** 675; **7,** 429; **8,** 451
Ghiringhelli, D., **11,** 283
Ghosal, S., **9,** 151
Ghosez, L., **1,** 222; **2,** 81, 118; **4,** 95, 135; **5,** 99, 138, 219, 684; **6,** 123; **7,** 66; **9,** 105, 396; **10,** 60; **11,** 51, 111, 176, 201, 407, 562; **12,** 125, 183
Ghosh, P. B., **1,** 1044; **2,** 376
Ghosh, R., **11,** 536
Ghosh, S., **7,** 387; **10,** 110, 363, 426; **11,** 143; **12,** 145
Ghosh, S. R., **3,** 65
Ghosh, S. S., **7,** 321
Ghribi, A., **12,** 350
Giacobbe, T. J., **2,** 406
Giacomelli, G., **11,** 563
Giacomini, E., **10,** 333
Giacomoni, J. C., **9,** 306
Giacopello, D., **3,** 171
Giam, C. S., **4,** 295; **8,** 267
Gianetti, J. P., **4,** 393
Giannangeli, M., **8,** 425
Giannini, D., **3,** 265
Giannis, A., **10,** 373
Giannoli, B., **6,** 607
Gianotti, R., **1,** 783
Giants, T. W., **4,** 282
Gibbons, C., **4,** 14; **5,** 441
Gibbons, E. G., **6,** 634; **11,** 342
Gibbs, C. G., **5,** 516
Gibbs, D. E., **1,** 1254; **7,** 170
Gibbs, J. J., **5,** 537
Gibian, H., **1,** 231
Gibson, A. R., **7,** 238
Gibson, D. H., **5,** 339
Gibson, D. T., **5,** 460
Gibson, G. W., **2,** 53
Gibson, H., **10,** 359
Gibson, J. A., **7,** 242
Gibson, M., **2,** 462
Gibson, M. S., **3,** 67; **5,** 499; **7,** 166
Gibson, T., **10,** 456
Gibson, T. W., **7,** 374
Gick, H., **1,** 62
Giddey, A., **12,** 492
Giddings, B., **1,** 1196
Giddings, P. J., **10,** 21
Gidley, G. C., **2,** 50; **3,** 165
Gierer, J., **5,** 580
Giering, W. P., **3,** 101; **5,** 686; **6,** 155, 539; **8,** 41, 455
Giersch, W., **5,** 424
Giesbrecht, E., **5,** 281
Giese, B., **10,** 254; **11,** 317; **12,** 303, 523
Giese, H. Y., **11,** 529
Giese, R. W., **3,** 20, 153
Gieseche, H., **7,** 23
Giesemann, G., **11,** 112

Giezynski, R., **7,** 374
Gifford, M., **9,** 293
Giga, A., **5,** 564, 705
Giger, R., **5,** 136
Gigg, J., **2,** 159
Gigg, R., **1,** 309; **2,** 159; **3,** 122; **5,** 740; **12,** 363
Giguere, R. J., **11,** 299
Gil, G., **6,** 223; **7,** 281; **8,** 302; **9,** 323; **12,** 464
Gilani, S. S. H., **1,** 850; **2,** 282, 325, 326, 445
Gil-Av, E., **1,** 1023
Gilbert, A., **1,** 465
Gilbert, D. G., **5,** 261
Gilbert, D. P., **5,** 389; **8,** 189
Gilbert, E. C., **2,** 229; **3,** 166
Gilbert, E. E., **1,** 667; **2,** 346, 353; **4,** 7, 206
Gilbert, J. C., **8,** 454; **9,** 181; **10,** 157; **12,** 187
Gilbert, K. E., **10,** 336
Gilchrist, T. L., **3,** 171; **5,** 370; **9,** 214
Giles, P. M., Jr., **5,** 100
Giles, R. G. F., **10,** 81
Gilham, P. T., **2,** 364; **3,** 122
Giliani, S. S. H., **3,** 224
Gilinsky, P., **9,** 393
Gill, G. B., **8,** 83
Gill, H. S., **11,** 582
Gill, J. C., **5,** 551, 667
Gill, M., **9,** 520
Gillard, B. K., **6,** 25
Gillard, J. W., **10,** 59; **12,** 101
Gillard, M., **9,** 105, 396
Gillard, R. D., **5,** 764
Gillen, M. F., **8,** 468; **9,** 446; **10,** 62
Gillespie, D. G., **6,** 341
Gillespie, H. B., **1,** 433, 450, 864; **4,** 239
Gillespie, J. S., Jr., **6,** 468
Gillespie, P. D., **4,** 536
Gillespie, R. J., **1,** 397; **2,** 200; **8,** 159, 435
Gillick, J. G., **11,** 294
Gillis, B., **1,** 1131
Gillis, B. T., **1,** 247, 309; **2,** 282, 326
Gillis, H. R., **9,** 270; **10,** 181
Gillis, R. G., **1,** 1073
Gillon, A., **5,** 718
Gilman, H., **1,** 52, 96, 253, 345, 348, 424, 573, 611, 619, 689, 731, 845, 1103, 1251, 1256; **2,** 153, 179; **4,** 163, 283; **5,** 505, 624; **6,** 40; **7,** 133; **8,** 218, 276; **10,** 77, 314; **11,** 209, 554; **12,** 568
Gilman, J. W., **12,** 11
Gilman, N. W., **2,** 263, 369; **3,** 113; **4,** 53, 295, 318; **5,** 381; **9,** 251
Gilman, S., **7,** 253; **8,** 24
Gilmont, P., **1,** 67, 228
Gilmore, D. W., **7,** 94; **8,** 45, 249
Gilmore, G. N., **6,** 621
Gilmore, J. R., **4,** 74, 318
Gilmore, W. F., **1,** 268
Gilon, C., **2,** 43; **9,** 180
Gilow, H. M., **11,** 343
Gilpin, J. R., **1,** 135; **2,** 33
Gilsdorf, R. T., **1,** 595
Gimbarzevsky, B. P., **6,** 118
Ginebreda, A., **8,** 391
Gingrich, H. L., **5,** 667; **9,** 3
Ginos, J., **3,** 314
Ginsberg, H., **1,** 720
Ginsberg, H. F., **1,** 502
Ginsburg, D., **1,** 93, 1004; **5,** 136, 205, 685; **6,** 107
Ginsburg, G. S., **9,** 291
Ginzburg, Y. I., **2,** 465
Giordano, C., **6,** 560; **11,** 471
Giorgetti, J. P., **5,** 117
Giorgio, P. A. D., **8,** 263
Giormani, V., **1,** 310
Gipp, R., **11,** 206
Girard, A., **1,** 411
Girard, C., **4,** 437; **5,** 589, 713; **6,** 523; **7,** 402
Girard, J. P., **7,** 37, 182
Girard, P., **6,** 671; **8,** 439; **10,** 344; **11,** 283, 465
Giraud, M., **5,** 507
Girgis, N. S., **7,** 356
Girijavallabhan, M., **4,** 162
Girodeau, J. M., **6,** 137
Girotra, N. N., **2,** 159, 252; **5,** 5; **7,** 40, 45; **11,** 498
Gish, D. T., **1,** 737, 1139
Gisin, B. F., **8,** 81
Gisler, M., **7,** 304
Gisvold, O., **7,** 37
Giube-Jampel, E., **4,** 173; **8,** 57
Giudicelli, R., **6,** 570
Giumanini, A. G., **2,** 148
Giusti, G., **4,** 316; **5,** 57
Giveli, C., **10,** 432
Givens, R. S., **6,** 554
Giza, C. A., **1,** 85, 503, 661; **5,** 433
Gladstone, W. A. F., **2,** 257
Gladys, C. L., **4,** 132
Gladysz, J. A., **5,** 673; **8,** 214, 254, 310; **9,** 286; **11,** 401
Glamkowski, E. J., **5,** 62
Glaser, R., **4,** 240
Glaser, S., **4,** 536
Glass, C. A., **12,** 36
Glass, D. S., **1,** 926
Glass, R. S., **5,** 323; **7,** 167; **10,** 359, 427
Glass, T. E., **4,** 100
Glasscock, K. G., **10,** 413
Glasser, A. C., **4,** 45
Glatthard, R., **1,** 56, 330
Glatz, B., **9,** 299, 337, 406
Glavis, F. J., **2,** 231
Glaze, W. H., **1,** 1274
Glazer, E. A., **5,** 677
Glazier, E. R., **1,** 162
Gleason, J. C., **2,** 207
Gleason, J. G., **3,** 36, 149; **4,** 242; **5,** 66; **6,** 76, 118; **7,** 40; **9,** 236
Gleason, R. W., **3,** 219
Gleicher, G. J., **2,** 21; **3,** 7
Gleim, R. D., **7,** 223; **8,** 417, 458; **9,** 233
Gleiter, R., **7,** 268; **9,** 341
Glemser, O., **6,** 186
Glenat, R., **2,** 234
Glenn, T. H., **1,** 339
Gless, R. D., **6,** 2
Glickman, S. A., **3,** 241
Glieze, P. A., **11,** 551
Glinskaya, O. V., **3,** 323
Glockling, F., **1,** 397

Glockner, P., **4,** 41
Gloede, J., **1,** 120; **2,** 63
Gloor, U., **1,** 567
Glosauer, O., **1,** 942
Glotter, E., **2,** 72; **7,** 361; **8,** 451
Glowinski, R., **7,** 74
Glukhovtsev, V. G., **10,** 356
Glynn, G. A., **2,** 382
Gmuender, J., **1,** 11
Gnad, J., **3,** 147
Gnoj, O., **1,** 150; **2,** 77, 232, 344, 433
Gnuechtel, A., **1,** 664
Goasdoue, N., **7,** 116
Gobao, R. A., **11,** 115
Goble, P. H., **2,** 203
Godat, J. P., **1,** 187
Godefroi, E. F., **1,** 257; **6,** 514
Godel, T., **11,** 165, 518; **12,** 251
Godet, J. Y., **3,** 294; **8,** 445
Godfrey, C. R. A., **11,** 276; **12,** 102
Godfrey, I. M., **5,** 127
Godfrey, J. D., **10,** 403
Godfrey, M., **2,** 125
Godin, D., **1,** 219
Godinho, L. S., **2,** 200
Godleski, S. A., **8,** 39; **9,** 458; **11,** 275, 514
Godoy, J., **11,** 271
Godschalx, J. P., **10,** 26; **12,** 46, 576
Godt, H. C., Jr., **1,** 1131, 1216
Goe, G. L., **6,** 436
Goebel, P., **1,** 23
Goedecke, E., **5,** 448
Goedken, V., **12,** 3
Goeknel, E., **5,** 254
Goel, A. B., **8,** 175, 311; **11,** 152
Goel, O. P., **5,** 667
Goel, R. G., **9,** 335, 501
Goeldner, M. P., **9,** 127
Goemann, M., **10,** 396
Goendoes, G., **5,** 308; **11,** 446
Goering, H. L., **1,** 101, 1143; **6,** 142; **7,** 122; **11,** 188
Goering, J., **3,** 220
Goerler, K., **1,** 1015
Goesl, R., **1,** 484
Goethlich, L., **1,** 1101, 1247
Goettschi, E., **5,** 39
Goetz, A., **11,** 587
Goetz, N., **7,** 338
Goetzen, T., **5,** 28
Goewecke, S., **1,** 85
Goff, D. L., **5,** 582
Gofstein, R., **1,** 62
Goggin, P., **1,** 827
Gogins, K. A. Z., **11,** 237, 452
Gogte, V. N., **5,** 731
Goh, S. H., **4,** 201, 248; **6,** 230, 326; **12,** 275
Gohda, N., **9,** 507
Goheen, D. W., **1,** 1095
Goheen, G. E., **1,** 754
Gohke, K., **5,** 571
Gokel, G. W., **4,** 31; **5,** 28, 155; **7,** 20, 79, 304; **8,** 130, 255; **9,** 127, 186, 361, 387; **10,** 323; **12,** 147, 381
Gokhale, P. D., **6,** 475
Gokhale, S. D., **2,** 329
Golborn, P., **5,** 32, 740
Gold, A., **9,** 71
Gold, A. H., **2,** 147
Gold, A. M., **1,** 562
Gold, E. H., **4,** 442
Gold, H., **1,** 1211; **5,** 653; **6,** 229; **11,** 195
Gold, J., **6,** 84
Gold, J. M., **10,** 240, 384
Gold, J. R., **6,** 632
Gold, M. H., **1,** 343, 1226
Gold, P. M., **9,** 365; **11,** 187
Goldberg, A. A., **5,** 117
Goldberg, G. M., **6,** 632; **8,** 263; **9,** 74
Goldberg, I. H., **1,** 189
Goldberg, M. W., **1,** 659, 660
Goldberg, N. N., **6,** 339
Goldberg, O., **12,** 156
Goldberg, P., **3,** 164
Goldblatt, L. A., **1,** 367
Goldblith, S. A., **4,** 191
Goldblum, A., **5,** 606
Golden, S., **2,** 336
Goldfarb, A. R., **1,** 671
Gol'dfarb, Y. L., **1,** 1103
Goldfrank, M., **1,** 324
Goldhill, J., **8,** 197; **9,** 38, 522; **11,** 76
Golding, B. T., **5,** 335; **6,** 79; **12,** 242
Goldkamp, A. H., **1,** 248
Goldmacher, J. E., **1,** 484, 905
Goldmacher, S., **3,** 81
Goldman, A., **11,** 460
Goldman, I. M., **2,** 263
Goldman, L., **1,** 310, 1185; **2,** 165
Goldman, N. L., **1,** 603; **3,** 182; **5,** 482, 667
Goldner, H., **6,** 547
Goldschmid, H. R., **1,** 562
Goldschmidt, B. M., **4,** 84
Goldschmidt, S., **1,** 348, 951, 1101
Goldschmidt, Z., **4,** 31
Goldsmith, B., **6,** 534
Goldsmith, D., **5,** 78; **9,** 31; **12,** 378
Goldsmith, D. J., **4,** 199; **5,** 631; **6,** 353
Goldsmith, L., **9,** 259
Goldstein, A., **1,** 156, 160
Goldstein, A. W., **5,** 127
Goldstein, J. E., **11,** 458
Goldstein, J. H., **1,** 13
Goldstein, S., **5,** 667; **11,** 111
Goldwhite, H., **5,** 326, 650
Goldyrev, L. N., **1,** 66
Golebiewski, W. M., **5,** 667
Golfier, M., **2,** 363; **3,** 249; **4,** 427, 431; **5,** 580; **6,** 514
Goliaszewski, A., **12,** 372
Golob, A. M., **7,** 303; **9,** 11
Gombatz, K., **9,** 79, 304
Gomberg, M., **5,** 420; **9,** 110
Gomez-Parra, F., **7,** 320
Gompper, R., **1,** 403, 494, 1252; **5,** 136, 649; **10,** 158; **11,** 197
Gonis, G., **1,** 139; **2,** 68, 242
Gontarev, B. H., **1,** 409
Gonzalez, A. G., **5,** 631; **7,** 40; **10,** 459
Gonzalez, D., **1,** 643; **10,** 180

Gonzalez, F. B., **12,** 256
Gonzalez, J. M., **12,** 306
Gonzalez, T., **5,** 155
Gonzalez-Gomez, J. A., **12,** 523
Gonzy, G., **6,** 574
Gooch, E. E., **10,** 212; **11,** 269, 489
Goodacre, J., **6,** 597
Goodbrand, H. B., **5,** 751; **6,** 667
Goodchild, J., **2,** 35
Goodgame, D. M. L., **9,** 148
Goodin, R., **7,** 324
Goodlett, V. W., **2,** 426; **4,** 172
Goodman, I. J., **6,** 9
Goodman, L., **1,** 271, 1045, 1220; **2,** 434; **3,** 229; **4,** 154
Goodman, M., **1,** 235; **2,** 220
Goodman, M. M., **1,** 1059; **2,** 381
Goodson, L. H., **3,** 95
Goodwin, G. T., **12,** 127
Goodwin, S., **1,** 562, 604
Goodwin, T. E., **8,** 461
Goodwin, T. W., **1,** 642, 951
Goosen, A., **1,** 95
Goosen, A. J., **6,** 675
Goossens, D., **10,** 60
Gopal, H., **2,** 238; **3,** 244; **4,** 421
Gopal, R., **4,** 398, 438
Gopalan, A., **9,** 372; **10,** 321
Gopalan, B., **7,** 385; **9,** 432
Gopalan, R., **8,** 535; **9,** 371
Gorbatenko, W. I., **5,** 668
Gordinier, A., **6,** 486
Gordon, A. J., **4,** 421
Gordon, B., III, **9,** 87
Gordon, J. E., **5,** 329
Gordon, J. T., **1,** 500
Gordon, M., **1,** 518
Gordon, M. E., **1,** 854, 1237
Gordon, N. B., **10,** 422
Gordon, S. B., **5,** 551
Gore, I. Y., **1,** 937
Gore, J., **5,** 389, 420; **6,** 223, 342; **7,** 164, 223; **9,** 119; **10,** 101, 149, 290, 434; **11,** 321, 360; **12,** 51, 307, 437, 502
Gore, P. H., **1,** 158, 500, 1295; **5,** 667; **9,** 249
Gore, T. S., **4,** 424
Gorecka, A., **8,** 170
Gorenstein, D., **5,** 707
Gorgues, A., **2,** 203; **5,** 29, 319, 639; **7,** 20, 355; **11,** 407; **12,** 137
Gorman, R. R., **8,** 419
Gormish, J. F., **1,** 1194
Gorrichon, J. P., **11,** 346
Gorski, R. A., **6,** 252; **7,** 207
Gortler, B., **2,** 289
Gortz, H. H., **10,** 94
Gorup, B., **1,** 385
Gorzkowska, I., **8,** 391
Gorzynski, J. D., **4,** 35
Gosden, A. F., **1,** 732
Goshorn, R. H., **1,** 864; **2,** 174; **3,** 128
Gossauer, A., **7,** 377
Gosselck, J., **5,** 258; **9,** 483
Gosselink, E. P., **2,** 397
Gosteli, J., **5,** 511
Goswami, J. C., **5,** 329
Gotfredsen, W. O., **1,** 1228
Goto, J., **10,** 390; **11,** 419
Goto, R., **2,** 222
Goto, S., **8,** 354
Goto, T., **1,** 357, 374, 796, 802, 1064, 1284; **4,** 86; **8,** 421, 516; **9,** 110, 311, 426; **10,** 33; **11,** 99; **12,** 325
Goto, Y., **11,** 293
Gotoh, Y., **12,** 33
Gottardi, W., **2,** 109; **10,** 122
Gottfried, S., **1,** 290, 1178
Gotthardt, H., **5,** 230
Gottlieb, H. B., **1,** 1249
Gottlieb, H. E., **12,** 546
Gottlieb, W. B., **3,** 87
Gottschalck, H., **3,** 249
Gottschalk, P., **12,** 263
Gottsegen, A., **11,** 188
Gottstein, W. J., **1,** 183; **2,** 294; **5,** 700
Gotzmer, C., Jr., **5,** 649
Gouedard, M., **5,** 413
Gouge, M., **2,** 412
Gougoutas, J. Z., **3,** 12; **7,** 367
Gould, D. H., **1,** 13, 75, 1101; **2,** 353
Gould, E. S., **3,** 331; **5,** 77; **8,** 24, 29
Gould, F. E., **1,** 692, 730, 977; **6,** 423
Gould, K. J., **6,** 208, 494
Gould, S. J., **9,** 139
Goure, W. F., **12,** 46
Goutarel, R., **2,** 210
Gouverneur, P., **7,** 35
Govindachari, T. R., **4,** 395; **5,** 540, 542; **9,** 308; **10,** 161
Gower, L. H., **1,** 1254
Gowland, B. D., **10,** 72
Gowland, F. N., **9,** 458
Gowling, E. W., **3,** 212
Goya, S., **3,** 325
Graber, R. P., **1,** 160
Grabiel, T. F., **9,** 415
Grabowich, P., **1,** 782
Grabowski, E. J. J., **12,** 381
Gracela, R., **1,** 966
Gracner, M., **8,** 483
Gradeff, P. S., **7,** 312
Grady, G. L., **3,** 294; **4,** 520
Graebe, C., **1,** 536, 713, 828, 951
Graef, H. D., **6,** 596
Graefe, J., **12,** 48
Graefing, R., **8,** 245; **11,** 473
Graf, E., **6,** 138
Graf, R., **1,** 118; **2,** 61, 70; **3,** 53; **5,** 93, 283
Graf, W., **9,** 210, 269; **10,** 413; **11,** 91
Grafen, P., **1,** 777
Graham, C., **4,** 531
Graham, C. L., **1,** 52
Graham, C. R., **8,** 36
Graham, J. C., **4,** 387; **5,** 322
Graham, J. H., **1,** 964
Graham, J. R., **1,** 44
Graham, R. S., **11,** 430; **12,** 196
Graham, S. H., **2,** 289, 313
Graham, S. L., **7,** 31
Graham, W. D., **5,** 10

Graham, W. H., **1,** 214, 254; **2,** 135
Grakauskas, V., **5,** 705
Gram, H. F., **1,** 754
Gramain, J. C., **2,** 243; **5,** 580
Gramatica, P., **12,** 422
Gramlich, W., **8,** 330
Grammaticakis, M. P., **6,** 526
Grams, G. W., **2,** 26
Gramstad, T., **5,** 564, 705; **6,** 521
Granath, K. A., **1,** 1002
Grandbois, E. R., **11,** 153
Grandguillot, J. C., **9,** 97
Grandi, R., **7,** 331
Grandmaison, J. L., **5,** 362; **9,** 264
Granger, R., **7,** 37, 182
Granito, C., **2,** 229
Granitzer, W., **9,** 172
Granoth, I., **5,** 688; **6,** 585
Granowitz, S., **1,** 1284
Grant, F. W., **1,** 952
Grant, L. R., **1,** 964, 966; **5,** 411
Grant, M. S., **1,** 1152
Grant, P. K., **5,** 563
Grantham, R. J., **4,** 191
Grantham, R. K., **2,** 290
Gras, J. L., **6,** 223; **7,** 229; **8,** 333; **9,** 136, 193, 225, 227, 318, 432, 467; **11,** 335; **12,** 490
Gras, M. A. M. P., **1,** 247
Graselli, P., **6,** 74
Graser, F., **1,** 755
Grasley, M. H., **1,** 613
Grasselli, P., **1,** 1055; **2,** 274; **3,** 189, 201
Grasshoff, H. J., **1,** 608
Grassmann, W., **1,** 974
Gratz, J. P., **3,** 215
Gravel, D., **4,** 425, 431, 522
Graven, N., **1,** 219
Graves, J. M. H., **1,** 309, 926; **2,** 29; **6,** 258
Gravestock, M. B., **3,** 269; **4,** 531; **5,** 700
Gray, A. P., **1,** 904
Gray, A. R., **1,** 999, 1269
Gray, C. J., **6,** 597
Gray, D., **2,** 368
Gray, J., **8,** 15
Gray, M. D. M., **10,** 62
Gray, S., **1,** 96
Gray, S. L., **2,** 203
Graybill, B. M., **2,** 336
Graymore, J., **8,** 305
Grayshan, R., **3,** 136
Grayson, D. H., **4,** 55
Grayson, J. I., **7,** 26
Graziano, M. L., **12,** 365
Gream, G. E., **4,** 405
Greaves, E. O., **5,** 649
Greaves, P. M., **4,** 255; **5,** 243
Greber, G., **2,** 438
Greck, C., **11,** 292
Greco, A., **9,** 257
Greco, C. C., **2,** 389; **9,** 436
Greco, C. V., **1,** 1123; **2,** 98
Gree, R., **10,** 121, 223; **11,** 231
Green, A. E., **9,** 135
Green, A. G., **1,** 1087
Green, B., **1,** 207, 599; **2,** 60, 271, 303; **3,** 40, 198
Green, D. P. L., **3,** 198
Green, D. T., **3,** 160
Green, E. E., **4,** 119
Green, F. R., III, **8,** 64; **9,** 417
Green, G. F. H., **1,** 104, 954; **5,** 90
Green, G. W., **4,** 14
Green, I. R., **10,** 81
Green, J. W., **1,** 658, 818; **6,** 360
Green, M., **1,** 851; **3,** 97; **9,** 163
Green, M. J., **2,** 316; **7,** 140; **8,** 167
Green, M. L. H., **6,** 48; **9,** 426
Green, W. J., **8,** 60
Greenberg, S., **8,** 4
Greene, A., **6,** 536
Greene, A. E., **4,** 123, 127; **5,** 187; **7,** 7, 86; **8,** 80; **9,** 99, 436; **10,** 140; **11,** 235; **12,** 178, 357
Greene, F. D., **1,** 235, 409, 927; **2,** 309, 424; **3,** 234; **4,** 282; **5,** 424, 470
Greene, R. M. E., **11,** 330
Greene, T. M., **8,** 466
Greene, W. R. N., **5,** 155
Greener, E., **5,** 740
Greenfield, H., **1,** 228, 892
Greenfield, S., **2,** 72
Greenhalgh, R., **1,** 518
Greenhorn, J. D., **6,** 345
Greenhouse, R., **6,** 496, 644; **10,** 93
Greenlee, K. W., **4,** 135; **5,** 107
Greenlee, R. B., **2,** 344
Greenlee, W. J., **9,** 305; **11,** 276; **12,** 150
Greenlimb, P. E., **6,** 262
Greenough, W. B., III, **1,** 118
Greenspan, F. P., **1,** 457, 954
Greenspan, G., **3,** 204
Greenstein, J. P., **1,** 129
Greenwald, R., **1,** 318, 1246; **3,** 142
Greenwald, R. B., **1,** 224; **4,** 189
Greenwood, F. L., **1,** 198, 777
Greenwood, G., **4,** 328
Greenwood, N. N., **9,** 483
Greer, F., **1,** 940
Greer, S., **10,** 321
Gregoire de Bellemont, E., **1,** 380
Gregor, I. K., **6,** 504
Gregorio, G., **5,** 172, 190
Gregory, C. D., **7,** 122
Gregory, G. I., **2,** 228
Gregory, H., **1,** 745, 1196
Gregory, P., **10,** 306
Gregson, M., **3,** 87, 153, 269, 293; **5,** 5, 678; **6,** 598
Gregson, R. P., **5,** 109
Grehn, L., **12,** 159
Greibrokk, T., **5,** 28, 491
Greidanus, J. W., **3,** 228
Greig, C. G., **1,** 110
Greiger, R. E., **12,** 180
Greijdanus, B., **6,** 575
Gremban, R. S., **12,** 150
Grenda, V. J., **1,** 139
Grenier-Loustalot, M. F., **6,** 562
Grenon, B. J., **2,** 382
Gresham, T. L., **1,** 531, 957
Greuter, H., **7,** 9; **9,** 112, 481

Greve, W., **5,** 259
Grevels, F. W., **4,** 550
Grewe, F., **1,** 487
Grewe, R., **1,** 180, 884, 926, 1055
Grey, R. A., **12,** 46
Grezemkovsky, R., **1,** 219, 249
Gribble, G. W., **2,** 462; **5,** 614; **6,** 534, 553; **8,** 451; **9,** 421; **11,** 479
Grice, P., **10,** 423
Grie, R., **10,** 81
Grieco, C., **6,** 327
Grieco, P. A., **3,** 153, 269, 293; **4,** 302, 327; **5,** 5, 94, 244, 246, 276, 277, 293, 315, 361, 379, 418, 436, 460, 516, 526, 678, 758; **6,** 238, 267, 359, 403, 422, 460, 464, 552, 598, 632; **7,** 137, 160, 213, 253, 280, 307, 314; **8,** 24, 101, 140, 362, 397, 428, 437, 486, 492; **9,** 81, 277, 325; **10,** 31, 313, 366; **12,** 204, 282, 295, 314, 508
Grieder, A., **7,** 122
Gries, K., **8,** 395
Griesbaum, K., **3,** 178; **6,** 441; **9,** 343
Grieshaber, P., **5,** 250; **6,** 243
Grieshaber, T., **4,** 251
Griesinger, A., **6,** 564
Griesser, H., **10,** 337
Grifagni, M. A., **11,** 117
Griffin, C. E., **1,** 518, 1022
Griffin, G. W., **3,** 160; **4,** 260; **6,** 244; **8,** 102
Griffin, I. M., **7,** 224
Griffin, R. W., Jr., **1,** 633
Griffin, T. S., **6,** 543; **8,** 310
Griffith, J. R., **8,** 396
Griffith, W. P., **9,** 391
Griffiths, J., **7,** 268; **10,** 306
Griffiths, J. S., **5,** 406
Griffiths, P. G., **9,** 264
Grigat, E., **2,** 320
Grigg, R., **5,** 464; **9,** 349; **11,** 130
Grignan, J., **11,** 16
Grignard, V., **1,** 424; **8,** 214
Grignon-Dubois, M., **7,** 84
Grigoryan, M. A., **5,** 478
Grillot, G. F., **10,** 164
Grim, S. O., **1,** 1246; **7,** 62
Grimaldi, J., **7,** 281; **11,** 126; **12,** 51, 305
Grime, R., **1,** 1102
Grimm, D., **1,** 1101; **6,** 122
Grimm, E. G., **6,** 399
Grimm, K. G., **8,** 352
Grimme, W., **10,** 381
Grimshaw, J., **2,** 233; **4,** 194
Grimshaw, J. T., **4,** 194
Grimwood, B. E., **2,** 224
Grinberg, S., **5,** 728; **7,** 406
Gringauz, A., **3,** 155; **7,** 413
Grisdale, E. E., **6,** 137
Grisley, D. W., Jr., **1,** 1247; **2,** 447
Griswold, P. H., Jr., **1,** 654
Gritter, R. L., **1,** 642
Grivsky, E. M., **1,** 444
Grob, C. A., **1,** 93, 139, 276, 367, 562, 581, 929, 1033; **3,** 172; **4,** 413; **5,** 336, 370, 436, 580; **6,** 403; **8,** 126
Grob, V. D., **6,** 676
Grochowski, E., **7,** 406; **11,** 589
Grodski, A., **5,** 639
Groebel, B. T., **4,** 285; **5,** 375; **6,** 55, 118, 322; **9,** 483
Groebel, W., **8,** 399
Groebke, W., **1,** 652
Groeger, C., **7,** 277
Groen, M. B., **6,** 525; **11,** 96
Groen, S. H., **5,** 620
Groenewegen, P., **8,** 415; **9,** 387
Groening, T., **1,** 158; **2,** 351; **5,** 148, 161; **9,** 486
Groeninger, K., **12,** 303
Groenwall, S., **1,** 84, 101, 197, 266
Groetsch, H., **1,** 894
Groger, R., **8,** 188
Grohman, K., **6,** 171
Gromelski, S., Jr., **6,** 546
Gronowitz, J. S., **3,** 329
Gronowitz, S., **3,** 329; **10,** 188
Gros, E. G., **2,** 104
Gros, S. K., **2,** 378
Gross, A. W., **10,** 228; **12,** 285
Gross, B., **5,** 326; **7,** 35; **9,** 297; **10,** 306; **12,** 44
Gross, E., **1,** 176, 505, 1002
Gross, H., **1,** 120, 220, 1209; **2,** 43, 63, 120; **3,** 278; **7,** 174
Gross, N., **1,** 491
Gross, P., **5,** 250; **9,** 182
Gross, R. A., **4,** 31
Grosse, A. V., **1,** 106
Grosse, D., **10,** 59
Grosser, J., **5,** 200; **6,** 171
Grossert, J. S., **4,** 508; **5,** 160, 659; **11,** 493, 494
Grossman, J., **1,** 511, 526
Grossman, L. I., **1,** 1247
Grote, J., **12,** 11
Groth, U., **10,** 152; **11,** 193; **12,** 526
Grotjahn, D. B., **10,** 381
Groton, V., **9,** 465
Groutas, W. C., **10,** 113, 369
Grovenstein, E., Jr., **1,** 563
Grover, J. N., **5,** 464
Grover, P. K., **1,** 219, 730, 882, 1041, 1292
Groves, I. F., **11,** 132
Groves, J. T., **2,** 424; **6,** 261, 534; **11,** 462
Grozinger, K., **7,** 428
Grubbs, E. J., **1,** 195
Grubbs, R. H., **10,** 88; **12,** 55, 111
Gruber, J. M., **7,** 188, 189; **8,** 102; **11,** 124; **12,** 272
Gruber, L., **6,** 645; **7,** 160
Gruber, P., **4,** 172; **10,** 151
Gruber, R. J., **1,** 1122
Gruber, W., **1,** 971
Grubmuller, P., **9,** 322; **11,** 355
Grudzinskas, C. V., **9,** 335
Gruen, H., **1,** 1080
Gruenblat, E. I., **8,** 226
Gruenhut, N. S., **1,** 324
Gruenter, K., **6,** 653
Gruenwald, G. L., **10,** 113

Gruenze, H., **2,** 354
Gruessner, A., **1,** 156
Gruetzmacher, G. D., **4,** 191; **5,** 115, 261
Grummitt, O., **1,** 402, 785, 846, 1150, 1284, 1292; **2,** 390; **11,** 255
Grund, G., **1,** 652
Grund, H., **7,** 207; **10,** 243
Grundemann, E., **1,** 212, 465
Grundke-Schroth, G., **1,** 885
Grundmann, C., **1,** 370, 890, 1216, 1235, 1272; **2,** 42, 384; **8,** 386, 465
Grundon, M. F., **1,** 444, 927; **5,** 266; **6,** 359
Grundy, J., **1,** 329; **4,** 189
Grunewald, G. L., **1,** 231, 971; **7,** 97; **8,** 148
Gruntz, U., **8,** 521
Grunwell, J. R., **7,** 101
Grunze, H., **1,** 971
Gruter, H. F. M., **10,** 220
Grutzner, J. B., **7,** 174
Gruuller, P., **11,** 355
Grynkiewicz, G., **6,** 280; **7,** 406; **12,** 420
Gryszkiewicz-Trochimowski, E., **7,** 247
Grzejszczak, S., **6,** 191, 193; **7,** 126
Grzeskowiak, R., **6,** 147
Grzonka, Z., **2,** 126; **6,** 78
Grzybowska, J., **9,** 506
Gschwend, H. W., **8,** 192, 279; **9,** 8, 89, 309; **10,** 71, 75; **11,** 402
Gsottschneider, W., **4,** 100
Guaciaro, M. A., **9,** 87
Gualtieri, J. A., **9,** 352
Guanti, G., **9,** 475; **10,** 409; **12,** 444, 509, 510
Guarneri, M., **9,** 151
Guay, V., **12,** 197
Gubisch, N., **1,** 444
Guder, H. J., **11,** 587
Gudi, M. N., **4,** 173
Guedin-Vuong, D., **12,** 118
Guedj, R., **11,** 454
Guegan, R., **1,** 676
Guenthard, H. H., **5,** 620
Guenther, D., **1,** 292; **5,** 595
Guenther, F., **1,** 219
Guenther, H., **2,** 344; **4,** 579
Guenther, H. J., **8,** 205, 260
Guenther, K., **5,** 127
Guenther, P., **1,** 1136
Guenther, W. H. H., **1,** 595; **8,** 459; **9,** 35
Gueritte, F., **7,** 390
Guerrero, A., **11,** 486
Guerrero, T. J., **1,** 450, 864, 1033
Guerrieri, F., **5,** 474; **6,** 451
Guertine, D. L., **11,** 67
Guesten, H., **1,** 712; **2,** 345
Guette, J. P., **5,** 557; **10,** 322; **11,** 251
Guggenheim, T. L., **11,** 150; **12,** 476
Guggisberg, A., **9,** 379
Guha, P. C., **2,** 105
Guibe, F., **10,** 32, 413; **11,** 551; **12,** 525
Guida, A. R., **9,** 483
Guida, W. C., **5,** 693; **6,** 565; **7,** 353; **8,** 501; **9,** 483; **10,** 112; **11,** 499
Guilhemat, R., **11,** 1
Guillaumet, G., **7,** 327; **10,** 194, 368
Guillemonat, A., **1,** 999
Guillemont, M., **9,** 46
Guillerm, D., **11,** 216
Guillot, J., **5,** 537
Guillot, J. P., **12,** 36
Guindon, Y., **10,** 427, 463; **12,** 101, 200, 576
Guingant, A., **9,** 165; **10,** 403
Guinsig, R., **2,** 372
Guiseley, K. B., **2,** 394
Guitart, J., **11,** 486
Guittet, E., **8,** 334
Gulati, K. C., **1,** 1292; **4,** 15
Gulland, J. M., **1,** 685
Gum, S. O., **5,** 454
Gumulka, J., **9,** 343
Gumulka, M., **4,** 507
Gunar, V. I., **2,** 267
Gunatilaka, A. A. L., **7,** 111
Gund, T. M., **3,** 7; **4,** 31, 140; **5,** 756; **6,** 17
Gunda, E. T., **5,** 182
Gundermann, K., **9,** 508
Gundermann, K. D., **5,** 456; **6,** 471
Gundu-Rao, C., **9,** 399
Gunn, P. A., **5,** 655
Gunneweg, J. J., **5,** 92
Gunnina, H. E., **8,** 222
Gunsher, J. A., **1,** 758; **2,** 300; **4,** 234
Gunstone, F. D., **1,** 169, 814, 1004, 1008, 1011; **2,** 36, 248, 363
Guo, M., **12,** 326
Gupta, A. S., **9,** 1
Gupta, B. G. B., **8,** 248, 263, 518; **9,** 236, 255, 256, 324, 326, 488; **10,** 81, 96, 115, 409; **11,** 355, 553, 587
Gupta, D., **11,** 519
Gupta, D. N., **2,** 378; **11,** 492
Gupta, G. S., **4,** 369
Gupta, I., **11,** 341
Gupta, K., **4,** 31; **12,** 568
Gupta, O. D., **12,** 106
Gupta, P., **1,** 271
Gupta, R. C., **12,** 319
Gupta, S. C., **10,** 214
Gupta, S. K., **3,** 98, 123, 329; **4,** 25, 69, 70; **5,** 13, 740; **6,** 34, 573, 597, 617; **7,** 53
Gupte, S. S., **3,** 224; **4,** 383
Gupton, J. T., **11,** 195
Gurevich, A. I., **1,** 318
Gurfinkel, E., **3,** 124
Gurien, H., **4,** 367
Gurria, G. M., **6,** 17; **7,** 7; **8,** 13
Gurst, J. E., **4,** 265
Gurudutt, K. N., **10,** 234
Gurusiddappa, S., **9,** 227
Guschlbauer, W., **8,** 412
Guseinov, M. M., **10,** 459
Gustafson, D. H., **2,** 210; **3,** 142
Gut, M., **2,** 75
Gut, R., **9,** 241
Gutbrod, H. D., **5,** 250; **9,** 182
Gutenstein, M., **2,** 428; **3,** 225

Guterstein, M., **1,** 859
Guthikonda, R. N., **4,** 323; **5,** 723, 747
Guthrie, A. E., **10,** 182
Guthrie, J. D., **1,** 859
Guthrie, R. D., **5,** 430, 595
Gutierrez, C. G., **10,** 413
Gutmann, H., **1,** 114
Gutmann, H. R., **1,** 465
Gutmann, V., **6,** 564
Gutowski, F. D., **7,** 324
Gutowsky, H. S., **1,** 952; **2,** 348; **12,** 466
Gutsche, C. D., **1,** 56, 144, 195, 367, 573, 1100; **2,** 84; **8,** 501
Gutsell, E. S., **1,** 97
Guttmann, S., **1,** 155, 383, 1220
Gutzwiller, J., **2,** 453; **4,** 323; **5,** 73; **8,** 424
Guy, A., **8,** 451; **10,** 322, 378; **11,** 251
Guy, R. G., **1,** 1153; **5,** 662
Guynn, R., **1,** 450
Guyot, A., **1,** 343
Guyot, M., **6,** 503
Guziec, F. S., Jr., **4,** 27; **5,** 23; **6,** 509, 644; **8,** 463; **10,** 30; **11,** 197
Guzik, H., **1,** 731
Guzman, A., **4,** 390; **5,** 244, 537, 762
Gygax, P., **4,** 82; **5,** 113; **8,** 135
Gymer, G. E., **5,** 370

Haab, F., **1,** 796
Haack, A., **1,** 289, 682
Haaf, A., **3,** 240
Haaf, W., **1,** 407; **11,** 243
Haag, A., **1,** 360, 1246; **9,** 387
Haag, W., **1,** 1247
Haag, W. O., **1,** 1023
Haage, K., **4,** 128
Haak, P., **5,** 705; **6,** 231
Haake, M., **3,** 106; **11,** 211
Haake, P., **7,** 138
Haas, G., **9,** 337
Haas, H. J., **3,** 218
Haas, M. A., **2,** 270
Haas, W. L., **1,** 15
Haase, E., **8,** 272
Hasse, F., **1,** 486
Haase, K., **11,** 28; **12,** 450
Habeeb, A. F. S. A., **1,** 739
Haber, A., **1,** 51
Haber, F., **1,** 281
Haber, R. G., **1,** 866
Haber, S. B., **6,** 567; **7,** 335
Haberfield, P., **1,** 309, 1145
Habermehl, G., **3,** 240; **5,** 330
Habich, D., **10,** 2
Habisch, D., **1,** 125; **2,** 66
Hach, C. C., **1,** 999; **2,** 94, 360
Hach, V., **4,** 16; **5,** 601
Hachem, K., **10,** 382
Hachey, D. L., **6,** 598
Hachihama, Y., **1,** 637
Hacker, N. P., **12,** 156
Hackett, S., **12,** 317
Hackles, R. E., **2,** 431
Hackley, B. E., Jr., **2,** 392; **3,** 161
Hackney, R. J., **3,** 24
Haddad, Y. M. Y., **1,** 132; **2,** 229
Haddadin, M. J., **1,** 255, 341, 521, 933, 971, 1150, 1246; **4,** 209; **5,** 280, 600
Hadfield, J. R., **3,** 196
Hadler, H. I., **1,** 389, 703
Hadley, S. B., **11,** 188
Haeberlein, H., **3,** 320; **4,** 550
Haede, W., **1,** 139, 488; **4,** 296; **5,** 119, 633
Haefliger, W., **4,** 577; **10,** 159; **11,** 167, 222
Haehnle, R., **1,** 1056
Haenssle, P., **6,** 373
Haerle, H., **1,** 573
Haertl, R., **3,** 320; **4,** 550
Haeseler, P. R., **1,** 767
Haeuser, H., **2,** 402; **4,** 510; **5,** 650
Haffner, H. E., **5,** 677; **6,** 534
Haffner, K., **1,** 203, 364, 396, 1023, 1118; **2,** 154; **4,** 173; **5,** 198, 448; **7,** 119; **8,** 184
Haga, T., **11,** 526; **12,** 491
Haga, Y., **12,** 44
Hagaman, E. W., **5,** 10
Hagan, C. P., **8,** 284
Hagarty, J. D., **2,** 282, 326
Hagedorn, I., **6,** 268
Hagelloch, G., **1,** 295
Hageman, H. A., **1,** 176
Hageman, H. J., **2,** 38; **6,** 507
Hagemann, C. T., **5,** 614
Hagemann, H., **5,** 110
Hagemeyer, H. J., Jr., **1,** 7, 8, 526, 703
Hagen, J. P., **8,** 9; **9,** 283; **11,** 341, 349
Hagenbach, A., **5,** 689
Hagendoorn, J. A., **2,** 287
Hager, F. D., **1,** 500, 505, 864
Hagerty, A. F., **10,** 348
Haggerty, J. G., **4,** 375
Hagihara, N., **5,** 173; **6,** 61; **8,** 152; **10,** 117; **11,** 162; **12,** 476
Haginiwa, J., **5,** 601
Hagitani, A., **1,** 4; **2,** 12
Hagiwara, D., **6,** 91, 106; **10,** 61; **11,** 258, 562; **12,** 199, 381
Hagiwara, H., **7,** 235
Hagiwara, I., **12,** 174
Hagiwara, K., **6,** 11
Haglid, F., **1,** 966
Hagopian, R. A., **12,** 251
Hague, M. S., **11,** 437
Hah, Y. I., **9,** 418
Hahn, C. S., **12,** 195
Hahn, E. H., **5,** 205
Hahn, G., **12,** 357
Hahn, R. C., **1,** 1293
Hahn, V., **4,** 191
Hahnfeld, J. L., **6,** 261
Haidukewych, D., **6,** 20; **8,** 279
Haigh, F. C., **6,** 360
Hainaut, D., **5,** 266
Haines, A. H., **6,** 35
Haines, L. M., **8,** 206
Haines, R. L., **7,** 289
Haines, S. R., **12,** 298
Haitko, D. A., **12,** 557
Hajek, M., **6,** 518
Hajos, A., **2,** 27, 57; **5,** 89

Hajos, Z. G., **6,** 267, 411, 542; **12,** 415
Hajra, A. K., **1,** 269
Hakamada, I., **12,** 228
Hakata, T., **11,** 184
Hakimelahi, G. H., **9,** 485; **10,** 325; **11,** 90, 284
Hakomori, S., **2,** 173
Hakomori, S. I., **10,** 329
Hala, S., **1,** 1167
Halasz, S. P. v., **6,** 186
Halazy, S., **9,** 87, 307; **10,** 233, 319; **11,** 301
Halberstadt, I. K., **10,** 462
Halberstadt-Kausch, I. K., **10,** 462
Halbrook, N. J., **2,** 190
Halcour, K., **4,** 497
Haldane, J., **1,** 340
Hale, R. L., **4,** 31
Hales, N. J., **6,** 524; **7,** 325
Haley, J. F., Jr., **5,** 15
Halford, M. D. A., **4,** 461
Hall, A. M., **2,** 336
Hall, C. D., **4,** 152
Hall, C. M., **5,** 656
Hall, D., **5,** 113
Hall, D. H., **1,** 309
Hall, D. M., **1,** 905
Hall, D. R., **5,** 335; **6,** 547; **12,** 242
Hall, D. T., **5,** 622
Hall, D. W., **3,** 167
Hall, H. K., Jr., **1,** 1185; **6,** 480
Hall, H. T., **6,** 28, 104
Hall, H. T., Jr., **9,** 119
Hall, J. H., **1,** 1055; **7,** 304
Hall, L. D., **2,** 365; **5,** 351; **6,** 620; **11,** 489
Hall, P. L., **5,** 306
Hall, R. H., **1,** 199, 905; **5,** 136
Hall, S. E., **9,** 270
Hall, S. S., **4,** 290; **5,** 381, 454; **6,** 323; **7,** 195
Hall, T. C., **7,** 277
Hall, T. K., **2,** 263
Hall, T. W., **4,** 74; **5,** 102; **6,** 563
Hall, W. A., **1,** 1145
Hall, W. L., **4,** 505
Hallada, D. P., **11,** 61
Hallam, B. F., **10,** 223
Hallas, G., **1,** 73; **3,** 285; **4,** 189, 477; **7,** 352
Haller, R., **4,** 126; **5,** 389
Hallett, A., **6,** 425; **7,** 91
Halleux, A., **2,** 134
Halliday, D. E., **4,** 352
Hallman, P. S., **7,** 60
Hallsworth, A. S., **1,** 581
Halperin, G., **5,** 55; **6,** 297
Halpern, B., **1,** 267; **2,** 203, 228; **5,** 119, 709; **6,** 38
Halpern, J., **3,** 251; **5,** 658; **6,** 172
Halpern, O., **1,** 500
Halpern, Y., **6,** 24
Halsall, T. G., **1,** 144, 526, 1054; **3,** 49
Halstenberg, M., **7,** 106
Haltiwanger, R. C., **8,** 229
Halton, B., **4,** 405; **5,** 587
Haluska, R. J., **3,** 163
Halvey, N., **6,** 407
Halweg, K. M., **11,** 28
Ham, G. E., **1,** 982
Hamachi, I., **12,** 146, 502
Hamada, A., **7,** 345
Hamada, T., **3,** 231, 303
Hamada, Y., **7,** 138; **9,** 164; **10,** 173, 324; **11,** 216, 222, 239; **12,** 188
Hamaguchi, H., **10,** 310; **11,** 599
Hamajima, R., **6,** 510
Hamalainen, C., **1,** 531
Hamalidis, C., **1,** 745
Hamamoto, I., **11,** 551; **12,** 350
Hamamura, E. K., **2,** 181
Hamamura, T., **9,** 91
Hamana, H., **11,** 159; **12,** 179
Hamanaka, E., **1,** 723; **2,** 293
Hamanaka, E. S., **9,** 477
Hamanaka, N., **11,** 172, 514
Hamann, P. R., **12,** 402
Hamaoka, T., **5,** 101, 347
Hamashima, Y., **9,** 519
Hambsch, E., **1,** 295
Hamdan, A., **8,** 279
Hamelin, J., **10,** 256
Hamer, J., **1,** 84
Hamer, N. K., **2,** 386
Hamersma, J. W., **1,** 258; **5,** 33, 760; **6,** 675
Hamill, B. J., **4,** 544; **8,** 188; **9,** 181
Hamilton, C. S., **1,** 876
Hamilton, D. E., **11,** 138
Hamilton, F. H., **1,** 654, 1284
Hamilton, G. A., **5,** 308; **8,** 248, 463
Hamilton, M., **10,** 332
Hamilton, R. J., **1,** 1011
Hamilton, R. W., **1,** 1090
Hamity, M., **8,** 130
Hamlet, J. C., **1,** 19, 642, 802, 846, 1090
Hamlin, K. E., **3,** 185; **6,** 526; **8,** 446
Hammar, W. J., **2,** 267
Hammer, I., **1,** 348, 536
Hammer, R. N., **2,** 369
Hammett, L. P., **1,** 495, 1137, 1143
Hamming, M. C., **6,** 20
Hammond, G. S., **1,** 45, 348, 1081; **2,** 286; **5,** 420, 470; **7,** 310
Hammond, J. J., **7,** 73
Hammond, L. M., **5,** 127
Hammond, W. A., **1,** 108
Hammond, W. B., **1,** 531; **2,** 104; **10,** 138
Hamoaka, T., **6,** 73
Hamon, D. P. G., **1,** 573; **5,** 680; **6,** 524
Hampel, C. A., **6,** 152
Hampel, G., **6,** 564
Hampson, N. A., **2,** 369; **3,** 16; **5,** 20
Hampton, A., **1,** 269, 332, 377
Hampton, K. G., **1,** 342; **2,** 178; **6,** 146
Hamrick, P. J., Jr., **3,** 183
Hamsan, A., **11,** 424
Hamuro, J., **3,** 217
Han, B. H., **11,** 292, 599
Han, G., **11,** 479
Han, G. R., **12,** 90

Han, G. Y., **3,** 146; **4,** 237
Han, W. T., **12,** 365
Han, Y. K., **7,** 358; **8,** 411; **9,** 237, 438
Hanack, M., **1,** 1015, 1056; **3,** 308; **4,** 405, 533; **5,** 308, 480, 551; **6,** 479, 620, 621; **8,** 53, 286
Hanafusa, T., **1,** 563; **5,** 167; **6,** 244; **11,** 434; **12,** 407
Hanaoka, M., **9,** 215; **11,** 430
Hanaya, K., **5,** 602; **9,** 421
Hance, C. R., **5,** 725; **10,** 434
Hancock, E. M., **1,** 685, 1049
Hancock, J. E. H., **1,** 772
Hancock, R. A., **9,** 463; **10,** 375
Hancock, W. S., **5,** 55; **6,** 149; **7,** 59
Hand, C. R., **5,** 322
Handa, M., **10,** 188
Handa, V. K., **9,** 54
Handel, H., **10,** 112
Handford, B. O., **1,** 486; **2,** 27
Handrick, R. G., **1,** 156, 494
Hands, A. R., **2,** 443
Hands, D., **6,** 467
Haneda, A., **11,** 431
Hanel, P. G., **2,** 150
Hanes, R. M., **7,** 54
Hanessian, S., **2,** 431; **4,** 167, 200, 555; **5,** 742; **6,** 81; **7,** 345; **8,** 192; **9,** 121, 141; **10,** 125, 216, 254, 300, 427; **11,** 196, 339, 373, 497; **12,** 79, 396
Hanford, W. E., **1,** 292
Hangauer, D. G., **12,** 150
Hangauer, D. G., Jr., **10,** 57
Hankin, H., **1,** 720
Hann, R. A., **5,** 249
Hann, R. M., **1,** 955
Hanna, I., **2,** 243
Hanna, J., **9,** 20
Hanna, R., **2,** 210; **5,** 397; **8,** 331
Hanna, W. J. W., **3,** 267; **5,** 20
Hannah, D. J., **9,** 462
Hannan, W., **6,** 504
Hannick, S. M., **12,** 568
Hannon, S. J., **6,** 640; **10,** 249
Hansbury, E., **1,** 894
Hansck, P., **11,** 232
Hanseke, G., **1,** 340
Hansel, R., **9,** 413; **10,** 351
Hansen, B., **2,** 447; **6,** 566; **7,** 97; **9,** 157
Hansen, H. J., **5,** 51, 212, 588, 700, 764; **9,** 237; **10,** 79; **11,** 559
Hansen, H. V., **2,** 154
Hansen, J., **1,** 1075
Hansen, J. F., **6,** 128
Hansen, K. C., **5,** 51
Hansen, L. D., **6,** 137
Hansen, M. M., **12,** 14
Hansen, R. T., **8,** 43
Hansler, J., **10,** 337
Hansley, V. L., **1,** 268, 712
Hanslick, R. S., **1,** 859
Hanson, E. M., **2,** 31
Hanson, E. R., **1,** 500, 652
Hanson, H. K., **2,** 96
Hanson, H. T., **1,** 309
Hanson, J. R., **2,** 77; **3,** 61; **5,** 144; **6,** 283; **12,** 72
Hanson, R. N., **10,** 211
Hanson, W. H., **11,** 108
Hanssen, H. W., **2,** 257
Hansske, F., **12,** 289
Hansson, C., **5,** 249; **7,** 342
Hantke, K., **7,** 215
Hantzsch, A., **1,** 501
Hanus, V., **12,** 503
Hanze, A. R., **1,** 93, 690
Hanzel, R. S., **1,** 1216; **4,** 311
Hanzlik, R. P., **2,** 415; **4,** 53; **8,** 125
Haque, K. E., **6,** 657
Hara, H., **6,** 316
Hara, K., **12,** 316, 402
Hara, N., **5,** 148
Hara, S., **6,** 222, 403; **7,** 367; **8,** 481; **9,** 362, 472; **10,** 403; **11,** 15, 294; **12,** 236, 502
Hara, T., **11,** 15
Hara, Y., **6,** 377
Harada, F., **10,** 277
Harada, H., **12,** 106
Harada, K., **4,** 252; **7,** 295; **8,** 386
Harada, N., **10,** 277
Harada, T., **7,** 207; **8,** 83, 127; **9,** 235, 309, 406; **10,** 6, 87, 374; **11,** 203, 489, 520, 522, 525; **12,** 465
Harakal, M. E. J., **11,** 109
Harano, K., **4,** 293; **5,** 13
Harano, Y., **12,** 336
Harayama, T., **8,** 330; **9,** 79, 304, 318; **11,** 352
Harbert, C. A., **2,** 382; **11,** 85
Harchema, S., **9,** 165
Hardegger, E., **1,** 34, 339, 660; **6,** 409
Hardenstein, R., **10,** 107
Harder, R. J., **1,** 426, 1125; **6,** 517
Hardie, B. A., **1,** 1018
Hardin, B. A., **1,** 741
Hardin, C. V., **9,** 245
Harding, D. R. K., **7,** 59
Harding, K., **2,** 40, 117
Harding, K. E., **5,** 319; **6,** 124; **12,** 70, 303
Harding, P. J. C., **8,** 47; **9,** 57; **10,** 48
Hardinger, S. A., **11,** 420
Hardstone, J. D., **1,** 500
Hardtmann, G., **2,** 196; **4,** 361; **9,** 337
Hardy, D. V. N., **1,** 843
Hardy, E. E., **1,** 371
Hardy, J. P., **5,** 336
Hardy, P. M., **9,** 439
Hardy, W. B., **1,** 169, 1128; **3,** 43; **5,** 503
Harfenist, M., **1,** 642, 1226
Harges, M. J. P., **11,** 222
Hargis, J. H., **6,** 576
Hargrove, R. J., **5,** 705; **6,** 620
Hargrove, W. W., **1,** 33, 34
Harirchian, B., **7,** 207; **8,** 281, 400; **12,** 394
Harkiss, D., **8,** 230
Harland, P. A., **12,** 319
Harley-Mason, J., **1,** 135, 933, 1211; **2,** 382
Harlow, R. L., **5,** 561
Harman, R. E., **6,** 597
Harmata, M. A., **11,** 437; **12,**

283, 451
Harmon, A. D., **5,** 446, 609; **6,** 538
Harmon, K. M., **1,** 1258; **4,** 568
Harmon, R. E., **3,** 98, 123, 329; **5,** 13, 740
Harms, R., **5,** 257; **7,** 215
Harms, W. M., **4,** 267; **5,** 141
Harney, D. W., **6,** 624
Harper, J. L., **1,** 172
Harpold, M. H., **2,** 313
Harpp, D. N., **2,** 207, 373; **3,** 36, 149, 299; **4,** 242; **5,** 66; **6,** 76, 118, 280, 482, 634; **7,** 40, 194; **8,** 33, 57; **9,** 129, 236, 497; **10,** 219; **11,** 493
Harre, M., **12,** 564
Harrington, J. K., **3,** 18
Harrington, K. I., **4,** 185
Harrington, K. J., **6,** 96
Harrington, R. W., **3,** 123
Harris, A. D., **1,** 295
Harris, B. D., **12,** 476
Harris, C. E., **1,** 1027
Harris, C. M., **1,** 1040
Harris, G. C., **1,** 37, 187
Harris, H., **1,** 966
Harris, H. P., **5,** 155
Harris, J. E., **2,** 447
Harris, J. F., Jr., **7,** 368
Harris, L. E., **4,** 378, 391; **5,** 561
Harris, L. S., **1,** 1163; **6,** 283
Harris, M. M., **5,** 717
Harris, N. D., **4,** 194
Harris, P. G., **8,** 524; **12,** 550
Harris, R. C., **1,** 389
Harris, R. F., **1,** 570, 574
Harris, R. L., **4,** 199
Harris, S. A., **1,** 637, 730, 731, 860, 873, 1049
Harris, S. J., **7,** 29
Harris, T. M., **1,** 342, 1040, 1081; **2,** 178, 197; **7,** 207; **10,** 75; **12,** 314
Harrison, A. K., **2,** 425
Harrison, C. R., **6,** 208, 453, 494; **7,** 279; **9,** 503; **11,** 195
Harrison, D., **5,** 342
Harrison, G. C., **1,** 874
Harrison, H. R., **1,** 531, 705; **2,** 287
Harrison, I. T., **1,** 642; **2,** 34, 75, 201, 381; **3,** 258, 265; **4,** 305; **8,** 211; **9,** 406
Harrison, J. J., **9,** 21; **11,** 432
Harrison, J. M., **5,** 280
Harrison, L. W., **12,** 288, 303
Harrison, M. J., **6,** 316
Harrison, P. G., **12,** 202
Harrison, S., **2,** 75; **9,** 406
Harrison, W. A., **1,** 754
Harrisson, R., **1,** 444
Harshe, S. N., **1,** 309, 685
Hart, A. J., **6,** 638; **12,** 542
Hart, C. K., **7,** 236
Hart, D. J., **5,** 9, 70; **6,** 94; **8,** 76, 196, 203, 412; **11,** 550; **12,** 523
Hart, D. W., **4,** 249; **6,** 179
Hart, F. A., **8,** 148
Hart, H., **1,** 827, 1095; **2,** 173, 316; **4,** 196, 254, 375; **5,** 311, 394, 486; **8,** 445; **11,** 355, 479
Hart, L. S., **5,** 337
Hart, P. A., **3,** 6
Hartemink, M., **4,** 111
Hartenstein, A., **1,** 364
Hartenstein, J., **6,** 514
Hartford, T. W., **6,** 649
Hartgerink, J., **10,** 455
Hartke, K., **2,** 398; **6,** 379
Hartle, R. J., **3,** 145, 201
Hartley, D., **2,** 29; **6,** 258
Hartley, F. R., **6,** 61
Hartley, W. M., **7,** 257
Hartman, B. C., **6,** 103
Hartman, J. L., **1,** 465
Hartman, J. S., **7,** 359
Hartman, L., **1,** 66
Hartman, M. E., **3,** 139; **8,** 501
Hartman, W. W., **1,** 109, 176, 500, 685, 695, 864, 1027, 1064, 1105, 1123, 1169, 1237, 1283; **3,** 272
Hartmann, A. A., **4,** 231
Hartmann, H., **7,** 177
Hartmann, J., **5,** 552; **6,** 638; **7,** 401; **12,** 542
Hartmann, W., **9,** 3
Hartner, F. W., **8,** 391
Hartner, F. W., Jr., **12,** 578
Hartog, F. A., **10,** 5
Hartough, H. D., **1,** 500
Hartshorn, M., **2,** 208
Hartshorn, M. P., **1,** 355; **2,** 173, 267, 362; **8,** 440; **12,** 425
Hartter, D. R., **4,** 157; **5,** 176, 220
Hartung, W. H., **1,** 97, 692
Hartwell, J. L., **1,** 448, 682, 881, 927, 1027
Hartwig, I., **6,** 390
Hartwig, W., **9,** 428; **10,** 152
Hartz, G., **11,** 200
Hartzell, G. E., **3,** 219
Hartzell, S. L., **5,** 372; **7,** 42, 51
Hartzler, H. D., **1,** 926; **5,** 460
Haruki, E., **6,** 95
Haruna, M., **9,** 32; **10,** 66
Harusawa, S., **9,** 164; **10,** 145; **11,** 181; **12,** 188
Haruta, J., **9,** 310; **11,** 279, 340; **12,** 59, 559
Haruta, R., **11,** 181, 571
Harvey, D. F., **12,** 313, 561
Harvey, G. R., **2,** 417
Harvey, R. G., **1,** 1216; **2,** 273; **4,** 201, 248, 489; **5,** 86; **6,** 230, 316, 326, 658; **7,** 331; **8,** 155, 383; **9,** 34, 102, 238; **10,** 200; **11,** 150
Harville, R., **3,** 36
Harwood, H. J., **1,** 837; **5,** 127
Harwood, L. M., **9,** 376; **11,** 559
Hasan, F., **5,** 538
Hasan, I., **11,** 75
Hasan, S. K., **3,** 244; **8,** 401; **9,** 466
Hasbrouck, R. W., **5,** 319
Hase, C., **6,** 563
Hase, S., **1,** 744
Hase, T., **4,** 43; **6,** 383; **9,** 66
Hase, T. A., **9,** 423; **10,** 21, 252, 296

Hasegawa, A., **5,** 226
Hasegawa, H., **11,** 1
Hasegawa, J., **11,** 494
Hasegawa, K., **6,** 424
Hasegawa, M., **1,** 444, 608; **12,** 193
Hasegawa, S., **12,** 562
Hasegawa, T., **10,** 198; **12,** 64
Hasek, R. H., **1,** 389, 518, 986
Hasek, W. R., **1,** 1125
Hasenfratz, H., **5,** 552
Hasenhuettl, G., **6,** 492
Hashiguchi, S., **9,** 370
Hashimoto, H., **3,** 258; **4,** 153; **6,** 194, 216; **8,** 96, 172, 486; **9,** 344, 352; **10,** 56, 90; **12,** 437
Hashimoto, I., **3,** 212
Hashimoto, K., **8,** 227; **9,** 38, 414; **11,** 235; **12,** 506
Hashimoto, M., **11,** 258, 562; **12,** 199, 381
Hashimoto, N., **4,** 187; **10,** 431; **11,** 574
Hashimoto, S., **7,** 189, 391; **8,** 164, 165, 272, 472, 476; **9,** 478; **10,** 115, 117, 230, 419; **12,** 420, 466
Hashimoto, S. I., **9,** 290
Hashimoto, T., **4,** 109; **7,** 332
Hashimoto, Y., **1,** 195
Hashino, S., **10,** 71
Hashio, S., **10,** 390; **11,** 419
Hashizume, A., **9,** 166
Hashizume, T., **2,** 30
Hasibullah, **1,** 1167
Haskelberg, L., **1,** 66
Haskell, B. E., **7,** 172
Haskins, W. M., **8,** 365
Haslanger, M. F., **6,** 215; **8,** 87
Haslouin, J., **8,** 138
Hass, H., **1,** 952
Hass, H. B., **1,** 1073, 1101; **4,** 461
Hassall, C. H., **1,** 536; **2,** 386; **11,** 410
Hassanali, A., **2,** 238
Hassel, T., **7,** 312; **8,** 494
Hasserodt, U., **2,** 179, 279
Hassid, A. I., **4,** 239, 451
Hassig, R., **12,** 279, 547
Hasskul, T., **12,** 303
Hassner, A., **1,** 40, 125, 501, 754, 757, 1055, 1284; **2,** 25, 38, 101, 107, 223, 224, 225, 299; **3,** 34, 161, 163, 215; **4,** 187, 262, 323, 411; **5,** 93, 136, 280, 713; **6,** 114, 122, 319; **7,** 66, 225; **8,** 156; **9,** 154, 157, 180; **12,** 100, 242
Hasty, N. M., **5,** 491
Haszeldine, R. N., **1,** 504, 1058, 1227; **2,** 226, 227; **5,** 326, 564, 705; **6,** 521, 621; **7,** 324; **9,** 480
Hata, H., **8,** 115
Hata, K., **4,** 571; **6,** 500
Hata, N., **6,** 70; **11,** 427
Hata, T., **1,** 208; **2,** 40; **3,** 49, 297; **6,** 602, 644; **9,** 166; **11,** 178; **12,** 419, 557
Hata, Y., **5,** 569
Hatakeyama, S., **9,** 308; **11,** 75, 267
Hatanaka, C., **2,** 54, 305
Hatanaka, I., **7,** 359
Hatanaka, K., **11,** 604
Hatanaka, N., **9,** 333
Hatanaka, Y., **11,** 114
Hatanake, N., **10,** 142
Hatayama, Y., **9,** 151
Hatch, C. E., **5,** 687
Hatch, L. F., **1,** 215; **2,** 390; **3,** 273; **5,** 107
Hatch, R. H., **9,** 16
Hatch, R. L., **10,** 321
Hatcher, A. S., **6,** 398
Hatfield, G. L., **9,** 74
Hatfield, M. R., **1,** 149
Hathaway, B. J., **6,** 133
Hatt, H. H., **1,** 38, 444; **2,** 222; **5,** 15
Hatterman, R. L., **11,** 431
Hattori, H., **7,** 221
Hattori, K., **4,** 52; **11,** 249, 542; **12,** 411, 452
Hattori, M., **9,** 146
Hatz, E., **1,** 282, 283
Haubein, A. H., **1,** 573, 619
Haubenstock, H., **1,** 1189; **2,** 352; **4,** 312; **5,** 392, 565
Hauck, H. F., Jr., **10,** 221
Hauck, P. R., **11,** 170
Haug, E., **4,** 568; **10,** 113
Haumesser, W., **5,** 631
Hauptmann, H., **1,** 956; **6,** 82
Hauptmann, S., **1,** 445
Hauptschein, M., **3,** 143
Hausberg, H. H., **8,** 395
Hauschild, K., **9,** 361
Hauser, A., **7,** 122; **8,** 374
Hauser, C., **2,** 285
Hauser, C. F., **1,** 1246; **5,** 534
Hauser, C. R., **1,** 69, 70, 83, 160, 251, 342, 371, 481, 524, 601, 629, 685, 909, 911, 1033, 1040, 1041, 1080, 1081, 1118, 1256, 1258, 1261, 1286, 1292; **2,** 53, 173, 178, 197, 374; **3,** 183; **4,** 112, 297, 397, 489, 510; **5,** 160, 406, 725; **6,** 84, 636; **7,** 106; **8,** 190; **9,** 129; **10,** 434; **12,** 403
Hauser, D., **10,** 159
Hauser, F. H., **10,** 95
Hauser, F. M., **8,** 76; **9,** 370; **11,** 79; **12,** 490
Hauser, T. R., **1,** 673
Haushalter, R. C., **11,** 462
Hauske, J. R., **8,** 356, 499
Hausler, J., **9,** 156
Haut, S. A., **6,** 606
Hauteville, M., **5,** 620
Hauth, H., **5,** 118
Haveaux, B., **4,** 95; **6,** 123; **12,** 125
Havel, J. J., **5,** 654
Haven, A. C., Jr., **1,** 662
Havens, J. L., **8,** 110, 252
Havinga, E., **2,** 38; **8,** 115
Havlicek, S. C., **5,** 551
Hawkins, C., **7,** 268
Hawkins, D. W., **5,** 472
Hawkins, E. G. E., **5,** 474
Hawkins, J. M., **12,** 60
Hawkins, L., **3,** 171
Hawks, G. H., **1,** 345; **2,** 411; **4,** 497, 527
Hawley, G. G., **6,** 152

Hawley, L. B., Jr., **2,** 84
Haworth, R. D., **1,** 41
Hawthorne, J. O., **1,** 782
Hawthorne, M. F., **1,** 207, 673, 964; **3,** 29; **6,** 607
Hawtrey, A. O., **2,** 424
Hay, A. S., **1,** 154, 169, 652; **7,** 277; **10,** 138
Hay, G. W., **12,** 420
Hay, N., **4,** 71
Hayakawa, K., **11,** 28
Hayakawa, N., **5,** 224
Hayakawa, S., **6,** 430
Hayakawa, T., **2,** 389
Hayakawa, Y., **4,** 158; **5,** 224; **6,** 198, 201; **8,** 498, 499; **9,** 290, 478
Hayama, H., **5,** 534
Hayama, N., **6,** 573; **9,** 322; **12,** 337
Hayama, T., **11,** 33, 468
Hayami, H., **12,** 352
Hayami, J., **5,** 556, 607
Hayano, K., **11,** 96
Hayao, S., **1,** 904; **5,** 193
Hayase, M., **4,** 147
Hayase, Y., **2,** 132, 427; **3,** 96; **8,** 444
Hayashi, H., **5,** 433; **12,** 276, 338
Hayashi, I., **6,** 317, 403, 649
Hayashi, J., **11,** 533; **12,** 263
Hayashi, K., **3,** 294; **12,** 187
Hayashi, M., **5,** 756; **6,** 596; **8,** 312; **9,** 171, 436; **10,** 332; **11,** 447; **12,** 420, 466
Hayashi, N., **4,** 35; **6,** 170
Hayashi, R., **1,** 754
Hayashi, S., **5,** 226; **9,** 166; **10,** 99, 448
Hayashi, T., **6,** 14, 398, 569; **8,** 39; **9,** 147, 201; **10,** 38, 194; **11,** 20, 47, 59, 239, 240; **12,** 146, 172, 374, 502
Hayashi, Y., **10,** 14; **12,** 36, 307, 405, 508
Hayden, J. L., **8,** 522
Hayes, F. N., **1,** 894; **4,** 252
Hayes, L. J., **4,** 389
Hayes, N. F., **6,** 502
Hayez, E., **5,** 572
Haymaker, C. R., **1,** 731
Haynes, G. R., **1,** 444, 1197
Haynes, N. B., **1,** 87
Haynes, R., **10,** 209
Haynes, R. K., **5,** 486; **6,** 596, 657; **10,** 198
Haynes, W. M., **1,** 705; **2,** 287
Haynie, E. C., **5,** 460
Haynie, S. L., **11,** 54
Hays, H. R., **1,** 310, 378
Hayward, R. C., **5,** 655; **8,** 260; **9,** 249; **10,** 395; **11,** 515
Haywood, D. J., **10,** 300
Haywood-Farmer, J., **5,** 127; **7,** 52
Hazdra, J. J., **1,** 581, 611; **5,** 379
Hazen, G. G., **1,** 665
Hazen, G. H., **11,** 536; **12,** 220
Hazen, R. K., **1,** 1049
Hazra, B. G., **8,** 529
Hazum, E., **6,** 306, 625
He, D. W., **12,** 290
He, Z., **12,** 314
Heacock, D. J., **11,** 275
Heacock, J. F., **3,** 18
Head, D. B., **11,** 191
Heal, H. G., **1,** 1236
Heaney, H., **1,** 77, 573; **4,** 336, 411; **5,** 266; **6,** 280; **7,** 325; **9,** 388
Heany, H., **6,** 524
Hearn, M. J., **5,** 677
Hearon, W. M., **2,** 323
Hearst, P. J., **1,** 664, 1163
Heath, M. J., **2,** 422; **3,** 161
Heath, R. R., **10,** 332
Heath, W. F., **11,** 256
Heathcock, C., **1,** 501, 754, 1055; **3,** 163
Heathcock, C. H., **2,** 173, 224; **3,** 196; **4,** 323, 411; **5,** 276, 482, 761; **6,** 205, 215, 339, 441; **7,** 64, 258, 271, 279; **8,** 112; **9,** 283, 314; **10,** 133, 162, 326; **11,** 341, 349; **12,** 121, 188, 418, 502
Heathcock, C. L., **8,** 299
Heaton, P. C., **4,** 74
Heaton, P. R., **4,** 79
Heavner, G. A., **6,** 115
Hebbelynck, M. F., **1,** 139
Hebenbrock, K. F., **1,** 127
Hebert, E., **9,** 224
Hebert, N. C., **2,** 368
Hecht, J. K., **2,** 273
Hecht, R., **1,** 116
Hecht, S. M., **5,** 182, 412; **6,** 159; **8,** 358; **9,** 323
Heck, L. L., **1,** 424
Heck, R. F., **1,** 155, 1058; **4,** 379; **5,** 413, 497; **6,** 61, 157; **7,** 5, 274; **8,** 39, 382; **9,** 351, 482; **12,** 370
Hecker, E., **1,** 736, 1226
Heckert, R. E., **1,** 1136
Heckl, B., **11,** 401
Heckman, R. A., **2,** 48
Hedaya, E., **5,** 230
Hedayatullah, M., **6,** 514; **8,** 451; **10,** 378
Hederich, V., **1,** 1018; **3,** 251
Hedge, J. A., **1,** 527
Hedge, S. G., **10,** 209
Hedgecock, H. C., Jr., **6,** 625; **7,** 181
Hedges, S. H., **10,** 347
Hedgley, E. J., **6,** 583
Hedin, P. A., **5,** 177
Hedrick, G. W., **1,** 37
Heeger, A. J., **7,** 165, 387
Heeren, J. K., **1,** 854, 1141; **2,** 456; **7,** 62
Heermann, D., **9,** 167
Heffe, W., **5,** 39
Heffer, J. P., **1,** 778
Heffer, T. P., **5,** 498
Heffler, J. P., **10,** 303
Hegarty, A. F., **10,** 269
Hegde, S. G., **11,** 260
Hegde, V. R., **12,** 554
Hege, B. P., **2,** 48
Hegedus, L. S., **3,** 212; **4,** 354; **5,** 46, 437; **6,** 93, 105, 361, 419; **7,** 22, 250; **8,** 39, 324; **10,** 302; **11,** 47, 401, 514; **12,** 576
Heggs, R. P., **9,** 245; **10,** 206

Hehemann, D. G., **9,** 447
Hei, S. K., **4,** 253
Heiba, E. I., **2,** 264; **4,** 71; **6,** 127, 356, 357
Heibl, C., **6,** 285
Heidberg, J., **1,** 536
Heidelberger, C., **10,** 249
Heider, J., **8,** 130; **11,** 337
Heiker, F. R., **10,** 270
Heilbron, I. M., **1,** 144, 360, 754, 1157; **3,** 39
Heilbronner, E., **1,** 1148
Heilmann, R., **2,** 42
Heilmann, S. M., **5,** 389; **8,** 409; **9,** 128; **12,** 235
Heim, P., **1,** 207; **3,** 77; **4,** 126, 176
Heim, S., **4,** 146
Heimann, M. R., **12,** 428
Heimbach, H., **12,** 181
Heimbach, K., **1,** 636
Heimbach, P., **5,** 510
Hein, D. W., **1,** 904
Hein, R. W., **1,** 518
Heincke, K. D., **5,** 25
Heine, H. G., **4,** 47; **6,** 34; **9,** 3
Heine, H. W., **1,** 484, 1295; **4,** 541
Heinemann, U., **10,** 158
Heiney, R. E., **1,** 13
Heininger, H. U., **4,** 172
Heininger, S. A., **1,** 160
Heinke, B., **1,** 330
Heinrich, G. R., **9,** 11
Heintzelman, R. W., **7,** 136; **10,** 167
Heinz, G., **1,** 1272
Heinz, K. J., **10,** 135
Heinzelman, R. V., **1,** 392, 1049
Heinzer, F., **9,** 355
Heinzman, S. W., **11,** 138
Heischkeil, R., **1,** 195
Heisig, G. B., **1,** 402, 1073
Heisler, R. Y., **1,** 1044
Heissler, D., **9,** 38; **10,** 24
Heiszwolf, G. J., **1,** 524; **2,** 431
Heitkaemper, P., **4,** 122
Heitmann, W. R., **9,** 117
Heitz, L. J., **5,** 105
Heitz, M. P., **12,** 245
Heitzer, H., **4,** 256
Heitzler, H., **11,** 174
Helali, S. E., **6,** 409
Helbig, W., **11,** 87
Helbling, A. M., **8,** 101
Helbling, R., **1,** 652
Helder, R., **4,** 141; **5,** 205; **6,** 501; **7,** 311; **8,** 431
Heldeweg, R. F., **7,** 60, 347; **8,** 375; **9,** 145
Heldt, W. Z., **5,** 727
Helferich, B., **1,** 655, 664, 838, 1027, 1256; **2,** 4, 365; **6,** 511
Helfert, H., **3,** 32
Helgeson, R. C., **1,** 195, 1004; **5,** 103; **9,** 361
Helgren, P. F., **1,** 982
Heller, J. E., **5,** 57, 257
Heller, M., **1,** 281, 671, 764, 881, 1000; **2,** 117, 281
Heller, S. R., **2,** 98
Hellerman, L., **1,** 510
Helling, J. F., **6,** 596
Hellman, H. M., **1,** 478
Hellman, S., **10,** 211
Hellmann, H., **2,** 23, 118
Hellmuth, R., **1,** 561
Hellwinkel, D., **4,** 476
Helmchen, G., **7,** 162; **9,** 299, 337, 406; **12,** 227, 241
Helmholt, R. B., **4,** 516
Helmkamp, G. K., **1,** 742, 754, 1233; **2,** 274, 439; **3,** 163
Helmut, E., **9,** 325
Helquist, P. M., **4,** 35, 354; **6,** 361, 419; **8,** 117, 458; **9,** 70, 143, 283, 306, 333; **10,** 77, 129; **11,** 51
Helt, A., **5,** 626
Hemesley, P., **3,** 191
Hemmi, K., **12,** 199, 381
Hemo, J. H., **4,** 239; **8,** 232
Hems, B. A., **1,** 372, 500, 595, 642, 951
Henbest, H. B., **1,** 19, 37, 132, 144, 444, 561, 581, 594, 642, 643, 696, 795, 802, 846, 927; **2,** 229, 263, 281, 359; **3,** 166; **5,** 77; **7,** 64, 354; **9,** 292
Henderson, A., **3,** 81; **11,** 110
Henderson, A. T., **1,** 268
Henderson, G. H., **4,** 393
Henderson, G. N., **10,** 5
Henderson, R., **4,** 98
Henderson, R. M., **5,** 505
Henderson, W. A., Jr., **6,** 137
Henderson, W. W., **1,** 177
Hendrick, M., **5,** 342
Hendrick, M. E., **4,** 336
Hendrickson, J. B., **1,** 44, 1055; **2,** 42, 62, 146, 159; **3,** 34, 221; **5,** 29, 530, 560, 564, 705; **6,** 36, 44, 225, 648; **7,** 390
Hendrickson, Y. G., **1,** 465
Hendriks, K. B., **6,** 569
Hendry, L. B., **7,** 287
Henecka, H., **1,** 392
Henegar, K., **10,** 359
Henery, J., **2,** 112, 140; **3,** 101
Henery-Logan, K. R., **1,** 235
Heng, K. K., **6,** 215
Hengartner, U., **4,** 481; **9,** 260
Henggeler, B., **11,** 441
Henin, F., **10,** 202
Henke, K., **11,** 415
Henkel, J., **1,** 361
Henkler, H., **1,** 1075
Henle, F., **1,** 22
Henn, B. R., **1,** 289
Henne, A. L., **1,** 426, 777, 1226, 1227; **5,** 107
Hennessy, D. J., **1,** 1127
Hennig, G. R., **6,** 269
Henning, R., **9,** 22; **12,** 38, 212
Henninger, M., **7,** 418; **8,** 113
Hennion, G. F., **1,** 377, 1184, 1191; **2,** 465; **3,** 199; **4,** 311; **7,** 155; **8,** 94
Henrich, P., **1,** 782
Henrick, C. A., **3,** 112, 118; **4,** 164; **5,** 244; **6,** 165, 224, 586
Henrikson, B. W., **1,** 374; **5,** 593
Henry, J. A., **3,** 54
Henry, P. M., **5,** 497

Henry, R., **2,** 226
Henry, R. A., **2,** 423; **5,** 10; **7,** 337
Henry, W., **11,** 322
Henry, Y., **5,** 580
Henry-Basch, E., **7,** 240
Henseke, G., **1,** 695, 1101
Henseling, K. O., **7,** 298
Hentges, S. G., **10,** 291, 445
Hentschel, P., **6,** 263
Henze, H. R., **1,** 388, 892
Henze, M., **10,** 145
Henzen, A. V., **12,** 182
Hepburn, D. R., **6,** 220; **7,** 424
Hepburn, S. P., **12,** 95
Hepworth, J. D., **4,** 189, 477; **5,** 693; **7,** 352
Herald, C. L., **2,** 131, 237; **3,** 96
Herald, D. L., Jr., **4,** 369
Herb, S. F., **6,** 675
Herbach, J., **5,** 182; **11,** 347
Herbert, R. B., **2,** 296; **4,** 529; **5,** 659; **8,** 529; **10,** 347; **11,** 533
Herbertz, T., **1,** 186
Herbst, D. M., **2,** 359
Herbst, M. D., **9,** 483
Herbst, R. M., **1,** 448; **7,** 337
Herczegh, P., **8,** 132
Hergenrother, W. L., **1,** 309, 310
Hergott, H. H., **10,** 96; **11,** 587
Hering, H., **1,** 654, 1289
Herkes, F. E., **1,** 1058, 1227; **3,** 308
Herlinger, H., **1,** 494
Herman, G., **5,** 282; **6,** 133
Hermanek, S., **1,** 608; **4,** 316
Hermann, R. B., **1,** 643
Hermansen, D. E., **1,** 619
Hermecz, I., **12,** 231
Hermes, M. E., **1,** 174, 859; **5,** 476
Hernandez, A., **5,** 719
Hernandez, D., **11,** 249
Hernandez, J. E., **11,** 452
Hernandez, L., Jr., **8,** 88
Hernandez, O., **7,** 274; **8,** 180; **9,** 180; **10,** 156
Hernandez, R., **7,** 40
Hernandz, E., **6,** 162
Herndon, J. W., **12,** 344
Herold, A., **2,** 241; **5,** 19; **6,** 23
Herold, T., **9,** 363
Herr, M. E., **1,** 777, 890, 974
Herr, R. W., **3,** 112; **5,** 78; **6,** 245
Herranz, E., **8,** 365
Herrick, E. C., **1,** 1177
Herrick, J. J., **12,** 276
Herring, K. H., **12,** 154
Herriott, A. W., **5,** 644; **6,** 406
Herrman, K., **10,** 1
Herrmann, J. L., **5,** 303, 305, 406, 407, 445, 456; **6,** 312; **7,** 412; **9,** 217; **10,** 57
Herron, D. K., **3,** 313; **5,** 293; **6,** 383; **8,** 498
Herschied, J. D. H., **6,** 605
Herschied, J. D. M., **5,** 675
Herscovici, J., **10,** 273
Hershberg, E. B., **1,** 74, 411, 424, 456, 562, 573, 682, 713, 785, 787, 802, 929, 966, 974, 976, 1000, 1008, 1023, 1029, 1073, 1101, 1121, 1284; **2,** 353; **9,** 380
Hershberger, S. S., **10,** 149
Hershkowitz, R. L., **1,** 1249
Herten, B. W., **6,** 99
Hertenstein, U., **5,** 722; **6,** 633; **11,** 150
Hertenstein, W., **9,** 129
Hertler, W. R., **1,** 16, 500, 1136, 1184
Hertrich, I., **3,** 308
Hertzberg, R. P., **10,** 249
Hertzberg, S., **2,** 295
Hertzler, D. V., **2,** 234
Herve, Y., **12,** 417
Hervens, F., **5,** 198
Herweh, J. E., **5,** 268
Herwig, W., **4,** 163
Herynk, J., **1,** 658
Herz, J. E., **1,** 367, 1284; **4,** 223, 557; **5,** 669
Herz, W., **1,** 617, 876; **2,** 238, 318, 344
Herzberg, H., **8,** 2
Herzberg-Minzly, Y., **2,** 76; **7,** 21
Herze, P. Y., **11,** 237
Herzig, J., **12,** 546
Herzig, P. T., **1,** 394
Herzog, H., **2,** 49, 54; **4,** 65
Herzog, H. L., **1,** 47, 74, 966, 1000; **2,** 232, 433; **4,** 370; **5,** 552
Herzog, J., **4,** 204
Hes, J., **6,** 168
Hesbain-Frisque, A. M., **6,** 123; **11,** 51, 201
Heseltine, D. W., **6,** 409
Heslinga, L., **1,** 220
Heslop, R. B., **2,** 354
Hess, G. P., **1,** 235
Hess, H., **10,** 273
Hess, H. J., **1,** 318
Hess, H. M., **3,** 10; **6,** 409
Hess, K., **1,** 333
Hess, O., **1,** 841
Hess, W. W., **2,** 75; **4,** 216
Hesse, A., **3,** 185; **5,** 416
Hesse, G., **1,** 1108
Hesse, H., **2,** 249
Hesse, M., **5,** 588; **9,** 379; **11,** 479; **12,** 462
Hesse, R., **2,** 172
Hesse, R. H., **1,** 149; **2,** 200; **3,** 147, 262; **4,** 37, 511, 568; **5,** 381; **6,** 124, 264, 598; **7,** 158
Hessel, L. W., **12,** 294
Hessler, E. J., **1,** 80; **4,** 53
Hesson, D. P., **6,** 567; **7,** 404
Hester, W. F., **1,** 470
Hettrick, G., **8,** 471
Hetzel, F. W., **2,** 174; **3,** 128; **4,** 202
Heuberger, C., **10,** 413; **12,** 537
Heuck, K., **10,** 254; **11,** 317
Heuman, P., **6,** 534
Heumann, A., **5,** 13; **9,** 353; **11,** 45
Heus, R., **10,** 321
Heusel, G., **6,** 422
Heusler, K., **1,** 87, 169, 389, 562, 608, 624, 1254; **3,**

188; **9,** 318
Heusmann, P. L., **10,** 436
Heusser, H., **1,** 3, 471, 595, 630, 652, 1284; **6,** 91
Hevesi, L., **7,** 35, 184, 268; **8,** 64, 320; **9,** 34; **11,** 225
Hewett, W. A., **1,** 1157
Hewgill, F. R., **9,** 385
Hewlett, A. P., **1,** 424
Hewlett, C., **2,** 344
Hewson, A. T., **9,** 502; **11,** 426; **12,** 479
Hexner, P. E., **1,** 11, 445
Hextall, P. J., **1,** 1237
Hey, D. G., **1,** 1004, 1008, 1019
Hey, D. H., **1,** 155, 264, 323, 754, 928; **3,** 21
Heyd, W. E., **5,** 308
Heydkamp, W. R., **1,** 125, 484
Heyl, D., **1,** 637
Heyl, F. W., **1,** 777, 890, 974
Heyman, M. L., **5,** 491
Heymann, H., **1,** 72, 1054
Heymes, G., **1,** 1256
Heyn, H., **4,** 93
Heyn, J., **4,** 209; **12,** 187
Heyn, W., **5,** 322
Heyne, H. U., **4,** 456
Heyne, T. R., **6,** 665
Heyns, K., **1,** 432, 652, 1018; **3,** 249; **4,** 86; **5,** 601; **7,** 290; **10,** 270
Heywood, D. L., **1,** 791; **5,** 506
Hiatt, J. D., **2,** 323
Hiatt, J. E., **5,** 205
Hiatt, R. R., **3,** 331
Hibbert, P. G., **5,** 148
Hibi, T., **9,** 344
Hibino, J., **12,** 344
Hibino, S., **8,** 351, 492; **9,** 470, 473; **10,** 404
Hickinbottom, W. J., **1,** 598, 689, 1081, 1132, 1231
Hida, M., **3,** 258
Hida, T., **11,** 71
Hider, R. C., **5,** 91
Hidy, P. H., **1,** 345
Hieber, W., **1,** 1058
Hiebert, H. I., **2,** 112
Hiemstra, H., **8,** 431
Hiers, G. S., **1,** 452, 481; **11,** 78
Higa, M. T., **9,** 256
Higa, T., **5,** 667
Higashi, F., **4,** 210
Higashiura, K., **10,** 28
Higginbottom, C., **8,** 397
Higgins, G. M. C., **3,** 320
Higgins, J., **3,** 322
Higgins, J. G., **1,** 1249; **2,** 446
Higgins, R., **6,** 436; **12,** 123
Higgins, S. D., **11,** 268
Higgins, W., **10,** 172
Hignite, C., **1,** 904
Higuchi, H., **10,** 85
Higuchi, K., **12,** 125
Higuchi, M., **4,** 390; **5,** 564; **6,** 185
Higuchi, S., **5,** 13, 178; **6,** 158
Higuchi, T., **11,** 220; **12,** 172
Hii, P., **10,** 216
Hiiragi, M., **3,** 268
Hilbert, P. H., **4,** 544; **9,** 181; **10,** 157
Hilcken, V., **1,** 561
Hildebrand, R. P., **1,** 1011
Hilgetag, G., **4,** 197
Hill, A. E., **4,** 328; **5,** 437; **6,** 364
Hill, A. J., **1,** 52
Hill, A. S., **1,** 484
Hill, D. T. **2,** 300; **3,** 78; **4,** 234
Hill, D. W., **9,** 462
Hill, G. L., **5,** 600
Hill, H. A. O., **1,** 170
Hill, H. W., Jr., **4,** 505
Hill, J. G., **12,** 94
Hill, J. H., **10,** 195
Hill, J. W., **1,** 952
Hill, K. A., **4,** 23; **5,** 454; **6,** 83
Hill, K. R., **5,** 285
Hill, M. E., **3,** 96; **5,** 477; **6,** 420
Hill, M. K., **1,** 904
Hill, P., **1,** 777
Hill, R. K., **1,** 8, 185, 904, 1000, 1184; **3,** 73, 127; **4,** 167; **7,** 225; **9,** 470
Hill, R. M., **1,** 413
Hill, T., Jr., **1,** 118
Hillard, R. L., III, **6,** 154; **8,** 147; **12,** 163
Hillenbrand, G. F., **12,** 158
Hiller, G., **1,** 707
Hiller, H., **4,** 414
Hillman, M. E. D., **1,** 1231
Hilpert, L., **8,** 102
Hilscher, J. C., **4,** 251; **6,** 529
Hilton, B. D., **11,** 589
Himelstein, N., **1,** 892, 986
Himizu, J., **4,** 146
Himmele, W., **5,** 148; **8,** 18; **12,** 426
Himmelsbach, F., **11,** 56
Himmen, E., **2,** 365
Hinckley, A. A., **1,** 1080, 1143
Hindersinn, R. R., **1,** 986
Hindley, F. M., **7,** 155
Hindley, N. C., **7,** 414
Hindriks, H., **2,** 228
Hine, J., **1,** 844; **2,** 70; **3,** 78; **5,** 49; **6,** 261; **11,** 306
Hine, K. E., **5,** 311
Hinemann, S. D., **5,** 620
Hines, J. N., **2,** 53; **5,** 399
Hingerty, B., **6,** 568
Hinman, C. W., **4,** 357, 397
Hinman, R. L., **1,** 503, 518
Hinney, H. R., **9,** 39; **11,** 420
Hino, J. B., **1,** 76
Hino, K., **6,** 445
Hino, T., **7,** 268
Hinrichs, I., **1,** 180
Hinsberg, O., **1,** 837
Hinshaw, J. C., **4,** 15, 263
Hinterhofer, O., **2,** 393
Hinz, G., **1,** 1189
Hinze, S., **7,** 247
Hiraga, K., **12,** 387
Hirahata, Y., **3,** 337
Hirai, H., **5,** 764; **10,** 274; **11,** 539; **12,** 151
Hirai, K., **5,** 448, 459; **6,** 399; **8,** 342
Hirai, S., **1,** 1198; **4,** 147; **5,** 370; **9,** 13
Hirakura, M., **1,** 1096
Hirama, M., **6,** 170; **9,** 76, 79,

132, 304; **10,** 119, 198
Hirana, M., **8,** 330
Hirano, M., **9,** 119
Hirano, S., **5,** 534; **8,** 112; **10,** 318
Hirao, A., **9,** 422; **12,** 31
Hirao, I., **12,** 70, 241
Hirao, K., **6,** 453; **11,** 312
Hirao, N., **5,** 194, 441
Hirao, T., **6,** 450; **8,** 381; **9,** 349; **10,** 298, 391; **11,** 514, 555; **12,** 187, 336
Hiraoka, H., **12,** 356
Hiraoka, T., **3,** 81; **5,** 406; **8,** 208
Hirasawa, H., **8,** 314
Hirashima, T., **6,** 281; **9,** 237; **10,** 460
Hirata, K., **9,** 479
Hirata, Y., **2,** 462; **5,** 421, 446, 756; **7,** 135; **8,** 354
Hirayama, F., **9,** 129
Hirayama, M., **8,** 337; **11,** 570
Hirayama, T., **5,** 171
Hiriart, J. M., **4,** 443
Hirobe, M., **4,** 257; **9,** 242; **11,** 220, 445; **12,** 414
Hirohisa, K., **2,** 30
Hiroi, K., **4,** 302; **5,** 94, 246, 315, 436, 460; **6,** 238, 267, 316, 403, 464; **7,** 188; **8,** 197, 486; **9,** 269, 429; **10,** 171; **12,** 421
Hirose, T., **10,** 56; **11,** 185
Hirota, H., **11,** 317
Hirota, K., **5,** 572; **12,** 272, 363
Hirota, Y., **4,** 354
Hirotsu, K., **9,** 46, 119; **12,** 172
Hirowatari, N., **5,** 651
Hirsch, A. F., **1,** 1118
Hirsch, C., **2,** 93
Hirsch, E., **10,** 416; **11,** 40, 299
Hirsch, L. K., **4,** 183
Hirsch, R., **3,** 123
Hirschhorn, A., **2,** 17
Hirschmann, R., **1,** 293, 664, 764, 911, 1006, 1187; **2,** 5
Hirsekorn, F. J., **6,** 15
Hirt, R., **1,** 504
Hiscock, B. F., **2,** 373; **6,** 634
Hiser, R. D., **1,** 377
Hiskey, C. F., **1,** 664, 1187
Hiskey, R. G., **1,** 49, 339, 1083; **2,** 313; **4,** 45; **8,** 386; **10,** 245
Hitzel, E., **6,** 653
Hitzler, F., **1,** 235
Hixon, R. M., **1,** 1101, 1141
Hiyama, T., **5,** 333, 370, 534, 742; **6,** 91, 220, 385; **7,** 93; **8,** 112, 151, 292; **9,** 68, 256, 395; **10,** 318; **11,** 68, 134, 310, 470; **12,** 137, 209, 263, 304, 462
Hiyoshi, T., **12,** 180
Hjeds, H., **1,** 671
Hjelte, N. S., **1,** 445
Hlasta, D. J., **9,** 428
Hlavka, J. J., **1,** 181
Hnevkovsky, O., **8,** 535
Ho, A. J., **4,** 112
Ho, C. T., **6,** 577
Ho, H. C., **4,** 74, 472
Ho, L., **10,** 157
Ho, L. K., **11,** 154
Ho, M. S., **5,** 419
Ho, P. T., **8,** 232; **10,** 263; **12,** 553
Ho, R. K. Y., **3,** 90
Ho, T. L., **4,** 15, 74, 85, 96, 119, 340, 472, 555; **5,** 102, 170, 207, 566, 632, 671, 686; **6,** 50, 229, 250, 251, 470, 497, 529, 540, 543, 553, 563; **7,** 56, 124, 248, 254, 341, 365, 369, 417, 418; **8,** 248, 263, 302, 361, 364, 383, 459, 497; **9,** 59, 99, 105, 256, 500; **10,** 202, 401
Ho, T. N. S., **7,** 377
Hoa, K., **6,** 504
Hobara, S., **11,** 213
Hobbs, F. W., Jr., **12,** 350, 467
Hobbs, J. J., **1,** 500
Hoberg, H., **1,** 195, 245, 1022; **4,** 122
Hoblitt, R. P., **3,** 163
Hobson, J. D., **2,** 319
Hochstein, F. A., **1,** 384, 594, 595, 1033
Hock, K., **1,** 677
Hocker, J., **2,** 417; **7,** 23; **10,** 405
Hocking, M. B., **3,** 290
Hocks, P., **2,** 211, 362
Hocquaux, M., **12,** 414
Hodder, D. J., **9,** 264
Hodes, H. D., **8,** 150
Hodge, J. E., **1,** 518
Hodge, P., **2,** 195, 381; **3,** 265; **6,** 453, 628, 645; **7,** 279; **9,** 135, 503; **11,** 492
Hodges, M. L., **4,** 60, 455; **5,** 78
Hodges, R. J., **2,** 182; **3,** 134
Hodgins, T., **4,** 529
Hodgkinson, L. C., **9,** 376
Hodgson, G. L., **4,** 489; **5,** 52, 756
Hodgson, H. H., **1,** 756, 1100, 1116; **3,** 267
Hodgson, K. O., **6,** 35
Hodgson, P. K. G., **8,** 416; **9,** 381
Hodgson, W. G., **1,** 211
Hodkova, J., **5,** 330
Hodson, D., **3,** 92
Hoefle, G., **3,** 119; **4,** 416; **5,** 219, 339; **7,** 337; **8,** 442; **9,** 180
Hoefle, G. A., **6,** 373
Hoeft, E., **1,** 120, 220; **2,** 120; **7,** 174
Hoeft, V. E., **6,** 36
Hoeg, D. F., **1,** 1141
Hoegberg, S. A. G., **6,** 137
Hoeger, E., **1,** 562
Hoehn, H. H., **1,** 345, 657
Hoekstra, M. S., **6,** 416; **8,** 318, 356, 363; **10,** 234
Hoelzel, C. B., **1,** 563
Hoenicke, J., **5,** 756
Hoentjen, G., **6,** 507
Hoerauf, W., **1,** 185
Hoerig, C., **1,** 484
Hoermann, H., **1,** 582
Hoernfeldt, A. B., **3,** 329
Hoerster, H., **10,** 359

Hoerster, H. G., **5,** 693
Hoesch, K., **1,** 1292
Hoever, H., **3,** 338
Hofbauer, P., **6,** 118
Hofer, P., **1,** 139; **3,** 275
Hoff, D. J., **12,** 334
Hoff, M. C., **4,** 135
Hoff, S., **2,** 344; **5,** 55; **7,** 226; **8,** 322; **12,** 310
Hoff, Y. E., **4,** 191
Hoffenberg, D. S., **1,** 69
Hoffman, A. S., **1,** 61
Hoffman, C. J., **12,** 466
Hoffman, E. J., **2,** 353
Hoffman, F., **1,** 87, 402
Hoffman, H. M. R., **4,** 80, 328; **5,** 760
Hoffman, J., **1,** 465
Hoffman, J. M., **5,** 722; **7,** 399
Hoffman, J. M., Jr., **5,** 19, 220
Hoffman, L., Jr., **10,** 165
Hoffman, N. W., **5,** 625
Hoffman, R. V., **5,** 574; **7,** 252; **10,** 107; **11,** 495
Hoffman, R. W., **2,** 53
Hoffman, W., **10,** 376
Hoffman, W. A., III, **5,** 728; **11,** 589
Hoffman, W. H., **6,** 645
Hoffmann, A. K., **1,** 211, 268, 688; **3,** 81; **4,** 31
Hoffmann, E., **1,** 882
Hoffmann, F. W., **1,** 1216; **2,** 155
Hoffmann, H., **1,** 118, 1192, 1247, 1249; **5,** 73, 250, 694; **6,** 243; **7,** 289
Hoffmann, H. M. R., **1,** 1018; **2,** 371; **5,** 224, 437; **6,** 364, 482, 544; **7,** 281; **8,** 6, 534; **10,** 11; **11,** 28, 299; **12,** 156, 450
Hoffmann, J. A., **6,** 307
Hoffmann, P., **2,** 49, 54; **4,** 65
Hoffmann, P. T., **7,** 20
Hoffmann, R., **3,** 88, 251
Hoffmann, R. W., **1,** 50, 832; **2,** 402; **3,** 123; **4,** 510; **5,** 650; **9,** 38, 363; **11,** 87, 103; **12,** 101, 397
Hoffmann, W., **1,** 608; **4,** 472; **6,** 441
Hoffmeister, H., **1,** 506
Hoffpauir, C. L., **1,** 379
Hoffsommer, R. D., **1,** 37, 59, 974; **2,** 159, 252, 434
Hofmann, A., **1,** 282, 351, 942, 1252; **4,** 577; **6,** 2
Hofmann, A. F., **2,** 398
Hofmann, A. W., **1,** 666, 844; **4,** 31
Hofmann, C. M., **10,** 376
Hofmann, H., **5,** 136
Hofmann, H. P., **6,** 596
Hofmann, J. E., **1,** 309, 318, 431, 705, 927
Hofmann, K., **1,** 46, 85, 97, 111; **11,** 587
Hofmann, K. A., **1,** 764
Hofmeister, H., **9,** 324; **11,** 193; **12,** 79
Hogan, H. P., **4,** 54
Hogeveen, H., **5,** 587; **6,** 24; **7,** 60, 347, 392; **8,** 375; **9,** 145
Hogg, J. A., **1,** 394, 509, 690, 787, 1118; **2,** 132
Hogg, J. L., **8,** 461
Hogrefe, F., **6,** 675
Hohenlohe-Oehringen, K., **1,** 1273
Hohermuth, M. K., **9,** 135
Hohl, T., **1,** 796
Hohlbrugger, R., **9,** 506
Hohler, W., **6,** 268
Hohlstein, G., **1,** 44
Hohorst, F. A., **4,** 37
Hoiness, C. M., **5,** 589
Hoinowski, A. M., **6,** 412
Hoizey, M. J., **8,** 233
Hojas, Z. G., **1,** 659
Hojo, K., **4,** 339; **8,** 225
Hojo, M., **6,** 510; **7,** 135, 350; **8,** 289, 345; **11,** 293; **12,** 275
Hokama, T., **1,** 329, 757
Hokari, H., **10,** 90
Hoke, D., **3,** 264; **4,** 444
Hol, C. M., **3,** 57
Hola, S., **3,** 6
Holah, D. G., **6,** 133
Holan, G., **4,** 37
Holand, S., **8,** 489; **9,** 89
Holberg, B., **1,** 1157
Holbert, G. W., **7,** 85, 88; **8,** 373, 505
Holden, J. L., **4,** 249
Holden, R. F., **1,** 295
Holder, D., **5,** 508
Holder, G. M., **7,** 274
Holder, R. W., **7,** 216
Holick, M. F., **7,** 273
Holker, J. S. E., **11,** 39
Holladay, M. W., **11,** 170
Holland, C. L., **5,** 51
Holland, D., **9,** 408
Holland, G. W., **3,** 39; **4,** 53; **5,** 486
Holland, H. L., **5,** 334; **6,** 201; **7,** 239; **11,** 492
Holleman, A. F., **1,** 465, 1033
Hollenberg, D. H., **9,** 399
Holley, A. D., **10,** 194
Holley, R. W., **10,** 194
Holliman, E. G., **4,** 529
Holliman, F. G., **2,** 296
Hollingshead, R. G. W., **1,** 1194
Hollins, R. A., **5,** 624
Hollinshead, D. M., **12,** 38
Hollinshead, J. H., **6,** 524; **7,** 325
Hollister, K. R., **4,** 174; **5,** 373; **8,** 140
Hollitzer, O., **5,** 219; **7,** 141; **9,** 203
Holly, F. W., **1,** 1157; **2,** 5; **10,** 61
Holm, A., **1,** 177
Holm, B., **10,** 188
Holm, M. J., **2,** 220
Holm, P., **2,** 215
Holm, R. H., **5,** 64
Holm, R. T., **1,** 478
Holman, N. J., **12,** 36
Holman, R. T., **1,** 1269
Holmberg, G. H., **6,** 415
Holmberg, K., **6,** 566; **9,** 157
Holmes, A. B., **7,** 277; **9,** 311; **10,** 272; **12,** 125
Holmes, C. P., **12,** 256
Holmes, H. L., **1,** 383, 489,

938, 1073, 1087; **3,** 185, 237
Holmes, J., **1,** 42; **3,** 330
Holmes, J. B., **1,** 1247
Holmes, R. R., **1,** 68, 465
Holmquist, H. E., **1,** 228
Holmquist, R. K., **3,** 59
Holmstead, R. L., **6,** 604
Holschneider, F., **6,** 32
Holsten, J. R., **1,** 951
Holt, D. A., **10,** 194; **12,** 275
Holt, G., **3,** 92
Holtermann, H., **3,** 41
Holton, P. G., **1,** 219
Holton, P. J., **1,** 564
Holton, R. A., **3,** 332; **8,** 94, 178, 527; **12,** 75, 265
Holtschmidt, H., **4,** 524; **5,** 439
Holtz, H. D., **1,** 81, 658
Holtz, W. J., **11,** 180
Holubek, J., **1,** 608
Holubka, J. W., **9,** 220, 324
Holum, J. R., **1,** 147; **4,** 216
Holum, L. B., **1,** 244
Holy, A., **4,** 185
Holy, N. L., **7,** 132; **9,** 352
Holysz, R. P., **1,** 609
Holzel, R., **3,** 194
Holzmann, H., **5,** 445
Homeyer, A. H., **1,** 248
Homfeld, E., **3,** 290
Hommes, H., **8,** 407; **10,** 73; **11,** 103
Homolka, B., **1,** 414
Honda, I., **2,** 25; **3,** 15
Honda, T., **10,** 245; **11,** 550
Honek, J. F., **11,** 574
Honeyman, J., **2,** 4
Hong, C. Y., **12,** 573
Honma, K., **7,** 345
Honma, S., **3,** 5
Honma, T., **3,** 251
Honnen, L. R., **1,** 1258
Honold, E., **1,** 141
Honsberg, U., **2,** 412
Honsberg, W., **2,** 412
Hontz, A. C., **1,** 870, 963
Honwad, V. K., **3,** 329
Honzl, J., **1,** 97, 755
Hoogeboom, T. J., **4,** 288
Hoogenboom, B. E., **4,** 516; **6,** 600; **8,** 494
Hoogwater, D. A., **5,** 92
Hoogzand, C., **1,** 228; **7,** 144
Hook, J. M., **10,** 236
Hooker, S. C., **1,** 392
Hoornaert, C., **6,** 123; **11,** 550
Hooten, K. A., **6,** 399
Hooton, S., **1,** 444, 445
Hoover, F. W., **1,** 172
Hoover, J. R. E., **5,** 455
Hooz, J., **2,** 195, 445; **3,** 73; **4,** 5, 122, 146, 187, 221, 230, 552; **5,** 86, 225, 296, 408; **6,** 645, 665; **7,** 379; **9,** 32; **10,** 244
Hope, A., **5,** 119
Hopf, H., **4,** 173; **5,** 165; **6,** 271; **7,** 428
Hopff, H., **1,** 44, 1171
Hopkins, P. B., **8,** 33; **9,** 242, 301; **10,** 115; **11,** 189, 209; **12,** 121
Hopkinson, A. C., **8,** 503
Hoppe, B., **6,** 570
Hoppe, D., **5,** 257; **12,** 83
Hoppe, I., **6,** 343; **7,** 211, 215; **8,** 395; **10,** 107
Hopper, C. R., **5,** 266
Hopper, R. J., **1,** 1194
Hopper, S. P., **3,** 222, 225; **4,** 385; **8,** 408
Hoppmann, A., **9,** 151
Hopps, H. B., **4,** 79; **6,** 64
Hopton, H. P., **5,** 680
Horak, I., **5,** 668
Horcher, L. H. M., II, **12,** 14
Horeau, A., **1,** 561; **2,** 318; **7,** 17
Horgan, S. W., **4,** 411
Hori, H., **7,** 33; **9,** 63
Hori, I., **6,** 398, 569; **12,** 288, 374
Hori, K., **10,** 176; **11,** 396
Hori, T., **7,** 318; **8,** 24, 25, 43, 359; **9,** 111, 367
Horie, T., **11,** 480
Horiguchi, Y., **12,** 502
Horiike, M., **6,** 145
Horiike, T., **8,** 133, 162
Horikawa, H., **10,** 306
Horikawa, M., **10,** 277
Horike, A., **2,** 273
Horiki, K., **7,** 126
Horino, H., **7,** 115
Horita, K., **12,** 175
Horiuchi, C. A., **10,** 211; **11,** 131, 267; **12,** 256
Horiuchi, S., **6,** 653
Horler, H., **12,** 303
Horley, D., **2,** 173
Horn, D. H. S., **1,** 155; **3,** 24
Horn, H. G., **6,** 640
Horn, K. A., **11,** 286
Horn, P., **5,** 73, 250; **6,** 243
Horn, U., **5,** 329; **7,** 171
Horn, W., **5,** 15
Hornby, J. L., **8,** 310
Horne, D., **12,** 537
Horne, D. A., **8,** 454
Horne, W. H., **1,** 859
Horner, C. J., **7,** 336
Horner, L., **1,** 80, 129, 563, 712, 787, 790, 1056, 1247, 1249; **2,** 433; **4,** 76, 251; **5,** 105, 694; **8,** 535
Hornesky, S., **1,** 255
Horng, A., **3,** 23; **4,** 36; **5,** 41, 565
Horning, D. E., **3,** 125; **5,** 700
Horning, E. C., **1,** 248, 267, 604, 872, 890, 904, 1073, 1294; **2,** 303; **3,** 148, 312
Horning, M. G., **1,** 267, 890; **3,** 148
Hornish, R. E., **5,** 79, 381, 495
Hornung, K. H., **1,** 289
Horovitch, S., **5,** 66; **7,** 40
Horrocks, W. D., Jr., **6,** 51; **11,** 168
Horrom, B. W., **1,** 281
Horswill, E. C., **1,** 1008
Hortmann, A. G., **1,** 20, 465, 608; **2,** 173; **3,** 49; **4,** 199; **6,** 25
Horton, D., **1,** 68, 1157, 1234; **2,** 165, 175, 359; **3,** 128; **11,** 285, 329; **12,** 318
Horton, H. R., **1,** 485
Horton, R. L., **1,** 39
Horton, W. J., **1,** 1253

Horvath, B., **1,** 999
Horwell, D. C., **3,** 171; **4,** 282; **6,** 5, 587; **7,** 86
Horwitz, J. P., **1,** 630; **6,** 221
Horwitz, S. B., **5,** 315
Hosaka, K., **5,** 658; **6,** 11; **8,** 478; **11,** 324
Hoshi, M., **9,** 250
Hoshino, M., **5,** 182, 649; **10,** 25; **12,** 453
Hoshino, O., **6,** 316
Hoshino, Y., **9,** 404; **10,** 44, 196
Hoskin, D. H., **8,** 308
Hoskins, J. A., **1,** 500, 1295
Hoskinson, R. M., **1,** 244; **3,** 122
Hosoda, H., **5,** 75; **7,** 32
Hosoda, S., **7,** 52
Hosoda, Y., **9,** 138
Hosogai, T., **7,** 352
Hosokawa, T., **4,** 129; **5,** 190, 510; **6,** 573; **9,** 349; **12,** 89, 373
Hosomi, A., **7,** 372; **8,** 181, 486; **9,** 47, 69, 255, 352, 446, 494; **10,** 8, 56, 219, 289, 433; **11,** 103, 130, 275, 533, 580; **12,** 263, 540, 546
Hossain, M. B., **5,** 725
Hostettler, F., **8,** 396
Hotta, Y., **8,** 339
Hotten, T. M., **4,** 523
Hou, K. C., **12,** 259
Houbiers, J. P. M., **3,** 254
Houge, C., **11,** 111
Hough, L., **1,** 1187; **2,** 365; **5,** 436; **8,** 405; **10,** 349
Houghten, R. A., **10,** 337
Houghton, E., **3,** 303
Houghton, L. E., **2,** 35, 107; **7,** 155
Houk, K. N., **5,** 182
Houlihan, W. J., **10,** 296
House, H. O., **1,** 34, 49, 69, 72, 150, 180, 195, 268, 470, 526, 594, 606, 630, 787, 926, 937, 1178, 1193, 1256, 1258; **2,** 84, 92, 117, 143, 153, 157, 278, 283, 307, 359, 439, 441, 462; **3,** 20, 112, 153, 185, 237, 258, 265, 278, 312, 315; **4,** 63, 183, 199, 290, 302; **5,** 142, 149, 175, 243, 427, 445, 764; **6,** 203, 215, 225, 676; **7,** 295; **8,** 299, 302, 343; **10,** 276; **12,** 52
Housmans, J. G. H. M., **5,** 268
Houston, B., **1,** 152
Houston, T. L., **10,** 198
Houwing, H. A., **7,** 246
Hovakeemian, G. H., **12,** 163
How, M. J., **4,** 461
Howard, B. M., **7,** 352
Howard, C. C., **1,** 742
Howard, E., Jr., **1,** 1247
Howard, H., **11,** 150
Howard, J. B., **1,** 48
Howard, J. F., Jr., **1,** 228
Howard, J. W., **1,** 911, 1121
Howard, K. L., **1,** 339, 657
Howard, R. B., **1,** 75
Howard, S. I., **11,** 153
Howard, T. J., **2,** 294
Howard, T. R., **12,** 55
Howard, W. L., **1,** 209, 269, 982
Howart, M., **2,** 310
Howarth, B. D., **6,** 147
Howarth, M., **5,** 290
Howarth, T. T., **8,** 37, 427
Howe, E. E., **1,** 1100
Howe, G. R., **3,** 331
Howe, R., **1,** 1284
Howe, R. S., **9,** 387
Howell, C. F., **1,** 412
Howell, I. V., **7,** 315
Howell, J. A. S., **7,** 111
Howell, S. C., **10,** 168; **12,** 38
Howells, P. N., **10,** 42
Howells, R. D., **8,** 504
Howk, B. W., **1,** 228, 1209, 1294
Howton, D. R., **1,** 818, 1011, 1157
Hoy, R. C, **4,** 351
Hoye, T. R., **9,** 296, 434, 512; **11,** 309
Hoyer, G. A., **6,** 443
Hoyle, G., **1,** 1144
Hoyle, J., **11,** 494
Hoyng, C. F., **6,** 32; **11,** 152
Hoz, T., **6,** 589
Hozumi, T., **7,** 14, 15
Hrabie, J., **9,** 73
Hrib, N. J., **7,** 237
Hrinyo, T., **10,** 209
Hrnciar, P., **8,** 433
Hrnjez, B. J., **8,** 382; **9,** 351
Hrubiec, R. T., **12,** 555
Hruby, V. J., **9,** 503
Hrutford, B. F., **1,** 909
Hryb, D., **5,** 479
Hsia, S. L., **6,** 254
Hsiao, C. N., **11,** 42
Hsieh, H. H., **4,** 122
Hsu, C. K., **7,** 244
Hsu, H. C., **11,** 489
Hsu, I. H. S., **6,** 228, 229
Hsu, M. H., **12,** 288
Hsu, R. Y., **6,** 102
Hsu, S. Y., **12,** 90
Hsu, Y. F. L., **2,** 76; **5,** 631
Hsue, H. C., **10,** 212
Hu, H., **10,** 214; **11,** 412
Hu, Q. Y., **11,** 486
Hua, D., **11,** 91
Hua, D. H., **10,** 407; **11,** 58, 567; **12,** 57, 182
Huang, B. S., **5,** 177; **6,** 501; **7,** 87
Huang, C. C., **4,** 346
Huang, C. T., **5,** 262; **6,** 248
Huang, G. F., **5,** 344
Huang, H., **12,** 117
Huang, J., **10,** 110
Huang, N. Z., **11,** 534, 575
Huang, P., **1,** 445
Huang, S. B., **12,** 111
Huang, S. J., **1,** 188, 341, 342
Huang, S. L., **7,** 136; **8,** 200; **9,** 192
Huang, S. P., **6,** 283; **7,** 329; **10,** 194
Huang, T. N., **8,** 153
Huang, W. Y., **1,** 141, 731, 1064, 1284
Huang, X., **11,** 50
Huang, Y., **11,** 5

Huang-Minlon, **1,** 444, 526, 1178
Hubacher, M. H., **1,** 745
Hubbard, J. L., **9,** 278, 287, 393; **10,** 49, 77; **11,** 154
Hubbard, J. S., **10,** 75
Hubbard, W. N., **1,** 235, 1178
Hubbs, J. C., **10,** 264
Huber, F., **3,** 226
Huber, G., **1,** 562
Huber, R., **3,** 178
Huber, U., **11,** 533
Huber, W., **1,** 703, 1163; **3,** 67
Hubert, A. J., **2,** 182; **5,** 497, 572; **7,** 76, 313; **8,** 435; **10,** 342; **11,** 460
Hubert, J. C., **7,** 331
Hubert, T. D., **12,** 60
Hubert-Habart, M., **3,** 229
Huckin, S. N., **5,** 454; **8,** 62
Huckle, D., **4,** 502
Hudak, N. J., **1,** 758
Hudec, J., **1,** 260, 723; **3,** 277; **5,** 55
Hudlicky, M., **1,** 8; **3,** 96
Hudlicky, T., **9,** 265; **11,** 23; **12,** 568
Hudnall, P. M., **4,** 316; **5,** 756; **7,** 218; **10,** 251
Hudrlik, A. M., **7,** 220; **11,** 432; **12,** 126
Hudrlik, O., **1,** 777
Hudrlik, P. F., **2,** 438; **5,** 763; **6,** 201, 245; **7,** 220; **8,** 203; **11,** 90, 432; **12,** 126
Hudson, B., **6,** 426
Hudson, B. E., **1,** 866, 1256, 1261, 1292
Hudson, C. S., **1,** 476, 818, 955
Hudson, D., **6,** 425
Hudson, H. R., **6,** 220; **7,** 424
Hudson, R. A., **2,** 95
Hudson, R. F., **1,** 49; **4,** 481; **8,** 79; **11,** 70
Hudspeth, J. P., **8,** 178, 415; **10,** 443
Huebel, W., **1,** 228, 260; **2,** 140; **7,** 144
Huebner, C. F., **1,** 814
Huebner, R., **1,** 119
Hueckel, W., **1,** 139
Huenig, S., **1,** 258, 357, 370, 371, 444, 472, 707, 708, 1178; **4,** 529, 535; **5,** 165, 722; **6,** 24, 633; **7,** 188; **9,** 128, 129; **10,** 416; **11,** 40, 150, 176
Huet, F., **7,** 164; **8,** 340; **10,** 93
Huet, J., **7,** 349
Huettel, R., **1,** 1015
Huettig, C. F., **8,** 535
Huffman, J. C., **9,** 306; **11,** 584
Huffman, J. W., **1,** 574, 604; **4,** 183; **5,** 537; **7,** 268; **8,** 524; **12,** 109, 158, 550
Hufford, D. L., **1,** 636
Hufnal, J. M., **9,** 186
Hug, R. P., **5,** 28; **9,** 462
Hug, W., **9,** 237
Hugel, G., **2,** 238
Hughes, G. A., **2,** 29; **6,** 258
Hughes, G. K., **1,** 158
Hughes, H., **1,** 1292
Hughes, L. R., **9,** 450
Hughes, N. W., **2,** 36
Hughes, R., **7,** 101
Hughes, R. J., **7,** 26
Hughes, W. B., **1,** 1141
Hughey, J. L., **10,** 16
Hughmark, P. B., **1,** 718
Hugl, H., **5,** 365
Hugo, J. M., **1,** 60
Hugos, H., **1,** 619
Hugueny, J. C., **10,** 378
Huh, G., **5,** 433
Hui, R. A. H. F., **9,** 34; **11,** 39
Hui, R. C., **12,** 4
Huie, E. M., **11,** 345
Huisgen, R., **1,** 364, 754, 827, 832; **2,** 9, 21, 81; **3,** 132; **4,** 383, 405; **5,** 230, 280, 339, 649; **6,** 17, 24, 612; **7,** 207; **9,** 135; **10,** 243
Huisman, H. O., **1,** 247, 445; **5,** 319, 560; **7,** 331; **8,** 2
Hulce, M., **10,** 9; **11,** 315; **12,** 298
Hulin, B., **12,** 532
Hull, C. J., **1,** 668
Hull, D. C., **1,** 8, 526
Hullot, P., **7,** 204
Hulshof, L. A., **3,** 254
Hultberg, H., **11,** 483
Hultner, G., **12,** 278
Hulton, R. F., **1,** 1083
Humber, L. G., **2,** 405; **3,** 171
Hummel, K., **4,** 533; **6,** 620
Hummelen, J. C., **7,** 311; **8,** 431
Humphlett, W. J., **1,** 859, 1163
Hung, H. K., **5,** 621; **8,** 199
Hung, M. H., **12,** 557
Hung, S. C., **11,** 88
Hung, W. M., **4,** 243; **5,** 325; **6,** 4
Hung, Y. Y., **1,** 1015
Hunger, K., **1,** 669; **2,** 279
Hungerbuhler, E., **11,** 529
Hunsberger, I. M., **1,** 310, 355, 952, 1123; **2,** 348
Hunt, C. A., **7,** 391
Hunt, D. F., **4,** 322, 323
Hunt, E., **4,** 218
Hunt, J. C., **1,** 1090
Hunt, J. D., **3,** 289; **4,** 497, 501; **5,** 657
Hunt, J. H., **1,** 481
Hunt, J. S., **1,** 954; **2,** 228
Hunt, K., **5,** 280
Hunt, P., **5,** 61
Hunt, R. H., **1,** 926; **5,** 552
Hunter, D. H., **6,** 160; **8,** 130
Hunter, G. D., **1,** 777, 937
Hunter, J. H., **1,** 444
Hunter, N. R., **7,** 221, 288
Hunter, R. F., **1,** 718
Huntington, A., **8,** 397
Huntress, E. H., **1,** 244, 273
Hunziker, F., **1,** 1235
Huong, T. T., **4,** 343
Hurd, C. D., **1,** 244, 329, 388, 531, 563, 904, 1184; **2,** 12; **5,** 193
Hurley, E., Jr., **3,** 167
Hurnaus, R., **4,** 529; **10,** 151
Hurni, B., **8,** 349
Hurst, G. D., **12,** 81
Hurst, J. J., **1,** 562; **2,** 69
Hurst, K. M., **10,** 226
Hursthouse, M. B., **4,** 262

Hurwitz, B. M., **1,** 1087
Hurwitz, M. D., **2,** 202
Husain, A., **11,** 355, 553, 587; **12,** 527
Husain, S., **3,** 303
Husbands, J., **1,** 132; **2,** 229
Husby, R. M., **2,** 93
Huselton, J. K., **6,** 137
Huser, D. L., **10,** 356
Husk, G. R., **10,** 149
Hussain, S. A., **8,** 204
Hussey, A. S., **1,** 754, 904, 1064
Hussey, B. J., **10,** 93; **12,** 574
Hussmann, G. P., **11,** 562
Husson, H. P., **3,** 46, 115; **6,** 361
Husstedt, U. J. U., **9,** 399
Huston, D. M., **4,** 152
Hutchings, M., **5,** 607
Hutchings, M. G., **4,** 447; **6,** 536; **7,** 223
Hutchings, S. D., **9,** 352
Hutchins, C. W., **9,** 426
Hutchins, M. G., **4,** 559
Hutchins, R. A., **11,** 483
Hutchins, R. O., **2,** 92, 229; **3,** 166, 264; **4,** 444, 450, 451, 559; **5,** 415, 479, 600, 609, 645; **6,** 534, 538; **7,** 292, 335; **9,** 383, 426; **10,** 33, 390; **11,** 305, 499; **12,** 288, 446
Hutchins, R. R., **9,** 16; **12,** 280
Hutchinson, A. J., **8,** 102
Hutchinson, C. R., **5,** 446, 609; **6,** 538; **7,** 390
Hutchinson, J., **12,** 77
Hutchinson, L. L., **7,** 201; **10,** 240; **12,** 290
Hutchinson, R. C., **8,** 338
Hutchinson, R. J., **9,** 361
Hutchison, J. J., **4,** 204, 437, 524
Huth, A., **4,** 172; **5,** 740
Huthmacher, K., **6,** 521, 618
Hutmacher, H. M., **5,** 532; **6,** 317
Huttemann, T. J., **6,** 469
Huttner, G., **11,** 401
Hutton, J., **3,** 54, 88, 155; **12,** 390
Hutton, R. S., **5,** 19
Hutton, T. W., **1,** 195, 1005
Huttrer, C. P., **1,** 601
Huurdeman, W. F. J., **4,** 75, 446
Huxol, R. F., **4,** 174
Huynh, C., **1,** 125, 219; **9,** 125
Huyser, E. S., **1,** 45, 81, 209, 212, 506, 1195, 1196; **7,** 41
Huyser, W. S., **1,** 209
Hwa, J. C. H., **1,** 874
Hwang, D., **7,** 241
Hwang, J. S., **4,** 389
Hwang, K. J., **9,** 490; **10,** 360
Hwu, J. R., **12,** 63
Hyatt, A. A., **1,** 689
Hyatt, J. A., **8,** 312; **9,** 194
Hyde, D. K. A., **1,** 631
Hyde, E. K., **1,** 603
Hylarides, M. D., **11,** 489
Hylton, T., **2,** 187
Hyman, H. H., **6,** 670
Hymans, W. E., **1,** 144, 755
Hytta, R., **5,** 320

Iacona, R. N., **1,** 309
Iacono, S., **12,** 523
Ibanez, L. C., **1,** 75, 127
Ibata, T., **9,** 51
Ibne-Rasa, K. M., **1,** 1167; **4,** 466
Ibrahim, B. E. D., **6,** 618
Ibuka, T., **7,** 28; **8,** 47; **10,** 289; **12,** 350
Ibuki, I., **11,** 157
Ichijima, S., **6,** 158
Ichikawa, J., **12,** 202
Ichikawa, K., **1,** 1054; **5,** 167; **7,** 361; **9,** 63, 258
Ichikawa, Y., **12,** 325
Ichikizaki, I., **6,** 174; **8,** 360
Ichimura, K., **3,** 322
Ichino, M., **4,** 247
Ichinohe, Y., **4,** 369
Ichinone, Y., **9,** 148
Ichinose, I., **7,** 352; **10,** 377
Icke, R. N., **1,** 402
Idacavage, M. J., **8,** 507; **9,** 482, 488
Ide, J., **1,** 336, 926; **3,** 124; **4,** 562; **12,** 131
Ide, W. S., **1,** 166, 471, 481, 1027, 1113, 1116; **7,** 299
Ide, Y., **9,** 519
Idei, M., **11,** 267
Idler, D. R., **1,** 1246
Idoux, J. P., **3,** 267
Iemura, S., **7,** 391; **8,** 472
Iesce, M. R., **12,** 365
Iffland, D. C., **1,** 561, 1250
Iflah, S., **10,** 79
Ifzal, S. M., **2,** 165
Igami, M., **6,** 552
Igano, K., **1,** 732
Igarashi, K., **3,** 251; **5,** 55
Igbal, J., **9,** 191
Igersheim, F., **8,** 206
Ignatiadou-Ragoussis, V., **5,** 60; **6,** 430
Ignatova, E., **9,** 209
Iguchi, H., **9,** 47; **11,** 580; **12,** 540
Iguchi, K., **6,** 11; **7,** 365; **8,** 478, 481; **12,** 130
Iguchi, M., **5,** 319; **9,** 11
Iguchi, S., **8,** 312; **9,** 171, 436
Ihara, M., **4,** 407; **6,** 283; **7,** 329; **10,** 194; **11,** 40; **12,** 44
Ihn, W., **3,** 71
Iida, H., **10,** 347; **11,** 243; **12,** 327, 328, 444
Iida, K., **7,** 430; **11,** 599; **12,** 356
Iida, S., **12,** 125, 555, 565
Iida, T., **11,** 445
Iihama, T., **12,** 328
Iijima, S., **10,** 131; **11,** 275; **12,** 356
Iimori, T., **10,** 449
Iino, K., **8,** 208
Iinuma, M., **8,** 156
Iio, H., **8,** 421; **9,** 110, 426; **10,** 346; **12,** 502
Iitaka, Y., **9,** 349
Iizuka, K., **12,** 106
Ikan, R., **4,** 31
Ikariya, T., **11,** 277
Ikawa, T., **9,** 167; **10,** 56

Ikeda, H., **6,** 526; **11,** 352; **12,** 439
Ikeda, I., **8,** 256
Ikeda, K., **5,** 280, 538; **6,** 472
Ikeda, M., **4,** 257; **5,** 433; **6,** 233; **8,** 319; **10,** 198; **11,** 95
Ikeda, N., **11,** 64, 181, 378, 571; **12,** 19, 356
Ikeda, S., **1,** 637; **5,** 332; **7,** 52; **8,** 96; **11,** 249
Ikeda, T., **4,** 422; **10,** 305; **11,** 324
Ikeda, Y., **5,** 167; **8,** 505; **11,** 378; **12,** 19, 407
Ikegami, S., **2,** 267; **3,** 186, 264; **9,** 393; **10,** 414; **11,** 456, 552; **12,** 109
Ikegami, T., **5,** 764
Ikegawa, A., **10,** 12
Ikehara, F., **11,** 431
Ikehara, M., **1,** 289; **2,** 177; **7,** 286
Ikejira, T., **8,** 83
Ikejiri, T., **12,** 465
Ikekawa, N., **5,** 55, 381, 427; **7,** 373; **10,** 149; **11,** 570
Ikemori, S., **9,** 492
Ikeno, M., **8,** 218; **9,** 170
Ikeshima, H., **5,** 294; **6,** 317, 403, 649
Ikeya, Y., **9,** 269
Ikota, N., **6,** 193; **8,** 56; **9,** 300; **11,** 208, 460
Ikura, K., **7,** 225
Ila, H., **6,** 284; **10,** 229
Iles, D. H., **4,** 79
Ilett, J. D., **1,** 248; **4,** 481
Iley, D. E., **10,** 213
Illi, V. O., **9,** 361
Ilton, M. A., **2,** 331, 464
Imada, Y., **12,** 89
Imagawa, D. K., **12,** 139
Imagawa, M., **7,** 12
Imai, H., **5,** 332; **7,** 109, 314; **11,** 100; **12,** 95
Imai, I., **8,** 188
Imai, J., **11,** 599
Imai, S., **6,** 244; **11,** 15
Imai, Z., **3,** 265
Imakura, Y., **11,** 419
Imamoto, T., **10,** 266, 437; **11,** 114, 428, 430; **12,** 108, 345, 430, 543
Imamura, T., **6,** 596
Imanaka, T., **6,** 450; **7,** 277, 314
Imanishi, M., **1,** 144
Imanishi, T., **9,** 215; **11,** 430; **12,** 411
Imoto, E., **1,** 9; **6,** 95
Imperiali, B., **10,** 119
Imuta, M., **6,** 436; **7,** 268; **9,** 126
Inaba, M., **12,** 64
Inaba, S., **6,** 633; **8,** 486; **10,** 403; **12,** 335
Inaba, T., **12,** 444
Inagaki, S., **6,** 436
Inagaki, Y., **6,** 560; **11,** 429
Inaishi, M., **8,** 83; **10,** 87
Inaki, H., **10,** 378
Inamoto, H., **11,** 496
Inamoto, N., **6,** 560; **9,** 442; **11,** 214
Inamoto, Y., **6,** 359, 526; **7,** 349
Inanaga, J., **7,** 320; **9,** 107, 479; **11,** 552
Inatome, M., **1,** 652
Inayama, S., **5,** 234
Inderbitzin, M., **9,** 180
Infante, R., **1,** 509
Ing, H. R., **1,** 445
Ingersoll, A. W., **1,** 433, 838; **2,** 273
Ingham, R. K., **1,** 1251
Ingleson, D., **9,** 256
Ingold, C. F., **12,** 576
Ingold, C. K., **4,** 115
Inhoffen, H. H., **1,** 155, 350, 669, 911, 1210
Inman, C. E., **1,** 808
Inokawa, S., **7,** 58; **9,** 322; **12,** 337
Inokuchi, T., **9,** 133
Inomata, K., **6,** 596; **8,** 223, 231; **11,** 514; **12,** 155
Inone, K., **9,** 479
Inone, T., **11,** 293; **12,** 275
Inoue, A., **2,** 37
Inoue, H., **10,** 28
Inoue, I., **5,** 752; **6,** 70
Inoue, K., **1,** 686; **9,** 442; **11,** 383
Inoue, M., **9,** 233; **10,** 114; **11,** 150
Inoue, N., **1,** 484; **7,** 115
Inoue, S., **1,** 40; **2,** 307; **4,** 86, 467; **6,** 510, 578; **10,** 400; **12,** 213
Inoue, T., **4,** 175; **7,** 92, 428; **8,** 44, 505; **9,** 132
Inoue, Y., **9,** 344; **11,** 480
Inouye, Y., **3,** 258; **6,** 145; **9,** 54; **12,** 439
Insalaco, M. A., **2,** 45, 46; **3,** 36
Inubushi, Y., **4,** 101; **5,** 581; **7,** 28; **8,** 47; **9,** 318; **11,** 352
Inui, S., **12,** 89
Inukai, N., **1,** 40, 745; **2,** 24, 219, 307; **7,** 220
Inukai, T., **2,** 23
Invergo, B. J., **10,** 359
Ioannou, P. V., **6,** 79
Ipaktschi, J., **5,** 9, 70
Ipatieff, V. N., **1,** 1157
Ipe, D. P., **10,** 85
Ippen, J., **6,** 33, 479, 657; **8,** 52
Iqbab, A. F. M., **4,** 420; **5,** 326
Iqbal, J., **12,** 320
Iranshahi, L., **12,** 450
Iratcabal, P., **6,** 562
Irelan, J. R. S., **2,** 438
Ireland, R. E., **1,** 87, 106, 144, 166, 625, 818, 926, 986; **2,** 54, 112, 453; **3,** 57, 98, 174, 199, 258; **4,** 167, 309, 481, 527, 562; **5,** 410, 689; **6,** 279; **7,** 210; **8,** 245, 284, 302; **9,** 286; **10,** 2, 35; **12,** 282
Iriarte, J., **1,** 624; **5,** 605
Irick, G., **2,** 263
Irie, H., **9,** 176; **10,** 5; **11,** 435
Irie, K., **1,** 894; **3,** 231
Irie, T., **9,** 129
Iriguchi, J., **12,** 106
Irikawa, H., **2,** 462
Irisawa, J., **3,** 251

Iriuchijima, S., **4,** 475, 514; **5,** 683; **6,** 6; **9,** 68
Irngartinger, H., **11,** 460
Iruichijima, S., **5,** 4
Irvine, D. S., **3,** 54
Irvine, J., **1,** 1112
Irvine, J. C., **1,** 685
Irvine, J. W., Jr., **1,** 1142
Irving, E. M., **10,** 48
Irving, G. C., **1,** 180
Irving, H. M., **5,** 764
Irving, K. C., **4,** 541
Irwin, K. C., **3,** 331
Isaac, R., **6,** 167; **7,** 94
Isaacs, N. S., **4,** 552
Isaacson, Y. A., **7,** 413
Isagawa, K., **6,** 601; **9,** 146
Isayama, K., **4,** 107
Isbell, H. S., **2,** 263
Iseki, Y., **11,** 91, 373
Iseli, R., **9,** 523
Iselin, B., **1,** 130, 1154; **2,** 177, 179; **4,** 360
Ishibashi, M., **4,** 514
Ishibashi, Y., **11,** 267
Ishida, A., **6,** 412; **7,** 373
Ishida, H., **10,** 161
Ishida, M., **8,** 348; **9,** 316
Ishida, N., **5,** 406; **8,** 208; **12,** 182, 245
Ishida, Y., **12,** 11
Ishido, Y., **7,** 171; **10,** 207
Ishidoya, M., **7,** 153
Ishigami, T., **4,** 35, 270; **6,** 232
Ishige, M., **5,** 743; **6,** 659; **7,** 417; **8,** 434
Ishiguro, H., **12,** 236
Ishiguro, M., **8,** 259; **9,** 71, 408, 468; **10,** 142, 149; **11,** 64, 181, 378, 571
Ishiguro, T., **5,** 28
Ishiguro, Y., **12,** 131
Ishihara, H., **6,** 596
Ishihara, T., **7,** 5
Ishihara, Y., **11,** 143, 373; **12,** 514
Ishii, A., **11,** 496
Ishii, H., **9,** 434, 497; **11,** 498
Ishii, K., **10,** 377
Ishii, T., **11,** 324
Ishii, Y., **6,** 187; **7,** 220; **9,** 46, 344; **10,** 428; **11,** 571; **12,** 562
Ishikawa, H., **6,** 596; **7,** 373; **10,** 131, 238; **11,** 164, 249
Ishikawa, K., **8,** 102, 156
Ishikawa, M., **5,** 89, 552; **8,** 218
Ishikawa, N., **5,** 620; **6,** 581; **9,** 166; **10,** 44, 448; **11,** 599
Ishikawa, R., **8,** 526
Ishikura, K., **9,** 519
Ishimoto, S., **6,** 165, 641; **8,** 533
Ishino, Y., **9,** 237
Ishisawa, N., **7,** 165
Ishizaki, M., **12,** 25
Ishizuma, K., **8,** 373
Ishratullah, K., **9,** 191
Isidor, J. L., **4,** 165, 522; **5,** 208
Iskander, G. M., **3,** 254
Islam, A. M., **1,** 66, 861, 1137; **9,** 395
Isler, O., **1,** 114, 385, 388, 567, 614, 703, 1178
Ismail, H., **9,** 191
Ismail, Z. M., **10,** 11; **11,** 28
Isner, W. G., **1,** 1108
Isobe, M., **4,** 293; **6,** 267; **8,** 421, 516; **9,** 110, 311, 426; **11,** 99; **12,** 325
Isobe, T., **12,** 256
Isoe, S., **5,** 319, 491; **11,** 560
Israel, M., **10,** 59
Israelstam, S. S., **1,** 518
Issidorides, C. H., **1,** 309, 452, 643
Issleib, K., **9,** 66; **11,** 373
Isso, T., **4,** 395
Itagaki, K., **10,** 107; **11,** 36, 581
Itahara, T., **6,** 430, 479; **7,** 260; **9,** 349; **10,** 298; **11,** 391
Itai, T., **9,** 396
Itakura, J., **3,** 156
Itakura, K., **5,** 309, 434, 517; **6,** 361, 622; **7,** 14; **11,** 602
Itani, H., **12,** 121
Itano, H. A., **2,** 94, 317
Itaya, N., **3,** 41
Itaya, T., **2,** 305; **5,** 691
Itazaki, H., **1,** 444
Ito, A., **6,** 201
Ito, E., **11,** 180
Ito, H., **3,** 156; **5,** 13, 178; **6,** 174; **8,** 360; **11,** 20, 602
Ito, K., **1,** 1231; **3,** 300; **6,** 73; **9,** 32, 406, 464; **10,** 66, 398; **11,** 184; **12,** 31
Ito, S., **6,** 90, 145, 170; **7,** 47; **9,** 292; **11,** 172, 383, 462; **12,** 231
Ito, T., **5,** 190; **6,** 95; **9,** 46, 344; **10,** 36; **11,** 445
Ito, Y., **3,** 167; **4,** 101; **5,** 150, 163, 572, 581; **6,** 129, 141, 450, 517; **7,** 155, 322; **8,** 127, 299, 381, 519; **9,** 349; **10,** 84, 106, 185, 289, 381; **11,** 12, 100, 117; **12,** 59, 95, 573
Itoh, A., **8,** 165, 472, 476; **9,** 471; **10,** 164
Itoh, H., **8,** 312; **9,** 436
Itoh, I., **6,** 352, 496
Itoh, K., **10,** 107; **11,** 36, 373, 452, 522, 533, 581; **12,** 384, 435, 523, 547, 562
Itoh, M., **2,** 24, 285; **3,** 5, 39; **4,** 53; **5,** 347, 486; **6,** 91, 106, 149, 215, 295, 308; **7,** 4, 75; **8,** 260; **9,** 341; **10,** 61, 257
Itoh, N., **7,** 326, 342; **9,** 426
Itoh, S., **5,** 45
Itoh, T., **7,** 45; **10,** 457; **11,** 514
Itohi, K., **6,** 187
Itsuno, S., **9,** 422; **12,** 31
Ittah, Y., **6,** 542; **7,** 106; **8,** 516
Ivanetich, K., **8,** 272
Ivanoff, C., **1,** 711
Ivanov, C., **2,** 289
Iversen, T., **11,** 44
Ives, D. A. J., **1,** 444, 802, 926, 957; **4,** 540
Ives, J. L., **7,** 41; **9,** 341; **11,** 387
Iwacha, D. J., **4,** 58

Iwahara, T., **12,** 245
Iwahashi, H., **5,** 136, 441
Iwai, I., **1,** 336, 652, 926; **3,** 124
Iwai, K., **6,** 464, 466, 662; **9,** 403
Iwakuma, T., **7,** 326, 342; **9,** 426; **11,** 447; **12,** 454
Iwakura, Y., **3,** 119
Iwamoto, H., **7,** 220
Iwamoto, M., **1,** 318
Iwamura, H., **2,** 30; **7,** 3
Iwamura, J., **5,** 194
Iwanami, M., **5,** 561
Iwao, M., **11,** 373
Iwasaki, T., **5,** 178
Iwasaki, Y., **11,** 542
Iwasawa, H., **5,** 526
Iwasawa, M., **5,** 210
Iwasawa, N., **11,** 231, 526; **12,** 4, 117, 491
Iwashige, T., **2,** 159
Iwashita, T., **9,** 256
Iwashita, Y., **4,** 420
Iwata, C., **8,** 119; **9,** 123; **12,** 411
Iwata, N., **5,** 667
Iyer, K. N., **1,** 73, 604
Iyer, R., **12,** 77
Iyer, R. P., **9,** 411
Iyer, S., **12,** 130
Iyoda, J., **3,** 294; **5,** 150
Iyoda, M., **6,** 138; **11,** 275
Izatt, R. M., **6,** 137
Izawa, M., **11,** 445
Izawa, T., **6,** 308, 596; **10,** 449; **11,** 267, 518; **12,** 440
Izawa, Y., **12,** 399
Izumi, G., **6,** 596
Izumi, H., **2,** 37
Izumi, T., **11,** 293; **12,** 275
Izumi, Y., **8,** 96, 172; **9,** 406; **10,** 262; **12,** 503, 541
Izumiya, N., **2,** 177; **3,** 137
Izydore, R. A., **6,** 253

Jabri, N., **10,** 289
Jackman, L. M., **1,** 127, 129, 219; **2,** 263
Jackman, M., **1,** 956
Jacknow, B. B., **1,** 93
Jackson, A., **1,** 257
Jackson, A. C., **11,** 12; **12,** 11
Jackson, A. G., **7,** 138
Jackson, A. H., **1,** 703, 933; **2,** 107; **6,** 419; **12,** 481
Jackson, B. G., **1,** 603; **3,** 41
Jackson, C. L., **1,** 129
Jackson, D. A., **10,** 260
Jackson, D. K., **9,** 40
Jackson, E. L., **1,** 818
Jackson, H. L., **1,** 1006
Jackson, L. L., **4,** 371; **10,** 419
Jackson, R. A., **9,** 56
Jackson, R. W., **1,** 531, 808; **2,** 132, 362
Jackson, W. J., Jr., **2,** 215
Jackson, W. P., **10,** 22, 313
Jackson, W. R., **1,** 696; **2,** 281; **6,** 450; **7,** 277, 354; **9,** 240; **10,** 198; **11,** 503
Jacob, E. J., **4,** 33
Jacob, K., **4,** 466
Jacob, L., **12,** 212
Jacob, P., III, **5,** 486; **7,** 30, 56; **8,** 6; **9,** 58, 199
Jacob, T. A., **1,** 652
Jacob, T. M., **1,** 236, 662
Jacobi, P. A., **5,** 701; **10,** 232; **11,** 285; **12,** 445
Jacobs, A. M., **1,** 19
Jacobs, H. J. C., **3,** 57
Jacobs, P. M., **2,** 282
Jacobs, R. L., **1,** 955, 1167, 1216
Jacobs, S. A., **11,** 150
Jacobs, T. L., **1,** 1080
Jacobs, W. A., **6,** 2
Jacobsen, E. J., **11,** 47
Jacobsen, M., **5,** 591
Jacobsen, N., **5,** 17; **8,** 18
Jacobsen, N. W., **4,** 19
Jacobson, E. C., **1,** 1030
Jacobson, H. I., **1,** 1216
Jacobson, M., **1,** 814
Jacobson, N., **1,** 348
Jacobson, R. M., **6,** 545; **8,** 323; **9,** 421; **10,** 155
Jacobson, T., **2,** 423
Jacobsson, U., **6,** 24
Jacobus, J., **3,** 264
Jacoby, A. L., **1,** 573
Jacot-Guillarmod, A., **6,** 11
Jacques, J., **1,** 377, 561, 652, 856; **4,** 386; **10,** 461; **12,** 49
Jacquesy, J. C., **5,** 310; **6,** 273; **7,** 167; **8,** 240; **10,** 195
Jacquesy, R., **5,** 310, 609; **6,** 273; **7,** 167, 294
Jacquet, B., **12,** 414
Jacquier, R., **2,** 226; **4,** 172; **12,** 350
Jadhav, P. K., **10,** 224; **11,** 188, 191, 351, 568; **12,** 18, 399, 401
Jaeger, R., **1,** 904; **2,** 331
Jaeger, V., **3,** 130; **7,** 207; **8,** 205, 260; **10,** 237, 243; **11,** 292
Jaeggi, K. A., **2,** 117
Jaekle, H., **5,** 740
Jaen, J. C., **12,** 141
Jaffari, G. A., **4,** 75
Jaffe, A., **5,** 19
Jaffe, H. H., **1,** 353
Jaffe, M., **5,** 345
Jagdmann, G. E., Jr., **11,** 348
Jager, H., **9,** 43
Jager, V., **12,** 400
Jagur-Grodzinski, J., **4,** 287; **6,** 651; **11,** 156
Jahngen, E. G. E., Jr., **5,** 23, 266, 295
Jahnke, D., **1,** 1198; **4,** 128
Jahnke, U., **2,** 234
Jain, P. C., **7,** 11
Jain, S., **12,** 126
Jain, T. C., **4,** 45, 430; **6,** 304; **7,** 119; **10,** 58
Jaisli, F., **9,** 355
Jakiela, D. J., **10,** 77
Jakobsen, H. J., **1,** 84, 87, 198
Jakobsen, P., **2,** 55
Jakubke, H. D., **8,** 537
Jakubowski, A. A., **8,** 463
Jalandoni, C. C., **7,** 66
Jallabert, C., **8,** 119
Jallali-Naini, M., **11,** 216
James, B. D., **4,** 316
James, B. G., **4,** 189
James, D. C., **12,** 71
James, D. E., **11,** 395

James, D. S., **2,** 146
James, F. L., **1,** 1178
James, G. H., **3,** 202
James, K., **3,** 57; **6,** 569
James, L. B., **1,** 267
James, P. N., **1,** 285, 942
James, R., **6,** 581
Jamieson, I. M., **2,** 201
Jamison, J. D., **1,** 904
Jamison, W. C. L., **11,** 483
Jampel-Costa, E., **10,** 37
Janda, M., **4,** 70
Jani, A., **7,** 211
Janiga, E. R., **4,** 174
Jankovic, J., **7,** 196
Jann, K., **1,** 195
Janoski, E. J., **2,** 21
Janot, M. M., **2,** 75
Janousek, Z., **4,** 138; **5,** 198
Janout, V., **11,** 415
Jansen, A. B. A., **1,** 595, 642, 643, 951
Jansen, A. C. A., **12,** 191
Jansen, A. P., **1,** 321
Jansen, B. J. M., **7,** 310; **10,** 260; **12,** 317
Jansen, D. K., **10,** 253
Jansen, J. E., **1,** 531, 957
Jansing, J., **3,** 53
Janssen, D. E., **1,** 500, 1019, 1220; **5,** 523; **7,** 324
Janssen, G., **9,** 212
Janssen, M. J., **7,** 378
Janssen, W. A. J., **5,** 495
Januszkiewicz, K. R., **12,** 173, 381
Janzen, A. F., **7,** 242
Janzen, E. G., **1,** 310
Jaouen, G., **6,** 126; **7,** 71; **8,** 110; **10,** 14; **12,** 36
Jaouhari, R., **11,** 101
Japp, R. F., **4,** 369
Jarboe, C. H., **6,** 292
Jardine, F. H., **1,** 140, 1252; **2,** 453; **5,** 700
Jardine, I., **2,** 121; **4,** 40
Jardine, R. V., **1,** 970
Jarglis, P., **10,** 207; **11,** 72
Jarrar, A., **1,** 643
Jarrar, A. A., **8,** 410
Jarreau, F. X., **2,** 210
Jarrell, H. C., **5,** 470
Jart, A., **1,** 169
Jarvi, E. T., **11,** 341
Jarvis, B. B., **11,** 453
Jas, G., **12,** 345
Jaselskis, B., **2,** 458
Jasor, Y., **8,** 194
Jasserand, D., **7,** 37, 182
Jaszberenyi, J. C., **5,** 182
Jauhal, G. S., **9,** 217
Jaul, E., **2,** 174; **3,** 128
Jaunin, R., **1,** 187
Jaurand, G., **10,** 21
Jautelat, M., **2,** 181, 324; **3,** 141; **5,** 448
Javaid, K. A., **3,** 247
Javet, P., **6,** 284
Jaw, J. Y., **10,** 313
Jawdosiuk, M., **7,** 20
JaxaChamiec, A. A., **8,** 347; **11,** 309
Jay, E., **11,** 359
Jayalakshmi, S., **8,** 66
Jeanloz, R. W., **1,** 9
Jeanne-Carlier, R., **6,** 124; **7,** 69
Jedziniak, E. J., **5,** 61
Jefferies, P. R., **1,** 1004
Jefferson, A., **4,** 577
Jeffery, B. A., **2,** 296
Jeffery, E. A., **6,** 624
Jeffery, T., **12,** 370
Jeffery-Luong, T., **12,** 474
Jefford, C. W., **1,** 758; **2,** 300; **3,** 78; **4,** 234, 291, 293; **5,** 31, 389; **6,** 436, 455; **7,** 268, 269, 428; **8,** 272
Jeffs, P. W., **4,** 135; **7,** 384; **8,** 156; **9,** 34; **11,** 170
Jefson, M., **11,** 178
Jeger, O., **1,** 471, 562, 652, 814, 1284
Jeletzky, N. P., **4,** 87
Jellal, A., **12,** 305
Jellinek, E., **2,** 21
Jellinek, G., **1,** 942
Jemilev, U. M., **4,** 20
Jemison, R. W., **2,** 84
Jempty, T. C., **10,** 185; **11,** 237
Jen, K. Y., **12,** 466
Jen, T., **5,** 455, 626
Jendralla, H., **12,** 437
Jendrichovsky, J., **5,** 668
Jenkins, C. L., **5,** 152; **6,** 133, 665
Jenkins, D. M., **9,** 182
Jenkins, G. L., **1,** 1210
Jenkins, J. A., **5,** 567
Jenkins, P. R., **9,** 241
Jenkins, R. H., Jr., **11,** 108, 109
Jenkins, R. L., **1,** 402, 870
Jenkins, T. C., **8,** 204
Jenkinson, D. S., **1,** 1121
Jenner, E. L., **1,** 476, 892, 904; **2,** 357
Jennings, C. A., **9,** 8, 89
Jennings, R. C., **9,** 71
Jennings, W. B., **11,** 330
Jennings-White, C. L. D., **9,** 311; **12,** 125
Jensen, B. L., **11,** 188
Jensen, C. B., **1,** 999
Jensen, E. V., **1,** 182, 808, 1216
Jensen, F. R., **1,** 1251; **6,** 82
Jensen, H., **3,** 53; **5,** 499
Jensen, H. B., **3,** 329; **5,** 563
Jensen, H. P., **5,** 576
Jensen, K. A., **1,** 177; **6,** 583
Jensen, N. P., **2,** 238
Jensen, R. B., **5,** 169
Jensen, S. L., **1,** 127; **2,** 295
Jenson, T. M., **12,** 96
Jenssen, B., **1,** 926
Jentsch, J., **1,** 524
Jentzsch, J., **1,** 707
Jerina, D. M., **2,** 17; **4,** 459; **5,** 713; **6,** 358, 477; **7,** 274; **12,** 555
Jerina, M. S., **5,** 23
Jering, H., **5,** 439
Jernow, J., **9,** 184
Jernow, J. L., **5,** 714
Jerris, P. J., **9,** 96; **10,** 55, 265
Jerussi, R. A., **1,** 478; **4,** 270
Jeschkeit, H., **1,** 408
Jesser, F., **10,** 289; **12,** 315
Jessup, P. J., **7,** 147; **8,** 35, 221; **11,** 43
Jesthi, P. K., **6,** 329

Jetuah, F. K., **6,** 5
Jew, S., **8,** 424; **9,** 393
Jew, S. S., **10,** 332
Jewell, J. S., **2,** 165, 359
Jewell, R. A., **4,** 359
Jewett-Bronson, J., **11,** 188
Jeyaraman, R., **12,** 413
Jha, H. C., **10,** 410
Ji, S., **2,** 289
Jiang, J. L., **11,** 30; **12,** 221
Jibril, I., **12,** 278
Jigajinni, V. B., **11,** 160
Jimenez, C., **11,** 318
Jimenez, M. H., **10,** 300
Jimpso, K., **12,** 508
Jinbo, T., **10,** 131; **11,** 164
Jiricny, J., **9,** 244
Jitsukawa, K., **7,** 45; **10,** 457; **11,** 91; **12,** 89, 94
Jo, S., **10,** 176
Jobin, R. A., **2,** 222
Jobson, R. B., **2,** 39, 229
Jochims, J. C., **1,** 1284; **3,** 93
Jodoin, M., **7,** 135
Joensson, L., **7,** 274
Joerg, J., **1,** 562
Johanson, R., **5,** 320
Johansson, R., **11,** 467; **12,** 555
John, A. M., **8,** 289
John, D. I., **5,** 91; **10,** 21
John, E. V. O., **1,** 518
John, K. C., **8,** 476
John, L. S., **11,** 270
John, R. A., **11,** 514
John, T. V., **10,** 447
Johns, G., **7,** 79
Johns, R. B., **1,** 370
Johns, W. F., **1,** 360, 625, 926, 1184; **2,** 17, 29, 351; **4,** 366
Johnson, A. P., **1,** 310; **5,** 464; **9,** 189
Johnson, A. W., **1,** 186, 360, 370, 933; **6,** 456
Johnson, B. F. G., **6,** 39, 552; **7,** 111
Johnson, B. J., **1,** 783; **2,** 282
Johnson, C. D., **1,** 331; **2,** 156
Johnson, C. R., **1,** 94, 329, 814; **2,** 313; **3,** 47, 61, 106, 112, 206; **4,** 79, 174, 183; **5,** 78, 211, 231, 243, 459, 528, 609; **6,** 245, 395, 585; **9,** 194; **11,** 211, 344, 373, 570
Johnson, D., **1,** 658; **6,** 195
Johnson, D. A., **9,** 477
Johnson, D. K., **7,** 365; **8,** 477, 481
Johnson, D. L., **1,** 383; **10,** 454; **11,** 401
Johnson, D. W., **4,** 194; **6,** 615; **11,** 559
Johnson, F., **2,** 344; **3,** 191, 338; **5,** 421; **6,** 541; **11,** 301
Johnson, F. B., **2,** 58
Johnson, F. H., **3,** 127
Johnson, F. P., **6,** 91
Johnson, G., **6,** 115; **8,** 83; **9,** 199
Johnson, G. D., **1,** 1112
Johnson, G. J., **4,** 255
Johnson, G. S., **1,** 692, 730, 977; **6,** 423
Johnson, H. E., **1,** 367, 1100, 1167
Johnson, H. L., **1,** 124
Johnson, H. W., Jr., **1,** 78, 310, 978; **2,** 273
Johnson, J., **8,** 365
Johnson, J. E., **1,** 595; **4,** 389
Johnson, J. L., **1,** 777, 890; **6,** 534
Johnson, J. M., **1,** 643
Johnson, J. R., **1,** 22, 33, 207, 266, 489, 511, 636, 834, 907, 1087, 1112, 1150; **4,** 508
Johnson, K. K., **9,** 260
Johnson, L., **9,** 380
Johnson, M., **11,** 421
Johnson, M. A., **9,** 207
Johnson, M. C., **8,** 245
Johnson, M. R., **2,** 196; **4,** 326; **6,** 565; **9,** 110
Johnson, M. W., **10,** 292
Johnson, N. A., **5,** 77
Johnson, P. Y., **5,** 34, 687
Johnson, R. A., **6,** 490; **7,** 307; **8,** 259, 419; **9,** 249; **12,** 94
Johnson, R. C., **1,** 74; **4,** 227
Johnson, R. E., **3,** 264; **4,** 521; **9,** 83
Johnson, R. P., **12,** 570
Johnson, S., **9,** 283
Johnson, S. E., **6,** 195
Johnson, S. J., **8,** 117
Johnson, T. A., **7,** 171
Johnson, T. F. N., **1,** 486
Johnson, T. H., **7,** 289
Johnson, W. L., **8,** 516; **10,** 450
Johnson, W. M., **7,** 52
Johnson, W. S., **1,** 54, 86, 144, 149, 195, 198, 211, 213, 235, 377, 383, 389, 396, 524, 609, 703, 764, 777, 814, 829, 856, 881, 906, 926, 964, 986, 1064, 1080, 1131, 1143, 1178, 1184, 1286; **2,** 87, 203, 238, 331, 382, 464; **3,** 188, 302, 307, 308; **4,** 23, 472, 531; **5,** 55, 290, 319, 532, 552, 601, 631, 700; **6,** 615; **7,** 160, 345, 389; **8,** 186, 299, 503; **10,** 321, 436; **11,** 85; **12,** 377, 378
Johnson, W. W., **1,** 502
Johnston, D. B. R., **1,** 273; **2,** 426; **9,** 408; **10,** 100, 349
Johnston, D. E., **4,** 14, 126, 393; **6,** 121
Johnston, D. N., **6,** 419
Johnston, F., **1,** 7
Johnston, H. W., **1,** 110, 859
Johnston, J., **9,** 373
Johnston, J. D., **1,** 1090
Johnston, K. M., **5,** 542; **6,** 618; **7,** 295
Johnston, R. B., **2,** 147
Johnstone, H. F., **1,** 654; **10,** 209
Johnstone, L. M., **11,** 277
Johnstone, R. A. W., **6,** 446, 488; **7,** 341; **8,** 451; **9,** 388; **10,** 93, 359; **11,** 393; **12,** 574
Joines, R. C., **5,** 631
Jokic, A., **3,** 171

Jolad, S. D., **1,** 402
Jolly, W. L., **2,** 329
Joly, G., **7,** 167
Joly, R., **1,** 361, 465, 608, 808
Jommi, G., **3,** 158
Jonas, D. A., **2,** 313
Jonczyk, A., **4,** 31; **6,** 43; **7,** 20, 353; **8,** 461
Jones, A., **7,** 333; **8,** 522
Jones, B. D., **2,** 274
Jones, D., **2,** 159
Jones, D. A., Jr., **5,** 741
Jones, D. C., **3,** 36
Jones, D. G., **2,** 42
Jones, D. H., **2,** 313
Jones, D. L., **2,** 365; **5,** 351
Jones, D. M., **1,** 248; **2,** 295
Jones, D. N., **1,** 561; **2,** 316; **5,** 26, 276; **6,** 47
Jones, D. S., **1,** 1229
Jones, D. W., **1,** 927; **5,** 585
Jones, E., **1,** 562
Jones, E. R. H., **1,** 12, 19, 144, 186, 281, 355, 360, 389, 424, 595, 642, 643, 668, 723, 754, 802, 846, 928, 1040, 1157; **2,** 19, 243, 295; **3,** 39; **5,** 86; **6,** 304, 451, 514
Jones, F. N., **3,** 329; **5,** 176
Jones, G., **2,** 213, 246; **3,** 183, 291; **5,** 6, **6,** 291
Jones, G. H., **2,** 163, 181; **6,** 137; **12,** 41
Jones, H., **8,** 167
Jones, J. B., **2,** 163; **3,** 136
Jones, J. E., **1,** 456, 682, 881
Jones, J. F., **6,** 618
Jones, J. H., **1,** 486; **3,** 229; **8,** 73; **10,** 11
Jones, J. K. N., **1,** 989, 1187; **2,** 226, 395; **3,** 244; **5,** 470; **7,** 67
Jones, J. R., **2,** 274
Jones, K., **2,** 147; **9,** 349
Jones, L. D., **4,** 35; **10,** 77; **11,** 51
Jones, L. H., **9,** 207
Jones, L. W., **9,** 42
Jones, M., **5,** 399; **12,** 116
Jones, M. B., **4,** 501
Jones, M. F., **12,** 415
Jones, M., Jr., **3,** 204; **4,** 336; **12,** 330
Jones, N., **3,** 219
Jones, N. R., **2,** 187; **6,** 493
Jones, P. F., **2,** 187; **5,** 374
Jones, P. G., **1,** 802, 1090
Jones, P. R., **5,** 457
Jones, R. A., **4,** 502; **7,** 69
Jones, R. C. F., **9,** 470
Jones, R. E., **1,** 74, 952
Jones, R. G., **1,** 96, 257, 414, 470, 573, 837, 845, 951, 970; **2,** 153; **10,** 77
Jones, R. W., **6,** 398
Jones, S. R., **6,** 317; **9,** 270
Jones, S. S., **9,** 77
Jones, T. H., **10,** 361
Jones, T. K., **11,** 83; **12,** 441
Jones, V. K., **1,** 49, 1193
Jones, W. A., **1,** 814
Jones, W. D., **8,** 469
Jones, W. E., **1,** 13, 1157
Jones, W. J., **1,** 310, 952, 1011, 1018
Jones, W. M., **1,** 339, 613, 1054, 1230; **3,** 173
Jong, T. T., **12,** 434
Jongejan, E., **8,** 187
Jonker, C., **5,** 208
Jonkers, F. L., **6,** 381; **9,** 112, 319, 450
Joos, R., **5,** 15, 329; **7,** 171
Joos, W., **1,** 643
Jordaan, A., **5,** 136
Jordaan, J. H., **1,** 755
Jordan, A., **8,** 454
Jordan, E. F., Jr., **1,** 652, 759, 1271
Jordens, P., **2,** 87
Jordis, U., **8,** 358; **9,** 323
Jorge, Z. D., **10,** 459
Jorgensen, K. A., **11,** 491
Jorgensen, M. J., **1,** 595
Jorgensen, P. M., **1,** 169
Jorgenson, M. J., **1,** 35; **2,** 143, 242; **3,** 176; **5,** 391; **6,** 368
Jorritsma, H., **6,** 24
Jorritsma, R., **4,** 108
Josan, J. S., **4,** 185; **7,** 386
Jose, F. L., **4,** 60
Joseph, J. P., **2,** 310
Joseph, J. T., **5,** 339
Josephson, S., **10,** 432; **11,** 419
Josey, A. D., **1,** 904; **2,** 144
Josey, J., **11,** 416
Joshi, B. N., **2,** 239
Joshi, B. S., **1,** 409, 933
Joshi, G. C., **4,** 31; **6,** 675
Joshua, C. P., **6,** 34
Josiah, B. M., **10,** 183; **11,** 427
Joska, J., **1,** 624, 625
Jost, E., **1,** 424, 494
Jost, K. H., **1,** 885
Josty, P. L., **7,** 111
Jouannetaud, M. P., **8,** 240; **10,** 195
Joubert, J. P., **6,** 665
Jouffret, M., **9,** 207
Jouitteau, C., **10,** 306
Joukhadar, L., **8,** 285; **9,** 378
Joule, J. A., **1,** 8
Joullie, M. M., **1,** 425, 904; **8,** 284, 289, 342; **9,** 434; **10,** 21, 67, 182
Jourdan, F., **1,** 841
Jovtscheff, A., **5,** 129
Joyce, M. A., **6,** 137
Joyce, R. J., **1,** 253
Joye, N. M., Jr., **2,** 156, 332
Jucker, E., **1,** 20, 478, 1046
Judge, J. M., **1,** 1203; **5,** 600
Judkins, B. D., **9,** 269
Judy, W. A., **4,** 570
Juenge, E. C., **1,** 1274; **2,** 427; **3,** 297
Juenger, A., **1,** 1256
Juergens, E., **1,** 790
Jujita, E., **11,** 324
Jukes, A. E., **5,** 375
Julia, M., **1,** 360, 676; **2,** 267; **4,** 123; **5,** 16, 107, 183, 427; **6,** 141; **7,** 50, 327; **8,** 316; **9,** 132; **11,** 59, 346, 475; **12,** 140, 212, 486
Julia, S., **1,** 5, 215, 676, 1096, 1253; **2,** 267; **5,** 246, 406, 460; **6,** 631; **8,** 94, 222, 334, 391; **9,** 95
Julia, S. A., **1,** 574

Julia, T. F., **3,** 222
Julian, P. L., **1,** 150
Juliano, B. O., **2,** 175
Juliano, P. C., **1,** 1108
Jun, Y. M., **8,** 119
Junanez, Z., **10,** 240
Jund, K., **10,** 248
Juneja, H. R., **4,** 194
Jung, A., **12,** 356, 490, 502
Jung, C. J., **5,** 144
Jung, F., **5,** 129; **9,** 387
Jung, G., **6,** 422
Jung, M., **11,** 85
Jung, M. E., **5,** 355; **7,** 402, 415; **8,** 56, 178, 263, 326, 415, 525; **9,** 74, 151, 250, 255, 256, 491; **10,** 154, 219, 443; **11,** 28
Jung, S. H., **12,** 147
Jung, W., **9,** 428
Junga, I. G., **4,** 161
Junggren, U., **6,** 565
Jungheim, L. N., **10,** 32, 383, 423
Jungnickel, J. L., **1,** 528
Junjappa, H., **6,** 456; **10,** 229
Juo, C. H., **1,** 1171
Jurayama, E., **11,** 108
Jurch, G. R., Jun., **2,** 159
Jurczak, J., **6,** 280; **7,** 406; **10,** 258
Jurd, L., **1,** 500; **3,** 87; **6,** 518; **8,** 443
Jurewicz, A. T., **5,** 359
Jurgeleit, W., **2,** 433
Jurion, M., **3,** 304; **4,** 341; **5,** 693
Jurjev, V. P., **4,** 20
Jurlina, J. L., **8,** 260; **9,** 249; **11,** 515
Just, G., **1,** 771; **5,** 667; **7,** 428; **9,** 485; **10,** 325
Justus, R., **9,** 121
Jutand, A., **10,** 194
Jutland, A., **10,** 133
Jutzi, P., **11,** 111
Juve, H. D., Jr., **11,** 124, 268; **12,** 272

Kabalka, G. W., **2,** 15, 193, 195; **3,** 5, 29, 39, 91; **4,** 362, 447; **5,** 735; **6,** 98, 625; **7,** 55, 181; **8,** 80, 507; **9,** 99, 443, 489; **10,** 32, 212; **11,** 269, 489; **12,** 65, 288
Kabasakalian, P., **1,** 411
Kabbe, H. J., **1,** 190, 1151; **2,** 19
Kabengele, T., **5,** 389
Kaberia, F., **8,** 463
Kabeta, K., **12,** 146, 502
Kablaoui, M. S., **6,** 560
Kabuto, K., **6,** 206; **10,** 369
Kachensky, D. F., **12,** 23
Kacher, M., **6,** 538
Kachinski, J. L. C., **8,** 109
Kacprowicz, A., **6,** 601
Kaczmarek, C. S. R., **12,** 445
Kadaba, P. K., **4,** 45
Kader, A. T., **1,** 1185; **6,** 56
Kadesch, R. G., **1,** 940
Kadin, S. B., **2,** 378; **5,** 601
Kadish, A. F., **10,** 187
Kadlec, S. B., **9,** 391
Kadokura, A., **9,** 343
Kadokura, M., **12,** 449
Kadonaga, J. T., **10,** 89
Kaegi, H., **1,** 130
Kaelicke, J., **1,** 347
Kaemmerer, H., **1,** 561
Kaestner, P., **4,** 122
Kagal, S. A., **2,** 220
Kagami, M., **6,** 37
Kagan, F., **1,** 808, 1125
Kagan, H. B., **1,** 377, 652, 856; **2,** 318; **4,** 273; **5,** 19, 58, 361; **6,** 23, 269, 671; **8,** 439; **9,** 260; **10,** 206, 344; **11,** 283, 465; **12,** 94, 430
Kagawa, R., **5,** 535
Kagei, K., **8,** 156
Kageyama, T., **12,** 445
Kagi, D. A., **4,** 14; **6,** 19
Kagotani, M., **8,** 385; **10,** 32; **11,** 239, 240, 395
Kahle, G. G., **10,** 194
Kahle, U., **2,** 423
Kahlert, E., **1,** 672
Kahn, M., **10,** 272, 328; **11,** 99, 337; **12,** 359
Kahn, P., **10,** 268
Kahn, S. A., **9,** 352
Kahne, D., **11,** 536
Kahne, D. E., **12,** 152
Kahr, E., **4,** 33
Kahr, K., **1,** 476
Kai, F., **11,** 322
Kai, Y., **11,** 500; **12,** 85
Kaiho, T., **11,** 223
Kaiser, C., **2,** 173; **4,** 228, 268, 441; **7,** 182
Kaiser, E., **2,** 285
Kaiser, E. M., **1,** 581, 609, 611; **2,** 242, 374; **3,** 175, 183; **4,** 33, 63, 112, 297; **5,** 379, 406; **6,** 84, 339, 526, 636; **9,** 94
Kaiser, E. T., **11,** 359
Kaiser, H., **1,** 1023, 1118
Kaiser, R., **5,** 448
Kaiser, S., **5,** 225
Kaito, M., **8,** 382; **9,** 173, 350; **10,** 150
Kaiya, T., **9,** 349
Kajfez, F., **6,** 474; **9,** 234; **11,** 124
Kaji, A., **5,** 556, 607; **7,** 2, 149; **8,** 141, 315; **9,** 133, 188, 216, 434; **10,** 39, 183, 261, 413; **11,** 551; **12,** 47, 233, 350, 523, 551
Kaji, K., **9,** 38; **10,** 289; **12,** 506
Kaji, S., **12,** 214
Kajigaeshi, S., **4,** 405; **5,** 620
Kajihara, Y., **11,** 174
Kajima, M., **11,** 373
Kajimoto, T., **5,** 165; **6,** 430; **7,** 81; **12,** 30
Kajitani, M., **6,** 659
Kakac, B., **1,** 764
Kakese, F., **4,** 149
Kakihana, T., **2,** 173
Kakis, F. J., **2,** 368; **5,** 60, 580; **6,** 430
Kakisawa, H., **2,** 151, 318; **4,** 177; **9,** 256
Kakiuchi, H., **10,** 198
Kakoi, H., **4,** 86
Kakui, T., **8,** 56, 101; **9,** 95
Kakushima, M., **7,** 345; **10,**

373
Kalabin, G. A., **11,** 440
Kalb, L., **1,** 561
Kale, A. V., **9,** 410
Kale, N., **2,** 334; **3,** 232
Kaleya, R., **10,** 216
Kalicky, P., **4,** 265; **5,** 353
Kalinowski, H. O., **6,** 578; **8,** 517; **9,** 48; **10,** 395; **11,** 326; **12,** 193, 480
Kalir, A., **5,** 688; **8,** 433
Kalish, J., **4,** 255
Kallenberg, H., **8,** 415; **9,** 387
Kalli, M., **5,** 243
Kallio, R. E., **1,** 1011
Kallmerten, J. L., **9,** 337
Kalm, M. J., **1,** 874, 926, 976, 1073
Kalman, J. R., **4,** 283; **6,** 581; **11,** 32
Kalnins, M. V., **1,** 281, 705
Kaloustian, M. K., **6,** 583; **9,** 429; **10,** 417; **11,** 487
Kalvoda, J., **1,** 562, 624; **2,** 77, 422; **7,** 146
Kalyan, Y. B., **11,** 163
Kalyanam, N., **11,** 142, 453
Kalyanasundaram, S. K., **8,** 460
Kamachi, H., **2,** 151; **4,** 177
Kamakshi, C., **12,** 71
Kamano, Y., **5,** 507, 743
Kamata, K., **4,** 345; **6,** 389, 629; **7,** 231; **11,** 233; **12,** 383
Kamata, S., **4,** 147; **6,** 192, 582
Kamath, P. M., **7,** 23
Kambe, N., **10,** 206; **11,** 220, 498
Kambi, E., **5,** 319
Kameda, N., **9,** 148
Kamekawa, K., **11,** 361
Kamemoto, K., **4,** 529; **5,** 693
Kameswaran, V., **4,** 330; **5,** 746; **6,** 660; **7,** 421
Kametani, T., **1,** 122; **2,** 433; **3,** 145, 268; **4,** 340, 407, 428; **5,** 157; **6,** 283; **7,** 329; **9,** 32; **10,** 21, 194, 245; **11,** 550; **12,** 44
Kameyama, K. L., **12,** 272
Kamienski, C. W., **1,** 97, 573; **2,** 241; **5,** 649
Kamijo, T., **12,** 106
Kamikawaji, Y., **11,** 28
Kamimura, A., **11,** 551
Kaminaga, M., **5,** 571
Kamino, K., **9,** 414
Kaminski, F. E., **2,** 374
Kamio, S., **5,** 80
Kamitori, Y., **11,** 293; **12,** 275
Kamiya, T., **6,** 91; **10,** 61
Kamm, O., **1,** 407, 452, 1064, 1284
Kamm, R. M., **1,** 1073
Kamma, K., **2,** 289
Kammann, K. P., Jr., **2,** 12
Kammereck, R., **2,** 368
Kammeyer, C. W., **6,** 254
Kammula, S., **5,** 654
Kamonah, F. S., **9,** 249
Kamoshida, A., **7,** 352
Kampe, K. D., **3,** 53
Kampe, W., **2,** 387
Kampmeier, J. A., **1,** 39, 41, 677
Kampouris, E. M., **1,** 48
Kan, K., **11,** 494
Kan, R. O., **1,** 34, 1055
Kanai, K., **12,** 204
Kanakarajan, K. K., **9,** 151
Kanaki, J., **8,** 319; **10,** 398
Kanakura, A., **9,** 68
Kanamara, H., **6,** 392
Kanamaru, H., **8,** 347
Kanaoka, Y., **1,** 894; **2,** 107, 334; **3,** 231, 303
Kanazawa, R., **6,** 529; **7,** 329
Kanbara, H., **12,** 171
Kanda, Y., **5,** 614
Kandall, C., **3,** 34
Kandasamy, D., **5,** 645; **6,** 538; **11,** 483
Kandil, S. A., **2,** 53
Kandutsch, A. A., **10,** 304
Kane, M. J., **5,** 266
Kane, P. D., **12,** 41
Kane, S. S., **1,** 772
Kane, V. V., **2,** 104; **7,** 247; **12,** 330
Kaneda, K., **6,** 450, 653; **7,** 45, 277, 314; **10,** 457; **11,** 91; **12,** 89, 94
Kanehira, K., **11,** 240
Kaneko, C., **1,** 89, 603, 1273; **5,** 89, 552; **6,** 586; **9,** 467
Kaneko, K., **7,** 365; **9,** 315, 462
Kaneko, T., **6,** 378; **7,** 268
Kaneko, Y., **12,** 440
Kanemasa, S., **12,** 316, 539, 543
Kanematsu, K., **6,** 43, 297; **11,** 28
Kanemoto, S., **11,** 533; **12,** 63, 352, 419
Kanetani, R., **12,** 245
Kan-Fan, C., **2,** 9, 75; **3,** 115, 308; **4,** 184
Kang, H. J., **12,** 252
Kang, J., **9,** 94; **11,** 339; **12,** 103, 324
Kang, J. W., **8,** 153
Kang, K., **5,** 205
Kang, M., **12,** 294
Kang, S. K., **9,** 413
Kang, Y. H., **11,** 408
Kanghae, W., **8,** 95
Kanillis, P., **11,** 216
Kanner, B., **1,** 212
Kano, S., **6,** 547; **8,** 351, 492; **9,** 381, 438, 470, 473; **10,** 404
Kanojia, R. M., **5,** 438
Kantlehner, W., **4,** 529; **5,** 73, 250, 653; **6,** 58, 243; **9,** 182; **10,** 113
Kantor, S. W., **1,** 909, 911
Kaplan, B. E., **2,** 79
Kaplan, F., **6,** 552
Kaplan, J., **1,** 1116
Kaplan, J. P., **3,** 20, 153
Kaplan, L., **2,** 448; **4,** 339
Kaplan, L. J., **4,** 206
Kaplan, M. L., **3,** 324
Kaplan, R. B., **1,** 757; **2,** 300
Kapnang, H., **8,** 232
Kapoor, A., **1,** 783
Kapoor, S. K., **4,** 149
Kapoor, V. M., **6,** 170; **8,** 186
Kappe, T., **2,** 372; **6,** 96
Kappeler, H., **1,** 85, 385

Kappler, F. E., **5,** 551
Kapur, J. C., **5,** 127, 713
Kar, S. K., **6,** 548
Karabatsos, G. J., **3,** 176
Karabinos, J. V., **1,** 310; **5,** 277
Karady, S., **6,** 412; **9,** 107; **11,** 189
Karanewsky, D. S., **7,** 2; **9,** 138, 174
Karapinka, G. L., **3,** 333
Karas, L. J., **8,** 464
Karczag, A., **1,** 34
Karg, E., **1,** 782
Karger, M. H., **2,** 15; **3,** 198; **4,** 8; **5,** 5
Karigomi, H., **11,** 32
Karim, A., **3,** 67
Kariyone, K., **3,** 247; **6,** 65
Karkhanis, Y. D., **1,** 485
Karl, H., **10,** 350
Karle, D. W., **7,** 299
Karlson, P., **1,** 506
Karmas, G., **1,** 730, 843, 1286
Karnes, H. A., **2,** 174
Karpeles, R., **10,** 329
Karpetsky, T. P., **5,** 672
Karpf, M., **11,** 28
Karpitschka, E. M., **11,** 222
Karras, M., **8,** 42; **9,** 53, 471; **10,** 181; **11,** 12
Karrenbrock, F., **9,** 160
Karrer, P., **1,** 414, 595, 666, 685, 796, 1083, 1284; **6,** 37
Karsch, U., **2,** 63
Karstashov, V. R., **6,** 232
Karstens, A., **1,** 992
Kartoon, I., **5,** 728
Karunaratne, V., **12,** 116
Kasahara, A., **12,** 313
Kasai, M., **2,** 23
Kasai, R., **12,** 282
Kasai, Y., **5,** 217; **6,** 193; **7,** 107
Kashdan, D. S., **5,** 295, 394; **6,** 91
Kashelikar, D. V., **1,** 372, 655
Kashima, C., **6,** 534
Kashitani, T., **6,** 138
Kashman, Y., **4,** 188
Kasina, S., **6,** 223
Kaska, W. C., **1,** 262
Kaslow, C. E., **1,** 266, 866, 1027, 1178, 1269
Kaspi, J., **9,** 320
Kastner, S. M., **5,** 470
Kasturi, T. R., **1,** 119, 207, 599; **2,** 117, 246, 303, 344; **5,** 186
Kasuga, K., **9,** 370, 386
Katagiri, N., **5,** 309, 434; **6,** 361, 622; **7,** 14; **11,** 190
Katagiri, T., **10,** 247; **12,** 57
Katakai, R., **5,** 100
Katakawa, J., **9,** 176; **11,** 435
Katano, K., **11,** 322
Kataoka, H., **9,** 431; **10,** 281, 298; **11,** 514; **12,** 370, 562
Kataoka, K., **6,** 618
Kataoka, M., **9,** 91
Kataoka, T., **9,** 13
Katayama, E., **12,** 288, 513
Katayama, I., **10,** 5
Katayama, S., **12,** 327
Katchalski, E., **1,** 110, 1015; **2,** 59; **4,** 49, 87
Katcher, M. L., **10,** 38
Katchian, H., **9,** 391
Kate, V. N., **10,** 98
Katekar, G. F., **3,** 206; **4,** 174; **9,** 194
Katerberg, G. J., **1,** 220
Kates, M., **1,** 847
Kathawala, F., **2,** 196; **9,** 337
Katner, A. S., **4,** 172
Kato, H., **10,** 274; **11,** 12, 100
Kato, J., **11,** 112, 521
Kato, K., **4,** 149, 555; **5,** 348; **7,** 135; **8,** 530; **9,** 15; **10,** 400
Kato, M., **5,** 55, 421, 446; **6,** 170, 482; **8,** 446; **9,** 87, 277; **10,** 71
Kato, N., **11,** 334
Kato, S., **2,** 344; **4,** 257; **12,** 167
Kato, T., **2,** 37, 177; **3,** 229; **5,** 226, 266; **6,** 554; **7,** 352; **9,** 404, 410; **10,** 377; **11,** 190; **12,** 167
Kato, Y., **10,** 365
Katoh, N., **8,** 345
Katori, E., **11,** 445
Katritzky, A. R., **1,** 1044; **2,** 47; **6,** 618; **7,** 10, 409; **8,** 456, 520, 521; **9,** 309, 508; **11,** 502
Katsaros, N., **9,** 513
Katsoyannis, P. G., **1,** 1139
Katsuki, S., **1,** 718
Katsuki, T., **7,** 320; **9,** 479; **10,** 65; **11,** 94, 95, 463, 552; **12,** 59, 435, 573
Katsumura, A., **11,** 240
Katsumura, S., **5,** 319, 491; **9,** 191; **11,** 560
Katsuro, Y., **10,** 194; **11,** 59
Katten, E., **4,** 74
Kattenberg, J., **5,** 560
Katterman, L. C., **9,** 286
Katz, A., **1,** 87
Katz, A. H., **12,** 483
Katz, C., **4,** 84
Katz, J. J., **4,** 176; **5,** 203; **6,** 172, 208
Katz, L., **1,** 1015; **4,** 123, 534
Katz, R., **2,** 224, 315
Katz, S., **4,** 358, 457
Katz, T. J., **1,** 9, 983; **3,** 329; **4,** 339, 419; **5,** 530
Katzenellenbogen, E. R., **1,** 526
Katzenellenbogen, J. A., **2,** 153, 241, 263; **3,** 112, 113, 174, 210; **4,** 53, 161, 183, 295, 424, 535; **5,** 285, 381; **6,** 79, 180; **7,** 83; **8,** 122, 316; **9,** 251; **10,** 106; **11,** 321, 533
Katzin, M. I., **6,** 316
Kau, D., **1,** 802; **2,** 17
Kauer, J. C., **1,** 177, 247, 1040; **3,** 171
Kauffman, G. B., **1,** 1192; **2,** 400; **3,** 81
Kauffmann, H., **1,** 407
Kauffmann, T., **1,** 1075; **2,** 92; **5,** 610; **8,** 211, 280; **9,** 509; **11,** 424; **12,** 353
Kaufman, B. L., **7,** 413
Kaufman, C. W., **1,** 619
Kaufman, D., **1,** 157

Kaufman, G., **2,** 422
Kaufman, G. B., **4,** 517
Kaufman, G. M., **1,** 1187
Kaufman, J. B., **2,** 283
Kaufman, K. L., **1,** 518
Kaufman, P. R., **1,** 1131
Kaufman, S., **1,** 371
Kaufmann, D., **6,** 568; **9,** 272
Kaufmann, H., **2,** 219, 422; **5,** 344; **7,** 146
Kaufmann, L., **1,** 22
Kaufmann, S., **1,** 3, 39, 1090
Kaufmann, W. E., **1,** 1209
Kaugars, G., **1,** 424
Kauser, W. M., **1,** 1040
Kautsky, H., **6,** 510
Kautzner, B., **6,** 179
Kavai, I., **8,** 167
Kawabata, K., **9,** 315; **11,** 519
Kawabata, N., **1,** 253; **3,** 141; **5,** 89, 442; **7,** 74
Kawabata, T., **10,** 51, 52; **11,** 29; **12,** 29
Kawabe, K., **9,** 208
Kawabe, N., **4,** 69
Kawada, M., **10,** 390; **11,** 419; **12,** 316, 402
Kawagishi, T., **10,** 289, 441
Kawaguchi, K., **12,** 150
Kawahara, S., **12,** 445
Kawahara, T., **1,** 756
Kawai, M., **6,** 464, 466
Kawai, N., **8,** 211
Kawai, S., **2,** 159; **9,** 434
Kawai, T., **8,** 421; **9,** 110, 426
Kawakami, J. H., **2,** 267; **3,** 186
Kawakami, S., **11,** 239
Kawakami, Y., **12,** 195
Kawakubo, H., **8,** 135
Kawamata, T., **5,** 234
Kawami, Y., **10,** 289
Kawamoto, F., **7,** 277; **10,** 185
Kawamoto, I., **6,** 667
Kawamura, K., **7,** 5
Kawamura, S., **7,** 251; **8,** 358; **12,** 256
Kawanami, Y., **11,** 552; **12,** 59, 435, 573
Kawanishi, Y., **12,** 94
Kawanisi, M., **1,** 404; **2,** 202; **11,** 49, 80
Kawano, Y., **6,** 534
Kawara, T., **10,** 289; **11,** 141, 249
Kawasaki, A., **11,** 514
Kawasaki, H., **11,** 49, 80
Kawasaki, K., **10,** 237
Kawasaki, M., **6,** 70; **12,** 322
Kawasaki, T., **8,** 519; **9,** 506, 507
Kawase, M., **11,** 483; **12,** 432
Kawase, Y., **12,** 291
Kawashima, A., **10,** 182
Kawashima, K., **5,** 28; **8,** 535
Kawashima, M., **10,** 289; **12,** 25
Kawashima, T., **6,** 500
Kawata, K., **5,** 370; **9,** 106
Kawata, M., **5,** 308
Kawata, T., **5,** 13
Kawatani, H., **2,** 48; **5,** 702
Kawate, T., **11,** 434; **12,** 407
Kawauchi, H., **6,** 45
Kawazoe, T., **7,** 282
Kawazoe, Y., **5,** 344; **6,** 556
Kawesaki, K., **2,** 36
Kay, I. T., **9,** 361
Kaya, R., **10,** 227; **11,** 249
Kaye, A. D., **12,** 257
Kaye, H., **2,** 313, 429
Kaye, I. A., **1,** 357, 500, 601, 982
Kaye, R. L., **2,** 84
Kaye, S., **1,** 1235
Kayser, M., **7,** 99
Kayser, R. H., **6,** 617
Kazama, Y., **1,** 601
Kazan, J., **1,** 235
Kazlauskas, R., **9,** 442
Kazubski, A., **11,** 229, 230
Keach, DeW. T., **1,** 52
Kealy, T. J., **1,** 93, 850; **2,** 332
Keana, J. F. W., **1,** 54; **2,** 192; **4,** 223; **5,** 677
Keana, S. B., **2,** 192
Kearns, D. R., **5,** 491
Keating, M., **3,** 47; **4,** 282
Keaton, O. D., **1,** 243, 906, 940
Keaveney, W. P., **2,** 156, 444
Keay, B. A., **11,** 450
Keck, G. E., **8,** 259; **9,** 136, 193, 318, 326, 432, 449, 467; **10,** 161; **11,** 16, 127, 292, 335, 363; **12,** 23, 146
Keck, H., **1,** 1252
Keehn, P., **11,** 311
Keehn, P. M., **5,** 15; **7,** 390; **11,** 107
Keeley, D. E., **5,** 373; **6,** 320; **7,** 190; **8,** 140
Keely, S. L., Jr., **2,** 285
Keen, G. W., **6,** 20, 304
Keene, B. R. T., **5,** 651
Keeney, M., **1,** 566
Kees, F., **5,** 641
Kees, K., **6,** 469
Kees, K. L., **8,** 483; **9,** 467; **11,** 527
Keese, R., **9,** 217
Keeseler, K., **11,** 378
Keeton, R., **2,** 366
Kehm, B. B., **1,** 1095, 1210
Kehoe, L. J., **4,** 87
Keii, T., **6,** 504, 510
Keil, E., **1,** 618; **3,** 184
Keilich, G., **1,** 608
Keim, W., **5,** 463; **7,** 67
Keinan, E., **6,** 441; **7,** 273; **8,** 230, 377, 461; **9,** 175, 458; **10,** 296, 390, 391; **11,** 551; **12,** 475, 546
Keisel, R. J., **10,** 413
Keiser, J. E., **2,** 313; **3,** 61, 106; **6,** 9
Keith, D. R., **9,** 197; **10,** 201
Kelleghan, W. J., **10,** 117
Kelleher, R. G., **10,** 246
Keller, H., **2,** 273
Keller, J. W., **3,** 69
Keller, L., **11,** 51, 358
Keller, O., **7,** 91; **10,** 122
Keller, W. E., **9,** 361; **10,** 122
Kellert, M. D., **1,** 198
Kellett, R. E., **9,** 50
Kelley, H. C., **1,** 355
Kelley, J. A., **6,** 114; **8,** 102
Kelley, M. G., **2,** 394
Kelley, R. B., **4,** 540
Kellner, H. A., **11,** 63
Kellog, C. K., **5,** 381

Kellog, M. S., **6,** 455
Kellogg, K. G., **2,** 200
Kellogg, M. S., **9,** 477
Kellogg, R. M., **3,** 320; **4,** 210; **5,** 160; **7,** 392; **8,** 458; **9,** 100; **11,** 115, 118; **12,** 108, 292
Kellom, D. B., **1,** 905
Kelly, A. G., **10,** 250
Kelly, D. J., **12,** 379
Kelly, D. P., **2,** 431
Kelly, D. R., **9,** 65, 239
Kelly, J. E., **4,** 419; **6,** 617
Kelly, J. F., **3,** 53
Kelly, K. P., **10,** 30; **11,** 45
Kelly, M. J., **12,** 104
Kelly, R., **5,** 290
Kelly, R. B., **1,** 385, 643, 926, 957; **5,** 621; **6,** 181
Kelly, R. C., **5,** 319; **7,** 160, 257
Kelly, R. W., **2,** 79, 355; **3,** 57, 113, 114
Kelly, T. R., **4,** 375; **5,** 296; **6,** 383; **8,** 305; **9,** 61
Kelly, W. J., **9,** 421; **11,** 441, 479
Kelm, H., **5,** 649
Kelsey, R., **2,** 68, 242
Kem, K., **4,** 521
Kemp, D. S., **1,** 364, 484; **2,** 192; **5,** 342; **6,** 32, 195, 254; **8,** 252
Kemp, G., **6,** 280
Kemp, W., **7,** 295
Kempe, T., **7,** 336
Kempe, U. M., **4,** 82
Kemper, B., **11,** 87
Kempf, D. J., **10,** 75
Kempter, F. E., **5,** 3
Kendall, D. L., **1,** 281
Kendall, E. C., **1,** 331, 452, 537, 566, 654, 1000, 1006, 1118, 1283; **8,** 442
Kendall, M. C. R., **6,** 573
Kendall, P. M., **6,** 105, 204
Kende, A. S., **1,** 39, 808; **2,** 219; **4,** 302; **5,** 639; **6,** 390, 654; **7,** 322, 421; **8,** 458, 481; **9,** 337, 480; **10,** 75, 136, 195, 243, 246, 354, 415, 423; **11,** 70, 281, 299, 301, 318, 331, 391, 553
Kendrick, D. A., **12,** 125
Kenehan, E. F., **6,** 114
Kennard, K. C., **1,** 876
Kenne, L., **6,** 406; **9,** 141
Kennedy, A., **1,** 254, 1140
Kennedy, C. D., **4,** 407
Kennedy, E., **6,** 353
Kennedy, E. R., **1,** 505, 506, 508, 509, 511
Kennedy, J. P., **5,** 708
Kennedy, R. J., **1,** 952; **11,** 442
Kennepohl, G. J. A., **6,** 190
Kenner, G. W., **1,** 53, 199, 235, 248, 253, 347, 703, 730, 828, 1041, 1126, 1229, 1237, 1238, 1256; **2,** 133, 166; **3,** 153; **6,** 425; **7,** 138
Kenner, J., **1,** 562; **3,** 74
Kennewell, P. D., **8,** 347; **10,** 181
Kenney, H. E., **2,** 353
Kenny, D. H., **8,** 521
Kent, F. W., **1,** 164, 1081
Kent, G. J., **2,** 225; **3,** 215
Kent, P. W., **2,** 130
Kent, R. E., **1,** 872
Kenyon, J., **4,** 183
Kenyon, W. G., **1,** 1040; **2,** 374
Kenyon, W. O., **1,** 48
Keogh, J., **3,** 264
Kerb, U., **2,** 211, 362; **10,** 216
Kerber, R. C., **6,** 419; **7,** 170; **10,** 129
Kerdesky, F. A. J., **10,** 119; **11,** 599
Kerekes, I., **5,** 171, 539, 577; **9,** 400
Kerekgyarto, J., **11,** 157
Keresztesy, J. C., **1,** 731
Kerfanto, M., **1,** 707
Kergomard, A., **1,** 90, 1084; **5,** 589; **6,** 359
Kerkmann, T., **10,** 243
Kern, J., **6,** 243
Kern, J. W., **1,** 892
Kern, W., **1,** 500
Kerr, G. T., **8,** 263; **9,** 74
Kerrin, S. L., **5,** 336
Kershner, L. D., **5,** 652
Kerton, N. A., **9,** 396
Kerwin, J. F., **1,** 989
Kerwin, J. F., Jr., **9,** 506; **10,** 448; **11,** 20, 334
Kerwood, J. E., **8,** 399
Keshing, H. S., **10,** 448
Kesling, H. S., **9,** 166
Kessar, S. V., **4,** 398
Kesseler, K., **12,** 490, 502
Kessler, H., **1,** 166; **2,** 91; **11,** 106; **12,** 416
Kessler, W., **4,** 360
Kester, M., **5,** 538; **6,** 472
Kestner, M. M., **7,** 170
Ketterman, K. J., **5,** 652
Kettle, S. F. A., **1,** 520; **3,** 212
Keul, H., **6,** 441; **9,** 343
Keumi, T., **9,** 244, 245; **10,** 420; **11,** 256
Keung, E. C. H., **3,** 167
Keutmann, E. H., **1,** 411
Kevill, D. N., **1,** 1137; **8,** 341
Kevon, G. C., **10,** 49
Key, J. M., **3,** 131
Keyton, D. J., **2,** 248
Kezar, H. S., III, **11,** 36
Keziere, R. J., **2,** 153; **3,** 112; **6,** 530
Khafaji, A. N., **1,** 318
Khai, B. T., **10,** 142
Khajavi, M. S., **10,** 115
Khalafinejad, A., **12,** 144
Khalil, H., **6,** 667
Khalil, M. H., **5,** 114
Khan, B. T., **1,** 966
Khan, F. A., **6,** 90
Khan, H. A., **11,** 602
Khan, J. A., **8,** 249; **11,** 551
Khan, M. A., **1,** 1015
Khan, M. N., **6,** 628
Khan, M. N. I., **12,** 333
Khan, N., **6,** 514
Khan, N. A., **1,** 392, 759, 952, 1040; **5,** 486
Khan, N. H., **4,** 369; **5,** 571
Khan, S. A., **2,** 359; **6,** 288
Khand, I. U., **12,** 167

Khanna, J. M., **6,** 326
Khanna, P. L., **10,** 216
Khanna, V. K., **6,** 546; **9,** 151
Kharasch, M. S., **1,** 56, 101, 155, 169, 197, 409, 771, 772, 933, 1131, 1216; **3,** 295; **6,** 450; **7,** 95
Kharasch, N., **1,** 320, 1131; **2,** 21; **3,** 43; **6,** 232
Khathing, D. T., **11,** 453
Khatri, H. N., **9,** 470
Khatri, N. A., **9,** 177
Khawaja, T. A., **1,** 838
Khomutova, E. D., **1,** 277, 675
Khor, T. C., **6,** 454
Khorana, H. G., **1,** 189, 199, 235, 236, 332, 347, 352, 519, 662, 705, 745, 905, 1137, 1138, 1148, 1228, 1229; **2,** 228; **4,** 535; **7,** 414
Khorlina, I. M., **1,** 262, 1030
Khoujah, A. M., **6,** 597
Khouri, F., **10,** 417
Khuddus, M. A., **4,** 194; **5,** 266
Khuong-Huu, F., **4,** 288
Khuong-Huu, Q., **2,** 210; **9,** 416; **11,** 317
Khwaja, T. A., **3,** 223
Kibler, C. J., **1,** 678, 963, 1177
Kice, J. L., **11,** 408
Kidd, D. A. A., **1,** 351
Kidd, J. M., **7,** 324
Kido, F., **4,** 282; **6,** 530, 662; **9,** 516; **11,** 594
Kidokoro, N., **5,** 567
Kidwai, A. R., **4,** 369; **5,** 571
Kidwell, R. L., **2,** 435, 445; **3,** 310
Kieboom, A. P. G., **6,** 149; **10,** 321
Kieczykowski, G. R., **5,** 303, 407, 609; **6,** 312; **8,** 62, 268
Kiedrowski, G. v., **11,** 37
Kiefer, E. F., **1,** 18; **2,** 117
Kiefer, H., **1,** 351
Kiehlmann, E., **6,** 521
Kiel, F., **5,** 639
Kiel, W. A., **8,** 170; **9,** 35; **10,** 21, 452
Kielbania, A. J., Jr., **5,** 574
Kielbasinski, P., **11,** 275
Kienan, E., **12,** 288
Kienzle, F., **3,** 43, 289; **4,** 497, 501; **5,** 657; **7,** 365
Kieper, G., **12,** 353
Kierstead, R. C., **2,** 112
Kierstead, R. W., **1,** 235, 764; **2,** 29
Kierzek, R., **11,** 602
Kiesch, Y., **11,** 139
Kiesel, R. J., **1,** 1193, 1251
Kieslich, K., **1,** 1284
Kiess, A. A., **1,** 1055
Kiessel, M., **1,** 370, 371; **4,** 529; **6,** 24
Kiessling, A. D. A., **5,** 319
Kigasawa, K., **3,** 268
Kii, N., **11,** 91
Kiji, J., **1,** 318; **2,** 304; **5,** 42; **6,** 451
Kikkawa, I., **1,** 1198; **9,** 519; **10,** 194
Kikuchi, K., **6,** 534; **7,** 331; **8,** 451
Kikuchi, S., **12,** 313
Kikuchi, T., **3,** 265; **12,** 548
Kikugawa, K., **4,** 247
Kikugawa, Y., **3,** 264; **5,** 601; **6,** 534; **7,** 325; **8,** 51; **9,** 59; **11,** 69, 483; **12,** 432
Kikukawa, K., **3,** 36; **5,** 498; **10,** 298
Kilbee, G. W., **9,** 161
Kiliani, H., **1,** 144
Kill, R. J., **7,** 16
Killough, J. M., **8,** 238, 406
Kilmer, G. W., **1,** 1121
Kilpper, G., **3,** 220
Kilroy, M., **1,** 892, 986
Kim, B., **6,** 190
Kim, B. H., **9,** 51
Kim, C., **12,** 126
Kim, C. K., **6,** 660; **7,** 421; **8,** 143; **9,** 202, 514
Kim, C. S., **6,** 494
Kim, C. S. Y., **1,** 81, 96
Kim, C. U., **4,** 90, 192, 200, 220; **5,** 132, 748
Kim, C. W., **12,** 272
Kim, D. H., **7,** 1
Kim, J. C., **7,** 236
Kim, J. H., **4,** 254; **5,** 433; **8,** 284
Kim, J. K., **11,** 206
Kim, K., **4,** 52, 520; **5,** 551
Kim, K. H., **7,** 97
Kim, K. S., **8,** 391; **12,** 195
Kim, L., **5,** 740; **9,** 245
Kim, S., **8,** 87; **9,** 242, 301; **10,** 156; **11,** 294, 296, 373; **12,** 12, 83, 102, 193, 252, 276, 350, 446, 479, 573
Kim, S. C., **5,** 565; **8,** 310
Kim, S. W., **11,** 159
Kim, Y. C., **4,** 522; **8,** 382; **12,** 12
Kim, Y. H., **4,** 252; **8,** 204; **9,** 92; **12,** 412
Kim, Y. J., **12,** 446
Kimber, R. W. L., **2,** 336
Kimel, W., **1,** 266
Kimling, H., **3,** 254
Kimmel, T., **12,** 401
Kimoto, S., **2,** 36; **6,** 70
Kimpenhaus, W., **5,** 299
Kimura, J., **4,** 555; **10,** 182
Kimura, K., **11,** 134; **12,** 137
Kimura, M., **5,** 307, 308, 340; **6,** 304; **7,** 385; **8,** 444; **10,** 347; **11,** 324
Kimura, N., **10,** 289
Kimura, S., **12,** 562
Kimura, T., **12,** 199
Kimura, Y., **6,** 601; **8,** 501; **9,** 361; **12,** 546
Kinast, G., **7,** 339
Kincaid, K., **7,** 188
Kincl, F. A., **1,** 219
Kindler, K., **1,** 730, 782, 1121
King, A. O., **7,** 113; **8,** 87, 475, 507; **10,** 243, 423; **11,** 299, 514
King, B. T., **10,** 201
King, F. D., **5,** 45; **6,** 57, 58, 185, 652
King, F. E., **1,** 452
King, G., **1,** 818
King, H., **1,** 413, 1074

King, H. S., **1,** 500, 685, 864
King, J. F., **5,** 478, **6,** 190; **9,** 212
King, L. C., **1,** 162, 504; **2,** 343
King, P. F., **8,** 245
King, R. B., **5,** 610, 744; **8,** 41; **10,** 382
King, R. R., **9,** 481
King, R. W., **1,** 695, 841; **5,** 276; **6,** 469
King, T. J., **1,** 933; **8,** 37
King, T. P., **2,** 257
King, W., **2,** 363
King, W. B., **1,** 1023, 1073
Kingsbury, C. A., **1,** 309, 1145
Kingsbury, W. D., **3,** 47; **4,** 79
Kingsley, R. B., **1,** 129
Kinloch, E. F., **5,** 142
Kinnear, A. M., **1,** 666; **2,** 281
Kinnear, A. N., **3,** 105
Kinney, C. R., **1,** 156; **4,** 163
Kinney, R. E., **3,** 69
Kinney, R. J., **8,** 469
Kinney, W. A., **11,** 475
Kinnick, M. D., **5,** 183; **6,** 468; **8,** 168
Kinomura, Y., **2,** 36
Kinoshita, H., **4,** 101; **6,** 597; **8,** 223, 231; **11,** 514; **12,** 155
Kinoshita, K., **10,** 289
Kinoshita, M., **5,** 717; **6,** 257, 561; **8,** 509; **9,** 382, 491
Kinsel, E., **4,** 302; **5,** 406
Kinsella, M. A., **11,** 203
Kinson, P. L., **4,** 101; **5,** 505
Kinugawa, Z., **1,** 52
Kinzig, C. M., **6,** 17
Kipphardt, H., **12,** 31
Kipping, F. S., **5,** 339
Kiprianov, A. I., **9,** 305
Kira, M., **5,** 620
Kirahonoki, K., **3,** 46
Kirby, G. W., **1,** 320, 394; **2,** 263; **3,** 332; **6,** 581; **9,** 449; **11,** 562
Kirby, J. E., **4,** 163
Kirby, R. H., **1,** 424, 573
Kirch, L., **1,** 155
Kircher, H. W., **5,** 66, 690, 740
Kirchhof, W., **1,** 262, 478, 790
Kirchhoff, R., **7,** 418
Kirchhoff, R. A., **5,** 459; **6,** 395
Kirchner, F. K., **1,** 672; **3,** 49
Kirchner, J. R., **1,** 503
Kirchoff, R. A., **9,** 194
Kirk, D. N., **1,** 219, 285, 355, 361; **3,** 48; **5,** 370
Kirk, K. L., **4,** 282
Kirk, T. C., **11,** 12; **12,** 11
Kirkemo, C. L., **7,** 406
Kirkham, L. H., **2,** 201, 381; **3,** 265
Kirkolev, V. P., **4,** 562
Kirkpatrick, D., **4,** 293, 552; **6,** 208
Kirms, M. A., **12,** 426
Kirmse, W., **2,** 422; **4,** 512; **5,** 593; **10,** 121
Kirner, W. R., **1,** 1163
Kirpal, A., **1,** 465
Kirrmann, A., **1,** 385
Kirrstetter, R. G. H., **8,** 306
Kirsanow, A. V., **6,** 186
Kirsch, G., **2,** 345
Kirsch, P., **5,** 143
Kirschke, K., **5,** 503
Kirschleger, B., **12,** 77
Kirshenbaum, K., **12,** 247
Kirson, I., **12,** 475
Kirst, H. A., **2,** 241; **3,** 174; **4,** 355
Kirtley, M., **1,** 13
Kis, Z., **10,** 159
Kisch, H., **10,** 274
Kise, M. A., **1,** 754, 1270
Kiseleva, O. A., **11,** 341
Kishi, K., **9,** 314
Kishi, Y., **1,** 374; **4,** 86; **5,** 183; **6,** 611; **7,** 2, 135; **8,** 1, 64, 102; **9,** 110, 138, 174; **10,** 346; **11,** 20, 75, 85, 95, 134; **12,** 359, 568
Kishida, Y., **4,** 199; **5,** 448, 459; **6,** 399; **8,** 342
Kishimura, K., **9,** 362; **10,** 403
Kisielowski, L., **7,** 408
Kiso, Y., **3,** 197; **5,** 87, 361; **6,** 51, 59; **9,** 297, 464; **10,** 398; **11,** 168
Kispersky, J. P., **1,** 905
Kiss, J., **1,** 76, 736
Kissel, W. J., **2,** 238
Kissinger, L. W., **1,** 50, 755
Kissman, H., **7,** 268
Kissman, H. M., **1,** 61, 904
Kita, K., **11,** 489
Kita, Y., **5,** 433; **8,** 519; **9,** 111, 310, 506, 507; **11,** 255, 279, 340; **12,** 59, 221, 449, 559
Kitada, C., **11,** 330
Kitade, K., **9,** 297
Kitagawa, I., **1,** 309
Kitagawa, K., **6,** 618; **9,** 297, 464
Kitagawa, T., **12,** 59, 573
Kitagawa, Y., **6,** 328; **7,** 391; **8,** 164, 165, 472, 476
Kitahara, T., **6,** 372; **7,** 103, 114, 132, 233; **8,** 330
Kitahara, Y., **2,** 37, 129; **4,** 113, 316; **5,** 182, 266, 649; **6,** 170, 554; **7,** 352
Kitahonoki, K., **1,** 595; **2,** 242; **5,** 713
Kitai, M., **6,** 281, 295
Kitajima, H., **10,** 420
Kitajima, K., **11,** 109
Kitajima, M., **10,** 398
Kitajima, T., **11,** 522; **12,** 523
Kitakana, T., **8,** 176
Kitamoto, M., **7,** 17
Kitamura, K., **12,** 445
Kitamura, M., **9,** 188, 311; **11,** 99; **12,** 325
Kitao, T., **2,** 9; **11,** 140
Kitaoka, Y., **2,** 9
Kitatani, K., **5,** 742; **6,** 91, 385; **7,** 93; **8,** 292
Kitatsuji, E., **6,** 403
Kitazuma, S., **5,** 332
Kitazume, T., **10,** 44; **11,** 599
Kitchin, J. P., **9,** 335; **11,** 587
Kitching, W., **1,** 778; **2,** 267; **4,** 322; **5,** 497
Kitoh, R., **7,** 361

Kitomoto, M., **8,** 197
Kiuchi, K., **5,** 614
Kivikari, R., **10,** 296
Kivity, S., **2,** 333
Kiyohara, Y., **6,** 187
Kiyoi, R., **1,** 668
Kiyokawa, H., **9,** 111
Kiyooka, S., **12,** 150
Kjaer, A., **5,** 169
Kjeldsen, G., **12,** 29
Kjell, D. P., **11,** 203
Kjonaas, R. A., **8,** 178; **12,** 1
Klabunde, K. J., **5,** 419; **7,** 52; **11,** 309
Klaerner, F. G., **2,** 99; **4,** 459
Klager, K., **1,** 185; 905
Klages, F., **1,** 1188
Klamann, D., **1,** 756, 966; **2,** 197; **3,** 140; **6,** 263
Klanderman, B. H., **1,** 1102; **5,** 600
Klapproth, W. S., **2,** 310
Klasinc, L., **11,** 124
Klass, D. L., **1,** 198, 1004, 1128
Klauenberg, G., **2,** 378
Klauke, E., **2,** 49, 54; **4,** 65
Klausner, Y., **9,** 180
Klausner, Y. S., **6,** 288; **7,** 79
Klayman, D. L., **4,** 8; **6,** 543; **8,** 310
Klebe, J. F., **1,** 61
Klee, W., **1,** 98
Klei, B., **8,** 40
Kleid, D. G., **7,** 414
Kleiderer, E. C., **1,** 731, 951
Kleijn, H., **8,** 322; **9,** 277, 334; **10,** 290, 350; **11,** 13
Klein, D. A., **4,** 241
Klein, E., **1,** 444; **9,** 391
Klein, G. W., **2,** 26; **4,** 239
Klein, H., **6,** 441; **7,** 164, 226; **9,** 11; **11,** 604; **12,** 576
Klein, H. A., **10,** 146; **12,** 189
Klein, H. C., **1,** 500
Klein, J., **1,** 465, 1005; **3,** 124; **4,** 126, 183; **5,** 86
Klein, K. P., **2,** 173; **6,** 611; **7,** 347
Klein, M. P., **1,** 894
Klein, M. W., **9,** 220
Klein, O., **1,** 1101, 1247
Klein, P., **10,** 172; **11,** 248
Klein, R. A., **1,** 689; **2,** 109, 119
Klein, R. S., **9,** 399; **11,** 62
Klein, S. A., **6,** 534
Klein, U., **6,** 304
Kleinberg, J., **1,** 182, 506, 1195; **2,** 369
Kleiner, H. J., **6,** 375; **10,** 333
Kleinfelter, D. C., **1,** 758; **3,** 238
Kleinschmidt, R. F., **1,** 103, 848; **4,** 240
Kleinstueck, R., **4,** 550, 552; **5,** 727
Kleis, A. A. B., **1,** 1167; **3,** 291
Klemka, A. J., **1,** 1029
Klemm, K., **1,** 1252
Klemm, W., **5,** 286
Klemmensen, P., **1,** 957
Klemperer, W. A., **9,** 448
Klender, G. J., **4,** 162
Klenk, H., **5,** 702
Kleschick, W. A., **8,** 299; **9,** 88; **10,** 133; **11,** 349
Klessen, C., **8,** 537
Kleveland, K., **5,** 454; **6,** 558
Kliebisch, U., **12,** 187
Klieger, E., **1,** 231
Klima, W. L., **8,** 87; **10,** 243
Klimov, E. M., **4,** 333
Klimstra, P. D., **1,** 1011, 1091
Klinck, J., **1,** 430
Kline, E. R., **1,** 35
Kline, G. B., **1,** 970
Kling, T. A., **4,** 14
Klingemann, F., **4,** 369
Klinger, H., **2,** 198
Kliss, R. M., **1,** 1022
Klobucar, W. D., **7,** 65
Kloc, K., **9,** 99; **10,** 348, 354
Kloesgen, G., **5,** 639
Kloetzel, M. C., **2,** 363
Kloetzer, W., **9,** 506; **11,** 222
Klohs, M. W., **1,** 367
Klonowski, R. S., **1,** 927
Kloosterziel, H., **1,** 1108; **2,** 431
Kloppner, E., **11,** 167, 222
Klosa, J., **1,** 1184
Klose, T. R., **6,** 264, 615
Klose, W., **9,** 90; **10,** 451; **11,** 281
Klos-Ringuet, M., **12,** 420
Kloster-Jensen, E., **7,** 428
Klostermeyer, H., **5,** 700
Klotz, I. M., **1,** 13
Kluba, M., **5,** 650; **9,** 444
Kluender, H. C., **4,** 194
Klug, H. H., **7,** 269
Klug, J. T., **10,** 184
Klug, W., **4,** 133
Kluge, A. F., **5,** 120, 360; **9,** 163; **11,** 182
Kluge, M., **1,** 445
Kluiber, R. W., **1,** 86, 211, 213, 524
Klumpp, G. W., **6,** 143
Klun, T. P., **9,** 458
Klunder, J. M., **12,** 94
Klundt, I. L., **7,** 58
Klutchko, S., **2,** 154
Klyne, W., **2,** 75
Knaack, D. F., **1,** 1132
Knabe, J., **1,** 652
Knaggs, J. A., **4,** 552
Knapp, D. R., **5,** 508
Knapp, K. H., **1,** 563
Knapp, S., **7,** 130; **10,** 16, 454; **11,** 260; **12,** 48, 78
Knaus, G., **6,** 204, 389; **7,** 231; **11,** 233; **12,** 383
Knauss, E., **1,** 343
Kneidl, F., **4,** 172
Knevel, A. M., **1,** 1210
Kniel, P., **1,** 270
Knifton, J. F., **6,** 655
Knight, D. W., **7,** 291
Knight, H. B., **1,** 759, 1142, 1269; **3,** 279
Knight, J. C., **1,** 207, 989; **3,** 275; **9,** 277
Knight, J. W., **8,** 387
Knight, M. K., **8,** 466
Knights, E. F., **2,** 31; **3,** 28; **5,** 47; **6,** 64; **8,** 6, 334
Knoche, H., **3,** 88
Knochel, P., **11,** 173, 361
Knoeber, M., Sr., **2,** 92
Knoevenagel, E., **6,** 260; **8,**

163
Knof, L., **1,** 679
Knoflach, J., **11,** 222
Knoll, F., **7,** 97
Knoll, F. M., **8,** 317; **9,** 44; **10,** 32
Knolle, J., **8,** 180
Knopka, W. N., **1,** 904
Knorr, E., **4,** 67
Knorr, L., **6,** 228
Knorr, R., **5,** 339
Knoth, W. H., **4,** 543; **6,** 81
Knothe, L., **2,** 67, 454
Knotnerus, J., **1,** 105
Knouzi, N., **12,** 551
Knowles, J. R., **9,** 394
Knowles, W. S., **1,** 94, 802, 957, 1235; **5,** 502; **11,** 31
Knox, G. A., **1,** 309, 926, 927; **10,** 223; **12,** 167
Knox, L. H., **1,** 109, 249, 802; **2,** 381; **3,** 265; **7,** 369
Knox, S. D., **6,** 47
Knudsen, J. S., **12,** 29
Knudsen, M. T., **12,** 167
Knupp, G., **11,** 411
Kny, H., **1,** 93, 276
Ko, A. I., **10,** 181
Ko, K. Y., **12,** 239
Ko, S. S., **11,** 95; **12,** 387
Kobal, V. M., **8,** 305
Kobayakawa, S., **9,** 257
Kobayashi, C. S., **6,** 642
Kobayashi, H., **10,** 8
Kobayashi, J., **6,** 152
Kobayashi, K., **8,** 119, 299; **11,** 109
Kobayashi, M., **5,** 210; **6,** 165, 641; **7,** 170; **8,** 533; **9,** 296; **10,** 391, 423; **11,** 131
Kobayashi, N., **9,** 403
Kobayashi, R., **4,** 158; **8,** 498
Kobayashi, S., **5,** 80, 360, 572; **6,** 291; **7,** 110; **8,** 96, 225, 253, 424, 501; **9,** 20; **10,** 449; **11,** 267, 330, 334; **12,** 256, 549
Kobayashi, S. K., **9,** 414
Kobayashi, T., **6,** 554; **7,** 352; **8,** 163; **9,** 208; **10,** 38, 236; **11,** 350, 373; **12,** 89, 172, 526
Kobayashi, Y., **2,** 136; **5,** 148, 621; **7,** 79; **8,** 127; **9,** 486; **10,** 110, 281, 298; **11,** 514; **12,** 171
Kober, W., **7,** 390; **11,** 587
Kobig, J., **10,** 186
Kobler, H., **6,** 569
Kobori, Y., **12,** 172
Kobrina, N. S., **5,** 424
Kobuke, Y., **9,** 96
Kobylecki, R. J., **6,** 147
Kocevar, M., **9,** 472
Koch, D., **7,** 260
Koch, G., **11,** 352
Koch, H., **1,** 407
Koch, H. P., **1,** 144
Koch, P., **6,** 509
Koch, S. D., **1,** 1022
Koch, T. H., **8,** 46
Koch, W. G., **1,** 619
Kocher, F. W., **1,** 74
Kocheshkov, Z., **11,** 323
Kochetkov, N. K., **1,** 277, 675; **4,** 333; **12,** 198
Kochhar, K. S., **12,** 1, 121
Kochhar, R., **3,** 243; **7,** 111
Kochi, J. K., **1,** 89, 101, 162, 563, 1081; **2,** 84, 87, 89, 238, 406; **3,** 59, 171, 239; **4,** 71, 164, 236, 281, 282; **5,** 146, 152, 370; **6,** 57, 133, 259, 624, 665; **7,** 411; **10,** 59, 140
Kochloefl, K., **3,** 261
Kochmann, E. L., **1,** 828
Koch-Pomeranz, U., **5,** 588
Kochs, P., **8,** 465
Kocienski, P. J., **4,** 405; **6,** 479; **7,** 301; **8,** 357; **9,** 446; **10,** 314; **11,** 67, 475, 576; **12,** 444, 502
Kock-van Dalen, A. C., **10,** 321
Kocor, M., **1,** 1000; **4,** 507; **5,** 353; **6,** 10
Kocovsky, P., **12,** 373
Kodama, H., **1,** 601; **8,** 289; **9,** 277
Kodama, M., **6,** 90; **7,** 47; **9,** 292
Kodama, S., **6,** 51; **11,** 168
Kodama, T., **7,** 235
Kodera, M., **11,** 430
Kodicek, E., **4,** 134
Koebernick, W., **6,** 545
Koebner, A., **1,** 904
Koebrich, G., **1,** 224, 1196; **2,** 119; **3,** 89; **4,** 139; **5,** 120, 200, 277; **6,** 171; **7,** 47; **8,** 359
Koechlin, B. A., **1,** 74, 1178
Koegler, H., **1,** 652
Koehl, W. I., Jr., **6,** 356
Koehl, W. J., Jr., **1,** 97; **2,** 264
Koehler, R. E., **4,** 60
Koelbl, H., **5,** 615
Koell, P., **4,** 86; **5,** 601, 702; **7,** 290
Koelliker, U., **4,** 105, 376
Koelsch, C. F., **1,** 57; **2,** 414
Koelsch, P. M., **8,** 196
Koeners, H. J., **10,** 231
Koeng, F. R., **5,** 33, 760; **6,** 675
Koenig, C., **1,** 364; **2,** 154
Koenig, F., **9,** 376, 439
Koenig, G., **6,** 618
Koenig, H., **1,** 318; **10,** 270
Koenig, K. E., **5,** 644; **6,** 35
Koenig, K. H., **1,** 561, 1015
Koenig, P. E., **1,** 618; **2,** 251
Koenig, W., **1,** 486; **2,** 134; **3,** 103, 156; **5,** 342; **6,** 288
Koenig, W. A., **5,** 701; **11,** 278
Koenigkramer, R. E., **10,** 145
Koenigs, W., **4,** 67
Koenigstein, O., **1,** 500
Koepke, B., **11,** 55
Koeppelmann, E., **8,** 211
Koerding, P., **1,** 992
Koermendy, K., **1,** 34
Koertge, T. E., **7,** 158
Koester, H., **1,** 1210; **3,** 41; **10,** 461; **11,** 12
Koester, R., **3,** 178, 187, 310; **4,** 405; **5,** 392, 544, 551, 593
Kofler, M., **1,** 567, 703
Kofod, H., **1,** 1167

Kofron, W. G., **3,** 232
Koga, K., **4,** 129, 270; **5,** 103; **6,** 307; **7,** 189, 325; **8,** 272, 424; **9,** 290, 393; **10,** 230, 238, 332; **11,** 447; **12,** 563
Koga, N., **2,** 178
Kogan, I. C., **1,** 601
Kogan, T. M., **11,** 315
Kogan, T. P., **12,** 298
Koganty, R. R., **5,** 249
Kogen, H., **10,** 230
Kogler, H., **12,** 416
Kogura, T., **4,** 563
Kogure, T., **5,** 740; **6,** 653; **8,** 59; **11,** 87, 131; **12,** 239
Koh, H. S., **2,** 263
Kohara, H., **9,** 51
Koharski, D., **3,** 264
Kohda, A., **12,** 231
Kohen, F., **1,** 281
Kohl, H., **4,** 296
Kohler, E. P., **1,** 195, 470, 713, 1049; **3,** 74
Kohler, H., **1,** 695
Kohli, V., **10,** 461
Kohn, E. J., **1,** 730
Kohn, H., **12,** 147
Kohn, P., **1,** 59
Kohno, M., **12,** 428
Kohno, S., **12,** 187
Kohno, T., **11,** 1
Koide, H., **5,** 370
Koike, S., **11,** 253
Koinuma, H., **10,** 274
Koizumi, M., **4,** 407
Koizumi, N., **7,** 143; **8,** 514; **10,** 149
Koizumi, T., **6,** 317, 403, 649; **7,** 282; **9,** 117; **11,** 539; **12,** 228
Koji, A., **11,** 155
Kojic-Prodic, B., **11,** 124
Kojima, H., **5,** 146; **7,** 146
Kojima, T., **2,** 23; **9,** 431
Kojima, Y., **4,** 222; **12,** 363
Kok, P., **9,** 116
Kokko, B. J., **11,** 323; **12,** 205
Kokubu, Y., **11,** 141
Kol, M., **10,** 1
Kolani, N., **11,** 64
Kolb, M., **4,** 285; **6,** 55, 322; **10,** 153
Kolbah, D., **6,** 474; **9,** 234
Kole, S. L., **12,** 267
Kolesnikov, V. V., **4,** 441
Kolind-Andersen, H., **6,** 565
Koll, W., **1,** 247
Kollek, L., **5,** 322
Koller, F., **1,** 664; **2,** 329
Kollmeier, H. J., **11,** 401
Kollonitsch, J., **1,** 603; **2,** 27, 57, 245, 412; **6,** 561; **8,** 251
Kolm, H., **1,** 1247
Kolobielski, M., **1,** 986
Kolodkina, I. I., **3,** 222
Kolomnikov, I. S., **4,** 562
Kolonitis, P., **2,** 142
Kolonko, K. J., **6,** 600; **8,** 492
Kolosov, M. N., **5,** 434
Kolsaker, P., **5,** 320, 495
Kolshorn, H., **9,** 410
Kolthoff, I. M., **1,** 951
Komamura, T., **9,** 415
Komar, D. A., **9,** 471
Komatsu, M., **6,** 158
Komatsu, T., **11,** 30; **12,** 15, 82
Komeno, T., **5,** 55
Komeshima, N., **9,** 290
Kometani, T., **7,** 160
Komin, J. B., **6,** 78
Komiyama, E., **6,** 547
Komiyama, M., **12,** 151
Komori, S., **5,** 336, 534
Komoto, R. G., **4,** 465
Kon, G. A. R., **1,** 182, 935
Konaka, R., **1,** 732
Kondo, A., **6,** 297
Kondo, F., **9,** 255; **11,** 203
Kondo, H., **11,** 494
Kondo, I., **5,** 667
Kondo, J., **11,** 213
Kondo, K., **3,** 219; **4,** 422; **5,** 341, 467, 575, 749; **6,** 436, 462, 509, 610, 662; **7,** 269; **8,** 173, 421, 454; **9,** 200, 356, 434, 458, 497; **10,** 165, 206; **11,** 180, 184, 466, 498
Kondo, M., **5,** 477
Kondo, S., **8,** 516
Kondo, T., **6,** 44
Kondratenko, N. V., **6,** 621
Konecky, M. P., **1,** 905
Konen, D. A., **4,** 428
Konieczny, M., **9,** 238; **10,** 200
Konig, W., **1,** 116
Konigsberg, M., **1,** 754
Konigsberg, N., **1,** 80
Koning, B. H., **8,** 225
Konishi, A., **8,** 40
Konishi, H., **3,** 212; **5,** 560
Konishi, M., **9,** 147; **10,** 38; **11,** 20, 239; **12,** 172
Kono, D. H., **5,** 731
Kono, H., **6,** 665
Konoike, T., **5,** 150; **6,** 141, 517; **7,** 322; **8,** 127
Konoishi, M., **11,** 240
Konopinska, D., **2,** 302
Konrad, P., **1,** 11
Konstantatos, J., **9,** 513
Konstas, S., **2,** 180
Kontowicz, A. D., **1,** 904
Konz, W. E., **4,** 383, 405
Koo, J., **1,** 890, 904; **2,** 216
Kooistra, D. A., **4,** 14, 530; **5,** 695
Koolpe, G. A., **9,** 374, 384; **10,** 324
Koonsvitsky, B. P., **5,** 86
Koopman, H., **1,** 519
Koos, E. W., **4,** 119
Kopecky, J., **2,** 210, 336; **3,** 96
Kopecky, K. R., **6,** 288; **8,** 45, 374
Kopet, R., **3,** 6
Kopka, I., **11,** 6
Kopp, R., **6,** 479
Koppe, V., **1,** 350
Koppel, G. A., **4,** 60; **5,** 183; **6,** 468; **8,** 2, 168; **9,** 75
Koppes, W. M., **5,** 6, 731; **6,** 646
Kopple, K. D., **1,** 274; **6,** 307
Kopylov-Shakmatov, N. N., **3,** 302
Korach, M., **1,** 182; **2,** 309
Koransky, W., **1,** 971; **2,** 354
Kordish, R. J., **11,** 551

Koreeda, M., **6,** 202; **7,** 143; **8,** 514; **10,** 259; **11,** 365, 543; **12,** 44, 205, 532
Korenowski, T. F., **8,** 120
Kormendy, M. F., **1,** 318, 1059, 1220; **2,** 381
Korn, A., **9,** 87
Kornblum, N., **1,** 281, 309, 310, 491, 519, 755, 952, 1011, 1018, 1027, 1211, 1250; **4,** 461; **5,** 566, 627; **6,** 229, 419; **7,** 170; **8,** 18, 74; **9,** 434; **10,** 331; **11,** 441, 551; **12,** 366
Kornet, M. J., **1,** 337, 519; **2,** 107; **5,** 23
Kornfeld, E. C., **1,** 573, 731, 970; **4,** 318
Korobko, V. G., **5,** 434
Korosec, P. S., **1,** 295
Korsakoff, L., **3,** 81
Korte, D. E., **7,** 250; **8,** 324
Korte, F., **2,** 279
Korte, W. D., **7,** 388
Korth, J., **5,** 709
Korth, T., **6,** 141
Korti, F., **2,** 179
Kortylewicz, Z., **11,** 495
Koruda, M., **11,** 4
Korytnyk, W., **1,** 206; **2,** 23; **8,** 167; **9,** 297
Korzeniowski, S. H., **5,** 763; **7,** 287; **8,** 130, 255; **9,** 361; **12,** 381
Kosak, A. I., **1,** 500
Kosarych, Z., **10,** 98; **11,** 182, 329
Kosel, C., **1,** 1075
Koshimies, J. K., **8,** 509
Koshino, J., **9,** 373
Koshland, D. E., Jr., **1,** 485
Koskinen, L., **10,** 359
Kosley, R. W., Jr., **6,** 610; **7,** 125
Kosma, S., **3,** 32
Kosolapoff, G. M., **1,** 334; **2,** 281; **4,** 350
Kosower, E., **1,** 531, 1073
Kosower, E. M., **1,** 518; **2,** 87; **4,** 439; **5,** 160
Kosswig, K., **1,** 790; **5,** 77
Kostanecki, S. von, **4,** 15
Koster, D. F., **4,** 231
Koster, J. B., **6,** 575
Koster, W., **7,** 244
Kostova, K., **12,** 462
Kosuda, S., **4,** 345
Kosugi, H., **6,** 464, 466, 662; **7,** 235
Kosugi, M., **11,** 16; **12,** 174
Koszalka, T. R., **1,** 636
Koszuk, J., **10,** 181
Kotake, H., **6,** 597; **8,** 231; **11,** 514; **12,** 155
Kotaki, H., **8,** 223
Kotani, E., **4,** 237
Koteel, C., **8,** 417
Kotera, K., **1,** 595; **2,** 242
Koto, S., **10,** 60
Kotodziejczyk, A. M., **12,** 222
Kotsuji, K., **10,** 171
Kotsuki, H., **6,** 529; **12,** 70
Kottenhahn, A. P., **1,** 928
Kottwitz, J., **6,** 443
Koul, V. K., **10,** 169
Koulkes, M., **1,** 652
Koutek, B., **3,** 115
Kouwenhoven, A. P., **5,** 19; **6,** 252; **7,** 146
Kouwenhoven, C. G., **5,** 230
Kovac, L. R., **1,** 935
Kovache, A., **1,** 407
Kovacic, P., **1,** 162, 264, 484, 1194; **2,** 55, 142, 425; **3,** 295; **4,** 14, 520; **5,** 103, 359; **6,** 146
Kovacs, G., **7,** 160; **8,** 260; **10,** 240
Kovacs, J., **1,** 783
Kovacs, K., **3,** 317
Kovar, R., **7,** 197
Kovelesky, A. C., **3,** 284; **4,** 484
Koveshnikova, G. M., **10,** 419
Kovitch, G. H., **6,** 190
Kow, R., **4,** 311; **5,** 417
Kowalski, C., **6,** 101; **8,** 299
Kowalski, C. J., **10,** 77, 312; **11,** 105
Kowalski, J., **9,** 372, 440
Kowanko, N., **1,** 977
Kowar, T., **5,** 568
Koyama, K., **11,** 429
Koyama, T., **11,** 131
Koyama, Y., **9,** 167
Koyanagi, M., **11,** 434
Kozar, L. G., **6,** 339
Kozarich, J. W., **5,** 182, 412; **6,** 159
Kozeschkow, K. A., **1,** 349
Kozikowski, A. P., **5,** 9, 70; **6,** 50, 292, 310; **7,** 161; **8,** 22, 183, 325, 415; **9,** 176; **10,** 66, 107, 110, 161, 232, 330; **11,** 20, 317, 345, 458; **12,** 305, 387, 546
Kozlowski, J., **11,** 373
Kozlowski, J. A., **12,** 350
Kozlowski, M. A., **2,** 377
Kozuka, A., **4,** 428
Kozuka, S., **2,** 101; **9,** 66; **10,** 399
Kozyrod, R. P., **11,** 32
Kozyukov, V. P., **10,** 432
Kraak, A., **2,** 87
Kraay, R. J., **1,** 970; **2,** 17, 351
Krabbenhoft, H. O., **7,** 161; **9,** 103
Kraegeloh, K., **10,** 441
Kraemer, G., **1,** 832, 1179
Kraetz, O., **1,** 966
Krafft, F., **5,** 281
Krafft, G. A., **11,** 209
Krafft, M. E., **12,** 75, 265
Kraft, H., **1,** 536
Kraft, L., **1,** 562
Kraiss, G., **5,** 118, 225
Krakower, G. W., **1,** 617; **2,** 173
Kramar, V., **1,** 1258
Kramer, B. D., **2,** 118
Kramer, D. N., **1,** 257
Kramer, G. W., **6,** 64; **8,** 49; **9,** 58; **10,** 77
Kramer, J. D., **7,** 118
Kramer, P. A., **5,** 19
Kramer, R., **8,** 117
Kramer, U., **9,** 379
Krantz, A., **5,** 639; **6,** 570

Krantz, A. B., **1,** 1002
Krantz, K. D., **1,** 1041
Krapcho, A. P., **1,** 56; **2,** 129, 382; **4,** 445, 448; **5,** 23, 266, 295; **6,** 91
Kratovil, E. J., **1,** 140
Kratzer, J., **1,** 756
Kratzer, O., **7,** 408
Kratzer, P., **1,** 882
Kratzer, R. H., **7,** 299
Kratzin, H., **7,** 97
Kratzl, K., **6,** 503
Kraus, C. A., **4,** 478
Kraus, G. A., **6,** 339; **7,** 199, 232; **8,** 59, 260, 291; **9,** 298; **10,** 1, 219, 351, 354, 418; **11,** 20, 232, 469, 481; **12,** 258, 263, 403
Kraus, J. G., **5,** 433
Kraus, J. L., **6,** 574
Kraus, M., **3,** 261
Kraus, W., **6,** 596
Krause, H. J., **1,** 155
Krause, J. F., **1,** 927
Krause, J. G., **6,** 456
Krause, M., **4,** 435
Krause, R. F., **1,** 849
Krauss, H. L., **1,** 151
Krausz, F., **2,** 229
Krausz, P., **6,** 658
Krauth, C. A., **1,** 477
Kray, W. C., Jr., **1,** 151
Kraychy, S., **1,** 389
Kreageloh, K., **11,** 587
Krebs, A., **1,** 343, 832, 1203; **3,** 254; **4,** 491
Krebs, A. W., **3,** 139
Krebs, E. P., **11,** 299
Krebs, L. S., **1,** 444
Kredel, J., **2,** 23
Kreevoy, M. M., **11,** 479
Kreft, A., **8,** 174
Kreft, A. F., III, **9,** 438
Kreider, E. M., **8,** 245
Kreiser, W., **4,** 577; **5,** 631; **10,** 243
Kreiter, C. G., **11,** 401
Kremer, C. B., **1,** 1158
Kremer, K. A. M., **10,** 129
Kremers, E., **1,** 1101
Krepski, L. R., **7,** 369; **8,** 156; **9,** 154, 180, 467; **12,** 235
Kresge, A. J., **2,** 310
Krespan, C. G., **1,** 426, 563, 1136; **2,** 125; **11,** 403
Kress, T. J., **2,** 391
Kresze, G., **1,** 98; **5,** 682; **7,** 50; **8,** 43; **9,** 75; **12,** 58
Kretchmer, R. A., **4,** 405, 479; **7,** 74
Kreutzberger, A., **2,** 423
Kreuz, K. L., **4,** 444
Kreuzbichler, L., **9,** 56
Kreyssler, H. A., **1,** 78
Kricheldorf, H. R., **2,** 438; **4,** 360; **5,** 534; **6,** 561; **7,** 378; **9,** 74; **10,** 59
Kricka, L. J., **4,** 502; **5,** 735
Krief, A., **5,** 66; **6,** 29, 90, 240, 362, 488; **7,** 35, 184, 195, 268, 367; **8,** 64, 199, 245, 299, 320, 402, 537; **9,** 32, 34, 87, 204, 235, 283, 307; **10,** 56, 174, 219, 233, 312, 319, 368, 454; **11,** 225, 301; **12,** 20, 219, 295
Krieger, J. K., **5,** 454; **6,** 495; **7,** 83, 141
Kriegesmann, R., **9,** 509; **11,** 424
Kriesel, D. C., **7,** 37
Krikorian, S. E., **1,** 1108
Krimen, L. I., **2,** 12; **3,** 4; **10,** 14
Krimer, M. Z., **11,** 163
Krinsky, P., **8,** 451
Krishnamurthy, H. G., **8,** 234
Krishnamurthy, S., **3,** 176; **4,** 312, 313, 314; **5,** 48, 565; **6,** 64, 349; **7,** 30, 216, 217; **8,** 49, 303; **9,** 287; **10,** 77, 237, 250; **11,** 292, 305
Krishnan, S., **8,** 248, 463
Krishna-Rao, G. S., **5,** 74; **6,** 498
Kristensen, J., **9,** 50, 283
Kritchevsky, T. H., **1,** 74, 1178
Krivun, S. V., **1,** 802
Kriz, J., **2,** 172, 241
Kroboth, K. C., **5,** 249
Kroehnke, E., **4,** 392
Kroehnke, F., **1,** 504, 747, 1188
Kroener, M., **4,** 35
Kroeninger, A., **4,** 364
Kroeper, H., **1,** 723
Krogh, J. A., **9,** 410
Krohn, K., **9,** 57, 228; **12,** 48
Krolikiewicz, K., **6,** 640; **9,** 22, 496; **11,** 588; **12,** 150, 239
Krolski, M. E., **11,** 232
Krommer, H., **5,** 168
Kronberger, K., **3,** 20, 153
Kropp, J. E., **2,** 225; **3,** 163, 215
Kropp, P. J., **1,** 1184; **9,** 171
Krow, G., **5,** 682; **7,** 395
Krow, G. R., **5,** 283; **6,** 383
Krsek, G., **1,** 228, 730
Kruber, O., **1,** 933
Krubiner, A., **5,** 225
Krubsack, A. J., **5,** 667
Kruck, P., **1,** 562
Krueck, K. H., **1,** 671, 1246
Krueerke, U., **1,** 228
Krueger, C. R., **1,** 1047; **5,** 719; **11,** 68
Krueger, D. S., **7,** 430; **8,** 513
Krueger, P. A., **1,** 502
Krueger, R. A., **2,** 422
Krueger, R. C., **12,** 1
Krueger, W., **6,** 511
Kruger, G., **2,** 344; **5,** 394
Kruger, J. A., **6,** 675
Kruizinga, R. M., **11,** 115
Kruizinga, W. H., **9,** 100; **11,** 118
Krull, I. S., **3,** 234
Krumkalns, E. V., **1,** 15
Kruse, C. G., **7,** 357; **8,** 90; **9,** 112, 450; **11,** 180
Kruse, C. W., **1,** 103, 527, 935
Kruse, L. I., **7,** 335; **12,** 111
Krutzsch, H. C., **2,** 444; **4,** 83
Krynitsky, J. A., **1,** 266, 531, 870, 1117, 1163
Kryshtal, G. V., **9,** 361
Krzikalla, H., **1,** 48, 1133
Ksander, G. M., **7,** 106; **9,**

207; **11,** 421
Ku, A. T., **5,** 310; **6,** 24
Ku, T., **6,** 230
Kubela, R., **10,** 55
Kubik, D. A., **1,** 195; **4,** 267; **5,** 356
Kubitzek, H., **9,** 167
Kubler, D. G., **11,** 469
Kubo, A., **8,** 115
Kubo, I., **11,** 281
Kubo, Y., **9,** 216
Kubota, H., **12,** 275
Kubota, I., **3,** 43
Kubota, K., **9,** 519
Kubota, M., **10,** 454
Kubota, T., **10,** 44
Kubota, Y., **12,** 384
Kucerovy, A., **12,** 420
Kuch, V. J., **5,** 19
Kucherov, V. F., **4,** 325; **5,** 424; **6,** 520; **9,** 361
Kuchinke, E., **1,** 938
Kucinski, P., **4,** 322
Kudo, H., **5,** 602; **9,** 421
Kudo, M., **8,** 40
Kudo, Y., **5,** 210
Kuebel, B., **8,** 396; **10,** 151
Kuehle, E., **1,** 1121; **4,** 497; **6,** 32
Kuehling, D., **6,** 563
Kuehlmann, R., **4,** 539
Kuehn, A., **7,** 54
Kuehne, M. E., **1,** 56, 411, 904; **2,** 406; **7,** 277
Kuendig, E. P., **11,** 309
Kueng, F. E., **1,** 531
Kueng, W., **1,** 135
Kuenzler, P., **9,** 93
Kuepfer, V., **5,** 155
Kuepper, F. W., **5,** 742
Kueppers, H., **1,** 672; **2,** 271
Kuga, T., **3,** 231
Kugel, R. L., **4,** 387
Kugel, W., **6,** 58
Kugita, H., **1,** 41
Kugler, H. J., **5,** 268
Kuhl, U., **11,** 200
Kuhla, D. E., **2,** 224; **3,** 163
Kuhlen, L., **1,** 802
Kuhlmann, D., **8,** 211
Kuhlmann, H., **6,** 39, 289; **10,** 27
Kuhn, B., **5,** 345
Kuhn, D., **5,** 282; **6,** 133
Kuhn, D. G., **8,** 248, 463
Kuhn, L. P., **1,** 562
Kuhn, M., **1,** 1152; **4,** 45
Kuhn, R., **1,** 22, 57, 281, 295, 309, 318, 340, 348, 350, 465, 536, 567, 685, 890, 1236, 1284; **3,** 218; **8,** 509, 532
Kuhn, S. J., **1,** 4, 15, 69, 266, 407, 693, 741, 743, 748, 1018, 1145; **2,** 12, 297; **8,** 6
Kuhn, W., **4,** 94
Kuhn, W. E., **1,** 1284
Kuhnen, F., **5,** 370
Kui, Y. T. Y., **8,** 362
Kuipers, J. A. M., **10,** 320
Kuivila, H. G., **1,** 349, 1193, 1251; **2,** 424; **3,** 134, 294; **4,** 520; **5,** 686; **9,** 501; **10,** 413; **11,** 224
Kukhar, V. P., **6,** 170
Kukkola, P., **10,** 21
Kukla, M. J., **11,** 302
Kukolja, S., **6,** 281; **8,** 360
Kuksis, A., **3,** 208; **4,** 577
Kulak, H. J., **3,** 123
Kulenovic, S. T., **8,** 464
Kulesha, I. D., **8,** 81; **10,** 400
Kulganek, V. V., **9,** 361
Kulharni, A. K., **12,** 126
Kulig, M. J., **5,** 491
Kulik, S., **6,** 419
Kulka, M., **1,** 34
Kulkarni, A. B., **1,** 595
Kulkarni, A. K., **11,** 90, 432
Kulkarni, B. D., **7,** 188
Kulkarni, G. H., **2,** 361
Kulkarni, S. B., **5,** 731
Kulkarni, S. U., **8,** 49, 50; **9,** 398, 399; **10,** 40, 95; **11,** 266, 267; **12,** 236
Kulkowit, S., **9,** 51; **10,** 21
Kull, H., **1,** 1033
Kulp, S. S., **2,** 248
Kulsa, P., **9,** 458
Kumada, M., **5,** 361; **6,** 51, 59; **8,** 56, 101, 218, 381, 523; **9,** 95, 147, 201; **10,** 38, 194; **11,** 20, 59, 168, 239, 240, 374; **12,** 29, 146, 172, 182, 245, 502
Kumadaki, I., **5,** 148, 621; **7,** 79; **9,** 486
Kumadaki, S., **5,** 614
Kumagai, M., **6,** 653; **8,** 510; **10,** 428
Kumagai, T., **10,** 377
Kumagai, Y., **11,** 48
Kumagawa, T., **12,** 440
Kumai, M., **1,** 329
Kumamoto, T., **1,** 1216; **11,** 1
Kumar, A., **7,** 298
Kumar, K., **12,** 439
Kumar, S., **8,** 279
Kumar, V., **12,** 502
Kumaravel, G., **12,** 258
Kumari, V. G., **4,** 406
Kumazawa, K., **5,** 210
Kumazawa, T., **10,** 279; **12,** 502
Kume, A., **9,** 166; **11,** 178
Kume, F., **12,** 239
Kume, H., **5,** 587
Kummerow, F. A., **4,** 505
Kumobayashi, H., **11,** 54; **12,** 57
Kumura, M., **11,** 518
Kunaver, M., **9,** 472
Kuncl, K., **2,** 351
Kundiger, D., **1,** 926; **5,** 362; **9,** 264
Kundu, N. G., **9,** 421; **10,** 249; **11,** 275
Kunert, D. M., **8,** 88; **9,** 103
Kunert, F., **5,** 714
Kunerth, D. C., **4,** 247; **6,** 279
Kung, F. E., **6,** 456
Kuniak, M. P., **8,** 415
Kunieda, N., **6,** 257, 561
Kunieda, T., **3,** 127; **11,** 220, 445
Kunii, T., **5,** 576
Kunitama, Y., **12,** 263
Kunkl, W., **4,** 123
Kunstmann, M. P., **5,** 675
Kuntz, I., **1,** 342
Kunz, H., **8,** 93; **9,** 506; **11,** 78; **12,** 421

Kunz, R. A., **5,** 747; **6,** 662; **9,** 108
Kunzek, H., **5,** 727
Kunzer, H., **12,** 150
Kunzmann, E., **8,** 286
Kuo, C. H., **1,** 536, 933, 1087; **2,** 159, 252
Kuo, Y. C., **11,** 472
Kuo, Y. N., **4,** 33, 539; **6,** 114; **9,** 310; **12,** 181, 268
Kupchan, S. M., **1,** 60, 1158, 1168; **4,** 330; **5,** 315, 746, 760; **6,** 660; **7,** 421; **8,** 143, 365; **9,** 202, 514
Kupchik, E. J., **1,** 1193, 1251; **10,** 413
Kupfer, D., **10,** 360
Kupiecki, F. P., **3,** 199
Kupper, R. J., **11,** 589
Kuramoto, M., **5,** 348
Kuraoka, S., **12,** 543
Kurath, P., **1,** 87; **5,** 127
Kurbanov, M., **4,** 325
Kurek, J. T., **3,** 196, 197; **4,** 322
Kurihara, H., **8,** 115; **9,** 277
Kurihara, T., **7,** 406; **12,** 188
Kuriki, H., **12,** 365
Kurimoto, M., **9,** 207
Kurino, K., **11,** 16
Kurita, A., **8,** 56
Kurita, K., **5,** 677
Kurita, M., **12,** 439
Kurita, Y., **12,** 565
Kurkov, V., **1,** 335
Kurobe, H., **9,** 32; **10,** 21
Kuroda, A., **12,** 279
Kuroda, K., **8,** 373; **9,** 341; **10,** 56; **11,** 47, 61, 128
Kuroda, N., **4,** 405
Kuroda, S., **4,** 316; **6,** 170; **8,** 530
Kuroda, T., **11,** 75; **12,** 137
Kuroda, Y., **6,** 187; **7,** 20; **8,** 135
Kuroki, Y., **5,** 93
Kuromizu, K., **6,** 445
Kurosaki, T., **8,** 252
Kurosawa, K., **3,** 87
Kurosawa, T., **6,** 258
Kurosky, J. M., **12,** 81
Kuroyama, Y., **10,** 117; **11,** 162
Kurozumi, S., **6,** 165, 238, 464, 641; **8,** 533
Kurpjun, C., **1,** 885
Kursanov, D. N., **5,** 695; **6,** 616
Kurth, M., **11,** 414
Kurth, M. J., **9,** 296, 512; **11,** 309
Kurtz, A. N., **2,** 344
Kurtz, D. W., **6,** 409
Kurtz, J., **1,** 1015
Kurtz, P., **1,** 169
Kurusu, Y., **12,** 89
Kurz, M. E., **1,** 264; **2,** 55, 142; **4,** 74, 255; **5,** 25
Kurz, W., **9,** 261
Kurzer, F., **1,** 170, 533, 843, 1163; **3,** 50, 93; **6,** 32, 250; **12,** 507
Kusabayashi, S., **11,** 592
Kusamran, K., **8,** 299; **9,** 68
Kuschner, M., **4,** 84
Kuser, P., **9,** 180
Kushida, T., **11,** 275
Kusomoto, T., **11,** 114; **12,** 108
Kusuhara, C., **10,** 60
Kusumi, T., **9,** 256
Kusumoto, T., **12,** 543
Kutepow, N. v., **3,** 83
Kuthan, J., **7,** 16
Kutner, A., **1,** 754
Kutney, J. P., **1,** 652, 1284; **2,** 257; **3,** 196; **6,** 97, 496, 644
Kutney, R., **9,** 415
Kutsuma, T., **8,** 198; **9,** 188
Kuwabata, N., **2,** 134
Kuwajima, I., **1,** 196; **3,** 81; **4,** 112, 196; **5,** 526; **6,** 44; **7,** 152, 199, 354; **8,** 52, 210, 220, 227, 446, 468; **9,** 25, 79, 87, 233; **10,** 65, 71; **11,** 124, 235, 402, 544; **12,** 11, 29, 222, 378, 462, 490, 502, 516
Kuwajima, L., **4,** 224
Kuzima, T., **2,** 318
Kuznicki, R. E., **5,** 6
Kuzuhara, H., **11,** 30
Kuzuya, M., **5,** 311
Kwan, T., **6,** 141
Kwant, P. W., **6,** 24
Kwart, H., **1,** 348, 1056; **5,** 379, 749
Kwiatowski, G. T., **2,** 155, 281; **3,** 118, 333; **5,** 528; **6,** 339
Kwitowski, P. T., **2,** 91; **5,** 322
Kwok, R., **1,** 624
Kwon, K. S., **12,** 279
Kwon, S., **6,** 601
Kwon, Y. C., **12,** 399
Kyba, E. P., **3,** 206; **8,** 289; **9,** 361
Kyburz, E., **1,** 471
Kyburz, R., **9,** 523
Kyler, K. S., **12,** 337, 350
Kynnersley, J. D., **2,** 296
Kyotani, Y., **8,** 354
Kyriacou, D., **8,** 79
Kyrides, L. P., **1,** 883, 1073, 1292
Kyrtopoulos, S. A., **7,** 252

Laane, R. W. P. M., **7,** 311
Laasch, P., **3,** 303; **6,** 250
Labadie, J. W., **12,** 46
Labadie, S. S., **12,** 46
Labana, S. S., **5,** 411
Labandibar, P., **4,** 247
Labandiniere, R., **10,** 81
Labar, D., **8,** 64, 320
Labaw, C. S., **5,** 571; **6,** 494
L'abbe, G., **3,** 138; **4,** 225, 262; **5,** 280
Labinger, J. A., **6,** 48, 179; **7,** 102
Labler, L., **1,** 139; **11,** 287
Labovitz, J., **6,** 436
Labovitz, J. N., **6,** 615
Lacadie, J. A., **5,** 455
Lacasse, G., **2,** 191; **3,** 125
Lacey, A. B., **1,** 1073
Lacey, R. N., **1,** 266, 531
Lach, D., **5,** 136, 649
Lachman, A., **1,** 481
Lachmann, H. H., **5,** 244
Lachowicz, D. R., **4,** 444

Lack, R. E., **1,** 1169; **2,** 36
Lacombe, S., **10,** 459
Lacombe, S. M., **11,** 454
Laconi, A., **12,** 36
Lacroix, A., **7,** 285
Ladbury, J. W., **1,** 951
Ladduwahetty, T., **12,** 143
Ladenburg, K., **1,** 1101; **5,** 607
Ladisch, C., **2,** 5
Ladjama, D., **9,** 387
Ladkany, D., **2,** 43
Ladlow, M., **12,** 523
Ladner, W., **11,** 87
Laemmle, J., **5,** 419
Laffer, M. H., **5,** 389
LaFlamme, P. M., **1,** 96, 688
Lafontaine, J., **5,** 19
LaForge, F. B., **1,** 134
Laganis, E. D., **12,** 414
Lagarias, J. C., **10,** 337
Lagerlund, I., **7,** 109
Lago, J. M., **12,** 384
Laguerre, M., **12,** 220
Laguzza, B. C., **10,** 99
Lahav, M., **6,** 166
Lahde, E., **1,** 235
Lahey, F. N., **2,** 334
Lahournere, J. C., **7,** 378
Lai, H. K., **9,** 161, 181, 406, 428; **11,** 154
Lai, R., **6,** 172
Lai, Y. H., **9,** 72; **10,** 340, 401; **11,** 162
Laidlaw, G. M., **5,** 542
Laidler, D. A., **6,** 137
Laing, S. B., **2,** 65
Laird, A. H., **1,** 745, 1196, 1238
Lajis, N. H., **8,** 56; **9,** 55, 78
Lajoie, G., **12,** 59
Lajsic, S. P., **6,** 201
Lake, D. E., **1,** 23
Lakeman, J., **1,** 247
Lakshmikantham, M. V., **3,** 231; **7,** 200, 336; **11,** 599
Lal, B., **5,** 728, 729; **6,** 645; **8,** 211
Lal, G. S., **11,** 14
Lal, J., **7,** 380
Lalancette, J. M., **3,** 264; **4,** 97, 397, 444; **5,** 19, 601; **7,** 296, 410
Lalanne-Casson, B., **8,** 150
Laliberte, M., **4,** 444; **5,** 601
Lallemand, J. Y., **5,** 16, 183; **6,** 141; **11,** 216
Lalloz, L., **6,** 526
LaLonde, J., **1,** 147, 1049
Lalonde, M., **12,** 180
Lam, B., **7,** 338
Lam, H. Y. P., **10,** 26
Lam, L. K. T., **8,** 405
Lamaitre, P., **11,** 216
LaMattina, J. L., **6,** 446; **8,** 90
Lamb, N., **9,** 115
Lamb, R. C., **1,** 57
Lambert, B. F., **1,** 56, 563
Lambert, C., **12,** 51
Lambert, J. B., **5,** 33, 760; **6,** 675
Lambert, P. H., **11,** 588
Lambert, R. F., **1,** 581
Lambert, R. L., Jr., **3,** 222, 223, 225; **6,** 91
Lambert, R. W., **5,** 323
Lambert, Y., **3,** 33
Lambropoulou, V., **12,** 44
Lamby, E. J., **6,** 486
Lam-Chi, Q., **9,** 102
Lamm, B., **6,** 41, 564, 565, 567; **10,** 378
Lammek, B., **6,** 78
Lammens, H., **1,** 392; **2,** 94
Lammer, O., **12,** 554
Lammerink, B. H. M., **8,** 466; **10,** 320
Lammert, S. R., **6,** 281; **8,** 360
Lamon, R. W., **4,** 224; **6,** 250
LaMonica, G., **9,** 217
LaMontagne, M. P., **5,** 200
Lamotte, A., **10,** 370
Lampe, J., **10,** 133; **11,** 349
Lampman, G. M., **2,** 92; **4,** 324, 437, 475
Lamson, D. W., **4,** 444; **5,** 479, 600
Lance, D. G., **2,** 300; **3,** 215
Lancelin, J. M., **12,** 285
Lanciano, F., **10,** 396
Landa, S., **1,** 1167; **3,** 6
Landauer, S. R., **1,** 1249, 1250
Landen, G. L., **10,** 42
Landesberg, J. M., **2,** 394; **4,** 201, 534
Landesman, H., **1,** 348, 707; **5,** 523
Landesman, H. K., **1,** 562, 974
Landgraf, F. W., **1,** 235; **2,** 39, 229
Landgrebe, J., **7,** 193
Landgrebe, K., **12,** 258
Landick, R. C., **6,** 466
Landini, D., **5,** 323; **6,** 175, 272, 406, 543; **7,** 166; **9,** 53, 361
Landis, M. E., **5,** 727; **11,** 77
Landis, P. W., **1,** 802; **2,** 381; **3,** 265
Landmesser, N. G., **8,** 222
Landolt, F., **1,** 1269
Landon, M. J., **4,** 529
Landor, P. D., **1,** 169, 1250, 1295; **2,** 465; **3,** 99; **4,** 255; **5,** 69, 220, 243; **6,** 144
Landor, S. R., **1,** 68, 169, 1249, 1250, 1295; **2,** 173, 245, 330, 446, 465; **3,** 99; **4,** 255; **5,** 69, 220, 243, 460; **6,** 144, 147
Lane, C. F., **3,** 160; **4,** 53; **5,** 47; **6,** 64, 208, 259, 349, 538; **7,** 55; **11,** 479
Lane, J. F., **11,** 105
Lane, R. M., **7,** 18
Lane, S., **12,** 263
Lanford, R. B., **1,** 320
Lang, D., **1,** 182
Lang, G., **5,** 362; **9,** 264; **10,** 322
Lang, J., **1,** 518
Lang, R. W., **10,** 79
Lang, W., **1,** 276, 756; **2,** 329
Langdon, J. M., **1,** 289
Lange, B., **7,** 337; **8,** 442
Lange, R. G., **1,** 34
Lange, R. M., **2,** 316; **11,** 355
Langemann, A., **1,** 567; **2,** 211, 362
Langen, P., **5,** 309

Langenbeck, W., **1,** 329
Langer, E., **3,** 329; **5,** 740
Langer, W., **9,** 48; **11,** 326; **12,** 193, 480
Langford, P. B., **3,** 78; **6,** 261
Langford, R. B., **1,** 1131; **6,** 232
Langhals, H., **6,** 82; **10,** 65
Langham, W., **1,** 573, 689
Langheld, K., **1,** 894
Langin-Lanteri, M. T., **7,** 349
Langler, R. F., **6,** 589; **7,** 58; **11,** 494
Langley, W. D., **1,** 77, 119, 370, 402
Langlois, D. P., **1,** 156
Langlois, N., **5,** 361, 600; **7,** 390; **9,** 459
Langlois, Y., **3,** 46; **7,** 390; **9,** 459; **11,** 117
Langone, J. J., **4,** 294
Langry, K. C., **10,** 52
Lankelma, H. P., **1,** 425
Lankin, D. C., **4,** 411; **5,** 562
Lanpher, E. J., **1,** 848
Lansard, J. P., **12,** 357
Lansbury, P. T., **1,** 147, 600, 737; **2,** 120; **5,** 191, 193; **7,** 99; **10,** 57; **11,** 292, 310, 452, 494
Lansinger, J. M., **9,** 65, 107
Lantos, I., **6,** 363
Lanz, P., **2,** 454
Lanzilotti, A. E., **1,** 1185
Laonigro, G., **7,** 182
Laos, I., **1,** 1059
Lap, B. V., **9,** 416
Lape, H. E., **5,** 542
Lapenue, M., **5,** 167
Lapidot, Y., **1,** 1137
Lapierre, J. C., **1,** 747
Lapin, H., **1,** 1141
Lapinski, R. L., **1,** 509
Lapkin, M., **1,** 935
Laporte, O., **10,** 98; **11,** 150
Laporterie, A., **5,** 749; **9,** 372
Lapouyade, R., **4,** 247; **5,** 167
Lappert, M. F., **1,** 68; **2,** 147; **4,** 46; **5,** 374
Lapporte, S. J., **1,** 731
LaPrade, J. E., **1,** 1108
Lapworth, A., **1,** 952, 1128
Larber, M. E., **1,** 1286
Larcheveque, M., **2,** 210; **3,** 153; **4,** 247; **5,** 325, 543; **6,** 279, 323, 333, 352; **7,** 204; **8,** 299; **9,** 209, 217; **11,** 216
Lardicci, L., **11,** 563
Lardon, A., **1,** 74, 407, 771, 1226
Large, R., **5,** 474
Laricchiuta, O., **11,** 442
Larkin, J. M., **1,** 40; **2,** 25; **4,** 444
Larkin, S., **4,** 84
LaRochelle, R., **3,** 134
Larock, R. C., **3,** 91; **4,** 70; **6,** 573; **7,** 9, 53, 115; **8,** 385, 470; **9,** 174, 292; **10,** 99, 149, 302; **12,** 288, 303, 483
Laroy, R., **10,** 321
Larrabee, R. B., **1,** 604
Larscheid, M. E., **9,** 260
Larsen, A. A., **2,** 252
Larsen, D. S., **11,** 486
Larsen, E. H., **1,** 84, 87, 198
Larsen, J. W., **9,** 483
Larsen, R. D., Jr., **11,** 224
Larsen, S., **11,** 382
Larsen, S. D., **9,** 432; **11,** 335; **12,** 462
Larson, E. R., **9,** 450; **11,** 327; **12,** 313
Larson, G. L., **5,** 719; **6,** 162; **9,** 58, 517; **10,** 91; **11,** 127, 249; **12,** 322
Larson, J. K., **2,** 157, 462
Larson, W. D., **6,** 15
Larsson, F. C. V., **4,** 405
LaRue, M., **8,** 192
Lasarov, D., **2,** 289
Lasch, I., **6,** 375
Lashley, L. K., **10,** 135
Laskovics, F. M., **8,** 239; **10,** 455
Laskowski, S. C., **1,** 678
Lasne, M. C., **5,** 250
Laszlo, P., **5,** 649; **11,** 237; **12,** 231, 325
Latham, H. G., **1,** 485
Latham, H. G., Jr., **1,** 329, 772
Latham, R. A., **2,** 92
Latif, N., **7,** 356
Latimer, L. H., **11,** 301
La Torre, F., **9,** 151, 352, 411, 443; **10,** 306
Latourette, H. K., **1,** 692
Latrofa, A., **5,** 556; **12,** 65
Latt, R. S., **10,** 136
Lattrell, R., **5,** 136; **10,** 329
Lau, C. K., **7,** 212; **11,** 373
Lau, H. H., **12,** 483
Lau, K. S. Y., **8,** 36, 222; **10,** 117
Lau, P. T. S., **8,** 399
Lau, P. W. K., **7,** 356; **8,** 274
Laub, R. J., **6,** 208
Laubach, G. D., **1,** 127, 1122
Laucher, D., **12,** 315
Lauener, H., **1,** 504
Lauer, R. F., **4,** 413; **5,** 276, 522; **10,** 312
Lauer, W. M., **1,** 402, 1027
Laugal, J. A., **12,** 103
Laughton, C. A., **12,** 36
Laurenco, C., **6,** 191
Laurent, A., **10,** 459; **11,** 602; **12,** 223
Laurent, A. J., **11,** 454
Laurent, H., **9,** 324; **11,** 193; **12,** 79
Lautenber, P. C., **3,** 101
Lautens, M., **12,** 333, 345
Lautenschlaeger, F., **1,** 1122; **2,** 392; **3,** 275; **4,** 469
Lauterbach-Keil, G., **1,** 562
Lauwers, M., **9,** 204
Lavagnino, E. R., **1,** 982
Lavallee, P., **4,** 200, 555; **6,** 81; **11,** 196
Lavanish, J. M., **5,** 356
Laves, F., **1,** 1167, 1269
Lavie, D., **1,** 60; **2,** 72
Lavielle, G., **7,** 170
Lavielle, S., **8,** 245, 437
Lavrik, P. B., **5,** 486
Law, H. D., **1,** 152, 385, 736, 1228
Law, K. W., **12,** 394
Lawesson, S. O., **1,** 84, 87, 89,

101, 197, 198, 266, 957, 1141; **2,** 55; **4,** 247, 405; **5,** 325; **6,** 279, 558; **8,** 327; **9,** 50, 283; **11,** 55, 491
Lawrence, R. F., **10,** 425; **12,** 11
Lawrence, R. V., **1,** 37; **2,** 24, 156, 190, 332, 347; **4,** 393; **8,** 385
Lawrence, W. W., Jr., **4,** 42
Lawrynowicz, W., **12,** 462
Lawson, A., **1,** 170
Lawson, D. F., **6,** 628
Lawson, D. N., **10,** 149
Lawson, J. A., **5,** 624
Lawson, J. E., **1,** 563
Lawson, J. K., **3,** 272
Lawson, J. P., **11,** 597
Lawson, R., **1,** 465
Lawson, W. B., **1,** 505; **9,** 191
Lawton, B. T., **3,** 244
Lawton, E. A., **1,** 289
Lawton, R. G., **9,** 483
Laycock, D. E., **9,** 426; **10,** 112, 455; **11,** 119
Layer, R. W., **11,** 174
Layman, D., **5,** 25
Layton, R. B., **4,** 146, 221, 230; **5,** 225
Lazare, S., **9,** 474
Lazier, W. A., **1,** 156, 642
Lazzell, C. L., **1,** 379
Le, T., **12,** 362
Leaback, D. H., **1,** 110
Leach, C. A., **11,** 359, 602
Leach, M., **5,** 63
Leader, G. R., **4,** 63
Leader, H., **1,** 1081
Leake, P. H., **2,** 215
Leal, J. R., **1,** 676
Leandri, G., **12,** 420
Leapheart, T., **11,** 402
Leardini, R., **8,** 122; **9,** 56; **12,** 194
Learn, K., **10,** 390; **11,** 305; **12,** 446
Lease, M. F., **1,** 258
Leatham, M. J., **4,** 126
Leavitt, J. J., **1,** 904
LeBel, N. A., **5,** 454; **7,** 241
Le Berre, A., **5,** 117
Lebibi, J., **11,** 132
Le Bigot, Y., **11,** 433
LeBihan, J. Y., **12,** 36
LeBlanc, T., **5,** 667
Leblanc, Y., **11,** 196; **12,** 396
Leboeuf, M., **2,** 210
Le Borgne, J. F., **7,** 204
Leboul, J., **5,** 642
Le Breton, G. C., **11,** 88
Lebreux, C., **9,** 429
Lecadet, D., **5,** 182, 257
Lechartier, J. P., **4,** 416
Lecher, H., **3,** 21, 183; **6,** 32
Lechevallier, A., **8,** 340; **10,** 93
Lechleiter, J. C., **8,** 336
LeClef, B., **5,** 198; **8,** 157; **11,** 407
Leclerc, G., **4,** 567; **6,** 458
Leclercq, D., **6,** 279; **11,** 515
Leclercq, J., **2,** 297; **4,** 31; **6,** 657
Lecomte, J., **1,** 74
Le Coq, A., **7,** 20, 355; **11,** 407; **12,** 137
LeCorre, M., **9,** 66
Lederer, E., **1,** 411; **2,** 173
Lederer, K., **1,** 349; **9,** 50
Lederman, I., **1,** 502
Ledger, R., **1,** 374; **2,** 317
Ledig, W. O., **4,** 119
Ledlie, D. B., **5,** 658
Ledlie, M. A., **7,** 315
Lednicer, D., **1,** 978, 982, 1147
Ledon, H., **5,** 246, 460; **6,** 631; **8,** 222; **11,** 383
Le Drian, C., **11,** 235
Ledwith, A., **3,** 74; **4,** 79, 502, **5,** 735; **9,** 510; **10,** 452
Lee, A. S. K., **6,** 555
Lee, A. W. M., **9,** 423; **11,** 94, 95
Lee, C., **1,** 618; **9,** 31, 391; **10,** 368
Lee, C. S., **5,** 407
Lee, D. G., **3,** 265; **5,** 563, 622; **6,** 506; **8,** 391; **9,** 361, 391; **11,** 142, 441
Lee, D. L., **6,** 2
Lee, D. P., **6,** 38, 645
Lee, F. T., **1,** 606
Lee, G., **4,** 521
Lee, G. A., **7,** 338
Lee, G. R., **6,** 493
Lee, H. D., **10,** 95
Lee, H. H., **2,** 19
Lee, H. M., **5,** 182; **8,** 155; **11,** 347
Lee, J., **1,** 289; **8,** 341
Lee, J. B., **1,** 1247; **2,** 369, 445; **3,** 16, 149, 290; **5,** 20; **12,** 55
Lee, J. G., **9,** 418
Lee, J. I., **11,** 373; **12,** 12, 102, 350
Lee, J. K., **6,** 383; **9,** 302
Lee, J. T., **5,** 535
Lee, K., **11,** 174
Lee, L. F., **4,** 54; **6,** 546
Lee, L. H., **8,** 531
Lee, M. K. T., **6,** 24
Lee, P. E., **11,** 173
Lee, R. A., **4,** 309; **5,** 412
Lee, S. D., **12,** 175, 279
Lee, S. H., **11,** 544
Lee, S. J., **6,** 390; **8,** 254
Lee, S. P., **9,** 187; **11,** 373
Lee, S. Y., **3,** 163
Lee, T. B. K., **4,** 203; **5,** 17; **6,** 412
Lee, T. J., **11,** 180
Lee, T. V., **8,** 302; **11,** 134; **12,** 320
Lee, V., **2,** 104; **3,** 74; **5,** 620; **6,** 360, 503
Lee, W., **6,** 160
Lee, W. H., **2,** 368
Lee, Y. J., **5,** 614
Leed, A. R., **10,** 122
Leeding, M. V., **2,** 311
Leeds, J. P., **12,** 130
Leeds, N. S., **1,** 1283
Leeming, M. R. G., **2,** 355; **3,** 204
Leeney, T. J., **1,** 1211
Leeper, F. J., **9,** 302
Leeper, R. W., **1,** 177; **8,** 132
Leermakers, P. A., **2,** 429; **7,** 310
Lee-Ruff, E., **7,** 307; **8,** 419,

503
Lees, R. G., **7,** 414
Leese, C. L., **1,** 911
Leese, R. M., **8,** 451
Leete, J. F., **1,** 1027
LeFaivre, M. H., **7,** 158
Lefar, M., **1,** 787
Lefebvre, C., **12,** 486
Lefebvre, P., **5,** 626
Lefferts, E. B., **1,** 96, 573
Leffingwell, J. C., **3,** 194; **4,** 318
Leffler, A. J., **6,** 414
Leffler, J. E., **1,** 802; **5,** 476
Leffler, M., **1,** 187
Lefort, D., **2,** 163; **7,** 174
Leftwick, A. P., **3,** 194
Legault, R., **7,** 284; **9,** 291
Leger, F., **1,** 502
Leger, S., **11,** 339
Leggatt, T., **2,** 344
Leggetter, B. E., **1,** 445; **5,** 186
Leginus, J. M., **12,** 566
LeGoaller, R., **5,** 589
LeGoff, E., **1,** 310, 341, 935, 954, 1022; **4,** 196, 209; **5,** 568; **6,** 198
LeGoff, M. T., **3,** 152; **7,** 299
LeGras, J., **2,** 300; **4,** 234
Legris, C., **5,** 187; **6,** 354
Legzdins, P., **6,** 650; **7,** 313
Lehman, P. G., **4,** 172; **5,** 381
Lehmann, G., **1,** 7; **2,** 42
Lehmann, H. G., **1,** 318; **2,** 173
Lehmann, K. A., **7,** 340
Lehmkuhl, H., **1,** 1189, 1198; **5,** 306; **9,** 361; **11,** 249
Lehn, J. M., **3,** 252; **5,** 156; **6,** 137, 138; **8,** 130
Lehnert, W., **3,** 291; **4,** 508; **6,** 596
Lehong, N., **10,** 254
Lehr, F., **9,** 446; **10,** 381
Lehrfeld, J., **2,** 453
Lehtonen, E. M. M., **9,** 391
Leibfritz, D., **11,** 106
Leicester, J., **1,** 377, 935
Leichsenring, G., **1,** 476
Leichter, L. M., **4,** 429; **6,** 243
Leiden, T. M., **9,** 438
Leigh, G. J., **7,** 289
Leigh, T., **1,** 424, 1141
Leighton, J. P., **5,** 69
Leighton, P., **6,** 144
Leimgruber, W., **1,** 927; **7,** 113; **11,** 200
Leinwetter, M., **8,** 125
Leiserowitz, L., **6,** 166
Leising, E., **3,** 143
Leister, H., **1,** 322
Leitch, L. C., **2,** 215; **5,** 542
Leitich, J., **7,** 251
Leitmann, O., **6,** 564
Lelli, M., **8,** 122
LeMahieu, R., **1,** 450
LeMahieu, R. A., **7,** 274
Lemaire, M., **11,** 251
Lemal, D. M., **1,** 260, 1018; **2,** 397; **3,** 123, 251; **6,** 454
Lemay, G., **11,** 249
Lemberg, M. S., **3,** 243
LeMen, J., **2,** 9, 75; **8,** 233
LeMenOlivier, L., **8,** 233
LeMerrer, Y., **5,** 132; **6,** 610
Lemieux, R. U., **1,** 73, 755, 814; **6,** 44, 569; **10,** 213
Lemin, A. J., **1,** 144
Lemmerman, L. V., **1,** 1040
Lemmers, J. W. F. M., **3,** 66
Lempert, K., **4,** 14
Lenard, J., **2,** 216
Lengyel, I., **1,** 927
Lenhard, R. H., **1,** 16, 120, 281, 634, 764, 767, 881; **2,** 117
Lenich, F. T., **5,** 165
Lenk, C. T., **1,** 511
Lennon, J., **7,** 220
Lennon, M., **5,** 639
Lennon, P., **8,** 455
Lenoir, D., **4,** 240; **8,** 482, 488
Lenox, R. S., **4,** 161
Lentz, C. M., **6,** 343; **9,** 124, 241; **11,** 315
Lenz, G. R., **4,** 428
Leon, C., **4,** 228
Leonard, D. R. A., **5,** 475
Leonard, J. A., **1,** 323
Leonard, N. J., **1,** 195, 618, 652, 814, 926; **2,** 59, 93; **5,** 569; **7,** 11; **9,** 364
Leone, R. E., **1,** 320
Leone Bay, A., **11,** 78
Leong, A. Y. W., **9,** 159
Leong, B. K. J., **6,** 137
Leonhardt, M., **1,** 713
Leopold, E. J., **11,** 226
Leovey, E. M. K., **6,** 19
LePerchec, P., **8,** 374
Lepine, F., **12,** 59
Lepine, M. C., **7,** 392
Leplattenier, F. A., **6,** 417
Leplawy, M., **1,** 88; **8,** 170
Leplawy, M. T., **1,** 1229
Lepley, A. R., **2,** 344
Lepore, U., **4,** 480
Leppert, E., **6,** 561; **7,** 378
LeQuan, M., **12,** 416
Lerch, A., **1,** 96
Lerch, B., **1,** 189
Lerdal, D., **4,** 363; **5,** 491; **9,** 341
Leresche, J. P., **6,** 598
Lergier, W., **2,** 177
Leriverend, M. L., **11,** 301
Leriverend, P., **1,** 374; **11,** 301
Lerman, O., **9,** 223; **10,** 1; **11,** 5; **12,** 4
Lerner, L. M., **1,** 59
Lerner, L. R., **2,** 316
Leroux, Y., **5,** 86, 512; **10,** 198; **11,** 309
Leroy, J., **5,** 312; **7,** 158; **9,** 224
Lesbre, M., **5,** 749; **9,** 372
Lesher, G. Y., **9,** 438
Lesniak, S., **11,** 602
Lesnikowski, Z., **7,** 286
Lespieau, R., **1,** 1040
Lessard, J., **3,** 61, 275; **5,** 23
Lesslie, T. E., **1,** 244, 273
Lester, D. J., **8,** 32; **9,** 34, 501; **10,** 24, 446; **11,** 587
Lester, R., **5,** 633; **6,** 547
LeSuer, W. M., **1,** 1015, 1131
Lesur, B., **11,** 176
Lesur, B. M., **12,** 350
Letendre, L. J., **10,** 264, 289; **11,** 373
Lethbridge, A., **5,** 370, 659
LeThuillier, G., **11,** 59

Leto, J. R., **1,** 1142
Leto, M. F., **1,** 1142
Letsinger, R. L., **1,** 320, 337, 519, 904; **2,** 26, 33, 47, 298, 335; **3,** 19; **5,** 23; **6,** 115, 254, 415
Lett, R., **8,** 87
Lett, R. M., **12,** 411
Letters, R., **1,** 1128
Leu, H. R., **9,** 217
Leube, H., **1,** 1247
Leubner, G. W., **1,** 430, 1049
Leung, H. W., **6,** 158
Leung, K. K., **6,** 466
Leung, T., **5,** 397; **9,** 280
Leupold, E., **1,** 710
Leupold, I., **5,** 622
Leute, R., **4,** 425
Leutert, T., **8,** 184
Leval, L., **2,** 57
Levas, E., **8,** 132
Levasseur, L. A., **1,** 231; **5,** 206
Levenberg, P. A., **11,** 216
Levene, P. A., **1,** 1284; **2,** 89
Levene, R., **4,** 126, 183
Lever, A. B. P., **7,** 307
Lever, J. R., **11,** 284
Lever, O. W., Jr., **5,** 491, 620; **6,** 205, 373; **7,** 234
Levering, D. R., **1,** 730
Levin, G., **4,** 287
Levin, J., **8,** 438
Levin, J. I., **11,** 121
Levin, N., **1,** 97
Levin, R. H., **1,** 93
Levine, I., **1,** 1294
Levine, R., **1,** 69, 160, 228, 703, 1040, 1064, 1084, 1227, 1258, 1261; **2,** 273; **6,** 339; **7,** 135
Levine, S. G., **1,** 13, 671, 1138; **10,** 328
Levinsky, J. A., **1,** 1194
Levinson, A. S., **1,** 603, 617, 974
Levinson, M. I., **11,** 55
Levintow, L., **1,** 129
Levis, W. W., Jr., **2,** 174; **3,** 128
Levisalles, J., **3,** 117
Levison, J. J., **5,** 332; **7,** 60, 109; **9,** 167; **11,** 184
Levitin, I., **3,** 264
Levitt, T. E., **8,** 522
Levitz, R., **9,** 421
Levy, A. B., **4,** 230, 377; **5,** 280, 296, 351; **6,** 325; **7,** 234; **8,** 238; **9,** 304
Levy, D., **1,** 1249
Levy, G., **1,** 1148
Levy, J., **8,** 233
Levy, L. A., **1,** 501; **2,** 101, 223
Levy, N., **1,** 328
Lew, G., **5,** 57
Lew, H. Y., **1,** 500
Lewars, E. G., **5,** 478
Lewbart, M. L., **1,** 161, 841; **2,** 7; **3,** 220
Lewellyn, M. E., **6,** 215
Lewenz, G. F., **1,** 1235
Lewicki, J. W., **8,** 459
Lewin, A. H., **4,** 103, 480; **5,** 163; **8,** 125; **9,** 156
Lewin, N., **1,** 633
Lewis, A., **1,** 999
Lewis, A. J., **12,** 425
Lewis, B., **3,** 265
Lewis, B. A., **1,** 392
Lewis, B. B., **1,** 989
Lewis, C. E., **9,** 376
Lewis, G. S., **1,** 274; **3,** 265
Lewis, J., **6,** 39, 552; **7,** 10, 111, 409; **8,** 521
Lewis, J. R., **1,** 536; **9,** 377
Lewis, J. W., **1,** 207, 598; **3,** 10
Lewis, M. D., **9,** 196; **11,** 20, 134
Lewis, N. J., **8,** 208
Lewis, P. L., **12,** 154
Lewis, R. G., **2,** 210; **3,** 142
Lewis, R. L., **1,** 257
Lewis, R. N., **1,** 17
Lewis, S. D., **3,** 272
Lewis, T. B., **1,** 581; **3,** 175; **9,** 520
Lewis, T. R., **1,** 956
Lewis, W., **8,** 105; **9,** 277; **11,** 188
Lewis, W. P. G., **1,** 1246
Lex, C. G., **1,** 500
Ley, K., **5,** 110
Ley, S. V., **5,** 266; **6,** 568; **7,** 139; **8,** 32, 267; **9,** 34, 50, 375, 388; **10,** 22, 24, 168, 313; **11,** 36, 39, 276, 418; **12,** 38, 394
Leyba, C., **9,** 238
Leyendecker, F., **8,** 238; **10,** 289; **12,** 315
Leyshon, D. M., **1,** 1171
Leznoff, C. C., **2,** 330; **9,** 3; **11,** 30; **12,** 221
Lhomme, J., **2,** 403
Li, C. H., **1,** 176, 385; **9,** 415
Li, G. S., **3,** 299; **4,** 526; **8,** 499
Li, M. P., **6,** 482; **7,** 356
Li, P. T. J., **11,** 308, 599
Li, T., **2,** 87, 464; **3,** 302; **4,** 23; **10,** 33
Li, Z., **11,** 5
Liaaen-Jensen, S., **6,** 542
Liak, T. J., **10,** 300
Liang, P., **1,** 734, 1148
Liang, P. H., **8,** 125
Liang, Y. F., **10,** 276
Liao, C. C., **7,** 268
Liard, J. L., **6,** 534; **7,** 332
Libbey, W. J., **3,** 101
Liberatore, F., **6,** 534
Liberek, B., **1,** 617; **2,** 126
Libermann, D., **1,** 676
Libert, V., **11,** 407
Libit, L., **4,** 473
Libman, J., **5,** 648; **8,** 359; **12,** 463
Lichman, K. V., **4,** 79
Lichtenberg, D. W., **10,** 38
Lichtenberg, F., **12,** 83
Lichtenthaler, F. W., **1,** 741; **3,** 304; **10,** 207; **11,** 72
Lichtenwalter, G. D., **7,** 133
Lick, C., **12,** 386
Liddell, H. G., **1,** 1287
Lidy, W., **5,** 722
Lie, T. S., **5,** 92; **7,** 349; **8,** 103
Liebenow, W., **1,** 340
Lieber, E., **1,** 730, 832, 963; **11,** 421
Lieberknecht, A., **10,** 337
Lieberman, S., **1,** 1178

Lieberman, S. V., **4,** 69
Liebermann, C., **2,** 139
Liebermann, L., **1,** 140
Liebeskind, L., **6,** 390, 406
Liebeskind, L. S., **6,** 654; **7,** 421; **8,** 481; **10,** 223; **12,** 3, 130, 178
Liebig, J., **2,** 207
Liebig, R., **2,** 123
Liebscher, J., **7,** 177
Lied, T., **8,** 184; **9,** 178
Liedhegener, A., **2,** 417; **3,** 292; **4,** 510; **6,** 597; **10,** 405
Liehr, J. G., **2,** 457
Liehr, W., **6,** 640
Lienert, J., **3,** 322; **4,** 555; **5,** 283
Lienhard, U., **8,** 349
Liepa, A. J., **5,** 746; **6,** 660; **8,** 529
Liepkalns, V. A., **1,** 1194
Lier, E. F., **2,** 238
Liew, W. F., **12,** 366
Liewen, M. B., **4,** 148; **7,** 43
Liggett, W. R., **1,** 643
Light, K. K., **1,** 983, 1275
Lightfoot, E. N., **1,** 1122
Lightner, D. A., **5,** 491
Ligon, R. C., **5,** 319
Lilburn, J., **5,** 148
Lillie, T. S., **12,** 245
Lillien, I., **1,** 310, 341, 954
Lillis, G. A., **1,** 754
Lim, D., **5,** 406
Lim, M. I., **11,** 62
Lim, R. M., **9,** 250
Limacher, J., **8,** 374
Limasset, J. C., **2,** 372, 386; **3,** 258; **5,** 593; **10,** 265
Limborg, F., **1,** 636
Limn, W., **5,** 538
Limon, D. C., **1,** 219, 802
Lin, C. H., **4,** 146, 545; **6,** 670; **7,** 425
Lin, C. Y., **5,** 477, 639
Lin, F. F. S., **4,** 152
Lin, G. M. L., **8,** 341
Lin, H. C., **5,** 21, 52, 477; **6,** 24, 103, 273; **7,** 240
Lin, H. J., **8,** 303
Lin, J. J., **6,** 327; **7,** 197; **8,** 175, 291, 311
Lin, J. W. P., **6,** 577
Lin, L. H. C., **2,** 248
Lin, L. J., **12,** 73
Lin, M. H., **5,** 591; **7,** 141
Lin, M. T., **12,** 280
Lin, W. H., **5,** 534
Lin, Y., **3,** 277
Lin, Y. J., **12,** 465
Lin, Y. T., **2,** 344
Linares, H., **1,** 5
Lincoln, D. N., **1,** 397
Lincoln, F. H., **1,** 787, 1118; **2,** 132, 134, 362; **3,** 296; **4,** 562; **8,** 259; **9,** 249
Lind, F. K., **1,** 859
Lind, H., **2,** 129
Lindberg, B., **2,** 165, 317; **6,** 406; **9,** 141
Lindell, S. D., **12,** 377
Lindemann, H., **1,** 742; **8,** 188
Lindenberg, E., **1,** 120
Lindenmann, A., **1,** 20, 478, 1046; **3,** 231
Linder, S., **3,** 177
Linder, V., **8,** 198
Linderman, R. J., **11,** 275; **12,** 11
Lindert, A., **3,** 310; **4,** 309; **5,** 551; **8,** 500
Lindgren, B. O., **5,** 603
Lindhard, M., **1,** 501
Lindhoult, J. C., **7,** 296
Lindlar, H., **1,** 385, 388, 567, 1178; **3,** 172; **6,** 319
Lindner, D. L., **11,** 524
Lindner, R., **9,** 66
Lindsay, B. G., **9,** 458
Lindsay, D. G., **1,** 976
Lindsay, J. K., **1,** 601; **9,** 129
Lindsay, R. O., **1,** 832, 1100
Lindsay, W. S., **2,** 405
Lindsey, R. L., **11,** 77
Lindsey, R. V., Jr., **1,** 18, 892; **2,** 357; **3,** 329; **4,** 543; **6,** 81
Lindstrom, E. G., **1,** 17
Lindwall, H. G., **5,** 297
Lindy, L. B., **4,** 174; **5,** 373; **8,** 140
Lineback, D. R., **1,** 658
Ling, N. C., **4,** 31
Linge, H., **1,** 1133
Lingens, F., **12,** 414
Lingham, D. A., **9,** 214
Lingmann, H., **5,** 329
Link, C. M., **10,** 253
Link, K. P., **1,** 489
Linke, K. H., **5,** 329, 389
Linke, S., **2,** 195
Linkies, A., **8,** 391
Linn, B. O., **1,** 1081
Linn, C. P., **6,** 577
Linn, W. S., **4,** 162
Linsert, O., **1,** 652, 795, 1284
Linstead, R. P., **1,** 129, 219, 348, 667, 772, 782, 859, 1090, 1197, 1269
Linstrumelle, G., **5,** 246, 460; **6,** 495, 631, 666; **7,** 80, 83, 141; **8,** 94, 122, 190, 222, 322, 475; **9,** 125, 458; **10,** 194; **11,** 14; **12,** 176, 467, 474, 476
Lion, C., **4,** 183; **5,** 244; **6,** 165, 215; **8,** 238
Liotta, C. L., **5,** 155, 266; **6,** 137; **8,** 409; **9,** 361
Liotta, D., **5,** 482, 667; **7,** 268; **8,** 448; **9,** 31, 32, 373; **10,** 21, 345, 373; **11,** 14, 36, 299, 466, 475; **12,** 131, 378, 390
Liotta, E. L., **9,** 488
Liotta, R., **7,** 30; **8,** 49; **9,** 58
Lipinski, C. A., **3,** 174
Lipinski, M., **1,** 383, 881; **2,** 51; **10,** 67
Lipisko, B., **10,** 161
Lipisko, B. A., **7,** 304; **9,** 387
Lipkin, D., **1,** 712; **2,** 322
Lipowitz, J., **4,** 394
Lipp, H. I., **4,** 371; **10,** 419
Lipp, M., **1,** 882
Lippard, S. J., **8,** 47; **10,** 33
Lippsmeier, B., **7,** 390; **8,** 515
Lipshes, Z., **4,** 562
Lipshutz, B. H., **6,** 430; **7,** 260; **8,** 325, 374; **10,** 294, 431; **11,** 373, 574; **12,** 350, 365

Lipsky, S. D., **4,** 290; **5,** 381, 454
Liptak, A., **11,** 157
Lipton, M. F., **6,** 600; **8,** 182, 492; **9,** 177; **10,** 5
Lisini, A., **11,** 124
Liskamp, R. M. J., **9,** 283
Liso, G., **6,** 534
Lissel, M., **8,** 130; **9,** 127, 361, 393; **10,** 112; **11,** 127, 407; **12,** 263, 414
Lister, J. H., **1,** 285
Listl, M., **8,** 499; **9,** 121
Liston, A. J., **2,** 310; **3,** 154; **5,** 290
Lisy, J. M., **9,** 194
Litle, R. L., **1,** 124, 445, 1187
Littell, R., **1,** 764, 881, 1000, 1054, 1178
Litterer, W. E., **10,** 289
Little, E. L., **1,** 1136
Little, R. D., **8,** 305; **10,** 170; **12,** 62
Littlewood, P. S., **3,** 59
Litvan, F., **1,** 41
Liu, C. L., **12,** 483
Liu, D., **12,** 329
Liu, H. J., **6,** 253; **7,** 69; **8,** 199; **9,** 161, 181, 187, 406, 428; **10,** 373; **11,** 154, 373, 479
Liu, J. C., **5,** 23; **8,** 519
Liu, J. H., **5,** 359
Liu, J. J., **11,** 373
Liu, J. S., **1,** 397, 563
Liu, K. H., **6,** 55
Liu, K. T., **3,** 325; **4,** 96; **7,** 365; **9,** 11
Liu, S., **5,** 353
Liu, S. Y., **4,** 265
Liu, T., **10,** 252, 449
Liu, T. M., **1,** 281, 705
Liu, W. K., **1,** 176
Liu, W. L., **10,** 335
Lively, D. L., **12,** 122
Livinghouse, T., **9,** 259; **12,** 317
Livingston, A. L., **1,** 296
Livingston, C. M., **8,** 408
Livingston, J. R., **1,** 309
Liwschitz, Y., **1,** 51, 477
Lizzi, M. J., **11,** 195
Ljunggvist, A., **9,** 14
Lloyd, D. J., **4,** 477
Lloyd, E. T., **1,** 1171
Lloyd, J. M., **11,** 502
Lloyd, P. F., **1,** 741; **2,** 175
Lloyd, W. D., **1,** 37; **4,** 412
Lo, K. W., **11,** 359
Lo, S. M., **5,** 46; **6,** 105
Lobert, B., **3,** 240
Locher, R., **9,** 8, 381; **11,** 288
Lochmann, L., **5,** 406
Lochte, H. R., **4,** 162
Lock, C. J. L., **5,** 649
Lock, J. D., **9,** 212
Lock, M. V., **1,** 685
Lockard, J. P., **4,** 174
Locke, D. M., **1,** 144, 253, 707, 1041
Locke, J. M., **1,** 563, 1154; **6,** 374
Lockhart, I. M., **4,** 502
Lockhart, L. B., Jr., **1,** 33, 876
Locksley, H. D., **3,** 31
Lockwood, P. A., **6,** 288; **8,** 374
Lockwood, R. F., **8,** 150
Loder, D. J., **4,** 241
Loder, J. W., **1,** 952, 1121
Lods, L., **2,** 238
Loebenstein, W. V., **4,** 478
Loeber, D. E., **6,** 514
Loebl, H., **1,** 476
Loeblich, V. M., **2,** 24, 156, 332; **4,** 393
Loechel, W., **1,** 187
Loeffler, A., **3,** 336; **5,** 357
Loeffler, L. J., **1,** 8
Loeken, B., **1,** 37, 39, 82
Loeliger, P., **5,** 39
Loening, K. L., **4,** 132
Loeppky, R. N., **3,** 214
Loev, B., **1,** 318, 333, 518, 667, 1059, 1220; **2,** 270, 381; **5,** 626; **6,** 363
Loew, G., **1,** 594
Loew, I., **1,** 685
Loewe, W., **9,** 522
Loewen, P. C., **9,** 77
Loewenfeld, R., **1,** 322
Loewenstein, P. L., **4,** 161
Loewenthal, H. J. E., **2,** 76; **7,** 21
Loffler, H. P., **4,** 520
Lofquist, R., **5,** 10
Logan, C. J., **8,** 230
Logan, R. T., **10,** 304
Logan, T. J., **2,** 30
Logbo, K. D., **8,** 150
Logemann, W., **1,** 531; **3,** 41
Login, R. B., **3,** 104
Logue, M. W., **6,** 84, 148
Logullo, F. M., **1,** 46, 1194; **2,** 26
Logun, A. A., **4,** 336
Logusch, E. W., **9,** 334; **10,** 236; **12,** 11
Loh, J. P., **11,** 527
Lohaus, G., **2,** 70; **3,** 53; **5,** 136; **10,** 329
Lohaus, H., **1,** 933
Loher, H., **12,** 251
Lohmar, E., **4,** 579
Lohmar, R., **8,** 272
Lohmar, R. L., **1,** 865
Lohri, B., **8,** 509
Lohrmann, R., **1,** 1138, 1229; **4,** 535
Loibner, H., **7,** 406; **12,** 233
Loim, N. M., **5,** 695; **6,** 616
Lok, M. T., **5,** 156
Lola, L. K., **2,** 455
Loman, A. A., **3,** 31
Lomas, D., **2,** 389
Lomas, J. S., **4,** 247
Lombard, R., **1,** 821; **8,** 521
Lombardi, P., **9,** 491
Lombardino, J. G., **1,** 1083
Lombardo, L., **4,** 477; **8,** 171; **10,** 422; **11,** 338, 559; **12,** 322
Long, A. G., **1,** 104, 500, 671, 954, 1090; **5,** 90
Long, D. A., **12,** 310
Long, F. A., **1,** 1121
Long, J. W., **1,** 75
Long, L. H., **1,** 206, 207; **2,** 378
Long, L. M., **1,** 430
Long, L., Jr., **1,** 309, 452; **2,** 269; **5,** 226

Long, M. A., **5,** 287; **6,** 504
Long, N. H., **1,** 1055
Long, R., **2,** 110, 305
Longeray, R., **6,** 557
Longheed, G. S., **1,** 1131
Longley, R. I., Jr., **1,** 156, 362, 412, 697, 730
Longo, F. R., **3,** 81
Longone, D. T., **1,** 1136; **3,** 175
Longstaff, P. A., **1,** 140; **2,** 453
Longuet-Higgins, H. C., **2,** 88
Lonngren, J., **6,** 406; **9,** 141
Lonsky, W., **6,** 503
Loo, S. N., **2,** 200
Look, M., **2,** 206
Looker, J. H., **2,** 220
Loomis, G. L., **4,** 302, 309; **6,** 215
Loosmore, S. M., **9,** 212
Loots, M. J., **8,** 87
Loozen, H. J. J., **6,** 514; **7,** 323
Lopez, H., **5,** 88
Lopez, I. C., **1,** 730
Lopez, L., **4,** 477; **5,** 24, 644; **6,** 73, 375; **7,** 35, 245; **8,** 53
Lopez, M. I., **5,** 589
Lopez-Ortiz, J. F., **9,** 465
Lopiekes, D. V., **1,** 1022
Lopresti, R. J., **9,** 420; **10,** 155
Lorber, M. E., **2,** 224, 422; **3,** 57, 163
Lord, K. E., **5,** 351
Lord, P. D., **6,** 534
Lord, W. M., **4,** 184
Lorenc, L., **3,** 171
Lorenz, D. H., **1,** 1251
Lorenz, H., **11,** 401
Lorenz, M., **1,** 990
Lorenz, O., **1,** 1247
Lorenz, R. R., **1,** 708
Lorenzi, G. P., **1,** 262
Lorenzini, A., **5,** 160
Loreto, M. A., **7,** 308; **10,** 333
Lorette, N. B., **1,** 209, 269, 518
Loring, H. S., **1,** 407
Lorne, R., **6,** 666; **8,** 122; **9,** 125
Lorthrop, R. C., **8,** 386
Loshua, S. C., **1,** 452
Loskot, S., **10,** 116
Losse, G., **1,** 250, 345, 408; **4,** 251
Lotspeich, F. J., **1,** 849
Lott, R. S., **9,** 256
Lottenbach, W., **9,** 269
Lotter, H., **11,** 243, 299
Lotts, K. D., **7,** 215; **8,** 308
Lotz, W. W., **5,** 258
Loubinoux, B., **5,** 593; **6,** 70; **10,** 194, 368
Loudon, G. M., **6,** 233; **9,** 54
Louer, M., **8,** 110
Loufer, D. A., **2,** 80
Lougheed, G. S., **2,** 394
Louis, J. M., **3,** 249; **5,** 580; **6,** 514; **7,** 320
Lounasmaa, M., **2,** 10
Loupy, A., **8,** 133; **9,** 127; **10,** 378
Lourens, G. J., **5,** 136
Louw, R., **6,** 256; **7,** 3
Lovald, R. A., **3,** 123
Love, G. M., **5,** 394, 693
Loveitt, M. E., **7,** 224; **8,** 173, 249; **11,** 551
Lovel, C. G., **11,** 503
Lovell, A. V., **11,** 249
Lovell, B. J., **1,** 19
Lovell, C. H., **2,** 331
Lovey, A. J., **4,** 445; **5,** 266
Lovich, S. F., **12,** 474
Loving, B. A., **9,** 480
Lovtsova, A. N., **1,** 854
Lowe, G., **1,** 129
Lowe, J. A., **8,** 326
Lowery, M. K., **2,** 425; **3,** 295
Lowery, S., **12,** 419
Lowry, B. R., **1,** 89
Lowry, N. N., **2,** 424
Loza, M. C., **1,** 564
Loza, R., **10,** 354
Lu, L. D. L., **9,** 96; **12,** 94
Lubosch, W., **7,** 190; **8,** 517
Lubs, H. J., **2,** 199, 263; **3,** 194
Lucas, C. R., **6,** 48
Lucas, G. B., **1,** 827
Lucas, H., **5,** 225
Lucas, H. J., **1,** 372, 505, 506, 508, 509, 511, 828, 1011; **2,** 464
Lucas, M., **11,** 284
Lucas, R. A., **1,** 904
Lucast, D. H., **7,** 160; **8,** 509
Lucchetti, J., **8,** 402
Lucchini, V., **6,** 375; **11,** 336, 495
Lucente, G., **5,** 336
Luche, J. J., **7,** 122
Luche, J. L., **5,** 19, 58; **6,** 23, 215, 269; **7,** 243; **8,** 452; **9,** 99, 211, 226, 422; **10,** 5, 186, 366; **11,** 198, 232, 373; **12,** 357, 400
Luche, M. J., **8,** 245; **12,** 178
Luchetti, J., **8,** 245; **9,** 235
Luchter, K., **5,** 606
Luchter, K. M., **4,** 529
Lucier, J. J., **1,** 295
Lucius, G., **1,** 652
Luck, J. M., **1,** 80
Luckenbach, R., **5,** 266
Luczak, J., **4,** 194; **5,** 266; **6,** 229, 295
Luddy, F. E., **6,** 675
Ludt, R. E., **4,** 489; **5,** 406
Ludwig, S. N., **5,** 98
Ludwig, W., **1,** 563
Ludwikow, M., **6,** 160
Luebbecke, H., **7,** 175
Luecke, B., **2,** 42
Luecke, E., **1,** 444, 707, 708, 1178
Luederitz, O., **1,** 45
Lueking, M., **1,** 116
Luengo, J. I., **12,** 205
Luening, B., **3,** 252
Luening, U., **10,** 254; **11,** 317; **12,** 303
Luetke, H., **12,** 388
Luetolf, J., **9,** 43
Luettke, W., **6,** 646
Luettringhaus, A., **1,** 573, 795, 1146, 1284
Lugada, A. C., **11,** 255
Luh, T. Y., **8,** 105, 498
Luhowy, R. R., **5,** 581, 582
Luijten, J. G. A., **1,** 424,

1112, 1251; **7,** 378
Luis, F. R., **10,** 459
Luisi, P. L., **9,** 361
Luk, K. C., **6,** 286
Lukacovic, M. F., **5,** 170
Lukas, G., **1,** 266, 1163
Lukas, J., **5,** 19
Lukas, J. H., **6,** 252; **7,** 146
Lukaszewski, H., **1,** 746
Lukes, R. M., **1,** 360, 926, 1006
Lumb, J. T., **2,** 344; **5,** 574
Lumma, W. C., **6,** 412, 577
Lund, H., **1,** 367, 629
Lundeen, A. J., **1,** 1167
Lundeen, J. T., **9,** 121
Lunk, H. E., **2,** 226
Lunn, G., **8,** 352
Lunn, W. H., **1,** 310; **5,** 508
Lunsford, W. B., **6,** 115
Lunthankar, S. V., **8,** 483
Luo, F. T., **11,** 575; **12,** 411, 474
Luo, T., **7,** 280; **8,** 454
Luong-Thi, N. T., **4,** 183
Lusch, L. J., **5,** 182
Lusch, M. J., **6,** 377; **8,** 343; **9,** 334
Lusinchi, X., **6,** 73; **7,** 241; **11,** 39; **12,** 395
Lusk, D. I., **1,** 1141
Lusskin, R. M., **5,** 639
Lutomski, K. A., **12,** 312, 383
Luttman, C., **11,** 538
Lutz, E. F., **1,** 378, 1112, 1171
Lutz, R. E., **1,** 196; **3,** 160
Lutz, W., **8,** 16, 374
Lutz, W. B., **1,** 1147
Luvisi, J. P., **1,** 758
Luzzio, F. A., **10,** 30; **11,** 197; **12,** 366
Lwowski, W., **1,** 364, 736; **2,** 296
Ly, M., **5,** 259
Lydiate, J., **2,** 290
Lygo, B., **11,** 418
Lyle, G. G., **1,** 1190; **7,** 254; **8,** 427
Lyle, P. A., **12,** 537
Lyle, R. E., **1,** 1190; **5,** 266; **6,** 446; **7,** 254; **8,** 427
Lyman, R. L., **1,** 296
Lynch, B. M., **1,** 1015
Lynch, D. M., **5,** 186
Lynch, G. J., **3,** 196; **4,** 322
Lynch, J. E., **12,** 239
Lynch, P. P., **1,** 598; **3,** 10
Lyncheski, A. M., **1,** 966
Lynd, R. A., **6,** 201; **7,** 113, 147; **8,** 292
Lynde, E. J., **1,** 219
Lynen, F., **1,** 971
Lynn, D. G., **9,** 34
Lynn, J. T., **6,** 660; **7,** 421
Lyon, G. D., **5,** 677; **6,** 534
Lyon, J. T., **12,** 217
Lyons, D. E., **6,** 254
Lyons, J. A., **1,** 348
Lyons, J. E., **1,** 1235; **3,** 304; **5,** 174; **6,** 655; **7,** 60
Lysenko, Z., **8,** 28, 34, 289; **9,** 434; **10,** 21, 67
Lyssy, T. M., **1,** 712
Lyster, M. A., **8,** 263; **9,** 256; **10,** 219
Lythgoe, B., **1,** 595, 634; **3,** 59; **4,** 218; **6,** 341; **8,** 171, 498; **9,** 156, 417

Ma, C. L., **7,** 71
Ma, K. W., **6,** 534
Ma, P., **10,** 147; **11,** 94, 95
Ma, S. Y., **1,** 566
Ma, V., **1,** 859
Maag, H., **4,** 187
Maalouf, G., **5,** 397
Maas, S. P. J., **1,** 147
Maasboel, A., **4,** 379
Maass, W. S. G., **4,** 199
Maassen, J. A., **4,** 428
Maat, L., **7,** 349; **8,** 103
Mabry, T. J., **2,** 318, 348
Macaione, D. P., **6,** 142
MacAkam, T., **12,** 570
MacAlpine, G. A., **10,** 55, 58
Macartney, J. H., **1,** 707
Macaulay, D. B., **5,** 102
MacBride, J. A. H., **1,** 219, 269
Maccagnani, G., **5,** 516
Macchia, B., **2,** 159
Macchia, F., **2,** 159; **3,** 73
Maccioni, A., **7,** 4
MacCorquodale, D. W., **1,** 140
Macdonald, C. G., **3,** 122
Macdonald, J. E., **9,** 490; **10,** 425
MacDonald, K. I., **5,** 20
Macdonald, T. L., **6,** 386, 607; **7,** 115; **8,** 272; **9,** 334; **10,** 87, 220
Macey, W. A. T., **1,** 377, 935
MacGregor, P., **1,** 808
Mach, K., **12,** 503
Machacek, J., **2,** 382; **3,** 261
Machacek, V., **1,** 1167
Machado-Araujo, F. W., **10,** 290
Macharia, B. W., **4,** 455
Machell, G., **1,** 1158, 1163
Machida, H., **12,** 453
Machida, Y., **6,** 319, 490
Machiguchi, T., **5,** 182, 649
Machii, Y., **8,** 256
Machinek, R., **6,** 646
Machleder, W. H., **2,** 309; **3,** 50
Machleidt, H., **1,** 808, 814
Mack, K. E., **5,** 705
Mack, M. P., **6,** 121; **9,** 506
Mack, W., **2,** 388
Mackawa, H., **5,** 45
Mackay, D., **7,** 288
MacKay, D. D., **1,** 1065; **3,** 272
MacKenzie, J. C. J., **1,** 247
Mackenzie, K., **1,** 834
Mackie, D. W., **4,** 201
Mackie, R. K., **1,** 1216; **2,** 433; **5,** 253
Mackor, A., **11,** 143
Maclagan, N. F., **1,** 334, 518
MacLaren, J. A., **12,** 222
MacLaurin, R. D., **1,** 129
Maclean, A. F., **1,** 531
MacLean, D. B., **5,** 334; **6,** 201
MacMillan, J., **7,** 189
MacMillan, J. H., **5,** 720; **6,** 632
Macnicol, D. D., **5,** 542
MacNicol, D. D., **6,** 550; **8,**

135; **9,** 441; **11,** 203
Macomber, R. S., **5,** 551; **6,** 229; **7,** 394
MacPeek, D. L., **1,** 790
MacPherson, D. T., **12,** 479
MacPhillamy, H. B., **1,** 904; **2,** 79
MacRae, D. M., **7,** 271
MacSweeney, D. F., **4,** 489; **5,** 756
Madar, K., **9,** 361
Madden, D. A., **1,** 69
Maddox, M. L., **12,** 41
Madeja, R., **5,** 51
Mader, W. J., **1,** 62
Madhavarao, M., **6,** 539; **8,** 455
Madigan, D. M., **6,** 675
Madjdabadi, F. A., **12,** 310
Madox, M. L., **8,** 211
Madson, M. A., **11,** 336
Madumelu, C. B., **8,** 451
Maeda, A., **9,** 462
Maeda, H., **11,** 255
Maeda, K., **4,** 129; **5,** 190, 510; **6,** 573; **12,** 245
Maeda, M., **11,** 373
Maeda, N., **12,** 29, 514
Maeda, S., **1,** 322
Maeda, T., **9,** 13; **12,** 288
Maeda, Y., **11,** 184
Maegerlein, H., **5,** 668
Maehr, H., **5,** 63
Maekawa, K., **5,** 9
Maeke, S., **3,** 260
Maemoto, K., **12,** 231
Maercker, A., **3,** 142; **11,** 147
Maerker, G., **1,** 53
Maerkl, G., **1,** 720; **2,** 441; **5,** 30, 647
Maestrone, T., **6,** 110
Magaha, H. S., **11,** 309
Magboul, B. I., **3,** 254
Magennis, S. A., **7,** 274
Magerlein, B. J., **1,** 690, 808, 1000; **4,** 296; **5,** 304
Maggerramov, A. M., **6,** 232
Maggi, D., **7,** 133
Maggiulli, C. A., **2,** 320
Magid, R. M., **8,** 516; **10,** 450
Magidman, P., **6,** 675
Magnane, R., **10,** 174, 319
Magnani, A., **1,** 150, 1073
Magnien, E., **5,** 447
Magnin, D., **10,** 369
Magnus, P. D., **4,** 43, 567; **5,** 116, 486, 734; **6,** 100, 109, 187, 241, 583, 657; **7,** 207; **8,** 32, 274, 278, 281, 400; **9,** 272, 284, 372, 495, 499; **10,** 8, 84, 141, 430, 444; **11,** 83, 129, 150, 331, 402, 465, 577, 584; **12,** 118, 167, 394, 483, 507
Magnusson, G., **5,** 395; **8,** 245; **10,** 265
Magolda, R. L., **8,** 259; **9,** 32, 237; **10,** 21
Magoon, E. F., **2,** 242
Magrath, D. I., **1,** 53
Mah, R. H. W., **1,** 84
Mah, T., **6,** 348
Mahadevan, A. P., **1,** 756, 1100
Mahadevan, V., **1,** 320, 519; **5,** 23
Mahaim, C., **12,** 121, 418
Mahajan, J. R., **6,** 124, 441, 542; **8,** 374
Mahal, H. S., **4,** 15, 424
Mahalingam, S., **10,** 220
Mahatantila, C. P., **12,** 333
Mahe, C., **11,** 49
Mahler, J. E., **3,** 243; **7,** 111
Mahon, J. J., **1,** 1145
Mahon, M., **10,** 168; **12,** 38
Mahoney, L. R., **1,** 45
Mahran, M. R., **2,** 384; **4,** 356
Mai, J., **1,** 491
Mai, K., **12,** 150, 319
Maienfisch, P., **12,** 81
Maier, G., **3,** 204
Maier, J., **5,** 21
Maier, W. F., **8,** 486; **9,** 322, 522; **10,** 243; **11,** 355; **12,** 278
Maillard, A., **5,** 61
Maillard, B., **11,** 101
Mains, G. J., **4,** 346
Maiolo, F., **11,** 203, 473
Maiorana, S., **4,** 469; **9,** 119
Maisey, R. F., **8,** 245
Maisto, K. D., **8,** 158, 487
Maitlis, P. M., **1,** 56; **5,** 649; **7,** 95; **8,** 153; **10,** 34
Maitra, U., **12,** 314
Maitte, M. P., **6,** 514
Maitte, P., **2,** 89
Majerski, K. M., **5,** 639
Majerski, Z., **5,** 639
Majert, H., **1,** 1189
Majerus, G., **2,** 255
Majestic, V. K., **10,** 249
Majetich, G., **6,** 359; **7,** 213; **8,** 24; **9,** 277; **12,** 25
Major, R. T., **1,** 318, 1081; **5,** 607
Majumdar, S. P., **6,** 253
Mak, C. P., **9,** 151
Mak, K. T., **4,** 472
Makabe, O., **10,** 273
Makaiyama, T., **1,** 1216
Makani, S., **10,** 195
Makar, S. M., **1,** 131
Makata, H., **2,** 333
Maki, Y., **2,** 411; **4,** 282; **6,** 534; **7,** 331; **8,** 451; **12,** 272, 363
Makin, M. I. H., **9,** 393
Makin, S. M., **1,** 636
Makino, H., **11,** 24
Makino, S., **4,** 158; **5,** 224; **6,** 198; **8,** 498, 499
Makisumi, S., **2,** 177
Makita, M., **1,** 427
Makofske, R., **8,** 81
Makofske, R. C., **9,** 415
Makosza, M., **3,** 19; **4,** 31; **5,** 28, 530; **6,** 43, 160, 601; **7,** 20, 353; **8,** 391, 461; **9,** 361
Makriyannis, A., **5,** 213; **8,** 20
Maksimovic, Z., **3,** 171
Mal, D., **12,** 490
Malacria, M., **11,** 321, 360; **12,** 51, 163, 307, 437
Malaitong, N., **10,** 394
Malaprade, L., **1,** 818
Malatesta, V., **6,** 628
Malaval, A., **6,** 441
Malchenko, S., **9,** 172
Maldonado, L., **4,** 302; **6,** 530

Maldonado, L. A., **10,** 154
Malecki, F. E., **1,** 1188
Malek, G., **2,** 191
Malek, J., **3,** 261; **4,** 442; **6,** 518, 529
Maleski, R. J., **9,** 324
Maletina, I. I., **6,** 301
Malfroot, T., **12,** 113
Malherbe, R., **9,** 154
Malhotra, R., **9,** 255, 256, 320, 335; **10,** 96; **11,** 355
Malhotra, S., **9,** 377; **11,** 491
Malhotra, S. K., **1,** 926; **2,** 344
Mali, R. S., **6,** 90
Malik, A., **12,** 532
Malinowski, M., **11,** 529
Malkus, H. L., **6,** 628
Mallamo, J. P., **10,** 194; **11,** 315
Mallan, J. M., **1,** 581
Mallano, J. P., **9,** 283
Malleron, J. L., **10,** 34; **11,** 53
Mallik, R., **5,** 217
Mallinson, P. R., **11,** 203
Mallon, C. B., **6,** 137
Mallory, F. B., **1,** 500, 1044, 1087
Malmberg, C., **9,** 274
Malone, G. R., **3,** 284; **4,** 484
Malone, T. C., **11,** 241
Maloney, J. R., **8,** 427
Malpass, J. R., **4,** 93; **8,** 106
Malte, A. M., **5,** 97
Maltes, K. C., **8,** 338
Mamada, Y., **10,** 145
Mamantov, A., **4,** 529
Mamedov, S. A., **10,** 459
Mamer, O. A., **2,** 222; **4,** 9
Mami, I. S., **6,** 581
Mamlok, L., **1,** 377, 856
Mammarella, R. E., **9,** 492
Man, E. H., **1,** 117; **7,** 391; **12,** 106
Manabe, O., **6,** 281
Manas, A. R. B., **6,** 651
Manasse, O., **1,** 1101
Manatt, S. L., **5,** 336
Mancera, O., **1,** 3, 150, 642, 1090; **2,** 259
Manchand, P. S., **4,** 42
Mancilla, J. M., **5,** 160
Mancini, M. L., **11,** 574
Manck, P., **1,** 928
Mancuso, A. J., **8,** 200; **9,** 192; **10,** 167
Mandai, T., **8,** 382; **9,** 173, 350; **10,** 150, 390; **11,** 419; **12,** 316, 402
Mandal, A. K., **8,** 50, 174; **9,** 317; **10,** 50, 224; **11,** 188, 351
Mandel, G., **12,** 362
Mandel, G. S., **9,** 11
Mandel, N., **12,** 362
Mandel, N. S., **9,** 11
Mandelbaum, A., **12,** 277
Mandell, L., **1,** 13, 50, 150
Mander, L. N., **3,** 65; **5,** 55; **6,** 615; **8,** 39; **10,** 236, 422; **11,** 559; **12,** 86, 321
Mandeville, W. H., **6,** 343
Mandolini, L., **6,** 479; **9,** 127
Mane, R. B., **6,** 498
Manecke, G., **5,** 148
Manescalchi, F., **6,** 304; **7,** 182
Manfre, R. J., **9,** 471
Mangeney, P., **9,** 459
Mangia, A., **8,** 248
Mangiaracina, P., **12,** 454
Mangold, D., **10,** 331
Mangold, H. K., **2,** 254, 313
Mangoni, L., **5,** 349; **6,** 296, 586; **7,** 181, 182; **11,** 128
Mangravite, J. A., **10,** 413
Mangum, M. G., **5,** 705; **6,** 231, 620
Manhas, M. S., **5,** 713, 728; **6,** 645; **8,** 211; **10,** 115, 448; **12,** 554, 568
Manis, P. A., **11,** 217
Manitto, P., **3,** 171; **5,** 370; **12,** 422
Maniwa, K., **5,** 4, 683; **6,** 6
Manley, N. C., **1,** 795
Mann, F. G., **1,** 345, 506, 510; **2,** 34, 179; **3,** 83; **5,** 410; **7,** 324; **10,** 39; **12,** 19
Mann, J., **5,** 441; **12,** 41
Mann, M. J., **1,** 257, 970
Mann, S., **1,** 240
Mann, T. A., **2,** 433
Mannering, G. C., **5,** 651
Manners, G. D., **6,** 518
Mannich, C., **2,** 7
Manning, D. T., **1,** 755
Manning, M. J., **7,** 192; **8,** 281
Manning, R. A., **6,** 500
Manning, R. E., **1,** 94
Mannschreck, A., **1,** 116
Manolopoulo, M., **1,** 165
Mansfield, G. H., **1,** 1040
Mansfield, K. T., **1,** 247; **2,** 87; **3,** 66
Manske, R., **9,** 483
Manske, R. F., **1,** 445
Manske, R. F. H., **6,** 201
Mansuy, D., **5,** 16, 183; **7,** 50
Mantecon, R. E., **4,** 223, 557
Manuel, G., **9,** 372
Manville, J. F., **2,** 365
Manyik, R. M., **1,** 70
Manzocchi, A., **8,** 110, 534; **9,** 376, 421; **10,** 382; **12,** 444
Mao, C., **5,** 160
Mao, C. L., **4,** 510
Mao-Chin, L., **3,** 161
Maquenne, M., **1,** 494
Marakowski, J., **5,** 682
Marans, N. S., **6,** 632
Marathe, K. G., **6,** 583
Marathey, M. G., **1,** 162
Marazza, F., **10,** 437
Marbet, R., **2,** 231; **4,** 235, 328
Marburg, S., **6,** 561
Marchand, A. P., **3,** 134, 322
Marchand, B., **1,** 562, 764; **9,** 207
Marchand, E., **6,** 472; **8,** 416
Marchand-Brynaert, J., **4,** 95; **5,** 138; **10,** 60
Marchelli, R., **9,** 190, 361
Marchese, G., **10,** 289; **12,** 171, 557
Marchese, L., **6,** 375; **7,** 245
Marchesini, A., **7,** 331
Marchetti, M., **11,** 125
Marchini, P., **6,** 534
Marchiori, F., **6,** 419
Marckwald, W., **1,** 734, 742; **5,** 286; **11,** 359

Marco, G. J., **1,** 257, 1246
Marcus, R., **1,** 223
Marcuzzi, F., **6,** 114; **11,** 336
Marecek, J. F., **6,** 8, 219
Marek, P. J., **7,** 237
Mares, F., **8,** 438
Maresca, L., **5,** 468
Maresca, L. M., **8,** 64
Marfat, A., **8,** 117; **9,** 70, 333; **10,** 115, 199
Margartha, P., **10,** 160
Margrave, J. L., **1,** 235; 1178
Marguerite, J., **11,** 127
Margulis, T. N., **4,** 252
Marhenke, R. L., **5,** 14
Marica, E., **2,** 112; **6,** 279
Maricich, T. J., **1,** 364, 736; **2,** 296
Mariella, R. P., **1,** 383, 1073, 1095
Marigliano, H. M., **2,** 326, 332
Maring, C., **12,** 561
Marini-Bettolo, G. B., **4,** 424
Marini-Bettolo, R., **11,** 325
Marinier, B., **1,** 180; **4,** 522
Marino, G., **5,** 624
Marino, J. P., **1,** 310; **5,** 68, 714, 744; **6,** 121, 209, 331, 378, 466, 556; **7,** 193, 199; **9,** 202, 286, 333; **10,** 290; **11,** 170, 224, 275; **12,** 108, 141, 178
Marino, M. L., **4,** 373
Marinovic, M. V., **12,** 523
Marinovic, N., **7,** 160, 314; **8,** 437; **9,** 81
Marioni, F., **5,** 568
Mark, C., **10,** 206
Mark, V., **1,** 432, 1216, 1229; **4,** 473
Markby, R., **1,** 228, 520, 723
Markees, D. G., **5,** 216
Marker, R. E., **1,** 119; **2,** 353
Markes, J. H. H., **1,** 60; **2,** 322; **5,** 512; **7,** 59
Markezich, R. L., **5,** 700
Markgraf, J. H., **4,** 254
Markham, J. L., **7,** 268
Markham, R., **2,** 228
Markiewicz, W., **8,** 448; **11,** 475
Markiewicz, W. T., **10,** 307
Markiewitz, K. H., **1,** 1103
Markov, P., **1,** 711; **2,** 289
Markovac, A., **3,** 161
Markovits-Kornis, R., **4,** 14
Markovskii, L. N., **6,** 184, 186
Markowitz, M., **10,** 33; **11,** 499
Markstein, J., **5,** 551
Marktscheffel, F., **1,** 562, 598
Markus, A., **4,** 31
Markwell, R. E., **7,** 158
Marlborough, D. I., **2,** 80
Marlewski, T. A., **5,** 6
Marman, T. H., **12,** 303
Marmor, R. M., **10,** 157
Marmor, R. S., **3,** 114; **4,** 544; **5,** 158; **6,** 189; **9,** 181
Marmor, S., **1,** 1087; **8,** 463; **10,** 209
Marolewski, T. A., **3,** 166; **4,** 266
Maroski, J. G., **5,** 560; **10,** 209
Marquarding, D., **7,** 20; **9,** 82; **11,** 352
Marquet, A., **1,** 377, 652, 856, 970; **4,** 386; **7,** 35; **8,** 194, 245, 437
Marquet, B., **10,** 459
Marquet, J., **10,** 102
Marquis, E. T., **1,** 689
Marquis, R., **1,** 636
Marr, D. H., **4,** 387
Marrero, R., **6,** 114; **8,** 272; **10,** 93; **12,** 272
Mars, S., **5,** 60
Marschalk, C., **8,** 458; **9,** 376, 439
Marschall, H., **6,** 74; **7,** 102, 394; **8,** 194
Marsden, C., **1,** 240
Marsden, J. C., **2,** 27
Marsh, F. D., **1,** 174; **2,** 93; **5,** 476; **11,** 119
Marsh, J. P., Jr., **1,** 1220
Marsh, M. M., **1,** 866
Marsh, W. C., **2,** 352
Marshall, C. W., **1,** 1059, 1178
Marshall, D., **1,** 669
Marshall, D. R., **3,** 199, 258
Marshall, F. J., **4,** 283
Marshall, G. R., **4,** 164; **6,** 149
Marshall, H., **5,** 319
Marshall, J. A., **1,** 161, 625, 703; **2,** 54, 153, 173, 307; **3,** 74, 77, 112; **4,** 183, 285, 413, 472; **5,** 601; **6,** 215, 454; **8,** 159, 181, 464; **10,** 134; **12,** 11, 96
Marshall, J. L., **2,** 144; **3,** 33; **9,** 416
Marshall, J. P., **2,** 362
Marshall, W. S., **1,** 718
Marsi, K. L., **5,** 523
Marsi, M., **11,** 401
Marsili, A., **3,** 240
Marsman, B., **10,** 338
Marsmann, H. C., **6,** 640
Martel, B., **4,** 443
Martel, J., **1,** 219, 465
Martel, R., **2,** 191
Martell, A. E., **1,** 536
Martelli, J., **10,** 81, 223
Marten, D. F., **5,** 243; **8,** 455
Martens, H. F., **10,** 220
Martens, T. F., **1,** 47
Martin, A., **11,** 445
Martin, B. D., **1,** 643
Martin, D., **2,** 63, 159; **3,** 51, 123; **9,** 127
Martin, D. G., **1,** 144, 609, 1064, 1125, 1131
Martin, D. J., **1,** 689; **2,** 278, 376, 389; **9,** 436
Martin, E. C., **1,** 754
Martin, E. J., **2,** 92
Martin, E. L., **1,** 485, 654, 718, 837, 1163, 1284, 1289; **2,** 61
Martin, G., **5,** 339
Martin, H., **1,** 1015
Martin, J., **1,** 1295; **3,** 191; **5,** 421; **9,** 172
Martin, J. A., **2,** 35
Martin, J. B., **1,** 66
Martin, J. C., **1,** 80, 814, 986;

4, 206, 207; **5,** 272; **6,** 239; **9,** 485; **10,** 71; **12,** 379
Martin, J. D., **5,** 631; **7,** 40
Martin, J. R., **5,** 127, 514
Martin, K. J., **1,** 254
Martin, K. R., **2,** 241
Martin, M., **12,** 538
Martin, M. G., **12,** 203, 444
Martin, M. M., **2,** 248; **6,** 166
Martin, P., **9,** 112, 481; **12,** 566
Martin, P. K., **2,** 263
Martin, R. H., **1,** 80, 139, 1116
Martin, R. J. L., **1,** 599; **2,** 243
Martin, R. L., **9,** 483
Martin, R. S., **10,** 295
Martin, S. F., **7,** 105, 411; **8,** 219, 316, 323; **9,** 162, 341; **11,** 179; **12,** 186
Martin, S. J., **11,** 206; **12,** 208
Martin, V. S., **11,** 94, 95, 463
Martin, W. B., **1,** 1081; **8,** 507
Martin, W. F., **1,** 1168
Martina, D., **12,** 245
Martina, V., **12,** 557
Martinelli, J. E., **3,** 49
Martinez, A. G., **12,** 338
Martinez, A. M., **5,** 140
Martinez, G. R., **9,** 100; **10,** 436
Martinez, J., **6,** 470
Martinez-Davila, C., **9,** 87; **10,** 89
Martinez-Utrilla, R., **10,** 224, 354
Martini, C. M., **1,** 956
Martins, F. J. C., **6,** 675
Martin-Smith, M., **5,** 436
Martius, C., **1,** 195
Martz, J. T., **12,** 113
Martz, M. D., **1,** 876; **4,** 18
Marullo, N. P., **10,** 245
Maruoka, K., **8,** 164; **9,** 478; **11,** 160, 188, 249, 542, 590; **12,** 11, 193, 344, 356, 378, 480
Maruta, R., **9,** 516
Maruyama, F., **11,** 445
Maruyama, H., **2,** 445
Maruyama, K., **8,** 335, 381; **9,** 8, 77, 334, 500; **10,** 55, 133, 289, 411, 415, 417, 431, 451; **11,** 12, 20, 22, 52, 75, 143, 365, 373, 415; **12,** 15, 20, 29, 82, 405, 514
Maruyama, M., **5,** 760
Marvel, C. S., **1,** 17, 47, 48, 134, 169, 251, 370, 402, 407, 412, 452, 481, 500, 718, 864, 870, 962, 963, 1023, 1056, 1073, 1081, 1184, 1284; **4,** 77; **11,** 78
Marvell, E. N., **1,** 866
Marx, A. F., **1,** 360
Marx, G. S., **4,** 562
Marx, J. L., **6,** 102
Marx, J. N., **5,** 144, 631; **9,** 159; **12,** 465
Marx, M., **3,** 269; **5,** 5; **8,** 408
Marxer, A., **8,** 184
Maryanoff, B. E., **4,** 444, 450, 451; **5,** 55, 609; **7,** 335; **8,** 46; **11,** 300, 483
Maryanoff, C. A., **7,** 292; **11,** 483
Masada, H., **8,** 441
Masaki, Y., **5,** 379; **6,** 267, 422; **9,** 38; **10,** 289; **12,** 506
Masamune, S., **3,** 38, 102; **4,** 339, 435; **5,** 168, 331; **6,** 56, 192, 582; **7,** 20, 61, 400; **8,** 444; **9,** 96, 132; **10,** 119; **11,** 94, 95, 223; **12,** 94, 249, 277
Masamune, T., **5,** 348; **7,** 208; **8,** 218; **9,** 198; **12,** 506
Mascarella, S. W., **12,** 350
Masci, B., **9,** 127
Mascitti, A., **1,** 859
Masdupuy, E., **1,** 618
Mashima, K., **11,** 174
Mashraqui, S., **11,** 311
Mashraqui, S. H., **12,** 292
Masilamani, D., **7,** 292; **8,** 466; **11,** 483
Maskimova, N. G., **11,** 83
Masler, W. F., **6,** 416
Maslin, D. N., **1,** 1030; **3,** 259
Mason, D., **1,** 954
Mason, H. L., **1,** 150
Mason, H. S., **1,** 34
Mason, J. L., **1,** 832
Mason, M. M., **4,** 291, 294
Mason, P. S., **1,** 618
Mason, R. F., **4,** 183
Mason, R. W., **9,** 158, 172, 438
Massa, W., **11,** 87
Masse, G. M., **6,** 667
Masse, J. P., **7,** 239
Massengale, J. T., **1,** 518
Massiot, G. S., **6,** 361; **8,** 392
Masson, S., **2,** 226; **3,** 165; **5,** 250; **6,** 300
Massuda, D., **6,** 100; **7,** 13
Massy-Westropp, R. A., **5,** 55
Mastagli, P., **6,** 596
Mastalerz, H., **11,** 121; **12,** 538
Mastre, T. A., **6,** 224, 586
Mastres, R., **10,** 310
Mastronardi, I. O., **2,** 104
Masuda, A., **6,** 55; **7,** 222
Masuda, R., **6,** 510; **7,** 135, 350; **8,** 289; **11,** 293; **12,** 275
Masuda, T., **6,** 369; **11,** 24; **12,** 279
Masuda, Y., **5,** 535; **6,** 259, 510; **7,** 156; **9,** 250
Masui, K., **5,** 42
Masui, Y., **12,** 199
Masumura, M., **11,** 480
Masunaga, T., **10,** 391; **11,** 555; **12,** 187
Masure, D., **5,** 406; **8,** 505
Masuri, D., **6,** 622
Masuya, H., **9,** 151
Masuyama, Y., **7,** 123; **12,** 89, 246
Matacz, Z., **9,** 361
Matar, A., **1,** 452
Matarasso-Tchiroukhine, E., **10,** 411; **11,** 143
Mateen, B., **8,** 255
Mateescu, G., **2,** 112; **6,** 279
Mateos, J. L., **1,** 125, 598, 624
Mathai, I. M., **2,** 92; **3,** 69

Matharu, S. S., **10,** 181
Matheny, N. P., **1,** 73
Mather, A. M., **8,** 331
Matheson, J. M., **1,** 127
Matheson, M. S., **6,** 670
Mathew, C. T., **4,** 369; **5,** 5
Mathew, K. K., **1,** 882, 1116, 1284
Mathey, F., **5,** 25, 465; **6,** 393, 412; **7,** 83; **9,** 224
Mathiaparanam, P., **8,** 33
Mathias, R., **4,** 31; **5,** 28
Mathieson, D. W., **5,** 646
Mathieu, D., **8,** 486
Mathieu, J., **1,** 23, 125, 215, 465
Mathre, D. J., **10,** 112; **11,** 381; **12,** 362
Mathur, N. K., **4,** 469
Mathur, S. S., **4,** 505
Mathy, A., **12,** 325
Matier, W. L., **5,** 700
Matiskella, J. D., **5,** 687
Matkovics, B., **5,** 308
Matl, V. G., **4,** 45
Matlin, S. A., **4,** 204, 570; **9,** 48
Matlock, P. L., **7,** 340
Matloubi, F., **10,** 407
Matloubi-Moghadam, F., **11,** 106, 538
Matough, M. F. S., **2,** 445
Matsubara, F., **9,** 519
Matsubara, I., **9,** 404; **11,** 253
Matsubara, S., **12,** 63, 64, 304, 344, 524
Matsuda, H., **5,** 459; **6,** 399; **9,** 94
Matsuda, I., **8,** 513, 515; **10,** 262, 428; **11,** 571; **12,** 503, 541
Matsuda, K., **12,** 539
Matsuda, M., **6,** 377
Matsuda, R., **9,** 326
Matsuda, S., **5,** 667
Matsuda, T., **5,** 325, 498; **6,** 59; **10,** 221, 298; **12,** 392
Matsuda, Y., **9,** 429
Matsueda, R., **2,** 445; **3,** 322
Matsugashita, S., **5,** 433
Matsugo, S., **6,** 436; **7,** 268
Matsuguma, H. J., **1,** 484
Matsui, K., **11,** 180
Matsui, M., **3,** 41, 65; **4,** 155; **5,** 553; **7,** 40; **8,** 44, 354
Matsui, S., **10,** 231
Matsui, T., **5,** 13, 178
Matsukawa, K., **11,** 140
Matsuki, Y., **6,** 90; **7,** 47; **9,** 292
Matsumara, S., **3,** 299
Matsumoto, A., **11,** 239
Matsumoto, G., **4,** 405
Matsumoto, H., **2,** 285; **5,** 741; **6,** 655; **8,** 56, 161, 240; **9,** 201, 404, 492; **10,** 44, 196, 245; **11,** 240, 253; **12,** 335
Matsumoto, K., **5,** 178, 214; **6,** 158; **11,** 500; **12,** 462
Matsumoto, K. K., **9,** 233
Matsumoto, M., **6,** 436, 610; **7,** 269; **8,** 373; **9,** 341; **11,** 47, 61, 172, 383, 462; **12,** 231, 428, 523
Matsumoto, S., **6,** 185
Matsumoto, T., **8,** 441; **9,** 138, 296; **11,** 96, 445; **12,** 513, 543, 546
Matsumoto, Y., **6,** 194
Matsumura, A., **5,** 89, 442
Matsumura, N., **3,** 156; **6,** 95
Matsumura, Y., **9,** 465; **10,** 310; **11,** 533, 542; **12,** 11
Matsuo, T., **9,** 151
Matsuoka, Y., **10,** 277
Matsushita, H., **9,** 205; **10,** 391; **11,** 514
Matsushita, Y., **10,** 99
Matsuura, A., **5,** 576
Matsuura, S., **5,** 576; **7,** 294; **8,** 156
Matsuura, T., **3,** 245; **6,** 430, 436, 479; **7,** 12, 260, 268; **9,** 140
Matsuyama, H., **10,** 361
Matsuyama, K., **8,** 83; **12,** 465
Matsuzaki, Y., **12,** 288
Matta, K. L., **9,** 54; **11,** 24
Mattenberger, A., **11,** 542
Mattenci, M. D., **10,** 461
Matter, M., **1,** 56, 330
Matteson, D. S., **1,** 84; **6,** 329; **9,** 1, 262; **10,** 239, 292; **11,** 573; **12,** 81
Matthew, C. P., **12,** 276
Matthew, J. A., **10,** 294
Matthews, A. O., **1,** 1064
Matthews, C. N., **2,** 447
Matthews, G. J., **3,** 161
Matthews, H. R., **2,** 263
Matthews, R. S., **1,** 982; **4,** 405; **8,** 101; **12,** 344
Matthews, R. W., **10,** 348
Matthews, W. S., **5,** 97; **6,** 577
Matthey, G., **1,** 220, 1209
Matthies, D., **1,** 730
Mattingly, P. G., **9,** 506; **10,** 342, 400, 448
Mattingly, T. W., Jr., **1,** 364, 736
Mattocks, A. R., **8,** 4; **10,** 330
Mattox, V. R., **1,** 161, 331, 566, 841, 1000, 1006; **8,** 442
Matturro, M. G., **7,** 216
Maturaba, S., **10,** 165
Matuszko, A., **6,** 58
Matz, J. R., **10,** 245, 256, 383; **11,** 527; **12,** 280
Maujean, A., **7,** 3
Maume, G. M., **3,** 312
Maumy, M., **4,** 123, 437; **7,** 324
Maurer, B., **5,** 481
Maurer, H., **7,** 428
Maurer, K., **1,** 329
Maurer, P. J., **12,** 431
Maurette, M. T., **12,** 414
Mauri, F., **3,** 158
Maurin, R., **6,** 223
Maurukas, J., **1,** 847
Mausner, M., **1,** 342
Mauthner, J., **1,** 1284
Mauz, O., **6,** 202
Mauze, B., **8,** 151
Mauzerall, D., **1,** 1083; **6,** 37
Mawhinney, T. P., **11,** 336
Maxson, R. N., **1,** 1131
Maxwell, C. E., **1,** 1121
May, E. L., **1,** 772
May, G. E., **1,** 1112

May, H. J., **1,** 94
May, K. D., **6,** 546
May, L. M., **6,** 124
May, P. J., **2,** 228
Maycock, C. D., **10,** 422; **11,** 378, 379
Mayer, F., **5,** 13
Mayer, G. D., **11,** 305
Mayer, H., **1,** 385; **2,** 198; **6,** 527; **12,** 194
Mayer, R., **1,** 281, 707, 933; **3,** 275; **6,** 583; **7,** 289
Mayer, R. L., **1,** 601
Mayer, R. P., **1,** 96
Mayer, U., **1,** 101
Mayer, W., **1,** 341
Mayerle, J. J., **10,** 33
Mayer-Shocket, N., **10,** 125
Maynard, J. A., **1,** 253
Maynard, J. R., **2,** 313
Maynez, S. R., **6,** 9
Mayo, F. R., **1,** 56, 197; **6,** 450; **7,** 95
Mayo, J., **1,** 85
Mayr, A., **9,** 66
Mayr, A. J., **11,** 156
Mayr, H., **6,** 17; **10,** 462; **11,** 604; **12,** 576
Mazenod, F., **7,** 74, 239
Mazharuddin, M., **5,** 13
Maziak, L., **12,** 59
Mazieres, C., **5,** 58
Mazuda, R., **8,** 345
Mazumdar, B. N. G., **1,** 518
Mazur, R. H., **1,** 486, 494; **2,** 464; **5,** 741
Mazur, S., **4,** 363; **5,** 491; **7,** 268; **9,** 341
Mazur, Y., **1,** 73, 104, 207, 604, 614, 999, 1000; **2,** 15, 132, 361; **3,** 198; **4,** 8; **5,** 5, 506; **6,** 441, 574; **7,** 273; **8,** 161, 230, 377, 461; **9,** 156; **10,** 185, 296; **11,** 237
Mazurek, M., **1,** 626
Mazurek, M. A., **9,** 250
Mazza, S., **6,** 364
Mazzu, A. L., Jr., **9,** 323
McAlees, A. J., **5,** 655
McAndrews, C., **4,** 309
McArthur, C. R., **9,** 3; **11,** 30; **12,** 221
McArthur, R. E., **1,** 619
McBee, E. J., **4,** 529
McBee, E. T., **1,** 270, 333, 511, 1131, 1132; **2,** 206, 227; **5,** 329
McBride, E., **5,** 756
McBride, J. J., Jr., **8,** 162
McBride, J. M., **1,** 89; **5,** 19, 603
McCabe, P. H., **8,** 408
McCain, G. H., **1,** 518
McCaleb, K. E., **1,** 389, 703
McCall, J. M., **6,** 161, 588
McCallum, J. S., **11,** 401
McCallum, K. S., **1,** 742, 1226
McCandlish, C. S., Jr., **6,** 676
McCann, E. L., III, **6,** 414
McCants, D., Jr., **1,** 94, 329, 814
McCarney, C. C., **6,** 675
McCarry, B. E., **4,** 531; **5,** 700
McCarthy, D. G., **10,** 269
McCarthy, E. P., **1,** 1131
McCarthy, J. R., **9,** 423
McCarthy, T. J., **8,** 198
McCarthy, W. C., **1,** 430
McCaskie, J. E., **5,** 561
McCasland, G. E., **1,** 1008; **2,** 364
McCaully, R. J., **3,** 13, 214
McCausland, J. H., **7,** 347
McChesney, J. D., **9,** 380
McClanahan, J. L., **1,** 172
McClaugherty, W. C., **4,** 389
McClenaghan, I., **2,** 355
McClory, M. R., **5,** 129
McCloskey, A. L., **1,** 86, 211, 213, 524
McCloskey, C. J., **9,** 468
McCloskey, C. M., **1,** 1005, 1163
McCloskey, J. E., **4,** 45, 430; **6,** 304; **7,** 119
McClure, D. E., **10,** 11
McClure, G. L., **5,** 117
McClure, J. D., **1,** 465, 827
McCluskey, J. G., **2,** 319
McCollum, G. J., **4,** 475; **6,** 577
McColm, E. M., **1,** 412
McCombe, K. M., **11,** 330
McCombie, H., **1,** 253; **7,** 106
McCombie, S. W., **8,** 460, 498
McCombs, C. A., **7,** 402
McComsey, D. F., **8,** 46
McConaghy, J. S., Jr., **2,** 296
McCorkindale, N. J., **2,** 117
McCormack, M. T., **10,** 269
McCormack, W. B., **1,** 509, 690, 695; **6,** 393
McCormick, J., **1,** 1121
McCormick, J. P., **3,** 112; **4,** 53; **5,** 132; **6,** 354; **9,** 334; **10,** 292
McCormick, J. R. D., **9,** 283
McCorquodale, D. J., **2,** 27
McCourt, D. W., **11,** 501
McCown, J. D., **8,** 504
McCoy, L. L., **1,** 80, 1081
McCrae, W., **1,** 159, 169
McCreadie, T., **4,** 435
McCready, R., **9,** 509; **10,** 205
McCready, R. M., **1,** 685
McCready, R. S., **11,** 292
McCreer, D., **4,** 163
McCrindle, R., **1,** 185; **5,** 655
McCullagh, L. N., **4,** 9
McCulloch, A. W., **3,** 9; **12,** 107
McCullough, J. D., **8,** 24, 29
McCullough, R. L., **9,** 207
McCurry, P., **6,** 364
McCurry, P. M., Jr., **7,** 293; **8,** 399; **11,** 327
McCusker, P. A., **1,** 1191
McCutcheon, J. W., **1,** 1287
McDaniel, R. L., Jr., **10,** 328
McDermott, J. R., **5,** 615
McDermott, J. X., **6,** 124
McDermott, M., **5,** 155
McDivitt, J. R., **4,** 215
McDonald, C., **2,** 353
McDonald, E., **6,** 481, 629; **8,** 110
McDonald, H., **11,** 486
McDonald, I. A., **5,** 532
McDonald, J. H., III, **8,** 323; **9,** 251, 476; **10,** 194

McDonald, R. M., **2,** 422
McDonald, R. N., **1,** 320, 337, 613, 772, 933, 1152, 1246; **2,** 69, 182, 373; **3,** 50, 259; **4,** 455; **5,** 299, 341
McDonough, L. M., **5,** 285
McDougal, P. G., **9,** 270; **11,** 96, 118; **12,** 99
McDowell, D. C., **10,** 321
McDowell, J. W., **1,** 273, 1230
McElhinney, R. S., **1,** 927; **9,** 292
McElvain, S. M., **1,** 169, 402, 536, 686, 872, 926, 933, 951, 967, 1073, 1080; **2,** 330; **5,** 119, 362, 620; **7,** 340; **9,** 264
McElwee, J., **5,** 66
McEnroe, F. J., **4,** 290
McEvoy, F. J., **1,** 671; **5,** 553
McEwen, W. E., **1,** 182, 904, 1044; **4,** 390, 510; **5,** 433, 728; **11,** 150
McEwen, W. L., **1,** 454, 870, 952
McFarland, J. W., **1,** 48, 872, 1178
McFarland, P. E., **4,** 183
McFarlin, R. F., **1,** 624
McGahren, W. J., **2,** 220
McGarrity, J. F., **5,** 349
McGarry, E. J., **5,** 675
McGarvey, B. R., **7,** 60
McGarvey, G., **8,** 12, 261
McGary, C. W., Jr., **2,** 310
McGee, J. J., **1,** 1002
McGee, L. R., **10,** 133
McGhie, J. F., **1,** 152, 731; **2,** 238; **3,** 171; **5,** 472; **6,** 5, 587; **8,** 285; **9,** 378
McGillivray, G., **3,** 289; **4,** 492, 501
McGinn, F. A., **1,** 860
McGirk, R. H., **4,** 110
McGlinchey, M. J., **5,** 654
McGlotten, J., **1,** 411
McGonigal, W. E., **1,** 310
McGovern, T. P., **1,** 643
McGowan, D. A., **8,** 358; **11,** 208
McGowan, L. S., **12,** 344
McGrath, J. P., **6,** 114; **8,** 102
McGrath, K. N., **5,** 406
McGreer, D. E., **1,** 374
McGregor, A. C., **1,** 98, 250
McGuckin, W. F., **1,** 150, 331
McGuire, M. A., **11,** 47, 401
McGuire, S. E., **1,** 904
McGuirk, P. R., **8,** 117; **9,** 333
McHenry, B. M., **12,** 280
McInnes, A. G., **3,** 9; **12,** 107
McIntosh, A. V., Jr., **1,** 93
McIntosh, J. M., **4,** 262; **5,** 751; **6,** 667
McIntyre, P. S., **3,** 157
McKay, A. F., **1,** 135, 141, 195, 666; **3,** 51; **6,** 585
McKay, F. C., **1,** 182
McKean, D. R., **5,** 508; **9,** 256, 272; **10,** 219; **11,** 275, 346; **12,** 263, 445
McKee, R., **7,** 132, 227
McKee, R. L., **11,** 184
McKeer, L. C., **12,** 123
McKelvey, R. D., **5,** 391
McKelvie, N., **1,** 223; **2,** 447; **4,** 552; **7,** 97
McKendrick, J. J., **5,** 542; **8,** 135; **11,** 203
McKenna, C., **3,** 243
McKenna, C. E., **9,** 74, 256
McKenna, J., **1,** 1284; **9,** 89
McKenna, J. F., **1,** 69
McKenna, M. C., **9,** 256
McKenna, M. G., **11,** 152
McKennis, J. S., **4,** 74; **7,** 56
McKenzie, B. F., **1,** 452, 537, 566, 1000, 1006, 1118, 1283
McKenzie, R. F., **8,** 442
McKenzie, S., **5,** 253
McKenzie, T. C., **8,** 152
McKeon, J. E., **3,** 194; **6,** 573; **8,** 36; **9,** 42
McKeon, T. F., Jr., **1,** 377
McKervey, A., **10,** 21
McKervey, M. A., **2,** 248; **3,** 200; **4,** 14, 126, 240, 393; **6,** 121, 473; **9,** 51, 322; **10,** 246; **11,** 355; **12,** 44, 425
McKillop, A., **1,** 345, 465; **2,** 406, 411; **3,** 286, 289; **4,** 492, 497, 501, 527; **5,** 28, 164, 429, 430, 475, 655, 657, 659; **6,** 144, 147, 206, 360, 581; **7,** 118, 361, 365; **8,** 316, 476, 477, 480, 481; **9,** 462, 463; **10,** 397; **11,** 110; **12,** 310, 452, 483
McKinley, J. W., **4,** 10
McKinley, S. V., **4,** 573
McKinney, M. A., **5,** 765
McKinnie, B. G., **8,** 69, 341
McKown, W. D., **5,** 620
McKusick, B. C., **1,** 276, 426, 785, 859, 1006, 1136
McLachlin, D. M., **1,** 963
McLamore, W. M., **1,** 389, 1254; **9,** 318
McLaren, A. B., **5,** 287
McLaren, F. R., **10,** 198
McLaughlin, K. C., **1,** 414, 837, 951
McLaughlin, T. G., **4,** 99
McLay, G. W., **2,** 411
McLean, E. A., **11,** 28
McLean, R., **5,** 626
McLean, S., **6,** 253; **11,** 188, 388
McLeod, D. J., **1,** 1163, 1203
McLeod, G. L., **1,** 1087
McLoughlin, B. J., **2,** 29; **6,** 258
McLoughlin, J. I., **12,** 399
McMahon, E. M., **1,** 389
McMahon, R. E., **1,** 1073, 1138; **3,** 285
McManus, S. P., **1,** 703
McMartin, K., **4,** 40
McMaster, I. T., **5,** 389
McMeekin, W., **1,** 329
McMeeking, J., **5,** 35, 649
McMillan, A., **1,** 520
McMinn, D. G., **7,** 387
McMorris, T. C., **1,** 841; **9,** 380; **10,** 265
McMullen, C. W., **1,** 966
McMurry, J., **6,** 450

McMurry, J. E., **2,** 151, 176; **3,** 72; **4,** 100, 232, 507; **5,** 169, 370, 392, 495, 528, 671; **6,** 255, 588, 589; **7,** 106, 209, 369; **8,** 79, 311, 375, 483; **9,** 207, 467; **10,** 110, 175, 278, 289; **11,** 281, 421, 527; **12,** 350, 373, 395, 493
McMurry, P., **6,** 364
McNab, H., **8,** 314
McNally, S., **1,** 68
McNamara, J. M., **11,** 75, 85
McNamee, P. I., **1,** 235
McNeil, D., **1,** 782; **5,** 230
McNinch, H. A., **1,** 52
McNutt, R. W., **6,** 383
McOmie, J. F. W., **1,** 67, 134, 430; **2,** 33; **3,** 31; **4,** 250; **5,** 700; **10,** 268; **12,** 156
McOsker, C. C., **7,** 9
McPhail, A. T., **7,** 140; **8,** 167
McPhee, J. A., **8,** 238
McPherson, S. P., **1,** 832
McPhillips, J. J., **6,** 9
McQuillin, F. J., **1,** 52, 703, 1284; **2,** 121; **3,** 49; **4,** 40; **8,** 69
McQuillin, P. J., **2,** 252
McRitchie, D. D., **1,** 289
McShane, H. F., **1,** 402, 679
McShane, L. J., **8,** 2
McSweeney, J. V., **1,** 318
McWherter, P. W., **8,** 169
McWhirter, J., **1,** 1163
Mead, E. J., **1,** 603, 1109
Meader, A. L., **1,** 195
Meadows, J. D., **12,** 411
Mea-Jacheet, D., **7,** 17
Meakins, G. D., **1,** 754, 865, 989, 1004, 1008, **2,** 243; **3,** 220; **4,** 204; **6,** 304, 514
Meana, M. C., **4,** 415
Means, G. E., **9,** 227
Meanwell, N. A., **11,** 344, 570
Mebane, A. D., **1,** 574, 730, 843, 1286
Mebane, R. C., **8,** 249
Mechoulam, R., **1,** 815, 926, 1178; **2,** 278; **3,** 191; **5,** 606; **12,** 121
Mecke, R., **1,** 1058, 1152
Medema, D., **1,** 1167; **3,** 291
Medici, A., **6,** 470; **7,** 335; **8,** 122; **9,** 426
Medlik, A., **4,** 126; **5,** 86
Meek, E. G., **1,** 1210
Meek, J. S., **1,** 658
Meerwein, H., **1,** 272, 282, 283, 1018, 1189, 1211, 1232; **3,** 33, 251, 303; **4,** 167; **5,** 639, 714; **6,** 250; **11,** 206
Meesschaert, B., **9,** 212
Meguriya, N., **11,** 378
Mehlenbacher, V. C., **1,** 4
Mehlose, K., **1,** 487
Mehrotra, A. K., **5,** 351, 476; **10,** 328; **12,** 527
Mehrotra, M. L., **4,** 547; **8,** 102
Mehta, A. M., **6,** 170
Mehta, A. S., **1,** 309
Mehta, C. V., **1,** 832
Mehta, G., **2,** 70; **4,** 149; **6,** 287, 297; **7,** 56; **8,** 106; **9,** 13
Mehta, N. B., **1,** 444
Mehta, S., **3,** 61
Mehta, S. M., **1,** 1102
Meidar, D., **9,** 320; **10,** 275
Meienhofer, J., **1,** 743; **6,** 445; **8,** 81; **9,** 415; **10,** 300
Meier, G. P., **11,** 121
Meier, H., **4,** 324; **8,** 272, 465; **9,** 410
Meier, J., **1,** 289
Meier, J. P., **2,** 140
Meier, M. M., **10,** 409
Meier, W., **11,** 287
Meijer, J., **5,** 95, 168; **6,** 144, 146, 270; **7,** 80, 164, 226; **9,** 95, 106, 334; **10,** 290; **11,** 13
Meijs, G. F., **11,** 161
Meimarogl, C., **1,** 970
Meiners, A. F., **2,** 227
Meinert, H., **5,** 309
Meinwald, J., **1,** 125, 144, 562, 563, 566, 754, 755, 758, 814, 999; **2,** 50, 69, 79, 208, 422; **3,** 45, 221; **4,** 424; **5,** 411, 614; **6,** 150; **10,** 110; **11,** 178
Meinwald, Y. C., **1,** 566, 754, 1101, 1136; **2,** 208
Meisels, A., **1,** 814
Meister, H., **1,** 186
Meisters, A., **5,** 708; **6,** 624
Meisters, G., **2,** 87; **5,** 160
Meites, L., **7,** 418
Mekler, A. B., **1,** 685, 730
Melamed, G., **1,** 730
Melby, L. R., **1,** 276, 377, 1136, 1269
Melchionne, S., **4,** 84
Meldrum, A. N., **1,** 527; **8,** 314
Melega, W. P., **7,** 118; **8,** 493
Melhuish, R. R., **9,** 462
Melillo, D. G., **4,** 63; **5,** 427; **10,** 252, 449
Melin, J., **5,** 19; **6,** 23
Melis, R., **9,** 353
Meller, J. M., **9,** 270
Mellor, D. P., **3,** 233
Mellor, J. M., **2,** 238; **4,** 318; **6,** 317; **8,** 393
Melmed, K. M., **8,** 47
Melmed, K. N., **10,** 33
Melnick, B., **7,** 268
Melpolder, J. B., **7,** 274
Mels, S. J., **5,** 330
Melton, J., **4,** 507; **5,** 495, 671; **8,** 375
Meltz, C. N., **12,** 70
Meltzer, R. I., **2,** 154
Melville, D. B., **1,** 731
Melvin, L. S., Jr., **5,** 68; **6,** 215, 247; **7,** 41; **8,** 87
Menachem, Y., **9,** 223
Menapace, L. W., **1,** 1193, 1251; **3,** 134
Menapace, S. A., **4,** 570
Menard, M., **2,** 191
Menchen, S. M., **8,** 170, 455, 523; **9,** 35, 325, 511; **10,** 21, 362, 452
Mende, U., **6,** 443
Mendel, A., **2,** 205
Mendelowitz, P. C., **8,** 408
Mendelson, L., **11,** 51
Mendenhall, G. D., **3,** 324

Mendoza, A., **9,** 262
Menefee, A., **11,** 56
Meneghini, F., **5,** 581, 582
Meney, J., **4,** 252
Mengel, R., **8,** 412; **9,** 506
Menger, F. M., **6,** 70; **9,** 391; **10,** 368
Mengler, H., **1,** 531
Menicagli, R., **8,** 232
Menini, E., **1,** 87
Menon, B., **7,** 302; **8,** 130, 524
Menotti, A. R., **1,** 1049; **12,** 139
Mentzen, B., **11,** 383
Menz, F., **3,** 292; **6,** 597
Menzel, H., **4,** 544; **10,** 431
Menzel, I., **9,** 410
Menzies, I. D., **4,** 567; **5,** 486; **6,** 657
Mercer, A. J. H., **2,** 443
Mercer, F., **8,** 88; **9,** 103
Mercer, G. D., **6,** 504
Merchant, J. R., **9,** 1
Mercier, C., **7,** 102
Mercier, D., **5,** 642
Mercier, F., **9,** 89
Merckling, N., **1,** 629
Mercury, M. L., **4,** 87
Merenyi, F., **1,** 526; **4,** 273
Merenyi, R., **2,** 134; **11,** 476
Merer, J. J., **1,** 885
Mergelsberg, I., **10,** 65
Mergen, W. W., **10,** 113
Merholz, C. A., **9,** 50
Meric, R., **10,** 265
Merijanian, A., **8,** 280
Merk, B., **1,** 144
Merk, F. H., **1,** 1121
Merk, W., **4,** 297
Merkel, C., **6,** 267
Merkel, D., **5,** 120
Merkle, H. R., **1,** 1196
Merkle, U., **8,** 272
Merkushev, E. B., **10,** 419
Merlin, J. C., **10,** 370
Merlini, L., **2,** 117; **3,** 87; **6,** 517; **9,** 413; **10,** 351
Merour, J. Y., **9,** 207
Merrer, Y. L., **6,** 331
Merrifield, R. B., **1,** 519; **2,** 228; **6,** 304, 307; **8,** 230; **11,** 256; **12,** 479
Merrifield, R. E., **1,** 1136
Merrill, R. E., **6,** 91, 570; **9,** 259; **10,** 38
Merrill, S. H., **1,** 1081
Merritt, C., Jr., **1,** 33
Merritt, R. F., **1,** 773; **2,** 278, 302
Mersch, R., **3,** 303; **6,** 250
Mersereau, M., **3,** 329
Mertel, H. E., **1,** 1006
Merten, R., **1,** 677; **7,** 23
Mertes, M. P., **6,** 168; **10,** 62
Mertz, C., **5,** 282
Merz, A., **5,** 30, 647; **6,** 43
Merz, I., **6,** 665
Merzoni, S., **6,** 451
Mesbergen, W. B., **6,** 209; **9,** 286
Meshishnek, M. J., **5,** 127
Messeguer, A., **5,** 205; **9,** 52; **11,** 124
Messer, L. A., **7,** 282
Messmer, A., **5,** 601
Mester, L., **1,** 695; **2,** 23
Mestres, R., **10,** 60; **11,** 416, 466
Meszaros, M., **1,** 34
Meszaros, Z., **12,** 231
Metcalf, B. W., **7,** 389; **8,** 186; **10,** 248
Metcalf, W. S., **1,** 214
Metcalfe, T. P., **1,** 976
Meteyer, T. E., **4,** 405
Meth-Cohn, O., **1,** 827; **3,** 194; **5,** 631; **6,** 73; **9,** 515
Metlin, S., **1,** 228
Metras, F., **6,** 562
Metrione, R. M., **2,** 147
Metsumoto, M., **10,** 56
Metz, F., **11,** 462
Metz, G., **4,** 397
Metzger, H., **1,** 318
Metzger, J., **5,** 429; **6,** 228; **7,** 124; **8,** 486; **9,** 207
Metzner, P., **11,** 340
Metzner, T. J., **10,** 303
Meuley, W. C., **1,** 1039
Meunier, A., **12,** 251
Meunier, B., **6,** 179
Meunier, J. M., **8,** 59
Meurling, P., **4,** 196
Meuwsen, A., **1,** 484
Meyborg, H., **5,** 702
Meyer, A., **5,** 320; **6,** 126; **7,** 71
Meyer, C., **4,** 541
Meyer, E., **1,** 1226
Meyer, E. W., **1,** 150
Meyer, F., **3,** 304
Meyer, G., **5,** 668
Meyer, H., **1,** 3
Meyer, H. E., **1,** 352
Meyer, H. H., **8,** 308
Meyer, I., **11,** 588
Meyer, J., **1,** 1247; **12,** 551
Meyer, K., **1,** 629, 802, 904
Meyer, K. H., **1,** 140, 562, 1169
Meyer, L. U., **7,** 163
Meyer, M. D., **12,** 111
Meyer, M. W., **9,** 43
Meyer, N., **7,** 378; **8,** 69; **9,** 75, 87
Meyer, R. B., **1,** 1040
Meyer, V., **1,** 510, 1011
Meyer, W., **7,** 299
Meyer, W. L., **1,** 603, 617, 777, 974; **6,** 500
Meyer, W. U., **11,** 535
Meyer-Dayan, M., **9,** 462
Meyers, A. I., **1,** 982; **3,** 15, 284, 314; **4,** 345, 418, 484, 541; **5,** 49, 700, 715; **6,** 20, 204, 368, 389, 404, 585, 629, 630; **7,** 17, 109, 231; **8,** 279, 341; **9,** 1, 22, 43, 293, 313, 323; **10,** 12; **11,** 233, 348, 597; **12,** 11, 32, 312, 383, 384, 564
Meyers, A. I. M., **4,** 552
Meyers, C. Y., **1,** 518, 938; **4,** 475; **5,** 97; **6,** 577
Meyers, H., **11,** 416
Meyers, W. E., **11,** 393
Meyr, R., **1,** 383, 407, 844, 881, 927; **2,** 51; **4,** 101; **10,** 67
Meystre, C., **1,** 562, 624, 818, 1000; **3,** 188
Mezaros, L., **2,** 233
Mhehe, G. L., **8,** 199

Miac, C. K., **3,** 297
Miarka, S. V., **6,** 291
Micetich, R. G., **7,** 290
Micev, I., **5,** 129
Michael, A., **1,** 674, 1269
Michael, D., **1,** 1247
Michael, K. W., **7,** 388
Michaelis, A., **1,** 425; **2,** 33; **4,** 556
Michaels, R. J., **1,** 281; **2,** 101
Michaels, R. J., Jr., **1,** 682
Michaelson, R. C., **5,** 77; **8,** 393
Micha-Screttas, M., **8,** 306; **9,** 285
Michaud, C., **5,** 312
Michaud, H., **5,** 168
Micheel, F., **1,** 990
Michejda, C. J., **11,** 589
Michel, R. E., **1,** 772; **6,** 419
Micheli, R. A., **1,** 827
Michelot, D., **5,** 460; **8,** 94; **11,** 14
Michelotti, E. L., **9,** 148
Michelotti, F. W., **1,** 1095
Michelson, A. M., **1,** 1256
Michener, E., **6,** 383; **7,** 395
Michet, A., **8,** 196
Michie, J. K., **12,** 1
Michishita, K., **5,** 427
Michl, R. J., **4,** 480
Michler, W., **1,** 1146
Michniewicz, J., **5,** 309
Michno, D. M., **8,** 427
Micovic, I. V., **6,** 201
Micovic, V. M., **1,** 180, 594
Middleton, E., **3,** 24
Middleton, W. J., **1,** 377, 563, 963, 1136, 1246, 1269; **6,** 184; **8,** 167; **9,** 54; **10,** 142, 421, 453; **11,** 590
Midfley, J. M., **6,** 467
Midgley, I., **5,** 381
Midgley, J. M., **4,** 79
Midland, M. M., **4,** 230, 362, 377; **5,** 296, 347, 486; **6,** 284, 325; **8,** 286, 334, 403; **10,** 49, 321; **11,** 173, 229, 230, 430, 431; **12,** 96, 196, 399
Midorikawa, H., **6,** 14, 398, 569
Midura, W., **6,** 191; **9,** 66
Mielert, A., **1,** 182, 1136
Mier, J. D., **1,** 1210
Mierzwa, S. A., Jr., **1,** 257, 386, 1095
Mies, W., **12,** 33
Miescher, K., **1,** 477, 690, 1203; **2,** 173
Migdalof, B. H., **3,** 47
Miginiac, P., **5,** 175; **9,** 279
Migita, T., **8,** 188; **11,** 16; **12,** 174
Migliorese, K. G., **6,** 670
Mignani, G., **9,** 258
Migron, Y., **9,** 414
Mihailovic, M. L., **1,** 562, 563, 594; **3,** 45, 171, 254; **5,** 61; **7,** 196; **12,** 272
Mihelich, E. D., **4,** 405; **5,** 715; **6,** 20, 368, 389, 629; **8,** 279; **9,** 82; **11,** 99; **12,** 344, 365
Mihina, J. S., **1,** 448
Mijs, W. J., **3,** 168; **6,** 507
Mika-Gibala, A., **11,** 355
Mikami, K., **11,** 103; **12,** 96, 143
Mikami, Y., **8,** 115
Mikeska, L. A., **7,** 160
Mikhail, G., **12,** 546
Mikhailopulo, I. A., **2,** 267; **5,** 6
Mikhaleva, A. I., **11,** 440
Miki, M., **8,** 130
Miki, T., **9,** 151
Miki, Y., **5,** 433
Miklasiewicz, E. J., **1,** 1203
Mikol, G. J., **1,** 318; **2,** 173
Mikolajczak, K. L., **2,** 139
Mikolajczyk, M., **4,** 194; **5,** 266; **6,** 191, 193, 229, 295; **7,** 126, 373, 389; **8,** 135, 198, 232, 488; **9,** 66, 244; **12,** 327
Mikulak, R., **9,** 139
Mikulec, R. A., **1,** 658; **5,** 741
Milano, F., **12,** 458
Milas, N. A., **1,** 89, 465, 477, 1058; **4,** 20
Milcent, G. A., **4,** 431
Milcent, R., **6,** 514
Miles, D., **1,** 430
Miles, D. E., **9,** 186
Miles, D. H., **4,** 119, 120, 531; **5,** 177, 513; **6,** 501; **7,** 87
Miles, G. B., **1,** 477
Miles, M. L., **1,** 1081, 1246; **5,** 160
Milewich, L., **1,** 139; **5,** 508
Milewski, C., **4,** 444
Milewski, C. A., **4,** 450, 451, 559; **5,** 609; **7,** 335
Militzer, H., **4,** 317
Militzer, W., **1,** 502
Milius, R. A., **11,** 500
Milkowski, J. D., **2,** 5
Milks, J. E., **1,** 470
Mill, P. J., **1,** 518
Millar, I. T., **1,** 77, 573
Millar, M. D., **4,** 139
Millard, A. A., **8,** 67, 415
Millard, B. J., **7,** 341
Miller, A. D., **6,** 339
Miller, A. E. G., **1,** 262
Miller, A. S., **5,** 280
Miller, A. W., **2,** 30
Miller, B., **1,** 293
Miller, B. J., **2,** 245
Miller, C. E., **1,** 221, 258, 481
Miller, C. H., **4,** 424; **6,** 369
Miller, D., **4,** 40
Miller, D. B., **12,** 444
Miller, D. C., **6,** 620
Miller, D. D., **12,** 493
Miller, D. G., **7,** 32
Miller, D. J., **11,** 514
Miller, E., **1,** 506, 1090
Miller, E. G., **1,** 125
Miller, G. E., **1,** 1105; **3,** 272
Miller, G. L., **1,** 634
Miller, G. W., **2,** 93
Miller, H. E., **2,** 318
Miller, H. W., **2,** 42
Miller, J., **1,** 309
Miller, J. A., **6,** 259, 308, 676; **10,** 423; **11,** 572; **12,** 1, 54, 171, 474
Miller, J. B., **1,** 339
Miller, J. G., **9,** 261
Miller, J. L., **10,** 185
Miller, J. M., **7,** 239; **8,** 264,

411; **9,** 448
Miller, J. N., **4,** 406
Miller, K. E., **1,** 500, 731
Miller, L. A., **1,** 652
Miller, L. E., **1,** 518
Miller, L. L., **6,** 419; **8,** 109; **10,** 63; **11,** 237, 253, 452
Miller, L. S., **1,** 619; **2,** 397
Miller, M. A., **1,** 643
Miller, M. J., **6,** 233; **9,** 506; **10,** 342, 400, 448
Miller, M. W., **3,** 247
Miller, N., **2,** 412
Miller, P. S., **2,** 26; **3,** 19
Miller, R., **1,** 911
Miller, R. B., **3,** 191; **5,** 315, 556; **6,** 492; **8,** 12, 194, 261; **9,** 304; **11,** 388
Miller, R. D., **2,** 326; **9,** 256, 272; **10,** 219; **11,** 275; **12,** 263, 279, 547
Miller, R. E., **1,** 767
Miller, R. G., **1,** 46
Miller, R. L., **4,** 161
Miller, R. W., **7,** 140
Miller, S. E., **1,** 1008, 1073
Miller, S. I., **1,** 1084; **2,** 92; **3,** 69; **7,** 395
Miller, S. J., **9,** 324; **10,** 219
Miller, W. T., Jr., **1,** 81, 96, 224, 1196
Millet, G. H., **8,** 456
Millett, W. H., **1,** 407, 837
Millidge, A. F., **1,** 782
Milliet, P., **7,** 241; **11,** 39
Milligan, B., **1,** 1004
Milligan, W. O., **9,** 237
Millon, J., **6,** 666; **7,** 80
Mills, A., **5,** 410
Mills, J. A., **2,** 23
Mills, J. E., **7,** 322; **8,** 458, 481
Mills, J. S., **1,** 107, 150, 808; **2,** 39, 267
Mills, M. T., **11,** 322
Mills, N. S., **9,** 87
Mills, R. H., **1,** 1203; **4,** 85
Mills, R. W., **5,** 591, 756
Mills, S., **11,** 70; **12,** 65
Mills, W. H., **2,** 441
Millward, B. B., **1,** 927
Milner, D. J., **9,** 408
Milovanoic, J., **7,** 196
Milsom, P. J., **4,** 405
Milstein, D., **8,** 36; **9,** 42; **11,** 395
Milstein, S. R., **8,** 440
Mimoun, H., **4,** 204; **6,** 430; **8,** 206; **10,** 203, 206, 299; **11,** 430
Mimura, H., **12,** 337
Mimura, T., **5,** 621; **8,** 201; **9,** 473; **11,** 537
Min, Y. F., **11,** 387
Minabe, M., **9,** 343
Minakata, H., **10,** 289
Minameda, H., **3,** 329
Minami, I., **11,** 514; **12,** 16, 370
Minami, K., **1,** 732; **2,** 295; **4,** 506
Minami, N., **7,** 199; **9,** 87; **11,** 95
Minami, T., **4,** 573; **8,** 348; **9,** 316; **12,** 36
Minamikawa, J., **5,** 433; **6,** 233; **8,** 319; **9,** 256
Minamoto, K., **4,** 52; **5,** 567
Minaskanian, G., **9,** 159
Minasz, R. J., **1,** 854
Minato, A., **6,** 59; **11,** 168
Minato, H., **1,** 89, 262, 1189; **2,** 142; **7,** 170
Mincher, D. J., **10,** 364
Mincione, E., **8,** 50, 385; **9,** 45; **10,** 396
Mincuzzi, A., **7,** 245
Mineo, S., **7,** 251; **8,** 358
Minieri, P. P., **4,** 358; **5,** 639
Minisci, F., **3,** 38; **4,** 237; **6,** 21
Minklei, A. O., **1,** 619
Minnis, R., **2,** 107
Minnis, W., **1,** 500
Minobe, M., **12,** 256
Minor, J. L., **2,** 220
Minot, M., **1,** 281
Minster, D. K., **8,** 358; **9,** 323
Minta, A., **10,** 188; **11,** 382
Minton, M. A., **12,** 426
Mintz, M. J., **2,** 50; **3,** 38
Mio, S., **11,** 99; **12,** 325
Mioduski, J., **6,** 150
Mioskowski, C., **5,** 210; **9,** 247; **10,** 407; **12,** 241, 551
Miranti, T. S., **12,** 145
Mirek, J., **5,** 617
Mirinov, V. F., **11,** 83
Mironov, V. F., **10,** 356, 432
Mirrington, R. N., **2,** 238; **3,** 115; **4,** 465; **5,** 109
Mirsch, M., **1,** 120
Misawa, H., **10,** 148
Misbach, P., **3,** 214; **4,** 256
Mise, T., **9,** 201; **11,** 239, 240
Miser, J. R., **8,** 125
Mishima, H., **8,** 115
Mishima, T., **6,** 220, 385
Mishrikey, M. M., **7,** 10; **8,** 456
Mishriky, N., **7,** 356
Misiti, D., **1,** 448; **2,** 213; **6,** 162, 607; **7,** 175; **8,** 250; **9,** 352, 411, 432, 443; **10,** 306
Miskowski, C., **11,** 538
Mislanker, D. G., **11,** 566
Mislow, K., **1,** 860; **2,** 57; **3,** 148; **4,** 35; **5,** 55
Mison, P., **4,** 240; **7,** 250; **12,** 223
Misra, R. N., **7,** 220; **11,** 437
Misra, S. C., **8,** 285; **9,** 378
Misterkiewicz, B., **7,** 366
Misu, K., **7,** 73
Misumi, S., **6,** 500; **8,** 441; **10,** 85
Mita, N., **9,** 458
Mitachi, S., **9,** 201
Mitamura, S., **4,** 467; **9,** 314; **11,** 109
Mitani, K., **8,** 319
Mitarashi, H., **1,** 842
Mitch, C. H., **10,** 71; **11,** 215
Mitchell, A. R., **6,** 307
Mitchell, D. L., **1,** 1055
Mitchell, F. W., Jr., **1,** 372
Mitchell, G., **10,** 320
Mitchell, J. R., **4,** 80
Mitchell, J. W., **2,** 267
Mitchell, M. A., **7,** 115
Mitchell, M. J., **1,** 343, 828, 846; **3,** 231; **4,** 9; **5,** 551; **7,** 260

Mitchell, P. R. K., **10,** 81
Mitchell, P. W. D., **1,** 219
Mitchell, R. H., **2,** 344; **4,** 115; **5,** 624, 641, 729; **6,** 90, 235; **9,** 72; **10,** 340
Mitchell, R. W., **6,** 650; **7,** 313
Mitchell, T. R. B., **1,** 132; **2,** 229; **3,** 166; **9,** 349
Mitchell, W., **1,** 1168
Mitgau, R., **1,** 595, 694
Mitin, Y. V., **3,** 323
Mitra, A., **9,** 251, 476; **12,** 233
Mitra, D. K., **4,** 322; **11,** 503
Mitra, M. N., **2,** 294
Mitra, P., **10,** 256
Mitra, R. B., **1,** 82, 318; **2,** 150; **7,** 339
Mitscher, L. A., **10,** 62
Mitschke, K. H., **9,** 224
Mitschler, A., **10,** 203, 299
Mitsudo, T., **4,** 465; **5,** 358, 625; **6,** 485, 486, 552; **7,** 183, 341; **8,** 420; **9,** 108; **10,** 148
Mitsue, Y., **12,** 51
Mitsuhashi, H., **2,** 108, 117, 453
Mitsuhashi, T., **8,** 120
Mitsui, H., **12,** 247
Mitsui, S., **5,** 210, 571
Mitsui, Y., **10,** 289
Mitsukuchi, M., **5,** 427
Mitsunobu, O., **1,** 384; **2,** 40; **4,** 149, 555; **7,** 406; **10,** 182, 449; **11,** 589; **12,** 553
Mitsuo, N., **11,** 480
Mittag, E., **1,** 1261
Mittag, R., **1,** 664
Mittelmeijer, M. C., **7,** 415; **10,** 456
Miura, C., **6,** 201
Miura, H., **9,** 269; **10,** 171
Miura, K., **10,** 194; **11,** 23, 315
Miura, M., **9,** 148; **10,** 194; **11,** 592
Miura, S., **6,** 165, 641
Miwa, M., **9,** 123, 279
Miwa, T., **6,** 403; **10,** 279; **12,** 338
Miwa, T. K., **1,** 865
Mixan, C. E., **8,** 79
Miyadera, T., **11,** 125
Miyagi, S., **9,** 349
Miyaji, Y., **3,** 265
Miyake, H., **4,** 374; **8,** 312; **9,** 436; **10,** 413; **11,** 155, 551; **12,** 233, 523, 551
Miyake, K., **6,** 279
Miyake, N., **5,** 361; **6,** 51, 59
Miyake, T., **11,** 164
Miyamoto, T., **11,** 599
Miyamoyo, M., **11,** 599
Miyano, H., **12,** 526
Miyano, M., **1,** 772; **10,** 21
Miyano, O., **8,** 231
Miyano, S., **1,** 982; **3,** 258; **4,** 153; **6,** 194; **8,** 96, 172; **10,** 90; **12,** 94
Miyasaka, T., **11,** 518, 519; **12,** 485
Miyashita, A., **10,** 36; **11,** 54; **12,** 57
Miyashita, K., **1,** 894; **3,** 231
Miyashita, M., **5,** 276; **6,** 552; **7,** 253, 307; **8,** 428; **9,** 410; **10,** 279; **12,** 502
Miyashita, S., **10,** 44
Miyata, N., **9,** 242; **12,** 555
Miyata, S., **10,** 381; **11,** 117
Miyata, T., **9,** 237
Miyataka, H., **11,** 480
Miyaura, M., **8,** 260
Miyaura, N., **4,** 53; **5,** 347, 486; **6,** 149, 215, 295, 308; **7,** 4, 75; **10,** 257; **11,** 393, 514
Miyazaki, M., **10,** 273, 307; **12,** 384
Miyazaki, T., **11,** 542; **12,** 193
Miyazaki, Y., **12,** 199, 381
Miyazawa, Y., **8,** 510
Miyoshi, H., **7,** 361; **9,** 443
Miyoshi, M., **3,** 15; **5,** 178, 214; **6,** 158
Miyoshi, N., **6,** 224; **8,** 5, 29; **9,** 200, 434, 497; **11,** 220
Mizoguchi, T., **3,** 36; **4,** 270; **6,** 307
Mizsak, S. A., **8,** 259
Mizuno, A., **10,** 324
Mizuno, K., **6,** 534; **9,** 361
Mizuno, S., **8,** 337
Mizuno, Y., **5,** 538; **6,** 152, 472; **9,** 176; **10,** 5; **11,** 435
Mizushima, K., **7,** 183
Mizutaki, S., **12,** 452
Mizutani, K., **12,** 282
Mizutani, M., **12,** 256
Mlinaric, K., **5,** 639
Mlochowski, J., **9,** 99; **10,** 348, 354
Mlotkiewicz, J. A., **9,** 65
Mlotkowska, B., **7,** 126; **9,** 429
Mo, Y. K., **5,** 311, 577
Moakley, D. F., **2,** 344
Mobbs, B. E., **12,** 36
Mobbs, D. B., **4,** 397
Mobilio, D., **11,** 260; **12,** 137
Mocadlo, P. E., **1,** 89; **3,** 59
Mochizuki, D., **9,** 422
Mochizuki, F., **9,** 257
Mochizuki, K., **9,** 391
Mock, G. A., **10,** 272
Mock, W. L., **1,** 42, 258; **2,** 392; **3,** 139; **7,** 347; **8,** 501; **11,** 175
Moczar, E., **1,** 695
Modaressi, S., **12,** 187
Modena, G., **6,** 375; **11,** 61, 336; **12,** 94
Moder, T. I., **6,** 82
Modest, E. J., **1,** 230; **3,** 9
Mody, N. V., **5,** 177; **9,** 426; **10,** 9; **11,** 24
Moe, N. S., **2,** 13, 153
Moedlhammer, U., **6,** 271
Moeens, L., **11,** 434
Moehizuki, A., **9,** 349
Moehle, K. D., **1,** 911
Moehle, H., **5,** 427
Moell, N., **3,** 190
Moeller, F., **1,** 190; **2,** 101; **4,** 18; **7,** 88
Moeller, G., **1,** 561, 1015
Moerck, R., **9,** 372; **10,** 444; **11,** 83, 584
Moerck, R. E., **8,** 400
Moerikofer, A. W., **1,** 207
Moerk, P. C., **1,** 926
Moersch, G. W., **1,** 772; **4,** 302

Moffat, J., **1,** 383
Moffatt, F., **11,** 414
Moffatt, J. G., **1,** 173, 236, 269, 310, 332, 518, 802, 1148, 1250; **2,** 163, 181, 365, 446; **4,** 451, 559; **5,** 188; **6,** 229; **7,** 407; **8,** 4; **12,** 41
Moffett, R. B., **1,** 182
Moffett, R. B., **1,** 444, 526, 594, 595, 730, 1073, 1253; **4,** 231
Mohacsi, E., **7,** 113
Mohanraj, S., **12,** 350
Mohnhaupt, M., **11,** 414
Mohri, K., **5,** 691; **11,** 312
Moine, G., **12,** 298
Moinet, G., **5,** 18, 706
Moir, R. Y., **1,** 155
Moiseenkov, A. M., **5,** 596
Moiseeva, L. V., **5,** 695
Moje, S., **2,** 78
Molander, G. A., **8,** 7; **11,** 514; **12,** 5, 54, 430
Molau, G. E., **1,** 20
Mole, M. L., **4,** 183
Mole, T., **1,** 444; **5,** 708; **6,** 624
Molho, D., **5,** 507; **8,** 407; **9,** 462
Molin, M., **5,** 129
Molina, G., **4,** 135; **11,** 170
Molina, G. A., **5,** 18
Molina, J., **8,** 156
Molina-Buendia, P., **7,** 10, 409; **8,** 521
Molinero, E. G., **11,** 110
Molines, H., **6,** 263; **9,** 314; **11,** 418
Molinet, C., **1,** 1131
Molinski, T. F., **12,** 195
Molko, D., **6,** 503
Mollan, R. C., **6,** 254
Mollet, P., **1,** 222; **4,** 135
Molloy, B. B., **1,** 129, 219
Mols, O., **12,** 318
Mommaerts, J., **5,** 198
Momose, D., **12,** 130
Monaco, P., **6,** 586
Monagle, J. J., **1,** 235, 338, 679, 695, 1231; **8,** 208
Monahan, A. S., **2,** 84
Monder, C., **2,** 426; **7,** 177
Mondon, A., **1,** 659; **5,** 370; **10,** 61
Mondron, P. J., **6,** 646
Money, T., **2,** 256; **4,** 489; **5,** 52, 756; **7,** 71; **9,** 115
Mong, G. M., **7,** 189
Mongelli, N., **8,** 521
Monier-Williams, G. W., **1,** 362
Moniz, W. B., **1,** 1087
Monkiewicz, J., **10,** 181
Monkovic, I., **8,** 143
Monneret, C., **9,** 416
Monpert, A., **10,** 223
Monson, R. S., **4,** 247; **5,** 325
Monssebois, C., **2,** 293
Montaigne, R., **1,** 222; **2,** 118; **4,** 135
Montanari, F., **3,** 165; **5,** 323; **6,** 175, 191; **9,** 361; **12,** 298, 509
Montavon, M., **1,** 114, 385, 388, 614, 1178
Montecalvo, D. F., **5,** 486
Monteiro, H. J., **5,** 693; **6,** 3, 374; **8,** 34
Monteux, R., **3,** 264
Montforts, F. P., **9,** 83, 349
Montgomery, H. A. C., **1,** 861
Montgomery, J. A., **1,** 244
Montgomery, L. K., **1,** 220; **2,** 118
Monthony, J., **2,** 238
Monthony, J. F., **3,** 59
Monti, D., **12,** 422
Monti, H., **2,** 372; **12,** 420
Montijn, P. P., **2,** 344, 462
Montillier, J. P., **3,** 299
Montury, M., **9,** 61; **10,** 149, 434
Montzka, T. A., **5,** 687
Mooberry, D. D., **5,** 566
Mooberry, J. B., **2,** 196; **9,** 337
Moody, C. J., **8,** 529
Moody, R. J., **6,** 329; **10,** 239
Moog, L., **1,** 1256
Mook, R., Jr., **12,** 523
Mookherjee, B. D., **4,** 119, 372
Moon, S., **10,** 9
Moon, Y. C., **11,** 294
Mooney, B., **1,** 954
Mooney, R. A., **1,** 855
Mooradian, A., **1,** 1131
Moore, B., **1,** 952
Moore, C. G., **3,** 320
Moore, D. R., **7,** 411; **9,** 341
Moore, F. W., **1,** 573, 689
Moore, G. A., **7,** 138
Moore, G. L., **6,** 148
Moore, H. M., **9,** 103
Moore, H. W., **1,** 285, 448; **2,** 165, 213; **4,** 57, 60, 107, 148, 199; **5,** 639; **6,** 78; **7,** 43; **8,** 88, 163; **12,** 112
Moore, J. A., **1,** 195; **4,** 419; **6,** 467, 617
Moore, J. L., **9,** 423
Moore, L. L., **1,** 1073
Moore, M., **1,** 782, 796, 1184
Moore, M. L., **4,** 451
Moore, M. W., **11,** 568
Moore, R. A., **2,** 273
Moore, R. E., **1,** 444
Moore, R. H., **8,** 374
Moore, S., **1,** 489, 519
Moore, S. S., **9,** 361
Moore, T. L., **2,** 222
Moore, W. R., **1,** 96, 688, 1108; **2,** 278; **3,** 200, 204; **4,** 162
Moormeier, L. F., **8,** 69
Moorthy, S. N., **4,** 438; **6,** 523
Moos, J., **5,** 195
Moosmueller, F., **2,** 126
Moppett, C. E., **2,** 462; **5,** 127
Moracci, F. M., **6,** 534
Morales, C., **4,** 316; **5,** 57
Moralioglu, E., **4,** 167
Moran, T. A., **6,** 341
Morand, P., **1,** 219; **7,** 99
Morand, P. F., **1,** 219, 562
Morandi, J. R., **3,** 329
Morandini, F., **10,** 37
Morber, G., **10,** 59
Mordenti, L., **11,** 486
Moreau, B., **8,** 245, 437
Moreau, C., **4,** 364; **6,** 441

Moreau, G., **5,** 580
Moreau, J., **4,** 455
Moreau, J. J. E., **4,** 562
Moreau, N., **4,** 427; **5,** 580
Moreau, P., **4,** 298
Moreau, S., **5,** 310
Morehouse, F. S., **1,** 149
Morehouse, S. M., **1,** 778; **5,** 498; **10,** 303
Morel, G., **8,** 416
Moreland, M., **8,** 60
Moreland, W. T., Jr., **1,** 370, 1203
Morella, A. M., **12,** 41
Morelli, I., **3,** 240; **5,** 568
Morely, J. O., **7,** 331
Moreno, G., **2,** 321
Moreno-Manas, M., **6,** 417; **10,** 102
Moret, E., **12,** 542
Moretti, I., **3,** 291
Moretti, R., **12,** 251
Morey, M. C., **11,** 574; **12,** 365
Morgan, A. R., **10,** 213
Morgan, G. L., **6,** 656
Morgan, G. T., **9,** 50
Morgan, K., **10,** 43
Morgan, K. D., **7,** 414
Morgan, L. R., Jr., **1,** 84
Morgan, P. H., **6,** 538
Morgan, R. L., **1,** 808
Morgan, T. A., **10,** 259
Morgan, V. L., **12,** 258
Morgans, D. J., Jr., **9,** 503; **10,** 405; **11,** 518
Morgenthau, J. L., Jr., **1,** 595
Mori, A., **5,** 641; **11,** 571; **12,** 193, 356, 378
Mori, F., **6,** 610, 662
Mori, H., **1,** 1131
Mori, I., **11,** 68, 542; **12,** 344, 462
Mori, K., **3,** 41, 65; **4,** 155; **5,** 553; **6,** 216; **8,** 138; **9,** 138, 160; **10,** 417
Mori, M., **9,** 349
Mori, S., **1,** 1284; **9,** 132, 519; **10,** 119; **11,** 574; **12,** 217
Mori, T., **2,** 248; **3,** 312; **11,** 221; **12,** 125, 201, 477, 565
Mori, Y., **7,** 28; **8,** 47; **9,** 277; **12,** 143, 160
Moriarity, R. M., **1,** 402
Moriarty, R. M., **2,** 238; **3,** 244; **4,** 421; **6,** 477; **7,** 268; **10,** 214, 400; **11,** 270, 412; **12,** 259, 386
Moriconi, E. J., **1,** 465; **3,** 53; **7,** 66
Morikam, M., **5,** 280
Morikawa, A., **6,** 613; **7,** 27, 40; **11,** 550
Morikawa, M., **2,** 304
Morikawa, T., **8,** 127; **10,** 110; **12,** 199
Morimoto, T., **9,** 119; **12,** 63
Morin, J. M., Jr., **9,** 87; **11,** 471
Morin, R. D., **2,** 305
Morin-Allory, L., **12,** 285
Morisaki, M., **5,** 55, 381, 427
Morisaki, Y., **12,** 402
Morishima, N., **10,** 60
Morita, K., **1,** 144, 289, 808; **5,** 28
Morita, S., **6,** 556; **10,** 206
Morita, T., **8,** 109; **9,** 255, 256; **10,** 8, 62, 98; **11,** 128, 587
Morita, Y., **3,** 310
Moritani, I., **1,** 926; **2,** 349; **3,** 212; **4,** 129, 365; **5,** 190, 467, 499, 510; **6,** 64, 377, 392, 443, 445, 573
Morito, N., **9,** 51
Moritoki, H., **9,** 464
Moritz, K. L., **1,** 364
Moriuti, S., **2,** 81
Moriwake, T., **1,** 629; **12,** 64
Moriya, H., **12,** 356
Moriya, O., **11,** 550; **12,** 523
Moriyama, H., **2,** 227
Moriyama, Y., **11,** 317
Morizawa, Y., **11,** 68, 533; **12,** 344, 352, 524
Mork, H. M., **1,** 808
Morlacchi, F., **7,** 315
Morley, B., **2,** 294
Morley, J. F., **1,** 999
Morley, J. O., **12,** 70
Morley, J. R., **2,** 369; **3,** 16
Morley, J. S., **1,** 745, 1196; **2,** 56
Morlock, G., **1,** 403
Morneweck, S. T., **1,** 264
Moroder, F., **11,** 324
Moroder, L., **7,** 91
Moro-oka, Y., **7,** 307; **9,** 167; **10,** 56
Moroz, E., **1,** 125
Morrell, D. G., **6,** 57; **10,** 59
Morris, D. E., **5,** 46
Morris, D. G., **6,** 515
Morris, D. L., **6,** 475
Morris, G. E., **8,** 137; **10,** 116
Morris, G. F., **1,** 906
Morris, G. J., **10,** 394
Morris, J. M., **1,** 618
Morris, L. J., **1,** 814, 1004, 1011; **2,** 363
Morris, L. R., **1,** 967
Morris, M. J., **9,** 439
Morris, R. N., **3,** 74
Morris, S. G., **6,** 675
Morrise, G. E., **12,** 152
Morrisey, M. M., **12,** 428
Morris-Natschke, S., **12,** 534
Morrison, D. C., **1,** 837
Morrison, D. E., **1,** 970
Morrison, G. C., **1,** 872
Morrison, G. F., **3,** 73; **4,** 5
Morrison, H., **1,** 643
Morrison, J. D., **6,** 416; **8,** 186; **9,** 48; **11,** 153
Morrison, M. A., **10,** 448
Morrison, W. H., III, **4,** 480, 505; **5,** 651
Morrow, C. J., **6,** 2, 141, 229, 416
Morrow, D. F., **1,** 772
Morrow, S. D., **11,** 420
Morse, A. T., **2,** 173
Mortensen, H. E., **1,** 378
Mortimer, R., **7,** 379
Mortimer, R. D., **9,** 32
Morton, A. A., **1,** 248
Morton, D. R., **5,** 290, 700; **8,** 503
Morton, F., **1,** 562
Morton, G. O., **5,** 675
Morton, H. E., **9,** 334; **10,**

290; **12,** 200
Morton, J. A., **10,** 313; **11,** 418; **12,** 455
Morton, J. R., **1,** 754
Morton, R. A., **1,** 642, 951
Morton, R. B., **2,** 35
Morton, T. H., **5,** 454; **6,** 589
Mortreux, A., **12,** 53
Mory, R., **1,** 289; **9,** 335
Morzycki, J. W., **11,** 39, 276; **12,** 102
Mosby, W. L., **1,** 445; **2,** 300
Mosettig, E., **1,** 309, 772, 782, 1049; **3,** 74; **4,** 369; **5,** 680; **7,** 277
Mosher, H. S., **1,** 17, 465, 787; **4,** 152; **5,** 231; **8,** 186; **9,** 48; **11,** 330; **12,** 233, 282
Mosher, R. A., **1,** 1216
Mosher, W. A., **1,** 147, 336
Moskal, J., **12,** 511
Moss, A., **9,** 145
Moss, A. M., **3,** 148
Moss, G. P., **12,** 194
Moss, R. A., **2,** 135; **4,** 529; **5,** 17, 155, 195, 693; **6,** 137
Mossman, A. B., **6,** 36; **9,** 256; **10,** 219
Mostowicz, D., **6,** 457
Motcock, G. T., **1,** 619
Motherwell, R. S. H., **11,** 551
Motherwell, W. B., **5,** 714; **8,** 131; **9,** 119, 335, 501; **10,** 393, 413, 446; **11,** 39, 276, 515, 551, 587; **12,** 102, 338, 417, 515, 523
Motohashi, S., **11,** 283
Motoki, S., **5,** 534
Motoyama, I., **4,** 571
Motoyoshiya, J., **9,** 216
Mott, L., **1,** 1249; **3,** 322; **4,** 555; **5,** 283
Mott, R. C., **7,** 430; **9,** 249; **11,** 268
Motto, M. G., **9,** 207
Mottram, E. N., **1,** 952
Moulineau, C., **4,** 183; **5,** 244
Moulines, F., **10,** 98; **11,** 150
Moulines, J., **6,** 279; **11,** 515
Mourad, M. S., **12,** 65
Mourgues, P., **4,** 427; **5,** 580
Mousseron, M., **2,** 132, 226
Mousseron-Canet, M., **2,** 226; **4,** 238
Moutardier, G., **6,** 596
Mowery, D. F., Jr., **1,** 518
Moy, D., **6,** 573
Moya-Portuguez, M., **10,** 60
Moye, C. J., **9,** 520
Moyer, C. E., Jr., **2,** 344
Moyer, J. R., **1,** 795
Moyer, W. W., Jr., **1,** 1081
Moyle, M., **1,** 1169
Mozdzen, E. C., **9,** 256
Mozingo, R., **1,** 402, 454, 730, 731, 782; **4,** 267, 369; **7,** 277
Mrani, M., **11,** 63
Mrowca, J. J., **1,** 983; **4,** 419; **9,** 217
Mubarak, A. M., **10,** 357
Muchmore, D. C., **4,** 183, 481
Muchowski, J. M., **2,** 422; **3,** 125; **5,** 700; **10,** 78, 93
Mucke, W., **3,** 51
Muckensturm, B., **5,** 397
Mudry, C. A., **5,** 266
Muehl, G., **4,** 510; **5,** 650
Muehle, G., **1,** 11
Muehle, H., **9,** 523
Muellen, K., **4,** 579
Mueller, A. K., **7,** 385
Mueller, B., **6,** 186
Mueller, B. W., **10,** 122
Mueller, D. C., **3,** 223; **4,** 377
Mueller, E., **1,** 166, 178, 195, 754, 933, 1103; **2,** 91, 405; **3,** 69, 329; **4,** 172, 318, 324; **5,** 740; **6,** 653
Mueller, G., **1,** 677
Mueller, G. C., **2,** 27
Mueller, H., **1,** 262, 346, 685, 847, 1075
Mueller, H. R., **1,** 258; **4,** 535
Mueller, J., **11,** 401
Mueller, K. F., **5,** 649
Mueller, L. G., **9,** 483
Mueller, N., **6,** 441; **10,** 376
Mueller, P., **4,** 74; **5,** 325, 581, 607; **6,** 19; **9,** 115, 180; **10,** 418; **11,** 271
Mueller, P. M., **6,** 286; **7,** 289; **10,** 103
Mueller, R., **12,** 414
Mueller, R. H., **1,** 837; **4,** 227, 309; **7,** 210; **10,** 34; **11,** 294
Mueller, R. K., **5,** 15, 329; **7,** 171
Mueller, W., **4,** 343, 397; **5,** 457, 542
Mueller, W. E., **6,** 304, 514
Mueller, W. M., **6,** 511
Mueller-Johnson, M., **11,** 194
Mueller-Starke, H., **12,** 150
Muellner, F. X., **1,** 1235
Muench, G., **1,** 971
Muench, W. C., **4,** 139
Muencher, G., **4,** 512
Muensterer, H., **12,** 58
Muenzenmaier, W., **6,** 47
Muerke, H., **1,** 652
Mueter, B., **4,** 550
Muetterties, E. L., **6,** 15; **7,** 391
Muffler, H., **5,** 653
Mugdan, M., **1,** 477
Mugge, E., **5,** 306
Mugrage, B., **11,** 566
Muhammad, N., **1,** 1167
Muhm, H., **5,** 740
Muhs, W. H., **8,** 4
Mui, J. Y. P., **1,** 854, 1237; **2,** 328, 442
Mukaiyama, T., **1,** 196, 208, 384, 686, 843, 1216; **2,** 40, 129, 323, 445; **3,** 21, 49, 183, 322; **5,** 9, 80, 160, 161, 321; **6,** 53, 118, 308, 376, 412, 596, 613; **7,** 27, 40, 92, 110, 260, 372, 373, 428; **8,** 20, 92, 96, 225, 253, 424, 505; **9,** 20, 42, 51, 106, 132, 187, 234, 246, 414; **10,** 6, 12, 33, 41, 84, 208, 266, 277, 306, 374, 458; **11,** 112, 165, 184, 185, 203, 231, 249, 361, 410, 518, 520, 521, 522, 525, 526, 550; **12,** 4, 48, 60, 117, 150, 173, 202, 351, 413, 491,

502, 549
Mukawa, F., **1,** 562
Mukherjee, S. L., **1,** 940, 1096
Mukherjee, T. K., **1,** 231; **5,** 206
Mukhopadhyay, T., **11,** 207, 500; **12,** 418
Muller, B., **5,** 474; **6,** 419
Muller, B. L., **11,** 437
Muller, G., **1,** 125, 219
Muller, H. J., **11,** 2
Muller, J. C., **4,** 123, 127; **5,** 187
Muller, J. F., **4,** 134
Muller, R. J., **1,** 926
Muller, W., **9,** 324
Mullican, M. D., **10,** 272; **11,** 281; **12,** 207
Mullins, D., **9,** 236
Mullins, M., **6,** 628
Mullins, M. J., **8,** 9
Mullis, J. C., **11,** 150
Mullock, E. B., **2,** 334
Mulvaney, J. E., **3,** 285
Mulvaney, M., **10,** 130
Mulzer, J., **8,** 517; **9,** 288, 506; **10,** 243; **11,** 200; **12,** 278, 554
Mumaw, C. E., **1,** 1254
Mumford, C., **6,** 288; **8,** 374
Munavalli, S., **1,** 1136
Munavu, R. M., **4,** 345; **6,** 27, 404
Munch-Petersen, J., **1,** 169, 1043
Mundy, B. P., **4,** 448; **9,** 257
Mundy, D., **5,** 26, 276
Muneyama, K., **2,** 177
Muneyuki, R., **6,** 502
Munjal, R. C., **5,** 563; **9,** 145
Munk, M. E., **1,** 345
Munro, D. J., **1,** 1102
Munroe, J. E., **10,** 115; **11,** 335
Mura, A. J., Jr., **6,** 133, 359, 606; **7,** 76, 410
Murahashi, S., **1,** 228; **12,** 247
Murahashi, S. I., **4,** 365; **5,** 190, 499, 510; **6,** 64, 392, 443, 445, 573; **7,** 31, 201, 237; **8,** 346, 347, 381, 475; **9,** 349, 356, 458; **10,** 239, 269; **11,** 88, 184, 295, 342, 514; **12,** 51, 89, 373, 446
Murai, A., **7,** 208; **8,** 218; **9,** 198; **12,** 506
Murai, S., **4,** 158; **5,** 93, 589; **6,** 73, 224, 424, 509; **8,** 5, 28, 29, 133, 162, 172, 173, 498; **9,** 145, 151, 200, 434, 497; **10,** 206, 276, 312; **11,** 220, 466; **12,** 106, 144, 167, 336
Murai, T., **12,** 167
Murai, Y., **10,** 273, 374
Murakami, M., **1,** 4; **4,** 147; **5,** 428; **7,** 220; **9,** 519; **12,** 48, 150, 549
Murakami, S., **7,** 79
Muraki, M., **3,** 36; **6,** 53; **12,** 60
Muralidharan, V. B., **11,** 208, 551
Muramatsu, I., **2,** 12
Muramatsu, L., **1,** 4
Muramatsu, S., **6,** 667
Muramatsu, T., **9,** 421
Muranaka, K., **12,** 432
Muraoka, M., **3,** 137
Murase, I., **5,** 150, 163
Murata, I., **2,** 129; **4,** 339; **5,** 587, 713; **7,** 5
Murata, S., **8,** 513; **9,** 497; **10,** 262, 428, 441, 444; **11,** 571, 587
Murawski, D., **8,** 342
Murayama, E., **11,** 206; **12,** 548
Murchison, J. T., **1,** 388
Murdoch, J. R., **12,** 251
Murdock, K. C., **1,** 39
Murdock, T. O., **7,** 52
Muren, J. F., **1,** 827
Murofushi, T., **7,** 152, 354
Murphy, A., **11,** 203
Murphy, C. J., **1,** 52
Murphy, D. P. H., **5,** 672
Murphy, G. J., **4,** 385; **5,** 189, 515; **8,** 151
Murphy, G. P., **7,** 207
Murphy, J. A., **9,** 283, 302
Murphy, M., **2,** 435; **3,** 310
Murphy, P. T., **9,** 442
Murphy, R., **7,** 321
Murphy, T. J., **1,** 695, 841
Murphy, W. S., **1,** 125, 484; **9,** 218; **10,** 165, 373; **11,** 472
Murray, A. M., **6,** 515
Murray, A. W., **12,** 226
Murray, H. H., **12,** 333
Murray, I. G., **3,** 31
Murray, J. I., **1,** 1095, 1269
Murray, K., **1,** 207
Murray, K. J., **1,** 207
Murray, M. A., **1,** 730
Murray, M. F., **1,** 257; **4,** 45
Murray, P. J., **11,** 418
Murray, R. D. H., **5,** 591
Murray, R. E., **8,** 292; **10,** 308; **11,** 188
Murray, R. K., Jr., **2,** 316; **5,** 132, 329
Murray, R. W., **1,** 85, 643, 844; **3,** 324; **4,** 559; **6,** 577; **7,** 271; **9,** 341; **12,** 413
Murray, T. F., **1,** 47, 1073; **8,** 385; **11,** 397
Murray, W. P., **6,** 494
Murray-Rust, J., **8,** 159, 435; **9,** 65
Murray-Rust, P., **8,** 159, 435; **9,** 65
Murrill, N., **1,** 966
Murtiashaw, C. W., **10,** 156, 443
Murto, M. L., **1,** 309
Musallam, H. A., **12,** 386
Muscio, O. I., Jr., **8,** 119
Musgrave, O. C., **1,** 127
Musgrave, T. R., **4,** 100
Musgrave, W. K. R., **1,** 827
Mushika, Y., **3,** 49; **6,** 207
Musliner, W. J., **2,** 320; **4,** 377
Musser, J. H., **6,** 255; **8,** 311; **9,** 207
Musso, H., **5,** 532; **6,** 317; **7,** 277
Mustafa, A., **2,** 384; **4,** 188, 356
Muth, C. L., **9,** 256
Muth, K., **1,** 846

Muth, R., **6,** 467
Muthard, J. L., **9,** 123
Muthukrishnan, R., **7,** 401
Muxfeldt, H., **1,** 1211; **2,** 196, 256, 333, 378; **4,** 361; **7,** 162; **9,** 299, 337, 406
Muzart, J., **10,** 302; **11,** 395
Muzika, K., **5,** 595
Myatt, H. L., **6,** 64
Mychajlowskij, W., **5,** 68, 376; **6,** 482; **7,** 194, 279; **8,** 57; **9,** 497
Mycz, D. M., **10,** 225
Myddleton, W. W., **1,** 652
Myers, A. G., **12,** 52
Myers, A. I., **4,** 327
Myers, G., **3,** 171
Myers, R. F., **4,** 531; **8,** 503
Myers, T. C., **1,** 182
Myerson, J., **8,** 260; **9,** 249
Myles, J. R., **1,** 248
Mylo, B., **1,** 636
Myong, S. O., **10,** 170
Myrboh, B., **10,** 229

Naae, D. G., **5,** 63; **9,** 166; **10,** 448
Nabeya, A., **3,** 284; **4,** 484
Nace, H. R., **1,** 309, 624; **3,** 72, 124, 128
Nachmias, G., **1,** 411
Nade, N., **10,** 273
Nadelson, J., **7,** 211
Nader, B., **11,** 113
Nader, R. B., **9,** 429
Naderi, M., **11,** 62
Nadir, U. K., **10,** 169
Nadjo, L., **11,** 255
Nador, K., **5,** 118, 225
Naef, F., **6,** 215, 479, 675; **9,** 361; **11,** 373
Naef, R., **11,** 299, 529
Naegely, P. C., **10,** 403
Naemura, K., **1,** 999
Nafti, A., **12,** 223
Nagabhushan, T. L., **1,** 755; **3,** 316
Nagai, H., **11,** 243
Nagai, S., **10,** 173
Nagai, T., **5,** 486; **10,** 427
Nagai, Y., **4,** 563; **5,** 740, 741; **6,** 613, 633, 655; **8,** 161, 240; **9,** 404, 492; **10,** 44, 196, 428; **11,** 253, 448, 514
Nagakura, I., **6,** 344, 648; **7,** 212; **11,** 373
Nagamatsu, T., **6,** 547; **7,** 130
Nagami, K., **9,** 215
Nagano, T., **11,** 445; **12,** 414
Nagao, K., **11,** 159
Nagao, S., **12,** 110
Nagao, Y., **7,** 365; **9,** 221, 315, 462; **11,** 1, 324, 518, 519; **12,** 485
Nagaoka, H., **8,** 444; **10,** 346; **11,** 134
Nagaoka, M., **12,** 263
Nagarajan, R., **5,** 428
Nagarkatti, J. P., **5,** 510
Nagasaki, T., **1,** 1189; **2,** 142
Nagasampagi, B. A., **6,** 3; **7,** 32
Nagasawa, H. T., **6,** 174
Nagasawa, K., **6,** 368
Nagasawa, N., **8,** 516
Nagase, H., **7,** 135; **8,** 354; **9,** 136
Nagase, K., **1,** 59
Nagashima, A., **4,** 405
Nagashima, H., **8,** 28; **10,** 176, 302; **11,** 391, 396
Nagashima, N., **11,** 239
Nagashima, T., **10,** 103; **12,** 447
Nagasuna, K., **11,** 174
Nagata, T., **6,** 296
Nagata, W., **1,** 244, 444, 1198; **2,** 128, 132, 427; **3,** 96; **4,** 526; **5,** 370; **9,** 13, 24, 328, 519; **12,** 121
Nagata, Y., **4,** 147, 526
Nagayama, I., **8,** 198; **9,** 188
Nagayama, M., **5,** 641
Nagayashi, K., **12,** 231
Nagel, A., **6,** 411
Nagel, C. J., **12,** 379
Nagendrappa, G., **3,** 101; **4,** 405; **10,** 356
Nagihara, N., **5,** 146
Nagira, K., **10,** 298
Nagler, M. J., **9,** 257
Nagura, K., **10,** 305
Nagy, P. L. I., **9,** 426
Nagy, S. M., **1,** 481
Nahm, S., **11,** 202
Naidu, M. V., **10,** 136
Naik, D. N., **1,** 1261
Naik, N. C., **7,** 386
Naik, N. N., **4,** 46
Naik, R. G., **12,** 428
Nainan, K. C., **11,** 502
Nair, M. D., **1,** 827
Nair, M. G., **2,** 344
Nair, P. G., **6,** 34
Naito, K., **12,** 286
Naito, T., **9,** 77
Naito, Y., **9,** 328; **10,** 281
Najasaki, T., **1,** 262
Najer, H., **6,** 570
Najera, C., **11,** 318
Naka, M., **7,** 74
Nakabayashi, J., **10,** 44, 196
Nakabayashi, S., **11,** 587
Nakagawa, I., **6,** 602
Nakagawa, K., **1,** 563, 732; **2,** 295; **4,** 506; **7,** 251; **8,** 358; **10,** 277; **12,** 449
Nakagawa, M., **6,** 138; **7,** 268; **9,** 91
Nakagawa, Y., **10,** 148
Nakagome, T., **12,** 508
Nakahama, S., **9,** 422; **12,** 31
Nakahara, A., **7,** 79
Nakahara, T., **3,** 312
Nakai, A., **11,** 430
Nakai, H., **9,** 171
Nakai, T., **6,** 14, 365, 540; **8,** 201; **9,** 166, 398, 473; **10,** 448; **11,** 103, 537; **12,** 96
Nakai, Y., **10,** 377
Nakaido, S., **8,** 188
Nakajima, I., **11,** 168
Nakajima, K., **1,** 500, 505
Nakajima, M., **2,** 263; **5,** 226; **7,** 20; **9,** 166
Nakajima, T., **12,** 29, 316, 402
Nakajima, Y., **7,** 406; **12,** 221
Nakajo, E., **12,** 51
Nakamori, S., **11,** 1
Nakamura, A., **3,** 329; **8,** 40; **10,** 303, 437; **11,** 174, 310

Nakamura, C. Y., **8,** 367
Nakamura, E., **6,** 44; **7,** 152, 354; **8,** 52, 220, 227, 468; **9,** 283; **11,** 235; **12,** 222, 462, 490, 502, 537
Nakamura, F., **9,** 188
Nakamura, H., **4,** 250; **6,** 65; **7,** 52, 428; **8,** 327; **10,** 277
Nakamura, K., **2,** 222; **8,** 509; **9,** 491; **11,** 445
Nakamura, M., **11,** 462
Nakamura, N., **4,** 562; **6,** 56; **12,** 131, 446
Nakamura, S., **9,** 464; **10,** 398; **12,** 188
Nakamura, T., **9,** 317
Nakane, M., **8,** 338
Nakanishi, A., **6,** 585
Nakanishi, K., **1,** 147; **2,** 227, 318; **6,** 202; **9,** 77, 198, 207; **10,** 307; **11,** 128
Nakanishi, N., **6,** 585
Nakanishi, S., **1,** 182, 808; **9,** 278
Nakanishi, T., **7,** 111
Nakano, A., **9,** 479
Nakano, J., **11,** 190
Nakano, K., **9,** 519
Nakano, M., **6,** 618
Nakano, S., **6,** 374
Nakano, T., **1,** 444, 608; **5,** 13, 178, 741; **6,** 158; **8,** 161, 240; **9,** 492; **10,** 44, 196; **11,** 253, 448
Nakao, R., **4,** 526; **11,** 127
Nakasuji, K., **4,** 339; **5,** 587; **7,** 5
Nakata, F., **7,** 79
Nakata, H., **1,** 989
Nakata, T., **1,** 732; **5,** 163; **9,** 77, 110; **10,** 461; **11,** 600; **12,** 288, 441, 573
Nakatani, M., **12,** 546
Nakatani, Y., **8,** 535
Nakatsubo, F., **4,** 86
Nakatsugawa, K., **6,** 633, 653
Nakatsuji, S., **9,** 91
Nakatsuka, M., **10,** 84, 106, 185, 381; **11,** 117
Nakatsuka, S., **5,** 183
Nakatsuka, T., **5,** 171; **6,** 51
Nakaya, T., **2,** 264
Nakayama, A., **7,** 165
Nakayama, H., **9,** 431
Nakayama, J., **3,** 294; **6,** 96; **8,** 34, 238; **10,** 25; **12,** 453
Nakayama, K., **6,** 129; **9,** 188, 341
Nakayama, M., **10,** 99; **11,** 480
Nakazaki, M., **1,** 999; **12,** 191
Nakazaki, N., **7,** 171; **10,** 207
Nakazawa, T., **5,** 667
Nakazima, K., **7,** 345
Nalley, E. A., **4,** 389
Namanworth, E., **2,** 200
Namba, T., **5,** 58
Nambara, T., **3,** 325; **7,** 32
Nambiar, K. P., **9,** 242, 301
Nambu, H., **1,** 1216; **3,** 29, 237
Nambudiry, M. E. N., **6,** 341
Namigata, F., **2,** 424
Namkung, M. J., **1,** 444, 500, 1141
Namy, J. L., **8,** 439; **10,** 344; **11,** 465; **12,** 430
Nanba, K., **11,** 480
Nanbu, A., **5,** 571
Nanda, R. K., **9,** 103
Nandi, U. S., **1,** 45
Nangia, A., **12,** 454
Nanninga, T. N., **11,** 260
Naoi, Y., **5,** 13, 178; **6,** 158
Naota, T., **11,** 184
Napier, D. R., **1,** 124, 1019, 1187
Napier, R., **1,** 101
Napier, R. P., **5,** 394
Naples, J. O., **5,** 356; **7,** 32
Naquib, Y. M. A., **10,** 225
Naqvi, S. M., **1,** 630
Nara, M., **8,** 320, 403
Narang, S. A., **1,** 662; **5,** 309, 434, 517; **6,** 361, 622; **7,** 14, 15
Narang, S. C., **8,** 263, 361, 518; **9,** 236, 255, 256, 320, 324, 326, 335; **10,** 96, 115, 275, 329, 409; **11,** 275, 355, 553, 587; **12,** 263
Naranjo, S. B., **9,** 361
Narasaka, K., **5,** 9, 160; **6,** 596; **7,** 372; **8,** 20; **10,** 410; **11,** 184, 185; **12,** 118, 275, 338, 444
Narasimhan, K., **3,** 177
Narasimhan, L., **9,** 8, 40
Narasimhan, N. S., **6,** 90; **12,** 175
Narasimhan, S., **11,** 69, 293
Narayan, B. A., **10,** 87
Narayan, K. V., **2,** 464
Narayana, M., **1,** 689; **2,** 376
Narayanan, C. R., **1,** 73, 604; **6,** 3; **7,** 32; **8,** 364
Narayanan, V. L., **3,** 231
Narayanaswamy, M., **3,** 126
Narbonne, C., **7,** 167
Nardin, G., **11,** 124
Narimatsu, S., **11,** 36
Narisada, M., **2,** 427; **3,** 81; **4,** 147; **8,** 39
Narisano, E., **9,** 475; **10,** 409; **12,** 444, 509, 510
Narita, K., **2,** 147
Naritomi, T., **12,** 435
Narnik, M., **1,** 389
Narula, A. P. S., **11,** 347; **12,** 272
Narula, A. S., **3,** 54
Naruse, K., **11,** 514
Naruse, M., **3,** 228; **5,** 435; **6,** 295
Naruse, N., **3,** 60, 214; **6,** 215
Naruta, Y., **9,** 8, 500; **10,** 411, 451; **11,** 22, 143; **12,** 405
Narwid, T. A., **5,** 49, 700
Naser-ud-din, **5,** 283
Nash, B. W., **1,** 735
Nash, D. R., **4,** 63
Nash, R. D., **6,** 492; **9,** 304
Nash, S. A., **12,** 275
Nashed, M. A., **9,** 54
Nasielski, J., **5,** 143
Nasipuri, D., **1,** 522, 599; **2,** 243; **5,** 141, 614; **11,** 71
Naso, F., **4,** 477; **5,** 556; **10,** 289
Nasutavicus, W. A., **6,** 541
Natale, N. R., **7,** 335; **9,** 383, 426; **11,** 465

Natalie, K. J., Jr., **9,** 494
Natarajan, S., **7,** 384; **8,** 344
Natelson, S., **1,** 290, 1128, 1178
Nathan, A. H., **1,** 690, 787, 1118
Nathan, E. C., **4,** 405; **6,** 479
Nath-Guthikonda, R., **4,** 577
Natile, G., **5,** 647
Nations, R. G., **1,** 986
Natsukawa, K., **6,** 509
Natsume, M., **5,** 344, 614
Natsume, S., **9,** 396
Natta, G., **1,** 228, 1171
Naumann, A., **11,** 2
Naumann, K., **3,** 148; **4,** 35
Naumann, M. O., **2,** 364
Naumann, R., **1,** 927
Navarette, B. J., **4,** 412
Navarre, J. M., **4,** 522
Navarro, M., **11,** 407
Nave, D. W., **8,** 461
Nave, P. M., **3,** 45
Naves, Y. R., **1,** 500
Nayak, U. R., **6,** 475
Naylor, C. A., Jr., **1,** 951, 1293
Naylor, M. A., **1,** 755
Naylor, P., **1,** 833
Naylor, R. D., **4,** 497
Nazarenko, N., **4,** 484
Nazareno, L., **4,** 489
Nazarov, I. N., **1,** 1272, 1273
Nazer, B., **12,** 406, 414, 485
Nazer, M. Z., **1,** 23, 187, 606; **2,** 17, 381
Ncube, S., **8,** 238; **9,** 5
Neal, G. W., **8,** 80
Neale, R. S., **1,** 139; **4,** 139
Nealy, D. L., **2,** 397
Neark, H. F., **1,** 1171
Neary, N., **8,** 438
Neber, P. W., **5,** 433
Nechvatal, A., **1,** 93
Neckers, D. C., **4,** 14; **5,** 13, 491; **6,** 19
Necoechea, E., **1,** 219
Nedelec, J. Y., **7,** 174
Nee, G., **9,** 502; **10,** 198
Needham, L. L., **4,** 378; **7,** 88
Needles, H. L., **1,** 78; **8,** 272
Neef, G., **10,** 5
Neelakantan, S., **1,** 235
Neely, L., **1,** 519
Neeman, M., **1,** 195, 396, 764, 808; **2,** 311
Neese, H. J., **11,** 378
Neeser, J. R., **11,** 489
Nef, J. U., **1,** 844
Neff, J. R., **8,** 299
Nefkens, G. H. L., **1,** 112, 135, 486
Negishi, A., **3,** 219
Negishi, E., **2,** 149; **3,** 43; **4,** 36, 53, 69, 176; **5,** 47, 57, 186, 233, 486; **6,** 91, 208, 399, 492, 570, 601; **7,** 26, 96, 113, 357, 379; **8,** 41, 87, 407, 475, 507; **9,** 482, 488; **10,** 133, 243, 391, 423; **11,** 299, 514, 568, 575; **12,** 54, 171, 411, 474
Negishi, E. I., **10,** 398
Negishi, Y., **11,** 214
Negoro, K., **11,** 343, 421
Negri, D. P., **8,** 64
Negron, G., **12,** 533
Neher, H. T., **1,** 652
Neher, W. T., **2,** 231
Nehring, R., **1,** 1137
Neidle, S., **4,** 262
Neilan, J. P., **7,** 5
Neilson, T., **1,** 1055; **5,** 627; **6,** 605
Neiman, L. A., **4,** 282
Neiman, M. B., **6,** 114
Neirabeyeth, M. A., **7,** 35
Neises, B., **8,** 163; **9,** 157; **11,** 360
Neish, A. C., **4,** 199
Neisser, M., **11,** 170
Neiswender, D. D., Jr., **1,** 1087
Nel, M., **9,** 132
Nelb, R. G., **1,** 610
Nelds, J. M., **8,** 52
Nelke, J. M., **4,** 490; **6,** 141; **8,** 109
Nelsen, E. F., **1,** 211; **2,** 32; **5,** 609
Nelsen, T. R., **6,** 102, 597
Nelson, A. B., **1,** 1100
Nelson, J. A., **1,** 164; **2,** 79
Nelson, J. H., **10,** 42
Nelson, J. S., **1,** 367
Nelson, J. V., **8,** 415; **9,** 140; **10,** 328
Nelson, N. A., **1,** 56, 268, 1041
Nelson, P., **6,** 554
Nelson, P. H., **3,** 87; **10,** 78
Nelson, R. B., **5,** 614
Nelson, S. J., **4,** 270
Nelson, V., **5,** 620; **6,** 146
Nematollahi, J., **6,** 223
Nemec, J., **10,** 125
Nemec, M., **4,** 70
Nemer, M. J., **11,** 124, 284
Nemo, T. E., **11,** 462
Nemoto, H., **9,** 32; **10,** 21, 182, 302; **12,** 447
Nencki, M., **1,** 1292
Nenitzescu, C. D., **1,** 191, 343, 657, 1027; **2,** 79, 112; **6,** 279; **7,** 94
Neoh, S. B., **6,** 471
Nerdel, F., **2,** 197; **3,** 140; **6,** 263
Nerger, W., **1,** 352
Neri, O., **11,** 128
Neri, R., **5,** 552
Nes, W. R., **1,** 157
Nesbit, S. L., **8,** 352
Nesbitt, B. F., **5,** 633; **6,** 547
Nesmejanow, A. N., **1,** 349, 654; **5,** 44
Nestler, G., **5,** 511
Nestrick, T. J., **3,** 243
Neth, F. T., **1,** 124
Nettleton, D. E., Jr., **1,** 703
Neubauer, D., **3,** 83
Neubauer, O., **1,** 954
Neuberg, C., **1,** 140
Neuberger, A., **1,** 531
Neubert, K., **8,** 537
Neubert, M. E., **8,** 366
Neubold, G. T., **8,** 265
Neuenschwander, K., **10,** 354, 418; **11,** 20, 469; **12,** 420
Neuenschwander, M., **8,** 184, 349; **9,** 323
Neugebauer, F. A., **1,** 57; **4,** 24

Neuklis, W. A., **1,** 772
Neukom, C., **10,** 155
Neuman, S. M., **7,** 411
Neumann, F., **1,** 1000
Neumann, F. W., **1,** 1027, 1283
Neumann, H., **3,** 116; **7,** 47; **8,** 318
Neumann, H. M., **5,** 419
Neumann, S. M., **6,** 259
Neumann, W. P., **1,** 262
Neumeister, J., **9,** 343
Neumeyer, J. L., **6,** 477
Neunhoeffer, H., **4,** 120; **5,** 329, 595
Neurath, G., **1,** 736
Neureiter, N. P., **1,** 1216
Neuss, N., **1,** 796
Neuwirth, H., **1,** 796
Nevell, T. P., **5,** 603
Nevestveit, C., **9,** 165
Neville, G. A., **4,** 431
Neville, R. G., **1,** 1002
Nevitt, T. D., **1,** 1143
Newall, G. E., **3,** 96
Newbold, B. T., **1,** 465
Newbould, J., **1,** 818
Newcomb, M., **8,** 299
Newcomer, J. S., **1,** 270, 1132
Newell, L. C., **1,** 849
Newell, R. G., **5,** 297
Newhall, W. F., **1,** 176
Newkirk, J. D., **9,** 182
Newkome, G. R., **1,** 290; **2,** 155; **3,** 117; **5,** 117, 249, 457; **6,** 244, 479; **10,** 249
Newman, H., **1,** 1226; **4,** 161; **5,** 45
Newman, H. M., **12,** 89
Newman, M. S., **1,** 22, 81, 106, 158, 281, 343, 424, 432, 489, 518, 659, 660, 678, 685, 696, 703, 730, 754, 777, 890, 904, 938, 952, 963, 974, 978, 982, 1005, 1019, 1023, 1064, 1081, 1087, 1147, 1178, 1203, 1226; **2,** 36, 101, 104, 165, 173, 174, 203, 270, 404, 455; **3,** 74, 118, 128, 183, 187; **4,** 31, 54, 122, 202, 243, 248, 296, 389, 564, 565; **5,** 107, 144, 304, 325, 359, 360, 427, 593, 620, 713; **6,** 4, 360, 406, 503, 546; **8,** 245, 279, 446; **9,** 151, 465
Newman-Evans, D. D., **11,** 551
Newmann, H., **3,** 57; **5,** 80
Newstead, R. C., **2,** 344
Newton, A., **1,** 1229
Newton, B. N., **7,** 170
Newton, D. J., **3,** 285
Newton, L. W., **1,** 7, 276, 452
Newton, M. V., **1,** 383
Newton, R. F., **9,** 65, 239; **10,** 94
Newton, R. J., Jr., **9,** 99, 489
Newton, T. A., **4,** 541
Newton, T. W., **12,** 212
Neya, M., **12,** 199, 381
Ng, K. S., **10,** 395; **11,** 119
Ng, Y. H., **1,** 982
Ngugen, L. T., **11,** 589
Nguyen, N. H., **11,** 430
Niaz, G. R., **3,** 285
Nice, F. A., **2,** 228
Nichisch, K., **9,** 90
Nicholas, K. M., **6,** 155; **8,** 150; **10,** 130; **12,** 167
Nicholas, R. D., **1,** 791; **2,** 17; **3,** 6
Nicholls, A. C., **9,** 439
Nicholls, B., **6,** 104, 126; **9,** 292
Nichols, G., **1,** 47, 1163
Nichols, J., **1,** 362, 367
Nichols, S. B., **4,** 63
Nicholson, D. G., **1,** 23
Nicholson, J., **1,** 833
Nicholson, R. T., **5,** 144
Nickel, D. L., **6,** 40
Nickell, D., **9,** 449
Nickisch, K., **10,** 451; **11,** 281
Nicklin, P. D., **5,** 646
Nickon, A., **1,** 484, 642, 952; **4,** 363; **5,** 491
Nickson, T. E., **11,** 452
Niclas, H. J., **2,** 159
Nicolaides, E. D., **1,** 1220
Nicolaides, N., **1,** 1167, 1269
Nicolaou, K. C., **5,** 286, 288; **6,** 247, 319, 490; **8,** 28, 34, 87, 259, 489; **9,** 24, 31, 32, 169, 237, 283, 367; **10,** 21, 147, 313; **11,** 559; **12,** 143, 185, 476, 536
Nicoletti, R., **7,** 20; **8,** 13
Nicolini, M., **12,** 502
Nidy, E. G., **6,** 490; **8,** 259, 419; **9,** 249
Nie, P. L., **7,** 10, 409; **8,** 520, 521
Niedballa, A., **11,** 524
Niedballa, U., **6,** 640
Niedenzu, K., **9,** 63, 152; **11,** 590; **12,** 179
Niederpruem, H., **1,** 1047; **3,** 262; **4,** 155, 409, 568; **5,** 480; **6,** 273
Nieduzak, T. R., **11,** 317
Nieh, M. T., **4,** 570
Nielsen, A. T., **1,** 178; **4,** 14, 510; **5,** 441; **10,** 296
Nielsen, D. R., **1,** 182; **2,** 309
Nielsen, M. L., **2,** 207
Nielsen, S. D., **1,** 999
Nielsen, S. F., **6,** 137
Niemann, C., **1,** 454
Nienhouse, E. S., **2,** 120
Nierth, A., **4,** 533; **6,** 620
Niethammer, K., **1,** 343
Nietzki, R., **3,** 156
Nieuwenhuis, H. J. W., **1,** 927, 1254
Nieuwenhuis, J., **1,** 360
Nieuwland, J. A., **1,** 12, 659, 1040
Nieves, I., **6,** 162
Niewoehner, C. B., **9,** 391
Niewohner, U., **10,** 308
Nigam, A., **10,** 198
Nigh, W. G., **9,** 121
Nihonyanagi, M., **5,** 740; **6,** 613, 653; **10,** 427
Niida, T., **9,** 317; **11,** 322
Niino, Y., **11,** 480
Niinomi, Y., **11,** 591
Nikaido, T., **6,** 655
Nikishin, G. I., **5,** 157; **6,** 316
Niklas, H., **1,** 445

Nikrad, P. V., **11,** 255
Nilles, G. P., **2,** 107; **6,** 98
Nilsen, B. P., **6,** 517
Nilsson, A., **6,** 23; **12,** 502
Nilsson, M., **1,** 526; **4,** 273
Nilsson, N. H., **8,** 327; **9,** 50
Nilsson, T., **5,** 603
Nimetz, A. A., **1,** 563
Nimgirawath, S., **9,** 68
Nimitz, J. S., **9,** 414; **12,** 233
Ninagawa, A., **9,** 94
Ninomiya, K., **4,** 211; **5,** 280; **7,** 138
Nishi, A., **5,** 13, 178; **6,** 158
Nishi, T., **12,** 429
Nishibuchi, T., **9,** 278
Nishida, A., **11,** 456
Nishida, I., **10,** 453
Nishida, M., **10,** 183
Nishida, R., **12,** 64
Nishida, S., **12,** 384
Nishida, T., **9,** 316
Nishide, H., **12,** 307
Nishide, K., **9,** 13; **10,** 11; **12,** 30
Nishigaki, S., **5,** 479
Nishigaki, Y., **7,** 12
Nishiguchi, I., **8,** 532; **9,** 415; **10,** 460; **11,** 599; **12,** 439
Nishiguchi, T., **5,** 332; **6,** 653; **7,** 109, 314
Nishihara, A., **3,** 43
Nishihata, K., **11,** 322
Nishikawa, K., **5,** 560
Nishikawa, M., **1,** 289
Nishikawa, N., **10,** 14
Nishiki, M., **11,** 480
Nishimura, J., **1,** 253; **2,** 134; **3,** 141; **5,** 311
Nishimura, K., **11,** 514
Nishimura, O., **5,** 360; **6,** 291
Nishimura, S., **2,** 215, 305; **5,** 42; **7,** 3, **8,** 434; **9,** 257, 408
Nishimura, T., **1,** 147; **3,** 101
Nishimura, Y., **10,** 240
Nishinaga, A., **3,** 245; **6,** 430, 479; **7,** 260; **9,** 140
Nishio, K., **9,** 256; **12,** 263
Nishio, T., **6,** 161; **10,** 93
Nishitani, K., **7,** 43; **10,** 21; **11,** 152, 158
Nishitani, Y., **9,** 13, 519
Nishiuchi, K., **11,** 419
Nishiwaki, K., **5,** 620
Nishiyama, H., **9,** 136; **10,** 107; **11,** 36, 373, 452, 522, 533; **12,** 435, 523, 547
Nishiyama, K., **11,** 32; **12,** 539
Nishiyama, N., **11,** 581
Nishiyama, S., **11,** 516
Nishizauka, T., **10,** 347
Nishizawa, K., **4,** 107
Nishizawa, M., **7,** 137, 160, 253, 314; **8,** 24, 492; **9,** 81, 170, 478; **10,** 149, 453; **12,** 191, 307, 405, 508
Nishizawa, Y., **8,** 122
Nist, B. J., **4,** 113
Nitasaka, T., **10,** 413
Nitecki, D. E., **1,** 274; **2,** 203
Nitta, M., **11,** 350; **12,** 526
Nitta, T., **2,** 129
Nitta, Y., **1,** 247
Nitz, T. J., **12,** 139
Nitzsche, M., **1,** 707
Niu, C. H., **7,** 294
Nivard, R. J. F., **1,** 112, 486; **5,** 614, 653; **8,** 537; **9,** 523; **11,** 71
Niwa, H., **9,** 243; **11,** 71
Niwa, I., **8,** 28; **10,** 312
Niwa, M., **5,** 319; **7,** 165; **9,** 11
Nixon, J., **7,** 94
Nixon, J. R., **8,** 249
Niznik, G. E., **3,** 280; **4,** 186, 480, 505; **5,** 651; **6,** 643; **8,** 464
Noall, W. I., **8,** 260; **10,** 455
Nobayashi, Y., **12,** 70
Nobile, A., **1,** 1000
Nobis, J. F., **8,** 69
Noble, E. G., **1,** 1197
Noble, W. R., **1,** 1269; **2,** 310
Nobuhara, Y., **1,** 668
Noda, A., **10,** 289; **11,** 249
Noda, S., **5,** 641
Noda, Y., **11,** 594
Node, M., **7,** 33; **9,** 13, 63; **10,** 11, 52; **11,** 29; **12,** 29, 30
Nodiff, E. A., **3,** 143
Noding, S. A., **9,** 46; **10,** 131
Noe, C. R., **12,** 339
Noe, J. L., **1,** 673
Noel, Y., **5,** 355
Noels, A., **5,** 626
Noels, A. F., **7,** 313; **8,** 435; **10,** 342; **11,** 460
Noeth, H., **1,** 16, 273; **6,** 576
Noftle, R. E., **12,** 173
Nogami, J., **5,** 42
Nogi, T., **2,** 305
Nogradi, M., **7,** 365; **11,** 188
Noguchi, H., **11,** 231
Noguchi, I., **3,** 145
Noguchi, M., **9,** 205
Noguchi, S., **1,** 144, 289
Noguchi, Y., **9,** 353; **10,** 221
Nogucki, J., **2,** 389
Noguez, J. A., **6,** 267, 403
Nohara, A., **5,** 253, 617
Nohira, H., **1,** 843; **12,** 313
Nojima, H., **6,** 106
Nojima, M., **5,** 171, 539, 577; **6,** 296; **11,** 592
Nojime, M., **9,** 400
Nokami, J., **6,** 257, 561; **11,** 419; **12,** 392, 486
Nolan, S. M., **9,** 512; **11,** 220
Nolan, T. J., **2,** 445
Noland, N. E., **8,** 461
Noland, W. E., **1,** 251, 741, 744; **4,** 468; **6,** 545
Nolen, R. L., **4,** 484; **5,** 49, 700, 715
Noller, C. R., **1,** 471, 500, 664, 730, 874, 881, 1163, 1232, 1293; **10,** 306
Noller, D. C., **11,** 576
Nolley, J. P., Jr., **5,** 497
Nolte, R. J. M., **11,** 567
Noltes, J. G., **1,** 1193, 1251
Nomine, G., **1,** 215, 361, 518, 608
Nomizu, S., **12,** 64
Nomoto, T., **12,** 399
Nomura, K., **11,** 494
Nomura, M., **6,** 17
Nomura, R., **9,** 94
Nomura, Y., **11,** 33, 417, 468

Nonaka, R., **4,** 390
Nonaka, T., **12,** 419
Nonhebel, D. C., **1,** 162, 163; **2,** 87; **5,** 160
Norcross, B. E., **1,** 1080; **7,** 301
Nord, F. F., **1,** 595, 767
Nordberg, R. E., **12,** 370
Nordin, I. C., **1,** 624, 652, 983; **4,** 167
Nordin, P., **1,** 1002
Nordlander, J. E., **8,** 299; **9,** 387
Norell, J. B., **5,** 337
Norin, T., **2,** 238; **7,** 336; **9,** 462
Norman, L. R., **5,** 631
Norman, R. O. C., **1,** 212, 476; **2,** 159; **3,** 171; **5,** 102, 370, 659; **6,** 70, 316, 443; **12,** 123
Normant, H., **1,** 424, 431, 712; **2,** 92, 210; **3,** 153; **4,** 247; **5,** 543; **6,** 279, 333; **7,** 204; **9,** 217; **10,** 198
Normant, J., **1,** 431; **6,** 143
Normant, J. F., **2,** 398; **3,** 153; **4,** 139; **5,** 94, 167, 168, 406, 411, 667; **6,** 189, 270, 331, 345, 577, 622, 664; **7,** 66, 83, 222; **8,** 313, 505; **9,** 138, 154, 249, 289; **10,** 1, 194, 289, 290, 326; **12,** 350
Normark, J. R., **5,** 160
Normington, J. B., **1,** 1095
Norris, E., **2,** 94
Norris, F., **5,** 357
Norris, F. A., **6,** 555
Norris, J. F., **1,** 928, 1029
Norris, K. E., **7,** 414
Norris, R. K., **5,** 734
Norton, A. A., **1,** 848
Norton, D. G., **1,** 292, 864, 1287; **4,** 577
Norton, J. A., **1,** 343
Norton, J. R., **6,** 451; **8,** 385; **11,** 397
Norton, K. B., **1,** 60
Norton, R. V., **4,** 122; **5,** 455
Norton, S. J., **5,** 538
Norton, S. T., **6,** 472
Norval, S., **2,** 334
Norymberski, J. K., **1,** 87, 1046, 1284
Nosaki, K., **2,** 173
Nose, Y., **11,** 494
Noszko, L., **1,** 748
Notani, J., **6,** 106
Noureldin, N. A., **11,** 142, 441
Novack, V. T., **10,** 445
Novak, L., **2,** 394; **12,** 508
Nowak, R. M., **1,** 703; **2,** 14
Noyes, W. A., **1,** 41, 97, 933; **4,** 162; **5,** 566; **7,** 49
Noyons, E. C., **1,** 321
Noyori, R., **2,** 81, 173, 248, 344; **3,** 20, 101, 155; **4,** 35, 39, 158, 270; **5,** 224; **6,** 45, 170, 198, 201; **8,** 468, 498, 499; **9,** 170, 458, 478, 497, 498, 520; **10,** 36, 149, 289, 391, 441, 444, 453; **11,** 48, 54, 67, 75, 172, 276, 514, 568, 587; **12,** 57, 191, 420, 462, 466
Nozaki, H., **1,** 601; **2,** 81, 248, 344; **3,** 69, 312; **4,** 139, 250, 322; **5,** 7, 64, 77, 200, 210, 293, 333, 370, 374, 398, 435, 534, 742, 749; **6,** 91, 163, 183, 220, 222, 255, 281, 295, 328, 385; **7,** 45, 83, 90, 93, 102, 243, 391; **8,** 112, 151, 164, 165, 174, 246, 292, 339, 472, 476; **9,** 68, 77, 256, 395, 471, 478; **10,** 114, 142, 164, 165, 293, 391; **11,** 68, 99, 114, 150, 194, 365, 470, 533, 542; **12,** 51, 63, 64, 137, 263, 304, 344, 352, 419, 462, 524
Nozaki, M., **8,** 44
Nozaki, N., **10,** 318; **11,** 134
Nozawa, S., **3,** 39
Nozoe, S., **1,** 703, 1283; **5,** 756; **6,** 675
Nudelman, A., **12,** 546
Nuerrenbach, A., **8,** 248
Nuessler, C., **9,** 48; **11,** 326
Nuetzel, K., **6,** 596
Nugent, W. A., **9,** 448; **12,** 171, 350, 467
Nukada, K., **1,** 754
Nukada, T., **10,** 417
Nukina, S., **1,** 452
Null, G., **1,** 1056
Numata, T., **4,** 6; **5,** 5; **7,** 225; **10,** 401
Numazawa, M., **12,** 263
Nunes, B. J., **6,** 441
Nunn, A. J., **4,** 75
Nunn, E. E., **7,** 23
Nunn, M. J., **6,** 259
Nunokawa, O., **8,** 486; **9,** 156
Nunomoto, S., **12,** 195
Nussbaum, A. L., **1,** 269, 1000; **2,** 462
Nussbaumer, T., **1,** 219
Nussim, M., **1,** 59, 104, 207
Nutile, A. N., **1,** 671
Nutting, W. H., **4,** 359
Nuvole, A., **12,** 46
Nuyans, L. J., **9,** 60
Nuzzo, R. G., **8,** 522; **11,** 54
Nwaukwa, S. O., **11,** 107
Nwokogu, G., **11,** 479
Nyberg, K., **6,** 52
Nybraaten, G., **6,** 542
Nye, M. J., **4,** 340, 411
Nyikos, S. J., **10,** 356
Nyitrai, J., **4,** 14
Nylen, P., **1,** 371
Nylund, E. L., **9,** 423
Nysted, L. N., **1,** 1145
Nystroem, J. E., **12,** 475
Nystrom, R. F., **1,** 594, 595, 598, 599, 884
Nyu, K., **3,** 145

Oae, K., **5,** 639
Oae, S., **2,** 9, 101; **3,** 36; **4,** 6, 175; **5,** 5, 57, 104, 639; **6,** 369; **7,** 225; **8,** 200, 204, 504; **9,** 66, 92, 191; **10,** 174, 194, 220, 401, 450; **11,** 428
Oakes, V., **1,** 208
Oakwood, T. S., **1,** 50; **5,** 644

Obana, M., **11,** 277
Obata, T., **2,** 40
Obayashi, M., **5,** 360; **6,** 291; **7,** 243; **8,** 246; **10,** 114; **11,** 180, 310; **12,** 462
Obenland, C. O., **1,** 73
Oberender, H., **5,** 503
Oberholzer, M. E., **7,** 62
Oberrauch, H., **1,** 87
Oberster, A. E., **1,** 87
Obinata, T., **10,** 289
O'Boyle, M., **10,** 130
O'Brien, D. H., **6,** 638; **12,** 542
O'Brien, E., **11,** 39
O'Brien, J., **4,** 43, 250, 330
O'Brien, J. P., **1,** 808; **7,** 64
O'Brien, R. E., **1,** 1203; **2,** 222
Obrzut, M. L., **12,** 426
Occolowitz, J. L., **2,** 433
Ochiai, E., **1,** 842, 876
Ochiai, H., **12,** 51
Ochiai, M., **7,** 365; **9,** 270, 462; **10,** 8; **11,** 270, 516, 544; **12,** 259, 523
Ochrymowycz, L. A., **10,** 409
Ochsner, P., **1,** 1058
Ochwat, P., **1,** 1029
Ockerman, L. T., **4,** 406
O'Connell, E. J., Jr., **3,** 160; **4,** 260; **6,** 244
O'Connor, B. R., **1,** 41; **5,** 57, 257
O'Connor, G. L., **1,** 624; **3,** 72
O'Connor, R. T., **1,** 37
O'Connor, U., **11,** 260
O'Connor, W. F., **1,** 465
Oda, M., **4,** 316, 520; **10,** 113; **11,** 24, 275
Oda, O., **4,** 562; **12,** 131
Oda, R., **3,** 217; **4,** 80
Oda, Y., **9,** 458; **11,** 514
Odagi, T., **3,** 20
Odaira, Y., **9,** 269
Ode, R. H., **7,** 135
Odeh, I. M. A., **10,** 232
O'Dell, D. E., **10,** 87
Odenigbo, G., **5,** 740
Odier, M. E., **1,** 140
Odinokov, V. N., **9,** 343
O'Doherty, J., **10,** 333
Odom, H. C., **3,** 219; **4,** 371
O'Donnell, M. J., **8,** 391; **11,** 407; **12,** 329
O'Donohue, A. M., **10,** 394
O'Dowd, M. L., **10,** 77
Odum, R. A., **1,** 832; **2,** 444
Odyck, O., **4,** 255
Oediger, H., **1,** 190, 1249; **2,** 101, 211, 448; **4,** 18; **7,** 88
Oehldrich, J., **11,** 194
Oehler, E., **3,** 336; **4,** 157; **8,** 246
Oehlschlager, A. C., **1,** 364; **3,** 18, 329; **5,** 351
Oei, H. A., **11,** 326
Oei, H. O., **9,** 48
Oeller, M., **11,** 150
Oelschlaeger, H., **1,** 1116
Oertel, G., **5,** 439
Oertel, H., **2,** 21
Oertle, K., **6,** 625; **7,** 142; **10,** 99
Oesch, F., **7,** 200; **11,** 167, 256
Oesterle, T., **11,** 587
Oesterling, R. E., **1,** 808
Offenbaecker, M., **1,** 796
Offenhauer, R. D., **2,** 32
Offermann, K., **1,** 859
Ofner, A., **1,** 567
Ofner, S., **9,** 83, 349
Ofstead, E. A., **4,** 570
Ogard, A. E., **1,** 1030
Ogasawara, H., **4,** 366
Ogasawara, K., **2,** 433; **4,** 340; **5,** 600; **9,** 308, 458; **11,** 188, 449; **12,** 527
Ogata, T., **7,** 58
Ogata, Y., **1,** 500, 505; **2,** 222; **4,** 372; **8,** 83, 248; **10,** 87, 305; **12,** 465
Ogawa, A., **10,** 378; **11,** 466; **12,** 144
Ogawa, H., **5,** 257, 702; **9,** 133; **11,** 24, 466
Ogawa, K., **8,** 360
Ogawa, M., **5,** 325; **6,** 316; **8,** 233
Ogawa, N., **11,** 119
Ogawa, S., **12,** 313
Ogawa, T., **5,** 28; **7,** 40; **8,** 44, 354; **11,** 587
Ogawa, Y., **9,** 59; **12,** 11, 36
Ogg, C. L., **2,** 316
Ogibin, Y. N., **6,** 316
Ogilvie, K. K., **2,** 26, 47, 298; **4,** 58; **5,** 75; **7,** 58, 354; **8,** 59, 82, 468; **9,** 77, 446; **10,** 62; **11,** 90, 124, 284
Ogino, T., **9,** 391
Ogiso, A., **1,** 652
Ogiso, H., **2,** 313
Ogliaruso, M., **1,** 33, 772
Ogloblin, K. A., **9,** 214
Ogoshi, H., **5,** 560
O'Grodnick, J. S., **5,** 444
Ogura, F., **10,** 85
Ogura, H., **8,** 163; **9,** 208; **10,** 173; **11,** 227; **12,** 219
Ogura, K., **4,** 342, 467, 475; **5,** 457; **6,** 392, 446; **8,** 289, 345; **9,** 277, 314; **10,** 323; **11,** 131, 243; **12,** 29, 327, 328, 444
Ogura, M., **9,** 414
Oguri, T., **7,** 178; **8,** 101; **9,** 81
Oguro, K., **9,** 233; **10,** 194
Oh, S., **7,** 177
O'Hara, E., **4,** 291, 294
Ohara, E., **10,** 21
Ohara, K., **1,** 329
Ohara, M., **6,** 152
Ohara, S., **5,** 163
Ohashi, M., **2,** 151; **4,** 177; **5,** 764
Ohashi, T., **11,** 494
Ohashi, Y., **11,** 514
Ohata, H., **7,** 155
Ohba, N., **9,** 34; **10,** 24
Ohdoi, K., **12,** 546
Ohfune, Y., **8,** 24, 140, 441; **9,** 277
Ohira, S., **10,** 99
Ohkata, K., **6,** 244
Ohkawa, K., **8,** 161, 240; **9,** 492
Ohkawa, S., **10,** 247
Ohki, E., **1,** 1283; **5,** 756; **6,** 675
Ohki, M., **4,** 155
Ohl, H., **5,** 595
Ohle, H., **1,** 1101
Ohlinger, H., **1,** 611, 1096

Ohlmstead, W., **5,** 19
Ohloff, G., **1,** 444; **2,** 422; **5,** 424, 481, 576; **6,** 479, 675; **8,** 245; **9,** 154, 462; **10,** 58; **11,** 519
Ohme, R., **1,** 125, 484, 1087; **2,** 219, 399, 430; **3,** 147; **4,** 60, 455, 456, 527; **8,** 342
Ohmizu, H., **11,** 542
Ohmori, M., **11,** 288
Ohnami, T., **11,** 258, 562
Ohnishi, H., **12,** 70
Ohnishi, K., **8,** 444
Ohnishi, S., **10,** 151
Ohnishi, T., **3,** 15
Ohnishi, Y., **2,** 101; **6,** 37; **11,** 407
Ohno, A., **6,** 37, 573; **7,** 252
Ohno, H., **8,** 424
Ohno, K., **1,** 140, 1252; **2,** 67, 453; **3,** 329; **9,** 136
Ohno, M., **1,** 258, 754; **2,** 177, 445; **3,** 60, 101, 214; **4,** 69; **6,** 145, 215; **7,** 79; **10,** 449; **11,** 267; **12,** 256
Ohno, Y., **9,** 200
Ohnuma, T., **6,** 312; **11,** 427; **12,** 429
Ohoi, F., **6,** 145
Ohoka, M., **5,** 336, 534
Ohowa, M., **9,** 422
Ohr, J., **9,** 383
Ohrui, H., **12,** 41
Ohsawa, A., **5,** 148; **7,** 79
Ohsawa, H., **12,** 456
Ohsawa, T., **11,** 431; **12,** 44
Ohse, H., **2,** 445
Ohshima, M., **12,** 48
Ohshiro, Y., **6,** 158; **9,** 216; **10,** 298, 391; **11,** 514, 555; **12,** 187, 336
Ohsugi, Y., **11,** 172
Ohsumi, T., **12,** 351
Ohta, A., **7,** 73; **11,** 542
Ohta, H., **6,** 260; **8,** 346; **12,** 214
Ohta, K., **3,** 229; **12,** 29, 30
Ohta, M., **1,** 250, 562, 1217; **9,** 156; **12,** 338
Ohta, S., **6,** 70; **11,** 155, 449
Ohta, T., **6,** 618; **11,** 172; **12,** 180, 373
Ohtaka, H., **5,** 55
Ohtani, K., **12,** 282
Ohtani, M., **9,** 318
Ohtomi, M., **6,** 138
Ohtsuka, H., **12,** 573
Ohtsuka, T., **11,** 96
Ohtsuka, Y., **9,** 305
Oikawa, A., **9,** 372
Oikawa, Y., **6,** 258; **8,** 314; **10,** 252; **11,** 167, 312; **12,** 175
Oine, I., **6,** 70
Oishi, T., **4,** 345, 529; **5,** 693; **6,** 312; **9,** 77, 305; **10,** 461; **11,** 431, 600; **12,** 288, 374, 441, 573
Ojima, I., **4,** 563; **5,** 740, 749; **6,** 613, 633, 653; **8,** 59, 486, 510; **10,** 142, 403, 427, 428; **11,** 87, 131; **12,** 333
Oka, H., **5,** 447
Oka, K., **6,** 403; **7,** 367; **8,** 481; **9,** 472; **11,** 127
Oka, S., **6,** 573; **7,** 252
Okabe, H., **12,** 109
Okabe, M., **11,** 136
Okada, A., **7,** 400
Okada, H., **8,** 119; **12,** 503, 541
Okada, I., **3,** 322
Okada, J., **6,** 659
Okada, K., **4,** 69; **9,** 24, 519
Okada, M., **11,** 421
Okada, S., **9,** 519
Okada, S. S., **11,** 23
Okada, T., **3,** 217, 265; **4,** 550
Okahara, M., **5,** 534; **8,** 130, 256
Okajima, H., **7,** 268
Okajima, T., **4,** 465
Okamoto, H., **6,** 534
Okamoto, M., **1,** 258, 754, 1216; **2,** 36; **3,** 101; **11,** 155, 449
Okamoto, T., **6,** 573, 618
Okamoto, Y., **2,** 349; **5,** 329; **7,** 424; **8,** 109; **9,** 255, 256; **10,** 8, 62, 98; **11,** 59, 70, 128 587
Okamura, H., **9,** 148; **10,** 194
Okamura, N., **6,** 43
Okamura, W. H., **3,** 59; **11,** 40
Okamura, Y., **6,** 500
Okano, K., **12,** 63
Okano, M., **4,** 21; **6,** 23, 141; **7,** 12, 361; **8,** 120; **9,** 35, 443; **10,** 21, 106, 368; **11,** 37, 397, 455, 604
Okano, T., **9,** 238
Okawa, K., **1,** 744
Okawara, M., **6,** 14, 34, 365, 540; **7,** 27, 123, 165; **8,** 44; **9,** 193, 225, 398, 477; **10,** 413; **11,** 550, 551; **12,** 445, 523
Okawara, R., **4,** 550; **12,** 486
Okawara, T., **4,** 252; **7,** 101; **9,** 353; **10,** 221
Okazaki, H., **6,** 8, 219; **8,** 120
Okazaki, K., **4,** 80; **10,** 207; **12,** 291
Okazaki, R., **6,** 560; **9,** 442; **11,** 214, 496; **12,** 86
Okazaki, T., **8,** 198; **9,** 188
O'Keefe, D. F., **4,** 37
O'Keefe, J. H., **2,** 182
Oki, M., **9,** 318
Okita, M., **9,** 349
Okita, T., **6,** 198
Oklobdzija, M., **6,** 474; **11,** 124
Okogun, J. I., **6,** 139
Okon, A. U., **12,** 221
Okonkwo, J. O., **12,** 320
Okonogi, T., **5,** 363
Oksme, S., **9,** 309
Oku, A., **2,** 173; **4,** 196; **5,** 470; **9,** 431, **11,** 489
Oku, M., **6,** 24
Okuda, T., **1,** 16, 289, 380, 1073
Okuda, Y., **12,** 344, 352
Okude, Y., **8,** 112
Okuhn, S., **7,** 93
Okukado, N., **8,** 87, 475, 507; **10,** 391; **11,** 514
Okukado, N. T., **12,** 474
Okukudo, N., **9,** 479
Okuma, K., **5,** 649; **12,** 214

Okumoto, H., **11,** 349, 392, 514
Okumura, H., **12,** 446
Okumura, K., **3,** 15; **5,** 178
Okumura, O., **5,** 641
Okumura, T., **1,** 1198
Okumura, Y., **1,** 1131, 1163
Okuniewicz, F. J., **7,** 280
Okura, I., **6,** 504
Okwute, K. S., **6,** 139
Olafson, R. A., **12,** 113
Olah, G., **1,** 748
Olah, G. A., **1,** 4, 15, 44, 69, 266, 397, 407, 693, 741, 743, 747, 748, 1018, 1145; **2,** 12, 200, 297; **3,** 202, 251; **4,** 329, 567; **5,** 21, 52, 116, 171, 232, 310, 311, 312, 477, 539, 577, 716; **6,** 24, 103 105, 109, 273, 286, 474; **7,** 22, 240, 248, 254, 271, 341, 369, 370, 416, 417, 418; **8,** 6, 113, 248, 263, 361, 364, 383, 518; **9,** 66, 67, 236, 244, 245, 255, 256, 320, 324, 326, 335, 396, 400, 488; **10,** 81, 96, 98, 115, 167, 202, 207, 251, 275, 329, 401, 409, 457; **11,** 244, 256, 275, 355, 454, 553, 587, 593; **12,** 263, 420, 527
Olah, J. A., **1,** 69, 748; **4,** 567; **5,** 310, 716; **6,** 105; **7,** 22; **9,** 320, 400
Olah, V. A., **11,** 157
Olander, C. P., **1,** 845
Olcott, H. S., **1,** 860, 873
Oldenziel, O. H., **4,** 497, 516; **5,** 685; **6,** 600; **8,** 494
Olesen, S. O., **9,** 50
Oliva, G., **6,** 430
Olivares, E., **1,** 808
Oliver, J. A., **9,** 169
Oliver, J. E., **5,** 693; **6,** 647; **7,** 338, 407
Oliver, J. P., **1,** 1193; **6,** 573
Oliver, L. K., **3,** 93
Oliver, M., **10,** 209
Oliver, R., **4,** 265
Oliveri-Mandala, E., **1,** 448
Oliverio, A., **2,** 117
Olivero, A. G., **10,** 422; **11,** 378
Oliveros, E. P., **12,** 414
Oliveto, E. P., **1,** 13, 74, 75, 150, 269, 526, 755, 787, 802, 808, 966, 1000, 1101, 1107; **2,** 77, 344, 355, 462; **5,** 552; **9,** 380
Olivier, L., **8,** 233
Ollapally, A., **9,** 183
Ollendorff, G., **1,** 476
Ollerenshaw, J., **5,** 677
Ollerenshaw, J. E., **1,** 354
Olli, L. K., **9,** 121
Ollinger, J., **3,** 123; **5,** 528; **6,** 238
Ollis, W. D., **2,** 117, 406; **3,** 87, 286, 320
Ollivier, J., **12,** 248
Olmstead, H. D., **3,** 312; **4,** 302; **5,** 764
Olofson, R. A., **1,** 310, 385; **2,** 198; **4,** 311; **5,** 374, 398, 714; **7,** 215; **8,** 190, 226, 308, 531; **10,** 29, 89
Olsen, C., **4,** 534
Olsen, D. J., **10,** 302
Olsen, D. O., **5,** 675; **6,** 1
Olsen, M. J., **1,** 139
Olsen, R. E., **1,** 1040
Olsen, R. K., **1,** 1153; **3,** 202; **6,** 193
Olson, C. E., **1,** 33, 102
Olson, D. R., **5,** 713
Olson, G. L., **2,** 87; **4,** 23; **6,** 656; **7,** 109, 414; **10,** 43
Olson, R. E., **9,** 87
Olsson, L. I., **5,** 86; **6,** 148
Olszewski, W. F., **1,** 1247
Oltvoort, J. J., **10,** 116
Omata, T., **5,** 104, 639; **10,** 174
O'Meara, D., **2,** 4
Omietanski, G. M., **1,** 124
Omote, Y., **6,** 161; **8,** 120; **10,** 93
Omura, H., **10,** 276; **12,** 336
Omura, K., **7,** 136; **8,** 200
On, H. P., **9,** 277; **11,** 188
Ona, H., **4,** 435; **9,** 519
Onaka, M., **6,** 596; **10,** 33, 277
Onaka, T., **4,** 232
Onan, K. D., **8,** 167
Onato, T., **2,** 40
Onda, M., **1,** 1101
Ondetti, M. A., **1,** 181
O'Neil, J., **5,** 19
O'Neill, G. J., **1,** 191, 854; **2,** 457
O'Neill, I. K., **1,** 755
O'Neill, R. C., **1,** 1131
O'Neill, T. H., **7,** 41
Ong, B. S., **7,** 67, 194, 279; **8,** 57; **9,** 497; **11,** 28
Ong, C. W., **12,** 267
Ong, H. H., **10,** 251
Ongler, P. A., **9,** 169
Onishi, T., **9,** 316
Onishi, Y., **9,** 66
Ono, H., **5,** 348
Ono, I., **6,** 70
Ono, M., **7,** 208; **8,** 218; **9,** 198; **11,** 352
Ono, N., **5,** 556; **7,** 2; **8,** 141, 315; **9,** 133, 216, 434; **10,** 183, 413; **11,** 155, 551; **12,** 47, 233, 350, 523, 551
Ono, T., **8,** 351, 492; **9,** 473; **12,** 392
Ono, Y., **6,** 510
Onoe, A., **6,** 23, 141; **7,** 12; **8,** 120
Onopchenko, A., **4,** 99
Onoue, H., **1,** 563, 732; **2,** 295; **4,** 365
Onozuka, J., **12,** 29
Ooi, Y., **11,** 480
Ookawa, A., **11,** 479; **12,** 276
Oomkes, P. G., **10,** 409
Ooms, P. H. J., **5,** 653; **8,** 537; **9,** 523
Oota, Y., **5,** 553
Opara, A. E., **3,** 336; **5,** 580
Opel, J., **9,** 483
Openshaw, H. T., **3,** 157; **6,** 43; **7,** 106, 353
Opheim, K., **4,** 424
Opie, J. W., **1,** 393
Opitz, G., **2,** 104; **6,** 564

Oppenauer, R. V., **1,** 24, 87; **2,** 58
Oppenheimer, E., **2,** 453
Opperman, M., **4,** 191
Oppolzer, W., **2,** 181; **8,** 534; **9,** 169; **10,** 84, 181, 437; **11,** 165, 249, 309, 414, 415, 557; **12,** 104, 251, 290, 326
Oppong, I. V., **11,** 154
Orahovats, A. S., **9,** 503
Orazi, O. O., **1,** 74, 259; **2,** 108; **3,** 4
Orban, J., **12,** 84
Orchin, M., **1,** 155, 228, 759, 992, 1034, 1237; **2,** 80
Orda, V. V., **6,** 301
Orena, M., **7,** 69, 71, 298; **9,** 257, 383; **10,** 211; **11,** 267; **12,** 87
Orere, D. M., **8,** 409; **9,** 244
Orfanopoulos, M., **7,** 388; **8,** 501; **10,** 144
Orgel, L. E., **1,** 520
Oribe, T., **9,** 117
Orito, K., **9,** 341
Oritz, B., **4,** 432
Oriyama, T., **12,** 150
Orlov, N. F., **7,** 413
Ormand, K. L., **2,** 406; **3,** 286
Ornaf, R. M., **10,** 454; **12,** 48
Orndorff, W. R., **3,** 157
Ornstein, P. L., **8,** 263; **11,** 166
Oroshnik, W., **1,** 574, 730, 843, 1286; **2,** 163
Orr, D. E., **2,** 58
Orr, J. C., **1,** 625; **3,** 329; **11,** 446
Orszulik, S. T., **9,** 463
Ortaggi, G., **8,** 385; **9,** 45
Ortar, G., **7,** 361; **10,** 248
Ortega, C., **11,** 467
Orten, J. M., **1,** 413
Ortiz, M., **11,** 127
Ortiz de Montellano, P. R., **5,** 243, 537; **7,** 244
Orton, G., **5,** 491
Orttmann, H., **1,** 767
Ortuno, R. M., **11,** 555
Orwoll, E. F., **1,** 158, 1269
Osaka, N., **11,** 581; **12,** 547
Osakada, K., **11,** 277
Osaki, K., **9,** 176
Osaki, M., **7,** 372; **9,** 187
Osata, R. L., **2,** 47
Osato, R. L., **1,** 86, 1229
Osawa, E., **3,** 7; **6,** 17
Osawa, T., **1,** 9
Osawa, Y., **1,** 808; **2,** 311
Osbond, J. M., **1,** 567
Osborn, C. L., **1,** 318, 927
Osborn, J. A., **1,** 140, 1252; **2,** 453; **5,** 113, 700; **6,** 105; **7,** 412; **8,** 333; **10,** 149
Osborne, A. G., **2,** 412
Osby, J. O., **12,** 444
Oscar, J. A., **1,** 518
Oscarson, S., **11,** 483
Oshima, A., **6,** 430
Oshima, K., **5,** 7, 293, 374, 749; **6,** 183, 328; **7,** 244, 256; **8,** 206, 339; **9,** 77, 471; **10,** 142, 164, 293, 391; **11,** 99, 114, 194, 318, 365, 533, 542; **12,** 63, 304, 344, 352, 419, 524
Oshima, T., **2,** 36
Oshima, U., **5,** 60
Osiecki, J., **2,** 150
Oskay, E., **1,** 170; **2,** 263
Osman, M. A., **11,** 379
Osowska, K., **7,** 104
Oster, T. A., **12,** 314
Osterberg, A. E., **1,** 654
Osterwinter, B., **10,** 365
Osteryoung, R. A., **9,** 435
Ostrow, R. W., **7,** 301
Ostrum, G. K., **1,** 162; **2,** 343
Osuch, C., **4,** 25; **6,** 339
Osuga, D. T., **1,** 658
Osuka, A., **10,** 174; **12,** 143, 160, 275, 449
O'Sullivan, M. J., **3,** 53
Oswald, A. A., **11,** 67
Otani, N., **7,** 47
Otera, J., **10,** 390; **12,** 316, 402, 486
Oth, J. F. M., **2,** 134; **4,** 216, 293
Otomasu, H., **11,** 231
Ototani, N., **2,** 37
Otsubo, R., **6,** 90
Otsubo, T., **10,** 85
Otsuji, Y., **6,** 95; **9,** 146, 278, 361
Otsuka, S., **3,** 329; **6,** 49; **8,** 40; **9,** 238; **11,** 47, 54; **12,** 57
Ott, A. C., **1,** 257; **4,** 45
Ott, E., **1,** 527, 870, 883; **4,** 405
Ott, G. H., **1,** 74
Ott, H., **1,** 827
Ott, W., **12,** 390
Ottenheijm, H. C. J., **6,** 172; **7,** 310; **9,** 156, 283
Ottens, B., **1,** 1101
Otter, B., **2,** 365; **10,** 349
Otterbacher, T. J., **1,** 401
Ottley, P. M., **12,** 379
Ottmann, G., **1,** 370
Otto, K., **1,** 411
Otto, S., **1,** 1231
Ottolenghi, A., **8,** 57; **11,** 347
Ottow, E., **12,** 411
Ottridge, A. P., **9,** 71
Ouannes, C., **1,** 377, 856; **2,** 101
Ouchida, S., **9,** 479
Oudenes, J., **10,** 244
Oudman, D., **6,** 525
Ouellette, D., **9,** 312; **11,** 340, 422
Ouellette, R. J., **1,** 563, 1151
Oughton, J. F., **1,** 150, 1090
Ourari, A., **11,** 454
Ourisson, G., **1,** 594, 999, 1011, 1136; **2,** 238, 255, 329, 403; **3,** 76; **4,** 123, 127, 183; **5,** 187, 631; **6,** 407
Ouroussoff, N., **9,** 376, 439
Ousset, J. B., **12,** 241, 551
Ovechkina, L. F., **2,** 267
Overbeck, H., **1,** 352
Overberger, C. G., **1,** 445, 834, 911, 1083, 1095; **2,** 313, 429; **3,** 147; 5, 406, 591; **7,** 23; **10,** 245
Overchuk, N. A., **1,** 747
Overend, W. G., **1,** 413, 989,

1090; **2,** 359
Overman, J. D., **5,** 727
Overman, L. E., **3,** 258; **5,** 429, 727; **6,** 605; **7,** 147, 382; **8,** 35, 221, 317; **9,** 44, 479; **10,** 32, 118; **11,** 43, 47, 241, 292; **12,** 14, 51, 193, 350
Overtani, M., **11,** 283
Overton, K. H., **1,** 185
Owada, H., **10,** 21, 368; **11,** 455
Owen, G. R., **3,** 198; **4,** 14, 451; **6,** 605; **8,** 362
Owen, L. N., **1,** 357, 667, 952, 955, 1090, 1137; **2,** 383
Owen, N. E. T., **7,** 292
Owen, T. C., **1,** 37, 643
Owens, R. A., **7,** 395
Owens, R. M., **6,** 406
Owings, F. F., **1,** 989
Owsley, D. C., **1,** 162; **4,** 102, 490; **5,** 44; **8,** 265; **9,** 298
Oxenrider, B. C., **7,** 347
Oxford, A. E., **1,** 563
Oxley, P., **1,** 1184
Oxman, M., **1,** 185
Oxman, M. A., **1,** 471
Oya, M., **5,** 100
Oyama, H., **11,** 496
Oyamada, H., **11,** 479; **12,** 444
Oyler, A. R., **6,** 366, 496; **7,** 207
Oyster, L., **1,** 828
Ozaki, K., **5,** 294; **6,** 317, 403, 649
Ozari, Y., **6,** 651
Ozawa, F., **11,** 165
Ozawa, K., **12,** 213
Ozawa, S., **10,** 164
Ozawa, T., **4,** 257; **9,** 257
Ozorio, A. A., **9,** 159
Ozretich, T. M., **2,** 278

Paaren, H. E., **10,** 400
Paatz, R., **2,** 216
Pabon, H. J. J., **7,** 296; **8,** 169; **10,** 68
Pabst, W. E., **8,** 168
Pac, C., **6,** 534
Pacevitz, H. A., **6,** 40
Pachlatko, J. P., **9,** 334
Pachler, K. G. R., **2,** 315
Pachter, I. J., **4,** 540; **6,** 629; **7,** 110
Pacifici, J. G., **1,** 57; **5,** 470
Paciorek, K. L., **7,** 299
Pacquer, D., **5,** 257
Pacsu, E., **1,** 818
Paddock, N. L., **2,** 207
Paddon-Row, M. N., **9,** 416
Padegimas, S. J., **2,** 336; **5,** 672
Padgett, H., **5,** 495
Padgett, H. C., **7,** 293
Padilla, A., **9,** 436
Padley, F. B., **1,** 1011
Padmanabhan, S., **12,** 167
Padmasani, R., **1,** 235
Padwa, A., **1,** 90; **2,** 423; **5,** 486; **6,** 137; **8,** 261; **10,** 354
Paessler, P., **5,** 319
Paetzold, P. I., **3,** 32
Paetzold, R., **8,** 198
Pagano, A. S., **1,** 323, 827; **2,** 316; **3,** 221; **8,** 505
Paganou, A., **2,** 178, 317
Page, G. A., **1,** 720, 787
Page-Lecuyer, A., **5,** 58
Paglialunga, S., **7,** 210
Paglietti, G., **5,** 230; **12,** 46
Pagni, R. M., **4,** 339; **5,** 391
Pagnoni, U. M., **7,** 1, 331
Pahler, L. F., **9,** 256
Pai, F. C., **12,** 444
Pai, G. G., **11,** 430; **12,** 399
Pai, H. C., **10,** 410
Paige, J. N., **3,** 219
Paik, H. N., **8,** 267, 391
Pailer, M., **2,** 154; **4,** 189
Paine, J. B., III, **7,** 56
Paine, R. A., **2,** 320
Pajetta, P., **6,** 419
Pajouhesh, H., **12,** 122
Pak, C. S., **5,** 48
Pakrashi, S. C., **1,** 104
Pal, B. C., **9,** 411
Pal, D., **4,** 398
Palaniswamy, V. A., **11,** 321
Pale, P., **11,** 395
Palecek, J., **7,** 16
Palermo, R. E., **8,** 393; **11,** 318
Paleveda, W. J., **10,** 61
Paliokas, A., **2,** 368
Pall, D. B., **4,** 478
Palladin, A. V., **1,** 1049
Palmer, C. S., **8,** 169
Palmer, F. S., **1,** 456
Palmer, J. R., **9,** 269
Palmer, K. J., **3,** 251, 252
Palmer, M. H., **3,** 157
Palmer, P. J., **1,** 152, 274
Palmer, R., **5,** 582
Palmer, W. G., **6,** 52
Palmertz, I., **6,** 564, 567; **10,** 378
Palmieri, G., **7,** 175; **8,** 250; **9,** 432
Palmisano, G., **6,** 203; **7,** 56
Palomo, A. L., **12,** 546
Palomo, C., **10,** 60, 310; **11,** 384, 410, 416, 582; **12,** 79, 122, 199, 204, 384
Palomo-Coll, A., **11,** 58
Palomo-Coll, A. L., **10,** 41; **11,** 58
Palton, D. S., **2,** 379
Palumbo, G., **6,** 586; **12,** 555
Palumbo, Z., **11,** 128
Pan, B. C., **11,** 58
Pan, H. L., **1,** 500, 1212; **2,** 159; **5,** 333; **6,** 17, 578; **10,** 346
Pan, T. P., **2,** 289
Pan, Y., **9,** 309
Pan, Y. G., **8,** 525; **9,** 151
Panajotova, B., **5,** 129
Pande, L. M., **4,** 31; **6,** 675
Pandey, G. D., **9,** 95
Pandey, P. N., **6,** 287, 297; **7,** 56; **11,** 393
Pandhare, E. D., **4,** 189
Pandit, A., **2,** 404
Pandit, U. K., **1,** 247, 445
Panek, E. J., **2,** 400; **3,** 81; **4,** 183; **5,** 325
Panek, J. S., **10,** 409; **12,** 216, 474, 529
Panetta, C. A., **3,** 299; **4,** 399; **6,** 607; **8,** 499

Pangon, G., **5,** 129
Panizzi, L., **1,** 289
Pankow, L. M., **10,** 391
Pannell, K. H., **9,** 66; **11,** 156
Panneman, H. J., **1,** 360
Pannwitz, P., **1,** 407
Panouse, J. J., **1,** 87
Panse, P., **1,** 49
Pant, B. C., **3,** 51
Panunto, T. W., **11,** 108
Panunzio, M., **5,** 315; **6,** 306, 486; **7,** 182, 300
Panzica, R. P., **10,** 450
Paoletti, P., **11,** 563
Paolucci, G., **4,** 512; **9,** 151
Papa, A. J., **2,** 404
Papa, D., **1,** 502, 720, 1163
Papadakis, I., **6,** 514
Papadimas, D., **10,** 329
Papadopoulos, E. P., **1,** 643; **4,** 224; **7,** 146
Papadopoulos, K., **12,** 31
Papahatjis, D. P., **12,** 185, 536
Papalambrou, P., **1,** 664
Papanastassiou, Z. B., **1,** 207, 520
Pape, P. G., **1,** 606
Papenmaier, G. J., **1,** 383
Papies, O., **10,** 381
Papoula, M. T. B., **9,** 501; **10,** 446; **11,** 587
Pappalardo, J. A., **1,** 12, 33
Pappalardo, P., **11,** 577; **12,** 507
Pappas, B. C. T., **4,** 287
Pappas, J. J., **1,** 926, 1131; **2,** 156, 444; **6,** 103
Pappas, S. P., **1,** 562; **4,** 282
Pappo, R., **1,** 180, 207, 814, 829, 926, 986, 1184; **5,** 144, 685; **9,** 269
Paproth, T. G., **5,** 667
Paquer, D., **5,** 182
Paquette, L. A., **1,** 125, 139, 290, 362, 1041, 1136, 1212; **2,** 66, 70, 104, 207, 210, 224; **3,** 50, 74, 163, 234, 251, 259; **4,** 94, 290, 375, 429, 435, 469, 520; **5,** 15, 97, 213, 230, 389; **6,** 24, 70, 279, 352, 519, 568; **7,** 65, 66, 118, 268; **8,** 245, 289, 290, 344, 400, 493; **9,** 123, 237, 341, 438, 499; **10,** 55, 96, 265, 316; **11,** 25, 42, 78, 236, 286, 427, 475; **12,** 127, 139, 237, 390, 462, 502
Paquin, R., **11,** 145
Paradisi, M. P., **4,** 294; **10,** 248
Paragamian, V., **1,** 34
Parameswaran, K. N., **1,** 563
Paraschiv, M., **8,** 521
Paraskewas, S., **6,** 450
Paraye, E. R., **7,** 239
Pardasani, R. T., **11,** 311
Pardo, S. N., **10,** 144, 363; **11,** 180
Paredes, M. C., **10,** 224, 354
Parello, J., **1,** 695
Parent, K., **1,** 281, 705
Parent, R. A., **1,** 573
Parente, A., **8,** 440
Parente, R. A., **9,** 291
Parham, J. C., **1,** 1145
Parham, M. E., **9,** 54
Parham, W. E., **1,** 257, 386, 680, 976, 1041, 1095, 1105, 1210; **2,** 328, 351; **4,** 539; **5,** 4, 620; **6,** 283; **10,** 77, 246; **11,** 281
Parikh, J. R., **2,** 166, 394
Parikh, V. M., **1,** 989
Paris, G., **1,** 412
Paris, J. M., **11,** 475
Parish, E. J., **4,** 119, 120; **5,** 177, 513; **6,** 501; **7,** 196; **10,** 304, 335, 340; **12,** 419
Parish, J. H., **1,** 1139; **2,** 403
Parish, R. C., **1,** 1226
Park, B. K., **11,** 39
Park, C. H., **1,** 562
Park, P., **12,** 546
Parker, A. J., **1,** 309; **2,** 145; **4,** 377, 477
Parker, C. O., **2,** 135
Parker, D. G., **7,** 271
Parker, D. W., **11,** 401
Parker, J. E., **11,** 216
Parker, K. A., **4,** 409; **6,** 610, 634; **7,** 125; **8,** 169; **9,** 337, 413
Parker, S. R., **9,** 156
Parker, T., **6,** 284; **8,** 1
Parker, T. L., **9,** 337
Parker, V. D., **4,** 532; **5,** 88; **6,** 23
Parker, W., **1,** 1295; **5,** 542, 639; **9,** 65, 172
Parker, W. E., **2,** 316
Parker, W. L., **5,** 421
Parkes, G. D., **1,** 1194
Parkham, W. E., **12,** 207
Parkin, C., **5,** 20
Parkinson, B., **4,** 520
Parks, C. R., **1,** 1247
Parks, D. B., **2,** 309
Parks, J. F., **5,** 64
Parlar, H., **11,** 24
Parliment, M. W., **2,** 348
Parliment, T. H., **2,** 348
Parlman, R. M., **6,** 349, 552; **8,** 217
Parnas, J., **1,** 536
Parnell, C. A., **12,** 163
Parnell, E. W., **1,** 1131
Parnes, Z. N., **5,** 695; **6,** 616
Parr, J. E., **4,** 174; **5,** 373; **8,** 140
Parrick, J., **1,** 331
Parrilli, M., **5,** 349; **6,** 296; **7,** 181, 182
Parrish, D. R., **6,** 267, 411; **9,** 352; **12,** 415
Parrish, F. W., **2,** 269; **5,** 226
Parry, R. J., **4,** 531; **5,** 700
Parshall, G. W., **3,** 171; **6,** 48; **8,** 84
Parsons, C. G., **2,** 182
Parsons, J. L., **3,** 329
Parsons, P. J., **9,** 371; **10,** 355
Parsons, T. G., **1,** 1081
Parsons, W. H., **9,** 486
Partch, R. E., **1,** 563; **2,** 238, 239; **3,** 45
Parthasarathy, P. C., **1,** 633
Partis, R. A., **5,** 358
Parton, B., **2,** 35
Partos, R. D., **5,** 329
Partridge, J. J., **3,** 74; **4,** 162;

5, 283; 8, 338; 11, 226
Partridge, M. W., 1, 33, 445, 1029, 1210
Partyka, R. A., 5, 687
Paryzek, Z., 12, 555
Pascali, V., 5, 9
Pasedach, H., 4, 472
Pashinnik, V. E., 6, 184, 186
Paskins, K. N., 4, 477; 7, 352
Paskovich, D. H., 1, 1018
Pasquet, C., 5, 649
Pasquini, M. A., 10, 112
Pasquori, I., 1, 1171
Pass, S., 11, 79
Passannanti, S., 7, 207
Passeron, S., 1, 236
Pasternak, V. I., 6, 170
Pasternak, V. Z., 2, 159
Pasto, D. J., 1, 139, 258; 4, 383; 7, 155; 11, 503
Pastuszak, J. J., 4, 360
Pastuszak, J. P., 10, 277
Pasynkiewicz, S., 7, 374
Patai, J., 1, 772
Pataki, J., 1, 1090; 2, 311
Pataki, S., 1, 802
Patchett, A. A., 1, 926, 957; 4, 540; 7, 40; 9, 305
Patchornik, A., 1, 110, 581, 859; 2, 59; 4, 49; 6, 319; 9, 191; 11, 79
Pate, C. T., 5, 511
Patel, A. D., 4, 135
Patel, A. N., 1, 1249, 1250; 2, 330, 446; 5, 243, 460
Patel, D. D., 12, 1
Patel, D. K., 1, 355
Patel, D. V., 12, 78
Patel, K. M., 4, 309; 6, 600; 9, 401
Patel, P. K., 5, 717
Patel, R. C., 9, 508
Patel, R. P., 1, 1163
Patel, R. R., 4, 119, 372
Patel, S. K., 10, 51; 12, 212
Patel, V., 8, 130
Patel, V. V., 7, 176
Patelli, B., 1, 72, 927
Pater, R. H., 11, 442
Paterson, I., 8, 486; 9, 38, 108, 151, 492, 522; 11, 228
Pathak, S. P., 2, 361
Patil, G., 12, 150, 319
Patil, S. R., 11, 253
Patin, H., 7, 111; 8, 437; 9, 258; 11, 49
Patney, H. K., 7, 224
Patoiseau, J. F., 5, 310; 8, 240
Paton, J. M., 3, 61
Patrick, D. W., 6, 642
Patrick, J. B., 7, 268
Patrick, T. B., 3, 118, 187; 4, 54; 5, 460; 6, 190; 7, 158; 9, 224
Patrick, T. M., Jr., 1, 198, 963
Patrie, W. J., 9, 136; 10, 122; 12, 72
Patrizi, R., 4, 439
Patronik, V. A., 6, 252
Pattabhiraman, T., 4, 31
Pattenden, G., 3, 191; 4, 189; 5, 567; 7, 291; 9, 396, 431; 11, 461, 491; 12, 523
Pattenden, G. E., 2, 190
Patter, L., 9, 317
Patterson, D. B., 1, 1251
Patterson, D. R., 9, 429
Patterson, E. L., 1, 1157
Patterson, I., 11, 602
Patterson, J. M., 1, 963
Patterson, J. W., 10, 278
Patterson, L. A., 1, 454
Patterson, W., 10, 243, 423
Pattison, F. L. M., 1, 574, 603, 808, 992; 2, 40, 53
Pattison, V. A., 4, 473
Patton, D. S., 3, 123
Patton, J. W., 1, 518
Patton, R., 2, 263
Patton, W., 2, 228
Patwardhan, A. V., 2, 98
Patwardhan, S. A., 4, 510; 5, 356
Paty, P. B., 9, 373
Patzelt, H., 1, 396
Pau, J. K., 11, 206
Paudler, W. W., 2, 391
Paufler, R. M., 1, 703, 971
Paukstelis, J. V., 1, 195; 2, 59; 4, 455; 5, 569
Paul, B., 9, 297
Paul, D. E., 1, 712
Paul, H., 1, 1116
Paul, J. G., 9, 377
Paul, J. M., 3, 128
Paul, K. G., 11, 301
Paul, M. A., 1, 495
Paul, R., 1, 116, 257, 524; 4, 437; 7, 324
Pauling, H., 6, 307, 656; 7, 414
Paulissen, R., 5, 497, 572
Paull, K. D., 2, 344; 3, 275
Pauls, H. W., 10, 213; 12, 527
Paulsen, H., 1, 432, 1018; 4, 218; 5, 259, 329, 702; 6, 545; 10, 270
Paulson, D. R., 2, 309
Paulsson, L., 3, 252
Paulus, H., 8, 234
Pauly, H., 1, 134; 9, 508
Pausacker, K. H., 1, 509
Pauson, P. L., 10, 223; 11, 163; 12, 167
Paust, J., 2, 99, 119, 126; 3, 338; 8, 18, 248
Pauzenga, U., 2, 21
Pavey, D. F., 6, 147
Pavia, M. R., 9, 283
Pavlath, A., 1, 1018
Pavlic, A. A., 1, 730
Pawelczak, K., 5, 87
Pawelzik, K., 11, 44
Pawlak, M., 6, 77
Pawlowicz, R., 9, 506
Pawson, B. A., 2, 273
Paxton, H. J., 11, 596
Payne, C. K., 1, 849
Payne, D. A., 5, 160
Payne, G. B., 1, 89, 465, 470, 471, 477, 478, 820; 2, 196
Payne, H. A. S., 1, 129, 219
Payne, L. B., 2, 281
Payne, M. J., 9, 387
Paynter, O. I., 11, 490; 12, 379
Pazenko, Z. N., 9, 305
Peace, B. W., 5, 245, 246, 718
Peach, M. E., 8, 354
Peacock, R. D., 4, 407
Peagram, M. J., 2, 53; 5, 399
Peake, S. L., 8, 24, 101

Pearce, A., **7,** 375; **9,** 151
Pearce, A. A., **1,** 207
Pearce, H. L., **8,** 259; **9,** 269, 430
Pearce, P. J., **4,** 287
Pearce, R. J., **5,** 639
Pearl, I. A., **1,** 937, 1015
Pearl, N. J., **5,** 544
Pearlman, B. A., **9,** 194; **11,** 75
Pearlman, M., **8,** 352
Pearlman, P. S., **10,** 48, 378
Pearlman, W. M., **1,** 353, 782; **3,** 136; **10,** 300
Pearn, J., **12,** 112
Pearson, A. J., **12,** 90, 267, 333
Pearson, B. D., **1,** 354
Pearson, D. E., **1,** 33, 34, 243, 375, 377, 424, 905, 906, 940; **2,** 37, 44, 429; **5,** 552; **7,** 393; **8,** 50; **12,** 403
Pearson, D. L., **1,** 730, 741, 1073, 1100, 1258; **4,** 568
Pearson, I., **5,** 662
Pearson, N. R., **10,** 95; **11,** 188
Pearson, R. E., **1,** 80
Pearson, R. G., **7,** 122; **8,** 245
Pearson, R. L., **8,** 361
Pearson, W. H., **10,** 14; **12,** 37, 313
Pechet, M. M., **1,** 149; **2,** 200; **3,** 147, 262; **4,** 37, 511, 568; **5,** 381; **6,** 124, 264, 598; **7,** 158
Pecord, A. J., **12,** 474
Pedersen, B. S., **8,** 327; **9,** 50
Pedersen, C., **6,** 583
Pedersen, C. J., **4,** 143
Pedersen, E. B., **4,** 247; **5,** 325; **6,** 279
Pedersen, K., **2,** 55
Pederson, R. L., **1,** 257; **4,** 45
Pederson, U., **11,** 55
Pednekar, P. R., **11,** 479
Pedone, C., **5,** 115
Pedulli, G. F., **12,** 194
Peeling, E. R. A., **1,** 892; **3,** 51
Peer, H. G., **2,** 389; **9,** 24
Peet, J. H. J., **6,** 454
Peet, N. P., **4,** 4, 510; **10,** 276
Peet, W. G., **12,** 344
Peevey, R. M., **12,** 121
Pegel, K. H., **5,** 675
Pegram, J. J., **10,** 431; **11,** 574
Peiffer, G., **12,** 53
Pelah, Z., **2,** 378
Pelavin, L., **6,** 9
Pelc, B., **1,** 608; **4,** 134; **5,** 330
Pelegrina, D. R., **6,** 114; **7,** 271
Pelister, Y., **9,** 306
Pellacani, L., **7,** 308; **8,** 425; **10,** 333
Pelle, G., **6,** 26
Pellegata, R., **6,** 203; **9,** 234
Pellegrini, J. P., **11,** 432
Pellet, M., **7,** 164; **8,** 340
Pelletier, S. W., **1,** 144, 253, 633, 707, 1041, 1184; **2,** 84, 203, 313; **4,** 369; **5,** 747; **6,** 201; **9,** 426; **10,** 9; **11,** 24
Pellicciari, R., **6,** 631; **9,** 408; **11,** 156
Pelster, H., **1,** 396
Pelter, A., **1,** 310; **4,** 447; **5,** 607; **6,** 208, 295, 494, 536, 554; **7,** 26, 89, 101, 320, 333; **8,** 238, 522; **9,** 5, 317, 413, 464; **10,** 25, 351; **11,** 70, 75, 306; **12,** 13, 65, 71, 195
Pena, A. E., **5,** 486
Pendse, A. D., **9,** 353
Peng, C., **8,** 445
Peng, T. Y., **7,** 293
Penman, D. R., **1,** 430
Penn, R. E., **7,** 380
Pennanen, S. I., **8,** 166
Penners, N. H. G., **8,** 225
Penney, C. L., **11,** 410
Pennings, M. L. M., **9,** 165
Penprase, W. G., **2,** 363
Penque, R., **6,** 414
Pentimalli, L., **3,** 119
Penton, H. R., Jr., **4,** 228, 344; **5,** 444; **8,** 334
Penzes, P., **5,** 308
Penzhorn, R. D., **5,** 127
Peone, J., Jr., **5,** 172
Pepe, J. P., **8,** 531
Peppard, D. J., **5,** 474; **6,** 419
Pepper, E. P., **2,** 465
Pepper, J. M., **1,** 165
Pepperman, A. B., **11,** 469
Perche, J. C., **5,** 542
Perchinunno, M., **4,** 237
Perchonock, C., **5,** 19
Percival, A., **7,** 397
Perelman, M., **1,** 970; **2,** 351
Perera, D., **1,** 51, 477
Pereyre, M., **2,** 448; **3,** 294; **7,** 380; **8,** 445; **11,** 1, 16
Perez, A. D., **12,** 178
Perez, C., **7,** 40
Perez, J. A., **12,** 565
Perez-Prieto, J., **11,** 319
Perfetti, R. B., **5,** 306; **7,** 7
Periasamy, M., **8,** 81; **9,** 289
Periasamy, M. P., **5,** 651; **6,** 643; **9,** 458
Pericas, M. A., **8,** 144; **9,** 52; **11,** 124
Peries, R., **8,** 222
Perilstein, W. L., **1,** 777
Perina, I., **5,** 281
Perkin, A. G., **1,** 1056
Perkin, F. M., **1,** 152
Perkins, E. G., **1,** 1286
Perkins, L. M., **6,** 561
Perkins, M. J., **3,** 21; **8,** 204; **9,** 35
Perkins, N. A., **2,** 47
Perkins, P., **9,** 143
Perkow, N., **3,** 304
Perl, C., **4,** 6
Perlin, A. S., **1,** 562, 626; **4,** 201; **11,** 410
Perlman, P. L., **1,** 1000
Perlstein, T., **1,** 500, 1203
Pernert, J. C., **1,** 865
Pernet, A. G., **4,** 200
Perold, G. W., **2,** 315
Perozzi, E. F., **4,** 206
Perraud, R., **2,** 372; **3,** 258
Perregaard, J., **6,** 558
Perren, E. A., **1,** 666; **2,** 281; **3,** 105
Perret, C., **11,** 309
Perrier, G., **1,** 33
Perrin, C. L., **3,** 302

Perrine, T. D., **1,** 933
Perriot, P., **6,** 189, 331; **9,** 138, 154
Perron, Y. G., **2,** 191
Perrotti, E., **6,** 509
Perry, B. J., **1,** 685, 876, 1216, 1229
Perry, D. A., **11,** 76; **12,** 85
Perry, D. H., **4,** 250; **5,** 760; **6,** 675; **7,** 365
Perry, F. M., **1,** 720
Perry, M. W. D., **12,** 87
Perry, R. A., **8,** 130
Perry, R. H., Jr., **1,** 215
Perst, H., **2,** 345
Perumal, P. T., **11,** 441; **12,** 18
Perun, T. J., **1,** 610; **5,** 514
Peruzzotti, G., **4,** 221, 517; **6,** 201
Perz, R., **10,** 84; **11,** 554; **12,** 109
Pesaro, M., **1,** 927; **6,** 215; **8,** 143
Pesce, G., **4,** 477; **5,** 644; **6,** 375; **7,** 35, 245
Peseckis, S. M., **12,** 513
Peska, J., **2,** 172, 241
Pesnelle, P., **1,** 1011; **5,** 474; **6,** 419
Petasis, N. A., **9,** 283
Pete, J. P., **7,** 170; **9,** 378, 416; **10,** 202, 302; **11,** 395
Peter, D., **2,** 390
Peter, H., **1,** 1018; **6,** 186
Peter, H. H., **1,** 56, 573
Peter, R., **10,** 138, 271; **11,** 378; **12,** 356
Peters, A. T., **4,** 184
Peters, D. A. V., **2,** 40
Peters, E. D., **1,** 89, 528
Peters, J. A., **4,** 369; **5,** 319
Peters, J. W., **4,** 412; **5,** 681; **6,** 241, 487
Peters, M. A., **1,** 339, 370, 1029, 1082, 1101
Peters, R. A., **5,** 517
Peterse, A. J. G. M., **8,** 225
Petersen, H., **9,** 410
Petersen, J. M., **1,** 94, 629, 1095
Petersen, M. R., **3,** 302; **4,** 331
Petersen, Q. R., **1,** 1178
Petersen, R. C., **1,** 281
Petersen, R. J., **2,** 211
Peterson, B., **12,** 333
Peterson, D., **6,** 201, 245; **8,** 203
Peterson, D. J., **4,** 53; **5,** 528, 725; **10,** 433; **11,** 209, 554
Peterson, D. S., **2,** 403
Peterson, E. A., **1,** 519
Peterson, J. O., **1,** 600
Peterson, J. R., **6,** 496; **12,** 33, 212
Peterson, J. S., **12,** 94
Peterson, K. B., **5,** 287
Peterson, L. H., **1,** 1081, 1157
Peterson, L. I., **2,** 84, 176
Peterson, P. E., **1,** 1220; **5,** 700; **6,** 381; **7,** 240; **10,** 9, 436
Peterson, R. A., **1,** 1203
Peterson, W. D., **1,** 837, 1101
Peterson, W. R., Jr., **5,** 720
Petiniot, N., **8,** 435; **10,** 342; **11,** 460
Petit, F., **12,** 53
Petit, Y., **5,** 609
Petraglia, S. P., **12,** 538
Petragnani, N., **1,** 114, 349; **6,** 339; **8,** 398
Petraitis, J. J., **7,** 125; **8,** 169
Petrakis, K. S., **12,** 335
Petrarca, A. E., **4,** 132
Petrier, C., **11,** 198, 373; **12,** 357, 400
Petrillo, E. W., Jr., **3,** 204
Petrini, M., **12,** 405
Petrissans, J., **6,** 562
Petrov, A. A., **2,** 438; **5,** 44; **11,** 341
Petrov, A. D., **10,** 356; **11,** 83
Petrova, J., **6,** 189, 191
Petrow, V., **1,** 147, 219, 285, 355, 361, 767, 1178; **2,** 42, 227
Petrow, V. A., **12,** 72
Petru, F., **1,** 882
Petry, R. C., **1,** 1140
Petrzilka, M., **5,** 113; **6,** 76; **9,** 32, 366; **10,** 312
Petrzilka, T., **1,** 276, 384
Pettee, J. M., **6,** 234
Petterson, R. C., **1,** 93
Pettit, G. R., **1,** 34, 119, 207, 281, 598, 599, 705, 730, 802, 1055; **2,** 60, 117, 131, 198, 237, 271, 303, 344, 433; **3,** 40, 96, 128, 198, 275; **4,** 83, 150; **5,** 186, 391, 507, 743; **9,** 277
Pettit, R., **1,** 42; **2,** 95, 112, 140; **3,** 101, 243, 330, 336; **4,** 74; **6,** 150; **7,** 56, 111; **8,** 150, 476
Pettitt, D. J., **1,** 1233; **2,** 274, 439
Pettus, J. A., Jr., **9,** 442
Petty, C. B., **8,** 35; **9,** 479; **11,** 43
Petty, E. H., **11,** 460
Petty, H. E., **5,** 299
Petty, J. D., **6,** 339
Pews, R. G., **2,** 69; **5,** 322; **8,** 79
Pfaeffli, P., **5,** 118
Pfaendler, H. R., **3,** 172
Pfaffenbach, P. I., **1,** 708
Pfaffenberger, C. D., **2,** 149
Pfaltz, A., **11,** 542; **12,** 568
Pfannenstiel, A., **1,** 1015
Pfau, A. St., **1,** 370
Pfeffer, P. E., **3,** 185; **4,** 247, 302; **5,** 406
Pfeiderer, W., **11,** 56
Pfeiffer, B., **12,** 212
Pfeiffer, F. R., **3,** 297
Pfeiffer, W., **8,** 340
Pfeil, E., **1,** 1211
Pfenninger, F., **1,** 657
Pfenninger, J., **10,** 413
Pfiffner, A., **1,** 34
Pfister, G., **3,** 98
Pfister, J. R., **12,** 529
Pfister, K., 3rd, **1,** 476, 1006
Pfitzner, H., **4,** 414
Pfitzner, K. E., **1,** 310
Pflaum, D. J., **1,** 379
Pfleger, H., **6,** 510
Pfleiderer, W., **5,** 3; **6,** 547
Pfluger, C. E., **5,** 561
Pfohl, S., **8,** 18
Pfohl, W., **1,** 1189, 1198

Phadke, R., **4,** 579
Phan-Tan-Luu, R., **8,** 486
Philip, G., **6,** 19
Philip, J. B., Jr., **8,** 119
Philipp, A., **1,** 614
Philippe, J. L., **5,** 129
Philips, H., **10,** 407
Philips, J. C., **2,** 210
Philips, K. D., **6,** 221
Phillippe, M., **4,** 183
Phillipps, G. H., **1,** 107, 561; **2,** 228; **3,** 96
Phillips, B., **1,** 790, 791; **5,** 506
Phillips, C., **6,** 416
Phillips, D. D., **1,** 157
Phillips, D. I., **5,** 198
Phillips, G. B., **11,** 12; **12,** 11
Phillips, G. W., **9,** 162; **11,** 179
Phillips, H., **12,** 160
Phillips, J. B., **1,** 392; **2,** 75
Phillips, J. G., **11,** 152
Phillips, L., **5,** 675
Phillips, L. R., **6,** 151; **9,** 306, 494; **10,** 329
Phillips, M., **1,** 66
Phillips, M. L., **11,** 275
Phillips, R., **1,** 321, 865
Phillips, R. J., **8,** 60
Phillips, R. R., **8,** 22
Phillips, W. D., **1,** 1136
Phillips, W. G., **5,** 609
Phillips, W. V., **8,** 299, 343
Phillipson, J. D., **4,** 395
Philpott, P. G., **1,** 567
Phoenix, F. H., **9,** 337
Photaki, I., **1,** 346, 859, 1256
Photis, J. M., **5,** 389; **8,** 289; **12,** 237
Piancatelli, G., **8,** 12, 427; **9,** 399; **10,** 335, 336, 366; **11,** 452; **12,** 256
Piantadosi, C., **1,** 518, 1118
Piatak, D. M., **1,** 34, 599, 1055; **2,** 359; **4,** 74
Piazza, G., **5,** 114
Picard, J. P., **6,** 323; **7,** 199
Picard, P., **6,** 279; **11,** 515
Piccolo, O., **10,** 37; **12,** 502
Pichat, L., **5,** 723
Piche, L., **1,** 737
Pichl, K. E., **1,** 140
Pickenhagen, W., **4,** 472; **10,** 152
Picker, D., **5,** 644; **6,** 406
Picker, D. H., **6,** 19
Pickett, J. E., **11,** 387
Pickholtz, Y., **4,** 254
Pickles, G. M., **6,** 581
Picot, A., **6,** 73; **7,** 241
Pidcock, A., **5,** 446
Piechocki, C., **5,** 167; **11,** 462
Piechucki, C., **6,** 567; **7,** 355; **10,** 1
Pienkowski, J. J., **5,** 6
Pienta, N. J., **9,** 171
Piepers, O., **9,** 100
Pier, S. M., **1,** 870
Pierce, A. C., **1,** 425
Pierce, D., **9,** 73
Pierce, J. B., **4,** 290; **6,** 91
Pierce, J. K., **2,** 344; **7,** 178
Pierce, J. V., **1,** 1157
Pierce, O. R., **2,** 227
Pierce, T. E., **10,** 295
Piercey, M. R., **1,** 756
Pierdet, A., **1,** 215
Pieronczyk, W., **9,** 49
Pierre, J. L., **2,** 372; **3,** 258; **5,** 589; **10,** 112
Piers, E., **1,** 158, 652, 1284; **2,** 131, 153, 453; **3,** 65, 112, 196; **4,** 103, 183; **6,** 344, 376, 530, 648; **7,** 212; **8,** 492; **9,** 181, 333, 334; **10,** 290; **11,** 373; **12,** 116
Piers, K., **1,** 970
Pierson, A. K., **6,** 283
Pierson, G. O., **5,** 424
Pieter, R., **5,** 86; **7,** 312; **9,** 17
Pietra, F., **4,** 417; **5,** 188; **6,** 27
Pietra, S., **1,** 142, 444, 986
Pietrasanta, Y., **4,** 69
Pietraszkiewicz, M., **7,** 50
Pietrusiewicz, K. M., **10,** 181
Pietrusza, E. W., **1,** 1218
Pietsch, H., **8,** 391
Pietta, P. G., **4,** 164
Pifferi, G., **9,** 234
Pigeon, P., **11,** 407
Pigiere, C., **12,** 350
Pigman, W., **4,** 67
Pignolet, L. H., **12,** 111
Pigott, F., **4,** 411; **9,** 260
Pigott, H. D., **6,** 238, 400
Pihlaja, K., **5,** 68
Pike, D., **6,** 141
Pike, J. E., **1,** 609, 808, 1125, 1184; **2,** 134, 362; **3,** 296; **4,** 562
Pike, R. A., **1,** 982; **2,** 368; **5,** 713
Pike, S., **10,** 326
Pilbeam, A., **1,** 1141
Pilette, J. F., **9,** 380
Pilgram, K., **2,** 445
Pilgrim, A., **2,** 222
Pilgrim, F. J., **1,** 1184
Pilgrim, W. R., **4,** 53
Pilkiewicz, F. G., **5,** 155, 195; **6,** 137
Pillai, C. N., **1,** 20
Pillai, M. D., **5,** 319
Pillay, M. K., **11,** 41, 146
Pilling, G. M., **2,** 330
Pillot, J. P., **7,** 9
Piloty, O., **7,** 207
Pinck, L. A., **1,** 170
Pincock, R. E., **3,** 69; **4,** 10, 63; **6,** 515; **7,** 52
Pinder, A. R., **1,** 13, 1157; **3,** 219; **4,** 371
Pine, S. H., **4,** 239, 451; **10,** 88
Pines, H., **1,** 20, 927, 1023
Pines, S. H., **2,** 377; **6,** 19
Pingert, F. P., **1,** 158
Ping-Lin, K., **8,** 130
Pinhas, A. R., **12,** 279
Pinhey, J. T., **1,** 827; **4,** 283; **6,** 581; **11,** 32
Pinkney, P. S., **1,** 1080; **2,** 159
Pinkus, A. G., **5,** 534
Pinnell, R. P., **1,** 1195
Pinnick, H. W., **4,** 461; **6,** 114; **7,** 170; **8,** 56, 299; **9,** 55, 78, 190, 274; **11,** 28, 219, 275, 481, 535; **12,** 1, 121, 411
Pino, P., **1,** 228, 262; **4,** 420; **5,** 331; **8,** 437
Pinson, R., Jr., **1,** 808, 1122
Pinsonnault, J., **8,** 486
Pinza, M., **9,** 234

Piotrowski, A., **12,** 507
Piozzi, F., **4,** 133, 373
Piper, J. U., **5,** 106, 292, 428
Piper, T. S., **9,** 426
Piraux, M., **9,** 380
Pirisi, F. M., **6,** 175
Pirkle, W. H., **1,** 258, 322; **2,** 282, 326; **6,** 416; **8,** 356, 357, 499; **9,** 321
Piroue, R. P., **1,** 140
Pirrung, M. C., **9,** 283, 471; **10,** 133, 162; **11,** 341, 349
Pisciotti, F., **7,** 9
Piskorz, C. F., **9,** 54
Piszkiewicz, L. W., **2,** 453
Pitchen, P., **12,** 94
Pitchforth, L. L., **1,** 1253
Piteau, M., **12,** 113
Pitt, B. M., **1,** 1131; **10,** 316
Pitt, C. G., **3,** 308
Pitteloud, R., **9,** 32
Pitteroff, W., **1,** 96; **2,** 271
Pittman, C. U., Jr., **1,** 397; **6,** 59; **7,** 54; **10,** 276
Pitts, E. H., Jr., **1,** 951
Pitts, J. N., Jr., **4,** 412; **5,** 511, 681; **6,** 487
Pitzele, B. S., **6,** 348
Pivawer, P. M., **2,** 25
Piwinski, J. J., **10,** 312
Pizey, J. S., **6,** 357
Pizzo, F., **9,** 13
Pizzolato, G., **4,** 323; **5,** 73; **8,** 383; **10,** 309
Pizzorno, M. T., **4,** 502; **7,** 362
Place, P., **5,** 420; **9,** 119; **10,** 101
Plachky, M., **7,** 133
Plaha, L., **1,** 764
Plamondon, J. E., **5,** 41
Plaquevent, J. C., **9,** 205
Platt, K. L., **7,** 200; **11,** 167, 256
Platt, T. E., **12,** 223
Plattner, E., **6,** 284
Plattner, J. J., **3,** 71, 325; **4,** 424
Plattner, J. T., **6,** 2
Plattner, P. A., **1,** 276, 370, 470, 595, 664, 747, 963, 1121; **2,** 329
Platz, M., **10,** 363
Plaumann, H. P., **10,** 50
Plaut, H., **12,** 212
Plenchette, A., **6,** 279
Plesek, J., **1,** 181; **4,** 316
Pless, J., **1,** 281, 1196; **4,** 46; **5,** 49, 645
Pletcher, D., **8,** 410
Pletcher, J., **5,** 394
Plettenberg, H., **7,** 181
Plieninger, H., **1,** 608; **4,** 172; **8,** 330
Ploetz, T., **3,** 79
Ploss, G., **2,** 154
Pluim, H., **9,** 403; **10,** 27
Plume, G., **1,** 486
Plummer, C. W., **1,** 1011
Pober, K. W., **3,** 139, 258, 338
Pobiner, H., **1,** 431, 705; **6,** 430
Pochat, F., **8,** 132
Pochini, A., **8,** 245; **9,** 489
Pockels, U., **1,** 573
Pocker, Y., **5,** 370
Podesta, J. C., **6,** 581
Poduska, K., **1,** 524
Poels, E. K., **9,** 450
Poeth, T., **4,** 103
Pogonowski, C. S., **5,** 526; **6,** 238, 460; **8,** 268
Pohl, R. J., **1,** 1019
Pohlke, R., **1,** 343, 828
Pohmakotr, M., **8,** 8; **11,** 438
Pohmer, L., **1,** 77, 255
Poindexter, G. S., **10,** 251
Pointner, A., **9,** 288, 506
Poirier, J. M., **8,** 72
Poirier, M., **11,** 554
Poje, A. J., **7,** 207
Poje, M., **6,** 486; **8,** 417; **10,** 214
Pojer, P. M., **2,** 214; **7,** 135; **9,** 190; **10,** 364
Pol, E. H., **2,** 228
Polacheck, J. W., **1,** 476
Polak, R. J., **1,** 935; **5,** 551
Polanc, S., **8,** 483; **10,** 271
Polenski, B., **1,** 408
Polgar, L., **2,** 323
Poli, G., **12,** 226
Politzer, I. R., **4,** 484
Politzer, J. M., **3,** 284
Polk, D., **11,** 195
Poll, T., **12,** 227
Polla, E., **12,** 231
Pollack, M. A., **1,** 244
Pollack, N. M., **4,** 9
Pollack, R. M., **6,** 617
Pollak, A., **5,** 512; **6,** 300, 670; **7,** 242
Pollaud, R., **1,** 730
Pollet, P., **8,** 311
Pollini, G. P., **4,** 490; **5,** 553; **6,** 497, 560, 566; **9,** 151; **11,** 324
Pollmann, W., **1,** 894
Poloni, M., **2,** 207
Polonovski, M., **2,** 9
Polovsky, S. B., **6,** 415
Polster, R., **1,** 909
Polston, N. L., **4,** 141; **5,** 467
Polzhofer, K. P., **3,** 36
Pomeroy, J. H., **1,** 61
Pommer, H., **4,** 472; **8,** 248
Pommier, J. C., **11,** 161
Ponaras, A. A., **7,** 148, 283
Pond, D. M., **2,** 159
Ponder, B. W., **1,** 566; **2,** 239
Ponpipom, M. M., **4,** 555
Pons, J. M., **11,** 534
Ponsati, O., **11,** 555
Ponsford, R. J., **6,** 597; **9,** 408
Ponsold, K., **1,** 318, 445, 671; **3,** 71
Ponsold, R., **1,** 501
Ponti, F., **8,** 110; **10,** 382
Ponzio, G., **1,** 329, 833
Pool, W. O., **1,** 837
Poonian, M. S., **3,** 101
Poos, G. I., **1,** 147, 176, 360, 703, 802, 926
Pope, B. M., **4,** 128; **8,** 57, 145
Pope, H. W., **1,** 33, 34
Popien, D., **4,** 15
Popjak, G., **1,** 147, 777, 814, 937
Popoff, I. C., **4,** 63
Popov, A. I., **1,** 1132
Popovitz-Biro, R., **6,** 166
Popp, F. D., **1,** 904; **3,** 166; **4,** 390, 510; **5,** 433, 728; **11,**

150
Popper, T. L., **1,** 1000; **2,** 232, 326, 332, 433, 462
Poppi, R. G., **5,** 633; **6,** 547
Popplestone, C. R., **5,** 339
Porath, J., **1,** 1001, 1002
Porcher, H., **9,** 299, 337, 406
Poredda, S., **3,** 312
Pornet, J., **10,** 303, 381; **11,** 64
Porskamp, P. A. T. W., **8,** 465, 466
Port, W., **1,** 53
Porta, O., **6,** 21; **12,** 492
Portella, C., **7,** 170; **9,** 416
Porter, A. E. A., **8,** 159, 435; **10,** 342
Porter, A. S., **5,** 658
Porter, C. C., **1,** 841
Porter, H. D., **1,** 842, 1101
Porter, H. K., **5,** 632
Porter, J., **10,** 156
Porter, M., **3,** 320
Porter, N. A., **6,** 167; **7,** 94; **8,** 45, 249; **10,** 203
Porter, P. K., **1,** 134, 718
Porter, R. R., **1,** 322
Portlock, D. E., **9,** 274
Portnoy, R., **4,** 484
Portnoy, R. C., **4,** 460
Portoghese, P. S., **5,** 370
Portz, W., **1,** 838
Porwoll, J. P., **12,** 434
Porzi, G., **9,** 383; **10,** 142, 211; **11,** 267
Posler, J., **2,** 195; **3,** 182
Posner, G. H., **2,** 153, 241; **3,** 113, 313; **4,** 63, 128, 183, 295, 302, 309; **5,** 188, 243, 244, 396, 415; **6,** 17, 215, 343, 465; **7,** 7, 211; **8,** 12, 13; **9,** 11, 124, 219, 241, 283, 470; **10,** 9, 194; **11,** 23, 24, 315; **12,** 298
Posner, J., **1,** 1203; **2,** 427
Pospisil, J., **6,** 656
Poss, M. A., **11,** 334
Possel, O., **6,** 567; **8,** 494
Post, H. W., **1,** 244, 1073, 1138; **3,** 285; **11,** 576
Post, M. E., **7,** 241
Posternak, T., **1,** 329, 1005, 1007
Posthault, M., **10,** 370
Posthumus, T., **1,** 1000
Postlethwaite, J. D., **6,** 133
Postovaskii, I. Y., **1,** 66
Postowskue, Y., **8,** 226
Posvic, H., **1,** 383
Potekhin, A. A., **9,** 214
Potempa, S. J., **1,** 320
Potier, P., **2,** 9, 75; **3,** 46, 115, 308; **4,** 184; **6,** 361; **7,** 390; **9,** 459; **12,** 417
Potoski, J. R., **2,** 328
Potrosow, W. I., **1,** 349
Potter, G. F., **10,** 99
Potter, G. J., **8,** 260
Potter, H., **1,** 195, 713, 1049; **3,** 74
Potter, N. H., **3,** 34
Potts, B., **12,** 390
Potts, I. W., Jr., **5,** 276
Potts, K. T., **1,** 372, 1040; **2,** 61, 93; **3,** 303; **4,** 209; **11,** 282
Pouet, M. J., **12,** 105
Pougny, J. R., **11,** 224
Poulin, J. C., **9,** 260
Poulose, A., **9,** 391
Poulter, C. D., **7,** 132; **8,** 194; **10,** 455
Poulton, G. A., **6,** 99, 583
Poupart, J., **2,** 36
Poupart, M. A., **10,** 125; **12,** 160
Pourcelot, G., **9,** 365
Poutsma, M. L., **2,** 226; **8,** 19
Povall, T. J., **6,** 446
Povarny, M., **5,** 225
Powell, A. R., **1,** 778; **5,** 498; **10,** 303
Powell, D. L., **6,** 24
Powell, D. W., **11,** 121
Powell, J. E., Jr., **4,** 65; **8,** 338
Powell, J. R., **1,** 170, 843; **6,** 32
Powell, J. W., **1,** 1139, 1187; **2,** 403
Powell, K. A., **9,** 391
Powell, R. G., **2,** 139
Powell, R. W., **4,** 378
Powell, V. H., **3,** 238
Powers, J. C., **1,** 1081
Powers, P. J., **2,** 166
Poyser, J. P., **4,** 262
Prabhakar, S., **2,** 203; **5,** 747
Pradhan, S. K., **1,** 219, 927; **9,** 377; **11,** 491
Prager, R. H., **4,** 314; **6,** 350; **7,** 321
Pragnell, J., **6,** 304, 514
Pragnell, M. J., **1,** 345; **2,** 34, 179; **5,** 410; **12,** 19
Prahash, S. R., **6,** 3
Praill, P. F. G., **1,** 716, 802
Prakasa-Rao, A. S. C., **6,** 475
Prakash, G., **11,** 426
Prakash, G. K. S., **9,** 66, 67; **10,** 202, 251; **11,** 454, 593; **12,** 420
Prakash, I., **12,** 386
Prakash, O., **12,** 386
Prakash, S. R., **7,** 32
Pramanik, B. N., **8,** 211
Prange, T., **5,** 580
Prange, U., **4,** 216, 293
Prantz, E., **8,** 246
Prasad, D., **1,** 834, 1191
Prasad, H. S., **9,** 335
Prasad, J. S., **8,** 234
Prasad, K. S. N., **6,** 411
Prasanna, S., **9,** 370; **11,** 79
Prashad, M., **11,** 407
Prather, J., **2,** 243
Prati, L., **12,** 298
Prato, M., **11,** 336
Pratt, A. J., **11,** 326
Pratt, B. C., **1,** 228
Pratt, D. S., **3,** 157
Pratt, E. F., **1,** 642, 643; **2,** 263
Pratt, R. D., **3,** 336
Pratt, T. M., **7** , 59
Pratt, W. L. C., **1,** 319
Pravda, Z., **1,** 859
Pray, B. O., **8,** 263; **9,** 74
Preckel, M. M., **5,** 73, 416; **6,** 243; **11,** 596
Preftitsi, S., **12,** 450
Pregaglia, G. F., **5,** 172, 190
Preitner, G., **1,** 110
Prelog, V., **1,** 976; **2,** 112, 213, 318; **4,** 191; **5,** 620; **8,** 509

Premasagar, V., **11,** 321
Premuzic, E., **2,** 77; **3,** 61
Preobrajensky, N. A., **1,** 360; **3,** 222
Press, J. B., **4,** 340; **9,** 61; **10,** 15, 376
Prestidge, R. L., **7,** 59
Preston, J., **1,** 371
Preston, P. N., **12,** 70
Prestwich, G. D., **6,** 615
Preuschhof, H., **4,** 60, 455, 456
Preuss, T., **8,** 377
Previtera, L., **6,** 586
Prevost, C., **1,** 1008
Prewo, R., **9,** 299, 337, 406
Prey, V., **1,** 23, 966
Price, A. W., **2,** 35
Price, C. C., **1,** 12, 33, 309, 377, 380, 472, 481, 818, 1073, 1105, 1203; **5,** 600
Price, J. A., **1,** 367, 629, 1203
Price, J. G., **2,** 378
Price, J. R., **1,** 471
Price, L. G., **11,** 228
Price, M. F., **12,** 70
Price, P., **1,** 518; **4,** 221, 517; **6,** 201
Price, R. T., **11,** 12
Price, S., **1,** 1163
Prichard, W. W., **1,** 489
Prickett, J. E., **7,** 100, 323
Priddy, D. B., **3,** 182
Priebe, W., **11,** 285
Priest, D. N., **4,** 247
Prietzel, H., **5,** 168
Prigge, H., **4,** 275
Prill, E. A., **1,** 933
Prince, T. L., **10,** 369
Prins, D. A., **1,** 818, 1033
Prinzbach, H., **2,** 67, 454; **5,** 693; **6,** 584
Prior, M., **6,** 417
Prisbe, E. J., **7,** 407
Prisbylla, M., **10,** 161, 403
Pritchard, N. D., **2,** 165
Pritzsche, A., **1,** 680
Priyono, W., **12,** 48
Proba, Z. A., **11,** 90, 284
Probst, W. J., **1,** 74; **4,** 54; **5,** 460
Proctor, B. E., **4,** 191
Proebster, M., **10,** 36
Proehl, G. S., **9,** 364
Profitt, J. A., **6,** 352; **10,** 251
Prokai, B., **1,** 689, 854
Prokipcak, J. M., **1,** 1261; **2,** 222; **6,** 456
Prokopiou, P. A., **8,** 285; **9,** 378; **10,** 322
Proksch, E., **8,** 377
Proops, W. R., **1,** 169
Prosad, H. S., **9,** 501
Proskauer, S., **1,** 1235
Proskow, S., **1,** 465
Pross, A., **4,** 260; **6,** 254
Prossel, G., **6,** 122
Prosser, T., **1,** 1163
Prostenik, M. V., **8,** 87
Protopapa, H. K., **1,** 230
Protschuk, G., **8,** 363; **10,** 234
Prout, F. S., **1,** 16, 413; **5,** 6
Provelenghiou, C., **9,** 296
Prox, A., **1,** 116, 486
Pruckner, A., **12,** 233
Pruett, R. L., **1,** 249
Pryke, J. M., **1,** 731
Pryor, W. A., **4,** 152
Psaume, B., **10,** 434
Pu, L. S., **5,** 332
Puar, M. S., **7,** 244
Puckette, T. A., **11,** 179
Pucknat, J., **3,** 336
Puda, J. M., **5,** 6
Puetter, R., **2,** 320
Pugin, A., **6,** 417
Puglia, G., **8,** 245
Pujol, F., **11,** 124
Puleo, R., **11,** 36
Pullukat, T. J., **2,** 412
Pulwer, M. J., **10,** 295
Pummerer, R., **1,** 755; **6,** 308
Pump, J., **11,** 68
Punja, N., **2,** 173
Puntambeker, S. V., **1,** 1229
Pura, J. L., **4,** 185
Purcell, K. F., **4,** 450
Purcell, T. A., **5,** 98
Purdie, D., **7,** 324
Purdie, T., **1,** 685
Purick, R., **5,** 62
Purkayastha, M. L., **11,** 393
Purkayastha, R., **2,** 27
Purohit, V. G., **9,** 361
Purrington, S. T., **11,** 284; **12,** 158
Purushothaman, K. K., **3,** 50; **8,** 432
Pushpananda, K., **11,** 422
Puskas, I., **5,** 508
Pusset, J., **7,** 299
Pustinger, J. V., **2,** 447
Puterbaugh, W. H., **1,** 524, 601
Puthenpurayil, J., **6,** 354
Putnam, R. E., **1,** 426, 1271
Putnam, S. T., **1,** 562
Putt, S. R., **11,** 524
Putze, B., **4,** 216, 293
Pye, W. E., **8,** 76
Pyman, F. L., **5,** 758
Pyne, S. G., **12,** 569
Pyrek, J. S., **5,** 209
Pyron, R. S., **3,** 173
Pyun, C., **12,** 276
Pyuskyulev, B., **5,** 64

Qazi, T. U., **6,** 285
Quabeck, G., **1,** 531
Quagliato, D. A., **11,** 584
Quail, J. W., **4,** 406
Quallich, G., **12,** 561
Quang, L. V., **1,** 1022
Quarterman, L. A., **6,** 670
Quast, H., **4,** 35; **5,** 641
Quayle, O. R., **1,** 1064
Quayle, P., **10,** 248; **11,** 536
Querry, M. V., **1,** 198; **10,** 187
Quesada, M. L., **7,** 24
Quezada, P., **4,** 312
Qui, N. T., **8,** 486; **9,** 156; **10,** 103
Quici, S., **6,** 406, 543; **9,** 11
Quick, J., **9,** 300; **11,** 500
Quick, S. J., **12,** 263
Quilico, A., **1,** 221, 289, 481, 952
Quilliam, M. A., **5,** 75
Quillinan, A. J., **6,** 90, 424
Quin, L. D., **2,** 279
Quinn, C. B., **4,** 405; **5,** 551
Quinn, H. A., **4,** 240, 241
Quinn, H. W., **1,** 1018; **3,** 251

Quinn, N. R., **12,** 538
Quinn, R. A., **1,** 952
Quinn, R. J., **9,** 442
Quinney, J. C., **6,** 100
Quint, F., **1,** 1150
Quintana, J., **5,** 163
Quintily, U., **11,** 336
Quirk, J. M., **10,** 116
Quo, S. G., **1,** 424
Qureski, A. K., **2,** 397; **9,** 449
Qureski, M. I., **1,** 1011; **2,** 300

Raab, W., **9,** 381; **10,** 323
Raasch, M. S., **1,** 1125
Rabe, J., **12,** 156
Rabeb, F., **9,** 361
Raber, D. J., **4,** 529; **5,** 693; **6,** 565; **7,** 353; **8,** 501; **9,** 483; **11,** 499
Rabiant, J., **1,** 845
Rabideau, P. W., **10,** 356
Rabinovitz, M., **6,** 670; **7,** 425; **12,** 400
Rabinowitz, J., **1,** 905
Rabinowitz, R., **1,** 223, 1216; **9,** 256
Rabinsohn, Y., **1,** 51, 477
Rabjohn, N., **1,** 111, 247, 361, 402, 999; **4,** 149, 334; **7,** 319
Rabone, K. L., **3,** 277
Raby, C., **4,** 263
Rachele, J. R., **1,** 269
Racherla, U. S., **12,** 70
Rachlin, A. I., **4,** 367
Radatus, B., **5,** 389; **7,** 196
Radau, M., **6,** 379
Radcliffe, C. D., **10,** 154
Rademacher, D. R., **6,** 554
Radhakrishna, A. S., **9,** 54; **12,** 259
Radhakrishnan, J., **6,** 546
Radhakrishnan, T. V., **9,** 377; **11,** 491
Radics, L., **6,** 645
Radin, N. S., **1,** 269, 1163
Radlick, P., **1,** 999, 1022; **2,** 236; **5,** 31, 447; **6,** 91
Radscheit, K., **1,** 350; **4,** 296; **5,** 119, 633
Raduechel, B., **6,** 443; **10,** 329
Radunz, H., **6,** 141
Radziszewski, B., **1,** 471, 494
Raecke, B., **1,** 711
Raeth, C., **5,** 106, 286
Raetz, R., **1,** 894
Raff, P., **1,** 618; **3,** 184
Raffelson, H., **1,** 802, 957
Rafikov, S. R., **4,** 20
Rafla, F. K., **7,** 10; **8,** 456
Raggio, M. L., **7,** 107
Raggon, J., **11,** 439
Raghavan, P. R., **9,** 2
Raghu, S., **8,** 455
Ragnarsson, U., **12,** 159
Raha, C., **1,** 211, 276
Rahal, S., **4,** 431, 522
Rahm, A., **11,** 1
Rahman, A., **4,** 466
Rahman, A. U., **4,** 6, 468
Rahman, H. U., **12,** 370
Rahman, M., **5,** 464
Rahman, W., **9,** 191
Rahrs, E. J., **1,** 1027
Rai, C., **1,** 904
Rai, H. S., **4,** 424
Rainer, G., **1,** 404
Raj, K., **10,** 236
Rajadhyaksha, S. N., **12,** 1
Rajagopal, S., **1,** 402
Rajagopalan, K., **1,** 176; **7,** 385
Rajagopalan, P., **2,** 178; **3,** 131; **6,** 556
Rajagopalan, S., **1,** 80, 367, 465, 819, 962, 1000, 1284; **4,** 416
RajanBabu, T. V., **12,** 559
Rajappa, S., **1,** 364; **2,** 192
Rajaram, J., **12,** 272
Raju, M. S., **10,** 426
Raju, N., **12,** 324, 350
Rakhit, S., **1,** 377
Rakiewicz, D. M., **12,** 422
Rakin, D., **11,** 327
Rakoneza, N., **1,** 986
Rakowski, M. C., **6,** 15
Rakshys, J. W., Jr., **4,** 573
Raleigh, D., **1,** 636; **2,** 144; **9,** 439
Raley, J. H., **2,** 376
Rall, G. J. H., **7,** 62; **8,** 391; **9,** 463; **10,** 397
Ralli, P., **11,** 282
Ralls, J. W., **1,** 1178; **5,** 290
Ralph, P. K., **1,** 235
Ralston, A. W., **1,** 837
Ramachandra, R., **9,** 300
Ramachandran, J., **1,** 385; **11,** 421
Ramachandran, L. K., **1,** 814
Ramachandran, S., **1,** 685, 696, 703, 730, 890, 974; **2,** 173, 252
Ramadan, N., **5,** 713
Ramage, R., **6,** 97, 425; **7,** 138; **8,** 79
Ramaiah, M., **8,** 300
Ramakrishnan, K., **4,** 526
Ramakrishnan, V. T., **5,** 632
Ramalakshmi, Y., **12,** 122
Ramamoorthy, B., **7,** 414
Raman, H., **5,** 511
Raman, K., **12,** 425
Raman, P. S., **3,** 252
Raman, P. V. A., **1,** 962
Ramana-Rav, V. V., **9,** 99
Ramanathan, H., **10,** 236; **12,** 523
Ramanathan, N., **1,** 1234; **2,** 98
Rama-Rao, A. V., **4,** 189; **5,** 347; **6,** 296; **8,** 325
Ramasseul, R., **7,** 384
Ramasubbu, A., **9,** 349
Ramaswamy, K., **8,** 460
Ramaswamy, P. S., **8,** 364
Ramaswamy, S., **10,** 38
Rambacher, P., **3,** 260
Rambaud, M., **12,** 77, 527
Ramer, R. M., **2,** 131
Ramey, K. C., **2,** 159, **5,** 682
Ramey, P. S., **2,** 226
Ramirez, F., **1,** 223, 1216, 1234, 1247; **2,** 98, 447; **4,** 536, 552; **5,** 268; **6,** 8, 219, 381; **7,** 97; **8,** 219
Ramirez, J. R., **12,** 322
Ramirez, J. S., **12,** 395
Ramirez, M. A., **7,** 40
Rammler, D. H., **1,** 189, 235, 802, 1137
Ramp, F. L., **1,** 207, 1190

Rampal, J. B., **11,** 123
Ramsay, O. B., **1,** 118
Ramsdell, P. A., **1,** 510
Ramsden, C. A., **7,** 10, 409; **8,** 520, 521
Ramsey, B., **2,** 200
Ramsey, J. B., **1,** 121
Ramsperger, H. C., **8,** 214
Ramuz, H., **2,** 263
Rana, S. S., **11,** 24
Ranade, A. C., **8,** 66
Rance, M. J., **3,** 69
Ranchschwalbe, G., **10,** 186
Rand, C. L., **11,** 568, 575; **12,** 54
Rand, L., **1,** 121, 333, 444, 511, 935, 1197
Randaccio, L., **11,** 124
Randall, A. A., **1,** 1203
Randall, J. L., **12,** 185
Rando, R. R., **4,** 336
Ranganathan, D., **5,** 351, 476; **8,** 102
Ranganathan, S., **5,** 351, 476; **6,** 548; **8,** 102
Rangarajan, T., **6,** 20; **7,** 171
Raninger, F., **7,** 385
Rank, B., **1,** 562
Rannikko, V., **1,** 662
Ransom, C. J., **6,** 229, 605; **7,** 79
Ransom, R., **2,** 139
Ranu, B. C., **5,** 161
Rao, A. S., **2,** 236; **3,** 126; **5,** 370; **7,** 188
Rao, A. S. C. P., **9,** 194; **12,** 217
Rao, B. V. G., **11,** 167
Rao, C. G., **6,** 283; **7,** 2; **9,** 398; **10,** 40; **12,** 259
Rao, C. N. R., **1,** 832, 963
Rao, C. T., **12,** 569
Rao, D. V., **1,** 563; **10,** 202
Rao, G. S. K., **10,** 29, 136
Rao, G. V., **2,** 368
Rao, J. M., **11,** 306
Rao, K. R. N., **12,** 481
Rao, P. N., **1,** 87, 802; **4,** 33
Rao, P. V., **2,** 252
Rao, U. V., **1,** 1193
Rao, V. V., **1,** 518
Rao, V. V. R., **9,** 398
Rao, Y. S., **1,** 1126; **4,** 479; **6,** 6, 228; **7,** 295; **8,** 59
Raoul, Y., **2,** 329
Rapala, R. T., **1,** 970, 982, 986; **2,** 17, 351
Raper, R., **6,** 411
Raphael, R. A., **1,** 66, 185, 392, 574, 659, 861, 911, 926, 1137, 1295; **3,** 286; **4,** 497; **5,** 57, 98, 257, 591; **7,** 277; **9,** 172, 395, 450; **10,** 272; **11,** 461
Raphalen, A., **6,** 574
Rapoport, H., **1,** 135, 927, 1055; **2,** 263, 331, 363; **3,** 71, 325; **4,** 359, 374, 424, 432, 505; **5,** 508; **6,** 2, 141, 229; **7,** 293; **9,** 99, 375; **10,** 337; **11,** 244; **12,** 431, 450
Rapoport, L., **2,** 427
Rapp, R. D., **2,** 68, 242; **3,** 207
Rappa, A., **12,** 103
Rappoport, Z., **1,** 778; **4,** 323; **5,** 497
Rasburn, J. W., **1,** 331
Rascher, L., **2,** 163
Raschig, F., **1,** 124; **5,** 329
Rasetti, V., **9,** 349
Rashid, A., **2,** 387
Rasmussen, J., **12,** 523
Rasmussen, J. K., **5,** 93, 136; **6,** 122; **7,** 66; **8,** 409; **9,** 128; **12,** 235
Rasmussen, J. R., **11,** 551
Rasmusson, G. H., **3,** 37; **5,** 640
Rassat, A., **5,** 639; **7,** 384
Rassweiler, C. F., **1,** 6
Rastall, M. H., **10,** 225
Rastetter, W. H., **9,** 196
Raston, C. L., **12,** 290
Ratajczak, A., **7,** 366
Ratananukul, P., **12,** 44
Ratcliffe, A. H., **6,** 97
Ratcliffe, B. E., **11,** 85
Ratcliffe, C. T., **9,** 245
Ratcliffe, N. M., **12,** 38
Ratcliffe, R., **3,** 57
Ratcliffe, R. W., **4,** 97; **5,** 21; **7,** 70; **9,** 162; **10,** 342
Rathfelder, P., **1,** 643
Rathke, M. W., **1,** 484, 1231; **2,** 15, 60, 193, 195, 253, 285, 344; **3,** 29, 91, 160, 173, 237, 310; **4,** 309, 311, 447; **5,** 371, 372, 417, 690; **6,** 259, 675; **7,** 42, 51; **8,** 67, 203, 415, 500, 525; **9,** 151; **11,** 6, 217, 604
Raths, R. A., **8,** 323
Ratier, M., **7,** 380; **8,** 445
Ratovelomana, V., **10,** 194; **12,** 176, 467, 476
Ratts, K. W., **1,** 223, 257, 1246
Ratzenhofer, M., **10,** 274
Ratzkin, H., **1,** 236
Raube, R., **1,** 1073
Rauch, E., **5,** 610
Raucher, S., **8,** 28; **9,** 374, 384, 490; **10,** 172, 324, 360, 425; **11,** 428; **12,** 11
Rauchfuss, T. B., **6,** 504
Rauchschwalbe, G., **6,** 262
Rauckman, E. J., **6,** 456
Raudenbusch, W., **5,** 436; **6,** 403
Raudnitz, H., **1,** 710
Raulins, N. R., **6,** 127; **7,** 2; **10,** 32, 425
Rausch, M. D., **6,** 28; **7,** 374; **10,** 422
Rausser, R., **1,** 966, 1101
Rautenstrauch, V., **3,** 87; **6,** 323; **8,** 245; **11,** 472; **12,** 51
Ravenscroft, P. D., **12,** 289
Ravindran, N., **4,** 130, 347, 377; **5,** 101, 191, 296, 347, 467; **7,** 96, 249; **8,** 143, 354
Ravindranath, B., **10,** 234
Ravindranathan, M., **11,** 142
Ravindranathan, T., **3,** 198; **4,** 77, 497
Ravn-Petersen, L. S., **12,** 29
Rawalay, S. S., **2,** 348
Rawlins, M. F., **8,** 393

Rawlinson, D. J., **4,** 66; **5,** 370
Rawson, R. J., **3,** 258
Ray, A. B., **4,** 395
Ray, F. E., **1,** 1064
Ray, P. S., **11,** 110
Ray, R., **10,** 292
Ray, R. E., **1,** 1059
Ray, T., **12,** 90
Ray, W. C., **5,** 495
Raychaudhuri, S. R., **10,** 154; **11,** 143; **12,** 145, 444
Raymond, K. N., **5,** 574
Raymond, M. A., **1,** 1246
Raymond, M. G., **5,** 602
Raymond, P., **2,** 105; **3,** 76
Rayner, L. S., **1,** 634
Raynolds, P. W., **7,** 192; **8,** 281, 308, 502
Raynolds, S., **1,** 1064
Razdan, R. K., **9,** 274
Raznikiewicz, T., **1,** 1284
Re, G. D., **11,** 563
Re, L., **5,** 481; **7,** 79; **9,** 335; **11,** 385
Read, C. M., **10,** 24
Read, G., **2,** 387; **3,** 336; **5,** 758
Read, R. R., **1,** 1289
Read, T. O., **1,** 1246; **5,** 454
Reagan, J., **11,** 388
Ream, B. C., **8,** 36; **9,** 42
Reamer, R. A., **11,** 189
Reames, D. C., **10,** 72
Reap, J. J., **5,** 277; **6,** 403
Rebek, J., Jr., **5,** 675; **6,** 36, 254; **9,** 509; **10,** 205
Reber, F., **1,** 407, 771
Rebert, N. W., **7,** 79
Rebsdat, S., **5,** 73, 250; **6,** 243
Recondo, E., **1,** 236
Reczek, J., **8,** 252
Reddy, A. V., **9,** 13
Reddy, G. S., **8,** 84
Reddy, J. S., **1,** 13
Reddy, K. V., **5,** 428
Redemann, C. E., **1,** 454, 828
Redlich, H., **11,** 535
Redman, B. T., **3,** 87
Redmond, J. W., **2,** 238
Redmond, W., **1,** 43, 1146
Redmore, D., **1,** 87
Reece, C. A., **2,** 187
Reed, D. E., **1,** 195
Reed, H. W. B., **1,** 61
Reed, J. N., **9,** 8, 466; **11,** 373
Reed, L. A., III, **12,** 249
Reed, L. J., **1,** 1210; **11,** 281; **12,** 207
Reed, R., **1,** 424, 1203
Reed, R. A., **1,** 208
Reed, S. F., Jr., **3,** 36
Reed, W. R., **2,** 207
Reeder, R. A., **6,** 137
Rees, C. W., **1,** 323, 484, 563; **2,** 67; **3,** 11, 46, 47, 171; **4,** 75, 282; **5,** 370
Rees, R., **2,** 146, 372
Reese, A. L., **4,** 40
Reese, C. B., **1,** 838, 976; **2,** 271; **3,** 198, 223, 285; **4,** 14, 435, 451, 535; **5,** 675, 707; **6,** 229, 367, 519, 605; **7,** 79, 298, 392; **8,** 128, 362, 409; **9,** 77, 110, 139, 244; **10,** 255, 423; **11,** 537, 566
Reese, P. B., **12,** 72
Reese, R. M., **9,** 421
Reetz, M. T., **3,** 223; **4,** 325; **7,** 133; **8,** 228, 486, 525, 537; **9,** 522; **10,** 138, 243, 271, 276, 373; **11,** 150, 378, 524, 533; **12,** 150, 181, 278, 356, 490, 502, 576
Reetz, T., **1,** 371
Reeve, W., **1,** 157, 977; **4,** 260; **6,** 649
Reeves, P. C., **4,** 439; **9,** 489
Reeves, W. A., **1,** 379, 859
Reeves, W P., **7,** 170, 246
Refn, S., **1,** 169
Regan, A. C., **12,** 318
Regan, C. M., **1,** 339
Regel, W., **10,** 59
Regen, S. L., **4,** 564; **5,** 741; **6,** 38, 645; **8,** 417; **9,** 11, 423; **10,** 198, 328; **11,** 415
Regenass, F. A., **1,** 293
Regenye, R., **12,** 94
Reggel, L., **1,** 570, 615, 1237
Regitz, M., **1,** 1179; **2,** 417; **3,** 292; **4,** 60, 510; **6,** 597; **10,** 405
Reglier, M., **9,** 353
Regnault, A., **3,** 33; **4,** 522
Regnier, B., **9,** 204
Regondi, V., **10,** 255
Rehberg, C. E., **1,** 407, 1177
Rehn, D., **9,** 82
Reho, A., **6,** 534; **12,** 65
Rehwinkel, H., **11,** 529
Rei, M. H., **2,** 215; **3,** 196
Reibenspies, J. H., **12,** 412
Reich, D. A., **1,** 222
Reich, H., **1,** 74, 331, 764
Reich, H. J., **4,** 281; **5,** 26, 522, 523; **6,** 235, 338, 339, 460, 549; **7,** 120, 287; **8,** 24, 25, 28, 101; **9,** 35, 87, 195, 283, 365, 430; **10,** 170; **11,** 60, 178; **12,** 11, 547
Reich, I. L., **5,** 522; **6,** 235, 460; **9,** 195; **12,** 547
Reich, P., **8,** 198
Reich, W. S., **1,** 838
Reichardt, C., **2,** 345; **5,** 251, 438
Reiche, D. M., **4,** 260
Reichel, C. J., **10,** 161
Reichen, W., **9,** 309
Reichenbach, T., **5,** 556; **8,** 12
Reichert, J. S., **1,** 12, 659
Reichle, W. R., **1,** 445, 1187
Reichlin, D., **11,** 165, 414; **12,** 251
Reichmanis, E., **5,** 127
Reichstein, T., **1,** 74, 156, 286, 329, 407, 411, 561, 764, 771, 802, 818, 1226; **2,** 7; **8,** 523
Reid, D. E., **1,** 1019
Reid, D. H., **1,** 129, 219, 1258; **3,** 128; **4,** 501; **5,** 253
Reid, E. E., **1,** 271, 454
Reid, G. R. **10,** 373
Reid, I., **4,** 253
Reid, J. D., **1,** 531
Reid, R. G., **12,** 226
Reid, W., **2,** 412; **11,** 170
Reif, D. J., **1,** 72, 787

Reif, E., **1,** 476
Reif, W., **4,** 472; **8,** 18
Reiff, H., **2,** 249
Reiffers, S., **3,** 312; **4,** 539; **5,** 295
Reifschneider, W., **6,** 465
Reike, R. D., **10,** 251; **12,** 140
Reilly, C. A., **7,** 67
Reimlinger, H., **1,** 195, 370, 643; **4,** 19, 122, 172, 248; **5,** 572
Rein, B. M., **1,** 1249
Reinarz, R. B., **5,** 454
Reindel, F., **1,** 952
Reindel, W., **6,** 308
Reinecke, M. G., **2,** 429; **5,** 687
Reinefeld, E., **5,** 25
Reinehr, D., **5,** 306
Reineke, C. E., **2,** 373; **3,** 259
Reingold, I. D., **6,** 573; **9,** 189
Reinhart, J., **1,** 1226
Reinheckel, H., **1,** 1198; **4,** 128
Reinheimer, J. D., **1,** 322
Reinhoudt, D. N., **5,** 92, 230; **9,** 165, 177; **11,** 68
Reininger, K., **3,** 336
Reinking, P., **10,** 216
Reinstein, J. A., **9,** 139
Reinstein, M., **1,** 677; **2,** 134
Reis, H., **3,** 83
Reischer, R. J., **9,** 194
Reischl, W., **9,** 441, 502; **10,** 447; **11,** 40
Reisdorf, D., **3,** 153
Reisner, D. B., **1,** 872
Reissenweber, G., **10,** 331
Reissig, H. U., **9,** 333; **10,** 416; **11,** 40, 299
Reist, E. J., **1,** 1045; **4,** 161
Reist, H. N., **2,** 363
Reist, H. R., **2,** 331
Reitano, M., **5,** 713
Reiter, P. L., **5,** 702
Reiter, U., **8,** 395
Reitsema, R. H., **1,** 60, 1064
Reitz, R. L., **2,** 336
Reitz, R. R., **4,** 455
Rejoan, A., **12,** 36
Rekers, J. W., **10,** 202
Relles, H. M., **6,** 647
Remar, J. F., **1,** 1107
Remers, W. A., **1,** 942; **2,** 348
Remick, M. N., **1,** 595
Remillard, B. D., **9,** 139
Remion, J., **7,** 367; **10,** 312
Rempel, G. L., **6,** 426, 650; **7,** 313
Rempfer, H., **1,** 1252
Remsen, I., **1,** 870
Remuson, R., **12,** 78
Remy, D. C., **1,** 1080; **7,** 391
Ren, W. Y., **11,** 62
Renaldo, A. F., **12,** 51
Renaud, D. J., **1,** 1203
Renaut, P., **11,** 579
Rene, L., **6,** 498
Renfrow, A., **1,** 926
Renfrow, W. B., **1,** 481, 909, 926, 1033, 1261
Renga, J. M., **5,** 26, 522; **6,** 235, 460; **7,** 120, 287; **8,** 9; **11,** 178; **12,** 329
Rengaraju, S., **6,** 283; **7,** 2
Renge, T., **6,** 73
Renger, B., **5,** 717; **7,** 312; **8,** 517
Renick, K. J., **6,** 307
Renk, E., **1,** 562
Renken, T. L., **7,** 380
Renn, K., **1,** 348
Renneboog, R. M., **10,** 225, 316
Rennecke, R. W., **7,** 290
Renner, R., **5,** 577
Rennhard, H. H., **1,** 4
Renold, W., **2,** 318; **9,** 462
Renoll, M., **1,** 1178
Rens, J., **2,** 278
Rens, M., **6,** 123; **12,** 125
Renwick, J. A. A., **1,** 808
Repic, O., **4,** 413; **11,** 472
Replogle, L. L., **2,** 313
Reppe, H. G., **1,** 1261
Reppe, W., **1,** 61, 185, 190, 520, 723; **3,** 83; **4,** 420
Rerick, M. N., **1,** 598, 599, 1055; **5,** 186
Res, M., **1,** 986
Reshef, D., **11,** 184
Reske, E., **3,** 97
Resnati, G., **10,** 409; **12,** 510
Respess, W. L., **2,** 153, 278, 283, 307; **3,** 278
Ressler, C., **1,** 236, 372, 1139; **5,** 170
Restelli, A., **12,** 298, 509
Rettig, M. F., **10,** 34
Reusch, W., **1,** 450; **2,** 245, 344; **3,** 182; **4,** 309; **6,** 600; **10,** 51
Reuschling, D., **8,** 391
Reuschling, D. B., **4,** 392
Reuss, R. H., **5,** 713; **6,** 114
Reuter, J. M., **8,** 161
Reutov, O. A., **1,** 854
Reutrakul, V., **8,** 95, 299; **9,** 68, 271
Reverdin, F., **1,** 1095
Revial, G., **10,** 267
Rey, P., **7,** 384
Reye, C., **10,** 84; **11,** 117, 554; **12,** 109
Reymond, D., **1,** 329
Reynen, W. A. P., **7,** 258
Reynolds, D. D., **1,** 33, 48, 383, 452, 454, 1256
Reynolds, D. P., **9,** 239
Reynolds, G. A., **1,** 629
Reynolds, G. F., **1,** 353; **3,** 37
Reynolds, H. H., **1,** 66
Reynolds, W. E., **1,** 832
Rezende, M. C., **8,** 521
Rhee, H. K., **6,** 70
Rhee, I., **10,** 276; **12,** 336
Rhee, J. U., **6,** 70
Rhee, R. P., **8,** 76; **9,** 370; **10,** 95
Rheinboldt, H., **5,** 281
Rhoads, S. J., **1,** 772; **6,** 127; **7,** 2; **10,** 32, 425
Rhodes, S. P., **3,** 29
Rhouati, S., **9,** 515
Rhyne, L. D., **9,** 122
Riba, M., **9,** 427; **11,** 486
Ribas, I., **5,** 540
Ricard, R., **12,** 416
Ricart, R., **9,** 52
Ricca, A., **1,** 221
Riccardi, G., **9,** 119
Ricci, A., **10,** 326; **11,** 117
Riccieri, F. M., **6,** 541

Ricciuti, C., **1,** 952; **2,** 316; **9,** 391
Rice, E. M., **12,** 538
Rice, J. E., **11,** 70
Rice, K. C., **1,** 310
Rice, K. E., **7,** 236
Rice, L. E., **12,** 568
Rice, R. G., **1,** 730
Rice, R. V., **1,** 503
Richard, T. J., **9,** 196
Richards, D. H., **4,** 287
Richards, D. J., **11,** 450
Richards, H. H., **1,** 1002
Richards, K. E., **2,** 173, 267
Richards, R., **6,** 143
Richards, R. L., **12,** 502
Richards, R. W., **9,** 520
Richardson, A. C., **3,** 153; **5,** 436; **8,** 405
Richardson, A., Jr., **2,** 93
Richardson, G., **6,** 645
Richardson, G. A., **10,** 145
Richer, J. C., **1,** 180; **2,** 352; **4,** 498
Richey, F. A., Jr., **3,** 315
Richheimer, S. L., **4,** 202
Richman, J. E., **3,** 60; **5,** 155, 303, 305, 407, 445; **8,** 268
Richmond, G. D., **2,** 173
Richmond, J. M., **5,** 341
Richter, H. J., **1,** 34, 563, 680
Richter, M., **1,** 329
Richter, P., **4,** 66
Richter, R., **2,** 42; **12,** 204
Richter, T. L., **7,** 323
Richtmyer, N. K., **1,** 470, 562, 955
Richwine, H. J., **5,** 601
Rick, E. A., **11,** 269
Rickards, R. W., **2,** 213; **12,** 70
Rickborn, B., **1,** 207, 309, 926, 1145; **4,** 183, 297, 326; **6,** 103, 565; **7,** 220; **10,** 245; **12,** 286, 421
Rickenbacher, H. R., **1,** 978
Rico, J. G., **12,** 99
Riddell, F. G., **9,** 65
Rideout, D. C., **11,** 151; **12,** 314
Rideout, W. H., **1,** 182; **2,** 309
Ridge, R. J., **8,** 255
Rieber, N., **3,** 254
Riebsomer, J. L., **1,** 1112
Rieche, A., **1,** 120, 212, 220, 361, 465, 472, 1209; **2,** 120; **7,** 174
Rieck, J. N., **7,** 139
Ried, W., **1,** 11, 62, 531, 977, 1187; **4,** 33, 94, 123, 258, 397, 547; **7,** 318
Rieder, W., **5,** 132; **6,** 118
Riediker, M., **9,** 210; **11,** 91, 321, 395
Riedl, J., **1,** 1015
Riefling, B., **7,** 53; **8,** 385
Riegel, B., **1,** 1059, 1178; **5,** 62, 290
Riegel, E. R., **1,** 1073
Riegel, J., **8,** 211
Riegl, J., **5,** 283
Riego, J. M., **11,** 466
Riehl, J. J., **9,** 38, 387; **10,** 24
Riehl, T., **5,** 693
Rieke, R. D., **4,** 316; **5,** 419, 756; **6,** 293, 675; **7,** 218; **9,** 122; **11,** 308, 599; **12,** 335
Rieker, A., **7,** 260
Rieker, W. F., **12,** 384
Riemann, U., **1,** 652
Riemer, J., **1,** 767; **9,** 337
Riemland, E., **11,** 492
Riemschneider, R., **1,** 1137
Rienaecker, R., **12,** 48
Riesner, D. B., **2,** 303
Riess, J., **1,** 861, 881
Riess, J. G., **5,** 527
Rietz, E., **7,** 333
Rieveschl, G., Jr., **1,** 1064
Rifi, M. R., **4,** 370
Rigamonti, J., **1,** 43
Rigassi, N., **6,** 409
Rigby, J. H., **7,** 137; **8,** 140, 210; **11,** 439; **12,** 127, 534
Rigby, J. R., **11,** 75
Rigby, W., **1,** 60, 939, 1046; **3,** 290; **4,** 40
Riggs, N. V., **1,** 509
Riggs, R. M., **9,** 54
Rihs, G., **11,** 362
Riiber, C. N., **4,** 189
Riley, H. A., **1,** 999
Riley, H. L., **1,** 999
Riley, J. G., **7,** 207
Riley, J. P., **1,** 165
Rimbault, C. Q., **7,** 269
Rimerman, R., **6,** 552
Rimmelin, P., **5,** 61
Rinaldi, P. L., **9,** 321
Rinderknecht, H., **1,** 859; **2,** 427; **3,** 225
Rindone, B., **5,** 64; **6,** 356, 518; **11,** 445
Rinehart, J. K., **2,** 328
Rinehart, K. L., Jr., **1,** 392, 1286; **3,** 131
Ring, H., **6,** 306
Ringold, H. J., **1,** 37, 39, 60, 82, 107, 127, 150, 219, 333, 500, 526, 562, 624, 705, 730, 741, 808, 926, 1000, 1073, 1090, 1100; **2,** 117, 132
Rings, M., **4,** 491
Ringsdorf, H., **8,** 471
Riniker, B., **1,** 471
Rinke, H., **2,** 389
Rinkes, I. J., **1,** 1116
Riobe, O., **4,** 437; **7,** 324
Riordan, J. F., **1,** 1148; **2,** 404
Riordan, P. D., **11,** 554
Riordan, R. C., **3,** 116
Rios, A., **4,** 254
Ripka, W. C., **5,** 640
Ripoll, J. L., **3,** 221; **5,** 593
Rischbieth, P., **1,** 566
Risius, A. C., **7,** 392; **10,** 255
Risler, W., **4,** 35
Rist, C. E., **1,** 518
Ritchie, B., **1,** 881
Ritchie, C. D., **2,** 159
Ritchie, E., **2,** 214; **4,** 282
Ritchie, R. G. S., **5,** 470
Ritscher, J. S., **4,** 339
Rittel, W., **1,** 130; **2,** 177
Ritter, A., **1,** 427, 1218, 1232, 1236; **2,** 412; **4,** 66; **7,** 400
Ritter, D. M., **1,** 1109
Ritter, E. J., **2,** 174; **3,** 128
Ritter, F. T., **11,** 131
Ritter, J. J., **3,** 284; **4,** 255, 358; **5,** 351, 639

Ritter, R. H., **11,** 340
Rittersdorf, W., **1,** 1195
Rittle, K. E., **7,** 391
Ritvay-Emandity, K., **2** 57
Rius, C., **5,** 762
Rivard, D. E., **9,** 139
Rivera, J., **5,** 205; **9,** 341
Rivers, G. T., **7,** 94, 376; **10,** 240, 384
Rivetti, F., **6,** 375
Riviere, H., **1,** 465; **8,** 119; **10,** 413
Rivlin, V. G., **1,** 562
Rizvi, S. Q. A., **11,** 108
Rizzardo, E., **5,** 381; **6,** 612
Rizzi, G. P., **4,** 274
Rizzi, J. J., **9,** 337; **11,** 301
Rizzi, J. P., **10,** 75
Rizzo, V., **9,** 361
Ro, R. S., **1,** 34
Roach, W. S., **3,** 234
Robb, C. M., **1,** 1131
Robben, W. M. M., **7,** 323
Robbiani, C., **10,** 181; **11,** 415
Robbins, C. M., **8,** 290
Robbins, F., **1,** 1102
Robbins, J. H., **4,** 387
Robbins, M. D., **4,** 53, 74; **6,** 548
Roberge, G., **10,** 384
Roberson, E. B., **1,** 808
Robert, D. U., **5,** 527
Roberts, A., **1,** 598
Roberts, B. P., **7,** 175
Roberts, B. W., **6,** 552; **11,** 539
Roberts, C. W., **1,** 1139; **4,** 42; **5,** 322, 406
Roberts, D. A., **9,** 169
Roberts, D. C., **1,** 1080; **6,** 195
Roberts, D. L., **2,** 48
Roberts, F. E., **1,** 720; **6,** 412; **9,** 67; **12,** 220
Roberts, G. P., **1,** 741
Roberts, H., **5,** 700
Roberts, J. C., **3,** 69
Roberts, J. D., **1,** 195, 339, 1073, 1137, 1138, 1143; **2,** 464; **3,** 285; **7,** 250
Roberts, J. L., **5,** 655; **7,** 132; **8,** 194; **10,** 395
Roberts, J. S., **9,** 65, 172; **10,** 250
Roberts, M. R., **8,** 62; **9,** 349, 486
Roberts, R. A., **12,** 502
Roberts, R. D., **9,** 449
Roberts, R. J., **3,** 87
Roberts, R. M., **1,** 1210
Roberts, S. M., **6,** 157; **7,** 174; **9,** 65, 239; **11,** 550
Roberts, T. G., **9,** 214
Roberts, W., **1,** 318
Robertson, A. K., **4,** 80
Robertson, A. V., **1,** 860
Robertson, D. A., **2,** 173; **6,** 25
Robertson, D. E., **2,** 329
Robertson, D. N., **1,** 257, 634
Robertson, G. B., **5,** 172
Robertson, G. R., **1,** 1064; **3,** 272
Robey, R. L., **4,** 497, 501; **6,** 546; **7,** 365; **10,** 71
Robin, J. P., **8,** 114
Robinet, F. G., **1,** 339
Robins, B. K., **1,** 257
Robins, M. J., **6,** 555; **8,** 4; **10,** 307; **12,** 289
Robinson, A. B., **2,** 216
Robinson, C. A., **2,** 127
Robinson, C. H., **1,** 13, 75, 139, 150, 269, 808; **2,** 77; **4,** 98
Robinson, D. B., **1,** 1284
Robinson, D. S., **9,** 167
Robinson, F. M., **1,** 360
Robinson, G. M., **1,** 952
Robinson, J. C., **5,** 441
Robinson, J. C., Jr., **1,** 838
Robinson, J. M., **5,** 249, 457
Robinson, L., **6,** 249; **7,** 239
Robinson, M., **6,** 5, 587
Robinson, M. J. T., **2,** 21
Robinson, M. L., **7,** 155
Robinson, R., **1,** 7, 34, 41, 166, 383, 433, 471, 703, 904, 952; **2,** 331; **3,** 41
Robinson, R. E., **1,** 581, 754
Robinson, R. M., **1,** 983
Robinson, S. A., **1,** 952
Robinson, S. D., **5,** 332; **7,** 5, 60, 109; **11,** 184
Robison, B. L., **1,** 491
Robison, M. M., **1,** 491, 847
Robson, J. H., **1,** 323, 1226
Robson, R., **2,** 30
Rocchi, R., **6,** 419
Rocchio, J. J., **4,** 290
Rocek, J., **2,** 75; **5,** 140, 538, 563; **7,** 293
Rocek, M. D., **4,** 74
Roche, L. K., **1,** 631
Rochefort, M. P., **9,** 413; **10,** 351
Roches, D. D., **8,** 266
Rochow, E. G., **1,** 705, 1047; **5,** 719
Rockell, C. J. M., **11,** 167
Rocket, B. W., **6,** 454
Rockwell, N., **10,** 209
Rodda, H. J., **1,** 731
Rode, L., **5,** 319
Rodebaugh, R., **5,** 682
Rodebaugh, R. M., **3,** 330
Rode-Gowal, H., **6,** 229
Rodeheaver, G. T., **4,** 322, 323
Rodehorst, R., **3,** 57
Rodewald, L. B., **2,** 21; **3,** 7
Rodewald, P. G., **6,** 356
Rodgman, A., **1,** 247
Rodin, J. O., **2,** 187
Rodin, R. L., **1,** 783; **5,** 667
Rodin, W. A., **12,** 73
Rodini, D., **8,** 15
Rodini, D. J., **9,** 13; **10,** 180, 187; **11,** 12
Rodionow, W., **1,** 855
Rodrigo, R., **10,** 50; **11,** 450
Rodrigo, R. G. A., **6,** 201
Rodrigue, A., **11,** 145
Rodrigues, R., **6,** 339
Rodriguez, A., **9,** 56
Rodriguez, H. R., **2,** 313; **9,** 8, 89; **10,** 75; **11,** 402
Rodriguez, J., **11,** 256
Rodriguez, M., **10,** 342
Rodriguez, O., **8,** 519
Rodriques, K. E., **12,** 48
Rodriquez, M. L., **5,** 631
Rodriquez de Roca, M., **11,**

127
Rodriquez-Hahn, L., **8,** 452
Roe, A., **1,** 44
Roe, A. S., **2,** 12
Roebke, H., **1,** 161; **2,** 153
Roeblich, V. M., **1,** 37
Roebuck, A., **1,** 465
Roeder, H., **1,** 476
Roedig, A., **1,** 720, 938; **4,** 9; **5,** 475
Roefols, W. L., **4,** 37
Roehle, G., **6,** 511
Roehnert, H., **1,** 842
Roeller, H., **2,** 442
Roelofsen, D. P., **2,** 287; **4,** 345
Roenneberg, H., **12,** 70
Roensch, H., **1,** 1090
Roepsch, A., **1,** 851
Roesch, L., **12,** 345
Roescheisen, G., **2,** 405
Roeske, R., **1,** 524
Roessler, F., **10,** 162; **12,** 212
Roessler, P., **6,** 228
Roffia, P., **5,** 190
Rogalski, W., **1,** 1211; **2,** 378
Rogers, C. U., **1,** 841
Rogers, D. Z., **6,** 17; **8,** 12; **9,** 11
Rogers, F. F., Jr., **2,** 427
Rogers, G. T., **3,** 197
Rogers, H. A. J., **2,** 344
Rogers, H. R., **6,** 124
Rogers, J. B., **7,** 367
Rogers, J. W., **1,** 85, 86
Rogers, P. E., **5,** 211
Rogers, R., **11,** 203
Rogerts, D. A., **9,** 156
Rogic, M. M., **2,** 15, 193, 195, 285, 344; **3,** 29, 160, 237; **4,** 311, 447; **6,** 259, 423, 611; **7,** 347; **8,** 364, 466; **10,** 138
Rogier, E. R., **1,** 389, 703, 986
Roginski, E., **1,** 983, 986
Rogozhin, S. V., **6,** 634
Rohner, J., **5,** 479
Rohr, W., **1,** 1228
Rohrmann, E., **2,** 353
Roitburd, G. V., **6,** 520
Rojahn, W., **9,** 391
Rojas, A. C., **5,** 86
Rokach, J., **1,** 403; **12,** 576
Rokos, H., **5,** 3
Rolajczyk, M. M., **11,** 275
Roland, D. M., **9,** 22
Roland, J. R., **1,** 1136; **5,** 505
Rolando, C., **11,** 59; **12,** 140, 212
Roll, L. J., **1,** 695, 1283
Rolla, F., **5,** 323; **6,** 272, 406, 543; **7,** 166; **9,** 53, 361; **11,** 479
Roller, P. P., **11,** 501
Rollett, A., **1,** 119
Rollin, A. J., **7,** 271; **8,** 299
Rollin, G., **4,** 97, 397
Rollin, P., **11,** 224, 452
Rollinson, S. W., **11,** 321
Rolls, J. W., **1,** 389
Roloff, H., **1,** 652
Roman, C., **5,** 86
Roman, S. A., **2,** 263; **3,** 113; **4,** 53, 295; **5,** 381; **9,** 251
Romanelli, M. G., **2,** 201
Romanet, R. F., **5,** 303, 445; **6,** 312
Romano, L. J., **6,** 414; **7,** 307
Romeo, A., **4,** 294; **7,** 361
Romero, A., **9,** 11
Romero, M. A., **1,** 367, 561, 1143
Romers, C., **12,** 294
Romo, J., **1,** 643
Rona, P., **2,** 153; **3,** 106
Rona, R. J., **6,** 245; **7,** 220
Ronald, R. C., **5,** 354; **6,** 109; **9,** 65, 107; **12,** 113, 245
Ronault, G. F., **5,** 614
Ronchetti, F., **5,** 370
Ronchi, A. U., **4,** 134
Ronco, A., **1,** 703
Rondestvedt, C. S., Jr., **1,** 1127; **2,** 91; **4,** 18; **7,** 171
Ronen-Braustein, I., **6,** 612
Ronlan, A., **6,** 23
Ronwin, E., **1,** 130
Ronzini, L., **5,** 556; **12,** 171, 557
Ronzio, A. R., **1,** 414, 533, 999, 1049
Rooks, W., **2,** 201
Rooney, J. J., **4,** 14, 126, 240, 241, 393; **6,** 121, 473; **11,** 355
Roos, H., **1,** 66
Roos, J., **11,** 43
Roos, L., **2,** 80
Roosen, P., **1,** 842
Roosevelt, C. S., **4,** 539
Roosz, M., **12,** 532
Root, W. G., **6,** 198
Roozpeikar, B., **4,** 535; **5,** 707; **7,** 392
Roper, J. N., Jr., **1,** 389
Rorig, K., **1,** 1090
Rosan, A., **6,** 539; **8,** 455
Rosati, R. L., **9,** 458
Roscher, N. M., **5,** 61
Rose, A. A., **10,** 126
Rose, B. F., **4,** 501; **8,** 143, 527
Rose, E. H., **9,** 417
Rose, F. L., **2,** 93
Rose, J. D., **1,** 165
Rose, M. E., **9,** 388
Rose, R. K., **10,** 9; **11,** 23
Rosegay, A., **2,** 412
Rosen, G. M., **6,** 456
Rosen, L. J., **5,** 551
Rosen, P., **1,** 603; **3,** 182; **6,** 430; **9,** 184
Rosen, S., **2,** 224
Rosen, W. E., **1,** 952
Rosenau, J. D., **1,** 176
Rosenberg, D. E., **12,** 381
Rosenberg, H. M., **2,** 23
Rosenberg, S. D., **1,** 1251
Rosenblum, L. D., **5,** 244
Rosenblum, M., **5,** 610; **6,** 155, 539; **8,** 41, 455; **10,** 362; **11,** 485
Rosenburg, D. W., **1,** 652, 665
Rosendahl, F., **1,** 383, 881; **2,** 51; **10,** 67
Rosenfeld, D. D., **1,** 431, 927; **2,** 344
Rosenfeld, M. N., **6,** 241; **7,** 139; **8,** 32; **9,** 400; **10,** 24
Rosenfeld, R. S., **1,** 1284
Rosenfeld, S. M., **8,** 198
Rosenfelder, W. J., **1,** 652

Rosenheim, O., **1,** 1000
Rosenkranz, G., **1,** 3, 37, 39, 60, 82, 150, 642, 643, 731, 1000, 1090; **2,** 132, 259
Rosenman, H., **2,** 453
Rosenmund, K. W., **1,** 782, 976; **8,** 2
Rosenmund, P., **12,** 36
Rosenquist, N. R., **7,** 89, 428
Rosenstein, F. U., **5,** 66, 690, 740
Rosenthal, A., **1,** 324
Rosenthal, A. F., **7,** 413
Rosenthal, D., **1,** 782; **3,** 215
Rosenthal, I., **4,** 412; **5,** 681; **6,** 490; **7,** 307
Rosewater, W. H., **4,** 103
Rosi, D., **5,** 542
Rosini, G., **2,** 207; **4,** 387; **5,** 532; **6,** 76, 470, 534; **7,** 335; **8,** 122; **9,** 426; **12,** 405, 418
Roskos, P. D., **2,** 425; **3,** 295
Rosowsky, A., **3,** 9; **5,** 591; **7,** 141
Ross, A., **4,** 369
Ross, D. A., **5,** 700
Ross, J. A., **1,** 876; **6,** 454
Ross, J. H., **1,** 928
Ross, N. C., **1,** 703
Ross, W. A., **1,** 731
Rossana, D. M., **11,** 151
Rossbach, F., **2,** 398; **6,** 379
Rosser, R. M., **11,** 70; **12,** 65
Rossi, D., **5,** 336
Rossi, J. C., **7,** 37, 182
Rossi, R., **8,** 289; **9,** 277
Rossi, R. A., **4,** 153, 398; **5,** 217
Rossi, S., **4,** 460
Rossini, G., **12,** 275
Rossiter, B. E., **9,** 82; **10,** 65; **11,** 463; **12,** 94
Rossy, P. A., **5,** 168; **10,** 376
Roster, A. M., **4,** 512
Rotard, W., **11,** 45
Rotermund, G. W., **3,** 187
Roth, B., **11,** 281, 391
Roth, B. D., **10,** 418
Roth, H. D., **3,** 87
Roth, J. S., **3,** 16, 116
Roth, K., **2,** 174; **9,** 158; **11,** 176
Roth, M., **5,** 39; **9,** 185
Roth, R. H., **2,** 348
Roth, R. J., **4,** 419
Roth, R. W., **5,** 624; **8,** 403
Roth, W. R., **1,** 351; **2,** 140; **3,** 69
Rothe, I., **2,** 397
Rothe, M., **2,** 397
Rothemund, P., **12,** 139
Rothman, E. S., **1,** 500, 1203; **2,** 353
Rothman, L. A., **1,** 1193, 1251
Rothrock, H. S., **1,** 172
Rothstein, R., **12,** 466
Rotman, A., **5,** 506
Rotstein, D., **7,** 335
Rott, U., **1,** 671
Rottenberg, M., **6,** 281
Rotter, U., **5,** 148
Rouessac, F., **3,** 93; **4,** 141; **8,** 138; **9,** 97, 442
Rougny, A., **7,** 177
Rouhi-Laridjani, M., **1,** 561
Roumestant, M. L., **5,** 420; **12,** 51
Roundhill, D. M., **6,** 504
Rouot, B., **6,** 458
Roush, D. M., **9,** 222; **10,** 180
Roush, W. R., **5,** 675; **8,** 337; **9,** 270; **10,** 181; **11,** 95, 218; **12,** 277, 350, 513
Rousseau, G., **1,** 626; **8,** 374; **10,** 460; **11,** 121; **12,** 74
Roussel, A., **4,** 135; **6,** 455
Roussel, M., **10,** 203
Roussel, R., **7,** 296
Roussi, G., **2,** 110; **7,** 299; **9,** 381; **12,** 533
Roustan, J. L., **9,** 207
Rouve, A., **10,** 150
Rouvier, E., **9,** 306
Roux, D. G., **7,** 62; **8,** 391
Rowe, B. A., **11,** 32
Rowe, C. A., Jr., **1,** 309, 926
Rowe, E. J., **1,** 518
Rowe, F. M., **1,** 1087
Rowe, K., **6,** 536; **7,** 101
Rowland, A. T., **1,** 309
Rowland, C. R., **11,** 132
Rowland, S. P., **1,** 1294
Rowley, H. H., **2,** 254
Rowley, R. J., **11,** 22
Roy, A., **3,** 4
Roy, G., **8,** 278; **9,** 284; **11,** 129, 331; **12,** 118
Roy, R., **9,** 141
Roy, S., **3,** 4
Royer, G. P., **9,** 227
Royer, R., **1,** 1187; **2,** 353; **4,** 416; **5,** 567; **6,** 498; **11,** 360
Royes, G. P., **11,** 393
Rozantzev, E. G., **6,** 114
Rozen, S., **4,** 335; **5,** 680; **7,** 158; **9,** 223; **10,** 1; **11,** 5; **12,** 4
Rozzell, J. D., **11,** 460
Rua, L., **4,** 444; **6,** 538
Rubenstein, M., **6,** 319
Rubezhov, A. Z., **8,** 412
Rubin, M., **1,** 802
Rubin, M. B., **1,** 777, 1184; **2,** 462
Rubinstein, R., **5,** 136
Rubio-Lightbourn, J., **5,** 381, 427
Rubottom, G. M., **4,** 247; **5,** 589, 614; **6,** 114; **7,** 188, 189, 271, 430; **8,** 102, 272, 513; **9,** 249; **10,** 93; **11,** 124, 268; **12,** 126, 272, 571
Ruby, P. R., **1,** 402
Ruch, E., **4,** 384
Ruchirawat, S., **8,** 367
Ruck, F., **1,** 66
Rudakova, I. P., **3,** 222
Ruddick, J. D., **7,** 313
Ruddick, J. N. R., **11,** 526
Ruddick, J. P., **6,** 650
Ruddy, A. W., **1,** 169
Ruden, R. A., **4,** 154, 183; **5,** 454, 751; **6,** 84, 188, 322, 635, 636; **8,** 170; **10,** 289; **12,** 221
Ruder, S. M., **12,** 113
Rudinger, J., **1,** 97, 755, 859
Rudler, H., **3,** 117; **4,** 268

Rudnick, L. R., **5,** 763
Rudolph, W., **1,** 150
Rudowski, A., **1,** 834, 1191
Ruechardt, C., **1,** 49, 116, 409, 723, 966; **6,** 82
Ruecker, C., **11,** 64, 91, 566, 567; **12,** 97
Ruecker, G., **10,** 359
Rueegg, R., **1,** 114, 385, 388, 567, 614, 1178
Ruehlmann, K., **2,** 438; **3,** 312; **5,** 727; **10,** 45; **12,** 502
Ruempler, K. D., **4,** 163
Rueppel, M. L., **4,** 374; **6,** 2
Ruest, L., **3,** 275; **5,** 234
Rueter, J., **3,** 292; **4,** 510
Ruettimann, A., **6,** 107, 222; **12,** 194
Ruetz, L., **4,** 512
Ruf, W., **10,** 454
Rufenact, R., **5,** 660
Ruff, J. K., **1,** 773; **6,** 607; **8,** 385
Ruff, O., **1,** 476
Ruffner, R. J., **7,** 76; **8,** 126; **11,** 329
Ruggli, P., **5,** 479
Ruhoff, J. R., **1,** 271, 454, 951
Ruiz, A., **2,** 201
Ruiz, V. M., **5,** 758
Rummens, F. H. A., **2,** 329
Rummert, G., **1,** 669
Rundberg, E. G. S., Jr., **7,** 301
Rundel, W., **1,** 195, 754; **3,** 69; **4,** 122; **5,** 507
Runeberg, J., **1,** 392; **2,** 366
Runge, F., **1,** 672
Runge, T. A., **10,** 32
Runquist, A. W., **6,** 17; **7,** 7; **8,** 12; **9,** 470
Ruoff, P. M., **2,** 394
Rupani, P., **10,** 25
Rupe, H., **2,** 203; **5,** 319
Rupp, H. D., **5,** 668
Ruppert, J. F., **5,** 756; **7,** 173
Ruppert, W., **6,** 187
Rusche, J., **1,** 120
Ruschig, H., **1,** 139, 488; **4,** 296
Ruse, M., **12,** 546
Rusek, J. J., **9,** 87
Rush, J. E., **4,** 132
Rush, R. V., **2,** 444
Ruske, W., **1,** 1272
Russ, M., **10,** 59, 98
Russ, W. K., **12,** 220
Russe, R., **9,** 246
Russel, G. A., **10,** 409
Russel, J. H., **1,** 286
Russel, K. E., **6,** 471
Russell, C. E., **12,** 576
Russell, C. G., **10,** 312, 452
Russell, C. R., **6,** 638; **12,** 542
Russell, G. A., **1,** 80, 296, 310, 318, 732, 1131; **2,** 173; **6,** 419, 628
Russell, J. A., **2,** 87
Russell, J. R., **1,** 195, 465, 1157
Russell, L. J., **12,** 258
Russell, M., **1,** 847
Russell, M. J. H., **8,** 153
Russell, P. B., **1,** 444
Russell, R. A., **9,** 61; **10,** 313
Russell, R. R., **1,** 1269
Russell, T. W., **4,** 351; **6,** 446
Russo, G., **5,** 370
Rustemeier, K., **10,** 194
Rusterholz, D. B., **11,** 407
Ruston, S., **6,** 341
Rutenberg, A. M., **1,** 62
Ruth, J. M., **5,** 285
Rutherford, D., **1,** 180
Rutherford, K. G., **1,** 43, 1146, 1226, 1261
Rutherford, R. G., **2,** 222
Rutkowski, A. J., **1,** 1191
Rutledge, P. S., **1,** 1008; **3,** 66; **5,** 157, 395, 655; **7,** 360; **8,** 260, 481; **9,** 249, 458; **10,** 395; **11,** 305, 486, 515
Rutschmann, J., **1,** 384
Ruyle, W. V., **1,** 652
Ruzicka, L., **1,** 119, 747, 933, 963; **2,** 213; **6,** 91
Ruzic-Toros, Z., **11,** 124
Ruzo, L. O., **11,** 445
Ruzziconi, R., **11,** 120
Ryabov, A. D., **11,** 516; **12,** 484
Ryan, G., **2,** 424
Ryan, K., **10,** 252, 449
Ryan, K. J., **1,** 1045
Ryan, P. W., **1,** 581
Ryan, R. W., Jr., **1,** 1150
Ryang, M., **3,** 212; **4,** 302, 354; **12,** 336
Rybalka, B., **5,** 649
Rychnovsky, S. D., **11,** 267, 600; **12,** 256, 523
Ryder, D. J., **9,** 317, 464
Rydon, H. N., **1,** 60, 208, 851, 1249, 1250; **2,** 322, 446; **4,** 557; **5,** 512; **7,** 59; **9,** 439
Ryerson, G. D., **1,** 72
Rylander, P. N., **1,** 892, 982, 983, 986, 989, 1116; **3,** 216; **5,** 406; **9,** 352
Ryntz, R. A., **9,** 220
Ryono, L. S., **5,** 463; **11,** 51
Ryschkewitsch, G. E., **11,** 502
Ryser, G., **1,** 114, 567
Ryskiewicz, E. E., **1,** 285
Ryu, I., **6,** 73; **8,** 28, 133, 162, 172; **9,** 151; **10,** 276, 312; **12,** 106, 144, 336
Rzeszotarska, B., **1,** 486; **4,** 66; **5,** 87

Saa, J., **5,** 540
Saa, J. M., **7,** 260
Saalfrank, R. W., **3,** 94; **5,** 209; **7,** 102; **11,** 176
Saavedra, J. E., **7,** 254; **8,** 427
Saba, A., **9,** 260
Sabacky, M. J., **5,** 502; **11,** 31
Sabesan, S. I., **11,** 479
Sabet, C. R., **1,** 52
Sabicri, M., **9,** 205
Sable, H. Z., **9,** 391
Sabo, E. F., **1,** 74, 150, 802
Sabol, M. A., **5,** 330
Sabol, M. R., **12,** 294
Saburi, M., **11,** 277
Saccomano, N. A., **12,** 449
Sacconi, L., **5,** 42; **11,** 138, 563
Sachdev, H. S., **2,** 326; **3,** 12, 13, 214; **6,** 233; **7,** 402
Sachdev, K., **7,** 402

Sachdeva, R., **9,** 374
Sachinvala, N. D., **12,** 444
Sachs, D. H., **1,** 392, 606, 1108
Sachs, F., **5,** 671
Sack, A., **1,** 1118
Sacki, H., **9,** 479
Sacks, C. E., **6,** 234; **7,** 32; **9,** 88
Sackur, O., **1,** 1131
Sadana, K. L., **5,** 75; **8,** 59; **9,** 77
Saddler, J., **10,** 209
Sadeh, S., **5,** 33
Sadhu, K. M., **12,** 81
Sadler, A. C., **10,** 5
Sadler, I. H., **3,** 87
Sadovaja, N. K., **6,** 232
Sadowski, J., **2,** 27
Sadowski, J. S., **5,** 148
Saegusa, T., **3,** 167; **4,** 101, 107, 109; **5,** 150, 163, 171, 572, 581; **6,** 129, 130, 141, 145, 450, 517; **7,** 74, 155, 322; **8,** 127, 299, 381, 501; **9,** 123, 279, 349; **10,** 84, 106, 185, 237, 381; **11,** 12, 100, 117; **12,** 95
Saeki, H., **2,** 159
Saeki, T., **7,** 135; **8,** 345
Sae-Lee, C., **3,** 161
Saeman, M., **9,** 119
Saenger, W., **3,** 12; **6,** 568; **7,** 367
Safe, S., **1,** 155
Saferstein, L., **2,** 305
Saffhill, R., **2,** 271; **3,** 198; **8,** 362
Safir, S. R., **10,** 376
Saft, M. S., **11,** 420
Sagami, H., **5,** 266
Sagatys, D. S., **4,** 247
Sage, J. M., **11,** 439
Sager, W. F., **1,** 414, 827, 1116; **2,** 263
Sagredos, A. N., **3,** 33
Sah, P. P. T., **1,** 566
Saha, C. R., **6,** 45
Saha, J. G., **2,** 352
Sahai, M., **12,** 475
Sahm, W., **2,** 92
Sahoo, S. P., **11,** 382
Sahraoui-Taleb, S., **12,** 125
Sahu, D. P., **10,** 448; **12,** 554
Sahyun, M. R. V., **1,** 309, 927
Sai, M., **9,** 13; **10,** 11
Saidi, M. R., **6,** 539; **11,** 533
Saigo, K., **6,** 118, 308, 376, 613; **7,** 27, 40, 92, 372; **8,** 20, 96; **11,** 550; **12,** 313
Saika, D., **6,** 115
Sailliant, R., **8,** 522
Saimoto, H., **9,** 256; **11,** 365, 470; **12,** 263, 304
Saindane, M., **10,** 345, 373; **11,** 14, 299, 466; **12,** 131, 378, 390
Sainsbury, M., **12,** 257
Sainte, F., **11,** 51
Saint-Ruf, G., **3,** 240; **5,** 542
Saito, E., **6,** 462
Saito, H., **5,** 571
Saito, I., **6,** 241, 436; **7,** 268
Saito, K., **8,** 51
Saito, M., **9,** 494; **10,** 433; **11,** 514; **12,** 540
Saito, S., **11,** 343; **12,** 64
Saito, T., **5,** 691; **6,** 453; **9,** 133, 471
Saito, Y., **11,** 12, 20; **12,** 20
Saitoh, Y., **11,** 317
Sajus, L., **2,** 252; **4,** 204; **5,** 537
Sakaguchi, R., **9,** 34; **10,** 24
Sakai, I., **2,** 445; **6,** 430; **7,** 81; **11,** 574
Sakai, K., **4,** 54, 562, 571; **5,** 13, 178; **6,** 158; **7,** 424; **10,** 302; **11,** 396; **12,** 131
Sakai, M., **2,** 238; **4,** 322, 435; **6,** 198; **11,** 157
Sakai, N., **5,** 64
Sakai, S., **5,** 601; **6,** 96, 187; **7,** 47; **8,** 233; **9,** 46, 344, 484
Sakai, T., **7,** 170; **10,** 306; **12,** 277, 365
Sakai, Y., **9,** 269; **10,** 252
Sakairi, N., **7,** 171; **10,** 207
Sakajira, T., **8,** 441
Sakakibara, J., **9,** 349
Sakakibara, S., **1,** 40, 668, 745; **2,** 24, 25, 216, 219, 307; **3,** 15; **9,** 297; **12,** 199
Sakakibara, T., **9,** 41, 349; **10,** 21
Sakakibara, Y., **6,** 573; **11,** 157
Sakakura, T., **12,** 89, 118
Sakamaki, H., **9,** 148
Sakamoto, H., **9,** 257
Sakamoto, T., **8,** 198; **9,** 188; **10,** 346; **11,** 480
Sakan, K., **12,** 11
Sakan, T., **2,** 387; **5,** 319, 491
Sakanishi, K., **8,** 317
Sakashita, T., **5,** 160
Sakasi, T., **10,** 346
Sakata, J., **8,** 468; **10,** 453; **12,** 462
Sakata, K., **12,** 440
Sakata, T., **2,** 263
Sakata, Y., **6,** 500; **10,** 85; **12,** 263, 546
Sakdarat, S., **12,** 112
Sakito, Y., **9,** 20
Sako, M., **12,** 272, 363
Sako, N., **4,** 113
Sakoh, H., **12,** 316
Sakrikar, S., **5,** 335; **12,** 242
Saksena, A. K., **3,** 59; **12,** 454
Sakuma, K., **9,** 38; **10,** 289
Sakuma, Y., **5,** 561; **9,** 160
Sakuraba, M., **4,** 420
Sakurago, H., **1,** 796
Sakurai, H., **1,** 409; **5,** 341, 428, 575, 620; **6,** 534; **7,** 372, 400; **8,** 109, 181, 486; **9,** 47, 69, 255, 256, 352, 446, 494; **10,** 8, 56, 62, 98, 219, 289, 433; **11,** 103, 128, 130, 203, 275, 533, 580, 587; **12,** 263, 502, 540, 546
Sakurai, T., **7,** 79; **11,** 30
Sakurai, Y., **4,** 369
Sakuta, K., **11,** 36; **12,** 435, 547
Sala, T., **8,** 468; **9,** 69
Saladin, E., **4,** 336
Salama, M. A., **5,** 731
Salamon, A., **1,** 119
Salauen, J. R., **4,** 251, 305,

472; **5,** 713; **11,** 127; **12,** 248
Salazar, J. A., **7,** 40
Salbaum, H., **6,** 95
Sale, A. A., **4,** 75
Saleh, S. A., **5,** 700; **11,** 434
Salem, A. Y., **1,** 131
Salem, G., **12,** 290
Salem, G. F., **9,** 324; **11,** 275, 587
Salerno, G., **9,** 489
Salim, N., **1,** 68
Saliman, F. M., **9,** 133
Salisbury, J. M., **1,** 170
Salisbury, L., **1,** 561
Salisbury, P., **9,** 115
Salisova, M., **9,** 127; **10,** 223
Salmi, E. J., **1,** 662
Salmond, W. G., **8,** 110, 158, 252, 448, 487
Salomon, M. F., **5,** 32; **6,** 665; **8,** 45; **10,** 144, 363
Salomon, R. G., **4,** 221, 517; **5,** 152; **6,** 201, 624, 665; **7,** 223; **8,** 45, 109, 161; **9,** 239; **10,** 110, 144, 154, 363; **11,** 143, 180; **12,** 145, 444
Salomone, R. A., **3,** 53
Salonen, K., **10,** 252
Saltzman, H., **1,** 508, 509, 511, 787
Salvadori, G., **1,** 49
Salzmann, T. N., **5,** 247, 277; **6,** 218, 238; **10,** 342
Sam, D. J., **4,** 143; **5,** 155, 176; **11,** 119
Samain, D., **7,** 100; **8,** 150, 176
Samanen, J. M., **5,** 744
Samaraj, L. I., **5,** 668
Samaritano, R. H., **1,** 59
Sambeth, J., **1,** 971
Sambur, V. P., **6,** 621
Sammes, P. G., **4,** 162, 204, 262, 570; **5,** 758; **7,** 268; **8,** 347, 402; **9,** 48, 285; **10,** 181; **11,** 157, 309; **12,** 554
Sammour, A. E. M. A., **1,** 444
Samokhavalov, G. I., **1,** 388
Sampathkumar, P. S., **9,** 73
Sample, T. E., Jr., **2,** 390; **3,** 273
Sampson, W. L., **1,** 470
Sams, J. R., **11,** 526
Samsel, E. G., **11,** 397
Samson, M., **8,** 131
Samuel, D., **1,** 1139
Samuels, S. B., **8,** 41
Samuelsson, B., **9,** 121, 508; **10,** 329, 350; **11,** 467; **12,** 71, 139, 366, 419, 555
Sanchez, B. L., **4,** 239, 451
Sanchez, E. L., **12,** 141
Sanchez, M. B., **1,** 1000
Sandberg, R., **1,** 84, 266
Sandborn, L. T., **1,** 1064, 1229
Sandefur, L. O., **5,** 244
Sander, J., **12,** 353
Sander, V., **7,** 298
Sanders, A., **5,** 686
Sanders, F., **1,** 1157
Sanders, H. J., **4,** 438
Sanders, H. P., **10,** 255; **11,** 537
Sanders, J. R., **5,** 14
Sanderson, A. R., **1,** 1128
Sanderson, J. J., **1,** 70, 1118
Sanderson, J. R., **6,** 577
Sanderson, T. F., **1,** 37, 187
Sandin, R. B., **1,** 502, 509, 872, 1116; **4,** 163
Sandler, S. R., **2,** 322, 363; **3,** 72; **4,** 165, 523
Sandmann, R. A., **3,** 6
Sandmeyer, T., **1,** 481
Sandner, M. R., **5,** 571
Sandri, J. M., **1,** 1095
Sandri, S., **5,** 86; **7,** 69, 71, 298; **9,** 257; **10,** 211; **11,** 267; **12,** 87
Sands, C. A., **1,** 964
Sands, D. R., **3,** 40
Sands, R. D., **7,** 9
Sandstrom, W. M., **1,** 1033
Sandulesco, G., **1,** 411
Sane, P. P., **5,** 370
Sanfer, P., **8,** 122
San Filippo, J., Jr., **2,** 400; **3,** 81, 112; **4,** 183; **5,** 600; **6,** 414, 490; **7,** 307; **8,** 112, 419, 522
Sanfilippo, P. J., **11,** 391
Sanfilippo, S. J., **1,** 331
Sanford, A., **3,** 329
Sangaiah, R., **10,** 29
Sanger, F., **1,** 322
Sangster, I., **1,** 834, 1191
Sanitra, R., **9,** 8
Sanjoh, H., **5,** 658; **7,** 365; **8,** 481
Sankaran, D. K., **2,** 105
Sankaran, V., **6,** 546
Sankarappa, S. K., **6,** 546
Sanna, P., **12,** 46
Sannie, C., **1,** 87
Sanno, Y., **5,** 253, 617
Sano, H., **5,** 150; **9,** 146; **10,** 413; **11,** 551; **12,** 344
Sano, K., **6,** 187
Sano, S., **5,** 13, 178; **6,** 158
Sano, T., **4,** 555; **8,** 453
Sansoulet, J., **11,** 407
Santamello, E., **6,** 518
Santaniello, E., **8,** 110, 534; **9,** 376, 421; **10,** 42, 324, 382; **11,** 157; **12,** 444, 458
Santelli, M., **5,** 389; **11,** 126, 534; **12,** 51, 70, 305
Santelli-Rouvier, C., **12,** 70
Santelli-Rouvier, E. J., **12,** 51
Santhanakrishan, T. S., **4,** 146
Santhanakrishnan, N. A., **2,** 29
Santi, W., **5,** 580
Santiago, M. L., **10,** 120, 211
Santiesteban, H., **8,** 448; **11,** 475
Santilli, A. A., **1,** 85
Santos, R., **7,** 368
Santosusso, T. M., **6,** 228
Santroch, J., **2,** 301
Santucci, E., **6,** 541
Santurri, P., **1,** 1102
Sanyal, B., **5,** 161; **6,** 70; **7,** 387; **10,** 426
Sapp, J., **1,** 1220; **2,** 434
Sapuppo, N., **7,** 236
Saraf, S. D., **5,** 663

Saraie, T., **5,** 28
Sard, H., **10,** 442
Sarda, P., **3,** 337; **4,** 577; **8,** 374, 535
Sardarian, A., **12,** 62
Sarel, S., **1,** 444, 815, 1081; **2,** 224, 315; **3,** 86; **5,** 358; **7,** 144
Sarett, L. H., **1,** 87, 147, 360, 402, 662, 685, 703, 764, 802, 881, 926, 1000, 1006
Sargent, G. D., **5,** 470
Sargent, M. V., **1,** 1284; **3,** 339; **4,** 508; **5,** 127, 269, 672; **8,** 468; **9,** 69; **10,** 201; **11,** 308; **12,** 65
Sarhangi, A., **8,** 499
Sarid, S., **1,** 581
Sarin, P. S., **1,** 730
Saris, L. E., **7,** 336
Sarkanen, K. V., **5,** 555
Sarkar, M., **10,** 426
Sarkar, T., **10,** 84
Sarkis, A., **1,** 124
Sarkisian, G. M., **5,** 290
Sarma, D. N., **12,** 305
Sarma, J. C., **11,** 128
Sarma, P. S., **1,** 814
Sarnowski, R., **9,** 138
Sartori, G., **8,** 245; **9,** 14; **10,** 462
Sartori, P., **7,** 390; **8,** 515
Sasaki, A., **9,** 106
Sasaki, I., **12,** 526
Sasaki, J., **11,** 580; **12,** 540
Sasaki, K., **1,** 161; **9,** 255; **10,** 219; **11,** 275; **12,** 263, 540, 548
Sasaki, M., **8,** 532; **9,** 415; **10,** 460; **11,** 373, 452, 599; **12,** 439
Sasaki, N., **6,** 215
Sasaki, O., **4,** 21
Sasaki, S., **5,** 89, 552
Sasaki, T., **3,** 260; **4,** 52, 93; **5,** 28, 567; **6,** 43, 297; **7,** 79
Sasakura, K., **9,** 63, 152; **12,** 179
Sasaoka, M., **12,** 110
Sasatani, S., **12,** 11, 193
Sasho, M., **11,** 255; **12,** 449
Sasse, K., **10,** 331
Sasse, W. H. F., **1,** 731, 977
Sasson, I., **6,** 436; **9,** 73
Sasson, Y., **4,** 564; **5,** 95, 96, 741; **6,** 426, 650, 655; **9,** 418; **10,** 79
Sastry, K. A. R., **11,** 489
Satake, M., **9,** 344
Satek, L., **5,** 419; **11,** 309
Sathe, V. M., **3,** 126
Satish, S., **11,** 453
Sato, A., **4,** 155
Sato, E., **1,** 894
Sato, F., **8,** 289; **9,** 233, 277; **10,** 131, 194, 238; **11,** 164, 175, 599; **12,** 171, 356, 449, 513
Sato, H., **5,** 621
Sato, K., **4,** 222; **8,** 337; **10,** 277; **11,** 391, 392; **12,** 213
Sato, M., **8,** 289; **9,** 233, 277; **10,** 131, 194, 238, 239; **11,** 164, 175, 599; **12,** 171, 344, 352, 356, 449, 513
Sato, N., **10,** 174
Sato, O., **11,** 227
Sato, S., **5,** 148; **8,** 289; **9,** 269, 393, 429; **10,** 171; **11,** 131; **12,** 541
Sato, T., **1,** 756; **6,** 500, 596; **7,** 199, 428; **8,** 253, 446; **9,** 87, 233, 246, 520; **10,** 289; **11,** 141, 221, 235, 249, 293, 358, 514; **12,** 33, 125, 201, 231, 282, 477, 548, 555, 565
Sato, Y., **1,** 329; **7,** 146, 165; **10,** 273; **11,** 591
Satoh, D., **7,** 332
Satoh, H., **5,** 433
Satoh, J. Y., **4,** 107; **8,** 53, 367; **10,** 211; **11,** 131, 267; **12,** 256
Satoh, M., **11,** 178
Satoh, T., **3,** 265; **5,** 602; **9,** 34; **10,** 24; **11,** 480; **12,** 344, 440
Satoh, Y., **11,** 294; **12,** 236
Satomi, M., **11,** 283
Satomi, N., **10,** 305
Satori, P., **4,** 264
Satsumabayashi, S., **5,** 534
Sattar, A., **9,** 65
Satyanarayana, P., **11,** 167
Satzinger, G., **6,** 514
Saucy, G., **1,** 652, 1178, 1284; **2,** 231; **4,** 235, 328; **6,** 656; **7,** 414; **8,** 192; **9,** 420; **10,** 43, 155
Sauer, G., **6,** 411
Sauer, H., **1,** 832
Sauer, J., **1,** 182, 1136; **2,** 23, 326; **5,** 663
Sauer, J. C., **1,** 228, 292, 1203, 1209, 1284, 1294; **4,** 414
Sauer, J. D., **5,** 117, 457; **6,** 244, 479; **10,** 249
Sauer, R. O., **1,** 427
Sauerbier, M., **5,** 740
Sauers, C. K., **1,** 367
Sauers, R. A., **4,** 294
Sauers, R. R., **1,** 573; **4,** 291
Sauerwald, M., **11,** 533; **12,** 356, 576
Saunders, B. C., **1,** 248, 253, 322; **2,** 441; **4,** 481; **5,** 650; **7,** 106
Saunders, J. H., **1,** 12, 872, 911, 1027; **5,** 280
Saunders, J. O., **10,** 156, 443
Saunders, M., **1,** 844; **5,** 19
Saunders, W. H., Jr., **7,** 170
Sausen, G. N., **1,** 986, 1136
Saussine, L., **9,** 132; **11,** 59
Sauter, F. J., **4,** 63; **12,** 52
Sauter, H., **5,** 693; **6,** 173; **7,** 100
Sauvage, J. F., **1,** 754
Sauvage, J. P., **5,** 156; **6,** 137
Sauve, D. M., **1,** 581, 754
Sauvetre, R., **6,** 622; **8,** 245, 505
Savard, J., **9,** 303; **12,** 196
Savas, E. S., **1,** 1216
Savelli, G., **5,** 624
Savides, C., **1,** 40, 927
Savige, W. E., **7,** 135; **8,** 51
Savignac, P., **5,** 512; **6,** 189, 191, 279; **7,** 83, 104; **8,** 168
Saville, B., **3,** 320

Saville, J. F., **7,** 69
Saville, N. M., **1,** 68
Savoia, D., **6,** 43; **8,** 406; **9,** 378, 477; **10,** 297, 322; **11,** 276, 356; **12,** 571
Savona, G., **4,** 373
Savu, P. A., **10,** 106
Saw, A. C., **9,** 446
Sawa, Y. K., **1,** 322
Sawachi, Y., **4,** 395
Sawada, H., **6,** 220
Sawada, I., **12,** 219
Sawada, K., **11,** 258, 562
Sawada, S., **3,** 258; **4,** 167; **12,** 439
Sawahata, T., **8,** 478
Sawaki, Y., **8,** 248
Saward, C. J., **6,** 154
Sawaya, H. S., **6,** 215
Sawaya, T., **5,** 308
Sawicki, E., **1,** 673
Sawyer, A. K., **1,** 349
Sawyer, R. L., **1,** 257
Sax, K. J., **1,** 80, 1178
Sax, M., **5,** 394
Saxena, M. K., **4,** 173
Saxena, M. P., **9,** 343
Saxton, J. E., **1,** 1040
Saxton, M. R., **1,** 230
Sayed, Y. A., **6,** 283; **10,** 77, 246
Sayer, T. S. B., **6,** 206; **7,** 118; **8,** 299
Sayigh, A., **1,** 536
Sayigh, A. A. R., **2,** 101
Sayo, N., **12,** 96
Sayre, J. L., **1,** 629
Scahill, T. A., **11,** 401
Scaife, C. W., **1,** 328; **7,** 253
Scala, A., **7,** 70, 428; **9,** 391
Scala, A. A., **1,** 424
Scalera, M., **1,** 1128
Scanio, C. J. V., **4,** 371
Scanlan, J. T., **1,** 465, 759, 992
Scanlon, B., **2,** 369; **3,** 16
Scarborough, R. M., Jr., **8,** 305, 307, 448; **9,** 24; **10,** 343, 356; **11,** 5
Scarpa, M., **10,** 413
Scarpati, R., **12,** 365
Scarponi, U., **6,** 110
Scartazzini, R., **10,** 122
Scartoni, V., **5,** 568
Scatturin, A., **5,** 647
Scerbo, L., **2,** 422
Scettri, A., **8,** 12, 427; **10,** 366; **11,** 452; **12,** 256
Schaaf, T. K., **3,** 67, 308; **4,** 105, 376; **5,** 639
Schaafsma, S. E., **4,** 108
Schaal, V., **1,** 720
Schaap, A. P., **3,** 324; **5,** 491, 727; **6,** 469
Schaap, L., **1,** 927
Schade, G., **12,** 388
Schadenberg, H., **6,** 525
Schaechter, O., **1,** 1144
Schaefer, A. D., **2,** 101
Schaefer, D., **3,** 254
Schaefer, G., **11,** 103
Schaefer, G. F., **1,** 1065
Schaefer, H. J., **9,** 43, 160, 399; **10,** 29
Schaefer, J. P., **1,** 999, 1000, 1169, 1249; **2,** 385, 446; **3,** 322
Schaefer, W., **1,** 67, 1121; **2,** 23; **4,** 425, 497
Schaefer-Ridder, M., **7,** 268, 269
Schaeffer, D. J., **4,** 472
Schaeffer, G. F., **3,** 272
Schaer, B., **5,** 562; **8,** 400
Schaerfe, E., **1,** 782
Schafer, D. J., **4,** 391
Schafer, T. R., **9,** 334
Schafer, W. R., **1,** 561
Schaffer, E., **1,** 966
Schafferman, R., **4,** 46
Schaffner, I. J., **1,** 703
Schaffner, K., **1,** 814, 827; **10,** 122
Schagerer, K., **5,** 251
Schalk, D. E., **7,** 415
Schallenberg, E. E., **3,** 143
Schaller, H., **1,** 189, 848, 1138, 1228
Schallhorn, C. H., **3,** 251, 252
Schallner, O., **5,** 681
Schambach, R. A., **3,** 65
Schamp, N., **6,** 118; **8,** 15, 105, 463; **9,** 182; **11,** 434
Schank, K., **2,** 63, 233; **3,** 190; **5,** 552; **9,** 192; **11,** 388; **12,** 386
Schank, R. L., **5,** 713
Schaper, U. A., **11,** 241
Schaper, W., **12,** 439
Schappell, F. G., **1,** 20
Schappert, R., **11,** 460
Scharf, D. J., **2,** 120
Scharf, G., **4,** 562
Scharf, H. D., **2,** 123; **6,** 172; **10,** 56
Scharp, J., **3,** 37
Schatz, P. F., **4,** 349
Schaub, B., **11,** 120
Schaub, F., **6,** 165
Schaub, R. E., **1,** 40, 198, 603; **4,** 517
Schauble, J. H., **9,** 477
Schauder, J. R., **12,** 219
Schaus, J. M., **8,** 72
Scheck, D., **11,** 485
Scheckenbach, F., **1,** 247
Scheel, F. M., **3,** 228; **4,** 389
Scheeren, H. W., **8,** 537; **9,** 523; **11,** 71
Scheeren, J. W., **5,** 614, 653; **11,** 71
Scheffer, A., **1,** 470
Scheffer, J. R., **9,** 181
Scheffler, K., **1,** 933
Scheffold, R., **4,** 336; **7,** 277; **10,** 359
Schegolev, A. A., **6,** 520; **11,** 163
Scheiber, P., **5,** 225
Scheibler, H., **2,** 330
Scheibye, S., **8,** 327; **9,** 50
Scheidegger, U., **1,** 814
Scheidl, F., **9,** 260; **11,** 29
Scheifele, H. J., **1,** 1184
Scheinbaum, M. L., **4,** 358, 360, 540; **6,** 629
Scheiner, P., **1,** 562
Scheinmann, F., **4,** 577; **5,** 32, 740; **6,** 90, 203, 424
Scheit, K. H., **2,** 188, 387
Scheithauer, S., **6,** 583
Schell, D. M., **4,** 393
Schell, R. A., **4,** 87

Schellenberg, W. D., **1,** 185
Schellenberger, H., **1,** 562
Schempf, R., **11,** 32
Schenck, F., **1,** 172
Schenck, G. O., **4,** 550; **5,** 491; **6,** 586
Schenk, G. H., **1,** 802; **2,** 13
Schenk, H. R., **1,** 471
Schenk, W. N., **8,** 416
Schenker, E., **1,** 1054
Schepers, R. G., **1,** 139
Schepp, H., **2,** 422
Scheraga, H. A., **10,** 398
Scherer, K. V., **5,** 155
Scherer, K. V., Jr., **2,** 144
Scherowsky, G., **3,** 140
Scherrer, R. A., **4,** 87
Scheuer, P. J., **4,** 31
Scheunemann, K. H., **7,** 215
Schevier, J. R., **8,** 267
Schexnayder, D. A., **1,** 320
Schick, A., **1,** 933
Schick, H., **1,** 318
Schickfluss, R., **5,** 170
Schiedt, U., **1,** 933
Schiehser, G. A., **10,** 167
Schiemann, G., **1,** 44, 396
Schiemenz, G. P., **2,** 179; **11,** 75, 147
Schiess, M., **11,** 378; **12,** 356, 502
Schiess, P. W., **1,** 144, 562, 581, 926, 986; **4,** 413; **8,** 126
Schiessler, R. W., **1,** 1167, 1269
Schiff-Shenhav, H., **1,** 882
Schifman, A. L., **8,** 59; **9,** 77
Schildknecht, H., **1,** 153
Schill, G., **1,** 114; **5,** 457; **6,** 267
Schiller, H., **1,** 977
Schilling, P., **5,** 577
Schilling, W., **6,** 582; **8,** 444
Schiloff, J. D., **7,** 288
Schiltz, L. R., **1,** 652
Schimelpfenig, C. W., **1,** 926; **2,** 303, 344
Schimke, H., **1,** 1196
Schimmelschmidt, K., **2,** 414
Schindlbauer, H., **2,** 154, 393; **4,** 248
Schindler, O., **1,** 286
Schinkowski, K., **2,** 66
Schinski, W., **4,** 291, 294
Schinz, H., **1,** 76, 172, 180, 377, 1015; **2,** 213
Schinzer, D., **8,** 525; **10,** 326; **12,** 11
Schipper, E., **1,** 367; **2,** 163
Schipperijn, A. J., **5,** 593
Schirch, P. F. T., **9,** 222
Schirmer, J. P., **1,** 187, 708
Schirp, H., **1,** 711
Schittenhelm, D., **5,** 427
Schlaefer, F. W., **1,** 1187
Schlager, J. J., **11,** 445
Schlatmann, J. L. M. A., **3,** 228
Schlecht, M. F., **11,** 373; **12,** 121, 418
Schlegel, D. C., **3,** 131
Schleich, K., **12,** 180
Schlenk, H., **1,** 1269
Schlenk, W., Jr., **1,** 1269
Schlenker, W., **10,** 27
Schlesinger, G., **3,** 90
Schlesinger, H. I., **1,** 206, 594, 603, 1054, 1109, 1230, 1251
Schlessinger, R. H., **5,** 220, 303, 305, 406, 407, 445, 456; **6,** 253, 312; **7,** 24, 280, 412, 421; **8,** 62, 268; **9,** 217, 349, 486; **10,** 57; **11,** 334
Schletter, I., **5,** 319
Schleussner, K., **1,** 536
Schleyer, P. von R., **1,** 758, 791, 1184; **2,** 17, 21; **3,** 6, 7; **4,** 31, 140, 240; **5,** 10, 639, 658, 680, 705, 756; **6,** 17; **9,** 322; **11,** 103, 355
Schlichting, O., **1,** 185
Schlierf, C., **5,** 615; **7,** 235
Schloeder, H., **1,** 1261
Schlosberg, R. H., **3,** 202
Schlossarczyk, H., **5,** 588
Schlosser, M., **1,** 671, 927; **2,** 344; **3,** 146; **5,** 120, 226, 552; **6,** 175, 262, 511, 638; **7,** 401; **8,** 222, 448; **9,** 89, 381, 420; **10,** 186; **11,** 120, 240; **12,** 397, 542
Schlubach, H. H., **2,** 23
Schlude, H., **4,** 248, 414, 425
Schluenz, R. W., **6,** 647
Schmand, H. L. K., **6,** 170; **7,** 97
Schmeiseer, M., **7,** 390
Schmeising, R., **9,** 176
Schmeisser, M., **4,** 264; **8,** 515
Schmeltz, I., **4,** 247
Schmelzer, H. G., **4,** 524
Schmerling, L., **1,** 758
Schmid, C. R., **12,** 480
Schmid, E., **5,** 212
Schmid, G., **9,** 138, 174; **10,** 346; **12,** 388
Schmid, G. H., **10,** 22
Schmid, H., **1,** 595, 1284; **5,** 51, 212, 588, 700, 764; **8,** 9; **9,** 379; **11,** 559
Schmid, H. H. O., **2,** 313
Schmid, HJ., **1,** 93, 139
Schmid, K. H., **10,** 120
Schmid, M., **1,** 289; **5,** 764; **9,** 335
Schmid, R., **8,** 9; **10,** 436
Schmidbaur, H., **2,** 173; **5,** 724; **9,** 224; **11,** 68
Schmidhammer, L., **1,** 116
Schmidhauser, J., **9,** 74
Schmidlin, J., **1,** 262, 477, 690, 802, 1015
Schmidt, A. H., **4,** 94; **10,** 59
Schmidt, C., **3,** 171; **7,** 145, 196
Schmidt, E., **1,** 235; **2,** 126; **12,** 3
Schmidt, E. A., **5,** 760
Schmidt, E. K. G., **3,** 336; **6,** 552
Schmidt, G., **9,** 220, 399; **10,** 336; **12,** 85
Schmidt, H., **11,** 29
Schmidt, H. J., **5,** 282; **9,** 43; **10,** 29
Schmidt, H. M., **2,** 438
Schmidt, J., **4,** 63; **7,** 239
Schmidt, K. F., **1,** 448
Schmidt, M., **1,** 140; **8,** 523; **9,** 48, 511; **10,** 50; **11,** 68,

326
Schmidt, O. T., **1,** 1226
Schmidt, P., **9,** 506
Schmidt, R., **11,** 87
Schmidt, R. D., **1,** 484
Schmidt, R. E., **1,** 66
Schmidt, R. R., **8,** 226; **9,** 130
Schmidt, T. E., **4,** 375
Schmidt, U., **1,** 777, 1217; **2,** 438; **3,** 336; **8,** 246; **9,** 55, 156, 167; **10,** 337; **11,** 75, 421
Schmidt, W., **8,** 18; **9,** 510
Schmidt, W. W., **6,** 676
Schmidtberg, G., **4,** 539
Schmidt-Thome, J., **1,** 139, 488, 881, 1284
Schmiegel, K. K., **1,** 1178
Schmiegel, W. W., **1,** 1178; **3,** 182
Schmiesing, R. J., **9,** 176; **10,** 66
Schmir, G. L., **1,** 494
Schmit, C., **12,** 125
Schmit, J. P., **9,** 380
Schmitt, A. H., **10,** 98
Schmitt, G., **6,** 201
Schmitt, H. J., **8,** 109
Schmitt, J., **1,** 51
Schmitt, J. L., **5,** 86
Schmitt, S. M., **9,** 408; **10,** 100, 349
Schmitz, E., **1,** 125, 212, 361, 465, 484, 1087; **2,** 66, 219, 399, 430; **3,** 147; **4,** 527; **8,** 342
Schmitz, F. J., **4,** 31
Schmitz, J., **1,** 494
Schmitz, R. F., **6,** 143
Schmoetzer, G., **4,** 157
Schmutz, J., **1,** 504
Schmutzler, R., **1,** 135; **3,** 215; **5,** 527; **11,** 543
Schnabel, E., **2,** 49, 54; **4,** 65; **5,** 700
Schnatter, W., **12,** 266
Schnautz, N. G., **5,** 381
Schnegg, U., **4,** 383
Schneider, A., **2,** 21
Schneider, D. F., **2,** 19, 274
Schneider, H. J., **1,** 157; **5,** 389; **9,** 324
Schneider, J., **1,** 396, 832, 1189; **2,** 402; **8,** 248
Schneider, J. A., **7,** 271; **10,** 194
Schneider, J. J., **3,** 220
Schneider, K., **1,** 1189
Schneider, L., **1,** 796
Schneider, M., **6,** 665
Schneider, N. M., **1,** 262
Schneider, O., **1,** 764
Schneider, P., **6,** 186; **12,** 290
Schneider, S., **9,** 269
Schneider, W., **1,** 1118
Schneider, W. P., **1,** 567, 690, 906, 926; **2,** 132; **3,** 296; **4,** 562; **9,** 249
Schneiderwind, R. G. K., **11,** 63
Schnell, B., **1,** 1103
Schnider, O., **1,** 518
Schniepp, C. E., **9,** 439
Schniepp, L. E., **1,** 257, 636; **2,** 144
Schnizer, A. W., **1,** 531
Schnoes, H. K., **7,** 273
Schnorrenberg, G., **9,** 203
Schnorrenberg, W., **1,** 465; **12,** 245
Schnur, R. C., **6,** 379; **8,** 531
Schnurrenberger, P., **11,** 529
Schnyder, J., **6,** 281
Schobert, R., **12,** 269
Schoberth, W., **4,** 405
Schoch, E., **5,** 639
Schock, R. U., **1,** 954
Schodder, H., **1,** 20
Schoeberl, A., **1,** 72
Schoeler, H., **11,** 170
Schoellkopf, U., **1,** 96, 672, 678, 697, 1246; **2,** 119, 271; **3,** 254; **4,** 271, 272, 325, 516; **5,** 234, 257, 267, 296, 681, 685; **6,** 187, 343, 373; **7,** 207, 211, 215, 216; **8,** 393; **9,** 428; **10,** 107, 152; **11,** 193, 232
Schoellkopf, W., **3,** 223
Schoemaker, H. E., **7,** 331; **8,** 232; **9,** 83
Schoen, G., **1,** 283
Schoen, K. L., **1,** 1095
Schoen, N., **1,** 272, 282; **4,** 167
Schoenberg, A., **1,** 444, 1258; **6,** 61
Schoene, D. L., **7,** 23
Schoeneck, W., **1,** 1075
Schoenecker, B., **1,** 671
Schoenefeld, J., **4,** 31; **5,** 28
Schoenheimer, R., **1,** 1090
Schoenwaelder, K. H., **12,** 291
Schoepf, C., **1,** 7
Schofield, J. A., **1,** 352
Schofield, K., **1,** 500
Schofield, P., **1,** 110
Scholes, G., **8,** 36
Scholl, H. J., **3,** 87
Scholl, R., **1,** 904; **6,** 525
Scholler, K. L., **1,** 350
Scholten, D. J., **4,** 548
Scholz, C. R., **1,** 601, 967, 970; **2,** 79
Scholz, D., **11,** 44, 591
Scholz, H. U., **7,** 216
Scholz, K. H., **9,** 3
Schommer, W., **1,** 834
Schonbaum, G. R., **1,** 153
Schonholzer, S., **9,** 523
Schoolenberg, J., **3,** 329
Schore, N. E., **12,** 167
Schori, E., **11,** 156
Schorno, K. S., **2,** 150
Schors, A., **2,** 80, 389
Schosser, H. P., **1,** 404
Schostarez, H., **10,** 55, 265; **11,** 236
Schott, H., **5,** 510
Schow, S. R., **9,** 380; **10,** 265
Schowen, R. L., **5,** 266
Schrader, E., **1,** 885
Schraepler, U., **2,** 438
Schramm, G., **1,** 894
Schramm, R. F., **12,** 467
Schramm, R. M., **2,** 310
Schramm, S., **1,** 484; **2,** 219
Schrank, W., **8,** 395
Schranzer, C. N., **11,** 136
Schranzer, G., **6,** 124
Schrauzer, G. N., **1,** 520, 723;

2, 124; **3,** 20, 90; **4,** 41; **11,** 48
Schreckenberg, M., **4,** 447; **5,** 607; **6,** 536
Schrecker, A. W., **1,** 524, 927
Schreiber, E. C., **1,** 1122
Schreiber, F. G., **8,** 124
Schreiber, J., **1,** 147, 274, 282, 283, 574, 617, 927, 1018; **2,** 422; **4,** 187; **5,** 4, 15, 320, 329; **7,** 171
Schreiber, K., **1,** 1090; **12,** 113
Schreiber, S. L., **10,** 104; **11,** 388; **12,** 305, 366
Schreibmann, A. A. P., **3,** 7
Schreiner, J. L., **8,** 272
Schrenk, H. M., **1,** 927; **2,** 344
Schreurs, H., **5,** 168; **6,** 270
Schreurs, P. H. M., **7,** 226
Schreyer, J. M., **4,** 406
Schriesheim, A., **1,** 309, 318, 348, 431, 705, 926, 927; **6,** 430
Schrobilgen, G. J., **7,** 359
Schrock, R. R., **7,** 403, 412; **8,** 333
Schroder, F. W., **11,** 111
Schroeck, C. W., **3,** 106; **5,** 231; **11,** 211
Schroedel, R., **1,** 1108
Schroeder, B., **2,** 326
Schroeder, E., **1,** 231
Schroeder, E. F., **1,** 952; **10,** 273
Schroeder, G., **1,** 821; **4,** 216, 293; **6,** 548
Schroeder, H. E., **1,** 456, 1096; **3,** 337
Schroeder, L. R., **1,** 658; **6,** 360
Schroeder, M., **9,** 391; **10,** 290
Schroeder, M. A., **5,** 143
Schroeder, R., **4,** 271, 272; **5,** 685; **6,** 187; **10,** 107
Schroeder, W., **1,** 1015, 1152
Schroepfer, G. J., Jr., **2,** 368; **7,** 196; **10,** 304, 335, 340
Schroeter, G., **1,** 119, 345, 1043
Schroll, G., **1,** 581, 957
Schroth, G., **5,** 649
Schrott, W., **11,** 222
Schubart, R., **1,** 190; **5,** 574
Schubert, H., **12,** 31
Schubert, H. W., **6,** 564
Schuda, P. F., **7,** 114, 132, 233; **8,** 330; **9,** 79, 304; **11,** 605; **12,** 428
Schudde, E. P., **8,** 375
Schudel, P., **1,** 388, 927; **3,** 243
Schue, F., **5,** 61
Schueler, P. E., **5,** 17
Schuessler, H., **1,** 235
Schuett, H., **8,** 2
Schuett, W. R., **1,** 731
Schuette, H., **5,** 593
Schuetz, G., **1,** 1258
Schuetz, R. D., **1,** 808, 955, 1167, 1216; **2,** 107
Schugar, H., **10,** 16
Schuhknecht, C., **9,** 192; **11,** 388
Schuijl, P. J. W., **5,** 293, 749
Schulenberg, J. W., **5,** 542
Schuller, W. H., **2,** 347
Schulman, E. M., **8,** 239
Schulman, M. F., **4,** 290
Schulte, G., **9,** 102
Schulte-Elte, K., **8,** 245
Schulte-Elte, K. H., **2,** 422; **5,** 576; **9,** 154; **11,** 437
Schulte-Frohlinde, D., **1,** 933; **2,** 345
Schulthess, A. H., **9,** 311
Schultz, A. G., **5,** 394; **7,** 62, 292; **8,** 392; **9,** 363; **10,** 403; **11,** 234
Schultz, G., **1,** 828
Schultz, H. L., **10,** 205
Schultz, H. P., **2,** 229
Schultz, H. S., **1,** 477
Schultz, J., **12,** 483
Schultz, T. H., **8,** 299
Schultze, E., **1,** 561
Schulz, B., **8,** 248
Schulz, H., **1,** 114, 1246; **2,** 98
Schulz, J., **1,** 1075; **5,** 610
Schulz, J. A., **9,** 269
Schulz, J. G. D., **4,** 99
Schulz, L., **2,** 429; **4,** 317
Schulze, C. C., **1,** 177; **8,** 132
Schulze, F., **1,** 424, 611
Schulze, G., **4,** 475
Schulze, H., **5,** 456
Schulze, P. E., **10,** 216
Schulze, W., **1,** 561, 1015
Schumacher, H., **2,** 119; **3,** 223; **4,** 325
Schumacher, W., **8,** 286
Schumaker, R. R., **2,** 159; **4,** 223; **8,** 305
Schumann, D., **11,** 2
Schumann, I., **1,** 110, 841
Schumb, W. C., **1,** 457
Schunn, R. A., **6,** 570
Schure, R. M., **2,** 119
Schurig, V., **10,** 206
Schurink, H. B., **1,** 874, 1090
Schuster, D. I., **1,** 606; **11,** 347
Schut, J., **5,** 472; **7,** 377
Schwab, L. S., **5,** 216
Schwab, W., **10,** 237; **11,** 292
Schwartz, A., **5,** 582; **6,** 121, 556; **7,** 361; **9,** 202
Schwartz, D. R., **10,** 206
Schwartz, J., **4,** 249; **5,** 332; **6,** 48, 179, 451; **7,** 102; **8,** 43, 87; **9,** 104; **11,** 321, 395; **12,** 372, 578
Schwartz, M. A., **1,** 80, 1171; **3,** 182, 332; **4,** 53, 501; **6,** 581; **8,** 143, 527; **9,** 463; **11,** 593
Schwartz, N., **1,** 536, 642
Schwartz, N. N., **1,** 139
Schwartz, N. Y., **3,** 275
Schwartz, R. D., **4,** 316; **5,** 14; **8,** 120
Schwartz, S. J., **7,** 234; **9,** 304
Schwartzman, L. H., **1,** 262
Schwartzman, S., **7,** 109
Schwartzman, S. M., **6,** 225, 648
Schwarz, G., **1,** 871
Schwarz, H., **1,** 235
Schwarz, M., **2,** 17; **4,** 247
Schwarz, R., **1,** 654, 1289
Schwarz, R. A., **8,** 190
Schwarz, V., **1,** 149
Schwarz, W. M., **4,** 439
Schwarzenbach, K., **1,** 195, 1022; **3,** 69

Schwebke, G. L., **1,** 52
Schweckendiek, W. J., **1,** 61
Schweiger, R., **1,** 476
Schweizer, D., **4,** 529
Schweizer, E. E., **1,** 191, 257, 386, 854, 976, 983, 1095, 1275; **2,** 96, 457; **3,** 333; **4,** 405, 573; **5,** 751; **6,** 494; **10,** 225
Schweizer, E. H., **1,** 1252
Schweizer, R., **1,** 561
Schweizer, W. B., **11,** 326
Schwellnus, K., **8,** 537; **11,** 524; **12,** 181
Schwenck, G. O., **1,** 247
Schwenk, E., **1,** 562, 707, 720, 802, 1000, 1163
Schwenk, R., **4,** 15
Schwenke, K. D., **2,** 397
Schwenter, J., **10,** 350
Schwerin, S., **1,** 125
Schwerzel, R. E., **4,** 129
Schwier, J. R., **11,** 351
Schwindeman, J. A., **10,** 444; **11,** 83
Schwindeman, J. A., **9,** 499; **11,** 150, 584
Schwochau, M., **2,** 438; **9,** 55
Schwyzer, R., **1,** 85, 130, 380, 385, 445, 736, 796, 1220, 1229; **2,** 177; **7,** 285
Sciacovelli, O., **10,** 289
Sciaky, R., **1,** 927
Sciaraffa, P. L., **3,** 292
Sciarro, R., **5,** 479
Scilly, N. F., **4,** 287; **5,** 377
Scoffone, E., **1,** 266, 310
Scola, D. A., **1,** 518
Scolastico, C., **5,** 64; **6,** 356, 518; **9,** 475; **10,** 409; **12,** 226, 444, 510
Scopes, P. M., **2,** 389
Scorrano, G., **11,** 336
Scotini, R., Jr., **1,** 1148
Scott, A., **5,** 627; **8,** 18
Scott, A. D., **12,** 419
Scott, A. I., **6,** 436
Scott, C. A., **7,** 245
Scott, C. B., **1,** 1216; **11,** 56
Scott, H., **6,** 570
Scott, J. E., **1,** 122, 718
Scott, J. W., **9,** 260, 352
Scott, K. W., **4,** 570
Scott, L. T., **5,** 165, 188, 220, 356, 756; **7,** 32, 81; **8,** 299; **12,** 426
Scott, M. D., **1,** 444, 927
Scott, N. D., **1,** 268, 712
Scott, R., **11,** 130
Scott, R. W., **1,** 870; **4,** 317
Scott, S. W., **3,** 332; **8,** 527
Scott, W. A., **8,** 374
Scott, W. B., **4,** 10, 63
Scott, W. J., **1,** 1058; **10,** 289; **12,** 395, 474
Scott, W. L., **4,** 77
Scott-Burden, T., **2,** 424
Scouten, C. G., **2,** 182, 289; **5,** 47; **6,** 64; **8,** 6, 334; **9,** 58
Scouten, C. J., **7,** 30
Screttas, C. G., **1,** 1204; **2,** 53; **4,** 349; **5,** 649; **8,** 306; **9,** 285
Screttri, A., **9,** 399; **10,** 335, 336
Scribner, R. M., **1,** 329; **5,** 640; **7,** 282
Scriesham, A., **2,** 344
Scripko, J. G., **11,** 317; **12,** 305
Scriven, E. F. V., **12,** 199
Scudder, P. H., **8,** 214
Scullard, P. W., **5,** 436; **6,** 403
Scully, C. N., **1,** 372
Scully, F. E., Jr., **8,** 419; **11,** 445
Seabright, C. A., **9,** 466
Seager, J. H., **1,** 652
Sealfon, S., **9,** 13
Seamans, R. E., **5,** 667
Searcy, A. M., **1,** 93
Searle, N. E., **1,** 370
Searle, R., **1,** 1125
Searle, R. J. G., **1,** 1216; **9,** 361
Searles, S., **1,** 378, 452, 1232; **5,** 299
Searles, S., Jr., **1,** 310
Sebulsky, SR. T., **4,** 393
Secci, M., **7,** 4
Seconi, G., **10,** 326
Secrist, J. A., III, **7,** 88; **8,** 87; **9,** 180
Seda, R., **12,** 322
Seden, T., **1,** 222
Sedergran, T. C., **10,** 205
Sedlak, J., **1,** 198
Sedmera, P., **12,** 503
Seebach, D., **2,** 101, 187, 324; **3,** 136, 330; **4,** 165, 216, 241, 285, 380, 564; **5,** 78, 80, 86, 267, 375, 700, 717; **6,** 29, 55, 84, 90, 118, 133, 238, 248, 267, 322, 339, 493, 578, 636, 651; **7,** 26, 47, 190, 312, 378, 412; **8,** 8, 69, 318, 363, 494, 517; **9,** 8, 48, 87, 381, 446, 483; **10,** 234, 381, 395, 422, 449; **11,** 173, 207, 288, 295, 299, 326, 361, 378, 379, 423, 438, 500, 529, 591; **12,** 85, 193, 303, 356, 358, 418, 480, 502
Seebeck, E., **1,** 802
Seebeck, S., **1,** 561
Seefelder, M., **1,** 289; **2,** 134; **7,** 424
Seeger, A., **9,** 324; **10,** 5
Seehafer, J., **4,** 54
Seekircher, R., **4,** 99
Seela, F., **4,** 407; **11,** 407
Seeler, F., **1,** 970
Seelert, K., **1,** 318
Seeley, D. A., **5,** 66
Seeliger, A., **1,** 685; **3,** 93; **4,** 552
Seer, C., **6,** 525
Seese, W. S., **1,** 125
Seewald, A., **7,** 141; **9,** 203
Sega, A., **11,** 124
Segal, H. L., **2,** 147
Segal, M., **8,** 395
Segall, Y., **5,** 688
Segawa, J., **9,** 310; **11,** 279, 340; **12,** 559
Segawa, T., **6,** 618
Segnitz, A., **3,** 329; **4,** 318; **5,** 740; **6,** 653
Segoe, K., **12,** 95
Segovia, R., **1,** 841
Sehm, E. J., **1,** 164, 814, 1081

Sei, K., **1,** 601
Seide, W., **2,** 292
Seidel, B., **1,** 220
Seidel, P. R., **5,** 370
Seidig, K. D., **1,** 445
Seidl, G., **1,** 849
Seidl, L., **1,** 125
Seidner, R., **1,** 1081
Seidner, R. T., **3,** 102; **4,** 339
Seifert, P., **1,** 180
Seifert, W. K., **1,** 329; **3,** 130
Seigel, W. O., **4,** 465
Seigle, L. W., **4,** 461
Seikel, M., **1,** 465
Seilo, M., **5,** 68
Seip, D., **2,** 67, 454
Seiss, H. J., **11,** 87
Seitz, A. H., **2,** 26
Seitz, B., **10,** 462
Seitz, D. E., **9,** 255; **10,** 211, 433; **11,** 500, 544, 581
Seitz, E. P., **7,** 303
Seitz, S. P., **8,** 28; **9,** 31, 283, 367; **11,** 58; **12,** 57, 182, 185, 536
Seka, R., **1,** 837
Sekera, V. C., **1,** 17, 962, 1184
Seki, F., **1,** 1096
Seki, K., **6,** 312
Seki, S., **9,** 317
Seki, Y., **6,** 73
Sekiguaki, S., **10,** 110
Sekine, M., **6,** 644; **9,** 166; **11,** 178; **12,** 419, 557
Sekiya, A., **6,** 581
Sekiya, M., **1,** 1231; **3,** 33, 300; **6,** 73; **8,** 254; **10,** 164; **12,** 63
Sekizaki, H., **11,** 75, 85
Seko, K., **12,** 432
Selby, W. M., **1,** 573
Selig, H., **6,** 670; **7,** 425
Seliger, H., **10,** 94
Seligman, A. M., **1,** 62, 198; **2,** 10, 464
Selikson, S. J., **5,** 524; **6,** 430
Selim, A., **12,** 241
Selke, E., **12,** 36
Sell, G., **7,** 318
Selle, R., **4,** 37
Selling, H. A., **2,** 344
Sellstedt, J. H., **1,** 744
Selman, L. H., **3,** 9, 329; **4,** 14, 337
Selover, J. C., **11,** 401
Selvarajan, R., **2,** 464
Selve, C., **5,** 259, 326; **6,** 35, 280; **7,** 14; **9,** 297
Selwitz, C. M., **11,** 432
Seman, J. J., **12,** 220
Semenovsky, A. V., **4,** 325
Semigran, M. J., **7,** 370
Semmelhack, M. F., **2,** 292, 293; **3,** 307; **4,** 35, 354, 358, 457; **5,** 35, 463; **6,** 28, 104, 144, 361; **7,** 429; **8,** 22, 120, 126; **9,** 21, 46, 119; **10,** 100, 126, 226; **11,** 51, 132, 358, 401; **12,** 266, 372, 480
Semmler, E. J., **7,** 273
Semon, W. L., **1,** 481
Semonsky, M., **9,** 205
Semple, B. M., **2,** 47
Semple, J. E., **10,** 67, 182
Sen, D., **6,** 45
Sen, J. N., **1,** 348
Senaratne, A., **11,** 422
Sencherey, B. B., **12,** 258
Senda, Y., **5,** 210, 571; **9,** 13
Sendoda, Y., **6,** 510
Senet, J. P., **12,** 113
Seng, F., **1,** 114, 1246
Seng, R., **2,** 147
Senga, K., **5,** 479
Sengupta, D., **12,** 201
Senior, R. G., **4,** 282
Senn, J., **9,** 299, 337, 406
Seno, K., **9,** 221; **11,** 518, 519; **12,** 485
Senoh, S., **2,** 387
Sepelak, D. J., **10,** 146; **12,** 189
Sephton, H. H., **1,** 562
Sepiol, J., **5,** 617; **6,** 553
Sepp, D. T., **5,** 155, 370
Sepulchre, A. M., **5,** 287
Serafin, B., **4,** 31
Seraglia, R., **12,** 94
Serban, N., **2,** 210
Serckx-Poncin, B., **11,** 51, 201
Serebryakov, E. P., **1,** 95, 562; **5,** 424
Serfontein, W. J., **1,** 755
Serini, A., **1,** 764, 1210
Serino, A. J., **11,** 47, 514
Serizawa, Y., **12,** 506
Sermon, P. A., **5,** 321
Serota, S., **1,** 814
Serratosa, F., **2,** 19; **3,** 63; **5,** 163, 205; **8,** 144
Servi, S., **6,** 625
Servis, K. L., **4,** 327
Seshadri, S., **1,** 904
Seshadri, T. R., **1,** 235, 730; **3,** 9
Set, L., **12,** 556
Seta, A., **8,** 505
Setescak, L., **3,** 231
Seth, A. K., **11,** 359
Seth, S. R., **1,** 1292
Sethi, D. S., **5,** 675
Sethi, S. C., **5,** 486
Sethi, S. P., **12,** 86, 321
Sethna, S., **4,** 579
Sethna, S. M., **1,** 33, 954
Setinek, K., **3,** 115
Seto, S., **6,** 534; **7,** 331; **8,** 451
Setoi, H., **9,** 225
Setoyama, O., **6,** 453
Setsune, J., **11,** 140
Sette, J., **6,** 570
Settlage, P. H., **1,** 667, 963
Setton, R., **6,** 23, 269; **7,** 296
Seuring, B., **5,** 86; **8,** 517; **10,** 449
Seus, E. J., **1,** 1095, 1216
Severin, R., **4,** 357
Severin, T., **5,** 614; **6,** 500
Sevestre, H., **12,** 415
Sevrin, M., **7,** 35, 184, 195; **8,** 199, 299, 537; **9,** 283; **10,** 56, 368
Sexton, H., **1,** 866
Seybold, G., **6,** 285
Seybold, W., **5,** 717
Seyden-Penne, J., **1,** 503; **8,** 133, 245; **9,** 127, 235; **10,** 198, 378; **12,** 105
Seyfarth, H. E., **1,** 361
Seyferth, D., **1,** 19, 52, 213, 424, 507, 689, 854, 1141, 1193, 1237, 1246, 1250,

1274; **2,** 31, 328, 442, 456; **3,** 89, 114, 175, 222, 223, 225, 278; **4,** 53, 139, 377, 385, 544; **5,** 189, 454, 514, 515, 724; **6,** 91, 189, 596, 638; **7,** 62, 193; **8,** 151; **9,** 181, 251, 492, 509; **10,** 146, 157, 303, 431; **11,** 16, 111; **12,** 4, 189
Seyler, R. C., **1,** 56; **6,** 450; **7,** 95
Seymour, E., **7,** 33, 284; **9,** 60
Sgoutas, D. S., **4,** 437, 505
Sha, C. K., **7,** 195
Shabana, R., **9,** 50
Shabica, A. C., **1,** 75, 802; **2,** 38, 79
Shabtai, J., **1,** 1023
Shadwell, J., **1,** 393
Shaefer, C. G., **7,** 277; **9,** 335
Shaefer, W. E., **1,** 828
Shafer, P. R., **2,** 195
Shaffer, E. T., **2,** 96
Shah, G. D., **1,** 882, 1292
Shah, N. M., **1,** 33, 295, 685
Shah, R. C., **1,** 33, 882, 1292
Shah, S. K., **6,** 338; **9,** 158, 172, 283, 430, 438
Shah, V. R., **1,** 882, 1292
Shahak, I., **1,** 371, 504; **4,** 63, 335; **5,** 95, 96, 680; **6,** 216, 542, 644; **7,** 106; **8,** 516; **9,** 418
Shaida, W. A., **9,** 191
Shaikh, I. N., **4,** 189
Shaikh, S., **6,** 456
Shain, I., **4,** 439
Shakovski, B. G., **2,** 438
Shalon, Y., **3,** 86
Sham, H. L., **9,** 257, 468; **12,** 204
Shama, S. A., **8,** 15
Shambhu, M. B., **5,** 23, 249
Shamblee, D. A., **6,** 468
Shamma, M., **1,** 1107; **2,** 313, 394; **8,** 391
Shani, A., **2,** 88; **6,** 203; **10,** 184
Shank, R. S., **1,** 1022
Shankaran, K., **12,** 99
Shanklin, J. R., **5,** 231, 459, 528; **6,** 395
Shanley, E. S., **1,** 457
Shanmugam, P., **8,** 460
Shannon, J. S., **1,** 129
Shannon, P. V. R., **6,** 419
Shanzer, A., **10,** 125; **11,** 177; **12,** 185, 463
Shapiro, B. L., **7,** 389
Shapiro, E. L., **1,** 926, 1101; **2,** 344; **5,** 552, 595
Shapiro, I., **5,** 416
Shapiro, M. J., **7,** 389
Shapiro, P., **1,** 335, 1251
Shapiro, R., **8,** 272
Shapiro, R. H., **2,** 422; **3,** 161; **4,** 512; **6,** 600; **7,** 376; **8,** 492; **10,** 5
Shapiro, S. H., **1,** 251
Shapiro, S. L., **1,** 386
Shapley, J. R., **5,** 113
Sharanina, L. G., **5,** 44
Sharefkin, J. G., **1,** 508, 509, 511, 787
Shariff, A., **11,** 188
Sharkey, W. H., **1,** 1246, 1271; **4,** 543; **6,** 81; **9,** 54
Sharma, A. K., **5,** 267; **6,** 230; **7,** 136
Sharma, C. S., **5,** 486
Sharma, G. K., **6,** 453
Sharma, M., **8,** 167
Sharma, N. D., **11,** 330
Sharma, N. K., **5,** 129
Sharma, R. K., **3,** 43; **12,** 462
Sharma, R. P., **4,** 336; **11,** 128; **12,** 126, 305, 550
Sharma, R. T., **9,** 449
Sharma, S., **9,** 466
Sharma, S. C., **9,** 19
Sharma, S. D., **5,** 713
Sharma, S. N., **2,** 278
Sharma, V. K., **11,** 255
Sharmout, A. R., **11,** 407
Sharp, D. W. A., **1,** 1018
Sharp, J. C., **1,** 1108; **2,** 389; **3,** 61
Sharp, J. T., **4,** 80, 172; **6,** 600
Sharpe, A. G., **1,** 504, 1018, 1227
Sharpe, V. V., III, **12,** 154
Sharples, G. M., **3,** 212
Sharpless, K. B., **2,** 415; **4,** 99, 204, 407, 413, 570; **5,** 77, 276, 351, 522, 563, 576; **6,** 127, 235, 422, 487, 589, 642, 658; **7,** 244, 255, 256, 318; **8,** 24, 25, 29, 43, 65, 112, 206, 359, 365, 393; **9,** 32, 79, 80, 82, 111, 367; **10,** 65, 291, 312, 405, 445; **11,** 94, 95, 318, 463; **12,** 60, 94, 247, 506
Sharrocks, D. N., **7,** 101
Sharts, C. M., **5,** 216; **6,** 146
Shasha, B. S., **5,** 247; **10,** 39
Shashak, I., **1,** 277
Shatzmiller, S., **5,** 113
Shavel, J., Jr., **1,** 329, 872; **5,** 345, 756
Shaw, A., **4,** 435; **6,** 519; **7,** 298
Shaw, B. L., **1,** 360, 668, 1040; **2,** 434; **7,** 5; **12,** 111
Shaw, C. K., **5,** 567
Shaw, D. L., **1,** 563, 1151
Shaw, G., **1,** 12, 266; **10,** 364
Shaw, I. M., **7,** 290
Shaw, J. E., **3,** 219; **4,** 84, 97, 247; **6,** 279
Shaw, K. R., **11,** 518
Shaw, M. F., **4,** 412
Shaw, M. J., **5,** 20; **6,** 670
Shaw, R., **5,** 330
Shaw, R. A., **2,** 207; **3,** 173
Shawky, M., **1,** 131
Shea, J. L., **1,** 808
Shea, K. J., **12,** 11
Shea, R. G., **12,** 121
Sheats, J. R., **6,** 105
Sheats, W. B., **9,** 121
Shechter, H., **1,** 131, 212, 281, 309, 328, 373, 392, 696, 757, 808, 873, 905, 952, 1022, 1064, 1141, 1187, 1293; **2,** 287, 300, 348, 422; **4,** 340; **11,** 42
Shedrinsky, A., **11,** 575
Sheehan, J. C., **1,** 4, 139, 181, 235, 274, 281, 371, 430, 744, 927, 956, 966, 1107,

1184, 1254; **2,** 127, 196; **4,** 27, 399; **5,** 23, 106, 428, 439; **6,** 506; **8,** 403
Sheehan, J. J., **2,** 465
Sheffy, F. K., **12,** 46, 56
Sheft, I., **1,** 1109
Shei, J. C., **11,** 562
Sheikh, H., **8,** 521
Sheikh, Y. M., **4,** 108, 265; **11,** 493
Shekhani, M. S., **9,** 283
Sheldon, J. C., **1,** 329
Sheldon, R. A., **4,** 282
Sheldrake, P. W., **8,** 87
Shellar, A. J., **11,** 355
Shellhamer, D. F., **5,** 127, 136, 441
Shelly, T. A., **5,** 207
Shelton, J. C., **2,** 50, 422
Shelton, J. R., **1,** 518
Shemin, D., **1,** 654
Shemyakin, M. M., **1,** 281, 318, 1231, 1246
Shen, C. M., **3,** 147; **5,** 539
Shen, M., **11,** 234
Shen, T. Y., **1,** 723; **6,** 451
Sheng, M. N., **2,** 287, 456; **3,** 331; **4,** 309, 346
Shenoi, S., **11,** 521
Shenoy, P. K., **1,** 1249; **2,** 446; **3,** 322
Shenstone, W. A., **1,** 754
Shenton, F. L., **6,** 675
Shenton, K. E., **6,** 451
Shepard, E. R., **1,** 982
Shepard, I. G., **1,** 6, 707, 734
Shepard, K. L., **4,** 204; **6,** 348
Shepard, M. S., **7,** 349
Shephard, K. P., **9,** 38
Shepherd, B. D., **1,** 1158, 1163
Shepherd, D. A., **1,** 777, 890
Shepherd, D. M., **2,** 4
Shepherd, J. P., **4,** 31
Shepherd, J. W., **1,** 40, 927
Shepherd, R. A., **7,** 207
Shepherd, T. M., **3,** 123
Sheppard, J. H., **9,** 317, 464
Sheppard, N., **1,** 506, 510
Sheppard, O. E., **1,** 1169
Sheppard, R. C., **1,** 828, 1229; **3,** 153; **6,** 425; **8,** 230
Sheppard, W. A., **1,** 117, 450, 849, 1006; **4,** 157; **5,** 176, 179, 216, 220, 505; **7,** 20; **10,** 304
Sher, F., **4,** 165
Sheradsky, T., **8,** 433
Sherbine, J. P., **12,** 280
Sheridan, J., **1,** 424
Sherman, C. S., **1,** 1289
Sherman, N., **1,** 139
Sherman, W. R., **1,** 1073, 1082, 1269
Sherred, A., **1,** 452
Sherrill, M. L., **4,** 87
Sherrington, D. C., **5,** 735; **9,** 510; **10,** 452
Sherry, J. J., **4,** 97, 247
Sherwin, P. F., **7,** 380
Sheth, J. P., **9,** 265
Sheverdina, N. J., **11,** 323
Shevlin, P. B., **5,** 654
Shew, D., **1,** 150
Shiba, S., **10,** 277
Shibaba, A., **8,** 481
Shibanuma, T., **8,** 96
Shibasaki, M., **4,** 472; **6,** 490; **8,** 180; **9,** 393; **10,** 414; **11,** 456, 552; **12,** 11, 36
Shibata, D., **12,** 485
Shibata, T., **11,** 206
Shibib, S. M., **10,** 251
Shibuya, M., **9,** 355
Shibuya, S., **4,** 407; **6,** 547; **8,** 351, 492; **9,** 381, 438, 470, 473
Shields, J. E., **1,** 452, 737
Shields, T. C., **1,** 318, 927; **2,** 344; **4,** 352, 405
Shigematsu, N., **1,** 53
Shih, H., **4,** 385; **5,** 514
Shih, J., **12,** 420
Shih, M. J., **11,** 124
Shih, T. L., **11,** 381
Shih, Y. S., **6,** 475
Shiihara, I., **3,** 294
Shildneck, P. R., **1,** 33, 1167
Shillington, J. K., **1,** 118
Shilov, E. A., **2,** 29
Shim, K. S., **1,** 260
Shim, S. C., **6,** 485, 486; **9,** 108
Shimabayashi, A., **11,** 155
Shimada, E., **6,** 376; **8,** 96
Shimada, J., **12,** 502
Shimada, K., **3,** 325; **11,** 292
Shimada, N., **12,** 506
Shimada, Y., **5,** 447, 518
Shimagaki, M., **5,** 163; **10,** 1; **12,** 288
Shimahara, M., **2,** 215, 305
Shimakawa, Y., **3,** 53
Shimamura, T., **5,** 499; **6,** 445
Shimano, Y., **7,** 225
Shimazu, H., **6,** 436
Shimizaki, M., **11,** 494
Shimizu, F., **9,** 478
Shimizu, H., **9,** 246; **10,** 289; **12,** 160
Shimizu, I., **7,** 278; **9,** 328; **10,** 281, 302, 417; **11,** 392, 514; **12,** 16, 370
Shimizu, K., **9,** 22, 208
Shimizu, M., **7,** 152; **8,** 210, 442, 468; **9,** 25, 79; **10,** 65, 71; **11,** 66, 124; **12,** 462, 491
Shimizu, N., **7,** 21; **11,** 470
Shimizu, T., **3,** 167; **5,** 572; **7,** 225; **9,** 140; **12,** 508
Shimizu, Y., **1,** 478; **2,** 117, 453; **5,** 436; **6,** 102, 403; **7,** 245
Shimoji, K., **5,** 7, 374; **6,** 255; **12,** 85, 247, 577
Shimomura, O., **3,** 127
Shimonishi, Y., **1,** 40, 668; **2,** 25, 216, 307
Shin, M., **1,** 40; **2,** 307
Shinagawa, S., **5,** 360; **6,** 291
Shine, H. J., **1,** 331
Shine, R. J., **4,** 8
Shiner, C. S., **6,** 490; **11,** 606; **12,** 11
Shiner, V. J., **1,** 266; **3,** 10
Shinhama, K., **9,** 92
Shinkai, I., **10,** 252, 449
Shinke, S., **10,** 99
Shinma, N., **9,** 484
Shinoda, M., **8,** 151, 417; **9,** 256, 395; **12,** 263, 304
Shinonaga, A., **8,** 133, 162

Shinoo, Y., **8,** 119; **9,** 123
Shinozuka, K., **9,** 160
Shinya, S., **7,** 165
Shioiri, T., **3,** 224; **4,** 211, 383; **5,** 217, 280; **6,** 193; **7,** 107, 138, 178; **8,** 211; **9,** 164; **10,** 145, 173, 324, 431; **11,** 44, 181, 216, 222, 472, 574; **12,** 188, 217, 538
Shiono, H., **6,** 14, 365, 540
Shiono, M., **6,** 596; **8,** 96; **12,** 502
Shiota, M., **2,** 215, 305; **5,** 743; **6,** 659; **7,** 417; **8,** 434
Shiota, T., **12,** 247
Shiotani, S., **7,** 160
Shiozaki, M., **3,** 41; **8,** 208
Shipkowski, E. R., **1,** 935
Shipman, J. J., **1,** 182
Shippey, M. A., **7,** 403; **8,** 514
Shirafuji, T., **3,** 69; **4,** 322
Shirahama, H., **8,** 441; **11,** 96, 445
Shirahata, A., **9,** 69, 255, 446; **10,** 289; **11,** 130
Shirai, H., **7,** 146
Shirai, I., **7,** 165
Shirai, K., **11,** 1
Shirai, N., **9,** 349
Shiraishi, M., **9,** 15; **12,** 190
Shirley, D. A., **1,** 742
Shirley, R. V., **1,** 465
Shiro, M., **11,** 324
Shiroyama, M., **8,** 248
Shishiyama, Y., **10,** 114; **11,** 150
Shiue, C. Y., **6,** 288
Shivers, J. C., **1,** 1292
Shmelev, L. V., **4,** 325
Shoaf, C. J., **1,** 624; **6,** 342
Shockravi, A., **8,** 92
Shoda, S., **8,** 92; **10,** 6, 84, 374; **11,** 203, 525; **12,** 549
Shoenberger, D. C., **11,** 499
Shoesmith, J. B., **1,** 340
Shoham, G., **11,** 524
Shoji, Y., **6,** 170
Shokina, V. V., **1,** 360
Shone, R. L., **3,** 176
Shonle, H. A., **1,** 470
Shono, T., **1,** 637; **8,** 532; **9,** 415, 465; **10,** 310; 460; **11,** 533, 599; **12,** 439
Shoppee, C. W., **1,** 19, 489, 624; **2,** 36
Shore, P. A., **1,** 476
Short, F. W., **5,** 266
Short, J. F., **3,** 233
Short, J. W., **12,** 131
Short, M. R., **5,** 585
Short, W. A., **1,** 872
Short, W. F., **1,** 1184
Shotter, R. G., **5,** 542
Shotter, R. Z., **7,** 295
Shoua, S., **5,** 760; **7,** 429
Shoulders, B. A., **1,** 318, 927
Showalter, H. D. H., **11,** 524
Showell, J. S., **1,** 195, 465, 1157; **2,** 310
Showler, A. J., **1,** 66, 1226
Shreeve, J. M., **4,** 37; **12,** 106
Shridhar, D. R., **1,** 166
Shriner, R. L., **1,** 267, 402, 472, 654, 668, 841, 859, 892, 951, 1015, 1027, 1073, 1106, 1283, 1292; **3,** 336; **4,** 577
Shu, B. Y., **9,** 489
Shu, J. S., **6,** 504
Shubart, R., **5,** 620
Shudo, K., **6,** 618
Shue, H. J., **7,** 140; **8,** 167
Shulgin, A. T., **7,** 56
Shull, D. W., **7,** 76; **8,** 126; **11,** 329
Shulman, J. I., **2,** 106, 441; **3,** 95, 97, 142; **4,** 151; **5,** 749
Shulman, R. S., **1,** 633
Shults, R. H., **7,** 155
Shum, W., **9,** 448
Shumaker, S. H., **1,** 718
Shundo, T., **9,** 278
Shunk, C. H., **1,** 389, 703, 771, 1081
Shupack, S. I., **1,** 1034
Shutt, R., **5,** 341; **6,** 422
Shvo, Y., **1,** 60; **3,** 22; **6,** 306, 625; **11,** 184
Sib, S., **5,** 194, 414
Sibi, M. P., **11,** 479; **12,** 99
Sibilia, J. P., **4,** 206
Sichel, E., **5,** 671
Sicher, J., **1,** 1258; **3,** 316; **4,** 440
Sicsic, S., **4,** 183
Sidani, A. R., **5,** 138; **6,** 123; **12,** 125
Siddall, G. B., **10,** 421
Siddall, J., **2,** 29; **6,** 258
Siddall, J. B., **1,** 309, 926; **2,** 201, 344, 362; **3,** 112, 278; **6,** 165
Siddappa, S., **5,** 339
Siddiqui, M. N. U., **5,** 148
Siderius, H., **4,** 516
Sidgwick, N. V., **1,** 172
Sidhu, R. S., **8,** 163
Sidky, M. M., **2,** 384; **4,** 188, 356
Sidwell, W. T. L., **5,** 436
Sieber, A., **3,** 96
Sieber, N., **1,** 1292
Sieber, P., **1,** 85, 130, 445, 736, 1220, 1229; **2,** 177, 179; **8,** 511
Sieber, W., **5,** 588
Siebert, E., **1,** 856
Siebert, H., **8,** 397
Siebert, W., **10,** 454
Sieburth, S. M., **10,** 328; **12,** 275
Sieczkowski, J., **2,** 394; **4,** 201
Siegel, E., **1,** 465, 667
Siegel, H., **8,** 18; **12,** 426
Siegel, S., **6,** 195
Siegemund, G., **5,** 640
Siegfried, B., **5,** 325, 607; **7,** 73
Sieja, J. B., **2,** 59
Sieloff, R. F., **8,** 343
Siemer, H., **1,** 911
Siemiatycki, M., **1,** 1258
Siemion, I. Z., **2,** 302
Sienkowski, K. J., **5,** 329
Sierakowski, A. F., **9,** 123
Siewert, G., **2,** 7; **4,** 457
Sifferd, R. H., **9,** 415
Siffert, O., **5,** 107
Sifniades, S., **7,** 3
Sigalov, M. V., **11,** 440
Sigaut-Titeux, F., **8,** 233
Sigel, H., **1,** 755

Sigfried, B., **7,** 49
Siggins, J. E., **2,** 252
Sigler, G. F., **5,** 477
Sigurdson, E. R., **9,** 115
Sih, C. J., **4,** 221, 517; **6,** 201; **7,** 32; **8,** 417, 458; **9,** 233
Sih, J. C., **5,** 326, 438
Siirala-Hansen, K., **6,** 93, 361; **7,** 22
Sikkel, B. J., **11,** 308
Sikorski, J. A., **10,** 95; **12,** 414
Silber, R. H., **1,** 841
Silberman, R. G., **2,** 70
Silbert, L. S., **1,** 89, 187, 465, 667; **3,** 185; **4,** 247, 302, 428; **5,** 406
Silhavy, P., **6,** 518
Silks, L. A., **12,** 350
Sillion, B., **10,** 85; **11,** 118
Silloway, H. L., **1,** 1105; **3,** 272
Silva, M. L., **2,** 300
Silveira, A., Jr., **6,** 399; **7,** 379; **8,** 475; **9,** 482; **10,** 243; **12,** 474
Silver, B. L., **1,** 1139
Silver, K., **4,** 119
Silver, M. S., **9,** 139
Silver, S. M., **5,** 725
Silverman, I. R., **12,** 378
Silverman, M., **2,** 206
Silverman, R. B., **11,** 170
Silverman, S. B., **8,** 501
Silversmith, E. F., **1,** 1143
Silverstein, R. M., **1,** 285, 1058; **2,** 187, 281
Silverwood, H. A., **1,** 759, 992
Silvestri, M., **6,** 588, 589; **11,** 337
Sim, S. K., **6,** 160
Simakhina, N. D., **10,** 419
Simalty, M., **5,** 194, 414
Simamura, O., **1,** 796; **3,** 294
Simchen, G., **1,** 283, 1252; **4,** 167; **5,** 73, 250, 254; **6,** 243, 569; **7,** 390; **8,** 515; **9,** 93; **10,** 1, 96, 441; **11,** 587; **12,** 204
Simeone, J. F., **3,** 20, 153, 182
Simington, R. M., **1,** 846
Simmonds, D. J., **11,** 490; **12,** 379
Simmons, D. P., **11,** 249, 557
Simmons, H. D., Jr., **1,** 854
Simmons, H. E., **1,** 507, 1022, 1293; **2,** 93, 270; **3,** 46, 141; **4,** 143, 411; **5,** 57, 155, 257, 589; **7,** 237; **12,** 221
Simmons, T. C., **1,** 1216
Simms, J. J., **12,** 197
Simon, E. S., **12,** 515
Simon, H. J., **8,** 522
Simon, M. S., **7,** 367
Simon, P., **1,** 5; **9,** 297
Simon, R. M., **10,** 146; 12, 189
Simone, R. A., **5,** 537
Simoneau, B., **12,** 407
Simoni, D., **11,** 324
Simonidesz, V., **7,** 160; **8,** 260
Simonneaux, G., **6,** 126; **8,** 110
Simonnin, M. P., **12,** 105
Simons, D. M., **1,** 1210
Simons, J. H., **1,** 456
Simons, J. K., **1,** 230
Simons, S. S., Jr., **4,** 282
Simonson, D. R., **4,** 21
Sim Ooi, N., **12,** 481
Simpkin, D. J., **9,** 480
Simpkins, N. S., **12,** 394
Simpson, M., **9,** 107
Simpson, P. L., **1,** 52
Sims, C. L., **5,** 406, 681
Sims, H., **1,** 874
Sims, J. J., **2,** 236; **3,** 9, 329; **4,** 14, 337; **9,** 442
Sims, P., **1,** 954
Sims, R. J., **10,** 112; **11,** 145
Sinay, P., **10,** 21; **11,** 224, 452; **12,** 285
Sinclair, J. A., **5,** 347; **7,** 181; **8,** 7; **10,** 127
Sing, A., **8,** 88
Sing, Y. L., **8,** 163
Singaram, B., **8,** 267; **9,** 278; **11,** 154, 351; **12,** 5, 13, 194, 334
Singaram, S., **9,** 434; **12,** 195
Singer, A. W., **1,** 402, 951
Singer, B., **6,** 43
Singer, C. P., **8,** 9
Singer, H., **8,** 109; **12,** 173
Singer, M. S., **3,** 18
Singer, S. P., **7,** 318; **8,** 43
Singh, A., **10,** 21, 452
Singh, B., **4,** 456; **9,** 438
Singh, B. B., **12,** 259
Singh, B. P., **12,** 527
Singh, G., **5,** 724; **6,** 392
Singh, J., **5,** 463; **9,** 377; **10,** 443
Singh, K., **10,** 26
Singh, M., **4,** 398
Singh, N., **4,** 31
Singh, P., **7,** 325; **11,** 337
Singh, P. P., **1,** 309, 685; **8,** 228
Singh, R., **10,** 26
Singh, R. K., **6,** 43, 217, 230; **7,** 123, 227, 293; **8,** 179, 330; **11,** 327
Singh, R. P., **2,** 348; **10,** 346
Singh, S. M., **11,** 146; **12,** 70
Singhal, G. H., **2,** 330
Singleton, D. A., **12,** 532
Singleton, D. M., **3,** 59; **5,** 204; **10,** 140
Singleton, F. G., **1,** 444
Singleton, V. D., Jr., **7,** 122
Sinha, N. D., **8,** 122; **11,** 12
Sinhababu, A. K., **12,** 284
Sinnige, H. J. M., **3,** 337; **7,** 349; **8,** 103
Sinnott, M. L., **3,** 202
Sinnreich, J., **5,** 160; **9,** 311
Sinnwell, V., **4,** 218
Sinoway, L., **6,** 465
Sinwel, F., **1,** 562
Sinz, A., **1,** 484
Sipido, V., **5,** 691
Sipio, W. J., **8,** 259; **9,** 31, 32; **10,** 21
Sipos, F., **1,** 1258; **4,** 223
Sircar, J. C., **1,** 982
Siret, P., **9,** 136, 193, 318, 432, 467; **11,** 335
Sirna, A., **8,** 385; **9,** 45
Siroi, T., **9,** 23
Sirrenberg, W., **9,** 402
Sisani, E., **9,** 408; **11,** 156

Sisido, K., **1,** 52, 601; **5,** 210; **8,** 340
Sisler, H. H., **1,** 124, 152
Sisti, A. J., **2,** 146; **3,** 36
Sit, S. Y., **11,** 158
Sita, G. E., **1,** 652
Sita, L. R., **12,** 94
Sitrin, R., **5,** 511
Sittig, M., **1,** 1034
Sivak, A., **4,** 84
Sivakumar, R., **12,** 288
Sivanandaiah, K. M., **6,** 288; **8,** 383; **9,** 227, 352
Sivaram, S., **5,** 615, 708; **11,** 142
Sivasankaran, K., **1,** 892
Sjoeberg, B., **1,** 183; **4,** 196; **5,** 643
Sjoeberg, K., **2,** 173; **4,** 196; **7,** 22, 250
Sjoeberg, S., **1,** 183
Sjoequist, J., **1,** 845
Sjolander, N. O., **9,** 283
Skarzewski, J., **12,** 107
Skattebol, L., **1,** 12, 281, 370, 389, 424, 688, 689, 926; **2,** 278, 295; **3,** 47, 204, 234; **4,** 5; **5,** 28, 283, 454; **6,** 558; **7,** 193
Skeean, R. W., **7,** 345
Skell, P. S., **1,** 80, 909; **2,** 211; **5,** 654; **6,** 190; **8,** 260
Skeludyakov, V. D., **10,** 432
Sket, B., **5,** 512
Skiles, J. W., **8,** 391
Skinner, G. A., **2,** 159
Skinner, G. S., **7,** 146
Skinner, W. A., **1,** 754
Sklarz, B., **2,** 313, 315, 397; **9,** 449
Skold, C. N., **7,** 116
Skorcz, J. A., **2,** 374
Skorna, G., **8,** 214
Skotnicki, J., **6,** 534
Skovlin, D. O., **3,** 173
Skuballa, W., **6,** 443
Skulan, T., **5,** 542
Skulnick, H. I., **10,** 182
Skulski, M., **6,** 585
Slack, D., **9,** 426
Slack, W. E., **5,** 667
Slade, J. S., **10,** 289
Slagel, R. C., **2,** 381
Slater, C. D., **2,** 92
Slater, G. P., **1,** 561
Slates, H. L., **1,** 152, 536, 642, 664, 933, 974, 1004, 1087, 1090, 1171; **2,** 159, 252, 354
Slaugh, L. H., **2,** 242, 346, 376
Slaunwhite, W. D., **3,** 4
Slaunwhite, W. R., Jr., **1,** 519
Slaven, R. W., **5,** 96
Slayden, S. W., **8,** 507
Slaytor, M., **1,** 598
Slebioda, M., **12,** 222
Slegeir, W. A. R., **5,** 582
Sleiter, G., **11,** 120
Slesar, L. N., **7,** 413
Slessor, K. N., **2,** 165, 317
Sletzinger, M., **1,** 1008; **5,** 62; **6,** 412; **9,** 107; **10,** 252, 449
Sliam, E., **6,** 279; **7,** 94
Slighter, R. G., **2,** 17
Slinger, C. J., **11,** 551
Sliwinski, W. F., **5,** 705
Sliwka, H. R., **9,** 237
Slix, J. A., **4,** 508
Sloan, A. D. B., **3,** 338
Slobbe, J., **6,** 323; **7,** 188
Slocombe, R. J., **5,** 280
Slocum, D. W., **5,** 86; **9,** 8, 89
Slomp, G., **1,** 75, 777, 890
Slooff, G., **1,** 796
Slorach, S. A., **1,** 1029, 1210
Slota, P. J., Jr., **1,** 432
Slotta, K. H., **1,** 176, 455
Slougui, N., **11,** 121; **12,** 74
Slusarchyk, W. A., **7,** 244
Sly, W. G., **1,** 228
Smadja, W., **10,** 343
Smael, P., **5,** 593
Small, V. R., Jr., **6,** 260
Smalley, R. K., **1,** 904
Smart, B. E., **11,** 119
Smart, J. C., **10,** 38
Smart, N. A., **1,** 97
Smart, W. D., **1,** 982
Smat, R. J., **1,** 906
Smejkal, J., **2,** 210, 336; **3,** 96; **7,** 407
Smentowski, F. J., **1,** 310
Smetana, R. D., **4,** 559
Smets, G., **4,** 225; **5,** 280
Smick, D., **4,** 74
Smick, R. L., **1,** 364
Smiles, S., **1,** 3, 472
Smiley, R. A., **1,** 17, 309
Smillie, R. D., **6,** 530; **8,** 492
Smissman, E. E., **1,** 185, 827; **5,** 213, 249; **6,** 646; **8,** 20
Smit, P. J., **2,** 87; **3,** 66
Smit, W. A., **4,** 325; **6,** 520; **11,** 163
Smit, W. C., **1,** 465
Smith, A. B., III, **4,** 512; **6,** 70, 142; **8,** 305, 307, 448; **9,** 24, 87, 96; **10,** 55, 56, 265, 289, 343, 356; **11,** 5, 109, 216
Smith, A. C. B., **5,** 342
Smith, A. E., **1,** 1167, 1269
Smith, A. K., **8,** 153
Smith, B. C., **2,** 207; **3,** 173
Smith, B. F., **5,** 315; **11,** 562
Smith, B. H., **1,** 1101
Smith, B. V., **9,** 35
Smith, C., **3,** 320
Smith, C. A., **7,** 174
Smith, C. D., **4,** 5
Smith, C. P., **2,** 98; **5,** 268
Smith, C. R., Jr., **1,** 865; **2,** 139
Smith, C. V., **5,** 587
Smith, C. W., **1,** 292, 478, 694, 864, 1287; **4,** 577
Smith, D. A., **12,** 11
Smith, D. G., **5,** 62; **6,** 393; **7,** 245
Smith, D. J. H., **4,** 573; **5,** 62; **6,** 393; **7,** 245; **10,** 62; **12,** 112
Smith, D. M., **5,** 148
Smith, D. S., **1,** 335, 445, 658
Smith, E. B., **1,** 57
Smith, E. H., **3,** 320
Smith, E. M., **4,** 484
Smith, E. W., **1,** 1116
Smith, G., **3,** 171; **4,** 567
Smith, G. A., **12,** 64
Smith, G. C., **12,** 127

Smith, G. F., **1,** 285, 1284; **4,** 473; **9,** 501
Smith, G. L., **1,** 1136; **4,** 215
Smith, G. P., **8,** 19
Smith, H., **1,** 56; **2,** 29, 372; **3,** 204; **6,** 258
Smith, H. A., **1,** 982; **8,** 284
Smith, H. G., **2,** 256; **9,** 302
Smith, H. L., **6,** 317
Smith, J., **2,** 344
Smith, J. A., **1,** 1187
Smith, J. C., **10,** 50
Smith, J. F., **3,** 320
Smith, J. G., **6,** 416; **9,** 196, 420; **11,** 51; **12,** 538
Smith, J. N., **4,** 45
Smith, J. R. L., **1,** 476; **6,** 70; **9,** 361; **12,** 123
Smith, J. S., **1,** 906
Smith, K., **4,** 447; **5,** 155; **6,** 295, 536; **7,** 26, 101, 333; **8,** 175, 238, 522; **9,** 5, 393
Smith, K. J., **1,** 1122
Smith, K. M., **10,** 52
Smith, L. A., **1,** 1123
Smith, L. H., **1,** 874
Smith, L. I., **1,** 339, 345, 362, 393, 445, 489, 657, 674, 777, 832, 1116, 1269; **5,** 614
Smith, L. L., **1,** 249; **2,** 245; **5,** 491
Smith, L. M., **3,** 45
Smith, L. R., **1,** 772; **2,** 426; **5,** 513; **6,** 36, 59, 634; **7,** 383
Smith, L. W., **10,** 330
Smith, M., **1,** 161, 189, 236; **2,** 126
Smith, M. A., **5,** 510
Smith, M. B., **12,** 555
Smith, N. L., **1,** 124
Smith, N. R., **1,** 156, 158, 962, 976
Smith, P. A. S., **1,** 34, 445, 1044, 1055, 1100; **4,** 272, 369; **5,** 280, 454
Smith, P. J., **4,** 406; **7,** 27
Smith, P. N., **1,** 357, 1137
Smith, P. V., **1,** 1283
Smith, P. W. G., **2,** 30
Smith, R. A., **6,** 76, 280, 651; **9,** 129
Smith, R. A. J., **6,** 215; **9,** 462; **12,** 208
Smith, R. D., **1,** 117, 507, 904, 1022, 1284, 1293; **3,** 141; **4,** 411
Smith, R. F., **1,** 290, 444; **2,** 217; **11,** 491
Smith, R. G., **3,** 123; **4,** 461; **7,** 170; **11,** 551
Smith, R. H., **1,** 89
Smith, R. K., **9,** 313
Smith, R. L., **2,** 198, 303, 433; **11,** 180
Smith, R. M., **2,** 213
Smith, R. S., **7,** 411
Smith, R. V., **6,** 9
Smith, S., **1,** 570; **11,** 157; **12,** 554
Smith, S. A., **9,** 438
Smith, S. G., **1,** 705
Smith, T. D., **1,** 163
Smith, T. N., **4,** 352
Smith, W. C., **1,** 1124, 1125; **2,** 136
Smith, W. C., Jr., **8,** 141
Smith, W. E., **3,** 299; **4,** 243; **11,** 554
Smith, W. N., **5,** 382, 649
Smith, W. T., **1,** 1087
Smith, W. T., Jr., **1,** 158; **6,** 40
Smithers, R. H., **9,** 130
Smith-Palmer, T., **7,** 360
Smolik, S., **1,** 764
Smolin, E. M., **1,** 1040, 1073; **2,** 427
Smolinsky, G., **1,** 567; **2,** 433; **5,** 61
Smoot, J., **5,** 727
Smorada, R. L., **9,** 8
Smrt, J., **1,** 4, 289, 388; **2,** 198
Smucker, L. D., **2,** 457
Smula, V., **5,** 334
Smyth, J., **1,** 345
Smyth, R. B., **2,** 57
Snader, K. M., **1,** 318
Snatzke, G., **1,** 147, 281, 989; **11,** 87
Sneed, C. N., **9,** 42
Sneen, R. A., **1,** 73
Snell, E. D., **11,** 502
Snider, B. B., **4,** 478; **5,** 13, 62, 530, 660; **6,** 300; **7,** 9, 237; **8,** 15, 42; **9,** 13, 53, 222, 471; **10,** 11, 180, 181, 187, 216; **11,** 12; **12,** 11, 71
Snieckus, V., **9,** 8; **10,** 75; **11,** 373; **12,** 99
Snoble, K. A. J., **4,** 304
Snoddy, C. S., Jr., **1,** 664, 1187
Snoussi, M., **8,** 168
Snover, J. A., **1,** 207
Snow, J. T., **2,** 142; **3,** 23, 219; **4,** 38, 110; **7,** 181
Snow, R. A., **7,** 65; **9,** 123
Snow, R. J., **12,** 492
Snowden, R. L., **11,** 437, 557
Snowdon, L. R., **3,** 31
Snyder, C. D., **4,** 432; **9,** 99
Snyder, C. E., **9,** 480
Snyder, C. H., **1,** 927, 937; **2,** 368; **4,** 194
Snyder, E. I., **1,** 258; **2,** 368; **4,** 552
Snyder, H. R., **1,** 156, 251, 285, 378, 433, 450, 471, 494, 527, 731, 838, 855, 864, 904, 905, 955, 1095, 1152, 1153, 1163, 1167; **5,** 441; **9,** 1; **11,** 61
Snyder, J. P., **2,** 207; **3,** 94; **5,** 209, 491; **11,** 176
Snyder, W. H., **1,** 309; **5,** 551
So, K. H., **8,** 264; **9,** 448
Soai, K., **6,** 596; **8,** 253; **9,** 246; **11,** 479; **12,** 25, 276, 444
Sobala, M. C., **8,** 158, 448
Sobel, A. E., **1,** 1128
Sober, H. A., **1,** 519
Sobieray, D. M., **11,** 187
Soccolni, F., **12,** 46
Soden, H. v., **4,** 556
Sodeoka, M., **12,** 36
Soderquist, J. A., **12,** 100
Soeder, R. W., **1,** 976; **2,** 351
Soehngen, B., **4,** 579
Soenke, H., **1,** 1189
Soerbo, B., **1,** 351
Sofuku, S., **2,** 12

Soga, K., **7,** 52
Sogadji, K., **8,** 133
Sogah, G. D. Y., **11,** 145
Soh, K. S., **2,** 334
Sohar, P., **4,** 376
Sohn, J. E., **9,** 283; **10,** 133; **11,** 349
Sohn, M. B., **4,** 336
Soja, P., **7,** 135
Sokol, P. E., **1,** 33, 772
Sokolovsky, M., **1,** 110; **2,** 404
Sokolowska, T., **1,** 83
Solar, S. L., **2,** 159
Solash, J., **5,** 587
Soldat, W. D., **5,** 601
Soleiman, M., **6,** 665
Solera, P. S., **8,** 223
Solladie, G., **5,** 210; **6,** 162; **9,** 364; **10,** 407; **11,** 106, 292, 315, 524, 538; **12,** 241, 298
Solladie-Cavallo, A., **9,** 364; **11,** 292
Solleder, J. B., **5,** 51
Sollman, P. B., **1,** 162; **9,** 269
Solmon, D. M., **5,** 315
Solms, U., **1,** 582
Solo, A. J., **2,** 326
Solodar, J., **6,** 569
Solomon, D. H., **1,** 561
Solomon, D. M., **3,** 258
Solomon, M. G., **1,** 1127
Solomon, P., **5,** 394
Solomon, P. H., **5,** 532
Solomon, S., **1,** 688, 926; **3,** 47, 204, 234
Solomonica, E., **1,** 191, 657, 1027
Soloski, E. J., **12,** 547
Soloway, A. H., **7,** 174
Soloway, S. B., **1,** 1011
Soltes, E. J., **2,** 180
Soltzberg, S., **1,** 1101
Soma, N., **4,** 199
Soman, R., **4,** 167; **11,** 519
Sombroek, W., **1,** 321
Somei, M., **5,** 344; **9,** 467; **10,** 400
Somekh, L., **11,** 177; **12,** 185
Somer, L. H., **9,** 74
Sommelet, M., **1,** 430
Sommer, H. Z., **4,** 371; **10,** 419
Sommer, J. M., **5,** 61
Sommer, L. H., **1,** 1218; **3,** 304; **4,** 58, 177; **6,** 84, 632; **7,** 388; **8,** 263
Sommer, N. B., **1,** 266, 1027
Sommer, P., **7,** 28, 53
Sommer, T. J., **12,** 305
Sommerfeld, C. D., **5,** 491
Sommers, A. H., **1,** 281
Somvichien, N., **2,** 35
Son, J. C., **12,** 276
Sonada, A., **9,** 349
Sonada, N., **11,** 220
Sondengam, B. L., **4,** 239; **8,** 232; **10,** 459; **12,** 570
Sonderquist, J. A., **11,** 68
Sondheimer, F., **1,** 3, 37, 39, 59, 82, 159, 207, 389, 614, 642, 802, 815, 926, 999, 1000, 1090, 1178, 1254, 1284; **2,** 88, 132, 241, 259, 330, 344, 361, 462; **3,** 113, 194, 339; **4,** 48, 105, 110, 196, 366, 405; **5,** 486; **6,** 138; **7,** 119, 324; **9,** 318
Sone, T., **12,** 421
Song, I. H., **1,** 49, 966; **2,** 353
Song, J., **1,** 927
Song, Y. H., **12,** 195
Songstad, J., **8,** 245
Sonneborn, H., III, **1,** 1284
Sonnenberg, F. M., **11,** 12
Sonnenburg, F., **1,** 1000
Sonnet, P. E., **5,** 693; **6,** 279, 647; **7,** 109, 338, 407; **8,** 504; **10,** 332, 419, 450
Sonnewald, U., **11,** 127
Sonntag, N. O. V., **3,** 177
Sonoda, A., **7,** 31, 201, 237; **8,** 346, 347, 381; **11,** 88, 295; **12,** 446
Sonoda, N., **3,** 156, 247; **4,** 422; **5,** 93, 341, 575, 589; **6,** 73, 224, 424, 509; **8,** 5, 28, 29, 133, 162, 172, 173, 498; **9,** 145, 151, 200, 434, 497; **10,** 206, 276, 312; **11,** 466, 498; **12,** 106, 144, 167, 336
Sonogashira, K., **6,** 61; **8,** 152; **10,** 117; **11,** 162; **12,** 476
Sonveaux, E., **5,** 684; **9,** 105
Sood, A., **6,** 70
Sood, H. R., **6,** 600
Sood, R., **4,** 221, 517; **6,** 201
Sood, V. K., **6,** 597
Sootome, N., **12,** 548
Soper, F. G., **1,** 214
Sorace, R., **6,** 467
Sorba, J., **7,** 174
Sorell, T. N., **10,** 175
Sorenson, C. M., **10,** 5
Sorgi, K. L., **10,** 66; **11,** 20
Sorm, F., **1,** 139, 289, 526, 624, 1002; **2,** 84
Sormova, Z., **1,** 1002
Sornay, R., **8,** 59
Soroos, H., **2,** 385
Sorrell, T. N., **8,** 47; **9,** 103; **10,** 48, 378
Sorrenti, P., **12,** 405
Sorrie, A. J. S., **1,** 1029
Sorriso, S., **5,** 114
Sosnovsky, G., **1,** 101, 1196; **4,** 66; **5,** 370; **9,** 410
Sotheeswaran, S., **1,** 562; **3,** 171
Soto, A. R., **1,** 927
Sotowicz, A. J., **7,** 223
Souchi, T., **10,** 36
Soucy, M., **6,** 376; **8,** 492
Soukup, M., **9,** 355
Soulen, R. L., **6,** 553; **7,** 97
Souma, Y., **5,** 150
Soumillion, J. P., **7,** 35
Souppe, J., **11,** 465; **12,** 430
Sousa, J. A., **1,** 1055
South, A., Jr., **1,** 1151
South, M. S., **10,** 223; **12,** 130
Southard, G. L., **6,** 234
Souther, S. K., **10,** 450
Southern, P. F., **12,** 390
Southgate, R., **9,** 408
Southwick, P. L., **1,** 503
Souto-Bachiller, F. A., **7,** 20, 61
Soverini, M., **9,** 426

Sowa, F. J., **1,** 69
Sowa, J. R., **6,** 486, 545
Sowa, W., **2,** 165
Sowden, J. C., **1,** 329
Sowerby, R. L., **4,** 65, 380
Sowers, E. E., **1,** 1178
Sowinski, A. F., **6,** 414; **9,** 221
Sowinski, F., **2,** 129; **5,** 213
Soyama, H., **11,** 165
Soysa, H. S. D., **8,** 240
Sozzani, P., **9,** 376; **10,** 324
Spadafino, L. P., **1,** 57
Spadaro, J., **6,** 56
Spaeth, E., **1,** 159
Spaeth, E. C., **1,** 35
Spahic, B., **6,** 175; **11,** 120
Spangenberg, R., **5,** 169
Spangler, C. W., **4,** 119; **6,** 649; **11,** 203
Spangler, F. W., **1,** 379, 890, 1177
Spangler, P. L., **3,** 297
Spangler, R. J., **4,** 254; **5,** 551
Sparke, M. B., **1,** 689
Sparrow, J. T., **6,** 304; **9,** 401
Spassky, N., **5,** 557
Spassow, A., **1,** 83, 160, 629, 1292
Spatz, S. M., **1,** 353
Spaude, S., **1,** 1075
Spaziano, V. T., **5,** 140
Speakman, P. R. H., **8,** 386
Specian, A. C., Jr., **8,** 192
Speck, D. H., **6,** 137
Speckamp, W. N., **5,** 389; **7,** 331; **8,** 232; **9,** 83
Spector, M. L., **1,** 782; **2,** 304
Speers, L., **1,** 790
Speeter, M., **12,** 568
Speh, P., **5,** 653
Spehar, A., **3,** 77, 123, 264
Speidel, P. E., **1,** 1256
Speier, A., **5,** 336
Speier, G., **9,** 393; **11,** 141, 445
Speier, J. L., **5,** 294
Speltz, L. M., **7,** 415
Spence, J. M., **1,** 782; **8,** 383
Spencer, A., **6,** 426, 650; **11,** 253
Spencer, C., **1,** 731
Spencer, C. F., **1,** 982; **2,** 368; **4,** 31
Spencer, D. A., **2,** 38
Spencer, H., **9,** 235
Spencer, H. K., **7,** 387; **9,** 430
Spencer, I. D., **1,** 481
Spencer, R. R., **1,** 1045
Spencer, T. A., **1,** 1178; **2,** 195, 462; **3,** 182; **4,** 197, 305; **5,** 294; **6,** 573; **9,** 449; **12,** 234
Speranza, G., **12,** 422
Sperber, N., **1,** 1033
Sperley, R. J., **2,** 328, 351
Sperling, R., **1,** 1127
Spero, G. B., **1,** 731; **2,** 134, 362
Speziale, A. J., **1,** 223, 253, 257, 357, 772, 1167, 1192, 1246, 1247; **2,** 425, 426; **5,** 513; **6,** 36, 252, 634; **7,** 383
Spiegel, B. I., **8,** 87, 507
Spiegelberg, H., **1,** 76
Spielman, M. A., **1,** 172, 1121; **4,** 497
Spier, J. L., **9,** 284
Spierenburg, J., **5,** 371
Spiers, D. B., **1,** 859
Spiess, E., **11,** 358
Spilker, A., **1,** 832
Spillane, L. J., **1,** 489
Spillane, R. J., **8,** 47
Spille, J., **3,** 303; **6,** 250
Spilners, I. J., **3,** 145, 201
Spina, K. P., **12,** 445
Spindell, D. K., **10,** 11, 180
Spitzer, U. A., **3,** 265; **6,** 506
Spitzner, D., **7,** 210; **12,** 228
Spletzer, E. G., **12,** 279
Spliethoff, B., **12,** 290
Spoerri, P. E., **1,** 1128; **3,** 177; **7,** 424
Spohn, R. J., **3,** 89, 278
Spokes, G. N., **2,** 26
Sponagel, P., **3,** 69
Sponlein, W., **12,** 400
Sporzynski, A., **7,** 247
Sprague, G. S., **2,** 268
Sprecher, M., **1,** 815, 1178
Spring, F. S., **1,** 885, 1284; **3,** 54; **8,** 265
Springer, J., **12,** 266
Springer, J. M., **4,** 357, 397; **6,** 546
Springer, J. P., **4,** 370; **10,** 330; **12,** 387
Springfield, J. R., **4,** 389
Sprinzak, Y., **1,** 737, 1253, 1254
Sprotte, U., **9,** 428
Spruegel, W., **4,** 90
Spur, B., **12,** 437
Spyroudis, S., **6,** 301; **9,** 54
Squires, T. G., **6,** 676; **11,** 562
Sraga, J., **8,** 433
Sramek, M., **11,** 197
Srebnik, M., **12,** 121
Sreckumar, C., **10,** 327, 445
Srinivasan, N. S., **9,** 391
Srinivasan, R., **1,** 169; **2,** 79
Srivastava, H. C., **1,** 309, 685; **8,** 228
Srivastava, R. K., **2,** 278
Srnak, T., **11,** 23
Srogl, J., **4,** 70; **10,** 188
Sroog, C. E., **1,** 594
Srzic, D., **11,** 124
Staab, H. A., **1,** 116, 340, 408, 424, 494, 595, 848, 1152, 1164, 1228; **2,** 61; **7,** 400; **8,** 155; **11,** 497
Staab, W., **1,** 1226
Stacey, F. W., **7,** 368
Stacey, G. J., **1,** 248, 253; **4,** 481; **7,** 106
Stacey, M., **1,** 68, 122, 413, 1226; **2,** 175, 274; **4,** 461, 533
Stache, U., **1,** 350; **4,** 296; **5,** 119, 633
Stacy, G. W., **1,** 472, 658, 1105
Stadler, P., **2,** 23; **4,** 218
Stadler, P. A., **1,** 942; **4,** 577; **5,** 4, 160, 161; **6,** 560; **8,** 186; **9,** 515
Stadnichuk, M. D., **2,** 438
Staehle, M., **9,** 89; **10,** 186
Staellberg-Stenhagen, S., **1,** 772
Staeudle, H., **1,** 802; **2,** 353

Stafforst, D., **6,** 187
Stahel, R., **1,** 340
Stahl, E., **1,** 1000
Stahl, I., **9,** 483; **10,** 426
Stahl, R. E., **1,** 1101
Stahman, M. A., **1,** 820
Stahnke, M., **10,** 216
Staiger, G., **1,** 942
Staires, S. K., **6,** 479
Stairs, R. A., **2,** 79
Staklis, A., **1,** 563; **6,** 317
Staley, S. W., **4,** 290; **5,** 544
Stalick, W. M., **5,** 107, 593; **8,** 446
Stam, M. F., **6,** 249; **7,** 239
Stamm, D., **1,** 736
Stamm, E., **9,** 217
Stamm, H., **5,** 756; **8,** 21
Stammer, C. H., **9,** 256; **10,** 136
Stamos, I. K., **5,** 561; **11,** 352
Stamper, W. E., **1,** 34
Stampfli, J. G., **1,** 109, 1169
Stancer, H. C., **1,** 847
St. Andre, A. F., **2,** 79; **3,** 57
Standring, D. N., **9,** 394
Stanescu, L., **4,** 57
Stanfield, C. F., **11,** 216
Stang, P., **3,** 101
Stang, P. J., **4,** 533; **5,** 705; **6,** 231, 620; **8,** 146, 412; **9,** 141; **10,** 123, 289; **12,** 532
Stange, H., **1,** 692, 730; **2,** 274
Stange, O., **1,** 141, 850
Stanger, H., **4,** 100
Stankovic, J., **3,** 254; **5,** 61
Stanley, T. W., **1,** 673
Stanovnik, B., **5,** 254; **8,** 483; **9,** 472; **10,** 274
Stansbury, H. A., Jr., **1,** 169, 755, 791; **5,** 506
Stansfield, F., **3,** 254
Stansfield, R. E., **9,** 387
Stanton, J. L., **6,** 32; **8,** 356; **9,** 321
Stapp, P. R., **4,** 240; **5,** 49
Stapp, R., **1,** 848
Staps, R. J. F. M., **5,** 614
Stararo, A., **11,** 336
Starcher, P. S., **1,** 790
Stare, F. J., **6,** 37
Staring, E. G. J., **11,** 135
Stark, A., **1,** 484
Stark, C. I., **10,** 204
Stark, C. J., Jr., **11,** 211
Stark, D. D., **1,** 618
Starkey, E. B., **1,** 44, 169, 396
Starks, C. M., **4,** 31; **5,** 323; **6,** 406; **9,** 361
Starling, W. W., **1,** 1000
Staros, J. V., **9,** 394
Staroscik, J., **4,** 183
Starr, D., **1,** 1141
Starr, L. D., **1,** 296
Starratt, A. N., **6,** 124
Starrett, R. M., **4,** 371
Stary, F. E., **7,** 271
Staschewski, D., **1,** 1289
Staskiewicz, S., **12,** 220
Staskun, B., **1,** 720, 730, 1087; **2,** 294; **4,** 351
Stassinopoulou, C. I., **9,** 513
Stathom, A. A., **10,** 126
Staub, A. P. A., **5,** 742
Staude, E., **9,** 56
Staudinger, G. K., **5,** 339
Staudinger, H., **1,** 345, 657, 767, 771, 1247; **2,** 105; **4,** 260; **5,** 155; **7,** 3; **11,** 588; **12,** 551
Stauffer, R. D., **5,** 35; **6,** 144; **8,** 120; **11,** 51
Staunton, J., **1,** 9; **6,** 9; **9,** 283, 302; **12,** 318
Stavber, S., **10,** 85; **11,** 118
Stavric, S., **1,** 219
Stawinski, J., **7,** 14, 15
Stearns, J. A., **1,** 1292
Stec, W., **1,** 88
Stec, W. J., **8,** 133
Stecher, O., **9,** 56
Steckhan, E., **9,** 510; **10,** 101
Stedman, R. J., **1,** 347, 1126; **2,** 397
Steele, D., **1,** 986
Steele, D. R., **3,** 216
Steele, E. S., **4,** 74
Steele, J., **11,** 441
Steele, R. B., **2,** 142, 249; **7,** 147
Steele, W. R. P., **3,** 22
Steen, K., **1,** 272, 283
Stefani, A., **5,** 331
Stefanovic, M., **3,** 171
Steffgen, F. W., **1,** 147
Steglich, W., **1,** 1220, 1271; **2,** 263; **3,** 119; **4,** 172, 411, 416; **5,** 219, 339; **7,** 141, 247; **8,** 163, 272, 396; **9,** 157, 180, 203; **10,** 151, 308; **11,** 360, 529
Stehower, K., **2,** 454
Steichen, D. S., **9,** 341
Steiger, M., **8,** 523
Steiger, R. E., **1,** 481, 533, 963
Steigman, J., **1,** 1137, 1143
Stein, C. A., **5,** 454
Stein, G., **1,** 476, 832; **8,** 47
Stein, P. D., **11,** 458
Stein, P. M., **9,** 322; **11,** 355
Stein, R. G., **3,** 50; **6,** 400
Stein, R. P., **3,** 204; **9,** 40
Stein, W. H., **1,** 519
Steinacker, K. H., **1,** 1210; **5,** 511
Steinbach, K., **11,** 87
Steinbach, R., **10,** 138, 271; **11,** 378
Steinbeck, K., **8,** 391; **10,** 365
Steinberg, A. M., **1,** 444
Steinberg, G. M., **1,** 248
Steinberg, H., **4,** 108; **8,** 187
Steindel, S. J., **5,** 381
Steiner, G., **5,** 649
Steiner, K., **1,** 34; **6,** 409
Steiner, R. I., **7,** 148
Steiner, R. P., **5,** 680
Steiner, U., **1,** 1015
Steinfatt, M., **6,** 471
Steinhardt, C. K., **1,** 195, 594, 618
Steinkopf, W., **1,** 1195
Steinman, D. H., **6,** 348
Steinmaus, H., **6,** 375
Steinmetz, A., **6,** 441
Steinpreis, R., **1,** 655
Steklenski, D. J., **5,** 105
Stelakatos, G. C., **2,** 178, 317
Stelander, B., **5,** 198; **8,** 157
Steliou, K., **9,** 236; **10,** 125; **11,** 63; **12,** 160
Stelzer, O., **11,** 543
Stemke, J. E., **6,** 600; **7,** 48; **9,**

488
Stemmle, B., **8,** 395
Stempel, A., **1,** 235; **5,** 63
Stempel, G. H., Jr., **1,** 50
Stenberg, J. F., **1,** 595
Stenberg, V. I., **1,** 195; **4,** 267; **5,** 356
Stenhagen, E., **2,** 166
Stening, T. C., **2,** 4
Stenlake, J. B., **1,** 643
Stepanov, F. N., **5,** 756
Stephan, D., **12,** 137
Stephan, W., **8,** 525; **10,** 276
Stephani, R., **5,** 527
Stephanon, S. E., **1,** 952
Stephen, E., **1,** 990
Stephen, H., **1,** 990
Stephen, J. P., **8,** 521
Stephens, C. R., **1,** 4, 1184
Stephens, E. R., **5,** 511
Stephens, J. R., **1,** 1084
Stephens, L. J., **5,** 667
Stephens, R. D., **1,** 151; **2,** 78; **6,** 628; **7,** 237; **12,** 144
Stephenson, E. F. M., **1,** 450, 864; **2,** 222
Stephenson, L., **1,** 609, 1090; **3,** 96
Stephenson, L. M., **8,** 534; **10,** 144
Stephenson, O., **1,** 140
Stephenson, T. A., **1,** 778; **4,** 421; **5,** 193, 498; **10,** 303
Steppan, W., **11,** 587
Steppel, R. N., **2,** 69; **3,** 50
Stercho, Y. P., **12,** 288, 446
Sterk, H., **2,** 372
Sterling, C. J. M., **6,** 56
Sterling, J., **1,** 1220
Sterling, J. J., **5,** 243, 244, 396, 415; **6,** 343
Stermitz, F. R., **2,** 226
Stern, A., **1,** 1251
Stern, E. W., **1,** 782; **2,** 304; **6,** 124
Stern, F., **1,** 531
Stern, P., **4,** 536; **5,** 551
Stern, R., **1,** 134; **2,** 252
Stern, R. L., **2,** 287
Sternbach, D., **5,** 705; **9,** 355
Sternbach, D. D., **6,** 36; **6,** 44; **11,** 151, 483
Sternbach, L. H., **7,** 255
Sternberg, E. D., **10,** 127; **12,** 163
Sternberg, H. W., **1,** 155, 228, 520, 723
Sternberg, J. A., **12,** 86
Sternhell, S., **1,** 927, 1203; **2,** 222; **4,** 260, 283; **6,** 254, 581; **7,** 188; **9,** 520; **11,** 32
Sterns, M., **9,** 61
Sterzycki, R., **9,** 400
Stetin, C., **11,** 161
Stetter, H., **1,** 23, 67, 1210; **2,** 17, 172; **3,** 19, 97; **4,** 435, 447; **5,** 167, 511, 607; **6,** 39, 289, 536; **7,** 17, 340; **8,** 454; **10,** 27
Steven, A. B., **1,** 1056
Stevens, C. L., **1,** 345, 1090, 1131; **2,** 330; **3,** 123, 161; **5,** 283
Stevens, H. C., **1,** 222
Stevens, I. D. R., **1,** 247, 850; **2,** 282, 325; **3,** 224; **4,** 383; **7,** 288
Stevens, J. I., **5,** 200
Stevens, K. L., **6,** 518
Stevens, R. M., **9,** 387
Stevens, R. V., **2,** 285; **4,** 161; **9,** 259, 423; **10,** 365; **11,** 488
Stevens, R. W., **11,** 526; **12,** 491
Stevens, T. E., **1,** 254, 323, 503, 1178; **3,** 130
Stevens, T. S., **1,** 1187; **7,** 331
Stevens, W., **1,** 4
Stevenson, G., **1,** 80
Stevenson, H. B., **1,** 18, 172
Stevenson, J. W. S., **11,** 53
Stevenson, P., **11,** 130
Stevenson, R., **1,** 236, 329, 561, 1249; **4,** 252; **8,** 124, 529
Steward, D. G., **7,** 295
Stewart, A., **8,** 408
Stewart, C. A., Jr., **1,** 1136
Stewart, D., **3,** 267; **5,** 266; **6,** 359
Stewart, F. H. C., **1,** 374; **2,** 220, 305, 317; **5,** 627
Stewart, J., **1,** 3
Stewart, J. A. G., **3,** 87
Stewart, J. C. M., **3,** 223
Stewart, J. E., **1,** 1011
Stewart, J. M., **1,** 378, 519, 955; **4,** 521
Stewart, O. J., **1,** 170; **7,** 252
Stewart, P., **10,** 25
Stewart, R. C., **5,** 191, 193
Stewart, R. F., **8,** 109; **10,** 63
Stewart, T. S., **2,** 377
Steyermark, P. R., **1,** 172
Stezowski, J. J., **9,** 299, 337, 406
Stibbard, J. H. A., **8,** 393
Stibor, I., **4,** 70; **10,** 188
Stick, R. V., **6,** 5, 569, 587; **7,** 380; **9,** 378
Sticker, R. E., **2,** 58, 239
Stickler, J. C., **2,** 282, 326
Stiff, J. F., **1,** 67
Stilbs, P., **7,** 359
Stiles, M., **1,** 46, 96, 360, 633; **2,** 26, 146
Still, I. W. J., **8,** 401; **9,** 103, 466; **10,** 320
Still, W. C., **6,** 386, 607; **7,** 115, 148, 400; **8,** 197, 415, 497; **9,** 81, 251, 334, 377, 476; **10,** 194, 327, 398, 445; **11,** 518, 542, 605; **12,** 137, 320, 484
Stille, J. K., **1,** 445, 777, 1011; **4,** 119; **6,** 59; **8,** 36; **9,** 42; **10,** 26, 133; **11,** 7, 395, 521; **12,** 46, 56, 474, 475, 576
Stille, J. R., **11,** 7; **12,** 55
Stillich, O., **1,** 1118
Stillings, M. R., **6,** 70
Stilmar, F. B., **1,** 456
Stipanovic, R. D., **2,** 382; **11,** 85
Stirling, C. J. M., **1,** 1185; **11,** 539
Stirling, I., **6,** 597; **8,** 427
Stitzel, R. E., **6,** 9
Stiverson, R. K., **5,** 437
Stjernstroem, N. E., **1,** 966
Stobart, S. R., **12,** 202

Stobie, A. A., **11,** 276, 587; **12,** 102
Stock, A. M., **1,** 952; **11,** 442
Stock, L. M., **1,** 81, 658, 1226; **8,** 498
Stockel, R. F., **1,** 905
Stocker, J. H., **1,** 982
Stockis, A., **6,** 306
Stockmann, H., **1,** 45
Stocky, T. P., **5,** 48
Stoddart, J. F., **6,** 137
Stodola, F. H., **1,** 370, 1073, 1283; **3,** 325
Stoeckel, K., **4,** 105, 366
Stoekel, K., **4,** 405
Stoermer, R., **4,** 189
Stoffey, D. G., **1,** 581
Stoilova, V., **9,** 503
Stojanac, N., **6,** 70
Stojanac, Z., **6,** 70; **10,** 55
Stojda, R. J., **9,** 299, 337, 406
Stojiljkovic, A., **1,** 180, 563
Stokes, B. G., **6,** 385
Stokes, C. S., **3,** 143
Stokes, P., **2,** 405
Stokstad, E. L. R., **1,** 1157
Stolar, S. M., **1,** 764, 881
Stolberg, U. G., **1,** 892
Stoll, A., **1,** 20, 351, 384, 478, 802, 1046; **6,** 2
Stoll, A. P., **6,** 309
Stoll, A. T., **12,** 474
Stoll, M., **1,** 119; **10,** 150
Stolle, W. T., **8,** 493; **9,** 473
Stoller, H., **12,** 180
Stolow, R. D., **1,** 581; **2,** 250; **4,** 282
Stonard, R. J., **11,** 128
Stone, B. E., **2,** 56
Stone, D. B., **2,** 247, 344
Stone, G. R., **1,** 982, 983, 986
Stone, H., **1,** 873, 905, 1141
Stone, L. C., **4,** 45
Stone, R. M., **1,** 905
Stoochnoff, B. A., **4,** 455; **5,** 615
Stoodley, R. J., **5,** 427; **10,** 260; **11,** 441
Stoodley, R. T., **12,** 319
Stopp, G., **1,** 272, 282, 283; **4,** 167
Storer, R., **5,** 567
Stork, G., **1,** 76, 268, 277, 360, 500, 562, 603, 642, 707, 814, 889, 890, 926, 974; **2,** 29, 84, 151, 438; **3,** 153, 182, 269, 293; **4,** 177, 302, 323, 409, 577; **5,** 5, 355, 375, 406, 463, 678, 723, 747; **6,** 101, 103, 267, 339, 341, 530, 598; **7,** 24, 130, 199, 204, 216, 232, 283; **8,** 193, 245, 408; **9,** 159, 261, 283, 334, 377, 421, 503; **10,** 236, 443; **11,** 103, 491, 550, 606; **12,** 11, 152, 272, 359, 449, 462, 523
Stork, P. J., **7,** 142
Storm, D. L., **2,** 195
Stormont, R. T., **1,** 5
Storni, A., **1,** 80; **4,** 53
Storr, R. C., **1,** 484; **2,** 67; **3,** 46, 47; **4,** 282
Story, P. R., **1,** 7, 101; **2,** 182, 263, 289; **3,** 123; **5,** 495, 649; **6,** 577
Stott, P. E., **7,** 79; **10,** 112
Stotter, P. L., **4,** 23; **5,** 79, 381, 454, 495; **6,** 83
Stoudt, G. S., **11,** 139
Stout, E. I., **5,** 247
Stout, E. J., **10,** 39
Stout, R., **9,** 121
Stowell, J. C., **1,** 927; **3,** 234, 251; **4,** 60, 375, 505; **5,** 672; **7,** 37; **9,** 197; **10,** 201, 221
Stoye, D., **5,** 329
Strachan, R. G., **1,** 293; **5,** 601
Strack, E., **1,** 536
Straley, J. M., **5,** 505
Strand, G., **6,** 319
Strand, J. W., **5,** 103
Strange, G. A., **11,** 305
Stransky, W., **6,** 279; **7,** 329; **8,** 234
Strassburger, P., **7,** 374
Stratford, M. J. W., **1,** 643; **2,** 263
Strathdee, R. S., **6,** 600
Strating, J., **2,** 370; **3,** 37, 312; **4,** 455, 459, 516, 539; **5,** 127, 295, 491, 495, 603; **11,** 44
Straub, H., **5,** 740; **6,** 47
Straus, D. B., **1,** 257
Straus, F., **1,** 970; **5,** 322
Strauss, H. E., **1,** 970
Strauss, H. F., **11,** 249
Strauss, J. U. G., **6,** 450
Strausz, O. P., **3,** 138; **6,** 190; **8,** 222
Strawson, C. J., **6,** 517
Streck, R., **5,** 742
Strecker, A., **9,** 164
Streckert, G., **4,** 567
Stredronsky, E. R., **3,** 154
Streeck, H., **1,** 682
Street, S. D. A., **12,** 444
Strege, P. A., **5,** 502
Strege, P. E., **6,** 395, 450; **7,** 327; **8,** 217
Streib, W. E., **5,** 152
Streichfuss, D., **5,** 740
Streith, J., **6,** 290; **9,** 246; **11,** 362
Streitwieser, A., Jr., **1,** 1143; **3,** 101; **6,** 521
Strekowski, L., **12,** 117, 229
Streltsova, S. G., **2,** 29
Strewe, W. S., **11,** 105
Strickland, R. C., **5,** 49, 700
Strickland, S. M. S., **10,** 443
Strijtveen, B., **11,** 118
Strike, D. P., **2,** 372
Stringer, O. D., **11,** 108, 109; **12,** 393
Stringham, R. A., **10,** 413
Strobach, D. R., **9,** 11
Strohmayer, H. F., **1,** 855
Strohmeier, W., **6,** 28, 104, 653; **7,** 54, 60
Stromberg, V. L., **1,** 904
Stromquist, P., **7,** 382
Stromqvist, M., **12,** 502
Strong, J. G., **5,** 394
Strouse, C. E., **9,** 259
Strub, H., **11,** 362
Strube, R. E., **1,** 524; **6,** 255; **10,** 182

Struble, D. L., **3,** 171
Struckmeyer, K., **4,** 345
Strung, G. M., **11,** 14
Strunk, R. J., **4,** 520; **5,** 686
Struve, A., **1,** 884, 926
Struve, W. S., **1,** 195; **4,** 544
Stryker, J. M., **12,** 272
Stuart, R. S., **2,** 173
Stubbings, R., **1,** 1102
Stubbs, C. A., **11,** 488
Stubbs, R. D., **1,** 1284
Stuber, F. A., **10,** 202
Stuber, H. A., **8,** 466
Studer, R. O., **2,** 177
Studnicka, B. J., **3,** 178
Studt, P., **8,** 405, 446
Stuebe, C., **1,** 425
Stueber, O., **1,** 1011
Stuehler, H., **5,** 724; **9,** 224
Stuessi, R., **8,** 215
Stuetz, A., **9,** 172; **12,** 233
Stuetz, P., **5,** 4, 160, 161
Stull, A., **3,** 272
Stump, B. L., **1,** 982
Stumpf, W., **1,** 478, 790
Sturm, G. P., **1,** 1254
Sturtz, G., **6,** 574; **10,** 6
Stylos, N., **1,** 383
Su, T. M., **5,** 705
Su, W. Y., **12,** 288
Suammons, R. E., **8,** 529
Suarato, A., **9,** 491
Suarez. E., **7,** 40; **11,** 445
Subba Rao, B. C., **1,** 206, 207, 603, 624, 1055
Subba Rao, G., **4,** 33, 135
Subba Rao, G. S. R., **2,** 331, 453; **10,** 236
Subbarao, H. N., **10,** 346
Subbarao, R., **2,** 368
Subbarow, Y., **1,** 80
Subo, E. F., **1,** 782
Subrahmanyam, C., **6,** 208; **9,** 317
Subramanian, L. R., **5,** 480; **12,** 338
Subramanian, R., **7,** 380; **9,** 361, 377, 378; **11,** 491
Subramanyam, G., **6,** 254
Subramanyam, V., **7,** 174
Suchanek, P., **11,** 103
Suckfuell, F., **1,** 652
Sucrow, W., **6,** 304
Suczak, J., **8,** 198
Suda, M., **4,** 435; **7,** 119; **10,** 448; **11,** 391
Suda, T., **5,** 89, 552
Sudakova, E. V., **7,** 413
Suder, B. J., **12,** 568
Sudo, K., **11,** 574; **12,** 538
Sudo, R., **3,** 322
Sudo, T., **12,** 486
Sudoh, R., **8,** 505; **9,** 41; **10,** 21
Sudol, J. J., **5,** 230
Sudweeks, W. B., **6,** 493
Sueess, R., **6,** 309
Suemitsu, R., **8,** 266; **9,** 207; **10,** 365; **12,** 449
Suemune, H., **12,** 131
Sues, O., **1,** 876
Sufrin, J. R., **5,** 560
Suga, K., **2,** 241, 289; **3,** 208; **4,** 240, 349, 350, 572; **5,** 470, 503; **6,** 415; **9,** 431
Suga, S., **10,** 142
Sugano, K., **8,** 314; **10,** 252
Sugasawa, S., **3,** 320
Sugasawa, T., **2,** 427; **9,** 63, 152; **11,** 159; **12,** 179
Sugawara, T., **9,** 151
Sugden, J. K., **5,** 14
Suggs, J. W., **5,** 473, 739; **6,** 380, 499, 604; **7,** 91, 309; **8,** 109; **9,** 114; **10,** 116
Sugi, Y., **8,** 333
Sugihara, J. M., **1,** 834
Sugihara, Y., **2,** 227; **5,** 713; **6,** 192
Sugimori, A., **6,** 659
Sugimoto, A., **5,** 89, 552; **6,** 586
Sugimoto, H., **8,** 59, 291
Sugimoto, K., **6,** 260; **7,** 155; **8,** 454
Sugimoto, N., **1,** 53
Sugimoto, T., **4,** 431; **8,** 83; **10,** 87, 176; **11,** 604
Sugimura, H., **10,** 194
Sugimura, Y., **4,** 199; **9,** 191
Sugino, E., **10,** 404
Suginome, H., **5,** 348; **11,** 393; **12,** 306
Sugita, J., **1,** 732
Sugita, T., **11,** 267
Sugiura, K., **12,** 291
Sugiura, M., **5,** 713
Sugiura, S., **4,** 86
Sugiura, Y., **12,** 345
Sugiyama, A., **6,** 534
Sugiyama, H., **7,** 331; **8,** 451
Sugiyama, K., **10,** 330
Sugiyama, N., **8,** 120
Sugiyama, T., **7,** 252; **12,** 130
Sugiyama, Y., **12,** 59
Sugizaki, M., **10,** 182
Suh, J. T., **1,** 176, 185
Suhl, K., **7,** 377
Suhr, H., **1,** 195; **2,** 91
Suhr, R. G., **10,** 159
Suida, W., **1,** 1284
Suijten, J. G. H., **1,** 1193
Sujeeth, P. K., **9,** 465
Sukenik, C. H., **10,** 253
Sukhi, L., **7,** 413
Sukhoverkhov, V. D., **5,** 756
Suksamrarn, A., **6,** 481; **8,** 110
Sukumar, D., **9,** 361
Sukumaran, K. B., **9,** 34, 102, 238
Sullivan, A. P., Jr., **1,** 476
Sullivan, D. F., **4,** 309; **5,** 371, 372, 690; **8,** 525; **9,** 151; **11,** 472
Sullivan, E. A., **1,** 1143; **4,** 450
Sullivan, G. R., **7,** 389
Sullivan, P. T., **5,** 538; **6,** 472
Sully, B. D., **1,** 1117
Sulston, J. E., **2,** 271; **3,** 198
Sultana, N., **10,** 360
Sultanbawa, M. U. S., **3,** 307
Sum, F. W., **9,** 334; **10,** 289
Sum, P. E., **9,** 65, 114
Sumaria, S., **9,** 470
Sumelius, S., **6,** 656
Sumi, K., **11,** 544
Sumi, S., **11,** 434; **12,** 407
Sumiki, Y., **3,** 41
Sumitani, K., **6,** 51; **11,** 168
Summerbell, R. K., **1,** 631; **2,** 226

Summers, G. H. R., **1,** 19, 35, 489, 624
Summers, R. M., **1,** 1178, 1269
Summers, S. T., **10,** 32
Sumoto, K., **5,** 433; **6,** 233
Sumrell, G., **1,** 926
Sun, R. C., **9,** 260
Sunaga, M., **5,** 363
Sunay, U., **11,** 475
Sundararaman, P., **8,** 374, 375
Sundberg, R. J., **1,** 1216
Sundelin, K. G. R., **6,** 554
Sundermeyer, W., **5,** 722
Sunder-Plassman, P., **2,** 271; **3,** 198; **4,** 146
Sunderwirth, S. G., **1,** 296
Sundin, C. E., **4,** 194
Sundin, S., **2,** 238
Sundri, S., **9,** 383
Sundstroem, G., **1,** 84
Sung, J. S., **10,** 312
Sunjic, V., **6,** 474; **9,** 234; **11,** 124
Sunkul, S., **8,** 367
Sunthankar, S. V., **6,** 19, 588; **7,** 333
Supanekar, C. V., **8,** 383
Supanekar, V. R., **1,** 782; **2,** 303
Sur, S. C., **8,** 106
Surber, B. W., **11,** 596
Surgenor, D. M., **1,** 89; **4,** 20
Surmatis, J. D., **1,** 567
Surrey, A. R., **1,** 295, 452, 471, 536, 881; **2,** 208
Surridge, J. H., **1,** 247; **2,** 139, 368; **4,** 107
Surtees, J. R., **1,** 643
Surya, G. K., **10,** 457
Surya Prakash, G. K., **7,** 248, 370
Suryawanshi, S. N., **11,** 442
Surzur, J. M., **4,** 123; **9,** 122
Suschitzky, H., **1,** 827, 904; **2,** 334; **3,** 194; **4,** 372, 397; **5,** 136
Susi, P. V., **1,** 17
Suskind, S. P., **2,** 263
Suss, D., **10,** 103
Sussman, S., **1,** 89, 477
Sustmann, R., **11,** 134
Sutbeyaz, Y., **12,** 139
Sutcliffe, F. K., **1,** 34
Suter, C., **5,** 474, 511; **6,** 419
Suter, C. M., **1,** 18, 1121, 1127, 1131, 1163
Sutherland, I. O., **2,** 117, 406; **3,** 87, 286, 320
Sutherland, J. K., **2,** 318, 462; **4,** 5; **9,** 87, 376; **11,** 108
Sutherland, M. D., **1,** 1011
Sutherland, R. G., **1,** 129, 219
Suthers, B. R., **1,** 34
Sutou, N., **11,** 514
Sutter, M., **1,** 764
Suzinome, H., **11,** 268
Suzukamo, G., **10,** 5
Suzuki, A., **2,** 285; **3,** 5, 39; **4,** 53; **5,** 347, 486; **6,** 25, 149, 215, 259, 295, 308; **7,** 4, 75, 153, 156; **8,** 240, 260; **9,** 362, 373; **10,** 257, 306, 403; **11,** 15, 252, 294, 393, 514; **12,** 219, 236, 502
Suzuki, E., **6,** 510, 578
Suzuki, F., **7,** 417; **8,** 417, 458; **9,** 32, 233
Suzuki, H., **2,** 222, 395; **3,** 119, 164; **4,** 84, 260; **5,** 167; **6,** 244; **8,** 160, 417; **9,** 167, 328; **10,** 56, 174, 281; **12,** 143, 160, 275, 449, 508
Suzuki, J., **6,** 73; **8,** 188
Suzuki, K., **1,** 247; **2,** 129; **3,** 33, 300; **6,** 73; **8,** 253, 254; **9,** 32, 107, 246, 343; **10,** 12, 21, 164, 208; **11,** 168, 466; **12,** 288, 413, 513
Suzuki, M., **1,** 16, 289; **3,** 322; **5,** 178, 214, 658; **6,** 158, 392; **9,** 458, 497, 498; **10,** 289, 391, 441; **11,** 67, 75, 172, 243, 514, 587; **12,** 546
Suzuki, N., **12,** 399
Suzuki, R., **1,** 52; **2,** 442; **11,** 395
Suzuki, S., **2,** 238; **3,** 265; **4,** 222; **5,** 602; **7,** 295, 428; **8,** 512; **10,** 258; **11,** 480, 568; **12,** 167
Suzuki, T., **2,** 241, 289; **3,** 208; **5,** 157; **10,** 289; **11,** 48, 99, 365; **12,** 437, 456
Suzuki, Y., **3,** 265; **6,** 244; **10,** 194; **11,** 175; **12,** 322
Svanholm, U., **4,** 532; **5,** 88
Svanholt, K. L., **5,** 357
Svedi, A., **1,** 228
Svensson, T., **2,** 463
Sviridov, A. F., **12,** 198
Svoboda, J. J., **4,** 329, 567; **5,** 116, 232, 310, 716; **6,** 24, 109
Swain, C. G., **3,** 157
Swain, C. J., **12,** 289
Swamer, F. W., **1,** 251, 371, 524, 1080; **4,** 397; **7,** 106; **12,** 403
Swaminathan, K., **8,** 175
Swaminathan, S., **1,** 685, 703, 730; **2,** 464; **7,** 385
Swan, G. A., **1,** 407, 927
Swan, J. M., **1,** 1139; **5,** 294
Swann, B. P., **3,** 289; **4,** 497; **5,** 657; **8,** 477
Swann, S., Jr., **1,** 1095
Swarts, F., **2,** 268; **3,** 78
Sweeley, C. C., **1,** 427
Sweeney, A., **4,** 293
Sweeny, J. C., **9,** 449
Swenson, J. S., **1,** 1203
Swenson, W., **9,** 480
Swenton, J. S., **2,** 118; **6,** 675; **7,** 192, 429; **8,** 281, 308, 502; **9,** 8, 40, 194, 520; **11,** 385
Swerdloff, M. D., **6,** 423; **8,** 364
Swern, D., **1,** 89, 187, 195, 465, 652, 667, 759, 952, 992, 1142, 1157, 1269, 1271; **2,** 122, 224, 310, 316; **3,** 279; **4,** 194; **5,** 266, 267; **6,** 115, 228, 230; **7,** 136; **8,** 161, 200, 438; **9,** 192, 391; **10,** 167
Swett, L. R., **1,** 281, 1081

Swierczewski, G., **2,** 110; **6,** 270
Swierczewski, S., **10,** 37
Swift, G., **2,** 224
Swigert, J., **5,** 693
Swindells, R., **1,** 865, 989
Swisher, J. V., **1,** 121, 935
Swithenbank, C., **1,** 999
Swoboda, J. A., **2,** 368
Swoboda, J. J., **5,** 631
Syhora, K., **2,** 372
Sykes, P. J., **2,** 65, 79, 355; **3,** 57
Sylvestre-Panthet, P., **7,** 281
Sym, G. J., **11,** 203
Synder, H. R., **4,** 437
Synerholm, M. E., **1,** 1141, 1292
Synnes, E., **2,** 173
Syper, L., **2,** 65, 369; **9,** 99; **10,** 348, 354
Syrdal, D. D., **4,** 531
Szabo, V., **5,** 499
Szantay, C., **2,** 142; **12,** 508
Szarek, W. A., **2,** 300; **3,** 215, 244; **5,** 470; **7,** 67, 238; **8,** 391; **12,** 420
Szarvasy, E., **1,** 633
Szary, A. C., **5,** 374
Szczepek, W. J., **9,** 343
Szczygielska-Nowosielska, A., **10,** 125
Szechner, B., **6,** 291
Szekely, I., **7,** 160; **8,** 259
Szelke, M., **1,** 748
Szeza, W., **9,** 361
Szilagyi, S., **6,** 454; **9,** 103
Szmant, H. H., **1,** 509, 1247; **2,** 211; **3,** 166; **6,** 27
Szmuszkovicz, J., **1,** 147, 180, 389, 703, 705, 707, 772, 974, 1122; **4,** 542
Szpilfogel, S., **1,** 1000
Szwarc, M., **4,** 287
Szyndler, R., **5,** 257

Taaffee, T. A., **9,** 324
Tabata, M., **6,** 25, 295; **7,** 101, 153
Taber, D. F., **7,** 21, 122; **8,** 408, 421; **9,** 377; **10,** 79; **11,** 434, 460; **12,** 425
Taber, T. R., **9,** 88
Tabern, D. L., **1,** 954
Taborsky, R. G., **1,** 444
Tabusa, R. W., **12,** 413
Tabushi, I., **3,** 217; **4,** 31, 80, 475; **5,** 257; **7,** 20; **8,** 135
Tachi, K., **6,** 653
Tachibana, K., **11,** 75
Tachibana, Y., **8,** 532
Tachimori, Y., **8,** 505
Tada, H., **2,** 177
Tada, M., **8,** 377; **9,** 372; **11,** 136
Tada, N., **12,** 110
Tadanier, J., **5,** 127
Taddia, R., **6,** 497, 560
Tadema, G., **7,** 80
Tadros, W., **6,** 503
Taeger, E., **1,** 672
Taemnefors, I., **6,** 148
Taft, D. D., **1,** 808
Taga, T., **9,** 176
Tagaki, K., **6,** 573
Tagaki, W., **2,** 101; **3,** 36; **5,** 363, 752; **9,** 66; **11,** 466
Tagami, H., **6,** 308
Tagliavini, E., **10,** 322; **11,** 276, 356; **12,** 571
Taguchi, H., **4,** 139; **5,** 200, 374, 398; **6,** 163, 222, 255; **7,** 90, 102; **8,** 1; **9,** 269
Taguchi, K., **4,** 345
Taguchi, T., **3,** 21, 183; **4,** 293; **5,** 13; **6,** 556; **8,** 127, 327; **10,** 110
Taguchi, Y., **6,** 207
Tahara, A., **1,** 603, 1273; **5,** 163
Tahir, A. R., **4,** 466
Tahk, F. C., **2,** 285
Tai, A., **9,** 406
Tai, W. T., **1,** 370
Taillefer, R. J., **9,** 429
Taimr, L., **6,** 656
Tait, S. J. D., **8,** 410
Taits, S. Z., **1,** 1103
Tajika, M., **9,** 201; **11,** 240
Tajima, K., **8,** 16; **12,** 282
Tajima, Y. A., **2,** 206
Takabe, K., **10,** 247; **12,** 57
Takacs, J. M., **10,** 332
Takada, H., **11,** 67
Takada, K., **3,** 15
Takagaki, T., **11,** 431
Takagi, K., **4,** 372; **9,** 322; **12,** 337
Takagi, M., **5,** 325, 498; **9,** 200
Takagi, S., **8,** 289; **10,** 318
Takagi, Y., **11,** 24
Takahagi, Y., **6,** 547
Takahashi, H., **1,** 250, 1217; **5,** 7, 165, 534, 749; **6,** 430; **7,** 81; **11,** 190, 231
Takahashi, K., **2,** 129, 241; **4,** 340, 407; **5,** 157, 190, 561; **6,** 170; **8,** 115; **10,** 107, 347; **12,** 327, 328, 444
Takahashi, M., **8,** 56, 381; **10,** 56, 281, 302; **12,** 89
Takahashi, N., **4,** 31; **6,** 504
Takahashi, R., **6,** 201
Takahashi, S., **5,** 146; **10,** 117; **11,** 162
Takahashi, T., **8,** 28, 377; **9,** 370, 386; **10,** 281, 298, 302; **11,** 317; **12,** 63, 447, 474, 562
Takahashi, T. T., **8,** 53, 367
Takahashi, Y., **4,** 240, 572; **5,** 13, 178; **6,** 573; **8,** 240; **9,** 46, 344; **11,** 243, 252, 431
Takahata, A., **5,** 441
Takahata, H., **11,** 407; **12,** 431
Takahatake, Y., **8,** 421, 454
Takai, I., **10,** 21
Takai, K., **6,** 106; **8,** 339; **10,** 142, 293, 391; **11,** 542; **12,** 63, 64, 137, 344
Takai, M., **12,** 199
Takai, Y., **8,** 29
Takaishi, N., **6,** 359, 526; **7,** 349
Takaki, K., **11,** 343, 421
Takaku, H., **5,** 447, 518
Takamatsu, N., **6,** 174
Takami, Y., **5,** 190; **6,** 95
Takamoto, T., **8,** 505

Takamura, N., **4,** 270; **6,** 307; **11,** 460
Takano, S., **3,** 145; **5,** 600; **9,** 308; **11,** 188, 267, 449; **12,** 527
Takano, T., **6,** 628
Takano, Y., **3,** 46; **5,** 713; **12,** 516
Takao, S., **11,** 518, 519
Taka-Oka, K., **12,** 449
Takasaka, N., **8,** 354
Takasaki, K., **9,** 296
Takase, K., **8,** 16, 155
Takase, M., **11,** 479
Takatani, M., **9,** 318; **11,** 94, 352
Takaya, H., **2,** 81; **3,** 20, 101; **4,** 35, 158; **5,** 224; **6,** 45; **8,** 498; **10,** 36; **11,** 48, 54; **12,** 57
Takaya, T., **1,** 9
Takayama, H., **9,** 341; **11,** 288; **12,** 456
Takayama, K., **4,** 293; **11,** 16
Takayanagi, H., **5,** 165; **6,** 430; **7,** 81; **8,** 119; **9,** 410; **10,** 103
Takazawa, O., **7,** 92
Takebayashi, M., **9,** 51
Takechi, H., **5,** 693
Takeda, A., **5,** 553; **7,** 260; **9,** 382; **10,** 306; **11,** 24; **12,** 279, 365, 432
Takeda, H., **2,** 347
Takeda, K., **1,** 104; **5,** 55, 713; **8,** 163; **9,** 208; **10,** 173; **11,** 227; **12,** 219
Takeda, M., **1,** 41; **4,** 90, 220; **5,** 748; **11,** 447; **12,** 454
Takeda, R., **9,** 25
Takeda, T., **9,** 187
Takeda, Y., **1,** 52
Takegami, Y., **4,** 465; **5,** 358, 625; **6,** 485, 486, 552; **7,** 183, 341; **8,** 42, 420; **9,** 108; **11,** 267
Takehara, H., **11,** 407
Takehira, Y., **6,** 146
Takei, H., **5,** 80, 321; **6,** 279; **8,** 327; **9,** 148, 414; **10,** 194
Takei, S., **6,** 534
Takemasa, Y., **12,** 563
Takemura, K. H., **10,** 187
Takenaka, H., **12,** 307
Takenaka, S., **12,** 543
Takenaka, Y., **6,** 216
Takenishi, T., **3,** 229
Takeshima, T., **4,** 395
Takeshita, H., **6,** 170
Takeshita, T., **5,** 381
Taketomi, T., **11,** 54; **12,** 57
Takeuchi, M., **11,** 235; **12,** 291
Takeuchi, Y., **11,** 33, 417, 468
Takeyama, M., **8,** 319
Takeyama, T., **12,** 430
Takigawa, T., **5,** 553; **6,** 216; **10,** 31
Takimoto, S., **7,** 320; **9,** 479
Takinami, S., **12,** 236
Takita, S., **10,** 24
Takiyama, N., **12,** 345
Takizawa, Y., **8,** 120
Takken, H. J., **7,** 124; **8,** 530
Takoya, H., **10,** 281
Talaty, C. N., **2,** 178; **3,** 131; **6,** 556
Talaty, E. R., **2,** 34; **10,** 391
Talbiersky, J., **8,** 226; **10,** 337
Taleb-Sahraoui, S., **11,** 562
Talies, G., **9,** 509
Talkowski, C. J., **7,** 20
Talley, B. G., **10,** 450
Talley, J. J., **9,** 128; **12,** 283
Talley, P., **8,** 238
Tam, C. C., **11,** 448
Tam, J. P., **11,** 256; **12,** 479
Tam, S. Y. K., **5,** 389; **7,** 196
Tam, W., **8,** 214; **11,** 401; **12,** 344
Tam, W. W., **11,** 488
Tamagaki, S., **7,** 359; **11,** 466
Tamaki, K., **12,** 452
Tamao, K., **5,** 361; **6,** 51, 59; **8,** 56, 101, 381, 523; **9,** 95, 201; **11,** 168, 240, 374; **12,** 29, 182, 245
Tamaoka, T., **12,** 392
Tamarkin, D., **12,** 400
Tamaru, Y., **4,** 475; **7,** 207; **8,** 105, 382, 385; **9,** 235, 309; **10,** 32, 299; **11,** 395; **12,** 51, 256
Tamblyn, W. H., **10,** 382; **11,** 502
Tambor, J., **4,** 15
Tamborski, C., **12,** 547
Tamm, C., **1,** 329, 624; **5,** 90; **11,** 359, 602
Tamura, C., **8,** 115
Tamura, F., **5,** 87
Tamura, M., **4,** 164, 236; **8,** 340, 386; **10,** 5
Tamura, N., **10,** 24, 39; **11,** 188
Tamura, R., **8,** 141; **10,** 413; **11,** 514, 551; **12,** 47, 266
Tamura, T., **7,** 220
Tamura, Y., **4,** 257; **5,** 433; **6,** 118, 233; **8,** 319, 519; **9,** 111, 310, 506, 507; **11,** 255, 279, 340; **12,** 59, 221, 449, 559
Tamura, Z., **2,** 159
Tamuro, Y., **10,** 238
Tamuru, Y., **6,** 396
Tan, S. I., **2,** 107
Tanabe, K., **1,** 754; **7,** 221; **8,** 231
Tanabe, M., **1,** 269; **2,** 344, 423; **3,** 262; **4,** 270; **5,** 394, 552; **8,** 167
Tanabe, T., **9,** 491
Tanabe, Y., **4,** 369; **12,** 173
Tanaka, H., **3,** 156; **9,** 23, 114; **10,** 31; **12,** 110
Tanaka, J., **12,** 57
Tanaka, K., **1,** 1002; **5,** 607; **6,** 65, 585; **7,** 89, 149; **8,** 315; **9,** 133, 166, 188, 216, 434; **10,** 39, 231, 261; **11,** 83
Tanaka, M., **4,** 465; **5,** 358, 486, 625; **6,** 552; **9,** 207, 435; **10,** 38, 126, 411; **11,** 269, 434, 470; **12,** 89, 172
Tanaka, N., **12,** 539
Tanaka, O., **12,** 282
Tanaka, S., **5,** 77, 200; **6,** 183, 586; **7,** 102; **12,** 241
Tanaka, T., **5,** 534; **6,** 165, 641; **10,** 71; **11,** 125, 600;

12, 175, 182, 245, 441, 573
Tanaka, Y., **7,** 221; **9,** 470; **10,** 404; **11,** 164, 365, 543; **12,** 171, 214, 449, 513
Tanba, Y., **6,** 443
Tancrede, J. M., **5,** 610; **6,** 539
Tane, J. P., **10,** 127
Taneja, H. R., **4,** 165; **11,** 445
Tanerede, J., **8,** 455
Tang, C. P., **6,** 166
Tang, C. S., **2,** 217
Tang, C. S. F., **6,** 2, 141, 229
Tang, P. C., **11,** 401; **12,** 136
Tang, P. W., **9,** 219, 283; **10,** 194
Tang, R., **5,** 55; **8,** 438
Tang, R. T., **5,** 146
Tang, W. P., **4,** 340, 411
Tangari, N., **6,** 74, 506; **7,** 315
Tanguy, L., **6,** 172
Tani, J., **6,** 70
Tani, K., **11,** 54; **12,** 57
Tanida, H., **6,** 502
Tanigawa, Y., **6,** 392; **8,** 346, 347; **9,** 269; **10,** 269; **11,** 342, 514
Taniguchi, H., **2,** 92; **3,** 69; **11,** 249
Taniguchi, N., **12,** 155
Taniguchi, Y., **12,** 59, 573
Tanigushi, R., **10,** 420
Tanikaga, R., **7,** 2, 149; **8,** 141; **9,** 188; **10,** 183; **11,** 155
Tanimoto, S., **10,** 176; **11,** 604
Tanimoto, Y., **9,** 170; **12,** 191
Tanino, H., **4,** 86
Tanio, M., **8,** 519
Tanioka, K., **7,** 252
Tanis, S. P., **9,** 77, 198; **11,** 191
Taniyama, E., **12,** 303
Tanizawa, K., **3,** 231
Tannenbaum, S., **1,** 1235
Tanner, D. D., **2,** 50; **3,** 165; **11,** 551
Tanner, H. A., **3,** 96
Tanno, N., **10,** 238; **12,** 508
Tao, E. V. P., **4,** 203
Tara, H., **7,** 361
Taraz, K., **4,** 230
Tarbell, D. S., **1,** 154, 164, 276, 367, 444, 610, 629, 636, 720, 787, 814, 1081, 1131, 1203; **2,** 45, 46, 378; **3,** 36, 261, 302; **4,** 128, 577; **8,** 57, 145; **9,** 193
Tarbin, J. A., **12,** 452
Tarchini, C., **6,** 175
Tardella, P. A., **3,** 299; **7,** 308; **8,** 425; **10,** 333
Tardivat, J. C., **6,** 359
Tarelli, E., **5,** 436
Tarhouni, T., **12,** 77
Tark, S. Y., **5,** 127
Tarka, S. M., **5,** 641
Tarlton, E. J., **1,** 952
Tarnow, H., **4,** 524
Tarnowski, B., **9,** 515
Tarnowski, T. L., **9,** 361
Tarrant, P., **2,** 389
Tarrat, H. J. F., **2,** 21
Tarzia, G., **3,** 147, 262; **4,** 37; **7,** 158
Tasai, H., **6,** 232
Tasaka, K., **1,** 208
Taschner, E., **1,** 83, 486, 617, 1131, 1184
Taschner, M. J., **8,** 260; **10,** 351, 418; **11,** 481
Tashiro, M., **1,** 329
Tashlick, I., **7,** 23
Tasker, P. A., **5,** 230
Tassel, M., **4,** 288
Taster, D. G., **5,** 276
Tatake, J. G., **8,** 483
Tatchell, A. R., **2,** 245
Tate, B. E., **1,** 7
Tate, D. P., **1,** 581
Tate, M. E., **1,** 180, 1081
Taticchi, A., **5,** 624; **9,** 13
Tatlow, C. E. M., **1,** 1226
Tatlow, J. C., **1,** 1226; **4,** 533
Tatsuda, M., **11,** 47
Tatsuno, T., **7,** 225, 338; **9,** 269; **11,** 283
Tatsuno, Y., **8,** 40; **11,** 47
Tatsuta, K., **9,** 382
Tatum, C. M., Jr., **5,** 470
Taub, B., **1,** 76, 1011
Taub, D., **1,** 37, 59, 160, 389, 536, 808, 933, 974, 1087, 1171, 1254; **2,** 159, 252, 434; **9,** 305, 318
Taub, W., **5,** 357
Taubert, R., **5,** 329
Taufer, I., **12,** 241
Tautz, W., **9,** 184
Tavares, D. F., **5,** 552
Tavernari, D., **8,** 122
Tavs, P., **1,** 151
Tawara, Y., **5,** 560
Tawarayama, Y., **11,** 114
Tawney, P. O., **1,** 169
Taya, K., **2,** 228; **11,** 487
Taya, S., **11,** 103
Tayal, S. R., **7,** 298
Taylor, A. R., **6,** 304
Taylor, B. J., **9,** 116
Taylor, B. S., **12,** 538
Taylor, D. A., **4,** 473
Taylor, E. A., **4,** 228, 344; **5,** 444; **8,** 334
Taylor, E. C., **1,** 54, 257, 345, 404, 465, 870, 1073, 1082, 1269; **2,** 129, 202, 406, 411; **3,** 22, 43, 286, 289; **4,** 282, 460, 492, 497, 501, 503, 527; **5,** 213, 253, 655, 657, 659, 701; **6,** 581; **7,** 361, 365; **8,** 90, 476, 477, 480, 481, 508; **9,** 462, 463, 493; **10,** 397; **11,** 123; **12,** 483
Taylor, E. G., **9,** 438
Taylor, E. J., **7,** 335
Taylor, E. R., **1,** 452, 876, 951; **4,** 309
Taylor, F. B., **7,** 118
Taylor, G. A., **11,** 350
Taylor, G. F., **6,** 524; **7,** 147; **8,** 35
Taylor, G. N., **2,** 26; **3,** 129; **4,** 239
Taylor, H. M., **10,** 159
Taylor, H. T., **3,** 219
Taylor, H. W., **4,** 189
Taylor, J. A., **2,** 295
Taylor, J. B., **10,** 181
Taylor, J. W., **1,** 563

Taylor, K. G., **5,** 693
Taylor, L. D., **1,** 134; **12,** 223
Taylor, M. D., **1,** 964, 966; **5,** 411
Taylor, M. E., **1,** 72
Taylor, R., **5,** 57, 257; **6,** 91
Taylor, R. J. K., **8,** 171; **10,** 94
Taylor, R. K., **12,** 263
Taylor, R. T., **6,** 70; **8,** 344, 493
Taylor, S. C., **3,** 51
Taylor, S. K., **10,** 135
Taylor, W. C., **2,** 214; **4,** 282
Taylor, W. I., **1,** 563
Taylor-Papadimitriou, J., **2,** 178
Taymaz, K., **5,** 391
Tazaki, M., **9,** 200; **12,** 213
Tazuma, J. J., **1,** 164
Tchistow, W. O., **2,** 39
Tchoubar, B., **1,** 1131; **2,** 210; **9,** 502
Teague, P. C., **1,** 872
Teague, S. J., **11,** 491
Tebbe, F. N., **6,** 48; **8,** 84
Tedder, J. M., **1,** 1179, 1226
Tedeschi, R. J., **1,** 186
Teeter, H. M., **1,** 93
Teeter, L. A., **3,** 81
Tefertiller, B. A., **2,** 359
Tegner, C., **1,** 689
Tehan, F. J., **5,** 156
Teicher, B. A., **7,** 143; **8,** 514
Teichmann, H., **4,** 197
Teige, B., **5,** 495
Teitel, S., **4,** 43, 250, 330; **7,** 64
Teixeira, H. L. S., **6,** 339
Tekling, C., **2,** 263, 295
Telang, S. G., **6,** 588
Telegina, N. I., **1,** 636
Telford, R. P., **8,** 451; **10,** 359
Telinski, T. J., **1,** 1090
Teller, D. M., **1,** 249
Telly, D. L., **7,** 11
Telschow, J. E., **7,** 136; **8,** 208
Temme, G. H., III, **5,** 495
Temple, A. F., **7,** 345
Temple, C., Jr., **1,** 244
Temple, D. L., **3,** 15, 314; **5,** 621, 715; **6,** 20; **8,** 279
Temple, J. S., **11,** 395
Temple, P., **5,** 399
Templeton, J. F., **1,** 927; **2,** 146
Tenaglia, A., **12,** 53
TenBrink, R. E., **6,** 161, 588
Tener, G. M., **1,** 173; **2,** 228
ten Haken, P., **5,** 465
Terabe, S., **2,** 273
Terada, A., **2,** 299
Teramoto, T., **8,** 252
Teramura, K., **12,** 508
Teranishi, A. Y., **4,** 99, 413; **5,** 522, 764; **8,** 112; **10,** 312
Teranishi, R., **1,** 37, 983
Teranishi, S., **6,** 450, 653; **7,** 45, 277, 314; **8,** 526; **10,** 457; **11,** 91; **12,** 89, 94
Terao, K., **12,** 41
Terao, S., **6,** 436; **9,** 15; **12,** 190
Terasaki, T., **8,** 59
Terasawa, I., **3,** 60, 214; **6,** 215
Terasawa, M., **6,** 450; **7,** 277
Terasawa, S., **12,** 538
Terasawa, T., **1,** 1198; **4,** 147
Terashima, M., **11,** 70
Terashima, S., **4,** 63, 77, 472, 479; **5,** 160; **7,** 17; **8,** 197, 320, 403, 424; **9,** 393; **10,** 238, 332; **11,** 447; **12,** 322, 546
Teratani, S., **11,** 24, 466
Terauchi, M., **10,** 261
Terawaki, Y., **9,** 353
Terem, B., **9,** 35
Terent'cv, A. P., **1,** 334
ter Horst, I., **1,** 834
Termine, E. J., **10,** 324; **11,** 75
Terpinski, J., **12,** 462
Terpstra, J. W., **12,** 224
Terrell, R., **1,** 707, 974
Terrett, N. K., **12,** 212
Terry, P. H., **1,** 85, 210
Terry, W. G., **1,** 703
Terzeau, H., **1,** 904
Teschner, M., **6,** 238
Tesser, G. I., **1,** 112, 486
Tessier, J., **1,** 215, 518
Testaferri, L., **11,** 166, 203, 473
Tetaz, J. R., **1,** 430, 1102
Teter, L. A., **1,** 1192; **2,** 283, 400; **4,** 517
Tetlow, A. J., **1,** 506, 510
Tette, J. P., **9,** 402
Teuben, J. H., **8,** 40
Teuber, H. J., **1,** 942; **2,** 348; **4,** 411
Teunissen, H. P., **1,** 566
Texier, F., **6,** 472
Texier-Boullet, F., **11,** 24
Teyssie, P., **5,** 497, 572; **7,** 76, 313; **8,** 435; **10,** 342; **11,** 460
Thacher, A. F., **1,** 595; **2,** 143, 242
Thain, E. M., **1,** 881
Thaisrivongs, S., **8,** 284; **12,** 303
Thaler, W., **1,** 93
Thaler, W. A., **1,** 247; **4,** 215
Thalmann, A., **6,** 247, 520, 625; **7,** 142
Thamnusan, P., **9,** 271
Thanh, T. N., **6,** 574
Thatcher, A. F., **3,** 176
Thatcher, D. N., **5,** 10
Thatte, S. D., **1,** 85
Thayer, A. L., **5,** 491; **6,** 469
Thayer, F. K., **1,** 452, 1027; **11,** 78
Thea, S., **12,** 509
Thebtaranonth, C., **10,** 394
Thebtaranonth, Y., **8,** 367; **9,** 21
Thedford, R., **1,** 600, 737
Theilacker, W., **1,** 125; **2,** 67
Theilig, G., **1,** 403
Theiling, L. F., **5,** 534
Theis, A. B., **10,** 106
Theis, R. J., **2,** 268
Theis, R. W., **5,** 507
Theisen, D., **3,** 19
Thenn, W., **5,** 537
Theobald, D. W., **1,** 526; **2,** 238; **3,** 49
Theobald, J. M., **1,** 296
Theodoridis, G., **11,** 282
Theodoropoulos, D., **1,** 737,

1256
Theodoropulos, S., **5,** 230
Theopold, K. H., **11,** 137
Theriault, N. Y., **8,** 59, 82; **9,** 77
Therien, M., **10,** 216
Therien, M. J., **12,** 251
Thesing, J., **9,** 402
Thewalt, I., **1,** 243, 1150
Thieffry, A., **9,** 141; **11,** 501; **12,** 548
Thiele, J., **1,** 22, 45, 147, 219, 444, 536, 850, 909; **2,** 9, 10; **3,** 339; **4,** 579
Thielecke, W., **4,** 140, 317
Thielke, D., **6,** 2
Thiem, J., **10,** 350
Thiem, K. W., **5,** 120
Thieme, H. K., **4,** 389
Thier, W., **1,** 258; **4,** 535; **5,** 433
Thierry, J., **12,** 417
Thies, I., **9,** 322; **11,** 355
Thies, R. W., **7,** 303
Thijs, L., **5,** 127, 495, 516
Thilo, E., **1,** 894
Thind, S. S., **11,** 502
Thio, J., **4,** 216; **5,** 587
Thio, P. A., **2,** 107
Thio, T., **4,** 293
Thirasasna, N., **8,** 367
Thiullier, A., **11,** 318
Thivolle-Cazat, J., **11,** 253
Thobe, J., **2,** 179
Thom, I. G., **10,** 223
Thoma, K., **5,** 39
Thomas, A., **1,** 643; **2,** 263
Thomas, A. B., **1,** 705
Thomas, A. F., **4,** 361; **5,** 631; **7,** 184
Thomas, A. M., **4,** 45
Thomas, B. R., **1,** 444, 802
Thomas, C. B., **2,** 159; **5,** 102, 370, 659; **6,** 443; **11,** 268
Thomas, D., **5,** 127
Thomas, D. A., **1,** 561
Thomas, E., **11,** 500; **12,** 418
Thomas, E. G., **1,** 281, 705; **2,** 270
Thomas, E. J., **5,** 399; **10,** 21; **11,** 326
Thomas, E. W., **7,** 320; **8,** 284, 442
Thomas, G. G., **1,** 518
Thomas, G. H., **1,** 511, 526
Thomas, G. H. S., **1,** 122; **2,** 165
Thomas, G. J., **11,** 410
Thomas, G. M., **2,** 263
Thomas, J. A. G., **2,** 190
Thomas, J. O., **2,** 126
Thomas, L. H., **4,** 25
Thomas, L. L., **5,** 601
Thomas, M. J., **2,** 309
Thomas, M. T., **5,** 614; **7,** 304
Thomas, O. H., **1,** 581, 754
Thomas, P. D., **1,** 652
Thomas, P. J., **12,** 425
Thomas, P. L., **1,** 181
Thomas, R., **5,** 740
Thomas, R. C., **7,** 181; **10,** 71
Thomas, S. L. S., **1,** 782
Thomas, S. P., **4,** 350
Thomas, T. M., **2,** 400
Thomas, W. R., **4,** 63
Thomason, S. C., **11,** 469
Thomassin, R., **5,** 669
Thome, C., **1,** 349
Thommen, W., **6,** 215, 675; **11,** 472
Thompson, A., **1,** 652
Thompson, A. R., **2,** 438
Thompson, C. M., **1,** 393
Thompson, D., **1,** 606; **2,** 397
Thompson, D. J., **6,** 552
Thompson, E. A., **5,** 75
Thompson, G., **1,** 1253
Thompson, G. W., **4,** 406
Thompson, H., **9,** 377; **11,** 491
Thompson, H. G., **10,** 164
Thompson, H. T., **1,** 195, 713, 1049; **3,** 74
Thompson, H. W., **2,** 133, 335
Thompson, J. G., **2,** 96, 457
Thompson, J. L., **1,** 93; **2,** 134, 362; **8,** 259; **9,** 249
Thompson, M., **10,** 34
Thompson, M. D., **6,** 137
Thompson, M. J., **2,** 368
Thompson, P., **1,** 1030
Thompson, P. A., **11,** 87; **12,** 397, 426
Thompson, P. G., **4,** 37
Thompson, Q. E., **1,** 957, 963, 1235; **3,** 324
Thompson, R., **1,** 570
Thompson, R. B., **1,** 12, 178; **11,** 2
Thompson, R. C., **10,** 85
Thompson, R. H., **2,** 174, 344
Thompson, W. J., **9,** 286; **12,** 537
Thomsen, I., **6,** 558; **9,** 50
Thomson, J. B., **2,** 208; **3,** 36, 140; **4,** 53
Thomson, P. J., **1,** 1238
Thomson, R. H., **1,** 685, 1029
Thomson, S. J., **1,** 1145
Thoren, S., **5,** 395
Thorne, M. P., **7,** 346
Thorpe, F. G., **4,** 126; **6,** 581
Thorpe, J. F., **1,** 182, 935
Thorpe, W. D., **7,** 138
Thorsen, M., **11,** 55
Thorsen, P. T., **11,** 479
Thorsett, E. D., **11,** 276
Threlfall, T., **1,** 927
Throckmorton, J. R., **2,** 351
Thuan, S. L. T., **4,** 427; **6,** 514
Thuillier, A., **2,** 226; **3,** 165; **5,** 182, 250, 257; **6,** 300; **9,** 95; **11,** 113
Thummel, R. P., **7,** 188; **10,** 245; **12,** 421
Thurman, J. C., **1,** 289
Thurman, N., **1,** 654
Thweatt, J. G., **4,** 172; **5,** 160
Thyagarajan, B. S., **1,** 176, 933; **4,** 174
Thyagarajan, G., **2,** 263
Tibbetts, F., **1,** 85
Tiberi, R., **1,** 269
Tichy, M., **1,** 1258; **3,** 316
Ticozzi, C., **2,** 173; **7,** 133; **12,** 510
Tideswell, J., **6,** 341
Tidwell, T. T., **6,** 589; **9,** 87
Tiecco, M., **11,** 166, 203, 473
Tieckelmann, H., **1,** 1073, 1138; **3,** 285
Tiedeman, T., **1,** 658

Tieman, C. H., **1,** 1226
Tien, J. M., **1,** 310, 355
Tiensripojamarn, A., **9,** 68
Tierney, P. A., **1,** 68, 206; **4,** 130
Tietze, B. M., **8,** 365
Tietze, L. F., **4,** 65; **7,** 339; **8,** 338; **11,** 37, 587
Tiffany, B. D., **1,** 611
Tigchelaar, M., **11,** 13
Tijhuis, M. W., **9,** 156
Tilak, B. D., **5,** 731
Tilak, M. A., **6,** 307
Tilden, W. A., **1,** 754
Tilley, J. N., **2,** 101
Tilley, J. W., **3,** 199, 258
Tillman, P., **1,** 364
Tillmanns, E. J., **3,** 284
Tillotson, A., **1,** 152
Timar, E., **8,** 132
Timberlake, J. F., **8,** 138
Timberlake, J. W., **4,** 60, 455; **5,** 78
Timko, J. M., **5,** 103; **8,** 356; **9,** 321
Timmer, E. C., **5,** 340
Timmermans, G. J., **6,** 575
Timmons, C. J., **1,** 87
Timmons, M. L., **3,** 308
Timmons, R. J., **1,** 258, 909; **4,** 215
Timms, G. H., **4,** 507
Timms, R. N., **7,** 277
Timpe, W., **12,** 444
Tin, K. C., **3,** 276
Tinapp, P., **5,** 607
Tiner-Harding, T., **11,** 152
Ting, P. C., **12,** 458
Tingoli, M., **11,** 166, 203, 473
Tinker, H. B., **5,** 46
Tinsley, S. W., Jr., **5,** 62
Tinti, D. S., **1,** 78
Tinucci, L., **12,** 502
Tio, C. O., **1,** 595
Tiollais, R., **4,** 283
Tipson, R. S., **1,** 140, 1090; **2,** 89; **4,** 411
Tipton, C. D., **3,** 131
Tirpak, J. G., **6,** 133; **10,** 110
Tirpak, R. E., **11,** 604
Tishler, M., **1,** 195, 470, 476, 642, 652, 713, 952, 962, 1000, 1033, 1049, 1073, 1178, 1284; **3,** 74; **8,** 463
Tisler, M., **1,** 370; **5,** 254; **8,** 483; **9,** 472; **10,** 274
Titov, M. I., **1,** 524
Titus, R. L., **2,** 107
Tius, M. A., **8,** 316; **9,** 296, 493; **10,** 169, 172; **11,** 533; **12,** 310
Tiwari, K. P., **9,** 95
Tkatchenko, I., **11,** 253
T'Kint, C., **9,** 105
Toba, M., **12,** 202
Tobe, Y., **9,** 269
Tobey, S. W., **1,** 1108; **2,** 34; **6,** 553
Tobias, M. A., **5,** 394
Tobiki, H., **12,** 508
Tobin, P. S., **10,** 396
Tobinaga, S., **4,** 237
Tobler, E., **1,** 353
Tocanne, J. F., **1,** 838
Tocco, G., **1,** 1123
Tochtermann, W., **3,** 254; **8,** 306
Toczek, J., **11,** 134
Toda, F., **3,** 119; **6,** 146; **7,** 89, 430; **11,** 83
Toda, H., **7,** 110
Toda, M., **3,** 337; **5,** 756
Toda, S., **11,** 544
Todd, A. H., **9,** 139
Todd, A. R., **1,** 53, 199, 235, 352, 881, 1256; **6,** 43; **7,** 106, 353; **12,** 492
Todd, C. W., **9,** 193
Todd, D., **1,** 444, 536
Todd, H. R., **1,** 267, 1073
Todd, L. J., **1,** 854
Todd, M., **11,** 67, 475
Todd, M. J., **1,** 1216
Toder, B. H., **6,** 70; **10,** 55, 356; **11,** 299
Todeschini, R., **12,** 226
Todesco, P. E., **4,** 477; **5,** 24, 556, 595, 644; **10,** 318
Toehl, A., **3,** 164
Toeke, L., **2,** 142
Toekes, L., **2,** 34, 153
Toeldte, W., **1,** 1256
Toemoeskoezi, I., **6,** 645; **7,** 160; **8,** 260
Toepel, T., **1,** 185
Toft, P., **3,** 154
Togo, H., **10,** 220, 450; **11,** 428
Toh, H. T., **4,** 37; **7,** 158
Tohda, Y., **6,** 61; **8,** 152; **12,** 476
Tohma, M., **5,** 307, 308, 340; **6,** 304
Toi, H., **6,** 64; **7,** 31, 201; **11,** 88, 295; **12,** 446
Toi, K., **2,** 94
Toki, T., **12,** 167
Tokita, S., **6,** 547
Tokitoh, N., **12,** 86
Tokoroyama, T., **6,** 529; **7,** 329; **12,** 502
Tokue, R., **8,** 163
Tokumaru, K., **1,** 796
Tokuno, E., **8,** 127; **10,** 110
Tokura, H., **5,** 58
Tokura, N., **1,** 756; **3,** 299; **5,** 486; **6,** 296
Tokuyama, T., **2,** 387; **11,** 292
Tolbert, B. M., **8,** 466
Tolgyesi, W. S., **1,** 44, 693; **8,** 6
Tolkiehn, K., **9,** 57
Tolkmith, H., **6,** 115
Tolksdorf, S., **1,** 1000; **5,** 552
Tollens, B., **1,** 566, 1171
Tolman, C. A., **6,** 654
Tolstikov, G. A., **4,** 20; **9,** 343
Tom, T., **8,** 446
Toma, C., **7,** 409
Tomari, M., **4,** 149
Tomasello, J. A. V., **8,** 254
Tomasi, G. E., **1,** 619
Tomasi, R. A., **8,** 218
Tomasik, W., **10,** 292
Tomasini, C., **12,** 87
Tomasz, J., **1,** 34
Tomasz, M., **1,** 360, 642, 1041
Tomboulian, P., **1,** 1256; **2,** 454
Tombret, F., **11,** 583
tom Dieck, H., **12,** 157
Tomer, K., **2,** 422
Tomesch, J. C., **8,** 126; **9,** 46

Tomi, K., **6,** 552
Tominaga, M., **5,** 553
Tomino, I., **9,** 170; **12,** 191
Tomioka, H., **10,** 142, 293; **11,** 99, 114, 194, 365
Tomioka, K., **7,** 325; **9,** 1; **10,** 230; **12,** 563
Tomisawa, S., **4,** 475
Tomita, H., **6,** 451
Tomita, K., **11,** 231
Tomita, M., **11,** 435
Tomita, S., **4,** 101; **5,** 150, 581
Tomita, T., **5,** 307, 340; **6,** 304
Tomiyoshi, N., **11,** 445
Tomizawa, K., **4,** 372; **10,** 305
Tomlin, J. E., **1,** 731
Tommila, E., **1,** 309
Tomoda, S., **10,** 226; **11,** 33, 417, 468
Tomoeda, M., **1,** 905
Tomoi, M., **10,** 198
Tong, Y. C., **9,** 11
Tonnesen, G. L., **10,** 211
Tonnis, J. A., **6,** 146
Tonooka, K., **1,** 686
Top, S., **10,** 14; **12,** 36
Topgi, R. S., **12,** 523
Topliss, G. B., **2,** 462
Topp, A., **3,** 69, 86
Tor, Y., **11,** 5
Tordeux, M., **6,** 70
Tori, K., **5,** 713
Torigoe, M., **1,** 731
Torii, I., **2,** 289
Torii, S., **5,** 42; **9,** 23, 114, 133; **12,** 110
Torimitsu, S., **3,** 214
Torisawa, Y., **9,** 393; **10,** 414; **11,** 552; **12,** 109
Toriumi, K., **10,** 36
Torkelson, S., **7,** 400; **8,** 299
Tornare, J. M., **12,** 61
Toromanoff, E., **1,** 215; **5,** 266
Torre, G., **3,** 291
Torrence, P. F., **11,** 599
Torres, L. E., **12,** 322
Torres, M., **5,** 163
Torri, G., **5,** 109
Torssell, K., **2,** 163; **9,** 19
Torssell, K. B. G., **12,** 29
Tortorella, V., **6,** 506; **7,** 315
Tortorello, A. J., **7,** 208
Toru, T., **3,** 260; **6,** 165, 247, 641; **8,** 533
Torupka, E. J., **2,** 257; **4,** 63
Toshida, Y., **7,** 81
Toshimitsu, A., **8,** 120; **9,** 35; **10,** 21, 106, 368; **11,** 397, 455; **12,** 41, 121
Toshimoto, A., **11,** 37
Tosk, E., **3,** 155
Toso, R., **11,** 124
Tosolini, G., **1,** 1226
Totter, J. R., **1,** 163, 402
Totzauer, W., **12,** 168
Tou, J. S., **10,** 51
Toube, T. P., **6,** 514
Toubiana, R., **2,** 79
Tour, J. J. M., **11,** 299
Touster, O., **6,** 423
Touzin, A. M., **5,** 219; **7,** 108; **8,** 222; **9,** 274; **12,** 183
Towles, H. V., **1,** 881, 1141
Townsend, C. A., **10,** 106; **11,** 589
Townsend, J. M., **4,** 204, 305; **6,** 48, 573; **9,** 260
Townsend, L. B., **9,** 306
Toy, A. D. F., **6,** 215; **9,** 187
Toy, M. S., **2,** 175
Toyama, S., **11,** 44
Toyama, T., **5,** 441; **8,** 188
Toye, J., **6,** 123; **12,** 125
Toyne, K. J., **1,** 562; **3,** 131, 171
Toyoda, T., **9,** 63, 152; **11,** 223; **12,** 179
Toyofuku, H., **11,** 75
Toyotani, S., **9,** 269
Tozune, S., **8,** 188
Traas, P. C., **5,** 253; **7,** 124; **8,** 530
Trabert, H., **5,** 286
Trachtenberg, E. N., **2,** 75
Traenckner, H. J., **2,** 271
Trahanovsky, W. S., **1,** 121; **2,** 65, 455; **3,** 45; **4,** 74; **5,** 102; **6,** 620; **11,** 596
Trail, M. M., **1,** 524
Trainor, J. T., **8,** 290
Traitler, H., **6,** 503
Trammell, G. L., **5,** 98; **7,** 345
Tramontano, A., **8,** 334, 403; **10,** 49, 321; **11,** 207
Trangle, K., **5,** 342
Trangle, M., **5,** 342
Trapani, G., **12,** 65
Trapasso, L. E., **1,** 207, 1190
Trapp, H., **5,** 120; **7,** 47
Trask, E. G., **2,** 282
Traub, H., **1,** 1252
Traube, W., **1,** 1128
Trautz, V., **4,** 270
Trave, R., **7,** 1, 331
Traxler, J. T., **8,** 532
Traxler, R. N., **1,** 350
Traylor, T. G., **1,** 652; **5,** 427; **6,** 640
Traynard, J. C., **1,** 952
Traynard, P., **5,** 669
Traynelis, V. J., **1,** 309, 310, 318; **5,** 641; **7,** 135, 139
Traynham, J. G., **2,** 247, 344
Traynor, S. G., **10,** 405
Trecker, D. J., **3,** 7, 194; **5,** 152, 571
Trefonas, L. M., **5,** 78
Trego, B. R., **3,** 320
Trehan, I. R., **2,** 426
Treiber, A. J. H., **1,** 854
Treibs, A., **4,** 466
Treibs, W., **1,** 181, 476, 652, 767, 842, 885; **2,** 430
Tremble, J., **2,** 153
Tremelling, M. J., **8,** 408
Tremont, S. J., **12,** 370
Tremper, A., **8,** 261
Tremper, A. W., **8,** 102; **9,** 428
Trenbeath, S., **7,** 32; **8,** 417, 458; **9,** 233
Trend, J. E., **7,** 120; **8,** 25
Trenkle, R. W., **4,** 372
Trenner, N. R., **1,** 293, 1081
Trepanier, D. L., **5,** 127
Treptow, W., **10,** 123, 289
Trescher, V. J., **2,** 21
Tresper, E., **5,** 167
Trevoy, L. W., **1,** 383, 595, 1073
Tribo, M., **4,** 45
Trickes, G., **8,** 465

Triebel, C. A., **11,** 203
Trifilova, L., **8,** 226
Trifonov, L. S., **9,** 503
Triggle, D. J., **3,** 224
Trimitsis, G. B., **5,** 652
Trinor, D. A., **11,** 128
Triplett, J., **4,** 45
Tripoulas, N. A., **9,** 387
Trippett, S., **1,** 1247; **4,** 573; **5,** 269, 294; **6,** 268; **9,** 481; **12,** 234
Trischmann, H., **1,** 295, 309, 318, 685, 1236; **4,** 24; **8,** 509
Trisler, J. C., **5,** 249
Trister, S. M., **1,** 818
Tristram, E. W., **1,** 491, 676
Trivedi, J. P., **3,** 266
Troesken, O., **1,** 243, 1150
Troffkin, H. J., **2,** 268
Trofimenko, S., **1,** 262
Trofimov, B. A., **11,** 440
Troger, J., **9,** 216
Troisi, L., **11,** 442
Trolliet, M., **6,** 557
Trombini, C., **8,** 406; **9,** 378, 477; **10,** 297, 322; **11,** 276, 356; **12,** 571
Trommer, W., **5,** 342
Trompenaars, W. P., **9,** 165, 177
Tronchet, J., **3,** 249
Tronchet, J. M. J., **3,** 249
Trong, A. N., **5,** 693
Tronich, W., **2,** 173; **5,** 514, 724
Trope, A. F., **10,** 454
Trost, B. M., **1,** 526, 1256, 1258; **2,** 173, 278, 439, 442; **3,** 134, 258; **4,** 63, 101, 214, 255, 370, 469; **5,** 68, 73, 247, 262, 277, 281, 373, 416, 502, 503, 505, 747; **6,** 32, 118, 155, 218, 238, 243, 316, 320, 369, 395, 396, 450, 464, 466, 662; **7,** 41, 122, 137, 188, 190, 233, 327, 358; **8,** 39, 52, 105, 140, 148, 210, 214, 217, 352, 356, 392, 476; **9,** 108, 175, 269, 270, 321, 458; **10,** 14, 32, 120, 165, 211, 240, 303, 328, 383, 384, 390, 391; **11,** 31, 66, 82, 96, 108, 118, 159, 166, 206, 260, 301, 514, 579; **12,** 29, 37, 61, 78, 208, 218, 246, 252, 288, 333, 344, 345, 475, 557
Trouliman, H. D., **1,** 430
Troxler, F., **1,** 942; **6,** 2
Trozzolo, A. M., **1,** 643
Trpin, J., **2,** 159
Truce, W. E., **1,** 102, 581, 720, 1127, 1131; **3,** 221; **4,** 174; **5,** 373, 478; **6,** 377; **7,** 237; **8,** 140, 245; **9,** 8, 67, 334; **11,** 219
Trucker, D. E., **1,** 188
Trudell, M. L., **12,** 224
Trueb, W., **5,** 538
Truempy, F., **1,** 951
Truesdale, L. K., **4,** 77, 543; **5,** 207, 722; **6,** 399, 642; **7,** 318; **8,** 352
Truett, M. L., **1,** 964
Trumbull, E. R., **1,** 611
Trumbull, P. A., **1,** 611
Trus, B., **5,** 689
Truscheit, E., **1,** 190; **2,** 19; **4,** 119
Trust, R. I., **4,** 527
Trybulski, E. J., **7,** 91; **8,** 87
Tsai, A., **5,** 561
Tsai, D. J. S., **11,** 573; **12,** 96
Tsai, L., **1,** 876
Tsai, M., **7,** 103
Tsai, M. Y., **8,** 176
Tsai, T. Y. R., **10,** 188; **11,** 325, 382
Tsai, Y. M., **11,** 550; **12,** 350, 523
Tsamo, E., **9,** 364
Tsang, W. G., **8,** 305
Tsangaris, J., **1,** 737
Tsao, J. H., **8,** 471
Tsaroom, S., **9,** 418
Tsay, Y., **7,** 322; **8,** 458
Tschesche, H., **5,** 439
Tschesche, R., **1,** 814; **2,** 139; **4,** 345
Tschitschibabin, A. E., **4,** 87; **7,** 139
Tschudi, G., **1,** 444, 927; **7,** 349
Tschudi, W., **5,** 580
Tseng, C., **5,** 319
Tseng, C. C., **11,** 188
Tseng, C. K., **5,** 537
Tseng, C. P., **11,** 123
Tsizin, Y. S., **8,** 115
Tsou, H. R., **11,** 175
Tsou, T. T., **10,** 59
Tsubata, K., **9,** 465; **10,** 310
Tsuboi, S., **5,** 553; **7,** 382; **9,** 382; **11,** 24; **12,** 279, 432
Tsuboniwa, N., **12,** 344, 524
Tsuchida, K., **5,** 538
Tsuchihashi, G., **4,** 467, 475, 514; **5,** 4, 457, 683; **6,** 6, 392, 446; **8,** 345; **9,** 68, 314; **10,** 323; **11,** 109, 243; **12,** 29, 288, 513
Tsuchihashi, G. I., **4,** 342
Tsuchihashi, K., **7,** 349
Tsuchimoto, M., **7,** 3
Tsuchiya, H. M., **4,** 366
Tsuchiya, T., **9,** 188
Tsuda, K., **1,** 1283; **5,** 756; **6,** 675
Tsuda, M., **10,** 304
Tsuda, T., **4,** 107, 109; **5,** 171; **6,** 130, 145; **7,** 74; **9,** 123, 279; **10,** 237
Tsuda, Y., **6,** 534; **8,** 453; **10,** 252
Tsuge, O., **12,** 316, 539, 543
Tsuge, S., **12,** 201
Tsugoshi, T., **12,** 446, 449
Tsui, F. P., **6,** 273
Tsuitani, K., **11,** 407
Tsuji, T., **1,** 140, 603, 1252; **2,** 67, 304, 305, 453; **3,** 182, 329; **5,** 165; **7,** 81, 278; **8,** 28, 119, 160, 382, 486, 512; **9,** 156, 173, 328, 350, 370, 386; **10,** 103, 150, 176, 258, 281, 298, 302, 417; **11,** 349, 391, 392, 396, 514; **12,** 16, 370, 447, 562
Tsuji, N., **1,** 322

Tsuji, T., **1,** 310, 732; **2,** 94, 163; **6,** 430; **9,** 13, 42, 106, 519; **11,** 514; **12,** 121
Tsuji, Y., **11,** 172; **12,** 180
Tsujihara, K., **5,** 639; **6,** 369
Tsujimoto, K., **9,** 207
Tsujino, Y., **9,** 205
Tsujitani, R., **8,** 40
Tsukamoto, A., **1,** 380, 595, 610, 625
Tsukamoto, M., **12,** 241, 502
Tsukanaka, M., **5,** 333; **6,** 220
Tsukanaka, T., **12,** 399
Tsukayama, M., **11,** 480
Tsukui, N., **11,** 551
Tsumaki, H., **11,** 489
Tsumiyama, T., **12,** 51
Tsuno, Y., **5,** 476; **11,** 470
Tsunoda, J. N., **2,** 147
Tsunoda, M., **6,** 118
Tsunoda, T., **9,** 498; **10,** 441
Tsurata, H., **10,** 236
Tsurugi, J., **4,** 526
Tsuruta, N., **3,** 212
Tsushima, T., **8,** 6
Tsutsui, H., **12,** 553
Tsutsui, M., **8,** 225
Tsutsumi, K., **6,** 662; **9,** 516
Tsutsumi, S., **2,** 293; **3,** 156, 212, 247; **4,** 158, 302, 354, 422; **5,** 93, 341; **8,** 345
Tsuyuki, T., **8,** 377
Tubiani, W., **4,** 67
Tuchinda, P., **9,** 161
Tucker, B., **12,** 204
Tucker, S. H., **1,** 169
Tudball, J. K., **1,** 1171
Tuddenham, D., **11,** 94
Tuddenham, R. M., **5,** 13
Tuezuen, C., **1,** 33, 772
Tufariello, J. J., **1,** 562; **2,** 69, 238; **6,** 597; **9,** 382, 402
Tuggle, R. M., **5,** 408
Tuinman, A., **5,** 244; **6,** 600
Tuladhar, S. M., **12,** 425
Tuley, W. F., **1,** 412, 892
Tulich, L., **10,** 456
Tulis, R. W., **6,** 506
Tull, R., **1,** 952, 1131
Tullar, B. F., **1,** 80, 198, 865
Tuller, F. N., **2,** 112
Tulley, A., **3,** 96
Tulloch, C. W., **12,** 106
Tullock, C. W., **1,** 117, 1124, 1125; **5,** 640
Tullock, G. W., **2,** 382
Tully, C. R., **9,** 78
Tuman, R. W., **3,** 123
Tummes, H., **5,** 205
Tumura, K., **9,** 117
Tuncay, A., **5,** 652
Tundo, A., **12,** 194
Tundo, P., **7,** 79, 164; **9,** 361
Tunemoto, D., **6,** 462; **8,** 421, 454; **10,** 165
Tung, C. C., **1,** 1247
Tung, J. S., **12,** 450
Tunick, A. A., **2,** 279
Tun-Kyi, A., **7,** 285; **8,** 360; **9,** 397
Tuppy, H., **2,** 147
Turchi, I. J., **9,** 493
Turecek, F., **12,** 503
Turley, R., **5,** 389
Turnbull, J. H., **1,** 861, 1210
Turnbull, K., **8,** 401; **9,** 466; **10,** 320
Turnbull, P., **2,** 372
Turner, A., **1,** 219
Turner, A. B., **2,** 117; **3,** 86, 87; **4,** 134, 240; **6,** 170; **7,** 97
Turner, A. F., **1,** 235, 745
Turner, E. E., **1,** 562, 1103; **8,** 280
Turner, E. S., **9,** 35
Turner, J. M., **1,** 1126, 1237, 1238
Turner, J. O., **1,** 397
Turner, J. V., **11,** 446, 559; **12,** 84, 195
Turner, L., **1,** 465, 787
Turner, L. M., **5,** 13; **6,** 19
Turner, M., **6,** 1
Turner, P. B., **1,** 76
Turner, R. B., **1,** 394, 703, 802, 1006, 1178; **8,** 442
Turner, R. W., **1,** 222; **8,** 245
Turner, S., **7,** 69
Turner, W. N., **1,** 1234
Turney, T. W., **8,** 153
Turnquest, B. W., **1,** 494
Turose, E., **11,** 121
Turrell, A. G., **8,** 480; **10,** 397
Turro, N. J., **1,** 531; **2,** 104, 286, 429; **6,** 482; **7,** 310; **8,** 519
Tursch, B. M., **1,** 707; **4,** 31, 265
Tuschka, T., **12,** 286
Tuszy-Maczka, M., **1,** 1000
Tuttle, N., **1,** 500
Tveten, J. L., **1,** 349
Twansato, R. T., **4,** 480
Tweddle, N. J., **4,** 93; **8,** 106
Tweeddale, H. J., **10,** 330
Tweig, R. J., **7,** 94
Tweit, R. C., **1,** 127, 1157
Twelves, R. R., **1,** 1253
Twitchin, B., **12,** 84
Twohig, M. F., **12,** 425
Tyagi, B. S., **1,** 570
Tyczkowswki, E. A., **1,** 808
Tykelar, Z., **11,** 141
Tyle, Z., **2,** 144
Tyler, L. J., **4,** 58, 177
Tyler, P. C., **11,** 373
Tyman, J. H. P., **1,** 731; **5,** 571; **6,** 517
Tyner, D. A., **1,** 389
Tyobeka, T. E., **10,** 375
Tyrlik, S., **5,** 671; **6,** 588, 590
Tyson, F. T., **1,** 1113
Tyssee, D. A., **7,** 138
Tzeng, D., **12,** 220
Tzodikov, N. R., **7,** 234
Tzougraki, C., **8,** 81; **9,** 415

Uang, B. J., **11,** 317; **12,** 561
Uba, K., **9,** 242
Ubhelohde, A. R. J. P., **9,** 289
Ucciani, E., **5,** 677; **6,** 172
Uchibayashi, M., **9,** 377; **11,** 491
Uchida, A., **5,** 667
Uchida, K., **7,** 83; **8,** 174
Uchida, M., **4,** 196
Uchida, S., **12,** 411
Uchida, Y., **10,** 399
Uchimaru, T., **8,** 505; **9,** 132; **11,** 184, 185; **12,** 118
Uchino, M., **8,** 254

Uchino, N., **11,** 157
Uchiyama, A., **3,** 51
Uchiyama, M., **12,** 94
Uchiyama, T., **7,** 277
Ucko, D. A., **8,** 47
Uda, H., **1,** 318, 1022; **4,** 282; **6,** 464, 466, 662; **7,** 235, 339
Udaka, S., **9,** 176
ud Din, Z., **5,** 427; **6,** 503, 546
Udding, A. C., **2,** 370
Udenfriedin, S., **7,** 268
Udenfriend, S., **1,** 476; **5,** 308
Udo, H., **2,** 150
Udodong, U. E., **12,** 445
Uebel, J. J., **1,** 814
Ueberwasser, H., **1,** 1203
Ueda, M., **6,** 25; **11,** 29
Ueda, N., **5,** 713
Ueda, S., **2,** 462; **12,** 508
Ueda, T., **6,** 659; **7,** 424
Ueda, Y., **6,** 76, 322
Uehara, H., **12,** 555
Uekama, K., **9,** 129
Ueki, M., **2,** 40, 445
Uematsu, T., **7,** 345
Uemura, M., **10,** 14; **11,** 292; **12,** 36
Uemura, S., **4,** 21; **5,** 167; **6,** 23, 141; **7,** 12, 361; **8,** 120; **9,** 35, 443; **10,** 21, 106, 368; **11,** 37, 253, 397, 455; **12,** 41, 121, 452
Ueno, H., **11,** 514
Ueno, K., **8,** 267; **12,** 131
Ueno, Y., **1,** 9; **6,** 34; **7,** 27, 123, 165; **8,** 44; **9,** 225, 477; **10,** 413; **11,** 550, 551; **12,** 445, 523
Uesugi, S., **7,** 286
Uevaya, S., **12,** 344
Uff, B. C., **3,** 101; **4,** 46
Ugelstad, J., **1,** 926
Ugi, I., **1,** 383, 407, 844, 859, 881, 927; **2,** 21, 49, 51, 54; **4,** 65, 101, 384, 536, 555; **6,** 8, 91; **7,** 20, 366; **8,** 214, 499; **9,** 82, 121; **10,** 67; **11,** 112
Ugi, I. K., **11,** 63
Ugo, R., **5,** 172, 190; **9,** 217
Uh, H. S., **5,** 285, 325, 481, 631; **6,** 582
Uher, M., **5,** 668
Uhl, J., **5,** 136
Uhle, F. C., **1,** 1163; **12,** 444
Uhle, K., **10,** 444
Uhlig, F., **1,** 904
Uhlmann, J. G., **3,** 103
Uhm, S. J., **5,** 756; **6,** 675
Uhm, S. T., **11,** 308, 599
Uibrig, C., **2,** 139
Uijttewaal, A. P., **6,** 381
Ukai, J., **12,** 19
Ukaji, Y., **12,** 275
Ukawa, K., **5,** 617; **9,** 464; **10,** 398
Ukita, T., **12,** 523
Ulbricht, T. L. V., **1,** 380, 1073, 1090; **3,** 197; **9,** 480
Uldall, I., **1,** 77
Ulich, L. H., **1,** 771
Ullerich, K., **3,** 304
Ullman, E. F., **1,** 470
Ullmann, F., **1,** 1058; **3,** 69
Ullrey, J. C., **4,** 247
Ullyott, G. E., **4,** 189
Ulman, A., **7,** 273
Ulrich, C., **1,** 955
Ulrich, P., **5,** 600; **6,** 367; **7,** 142, 229; **9,** 68
Ultee, W. J., **7,** 3
Umani-Ronchi, A., **2,** 173; **3,** 125; **4,** 197; **5,** 9, 315; **6,** 43, 74, 306, 486; **7,** 300; **8,** 406; **9,** 378, 477; **10,** 297, 322; **11,** 276, 356; **12,** 571
Umbreit, G. R., **1,** 643
Umbreit, M. A., **4,** 570; **6,** 658; **8,** 65
Umeda, I., **4,** 270; **6,** 45
Umeda, Y., **10,** 90
Umemoto, T., **6,** 500; **8,** 421
Umen, M. J., **5,** 149, 243; **6,** 225
Umetani, T., **5,** 253
Umeyama, K., **11,** 160; **12,** 171
Umezawa, B., **6,** 316
Umezawa, K., **5,** 439
Umezawa, O., **1,** 122
Umezawa, S., **9,** 188
Umezu, M., **7,** 12
Umhoefer, R. R., **1,** 631
Umhoefer, S. G., **9,** 410
Umino, N., **7,** 326, 342
Unde, G., **8,** 245
Underwood, G. R., **6,** 628
Underwood, H. W., **1,** 39, 828, 1058
Undheim, K., **1,** 208; **6,** 517; **10,** 326
Uneme, H., **8,** 315; **10,** 231
Ungaro, R., **8,** 245; **9,** 489
Ungefug, G. A., **4,** 42
Unger, L. R., **5,** 491; **8,** 374
Unger, P. L., **10,** 159
Ungnade, H. E., **1,** 50, 158, 270, 755, 1011, 1269; **2,** 206
Unni, M. K., **1,** 206
Uno, F., **2,** 422
Uno, H., **1,** 289; **11,** 22; **12,** 405
Uno, T., **9,** 382
Unrau, A. M., **3,** 329; **5,** 339
Unser, M. J., **3,** 199
Untch, K. G., **1,** 689; **2,** 278, 376; **3,** 87; **5,** 120, 360; **9,** 261; **11,** 481
Unterweger, W. D., **7,** 162
Upton, C. J., **6,** 348
Upton, R. M., **9,** 48
Urabe, H., **10,** 65; **11,** 402, 544; **12,** 29, 378, 516
Urasaki, I., **10,** 305
Urbach, H., **12,** 38
Urban, F. J., **4,** 60
Urban, G., **7,** 358
Urban, R. S., **1,** 927
Urbanski, T., **2,** 176; **3,** 213, 284
Urenovitch, J. V., **5,** 640
Uribe, G., **11,** 219
Urion, E., **1,** 424
Urpilainen, P., **10,** 359
Urquhart, G. G., **1,** 1167
Urquiza, M., **1,** 642
Urquiza, R., **1,** 564
Urry, G., **2,** 412
Urso, F., **12,** 112
Ursprung, J. J., **1,** 465

Urushibara, Y., **6,** 659
Uschakow, M. I., **2,** 39
Usgaonkar, R. N., **7,** 424
Ushida, S., **9,** 8
Usieli, V., **7,** 144
Usinger, R. P., Jr., **1,** 1216
Uskokovic, M. R., **1,** 764; **2,** 75; **4,** 162, 323; **5,** 73, 283; **8,** 338, 383; **10,** 180, 309, 400; **11,** 12, 226
Usselman, M. C., **5,** 176
Usui, M., **6,** 376; **8,** 96
Usui, T., **9,** 188
Utaka, M., **12,** 365
Utawanit, T., **6,** 180; **12,** 121, 418
Utermohlen, W. P., Jr., **1,** 389; **2,** 391
Utimoto, K., **5,** 64, 435; **6,** 281, 295; **7,** 83, 243; **8,** 174, 246, 340; **10,** 114; **11,** 150; **12,** 51
Utne, T., **1,** 367, 676; **2,** 39, 229
Uttley, M. F., **7,** 60
Utzinger, G. E., **1,** 293
Uyeda, S., **7,** 135
Uyehara, T., **4,** 113; **8,** 197; **9,** 410
Uyeo, S., **9,** 519
Uzan, R., **11,** 255
Uzarewicz, A., **9,** 138
Uznanski, B., **8,** 133

Vacca, J. P., **10,** 57; **11,** 292, 310, 494
Vagberg, J., **12,** 370
Vaidya, A. S., **2,** 236
Vaidya, V. M., **2,** 80
Vaidyanathaswamy, R., **2,** 278, 376; **6,** 523
Vail, P. D., **11,** 420
Vaille, J., **11,** 318
Vajs, V., **9,** 189
Vakatkar, V. V., **7,** 333; **8,** 483
Vakilwala, M. V., **1,** 1102
Valade, J., **2,** 448; **7,** 378
Valcho, J. J., **6,** 91; **9,** 364
Valenta, Z., **1,** 1284; **6,** 70; **7,** 345; **10,** 55, 373
Valente, L. F., **8,** 507; **9,** 488; **10,** 391, 423
Valente, V. R., **2,** 69
Valentine, D., Jr., **9,** 260
Valentine, J. S., **6,** 490; **7,** 307
Valentino, D. S., **8,** 53
Valenty, S., **6,** 190
Valery, J. M., **8,** 260
Valette, G., **8,** 299
Valiaveedan, G. D., **1,** 655; **4,** 366
Valicenti, J. A., **1,** 309
Valkanas, G., **1,** 44
Vallee, B. L., **1,** 1148; **2,** 404
Valls, J., **3,** 63
Valnot, J. Y., **5,** 58, 377; **8,** 193
Valoti, E., **12,** 502
Valpey, R. S., **11,** 514; **12,** 127
VanAllen, J. A., **1,** 39, 373, 402, 837, 1040, 1041, 1049, 1064, 1073, 1106, 1150, 1167, 1209; **9,** 159
van Alphen, J., **9,** 362
Van Ammers, M., **1,** 827
Vanasse, B. J., **10,** 300
van Asten, J. J. A., **6,** 256; **7,** 3
Van Audenhove, M., **10,** 45
Van Auken, T. V., **1,** 978; **2,** 273
van Bac, N., **1,** 730
van Balen, H. C. J. G., **11,** 71
van Beelen, D. C., **6,** 458
van Bekkum, H., **1,** 1167; **2,** 287; **3,** 291, 329; **4,** 345, 369; **5,** 319; **6,** 149, 575; **7,** 188; **10,** 321
van Bergen, T. J., **4,** 191; **5,** 115, 261, 389; **8,** 458
van Berkel, J., **8,** 103
van Boeckel, C. A. A., **10,** 116
van Boom, J. H., **2,** 344; **5,** 675; **6,** 605; **8,** 362; **10,** 116, 231
Van Brussel, W., **6,** 201
Van Camp, A., **10,** 60
Van Campen, M. G., Jr., **1,** 207
Van Catledge, F. A., **6,** 554
van Dam, P. B., **7,** 415; **10,** 456
Van de Castle, J. F., **1,** 642
Vandenberg, G. E., **1,** 808
van den Bosch, C. B., **7,** 188
van den Brink, W. M., **3,** 157
Vandenbulcke-Coyette, B., **11,** 476
Van Den Elzen, R., **5,** 129
van den Goorbergh, J. A. M., **10,** 181
van den Ham, D. M. W., **7,** 241
Vandenheuvel, F. A., **1,** 892
van de Putte, T., **3,** 329
Van der Koot, J. P., **4,** 119
van der Baan, J. L., **9,** 55
Vanderbilt, D. P., **7,** 287
van der Brink, W. M., **1,** 1188; **2,** 423
van der Gen, A., **1,** 247; **5,** 371, 631; **6,** 381; **7,** 357; **8,** 90, 415; **9,** 112, 319, 387, 450; **10,** 181; **11,** 180, 353; **12,** 182, 334
Vanderhaeghe, H., **1,** 526
van der Helm, D., **5,** 725; **9,** 488
Van Der Kerk, G. J. M., **1,** 424, 1112, 1193, 1251; **7,** 378
Van der Kooi, H. O., **6,** 458
van der Leij, M., **8,** 465, 466
van der Meer, D., **7,** 241
van der Plas, H. C., **1,** 1163; **8,** 419
Van Der Puy, M., **6,** 261
Vanderslice, C. W., **3,** 77, 123, 264
van der Steen, J., **5,** 340
Vander Stouw, G. G., **1,** 1187
Van Derveen, D., **11,** 349
Van der Ven, S., **1,** 1108
Vanderwaart, B. E., **5,** 551
Vanderwerf, C. A., **1,** 1040, 1044, 1269
van der Wolf, L., **8,** 169
van der Zeeuw, A. J., **1,** 268
Vander Zwan, M. C., **5,** 360, 427; **8,** 245, 391
Vandewalle, J. J. M., **4,** 122,

248
Vandewalle, M., **5,** 691; **6,** 201; **8,** 131; **10,** 45
Vandewalle, M. E., **9,** 116
van Dijck, L. A., **10,** 350
Van Dine, H. A., **2,** 173
Van Dongen, J. P. C. M., **2,** 344
Van Dorp, D. A., **1,** 140, 360, 689, 1000; **2,** 127, 374; **3,** 93
Van Dort, H. M., **2,** 360; **3,** 168
van Driel, H., **4,** 267
Van Duuren, B. L., **1,** 787; **4,** 84
Van Dyke, M., **2,** 165
Van Dyke Tiers, G., **1,** 870
Van Eenoo, M., **9,** 204; **10,** 174, 319
van Elburg, P., **12,** 334
Van Ende, D., **5,** 66; **6,** 362, 488; **7,** 195
Van Es, A., **1,** 4
Van Es, T., **1,** 407, 720; **2,** 217; **4,** 351
van Es, T., **9,** 245
Van Ess, P. R., **1,** 689
Vangedal, S., **1,** 1228
van Gelderen, M. J., **4,** 105
van Gennep, H. E., **4,** 516
Van Hecke, G. R., **6,** 51; **11,** 168
van Heerden, F. R., **8,** 391
van Hengstum, A. J., **10,** 321
Van Hoozer, R., **1,** 1167
Van Horn, D. E., **7,** 113; **8,** 87, 507; **9,** 132, 488; **11,** 568; **12,** 54
Van Horn, W. F., **3,** 299
Van Hulle, F., **5,** 691
Van Hulle, M. T., **11,** 49
Vanino, L., **1,** 46
Vankar, Y. D., **9,** 66, 67, 396, 400, 488; **10,** 98, 167, 202; **12,** 258, 569
van Keulen, B. J., **9,** 100
van Klaveren, W., **1,** 465
van Leeuwen, P. M., **8,** 225
van Leusen, A. M., **4,** 516; **5,** 685; **6,** 567, 600; **7,** 246, 377; **8,** 343, 494; **10,** 409; **11,** 44; **12,** 511
van Leusen, D., **8,** 343, 494; **10,** 409; **11,** 50; **12,** 511
Vanlierde, H., **4,** 135
van Look, G., **10,** 122
Van Mater, H. L., **1,** 1295
Van Meerssche, M., **11,** 111
VanMeter, J. P., **3,** 133, 228, 273
van Minnen-Pathuis, G., **7,** 188
van Mourik, G. L., **7,** 296
van Muijlwijk, A. W., **6,** 149
Vanneman, C. R., **1,** 1226
Vannes, P., **11,** 111
van Norden, J. J., **8,** 340
Van Order, R. B., **5,** 297
van Overstraeten, A., **1,** 258; **4,** 122
van Peppen, J. F., **6,** 611
van Rantwijk, F., **3,** 329
van Reijendam, J. W., **4,** 267; **5,** 330
Van Rens, E. M. M., **5,** 516
Van Rheenan, V., **9,** 38
Van Rheenen, V., **2,** 256; **3,** 66; **5,** 290, 319; **7,** 160, 257
van Rijn, P. E., **11,** 103
Van Royen, L. A., **9,** 23, 223
van Schaik, T. A. M., **8,** 340; **12,** 182
Van Schooten, J., **1,** 105
Vanstone, A. E., **2,** 72
van Straten, J. W., **8,** 340
Van Strien, R. E., **6,** 84
van Tamelen, E. E., **1,** 80, 258, 562, 730, 909, 1171; **2,** 336, 415; **3,** 93, 101, 112, 113, 191; **4,** 53, 268, 282, 287, 448, 535; **5,** 319, 351, 673; **6,** 379, 589; **9,** 438; **11,** 200, 214
Vanterpool, A., **3,** 123
Van Verth, J. E., **5,** 491
van Vyve, T., **4,** 138; **5,** 198
Van Wazer, J. R., **2,** 334
van Wijk, A. M., **7,** 188
van Zutphen, L., **5,** 13
van Zyl, J. J., **8,** 391
Varaprath, S., **12,** 483
Varescon, F., **11,** 383
Vargas, L. A., **7,** 413
Varie, D. L., **12,** 85
Varimbi, S. P., **1,** 1087
Varkony, T. H., **6,** 441; **7,** 273; **10,** 296
Varland, R. H., **1,** 383
Varley, J. H., **7,** 341
Varma, K. R., **1,** 484
Varma, R. K., **4,** 105, 310, 376; **6,** 202; **12,** 259
Varma, R. S., **12,** 65, 288
Varma, V., **1,** 263; **5,** 41; **8,** 385; **11,** 397
Varney, M. D., **12,** 282
Varrentrapp, F., **1,** 937
Varshavskaya, L. S., **3,** 222
Vartanian, P. F., **6,** 604
Varvoglis, A., **6,** 301; **9,** 54; **11,** 411; **12,** 386
Vas, S., **2,** 458
Vasella, A., **5,** 576
Vasilevskis, J., **9,** 352
Vaska, L., **2,** 434; **5,** 172
Vass, G., **5,** 287; **7,** 392
Vatakencherry, P. A., **1,** 989
Vatele, J. M., **11,** 497; **12,** 316
Vaughan, J. R., Jr., **1,** 86, 1203, 1229; **2,** 47
Vaughan, W. R., **1,** 158, 927, 1095, 1163, 1286; **3,** 171; **4,** 21
Vaughn, H. L., **6,** 548
Vaughn, T. H., **1,** 1040
Vaughn, W. L., **1,** 1287
Vaultier, M., **10,** 120, 211; **11,** 588; **12,** 551
Vavasour, G. R., **1,** 135, 141
Vavon, G., **8,** 8
Vaziri, C., **4,** 431
Vazquez, M. A., **6,** 114; **7,** 271
Veale, H. S., **8,** 438
Veazey, R. L., **12,** 154
Veber, D. F., **2,** 5; **10,** 61
Vechio, R. L., **4,** 74
Vedejs, E., **2,** 99, 126, 196; **3,** 136, 338; **4,** 304; **5,** 32, 270, 445, 661, 680; **6,** 15, 384, 573, 628; **7,** 136,

150, 207; **8,** 9, 208, 493; **9,** 100, 337, 473; **10,** 436; **11,** 113, 121, 209, 254, 382; **12,** 85, 178, 462
Vederas, J. C., **4,** 261
Veer, W. L. C., **1,** 1101
Veeraraghavan, S., **11,** 150
Veeravagu, P., **1,** 927
Veibel, S., **2,** 62
Vela, F., **4,** 573
Vela, V., **8,** 87
Velarde, E., **1,** 249, 802; **2,** 381; **3,** 265; **4,** 150; **5,** 216, 605
Velasco, M., **1,** 82
Veliev, M. G., **10,** 459
Velluz, L., **1,** 80, 215, 465, 495, 626, 978, 1256
Veltmann, H., **7,** 97
Venakis, T., **9,** 180; **10,** 418
Venanzi, L. M., **2,** 110, 403
Venegas, M. G., **8,** 305; **12,** 62
Venier, C. G., **4,** 122; **5,** 516; **6,** 457; **8,** 465; **11,** 562
Venit, J., **10,** 141
Venkataraman, K., **1,** 206, 1292; **4,** 15, 189, 424; **10,** 115
Venkataramani, P. S., **2,** 245
Venkatasubramanian, G. B., **3,** 9
Venkatesawaran, R. V., **12,** 201
Venkateswarlu, A., **4,** 177, 478; **5,** 394
Venkov, A. P., **9,** 426; **10,** 9
Venner, H., **4,** 390
Vennstra, G. E., **6,** 567
Venton, D. L., **11,** 88
Ventura, J. J., **3,** 294
Venturello, C., **6,** 451
Venturello, P., **4,** 133; **9,** 361
Vera, M., **5,** 762
Veracini, C. A., **6,** 27
Verardo, G., **10,** 297
Verbicky, J. W., Jr., **11,** 393
Verbit, L., **3,** 101
Verboom, V., **11,** 68
Verboom, W., **9,** 106, 111
Verbrugge, C., **2,** 368
Vercellone, A., **1,** 72
Vercellotti, J. R., **1,** 122; **6,** 676
Verdegaal, C. H. M., **10,** 231
Verdone, J. A., **10,** 413
Verhe, R., **6,** 118; **8,** 15, 105, 463; **9,** 182; **11,** 434
Verheyden, J. P. H., **1,** 1250; **2,** 365, 446; **4,** 451, 559; **5,** 188; **7,** 407
Verhoeven, J., **10,** 231
Verhoeven, T. R., **7,** 327; **8,** 476; **9,** 79, 82, 458; **11,** 514
Verkade, J. G., **6,** 469; **7,** 170
Verkade, P. E., **1,** 1167; **3,** 291
Verkruyesee, H. D., **2,** 374; **10,** 73; **11,** 103, 432
Verlade, E., **4,** 577
Verley, M. A., **1,** 1040
Vermeer, P., **5,** 95, 168; **6,** 144, 146, 270; **7,** 80, 164, 226; **8,** 322; **9,** 95, 111, 277, 334; **10,** 290, 350, 434; **11,** 13; **12,** 350
Vermeeren, H. P. W., **7,** 3
Vermes, J. P., **3,** 61
Verniere, C., **10,** 101
Vernon, C. A., **1,** 1107
Vernon, I. H., **1,** 329
Verpeaux, J. N., **11,** 59; **12,** 140
Vertalier, S., **1,** 87
Verzele, M., **1,** 392; **2,** 94
Veschambre, H., **5,** 589
Vesley, G. F., **2,** 429; **4,** 267
Vessal, B., **11,** 62
Vessel, C. R., **10,** 187
Vessiere, R., **8,** 196; **10,** 174
Vessieres, A., **12,** 36
Vessman, J., **3,** 148
Vest, R. D., **7,** 237
Vesterager, N. O., **4,** 247; **5,** 325
Vestergaard, J. H., **2,** 62
Vetter, H., **1,** 520
Vetter, H. J., **6,** 576
Veyrieres, A., **11,** 501
Veysoglu, T., **5,** 127; **10,** 62
Viala, J., **8,** 302
Vialle, J., **5,** 182
Viallefont, P., **12,** 350
Viau, R., **5,** 129
Vib, J., **2,** 382
Vibuljan, P., **10,** 246
Vicens, J., **9,** 32
Vickers, E., **4,** 461
Vickery, B., **8,** 463
Vickery, E. H., **9,** 256
Victor, R., **5,** 358; **7,** 144
Victor, R. R., **2,** 344
Vida, J. A., **3,** 250; **4,** 228
Vidal, M., **2,** 372
Vidari, G., **10,** 366
Vidril-Robert, D., **12,** 414
Vidrine, D. W., **7,** 240
Viehbeck, A., **9,** 387
Viehe, H. G., **1,** 677; **2,** 134; **3,** 130; **4,** 95, 138; **5,** 198; **7,** 124; **8,** 157; **11,** 476
Vieira, E., **11,** 140
Vieregge, H., **1,** 360
Vierhapper, F. W., **6,** 615
Vietmeyer, N. D., **2,** 422
Vietti, D. E., **6,** 128
Vigneron, J. P., **9,** 308; **10,** 265
Vila, J., **1,** 141
Vilkas, E., **2,** 173
Vilkas, M., **1,** 158, 561, 743, 795; **8,** 311
Vill, J. J., **1,** 114
Villacorta, G. M., **11,** 373
Villani, F., **7,** 246
Villani, F. J., **2,** 433
Villanueva, P., **4,** 432
Villarica, R. M., **2,** 195; **3,** 182
Ville, G., **10,** 343
Villemin, D., **10,** 413
Villenave, J. J., **11,** 101
Villiani, F. J., Jr., **8,** 475
Villieras, J., **4,** 139, 234; **5,** 94, 187, 406; **6,** 143, 189, 270, 331, 345, 354, 577, 622, 664; **7,** 66, 83, 170; **9,** 138, 154; **12,** 77, 527
Villiger, V., **1,** 17, 119
Villotti, R., **1,** 526
Vilsmaier, E., **4,** 90; **6,** 118
Vilsmeier, A., **1,** 289, 682
Vincent, B. F., Jr., **1,** 207; **3,** 77

Vincent, J. E., **11,** 260, 579
Vincenti, C. B., **9,** 151
Vincke, P., **1,** 1272
Vinczer, P., **12,** 508
Vineyard, B. D., **5,** 502; **11,** 31
Vingiello, F. A., **1,** 424
Vinick, F. J., **9,** 309; **10,** 295
Vinion, G., **12,** 86
Vinogradov, M. G., **5,** 157
Vinogradova, L. P., **1,** 465
Vinokur, E., **5,** 725; **6,** 323
Vinson, J. W., **6,** 620
Vinter, J. G., **8,** 534
Viola, H., **7,** 289
Viola, J., **9,** 323
Viout, P., **5,** 266
Vipond, H. J., **1,** 1029, 1210
Viriot-Villaume, M. L., **6,** 70
Virk, K., **1,** 309, 685
Viscogliosi, L. A., **11,** 145
Viscontini, M., **1,** 289, 629, 666, 1221
Vishnuvajjala, B., **4,** 501; **8,** 143, 527
Vishwakarma, L. C., **12,** 393
Vismara, E., **11,** 529
Visnick, M., **12,** 117, 229
Visser, C. W., **11,** 68
Visser, G. W., **9,** 177
Visser, R. G., **10,** 96
Viswanatha, V., **5,** 74
Viswanathan, M., **12,** 538
Visweswariah, S., **11,** 426
Vit, J., **3,** 260, 261
Vitale, W., **2,** 27
Vitali, R., **1,** 1210
Viti, S. M., **11,** 94, 95, 477; **12,** 94
Vitrone, J., **6,** 423; **8,** 364
Vittemberga, B., **1,** 904
Vizsolyi, J. P., **1,** 235
Vladuchick, S. A., **5,** 589
Vladuchick, W. C., **8,** 52
Vlahov, R., **5,** 595
Vlattas, I., **2,** 34, 40, 99, 117, 126; **3,** 338; **5,** 527
Vliet, E. B., **1,** 144, 1064
Vlietstra, E. J., **11,** 567
Voaden, D. J., **7,** 67
Vocelle, D., **2,** 310
Vodoz, C. A., **1,** 41
Voelkel, M. B., **1,** 322
Voelter, W., **3,** 329; **5,** 701; **12,** 532
Voepel, K. H., **2,** 154
Voett, V., **3,** 308
Voetter, H. D., **5,** 595
Vofsi, D., **9,** 311
Vogel, A., **2,** 344
Vogel, A. I., **1,** 377, 935
Vogel, A. L., **4,** 100
Vogel, D. E., **12,** 107
Vogel, E., **1,** 190, 351, 377; **2,** 99, 344; **3,** 87; **4,** 133, 459, 579; **5,** 491; **6,** 33, 479, 657, 675; **7,** 268, 269; **9,** 140
Vogel, F., **7,** 102; **8,** 194, 303
Vogel, M. K., **10,** 209
Vogel, P., **5,** 269; **10,** 392; **11,** 140; **12,** 61
Vogel, P. W., **10,** 172
Vogeler, K., **1,** 330, 494, 1164
Vogeli, U., **10,** 359
Vogl, O., **1,** 1210
Vogler, K., **2,** 454
Vogt, B. R., **1,** 93, 125; **2,** 66; **5,** 230; **11,** 341
Vogt, C. M., **1,** 329
Vogt, L. H., Jr., **3,** 245
Vogt, P. F., **5,** 552
Vogt, P. J., **1,** 1210
Vogt, R. R., **1,** 1040
Vogt, S., **11,** 472
Voight, E., **9,** 410
Voigt, A., **1,** 367, 629
Voigt, B., **5,** 424
Volante, R. P., **11,** 589
Volger, H. C., **1,** 164; **2,** 303; **3,** 66
Volkenburgh, R. V., **7,** 195
Volkmann, R. A., **5,** 315, 631; **6,** 615; **12,** 70
Volkmer, F., **9,** 216
Vollbracht, L., **3,** 168
Vollering, M. C., **11,** 423
Vollhardt, K. P. C., **4,** 119; **5,** 173; **6,** 154, 668; **7,** 84, 95, 218; **8,** 147; **9,** 143; **10,** 127; **11,** 162; **12,** 163, 526
Vollmann, H., **1,** 682
Volpers, F., **1,** 20
Vol'pin, M. E., **3,** 264; **4,** 562
Volz, W. E., **7,** 66
von Auwers, K., **1,** 1101
von Beroldingen, L. A., **5,** 528
von Bonin, W., **1,** 1055
von Braun, J., **1,** 150, 176, 448, 874, 889
von Bredow, K., **6,** 173
von Buelow, B. G., **2,** 422
Voncken, W. G., **5,** 707
Vonderheid, C., **9,** 381
von Euw, J., **1,** 818
von Gizycki, U., **4,** 361
von Grote, A., **1,** 566
von Gustorf, E. K., **4,** 550
von Hinrichs, E., **4,** 555; **11,** 112
von Hofmann, A. W., **1,** 1138
von Imhof, A., **11,** 401
von Mutzenbecker, G., **1,** 444
von Pechmann, H., **1,** 195, 518, 928; **2,** 216, 377
von Philipsborn, W., **1,** 1284
von Planta, C., **1,** 567
von Rudloff, E., **1,** 20, 814, 962; **4,** 194
von Saltza, M. H., **1,** 1157
von Schuh, H. G., **1,** 403, 842
von Strandtmann, M., **5,** 756
von Stryk, F., **1,** 664
von Wartburg, A., **4,** 45
von Werder, F., **1,** 172
Voong, S. T., **1,** 481
Voorhees, K. J., **5,** 156, 590
Voorhees, V., **1,** 892
Vo-Quang, L., **1,** 268
Vo-Quang, Y., **1,** 268
Vorbrueggen, H., **1,** 104, 282, 283, 814; **4,** 416; **5,** 260; **6,** 443, 640; **9,** 22, 180, 496; **10,** 441; **11,** 524, 588; **12,** 150, 239
Vorlaender, D., **1,** 243, 680, 856, 1073, 1261
Vorster, W., **5,** 281
Vose, C. E., **1,** 870; **4,** 317
Voser, W., **1,** 1000
Voskuil, W., **5,** 268

Voss, G., **1,** 938
Voss, J., **5,** 535; **11,** 55
Voss, P., **5,** 480; **6,** 273
Vossen, D., **3,** 33
Vostrowsky, O., **6,** 279; **7,** 329; **8,** 234
Voswinckel, H., **1,** 710
Votral, R. L., **2,** 457
Vowinkel, E., **1,** 235; **5,** 170, 207
Vrachnou-Astra, E., **9,** 513
Vranesic, B., **8,** 64
Vrba, L., **9,** 205
Vredenburgh, W. A., **1,** 609
Vrielink, J. J., **6,** 143
Vriesen, C. W., **3,** 221
Vrijland, M. S. A., **8,** 132
Vroman, H. E., **2,** 368
Vuillerme, J. P., **6,** 359
Vuitel, L., **6,** 11
Vukov, R., **3,** 38
Vyas, D. M., **7,** 238
Vyas, G. N., **1,** 295, 685

Wachter, M., **6,** 511
Wachter, W., **1,** 506, 510
Wacker, W. E. C., **1,** 1148
Wackerle, L., **6,** 91; **9,** 247
Wada, A., **11,** 255
Wada, E., **4,** 405; **9,** 398; **12,** 316
Wada, F., **10,** 298
Wada, M., **4,** 555; **8,** 327; **10,** 119; **11,** 91, 373
Waddington, G., **8,** 214
Waddington, T. C., **6,** 301
Waddington-Feather, S. M., **1,** 595
Wade, K., **3,** 128
Wade, L. G., Jr., **6,** 497; **8,** 94; **9,** 107, 435
Wade, P. A., **5,** 566; **7,** 170; **9,** 39; **11,** 41, 146, 420
Wade, R., **4,** 391
Wade, R. C., **4,** 450
Wade, R. H., **1,** 730, 741, 1073, 1100
Wade, R. S., **3,** 59; **4,** 97
Wade, T. N., **11,** 454
Wade, W. S., **11,** 208
Wadia, M. S., **8,** 364
Wadsworth, A. H., **11,** 426
Wadsworth, D. H., **2,** 155; **10,** 9
Wadsworth, H. J., **12,** 72
Wadsworth, W. S., **10,** 112
Wadsworth, W. S., Jr., **1,** 250, 253, 1095, 1138, 1216, 1217, 1218; **3,** 144; **8,** 337
Waegell, B., **2,** 300; **4,** 67, 234; **5,** 13; **9,** 353; **11,** 45; **12,** 53
Waelchli, R., **11,** 479
Waescher, K., **1,** 134
Waetjen, F., **11,** 147
Wagatsuma, M., **5,** 160
Wagatsuma, N., **3,** 268
Wagatsuma, S., **5,** 13, 178
Wagenaar, A., **5,** 127, 516
Wagenknecht, J. H., **4,** 85
Wager, J. S., **4,** 430
Wagland, A. A., **1,** 802
Wagle, D. R., **10,** 115
Wagner, A. F., **1,** 1081, 1157
Wagner, B. E., **5,** 64
Wagner, C. D., **1,** 89
Wagner, D., **5,** 188
Wagner, D. P., **4,** 367
Wagner, E. C., **1,** 5, 407, 837, 859, 870, 963
Wagner, E. R., **5,** 10
Wagner, G. W., **1,** 633
Wagner, K., **4,** 524
Wagner, R., **4,** 258; **5,** 136
Wagner, R. I., **4,** 191
Wagner, R. M., **5,** 49
Wagner, S. D., **6,** 93, 361
Wagner, W., **1,** 935
Wagner, W. J., **3,** 50; **6,** 400
Wagner, W. M., **1,** 1108
Wagner, W. S., **3,** 272
Wagner-Jauregg, T., **9,** 325
Waheed, N., **7,** 331
Wahl, G. M., **5,** 296
Wahl, R., **5,** 73, 250; **6,** 243
Wahlborg, H. J., **1,** 617
Wahren, R., **7,** 340
Waight, E. S., **5,** 675
Wailes, P. C., **1,** 430; **6,** 179
Waite, J. A., **1,** 1210
Waitkins, G. R., **5,** 341; **6,** 422
Waits, H. P., **4,** 439
Wakabayashi, M., **6,** 500
Wakabayashi, N., **1,** 562; **5,** 727
Wakabayashi, S., **11,** 419; **12,** 392, 486
Wakabayashi, T., **2,** 427; **4,** 147; **11,** 487
Wakabayashi, Y., **11,** 150; **12,** 33
Wakamatsu, T., **6,** 628; **7,** 225; **9,** 414; **10,** 378; **11,** 213; **12,** 429
Wakatsuki, Y., **5,** 173; **6,** 128
Wakayama, S., **5,** 45
Wakefield, B. J., **1,** 629; **2,** 254
Wakeford, D. H., **4,** 390
Wakelin, R. W., **5,** 517
Wakeman, N., **1,** 1101
Wakemoto, H., **11,** 140
Wakimasu, M., **11,** 330
Wakisaka, K., **3,** 268; **5,** 551
Wakita, Y., **11,** 373
Wakselman, C., **5,** 219; **6,** 70, 263; **7,** 158; **9,** 224, 314
Wakselman, M., **4,** 173; **8,** 57; **10,** 91; **11,** 359
Walba, D. M., **4,** 324; **5,** 410; **8,** 302; **9,** 391; **11,** 139, 302
Walborsky, H. M., **1,** 818, 1011, 1157; **3,** 280, 329; **4,** 186, 290, 480, 505, 562; **5,** 651; **6,** 91, 145, 643; **7,** 238; **8,** 464; **9,** 458; **11,** 527; **12,** 121
Wald, K., **6,** 76
Walde, A., **10,** 186
Walden, P., **1,** 955
Waldmann, H., **1,** 882
Waldmeier, F., **11,** 359, 602
Waldron, J. T., **5,** 551
Waldschmidt-Leitz, E., **6,** 445
Waldvogel, G., **2,** 211, 362
Walia, J., **4,** 282
Walinsky, S. W., **4,** 218; **5,** 714
Walker, A., **10,** 248; **11,** 536
Walker, B. H., **2,** 72, 213, 422
Walker, B. J., **6,** 341

Walker, C. C., **2,** 287
Walker, D., **1,** 219; **2,** 112; **6,** 234, 453
Walker, D. G., **10,** 232; **11,** 285
Walker, D. L., **8,** 331
Walker, D. M., **1,** 1247; **5,** 269, 294; **6,** 268; **11,** 194; **12,** 234
Walker, D. R., **2,** 239
Walker, E. C., **8,** 375
Walker, F. H., **5,** 537
Walker, F. J., **9,** 304; **11,** 94, 95
Walker, F. W., **5,** 419
Walker, G. L., **11,** 343
Walker, G. N., **1,** 890, 1253
Walker, H. G., Jr., **1,** 685
Walker, J. A., **4,** 577; **7,** 181
Walker, J. F., **1,** 268, 712
Walker, J. F., Jr., **2,** 344
Walker, K. A. M., **2,** 453; **8,** 399
Walker, L. E., **1,** 290; **4,** 473
Walker, L. F., **2,** 344
Walker, P. G., **1,** 110
Walker, R., **1,** 764
Walker, R. W., **1,** 664
Walker, T., **1,** 595, 609, 642, 652, 802, 951, 1090
Walker, W. E., **4,** 352
Wall, A., **12,** 77
Wall, D. K., **3,** 92
Wall, G., **7,** 166
Wall, M. E., **1,** 13, 318, 500, 814, 1203; **2,** 173, 216, 353; **3,** 308
Wallace, B., **8,** 83
Wallace, I. H. M., **11,** 533
Wallace, R. A., **9,** 463
Wallace, R. G., **10,** 207
Wallace, T. J., **1,** 309, 310, 348, 431, 642, 705, 927, 1145; **6,** 430
Wallace, T. W., **8,** 272; **9,** 123
Wallach, D., **9,** 400
Wallach, O., **1,** 250; **6,** 357
Wallbillich, G. E. H., **2,** 118
Wallbridge, M. G. H., **9,** 103
Wallenfels, K., **1,** 295, 350, 685; **8,** 163
Walley, D. R., **12,** 263, 282
Wallin, S., **11,** 483
Walling, C., **1,** 17, 45, 80, 89, 90, 93, 158, 197, 212, 1157, 1216, 1228; **2,** 50; **3,** 38; **5,** 160
Wallingford, V. H., **1,** 248, 502
Wallis, A. F. A., **5,** 555
Wallis, C. J., **7,** 229; **9,** 301
Wallis, E. S., **1,** 506; **11,** 105
Wallis, T. G., **7,** 268
Walls, F., **4,** 432
Walpole, A. L., **1,** 782
Walrond, R. E., **5,** 136
Walsh, E. J., Jr., **1,** 1193
Walsh, E. N., **6,** 215; **9,** 187
Walsh, H. G., **2,** 238
Walsh, M. R., **1,** 139
Walsh, W. L., **1,** 39, 1058
Walshaw, K. B., **1,** 526; **2,** 389
Walshaw, R. B., **3,** 49
Walshe, N. D. A., **12,** 127
Walsingham, R. W., **9,** 56
Walter, D., **4,** 35
Walter, L. A., **3,** 95
Walter, M., **1,** 518
Walter, R. I., **1,** 754
Waltermire, R. E., **11,** 502
Walters, D. R., **1,** 110
Walters, M. E., **8,** 482; **9,** 467
Walters, P. M., **1,** 686
Walters, R. M., **5,** 727
Walther, G., **1,** 1152, 1164, 1228
Walther, R., **1,** 842
Walther, W., **1,** 562
Walton, D. R. M., **2,** 438; **4,** 265; **5,** 45, 69, 375; **6,** 57, 58, 185, 652; **7,** 29; **9,** 311; **10,** 183, 326; **11,** 427
Walton, E., **1,** 986, 1157
Walton, R. A., **10,** 348
Wambsgans, A., **1,** 93
Wamhoff, H., **4,** 376; **6,** 76
Wamser, C. C., **8,** 15
Wan, C. N., **7,** 220
Wan, C. S. K., **12,** 416
Wancowicz, D. J., **8,** 531
Wand, M. D., **9,** 391; **11,** 302
Wang, A., **8,** 341
Wang, B. C., **12,** 387
Wang, C. H., **5,** 430
Wang, C. L. J., **6,** 464, 632; **7,** 213, 280; **9,** 110, 138, 174, 277
Wang, D., **12,** 220
Wang, E. J., **4,** 339
Wang, I. C., **5,** 394
Wang, J. L., **4,** 570
Wang, L. L., **1,** 904
Wang, N., **6,** 368
Wang, P. C., **8,** 289; **9,** 434; **10,** 67; **12,** 329
Wang, S., **1,** 332, 377; **5,** 127
Wang, S. S., **1,** 1015; **8,** 81; **9,** 415
Wang, S. W., **6,** 254
Wang, T. F., **12,** 439
Wang, T. S., **1,** 281
Wang, W. L., **12,** 4
Wang, Y., **3,** 49
Wang, Y. F., **7,** 132; **11,** 267
Wangen, L. E., **2,** 92
Wankel, R. A., **1,** 536
Wann, R. E., **1,** 1131, 1216
Wann, S. R., **11,** 479
Wannagat, U., **1,** 44, 1047; **3,** 262; **4,** 155, 409, 568; **5,** 45; **6,** 640; **11,** 68
Wannowius, H., **1,** 562, 764
Wanzlick, H. W., **1,** 187; **2,** 234, 430; **6,** 375
Warawa, E. J., **1,** 9
Warburton, W. K., **1,** 609, 652, 782
Ward, A. D., **12,** 41
Ward, D. I., **7,** 67
Ward, E. R., **1,** 756, 1100; **3,** 267
Ward, H. R., **1,** 688; **2,** 53, 278; **3,** 204; **4,** 339
Ward, J. A., **2,** 250
Ward, J. P., **2,** 127, 374; **3,** 93
Ward, J. W., **5,** 495
Ward, M. A., **6,** 457
Ward, M. G., **1,** 625
Ward, M. L., **1,** 1294
Ward, N. D., **3,** 101
Ward, P. J., **5,** 102
Ward, R. S., **6,** 675; **7,** 320;

11, 167; **12,** 71
Wardlaw W., **11,** 185
Ware, J. C., **1,** 162, 163
Wareing, J., **5,** 564
Warin, R., **7,** 76; **10,** 342
Waring, A., **2,** 289
Waring, A. J., **1,** 827
Waring, A. M., **2,** 217
Waring, D. R., **3,** 285
Warkentin, J., **10,** 58
Warnant, J., **1,** 465, 608, 808
Warne, T. M., Jr., **2,** 307
Warneke, J., **8,** 534
Warner, C. D., **6,** 526
Warner, C. R., **1,** 1193, 1251
Warner, P. O., **1,** 935
Warner, R. W., **11,** 514
Warnhoff, E. W., **1,** 144, 370, 609, 1064, 1131; **5,** 282; **6,** 284; **7,** 304; **10,** 342; **12,** 224
Warnock, W. D. C., **2,** 257
Warren, C. D., **1,** 309; **2,** 159; **3,** 122
Warren, C. K., **1,** 127
Warren, C. L., **3,** 171
Warren, J. D., **6,** 632; **10,** 15
Warren, M. E., Jr., **2,** 331
Warren, R. W., **2,** 21
Warren, S., **6,** 341; **7,** 26, 229, 375; **9,** 5, 301; **10,** 2; **11,** 425, 596; **12,** 477
Warrener, R. N., **2,** 295; **9,** 61; **10,** 313
Warthen, J. D., Jr., **5,** 591
Wartski, L., **9,** 235
Warwel, S., **6,** 201
Warwick, G. P., **1,** 430
Warzynski, R. J., **1,** 630
Washburn, L. C., **7,** 393
Washburn, W., **5,** 353
Washburn, W. H., **2,** 101
Washburn, W. N., **4,** 265
Washburne, S. S., **5,** 514, 720; **6,** 632
Washecheck, P. H., **5,** 725
Washita, H., **9,** 123
Wasielewski, C., **1,** 83, 1131
Wasmuth, D., **11,** 295
Wassenaar, S., **3,** 320
Wasserman, E., **5,** 19
Wasserman, H. H., **1,** 360, 465, 668, 1022; **2,** 202; **4,** 251, 430; **5,** 15, 292, 491, 677; **6,** 241, 430, 436; **7,** 41, 260; **8,** 102, 374; **9,** 341, 428; **10,** 294, 295, 361; **11,** 387; **12,** 365
Wassmundt, F. W., **2,** 336
Wasson, R. L., **1,** 72, 470
Wat, E. K. W., **2,** 292
Watabe, T., **10,** 113
Watabe, Y., **9,** 41
Watanabe, A., **10,** 391
Watanabe, F., **8,** 39
Watanabe, H., **5,** 330, 601; **8,** 453; **10,** 148, 194; **11,** 164, 428, 514; **12,** 171
Watanabe, I., **9,** 188
Watanabe, J., **11,** 243
Watanabe, K., **3,** 245; **6,** 152, 596; **9,** 123, 242, 462; **10,** 148; **11,** 494; **12,** 29, 502
Watanabe, M., **5,** 569; **10,** 75, 378; **11,** 1, 550
Watanabe, N., **12,** 428
Watanabe, S., **2,** 241, 289; **3,** 208; **4,** 240, 349, 350, 572; **5,** 470, 503; **6,** 415; **9,** 431; **10,** 87
Watanabe, T., **7,** 73; **12,** 327
Watanabe, W. H., **1,** 652; **3,** 302
Watanabe, Y., **4,** 465; **5,** 358, 625; **6,** 485, 486, 552; **7,** 183, 341; **8,** 92, 420, 478; **9,** 42, 51, 106, 108; **10,** 41, 148, 306, 401; **11,** 172, 361; **12,** 180
Waterhouse, A., **7,** 155
Waterhouse, I., **8,** 171, 498; **9,** 156, 417
Waterman, E. L., **4,** 354; **6,** 93, 361; **7,** 22, 250
Waters, J. A., **2,** 387
Waters, R. M., **4,** 247; **7,** 67
Waters, W. A., **1,** 159, 212, 562, 1015, 1131
Waters, W. L., **4,** 162; **5,** 427; **7,** 271
Waterson, D., **12,** 212
Watkins, G. L., **6,** 284
Watkins, J. J., **11,** 373
Watkins, R. J., **2,** 309
Watson, A. A., **3,** 320
Watson, C. R., **4,** 339; **5,** 391
Watson, H. G., **1,** 248; **4,** 481
Watson, J., **2,** 317
Watson, T. R., **5,** 436
Watson, W. D., **7,** 350
Watson, W. F., **1,** 348
Watson, W. H., **10,** 347; **11,** 77, 518
Watt, A. N., **2,** 145
Watt, C. T., **1,** 247
Watt, D. S., **2,** 462; **3,** 182; **4,** 151, 196, 473; **5,** 75, 412, 524; **6,** 114, 352, 430; **7,** 107, 260; **8,** 168, 229; **12,** 294, 337, 412
Watt, G. W., **1,** 56
Wattanasin, S., **8,** 299; **9,** 218; **10,** 165, 373
Watthey, J. W. H., **1,** 926
Watts, C. T., **3,** 224; **4,** 383; **7,** 288
Watts, L., **6,** 150
Watts, M. L., **1,** 67; **2,** 33
Watts, P. C., **3,** 191; **5,** 421
Watts, W. E., **5,** 266; **6,** 359; **12,** 167
Waugh, F., **4,** 265; **5,** 45, 69
Waugh, T. D., **1,** 80, 219, 414, 533, 999, 1049
Wawrzyniewicz, M., **4,** 31
Wawzonek, S., **1,** 452, 1011, 1040, 1073; **3,** 178; **5,** 254, 389
Wayaku, M., **6,** 653; **7,** 314
Wayda, A. L., **8,** 69
Waykole, L., **12,** 378
Wayland, B. B., **12,** 467
Wayne, W., **1,** 24, 654
Weaver, J. D., **4,** 378; **8,** 50
Weaver, T. D., **2,** 195; **3,** 182
Weavers, R. T., **5,** 563
Webb, F. J., **10,** 314
Webb, G., **5,** 321
Webb, G. B., **1,** 609, 652
Webb, I. D., **2,** 292
Webb, J. L., **5,** 337
Webb, M. J., **4,** 265
Webb, R. A., **5,** 510

Webb, R. F., **1,** 667, 1090, 1256
Webb, R. G., **1,** 57
Webb, R. L., **5,** 694
Webb, R. R., **10,** 161; **11,** 292, 363; **12,** 303, 391
Webb, T. E., **4,** 391
Webb, T. R., **8,** 64
Webber, G. M., **1,** 56, 612
Webber, J. M., **1,** 68
Webber, S. E., **12,** 476
Weber, A., **5,** 552; **10,** 11
Weber, J., **5,** 205
Weber, K. H., **1,** 129, 1056
Weber, L., **1,** 1101; **2,** 344; **5,** 552; **6,** 450; **8,** 217
Weber, R., **1,** 1075
Weber, S., **2,** 159, 252, 354
Weber, W., **11,** 192
Weber, W. P., **4,** 31; **5,** 28, 155, 644; **6,** 35; **7,** 20, 79; **8,** 240; **9,** 361; **11,** 150; **12,** 220, 321
Webster, B., **1,** 1179
Webster, B. R., **2,** 213
Webster, C. J., **1,** 127
Webster, D. E., **1,** 892; **6,** 426
Webster, E. R., **1,** 379
Webster, J. A., **5,** 294
Webster, O. W., **4,** 157; **5,** 176, 179
Webster, R. G., **4,** 501; **5,** 253
Wechter, W. J., **1,** 75, 207
Weddige, H., **1,** 250
Weedon, A. C., **9,** 237, 477; **12,** 416
Weedon, B. C. L., **1,** 127, 144, 155, 165, 170, 360, 937; **2,** 19, 263, 274; **3,** 39, 194; **6,** 514; **12,** 194
Weeks, J. L., **6,** 670
Weeks, P. D., **5,** 32; **6,** 573
Weerasooriya, U., **9,** 181; **10,** 157
Wege, G., **4,** 477
Wege, P. M., **4,** 254
Wegener, J., **6,** 2
Wegler, R., **1,** 487, 1121; **3,** 88; **4,** 497
Wegner, E., **1,** 125
Wegner, P., **1,** 1236; **3,** 316
Wehl, J., **1,** 281; **3,** 275
Wehlacz, J. T., **6,** 165
Wehle, D., **12,** 187
Wehman, A. T., **4,** 573; **10,** 225
Wehner, G., **6,** 633; **7,** 188; **9,** 128, 129; **10,** 416
Wehrli, H., **10,** 122
Wehrli, P. A., **4,** 411; **5,** 49; **8,** 400
Wehrmeister, H. L., **1,** 320; **2,** 329; **3,** 314
Weichert, R., **11,** 193
Weichet, J., **1,** 764
Weidemann, P., **4,** 397
Weider, P., **9,** 449
Weidig, C., **4,** 389
Weidinger, H., **1,** 289; **2,** 134; **7,** 424
Weidmann, B., **10,** 422; **11,** 378, 379; **12,** 356, 358
Weidmann, C., **1,** 561
Weidmann, H., **9,** 64; **12,** 444
Weiffen, J., **11,** 127
Weigel, H., **1,** 1055; **2,** 378; **10,** 375
Weigel, L. O., **7,** 277; **8,** 319; **9,** 335; **10,** 407
Weigele, M., **2,** 256
Weigert, F. J., **5,** 176, 503
Weigold, H., **6,** 179
Weijlard, J., **1,** 98, 476, 562
Weil, E. D., **1,** 1122
Weil, J. A., **1,** 536; **6,** 670
Weil, T., **2,** 417
Weil, T. A., **5,** 205
Weiland, J. H. S., **3,** 97
Weiler, L., **5,** 454; **8,** 62; **9,** 65, 114, 334; **10,** 289; **11,** 34
Weimann, G., **1,** 189, 1138
Weimar, W. R., Jr., **3,** 134, 322
Weinberg, D. S., **1,** 1249
Weinberg, M. L. D., **8,** 199
Weinberg, N. L., **2,** 191
Weindlein, J., **9,** 224
Weiner, M. A., **1,** 19, 1250, 1274; **4,** 53; **11,** 16
Weiner, N., **1,** 357, 674, 1073
Weiner, S. A., **1,** 318
Weinert, J., **7,** 52
Weingarten, H., **2,** 400, 415; **4,** 430; **9,** 470; **12,** 502
Weinges, K., **8,** 395
Weingold, D. H., **11,** 502
Weinheimer, A. J., **1,** 909, 911; **4,** 31; **5,** 725
Weinkauff, O. J., **1,** 347, 1203; **4,** 85
Weinmayr, V., **3,** 337
Weinreb, S. M., **4,** 579; **5,** 713; **8,** 182, 507; **9,** 16, 177; **11,** 115, 121, 202; **12,** 420, 455
Weinshenker, N. M., **3,** 67; **5,** 539; **6,** 92, 582
Weinstein, B., **1,** 767; **4,** 54; **5,** 78; **7,** 247, 377
Weinstein, R. M., **11,** 111; **12,** 4
Weinstein-Lanse, F., **5,** 667
Weinstock, J., **1,** 518, 685, 1043, 1284; **2,** 173; **3,** 19; **4,** 228, 268, 441; **6,** 83; **7,** 182
Weinstock, L. M., **6,** 412; **9,** 107; **11,** 189, 249, 536
Weintraub, P. M., **1,** 124
Weis, C. D., **1,** 231, 1247; **6,** 173; **7,** 100
Weis, K., **1,** 655
Weisbach, J. A., **3,** 297
Weisblat, D. I., **1,** 526, 754, 1209
Weischedel, F., **5,** 329
Weise, A., **2,** 159; **3,** 123
Weisenborn, F. L., **1,** 1080; **11,** 341
Weisenfeld, R. B., **8,** 306; **9,** 262, 285; **11,** 47; **12,** 61
Weiser, H. B., **9,** 237
Weisgerber, C. A., **1,** 50; **5,** 644
Weisgerber, G., **2,** 216
Weisheit, W., **5,** 205
Weisman, G. R., **5,** 609; **11,** 153
Weiss, A., **1,** 151, 562; **5,** 186
Weiss, C., **8,** 228
Weiss, E., **1,** 260; **2,** 7
Weiss, G., **4,** 475
Weiss, H. D., **2,** 155

Weiss, H. G., **5,** 416
Weiss, J., **1,** 1144
Weiss, K., **1,** 247
Weiss, L. B., **7,** 88; **8,** 373
Weiss, M., **1,** 401
Weiss, M. J., **1,** 40, 61, 150, 603, 942; **2,** 219, 348; **4,** 517
Weiss, R., **1,** 1027, 1121; **5,** 615; **7,** 235; **10,** 203, 299
Weiss, R. H., **6,** 192
Weiss, S., **5,** 168
Weiss, U., **1,** 601; **11,** 194
Weiss, W., **1,** 476; **4,** 260
Weissberger, A., **1,** 481, 678, 842, 1177; **4,** 497
Weissberger, E., **6,** 306
Weissenberg, M., **8,** 451
Weissenfels, M., **1,** 652
Weissgerber, R., **1,** 933
Weisshaus, S. Z., **4,** 239
Weissman, P. M., **1,** 624, 625
Weissman, S. I., **1,** 712
Weisz, A., **12,** 277
Weisz, I., **3,** 317
Weitemeyer, C., **5,** 681
Weitz, E., **1,** 470
Weizmann, C., **1,** 66
Weizmann, M., **1,** 772
Welch, A. D., **10,** 125
Welch, E., **1,** 841, 1292
Welch, F. J., **11,** 596
Welch, J. T., **6,** 286, 474; **7,** 248, 416, 417; **8,** 113; **9,** 400
Welch, R. W., **1,** 758
Welch, S. C., **5,** 410, 689; **8,** 284, 482; **9,** 194, 438, 467; **11,** 527; **12,** 217
Welch, W. M., **8,** 425
Welcher, A. D., **1,** 248, 253, 1139
Welcher, R. P., **4,** 503
Welford, M., **2,** 389
Welker, M. E., **12,** 3
Wellander, L. L., **11,** 203
Wellard, N. K., **7,** 277
Weller, H. N., **10,** 365
Weller, T., **10,** 234, 381
Welling, L. L., **6,** 76
Wellman, G. R., **7,** 338
Wellman, J., **10,** 101
Wellman, W. F., **1,** 905
Wells, A. F., **7,** 324
Wells, A. G., **6,** 35
Wells, D., **5,** 406; **6,** 32; **8,** 455
Wells, G. J., **11,** 286; **12,** 462
Wells, I. C., **5,** 700
Wells, J. I., **3,** 69
Wells, P. B., **5,** 321; **6,** 426
Wells, P. R., **9,** 442
Wells, R. J., **9,** 442
Wells, W. W., **1,** 427
Welstead, W. J., Jr., **1,** 695, 841; **2,** 59
Welter, H., **10,** 421
Welter, M., **2,** 321
Wempen, I., **2,** 412
Wemple, J., **6,** 252, 338, 542; **7,** 160, 207; **8,** 509
Wemple, J. N., **1,** 191
Wendel, K., **1,** 116, 1164; **2,** 61
Wendelborn, D. F., **8,** 25
Wender, I., **1,** 155, 228, 520, 570, 615, 723, 1000; **5,** 205
Wender, P. A., **4,** 334; **7,** 193, 236; **8,** 72, 336; **9,** 273, 398; **10,** 71, 264, 289, 328; **11,** 373; **12,** 275
Wenderoth, B., **10,** 138; **11,** 378; **12,** 356
Wendisch, D., **5,** 574
Wendland, R., **1,** 147, 1049
Wendler, N. L., **1,** 37, 59, 152, 160, 536, 642, 664, 911, 927, 933, 974, 1004, 1087, 1090, 1171, 1178, 1187; **2,** 159, 252, 354, 434; **7,** 40, 45; **11,** 498
Wendschuh, P. H., **5,** 511
Wenis, E., **1,** 289
Wenkert, E., **1,** 89, 273, 603, 652, 1178, 1273; **2,** 9; **3,** 41, 199, 258; **5,** 10; **8,** 188; **9,** 13, 148, 408; **10,** 426; **11,** 166, 358; **12,** 141, 228
Wenthe, A. M., **2,** 101
Wentland, M. P., **2,** 285
Wentrup, C., **9,** 309
Wentworth, R. A. D., **5,** 742; **8,** 522
Wentworth, S. E., **3,** 292; **6,** 142
Wentworth, W. A., **7,** 225
Wepplo, P. J., **5,** 303; **6,** 312
Wepster, B. M., **1,** 1167; **3,** 291
Werber, F. X., **1,** 904
Wermuth, C. G., **5,** 4, 320, 601, 609; **6,** 458; **8,** 166
Werner, A., **5,** 717
Werner, E., **2,** 238; **4,** 550
Werner, G., **5,** 170
Werner, H., **11,** 401
Werner, J. J., **3,** 336
Werner, N. W., **2,** 376
Werner, U., **10,** 444
Werner, W., **3,** 89; **5,** 200
Wernsmann, P., **10,** 78
Wersin, G., **7,** 91; **10,** 122
Werstiuk, E. S., **5,** 627; **6,** 605
Wertheimer, H. C., **4,** 191
Werthemann, L., **3,** 302
Weselowsky, V. W., **5,** 596
Wessel, H., **2,** 67
Wesseler, E. P., **4,** 529
Wessels, J. C. M., **8,** 103
Wessely, F., **1,** 562
West, C. T., **5,** 695; **6,** 616; **7,** 9
West, D. E., **2,** 33; **3,** 31, 171
West, D. L., **10,** 320
West, E. S., **1,** 295
West, H. D., **1,** 652
West, P. J., **6,** 583
West, R., **1,** 963, 1108, 1274; **2,** 34, 91; **3,** 119; **5,** 86, 322; **6,** 24, 58, 553; **12,** 37, 86
West, R. C., **9,** 352
West, R. W., **1,** 208
West, T. S., **1,** 528
West, W., **9,** 93; **11,** 587
Westall, S., **7,** 28
Westberg, H. H., **2,** 236; **4,** 435
Westcott, N. D., **4,** 37
Westcott, P. A., **4,** 100
Westdorp, I., **11,** 529
Westerbeek, J. M. M., **11,** 423

Westerlund, C., **11,** 228
Westermann, J., **10,** 138, 271
Westfelt, L., **2,** 366
Westheimer, F. H., **1,** 328, 1083; **2,** 75, 310; **4,** 345; **5,** 707; **6,** 37
Westmijze, H., **7,** 164, 226; **8,** 322; **9,** 111, 277, 334; **10,** 290, 350, 434; **12,** 350
Westmore, J. B., **5,** 75
Weston, A. F., **4,** 276
Weston, A. W., **1,** 682; **3,** 185; **6,** 526; **8,** 446
Weston, G., **1,** 219
Weston, G. O., **1,** 219
Westphal, D., **6,** 318
Westphal, O., **4,** 457
Westphal, U., **1,** 45
Westphalen, K. O., **11,** 193
Westra, J. G., **5,** 340
Wetter, H., **6,** 107; **9,** 241
Wetter, W. P., **1,** 1163
Wettlaufer, D. G., **12,** 383
Wettstein, A., **1,** 50, 87, 262, 562, 608, 624, 802, 1000, 1015
Wetzel, W. H., **1,** 500
Weuster, P., **9,** 184; **12,** 205
Wexler, B. A., **10,** 289
Wexler, S., **6,** 436
Wey, J. E., **12,** 571
Weyell, D. J., **1,** 274
Weyenberg, D. R., **3,** 103
Weyerstahl, P., **1,** 756; **2,** 197; **3,** 45; **4,** 31; **5,** 28; **6,** 74, 263; **7,** 102, 298, 394; **8,** 194; **9,** 97
Weygand, F., **1,** 322, 347, 414, 486, 494, 595, 669, 694, 851, 1188, 1220, 1271; **2,** 134, 187; **3,** 143; **7,** 247; **12,** 44
Weygand, F. W., **1,** 116
Weyler, W., Jr., **4,** 57, 148; **5,** 639; **6,** 78; **7,** 43
Weymouth, F. J., **1,** 199
Whalen, D. L., **3,** 251, 252
Whalen, D. M., **1,** 224, 1196
Whaley, A. M., **1,** 1292
Whaley, W. M., **4,** 395; **5,** 540, 542; **9,** 308; **10,** 161
Whalley, M., **1,** 859, 1269
Whalley, W. B., **1,** 322; **4,** 79; **6,** 467
Whang, J. J., **7,** 241
Wharton, P. S., **1,** 360, 444, 668; **4,** 194
Whatley, L. S., **6,** 24
Wheaton, G. A., **7,** 60
Wheeler, A. S., **1,** 444
Wheeler, D. M. S., **9,** 462; **10,** 396
Wheeler, J. W., **1,** 563
Wheeler, M. M., **10,** 396
Wheeler, O. H., **1,** 411, 598, 624, 643; **5,** 667
Wheeler, T. N., **3,** 45
Wheeler, T. S., **1,** 392, 962, 963
Wherli, P. A., **5,** 562
Whimp, P. O., **5,** 172
Whistler, R. L., **1,** 476; **2,** 13, 180, 323
Whitcombe, M. J., **10,** 48
White, A. M. S., **1,** 598; **5,** 391
White, A. W., **10,** 71
White, C., **8,** 153
White, C. T., **11,** 349
White, D. A., **5,** 685; **6,** 475
White, D. L., **12,** 333
White, D. M., **1,** 61
White, D. R., **6,** 103; **7,** 170; **9,** 75, 299, 337, 406
White, E., **7,** 338
White, E. H., **1,** 329, 697; **2,** 176, 282; **5,** 672
White, J., **1,** 755
White, J. D., **1,** 652; **2,** 378; **3,** 71; **5,** 42, 98, 756; **6,** 497; **7,** 64, 173, 200, 345, 406; **9,** 517; **10,** 167; **11,** 319
White, J. E., **1,** 885
White, J. F., **7,** 365
White, J. G., **3,** 53
White, J. W., **1,** 1254
White, K., **5,** 756
White, K. B., **10,** 214
White, M. A., **10,** 116
White, M. R., **7,** 170; **12,** 532
White, P. D., **6,** 338
White, P. S., **10,** 373
White, R., **5,** 113
White, R. E., **4,** 14
White, R. W., **1,** 819
White, S., **10,** 12
White, W., **12,** 502
White, W. A., **2,** 400, 415; **9,** 470
White, W. N., **1,** 889; **2,** 29; **6,** 341
Whited, E. A., **5,** 649
Whitehead, C. W., **1,** 1095, 1210
Whitehouse, M. L., **9,** 293
Whitehouse, N. R., **5,** 427
Whitehouse, P. A., **2,** 60; **3,** 40
Whitehouse, R. D., **5,** 26, 276
Whitehurst, J. S., **2,** 72
Whitelaw, S. P., **1,** 667
Whiteley, M. A., **1,** 208
Whiter, P. F., **1,** 1249; **2,** 330, 446; **5,** 243, 460
Whitesell, J. K., **5,** 511; **6,** 338; **8,** 101, 197; **10,** 245; **11,** 415; **12,** 389
Whitesides, G. M., **2,** 153, 283, 307, 400; **3,** 81, 112, 153, 154, 278; **4,** 183, 564; **5,** 148, 175, 454, 600; **6,** 124, 343, 495; **7,** 83, 141, 324; **9,** 147, 221; **11,** 54
Whitesides, T., **2,** 236
Whitesides, T. H., **5,** 96
Whitesitt, C. A., **8,** 498
Whitfield, G. H., **2,** 305
Whitham, G. H., **1,** 101, 723; **2,** 53; **4,** 204; **5,** 399, 410; **7,** 333; **8,** 290; **9,** 82; **10,** 183
Whithan, G. H., **6,** 628
Whiting, D. A., **5,** 257; **7,** 197; **8,** 238
Whiting, M. C., **1,** 12, 186, 281, 318, 360, 389, 424, 444, 595, 643, 668, 669, 723, 833, 1040, 1139, 1145, 1187; **2,** 19, 105, 295, 403; **3,** 76, 153, 202; **5,** 86; **6,** 104, 126, 451; **11,** 490; **12,** 379

Whitlock, B. J., **6,** 354; **8,** 66
Whitlock, H. W., Jr., **1,** 1081, 1258; **3,** 258; **4,** 349; **6,** 354; **8,** 66
Whitman, B., **1,** 720, 802
Whitman, G. H., **8,** 290
Whitman, G. M., **1,** 228
Whitmore, F. C., **1,** 500, 652, 654, 741, 1100, 1218, 1284; **6,** 84; **8,** 263; **9,** 74
Whitmore, M. G., **1,** 741, 1100
Whitney, C. C., **2,** 142; **4,** 38, 110, 161; **5,** 347, 467; **7,** 181; **11,** 266
Whitney, D. B., **12,** 479
Whittaker, D., **3,** 10
Whittaker, G., **4,** 172
Whittaker, M., **8,** 260
Whittaker, N., **1,** 634; **3,** 157
Whitten, C. E., **3,** 113; **4,** 128, 183; **5,** 243, 396, 415; **6,** 343, 389; **7,** 211, 231; **9,** 313
Whittier, G. L., **9,** 480
Whittle, A. J., **10,** 22
Whittle, B. J., **11,** 36
Whittle, C. W., **1,** 979
Wiartalla, R., **9,** 11
Wiaux-Zamar, C., **7,** 66; **11,** 111
Wiberg, E., **1,** 35; **9,** 56
Wiberg, K. B., **1,** 89, 195, 402, 430, 471, 999, 1005; **2,** 79; **3,** 129, 143, 178; **4,** 113, 350; **5,** 205, 507, 563, 574
Wiberg, N., **4,** 155
Wiberley, S. E., **2,** 360
Wicha, J., **2,** 243, 359; **3,** 220; **5,** 106, 353; **6,** 74
Wichelhaus, H., **1,** 392, 1210
Wichers, E., **1,** 892
Wichman, D., **11,** 194
Wicholas, M., **4,** 107
Wichterle, O., **1,** 8
Wick, A., **6,** 222
Wick, A. E., **1,** 272
Wick, A. K., **2,** 351
Wick, E., **4,** 87
Wick, E. H., **4,** 191
Wickberg, B., **2,** 366; **5,** 249; **7,** 342
Wickens, J. C., **1,** 567
Wickham, P. P., **4,** 40
Widdowson, D. A., **1,** 642; **3,** 224; **4,** 383; **5,** 349, 437; **7,** 16, 111, 421; **9,** 400; **11,** 604
Wideman, L. G., **2,** 72
Widequist, S., **2,** 92
Widera, R. P., **5,** 28; **7,** 20
Widiger, G. N., **6,** 11
Widler, L., **10,** 422; **11,** 378
Widmer, E., **1,** 34
Widmer, J., **9,** 434
Widmer, U., **5,** 700; **11,** 559
Wiebecke, G. H., **11,** 604
Wiechert, K., **1,** 456
Wiechert, R., **2,** 211, 362; **3,** 126; **6,** 411; **9,** 324; **12,** 79
Wiechman, B., **10,** 157
Wieckert, R., **10,** 216
Wieczorek, J. J., **1,** 612
Wiedhaup, K., **2,** 87; **4,** 23; **5,** 631
Wiehler, G., **1,** 72
Wiejak, S., **4,** 66; **5,** 87
Wiel, J. ter, **3,** 155
Wieland, A. A., **2,** 59
Wieland, D. M., **3,** 112; **4,** 183; **6,** 245
Wieland, H., **1,** 141, 796, 832; **2,** 174
Wieland, P., **1,** 562, 608, 624, 1203; **2,** 173, 219, 250, 422; **3,** 188, 293; **5,** 344
Wieland, T., **1,** 86, 97, 330, 494, 755, 1101, 1164, 1203; **2,** 193, 279, 423; **3,** 96; **4,** 552
Wieleseck, R., **1,** 1084
Wielogorski, Z., **9,** 343
Wiemann, J., **4,** 427
Wiemer, D. F., **12,** 60
Wierenga, W., **10,** 182
Wiering, J. S., **7,** 311
Wieringa, J. H., **5,** 491
Wiersema, A. K., **2,** 87
Wies, R. A., **2,** 368
Wieschollek, R., **12,** 353
Wiesner, K., **1,** 219; **2,** 301; **5,** 595; **10,** 188; **11,** 325, 382
Wiessler, M., **8,** 272
Wiest, H., **1,** 1136
Wife, R. L., **4,** 48; **5,** 585
Wigfield, D. C., **2,** 163; **5,** 391
Wightman, R. H., **1,** 310, 859; **5,** 517; **6,** 622; **7,** 119; **10,** 335; **11,** 160
Wihler, H. D., **7,** 404
Wijers, H. E., **5,** 208
Wijneberg, J. B. P. A., **7,** 331; **9,** 83
Wijsman, A., **8,** 90; **11,** 180
Wikholm, R. J., **2,** 165; **4,** 107, 199
Wilberg, N., **10,** 120
Wilby, A. H., **11,** 393
Wilchek, M., **1,** 110, 581, 859
Wilcox, C. S., **8,** 245, 284; **12,** 528
Wilcox, H. W., **1,** 754
Wilcox, M. E., **2,** 348
Wild, D., **4,** 172
Wild, F., **1,** 77, 319
Wilde, H., **1,** 445
Wildeman, J., **6,** 600; **7,** 246; **8,** 494
Wilding, I. G. E., **1,** 248; **4,** 481
Wildman, W. C., **1,** 389; **3,** 332
Wilds, A. L., **1,** 37, 56, 195, 251, 268, 389, 500, 690, 703, 771
Wildsmith, E., **4,** 507
Wiley, D. W., **1,** 273, 963, 1136
Wiley, G. A., **1,** 1249
Wiley, P., **4,** 497
Wiley, R. A., **6,** 554
Wiley, R. H., **1,** 11, 12, 156, 158, 445, 465, 658, 962, 963, 976; **2,** 263; **4,** 497; **6,** 292
Wilgus, H. S., III, **2,** 10
Wilhalm, B., **6,** 675
Wilhelm, R. S., **11,** 373
Wilka, E. M., **12,** 193, 480
Wilke, G., **1,** 262, 1154; **2,** 422; **3,** 101; **4,** 35; **5,** 471, 510; **8,** 69

Wilkes, M. C., **9,** 391; **11,** 596
Wilkins, A. L., **6,** 304, 514
Wilkins, C. K., Jr., **1,** 677
Wilkins, C. L., **6,** 521
Wilkins, J. M., **6,** 225; **8,** 302
Wilkins, S. W., **5,** 571
Wilkinson, A. J., **2,** 110
Wilkinson, G., **1,** 140, 182, 392, 778, 1142, 1252; **2,** 453; **4,** 421; **5,** 193, 331, 498, 700; **6,** 105, 426, 454, 650; **7,** 60, 313; **8,** 109; **9,** 426; **10,** 149, 303; **11,** 253; **12,** 202
Wilkinson, J. M., **10,** 160, 424
Wilkinson, M., **3,** 93
Wilkinson, P. A., **1,** 754
Wilkinson, R. G., **1,** 39
Willard, A. K., **6,** 279; **7,** 210
Willard, C., **1,** 285
Willard, J. J., **2,** 27, 274
Willems, A. G. M., **1,** 445
Willenberg, E., **1,** 408
Willet, J. D., **5,** 351
Willett, J. E., **1,** 851
Willey, K., **11,** 407
Willey, P. R., **12,** 280
Willfang, G., **1,** 1211
Willgerodt, C., **1,** 506, 1121
Willholm, B., **11,** 472
Williams, A. J. S., **2,** 289
Williams, A. L., **3,** 69
Williams, B., **1,** 321
Williams, B. J., **8,** 230
Williams, C. C., **4,** 218
Williams, D. H., **1,** 104, 444, 731; **2,** 378
Williams, D. J., **6,** 536; **11,** 39, 604
Williams, D. K., **6,** 660
Williams, D. M., **1,** 967; **3,** 128
Williams, D. R., **4,** 204, 413; **6,** 487; **7,** 17; **9,** 43; **10,** 12, 21; **11,** 152, 158; **12,** 112
Williams, E., **8,** 397
Williams, E. B., Jr., **4,** 45
Williams, E. H., **3,** 244
Williams, F. T., Jr., **1,** 952
Williams, G. J., **2,** 68, 242; **7,** 221
Williams, G. M., **11,** 401
Williams, H. J., **5,** 542; **8,** 344
Williams, H. R., **1,** 17; **4,** 152
Williams, H. W. R., **7,** 312
Williams, J. A., **1,** 198
Williams, J. B., **5,** 206
Williams, J. F. A., **1,** 80
Williams, J. H., **1,** 1178
Williams, J. K., **1,** 18, 1136
Williams, J. L., **4,** 437
Williams, J. L. R., **1,** 33
Williams, J. R., **5,** 290, 491, 727; **8,** 374
Williams, J. W., **1,** 266, 531, 870, 1117; **6,** 40
Williams, K., **6,** 38
Williams, K. I. H., **1,** 164, 814, 1081
Williams, L., **11,** 393; **12,** 13
Williams, M. W., **1,** 97, 296; **9,** 51
Williams, N. R., **1,** 248, 253, 1041; **2,** 133
Williams, P. H., **1,** 465, 471, 477, 827
Williams, R. E., **4,** 223; **11,** 47
Williams, R. G., **5,** 391
Williams, R. M., **5,** 233; **10,** 170; **11,** 416; **12,** 437
Williams, R. O., **1,** 260, 723
Williams, R. V., **9,** 72; **11,** 42
Williams, S. B., **9,** 136
Williams, T. H., **9,** 260
Williams, V. Z., Jr., **2,** 21; **3,** 7; **4,** 350; **6,** 17
Williams, W. M., **4,** 333, 344; **5,** 93; **8,** 334
Williamson, D. H., **7,** 68
Williamson, K. L., **1,** 149, 856, 1143; **2,** 76; **5,** 631
Williamson, S. M., **1,** 909, 1041, 1044
Williard, P. G., **8,** 245; **10,** 50; **12,** 434
Willimann, L., **1,** 76
Willis, A. C., **12,** 333
Willis, B. J., **3,** 320; **4,** 550; **10,** 240
Willis, C., **9,** 147
Willis, C. J., **1,** 1237
Willis, J. P., **11,** 253, 452
Willis, R. G., **1,** 926
Willis, W. W., Jr., **11,** 60
Willmott, W. E., **8,** 159, 435
Willner, D., **1,** 60
Willstaetter, R., **1,** 536, 561, 1015; **2,** 25; **6,** 445
Willy, W. E., **6,** 224, 586; **11,** 346; **12,** 445
Wilmarth, W. K., **1,** 536
Wilms, H., **1,** 832
Wilputte-Steinert, L., **5,** 143
Wilshire, J. F. K., **2,** 178
Wilsmore, N. T. M., **1,** 7
Wilson, A. N., **1,** 637, 860, 873, 1157
Wilson, B. M., **1,** 1090
Wilson, C. A., II, **7,** 383
Wilson, C. V., **1,** 500, 595, 859, 963, 1008, 1019, 1095, 1216, 1220; **5,** 523, 655; **7,** 324
Wilson, C. W., **1,** 66
Wilson, D. A., **1,** 735; **2,** 165
Wilson, D. R., **4,** 535; **8,** 135; **11,** 203
Wilson, E., **1,** 962, 1000, 1178, 1284
Wilson, E. M., **6,** 453
Wilson, E. R., **3,** 178
Wilson, G. E., Jr., **5,** 115; **7,** 207; **8,** 393
Wilson, H. L., **5,** 339
Wilson, I. B., **1,** 1157
Wilson, J. G., **1,** 430, 1102
Wilson, J. M., **12,** 312, 383
Wilson, J. S., **10,** 307
Wilson, J. W., **1,** 154, 276, 1131; **12,** 13
Wilson, J. Z., **12,** 127, 534
Wilson, K. E., **3,** 102
Wilson, M. E., **11,** 54
Wilson, N., **9,** 304
Wilson, R., **5,** 470
Wilson, R. A. L., **1,** 594; **5,** 77; **7,** 64
Wilson, R. M., **10,** 202
Wilson, S. E., **7,** 113, 373
Wilson, S. R., **6,** 151; **7,** 415; **9,** 306, 494; **10,** 71; **11,** 437, 575; **12,** 70

Wilson, T., **2,** 101
Wilson, T. L., **1,** 865
Wilson, W., **1,** 928, 1210; **5,** 345
Wilson, W. C., **1,** 1269
Wilt, J. W., **3,** 50; **4,** 112; **6,** 400
Wilt, M. H., **1,** 782
Wiltshire, C., **4,** 568
Wilzbach, K. E., **1,** 1251; **4,** 339
Winans, R. E., **10,** 85
Windaus, A., **1,** 141, 172, 652, 795, 1284; **2,** 139; **3,** 61
Windholz, T. B., **2,** 426
Windus, W., **1,** 1163, 1167
Wineman, R. J., **1,** 1022
Winer, A. M., **6,** 487
Wing, R. M., **9,** 442
Wingler, F., **1,** 1022; **3,** 258
Wingrove, A. S., **2,** 118, 393
Winkeler, H. D., **11,** 407
Winkelmann, E. H., **1,** 80, 563; **9,** 393
Winkelmueller, W., **1,** 44, 396
Winkhaus, G., **12,** 173
Winkler, H. J. S., **1,** 978; **2,** 273; **3,** 200
Winkler, J., **7,** 93
Winkler, J. D., **11,** 606; **12,** 449
Winslow, E. C., **1,** 231
Winstead, M. B., **1,** 484; **5,** 601
Winstein, S., **1,** 563, 606, 705, 778, 926, 1004, 1011, 1022, 1137, 1143, 1151; **2,** 19, 397; **3,** 277; **4,** 154, 323; **5,** 319, 497
Winston, A., **1,** 1108; **2,** 389
Winter, D. P., **8,** 81
Winter, E., **1,** 147; **2,** 10
Winter, G., **5,** 195
Winter, H. W., **9,** 309
Winter, M., **1,** 281; **2,** 422
Winter, P. K., **1,** 23
Winter, R., **1,** 989
Winter, R. A. E., **1,** 1234; **2,** 441
Winter, R. E. K., **2,** 99, 126; **3,** 338
Winter, S. R., **4,** 465
Winterfeldt, E., **6,** 2, 141, 201
Winterman, D. R., **3,** 22
Winternitz, F., **2,** 132; **6,** 470
Wintersberger, E., **2,** 147
Winterstein, A., **1,** 22, 350, 536, 1284; **8,** 532
Wintersteiner, O., **1,** 110, 782, 796, 1184
Wintner, C., **5,** 329; **7,** 171
Winwick, T., **12,** 70
Wippel, H. G., **1,** 1247; **5,** 694
Wiriyachitra, P., **8,** 359
Wirkkala, R. A., **1,** 1221
Wirth, D. D., **10,** 452
Wirth, H. O., **1,** 500
Wirth, J. G., **3,** 245
Wirth, R. K., **8,** 324
Wirth, R. P., **8,** 94
Wirz, J., **7,** 428
Wis, M. L., **8,** 232
Wisansky, W. A., **1,** 169, 500
Wise, H. E., Jr., **1,** 1054
Wise, P. H., **5,** 325
Wisegarver, B. B., **1,** 402
Wiselogle, F. Y., **1,** 1284
Wiseman, J. R., **2,** 173; **3,** 103; **4,** 405, 542; **5,** 551; **8,** 336
Wishnok, J. S., **4,** 339; **5,** 639
Wiskott, E., **5,** 681
Wislicenus, H., **1,** 22
Wislicenus, J., **1,** 970
Wissinger, J. E., **12,** 411
Wissmann, H., **10,** 333
Wissner, A., **5,** 614; **6,** 552; **8,** 523; **9,** 335, 512; **11,** 281
Wistrand, L. G., **6,** 52
Witherell, D. R., **1,** 1011
Withers, G. P., **7,** 220; **9,** 325
Withrow, J. R., **1,** 108
Witiak, D. T., **11,** 493
Witkop, B., **1,** 176, 274, 309, 505, 562, 860, 904, 1002; **2,** 107, 387; **3,** 127; **7,** 268; **9,** 191
Witnauer, L. P., **1,** 1269
Witt, K. E., **9,** 468
Wittcoff, H., **1,** 401, 1008
Wittekind, R. R., **5,** 345
Witten, C. H., **1,** 870, 1117
Wittenberg, D., **1,** 345, 1261; **5,** 361; **11,** 554
Wittenbrook, L. S., **2,** 104; **3,** 74; **4,** 215
Wittenburg, E., **2,** 370
Witter, A., **2,** 147
Wittig, G., **1,** 50, 77, 96, 195, 255, 268, 343, 573, 618, 643, 671, 678, 697, 832, 845, 909, 1022, 1141, 1246, 1247, 1261; **2,** 179, 249, 368; **3,** 69, 141, 184, 185, 258; **4,** 68, 163, 204, 209, 354, 356, 437, 491, 524; **5,** 120, 361, 416; **9,** 476; **11,** 103; **12,** 187
Wittreich, P. E., **1,** 1081
Wittstruck, T., **5,** 444
Witzel, T., **12,** 523
Witzemann, E. J., **1,** 952
Witzmann, E. J., **10,** 273
Wlos, R., **2,** 311
Wlostowska, J., **7,** 353; **8,** 461
Wluka, D. J., **1,** 34
Wnuk, J., **7,** 247
Woerffel, V., **1,** 139
Woessner, W., **7,** 191
Woessner, W. D., **4,** 218; **8,** 223
Wogan, G. N., **7,** 210
Woggon, W. D., **9,** 349
Wohl, A., **1,** 636; **4,** 48
Wohl, R. A., **5,** 20, 476, 639, 714
Wohler, F., **8,** 320
Wohllebe, J., **4,** 441, 459
Wojcicki, A., **12,** 3
Wojciechowska, H., **9,** 506
Wojciechowski, K., **9,** 361; **11,** 407; **12,** 329
Wojtowicz, J. A., **5,** 551
Wolak, R. A., **9,** 387
Wolak, R. P., Jr., **10,** 323
Wolber, G. J., **7,** 207
Wolf, C. N., **1,** 472, 1084
Wolf, D., **1,** 1075
Wolf, D. E., **1,** 730, 731
Wolf, F. J., **1,** 98, 562
Wolf, H. R., **5,** 734
Wolf, I., **2,** 23
Wolf, N., **12,** 444

Wolf, P. F., **3,** 279
Wolf, R. E., **4,** 290
Wolf, S. F., **6,** 36
Wolf, W. A., **2,** 62
Wolfe, J. F., **1,** 70
Wolfe, J. K., **1,** 411
Wolfe, J. R., Jr., **1,** 1143
Wolfe, S., **1,** 207; **2,** 39, 226, 395; **3,** 241, 244; **4,** 48, 53; **6,** 345; **12,** 576
Wolfel, G., **12,** 441
Wolff, C., **5,** 207
Wolff, G., **4,** 183
Wolff, H., **1,** 448
Wolff, I. A., **1,** 865; **2,** 139
Wolff, K., **1,** 956
Wolff, L., **1,** 832
Wolff, M., **12,** 416
Wolff, M. E., **1,** 624, 989; **5,** 639; **7,** 385
Wolff, R., **2,** 372
Wolff, R. E., **1,** 9
Wolff, S., **4,** 108; **7,** 278
Wolffenstein, R., **1,** 486
Wolfhagen, J. L., **2,** 69
Wolflehner, W., **9,** 64
Wolfmeier, U., **4,** 86
Wolfram, J., **11,** 170
Wolfrom, M. L., **1,** 66, 122, 324, 450, 652, 654, 658, 754, 755, 814, 1101, 1157, 1226; **2,** 13, 175, 180, 323; **3,** 143
Wolfstirn, K. B., **1,** 158
Wolinsky, J., **1,** 926, 1157, 1217; **2,** 238; **3,** 104; **5,** 14, 551; **10,** 209; **11,** 260
Wolinsky, L. E., **7,** 36
Wollenberg, R. H., **5,** 44; **6,** 56, 603; **7,** 229; **8,** 222, 279; **9,** 324, 414; **10,** 144; **11,** 458
Wollensak, J., **1,** 889
Wollmann, T., **12,** 257
Wollowitz, S., **8,** 25; **9,** 195
Wolman, Y., **2,** 43, 333; **6,** 78
Wolnak, B., **1,** 39, 320
Wolochowicz, I., **5,** 671; **6,** 588, 590
Wolovsky, R., **1,** 159, 1011, 1141; **2,** 88
Wolter, A., **8,** 188
Woltermann, A., **8,** 280; **9,** 509
Wolters, E. T., **4,** 45
Wolters, J., **5,** 371; **6,** 458
Wolthan, H., **1,** 1261
Wolthuis, E., **1,** 127
Womack, E. B., **1,** 255, 1100, 1163
Wonchoba, E., **10,** 304
Wong, C. K., **8,** 170, 522; **9,** 32, 35, 511; **10,** 21, 452
Wong, C. M., **4,** 15, 74, 85, 340, 472; **5,** 102, 170, 566, 632, 671, 686; **6,** 50, 229, 470, 497, 529, 543, 563; **7,** 365; **10,** 26
Wong, D. H., **10,** 216
Wong, E. L., **1,** 1081
Wong, G. B., **6,** 450
Wong, H., **8,** 143
Wong, H. N. C., **4,** 472; **7,** 324
Wong, J. Y., **6,** 92
Wong, L. T. L., **3,** 277; **4,** 425
Wong, P. K., **6,** 59; **8,** 36, 455
Wong, R. Y., **8,** 401
Woo, E. P., **3,** 113, 194; **4,** 472
Woo, S. H., **12,** 403
Woo, S. L., **7,** 279
Wood, A., **1,** 424
Wood, C. S., **1,** 500, 1087
Wood, G. W., **1,** 19, 144
Wood, H. B., **1,** 755; **2,** 163
Wood, H. C. S., **1,** 1055
Wood, J. F., **2,** 54
Wood, J. L., **1,** 1152
Wood, J. P., **4,** 570
Wood, J., Jr., **1,** 1289
Wood, K. H., **1,** 253
Wood, K. R., **2,** 130
Wood, N. F., **1,** 927; **3,** 188
Wood, R. D., **11,** 76
Wood, R. H., **4,** 406
Wood, S. E., **1,** 207
Woodard, S. S., **11,** 94, 95
Woodbridge, D. T., **6,** 224, 241; **8,** 29
Woodburn, H. M., **1,** 594, 954
Woodbury, R. P., **8,** 203, 525; **9,** 151
Woodcock, S. J., **1,** 248; **4,** 481
Woodgate, P. D., **7,** 360; **8,** 260; **9,** 249, 458; **10,** 395; **11,** 305, 428, 486, 515
Woodruff, E. H., **1,** 33
Woodruff, R. A., **3,** 183; **4,** 297; **5,** 189
Woods, G. F., **1,** 257; **4,** 397
Woods, G. F., Jr., **1,** 257
Woods, L. A., **2,** 153
Woods, L. L., **1,** 1220; **2,** 434
Woods, T. S., **6,** 543
Woodward, F. E., **12,** 127
Woodward, R. B., **1,** 9, 37, 235, 351, 364, 385, 389, 484, 573, 764, 890, 926, 927, 957, 970, 1004, 1116, 1254; **2,** 39, 48, 198; **3,** 49, 88, 251, 296; **4,** 82, 173, 540; **5,** 141, 334, 511; **6,** 629; **8,** 226; **9,** 318, 363; **11,** 268, 324, 524; **12,** 79, 94
Wooldridge, K. R., **1,** 129, 219
Wooley, D. W., **1,** 519
Woolfolk, E. O., **1,** 570, 832, 1237
Woolhouse, R. A., **1,** 1171
Woollett, G. H., **1,** 502
Woolsey, N. F., **1,** 95, 562; **5,** 114
Wooten, W. C., **11,** 184
Word, J. L., **9,** 177
Worden, L. R., **1,** 209, 581, 611; **2,** 6; **3,** 175
Work, J. D., **1,** 860
Wormald, P. C., **11,** 533
Worrall, D. E., **1,** 741
Worrall, I. J., **3,** 10
Worrall, R., **1,** 1226
Worrall, W. S., **1,** 926; **6,** 103
Worster, P. M., **9,** 3; **11,** 30; **12,** 221
Worth, B. R., **7,** 111
Worthington, P. A., **10,** 168; **12,** 38
Wotiz, J. H., **1,** 228; **2,** 182; **4,**

231
Wotring, R. W., Jr., **2,** 404
Wovkulich, P. M., **9,** 87, 96; **11,** 12
Woznow, R. J., **6,** 70
Wrede, F., **1,** 536
Wren, D., **4,** 490
Wriede, P., **1,** 712; **2,** 289
Wright, G. F., **1,** 195, 247, 1033; **3,** 225; **6,** 416
Wright, J. B., **1,** 97, 276
Wright, M., **2,** 53; **4,** 502
Wright, M. E., **12,** 46
Wright, M. J., **6,** 488
Wright, N. D., **5,** 230
Wright, P. W., **6,** 341
Wright, S. H. B., **5,** 294
Wright, S. W., **12,** 247
Wright, W. B., Jr., **2,** 61
Wright, W. D., **3,** 175
Wrighton, M., **5,** 143
Wrigley, T. I., **1,** 581, 795, 802
Wrobel, J., **10,** 296
Wrobel, J. E., **11,** 483; **12,** 70
Wrobel, S. J., Jr., **6,** 254
Wroble, R. R., **7,** 260
Wroebel, J. T., **5,** 667
Wu, C. S., **5,** 420, 470
Wu, D. K., **6,** 103
Wu, E. S. C., **5,** 661; **6,** 384; **7,** 429
Wu, G. S., **2,** 87; **5,** 160
Wu, J. S., **8,** 374; **9,** 428
Wu, M. T., **1,** 236, 904
Wu, S. M., **5,** 249
Wu, T. C., **5,** 319
Wucherpfenning, W., **8,** 43
Wudl, F., **4,** 203; **6,** 121
Wueest, H., **3,** 247; **4,** 283, 427; **5,** 9, 254; **6,** 77, 514; **8,** 125, 375, 488; **9,** 1, 70; **10,** 36
Wuensch, E., **1,** 524; **5,** 169; **7,** 91
Wuerthwein, E. U., **5,** 438
Wuest, H. H., **11,** 527
Wulff, C., **1,** 67
Wulff, G., **2,** 139; **4,** 345; **6,** 511
Wulff, J., **6,** 612
Wulff, W., **11,** 132
Wulff, W. D., **11,** 401; **12,** 136
Wulfman, D. S., **5,** 245, 246, 718, 742
Wulfman, D. W., **4,** 9
Wunderlich, K., **1,** 1018; **3,** 251; **5,** 714
Wunderly, S. W., **8,** 310
Wuonola, M. A., **5,** 141
Wursthorn, K. R., **9,** 492, 501
Wuts, P. G. M., **8,** 181; **10,** 106; **11,** 87; **12,** 397, 426
Wyatt, M., **5,** 471
Wychuyse, F., **9,** 182
Wycpalek, A. F., **9,** 380
Wydra, R., **12,** 555
Wyler, H., **2,** 348; **5,** 614
Wylie, A. G., **1,** 1055
Wyman, D. P., **1,** 1131
Wyman, L. J., **1,** 107, 954
Wymann, W. E., **12,** 529
Wynberg, H., **1,** 389, 703, 1041, 1105, 1178; **2,** 87, 238, 370; **3,** 37, 124, 254, 275, 312, 337; **4,** 75, 141, 267, 446, 539; **5,** 205, 295, 472, 491, 603, 731; **6,** 501, 525, 575; **7,** 301, 311; **8,** 115, 431; **9,** 403; **10,** 27, 338; **11,** 135
Wynn, W. K., **1,** 808
Wysocki, A. J., **1,** 1126
Wysong, R. D., **2,** 37, 44, 429
Wyss, G., **10,** 350
Wyss, H., **10,** 359
Wyvratt, M. J., **5,** 213, 230

Xian, Y. T., **12,** 525
Xu, Y., **11,** 5
Xu, Z. B., **12,** 387
Xuong, N. D., **1,** 730, 731

Yabuki, Y., **7,** 2
Yadani, N., **11,** 157
Yagen, B., **6,** 202
Yagi, H., **4,** 459; **6,** 358; **7,** 274; **12,** 555
Yagi, K., **1,** 476; **5,** 470
Yagi, M., **11,** 1, 324
Yagi, N., **11,** 430
Yagi, S., **12,** 188
Yagihara, Y., **8,** 188
Yagupolskii, L. M., **6,** 301, 621
Yagupsky, G., **5,** 331
Yahata, N., **12,** 327
Yahner, Y. A., **4,** 539
Yajima, H., **2,** 36, 48; **3,** 197; **5,** 87, 702; **6,** 618; **8,** 319; **9,** 297; **10,** 398
Yamachika, N. J., **3,** 62
Yamada, F., **9,** 467
Yamada, G., **11,** 324
Yamada, H., **4,** 93; **12,** 508
Yamada, K., **5,** 421, 446; **6,** 295; **7,** 135; **8,** 260, 354; **9,** 426; **11,** 71, 447; **12,** 454
Yamada, M., **8,** 119; **9,** 41, 123; **10,** 149; **11,** 421; **12,** 191
Yamada, S., **2,** 317; **3,** 264; **4,** 211, 270, 472; **5,** 89, 160, 217, 280, 470, 552, 601; **6,** 193, 307; **7,** 17, 107, 138, 178, 189, 325; **8,** 51, 197, 320, 403; **9,** 341, 421; **10,** 173, 230; **11,** 222, 268, 288; **12,** 306, 421, 456
Yamada, S. I., **6,** 193
Yamada, T., **5,** 148; **9,** 133, 173; **10,** 150; **12,** 57
Yamada, Y., **5,** 658; **6,** 11; **7,** 365; **8,** 382, 444, 478, 481; **10,** 299; **11,** 95; **12,** 51, 130
Yamagata, H., **11,** 178
Yamagata, T., **11,** 54; **12,** 57
Yamagishi, N., **8,** 315
Yamaguchi, H., **4,** 435; **10,** 85; **11,** 270, 492; **12,** 259
Yamaguchi, M., **7,** 320; **9,** 107, 479, 519; **11,** 112, 410, 521, 552; **12,** 59, 70, 241, 435, 573
Yamaguchi, R., **11,** 49, 80
Yamaguchi, S., **5,** 231; **10,** 369; **12,** 291
Yamaguchi, T., **8,** 419
Yamaguchi, Y., **4,** 572
Yamaichi, A., **9,** 235; **11,** 431

Yamaji, T., **12,** 171
Yamakado, Y., **11,** 64
Yamakawa, K., **7,** 43; **9,** 34; **10,** 24; **12,** 344, 440
Yamakawa, M., **11,** 48
Yamakawa, T., **8,** 382; **9,** 350
Yamamato, Y., **12,** 446
Yamamori, T., **10,** 231
Yamamoto, A., **5,** 332; **7,** 43; **11,** 165, 522
Yamamoto, B. R., **10,** 198
Yamamoto, G., **6,** 509
Yamamoto, H., **3,** 142, 313; **4,** 139, 250; **5,** 7, 77, 200, 293, 374, 381, 398, 749; **6,** 141, 163, 183, 222, 255, 328, 582; **7,** 45, 90, 102, 268, 391; **8,** 56, 164, 165, 292, 472, 476; **9,** 68, 171, 201, 256, 257, 468, 478; **11,** 64, 160, 181, 188, 249, 378, 542, 571, 590; **12,** 11, 19, 193, 263, 344, 356, 378, 407, 480
Yamamoto, I., **6,** 644
Yamamoto, J., **7,** 12
Yamamoto, K., **4,** 110, 465; **5,** 361, 514; **6,** 51, 451, 654; **7,** 278; **8,** 486, 512; **9,** 156, 201, 486; **10,** 240, 258, 417; **11,** 239, 428, 514; **12,** 191
Yamamoto, M., **5,** 470; **8,** 316; **9,** 108
Yamamoto, S., **10,** 289; **11,** 373; **12,** 121
Yamamoto, T., **9,** 193; **11,** 165, 220, 514
Yamamoto, Y., **2,** 173; **3,** 69, 212; **4,** 10, 53, 128; **5,** 57, 182, 467; **6,** 64, 322, 377, 534; **7,** 31, 201, 237; **8,** 57, 145, 335, 381; **9,** 77, 334; **10,** 55, 133, 151, 289, 299, 411, 415, 417, 431, 451; **11,** 12, 20, 52, 75, 88, 143, 233, 295, 365, 373; **12,** 15, 20, 29, 82, 514, 546
Yamamoto, Y. S., **8,** 531
Yamamura, K., **8,** 475
Yamamura, M., **6,** 443, 573; **9,** 458
Yamamura, S., **2,** 462; **3,** 337; **5,** 319, 756; **9,** 11, 207; **11,** 516
Yamamura, Y., **11,** 160
Yamamuro, A., **10,** 113
Yamana, Y., **12,** 336
Yamanaka, A., **11,** 352
Yamanaka, E., **5,** 601
Yamanaka, H., **1,** 1216; **2,** 273, 424; **10,** 346
Yamane, M., **6,** 436
Yamane, S., **11,** 533
Yamane, T., **5,** 498
Yamanishi, T., **4,** 191
Yamartino, R. L., **1,** 228
Yamasaki, M., **3,** 325; **4,** 559
Yamashina, I., **1,** 718
Yamashita, A., **6,** 476; **7,** 296; **8,** 120; **9,** 46; **10,** 100; **11,** 401
Yamashita, K., **7,** 32
Yamashita, M., **5,** 358, 625; **6,** 392, 446, 485, 486, 552; **7,** 183, 341; **8,** 266, 345, 420; **9,** 207; **10,** 365; **11,** 243; **12,** 29, 333, 449
Yamashita, S., **5,** 89, 442; **7,** 74
Yamashita, T., **2,** 147
Yamashita, Y., **5,** 571; **12,** 195
Yamashito, M., **8,** 42
Yamato, E., **3,** 320
Yamato, H., **11,** 240
Yamato, Y., **11,** 415
Yamauchi, K., **5,** 717; **8,** 509; **9,** 491
Yamauchi, M., **10,** 303, 437; **12,** 327
Yamauchi, O., **7,** 79
Yamauchi, T., **12,** 452
Yamaura, M., **12,** 437
Yamawaki, J., **10,** 326; **11,** 434; **12,** 407
Yamaya, M., **8,** 174
Yamazaki, H., **1,** 1193; **3,** 222; **5,** 173; **6,** 128; **7,** 193
Yamazaki, K., **5,** 308, 764
Yamazaki, N., **4,** 210; **9,** 422
Yamazaki, T., **11,** 407
Yamazaki, Y., **6,** 55; **9,** 207
Yan, C. C., **8,** 299
Yan, C. F., **8,** 397, 399; **10,** 272; **11,** 327
Yan, T. H., **11,** 286
Yanagawa, H., **5,** 266
Yanagi, K., **12,** 256
Yanagi, T., **11,** 514; **12,** 402
Yanagida, S., **5,** 336, 534
Yanagisawa, A., **11,** 75, 568
Yanagisawa, I., **7,** 220
Yanagisawa, K., **9,** 458
Yanagita, M., **1,** 637, 1096
Yanagiya, M., **11,** 445
Yanai, M., **2,** 318
Yanami, T., **6,** 482; **7,** 253; **9,** 410; **12,** 502
Yanase, R., **3,** 139, 258, 338
Yaney, D., **1,** 124
Yang, C. O., **12,** 167
Yang, D. D. H., **1,** 4, 744
Yang, D. T. C., **6,** 201; **7,** 55
Yang, I. W., **9,** 361
Yang, L. T. A., **5,** 25
Yang, N. C., **1,** 89, 101; **3,** 166; **4,** 266; **5,** 648
Yang, S., **10,** 156; **12,** 252, 573
Yang, S. S., **8,** 167
Yang, Y. L., **12,** 551
Yang-Chung, G., **4,** 200
Yang-Lan, S., **11,** 194
Yankee, E. W., **4,** 174, 545
Yan-Kui, Y. T., **6,** 367
Yano, T., **6,** 445; **10,** 239
Yano, Y., **5,** 363, 752
Yanovskaya, L. A., **1,** 334; **9,** 361
Yanuka, Y., **1,** 444, 815; **2,** 224, 315; **3,** 86; **5,** 55; **6,** 297
Yany, F., **8,** 519
Yaouane, J. J., **10,** 6
Yarbro, C. L., **1,** 1118
Yardley, J. P., **2,** 131, 237; **3,** 40, 96; **5,** 31; **7,** 116
Yardley, Y. P., **1,** 219
Yardly, G. V., **2,** 60
Yarnell, T. M., **8,** 186, 503
Yaroslavsky, C., **4,** 49, 87; **6,**

510
Yashima, H., **10,** 56
Yashuhara, F., **10,** 369
Yashunsky, D. V., **12,** 198
Yasuda, A., **6,** 183; **7,** 45; **10,** 36, 281
Yasuda, D. M., **8,** 167
Yasuda, H., **10,** 303, 437; **11,** 174, 279, 340; **12,** 59, 221, 559
Yasuda, K., **1,** 1131
Yasuda, M., **6,** 21; **10,** 149
Yasufuku, K., **9,** 431
Yasuhara, T., **4,** 422
Yasui, B., **5,** 441
Yasumura, M., **11,** 343
Yasunami, M., **8,** 155
Yasunobu, K. T., **2,** 147
Yasuoka, N., **7,** 170
Yasuyuki, K., **11,** 279
Yatagai, H., **6,** 377; **7,** 237; **8,** 381; **9,** 77, 334; **10,** 289, 411, 415, 417, 431, 451; **11,** 143, 365, 373; **12,** 514
Yates, D. H., **6,** 675
Yates, J. B., **9,** 326; **11,** 16, 363
Yates, P., **1,** 34, 169, 643, 927; **5,** 675; **7,** 188; **11,** 341
Yates, S. W., **12,** 258
Yatsimirsky, A. K., **11,** 516; **12,** 484
Yavarzadeh, R., **12,** 257
Yax, E., **2,** 255
Yazawa, H., **3,** 247; **6,** 65; **8,** 1
Yeats, R. B., **2,** 252
Yee, K. C., **6,** 267
Yee, Y. K., **7,** 292; **9,** 363; **12,** 462
Yeh, H. J. C., **7,** 274
Yelm, K. E., **12,** 547
Yen, C. C., **7,** 352
Yeong, Y. C., **5,** 55
Yeung, B. W. A., **12,** 116
Yiannios, C. N., **1,** 310; **5,** 277
Ykman, P., **4,** 225; **6,** 480
Yoakim, C., **12,** 101, 200
Yoda, Y., **11,** 87
Yogai, S., **10,** 279
Yogo, T., **9,** 373
Yokoe, M., **5,** 534
Yokohama, S., **10,** 194
Yokomatsu, T., **6,** 547; **8,** 492; **9,** 381, 473
Yokommatsu, T., **8,** 351
Yokoo, C., **5,** 621
Yokoo, K., **11,** 249
Yokoohji, K., **12,** 414
Yokota, K., **10,** 38; **12,** 172
Yokota, Y., **2,** 323
Yokoyama, H., **10,** 437; **11,** 428
Yokoyama, K., **4,** 158; **5,** 224; **8,** 468, 499; **12,** 462
Yokoyama, M., **4,** 395; **10,** 437; **11,** 114, 428, 430; **12,** 108, 430, 543
Yokoyama, T., **6,** 141; **8,** 501
Yokoyama, Y., **6,** 193; **7,** 107; **8,** 24, 101, 362, 397
Yoneda, F., **1,** 247; **2,** 129; **4,** 390; **5,** 479, 561, 564; **6,** 185, 547; **7,** 130; **9,** 160; **12,** 350
Yoneda, N., **7,** 271; **8,** 240; **11,** 243, 252
Yoneda, R., **12,** 188
Yoneda, T., **1,** 4
Yonekura, M., **12,** 44
Yonemitsu, O., **1,** 894; **3,** 231, 303; **6,** 258, 453; **8,** 314; **10,** 252; **11,** 167, 312; **12,** 175
Yonezawa, K., **4,** 101; **5,** 150; **6,** 129
Yonezawa, T., **5,** 764
Yoo, B. K., **12,** 195
Yoo, S., **8,** 87; **9,** 242, 301
Yoo, S. C., **5,** 394
Yoon, N. M., **1,** 35, 625; **2,** 24, 108; **3,** 23, 77; **4,** 176; **5,** 48; **7,** 30; **8,** 267; **11,** 351; **12,** 276, 485
Yorifuji, T., **10,** 194
Yorka, K. V., **1,** 964
Yoshi, E., **6,** 317
Yoshida, H., **7,** 58
Yoshida, J., **8,** 56, 381, 523; **12,** 345
Yoshida, J. I., **11,** 374
Yoshida, K., **5,** 427; **8,** 486; **12,** 314
Yoshida, M., **3,** 294; **9,** 188, 343
Yoshida, N., **10,** 182
Yoshida, S., **9,** 256
Yoshida, T., **5,** 57, 172, 233; **6,** 399, 492; **7,** 26, 379; **9,** 238; **10,** 133, 239; **11,** 343; **12,** 54, 171
Yoshida, Y., **9,** 96
Yoshida, Z., **4,** 31, 475; **5,** 560; **7,** 20, 207; **8,** 382, 385; **9,** 235, 309; **10,** 32, 299; **11,** 395; **12,** 51, 256
Yoshidawa, S., **9,** 205
Yoshifuji, M., **6,** 28; **8,** 87; **9,** 119
Yoshifuji, S., **5,** 427
Yoshigi, M., **12,** 221
Yoshii, E., **3,** 325; **5,** 294; **6,** 403, 649; **7,** 282; **9,** 117; **11,** 539; **12,** 228
Yoshikawa, K., **7,** 268
Yoshikawa, M., **3,** 229
Yoshikawa, S., **5,** 42; **11,** 277
Yoshikawa, Y., **5,** 641
Yoshikoshi, A., **4,** 282; **6,** 482, 662; **7,** 253; **8,** 428; **9,** 277, 410, 516; **10,** 279; **11,** 594; **12,** 502
Yoshimoto, M., **5,** 406
Yoshimura, I., **8,** 345; **11,** 159
Yoshimura, J., **12,** 437
Yoshimura, M., **5,** 58
Yoshimura, N., **5,** 499; **6,** 445
Yoshimura, T., **5,** 639; **8,** 141; **10,** 174
Yoshinari, T., **10,** 257
Yoshino, A., **5,** 764
Yoshioka, H., **9,** 296
Yoshioka, M., **1,** 244, 1198; **2,** 128, 427; **4,** 147, 526; **9,** 13, 328, 519
Yoshioka, T., **10,** 252; **11,** 167; **12,** 175
Yoshisato, E., **2,** 293
Yoshiska, T., **7,** 89
Yoshitake, J., **8,** 486; **9,** 156
Yoshitome, T., **11,** 49
Yoshizawa, T., **11,** 75
Yosioka, I., **9,** 269

Yost, D. M., **1,** 773
Yosuda, M. S., **12,** 529
Youlus, J., **1,** 386
Young, D. A., **5,** 160
Young, D. P., **1,** 477
Young, D. W., **8,** 316; **9,** 462; **10,** 397
Young, E. I., **2,** 76
Young, F. G., **1,** 1118
Young, G. T., **1,** 76, 97, 296, 486; **2,** 27, 389; **4,** 391; **7,** 139; **9,** 51
Young, H., **2,** 238
Young, J. F., **1,** 140, 1252; **2,** 453; **5,** 700
Young, L. B., **1,** 121; **2,** 65
Young, M., **7,** 244
Young, M. G., **3,** 45
Young, M. V., **8,** 112
Young, M. W., **5,** 276; **6,** 235, 422
Young, R. J., **1,** 509, 562
Young, R. N., **5,** 116; **6,** 109, 524; **8,** 432; **10,** 463; **12,** 530
Young, R. V., **1,** 1103
Young, R. W., **1,** 248, 253
Young, S. D., **7,** 376; **9,** 283; **11,** 208, 341, 460; **12,** 188
Young, S. I., **11,** 349
Young, T. E., **5,** 105
Young, W. G., **1,** 778, 1137, 1143; **4,** 323; **5,** 497
Youssefyeh, R. D., **1,** 73, 102, 604
Yovanidis, C., **2,** 178
Yu, L. C., **9,** 306; **11,** 422
Yu, M., **11,** 471
Yu, R. J., **2,** 315
Yu, S. H., **3,** 160; **6,** 105
Yu, S. W., **4,** 260
Yu, Y. K., **8,** 392
Yu, Y. S., **8,** 254
Yuan, C., **1,** 357
Yuan, E., **1,** 269
Yuasa, Y., **9,** 438; **12,** 392
Yudis, M. D., **1,** 342; **2,** 326, 332
Yui, S., **5,** 45
Yukawa, T., **2,** 238
Yukimoto, Y., **11,** 208
Yukizaki, H., **12,** 125
Yunker, M., **5,** 389
Yura, Y., **6,** 667
Yurkevich, A. M., **3,** 222
Yurtanov, A. I., **6,** 634
Yus, M., **9,** 233; **11,** 303, 318
Yuyama, M., **6,** 161

Zabel, V., **11,** 77
Zabkiewicz, J. A., **1,** 926
Zabriskie, J. L., **1,** 322
Zabrocki, J., **8,** 170
Zador, M., **4,** 498
Zadrazil, S., **1,** 1002
Zaffaroni, A., **1,** 411
Zahler, R., **10,** 88
Zahler, R. E., **8,** 245
Zahn, H., **1,** 235
Zahn, K., **1,** 1029
Zahora, E. P., **5,** 25
Zaidlewicz, M., **7,** 249; **9,** 138; **11,** 267
Zaitseva, G. V., **5,** 6
Zajac, W. W., Jr., **12,** 531
Zajacek, J. G., **2,** 287, 456; **3,** 331; **4,** 346
Zakharkin, L. I., **1,** 262, 1030; **3,** 259
Zaki, M., **5,** 663
Zalar, F. V., **1,** 927; **2,** 344
Zalikowski, J. A., **10,** 336
Zalkow, L. H., **1,** 364, 561; **3,** 18; **5,** 381; **8,** 409
Zally, W. J., **7,** 255
Zalukaev, L. P., **5,** 695
Zamarlik, H., **9,** 442
Zambito, A. J., **1,** 1100
Zambone, R., **10,** 272
Zamboni, R., **11,** 99
Zamecnik, J., **6,** 181
Zamojski, A., **6,** 280, 291
Zanardi, G., **12,** 194
Zanarotti, A., **6,** 517; **9,** 413; **10,** 351
Zander, H., **1,** 1128
Zander, M., **1,** 127; **2,** 321; **4,** 316
Zanirato, V., **11,** 324
Zanotti, V., **12,** 275
Zaoral, M., **1,** 289, 1229
Zapata, A., **10,** 433; **11,** 581
Zapf, L., **11,** 355
Zapp, J. A., Jr., **6,** 278
Zappelli, P., **5,** 495
Zappi, E. V., **1,** 177
Zard, S. Z., **12,** 102, 338, 515
Zarecki, A., **5,** 106, 353
Zaretzkii, Z., **6,** 166
Zart, A., **1,** 842
Zaslona, A. T., **11,** 425
Zatorski, A., **6,** 191, 193; **7,** 126
Zaugg, H. E., **1,** 281, 309, 531, 1033, 1073, 1081; **2,** 101; **7,** 57; **8,** 507; **10,** 403
Zavgorodnii, V. S., **5,** 44
Zavialov, S. I., **2,** 267
Zavist, A. F., **3,** 295
Zav'yalov, S. I., **1,** 465, 1272, 1273
Zawadzki, S., **7,** 104
Zayed, M. F., **4,** 188
Zayed, S. M. A. D., **1,** 45; **4,** 188, 356
Zbaida, D., **8,** 433
Zbikowski, I., **2,** 310
Zbiral, E., **1,** 562; **2,** 238; **4,** 278, 550; **5,** 354, 365; **6,** 297, 318; **7,** 406; **9,** 441, 502; **10,** 447
Zbirovsky, M., **5,** 668
Zdanovich, V. I., **5,** 695
Zderic, J. A., **1,** 104, 219, 624, 802
Zderic, S. A., **8,** 403; **10,** 49, 321
Zecchini, G. P., **10,** 248
Zecher, W., **4,** 524
Zechmeister, L., **1,** 500
Zee-Cheng, K. Y., **2,** 361
Zeeh, B., **1,** 195; **2,** 51; **7,** 338
Zeelen, F. J., **11,** 96
Zefirov, N. S., **4,** 441, 472; **6,** 232
Zeftel, L., **1,** 610
Zehavi, U., **6,** 420
Zeidler, D., **4,** 251
Zeidler, U., **5,** 555
Zeile, K., **1,** 3
Zelawski, Z. S., **7,** 45
Zelesko, M. J., **2,** 411; **3,** 289; **4,** 501

Zelle, R. E., **12,** 204, 282
Zeller, H. G., **1,** 942
Zeller, P., **1,** 114, 385, 388, 614, 1178
Zelonka, R. A., **6,** 665
Zembayashi, M., **8,** 523; **11,** 168
Zemlicka, J., **1,** 4, 289
Zemplen, G., **2,** 23
Zen, S., **10,** 60
Zenarosa, C. V., **3,** 98, 123
Zenbayashi, M., **5,** 581
Zengierski, L., **4,** 473
Zenki, S., **12,** 247
Zenner, K. F., **5,** 439; **11,** 206
Zera, R. T., **6,** 174
Zerby, G. A., **9,** 186
Zergenyi, J., **5,** 436; **6,** 403
Zerner, B., **1,** 153; **5,** 617
Zervas, L., **1,** 110, 321, 346, 745, 859, 1006, 1256, 1261; **2,** 163, 178, 180, 317, 389
Zetterberg, K., **7,** 22
Zetzsche, F., **1,** 352, 976
Zeyfang, D., **4,** 173
Zhang, B. W., **11,** 387
Zharkov, V. V., **5,** 6
Zhdanov, Y. A., **1,** 802
Zhou, Y. F., **11,** 575
Zhukova, S. V., **4,** 282
Ziegenbein, W., **1,** 262, 285, 289, 756
Zieger, H. E., **5,** 468
Ziegler, C. B., **8,** 382; **9,** 351
Ziegler, D., **2,** 263, 349
Ziegler, E., **2,** 372; **6,** 96; **7,** 385
Ziegler, F. E., **4,** 334; **5,** 227; **7,** 236; **8,** 122, 299; **10,** 34, 312; **11,** 47, 448; **12,** 143, 439
Ziegler, G. R., **1,** 1084
Ziegler, J. B., **1,** 75, 378, 802, 955; **2,** 38
Ziegler, J. C., **7,** 14, 35; **9,** 297
Ziegler, K., **1,** 262, 455, 573, 611, 618, 832, 1096, 1103, 1189, 1198, 1261; **3,** 297
Ziehn, K. D., **4,** 550, 552; **5,** 727
Zielinski, W. S., **7,** 286
Ziemann, H., **5,** 574
Ziffer, H., **6,** 206; **9,** 126
Zigler, S., **12,** 37
Zigman, A. R., **3,** 178
Zika, R. G., **4,** 18
Zikra, N., **11,** 373
Zilkha, A., **8,** 57; **11,** 347
Zilliken, F., **10,** 410
Zima, G., **9,** 31, 32, 373; **10,** 21, 345
Ziman, S. D., **4,** 63, 469
Zimmer, G., **1,** 1116
Zimmer, H., **4,** 411; **5,** 562; **6,** 392; **10,** 145
Zimmer, N., **7,** 407
Zimmer, W. F., **1,** 1227
Zimmerman, C. A., **4,** 14; **5,** 441
Zimmerman, D. M., **10,** 71
Zimmerman, H. E., **1,** 231, 703, 971, 1018; **4,** 174
Zimmerman, J. E., **1,** 487, 524; **2,** 47
Zimmerman, R., **3,** 123
Zimmerman, W. T., **6,** 310; **7,** 94, 376
Zimmermann, F. J., **1,** 870
Zimmermann, G., **10,** 260
Zimmermann, J. F., **2,** 279
Zimmermann, M., **2,** 112; **3,** 146
Zincke, T., **1,** 970
Zingaro, R. A., **8,** 280
Zink, M. P., **5,** 734
Zinner, G., **1,** 1123
Zinner, H., **1,** 935
Zinser, D., **1,** 364
Zioudrou, C., **12,** 44
Zipkin, R. E., **11,** 559; **12,** 476
Zitsman, J., **5,** 34
Zletz, A., **1,** 432
Zlotogorski, C., **6,** 108
Zobel, H., **1,** 865
Zobrist, F., **1,** 172
Zoda, M. F., **11,** 593
Zoebelein, A., **1,** 951
Zoellner, E. A., **1,** 424, 573, 1229
Zoellner, R., **4,** 367
Zoerner, A., **8,** 535
Zoller, L. W., III, **11,** 28
Zoller, U., **6,** 290
Zollinger, H., **1,** 289; **9,** 335
Zoltewicz, J. A., **4,** 503
Zombeck, A., **11,** 138
Zon, G., **3,** 148; **4,** 35, 435; **6,** 273
Zook, H. D., **5,** 49
Zoran, A., **6,** 108
Zorbach, W. W., **1,** 595, 655
Zoretic, P. A., **6,** 110, 229; **7,** 135, 310
Zorko, F., **8,** 87
Zorn, H., **2,** 393
Zosel, K., **1,** 1189, 1198
Zovko, M. J., **4,** 103
Zschokke, H., **1,** 156
Zsindely, J., **5,** 700; **11,** 559
Zubarovskii, V. M., **5,** 24
Zubiani, A., **6,** 74
Zubiani, G., **2,** 274; **3,** 189, 201
Zubrick, J. W., **5,** 609; **6,** 137
Zuckerman, J. J., **11,** 209
Zueger, M. F., **11,** 529
Zuehlsdorff, G., **1,** 652
Zuercher, A., **1,** 652
Zugaza-Bibao, A., **10,** 41
Zupan, M., **5,** 512; **6,** 300, 670; **7,** 242; **10,** 85; **11,** 118
Zupancic, B., **2,** 159
Zur, Z., **6,** 471
Zurflueh, R., **5,** 90
Zurqiyah, A., **2,** 78; **3,** 62
Zurr, D., **4,** 567
Zutterman, Z., **12,** 20
Zvacek, J., **1,** 764
Zvilichovsky, G., **5,** 214
Zwanenburg, B., **3,** 155; **5,** 127, 495, 516; **6,** 567; **8,** 465, 466; **10,** 320
Zwanenburg, D. J., **7,** 258
Zweifel, G., **1,** 59, 72, 206, 207, 255, 263, 276, 1141, 1191, 1228; **2,** 142, 249; **3,** 23, 91; **4,** 36, 38, 110, 141, 161; **5,** 41, 347, 397, 467, 565; **6,** 201, 295; **7,** 113, 147, 181; **8,** 105,

162, 292; **9,** 277, 280; **10,** 95, 308, 423; **11,** 188, 266, 572; **12,** 357, 474
Zweig, J., **6,** 215
Zweirzchowska, Z., **6,** 291
Zwick, W., **12,** 303
Zwierzak, A., **5,** 650; **6,** 43; **7,** 20, 104, 353; **8,** 146, 170, 391; **9,** 429, 444; **11,** 407
Zwikker, J. W., **11,** 567
Zwilgmeyer, F., **1,** 1073

REAGENTS INDEX

Acetaldehyde, 1, 3; **10,** 177; **12,** 8
Acetaldehyde 3-bromopropyl ethyl acetal, **4,** 226
Acetaldehyde N-*t*-butylimine, **9,** 1
Acetaldehyde cyclohexylimine, **2,** 249
Acetaldehyde dimethyl acetal (*see* 1,1-Dimethoxyethane)
Acetaldoxime, **1,** 3
Acetamide, **1,** 3
p-Acetamidobenzenesulfonyl chloride, **1,** 3
Acetamidomethanol, **2,** 4
Acetanilide, **4,** 79
Acetatohydridotris(triphenylphosphine)-ruthenium, **11,** 171
(Acetato)phenylmercury, **6,** 458
(Acetato)phenylpalladium, **4,** 379
Acetic acid, **1,** 898, 1026, 1223, 1291; **2,** 5, 6; **5,** 3; **6,** 16, 355; **7,** 1; **8,** 1, 221; **9,** 10; **12,** 415
Acetic acid–Acetic anhydride (*see* Acetic anhydride–Acetic acid)
Acetic anhydride (*see also* Acetic anhydride + catalysts), **1,** 3, 479, 528, 1228; **2,** 7; **4,** 234; **5,** 3; **6,** 1; **7,** 1; **8,** 1; **9,** 1; **11,** 1, 11
Acetic anhydride–Acetic acid, **1,** 797; **2,** 5; **6,** 1, 4; **8,** 2
Acetic anhydride–Acetyl chloride, **7,** 1
Acetic anhydride–Boron trifluoride–Acetic acid, **1,** 70
Acetic anhydride–Boron trifluoride etherate, **1,** 69, 71, 72, 604; **2,** 7, 8, 35, 36; **3,** 33; **6,** 3; **11,** 1
Acetic anhydride–Calcium hydride, **2,** 58
Acetic anhydride–Chromium(II) acetate, **6,** 4
Acetic anhydride–4-Dimethylaminopyridine, **3,** 119; **9,** 178
Acetic anhydride–Ferric chloride, **6,** 260; **8,** 228; **12,** 1
Acetic anhydride–Ion-exchange resin, **1,** 514
Acetic anhydride–Magnesium bromide, **6,** 352
Acetic anhydride–Magnesium perchlorate, **8,** 14
Acetic anhydride–Manganese(III) acetate, **6,** 355
Acetic anhydride–Methanesulfonic acid, **6,** 3
Acetic anhydride–Nitric acid, **5,** 475
Acetic anhydride–Perchloric acid, **1,** 798; **2,** 9, 309
Acetic anhydride–Phosphoric acid, **1,** 898; **6,** 3
Acetic anhydride–Potassium acetate, **1,** 906; **7,** 2
Acetic anhydride–Pyridine, **1,** 958, 1174; **6,** 4; **7,** 66; **9,** 1
Acetic anhydride–Pyridine N-oxide, **5,** 567
Acetic anhydride–Pyridinium chloride, **2,** 353; **4,** 4
Acetic anhydride–Sodium acetate, **1,** 1024; **2,** 9; **5,** 4, 470; **6,** 5; **7,** 2; **8,** 1
Acetic anhydride–Sulfuric acid, **2,** 8; **6,** 560
Acetic anhydride–Titanium(III) acetate, **6,** 4
Acetic anhydride–*p*-Toluenesulfonic acid, **1,** 1174; **4,** 509; **5,** 674; **11,** 1
Acetic anhydride–Triethylamine, **5,** 3, 4; **6,** 6
Acetic anhydride–Trifluoroacetic anhydride, **7,** 2
Acetic anhydride–Zinc chloride, **1,** 1289; **4,** 4
Acetic-formic anhydride, **1,** 4; **2,** 10; **3,** 4; **4,** 5
Acetic-phosphoric anhydride, **2,** 12
Acetoacetic acid, **6,** 21; **12,** 1
Acetoacetyl fluoride, **1,** 4
Aceto(carbonyl)cyclopentadienyl-(triphenylphosphine)iron, **12,** 1
Acetoinenediolcyclopyrophosphate [Bis(4,5-dimethyl-2-oxo-1,3,2-dioxaphospholenyl)oxide], **6,** 6, 218
Acetone, **1,** 332, 512, 528, 1173; **2,** 13, 89; **3,** 4; **6,** 9, 518; **8,** 228
Acetone-1,3-bis(triphenylphosphonium) dichloride, **1,** 925
Acetone cyanohydrin, **1,** 5; **8,** 409; **11,** 1
Acetone cyanohydrin nitrate, **1,** 5, 706
Acetone cyanohydrin trimethylsilyl ether, **5,** 207
Acetone cyclohexylimine, **6,** 332
Acetone dibutylketal, **1,** 209
1,3-Acetonedicarboxylic acid, **1,** 6, 411; **11,** 2
Acetone diethyl ketal (*see* 2,2-Diethoxypropane)

Acetone diethyl thioketal [*see* 2,2-Bis-(ethylthio)propane]
Acetone N,N-dimethylhydrazone, **8,** 276, 511; **9,** 185
Acetone dimethyl ketal (*see* 2,2-Dimethoxypropane)
Acetone hydrazone, **2,** 105; **3,** 74, 153
Acetone phenylhydrazone, **1,** 1290
Acetone–Potassium peroxomonosulfate, **11,** 442; **12,** 413
Acetonitrile, **1,** 1291; **2,** 13, 209; **3,** 266; **5,** 621, 634; **6,** 69, 84, 636; **7,** 303; **9,** 324; **10,** 14, 460
Acetonitrile oxide, **11,** 289, 457
Acetophenone, **4,** 5
Acetophenone cyclohexylimine, **6,** 332
Acetophenone, methoxymethyl enol ether, **12,** 99
Acetophenone phenylhydrazone, **4,** 172
ω-Acetophenonesulfonyl chloride, **3,** 221
2-Acetoxyacrylonitrile, **1,** 7; **2,** 13; **11,** 139
1-Acetoxybutadiene, **1,** 7; **3,** 104; **9,** 56
3-Acetoxy-2-ethoxy-1-propene, **10,** 386
3β-Acetoxy-Δ^5-etienic acid, **1,** 9
2-Acetoxyisobutyryl chloride, **8,** 3
2-Acetoxy-3-*p*-methoxyphenylthio-1,3-butadiene, **8,** 52
2-Acetoxy-1-methoxy-3-trimethylsilyloxy-1,3-butadiene, **11,** 2
2-Acetoxy-2-methylbutyryl chloride, **8,** 3
5-Acetoxymethyl-4-methoxy-*o*-benzoquinone, **6,** 363
Acetoxymethyl methyl selenide, **8,** 5
2-Acetoxymethyl-3-trimethylsilyl-1-propene, **9,** 11; **11,** 578
N-Acetoxyphthalimide, **1,** 9
3-Acetoxypropyltriphenylphosphonium bromide, **4,** 401
2- or 3-Acetoxypyridine, **1,** 9
(1E,3E)-1-Acetoxy-4-trimethylsilyl-1,3-butadiene, **11,** 3
3-Acetoxy-1-trimethylsilyl-1,3-butadiene, **9,** 266
3-Acetoxy-2-trimethylsilylmethyl-1-propene, **9,** 454; **11,** 578
3-Acetoxy-1-trimethylsilyl-1-propene, **11,** 509
3-Acetoxy-3-trimethylsilyl-1-propene, **11,** 509
(R)-2-Acetoxy-1,1,2-triphenylethanol, **12,** 3
Acetylacetone (*see* 2,4-Pentanedione)
Acetylacetylene (*see* 3-Butyn-2-one)
Acetyl bromide, **6,** 9; **9,** 523
Acetyl chloride (*see also* Acetyl chloride + co-reactants), **1,** 11, 263, 771; **2,** 383, 408, 448; **4,** 5, 346; **5,** 4, 237, 332, 410; **6,** 164; **7,** 3; **8,** 1, 86; **9,** 448; **11,** 11; **12,** 471
Acetyl chloride–Aluminum chloride, **1,** 27, 131, 677; **3,** 8
Acetyl chloride–4-Dimethylaminopyridine, **9,** 387
Acetyl chloride–N,N-Dimethylaniline, **1,** 275
Acetyl chloride–Magnesium, **1,** 629
Acetyl chloride–Silver cyanide, **9,** 498
Acetyl chloride–Tetrabutylammonium hydrogen sulfate, **9,** 357
Acetyl chloride–Tin(IV) chloride, **1,** 1111
Acetyl chloride–2-Trimethylsilylethanol, **11,** 4
Acetyl chloride–Zinc halides, **9,** 521, 523
Acetyl chloride–Zirconium(IV) chloride, **1,** 1295
Acetyl cyanide, **10,** 1
Acetylene, **1,** 11, 227, 389, 712; **5,** 173; **6,** 61, 576; **10,** 284
Acetylenedicarbononitrile (*see* Dicyanoacetylene)
Acetylenedicarboxylic acid, **1,** 12, 343
Acetylenedimagnesium bromide, **1,** 634
α-Acetyl-β-ethoxy-N-ethoxycarbonyl-acrylamide, **1,** 12
Acetyl fluoride, **2,** 408; **11,** 6
Acetyl hexachloroantimonate, **8,** 5
Acetyl hexafluoroantimonate, **1,** 692
Acetyl hypobromite, **1,** 12
Acetyl hypofluorite, **10,** 1; **11,** 5; **12,** 3
Acetyl hypoiodite, **1,** 496, 504
N-Acetylimidazole, **1,** 13; **7,** 51
S-Acetylmercaptosuccinic anhydride, **1,** 13
Acetyl methanesulfonate, **5,** 5; **9,** 2
Acetylmethylene(triphenyl)arsorane, **11,** 5
Acetylmethylene(triphenyl)phosphorane, **2,** 306; **5,** 375
1-Acetyl-1-methylhydrazine, **4,** 7
Acetyl nitrate, **1,** 13; **7,** 3; **12,** 530
3-Acetyl-4-oxazoline-2-one, **9,** 2
N-Acetylpyrrole, **8,** 414
2-Acetylsalicylic acid chloride, **5,** 6
Acetyl sulfuric acid, **4,** 7
Acetyl tetrafluoroborate, **11,** 6
3-Acetylthiazolidine-2-thione, **12,** 490
1-Acetyl-2-thiourea, **4,** 7
Acetyl *p*-toluenesulfonate, **2,** 14; **4,** 8; **6,** 10
3-Acetyl-1,5,5-trimethylhydantoin, **3,** 4

Acrolein, **2,** 15; **3,** 5; **10,** 2; **11,** 11
Acrolein diethyl acetal
(*see* 3,3-Diethoxy-1-propene)
Acrylic acid, **4,** 18; **6,** 117
Acrylonitrile, **1,** 14, 973, 1200; **4,** 40; **7,** 4
(2R,4R)- and (2S,4S)-N-Acryloyl-4-(diphenylphosphine)-2-[(diphenylphosphine)methyl]pyrrolidine, **11,** 6
2-Acyl-1,3-dithianes, **4,** 216
Acyl fluorides, **1,** 14, 407
Acyllithium reagents, **5,** 233; **11,** 111; **12,** 4, 105
3-Acylthiazolidine-2-thiones, **11,** 323, 518; **12,** 4, 490
Adamantane, **3,** 5
Adamantane-1-carbonyl chloride, **2,** 15
1-Adamantyl chloroformate, **1,** 15
1-Adamantyl hexafluoroantimonate, **5,** 310
Adams catalyst (*see* Platinum oxide)
Adipic acid, **1,** 15, 766
Adkins catalyst, **1,** 156, 157
Adogen 464, **6,** 10; **7,** 4; **8,** 268, 387; **9,** 357, 426; **11,** 486
Alane (*see* Aluminum hydride, Lithium aluminum hydride–Aluminum chloride)
β-Alanine, **1,** 16; **5,** 6; **6,** 10
Aliquat 336, **4,** 28, 30; **5,** 460; **6,** 404; **7,** 239, 246, 380; **8,** 387; **9,** 358; **10,** 305; **11,** 126, 403; **12,** 105, 380
Alkanoylketene dithioacetals, **11,** 282
B-1-Alkenyl-9-borabicyclo[3.3.1]nonanes, **7,** 30; **8,** 6; **9,** 57; **10,** 105; **12,** 5
2-Alkoxy-1,3-benzodithiolanes, **8,** 236
(S)-1-Alkoxymethyl-2-aminopropane, polymeric, **9,** 3
Alkoxymethyllithium reagents, **12,** 351
Alkylaluminum halides (*see also* specific names), **1,** 245; **4,** 144; **6,** 251; **7,** 146; **8,** 42, 164; **9,** 11; **10,** 9, 164, 177; **11,** 7, 12, 539; **12,** 2, 5, 28, 104, 252, 261, 512
Alkylamine-boranes, **1,** 16
N-Alkyl-β-aminoethyltriphenylphosphonium bromides, **12,** 11
2-Alkyl-1,3-benzodithiolanes, **8,** 34, 236; **9,** 4
Alkyl chloroformates (*see also* Ethyl chloroformate, Methyl chloroformate, other specific names), **1,** 40, 86, 109, 182, 1185; **2,** 24, 47, 48, 59, 307, 423; **3,** 15; **5,** 117, 359; **6,** 307; **8,** 499; **10,** 431; **11,** 419; **12,** 12, 15
Alkylcopper reagents (*see also* Organocopper reagents, specific names), **11,** 13
Alkylcopper reagents–Aluminum chloride, **10,** 286
Alkylcopper reagents–Boron trifluoride complex, **8,** 334, 335; **9,** 333; **10,** 54, 282; **11,** 368, 413; **12,** 250, 347
Alkyldimesitylboranes, **12,** 12
Alkyldiphenylphosphine oxides, **10,** 2; **12,** 477
Alkyl disulfides, *see* Dialkyl disulfides
Alkyl ethers, *see* Dialkyl ethers
N-Alkyl-N-fluoro-*p*-toluenesulfonamides, **12,** 231
Alkyl hydroperoxides (*see also* specific names), **1,** 17
N-Alkylhydroxylamines, **4,** 80; **5,** 111, 123; **6,** 107; **7,** 241; **8,** 135; **9,** 42; **12,** 13, 95, 322
Alkyllithium reagents (*see also* Organolithium agents, specific names), **9,** 5; **10,** 3; **11,** 13; **12,** 14
Alkyl orthotitanates [*see* Titanium(IV) alkoxides]
1-(Alkylthio)vinyllithium (*see also* 1-Phenylthiovinyllithium), **5,** 6
Alkyltris(dialkylamino)titanium derivatives, **11,** 375
B-1-Alkynyl-9-borabicyclo[3.3.1]nonanes, **8,** 6
Allene, **1,** 18; **2,** 388; **3,** 327; **5,** 501; **6,** 251; **11,** 512; **12,** 55
Allenylboronic acid, **11,** 181
Allylacetophenone, **8,** 7
Allyl alcohol, **1,** 268; **5,** 80; **9,** 10
Allyl alcohol, 0,2-dilithio derivative, **6,** 11
B-Allyl-9-borabicyclo[3.3.1]nonane, **12,** 15
Allyl bromide, **1,** 1287; **6,** 199; **12,** 106
Allyl-*t*-butyldimethylsilane, **10,** 8, 62
Allyl chloride, **1,** 1040; **10,** 87
Allyl chloroformate, **12,** 15
Allylcopper, **4,** 220
Allyl diazoacetate, **8,** 222
Allyldifluoroiodotin, **10,** 6, 374
Allyldiisobutylaluminum, **12,** 169
B-Allyldiisocaranylborane, **12,** 18
B-Allyldiisopinocampheylborane, **12,** 17
Allyldimesitylborane, **12,** 12
S-Allyl N,N-dimethyldithiocarbamate, **6,** 11, 539
(Allyldimethylsilyl)methylmagnesium chloride, **12,** 244
Allyldiphenylphosphine oxide, **12,** 19
Allyl ethyl sulfide, **11,** 377

O-Allylhydroxylamine, **10,** 5
Allylidenecyclopropane, **12,** 19
Allylidenetriphenylphosphorane, **5,** 7, 70; **6,** 14
Allyl iodide, **3,** 253
Allyl isopropyl sulfide, **9,** 76
Allyllithium, **1,** 18, 1250; **5,** 82, 175
Allylmagnesium bromide, **1,** 415; **9,** 229; **12,** 246
Allylmagnesium bromide–Copper(I) iodide, **8,** 121; **12,** 496
Allylmagnesium chloride, **1,** 420; **11,** 436; **12,** 15
Allylmercaptan, **5,** 84
π-Allylnickel bromide [*see* Di-μ-bromobis-(allyl)dinickel]
π-Allylnickel halides (*see also* specific names), **2,** 291; **3,** 210; **4,** 33, 353, 354, 355; **5,** 437; **6,** 93, 361; **7,** 250; **8,** 181, 324
Allyloxytriethylsilane, **6,** 607
π-Allylpalladium acetate dimer, **7,** 4
π-Allylpalladium chloride dimer [*see* Di-μ-chlorobis(allyl)dipalladium]
π-Allylpalladium complexes (*see also* specific names), **2,** 109; **5,** 189; **6,** 394; **7,** 4, 5; **9,** 132; **12,** 371
Allyl phenyl ether, **1,** 19; **4,** 365; **5,** 175; **11,** 14
Allyl phenyl selenide, **11,** 17; **12,** 20
Allyl phenyl sulfide **10,** 197; **11,** 28
Allyl phenyl sulfone, **8,** 405; **10,** 196
P-Allyl-N,N,N′,N′-tetramethylphosphonic diamide, **3,** 105
Allyl tetramethylphosphorodiamidate, **10,** 5
Allyltin difluoroiodide, **10,** 6, 374
Allyltributyltin, **11,** 15; **12,** 21, 391
Allyl trifluoromethanesulfonate, **5,** 704; **8,** 8
Allyltrimethylgermanium, **11,** 516
Allyltrimethylsilane, **7,** 8, 370; **8,** 273; **9,** 155, 372, 445; **10,** 1, 6, 8, 130, 430, 439; **11,** 16, 117, 228, 516, 529, 573, 584; **12,** 23, 260, 375, 434, 495, 544
Allyltrimethylsilylzinc chloride, **10,** 8
Allyltrimethyltin, **9,** 8, 47; **11,** 20, 516
Allyltriphenylphosphonium bromide, **5,** 633
Allyltriphenyltin, **1,** 18, 1250
trihapto-Allyltris(trimethylphosphite)-cobalt(I), **6,** 15
Allyl vinyl sulfide, **5,** 749
Allylzinc bromide, **12,** 54
Alpine borane (*see* B-3-Pinanyl-9-borabicyclo[3.3.1]nonane)
Alumina, **1,** 19, 20, 755, 828, 958, 1167; **2,** 17, 18, 299, 418; **3,** 6; **4,** 8; **5,** 348, 432; **6,** 16; **7,** 5; **8,** 9; **9,** 8; **10,** 8, 347; **11,** 22; **12,** 508
Aluminum, **2,** 19; **3,** 7; **4,** 9
Aluminum amalgam, **1,** 20, 653; **2,** 255; **3,** 7; **5,** 9; **9,** 449; **11,** 24
Aluminum amalgam–Trifluoroacetic acid–*t*-Butyl alcohol, **9,** 518
Aluminum azide, **5,** 9
Aluminum bromide, **1,** 22; **2,** 19, 125; **3,** 7; **4,** 10; **5,** 10; **6,** 17; **9,** 11; **12,** 26
Aluminum bromide–Ethanethiol, **9,** 13
Aluminum *t*-butoxide, **1,** 23; **2,** 21
Aluminum *t*-butoxide–Raney nickel, **8,** 13
Aluminum chloride, **1,** 24, 677; **2,** 21; **3,** 7; **4,** 10, 357; **5,** 10, 763; **6,** 17; **7,** 7, 252, 371; **8,** 13, 86; **9,** 11; **10,** 9, 180, 262; **11,** 8, 25; **12,** 26, 501
Aluminum chloride–Dialkyl sulfides, **10,** 11
Aluminum chloride–Ethanethiol, **9,** 13; **10,** 11; **11,** 28; **12,** 29
Aluminum chloride–Ethanol, **8,** 15
Aluminum chloride–N-Methylaniline, **9,** 470
Aluminum chloride–Potassium chloride–Sodium chloride, **9,** 13
Aluminum chloride–Potassium phenoxide, **9,** 14
Aluminum chloride–Sodium iodide, **12,** 29
Aluminum cyclohexoxide, **1,** 34; **4,** 15; **6,** 20
Aluminum ethoxide, **6,** 180
Aluminum hydride (*see also* Lithium aluminum hydride–Aluminum chloride), **1,** 34, 595; **2,** 23, 243; **3,** 9, 102; **4,** 233; **5,** 13; **6,** 19
Aluminum iodide, **1,** 35; **12,** 30
Aluminum isopropoxide, **1,** 35, 50, 983; **3,** 10; **4,** 15; **5,** 14; **6,** 19; **8,** 15; **9,** 14; **11,** 29, 296
Aluminum oxide (*see* Alumina)
Aluminum telluride, **10,** 205
Aluminum thiophenoxide, **9,** 15
Amberlite ion-exchange resin, **1,** 512; **2,** 227; **4,** 266, 268; **5,** 356; **6,** 302; **8,** 263; **11,** 276; **12,** 87
Amberlyst ion-exchange resin, **5,** 355; **6,** 302; **7,** 182; **8,** 263; **9,** 256; **11,** 263, 266, 276
Amidines, bicyclic (*see also* specific names), **4,** 16
p-Aminoacetophenone, **4,** 18
p-Aminoazobenzene, **1,** 37
p-Aminobenzoic acid, **2,** 24
1-Aminobenzotriazole, **1,** 37; **3,** 10

(S)-(−)-1-Amino-2-(*t*-butyldialkylsilyloxymethyl)pyrrolidine, **12,** 32
1-Amino-2,3-*trans*-diphenylaziridine, **5,** 328
N-Amino-4,6-diphenylpyridone, **7,** 10
2-Aminoethanol, **1,** 357; **2,** 189; **4,** 222; **6,** 225
4-Amino-3-hydrazino-5-mercapto-1,2,4-triazole, **4,** 19
1-Amino-(S)-2-[(R)-1-hydroxyethyl]indoline, **3,** 11
(S)-1-Amino-2-hydroxymethylindoline, **3,** 12
Aminoiminomethanesulfinic acid (*see* Formamidinesulfinic acid)
Aminomalononitrile, **4,** 459
(R)-1-Amino-2-methoxymethyl-1-pyrrolidine (RAMP), **9,** 17; **12,** 30
(S)-1-Amino-2-methoxymethyl-1-pyrrolidine (SAMP), **7,** 10; **8,** 16; **9,** 17; **12,** 30
(S)-(−)-2-Amino-1-methoxy-3-phenylpropane, **10,** 11
(1S,2S)-2-Amino-3-methoxy-1-phenyl-1-propanol, **6,** 204, 388; **7,** 229; **9,** 312; **12,** 310, 381
(S)-(−)-2-Amino-3-methyl-1,1-diphenyl-1-butanol, **12,** 31
(S)-15-Aminomethyl-14-hydroxy-5,5-dimethyl-2,8-dithia[9](2,5)-pyridinophane, **11,** 29
(+)-3-Aminomethylpinane, **8,** 17
2-Amino-2-methyl-1,3-propanediol, **2,** 24
2-Amino-2-methyl-1-propanol, **1,** 37; **3,** 14, 314; **4,** 540; **6,** 20, 368
1-Amino-2-phenylaziridine, **5,** 328
(1S,2S)-2-Amino-1-phenyl-1,3-propanediol, **6,** 386; **12,** 32, 383
N-Aminophthalimide, **1,** 38; **5,** 14
(S)-2-Aminopropyl benzyl ether, **11,** 30
2-Aminopyridine, **9,** 18
(S)-(−)-1-Amino-2-(silyloxymethyl)-pyrrolidines (SASP), **12,** 32
5-Amino-1*H*-tetrazole, **5,** 653
2-Aminothiazole, **4,** 19
N-Aminotriphenylphosphinimine, **2,** 286
N-Amino-2,4,6-triphenylpyridinium perchlorate, **7,** 409
Ammonia, **1,** 1290; **3,** 288; **4,** 420; **5,** 15
Ammonium acetate, **1,** 38; **5,** 6; **7,** 11
Ammonium azide, **5,** 9
Ammonium bromide, **6,** 473
Ammonium cerium(IV) nitrate or sulfate [*see* Cerium(IV) ammonium nitrate or sulfate]
Ammonium formate, **1,** 38; **12,** 369
Ammonium metavanadate, **1,** 39, 1057
Ammonium nitrate, **1,** 39
Ammonium peroxydisulfate (*see* Ammonium persulfate)
Ammonium persulfate, **1,** 952; **3,** 238; **5,** 15; **6,** 20; **12,** 33
Ammonium persulfate–Silver nitrate, **3,** 15; **5,** 16; **8,** 18; **9,** 18
Ammonium polysulfide, **1,** 1120; **6,** 21
Ammonium sulfide, **3,** 270
Ammonium tetrafluoroborate, **5,** 17; **6,** 132
Amphetamine, **5,** 18
t-Amyl azidoformate, **2,** 24
t-Amyl chloroformate (*see also* Alkyl chloroformates), **1,** 40; **2,** 24, 307; **3,** 15
t-Amyl diazoacetate, **8,** 40
t-Amyl hydroperoxide, **4,** 20
Amyl nitrite, **1,** 41, 923, 1098; **5,** 18
t-Amylquinolyl-8-carbonate, **4,** 20
Amylsodium (*see* Pentylsodium)
Aniline, **1,** 41; **6,** 21, 158; **8,** 105
(S)-4-Anilino-3-methylamino-1-butanol, **11,** 292; **12,** 33
(2S,4S)-(Anilinomethyl)-1-ethyl-4-hydroxypyrrolidine, **10,** 12
(S)-2-(Anilinomethyl)pyrrolidine, **9,** 19
Anisole, **1,** 41
p-Anisolesulfonyl chloride, **9,** 297
(R)-(+)-*o*-Anisyl(cyclohexyl)methylphosphine (CAMP), **5,** 500; **11,** 31
(S)-(+)-2-(*p*-Anisylsulfinyl)-2-cyclopentenone, **12,** 296
Anthracene, **1,** 539; **5,** 662; **6,** 173
Anthracene-9,10-diimine, **1,** 42
Anthranilic acid, **1,** 46, 268, 1064, 1194; **3,** 272; **5,** 359; **6,** 96
Anthraquinone, **4,** 20
9-Anthrylmethyl *p*-nitrophenyl carbonate, **8,** 18
Antifoam compounds, **1,** 42
Antimony(III) chloride, **1,** 423; **8,** 19, 536; **11,** 602
Antimony(V) chloride, **1,** 42; **4,** 21; **5,** 18, 20, 115; **6,** 22; **7,** 12; **8,** 500
Antimony(III) fluoride, **6,** 553
Antimony(V) fluoride, **5,** 19, 20; **6,** 23, 103; **9,** 20
Antimony(V) fluoride–Sulfur dioxide, **3,** 202
Arenechromium tricarbonyls [*see* Arene-(tricarbonyl)chromium complexes]

Arenediazonium halides, **1,** 3, 166, 402, 490, 653, 743, 1177; **2,** 62, 91; **4,** 18, 24; **10,** 306
Arenediazonium hexafluorophosphates, **1,** 42; **12,** 380
Arenediazonium tetrahaloborates, **1,** 43, 349, 394, 722, 1052; **2,** 25; **3,** 16, 294; **4,** 562; **5,** 147, 697; **8,** 22, 130, 255; **9,** 102, 360, 465; **10,** 298; **12,** 380
Arenesulfonyl azides, **2,** 62; **3,** 17; **4,** 24; **5,** 19, 475; **10,** 422; **11,** 535, 536; **12,** 220
Arenesulfonyl halide–Antimony(V) halide complexes, **5,** 20
Arene(tricarbonyl)chromium complexes, **6,** 27, 103, 125; **7,** 71; **8,** 21, 110; **9,** 21, 117, 364; **10,** 13, 100; **11,** 131, 398; **12,** 34, 575
Argentic, *see* Silver(II)
Arsenic trihalides, **3,** 16
Aryllead triacetates, **11,** 31
2-Arylsulfonyl oxaziridines, **12,** 392
Arylthallium bis(trifluoroacetates), **3,** 287, **4,** 498; **5,** 658; **6,** 579; **8,** 480; **9,** 462; **10,** 300; **12,** 483
Ascarite, **4,** 21
Azidoacetyl chloride, **5,** 21
Azido(methoxy)acetyl chloride, **5,** 21
Azidomethyl phenyl sulfide, **10,** 14; **12,** 37
α-Azidostyrene, **6,** 24
Azidotributyltin, **7,** 377
Azidotrimethylsilane, **1,** 1236; **3,** 316; **4,** 542; **5,** 207, 282, 719; **6,** 632; **7,** 394; **8,** 287; **9,** 21; **10,** 14, 113, 120; **11,** 32; **12,** 37, 464
Azidotris(dimethylamino)phosphonium hexafluorophosphate, **6,** 25
Aziridine, **1,** 378; **4,** 22; **10,** 348; **11,** 23
Azobenzene-4-carbohydrazide, **1,** 44
Azobisisobutyronitrile, **1,** 45
Azodicarbonamide, **1,** 45
1,1′-(Azodicarbonyl)dipiperidine, **8,** 19; **9,** 22

Barium–Ammonia, **1,** 107
Barium hydroxide, **4,** 23; **9,** 126; **12,** 38
Barium manganate, **8,** 21; **9,** 23; **10,** 16, 168; **12,** 38
Barium oxide, **1,** 683; **9,** 23
Barium permanganate, **1,** 46
Bate's reagent {μ-Oxo-bis[tris-(dimethylamino)phosphonium] bis(tetrafluoroborate)}, **6,** 425
9-BBN (*see* 9-Borabicyclo[3.3.1]nonane)
Benzal bromide or chloride, **5,** 155, 453
Benzaldehyde, **1,** 46; **5,** 398; **6,** 26; **7,** 6
Benzaldehyde ethylene dithioacetal (*see* 2-Phenyl-1,3-dithiolane)
Benzaldehyde phenylhydrazone, **4,** 172
Benzamidine, **6,** 27
μ-[Benzenaminato(2-)]dibromodimagnesium, **4,** 33
Benzene, **2,** 21, 125; **3,** 288; **6,** 364
Benzeneboronic acid (*see also* Boronic acid resins), **1,** 833; **2,** 317; **3,** 221; **5,** 513; **7,** 284; **9,** 23; **11,** 408, 513
2-Benzenecarbothio-1-methylpyridinium chloride, **9,** 40
Benzenechromium tricarbonyl [*see also* Arene(tricarbonyl)chromium complexes], **6,** 27; **8,** 21
Benzenediazonium 2-carboxylate, **1,** 46, 268, 360, 1064, 1194; **2,** 26, 390; **3,** 272; **5,** 359; **6,** 96
Benzenediazonium chloride (*see also* Arenediazonium halides), **4,** 24
Benzenediazonium tetrafluoroborate (*see also* Arenediazonium tetrahaloborates), **1,** 43; **8,** 22
Benzeneperoxyseleninic acid, **8,** 22
Benzeneselenenic acid, **8,** 24, 320; **9,** 24
Benzeneselenenic anhydride, **9,** 25, 79
Benzeneselenenyl acetate, **5,** 520; **8,** 28; **9,** 200
Benzeneselenenyl bromide (*see also* Benzeneselenenyl halides), **5,** 274, 518; **6,** 336, 459; **7,** 286; **8,** 25; **9,** 25; **10,** 16, 296; **11,** 34, 468; **12,** 241
Benzeneselenenyl bromide–Silver nitrate, **11,** 33
Benzeneselenenyl bromide–Silver trifluoroacetate (*see also* Benzene-selenenyl trifluoroacetate), **5,** 522
Benzeneselenenyl chloride (*see also* Benzeneselenenyl halides), **5,** 518; **6,** 459; **7,** 286; **8,** 25; **9,** 25, 367; **10,** 16, 170; **11,** 34, 266; **12,** 39
Benzeneselenenyl chloride–Aluminum chloride, **11,** 33
Benzeneselenenyl chloride–Silver trifluoroacetate (*see also* Benzene-selenenyl trifluoroacetate), **5,** 522
Benzeneselenenyl halides, **5,** 274, 518; **6,** 336, 459; **7,** 286; **8,** 25; **9,** 25, 367; **10,** 16, 170, 296; **11,** 34, 36, 266, 468; **12,** 39, 241
Benzeneselenenyl hexafluoroantimonate or

phosphate, **10,** 21
Benzeneselenenyl iodide (*see also* Benzeneselenenyl halides), **11,** 36
Benzeneselenenyl trifluoroacetate, **5,** 518, 522; **6,** 459
Benzeneseleninic acid, **5,** 520; **8,** 28; **9,** 32, 79; **10,** 22
Benzeneseleninic anhydride, **6,** 240; **7,** 139; **8,** 29; **9,** 32; **10,** 22; **11,** 37, 275
Benzeneselenocyanate (*see* Phenyl selenocyanate)
Benzeneselenol, **6,** 28, 85, 362; **8,** 447, 536; **9,** 35, 432; **10,** 412; **11,** 537
Benzenesulfenyl chloride, **5,** 523; **6,** 30; **8,** 32; **9,** 35, 429; **10,** 3, 24; **11,** 34, 39; **12,** 42
Benzenesulfinic acid, **9,** 132
Benzenesulfonic anhydride, **1,** 46
Benzenesulfonyl azide (*see also* Arenesulfonyl azides), **3,** 17; **4,** 24
Benzenesulfonyl bromide or chloride, **1,** 46; **3,** 18; **4,** 106; **5,** 22
Benzenesulfonyl fluoride, **11,** 40
N-Benzenesulfonylformimidic acid ethyl ester, **3,** 18
Benzenesulfonylhydrazide, **4,** 535; **7,** 13
Benzenesulfonyl isocyanate, **1,** 48
Benzenesulfonylnitrile oxide, **9,** 39; **11,** 40
Benzenesulfonyltetrazole, **7,** 13
1-Benzenesulfonyl-2-trimethylsilylethane, **11,** 41
(E)-1-Benzenesulfonyl-2-(trimethylsilyl)-ethylene, **11,** 42
(Benzenethiolato)diethylaluminum, **10,** 281; **11,** 542; **12,** 343
Benzhydryl azidoformate, **1,** 48
Benzil monohydrazone, **1,** 344, 655
5-Benzisoxazolemethylene chloroformate, **6,** 32
Benzocyclobutenedione monoethylene ketal, **9,** 40
Benzocyclobutenylmethylmagnesium bromide, **11,** 308
Benzocyclopropene, **4,** 402; **6,** 33
1,3,2-Benzodioxaborole (*see* Catecholborane)
1,3-Benzodithiole-2-carbene, **6,** 96
1,3-Benzodithiolylium perchlorate or tetrafluoroborate, **8,** 34; **10,** 24, 25
Benzohydroxamic acid, **9,** 449
Benzoic acid, **1,** 49; **3,** 231; **7,** 405
Benzoic anhydride, **1,** 49; **3,** 262; **5,** 23, 211
S-Benzoic O, O-diethyl phosphorodithioic anhydride, **6,** 34
Benzoin, **6,** 34
Benzoin cyclic carbonate, **4,** 25; **5,** 23
Benzonitrile, **1,** 520; **2,** 404; **5,** 533, 742; **6,** 69
Benzonitrile oxide, **2,** 383
Benzophenone hydrazone, **1,** 338, 655; **2,** 294; **6,** 452
Benzophenone N-methylimine, **8,** 210
1,4-Benzoquinone, **1,** 49, 624
Benzo-1,2,3-thiadiazole 1, 1-dioxide, **1,** 50
1,3-Benzothiazole, **8,** 274
Benzo-1,3-thiazole-2-carboxaldehyde, **5,** 24
2-Benzothiazolyllithium, **8,** 274
1-Benzotriazolyl *t*-butyl carbonate, **12,** 83
Benzotriazolyl-N-hydroxytris(dimethyl-amino)phosphonium hexafluoro-phosphate [*see* Benzotriazolyloxytris-(dimethylamino)phosphonium hexafluorophosphate]
Benzotriazolyloxytris(dimethylamino)-phosphonium hexafluorophosphate (BOP), **6,** 34; **7,** 14
O-Benzotriazolyl-N,N,N′,N′-tetramethyl-uronium hexafluorophosphate, **12,** 44
Benzotrichloride, **1,** 1140
Benzoyl bromide, **5,** 249
Benzoylcarbene, **2,** 81
Benzoyl chloride (*see also* Benzoyl chloride + catalysts), **1,** 30, 50, 1258; **4,** 311, 346, 464; **5,** 24, 45, 145, 403, 410; **6,** 171; **8,** 86, 151, 313, 318, 510
Benzoyl chloride–Pyridine, **1,** 20, 959
Benzoyl chloride–Zinc chloride, **1,** 1290
Benzoyl chloride–Zirconium(IV) chloride, **1,** 1295
Benzoyl cyanide, **6,** 35
Benzoyl fluoride, **1,** 15
N-Benzoylimidazole, **1,** 424, 493
Benzoyl isothiocyanate, **1,** 50
Benzoylmethylenetriphenylphosphorane, **2,** 416
Benzoyl nitrate, **5,** 25; **7,** 3
2-Benzoyloxymethylbenzoyl chloride, **7,** 14
N-Benzoylperoxycarbamic acid, **6,** 35
N-Benzoyl-N′-phenyldiazene, **1,** 833
N-Benzoylpiperidine, **1,** 886
3-Benzoylpropionic acid, **2,** 26; **3,** 19
Benzoyltetrazole, **7,** 15
2-Benzoylthio-1-methylpyridinium chloride, **9,** 40
Benzoyl trifluoromethanesulfonate, **12,** 44

N-Benzoyl-N′-trifluoromethanesulfonylhydrazide, **6,** 36
Benzyl acetate, **9,** 82
Benzylamine, **1,** 51; **5,** 479; **8,** 70; **10,** 26
Benzyl bromide, **3,** 253; **5,** 25
Benzyl bromomethyl sulfide, **5,** 25
Benzyl *trans*-1,3-butadiene 1-carbamate, **8,** 34; **9,** 479; **11,** 43
Benzylbutyldimethylammonium chloride, polymeric, **6,** 38
Benzyl carbazate, **1,** 110
Benzyl chloride, **1,** 1080; **2,** 158; **3,** 19; **4,** 310; **8,** 110
Benzylchlorobis(triphenylphosphine)palladium(II), **8,** 35; **9,** 41; **10,** 26; **12,** 44
Benzyl chloroformate, **1,** 109; **2,** 59
Benzyl chloromethyl ether, **1,** 52; **2,** 26; **12,** 430
Benzyl chloromethyl sulfide, **2,** 27
N-Benzylcinchoninium chloride, **12,** 380
3-O-Benzyl-1, 2-O-cyclohexylidene-α-D-glucofuranose, **2,** 243
Benzyl diazoacetate, **8,** 222
1-Benzyl-1,2-dihydroisonicotinamide, **12,** 46
1-Benzyl-1,4-dihydronicotinamide, **6,** 36; **7,** 15; **11,** 419; **12,** 47
Benzyldimethylanilinium hydroxide, **6,** 37
Benzyldodecyldimethylammonium bromide, **7,** 16
(+)-2-Benzylethanamine, **1,** 978
Benzyl(2-hydroxyethyl)dimethylammonium hydroxide, **6,** 220
3-Benzyl-5-(2-hydroxyethyl)-4-methyl-1,3-thiazolium chloride, **6,** 38, 289; **7,** 16; **10,** 27
N-Benzylhydroxylamine, **9,** 42; **12,** 13
O-Benzylhydroxylamine, **9,** 504; **12,** 273
Benzylideneacetone(tricarbonyl)iron, **6,** 39; **8,** 36
N-Benzylideneaniline, **12,** 194
Benzylidenetriphenylphosphorane, **5,** 30, 299; **11,** 147
Benzyl lithioacetate, **9,** 82
Benzyllithium, **1,** 52; **2,** 343; **5,** 552
Benzylmagnesium bromide, **9,** 53, 229
Benzylmagnesium chloride, **1,** 253, 419
(Benzylmethoxymethyl)methylamine, **7,** 17; **9,** 42
(−)-N-Benzyl-N-methylephedrinium bromide, **6,** 39
Benzyl methyl ether, **1,** 52
3-Benzyl-4-methylthiazolium chloride, **7,** 18
O-Benzylmonoperoxycarbonic acid, **6,** 40
Benzyl nitrile, **1,** 1066
Benzyl *p*-nitrophenyl carbonate, **7,** 78
3-(2-Benzyloxyacetyl)thiazolidine-2-thione, **12,** 491
9-(Benzyloxy)anthracene, **12,** 47
Benzyloxycarbonylhydrazine, **1,** 110
(S)-N-Benzyloxycarbonylproline, **11,** 447; **12,** 454
5-Benzyloxymethyl-1,3-cyclopentadiene, **4,** 77
2-Benzyloxy-1-propene, **12,** 48
Benzyl phenyl sulfide, **6,** 131
Benzyl phenyl sulfoxide, **6,** 394
O-Benzyl phosphorus O,O-diphenylphosphoric anhydride, **1,** 52
N-Benzylquinidinium chloride, **7,** 311
(−)-Benzylquininium chloride, **7,** 311; **8,** 430; **10,** 27, 338
(−)-Benzylquininium fluoride, **8,** 430
Benzylsodium, **6,** 40
Benzylsulfonyldiazomethane, **11,** 43, 472
3-Benzylthiazolium bromide, **10,** 27
Benzylthiocarbonyl chloride, **2,** 26
Benzylthiomethyl halides (*see* Benzyl bromomethyl sulfide or Benzyl chloromethyl sulfide)
Benzyl *p*-toluenesulfonate, **11,** 44
Benzyltributylammonium chloride, **6,** 41
Benzyl 3-tributylstannylacrylate, **12,** 56
Benzyl trichloroacetimidate, **11,** 44
Benzyltriethylammonium chloride (TEBA), **3,** 19; **4,** 27; **5,** 26; **6,** 41, 566; **7,** 18, 78, 239, 353; **8,** 248, 388; **9,** 357; **10,** 293, 305
Benzyl(triethyl)ammonium permanganate, **9,** 43; **10,** 28; **11,** 44; **12,** 48
N-Benzyl trifluoromethanesulfonamide, **5,** 29; **6,** 43
Benzyl trifluoromethanesulfonate, **6,** 44
Benzyltrimethylammonium chloride, **1,** 53
Benzyltrimethylammonium cyanide, **1,** 53
Benzyltrimethylammonium fluoride, **6,** 44; **7,** 353; **8,** 295; **9,** 128; **10,** 29; **12,** 461
Benzyltrimethylammonium hydroxide (Triton B), **1,** 1252; **5,** 29; **7,** 20; **8,** 36; **12,** 48, 478
Benzyltrimethylammonium iodide, **1,** 53
Benzyltrimethylammonium mesitoate, **1,** 54
Benzyltrimethylsilane, **6,** 637
Benzyltriphenylphosphonium chloride, **1,** 279; **5,** 30

Benzyltris(dimethylamino)phosphonium bromide, **2,** 210
Biacetyl (*see* 2,3-Butanedione)
BINAP [*see* (R)-(+)- and (S)-(−)-2,2′-Bis-(diphenylphosphine)-1,1′-binaphthyl]
(S)-(+)- and (R)-(−)-1,1′-Binaphthyl-2,2′-diyl hydrogen phosphate, **12,** 49
4-Biphenylcarbonyl chloride, **4,** 376; **7,** 282
p-Biphenyl isocyanate, **4,** 105
[2-(*p*-Biphenyl)isopropyl]phenyl carbonate, **2,** 179
2,2′-Bipyridinium chlorochromate, **10,** 30; **11,** 44
Birch reduction (*see also* Lithium–Ammonia, Sodium–Ammonia), **1,** 54; **2,** 27; **3,** 19; **4,** 31; **5,** 30, 378; **7,** 21; **8,** 38
Bis[μ-(acetato)]bis(2-propenyl)dipalladium, **7,** 4
Bis(acetonitrile)chloronitropalladium(II), **10,** 30; **11,** 45
Bis(acetonitrile)dichloropalladium(II), **7,** 21; **8,** 39; **9,** 44, 456; **10,** 30; **11,** 46; **12,** 50
Bis(acetylacetonate)oxovanadium (*see* Vanadyl acetylacetonate)
Bis(acetylacetonate)palladium(II) (*see* Palladium(II) acetylacetonate)
Bis(acrylonitrile)nickel(0), **3,** 20; **6,** 45; **11,** 48
2,3-Bis(azidomethyl)-1,3-butadiene, **5,** 33
Bis(benzonitrile)dibromopalladium(II), **11,** 80
Bis(benzonitrile)dichloropalladium(II), **1,** 56; **4,** 129; **5,** 31, 502, 574; **6,** 45; **7,** 95, 277, 371; **9,** 44; **10,** 31; **11,** 48; **12,** 51
Bis(benzonitrile)dichloroplatinum(II), **11,** 49
Bis(benzonitrile)palladium(II) chloride [*see* Bis(benzonitrile)dichloropalladium(II)]
Bis(benzoyloxy)borane, **10,** 32
Bis(benzyltriethylammonium) dichromate, **11,** 50
Bis[1,2-bis(diphenylphosphine)ethane]-palladium(0), **9,** 344; **10,** 32; **11,** 508; **12,** 473
Bis[1,3-bis(diphenylphosphine)propane]-chlororhodium(I), **12,** 111
Bis(α-bromobenzyl)phenylphosphine oxide, **4,** 116
N,N-Bis(bromomagnesio)anilide, **4,** 33
1,4-Bis(bromomagnesio)butane, **1,** 415
1,5-Bis(bromomagnesio)pentane, **9,** 355
2,3-Bis(bromomethyl)-1,3-butadiene, **5,** 32
Bis(*t*-butoxycarbonyl)amine, **1,** 210
1,3-Bis(*t*-butyldimethylsilyloxy)-2-aza-1,3-diene, **11,** 50
Bis(N-*t*-butylsalicylaldiminato)copper(II), **12,** 52
Bis[(−)-camphorquinone-α-dioximato]-cobalt(II) hydrate, **8,** 39
Bis(3-carboxy-4-nitrophenyl) disulfide, **1,** 351
Bis(chlorodicarbonyl)rhodium (*see* Tetracarbonyldi-μ-chlorodirhodium)
Bis(chlorodimethylsilyl)ethane, **10,** 140; **12,** 179
1,4-Bis(chloromethoxy)butane, **6,** 105; **7,** 22
Bis(chloromethyl) ether, **5,** 34
Bis(chloromethyl) sulfoxide, **7,** 380
μ-Bis(cyanotrihydroborato)tetrakis-(triphenylphosphine)dicopper(I), **10,** 33
Bis(1,5-cyclooctadiene)nickel(0), **4,** 33; **5,** 34; **6,** 45; **7,** 428; **9,** 45; **10,** 33; **11,** 51; **12,** 52
Bis(1,5-cyclooctadiene)nickel(0)–(−)-Benzylmethylphenylphosphine, **4,** 35
Bis(cyclopentadienyl)bis(μ-nitrosyl-N:N)-dicobalt, **10,** 116
Bis(cyclopentadienyl)dichlorozirconium(II) [*see* Dichlorobis(cyclopentadienyl)-zirconium(II)]
Bis(cyclopentadienyl)(diiodozinc)-μ-methylenetitanium, **12,** 506
Bis(cyclopentadienyl)diiodozirconium, **12,** 53
Bis(cyclopentadienyl)isoprenezirconium, **11,** 174
Bis(cyclopentadienyl)methyltitanium, **8,** 40
Bis(cyclopentadienyl)niobium trihydride, **6,** 47
Bis(cyclopentadienyl)titanacyclobutanes, **12,** 54, 110
Bis(cyclopentadienyl)titanium-μ-chloro-μ-methylidenedimethylaluminum [*see* μ-Chlorobis(cyclopentadienyl)(dimethyl-aluminum)-μ-methylenetitanium]
Bis(cyclopentadienyl)titanium dichloride [*see* Dichlorobis(cyclopentadienyl)titanium]
Bis(cyclopentadienyl)trihydridoniobium, **6,** 47
Bis(cyclopentadienyl)(η^3-1-trimethyl-silylallyl)titanium, **11,** 174
Bis(dialkylamino)alanes–Dichlorobis-(cyclopentadienyl)titanium, **9,** 46
Bis(1-diazo-2-ethoxy-2-oxoethyl)mercury, **6,** 190
Bis(dibenzylideneacetone)palladium(0), **9,** 46, 344; **10,** 34; **11,** 53; **12,** 56, 575
Bis(dibutylacetoxytin) oxide, **4,** 393

Bis(dicarbonylcyclopentadienyl)iron [*see* Di-μ-carbonyldicarbonylbis-(cyclopentadienyl)diiron]
Bis-3,4-dichlorobenzoyl peroxide, **3,** 21
Bis(dichlorotrimethylsilylmethyl)mercury, **2,** 30
Bis(diethoxyphosphinyl)methane, **6,** 191
Bis(diethoxyphosphinyl)methyl(dimethyl)-amine (*see* Tetraethyl dimethyl-aminomethylenediphosphonate)
Bis(diethylaluminum) sulfate, **9,** 46
Bis(2,6-diethylphenyl)carbodiimide, **1,** 56
Bis(diethylphosphono)methane, **6,** 191
Bis(diisobutylaluminum) 1,2-ethane-dithiolate, **12,** 342
Bis(diisopropylamino)aluminum hydride, **7,** 196
1,3-Bis(dimethoxyphosphinyl)-2-propanone, **10,** 35
Bis(dimethylaluminum) 1,2-ethanedithiolate, **5,** 35
Bis(dimethylaluminum) 1,3-propane-dithiolate, **6,** 49
4,4′-Bis(dimethylamino)benzophenone (Michler's ketone), **2,** 285
(S,S)-(+)-1,4-Bis(dimethylamino)-2,3-dimethoxybutane, **9,** 47
Bis(dimethylamino)methane, **3,** 21; **7,** 359
Bis(dimethylamino)methoxymethane [*see also t*-Butoxybis(dimethylamino)-methane], **10,** 34
1,8-Bis(dimethylamino)naphthalene (Proton sponge), **3,** 22; **4,** 35; **6,** 50, 59, 596; **7,** 372
Bis(dimethylamino) phosphorochloridate, **4,** 480; **11,** 372
Bis(dimethylamino) phosphorochloridite, **6,** 575
Bis(3-dimethylaminopropyl)-phenylphosphine, **5,** 36; **10,** 34
1,4-Bis(dimethylamino)-1,1,4,4-tetramethyldisilethylene, **12,** 476
Bis(3,6-dimethyl)borepane, **4,** 35
Bis(N,N-dimethylformamido)-oxodiperoxymolybdenum(VI), **9,** 48
Bis(4,5-dimethyl-2-oxo-1,3,2-dioxaphospholenyl)oxide, **6,** 6, 218
1,3-Bis(dimethylphosphono)-2-propanone, **10,** 35
Bis(2,4-dinitrophenyl) carbonate, **1,** 56; **2,** 29
[Bis(1,3,2-dioxaborolan-2-yl)methyl]lithium, **6,** 328; **10,** 239
α,γ-Bis(diphenylene)-β-phenylallyl, **1,** 57
Bis(1,3-diphenyl-2-imidazolidinylidene), **7,** 22
trans-2,3-Bis(diphenylphosphine)bicyclo-[2.2.1]hept-5-ene (Norphos), **9,** 48; **10,** 36
[*trans*-2,3-Bis(diphenylphosphine)bicyclo-[2.2.1]hept-5-ene]dichloronickel(II), **10,** 36
(R)-(+)- and (S)-(−)-2,2′-Bis(diphenyl-phosphine)-1,1′-binaphthyl (BINAP), **10,** 36; **11,** 53; **12,** 56
[2,2′-Bis(diphenylphosphine)-1,1′-binaphthyl](cyclooctadiene)rhodium(I) perchlorate, **11,** 54; **12,** 57
[2,2′-Bis(diphenylphosphine)-1,1′-binaphthyl](norbornadiene)rhodium(I) perchlorate, **10,** 36
1,4-Bis(diphenylphosphine)butane (dppb), **12,** 473
[1,4-Bis(diphenylphosphine)butane]-(1,5-cyclooctadiene)iridium(I) tetrafluoroborate, **12,** 427
1,4-[Bis(diphenylphosphine)butane]-(norbornadiene)rhodium(I) tetrafluoroborate, **12,** 426
1,2-Bis(diphenylphosphine)ethane (dppe, Diphos), **6,** 447; **9,** 451; **10,** 34, 389; **11,** 53, 81, 391; **12,** 16, 129, 473
[1,2-Bis(diphenylphosphine)ethane]bis-(triphenylphosphine)palladium(0), **10,** 133
[1,2-Bis(diphenylphosphine)ethane]-cyclohexyneplatinum(0), **5,** 171
[1,2-Bis(diphenylphosphine)ethane]-(dichloro)nickel(II), **7,** 250; **9,** 147; **10,** 36; **12,** 146, 171
[1,2-Bis(diphenylphosphine)ethane]-dichloropalladium(II), **6,** 447
[1,2-Bis(diphenylphosphine)ethane]nickel(II) chloride {*see* [1,2-Bis(diphenyl-phosphine)ethane](dichloro)nickel(II)}
Bis[2-(diphenylphosphine)ethyl]amine hydrochloride, **11,** 54
1,1′-Bis(diphenylphosphine)ferrocene (dppf), **10,** 37
[1,1′-Bis(diphenylphosphine)ferrocene]-dichloronickel(II), **10,** 37; **12,** 172
[1,1′-Bis(diphenylphosphine)ferrocene]-dichloropalladium(II), **9,** 147; **10,** 37; **12,** 46, 171
(R)-1-(S)-1′,2-Bis(diphenylphosphine)-

ferrocenylethanol, **11,** 237
1,3-Bis(diphenylphosphine)propane, **8,** 474
[1,3-Bis(diphenylphosphine)propane]-dichloronickel(II), **6,** 50; **10,** 191; **11,** 167; **12,** 244
[1,2-Bis(diphenylphosphine)propane]-(norbornadiene)rhodium perchlorate, **9,** 452
Bis[2-(diphenylphosphinyl)ethyl]amine hydrochloride, **11,** 54
Bis(2,2-dipyridyl)silver(II) peroxydisulfate, **6,** 51
1,2-Bis(1,3-dithian-2-yl)ethane, **6,** 493; **8,** 223
Bis(ethenyl) sulfone (*see* Divinyl sulfone)
Bis(ethyl diazoacetate)mercury, **6,** 190
Bis(ethylenediamine)chromium(II) perchlorate, **3,** 58; **4,** 97
Bis(ethylthio)acetic acid, **9,** 228; **10,** 38
2,2-Bis(ethylthio)propane, **10,** 373
Bis(fluoroxy)difluoromethane, **4,** 36
Bis(iodomethyl)zinc–Zinc iodide (*see* Simmons–Smith reagent)
Bis(isopropylthio)boron bromide [*see* Bromobis(isopropylthio)boron]
Bis(methoxycarbonyl)carbene, **8,** 159
Bis(methoxycarbonyl)sulfodiimide, **12,** 57
trans-2,5-Bis(methoxymethoxymethyl)-pyrrolidine, **12,** 573
2,3-Bis(methoxymethyl)-1,3-butadiene, **5,** 33
Bis(1-methoxy-2-methyl-1-propenyloxy)-dimethylsilane, **12,** 58
trans-2,5-Bis(methoxymethyl)pyrrolidine, **12,** 59
2,4-Bis(4-methoxyphenyl)-1,3-dithia-2,4-diphosphetane-2,4-disulfide (Lawesson's reagent), **8,** 327; **9,** 49; **10,** 39; **11,** 54; **12,** 59
Bis(*p*-methoxyphenyl) telluroxide, **9,** 50; **11,** 55
Bis[methoxy(thiocarbonyl)] disulfide, **10,** 39
Bis(*o*-methylaminophenyl) disulfide, **9,** 305
(S,S)- or (R,R)-N,N′-Bis(α-methylbenzyl)-sulfamide, **12,** 60
Bis(3-methyl-2-butyl)borane (*see* Disiamylborane)
Bis(*trans*-2-methylcyclohexyl)borane, **4,** 37
Bis(methylcyclopentadienyl)tin(II), **12,** 201
Bis(N-methyl-N-nitroso)terephthalamide, **1,** 59, 191
Bis(N-methylpiperazinyl)aluminum hydride, **6,** 52; **12,** 60
1,1-Bis(methylseleno)cyclobutane, **10,** 232
1,1-Bis(methylseleno)cyclopropane, **9,** 85
1,3-Bis(methylthio)allyllithium, **4,** 38; **6,** 53
1,2-Bis(methylthio)-1,3-butadiene, **10,** 40
2,4-Bis(methylthio)-2,4-dithioxo-cyclodiphosphathiane, **11,** 56
Bis(methylthio)methane, **7,** 24
Bis(methylthio)methyllithium, **7,** 24
Bis(methylthio)(trimethylsilyl)methane, **6,** 53
Bis(methylthio)(trimethylsilyl)methyllithium, **6,** 53
Bis(methylthio)(trimethylstannyl)methane, **6,** 54
Bis(methylthio)(trimethylstannyl)-methyllithium, **6,** 54
Bismuth(III) acetate, **4,** 40
Bismuth(III) chloride, **7,** 24
Bismuth(III) oxide, **1,** 59
Bismuth(III) oxide–Tin(IV) oxide, **6,** 55
Bis(*p*-nitrophenethyl) phosphorochloridate, **11,** 56
Bis(4-nitrophenyl) carbonate, **1,** 330
Bis(*p*-nitrophenyl) hydrogen phosphate, **1,** 331; **2,** 29
Bis(*o*-nitrophenyl) phenylphosphonate, **9,** 50; **10,** 41
Bis(*p*-nitrophenyl) phenylphosphonate, **9,** 50
2,5-Bis[(Z)-(2-nitrophenylsulfenyl)-methylene]-3,6-dimethylene-7-oxabicyclo[2.2.1]heptane, **12,** 60
Bis(norbornadiene)rhodium(I) perchlorate–(R)-1-(S)-1′,2-Bis(diphenylphosphine)-ferrocenylethanol, **11,** 239
N,N-Bis(2-oxo-3-oxazolidinyl)phosphoro-diamidic chloride, **10,** 41; **11,** 57
Bis(2,4-pentanedionato)copper (*see* Copper (II) acetylacetonate)
Bis(2,4-pentanedionato)nickel(II) (*see* Nickel(II) acetylacetonate)
Bis(2,4-pentanedionato)palladium(II) [*see* Palladium(II) acetylacetonate]
Bis-*o*-phenylene pyrophosphite, **1,** 60; **2,** 30
1,3-Bis(phenylseleno)propene, **11,** 59
1,3-Bis(phenylseleno)-3-trimethylsilyl-1-propene, **11,** 60
Bis(phenylsulfinyl)methane, **6,** 191, 394
1,1-Bis(phenylsulfonyl)cyclopropane, **12,** 61
(Z)-1,2-Bis(phenylsulfonyl)ethylene, **11,** 60
Bis(phenylsulfonyl)methane, **9,** 457
N,N-Bis(phenylthio)benzenesulfenamide, **5,** 732

1,1-Bis(phenylthio)cyclopropane, **8,** 306
Bis(phenylthio)methane, **6,** 131, 267; **7,** 25, 75; **8,** 126; **10,** 42; **11,** 303
Bis(phenylthio)methyllithium, **6,** 131, 267; **7,** 25; **8,** 126; **10,** 42
1,3-Bis(phenylthio)propane, **10,** 231
Bis(propenyl)mercury, **6,** 573
Bis(N-propylsalicylideneaminate)cobalt(II), **11,** 61
Bis(N-propylsalicylideneaminate)copper(II), **5,** 41
Bis(pyridine)dimethylformamido-dichlororhodium borohydride, **4,** 40
Bispyridinesilver permanganate, **11,** 61; **12,** 62
Bis(3-salicylideneaminopropyl)amine-cobalt(II), **3,** 245
Bis(salicylidene)ethylenediiminocobalt(II) (*see* Salcomine)
[Bis(salicylidene-γ-iminopropyl)]-methylaminecobalt(II), **11,** 138
Bis(tetrabutylammonium) dichromate, **9,** 53; **10,** 42
Bis(*p*-toluenesulfonato)zinc, **11,** 605
Bis(*p*-toluenesulfonyl)seleniumdiimide, **7,** 316
Bis(N-*p*-toluenesulfonyl)sulfodiimide, **7,** 317; **8,** 43
1,2-Bis(β-tolylsulfonylethoxycarbonyl)-diazene, **6,** 55
Bis(tribromomethyl)mercury, **2,** 30
Bis(tributylphosphine)nickel dibromide [*see* Dibromobis(tributylphosphine)-nickel(II)]
trans-1,2-Bis(tributylstannyl)ethylene, **5,** 43; **6,** 56
Bis(tributyltin) oxide, **6,** 56; **7,** 26; **8,** 43; **9,** 53; **11,** 62
Bis(tributyltin) peroxide, **8,** 44
Bis(tricaprylylmethyl)ammonium nitrosodisulfonate, **10,** 42
Bis(2,2,2-trichloro-1,1-dimethylethyl) phosphorochloridate, **11,** 62
Bis(2,2,2-trichloroethyl) azodicarboxylate, **12,** 62
Bis(2,2,2-trichloroethyl) phosphoro-chloridate, **2,** 187; **3,** 136
Bis(trichloromethyl)mercury, **2,** 30
Bis(tricyclohexyltin) selenide or sulfide–Boron trichloride, **11,** 63
Bis(triethylphosphine)nickel(II) bromide, **5,** 42
Bis(trifluoroacetoxy)borane, **8,** 45
[Bis(trifluoroacetoxy)iodo]benzene [*see* Phenyliodine(III) bis(trifluoroacetate)]
O,N-Bis(trifluoroacetyl)hydroxylamine, **1,** 60
Bis(trifluoroacetyl) sulfate, **10,** 375
Bis(2,2,2-trifluoroethoxy)triphenyl-phosphorane, **10,** 43
Bis(2,2,2-trifluoroethyl) methoxycarbonyl-methylphosphonate, **12,** 320
Bis(trifluoromethyl)diazomethane, **1,** 561
Bis(trifluoromethyl)methylenetriphenyl-phosphorane, **9,** 54
Bis(trifluoromethylthio)mercury, **7,** 391
1,3-Bis(triisopropylsilyl)propyne, **11,** 63
(2R,3R)-Bis(trimethylacetyl)tartaric acid, **9,** 204
Bis(trimethylsilyl)acetamide (BSA), **1,** 61; **2,** 30; **3,** 23; **6,** 634; **9,** 234; **12,** 330, 473
Bis(trimethylsilyl)acetylene, **5,** 44; **6,** 153; **7,** 8, 84, 94; **8,** 146; **9,** 142, 311; **10,** 126, 183; **11,** 162, 426; **12,** 162
Bis(trimethylsilyl)amidocopper (*see* Copper(I) hexamethyldisilazide)
1,4-Bis(trimethylsilyl)butadiyne, **5,** 624
1,4-Bis(trimethylsilyl)-2-butene, **7,** 8
1,4-Bis(trimethylsilyl)-2-butyne, **11,** 64
N,O-Bis(trimethylsilyl) carbamate, **7,** 28
Bis(trimethylsilyldichloromethyl)mercury, **2,** 30
Bistrimethylsilyl ether (*see* Hexamethyldisiloxane)
1,1-Bis(trimethylsilyl)ethylene, **5,** 374
Bis(trimethylsilyl)ethyne [*see* Bis-(trimethylsilyl)acetylene]
Bis(trimethylsilyl)formamide, **6,** 58
O,N-Bis(trimethylsilyl)hydroxylamine, **6,** 58
C,O-Bis(trimethylsilyl)ketene methyl ketal, **10,** 261
Bis(trimethylsilyl) lithiomalonate, **9,** 55
Bis(trimethylsilyl) malonate, **7,** 53; **9,** 55
Bis(trimethylsilyl)mercury, **9,** 55
Bis(trimethylsilyl)methane, **5,** 617; **7,** 396
N,N-Bis(trimethylsilyl)methoxymethylamine, **12,** 62
2,3-Bis(trimethylsilylmethyl)-1,3-butadiene, **11,** 64; **12,** 77
Bis(trimethylsilylmethyl)mercury, **6,** 637
Bis(trimethylsilyl) monoperoxysulfate, **9,** 56
1,3-Bis(trimethylsilyloxy)-1,3-butadiene, **7,** 27; **9,** 56; **10,** 44
2,3-Bis(trimethylsilyloxy)-1,3-butadiene, **4,** 436; **8,** 46

1,2-Bis(trimethylsilyloxy)cyclobutene, **10,** 45
1,3-Bis(trimethylsilyloxy)-1,3-cyclohexadiene, **8,** 46, 297
2,3-Bis(trimethylsilyloxy)-1,3-cyclohexadiene, **8,** 297
2,5-Bis(trimethylsilyloxy)furans, **10,** 45
1,3-Bis(trimethylsilyloxy)-1-methoxy-1,3-butadiene [*see* 1-Methoxy-1,3-bis-(trimethylsilyloxy)-1,3-butadiene]
1,1-Bis(trimethylsilyloxy)-1-propene, **11,** 600
Bis(trimethylsilyl) peroxide, **11,** 67; **12,** 63
Bis(trimethylsilyl) peroxide–Vanadyl acetylacetonate, **12,** 63
1,3-Bis(trimethylsilyl)propyne, **10,** 126
N,O-Bis(trimethylsilyl) sulfamate, **7,** 28
Bis(trimethylsilyl) sulfate, **10,** 176; **11,** 67
Bis(trimethylsilyl)thioketene, **7,** 28
Bis(trimethylsilyl) trimethylsilyloxymethylphosphonite, **7,** 412
Bis(trimethylsilyl)urea, **9,** 113; **11,** 68
2,3-Bis(trimethylstannyl)-1,3-butadiene, **12,** 547
Bis(triphenylphosphine)carbonylphenylrhodium(I), **5,** 45
Bis(triphenylphosphine)copper(I) borohydride, **8,** 47; **9,** 57; **10,** 47
Bis(triphenylphosphine)copper(I) cyanoborohydride, **10,** 33
Bis(triphenylphosphine)dicarbonylnickel [*see* Dicarbonylbis(triphenylphosphine)-nickel]
Bis(triphenylphosphine)dicyanonickel(0), **4,** 40
Bis(triphenylphosphine)nickel(0), **4,** 41
Bis(triphenylphosphine)nickel(II) bromide [*see* Dibromobis(triphenylphosphine)-nickel(II)]
Bis(triphenylphosphine)nickel(II) chloride [*see* Dichlorobis(triphenylphosphine)-nickel(II)]
Bis(triphenylphosphine)nickel dicarbonyl [*see* Dicarbonylbis(triphenylphosphine)-nickel]
Bis(triphenylphosphine)palladium(II) halides [*see* Dichloro- or Dihalobis(triphenyl-phosphine)palladium(II)]
Bis(triphenylphosphine)rhodium carbonyl chloride [*see* Carbonylchlorobis-(triphenylphosphine)rhodium(I)]
Blue tetrazolium, **1,** 61
9-Borabicyclo[3.3.1]nonanates, **6,** 62; **7,** 30, 201; **11,** 88, 295; **12,** 446
9-Borabicyclo[3.3.1]nonane (9-BBN), **2,** 31; **3,** 24; **4,** 41; **5,** 46, 201; **6,** 62; **7,** 29; **8,** 47, 406; **9,** 57; **10,** 48; **11,** 68
9-Borabicyclo[3.3.1]nonane–Pyridine, **8,** 49
9-Borabicyclo[3.3.1]nonyl halides [*see* B-Bromo(B-Chloro-, B-Iodo-)-9-borabicyclo[3.3.1]nonane]
9-Borabicyclononyl trifluoromethanesulfonate (*see also* Dialkylboryl trifluoromethanesulfonates), **8,** 504; **9,** 131; **10,** 119; **11,** 159, 232
Borane–Alkylamines (*see also* specific reagents), **1,** 16
Borane–(S)-(–)-2-Amino-3-methyl-1,1-diphenyl-1-butanol, **12,** 31
Borane–Ammonia, **10,** 12
Borane–*t*-Butylamine, **1,** 16; **11,** 87
Borane–Diisopropylamine, **11,** 188
Borane–Dimethylamine, **1,** 16, 273
Borane–Dimethyl sulfide (BMS), **4,** 124, 191; **5,** 47; **6,** 64, 294; **8,** 46, 49; **9,** 398; **10,** 49, 236, 423; **11,** 69, 187, 517; **12,** 64
Borane–Diphenylamine, **11,** 219
Borane–Ethylamine, **1,** 16
Borane–Isopropylamine, **10,** 12
Borane–Methylamine, **1,** 16
Borane–1,4-Oxathiane, **10,** 50
Borane–Pyridine, **1,** 16, 963; **8,** 46, 50; **9,** 59; **11,** 69, 257; **12,** 65
Borane–Tetrahydrofuran (*see also* Diborane), **1,** 57, 199, 483; **2,** 106, 192; **3,** 42, 76; **4,** 124, 175; **5,** 48, 184; **6,** 161, 580; **7,** 89, 321, 333; **8,** 141; **9,** 136, 355; **11,** 69, 156, 291, 487; **12,** 65
Borane–Triethylamine, **3,** 299
Borane–Trimethylamine, **1,** 16, 1229; **5,** 708; **12,** 65
Borane–Triphenylphosphine, **11,** 70
Borane–Triphenyl phosphite, **11,** 70
Boric acid, **1,** 63; **2,** 32; **3,** 29; **4,** 41; **5,** 48; **9,** 23, 59; **11,** 70
(–)-Borneol, **10,** 101
Bornyloxyaluminum dichloride, **11,** 70
Boronic acid resins, **7,** 33, 284; **9,** 59
p-Boronobenzoic acid, **2,** 32
Boron oxide, **2,** 32; **3,** 33
Boron thiophenoxide [*see* Tris(phenylthio)-borane]
Boron triacetate, **9,** 60
Boron tribromide, **1,** 66, 68; **2,** 33; **3,** 30; **4,** 42;

5, 49; **6,** 64; **9,** 61, 255; **10,** 50
Boron tribromide–Sodium iodide–15-Crown-5, **11,** 71
Boron trichloride, **1,** 67, 68; **2,** 34; **3,** 31; **4,** 42; **5,** 50; **6,** 65; **9,** 62, 448; **10,** 50; **12,** 65, 497
Boron trifluoride (*see also* Boron trifluoride etherate), **1,** 31, 68; **3,** 32; **5,** 51; **7,** 31; **10,** 50; **11,** 71; **12,** 28, 494
Boron trifluoride–Acetic acid, **1,** 69; **9,** 492; **10,** 51; **12,** 68
Boron trifluoride dibutyl etherate, **4,** 43
Boron trifluoride–Dimethyl sulfide, **10,** 51
Boron trifluoride–Ethanethiol (*see also* Boron trifluoride etherate–Ethanethiol), **9,** 63; **10,** 51
Boron trifluoride etherate (*see also* Boron trifluoride), **1,** 70; **2,** 35, 160; **3,** 33; **4,** 44; **5,** 52; **6,** 65, 200, 245; **7,** 31; **8,** 51; **9,** 9, 20, 60, 64; **10,** 52, 370; **11,** 72; **12,** 21, 66, 115, 501
Boron trifluoride etherate–Acetic anhydride (*see* Acetic anhydride–Boron trifluoride etherate)
Boron trifluoride etherate–Ethanethiol (*see also* Boron trifluoride–Ethanethiol), **7,** 33
Boron trifluoride–Hydrogen fluoride (*see also* Tetrafluoroboric acid), **2,** 37; **4,** 253
Boron trifluoride–Methanol, **1,** 73; **9,** 64
Boron trifluoride–Trifluoroacetic anhydride, **4,** 45
Boron triiodide, **1,** 68; **9,** 65
Boron tris(trifluoroacetate), **4,** 46; **5,** 55
Boron trisulfide, **5,** 55; **11,** 63
Brederick's reagent [*see t*-Butoxybis-(dimethylamino)methane]
Brij surfactants, **1,** 892; **6,** 70
Bromanil, **1,** 127
Bromine, **1,** 355, 499, 927, 1159; **2,** 142, 221; **3,** 34, 131, 160; **4,** 46, 306, 537; **5,** 55, 97, 100, 188, 452, 556, 568, 712; **6,** 70, 259; **7,** 33, 55; **8,** 52; **9,** 65, 277, 399; **10,** 56; **11,** 75, 187; **12,** 70
Bromine–Acetic acid, **1,** 519, 1025
Bromine–Aluminum halides, **1,** 32; **6,** 17; **9,** 11
Bromine–Antimony(V) chloride–Sulfur dioxide, **5,** 57
Bromine azide, **2,** 37; **3,** 34; **5,** 66
Bromine–Boron trifluoride etherate, **5,** 54
Bromine–*t*-Butylamine, **2,** 429
Bromine chloride, **1,** 73, 74; **2,** 38; **6,** 102
Bromine–Chlorosulfuric acid, **10,** 86
Bromine–Cobalt(II) alkanoates, **12,** 72
Bromine–Dibenzo-18-crown-6, **9,** 66; **11,** 156
Bromine–Dioxane, **1,** 333; **2,** 177; **3,** 130; **5,** 58
Bromine dipyridine nitrate, **2,** 38
Bromine fluoride, **1,** 75; **2,** 39, 365; **4,** 48; **5,** 351; **8,** 53; **11,** 454
Bromine–Hexamethylphosphoric triamide, **7,** 35
Bromine–Mercury(II) oxide (*see* Mercury(II) oxide–Bromine)
Bromine–Nickel(II) alkanoates, **11,** 358; **12,** 72
Bromine on Graphite, **5,** 58
Bromine–Phosphorus(III) bromide(chloride), **1,** 874, 875
Bromine–Pyridine, **1,** 960; **2,** 350; **8,** 104
Bromine–Silver(I) oxide, **1,** 73; **3,** 253
Bromine–Silver salts, **1,** 73, 1015; **2,** 365; **3,** 253; **5,** 59, 60, 61, 351; **7,** 36; **12,** 72
Bromine–Sulfur, **1,** 1120
Bromine–Thallium(III) acetate (*see* Thallium(III) acetate–Bromine)
Bromine–Titanium(IV) chloride, **4,** 508; **5,** 672
Bromoacetaldehyde diethyl acetal, **1,** 883, 1105
N-Bromoacetamide (NBA), **1,** 74; **2,** 39, 330; **4,** 47; **5,** 61
N-Bromoacetamide–Hydrogen chloride, **1,** 74
N-Bromoacetamide–Hydrogen fluoride, **1,** 75; **2,** 39; **4,** 48
N-Bromoacetamide–Pyridine–Sulfur dioxide, **1,** 75
N-Bromoacetamide–Silver acetate–Acetic acid, **7,** 36
Bromoacetone, **3,** 235
α-Bromoacetophenone (*see* Phenacyl bromide)
Bromoacetyl halides, **6,** 103; **12,** 549
Bromoacetylium hexafluoroantimonate, **6,** 103
Bromoacetylmethylenetriphenylphosphorane, **9,** 66
2-Bromoacrolein, **2,** 15; **5,** 62
Bromoallene, **8,** 288
2-Bromoallyl alcohol, **6,** 11
(2-Bromoallyl)trimethylsilane, **11,** 64, 80

Bromo(1-allyl)zinc, **12,** 54
Bromobenzene, **6,** 347; **10,** 298
p-Bromobenzyl chloroformate, **1,** 76
Bromobis(isopropylthio)borane, **12,** 57
B-Bromo-9-borabicyclo[3.3.1]nonane, **12,** 200, 236
4-Bromo-2-butanone ethylene ketal, **1,** 75; **7,** 148
Bromo-*t*-butylmethoxyphenylsilane, **12,** 100
(−)-α-Bromocamphor-10-sulfonic acid, **2,** 40
N-Bromocaprolactam, **1,** 76
Bromocarbamide, **1,** 76
Bromocarbene, **4,** 442
p-Bromocarbobenzoxy chloride, **1,** 76
1-Bromo-3-chloro-2-butene, **10,** 235
Bromochlorocarbene, **1,** 852; **3,** 222; **8,** 129
1-Bromo-2-chlorocyclopropene, **12,** 72
(1-Bromo-1-chloro-2,2,2-trifluoroethyl)-phenylmercury, **3,** 223; **4,** 376
(E)-Bromocrotylbis(cyclopentadienyl)-titanium, **12,** 354
α-Bromocyanoacetamide, **2,** 40
2-Bromo-2-cyano-N,N-dimethylacetamide, **6,** 73
Bromocyanoketene, **8,** 88
5-Bromocyclopentadiene, **5,** 19
1-Bromo-1-cyclopropyllithium, **5,** 78
(Bromodichloromethyl)phenylmercury, **1,** 851; **2,** 326; **4,** 385; **5,** 514; **7,** 282
2-Bromo-1,1-diethoxyethane, **1,** 883, 1105
Bromodiethylborane, **11,** 261
Bromodifluoromethyltriphenylphosphonium bromide, **5,** 62
3-Bromo-1,1-difluoro-1-propene, **10,** 146
3-Bromo-3,3-difluoro-1-propene, **10,** 146
3-Bromo-4,5-dihydro-5-hydroperoxy-4,4-dimethyl-3,5-diphenyl-3*H*-pyrazole, **11,** 76
1-Bromo-1,2-dimethoxyethylene, **7,** 116
Bromodimethylborane, **12,** 199
Bromodimethylsulfonium bromide, **4,** 174; **5,** 114; **9,** 66; **11,** 198
α-Bromodiphenylacetyl bromide, **2,** 444
5-Bromo-1,4-diphenyl-3-methylthio-*s*-triazolium bromide, **8,** 53
α-Bromoethanesulfonyl bromide, **12,** 77
2-Bromoethanol, **5,** 117; **11,** 77; **12,** 68
1-Bromo-1-ethoxycyclopropane, **12,** 73
1-Bromo-2-ethoxycyclopropyllithium, **9,** 67
(Z)-1-Bromo-2-ethoxyethylene, **8,** 221
3-Bromo-2-ethoxy-1-propene, **5,** 292
(α-Bromoethyl)diethylborane, **4,** 10
2-(2-Bromoethyl)-1,3-dioxane, **7,** 37; **11,** 78
2-(2-Bromoethyl)-1,3-dioxolane, **6,** 512; **7,** 37; **8,** 217; **9,** 70
2-(2-Bromoethyl)-2-methyl-1,3-dioxolane, **1,** 75; **7,** 148
(2-Bromoethyl)trimethylsilane, **11,** 574
o-Bromofluorobenzene, **1,** 77, 343; **2,** 45; **3,** 177
Bromofluorocarbene, **3,** 77
6-Bromo-4-fluoro-1,3-dinitrobenzene, **5,** 63
α-Bromo-α-fluorotoluene, **6,** 136
Bromoform, **1,** 425; **3,** 140, 165; **5,** 398; **7,** 66; **9,** 91, 357
Bromoform–Diethylzinc (*see* Diethylzinc–Bromoform)
Bromoform–Potassium *t*-butoxide, **1,** 96, 686, 758, 916; **3,** 174, 234; **4,** 438; **5,** 450; **6,** 468; **8,** 343
Bromoform–Potassium carbonate–18-Crown-6, **11,** 144
Bromoform–Potassium hydroxide–Alumina, **2,** 18
Bromoform–Sodium hydroxide–Benzyl-triethylammonium chloride, **5,** 26; **6,** 42
Bromoform–Sodium hydroxide–(−)-N,N-Ethylmethylephedrinium bromide, **6,** 220
Bromoform–Sodium hydroxide–Tetraalkylammonium halide, **12,** 517
3-Bromofuran, **10,** 188
o-Bromoiodobenzene, **1,** 77
2-Bromoisobutyryl bromide, **1,** 290, 863, 1287
1-Bromo-1-lithiocyclopropanes, **5,** 78
2-Bromo-6-lithiopyridine [*see* 2-(6-Bromopyridyl)lithium]
2-(2-Bromomagnesioethyl)-2-methyl-1,3-dioxolane, **7,** 148
Bromomagnesium alkyl(methyl)cuprates, **8,** 237
Bromomagnesium diethylamide, **5,** 209
Bromomagnesium diisopropylamide, **9,** 173; **12,** 74
Bromomagnesium hexamethyldisilazide, **12,** 75
Bromomagnesium isopropylcyclohexylamide, **10,** 115; **12,** 75
Bromomagnesium N-methylanilide, **4,** 11; **5,** 440
Bromomalononitrile, **1,** 1202; **2,** 428; **3,** 237
2-Bromomesitylene, **1,** 419; **8,** 317; **10,** 234,

299; **12,** 12
Bromomethane, **1,** 673
Bromomethanesulfonyl bromide, **12,** 75
2-Bromo-4′-methoxyacetophenone, **5,** 279
Bromomethoxycarbonylcarbene, **3,** 222
1-Bromo-2-methoxycyclopropane, **6,** 366
4-Bromo-2-methoxyfuran, **11,** 325
3-Bromo-2-methoxy-1-propene (*see* 2-Methoxyallyl bromide)
2-Bromomethyl-1,3-butadiene, **2,** 186
1-Bromo-3-methyl-1,2-butadiene, **5,** 460
1-Bromo-3-methyl-2-butene (Prenyl bromide), **3,** 355; **4,** 355, 487; **5,** 64, 84; **6,** 151
(Bromomethyl)chlorodimethylsilane, **12,** 521
Bromomethylenetriphenylphosphorane, **10,** 56
Bromomethyllithium, **3,** 201; **6,** 74; **12,** 77
Bromomethyl methyl ether, **10,** 56
3-Bromo-4-methyl-3-penten-2-one, **4,** 10
Bromomethyl phenyl sulfoxide, **9,** 68
3-Bromo-2-methylpropene (*see* Methallyl bromide)
2-Bromo-2-methylpropionyl bromide, **1,** 290, 863, 1287
2-Bromo-1-methylpyridinium iodide (*see also* 2-Halopyridinium salts), **6,** 375
5-Bromo-3-methylthio-1,4-diphenyl-5-triazolium bromide, **8,** 53
Bromomethyl *p*-tolyl sulfone, **6,** 41
2-Bromomethyl-3-(trimethylsilylmethyl)-1,3-butadiene, **12,** 77
3-Bromo-3-methyl-2-trimethylsilyloxy-1-butene, **9,** 68; **10,** 286
5-Bromo-7-nitroindoline, **11,** 79
Bromonium dicollidine perchlorate, **12,** 78
5-Bromo-2-pentanone ethylene ketal, **6,** 74
p-Bromophenacyl bromide, **1,** 77; **3,** 34; **5,** 154
Bromophenylcarbene, **5,** 155
(Bromo)phenylmercury, **4,** 353
3-Bromo-1-(phenylsulfonyl)-1-propene, **9,** 230
Bromo(phenyl)zinc, **5,** 753
N-Bromopolymaleimide, **4,** 49
3-Bromopropanal ethylene acetal [*see* 2-(2-Bromoethyl)-1,3-dioxolane]
2-Bromopropane, **9,** 69
3-Bromopropane-1,2-diol, **4,** 154
1-Bromo-1-propene, **4,** 236
3-Bromo-1-propene (*see* Allyl bromide)
2-Bromo-2-propen-1-ol, **6,** 11
2-Bromo-2-propenyltrimethylsilane, **11,** 64, 80
β-Bromopropionaldehyde ethylene acetal [*see* 2-(2-Bromoethyl)-1,3-dioxolane]
3-Bromopropionaldehyde trimethylene acetal, **7,** 37; **11,** 78
3-Bromopropionic acid, **8,** 303
3-Bromopropionitrile, **1,** 77
3-Bromopropionyl isocyanate, **1,** 78
3-Bromopropyl 1-ethoxyethyl ether, **4,** 226
2-(3-Bromopropyl)-2-methyl-1,3-dioxolane, **6,** 74
3-Bromopropyltriphenylphosphonium bromide, **6,** 479
3-Bromopropyne (*see* Propargyl bromide)
2-(6-Bromopyridyl)lithium, **4,** 490; **5,** 63
N-Bromosuccinimide (NBS), **1,** 78, 197; **2,** 40; **3,** 34, 220; **4,** 49; **5,** 65, 127; **6,** 71, 74, 117, 178, 474; **7,** 33, 37, 102; **8,** 54, 235, 256; **9,** 70, 387, 443; **10,** 57, 94; **11,** 79; **12,** 71, 79, 253
N-Bromosuccinimide–Dimethylformamide, **9,** 72
N-Bromosuccinimide–Dimethyl sulfide, **4,** 90
N-Bromosuccinimide–Dimethyl sulfoxide (*see* Dimethyl sulfoxide–N-Bromosuccinimide)
N-Bromosuccinimide–Silica, **6,** 510
N-Bromosuccinimide–Silver salts, **4,** 216; **7,** 182
N-Bromosuccinimide–Sodium azide, **5,** 66
(1-Bromo-1,2,2,2-tetrafluoroethyl)-phenylmercury, **5,** 515
1-Bromo-3-tetrahydropyranyloxy-1-propyne, **6,** 576
1-Bromo-3,3,6,6-tetramethoxy-1,4-cyclohexadiene, **7,** 191
Bromotitanium triisopropoxide, **11,** 375
Bromo-*o*-tolylbis(triethylphosphine)-nickel(II), **10,** 58
Bromotrichloromethane, **1,** 80, 90, 224; **2,** 42; **7,** 40; **11,** 131; **12,** 417
Bromotrifluoromethane, **1,** 81; **2,** 42; **5,** 148
Bromotriisopropoxytitanium, **11,** 375
2-Bromo-1,3,5-trimethylbenzene (*see* 2-Bromomesitylene)
5-Bromo-2,2,5-trimethyl-1,3-dioxane-4,6-dione, **5,** 66; **7,** 41
Bromotrimethylsilane, **8,** 262; **9,** 73, 255; **10,** 59, 96, 216; **12,** 79, 463
Bromotrimethylsilane–Cobalt(II) bromide,

10, 59
(Z)-2-Bromo-1-(trimethylsilyloxy)ethylene, **11,** 582
1-Bromo-3-trimethylsilyloxy-1-propyne, **6,** 576
1-Bromo-1-trimethylsilyl-1(Z),4-pentadiene, **11,** 80
3-Bromo-1-trimethylsilyl-1-propene, **12,** 487
2-Bromo-3-trimethylsilyl-1-propene, **11,** 64, 80
4-Bromo-2,5,7-trinitrofluorenone, **1,** 81
Bromotriphenylmethane, **1,** 1254
Bromotriphenylphosphonium bromide (*see* Triphenylphosphine dibromide)
Bromourea, **1,** 76
1-Bromo-2-vinylcyclopropane, **7,** 192, 199, 212
(1-Bromovinyl)trimethylsilane, **5,** 374; **7,** 193; **8,** 56, 515; **10,** 442; **11,** 286; **12,** 72, 198
(2-Bromovinyl)trimethylsilane, **11,** 82
(1-Bromovinyl)triphenylsilane, **5,** 68, 375
Brown and Brown catalyst [*see* Hydrogen hexachloroplatinate(IV)–Sodium borohydride]
Brucine, **11,** 83
Burgess reagent [*see* Ethyl(carboxysulfamoyl)triethylammonium hydroxide inner salt]
1,3-Butadiene, **1,** 31, 81, 426, 461, 539, 624; **2,** 22, 217; **4,** 351; **5,** 159, 471; **6,** 173; **9,** 13
1,3-Butadiene monoxide, **4,** 53; **5,** 483
Butadiene sulfone (*see* 3-Sulfolene)
(Butadiene)tricarbonyliron, **10,** 222
1-Butadienyldimethylsulfonium tetrafluoroborate, **9,** 74
1,3-Butadienyl-2-magnesium chloride, **12,** 195
trans-1-Butadienyltriphenylphosphonium bromide, **6,** 76
Butadiyne, **1,** 185
Butadiynyl(trimethyl)silane, **5,** 68; **6,** 143
Butane-1,4-bis(triphenylphosphonium) dibromide, **10,** 60
Butane 1,4-(dimagnesium bromide), **1,** 415
D-(–)- and L-(+)-2,3-Butanediol, **1,** 81; **7,** 242; **11,** 84; **12,** 80, 480
2,3-Butanedione, **7,** 21; **11,** 193
L-(+)-Butane-2,3-dithiol, **1,** 82
1-Butanethiol, **2,** 53; **4,** 64, 196, 458; **7,** 411
2-Butanone, **1,** 678
Butanone ethylene ketal, **1,** 82
2-Butene-1,4-diol, **6,** 444
1-Buten-3-one (*see* Methyl vinyl ketone)
2-Butenyl-, *see also* Crotyl-
2-(2-Butenyl)-1,3-dithiane, **11,** 447
2-Butenylidenetriphenylphosphorane, **5,** 69
2-Butenyltributyltin (*see* Tributylcrotyltin)
3-Butenyltributyltin, **4,** 53
3-(1-Butenyl)trimethylsilane, **9,** 446
t-Butoxyallene, **8,** 322
t-Butoxybis(dimethylamino)methane [Brederick's reagent, *see also* Bis(dimethylamino)methoxymethane], **5,** 71; **6,** 242; **7,** 41; **9,** 76, 246
t-Butoxycarbonyl azide (*see* *t*-Butyl azidoformate)
2-*t*-Butoxycarbonyl-1,3-butadiene, **12,** 155
t-Butoxycarbonyl chloride (*see* *t*-Butyl chloroformate)
1-*t*-Butoxycarbonyl-4-dimethylaminopyridinium chloride, **4,** 173
1-*t*-Butoxycarbonyl-4-dimethylaminopyridinium tetrafluoroborate, **8,** 57
(2S,4S)-N-(*t*-Butoxycarbonyl)-4-(diphenylphosphine)-2-[(diphenylphosphine)methyl]pyrrolidine, **8,** 57; **11,** 87
[N-(*t*-Butoxycarbonyl)-4-diphenylphosphine]-2-[(diphenylphosphine)methyl]pyrrolidine(cyclooctadiene)rhodium chloride or perchlorate, **11,** 87
t-Butoxycarbonyl fluoride (*see* *t*-Butyl fluoroformate)
t-Butoxycarbonylhydrazine (*see* *t*-Butyl carbazate)
t-Butoxycarbonyl-N-hydroxysuccinimide ester, **2,** 42
t-Butoxycarbonylimidazole, **1,** 98
2-*t*-Butoxycarbonyl-1-methoxycarbonylethylidenetriphenylphosphorane, **8,** 77
(*t*-Butoxycarbonylmethyl)diethylaluminum, **7,** 103
N-Butoxycarbonylmethylene-*p*-toluenesulfonamide, **1,** 98
1-(*t*-Butoxycarbonyloxy)benzotriazole, **12,** 83
2-*t*-Butoxycarbonyloxyimino-2-phenylacetonitrile, **6,** 91; **10,** 61
N-(*t*-Butoxycarbonyloxy)phthalimide, **2,** 43
N-(*t*-Butoxycarbonyloxy)succinimide, **2,** 42
N-*t*-Butoxycarbonylpyrrole, **2,** 45
1-*t*-Butoxycarbonyl-1,2,4-triazole, **4,** 65
2-*t*-Butoxyfuran, **8,** 59
t-Butoxyketene, **8,** 144
t-Butoxymagnesium bromide, **8,** 20

t-Butoxymethyllithium, **12,** 350
t-Butoxytrimethylsilane, **10,** 444
t-Butyl acetate, **1,** 82; **5,** 371; **6,** 84, 336, 636; **9,** 82; **10,** 67
t-Butyl acetoacetate, **1,** 83; **3,** 47; **8,** 60
t-Butylacetylene, **5,** 547, 705; **6,** 478, 486
t-Butyl acrylate, **12,** 155
Butylamine, **2,** 286; **4,** 53
t-Butylamine, **1,** 84; **2,** 43; **5,** 117; **10,** 62, 248; **12,** 383
t-Butylaminoborane (*see* Borane–*t*-Butylamine)
Butyl azide, **1,** 84; **5,** 278
t-Butyl azidoformate, **1,** 84; **2,** 44; **3,** 36; **4,** 54; **6,** 77; **7,** 41
t-Butyl benzotriazol-1-yl carbonate, **12,** 83
t-Butyl bis(trimethylsilyl)acetate, **7,** 42
t-Butyl bis(trimethylsilyl)lithioacetate [*see* *t*-Butyl lithiobis(trimethylsilyl)acetate]
B-Butyl-9-borabicyclo[3.3.1]nonane, **9,** 76; **11,** 88
t-Butyl carbazate, **1,** 85; **2,** 46; **3,** 36; **11,** 88
t-Butylcarbonic diethylphosphoric anhydride, **2,** 46; **3,** 36
t-Butyl chloride, **1,** 86; **2,** 46
t-Butyl chloroacetate, **1,** 86; **3,** 36
t-Butylchlorodimethylsilane (*see* *t*-Butyldimethylchlorosilane)
t-Butylchlorodiphenylsilane (*see* *t*-Butyldiphenylchlorosilane)
sec-Butyl chloroformate, **1,** 86; **2,** 47
t-Butyl chloroformate, **1,** 86; **2,** 48
t-Butyl 2-chloro-2-lithiotrimethylsilylacetate, **8,** 60
t-Butyl chloromethyl ether, **9,** 77
t-Butyl 2-chloro-2-trimethylsilylacetate, **8,** 60
t-Butyl chromate, **1,** 86; **2,** 48; **3,** 37; **4,** 54; **5,** 73, 696; **7,** 42; **9,** 397
Butylcopper, **5,** 748; **6,** 163; **9,** 333; **11,** 369
Butylcopper–Boron trifluoride (*see also* Alkylcopper reagents–Boron trifluoride complex), **8,** 335; **9,** 333; **10,** 54; **11,** 368, 413
Butylcopper–Trialkylboranes, **9,** 333
t-Butyl cyanoacetate, **1,** 87
t-Butylcyanodimethylsilane, **6,** 80
t-Butyl cyanoformate, **1,** 87
t-Butylcyanoketene, **4,** 55, 148; **6,** 78; **7,** 43
t-Butyl diazoacetate, **2,** 416; **7,** 313; **9,** 20, 483; **11,** 458
t-Butyl diazomethyl ketone, **5,** 162
t-Butyl dilithioacetoacetate, **8,** 60
Butyl (dimethylamino)methyl ether, **11,** 274
t-Butyldimethylchlorosilane, **4,** 57, 176; **5,** 74; **6,** 78, 277; **7,** 210; **8,** 59, 243; **9,** 77, 444; **10,** 62; **11,** 88; **12,** 83, 281
t-Butyldimethylcyanosilane, **6,** 80
t-Butyldimethyliodosilane, **10,** 63
t-Butyldimethylsilyl chloride (*see* *t*-Butyldimethylchlorosilane)
t-Butyldimethylsilyl cyanide, **6,** 80
t-Butyldimethylsilyl ethylnitronate, **12,** 84
t-Butyldimethylsilyl hydroperoxide–Mercury(II) trifluoroacetate, **12,** 85
O-*t*-Butyldimethylsilylhydroxylamine, **12,** 85
t-Butyldimethylsilyl iodide, **10,** 63
N-(*t*-Butyldimethylsilyl)-N-methylacetamide, **11,** 336
N-(*t*-Butyldimethylsilyl)-N-methyltrifluoroacetamide, **11,** 336
4-*t*-Butyldimethylsilyloxy-3-penten-2-one, **10,** 62
t-Butyldimethylsilyl perchlorate, **9,** 78
t-Butyldimethylsilyl phenyl selenide, **10,** 63
t-Butyldimethylsilyl trifluoromethanesulfonate, **10,** 63; **11,** 90; **12,** 86
3-(*t*-Butyldimethylsilyl)-1-(trimethylsilyl)-propyne, **11,** 63
t-Butyldiphenylchlorosilane, **6,** 81; **12,** 87
t-Butyldiphenylsilyl chloride (*see* *t*-Butyldiphenylchlorosilane)
t-Butyldiphenylsilyllithium, **11,** 553
t-Butyl ethyl malonate, **1,** 522
t-Butyl fluoroformate (BOC-F), **2,** 48, 54; **4,** 65; **6,** 91
N-*t*-Butylformamide, **1,** 212
t-Butylhydrazine, **12,** 87
Butyl hydroperoxide, **1,** 17
sec-Butyl hydroperoxide, **1,** 17
t-Butyl hydroperoxide (TBHP), **1,** 88; **2,** 49; **3,** 37; **5,** 75; **6,** 81, 179; **7,** 43; **8,** 24, 62; **9,** 78; **10,** 64; **11,** 397; **12,** 88
t-Butyl hydroperoxide–Alumina, **8,** 64
t-Butyl hydroperoxide–(–)-Benzylquininium chloride, **10,** 338
t-Butyl hydroperoxide–Benzyltrimethylammonium hydroxide, **1,** 88; **5,** 338; **9,** 80
t-Butyl hydroperoxide–Benzyltrimethylammonium tetrabromooxomolybdate, **12,** 89
t-Butyl hydroperoxide–Chromium carbonyl,

12, 89
t-Butyl hydroperoxide–Cobalt(II) acetylacetonate, **5,** 77
t-Butyl hydroperoxide–Dialkyl tartrate–Dichlorodiisopropoxytitanium(IV), **12,** 91
t-Butyl hydroperoxide–Dialkyl tartrate–Titanium(IV) *t*-butoxide, **12,** 91
t-Butyl hydroperoxide–Dialkyl tartrate–Titanium(IV) isopropoxide, **10,** 64; **11,** 92; **12,** 90
t-Butyl hydroperoxide–Diaryl diselenides, **9,** 25, 79; **10,** 65
t-Butyl hydroperoxide–Dichlorotris-(triphenylphosphine)ruthenium(II), **12,** 88
t-Butyl hydroperoxide–Dimesityl diselenide, **9,** 79; **10,** 65
t-Butyl hydroperoxide–Diphenyl diselenide, **9,** 25, 79
t-Butyl hydroperoxide–Ferrous sulfate, **3,** 38
t-Butyl hydroperoxide–Manganese(II) acetylacetonate, **3,** 194
t-Butyl hydroperoxide–Manganese(III) acetylacetonate, **5,** 77
t-Butyl hydroperoxide–Molybdenum carbonyl , **2,** 287; **3,** 206; **4,** 346; **5,** 75; **8,** 62; **9,** 81; **11,** 95
t-Butyl hydroperoxide–Molybdenyl acetylacetonate, **11,** 91; **12,** 89
t-Butyl hydroperoxide–Oxomethoxy-molybdenum(V) 5,10,15,20-tetraphenyl-porphyrin, **11,** 383
t-Butyl hydroperoxide–Palladium *t*-butyl peroxide trifluoroacetate, **10,** 299
t-Butyl hydroperoxide–Potassium hydride, **9,** 81
t-Butyl hydroperoxide–Ruthenium(III) acetylacetonate, **12,** 88
t-Butyl hydroperoxide–Ruthenium(III) chloride, **12,** 88
t-Butyl hydroperoxide–Selenium dioxide, **8,** 64; **9,** 79; **10,** 65; **11,** 96
t-Butyl hydroperoxide–Titanium(III) acetylacetonate, **5,** 76
t-Butyl hydroperoxide–Titanium(IV) chloride, **12,** 94
t-Butyl hydroperoxide–Titanium(IV) isopropoxide, **5,** 76
t-Butyl hydroperoxide–Triton B, **1,** 88; **5,** 338; **9,** 80
t-Butyl hydroperoxide–Vanadyl acetylacetonate, **2,** 456; **3,** 331; **5,** 75; **7,** 43, 44, 63; **8,** 62; **9,** 81; **10,** 66; **11,** 95, 97; **12,** 94
t-Butyl hydroperoxide–Vanadyl acetylacetonate–N-Phenylcampholyl-hydroxamic acid, **8,** 393
t-Butyl hydroperoxide–Vanadyl acetylacetonate–Titanium(IV) isopropoxide, **11,** 99
N-*t*-Butylhydroxylamine, **5,** 123; **12,** 95
t-Butyl hypobromite, **1** 90
t-Butyl hypochlorite, **1,** 90; **2,** 50; **3,** 38, 223; **4,** 58; **5,** 77; **6,** 82; **7,** 79; **10,** 66, 209
t-Butyl hypochlorite–Dialkyl sulfides, **6,** 118
t-Butyl hypoiodite, **1,** 94; **2,** 50
(*t*-Butylimido)trioxidoosmium(VIII), **6,** 641; **10,** 445
t-Butyliododimethylsilane, **10,** 63
t-Butyl 4-iodo-2-methyl-2-butenoate, **6,** 82
t-Butyl γ-iodotiglate, **6,** 82
t-Butyl isobutyrate, **6,** 84
Butyl isocyanate, **5,** 532
t-Butyl isocyanide, **2,** 50; **5,** 467, 581; **7,** 20; **9,** 82, 121, 123; **10,** 67; **11,** 99, 420; **12,** 95
O,O-*t*-Butyl O-isopropenyl peroxycarbonate, **11,** 100
t-Butyl lithioacetate, **4,** 306; **5,** 303, 371; **6,** 84, 636; **7,** 103; **8,** 21, 66; **9,** 82; **10,** 67
t-Butyl lithiobis(trimethylsilyl)acetate, **7,** 42
t-Butyl α-lithioisobutyrate, **6,** 84; **8,** 21
t-Butyl lithio(trimethylsilyl)acetate, **5,** 371; **7,** 51; **12,** 228
Butyllithium, **1,** 95, 572, 687; **2,** 51; **4,** 60; **5,** 78, 80, 399; **6,** 85, 189, 643, 666; **7,** 45, 132, 206; **8,** 65, 71, 201, 347, 524; **9,** 83, 154, 183, 250; **10,** 68, 72, 462; **11,** 14, 101; **12,** 96, 142, 324, 351, 410
sec-Butyllithium, **3,** 280; **5,** 78; **6,** 570; **9,** 6, 87, 487; **10,** 75; **11,** 102; **12,** 97
t-Butyllithium, **1,** 96; **5,** 79, 676; **6,** 109, 642; **7,** 47; **8,** 70; **9,** 85, 89; **10,** 76; **11,** 14, 103, 187; **12,** 100, 533
Butyllithium–(S,S)-(+)-1,4-Bis(dimethyl-amino)-2,3-dimethoxybutane, **9,** 47
Butyllithium–Boron trifluoride etherate, **12,** 68
Butyllithium–Cryptates, **5,** 85
Butyllithium–1,4-Diazabicyclo[2.2.2]octane, **6,** 87; **7,** 46; **9,** 473
Butyllithium–Diisobutylaluminum hydride,

12, 96
Butyllithium–Hexamethylphosphoric triamide, **9,** 87; **10,** 197; **11,** 14
t-Butyllithium–Hexamethylphosphoric triamide, **8,** 72; **9,** 139
Butyllithium–Magnesium bromide, **10,** 7
Butyllithium–Nickel(II) bromide, **8,** 66
Butyllithium–Potassium *t*-butoxide, **2,** 343; **5,** 552; **7,** 400; **8,** 67; **9,** 87; **10,** 72; **11,** 103; **12,** 97
sec-Butyllithium–Potassium *t*-butoxide, **6,** 262; **12,** 99
Butyllithium–Pyridine, **2,** 351
Butyllithium–Tetramethylethylenediamine, **1,** 1203; **2,** 403; **3,** 284; **4,** 485; **5,** 80; **7,** 47; **8,** 67; **9,** 5, 86, 486; **10,** 73
sec-Butyllithium–Tetramethylethylenediamine, **5,** 79; **8,** 69; **9,** 16, 88; **10,** 73, 75; **12,** 97, 533
t-Butyllithium–Tetramethylethylenediamine, **3,** 284
Butylmagnesium bromide, **1,** 377, 415; **3,** 150; **7,** 283; **10,** 193
t-Butylmagnesium bromide, **2,** 205; **10,** 406; **12,** 298, 509
Butylmagnesium bromide–Copper(I) halides, **1,** 168; **6,** 270; **7,** 149; **10,** 190
Butylmagnesium bromide–Iron(III) reagents, **7,** 155
Butylmagnesium bromide–Methylcopper, **8,** 237
Butylmagnesium bromide–Nickel(II) reagents, **6,** 50
Butylmagnesium chloride, **9,** 124
t-Butylmagnesium chloride, **1,** 415, 421; **4,** 63
Butylmagnesium chloride–Copper(I) iodide, **9,** 124
Butyl mercaptan (*see* 1-Butanethiol)
N-*t*-Butylmethacrylamide, **10,** 74
t-Butylmethoxyphenylsilyl bromide, **12,** 100
t-Butyl methyl ether, **4,** 333; **12,** 350
t-Butyl methyl iminodicarboxylate, potassium salt, **8,** 72
Butyl nitrile, **1,** 17
Butyl nitrite, **1,** 97; **6,** 127
t-Butyl nitrite, **1,** 97; **7,** 48
t-Butyl *p*-nitrophenyl carbonate, **1,** 97
N-*t*-Butyl osmiumate, **6,** 641; **10,** 445
t-Butyl pentachlorophenyl carbonate, **2,** 54
t-Butyl perbenzoate, **1,** 98; **2,** 54; **4,** 66; **7,** 49; **9,** 90
2-*t*-Butylperoxy-1,3,2-dioxaborolane, **12,** 101
t-Butylperoxy isopropyl carbonate, **2,** 55
t-Butyl phenyl carbonate, **1,** 85; **5,** 86
O-*t*-Butyl S-phenyl thiocarbonate, **7,** 41
Butylphenyltin dihydride, polymeric, **6,** 92
N-Butylpiperidine, **6,** 601
t-Butyl 2-pyridyl carbonate, **12,** 102
t-Butyl 8-quinolyl carbonate, **4,** 66; **5,** 87
N-*t*-Butylsulfamoyl chloride, **5,** 699
N^1-*t*-Butyl-N^2,N^2,N^3,N^3-tetramethylguanidine (*see also* Guanidines), **11,** 105, 249; **12,** 102
N″-(*t*-Butyl)-N,N,N′,N′-tetramethylguanidinium *m*-iodylbenzoate, **12,** 102
trans-4-(*t*-Butylthio)-3-buten-2-one, **9,** 90
S-*t*-Butyl thiolacetate, **9,** 92
S-*t*-Butylthio lithioacetate, **9,** 92
(4S)-*t*-Butylthio-(2S)-methoxymethyl-N-pivaloylpyrrolidine, **12,** 315
t-Butyl thionitrate, **9,** 91
1-*t*-Butylthio-1-trimethylsilyloxyethylene, **9,** 92
Butyl N-(*p*-toluenesulfonyl)iminoacetate, **7,** 50
(R)-(+)-*t*-Butyl (*p*-tolylsulfinyl)acetate, **10,** 407; **11,** 106, 538
(R)-(+)-*t*-Butyl 2-(*p*-tolylsulfinyl)propionate, **11,** 538
t-Butyl 2,4,5-trichlorobenzoate, **2,** 55
t-Butyl trifluoroacetate, **5,** 87
t-Butyl trimethylsilylacetate, **5,** 371; **7,** 51
Butyl vinyl ether, **5,** 28; **6,** 609
2-Butyne, **1,** 520; **2,** 23
3-Butyn-1-ol, **3,** 28
3-Butyn-2-ol, **4,** 423
3-Butyn-2-one, **3,** 38
2-Butynoic acid, **6,** 345; **8,** 308
Butyric anhydride, **5,** 88
γ-Butyrolactone, **1,** 27, 101; **5,** 234, 403, 456, 667; **7,** 199; **12,** 505

Cadmium, **7,** 52
Cadmium acetate–Zinc acetate, **1,** 103
Cadmium borohydride–Dimethylformamide, **12,** 574
Cadmium carbonate, **4,** 67
Cadmium chloride, **1,** 422; **4,** 485; **8,** 69
Calcium, **5,** 89
Calcium amalgam, **8,** 74
Calcium–Amines, **9,** 94; **12,** 103
Calcium–Ammonia, **1,** 104, 106; **2,** 58; **4,**

160; **9,** 94, 273
Calcium borohydride, **2,** 57; **5,** 89; **11,** 293
Calcium carbide, **2,** 58; **3,** 40
Calcium carbonate, **1,** 103; **2,** 57, 246; **4,** 67; **5,** 89
Calcium chloride, **1,** 104, 354; **4,** 368; **10,** 251
Calcium ethoxide, **7,** 52
Calcium hexamine, **1,** 104
Calcium hydride, **1,** 105; **2,** 58
Calcium hypochlorite, **1,** 488; **11,** 107
Calcium oxide, **9,** 23
Calcium sulfate, **1,** 107, 354; **4,** 266; **5,** 356
CAMP [*see* (R)-(+)-*o*-Anisyl(cyclohexyl)-methylphosphine]
(+)-Camphanic acid, **8,** 74
Camphor-10-sulfonic acid (CSA), **1,** 108; **2,** 58; **4,** 68; **6,** 611; **11,** 107; **12,** 103, 178, 198, 253
d- and *l*-10-Camphorsulfonyl chloride, **1,** 109; **11,** 109; **12,** 68
[(−)-Camphor-10-ylsulfonyl]-3-aryloxaziridines, **11,** 108
CAN [*see* Cerium(IV) ammonium nitrate]
ε-Caprolactam, **9,** 316
Capryl alcohol, **1,** 109
Carbobenzoxy chloride (*see* Benzyl chloroformate)
Carbobenzoxyhydrazide, **1,** 110
(Carbo-*t*-butoxymethyl)diethylaluminum, **7,** 103
Carboethoxy-, *see also* Ethoxycarbonyl-
Carboethoxyformimidate, **11,** 109
Carbohydrazide, **3,** 92
Carbon, activated, **7,** 52
Carbon dioxide, **1,** 421, 1037, 1102, 1256; **2,** 141, 184, 273; **3,** 40, 180, 280; **4,** 41, 480, 551; **5,** 93, 295; **6,** 94, 130; **7,** 52; **8,** 150, 163; **9,** 52, 94, 122, 146, 344, 487; **10,** 274; **11,** 589
Carbon disulfide, **1,** 114; **2,** 63; **3,** 92; **4,** 195, 404; **5,** 94, 173; **6,** 95, 225; **8,** 494; **9,** 333; **11,** 245
Carbon disulfide–Methyl iodide, **5,** 242; **9,** 94; **12,** 316
Carbon disulfide–Methyl iodide–Sodium hydride, **5,** 95
Carbonic difluoride, **1,** 116; **12,** 106
Carbon monoxide, **1,** 167, 226, 268, 782, 1058; **2,** 60, 204, 253, 304; **3,** 25, 41, 82; **4,** 69, 87, 107, 175, 420, 422, 463; **5,** 58, 96, 326, 472, 502, 503, 654; **6,** 59, 60, 177, 306, 451, 508; **7,** 53, 60, 101, 355; **8,** 76, 149, 385, 437; **9,** 95, 144, 348, 393; **10,** 38, 126, 133, 298, 302; **11,** 111, 163, 252, 269, 392, 465, 498; **12,** 45, 54, 105, 112, 165, 172, 425, 426, 470, 474
Carbon oxysulfide, **11,** 112
Carbon suboxide, **6,** 96; **7,** 53
Carbon tetrabromide, **2,** 119, 275
Carbon tetrabromide–Methyllithium, **6,** 523
Carbon tetrabromide–Tin(II) fluoride, **11,** 111
Carbon tetrachloride, **3,** 43; **4,** 69; **5,** 355; **6,** 596; **11,** 131; **12,** 151, 417
Carbon tetrachloride–Potassium hydroxide (*see* Potassium hydroxide–Carbon tetrachloride)
1,1′-Carbonylbis(3-imidazolium) bismethanesulfonate, **11,** 112
Carbonyl bromide fluoride, **5,** 640
Carbonyl chloride (*see* Phosgene)
Carbonyl chloride fluoride, **5,** 640; **6,** 91
Carbonylchlorobis(triphenylphosphine)-iridium(I), **6,** 652; **7,** 68; **10,** 79
Carbonylchlorobis(triphenylphosphine)-rhodium(I), **2,** 67; **4,** 562; **5,** 46; **6,** 105; **7,** 100; **10,** 79
Carbonyl cyanide, **2,** 60
Carbonylcyclopropane, **6,** 96
Carbonyl difluoride, **1,** 116; **12,** 106
Carbonyldihydridotris(triphenylphosphine)-ruthenium, **7,** 109
N,N′-Carbonyldiimidazole, **1,** 114, 181, 591; **2,** 61; **5,** 22, 97, 421; **6,** 97; **8,** 77; **9,** 85, 96; **10,** 252, 318; **11,** 125, 155; **12,** 106
2-Carbonyl-1,3-dithiane, **5,** 99
1,1-Carbonyldi-1,2,4-triazole, **1,** 116; **2,** 61; **8,** 77
1,1-Carbonyldi-1,2,4-triazole–Hydrogen peroxide, **6,** 35
Carbonyl fluoride, **1,** 116; **12,** 106
Carbonylhydridotris(triphenylphosphine)-rhodium(I), **4,** 249; **5,** 331, 332; **7,** 53; **8,** 437
Carbonylhydridotris(triphenylphosphine-rhodium)–2,3-O-Isopropylidene-2,3-dihydroxy-1,4-bis(diphenylphosphine)-butane, **9,** 259
Carbonylisocyanatidic chloride, **5,** 109
Carbonylphenylbis(triphenylphosphine)-rhodium(I), **5,** 45
N-Carbonylsulfamic acid chloride (*see*

Chlorosulfonyl isocyanate)
N-Carbonylsulfamoyl chloride (*see* Chlorosulfonyl isocyanate)
Carbonyl sulfide, **11,** 112
p-Carboxybenzenediazonium chloride, **2,** 62
p-Carboxybenzenesulfonyl azide, **2,** 62
(4-Carboxybutyl)triphenylphosphonium bromide, **11,** 300
2-Carboxy-1,3-dithiane, **5,** 99
2-(2-Carboxyethyl)-2-methyl-1,3-dithiolane, **10,** 230
(2-Carboxyethyl)triphenylphosphonium perbromide, **6,** 97
N-Carboxyglycine anhydride (*see* 2,5-Oxazolidinedione)
2-Carboxy-1-methoxycarbonylethylidene-triphenylphosphorane, **8,** 77
6-Carboxy-5-methoxycarbonyl-7-oxabicyclo-[2.2.1]hept-2-ene, **9,** 97
1-(Carboxymethyl)pyridinium chloride hydrazide (Girard reagent P), **1,** 410
(Carboxymethyl)trimethylammonium chloride hydrazide (Girard reagent T), **1,** 410; **10,** 346
o-Carboxyphenyl *o*-carboxybenzenethio-sulfonate, **5,** 100
(*p*-Carboxyphenyl)dihydroxyborane, **2,** 32
(*o*-Carboxyphenyl)phenyliodonium hydroxide inner salt, **1,** 341, 953; **4,** 207
4-(4-Carboxyphenyl)semicarbazide, **1,** 118
(2-Carboxy-2-propenyl)triphenylphos-phonium bromide, **9,** 97
(E)-(Carboxyvinyl)trimethylammonium betaine, **12,** 106
Caro's acid, **1,** 118; **3,** 43; **6,** 97; **8,** 79
CAS (*see* Cerium(IV) ammonium sulfate)
Catecholborane, **4,** 25, 69; **5,** 100; **6,** 33, 98; **7,** 54; **8,** 46, 79; **9,** 97, 443, 455; **10,** 79
Catechyl phosphorus trihalide (*see* 2,2,2-Tribromo- or 2,2,2-Trichlorobenzo-1,3,2-dioxaphosphole)
Cathyl chloride (*see* Ethyl chloroformate)
Celite, **1,** 120; **9,** 115; **11,** 22, 237
Cellulose, **11,** 22
Ceric, see Cerium(IV)
Cerium(IV) acetate, **4,** 71
Cerium amalgam, **11,** 114
Cerium(IV) ammonium nitrate (CAN), **1,** 120; **2,** 63; **3,** 44; **4,** 71; **5,** 101; **6,** 99, 197; **7,** 55, 56; **9,** 99; **10,** 79, 143; **12,** 71
Cerium(IV) ammonium nitrate–Sodium bromate, **10,** 81; **11,** 114
Cerium(IV) ammonium sulfate (CAS), **7,** 56; **8,** 80; **12,** 107
Cerium(III) chloride, **9,** 99, 211; **12,** 107, 345
Cerium(III) hydroxide, **1,** 121
Cerium(III) iodide, **9,** 99; **11,** 114; **12,** 107
Cerium(IV) oxide–Hydrogen peroxide, **6,** 99
Cerium(IV) potassium nitrate, **4,** 72
Cerium(IV) pyridinium chloride, **11,** 184
Cerium(III) sulfate, **9,** 99
Cerium(IV) sulfate, **2,** 65
Cerium(IV) sulfate–Sodium bromate, **11,** 114
Cerium(IV) trifluoroacetate, **5,** 102
Cerous, *see* Cerium(III)
Cesium bicarbonate, **7,** 57; **8,** 81
Cesium carbonate, **7,** 57; **8,** 81; **9,** 99; **11,** 114; **12,** 108
Cesium fluoride, **1,** 121, 933; **2,** 346; **6,** 100; **7,** 57; **8,** 81; **9,** 100; **10,** 81, 434; **11,** 115, 434, 554; **12,** 86, 108, 407
Cesium fluoride–Tetraalkoxysilanes, **10,** 82; **11,** 117; **12,** 108
Cesium fluoroxysulfate, **10,** 84; **11,** 118
Cesium 1,1,1,2,2,2-hexachloro-μ-(hydrido)-di-μ-(chloro)dimolybdenum(III), **8,** 522
Cesium hydrogen carbonate, **7,** 57; **8,** 81
Cesium propionate, **11,** 118; **12,** 109
Cetyltrimethylammonium borohydride, **1,** 1143
Cetyltrimethylammonium bromide, **1,** 121; **6,** 70; **8,** 389
Chloral, **1,** 122, 480; **3,** 45; **5,** 429; **6,** 100; **7,** 6; **8,** 12, 82; **9,** 121
Chloramine, **1,** 122; **2,** 65; **3,** 45; **4,** 74; **5,** 103
Chloramine-T, **4,** 74, 75, 445; **5,** 104; **7,** 58; **8,** 83; **9,** 101, 367; **10,** 85; **11,** 118; **12,** 121
Chloramine-T–Diphenyl diselenide or disulfide, **9,** 101
Chloranil, **1,** 125; **2,** 66; **3,** 46; **4,** 75, 132; **10,** 85, 330
o-Chloranil, **1,** 127, 128; **2,** 345; **4,** 76; **5,** 104; **7,** 355; **9,** 102
Chlorauric acid [*see* Hydrogen tetra-chloroaurate(III)]
Chloreal (*see* Trichloroisocyanuric acid)
Chlorine, **5,** 105, 556; **6,** 101; **7,** 58; **9,** 182; **10,** 94, 209
Chlorine bromide (*see* Bromine chloride)
Chlorine–Chlorosulfuric acid, **8,** 83; **10,** 86
Chlorine dipyridine nitrate, **2,** 38
Chlorine–Hexamethylphosphoric triamide, **7,**

35
Chlorine oxide, **11,** 119; **12,** 110
Chlorine–Pyridine, **5,** 106
Chloroacetaldehyde, **5,** 106; **6,** 107
Chloroacetaldehyde diethyl acetal, **5,** 107, 592
Chloroacetic anhydride, **1,** 129
γ-Chloroacetoacetyl chloride, **4,** 459
Chloroacetone, **1,** 870, 1166; **3,** 147
Chloroacetonitrile, **1,** 129; **3,** 235; **4,** 30; **6,** 102, 159; **10,** 48
Chloroacetyl chloride, **1,** 130; **3,** 46
Chloroacetyl fluoride, **6,** 103
Chloroacetylhydrazine, **1,** 130, 258
Chloroacetyl isocyanate, **6,** 634
Chloroacetylium hexafluoroantimonate, **6,** 103
2-Chloroacrylonitrile, **3,** 66; **4,** 76; **5,** 107; **6,** 142; **8,** 444; **9,** 75; **11,** 139
2-Chloroacrylyl chloride, **4,** 77
Chloroalane (*see* Monochloroalane)
α-Chloroallyllithium, **10,** 87
Chlorobenzene, **4,** 311; **6,** 346
π-(Chlorobenzene)chromium tricarbonyl [*see also* Arene(tricarbonyl)chromium complexes], **6,** 103
4-Chlorobenzenediazonium tetrafluoroborate (*see also* Arenediazonium tetrahaloborates), **9,** 102
p-Chlorobenzeneselenenyl bromide, **9,** 26
3-Chloro-1,2-benzisothiazole 1,1-oxide (γ-Saccharin chloride), **1,** 990
2-Chloro-1,3,2-benzodioxaphosphole, **2,** 321; **5,** 516; **7,** 58
1-Chlorobenzotriazole, **2,** 67; **3,** 46; **4,** 78; **5,** 109
p-Chlorobenzoyl nitrite, **4,** 79
p-Chlorobenzoyl alcohol, **7,** 59
Chlorobis[1,3-bis(diphenylphosphine)-propane]rhodium, **12,** 111
μ-Chlorobis(cyclopentadienyl)(dimethyl-aluminum)-μ-methylenetitanium (Tebbe reagent), **8,** 83; **10,** 87; **11,** 52; **12,** 54, 110
Chlorobis(cyclopentadienyl)hydrido-zirconium(IV) (Schwartz's reagent), **6,** 175; **7,** 101; **8,** 84; **9,** 104; **11,** 119
Chlorobis(cyclopentadienyl)tetrahydro-boratozirconium(IV), **9,** 103
B-Chloro-9-borabicyclo[3.3.1]nonane, **12,** 236
1-Chloro-4-bromomethoxybutane, **6,** 104
4-Chloro-2-butanone, **1,** 698
3-Chloro-1-butene (α-Methallyl chloride), **5,** 392
1-Chloro-2-butene (*see* Crotyl chloride)
4-Chloro-1-butenyl-2-lithium, **12,** 113
1-Chloro-2-butenyllithium, **10,** 87
4-Chloro-1-butenyl-2-magnesium bromide, **12,** 115
Chlorocarbene, **1,** 95, 130; **6,** 171
1-Chlorocarbonylbenzotriazole, **8,** 87
Chloro(carbonyl)bis(triphenylphosphine)-rhodium(I) [*see* Carbonylchlorobis-(triphenylphosphine)rhodium(I)]
N-Chlorocarbonyl isocyanate, **5,** 109
Chlorocarbonylrhodium(I) dimer (*see* Tetracarbonyldi-μ-chlorodirhodium)
B-Chlorocatecholborane, **9,** 98
6-Chloro-1-*p*-chlorobenzenesulfonyl-oxybenzotriazole, **6,** 106
1-Chloro-4-(chloromethoxy)butane, **6,** 104; **7,** 22
3-Chloro-2-chloromethyl-1-propene, **2,** 290; **3,** 101
α-Chlorocrotyllithium, **10,** 87
Chlorocyanoketene, **8,** 88; **9,** 103; **12,** 111
α-Chloro-N-cyclohexylacetaldonitrone, **4,** 80; **6,** 107
N-Chloro-N-cyclohexylbenzenesulfonamide, **2,** 67
α-Chloro-N-cyclohexylpropanaldonitrone, **4,** 80; **5,** 110; **6,** 106
Chloro-1,5-cyclooctadieneiridium(I) dimer [*see* Di-μ-chlorobis(1,5-cyclooctadiene)-diiridium]
1-Chloro-3-diazoacetone, **5,** 114
Chlorodiazomethane, **1,** 130
1-Chloro-3-diazo-2-propanone, **5,** 114
Chlorodicarbonylrhodium(I) dimer (*see* Tetracarbonyldi-μ-chlorodirhodium)
2-Chloro-1,1-diethoxyethane, **5,** 107, 592
2-Chloro-1,1-diethoxyethylene, **9,** 522
2-Chloro-2-diethoxyphosphinylacetic acid, **8,** 168
Chlorodifluoromethane, **4,** 314; **7,** 60
4-Chloro-2,3-dihydrofuran, **11,** 119
3-Chloro-4,5-dihydrofuryl-2-copper, **11,** 119
Chlorodiiodomethane, **5,** 27; **7,** 74
2-Chloro-1,1-dimethoxyethylene, **5,** 119
(Z)-1-Chloro-1,2-dimethoxyethylene, **5,** 204
1-Chloro-1-dimethylaminoisoprene, **9,** 104
Chloro(dimethylamino)methoxyphosphine, **10,** 88

1-Chloro-1-dimethylamino-2-methyl-1,3-butadiene, **9,** 104
2-Chloro-4,5-dimethyl-2-oxo-1,3,2-dioxaphosphole, **6,** 7
Chlorodimethylphenylsilane, **2,** 278
Chlorodimethylsulfonium chloride (Corey–Kim reagent), **4,** 191; **5,** 114, 130, 132; **9,** 67, 105
Chlorodimethylsulfoxonium chloride, **4,** 200; **5,** 130; **6,** 229
Chlorodimethyl(2,4,6-tri-*t*-butylphenoxy)-silane, **11,** 217
2-Chloro-1,3,2-dioxaborolane, **12,** 101, 224
2-Chloro-1,3,2-dioxaphospholane, **1,** 372
Chlorodiphenylcarbenium hexachloroantimonate, **5,** 115; **6,** 108
Chlorodiphenylmethylene hexachloro-antimonate, **5,** 115; **6,** 108
Chlorodiphenylmethylsilane, **10,** 91; **12,** 321
Chlorodiphenylphosphine, **4,** 303; **11,** 120, 564
Chlorodiphenylphosphine oxide (*see* Diphenylphosphinic chloride)
2-Chloro-1,3,2-dithiaborolane, **12,** 112
2-Chloro-1,3-dithiane, **8,** 88
2-Chloroethanesulfonyl chloride, **5,** 116
2-Chloroethanol, **5,** 117; **8,** 177
2-(Chloroethoxy)carbene, **2,** 118; **7,** 214; **9,** 83; **10,** 89
Chloro(ethoxycarbonyl)carbene, **3,** 222
2-Chloro-3-ethylbenzoxazolium tetrafluoroborate, **8,** 90; **9,** 105; **10,** 204
N-Chloroethyl carbamate (*see* N-Chlorourethane)
α-Chloroethyl chloroformate, **12,** 112
2-Chloroethyl chloromethyl ether, **7,** 214; **10,** 89
2-Chloroethyl(dimethyl)sulfonium iodide, **12,** 113
α-Chloroethyl ethyl ether, **8,** 495
3-Chloro-2-ethyl-1-propene, **5,** 544
2-Chloroethyl thiocyanate, **9,** 106
α-Chloroethyltrimethylsilane, **8,** 277; **11,** 128
β-Chloroethyl vinyl ketone, **3,** 47
Chloroethynyllithium, **6,** 294
Chlorofluorocarbene, **1,** 254; **2,** 273; **6,** 175, 511
Chlorofluoromethylenetriphenyl-phosphorane, **2,** 443; **4,** 83
Chloroform (*see also* Chloroform + co-reactants), **1,** 29, 130, 131, 223, 425; **3,** 140; **4,** 22, 114, 314; **5,** 28, 398; **8,** 92, 388; **12,** 517
Chloroform–Dibenzoyl peroxide, **1,** 131
Chloroform–Di-*t*-butyl peroxide, **1,** 131
Chloroform–Ethylene oxide–Tetraethylammonium bromide, **2,** 197
Chloroform–Lithium triethylmethoxide, **6,** 349
Chloroform–Potassium *t*-butoxide, **2,** 365; **3,** 85; **4,** 130, 402; **7,** 66
Chloroform–Potassium hydroxide–Alumina, **2,** 18
Chloroform–Sodium hydroxide, **1,** 844
Chloroform–Sodium hydroxide–Benzyltriethylammonium chloride, **4,** 27; **5,** 27; **6,** 38, 42; **7,** 19; **9,** 360
Chloroform–Sodium hydroxide–β-Cyclodextrin, **12,** 151
Chloroform–Sodium hydroxide–Dibenzo-18-crown-6, **6,** 159
Chloroform–Sodium hydroxide–(−)-N,N-Dimethylephedrinium bromide, **6,** 220
Chloroform–Sodium hydroxide–Tributylamine, **6,** 601
1-(Chloroformyl)benzotriazole, **8,** 87
2-(Chloroformyl)-1,3-dithiane, **5,** 99
2-[(Chloroformyl)oxy]ethyltriphenyl-phosphonium chloride, **8,** 92
Chloro(hexamethylbenzene)hydrido-triphenylphosphinerhodium, **8,** 153
Chlorohydridobis(triethylphosphine)-platinum, **5,** 446
Chlorohydridotris(triphenylphosphine)-ruthenium(II), **6,** 425; **7,** 60
5-Chloro-2-hydroxybenzophenone, **5,** 118
3-Chloro-2-hydroxyiminopropanal, **9,** 213
1-Chloro-2-iodoethane, **9,** 106
(E)-1-Chloro-2-iodoethylene, **12,** 468
Chloroiodomethane, **10,** 89
Chloroiodomethane–Zinc-copper couple, **3,** 256
Chloroiridic acid [*see* Hydrogen hexachloroiridate(IV)]
Chloroketene dialkyl acetals, **5,** 119; **9,** 522
1-Chloro-1-lithio-2-butene, **10,** 87
4-Chloro-2-lithio-1-butene, **12,** 113
3-Chloro-1-lithio-3-methyl-1-butyne, **8,** 93
3-Chloro-3-lithio-2-methyl-1-propene, **10,** 87
5-Chloro-2-lithio-1-pentene, **12,** 115
Chloromagnesium dimethylcuprate, **10,** 287
Chloromaleic anhydride, **3,** 104; **5,** 362

α-Chloromethallyllithium, **10,** 87
Chloromethane, **1,** 674
Chloromethanesulfonyl chloride, **2,** 103; **3,** 74
Chloro(methoxycarbonyl)ketene, **11,** 110
Chloro(methoxy)methylalane, **5,** 438
[Chloro(*p*-methoxyphenyl)methylene]-diphenylammonium chloride, **11,** 220
Chloro(methyl)aluminum amides, **11,** 121
2-Chloro-3-methyl-1,3-benzothiazolium trifluoromethanesulfonate, **7,** 61
1-Chloro-3-methyl-2-butene (Prenyl chloride), **9,** 52
3-Chloro-3-methyl-1-butyne, **5,** 27, 460; **7,** 77; **8,** 93
(3-Chloro-3-methyl-1-butynyl)lithium, **8,** 93
Chloromethylcarbene, **6,** 530; **10,** 90; **11,** 121
4-Chloromethyl-3,5-dimethylisoxazole, **1,** 276; **2,** 150; **4,** 177
(Chloromethyl)dimethylsilane, **12,** 27
Chloro(methyl)diphenylsilane, **10,** 91; **12,** 321
(Chloromethylene)dimethylammonium halides (*see* Vilsmeier reagent)
Chloromethyleneoxomolybdenum, **12,** 352
Chloromethylenetriphenylphosphorane, **1,** 1239; **5,** 119, 449; **7,** 62; **8,** 96
Chloromethyl ethyl ether, **7,** 112, 213; **12,** 262
Chloromethyllithium, **12,** 77
Chloromethylmaleic anhydride, **1,** 920
(−)-2-Chloromethyl-4-methoxymethyl-5-phenyloxazoline, **6,** 204
Chloromethyl methyl ether, **1,** 132, 672; **3,** 150; **4,** 83, 328; **5,** 120; **6,** 18, 109, 304; **7,** 61, 111; **8,** 387; **9,** 107, 520; **10,** 460; **12,** 116
4-Chloromethyl-3-methylisoxazole, **1,** 675
Chloromethyl methyl sulfide, **4,** 84; **6,** 109; **8,** 94; **9,** 107; **12,** 117, 262
2-Chloromethyl-4-nitrophenyl phosphorodichloridate, **3,** 48
Chloromethyl phenyl sulfide, **6,** 510; **9,** 107, 521
Chloromethyl phenyl sulfone, **12,** 412
Chloromethyl phenyl sulfoxide, **8,** 94; **9,** 271
Chloromethylphosphonic dichloride, **1,** 135
Chloromethyl phosphonothioic dichloride, **1,** 135
Chloromethylphosphoric dichloride, **1,** 135
N-Chloromethylphthalimide, **1,** 135
3-Chloro-2-methylpropene (*see* Methallyl chloride)
3-Chloro-2-methyl-1-propenyl-3-lithium, **10,** 87
4-(Chloromethyl)pyridine hydrochloride, **4,** 391
2-(Chloromethyl)pyridine N-oxide, **5,** 538; **6,** 472
2-Chloro-1-methylpyridinium iodide (*see also* 2-Halopyridinium salts), **7,** 110; **8,** 95; **10,** 230; **11,** 518; **12,** 116
2-Chloromethyl-4,4,5,5-tetramethyl-1,3,2-dioxaborolane, **11,** 86; **12,** 397
1-Chloro-2-methyl-N,N-tetramethylene-propenylamine, **12,** 124
Chloro(methylthio)carbene, **5,** 194; **6,** 136
Chloromethyl *p*-tolyl sulfide, **5,** 550
Chloromethyl *p*-tolyl sulfone, **5,** 550; **6,** 41; **8,** 340
(Chloromethyl)trimethylsilane, **6,** 635; **8,** 277, 307; **12,** 118
3-Chloro-3-methyl-2-trimethylsilyloxy-1-butene, **11,** 470
1-Chloro-2-methyl-1-trimethylsilyl-1-propene, **8,** 410
2-Chloromethyl-3-trimethylsilyl-1-propene, **11,** 259
(Chloromethyl)trimethyltin, **8,** 308
(Chloromethyl)triphenylphosphonium halides, **7,** 62; **8,** 96
Chloro(norbornadiene)rhodium dimer [*see* Dichlorobis(norbornadiene)-dirhodium(I)]
2-Chloro-2-oxobenzo-1,3,2-dioxaphosphole, **1,** 837; **2,** 321; **3,** 223; **10,** 91
1-Chloro-3-pentanone, **3,** 49; **6,** 110
5-Chloro-1-penten-3-one, **3,** 47
5-Chloro-1-pentenyl-2-lithium, **12,** 115
m-Chloroperbenzoic acid (MCPBA), **1,** 135; **2,** 68, 309; **3,** 49; **4,** 85; **5,** 3, 120; **6,** 46, 110, 157, 459; **7,** 44, 56, 62; **8,** 97, 290; **9,** 81, 108; **10,** 92; **11,** 71, 95, 97, 99, 122; **12,** 118, 210, 243, 452
m-Chloroperbenzoic acid–Potassium fluoride, **11,** 124
m-Chloroperbenzoic acid–2,2,6,6-Tetramethylpiperidine, **6,** 110; **8,** 99
1-Chloro-2-phenylacetylene, **11,** 553
Chlorophenylcarbene, **5,** 155
p-Chlorophenyl chlorothionoformate, **3,** 50
2-Chlorophenyl dichlorophosphite, **6,** 114
p-Chlorophenyl isocyanate, **12,** 386
Chlorophenylketene, **5,** 279
p-Chlorophenylmagnesium bromide, **1,** 415

(Chloro)phenylmercury (*see* Phenylmercuric chloride)
7-Chloro-5-phenyl-1-[(S)-α-phenylethyl]-1,3-dihydro-2*H*-1,4-benzodiazepine-2-one, **11,** 124
o-Chlorophenyl phosphorodichloridite, **6,** 114
4-Chloro-2-phenylquinazoline, **4,** 86
Chloro(phenylsulfinyl)methyllithium, **8,** 94; **9,** 271
5-Chloro-1-phenyltetrazole, **2,** 319; **4,** 377; **7,** 64; **8,** 102; **10,** 93
1-Chloro-2-phenylthioacetylene, **11,** 553
3-Chloro-1-phenylthio-1-propene, **6,** 605
9-Chloro-9-phenylxanthene, **9,** 110
Chloro(phenyl)zinc (*see* Phenylzinc chloride)
Chloropicrin, **1,** 139
N-Chloropiperidine, **12,** 123
Chloroplatinic acid [*see* Hydrogen hexachloroplatinate(IV)]
N-Chloropolymaleimide, **4,** 87
2-Chloropropanal, **4,** 80; **6,** 107
3-Chloro-1-propene (*see* Allyl chloride)
2-Chloropropionitrile, **1,** 912
3-Chloropropyne, **5,** 565
3-Chloropropynyllithium (Lithium 3-chloropropargylide), **5,** 397
Chloro(pyridine)cobaloxime(III), **11,** 136
5-Chlorosalicylaldehyde, **1,** 139
N-Chloro-N-sodiourethane, **6,** 115
N-Chlorosuccinimide (NCS), **1,** 139; **2,** 69; **3,** 142; **5,** 127; **6,** 75, 115, 178; **7,** 39, 65; 368; **8,** 103, 235, 463; **9,** 111, 276, 367; **10,** 67, 94; **12,** 121
N-Chlorosuccinimide–Benzeneselenenyl chloride, **9,** 111
N-Chlorosuccinimide–Dialkyl sulfides, **6,** 118
N-Chlorosuccinimide–Diisopropyl sulfide, **12,** 195
N-Chlorosuccinimide–Dimethyl sulfide (Corey–Kim reagent), **4,** 87, 192; **5,** 129, 130; **6,** 118, 120, 608; **8,** 200; **9,** 111, 365
N-Chlorosuccinimide–Diphenyl diselenide, **9,** 111, 367
N-Chlorosuccinimide–1,3-Dithiane, **6,** 119
N-Chlorosuccinimide–Silica, **6,** 510
N-Chlorosuccinimide–Silver nitrate, **4,** 216
N-Chlorosuccinimide–Triethylamine, **6,** 120
N-Chlorosuccinimide–Trifluoroacetic acid, **10,** 419
1-Chlorosulfinyl-4-dimethylaminopyridinium chloride, **12,** 122
Chlorosulfonic acid, **1,** 139; **2,** 70; **4,** 436; **6,** 121; **7,** 12
Chlorosulfonyl isocyanate (CSI), **1,** 117; **2,** 61, 70; **3,** 51; **4,** 90, 382; **5,** 132, 283; **6,** 122; **7,** 65; **8,** 105; **10,** 94; **11,** 125; **12,** 122
Chlorosulfuric acid (*see* Chlorosulfonic acid)
2-Chlorotetrahydrofuran, **9,** 112
Chlorothexylborane, **10,** 95
Chlorothexylborane–Dimethyl sulfide, **12,** 485
4-Chlorothiophenol, **6,** 208
Chlorotitanium triisopropoxide, **11,** 375
Chlorotitanium triphenoxide, **11,** 376
Chlorotribenzylsilane, **5,** 338
Chloro(trichloroethyl)ketene, **9,** 112
N-Chlorotriethylammonium chloride, **12,** 122
N-(2-Chloro-1,1,2-trifluoroethyl)diethylamine, **1,** 249; **2,** 130; **3,** 95; **4,** 149; **5,** 214, 436; **6,** 66, 186; **9,** 224
Chlorotrifluoroethylene, **1,** 1072; **4,** 94
Chloro(trifluoromethyl)carbene, **3,** 223; **4,** 376
Chlorotriisopropoxytitanium, **11,** 375
Chlorotriisopropylsilane, **5,** 75
1-Chloro-N,N,2-trimethylpropenylamine, **4,** 94; **5,** 136; **6,** 122; **7,** 66; **12,** 123
Chlorotrimethylsilane, **1,** 1232; **2,** 207, 435; **3,** 310; **4,** 214, 308, 537; **5,** 81, 417, 462, 709; **6,** 522, 626, 675; **7,** 13, 66, 128; **8,** 107, 492; **9,** 112, 255, 387, 444, 487; **10,** 96; **11,** 125; **12,** 126, 263, 264, 280, 463
Chlorotrimethylsilane–Acetic anhydride, **12,** 126
Chlorotrimethylsilane–Lithium, **5,** 711; **6,** 323
Chlorotrimethylsilane–Lithium bromide, **10,** 96
Chlorotrimethylsilane–Magnesium amalgam, **6,** 352
Chlorotrimethylsilane–Magnesium–Hexamethylphosphoric triamide, **5,** 713
Chlorotrimethylsilane–Sodium, **2,** 435; **3,** 311; **4,** 32, 537; **5,** 711; **6,** 627; **8,** 107, 446; **10,** 356; **11,** 127; **12,** 126
Chlorotrimethylsilane–Sodium bromide, **10,** 96
Chlorotrimethylsilane–Sodium iodide (*see also* Iodotrimethylsilane), **8,** 108; **9,** 251, 255; **10,** 97; **11,** 127, 274; **12,** 127
Chlorotrimethylsilane–Sodium iodide–Zinc,

11, 128
Chlorotrimethylsilane–Tetraethylammonium bromide, **12,** 79
Chlorotrimethylsilane–Zinc, **5,** 714; **6,** 628; **7,** 429; **8,** 532; **10,** 98; **12,** 568
(E)-1-Chloro-3-trimethylsilyl-2-butene, **5,** 355
Chlorotrimethylsilylcarbene, **2,** 31
1-Chloro-1-(trimethylsilyl)ethyllithium, **8,** 277; **11,** 128
Chloro[(trimethylsilyl)methyl]ketene, **12,** 127
Chloro(trimethylsilyl)methyllithium, **8,** 277; **12,** 118
1-Chloro-1-trimethylsilyl-2-methylpropene, **7,** 358
3-Chloro-2-trimethylsilylmethylpropene, **11,** 259
3-Chloro-2-trimethylsilyloxy-1-propene, **10,** 286; **11,** 129
4-Chloro-2-trimethylstannyl-1-butene, **12,** 113
Chlorotrimethyltin (*see* Trimethyltin chloride)
Chlorotriphenoxytitanium, **11,** 376
Chlorotriphenylmethane (*see* Triphenylmethyl chloride)
Chlorotriphenylphosphonium chloride (*see* Triphenylphosphine dichloride)
Chlorotriphenyltin (*see* Triphenyltin chloride)
Chlorotris(methyldiphenylphosphine)-rhodium(I), **7,** 67
Chlorotris(neomenthyldiphenylphosphine)-rhodium(I), **6,** 416
Chlorotris(triphenylphosphine)cobalt, **10,** 274; **12,** 128
Chlorotris(triphenylphosphine)rhodium(I) (Wilkinson's catalyst), **1,** 140, 1252; **2,** 448; **3,** 89, 325; **4,** 559; **5,** 496, 572, 736; **6,** 652; **7,** 68; **8,** 109; **9,** 113, 257; **10,** 79, 98; **11,** 130; **12,** 130, 182
Chlorotris(triphenylphosphine)rhodium(I)–Hydrosilanes, **4,** 562; **5,** 739; **6,** 652; **7,** 282; **11,** 130
N-Chlorourethane, **4,** 139; **6,** 115
β-Chlorovinyl ethyl ether, **1,** 140, 613
β-Chlorovinyl methyl ether, **4,** 549
(2-Chlorovinyl)trimethylsilane, **5,** 556
Chloro(vinyl)zinc, **11,** 505
Chlorozirconium tributoxide, **11,** 379
Cholesterol, **1,** 141
Cholesteryl chloroformate, **1,** 141
Choline, **1,** 142
Chromic acid (*see* Sodium dichromate)
Chromic anhydride [*see* Chromium(VI) oxide]
Chromium(II) acetate, **1,** 147; **2,** 75; **3,** 59, 60; **4,** 97; **5,** 143
Chromium(II) acetate–Ethylenediamine (*see also* Chromium(II)–Amine complexes), **5,** 142
Chromium(II)–Amine complexes, **3,** 57; **4,** 97; **5,** 121, 142; **9,** 117
Chromium carbene complexes, **11,** 397; **12,** 132
Chromium carbonyl, **5,** 142; **6,** 27, 103, 125; **7,** 71; **8,** 110; **9,** 117; **10,** 100; **11,** 131, 397; **12,** 132
Chromium(II) chloride [*see also* Chromium (III) chloride–Lithium aluminum hydride], **1,** 149; **2,** 76; **3,** 60; **5,** 144; **7,** 73; **9,** 511; **11,** 132; **12,** 136
Chromium(III) chloride, **11,** 232
Chromium(III) chloride–Lithium aluminum hydride (*see also* Chromium(II) chloride), **8,** 110; **9,** 119; **10,** 101
Chromium hexacarbonyl (*see* Chromium carbonyl)
Chromium(VI) oxide, **1,** 144; **2,** 72; **3,** 54; **4,** 96; **5,** 140; **7,** 70; **9,** 115, 289; **10,** 99, 306
Chromium(VI) oxide–Acetic acid (*see* Fieser reagent)
Chromium(VI) oxide–Acetic anhydride (Thiele reagent), **1,** 146; **7,** 70
Chromium(VI) oxide–Amberlyst, **7,** 69
Chromium(VI) oxide–2,2′-Bipyridine, **10,** 30
Chromium(VI) oxide–Diethyl ether, **9,** 115
Chromium(VI) oxide–Dimethylformamide (Snatzke's reagent), **1,** 147, 278; **3,** 56; **11,** 131
Chromium(VI) oxide–3,5-Dimethylpyrazole complex, **5,** 142; **8,** 110; **9,** 116
Chromium(VI) oxide–Graphite, **4,** 97; **5,** 732
Chromium(VI) oxide–Hexamethylphosphoric triamide, **3,** 150; **7,** 71; **9,** 116
Chromium(VI) oxide–Pyridine (*see also* Collin's reagent, Sarett reagent, Ratcliff's reagent), **1,** 145; **2,** 74; **3,** 55; **4,** 96, 215; **7,** 70; **9,** 121; **10,** 99; **11,** 139; **12,** 139
Chromium(II) perchlorate, **10,** 139
Chromium(II) perchlorate–Butanethiol, **10,** 101
Chromium(II) perchlorate–Ethylenediamine

[*see also* Chromium(II)–Amine complexes], **3,** 58; **4,** 97
Chromium(II) sulfate, **1,** 150; **2,** 77; **3,** 62; **5,** 144
Chromium trioxide (*see* Chromium(VI) oxide)
Chromous, *see* Chromium(II)
Chromyl acetate, **1,** 147, 151; **2,** 78
Chromyl chloride, **1,** 151; **2,** 79; **3,** 62; **4,** 98; **5,** 144; **6,** 126; **8,** 112; **11,** 134
Chromyl trichloroacetate, **1,** 152
Cinchona alkaloids (*see also* specific names), **6,** 501; **7,** 311; **8,** 430; **9,** 403; **10,** 338; **11,** 134, 374, 456; **12,** 380
Cinchonidine (*see also* Cinchona alkaloids), **11,** 456
Cinchonine (*see also* Cinchona alkaloids), **6,** 501; **11,** 374
N-*trans*-Cinnamoylimidazole, **1,** 153
Cinnamyltributyltin, **11,** 542
Cinnamyltrimethylsilane, **11,** 228; **12,** 146, 497
Cinnamyltriphenylphosphonium chloride, **1,** 613, 1238
Cinnamyltriphenyltin, **11,** 543
Claisen's alkali, **1,** 153; **6,** 127
Claycop, **12,** 231
Clayfen, **11,** 237; **12,** 231
Cobaloxime(I), **11,** 135; **12,** 520
Cobalt(II) acetate, **2,** 80; **4,** 99
Cobalt(III) acetate, **5,** 157; **6,** 127; **9,** 119
Cobalt(II) acetate–Hydrogen bromide, **1,** 154
Cobaltacyclopentan-2-ones, **11,** 136
Cobalt(II) bis(salicylidene-γ-iminopropyl)-methylamine, **11,** 138
Cobalt boride–Borane–*t*-Butylamine, **11,** 138
Cobalt(II) chloride, **1,** 155; **6,** 127, 145; **10,** 101
Cobalt(III) fluoride, **8,** 113; **10,** 397
Cobalt hydridocarbonyl (*see* Tetracarbonylhydridocobalt)
Cobaltocene (*see* Dicyclopentadienylcobalt)
Cobaltous, *see* Cobalt(II)
Cobalt(II) phthalocyanine, **8,** 499; **9,** 119; **10,** 102; **11,** 63, 138
Cobalt *meso*-tetraphenylporphine, **12,** 138
Cobalt(III) trifluoroacetate, **5,** 146
Collidine, **1,** 155; **4,** 61, 117; **11,** 105
2,4,6-Collidinium *p*-toluenesulfonate, **12,** 139
Collins reagent (*see also* Chromium(VI) oxide–Pyridine, Sarett reagent), **2,** 74; **3,** 55; **4,** 215; **5,** 74, 285; **6,** 124; **8,** 496, 512; **9,** 121, 397, 398; **10,** 99; **11,** 139, 215, 216; **12,** 38, 139
Collman's reagent (*see* Disodium tetracarbonylferrate)
Copoly(ethylene–N-hydroxymaleimide), **2,** 80
Copper, **1,** 157; **2,** 81, 82; **3,** 63; **4,** 102; **5,** 146; **6,** 184; **7,** 73; **8,** 113; **9,** 122; **11,** 553; **12,** 140
Copper(I) acetate, **2,** 89; **6,** 142; **7,** 80
Copper(II) acetate, **1,** 157, 159; **2,** 18, 84; **3,** 65; **4,** 105; **5,** 156; **6,** 138; **7,** 126; **10,** 103; **12,** 140
Copper(II) acetate–Amine complexes [*see also* Copper(II)–Amine complexes], **3,** 65; **5,** 157; **8,** 115
Copper(II) acetate–2,2′-Bipyridyl, **3,** 65
Copper(I) acetate–*t*-Butyl isocyanide (*see also* Copper(0)–Isonitrile complexes, Copper(I) oxide–*t*-Butyl isocyanide), **5,** 163
Copper(II) acetate–Copper(II) tetrafluoroborate, **11,** 139
Copper(II) acetate–Ferrous(II) sulfate, **10,** 103
Copper(II) acetate–Morpholine, **8,** 115
Copper(II) acetate–Oxygen, **6,** 429
Copper(II) acetate–1,10-Phenanthroline, **3,** 65
Copper(II) acetate–Pyridine, **3,** 65
Copper(II) acetate–Sodium hydride–Sodium *t*-amyloxide, **10,** 365
Copper(I) acetylacetonate, **4,** 100
Copper(II) acetylacetonate, **2,** 81; **3,** 62; **5,** 244; **8,** 159; **9,** 51; **10,** 416; **12,** 422
Copper(I) alkoxides, **4,** 109; **5,** 148, 395; **6,** 144; **9,** 122
Copper(II)–Amine complexes, **1,** 168; **2,** 400; **3,** 65; **5,** 157; **8,** 114, 115
Copper–Ascorbic acid, **1,** 155
Copper–Benzoic acid, **1,** 158
Copper(I) bis(trimethylsilylamide), **5,** 45; **6,** 57
Copper(I) bromide, **1,** 165; **2,** 90; **3,** 67; **4,** 108; **5,** 69, 163; **6,** 143, 662; **7,** 79; **8,** 116; **10,** 186; **11,** 140
Copper(II) bromide, **1,** 161; **2,** 84, 86; **5,** 158, 159; **6,** 138, 227; **8,** 119; **10,** 106
Copper(I) bromide–Dimethyl sulfide, **6,** 225; **8,** 117, 235; **9,** 70; **10,** 104; **12,** 114, 222

Copper(I) bromide–Lithium carbonate, **6,** 18
Copper(I) bromide–Lithium trimethoxyaluminum hydride (*see also* Copper hydride ate complexes), **7,** 80; **8,** 120
Copper(II) bromide–Nitric oxide, **2,** 87
Copper(I) bromide–Sodium bis-(2-methoxyethoxy)aluminum hydride (*see also* Copper hydride ate complexes), **6,** 144; **8,** 120
Copper bronze, **1,** 155; **3,** 62; **4,** 100; **11,** 49; **12,** 140
Copper(I) *t*-butoxide (*see also* Copper(I) alkoxides), **4,** 109; **5,** 395; **6,** 144; **9,** 122
Copper *t*-butylacetylide, **5,** 148; **6,** 225, 329; **8,** 300
Copper(0)–*t*-Butyl isocyanide [*see also* Copper(0)–Isonitrile complexes], **6,** 129
Copper carbonate, basic, **1,** 158, 163; **4,** 101
Copper(I) carbonyl ion, **5,** 149
Copper(I) chloride, **1,** 166; **2,** 91; **3,** 67; **4,** 109, 158; **5,** 159, 164, 450, 581; **6,** 143, 145; **7,** 80; **8,** 87, 118; **9,** 123; **10,** 197, 355; **11,** 49, 140; **12,** 141
Copper(II) chloride, **1,** 161, 163; **2,** 84; **3,** 66; **4,** 105; **5,** 158; **6,** 139, 259; **7,** 79, 126; **8,** 119; **9,** 123; **10,** 103, 106
Copper(I) chloride–Ammonium chloride, **1,** 165
Copper(II) chloride–Copper(II) acetate, **6,** 140
Copper(II) chloride–Copper(II) oxide, **5,** 160; **9,** 34
Copper(II) chloride–Iodine, **6,** 140
Copper(II) chloride–Lithium chloride [*see* Dilithium tetrachlorocuprate(II)]
Copper(II) chloride–Nitric oxide, **2,** 87
Copper(I) chloride–Oxygen, **4,** 110; **5,** 165; **6,** 429; **7,** 80; **8,** 118; **11,** 141
Copper(II) chloride–Oxygen, **8,** 120
Copper(I) chloride–Tetrabutylammonium chloride, **8,** 119
Copper(II) chloride–Tetramethylethylenediamine, **1,** 168; **2,** 400
Copper(II) chloride–Zinc chloride–Lithium aluminum hydride, **5,** 161
Copper chromite (Lazier catalyst), **1,** 156; **2,** 82
Copper chromite–Quinoline, **1,** 975
Copper–Chromium oxide (Adkin's catalyst), **1,** 156, 157
Copper(I) cyanide, **1,** 278, 391, 696, 960; **2,** 81; **5,** 166; **6,** 143, 146; **7,** 333; **9,** 124; **10,** 1, 162; **11,** 366; **12,** 568
Copper(II) cyanide, **5,** 166
Copper(I) cyanoacetate, **5,** 171; **7,** 74
Copper(II) cyanoacetate, **5,** 171
Copper halide nitrosyls, **7,** 73
Copper(II) hexafluoroacetylacetonate, **6,** 664
Copper(I) hexamethyldisilazide, **5,** 45; **6,** 57
Copper hydride ate complexes, **3,** 154; **5,** 168, 330; **6,** 144, 492; **7,** 80; **8,** 120; **10,** 237; **12,** 286
Copper(II) hydroxide, **10,** 103
Copper(I) iodide, **1,** 169; **2,** 82, 92; **3,** 69; **5,** 167; **6,** 147; **7,** 81, 238; **8,** 121; **9,** 124; **10,** 107; **11,** 141; **12,** 141
Copper(I) iodide–Dibutyl sulfide, **5,** 148
Copper(I) iodide–Diisopropyl sulfide, **4,** 219; **5,** 747; **7,** 129
Copper(I) iodide–Dimethyl sulfide, **11,** 141
Copper(I) iodide–Grignard reagents [*see* Grignard reagents–Copper(I) halides]
Copper(I) iodide–Lithium trimethoxyaluminum hydride (*see also* Copper hydride ate complexes), **5,** 168
Copper(I) iodide–Potassium naphthalenide, **9,** 122
Copper(I) iodide–Tributylphosphine, **2,** 282, 400, 441; **3,** 79, 278; **4,** 221, 516; **6,** 266
Copper(I) iodide–Tributyl phosphite, **2,** 441
Copper(I) iodide–Triethylphosphine–Lithium naphthalenide, **12,** 140
Copper(I) iodide–Triethyl phosphite, **8,** 122; **12,** 143
Copper(I) iodide–Trimethyl phosphite, **3,** 278; **5,** 244, 717; **6,** 630
Copper(0)-Isonitrile complexes [*see also* Copper(I) oxide–*t*-Butyl isocyanide, Copper(I) acetate–*t*-Butyl isocyanide], **6,** 128; **9,** 122
Copper(I) isoprenylacetylide (*see* 3-Methyl-3-buten-1-ynylcopper)
Copper(II) methoxide, **4,** 107
Copper(I) methyltrialkylborates, **7,** 75
Copper(II) nitrate–Acetic anhydride, **1,** 163; **2,** 87
Copper(II) nitrate/K10 Bentonite (Claycop), **12,** 231
Copper(II) nitrate–Pyridine, **1,** 164; **4,** 107
Copper(I) oxide, **1,** 169; **2,** 83; **5,** 149; **10,** 107
Copper(II) oxide, **4,** 108; **5,** 161
Copper(I) oxide–*t*-Butyl isocyanide (*see also*

Copper(0)–Isonitrile complexes, Copper(I) acetate–*t*-Butyl isocyanide), **4,** 101; **5,** 150
Copper(I) oxide–Copper(II) nitrate, **8,** 124
Copper(II) perchlorate, **7,** 79
Copper(II) permanganate, **9,** 389; **11,** 142
Copper(I) phenoxides, **5,** 148
Copper(I) phenylacetylide, **4,** 101; **7,** 237; **12,** 143
Copper(I) phenylacetylide–Tributylphosphine, **6,** 130
Copper(II) sulfate, **1,** 157, 164; **2,** 82, 89; **3,** 63; **4,** 528; **5,** 162, 718; **6,** 141; **8,** 125; **10,** 107
Copper(II) sulfate–Pyridine, **9,** 125
Copper(II) sulfate–Sodium borohydride, **11,** 142
Copper(II) tetrafluoroborate, **3,** 66; **5,** 722; **6,** 142; **12,** 144
Copper(I) trifluoromethanesulfonate, **5,** 151; **6,** 130; **7,** 75; **8,** 125; **10,** 108, 340; **11,** 113, 142; **12,** 144
Copper(II) trifluoromethanesulfonate, **5,** 152; **6,** 664; **7,** 76; **8,** 126; **10,** 110; **11,** 58, 460
Copper(I) trimethylsilylacetylide, **6,** 148
Corey–Kim reagent (*see* Chlorodimethylsulfonium chloride, N-Chlorosuccinimide–Dimethyl sulfide)
Corey's reducing agent, **4,** 103
Cornforth reagent (*see also* Chromium(VI) oxide–Pyridine), **1,** 146; **5,** 141
Crotonaldehyde, **1,** 697; **2,** 391; **3,** 5
Crotonic acid, **4,** 348
Crotonic anhydride, **8,** 127; **11,** 596
Crotylbis(cyclopentadienyl)halotitanium reagents, **12,** 354
B-Crotyl-9-borabicyclo[3.3.1]nonane, **9,** 58; **12,** 81
Crotyl carbamates, **12,** 82
Crotyl chloride, **2,** 290; **3,** 178; **5,** 392; **10,** 87
Crotyldimethoxyborane, **11,** 240
Crotyldimethylaluminum, **12,** 170
Crotyllithium, **10,** 416
Crotylmagnesium bromide, **8,** 122
Crotylmagnesium chloride, **9,** 94; **11,** 308; **12,** 28, 371
Crotylmagnesium chloride–Dichlorobis-(cyclopentadienyl)titanium, **6,** 48
Crotylpalladium chloride [*see* Dichlorobis-(crotyl)dipalladium]
Crotyl phenyl selenide, **6,** 338
2-Crotyl-4,4,5,5-tetramethyl-1,3,2-dioxaborolane, **11,** 86
Crotyltitanium triisopropoxide, **12,** 354
Crotyltributyltin (*see* Tributylcrotyltin)
Crotyltriisopropoxytitanium, **12,** 354
Crotyltrimethylsilane, **12,** 146, 434, 497
12-Crown-4, **6,** 134
15-Crown-5, **6,** 134; **7,** 77; **8,** 128; **11,** 144
15-Crown-6, **10,** 111
18-Crown-6, **5,** 152; **6,** 135; **7,** 76; **8,** 128, 390; **9,** 126, 359; **10,** 111; **11,** 144, 403
Crown ethers (*see also* specific names), **4,** 142; **5,** 152; **6,** 133; **7,** 76; **8,** 128, 389; **9,** 126, 359; **10,** 110; **11,** 143; **12,** 147, 380
Cryptates, **5,** 156; **6,** 137; **8,** 130; **9,** 127; **10,** 112, 378; **12,** 404
Cryptate[2.2.1], **10,** 378
CSA (*see* Camphor-10-sulfonic acid)
CSI (*see* Chlorosulfonyl isocyanate)
Cumene hydroperoxide–Titanium(III) acetylacetonate, **5,** 76
Cumene hydroperoxide–Titanium(IV) isopropoxide, **5,** 76
Cupric, *see* Copper(II)
Cuprous, *see* Copper(I)
Cyamelide, **1,** 170
Cyanamide, **1,** 170; **3,** 71; **5,** 168; **12,** 147
Cyanic acid, **1,** 170
Cyanoacetic acid, **1,** 39, 515, 959; **5,** 248; **11,** 145
Cyano-*t*-butoxychloroformate, **5,** 168
Cyano-*t*-butyldimethylsilane, **6,** 80
Cyanodiethylaluminum–Chlorotrimethylsilane, **8,** 130
2-Cyanoethyl phosphate, **1,** 172
Cyanogen, **2,** 142
Cyanogen azide, **1,** 173; **2,** 92; **3,** 71; **5,** 169
Cyanogen bromide, **1,** 174, 176; **2,** 93; **4,** 110; **5,** 57, 169; **6,** 148; **8,** 131; **9,** 127
Cyanogen chloride, **1,** 176; **4,** 110; **8,** 132
Cyanogen chloride N-oxide, **11,** 146
Cyanogen iodide, **8,** 132
N-Cyanoguanidine, **1,** 229
Cyanomethylcopper, **4,** 111; **5,** 171
Cyanomethyldiphenylphosphine oxide, **8,** 133
Cyanomethylenetriphenylphosphorane, **4,** 549; **11,** 146
Cyanomethylidenebis(triphenylphosphonium) dibromide, **4,** 112
Cyanomethyllithium, **6,** 84, 636
Cyanotributyltin, **10,** 411

Cyanotrimethylsilane, **4,** 542; **5,** 720; **6,** 632; **7,** 397; **8,** 133; **9,** 127; **10,** 1, 112; **11,** 147, 206; **12,** 79, 148, 235, 319, 376, 545, 568
Cyanotrimethylsilane–Triethylaluminum, **10,** 113
Cyano(trimethylsilyl)methyllithium, **10,** 428
2-Cyano-2-trimethylsilyloxypropane, **5,** 207
1-Cyanovinyl acetate (*see* 2-Acetoxyacrylonitrile)
Cyanuric acid, **1,** 170
Cyanuric chloride (*see also* Trichloroisocyanuric acid), **1,** 170; **3,** 72; **4,** 522; **5,** 687; **6,** 149; **10,** 114, 115
Cyanuric fluoride, **5,** 171; **10,** 115
Cyclic polyethers, chiral, **5,** 103; **6,** 134; **11,** 144
Cyclobutadieneiron tricarbonyl [*see* Tricarbonyl(1,3-cyclobutadiene)iron]
1-Cyclobutenylmethyllithium, **6,** 151
Cyclobutylidene, **12,** 150
α-Cyclodextrin, **8,** 135; **9,** 129
β-Cyclodextrin, **6,** 151; **8,** 133; **9,** 129; **11,** 150; **12,** 151
γ-Cyclodextrin, **9,** 129
Cyclodextrins, **6,** 151; **8,** 133; **9,** 129; **11,** 150; **12,** 151
Cycloheptaamylose (*see* β-Cyclodextrin)
Cycloheptatriene, **1,** 177
Cyclohexaamylose (*see* α-Cyclodextrin)
1,2-Cyclohexadiene, **2,** 178
1,2-Cyclohexanedione, **2,** 93
1,3-Cyclohexanedione, **1,** 178
1,4-Cyclohexanedione, **1,** 178
Cyclohexanone, **1,** 178
Cyclohexanone cyclohexylimine, **6,** 333
Cyclohexene oxide, **7,** 33
2-Cyclohexenone, **1,** 180
Cyclohexylamine, **1,** 252; **2,** 249; **4,** 300; **6,** 332; **7,** 201; **8,** 463
Cyclohexylboron dichloride, **4,** 377
2-Cyclohexylcyclohexanone, **1,** 180
N-Cyclohexylformamide, **1,** 382
N-Cyclohexylhydroxylamine, **4,** 80; **5,** 111; **6,** 107; **8,** 135
N-Cyclohexylidenecyclohexylamine, **6,** 333
Cyclohexyl isocyanide, **6,** 152
Cyclohexylisopropylaminomagnesium bromide (*see* Bromomagnesium isopropylcyclohexylamide)
Cyclohexyl metaborate trimer, **3,** 72; **4,** 112
(R)-(+)-Cyclohexyl(*o*-methoxyphenyl)(methyl)phosphine, **5,** 500; **11,** 31
1-Cyclohexyl-3-(2-morpholinoethyl)carbodiimide, **1,** 181
1-Cyclohexyl-3-(2-morpholinoethyl)carbodiimide metho-*p*-toluenesulfonate, **11,** 151
(Cyclohexyne)platinum(diphos), **5,** 171
Cyclooctaamylose (*see* γ-Cyclodextrin)
(1,5-Cyclooctadiene)bis(methyldiphenylphosphine)iridium(I) hexafluorophosphate, **8,** 135; **10,** 115
(1,5-Cyclooctadiene)(pyridine)(tricyclohexylphosphine)iridium(I) hexafluorophosphate, **12,** 151, 427
(1,5-Cyclooctadiene)(pyridine)(triphenylphosphine)iridium(I) hexafluorophosphate, **10,** 116
Cyclopentadiene, **1,** 181; **2,** 22; **4,** 130, 328; **5,** 99, 268, 476; **6,** 291; **8,** 137; **10,** 112
Cyclopentadienone dialkyl ketals, **1,** 270, 605; **2,** 94, 143, 396; **3,** 103; **8,** 178; **10,** 153
Cyclopentadienylcobalt (*see* Dicyclopentadienylcobalt)
Cyclopentadienylcobalt dicarbonyl (*see* Dicarbonylcyclopentadienylcobalt)
Cyclopentadienyl(diiodo)triphenylphosphinecobalt, **5,** 173
Cyclopentadienyliron dicarbonyl dimer [Di-μ-carbonyldicarbonylbis(cyclopentadienyl)diiron], **8,** 41
π-(Cyclopentadienyl)(isobutenyl)iron dicarbonyl tetrafluoroborate [*see* Dicarbonylcyclopentadienyl(2-methyl-1-propene)iron tetrafluoroborate]
Cyclopentadienylmagnesium hydride, **11,** 152
Cyclopentadienylnitrosylcobalt dimer, **10,** 116
Cyclopentadienylthallium(I), **8,** 138
Cyclopentadienyltitanium trichloride–Lithium aluminum hydride (*see* Trichlorocyclopentadienyltitanium–Lithium aluminum hydride)
Cyclopentadienyltribenzyltitanium, **4,** 112
π-Cyclopentadienyl(triphenylphosphine)cobaltatetraphenylcyclopentadiene, **5,** 173
π-Cyclopentadienylvanadium carbonyl [*see* Tetracarbonyl(cyclopentadienyl)vanadium]
Cyclopentanol, **1,** 182

Cyclopentanone, **1,** 182
Cyclopentyl chloroformate, **1,** 182
Cyclopentylidenetriphenylphosphorane, **7,** 183
Cyclopentylthexylboryl trifluoromethanesulfonate (*see also* Dialkylboryl trifluoromethanesulfonates)
Cyclopropene, **4,** 113
Cyclopropenone 1,3-propanediyl ketal, **12,** 152
B-Cyclopropyl-9-borabicyclo[3.3.1]nonane, **3,** 28
Cyclopropyldiphenylsulfonium tetrafluoroborate, **4,** 213; **6,** 155
Cyclopropylidenemethanone, **6,** 96
Cyclopropylidenetriphenylphosphorane, **2,** 96
Cyclopropylmagnesium bromide, **5,** 307
Cyclopropyl methyl ketone, **1,** 676
Cyclopropylmethylmagnesium bromide, **11,** 308
1-Cyclopropyl-1-nitrosourea, **5,** 591
Cyclopropyl phenyl sulfide, **5,** 372; **8,** 138; **11,** 284
Cyclopropyltrimethylsilane, **7,** 84
1-Cyclopropyl-1-trimethylsilyloxyethylene, **8,** 140
Cyclopropyltriphenylphosphonium bromide, **2,** 95
Cysteamine, **3,** 266

DABCO (*see* 1,4-Diazabicyclo[2.2.2]octane)
Danishefsky's diene (*see trans*-1-Methoxy-3-trimethylsilyloxy-1,3-butadiene)
Darvon alcohol, **5,** 231; **6,** 206; **8,** 184; **10,** 154; **11,** 153
DAST [*see* (Diethylamino)sulfur trifluoride]
DATMP (*see* Diethylaluminum 2,2,6,6-tetramethylpiperidide)
DBN (*see* 1,5-Diazabicyclo[4.3.0]nonene-5)
DBU (*see* 1,8-Diazabicyclo[5.4.0]undecene-7)
DCC (*see* Dicyclohexylcarbodiimide)
DDQ (*see* 2,3-Dichloro-5,6-dicyano-1,4-benzoquinone)
DEAD (*see* Diethyl azodicarboxylate)
DEAE–Cellulose ion-exchange resin, **1,** 518; **2,** 227
Dehydroabietylamine, **1,** 183; **2,** 97; **3,** 73; **6,** 575; **10,** 117
Dehydro-N,N′N″-tricyclohexylguanidinohexacarbonyldiiron(0), **4,** 114
Dessicants (*see also* Molecular sieves, Drying agents), **9,** 130
DET (*see* Diethyl tartrate)
Diacetatobis(triphenylphosphine)palladium(II) (*see also* Palladium(II) acetate–Triphenylphosphine), **6,** 156; **7,** 274; **8,** 380; **10,** 117
Diacetatobis(tri-*o*-tolylphosphine)palladium(II), **8,** 380
(E,E)-1,4-Diacetoxybutadiene, **1,** 183; **3,** 73; **7,** 85; **9,** 130
1,4-Diacetoxy-2-butene, **4,** 365
Diacetoxydimethylsilane, **3,** 113
Diacetyl chromate (*see* Chromyl acetate)
Diacetylene, **1,** 185
Diacetyl–Trimethyl phosphite adduct (*see* 2,2,2-Trimethoxy-4,5-dimethyl-1,3-dioxaphospholene)
Diacyl peroxides, **1,** 186
Dialkoxycarbenium hexafluorophosphates, **5,** 716
Dialkoxycarbenium tetrafluoroborates, **3,** 303; **4,** 114, 144; **5,** 623, 714; **11,** 175
Dialkylaluminum amides (*see also* Diethylaluminum amides, Dimethylaluminum amides, specific names), **6,** 181; **8,** 164, 182; **9,** 177; **10,** 117
Dialkylboryl trifluoromethanesulfonates (*see also* 9-Borabicyclononyl trifluoromethanesulfonate), **7,** 91; **8,** 504; **9,** 131, 140; **10,** 118; **11,** 159, 232, 233, 379; **12,** 155
Dialkylcopperlithiums (*see* Lithium dialkylcuprates)
Diallylamine, **12,** 505
Diallylcopperlithium (*see* Lithium diallylcuprate)
Diallyldibromotin, **11,** 520
Di-π-allylpalladium chloride [*see* Di-μ-chlorobis(allyl)dipalladium]
Diallyltin dibromide, **11,** 520
Diallylzinc, **12,** 54
Diaminomaleonitrile, **4,** 155; **5,** 175, 220
4,4′-Di(2-amino-6-methylpyrimidyl) disulfide, **9,** 413
2,4-Diamino-1,3-thiazole hydrotribromide, **10,** 119
Diammonium hexachloroplatinate(IV), **1,** 890
Diammonium tetrachloroplatinate(II), **3,** 81
Diamylamine (*see* Dipentylamine)

dl-Dianilino-1,2-diphenylethane, **1,** 187
1,2-Dianilinoethane, **1,** 187; **2,** 98, 430
3,3'-Dianisole bis-4,4'-(3,5-diphenyltetrazolium) chloride, **1,** 61
2,5-Di-*p*-anisyl-3,4-diphenylcyclopentadienone, **1,** 188
Di-*p*-anisylphenylmethyl chloride, **1,** 188
1,5-Diazabicyclo[4.3.0]nonene-5 (DBN), **1,** 189; **2,** 98; **4,** 16, 116, 459; **5,** 176; **6,** 157, 411; **7,** 86; **8,** 256; **12,** 156
1,4-Diazabicyclo[2.2.2]octane (DABCO), **1,** 1203; **2,** 99, 429; **4,** 119; **5,** 176, 487; **6,** 157, 278, 534; **7,** 86; **11,** 153; **12,** 155
1,4-Diazabicyclo[2.2.2]octane–Bromine complex, **2,** 99, 429; **8,** 141
1,8-Diazabicyclo[5.4.0]undecene-7 (DBU), **2,** 101; **4,** 16, 119; **5,** 177, 532; **6,** 94, 158, 411; **7,** 87; **8,** 141, 256; **9,** 132, 497; **11,** 155; **12,** 156, 479
Diazadieneiron(0) complexes, **12,** 156
Diazald (*see p*-Tolylsulfonylmethylnitrosoamide)
p-Diazanylbenzenesulfonic acid, **1,** 842
2,5-Diazido-3,6-di-*t*-butyl-1,4-benzoquinone, **4,** 55
Diazidotin dichloride, **10,** 120
Diazoacetaldehyde, **2,** 101; **3,** 73; **4,** 5, 120
Diazoacetaldehyde dimethyl acetal, **10,** 120
Diazoacetic acid azide, **4,** 120
Diazoacetone, **10,** 244; **11,** 155; **12,** 42
Diazoacetonitrile, **2,** 195
α-Diazoacetophenone, **2,** 81; **3,** 253; **6,** 473, 577; **7,** 157; **8,** 107; **9,** 20; **10,** 274
Diazoacetyl azide, **4,** 120
Diazoacetyl chloride, **9,** 133
1-Diazo-2-butanone, **4,** 100
3-Diazo-2-butanone, **4,** 168
Diazodicyanoimidazole, **5,** 178
2-Diazo-1,1-dimethoxyethane, **10,** 120
Diazoethane, **3,** 73
9-Diazofluorene, **1,** 190; **6,** 577
1-Diazolithioacetone, **11,** 155
Diazomethane, **1,** 191, 531; **2,** 102; **3,** 74, 91; **4,** 120; **5,** 155, 179, 282, 569; **6,** 159, 250, 442, 554, 577; **7,** 88; **9,** 133; **10,** 9, 79; **11,** 391; **12,** 157
Diazomethane–Alumina, **3,** 6; **9,** 135
Diazomethane–Copper(I) halide, **1,** 165; **2,** 90; **3,** 68; **5,** 591
Diazomethane–Silica or Zeolite, **9,** 135
(Diazomethyl)trimethylsilane (*see* Trimethylsilyldiazomethane)
(Diazomethyl)trimethylsilyllithium [*see* Diazo(trimethylsilyl)methyllithium]
2-Diazopropane, **2,** 105; **3,** 74; **5,** 516
Diazo(trimethylsilyl)methyllithium, **7,** 216; **11,** 573
DIBAH, DIBAL (*see* Diisobutylaluminum hydride)
Dibenzenesulfenamide, **3,** 20, 182
Dibenzo-15-crown-5, **6,** 174
Dibenzo-18-crown-6, **4,** 142; **6,** 159, 174; **8,** 129; **9,** 127
Dibenzo-18-crown-6–Bromine, **9,** 66; **11,** 156
trans-1,2-Dibenzoylethylene, **1,** 195
Dibenzoyl peroxide, **1,** 196; **4,** 122, 169; **5,** 15, 182; **6,** 160; **7,** 89; **12,** 157
Dibenzoyl peroxide–Nickel(II) bromide, **9,** 136; **10,** 121
3-Dibenzylamino-2-methyl-1-propanol, **10,** 155
Dibenzylammonium trifluoroacetate, **9,** 136, 318
Dibenzyl carboxymethylphosphonate, **5,** 183; **8,** 168
Dibenzyl chlorophosphate, **1,** 198
Dibenzyl chlorophosphonate (*see* Dibenzyl phosphorochloridate)
Dibenzyl ketone, **1,** 198; **4,** 123
Dibenzyl malonate, **1,** 198
1,3-Dibenzyl-2-methyl-1,3,2-diazaphospholidine, **2,** 105
Dibenzyloxyphosphinylacetic acid, **5,** 183; **8,** 168
Dibenzyl phosphite, **1,** 198
α-Dibenzylphosphonoacetic acid, **5,** 183; **8,** 163
Dibenzyl phosphorochloridate, **1,** 198
N,N'-Dibenzyltartardiamide, **12,** 91
Diborane (*see also* Borane–Tetrahydrofuran), **1,** 199; **2,** 71, 106, 117; **3,** 76; **4,** 124; **5,** 184; **6,** 161; **7,** 89; **8,** 141; **9,** 136; **11,** 156
Diborane–Boron trifluoride, **2,** 107
Diborane–Lithium borohydride, **2,** 108
Diborane–Sodium borohydride, **2,** 108
Dibromantin (*see* 1,3-Dibromo-5,5-dimethylhydantoin)
1,3-Dibromoacetone, **5,** 240
Dibromoacetonitrile, **5,** 186
Dibromoalane, **12,** 192, 377
1,1-Dibromoalkyllithium reagents, **5,** 398,

403; **6,** 162; **7,** 90; **9,** 138; **11,** 104, 158
Di-μ-bromobis(allyl)dinickel (*see also* π-Allylnickel halides), **3,** 210; **4,** 354
Dibromobis(cyclopentadienyl)zirconium, **12,** 53
Di-μ-bromobis[(2-ethoxycarbonyl)-2-propenyl]dinickel, **4,** 33; **6,** 93
Di-μ-bromobis(2-methallyl)dinickel (*see also* π-Allylnickel halides), **4,** 353; **6,** 361; **7,** 250
Di-μ-bromobis(2-methoxyallyl)dinickel, **5,** 437; **8,** 324
Di-μ-bromobis(3-methyl-2-butenyl)dinickel (*see also* π-Allylnickel halides), **2,** 291; **4,** 355; **8,** 181
Dibromobis(tributylphosphine)nickel(II), **5,** 42
Dibromobis(tributylphosphine)nickel(II)–Sodium borohydride, **5,** 42
Dibromobis(triethylphosphine)nickel(II), **6,** 57
Dibromobis(triphenylphosphine)cobalt(II), **3,** 89; **4,** 126
Dibromobis(triphenylphosphine)nickel(II), **6,** 58
Dibromobis(triphenylphosphine)nickel(II)–Zinc, **11,** 157
Dibromobis(triphenylphosphine)-palladium(II), **6,** 60
Dibromoborane–Dimethyl sulfide, **8,** 143; **11,** 262
1,4-Dibromobutane, **5,** 404
Dibromocarbene, **1,** 96, 606, 686, 758, 852, 916; **2,** 18, 30, 276, 379; **3,** 174, 234; **4,** 438; **5,** 26, 450; **6,** 42, 68, 136, 220, 468, 601; **8,** 343, 388; **11,** 144; **12,** 517
Dibromochloromethane, **8,** 129
Dibromochloromethyl(phenyl)mercury, **1,** 851; **3,** 222
1,1-Dibromocyclobutane, **12,** 150
3,4-Dibromo-1-cyclobutene, **3,** 335
2,6-Dibromo-1-cyclohexanone, **5,** 223, 240
Dibromodifluoromethane, **1,** 81, 207; **12,** 158
1,3-Dibromo-1,1-difluoropropane, **12,** 188
5,5-Dibromo-2,2-dimethyl-1,3-dioxane-4,6-dione, **8,** 143
1,3-Dibromo-5,5-dimethylhydantoin, **1,** 208, **2,** 108; **12,** 158
2,4-Dibromo-2,4-dimethyl-3-pentanone, **4,** 158; **5,** 222, 759
Dibromodiphenylmethane, **1,** 520
1,2-Dibromoethane, **1,** 374; **4,** 126; **5,** 187; **6,** 42
1,1-Dibromo-2-ethoxycyclopropane, **9,** 67
1,2-Dibromoethylene, **6,** 341
Dibromofluoromethane, **3,** 77; **4,** 29
3,5-Dibromo-4-heptanone, **5,** 221
Dibromoisocyanuric acid, **2,** 109; **10,** 122
Dibromoketene, **2,** 462
Dibromomalonamide, **1,** 208
Dibromomethane, **2,** 273; **3,** 189; **4,** 442; **5,** 398, 403; **6,** 10, 162, 274; **7,** 4, 90; **8,** 387; **11,** 157, 158
Dibromomethane–Cesium or Potassium fluoride, **7,** 239
Dibromomethane–Lithium, **3,** 201
Dibromomethane–Zinc-copper couple (*see* Simmons–Smith reagent)
Dibromomethane–Zinc–Titanium(IV) chloride, **8,** 339; **11,** 337; **12,** 322
Dibromomethoxycarbonylmethyl(phenyl) mercury, **3,** 222
(2-Dibromomethyl)benzoyl chloride, **9,** 138
(Dibromomethylene)cyclobutane, **12,** 186
Dibromo-μ-methylenedimagnesium, **2,** 273; **3,** 189, 201
Dibromomethylenetriphenylphosphorane, **6,** 211
Dibromomethyllithium (*see also* 1,1-Dibromoalkyllithium reagents), **5,** 398, 403; **6,** 162; **7,** 90; **11,** 104, 158
2,4-Dibromo-3-pentanone, **4,** 158; **5,** 222
1,3-Dibromo-2-pentene, **7,** 90
1,4-Dibromo-2-pentene, **9,** 358
2,2-Dibromopropane, **11,** 102
2,2-Dibromopropane-1,3-diol, **7,** 91
2,3-Dibromopropene, **1,** 420
2,6-Dibromopyridine, **4,** 490; **5,** 63
1,2-Dibromotetrachloroethane, **1,** 209
Dibromotriphenylantimony, **9,** 94
Dibromotriphenylphosphorane (*see* Triphenylphosphine dibromide)
Dibromotris(triphenylphosphine)-ruthenium(II), **11,** 171
Di-*t*-butoxyacetylene, **8,** 144
1,3-Dibutoxy-1-lithio-1-propene, **9,** 139
2,2-Dibutoxypropane, **1,** 209
1,3-Dibutoxypropene, **9,** 139
1,3-Dibutoxy-1-propenyllithium, **9,** 139
Di-*t*-butyl acetamidomalonate, **1,** 523
Dibutylamine, **1,** 209
Di-*t*-butyl azodiformate, **1,** 209

3,5-Di-*t*-butyl-1,2-benzoquinone, **3,** 78
2,6-Di-*t*-butyl-*p*-benzoquinone, **1,** 218; **9,** 139
Dibutylboron chloride, **5,** 201
Dibutylboryl trifluoromethanesulfonate, **7,** 91; **9,** 131, 140; **10,** 119; **11,** 159, 379; **12,** 155
Di-*t*-butyl bromomalonate, **8,** 213
Di-*t*-butyl bromophosphate, **7,** 20; **8,** 146
Di-*t*-butyl(*t*-butylthio)borane, **9,** 132
Dibutylchloroborane, **5,** 201
Di-*t*-butyl chlorophosphate, **7,** 20
Di-*n*- (or *t*-)butylcopperlithium [*see* Lithium di-*n*- (or *t*-)butylcuprate]
2,6-Di-*t*-butyl-*p*-cresol, **5,** 487
Di-*t*-butyl dicarbonate, **4,** 128; **7,** 91; **8,** 145; **10,** 122; **12,** 159
Di-*t*-butyldichlorosilane, **11,** 159
Dibutyldichlorotin, **1,** 213, 423; **11,** 161
2,6-Di-*t*-butyl-α-dimethylamino-*p*-cresol (Ethyl antioxidant 703), **4,** 128
Di-*t*-butyl diperoxycarbonate, **6,** 166
Dibutyldivinyltin, **1,** 213
Dibutyl(ethoxycarbonylmethyl)telluronium bromide, **12,** 159
2,6-Di-*t*-butyl-4-hydroxymethylphenol (Ethyl antioxidant 754), **4,** 257
Di-*t*-butyl iminodicarboxylate, **1,** 210
Dibutylmagnesium, **8,** 253
Di-*t*-butyl malonate, **1,** 210, 523
2,6-Dibutyl-4-methoxyphenyl propionate, **10,** 162
(2,6-Di-*t*-butyl-4-methylphenoxy)-methylaluminum trifluoromethane-sulfonate, **11,** 159
2,6-Di-*t*-butyl-4-methylpyridine, **6,** 406; **8,** 145; **9,** 141, 485; **10,** 123; **11,** 160
Di-*t*-butylmethylsilyl perchlorate, **9,** 78
2,6-Di-*t*-butyl-4-nitrophenol, **10,** 182
Di-*t*-butyl nitroxide, **1,** 211; **3,** 81; **4,** 129; **11,** 160
Di-*t*-butyl peroxide, **1,** 211, 280; **11,** 131; **12,** 275
Di-*t*-butylperoxide–Titanium(IV) chloride, **12,** 94
Di-*t*-butylperoxyoxalate, **6,** 166; **7,** 93
Di(2-*t*-butylphenyl) phosphorochloridate, **6,** 167
Dibutyl phosphite, **7,** 353
Di-*t*-butyl phosphorobromidate, **7,** 20; **8,** 146
Di-*t*-butyl phosphorochloridate, **7,** 20
N,N-Dibutylpiperidinium iodide, **6,** 601
2,6-Di-*t*-butylpyridine, **1,** 212; **2,** 109; **6,** 406; **10,** 282, 440
Di-*t*-butylsilyl bis(trifluoromethane-sulfonate), **11,** 189
Di-*t*-butyl succinate, **1,** 212
Dibutyltelluronium carboethoxymethylide, **12,** 159
Dibutyl (tetrahydropyranyloxymethyl)-phosphonate, **9,** 163
Dibutyltin dichloride, **1,** 213, 423; **11,** 161
Dibutyltin dilaurate, **7,** 346
Dibutyltin oxide, **5,** 188; **9,** 141; **10,** 123; **11,** 501; **12,** 160
Dibutylzinc, **5,** 753
Dicarbonylbis(triphenylphosphine)nickel, **1,** 61; **4,** 40; **7,** 94; **8,** 147, 498; **10,** 125; **11,** 80
Dicarbonylbis(triphenyl phosphite)nickel, **1,** 18
Dicarbonylcyclopentadienylcobalt, **5,** 172; **6,** 153; **7,** 84, 94; **8,** 146; **9,** 142; **10,** 126; **11,** 136, 161; **12,** 160, 180
Dicarbonyl(cyclopentadienyl)[(dimethyl-sulfonium)methyl]iron tetrafluoroborate, **9,** 143
Dicarbonylcyclopentadienylironethyl-(methylphenyl)sulfonium fluoro-sulfonate, **10,** 128
Dicarbonylcyclopentadienylironmethyl-dimethylsulfonium tetrafluoroborate, **9,** 143
Dicarbonyl(cyclopentadienyl)[(methylphenyl-sulfonium)ethyl]iron trifluoromethane-sulfonate, **10,** 128
Dicarbonylcyclopentadienyl(2-methyl-1-propene)iron tetrafluoroborate, **6,** 154
Di-μ-carbonyldecacarbonyltri-*triangulo*-iron, **4,** 534; **5,** 357, 706; **7,** 144; **8,** 130, 390
Di-μ-carbonyldicarbonylbis(cyclo-pentadienyl)diiron, **8,** 41, 390; **10,** 455
Dicarbonyldichlorobis(triphenylphosphine)-ruthenium(II), **4,** 421; **5,** 193
Di-μ-carbonylhexacarbonyldicobalt, **1,** 154, 224; **3,** 89; **4,** 139; **5,** 204; **6,** 172; **7,** 68, 99; **8,** 148; **9,** 144, 286; **10,** 129; **11,** 162; **12,** 163
Dichloramine, **1,** 213
1,3-Dichloroacetone, **2,** 109
Dichloroacetonitrile, **3,** 236
Dichloroacetyl chloride, **1,** 222; **3,** 87
Dichloroacetylene, **11,** 552

Dichloroalane, **1,** 595; **2,** 243; **12,** 192, 333
Di-μ-chloro-π-allyldipalladium [*see* Di-μ-chlorobis(allyl)dipalladium]
gem-Dichloroallyllithium, **5,** 188; **8,** 150; **9,** 145
(3,3-Dichloroallyl)triphenyllead, **5,** 188
2,2-Dichloro-1,3-benzodioxole, **1,** 119; **4,** 70
α,α-Dichlorobenzyllithium, **1,** 1140
Di-μ-chlorobis(allyl)dipalladium, **2,** 109; **5,** 189; **6,** 394; **7,** 5; **9,** 132
Dichlorobis(benzonitrile)palladium(II) [*see* Bis(benzonitrile)dichloropalladium(II)]
Dichlorobis(crotyl)dipalladium, **12,** 371
Dichlorobis(cyclohexene)palladium(II), **5,** 502
Di-μ-chlorobis(1,5-cyclooctadiene)diiridium, **5,** 113
Dichlorobis(1,5-cyclooctadiene)dirhodium–N-Acryloyl-4-(diphenylphosphine)-2-[(diphenylphosphine)methyl]pyrrolidine, **11,** 6
Dichlorobis(1,5-cyclooctadiene)dirhodium–2,3-Bis(diphenylphosphine)bicyclo-[2.2.1]-hept-5-ene, **9,** 49
Dichlorobis(1,5-cyclooctadiene)dirhodium–2,3-O-Isopropylidene-2,3-dihydroxy-1,4-bis(diphenylphosphine)butane, **9,** 259
Dichlorobis(cyclopentadienyl)titanium, **6,** 48, 179; **9,** 146; **10,** 130; **11,** 163, 174; **12,** 168
Dichlorobis(cyclopentadienyl)titanium–Bis(dialkylamino)alanes, **9,** 46
Dichlorobis(cyclopentadienyl)titanium–Dialkylaluminum chloride, **9,** 471
Dichlorobis(cyclopentadienyl)titanium–Lithium borohydride, **9,** 146
Dichlorobis(cyclopentadienyl)titanium–Magnesium, **6,** 596
Dichlorobis(cyclopentadienyl)titanium–Trialkylaluminums, **8,** 507; **9,** 146, 471, 488
Dichlorobis(cyclopentadienyl)zirconium(II), **8,** 86; **10,** 131; **11,** 51, 52; **12,** 53, 297
Dichlorobis(cyclopentadienyl)zirconium–*t*-Butylmagnesium chloride, **12,** 171
Dichloro[1,2-bis(diphenylphosphine)ethane]-nickel(II), **7,** 250; **9,** 147; **10,** 36; **12,** 146, 171
Dichloro[1,1′-bis(diphenylphosphino)ferrocene]palladium(II), **9,** 147; **10,** 37; **12,** 46, 171
Dichloro[1,3-bis(diphenylphosphine)-propane]nickel(II), **6,** 50; **10,** 191; **11,** 167; **12,** 244
Di-μ-chlorobis(1,5-hexadiene)dirhodium, **12,** 172
Di-μ-chlorobis(1,5-hexadiene)dirhodium–N-Acryloyl-4-(diphenylphosphine)-2-[(diphenylphosphine)methyl]pyrrolidine, **11,** 6
Di-μ-chlorobis(1,5-hexadiene)dirhodium–(2S,4S)-N-*t*-Butoxycarbonyl-4-(diphenylphosphine)-2-[(diphenylphosphine)methyl]pyrrolidine, **8,** 58
Di-μ-chlorobis(1,5-hexadiene)dirhodium–(S)-α-(R)-2-Diphenylphosphine-ferrocenyl ethyldimethylamine, **9,** 200
Dichlorobis(isopropoxy)titanium(IV), **11,** 164; **12,** 104
Dichlorobis(methyldiphenylphosphine)-palladium(II), **11,** 165
Dichlorobis(norbornadiene)dirhodium(I), **9,** 108
Dichlorobis(norbornadiene)dirhodium–1,2-Ethanediylbis[(*o*-methoxyphenyl)-phenylphosphine], **12,** 364
Dichlorobis(norbornadiene)dirhodium–Methyl 2,3-bis-O-diphenylphosphino-4,6-O-benzylidene-α-glucopyranoside, **8,** 333
Dichlorobis(tributylphosphine)nickel(II), **5,** 42
Dichlorobis(trifluoromethanesulfonato)-titanium(IV), **12,** 173
Dichlorobis(triphenylphosphine)nickel(II), **2,** 110; **6,** 59; **8,** 523; **9,** 147; **10,** 190; **11,** 165
Dichlorobis(triphenylphosphine)palladium(II), **3,** 81; **5,** 190; **6,** 59, 60; **7,** 95; **8,** 151; **10,** 133; **11,** 511; **12,** 525
Dichlorobis(triphenylphosphine)palladium(II)–Copper(I) iodide, **12,** 475
Dichlorobis(triphenylphosphine)palladium(II)–Diisobutylaluminum hydride, **7,** 95
Dichlorobis(triphenylphosphine)palladium(II)–Zinc, **11,** 514
Dichlorobis(triphenylphosphine)platinum(II)–Tin(II) chloride, **9,** 148
Dichlorobis(tri-*o*-tolylphosphine)palladium(II), **12,** 173, 516
Dichloroborane, **4,** 129
Dichloroborane diethyl etherate, **4,** 130; **5,** 191; **7,** 96
Dichloroborane–Dimethyl sulfide, **8,** 354

1,1-Dichloro-3-bromopropene, **5,** 191
1,4-Dichloro-2-butanone, **9,** 75
1,3-Dichloro-2-butene, **1,** 214, 1096; **2,** 111; **3,** 179; **4,** 472; **7,** 203; **8,** 152; **9,** 294; **10,** 134; **12,** 472
1,4-Dichloro-2-butene, **7,** 88, 148; **10,** 134
1,1-Dichloro-1-buten-4-ol, **5,** 192
Dichlorocarbene, **1,** 95, 224, 386, 425, 605, 844, 852, 1107; **2,** 18, 30, 197, 326, 365, 380, 388, 411; **3,** 45, 85, 143; **4,** 27, 114, 130, 233, 385, 402; **5,** 27, 332, 514; **6,** 38, 41, 136, 159, 219, 349, 553, 601, 621; **7,** 18, 282, 341; **8,** 144, 388; **9,** 360; **11,** 406; **12,** 151, 517
3,4-Dichlorocyclobutene, **2,** 112, 140
Dichlorocyclohexylborane, **4,** 377
α,β-Dichloro-N-cyclohexylpropanaldonitrone, **5,** 111
Dichloro(1,5-cyclooctadiene)palladium(II), **12,** 48
Dichlorodicarbonylbis(triphenylphosphine)-ruthenium(II), **4,** 421; **5,** 193
Di-μ-chlorodichlorobis(pentamethylcyclo-pentadienyl)diiridium or -rhodium, **8,** 152
2,3-Dichloro-5,6-dicyano-1,4-benzoquinone (DDQ), **1,** 127, 128, 215; **2,** 112; **3,** 83; **4,** 130; **5,** 193; **6,** 168, 511; **7,** 96; **8,** 153, 442, 512, 519; **9,** 148; **10,** 135; **11,** 166; **12,** 174
Dichloro(diethoxyphosphinyl)methyllithium, **6,** 188
Dichloro(diethylamino)borane, **9,** 63, 152
1,1-Dichloro-2,2-difluoroethylene, **1,** 220; **2,** 117
Dichlorodiisopropoxytitanium(IV), **11,** 164; **12,** 104
Di-μ-chlorodimethoxybis(pyridine)dicopper, **10,** 136
1,4-Dichloro-1,4-dimethoxybutane, **12,** 175
1,1-Dichlorodimethyl ether (*see* Dichloromethyl methyl ether)
Dichloro(dimethylformamido)bis(pyridine)-rhodium borohydride, **4,** 40
Dichlorodimethylmolybdenum, **12,** 353
Dichlorodimethylsilane, **3,** 114; **4,** 183; **12,** 220
Dichlorodimethylsilane–Magnesium amalgam, **4,** 183
Dichlorodimethyltitanium, **10,** 138, 270
4,4′-Dichlorodiphenyl diselenide, **6,** 421
Dichlorodiphenylmethane, **1,** 520
1,1-Dichloroethane, **6,** 530; **10,** 90; **11,** 121
5,7-Dichloro-N-ethylbenzisoxazolium tetrafluoroborate, **2,** 192
1,1-Dichloroethylene, **1,** 220; **2,** 118
1,2-Dichloroethylene, **6,** 294, 341; **12,** 175
(−)-*trans*-Dichloro(ethylene)[α-methyl-benzylamine]platinum(II), **2,** 272; **3,** 199
1,1-Dichloroethyl ethyl ether, **1,** 219
Dichlorofluoromethane, **6,** 175, 511
Dichloroformaldoxime, **1,** 221
Dichloro[2,3-O-isopropylidene-2,3-dihydroxy-1,4-bis(diphenylphosphine)-butane]nickel(II), **5,** 360; **11,** 563
Dichloroketene, **1,** 221, 1287; **2,** 118; **3,** 87; **4,** 134; **8,** 156; **9,** 103, 152; **10,** 139; **11,** 168; **12,** 176
Dichloromaleic anhydride, **4,** 135; **5,** 195, 439
Dichloromethane (*see* Methylene chloride)
Dichloromethaneboronic acid, **11,** 86; **12,** 80, 397
[Dichloro(methoxycarbonyl)methyl]-phenylmercury, **3,** 222
1,3-Dichloro-3-methylbutane, **9,** 51
Dichloromethyl 2-chloroethyl ether, **2,** 118
(Dichloromethyl)dihydroxyborane (*see* Dichloromethaneboronic acid)
2-Dichloromethyl-4,5-dimethyl-1,3,2-dioxaborolane, **12,** 80
Dichloromethylenedimethylammonium chloride, **4,** 135; **5,** 195; **6,** 170; **8,** 156
Dichloromethylenetriphenylphosphorane, **1,** 222, 1239; **7,** 97
Dichloromethylenetris(dimethylamino)-phosphorane, **8,** 157
Dichloromethyllithium, **1,** 223; **2,** 119; **3,** 89; **4,** 138; **5,** 199, 398; **6,** 170, 344; **9,** 154
Dichloromethyl methyl ether, **1,** 220; **2,** 120; **5,** 200, 671; **6,** 171; **7,** 98, 249; **8,** 485; **9,** 154
Dichloromethyl methyl sulfide, **5,** 194; **6,** 136
Dichloromethyl phenyl sulfone, **6,** 41
1,3-Dichloro-3-methyl-2-trimethyl-silyloxy-1-butene, **11,** 470
Dichlorophenylborane, **4,** 377; **12,** 178
Dichlorophenylcarbenium hexachloroantimonate, **5,** 116
(Z)-1,2-Dichloro-4-phenylthio-2-butene, **7,** 98
2,3-Dichloropropanal, **5,** 111
1,3-Dichloro-1-propene, **6,** 129; **7,** 326
2,3-Dichloro-1-propene, **2,** 120; **6,** 332; **7,** 326; **8,** 158; **12,** 472

3,3-Dichloro-1-propene, **6,** 129; **9,** 145
Dichlorosilane, **4,** 139
Di-μ-chlorotetra(ethylene)dirhodium, **5,** 203
Di-μ-chlorotetra(ethylene)dirhodium–2,3-O-Isopropylidene-2,3-dihydroxy-1,4-bis(diphenylphosphine)butane, **4,** 273
Di-μ-chlorotetrakis(cyclooctene)diiridium, **5,** 113
Di-μ-chlorotetrakis(cyclooctene)dirhodium–2,3-O-Isopropylidene-2,3-dihydroxy-1,4-bis(diphenylphosphine)butane, **4,** 273; **6,** 309
Dichlorotetrakis(triphenylphosphine)-ruthenium(II), **5,** 741
1,4-Dichloro-1,1,4,4-tetramethyl-disilylethylene [*see* Bis(chloro-dimethylsilyl)ethane]
2,5-Dichlorothiophenium bismethoxy-carbonylmethylide, **8,** 159
Dichlorotitanium bis(isopropoxide) [*see* Dichlorodiisopropoxytitanium(IV)]
Dichlorotriphenylphosphorane (*see* Triphenylphosphine dichloride)
Dichlorotris(dimethylamino)phosphorane, **11,** 170
Dichlorotris(triphenylphosphine)ruthenium(II), **2,** 121; **4,** 562, 564; **5,** 740; **6,** 654; **7,** 99, 244; **8,** 159; **9,** 491; **10,** 79, 141, 274; **11,** 47, 171, 223, 270, 448; **12,** 63, 88, 179
N,N-Dichlorourethane, **2,** 121; **4,** 139; **8,** 161; **9,** 156
Dichlorovinylene carbonate, **2,** 122; **6,** 172
Dicobalt octacarbonyl (Di-μ-carbonylhexa-carbonyldicobalt), **7,** 99
Di(cobalttetracarbonyl)zinc (*see* Zinc tetracarbonylcobaltate)
Dicyandiamide, **1,** 229
Dicyanoacetylene, **1,** 230; **2,** 21, 124; **4,** 140, 547; **5,** 205
9,10-Dicyanoanthracene, **9,** 336
Dicyanobis(triphenylphosphine)nickel(0), **4,** 40
2,3-Dicyano-1,3-butadiene, **6,** 173; **7,** 100
Dicyanocarbene, **1,** 1202
1,2-Dicyanocyclobutene, **6,** 172
Dicyanodiazomethane, **1,** 561; **2,** 125
Dicyanodimethylsilane, **8,** 161
9-Dicyanomethylene-2,4,7-trinitrofluorene, **1,** 231; **5,** 206
(Dicyclohexylamido)copper(I), **11,** 368
Dicyclohexylamine, **1,** 231
Dicyclohexylborane, **1,** 205; **3,** 90; **4,** 141; **5,** 39; **6,** 62; **7,** 82; **8,** 162; **11,** 172, 269
Dicyclohexylcarbodiimide (DCC), **1,** 231, 517; **2,** 126; **3,** 91; **4,** 141; **5,** 22, 206, 532; **6,** 174, 291; **7,** 100; **8,** 162; **9,** 156; **10,** 142; **11,** 125, 173
Dicyclohexylcarbodiimide–Copper(I) chloride, **9,** 184
Dicyclohexylcarbodiimide–4-Dimethyl-aminopyridine, **8,** 163; **9,** 156, 179; **11,** 360
Dicyclohexylcarbodiimide–4-Pyrrolidino-pyridine, **9,** 156
Dicyclohexyl-18-crown-6, **4,** 142; **5,** 153, 207; **6,** 174, 405; **7,** 77
Dicyclohexylethylamine, **1,** 370, 1058; **2,** 195; **3,** 325
Dicyclohexyl(methylthio)borane, **7,** 101
N,N′-Dicyclohexyl-4-morpholino-carboxamidine, **1,** 236; **2,** 126
Di(cyclopentadienyl)(chloro)hydrido-zirconium(IV) [*see* Chlorobis(cyclo-pentadienyl)hydridozirconium(IV)]
Dicyclopentadienylcobalt (Cobaltocene), **5,** 145; **6,** 128; **12,** 180
Dicyclopentadienylcobalt–Oxygen adduct, **5,** 145
Dicyclopentadienyl(isoprene)zirconium, **11,** 174
Dicyclopentadienyltitanium, **5,** 672; **6,** 596
η^5-Dicyclopentadienyltitanium dichloride [*see* Dichlorobis(cyclopentadienyl)-titanium]
Dicyclopentadienyl(1-trimethylsilylallyl)-titanium, **11,** 174
Dicyclopentadienylvanadium, **5,** 744
Dicyclopentylboryl trifluoromethane-sulfonate, **9,** 131; **10,** 118
Diethoxyaluminum chloride, **6,** 180
2,5-Diethoxy-1,4-benzoquinone, **8,** 163
1,2-Diethoxy-1,2-bis(trimethylsilyloxy)-ethylene, **12,** 181
Diethoxycarbenium tetrafluoroborate (*see also* Dialkoxycarbenium tetrafluoroborates), **4,** 144; **11,** 175
1,3-Diethoxycarbonylallene, **8,** 414
3,4-Diethoxycarbonyl-2,5-dimethylfuran, **5,** 359
5,5-Diethoxycyclopentadiene, **2,** 94
1,1-Diethoxy-N,N-dimethylethylamine (*see* N,N-Dimethylacetamide dialkyl acetals)

1,2-Diethoxy-1,2-disilyloxyethylene, **12,** 181
2,2-Diethoxyethanethiol, **1,** 1105
1,1-Diethoxyethylene (*see* Ketene diethyl acetal)
2,2-Diethoxyethylidenetriphenylphosphorane, **9,** 157
3,3-Diethoxy-2-lithio-1-propene (*see* 3,3-Diethoxy-1-propenyl-2-lithium)
3,3-Diethoxy-1-lithio-1-propyne, **5,** 207
Diethoxymethane, **11,** 241
Diethoxymethyl acetate, **1,** 243
(Diethoxymethyl)diphenylphosphine oxide, **12,** 181
Diethoxymethylsilane, **12,** 182
3,3-Diethoxy-1-(methylthio)propyne, **5,** 207
Diethoxyphosphinylacetaldehyde, **2,** 131
Diethoxyphosphinylacetaldehyde diethyl acetal, **2,** 131
Diethoxyphosphinylacetic acid, **8,** 168; **11,** 281
Diethoxyphosphinylacetic acid, anhydride with trifluoroacetic acid, **11,** 281
1-Diethoxyphosphinyl-2-butanone, **12,** 327
(Diethoxyphosphinyl)difluoromethyllithium, **11,** 180
2-Diethoxyphosphinyl-1,3-dithiane, **8,** 89; **11,** 179
2-(Diethoxyphosphinyl)propionic acid, **11,** 560
2-(Diethoxyphosphinyl)propionic acid, anhydride with trichloroacetic acid, **11,** 560
2-(Diethoxyphosphinyl)propionitrile, **7,** 106
2,2-Diethoxypropane (Acetone diethyl acetal), **1,** 244; **3,** 120; **5,** 208
1,1-Diethoxy-1-propene, **1,** 685
3,3-Diethoxy-1-propene (Acrolein diethyl acetal), **6,** 270; **9,** 158
3,3-Diethoxy-1-propenyl-2-copper, **7,** 213
3,3-Diethoxy-1-propenyl-2-lithium, **5,** 131; **6,** 330; **7,** 212; **8,** 300
3,3-Diethoxy-1-propyne, **2,** 126; **3,** 93; **7,** 80
3,3-Diethoxy-1-propynyllithium, **5,** 207
3,3-Diethoxypropynylmagnesium bromide, **2,** 127; **3,** 93
2,2-Diethoxyvinylidenetriphenylphosphorane, **3,** 93; **5,** 209; **7,** 102; **11,** 175
N,N-Diethylacetamide, **9,** 220
Diethyl acetylenedicarboxylate, **1,** 244; **2,** 127
Diethyl acetylmethylmalonate, **7,** 102
Diethyl acetylmethylphosphonate, **12,** 527
Diethyl acetylphosphonate, **11,** 178
Diethyl allylthiomethylphosphonate, **3,** 94
Diethylaluminum amides (*see also* Dialkylaluminum amides, Dimethylaluminum amides), **6,** 181; **8,** 164, 165; **10,** 117, 118
Diethylaluminum anilide, **10,** 118
Diethylaluminum benzeneacetylide, **11,** 364
Diethylaluminum benzenethiolate, **10,** 281; **11,** 542; **12,** 343
Diethylaluminum chloride, **4,** 144; **8,** 164; **10,** 178; **11,** 7; **12,** 5, 512
Diethylaluminum cyanide (*see also* Organoaluminum cyanides), **1,** 244; **2,** 127; **4,** 146; **6,** 180; **8,** 130; **9,** 158, 328; **12,** 182
Diethylaluminum ethoxyacetylide, **7,** 103
Diethylaluminum iodide, **1,** 245; **12,** 5
Diethylaluminum N-methylanilide, **8,** 165
Diethylaluminum phenylacetylide, **4,** 146
Diethylaluminum propylamide, **10,** 118
Diethylaluminum 2,2,6,6-tetramethylpiperidide (DATMP), **6,** 181; **8,** 164
Diethylaluminum tributylplumbate, **12,** 340
Diethylaluminum tributyltin, **12,** 339
Diethylaluminum trimethylsilylacetylide, **8,** 42
Diethylamidosulfurous acid, methyl ester, **1,** 1123; **2,** 392
Diethylamine, **1,** 388, 397, 1127; **4,** 148; **5,** 446, 469
N,N-Diethylaminoacetonitrile, **9,** 159
4-Diethylamino-2-butanone, **1,** 698
(Diethylamino)dimethyloxosulfonium tetrafluoroborate, **5,** 210
Diethylaminomagnesium bromide, **5,** 209
(N,N-Diethylamino)methyloxosulfonium methylide, **5,** 210
1-Diethylamino-3-pentanone, **1,** 388
Diethylaminophenylacetylene, **5,** 684
2-Diethylamino-4-phenylsulfonyl-2-butenenitrile, **12,** 182
2-Diethylamino-4-phenylthio-2-butenenitrile, **11,** 176
(Diethylamino)propyne, **2,** 133; **3,** 98; **4,** 225; **5,** 217, 477; **7,** 107; **8,** 165; **9,** 164; **12,** 183
(Diethylamino)sulfur trifluoride (DAST), **6,** 183; **8,** 166; **9,** 149, 224; **10,** 142; **11,** 177; **12,** 183
Diethylamino(trimethyl)silane, **3,** 317; **4,** 544; **6,** 184, 634
Diethylaniline, **5,** 212

Diethyl azodicarboxylate (DEAD), **1,** 245; **2,** 128; **4,** 148; **5,** 212; **6,** 185; **9,** 160; **11,** 177
Diethylbenzeneselenenamide, **11,** 178
Diethyl benzoylphosphonate, **11,** 178
Diethyl N-benzylideneaminomethylphosphonate, **9,** 161; **11,** 178; **12,** 185
Diethyl benzylphosphonate, **1,** 1093, 1212; **2,** 432; **10,** 111
Diethylboron bromide, **11,** 261
Diethylboryl trifluoromethanesulfonate, **12,** 155
Diethyl bromophosphate, **8,** 170
Diethyl(*t*-butoxycarbonylmethyl)alane, **7,** 103
Diethyl *t*-butoxy(cyano)methylphosphonate, **8,** 167
Diethylcarbo-*t*-butoxymethylaluminum, **7,** 103
Diethyl 3-carboethoxy-2-oxopropylphosphonate, **10,** 181
Diethyl carboethoxyphosphonate, **1,** 370
Diethyl carbonate, **1,** 247, 420, 1069; **2,** 129; **3,** 95, 232; **5,** 213, 234, 294
Diethyl carboxychloromethylphosphonate, **8,** 168
Diethyl 1-carboxyethylphosphonate, **11,** 560
Diethyl carboxymethylphosphonate, **8,** 168; **11,** 281
Diethyl chloromethylphosphonate, **6,** 188
Diethyl chlorophosphate (*see* Diethyl phosphorochloridate)
Diethyl chlorophosphite, **1,** 248
Diethyl chlorophosphonite (*see* Diethyl phosphorochloridite)
Diethyl(2-chloro-1,1,2-trifluoroethyl)amine, **1,** 249; **2,** 130; **3,** 95; **4,** 149; **5,** 214, 436; **6,** 66, 186; **9,** 224
Diethylcopperlithium (*see* Lithium diethylcuprate)
Diethylcyanamide, **1,** 249
Diethyl cyanomethylphosphonate, **1,** 250; **2,** 130; **3,** 96; **4,** 150; **6,** 566
Diethyl β-(cyclohexylimino)ethylphosphonate, **2,** 131; **3,** 96
Diethyl diazomalonate, **2,** 368; **5,** 244; **6,** 631
Diethyl (diazomethyl)phosphonate, **12,** 186
Diethyl dibromomalonate, **8,** 169
Diethyl dibromomethylphosphonate, **7,** 104
Diethyl dibromophosphoramidate, **7,** 103
Diethyl 2,2-dichloro-1-ethoxyethenyl phosphate, **9,** 160
Diethyl dichlorolithiomethylphosphonate, **6,** 188
Diethyl dichloromethylphosphonate, **6,** 188
Diethyl 2,2-diethoxyethylphosphonate, **2,** 131
Diethyl diethoxyphosphinylsuccinate, **1,** 1138
Diethyl difluoromethylphosphonate, **11,** 180
Diethyl difluorotrimethylsilylmethylphosphonate, **11,** 180
Diethyl[dimethyl(phenyl)silyl]aluminum, **12,** 342
Diethyl (1,3-dithian-2-yl)phosphonate, **8,** 89; **11,** 179
S,S′-Diethyl dithiomalonate, **9,** 160
Diethylene glycol dimethyl ether, **1,** 255
Diethylene orthocarbonate, **6,** 186
Diethylene tetrathioorthocarbonate, **6,** 187
N,N-(Diethylethanamine)trihydridoaluminum, **6,** 607
Diethyl ethoxycarbonyl(iodo)methylphosphonate, **1,** 1218
Diethyl ethoxycarbonylmethylphosphonate (*see* Triethyl phosphonoacetate)
Diethyl(ethoxyethynyl)aluminum, **7,** 103
Diethyl ethoxymethylenemalonate, **5,** 216
Diethyl ethylenepyrophosphite (*see* Ethyl ethylene pyrophosphite)
Diethyl ethylidenemalonate, **1,** 1252; **9,** 161
Diethyl ethylmalonate, **1,** 197
Diethyl ethylthiomethylphosphonate, **3,** 97
Diethylformamide diethyl acetal, **8,** 169
Diethyl formylmethylphosphonate, **2,** 131
Diethyl hydrogen phosphate, **1,** 208
Diethyl hydrogen phosphite (*see* Diethyl phosphite)
O,O-Diethyl phosphoroselenoate, **8,** 169
N,N-Diethylhydroxylamine, **6,** 187
Diethyl(iodomethyl)aluminum, **1,** 245
Diethyl isocyanomethylphosphonate, **4,** 271; **6,** 187
Diethylketene, **9,** 85
Diethyl ketomalonate (*see* Diethyl oxomalonate)
Diethyl lithio-N-benzylideneaminomethylphosphonate, **9,** 161; **11,** 179; **12,** 185
Diethyl lithiomorpholinomethylphosphonate, **7,** 105
Diethyl lithiosuccinate, **8,** 296
Diethylmagnesium, **8,** 253
Diethyl maleate, **1,** 197
Diethyl malonate, **1,** 627, 887, 1069, 1251, 1268; **4,** 369, 467, 507; **5,** 157, 500, 501; **6,**

42, 418, 449; **7,** 277
Diethyl methoxyethoxymethylphosphonate, **9,** 162
Diethyl methoxymethylphosphonate, **9,** 162
Diethyl methylenemalonate, **3,** 96
Diethyl methylmalonate, **5,** 216
Diethylmethyl(methylsulfonyl)ammonium fluorosulfonate (Easy Mesyl), **6,** 190
Diethylmethylsilane, **9,** 145; **12,** 166
Diethyl methylsulfinylmethylphosphonate, **6,** 193
Diethyl methylsulfonylmethylphoshonate, **4,** 61
Diethyl (1-methylthio)ethylphosphonate, **4,** 151
Diethyl methylthiomethylphosphonate, **3,** 97; **4,** 151
Diethyl morpholinomethylphosphonate, **7,** 105
Diethyl oxalate, **1,** 250, 929, 1067, 1268; **2,** 132; **6,** 597; **7,** 106; **11,** 247; **12,** 34
Diethyl 2-oxobutylphosphonate, **12,** 327
Diethyl oxomalonate, **1,** 1112; **5,** 244; **6,** 188; **7,** 291, 396; **8,** 170; **10,** 143; **11,** 180
Diethyl (2-oxopropyl)phosphonate, **12,** 527
Diethyl peroxide, **4,** 152
Diethyl N-phenylamidophosphate, **7,** 106
Diethyl phenyl orthoformate, **3,** 97
Diethyl N-phenylphosphoramidate, **7,** 106
Diethyl phenylsulfinylmethylphosphonate, **6,** 191
Diethyl phenylthiomethylphosphonate, **6,** 191
Diethyl 1-phenyl-1-trimethylsilyloxymethylphosphonate, **10,** 145
Diethyl phosphate, **1,** 208
Diethyl phosphite, **1,** 251; **2,** 132; **6,** 192; **7,** 353; **8,** 389; **12,** 187
Diethylphosphonoacetaldehyde, **2,** 131
Diethylphosphonoacetaldehyde diethyl acetal, **2,** 131
Diethylphosphonoacetic acid, **8,** 168; **11,** 281
2-Diethylphosphono-1,3-dithiane, **8,** 89; **11,** 179
2-(Diethylphosphono)propionic acid, **11,** 560
2-(Diethylphosphono)propionitrile, **7,** 106
Diethyl phosphorobromidate, **8,** 170
Diethyl phosphorochloridate, **1,** 248; **3,** 98; **4,** 152, 398, 481; **5,** 217; **6,** 192, 336; **8,** 171; **9,** 1, 332, 467; **10,** 243; **11,** 297; **12,** 352
Diethyl phosphorochloridite, **1,** 248
Diethyl phosphorocyanidate (DEPC), **5,** 217; **6,** 192; **7,** 107; **9,** 163; **10,** 145; **11,** 181; **12,** 187
Diethylphosphorylmethyl methyl sulfoxide, **6,** 193
N,N-Diethyl-1-propynylamine (*see* N,N-Diethylaminopropyne)
Diethylsilane–Chlorotris(triphenylphosphine)rhodium(I) [*see* Chlorotris-(triphenylphosphine)rhodium(I)–Hydrosilanes]
Diethyl succinate, **1,** 911, 1067, 1076; **8,** 296
Diethyl sulfate, **1,** 18, 253, 370, 419; **2,** 153; **3,** 325; **5,** 81
Diethyl sulfoxide, **6,** 257
Diethyl tartrate (DET), **10,** 64; **11,** 92, 181; **12,** 90
Diethyl [(2-tetrahydropyranyloxy)methyl]-phosphonate, **11,** 181; **12,** 188
Diethylthallium *t*-butoxide, **7,** 108
Diethyl(tributylstannyl)aluminum, **12,** 339
Diethyl trichloromethylphosphonate, **6,** 189
N,N-Diethyl-1,2,2-trichlorovinylamine, **1,** 253; **2,** 134
Diethyltrimethylsilylamine, **3,** 317; **4,** 544; **6,** 184, 634
Diethyl(trimethylsilylethynyl)aluminum, **8,** 42
Diethyl trimethylsilyloxycarbonylmethylphosphonate, **8,** 171
Diethyl 1-trimethylsilyloxyethylphosphonate, **9,** 165
Diethyl trimethylsilyl phosphite, **9,** 166
Diethylzinc, **3,** 257; **5,** 219
Diethylzinc–Benzal iodide, **2,** 134
Diethylzinc–Bromoform, **6,** 194; **12,** 74
Diethylzinc–Bromoform–Oxygen, **9,** 194
Diethylzinc–1,1-Diiodoethane, **3,** 141; **11,** 186
Diethylzinc–Diiodomethane, **1,** 253; **2,** 134; **4,** 153; **7,** 154; **8,** 172; **11,** 182
Diethylzinc–Iodoform, **4,** 153; **6,** 194
Difluoramine, **1,** 213, 253; **2,** 134
gem-Difluoroallyllithium, **10,** 146; **12,** 188
Difluorocarbene, **1,** 1237; **2,** 42, 193, 379, 411; **3,** 224, 265; **4,** 384; **5,** 63, 603; **7,** 60; **12,** 158
Difluoromethylenetriphenylphosphorane, **1,** 1058, 1227; **2,** 281, 379; **9,** 166; **10,** 448
sym-Difluorotetrachloroacetone, **1,** 254; **2,** 135
2,2-Difluoro-1-tolylsulfonyloxyvinyllithium, **9,** 166

Difluorotriphenylphosphorane (*see* Triphenylphosphine difluoride)
Digitonin, **2,** 136
Diglyme, **1,** 255
Dihalobis(triphenylphosphine)palladium(II), **3,** 81; **5,** 190; **6,** 59, 60; **7,** 95; **8,** 151; **10,** 133; **11,** 511; **12,** 525
Dihalocarbene (*see also* specific names), **4,** 231; **5,** 656
Dihalo(methoxycarbonyl)methyl(phenyl)-mercury, **3,** 222
Dihydridotetrakis(tributylphosphine)-ruthenium(II), **10,** 148
Dihydridotetrakis(triphenylphosphine)-ruthenium(II), **7,** 109; **9,** 167; **10,** 55; **11,** 182
Dihydridotris(triphenylphosphine)ruthenium, **11,** 182
1,3-Dihydrobenzo[c]thiophene 2,2-dioxide, **9,** 168; **10,** 146
1,2-Dihydro-4,6-dimethyl-2-thioxo-3-pyridinecarbonitrile, **9,** 167
3,4-Dihydro-2-ethoxy-2*H*-pyran, **1,** 362
2,3-Dihydrofuran, **8,** 70; **12,** 189
Dihydrogen hexachloroiridate(IV) [*see* Hydrogen hexachloroiridate(IV)]
Dihydrogen hexachloroplatinate(IV) + co-reactants (*see* Hydrogen hexachloroplatinate(IV) + co-reactants)
1,3-Dihydroisothianaphthalene-2,2-dioxide, **9,** 168; **10,** 146
5,6-Dihydro-4-methoxy-2*H*-pyran, **2,** 271; **3,** 197; **5,** 438, 674; **6,** 367
4,5-Dihydro-5-methyl-1,3,5-dithiazine, **8,** 305
4,5-Dihydro-5-methyl-1,3,5-dithiazin-2-yllithium, **8,** 305; **11,** 302
4,5-Dihydro-2-methylthiazole, **4,** 344; **6,** 403
1,4-Dihydronaphthalene-1,4-*endo*-oxide, **1,** 255
5,6-Dihydrophenanthridine, **11,** 184
3,4-Dihydro-2*H*-pyran, **1,** 256; **2,** 52; **3,** 99; **5,** 220, 674; **7,** 109; **8,** 70; **9,** 256; **10,** 147
3,4-Dihydro-2*H*-pyrido[1,2-a]pyrimidine-2-one, **7,** 110
Dihydroquinidine acetate, **10,** 291
Dihydroquinine acetate, **10,** 291
Dihydroresorcinol, **1,** 178
5,6-Dihydro-2,4,4,6-tetramethyl-4*H*-1,3-oxazine (*see* 2,4,4,6-Tetramethyl-5,6-dihydro-1,3-(4*H*)-oxazine)
2,5-Dihydrothiophene-3,4-dicarboxylic anhydride, **10,** 187
2,5-Dihydrothiophene-1,1-dioxide (*see* 3-Sulfolene)
2,2-Dihydro-2,2,2-triphenyl-1,2-oxaphospholane
2,2′-Dihydroxy-1,1′-binaphthyl, **9,** 169; **10,** 148; **11,** 374; **12,** 49, 190
(R)- and (S)-2,2′-Dihydroxy-1,1′-binaphthyl–Lithium aluminum hydride, **9,** 169; **10,** 148; **12,** 190
(S)-(−)-10,10′-Dihydroxy-9,9′-biphenanthryl, **12,** 191
p-Dihydroxyborylbenzyloxycarbonyl chloride, **6,** 194
1,2-Dihydroxy-3-bromopropane, **4,** 154
3,4-Dihydroxy-3-cyclobutene-1,2-dione, **4,** 466
Di(2-hydroxyethyl) sulfide, **9,** 241
Diimide, **1,** 42, 257, 482, 909; **2,** 139; **3,** 99; **4,** 154, 535; **5,** 220, 707; **6,** 195; **7,** 121, 392; **8,** 172
Diiminobutanedinitrile(Diiminosuccinonitrile), **4,** 155; **5,** 220
Diiodoborane, **7,** 196
3,4-Diiodo-1-cyclobutene, **3,** 335
1,3-Diiodo-5,5-dimethylhydantoin, **1,** 258
Diiododimethylsilane, **9,** 170
1,1-Diiodoethane, **6,** 254; **11,** 182
1,1-Diiodoethane–Diethylzinc (*see* Diethylzinc–1,1-Diiodoethane)
Diiodomethane, **1,** 506; **2,** 273; **3,** 189; **5,** 398; **7,** 74; **9,** 170
Diiodomethane–Zinc–Copper- or Silver-couple (*see* Simmons-Smith reagent)
Diiodomethane–Zinc–Trimethylaluminum, **8,** 339
Diiodo(methyl)bis(triphenylphosphine)-rhodium, **10,** 149
Diiodo-μ-methylenedimagnesium, **2,** 273; **3,** 189, 201
Diiodomethyllithium, **5,** 398
1,3-Diiodopropane, **6,** 129
Diiodotriphenylphosphorane (*see* Triphenylphosphine–Iodine)
Diiodotris(triphenylphosphine)cobalt, **3,** 89
Diiron nonacarbonyl (*see* Tri-μ-carbonyl-hexacarbonyldiiron)
Diisobutylaluminum chloride, **11,** 12, 539; **12,** 2
Diisobutylaluminum 2,6-di-*t*-butyl-4-methylphenoxide, **9,** 171

Diisobutylaluminum hydride (DIBAH, DIBAL), **1,** 209, 260, 1189; **2,** 140; **3,** 101, 174; **4,** 158; **5,** 224; **6,** 198; **7,** 111, 147; **8,** 104, 173; **9,** 171, 274; **10,** 149; **11,** 185, 291, 518; **12,** 191, 239, 512
Diisobutylaluminum phenoxide, **9,** 172; **10,** 150
Diisopinocampheylborane, **1,** 205, 262; **4,** 161; **5,** 39, 283; **6,** 202; **7,** 180; **8,** 174; **11,** 188, 225; **12,** 193
Diisopropoxydimethyltitanium, **12,** 460
(Diisopropoxymethylsilyl)methylmagnesium chloride, **12,** 244
Diisopropylamine–borane (*see* Borane–Diisopropylamine)
Diisopropylaminomagnesium bromide (*see* Bromomagnesium diisopropylamide)
Diisopropyl benzoylphosphonate, **11,** 178
Diisopropylbis(trifluoromethanesulfonato)-silane, **11,** 189
2,2′-Di(1-isopropyl-4-*t*-butylimidazolyl) disulfide, **7,** 141
Diisopropylcarbamoyllithium, **8,** 174
Diisopropylcarbodiimide, **7,** 76
Diisopropyl chlorophosphate, **9,** 312
Diisopropylethylamine (Hunig's base), **1,** 371; **4,** 230; **7,** 148; **10,** 151; **11,** 175; **12,** 225
N,N-Diisopropylformamide, **8,** 174
N,N′-Diisopropylhydrazine, **4,** 162
(−)-2,3;4,6-Di-O-isopropylidene-2-keto-L-gulonic acid hydrate, **7,** 113
Diisopropyl methylphosphonate, **1,** 1229
Diisopropyl peroxydicarbonate, **1,** 263; **2,** 142; **12,** 194
2,6-Diisopropylphenoxymagnesium hydride, **8,** 175
Diisopropyl phosphite, **10,** 390
Diisopropyl phosphorochloridate, **9,** 312
Diisopropylsilyl bis(trifluoromethane-sulfonate), **11,** 189
Diisopropylsilyl ditriflate, **11,** 189
Diisopropyl sulfide–N-Chlorosuccinimide (*see* N-Chlorosuccinimide–Diisopropyl sulfide)
Diisopropyl tartrate, **11,** 94, 181; **12,** 90
Diketene, **1,** 264, 528; **5,** 225; **6,** 202, 592; **11,** 189
Diketene–acetone adduct (*see* 2,2,6-Trimethyl-1,3-dioxolene)
Diketene–acetophenone adduct, **1,** 266
Diketene–Iodotrimethylsilane, **10,** 151
2,4-Diketo(16-crown-5), **6,** 133
2,4-Diketo(19-crown-6), **6,** 133
Dilithioacetate, **3,** 208; **7,** 113; **8,** 175; **11,** 198
Dilithioacetylene (*see* Dilithium acetylide)
2,2′-Dithiobiphenyl, **4,** 162
1,3-Dilithiopropyne, **6,** 202
Dilithium acetylide, **6,** 148, 324
Dilithium dialkyl(cyano)cuprates, **11,** 366; **12,** 349
Dilithium pentamethyltricuprate, **11,** 365
Dilithium tetrabromonickelate(II), **12,** 195
Dilithium tetrachlorocuprate(II), **2,** 85; **4,** 163; **5,** 159, 226; **6,** 139, 203; **7,** 79, 114; **8,** 120, 176; **11,** 116, 190; **12,** 195, 524
Dilithium tetrachloropalladate(II), **5,** 413; **7,** 114; **8,** 176, 470; **9,** 174; **12,** 482, 566
Dilithium tributylcuprate, **7,** 115
Dilithium trimethylcuprate, **6,** 386; **7,** 115; **11,** 365
Dilithium tris(1-pentynyl)cuprate, **6,** 203
Dilongifolylborane, **11,** 191
Dimedone, **1,** 266
Dimenthyl fumarate, **12,** 294
1,4-Dimercapto-2,3-butanediol, **9,** 394
1,2-Di(mercaptomethyl)-4,5-dimethylbenzene, **1,** 277; **2,** 153
Di-2-mesitylborane, **5,** 408; **12,** 195
Dimesityl(methyl)borane, **12,** 12
Dimesityl(trimethylsilylmethyl)borane, **12,** 12
Dimethoxyacetylene, **5,** 204
4,4′-Dimethoxybenzhydrol, **3,** 102
4,4′-Dimethoxybenzhydrylamine, **9,** 174; **12,** 149, 287
2,4-Dimethoxybenzylamine, **4,** 164
3,5-Dimethoxybenzyl *p*-nitrophenyl carbonate, **2,** 142
1(E),3-Dimethoxybutadiene, **11,** 192
3,3-Dimethoxy-2-butanone, **2,** 255
Dimethoxycarbene, **5,** 714
Dimethoxycarbenium tetrafluoroborate (*see also* Dialkoxycarbenium tetrafluoro-borates), **4,** 114; **5,** 623, 714
5,5-Dimethoxycyclopentadiene, **2,** 94
3,3-Dimethoxycyclopropene, **12,** 152
2,5-Dimethoxy-2,5-dihydrofuran, **1,** 635
(3S,6S)-(+)-2,5-Dimethoxy-3,6-dimethyl-3,6-dihydropyrazine, **10,** 151; **11,** 192
1,1-Dimethoxy-N,N-dimethylethylamine (*see* N,N-Dimethylacetamide dialkyl acetals)
1,1-Dimethoxyethane, **2,** 4
1,2-Dimethoxyethane (DME), **1,** 267; **2,** 143

2-(2,2-Dimethoxyethyl)-1,3-dithiane, **4,** 164
2-(2′,2′-Dimethoxyethyl)-1,3-dithianyl-2-lithium, **4,** 165
1,1-Dimethoxyethylene (*see* Ketene dimethyl acetal)
cis-1,2-Dimethoxyethylene, **4,** 156
Dimethoxymethane, **1,** 671; **2,** 271; **6,** 18, 374; **7,** 115; **9,** 521; **11,** 193, 354
(S)-1-(Dimethoxymethyl-2-methoxymethyl)-pyrrolidine, **10,** 152
trans-2,4-Dimethoxymethyl-5-phenyl-oxazoline, **6,** 204
1,1-Dimethoxy-2-methylthioethane, **4,** 191
(Dimethoxymethyl)trimethoxysilane, **3,** 103
Dimethoxyphosphinyldiazomethane (*see* Dimethyl diazomethylphosphonate)
(Dimethoxyphosphinyl)methyllithium, **6,** 339
R-(−)-(Dimethoxyphosphinyl)methyl *p*-tolyl sulfoxide, **13,** 295
2,2-Dimethoxypropane, **1,** 268; **2,** 177; **4,** 157; **5,** 226; **10,** 438
1,1-Dimethoxy-1-propene, **9,** 523
4,6-Dimethoxy-2-pyrone, **9,** 175
5,5-Dimethoxy-1,2,3,4-tetrachlorocyclo-pentadiene, **1,** 270, 605; **2,** 143; **3,** 103; **8,** 178; **10,** 153
2,5-Dimethoxytetrahydrofuran, **2,** 144
2,3-Dimethoxy-N,N,N′,N′-tetramethyl-1,4-butanediamine, **9,** 47
1,3-Dimethoxy-1-trimethylsilyloxy-1,3-butadiene, **12,** 196
1,1-Dimethoxy-1-trimethylsilyloxy-1,3-butadiene, **8,** 178; **12,** 196
1,2-Dimethoxy-1-trimethylsilyloxy-1,3-pentadiene, **12,** 197
4,4′-Dimethoxytriphenylmethyl chloride, **1,** 188
Di(α-methoxyvinyl)copperlithium [*see* Lithium di(α-methoxyvinyl)cuprate]
1,2-Dimethoxyvinyllithium, **7,** 116
N,N-Dimethylacetamide, **1,** 270; **2,** 144; **4,** 165; **6,** 84, 636
N,N-Dimethylacetamide dialkyl acetals, **1,** 271; **4,** 166; **5,** 226; **8,** 179, 180; **10,** 151, 154
Dimethyl acetamidomalonate, **11,** 505
Dimethyl 1,3-acetonedicarboxylate, **9,** 176; **11,** 193
Dimethyl acetylenedicarboxylate (DMAD), **1,** 236, 272, 343, 1149; **2,** 145; **3,** 9, 103; **4,** 41, 168; **5,** 227; **6,** 206, 584; **7,** 117, 161; **9,** 176; **11,** 371
1,1-Dimethylallene, **5,** 279
3,3-Dimethylallyl-, *see also* 3-Methyl-2-butenyl-
3,3-Dimethylallyl bromide (*see* 1-Bromo-3-methyl-2-butene)
N,N′-Dimethyl-2-allyl-1,3,2-diazaphospholidine 2-oxide, **3,** 104
3,3-Dimethylallylnickel bromide [*see* Di-μ-bromobis(3-methyl-2-butenyl)nickel]
(α,α-Dimethylallyl)trimethylsilane, **3,** 446; **8,** 181
Dimethylaluminum acetylide, **8,** 385
Dimethylaluminum amide, **8,** 182; **9,** 177
Dimethylaluminum amides (*see also* Dialkyl-aluminum amides, Diethylaluminum amides), **8,** 182; **9,** 177; **10,** 118
Dimethylaluminum benzenethiolate, **5,** 36; **9,** 15; **10,** 163; **11,** 194
Dimethylaluminum benzylamide, **8,** 182
Dimethylaluminum benzylthiolate, **5,** 36
Dimethylaluminum chloride, **8,** 42; **10,** 177; **11,** 7; **12,** 5
Dimethylaluminum cyanide (*see also* Organoaluminum cyanides), **12,** 184
Dimethylaluminum iodide, **10,** 164
Dimethylaluminum 2-methyl-2-propanethiolate, **9,** 15
Dimethylaluminum methylselenolate, **8,** 182; **11,** 194; **12,** 197
Dimethylaluminum phenylmethanethiolate, **5,** 36
(Dimethylamido)phosphoryl dichloride (*see* N,N-Dimethylphosphoramidic dichloride)
Dimethylamine, **1,** 117; **5,** 107, 609; **6,** 554; **7,** 119
Dimethylamine–Borane (*see* Borane–Dimethylamine)
3-(Dimethylamino)-2-azaprop-2-en-1-ylidene dimethylammonium chloride (Gold's reagent), **11,** 194
(*p*-Dimethylamino)benzaldehyde, **1,** 273; **2,** 146
4-Dimethylamino-1-*t*-butoxycarbonyl-pyridinium chloride, **4,** 173
4-Dimethylamino-3-butyn-2-one, **8,** 183; **9,** 177; **12,** 198
N,N-Dimethylaminocyclopropylphenyl-oxosulfonium tetrafluoroborate, **4,** 173
(Dimethylamino)dimethyloxosulfonium

tetrafluoroborate, **5,** 210
N-(4-Dimethylamino-3,5-dinitrophenyl)-maleimide, **2,** 146
2-Dimethylamino-1,3-dioxolane, **1,** 283
2-Dimethylamino-N,N'-diphenyl-1,3,2-diazaphospholidine, **11,** 196
2-(Dimethylamino)ethanol, **10,** 244
[R-(R*,S*)]-α-[1-(Dimethylamino)ethyl]-benzenemethanol (*see* N-Methylephedrine)
2-Dimethylaminoethylidenetriphenylphosphorane, **8,** 184
(2-Dimethylaminoethyl)triphenylphosphonium bromide, **8,** 184
(2S,3R)-(+)-4-Dimethylamino-3-methyl-1,2-diphenyl-2-butanol (*see* Darvon alcohol)
(Dimethylamino)methyl methyl ether, **11,** 274
(Dimethylamino)methyloxosulfonium methylide, **5,** 210
3-Dimethylamino-2-methyl-1-propanol, **10,** 155
1-Dimethylamino-2-nitroethylene, **4,** 357
2-(Dimethylamino)-4-nitrophenyl phosphate, **6,** 207
2-(Dimethylamino)-3-pentenonitrile, **10,** 155
4-Dimethylaminophenylmagnesium bromide, **2,** 205
(Dimethylamino)phenyloxosulfonium methylide, **3,** 105; **5,** 231
(E)-(Dimethylamino)phenyl(2-phenylethenyl)sulfoxonium tetrafluoroborate, **4,** 174
3-Dimethylaminopropylamine, **1,** 274
1-(3-Dimethylaminopropyl)-3-ethylcarbodiimide, **1,** 274; **12,** 199
4-Dimethylaminopyridine (DMAP), **3,** 118; **5,** 260; **9,** 178; **10,** 155; **12,** 57, 199
4-(Dimethylamino)pyridinium chlorochromate, **11,** 196
4-(Dimethylamino)pyridinium chlorosulfite chloride, **12,** 199
4-Dimethylamino-1,1,2-trimethoxy-1,3-butadiene, **11,** 197
Dimethylaminotrimethyltin, **2,** 147
4-Dimethylamino-N-triphenylmethylpyridinium chloride, **10,** 156
N,N-Dimethylaniline, **1,** 274; **2,** 147
N,N-Dimethylbenzeneselenenamide, **7,** 119
2,6-Dimethylbenzoquinone, **6,** 65
N,N-Dimethylbenzylamine, **10,** 26
(S,R)-N,N-Dimethyl-1-[1',2-bis-(diphenylphosphine)ferrocenyl]-ethylamine, **11,** 514
Dimethyl bis(methylthio)methylphosphonate, **7,** 125
5,5-Dimethyl-1,3-bis(trimethylsilyloxy)-1,3-cyclohexadiene, **8,** 47
Dimethylboron bromide, **12,** 199
Dimethyl 3-bromo-2-ethoxypropenylphosphonate, **9,** 180
Dimethyl bromomalonate, **12,** 200
Dimethylbromomethyleneammonium bromide (*see also* Vilsmeier reagent), **6,** 220; **7,** 422
Dimethylbromonium hexafluoroantimonate, **5,** 231
Dimethylbromosulfonium bromide (*see* Bromodimethylsulfonium bromide)
2,3-Dimethylbutadiene, **1,** 276; **5,** 471; **12,** 134
2,3-Dimethyl-2-butylborane (*see* Thexylborane)
B-3,3-Dimethyl-1-butynyl-9-borabicyclo-[3.3.1]nonane, **8,** 6
3,3-Dimethyl-1-butynylcopper (*see* Copper *t*-butylacetylide)
Dimethylcadmium, **2,** 245
Dimethylcarbamoyl chloride, **8,** 151
Dimethylcarbamoyl fluoride, **10,** 29
(Dimethylcarbamoyl)lithium, **5,** 233
Dimethylcarbene, **11,** 102
Dimethyl carbonate, **2,** 149; **5,** 234; **12,** 201
Dimethylchloroforminium chloride (*see* Vilsmeier reagent)
Dimethylchloromethyleneammonium chloride (*see* Vilsmeier reagent)
3,5-Dimethyl-4-chloromethylisoxazole (*see* 4-Chloromethyl-3,5-dimethylisoxazole)
N,N-Dimethyl-1-chloro-2-methyl-1-propenylamine (*see* 1-Chloro-N,N,2-trimethylpropenylamine)
Dimethylchloronium chloropentafluoroantimonate, **8,** 186
2,2-Dimethyl-6-(*p*-chlorophenylthiomethylene)cyclohexanone, **6,** 208
Dimethylchlorosulfonium chloride (*see* Chlorodimethylsulfonium chloride)
Dimethylcopperlithium (*see* Lithium dimethylcuprate)
N,N-Dimethylcyanamide, **4,** 137; **5,** 196
5,5-Dimethyl-1,3-cyclohexanedione, **1,** 266
1,1'-Dimethyl(cyclopentadienyl)tin(II), **12,**

201
3,3-Dimethylcyclopropene, **5,** 35
Dimethyldiacetoxysilane, **3,** 113
Dimethyl diazomalonate, **5,** 244, 717; **8,** 187, 435; **11,** 208; **12,** 203
Dimethyldiazomethane (*see* 2-Diazopropane)
Dimethyl diazomethylphosphonate, **3,** 113; **4,** 544; **8,** 188; **9,** 181; **10,** 157
L-α,α'-Dimethyldibenzylamine, **10,** 245
N,N-Dimethyl dichlorophosphoramide, **6,** 215; **9,** 187
Dimethyldichlorosilane (*see* Dichlorodimethylsilane)
7,8-Dimethyl-1,5-dihydro-2,4-benzodithiepin, **6,** 216
1,2-Dimethyl-4,5-di(mercaptomethyl)-benzene, **1,** 277; **2,** 153
S,S-Dimethyl-N-(2,4-dinitrophenyl)-sulfilimine, **9,** 193
2,2-Dimethyl-1,3-dioxane-4,6-dione (*see* Meldrum's acid)
1-(4,5-Dimethyl-1,3,2-dioxaphosphol-2-yl)-1*H*-imidazole, **6,** 218
6,6-Dimethyl-5,7-dioxaspiro[2.5]octane-4,8-dione, **6,** 96, 216; **7,** 122
Dimethyldioxirane (*see* Acetone–Potassium peroxomonosulfate)
trans-[2,2-Dimethyl-1,3-dioxolane-4,5-diyl)-bis(methylene)]bis(diphenylphosphine) [*see* 2,3-O-Isopropylidene-2,3-dihydroxy-1,4-bis(diphenylphosphine)-butane]
Dimethyl disulfide, **5,** 79, 246, 698; **6,** 217; **7,** 127; **8,** 103, 189, 416; **9,** 91, 281, 473
2,2-Dimethyl-1,3-dithiane, **10,** 81
Dimethyl dithiobis(thioformate), **5,** 247
N,N-Dimethyldithiocarbamoylacetonitrile, **7,** 123
S,S'-Dimethyl dithiocarbonate, **9,** 181
3,4-Dimethylenepyrrolidine, **5,** 32
3,4-Dimethylenetetrahydrofuran, **5,** 32
3,4-Dimethylenethiolane, **5,** 33
(−)-N,N-Dimethylephedrinium bromide, **6,** 219
O,O-Dimethyl ethanebisthioate, **5,** 247
N-(1,2-Dimethylethenylenedioxy-phosphoryl)imidazole, **6,** 218
Dimethylethylidenecarbene (*see* Isopropylidene carbene)
N,N-Dimethyl-O-ethylphenylpropiol-amidium tetrafluoroborate, **7,** 124
Dimethylethynylaluminum, **8,** 385
N,N-Dimethylformaldimmonium trifluoroacetate [*see* Dimethyl-(methylene)ammonium trifluoroacetate]
N,N-Dimethylformamide (DMF), **1,** 212, 278, 1146; **2,** 153; **3,** 115; **4,** 184; **5,** 221, 233, 247, 376; **6,** 276; **7,** 124; **8,** 189; **9,** 182; **11,** 198, 346; **12,** 98, 203
Dimethylformamide–Chlorine, **5,** 249
Dimethylformamide–N-Chlorosuccinimide, **5,** 129
Dimethylformamide dialkyl acetals (*see also* specific names), **1,** 281, 283; **2,** 154; **3,** 115; **4,** 184; **5,** 253, 254; **6,** 221, 222, 427; **7,** 125, 343; **8,** 191; **9,** 182, 183; **10,** 157, 158; **11,** 198; **12,** 204
Dimethylformamide dibenzyl acetal, **6,** 221; **7,** 343
Dimethylformamide diethyl acetal, **1,** 281; **2,** 154; **3,** 115; **5,** 253; **7,** 125; **9,** 182; **10,** 157
Dimethylformamide dimethyl acetal, **3,** 115; **4,** 184; **5,** 254; **6,** 221, 427; **7,** 343; **8,** 191; **10,** 158; **11,** 198
Dimethylformamide–Dimethyl sulfate adduct, **1,** 282; **2,** 154; **5,** 250; **9,** 182
Dimethylformamide dineopentyl acetal, **1,** 283; **6,** 222; **9,** 183; **12,** 204
Dimethylformamide ethylene ketal, **1,** 283
Dimethylformamide–Methanesulfonyl chloride, **7,** 423
Dimethylformamide–Oxalyl chloride (*see also* Vilsmeier reagent), **1,** 286; **5,** 251; **8,** 186, 529; **9,** 335
Dimethylformamide–Phosgene (*see also* Vilsmeier reagent), **1,** 284, 286; **5,** 251
Dimethylformamide–Phosphorus(III) chloride (*see also* Vilsmeier reagent), **1,** 286; **7,** 292
Dimethylformamide–Phosphorus(V) chloride (*see also* Vilsmeier reagent), **6,** 220
Dimethylformamide–Phosphoryl chloride (*see also* Vilsmeier reagent), **1,** 284; **2,** 154; **3,** 228; **5,** 251
Dimethylformamide–Sodium acetate, **1,** 285
Dimethylformamide–Thionyl bromide (*see also* Vilsmeier reagent), **3,** 16
Dimethylformamide–Thionyl chloride (*see also* Vilsmeier reagent), **1,** 286; **3,** 16, 116; **4,** 186; **5,** 652; **9,** 465; **12,** 204
Dimethylformamide–Triphenylphosphine dibromide (*see also* Vilsmeier reagent),

3, 321; **7,** 422
O,O-Dimethyl formylphosphonate S,S-dimethyl thioacetal, **7,** 125
2,5-Dimethylfuran, **6,** 346
2,2-Dimethyl-3(2*H*)-furanone, **10,** 159
N,N-Dimethylglycinehydrazine, **1,** 289
Dimethylglyoxime, **8,** 385
5,5-Dimethylhydantoin, **7,** 126
N,N′-Dimethylhydrazine, **12,** 489
N,N-Dimethylhydrazine, **1,** 289; **2,** 154; **3,** 117; **5,** 254; **6,** 223; **7,** 126, 201, 283, 416; **8,** 192, 370; **9,** 184; **11,** 200, 491; **12,** 205
N,O-Dimethylhydroxylamine, **11,** 201; **12,** 205
1,3-Dimethyl-2-imidazolidinone (DMI), **10,** 6, 374; **11,** 118, 202, 521; **12,** 477
N,N-Dimethylisobutenylamine, **4,** 170
3,5-Dimethylisoxazole, **6,** 533; **11,** 257
Dimethylketene, **1,** 290, 863, 1287; **5,** 291; **6,** 223; **9,** 185
Dimethylketene methyl trimethylsilyl ketal (*see also* Ketene alkyl trialkylsilyl acetals), **4,** 32; **8,** 484; **10,** 401
Dimethyl 3-ketoglutarate (*see* Dimethyl 1,3-acetonedicarboxylate)
Dimethyl lithiomethylphosphonate, **6,** 339
Dimethylmagnesium, **1,** 292; **6,** 386
Dimethyl maleate, **7,** 161
2,3-Dimethylmaleic anhydride, **11,** 439
Dimethyl malonate, **5,** 302, 685; **6,** 140, 447, 595; **8,** 193, 475; **9,** 453; **12,** 330
Dimethylmalonic acid, **1,** 290
Dimethylmalonic acid dimethyl ketene acylal, **1,** 291
N,N-Dimethylmethanesulfonamide, **1,** 611
Dimethyl methoxycarbonylmethylphosphonate (*see* Trimethyl phosphonoacetate)
Dimethyl methoxymethylenemalonate, **11,** 279; **12,** 206
N,N-Dimethyl-α-methylbenzylamine, **11,** 134
Dimethyl(methylene)ammonium halides [*see also* Dimethyl(methylene)ammonium salts], **4,** 186; **7,** 130; **8,** 194; **10,** 160; **11,** 208, 310; **12,** 260
Dimethyl(methylene)ammonium iodide (Eschenomoser's salt), **4,** 186; **7,** 130; **8,** 194; **10,** 160; **11,** 208, 310; **12,** 260
Dimethyl(methylene)ammonium salts, **3,** 114, 308; **4,** 184, 186; **7,** 130; **8,** 194; **10,** 160; **11,** 208, 310; **12,** 260
Dimethyl(methylene)ammonium trifluoroacetate, **3,** 114, 308; **4,** 184; **7,** 132; **8,** 194
Dimethyl methylmalonate, **12,** 331
Dimethyl methylphosphonate, **3,** 117; **6,** 339; **11,** 203
Dimethyl methylphosphonothioate, **2,** 155
4,4-Dimethyl-2-methylthio-2-oxazoline, **6,** 584
Dimethyl(methylthio)sulfonium tetrafluoroborate, **11,** 204; **12,** 207
Dimethylmolybdenum dichloride, **12,** 358
N,N-Dimethyl-4-nitrosoaniline, **1,** 746
5,5-Dimethyl-N-nitroso-2-oxazolidinone, **3,** 118
Dimethyl oxalate, **1,** 251
4,4-Dimethyl-1,3-oxathiolane-3,3-dioxide, **9,** 186
Dimethyl 2-oxoglutaconate, **11,** 206
Dimethyl 3-oxoglutarate (*see* Dimethyl 1,3-acetonedicarboxylate)
1,3-Dimethyl-2-oxohexahydropyrimidine (DMPU), **11,** 207
Dimethyl oxomalonate, **5,** 244; **11,** 207
Dimethyloxosulfonium methylide (*see* Dimethylsulfoxonium methylide)
Dimethylphenacylselenonium ylide, **5,** 257
Dimethylphenylchlorosilane, **2,** 278
1,3-Dimethyl-2-phenyl-1,3,2-diazaphospholidine, **11,** 208
N,N-Dimethyl-*p*-phenylenediamine, **1,** 293
2,6-Dimethylphenyl propionate, **10,** 162
Dimethyl(phenyl)selenonium methyl sulfate, **11,** 343
Dimethylphenylsilane, **12,** 209
Dimethylphenylsilyllithium, **7,** 133; **8,** 196; **10,** 162; **11,** 209; **12,** 210
Dimethylphenylsilyl(methyl)magnesium, **12,** 342, 352
N,S-Dimethyl-S-phenylsulfoximine, **5,** 458; **6,** 395; **11,** 210, 344
Dimethyl(phenylthio)aluminum (*see* Dimethylaluminum benzenethiolate)
Dimethyl(phenylthiomethyl)amine, **10,** 164
Dimethyl phosphite, **1,** 293; **4,** 188; **5,** 258
Dimethyl phosphite–Chloroacetic acid, **11,** 211
N,N-Dimethylphosphoramidic dichloride, **6,** 215; **9,** 187
R-(−)-(Dimethylphosphoryl)methyl *p*-tolyl sulfoxide, **12,** 295
2,6-Dimethylpiperidine, **2,** 156

2-(2,6-Dimethylpiperidino)acetonitrile, **11,** 212
N,N-Dimethyl-1,3-propanediamine, **1,** 274
2,2-Dimethyl-1,3-propanediol, **5,** 259
N,N-Dimethylpropionamide dimethyl acetal, **9,** 241
1,3-Dimethylpropyleneurea, **11,** 207
N,N-Dimethyl-4-pyridinamine (*see* 4-Dimethylaminopyridine)
2,4-Dimethylpyridine, **6,** 224
2,6-Dimethylpyridine, **1,** 626; **6,** 606
1,2-Dimethylpyrrolidine, **11,** 134
(+)-*trans*-2,5-Dimethylpyrrolidine, **8,** 197
Dimethyl selenide, **6,** 240
Dimethylselenonium methylide, **6,** 240
Dimethyl selenoxide, **6,** 224; **8,** 197
Dimethylsilylene, **8,** 218; **12,** 220
N,N-Dimethylsulfamoyl chloride, **9,** 187
Dimethylsulfamoylmethyllithium, **1,** 611
Dimethyl sulfate, **1,** 18, 293, 370, 419; **3,** 325; **4,** 188; **5,** 81, 647; **9,** 445
Dimethyl sulfate–Barium oxide, **1,** 295
Dimethyl sulfate–Potassium carbonate, **1,** 294, 295; **2,** 153; **4,** 188, 189
Dimethyl sulfate–Sodium bicarbonate, **1,** 295
Dimethyl sulfide, **2,** 156, 403; **4,** 190; **5,** 260; **6,** 398
Dimethyl sulfide–Borane (*see* Borane–Dimethyl sulfide)
Dimethyl sulfide–N-Bromosuccinimide (*see* N-Bromosuccinimide–Dimethyl sulfide)
Dimethyl sulfide–Chlorine (*see* Chlorodimethylsulfonium chloride)
Dimethyl sulfide–N-Chlorosuccinimide (*see* N-Chlorosuccinimide–Dimethyl sulfide)
Dimethyl sulfide–Cuprous bromide (*see* Copper(I) bromide-Dimethyl sulfide)
Dimethyl sulfide ditriflate (*see* Dimethyl sulfide–Trifluoroacetic anhydride)
Dimethyl sulfide–Methanesulfonic acid, **12,** 212
Dimethyl sulfide–Trifluoroacetic anhydride, **6,** 225; **9,** 188
Dimethyl sulfite, **1,** 296
3,4-Dimethyl-3-sulfolene, **8,** 287
Dimethyl sulfone, **1,** 296; **2,** 157; **7,** 78
Dimethylsulfonium ethoxycarbonylmethylide, **10,** 164
Dimethylsulfonium 2-ketotetrahydrofuranyl-3-ylide, **5,** 262
Dimethylsulfonium methylide, **1,** 314; **2,** 169; **3,** 124; **4,** 196; **5,** 646; **6,** 219; **8,** 198; **9,** 188; **11,** 213
Dimethyl sulfoxide (DMSO), **1,** 296; **2,** 157; **3,** 119; **4,** 192; **5,** 263; **6,** 225; **7,** 133; **8,** 198; **9,** 189; **10,** 166; **11,** 214; **12,** 212
Dimethyl sulfoxide–Acetic anhydride, **2,** 163; **3,** 121; **4,** 199; **7,** 135; **8,** 464; **9,** 190
Dimethyl sulfoxide–Boron trifluoride etherate, **1,** 304; **2,** 94
Dimethyl sulfoxide–Bromine, **1,** 308; **6,** 229
Dimethyl sulfoxide–N-Bromosuccinimide, **2,** 159; **3,** 34; **4,** 51, 199; **9,** 71, 73
Dimethyl sulfoxide–*t*-Butyl bromide, **9,** 190
Dimethyl sulfoxide–*t*-Butyl hypochlorite, **7,** 135; **10,** 167
Dimethyl sulfoxide–Chlorine (*see* Chlorodimethylsulfonium chloride)
Dimethyl sulfoxide–N-Chlorosuccinimide, **4,** 199
Dimethyl sulfoxide–Chlorosulfonyl isocyanate, **10,** 167
Dimethyl sulfoxide–Chlorotrimethylsilane, **9,** 190
Dimethyl sulfoxide–Cyanuric chloride, **5,** 266
Dimethyl sulfoxide–1-Cyclohexyl-3-(2-morpholinoethyl)carbodiimide metho-*p*-toluenesulfonate, **6,** 227
Dimethyl sulfoxide–Dicyclohexylcarbodiimide (Pfitzer–Moffatt reagent), **1,** 304; **2,** 162; **3,** 121; **5,** 132; **6,** 228; **8,** 199; **11,** 223
Dimethyl sulfoxide–Diethylcarbodiimide, **2,** 162
Dimethyl sulfoxide–Diphenylketene-*p*-tolylimine, **3,** 122
Dimethyl sulfoxide–Iodine, **3,** 161; **5,** 347; **6,** 227, 295; **8,** 200; **9,** 190
Dimethyl sulfoxide–Methanesulfonic anhydride, **5,** 266; **9,** 191
Dimethyl sulfoxide–4-Nitrobenzenesulfonyl chloride, **4,** 193
Dimethyl sulfoxide–Oxalyl chloride (Swern reagent), **8,** 200; **9,** 192; **10,** 166, 167; **11,** 215
Dimethyl sulfoxide–Phosgene, **1,** 856; **6,** 226
Dimethyl sulfoxide–3-Sulfopropanoic anhydride complex, **9,** 192
Dimethyl sulfoxide–Sulfuric acid, **6,** 227
Dimethyl sulfoxide–Sulfur trioxide, **2,** 165; **4,** 200; **6,** 229; **7,** 159; **8,** 200; **10,** 167; **11,** 216

Dimethyl sulfoxide–Sulfur trioxide–Pyridine, **2,** 394; **4,** 201
Dimethyl sulfoxide–Sulfuryl chloride, **8,** 200
Dimethyl sulfoxide–Thionyl chloride, **8,** 200; **12,** 213
Dimethyl sulfoxide–*p*-Toluenesulfonic acid, **6,** 227
Dimethyl sulfoxide–*p*-Toluenesulfonyl chloride, **5,** 266
Dimethyl sulfoxide–Trichloroacetic acid, **6,** 227
Dimethyl sulfoxide–Trichloroacetic anhydride, **9,** 192
Dimethyl sulfoxide–Trifluoroacetic acid, **6,** 226
Dimethyl sulfoxide–Trifluoroacetic anhydride (Swern reagent), **5,** 266; **6,** 230; **7,** 136; **8,** 200; **10,** 166, 168
Dimethyl sulfoxide–2,4,6-Trinitrobenzenesulfonic acid, **4,** 193; **5,** 265
Dimethylsulfoxonium methylide, **1,** 315; **2,** 171; **3,** 125; **4,** 197; **5,** 254, 646; **6,** 41, 378; **7,** 133; **8,** 194; **9,** 186; **10,** 168; **12,** 213
N,N′-Dimethyltartaramide, **12,** 192, 480
Dimethyl tartrate, **11,** 298
Dimethyl(tetrahydro-2-oxo-3-furanyl)-sulfonium tetrafluoroborate, **5,** 262
Dimethyl 1,2,4,5-tetrazine-3,6-dicarboxylate, **12,** 214
2,4-Dimethylthiazole, **4,** 202
N,N-Dimethylthiocarbamoyl chloride, **2,** 173; **3,** 127; **4,** 202; **8,** 200
Dimethylthiocarbamoyllithium, **5,** 234, 267
N,N-Dimethylthioformamide, **3,** 128; **5,** 234, 267
Dimethylthioformamide–Acetic anhydride, **3,** 128
N,N-Dimethylthiopivalamide, **7,** 190
Dimethyltitanium bis(isopropoxide), **12,** 460
Dimethyltitanium dichloride, **10,** 138, 270
S,S-Dimethyl-N-*p*-toluenesulfonylsulfilimine, **9,** 193
Dimethyl N-(*p*-toluenesulfonyl)sulfoximine, **3,** 205; **9,** 193; **12,** 216
Dimethyl-N-*p*-tolylketenimine, **5,** 282
N,N-Dimethyl-α-*p*-tolylsulfinylacetamide, **12,** 298, 509
S,S-Dimethyl-N-tosylsulfilimine, **9,** 193
Dimethyl N-(*p*-tosyl)sulfoximine, **3,** 205; **9,** 193; **12,** 216
Dimethyl(2,4,6-tri-*t*-butylphenoxy)-chlorosilane, **11,** 217
4,5-Dimethyl-2,2,2-trimethoxy-1,3-dioxaphospholene, **2,** 97; **4,** 535; **5,** 267; **7,** 392
N,N-Dimethyl-α-trimethylsilylacetamide, **8,** 201
Dimethyltrimethylstannylamine, **2,** 147
1,3-Dimethyluracil, **9,** 194
6,6-Dimethyl-2-vinyl-5,7-dioxaspiro[2.5]-octane-4,8-dione, **6,** 230
Dimethylvinylidenecarbene, **5,** 27, 460; **8,** 93
2,2-Dimethylvinyl triflate, **6,** 230
4,5-Dimethyl-*o*-xylene-α,α′-dithiol, **1,** 277; **2,** 153
Dimethylzinc, **3,** 128; **12,** 357
Dimorpholinophosphinyl chloride, **1,** 861
Dimsyllithium (*see* Lithium methylsulfinylmethylide)
Dimsylpotassium (*see* Potassium methylsulfinylmethylide)
Dimsylsodium (*see* Sodium methylsulfinylmethylide)
2,4-Dinitrobenzaldehyde, **1,** 318
2,4-Dinitrobenzeneseleninic acid, **8,** 359
2,4-Dinitrobenzenesulfenamide, **9,** 267
2,4-Dinitrobenzenesulfenyl chloride, **1,** 319; **2,** 21; **6,** 231; **9,** 194
3,5-Dinitrobenzenesulfonyl chloride, **3,** 307
2,4-Dinitrobenzenesulfonylhydrazide, **6,** 232
3,4-Dinitrobenzoic acid, **1,** 320
3,5-Dinitrobenzoyl *t*-butyl nitroxyl, **8,** 204
3,5-Dinitrobenzoyl chloride, **1,** 320
Di-*p*-nitrobenzyl phosphorochloridate, **1,** 321
4,4′-Dinitrodiphenylnitroxide, **2,** 174
2,4-Dinitrofluorobenzene (DNF, Sanger's reagent), **1,** 321; **2,** 175; **6,** 233
Dinitrogen pentoxide, **1,** 322
Dinitrogen tetroxide (*see* Nitrogen dioxide)
Dinitrogen trioxide–Boron trifluoride, **1,** 329
2,4-Dinitroiodobenzene, **1,** 329
3,5-Dinitromesitylglyoxal, **3,** 79
3,5-Dinitroperbenzoic acid, **9,** 196; **11,** 217
Di(*p*-nitrophenyl) carbonate, **1,** 330
2,4-Dinitrophenylhydrazine, **1,** 330, 565, 1114; **2,** 176; **6,** 587
Di-*p*-nitrophenyl hydrogen phosphate, **1,** 331; **2,** 29
O-(2,4-Dinitrophenyl)hydroxylamine, **6,** 233
2,4-Dinitrophenyl phenylmethanesulfonate, **5,** 478
Di-*p*-nitrophenyl sulfite, **2,** 176

N,N′-Dinitroso-N,N′-dimethyloxamide, **1,** 192
DIOP [*see* 2,3-O-Isopropylidene-2,3-dihydroxy-1,4-bis(diphenylphosphine)-butane]
Dioxane, **1,** 333
Dioxane–Bromine, **1,** 333; **2,** 177; **3,** 130; **5,** 58
Dioxane diphosphate, **1,** 334; **3,** 131
2-(1,3-Dioxan-2-ylethylidene)triphenyl-phosphorane, **9,** 196
[2-(1,3-Dioxan-2-yl)ethyl]triphenyl-phosphonium bromide, **9,** 196
1,5-Dioxaspiro[4.5]deca-7,9-diene, **2,** 94
1,4-Dioxaspiro[4.4]nona-6,8-diene, **2,** 94
Dioxobis(*t*-butylimido)osmium(VIII), **8,** 206
1,3-Dioxolan-2-ylmethylenetriphenyl-phosphorane, **5,** 269
1,3-Dioxolan-2-ylmethyltriphenyl-phosphonium bromide, **5,** 269
Dioxygentetrakis(triphenylarsine)rhodium(I) hexafluorophosphate or perchlorate, **8,** 206
Dipentylamine, **1,** 187
Diperoxo-oxohexamethylphosphoramido-molybdenum(VI) [*see* Oxodiperoxy-molybdenum(pyridine)(hexamethyl-phosphoric triamide)]
Diphenoxyphosphinyl isothiocyanate, **1,** 346
1,3-,Diphenylacetone toluenesulfonyl-hydrazone, **10,** 3
Diphenylacetyl chloride, **1,** 344; **4,** 527
Diphenylacetylene, **1,** 241, 335; **5,** 738
2-Diphenylacetyl-1,3-indanedione-1-hydrazone, **1,** 335
Diphenylamine–Borane (*see* Borane–Diphenylamine)
(Diphenylarsinyl)methyllithium, **8,** 280
N,N-Diphenylbenzamide–Oxalyl chloride, **12,** 565
2,5-Diphenyl-3,4-benzofuran (*see* 1,3-Diphenylisobenzofuran)
Diphenylborane, **9,** 198
2,3-Diphenylbutadiene, **1,** 336
(E,E)-1,4-Diphenylbutadiene, **1,** 337
Diphenylcarbamoyl chloride, **1,** 337; **2,** 177
N,N-Diphenylcarbamoylpyridinium chloride, **4,** 204
Diphenylcarbene, **2,** 81
Diphenylcarbodiimide, **1,** 337; **2,** 178; **3,** 131; **8,** 208
Diphenyl chlorophosphate (*see* Diphenyl phosphorochloridate)
S,S-Diphenyl N-chlorosulfilimine, **5,** 639
Diphenylcopperlithium (*see* Lithium diphenylcuprate)
Diphenyldiazomethane, **1,** 338, 655, 1258; **2,** 81, 178, 294; **4,** 204; **6,** 234, 452, 508, 643
Diphenyldi(1,1,1,3,3,3-hexafluoro-2-phenyl-2-propoxy)sulfurane, **4,** 205; **5,** 270; **6,** 239; **8,** 208
Diphenyl diselenide, **5,** 272; **6,** 235; **7,** 136; **9,** 199; **10,** 170
Diphenyl diselenide–Benzeneseleninic acid (*see also* Benzeneselenenic acid), **8,** 24
Diphenyl diselenide–Benzeneseleninic anhydride (*see also* Benzeneselenenic anhydride), **9,** 25
Diphenyl diselenide–Bromine (*see also* Benzeneselenenyl bromide), **8,** 28
Diphenyl diselenide–Bromine–Hexabutyldistannoxane, **8,** 209
Diphenyl diselenide–Copper(II) acetate, **9,** 199
Diphenyl diselenide–Iodylbenzene (*see also* Benzeneseleninic anhydride), **11,** 37
Diphenyl diselenide–*m*-Iodylbenzoic acid (*see also* Benzeneseleninic anhydride), **11,** 37
Diphenyl diselenide–Lead tetraacetate, **9,** 199
Diphenyl diselenide–Selenium(IV) oxide, **11,** 219
Diphenyl disulfide, **5,** 246, 276; **6,** 235, 336; **7,** 47, 137; **8,** 210, 416; **9,** 386, 429; **10,** 171; **11,** 220
Diphenyl disulfide–Tributylphosphine (*see* Tributylphosphine–Diphenyl disulfide)
Diphenyl 3-ethoxycarbonyl-2-oxopropylphosphonate, **10,** 181
Diphenylethylene carbonate, **4,** 25; **5,** 23
N,N-Diphenylethylenediamine (*see* 1,2-Dianilinoethane)
Diphenylethylsulfonium tetrafluoroborate, **2,** 180
N,N′-Diphenylformamidine, **1,** 339
1,1-Diphenylhexane, **5,** 277
1,1-Diphenylhexyllithium, **5,** 277
1,1-Diphenylhydrazine, **1,** 340; **2,** 178
Diphenyliodonium bromide, **1,** 340
Diphenyliodonium-2-carboxylate monohydrate [*see* (*o*-Carboxyphenyl)-phenyliodonium hydroxide inner salt]
Diphenyliodonium chloride, **1,** 341; **2,** 178

S,S-Diphenyl N-iodosulfilimine, **5,** 639
1,3-Diphenylisobenzofuran, **1,** 342, 831, 918; **2,** 178; **3,** 273; **4,** 209; **6,** 100, 347, 471
1,3-Diphenyl-3,4-isobenzofuran, **2,** 439
Diphenylketene, **1,** 343, 359, 655, 1240; **2,** 444; **3,** 131; **4,** 210, 527; **5,** 278
Diphenylketene *p*-tolylimine (*see* Diphenyl-N-*p*-tolylketenimine)
Diphenyl 2-keto-3-oxazolinylphosphonate, **11,** 220
Diphenylmercury, **1,** 507
Diphenylmethane, **5,** 156, 400; **7,** 302
Diphenylmethoxymethylphosphine, **10,** 172
N,N-Diphenyl-*p*-methoxyphenyl-chloromethyleniminium chloride, **11,** 220
Diphenylmethyl chloride, **2,** 178
N-(Diphenylmethylene)methylamine, **8,** 210
endo, cis-(Diphenylmethyl)isoborneol, **11,** 412
Diphenylmethyllithium, **2,** 343; **5,** 400
Diphenylmethylpotassium, **7,** 302
Diphenyl(methyl)silyl chloride [*see* Chloro-(methyl)diphenylsilane]
Diphenylmethylsodium, **5,** 156
1,3-Diphenylnaphtho[2.3-c]furan, **3,** 132
Diphenyl-1-(phenylthiovinyl)phosphine oxide, **11,** 424
Diphenylphosphine, lithium salt [*see* (Diphenylphosphine)lithium]
Diphenylphosphidocopper(I), **11,** 368
Diphenylphosphine, **1,** 345; **4,** 505; **5,** 408
(S)-α-(R)- or (R)-α-(S)-2-Diphenylphosphine-ferrocenyl ethyldimethylamine, **9,** 200; **11,** 19, 237
(R)-α-(S)-2-Diphenylphosphineferrocenyl ethyldimethylamine, **11,** 19
(Diphenylphosphine)lithium (LDP), **1,** 345; **2,** 179; **4,** 303; **5,** 408; **6,** 340; **8,** 302; **11,** 368
Diphenylphosphine oxide, **11,** 565
Diphenylphosphinic chloride, **6,** 340; **7,** 138
p-Diphenylphosphinobenzoic acid, **2,** 179
Diphenylphosphinodithioic acid, **10,** 171
1-Diphenylphosphino-1-methoxymethyllithium, **10,** 172
Diphenylphosphinyl chloride, **6,** 340; **7,** 138
Diphenylphosphinylhydrazine, **9,** 429
O-(Diphenylphosphinyl)hydroxylamine, **11,** 221
Diphenyl phosphite, **4,** 188, 210
Diphenyl phosphoroazidate (DPPA), **4,** 210; **5,** 280; **6,** 193; **7,** 138, 394; **8,** 211; **10,** 173; **11,** 222; **12,** 217
Diphenyl phosphorochloridate, **1,** 345; **2,** 180; **3,** 133; **7,** 93; **10,** 310, 390; **11,** 223; **12,** 409
Diphenylphosphoroisothiocyanatidate, **1,** 346
Diphenylphosphoryl azide (*see* Diphenyl phosphoroazidate)
Diphenyl phthalate, **1,** 347
Diphenylpicrylhydrazyl, **1,** 347
Diphenyl 2-(N-piperidinyl)ethylphosphine oxide, **11,** 595
Diphenyl-4-pyridylmethanol, **7,** 139
Diphenyl-4-pyridylmethylchloride, **7,** 139
Diphenyl-4-pyridylmethylpotassium, **9,** 384
Diphenyl selenide, **6,** 240
Diphenylseleninic anhydride (*see* Benzeneseleninic anhydride)
Diphenylselenium bis(trifluoroacetate), **11,** 223
Diphenyl selenoxide, **5,** 280; **9,** 201
Diphenylsilane, **1,** 348; **5,** 361
Diphenylsilane–Chlorotris(triphenyl-phosphine)rhodium(I) [*see* Chlorotris-(triphenylphosphine)rhodium(I)–Hydrosilanes]
Diphenyl N-succinimidylphosphate, **10,** 173
Diphenyl sulfide, **6,** 241
S,S-Diphenylsulfilimine, **5,** 639; **10,** 174
Diphenylsulfonium cyclopropylide, **3,** 133; **4,** 211; **5,** 281, 416; **6,** 242; **7,** 140; **8,** 212; **12,** 217
Diphenylsulfonium isopropylide, **2,** 180
Diphenylsulfonium methylide, **5,** 281
Diphenyl sulfoxide, **1,** 348; **9,** 396
S,S-Diphenylsulfoximine, **5,** 432
Diphenyltelluride, **1,** 348
4,6-Diphenylthieno[3.4-*d*][1,3]dioxol-2-one 5,5-dioxide, **7,** 140; **9,** 202
Diphenyltin dihydride, **1,** 349; **3,** 134; **11,** 224
Diphenyl N-(*p*-toluenesulfonyl)sulfoximine, **5,** 432
Diphenyl-N-*p*-tolylketenimine, **1,** 345; **3,** 98; **5,** 282
Diphenyl triphenylphosphoranylidene-methylphosphonate, **2,** 181
Diphenylvinylphosphine oxide, **11,** 595
Diphenylzinc, **1,** 349
DIPHOS [*see* 1,2-Bis(diphenylphosphine)-ethane]
Diphosgene, **8,** 214
Diphosphorus tetraiodide, **1,** 349; **6,** 243; **9,** 203; **10,** 174, 318; **11,** 224; **12,** 218

(+)-Di-3-pinanylborane (*see* Diisopinocampheylborane)
(2R,3R)-Dipivaloyltartaric acid, **9,** 204
Dipotassium acetylide, **9,** 35
Dipotassium azodicarboxylate, **1,** 257
Dipotassium cyanodithioimidocarbonate, **4,** 215
Dipotassium hexachloroosmiate(IV)–Zinc, **8,** 412
Dipotassium hexachlorotungstate(IV), **4,** 407
Dipotassium platinum tetrachloride [*see* Dipotassium tetrachloroplatinate(II)]
Dipotassium tetracarbonylferrate, **8,** 214
Dipotassium tetrachloroplatinate(II), **2,** 182; **4,** 215
Dipotassium tetramethylosmiate, **1,** 955
Di-(E)-propenylcopperlithium {*see* Lithium di[(E)-1-propenyl]cuprate}
Dipropylboryl trifluoromethanesulfonate, **12,** 155
Dipropylcopperlithium (*see* Lithium dipropylcuprate)
Dipyridine chromium(VI) oxide (*see* Collins reagent)
2,2′-Dipyridyl diselenide, **10,** 368
2,2′-Dipyridyl disulfide, **5,** 285; **6,** 246; **8,** 214
2,2′-Dipyridyl disulfide–Triphenylphosphine (*see* Triphenylphosphine–2,2′-Dipyridyl disulfide)
Disiamylborane, **1,** 57, 201; **2,** 29; **3,** 22, 160; **4,** 37, 175; **5,** 39, 417; **6,** 62; **8,** 41; **9,** 455; **10,** 40; **11,** 226; **12,** 194
Disodium acetylide, **2,** 182
Disodium hexachloroiridate, **5,** 287
Disodium phenanthrenide, **1,** 350; **2,** 182
Disodium platinum tetrachloride, **3,** 134
Disodium tetracarbonylferrate (Collman's reagent), **3,** 267; **4,** 461; **5,** 624; **6,** 550; **7,** 341; **8,** 216; **9,** 205; **10,** 174
Disodium tetracarbonylferratesesqui-dioxanate, **10,** 175
Disodium tetrachloropalladate(II), **4,** 369; **5,** 500, 626; **6,** 45, 447; **8,** 217, 475; **9,** 207; **10,** 175
Disodium tetrachloroplatinate(II), **3,** 134
Disodium tetracyanonickelate(II), **9,** 207
N,N′-Disuccinimidyl carbonate, **9,** 208; **11,** 226
Di(N-succinimidyl)oxalate, **12,** 219
1,3-Dithiane (*see also* 2-Lithio-1,3-dithianes), **2,** 182; **3,** 135; **4,** 216, 241; **5,** 287, 660; **6,** 119, 248; **7,** 142; **8,** 217, 346; **9,** 48, 235; **10,** 231; **11,** 285; **12,** 573
1,3-Dithiane-2-thione, **2,** 441
1,3-Dithian-2-ylidenemethanone, **5,** 99
1,3-Dithian-2-ylidenetriphenylphosphorane, **8,** 89
1,3-Dithian-2-yllithium (*see* 2-Lithio-1,3-dithianes)
1,3-Dithienium tetrafluoroborate, **4,** 218; **11,** 227
5,5′-Dithiobis(2-nitrobenzoic acid), **1,** 351
1,3-Dithiolan-2-yltriphenylphosphonium tetrafluoroborate, **10,** 176
Dithiothreitol, **9,** 394
Di-*p*-toluoyl-D-tartrate, **1,** 351
Di-*p*-tolylcarbodiimide, **1,** 351
Di-*p*-tolylmethane, **7,** 302
Di-*p*-tolylmethylpotassium, **7,** 302
Di(2,2,2-trichloroethyl) phosphorochloridate, **2,** 187; **3,** 136
Di(trimethylsilyl) sulfate, **10,** 176; **11,** 67
Divinylallene (*see* 1,3,4,6-Heptatetraene)
Divinylcopperlithium (*see* Lithium divinylcuprate)
Divinylmercury, **1,** 352
Divinyl sulfone, **7,** 23
DMAD (*see* Dimethyl acetylenedi-carboxylate)
DMAP (*see* 4-Dimethylaminopyridine)
DME (*see* 1,2-Dimethoxyethane)
DMF (*see* N,N-Dimethylformamide)
DMI (*see* 1,3-Dimethyl-2-imidazolidinone)
DMPU (*see* 1,3-Dimethyl-2-oxohexahydropyrimidine)
DMSO (*see* Dimethyl sulfoxide)
DNF (*see* 2,4-Dinitrofluorobenzene)
Dodecacarbonyltriiron (*see* Di-μ-carbonyldecacarbonyltri-*triangulo*-iron)
Dodecacarbonyltri-*triangulo*-ruthenium, **7,** 244, 315
Dodecamethylcyclohexasilane, **8,** 218; **12,** 219
p-Dodecylbenzenesulfonyl azide, **11,** 535; **12,** 220
(−)-N-Dodecyl-N-methylephedrinium bromide, **6,** 249; **7,** 144, 238; **11,** 337
Dowex ion-exchange resin, **1,** 512; **8,** 263; **9,** 256
Dowtherm A., **1,** 353
Dppb [*see* 1,4-Bis(diphenylphosphine)butane]
Dppe [*see* 1,2-Bis(diphenylphosphine)ethane]
Dppf [*see* 1,1′-Bis(diphenylphosphine)-

ferrocene]
Drying agents (*see also* Molecular sieves, Dessicants), **1,** 353
Duolite ion-exchange resin, **10,** 220

Easy Mesyl, **6,** 190
EDA (*see* Ethylenediamine)
EDTA (*see* Ethylenediaminetetraacetic acid tetrasodium salt)
EEDQ (*see* N-Ethoxycarbonyl-2-ethoxy-1,2-dihydroquinoline)
NB-Enantrane, **11,** 229
NB-Enantride, **11,** 229
Ephedrine, **5,** 289; **9,** 209; **11,** 230
Epichlorohydrin (1-Chloro-2,3-epoxypropane), **1,** 355; **4,** 222; **5,** 290
3,4-Epoxy-1-butyne, **5,** 484
1,2-Epoxycyclohexane, **7,** 33
1,2-Epoxy-3-phenoxypropane, **1,** 355
N-(2,3-Epoxypropylidene)cyclohexylamine N-oxide, **9,** 209
Erbium(III) chloride, **9,** 99, 210; **11,** 232
Eschenmoser's salt [*see* Dimethyl(methylene)-ammonium iodide]
Ethane-1,2-diaminoborane, **1,** 355
1,2-Ethanedithiol, **1,** 356, 579, 728, 902, 1174; **2,** 189, 433; **5,** 290, 533; **8,** 367; **10,** 275; **12,** 501, 577
1,2-Ethanedithiol bis(*p*-toluenesulfonate), **4,** 540; **6,** 628
1,2-Ethanediylbis[chlorodimethylsilane], **10,** 140; **12,** 179
(R,R)-1,2-Ethanediylbis[(*o*-methoxyphenyl)-phenylphosphine], **9,** 259
1,2-Ethanediylbis(tributylstannane), **5,** 43; **6,** 56
S,S'-Ethanediyl *p*-toluenesulfonothioate, **5,** 540; **6,** 628
Ethanethiol, **2,** 209; **6,** 16
Ethanethiol–Methanesulfonic acid, **12,** 212
Ethanol, **1,** 1173; **12,** 305, 545
Ethanolamine, **1,** 357; **2,** 189; **4,** 222; **6,** 225
Ethenesulfonyl chloride–Trimethylamine, **9,** 211
Ethoxyacetyl chloride, **3,** 137
Ethoxyacetylene, **1,** 357; **2,** 190, 270; **4,** 225; **5,** 290
Ethoxyacetylene–Magnesium bromide, **7,** 145
2-Ethoxyallyl bromide, **5,** 292
2-Ethoxyallylidenetriphenylphosphorane, **8,** 219
(2-Ethoxyallyl)triphenylphosphonium iodide, **8,** 219
2-Ethoxyallyl vinyl sulfide, **5,** 292
4-Ethoxy-1,3-butadienyllithium, **8,** 278
1-Ethoxy-1-buten-3-yne, **1,** 360, 669
4-Ethoxy-3-buten-1-ynyllithium, **1,** 669
1-Ethoxy-1-butyne, **5,** 592; **8,** 446
Ethoxycarbene, **7,** 213
2-Ethoxycarbonylallyl bromide, **4,** 33; **6,** 93
π-(2-Ethoxycarbonylallyl)nickel bromide, **4,** 33; **6,** 93
o-Ethoxycarbonylbenzyl phenyl sulfoxide, **8,** 75
Ethoxycarbonylcyclopropyltriphenyl-phosphonium tetrafluoroborate, **5,** 90; **6,** 93; **8,** 76; **10,** 88
2-Ethoxycarbonyl-1,3-dithiane, **4,** 230; **11,** 405
2-Ethoxycarbonyl-1,3-dithiolane, **5,** 304
N-Ethoxycarbonyl-2-ethoxy-1,2-dihydroquinoline (EEDQ), **2,** 191; **3,** 137; **4,** 223; **11,** 283
Ethoxycarbonylformimidate, **11,** 109
Ethoxycarbonylformonitrile oxide, **7,** 145
Ethoxycarbonylhydrazine (*see* Ethyl carbazate)
Ethoxycarbonyl isothiocyanate, **4,** 223; **5,** 512; **6,** 250; **7,** 146
Ethoxycarbonylmethylcopper, **4,** 224
Ethoxycarbonylmethylenetriphenyl-phosphorane, **1,** 49, 111, 112, 361, 1241; **2,** 416; **4,** 549; **5,** 299; **6,** 635; **10,** 78; **11,** 24
Ethoxycarbonylmethyltriphenyl-phosphonium bromide, **1,** 111
N-(Ethoxycarbonyl)phthalimide, **1,** 111; **5,** 91; **9,** 212; **12,** 221
N-Ethoxycarbonylpyridinium chloride, **11,** 369
2-Ethoxycarbonyl-3-trimethylsilylpropene [*see* Ethyl 2-(trimethylsilylmethyl)-acrylate]
1-Ethoxycyclopropyllithium, **12,** 73
Ethoxy(diethylaluminum)acetylide, **7,** 103
Ethoxydiethylsilane, **5,** 293
2-Ethoxy-1,2-dihydro-1-methyl-6,8-dinitroquinoline, **1,** 361
2-Ethoxy-3,4-dihydro-1,2-pyran, **1,** 362
Ethoxydiphenylphosphine, **3,** 139
2-Ethoxy-1,3-dithiolane, **11,** 604

1-(1-Ethoxyethoxyallyl)trimethylsilane, **12,** 506
1-Ethoxy-2-ethyl-3-trimethylsilyloxy-1,3-butadiene, **11,** 71
Ethoxyethynyllithium, **1,** 359, 613
Ethoxyethynylmagnesium bromide, **1,** 358
Ethoxyethynylsodium, **5,** 107
Ethoxyketene, **3,** 137
1-Ethoxy-1-lithiocyclopropane, **12,** 73
Ethoxymethyleneaniline, **1,** 362
1-Ethoxy-1-propene, **1,** 385
(2-Ethoxy-1-propenyl)triphenylphosphonium iodide, **8,** 219
1-Ethoxy-1-propyne, **5,** 107
2-Ethoxypyridine N-oxide, **1,** 362
1-Ethoxy-4-tributylstannyl-1,3-butadiene, **8,** 278
1-Ethoxy-2-tributylstannylethylene, **8,** 221
Ethoxy(trimethylsilyl)acetylene, **12,** 221
1-Ethoxy-1-trimethylsilyloxycyclopropane, **8,** 219; **12,** 221, 499
1-Ethoxyvinyllithium, **6,** 205; **11,** 232; **12,** 325
(Z)-2-Ethoxyvinyllithium, **8,** 221; **11,** 342
2-Ethoxyvinyltriphenylphosphonium iodide, **5,** 294
1-Ethoxyvinylzinc chloride, **12,** 468
Ethyl acetate, **1,** 1066, 1075; **3,** 172; **7,** 198; **8,** 121, 225
Ethyl acetoacetate, **1,** 29, 278, 514, 887, 1166, 1268; **4,** 467, 507, 555; **6,** 264, 594; **7,** 293; **9,** 438; **12,** 222
N-Ethylacetonitrilium tetrafluoroborate, **6,** 250
Ethyl acrylate, **2,** 22; **6,** 208, 251; **7,** 4
Ethylal, **11,** 241
Ethylaluminum dichloride, **6,** 251; **7,** 146; **9,** 11; **10,** 9, 177; **11,** 7; **12,** 5, 28, 104, 252, 261
Ethylamine, **1,** 372
Ethylamine–Borane (*see* Borane–Ethylamine)
Ethyl antioxidants, **4,** 128, 257, 337, 503, 545
Ethyl azidoformate, **1,** 363; **2,** 191; **3,** 138; **4,** 225; **7,** 146
Ethyl benzenesulfonylformimidate, **3,** 18
N-Ethylbenzisoxazolium tetrafluoroborate, **1,** 364; **2,** 192
Ethyl benzoylacetate, **12,** 332
Ethyl bis(2-trimethylsilylallyl) phosphate, **9,** 497
Ethyl bromoacetate, **2,** 192, 197; **3,** 26, 172, 236; **6,** 293
Ethyl 4-bromocrotonate, **3,** 236
Ethyl bromocyanoacetate, **3,** 237
Ethyl 3-bromo-2-hydroxyiminopropanoate, **9,** 212
Ethyl 2-bromomethylacrylate, **4,** 33; **6,** 93
Ethyl (E)-1,3-butadiene-1-carbamate, **7,** 147; **8,** 220
Ethyl 2-(*t*-butylsulfinyl)acetate, **6,** 561
Ethyl carbamate (Urethane), **1,** 1270
Ethyl carbazate, **1,** 360; **2,** 191
1-Ethyl-4-carbomethoxypyridinyl, **4,** 438
S-Ethyl carbonochloridothioate, **6,** 252
Ethyl carbonoperoxoate, **9,** 220
Ethyl(carboxysulfamoyl)triethylammonium hydroxide inner salt (Burgess reagent), **4,** 227; **5,** 442; **8,** 139
Ethyl chloroacetate, **1,** 911, 1039, 1077, 1094, 1166; **2,** 192; **4,** 234
Ethyl chlorocarbonate (*see* Ethyl chloroformate)
Ethyl chlorodiazoacetate, **3,** 222
Ethyl chloroformate (*see also* Alkyl chloroformates, Methyl chloroformate), **1,** 364, 1199; **2,** 193, 319; **4,** 228; **5,** 117, 294; **6,** 307, 336, 551; **7,** 147; **8,** 307; **9,** 214; **12,** 223
Ethyl chlorothiolformate, **6,** 252
Ethyl crotonate, **4,** 306
Ethyl cyanoacetate, **1,** 16, 39, 121, 887, 1071, 1268; **4,** 122, 555; **6,** 42, 140; **7,** 406; **12,** 332
O-Ethyl S-cyanomethyl dithiocarbonate, **9,** 215
Ethyl diazoacetate, **1,** 367, 1246, 1258; **2,** 109, 193, 464; **3,** 138, 210, 253, 255; **4,** 228; **5,** 151, 295, 496, 571; **6,** 252, 474, 672; **7,** 5, 313; **8,** 40, 222, 501; **9,** 20, 406; **10,** 274, 341, 382; **11,** 461; **12,** 223
Ethyl diazolithioacetate, **5,** 295; **7,** 5
Ethyl diazopyruvate, **12,** 140
Ethyl dibromoacetate, **2,** 195; **3,** 26, 236
Ethyl dichloroacetate, **2,** 195; **3,** 26; **6,** 331
Ethyl diethoxyacetate, **6,** 253
Ethyl 4-diethoxyphosphinylcrotonate, **8,** 337
Ethyl diethoxyphosphinylformate, **1,** 370
N-Ethyl(diethoxyphosphinyl)-methylketenimine, **9,** 216
Ethyl 4-diethoxyphosphinyl-3-oxobutanoate, **10,** 181
Ethyl 2-diethoxyphosphinylpropionate, **5,**

693
Ethyl 4-diethylphosphonocrotonate, **8,** 337
N-Ethyl(diethylphosphono)-methylketenimine, **9,** 216
Ethyl 2-diethylphosphonopropionate, **5,** 693
Ethyldiisopropylamine (*see* Diisopropylethylamine)
Ethyl 4,4-dimethoxy-2-phenylthiobutyrate, **8,** 222
1-Ethyl-3-(3′-dimethylaminopropyl)-carbodiimide (or hydrochloride salt), **1,** 371; **2,** 196; **9,** 504
Ethyl (dimethylsulfuranylidine)acetate, **2,** 196
Ethyl diphenylphosphinite, **3,** 139
Ethyl 4-diphenylphosphinyl-3-oxobutanoate, **10,** 181
Ethyldiphenylsulfonium tetrafluoroborate, **2,** 180
Ethyl 1,3-dithiane-2-carboxylate, **4,** 230; **11,** 405
Ethyl 1,3-dithiolane-2-carboxylate, **5,** 304
Ethylene, **1,** 605, 1150; **2,** 144; **3,** 8; **4,** 11, 337; **5,** 159
2,2′-Ethylenebis(1,3-dithiane), **6,** 493; **8,** 223
Ethylenebis(*p*-toluenethiosulfonate), **4,** 540; **6,** 628
Ethylenebis(triphenylphosphine)platinum(0), **9,** 216
1,2-Ethylenebis(triphenylphosphonium) dibromide, **6,** 665
Ethylene carbonate, **1,** 372; **5,** 696
Ethylene chloroboronate, **12,** 101, 224
Ethylene chlorophosphite, **1,** 372
Ethylenediamine (EDA), **1,** 372, 567; **4,** 231; **10,** 297; **12,** 338
Ethylenediaminediborane, **1,** 355
Ethylenediaminetetraacetic acid tetrasodium salt (EDTA), **1,** 373; **5,** 296
Ethylene dibromide (*see* 1,2-Dibromoethane)
Ethylene diisocyanate, **2,** 100
3,3-Ethylenedioxybutylmagnesium bromide, **7,** 148
Ethylene dithiotosylate, **4,** 540; **6,** 628
Ethylene glycol, **1,** 15, 375, 818, 966, 1173; **5,** 296, 319, 481, 718; **6,** 27, 302, 381; **9,** 217, 257
Ethylene glycol dimethyl ether (*see* 1,2-Dimethoxyethane)
Ethylene glycol–*p*-Toluenesulfonic acid, **10,** 181
Ethylene oxide, **1,** 355, 377, 420; **2,** 196; **3,** 140; **4,** 231; **5,** 297; **6,** 199; **8,** 150
Ethylene sulfide, **1,** 378
Ethylenesulfonyl chloride–Trimethylamine, **9,** 211
Ethylenimine (*see* Aziridine)
O-Ethyl-S-ethoxycarbonylmethyl dithiocarbonate, **7,** 148
Ethyl ethoxyiminoacetate, **11,** 109
Ethyl ethoxymagnesiomalonate, **1,** 365, 627, 1199
Ethyl ethoxymethylenecyanoacetate, **1,** 380
Ethyl ethylene pyrophosphite, **1,** 250
Ethyl 3-ethyl-5-methyl-4-isoxazolecarboxylate, **4,** 232
Ethyl ethylsulfinylmethyl sulfide, **5,** 299
Ethyl ethylthiomethyl sulfoxide, **5,** 299
Ethyl fluoroacetate, **5,** 304
3-Ethyl-2-fluorobenzothiazolium tetrafluoroborate, **8,** 223
Ethyl fluoroformate, **10,** 29
Ethyl formate, **1,** 380, 420, 1068, 1077, 1094; **2,** 197; **3,** 185; **4,** 233; **5,** 608; **12,** 226
Ethyl glyoxylate ethylene thioacetal, **5,** 304
Ethyl hydroperoxide, **1,** 17
2-Ethyl-7-hydroxybenzisoxazolium tetrafluoroborate, **2,** 192; **6,** 253
3-Ethyl-5-(2-hydroxyethyl)-4-methyl-1,3-thiazolium chloride, **7,** 16
N-Ethylidene-*t*-butylamine, **9,** 1
Ethylidene chloride (*see* 1,1-Dichloroethane)
N-Ethylidenecyclohexylamine, **2,** 249
Ethylidene iodide (*see* 1,1-Diiodoethane)
Ethylidenetriphenylphosphorane, **1,** 310, 1240; **3,** 141; **6,** 275; **8,** 339; **10,** 326
Ethyl iodide, **1,** 17; **5,** 692
Ethyl iodide–Zinc-copper couple, **3,** 257
Ethyl(iodomethyl)zinc, **3,** 257
Ethyl isocyanate, **1,** 383
Ethyl (S)-lactate, **8,** 298; **12,** 226
Ethyl lithioacetate, **3,** 172; **6,** 255; **7,** 198; **8,** 121, 225, 258
Ethyl lithiopropiolate, **9,** 217
Ethyl lithio(trimethylsilyl)acetate, **5,** 373; **6,** 255; **11,** 234; **12,** 227, 228
Ethyllithium, **5,** 306, 324; **9,** 84
Ethyllithium–Tetramethylethylenediamine, **5,** 306
Ethylmagnesium bromide (*see also* Grignard reagents), **1,** 415; **2,** 205; **5,** 324; **7,** 206; **9,** 218, 229; **10,** 189, 192; **11,** 247
Ethylmagnesium bromide–Copper(I) iodide

(*see also* Grignard reagents–Copper(I) halides), **6,** 148; **7,** 149; **9,** 219
Ethylmagnesium bromide–Dichlorobis-(cyclopentadienyl)titanium, **9,** 146
Ethylmagnesium chloride (*see also* Grignard reagents), **1,** 415
Ethyl malonate, **6,** 255; **10,** 182
Ethyl mandelate, **8,** 298
Ethylmanganese chloride, **9,** 289
Ethyl mercaptan (*see* Ethanethiol)
Ethyl 2-mercaptoethyl carbonate, **1,** 383
Ethyl metaphosphate (*see* Polyphosphate ester)
Ethyl-N-methoxyacetimidate, **12,** 297
(4S,5S)-2-Ethyl-4-(methoxymethyl)-5-phenyloxazoline, **11,** 232
2-Ethyl-2-methyl-1,3-dioxolane, **1,** 82
Ethyl 2-(methyldiphenylsilyl)propionate, **12,** 322
(−)-N,N-Ethylmethylephedrinium bromide, **6,** 219
1-Ethyl-3-methyl-3-phospholene-1-oxide, **1,** 679
Ethyl 2-methylsulfinylacetate, **6,** 255
Ethyl (methylthio)acetate, **5,** 260
Ethyl (E)-3-methylthio-2-methyl-2-butenoate, **9,** 219
N-Ethylmorpholine, **1,** 383
Ethyl nitroacetate, **4,** 507
Ethyl nitroacetate–Diethyl azodicarboxylate–Triphenylphosphine, **10,** 182
Ethyl 4-nitrobutanoate, **7,** 11
Ethyl orthoformate (*see* Triethyl orthoformate)
Ethyl orthosilicate, **1,** 387
O-Ethylperoxycarbonic acid, **9,** 220
Ethyl N-phenyliminophosphite, **1,** 384
N-Ethyl-5-phenylisoxazolium-3′-sulfonate (Woodward's reagent K), **1,** 181, 384; **2,** 197; **3,** 137; **5,** 306; **8,** 226
Ethyl 2-phenylsulfinylacetate, **6,** 256; **10,** 183
Ethyl phenyl sulfone, **9,** 248
Ethyl β-phenylsulfonylpropiolate, **11,** 233
Ethyl phenyl sulfoxide, **6,** 257
Ethyl phospenimidoate, **1,** 384
Ethylphosphonic acid bis(dimethylamide), **2,** 280
N-Ethylpiperidine, **12,** 4
Ethyl potassio(trimethylsilyl)acetate, **12,** 228
Ethyl propenyl ether, **1,** 385
Ethyl propiolate, **1,** 1142; **7,** 7; **8,** 259; **9,** 217
Ethyl propionate, **1,** 385
Ethyl β-(1-pyrrolidinyl)acrylate, **8,** 226
N-Ethyl-2-pyrrolidone (NEP), **9,** 220
1-Ethylsulfinyl-3-pentanone, **9,** 221
P-Ethyl-N,N,N′,N′-tetramethylphosphonic diamide, **2,** 280
Ethylthiocyanate, **5,** 28
O-Ethyl thioformate, **11,** 495
Ethyl thioltrifluoroacetate, **3,** 143
N-Ethylthiophthalimide, **11,** 493
Ethylthiotributyltin, **8,** 354; **11,** 584
Ethylthiotrimethylsilane, **11,** 586
(R)-Ethyl *p*-tolylsulfinylmethylene-propionate, **12,** 228
Ethyl tributylstannyl sulfide, **8,** 354; **11,** 584
Ethyl trichloroacetate, **1,** 386, 852; **2,** 300; **3,** 143; **4,** 233
Ethyl α-trifluoromethylsulfonyloxyacetate, **7,** 149
S-Ethyl trifluorothioacetate, **3,** 143
Ethyltrihexylammonium hydroxide, **1,** 1228
Ethyl trimethylsilylacetate, **5,** 373; **7,** 150; **8,** 226; **11,** 234, 247; **12,** 227, 228, 459
Ethyl N-trimethylsilylcarbamate, **6,** 412
Ethyl trimethylsilyl malonate, **5,** 723
Ethyl 2-(trimethylsilylmethyl)acrylate, **10,** 55
Ethyl trimethylsilyl sulfide, **11,** 586
4-Ethyl-2,6,7-trioxa-1-phosphabicyclo-[2.2.2]octane, **3,** 143
Ethyltriphenylphosphonium salts, **1,** 1242; **10,** 326
Ethyl vinyl ether, **1,** 386; **2,** 198; **3,** 300; **4,** 234; **6,** 529; **8,** 427, 428; **10,** 184; **11,** 232, 235
Ethyl vinyl ketone, **1,** 388
Ethylzinc iodide, **3,** 257
Ethynyllithium (*see* Lithium acetylide)
Ethynylmagnesium bromide, **1,** 11, 389, 415
β-Ethynyl-β-propiolactone, **10,** 283
Ethynyl *p*-tolyl sulfone, **10,** 183
Ethynyltributyltin, **5,** 43
Ethynyltrimethylsilane (*see* Trimethylsilylacetylene)
Eu(fod)$_3$, **4,** 211; **12,** 196, 312, 559
Eu(hfc)$_3$, **12,** 196, 561

Fehling solution, **1,** 390
Fenchone, **8,** 228
Fenton's reagent (Hydrogen peroxide–Iron salts), **1,** 472
Ferric, *see also* Iron(III)
Ferric acetylacetonate–Hydrogen peroxide

(*see* Hydrogen peroxide–Iron(III) acetylacetonate)
Ferric azide, **7,** 153
Ferric chloride, **1,** 31, 390, 392; **2,** 199; **3,** 145; **4,** 236; **5,** 307, 555; **6,** 259, 555; **7,** 153; **8,** 228, 229; **9,** 151, 222, 489; **10,** 106, 185, 373; **11,** 137, 237; **12,** 26, 230
Ferric chloride–Acetic anhydride (*see* Acetic anhydride–Ferric chloride)
Ferric chloride–Butyllithium, **6,** 260
Ferric chloride–Dimethylformamide, **4,** 236
Ferric chloride–Isopropylmagnesium chloride (*see* Isopropylmagnesium chloride–Iron(III) reagents)
Ferric chloride–Sodium hydride, **7,** 155
Ferric ethoxide, **6,** 95
Ferric nitrate, **4,** 107
Ferric nitrate/K10 Bentonite (Clayfen), **11,** 237; **12,** 231
Ferric selenocyanate, **7,** 155
Ferric sulfate, **10,** 223
Ferric thiocyanate, **6,** 259; **7,** 155
Ferrocenylphosphines, **9,** 200; **10,** 37; **11,** 19, 237, 514
Ferrous chloride, **3,** 145, 200; **6,** 555
Ferrous chloride–Oxygen, **6,** 428
Ferrous perchlorate, **6,** 260; **9,** 258
Ferrous sulfate, **1,** 393; **4,** 237; **11,** 276
Ferrous sulfate–Oxygen, **5,** 308
Fetizon's reagent (*see* Silver carbonate–Celite)
Fieser reagent (Chromium(VI) oxide–Acetic acid), **1,** 145; **2,** 72; **3,** 56
Fieser's solution, **1,** 393
Fischer reagent, **1,** 528
Florisil, **1,** 393; **8,** 221; **9,** 223; **11,** 35
Fluorene-9-carboxylic acid, **1,** 394
9-Fluorenylmethanol, **5,** 308
9-Fluorenylmethyl azidoformate, **3,** 145
9-Fluorenylmethyl chloroformate, **3,** 145; **4,** 237; **8,** 230
N-Fluorenyltrifluoromethane-sulfonamide, **6,** 43
Fluorine, **5,** 309
Fluorine–Sodium trifluoroacetate, **9,** 223
N-Fluoro-N-alkyl-*p*-toluenesulfonamides, **12,** 231
Fluoroantimonic acid (*see* Hexafluoro-antimonic acid)
Fluorobenzene, **1,** 845
Fluoroboric acid (*see* Tetrafluoroboric acid)
2-Fluoro-1-buten-3-one, **6,** 263
Fluorodiiodomethane, **6,** 261
B-Fluorodimesitylborane, **5,** 408; **12,** 12, 195
Fluorodimethoxyborane, **6,** 261; **10,** 186; **11,** 240
2-Fluoro-2,2-dinitroethanol, **5,** 704
2-Fluoro-1-ethylpyridinium tetrafluoroborate (*see also* 2-Halopyridinium salts), **10,** 83
Fluoroform, **1,** 425
Fluoroiodocarbene, **4,** 29
Fluoromethylenetriphenylphosphorane, **3,** 146; **6,** 262
2-Fluoro-1-methylpyridinium *p*-toluene-sulfonate (*see also* 2-Halopyridinium salts), **8,** 96; **9,** 42
Fluorophenylcarbene, **6,** 136
Fluorosulfuric acid, **1,** 396; **2,** 70, 199; **4,** 436; **5,** 310; **6,** 262
Fluorosulfuric acid–Antimony(V) fluoride (Magic acid), **1,** 396; **2,** 199; **5,** 309, 312
Fluorotributylmethylphosphorane, **9,** 223
Fluorotributyltin (*see* Tributyltin fluoride)
Fluoro(trifluoromethyl)carbene, **5,** 515
Fluorotrimethylsilane, **7,** 151; **12,** 459
2-Fluoro-1,3,5-trinitrobenzene, **8,** 230; **10,** 155
Fluorotriphenylphosphonium fluoride, **2,** 135
1-Fluorovinyl methyl ketone, **6,** 263
Fluoroxytrifluoromethane (*see* Trifluoromethyl hypofluorite)
Formaldehyde, **1,** 309, 397, 420, 530, 746, 1252; **2,** 200, 399; **3,** 141, 238, 301, 450, 488; **5,** 48, 312, 609, 673; **6,** 264, 305, 485, 590; **7,** 126, 158; **8,** 231, 242; **9,** 23, 52, 224, 487; **10,** 186; **11,** 8, 114, 240; **12,** 6, 35, 232, 511
Formaldehyde diethyl acetal, **11,** 241
Formaldehyde–Diethylamine, **3,** 190; **5,** 94
Formaldehyde diethyl thioacetal monoxide, **5,** 299
Formaldehyde–Diisopropylamine–Copper(I) bromide, **9,** 225
Formaldehyde–Dimethylaluminum chloride, **10,** 186
Formaldehyde–Dimethylamine, **1,** 398, 765; **5,** 314; **8,** 231
Formaldehyde dimethyl dithioacetal S,S-dioxide, **11,** 242
Formaldehyde dimethyl thioacetal, **7,** 24
Formaldehyde dimethyl thioacetal monoxide, **4,** 341; **5,** 456; **6,** 390, 446; **8,** 344; **9,** 314

Formaldehyde diphenyl thioacetal [*see* Bis(phenylthio)methane]
Formaldehyde–Hydrochloric acid, **1,** 399; **4,** 238; **6,** 18; **10,** 161
Formaldoxime, **1,** 402
Formamide, **1,** 212, 402; **2,** 201; **3,** 147
Formamidine acetate, **1,** 403; **2,** 202; **3,** 147
Formamidinesulfinic acid, **4,** 506; **5,** 668; **6,** 586; **8,** 232
Formamidinoyl isothiocyanates, **5,** 315
Formic acid, **1,** 404, 706, 757; **2,** 126, 153, 202; **3,** 147; **4,** 239; **5,** 191, 192, 316, 698; **7,** 160; **8,** 232; **9,** 226; **10,** 189; **11,** 243; **12,** 179
Formic acid–Formamide, **1,** 407; **8,** 232
2-Formyl-1,3-benzothiazole, **5,** 24
Formylcarbene, **2,** 102; **4,** 5
Formyl chloride, **1,** 167
(α-Formylethylidene)triphenylphosphorane, **12,** 234
Formyl fluoride, **1,** 14, 407, 1018; **2,** 12
Formylhydrazine, **1,** 382
N-Formylimidazole, **1,** 407; **8,** 233
(S)-N-Formyl-2-methoxymethylpyrrolidine, **11,** 243
2-(N-Formyl-N-methyl)aminopyridine, **8,** 341; **10,** 265
Formylmethylenetriphenylphosphorane, **5,** 269; **6,** 267; **8,** 234; **9,** 157
4-Formyl-1-methylpyridinium benzenesulfonate, **11,** 244
N-Formylpiperidine, **11,** 244
Fremy's salt (*see* Potassium nitrosodisulfonate)
Fumaric acid, **5,** 319
Furan, **1,** 255; **4,** 328; **5,** 222; **6,** 173, 195; **7,** 161; **9,** 227, 519; **10,** 187; **11,** 139; **12,** 69
Furfuryl chloroformate, **1,** 408
β-Furyllithium, **10,** 188; **12,** 465

Gallium oxide, **4,** 241
Galvinoxyl, **1,** 409
Geranyltrimethyltin, **9,** 500
Girard reagent P, **1,** 410
Girard reagent T, **1,** 410; **10,** 346
Glutaraldehyde, **1,** 411, 740
Glutaric acid, **1,** 412, 1029
Glutaronitrile, **4,** 241
Glycerol, **2,** 391
Glycine, **1,** 412
Glyme (*see* 1,2-Dimethoxyethane)
Glyoxal, **1,** 413, 740; **11,** 194
Glyoxylic acid, **5,** 320; **7,** 162; **9,** 228
Glyoxylic acid dithioethyl acetal, **9,** 228; **10,** 38
Glyoxyloyl chloride *p*-toluenesulfonylhydrazone, **12,** 52
Gold, **5,** 321; **7,** 73, 162
Gold chloride [*see* Hydrogen tetrachloroaurate(III)]
Gold's reagent, **11,** 194
Graphite bisulfate, **6,** 269
Grignard reagents (*see also* specific name), **1,** 374, 415, 572, 677, 1202; **2,** 205; **3,** 159; **4,** 260; **5,** 321, 419; **6,** 269; **7,** 163; **8,** 235; **9,** 229; **10,** 189; **11,** 245; **12,** 235, 290
Grignard reagents–(S,S)-(+)-1,4-Bis(dimethylamino)-2,3-dimethoxybutane, **9,** 47
Grignard reagents–Cobalt(II) chloride, **1,** 155
Grignard reagents–Copper(I) halides, **1,** 167; **2,** 91, 127, 400; **3,** 278; **4,** 517; **5,** 167; **6,** 143, 146, 147, 269, 576; **7,** 80, 82, 163, 226; **8,** 117, 122, 235, 322; **9,** 124; **10,** 54, 190; **11,** 141, 248; **12,** 124, 143, 564
Grignard reagents–Dichlorobis(cyclopentadienyl)titanium, **9,** 146; **10,** 130
Grignard reagents–Dilithium tetrachlorocuprate(II), **4,** 163; **5,** 226; **6,** 203; **7,** 114; **8,** 176
Grignard reagents–Iron(III) reagents, **4,** 236; **6,** 259; **7,** 155, 411; **9,** 222
Grignard reagents–Manganese(II) chloride, **7,** 222
Grignard reagents–Methylcopper, **8,** 237
Grignard reagents–Nickel(0) reagents, **12,** 467
Grignard reagents–Nickel(II) reagents, **2,** 110; **6,** 50; **7,** 250; **9,** 201; **10,** 36, 37, 189, 276; **11,** 59, 239; **12,** 171, 172, 244
Grignard reagents–Palladium(0) reagents, **6,** 571; **8,** 472; **9,** 455
Grignard reagents–Palladium(II) reagents, **9,** 147; **12,** 172
Grignard reagents–Thallium(I) bromide, **3,** 286
Grignard reagents–Titanium(IV) chloride, **1,** 1170; **6,** 594; **11,** 245
Guaiacylmethyl chloride, **10,** 194
Guanidines, **1,** 1071, 1145; **4,** 489; **6,** 254; **7,** 246; **9,** 179; **11,** 105, 249; **12,** 241, 477
B-Halo-9-borabicyclo[3.3.1]nonane, **12,** 236

Haloform, **4,** 231; **5,** 656
Haloform test, **1,** 425
2-Halopyridinium salts, **6,** 375; **7,** 110; **8,** 95, 96; **9,** 42, 234; **10,** 83, 230; **11,** 184, 518; **12,** 116
N-Heptafluorobutyrylimidazole, **3,** 148
Heptafluoro-1-iodopropane, **2,** 226
Heptafluoro-1-methylethyl phenyl ketone, **7,** 165
Heptamethyldisilazane, **4,** 539
1,3,4,6-Heptatetraene, **6,** 271
Hexabromocyclopentadiene, **5,** 322
Hexabutylditin, **7,** 165
Hexacarbonylchromium (*see* Chromium carbonyl)
Hexacarbonylmolybdenum (*see* Molybdenum carbonyl)
Hexacarbonyltetra-μ-chlorodiruthenium, **11,** 49
Hexacarbonyl[μ-(N,N′,N″-tricyclohexylguanidinato(2-)]-N,N′;N,N′-diiron, **4,** 114
Hexacarbonyltungsten (*see* Tungsten carbonyl)
Hexachloroacetone, **8,** 239
Hexachloro-1,3-butadiene, **10,** 195
2,3,4,5,6,6- or 2,3,4,4,5,6-Hexachloro-2,4-cyclohexadien-1-one, **11,** 251
Hexachlorocyclopentadiene, **1,** 425; **2,** 206; **5,** 322
Hexachlorocyclotriphosphazene (*see* Phosphonitrilic chloride trimer)
Hexachlorodisilane, **3,** 148; **8,** 431; **10,** 195
L-N-Hexadecyl-N-methylephedrinium bromide, **7,** 239
Hexadecyltributylphosphonium bromide (*see* Tributylhexadecylphosphonium bromide)
Hexaethylphosphorous triamide [Tris-(diethylamino)phosphine], **1,** 425; **2,** 207; **3,** 148; **4,** 242; **5,** 323
Hexafluoroacetone–Hydrogen peroxide, **3,** 149; **9,** 244; **10,** 206
Hexafluoroantimonic acid, **2,** 216; **5,** 231, 309; **6,** 272; **7,** 166; **8,** 239; **10,** 195; **11,** 252
Hexafluoro-2-butyne, **1,** 425; **2,** 207
Hexafluorophosphoric acid, **1,** 42, 426; **2,** 402; **4,** 252
cis-3a,4,5,6,7,7a-Hexahydrobenzothiazolidine-2-thione, **11,** 304
1,3,3a,4,7,7a-Hexahydro-4,7-methanobenzo-[c]thiophene 2,2-dioxide, **12,** 236
Hexahydro-4,4,7-trimethyl-4*H*-1,3-benzothiin, **12,** 237
Hexakis(acetato)trihydrato-μ_3-oxotrisrhodium acetate, **11,** 252
Hexamethylbenzene, **1,** 426
Hexamethylcyclotrisilathiane, **8,** 240
Hexamethyldisilane, **11,** 253, 511; **12,** 461
Hexamethyldisilane–Pyridinium bromide perbromide, **10,** 96
Hexamethyldisilathiane, **8,** 240
Hexamethyldisilazane, **1,** 427; **2,** 159, 207; **5,** 323; **6,** 273; **7,** 167; **8,** 29; **9,** 234; **11,** 38; **12,** 239
Hexamethyldisiloxane, **6,** 226; **8,** 240; **9,** 55; **10,** 44, 196
Hexamethylditin, **6,** 273; **9,** 500
Hexamethylenetetramine, **1,** 427, 898; **2,** 208; **4,** 243; **9,** 234
Hexamethylphosphoric triamide (HMPA, HMPT), **1,** 430; **2,** 208; **3,** 149; **4,** 244; **5,** 323; **6,** 273; **7,** 168, 210; **8,** 240; **9,** 220, 235; **10,** 101, 196; **11,** 64, 253, 448; **12,** 216, 239, 505
Hexamethylphosphoric triamide–Thionyl chloride, **3,** 153
Hexamethylphosphorous triamide, **1,** 431; **2,** 210; **3,** 149; **4,** 247; **5,** 725; **6,** 279; **7,** 169; **9,** 235; **10,** 199
Hexamethylphosphorous triamide–Carbon tetrachloride, **2,** 445; **5,** 259, 325
Hexamethylphosphorous triamide–Dialkyl azodicarboxylates, **6,** 280; **10,** 448
Hexamethylphosphorous triamide–Dibromodifluoromethane, **10,** 448
Hexamethylphosphorous triamide–Iodine, **9,** 236
Hexamine (*see* Hexamethylenetetramine)
Hexaminecalcium, **1,** 104
Hexarhodium hexadecacarbonyl (*see* Tetra-μ_3-carbonyldodecacarbonylhexarhodium)
Hexylamine, **7,** 170
B-Hexyl-9-borabicyclo[3.3.1]nonane, **10,** 257
B-1-Hexynyl-9-borabicyclo[3.3.1]nonane, **8,** 7
1-Hexynylcopper, **6,** 143
Heyn's catalyst, **1,** 432; **5,** 326; **7,** 320; **8,** 466
Hippuric acid, **1,** 432
(S)-Histidine, **9,** 393

HMPA, HMPT (*see* Hexamethylphosphoric triamide)
Homophthalic anhydride, **11,** 254
Hunig's base (*see* Diisopropylethylamine)
Hyamine germicides, **1,** 433
Hydrazine, **1,** 257, 290, 434, 543, 1074, 1097; **2,** 139, 211, 327; **3,** 100, 153; **4,** 248, 496; **5,** 220, 244, 327, 693, 726; **6,** 280; **7,** 170; **8,** 127, 245; **9,** 236; **11,** 255, 284; **12,** 241
Hydrazine acetate, **1,** 445; **10,** 207
Hydrazine sulfate, **1,** 445
p-Hydrazinobenzenesulfonic acid, **1,** 842
Hydrazoic acid, **1,** 446; **2,** 211; **5,** 329; **8,** 287; **12,** 242
Hydrido(butyl)copperlithium, **5,** 330
Hydridocarbonyltris(triphenylphosphine)-rhodium(I) [*see* Carbonylhydridotris-(triphenylphosphine)rhodium(I)]
Hydridodinitrogentris(triphenylphosphine)-cobalt(I), **5,** 331
Hydridotetrakis(triphenylphosphine)-rhodium(I), **5,** 332; **11,** 255; **12,** 540
Hydrido(tributylphosphine)copper(I) (*see also* Copper hydride ate complexes), **3,** 154
Hydridotris(triisopropylphosphine)-rhodium(I), **9,** 238
Hydridotris(triphenylphosphine)carbonyl-rhodium(I) [*see* Carbonylhydridotris-(triphenylphosphine)rhodium(I)]
Hydriodic acid (*see also* Hydrogen iodide), **1,** 449; **2,** 213, 221; **6,** 281; **8,** 246; **9,** 238; **10,** 199; **11,** 255
Hydriodic acid–Red phosphorus, **1,** 449, 862, 865; **7,** 171
Hydrobromic acid (*see also* Hydrogen bromide), **1,** 450; **2,** 214; **3,** 154; **4,** 249; **5,** 332; **6,** 282; **9,** 359
Hydrobromic acid–Dimethyl sulfoxide, **5,** 333
Hydrochloric acid (*see also* Hydrogen chloride), **1,** 308; **4,** 250; **5,** 322, 333, 507; **6,** 283
Hydrofluoric acid (*see also* Hydrogen fluoride), **6,** 284; **9,** 238
Hydrogen bromide (*see also* Hydrobromic acid), **1,** 3, 196, 453; **4,** 251; **6,** 284; **7,** 99; **8,** 488; **10,** 200; **11,** 75
Hydrogen bromide–Acetic acid, **1,** 452, 965; **2,** 305; **5,** 335; **7,** 171; **10,** 317; **12,** 242
Hydrogen bromide–Trifluoroacetic acid, **4,** 251
Hydrogen chloride (*see also* Hydrochloric acid), **2,** 215; **4,** 252; **5,** 335; **6,** 278, 285; **7,** 172; **8,** 246; **9,** 239
Hydrogen chloride–Acetic acid, **6,** 389
Hydrogen chloride–Titanium(IV) chloride, **10,** 201
Hydrogen cyanide, **1,** 226, 454, 1197; **2,** 261; **3,** 193; **4,** 252, 512, 538; **9,** 239, 293
Hydrogen cyanide–Diethylaluminum chloride, **2,** 427
Hydrogen cyanide–Triethylaluminum, **1,** 1198; **4,** 146, 526; **9,** 328
Hydrogen cyanide–Triethylamine, **5,** 689
Hydrogen cyanide–Trimethylaluminum, **2,** 427
Hydrogen fluoride (*see also* Hydrofluoric acid), **1,** 455; **2,** 215; **4,** 252; **5,** 336; **6,** 285; **7,** 173; **9,** 240; **11,** 453
Hydrogen fluoride–Antimony pentafluoride (*see* Hexafluoroantimonic acid)
Hydrogen fluoride–Boron trifluoride (*see* Boron trifluoride–Hydrogen fluoride)
Hydrogen fluoride–Dimethyl sulfide, **11,** 256
Hydrogen fluoride–Pyridine [*see* Pyridinium poly(hydrogen fluoride)]
Hydrogen hexachloroiridate(IV), **1,** 131; **2,** 67; **3,** 47; **4,** 83; **5,** 119
Hydrogen hexachloroplatinate(IV)–Dichlorosilane, **4,** 139
Hydrogen hexachloroplatinate(IV)–Diethoxymethylsilane, **12,** 243
Hydrogen hexachloroplatinate(IV)–Ethoxydiethylsilane, **5,** 293
Hydrogen hexachloroplatinate(IV)–Sodium borohydride, **1,** 890; **2,** 333
Hydrogen hexachloroplatinate(IV)–Tin(IV) chloride, **4,** 87
Hydrogen hexachloroplatinate(IV)–Tribenzylsilane, **1,** 891
Hydrogen hexachloroplatinate(IV)–Trichlorosilane, **8,** 54, 98, 378
Hydrogen hexachloroplatinate(IV)–Triethylsilane, **3,** 51
Hydrogen iodide (*see also* Hydriodic acid), **7,** 173; **10,** 217
Hydrogen peroxide, **1,** 202, 456, 457, 466, 471, 529, 800; **2,** 216, 217; **3,** 154, 155; **4,** 253, 254; **5,** 337, 738; **6,** 286, 459, 486; **7,** 174; **8,** 247; **9,** 241; **10,** 201; **12,** 242
Hydrogen peroxide–Acetic acid (*see also*

Peracetic acid), **1,** 459; **3,** 155; **6,** 286; **8,** 248; **10,** 331
Hydrogen peroxide–Acetoxymethyl methyl selenide, **8,** 5
Hydrogen peroxide–Alumina, **8,** 64
Hydrogen peroxide–Aluminum chloride, **4,** 255
Hydrogen peroxide–Ammonium heptamolybdate, **12,** 245
Hydrogen peroxide–Benzeneseleninic acid, **8,** 22
Hydrogen peroxide–Benzoyl isocyanate, **6,** 35
Hydrogen peroxide–Bis(2,4-dinitrophenyl) diselenide, **8,** 359
Hydrogen peroxide–Bis(*o*-nitrophenyl) diselenide, **8,** 359
Hydrogen peroxide–Boron trifluoride, **1,** 463
Hydrogen peroxide–N-Bromosuccinimide, **5,** 618
Hydrogen peroxide–N,N′-Carbonyldiimidazole, **9,** 242
Hydrogen peroxide–1,1-Carbonyldi-1,2,4-triazole, **6,** 35
Hydrogen peroxide–Cerium(IV) oxide, **6,** 99
Hydrogen peroxide–*p*-Chlorophenyl isocyanate, **3,** 156
Hydrogen peroxide–1,3-Dibromo-5,5-dimethylhydantoin, **5,** 618; **6,** 287
Hydrogen peroxide–Dicyclohexylcarbodiimide, **1,** 231
Hydrogen peroxide–Diisopropylcarbodiimide, **8,** 248
Hydrogen peroxide–Ferric acetate, **1,** 473
Hydrogen peroxide–Ferric acetylacetonate, **5,** 307; **6,** 304
Hydrogen peroxide–Ferric sulfate, **1,** 473
Hydrogen peroxide–Ferrous chloride, **1,** 473
Hydrogen peroxide–Ferrous sulfate, **1,** 473; **5,** 340
Hydrogen peroxide–Formic acid, **8,** 248
Hydrogen peroxide–Hexafluoroacetone, **3,** 149; **9,** 244; **10,** 206
Hydrogen peroxide–Hydrobromic acid, **8,** 248
Hydrogen peroxide–Hydrochloric acid, **1,** 459; **7,** 174; **8,** 248
Hydrogen peroxide–Hydrogen fluoride–Boron trifluoride etherate, **2,** 402; **11,** 256
Hydrogen peroxide–Iron(III) acetylacetonate, **5,** 307; **6,** 304
Hydrogen peroxide–Iron salts (Fenton's reagent), **1,** 472; **5,** 340
Hydrogen peroxide–Lead dioxide, **1,** 473
Hydrogen peroxide–Mercury(II) trifluoroacetate, **8,** 248
Hydrogen peroxide–N-Methylmorpholine oxide, **1,** 690; **3,** 204
Hydrogen peroxide–Osmium tetroxide, **1,** 475
Hydrogen peroxide–Palladium acetate, **10,** 203
Hydrogen peroxide–Perchloric acid, **7,** 174
Hydrogen peroxide–Phenyl isocyanate, **3,** 156
Hydrogen peroxide–Potassium carbonate, **9,** 243
Hydrogen peroxide–Pyridinium polyhydrogen fluoride, **9,** 244
Hydrogen peroxide–Salt and oxide catalysts, **1,** 472
Hydrogen peroxide–Selenium(IV) oxide, **1,** 477; **2,** 362; **3,** 156; **8,** 440; **9,** 244
Hydrogen peroxide–Silver nitrate, **8,** 247
Hydrogen peroxide–Silver trifluoroacetate, **7,** 175; **8,** 247, 249; **10,** 203
Hydrogen peroxide–Sodium peroxide, **7,** 175; **8,** 250
Hydrogen peroxide–Sodium tungstate, **1,** 51, 475; **10,** 205; **12,** 246
Hydrogen peroxide–1,1,3,3-Tetrachloroacetone, **10,** 204
Hydrogen peroxide–Tetrafluoroboric acid, **1,** 1139
Hydrogen peroxide–Thallium(I) ethoxide, **6,** 286
Hydrogen peroxide–Triethyl orthoacetate, **10,** 204
Hydrogen peroxide–Tris(acetylacetonate)-iron(III), **5,** 307; **6,** 304
Hydrogen peroxide–Tungstic acid, **1,** 51, 475; **10,** 205; **12,** 246
Hydrogen peroxide–Vanadium(V) oxide, **1,** 474
Hydrogen peroxide–Vilsmeier reagent, **11,** 256
Hydrogen selenide, **6,** 288; **7,** 175
Hydrogen selenide–Triethylamine, **5,** 341
Hydrogen sulfide, **1,** 707, 962, 1026; **4,** 256; **5,** 693, 726; **8,** 459; **12,** 247
Hydrogen telluride, **10,** 205
Hydrogen tetrachloroaurate(III), **1,** 414; **5,**

647; **7,** 277
2-Hydroperoxyhexafluoro-2-propanol, **3,** 149; **9,** 244; **10,** 206
2-Hydroperoxy-2-nitropropane, **4,** 256
2-Hydroperoxy-1,1,3,3-tetrachloro-2-propanol, **10,** 204
Hydroquinone, **5,** 341
1-Hydroxybenzotriazole, **3,** 156; **5,** 342; **6,** 288, 425; **11,** 159; **12,** 199, 421
10-Hydroxy-10,9-boroxarophenanthrene, **2,** 32
4-Hydroxy-2-butenolide, **10,** 290
1-Hydroxycyclopropanecarboxaldehyde, **12,** 247
4-Hydroxy-5,5-dimethyl-1-hexen-3-one, **12,** 248
(S)-(2-Hydroxy-N,N-dimethylpropanamide–O,O′)-oxodiperoxymolybdenum(VI), **10,** 206
(S)-(+)-α-Hydroxy-β,β-dimethylpropyl vinyl ketone, **12,** 248
5-(2′-Hydroxyethyl)-4-methyl-3-benzylthiazolium chloride [*see* 3-Benzyl-5-(2-hydroxyethyl)-4-methyl-1,3-thiazolium chloride]
α-Hydroxyhippuric acid, **6,** 289
(1R)-(−)-*cis*-3-Hydroxyisobornyl neopentyl ether, **12,** 249
3-Hydroxy-4-keto-3,4-dihydrobenzo-1,2,3-triazine, **5,** 342
Hydroxylamine, **1,** 478, 565, 903, 939; **2,** 217; **5,** 206; **6,** 400, 533, 538; **7,** 176, 225; **9,** 245, 409, 438; **10,** 206; **11,** 257; **12,** 67, 251
Hydroxylamine–Potassium cyanide, **5,** 342
Hydroxylamine sulfate, **1,** 481
Hydroxylamine-O-sulfonic acid, **1,** 124, 258, 481; **2,** 217; **3,** 99, 156; **4,** 256; **5,** 343, 433; **6,** 290; **8,** 250; **9,** 245; **10,** 207
Hydroxylaminium acetate, **10,** 207
2-Hydroxymethylanthraquinone, **8,** 252
3-Hydroxy-3-methylbutylidenetriphenylphosphorane, lithium salt, **8,** 251
4-Hydroxymethyl-2,6-di-*t*-butylphenol, **4,** 257
(−)-3-Hydroxy-5-methylhydantoin, **8,** 252
(S)-2-Hydroxymethyl-1-methylpyrrolidine, **10,** 266
(2S,2′S)-2-Hydroxymethyl-1-[(1-methylpyrrolidin-2-yl)methyl]-pyrrolidine, **8,** 253; **9,** 246; **10,** 207
N-Hydroxymethylphthalimide, **1,** 484; **8,** 253
1-Hydroxymethyl-2-tributylstannylcyclopropane, **12,** 525
2-Hydroxymethyl-3-trimethylsilyl-1-propene, **9,** 454; **11,** 258, 579; **12,** 252
2-Hydroxy-1,4-naphthoquinone, **1,** 484
2-Hydroxy-5-nitrobenzyl bromide, **1,** 485
N-Hydroxy-5-norbornene-2,3-dicarboximide, **6,** 290; **7,** 177
2-Hydroxy-5-oxo-5,6-dihydro-2*H*-pyran, **6,** 291
9-Hydroxy-9-phenylanthrone, **4,** 378
N-Hydroxyphthalimide, **1,** 485; **3,** 157; **7,** 177, 406
(1S,2S,5S)-2-Hydroxypinan-3-one, **7,** 177
N-Hydroxypiperidine, **1,** 486
2-Hydroxypyridine (*see* 2-Pyridone)
3-Hydroxypyridine, **1,** 486
3-Hydroxy-2-pyrone, **6,** 291; **8,** 254
8-Hydroxyquinolinedihydroboronite, **12,** 252
4-Hydroxysalcomine, **7,** 316
N-Hydroxysuccinimide, **1,** 487; **5,** 342; **6,** 425; **8,** 536; **9,** 246
N-Hydroxysuccinimide trifluoroacetate, **2,** 219; **3,** 157
4-Hydroxy-2,2,6,6-tetramethylpiperidinooxyl, **9,** 247
Hypobromous acid, **1,** 74, 80, 800; **4,** 48, 49; **5,** 61; **6,** 75; **10,** 58; **11,** 79
Hypochlorous acid, **1,** 245, 487; **10,** 208; **11,** 260
Hypofluorous acid, **8,** 254
Hypohalite solution, **1,** 488
Hypophosphorous acid, **1,** 489; **3,** 157; **8,** 255

Imidazole, **1,** 492; **2,** 220; **3,** 323; **4,** 176; **5,** 532; **12,** 416
1-(2-Imidazoline-2-yl)-2-methylthio-2-imidazoline hydroiodide, **5,** 345
N-Imino-N,N-dimethyl-2-hydroxypropanaminium ylide, **8,** 256
Indane-1,2,3-trione (*see* Ninhydrin)
Indanocyclone, **4,** 258
Indicators, **1,** 494
Indicators, Hammett acidity, **1,** 495
Indium, **6,** 293
Iodine, **1,** 495; **2,** 220, 321; **3,** 159; **4,** 105, 258, 295, 306; **5,** 97, 346, 516; **6,** 75, 117, 178, 259, 293, 662; **7,** 179; **8,** 104, 180, 235, 256; **9,** 52, 91, 199, 248, 386, 498; **10,** 210; **11,** 261, 289, 600; **12,** 253, 478
Iodine–Aluminum chloride–Copper(II)

chloride, **11,** 267
Iodine azide, **1,** 500; **2,** 100, 222; **3,** 160; **4,** 262; **5,** 350; **6,** 297; **8,** 260; **10,** 211
Iodine–Chlorosulfuric acid, **10,** 86
Iodine–Copper(II) acetate, **8,** 260; **10,** 211; **11,** 267; **12,** 256
Iodine–Copper(I) chloride–Copper(II) chloride, **12,** 256
Iodine–Dimethyl sulfoxide (*see* Dimethyl sulfoxide–Iodine)
Iodine fluoride, **2,** 365; **5,** 351
Iodine isocyanate, **1,** 501; **2,** 223; **3,** 161; **4,** 410
Iodine–Lead tetraacetate, **1,** 552
Iodine–Mercury(II) oxide (*see* Mercury(II) oxide–Iodine)
Iodine monobromide, **1,** 501; **2,** 224; **6,** 297
Iodine monochloride, **1,** 502; **8,** 261; **9,** 277; **10,** 212; **11,** 268, 488
Iodine–Morpholine complex, **1,** 502
Iodine nitrate, **2,** 224
Iodine pentafluoride, **1,** 503
Iodine–Periodic acid dihydrate, **2,** 220; **3,** 163; **4,** 259
Iodine–Potassium *t*-butoxide, **5,** 349
Iodine–Potassium iodate, **5,** 349; **6,** 296; **7,** 180, 182
Iodine–Potassium iodide, **4,** 260; **9,** 249
Iodine–Pyridine, **1,** 503
Iodine–Pyridine N-oxide, **6,** 296
Iodine–Pyridine–Sulfur dioxide, **6,** 296
Iodine–Pyridinium chlorochromate, **12,** 256
Iodine–Silver acetate, **1,** 1003; **4,** 261; **7,** 37; **9,** 249; **12,** 256
Iodine–Silver benzoate, **1,** 1007; **3,** 251
Iodine–Silver carboxylates, **1,** 495, 504, 1003, 1007, 1019, 1220; **3,** 251; **4,** 261; **5,** 555; **7,** 37; **9,** 249; **11,** 268; **12,** 256
Iodine–Silver fluoride, **1,** 504; **2,** 365; **5,** 351
Iodine–Silver nitrate, **2,** 224; **11,** 268; **12,** 257
Iodine–Silver oxide, **7,** 180
Iodine–Silver salts, **1,** 504
Iodine–Silver trifluoroacetate, **1,** 495, 504, 1019, 1220; **5,** 555
Iodine–Thallium(I) acetate (*see* Thallium(I) acetate–Iodine)
Iodine thiocyanate, **4,** 263
Iodine tris(trifluoroacetate), **4,** 263; **7,** 181
Iodoacetamide, **1,** 504
Iodoacetonitrile, **6,** 494
Iodoacetoxy(tributyl)tin, **12,** 258
Iodoalane (*see* Monoiodoalane)
Iodobenzene, **1,** 505
Iodobenzene bis(trifluoroacetate) [*see* Phenyliodine(III) bis(trifluoroacetate)]
Iodobenzene diacetate (*see* Phenyliodine(III) diacetate)
Iodobenzene dihalide (*see* Phenyliodine(III) dihalide)
Iodobenzene ditrifluoroacetate [*see* Phenyliodine(III) bis(trifluoroacetate)]
o-Iodobenzoic acid, **1,** 506
m-Iodobenzoic anhydride, **10,** 215
B-Iodo-9-borabicyclo[3.3.1]nonane, **12,** 236
4-Iodobutyltrimethyltin, **10,** 219
Iodocyanoketene, **8,** 88
Iodo-2,4-dinitrobenzene, **1,** 329
5-Iodo-2,4-dinitrophenylhydrazine, **1,** 506
1-Iodoethyl benzoate, **3,** 141
Iodoethynyl(trimethyl)silane, **4,** 265
Iodoform, **1,** 425; **3,** 165; **4,** 29, 266; **5,** 27; **12,** 417
1-Iodoheptafluoropropane, **2,** 226
Iodo(iodomethyl)mercury, **1,** 506
Iodo(iodomethyl)zinc (*see* Simmons–Smith reagent)
Iodomethane (*see* Methyl iodide)
Iodo(methyl)calcium, **5,** 442
Iodomethyl methyl ether, **9,** 250
Iodomethyl methyl sulfide, **5,** 353; **6,** 301
3-Iodo-2-methylpropene (*see* Methallyl iodide)
Iodomethyltributyltin, **9,** 250, 475; **11,** 543
Iodomethyltrimethylsilane, **9,** 445
2-Iodomethyl-3-trimethylsilyl-1-propene, **11,** 579
Iodomethyltriphenyltin, **9,** 509
Iodomethylzinc iodide (*see* Simmons–Smith reagent)
Iodonium di-*sym*-collidine perchlorate, **10,** 212; **11,** 269; **12,** 526
Iodophenylbis(triphenylphosphine)-palladium, **11,** 269
Iodo(phenyl)ytterbium, **11,** 249
2-Iodopropane, **5,** 661
3-Iodo-1-propene, **3,** 253
(3-Iodopropyl)trimethylsilane, **12,** 26
Iodoso, *see also* Iodosyl
Iodosobenzene diacetate or dihalides, *see* Phenyliodine(III)
N-Iodosuccinimide (NIS), **1,** 510; **2,** 40; **10,** 216, 349; **12,** 79, 258

N-Iodosuccinimide–Acetic acid, **7,** 182
N-Iodosuccinimide–Carboxylic acids, **7,** 182
N-Iodosuccinimide–Tetrabutylammonium iodide, **10,** 216
Iodosylbenzene, **1,** 507; **10,** 213; **11,** 270; **12,** 258
Iodosylbenzene–Ruthenium catalysts, **11,** 270
o-Iodosylbenzoic acid, **1,** 509; **12,** 259
Iodotri(butylphosphine)copper, **2,** 282, 400, 441; **3,** 79, 278; **4,** 221, 516; **6,** 266
Iodotrichlorosilane, **11,** 466
Iodotriethylsilane, **12,** 259
Iodotrifluoromethane (*see* Trifluoroiodomethane)
Iodo(trimethyl phosphite)copper, **3,** 278; **5,** 244, 717; **6,** 630
Iodotrimethylsilane, **8,** 261; **9,** 74, 251; **10,** 59, 97, 113, 216; **11,** 127, 253, 271, 481; **12,** 127, 259, 317, 463, 527, 544
(E)-1-Iodo-3-trimethylsilyl-2-butene, **5,** 355
3-Iodo-2-trimethylsilylmethyl-1-propene, **11,** 579
3-Iodo-1-trimethylsilylpropyne, **11,** 522
1-Iodo-4-(trimethylstannyl)butane, **10,** 219
Iodotriphenylphosphonium iodide (*see* Triphenylphosphine–Iodine)
Iodoxy, *see* Iodyl
Iodylbenzene, **1,** 511; **11,** 275
m-Iodylbenzoic acid, **12,** 102
Ion-exchange resins (*see also* specific names), **1,** 511; **2,** 227; **4,** 266, 267; **5,** 355; **6,** 302; **7,** 182; **8,** 263; **9,** 256; **10,** 220; **11,** 276; **12,** 87
Iridium catalysts, **2,** 228, 434; **8,** 135; **9,** 257
Iridium(III) chloride, **1,** 132; **2,** 67; **5,** 572
Iridium(IV) chloride, **2,** 228; **3,** 166
Iridium–Silica, **10,** 342
Iridium tetrachloride [*see* Iridium(IV) chloride]
Iridium trichloride [*see* Iridium(III) chloride]
Iron, **1,** 519; **2,** 229; **3,** 167; **5,** 357; **9,** 257; **12,** 263
Iron(II), *see* Ferrous
Iron(III), *see also* Ferric
Iron–Acetic acid, **8,** 264
Iron(III) acetylacetonate, **4,** 268; **6,** 259, 304; **12,** 557
Iron–carbene complexes, **12,** 265
Iron carbonyl, **1,** 519; **2,** 229; **3,** 167, 207, 242; **4,** 268, 420; **5,** 96, 357; **6,** 304, 625; **7,** 183; **8,** 13, 265, 390, 533; **9,** 67, 118; **10,** 221; **11,** 570; **12,** 266
Iron ennacarbonyl (*see* Tri-μ-carbonyl-hexacarbonyldiiron)
Iron–Graphite, **11,** 276
Iron pentacarbonyl (*see* Iron carbonyl)
Isoamyl nitrate, **1,** 40; **2,** 25
Isoamyl nitrite, **1,** 40, 46, 268, 751; **4,** 270; **5,** 358, 653; **6,** 307, 547; **9,** 258; **12,** 450
Isoamyl nitrite–Trifluoroacetic acid, **1,** 520
Isobenzofuran, **1,** 521
Isobornylamine, **9,** 62
Isobornyl chloroformate, **5,** 359
Isobornyloxyaluminum dichloride, **1,** 522
Isobornyloxycarbonyl chloride, **5,** 359
Isobutene, **1,** 522; **5,** 82; **9,** 87
Isobutenyllithium (*see* Methallyllithium)
Isobutylaluminum dichloride, **6,** 307
Isobutyl chloroformate, **1,** 86; **2,** 47; **6,** 307; **7,** 307
Isobutylmagnesium bromide, **12,** 168
Isobutyronitrile, **10,** 110
Isocyanic acid, **1,** 170
Isocyanomethane phosphonic acid diethyl ester, **4,** 271; **6,** 187
Isocyanomethyllithium, **4,** 272; **10,** 231; **11,** 285
Isocyanomethyl *p*-tolyl sulfone (*see* Tosylmethyl isocyanide)
Isocyanotolylthiomethyllithium, **4,** 516
B-Isopinocampheyl-9-borabicyclo[3.3.1]-nonane (see B-3-Pinanyl-9-borabicyclo-[3.3.1]nonane)
Isopinocampheylborane (*see* Mono-isopinocampheylborane)
Isoprene, **1,** 1112; **4,** 348, 350; **5,** 469, 630; **6,** 48
Isoprene epoxide, **6,** 307
Isoprene hydrobromide (*see* 1-Bromo-3-methyl-2-butene)
Isoprenyltrimethylsilane (*see* 2-Trimethyl-silylmethyl-1,3-butadiene)
Isopropanol (*see* 2-Propanol)
Isopropenyl acetate, **1,** 524; **2,** 191, 230; **3,** 103; **4,** 100, 273; **5,** 237; **6,** 308, 356
Isopropenyl ether (*see* 2-Methoxypropene)
Isopropenyllithium, **2,** 456
Isopropenylmagnesium bromide, **2,** 92; **9,** 150
Isopropenyl methyl ether (*see* 2-Methoxy-propene)
Isopropenyl phenyl sulfone, **10,** 316

2-Isopropenyl-4,4,6-trimethyl-5,6-dihydro-1,3-(4*H*)-oxazine, **3,** 283
Isopropenyltriphenylphosphonium bromide, **6,** 667; **10,** 225
1-Isopropoxy-1-trimethylsilyloxycyclopropane, **12,** 499
Isopropylamine, **2,** 44, 414
Isopropylamine–Borane (*see* Borane–Isopropylamine)
Isopropyl bromide, **9,** 69
Isopropylcyclohexylaminomagnesium bromide, **10,** 115; **12,** 75
Isopropyl dichloroacetate, **6,** 331
(3S,6R)-(+)-6-Isopropyl-2,5-dimethoxy-3-methyl-3,6-dihydropyrazine, **11,** 193
Isopropyl glyoxylate, **12,** 514
Isopropylidenecarbene, **3,** 118, 187; **6,** 230, 619; **7,** 358
Isopropylidene diazomalonate, **9,** 258
2,3-O-Isopropylidene-2,3-dihydroxy-1,4-bis-(diphenylphosphine)butane (DIOP), **4,** 273; **5,** 360, 500; **6,** 309; **9,** 259; **10,** 387; **11,** 31, 277, 563
[2,3-O-Isopropylidene-2,3-dihydroxy-1,4-bis-(diphenylphosphine)butane]palladium, **9,** 239
2,3-O-Isopropylidene-2,3-dihydroxy-1,4-bis-(di-*m*-tolylphosphine)butane, **9,** 259
Isopropylidenediphenylsulfurane, **7,** 183
Isopropylidene isopropylidenemalonate, **6,** 309
Isopropylidene malonate (*see* Meldrum's acid)
Isopropylidenetriphenylphosphorane, **1,** 1239; **5,** 361; **7,** 183; **8,** 339; **11,** 147; **12,** 294
Isopropyl iodide, **5,** 661
Isopropyl isocyanate, **11,** 277
Isopropylmagnesium bromide or chloride, **6,** 179; **8,** 194; **12,** 245
Isopropylmagnesium chloride–Iron(III) reagents, **9,** 257
4-Isopropyl-2-oxazolidinone, **11,** 379; **12,** 359
Isopropyl phenylsulfonylacetate, **9,** 454
Isopropylphosphonic acid bis(dimethylamide), **2,** 280
2-Isopropylpyrrolidine, **4,** 471
P-Isopropyl-N,N,N′,N′-tetramethylphosphonic diamide, **2,** 280
Isovaline, **4,** 274
Itaconic anhydride, **8,** 137

Jones reagent, **1,** 142; **2,** 70; **3,** 54; **4,** 95; **6,** 123; **7,** 68; **12,** 131

KAPA (see Potassium 3-aminopropylamide)
Karl Fischer reagent, **1,** 528
KDA (see Potassium diisopropylamide)
Ketene, **1,** 359, 528; **2,** 104, 232; **3,** 32; **4,** 275; **5,** 291; **6,** 11
Ketene alkyl trialkylsilyl acetals(ketals), **4,** 32, 538; **5,** 689; **7,** 371; **8,** 369, 484; **9,** 310; **10,** 92, 261, 401; **11,** 279, 340; **12,** 62, 74, 271, 500, 558
Ketene bis(methylthio)ketal monoxide (*see* Ketene dimethyl thioacetal monoxide)
Ketene bis(phenylthio)ketal, **9,** 261
Ketene bis(trimethylsilyl)acetals(ketals), **11,** 600; **12,** 268
Ketene *t*-butyldimethylsilyl methyl acetal, **11,** 279
Ketene diethyl acetal, **1,** 165, 917; **5,** 362; **9,** 262; **10,** 226
Ketene diethyl thioacetal monoxide, **11,** 155
Ketene dimethyl acetal, **9,** 263; **11,** 279; **12,** 206
Ketene dimethyl thioacetal monoxide, **5,** 302; **6,** 311; **8,** 268; **9,** 261; **10,** 227
Ketene diphenyl thioacetal, **9,** 261
Ketene methylthio phenylthio acetal, **8,** 281
Ketenylidenetriphenylphosphorane, **2,** 446; **10,** 450; **11,** 281; **12,** 268
3-Ketoglutaric acid (*see* 1,3-Acetonedicarboxylic acid)
α-Ketoglutaric acid, **1,** 531
α-Ketoketene dithioacetals, **11,** 282
Kiliani reagent, **1,** 144

Lanthanum chloride, **9,** 211
Lanthanum nitrate, **11,** 283
Lauryldimethylsulfonium chloride, **5,** 363
Lawesson's reagent [*see* 2,4-Bis-(4-methoxyphenyl)-1,3-dithia-2,4-diphosphetane-2,4-disulfide]
Lazier catalyst (*see* Copper chromite)
LDA (*see* Lithium diisopropylamide)
LDMAN [*see* 1-(Dimethylamino)-naphthalenide]
LDP [*see* (Diphenylphosphine)lithium]
Lead, **2,** 233
Lead(II) acetate, **1,** 532; **2,** 233; **4,** 276

Lead(IV) acetate azides, **4,** 276; **5,** 363
Lead carbonate, basic, **9,** 265
Lead dioxide (*see* Lead(IV) oxide)
Lead(IV) fluoride, **1,** 563
Lead(II) oxide, **1,** 536
Lead(IV) oxide, **1,** 533; **2,** 233, 260, 345; **3,** 168; **5,** 583; **7,** 55
Lead(IV) oxide–Boron trifluoride etherate, **11,** 283
Lead tetraacetate (LTA), **1,** 537; **2,** 234; **3,** 168, 287; **4,** 278, 321; **5,** 14, 43, 365, 726; **6,** 4, 313, 426; **7,** 55, 185; **8,** 118, 269; **9,** 265; **10,** 214, 228, 281, 397; **12,** 270
Lead tetraacetate–Boron trifluoride etherate, **1,** 541; **2,** 234; **6,** 316; **10,** 228
Lead tetraacetate–Copper(II) acetate, **2,** 84, 236; **3,** 168; **5,** 365; **6,** 313; **7,** 186
Lead tetraacetate–N-Halosuccinimides, **2,** 238; **5,** 370; **6,** 317
Lead tetraacetate–Metal halides, **1,** 557; **4,** 278; **5,** 370; **11,** 283
Lead tetraacetate–Trifluoroacetic acid, **6,** 317
Lead tetrabenzoate, **1,** 563; **7,** 186, 189
Lead tetrakis(trifluoroacetate) (LTTFA), **2,** 238; **4,** 282; **5,** 370; **6,** 318
Lead tetrakis(trifluoroacetate)–Diphenyl disulfide, **9,** 269
Lead(II) thiocyanate, **1,** 30
LEDA (*see* Lithium–Ethylenediamine, N-Lithioethylenediamine)
Lemieux–Johnson reagent (*see* Sodium periodate–Osmium tetroxide)
Lemiuex–von Rudloff reagent (*see* Sodium periodate–Potassium permanganate)
Leucine *t*-butyl ester, **7,** 189; **8,** 272; **10,** 229
Levulinic acid, **1,** 564; **2,** 239; **10,** 230; **11,** 283
Levulinic anhydride, **6,** 318; **11,** 283
LiICA (*see* Lithium N-isopropylcyclohexylamide)
Lindlar catalyst, **1,** 566; **3,** 171; **4,** 283; **6,** 319; **9,** 270; **11,** 553; **12,** 272
Lithioacetonitrile, **6,** 84, 636
Lithioallyltrimethylsilane {*see* [(Trimethylsilyl)allyl]lithium}
2-Lithiobenzothiazole, **8,** 274
α-Lithiobenzyl phenyl sulfide, **5,** 528
Lithio *t*-butyl acetate (*see* *t*-Butyl lithioacetate)
Lithio *t*-butyl trimethylsilylacetate, **5,** 371; **7,** 51; **12,** 228
α-Lithio-α-chloroethyltrimethylsilane, **8,** 277; **11,** 128
Lithio chloromethylphenyl sulfoxide, **8,** 94; **9,** 271
α-Lithio-α-chloromethyltrimethylsilane, **8,** 277; **12,** 118
1-Lithiocyclopropyl methyl selenide, **9,** 85; **10,** 318
1-Lithiocyclopropyl phenyl sulfide, **5,** 372; **6,** 319; **7,** 190; **8,** 138, 139; **9,** 271, 284; **10,** 231; **11,** 284; **12,** 19, 280
5-Lithio-2,3-dihydrofuran, **8,** 70
2-Lithio-4,5-dihydro-5-methyl-[4*H*]-1,3,5-dithiazine, **8,** 305; **11,** 302
6-Lithio-3,4-dihydro-(2*H*)-pyran, **8,** 70
2-Lithio-2-(2′,2′-dimethoxyethyl)-1,3-dithiane, **4,** 165
Lithio-N,N-dimethylacetamide, **6,** 84, 636
Lithio-N,N-dimethylthiopivalamide, **7,** 190
2-Lithio-1,3-dithianes (*see also* 1,3-Dithiane), **2,** 183; **3,** 135; **4,** 165; **5,** 287; **6,** 248, 667; **7,** 143; **8,** 217, 346; **9,** 48, 235; **10,** 231; **11,** 285; **12,** 573
1-Lithio-4-ethoxybutadiene, **8,** 278
Lithio ethyl acetate (*see* Ethyl lithioacetate)
N-Lithioethylenediamine (LEDA, *see also* Lithium–Ethylenediamine), **1,** 567; **2,** 239; **11,** 296
Lithio ethyl trimethylsilylacetate [*see* Ethyl lithio(trimethylsilyl)acetate]
2-Lithioisobutyronitrile, **8,** 22; **9,** 21
α-Lithio-α-methoxyallene, **7,** 226; **8,** 320; **9,** 272; **10,** 257; **12,** 310, 420
1-Lithio-1-methoxycyclopropane, **10,** 383
1-Lithio-2-methoxycyclopropane, **6,** 366
2-(2-Lithio-4-methoxyphenyl)-4,4-dimethyl-2-oxazoline, **8,** 279
2-Lithio-2-methoxypropane, **10,** 244
1-Lithio-3-methoxy-3-trimethylsilylallene, **11,** 577
Lithiomethyl(diphenylarsine) oxide, **8,** 280
2-Lithio-2-methyl-1,3-dithiane, **3,** 136
α-Lithiomethylenetriphenylphosphorane, **11,** 338
Lithiomethylisocyanide, **4,** 272; **10,** 231; **11,** 285
2-Lithio-2-methylpropionitrile, **8,** 22; **9,** 21
α-Lithiomethylselenocyclobutane, **10,** 232
2-Lithio-2-methylthio-1,3-dithiane, **4,** 217; **7,** 191
1-Lithio-1-methyl-2-vinylcyclopropane, **9,** 273

2-Lithiopropionitrile, **9,** 84
2-Lithio-2-propyl-1,3-benzodithiolane, **9,** 4
Lithiopropylidene-*t*-butylimine, **4,** 283
1-Lithio-3-tetrahydropyranyloxy-1-propyne, **7,** 320
1-Lithio-3,3,6,6-tetramethoxy-1,4-cyclohexadiene, **7,** 191; **9,** 40
2-Lithio-3,3,6,6-tetramethoxy-1,4-cyclohexadiene, **8,** 280
2-Lithiothiomethoxy-1,3-thiazoline, **6,** 399
6-Lithio-2H-thiopyran, **8,** 242
α-Lithio-*p*-tolylthiomethylisocyanide, **4,** 516
1-Lithio-2-tributylstannylethylene, **5,** 43
3-Lithio-1-triisopropylsilyl-1-propyne, **11,** 566
1-Lithio-2,4,6-trimethylbenzene (*see* Mesityllithium)
Lithiotrimethylsilylacetonitrile, **10,** 428
2-Lithio-2-trimethylsilyl-1,3-dithiane, **2,** 184; **4,** 284; **5,** 374; **6,** 320; **10,** 380
1-Lithio-1-trimethylsilylethylene [*see* (1-Lithiovinyl)trimethylsilane]
1-Lithio-3-trimethylsilyl-1-propyne, **11,** 63
3-Lithio-1-trimethylsilyl-1-propyne, **2,** 239; **3,** 173; **6,** 638; **11,** 566
Lithio triphenylphosphinoacetonide, **5,** 375
2-Lithio-1,3,5-trithiane, **3,** 329
1-Lithio-2-vinylcyclopropane, *cis-* and *trans-*, **7,** 192, 212; **9,** 273
α-Lithiovinyl phenyl sulfide, **8,** 281; **9,** 371
2-Lithiovinyltributyltin, **5,** 43
(1-Lithiovinyl)trimethylsilane, **5,** 374; **7,** 193, 279; **8,** 57; **9,** 498; **10,** 442; **11,** 286
2-(E)-Lithiovinyltrimethylsilane, **10,** 443
(1-Lithiovinyl)triphenylsilane, **5,** 68, 375
Lithium, **1,** 570, 638; **3,** 174; **4,** 286; **5,** 376; **6,** 666; **10,** 234; **11,** 126
Lithium acetylide, **1,** 573; **2,** 241, 288; **5,** 346; **6,** 324, 451; **8,** 285, 472; **12,** 272, 351, 410
Lithium acetylide–Ethylenediamine, **1,** 574; **2,** 208; **5,** 382; **6,** 324; **7,** 195
Lithium acetylides, **2,** 166; **4,** 551; **5,** 346; **7,** 101; **9,** 111; **11,** 288; **12,** 70, 207
Lithium–Alkylamines (*see also* specific names, specific Lithium dialkylamides), **1,** 574; **2,** 241; **3,** 175; **4,** 287; **5,** 377; **6,** 322; **7,** 194
Lithium alkyl(*t*-butoxy)cuprates, **5,** 395, 415
Lithium alkyl(*t*-butylthio)cuprates, **5,** 415
Lithium alkyl(cyano)cuprates, **9,** 329; **10,** 287
Lithium alkyl(diethylamino)cuprates, **5,** 415
Lithium alkyl(phenoxy)cuprates, **5,** 415
Lithium alkyl(phenylthio)cuprates, **5,** 414; **6,** 343, 465; **7,** 211, 212; **9,** 329; **11,** 372
Lithium aluminum hydride, **1,** 383, 581, 598, 677; **2,** 242, 378; **3,** 176, 261, 275; **4,** 160, 291, 312, 473; **5,** 326, 382; **6,** 325; **7,** 196; **8,** 286, 290, 450; **9,** 274, 444; **10,** 236, 357; **11,** 289; **12,** 272
Lithium aluminum hydride–Aluminum chloride (*see also* Aluminum hydride), **1,** 34, 595, 620; **2,** 243; **3,** 176, 275; **4,** 293; **5,** 186, 389; **7,** 372; **8,** 289
Lithium aluminum hydride–(S)-4-Anilino-3-methylamino-1-butanol, **11,** 292; **12,** 33
Lithium aluminum hydride–(S)-2-(Anilinomethyl)pyrrolidine, **8,** 423
Lithium aluminum hydride–3-O-Benzyl-1,2-O-cyclohexylidene-α-D-glucofuranose complex, **2,** 243
Lithium aluminum hydride–Berylium chloride, **5,** 14
Lithium aluminum hydride–(S,S)- or (R,R)-N,N′-Bis(methylbenzyl)sulfamide, **12,** 60
Lithium aluminum hydride–Boron trifluoride etherate, **1,** 599; **9,** 277
Lithium aluminum hydride–Cobalt(II) chloride, **8,** 291
Lithium aluminum hydride–Copper(I) *t*-butoxide, **10,** 237
Lithium aluminum hydride–Copper(II) chloride, **5,** 9
Lithium aluminum hydride–Copper(I) iodide, **6,** 326; **7,** 196; **10,** 237
Lithium aluminum hydride–Cryptate[2.2.1], **9,** 127
Lithium aluminum hydride–Darvon alcohol, **5,** 231; **6,** 206; **8,** 184; **11,** 153
Lithium aluminum hydride–1,4-Diazabicyclo[2.2.2]octane, **9,** 278
Lithium aluminum hydride–Di-μ-carbonyldecacarbonyltri-*triangulo*-iron, **9,** 278
Lithium aluminum hydride–Dichlorobis-(cyclopentadienyl)zirconium, **10,** 237
Lithium aluminum hydride–(R)- and (S)-2,2′-Dihydroxy-1,1′-binaphthyl (Noyori's reagent), **9,** 169; **10,** 148; **12,** 190
Lithium aluminum hydride–(S)-(−)-10,10′-Dihydroxy-9,9′-biphenanthryl, **12,** 191
Lithium aluminum hydride–Diphosphorus tetraiodide, **12,** 275

Lithium aluminum hydride–Iron carbonyl, **9,** 278
Lithium aluminum hydride–Mesitylcopper, **10,** 237
Lithium aluminum hydride–(–)-N-Methylephedrine, **9,** 308; **10,** 238; **12,** 322
Lithium aluminum hydride–(–)-N-Methylephedrine-3,5-Dimethylphenol, **10,** 265
Lithium aluminum hydride–Nickel(II) chloride, **8,** 291
Lithium aluminum hydride–(–)-*cis*-Pinanediol, **5,** 387
Lithium aluminum hydride–Pyridine, **1,** 599, 600; **4,** 294
Lithium aluminum hydride–Silica, **11,** 293; **12,** 275
Lithium aluminum hydride–Sodium methoxide, **4,** 295
Lithium aluminum hydride–Sulfuric acid, **5,** 14
Lithium aluminum hydride–Titanium(III) chloride (*see* Titanium(III) chloride–Lithium aluminum hydride)
Lithium aluminum hydride–Tri-μ-carbonylhexacarbonyldiiron, **9,** 278
Lithium aluminum hydride–(S)-2-(2,6-Xylidinomethyl)pyrrolidine, **10,** 458
Lithium aluminum hydride–Zinc chloride, **5,** 14
Lithium aluminum trialkoxyhydrides (*see* Lithium trialkoxyaluminum hydrides)
Lithium amalgam, **3,** 177; **4,** 287
Lithium amide, **1,** 573, 600; **3,** 178; **5,** 392; **9,** 278
Lithium–Ammonia (*see also* Birch reduction), **1,** 54, 601; **2,** 28, 245; **3,** 19, 179; **4,** 288; **5,** 30, 240, 379, 462; **6,** 235, 322; **7,** 21, 195; **8,** 189, 282; **9,** 273; **10,** 234; **11,** 286, 472, 473
Lithium *t*-amyl-*t*-butylamide, **4,** 311
Lithium anilidocyanocuprates, **11,** 367
Lithium azide, **6,** 327
Lithium benzeneselenoate, **8,** 183, 306; **10,** 238; **11,** 475
Lithium benzenethiolate, **4,** 311; **5,** 418; **8,** 183, 304
Lithium biphenylide, **1,** 612; **2,** 249
Lithium bis(alkylamino)cuprates, **6,** 327
Lithium bisbenzenesulfenimide, **3,** 20, 182
Lithium bis(3,3-diethoxy-1-propen-2-yl)cuprate, **6,** 331
Lithium bis(N,N-diethylcarbamoyl)cuprate, **9,** 279
Lithium bis(dimethylphenylsilyl)cuprate, **8,** 196; **12,** 210
Lithium bis(ethylenedioxyboryl)methide, **6,** 328; **10,** 239
Lithium bis(N-methyl-N-phenylcarbamoyl)cuprate, **11,** 373
Lithium bis(trimethylsilyl)amide (*see* Lithium hexamethyldisilazide)
Lithium 1,1-bis(trimethylsilyl)ethoxide, **7,** 198
Lithium 1,1-bis(trimethylsilyl)-3-methylbutoxide, **7,** 198
Lithium bis(3-trimethylsilyl-1-propen-2-yl)cuprate, **11,** 65
Lithium bis(2-vinylcyclopropyl)cuprate, **7,** 199
Lithium 9-boratabicyclo[3.3.1]nonane, **12,** 275
Lithium borohydride, **1,** 200, 603; **4,** 296; **10,** 378; **11,** 293; **12,** 276
Lithium bromide, **1,** 604; **2,** 245; **3,** 75, 95, 186, 306; **4,** 297; **5,** 326; **8,** 223; **12,** 100
Lithium bromide–Boron trifluoride etherate, **1,** 604
Lithium bromide–Dimethylformamide, **1,** 609
Lithium bromide–Dimethyl sulfoxide, **4,** 193
Lithium bromide–Hexamethylphosphoric triamide, **4,** 297; **5,** 395; **7,** 200; **11,** 300
Lithium bromide–Zinc, **7,** 338
Lithium bromo(methyl)cuprate–Diisobutylamine complex, **6,** 213
Lithium bronze, **11,** 293
Lithium 1,3-butadiene-1-olate, **8,** 291
Lithium–*t*-Butanol–Tetrahydrofuran, **1,** 604; **4,** 291
Lithium butoxide, **10,** 4
Lithium *t*-butoxide, **3,** 183; **4,** 297; **6,** 116
Lithium *t*-butoxy(*t*-butyl)cuprate, **5,** 396
Lithium *t*-butylbis(2-methylpropyl)aluminate, **10,** 239; **11,** 296
Lithium butylborohydride, **11,** 294
Lithium butyl(cyano)cuprate, **9,** 330
Lithium butyldiisobutylaluminum hydride, **10,** 239; **12,** 191, 276
Lithium *t*-butyldiisobutylaluminum hydride, **10,** 239; **11,** 296
Lithium (*t*-butylethynyl)methylcuprate, **5,** 149
Lithium *t*-butylethynyl(vinyl)cuprate, **5,** 149

Lithium butyl(hydrido)cuprate, **5,** 330
Lithium *t*-butylmercaptide, **5,** 412; **6,** 345; **7,** 210
Lithium *n*- or *t*-butyl(3-methoxy-3-methylbutynyl)cuprate, **8,** 325
Lithium *t*-butylneopentylamide, **4,** 310
Lithium *t*-butyl-*t*-octylamide, **12,** 285
Lithium *t*-butyl(phenylthio)cuprate, **5,** 414; **6,** 343, 465; **7,** 211
Lithium carbonate, **1,** 606; **3,** 183; **5,** 396
Lithium carbonate–Lithium halides, **1,** 606; **2,** 245; **4,** 298; **5,** 395, 452; **6,** 17; **7,** 200; **8,** 244; **12,** 203, 277
Lithium chlorate, **8,** 212
Lithium chloride, **1,** 609; **2,** 246; **3,** 95, 186; **8,** 223
Lithium chloride–1,5-Diazabicyclo[5.4.0]-undecene-5, **12,** 277
Lithium chloride–Diisopropylethylamine, **12,** 277
Lithium chloride–Dimethylformamide, **1,** 606
Lithium chloride–Dimethyl sulfoxide–Hexamethylphosphoric triamide, **7,** 169
Lithium chloride–Hexamethylphosphoric triamide, **6,** 233, 278; **10,** 240; **12,** 277
Lithium chloride–Hydrogen peroxide, **7,** 220
Lithium chloroacetylide, **6,** 294
Lithium 3-chloropropargylide, **5,** 397; **9,** 279; **11,** 294
Lithium–Chlorotrimethylsilane (*see* Chlorotrimethylsilane–Lithium)
Lithium cryptate[2.1.1], **12,** 405
Lithium cryptate[2.2.1], **9,** 127
Lithium cyanoborohydride, **1,** 618; **3,** 183
Lithium cyano(dimethylphenylsilyl)cuprate, **10,** 163
Lithium cyano(isohexyl)cuprate, **10,** 288
Lithium cyano(methyl)cuprate, **9,** 329
Lithium cyano(phenyl)cuprate, **9,** 329
Lithium cyano(trimethylsilyl)cuprate, **11,** 80
Lithium N-cyclohexylisopropylamide (*see* Lithium N-isopropylcyclohexylamide)
Lithium cyclopropyl(phenylthio)cuprate, **7,** 211; **11,** 372
Lithium diallylcuprate, **5,** 175; **7,** 86
Lithium 4,4′-di-*t*-butylbiphenylide, **7,** 200; **10,** 240
Lithium 9,9-dibutyl-9-borabicyclo[3.3.1]-nonanate, **6,** 62; **7,** 30, 201; **11,** 295; **12,** 446
Lithium dibutylcuprate, **2,** 152; **3,** 79, 112; **4,** 127; **5,** 187, 240; **6,** 163, 460; **7,** 92; **8,** 291; **10,** 191, 285, 286, 390; **11,** 366; **12,** 345
Lithium di-*t*-butylcuprate, **6,** 164, 210
Lithium 2,6-di-*t*-butyl-4-methylphenoxide, **5,** 242, 412; **6,** 95
Lithium dicyclohexylamide, **2,** 246, 337; **4,** 311; **5,** 397; **12,** 12
Lithium diethenyl(tributylphosphine)cuprate, **4,** 221
Lithium diethoxyaluminum hydride, **1,** 610
Lithium [(α-diethoxymethyl)vinyl]-phenylthiocuprate, **7,** 212
Lithium (1,1-diethoxy-2-propenyl)(3,3-dimethyl-1-butynyl)cuprate, **8,** 300
Lithium (3,3-diethoxy-1-propen-2-yl)-phenylthiocuprate, **7,** 212
Lithium di(1-ethoxyvinyl)cuprate, **6,** 205
Lithium di(2-ethoxyvinyl)cuprate, **8,** 222
Lithium diethylamide, **1,** 610; **2,** 247; **4,** 298; **5,** 398; **6,** 189, 331, 334; **11,** 295
Lithium diethylamide–Hexamethyl-phosphoric triamide, **6,** 332; **7,** 201
Lithium diethylcuprate, **2,** 152; **3,** 112; **4,** 295; **5,** 237, 324; **9,** 332
Lithium diethylcuprate-Tributylphosphine, **5,** 237
Lithium diisobutylmethylaluminum hydride, **2,** 248; **8,** 292
Lithium diisopropenylcuprate, **6,** 212
Lithium diisopropylamide (LDA), **2,** 249; **3,** 184; **4,** 61, 298; **5,** 400; **6,** 85, 334; **7,** 204, 209; **8,** 189, 292, 387, 489; **9,** 280, 429; **10,** 241; **11,** 102, 296, 602; **12,** 277, 448
Lithium diisopropylamide–Chlorotri-methylsilane, **12,** 279
Lithium diisopropylamide–Hexamethyl-phosphoric triamide, **5,** 402, 406; **7,** 207
Lithium dimesitylborohydride bis-(dimethoxymethane), **5,** 408
Lithium dimethoxyphosphinylmethylide [*see* (Dimethoxyphosphinyl)methyllithium]
Lithium di(α-methoxyvinyl)cuprate, **6,** 204
Lithium–Dimethylamine, **1,** 574
Lithium 2-dimethylaminoethoxide, **10,** 244
Lithium 1-(dimethylamino)naphthalenide (LDMAN), **10,** 244, 383; **11,** 422; **12,** 279
Lithium dimethylaminosulfonylmethide (*see* Dimethylsulfamoylmethyllithium)
Lithium (3,3-dimethyl-1-butynyl)-1,1-diethoxy-2-propenylcuprate, **8,** 300

Lithium dimethylcarbamoylnickel tricarbonylate, **4,** 302
Lithium dimethylcuprate, **2,** 151; **3,** 98, 106; **4,** 65, 177, 295; **5,** 187, 234, 462; **6,** 209, 225, 622; **7,** 120, 132; **8,** 301; **9,** 328, 507; **10,** 139, 193, 282; **11,** 154, 371; **12,** 198, 216
Lithium dimethylcuprate–Boron trifluoride, **10,** 53; **12,** 348
Lithium L-α,α′-dimethyldibenzylamide, **10,** 245
Lithium dimsylate (*see* Lithium methylsulfinylmethylide)
Lithium–Diphenyl, **1,** 612; **2,** 249
Lithium diphenylcuprate, **3,** 106; **4,** 108; **5,** 521; **6,** 234; **9,** 268; **10,** 286
Lithium diphenylphosphine [*see* (Diphenylphosphine)lithium]
Lithium di(2-propenyl)cuprate, **6,** 212
Lithium di[(E)-1-propenyl]cuprate, **2,** 400; **6,** 495; **7,** 141; **8,** 302
Lithium dipropylcuprate, **5,** 283; **6,** 245
Lithium diselenide, **8,** 310
Lithium di(*p*-tolyl)cuprate, **11,** 314
Lithium divinylcuprate, **4,** 219; **5,** 288; **6,** 225; **8,** 190; **12,** 345
Lithium divinylcuprate–Tributylphoshine complex, **4,** 221
Lithium ethanethiolate, **9,** 455
Lithium ethenethiolate, **5,** 292, 749
Lithium ethoxide, **1,** 612; **3,** 187
Lithium ethoxyacetylide, **1,** 359, 613
Lithium α-ethoxycarbonylvinyl(1-hexynyl)-cuprate, **6,** 329
Lithium–Ethylamine, **1,** 574; **2,** 241; **3,** 175; **4,** 288; **5,** 377, 381; **7,** 194; **8,** 284; **9,** 377; **10,** 236; **11,** 287; **12,** 219
Lithium–Ethylamine; Raney nickel, **5,** 381
Lithium (1-ethylcyclohexyl)-1,1,3,3-tetramethylbutylamide, **4,** 311
Lithium–Ethylenediamine (LEDA, *see also* N-Lithioethylenediamine), **1,** 614; **2,** 250; **3,** 186; **4,** 291, 488; **9,** 274
Lithium ethylenethiolate, **5,** 292, 749
Lithium ethylmercaptide, **5,** 292
Lithium hexamethyldisilazide, **4,** 296, 409, 568; **5,** 393; **6,** 542; **7,** 197; **9,** 283; **11,** 300; **12,** 257, 280
Lithium–Hexamethylphosphoric triamide, **1,** 430; **2,** 208; **3,** 19, 150; **4,** 246; **5,** 543; **6,** 274, 323
Lithium (1-hexynyl)(2-tributylstannylvinyl)-cuprate, **12,** 470
Lithium hydride, **2,** 251; **4,** 304; **10,** 400
Lithium hydroxide, **7,** 208; **8,** 268; **12,** 226, 282
Lithium iodide, **1,** 279, 615, 617; **4,** 304; **5,** 123, 410; **7,** 208; **9,** 283; **10,** 245; **11,** 300; **12,** 282
Lithium iodide–Boron trifluoride, **4,** 305
Lithium iodide–Dimethylformamide, **1,** 617; **5,** 323; **7,** 209; **12,** 282
Lithium iodide–Hydrogen peroxide, **7,** 220
Lithium B-isopinocampheyl-9-borabicyclo-[3.3.1]nonylhydride, **8,** 303
Lithium isopropoxide, **9,** 283
Lithium–Isopropylamine, **1,** 574
Lithium N-isopropylcyclohexylamide (LiICA), **4,** 306; **5,** 398, 411; **7,** 209; **10,** 67; **12,** 240
Lithium isopropylmercaptide, **6,** 345
Lithium lithioacetate, **3,** 208; **7,** 113; **8,** 175; **11,** 198
Lithium *o*-lithiobenzoate, **10,** 246
Lithium *o*-lithiophenoxide, **12,** 283
Lithium α-lithiophenoxyacetate, **8,** 294
Lithium β-lithiopropionate, **8,** 303
Lithium methanethiolate, **8,** 303; **12,** 317
Lithium methoxide, **5,** 617
Lithium methoxyaluminum hydride, **5,** 384; **6,** 341
Lithium methoxycarbonylvinyl(1-hexynyl)-cuprate, **6,** 330
Lithium 2-methoxyethoxide, **3,** 186
Lithium 3-methoxypropyl(α-methylbenzyl)-amide, **12,** 318
Lithium (γ-methoxypropyl)-α-phenylethylamide, **12,** 318
Lithium methoxy(trimethylsilyl)methylide, **9,** 284; **10,** 246; **11,** 331
Lithium–Methylamine, **1,** 574; **3,** 175; **4,** 288; **5,** 378; **6,** 322
Lithium–N-Methylaniline, **1,** 618
Lithium N-methylanilinoacetylide, **10,** 267
Lithium methylmercaptide, **8,** 303; **12,** 317
Lithium methyl(phenylseleno)cuprate, **12,** 394
Lithium methyl(phenylthio)cuprate, **6,** 343
Lithium 2-methylpropane-2-thiolate, **5,** 412; **6,** 345; **7,** 210
Lithium [N-methyl-N-(2-pyridyl)]amide, **12,** 284

Lithium methylsulfinylmethylide, **2,** 166; **4,** 195; **6,** 342, 369; **12,** 283, 451
Lithium methylsulfinylmethylide–Tributylborane, **11,** 302
Lithium methylthioformaldine, **8,** 305; **11,** 302
Lithium methyl(vinyl)cuprate, **5,** 236; **6,** 342
Lithium morpholide, **1,** 574; **12,** 284
Lithium naphthalenide, **2,** 288; **3,** 81, 208; **4,** 348; **5,** 468; **6,** 415; **8,** 303, 305; **9,** 284; **10,** 247; **11,** 302; **12,** 284
Lithium nitride, **1,** 618; **2,** 251
Lithium *t*-octyl-*t*-butylamide, **12,** 285
Lithium orthophosphate (*see* Lithium phosphate)
Lithium palladium chloride [*see* Dilithium tetrachloropalladate(II)]
Lithium (1-pentynyl)(3-tetrahydro-pyranyloxy-1-propenyl)cuprate, **6,** 603
Lithium (1-pentynyl)(2-tributylstannylvinyl)-cuprate, **5,** 43
Lithium perchlorate, **4,** 212; **5,** 413; **6,** 103; **12,** 549
Lithium perhydro-9b-boraphenalylhydride, **3,** 187; **4,** 309, 517
Lithium phenylethynolate, **6,** 343; **7,** 210; **10,** 247
Lithium phenylthiolate (*see* Lithium thiophenoxide)
Lithium phenylthio(trimethylstannyl)cuprate, **9,** 330; **10,** 289
Lithium phenylthio(2-vinylcyclopropyl)-cuprate, **7,** 212; **9,** 329
Lithium phosphate, **4,** 309; **6,** 334
Lithium phthalide, **9,** 285
Lithium piperidide, **6,** 344; **12,** 498
Lithium 1-propanethiolate, **3,** 188; **5,** 415; **6,** 345, 496
Lithium 2-propanethiolate, **6,** 345
Lithium–Propylamine, **1,** 574, 618; **5,** 378
Lithium propylmercaptide, **3,** 188; **5,** 415; **6,** 345, 496
Lithium pyrrolidide, **5,** 416; **6,** 327
Lithium 2-(1-pyrrolidinyl)methylpyrrolidide, **12,** 421
Lithium selenide, **8,** 310
Lithium selenophenolate, **8,** 183, 306; **10,** 238; **11,** 475
Lithium–Sodium alloy, **1,** 618
Lithium tetrabutylaluminate–(–)-N-Methylephedrine, **7,** 240
Lithium tetrafluoroborate, **5,** 416; **10,** 248; **11,** 574
Lithium tetrakis(N-dihydropyridyl)aluminate (*see* Lithium aluminum hydride–Pyridine)
Lithium 2,2,6,6-tetramethylpiperidide (LiTMP), **4,** 310; **5,** 417; **6,** 345; **7,** 213; **8,** 307; **9,** 83, 281, 285, 362; **10,** 243, **11,** 243, 297; **12,** 285
Lithium thioethoxide, **9,** 455
Lithium thiomethoxide, **8,** 303; **12,** 317
Lithium thiophenoxide, **4,** 311; **5,** 418; **8,** 183, 304
Lithium *p*-toluenesulfonate, **11,** 605
Lithium *p*-tolylsulfinylmethylide, **4,** 513; **5,** 682; **11,** 292
Lithium trialkoxyaluminum hydrides (*see also* specific names), **1,** 620, 625; **2,** 251, 252; **3,** 188; **4,** 312; **5,** 392; **10,** 248; **11,** 305; **12,** 286
Lithium tri-*t*-butoxyaluminum hydride, **1,** 620, 983; **2,** 251; **3,** 25, 188; **4,** 292, 312; **7,** 329; **10,** 248; **12,** 201
Lithium tri-*t*-butoxyaluminum hydride–Copper(I) bromide, **12,** 286
Lithium tri-*t*-butoxyaluminum hydride–Triethylborane, **4,** 312
Lithium tri-*sec*-butylborohydride (L-Selectride®), **4,** 104, 312; **5,** 557; **6,** 348; **7,** 307; **8,** 308, 309; **10,** 248; **11,** 445, 483; **12,** 239, 286, 512
Lithium tributylmanganate, **3,** 112
Lithium tri-*t*-butylzincate, **8,** 516
Lithium tricarbonyl(dimethylcarbamoyl)-nickelate, **4,** 302
Lithium trichloromethide (*see* Trichloromethyllithium)
Lithium trichloropalladate(II), **12,** 288
Lithium triethoxyaluminum hydride, **1,** 625; **2,** 252
Lithium triethylborohydride (Super hydride®), **4,** 313; **6,** 348; **7,** 196, 215; **8,** 309, 450; **9,** 286, 297; **10,** 169, 249; **11,** 288, 304; **12,** 216, 288, 289
Lithium triethylborohydride–Aluminum *t*-butoxide complex, **9,** 287
Lithium triethylcarboxide (*see* Lithium triethylmethoxide)
Lithium triethylmanganate, **3,** 112
Lithium triethylmethoxide, **4,** 314; **5,** 200; **6,** 349; **7,** 98

Lithium trifluoroacetate, **6,** 103
Lithium trimethoxyaluminum hydride, **1,** 625; **2,** 252; **3,** 25
Lithium trimethylcobaltate, **3,** 112
Lithium trimethylferrate, **3,** 112, 312
Lithium trimethylmanganate, **3,** 112
Lithium trimethylsilanolate, **12,** 414
Lithium trimethylsilylacetylide, **9,** 106, 246, 311; **10,** 208
Lithium trimethylsilyldiazomethane [*see* Diazo(trimethylsilyl)methyllithium]
Lithium trimethylzincate, **8,** 515
Lithium triphenylmethide (*see* Triphenylmethyllithium)
Lithium tris(3-ethyl-3-pentyloxy)aluminum hydride, **11,** 305
Lithium trisiamylborohydride (LS-Selectride®), **7,** 216
Lithium tris[(E)-2-methylcyclopentyl]-borohydride, **7,** 216
Lithium tris(phenylthio)methanide [*see* Tris-(phenylthio)methyllithium]
Lithium vinylmercaptide, **5,** 292, 749
LiTMP (*see* Lithium 2,2,6,6-tetramethyl-piperidide)
LTA (see Lead tetraacetate)
LTTFA [*see* Lead tetrakis(trifluoroacetate)]
2,4- or 2,6-Lutidine (*see* 2,4- or 2,6-Dimethylpyridine)
2,6-Lutidine-3,5-dicarboxylic hydrazide, **1,** 626
2,6-Lutidinium *p*-toluenesulfonate, **12,** 139

Magic Acid (*see* Fluorosulfuric acid–Antimony(V) fluoride)
Magic Methyl (*see* Methyl fluorosulfonate)
Magnesium, **1,** 627, 686; **2,** 205, 254; **3,** 159, 189; **4,** 315; **5,** 419; **6,** 351; **7,** 218, 398; **8,** 464; **9,** 70; **10,** 251; **11,** 126, 246, 307; **12,** 290
Magnesium acetate, **9,** 96
Magnesium aluminum hydride (*see* Tetra-μ-hydridotetrahydro-aluminummagnesium)
Magnesium amalgam, **1,** 627, 653; **3,** 189
Magnesium–Anthracene, **12,** 290
Magnesium bromide, **2,** 254; **5,** 763; **6,** 595; **8,** 273; **9,** 310; **12,** 21, 184
Magnesium bromide–Acetic anhydride (*see* Acetic anhydride–Magnesium bromide)
Magnesium bromide etherate, **1,** 629; **7,** 218; **9,** 288
Magnesium bromide etherate–Hydrogen peroxide, **7,** 220
Magnesium carbonate, **4,** 579
Magnesium chloride, **10,** 251
Magnesium chloride–Potassium (*see also* Magnesium), **4,** 315; **5,** 419
Magnesium chloride–Sodium (*see also* Magnesium), **4,** 315
Magnesium chloride–Sodium naphthalenide (*see also* Magnesium), **8,** 464
Magnesium ethoxide, **9,** 96
Magnesium ethyl malonate, **8,** 311
Magnesium hydride, **8,** 175
Magnesium hydride–Copper(I) *t*-butoxide, **8,** 311
Magnesium hydride–Copper(I) iodide, **8,** 311
Magnesium iodide, **1,** 630; **5,** 420; **6,** 353, 595; **8,** 246; **9,** 60
Magnesium iodide etherate, **1,** 630; **12,** 290
Magnesium–Mercury(II) chloride, **2,** 255
Magnesium methoxide, **2,** 255; **3,** 189; **6,** 351; **7,** 220; **10,** 251; **11,** 309
Magnesium methyl carbonate (MMC), **1,** 631; **2,** 256; **3,** 190; **5,** 420; **6,** 354; **11,** 310
Magnesium naphthalenide, **1,** 711; **2,** 289
Magnesium oxide, **1,** 633; **2,** 256; **7,** 221
Magnesium permanganate, **12,** 576
Magnesium sulfate, **1,** 354, 634; **5,** 421
Magnesium trifluoromethanesulfonate, **12,** 577
Malealdehyde, **1,** 634
Maleic anhydride, **1,** 238, 1027, 1149, 1290; **4,** 316; **5,** 422; **7,** 161
(Maleic anhydride)bis(triphenylphosphine)-palladium(0), **5,** 502
Malonaldehyde, **2,** 257
Malonaldehyde bis(diethyl acetal), **1,** 1167; **2,** 257
Malonic acid, **1,** 887, 959, 1196; **2,** 433; **5,** 248
Malononitrile, **1,** 16, 413, 962, 1135; **4,** 317, 555; **5,** 6, 501; **6,** 11
Malonyl dichloride, **12,** 291
(S)-(+)- and (R)-(−)-Mandelic acid, **12,** 291
Manganese(III) acetate, **2,** 263; **4,** 318; **5,** 157; **6,** 355; **7,** 221; **12,** 292
Manganese(III) acetylacetonate, **2,** 264; **3,** 194
Manganese(II) chloride, **7,** 222; **9,** 288
Manganese(II) chloride–Lithium aluminum hydride, **11,** 310

Manganese dioxide (*see* Manganese(IV) oxide)
Manganese(II) iodide, **7,** 222; **8,** 312; **9,** 289; **10,** 290
Manganese(IV) oxide, **1,** 636, 637, 828; **2,** 257; **3,** 191; **4,** 317; **5,** 422; **6,** 357; **8,** 312, 357, 412, 443; **9,** 151; **10,** 330; **11,** 311, 451
Manganese(III) sulfate, **9,** 289
Manganic, *see* Manganese(III)
McMurry's Reagent (*see* Titanium(III) chloride–Lithium aluminum hydride)
MCPBA (*see* m-Chloroperbenzoic acid)
Meldrum's acid, **1,** 526; **6,** 309; **8,** 313, 484; **9,** 192; **10,** 252; **11,** 311
MEM chloride (*see* β-Methoxyethoxymethyl chloride)
d- and *l*-Menthol, **10,** 101; **11,** 374; **12,** 294
(−)-Menthoxyacetyl chloride, **6,** 357
Menthoxyaluminum dichloride, **9,** 289
(S)-(−)-Menthyl *p*-methoxybenzenesulfinate, **12,** 296
(S)- or (R)-Menthyl *p*-toluenesulfinate, **10,** 405; **11,** 312; **12,** 295, 509
α-Mercaptoacetaldehyde diethyl acetal, **1,** 1105
Mercaptoacetic acid (*see* Thioglycolic acid)
α-Mercapto-γ-butyrolactone, **8,** 314
2-Mercaptoethanol, **1,** 356, 643; **5,** 718; **6,** 381
(−)-10-Mercaptoisoborneol, **9,** 290
2-Mercaptopyridine, **8,** 360
Mercuric, *see* Mercury(II)
Mercurous, *see* Mercury(I)
Mercury, **9,** 291
Mercury(II) acetate, **1,** 497, 542, 644, 1150; **2,** 234, 264, 310; **3,** 90, 194; **4,** 319; **5,** 370, 424; **6,** 123, 198, 358; **7,** 222; **8,** 315; **9,** 291, 411, 461; **10,** 252; **11,** 315, 469, 471; **12,** 298, 434
Mercury(II) acetate–Ethylenediaminetetraacetic acid, **5,** 427
Mercury(II) acetate–Nafion-H, **12,** 303
Mercury(II) acetate–Sodium trimethoxyborohydride, **10,** 253
Mercury(II) azide, **3,** 196; **4,** 323
Mercury bis(ethyl diazoacetate), **4,** 325
Mercury bis(*p*-toluenesulfonamide), **1,** 660
Mercury(II) bromide, **6,** 663
Mercury(II) chloride, **1,** 497, 652; **5,** 427; **6,** 259, 359; **7,** 53; **9,** 291; **12,** 304
Mercury(II) chloride–Cadmium carbonate, **6,** 581
Mercury(II) cyanide, **1,** 655; **3,** 197
Mercury(II) iodide, **3,** 70
Mercury(II) methanesulfonate, **8,** 444
Mercury(I) nitrate, **5,** 430; **6,** 360
Mercury(II) nitrate, **3,** 197; **4,** 495; **5,** 428; **7,** 223, 382; **10,** 254; **11,** 317
Mercury(II) nitrate–Hydrogen peroxide, **7,** 224
Mercury(II) nitrite, **9,** 292
Mercury(II) oxide, **1,** 655, 831; **2,** 267; **4,** 323; **5,** 428; **6,** 360; **7,** 224; **8,** 316; **9,** 293; **12,** 305, 434
Mercury(II) oxide–Boron trifluoride, **11,** 318
Mercury(II) oxide–Bromine, **1,** 657; **4,** 323, 324; **5,** 428
Mercury(II) oxide–Iodine, **1,** 657, 658; **3,** 44; **5,** 347; **11,** 267; **12,** 305
Mercury(II) oxide–Mercury(II) bromide, **6,** 360
Mercury(II) oxide–Tetrafluoroboric acid, **9,** 293; **10,** 254; **11,** 318; **12,** 306
Mercury(II) perchlorate, **5,** 428; **9,** 315
Mercury(II) pivalate, **11,** 319
Mercury(II) propionate, **6,** 123
Mercury(II) sulfate, **1,** 514, 658; **5,** 428
Mercury(II) tetrafluoroborate, **5,** 428
Mercury(II) thiocyanate, **9,** 293
Mercury(I) trifluoroacetate, **2,** 268
Mercury(II) trifluoroacetate, **1,** 659; **3,** 195; **4,** 319, 325; **5,** 429; **6,** 360, 581; **7,** 382; **8,** 316, 355; **9,** 294, 320; **10,** 373; **11,** 320; **12,** 198, 301, 306
Mercury(II) trifluoromethanesulfonate–N,N-Dimethylaniline, **12,** 307
O-Mesitoylhydroxylamine, **1,** 660
2-Mesitylenesulfonyl chloride (Trimsyl chloride), **1,** 235, 661; **5,** 430; **6,** 625; **8,** 318
(Mesitylenesulfonyl)hydrazide (MSH), **10,** 255
(O-Mesitylenesulfonyl)hydroxylamine, **5,** 430; **6,** 320; **8,** 319; **10,** 256
(Mesitylenesulfonyl)imidazole, **5,** 434
1-(Mesitylenesulfonyl)-1,2,4-triazole, **5,** 434; **6,** 361
Mesitylglyoxal, **3,** 79
Mesityllithium, **1,** 610; **8,** 317; **10,** 234; **12,** 12
2-Mesitylmagnesium bromide, **1,** 419; **6,** 59
Mesityl oxide, **4,** 11
Mesityl oxide ethylene ketal, **1,** 662

Mesityloxytris(dimethylamino)phosphonium azide, **9,** 296
Mesoxalonitrile, **2,** 60
Mesyl chloride (*see* Methanesulfonyl chloride)
Metaphosphoric acid, **1,** 665
Methacrolein, **2,** 15
Methacrolein dimethylhydrazone, **11,** 201
Methallyl alcohol, **8,** 67; **9,** 371
Methallyl bromide, **4,** 353; **6,** 361
Methallyl chloride, **2,** 181; **3,** 178, 310; **4,** 305, 403; **5,** 592; **10,** 87
α-Methallyl chloride, **5,** 392
Methallyl dichloride, **2,** 290; **3,** 101
B-Methallyldiisopinocampheylborane, **12,** 18
S-Methallyl N,N-dimethyldithiocarbamate, **6,** 11, 539
Methallyl iodide, **1,** 915
Methallyllithium, **5,** 82, 552
Methallylmagnesium chloride–Dichlorobis(cyclopentadienyl)titanium, **6,** 49
Methallylmagnesium halides, **9,** 94; **11,** 308, 436
π-2-Methallylnickel bromide, **4,** 353; **6,** 361; **7,** 250
Methallyl phenyl sulfoxide, **5,** 401
Methallyltributyltin, **10,** 387; **12,** 23
Methallyltrimethylsilane, **10,** 7; **12,** 375
Methanephosphonyl dichloride, **1,** 666
Methaneselenenic acid, **8,** 319
Methaneselenenyl acetate, **8,** 5
Methaneselenol, **6,** 361
Methanesulfinyl chloride, **5,** 434; **7,** 224
N-Methanesulfinyl-*p*-toluidide, **2,** 269
Methanesulfonic acid, **1,** 666; **2,** 270; **4,** 326; **10,** 256; **11,** 321
Methanesulfonic anhydride, **1,** 667
Methanesulfonyl chloride, **1,** 17, 662; **2,** 268; **4,** 326; **5,** 435; **6,** 265, 362, 400, 598; **7,** 225; **10,** 311; **11,** 322
Methanesulfonyl chloride–Lithium chloride, **4,** 327
Methanesulfonyl chloride–Sulfur dioxide, **1,** 664
(S)-N-Methanesulfonylphenylalanyl chloride, **8,** 320
Methanethiol, **5,** 698
Methanol, **1,** 678; **2,** 36; **5,** 481; **6,** 16, 568, 592, 645; **8,** 10; **12,** 305
3% Methanolic hydrogen chloride, **1,** 668
Methanol–Sodium tetraborate, **11,** 322
Methoxyacetonitrile, **1,** 422; **5,** 436
4-Methoxy-5-acetoxymethyl-*o*-benzoquinone, **6,** 363
Methoxyacetyl chloride, **4,** 328; **12,** 310
Methoxyacetylene, **1,** 668; **2,** 270
Methoxyallene, **6,** 146; **7,** 163, 225; **8,** 320; **9,** 272; **10,** 257; **12,** 310, 420
1-Methoxyallenyllithium, **7,** 226; **8,** 320; **9,** 272; **10,** 257; **12,** 310, 420
2-Methoxyallyl bromide, **4,** 327; **5,** 437; **6,** 364; **8,** 107, 322
S-(2-Methoxyallyl)-N,N-dimethyldithiocarbamate, **6,** 364
2-(2-Methoxy)allylidene-1,3-dithiane, **7,** 226
3-Methoxyallylidenetriphenylphosphorane, **8,** 323
π-(2-Methoxyallyl)nickel bromide, **5,** 437; **8,** 324
Methoxyamine, **1,** 670; **11,** 322
Methoxybenzene, **1,** 41
p-Methoxybenzenesulfonyl chloride, **9,** 297
p-Methoxybenzylazidoformate, **1,** 668; **3,** 197
p-Methoxybenzyl chloride, **1,** 668
p-Methoxybenzyl itaconate, **6,** 365
p-Methoxybenzylmagnesium bromide, **8,** 116
p-Methoxybenzyloxycarbonyl azide, **1,** 668; **3,** 197
1-Methoxy-1,3-bis(trimethylsilyloxy)-1,3-butadiene, **9,** 54; **10,** 46, 257; **11,** 66
4-Methoxy-1,2-bis(trimethylsilyloxy)-1,3-butadiene, **11,** 332
1-Methoxy-1,3-butadiene, **10,** 258
4-Methoxy-2-butanone, **1,** 702
1-Methoxy-3-buten-2-one, **3,** 198
4-Methoxy-3-buten-2-one, **8,** 7; **10,** 258; **12,** 5
1-Methoxy-1-buten-3-yne, **1,** 669; **9,** 297
4-Methoxy-3-buten-1-ynylcopper, **9,** 297
4-Methoxy-3-buten-1-ynylmagnesium bromide, **1,** 669
Methoxycarbene, **3,** 103
Methoxycarbenium hexafluoroantimonate, **4,** 328
Methoxycarbonylbenzoyl chloride, **5,** 91
Methoxycarbonyl-1,3-dithiane, **12,** 282
Methoxycarbonylethyl(diphenyl)phosphine oxide, **12,** 105
Methoxycarbonylhydrazine (*see* Methyl carbazate)
Methoxycarbonylmethylenetriphenylphosphorane, **1,** 112, 1240; **2,** 60, 98; **3,** 40

Methoxycarbonylmethyl vinyl ketone (*see* Methyl 3-oxo-4-pentenoate)
Methoxycarbonyl-1-penten-3-one, **1,** 1271
p-Methoxycarbonylperbenzoic acid, **4,** 68
Methoxycarbonylpyrrole, **3,** 9; **5,** 223
N-Methoxycarbonylsulfamyl chloride, **4,** 332; **5,** 92
4(R)-Methoxycarbonyl-1,3-thiazolidine-2-thione, **11,** 323
4-Methoxycarbonylthiolane-3-one, **11,** 324
3-Methoxycatecholborane, **9,** 98
4-Methoxycrotonic anhydride, **8,** 127
1-Methoxycyclopropanol, **4,** 107
1-Methoxy-1-cyclopropyllithium, **10,** 383
2-Methoxycyclopropyllithium, *cis*- and *trans*-, **6,** 366
4-Methoxy-5,6-dihydro-2*H*-pyran, **2,** 271; **3,** 197; **5,** 438, 674; **6,** 367
Methoxydimethylsulfonium hexachloroantimonate, **6,** 377
2-Methoxy-1,3-dioxolane, **9,** 298
erythro-2-Methoxy-1,2-diphenylethylamine, **12,** 313
Methoxydiphenylphosphine, **6,** 265
2-Methoxyethanol, **12,** 305
β-Methoxyethoxymethyl chloride (MEMCl), **7,** 227; **9,** 299
1-Methoxyethyldiphenylphosphine oxide, **7,** 229
α-Methoxyethylidenetriphenylphosphorane, **1,** 669
2-Methoxy-4-furyllithium, **11,** 325
6-Methoxy-(E,E)-3,5-hexadienoic acid, **12,** 314
6-Methoxy-7-hydroxy-3,4-dihydroisoquinolinium methiodide, **4,** 329
3-Methoxyisoprene, **3,** 300; **4,** 330
1-(Methoxymethoxy)styrene, **12,** 99
N-Methoxy-N-methylacetamide, **12,** 314
Methoxymethylaluminum chloride, **5,** 438
4-Methoxy-5-methyl-*o*-benzoquinone, **6,** 363
(Methoxymethyl)bis(trimethylsilyl)amine, **12,** 62
Methoxymethyl bromide, **10,** 56
1-Methoxy-3-methyl-1,3-butadiene, **9,** 60
2-Methoxy-3-methyl-1,3-butadiene (*see* 3-Methoxyisoprene)
(2S)-Methoxymethyl-(4S)-*t*-butylthio-N-pivaloylpyrroline, **12,** 315
3-Methoxy-3-methylbutynylcopper, **8,** 324; **9,** 300
Methoxymethylchloroalane, **5,** 438
5-Methoxymethylcyclopentadiene, **3,** 66; **5,** 476; **6,** 117
Methoxymethyl(diphenyl)phosphine oxide, **7,** 229; **9,** 301
3-Methoxymethylene-2,4-bis(trimethylsilyloxy)-1,4-pentadiene, **12,** 315
Methoxymethylenetriphenylphosphorane, **1,** 671; **2,** 237, 271; **3,** 198; **5,** 438; **6,** 368; **8,** 371
Methoxymethyl isocyanate, **5,** 439
Methoxymethyllithium, **1,** 672
Methoxymethylmagnesium chloride, **1,** 672
Methoxymethyl methanesulfonate, **3,** 198
(4S,5S)-4-Methoxymethyl-2-methyl-5-phenyl-2-oxazoline, **6,** 386; **7,** 229; **9,** 312
2-Methoxy-6-methyl-1,4,3,5-oxathiadiazine 4,4-dioxide, **4,** 331
Methoxymethyl phenyl sulfide, **6,** 369; **10,** 260; **12,** 316
Methoxymethyl phenyl sulfone, **12,** 401
2-Methoxy-3-methyl-1-phenylthio-1,3-butadiene, **11,** 328
2-Methoxy-6-methyl-4-pyranone, **9,** 302
4-Methoxy-6-methyl-2-pyranone, **9,** 302
6-Methoxy-4-methyl-2-pyranone, **8,** 325
(S)-(+)-2-Methoxymethylpyrrolidine, **10,** 259; **11,** 326
Methoxymethylsulfonylbenzene, **12,** 401
[(Methoxymethyl)thio]benzene, **6,** 369; **10,** 260; **12,** 316
3-Methoxy-1-methylthio-1-propyne, **6,** 397
(Methoxymethyl)trimethylsilane, **9,** 284; **11,** 331
1-Methoxy-2-methyl-3-trimethylsilyloxy-1,3-butadiene, **12,** 390
1-Methoxy-3-methyl-1-trimethylsilyloxy-1,3-butadiene, **9,** 302
1-Methoxy-2-methyl-3-trimethylsilyloxy-1,3-pentadiene, **11,** 326, 333; **12,** 312
Methoxymethyltriphenylphosphonium iodide, **9,** 250
Methoxymethyl vinyl ketone, **3,** 198
4-Methoxy-3-penten-2-one, **7,** 231
p-Methoxyphenacyl bromide, **5,** 439
2-(4-Methoxyphenyl)-4,4-dimethyl-2-oxazoline, **8,** 279
2-(*o*-Methoxyphenyl)-4,4-dimethyl-2-oxazoline, **6,** 368
4-Methoxy-1-phenylseleno-2-trimethylsilyloxy-1,3-butadiene, **8,** 398

(S)-(+)-2-(*p*-Methoxyphenylsulfinyl)-2-cyclopentenone, **12,** 296
(Z)-2-Methoxy-1-(phenylthio)-1,3-butadiene, **7,** 76; **11,** 328
2-Methoxy-3-phenylthio-1,3-butadiene, **7,** 232
1-Methoxy-1-phenylthiocyclopropane, **10,** 383
Methoxy(phenylthio)methane, **6,** 369; **10,** 260; **12,** 316
Methoxy(phenylthio)methyllithium, **6,** 369; **9,** 473; **10,** 260; **12,** 316
p-Methoxyphenylthionophosphine sulfide dimer [*see* 2,4-Bis(4-methoxyphenyl)-1,3-dithia-2,4-diphosphetane-2,4-disulfide]
2-Methoxy-2-phenylthiopropane, **10,** 244
3-Methoxy-1-phenylthio-1-propene, **8,** 327
Methoxy(phenylthio)trimethylsilylmethane, **12,** 317
Methoxy(phenylthio)trimethylsilyl-methyllithium, **12,** 317
2-Methoxypropene, **2,** 230; **5,** 360; **9,** 304; **11,** 329; **12,** 291, 317
2-Methoxy-2-propyllithium, **10,** 244
(3-Methoxypropyl)-α-methylbenzylamine, **12,** 318
(γ-Methoxypropyl)-α-phenylethylamine, **12,** 318
2-Methoxypyridine, **9,** 180
Methoxytributyltin, **9,** 141; **12,** 174
(R)- and (S)-α-Methoxy-α-trifluoro-methylphenylacetic acid, **11,** 329
4-Methoxy-2,3,6-trimethylbenzenesulfonyl chloride, **11,** 330
Methoxytrimethylsilane, **10,** 439; **12,** 466
1-Methoxy-2-trimethylsilylacetylene, **9,** 142; **12,** 161
1-Methoxy-1-trimethylsilylallene, **11,** 577
3-Methoxy-3-trimethylsilyl-1-allenyllithium, **11,** 577
1-Methoxy-1-trimethylsilylethylene, **12,** 100
Methoxy(trimethylsilyl)methyllithium, **9,** 284; **10,** 246; **11,** 331
(E)-1-Methoxy-3-trimethylsilyloxy-1,3-butadiene (Danishefsky's diene), **6,** 370; **7,** 131, 233; **8,** 328; **9,** 303; **10,** 260; **11,** 332, 475, 548; **12,** 318, 560
1-Methoxy-1-trimethylsilyloxycyclopropane, **12,** 499
1-Methoxy-1-trimethylsilyloxy-1-propene, **9,** 310; **11,** 340
3-Methoxy-1-trimethylsilyl-1-propyne, **6,** 154
α-Methoxyvinyllithium, **6,** 204, 372; **7,** 233; **8,** 276, 331; **9,** 304; **12,** 100
1-(Methoxyvinyl)trimethylsilane, **12,** 100
Methyl acetoacetate, **5,** 503; **6,** 311
Methyl 2-acetylacrylate, **11,** 334
Methyl acrylate, **5,** 11, 411, 413, 439; **6,** 61; **7,** 7; **11,** 7
Methylal (*see* Dimethoxymethane)
Methyl (allylthio)acetate, **10,** 261
Methylaluminum bis(trifluoroacetate), **8,** 471
Methylaluminum dichloride, **11,** 7; **12,** 5
Methylamine, **1,** 6; **5,** 15, 598
Methylamine–Borane (*see* Borane–Methylamine)
(1R,2S)-(−)-α-(1-Methylaminoethyl)benzyl alcohol (*see* Ephedrine)
o-(Methylamino)phenyl disulfide, **9,** 305
N-Methylanilinium trifluoroacetate, **8,** 331; **9,** 224
Methylanilinomagnesium bromide, **4,** 11; **5,** 440
Methyl azidoformate, **1,** 363; **5,** 441
Methyl α-benzalaminocrotonate, **9,** 305
Methyl benzenesulfinate, **6,** 374
3-Methylbenzothiazole-2-thione, **6,** 374
3-Methyl-2-benzothiazolinone hydrazone, **1,** 672
α-Methylbenzylamine, **1,** 838; **2,** 271; **3,** 199; **5,** 441; **6,** 162, 457; **8,** 393; **9,** 363; **10,** 67; **11,** 411; **12,** 319
N-Methylbenzylidenecyclohexylamine, **6,** 332
Methyl 2,3-bis-O-diphenylphosphine-4,6-O-benzylidene-α-D-glucopyranoside, **8,** 333
Methyl bis(ethylthio)acetate, **5,** 444
Methyl bis(methylthio)sulfonium hexachloroantimonate, **6,** 375; **7,** 235; **11,** 335
Methyl bis(trifluoroethoxy)phosphinyl-acetate, **12,** 320
Methyl 2-bis(2,2,2-trifluoroethoxy)-phosphinylpropionate, **12,** 320
Methyl bis(trifluoroethyl)phosphonoacetate, **12,** 320
Methyl 2-bis(trifluoroethyl)phosphono-propionate, **12,** 320
5-Methyl-1,3-bis(trimethylsilyl)-1,3-cyclohexadiene, **7,** 27
O-Methyl-C,O-bis(trimethylsilyl)ketene acetal, **10,** 261

5-Methyl-1,3-bis(trimethylsilyloxy)-1,3-cyclohexadiene, **8,** 47
B-Methyl-9-borabicyclo[3.3.1]nonane, **4,** 310
Methyl borate (*see* Trimethyl borate)
Methyl bromide, **1,** 673
Methyl bromide–Lithium bromide, **10,** 262
Methyl bromoacetate, **5,** 398
Methyl 2-bromocrotonate, **7,** 235
Methyl 4-bromocrotonate, **6,** 389
1-Methyl-2-bromopyridinium iodide, **6,** 375
Methyl β′-bromotiglate, **3,** 335
2-Methyl-1,3-butadiene (*see* Isoprene)
3-Methyl-1,2-butadiene, **5,** 279
3-Methylbuta-1,2-dienylidene (*see* Dimethylvinylidenecarbene)
3-Methyl-1,3-butadienyl phenyl sulfoxide, **8,** 333
3-Methyl-3-buten-1-ol, **5,** 84
3-Methyl-2-butenyl bromide (*see* 1-Bromo-3-methyl-2-butene)
B-(3-Methyl-2-butenyl)-diisopinocampheylborane, **12,** 17
3-Methyl-2-butenylidenetriphenyl-phosphorane, **5,** 69
3-Methyl-2-butenylmagnesium chloride, **9,** 125
3-Methyl-2-butenylmagnesium halides–Copper(I) iodide, **8,** 121; **9,** 125
3-Methyl-2-butenylmanganese chloride, **9,** 289
3-Methyl-2-butenylnickel bromide [*see* Di-μ-bromobis(3-methyl-2-butenyl)-dinickel]
3-(3-Methyl-1-butenyl)trimethylsilane, **3,** 446; **8,** 181
3-Methyl-2-butenyltrimethylsilane, **7,** 9
3-Methyl-2-butenyltrimethyltin, **9,** 47
3-Methyl-3-buten-1-ynylcopper, **8,** 123
B-(3-Methyl-2-butyl)-9-borabicyclo[3.3.1]-nonane, **8,** 334
N-Methyl-N-(*t*-butyldimethylsilyl)acetamide, **11,** 336
Methyl 3-*t*-butyldimethylsilyloxycrotonate, **10,** 62
N-Methyl-N-(*t*-butyldimethylsilyl)-trifluoroacetamide, **11,** 336
Methyl *t*-butyl ether (*see* *t*-Butyl methyl ether)
3-Methyl-1-butyn-3-ol, **1,** 674; **2,** 288
Methylcalcium iodide, **5,** 442
Methyl carbazate, **4,** 333; **7,** 235
Methyl(carboxysulfamoyl)triethylammonium hydroxide inner salt, **5,** 442
Methyl chloride, **1,** 674
Methyl chloroacetate, **3,** 150
Methyl 2-chloroacrylate, **11,** 7
Methyl chloroformate (*see also* Alkyl chloroformates, Ethyl chloroformate), **2,** 100; **5,** 611; **6,** 376; **7,** 147, 236; **9,** 306; **11,** 336
3-Methyl-4-chloromethylisoxazole, **1,** 675
Methyl α-chloro-α-phenylthioacetate, **12,** 320
Methyl chlorosulfinate, **1,** 675
Methyl chlorosulfinate–Dimethyl sulfoxide, **6,** 377
Methyl chlorothioformate, **3,** 51
Methylcopper, **4,** 334; **5,** 148; **6,** 377; **7,** 236; **8,** 117; **9,** 333
Methylcopper–Boron trifluoride complex, **8,** 334; **12,** 250
Methylcopper–Tributylphosphine, **2,** 282
Methylcopper–Trimethyl phosphite, **2,** 441
Methyl cyanoacetate, **5,** 178; **11,** 222
Methyl cyanodithioformate, **7,** 237
Methyl cyanoformate, **12,** 321
Methyl cyclobutenecarboxylate, **8,** 335; **9,** 306; **10,** 263
(R)-4-Methylcyclohexylidenemethylcopper, **7,** 238
Methylcyclopentadiene, **3,** 200
2-Methylcyclopentenone-3-dimethyl-sulfoxonium methylide, **6,** 378
Methyl cyclopropyl ketone, **1,** 676
Methyl diazoacetate, **2,** 82; **7,** 76; **8,** 434; **10,** 340, 416
Methyl diazopropionate, **4,** 335
O-Methyldibenzofuranium tetrafluoroborate, **4,** 336
N-Methyl-N,N′-di-*t*-butylcarbodiimium tetrafluoroborate, **6,** 378
Methyl dichloroacetate, **5,** 398
Methyl dichlorofluoroacetate, **2,** 273
Methyl dichlorophosphite, **10,** 88
N-Methyl-N,N′-dicyclohexylcarbodiimidium iodide, **4,** 336; **7,** 238
Methyl 4-diethoxyphosphinylcrotonate, **8,** 336
Methyl 4-diethylphosphonocrotonate, **8,** 336
Methyl diethylthioacetate, **5,** 444
Methyl diformylacetate, **8,** 338
Methyl 3-(N,N-dimethylamino)propionate, **9,** 306
Methyl(diphenyl)arsane oxide, **8,** 280

Methyldiphenylchlorosilane, **10,** 91; **12,** 321
S-Methyl-1,4-diphenylisothiosemicarbazide, **5,** 445
Methyl diphenylphosphinite, **6,** 265
Methyl 1,3-dithiane-2-carboxylate, **12,** 282
2-(2-Methyl-1,3-dithianyl)lithium, **3,** 136
Methyl dithioacetate, **11,** 340
N-Methyl-N-dodecylephedrinium bromide [*see* (−)-N-Dodecyl-N-methylephedrinium bromide]
4,4′-Methylenebis(2,6-di-*t*-butylphenol) (Ethyl antioxidant 702), **4,** 337
Methylenebis[magnesium bromide(iodide)], **2,** 273; **3,** 189, 201
Methylenebis(urethane), **1,** 677
Methylene bromide (*see* Dibromomethane)
Methylene chloride, **1,** 95, 223, 676; **2,** 273; **4,** 337; **5,** 322, 398; **6,** 565; **7,** 239
Methylene chloride–Butyllithium, **4,** 337; **7,** 239
Methylene chloride–Cesium or Potassium fluoride, **7,** 239
Methylene chloride–Methyllithium, **7,** 239
Methylenecyclobutane, **6,** 151
Methylenecyclopropane, **3,** 20; **4,** 35; **5,** 35, 123; **9,** 46
Methylene iodide (*see* Diiodomethane)
2-Methylene-1,3-propanediol, **6,** 379
Methylenetriphenylphosphorane, **1,** 95, 310, 678, 1238; **5,** 299, 589; **6,** 380, 612; **8,** 244, 339, 491; **9,** 307; **10,** 264; **11,** 147, 338
(−)-N-Methylephedrine, **7,** 240; **8,** 340, 430; **9,** 308; **10,** 238; **12,** 322
(−)-N-Methylephedrine–Lithium aluminum hydride–3,5-Dimethylphenol, **10,** 265
(−)-N-Methylephedrine–Lithium tetrabutylaluminate, **7,** 240
N-Methylephedrinium chloride, **8,** 340
Methyl ethanedithioate, **11,** 340
Methyl ethyl ketone, **1,** 678
3-Methyl-1-ethyl-3-phospholene 1-oxide, **1,** 679
Methyl fluorene-9-carboxylate, **1,** 679
Methyl fluoride–Antimonyl(V) fluoride, **3,** 201; **6,** 381; **7,** 240
Methyl fluorosulfonate (Magic Methyl), **3,** 202; **4,** 339; **5,** 445, 608, 691; **6,** 381; **7,** 240, **8,** 340, **9,** 307
N-Methylformamide, **2,** 202
N-Methylformanilide, **1,** 680; **6,** 86
N-Methylformanilide–Phosphoryl chloride (see also Vilsmeier reagent), **1,** 681, 878
Methyl formate, **5,** 446; **8,** 150
2-(N-Methyl-N-formyl)aminopyridine, **8,** 341; **10,** 265
2-Methylfuran, **1,** 682
Methyl glyoxylate, **12,** 514
Methylhydrazine, **4,** 340
Methyl hydroperoxide, **1,** 17
N-Methylhydroxylamine, **7,** 241; **12,** 322
O-Methylhydroxylamine (*see* Methoxyamine)
N-Methylhydroxylamine-O-sulfonic acid, **8,** 341
3-Methyl-3-hydroxymethyloxetane, **12,** 323
(S)-1-Methyl-2-hydroxymethylpyrrolidine, **10,** 266
Methyl 2-hydroxypropionate, **5,** 402
Methyl hypobromite, **2,** 331; **5,** 446
N,N′,N″-Methylidynetrisformamide, **1,** 1251
N-Methylimidazole, **5,** 447
Methyl iodide, **1,** 17, 682; **2,** 242, 274; **3,** 202; **4,** 341; **5,** 447, 459; **6,** 384; **7,** 242; **8,** 342; **9,** 308; **12,** 106
Methyl iodide–1,3-Propanedithiol, **9,** 308
Methyliodine(III) difluoride, **7,** 242
Methyl isocyanate, **2,** 320; **4,** 341
Methyl isocyanide, **10,** 231; **11,** 11, 285; **12,** 354, 498
Methyl isothiocyanate, **5,** 173
5-Methylisoxazole, **9,** 309
3-Methyl-5(4*H*)-isoxazolone, **9,** 309
Methylketene bis(trimethylsilyl) acetal, **11,** 600
Methylketene diethyl acetal, **1,** 685
Methylketene dimethyl acetal, **9,** 523
Methylketene methyl trimethylsilyl acetal, **9,** 310; **11,** 340
Methyl lithiodithioacetate, **11,** 340
Methyl α-lithio-α-methyldiphenylsilylacetate, **12,** 326
Methyl 1-lithio-1-methylselenoacetate, **8,** 241
Methyl 1-lithio-1-phenylselenoacetate, **8,** 240
Methyllithium, **1,** 686, 1141; **2,** 274; **3,** 202; **5,** 187, 448, 676, 722; **6,** 57, 90, 384, 386; **7,** 242; **8,** 342; **9,** 7, 310, 364; **10,** 3, 282; **11,** 14; **12,** 126, 324
Methyllithium–Dimethylcopperlithium (*see* Dilithium trimethylcuprate)
Methyllithium–Lithium bromide, **9,** 311
Methyllithium–Tetramethylethylenediamine, **9,** 311; **11,** 13, 340
Methyllithium–Titanium(III) chloride, **4,** 506

Methylmagnesium bromide, **1,** 167, 415, 688, 689; **5,** 419; **10,** 192, 266; **11,** 13; **12,** 332, 360
Methylmagnesium bromide–Copper(I) halides, **1,** 166; **6,** 147; **8,** 117
Methylmagnesium bromide–Iron(III) reagents, **4,** 236; **7,** 155
Methylmagnesium bromide–Nickel(II) acetylacetonate–Trimethylaluminum, **8,** 42; **9,** 52
Methylmagnesium bromide–Nickel(II) reagents, **2,** 110; **6,** 50
Methylmagnesium bromide–Titanium(IV) chloride, **12,** 376
Methylmagnesium chloride, **1,** 292, 415; **5,** 187
Methylmagnesium iodide, **1,** 415, 689; **2,** 278; **3,** 204; **6,** 386; **8,** 508; **9,** 230, 365; **10,** 191
Methylmagnesium iodide–Copper(I) chloride, **1,** 168; **2,** 91
Methyl mercaptan, **5,** 698
Methyl methanethiosulfonate, **5,** 454; **7,** 243; **12,** 325
Methyl methoxymagnesium carbonate (*see* Magnesium methyl carbonate)
(4S,5S)-2-Methyl-4-methoxymethyl-5-phenyl-2-oxazoline, **6,** 386; **7,** 229; **9,** 312
Methyl 5-methoxy-3-oxopentanoate (Nazarov's reagent), **9,** 314
Methyl methoxypropiolate, **11,** 340
Methyl α-(methyldiphenylsilyl)acetate, **12,** 325
Methyl O-methyllactate, **11,** 341
2-Methyl-2-(2-methyl-1-propenyl)-1, 3-dioxolane, **1,** 662
Methyl methylselenoacetate, **8,** 241
Methyl methylsulfinylmethyl sulfide (*see* Methyl methylthiomethyl sulfoxide)
Methyl methylsulfonylacetate, **4,** 369
Methyl methylsulfonylmethyl sulfide, **11,** 242
Methyl 2-methylthioacrylate, **5,** 455
Methyl 4-methylthiocrotonate, **6,** 389
Methyl methylthiomethyl sulfone, **11,** 242
Methyl methylthiomethyl sulfoxide, **4,** 341; **5,** 456; **6,** 390, 446; **8,** 344; **9,** 314
N-Methylmorpholine, **1,** 690; **2,** 278
N-Methylmorpholine N-oxide, **7,** 244
N-Methylmorpholine N-oxide–Hydrogen peroxide (*see* Hydrogen peroxide–N-Methylmorpholine oxide)
Methylneopentylphenylsilane–Boron trifluoride, **7,** 388
Methyl nitrate, **1,** 691; **5,** 51
Methyl nitrite, **1,** 691
N-Methyl-N′-nitro-N-nitrosoguanidine, **1,** 192
Methyl 2-nitrophenyl disulfide, **9,** 314
N-Methyl-N-nitroso-*p*-toluenesulfonamide, **1,** 191, 1187; **2,** 102
N-Methyl-N-nitrosourea (*see* N-Nitroso-N-methylurea)
Methyl orthosilicate, **12,** 260
4-Methyloxazole, **10,** 270
Methyl oxocarbonium hexafluoroantimonate, **1,** 692
Methyl 3-oxo-4-pentenoate, **1,** 1272; **5,** 746; **6,** 662
Methyl (E)-2,4-pentadienoate, **2,** 279
2-Methyl-2,4-pentanediol, **1,** 693
4-Methyl-1-penten-3-one, **5,** 464
Methyl perchlorate, **8,** 340
Methyl phenacyl sulfide, **4,** 190
N-Methyl-N-phenylaminoacetylene, **10,** 266
N-Methyl-N-phenylaminoethynyllithium, **10,** 267
N,N-Methylphenylamino(tributylphosphonium) iodide, **8,** 345; **10,** 268; **11,** 342
N,N-Methylphenylaminotriphenylphosphonium iodide, **6,** 392; **8,** 346
N-Methyl-N-phenylbenzohydrazonyl bromide, **10,** 269
N-Methyl-N-phenylcarbamoyl chloride, **1,** 694
O-4-Methylphenyl chlorothioformate, **4,** 342; **5,** 457; **9,** 474
Methyl α-phenylglycinate, **8,** 395; **10,** 308
1-Methyl-1-phenylhydrazine, **1,** 694
3-Methyl-2-phenyl-1,3,2-oxazaphospholine, **2,** 323
4-Methyl-5-phenyl-2-oxazolidinone, **11,** 379; **12,** 359
3-Methyl-1-phenyl-2-phospholene, **6,** 392; **7,** 244
3-Methyl-1-phenyl-3-phospholene 1-oxide, **1,** 695; **2,** 279; **4,** 343
(R)-(−)-Methylphenylpropylphosphine, **10,** 387
Methyl phenyl selenide, **6,** 86
Methyl phenylselenoacetate, **8,** 240
Methyl(phenyl)selenoniomethanide, **11,** 343
Methyl phenyl selenoxide, **10,** 291

Methyl phenylsulfinylacetate, **6,** 393; **8,** 347; **11,** 309, 343
(R)-(4-Methylphenylsulfinyl)ethyl-N-methoxyacetimidate, **12,** 297
Methyl phenyl sulfone, **6,** 43
N-Methylphenylsulfonimidoyl-1-ethyllithium, **5,** 459
N-Methylphenylsulfonimidoylmethyllithium, **5,** 458; **6,** 395; **11,** 210, 343
Methyl phenylsulfonylacetate, **6,** 447
2-Methyl-2-(2-phenylsulfonylethyl)-1,3-dioxolane, **6,** 462
Methyl(phenylthio)ketene, **8,** 348; **9,** 315
Methyl (phenylthio)propiolate, **11,** 341
(Z)-3-Methyl-1-phenylthio-2-trimethylsilyloxy-1,3-butadiene, **11,** 344
S-Methyl-S-phenyl-N-*p*-toluenesulfonylsulfoximide, **3,** 204
Methylphosphonic acid bis(dimethylamide), **2,** 280
Methylphosphonic dichloride, **1,** 666
Methyl phosphorodichloridite, **10,** 88
4-(4′-Methyl-1′-piperazinyl)-3-butyn-2-one, **8,** 348
4-Methyl-1-piperidone, **7,** 107
(S)-N-Methylprolinol, **12,** 315
2-Methyl-2-propene (*see* Isobutene)
2-Methyl-2-propen-1-ol (*see* Methallyl alcohol)
2-Methyl-2-propenyl-, *see also* Methallyl-
(2-Methyl-2-propenyl)tributyltin (*see* Tributylmethallyltin)
2-Methyl-1-propenyl trifluoromethanesulfonate, **6,** 230
β-Methyl-β-propiolactone, **10,** 283
Methyl propiolate, **1,** 1142
1-Methyl-2-(1*H*)-pyridothione–Benzoyl chloride, **9,** 40
N-Methyl-N-(2-pyridyl)formamide, **8,** 341; **10,** 265
Methyl α-pyrone-6-carboxylate, **1,** 272
N-Methylpyrrole, **5,** 223
2-Methylpyrrolidine, **4,** 471
(S)-1-[1-Methyl-2-pyrrolidinyl]methylpiperidine, **11,** 525; **12,** 490
N-Methyl-2-pyrrolidone (NMP), **1,** 696; **2,** 281; **9,** 316; **11,** 346
1-Methylselenocyclobutyllithium, **10,** 232
1-Methylseleno-1-cyclopropyllithium, **9,** 85; **10,** 318
Methylselenomethyllithium, **9,** 307
3-Methyl-2-selenoxo-1,3-benzothiazole, **7,** 245; **10,** 270
Methyl sorbate, **5,** 471
3-Methylsulfonyl-2,5-dihydrofuran, **10,** 270
1-Methylsulfonyloxybenzotriazole, **6,** 106
(S)-N-Methylsulfonylphenylalanyl chloride, **8,** 320
Methyl N-sulfonylurethane, **4,** 332, 343; **5,** 92
2-Methyl-2-thiazoline, **4,** 344; **6,** 403
Methylthioacetaldehyde, **4,** 190
Methylthioacetaldehyde dimethyl acetal, **4,** 191
Methylthioacetic acid, **6,** 395
Methylthioacetone, **4,** 190; **5,** 303
Methylthioacetone–Dimethyl sulfide, **6,** 118
Methylthioacetonitrile, **8,** 349
α-Methylthioacetophenone, **4,** 190
α-Methylthio-γ-butyrolactone, **5,** 262
3-Methylthio-1,4-diphenyl-5-triazolium iodide, **6,** 396; **8,** 351
2-Methylthio-1,3-dithiane, **7,** 191
2-(2-Methylthio-1,3-dithianyl)lithium, **4,** 217; **7,** 191
2-Methylthiomaleic anhydride, **8,** 351
1-Methylthio-3-methoxypropyne, **6,** 397
Methylthiomethyl-N,N-dimethyldithiocarbamate, **6,** 398
Methylthiomethyllithium, **2,** 403; **6,** 398; **7,** 379
N-Methylthiomethylpiperidine, **12,** 326
Methylthiomethyl *p*-tolyl sulfone, **12,** 327
4-(Methylthio)phenol, **2,** 281
N-Methylthiophthalimide, **11,** 493
1-Methylthio-2-propanone, **4,** 190; **5,** 303; **6,** 118
3-Methylthio-2-propenyl *p*-tolyl sulfone, **12,** 327
2-Methylthio-2-thiazoline, **5,** 459; **6,** 585
Methyl thiotosylate, **5,** 456, 460; **6,** 400; **7,** 245
Methylthiotributyltin, **8,** 352
1-Methylthio-3-triethylsilyloxy-1,4-pentadiene, **11,** 557
1-(Methylthio)-3-triethylsilyloxypentadienyllithium, **11,** 557
Methylthiotrimethylsilane, **6,** 399; **8,** 352; **10,** 113, 426
3-(Methylthio)-3-(trimethylsilyl)-1-propene, **12,** 336
1-Methylthiovinyl(triphenyl)phosphonium chloride, **11,** 425
3-Methyl-2-thioxobenzothiazole, **7,** 245

Methyltitanium trichloride, **10,** 270; **12,** 355
Methyltitanium triisopropoxide, **10,** 422; **12,** 297
Methyl *p*-toluenesulfinate, **6,** 374, 400
N,N-Methyltoluenesulfonylhydrazide, **9,** 411
Methyl N-(*p*-toluenesulfonylsulfonimidoyl)-methylsodium, **9,** 193; **12,** 217
S-Methyl *p*-toluenethiosulfonate, **5,** 456, 460; **6,** 400; **7,** 245
Methyl N-tolylsulfonylmethyl-thiobenzimidate, **7,** 245
N-Methyl-N-tolylsulfonylpyrrolidinium perchlorate, **4,** 345
(R)-(+)-Methyl *p*-tolyl sulfoxide, **4,** 466, 513; **5,** 682; **11,** 292
1-Methyl-3-*p*-tolyltriazene, **1,** 696; **2,** 282; **4,** 345
Methyl N-tosylmethylthiobenzimidate, **7,** 245
N-Methyl-N-tosylpyrrolidinium perchlorate, **4,** 345
4-Methyl-1,2,4-triazoline-3,5-dione, **2,** 282
Methyl(tributylphosphine)copper complex, **2,** 282
Methyltributylphosphonium fluoride, **9,** 223
Methyl tributylstannyl sulfide, **8,** 352
Methyltricaprylylammonium chloride (*see* Tricaprylylmethylammonium chloride)
Methyl trichloroacetate, **12,** 329
Methyltrichlorosilane, **11,** 553
N-Methyl-C-(trichlorotitanio)formimidoyl chloride, **12,** 354
Methyl α-triethylsilylvinyl ketone, **5,** 461
Methyl trifluoroacetate, **7,** 246
Methyl trifluoromethanesulfonate, **5,** 445; **6,** 406; **8,** 340; **12,** 329
Methyltriisopropoxytitanium (*see* Triisopropoxymethyltitanium)
(S)-(−)-Methyltri(2-methyl-1-butoxy)-titanium, **10,** 422
Methyl trimethylsilylacetate, **12,** 326
Methyl 2-trimethylsilylacrylate, **5,** 180; **11,** 346
3-Methyl-1-trimethylsilyl-1,2-butadiene, **9,** 372
3-Methyl-1-trimethylsilyl-2-butene, **7,** 9
3-Methyl-3-trimethylsilyl-1-butene, **3,** 446; **8,** 181
3-Methyl-1-trimethylsilyl-3-buten-2-one, **11,** 85
N-Methyl-N-trimethylsilylmethyl-N′-*t*-butylformamidine, **11,** 347
2-Methyl-1-(trimethylsilylmethyl)pyridinium trifluoromethanesulfonate, **12,** 542
2-Methyl-2-trimethylsilyloxypentan-3-one, **11,** 348
2-Methyl-3-trimethylsilyl-1-propene, **10,** 7; **12,** 375
Methyl trimethylsilyl sulfide, **6,** 399; **8,** 352; **10,** 113, 426
Methyl α-trimethylsilylvinyl ketone, **5,** 461; **6,** 406; **11,** 349
Methyltrioctylammonium chloride (*see* Tricaprylylmethylammonium chloride)
Methyltriphenoxyphosphonium iodide, **1,** 1249; **2,** 283, 446; **4,** 448, 556, 557; **5,** 421; **6,** 649; **8,** 354; **11,** 202
Methyltriphenylphosphonium bromide, **1,** 697, 1238
Methyltriphenylphosphonium permangate, **10,** 446
Methyltris(diethylamino)titanium, **11,** 375
Methyltris(trimethyl phosphite)copper, **2,** 441
Methyltris(triphenylphosphine)rhodium(I), **5,** 463
1-Methyl-2-vinyl-1-cyclopropyllithium, **9,** 273
Methyl vinyl ether, **1,** 697; **6,** 372; **8,** 276; **12,** 100
Methyl vinyl ketone, **1,** 697, 1112; **2,** 283, 391; **5,** 464; **6,** 38, 407; **7,** 71, 247; **10,** 272; **11,** 10; **12,** 329
2-Methyl-6-vinylpyridine, **6,** 409
Mevalonolactone, **1,** 809; **3,** 247; **4,** 345; **6,** 255, 512
Michler's ketone, **2,** 285
Mitsunobu reagent (*see* Triphenylphosphine–Diethyl azodicarboxylate)
MMC (*see* Magnesium methyl carbonate)
Molecular sieves (*see also* Dessicants, Drying agents), **1,** 354, 703; **2,** 286; **3,** 206; **4,** 345, 450; **5,** 465; **6,** 411; **8,** 486; **9,** 316; **10,** 273; **11,** 350
Molybdenum carbonyl, **2,** 287; **3,** 206; **4,** 346; **7,** 165, 247; **9,** 317; **10,** 273; **11,** 350, 383; **12,** 330, 344
Molybdenum(V) chloride, **6,** 412; **12,** 352
Molybdenum(V) chloride–Triphenyl-phosphine, **10,** 274
Molybdenum(V) chloride–Zinc, **10,** 274
Molybdenum(VI) fluoride, **5,** 465; **6,** 412; **7,** 248
Molybdenum hexacarbonyl (*see* Molybdenum carbonyl)

Molybdenum(V) trichloride oxide, **7,** 248
Molybdenum(V) trichloride oxide–Zinc, **7,** 248
Molybdenyl chloride, **7,** 248
Monochloroalane, **1,** 595; **2,** 243; **3,** 9; **8,** 290; **12,** 333
Monochloroborane diethyl etherate, **4,** 346; **5,** 465; **7,** 249
Monochloroborane–Dimethyl sulfide, **8,** 354; **10,** 95
Monoiodoalane, **6,** 327; **7,** 196
Monoisopinocampheylborane, **8,** 267; **9,** 317; **10,** 224; **11,** 350; **12,** 333
Montmorillonite clay K-10, **10,** 184
MoOPH [*see* Oxodiperoxymolybdenum-(pyridine)(hexamethylphosphoric triamide)]
Morpholine, **1,** 197, 705, 972, 1175; **4,** 351, 393; **6,** 429; **8,** 486; **11,** 352
Morpholine–Camphoric acid, **9,** 317; **11,** 352
1-Morpholino-1-cyclohexene, **1,** 707; **6,** 433
2-Morpholinoethyl isocyanide, **11,** 352
N-Morpholinomethyldiphenylphosphine oxide, **9,** 318; **11,** 352; **12,** 334
Mucochloric acid, **8,** 88

Nafion-H, **9,** 320; **10,** 275; **11,** 354; **12,** 335
Nafion-H–Palladium on carbon, **10,** 275
Naphthacene-9,10,11,12-diquinone, **1,** 709
Naphthalene-2,3-dicarboxylic acid, **1,** 710
Naphthalene-2,6-dicarboxylic acid, **1,** 710
Naphthalene–Lithium (*see* Lithium naphthalenide)
Naphthalene–Magnesium (*see* Magnesium naphthalenide)
Naphthalene–Sodium (*see* Sodium naphthalenide)
Naphthalene-2-sulfonic acid, **1,** 712
Naphthalene-2-sulfonyl azide, **11,** 536
1,2-Naphthalic anhydride, **1,** 713
1,8-Naphthalic anhydride, **1,** 713
2-Naphthol-6-sulfonic acid, **5,** 470
1,2-Naphthoquinone, **1,** 713
1,4-Naphthoquinone, **1,** 714
1,2-Naphthoquinone-4-sulfonate, ammonium and potassium salts, **1,** 717
α-Naphthyldiphenylmethyl chloride, **6,** 415
(S)-1-α-(Naphthyl)ethylamine, **8,** 355; **9,** 320
(R)-1-(1-Naphthyl)ethyl isocyanate, **6,** 416; **8,** 356; **9,** 321
α-Naphthyl isocyanate, **1,** 718
α-Naphthyl isothiocyanate, **1,** 718
Nazarov's reagent (*see* Methyl 5-methoxy-3-oxopentanoate)
NBA (*see* N-Bromoacetamide)
NBS (*see* N-Bromosuccinimde)
NCS (*see* N-Chlorosuccinimide)
Neomenthyldiphenylphosphine, **6,** 416
NEP (*see* N-Ethyl-2-pyrrolidone)
Nickel, **7,** 73; **12,** 335
Nickel(II) acetate, **1,** 718; **5,** 470; **6,** 417; **12,** 336
Nickel(II) acetate–Sodium hydride–*t*-Amyl oxide, **10,** 365
Nickel(II) acetylacetonate, **5,** 471; **6,** 417; **7,** 250; **8,** 42, 84; **9,** 51; **10,** 42, 189; **11,** 58; **12,** 336, 357
Nickel(II) acetylacetonate–Diethylaluminum ethoxide, **12,** 52
Nickel(II) acetylacetonate–Diisobutyl-aluminum hydride, **7,** 95; **8,** 42
Nickel(II) acetylacetonate–Diisopropoxy-phenylphosphine–Sodium borohydride, **4,** 352
Nickel(II) acetylacetonate–Trialkyl-aluminums, **6,** 623; **8,** 42; **9,** 52; **11,** 59
Nickel(II) acetylacetonate–Triphenyl-phosphine–Triethylaluminum, **5,** 471
Nickel–Alumina, **9,** 321
Nickel–aluminum alloy (*see* 1:1 Raney nickel alloy)
Nickel boride, **1,** 720; **2,** 289; **3,** 208; **4,** 351; **5,** 471; **11,** 37; **12,** 335
Nickel(II) bromide, **4,** 351; **8,** 119; **10,** 106; **11,** 59
Nickel(II) bromide–Diisopropoxyphenyl-phosphine, **4,** 351
Nickel(II) bromide–Zinc, **9,** 322
Nickel carbonyl, **1,** 44, 720; **2,** 181, 290; **3,** 210; **4,** 353; **5,** 472; **6,** 93, 417, 451; **7,** 250; **8,** 515; **10,** 125, 276; **11,** 356; **12,** 335
Nickel carbonyl–Butyllithium, **3,** 211
Nickel catalysts, **1,** 440, 718, 720, 723, 731; **2,** 289, 293; **3,** 208; **4,** 267, 350, 351; **5,** 263, 471, 570, 736; **6,** 502; **7,** 312; **8,** 433; **9,** 321, 405; **10,** 339; **11,** 37, 355, 356, 457; **12,** 175, 218, 335, 338, 422
Nickel(II) chloride, **8,** 119; **10,** 106; **12,** 336
Nickel(II) chloride–*trans*-2,3-Bis-(diphenylphosphine)bicyclo[2.2.1]-hept-5-ene, **10,** 36
Nickel(II) chloride–1,2-Bis(diphenyl-

phosphine)ethane, **7,** 250; **9,** 147; **10,** 36; **12,** 146, 171
Nickel(II) chloride–1,1′-Bis(diphenylphosphine)ferrocene, **10,** 37; **12,** 172
Nickel(II) chloride–2-Diphenylphosphineferrocenyl ethyldimethylamine, **9,** 200; **11,** 237
Nickel(II) chloride–Ethylene, **10,** 276
Nickel(II) chloride–2,3-O-Isopropylidene-2,3-dihydroxy-1,4-bis(diphenylphosphine)butane, **5,** 360; **11,** 563
Nickel(II) chloride–Triphenylphosphine, **11,** 358
Nickel chloride–Zinc, **10,** 277
Nickel cyanide, **4,** 356
Nickel(II) 2-ethylhexanoate (*see also* Bromine–Nickel(II) alkanoates), **11,** 358
Nickel–Graphite, **11,** 356
Nickel on alumina, **11,** 355
Nickel peroxide, **1,** 642, 731; **2,** 260, 294; **5,** 474; **7,** 250; **8,** 118, 357, 443; **9,** 322; **10,** 277
Nickel(II) sulfate, **2,** 295
Nickel tetracarbonyl (*see* Nickel carbonyl)
Nickel tetraphenylporphine, **3,** 81
Ninhydrin, **1,** 732; **4,** 356
NIS (*see* N-Iodosuccinimide)
Nishimura catalyst (*see* Rhodium oxide–Platinum oxide)
Nitric acid, **1,** 733; **3,** 212; **4,** 356; **5,** 474; **6,** 284
Nitric acid–Acetic anhydride (*see* Acetic anhydride–Nitric acid)
1,1′,1″-Nitrilotri-2-propanol [*see* Tris-(2-hydroxypropyl)amine]
4′-Nitroazobenzene-4-carbonyl chloride, **1,** 735; **4,** 359
Nitrobenzene, **2,** 295, 400; **7,** 251; **8,** 358
4-Nitrobenzenediazonium chloride, **10,** 306
2-Nitrobenzeneseleninic acid, **8,** 359
o-Nitrobenzene selenocyanate, **6,** 420; **7,** 252; **8,** 362; **10,** 278
p-Nitrobenzene selenocyanate, **9,** 325
2-Nitrobenzenesulfenamide, **9,** 267
o-Nitrobenzenesulfenyl chloride, **1,** 745; **4,** 359; **6,** 419; **8,** 360; **10,** 277; **11,** 360
p-Nitrobenzenesulfonoxyurethane, **1,** 736; **2,** 296
o-Nitrobenzenesulfonyl azide, **5,** 475
o-Nitrobenzenesulfonyl chloride, **6,** 419
p-Nitrobenzenesulfonyl chloride, **6,** 531
p-Nitrobenzenesulfonyl-4-nitroimidazole, **11,** 359
m-Nitrobenzenesulfonyl peroxide, **2,** 296
p-Nitrobenzenesulfonyl peroxide, **7,** 251
5-Nitro-[3*H*]-1,2-benzoxathiole S,S-dioxide, **11,** 359
o-Nitrobenzoyl chloride, **6,** 288
p-Nitrobenzoyl chloride, **1,** 736; **6,** 288
p-Nitrobenzyl bromide, **1,** 736; **8,** 360
p-Nitrobenzyl chloride, **1,** 736
p-Nitrobenzyl chloroformate, **1,** 737
p-Nitrobenzyl *p*-toluenesulfonate, **1,** 736
p-Nitrocarbobenzoxy chloride, **1,** 737
4-Nitrocatecholborane, **9,** 98
N-Nitrocollidinium tetrafluoroborate, **8,** 361
3-Nitro-2-cyclohexenone, **10,** 277
3-Nitro-2-cyclopentenone, **10,** 278
2-Nitro-1-dimethylaminoethylene, **4,** 357
Nitroethane, **4,** 357; **12,** 84
Nitroethane–Pyridinium chloride, **11,** 359
Nitroethylene, **5,** 476; **6,** 481
Nitrogen, **1,** 737
Nitrogen dioxide, **1,** 324, 329, 737; **2,** 175; **3,** 130; **4,** 202; **5,** 268; **6,** 284; **7,** 252; **8,** 204, 361
Nitrogen dioxide–Iodine, **2,** 225; **3,** 130; **8,** 205
Nitrogen oxides, **1,** 322, 324, 737; **2,** 175; **3,** 130; **4,** 202; **5,** 268; **7,** 252; **8,** 204, 361
Nitrogen trioxide, **1,** 737; **6,** 548; **7,** 252
1-Nitroguanyl-3,5-dimethylpyrazole, **1,** 738
3-Nitromesityglyoxal, **3,** 79
Nitromethane, **1,** 515, 739, 903; **4,** 357, 489, 507; **5,** 685; **8,** 129, 430; **9,** 323, 447; **10,** 330; **12,** 337, 462, 478
3-Nitro-N-nitrosocarbazole, **1,** 741
Nitronium hexafluorophosphate, **5,** 476
Nitronium tetrafluoroborate, **1,** 742, 1144; **2,** 296; **4,** 358; **5,** 476; **7,** 253; **8,** 361; **9,** 324
Nitronium trifluoromethanesulfonate, **5,** 477; **6,** 420
p-Nitroperbenzoic acid, **1,** 743; **9,** 324; **11,** 360
o-Nitrophenol, **5,** 477; **6,** 470
p-Nitrophenol, **1,** 743; **2,** 297; **6,** 470
p-Nitrophenyl acetate, **2,** 297
4-(4-Nitrophenylazo)benzoyl chloride, **1,** 735; **4,** 359
1-*o*-Nitrophenyl-1,3-butadiene, **1,** 743
p-Nitrophenyl chloroformate, **2,** 297; **5,** 118
p-Nitrophenyl dichlorophosphate, **1,** 744; **3,** 213

p-Nitrophenyl (1,2-dimethylethylenedioxy)-phosphate, **6,** 8
p-Nitrophenyl formate, **1,** 744
p-Nitrophenylhydrazine, **5,** 349
4-Nitrophenyl phenyl phosphorochloridate, **8,** 362
p-Nitrophenyl phosphorodichloridate, **1,** 744; **3,** 213
o-Nitrophenyl selenocyanate, **6,** 420; **7,** 252; **8,** 362; **10,** 278
p-Nitrophenyl selenocyanate, **9,** 325
o-Nitrophenylsulfenyl chloride (*see* *o*-Nitrobenzenesulfenyl chloride)
o-Nitrophenyl thiocyanate, **9,** 325
1-Nitro-1-(phenylthio)propene, **10,** 279
p-Nitrophenyl α-toluenesulfonate, **5,** 477
p-Nitrophenyl trichloroacetate, **1,** 745
p-Nitrophenyl trifluoroacetate, **1,** 745
2-Nitro-3-pivaloyloxypropene, **11,** 360
1-Nitropropane, **1,** 745
2-Nitropropane, **1,** 1101, 1252; **5,** 685
1-Nitropropene, **10,** 197
2-Nitropropene, **6,** 481; **7,** 253
β-Nitropropionyl chloride, **8,** 363
2-Nitroprop-2-yl hydroperoxide, **3,** 213
3-Nitro-2-pyridinesulfenyl chloride, **9,** 325
(5-Nitropyridyl)diphenyl phosphinate, **11,** 361
1-(N-Nitrosoacetylaminomethyl)-cyclohexanol, **4,** 30
Nitrosobenzene, **5,** 478; **11,** 362
1-Nitrosobenzotriazole, **3,** 139
Nitrosocarbonylmethane, **9,** 326; **11,** 362
N-Nitrosodimethylamine, **5,** 479; **6,** 337
p-Nitroso-N,N-dimethylaniline, **1,** 746
(S)-1-Nitroso-2-methylindoline-2-carboxylic acid, **3,** 214
N-Nitroso-N-methylurea (*see also* Diazomethane), **1,** 192, 747; **5,** 179; **6,** 159
Nitrosonium hexafluorophosphate, **1,** 747; **9,** 326
Nitrosonium hydrogen sulfate, **9,** 324
Nitrosonium tetrafluoroborate, **1,** 43, 747; **4,** 360; **6,** 226; **7,** 253; **8,** 364
β-Nitrostyrene, **6,** 422
Nitrosyl chloride, **1,** 43, 97, 748; **2,** 239, 298; **3,** 214; **5,** 658; **6,** 422; **7,** 254; **8,** 364
Nitrosyl fluoride, **1,** 755; **2,** 299; **3,** 214
Nitrosyl hexafluorophosphate, **1,** 747; **9,** 326
Nitrosylsulfate, **1,** 755; **2,** 299
Nitrosyl tetrafluoroborate (*see* Nitrosonium tetrafluoroborate)
Nitryl chloride, **1,** 756, 1018; **2,** 300
Nitryl iodide, **1,** 757; **2,** 300; **3,** 215
NMP (*see* N-Methyl-2-pyrrolidone)
Nonacarbonyldiiron (*see* Tri-μ-carbonyl-hexacarbonyldiiron)
Nonafluorobutanesulfonic anhydride, **5,** 480
Nonafluorobutanesulfonyl fluoride, **5,** 479; **11,** 40; **12,** 338
Norbornene, **1,** 757; **2,** 300
1-Norbornylmagnesium chloride, **10,** 251
(−)-Norephedrine, **12,** 338
Norphos {*see* (E)-2,3-Bis(diphenylphosphine)bicyclo[2.2.1]hept-5-ene}
Noyori's reagent (*see* Lithium aluminum hydride–2,2′-Dihydroxy-1,1′-binaphthyl)

Octacarbonyldicobalt (*see* Di-μ-carbonyl-hexacarbonyldicobalt)
Octacarbonyl(zinc)dicobalt, **2,** 123; **3,** 89; **4,** 139
1,7-Octadien-3-one, **9,** 327; **10,** 280
[(2S)-(2α,3aα,4α,7α,7aα)]-2,3,3a,4,5,6,7,7a-Octahydro-7,8,8-trimethyl-4,7-methanobenzofuran-2-ol, **12,** 339
2-Octanol, **1,** 109
Olah's reagent [*see* Pyridinium poly-(hydrogen fluoride)]
Oleic acid, **1,** 759
Organoaluminum cyanides (*see also* specific names), **1,** 244; **2,** 127; **4,** 146; **6,** 180; **9,** 158, 328; **12,** 182
Organoaluminum reagents (*see also* Alkylaluminum halides, Dialkylaluminum halides, Organoaluminum cyanides, specific names), **9,** 276; **10,** 131, 281; **11,** 17, 25, 246, 364, 506; **12,** 82, 143, 176, 207, 339, 468
Organocerium reagents, **12,** 345
Organocopper reagents (*see also* specific names), **4,** 265; **6,** 376; **7,** 93, 194; **8,** 356; **9,** 240, 328, 498; **10,** 268, 282; **11,** 13, 365, 425; **12,** 69, 315, 345
Organolithium reagents (*see also* specific names), **1,** 207, 280, 571, 619; **3,** 283; **5,** 379, 641, 650; **6,** 105, 203, 368, 443, 510, 570, 571; **7,** 67, 80; **8,** 190, 194, 346; **9,** 5, 94, 217, 246, 313; **10,** 3, 290; **11,** 13; **12,** 14, 101, 175, 205, 217, 310, 350, 382, 397, 466

Organolithium reagents–Boron trifluoride etherate, **12,** 68
Organolithium reagents–Copper(I) halides, **12,** 143
Organolithium reagents–Palladium(0) reagents, **9,** 455
Organolithium reagents–Titanium(IV) chloride, **12,** 376
Organomagnesium reagents (*see also* Grignard reagents, specific names), **1,** 292; **6,** 386; **8,** 253; **9,** 246; **11,** 314; **12,** 352
Organomanganese(II) chlorides, **9,** 288
Organomanganese(II) iodides, **7,** 222; **8,** 312; **9,** 289; **10,** 290
Organomolybdenum reagents, **12,** 352
Organopentafluorosilicates, **8,** 54, 98, 378; **9,** 95; **11,** 373
Organotitanium reagents (*see also* specific names), **4,** 112; **8,** 40, 83; **10,** 87, 138, 270, 422; **11,** 52, 174, 314, 374; **12,** 54, 110, 353
Organozinc reagents (*see also* specific names), **1,** 349; **3,** 128, 257, 257; **5,** 219, 753; **7,** 52; **8,** 472, 507; **9,** 450; **10,** 8, 389; **11,** 25, 468, 505; **12,** 46, 53, 54, 356, 468
Organozirconium reagents, **11,** 378, 394; **12,** 358
Orthophosphoric acid (*see* Phosphoric acid)
Osmium-on-carbon, **3,** 216
Osmium tetroxide, **1,** 458, 509, 690, 759; **2,** 301, 358; **9,** 459; **10,** 290; **12,** 211, 358
Osmium tetroxide–Barium chlorate, **1,** 764
Osmium tetroxide–*t*-Butyl-N-chloro-N-argentocarbamate, **8,** 365
Osmium tetroxide–*t*-Butyl hydroperoxide, **1,** 88; **7,** 255
Osmium tetroxide–Chloramine-T, **7,** 256; **9,** 101
Osmium tetroxide–Dihydroquinidine or Dihydroquinine acetate, **10,** 291
Osmium tetroxide–Diphenyl selenoxide, **10,** 291
Osmium tetroxide–Hydrogen peroxide, **1,** 475
Osmium tetroxide–N-Methylmorpholine N-oxide, **7,** 256; **9,** 334; **10,** 291
Osmium tetroxide–Potassium chlorate, **1,** 759; **4,** 361; **9,** 334
Osmium tetroxide–Silver chlorate, **2,** 363
Osmium tetroxide–Sodium chlorate, **1,** 759; **2,** 301; **6,** 424
Osmium tetroxide–Sodium periodate (*see* Sodium periodate–Osmium tetroxide)
Osmium tetroxide–Trimethylamine N-oxide–Pyridine, **10,** 292
Oxalacetic acid, **4,** 461
Oxalic acid, **1,** 764; **2,** 301; **5,** 481, 538; **9,** 85; **12,** 408
Oxalyl bromide, **1,** 767
Oxalyl chloride, **1,** 28, 767; **2,** 301; **3,** 216; **4,** 361; **5,** 481, 720; **6,** 424; **7,** 211, 257; **8,** 365; **11,** 379
Oxalyl chloride–Aluminum chloride, **8,** 366
Oxalyl chloride–Dimethylformamide (*see* Dimethylformamide–Oxalyl chloride)
Oxalyl chloride–Sodium iodide, **9,** 335
Oxalyl cyanide, **4,** 155
2,5-Oxazolidinedione, **5,** 99
2-Oxazolidones, chiral, **11,** 379; **12,** 359
Oxetane (*see* Trimethylene oxide)
μ-Oxo-bis[tris(dimethylamino)phosphonium] bis(tetrafluoroborate) (Bate's reagent), **6,** 425
Oxodiperoxymolybdenum(pyridine)-(hexamethylphosphoric triamide) (MoOPH), **4,** 203; **5,** 269; **7,** 136; **8,** 206; **9,** 197; **10,** 169; **11,** 218, 382; **12,** 393
α-Oxoglutaric acid, **1,** 531
3-Oxoglutaric acid (*see* 1,3-Acetonedicarboxylic acid)
Oxohexakis(μ-trimethylacetato)trimethanoltriiron(III) chloride, **11,** 382
Oxomethoxymolybdenum(V) 5,10,15,20-tetraphenylporphyrin, **11,** 383
Oxone® (*see* Potassium peroxomonosulfate)
Oxoperoxobis(N-phenylbenzohydroxamato)-molybdenum(VI), **10,** 292
(2-Oxo-2-phenylethylide)dimethylselenonium ylide, **5,** 257
2-Oxo-3-trimethylsilyltetrahydro-1,3-oxazole (*see* 3-Trimethylsilyl-2-oxazolidone)
2-Oxo-3-(triphenylphosphoranylidene)-propyllithium, **5,** 375
Oxotriruthenium acetate complex [*see* Tris-(aquo)hexa-μ-acetato-μ_3-oxotriruthenium(III,III,III) acetate]
Oxotris(*t*-butylimido)osmium(VIII), **8,** 206
Oxotris(triphenylsilanolato)vanadium, **6,** 655; **7,** 413
μ-Oxybis(chlorotriphenylbismuth), **9,** 335; **11,** 381

Oxybis(2,4-pentanedionato)vanadium (*see* Vanadyl acetylacetonate)
Oxygen, **1,** 431, 921, 1253; **2,** 80, 161, 341, 360, 400; **3,** 174; **4,** 362; **5,** 482; **6,** 426, 477, 663; **7,** 258; **8,** 366, 438; **9,** 335, 393, 435; **10,** 293; **11,** 384; **12,** 362
Oxygen fluoride, **1,** 772; **2,** 302
Oxygen–Palladium, **5,** 491
Oxygen, singlet, **2,** 100, 201; **3,** 144, 323; **4,** 362, 412; **5,** 425, 486, 576, 619; **6,** 241, 431, 487; **7,** 21, 41, 261; **8,** 173, 367; **9,** 9, 338; **10,** 294; **11,** 385; **12,** 363
Ozone, **1,** 773, 1052, 1090, 1234; **2,** 156, 444; **3,** 144, 172; **4,** 363; **5,** 261, 320, 491, 647; **6,** 436, 459; **7,** 269; **8,** 64, 374; **9,** 341; **10,** 295; **11,** 387; **12,** 365, 385
Ozone–Silica gel, **6,** 440; **7,** 271; **8,** 375; **9,** 343; **10,** 296

Palladium(II) acetate, **1,** 778; **2,** 303; **4,** 319, 365; **5,** 496; **6,** 442; **7,** 274; **8,** 378; **9,** 344; **10,** 297, 340; **11,** 389; **12,** 88, 177, 367
Palladium(II) acetate–1,2-Bis(diphenylphosphine)ethane, **11,** 391; **12,** 16
Palladium(II) acetate–(S,R)-N,N-Dimethyl-1-[1′,2-bis(diphenylphosphine)-ferrocenyl]ethylamine, **11,** 508
Palladium(II) acetate–Lithium chloride, **10,** 298; **12,** 367
Palladium(II) acetate–Sodium chloride, **6,** 443
Palladium(II) acetate–Sodium hydride, **12,** 371
Palladium(II) acetate–Sodium hydride–Sodium *t*-amyloxide, **10,** 365; **12,** 370
Palladium(II) acetate–Tributylphosphine, **12,** 369
Palladium(II) acetate–Triphenylphosphine [*see also* Diacetatobis(triphenylphosphine)palladium(II)], **4,** 365; **5,** 497; **6,** 60; **8,** 381; **9,** 95, 348, 349; **10,** 298; **11,** 392
Palladium(II) acetate–Tri-*o*-tolylphosphine, **5,** 497; **9,** 350; **11,** 390; **12,** 342
Palladium(II) acetylacetonate, **6,** 45; **9,** 348
Palladium(II) acetylacetonate–Triphenylphosphine, **5,** 190; **12,** 473
Palladium black (*see also* Palladium catalysts), **1,** 778; **4,** 365; **5,** 498; **6,** 443; **7,** 274; **8,** 383; **9,** 351; **10,** 299; **11,** 393
Palladium *t*-butyl peroxide trifluoroacetate, **10,** 203, 299; **12,** 88
Palladium catalysts (*see also* specific names), **1,** 440, 778, 975, 982; **2,** 214, 303, 305, 356, 429; **3,** 218, 274; **4,** 366, 367, 368; **5,** 499; **6,** 445, 615; **7,** 275; **8,** 382, 385; **9,** 351, 353; **10,** 297, 299, 336; **11,** 392; **12,** 338
Palladium(II) chloride, **1,** 782; **2,** 303; **4,** 369; **5,** 496, 500; **6,** 447; **7,** 22, 277; **8,** 384; **9,** 95, 352; **10,** 31, 300, 355; **11,** 61, 393, 393; **12,** 177, 336, 371, 399, 482
Palladium(II) chloride–(+)-*o*-Anisylcyclohexylmethylphosphine, **5,** 500
Palladium(II) chloride–*o*-Benzoquinone, **10,** 302
Palladium(II) chloride–1,2-Bis(diphenylphosphine)ethane, **6,** 447
Palladium(II) chloride–1,1′-Bis(diphenylphosphine)ferrocene, **9,** 147; **10,** 37; **12,** 46, 171
Palladium(II) chloride–Chloranil, **7,** 278
Palladium(II) chloride–Copper(I) chloride, **7,** 278; **10,** 301; **11,** 396; **12,** 372
Palladium(II) chloride–Copper(II) chloride, **2,** 304; **5,** 496; **8,** 384; **9,** 353; **10,** 302; **12,** 372, 373, 380
Palladium(II) chloride–Diphenylbenzylphosphine, **6,** 450; **7,** 277
Palladium(II) chloride–2-Diphenylphosphineferrocenyl ethyldimethylamine, **11,** 19, 237
Palladium(II) chloride–Hexamethylphosphorous triamide, **6,** 448
Palladium(II) chloride–2,3-O-Isopropylidene-2,3-dihydroxy-1,4-bis(diphenylphosphine)butane, **5,** 500
Palladium(II) chloride–Manganese(IV) oxide, **5,** 496
Palladium(II) chloride–P-Phenyl-1-phospha-3-methyl-3-cyclopentene, **5,** 503
Palladium(II) chloride–Potassium dichromate, **5,** 496
Palladium(II) chloride–Silver(I) acetate, **11,** 396
Palladium(II) chloride–(−)-Sparteine, **5,** 500
Palladium(II) chloride–Tetrabutylammonium chloride, **9,** 352
Palladium(II) chloride–Thiourea, **6,** 450; **7,** 278; **8,** 385; **11,** 397
Palladium(II) chloride–Tin(II) chloride–Triphenylphosphine, **11,** 397

Palladium(II) chloride–Triethylsilane (*see* Triethylsilane–Palladium(II) chloride)
Palladium(II) chloride–Triphenylphosphine, **5,** 503
Palladium(II) chloride–Tri-*o*-tolylphosphine, **6,** 448
Palladium–Graphite, **10,** 297
Palladium hydroxide, **1,** 782; **2,** 305; **8,** 385; **9,** 353
Palladium hydroxide on barium sulfate (*see also* Palladium catalysts), **3,** 218
Palladium nitrate, **6,** 451
Palladium on barium sulfate (*see also* Palladium catalysts), **1,** 778; **7,** 275
Palladium on calcium carbonate (*see also* Palladium catalysts), **4,** 366
Palladium on calcium carbonate, lead poisoned (*see* Lindlar catalyst)
Palladium on carbon (*see also* Palladium catalysts), **1,** 778, 982; **2,** 303; **4,** 367, 368; **5,** 499; **6,** 445; **7,** 275; **8,** 382; **9,** 351; **10,** 299; **11,** 393; **12,** 338
Palladium on charcoal (*see also* Palladium catalysts), **1,** 440, 778; **2,** 303; **3,** 274; **4,** 368; **5,** 499; **6,** 446; **7,** 275; **8,** 382; **10,** 336
Palladium(0)–Phosphines, **9,** 344
Palladium, poisoned (*see also* Lindlar catalyst, Rosenmund catalyst), **1,** 566, 975; **2,** 356; **4,** 367; **7,** 275
Palladium–Poly(ethylenimine), **11,** 393
Palladium(II) trifluoroacetate, **10,** 302; **12,** 373
Payne's reagent (*see* Peroxybenzimidic acid)
PCC (*see* Pyridinium chlorochromate)
PDC (*see* Pyridinium dichromate)
PEG (*see* Polyethylene glycols)
Pentacarbonyliron (*see* Iron carbonyl)
Pentacarbonyl(methoxyphenylmethylene)-chromium(0), **11,** 397
Pentacarbonyl(methoxyphenylmethylene)-tungsten(0), **4,** 378
Pentacarbonyl(trimethylsilyl)manganese, **11,** 401
Pentacarbonyl(triphenylphosphine)tungsten, **8,** 525
Pentachlorophenol, **1,** 782; **2,** 305
Pentachlorophenyl dichloroacetate, **2,** 305
Pentachlorophenyl trichloroacetate, **2,** 305
1,3-Pentadiene (Piperylene), **1,** 890; **4,** 10; **6,** 65
1,4-Pentadiene, **1,** 783
3,4-Pentadien-2-one, **9,** 227
Pentadienylllithium, **10,** 303
Pentadienyl-1-pyrrolidinecarbodithioate, **12,** 374
(2,4-Pentadienyl)trimethylsilane, **10,** 303, 436
Pentaerythrityl tetrabromide, **4,** 370
Pentaethoxyphosphorane, **2,** 305
Pentafluoro(fluoromethane)antimony (*see* Methyl fluoride–Antimony(V) fluoride)
Pentafluorophenylcopper, **5,** 504; **6,** 451; **10,** 303
Pentaisopropylguanidine (*see also* Guanidines), **11,** 249
Pentamethoxyphosphorane, **9,** 353
Pentamethylbenzyl chloride, **2,** 305
3,3,6,9,9-Pentamethyl-2,10-diazabicyclo-[4.4.0]decene, **9,** 354
Pentamethylenedimagnesium bromide, **9,** 355
Pentamethylphosphonic diamide, **2,** 280
1,2,2,6,6-Pentamethylpiperidine, **4,** 370; **10,** 419
Pentane-1,5-di(magnesium bromide), **9,** 355
(2R,4R)-Pentanediol, **12,** 192, 355, 375
(2,4-Pentanedionato)copper(I) (*see* Copper(I) acetylacetonate)
2,4-Pentanedione, **1,** 10; **5,** 302, 501, 685; **6,** 629
S,S′-1,5-Pentanediyl *p*-toluenesulfono-thioate, **4,** 539; **5,** 71; **6,** 628
Pentaphenylantimony, **9,** 94
1-Penten-3-one, **1,** 388
trans-3-Penten-2-one, **2,** 306; **3,** 218; **4,** 371
t-Pentyl, *see t*-Amyl
Pentyl nitrite (*see* Amyl nitrite)
Pentylsodium, **5,** 652; **7,** 358, 400
1-Pentynylcopper, **6,** 603
1-Pentynylcopper–Hexamethylphosphoric triamide, **11,** 370
Peracetic acid (*see also* Hydrogen peroxide–Acetic acid), **1,** 459, 785, 787, 795; **2,** 220, 307; **3,** 155, 219; **4,** 372; **5,** 505; **6,** 286, 452; **8,** 386; **11,** 402; **12,** 378
Peracetic acid–Boron trifluoride etherate, **4,** 372
Peracetic acid–Manganese(II) acetylacetonate, **3,** 194
Peracetic acid–Sodium acetate, **7,** 279
Peracetic acid–Sulfuric acid, **4,** 372
Perbenzoic acid, **1,** 458, 666, 791; **3,** 219; **5,** 76; **6,** 453; **7,** 265, 279; **8,** 386
Perbenzoic acid–2,2,6,6-Tetramethyl-

piperidine, **6,** 111
Perchloric acid, **1,** 796; **2,** 70, 221, 309; **3,** 220; **5,** 506, 637; **6,** 453; **7,** 279; **8,** 14; **11,** 402
Perchloryl fluoride, **1,** 802; **2,** 310; **6,** 454; **7,** 39, 280; **10,** 304
Perdichloromaleic acid, **5,** 511
Perfluoroallyl fluorosulfonate, **11,** 403
Perfluorotetramethylcyclopentadienone, **6,** 454
Performic acid, **1,** 457, 826; **4,** 373
Periodates, **1,** 809; **2,** 311; **4,** 373; **5,** 507; **6,** 455
Periodic acid [*see also* Potassium or Sodium periodate], **1,** 815; **2,** 313; **3,** 220; **4,** 374; **5,** 508; **10,** 304
Periodic acid–Chromium(VI) oxide, **2,** 315
Periodinane, **12,** 378
Perlauric acid, **2,** 315; **3,** 220
Permaleic acid, **1,** 819; **12,** 379
Permonophosphoric acid, **10,** 304
Permutit ion-exchange resin, **1,** 512
Peroxyacetimidic acid, **1,** 470; **8,** 387
Peroxyacetyl nitrate, **5,** 510
Peroxybenzimidic acid (Payne's reagent), **1,** 469; **4,** 375; **5,** 121, 125, 511; **6,** 455; **7,** 281
Peroxybis(triphenylphosphine)palladium, **5,** 510
Peroxydichloromaleic acid (*see* Perdichloromaleic acid)
Peroxysulfuric acid (*see* Caro's acid)
Peroxytetrakis(triphenylarsine)rhodium hexafluorophosphate or perchlorate, **8,** 206
Peroxytrichloroacetimidic acid, **12,** 379
Peroxytrifluoroacetic acid (*see* Trifluoroperacetic acid)
Perphthalic acid, **1,** 794, 819; **2,** 69; **3,** 220
Persuccinic acid, **1,** 820
Pertrifluoroacetic acid (*see* Trifluoroperacetic acid)
PFC (*see* Pyridinium fluorochromate)
Pfitzer–Moffatt reagent (*see* Dimethyl sulfoxide–Dicyclohexylcarbodiimide)
Phase-transfer catalysts (*see also* specific names), **8,** 387; **9,** 356; **10,** 305; **11,** 403; **12,** 379
Phenacyl bromide, **1,** 370; **2,** 316; **3,** 235; **5,** 758; **12,** 381
Phenacyldimethylsulfonium bromide, **6,** 456
Phenacylidenedimethylsulfurane, **6,** 456
Phenacylsulfonyl chloride, **3,** 221
Phenacyltriphenylphosphonium bromide, **1,** 1245
Phenanthrenequinone, **1,** 827; **2,** 317
Phenanthrene–Sodium, **4,** 375
Phenetole, **1,** 828
Phenol, **1,** 828; **6,** 307
Phenoxyacetic acid, **5,** 17; **8,** 294; **9,** 361
2-Phenoxy-1,3,2-benzodioxaphosphole, **5,** 512
Phenoxycarbene, **1,** 96
Phenoxycarbonyl isocyanate, **5,** 512
Phenoxymagnesium bromide, **8,** 242
Phenoxymethylenetriphenylphosphorane, **1,** 1240
Phenyl acetate, **1,** 829
(S)-Phenylalanine, **6,** 411; **9,** 393; **12,** 415
(1S,2S)-(+)-1-Phenyl-2-amino-3-methoxy-1-propanol [*see* (1S,2S)-2-Amino-3-methoxy-1-phenyl-1-propanol]
(1S,2S)-1-Phenyl-2-amino-1,3-propanediol [*see* (1S,2S)-2-Amino-1-phenyl-1,3-propanediol]
Phenyl azide, **1,** 829; **2,** 45; **3,** 17; **4,** 375; **5,** 513; **9,** 134
p-Phenylazobenzenesulfonyl chloride, **1,** 832
p-Phenylazobenzoyl chloride, **1,** 20, 832
p-Phenylazomaleinanil, **1,** 833; **2,** 317
2-(*p*-Phenylazophenyl)isopropyl azidoformate, **7,** 285
N-(*p*-Phenylazophenyl)maleimide, **1,** 833; **2,** 317
N-Phenylbenzamidine, **5,** 513
p-Phenylbenzeneselenol, **9,** 35
Se-Phenyl benzeneselenosulfonate (*see* Phenylselenenyl benzenesulfonate)
Phenyl benzenethiosulfonate, **8,** 391; **9,** 362, 429
2-Phenylbenzothiazoline, **12,** 384
p-Phenylbenzoyl chloride, **4,** 376; **7,** 282
Phenylbenzoyldiazomethane, **1,** 655
N-Phenyl-N′-benzoyldiimide, **1,** 833
Phenylboric acid (*see* Benzeneboronic acid)
(+)-3-Phenylbornane-2,3-diol, **9,** 363; **11,** 86
Phenylboron dichloride, **4,** 377; **12,** 178
Phenylboronic acid (*see* Benzeneboronic acid)
Phenyl(1-bromo-1-chloro-2,2,2-trifluoroethyl)mercury, **3,** 223; **4,** 376
Phenyl(bromodichloromethyl)mercury, **2,** 326; **4,** 385; **5,** 514; **7,** 282
Phenyl(1-bromo-1,2,2,2-tetrafluoroethyl)-

mercury, **5,** 515
(S)-2-Phenylbutanoyl chloride, **4,** 124
Phenyl *t*-butyldimethylsilyl selenide, **10,** 63
α-Phenylbutyric anhydride, **2,** 317
S-(+)-2-Phenylbutyryl chloride, **4,** 124
N-Phenylcampholylhydroxamic acid, **8,** 392
Phenylcarbene, **3,** 336; **8,** 110
Phenylcarbonimidic dichloride, **1,** 843; **6,** 458
Phenyl chloroformate, **2,** 318; **5,** 248
Phenyl chlorosulfate, **11,** 410
Phenyl chlorosulfite, **6,** 456
1-Phenyl-5-chlorotetrazole (*see* 5-Chloro-1-phenyltetrazole)
Phenyl chlorothionocarbonate, **10,** 306
Phenylcopper, **6,** 143, 234; **7,** 282; **9,** 333; **12,** 140
Phenylcopper–Boron trifluoride, **12,** 347
Phenyl cyanate, **2,** 320; **10,** 307
Phenylcyclone, **1,** 834
Phenyldiazomethane, **1,** 366, 834; **2,** 320; **4,** 73; **5,** 515; **6,** 473; **10,** 308
Phenyl diazomethyl sulfoxide, **6,** 457
Phenyl(dibromochloromethyl)mercury, **3,** 222
Phenyldichloroborane (*see* Dichlorophenylborane)
Phenyl dichlorophosphate, **1,** 847; **9,** 187; **11,** 478; **12,** 384
Phenyl dichlorophosphate–Dimethylformamide, **11,** 410
Phenyl(dihalocarbomethoxymethyl)mercury, **3,** 222
2-Phenyl-1,3-dioxol-4-ene, **5,** 662
2-Phenyl-1,3-dithiolane, **6,** 26
o-Phenylenediamine, **1,** 834; **8,** 393
Phenylene orthosulfite, **8,** 393
o-Phenylene phosphorochloridite (*see* 2-Chloro-1,3,2-benzodioxaphosphole)
α-Phenylethylamine (*see* α-Methylbenzylamine)
(S)-(−)-5-(α-Phenylethyl)semioxamazide, **9,** 364
(Phenylethynyl)copper [*see* Copper(I) phenylacetylide]
Phenylethynylmagnesium bromide, **5,** 664
α-Phenylglycine methyl ester, **8,** 395; **10,** 308
Phenylhydrazine, **1,** 838, 899; **2,** 322; **4,** 138; **6,** 457; **7,** 284; **8,** 424; **10,** 318
Phenylhydrazine-*p*-sulfonic acid, **1,** 842
2-Phenylhydrazonopropanal, **6,** 500
9-Phenyl-9-hydroxyanthrone, **4,** 378
Phenyliodine(III) bis(trifluoroacetate), **6,** 301; **9,** 54
Phenyliodine(III) diacetate, **1,** 508; **3,** 166; **4,** 266; **8,** 351; **10,** 214; **11,** 271, 411; **12,** 384
Phenyliodine(III) diacetate–Azidotrimethylsilane, **4,** 278; **5,** 354; **6,** 297
Phenyliodine(III) dichloride, **1,** 505; **2,** 225; **3,** 142, 164; **4,** 264; **5,** 352; **6,** 298; **9,** 249; **10,** 215
Phenyliodine(III) difluoride, **6,** 300
Phenyliodo dichloride [*see* Phenyliodine(III) dichloride]
Phenyliodoso diacetate [*see* Phenyliodine(III) diacetate]
Phenyl isocyanate, **1,** 252, 842; **2,** 322; **4,** 378; **5,** 94, 211; **7,** 52, 284; **8,** 205, 396; **10,** 309; **12,** 386
Phenyl isocyanide, **1,** 843; **10,** 309
Phenyl isocyanide dichloride, **1,** 893; **6,** 458
Phenyl isothiocyanate, **1,** 844; **2,** 323
N-Phenylketeniminyl(triphenyl)phosphorane, **12,** 387
Phenyllithium, **1,** 845, 908, 1232; **2,** 343; **6,** 307, 636, 643; **8,** 396; **9,** 364; **10,** 5
Phenyllithium–Boron trifluoride etherate, **12,** 68
Phenyllithium–Tetramethylethylenediamine, **3,** 284
Phenylmagnesium bromide, **1,** 98, 415, 673, 992; **2,** 11, 205; **3,** 14; **4,** 315; **5,** 722; **7,** 132, 283; **9,** 106, 230
Phenylmagnesium bromide–Cobalt(II) chloride, **1,** 155
Phenylmagnesium bromide–Copper(I) halides, **9,** 124
Phenylmagnesium bromide–Nickel(II) reagents, **6,** 50
Phenylmagnesium fluoride, **4,** 315
N-Phenylmaleimide, **1,** 845
8-Phenylmenthol, **10,** 178; **11,** 412; **12,** 389
8-Phenylmenthyl acrylate, **10,** 402; **11,** 412
8-Phenylmenthyl crotonate, **11,** 412
8-Phenylmenthyl glyoxylate, **11,** 414; **12,** 514
8-Phenylmenthyl pyruvate, **12,** 389
2-Phenylmercaptoethanol, **5,** 516
Phenylmercuric acetate, **6,** 458
Phenylmercuric bromide, **4,** 353
Phenylmercuric chloride, **4,** 353; **9,** 352
Phenylmercuric hydroxide, **11,** 415
Phenylmercuric perchlorate, **11,** 415
Phenylmethanesulfonyl fluoride, **2,** 323

Phenylmethoxycarbenepentacarbonyltungsten(0), **4,** 378
2-Phenyl-3-methyl-1,3,2-oxaphospholine, **2,** 323
N-Phenylmorpholine, **1,** 846
Phenyl-β-naphthylamine, **1,** 846
4-Phenyloxazole, **12,** 389
Phenyl 2-oxopropyl sulfoxide, **5,** 524
Phenylpalladium acetate, **4,** 379
1-Phenyl-4-penten-1-one, **8,** 7
Phenyl [2-(*p*-phenylazophenyl)isopropyl] carbonate, **7,** 285
Phenyl phenylmethanesulfonate, **5,** 478
Phenyl N-phenylphosphoramidochloridate, **7,** 286; **10,** 310; **11,** 416
3-Phenyl-2-(phenylsulfonyl)oxaziridine, **12,** 392
Phenyl(phenylthio)methyllithium, **5,** 528
Phenyl phenylthiomethyl sulfoxide, **6,** 394
Phenyl(phenylthio)trimethylsilylmethane, **10,** 314
Phenyl(phenylthio)trimethylsilylmethyllithium, **10,** 314
Phenylphosphonic dichloride, **1,** 846
Phenyl phosphorodiamidate, **5,** 517
Phenyl phosphorodichloridate (*see* Phenyl dichlorophosphate)
Phenylphosphorodi(1-imidazolidate), **1,** 847
Phenylpotassium, **1,** 848
Phenyl propargyl selenide, **9,** 365
(E)-1-Phenyl-1-propene oxide, **11,** 337
Phenyl 2-propenyl sulfone, **10,** 316
(3-Phenyl-2-propenyl)triphenylphosphonium chloride, **1,** 613, 1238
(+)-1-Phenyl-2-propylamine, **1,** 978
Phenyl 8-quinolyl phosphate, **5,** 517
Phenylselenenyl acetate (*see* Benzeneselenenyl acetate)
Phenylselenenyl benzenesulfonate, **11,** 407; **12,** 390
Phenylselenenyl halides (*see* Benzeneselenenyl bromide or chloride)
Phenylselenenyl *p*-toluenesulfonate, **10,** 315; **12,** 394
Phenylselenenyl trifluoroacetate (*see* Benzeneselenenyl trifluoroacetate)
Phenylseleninyl chloride, **6,** 459
Phenylselenoacetaldehyde, **9,** 365; **10,** 310
Phenylselenoacetone, **10,** 311
Phenylseleno(*t*-butyldimethyl)silane, **10,** 63
Phenylselenocopper, **12,** 394
Phenyl selenocyanate, **7,** 253; **8,** 119, 396; **9,** 34; **11,** 33, 416
Phenyl selenocyanate–Copper(II) chloride, **9,** 34
Phenylselenoethylene (*see* Phenyl vinyl selenide)
Phenylselenomethyllithium, **6,** 549
4-Phenylseleno-1-penten-3-one, **8,** 397
1-Phenylseleno-1-phenylthioethylene, **8,** 281
N-Phenylselenophthalimide, **9,** 366; **10,** 312; **11,** 417; **12,** 390
2-Phenylselenopropionic acid, **8,** 397
N-Phenylselenosuccinimide, **9,** 366
Phenylselenotrimethylsilane, **9,** 373, 432, 496; **10,** 436
Phenylselenotrimethylsilylmethyllithium, **7,** 402
1-Phenylseleno-2-trimethylsilyloxy-4-methoxy-1,3-butadiene, **8,** 398
Phenylsilane, **5,** 523
Phenylsodium, **1,** 848; **8,** 69
Phenylsulfenyl chloride (*see* Benzenesulfenyl chloride)
Phenylsulfinylacetone, **5,** 524
α-(Phenylsulfinyl)acetonitrile, **6,** 395; **11,** 418; **12,** 391
3-Phenylsulfinyl-2-butanone, **5,** 526
Phenylsulfinylcarbene, **6,** 457
Phenylsulfinyldiazomethane, **6,** 457
Phenylsulfinylethylene, **6,** 468; **8,** 399; **11,** 427, 547
Phenylsulfinylmethyl phenyl sulfide, **6,** 394
p-Phenylsulfonylbenzoyl chloride, **1,** 849
4-Phenylsulfonyl-2-butanone ethylene ketal, **6,** 462
2-Phenylsulfonylethyl chloroformate, **11,** 419
2-(2-Phenylsulfonylethyl)-1,3-dioxolane, **6,** 461
Phenylsulfonylethylene, **10,** 170, 315; **11,** 427, 475
α-Phenylsulfonylethyllithium, **9,** 84, 248
(2-Phenylsulfonylethyl)trimethylsilane, **9,** 445
3-Phenylsulfonyl-1(3*H*)-isobenzofuranone, **8,** 76; **9,** 368; **10,** 313
3-(Phenylsulfonyl)isobutyraldehyde ethylene acetal, **6,** 461
Phenylsulfonylmethylmagnesium bromide, **2,** 166
(Phenylsulfonyl)nitromethane, **11,** 419
2-(Phenylsulfonyl)-3-phenyloxaziridine, **12,** 392
γ-Phenylsulfonylphthalide (*see* 3-Phenyl-

sulfonyl-1(3*H*)-isobenzofuranone)
β-(Phenylsulfonyl)propionaldehyde ethylene acetal, **6,** 461
2-(1-Phenylsulfonyl-2-propyl)-1,3-dioxolane, **6,** 461
1-Phenylsulfonyl-2-trimethylsilylethane, **11,** 41
(E)-1-Phenylsulfonyl-2-trimethylsilylethylene, **11,** 42
Phenylsulfonyl(trimethylsilyl)methane, **12,** 393
Phenylsulfonyl(trimethylsilyl)methyllithium, **12,** 393
Phenylsulfur trifluoride, **1,** 849
Phenyltetrafluorophosphorane, **5,** 526
1-Phenyl-2-tetrazoline-5-thione, **11,** 420
Phenylthioacetic acid, **6,** 463
Phenylthioacetyl chloride, **9,** 370
Phenylthioacetylene, **9,** 370
Phenylthiobenzyllithium, **6,** 131
1-Phenylthio-1,3-butadiene, **7,** 75
3-Phenylthio-3-buten-2-one, **11,** 421
Phenylthiocopper, **5,** 414; **6,** 465; **7,** 211, 212
2-Phenylthiocyclobutanone, **11,** 422
1-Phenylthio-1-cyclopropyllithium (*see* 1-Lithiocyclopropyl phenyl sulfide)
1-Phenylthiocyclopropyltriphenylphosphonium tetrafluoroborate, **6,** 465
2-Phenylthioethanol, **5,** 516
Phenylthioethylene, **8,** 281; **9,** 371
Phenylthiomethyl azide, **10,** 14; **12,** 37
O-(Phenylthiomethyl)hydroxylamine, **5,** 527
Phenylthiomethyllithium, **2,** 99, 324; **4,** 379; **5,** 527; **6,** 596; **11,** 422
Phenylthiomethyl(trimethyl)silane (*see* Phenylthiotrimethylsilylmethane)
Phenylthiomethyl(triphenyl)tin, **11,** 424
N-Phenylthiophthalimide, **11,** 493
2-Phenylthiopropionic acid, **6,** 466
2-(Phenylthio)propionyl chloride, **9,** 315
N-Phenylthiosuccinimide, **8,** 399
Phenylthiotriethylsilane, **6,** 613
Phenylthiotrimethylsilane, **6,** 399; **10,** 426, 462; **12,** 63, 535
1-Phenylthio-1-trimethylsilylcyclopropane, **12,** 280
1-Phenylthio-1-trimethylsilylethylene, **8,** 281; **11,** 83, 583; **12,** 394
1-Phenylthio-2-trimethylsilylethylene, **11,** 583
Phenylthiotrimethylsilylmethane, **7,** 402; **10,** 313; **11,** 423
Phenylthio(trimethylsilyl)methyllithium, **10,** 314; **11,** 423
2-Phenylthio-2-trimethylsilylpropane, **12,** 280
Phenylthio(triphenylstannyl)methane, **11,** 424
Phenylthio(triphenylstannyl)methyllithium, **11,** 424
1-(Phenylthiovinyl)diphenylphosphine oxide, **11,** 424
1-Phenylthiovinyllithium, **8,** 281; **9,** 371
1-Phenylthiovinyl(triphenyl)phosphonium iodide, **9,** 502; **11,** 425
Se-Phenyl *p*-tolueneselenosulfonate, **10,** 315; **12,** 394
Phenyl α-toluenesulfonate, **5,** 478
(R)-(+)-Phenyl (*p*-tolylsulfinyl)acetate, **10,** 405
N-Phenyl-1,2,4-triazoline-3,5-dione (PTAD), **1,** 849; **2,** 324; **3,** 223; **4,** 381; **5,** 528; **6,** 253, 299, 433, 467; **7,** 287; **9,** 372; **12,** 395
Phenyl(tribromomethyl)mercury, **1,** 851; **2,** 276; **6,** 468
Phenyl(trichloromethyl)mercury, **1,** 851; **2,** 326
Phenyl trifluoroacetate, **1,** 850
N-Phenyltrifluoromethanesulfonimide, **5,** 530; **12,** 395
Phenyltrifluoromethylketene, **4,** 383
Phenyl(trifluoromethyl)mercury, **3,** 224; **4,** 384
Phenyl(trihalomethyl)mercury (*see also* specific names), **1,** 851; **2,** 326; **3,** 222, 224, 225; **4,** 384, 385; **5,** 514, 530; **6,** 468; **7,** 282
N-Phenyltrimethylacetimidoyl chloride, **1,** 854
Phenyltrimethylammonium ethoxide, **1,** 854
Phenyltrimethylammonium perbromide (PTAB), **1,** 855, 970; **2,** 328; **4,** 386; **5,** 531; **11,** 75, 426
1-Phenyl-2-trimethylsilylacetylene, **7,** 354
Phenyl 2-(trimethylsilyl)ethynyl sulfone, **11,** 426
Phenyl trimethylsilylmethyl sulfide, **7,** 402; **10,** 313; **11,** 423
Phenyl trimethylsilylpropargyl ether, **9,** 373
Phenyl trimethylsilyl selenide, **9,** 373, 432, 496; **10,** 439
Phenyltungsten trichloride–Aluminum chloride, **7,** 288
Phenyl vinyl selenide, **8,** 294; **9,** 282, 374
Phenyl vinyl sulfide, **8,** 281; **9,** 371

Phenyl vinyl sulfone, **10,** 170, 315; **11,** 427, 475
Phenyl vinyl sulfoxide, **6,** 468; **8,** 399; **11,** 427, 547
Phenylytterbium iodide, **11,** 249
Phenylzinc bromide, **5,** 753
Phenylzinc chloride, **11,** 505
Phosgene, **1,** 856; **2,** 328; **3,** 225; **4,** 157; **5,** 532, 663; **7,** 289
Phosgene immonium chloride (*see* Dichloromethylenedimethylammonium chloride)
Phosphatolead(IV) acids, **3,** 225
1-Phospha-2,8,9-trioxaadamantane ozonide, **6,** 469
Phosphine, **1,** 859; **2,** 329
Phosphonamides, chiral, **12,** 396
Phosphonitrilic chloride trimer, **2,** 206; **4,** 386; **5,** 322; **6,** 469
Phosphonium iodide, **1,** 859
Phosphoric acid, **1,** 860, 1195; **4,** 387; **7,** 255, 348; **10,** 317
Phosphoric acid–Boron trifluoride complex, **7,** 289
Phosphoric acid–Formic acid, **1,** 860; **5,** 534; **10,** 317
Phosphoric acid–Methanesulfonic acid, **10,** 256
Phosphorodimorpholidic chloride, **1,** 861
Phosphorous acid, **5,** 447
Phosphorus(III) bromide, **1,** 862, 873, 874; **2,** 330, 335; **7,** 292
Phosphorus(V) bromide, **1,** 865
Phosphorus(III) bromide–Copper(I) bromide, **6,** 183
Phosphorus(III) chloride, **1,** 28, 875; **8,** 400; **10,** 318; **12,** 1
Phosphorus(V) chloride, **1,** 771, 866; **4,** 388, 505; **5,** 534; **6,** 470, 497; **7,** 290; **10,** 318
Phosphorus(V) chloride–Sodium iodide, **9,** 335
Phosphorus(V) chloride–Zinc chloride, **1,** 870
Phosphorus diiodide (*see* Diphosphorus tetraiodide)
Phosphorus heptasulfide, **1,** 864
Phosphorus–Hydriodic acid (*see* Hydriodic acid–Red phosphorus)
Phosphorus(III) iodide, **1,** 497; **9,** 85, 203; **10,** 174, 318; **11,** 224; **12,** 218
Phosphorus(V) oxide, **1,** 871; **2,** 329; **7,** 291; **10,** 319
Phosphorus(V) oxide–Hexamethyldisiloxane (*see* Trimethylsilyl polyphosphate)
Phosphorus(V) oxide–Methanesulfonic acid, **5,** 535; **7,** 291; **8,** 400; **11,** 428
Phosphorus(V) oxide–Phosphoric acid, **1,** 872; **5,** 535
Phosphorus(V) oxide–Pyridine, **2,** 330
Phosphorus(V) oxide–Triethylamine, **2,** 330
Phosphorus oxychloride (*see* Phosphoryl chloride)
Phosphorus penta-, *see* Phosphorus(V)
Phosphorus pentoxide (*see* Phosphorus(V) oxide)
Phosphorus, red, **1,** 861
Phosphorus(red)–Iodine, **1,** 862; **6,** 470
Phosphorus(V) sulfide, **1,** 135, 333, 870; **3,** 226; **4,** 389; **5,** 534, 653; **6,** 470; **7,** 290; **8,** 401; **9,** 374, 466; **10,** 320; **11,** 428
Phosphorus tri-, *see also* Phosphorus(III)
Phosphorus trichloride–Dimethylformamide [*see* Dimethylformamide–Phosphorus(III) chloride]
Phosphoryl chloride (Phosphorus oxychloride), **1,** 609, 771, 876, 924; **2,** 63, 330, 335; **3,** 228; **4,** 389, 390; **5,** 535; **7,** 292; **8,** 401; **9,** 374; **11,** 429; **12,** 48
Phosphoryl chloride–N,N-Dimethylformamide (*see* Dimethylformamide–Phosphoryl chloride)
Phosphoryl chloride–Sodium iodide, **9,** 335
Phosphoryl chloride–Tin(II) chloride, **1,** 881
Phosphoryl chloride–Trimethyl phosphate, **3,** 229
Phosphoryl chloride–Zinc chloride, **1,** 880
Phosphotungstic acid, **5,** 537
1,4-Phthalazinedione, **2,** 331
Phthalic anhydride, **1,** 882, 1200
Phthalide, **5,** 537; **8,** 402
Phthalimide, **4,** 553; **11,** 597
Phthalimide–N-sulfenyl chloride, **9,** 375
Phthaloyl chloride, **1,** 882
Phthaloyl peroxide, **6,** 471
(S)-N-Phthaloylphenylalanyl chloride, **8,** 402
Phthalyl alcohol, **1,** 883
3-Picoline, **6,** 336
Picolinic acid, **5,** 538; **7,** 293
4-Picolyl chloride hydrochloride, **4,** 391
2-Picolyl chloride N-oxide, **5,** 538
2-Picolyl chloride 1-oxide hydrochloride, **6,** 472
2- or 4-Picolyllithium, **5,** 649
Picric acid, **1,** 884; **4,** 391

Picryl azide, **1,** 885
Picryl chloride, **1,** 885; **4,** 391
Pinacol, **11,** 86; **12,** 397
Pinacol chloromethaneboronate (*see* 2-Chloromethyl-4,4,5,5-tetramethyl-1,3,2-dioxaborolane)
Pinanediol, **5,** 387; **12,** 80
B-3-Pinanyl-9-borabicyclo[3.3.1]nonane, **8,** 403; **10,** 320; **11,** 429; **12,** 397
Piperazine, **4,** 392
Piperidine, **1,** 886, 972, 1039, 1044; **2,** 28, 332; **4,** 393; **6,** 472; **7,** 293; **12,** 330
Piperidinium acetate, **1,** 887; **9,** 318, 376; **11,** 352
1-Piperidinyl-1-cyclohexene, **4,** 365
Piperylene (*see* 1,3-Pentadiene)
Pivalic acid, **1,** 1229; **12,** 505
Pivaloyl chloride, **1,** 1229; **8,** 404; **12,** 282
Platinum, **7,** 73
Platinum–Alumina catalyst (*see also* Platinum catalysts), **4,** 393
Platinum black (*see also* Platinum catalysts), **1,** 890; **10,** 321
Platinum *t*-butyl peroxide trifluoroacetate, **11,** 430
Platinum catalysts, **1,** 432, 440, 890, 892, 982, 1220; **2,** 332; **4,** 368, 393; **5,** 326, 446; **6,** 472, 615; **8,** 382, 405; **9,** 352; **10,** 321, 342, **12,** 338
Platinum on carbon (*see also* Platinum catalysts), **1,** 982; **10,** 321
Platinum oxide (Adam's catalyst), **1,** 890; **2,** 332; **5,** 446; **6,** 615; **7,** 275
Platinum–Silica (*see also* Platinum catalysts), **1,** 891; **6,** 472; **10,** 342
Platinum sulfide-on-carbon (*see also* Platinum catalysts), **1,** 892
Platinum–Tin(II) chloride, **1,** 891
Polyethylene glycols (PEG), **1,** 17; **9,** 360, 376; **12,** 399
Polyhexamethylenecarbodiimide, **2,** 333
Polyhydrogen fluoride–Pyridine [*see* Pyridinium poly(hydrogen fluoride)]
Polymeric carbodiimide reagent, **5,** 539
Polymethylhydrosiloxane, **4,** 393
Polyoxyethylated lauryl alcohol, **1,** 892
Polyphosphate ester (PPE), **1,** 892; **2,** 333; **3,** 229; **4,** 394; **5,** 539; **6,** 474; **9,** 376; **11,** 427, 430
Polyphosphoric acid (PPA), **1,** 894; **2,** 334, 402; **3,** 231; **4,** 395; **5,** 536, 540; **6,** 474; **7,** 294; **10,** 321; **12,** 399
Potassium, **1,** 905; **5,** 559; **7,** 296; **8,** 69; **12,** 400
Potassium acetate, **1,** 906; **5,** 154; **6,** 137, 278, 475, 640; **8,** 130
Potassium acetate–Potassium iodide, **5,** 325
Potassium acetylenedicarboxylate, **1,** 907
Potassium–Alumina, **10,** 322
Potassium amide, **1,** 907; **2,** 336; **3,** 232; **4,** 61, 153, 398; **5,** 400, 543
Potassium 3-aminopropylamide (KAPA), **6,** 476; **7,** 296; **8,** 406; **9,** 378; **11,** 431; **12,** 400
Potassium–Ammonia, **8,** 38; **9,** 273, 377
Potassium *t*-amyloxide, **1,** 905, 939; **5,** 551; **9,** 379
Potassium azodicarboxylate, **1,** 909; **2,** 139; **3,** 100
Potassium benzeneselenoate, **9,** 373, 433
Potassium bis(trimethylsilyl)amide (*see* Potassium hexamethyldisilazide)
Potassium bisulfate, **1,** 909
Potassium borohydride, **4,** 398; **6,** 476; **8,** 407
Potassium borohydride–(−)-N-Dodecylmethylephedrinium bromide, **6,** 249
Potassium bromate, **11,** 431
Potassium bromide–Dicyclohexyl-18-crown-6, **5,** 154
Potassium bromide–Tricaprylylmethylammonium chloride, **6,** 405
Potassium 1-butanethiolate, **1,** 729
Potassium *t*-butoxide, **1,** 103, 905, 911; **2,** 336; **3,** 233; **4,** 358, 399; **5,** 544; **6,** 334, 477; **7,** 296; **8,** 71, 293, 407; **9,** 380; **10,** 56, 323; **11,** 432; **12,** 401
Potassium *t*-butoxide–Benzophenone, **9,** 382
Potassium *t*-butoxide–*t*-Butyl alcohol complex, **11,** 432; **12,** 402
Potassium *t*-butoxide–Butyllithium (*see* Butyllithium–Potassium *t*-butoxide)
Potassium *t*-butoxide–18-Crown-6, **6,** 135; **8,** 128; **9,** 126; **10,** 111, 323
Potassium *t*-butoxide–Dimethyl sulfoxide, **1,** 298, 914; **2,** 336; **3,** 233; **4,** 401; **5,** 544, 591; **6,** 228, 478, 479; **7,** 298; **9,** 307, 380; **11,** 432
Potassium *t*-butoxide–Hexamethylphosphoric triamide, **1,** 431; **7,** 297
Potassium–*t*-Butylamine, **8,** 285; **9,** 377
Potassium butylmercaptide, **1,** 729
Potassium carbonate, **1,** 1014; **5,** 552; **8,** 408;

9, 382; **10,** 323; **11,** 433; **12,** 406
Potassium carbonate–18-Crown-6, **6,** 135; **9,** 358; **12,** 403
Potassium carbonate–Dimethyl sulfoxide, **6,** 479
Potassium carbonate–Hexamethylphosphoric triamide, **6,** 274
Potassium chlorate, **1,** 927
Potassium chloride–Dicyclohexyl-18-crown-6, **6,** 480
Potassium chloride–Dimethylformamide, **9,** 423
Potassium chloride–Tricaprylylmethylammonium chloride, **6,** 405
Potassium chromate, **7,** 298; **9,** 357
Potassium chromate–Dicyclohexyl-18-crown-6, **7,** 298
Potassium–Crown ether, **10,** 322; **11,** 431
Potassium cryptate[2.2.2], **5,** 156; **6,** 137; **9,** 127; **10,** 112; **12,** 404
Potassium cyanide, **1,** 401, 878, 1047, 1197; **3,** 288; **4,** 512; **5,** 553; **7,** 299; **8,** 409, 453, 473; **9,** 328; **10,** 145, 324, 398; **11,** 269, 433; **12,** 405
Potassium cyanide–Acetone cyanohydrin (*see also* Acetone cyanohydrin), **8,** 409
Potassium cyanide–Chlorotrimethylsilane, **8,** 409
Potassium cyanide–Crown ethers, **6,** 135; **7,** 77, 299; **10,** 324; **11,** 433; **12,** 148, 406
Potassium cyanide–Silica, **10,** 347
Potassium cyanide–Tricaprylylmethylammonium chloride, **6,** 406
Potassium decacarbonylhydridochromate (*see* Potassium di-μ-carbonyloctacarbonyl-μ-hydridochromate)
Potassium diazomethanedisulfonate, **1,** 928
Potassium 2,6-di-*t*-butyl-4-methylphenoxide, **10,** 48
Potassium 2,6-di-*t*-butylphenoxide, **3,** 234; **5,** 557
Potassium di-μ-carbonyloctacarbonyl-μ-hydridochromate, **7,** 299
Potassium dichromate, **6,** 124; **7,** 68; **8,** 410; **9,** 383; **10,** 324; **12,** 405
Potassium dichromate–Polyethylene glycol, **9,** 376
Potassium dichromate–Tetrabutylammonium bisulfate, **8,** 410
Potassium diisopropylamide (KDA), **10,** 324
Potassium diisopropylamide–Lithium *t*-butoxide, **9,** 383
Potassium 9-(2,3-dimethyl-2-butoxy)-9-boratabicyclo[3.3.1]nonane, **12,** 406
Potassium–Dimethyl sulfoxide, **2,** 158
Potassium diphenyl-4-pyridylmethide, **9,** 384
Potassium ethanethiolate, **1,** 1067
Potassium ethoxide, **1,** 928
Potassium–Ethylenediamine, **4,** 409
Potassium ethylmercaptide, **1,** 1067
Potassium ethylxanthate, **5,** 554
Potassium fencholate, **8,** 228
Potassium ferrate(VI), **4,** 405
Potassium ferricyanide, **1,** 929; **2,** 345; **4,** 406; **5,** 427, 554; **6,** 480; **7,** 300; **8,** 410; **9,** 385
Potassium fluoride, **1,** 121, 375, 933; **2,** 346; **5,** 376, 555; **6,** 481; **8,** 129, 410, 480; **9,** 224; **10,** 325; **11,** 434, 554; **12,** 86, 406
Potassium fluoride–Alumina, **12,** 407
Potassium fluoride–Celite, **10,** 326
Potassium fluoride–Crown ethers, **5,** 153; **6,** 480; **7,** 78, 152; **8,** 411, 467; **9,** 323; **10,** 325; **11,** 441, 525
Potassium fluoride–Dimethylformamide, **12,** 406
Potassium fluoride–Dimethyl sulfone, **1,** 296
Potassium fluoride–Dimethyl sulfoxide, **1,** 303, 933; **5,** 555; **6,** 482; **10,** 325
Potassium fluoride–Hexamethylphosphoric triamide, **5,** 556; **10,** 325
Potassium fluoride–N-Methyl-2-pyrrolidone, **5,** 556
Potassium fluoride–Oxygen, **11,** 434
Potassium fluoride–Tributylhexadecylphosphonium bromide, **5,** 322
Potassium fluoride–Tricaprylylmethylammonium chloride, **6,** 405; **8,** 390
Potassium formate, **5,** 556; **10,** 329
Potassium–Graphite, **4,** 397; **7,** 296; **8,** 405, 445; **9,** 378, 476; **10,** 326; **12,** 400
Potassium *t*-heptoxide (*see* Potassium triethylmethoxide)
Potassium hexacyanodinickelate(4-), **3,** 212
Potassium hexacyanoferrate(III) (*see* Potassium ferricyanide)
Potassium hexamethyldisilazide, **4,** 407; **5,** 557; **6,** 482; **9,** 229, 370, 385; **10,** 326; **12,** 320, 392, 407, 472
Potassium–Hexamethylphosphoric triamide, **1,** 430; **2,** 208; **4,** 245; **5,** 324, 543; **6,** 274
Potassium hydride, **1,** 935; **2,** 346; **4,** 409; **5,** 557; **6,** 199, 245, 482; **7,** 301; **8,** 412; **9,**

386, 482; **10,** 327; **11,** 435; **12,** 48, 114, 407
Potassium hydride–*sec*-Butyllithium–Tetramethylethylenediamine, **8,** 7; **11,** 437
Potassium hydride–Crown ethers, **7,** 302; **8,** 129, 412; **9,** 272, 387; **11,** 436, 438; **12,** 405
Potassium hydride–Hexamethylphosphoric triamide, **7,** 302; **11,** 435; **12,** 408
Potassium hydridotetracarbonylferrate (*see* Potassium tetracarbonylhydridoferrate)
Potassium hydrogen sulfate, **1,** 909
Potassium hydrogen persulfate, **1,** 952; **10,** 328; **11,** 442; **12,** 413
Potassium hydrogen sulfide, **1,** 935; **6,** 103
Potassium hydroxide, **1,** 153, 279, 375, 828, 935; **2,** 346; **3,** 238; **4,** 246, 409; **5,** 557; **6,** 486; **7,** 303; **8,** 415; **9,** 381, 387; **11,** 439; **12,** 411
Potassium hydroxide–Acetone, **1,** 938
Potassium hydroxide–Alumina, **2,** 18; **10,** 328
Potassium hydroxide–Carbon tetrachloride, **5,** 96, 560
Potassium hydroxide–Crown ethers, **5,** 154; **7,** 303; **8,** 128
Potassium hydroxide–Dimethyl sulfoxide, **2,** 158; **4,** 410; **5,** 108, 263, 558; **6,** 486; **7,** 303; **9,** 388; **11,** 439
Potassium hydroxide–Hexamethylphosphoric triamide, **1,** 431
Potassium hydroxide–Tricaprylylmethylammonium chloride, **9,** 358
Potassium hypochlorite, **1,** 938; **6,** 486
Potassium iodide, **1,** 873, 904; **2,** 214, 384; **3,** 287; **4,** 498; **5,** 327; **6,** 279, 473, 531
Potassium iodide–Boron triiodide, **10,** 329
Potassium iodide–Dicyclohexyl-18-crown-6, **5,** 154; **6,** 174
Potassium iodide–Dimethyl sulfoxide, **7,** 134; **9,** 423; **11,** 440
Potassium iodide–Tetrabutylphosphonium iodide, **9,** 357
Potassium iodide–Tricaprylylmethylammonium chloride, **6,** 405
Potassium iodide–Zinc-copper couple, **4,** 411
Potassium iodide–Zinc–Phosphorus(V) oxide, **10,** 329
Potassium manganate, **1,** 938
Potassium methoxide–Dicyclohexyl-18-crown-6, **5,** 154
Potassium methoxide–Dimethyl sulfoxide, **1,** 298
Potassium N-methylanilide, **6,** 545
Potassium 2-methyl-2-butoxide (*see* Potassium *t*-amyloxide)
Potassium 2-methylcyclohexoxide, **5,** 560
Potassium methylsulfinylmethylide, **1,** 311; **2,** 169; **4,** 409; **7,** 339; **10,** 329
Potassium naphthalenide, **9,** 431
Potassium nitrate, **5,** 561
Potassium nitrite–Crown ether, **6,** 135
Potassium nitrite–Dimethyl sulfoxide, **10,** 329
Potassium nitrite–Tricaprylylmethylammonium chloride, **8,** 390
Potassium nitrobenzeneperoxysulfonate, **12,** 412
Potassium nitrosodisulfonate (Fremy's salt), **1,** 940; **2,** 347; **3,** 238; **4,** 411; **5,** 562; **10,** 329; **12,** 131
Potassium nonaflate, **10,** 441
Potassium osmiate, **4,** 412
Potassium perchromate, **4,** 412; **6,** 487
Potassium periodate–Potassium permanganate, **6,** 455
Potassium permanganate, **1,** 942; **2,** 348; **5,** 562, 583; **8,** 416; **9,** 388; **10,** 330; **11,** 420, 440; **12,** 413
Potassium permanganate–Acetic anhydride, **4,** 412; **5,** 563; **6,** 487
Potassium permanganate–Adogen 464, **8,** 390; **9,** 390
Potassium permanganate–Alumina, **9,** 11
Potassium permanganate–Bentonite, **11,** 441
Potassium permanganate–Benzyltriethylammonium chloride, **5,** 644
Potassium permanganate–Cetyltrimethylammonium bromide, **5,** 644; **6,** 70
Potassium permanganate–Clay, **8,** 416
Potassium permanganate–Copper sulfate, **9,** 389; **11,** 441
Potassium permanganate–Dicyclohexyl-18-crown-6, **4,** 143; **6,** 175; **9,** 389
Potassium permanganate–Molecular sieves, **8,** 416
Potassium permanganate–Polyethylene glycol, **9,** 360
Potassium permanganate–Silica, **8,** 416; **11,** 440
Potassium permanganate–Sodium periodate (*see* Sodium periodate–Potassium

permanganate)
Potassium permanganate–Tetrabutylammonium bromide, **5,** 644
Potassium permanganate–Tricaprylylmethylammonium chloride, **4,** 30; **5,** 644
Potassium peroxodisulfate (*see* Potassium persulfate)
Potassium peroxomonosulfate (Oxone®), **1,** 952; **10,** 328; **11,** 442; **12,** 413
Potassium persulfate, **1,** 952; **2,** 348; **3,** 238; **5,** 563; **7,** 274; **8,** 417; **10,** 331; **11,** 441
Potassium persulfate–Silver(I) acetate–2,2′-Dipyridine, **6,** 52
Potassium persulfate–Silver salts, **1,** 954; **2,** 348; **3,** 239
Potassium phenylacetate, **6,** 137
Potassium phenylthiolate, **8,** 420
Potassium phthalimide, **7,** 166
Potassium pyrosulfite, **5,** 563
Potassium ruthenate, **9,** 391
Potassium selenocyanate, **6,** 487; **8,** 417
Potassium selenophenylate, **9,** 373, 433
Potassium superoxide, **6,** 488; **7,** 304; **8,** 417; **9,** 391; **11,** 442; **12,** 413
Potassium superoxide–18-Crown-6, **6,** 488; **7,** 304; **8,** 417; **11,** 443; **12,** 413
Potassium superoxide–Dimethyl sulfoxide, **6,** 488; **12,** 413
Potassium superoxide–Tricaprylylmethylammonium chloride, **9,** 391
Potassium tetracarbonylferrate (*see* Dipotassium tetracarbonylferrate)
Potassium tetracarbonylhydridoferrate, **5,** 314, 358; **6,** 305, 483; **8,** 266, 419
Potassium thiocyanate, **1,** 954; **6,** 137, 487; **8,** 390
Potassium thiolacetate, **1,** 955; **10,** 325
Potassium thiophenoxide, **8,** 420
Potassium triacetylosmate, **1,** 955
Potassium tri-*sec*-butylborohydride (K-Selectride®), **5,** 557; **6,** 490; **7,** 307; **11,** 445
Potassium tri-*sec*-butylborohydride–Copper(II) iodide, **6,** 492
Potassium triethylmethoxide, **2,** 349; **5,** 557
Potassium trifluoromethanesulfinate, **5,** 564
Potassium triisopropoxyborohydride, **5,** 565; **9,** 393; **10,** 95; **12,** 389, 414
Potassium trimethylsilanolate, **12,** 414
Potassium triphenylmethide (*see* Triphenylmethylpotassium)
PPA (*see* Polyphosphoric acid)
PPE (*see* Polyphosphate ester)
PPSE (*see* Trimethylsilyl polyphosphate)
Prenyl halide (*see* 1-Bromo-3-methyl-2-butene)
Prevost reagent (*see* Silver iododibenzoate)
(S)-(−)-Proline, **1,** 973; **6,** 411, 492; **7,** 307; **8,** 421; **9,** 318, 393; **10,** 331; **11,** 446; **12,** 414
(S)-Prolinol, **10,** 332
1,3-Propanediamine, **6,** 445
1,3-Propanediol, **5,** 567, 718; **6,** 27
1,3-Propanedithiol, **1,** 956; **4,** 413; **6,** 493; **7,** 368; **9,** 394; **11,** 404; **12,** 501
1,3-Propanedithiol bis(*p*-toluenesulfonate) [*see* Trimethylene bis(thiotosylate)]
1-Propanol, **8,** 13
2-Propanol, **4,** 345; **6,** 17; **8,** 11; **12,** 505
Propargyl alcohol, **9,** 394
Propargyl aldehyde, **4,** 414
Propargyl aldehyde diethyl acetal (*see* 3,3-Diethoxy-1-propyne)
Propargyl bromide, **1,** 1285; **2,** 19, 91, 127; **3,** 28; **6,** 493; **8,** 276; **10,** 332
Propargyl chloride, **5,** 565
Propargyl trifluoromethanesulfonate, **5,** 704
Propargyltrimethylsilane, **10,** 380; **11,** 575
Propargyltriphenylphosphonium bromide, **2,** 447; **6,** 494
Propargyltriphenyltin, **12,** 415
Propene, **5,** 82
2-Propen-1-ol (*see* Allyl alcohol)
2-Propene-1-thiol, **5,** 84
2-Propenyl acetate (*see* Isopropenyl acetate)
(E)-1-Propenyllithium, **6,** 308, 495; **7,** 141
2-Propenyllithium, **2,** 456
2-Propenylmagnesium bromide, **9,** 150
2-Propenylmagnesium bromide–Copper(I) iodide, **2,** 92
2-Propenyltriphenylphosphonium bromide, **6,** 667; **10,** 225
β-Propiolactone, **1,** 957; **10,** 283
Propiolaldehyde, **4,** 414
Propiolic acid, **6,** 495; **7,** 295
Propionaldehyde *t*-butylimine, **4,** 283
Propionic acid, **12,** 415
Propionitrile, **10,** 460
3-Propionylthiazolidine-2-thione, **12,** 4, 490
2-Propylamine, **2,** 44, 414
2-Propyl-1,3-benzodithiolane, **9,** 4
2-Propyl dichloroacetate, **6,** 331
Propylene oxide, **1,** 355; **6,** 333; **12,** 69

Propyl hydroperoxide, **1,** 17
Propylidene–*t*-butylamine, **4,** 283
N-(2-Propylidene)cyclohexylamine, **6,** 332
2-Propyl isocyanate, **11,** 277
2-Propylmagnesium halides (*see* Isopropylmagnesium bromide or chloride)
Propyl nitrate, **3,** 185
Propyl nitrite–Sodium nitrite, **5,** 565
Propylphosphonic anhydride, **10,** 333
Propylthiolithium (*see* Lithium 1-propanethiolate)
Propyne, **6,** 202
2-Propyn-1-ol, **9,** 394
Propynylmagnesium bromide, **3,** 173
1-Propynylsodium, **1,** 1037
1-Propynyltrimethylsilane, **2,** 239
2-Propynyltriphenyltin, **12,** 415
Proton sponge [*see* 1,8-Bis(dimethylamino)-naphthalene]
PTAB (*see* Phenyltrimethylammonium perbromide)
PTAD (*see* N-Phenyl-1,2,4-triazoline-3,5-dione)
Py (*see* Pyridine)
Pyrazole, **1,** 958, 1188
Pyrene, **4,** 414
Pyridazine 1-oxide, **9,** 395
Pyridine (Py), **1,** 279, 958; **2,** 349; **4,** 414; **6,** 497; **11,** 448; **12,** 416
Pyridine–Aluminum chloride, **2,** 351
Pyridine borane (*see* Borane–Pyridine)
Pyridine–Butyllithium (*see* Butyllithium–Pyridine)
4-Pyridinecarboxaldehyde, **11,** 448
2-Pyridineethanol, **12,** 416
Pyridine hydrochloride (*see* Pyridinium chloride)
Pyridine N-oxide, **1,** 966; **2,** 353; **5,** 567; **9,** 396
Pyridine perbromide, **1,** 966; **5,** 568
2-Pyridineselenenyl bromide, **11,** 455
Pyridine–Sodium dithionite, **5,** 566
2-Pyridinesulfenyl bromide or chloride, **8,** 214; **9,** 397
Pyridine–Sulfur trioxide complex (*see* Sulfur trioxide–Pyridine)
2-Pyridinethiol, **5,** 321
2-Pyridinethiol 1-oxide, **12,** 417
2,4,6-Pyridinetricarboxylic acid, **10,** 352
Pyridinium bromide perbromide, **1,** 967; **2,** 94, 350; **5,** 568; **6,** 499; **9,** 399; **10,** 333
Pyridinium (*d*)-camphor-10-sulfonate, **11,** 449
Pyridinium chloride, **1,** 964; **2,** 352; **3,** 239; **4,** 415; **5,** 566; **6,** 497; **7,** 308; **8,** 424; **10,** 333; **11,** 449
Pyridinium chlorochromate (PCC), **6,** 498; **7,** 308; **8,** 425; **9,** 72, 121, 397; **10,** 40, 273, 334; **11,** 71, 139, 235, 237, 450; **12,** 417
Pyridinium chlorochromate–Alumina, **10,** 335
Pyridinium chlorochromate–3,5-Dimethylpyrazole, **12,** 419
Pyridinium chlorochromate–Hydrogen peroxide, **10,** 335
Pyridinium chlorochromate–Pyrazole, **12,** 419
Pyridinium dichromate (PDC), **3,** 239; **9,** 399; **10,** 273, 335; **11,** 216, 453; **12,** 63
Pyridinium dichromate–Acetic anhydride, **12,** 419
Pyridinium fluorochromate (PFC), **11,** 453; **12,** 419
Pyridinium hydrobromide perbromide (*see* Pyridinium bromide perbromide)
Pyridinium hydrochloride (*see* Pyridinium chloride)
Pyridinium *m*-nitrobenzenesulfonate, **12,** 419
Pyridinium poly(hydrogen fluoride) (Olah's reagent), **5,** 538; **6,** 285, 473; **7,** 294; **9,** 399; **11,** 453; **12,** 183, 419
Pyridinium poly(hydrogen fluoride)–N-Bromosuccinimide, **5,** 539; **11,** 454
Pyridinium poly(hydrogen fluoride)–N-Iodosuccinimide, **5,** 539
Pyridinium poly(hydrogen fluoride)–Nitronium tetrafluoroborate, **5,** 539
Pyridinium-1-sulfonate (*see* Sulfur trioxide–Pyridine)
Pyridinium tetrafluoroboratochromate, **9,** 397
Pyridinium *p*-toluenesulfonate, **8,** 427; **9,** 55, 400; **12,** 139, 420
Pyridinium trifluoromethanesulfonate, **10,** 196
2-Pyridone, **1,** 1188; **2,** 207; **3,** 157
α-(4-Pyridyl)benzhydrol, **7,** 139
2-Pyridylbenzoate, **10,** 33
2-(2-Pyridyl)ethyl-*p*-nitrophenyl carbonate, **12,** 420
2-Pyridylselenenyl bromide, **11,** 455
α-Pyrone, **1,** 702, 970
Pyrophosphoryl chloride, **1,** 971; **2,** 353

Pyrrolidine, **1,** 972; **2,** 354; **3,** 240; **4,** 470; **7,** 309; **8,** 428
1-Pyrrolidino-1-cyclohexene, **2,** 262; **4,** 170; **5,** 191, 302, 455; **6,** 311, 409, 628
1-Pyrrolidino-1-cyclopentene, **2,** 279
4-Pyrrolidinopyridine, **4,** 416; **9,** 400; **10,** 336
(S)-(+)-2-(1-Pyrrolidinyl)methylpyrrolidine, **12,** 421
L-Pyrrolidonecarboxylic acid, **1,** 977
Pyrrolidone-2 hydrotribromide, **3,** 240
1-Pyrroline-1-oxide, **9,** 401
3-Pyrrolinium perchlorate, **5,** 569
Pyruvaldehyde 2-phenylhydrazone, **6,** 500
Pyruvic acid, **1,** 974
Pyruvonitrile, **10,** 1
Pyruvyl chloride, **7,** 310

Quinine, **6,** 81, 501; **7,** 311; **8,** 430; **9,** 403; **10,** 338; **11,** 134, 374, 456
Quinoline, **1,** 975; **2,** 356; **9,** 403
Quinoline–Acetic acid, **6,** 501
Quinuclidine, **1,** 976, 1203; **4,** 417; **10,** 445
Quinuclidine N-oxide, **8,** 431
3-Quinuclidinol, **6,** 501; **7,** 86

RAMP [*see* (R)-1-Amino-2-methoxy-methyl-1-pyrrolidine]
Raney cobalt catalyst, **1,** 977
Raney nickel (*see also* Nickel catalysts), **1,** 440, 723; **2,** 289, 293; **4,** 267; **5,** 263, 381, 570, 736; **6,** 502; **7,** 312; **8,** 433; **9,** 405; **10,** 339; **11,** 457; **12,** 175, 218, 338, 422
Raney nickel alloy, **1,** 718; **2,** 289; **4,** 350
Ratcliff's reagent [Chromium(VI) oxide–Pyridine], **7,** 70
Red-Al® [*see* Sodium bis(2-methoxyethoxy)-aluminum hydride]
Reformatsky reagent, **1,** 499, 629, 1285; **3,** 310, 334; **5,** 34, 753; **6,** 674; **8,** 533; **9,** 47; **10,** 147, 459; **11,** 511, 598; **12,** 126, 422, 567, 570
Reinecke salt, **1,** 977
Resolving agents, **1,** 977
Rexyn ion-exchange resin, **1,** 512; **10,** 220
Rhenium catalysts, **1,** 978
Rhenium heptaselenide, **1,** 979
Rhodium(II) acetate, **5,** 571; **7,** 76, 313; **8,** 159, 434; **9,** 406, 483; **10,** 340; **11,** 458; **12,** 203, 423
Rhodium(II) butanoate, **7,** 313
Rhodium carbonyl (*see* Tetra-μ_3-carbonyl-dodecacarbonylhexarhodium)
Rhodium(II) carboxylates, **5,** 571; **7,** 76, 313; **8,** 159, 434; **9,** 406, 483; **10,** 340; **11,** 458; **12,** 203, 423
Rhodium catalysts, **1,** 979, 982, 1220; **4,** 418; **5,** 574; **6,** 503, 504, 615; **8,** 433; **10,** 321; **11,** 460; **12,** 408, 426
Rhodium(III) chloride, **2,** 357; **3,** 242; **5,** 496, 501, 572; **6,** 504; **7,** 313; **8,** 435; **9,** 257, 499; **11,** 460
Rhodium(III) chloride–Lithium chloride, **8,** 470
Rhodium(III) chloride–Silica, **6,** 504
Rhodium(III) chloride–Triphenylphosphine, **7,** 314
Rhodium dicarbonyl chloride dimer (*see* Tetracarbonyldi-μ-chlorodirhodium)
Rhodium(II) methoxyacetate, **7,** 313
Rhodium on alumina (*see also* Rhodium catalysts), **1,** 979; **4,** 418; **6,** 503; **12,** 408
Rhodium on carbon (*see also* Rhodium catalysts), **1,** 982; **4,** 418; **5,** 574; **6,** 503; **10,** 321; **11,** 460
Rhodium(III) oxide, **4,** 420; **8,** 437
Rhodium oxide–Platinum oxide (Nishimura catalyst), **9,** 408, 468
Rhodium(II) pivaloate, **7,** 313; **9,** 406
Rhodium(III) porphyrins, **11,** 461
Rhodium trichloride (*see* Rhodium(III) chloride)
Rhodium(II) trifluoroacetate, **7,** 313; **10,** 340
Rochelle salt, **1,** 983
Rongalite®, **9,** 430
Rosenmund catalyst, **1,** 975; **4,** 367
Rubidinium carbonate, **12,** 108
Rubidium fluoride, **1,** 121, 933, 983
Ruthenium(III) acetylacetonate, **12,** 88
Ruthenium(III) acetylacetonate–Alumina, **7,** 315
Ruthenium catalysts, **1,** 441, 982, 983; **5,** 574; **7,** 314; **10,** 342
Ruthenium(III) chloride, **4,** 421; **5,** 572; **7,** 244; **8,** 437; **10,** 238, 343; **11,** 252, 462; **12,** 88
Ruthenium(III) chloride–Alumina, **7,** 315
Ruthenium(III) chloride–Sodium bromate (*see also* Ruthenium tetroxide), **1,** 986
Ruthenium(III) chloride–Sodium hypochlorite (*see also* Ruthenium tetroxide), **3,** 243
Ruthenium(III) chloride–Sodium periodate

(*see also* Ruthenium tetroxide), **5,** 508; **6,** 505; **11,** 462
Ruthenium(III) chloride–Triphenylphosphine, **10,** 141
Ruthenium dioxide [*see* Ruthenium(IV) oxide]
Ruthenium hydrogenation catalyst, **1,** 983; **5,** 574
Ruthenium on alumina, **1,** 983; **7,** 314
Ruthenium on carbon, **1,** 982, 983
Ruthenium(IV) oxide, **1,** 984; **12,** 428
Ruthenium(IV) oxide–Potassium periodate (*see also* Ruthenium tetroxide), **3,** 243
Ruthenium(IV) oxide–Sodium hypochlorite (*see also* Ruthenium tetroxide), **4,** 421; **6,** 506; **11,** 462
Ruthenium(IV) oxide–Sodium periodate (*see also* Ruthenium tetroxide), **1,** 986; **2,** 358; **3,** 243; **5,** 508; **6,** 504, 545; **8,** 438; **10,** 250; **11,** 462; **12,** 428
Ruthenium–Silica, **10,** 342
Ruthenium tetroxide, **1,** 986; **2,** 357; **3,** 243; **4,** 420; **5,** 508; **6,** 504; **7,** 315; **8,** 438; **10,** 334, 343; **11,** 462; **12,** 428
Ruthenium tetroxide–Sodium periodate, **1,** 813
Ruthenium trichloride hydrate [*see* Ruthenium(III) chloride]

γ-Saccharin chloride, **1,** 990
Salcomine, **2,** 360; **3,** 245; **6,** 507; **7,** 316; **11,** 61; **12,** 429
Samarium(II) iodide, **8,** 439; **10,** 344; **11,** 464; **12,** 429
SAMP [*see* (S)-1-Amino-2-methoxymethyl-1-pyrrolidine]
Sanger's reagent (*see* 2,4-Dinitrofluorobenzene)
Sarett reagent (Chromium(VI) oxide–Pyridine), **1,** 145; **2,** 74; **3,** 55
SASP [*see* (S)-(–)-1-Amino-2-(silyloxymethyl)pyrrolidine]
Schardinger dextrins (*see* Cyclodextrins)
Schwartz's reagent [*see* Chlorobis(cyclopentadienyl)hydridozirconium(IV)]
K-Selectride® (*see* Potassium tri-*sec*-butylborohydride)
L-Selectride® (*see* Lithium tri-*sec*-butylborohydride)
LS-Selectride® (*see* Lithium trisiamylborohydride)
Selenious acid (*see* Selenium(IV) oxide)
Selenium, **1,** 423, 990; **4,** 422; **5,** 575; **6,** 507; **8,** 439; **10,** 345; **11,** 465
Selenium–Chloramine-T, **7,** 316
Selenium(I) chloride, **7,** 318
Selenium(IV) chloride–*p*-Toluenesulfonamide, **7,** 316
Selenium(IV) fluoride, **5,** 576
Selenium(IV) oxide, **1,** 532, 992; **2,** 360; **3,** 245; **4,** 422; **5,** 575; **6,** 227, 509; **7,** 319; **8,** 439; **9,** 409; **10,** 345
Selenium(IV) oxide–Hydrogen peroxide (*see* Hydrogen peroxide–Selenium(IV) oxide)
Selenium oxychloride, **1,** 1000
Semicarbazide, **1,** 437, 1000; **2,** 98; **9,** 409; **10,** 346
Sephadex, **1,** 1000
L-(–)-Serine, **12,** 430
Silica, **2,** 280, 453; **4,** 82; **6,** 510; **8,** 221, 339; **9,** 37, 157, 410; **10,** 9, 346; **11,** 22, 466; **12,** 431, 523, 548
Silicic acid (*see* Silica)
Silicon(IV) chloride (*see* Tetrachlorosilane)
Silicon(IV) ethoxide, **1,** 387
Silicon(IV) fluoride, **12,** 466
Silicon(IV) methoxide, **12,** 260
Silicon tetraisocyanate, **1,** 1002
Siloxene, **6,** 510
Silver, **4,** 164; **7,** 73
Silver acetate, **1,** 1002; **2,** 362; **4,** 431; **6,** 90, 287, 511; **8,** 45, 440
Silver antimonyl hexafluoride, **5,** 577; **7,** 320; **11,** 467, 471
Silver azide, **9,** 296
Silver benzoate, **1,** 1004
Silver carbonate, **1,** 657, 1005; **2,** 363; **4,** 425; **6,** 360, 511; **8,** 441; **12,** 432
Silver carbonate–Celite (Fetizon's reagent), **2,** 363; **3,** 247; **4,** 425; **5,** 577; **6,** 511; **7,** 319; **8,** 441, **10,** 168; **11,** 492; **12,** 38, 433
Silver chlorate, **1,** 1005; **2,** 363
Silver chloride, **5,** 580; **10,** 347
Silver chlorodifluoroacetate, **3,** 249
Silver chromate, **1,** 1005
Silver chromate–Iodine, **8,** 442
Silver crotonate, **12,** 41
Silver cyanide, **1,** 1006; **5,** 581; **7,** 320; **8,** 442
Silver dibenzyl phosphate, **1,** 1006
Silver diethyl phosphate, **4,** 428
Silver difluoride, **1,** 1006
Silver difluorochloroacetate, **3,** 249

Silver diphenylphosphate, **1,** 1006
Silver(II) dipicolinate, **10,** 348
Silver fluoride, **2,** 364; **5,** 581, 721; **6,** 514
Silver(II) fluoride, **1,** 1006
Silver fluoride–Pyridine, **10,** 348
Silver heptafluorobutanoate, **6,** 515
Silver hexafluoroantimonate, **5,** 577; **7,** 320; **11,** 467, 471
Silver imidazolate, **10,** 349; **11,** 467
Silver iododibenzoate (Prévost reagent), **1,** 1007; **3,** 251; **9,** 411
Silver(I) nitrate, **1,** 1008, 1009; **2,** 362, 366; **3,** 252; **4,** 429; **5,** 582; **6,** 171; **7,** 321; **9,** 411; **10,** 254, 350; **12,** 433
Silver nitrite, **1,** 1011; **6,** 515
Silver nitrite–Mercury(II) chloride, **11,** 467
Silver(I) oxide, **1,** 483, 657, 1011; **2,** 368; **3,** 252; **4,** 430; **5,** 25, 64, 74, 560, 583; **6,** 286, 360, 515; **7,** 321; **8,** 118, 441, 442; **10,** 350; **11,** 468
Silver(II) oxide, **2,** 369; **4,** 431; **6,** 518; **7,** 322; **9,** 99, 412; **10,** 352; **11,** 469
Silver perchlorate, **2,** 369; **3,** 121, 250; **4,** 432; **5,** 585, 698, 762; **6,** 69, 518, 519; **7,** 142, 320, 322; **9,** 413; **10,** 354; **11,** 469
Silver(II) picolinate, **3,** 16; **4,** 432; **5,** 20
Silver sulfate, **1,** 1015; **3,** 254; **4,** 435
Silver tetrafluoroborate, **1,** 1015; **2,** 365; **3,** 250; **4,** 81, 428; **5,** 587; **6,** 246, 519; **8,** 443; **9,** 295, 414, 429; **11,** 471; **12,** 129, 144, 434, 458
Silver tetrafluoroborate–Dimethyl sulfoxide, **5,** 264
Silver *p*-toluenesulfonate, **1,** 303, 1018; **2,** 370; **3,** 254; **7,** 323
Silver(I) trifluoroacetate, **1,** 1018; **4,** 327; **6,** 364; **7,** 323; **8,** 444; **10,** 355; **11,** 471
Silver(II) trifluoroacetate, **8,** 527
Silver(I) trifluoromethanesulfonate, **4,** 533; **6,** 520; **7,** 142, 324; **9,** 414; **12,** 435
Simmons–Smith reagent, **1,** 1019; **2,** 371; **3,** 255; **4,** 159, 436; **5,** 588, 758; **6,** 521; **8,** 445; **9,** 415; **11,** 472, 598; **12,** 437
SMEAH [*see* Sodium bis(2-methoxyethoxy)-aluminum hydride]
Snatzke's reagent [*see* Chromium(VI) oxide–Dimethylformamide]
Sodium, **1,** 886, 1022; **4,** 437; **5,** 589; **7,** 324
Sodium acetanilidoborohydride, **7,** 325
Sodium acetate, **1,** 1024; **5,** 591; **8,** 390
Sodium acetoxyborohydride (*see also* Sodium acyloxyborohydrides), **7,** 325
Sodium acetylides, **1,** 11, 573, 1027, 1037; **2,** 166, 398
Sodium acyloxyborohydrides (*see also* specific names), **6,** 553; **7,** 326, 342; **11,** 447; **12,** 453
Sodium–Alcohol (*see also* Sodium–Ethanol), **1,** 828; **9,** 416; **10,** 236; **11,** 472
Sodium–Alumina, **1,** 1023
Sodium aluminum chloride (*see* Sodium tetrachloroaluminate)
Sodium aluminum diethyl dihydride (*see* Sodium diethylaluminum hydride)
Sodium aluminum hydride, **1,** 1030; **3,** 259
Sodium amalgam, **1,** 21, 267, 1030, 1033; **2,** 373; **3,** 259; **4,** 181, 438; **7,** 326; **8,** 171, 296; **9,** 416, 449; **10,** 236; **11,** 473; **12,** 439
Sodium amide, **1,** 573, 1034; **2,** 337, 373; **4,** 61, 439; **5,** 400, 544, 591; **6,** 525; **8,** 446
Sodium amide–Sodium *t*-butoxide, **4,** 439; **5,** 593; **6,** 526; **7,** 327; **9,** 417
Sodium *o*-aminobenzenesulfinate, **1,** 831
Sodium–Ammonia (*see also* Birch reduction), **1,** 54, 1041; **2,** 27, 374; **3,** 259; **4,** 32, 290, 438; **5,** 31, 589; **6,** 523; **7,** 324; **9,** 26, 273, 415; **10,** 355; **11,** 472
Sodium–Ammonia–Ethanol, **1,** 1041
Sodium *t*-amyloxide, **1,** 1096; **2,** 386; **6,** 546; **9,** 379
Sodium anilide, **1,** 157
Sodium anthracenide, **6,** 415
Sodium azide, **1,** 901, 1041, 1225; **2,** 209, 376; **3,** 259; **4,** 440; **5,** 326, 593; **6,** 279, 676; **7,** 246; **8,** 258, 433; **9,** 28
Sodium azide–18-Crown-6, **8,** 287
Sodium azide–Triphenylphosphine, **9,** 418
Sodium benzeneselenoate, **5,** 272, 273, 519; **6,** 235, 548; **7,** 341; **8,** 212, 447; **9,** 430, 432; **10,** 356, 368; **11,** 475; **12,** 439, 452
Sodium benzenesulfinate, **6,** 526; **9,** 462
Sodium benzenesulfonate, **1,** 1044
Sodium benzenethiolate (*see* Sodium thiophenoxide)
Sodium benzoate, **1,** 1044; **2,** 377; **12,** 141
Sodium benzylate, **1,** 1045
Sodium bicarbonate, **1,** 299; **3,** 260; **5,** 595; **9,** 357; **11,** 476
Sodium bis(2-methoxyethoxy)aluminum hydride (SMEAH, Red-Al®), **2,** 382; **3,** 260; **4,** 293, 360, 441; **5,** 384, 389, 596; **6,** 528; **7,** 327; **8,** 448; **9,** 166, 418; **10,** 357;

11, 93, 476; **12,** 217, 287, 440, 572
Sodium bis(2-methoxyethoxy)aluminum hydride–N-Methylpiperazine, **7,** 329
Sodium bis(2-methoxyethoxy)aluminum hydride–Morpholine, **7,** 329
Sodium bismuthate, **1,** 1045
Sodium bistrimethylsilylamide (*see* Sodium hexamethyldisilazide)
Sodium bisulfite, **2,** 377; **11,** 487
Sodium borohydride, **1,** 44, 200, 1049; **2,** 377; **3,** 262; **4,** 238, 443; **5,** 597; **6,** 326, 530; **7,** 329; **8,** 309, 449; **9,** 274, 420; **10,** 357, 410; **11,** 477, 499, 518; **12,** 46, 175, 287, 441, 453
Sodium borohydride–Acetic acid, **5,** 599, 601; **6,** 531
Sodium borohydride–Alumina, **9,** 421; **10,** 358
Sodium borohydride–Aluminum chloride, **1,** 1053
Sodium borohydride–Amberlite, **10,** 358
Sodium borohydride–(–)-Benzylquininium chloride, **8,** 431
Sodium borohydride–Bis(2,4-pentanedionato)copper, **9,** 421
Sodium borohydride–Bis(triphenylphosphine)nickel dihalides, **5,** 602
Sodium borohydride–Boron trifluoride, **1,** 1053
Sodium borohydride–1-Butanethiol, **8,** 449
Sodium borohydride–Cadmium chloride–DMF, **8,** 451; **10,** 359
Sodium borohydride–Cerium(III) chloride, **8,** 451; **9,** 422; **10,** 359
Sodium borohydride–Cobalt boride, **11,** 138
Sodium borohydride–Cobalt(II) chloride, **3,** 264; **10,** 359; **11,** 479
Sodium borohydride–Cobalt(II) phthalocyanine, **11,** 138
Sodium borohydride–Copper(II) acetylacetonate, **9,** 421
Sodium borohydride–Dicyclohexyl-18-crown-6, **6,** 175
Sodium borohydride–Dihalobis(triphenylphosphine)nickel(II), **5,** 602
Sodium borohydride–1,2;5,6-Di-*O*-isopropylidene-α-D-glucofuranose, **9,** 422
Sodium borohydride–Dimethylformamide, **3,** 262; **5,** 599; **6,** 533; **7,** 329; **8,** 450; **10,** 358; **11,** 448; **12,** 565
Sodium borohydride–Dimethyl sulfoxide, **3,** 120, 262; **4,** 443; **5,** 597; **12,** 443
Sodium borohydride–(–)-N-Dodecyl-N-methylephedrinium bromide, **6,** 249; **7,** 144, 239
Sodium borohydride–Ethanedithiol, **6,** 532; **7,** 330
Sodium borohydride–Ethanethiol, **8,** 449
Sodium borohydride–L-N-Hexadecyl-N-methylephedrinium bromide, **7,** 239
Sodium borohydride–Hexamethylphosphoric triamide, **6,** 531; **10,** 198
Sodium borohydride–Lithium chloride, **11,** 216
Sodium borohydride–Methanesulfonic acid, **11,** 479
Sodium borohydride–Nickel chloride, **7,** 332
Sodium borohydride–Palladium(II) chloride, **5,** 602; **11,** 479; **12,** 399
Sodium borohydride–Palladium on charcoal, **1,** 1054
Sodium borohydride–Polyethylene glycol, **9,** 376; **12,** 444
Sodium borohydride–Praseodymium chloride, **12,** 444
Sodium borohydride–Pyridine, **3,** 263; **10,** 360
Sodium borohydride–Rhodium(III) chloride, **11,** 480
Sodium borohydride–Sulfolane, **4,** 443
Sodium borohydride, sulfurated, **3,** 264; **4,** 444; **5,** 399, 601; **6,** 534; **7,** 331
Sodium borohydride–Tetraoctylammonium bromide, **11,** 478
Sodium borohydride–Thiophenol, **6,** 532; **7,** 330
Sodium borohydride–Tin(II) chloride, **11,** 480
Sodium borohydride–Tin(IV) chloride etherate, **8,** 451; **9,** 438
Sodium borohydride–Tributylhexadecylphosphonium bromide, **11,** 478
Sodium borohydride–Trifluoroacetic acid, **8,** 451; **9,** 421; **11,** 478
Sodium borohydride–Zinc perchlorate, **12,** 572
Sodium bromate, **1,** 1055
Sodium bromide, **6,** 285; **12,** 445
Sodium bromide–Chloramine-T, **11,** 489
Sodium bromide–Dimethylformamide, **6,** 534; **12,** 445

Sodium bromide–Dimethyl sulfoxide, **6,** 534
Sodium bromite, **12,** 445
Sodium–*t*-Butyl alcohol–Tetrahydrofuran, **1,** 1056; **2,** 378; **6,** 523; **7,** 325
Sodium (*t*-butylamino)borohydride, **12,** 446
Sodium 2-butylcyclohexoxide, **1,** 1056
Sodium *t*-butyl peroxide, **5,** 602
Sodium carbonate, **7,** 332; **9,** 358
Sodium chlorate, **1,** 39, 1056
Sodium chloride, **6,** 285, 473
Sodium chloride–Dimethylformamide, **4,** 445; **6,** 534; **9,** 422
Sodium chloride–Dimethyl sulfoxide, **4,** 445; **5,** 265; **6,** 534
Sodium chlorite, **5,** 603; **9,** 423; **11,** 481
Sodium 4-chlorobenzeneselenoate, **6,** 421
Sodium chlorodifluoroacetate, **1,** 1058, 1227; **2,** 193, 281, 379; **3,** 265; **5,** 603
Sodium N-chloro-*p*-toluenesulfonamide (*see* Chloramine-T)
Sodium–Chlorotrimethylsilane (*see* Chlorotrimethylsilane–Sodium)
Sodium chromate, **1,** 1059; **4,** 96; **5,** 605
Sodium cobalt carbonylate, **1,** 1058
Sodium cryptate[2.2.2], **5,** 156
Sodium cyanate, **1,** 1059; **2,** 381
Sodium cyanide, **1,** 17, 400, 442, 516, 1088; **2,** 261; **4,** 317, 446; **5,** 606; **6,** 529, 535; **7,** 109, 333; **8,** 453
Sodium cyanide–Alumina, **9,** 423
Sodium cyanide–Benzylbutyldimethylammonium chloride, **6,** 38
Sodium cyanide–(–)-Benzylquininium chloride, **8,** 430
Sodium cyanide–Chlorotrimethylsilane–Sodium iodide, **11,** 481
Sodium cyanide–Dimethylformamide, **4,** 447; **6,** 535; **8,** 453
Sodium cyanide–Dimethyl sulfoxide, **1,** 17, 297; **2,** 381, 445; **4,** 447; **5,** 265; **8,** 453; **9,** 423
Sodium cyanide–Hexamethylphosphoric triamide, **5,** 323, 606; **6,** 536; **9,** 423
Sodium cyanide–Hexylamine, **7,** 170
Sodium cyanide–Tetrabutylammonium bromide, **5,** 644
Sodium cyanoborohydride, **4,** 448; **5,** 607; **6,** 537; **7,** 334; **8,** 454; **9,** 353, 424; **10,** 33, 237, 360; **11,** 481; **12,** 445
Sodium cyanoborohydride–Acetic acid, **10,** 361
Sodium cyanoborohydride–Hexamethylphosphoric triamide, **4,** 448; **6,** 538; **9,** 424
Sodium cyanoborohydride–Tricaprylylmethylammonium chloride, **11,** 499
Sodium cyanoborohydride–Trifluoroacetic acid, **8,** 46; **10,** 168
Sodium cyanoborohydride–Zinc chloride, **12,** 446
Sodium 9-cyano-9-hydrido-9-borabicyclo[3.3.1]nonane, **11,** 483
Sodium cyclopentadienyldicarbonylferrate [*see* Sodium dicarbonyl(cyclopentadienyl)ferrate]
Sodium decacarbonyl-μ-hydridodichromate, **7,** 299
Sodium dicarbonyl(cyclopentadienyl)ferrate, **5,** 610; **6,** 538; **8,** 41, 454; **9,** 426; **10,** 362; **11,** 483
Sodium di-μ-carbonylhexacarbonyl-μ-hydridoferrate, **7,** 340
Sodium dichlorofluoroacetate, **2,** 443; **4,** 83
Sodium dichromate (Chromic acid), **1,** 74, 142, 678, 1059; **2,** 70, 72; **3,** 54, 265; **4,** 95; **5,** 138, 538; **6,** 123, 124, 225; **7,** 68, 293; **9,** 114; **12,** 131
Sodium dichromate–Cerium(III) acetate, **1,** 1062
Sodium dichromate–Cerium(IV) ammonium nitrate, **3,** 54
Sodium dichromate–Manganese(II) nitrate, **2,** 72
Sodium dichromate–Silica, **8,** 110; **11,** 216
Sodium dicyanocuprate, **3,** 265
Sodium diethylaluminum hydride, **4,** 438
Sodium O,O-diethyl phosphoroselenolate, **8,** 455
Sodium O,O-diethyl phosphorotelluroate, **8,** 455; **10,** 362; **11,** 485
Sodium dihydrogen phosphate, **1,** 395; **4,** 451
Sodium diisopropylamide, **1,** 1064
Sodium (dimethylamino)borohydride, **12,** 446
Sodium α-(N,N-dimethylamino)-naphthalenide, **10,** 362
Sodium N,N-dimethyldithiocarbamate, **6,** 364, 539
Sodium–Dimethyl sulfoxide, **2,** 158; **9,** 431
Sodium–Diphenyl diselenide (*see* Sodium benzeneselenoate)
Sodium 4,6-diphenyl-1-oxido-2-pyridone, **8,** 456

Sodium diselenide, **9,** 430
Sodium, dispersed on charcoal or graphite, **8,** 445
Sodium disulfide, **1,** 1064
Sodium dithionite, **1,** 1081; **5,** 615; **7,** 336; **8,** 456; **9,** 426; **10,** 363; **11,** 485; **12,** 46
Sodium 4-dodecylbenzenesulfonate, **10,** 306
Sodium dodecyl sulfate, **12,** 107
Sodium ethanedithiolate, **6,** 540
Sodium ethanethiolate, **3,** 115; **4,** 465; **11,** 492; **12,** 450
Sodium–Ethanol, **1,** 828; **9,** 416; **10,** 236; **11,** 472
Sodium ethoxide, **1,** 1065; **3,** 265; **4,** 451; **6,** 540; **8,** 256
Sodium ethoxyacetylide, **5,** 107
Sodium–Ethylenediamine, **4,** 231
Sodium ethylmercaptide, **3,** 115; **4,** 465; **11,** 492; **12,** 450
Sodium fluoride, **1,** 1073; **2,** 382; **6,** 473
Sodium formate, **8,** 458
Sodium hexamethyldisilazide, **1,** 1046; **3,** 261; **4,** 442, 568; **6,** 529, 545; **7,** 329; **8,** 490; **11,** 380; **12,** 441, 446
Sodium–Hexamethylphosphoric triamide, **1,** 430; **3,** 19, 151; **4,** 246; **5,** 324, 543, 613; **9,** 416; **11,** 473
Sodium hydrazide, **1,** 1074; **5,** 610
Sodium hydride, **1,** 1075; **2,** 337, 382; **4,** 452; **5,** 610, 624; **6,** 541, 566; **7,** 335; **8,** 458; **9,** 283, 427; **10,** 242; **12,** 410, 447
Sodium hydride–*t*-Butyl hypochlorite, **4,** 455
Sodium hydride–Butyllithium, **5,** 526; **8,** 60; **9,** 429
Sodium hydride–Copper(II) acetate–Sodium *t*-amyloxide, **10,** 365
Sodium hydride–N,N-Dimethylacetamide, **4,** 456
Sodium hydride–Dimethylformamide, **1,** 1077; **2,** 383; **4,** 453; **6,** 541, 542; **7,** 335; **9,** 428; **12,** 405
Sodium hydride–Dimethyl sulfoxide (*see also* Sodium methylsulfinylmethylide), **2,** 382; **4,** 452, 456; **10,** 364
Sodium hydride–Hexamethylphosphoric triamide, **4,** 246, 455, 456; **5,** 610; **6,** 200; **12,** 448
Sodium hydride–Methyl iodide, **1,** 1079; **2,** 383; **4,** 452; **5,** 614; **6,** 541
Sodium hydride–Nickel acetate–Sodium *t*-amyloxide, **10,** 365
Sodium hydride–Palladium(II) acetate–Sodium *t*-amyloxide, **10,** 365; **12,** 370
Sodium hydride–Sodium *t*-amyloxide–Zinc chloride, **11,** 486
Sodium hydridotetracarbonylferrate (*see* Sodium tetracarbonylhydridoferrate)
Sodium hydrogen carbonate, **1,** 299; **3,** 260; **5,** 595; **9,** 357; **11,** 476
Sodium hydrogen selenide, **6,** 542; **8,** 459
Sodium hydrogen sulfide, **3,** 266; **5,** 251; **9,** 429; **10,** 417; **11,** 487; **12,** 453
Sodium hydrogen sulfite, **2,** 377; **11,** 487
Sodium hydrogen telluride, **8,** 459; **10,** 365; **12,** 449
Sodium hydrosulfide (*see* Sodium hydrogen sulfide)
Sodium hydrosulfite (*see* Sodium dithionite)
Sodium hydroxide, **1,** 375, 394, 1083; **5,** 558, 616; **7,** 336; **8,** 460; **11,** 432
Sodium hydroxide–(−)-Benzylquininium chloride, **8,** 430
Sodium hydroxide–Dimethyl sulfoxide, **3,** 120
Sodium hydroxide–Hexadecyltrimethylammonium bromide, **7,** 336
Sodium hydroxide–Hexamethyiphosphoric triamide, **4,** 247; **6,** 274
Sodium hydroxide–Potassium carbonate, **9,** 429
Sodium hydroxide–Silica, **8,** 461
Sodium hydroxide–Tetrabutylammonium iodide, **5,** 646
Sodium hydroxymethyl sulfoxylate (Rongalite®), **9,** 430
Sodium hypobromite, **1,** 488, 1083; **2,** 383; **5,** 617; **7,** 336
Sodium hypochlorite, **1,** 425, 488, 1084; **2,** 67; **3,** 45, 243; **4,** 456; **5,** 617; **6,** 543; **7,** 337; **8,** 430, 461; **9,** 430, 519; **10,** 365; **11,** 487
Sodium hypoiodite, **1,** 425, 489
Sodium hypophosphite, **1,** 1087
Sodium iodate, **1,** 1055
Sodium iodide, **1,** 1087; **2,** 384; **3,** 186, 267; **4,** 456, 529; **6,** 285, 543; **7,** 338; **10,** 365, 419; **12,** 449
Sodium iodide–Acetic acid–Propionic acid or Sodium acetate, **1,** 881, 1116; **4,** 250; **5,** 595; **6,** 544
Sodium iodide–Acetone, **1,** 1088; **3,** 267; **4,** 457; **8,** 223
Sodium iodide–Acetonitrile, **12,** 450

Sodium iodide–Chloramine-T, **11,** 488
Sodium iodide–Copper, **6,** 544
Sodium iodide–Dimethylformamide, **1,** 329, 1087; **4,** 456; **7,** 338
Sodium iodide–Hydrochloric acid, **10,** 366
Sodium iodide–Methyl iodide, **2,** 324
Sodium iodide–Pivaloyl chloride, **11,** 489
Sodium iodide–Tributylhexadecylphosphonium halides, **5,** 323; **6,** 543
Sodium iodide–Trimethylacetyl chloride, **11,** 489
Sodium iodide–Trimethylsilyl polyphosphate, **12,** 543
Sodium iodide–Zinc, **7,** 338; **12,** 142
Sodium isopropanethiolate, **11,** 473
Sodium isopropoxide, **2,** 385
Sodium isopropylmercaptide, **11,** 473
Sodium lauryl sulfate (SLS), **10,** 253
Sodium–lead alloy, **2,** 385
Sodium metabisulfite, **1,** 1090
Sodium methaneselenoate, **10,** 368; **12,** 450
Sodium methanethiolate, **5,** 626; **8,** 19; **9,** 281; **11,** 547
Sodium methoxide, **1,** 1091; **4,** 457, 511; **5,** 617, 620; **6,** 147, 545; **8,** 9, 213, 463; **9,** 430; **11,** 187, 262, 489; **12,** 282, 448
Sodium methoxide–Dimethyl sulfoxide, **1,** 298; **2,** 385; **4,** 458; **5,** 394; **6,** 545; **10,** 366
Sodium methoxide–Hexamethylphosphoric triamide, **5,** 617
Sodium methoxide–Silica, **8,** 461
Sodium 2-methoxyethoxide, **5,** 620
Sodium *p*-methoxythiophenoxide, **8,** 421
Sodium N-methylanilide, **1,** 1095; **4,** 459; **6,** 545
Sodium N-methylanilide–Hexamethylphosphoric triamide, **10,** 367
Sodium 2-methyl-2-butoxide (*see* Sodium *t*-amyloxide)
Sodium methylmercaptide, **5,** 626; **8,** 19; **9,** 281; **11,** 547
Sodium methyl phosphorazidate, **2,** 386
Sodium methylsulfinylmethylide (*see also* Sodium hydride–Dimethyl sulfoxide), **1,** 310, 1242; **2,** 166; **3,** 123; **4,** 195; **5,** 394, 621; **6,** 6, 546; **7,** 338; **9,** 431; **10,** 329, 371; **11,** 489; **12,** 451
Sodium–Molecular sieves, **5,** 465
Sodium naphthalenide, **1,** 711; **2,** 289; **4,** 349; **5,** 468; **7,** 340; **8,** 406, 464; **9,** 377, 431; **11,** 431, 490
Sodium nitrate, **7,** 393; **9,** 324; **11,** 283
Sodium nitrite, **1,** 298, 395, 1097; **2,** 386; **4,** 459; **6,** 547; **7,** 169, **9,** 324, 432; **11,** 491; **12,** 419
Sodium nitrite–Acetic acid, **6,** 548
Sodium 2-nitropropanenitronate, **1,** 1101
Sodium nitrosodisulfonate (Fremy's salt, *see also* Potassium nitrosodisulfonate), **4,** 411
Sodium octacarbonylhydridodiiron (*see* Sodium di-μ-carbonylhexacarbonyl-μ-hydridoferrate)
Sodium octanolate, **5,** 621
Sodium palladous chloride [*see* Disodium tetrachloropalladate(II)]
Sodium perborate, **1,** 1102; **2,** 387; **4,** 59; **12,** 452
Sodium periodate, **1,** 809; **2,** 311; **3,** 219; **4,** 373; **5,** 508; **6,** 459; **7,** 126; **8,** 103; **10,** 143; **11,** 492
Sodium periodate–Osmium tetroxide (Lemieux–Johnson reagent), **1,** 812; **2,** 312; **11,** 387
Sodium periodate–Potassium permanganate (Lemieux–von Rudloff reagent), **1,** 810; **2,** 312; **5,** 507; **6,** 455
Sodium periodate–Ruthenium(IV) oxide (*see* Ruthenium(IV) oxide–Sodium periodate)
Sodium periodate–Ruthenium tetroxide (*see* Ruthenium tetroxide–Sodium periodate)
Sodium permanganate monohydrate, **10,** 368
Sodium peroxide, **6,** 548; **7,** 341; **11,** 492
Sodium persulfate, **1,** 1102; **3,** 267
Sodium phenanthrenide, **4,** 375
Sodium phenoxide, **1,** 1102
Sodium phosphate, monobasic, **1,** 395; **4,** 451
Sodium pivalate, **5,** 16
Sodium polysulfide (*see* Sodium sulfide–Sulfur)
Sodium–Potassium alloy, **1,** 1102; **2,** 436; **6,** 524; **11,** 431
Sodium potassium tartrate tetrahydrate (Rochelle salt), **1,** 983
Sodium 2-propanethiolate, **11,** 473
Sodium pyridylselenate, **10,** 368
Sodium ruthenate, **5,** 622; **9,** 132
Sodium selenide, **9,** 430
Sodium selenophenolate (*see* Sodium benzeneselenoate)
Sodium sulfate, **1,** 354, 1103

Sodium sulfide, **1,** 1104; **3,** 269; **4,** 77, 460; **5,** 623; **6,** 271; **7,** 380; **10,** 369; **11,** 492; **12,** 453

Sodium sulfide–Sulfur (Sodium polysulfide), **1,** 1064; **3,** 270, 272; **9,** 434

Sodium sulfide–Thiophenol, **9,** 434

Sodium sulfite, **2,** 387

Sodium superoxide, **9,** 434

Sodium telluride, **2,** 388; **5,** 624; **6,** 550; **9,** 430

Sodium tetraborate, **4,** 461; **11,** 322

Sodium tetracarbonylcobaltate, **1,** 1058

Sodium tetracarbonylferrate(II) (*see* Disodium tetracarbonylferrate)

Sodium tetracarbonylhydridoferrate, **4,** 269; **6,** 305, 484; **8,** 419; **9,** 435

Sodium tetrachloroaluminate, **1,** 1027; **2,** 372; **4,** 438; **6,** 524; **8,** 446; **9,** 435

Sodium tetraphenylcyclopentadienide, **2,** 66

Sodium *p*-thiocresolate, **7,** 342

Sodium thiocyanate, **1,** 1105; **5,** 626

Sodium thioethoxide, **3,** 115; **4,** 465; **11,** 492; **12,** 450

Sodium thiolacetate, **8,** 312

Sodium thiomethoxide, **5,** 626; **8,** 19; **9,** 281; **11,** 547

Sodium thiophenoxide, **1,** 1106; **3,** 268; **4,** 460; **6,** 272, 404, 552; **11,** 93; **12,** 216, 440, 505

Sodium thiosulfate, **1,** 1107; **4,** 466

Sodium *p*-toluenesulfinate, **5,** 377; **11,** 512

Sodium triacetoxyborohydride (*see also* Sodium acyloxyborohydrides), **6,** 553; **12,** 453

Sodium tricarbonyl(cyclopentadienyl)-vanadate(2-), **8,** 469

Sodium trichloroacetate, **1,** 1107; **2,** 388; **4,** 385; **6,** 553; **7,** 341

Sodium triethoxyaluminum hydride, **1,** 1108

Sodium triethylborohydride–Cobalt(II) chloride, **10,** 369

Sodium triethylborohydride–Iron(II) chloride, **10,** 369

Sodium triethylborohydride–Vanadium(III) chloride, **10,** 369

Sodium trifluoroacetoxyborohydride (*see also* Sodium acyloxyborohydrides), **7,** 326, 342

Sodium triisopropoxyborohydride, **5,** 565

Sodium–Trimesitylborane, **3,** 309; **4,** 349

Sodium trimethoxyborohydride, **1,** 1108; **3,** 268; **11,** 315; **12,** 300

Sodium trimethylacetate, **5,** 16

Sodium trimethylsilanolate, **12,** 414

Sodium triphenylmethide (*see* Triphenylmethylsodium)

Sodium triphenylmethoxyacetate, **5,** 627

Sodium tris[(S)-N-benzyloxycarbonylproly-oxy]hydroborate, **11,** 447; **12,** 454

Sodium tris(3,5-di-*t*-butylphenoxy)-borohydride, **10,** 369

Sodium tris(3,5-dimethylphenoxy)-borohydride, **10,** 369

Sodium trithiocarbonate, **2,** 389; **6,** 553; **9,** 435; **10,** 369; **11,** 493

Sodium tungstate, **6,** 455

Solvents, **1,** 236, 1109; **2,** 126; **4,** 466

(–)-Sparteine, **5,** 500

Spiro{bicyclo[4.2.0]octa-1,3,5-triene-7,2′-[1,3]-dioxolan}-8-one, **9,** 40

Squaric acid, **4,** 466

Stannic, *see* Tin(IV)

Stannous, *see* Tin(II)

Stearic anhydride, **1,** 1117

trans-Styrylmagnesium bromide, **5,** 664

(+)-(R)-trans-β-Styryl *p*-tolyl sulfoxide, **4,** 466

Succindialdehyde, **1,** 6; **9,** 439

Succinic acid, **1,** 1029

Succinic anhydride, **4,** 468, 479

Succinimide–N-sulfenyl chloride, **9,** 375

Succinimidodimethylsulfonium tetrafluoroborate, **6,** 555

Succinonitrile, **4,** 241

Succinoyl peroxide, **1,** 1117

Sulfamic acid, **1,** 1117

Sulfene, **11,** 494; **12,** 454

N-Sulfinylaniline, **6,** 556

N-Sulfinylbenzenesulfonamide, **9,** 439

N-Sulfinyltoluenesulfonamide, **12,** 455

Sulfoacetic acid, **1,** 1117; **2,** 389; **10,** 374

Sulfolane (*see* Tetramethylene sulfone)

3-Sulfolene, **2,** 389; **3,** 272; **4,** 468; **12,** 455

Sulfo-mix, **2,** 391

Sulfomonoperacid (*see* Caro's acid)

o-Sulfoperbenzoic acid, **4,** 469

Sulfosalicylic acid, **1,** 1118

Sulfur, **1,** 279, 423, 1118; **3,** 273; **5,** 632; **6,** 556

Sulfur dichloride, **1,** 1121; **2,** 391; **4,** 157, 469; **8,** 408

Sulfur dioxide, **1,** 1122; **2,** 392; **4,** 469; **5,** 633; **6,** 558; **7,** 346; **8,** 464; **9,** 440; **11,** 495

Sulfur dioxide–Copper(I) chloride, **11,** 495

Sulfur–Hexamethylphosphoric triamide, **6,** 557
Sulfuric acid, **1,** 71, 678, 900; **4,** 470; **5,** 633, 762; **6,** 200, 558; **7,** 12, 295, 347; **8,** 298; **9,** 294, 441
Sulfuric acid–Trifluoroacetic anhydride, **10,** 375
Sulfur monochloride, **1,** 1122; **3,** 275; **6,** 470, 560; **9,** 442; **11,** 495
Sulfurous acid monomethyl ester N,N-diethylamide, **1,** 1123; **2,** 392
Sulfur sesquioxide, **1,** 1123
Sulfur tetrafluoride, **1,** 426, 1123; **2,** 392; **5,** 640; **6,** 560
Sulfur trioxide, **1,** 1125; **4,** 422, 473; **5,** 640
Sulfur trioxide–Dimethylformamide, **1,** 1125
Sulfur trioxide–Dioxane, **1,** 1126; **2,** 393; **5,** 641; **6,** 561
Sulfur trioxide–Pyridine, **1,** 1127; **2,** 393; **3,** 275; **4,** 473; **9,** 396
Sulfur trioxide–Triethylamine, **9,** 396
Sulfur trioxide–Trimethylamine, **1,** 1128; **8,** 466
Sulfuryl chloride, **1,** 196, 1128; **2,** 394; **3,** 276; **4,** 215, 474; **5,** 128, 641; **6,** 298, 561; **7,** 55, 349; **8,** 345; **9,** 128; **10,** 94, 375
Sulfuryl chloride–Dialkyl sulfides, **6,** 118
Sulfuryl chloride isocyanate (*see* Chlorosulfonyl isocyanate)
Sulfuryl chloride–Pyridine, **5,** 642
Sulfuryl chloride–Silica, **6,** 510; **7,** 350
Sulfuryl chloride–Tetrahydrofuran, **7,** 356
Sulfuryl chlorofluoride, **6,** 562
N,N′-Sulfuryldiimidazole, **11,** 496
Super Hydride® (*see* Lithium triethylborohydride)
Swern reagent (*see* Dimethyl sulfoxide–Oxalyl chloride, Dimethyl sulfoxide–Trifluoroacetic anhydride)

Tartaraldehyde, **1,** 740
Tartaric acid, **9,** 406
Tartrate esters, **1,** 351; **10,** 64; **11,** 92, 94, 181, 298; **12,** 90
TASF [*see* Tris(diethyl- or dimethylamino)-sulfonium difluorotrimethylsilicate]
TBAF (*see* Tetrabutylammonium fluoride)
TBHP (*see* *t*-Butyl hydroperoxide)
TCNE (*see* Tetracyanoethylene)
Tebbe reagent [*see* μ-Chlorobis(cyclopentadienyl)(dimethylaluminum)-μ-methylenetitanium]
Tellurium, **11,** 498
Tellurium(IV) chloride, **4,** 476; **5,** 643; **9,** 443; **10,** 377; **12,** 457
Tellurium(IV) oxide, **11,** 396
1,3,4,6-Tetraacetylglycouril, **6,** 563
Tetraallyltin, **1,** 19; **10,** 26; **12,** 342
1,1,3,3-Tetrabromoacetone, **5,** 223; **6,** 195; **9,** 519
2,4,4,6-Tetrabromo-2,5-cyclohexadienone, **4,** 476; **5,** 643; **6,** 563; **7,** 351; **10,** 377; **11,** 498; **12,** 457
Tetrabromodiazocyclopentadiene, **5,** 328
Tetrabromomethane (*see* Carbon tetrabromide)
Tetrabutylammonium acetate, **3,** 277; **9,** 443
Tetrabutylammonium azide, **6,** 563; **10,** 377
Tetrabutylammonium borohydride, **6,** 564; **7,** 352; **10,** 378; **11,** 499
Tetrabutylammonium bromide, **4,** 477; **5,** 460, 644; **6,** 601; **7,** 353; **8,** 387; **10,** 305; **11,** 403
Tetrabutylammonium chloride, **6,** 565; **7,** 353; **9,** 360; **10,** 302
Tetrabutylammonium chlorochromate, **12,** 458
Tetrabutylammonium chromate, **9,** 443
Tetrabutylammonium cyanoborohydride, **5,** 645; **11,** 499
Tetrabutylammonium di-*t*-butylphosphate, **9,** 443
Tetrabutylammonium fluoride (TBAF), **4,** 477; **5,** 645; **7,** 57, 151, 353; **8,** 81, 282, 467; **9,** 444; **10,** 83, 378; **11,** 386, 499; **12,** 25, 458
Tetrabutylammonium fluoride, supported, **9,** 446; **11,** 117; **12,** 407
Tetrabutylammonium formate, **4,** 478
Tetrabutylammonium halides (*see also* specific names), **4,** 477; **5,** 460, 644, 645, 646; **6,** 565, 566, 601; **7,** 57, 151, 353, 355; **8,** 81, 282, 387, 467; **9,** 357, 360, 444, 447; **10,** 83, 302, 305, 378; **11,** 117, 386, 403, 499, 501; **12,** 25, 458
Tetrabutylammonium hydrogen sulfate, **6,** 565; **7,** 354; **8,** 461; **9,** 356, 429; **10,** 41, 305; **11,** 403; **12,** 380
Tetrabutylammonium hydroxide, **5,** 645; **11,** 500
Tetrabutylammonium hypochlorite, **10,** 209
Tetrabutylammonium iodide, **5,** 646; **6,** 565,

566; **7,** 355; **9,** 357, 447; **11,** 501
Tetrabutylammonium iodotetrachloride, **1,** 1132
Tetrabutylammonium octahydrotriborate, **11,** 501
Tetrabutylammonium octamolybdate, **9,** 447
Tetrabutylammonium oxalate, **4,** 478
Tetrabutylammonium periodate, **10,** 381
Tetrabutylammonium permanganate, **8,** 468
Tetrabutylammonium thiocyanate, **9,** 293
Tetrabutylammonium *p*-toluenesulfinate, **6,** 567
Tetrabutylammonium trimethylsilylate, **12,** 461
Tetrabutyldiacetoxytin oxide dimer, **4,** 393
1,1,6,6-Tetrabutyl-1,6-distanna-2,5,7,10-tetraoxycyclodecane, **12,** 462
Tetracarbonyl(cyclopentadienyl)vanadium, **5,** 174; **8,** 468
Tetracarbonyldi-μ-chlorodirhodium, **5,** 572, 585; **6,** 108; **7,** 59, 100; **8,** 469; **10,** 382; **11,** 49; **12,** 112
Tetracarbonyldi-μ-chlorodirhodium–Lithium chloride, **8,** 470
Tetracarbonyldodecacarbonylhexarhodium, **5,** 326; **6,** 504; **10,** 382
Tetracarbonyl(ethoxyphenylmethylidene)-iron(0), **12,** 265
Tetracarbonylhydridocobalt, **1,** 154, 225; **2,** 80
N,2,4,6-Tetrachloroacetanilide, **1,** 1132
Tetrachloroauric(III) acid [*see* Hydrogen tetrachloroaurate(III)]
Tetrachloro-1,2-benzoquinone (*see* *o*-Chloranil)
Tetrachloro-1,4-benzoquinone (*see* Chloranil)
Tetrachlorocyclopentadicnone, **1,** 1132
Tetrachlorocyclopentadienone ethylene ketal, **2,** 396
sym-Tetrachlorodifluoroacetone, **1,** 254; **2,** 135
6,7,8,9-Tetrachloro-1,4-dioxaspiro[4.4]-nona-6,8-diene, **2,** 396
Tetrachloroethane–Nitrobenzene, **4,** 479
Tetrachlorofuran, **1,** 1132
Tetrachlorooxomolybdenum, **5,** 745
Tetrachlorosilane, **3,** 277; **4,** 424; **10,** 347; **11,** 532; **12,** 463
Tetrachlorosilane–Sodium iodide, **11,** 466
Tetracyanoethylene (TCNE), **1,** 1133; **2,** 397; **5,** 647; **6,** 567
7,7,8,8-Tetracyanoquinodimethane, **1,** 1136; **12,** 464
Tetracyclone (*see* Tetraphenylcyclopentadienone)
1,1,3,3-Tetraethoxypropane [*see* Malonaldehyde bis(diethyl acetal)]
Tetraethoxysilane, **1,** 387
Tetraethylammonium acetate, **1,** 1136; **2,** 397
Tetraethylammonium borohydride, **1,** 1143; **11,** 499
Tetraethylammonium bromide, **3,** 140; **6,** 568
Tetraethylammonium chloride, **1,** 1137; **6,** 480; **8,** 91
Tetraethylammonium cyanide, **6,** 569
Tetraethylammonium fluoride, **5,** 68, 376, 555; **6,** 482; **7,** 356; **8,** 470; **9,** 448
Tetraethylammonium formate, **1,** 1137
Tetraethylammonium hydroxide, **1,** 1138
Tetraethylammonium iodide, **12,** 465
Tetraethylammonium periodate, **2,** 397; **9,** 448
Tetraethyl dimethylaminomethylenediphosphonate, **3,** 277; **11,** 502
Tetraethyl orthocarbonate, **1,** 1138
Tetraethyl orthosilicate, **1,** 387
Tetraethyl phosphonosuccinate, **1,** 1138
Tetraethyl pyrophosphite, **1,** 1138; **2,** 397
Tetraethylthiuram disulfide, **6,** 569
Tetrafluoroboric acid (*see also* Boron trifluoride–Hydrogen fluoride), **1,** 44, 394, 1139; **2,** 37, 397, 402; **4,** 253; **5,** 96, 372; **8,** 139; **10,** 382; **12,** 73, 210, 465
Tetrafluoroethylidene, **5,** 515
Tetrafluorohydrazine, **1,** 1139; **5,** 649
Tetrafluoropyrrolidine, **9,** 449
Tetrafluorosilane, **12,** 466
Tetraheptylammonium chloride, **2,** 398
Tetrahexylammonium bromide, **10,** 383
Tetra-μ-hydridotetrahydroaluminum-magnesium, **4,** 316
Tetrahydrofuran (THF), **1,** 1140; **2,** 398; **3,** 278; **4,** 346; **5,** 229, 277, 649; **6,** 347, 352, 568, 570
Tetrahydrofuran–Sulfuryl chloride (*see* Sulfuryl chloride–Tetrahydrofuran)
Tetrahydro-2*H*-1,3-oxazine-2-one, **6,** 570
5,6,8,9-Tetrahydro-7-phenyldibenzo[*c,h*]-xanthylium tetrafluoroborate, **11,** 502
Tetrahydrophthalazine, **2,** 398
Tetrahydropyran, **8,** 427

1-Tetrahydropyranyloxycyclopropanecarboxaldehyde, **12,** 247
3-Tetrahydropyranyloxy-1-propyne, **6,** 604; **7,** 320; **9,** 306
3-Tetrahydropyranyloxy-1-propynyllithium, **7,** 320
3-Tetrahydropyranyloxy-1-tributylstannyl-1-propene, **6,** 602, 604
3-Tetrahydropyranyloxy-1-trimethylsilyl-1-propyne, **12,** 353
Tetrahydro-4-pyrone, **5,** 438
2-Tetrahydrothienyl diphenylacetate, **9,** 450
Tetrahydrothiophene, **4,** 190
Tetraisobutyldialuminoxane, **8,** 471
Tetraisopropylthiuram disulfide, **12,** 466
Tetrakis(acetonitrile)copper(I) perchlorate, **6,** 130; **10,** 453
Tetrakis(acetonitrile)copper(I) tetrafluoroborate, **10,** 453
Tetrakis(acetonitrile)palladium tetrafluoroborate, **12,** 467
Tetrakis(butyl)distannoxane, **4,** 393
Tetrakis(diethylamino)titanium, **12,** 354
Tetrakis(N-dihydropyridyl)aluminate (*see* Lithium aluminum hydride–Pyridine)
Tetrakis(dimethylamino)titanium, **2,** 399; **12,** 354
Tetrakis[iodo(tributylphosphine)copper(I)] (*see* Copper(I) iodide–Tributylphosphine)
Tetrakis(2-methylpropyl)-μ-oxodialuminum, **8,** 471
Tetrakis(phenylthio)methane, **6,** 650
Tetrakis(phosphorus trichloride)nickel(0), **1,** 1142
Tetrakis(pyridine)copper(I) perchlorate, **4,** 479
Tetrakis(trichlorophosphine)nickel(0), **1,** 1142
Tetrakis(triisopropyl phosphite)palladium(0), **11,** 578
1,1,2,3-Tetrakis(trimethylsilyloxy)-1,3-butadiene, **10,** 384
Tetrakis(triphenylphosphine)nickel(0), **6,** 570; **7,** 357; **8,** 85; **9,** 450; **10,** 189; **11,** 51; **12,** 175, 467
Tetrakis(triphenylphosphine)palladium(0), **5,** 501, 649; **6,** 571; **7,** 355, 357; **8,** 85, 458, 472; **9,** 174, 344, 451; **10,** 26, 32, 285, 298, 384, 411; **11,** 59, 304, 392, 393, 503, 548, 575; **12,** 46, 54, 175, 339, 344, 467, 468, 524, 525, 538, 575
Tetrakis(triphenylphosphine)palladium(0)–1,4-Bis(diphenylphosphine)butane, **12,** 473
Tetrakis(triphenylphosphine)palladium(0)–1,4-Bis(diphenylphosphine)ethane, **9,** 454; **10,** 389; **11,** 81
Tetrakis(triphenylphosphine)palladium(0)–1,4-Bis(diphenylphosphine)propane, **8,** 474
Tetrakis(triphenylphosphine)palladium(0)–Copper(I) iodide, **12,** 475
Tetrakis(triphenylphosphine)palladium(0)–Triphenylphosphine, **8,** 475
Tetrakis(triphenylphosphine)palladium(0)–Zinc, **11,** 514
Tetrakis(triphenyl phosphite)nickel(0), **11,** 503
Tetralin, **1,** 1142
Tetralin hydroperoxide, **3,** 278
Tetralin hydroperoxide–Cyclohexyl metaborate, **3,** 279; **5,** 125
(S,S)-(–)-1,2,3,4-Tetramethoxybutane, **9,** 47
Tetramethoxyethylene, **2,** 401; **5,** 649
1,1,1,3-Tetramethoxypropane, **10,** 424
Tetramethoxysilane, **12,** 260
N,N,N′,N′-Tetramethylallylphosphonodiamide, **3,** 105
N,N,N′,N′-Tetramethylaminomethane, **3,** 21; **7,** 359
Tetramethylammonium acetate, **1,** 1142
Tetramethylammonium borohydride, **1,** 1143
Tetramethylammonium bromide, **1,** 1143; **6,** 534; **9,** 2
Tetramethylammonium *t*-butyl hydrogen phosphite, **5,** 650
Tetramethylammonium chloride, **11,** 403
Tetramethylammonium dimethyl phosphate, **6,** 573
Tetramethylammonium fluoride, **7,** 358; **8,** 410
Tetramethylammonium hydroxide, **11,** 514
Tetramethylammonium iodide, **9,** 2
Tetramethylammonium tetracarbonylhydridoferrate, **8,** 476
Tetramethylammonium tribromide, **1,** 1144
1,1,4,4-Tetramethyl-1,4-bis(N,N-dimethylamino)disilethylene, **12,** 476
1,1,3,3-Tetramethylbutyl isocyanide, **4,** 480; **5,** 650; **9,** 458
N,N,N′,N′-Tetramethylchloroform-

amidinium chloride, **12,** 477
Tetramethylcyclobutadiene–Aluminum chloride complex, **6,** 574
Tetramethylcyclobutane-1,3-dione, **1,** 290
N,N,N′,N′-Tetramethyldiamidophosphorochloridate, **4,** 480; **11,** 372
N,N,N′,N′-Tetramethyldiaminomethane, **3,** 21; **7,** 359
N,N,N′,N′-Tetramethyldiaminophosphorochloridite, **6,** 575
N,N,N′,N′-Tetramethyl-1,3-dichloro-2-azatrimethinecyanine, **5,** 196
2,4,4,6-Tetramethyl-5,6-dihydro-1,3-(4*H*)-oxazine, **3,** 280; **4,** 481; **5,** 651; **10,** 391
Tetramethylenebromonium hexafluoroantimonate, **5,** 231
Tetramethylenedimagnesium bromide, **1,** 415
Tetramethylene sulfone (Sulfolane), **1,** 1144; **2,** 402; **5,** 651; **11,** 429
Tetramethylene sulfoxide, **1,** 1145
N,N,N′,N′-Tetramethylethylenediamine (TMEDA, *see also* specific organolithiums + TMEDA), **2,** 403; **3,** 284; **4,** 485; **5,** 652; **6,** 278, 576; **7,** 358; **10,** 196; **12,** 138, 477, 491
N,N,N′,N′-Tetramethylformamidinium chloride (*see* Dimethylformamide–Thionyl chloride)
N,N,N′,N′-Tetramethylglutaramide, **4,** 490
1,1,3,3-Tetramethylguanidine (*see also* Guanidines), **1,** 1145; **4,** 489; **6,** 254; **7,** 246; **12,** 477
Tetramethylguanidinium azide, **2,** 403
2,3,5,6-Tetramethylidene-7-oxanorbornane, **10,** 391; **12,** 60
Tetramethylketenimmonium tetrafluoroborate, **4,** 95; **5,** 137
N,N,N′,N′-Tetramethylmethanediamine, **3,** 21; **7,** 359
Tetramethyl orthocarbonate, **3,** 285; **11,** 73
Tetramethyl orthosilicate, **12,** 260
Tetramethylphosphorodiamidous chloride, **6,** 575
2,2,6,6-Tetramethylpiperidine (TMP), **6,** 110
2,2,6,6-Tetramethylpiperidinium chloride, **8,** 99
2,2,6,6-Tetramethylpiperidinyl-1-oxy, **6,** 110; **11,** 160; **12,** 479
1,1,3,3-Tetramethyl-4-(4-pyridyl)guanidine, **9,** 179
N,N,N′,N′-Tetramethylsuccinamide, **4,** 490
(R,R)- and (S,S)-Tetramethyltartaric acid diamide, **12,** 192, 480
Tetramethylthiourea, **1,** 1145; **7,** 359
Tetramethyltin, **9,** 41; **10,** 126
Tetramethylurea, **1,** 1146; **2,** 455; **12,** 204
2,4,5,7-Tetranitrofluorenone, **1,** 1146
(+)- and (−)-α-(2,4,5,7-Tetranitro-9-fluorenylideneaminooxy)propionic acid, **1,** 1147; **2,** 404; **6,** 577
Tetranitromethane, **1,** 1147; **2,** 404; **10,** 392
Tetra-*p*-nitrophenyl pyrophosphate, **1,** 1148
Tetraoctylammonium bromide, **9,** 359; **11,** 403
1,4,6,9-Tetraoxaspiro[4.4]nonane, **6,** 186
Tetraphenylbismuth trifluoroacetate, **10,** 393; **11,** 515
Tetraphenylcyclopentadiene, **3,** 176
Tetraphenylcyclopentadienone (Tetracyclone), **1,** 236, 953, 1149; **2,** 367; **3,** 11, 139, 170, 176; **4,** 79, 207, 354, 490; **6,** 471, 577; **7,** 305
Tetraphenylcyclopentadienylsodium, **2,** 66
Tetraphenylethylene, **2,** 404; **6,** 577
Tetraphenylphosphonium bromide, **2,** 456
Tetraphenyltin, **9,** 41, 352
Tetraphosphorus decasulfide [*see* Phosphorus(V) sulfide]
Tetra-3-pinanyldiborane (*see* Diisopinocampheylborane)
1,4,6,9-Tetrathiaspiro[4.4]nonane, **6,** 186
Tetra-*p*-tolylphosphonium bromide, **2,** 456
Tetravinyllead, **1,** 1273
Tetravinyltin, **1,** 1273
5-Tetrazolyldiazonium chloride, **5,** 653
TFA (*see* Trifluoroacetic acid)
TFAA (*see* Trifluoroacetic anhydride)
TFSA (*see* Trifluoromethanesulfonic anhydride)
Thallium, **2,** 407; **3,** 285
Thallium(I) acetate, **5,** 654
Thallium(III) acetate, **1,** 1150; **2,** 406; **3,** 286; **4,** 320, 492, 497; **5,** 370, 655; **7,** 360; **9,** 459; **10,** 393
Thallium(III) acetate–Bromine, **3,** 286; **4,** 492
Thallium(I) acetate–Iodine, **5,** 654; **7,** 359; **8,** 260; **9,** 458; **10,** 394
Thallium(I) benzoate–Iodine, **5,** 654
Thallium(I) bromide, **2,** 405; **3,** 286; **5,** 655; **7,** 361; **10,** 134
Thallium *t*-butylmercaptide, **6,** 192, 581; **8,** 444

Thallium(I) carbonate, **4,** 502; **11,** 515
Thallium chlorodifluoroacetate, **2,** 411
Thallium(I) cyanide, **8,** 476
Thallium cyclopentadienide, **4,** 502; **5,** 19
Thallium(I) ethanethiolate, **9,** 460
Thallium(I) ethoxide, **2,** 407; **4,** 497, 501; **5,** 656; **6,** 577; **7,** 362; **10,** 395
Thallium(I) hydroxide, **6,** 578
Thallium(I) 2-methyl-2-propanethiolate, **6,** 192, 581; **8,** 444
Thallium(III) nitrate (TTN), **4,** 492; **5,** 656, 657; **6,** 578, 671; **7,** 362; **8,** 211, 476; **9,** 315, 347, 460; **10,** 395; **11,** 516; **12,** 131, 481
Thallium(III) nitrate–Celite, **10,** 396
Thallium(III) nitrate–Clay, **7,** 364; **12,** 481
Thallium(III) oxide, **1,** 1151
Thallium(III) perchlorate, **5,** 657; **7,** 365; **8,** 478
Thallium 1-propanethiolate, **6,** 192
Thallium 2-propanethiolate, **6,** 192
Thallium(III) sulfate, **4,** 498; **5,** 658
Thallium triacetate (*see* Thallium(III) acetate)
Thallium(I) trichloroacetate, **2,** 411
Thallium(III) trifluoroacetate (TTFA), **3,** 286; **4,** 497, 498; **5,** 658; **6,** 579; **7,** 365; **8,** 478; **9,** 462; **10,** 397; **11,** 515; **12,** 481
Thallium(III) trifluoroacetate–Iodine, **6,** 579
Thallium(III) trifluoroacetate–Palladium(II) acetate, **11,** 516; **12,** 484
Thallous, *see* Thallium(I)
Thexylborane, **1,** 205, 276; **2,** 148; **4,** 175; **5,** 56, 232; **6,** 207; **7,** 333; **8,** 76; **10,** 397; **11,** 516; **12,** 484
Thexylborane–N,N-Diethylaniline, **9,** 464
Thexylchloroborane–Dimethyl sulfide, **12,** 485
THF (*see* Tetrahydrofuran)
1,3,4-Thiadiazoline-2,5-dione, **5,** 659
1,3-Thiazolidine-2-thione, **11,** 323, 518; **12,** 485
2-(1,3-Thiazolinyl)thiomethyllithium, **6,** 399
Thiele reagent (*see* Chromium(VI) oxide–Acetic anhydride)
Thioacetamide, **1,** 870; **4,** 502
Thioanisole, **2,** 324; **4,** 380; **5,** 527, 660; **6,** 582; **11,** 422
Thioanisole–Chlorine, **4,** 89; **5,** 130; **6,** 582; **7,** 159
Thioanisole–N-Chlorosuccinimide, **5,** 130
Thioanisole–Trifluoroacetic acid, **10,** 398
Thioanisole–Trifluoromethanesulfonic acid, **9,** 464; **10,** 398
Thiobenzoyl chloride, **6,** 582
4,4′-Thiobis(6-*t*-butyl-*o*-cresol) (Ethyl antioxidant 736), **4,** 503
4,4′-Thiobis(6-*t*-butyl-3-methylphenol), **4,** 85
N,N′-Thiocarbonyldiimidazole, **1,** 1151, 1233; **2,** 411; **3,** 316; **5,** 661; **6,** 583; **11,** 549; **12,** 556
Thiocyanogen, **1,** 1152; **2,** 105, 440; **4,** 503
Thiocyanogen chloride, **1,** 1153; **5,** 661
Thioglycolic acid, **1,** 1153; **7,** 366
Thiolacetic acid, **1,** 1154; **5,** 326; **9,** 435
2-Thiomethyl-4,4-dimethyl-2-oxazoline, **6,** 584
Thionyl bromide, **1,** 1157; **4,** 245; **5,** 663
Thionyl chloride, **1,** 394, 879, 1158; **2,** 395, 410, 412; **3,** 290; **4,** 245, 388, 503; **5,** 533, 663; **6,** 497, 585, 676; **7,** 193, 366; **8,** 481; **9,** 203, 465; **10,** 399
Thionyl chloride–Dimethylformamide (*see* Dimethylformamide–Thionyl chloride)
Thionyl chloride–Sodium iodide, **9,** 335
Thionyl chloride–Triethylamine, **7,** 367
Thionyl chloride–Zinc chloride, **5,** 667
N,N′-Thionyldiimidazole, **1,** 115, 1163
Thiophene, **1,** 872; **5,** 223
Thiophenol, **4,** 505; **6,** 279, 392, 585; **7,** 75, 100, 354, 367; **8,** 431; **9,** 312, 447, 465; **10,** 281, 399; **11,** 537; **12,** 485, 505, 545
2-Thiophenylmagnesium bromide, **1,** 423
Thiophosgene, **5,** 667; **9,** 466
Thiophosphoryl bromide, **9,** 466; **10,** 320
2*H*-Thiopyran, **8,** 242
2-Thiopyridone, **8,** 360
2-Thiopyridyl chloroformate, **9,** 466
Thiosemicarbazide, **1,** 1164
Thiourea, **1,** 1164; **2,** 157, 412; **3,** 290; **6,** 586; **11,** 519
Thiourea dioxide (*see* Formamidinesulfinic acid)
2-Thioxo-1,3-dioxol-4-ene, **5,** 662
Thorium oxide (Thoria), **1,** 1167; **5,** 669
Tigloyl chloride, **1,** 1167
Tin, **1,** 1168; **2,** 414, 459; **11,** 519
Tin–Aluminum, **12,** 486
Tin amalgam, **1,** 1169
Tin(II) bromide, **1,** 1112; **10,** 374
Tin(II) chloride, **1,** 1113, 1116; **2,** 389, 414; **3,** 271; **4,** 529; **5,** 631; **6,** 554; **10,** 374; **11,**

521; **12,** 2, 338
Tin(IV) chloride, **1,** 31, 424, 1111, 1112; **3,** 269; **5,** 373, 627; **6,** 553, 639; **7,** 342; **8,** 139; **9,** 218, 436; **10,** 50, 370; **11,** 271, 522; **12,** 389, 486, 494, 545
Tin(IV) chloride–Dicyclohexylcarbodiimide, **5,** 631
Tin(II) chloride–Silver perchlorate, **10,** 374
Tin(II) chloride–Sodium borohydride, **11,** 480
Tin(IV) chloride–Sodium borohydride, **8,** 452; **9,** 438
Tin(II) fluoride, **10,** 6, 374; **11,** 202, 524
Tin(IV) fluoride, **6,** 555
Tin(II) octoate, **7,** 346
Tin(II) oleate, **7,** 346
Tin(II) trifluoromethanesulfonate, **11,** 525; **12,** 4, 490
Titanium(0), **6,** 587; **7,** 368; **8,** 481; **9,** 466; **10,** 401; **11,** 526, 529
Titanium(III) acetate, **6,** 586
Titanium(IV) alkoxides, **6,** 11; **10,** 404; **11,** 374, 528; **12,** 19, 353, 504
Titanium(IV) butoxide, **6,** 11
Titanium(III) chloride, **2,** 415; **4,** 506; **5,** 143, 669; **6,** 587; **7,** 369, 418; **8,** 482; **9,** 467; **10,** 400, 447; **11,** 529; **12,** 492
Titanium(IV) chloride, **1,** 31, 1169; **2,** 414; **3,** 291; **4,** 507; **5,** 80, 671; **6,** 590; **7,** 370; **8,** 83, 483; **9,** 468, 521; **10,** 180, 262, 347, 380, 401; **11,** 161, 428, 529; **12,** 21, 104, 354, 488, 494, 545
Titanium(IV) chloride–Diethylaluminum chloride, **12,** 502
Titanium(III) chloride–Diisobutylaluminum hydride, **9,** 467
Titanium(III) chloride–Hydrogen peroxide, **10,** 401
Titanium(III) chloride–Lithium [*see also* Titanium(0)], **7,** 368; **9,** 466; **10,** 401; **11,** 529
Titanium(III) chloride–Lithium aluminum hydride (McMurry's reagent), **5,** 391; **6,** 588; **7,** 369; **9,** 468; **10,** 401; **11,** 527, 529; **12,** 493
Titanium(IV) chloride–Lithium aluminum hydride, **6,** 596; **7,** 372; **8,** 288, 486; **9,** 276; **10,** 401; **11,** 385, 534
Titanium(III) chloride–Magnesium (*see also* Titanium(0), Titanium(III) chloride–Tetrahydrofuran–Magnesium, **6,** 587; **11,** 529
Titanium(IV) chloride–Magnesium amalgam, **7,** 373; **11,** 534; **12,** 503
Titanium(IV) chloride–N-Methylaniline, **9,** 470
Titanium(III) chloride–Methyllithium, **2,** 415
Titanium(III) chloride–Potassium [*see also* Titanium(0)], **7,** 368; **8,** 481; **9,** 466; **10,** 401; **11,** 529
Titanium(IV) chloride–Sodium borohydride, **10,** 404
Titanium(III) chloride–Tetrahydrofuran–Magnesium (*see also* Titanium(III) chloride–Magnesium), **5,** 671; **6,** 590
Titanium(IV) chloride–2,2,6,6-Tetramethylpiperidine, **9,** 470
Titanium(IV) chloride–Titanium(IV) isopropoxide, **7,** 371; **9,** 421; **11,** 164; **12,** 375, 503
Titanium(IV) chloride–Triethylaluminum, **1,** 1170
Titanium(IV) chloride–Trifluoroperacetic acid, **4,** 508
Titanium(IV) chloride–Zinc, **8,** 487
Titanium(III) chloride–Zinc–copper couple [*see also* Titanium(0)], **8,** 483; **9,** 466; **11,** 526
Titanium(IV) ethoxide, **11,** 528
Titanium(IV) fluoride, **1,** 426; **9,** 20
Titanium(IV) isopropoxide, **6,** 11; **10,** 404; **11,** 374, 528; **12,** 19, 353, 504
Titanium(IV) methoxide, **11,** 528
Titanium(IV) propoxide, **6,** 11
Titanium tetra-, *see* Titanium(IV)
Titanium trichloride (*see* Titanium(III) chloride)
Titanocene dichloride [*see* Dichlorobis-(cyclopentadienyl)titanium]
Titanocene methylene–Zinc iodide complex, **12,** 506
Titanocene (*see* Dicyclopentadienyltitanium)
TMEDA (*see* N,N,N′,N′-Tetramethylethylenediamine)
TMP (*see* 2,2,6,6-Tetramethylpiperidine)
Tollens reagent, **1,** 1171
Toluene diisocyanate, **1,** 1171
p-Tolueneselenol, **9,** 403
p-Toluenesulfenyl chloride–Silica, **6,** 510
p-Toluenesulfinic acid, **7,** 374; **8,** 488
p-Toluenesulfinyl chloride, **6,** 585; **11,** 8; **12,** 507

p-Toluenesulfinylimidazolide, **11,** 535
p-Toluenesulfonamide, **11,** 317
p-Toluenesulfonic acid (TsOH), **1,** 1172; **4,** 508; **5,** 673, 762; **6,** 597; **7,** 172, 374; **8,** 139, 488; **9,** 471; **11,** 535; **12,** 48, 507
p-Toluenesulfonyl azide, **1,** 1178; **2,** 415; **3,** 291; **4,** 510; **5,** 244, 460, 650, 675; **6,** 597; **7,** 19; **9,** 472; **10,** 405; **11,** 535
p-Toluenesulfonyl chloride, **1,** 235, 958, 1079, 1179; **3,** 292; **4,** 390, 510; **5,** 98, 333, 676; **6,** 401, 531, 598; **7,** 225; **8,** 489; **9,** 297, 472; **11,** 536
p-Toluenesulfonyl chloride–Dimethylformamide, **1,** 1184
p-Toluenesulfonyl chloride–Lithium chloride, **3,** 152, 293; **5,** 677; **6,** 598
p-Toluenesulfonyl cyanide, **11,** 536
p-Toluenesulfonylhydrazide, **1,** 257, 1051, 1185; **2,** 377, 417; **3,** 293; **4,** 154, 443, 449, 511, 535; **5,** 197, 423, 532, 607, 676, 678; **6,** 98, 161, 234, 537, 587, 598; **7,** 31, 48, 54, 124, 282, 329, 334, 375, 416; **8,** 80, 290, 460, 489; **9,** 98, 472; **10,** 79, 400; **11,** 537; **12,** 274, 508
p-Toluenesulfonyl isocyanate, **6,** 633; **7,** 377; **10,** 217
p-Toluenesulfonyl-S-methylcarbazate, **5,** 681
p-Toluenesulfonylmethylnitrosoamide, **1,** 191, 1187; **2,** 102
p-Toluenesulfonyl perchlorate, **1,** 1187
N-*p*-Toluenesulfonyltrifluoromethylimine, **5,** 681
p-Tolyldiazomethane, **5,** 515
o-Tolyl isocyanide, **8,** 293
p-Tolyllithium, **9,** 313
p-Tolylmagnesium bromide or chloride, **2,** 405; **5,** 419, 655
p-Tolylmethyl isocyanide, **7,** 377
p-Tolylsulfinylacetic acid, **10,** 405; **11,** 538; **12,** 508
(S)-(+)-2-(*p*-Tolylsulfinyl)-2-buten-4-olide, **12,** 295
p-Tolylsulfinylcarbanion lithium salt [*see p*-(Tolylsulfinyl)methyllithium]
(S)-2-(*p*-Tolylsulfinyl)-2-cyclopentenone, **11,** 312
(R)-(+)-(α-*p*-Tolylsulfinyl)-N,N-dimethylacetamide, **12,** 298, 509
(R)-(4-Tolylsulfinyl)-ethyl-N-methoxyacetimidate, **12,** 297
(*p*-Tolylsulfinyl)methyllithium, **4,** 513; **5,** 682; **11,** 292
(R)-(+)-3-(*p*-Tolylsulfinyl)propionic acid, **12,** 509
2-*p*-Tolylsulfonylethyl chloroformate, **1,** 1185
(*p*-Tolylsulfonyl)methyl isocyanide (*see* Tosylmethyl isocyanide)
p-Tolylsulfonylmethylnitrosoamide (Diazald), **1,** 191, 1187; **2,** 102
p-Tolylthiomethylisocyanide, **4,** 515
p-Tolylthionochlorocarbonate, **4,** 342; **5,** 457; **9,** 474
(S)-(+)-*p*-Tolyl *p*-tolylthiomethyl sulfoxide, **9,** 474; **10,** 408; **12,** 510
N-*p*-Tolylvinylmethylketenimine, **5,** 683
p-Tolyl vinyl sulfoxide, **4,** 467; **11,** 538
Tosic acid (*see p*-Toluenesulfonic acid)
TosMIC (*see* Tosylmethyl isocyanide)
Tosyl, *see* also *p*-Toluenesulfonyl
Tosylmethyl isocyanide, **4,** 272, 514; **5,** 684; **6,** 566, 600; **7,** 377; **8,** 493; **10,** 409; **11,** 539; **12,** 511
Trialkoxycarbenium hexafluorophosphates, **5,** 716
Trialkylaluminums (*see also* specific names), **11,** 539; **12,** 512
Trialkynylalanes, **6,** 600; **12,** 165, 341
Triaryl phosphates, **11,** 542
1,2,4-Triazine, **10,** 409
1,3,5-Triazine, **2,** 423; **10,** 410
1,2,4-Triazole, **1,** 1188; **2,** 423
1,2,4-Triazoline-3,5-dione, **5,** 268
Tribenzylchlorosilane, **5,** 338
Tribenzylchlorotin, **1,** 52
Tribenzylcyclopentadienyltitanium, **4,** 112
Tribenzyltin chloride, **1,** 52
ω-Tribromoacetophenone, **1,** 1188
Tribromoacetyl bromide, **2,** 462
2,2,2-Tribromobenzo-1,3,2-dioxaphosphole, **2,** 63
2,2,2-Tribromoethyl chloroformate, **2,** 423
Tribromofluoromethane, **5,** 640
Tribromomethyllithium, **2,** 119; **5,** 398; **9,** 138
(Tribromomethyl)phenylmercury, **1,** 851; **2,** 276; **6,** 468
2,4,6-Tribromophenol, **2,** 424
Tributoxychlorozirconium, **11,** 379
Tributoxyethyl phosphate, **1,** 42
Tri-*t*-butylaluminum, **10,** 133
Tributylamine, **1,** 1189; **6,** 596, 601
Tributylborane, **1,** 202, 1190; **3,** 39, 235; **4,** 52, 314; **6,** 398; **10,** 244, 410; **11,** 163; **12,** 443

Tributylboroxine, **1,** 1190
Tributyl(3-butenyl)tin, **4,** 53
Tributylchlorotin (*see* Tributyltin chloride)
Tributylcinnamyltin, **11,** 543
Tributylcrotyltin, **10,** 411, 416, 451; **11,** 143, 542; **12,** 21, 28, 81, 146, 513
Tributylcyanotin, **10,** 411
Tributyl(diethylaluminum)plumbate, **12,** 340
Tributyl(diethylaluminum)tin, **12,** 339
Tributyl(ethylthio)tin, **8,** 354; **11,** 584
Tributyl(ethynyl)tin, **5,** 43
Tributylfluorophosphonium bromide, **11,** 543
Tributylfluorotin, **11,** 544; **12,** 515
Tributylhexadecylphosphonium bromide, **5,** 322; **6,** 249, 271; **7,** 166; **9,** 359
Tributyl(iodoacetoxy)tin, **12,** 258
Tributyl(iodomethyl)tin, **9,** 250, 475; **11,** 543
Tributylmanganeselithium, **3,** 112
Tributylmethallyltin, **10,** 387; **12,** 23
Tributyl(1-methoxymethoxy-2-butenyl)tin, **11,** 325
Tributylmethoxytin, **9,** 141; **12,** 174
Tributyl(methylphenylamino)phosphonium iodide, **8,** 345; **10,** 268; **11,** 342
Tributylmethylphosphonium fluoride, **9,** 223
Tributyl(methylthio)tin, **8,** 352
(E)-Tributyl(3-penten-2-yl)tin, **11,** 542
2,4,6-Tri-*t*-butylphenoxy-N,N-dimethylcarbamate, **8,** 494
2,4,6-Tri-*t*-butylphenyllithium, **11,** 495
Tributyl phosphate, **1,** 42
Tributylphosphine, **1,** 1191, 1243; **2,** 443; **5,** 685; **6,** 643; **9,** 122; **12,** 365
Tributylphosphine–Copper(I) iodide complex (*see* Copper(I) iodide–Tributylphosphine)
Tributylphosphine–Dialkyl disulfides, **6,** 602
Tributylphosphine–Diphenyl disulfide, **9,** 200; **12,** 514
(Tributylphosphine)iodocopper(I) [*see* Copper(I) iodide–Tributylphosphine]
Tributylphosphine oxide, **1,** 1192
Tributylphosphine sulfide, **5,** 732
Tri-*t*-butylsilyl perchlorate, **9,** 78
Tributylstannylacetylene, **5,** 43
Tributylstannylcopper, **11,** 544
Tributylstannyl 2-iodopropionate, **12,** 258
Tributylstannyllithium (*see* Tributyltinlithium)
Tributylstannylmagnesium bromide, **12,** 340
Tributylstannylmethyl(trimethyl)silane, **10,** 433; **11,** 581
(E)-4-Tributylstannyl-2-pentene, **11,** 542
3-Tributylstannyl-2-propen-1-ol, **12,** 524
trans-1-Tributylstannyl-1-propene-3-tetrahydropyranyl ether, **6,** 602, 604
(E)-1-Tributylstannyl-2-trimethylsilylethylene, **10,** 443
(E)-2-Tributylstannylvinyllithium, **5,** 43
Tributyltin azide, **7,** 377
Tributyltincarbinol, **7,** 378
Tributyltin chloride, **6,** 604; **7,** 378; **10,** 210, 386, 451; **11,** 520
Tributyltin chloride–Boron trifluoride, **11,** 12
Tributyltincopper, **11,** 544
Tributyltin cyanide, **10,** 411
Tributyltin fluoride, **11,** 544; **12,** 515
Tributyltin hydride, **1,** 1192; **2,** 424; **3,** 294, 299; **4,** 518; **5,** 685; **6,** 604; **7,** 224, 379; **8,** 495, 497; **9,** 476; **10,** 411; **11,** 545; **12,** 219, 417, 474, 477, 516, 525
Tributyltinlithium, **8,** 495; **10,** 411, 413; **11,** 143, 325, 551; **12,** 523
Tributyltinlithium–Diethylaluminum chloride, **12,** 523
Tributyltinmagnesium bromide, **12,** 340
Tributyltinmethanol, **7,** 378
Tributyltin methoxide, **9,** 141; **12,** 174
Tributyltin trifluoromethanesulfonate, **12,** 23, 524
Tributyl(trimethylsilylmethyl)tin, **10,** 433; **11,** 581
Tributylvinyltin, **9,** 42; **12,** 45
Tri-*t*-butylzinclithium, **8,** 516
Tricaprylylmethylammonium borohydride, **1,** 1143
Tricaprylylmethylammonium chloride (Aliquat 336), **4,** 28, 30; **5,** 460; **6,** 404; **7,** 239, 246, 380; **8,** 387; **9,** 358; **10,** 305; **11,** 126, 403; **12,** 105, 380
Tricarbonyl(1,3-cyclobutadiene)iron, **2,** 95, 140; **3,** 101; **4,** 72; **6,** 55, 149; **7,** 55
Tri-μ-carbonylhexacarbonyldiiron, **1,** 259; **2,** 139; **3,** 101; **4,** 157; **5,** 204, 221; **6,** 195, 552; **7,** 110; **8,** 390, 498; **9,** 477; **12,** 525
Tricarbonyl(*p*-methoxybenzylideneacetone)-iron, **7,** 111
Tri-μ-carbonylnonacarbonyltetrarhodium, **8,** 437
Trichloramine, **1,** 1193; **2,** 424; **3,** 295; **4,** 520
Trichloroacetaldehyde (*see* Chloral)
Trichloroacetic acid, **1,** 1194; **2,** 425; **4,** 520; **8,**

224
Trichloroacetic anhydride, **7,** 380
Trichloroacetonitrile, **1,** 1194; **5,** 686; **6,** 604; **7,** 381; **11,** 265; **12,** 379, 526
ω-Trichloroacetophenone, **3,** 45
Trichloroacetyl bromide, **1,** 1287; **3,** 87
Trichloroacetyl chloride, **4,** 134, 521; **8,** 156; **9,** 152; **11,** 168
Trichloroacetyl isocyanate, **2,** 425; **7,** 382
2,2,2-Trichloro-1,3,2-benzodioxaphosphole, **1,** 120; **2,** 63
2,4,5-Trichlorobenzoyl chloride, **6,** 288
2,4,6-Trichlorobenzoyl chloride, **9,** 478; **11,** 552
2,2,2-Trichloro-*t*-butyl chloroformate, **8,** 499; **9,** 120
2,2,2-Trichloro-*t*-butyloxycarbonyl chloride, **8,** 499; **9,** 120
Trichlorocyclopentadienyltitanium–Lithium aluminum hydride, **7,** 374
2,2,2-Trichloro-1,1-dimethylethyl chloroformate, **8,** 499; **9,** 120
2,2,2-Trichloroethanol, **3,** 295; **4,** 521; **6,** 605
2,2,2-Trichloroethyl (E)-1,3-butadiene-1-carbamate, **9,** 479
2,2,2-Trichloroethyl chloroformate, **2,** 426; **3,** 296; **4,** 522; **5,** 686; **7,** 383
2,2,2-Trichloroethyl dichlorophosphate, **8,** 59
2,2,2-Trichloroethyl dihydrogen phosphate, **6,** 605
Trichloroethylene, **4,** 400; **9,** 479; **10,** 414; **11,** 552
Tri(2-chloroethyl) orthoformate, **3,** 297
2,2,2-Trichloroethyl phosphorodichloridate, **8,** 59
Trichlorofluoromethane, **5,** 640; **6,** 91
Trichloro(iodo)silane, **11,** 466
Trichloroisocyanuric acid (Chloreal), **2,** 426; **3,** 297; **4,** 522; **5,** 687; **6,** 605; **9,** 512
Trichloromethanesulfenyl chloride, **5,** 721
Trichloromethanesulfonyl halides, **1,** 1195
(Trichloromethyl)carbonimidic dichloride, **4,** 523
Trichloromethyl chloroformate, **8,** 214
Trichloromethylenemolybdenum, **12,** 352
Trichloromethylisocyanide dichloride, **4,** 523
Trichloromethyllithium, **1,** 223, 1196; **2,** 119; **4,** 324; **5,** 199, 398
Trichloro(methyl)silane, **11,** 553
Trichloro(methyl)silane–Sodium iodide, **11,** 553; **12,** 527
Trichloromethylsodium, **1,** 1108
Trichloromethyltitanium, **10,** 270; **12,** 355
2,4,5-Trichlorophenol, **1,** 1196
2,4,6-Trichlorophenylcyanoacetate, **7,** 384
Trichloro(phenyl)tungsten–Aluminum chloride, **7,** 288
3,3,3-Trichloro-1-propene, **5,** 188
Trichlorosilane, **3,** 298; **4,** 525; **5,** 687; **6,** 606; **9,** 320
Trichlorosilane–*t*-Amines, **3,** 298; **4,** 525; **5,** 688; **8,** 499; **11,** 553
Trichlorosilane–Tributylamine, **3,** 298
Trichlorosilane–Triethylamine, **11,** 553
Trichlorosilane–Tripropylamine, **3,** 298
α,α,α-Trichlorotoluene, **1,** 1140
Trichloro-1,2,4-triazine, **9,** 480
1,1,1-Trichloro-3,3,3-trifluoroacetone, **3,** 299; **6,** 606; **8,** 499
Trichlorovinylsilane, **5,** 749
Tricyclohexylborane, **2,** 60; **3,** 39; **5,** 397
Tricyclopentylborane, **2,** 15; **6,** 25; **9,** 5
Triethanolamine, **1,** 1196
Triethanolamine borate, **8,** 500
Triethoxydiiodophosphorane, **9,** 480
Triethyoxysilane, **11,** 554
Triethylallyloxysilane, **6,** 607
Triethylaluminum, **1,** 1197; **2,** 427; **3,** 299; **4,** 526; **5,** 688; **9,** 328; **10,** 415; **11,** 539; **12,** 512
Triethylamine, **1,** 1198, 1203; **2,** 427; **4,** 527; **5,** 689; **6,** 101; **7,** 88, 385; **9,** 481
Triethylamine–Borane (*see* Borane–Triethylamine)
Triethylamine–Hydrogen fluoride, **11,** 454
Triethylaminoaluminum hydride, **6,** 607
Triethylammonium benzoate, **1,** 1201
Triethylammonium dimethylphosphate, **6,** 574
Triethylammonium fluoride, **10,** 415
Triethylammonium formate, **3,** 300
Triethylammonium formate–Palladium, **9,** 481
Triethylammonium hydrofluoride, **10,** 415
Triethylborane, **2,** 195, 343; **3,** 5, 26, 235; **4,** 51, 53, 187; **9,** 482; **10,** 416; **12,** 407, 472
Triethylenediamine (*see* 1,4-Diazabicyclo-[2.2.2]octane)
Triethylene glycol dimethyl ether, **1,** 1228
Triethylenetetramine, **1,** 1204
Tri(2-ethylhexyl) phosphate, **1,** 42
Triethyl(iodo)silane, **12,** 259

Triethylmanganeselithium, **3,** 112
Triethyl(methoxy)tin, **6,** 613
Triethyl nitrilotricarboxylate, **1,** 1218
Triethyl orthoacetate, **3,** 300; **4,** 527; **6,** 607; **10,** 417
Triethyl orthoformate, **1,** 39, 376, 422, 1204, 1294; **2,** 430; **4,** 527; **5,** 356, 690; **6,** 594, 610; **7,** 385; **10,** 275, 439; **11,** 555
Triethyloxonium hexafluorophosphate, **5,** 716
Triethyloxonium tetrafluoroborate, **1,** 1210; **2,** 430; **3,** 303; **4,** 527; **5,** 691; **6,** 611; **7,** 386; **8,** 500; **9,** 482; **10,** 417
Triethylpentylammonium bromide, **9,** 360
Triethyl(phenylthio)silane, **6,** 613
Triethyl phosphate, **1,** 1212
Triethylphosphine, **1,** 1243
Triethyl phosphite, **1,** 208, 1212, 1234; **2,** 341, 432; **3,** 304, 311; **4,** 356, 529; **5,** 198, 693; **6,** 612; **7,** 387; **8,** 501; **11,** 555, 589
Triethyl phosphite–Copper(I) iodide [*see* Copper(I) iodide–Triethyl phosphite]
Triethyl phosphoenol pyruvate, **1,** 1216
Triethyl phosphonoacetate, **1,** 250, 1217; **6,** 566, 612; **12,** 229, 527
Triethyl phosphonocrotonate, **8,** 337
Triethyl phosphonoiodoacetate, **1,** 1218
Triethyl α-phosphonopropionate, **5,** 693
Triethylsilane, **1,** 1218; **2,** 433; **3,** 187, 304; **4,** 530; **5,** 694; **7,** 8; **9,** 498; **11,** 587; **12,** 166
Triethylsilane–Boron trifluoride, **7,** 387; **8,** 501; **9,** 483; **10,** 418
Triethylsilane–Methanesulfonic acid, **6,** 321
Triethylsilane–Palladium(II) chloride, **1,** 1218; **2,** 433; **5,** 694, 739
Triethylsilane–Palladium on carbon, **3,** 304; **5,** 694
Triethylsilane–Trifluoroacetic acid (*see also* Trifluoroacetic acid–Alkylsilanes), **5,** 695; **8,** 202, 502; **9,** 309, **11,** 482
2-Triethylsilyl-1,3-butadiene, **8,** 502
3-Triethylsilyl-3-buten-2-one, **5,** 461
1-Triethylsilyloxy-2-butene, **10,** 38; **12,** 172
3-Triethylsilyloxy-1-butene, **10,** 38; **12,** 172
3-Triethylsilyloxy-1,4-pentadiene, **11,** 556
3-Triethylsilyloxypentadienyllithium, **11,** 556
3-Triethylsilyloxy-1-propene, **6,** 607
Triethylsilyl perchlorate, **12,** 527
Triethylsilyl thiophenoxide, **6,** 613
Triethyltin methoxide, **6,** 613
Triethyl 1,2,4-triazine-3,5,6-tricarboxylate, **12,** 528
Triethyl N-tricarboxylate, **1,** 1218
Triflic acid (*see* Trifluoromethanesulfonic acid)
Triflic anhydride (*see* Trifluoromethanesulfonic anhydride)
Trifluoroacetaldehyde *p*-toluenesulfonylimine, **5,** 681
Trifluoroacetic acid (TFA), **1,** 1219; **2,** 433; **3,** 305; **4,** 530; **5,** 695; **6,** 613; **7,** 93, 109, 388; **8,** 1, 51, 488, 503; **9,** 300, 483; **10,** 370, 418; **11,** 557; **12,** 424, 529
Trifluoroacetic acid–Alkylsilanes (*see also* Triethylsilane–Trifluoroacetic acid), **5,** 695; **6,** 616; **8,** 502, 309
Trifluoroacetic acid trianhydride, with iodous acid [Iodine tris(trifluoroacetate)], **4,** 263; **7,** 181
Trifluoroacetic acid–Trifluoroacetic anhydride, **5,** 700
Trifluoroacetic anhydride (TFAA), **1,** 1221; **3,** 308; **5,** 697, 701; **6,** 616; **7,** 389; **8,** 511; **9,** 484; **12,** 530
Trifluoroacetic anhydride–Pyridine, **8,** 504; **9,** 484
Trifluoroacetic anhydride–Sodium iodide, **8,** 504
ω-Trifluoroacetophenone, **1,** 1226; **3,** 308
Trifluoroacetyl chloride, **10,** 419; **11,** 560
Trifluoroacetyl hypohalites, **1,** 1227; **10,** 419
N-Trifluoroacetylimidazole, **1,** 1227
Trifluoroacetyl iodide, **8,** 504
Trifluoroacetyl nitrate, **12,** 530
Trifluoroacetyl trifluoromethanesulfonate, **9,** 485
Trifluorochloroethylene, **6,** 622
2,2,2-Trifluoroethanesulfonyl chloride, **6,** 578
2,2,2-Trifluoroethylamine, **6,** 617
N-Trifluoroethylidene-*p*-toluenesulfonamide, **5,** 681
2,2,2-Trifluoroethyl *p*-toluenesulfonate, **9,** 166
Trifluoroiodomethane, **1,** 1228; **5,** 148; **11,** 599
Trifluoromethanesulfonic acid, **4,** 532; **5,** 701, 762; **6,** 617; **8,** 504; **9,** 485; **10,** 441; **12,** 531
Trifluoromethanesulfonic acid trimethylsilyl ester (*see* Trimethylsilyl trifluoromethanesulfonate)
Trifluoromethanesulfonic–alkanesulfonic

anhydrides, **6,** 618
Trifluoromethanesulfonic anhydride (TFSA), **4,** 532, 533; **5,** 702; **6,** 618; **7,** 390; **10,** 123, 282, 419; **11,** 160, 560; **12,** 439, 532
Trifluoromethanesulfonic–carboxylic anhydrides, **4,** 533; **10,** 420
Trifluoromethanesulfonyl chloride, **9,** 485
Trifluoromethanesulfonylimidazole, **5,** 705; **6,** 620
N-*p*-Trifluoromethylbenzylcinchoninium bromide, **12,** 379
Trifluoromethylcopper, **9,** 486
Trifluoromethyl hypofluorite, **2,** 200; **3,** 146, 262; **4,** 237; **5,** 312; **6,** 263; **7,** 156; **9,** 224; **10,** 420
Trifluoromethyl iodide, **1,** 1228; **5,** 148; **11,** 599
α-Trifluoromethylmandelic acid, **4,** 383
(Trifluoromethyl)phenylmercury, **3,** 224; **4,** 384
9-Trifluoromethylsulfonyl-9-borabicyclo-[3.3.1]nonane (*see* 9-Borabicyclononyl trifluoromethanesulfonate)
1-Trifluoromethylsulfonyl-2-methyl-1-trimethylsilyl-1-propene, **8,** 410
Trifluoromethylthiocopper, **6,** 621; **7,** 391
(Trifluoromethyl)trimethyltin, **1,** 1236; **2,** 442
Trifluoroperacetic acid, **1,** 463, 821; **2,** 316; **3,** 221; **4,** 375; **5,** 658; **7,** 281; **8,** 505; **10,** 421; **11,** 562
Trifluoroperacetic acid–Boron trifluoride, **2,** 316; **4,** 253
Trifluoroperacetic acid–Sulfuric acid, **4,** 372
Trifluorophenylsulfur, **1,** 849
Trifluoro(trichloromethyl)silane, **6,** 621
Trifluorovinyllithium, **6,** 622; **8,** 505
Triglyme, **1,** 1228
Trihexylborane, **2,** 15, 253, 368; **9,** 249
Trihexylethylammonium hydroxide, **1,** 1228
Trihydridobis(triphenylphosphine)-iridium(III), **2,** 434
Triiron dodecacarbonyl (*see* Di-μ-carbonyl-decacarbonyltri-*triangulo*-iron)
Triisobutylaluminum, **1,** 260, 1188; **4,** 535; **6,** 198; **7,** 391; **9,** 486; **10,** 133; **11,** 539
Triisobutylaluminum–Bis(N-methyl-salicylaldimine)nickel, **11,** 562
Triisobutylborane, **1,** 1190; **12,** 444
Triisooctyl phosphite, **2,** 439
Triisopinocampheylborane, **1,** 1228
Triisopropanolamine, **1,** 370; **3,** 325
Triisopropoxymethyltitanium, **10,** 422; **12,** 297
2,4,6-Triisopropylbenzenesulfonyl azide, **10,** 422
2,4,6-Triisopropylbenzenesulfonyl chloride (TPS), **1,** 1228; **3,** 308; **6,** 622
2,4,6-Triisopropylbenzenesulfonyl chloride–Tetrazole, **11,** 359
2,4,6-Triisopropylbenzenesulfonylhydrazide, **4,** 535; **5,** 706; **7,** 392; **8,** 409; **9,** 486; **10,** 392, 422; **11,** 563; **12,** 533
Triisopropyl phosphite, **1,** 1229; **4,** 356
1-Triisopropylsilyl-1-propyne, **11,** 566
3-Triisopropylsilyl-2-propynyllithium, **11,** 566
Triisopropylsilyl trifluoromethanesulfonate, **11,** 63, 567
Trimesitylborane, **3,** 308
1,3,3-Trimethoxybutane, **1,** 702
2,2,2-Trimethoxy-4,5-dimethyl-1,3-dioxa-phospholene, **2,** 97; **4,** 535; **5,** 267; **7,** 392
1,1,2-Trimethoxyethylene, **12,** 206
2,2,2-Trimethoxy-5-methyl-Δ^4-oxaphospholene, **5,** 707
1,1,1-Trimethoxy-3-phenylselenopropane, **9,** 489
Trimethylacetaldehyde phenylhydrazone, **4,** 24
Trimethylacetic acid (Pivalic acid), **1,** 1229; **12,** 505
Trimethylacetic formic anhydride, **11,** 567
Trimethylacetyl chloride (Pivaloyl chloride), **1,** 1229; **8,** 404; **12,** 282
Trimethylaluminum, **5,** 707; **6,** 622; **8,** 506; **10,** 390; **11,** 7, 539; **12,** 184, 192, 343, 480, 512
Trimethylaluminum–Dichlorobis(cyclo-pentadienyl)zirconium, **8,** 506; **10,** 388, 423; **11,** 567, 575
Trimethylaluminum–Titanium(IV) chloride, **10,** 138
Trimethylaluminum–Titanocene dichloride [*see* Dichlorobis(cyclopentadienyl)-titanium–Trialkylaluminums]
Trimethylamine, **2,** 278
Trimethylamine–Borane (*see* Borane–Trimethylamine)
Trimethylamine N-oxide, **1,** 1230; **2,** 434; **3,** 309; **6,** 624; **7,** 392; **8,** 507; **9,** 31, 489; **10,** 423; **12,** 533
Trimethylamine–Sulfur dioxide, **9,** 488

Trimethylaminoacetohydrazide chloride (Girard reagent T), **1,** 410; **10,** 346
Trimethylammonium formate, **1,** 1231; **2,** 435
Trimethylanilinium hydroxide, **5,** 708
2,4,6-Trimethylbenzenesulfonyl, *see* 2-Mesitylenesulfonyl
2,4,6-Trimethylbenzoic acid, **10,** 424
Trimethyl borate, **1,** 673; **2,** 435; **3,** 310; **9,** 398; **11,** 57, 457, 573
Trimethylboroxine, **1,** 1231
Trimethylcarbenium hexafluoroantimonate, **5,** 310; **6,** 24
Trimethylchlorosilane (*see* Chlorotrimethylsilane)
Trimethylchlorosilane–Magnesium–Hexamethylphosphoric triamide, **5,** 713
Trimethylchlorosilane–Zinc (*see* Chlorotrimethylsilane–Zinc)
Trimethylcobaltlithium, **3,** 312
2,2,6-Trimethyl-1,3-dioxolenone, **1,** 266; **10,** 424
Trimethylene bis(thiotosylate), **1,** 956; **4,** 539; **5,** 71; **6,** 242, 628; **7,** 186
Trimethylene glycol (see 1,3-Propanediol)
Trimethylene oxide, **1,** 1232; **5,** 229; **7,** 80
Trimethylironlithium, **3,** 112, 312
Trimethylmanganeselithium, **3,** 112
Trimethyl β-(methoxy)orthopropionate, **10,** 424
Trimethyl(3-methyl-2-butenyl)silane, **7,** 9
Trimethyl[3-(3-methyl-1-butenyl)]silane, **3,** 446; **8,** 181
Trimethyl(3-methyl-2-butenyl)tin, **9,** 47
Trimethyl(methylthio)silane, **6,** 399; **8,** 352; **10,** 113, 426
Trimethyl orthoacetate, **5,** 709; **10,** 47
Trimethyl orthoformate, **2,** 418; **3,** 313; **4,** 540; **5,** 714; **7,** 362; **8,** 180, 507; **10,** 425, 438; **11,** 47, 73, 568
(−)-4,6,6-Trimethyl-1,3-oxathiane, **8,** 508
4,4,6-Trimethyl-1,3-oxathiane, **12,** 534
2,4,4-Trimethyl-2-oxazoline, **3,** 313; **5,** 714; **6,** 629
N,4,4-Trimethyl-2-oxazolinium iodide, **4,** 540; **6,** 630
Trimethyloxonium hexafluoroantimonate, **3,** 202
Trimethyloxonium hexafluorophosphate, **5,** 716
Trimethyloxonium tetrafluoroborate, **1,** 1232; **2,** 431, 438; **3,** 314; **4,** 541; **5,** 693, 716; **6,** 464, 630; **8,** 340; **10,** 426; **12,** 24
Trimethyloxonium 2,4,6-trinitrobenzenesulfonate, **1,** 1232; **2,** 438
Trimethyloxosulfonium hydroxide, **8,** 509
2,4,6-Trimethylphenoxymagnesium bromide, **9,** 489
Trimethylphenylammonium ethoxide, **1,** 854
Trimethylphenylammonium perbromide (*see* Phenyltrimethylammonium perbromide)
4,4,6-Trimethyl-2-phenyl-5,6-dihydro-1,3-(4*H*)-oxazine, **3,** 282
N,N,P-Trimethyl-P-phenylphosphinothioic amide, **11,** 569; **12,** 534
Trimethyl 3-(phenylseleno)orthopropionate, **9,** 489
Trimethyl(phenylseleno)silane, **9,** 373, 432, 496; **10,** 439
Trimethylphenylsilane–Iodine, **9,** 252
Trimethyl(phenylthiomethyl)silane, **7,** 402; **10,** 313; **11,** 423
Trimethyl(phenylthio)silane (*see* Phenylthiotrimethylsilane)
Trimethyl phosphate, **5,** 716; **7,** 393
Trimethyl phosphite, **1,** 131, 1233; **2,** 439; **3,** 315; **4,** 356, 541; **5,** 490, 717; **6,** 30; **7,** 393; **8,** 468; **9,** 490; **11,** 570; **12,** 536
Trimethyl phosphite–Copper(I) iodide (*see* Copper(I) iodide–Trimethyl phosphite)
Trimethyl phosphite–Methylcopper complex (*see* Methyl copper–Trimethyl phosphite)
Trimethyl phosphonoacetate, **2,** 442; **7,** 394; **10,** 384
Trimethylphosphonoglycolate, **12,** 537
4,4,6-Trimethyl-2-(2-propenyl)-5,6-dihydro-1,3-(4H)-oxazine, **3,** 283
Trimethyl(1-propenyl)tin, **12,** 470
2,4,6-Trimethylpyridine (*see* Collidine)
2,4,6-Trimethylpyrylium sulfoacetate, **11,** 571
Trimethylselenonium hydroxide, **9,** 491
Trimethylsilane, **1,** 1235; **9,** 498; **12,** 166
N-Trimethylsilylacetamide, **1,** 1232, 1235; **9,** 113
N-Trimethylsilylacetanilide, **7,** 372
Trimethylsilylacetic acid, **6,** 631
Trimethylsilylacetone, **6,** 84; **12,** 377
Trimethylsilylacetonitrile, **10,** 427; **11,** 571
Trimethylsilylacetyl chloride, **8,** 509
Trimethylsilylacetylene, **6,** 148; **7,** 276; **9,** 106, 246; **10,** 117, 127, 156, 208; **11,** 80; **12,** 162

(Trimethylsilyl)acetyltrimethylsilane, **11,** 572
Trimethylsilylallene, **10,** 428
2-Trimethylsilylallylalcohol, **9,** 497
[(Trimethylsilyl)allyl]lithium, **8,** 273; **9,** 155; **10,** 8, 430; **11,** 12, 17, 174, 531, 572
Trimethylsilyl azide (*see* Azidotrimethylsilane)
2-Trimethylsilyl-1,3-benzothiazole, **11,** 117
Trimethylsilylbenzyllithium, **6,** 637
α-(Trimethylsilyl)benzylmagnesium bromide, **11,** 19
Trimethylsilylbromoketene, **7,** 395
1-Trimethylsilyl-1,3-butadiene, **7,** 395; **9,** 491
3-Trimethylsilyl-1,2-butadiene, **10,** 429; **12,** 345
Trimethylsilylbutadiyne, **5,** 68; **6,** 143
3-Trimethylsilyl-1-butene, **9,** 446
1-Trimethylsilyl-1-buten-3-ol, **9,** 241
(E)-3-Trimethylsilyl-2-buten-1-ol, **5,** 355
4-Trimethylsilyl-2-buten-1-ol, **12,** 538
3-Trimethylsilyl-3-buten-2-one, **5,** 461; **6,** 406; **11,** 349
Trimethylsilylcarbene, **4,** 544; **8,** 307
Trimethylsilyl chloride (*see* Chlorotrimethylsilane)
Trimethylsilylcopper, **12,** 538
α-Trimethylsilylcrotyl-9-borabicyclo[3.3.1]-nonane, **10,** 430; **12,** 82
Trimethylsilyl cyanide (*see* Cyanotrimethylsilane)
3-Trimethylsilyl-1-cyclopentene, **8,** 509
Trimethylsilyldiazomethane, **4,** 543; **7,** 216; **8,** 188; **10,** 431; **11,** 573; **12,** 538
Trimethylsilyl dichloroacetate, **9,** 491
Trimethylsilyl 2,2-dichloropropionate, **9,** 492
Trimethylsilyldiethylamine (*see* Diethyltrimethylsilylamine)
2-Trimethylsilyl-1,3-dithiane, **4,** 284; **6,** 321; **10,** 380
2-(2-Trimethylsilyl-1,3-dithianyl)lithium (*see* 2-Lithio-2-trimethylsilyl-1,3-dithiane)
2-Trimethylsilylethanol, **8,** 510; **11,** 574
2-Trimethylsilylethenyllithium, **10,** 443
[1-(Trimethylsilyl)ethenyl]lithium [*see* (1-Lithiovinyl)trimethylsilane]
β-(Trimethylsilyl)ethoxymethyl chloride, **10,** 431
2-(Trimethylsilyl)ethyl azidoformate, **8,** 471
2-(Trimethylsilyl)ethyl bromide, **11,** 574
2-(Trimethylsilyl)ethyl chloroformate, **8,** 471; **10,** 431
β-Trimethylsilylethylidenetriphenylphosphorane, **9,** 492; **11,** 575
α-[(Trimethylsilyl)ethyl]lithium, **12,** 106
β-[(Trimethylsilyl)ethyl]lithium, **11,** 574
(Trimethylsilyl)ethynylcopper, **6,** 148
Trimethylsilylethynyllithium, **9,** 106, 246, 311; **10,** 208
2-Trimethylsilylfuran, **11,** 402; **12,** 378
1-Trimethylsilylhexyllithium, **12,** 540
N-(Trimethylsilyl)imidazole, **7,** 399; **11,** 575
Trimethylsilyl isocyanate, **6,** 634
Trimethylsilylketene, **6,** 635; **11,** 500
Trimethylsilyl 2-lithio-2-ethoxycarbonylacetate, **5,** 722; **9,** 493
Trimethylsilyllithium, **7,** 400; **8,** 514; **9,** 493; **11,** 575; **12,** 538
Trimethylsilylmagnesium chloride, **11,** 58
Trimethylsilylmethanethiol, **11,** 576
1-Trimethylsilyl-1-methoxyallene, **11,** 577
2-(Trimethylsilylmethyl)-3-acetoxy-1-propene, **9,** 454; **11,** 578
2-(Trimethylsilylmethyl)allyl halides, **11,** 259, 579
Trimethylsilylmethyl azide, **12,** 538
α-Trimethylsilyl-2-methylbenzothiazole, **8,** 511
2-Trimethylsilylmethyl-1,3-butadiene, **9,** 493; **10,** 432; **11,** 580; **12,** 24, 539
3-Trimethylsilylmethyl-3-butenoic acid, **10,** 107
Trimethylsilylmethylcopper, **12,** 546
Trimethylsilylmethyldimesitylborane, **12,** 12
2-Trimethylsilylmethylenecyclobutane, **9,** 494
Trimethylsilylmethylenedimethylsulfurane, **9,** 495
Trimethylsilylmethylenetrimethylphosphorane, **5,** 724
Trimethylsilylmethylenetriphenylphosphorane, **5,** 723; **10,** 434
Trimethylsilylmethyllithium, **6,** 635, 637; **9,** 495; **10,** 433; **11,** 581; **12,** 105
Trimethylsilylmethylmagnesium chloride, **5,** 724; **6,** 631, 636; **10,** 433; **11,** 574, 575; **12,** 146, 540, 546
Trimethylsilylmethylpotassium, **6,** 637; **7,** 400; **12,** 541
2-(Trimethylsilylmethyl)-2-propen-1-ol, **9,** 454; **11,** 258, 579; **12,** 252
Trimethylsilylmethylsodium, **5,** 617
Trimethylsilylmethyl trifluoromethanesulfonate, **9,** 100; **10,** 434; **11,** 582; **12,**

542
2-Trimethylsilylmethyl-3-trimethylsilyl-1-propene, **10,** 7
Trimethylsilylmethyltriphenylphosphonium trifluoromethanesulfonate, **10,** 434
Trimethylsilyl nonaflate, **9,** 496
N-Trimethylsilyl-2-oxazolidinone, **11,** 582
3-Trimethylsilyl-2-oxazolidone, **11,** 383
2-Trimethylsilyloxyallyl chloride, **10,** 286; **11,** 29
1-Trimethylsilyloxy-1,3-butadiene, **8,** 291, 512
2-Trimethylsilyloxy-1,3-butadiene, **6,** 522; **7,** 401; **12,** 136
2-Trimethylsilyloxy-1,3-cyclohexadiene, **8,** 512
1-(Trimethylsilyloxy)cyclohexene, **5,** 718
3-Trimethylsilyloxy-1,3-pentadiene, **10,** 271
3-Trimethylsilyloxy-1,4-pentadiene, **11,** 556
2-Trimethylsilyloxy-1-propene, **5,** 718; **8,** 485
4-Trimethylsilyloxyvaleronitrile, **8,** 513
1-Trimethylsilyloxy-1-vinylcyclopropane, **6,** 522
(Z)-(Trimethylsilyloxy)vinyllithium, **11,** 582
1-Trimethylsilyl(pentadienyl)lithium, **10,** 436
Trimethylsilyl perchlorate, **6,** 639; **10,** 441
Trimethylsilyl phenyl selenide (*see* Phenyl trimethylsilyl selenide)
Trimethylsilylphenylselenomethyllithium, **7,** 402
1-(Trimethylsilyl)-1-(phenylthio)ethylene, **8,** 281; **11,** 83, 583; **12,** 394
1-(Trimethylsilyl)-2-(phenylthio)ethylene, **11,** 583
Trimethylsilyl(phenylthio)methane, **7,** 402; **10,** 313; **11,** 423
Trimethylsilyl polyphosphate (PPSE), **10,** 437; **11,** 427; **12,** 543
Trimethylsilylpotassium, **7,** 402; **8,** 513
2-Trimethylsilyl-2-propen-1-ol, **9,** 497
3-Trimethylsilyl-2-propen-1-ol, **11,** 164, 342; **12,** 440
2-(3-Trimethylsilyl-1-propenyl)magnesium bromide, **11,** 81
2-(3-Trimethylsilyl-1-propenyl)magnesium bromide–Copper(I) iodide, **11,** 81
3-(Trimethylsilyl)propyl iodide, **12,** 26
1-Trimethylsilyl-1-propyne, **2,** 239
3-Trimethylsilyl-1-propyne, **10,** 380; **11,** 575
3-Trimethylsilyl-2-propyn-1-ol, **5,** 355; **12,** 169, 440
1-Trimethylsilylpropynylcopper, **6,** 638
3-Trimethylsilylpropynyllithium, **11,** 63
3-Trimethylsilyl-3-propynyllithium, **2,** 239; **3,** 173; **6,** 638; **11,** 566
N-Trimethylsilylpyrrole, **6,** 347
Trimethylsilylsodium, **8,** 514
Trimethylsilyl tribromoacetate, **4,** 549
Trimethylsilyl trichloroacetate, **9,** 492
Trimethylsilyl trifluoroacetamide, **6,** 412
Trimethylsilyl trifluoromethanesulfonate, **6,** 639; **7,** 390; **8,** 514; **9,** 210, 497; **10,** 438; **11,** 274, 530, 584; **12,** 62, 127, 218, 262, 466, 532, 543
Trimethylsilyl trimethylsilylamidosulfonate, **7,** 28
3-Trimethylsilyl-3-trimethylsilyloxy-1-propene, **8,** 485
1-Trimethylsilyl-2-trimethylstannylethylene, **12,** 469
Trimethylsilylvinylketene, **10,** 441
α-Trimethylsilylvinyllithium, **5,** 374; **7,** 193, 279; **8,** 57; **9,** 498; **10,** 442; **11,** 286
(E)-β-(Trimethylsilyl)vinyllithium, **10,** 443
α-Trimethylsilylvinylmagnesium bromide, **8,** 515
β-Trimethylsilylvinylmagnesium bromide, **11,** 83
α-Trimethylsilylvinylmagnesium bromide–Copper(I) iodide, **8,** 515
Trimethylstannylcarbene, **8,** 307
Trimethylstannylcopper–Dimethyl sulfide, **12,** 113
Trimethylstannyldimethylamine, **2,** 147
2-Trimethylstannylethylidenetriphenylphosphorane, **6,** 640
Trimethylstannyllithium, **8,** 412; **9,** 330, 499; **12,** 547
2-Trimethylstannylmethyl-1,3-butadiene, **10,** 432; **12,** 574
Trimethylstannylmethyllithium, **12,** 547
Trimethylstannylsodium, **9,** 500
4,4,6-Trimethyl-2-styryl-5,6-dihydro-1,3-(4*H*)-oxazine, **3,** 284
Trimethylsulfonium bromide, **1,** 314
Trimethylsulfonium hexafluoroantimonate, **3,** 202
Trimethylsulfonium hydroxide, **9,** 491
Trimethylsulfonium iodide, **1,** 314, 1236; **5,** 646; **6,** 219
Trimethylsulfonium perchlorate, **1,** 314
Trimethylsulfoxonium chloride, **6,** 41

Trimethylsulfoxonium hydroxide, **8,** 509
Trimethylsulfoxonium iodide, **1,** 315, 1236; **5,** 646; **6,** 41
Trimethyltin chloride, **10,** 392, 451
Trimethyltin hydride, **7,** 380
Trimethyltinlithium, **8,** 412; **9,** 330, 499; **12,** 547
Trimethyltinsodium, **9,** 500
Trimethyl(trifluoromethyl)tin, **1,** 1236; **2,** 442
Trimethyl(triphenylmethoxy)silane, **10,** 444; **11,** 586
1,3,5-Trimethyl-2,4,6-tris[3,5-di-*t*-butyl-4-hydroxybenzyl]benzene (Ethyl antioxidant 330), **4,** 545
Trimethyltrityloxysilane, **10,** 444; **11,** 586
Trimethylvinylsilane (*see* Vinyltrimethylsilane)
Trimethylzinclithium, **8,** 515
Trimethyvinyltin, **12,** 45
Trimsyl chloride (see 2-Mesitylenesulfonyl chloride)
Trineopentylneopentylideneniobium or -tantalum, **7,** 403
1,3,5-Trinitrobenzene, **1,** 1237; **6,** 158
2,4,7-Trinitrofluorenone, **1,** 1237
(2,4,7-Trinitro-9-fluorenylidene)malononitrile, **1,** 231; **5,** 206
Tri-*p*-nitrophenyl phosphorotrithioite, **1,** 1237
Tri-2-norbornylborane, **2,** 15
Trioctylphosphine–Carbon tetrachloride, **2,** 445
Trioctylpropylammonium chloride, **6,** 640
s-Trioxane, **8,** 331
Trioxo(*t*-butylimido)osmium(VIII), **6,** 641; **10,** 445
1,2,3-Trioxo-2,3-dihydrophenalene, **4,** 546
Triphenylacetonitrile oxide, **7,** 146
Triphenylaluminum, **1,** 1260
Triphenylantimonyl dibromide, **9,** 94
Triphenylarsine oxide, **1,** 424; **4,** 547
Triphenylarsonium cinnamylide, **12,** 240
Triphenylarsonium ethylide, **10,** 445
Triphenylbismuth, **11,** 587
Triphenylbismuth carbonate, **9,** 501; **10,** 446; **11,** 587
Triphenylbismuth diacetate, **12,** 548
Triphenylbis(2,2,2-trifluoroethoxy)-phosphorane, **10,** 43
Triphenylborane, **1,** 1259; **8,** 524
Triphenylcarbenium hexachloroantimonate, **1,** 42; **3,** 330; **4,** 567
Triphenylcarbenium hexafluoroantimonate, **4,** 548, 565
Triphenylcarbenium hexafluoroarsenate, **4,** 548
Triphenylcarbenium hexafluorophosphate, **4,** 548, 565
Triphenylcarbenium perchlorate, **1,** 1256; **5,** 506; **12,** 549
Triphenylcarbenium tetrafluoroborate, **1,** 1256; **2,** 454; **4,** 565; **6,** 657; **7,** 414; **8,** 524; **9,** 426, 512; **10,** 455; **11,** 217; **12,** 72, 550
Triphenylcarbenium trifluoroacetate, **6,** 657
Triphenyldibromophosphorane (*see* Triphenylphosphine dibromide)
Triphenylmethane, **1,** 1256; **2,** 454; **5,** 156; **7,** 302
Triphenylmethanesulfenyl chloride, **1,** 1261
Triphenylmethyl bromide, **1,** 1254
Triphenylmethyl chloride, **1,** 1254; **2,** 453; **4,** 565; **5,** 468, 741; **6,** 656; **9,** 179
Triphenylmethyl isocyanide, **5,** 650; **6,** 642
Triphenylmethyllithium, **1,** 600, 1256; **2,** 343, 454; **4,** 308, 567; **12,** 117
Triphenylmethyllithium–Tetramethylethylenediamine, **4,** 127
2-[*p*-(Triphenylmethyl)phenyl]sulfonylethanol, **7,** 414
2-(*p*-Triphenylmethylphenyl)thioethanol, **7,** 414
Triphenylmethylphosphonium permanganate, **10,** 446
Triphenylmethylpotassium, **1,** 1103, 1258; **7,** 302; **8,** 524; **9,** 502; **12,** 550
Triphenylmethylpyridinium tetrafluoroborate, **5,** 741
Triphenylmethylsodium, **1,** 1030, 1259; **3,** 40; **5,** 156
Triphenyl orthothioformate, **6,** 650; **11,** 305, 591; **12,** 208
2,2,2-Triphenyl-1,2-oxaphosph(V)olane, **2,** 442
Triphenyl(phenylthiomethyl)tin, **11,** 424
Triphenyl(1-phenylthiovinyl)phosphonium iodide, **9,** 502; **11,** 425
Triphenyl phosphate, **7,** 57
Triphenylphosphine, **1,** 1238; **2,** 443; **3,** 317; **4,** 548; **5,** 38, 408, 685, 725; **6,** 643; **7,** 403; **11,** 588; **12,** 365, 550
Triphenylphosphine bis(trifluoromethanesulfonate), **6,** 648

Triphenylphosphine–Bromotrichloromethane, **9,** 503
Triphenylphosphine–Carbon tetrabromide, **2,** 445; **4,** 550; **6,** 644; **12,** 186
Triphenylphosphine–Carbon tetrabromide–Lithium azide, **6,** 644
Triphenylphosphine–Carbon tetrachloride, **1,** 1247; **2,** 445; **3,** 320; **4,** 551; **5,** 727; **6,** 644; **7,** 404; **8,** 516; **9,** 503; **10,** 447; **11,** 588; **12,** 551, 554
Triphenylphosphine dibromide, **1,** 1247; **2,** 446; **3,** 320; **4,** 555; **5,** 729; **6,** 645, 646; **7,** 109, 407; **9,** 112; **10,** 60; **12,** 554
Triphenylphosphine dibromide–Dimethylformamide, **5,** 732
Triphenylphosphine–Dibromodifluoromethane, **10,** 448
Triphenylphosphine dichloride, **1,** 1247; **5,** 730; **6,** 646; **7,** 407; **12,** 554
Triphenylphosphine–2,2′-Di(3-cyano-4,6-dimethylpyridyl) disulfide, **9,** 167
Triphenylphosphine–Diethyl azodicarboxylate (Mitsunobu reagent), **4,** 553; **5,** 727; **6,** 645; **7,** 404; **8,** 517; **9,** 423, 504, 506; **10,** 447, 448; **11,** 541, 589; **12,** 552
Triphenylphosphine–Diethyl azodicarboxylate–Methyl iodide (*see also* Triphenylphosphine–Diethyl azodicarboxylate), **9,** 506
Triphenylphosphine–Diethyl oxomalonate, **10,** 449
Triphenylphosphine difluoride, **2,** 135
Triphenylphosphine dihalides (*see also* specific names), **6,** 647; **7,** 407; **12,** 554
Triphenylphosphine diiodide (*see* Triphenylphosphine–Iodine)
Triphenylphosphine–2,2′-Dipyridyl disulfide, **3,** 322; **5,** 285; **6,** 246, 519; **7,** 141; **8,** 215; **10,** 449
Triphenylphosphine ditriflate, **6,** 648
Triphenylphosphine fluoromethylene, **3,** 146; **6,** 262
Triphenylphosphine–N-Halosuccinimides, **4,** 555; **5,** 728
Triphenylphosphine–Hexachloroacetone, **8,** 516; **10,** 449
Triphenylphosphine hydroiodide–Triphenylphosphine diiodide, **10,** 450
Triphenylphosphine–Iodine, **6,** 647; **7,** 407; **8,** 518; **9,** 236; **10,** 220, 450; **12,** 554
Triphenylphosphine–Iodine–Imidazole, **9,** 508
Triphenylphosphine–Iodoform-Imidazole, **10,** 450
Triphenylphosphine–Potassium *t*-butoxide, **5,** 729
Triphenylphosphine selenide, **4,** 556
Triphenylphosphine–Selenocyanogen, **9,** 506
Triphenylphosphine sulfide, **5,** 732
Triphenylphosphine–Thiocyanogen, **8,** 518; **9,** 507
Triphenylphosphine–Triiodoimidazole, **9,** 507
Triphenylphosphirene oxide, **4,** 116
Triphenyl phosphite, **1,** 1249; **2,** 446; **3,** 322; **4,** 556
Triphenyl phosphite dibromide, **1,** 1249; **2,** 446
Triphenyl phosphite methiodide (*see* Methyltriphenoxyphosphonium iodide)
Triphenyl phosphite ozonide, **3,** 323; **4,** 559; **7,** 408; **8,** 370, 519
Triphenyl phosphite–Trifluoromethanesulfonic anhydride, **10,** 449
Triphenylphosphonium iodide–Triphenylphosphine diiodide, **10,** 450
(Triphenylphosphoranylidene)acetaldehyde (*see* Formylmethylenetriphenylphosphorane)
Triphenylphosphoranylideneoxalacetonitrile, **4,** 547
(Triphenylphosphorylidene)ketene, **2,** 446; **10,** 450; **11,** 281; **12,** 268
Triphenylpropargylphosphonium bromide, **2,** 447; **6,** 494
Triphenyl(propargyl)tin, **12,** 415
Triphenyl(2-propenyl)phosphonium bromide, **6,** 667; **10,** 225
Triphenyl(2-propynyl)phosphonium bromide, **2,** 447; **6,** 494
2,4,6-Triphenylpyrylium fluoride, **9,** 508
2,4,6-Triphenylpyrylium iodide, **8,** 519
2,4,6-Triphenylpyrylium perchlorate, **7,** 408; **10,** 25
2,4,6-Triphenylpyrylium tetrafluoroborate, **8,** 520
2,4,6-Triphenylpyrylium thiocyanate, **8,** 521
Triphenylsilane, **12,** 209
[1-(Triphenylsilyl)ethenyl]lithium, **5,** 68, 375
Triphenylsilyl hydroperoxide, **9,** 509
α-Triphenylsilylvinyllithium, **5,** 68, 375
Triphenylstannane (*see* Triphenyltin hydride)

Triphenylstannylmethyllithium, **9,** 509
3-Triphenylstannyl-1-propyne, **12,** 415
Triphenylstibine oxide, **4,** 547
Triphenylsulfinamide, **5,** 732
Triphenyl thioborate, **7,** 409; **9,** 15
Triphenyltin chloride, **1,** 1250; **10,** 451
Triphenyltin hydride, **1,** 1250; **2,** 448; **3,** 324; **4,** 559; **5,** 734; **6,** 649; **8,** 521; **9,** 511; **10,** 451; **12,** 41, 555
Triphenyl(trimethylsilylmethyl)phosphonium trifluoromethanesulfonate, **10,** 434
Triphenylvinylphosphonium bromide (*see* Vinyltriphenylphosphonium bromide)
Triphenylvinylsilane, **6,** 637
Tripotassium hexachloromolybdate(III), **8,** 522
Tripotassium tri-μ-chlorohexachloroditungstate, **8,** 522
Tripropylaluminum, **5,** 734; **11,** 539
Tripropylamine, **9,** 360
Tris(acetonitrile)tricarbonyltungsten, **12,** 556
Tris(acetylacetonate)iron(III) [*see* Iron(III) acetylacetonate]
Tris(acetylacetonate)iron(III)–Hydrogen peroxide (*see* Hydrogen peroxide–Iron(III) acetylacetonate)
Tris(aquo)hexa-μ-acetato-μ_3-oxotriruthenium(III,III,III) acetate, **6,** 425, 650
Tris(*p*-bromophenyl)aminium hexachloroantimonate, **5,** 735; **9,** 510; **10,** 452; **12,** 532
Tris(dibenzoylmethide)iron(III), **6,** 259; **7,** 410; **12,** 429
Tris(dibenzylideneacetone)dipalladium-(chloroform), **11,** 508; **12,** 16
4,4′,4″-Tris(4,5-dichlorophthalamido)trityl bromide, **12,** 557
Tris(diethylamino)methyltitanium, **11,** 375
Tris(diethylamino)phosphine (*see* Hexaethylphosphorus triamide)
Tris(diethylamino)sulfonium difluorotrimethylsilicate (TASF), **10,** 452
Tris(dimethoxyboryl)methane, **6,** 329
Tris(dimethylamino)arsine, **2,** 399
Tris(dimethylamino)borane, **6,** 554; **11,** 590
Tris(dimethylamino)mesityloxyphosphonium azide, **9,** 296
Tris(dimethylamino)methane, **5,** 653; **7,** 411; **9,** 341
Tris(dimethylamino)phosphine (*see* Hexamethylphosphorous triamide)
Tris(dimethylamino)sulfonium difluorotrimethylsilicate (TASF), **11,** 590; **12,** 558
Tris(dimethylphenylphosphine)(norbornadiene)rhodium(I) hexafluorophosphate, **7,** 411
Tris(2,2-dimethylpropyl)(2,2-dimethylpropylidene)niobium or -tantalum, **7,** 403
Tris(1,3,2-dioxaborolan-2-yl)methane, **6,** 328; **10,** 239
Tris(1,3-diphenyl-1,3-propanedionato-O,O′)-iron [*see* Tris(dibenzoylmethide)iron(III)]
Tris(ethylenedioxyboryl)methane, **6,** 328; **10,** 239
Tris(ethylthio)borane, **8,** 522
Tris(formamino)methane, **1,** 1251
Tris(6,6,7,7,8,8,8-heptafluoro-2,2-dimethyl-3,5-octanedionato)europium [$Eu(fod)_3$], **4,** 211; **12,** 196, 312, 559
Tris(6,6,7,7,8,8,8-heptafluoro-2,2-dimethyl-3,5-octanedionato)ytterbium [$Yb(fod)_3$], **12,** 561
Tris[3-(heptafluoropropylhydroxymethylene)-*d*-camphorato]europium(III) [$Eu(hfc)_3$], **12,** 196, 561
Tris(1-hexynyl)aluminum, **12,** 165
Tris(2-hydroxypropyl)amine, **1,** 370; **3,** 325
Tris(methylseleno)borane, **10,** 454
Tris(methylthio)methane, **7,** 412; **10,** 453
Tris(methylthio)methyllithium, **7,** 280, 412; **10,** 453
Tris(2,4-pentanedionato)iron(III) (*see* Iron(III) acetylacetonate)
Tris(2,4-pentanedionato)manganese(III) (*see* Manganese(III) acetylacetonate)
Tris(2,4-pentanedionato)ruthenium(*see* Ruthenium(III) acetylacetonate)
Tris(phenylethynyl)aluminum, **6,** 601
Tris(phenylseleno)borane, **8,** 522; **9,** 511; **10,** 454
Tris(phenylthio)borane, **7,** 409; **9,** 15
Tris(phenylthio)methane, **6,** 650; **11,** 305, 591; **12,** 208
Tris(phenylthio)methyllithium, **6,** 650; **9,** 511; **11,** 305, 591
Tris(phenylthio)methylsodium, **6,** 651
Tris(2-picoline)copper(I) perchlorate, **4,** 480
Tris(tetrabutylammonium)hexacyanoferrate(III), **6,** 656
Tris(tetrabutylammonium)hydrogen pyrophosphate, **10,** 455

N,N′,N″-Tris(tetramethylene)-
phosphoroamide, **6,** 651
Tris(tribenzylideneacetylacetone)-
tripalladium(chloroform), **12,** 561
Tris(trimethylsilyl)aluminum, **12,** 343
Tris(trimethylsilylethynyl)aluminum, **12,** 341
Tris(trimethylsilyl)hydrazidocopper, **6,** 651
Tris(trimethylsilyl)hydrazidolithium, **6,** 651
Tris(trimethylsilyl)ketenimine, **11,** 591
Tris(trimethylsilyl)methane, **5,** 618
Tris(trimethylsilyl)methyllithium, **5,** 374, 618
Tris(trimethylsilyloxy)ethylene, **8,** 523; **9,** 512
Tris(trimethysilyl) phosphite, **7,** 412
Tris(triphenylphosphine)chlororhodium [*see* Chlorotris(triphenylphosphine)-rhodium(I)]
Tris(triphenylphosphine)nickel(0), **6,** 653; **8,** 523
Tris(triphenylphosphine)ruthenium dichloride [*see* Dichlorotris(triphenyl-phosphine)ruthenium(II)]
Tris(triphenylsilyl)vanadate, **6,** 655; **7,** 413
s-Trithiane, **2,** 103; **3,** 329; **4,** 564
2-(1,3,5-Trithianyl)lithium, **3,** 329
Triton B (*see* Benzyltrimethylammonium hydroxide)
Trityl, *see* Triphenylmethyl or Triphenylcarbenium
Tritylone alcohol (*see* 9-Hydroxy-9-phenylanthrone)
Tropylidene, **1,** 177
Tropylium bromide, **2,** 66
Tropylium perchlorate, **1,** 216; **5,** 742
Tropylium salts, **1,** 177, 216, 395, 869, 1257, 1261; **2,** 46; **4,** 568; **5,** 615, 742; **7,** 369
Tropylium tetrafluoroborate, **1,** 177, 395, 869, 1257, 1261; **7,** 369
TsOH (*see p*-Toluenesulfonic acid)
TTFA [*see* Thallium(III) trifluoroacetate]
TTN [*see* Thallium(III) nitrate]
Tungsten carbonyl, **6,** 658; **8,** 525
Tungsten carbonyl–Aluminum chloride, **8,** 525
Tungsten(VI) chloride, **4,** 569; **5,** 742; **6,** 658
Tungsten(VI) chloride–Butyllithium, **4,** 569; **6,** 658
Tungsten(VI) chloride–Ethylaluminum dichloride, **4,** 570; **5,** 742
Tungsten(VI) chloride–Lithium aluminum hydride, **4,** 570; **8,** 525
Tungsten(VI) chloride–Lithium iodide, **4,** 569
Tungsten(VI) chloride–Tetramethyltin, **7,** 415; **10,** 455
Tungsten(VI) fluoride, **7,** 416
Tungsten hexa-, *see also* Tungsten(VI)
Tungsten hexacarbonyl (*see* Tungsten carbonyl)
Tungsten(VI) oxide–Chlorosulfonic acid, **11,** 591
L-Tyrosine hydrazide, **2,** 454; **3,** 330

Uranium, **9,** 122
Uranium(IV) chloride–Sodium–Potassium alloy, **9,** 122
Uranium(VI) fluoride, **7,** 417
Urea, **1,** 1262; **2,** 455; **7,** 295
Urethane, **1,** 1270
Urushibara catalysts, **4,** 571; **5,** 743; **6,** 659; **7,** 417

(S)-Valine *t*-butyl ester, **12,** 563
L-Valinol, **12,** 563
Vanadium(II) chloride, **7,** 369, 418; **10,** 457; **11,** 593
Vanadium(III) chloride, **7,** 248
Vanadium(IV) chloride, **3,** 332
Vanadium(III) chloride–Lithium aluminum hydride, **10,** 401
Vanadium(IV) chloride–N-Methylaniline, **9,** 470
Vanadium(II) chloride-tetrapyridine complex, **5,** 744
Vanadium(V) oxide, **1,** 733, 1057
Vanadium oxyacetylacetonate (*see* Vanadyl acetylacetonate)
Vanadium oxytrihalides (*see* Vanadyl trichloride or trifluoride)
Vanadium oxytrifluoride (*see* Vanadyl trifluoride)
Vanadium(II) perchlorate, **9,** 513
Vanadium tetrachloride (*see* Vanadium(IV) chloride)
Vanadocene (*see* Dicyclopentadienyl-vanadium)
Vanadyl acetylacetonate, **2,** 456; **3,** 331; **11,** 47
Vanadyl acetylacetonate–Azobisiso-butyronitrile, **10,** 457
Vanadyl trichloride, **3,** 331; **5,** 744; **8,** 527; **9,** 463; **11,** 593
Vanadyl trifluoride, **5,** 745; **6,** 660; **7,** 418; **8,** 480, 528; **9,** 385, 513

Vilsmeier reagent (*see also* specific names), **1,** 284, 286, 681, 878, 1271; **2,** 154; **3,** 16, 116, 128, 228, 321; **4,** 186; **5,** 129, 249, 251, 479; **6,** 220, 279; **7,** 292, 422; **8,** 186, 529; **9,** 335, 465, 514; **10,** 457; **12,** 204, 564
Vilsmeier reagent–Lithium tri-*t*-butoxyaluminum hydride, **12,** 201
Vinyl acetate, **1,** 649, 1271; **10,** 175; **12,** 565
Vinyl bromide, **4,** 236; **12,** 469
β-Vinylbutenolide, **6,** 661; **9,** 515; **11,** 594
Vinyl β-carbomethoxyethyl ketone, **1,** 1271
Vinyl chloride, **1,** 1274
Vinyl chloroformate, **7,** 384; **8,** 530
Vinylcopper, **4,** 220; **5,** 747
Vinylcopper reagents, **4,** 517; **5,** 93; **6,** 266, 342, 576, 662; **7,** 82; **8,** 117, 235, 237; **9,** 248; **10,** 423; **11,** 367; **12,** 113, 349
2-Vinyl-1-cyclopropyllithium, **7,** 192, 212; **9,** 273
Vinyldiazomethane, **4,** 469; **6,** 664
Vinyldiphenylphosphine oxide, **11,** 595
1,2-Vinylenebis(triphenylphosphonium) dibromide, **6,** 665
Vinylene carbonate, **12,** 206
Vinylidene chloride, **1,** 220; **2,** 118
Vinylketene, **11,** 596
Vinyllithium, **1,** 1250, 1273; **2,** 456; **5,** 748; **6,** 325, 665; **7,** 47; **8,** 275; **10,** 190; **12,** 351, 410
Vinylmagnesium bromide, **1,** 213, 415; **8,** 201; **9,** 83, 366; **10,** 190; **12,** 410
Vinylmagnesium bromide–Copper(I) iodide, **12,** 142
Vinylmagnesium bromide–Methylcopper, **8,** 237
Vinylmagnesium chloride, **1,** 415; **4,** 572
Vinylmagnesium chloride–Copper(I) chloride, **2,** 92; **4,** 572
Vinyl methoxycarbonylmethyl ketone, **1,** 1272; **5,** 746; **6,** 662
Vinylsulfonyl chloride–Trimethylamine, **9,** 211
1-Vinylthioallyllithium, **5,** 749
Vinyltributyltin, **9,** 42; **12,** 45
Vinyltrichlorosilane, **5,** 749
Vinyltrimethylsilane, **6,** 100; **9,** 498; **10,** 444; **11,** 41, 42; **12,** 26, 540, 566
Vinyltrimethyltin, **12,** 45
Vinyltriphenylphosphonium bromide, **1,** 1274; **2,** 456; **3,** 333; **4,** 572; **5,** 750; **6,** 666; **7,** 235; **10,** 316; **11,** 596; **12,** 11
Vinyltriphenylsilane, **6,** 637
Vinylzinc chloride, **11,** 505
Vitamin B_{12}, **10,** 103; **11,** 139

Wilkinson's catalyst [*see* Chlorotris(triphenylphosphine)rhodium(I)]
Wittig reaction (*see also* WITTIG REAGENTS in "TYPE OF COMPOUND INDEX"), **4,** 573; **5,** 752; **6,** 668
Woodward's Reagent K (*see* N-Ethyl-5-phenylisoxazolium-3′-sulfonate)

Xenic acid, **2,** 458
Xenon(II) fluoride, **6,** 669; **7,** 425; **10,** 85
Xenon(VI) fluoride, **7,** 425
(S)-2-(2,6-Xylidinomethyl)pyrrolidine, **10,** 458

$Yb(fod)_3$, **12,** 561
Ytterbium, **9,** 517; **11,** 249
Ytterbium–Ammonia, **9,** 517
Ytterbium(II) chloride, **9,** 211
Ytterbium(II) iodide, **8,** 439; **12,** 429
Ytterbium(III) nitrate, **6,** 671

Zeo–Karb® ion-exchange resin, **1,** 513
Zeolite, **10,** 347
Zinc, **1,** 44, 373, 865, 1276, 1285, 1286; **2,** 459, 462; **3,** 334; **4,** 315, 368, 574; **5,** 753; **6,** 672; **7,** 52, 73, 369, 426, 428; **8,** 464, 532; **10,** 332, 459; **11,** 598; **12,** 567
Zinc–Acetic acid, **1,** 1276; **2,** 459; **4,** 574; **5,** 380, 754, 757; **8,** 282; **12,** 568
Zinc–Acetic anhydride–Tetramethylammonium bromide, **1,** 1143
Zinc–Acid anhydride–Catalyst, **1,** 1143; **4,** 577
Zinc amalgam, **1,** 652, 1287; **2,** 462; **4,** 574; **7,** 426; **8,** 534; **10,** 147
Zinc–Ammonium chloride, **5,** 758
Zinc borohydride, **3,** 337; **5,** 761; **10,** 460; **11,** 599; **12,** 287, 360, 440, 572
Zinc borohydride–Dimethylformamide, **12,** 574
Zinc bromide, **2,** 463; **4,** 250; **5,** 756, 762; **8,** 535; **9,** 520; **10,** 285, 461; **11,** 314, 600; **12,** 63
Zinc carbonate, **4,** 579; **8,** 535
Zinc chloride, **1,** 1289, 1294; **2,** 463, 464; **3,**

338; **5,** 763; **6,** 676; **7,** 430; **8,** 120, 536; **9,** 272, 522; **10,** 461; **11,** 602; **12,** 525, 552, 574
Zinc chloride–Benzylchlorobis(triphenylphosphine)palladium(II), **10,** 26
Zinc chloride–Phosphoryl chloride, **1,** 1292
Zinc chloride–Sodium naphthalenide, **8,** 464
Zinc–Chlorotrimethylsilane (*see* Chlorotrimethylsilane–Zinc)
Zinc–copper couple, **1,** 1292; **2,** 465; **3,** 129; **5,** 121, 221, 758; **6,** 673; **7,** 428; **8,** 533; **9,** 45; **10,** 459; **11,** 599; **12,** 569
Zinc/Copper–Isopropyl iodide, **10,** 460
Zinc cyanide, **1,** 1293
Zinc–1,2-Dibromoethane, **12,** 570
Zinc–Dimethylformamide, **3,** 338
Zinc dithionite, **1,** 1294
Zinc fluoride, **6,** 514
Zinc–Graphite, **12,** 570
Zinc–Hydrochloric acid–*t*-Butyl alcohol, **9,** 518
Zinc hydrosulfite, **1,** 1294
Zinc iodide, **1,** 1294; **5,** 765; **10,** 462; **11,** 604; **12,** 21, 69, 148, 576
Zinc nitrate, **1,** 1294
Zinc oxide, **1,** 1294; **5,** 622
Zinc perchlorate–(S)-15-Amino-14-hydroxy-5,5-dimethyl-2,8-dithia[9]-(2,5)pyridinophane, **11,** 29
Zinc permanganate, **12,** 576
Zinc–Potassium cyanide, **6,** 674; **8,** 534
Zinc–Silver couple, **4,** 436; **5,** 760; **9,** 519; **10,** 460; **12,** 571
Zinc tetracarbonylcobaltate, **2,** 123; **3,** 89; **4,** 139
Zinc–Titanium(IV) chloride, **11,** 599
Zinc *p*-toluenesulfonate, **11,** 605
Zinc trifluoromethanesulfonate, **12,** 577
Zinc–Zinc chloride, **6,** 675; **7,** 430
Zirconium(IV) butoxide, **11,** 379
Zirconium carbene complexes, **12,** 577
Zirconium(III) chloride, **12,** 497
Zirconium(IV) chloride, **1,** 1295; **10,** 237
Zirconium(IV) chloride–Lithium aluminum hydride, **9,** 276
Zirconium(IV) propoxide, **11,** 605